Handbook of
NANOSCIENCE, ENGINEERING, and TECHNOLOGY

SECOND EDITION

The Electrical Engineering Handbook Series

Series Editor
Richard C. Dorf
University of California, Davis

Titles Included in the Series

The Handbook of Ad Hoc Wireless Networks, Mohammad Ilyas
The Avionics Handbook, Second Edition, Cary R. Spitzer
The Biomedical Engineering Handbook, Third Edition, Joseph D. Bronzino
The Circuits and Filters Handbook, Second Edition, Wai-Kai Chen
The Communications Handbook, Second Edition, Jerry Gibson
The Computer Engineering Handbook, Vojin G. Oklobdzija
The Control Handbook, William S. Levine
The CRC Handbook of Engineering Tables, Richard C. Dorf
The Digital Avionics Handbook, Second Edition Cary R. Spitzer
The Digital Signal Processing Handbook, Vijay K. Madisetti and Douglas Williams
The Electrical Engineering Handbook, Second Edition, Richard C. Dorf
The Electric Power Engineering Handbook, Second Edition, Leonard L. Grigsby
The Electronics Handbook, Second Edition, Jerry C. Whitaker
The Engineering Handbook, Third Edition, Richard C. Dorf
The Handbook of Formulas and Tables for Signal Processing, Alexander D. Poularikas
The Handbook of Nanoscience, Engineering, and Technology, Second Edition
William A. Goddard, III, Donald W. Brenner, Sergey E. Lyshevski, and Gerald J. Iafrate
The Handbook of Optical Communication Networks, Mohammad Ilyas and
Hussein T. Mouftah
The Industrial Electronics Handbook, J. David Irwin
The Measurement, Instrumentation, and Sensors Handbook, John G. Webster
The Mechanical Systems Design Handbook, Osita D.I. Nwokah and Yidirim Hurmuzlu
The Mechatronics Handbook, Robert H. Bishop
The Mobile Communications Handbook, Second Edition, Jerry D. Gibson
The Ocean Engineering Handbook, Ferial El-Hawary
The RF and Microwave Handbook, Mike Golio
The Technology Management Handbook, Richard C. Dorf
The Transforms and Applications Handbook, Second Edition, Alexander D. Poularikas
The VLSI Handbook, Second Edition, Wai-Kai Chen

Handbook of
NANOSCIENCE,
ENGINEERING,
and TECHNOLOGY

SECOND EDITION

William A. Goddard III
California Institute of Technology
Pasadena, California, USA

Donald W. Brenner
North Carolina State University
Raleigh, North Carolina, USA

Sergey E. Lyshevski
Rochester Institute of Technology
Rochester, New York, USA

Gerald J. Iafrate
North Carolina State University
Raleigh, North Carolina, USA

CRC Press
Taylor & Francis Group
Boca Raton London New York

CRC Press is an imprint of the
Taylor & Francis Group, an **informa** business

CRC Press
Taylor & Francis Group
6000 Broken Sound Parkway NW, Suite 300
Boca Raton, FL 33487-2742

© 2007 by Taylor & Francis Group, LLC
CRC Press is an imprint of Taylor & Francis Group, an Informa business

Library of Congress Cataloging-in-Publication Data

Handbook of nanoscience, engineering, and technology / editors, William A. Goddard, III ... [et al.].
    p. cm. -- (The Electrical engineering handbook series)
    Includes bibliographical references and index.
    ISBN 0-8493-7563-0 (alk. paper)
    1. Nanotechnology--Handbooks, manuals, etc. I. Goddard, William A., 1937- II. Title. III. Series.

T174.7.H36 2007
620'.5--dc22   2006102977

**Visit the Taylor & Francis Web site at
http://www.taylorandfrancis.com**

**and the CRC Press Web site at
http://www.crcpress.com**

Dedication

For my wife Karen, for her dedication and love, and for Sophie and Maxwell.

Donald W. Brenner

For my dearest wife Marina, and for my children Lydia and Alexander.

Sergey E. Lyshevski

To my wife, Kathy, and my family for their loving support and patience.

Gerald J. Iafrate

Preface

The first edition of the *Handbook of Nanoscience, Engineering, and Technology* was published in early 2003, reflecting many of the nanoscience possibilities envisioned by Richard Feynman in his 1959 address, "There is Plenty of Room at the Bottom." In his address, Feynman speculated about what might be on the molecular scale, and challenged the technical community "to find ways of manipulating and controlling things on a small scale." Inspired by the vision of Feynman, nanoscience is today defined as the study of material manipulation and control at the molecular scale, that is, a spatial scale of the order of a few hundred angstroms, less than one-thousandth of the width of a human hair. The extraordinary feature of nanoscience is that it allows for the tailoring and combining of the physical, biological, and engineering properties of matter at a very low level of nature's architectural building blocks. Critical to progress in nanoscience has been the stunning new achievements in fabrication, chemical processing, and nano resolution tool development in the last five decades, driven in large part by the microelectronics revolution. These developments today allow for molecular level tailoring and control of materials not heretofore possible except through naturally occurring atomic processes.

Over 40 years later, driven by federal executive orders of Presidents W.J. Clinton and G.W. Bush, and a recently enacted Twenty-First Century Nanotechnology Research and Development Act, the visionary challenge put forth by Feynman in 1959 is well on its way to becoming a reality. As a testimonial to this reality, the first edition of the *Handbook* included broad categories of innovative nanoscience, engineering, and technology that was emerging in the 2003 timeframe. The present 2007 second edition extends the portfolio of innovative nano areas further, including additional chapters on textiles, nanomanufacturing, spintronics, molecular electronics, aspects of bionanotechnology, and nanoparticles for drug delivery; as well, this edition updates select chapters which appeared in the first edition.

Acknowledgments

Dr. Brenner would like to thank his current and former colleagues for their intellectual stimulation and personal support. Professor Brenner also wishes to thank the Office of Naval Research, the Army Research Office, the National Science Foundation, the Air Force Office of Scientific Research, the NASA–Ames and NASA–Langley Research Centers, and the Department of Energy for supporting his research group over the last 13 years.

Dr. Lyshevski acknowledges the many people who contributed to this book. First, thanks go to all contributors, to whom I would like to express my sincere gratitude. It gives me great pleasure to acknowledge the help the editors received from many people in the preparation of this handbook. The outstanding CRC Press team, especially Nora Konopka and Helena Redshaw, helped tremendously by providing valuable feedback. Many thanks to all of you.

Dr. Iafrate acknowledges the career support and encouragement from colleagues, the Department of Defense, the University of Notre Dame, and North Carolina State University.

Editors

William A. Goddard, III obtained his Ph.D. in engineering science (minor in physics) from the California Institute of Technology, Pasadena, in October 1964, after which he joined the faculty of the chemistry department at Caltech and became a professor of theoretical chemistry in 1975. In November 1984, Goddard was honored as the first holder of the Charles and Mary Ferkel Chair in Chemistry and Applied Physics. He received the Badger Teaching Prize from the Chemistry and Chemical Engineering Division for Fall 1995. Goddard is a member of the National Academy of Sciences (U.S.) and the International Academy of Quantum Molecular Science. He was a National Science Foundation (NSF) Predoctoral Fellow (1960–1964) and an Alfred P. Sloan Foundation Fellow (1967–1969). In 1978 he received the Buck–Whitney Medal (for major contributions to theoretical chemistry in North America). In 1988 he received the American Chemical Society Award for Computers in Chemistry. In 1999 he received the Feynman Prize for Nanotechnology Theory (shared with Tahir Cagin and Yue Qi). In 2000 he received a NASA Space Sciences Award (shared with N. Vaidehi, A. Jain, and G. Rodriquez).

He is a fellow of the American Physical Society and of the American Association for the Advancement of Science. He is also a member of the American Chemical Society, the California Society, the California Catalysis Society (president for 1997–1998), the Materials Research Society, and the American Vacuum Society. He is a member of Tau Beta Pi and Sigma Xi. His activities include serving as a member of the board of trustees of the Gordon Research Conferences (1988–1994), the Computer Science and Telecommunications Board of the National Research Council (1990–1993), and the Board on Chemical Science and Technology (1980s), and a member and chairman of the board of advisors for the Chemistry Division of the NSF (1980s).

In addition, Dr. Goddard serves or has served on the editorial boards of several journals (*Journal of the American Chemical Society, Journal of Physical Chemistry, Chemical Physics, Catalysis Letters, Langmuir,* and *Computational Materials Science*). He is director of the Materials and Process Simulation Center (MSC) of the Beckman Institute at Caltech. He was the principal investigator of an NSF Grand Challenge Application Group (1992–1997) for developing advanced methods for quantum mechanics and molecular dynamics simulations optimized for massively parallel computers. He was also the principal investigator for the NSF Materials Research Group at Caltech (1985–1991). He is a cofounder (1984) of Molecular Simulations Inc., which develops and markets state-of-the-art computer software for molecular dynamics simulations and interactive graphics for applications to chemistry, biological, and materials sciences. He is also a cofounder (1991) of Schrödinger, Inc., which develops and markets state-of-the-art computer software using quantum mechanical methods for applications to chemical, biological, and materials sciences. In 1998 he cofounded Materials Research Source LLC, dedicated to development of new processing techniques for materials with an emphasis on nanoscale processing of semiconductors. In 2000 he cofounded BionomiX Inc., dedicated to predicting the structures and functions of all molecules for all known gene sequences. Goddard's research activities focus on the use of quantum mechanics and of molecular dynamics to study reaction mechanisms in catalysis (homogeneous and heterogeneous);

the chemical and electronic properties of surfaces (semiconductors, metals, ceramics, and polymers); biochemical processes; the structural, mechanical, and thermodynamic properties of materials (semiconductors, metals, ceramics, and polymers); mesoscale dynamics; and materials processing. He has published over 440 scientific articles.

Donald W. Brenner is currently a professor in the Department of Materials Science and Engineering at North Carolina State University. He received his B.S. from the State University of New York College at Fredonia in 1982 and his Ph.D. from Pennsylvania State University in 1987, both in chemistry. He joined the Theoretical Chemistry Section at the United States Naval Research Laboratory as a staff scientist in 1987, and joined the North Carolina State University faculty in 1994. His research interests focus on using atomic and mesoscale simulation and theory to understand technologically important processes and materials. Recent research areas include first-principles predictions of the mechanical properties of polycrystalline ceramics; crack dynamics; dynamics of nanotribology, tribochemistry and nanoindentation; simulation of the vapor deposition and surface reactivity of covalent materials; fullerene-based materials and devices; self-assembled monolayers; simulations of shock and detonation chemistry; and potential function development. He is also involved in the development of new cost-effective virtual-reality technologies for engineering education. Professor Brenner's awards include the 2002 Feynman Award for Research in Nanotechnology (theory), the Alcoa Foundation Engineering Research Achievement Award (2000), the Veridian Medal Paper (coauthor) (1999), an Outstanding Teacher Award from the North Carolina State College of Engineering (1999), an NSF Faculty Early Career Development Award (1995), the Naval Research Laboratory Chemistry Division Young Investigator Award (1991), the Naval Research Laboratory Chemistry Division Berman Award for Technical Publication (1990), and the Xerox Award from Penn State for the best materials-related Ph.D. thesis (1987). He was the scientific cochair for the Eighth (2000) and Ninth (2001) Foresight Conferences on Molecular Nanotechnology, and is a member of the editorial boards for the journals *Molecular Simulation* and the *Journal of Computational and Theoretical Nanoscience*, as well as a member of the North Carolina State University Academy of Outstanding Teachers.

Sergey Edward Lyshevski was born in Kiev, Ukraine. He received his M.S. (1980) and Ph.D. (1987) degrees from Kiev Polytechnic Institute, both in electrical engineering. From 1980 to 1993, Dr. Lyshevski held faculty positions at the Department of Electrical Engineering at Kiev Polytechnic Institute and the Academy of Sciences of Ukraine. From 1989 to 1993, he was the Microelectronic and Electromechanical Systems Division Head at the Academy of Sciences of Ukraine. From 1993 to 2002, he was with Purdue School of Engineering as an associate professor of electrical and computer engineering. In 2002, Dr. Lyshevski joined Rochester Institute of Technology as a professor of electrical engineering. Dr. Lyshevski serves as a Full Professor Faculty Fellow at the U.S. Air Force Research Laboratories and Naval Warfare Centers. He is the author of ten books (including *Logic Design of NanoICs*, coauthored with S. Yanushkevich and V. Shmerko, CRC Press, 2005; *Nano- and Microelectromechanical Systems: Fundamentals of Micro- and Nanoengineering*, CRC Press, 2004; *MEMS and NEMS: Systems, Devices, and Structures*, CRC Press, 2002) and is the author or coauthor of more than 300 journal articles, handbook chapters, and regular conference papers. His current research activities are focused on molecular electronics, molecular processing platforms, nanoengineering, cognitive systems, novel organizations/architectures, new nanoelectronic devices, reconfigurable super-high-performance computing, and systems informatics. Dr. Lyshevski has made significant contributions in the synthesis, design, application, verification, and implementation of advanced aerospace, electronic, electromechanical, and naval systems. He has made more than 30 invited presentations (nationally and internationally) and serves as an editor of the CRC Press book series on *Nano- and Microscience, Engineering, Technology, and Medicine.*

Gerald J. Iafrate joined the faculty of North Carolina State University in August 2001. Previously, he was a professor at the University of Notre Dame; he also served as Associate Dean for Research in the College of Engineering, and as director of the newly established University Center of Excellence in Nanoscience and Technology. He has extensive experience in managing large interdisciplinary research programs. From 1989 to 1997, Dr. Iafrate served as the director of the U.S. Army Research Office (ARO). As director, he was the army's key executive for the conduct of extramural research in the physical and engineering sciences in response to Department of Defense objectives. Prior to becoming director of ARO, Dr. Iafrate was the director of electronic devices research at the U.S. Army Electronics Technology and Devices Laboratory (ETDL). Working with the National Science Foundation, he played a key leadership role in establishing the first-of-its-kind army–NSF–university consortium. He is currently a research professor of electrical and computer engineering at North Carolina State University, Raleigh, where his current interests include quantum transport in nanostructures, spontaneous emission from Bloch electron radiators, and molecular-scale electronics. Dr. Iafrate is a fellow of the IEEE, APS, and AAAS.

Contributors

S. Adiga
Department of Materials
    Science and Engineering
North Carolina State University
Raleigh, NC

Damian G. Allis
Department of Chemistry
Syracuse University
Syracuse, NY

Narayan R. Aluru
Department of Mechanical
    and Industrial Engineering
and Beckman Institute for
    Advanced Science and
    Technology
University of Illinois
Urbana, IL

D.A. Areshkin
Department of Materials
    Science and Engineering
North Carolina State University
Raleigh, NC

Supriyo Bandyopadhyay
Department of Electrical and
    Computer Engineering
Virginia Commonwealth
    University
Richmond, VA

Carola Barrera
Department of Chemical
    Engineering
University of Puerto Rico-
    Mayaguez
Mayaguez, Puerto Rico

R. Bashir
Birck Nanotechnology
    Center
School of Electrical and
    Computer Engineering
Weldon School of Biomedical
    Engineering
Purdue University
West Lafayette, IN

K. Bloom
Department of Biology
University of North Carolina
Chapel Hill, NC

Youssry Botros
Intel Corporation

A.M. Bratkovsky
Hewlett-Packard Laboratories
Palo Alto, CA

Adam B. Braunschweig
California Nanosystems
    Institute and the Department
    of Chemistry and
    Biochemistry
University of California
Los Angeles, CA

Donald W. Brenner
Department of Materials
    Science and Engineering
North Carolina State University
Raleigh, NC

Ahmed Busnaina
Northeastern University
Boston, MA

Marc Cahay
Department of Electrical and
    Computer Engineering and
    Computer Science
University of Cincinnati
Cincinnati, OH

Caihua Chen
Department of Electrical and
    Computer Engineering
University of Delaware
Newark, DE

Saurabh Chhaparwal
College of Textiles
North Carolina State
    University
Raleigh, NC

Petersen F. Curt
EM Photonics, Inc.
University of Delaware
Newark, DE

Supriyo Datta
School of Electrical and
    Computer Engineering
Purdue University
West Lafayette, IN

C.W. Davis
Medicine/Cystic Fibrosis
  Center
University of North
  Carolina
Chapel Hill, NC

M.S. Diallo
Materials and Process
  Simulation Center
Beckman Institute
California Institute of
  Technology
Pasadena, CA
and Department of Civil
  Engineeering
Howard University
Washington, DC

William R. Dichtel
California Nanosystems
  Institute and the Department
  of Chemistry and
  Biochemistry
University of California
Los Angeles, CA

William Dondero
College of Textiles
North Carolina State University
Raleigh, NC

James P. Durbano
EM Photonics, Inc.
University of Delaware
Newark, DE

N. Fedorova
College of Textiles
North Carolina State University
Raleigh, NC

Richard P. Feynman
(Deceased)
California Institute of
  Technology
Pasadena, CA

J.K. Fisher
Department of Biomedical
  Engineering
University of North Carolina
Chapel Hill, NC

Kosmas Galatsis
Department of Electrical
  Engineering
University of California
Los Angeles, CA

Tushar Ghosh
College of Textiles
North Carolina State
  University
Raleigh, NC

Russell E. Gorga
College of Textiles
North Carolina State
  University
Raleigh, NC

Stephen A. Habay
Department of Chemistry
University of California
Irvine, CA

Meredith L. Hans
Department of Materials
  Science and Engineering
Drexel University
Philadelphia, PA

J.A. Harrison
Chemistry Department
U.S. Naval Academy
Annapolis, MD

S.A. Henderson
Starpharma Limited
Melbourne, Victoria,
  Australia

Karl Hess
Beckman Institute for
  Advanced Science and
  Technology
and Department of Electrical
  and Computer Engineering
University of Illinois
Urbana, IL

Juan P. Hinestroza
Department of Fiber Science
Cornell University
Ithaca, NY
and College of Textiles
North Carolina State
  University
Raleigh, NC

Yanhong Hu
Department of Materials
  Science and Engineering
North Carolina State University
Raleigh, NC

Zushou Hu
Department of Materials
  Science and Engineering
North Carolina State University
Raleigh, NC

**Michael Pycraft
  Hughes**
School of Engineering
University of Surrey
Guildford, Surrey, England

Dustin K. James
Department of Chemistry
Rice University
Houston, TX

Jean-Pierre Leburton
Beckman Institute for
  Advanced Science and
  Technology
University of Illinois
Urbana, IL

S.W. Lee
Department of Biomedical
    Engineering
Yonsei University
Won-Ju, Kang-Won, Korea

Kostantin Likharev
Stony Brook University
Stony Brook, NY

Wing Kam Liu
Department of Mechanical
    Engineering
Northwestern University
Evanston, IL

Anthony M. Lowman
Department of Materials
    Science and Engineering
Drexel University
Philadelphia, PA

**Sergey Edward
    Lyshevski**
Department of Electrical
    Engineering
Rochester Institute of
    Technology
Rochester, NY

Joshua S. Marcus
Department of Applied Physics
California Institute of
    Technology
Pasadena, CA

William McMahon
Beckman Institute for Advanced
    Science and Technology
University of Illinois
Urbana, IL

Paula M. Mendes
California Nanosystems
    Institute and the Department
    of Chemistry and
    Biochemistry
University of California
Los Angeles, CA

M. Meyyappan
NASA Ames Research Center
Moffett Field, CA

Stephen Michielsen
College of Textiles
North Carolina State
    University
Raleigh, NC

Vladimiro Mujica
Department of Chemistry
Northwestern University
Evanston, IL

Brian H. Northrop
California Nanosystems
    Institute and the Department
    of Chemistry and
    Biochemistry
University of California
Los Angeles, CA

E. Timothy O'Brien
Department of Physics and
    Astronomy
University of North Carolina
Chapel Hill, NC

Fernando E. Ortiz
EM Photonics, Inc.
University of Delaware
Newark, DE

Roman Ostroumov
Department of Electrical
    Engineering
University of California
Los Angeles, CA

Mihri Ozkan
Department of Electrical
    Engineering
University of California
Riverside, CA

Clifford W. Padgett
Department of Materials
    Science and Engineering
North Carolina State University
Raleigh, NC

Gregory N. Parsons
Department of Chemical
    Engineering
North Carolina State University
Raleigh, NC

Magnus Paulsson
School of Electrical and
    Computer Engineering
Purdue University
West Lafayette, IN

Wolfgang Porod
Department of Electrical
    Engineering
University of Notre Dame
Notre Dame, IN

B. Pourdeyhimi
College of Textiles
North Carolina State University
Raleigh, NC

Dennis W. Prather
Department of Electrical and
    Computer Engineering
University of Delaware
Newark, DE

Dong Qian
Department of Mechanical
    Engineering
Northwestern University
Evanston, IL

Mark A. Ratner
Department of Chemistry
Northwestern University
Evanston, IL

Umberto Ravaioli
Beckman Institute for
  Advanced Science and
  Technology
University of Illinois
Urbana, IL

Carlos Rinaldi
Department of Chemical
  Engineering
University of Puerto
  Rico-Mayaguez
Mayaguez, Puerto Rico

Mihail C. Roco
National Science Foundation
and National Nanotechnology
  Initiative
Washington, DC

Slava V. Rotkin
Physics Department
Lehigh University
Bethlehem, PA

Rodney S. Ruoff
Department of Mechanical
  Engineering
Northwestern University
Evanston, IL

Melinda Satcher
Department of Fiber Science
Cornell University
Ithaca, NY

**Christian E.
  Schafmeister**
Chemistry Department
University of Pittsburgh
Pittsburgh, PA

J.D. Schall
Department of Materials
  Science and Engineering
North Carolina State University
Raleigh, NC

Ahmed S. Sharkawy
Department of Electrical and
  Computer Engineering
University of Delaware
Newark, DE

Olga A. Shenderova
International Technology
  Center
Research Triangle Park, NC
  and Department of Materials
  Science and Engineering
North Carolina State
  University
Raleigh, NC

Shouyuan Shi
Department of Electrical and
  Computer Engineering
University of Delaware
Newark, DE

James T. Spencer
Department of Chemistry
Syracuse University
Syracuse, NY

Deepak Srivastava
NASA Ames Research Center
Moffett Field, CA

Martin Staedele
Infineon Technologies
Corporate Research
Munich, Germany

J. Fraser Stoddart
California Nanosystems
  Institute and the Department
  of Chemistry and
  Biochemistry
University of California
Los Angeles, CA

S.J. Stuart
Department of Chemistry
Clemson University
Clemson, SC

R. Superfine
Department of Physics
  and Astronomy
University of North Carolina
Chapel Hill, NC

R.M. Taylor, II
Department of Computer
  Science
University of North Carolina
Chapel Hill, NC

Todd Thorsen
Department of Mechanical
  Engineering
Massachusetts Institute
  of Technology
Cambridge, MA

D.A. Tomalia
Dendritic Nanotechnologies,
  Inc. and Central Michigan
  University
Mt. Pleasant, MI

James M. Tour
Center for Nanoscale Science
  and Technology
Rice University
Houston, TX

Blair R. Tuttle
Pennsylvania State University
Behrend College
Erie, PA

Trudy van der Straaten
Beckman Institute for
  Advanced Science and
  Technology
University of Illinois
Urbana, IL

L. Vicci
Department of Computer
  Science
University of North Carolina
Chapel Hill, NC

Gregory J. Wagner
Department of Mechanical
  Engineering
Northwestern University
Evanston, IL

Kang Wang
Department of Electrical
  Engineering
University of California
Los Angeles, CA

Min-Feng Yu
Department of Mechanical and
  Industrial Engineering
University of Illinois
Urbana, IL

Ferdows Zahid
School of Electrical and
  Computer Engineering
Purdue University
West Lafayette, IN

Contents

Section 1 Nanotechnology Overview

1 There's Plenty of Room at the Bottom: An Invitation to Enter a New Field of Physics *Richard P. Feynman* 1-1

2 Room at the Bottom, Plenty of Tyranny at the Top *Karl Hess*............................2-1

3 National Nanotechnology Initiative — Past, Present, Future *Mihail C. Roco* 3-1

Section 2 Molecular and Nanoelectronics

4 Engineering Challenges in Molecular Electronics *Gregory N. Parsons* 4-1

5 Molecular Electronic Computing Architectures *James M. Tour and Dustin K. James*...5-1

6 Nanoelectronic Circuit Architectures *Wolfgang Porod* 6-1

7 Molecular Computing and Processing Platforms *Sergey Edward Lyshevski* 7-1

8 Spin Field Effect Transistors *Supriyo Bandyopadhyay and Marc Cahay*......................8-1

9 Electron Charge and Spin Transport in Organic and Semiconductor Nanodevices: Moletronics and Spintronics *A.M. Bratkovsky*................................9-1

10 Nanoarchitectonics: Advances in Nanoelectronics *Kang Wang, Kosmas Galatsis, Roman Ostroumov, Mihri Ozkan, Kostantin Likharev, and Youssry Botros*....................10-1

11 Molecular Machines   *Brian H. Northrop, Adam B. Braunschweig,*
Paula M. Mendes, William R. Dichtel, and J. Fraser Stoddart **11**-1

Section 3   Molecular Electronics Devices

12 Molecular Conductance Junctions: A Theory and Modeling
Progress Report   *Vladimiro Mujica and Mark A. Ratner*..**12**-1

13 Modeling Electronics at the Nanoscale   *Narayan R. Aluru, Jean-Pierre Leburton,*
William McMahon, Umberto Ravaioli, Slava V. Rotkin, Martin Staedele,
Trudy van der Straaten, Blair R. Tuttle, and Karl Hess **13**-1

14 Resistance of a Molecule   *Magnus Paulsson, Ferdows Zahid, and Supriyo Datta* **14**-1

Section 4   Manipulation and Assembly

15 Magnetic Manipulation for the Biomedical Sciences   *J.K. Fisher,*
L. Vicci, K. Bloom, E. Timothy O'Brien, C.W. Davis, R.M. Taylor, II, and R. Superfine...............**15**-1

16 Nanoparticle Manipulation by Electrostatic Forces
Michael Pycraft Hughes..**16**-1

17 Biological- and Chemical-Mediated Self-Assembly of Artificial
Micro- and Nanostructures   *S.W. Lee and R. Bashir*..**17**-1

18 Nanostructural Architectures from Molecular Building Blocks
Damian G. Allis and James T. Spencer..**18**-1

19 Building Block Approaches to Nonlinear and Linear Macromolecules
Stephen A. Habay and Christian E. Schafmeister **19**-1

20 Introduction to Nanomanufacturing   *Ahmed Busnaina* **20**-1

21 Textile Nanotechnologies   *B. Pourdeyhimi, N. Fedorova, William Dondero,*
Russell E. Gorga, Stephen Michielsen, Tushar Ghosh, Saurabh Chhaparwal, Carola Barrera,
Carlos Rinaldi, Melinda Satcher, and Juan P. Hinestroza 21-1

Section 5   Functional Structures

22 Carbon Nanotubes   *M. Meyyappan and Deepak Srivastava* 22-1

23 Mechanics of Carbon Nanotubes   *Dong Qian, Gregory J. Wagner,*
Wing Kam Liu, Min-Feng Yu, and Rodney S. Ruoff 23-1

24 Dendrimers — an Enabling Synthetic Science to Controlled Organic
Nanostructures   *D.A. Tomalia, S.A. Henderson, and M.S. Diallo* 24-1

25 Design and Applications of Photonic Crystals   *Dennis W. Prather,*
Ahmed S. Sharkawy, Shouyuan Shi, and Caihua Chen 25-1

26 Progress in Nanofluidics for Cell Biology   *Todd Thorsen*
and Joshua S. Marcus 26-1

27 Carbon Nanostructures and Nanocomposites   *Yanhong Hu, Zushou Hu,*
Clifford W. Padgett, Donald W. Brenner, and Olga A. Shenderova 27-1

28 Contributions of Molecular Modeling to Nanometer-Scale Science
and Technology   *Donald W. Brenner, Olga A. Shenderova, J.D. Schall, D.A. Areshkin,*
S. Adiga, J.A. Harrison, and S.J. Stuart 28-1

29 Accelerated Design Tools for Nanophotonic Devices and Applications
James P. Durbano, Ahmed S. Sharkawy, Shouyuan Shi, Fernando E. Ortiz, Petersen F. Curt,
and Dennis W. Prather 29-1

30 Nanoparticles for Drug Delivery   *Meredith L. Hans and Anthony M. Lowman* 30-1

Index I-1

1

Nanotechnology Overview

1 There's Plenty of Room at the Bottom: An Invitation to
   Enter a New Field of Physics *Richard P. Feynman* 1-1
   Transcript

2 Room at the Bottom, Plenty of Tyranny at the Top *Karl Hess*..2-1
   Rising to the Feynman Challenge • Tyranny at the Top • New Forms of Switching and
   Storage • New Architectures • How Does Nature Do It?

3 National Nanotechnology Initiative — Past, Present, Future *Mihail C. Roco* 3-1
   Introduction • Identifying a Megatrend in Science and Technology • Key Factors in
   Establishing NNI (about 2000) • The Beginning of NNI • NNI at Five Years (2001–2005)
   Technical Challenges in 2005–2020 • New Science and Engineering, New Governance
   Approach

1

There's Plenty of Room at the Bottom: An Invitation to Enter a New Field of Physics

1.1 Transcript..1-1
How Do We Write Small? • Information on a Small Scale •
Better Electron Microscopes • The Marvelous Biological
System • Miniaturizing the Computer • Miniaturization by
Evaporation • Problems of Lubrication • A Hundred Tiny
Hands • Rearranging the Atoms • Atoms in a Small World •
High School Competition

Richard P. Feynman
California Institute of Technology

This transcript of the classic talk that Richard Feynman gave on December 29, 1959, at the annual meeting of the American Physical Society at the California Institute of Technology (Caltech) was first published in the February 1960 issue (Volume XXIII, No. 5, pp. 22–36) of Caltech's *Engineering and Science*, which owns the copyright. It has been made available on the web at http://www.zyvex.com/nanotech/ feynman.html with their kind permission.

For an account of the talk and how people reacted to it, see Chapter 4 of *Nano!* by Ed Regis. An excellent technical introduction to nanotechnology is *Nanosystems: Molecular Machinery, Manufacturing, and Computation* by K. Eric Drexler.

1.1 Transcript

I imagine experimental physicists must often look with envy at men like Kamerlingh Onnes, who discovered a field like low temperature, which seems to be bottomless and in which one can go down and down. Such a man is then a leader and has some temporary monopoly in a scientific adventure. Percy Bridgman, in designing a way to obtain higher pressures, opened up another new field and was able to move into it and to lead us all along. The development of ever-higher vacuum was a continuing development of the same kind.

I would like to describe a field in which little has been done but in which an enormous amount can be done in principle. This field is not quite the same as the others in that it will not tell us much of fundamental physics (in the sense of "what are the strange particles?"); but it is more like solid-state physics in the sense that it might tell us much of great interest about the strange phenomena that occur in complex situations. Furthermore, a point that is most important is that it would have an enormous number of technical applications.

What I want to talk about is the problem of manipulating and controlling things on a small scale.

As soon as I mention this, people tell me about miniaturization, and how far it has progressed today. They tell me about electric motors that are the size of the nail on your small finger. And there is a device on the market, they tell me, by which you can write the Lord's Prayer on the head of a pin. But that's nothing; that's the most primitive, halting step in the direction I intend to discuss. It is a staggeringly small world that is below. In the year 2000, when they look back at this age, they will wonder why it was not until the year 1960 that anybody began seriously to move in this direction.

Why cannot we write the entire 24 volumes of the *Encyclopaedia Britannica* on the head of a pin?

Let's see what would be involved. The head of a pin is a sixteenth of an inch across. If you magnify it by 25,000 diameters, the area of the head of the pin is then equal to the area of all the pages of the *Encyclopaedia Britannica*. Therefore, all it is necessary to do is to reduce in size all the writing in the encyclopedia by 25,000 times. Is that possible? The resolving power of the eye is about 1/120 of an inch — that is roughly the diameter of one of the little dots on the fine half-tone reproductions in the encyclopedia. This, when you demagnify it by 25,000 times, is still 80 angstroms in diameter — 32 atoms across, in an ordinary metal. In other words, one of those dots still would contain in its area 1000 atoms. So, each dot can easily be adjusted in size as required by the photoengraving, and there is no question that there is enough room on the head of a pin to put all of the *Encyclopaedia Britannica*. Furthermore, it can be read if it is so written. Let's imagine that it is written in raised letters of metal; that is, where the black is in the encyclopedia, we have raised letters of metal that are actually 1/25,000 of their ordinary size. How would we read it?

If we had something written in such a way, we could read it using techniques in common use today. (They will undoubtedly find a better way when we do actually have it written, but to make my point conservatively I shall just take techniques we know today.) We would press the metal into a plastic material and make a mold of it, then peel the plastic off very carefully, evaporate silica into the plastic to get a very thin film, then shadow it by evaporating gold at an angle against the silica so that all the little letters will appear clearly, dissolve the plastic away from the silica film, and then look through it with an electron microscope!

There is no question that if the thing were reduced by 25,000 times in the form of raised letters on the pin, it would be easy for us to read it today. Furthermore, there is no question that we would find it easy to make copies of the master; we would just need to press the same metal plate again into plastic and we would have another copy.

How Do We Write Small?

The next question is, how do we write it? We have no standard technique to do this now. But let me argue that it is not as difficult as it first appears to be. We can reverse the lenses of the electron microscope in order to demagnify as well as magnify. A source of ions, sent through the microscope lenses in reverse, could be focused to a very small spot. We could write with that spot like we write in a TV cathode ray oscilloscope, by going across in lines and having an adjustment that determines the amount of material which is going to be deposited as we scan in lines.

This method might be very slow because of space charge limitations. There will be more rapid methods. We could first make, perhaps by some photo process, a screen that has holes in it in the form of the letters. Then we would strike an arc behind the holes and draw metallic ions through the holes; then we could again use our system of lenses and make a small image in the form of ions, which would deposit the metal on the pin.

A simpler way might be this (though I am not sure it would work): we take light and, through an optical microscope running backwards, we focus it onto a very small photoelectric screen. Then electrons come away from the screen where the light is shining. These electrons are focused down in size by the electron microscope lenses to impinge directly upon the surface of the metal. Will such a beam etch away the metal if it is run long enough? I don't know. If it doesn't work for a metal surface, it must be possible to find some surface with which to coat the original pin so that, where the electrons bombard, a change is made which we could recognize later.

There is no intensity problem in these devices — not what you are used to in magnification, where you have to take a few electrons and spread them over a bigger and bigger screen; it is just the opposite. The light which we get from a page is concentrated onto a very small area so it is very intense. The few electrons which come from the photoelectric screen are demagnified down to a very tiny area so that, again, they are very intense. I don't know why this hasn't been done yet!

That's the *Encyclopedia Britannica* on the head of a pin, but let's consider all the books in the world. The Library of Congress has approximately 9 million volumes; the British Museum Library has 5 million volumes; there are also 5 million volumes in the National Library in France. Undoubtedly there are duplications, so let us say that there are some 24 million volumes of interest in the world.

What would happen if I print all this down at the scale we have been discussing? How much space would it take? It would take, of course, the area of about a million pinheads because, instead of there being just the 24 volumes of the encyclopedia, there are 24 million volumes. The million pinheads can be put in a square of a thousand pins on a side, or an area of about 3 square yards. That is to say, the silica replica with the paper-thin backing of plastic, with which we have made the copies, with all this information, is on an area approximately the size of 35 pages of the encyclopedia. That is about half as many pages as there are in this magazine. All of the information which all of mankind has ever recorded in books can be carried around in a pamphlet in your hand — and not written in code, but a simple reproduction of the original pictures, engravings, and everything else on a small scale without loss of resolution.

What would our librarian at Caltech say, as she runs all over from one building to another, if I tell her that, 10 years from now, all of the information that she is struggling to keep track of — 120,000 volumes, stacked from the floor to the ceiling, drawers full of cards, storage rooms full of the older books — can be kept on just one library card! When the University of Brazil, for example, finds that their library is burned, we can send them a copy of every book in our library by striking off a copy from the master plate in a few hours and mailing it in an envelope no bigger or heavier than any other ordinary airmail letter. Now, the name of this talk is "There Is *Plenty* of Room at the Bottom" — not just "There Is Room at the Bottom." What I have demonstrated is that there is room — that you can decrease the size of things in a practical way. I now want to show that there is *plenty* of room. I will not now discuss how we are going to do it, but only what is possible in principle — in other words, what is possible according to the laws of physics. I am not inventing antigravity, which is possible someday only if the laws are not what we think. I am telling you what could be done if the laws are what we think; we are not doing it simply because we haven't yet gotten around to it.

Information on a Small Scale

Suppose that, instead of trying to reproduce the pictures and all the information directly in its present form, we write only the information content in a code of dots and dashes, or something like that, to represent the various letters. Each letter represents six or seven "bits" of information; that is, you need only about six or seven dots or dashes for each letter. Now, instead of writing everything, as I did before, on the surface of the head of a pin, I am going to use the interior of the material as well.

Let us represent a dot by a small spot of one metal, the next dash by an adjacent spot of another metal, and so on. Suppose, to be conservative, that a bit of information is going to require a little cube of atoms $5 \times 5 \times 5$ — that is 125 atoms. Perhaps we need a hundred and some odd atoms to make sure that the information is not lost through diffusion or through some other process.

I have estimated how many letters there are in the encyclopedia, and I have assumed that each of my 24 million books is as big as an encyclopedia volume, and have calculated, then, how many bits of information there are (10^{15}). For each bit I allow 100 atoms. And it turns out that all of the information that man has carefully accumulated in all the books in the world can be written in this form in a cube of material 1/200 of an inch wide — which is the barest piece of dust that can be made out by the human eye. So there is plenty of room at the bottom! Don't tell me about microfilm! This fact — that enormous amounts of information can be carried in an exceedingly small space — is, of course, well known to the biologists and resolves the mystery that existed before we understood all this clearly — of how it could be that, in the tiniest cell, all of the information for the organization of a complex creature such as ourselves can be stored. All this information — whether we have brown eyes, or whether we think at all, or that in the embryo the jawbone should first develop with a little hole in the side so that later a nerve can grow through it — all this information is contained in a very tiny fraction of the cell in the form of long-chain DNA molecules in which approximately 50 atoms are used for one bit of information about the cell.

Better Electron Microscopes

If I have written in a code with $5 \times 5 \times 5$ atoms to a bit, the question is, how could I read it today? The electron microscope is not quite good enough — with the greatest care and effort, it can only resolve about 10 angstroms. I would like to try and impress upon you, while I am talking about all of these things on a small scale, the importance of improving the electron microscope by a hundred times. It is not impossible; it is not against the laws of diffraction of the electron. The wavelength of the electron in such a microscope is only 1/20 of an angstrom. So it should be possible to see the individual atoms. What good would it be to see individual atoms distinctly? We have friends in other fields — in biology, for instance. We physicists often look at them and say, "You know the reason you fellows are making so little progress?" (Actually I don't know any field where they are making more rapid progress than they are in biology today.) "You should use more mathematics, like we do." They could answer us — but they're polite, so I'll answer for them: "What *you* should do in order for us to make more rapid progress is to make the electron microscope 100 times better."

What are the most central and fundamental problems of biology today? They are questions like, what is the sequence of bases in the DNA? What happens when you have a mutation? How is the base order in the DNA connected to the order of amino acids in the protein? What is the structure of the RNA; is it single-chain or double-chain, and how is it related in its order of bases to the DNA? What is the organization of the microsomes? How are proteins synthesized? Where does the RNA go? How does it sit? Where do the proteins sit? Where do the amino acids go in? In photosynthesis, where is the chlorophyll; how is it arranged; where are the carotenoids involved in this thing? What is the system of the conversion of light into chemical energy?

It is very easy to answer many of these fundamental biological questions; you just look at the thing! You will see the order of bases in the chain; you will see the structure of the microsome. Unfortunately, the present microscope sees at a scale which is just a bit too crude. Make the microscope one hundred times more powerful, and many problems of biology would be made very much easier. I exaggerate, of course, but the biologists would surely be very thankful to you — and they would prefer that to the criticism that they should use more mathematics.

The theory of chemical processes today is based on theoretical physics. In this sense, physics supplies the foundation of chemistry. But chemistry also has analysis. If you have a strange substance and you want to know what it is, you go through a long and complicated process of chemical analysis. You can analyze almost anything today, so I am a little late with my idea. But if the physicists wanted to, they could also dig under the chemists in the problem of chemical analysis. It would be very easy to make an analysis of any complicated chemical substance; all one would have to do would be to look at it and see where the atoms are. The only trouble is that the electron microscope is 100 times too poor. (Later, I would like to ask the question: can the physicists do something about the third problem of chemistry — namely, synthesis? Is there a physical way to synthesize any chemical substance?)

The reason the electron microscope is so poor is that the f-value of the lenses is only 1 part to 1000; you don't have a big enough numerical aperture. And I know that there are theorems which prove that it is impossible, with axially symmetrical stationary field lenses, to produce an f-value any bigger than so and so; and therefore the resolving power at the present time is at its theoretical maximum. But in every theorem there are assumptions. Why must the field be symmetrical? I put this out as a challenge: is there no way to make the electron microscope more powerful?

The Marvelous Biological System

The biological example of writing information on a small scale has inspired me to think of something that should be possible. Biology is not simply writing information; it is doing something about it. A biological system can be exceedingly small. Many of the cells are very tiny, but they are very active; they manufacture various substances; they walk around; they wiggle; and they do all kinds of marvelous things — all on a very small scale. Also, they store information. Consider the possibility that we too can make a thing very small which does what we want — that we can manufacture an object that maneuvers at that level!

There may even be an economic point to this business of making things very small. Let me remind you of some of the problems of computing machines. In computers we have to store an enormous amount of information. The kind of writing that I was mentioning before, in which I had everything down as a distribution of metal, is permanent. Much more interesting to a computer is a way of writing, erasing, and writing something else. (This is usually because we don't want to waste the material on which we have just written. Yet if we could write it in a very small space, it wouldn't make any difference; it could just be thrown away after it was read. It doesn't cost very much for the material).

Miniaturizing the Computer

I don't know how to do this on a small scale in a practical way, but I do know that computing machines are very large; they fill rooms. Why can't we make them very small, make them of little wires, little elements — and by little, I mean *little*. For instance, the wires should be 10 or 100 atoms in diameter, and the circuits should be a few thousand angstroms across. Everybody who has analyzed the logical theory of computers has come to the conclusion that the possibilities of computers are very interesting — if they could be made to be more complicated by several orders of magnitude. If they had millions of times as many elements, they could make judgments. They would have time to calculate what is the best way to make the calculation that they are about to make. They could select the method of analysis which, from their experience, is better than the one that we would give to them. And in many other ways, they would have new qualitative features.

If I look at your face I immediately recognize that I have seen it before. (Actually, my friends will say I have chosen an unfortunate example here for the subject of this illustration. At least I recognize that it is a man and not an apple.) Yet there is no machine which, with that speed, can take a picture of a face and say even that it is a man; and much less that it is the same man that you showed it before — unless it is exactly the same picture. If the face is changed; if I am closer to the face; if I am further from the face; if the light changes — I recognize it anyway. Now, this little computer I carry in my head is easily able to do that. The computers that we build are not able to do that. The number of elements in this bone box of mine are enormously greater than the number of elements in our "wonderful" computers. But our mechanical computers are too big; the elements in this box are microscopic. I want to make some that are submicroscopic.

If we wanted to make a computer that had all these marvelous extra qualitative abilities, we would have to make it, perhaps, the size of the Pentagon. This has several disadvantages. First, it requires too much material; there may not be enough germanium in the world for all the transistors which would have to be put into this enormous thing. There is also the problem of heat generation and power consumption; TVA would be needed to run the computer. But an even more practical difficulty is that the computer would be limited to a certain speed. Because of its large size, there is finite time required to get the information from one place to another. The information cannot go any faster than the speed of light — so, ultimately, when our computers get faster and faster and more and more elaborate, we will have to make them smaller and smaller. But there is plenty of room to make them smaller. There is nothing that I can see in the physical laws that says the computer elements cannot be made enormously smaller than they are now. In fact, there may be certain advantages.

Miniaturization by Evaporation

How can we make such a device? What kind of manufacturing processes would we use? One possibility we might consider, since we have talked about writing by putting atoms down in a certain arrangement, would be to evaporate the material, then evaporate the insulator next to it. Then, for the next layer, evaporate another position of a wire, another insulator, and so on. So, you simply evaporate until you have a block of stuff which has the elements — coils and condensers, transistors and so on — of exceedingly fine dimensions.

But I would like to discuss, just for amusement, that there are other possibilities. Why can't we manufacture these small computers somewhat like we manufacture the big ones? Why can't we drill holes, cut things, solder things, stamp things out, mold different shapes all at an infinitesimal level? What are the limitations as to how small a thing has to be before you can no longer mold it? How many times when you are working on something frustratingly tiny, like your wife's wristwatch, have you said to yourself, "If I could only train an ant to do this!" What I would like to suggest is the possibility of training an ant to train a mite to do this. What are the possibilities of small but movable machines? They may or may not be useful, but they surely would be fun to make.

Consider any machine — for example, an automobile — and ask about the problems of making an infinitesimal machine like it. Suppose, in the particular design of the automobile, we need a certain precision of the parts; we need an accuracy, let's suppose, of 4/10,000 of an inch. If things are more inaccurate than that in the shape of the cylinder and so on, it isn't going to work very well. If I make the thing too small, I have to worry about the size of the atoms; I can't make a circle of "balls" so to speak, if the circle is too small. So if I make the error — corresponding to 4/10,000 of an inch — correspond to an error of 10 atoms, it turns out that I can reduce the dimensions of an automobile 4000 times, approximately, so that it is 1 mm across. Obviously, if you redesign the car so that it would work with a much larger tolerance, which is not at all impossible, then you could make a much smaller device.

It is interesting to consider what the problems are in such small machines. Firstly, with parts stressed to the same degree, the forces go as the area you are reducing, so that things like weight and inertia are of relatively no importance. The strength of material, in other words, is very much greater in proportion. The stresses and expansion of the flywheel from centrifugal force, for example, would be the same proportion only if the rotational speed is increased in the same proportion as we decrease the size. On the other hand, the metals that we use have a grain structure, and this would be very annoying at small scale because the material is not homogeneous. Plastics and glass and things of this amorphous nature are very much more homogeneous, and so we would have to make our machines out of such materials.

There are problems associated with the electrical part of the system — with the copper wires and the magnetic parts. The magnetic properties on a very small scale are not the same as on a large scale; there is the "domain" problem involved. A big magnet made of millions of domains can only be made on a small scale with one domain. The electrical equipment won't simply be scaled down; it has to be redesigned. But I can see no reason why it can't be redesigned to work again.

Problems of Lubrication

Lubrication involves some interesting points. The effective viscosity of oil would be higher and higher in proportion as we went down (and if we increase the speed as much as we can). If we don't increase the speed so much, and change from oil to kerosene or some other fluid, the problem is not so bad. But actually we may not have to lubricate at all! We have a lot of extra force. Let the bearings run dry; they won't run hot because the heat escapes away from such a small device very, very rapidly.

This rapid heat loss would prevent the gasoline from exploding, so an internal combustion engine is impossible. Other chemical reactions, liberating energy when cold, can be used. Probably an external supply of electrical power would be most convenient for such small machines.

What would be the utility of such machines? Who knows? Of course, a small automobile would only be useful for the mites to drive around in, and I suppose our Christian interests don't go that far. However, we did note the possibility of the manufacture of small elements for computers in completely automatic factories, containing lathes and other machine tools at the very small level. The small lathe would not have to be exactly like our big lathe. I leave to your imagination the improvement of the design to take full advantage of the properties of things on a small scale, and in such a way that the fully automatic aspect would be easiest to manage.

A friend of mine (Albert R. Hibbs) suggests a very interesting possibility for relatively small machines. He says that although it is a very wild idea, it would be interesting in surgery if you could swallow the surgeon. You put the mechanical surgeon inside the blood vessel and it goes into the heart and "looks" around. (Of course the information has to be fed out.) It finds out which valve is the faulty one and takes a little knife and slices it out. Other small machines might be permanently incorporated in the body to assist some inadequately functioning organ.

Now comes the interesting question: how do we make such a tiny mechanism? I leave that to you. However, let me suggest one weird possibility. You know, in the atomic energy plants they have materials and machines that they can't handle directly because they have become radioactive. To unscrew nuts and put on bolts and so on, they have a set of master and slave hands, so that by operating a set of levers here, you control the "hands" there, and can turn them this way and that so you can handle things quite nicely.

Most of these devices are actually made rather simply, in that there is a particular cable, like a marionette string, that goes directly from the controls to the "hands." But, of course, things also have been made using servo motors, so that the connection between the one thing and the other is electrical rather than mechanical. When you turn the levers, they turn a servo motor, and it changes the electrical currents in the wires, which repositions a motor at the other end.

Now, I want to build much the same device — a master–slave system which operates electrically. But I want the slaves to be made especially carefully by modern large-scale machinists so that they are 1/4 the scale of the "hands" that you ordinarily maneuver. So you have a scheme by which you can do things at 1/4 scale anyway — the little servo motors with little hands play with little nuts and bolts; they drill little holes; they are four times smaller. Aha! So I manufacture a 1/4-size lathe; I manufacture 1/4-size tools; and I make, at the 1/4 scale, still another set of hands again relatively 1/4 size! This is 1/16 size, from my point of view. And after I finish doing this I wire directly from my large-scale system, through transformers perhaps, to the 1/16-size servo motors. Thus I can now manipulate the 1/16 size hands.

Well, you get the principle from there on. It is rather a difficult program, but it is a possibility. You might say that one can go much farther in one step than from one to four. Of course, this all has to be designed very carefully, and it is not necessary simply to make it like hands. If you thought of it very carefully, you could probably arrive at a much better system for doing such things.

If you work through a pantograph, even today, you can get much more than a factor of four in even one step. But you can't work directly through a pantograph which makes a smaller pantograph which then makes a smaller pantograph — because of the looseness of the holes and the irregularities of construction. The end of the pantograph wiggles with a relatively greater irregularity than the irregularity with which you move your hands. In going down this scale, I would find the end of the pantograph on the end of the pantograph on the end of the pantograph shaking so badly that it wasn't doing anything sensible at all.

At each stage, it is necessary to improve the precision of the apparatus. If, for instance, having made a small lathe with a pantograph, we find its lead screw irregular — more irregular than the large-scale one —we could lap the lead screw against breakable nuts that you can reverse in the usual way back and forth until this lead screw is, at its scale, as accurate as our original lead screws, at our scale.

We can make flats by rubbing unflat surfaces in triplicates together — in three pairs — and the flats then become flatter than the thing you started with. Thus, it is not impossible to improve precision on a small scale by the correct operations. So, when we build this stuff, it is necessary at each step to improve the accuracy of the equipment by working for a while down there, making accurate lead screws, Johansen blocks, and all the other materials which we use in accurate machine work at the higher level. We have to stop at each level and manufacture all the stuff to go to the next level — a very long and very difficult program. Perhaps you can figure a better way than that to get down to small scale more rapidly.

Yet, after all this, you have just got one little baby lathe 4000 times smaller than usual. But we were thinking of making an enormous computer, which we were going to build by drilling holes on this lathe to make little washers for the computer. How many washers can you manufacture on this one lathe?

A Hundred Tiny Hands

When I make my first set of slave "hands" at 1/4 scale, I am going to make ten sets. I make ten sets of "hands," and I wire them to my original levers so they each do exactly the same thing at the same time in parallel. Now, when I am making my new devices 1/4 again as small, I let each one manufacture ten copies, so that I would have a hundred "hands" at the 1/16 size.

Where am I going to put the million lathes that I am going to have? Why, there is nothing to it; the volume is much less than that of even one full-scale lathe. For instance, if I made a billion little lathes, each 1/4000 of the scale of a regular lathe, there are plenty of materials and space available because in the billion little ones there is less than 2% of the materials in one big lathe.

It doesn't cost anything for materials, you see. So I want to build a billion tiny factories, models of each other, which are manufacturing simultaneously, drilling holes, stamping parts, and so on.

As we go down in size, there are a number of interesting problems that arise. All things do not simply scale down in proportion. There is the problem that materials stick together by the molecular (Van der Waals) attractions. It would be like this: after you have made a part and you unscrew the nut from a bolt, it isn't going to fall down because the gravity isn't appreciable; it would even be hard to get it off the bolt. It would be like those old movies of a man with his hands full of molasses, trying to get rid of a glass of water. There will be several problems of this nature that we will have to be ready to design for.

Rearranging the Atoms

But I am not afraid to consider the final question as to whether, ultimately — in the great future — we can arrange the atoms the way we want; the very atoms, all the way down! What would happen if we could arrange the atoms one by one the way we want them (within reason, of course; you can't put them so that they are chemically unstable, for example).

Up to now, we have been content to dig in the ground to find minerals. We heat them and we do things on a large scale with them, and we hope to get a pure substance with just so much impurity, and so on. But we must always accept some atomic arrangement that nature gives us. We haven't got anything, say, with a "checkerboard" arrangement, with the impurity atoms exactly arranged 1000 angstroms apart, or in some other particular pattern. What could we do with layered structures with just the right layers? What would the properties of materials be if we could really arrange the atoms the way we want them? They would be very interesting to investigate theoretically. I can't see exactly what would happen, but I can hardly doubt that when we have some control of the arrangement of things on a small scale we will get an enormously greater range of possible properties that substances can have, and of different things that we can do.

Consider, for example, a piece of material in which we make little coils and condensers (or their solid state analogs) 1,000 or 10,000 angstroms in a circuit, one right next to the other, over a large area, with little antennas sticking out at the other end — a whole series of circuits. Is it possible, for example, to emit light from a whole set of antennas, like we emit radio waves from an organized set of antennas to beam the radio programs to Europe? The same thing would be to beam the light out in a definite direction with very high intensity. (Perhaps such a beam is not very useful technically or economically.)

I have thought about some of the problems of building electric circuits on a small scale, and the problem of resistance is serious. If you build a corresponding circuit on a small scale, its natural frequency goes up, since the wavelength goes down as the scale; but the skin depth only decreases with the square root of the scale ratio, and so resistive problems are of increasing difficulty. Possibly we can beat resistance through the use of superconductivity if the frequency is not too high, or by other tricks.

Atoms in a Small World

When we get to the very, very small world — say circuits of seven atoms — we have a lot of new things that would happen that represent completely new opportunities for design. Atoms on a small scale behave like nothing on a large scale, for they satisfy the laws of quantum mechanics. So, as we go down and fiddle around with the atoms down there, we are working with different laws, and we can expect to do different things. We can manufacture in different ways. We can use not just circuits but some system involving the quantized energy levels, or the interactions of quantized spins, etc.

Another thing we will notice is that, if we go down far enough, all of our devices can be mass produced so that they are absolutely perfect copies of one another. We cannot build two large machines so that the dimensions are exactly the same. But if your machine is only 100 atoms high, you only have to get it correct to 1/2% to make sure the other machine is exactly the same size — namely, 100 atoms high!

At the atomic level, we have new kinds of forces and new kinds of possibilities, new kinds of effects. The problems of manufacture and reproduction of materials will be quite different. I am, as I said, inspired by the biological phenomena in which chemical forces are used in repetitious fashion to produce all kinds of weird effects (one of which is the author).

The principles of physics, as far as I can see, do not speak against the possibility of maneuvering things atom by atom. It is not an attempt to violate any laws; it is something, in principle, that can be done; but in practice, it has not been done because we are too big.

Ultimately, we can do chemical synthesis. A chemist comes to us and says, "Look, I want a molecule that has the atoms arranged thus and so; make me that molecule." The chemist does a mysterious thing when he wants to make a molecule. He sees that it has that ring, so he mixes this and that, and he shakes it, and he fiddles around. And, at the end of a difficult process, he usually does succeed in synthesizing what he wants. By the time I get my devices working, so that we can do it by physics, he will have figured out how to synthesize absolutely anything, so that this will really be useless. But it is interesting that it would be, in principle, possible (I think) for a physicist to synthesize any chemical substance that the chemist writes down. Give the orders and the physicist synthesizes it. How? Put the atoms down where the chemist says, and so you make the substance. The problems of chemistry and biology can be greatly helped if our ability to see what we are doing, and to do things on an atomic level, is ultimately developed — a development which I think cannot be avoided.

Now, you might say, "Who should do this and why should they do it?" Well, I pointed out a few of the economic applications, but I know that the reason that you would do it might be just for fun. But have some fun! Let's have a competition between laboratories. Let one laboratory make a tiny motor which it sends to another lab which sends it back with a thing that fits inside the shaft of the first motor.

High School Competition

Just for the fun of it, and in order to get kids interested in this field, I would propose that someone who has some contact with the high schools think of making some kind of high school competition. After all, we haven't even started in this field, and even the kids can write smaller than has ever been written before. They could have competition in high schools. The Los Angeles high school could send a pin to the Venice high school on which it says, "How's this?" They get the pin back, and in the dot of the "i" it says, "Not so hot."

Perhaps this doesn't excite you to do it, and only economics will do so. Then I want to do something, but I can't do it at the present moment because I haven't prepared the ground. It is my intention to offer a prize of $1000 to the first guy who can take the information on the page of a book and put it on an area 1/25,000 smaller in linear scale in such manner that it can be read by an electron microscope.

And I want to offer another prize — if I can figure out how to phrase it so that I don't get into a mess of arguments about definitions — of another $1000 to the first guy who makes an operating electric motor — a rotating electric motor which can be controlled from the outside and, not counting the lead-in wires, is only a 1/64-inch cube.

I do not expect that such prizes will have to wait very long for claimants.

2

Room at the Bottom, Plenty of Tyranny at the Top

	2.1	Rising to the Feynman Challenge	2-1
	2.2	Tyranny at the Top	2-2
	2.3	New Forms of Switching and Storage	2-3
Karl Hess	2.4	New Architectures	2-5
University of Illinois	2.5	How Does Nature Do It?	2-6

2.1 Rising to the Feynman Challenge

Richard Feynman is generally regarded as one of the fathers of nanotechnology. In giving his landmark presentation to the American Physical Society on December 29, 1959, at Caltech, his title line was, "There's Plenty of Room at the Bottom." At that time, Feynman extended an invitation for "manipulating and controlling things on a small scale, thereby entering a new field of physics which was bottomless, like low-temperature physics." He started with the question, can we "write the Lord's prayer on the head of a pin," and immediately extended the goal to the entire 24 volumes of the *Encyclopaedia Britannica*. By following the Gedanken Experiment, Feynman showed that there is no physical law against the realization of such goals: if you magnify the head of a pin by 25,000 diameters, its surface area is then equal to that of all the pages in the *Encyclopaedia Britannica*.

Feynman's dreams of writing small have all been fulfilled and even exceeded in the past decades. Since the advent of scanning tunneling microscopy, as introduced by Binnig and Rohrer, it has been repeatedly demonstrated that single atoms can not only be conveniently represented for the human eye but manipulated as well. Thus, it is conceivable to store all the books in the world (which Feynman estimates to contain 10^{15} bits of information) on the area of a credit card! The encyclopedia, having around 10^9 bits of information, can be written on about 1/100 the surface area of the head of a pin.

One need not look to atomic writing to achieve astonishing results: current microchips contain close to 100 million transistors. A small number of such chips could not only store large amounts of information (such as the *Encyclopaedia Britannica*); they can process it with GHz speed as well. To find a particular word takes just a few nanoseconds. Typical disk hard drives can store much more than the semiconductor chips, with a trade-off for retrieval speeds. Feynman's vision for storing and retrieving information on a small scale was very close to these numbers. He did not ask himself what the practical difficulties were in achieving these goals, but rather asked only what the principal limitations were. Even he could not possibly foresee the ultimate consequences of writing small and reading fast: the creation of the Internet. Sifting through large databases is, of course, what is done during Internet browsing. It is not only the microrepresentation of information that has led to the revolution we are witnessing but also the ability to browse through this information at very high speeds.

Can one improve current chip technology beyond the achievements listed above? Certainly! Further improvements are still expected just by scaling down known silicon technology. Beyond this, if it were possible to change the technology completely and create transistors the size of molecules, then one could fit hundreds of billions of transistors on a chip. Changing technology so dramatically is not easy and less likely to happen. A molecular transistor that is as robust and efficient as the existing ones is beyond current implementation capabilities; we do not know how to achieve such densities without running into problems of excessive heat generation and other problems related to highly integrated systems. However, Feynman would not be satisfied that we have exhausted our options. He still points to the room that opens if the third dimension is used. Current silicon technology is in its essence (with respect to the transistors) a planar technology. Why not use volumes, says Feynman, and put all books of the world in the space of a small dust particle? He may be right, but before assessing the chances of this happening, I would like to take you on a tour to review some of the possibilities and limitations of current planar silicon technology.

2.2   Tyranny at the Top

Yes, we do have plenty of room at the bottom. However, just a few years after Feynman's vision was published, J. Morton from Bell Laboratories noticed what he called the *tyranny of large systems*. This tyranny arises from the fact that scaling is, in general, not part of the laws of nature. For example, we know that one cannot hold larger and larger weights with a rope by making the rope thicker and thicker. At some point the weight of the rope itself comes into play, and things may get out of hand. As a corollary, why should one, without such difficulty, be able to make transistors smaller and smaller and, at the same time, integrate more of them on a chip? This is a crucial point that deserves some elaboration.

It is often said that all we need is to invent a new type of transistor that scales to atomic size. The question then arises: did the transistor, as invented in 1947, scale to the current microsize? The answer is no! The point-contact transistor, as it was invented by Bardeen and Brattain, was much smaller than a vacuum tube. However, its design was not suitable for aggressive scaling. The field-effect transistor, based on planar silicon technology and the hetero-junction interface of silicon and silicon dioxide with a metal on top (MOS technology), did much better in this respect. Nevertheless, it took the introduction of many new concepts (beginning with that of an inversion layer) to scale transistors to the current size. This scalability alone would still not have been sufficient to build large integrated systems on a chip. Each transistor develops heat when operated, and a large number of them may be better used as a soldering iron than for computing. The saving idea was to use both electron and hole-inversion layers to form the CMOS technology. The transistors of this technology create heat essentially only during switching operation, and heat generation during steady state is very small. A large system also requires interconnection of all transistors using metallic "wires." This becomes increasingly problematic when large numbers of transistors are involved, and many predictions have been made that it could not be done beyond a certain critical density of transistors. It turned out that there never was such a critical density for interconnection, and we will discuss the very interesting reason for this below. Remember that Feynman never talked about the tyranny at the top. He only was interested in fundamental limitations. The exponential growth of silicon technology with respect to the numbers of transistors on a chip seems to prove Feynman right, at least up to now. How can this be if the original transistors were not scalable? How could one always find a modification that permitted further scaling?

One of the reasons for continued miniaturization of silicon technology is that its basic idea is very flexible: use solids instead of vacuum tubes. The high density of solids permits us to create very small structures without hitting the atomic limit. Gas molecules or electrons in tubes have a much lower density than electrons or atoms in solids typically have. One has about 10^{18} atoms in a cm^3 of gas but 10^{23} in a cm^3 of a solid. Can one therefore go to sizes that would contain only a few hundred atoms with current silicon technology? I believe not. The reason is that current technology is based on the doping of silicon with donors and acceptors to create electron- and hole-inversion layers. The doping densities are much lower than the densities of atoms in a solid, usually below 10^{20} per cm^3. Therefore, to go to the ultimate limits of atomic size, a new type of transistor, without doping, is needed. We will discuss such possibilities below. But even if we have such transistors, can they be interconnected? Interestingly enough, interconnection problems have always been overcome in the past. The reason was that use of the third dimension has been made for interconnects. Chip designers have used the third dimension — not to overcome the limitations that two dimensions place on the number of transistors, but to overcome the limitations that two dimensions present for interconnecting the transistors. There is an increasing number of stacks of metal interconnect layers on chips — 2, 5, 8. How many can we have? (One can also still improve the conductivity of the metals in use by using, for example, copper technology.)

Pattern generation is, of course, key for producing the chips of silicon technology and represents another example of the tyranny of large systems. Chips are produced by using lithographic techniques. Masks that contain the desired pattern are placed above the chip material, which is coated with photosensitive layers that are exposed to light to engrave the pattern. As the feature sizes become smaller and smaller, the wavelength of the light needs to be reduced. The current work is performed in the extreme ultraviolet, and future scaling must overcome considerable obstacles. Why can one not use the atomic resolution of scanning tunneling microscopes? The reason is, of course, that the scanning process takes time; and this would make efficient chip production extremely difficult. One does need a process that works "in parallel" like photography. In principle there are many possibilities to achieve this, ranging from the use of X-rays to electron and ion beams and even self-organization of patterns in materials as known in chemistry and biology. One cannot see principal limitations here that would impede further scaling. However, efficiency and expense of production do represent considerable tyranny and make it difficult to predict what course the future will take. If use is made of the third dimension, however, optical lithography will go a long way.

Feynman suggested that there will be plenty of room at the bottom only when the third dimension is used. Can we also use it to improve the packing density of transistors? This is not going to be so easy. The current technology is based on a silicon surface that contains patterns of doping atoms and is topped by silicon dioxide. To use the third dimension, a generalization of the technology is needed. One would need another layer of silicon on top of the silicon dioxide, and so forth. Actually, such technology does already exist: silicon-on-insulator (SOI) technology. Interestingly enough, some devices that are currently heralded by major chip producers as devices of the future are SOI transistors. These may be scalable further than current devices and may open the horizon to the use of the third dimension. Will they open the way to unlimited growth of chip capacity? Well, there is still heat generation and other tyrannies that may prevent the basically unlimited possibilities that Feynman predicted. However, billions of dollars of business income have overcome most practical limitations (the tyranny) and may still do so for a long time to come. Asked how he accumulated his wealth, Arnold Beckman responded: "We built a pH-meter and sold it for three hundred dollars. Using this income, we built two and sold them for $600 … and then 4, 8, … ." This is, of course, the well-known story of the fast growth of a geometric series as known since ages for the rice corns on the chess board. Moore's law for the growth of silicon technology is probably just another such example and therefore a law of business rather than of science and engineering. No doubt, it is the business income that will determine the limitations of scaling to a large extent. But then, there are also new ideas.

2.3 New Forms of Switching and Storage

Many new types of transistors or switching devices have been investigated and even mass fabricated in the past decades. Discussions have focused on GaAs and III-V compound materials because of their special properties with respect to electron speed and the possibility of creating lattice-matched interfaces and layered patterns of atomic thickness. Silicon and silicon-dioxide have very different lattice constants (spacing between their atoms). It is therefore difficult to imagine that the interface between them can be electronically perfect. GaAs and AlAs on the other side have almost equal lattice spacing, and two crystals can be perfectly placed on top of each other. The formation of superlattices of such layers of semiconductors has, in fact, been one of the bigger achievements of recent semiconductor technology and was made possible by new techniques of crystal growth (molecular beam epitaxy, metal organic chemical vapor deposition, and the like). Quantum wells, wires, and dots have been the subject of extremely interesting research and have enriched quantum physics for example, by the discovery of the Quantum Hall Effect and the Fractional Quantum Hall effect. Use of such layers has also brought significant progress to semiconductor electronics. The concept of modulation doping (selective doping of layers, particularly involving pseudomorphic InGaAs) has led to modulation-doped transistors that hold the current speed records and are used for microwave applications. The removal of the doping to neighboring layers has permitted the creation of the highest possible electron mobilities and velocities. The effect of resonant tunneling has also been shown to lead to ultrafast devices and applications that reach to infrared frequencies, encompassing in this way both optics and electronics applications. When it comes to large-scale integration, however, the tyranny from the top has favored silicon technology. Silicon dioxide, as an insulator, is superior to all possible III-V compound materials; and its interface with silicon can be made electronically perfect enough, at least when treated with hydrogen or deuterium.

When it comes to optical applications, however, silicon is inefficient because it is an indirect semiconductor and therefore cannot emit light efficiently. Light generation may be possible by using silicon. However, this is limited by the laws of physics and materials science. It is my guess that silicon will have only limited applications for optics, much as III-V compounds have for large-scale integrated electronics. III-V compounds and quantum well layers have been successfully used to create efficient light-emitting devices including light-emitting and semiconductor laser diodes. These are ubiquitous in every household, e.g., in CD players and in the back-lights of cars. New forms of laser diodes, such as the so-called vertical cavity surface emitting laser diodes (VCSELs), are even suitable to relatively large integration. One can put thousands and even millions of them on a chip. Optical pattern generation has made great advances by use of selective superlattice intermixing (compositionally disordered III-V compounds and superlattices have a different index of refraction) and by other methods. This is an area in great flux and with many possibilities for miniaturization. Layered semiconductors and quantum well structures have also led to new forms of lasers such as the quantum cascade laser. Feynman mentioned in his paper the use of layered materials. What would he predict for the limits of optical integration and the use of quantum effects due to size quantization in optoelectronics?

A number of ideas are in discussion for new forms of ultrasmall electronic switching and storage devices. Using the simple fact that it takes a finite energy to bring a single electron from one capacitor plate to the other (and using tunneling for doing so), single-electron transistors have been proposed and built. The energy for this single-electron switching process is inversely proportional to the area of the capacitor. To achieve energies that are larger than the thermal energy at room temperature (necessary for robust operation), extremely small capacitors are needed. The required feature sizes are of the order of one nanometer. There are also staggering requirements for material purity and perfection since singly charged defects will perturb operation. Nevertheless, Feynman may have liked this device because the limitations for its use are not due to physical principles. It also has been shown that memory cells storing only a few electrons do have some very attractive features. For example, if many electrons are stored in a larger volume, a single material defect can lead to unwanted discharge of the whole volume. If, on the other hand, all these electrons are stored in a larger number of quantum dots (each carrying few electrons), a single defect can discharge only a single dot, and the remainder of the stored charge stays intact.

Two electrons stored on a square-shaped "quantum dot" have been proposed as a switching element by researchers at Notre Dame. The electrons start residing in a pair of opposite corners of the square and are switched to the other opposite corner. This switching can be effected by the electrons residing in a neighboring rectangular dot. Domino-type effects can thus be achieved. It has been shown that architectures of cellular neural networks (CNNs) can be created that way as discussed briefly below.

A new field referred to as *spintronics* is developing around the spin properties of particles. Spin properties have not been explored in conventional electronics and enter only indirectly, through the Pauli principle, into the equations for transistors. Of particular interest in this new area are particle pairs that exhibit quantum entanglement. Consider a pair of particles in a singlet spin-state sent out to detectors or spin analyzers in opposite directions. Such a pair has the following remarkable properties: measurements of the spin on each side separately give random values of the spin (up/down). However, the spin of one side is always correlated to the spin on the other side. If one is up, the other is down. If the spin analyzers are rotated relative to each other, then the result for the spin pair correlation shows rotational symmetry. A theorem of Bell proclaims such results incompatible with Einstein's relativity and suggests the necessity of instantaneous influences at a distance. Such influences do not exist in classical information theory and are therefore considered a quantum addition to classical information. This quantum addition provides part of the novelty that is claimed for possible future quantum computers. Spintronics and entanglement are therefore thought to open new horizons for computing.

Still other new device types use the wave-like nature of electrons and the possibility to guide these waves by externally controllable potential profiles. All of these devices are sensitive to temperature and defects, and it is not clear whether they will be practical. However, new forms of architectures may open new possibilities that circumvent the difficulties.

2.4 New Architectures

Transistors of the current technology have been developed and adjusted to accommodate the tyranny from the top, in particular the demands set forth by the von Neuman architecture of conventional computers. It is therefore not surprising that new devices are always looked at with suspicion by design engineers and are always found wanting with respect to some tyrannical requirement. Many regard it extremely unlikely that a completely new device will be used for silicon chip technology. Therefore, architectures that deviate from von Neuman's principles have received increasing attention. These architectures invariably involve some form of parallelism. Switching and storage is not localized to a single transistor or small circuit. The devices are connected to each other, and their collective interactions are the basis for computation. It has been shown that such collective interactions can perform some tasks in ways much superior to von Neuman's sequential processing.

One example for such new principles is the cellular neural network (CNN) type of architectures. Each cell is connected by a certain coupling constant to its nearest neighbors, and after interaction with each other, a large number of cells settle on a solution that hopefully is the desired solution of a problem that cannot easily be done with conventional sequential computation. This is, of course, very similar to the advantages of parallel computation (computation by use of more than one processor) with the difference that it is not processors that interact and compute in parallel but the constituent devices themselves. CNNs have advantageously been used for image processing and other specialized applications and can be implemented in silicon technology. It appears that CNNs formed by using new devices, such as the coupled square quantum dots discussed above, could (at least in principle) be embedded into a conventional chip environment to perform a certain desired task; and new devices could be used that way in connection with conventional technology. There are at least three big obstacles that need to be overcome if this goal should be achieved. The biggest problem is posed by the desire to operate at room temperature. As discussed above, this frequently is equivalent to the requirement that the single elements of the CNN need to be extremely small, on the order of one nanometer. This presents the second problem — to create such feature sizes by a lithographic process. Third, each element of the CNN needs to be virtually perfect and free of defects that would impede its operation. Can one create such a CNN by the organizing and self-organizing principles of chemistry on semiconductor surfaces? As Dirac once said (in connection with difficult problems), "one must try." Of course, it will be tried only if an important problem exists that defies conventional solution. An example would be the cryptographically important problem of factorizing large numbers. It has been shown that this problem may find a solution through quantum computation.

The idea of quantum computation has, up to now, mainly received the attention of theoreticians who have shown the superior power of certain algorithms that are based on a few quantum principles. One such principle is the unitarity of certain operators in quantum mechanics that forms a solid basis for the possibility of quantum computing. Beyond this, it is claimed that the number of elements of the set of parameters that constitutes quantum information is much larger than the comparable set used in all of classical information. This means there are additional quantum bits (qubits) of information that are not covered by the known classical bits. In simpler words, there are instantaneous action at a distance and connected phenomena, such as quantum teleportation, that have not been used classically but can be used in future quantum information processing and computation. These claims are invariably based on the theorem of Bell and are therefore subject to some criticism. It is well known that the Bell theorem has certain loopholes that can be closed only if certain time dependencies of the involved parameters are excluded. This means that even if the Bell theorem were general otherwise, it does not cover the full classical parameter space. How can one then draw conclusions about the number of elements in parameter sets for classical and quantum information? In addition, recent work has shown that the Bell theorem excludes practically all time-related parameters — not only those discussed in the well-known loopholes. What I want to say here is that the very advanced topic of quantum information complexity will need further discussion even about its foundations. Beyond this, obstacles exist for implementation of qubits due to the tyranny from the top. It is necessary to have a reasonably large number of qubits in order to implement the quantum computing algorithms and make them applicable to large problems. All of these qubits need to be connected in a quantum mechanical coherent way. Up to now, this coherence has always necessitated the use of extremely low temperatures, at least when electronics (as opposed to optics) is the basis for implementation. With all these difficulties, however, it is clear that there are great opportunities for solving problems of new magnitude by harnessing the quantum world.

2.5   How Does Nature Do It?

Feynman noticed that nature has already made use of nanostructures in biological systems with greatest success. Why do we not copy nature? Take, for example, biological ion channels. These are tiny pores formed by protein structures. Their opening can be as small as a few one-tenths of a nanometer. Ion currents are controlled by these pores that have opening and closing gates much as transistors have. The on/off current ratio of ion channels is practically infinite, which is a very desirable property for large systems. Remember that we do not want energy dissipation when the system is off. Transistors do not come close to an infinite on/off ratio, which represents a big design problem. How do the ion channels do it? The various gating mechanisms are not exactly understood, but they probably involve changes in the aperture of the pore by electrochemical mechanisms. Ion channels do not only switch currents perfectly. They also can choose the type of ions they let through and the type they do not. Channels perform in this way a multitude of functions. They regulate our heart rate, kill bacteria and cancer cells, and discharge and recharge biological neural networks, thus forming elements of logic and computation. The multitude of functions may be a great cure for some of the tyranny from the top as Jack Morton has pointed out in his essay "From Physics to Function." No doubt, we can learn in this respect by copying nature. Of course, proteins are not entirely ideal materials when it comes to building a computer within the limits of a preconceived technology. However, nature does have an inexpensive way of pattern formation and replication — a self-organizing way. This again may be something to copy. If we cannot produce chip patterns down to nanometer size by inexpensive photographic means, why not produce them by methods of self-organization? Can one make ion channels out of materials other than proteins that compare more closely to the solid-state materials of chip technology? Perhaps carbon nanotubes can be used. Material science has certainly shown great inventiveness in the past decades.

Nature also has no problems in using all three dimensions of space for applying its nanostructures. Self-organization is not limited to a plane as photography is. Feynman's ultimate frontier of using three dimensions for information storage is automatically included in some biological systems such as, for example, neural networks. The large capacity and intricate capability of the human brain derives, of course, from this fact.

The multitude of nanostructure functionalities in nature is made possible because nature is not limited by disciplinary boundaries. It uses everything, whether physics or chemistry, mechanics or electronics — and yes, nature also uses optics, e.g., to harvest energy from the sun. I have not covered nanometer-size mechanical functionality because I have no research record in this area. However, great advances are currently made in the area of nanoelectromechanical systems (NEMS). It is no problem any more to pick up and drop atoms, or even to rotate molecules. Feynman's challenge has been far surpassed in the mechanical area, and even his wildest dreams have long since become reality. Medical applications, such as the insertion of small machinery to repair arteries, are commonplace. As we understand nature better, we will not only be able to find new medical applications but may even improve nature by use of special smart materials for our bodies. Optics, electronics, and mechanics, physics, chemistry, and biology need to merge to form generations of nanostructure technologies for a multitude of applications.

However, an area exists in which man-made chips excel and are superior to natural systems (if man-made is not counted as natural). This area relates to processing speed. The mere speed of a number-crunching machine is unthinkable for the workings of a biological neural network. To be sure, nature has developed fast processing; visual evaluations of dangerous situations and recognition of vital patterns are performed with lightening speed by some parallel processing of biological neural networks. However, when it comes to the raw speed of converting numbers, which can also be used for alphabetical ordering and for a multitude of algorithms, man-made chips are unequaled. Algorithmic speed and variability is a very desirable property, as we know from browsing the Internet, and represents a great achievement in chip technology.

Can we have both —the algorithmic speed and variability of semiconductor-based processors and, at the same time, three-dimensional implementations and the multitude of functionality as nature features it in her nanostructure designs? I would not dare to guess an answer to this question. The difficulties are staggering! Processing speed seems invariably connected to heat generation. Cooling becomes increasingly difficult when three-dimensional systems are involved and the heat generation intensifies.

But then, there are always new ideas, new materials, new devices, new architectures, and altogether new horizons. Feynman's question as to whether one can put the *Encyclopaedia Britannica* on the head of a pin has been answered in the affirmative. We have proceeded to the ability to sift through the material and process the material of the encyclopedia with lightning speed. We now address the question of whether we can process the information of three-dimensional images within the shortest of times, whether we can store all the knowledge of the world in the smallest of volumes and browse through gigabits of it in a second. We also proceed to the question of whether mechanical and optical functionality can be achieved on such a small scale and with the highest speed. Nature has shown that the smallest spatial scales are possible. We have to search for the greatest variety in functionality and for the highest possible speed in our quest to proceed in science from what is possible in principle to a function that is desirable for humanity.

3

National Nanotechnology Initiative — Past, Present, Future

3.1	Introduction	3-1
3.2	Identifying a Megatrend in Science and Technology	3-2
	What Is Nanotechnology? • Nanotechnology — A Key Component of Converging Technologies	
3.3	Key Factors in Establishing NNI (about 2000)	3-5
3.4	The Beginning of NNI	3-6
3.5	NNI at Five Years (2001–2005)	3-10
3.6	Technical Challenges in 2005–2020	3-16
3.7	New Science and Engineering, New Governance Approach	3-19
	Closing Remarks	
	Acknowledgments	3-21
	References	3-21
	Appendices A–D	3-22

Mihail C. Roco

National Science Foundation and
National Nanotechnology Initiative

3.1 Introduction

This chapter presents the genesis of the National Nanotechnology Initiative, its long-term view (2000–2020), its current status, and its likely evolution.

Nanoscale science and engineering activities are flourishing in the United States. The National Nanotechnology Initiative (NNI) is a long-term research and development (R&D) program that began in fiscal year (FY) 2001, and today coordinates 25 departments and independent agencies, including the National Science Foundation, the Department of Defense, the Department of Energy, the National Institutes of Health, the National Institute of Standards and Technology, and the National Aeronautical and Space Administration. The total R&D investment in FYs 2001–2006 was over $5 billion, increasing from the annual budget of $270 million in 2000 to $1.3 billion including congressionally directed projects in FY 2006. An important outcome is the formation of an interdisciplinary nanotechnology community with about 50,000 contributors A flexible R&D infrastructure with over 60 large centers, networks, and user facilities has been established since 2000. This expanding industry consists of more than 1500 companies with nanotechnology products with a value exceeding $40 billion at an annual rate of growth estimated at about 25%. With such growth and complexity, participation of a coalition of academic organizations, industry, businesses, civil organizations, government, and NGOs in nanotechnology development becomes essential as an alternative to the centralized approach. The role of government continues in basic research and education but its emphasis is changing, while the private sector becomes increasingly dominant in funding nanotechnology applications.

The NNI plan proposed in 1999 has led to a synergistic, accelerated, and interdisciplinary development of the field, and has motivated academic and industrial communities at the national and global levels. A key factor that contributed to establishing NNI in 2000 was the preparation work for identifying the core nanotechnology concepts and challenges. Secondly, the orchestrated effort to assemble fragmented disciplinary contributions and application–domain contributions has led to broad support from various stakeholders. Thirdly, the long-term view in planning and setting priorities was essential in the transformative governance of nanotechnology. Nanotechnology holds the promise of increasing efficiency in traditional industries and bringing radically new applications through emerging technologies. Several potential R&D targets by 2015–2020 are presented in this chapter.

The promise of nanotechnology, however, will not be realized by simply supporting research. A specific governing approach is necessary for emerging technologies and in particular for nanotechnology by considering its fundamental and broad implications. Optimizing societal interactions, R&D policies, and risk governance for nanotechnology development can enhance economic competitiveness and democratization.

3.2 Identifying a Megatrend in Science and Technology

3.2.1 What Is Nanotechnology?

Nanotechnology operates at the first level of organization of atoms and molecules for both living and anthropogenic systems. This is where the properties and functions of all systems are defined. Such fundamental control promises a broad and revolutionary technology platform for industry, biomedicine, environmental engineering, safety and security, food, water resources, energy conversion, and countless other areas.

The first definition of nanotechnology to achieve some degree of international acceptance was developed after consultation with experts in over 20 countries in 1987–1898 (Siegel et al., 1999; Roco et al., 2000). However, despite its importance, there is no globally recognized definition. Any nanotechnology definition would include three elements:

1. The size range of the material structures under consideration — the intermediate length scale between a single atom or molecule, and about 100 molecular diameters or about 100 nm. Here we have the transition from individual to collective behavior of atoms. This length scale condition alone is not sufficient because all natural and manmade systems have a structure at the nanoscale.
2. The ability to measure and restructure matter at the nanoscale; without it we do not have new understanding and a new technology; such ability has been reached only partially so far, but significant progress was achieved in the last five years.
3. Exploiting properties and functions specific to nanoscale as compared to the macro- or microscales; this is a key motivation for researching nanoscale.

According to the National Science Foundation and NNI, nanotechnology is the ability to understand, control, and manipulate matter at the level of individual atoms and molecules, as well as at the "supramolecular" level involving clusters of molecules (in the range of about 0.1 to 100 nm), in order to create materials, devices, and systems with fundamentally new properties and functions because of their small structure. The definition implies using the same principles and tools to establish a unifying platform for science and engineering at the nanoscale, and employing the atomic and molecular interactions to develop efficient manufacturing methods.

There are at least three reasons for the current interest in nanotechnology. First, the research is helping us fill a major gap in our *fundamental knowledge of matter*. At the small end of the scale — single atoms and molecules — we already know quite a bit from using tools developed by conventional physics and chemistry. And at the large end, likewise, conventional chemistry, biology, and engineering have taught us about the bulk behavior of materials and systems. Until now, however, we have known much less about the intermediate nanoscale, which is the natural threshold where all living and manmade systems work. The basic properties and functions of material structures and systems are defined here and, even more importantly, can be changed as a function of the organization of matter via "weak" molecular interactions (such as hydrogen bonds, electrostatic dipole, van der Waals forces, various surface forces, electro-fluidic forces, etc.). The intellectual drive toward smaller dimensions was accelerated by the discovery of size-dependent novel properties and phenomena. Only since 1981 have we been able to measure the size of a cluster of atoms on a surface (IBM, Zurich), and begun to provide better models for chemistry and biology self-organization and self-assembly. Ten years later, in 1991, we were able to move atoms on surfaces (IBM, Almaden). And after ten more years, in 2002, we assembled molecules by physically positioning the component atoms. Yet, we cannot visualize or model with proper spatial and temporal accuracy a chosen domain of engineering or biological relevance at the nanoscale. We are still at the beginning of this road.

A second reason for the interest in nanotechnology is that nanoscale phenomena hold the promise for *fundamentally new applications*. Possible examples include chemical manufacturing using designed molecular assemblies, processing of information using photons or electron spin, detection of chemicals or bio-agents using only a few molecules, detection and treatment of chronic illnesses by subcellular interventions, regenerating tissue and nerves, enhancing learning and other cognitive processes by understanding the "society" of neurons, and cleaning contaminated soils with designed nanoparticles. Using input from industry and academic experts in the U.S., Asia Pacific countries, and Europe between 1997 and 1999, we have projected that $1 trillion in products incorporating nanotechnology and about 2 million jobs world-wide will be affected by nanotechnology by 2015 (Roco and Bainbridge, 2001). Extrapolating from information technology, where for every worker, another 2.5 jobs are created in related areas, nanotechnology has the potential to create 7 million jobs overall by 2015 in the global market. Indeed, the first generation of nanostructured metals, polymers, and ceramics have already entered the commercial marketplace.

Finally, a third reason for the interest is the *beginning of industrial prototyping and commercialization* and that governments around the world are pushing to develop nanotechnology as rapidly as possible. Coherent, sustained R&D programs in the field have been announced by Japan (April 2001), Korea (July 2001), EC (March 2002), Germany (May 2002), China (2002), and Taiwan (September 2002). However, the first and largest such program was the U.S. NNI, announced in January 2000.

3.2.2   Nanotechnology — A Key Component of Converging Technologies

Science and engineering are the primary drivers of global technological competition. Unifying science based on the unifying features of nature at the nanoscale provides a new foundation for knowledge, innovation, and integration of technology. Revolutionary and synergistic advances at the interfaces between previously separated fields of science, engineering, and areas of relevance are poised to create nano-bio-info-cogno (NBIC) transforming tools, products, and services.

There is a longitudinal process of convergence and divergence in major areas of science and engineering (Roco, 2002; Roco and Bainbridge, 2003). For example, the convergence of sciences at the macroscale was proposed during the Renaissance, and it was followed by narrow disciplinary specialization in science and engineering in the 18th to 20th centuries. The convergence at the nanoscale reached its strength in about 2000, and one may estimate a divergence of the nanosystem architectures in the following decades. Current convergence at the nanoscale is happening because of the use of the same elements of analysis (that is, atoms and molecules) and of the same principles and tools, as well as the ability to make cause-and-effect connections from simple components to higher-level architectures. In nano realms, the phenomena/processes cannot be separated, and there is no need for discipline-specific averaging methods. In 2000, convergence had been reached at the nanoworld (Figure 3.1a) because typical phenomena in material nanostructures could be measured and understood with a new set of tools, and nanostructures have been identified as the foundation of biological systems, nanomanufacturing, and communications. A new challenge is building systems from the nanoscale that will require the combined use of nanoscale laws, biological principles, information technology, and system integration. Then, after 2020, one may expect divergent trends as a function of the system architecture. Several possible divergent trends are system architectures based on: guided molecular and macromolecular assembling; robotics; biomimetics; and evolutionary approaches. While in 2000 we assumed to have been at the beginning of the "S" development curve, we also may estimate that in 2020 to be in the fast ascend.

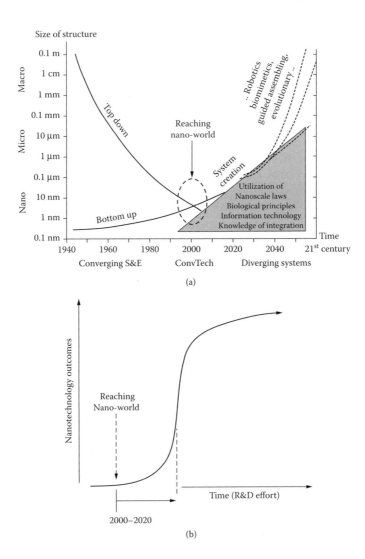

(a)

(b)

FIGURE 3.1 (a) Reaching the nanoworld (about 2000) and "converging technologies" approach for system creation from the nanoscale (2000–2020) towards new paradigms for nanosystem architectures in applications (after 2020); (b) Suggested nanotechnology R&D "S" development curve (schematic).

The transforming effect of converging technologies and particularly of emerging NBIC on society is expected to be large, not only because of the high rate of change in each domain and their synergism with global effect on science and engineering, but also because we are reaching qualitative thresholds in the advancement of each of the four domains. In the U.S., we have started two national initiatives on Information Technology Research (ITR, in 1999, about $2 billion in FY 2006) and National Nanotechnology Research (NNI, in 2000, reaching about $1.3 billion in FY 2006) (Figure 3.2). ITR and NNI provide the technological "push" with broad science and engineering platforms. Realizing the human potential, "the pull," would include the biotechnology and cognitive technologies. Several topical, agency-specific programs have been initiated in the field of biotechnology, such as NIH Roadmaps (including genome), NSF's Biocomplexity, and USDA's Roadmap. There was no national initiative on biotechnology and no large-scale programs on cognition, except for the core research programs in social, behavioral and economic sciences and centers for science or learning at NSF. There was a need to balance this situation, and a partial response was, in 2003, the launch of the Human and Social Dynamics NSF priority area. No special interagency program was established on system approach and cognitive sciences.

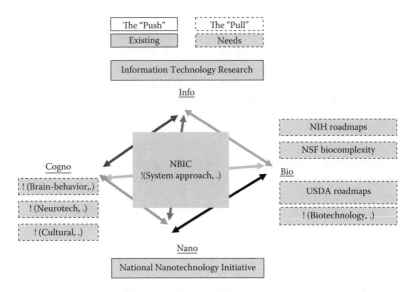

FIGURE 3.2 Converging technologies (NBIC) transforming tools: a survey of national R&D programs in 2000.

With proper attention to ethical issues and societal needs, these converging technologies could determine a tremendous improvement in human abilities, societal outcomes, the nation's productivity, and the quality of life of revolutionary products and services.

3.3 Key Factors in Establishing NNI (about 2000)

Key factors that have contributed to establishing NNI around the year 2000 and the rapid growth of nanotechnology are:

- The preparatory work for identifying core nanotechnology concepts encompassing all disciplines, including the definition of nanotechnology and what are the key research directions.
- The orchestrated effort to assemble fragmented disciplinary contributions and application-domain contributions and get broad and bottom-up support from various contributing communities and other stakeholders.
- Preparing the initiative as a science project. This included the long-term view (2000 to 2020) in planning and setting priorities on three time scales (5 years for strategic planning, 1 year for annual budgets, 1 month for interagency meetings on programs and collaborations decisions) and three levels (national, agency, and R&D program). The initial R&D focus in the first strategic plan (2001 to 2005) has been on fundamental research and "horizontal" multidisciplinary R&D with relevance to multiple application areas. A transition to "vertical" industrial development from the basic concepts is a focus for the second plan (2006 to 2010). A policy of inclusion and partnerships has been promoted, including international collaboration. The R&D projects have been aligned with societal needs and aspirations from the beginning, with a proactive role in the political and international context. The governing approach was data-driven and transformative (Roco, 2005a).

Fundamental changes envisioned through nanotechnology have required a *long-term R&D vision*. A two-decade timescale was planned for transitioning from the focus on passive nanostructures in 2001 to 2005 to molecular active nanosystems after 2015. To meet the challenges, the NNI was created, and a new approach for its governance was attempted (Roco, 2006).

The NNI bottom-up approach solidified the community of scientists to support the redirection of funds toward the new area of nanotechnology. Doing it top-down might have fractured our community and led to in-fighting. The orchestration of all the government agencies behind this concept was critical in getting the "jump start" that was achieved.

NNI was conceived as an *inclusive process* where various stakeholders would be involved. In 1999 we envisioned "a grand coalition" of academia, industry, government, states, local organizations, and the public that would advance nanotechnology. Twenty-five agencies covered most relevant areas of national interest in NNI in 2006, and industry already is investing more than the federal government for nanotechnology R&D in the U.S.

Personal observations made during research and interactions with the community in the 1980s helped me to pose the right questions. We identified nanotechnology as a "dormant" S&E opportunity, but with an "immense" potential. Creating a chorus to support nanotechnology, from 1990 to March 1999, was an important preliminary step.[1]

In the decade before the NNI, between 1990 and 2000, a main challenge was the search for the relevance of nanotechnology. We had to overcome three waves of skepticism. First, a concern was the limited relevance of the field and "pseudoscientific" claims. Then, it was the concern of large and unexpected consequences culminating with the risk of the so-called "grey-goo" scenarios. The third wave of concerns has been on environmental, health, and safety (EHS) implications; it arrived only later in 2002 to 2003, when industrial participation increased. And yet, the main issues related to long-term societal implications and human development reached the public and media by 2006.

The participation of multiple agencies is necessary because of the wide spectrum of the relevance of nanotechnology to the society. In November 1996, I organized a small group of researchers and experts from the government, including Stan Williams (Hewlett Packard), Paul Alivisatos (University of California, Berkeley), and Jim Murday (Naval Research Laboratory), and we started to do our homework in setting a vision for nanotechnology. We began with preparing supporting publications, including a report on research directions in 10 areas of relevance, despite low expectations of additional funding at that moment. In 1997–1998, we ran a program solicitation titled "Partnership in Nanotechnology: Functional Nanostructures" at NSF and we received feedback from the academic community. Also, we completed a worldwide study in academe, industry, and governments, together with a group of experts including Richard Siegel (Rensselaer Polytechnic Institute, then at Argonne National Laboratory) and Evelyn Hu (University of California, Santa Barbara). By the end of 1998, we had an understanding of the possibilities at the international level. The visits performed during that time interval were essential in developing an international acceptance for nanotechnology, and defining its place among existing disciplines.

NNI was prepared with the same rigor as a science project, between 1997 and 2000; we prepared a long-term vision for research and development (Roco et al., 2000) and we completed an international benchmarking of nanotechnology in academe, government, and industry (Siegel et al., 1999). Other milestones included a plan for the U.S. government investment (NSTC, 2000), a brochure explaining nanotechnology for the public (NSTC, 1999), and a report on the societal implications of nanoscience and nanotechnology (Roco and Bainbridge, 2001). More than 150 experts, almost equally distributed between academe, industry, and government, contributed in setting nanotechnology research directions, bringing into the dialogue experts like Richard Smalley (Rice University), Herb Goronkin (Motorola), and Meyya Mayyapan (NASA Ames). We distributed the brochure for the public to 30,000 organizations, proactively prepared a report for societal implications of nanoscience and nanotechnology, and tested the interest and capacity of the scientific community. This was the time to rally the interest of science and engineering communities and international support, as well as key industrial sectors.

3.4 The Beginning of NNI

On behalf of the interagency group, on March 11, 1999, in the historic Indian Hall at the White House's Office of Science and Technology Policy (OSTP), I proposed the NNI with a budget of half a billion dollars

---

[1]In 1990, I proposed nanoparticles synthesis and processing at high rates in the Emerging Technologies competition for a new programmatic topic at NSF. It was selected for funding, and became the first program in a federal agency dedicated to nanoscale science and engineering. The awards were for interdisciplinary groups of at least three coprincipal investigators in different disciplines. The program supported by four divisions at NSF was largely successful (the results were presented at grantees meetings, in 1994 and 1997).

for FY 2001. I was given 10 minutes to make the case. While two other topics were on the agenda of that meeting, nanotechnology captured the imagination of those present and discussions reverberated for about two hours. It was the first time that a forum at this level with representatives from the major federal R&D departments reached a decision to consider exploration of nanotechnology as a national priority. In parallel, over two dozen other competing topics were under consideration by OSTP for priority funding in FY 2001. We had the attention of Neal Lane, the presidential science advisor, and Tom Kalil, economic assistant to the president. However, few experts gave even a small chance to nanotechnology of becoming a national priority program. Despite this, after a long series of evaluations, NNI was approved and had a budget of $489 million in FY 2001 ($464 million proposed by six agencies and $25 million congressionally directed, see Table 3.1).

After that presentation, our focus changed. Because nanotechnology was not known to Congress or the Administration, establishing a clear definition of nanotechnology and communicating the vision to large communities and organizations took the center stage. Indeed, the period from March 1999 through to the end of the year was a time of very intense activity. Few experts gave even a small chance to nanotechnology for special funding by the White House. Nevertheless, with this proposal and the "homework" of studies completed, we focused our attention on the six major federal departments and agencies — the National Science Foundation (NSF), Department of Defense (DOD), Department of Energy (DOE), NASA, National Institutes of Health (NIH), and the National Institute of Standards and Technology (NIST) — that would place nanotechnology as a top priority during the summer of 1999.

We provided detailed technical input for two hearings in the Congress, in both the Subcommittee on Basic Science, Committee on Science, U.S. House of Representatives (June 22, 1999) and the Senate, and support was received from both parties. The preparatory materials included a full 200-page benchmarking report, 10-page research directions, and 1-page summary on immediate goals. After the hearing in the House, Nick Smith, the chair of the first public hearing in preparation of NNI, said "Now we have sufficient information to aggressively pursue nanotechnology funding." Rick Smalley came and testified despite his illness.

Then, the approval process moved to the Office of Management and Budget (OMB) (November 1999), Presidential Council of Advisors in Science and Technology (PCAST) (December 1999), and the Executive Office of the President (EOP, White House) (January 2000), and had supporting hearings in the House and Senate of the U.S. Congress (Spring 2000). In November 1999, the OMB recommended nanotechnology as the only new R&D initiative for FY 2001. On December 14, 1999, the PCAST highly recommended that the president fund nanotechnology R&D. Thereafter, it was a quiet month — we had been advised by the Executive Office of the President to restrain from speaking to the media about the topic because a White House announcement would be made. We prepared a draft statement. A video was being produced for the planned multimedia presentation, but we did not have time to complete it.

President Clinton announced the NNI at Caltech in January 2000 beginning with words such as "Imagine what could be done." He used only slides. After that speech, we moved firmly in preparing the federal plan for R&D investment, to identify the key opportunities and convincing potential contributors to be proactive. House and Senate hearings brought the needed recognition and feedback from Congress.

The selection of NNI at OMB, OSTP, and PCAST was in competition with other science and technology priorities for FY 2001, and only one topic — nanotechnology — was selected in the process.[2]

A challenge in the first years of the initiative with so many new developments was maintaining consistency, coherence, and original thinking.

Three names (nanotechnology definition, the name of the initiative — NNI, and of the National Nanotechnology Coordinating Office [NNCO]) were decided in the same time interval, 1999 to 2000. The name NNI was proposed on March 11, 1999, but it was under "further consideration" until the presidential announcement because of concerns from several professional societies and committees that the title did not include explicitly "science." We explained that we selected a simple name to show its relevance to society.

---

[2]I spoke to major professional societies (initially the American Chemical Society, then the Institute for Electric and Electronics Engineering, American Society of Mechanical Engineering, and American Institute of Chemical Engineering, and attended national meetings for the introduction of nanotechnology in about 20 countries.

TABLE 3.1 Contribution of Key Federal Departments and Agencies to NNI Investment 2000–2008

Federal Department or Agency	FY 2000 Actual ($M)	FY 2001 Actual ($M)	FY 2002 Actual ($M)	FY 2003 Actual ($M)	FY 2004 Actual ($M)	FY 2005 Actual ($M)	FY 2006 Actual ($M)	FY 2007 Estimate ($M)	FY 2008 Request ($M)
National Science Foundation (NSF)	97	150	204	221	256	335	360	373	390
Department of Defense (DOD)	70	125	224	322	291	252	324	317	375
Department of Energy (DOE)	58	88	89	134	202	208	231	293	331
National Institutes of Health (NIH)	32	40	59	78	106	165	192	170	203
National Institute of Standards and Technology (NIST)	8	33	77	64	77	79	78	89	97
National Aeronautics and Space Administration (NASA)	5	22	35	36	47	45	50	25	24
National Institute for Occupational Safety and Health (NIOSH)	—	—	—	—	—	3	4	5	5
Environmental Protection Agency (EPA)	—	5	6	5	5	7	5	9	10
Homeland Security (TSA)	—	—	2	1	1	1	2	2	1
Department of Agriculture (USDA: CSREES)	—	1.5	0	1	2	3	4	4	3
Department of Agriculture (USDA: Forest Service)							2	3	5
Department of Justice (DOJ)	—	1.4	1	1	2	2	1	1	1
Department of Transportation (DOT: FHVA)							1	1	1
Total, without Congr.-directed	270	464	697	862	989	1100	1241	1292	1445
(% of FY 2000 budget)	(100%)	(172%)	(258%)	(319%)	(366%)	(407%)	(%)	(%)	(%)
Congressionally-directed	—	25 (DOD)	40 (DOD)	80 (DOD)	103 (DOD)	~100 (DOD)	100 (DOD) 10 (NASA)	100 (DOD)	
Total, with Congr.-directed	270	489	737	942	1092	1200	1351	1392	1445
(% of FY 2000 budget)	(100%)	(181%)	(273%)	(349%)	(404%)	(444%)	(481%)	(515%)	(535%)

Note: Each FY begins on October 1 of the previous year and ends September 30 of the respective year.

The "Interagency Working Group on Nanoscale Science, Engineering and Technology" (IWGN) was established in 1998 (October 1998 to July 2000) as a crosscut working group in the National Science and Technology Council (NSTC), White House. In 1997, I was contacted by T. Khalil, who saw my publications on nanotechnology. Through NSF, I proposed an interagency group working across various topical committees. This was an important victory because the initial proposal in March 1998 was to have a working group under the NSTC's Materials Activities. Also, until March 11, 1999, we were advised to be cautious about interacting with media on nanotechnology risks. President Clinton announced the NNI at Caltech in January 2000. In July 2000, the White House elevated the IWGN to the level of "Nanoscale Science, Engineering and Technology" (NSET) Subcommittee of the NSTC's Committee on Technology with the role "to implement NNI" and I was appointed chairman. The Memorandum of Understanding to fund NNCO was signed by the last participating agency on January 17, 2001 on the last day of the Clinton Administration.

In the first year, the six agencies of the NNI invested about $490 million (including congressionally directed funding), only a few percentage points less than the tentative budget proposed on March 11, 1999. In FYs 2002 and 2003, NNI increased significantly, from 6 to 16 departments and agencies. The presidential announcement of NNI with its vision and program partially motivated or stimulated the international community. About 60 other countries have announced priority nanotechnology programs since the NNI announcement. It was as if nanotechnology had gone through a phase transition. What had once been perceived as blue-sky research of limited interest (or in the view of several groups, science fiction, or even pseudoscience), was now being seen as a key technology of the 21st century. The Bush Administration has increased the support for NNI, with higher presidential annual "budget requests" each year. The average annual rate of increase of the NNI budgets was over 35% (including congressionally directed funding) in the first five years. The structure of NNI programs in the first (FYs 2001–2005) and second (FYs 2006–2010) strategic plans is given in Appendix A. The list of NNI participating agencies in January 2006 is provided in Appendix B. Industry and the medical community embraced nanotechnology after 2002–2003 as a key competitive factor.

In 2000, we estimated a $1 trillion nanotechnology-related market of nanoproducts incorporating nanotechnology (Figure 3.3), and the demand for 2 million workers worldwide by 2015 — using input from industry and experts around the world. We also saw the increasing convergence of nanotechnology with modern biology, the digital revolution, and cognitive sciences in the first half of the 21st century.

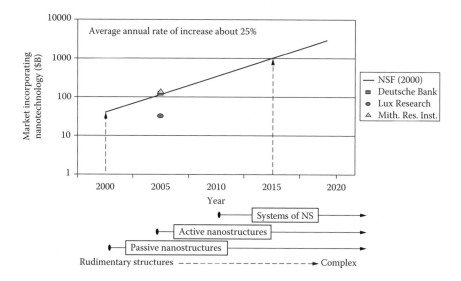

FIGURE 3.3 Estimation of the worldwide market incorporating nanotechnology (estimation made in 2000, at NSF; Roco and Bainbridge, 2001). These estimations were based on direct contacts with leading experts in large companies with related R&D programs in the U.S., Japan, and Europe, and the international study completed between 1997 and 1999 (Siegel et al., 1999).

After initially passing the House with a vote of 405-19 (H.R. 766), and then the Senate with unanimous support (S. 189) in November 2003, the "21st Century Nanotechnology R&D Act" was signed by the president on December 3, 2003. Bipartisan support is strong because nanotechnology progress is seen as "a higher purpose" beyond party affiliation. The Bush administration provided support and has increased the level of investment of NNI to about $1.3 billion in 2006. In 2007, congressional staff is planning to re-authorize the December 2003 Nanotechnology R&D Act.

3.5 NNI at Five Years (2001–2005)

There are major outcomes after five year (FYs 2001–2005) of the NNI. The R&D landscape for nanotechnology research and education has changed, advancing it from fragmented fields and questions such as "what is nanotechnology?" and "could it ever be developed?" to a highly competitive domain where the main question is "how can industry and medicine take advantage of it faster?" In only five years, nanoscience and nanotechnology have opened an era of integration of fundamental research and engineering from the atomic and molecular levels, increased technological innovation for economic manufacturing of products, and an enabling base for improving human health and cognitive abilities in the long term. For this reason, government investments worldwide for nanotechnology R&D have increased over fivefold in five years, rising to about $4.3 billion in 2005 from about $825 million in 2000, and all Fortune 500 companies in materials, electronics, and pharmaceuticals have made investments in nanotechnology after 2002. Of 30 Dow Jones companies, 19 have initiatives on application of nanotechnology in 2005. The NNI fuels these developments. By creating a "power house" of discoveries and innovations, the NNI has been the major driver for nanoscience and nanotechnology developments and applications in the U.S. and in the world. In 2005, NNI supported over 4000 projects and 60 new centers, networks, and user facilities in the U.S. About $10 billion were invested worldwide in 2006 by governments and industry for nanotechnology R&D. The vision of a decade ago has taken place.

Figure 3.4 shows the increase in the relative number of NSF awards in nanoscale science and engineering of the total NSF awards, which reached about 11% in 2005. On the same graph, one may see the increase, with some delay, in the relative number of nanotechnology-related papers in the top 20 cited journals (according to ISO), number of nanotechnology-related papers in three leading journals

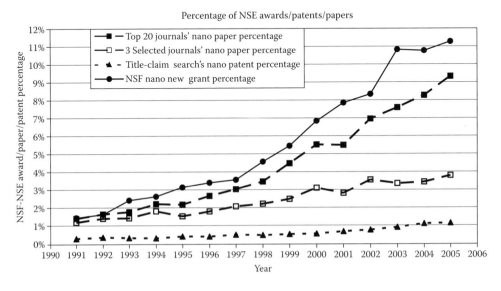

FIGURE 3.4 Timeline for the number of NSF awards, number of journal articles (ISO), and number of patents (USPTO) on nanoscale science and engineering published between 1991 and 2005. All documents were searched by the same keywords (as described in Huang et al., 2004) in the title and abstract, as well as in the "claims" for patents.

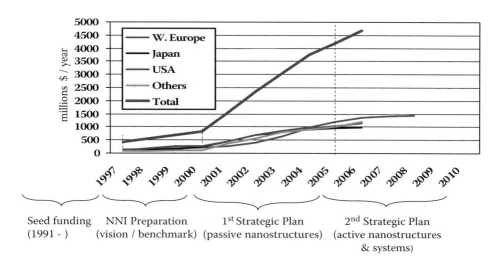

FIGURE 3.5 National government investments in nanotechnology R&D in the world and the U.S. in the past nine years (1997–2005) (see Table 3.3).

(*Science, Nature,* and *Proceedings of the National Academy of Science*), and number of nanotechnology patents at the U.S. Patent Office (USPTO). All documents were searched using the same method (searching for keywords in the title, abstract, and claims).

The NNI R&D program had a total investment of about $1.2 billion in FY 2005 (Table 3.1). Comparative budgets and investments per capita are given in Figure 3.5 and Table 3.2 to Table 3.4. About 65% of funds were dedicated to academic R&D institutions, 25% to government laboratories, and 10% to industry, of which 7% was for SBIR/STTR awards. The first five years (FYs 2001–2005) of NNI have led to significant science and engineering advances, have increased the confidence that nanotechnology development is one of the key technologies at the beginning of the 21st century, and have raised the challenges of responsible development, including EHS (Roco, 2003). The major achievements of NNI in the first five years are listed below.

(1) First, *NNI developed foundational knowledge in nanotechnology* with about 4000 projects funded during 2001 to 2005 in over 500 institutions (about 300 academic and over 200 small businesses and government laboratories) in all 50 states. Research is advancing toward systematic control of matter at the nanoscale faster than envisioned in 2000, when NNI was introduced with words like "Imagine what could be done in 20 to 30 years from now." In 2005, about 400 projects addressed molecular assembling and nanoscale devices and systems, tailoring molecules, and manipulating individual atoms. Such R&D projects were not feasible just five years ago. For example, Sam Stupp of Northwestern University has designed molecules for hierarchical self-assembling in desired materials. Alex Zettl of University of California at Berkeley has built the smallest nanomotor with an axis of a few nanometers, and Jim Heath at Caltech has analyzed and sensed cells as complex nanosystems. The time for reaching commercial

TABLE 3.2 Specific NNI R&D Expenditures Per Capita and Per GDP between 2000 and 2005

Index	FY 2000	FY 2001	FY 2002	FY 2003	FY 2004	FY 2005	FY 2006	FY 2007	FY 2008
GDP (trillion)	9.8	10.1	10.5	11.0	11.7	12.5			
Specific NNI R&D ($/capita) (assume 293 million of 2004)	0.9	1.7	2.5	3.2	3.7	4.2	4.6	4.8	4.9
Specific NNI R&D ($/$M GDP)	27.6	48.4	70.2	85.6	93.3	98.2			

Note: The NNI budgets include congressionally directed supplements.

TABLE 3.3 Estimated Government Nanotechnology R&D Expenditures, 1997–2006 ($Millions/Year)

Region	1997	1998	1999	2000	2001	2002	2003	2004	2005	2006	2007 Estimate	2008 Request
E.U.+	126	151	179	200	~225	~400	~650	~950	~1050	1150	—	—
Japan	120	135	157	245	~465	~720	~800	~900	~950	980	—	—
U.S.*	116	190	255	270	464	697	862	989	1200	1351	1397	1445
Others	70	83	96	110	~380	~550	~800	~900	~1000	1200	—	—
Total (% of 1997)	432	559	687	825	1534	2367	3112	3739	4200	4681		
	(100%)	(129%)	(159%)	(191%)	(355%)	(547%)	(720%)	(866%)	(972%)	(1083%)		

Note: National and E.U. funding is included. The E.U.+ includes countries in E.U. (15)/E.U.(25) and Switzerland (CH): the rate of exchange $1 = 1.1 euro until 2002, = 0.9 euro in 2003, and = 0.8 euro in 2004–2006; rate of exchange for Japan is $1 = 120 yen until 2002, = 110 yen in 2003, = 105 yen in 2004–2005; "Others" includes Australia, Canada, China, Eastern Europe, FSU, Israel, Korea, Singapore, Taiwan, and other countries with nanotechnology R&D; estimates use the nanotechnology definition as defined in the NNI (this definition does not include MEMS, microelectronics, or general research on materials) (see Roco, Williams and Alivisatos, 2000, Springer, also on http://nano.gov), and include the publicly reported government allocations spent in the respective financial years.

Note: An FY begins in the U.S. on October 1, and in most other countries, six month later around April 1; denotes the actual budget recorded at the end of the respective FY except for "Estimate" and "Request," without including congressionally directed budget.

TABLE 3.4   Specific NNI R&D Expenditure Per Capita and Per GDP between 2000 and 2005

Country/Region	Government Nanotech R&D, 2004 ($M)	Specific Nanotech R&D, 2004 ($/Capita)	Specific Nanotech R&D, 2004 ($/$M GDP)
U.S.	1100	3.7	93
E.U.–25	~1050	2.3	86
Japan	~950	7.4	250
China	~250	0.2	31
Korea	~300	6.2	350
Taiwan	~110	4.7	208

Note: The NNI budgets include congressionally directed supplements.
Source: From: Roco, 2005b.

prototypes has been reduced by at least a factor of two for key application areas such as detection of cancer, molecular electronics, and nanostructure-reinforced composites.

(2) NNI has been recognized for *creating an interdisciplinary nanotechnology community in the U.S.* According to an external review committee at NSF (NSF COV, 2004), "Two significant and enduring results have emerged from this investment: They are the creation of a nanoscale science and engineering community, and the fostering of a strong culture of interdisciplinary research."

(3) *Nanotechnology education and outreach* has impacted over 10,000 graduate students and teachers in 2005. Systemic changes are in preparation for education, by earlier introduction of nanoscience and reversing the "pyramid of science" with understanding of the unity of nature at the nanoscale from the beginning (Roco, 2003). Nanotechnology education has been expanded systematically to earlier education, including undergraduate (Nanotechnology Undergraduate Education program with over 80 awards since 2002) and high schools (since 2003), as well as informal education, science museums and the public. All major science and engineering colleges in the U.S. have introduced courses related to nanoscale science and engineering in the last five years. NSF has established recently three other networks with national outreach addressing education and societal dimensions: (1) The Nanoscale Center for Learning and Teaching aims to reach 1 million students in all 50 states in the next five years; (b) The Nanoscale Informal Science Education network will develop, among others, about 100 nanoscale science and technology museum sites in the next five years; (c) The Network on Nanotechnology in Society was established in September 2005 with four nodes at Arizona State University, University of California at Santa Barbara, University of South Carolina, and Harvard University. The Network will address both short-term and long-term societal implications of nanotechnology, as well as public engagement. All 15 Nanoscale Science and Engineering Centers sponsored by NSF have strong education and outreach activities. The exhibitions organized by the Cornell Nanobiotechnology Center are good illustrations of outreach efforts. The 3000-sq. ft. exhibition, "It's a Nano World," first opened at Ithaca, NY, in 2003, and had reached 3 million people by the end of 2005. It traveled to Epcot center in Florida, and science museums in Ohio, South Carolina, Louisiana, Michigan, Virginia, and Texas. The exhibition is aimed at 5- to 8-year-olds and their parents. In 2006, a new 5000-sq. ft. traveling exhibition, "Too Small to See," aims to explain to middle school students how nanotechnologists create and use devices on a molecular scale.

(4) *R&D and innovation results*: With about 25% of global government investments ($1 billion of $4 billions), the U.S. accounts for about 50% of highly cited papers (Zucker and Darby, 2005), ~60% of USPTO patents (Huang et al, 2004; Figure 3.4), and about 70% of startup companies (NanoBusiness, 2004) in nanotechnology worldwide. Industry investment in the U.S. has exceeded the NNI in R&D, and almost all major companies in traditional and emerging fields have nanotechnology groups at least to survey the competition. *Small Times* reported 1455 U.S. nanotechnology companies in March 2005, with roughly half being small businesses, and 23,000 new jobs were created in small startup "nano" companies. The NNI SBIR investment was about $80 million in FY 2005. More than 200 small businesses, with a total budget of approximately $60 million, have received support from NSF alone since 2001. Many of these are among the 600 "pure play" nanotechnology companies formed in the U.S. since 2001, identified

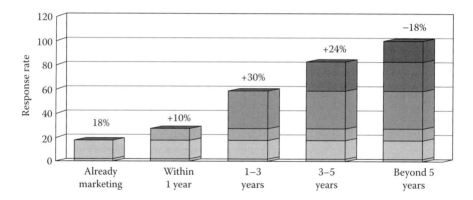

FIGURE 3.6 Commercialization timelines indicate many new nanoproduct introductions in 2007–2011, and the high level of expectations in the long-term (after NCMS, 2006).

in a survey by *Small Times*. All Fortune 500 companies in emerging materials, electronics, and pharmaceuticals have had nanotechnology activities since 2003.

In 2000, only a handful of companies had corporate interest in nanotechnology (under 1% of the companies). A survey performed by the National Center for Manufacturing Sciences (NCMS, 2006) at the end of 2005 showed that 18% of surveyed companies were already marketing nanoproducts. Also, a broad spectrum of new applications in advanced nanoparticles and nanocoatings, as well as durable goods, consumer electronics, and medical products are in the pipeline: 80% of the companies are expected to have nanoproducts by 2010 (within five years) and 98% in the longer term (Figure 3.6). Even if the survey is limited to about 600 manufacturing companies, it shows a strong vote of confidence from a variety of industrial sectors.

Figure 3.7 shows the number of USPTO patents searched by keywords in the "title-claims" (Huang et al., 2004). Although the U.S. has a commanding lead of about 61% that is similar in the range of 10%

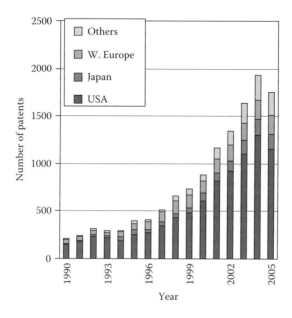

FIGURE 3.7 The U.S. has over 60% of world nanotechnology patents in the USPTO database (by searching in "title-abstract-claims" for nanotechnology keywords with an intelligent search engine; Huang et al., 2004).

with other major databases from Europe and Japan, the decrease in 2005 raises questions. In 2005, the overall number of USPTO patents decreased by about the same proportion.

An interesting result is the higher citation of patents authored by NSF grantees (see Figure 3.7, Huang et al., 2005). Statistical analyses show that the NSF-funded researchers and their patents have higher impact factors than other private and publicly funded reference groups. This suggests the importance of fundamental research and engineering on nanotechnology development. The number of cites per NSF-funded inventor is about 10 as compared to 2 for all inventors of NSE-related patents recorded at USPTO, and the corresponding Authority Score is 20 as compared to 1.

(5) *Significant infrastructure has been established in over 70 universities and 12 government laboratories with nanotechnology user capabilities.* About 60 large centers, networks, and user facilities have been created by NSF (24), NIH (19), DOE (5), NASA (4), DOD (3), NIST (2), and CDC/NIOSH (2). Two user networks established by NSF, the Network on Computational Nanotechnology (established in 2002) and the National Nanotechnology Infrastructure Network (established in 2003) have attracted over 12,000 academic, industrial, and government users in 2005 (see list of NSF centers in Appendix C). The DOE user facilities are located at five national laboratories, taking advantage of the existing large facilities there. NASA has established four academic-based centers.

(6) The NNI's vision of a *grand coalition* of academe, government, industry, and professional groups is taking shape. Over 22 regional alliances have been established throughout the U.S. that develop local partnerships and support commercialization and education. Professional societies have established specialized divisions, organized workshops, and continuing education programs, among them the American Association for the Advancement of Science, American Chemical Society, American Physics Society, Materials Research Society, American Society of Mechanical Engineers, American Institute of Chemical Engineers, Institute of Electrical and Electronics Engineers, and the American Vacuum Society. While federal R&D investment is increasing, attention is extending to the legislative and even judiciary branches of the U.S. government. Partnerships have been created between NNI and industry sectors (Consultative Boards for Advancing Nanotechnology [CBAN], including the electronic industry sector, chemical industry, and Industrial Research Institute). International agreements have been signed with over 25 countries. An example of partnerships is the International Institute for Nanotechnology (IIN) at Northwestern University in Illinois. With support from NSF, NIH, DOE, and NASA, this institute has developed partnerships with the state of Illinois, the city of Chicago, and private foundations to create a new kind of science-and-technology-driven regional coalition. With $300 million in funding for nanotechnology research, educational programs, and infrastructure, IIN has established a large pre-competitive nanoscale science and engineering platform for developing applications, demonstrating manufacturability, and training skilled researchers.

Example of partnerships are the industry–NNI collaborative boards for advancing nanotechnology. The NNI has established a new approach for interaction with industry sectors, besides the previous models in supporting academe–industry government collaboration and encouraging technological innovation. Consultative Boards for Advancing Nanotechnology (CBAN) are representing various industry sectors broadly and help coordinate interactions with electronics, chemical, business, medical/pharmaceutical, and car manufacturing sectors. They provide input to R&D planning for the short- and long-term and to EHS needs, and contribute to nanotechnology R&D and education. The CBAN with the electronics industry was established in October 2003, and five working groups have prepared various reports, and several collaborative activities in long-term R&D planning and funding of research have been completed. The main objectives of CBAN are:

- Jointly plan and support collaborative activities in key R&D areas
- Disseminate the NNI R&D results to industry and encourage technology transfer and technological innovation and industrial use
- Identify and promote new R&D for exploratory areas or gaps in current programs, including those with potential in niche markets
- Expand nanotechnology R&D in industry and long-term topics in academe

- Conduct periodic joint meetings and joint reports
- Exchange information via public hearings to provide opportunity for industrial partners to propose R&D topics

(7) *Societal implications* were addressed from the start of the NNI, beginning with the first research and education program on environmental and societal implications, issued by NSF in July 2000. In September 2000, the report on "Societal Implications of Nanoscience and Nanotechnology" was issued. In 2003, the number of projects in the area had grown significantly, funded by the NSF, EPA, NIH, NIOSH, DOE, and other agencies. Awareness of potential unexpected consequences of nanotechnology has increased, and federal agencies meet periodically to discuss those issues in the NSET Working Group Nanomaterials Environmental and Health Issues (NEHI). In the crosscut of all programs, societal implications and applications (addressing EHS; educational; ethical, legal, and other social implications) may be identified in about 10% of all NNI projects. In the first five years of NNI, the NSF investment with relevance to fundamental research supporting EHS aspects of nanotechnology is about $82 million, or about 7% of the NSF nanoscale science and engineering investment. Research has addressed the sources of nanoparticles and nanostructured materials in the environment (in air, water, soil, biosystems, and work environment), as well as the nonclinical biological implications. The safety of manufacturing nanomaterials is investigated in four NSF centers/networks including Rice University, Northeastern University, University of Pennsylvania, and NNIN (nodes at the University of Minnesota and Arizona State University are focused on nanoparticle measurement).

(8) *Leadership*: As a result of the NNI, the U.S. is recognized as the world leader in this area of science, technology, and economic opportunity. The NNI has catalyzed global activities in nanotechnology and served as a model for other programs. Several recognitions for contributions to NNI are given in Appendix D.

3.6   Technical Challenges in 2005–2020

Nanotechnology has the potential to change our comprehension of nature and life, develop unprecedented manufacturing tools and medical procedures, and even affect societal and international relations. Nanotechnology holds the promise of increasing the efficiency in traditional industries and bringing radically new applications through emerging technologies. The first set of nanotechnology grand challenges was established in 1999–2000, and NSET plans to update it in 2004. Let us imagine again what could be done. Ten potential developments by 2015 are:

1. *At least half of the newly designed advanced materials and manufacturing processes are built using control at the nanoscale at least in one of the key components.* Even if this control may be rudimentary as compared to the long-term potential of nanotechnology, this will mark a milestone towards the new industrial revolution as outlined in 2000. By extending the experience with information technology in the 1990s, I would estimate an overall increase of social productivity by at least 1% per year because of these changes. Silicon transistors will reach dimensions smaller than 10 nm and will be integrated with molecular or other kinds of nanoscale systems. Alternative technologies for replacing the electronic charge as information carrier with electron spin, phase, polarization, magnetic flux quanta, and/or dipole orientation are under consideration. Technologies will be developed for directed assembly of molecules and molecular modules into nonregular, hierarchically organized, device-oriented structures, and for creation of functional, nanoscale-building blocks. Lighter composite nanostructured materials, nanoparticle-laden more reactive and less pollutant fuels, and automated systems enabled by nanoelectronics will dominate automotive, aircraft, and aerospace industries. Top-down manufacturing is expected to integrate with bottom-up molecular assemblies using modular approaches. Nanoscale designed catalysts will expand the use in "exact" chemical manufacturing to cut and assemble molecular assemblies, with

minimal waste. Measurement and imaging of large domains of biological and engineering interest are expected to reach the atomic precision and time resolution of chemical reactions. Visualization and numerical simulation of 3-D domains with nanometer resolution will be necessary for engineering applications.

2. *Suffering from chronic illnesses is being sharply reduced.* It is conceivable that by 2015, our knowledge in detecting and treating tumors in their first year would be advanced and will have the ability to reduce suffering and death from cancer. In 2000, we aimed for earlier detection of cancer within 20 to 30 years. Today, based on the results obtained during 2001–2005 in understanding the cell and new instrumentation, we are trying to eliminate cancer as a cause of death if treated in a timely manner. Pharmaceutical synthesis, processing, and delivery will be enhanced by nanoscale control, and about half of the pharmaceuticals will use nanotechnology in a key component. Modeling the brain based on neuron-to-neuron interactions will be possible by using advances in nanoscale measurement and simulation.

3. *Converging science and engineering from the nanoscale* will establish a mainstream pattern for applying and integrating nanotechnology with biology, electronics, medicine, learning, and other fields (Roco and Bainbridge, 2003). It includes hybrid manufacturing, neuromorphic engineering, artificial organs, expanding the life span, and enhancing learning and sensorial capacities. Science and engineering of nanobiosystems will become essential to human health care and biotechnology. The brain and nervous system functions are expected to be measured with relevance to cognitive engineering.

4. *Life-cycle sustainability and biocompatibility will be pursued in the development of new products.* Knowledge development in nanotechnology will lead to reliable safety rules for limiting unexpected environmental and health consequences of nanostructures. Control of contents of nanoparticles will be performed in air, soils, and waters using a national network. International agreements will address the nomenclature, standards, and risk governance of nanotechnology.

5. *Knowledge development and education will originate from the nanoscale instead of the microscale.* A new education paradigm not based on disciplines but on unity of nature and education–research integration will be tested for K-16 (reversing the pyramid of learning [Roco, 2003]). Science and education paradigm changes will be at least as fundamental as those during the "microscale S&E transition" that originated in the 1950s, where microscale analysis and scientific analysis were stimulated by the space race and digital revolution. The new "nanoscale S&E transition" will change the foundation of analysis and the language of education, stimulated by nanotechnology products. This new "transition" originated at the threshold of the third millennium.

6. *Nanotechnology businesses and organizations* will restructure towards integration with other technologies, distributed production, continuing education, and forming consortia of complementary activities. Traditional and emerging technologies will be equally affected. An important development will be the creation of nanotechnology R&D platforms to serve various areas of applications with the same investigative and productive tools. Two examples are the nanotechnology platform created at a newly built laboratory by General Electric and the discovery instrumentation platform developed at the Sandia National Laboratory.

7. *The capabilities of nanotechnology for systematic control and manufacture at the nanoscale are envisioned to evolve in four overlapping generations of new nanotechnology products* (Roco, 2004b) (Figure 3.8). Each generation of products is marked here by the creation of first commercial prototypes using systematic control of the respective phenomena and manufacturing processing:

   (a) *The first generation of products (~2001–) is "passive nanostructures"* and is typically used to tailor macroscale properties and functions of materials. The specific behavior is stable in time. Illustrations are nanostructured coatings, dispersion of nanoparticles, and bulk materials — nanostructured metals, polymers, and ceramics.

   (b) *The second generation of products (~2005–) is "active nanostructures"* for mechanical, electronic, magnetic, photonic, biological, and other effects. It is typically integrated into microscale devices and systems. New transistors, components of nanoelectronics beyond CMOS,

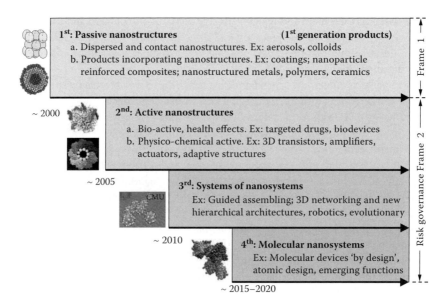

FIGURE 3.8 Timeline for beginning of industrial prototyping and nanotechnology commercialization: four generations of nanoproducts.

amplifiers, targeted drugs and chemicals, actuators, artificial "muscles," self-assembling materials during their use, and adaptive structures illustrate this.

(c) *The third generation (~2010–) is "systems of nanosystems with 3-D nanosystems"* using various syntheses and assembling techniques such as bio-assembling, robotics with emerging behavior, and evolutionary approaches. A key challenge is component networking at the nanoscale and hierarchical architectures. The research focus will shift towards heterogeneous nanostructures and supramolecular system engineering. This includes directed multiscale self-assembling, artificial tissues and sensorial systems, quantum interactions within nanoscale systems, processing of information using photons or electron spin, assemblies of nanoscale electromechanical systems (NEMS), and converging technologies (nano-bio-info-cogno) platforms integrated from the nanoscale.

(d) *The fourth generation (~2015–) will bring "heterogeneous molecular nanosystems,"* where the components of the nanosystems are reduced to molecules and macromolecules. Molecules will be used as devices and from their engineered structures and architectures will emerge fundamentally new functions. Designing new atomic and molecular assemblies is expected to increase in importance, including macromolecules "by design," nanoscale machines, and directed and multiscale self-assembling, exploiting quantum control, nanosystem biology for health care, human–machine interface at the tissue and nervous system level. Research will include topics such as: atomic manipulation for design of molecules and supramolecular systems, controlled interaction between light and matter with relevance to energy conversion among others, exploiting quantum control mechanical–chemical molecular processes, nanosystem biology for health care and agricultural systems, and human–machine interface at the tissue and nervous system level.

8. *Energy conversion* is a main objective of nanotechnology development, and exploratory projects in areas such as photovoltaic conversion and direct conversion of thermal to electric energy are expected to be developed.

9. *Water filtration and desalinization* using nanotechnology has high promise despite only scarce efforts being under way.

10. *Nanoinformatics:* Specific databases and methods to use them will be developed for characterization of nanocomponents, on materials and processes integrated at the nanoscale. Such databases will interface with the existing ones, such as bioinformatics, human, and plant genome databases.

3.7 New Science and Engineering, New Governance Approach

The promise of nanotechnology, however, will not be realized by simply supporting research. Just as nanotechnology is changing how we think about unity of matter at the nanoscale and manufacturing, it is also changing how we think about the management of the research enterprise.

This switch can be seen as the specialization of scientific disciplines has migrated to more unifying concepts for scientific research and system integration in engineering and technology.

Most of the major U.S. science and technology programs in the 20th century — such as space exploration and energy and environmental programs — have been "pulled" primarily by external factors. The economy, natural resources, national security, and international agreements and justifications have initiated top-down R&D funding decisions.

In contrast, nanotechnology development was initially "pushed" by fundamental knowledge (nanoscience and nanoengineering) and the long-term promise of its transformative power. For this reason, we have done the preparation and governance of nanotechnology differently. For nano, research policies have been motivated by long-term vision rather than short-term economic and political decisions.

Transforming and responsible development has guided many NNI decisions. Investments must have returns, the benefit-to-risk ratio must be justifiable, and societal concerns must be addressed. We have introduced nanomanufacturing as a grand challenge since 2002, and we have established a research program at NSF with the same name. NSF awarded three nanomanufacturing centers, and NSF, DOD, and NIST will create a network in 2006. In another example, in 2004–2005, NSF established a new kind of network with national goals and outreach. The six networks are in high school and undergraduate nanotechnology education, nanotechnology in society, informal nanotechnology science education, user facilities, nanotechnology computing, and hierarchical manufacturing. The NEHI (Nanotechnology Environmental and Health Issues) and NILI (Nanotechnology Innovation and Liaison with Industry) working groups were established by NSET (NSTC's Nanoscale Science, Engineering and Technology Subcommittee that coordinated NNI). The NNI has established a new approach for interaction with various industry sectors besides the previous models: CBAN as described earlier.

Improving technological innovation has been another strategy. While R&D activities have spread rapidly in the last five years (they now are in over 60 countries), and nanotechnology has been recognized as a key R&D domain, the economic potential and societal benefits of nanotechnology basically remain in the exploratory phase. Rather, they are seen as promising, and the national investment policies do not generally recognize nanotechnology as being as important as information technology and biotechnology. This may be explained by the relatively recent developments in nanoscale knowledge and the *limited economic understanding of its implications.* There is an apparent gap between the accelerated accumulation of scientific data and ways to apply the results safely and economically. The promise for future economic benefits is a key driver for any emerging technology, but it is generally difficult to document it.

Proactive actions have been taken for addressing societal implications. Immediate and long-term issues are addressed in parallel. We combined formal and informal approaches of interaction in order to receive better input from stakeholders, and encouraged push–pull dynamics (such as input of academic and industry perspectives) in setting research priorities. The speed and scope of nanotechnology R&D exceeds for now the capacity of researchers and regulators to fully assess human and environmental implications. A *specific framework for risk governance* is needed because nanotechnology developments are fundamental in the long-term, operating as an open and complex system. One may need to connect the governance at the national and the international levels. Interaction with industry, civil organizations, and the international community is essential for the responsible development of nanotechnology.

Adopting an *inclusive approach* in governance has been an initial strength of NNI. We developed partnerships with academic organizations, industry groups, various funding agencies, state and local governments, as well as international organizations. The International Dialog on Responsible Nanotechnology R&D (June 17–18, 2004 in Alexandria, VA) was the first meeting of government representatives from over 25 countries and the E.U. dedicated to broad societal issues that cannot be addressed by any single country. This activity may yield a set of principles, structured priorities, and mechanisms of interaction, including sharing data on responsible research and development of nanotechnology.

Long-term view has driven our *visionary approach* in planning, creating infrastructure, and addressing societal concerns. We need to develop anticipatory, deliberate, and proactive measures in order to accelerate the benefits of nanotechnology and its applications. Adaptive and corrective approaches in government organizations are to be established in the complex societal system with the goal of improved long-term risk governance. User- and civic-group involvement is essential for taking better advantage of the technology and developing a complete picture of its societal implications. A multidisciplinary, international forum is needed in order to better address the nanotechnology scientific, technological, and infrastructure development challenges. Optimizing societal interactions, R&D policies, and risk governance for the converging new technologies can enhance economical competitiveness and democratization. The International Risk Governance Council (IRGC, 2006) may provide an independent framework for identification, assessment, and mitigation of risk.

"NNI is a new way to run a national priority," said Charles West at the March 23, 2005 PCAST meeting while reviewing the NNI for Congress.

The Presidential Council of Advisors in Science and Technology (PCAST, 2005) endorsed the governing approach adopted by NNI: the Council "supports the NNI's high-level vision and goals, and the investment strategy by which those are to be achieved."

It was clear, however, that nanotechnology could not advance through the guidance of nanotechnologists or public policy administrators alone. The directions in which research was traveling were too complex and required more than a top-down management system.

Rather, it would require the efforts of all interested stakeholders. It would also require visionary, nanotechnology-specific, and multi-tier management to develop "higher purpose" goals. Key stakeholders needed to be involved from the beginning for a successful project. Furthermore, the introduction of nanotechnology was a global process.

The most recent NNI developments are with the Department of Labor for an anticipatory approach in training of workers for emerging nanotechnology application areas. We have support for nanotechnology from Congress on both sides of the aisle. The 21st century Nanotechnology R&D Act received 100% approval from the Senate in December 2003 (Congress, 2003), and the NSET, which coordinates NNI, passed unchanged from the Democratic to Republican administrations, receiving strong support from the White House. We have justified the development of nanotechnology as beyond the interest of a single political party. In January 2000, then President Clinton announced the NNI at the California Institute of Technology and in the State of the Union address and in January 2006, in the State of the Union address, President Bush listed nanotechnology as a top technological opportunity for national competitiveness.

3.7.1 Closing Remarks

The NNI has been the major driver for nanoscience and nanotechnology developments and applications in the United States and in the world.

Besides products, tools, and healthcare, nanotechnology also implies learning, imagination, infrastructure, inventions, public acceptance, culture, anticipatory laws, and S&E governance among other factors. In 1997–2000, we developed a vision, and in the first five years, 2001–2005, the vision has become a R&D reality. A main reason for the development of NNI has been the vision based on intellectual drive towards exploiting new phenomena and processes, developing a unified science and engineering platform from the nanoscale, and using the molecular and nanoscale interactions for efficient manufacturing. Another main reason has been the promise of broad societal implications, including $1 trillion per year by 2015 of products where nanotechnology plays a key role, which would require 2 million workers.

Nanotechnology is entering new S&E challenges (such as active nanostructures, nanosystems, and nanobiomedicine, advanced tools, environmental and societal implication studies, etc.) in 2006. All trends for journal papers, patents, and worldwide investments are still in exponential growth, with potential inflexion points in several years. There is a need for continuing long-term planning, interdisciplinary activities, and anticipatory measures involving interested stakeholders.

In the next 5 to 10 years, the challenges of nanotechnology will increase in new directions because there is:

A transition from investigating single phenomena and creating single nanoscale components to complex systems, active nanostructures, and molecular nanosystems

A transition from scientific discovery to technological innovation in advanced materials, nanostructured chemicals, electronics, and pharmaceuticals

Expansion into new areas of relevance such as energy, food and agriculture, nanomedicine, and engineering simulations from the nanoscale

Accelerating development, where the rate of discovery remains high and significant changes occur within intervals of years.

While expectations from nanotechnology may be overestimated in the short term, the long-term implications on health care, productivity, water and energy resources, and environment appear to be underestimated, provided proper consideration is given to educational and social implications.

Acknowledgments

This paper is based on my experience in coordinating the NNI. Opinions expressed here are those of the author and do not necessarily reflect the position of NSTC/NSET or NSF.

References

Congress (U.S.), 2003. *21st Century Nanotechnology Research and Development Act*, S.189. in November 2003; Public Law 108–153 on December 3, 2003, Washington, D.C.

Huang, Z., Chen, H., Chen, Z.K., and Roco, M.C., 2004. Longitudinal patent analysis for nanoscale science and engineering in 2003: country, institution and technology field analysis based on USPTO patent database, *J. Nanoparticle Res.*, Vol. 6, No. 4, pp. 325–354.

Huang, Z., Chen, H., Yan, L., and Roco, M.C., 2005. Longitudinal nanotechnology development (1990–2002): the National Science Foundation funding and its impact on patents, *J. Nanoparticle Res.*, Vol. 7(4/5), pp. 343–376.

NCMS (National Center for Manufacturing Science), 2006. 2005 NCMS-NSF cross-industry survey of the U.S. nanomanufacturing industry, NCMS, Detroit, MI.

NSF COV (Committee of Visitors), 2004. Report on Nanoscale Science and Engineering Program, NSF, Arlington, VA.

NSTC (M.C. Roco et al.), 2000, National Nanotechnology Initiative: the initiative and its implementation plan, National Science and Technology Council (NSTC), White House, Washington, D.C., July 2000.

NSTC (I. Amato, M.C. Roco et al.), 1999. Nanotechnology – shaping the world atom by atom, National Science and Technology Council (NSTC), White House, Washington, D.C., September 1999.

NSTC (National Science and Technology Council), 2004b. NNI Strategic Plan, Washington, D.C.

PCAST (Presidential Council of Advisors on Science and Technology), 2005. The National Nanotechnology Initiative at five years: assessment and recommendations of the National Nanotechnology Advisory Panel, OSTP, White House, Washington, D.C., May 2005.

Roco, M.C., Williams, R.S., and Alivisatos, P., eds., 2000. *Nanotechnology Research Directions*, U.S. National Science and Technology Council, Washington, D.C., Springer, 316 pages; http://www.wtec.org/loyola/nano/IWGN.Research.Directions/

Roco, M.C., 2002. Coherence and divergence in science and engineering megatrends, *J. Nanoparticle Res.*, Vol. 4, No. 1–2, pp. 9–19.

Roco, M.C., 2003. Converging science and technology at the nanoscale: opportunities for education and training, *Nature Biotechnol.*, Vol. 21, No. 10, pp. 1247–1249.

Roco, M.C., 2004. Nanoscale science and engineering: unifying and transforming tools, *AIChE J.*, Vol. 50, No. 5, pp. 890–897.

Roco, M.C., 2005a. The vision and strategy of the U.S. National Nanotechnology Initiative, in Schulte, J., ed., *Nanotechnology: Global Strategies, Industry Trends and Applications*, John Wiley & Sons, Ltd., 2005, pp. 79–94.

Roco, M.C., 2005b. International perspective on government nanotechnology funding in 2005, *J. Nanoparticle Res.*, Vol. 7, No. 6, pp. 707–712.

Roco, M.C., 2006. Governance of converging technologies integrated from the nanoscale, in *Progress in Converging Technologies*, W.S. Bainbridge and M.C. Roco (eds.), Ann. *NY acad. Sci.*, 1–23.

Roco, M.C. and Bainbridge, W.S., eds., 2001. *Societal Implications of Nanoscience and Nanotechnology*, Springer, Boston, 350 pages.

Roco, M.C. and Bainbridge, W.S., eds., 2003. *Converging Technologies for Improving Human Performance*, Springer, Boston, 468 pages.

Siegel, R., Hu, E., and Roco, M.C., eds., 1999. *Nanostructure Science and Technology*, National Science and Technology Council (NSTC), White House, Washington, D.C.; also published by Springer, Boston and Dordrecht.

Appendix A. The Structure of NNI in the First (2001–2005) and Second (2006–2010) Strategic Plans

Vision: A future in which the ability to understand and control matter on the nanoscale — 1 to 100 nm — leads to a revolution in technology and industry.

A1   NNI Modes of Support in FYs 2001–2005 (1st NNI Strategic Plan)

The funding strategy for the NNI was based on five modes of investment.

The first mode supports a balanced investment in fundamental research across the entire breadth of science and engineering, and it is led by the National Science Foundation.

The second mode, collectively known as the "grand challenges," focuses on nine specific R&D areas that are more directly related to applications of nanotechnology. They also are identified as having the potential to realize significant economic, governmental, and societal impact in about a decade. These challenges are:

1. Nanostructured materials by design
2. Manufacturing at the nanoscale
3. Chemical–biological–radiological–explosive detection, and protection
4. Nanoscale instrumentation, and metrology
5. Nanoelectronics, -photonics, and -magnetics
6. Health care, therapeutics, and diagnostics
7. Efficient energy conversion and storage
8. Microcraft and robotics
9. Nanoscale processes for environmental improvement

The third mode of investment supports centers of excellence that conduct research within host institutions. These centers pursue projects with broad multidisciplinary research goals that are not supported by more traditionally structured programs. These centers also promote education of future researchers and innovators, as well as training of a skilled technical workforce for the growing nanotechnology industry.

The fourth mode funds the development of infrastructure, instrumentation, standards, computational capabilities, and other research tools necessary for nanoscale R&D.

The fifth mode recognizes and funds research on the societal implications, and addresses educational needs associated with the successful development of nanoscience and nanotechnology. Besides graduate and postgraduate education activities, NSF supports nanoscale science and engineering programs for earlier nanotechnology education for undergraduates, high schools, and public outreach.

A2   NNI Goals in FYs 2006–2010 (2nd NNI Strategic Plan) (NSTC, 2004)

This plan describes the goals of the NNI as well as the strategy by which those goals are to be achieved. The goals are as follows:

Maintain a world-class research and development program aimed at realizing the full potential of nanotechnology

Facilitate transfer of new technologies into products for economic growth, jobs, and other public benefit

Develop educational resources, a skilled workforce, and the supporting infrastructure and tools to advance nanotechnology

Support responsible development of nanotechnology

The investment strategy includes the major subject categories of investment, or program component areas (PCAs), cutting across the interests and needs of the participating agencies:

1. Fundamental nanoscale phenomena and processes
2. Nanomaterials
3. Nanoscale devices and systems
4. Instrumentation research, metrology, and standards for nanotechnology
5. Nanomanufacturing
6. Major research facilities and instrumentation acquisition
7. Societal dimensions, including: environmental, health and safety issues; education; and ethical, legal, and other social issues.

Appendix B. NNI Members

List of federal agencies participating in the NNI (January 2006):

Federal agencies with budgets dedicated to nanotechnology research and development:

Department of Agriculture, Cooperative State Research, Education, and Extension Service (USDA/CSREES)

Department of Agriculture, Forest Service (USDA/FS)

Department of Defense (DOD)

Department of Energy (DOE)

Department of Homeland Security (DHS)

Department of Justice (DOJ)

Department of Transportation (DOT)

Environmental Protection Agency (EPA)

National Aeronautics and Space Administration (NASA)

National Institute of Standards and Technology (NIST, Department of Commerce)

National Institute for Occupational Safety and Health (NIOSH, Department of Health and Human Services/Centers for Disease Control and Prevention)

National Institutes of Health (NIH, Department of Health and Human Services)

National Science Foundation (NSF)

Other participating agencies:

Bureau of Industry and Security (BIS, Department of Commerce)

Consumer Product Safety Commission (CPSC)

Department of Education (ED)

Department of Labor (DOL)

Department of State (DOS)

Department of the Treasury (DOTreas)

Food and Drug Administration (FDA, Department of Health and Human Services)

International Trade Commission (ITC)
Intelligence Technology Innovation Center, representing the Intelligence Community (IC)
Nuclear Regulatory Commission (NRC)
Technology Administration (TA, Department of Commerce)
U.S. Patent and Trademark Office (USPTO, Department of Commerce)

Appendix C. List of NSF Centers and Networks in the Field of Nanoscale Science and Engineering Established Since 2000

University	Name of the Center, Network, or User Facility
Nanoscale Science and Engineering Centers (NSECs)	
Columbia University	Center for Electron Transport in Molecular Nanostructures
Cornell University	Center for Nanoscale Systems
Rensselaer Polytechnic Institute	Center for Directed Assembly of Nanostructures
Harvard University	Science for Nanoscale Systems and Their Device Applications
Northwestern University	Institute for Nanotechnology
Rice University	Center for Biological and Environmental Nanotechnology
University of California, Los Angeles	Center for Scalable and Integrated Nanomanufacturing
University of Illinois at Urbana-Champaign	Center for Nanoscale Chemical, Electrical, Mechanical, and Manufacturing Systems
University of California at Berkeley	Center for Integrated Nanomechanical Systems
Northeastern University	Center for High Rate Nanomanufacturing
Ohio State University	Center for Affordable Nanoengineering
University of Pennsylvania	Center for Molecular Function at the Nanoscale
Stanford University	Center for Probing the Nanoscale
University of Wisconsin	Center for Templated Synthesis and Assembly at the Nanoscale
Arizona State University, University of California, Santa Barbara, University of Southern California, Harvard University	Nanotechnology in Society Network
University of Massachusetts–Amherst	Hierarchical Nanomanufacturing
Centers from the Nanoscale Science and Engineering Education Solicitation	
Northwestern University	Nanotechnology Center for Learning and Teaching
Boston Museum of Science	Nanoscale Informal Science Education
NSF Networks and Centers that Complement the NSECs	
Cornell University and 12 other nodes	National Nanotechnology Infrastructure Network
Purdue University and 6 other nodes	Network for Computational Nanotechnology
Oklahoma University, Oklahoma State University	Oklahoma Nano Net
Cornell University	STC: The Nanobiotechnology Center
6 new MRSECs	Network of Materials Research Science and Engineering Centers (MRSEC)

Appendix D. Recognitions for NNI Contribution

The author of this chapter received several recognitions for contributing to NNI partially described in this chapter:

From U.S. interagency committees after announcing NNI (2000) and after the first strategic plan (2006):

—Interagency WGN/National Science and Technology Council Plaque:

"To Dr. M. Roco, who with big ideas has created a National Nanotechnology Program focused on things small"

Washington, D.C, February 16, 2000

—Nanoscale Science, Engineering and Technology, NSTC Plaque:

"To Mihail C. Roco, Dr Nano, Founding Father of the U.S. National Nanotechnology Initiative; For his vision, dedication and energy in advancing the field of nanotechnology in the U.S. and across the world"

Washington, D.C., January 19, 2006

FIGURE 3.D1 This nanograph of Dr. M. Roco was recorded at Oak Ridge National Laboratory using piezoresponse-force microscopy, one of the members of the family of techniques known as scanning probe microscopy, which can image and manipulate materials on the nanoscale. Each picture element is approximately 50 nm in diameter; the distance from chin to eyebrow is approximately 2.5 μm.

From professional organizations:

Dr. Roco is a Correspondent Member of the Swiss Academy of Engineering Sciences, a Fellow of the American Society of Mechanical Engineers, a Fellow of the Institute of Physics, and a Fellow of the American Institute of Chemical Engineers. The "Engineer of the Year" (two times, in 1999 and 2004) by the U.S. National Society of Professional Engineers and NSF, Distinguished Service Award of the NSF (2001); AIChE Nanoscale Science and Engineering Forum Award for "Leadership and service to the national science and engineering community through initiating and bringing to fruition the National Nanotechnology Initiative," New York, September 2005.

From national and international surveys:

Small Times Magazine: "Best of Small Tech Awards" ("Leader of the American nanotech revolution," Nov. 2002)

Forbes Nanotechnology: "First in Nanotechnology's Power Brokers," Forbes, March 2003 ("Leading Architect of NNI")

Scientific American: World Technology Leader — Scientific American Top 50 in 2004 (for "Led nearly $1-billion-a-year U.S. government effort in nanotechnology"; "building a solid consensus in the scientific and non-technical communities that nanotechnology is important for future scientific and economic development, and he has furthered public acceptance of nanotechnology"), Dec. 2004, New York

Nanotech Briefs, ABP International, Boston: 2005 Innovator Award, "For pioneering achievements in advancing the state of the art in nanotechnology," Nano 50, August 2005.

2

Molecular and Nanoelectronics

4 Engineering Challenges in Molecular Electronics *Gregory N. Parsons* 4-1
Introduction • Silicon-Based Electrical Devices and Logic Circuits • CMOS Device
Parameters and Scaling • Memory Devices • Opportunities and Challenges for Molecular
Circuits • Summary and Conclusions

5 Molecular Electronic Computing Architectures *James M. Tour, Dustin K. James*.........5-1
Present Microelectronic Technology • Fundamental Physical Limitations
of Present Technology • Molecular Electronics • Computer Architectures Based
on Molecular Electronics • Characterization of Switches and Complex
Molecular Devices • Conclusion

6 Nanoelectronic Circuit Architectures *Wolfgang Porod* 6-1
Introduction • Quantum-Dot Cellular Automata • Single-Electron Circuits • Molecular
Circuits • Summary

7 Molecular Computing and Processing Platforms *Sergey Edward Lyshevski* 7-1
Introduction • Data and Signal Processing Platforms • Microelectronics and Nanoelectronics:
Retrospect and Prospects • Performance Estimates • Synthesis Taxonomy in Design of ᴹICs and
Processing Platforms • Biomolecular Processing and Fluidic Molecular Electronics: Neuro-
biomimetics, Prototyping, and Cognition • Design of Three-Dimensional Molecular Integrated
Circuits: Data Structures, Decision Diagrams, and Hypercells • Three-Dimensional Molecular
Signal/Data Processing and Memory Platforms • Hierarchical Finite-State Machines and Their
Use in Hardware and Software Design • Adaptive Defect-Tolerant Molecular Processing-and-
Memory Platforms • Hardware–Software Design • Modeling and Analysis of Molecular
Electronic Devices • Conclusions

8 Spin Field Effect Transistors *Supriyo Bandyopadhyay, Marc Cahay*................................8-1
Introduction: The Spin Field Effect Transistor • Another Spinfet • Non-Idealities • Other
Types of Spinfets • What Motivates the Research in Spinfets? • Can "Spin" Ever Produce a
Power Advantage Over "Charge"? • Conclusion

**9 Electron Charge and Spin Transport in Organic and Semiconductor
Nanodevices: Moletronics and Spintronics** *A.M. Bratkovsky* 9-1
Introduction • Electron Transport in Molecular Devices: Molecular Electronics • Spin
Injection: Processes and Devices • Conclusions

10 Nanoarchitectonics: Advances in Nanoelectronics *Kang Wang, Kosmas Galatsis,*
Roman Ostroumov, Mihri Ozkan, Kostantin Likharev, Youssry Botros **10**-1
Introduction • The Nanoelectronics Landscape • Fabrication of Nanostructures
•Nanodevices • Nanoarchitectures • Conclusions

11 Molecular Machines *Brian H. Northrop, Adam B. Braunschweig, Paula M. Mendes,*
William R. Dichtel, J. Fraser Stoddart **11**-1
Introduction • Transferring Molecular Movement to the Solid State • Molecular Muscle
Systems • Molecular Nanovalves • Molecular Rotors • Surfaces with Controllable
Wettability • Conclusions

4

Engineering Challenges in Molecular Electronics

4.1 Introduction 4-1
4.2 Silicon-Based Electrical Devices and Logic Circuits.........4-2
   Two-Terminal Diode and Negative Differential Resistance
   Devices • Three-Terminal Bipolar, MOS, and CMOS Devices
   Basic Three-Terminal Logic Circuits • The Importance of Gain
4.3 CMOS Device Parameters and Scaling............................4-6
   Mobility and Subthreshold Slope • Constant Field Scaling and
   Power Dissipation • Interconnects and Parasitics • Reliability
   • Alternate Device Structures for CMOS
4.4 Memory Devices..4-11
   DRAM, SRAM, and Flash • Passive and Active Matrix Addressing
4.5 Opportunities and Challenges for Molecular
   Circuits...4-12
   Material Patterning and Tolerances • Reliability • Interconnects,
   Contacts, and the Importance of Interfaces • Power Dissipation
   and Gain • Thin-Film Electronics • Hybrid Silicon/Molecular
   Electronics
4.6 Summary and Conclusions 4-17
Acknowledgments...4-17
References..4-17

Gregory N. Parsons
North Carolina State University

4.1 Introduction

Manufacturing practices for complementary metal oxide semiconductor (CMOS) devices are arguably the most demanding, well developed, and lucrative in history. Even so, it is well recognized that historic trends in device scaling that have continued since the 1960s are going to face serious challenges in the next several years. Current trends in Moore's Law scaling are elucidated in detail in the Semiconductor Industry Association's *The International Technology Roadmap for Semiconductors.*[1] The 2001 roadmap highlights significant fundamental barriers in patterning, front-end processes, device structure and design, test equipment, interconnect technology, integration, assembly and packaging, etc.; and there are significant industry and academia research efforts focused on these challenges. There is also significant growing interest in potential leapfrog technologies, including quantum-based structures and molecular electronics, as possible means to redefine electronic device and system operation. The attention (and research funds) applied to potential revolutionary technologies is small compared with industrial efforts on silicon. This is primarily because of the tremendous manufacturing infrastructure built for silicon technology and the fact that there is still significant room for device performance improvements in silicon — even though many of the challenges described in the roadmap still have "no known solution."

Through continued research in leapfrog approaches, new materials and techniques are being developed that could significantly impact electronic device manufacturing. However, such transitions are not likely to be realized in manufacturing without improved insight into the engineering of current high-performance electronic devices.

Silicon devices are highly organized inorganic structures designed for electronic charge and energy transduction.[2–4] Organic molecules are also highly organized structures that have well-defined electronic states and distinct (although not yet well-defined) electronic interactions within and among themselves. The potential for extremely high device density and simplified device fabrication has attracted attention to the possibility of using individual molecules for advanced electronic devices (see recent articles by Ratner[5]; Kwok and Ellenbogen[6]; and Wada[7]). A goal of molecular electronics is to use fundamental molecular-scale electronic behavior to achieve electronic systems (with functional logic and/or memory) composed of individual molecular devices. As the field of molecular electronics progresses, it is important to recognize that current silicon circuits are likely the most highly engineered systems in history, and insight into the engineering driving forces in silicon technology is critical if one wishes to build devices more advanced than silicon. The purpose of this chapter, therefore, is to give a brief overview of current semiconductor device operation, including discussion of the strengths and weaknesses of current devices and, within the context of current silicon device engineering, to present and discuss possible routes for molecular electronics to make an impact on advanced electronics engineering and technology.

4.2 Silicon-Based Electrical Devices and Logic Circuits

4.2.1 Two-Terminal Diode and Negative Differential Resistance Devices

The most simple silicon-based solid-state electronic device is the p/n junction diode, where the current through the two terminals is small in the reverse direction and depends exponentially on the applied voltage in the forward direction. Such devices have wide-ranging applications as rectifiers and can be used to fabricate memory and simple logic gates.[8,9] A variation on the p/n diode is a resonant tunneling diode (RTD) where well-defined quantum states give rise to negative differential resistance (NDR). A schematic current vs. voltage trace for an NDR device is shown in Figure 4.1. Such devices can be made with inorganic semiconductor materials and have been integrated with silicon transistors[10–13] for logic devices with multiple output states to enhance computation complexity.

An example circuit for an NDR device with a load resistor is shown in Figure 4.1. This circuit can act as a switch, where V_{out} is determined by the relative voltage drop across the resistor and the diode. The resistance of the diode is switched from high to low by applying a short voltage pulse in excess of V_{dd} across the series resistor and diode, and the smaller resistance results in a small V_{out}. These switching circuits may be useful for molecular logic gates using RTD molecules, but several important issues need to be considered for applications involving two-terminal logic. One concern is the size of the output impedance. If the outlet voltage node is connected to a resistance that is too small (i.e., similar in magnitude to the RTD impedance), then the outlet voltage (and voltage across the RTD) will shift from the expected value; and this error will propagate through the circuit network. Another concern is that full logic gates fabricated with RTDs require an additional clock signal, derived from a controlled oscillator circuit. Such oscillators are readily fabricated using switching devices with gain, but to date they have not been demonstrated with molecular devices. Possibly the most serious concern is the issue of power dissipation. During operation, the current flows continuously through the RTD device, producing significant amounts of thermal energy that must be dissipated. As discussed below in detail, power dissipation in integrated circuits is a long-standing problem in silicon technology, and methodologies to limit power in molecular circuits will be a critical concern for advanced high-density devices.

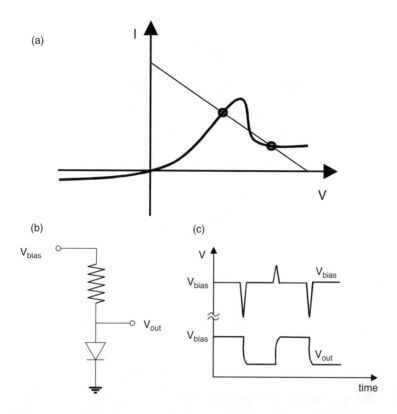

FIGURE 4.1 Schematic of one possible NDR device. (a) Schematic current vs. voltage curve for a generic resonant tunneling diode (RTD) showing negative differential resistance (NDR). The straight line is the resistance load line, and the two points correspond to the two stable operating points of the circuit. (b) A simple circuit showing an RTD loaded with a resistor. (c) Switching behavior of resistor/RTD circuit. A decrease in V_{bias} leads to switching of the RTD device from high to low impedance, resulting in a change in V_{out} from high to low state.

4.2.2 Three-Terminal Bipolar, MOS, and CMOS Devices

The earliest solid-state electronic switches were bipolar transistors, which in their most simple form consisted of two back-to-back p/n junctions. The devices were essentially solid-state analogs of vacuum tube devices, where a current on a base (or grid) electrode modulated the current between the emitter and collector contacts. Because a small change in the base voltage, for example, could enable a large change in the collector current, the transistor enabled signal amplification (similar to a vacuum tube device) and, therefore, current or voltage gain. In the 1970s, to reduce manufacturing costs and increase integration capability, industry moved away from bipolar and toward metal-oxide-semiconductor field effect transistor (MOSFET) structures, shown schematically in Figure 4.2. For MOSFET device operation, voltage applied to the gate electrode produces an electric field in the semiconductor, attracting charge to the silicon/dielectric interface. A separate voltage applied between the source and drain then enables current to flow to the drain in a direction perpendicular to the applied gate field. Device geometry is determined by the need for the field in the channel to be determined primarily by the gate voltage and not by the voltage between the source and drain. In this structure, current flow to or from the gate electrode is limited by leakage through the gate dielectric. MOS devices can be either NMOS or PMOS, depending on the channel doping type (p- or n-type, respectively) and the charge type (electrons or holes, respectively) flowing in the inversion layer channel. Pairing of individual NMOS and PMOS transistors results in a complementary MOS (CMOS) circuit.

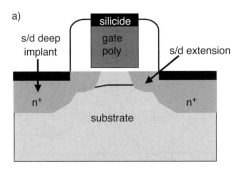

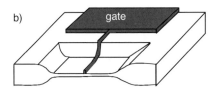

FIGURE 4.2 (a) Cross section of a conventional MOS transistor. (b) A three-dimensional representation of a MOS transistor layout. Two transistors, one NMOS and one PMOS, can be combined to form a complementary MOS (CMOS) device.

4.2.3 Basic Three-Terminal Logic Circuits

A basic building block of MOS logic circuits is the signal inverter, shown schematically in Figure 4.3. Logic elements, including, for example, NOR and NAND gates, can be constructed using inverters with multiple inputs in parallel or in series. Early MOS circuits utilized single-transistor elements to perform the inversion function utilizing a load resistor as shown in Figure 4.3a. In this case, when the NMOS is off (V_{in} is less than the device threshold voltage V_{th}), the supply voltage (V_{dd}) is measured at the outlet. When V_{in} is increased above V_{th}, the NMOS turns on and V_{dd} is now dropped across the load resistor; V_{out} is now in common with ground, and the signal at V_{out} is inverted relative to V_{in}. The same behavior is observed in enhancement/depletion mode circuits (Figure 4.3b) where the load resistor is replaced with another NMOS device. During operation of these NMOS circuits, current is maintained between V_{dd} and ground in either the high- or low-output state. CMOS circuits, on the other hand, involve combinations of NMOS and PMOS devices and result in significantly reduced power consumption as compared with NMOS-only circuits. This can be seen by examining a CMOS inverter structure as shown in Figure 4.3c. A positive input voltage turns on the NMOS device, allowing charge to flow from the output capacitance load to ground and producing a low V_{out}. A low-input voltage likewise enables the PMOS to turn on, and the output to go to the level of the supply voltage, V_{dd}. During switching, current is required to charge and discharge the channel capacitances, but current stops flowing when the channel and output capacitances are fully charged or discharged (i.e., when V_{out} reaches 0 or V_{dd}). In this way, during its static state, one of the two transistors is always off, blocking current from V_{dd} to ground. This means that the majority of the power consumed in an array of these devices is determined by the rate of switching and not by the number of inverters in the high- or low-output state within the array. This is a tremendously important outcome of the transition in silicon technology from NMOS to CMOS: *the power produced per unit area of a CMOS chip can be maintained nearly constant as the number density of individual devices on the chip increases.* The implications for this in terms of realistic engineered molecular electronic systems will be discussed in more detail below.

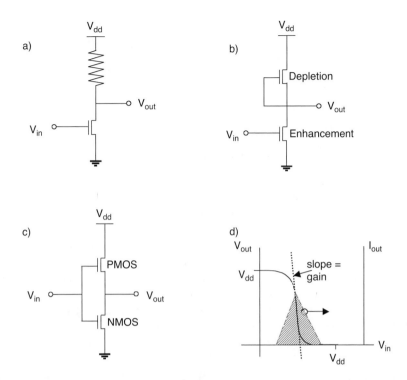

FIGURE 4.3 Inverter circuit approaches for NMOS and CMOS devices. (a) A simple inverter formed using an NMOS device and a load resistor, commonly used in the 1970s. (b) An inverter formed using an NMOS device and a depletion-mode load transistor. (c) Complementary MOS (CMOS) inverter using an NMOS and a PMOS transistor, commonly used since the 1980s. (d) A schematic voltage inverter trace for a CMOS inverter. The slope of the V_{out} vs. V_{in} gives the inverter gain. Also shown is the net current through the device as a function of V_{in}. Note that for the CMOS structure, current flows through the inverter only during the switching cycle, and no current flows during steady-state operation.

4.2.4   The Importance of Gain

Gain in an electronic circuit is generally defined as the ratio of output voltage change to input voltage change (i.e., voltage gain) or ratio of output current change to input current change (current gain). For a simple inverter circuit, therefore, the voltage gain is the slope of the V_{out} vs. V_{in} curve, and the maximum gain corresponds to the value of the maximum slope. A voltage gain in excess of one indicates that if a small-amplitude voltage oscillation is placed on the input (with an appropriate dc bias), a larger oscillating voltage (of opposite phase) will be produced at the output. Typical silicon devices produce gain values of several hundred. Power and voltage gain are the fundamental principles behind amplifier circuits used in common electronic systems, such as radios and telephones.

Gain is also critically important in any electronic device (such as a microprocessor) where voltage or current signals propagate through a circuit. Without gain, the total output power of any circuit element will necessarily be less than the input; the signal would be attenuated as it moved through the circuit, and eventually *high* and *low* states could not be differentiated. Because circuit elements in silicon technology produce gain, the output signal from any element is "boosted" back up to its original input value; and the signal can progress without attenuation.

In any system, if the power at the output is to be greater than the power at an input reference, then an additional input signal will be required. Therefore, at least three contact terminals are required to achieve gain in any electronic element: high voltage (or current) input, small voltage or current input, and large signal output. The goal of demonstrating molecular structures with three terminals that produce gain continues to be an important challenge.

4.3   CMOS Device Parameters and Scaling

4.3.1   Mobility and Subthreshold Slope

The speed of a circuit is determined by how fast a circuit output node is charged up to its final state. This is determined by the transistor drive current, which is related to the device dimension and the effective charge mobility, μ_{eff} (or transconductance), where mobility is defined as charge velocity per unit field. In a transistor operating under sufficiently low voltages, the velocity will increase in proportion to the lateral field (V_{sd}/L_{ch}), where V_{sd} is the source/drain voltage and L_{ch} is the effective length of the channel. At higher voltages, the velocity will saturate, and the lateral field becomes nonuniform. This *saturation velocity* is generally avoided in CMOS circuits but becomes more problematic for very short device lengths. Saturation velocity for electrons in silicon at room temperature is near 10^7 cm/s, and it is slightly smaller for holes. For a transistor operating with low voltages (i.e., in the linear regime), the mobility is a function of the gate and source/drain voltages (V_{gs} and V_{sd}), the channel length and width (L_{ch} and W_{ch}) and the gate capacitance per unit area (C_{ox}; Farads/cm^2):

$$\mu_{eff} = \frac{I_{sd}L_{ch}}{C_{ox}W_{ch}(V_{gs} - V_{th})V_{sd}} \tag{4.1}$$

The mobility parameter is independent of device geometry and is related to the current through the device. Because the current determines the rate at which logic signals can move through the circuit, the effective mobility is an important figure of merit for any electronic device.

Another important consideration in device performance is the subthreshold slope, defined as the inverse slope of the log (I_{sd}) vs. V_g curve for voltages below V_{th}. A typical current vs. voltage curve for an NMOS device is shown in Figure 4.4. In the subthreshold region, the current flow is exponentially dependent on voltage:

$$I_{sd} \propto \exp(qV/nkT) \tag{4.2}$$

where n is a number typically greater than 1. At room temperature, the ideal case (i.e., minimal charge scattering and interface charge trapping) results in n = 1 and an inverse slope $(2.3 \cdot kT)/q \approx 60$ mV/decade. The subthreshold slope of a MOS device is a measure of the rate at which charge diffuses from the channel region when the device turns off. Because the rate of charge diffusion does not change with device dimension, the subthreshold slope will not change appreciably as transistor size decreases. The nonscaling of subthreshold slope has significant implications for device scaling limitations. Specifically, it puts a limit on how much V_{th} can be reduced because the current at zero volts must be maintained low to control off-state current, I_{off}. This in turn puts a limit on how much the gate voltage can be decreased, which means that the ideal desired constant field scaling laws, described below, cannot be precisely followed.

Neither the mobility nor subthreshold slope is significantly affected by reduction in transistor size. Under ideal constant field scaling, the channel width, length, and thickness all decrease by the same factor (κ). So how is it that smaller transistors with less current are able to produce faster circuits? The speed of a transistor circuit is determined by the rate at which a logic element (such as an inverter) can change from one state to another. This switching requires the charging or discharging of a capacitor at the output node (i.e., another device in the circuit). Because capacitance $C = (W_{ch} \cdot L_{ch} \cdot \varepsilon\varepsilon_o)/t_{ox}$, the total gate capacitance will decrease by the scaling factor κ. The charge required to charge a capacitor is $Q = CV$, which

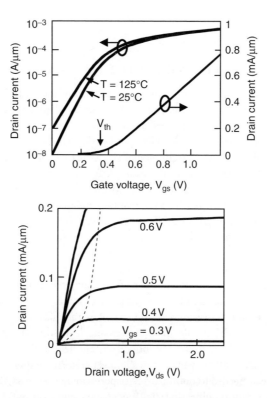

FIGURE 4.4 Operating characteristics for a typical MOS device. (top) Current measured at the drain as a function of voltage applied to the gate. The same data is plotted on linear and logarithmic scales. An extrapolation of the linear data is an estimate of the threshold voltage V_{th}. The slope of the data on the logarithmic scale for $V < V_{th}$ gives the subthreshold slope. Note that the device remains operational at temperatures $>100°C$, with an increase in inverse subthreshold slope and an increase in off-current (I_{off}) as temperature is increased. (bottom) Current measured at the drain as a function of source/drain voltage for various values of gate voltage for the same device as in (a). The dashed line indicates the transition to current saturation.

means that a reduction in C and a reduction in V by the factor κ results in a decrease in the total charge required by a factor of $κ^2$. Therefore, the decrease in current by κ still results in an increase in charging rate by the scaling factor κ, enabling the circuit speed to increase by a similar factor.

Another way to think about this is to directly calculate the time that it takes to charge the circuit output node. The charging rate of a capacitor is:

$$\frac{\partial V}{\partial T} = \frac{I}{C} \tag{4.3}$$

where C is the total capacitance in Farads. Because I and C both decrease with size, the charging rate is also independent of size. However, because the smaller device will operate at a smaller voltage, then the time that it takes to charge will decrease (and the circuit will become faster) as size decreases. An expression for charging time is obtained by integrating Equation 4.3, where $V = V_{sd}$, $I = I_{sd}$, and $C = (C_{ox} \cdot L_{ch} \cdot W_{ch})$; and substituting in for current from the mobility expression (Equation 4.1):

$$t = \frac{L_{ch}^2}{\mu_{eff}(V_{gs} - V_{th})} \tag{4.4}$$

This shows that as V_{gs} and L_{ch} decrease by κ, the charging time will also decrease by the same factor, leading to an increase in circuit speed.

4.3.2 Constant Field Scaling and Power Dissipation

As discussed in the SIA Roadmap, significant attention is currently paid to issues in front-end silicon processing. One of the most demanding challenges is new dielectric materials to replace silicon dioxide to achieve high gate capacitance with low gate leakage.[14] The need for higher capacitance with low leakage is driven primarily by the need for improved device speed while maintaining low power operation. Power has been an overriding challenge in MOS technologies since the early days of electronics, long before the relatively recent interest in portable systems. For example, reduced power consumption was one of the important problems that drove replacement of vacuum tubes with solid-state electronics in the 1950s and 1960s. Low power is required primarily to control heat dissipation, since device heating can significantly reduce device performance (especially if the device gets hot enough to melt). For current technology devices, dissipation of several watts of heat from a logic chip can be achieved using air cooling; and inexpensive polymer-based packaging approaches can be used. The increase in portable electronics has further increased the focus on problems associated with low power operation.

There are several possible approaches to consider when scaling electronic devices, including constant field and constant voltage scaling. Because of the critical need for low heat generation and power dissipation, current trends in shrinking CMOS transistor design are based on the rules of constant field scaling. Constant field scaling is an idealized set of scaling rules devised to enable device dimensions to decrease while output power density remains fixed. As discussed briefly above, constant field scaling cannot be precisely achieved, primarily because of nonscaling of the voltage threshold. Therefore, modifications in the ideal constant field scaling rules are made as needed to optimize device performance. Even so, the trends of constant field scaling give important insight into engineering challenges facing any advanced electronic device (including molecular circuits); therefore, the rules of constant field scaling are discussed here.

Heat generation in a circuit is related to the product of current and voltage. In a circuit operating at frequency f, the current needed to charge a capacitor C_T in half a cycle time is $(f \cdot C_T \cdot V_{dd})/2$, where V_{dd} is the applied voltage. Therefore, the power dissipated in one full cycle of a CMOS switching event (the *dynamic power*) is proportional to the total capacitance that must be charged or discharged during the switching cycle, C_T (i.e., the capacitance of the output node of the logic gate), the power supply voltage squared, V_{dd}^2, and the operation frequency, f:

$$P_{switch} = f C_T V_{dd}^2 \tag{4.5}$$

The capacitance of the output node is typically an input node of another logic gate (i.e., the gate of another CMOS transistor). If a chip has $\sim 10^8$ transistors/cm^2, a gate length of ~ 130 nm, and width/length ratio of 3:1, then the total gate area is $\sim 5\%$ of the chip area. Therefore, the total power consumed by a chip is related to the capacitance density of an individual transistor gate ($C/A = \varepsilon\varepsilon_o/t_{ox}$).

$$\frac{P_{switch}}{A_{gate}} = \frac{f\varepsilon\varepsilon_o V_{dd}^2}{t_{ox}}$$

$$\frac{P_{chip}}{A_{chip}} \approx 0.05 \left(\frac{P_{switch}}{A_{gate}} \right) \tag{4.6}$$

where t_{ox} is the thickness and ε is the dielectric constant of the gate insulator. This is a highly simplified analysis (a more complete discussion is given in Reference 15). It does not include the power loss associated with charging and discharging the interconnect capacitances, and it does not include the fact that in CMOS circuits, both NMOS and PMOS transistors are partially on for a short time during the switching transition, resulting in a small current flow directly to ground, contributing to additional power loss. It also assumes that there is no leakage through the gate dielectric and that current flow in the off state is negligible. Gate leakage becomes a serious concern as dielectric thickness decreases and tunneling increases, and off-state leakage becomes more serious as the threshold voltage decreases. These two processes result in an additional *standby power* term ($P_{standby} = I_{off} \cdot V_{dd}$) that must be added to the power loss analysis.

Even with these simplifications, the above equation can be used to give a rough estimate of power consumption by a CMOS chip. For example, for a 1GHz chip with V_{dd} of ~1.5V and gate dielectric thickness of 2 nm, the above equation results in ~150 W/cm^2, which is within a factor 3 of the ~50W/cm^2 dissipated for a 1GHz chip.[16] The calculation assumes that all the devices are switching on each cycle. Usually, only a fraction of the devices will switch per cycle, and the chip will operate with maximum output for only short periods, so the total power output will be less than the value calculated. A more complete calculation must also consider additional capacitances related to fan-out and loading of interconnect transmission lines, which will lead to power dissipation in addition to that calculated above. It is clear that heat dissipation problems in high-density devices are significant, and most high-performance processors require forced convection cooling to maintain operating temperatures within the maximum operation range of 70–80°C.[16]

The above relation for power consumption (Equation 4.5) indicates that in order to maintain power density, increasing frequency requires a scaled reduction in supply voltage. The drive to reduce V_{dd} in turn leads to significant challenges in channel and contact engineering. For example, the oxide thickness must decrease to maintain sufficient charge density in the channel region, but it must not allow significant gate leakage (hence the drive toward high dielectric constant insulators). Probably the most challenging problem in V_{dd} reduction is engineering of the device threshold voltage, V_{th}, which is the voltage at which the carrier velocity approaches saturation and the device turns on. If V_{th} is too small, then there is significant off-state leakage; and if it is too close to V_{dd}, then circuit delay becomes more problematic.

The primary strength of silicon device engineering over the past 20 years has been its ability to meet the challenges of power dissipation, enabling significant increases in device density and speed while controlling the temperature increases associated with packing more devices into a smaller area. To realize viable molecular electronic devices and systems, technologies for low-power device operation and techniques that enable power-conscious scaling methodologies must be developed. Discussion of power dissipation in molecular devices — and estimations of power dissipation in molecular circuits in comparison with silicon — have not been widely discussed, but are presented in detail below.

The steady-state operating temperature at the surface of a chip can be roughly estimated from Fourier's law of heat conduction:

$$Q = U \, \Delta T \qquad\qquad (4.7)$$

where Q is the power dissipation per unit area, U is the overall heat transfer coefficient, and ΔT is the expected temperature rise in the system. The heat generated per unit area of the chip is usually transferred by conduction to a larger area where it is dissipated by convection. If a chip is generating a net ~100mW/cm^2 and cooling is achieved by natural convection (i.e., no fan), then U~20W/(m^{2}K),[17] and a temperature rise of 50°C can be expected at the chip surface. (Many high-performance laptops are now issued with warnings regarding possible burns from contacting the hot casing surface.) A fan will increase U to 50W/(m^{2}K) or higher. Because constant field scaling cannot be precisely achieved in CMOS, the power dissipation in silicon chips is expected to increase as speed increases; and there is significant effort under way to address challenges specific to heat generation and dissipation in silicon device engineering. Organic materials will be much more sensitive to temperature than current inorganic electronics, so *the ability to control power consumption and heat generation will be one of the overriding challenges that must be addressed to achieve viable high-density and high-speed molecular electronic systems.*

4.3.3 Interconnects and Parasitics

For a given supply voltage, the signal delay in a CMOS circuit is determined by the charge mobility in the channel and the capacitance of the switch. Several other *parasitic* resistance and capacitance elements can act to impede signal transfer in the circuit. The delay time of a signal moving through a circuit (τ) is given by the product of the circuit resistance and capacitance: $\tau = RC$, and parasitic elements add resistance and capacitance on top of the intrinsic R and C in the circuit. Several parasitic resistances and capacitances exist within the MOS structure itself, including contact resistance, source/drain and "spreading" resistance, gate/source overlap capacitance, and several others.

Also important are the resistance and capacitance associated with the lines that connect one circuit to another. These interconnects can be local (between devices located close to each other on the chip) or global (between elements across the length of the chip). As devices shrink, there are significant challenges in interconnect scaling. Local interconnects generally scale by decreasing wire thickness and decreasing (by the same factor) the distance between the wires. When the wires are close enough that the capacitance between neighboring wires is important (as it is in most devices), the total capacitance per unit length C_L does not depend on the scaling factor. Because the resistance per unit length R_L increases as the wire diameter squared and the wire length decreases by L, then the delay time $\tau \sim R_L C_L L^2$ is not changed by device scaling. Moreover, using typical materials (Cu and lower-k dielectrics) in current device generations, local interconnect RC delay times do not significantly affect device speed. In this way, local interconnect signal transfer rates benefit from the decreasing length of the local interconnect lines. Global interconnects, on the other hand, generally increase in length as devices shrink due to increasing chip sizes and larger numbers of circuits per chip. This leads to significant signal delay issues across the chip. Solving this problem requires advanced circuit designs to minimize long interconnects and to reduce R_L and C_L by advanced materials such as high-conductivity metals and low dielectric constant insulators. This also implies that if sufficient function could be built into very small chips (much less than a few centimeters), using molecular components for example, issues of signal delay in global interconnects could become less of a critical issue in chip operation. However, sufficient current is needed in any network structure to charge the interconnecting transfer line. If designs for molecular devices focus on low current operation, then there may not be sufficient currents to charge the interconnect in the cycle times needed for ultrafast operation.

In addition to circuit performance, another concern for parasitic resistance and capacitance is in structures developed for advanced device testing. This will be particularly important as new test structures are developed that can characterize small numbers of molecules. As device elements decrease in size, intrinsic capacitances will increase. If the test structure contains any small parasitic capacitance in series or large parasitic capacitance in parallel with the device under testing, the parasitic can dominate the signal measured. Moreover, in addition to difficulties associated with small current measurements, there are some significant problems associated with measuring large-capacitance devices (including ultrathin dielectric films), where substantial signal coupling between the device and the lead wires, for example, can give rise to spurious parasitic-related results.

4.3.4 Reliability

A hallmark attribute of solid-state device technology recognized in the 1940s was that of reliability. Personal computer crashes may be common (mostly due to software problems), but seldom does a PC processor chip fail before it is upgraded. Mainframe systems, widely used in finance, business, and government applications, have reliability requirements that are much more demanding than PCs; and current silicon technology is engineered to meet those demands. One of the most important modes of failure in CMOS devices is gate dielectric breakdown.[14] Detailed mechanisms associated with dielectric breakdown are still debated and heavily studied, but most researchers agree that charge transport through the oxide, which occurs in very small amounts during operation, helps create defects which eventually create a shorting path (breakdown) across the oxide. Defect generation is also enhanced by other factors,

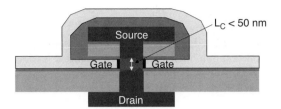

FIGURE 4.5 Schematic diagram of an example vertical field-effect transistor. For this device, the current between the source and drain flows in the vertical direction. The channel material is formed by epitaxial growth, where the channel length (L_c) is controlled by the film thickness rather than by lithography.

such as high operation temperature, which links reliability to the problem of power dissipation. Working with these restrictions, silicon devices are engineered to minimize oxide defect generation; and systems with reliable operation times exceeding 10 years can routinely be manufactured.

4.3.5　Alternate Device Structures for CMOS

It is widely recognized that CMOS device fabrication in the sub-50 nm regime will put significant pressure on current device designs and fabrication approaches. Several designs for advanced structures have been proposed, and some have promising capabilities and potential to be manufacturable.[18–22] Some of these structures include dual-gate designs, where gate electrodes on top and bottom of the channel (or surrounding the channel) can increase the current flow by a factor of 2 and reduce the charging time by a similar factor, as compared with the typical single-gate structure. Such devices can be partially depleted or fully depleted, depending on the thickness of the channel layer, applied field, and dielectric thickness used. Many of these devices rely on silicon-on-insulator (SOI) technology, where very thin crystalline silicon layers are formed or transferred onto electrically insulating amorphous dielectric layers. This electrical isolation further reduces capacitance losses in the device, improving device speed.

　　Another class of devices gaining interest is vertical structures. A schematic of a vertical device is shown in Figure 4.5. In these devices, the source and drain are on top and bottom of the channel region; and the thickness of the channel region is determined relatively easily by controlling the thickness of a deposited layer rather than by lithography. Newly developed thin film deposition approaches, such as atomic layer deposition, capable of highly conformal coverage of high dielectric constant insulators, make these devices more feasible for manufacturing with channel length well below 50 nm.[19]

4.4　Memory Devices

4.4.1　DRAM, SRAM, and Flash

The relative simplicity of memory devices, where in principle only two contact terminals are needed to produce a memory cell, makes memory an attractive possible application for molecular electronic devices. Current computer random access memory (RAM) is composed of dynamic RAM (DRAM), static RAM (SRAM), and flash memory devices. SRAM, involving up to six transistors configured as cross-coupled inverters, can be accessed very quickly; but it is expensive because it takes up significant space on the chip. SRAM is typically small (~1MB) and is used primarily in processors as cache memory. DRAM uses a storage capacitor and one or two transistors, making it more compact than SRAM and less costly to produce. DRAM requires repeated refreshing, so cycle times for data access are typically slower than with SRAM. SRAM and DRAM both operate at typical supply voltage, and the issue of power consumption and heat dissipation with increased memory density is important, following a trend similar to that for processors given in Equations 4.5 and 4.6. Overall power consumption per cm^2 for DRAM is smaller than that for processors. Flash memory requires higher voltage, and write times are slower than for DRAM; but it has the advantage of being nonvolatile. Power consumption is important for flash since it is widely used in portable devices.

4.4.2 Passive and Active Matrix Addressing

It is important to note that even though DRAM operates by storing charge in a two-terminal capacitor, it utilizes a three-terminal transistor connected to each capacitor to address each memory cell. This *active matrix* approach can be contrasted with a *passive matrix* design, where each storage capacitor is addressed by a two-terminal diode. The passive addressing approach is much more simple to fabricate, but it suffers from two critical issues: cross-talk and power consumption. Cross-talk is associated with fringing fields, where the voltage applied across a cell results in a small field across neighboring cells; and this becomes more dominant at higher cell density. The fringing field is a problem in diode-addressed arrays because of the slope of the diode current vs. voltage (IV) curve and the statistical control of the diode turn-on voltage. Because the diode IV trace has a finite slope, a small voltage drop resulting from a fringing field will give rise to a small current that can charge or discharge neighboring cells. The diode-addressing scheme also results in a small voltage drop across all the cells in the row and column addressed. Across thousands of cells, this small current can result in significant chip heating and power dissipation problems.

An active addressing scheme helps solve problems of cross-talk and power dissipation by minimizing the current flow in and out of cells not addressed. The steep logarithmic threshold of a transistor (Equation 4.2) minimizes fringing field and leakage problems, allowing reliable operation at significantly higher densities. The importance of active addressing schemes is not limited to MOS memory systems. The importance of active matrix addressing for flat panel displays has been known for some time,[23] and active addressing has proven to be critical to achieve liquid crystal displays with resolution suitable for most applications. Well-controlled manufacturing has reduced costs associated with active addressing, and many low-cost products now utilize displays with active matrix addressing.

Molecular approaches for matrix-array memory devices are currently under study by several groups. Challenges addressed by silicon-based memory, including the value and statistical control of the threshold slope, will need to be addressed in these molecular systems. It is likely that, at the ultrahigh densities proposed for molecular cross-bar array systems, some form of active matrix addressing will be needed to control device heating. This points to the importance of three-terminal switching devices for molecular systems.

4.5 Opportunities and Challenges for Molecular Circuits

The above discussion included an overview of current CMOS technology and current directions in CMOS scaling. The primary challenges in CMOS for the next several generations include lithography and patterning, tolerance control, scaling of threshold and power supply voltages, controlling high-density dopant concentration and concentration profiles, improving contact resistance and capacitances, increasing gate capacitance while reducing gate dielectric tunneling leakage, and maintaining device performance (i.e., mobility and subthreshold slope). These issues can be summarized into (at least) six distinct engineering challenges for any advanced electronic system:

- Material patterning and tolerance control
- Reliability
- Interconnects and parasitics
- Charge transport (including device speed and the importance of contacts and interfaces)
- Power and heat dissipation
- Circuit and system design and integration (including use of gain)

These challenges are not unique to CMOS or silicon technology, but they will be significant in any approach for high-density, high-speed electronic device technology (including molecular electronics). For molecular systems, these challenges are in addition to the overriding fundamental material challenges associated with design and synthesis, charge transport mechanisms, control of electrostatic and contact potentials, etc.

It is possible that alternate approaches could be developed to circumvent some of these challenges. For example, high-density and highly parallel molecular computing architectures could be developed such that the speed of an individual molecular device may not need to follow the size/speed scaling rules. It is important to understand, however, that the engineering challenges presented above must be addressed together. An increase in parallelism may enable lower device speeds, but it puts additional demands on interconnect speed and density, with additional problems in parasitics, heat and power dissipation, contacts, etc.

Lithography-based approaches to form CMOS device features less than 20 nm have been demonstrated[24,25] (but not perfected), and most of the other engineering challenges associated with production of 10–20 nm CMOS devices have yet to be solved. It is clear that Moore's law cannot continue to atomic-scale silicon transistors. This raises some natural questions:

1. At what size scale will alternate material technologies (such as quantum or molecular electronics) have a viable place in engineered electronic systems?
2. What fundamental material challenges should be addressed now to enable required engineering challenges to be met?

The theoretical and practical limits of silicon device speed and size have been addressed in several articles,[2–4,26–29] and results of these analyses will not be reviewed here. In this section, several prospective molecular computing architectures will be discussed in terms of the six engineering challenges described above. Then a prospective hybrid silicon/molecular electronics approach for engineering and implementing molecular electronic materials and systems will be presented and discussed.

4.5.1 Material Patterning and Tolerances

As devices shrink and numbers of transistor devices in a circuit increase, and variations occur in line width, film thickness, feature alignment, and overlay accuracy across a chip, significant uncertainty in device performance within a circuit may arise. This uncertainty must be anticipated and accounted for in circuit and system design, and significant effort focuses on statistical analysis and control of material and pattern tolerances in CMOS engineering. Problems in size and performance variations are expected to be more significant in CMOS as devices continue to shrink. Even so, alignment accuracy and tolerance currently achieved in silicon processing are astounding (accuracy of ~50 nm across 200 mm wafers, done routinely for thousands of wafers). The attraction of self-assembly approaches is in part related to prospects for improved alignment and arrangement of nanometer-scale objects across large areas. Nanometer scale arrangement has been demonstrated using self-assembly approaches, but reliable self-assembly at the scale approaching that routinely achieved in silicon manufacturing is still a significant challenge. Hopefully, future self-assembly approaches will offer capabilities for feature sizes below what lithography can produce at the time. What lithography will be able to achieve in the future is, of course, unknown.

4.5.2 Reliability

As discussed above, reliable operation is another hallmark of CMOS devices; and systems with reliable operation times exceeding 10 years can be routinely manufactured. Achieving this level of reliability in molecular systems is recognized as a critical issue, but it is not yet widely discussed. This is because molccular technology is not yet at the stage where details of various approaches can be compared in terms of reliability, and fundamental mechanisms in the failure of molecular systems are not yet discernible. Even so, some general observations can be made. Defect creation energies in silicon-based inorganic materials are fairly well defined. Creating a positive charged state within the silicon band gap, for example, will require energies in excess of silicon's electron affinity (> 4.1 eV). Ionization energies for many organic electronic materials are near 4–5 eV, close to that for silicon; but deformation energies are expected to be smaller in the organics, possibly leading to higher energy defect structures where less excess energy is needed to create active electronic defects. Therefore, reliability issues are expected to be more problematic in molecular systems as compared with silicon. Other factors such as melting temperatures and heat capacity also favor inorganic materials for reliable and stable operation. *Defect tolerant* designs are being considered that could overcome some of the problems of reliably interconnecting large numbers of molecular scale elements.[30] However, it is not clear how, or if, such an approach would manage a system with a defect density and distribution that changes relatively rapidly over time.

4.5.3   Interconnects, Contacts, and the Importance of Interfaces

Interconnection of molecular devices is also recognized as a critical problem. Fabrication and manipulation of molecular scale wires is an obvious concern. The size and precision for manipulating the wires must reach the same scale as the molecules. Otherwise, the largest device density would be determined by the density of the interconnect wire packing (which may not be better than future silicon devices) — not by the size of the molecules. Multilevel metallization technology is extremely well developed and crucial for fully integrated silicon systems; but as yet, no methodologies for multilevel interconnect have been demonstrated. Approaches such as metal nanocluster-modified viral particles[6] under study at the Naval Research Lab are being developed, in part, to address this issue. Also, as discussed above in relation to silicon technology, approaches will be needed to isolate molecular interconnections to avoid crosstalk and RC signal decay during transmission across and among chips. These problems with interconnect technology have the potential to severely limit realistic implementation of molecular-scale circuits.

The need to connect wires to individual molecules presents another set of problems. Several approaches to engineer linker elements within the molecular structure have been successful to achieve high-quality molecular monolayers on metals and other surfaces, and charge transport at molecule/metal surfaces is well established. However, the precise electronic structure of the molecule/metal contact and its role in the observed charge transport are not well known. As more complex material designs are developed to achieve molecular-scale arrays, other materials will likely be needed for molecular connections; and a more fundamental understanding of molecule/solid interfaces will be crucial. One specific concern is molecular conformational effects at contacts. Molecules can undergo a change in shape upon contact with a surface, resulting in a change in the atomic orbital configurations and change in the charge transfer characteristics at the interface. The relations among adsorption mechanisms, interface bond structure, configuration changes, and interface charge transfer need to be more clearly understood.

4.5.4   Power Dissipation and Gain

4.5.4.1   Two-Terminal Devices

The simplest device structures to utilize small numbers of molecules and to take advantage of self-assembly and "bottom-up" device construction involve linear molecules with contacts made to the ends. These can operate by quantum transport (i.e., conduction determined by tunneling into well-defined energy levels) or by coulomb blockade (i.e., conduction is achieved when potential is sufficient to overcome the energy of charge correlation). Molecular "shuttle" switches[31,32] are also interesting two-terminal structures. Two-terminal molecules can be considered for molecular memory[33–35] and for computation using, for example, massively parallel crossbar arrays[30] or nanocells.[36] All of the engineering challenges described above will need to be addressed for these structures to become practical, but the challenges of gain and power dissipation are particularly demanding.

The lack of gain is a primary problem for two-terminal devices, and signal propagation and fan-out must be supported by integration of other devices with gain capability. This is true for computation devices and for memory devices, where devices with gain will be needed to address and drive a memory array. Two-terminal logic devices will also suffer from problems of heat dissipation. The power dissipation described above in Equations 4.5 and 4.6 corresponds only to power lost in capacitive charging and discharging (dynamic power dissipation) and assumes that current does not flow under steady-state operation. The relation indicates that the dynamic power consumption will scale with the capacitance (i.e., the number of charges required to change the logic state of the device), which in principle could be small for molecular devices. However, quantum transport devices can have appreciable current flowing at steady state, adding another *standby power* term:

$$P_{standby} = I_{off} \cdot V_{dd} \qquad (4.8)$$

where I_{off} is the integrated current flow through the chip per unit area during steady-state operation. As discussed in detail above, complementary MOS structures are widely used now primarily because they can be engineered to enable negligible standby current (I_{off}). Standby power will be a serious concern if molecular devices become viable, and approaches that enable complementary action will be very attractive.

The relations above show that operating voltage is another important concern. Most molecular devices demonstrated to date use fairly high voltages to produce a switching event, and the parameters that influence this threshold voltage are not well understood. Threshold parameters in molecular devices must be better understood and controlled in order to manage power dissipation. Following the heat transfer analysis in Equation 4.7, generation of only ~200 mW/cm^2 in a molecular system cooled by natural convection would likely be sufficient to melt the device!

4.5.4.2 Multiterminal Structures

There are several examples of proposed molecular-scale electronic devices that make use of three or more terminals to achieve logic or memory operation. Approaches include the single electron transistor (SET),[11] the nanocell,[36] quantum cellular automata (QCA),[37] crossed or gated nanowire devices,[38–41] and field-effect devices with molecular channel regions.[42,43] Of these, the SET, nanowire devices, and field-effect devices are, in principle, capable of producing gain.

The nanocell and QCA structures are composed of sets of individual elements designed and organized to perform as logic gates and do not have specific provisions for gain built into the structures. In their simplest operating forms, therefore, additional gain elements would need to be introduced between logic elements to maintain signal intensity through the circuit. Because the QCA acts by switching position of charge on quantum elements rather than by long-range charge motion, the QCA approach is considered attractive for low power consumption. However, power will still be consumed in the switching events, with a value determined by the dynamic power equation (Equation 4.6). This dynamic power can be made small by reducing the QCA unit size (i.e., decreasing capacitance). However, reducing current will also substantially affect the ability to drive the interconnects, leading to delays in signal input and output. Also, the switching voltage must be significantly larger than the thermal voltage at the operating temperature, which puts a lower limit on V_{dd} and dynamic power consumption for these devices. The same argument for dynamic power loss and interconnect charging will apply to the switching processes in the nanocell device. The nanocell will also have the problem of standby power loss as long as NDR molecules have significant off-state leakage (i.e., low output impedance and poor device isolation). Output currents in the high impedance state are typically 1 pA, with as much as 1000 pA in the low impedance state. If a nanocell 1 µm × 1 µm contains 10^4 molecules,[36] all in the off state (I_{off} = 1pA) with V_{dd} = 2V, then the power dissipation is expected to be ~1 W/cm^2 (presuming 50% of the chip area is covered by nanocells). This power would heat the chip and likely impair operation substantially (from

Equation 4.7, ΔT would be much greater than 100°C under natural convection cooling). Stacking devices in three-dimensional structures could achieve higher densities, but it would make the heat dissipation problem substantially worse. Cooling of the center of a three-dimensional organic solid requires conductive heat transfer, which is likely significantly slower than cooling by convection from a two-dimensional surface. To address these problems, molecular electronic materials are needed with smaller operating voltage, improved off-state leakage, better on/off ratios, and sharper switching characteristics to enable structures with lower load resistances. It is important to note one aspect of the nanocell design: it may eventually enable incorporation of pairs of RTD elements in "Goto pairs," which under a limited range of conditions can show elements of current or voltage gain.[36,44]

Some devices, including the single-electron transistor (SET),[45,46] crossed and gated nanowire devices,[38–41] and field-effect devices with molecular channel regions[42,43] show promise for gain at the molecular scale; but none has yet shown true molecular-scale room temperature operation. The nanowire approach is attractive because of the possible capability for complementary operation (i.e., possibly eliminating standby power consumption). Field-effect devices with molecular channel regions show intriguing results,[42,43] but it is not clear how a small voltage applied at a large distance (30 nm) can affect transport across a small molecule (2 nm) with a larger applied perpendicular field. Even so, approaches such as these, and others with the prospect of gain, continue to be critical for advanced development of fully engineered molecular-scale electronic devices and systems.

4.5.5 Thin-Film Electronics

Thin-film electronics, based on amorphous silicon materials, is well established in device manufacturing. Charge transport rates in organic materials can challenge those in amorphous silicon; and organic materials may eventually offer advantages in simplicity of processing (including solution-based processing, for example), enabling very low-cost large-area electronic systems. The materials, fabrication, and systems engineering challenges for thin-film electronics are significantly different from those of potential single-molecule structures. Also, thin-film devices do not address the ultimate goal of high-density, high-speed electronic systems with molecular-scale individual elements. Therefore, thin-film electronics is typically treated as a separate topic altogether. However, some thin-film electronic applications could make use of additional functionality, including local computation or memory integrated within the large-area system. Such a system may require two distinctly different materials (for switching and memory, for example), and there may be some advantages of utilizing silicon and molecules together in hybrid thin-film inorganic/molecular/organic systems.

4.5.6 Hybrid Silicon/Molecular Electronics

As presented above, silicon CMOS technology is attractive, in large part, because of its ability to scale to very high densities and high speed while maintaining low power generation. This is achieved by use of: (1) complementary device integration (to achieve minimal standby power) and (2) devices with capability of current and voltage gain (to maintain the integrity of a signal as it moves within the circuit). Molecular elements are attractive for their size and possible low-cost chemical routes to assembly. However, the discussion above highlights many of the challenges that must be faced before realistic all-molecular electronic systems can be achieved.

A likely route to future all-molecular circuits is through engineered hybrid silicon/molecular systems. Realizing such hybrid systems presents additional challenges that are generally not addressed in studies of all-molecular designs, such as semiconductor/molecule chemical and electrical coupling, semiconductor/organic processing integration, and novel device, circuit, and computational designs. However, these additional challenges of hybrid devices are likely more surmountable in the near term than those of all-molecule devices and could give rise to new structures that take advantage of the benefits of molecules and silicon technology. For example, molecular RTDs assembled onto silicon CMOS structures could enable devices with multiple logic states, so more complex computation could be performed within the achievable design rules of silicon devices. Also, molecular memory devices integrated with CMOS transistors could enable ultrahigh density and ultrafast memories close-coupled with silicon to challenge SRAM devices in cost and performance for cache applications in advanced computing. Coupling molecules with silicon could also impact the problem of molecular characterization. As discussed above, parasitic effects in device characterization are a serious concern; and approaches to intimately couple organic electronic elements with silicon devices would result in structures with well-characterized parasitics, leading to reliable performance analysis of individual and small ensembles of molecules — critically important for the advance of any molecular-based electronic technology.

4.6 Summary and Conclusions

Present-day silicon technology is a result of many years of tremendously successful materials, device, and systems engineering. Proposed future molecular-based devices could substantially advance computing technology, but the engineering of proposed molecular electronic systems will be no less challenging than what silicon has overcome to date. It is important to understand the reasons why silicon is so successful and to realize that silicon has and will continue to overcome many substantial "show-stoppers" to successful production. Most of the engineering issues described above can be reduced to challenges in materials and materials integration. For example, can molecular switches with sufficiently low operating voltage and operating current be realized to minimize heat dissipation problems? Or, can charge coupling and transport through interfaces be understood well enough to design improved electrical contacts to molecules to control parasitic losses? It is likely that an integrated approach, involving close-coupled studies of fundamental materials and engineered systems including hybrid molecular/silicon devices, will give rise to viable and useful molecular electronic elements with substantially improved accessibility, cost, and capability over current electronic systems.

Acknowledgments

The author acknowledges helpful discussions with several people, including Veena Misra, Paul Franzon, Chris Gorman, Bruce Gnade, John Hauser, David Nackashi, and Carl Osburn. He also thanks Bruce Gnade, Carl Osburn, and Dong Niu for critical reading of the manuscript.

References

1. Semiconductor Industry Association, *The International Technology Roadmap for Semiconductors* (Austin, TX, 2001) (http://public.itrs.net).
2. R.W. Keyes, Fundamental limits of silicon technology, *Proc. IEEE* 89, 227–239 (2001).
3. J.D. Plummer and P.B. Griffin, Material and process limits in silicon VLSI technology, *Proc. IEEE* 89, 240–258 (2001).
4. R.L. Harriott, Limits of lithography, *Proc. IEEE* 89, 366–374 (2001).
5. M.A. Ratner, Introducing molecular electronics, *Mater. Today* 5, 20–27 (2002).
6. K.S. Kwok and J.C. Ellenbogen, Moletronics: future electronics, *Mater. Today* 5, 28–37 (2002).
7. Y. Wada, Prospects for single molecule information processing devices, *Proc. IEEE* 89, 1147–1173 (2001).
8. J.F. Wakerly, *Digital Design Principles and Practices* (Prentice Hall, Upper Saddle River, NJ, 2000).
9. P. Horowitz and W. Hill, *The Art of Electronics* (Cambridge University Press, New York, 1980).
10. L.J. Micheel, A.H. Taddiken, and A.C. Seabaugh, Multiple-valued logic computation circuits using micro- and nanoelectronic devices, *Proc. 23rd Intl. Symp. Multiple-Valued Logic*, 164–169 (1993).
11. K. Uchida, J. Koga, A. Ohata, and A. Toriumi, Silicon single-electron tunneling device interfaced with a CMOS inverter, *Nanotechnology* 10, 198–200 (1999).

12. R.H. Mathews, J.P. Sage, T.C.L.G. Sollner, S.D. Calawa, C.-L. Chen, L.J. Mahoney, P.A. Maki, and K.M. Molvar, A new RTD-FET logic family, *Proc. IEEE* 87, 596–605 (1999).

13. D. Goldhaber–Gordon, M.S. Montemerlo, J.C. Love, G.J. Opiteck, and J.C. Ellenbogen, Overview of nanoelectronic devices, *Proc. IEEE* 85, 521–540 (1997).

14. D.A. Buchanan, Scaling the gate dielectric: materials, integration, and reliability, *IBM J. Res. Dev.* 43, 245 (1999).

15. Y. Taur and T.H. Ning, *Fundamentals of Modern VLSI Devices* (Cambridge University Press, Cambridge, U.K., 1998).

16. Intel Pentium III Datasheet (ftp://download.intel.com/design/PentiumIII/datashts/24526408.pdf).

17. C.O. Bennett and J.E. Myers, *Momentum, Heat, and Mass Transfer* (McGraw-Hill, New York, 1982).

18. H. Takato, K. Sunouchi, N. Okabe, A. Nitayama, K. Hieda, F. Horiguchi, and F. Masuoka, High performance CMOS surrounding gate transistor (SGT) for ultra high density LSIs, *IEEE IEDM Tech. Dig.*, 222–226 (1988).

19. J.M. Hergenrother, G.D. Wilk, T. Nigam, F.P. Klemens, D. Monroe, P.J. Silverman, T.W. Sorsch, B. Busch, M.L. Green, M.R. Baker, T. Boone, M.K. Bude, N.A. Ciampa, E.J. Ferry, A.T. Fiory, S.J. Hillenius, D.C. Jacobson, R.W. Johnson, P. Kalavade, R.C. Keller, C.A. King, A. Kornblit, H.W. Krautter, J.T.-C. Lee, W.M. Mansfield, J.F. Miner, M.D. Morris, O.-H. Oh, J.M. Rosamilia, B.T. Sapjeta, K. Short, K. Steiner, D.A. Muller, P.M. Voyles, J.L. Grazul, E.J. Shero, M.E. Givens, C. Pomarede, M. Mazanec, and C. Werkhoven, 50nm vertical replacement gate (VRG) nMOSFETs with ALD HfO₂ and Al₂O₃ gate dielectrics, *IEEE IEDM Tech. Dig.*, 3.1.1–3.1.4 (2001).

20. D. Hisamoto, W.-C. Lee, J. Kedzierski, H. Takeuchi, K. Asano, C. Kuo, E. Anderson, T.-J. King, J. Bokor, and C. Hu, FinFET — a self-aligned double-gate MOSFET scalable to 20 nm, *IEEE Trans. Electron Devices* 47, 2320–2325 (2000).

21. C.M. Osburn, I. Kim, S.K. Han, I. De, K.F. Yee, J.R. Hauser, D.-L. Kwong, T.P. Ma, and M.C. Öztürk, Vertically-scaled MOSFET gate stacks and junctions: how far are we likely to go?, *IBM J. Res. Dev.*, (accepted) (2002).

22. T. Schulz, W. Rösner, L. Risch, A. Korbel, and U. Langmann, Short-channel vertical sidewall MOSFETs, *IEEE Trans. Electron Devices* 48 (2001).

23. P.M. Alt and P. Pleshko, Scanning limitations of liquid crystal displays, *IEEE Trans. Electron Devices* ED-21, 146–155 (1974).

24. R. Chau, J. Kavalieros, B. Doyle, A. Murthy, N. Paulsen, D. Lionberger, D. Barlage, R. Arghavani, B. Roberds, and M. Doczy, A 50nm depleted-substrate CMOS transistor (DST), *IEEE IEDM Tech. Dig.* 2001, 29.1.1–29.1.4 (2001).

25. M. Fritze, B. Tyrrell, D.K. Astolfi, D. Yost, P. Davis, B. Wheeler, R. Mallen, J. Jarmolowicz, S.G. Cann, H.-Y. Liu, M. Ma, D.Y. Chan, P.D. Rhyins, C. Carney, J.E. Ferri, and B.A. Blachowicz, 100-nm node lithography with KrF? *Proc. SPIE* 4346, 191–204 (2001).

26. D.J. Frank, S.E. Laux, and M.V. Fischetti, Monte Carlo simulation of 30nm dual-gate MOSFET: how short can Si go?, *IEEE IEDM Tech. Dig.*, 21.1.1–21.1.3 (1992).

27. J.R. Hauser and W.T. Lynch, Critical front end materials and processes for 50 nm and beyond IC devices, *(unpublished)*.

28. T.H. Ning, Silicon technology directions in the new millennium, *IEEE 38 Intl. Reliability Phys. Symp.* (2000).

29. J.D. Meindl, Q. Chen, and J.A. Davis, Limits on silicon nanoelectronics for terascale integration, *Science* 293, 2044–2049 (2001).

30. J.R. Heath, P.J. Kuekes, G.S. Snider, and S. Williams, A defect-tolerant computer architecture: opportunities for nanotechnology, *Science* 280, 1716–1721 (1998).

31. P.L. Anelli et al., Molecular meccano I. [2] rotaxanes and a [2]catenane made to order, *J. Am. Chem. Soc.* 114, 193–218 (1992).

32. D.B. Amabilino and J.F. Stoddard, Interlocked and intertwined structures and superstructures, *Chem. Revs.* 95, 2725–2828 (1995).

33. D. Gryko, J. Li, J.R. Diers, K.M. Roth, D.F. Bocian, W.G. Kuhr, and J.S. Lindsey, Studies related to the design and synthesis of a molecular octal counter, *J. Mater. Chem.* 11, 1162 (2001).

34. K.M. Roth, N. Dontha, R.B. Dabke, D.T. Gryko, C. Clausen, J.S. Lindsey, D.F. Bocian, and W.G. Kuhr, Molecular approach toward information storage based on the redox properties of porphyrins in self-assembled monolayers, *J. Vacuum Sci. Tech. B* 18, 2359–2364 (2000).

35. M.A. Reed, J. Chen, A.M. Rawlett, D.W. Price, and J.M. Tour, Molecular random access memory cell, *Appl. Phys. Lett.* 78, 3735–3737 (2001).

36. J.M. Tour, W.L.V. Zandt, C.P. Husband, S.M. Husband, L.S. Wilson, P.D. Franzon, and D.P. Nackashi, Nanocell logic gates for molecular computing, *(submitted)* (2002).

37. G. Toth and C.S. Lent, Quasiadiabatic switching for metal-island quantum-doc cellular automata, *J. Appl. Phys.* 85, 2977–2181 (1999).

38. X. Liu, C. Lee, C. Zhou, and J. Han, Carbon nanotube field-effect inverters, *Appl. Phys. Lett.* 79, 3329–3331 (2001).

39. Y. Cui and C.M. Lieber, Functional nanoscale electronic devices assembled using silicon nanowire building blocks, *Science* 291, 851–853 (2001).

40. Y. Huang, X. Duan, Y. Cui, L.J. Lauhon, K.-H. Kim, and C.M. Lieber, Logic gates and computation from assembled nanowire building blocks, *Science* 294, 1313–1317 (2001).

41. A. Bachtold, P. Hadley, T. Nakanishi, and C. Dekker, Logic circuits with nanotube transistors, *Science* 294, 1317–1320 (2001).

42. J.H. Schön, H. Meng, and Z. Bao, Self-assembled monolayer organic field-effect transistors, *Nature* 413, 713–716 (2001).

43. J.H. Schön, H. Meng, and Z. Bao, Field effect modulation of the conductance of single molecules, *Science* 294, 2138–2140 (2001).

44. J.C. Ellenbogen and J.C. Love, Architectures for molecular electronic computers: 1. logic structures and an adder designed from molecular electronic diodes, *Proc. IEEE* 88, 386–426 (2000).

45. M.A. Kastner, The single electron transistor, *Rev. Mod. Phys.* 64, 849–858 (1992).

46. H. Ahmed and K. Nakazoto, Single-electronic devices, *Microelectron. Eng.* 32, 297–315 (1996).

5

Molecular Electronic Computing Architectures

5.1   Present Microelectronic Technology.................................5-1
5.2   Fundamental Physical Limitations
       of Present Technology...5-2
5.3   Molecular Electronics 5-3
5.4   Computer Architectures Based
       on Molecular Electronics...5-4
       Quantum Cellular Automata (QCA) • Crossbar Arrays • The
       Nanocell Approach to a Molecular Computer: Synthesis • The
       Nanocell Approach to a Molecular Computer:
       The Functional Block
5.5   Characterization of Switches and Complex
       Molecular Devices 5-23
5.6   Conclusion...5-24
Acknowledgments..5-25
References..5-25

James M. Tour
Dustin K. James
Rice University

5.1   Present Microelectronic Technology

Technology development and industrial competition have been driving the semiconductor industry to produce smaller, faster, and more powerful logic devices. That the number of transistors per integrated circuit will double every 18–24 months due to advancements in technology is commonly referred to as *Moore's Law*, after Intel founder Gordon Moore, who made the prediction in a 1965 paper with the prophetic title "Cramming More Components onto Integrated Circuits."[1] At the time he thought that his prediction would hold until at least 1975; however, the exponentially increasing rate of circuit densification has continued into the present (Graph 5.1). In 2000, Intel introduced the Pentium 4 containing 42 million transistors, an amazing engineering achievement. The increases in packing density of the circuitry are achieved by shrinking the line widths of the metal interconnects, by decreasing the size of other features, and by producing thinner layers in the multilevel device structures. These changes are only brought about by the development of new fabrication techniques and materials of construction. As an example, commercial metal interconnect line widths have decreased to 0.13 μm. The resistivity of Al at 0.13 μm line width, combined with its tendency for electromigration (among other problems), necessitated the substitution of Cu for Al as the preferred interconnect metal in order to achieve the 0.13 μm line width goal. Cu brings along its own troubles, including its softness, a tendency to migrate into silicon dioxide (thus requiring a barrier coating of Ti/TiN), and an inability to deposit Cu layers via the

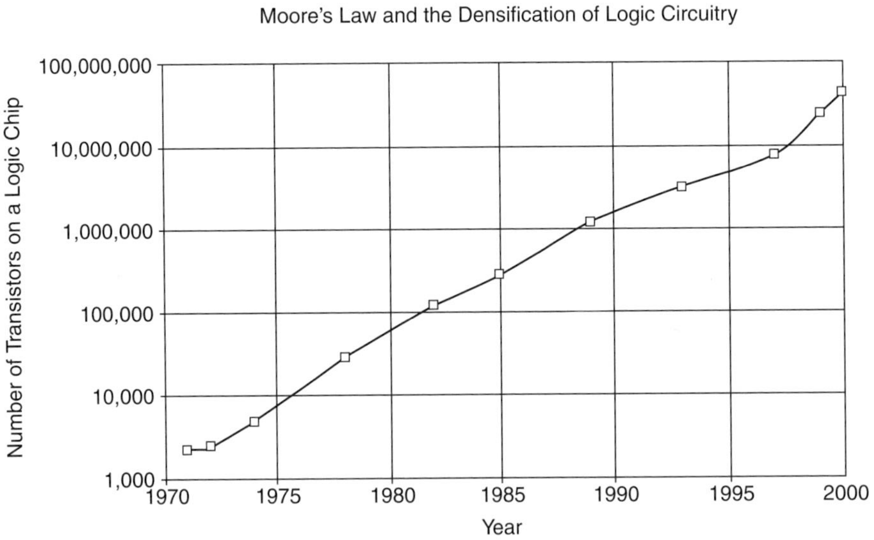

Moore's Law and the Densification of Logic Circuitry

GRAPH 5.1   The number of transistors on a logic chip has increased exponentially since 1972. (Courtesy of Intel Data.)

vapor phase. New tools for depositing copper using electroless electroplating and new technologies for removing the metal overcoats — because copper does not etch well — had to be developed to meet these and other challenges. To integrate Cu in the fabrication line, innovations had to be made all the way from the front end to the back end of the process. These changes did not come without cost, time, and Herculean efforts.

5.2   Fundamental Physical Limitations of Present Technology

This top-down method of producing faster and more powerful computer circuitry by shrinking features cannot continue because there are fundamental physical limitations, related to the material of construction of the solid-state-based devices, that cannot be overcome by engineering. For instance, charge leakage becomes a problem when the insulating silicon oxide layers are thinned to about three silicon atoms deep, which will be reached commercially by 2003–2004. Moreover, silicon loses its original band structure when it is restricted to very small sizes. The lithography techniques used to create the circuitry on the wafers has also neared its technological limits, although derivative technologies such as e-beam lithography, extreme ultraviolet lithography (EUV),[2] and x-ray lithography are being developed for commercial applications. A tool capable of x-ray lithography in the sub-100 nm range has been patented.[3]

Financial roadblocks to continued increases in circuit density exist. Intel's Fab 22, which opened in Chandler, Arizona, in October 2001, cost $2 billion to construct and equip; and it is slated to produce logic chips using copper-based 0.13 μm technology on 200 mm wafers. The cost of building a Fab is projected to rise to $15–30 billion by 2010[4] and could be as much as $200 billion by 2015.[5] The staggering increase in cost is due to the extremely sophisticated tools that will be needed to form the increasingly small features of the devices. It is possible that manufacturers may be able to take advantage of infrastructure already in place in order to reduce the projected cost of the introduction of the new technologies, but much is uncertain because the methods for achieving further increases in circuit density are unknown or unproven.

As devices increase in complexity, defect and contamination control become even more important as defect tolerance is very low — nearly every device must work perfectly. For instance, cationic metallic impurities in the wet chemicals such as sulfuric acid used in the fabrication process are measured in the part per billion (ppb) range. With decreases in line width and feature size, the presence of a few ppb of metal contamination could lead to low chip yields. Therefore, the industry has been driving suppliers to produce chemicals with part per trillion (ppt) contamination levels, raising the cost of the chemicals used.

Depending on the complexity of the device, the number of individual processing steps used to make them can be in the thousands.[6] It can take 30–40 days for a single wafer to make it through the manufacturing process. Many of these steps are cleaning steps, requiring some fabs to use thousands of gallons of ultra-pure water per minute.[7] The reclaim of waste water is gaining importance in semiconductor fab operations.[8] The huge consumption of water and its subsequent disposal can lead to problems where aquifers are low and waste emission standards require expensive treatment technology.

A new technology that addressed only one of the potential problems we have discussed would be of interest to the semiconductor industry. A new technology would be revolutionary if it produced faster and smaller logic and memory chips, reduced complexity, saved days to weeks of manufacturing time, and reduced the consumption of natural resources.

5.3 Molecular Electronics

How do we overcome the limitations of the present solid-state electronic technology? Molecular electronics is a fairly new and fascinating area of research that is firing the imagination of scientists as few research topics have.[9] For instance, *Science* magazine labeled the hook-up of molecules into functional circuits as the breakthrough of the year for 2001.[10] Molecular electronics involves the search for single molecules or small groups of molecules that can be used as the fundamental units for computing, i.e., wires, switches, memory, and gain elements.[11] The goal is to use these molecules, designed from the bottom up to have specific properties and behaviors, instead of present solid-state electronic devices that are constructed using lithographic technologies from the top down. The top-down approach is currently used in the silicon industry, wherein small features such as transistors are etched into silicon using resists and light; the ever-increasing demand for densification is stressing the industry. The bottom-up approach, on the other hand, implies the construction of functionality into small features, such as molecules, with the opportunity to have the molecules further self-assemble into the higher ordered structural units such as transistors. Bottom-up methodologies are quite natural in that all systems in nature are constructed bottom-up. For example, molecules with specific features assemble to form higher order structures such as lipid bilayers. Further self-assembly, albeit incomprehensibly complex, causes assembly into cells and further into high life forms. Hence, utilization of a diversity of self-assembly processes could lead to enormous advances in future manufacturing processes once scientists learn to further control specific molecular-level interactions.

Ultimately, given advancements in our knowledge, it is thought by the proponents of molecular electronics that its purposeful bottom-up design will be more efficient than the top-down method, and that the incredible structure diversity available to the chemist will lead to more effective molecules that approach optional functionality for each application. A single mole of molecular switches, weighing about 450 g and synthesized in small reactors (a 22-L flask might suffice for most steps of the synthesis), contains 6×10^{23} molecules — more than the combined number of transistors ever made in the history of the world. While we do not expect to be able to build a circuit in which each single molecule is addressable and is connected to a power supply (at least not in the first few generations), the extremely large numbers of switches available in a small mass illustrate one reason molecular electronics can be a powerful tool for future computing development.

The term *molecular electronics* can cover a broad range of topics. Petty, Bryce, and Bloor recently explored molecular electronics.[12] Using their terminology, we will focus on molecular-scale electronics instead of molecular materials for electronics. Molecular materials for electronics deal with films or crystals (i.e., thin-film transistors or light-emitting diodes) that contain many trillions of molecules per functional unit, the properties of which are measured on the macroscopic scale, while molecular-scale electronics deals with one to a few thousand molecules per device.

5.4 Computer Architectures Based on Molecular Electronics

In this section we will initially discuss three general architectural approaches that researchers are considering to build computers based on molecular-scale electronics and the advances made in these three areas in the years 1998–2001. In addition, we will touch upon progress made in measuring the electrical characteristics of molecular switches and in designing logic devices using molecular electronics components.

The first approach to molecular computing, based on quantum cellular automata (QCA), was briefly discussed in our prior review.[11] This method relies on electrostatic field repulsions to transport information throughout the circuitry. One major benefit of the QCA approach is that heat dissipation is less of an issue because only one to fractions of an electron are used rather than the 16,000 to 18,000 electrons needed for each bit of information in classical solid-state devices.

The second approach is based on the massively parallel solid-state Teramac computer developed at Hewlett-Packard (HP)[4] and involves building a similarly massively parallel computing device using molecular electronics-based crossbar technologies that are proposed to be very defect tolerant.[13] When applied to molecular systems, this approach is proposed to use single-walled carbon nanotubes (SWNT)[14–18] or synthetic nanowires[14,19–22] for crossbars. As we will see, logic functions are performed either by sets of crossed and specially doped nanowires or by molecular switches placed at each crossbar junction.

The third approach uses molecular-scale switches as part of a nanocell, a new concept that is a hybrid between present silicon-based technology and technology based purely on molecular switches and molecular wires (in reality, the other two approaches will also be hybrid systems in their first few generations).[23] The nanocell relies on the use of arrays of molecular switches to perform logic functions but does not require that each switching molecule be individually addressed or powered. Furthermore, it utilizes the principles of chemical self-assembly in construction of the logic circuitry, thereby reducing complexity. However, programming issues increase dramatically in the nanocell approach.

While solution-phase-based computing, including DNA computing,[24] can be classified as molecular-scale electronics, it is a slow process due to the necessity of lining up many bonds, and it is wedded to the solution phase. It may prove to be good for diagnostic testing, but we do not see it as a commercially viable molecular electronics platform; therefore, we will not cover it in this review.

Quantum computing is a fascinating area of theoretical and laboratory study,[25–28] with several articles in the popular press concerning the technology.[29,30] However, because quantum computing is based on interacting quantum objects called *qubits*, and not molecular electronics, it will not be covered in this review. Other interesting approaches to computing such as "spintronics"[31] and the use of light to activate switching[32] will also be excluded from this review.

5.4.1 Quantum Cellular Automata (QCA)

Quantum dots have been called *artificial atoms* or *boxes for electrons*[33] because they have discrete charge states and energy-level structures that are similar to atomic systems and can contain from a few thousand to one electron. They are typically small electrically conducting regions, 1 μm or less in size, with a variety of geometries and dimensions. Because of the small volume, the electron energies are quantized. No shell structure exists; instead, the generic energy spectrum has universal statistical properties associated with quantum chaos.[34] Several groups have studied the production of quantum dots.[35] For example, Leifeld and coworkers studied the growth of Ge quantum dots on silicon surfaces that had been precovered with 0.05–0.11 monolayer of carbon,[36] i.e., carbon atoms replaced about five to ten of every 100 silicon atoms at the surface of the wafer. It was found that the Ge dots grew directly over the areas of the silicon surface where the carbon atoms had been inserted.

Heath discovered that hexane solutions of Ag nanoparticles, passivated with octanethiol, formed spontaneous patterns on the surface of water when the hexane was evaporated;[37] and he has prepared superlattices of quantum dots.[38,39] Lieber has investigated the energy gaps in "metallic" single-walled

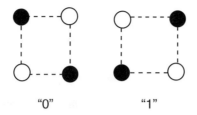

FIGURE 5.1 The two possible ground-state polarizations, denoted "0" and "1," of a four-dot QCA cell. Note that the electrons are forced to opposite corners of the cells by Coulomb repulsion.

carbon nanotubes[16] and has used an atomic-force microscope to mechanically bend SWNT in order to create quantum dots less than 100 nm in length.[18] He found that most metallic SWNT are not true metals and that, by bending the SWNT, a defect was produced that had a resistance of 10 to 100 kΩ. Placing two defects less than 100 nm apart produced the quantum dots.

One proposed molecular computing structural paradigm that utilizes quantum dots is termed a *quantum cellular automata* (QCA) wherein four quantum dots in a square array are placed in a cell such that electrons are able to tunnel between the dots but are unable to leave the cell.[40] As shown in Figure 5.1, when two excess electrons are placed in the cell, Coulomb repulsion will force the electrons to occupy dots on opposite corners. The two ground-state polarizations are energetically equivalent and can be labeled logic "0" or "1." Flipping the logic state of one cell, for instance by applying a negative potential to a lead near the quantum dot occupied by an electron, will result in the next-door cell flipping ground states in order to reduce Coulomb repulsion. In this way, a line of QCA cells can be used to do computations. A simple example is shown in Figure 5.2, the structure of which could be called a *binary wire*, where a "1" input gives a "1" output. All of the electrons occupy positions as far away from their neighbors as possible, and they are all in a ground-state polarization. Flipping the ground state of the cell on the left end will result in a domino effect, where each neighboring cell flips ground states until the end of the wire is reached. An inverter built from QCA cells is shown in Figure 5.3 — the output is "0" when the input is "1." A QCA topology that can produce *AND* and *OR* gates is called a *majority gate*[41] and is shown in Figure 5.4, where the three input cells "vote" on the polarization of the central cell. The polarization of the central cell is then propagated as the output. One of the inputs can be

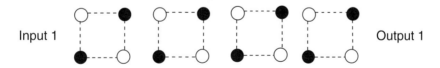

FIGURE 5.2 Simple QCA cell logic line where a logic input of 1 gives a logic output of 1.

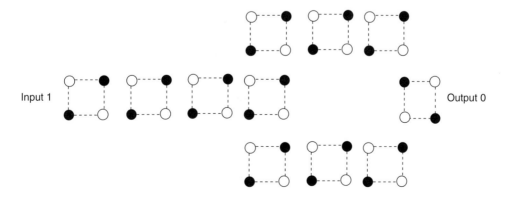

FIGURE 5.3 An inverter built using QCA cells such that a logic input of 1 yields a logic output of 0.

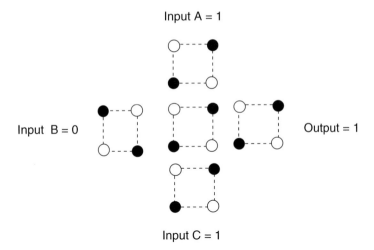

FIGURE 5.4 A QCA majority cell in which the three input cells A, B, and C determine the ground state of the center cell, which then determines the logic of the output. A logic input of 0 gives a logic output of 1.

designated a programming input and determines whether the majority gate produces an AND or an OR. If the programming gate is a logic 0, then the result shown in Figure 5.4 is OR while a programming gate equal to logic 1 would produce a result of AND.

A QCA fan-out structure is shown in Figure 5.5. Note that when the ground state of the input cell is flipped, the energy put into the system may not be enough to flip all the cells of both branches of the structure, producing long-lived metastable states and erroneous calculations. Switching the cells using a quasi-adiabatic approach prevents the production of these metastable states.[42]

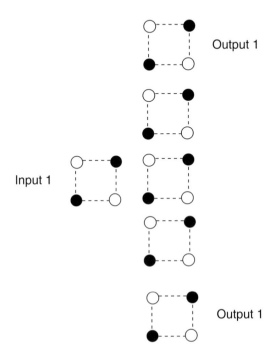

FIGURE 5.5 A fan-out constructed of QCA cells. A logic input of 1 produces a logic output of 1 at both ends of the structure.

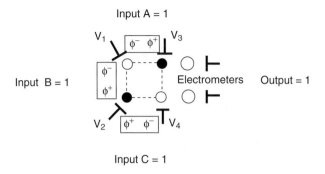

Input A = 1

Input B = 1 V_1 ... V_3 ... Electrometers   Output = 1

V_2 ... V_4

Input C = 1

FIGURE 5.6   A QCA majority cell as set up experimentally in a nonmolecular system.

Amlani and coworkers have demonstrated experimental switching of 6-dot QCA cells.[43–45] The polarization switching was accomplished by applying biases to the gates of the input double-dot of a cell fabricated on an oxidized Si surface using standard Al tunnel junction technology, with Al islands and leads patterned by e-beam lithography, followed by a shadow evaporation process and an *in situ* oxidation step. The switching was experimentally verified in a dilution refrigerator using the electrometers capacitively coupled to the output double-dot.

A functioning majority gate was also demonstrated by Amlani and coworkers,[46] with logic AND and OR operations verified using electrometer outputs after applying inputs to the gates of the cell. The experimental setup for the majority gate is shown in Figure 5.6, where the three input tiles A, B, and C were supplanted by leads with biases that were equivalent to the polarization states of the input cells. The negative or positive bias on a gate mimicked the presence or absence of an electron in the input dots of the tiles A, B, and C that were replaced. The truth table for all possible input combinations and majority gate output is shown in Figure 5.7. The experimental results are shown in Figure 5.8. A QCA binary wire has been experimentally demonstrated by Orlov and coworkers,[47] and Amlani and coworkers have demonstrated a leadless QCA cell.[48] Bernstein and coworkers have demonstrated a latch in clocked QCA devices.[49]

While the use of quantum dots in the demonstration of QCA is a good first step in reduction to practice, the ultimate goal is to use individual molecules to hold the electrons and pass electrostatic potentials down QCA wires. We have synthesized molecules that have been shown by *ab initio* computational methods to have the capability of transferring information from one molecule to another through electrostatic potential.[50] Synthesized molecules included three-terminal molecular junctions, switches, and molecular logic gates.

The QCA method faces several problems that need to be resolved before QCA-based molecular computing can become a reality. While relatively large quantum-dot arrays can be fabricated using existing methods, a major problem is that placement of molecules in precisely aligned arrays at the nanoscopic level is very difficult to achieve with accuracy and precision. Another problem is that degradation of only one molecule in the array can cause failure of the entire circuit. There has also

A	B	C	Output
0	0	0	0
0	0	1	0
0	1	1	1
0	1	0	0
1	1	0	1
1	1	1	1
1	0	1	1
1	0	0	0

FIGURE 5.7   The logic table for the QCA majority cell.

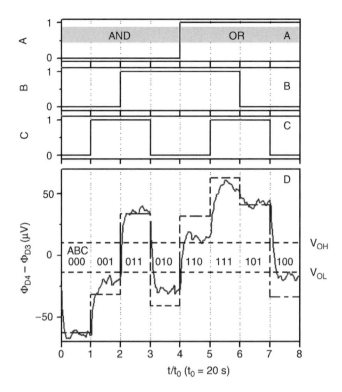

FIGURE 5.8   Demonstration of majority gate operation, where A to C are inputs in Gray code. The first four and last four inputs illustrate AND and OR operations, respectively. (D) Output characteristic of majority gate where $t_0 = 20$ s is the input switching period. The dashed stair-step-like line shows the theory for 70 mK; the solid line represents the measured data. Output high (V_{OH}) and output low (V_{OL}) are marked by dashed horizontal lines. (Reprinted from Amlani, I., Orlov, A.O., Toth, G., Bernstein, G.H., Lent, C.S., and Snider, G.L. *Science*, 284, 289, 1999. ©1999 American Association for the Advancement of Science. With permission.)

been some debate about the unidirectionality (or lack thereof) of QCA designs.[47,51–52] Hence, even small examples of 2-dots have yet to be demonstrated using molecules, but hopes remain high and researchers are continuing their efforts.

5.4.2   Crossbar Arrays

Heath, Kuekes, Snider, and Williams recently reported on a massively parallel experimental computer that contained 220,000 hardware defects yet operated 100 times faster than a high-end single processor workstation for some configurations.[4] The solid-state-based (not molecular electronic) Teramac computer built at HP relied on its fat-tree architecture for its logical configuration. The minimum communication bandwidth needed to be included in the fat-tree architecture was determined by utilizing Rent's rule, which states that the number of wires coming out of a region of a circuit should scale with the power of the number of devices (n) in that region, ranging from $n^{1/2}$ in two dimensions to $n^{2/3}$ in three dimensions. The HP workers built in excess bandwidth, putting in many more wires than needed. The reason for the large number of wires can be understood by considering the simple but illustrative city map depicted in Figure 5.9. To get from point A to point B, one can take local streets, main thoroughfares, freeways, interstate highways, or any combination thereof. If there is a house fire at point C, and the local streets are blocked, then by using the map it is easy to see how to go around that area to get to point B. In the Teramac computer, *street blockages* are stored in a defect database; when one device needs to communicate with another device, it uses the database and the map to determine how to get there. The Teramac design can therefore tolerate a large number of defects.

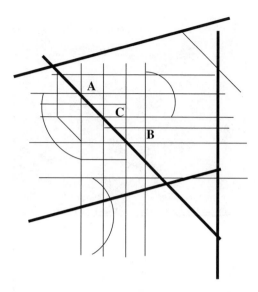

FIGURE 5.9   A simple illustration of the defect tolerance of the Teramac computer. In a typical city, many routes are available to get from point A to point B. One who dislikes traffic might take only city streets (thin lines) while others who want to arrive faster may take a combination of city streets and highways (thick lines). If there were a house fire at point C, a traveler intent on driving only on city streets could look at the map and determine many alternate routes from A to B.

In the Teramac computer (or a molecular computer based on the Teramac design), the wires that make up the address lines controlling the settings of the configuration switches and the data lines that link the logic devices are the most important and plentiful part of the computer. It is logical that a large amount of research has been done to develop nanowires (NW) that could be used in the massively parallel molecular computer. Recall that nanoscale wires are needed if we are to take advantage of the smallness in size of molecules.

Lieber has reviewed the work done in his laboratory to synthesize and determine the properties of NW and nanotubes.[14] Lieber used Au or Fe catalyst nanoclusters to serve as the nuclei for NW of Si and GeAs with 10 nm diameters and lengths of hundreds of nm. By choosing specific conditions, Lieber was able to control both the length and the diameter of the single crystal semiconductor NW.[20] Silicon NW doped with B or P were used as building blocks by Lieber to assemble semiconductor nanodevices.[21] Active bipolar transistors were fabricated by crossing n-doped NW with p-type wire base. The doped wires were also used to assemble complementary inverter-like structures.

Heath reported the synthesis of silicon NW by chemical vapor deposition using SiH_4 as the Si source and Au or Zn nanoparticles as the catalytic seeds at 440°C.[22,53] The wires produced varied in diameter from 14 to 35 nm and were grown on the surface of silicon wafers. After growth, isolated NW were mechanically transferred to wafers; and Al contact electrodes were put down by standard e-beam lithography and e-beam evaporation such that each end of a wire was connected to a metallic contact. In some cases a gate electrode was positioned at the middle of the wire (Figure 5.10). Tapping AFM indicated the wire in this case was 15 nm in diameter.

Heath found that annealing the Zi-Si wires at 550°C produced increased conductance attributed to better electrode/nanowire contacts (Figure 5.11). Annealing Au-Si wires at 750°C for 30 min increased current about 10^4, as shown in Figure 5.12 — an effect attributed to doping of the Si with Au and lower contact resistance between the wire and Ti/Au electrodes.

Much research has been done to determine the value of SWNT as NW in molecular computers. One problem with SWNT is their lack of solubility in common organic solvents. In their synthesized state, individual SWNT form ropes[54] from which it is difficult to isolate individual tubes. In our laboratory some solubility of the tubes was seen in 1,2-dichlorobenzene.[55] An obvious route to better solubilization

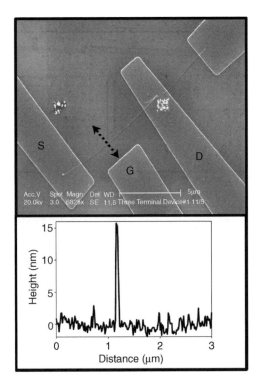

FIGURE 5.10   (Top) SEM image of a three-terminal device, with the source (S), gate (G), and drain (D) labeled. (Bottom) Tapping mode AFM trace of a portion of the silicon nanowire (indicated with the dashed arrow in the SEM image), revealing the diameter of the wire to be about 15 nm. (Reprinted from Chung, S.-W., Yu, J.-Y, and Heath, J.R., *Appl. Phys. Lett.*, 76, 2068, 2000. ©2000 American Institute of Physics. With permission.)

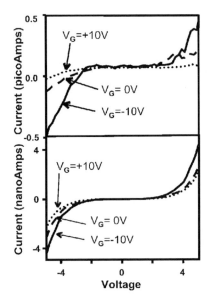

FIGURE 5.11   Three-terminal transport measurements of an as-prepared 15 nm Si nanowire device contacted with Al electrodes (top) and the same device after annealing at 550°C (bottom). In both cases, the gating effect indicates *p*-type doping. (Reprinted from Chung, S.-W., Yu, J.-Y, and Heath, J.R., *Appl. Phys. Lett.*, 76, 2068, 2000. ©2000 American Institute of Physics. With permission.)

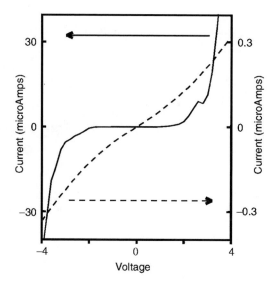

FIGURE 5.12 I(V) characteristics of Au-nucleated Si nanowires contacted with Ti/Au electrodes, before (solid line, current axis on left) and after (dashed line, current axis on right) thermal treatment (750°C, 1 h). After annealing the wire exhibits metallic-like conductance, indicating that the wire has been heavily doped. (Reprinted from Chung, S.-W., Yu, J.-Y, and Heath, J.R., *Appl. Phys. Lett.*, 76, 2068, 2000. ©2000 American Institute of Physics. With permission.)

is to functionalize SWNT by attachment of soluble groups through covalent bonding. Margrave and Smalley found that fluorinated SWNT were soluble in alcohols,[56] while Haddon and Smalley were able to dissolve SWNT by ionic functionalization of the carboxylic acid groups present in purified tubes.[57]

We have found that SWNT can be functionalized by electrochemical reduction of aryl diazonium salts in their presence.[58] Using this method, about one in 20 carbon atoms of the nanotube framework are reacted. We have also found that the SWNT can be functionalized by direct treatment with aryl diazonium tetrafluoroborate salts in solution or by *in situ* generation of the diazonium moiety using an alkyl nitrite reagent.[59] These functional groups give us handles with which we can direct further, more selective derivatization.

Unfortunately, fluorination and other sidewall functionalization methods can perturb the electronic nature of the SWNT. An approach by Smalley[54,60] and Stoddart and Heath[17] to increasing the solubility without disturbing the electronic nature of the SWNT was to wrap polymers around the SWNT to break up and solubilize the ropes but leave individual tube's electronic properties unaffected. Stoddart and Heath found that the SWNT ropes were not separated into individually wrapped tubes; the entire rope was wrapped. Smalley found that individual tubes were wrapped with polymer; the wrapped tubes did not exhibit the roping behavior. While Smalley was able to demonstrate removal of the polymer from the tubes, it is not clear how easily the SWNT can be manipulated and subsequently used in electronic circuits. In any case, the placement of SWNT into controlled configurations has been by a top-down methodology for the most part. Significant advances will be needed to take advantage of controlled placement at dimensions that exploit a molecule's small size.

Lieber proposed a SWNT-based nonvolatile random access memory device comprising a series of crossed nanotubes, wherein one parallel layer of nanotubes is placed on a substrate and another layer of parallel nanotubes, perpendicular to the first set, is suspended above the lower nanotubes by placing them on a periodic array of supports.[15] The elasticity of the suspended nanotubes provides one energy minima, wherein the contact resistance between the two layers is zero and the switches (the contacts between the two sets of perpendicular NW) are OFF. When the tubes are transiently charged to produce attractive electrostatic forces, the suspended tubes flex to meet the tubes directly below them; and a contact is made, representing the ON state. The ON/OFF state could be read by measuring the resistance

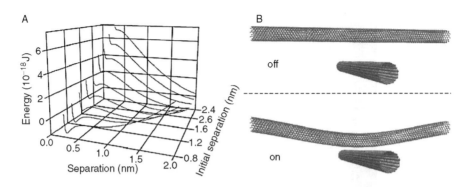

FIGURE 5.13 Bistable nanotubes device potential. (A) Plots of energy, $E_t = E_{vdW} + E_{elas}$, for a single 20 nm device as a function of separation at the cross point. The series of curves corresponds to initial separations of 0.8, 1.0, 1.2, 1.4, 1.6, 1.8, 2.0, 2.2, and 2.4 nm, with two well-defined minima observed for initial separations of 1.0 to 2.0 nm. These minima correspond to the crossing nanotubes being separated and in cdW contact. (B) Calculated structures of the 20 nm (10, 10) SWNT device element in the OFF (top) and ON (bottom) states. The initial separation for this calculation was 2.0 nm; the silicon support structures (elastic modulus of 168 Gpa) are not shown for clarity. (Reprinted from Rueckes, T., Kim, K., Joselevich, E., Tseng, G.Y., Cheung, C.–L., and Lieber, C.M., *Science*, 289, 94, 2000. © 2000 American Association for the Advancement of Science. With permission.)

at each junction and could be switched by applying voltage pulses at the correct electrodes. This theory was tested by mechanically placing two sets of nanotube bundles in a crossed mode and measuring the I(V) characteristics when the switch was OFF or ON (Figure 5.13). Although they used nanotube bundles with random distributions of metallic and semiconductor properties, the difference in resistance between the two modes was a factor of 10, enough to provide support for their theory.

In another study, Lieber used scanning tunneling microscopy (STM) to determine the atomic structure and electronic properties of intramolecular junctions in SWNT samples.[16] Metal–semiconductor junctions were found to exhibit an electronically sharp interface without localized junction states while metal–metal junctions had a more diffuse interface and low-energy states.

One problem with using SWNT or NW as wires is how to guide them in formation of the device structures — i.e., how to put them where you want them to go. Lieber has studied the directed assembly of NW using fluid flow devices in conjunction with surface patterning techniques and found that it was possible to deposit layers of NW with different flow directions for sequential steps.[19] For surface patterning, Lieber used NH_2-terminated surface strips to attract the NW; in between the NH_2- terminated strips were either methyl-terminated regions or bare regions, to which the NW had less attraction. Flow control was achieved by placing a poly(dimethylsiloxane) (PDMS) mold, in which channel structures had been cut into the mating surface, on top of the flat substrate. Suspensions of the NW (GaP, InP, or Si) were then passed through the channels. The linear flow rate was about 6.40 mm/s. In some cases the regularity extended over mm-length scales, as determined by scanning electron microscopy (SEM). Figure 5.14 shows typical SEM images of their layer-by-layer construction of crossed NW arrays.

While Lieber has shown that it is possible to use the crossed NW as switches, Stoddart and Heath have synthesized molecular devices that would bridge the gap between the crossed NW and act as switches in memory and logic devices.[61] The UCLA researchers have synthesized catenanes (Figure 5.15 is an example) and rotaxanes (Figure 5.16 is an example) that can be switched OFF and ON using redox chemistry. For instance, Langmuir–Blodgett films were formed from the catenane in Figure 5.15, and the monolayers were deposited on polysilicon NW etched onto a silicon wafer photolithographically. A second set of perpendicular titanium NW was deposited through a shadow mask, and the I(V) curve was determined. The data, when compared to controls, indicated that the molecules were acting as solid-state molecular switches. As yet, however, there have been no demonstrations of combining the Stoddart switches with NW.

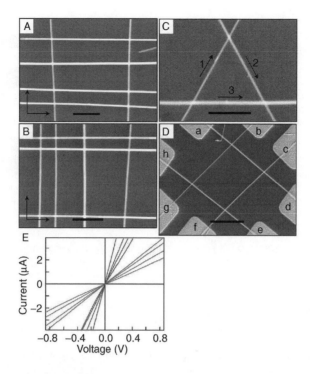

FIGURE 5.14 Layer-by-layer assembly and transport measurements of crossed NW arrays. (A and B) Typical SEM images of crossed arrays of InP NW obtained in a two-step assembly process with orthogonal flow directions for the sequential steps. Flow directions are highlighted by arrows in the images. (C) An equilateral triangle of GaP NW obtained in a three-step assembly process, with 60° angles between flow directions, which are indicated by numbered arrows. The scale bars correspond to 500 nm in (A), (B), and (C). (D) SEM image of a typical 2-by-2 cross array made by sequential assembly of n-type InP NW with orthogonal flows. Ni/In/Au contact electrodes, which were deposited by thermal evaporation, were patterned by e-beam lithography. The NW were briefly (3 to 5 s) etched in 6% HF solution to remove the amorphous oxide outer layer before electrode deposition. The scale bar corresponds to 2 μm. (E) Representative I(V) curves from two terminal measurements on a 2-by-2 crossed array. The solid lines represent the I(V) of four individual NW (ad, by, cf, eh), and the dashed lines represent I(V) across the four n–n crossed junctions (ab, cd, ef, gh). (Reprinted from Huang, Y., Duan, X., Wei, Q., and Lieber, C.M., *Science*, 291, 630, 2001. © 2001 American Association for the Advancement of Science. With permission.)

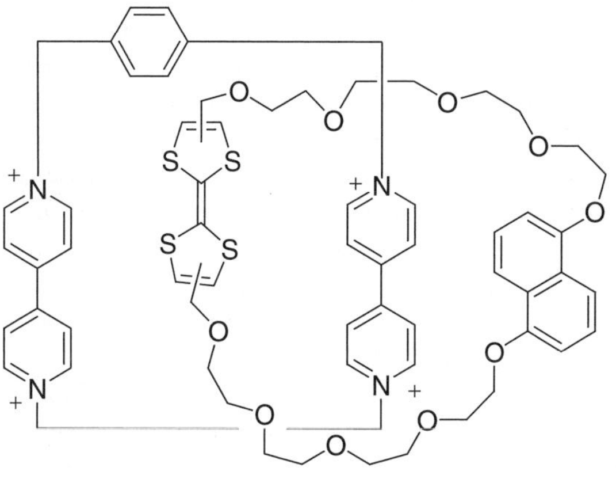

FIGURE 5.15 A catenane. Note that the two ring structures are intertwined.

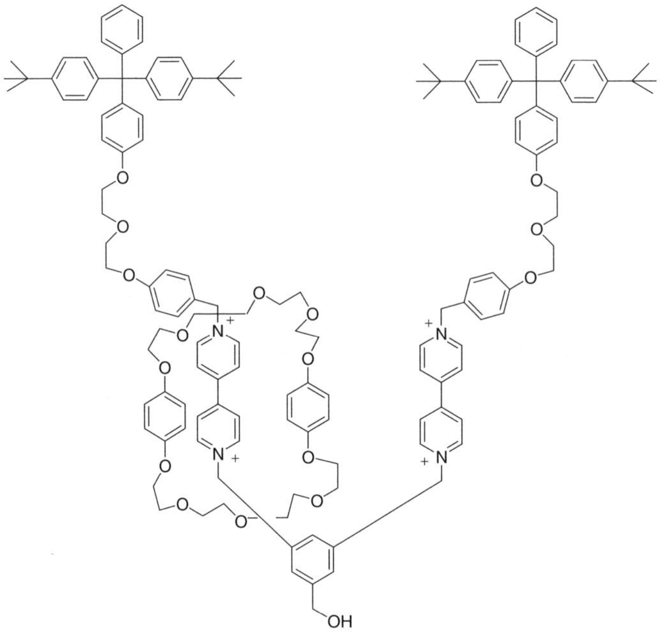

FIGURE 5.16 A [2] rotaxane. The two large end groups do not allow the ring structure to slip off either end.

Carbon nanotubes are known to exhibit either metallic or semiconductor properties. Avouris and coworkers at IBM have developed a method of engineering both multiwalled nanotubes (MWNT) and SWNT using electrical breakdown methods.[62] Shells in MWNT can vary between metallic or semiconductor character. Using electrical current in air to rapidly oxidize the outer shell of MWNT, each shell can be removed in turn because the outer shell is in contact with the electrodes and the inner shells carry little or no current. Shells are removed until arrival at a shell with the desired properties.

With ropes of SWNT, Avouris used an electrostatically coupled gate electrode to deplete the semiconductor SWNT of their carriers. Once depleted, the metallic SWNT can be oxidized while leaving the semiconductor SWNT untouched. The resulting SWNT, enriched in semiconductors, can be used to form nanotube-based field-effect transistors (FETs) (Figure 5.17).

The defect-tolerant approach to molecular computing using crossbar technology faces several hurdles before it can be implemented. As we have discussed, many very small wires are used in order to obtain the defect tolerance. How is each of these wires going to be accessed by the outside world? Multiplexing, the combination of two or more information channels into a common transmission medium, will have to be a major component of the solution to this dilemma. The directed assembly of the NW and attachment to the multiplexers will be quite complicated. Another hurdle is signal strength degradation as it travels along the NW. Gain is typically introduced into circuits by the use of transistors. However, placing a transistor at each NW junction is an untenable solution. Likewise, in the absence of a transistor at each cross point in the crossbar array, molecules with very large ON:OFF ratios will be needed. For instance, if a switch with a 10:1 ON:OFF ratio were used, then ten switches in the OFF state would appear as an ON switch. Hence, isolation of the signal via a transistor is essential; but presently the only solution for the transistor's introduction would be for a large solid-state gate below each cross point, again defeating the purpose for the small molecules.

Additionally, if SWNT are to be used as the crossbars, connection of molecular switches via covalent bonds introduces sp^3 linkages at each junction, disturbing the electronic nature of the SWNT and possibly obviating the very reason to use the SWNT in the first place. Noncovalent bonding will not provide the conductance necessary for the circuit to operate. Therefore, continued work is being done to devise and construct crossbar architectures that address these challenges.

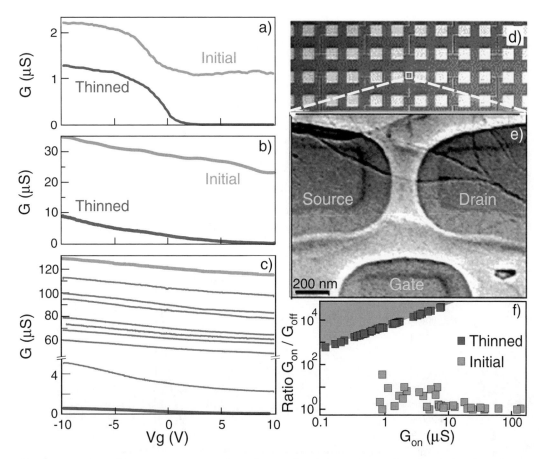

FIGURE 5.17 (a and b) Stressing a mixture of s- and m-SWNT while simultaneously gating the bundle to deplete the semiconductors of carriers resulted in the selective breakdown of the m-SWNT. The G(Vg) curve rigidly shifted downward as the m-SWNT were destroyed. The remaining current modulation is wholly due to the remaining s-SWNTs. (c) In very thick ropes, some s-SWNT must also be sacrificed to remove the innermost m-SWNT. By combining this technique with standard lithography, arrays of three-terminal, nanotube-based FETs were created (d and e) out of disordered bundles containing both m- and s-SWNT. Although these bundles initially show little or no switching because of their metallic constituents, final devices with good FET characteristics were reliably achieved (f). (Reprinted from Collins, P.G., Arnold, M.S., and Avouris, P., *Science*, 292, 706, 2001. © 2001 American Association for the Advancement of Science. With permission.)

5.4.3 The Nanocell Approach to a Molecular Computer: Synthesis

We have been involved in the synthesis and testing of molecules for molecular electronics applications for some time.[11] One of the synthesized molecules, the nitro aniline oligo(phenylene ethynylene) derivative (Figure 5.18), exhibited large ON:OFF ratios and negative differential resistance (NDR) when placed in a nanopore testing device (Figure 5.19).[63] The peak-to-valley ratio (PVR) was 1030:1 at 60 K.

The same nanopore testing device was used to study the ability of the molecules to hold their ON states for extended periods of time. The performance of molecules 1–4 in Figure 5.20 as molecular memory devices was tested, and in this study only the two nitro-containing molecules 1 and 2 were found to exhibit storage characteristics. The write, read, and erase cycles are shown in Figure 5.21. The I(V) characteristics of the Au-(1)-Au device are shown in Figure 5.22. The characteristics are repeatable to high accuracy with no degradation of the device noted even after 1 billion cycles over a one-year period.

The I(V) characteristics of the Au-(2)-Au were also measured (Figure 5.23, A and B). The measure logic diagram of the molecular random access memory is shown in Figure 5.24.

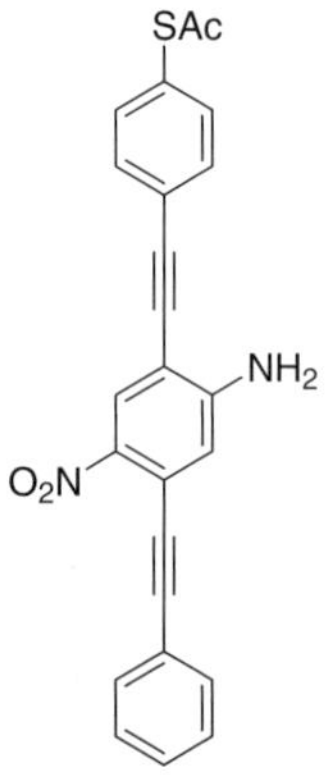

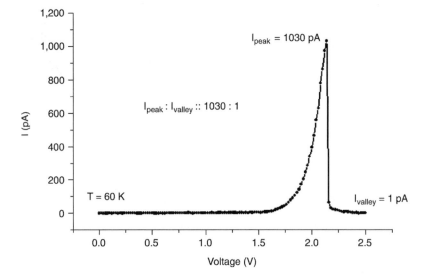

FIGURE 5.18   The protected form of the molecule tested in Reed and Tour's nanopore device.

FIGURE 5.19   I(V) characteristics of an Au-(2′-amino-4-ethynylphyenyl-4′-ethynylphenyl-5′-nitro-1-benzenethi-olate)-Au device at 60 K. The peak current density is ~50 A/cm², the NDR is ~− 400 μohm•cm², and the PVR is 1030:1.

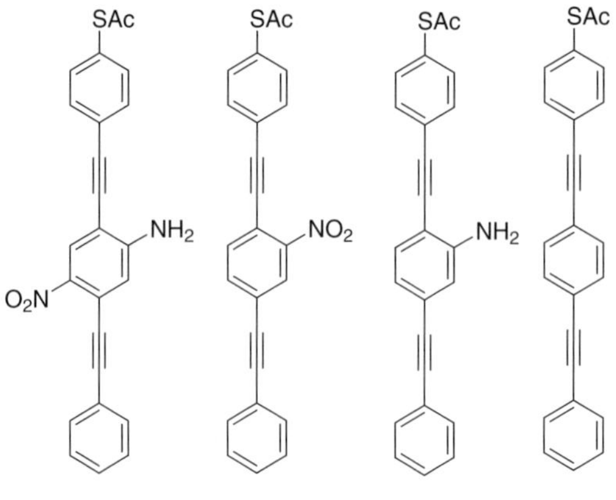

FIGURE 5.20   Molecules 1–4 were tested in the nanopore device for storage of high- or low-conductivity states. Only the two nitro-containing molecules 1 and 2 showed activity.

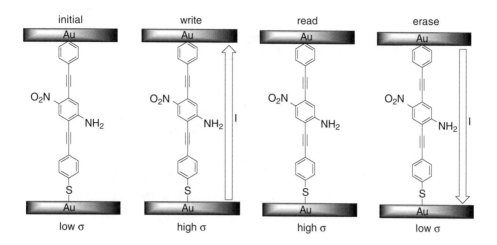

FIGURE 5.21 The memory device operates by the storage of a high- or low-conductivity state. An initially low-conductivity state (low σ) is changed into a high-conductivity state (high σ) upon application of a voltage. The direction of current that flows during the write and erase pulses is diagrammed by the arrows. The high σ state persists as a stored bit. (Reprinted from Reed, M.A., Chen, J., Rawlett, A.M., Price, D.W., and Tour, J.M. *Appl. Phys. Lett.*, 78, 3735, 2001. © 2001 American Institute of Physics. With permission.)

Seminario has developed a theoretical treatment of the electron transport through single molecules attached to metal surfaces[64] and has subsequently done an analysis of the electrical behavior of the four molecules in Figure 5.20 using quantum density functional theory (DFT) techniques at the B3PW91/6–31G* and B3PW91/LAML2DZ levels of theory.[65] The lowest unoccupied molecular orbit (LUMO) of nitro-amino functionalized molecule 1 was the closest orbital to the Fermi level of the Au. The LUMO of neutral 1 was found to be localized (nonconducting). The LUMO became delocalized (conducting) in the –1 charged state. Thus, ejection of an electron from the Au into the molecule to form a radical anion leads to conduction through the molecule. A slight torsional twist of the molecule allowed the orbitals to line up for conductance and facilitated the switching.

Many new molecules have recently been synthesized in our laboratories, and some have been tested in molecular electronics applications.[66–69] Since the discovery of the NDR behavior of the nitro aniline derivative, we have concentrated on the synthesis of oligo(phenylene ethynylene) derivatives. Scheme 5.1 shows the synthesis of a dinitro derivative. Quinones, found in nature as electron acceptors, can be easily reduced and oxidized, thus making them good candidates for study as molecular switches. The synthesis of one such candidate is shown in Scheme 5.2.

The acetyl thiol group is called a protected *alligator clip*. During the formation of a self-assembled monolayer (SAM) on a gold surface, for instance, the thiol group is deprotected *in situ*, and the thiol forms a strong bond (~2 eV, 45 kcal/mole) with the gold.

Seminario and Tour have done a theoretical analysis of the metal–molecule contact[70] using the B3PW91/LANL2DZ level of theory as implemented in Gaussian-98 in conjunction with the Green function approach that considers the "infinite" nature of the contacts. They found that Pd was the best metal contact, followed by Ni and Pt; Cu was intermediate, while the worst metals were Au and Ag. The best alligator clip was the thiol clip, but they found it was not much better than the isonitrile clip.

We have investigated other alligator clips such as pyridine end groups,[68] diazonium salts,[67] isonitrile, Se, Te, and carboxylic acid end groups.[66] Synthesis of an oligo(phenylene ethynylene) molecule with an isonitrile end group is shown in Scheme 5.3.

We have previously discussed the use of diazonium salts in the functionalization of SWNT. With modifications of this process, it might be possible to build the massively parallel computer architecture

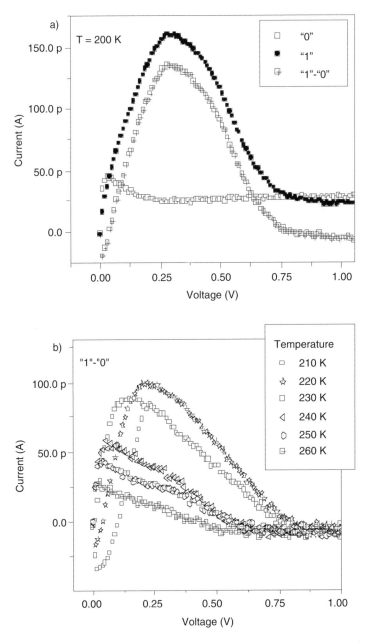

FIGURE 5.22   (a) The I(V) characteristics of an Au-(1)-Au device at 200 K. 0 denotes the initial state, 1 the stored written state, and 1–0 the difference of the two states. Positive bias corresponds to hole injection from the chemisorbed thiol-Au contact. (b) Difference curves (1–0) as a function of temperature. (Reprinted from Reed, M.A., Chen, J., Rawlett, A.M., Price, D.W., and Tour, J.M. *Appl. Phys. Lett.*, 78, 3735, 2001. © 2001 American Institute of Physics. With permission.)

using SWNT as the crosswires and oligo(phenylene ethynylene) molecules as the switches at the junctions of the crosswires, instead of the catenane and rotaxane switches under research at UCLA (see Figure 5.25). However, the challenges of the crossbar method would remain as described above. The synthesis of one diazonium switch is shown in Scheme 5.4. The short synthesis of an oligo(phenylene ethynylene) derivative with a pyridine alligator clip is shown in Scheme 5.5.

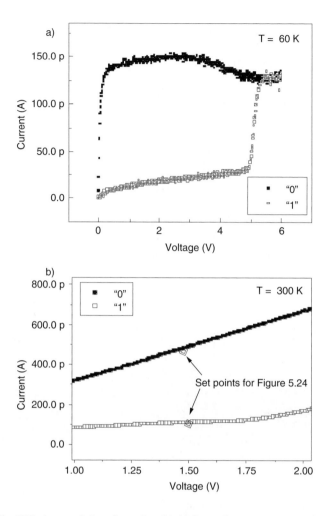

FIGURE 5.23 (a) The I(V) characteristics of stored and initial/erased states in an Au-(2)-Au device at 60 K and (b) ambient temperatures (300 K). The set points indicated are the operating point for the circuit of Figure 5.24. (Reprinted from Reed, M.A., Chen, J., Rawlett, A.M., Price, D.W., and Tour, J.M. *Appl. Phys. Lett.*, 78, 3735, 2001. ©2001 American Institute of Physics. With permission.)

5.4.4 The Nanocell Approach to a Molecular Computer: The Functional Block

In our conceptual approach to a molecular computer based on the nanocell, a small 1 μm^2 feature is etched into the surface of a silicon wafer. Using standard lithography techniques, 10 to 20 Au electrodes are formed around the edges of the nanocell. The Au leads are exposed only as they protrude into the nanocell's core; all other gold surfaces are nitride-coated. The silicon surface at the center of the nanocell (the molehole — the location of "moleware" assembly) is functionalized with $HS(CH_2)_3SiO_x$. A two-dimensional array of Au nanoparticles, about 30–60 nm in diameter, is deposited onto the thiol groups in the molehole. The Au leads (initially protected by alkane thiols) are then deprotected using UV/O_3; and the molecular switches are deposited from solution into the molehole, where they insert themselves between the Au nanoparticles and link the Au nanoparticles around the perimeter with the Au electrodes. The assembly of nanoparticles combined with molecular switches in the molehole will form hundreds to thousands of complete circuits from one electrode to another; see Figure 5.26 for a simple illustration. By applying voltage pulses to selected nanocell electrodes, we expect to be able to turn interior switches

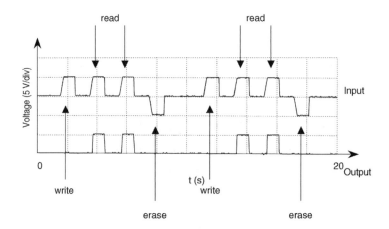

FIGURE 5.24   Measured logic diagram of the molecular random access memory. (Reprinted from Reed, M.A., Chen, J., Rawlett, A.M., Price, D.W., and Tour, J.M. *Appl. Phys. Lett.*, 78, 3735, 2001. © 2001 American Institute of Physics. With permission.)

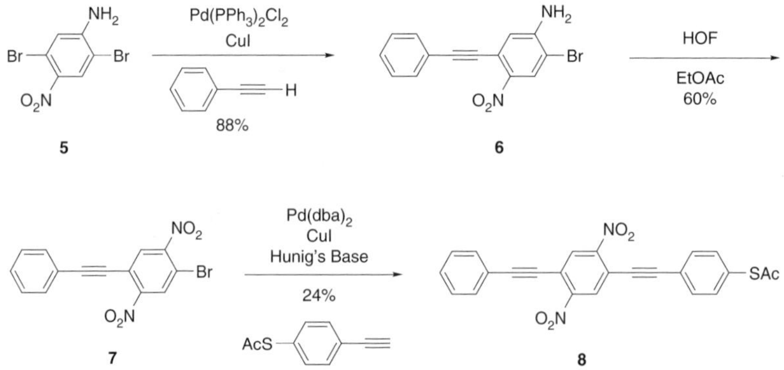

SCHEME 5.1   The synthesis of a dinitro-containing derivative. (Reprinted from Dirk, S.M., Price, D.W. Jr., Chanteau, S., Kosynkin, D.V., and Tour, J.M., *Tetrahedron*, 57, 5109, 2001. © ©2001 Elsevier Science. With permission.)

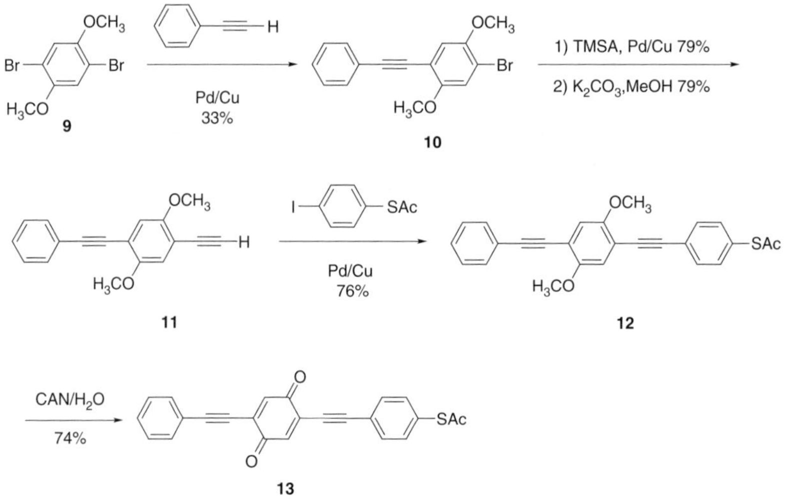

SCHEME 5.2   The synthesis of a quinone molecular electronics candidate. (Reprinted from Dirk, S.M., Price, D.W. Jr., Chanteau, S., Kosynkin, D.V., and Tour, J.M., *Tetrahedron*, 57, 5109, 2001. ©2001 Elsevier Science. With permission.)

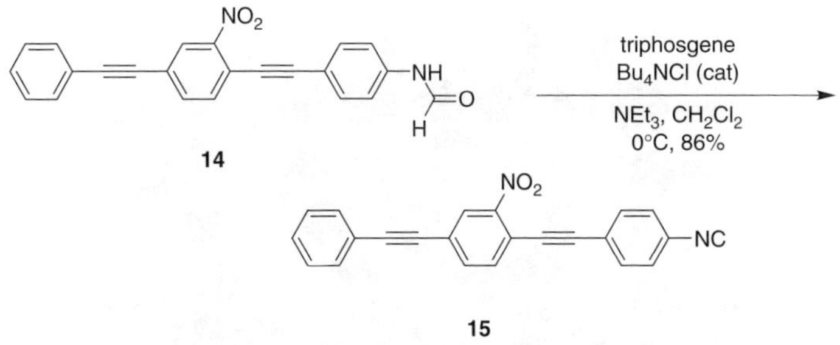

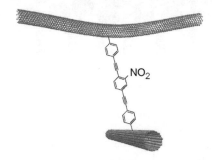

SCHEME 5.3 The formation of an isonitrile alligator clip from a formamide precursor.

FIGURE 5.25 Reaction of a bis-diazonium-derived nitro phenylene ethynylene molecule with two SWNT could lead to functional switches at cross junctions of SWNT arrays.

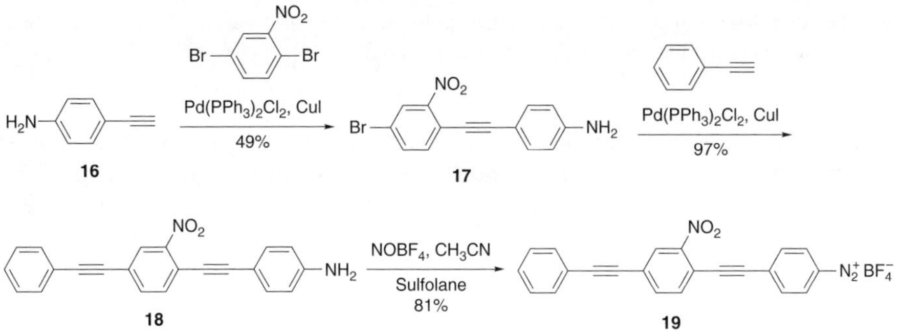

SCHEME 5.4 The synthesis of a diazonium-containing molecular electronics candidate. (Reprinted from Dirk, S.M., Price, D.W. Jr., Chanteau, S., Kosynkin, D.V., and Tour, J.M., *Tetrahedron*, 57, 5109, 2001. © 2001 Elsevier Science. With permission.)

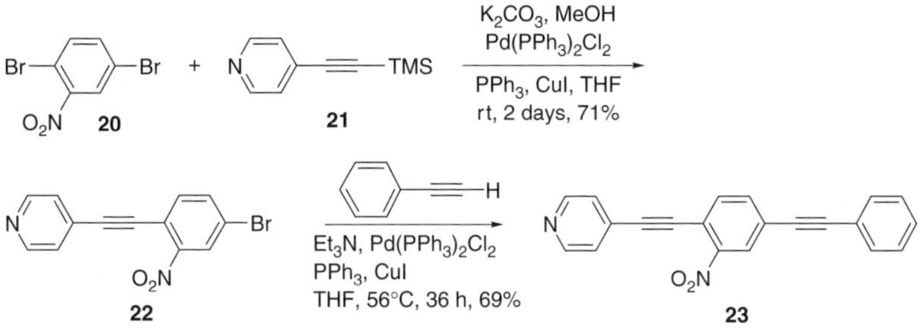

SCHEME 5.5 The synthesis of a derivative with a pyridine alligator clip. (Reprinted from Chanteau, S. and Tour, J.M., Synthesis of potential molecular electronic devices containing pyridine units, *Tet. Lett.*, 42, 3057, 2001. ©2001 Elsevier Science. With permission.)

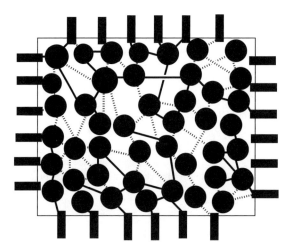

FIGURE 5.26   The proposed nanocell, with electrodes (black rectangles) protruding into the square molehole. Our simulations involve fewer electrodes. The metallic nanoparticles, shown here as black circles with very similar sizes, are deposited into the molehole along with organic molecular switches, not all of which are necessarily the same length or contain the same functionality. The molecular switches, with alligator clips on both ends, bridge the nanoparticles. Switches in the ON state are shown as solid lines while switches in the OFF state are shown as dashed lines. Because there would be no control of the nanoparticle or switch deposition, the actual circuits would be unknown. However, thousands to millions of potential circuits would be formed, depending on the number of electrodes, the size of the molehole, the size of the nanoparticles, and the concentration and identity of the molecular switches. The nanocell would be queried by a programming module after assembly in order to set the particular logic gate or function desired in each assembly. Voltage pulses from the electrodes would be used to turn switches ON and OFF until the desired logic gate or function was achieved.

ON or OFF, especially with the high ON:OFF ratios we have achieved with the oligo(phenylene ethynylene)s. In this way we hope to train the nanocell to perform standard logic operations such as AND, NAND, and OR. The idea is that we construct the nanocell first, with no control over the location of the nanoparticles or the bridging switches, and train it to perform certain tasks afterwards. Training a nanocell in a reasonable amount of time will be critical. Eventually, trained nanocells will be used to teach other nanocells. Nanocells will be tiled together on traditional silicon wafers to produce the desired circuitry. We expect to be able to make future nanocells 0.1 μm^2 or smaller if the input/output leads are limited in number, i.e., one on each side of a square.

While we are still in the research and development phase of the construction of an actual nanocell, we have begun a program to simulate the nanocell using standard electrical engineering circuit simulation programs such as SPICE and HSPICE, coupled with genetic algorithm techniques in three stages:[23]

1. With complete omnipotent programming, wherein we know everything about the interior of the constructed nanocell such as the location of the nanoparticles, how many switches bridge each nanoparticle pair, and the state of the conductance of the switches, and that we have control over turning specific switches ON or OFF to achieve the desired outcome without using voltage pulses from the outside electrodes
2. With omniscient programming, where we know what the interior of the nanocell looks like and know the conductance state of the switches, but we have to use voltage pulses from the surrounding electrodes to turn switches ON and OFF in order to achieve the desired outcome
3. With mortal programming, where we know nothing about the interior of the nanocell and have to guess where to apply the voltage pulses, we are just beginning to simulate mortal programming; however, it is the most critical type since we will be restricted to this method in the actual physical testing of the nanocell

Our preliminary results with omnipotent programming show that we can simulate simple logic functions such as AND, OR, and half-adders.

The nanocell approach has weaknesses and unanswered questions just as do the other approaches. Programming the nanocell is going to be our most difficult task. While we have shown that in certain circumstances our molecular switches can hold their states for extended periods of time, we do not know if that will be true for the nanocell circuits. Will we be able to apply voltage pulses from the edges that will bring about changes in conductance of switches on the interior of the nanocell, through extended distances of molecular arrays? Deposition of the SAMs and packaging the completed nanocells will be monumental development tasks. However, even with these challenges, the prospects for a rapid assembly of molecular systems with few restrictions to fabrication make the nanocell approach enormously promising.

5.5   Characterization of Switches and Complex Molecular Devices

Now that we have outlined the major classes of molecular computing architectures that are under consideration, we will touch upon some of the basic component tests that have been done. The testing of molecular electronics components has been recently reviewed.[11,71] Seminario and Tour developed a density functional theory calculation for determination of the I(V) characteristics of molecules, the calculations from which corroborated well with laboratory results.[72]

Stoddart and Heath have formed solid-state, electronically addressable switching devices using bistable [2] catenane-based molecules sandwiched between an n-type polycrystalline Si bottom electrode and a metallic top electrode.[73] A mechanochemical mechanism, consistent with the temperature-dependent measurements of the device, was invoked for the action of the switch. Solid-state devices based on [2] or [3] rotaxanes were also constructed and analyzed by Stoddart and Heath.[74,75]

In collaboration with Bard, we have shown that it is possible to use tuning-fork-based scanning probe microscope (SPM) techniques to make stable electrical and mechanical contact to SAMs.[76] This is a promising technique for quick screening of molecular electronics candidates. Frisbie has used an Au-coated atomic-force microscope (AFM) tip to form metal–molecule–metal junctions with Au-supported SAMs. He has measured the I(V) characteristics of the junctions, which are approximately 15 nm^2, containing about 75 molecules.[77] The I(V) behavior was probed as a function of the SAM thickness and the load applied to the microcontact. This may also prove to be a good method for quick screening of molecular electronics candidates.

In collaboration with Allara and Weiss, we have examined conductance switching in molecules 1, 2, and 4 (from Figure 5.20) by scanning tunneling microscopy (STM).[78] Molecules 1 and 2 have shown NDR effects under certain conditions, while molecule 4 did not.[63] SAMs made using dodecanethiol are known to be well packed and to have a variety of characteristic defect sites such as substrate step edges, film-domain boundaries, and substrate vacancy islands where other molecules can be inserted. When 1, 2, and 4 were separately inserted into the dodecanethiol SAMs, they protruded from the surrounding molecules due to their height differences. All three molecules had at least two states that differed in height by about 3 Å when observed by STM over time. Because topographic STM images represent a combination of the electronic and topographic structure of the surface, the height changes observed in the STM images could be due to a change in the physical height of the molecules, a change in the conductance of the molecules, or both. The more conductive state was referred to as ON, and the less conductive state was referred to as OFF. SAM formation conditions can be varied to produce SAMs with lower packing density. It was found that all three molecules switched ON and OFF more often in less ordered SAMs than in more tightly packed SAMs. Because a tightly packed SAM would be assumed to hinder conformational changes such as rotational twists, it was concluded that conformational changes controlled the conductance switching of all three molecules.

McCreery has used diazonium chemistry to form tightly packed monolayers on pyrolyzed photoresist film (PPF), a form of disordered graphitic material similar to glassy carbon.[79] Electrochemical reduction of stilbene diazonium salt in acetonitrile solvent in the presence of PPF forms a strong C–C bond between the stilbene molecule and carbons contained in the PPF. The I(V) characteristics of the stilbene junction were measured using Hg-drop electrode methods.

Lieber and coworkers constructed logic gates using crossed NW, which demonstrated substantial gain and were used to implement basic computations.[80] Avouris used SWNT that had been treated to prepare both p- and n-type nanotube transistors to build voltage inverters, the first demonstration of nanotube-based logic gates.[81] They used spatially resolved doping to build the logic function on a single bundle of SWNT. Dekker and coworkers also built logic circuits with doped SWNT.[82] The SWNT were deposited from a dichloroethane suspension, and those tubes having a diameter of about 1 nm and situated atop preformed Al gate wires were selected by AFM. Schön and coworkers demonstrated gain for electron transport perpendicular to a SAM by using a third gate electrode.[83] The field-effect transistors based on SAMs demonstrate five orders of magnitude of conductance modulation and gain as high as six. In addition, using two-component SAMs, composed of both insulating and conducting molecules, three orders of magnitude changes in conductance can be achieved.[84]

5.6 Conclusion

It is clear that giant leaps remain to be made before computing devices based on molecular electronics are commercialized. The QCA area of research, which has seen demonstrations of logic gates and devices earlier than other approaches, probably has the highest hurdle due to the need to develop nanoscopic quantum dot manipulation and placement. Molecular-scale quantum dots are in active phases of research but have not been demonstrated. The crossbar-array approach faces similar hurdles since the advances to date have only been achieved by mechanical manipulation of individual NWs, still very much a research-based phenomenon and nowhere near the scale needed for commercialization. Pieces of the puzzle, such as flow control placement of small arrays, are attractive approaches but need continued development. To this point, self-assembly of the crossbar arrays, which would simplify the process considerably, has not been a tool in development. The realization of mortal programming and development of the overall nanocell assembly process are major obstacles facing those working in the commercialization of the nanocell approach to molecular electronics. As anyone knows who has had a computer program crash for no apparent reason, programming is a task in which one must take into account every conceivable perturbation while at the same time not knowing what every possible perturbation is — a difficult task, to say the least. Many cycles of testing and feedback analysis will need to occur with a working nanocell before we know that the programming of the nanocell is successful.

Molecular electronics as a field of research is rapidly expanding with almost weekly announcements of new discoveries and breakthroughs. Those practicing in the field have pointed to Moore's Law and inherent physical limitations of the present top-down process as reasons to make these discoveries and breakthroughs. They are aiming at a moving target, as evidenced by Intel's recent announcements of the terahertz transistor and an enhanced 0.13 μm process.[85–87] One cannot expect that companies with "iron in the ground" will stand still and let new technologies put them out of business. While some may be kept off the playing field by this realization, for others it only makes the area more exciting. Even as we outlined computing architectures here, the first insertion points for molecular electronics will likely not be for computation. Simpler structures such as memory arrays will probably be the initial areas for commercial molecular electronics devices. Once simpler structures are refined, more precise methods for computing architecture will be realized. Finally, by the time this review is published, we expect that our knowledge will have greatly expanded, and our expectations as to where the technology is headed will have undergone some shifts compared with where we were as we were writing these words. Hence, the field is in a state of rapid evolution, which makes it all the more exciting.

Acknowledgments

The authors thank DARPA administered by the Office of Naval Research (ONR); the Army Research Office (ARO); the U.S. Department of Commerce, National Institute of Standards and Testing (NIST); National Aeronautics and Space Administration (NASA); Rice University; and the Molecular Electronics Corporation for financial support of the research done in our group. We also thank our many colleagues for their hard work and dedication. Dustin K. James thanks David Nackashi for providing some references on semiconductor manufacturing. Dr. I. Chester of FAR Laboratories provided the trimethylsilylacetylene used in the synthesis shown in Scheme 5.2.

References

1. Moore, G.E., Cramming more components onto integrated circuits, *Electronics*, 38, 1965.
2. Hand, A., EUV lithography makes serious progress, *Semiconductor Intl.*, 24(6),15, 2001.
3. Selzer, R.A. et al., Method of improving X-ray lithography in the sub-100 nm range to create high-quality semiconductor devices, U.S. patent 6,295,332, 25 September 2001.
4. Heath, J.R., Kuekes, P.J. Snider, G.R., and Williams, R.S., A defect-tolerant computer architecture: opportunities for nanotechnology, *Science*, 280, 1716, 1998.
5. Reed, M.A. and Tour, J.M., Computing with molecules, *Sci. Am.*, 292, 86, 2000.
6. Whitney, D.E., Why mechanical design cannot be like VLSI design, *Res. Eng. Des.*, 8, 125, 1996.
7. Hand, A., Wafer cleaning confronts increasing demands, *Semiconductor Intl.*, 24 (August), 62, 2001.
8. Golshan, M. and Schmitt, S., Semiconductors: water reuse and reclaim operations at Hyundai Semiconductor America, *Ultrapure Water*, 18 (July/August), 34, 2001.
9. Overton, R., Molecular electronics will change everything, *Wired*, 8(7), 242, 2000.
10. Service, R.F., Molecules get wired, *Science*, 294, 2442, 2001.
11. Tour, J.M., Molecular electronics, synthesis and testing of components, *Acc. Chem. Res.*, 33, 791, 2000.
12. Petty, M.C., Bryce, M.R., and Bloor, D., *Introduction to Molecular Electronics*, Oxford University Press, New York, 1995.
13. Heath, J.R., Wires, switches, and wiring: a route toward a chemically assembled electronic nano-computer, *Pure Appl. Chem.*, 72, 11, 2000.
14. Hu, J., Odom, T.W., and Lieber, C.M., Chemistry and physics in one dimension: synthesis and properties of nanowires and nanotubes, *Acc. Chem. Res.*, 32, 435, 1999.
15. Rueckes, T., Kim, K., Joselevich, E., Tseng, G.Y., Cheung, C.–L., and Lieber, C.M., Carbon nanotubes-based nonvolatile random access memory for molecular computing, *Science*, 289, 94, 2000.
16. Ouyang, M., Huang, J.–L., Cheung, C.-L., and Lieber, C.M., Atomically resolved single-walled carbon nanotubes intramolecular junctions, *Science*, 291, 97, 2001.
17. Star, A. et al., Preparation and properties of polymer-wrapped single-walled carbon nanotubes, *Angew. Chem. Intl. Ed.*, 40, 1721, 2001.
18. Bozovic, D. et al., Electronic properties of mechanically induced kinks in single-walled carbon nanotubes, *App. Phys. Lett.*, 78, 3693, 2001.
19. Huang, Y., Duan, X., Wei, Q., and Lieber, C.M., Directed assembly of one-dimensional nanostructures into functional networks, *Science*, 291, 630, 2001.
20. Gudiksen, M.S., Wang, J., and Lieber, C.M., Synthetic control of the diameter and length of single crystal semiconductor nanowires, *J. Phys. Chem. B*, 105, 4062, 2001.
21. Cui, Y. and Lieber, C.M., Functional nanoscale electronic devices assembled using silicon nanowire building blocks, *Science*, 291, 851, 2001.
22. Chung, S.-W., Yu, J.-Y, and Heath, J.R., Silicon nanowire devices, *App. Phys. Lett.*, 76, 2068, 2000.
23. Tour, J.M., Van Zandt, W.L., Husband, C.P., Husband, S.M., Libby, E.C., Ruths, D.A., Young, K.K., Franzon, P., and Nackashi, D., A method to compute with molecules: simulating the nanocell, submitted for publication, 2002.

24. Adleman, L.M., Computing with DNA, *Sci. Am.*, 279, 54, 1998.
25. Preskill, J., Reliable quantum computing, *Proc. R. Soc. Lond. A*, 454, 385, 1998.
26. Preskill, J., Quantum computing: pro and con, *Proc. R. Soc. Lond. A*, 454, 469, 1998.
27. Platzman, P.M. and Dykman, M.I., Quantum computing with electrons floating on liquid helium, *Science*, 284, 1967, 1999.
28. Kane, B., A silicon-based nuclear spin quantum computer, *Nature*, 393, 133, 1998.
29. Anderson, M.K., Dawn of the QCAD age, *Wired*, 9(9), 157, 2001.
30. Anderson, M.K., Liquid logic, *Wired*, 9(9), 152, 2001.
31. Wolf, S.A. et al., Spintronics: a spin-based electronics vision for the future, *Science*, 294, 1488, 2001.
32. Raymo, F.M. and Giordani, S., Digital communications through intermolecular fluorescence modulation, *Org. Lett.*, 3, 1833, 2001.
33. McEuen, P.L., Artificial atoms: new boxes for electrons, *Science*, 278, 1729, 1997.
34. Stewart, D.R. et al., Correlations between ground state and excited state spectra of a quantum dot, *Science*, 278, 1784, 1997.
35. Rajeshwar, K., de Tacconi, N.R., and Chenthamarakshan, C.R., Semiconductor-based composite materials: preparation, properties, and performance, *Chem. Mater.*, 13, 2765, 2001.
36. Leifeld, O. et al., Self-organized growth of Ge quantum dots on Si(001) substrates induced by sub-monolayer C coverages, *Nanotechnology*, 19, 122, 1999.
37. Sear, R.P. et al., Spontaneous patterning of quantum dots at the air–water interface, *Phys. Rev. E*, 59, 6255, 1999.
38. Markovich, G. et al., Architectonic quantum dot solids, *Acc. Chem. Res.*, 32, 415, 1999,
39. Weitz, I.S. et al., Josephson coupled quantum dot artificial solids, *J. Phys. Chem. B*, 104, 4288, 2000.
40. Snider, G.L. et al., Quantum-dot cellular automata: review and recent experiments (invited), *J. Appl. Phys.*, 85, 4283, 1999.
41. Snider, G.L. et al., Quantum-dot cellular automata: line and majority logic gate, *Jpn. J. Appl. Phys. Part I*, 38, 7227, 1999.
42. Toth, G. and Lent, C.S., Quasiadiabatic switching for metal-island quantum-dot cellular automata, *J. Appl. Phys.*, 85, 2977, 1999.
43. Amlani, I. et al., Demonstration of a six-dot quantum cellular automata system, *Appl. Phys. Lett.*, 72, 2179, 1998.
44. Amlani, I. et al., Experimental demonstration of electron switching in a quantum-dot cellular automata (QCA) cell, *Superlattices Microstruct.*, 25, 273, 1999.
45. Bernstein, G.H. et al., Observation of switching in a quantum-dot cellular automata cell, *Nanotechnology*, 10, 166, 1999.
46. Amlani, I., Orlov, A.O., Toth, G., Bernstein, G.H., Lent, C.S., and Snider, G.L., Digital logic gate using quantum-dot cellular automata, *Science*, 284, 289, 1999.
47. Orlov, A.O. et al., Experimental demonstration of a binary wire for quantum-dot cellular automata, *Appl. Phys. Lett.*, 74, 2875, 1999.
48. Amlani, I. et al., Experimental demonstration of a leadless quantum-dot cellular automata cell, *Appl. Phys. Lett.*, 77, 738, 2000.
49. Orlov, A.O. et al., Experimental demonstration of a latch in clocked quantum-dot cellular automata, *Appl. Phys. Lett.*, 78, 1625, 2001.
50. Tour, J.M., Kozaki, M., and Seminario, J.M., Molecular scale electronics: a synthetic/computational approach to digital computing, *J. Am. Chem. Soc.*, 120, 8486, 1998.
51. Lent, C.S., Molecular electronics: bypassing the transistor paradigm, *Science*, 288, 1597, 2000.
52. Bandyopadhyay, S., Debate response: what can replace the transistor paradigm?, *Science*, 288, 29, June, 2000.
53. Yu, J.-Y., Chung, S.-W., and Heath, J.R., Silicon nanowires: preparation, devices fabrication, and transport properties, *J. Phys. Chem. B*, 104, 11864, 2000.
54. Ausman, K.D. et al., Roping and wrapping carbon nanotubes, *Proc. XV Intl. Winterschool Electron. Prop. Novel Mater.*, Euroconference Kirchberg, Tirol, Austria, 2000.

55. Bahr, J.L. et al., Dissolution of small diameter single-wall carbon nanotubes in organic solvents? *Chem. Commun.*, 2001, 193, 2001.

56. Mickelson, E.T. et al., Solvation of fluorinated single-wall carbon nanotubes in alcohol solvents, *J. Phys. Chem. B.*, 103, 4318, 1999.

57. Chen, J. et al., Dissolution of full-length single-walled carbon nanotubes, *J. Phys. Chem. B.*, 105, 2525, 2001.

58. Bahr, J.L. et al., Functionalization of carbon nanotubes by electrochemical reduction of aryl diazonium salts: a bucky paper electrode, *J. Am. Chem. Soc.*, 123, 6536, 2001.

59. Bahr, J.L. and Tour, J.M., Highly functionalized carbon nanotubes using *in situ* generated diazonium compounds, *Chem. Mater.*, 13, 3823, 2001,

60. O'Connell, M.J. et al., Reversible water-solubilization of single-walled carbon nanotubes by polymer wrapping, *Chem. Phys. Lett.*, 342, 265, 2001.

61. Pease, A.R. et al., Switching devices based on interlocked molecules, *Acc. Chem. Res.*, 34, 433, 2001.

62. Collins, P.G., Arnold, M.S., and Avouris, P., Engineering carbon nanotubes and nanotubes circuits using electrical breakdown, *Science*, 292, 706, 2001.

63. Chen, J., Reed, M.A., Rawlett, A.M., and Tour, J.M., Large on-off ratios and negative differential resistance in a molecular electronic device, *Science*, 286, 1550, 1999.

64. Derosa, P.A. and Seminario, J.M., Electron transport through single molecules: scattering treatment using density functional and green function theories, *J. Phys. Chem. B.*, 105, 471, 2001.

65. Seminario, J.M., Zacarias, A.G., and Derosa, P.A., Theoretical analysis of complementary molecular memory devices, *J. Phys. Chem. A.*, 105, 791, 2001.

66. Tour, J.M. et al., Synthesis and testing of potential molecular wires and devices, *Chem. Eur. J.*, 7, 5118, 2001.

67. Kosynkin, D.V. and Tour, J.M., Phenylene ethynylene diazonium salts as potential self-assembling molecular devices, *Org. Lett.*, 3, 993, 2001.

68. Chanteau, S. and Tour, J.M., Synthesis of potential molecular electronic devices containing pyridine units, *Tet. Lett.*, 42, 3057, 2001.

69. Dirk, S.M., Price, D.W. Jr., Chanteau, S., Kosynkin, D.V., and Tour, J.M., Accoutrements of a molecular computer: switches, memory components, and alligator clips, *Tetrahedron*, 57, 5109, 2001.

70. Seminario, J.M., De La Cruz, C.E., and Derosa, P.A., A theoretical analysis of metal–molecule contacts, *J. Am. Chem. Soc.*, 123, 5616, 2001.

71. Ward, M.D., Chemistry and molecular electronics: new molecules as wires, switches, and logic gates, *J. Chem. Ed.*, 78, 321, 2001.

72. Seminario, J.M., Zacarias, A.G., and Tour, J.M., Molecular current–voltage characteristics, *J. Phys. Chem.*, 103, 7883, 1999.

73. Collier, C.P. et al., A [2]catenane-based solid-state electronically reconfigurable switch, *Science*, 289, 1172, 2000.

74. Wong, E.W. et al., Fabrication and transport properties of single-molecule thick electrochemical junctions, *J. Am. Chem. Soc.*, 122, 5831, 2000.

75. Collier, C.P., Molecular-based electronically switchable tunnel junction devices, *J. Am. Chem. Soc.*, 123, 12632, 2001.

76. Fan, R.-F.F. et al., Determination of the molecular electrical properties of self-assembled monolayers of compounds of interest in molecular electronics, *J. Am. Chem. Soc.*, 123, 2424, 2001.

77. Wold, D.J. and Frisbie, C.D., Fabrication and characterization of metal–molecule–metal junctions by conducting probe atomic force microscopy, *J. Am. Chem. Soc.*, 123, 5549, 2001.

78. Donahauser, Z.J. et al., Conductance switching in single molecules through conformational changes, *Science*, 292, 2303, 2001.

79. Ranganathan, S., Steidel, I., Anariba, F., and McCreery, R.L., Covalently bonded organic monolayers on a carbon substrate: a new paradigm for molecular electronics, *Nano Lett.*, 1, 491, 2001.

80. Huang, Y. et al., Logic gates and computation from assembled nanowire building blocks, *Science*, 294, 1313, 2001.

81. Derycke, V., Martel, R., Appenzeller, J., and Avouris, P., Carbon nanotubes inter- and intramolecular logic gates, *Nano Lett.*, 1, 453, 2001.

82. Bachtold, A., Hadley, P., Nakanishi, T., and Dekker, C., Logic circuits with carbon nanotubes transistors, *Science*, 294, 1317, 2001.

83. Schön, J.H., Meng, H., and Bao, Z., Self-assembled monolayer organic field-effect transistors, *Nature*, 413, 713, 2001.

84. Schön, J.H., Meng, H., and Bao, Z., Field-effect modulation of the conductance of single molecules, *Science*, 294, 2138, 2001.

85. Chau, R. et al., A 50 nm Depleted-Substrate CMOS Transistor (DST), International Electron Devices Meeting, Washington, D.C., December 2001.

86. Barlage, D. et al., High-Frequency Response of 100 nm Integrated CMOS Transistors with High-K Gate Dielectrics, International Electron Devices Meeting, Washington, D.C., December 2001.

87. Thompson, S. et al., An Enhanced 130 nm Generation Logic Technology Featuring 60 nm Transistors Optimized for High Performance and Low Power at 0.7–1.4 V, International Electron Devices Meeting, Washington, D.C., December 2001.

6

Nanoelectronic Circuit Architectures

6.1 Introduction..**6**-1
6.2 Quantum-Dot Cellular Automata...................................**6**-2
  Quantum-Dot Cell • QCA Logic • Computing with QCA
  • QCA Implementations
6.3 Single-Electron Circuits...**6**-8
6.4 Molecular Circuits **6**-9
6.5 Summary...**6**-11
Acknowledgments **6**-11
References **6**-11

Wolfgang Porod
University of Notre Dame

6.1 Introduction

The integrated circuit (IC), manufactured by optical lithography, has driven the computer revolution for more than four decades. Silicon-based technology allows for the fabrication of electronic devices with high reliability and of circuits with near-perfect precision. In fact, the main challenges facing conventional IC technology are not so much in making the devices, but in interconnecting them and in managing power dissipation.

IC miniaturization has provided the tools for imaging, manipulating, and modeling on the nanometer scale. These new capabilities have led to the discovery of new physical phenomena, which have been the basis for new device proposals; for a review, see Refs. [1–4]. Advantages of nanodevices include low power, high-packing densities, and speed. While there has been significant attention paid to the physics and chemistry of nanometer-scale device structures, there has been less appreciation of the need for new interconnection strategies for these new kinds of devices. In fact, the key problem is not so much how to make individual devices, but how to interconnect them in appropriate circuit architectures.

Nanotechnology holds the promise of putting a trillion molecular-scale devices in a square centimeter. How does one assemble a trillion devices per square centimeter? Moreover, this needs to be done quickly, inexpensively, and sufficiently reliably. What does one do with a trillion devices? If we assume that one can make them (and they actually work), how can this massive amount of devices be harnessed for useful computation? These questions highlight the need for innovative nanoelectronic circuit architectures.

In recent years, there has been significant progress in addressing the above issues [5]. To wit, "Nanocircuits" have been featured as the "Breakthrough of the Year 2001" in *Science* magazine [6]. Recent accomplishments include the fabrication of molecular circuits that are capable of performing logic operations.

The focus of this chapter is on the architectural aspects of nanometer-scale device structures. Device and fabrication issues will be referred to the literature. As a note of caution, this chapter attempts to survey an area that is under rapid development. Some of the architecture ideas described here have not yet been realized due to inherent fabrication difficulties. We attempt to highlight ideas that are at the forefront of the development of circuit architectures for nanoelectronic devices.

6.2 Quantum-Dot Cellular Automata

As device sizes shrink and packing densities increase, device–device interactions are expected to become ever more prominent [7,8]. While such parasitic coupling represents a problem for conventional circuitry, and efforts are being made to avoid it, such interactions may also represent an opportunity for alternate designs that utilize device–device coupling. Such a scheme appears to be particularly well suited for closely spaced quantum-dot structures, and the general notion of single-electrons switching on interacting quantum dots was first formulated in Ref. [9].

Based upon the emerging technology of quantum-dot fabrication, the Notre Dame NanoDevices group has proposed a scheme for computing with cells of coupled quantum dots [10], which has been termed "quantum-dot cellular automata" (QCA). To our knowledge, this is the first concrete proposal to utilize quantum dots for computing. There had been earlier suggestions that device–device coupling might be employed in a cellular-automaton-like scheme, but without an accompanying concrete proposal for a specific implementation [9,11–13].

The QCA cellular architecture is similar to other cellular arrays such as cellular neural/nonlinear networks (CNN) [14,15], in that they repeatedly employ the same basic cell with its associated near-neighbor interconnection pattern. The difference is that CNN cells have been realized by conventional CMOS circuitry, and the interconnects are provided by wires between cells in a local neighborhood [16,17]. For QCA, on the other hand, the coupling between cells is given by their direct physical interactions (and not by wires), which naturally takes advantage of the fringing fields between closely spaced nanostructures. The physical mechanisms available for interactions in such field-coupled architectures are electric (Coulomb) or magnetic interactions, in conjunction with quantum mechanical tunneling.

6.2.1 Quantum-Dot Cell

The original QCA proposal is based on a quantum-dot cell containing five dots [18], as schematically shown in Figure 6.1(a). The quantum dots are represented by the open circles, which indicate the confining electronic potential. In the ideal case, each cell is occupied by two electrons (shown as solid dots). The electrons are allowed to tunnel between the individual quantum dots inside an individual cell, but they are not allowed to leave the cell, which may be controlled during fabrication by the physical dot–dot and cell–cell distances.

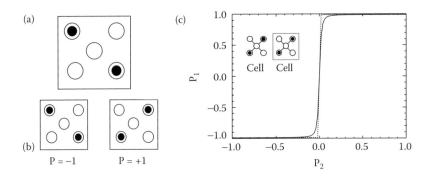

FIGURE 6.1 (a) Schematic diagram of a QCA cell consisting of five quantum dots and occupied by two electrons. (b) The two basic electronic arrangements in the cell, which can be used to represent binary information, $P = +1$ and $P = -1$. (c) Cell–cell response, which indicates that cell 1 abruptly switches when "driven" by only a small charge asymmetry in cell 2.

This quantum-dot cell represents an interesting dynamic system. The two electrons experience their mutual Coulombic repulsion, yet they are constrained to occupy the quantum dots inside the cell. If left alone, they will seek, by hopping between the dots, that configuration which corresponds to the physical ground state of the cell. It is clear that the two electrons will tend to occupy different dots on opposing corners of the cell because of the Coulomb energy cost associated with having them on the same dot or bringing them together closer. It is easy to see that the ground state of the system will be an equal superposition of the two basic configurations, as shown in Figure 6.1(b).

We may associate a "polarization" ($P=+1$ or $P=-1$) with either basic configuration of the two electrons in each cell. Note that this polarization is not a dipole moment but a measure of the alignment of the charge along the two cell diagonals. These two configurations may be interpreted as binary information, thus encoding bit values in the electronic arrangement inside a single cell. Any polarization between these two extreme values is possible, corresponding to configurations where the electrons are more evenly "smeared out" over all dots. The ground state of an isolated cell is a superposition with equal weight of the two basic configurations, and therefore has a net polarization of zero.

As described in the literature [19–21], this cell has been studied by solving the Schrödinger equation using a quantum-mechanical Hamiltonian model. Without going into the details, the basic ingredients of the theory are:

1. The quantized energy levels in each dot
2. The coupling between the dots by tunneling
3. The Coulombic charge cost for a doubly occupied dot
4. The Coulomb interaction between electrons in the same cell and also with those in neighboring cells

Numerical solutions of the Schrödinger equation confirm the intuitive understanding that the ground state is a superposition of the $P=+1$ and $P=-1$ states. In addition to the ground state, the Hamiltonian model yields excited states and cell dynamics.

The properties of an isolated cell were discussed above. Figure 6.1(c) shows how one cell is influenced by the state of its neighbor. As schematically depicted in the inset, the polarization of cell 2 is presumed to be fixed at a given value P_2, corresponding to a specific arrangement of charges in cell 2, and this charge distribution exerts its influence on cell 1, thus determining its polarization P_1. Quantum-mechanical simulations of these two cells yield the polarization response function shown in the figure. The important finding here is the strongly nonlinear nature of this cell–cell coupling. As can be seen, cell 1 is almost completely polarized even though cell 2 might only be partially polarized, and a small asymmetry of charge in cell 2 is sufficient to break the degeneracy of the two basic states in cell 1 by energetically favoring one configuration over the other.

6.2.2 QCA Logic

This bistable saturation is the basis for the application of such quantum-dot cells for computing structures. The above conclusions regarding cell behavior and cell–cell coupling are not specific to the five-dot cell discussed so far, but generalize to other cell configurations. Similar behavior is found for alternate cell designs, such as cells with only four dots in the corners (no central dot) or even cells with only two dots (molecular dipole). Based upon this bistable behavior of the cell–cell coupling, the cell polarization can be used to encode binary information. It has been shown that arrays of such physically interacting cells may be used to realize any Boolean logic functions [19–21].

Figure 6.2 shows examples of some basic QCA arrays. In each case, the polarization of the cell at the edge of the array is kept fixed; this so-called driver cell represents the input polarization that determines the state of the whole array. Each figure shows the cell polarizations that correspond to the physical ground state configuration of the whole array. Figure 6.2(a) shows that a line of cells allows the propagation of information, thus realizing a binary wire. Note that only information but no electric current flows down the line. Information can also flow around corners, as shown in Figure 6.2(b), and fan-out is possible (compare to

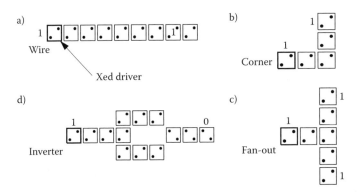

FIGURE 6.2 Examples of simple QCA structures showing (a) a binary wire, (b) signal propagation around a corner, (c) wire splitting and fan-out, and (d) an inverter.

Figure 6.2[c]). A specific arrangement of cells, such as the one shown in Figure 6.2(d), may be used to realize an inverter. In each case, electronic motion is confined within a given cell, but not between different cells. Only information, but not charge, is allowed to propagate over the whole array. This absence of current flow is the basic reason for the low-power dissipation in QCA structures.

The basic logic function, which is "native" to the QCA system, is majority logic [19–21]. Figure 6.3 shows a majority logic gate, which simply consists of an intersection of lines, and the "device cell" is just the one in the center. If we view three of the neighbors as inputs (kept fixed), then the polarization of the output cell is the one that "computes" the majority votes of the inputs. The figure also shows the majority logic truth table, which was computed (using the quantum-mechanical model) as the physical ground state polarizations for a given combination of inputs. The design of a majority logic gate using conventional circuitry would be significantly more complicated. The new physics of quantum mechanics gives rise to a new functionality, which allows a rather compact realization of majority logic. Note that conventional AND and OR gates are hidden in the majority logic gate. Inspection of the majority logic truth table reveals that if input *A* is kept fixed at 0, the remaining two inputs *B* and *C* realize an AND gate. Conversely, if *A* is held at 1, inputs *B* and *C* realize a binary OR gate. In other words, majority logic gates may be viewed as programmable AND and OR gates. This opens up the interesting possibility that the functionality of the gate may be determined by the state of the computation itself.

One may conceive of larger arrays representing more complex logic functions. The largest structure simulated so far (containing some 200 cells) is a single-bit full adder, which may be designed by taking advantage of the QCA majority logic gate as a primitive [19].

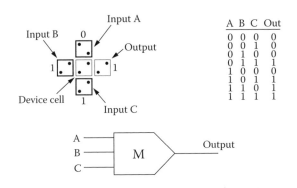

A	B	C	Out
0	0	0	0
0	0	1	0
0	1	0	0
0	1	1	1
1	0	0	0
1	0	1	1
1	1	0	1
1	1	1	1

FIGURE 6.3 Majority logic gate, which basically consists of an intersection of lines. Also shown are the computed majority-logic truth table and the logic symbol.

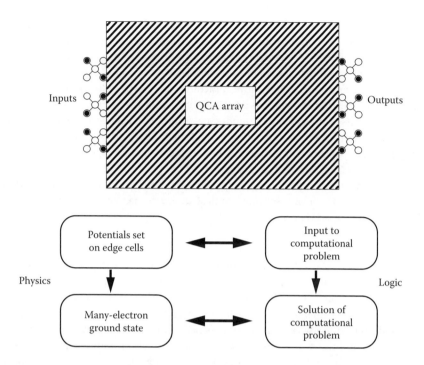

FIGURE 6.4 Schematic representation of computing with QCA arrays. The key concepts are "computing with the ground state" and "edge-driven computation." The physical evolution of the QCA structure is designed to mimic the logical solution path from input to output.

6.2.3 Computing with QCA

In a QCA array, cells interact with their neighbors, and neither power nor signal wires are brought to each cell. In contrast to conventional circuits, one does not have external control over each and every interior cell. Therefore, a new way is needed of using such QCA arrays for computing. The main concept is that the information in a QCA array is contained in the physical ground state of the system. The two key features that characterize this new computing paradigm are "computing with the ground state" and "edge-driven computation," which will be discussed in further detail below. Figure 6.4 schematically illustrates the main idea.

6.2.3.1 Computing with the Ground State

Consider a QCA array before the start of a computation. The array, left to itself, will have assumed its physical ground state. Presenting the input data, i.e., setting the polarization of the input cells, will deliver energy to the system, thus promoting the array to an excited state. The computation consists in the array reaching the new ground state configuration, compatible with the boundary conditions given by the fixed input cells. Note that the information is contained in the ground state itself, and not in how the ground state is reached. This relegates the question of the dynamics of the computation to one of secondary importance, although it is of significance, of course, for actual implementations. In the following, we will discuss two extreme cases for this dynamic, one where the system is completely left to itself, and another where exquisite external control is exercised.

- *Let physics do the computing*: The natural tendency of a system to assume the ground state may be used to drive the computation process. Dissipative processes due to the unavoidable coupling to the environment will relax the system from the initial excited state to the new ground state. The actual dynamics will be tremendously complicated since all the details of the system–environment coupling are unknown and uncontrollable. However, we do not have to concern ourselves with the detailed path in which the ground state is reached, as long as the ground state is reached. The attractive feature of this relaxation computation is that no external control is needed. However, there also are drawbacks in that the system may get "stuck" in metastable states and that there is no fixed time in which the computation is completed.

- *Adiabatic computing:* Due to the above difficulties associated with metastable states, Lent and coworkers have developed a clocked adiabatic scheme for computing with QCAs. The system is always kept in its instantaneous ground state, which is adiabatically transformed during the computation from the initial state to the desired final state. This is accomplished by lowering or raising potential barriers within the cells in concert with clock signals. The modulation of the potential barriers allows or inhibits changes of the cell polarization. The presence of clocks makes synchronized operation possible, and pipelined architectures have been proposed [19]. As an alternative to wired clocking schemes, optical pumping has been investigated as a means of providing power for signal restoration [22].

6.2.3.2 Edge-Driven Computation

Edge-driven computation means that only the periphery of a QCA array can be contacted, which is used to write the input data and to read the output of the computation. No internal cells may be contacted directly. This implies that no signals or power can be delivered from the outside to the interior of an array. All interior cells only interact within their local neighborhood. The absence of signal and power lines to each and every interior cell has obvious benefits for the interconnect problem and the heat dissipation. The lack of direct contact to the interior cells also has profound consequences for the way such arrays can be used for computation. Since no power can flow from the outside, interior cells cannot be maintained in a far-from-equilibrium state. Since no external signals are brought to the inside, internal cells cannot be influenced directly. These are the reasons why the ground state of the whole array is used to represent the information, as opposed to the states of each individual cell. In fact, edge-driven computation necessitates computing with the ground state! Conventional circuits, on the other hand, maintain devices in a far-from-equilibrium state. This has the advantage of noise immunity, but the price to be paid comes in the form of the wires needing to deliver the power (contributing to the wiring bottleneck) and the power dissipated during switching (contributing to the heat dissipation problem).

It should also be mentioned here that a formal link has been established between the QCA and CNN paradigms, which share the common feature of near-neighbor coupling. While CNN arrays obey completely classical dynamics, QCAs are mixed classical/quantum-mechanical systems. We refer to the literature for the details on such quantum cellular neural network systems [23,24].

6.2.4 QCA Implementations

The first QCA cell was demonstrated using Coulomb-coupled metallic islands [25]. These experiments showed that the position of one single electron can be used to control the position (switching) of a neighboring, single electron. This demonstrated the proof-of-principle of the QCA paradigm, namely that information can be encoded in the arrangements of electronic charge configurations. In these experiments, aluminum Coulomb-blockade islands represented the dots, and aluminum tunnel junctions provided the coupling. Electron-beam lithography and shadow evaporation were used for the fabrication. In a similar fashion, a binary QCA wire was realized [26]. The binary wire consisted of capacitively coupled double-dot cells charged with single electrons. The polarization switch caused by an applied input signal in one cell led to the change in polarization of the adjacent cell and so on down the line, as in falling dominos. Wire polarization was measured using single islands as electrometers. In addition, a functioning logic gate was also realized [27], where digital data was encoded in the positions of only two electrons. The logic gate consisted of a cell composed of four dots connected in a ring by tunnel junctions and two single-dot electrometers. The device operated by applying voltage inputs to the gates of the cell. Logic AND and OR operations have been verified using the electrometer outputs. Recently, a QCA shift register was demonstrated based on the same fabrication technology [28]. A drawback of these electrostatic QCA realizations using metallic Coulomb-blockade islands is operating at cryogenic temperatures. Recent experimental progress in this area is summarized in Refs. [29,30].

Molecular-scale QCA implementations hold the promise of room-temperature operation. The small size of molecules means that Coulomb energies are much larger than for metallic dots, and so operation at higher temperatures is possible. QCA molecules must have several redox centers that act as "quantum dots" arranged in the proper geometry. Furthermore, these redox centers must be able to respond to the local field created by another nearby QCA molecule. Several classes of molecules have been identified as candidates for possible molecular QCA operation [31,32]. It should be emphasized here that QCA implementations for molecular electronics represent an alternate viewpoint to the conventional approaches taken in the field of molecular electronics [33], which commonly use molecules as wires or switches. QCA molecules are not used to conduct electronic charge, but they represent structured "charge containers" that communicate with neighboring molecules through Coulombic interactions generated by particular charge arrangements inside the molecule.

In recent years, magnetic implementations have emerged as a promising alternative to QCA realizations for room-temperature operations. The advantage, compared to the above-mentioned electrostatic devices, is that logic gates featuring single-domain magnets in the size scale of 10 to 100 nm (above the superparamagnetic limit, but within the single-domain limit) are expected to operate at room temperature because of the relatively large magnetic coupling energies involved. Recent work demonstrated that QCA-like arrays of interacting nanometer-scale magnetic dots can be used to perform logic operations and to propagate information at room temperature [34,35]. Cowburn and coworkers [34] demonstrated QCA-like coupling in a line of circular Permalloy islands, while the Notre Dame team concentrated on elongated Permalloy islands, taking advantage of shape-induced magnetic anisotropy [35–37].

Nanometer-scale, single-domain magnets with elongated shapes are strongly bistable, as their remnant magnetization (magnetization at zero external magnetic field) always points along their long axis due to shape-induced magnetic anisotropy ("up" or "down"). This bistabilty is the basis for the use of such elongated islands in MRAM data storage technology. Even though a magnetizing force can rotate the magnetization away from the long axis, when the external force is removed, the magnet switches to either one of the two remnant stable states. The process of magnetizing perpendicular to the long axis can be pictured as the magnetizing force pulling down and releasing the energy barrier between the remanent ground states.

For the one-domain state, unlike the two-domain or vortex configurations, the magnetic flux lines close outside of the magnets. This creates strong magnetic stray fields that can be used to couple elements in proximity through their dipole–dipole interaction. The resulting magnetization pattern for an array of nanomagnets depends on their physical arrangement [38]. For example, lining several of these magnets along their long axes results in a line of magnets favoring their magnetization to point in the same direction (parallel alignment of dipoles), which is called a ferromagnetically ordered state. Placing them side-by-side and parallel to their long axes results in a line that favors antiparallel alignment of the magnetic dipoles, called an antiferromagnetically ordered state. In MQCA, these coupling-induced ordering phenomena are used to drive the computation.

Based on these magnetic-coupling phenomena, we have theoretically explored appropriate MQCA architectures [39], and we have experimentally demonstrated the basic MQCA logic gate, i.e., the three-input majority logic gate [40]. Here, similar to the process in the work of Parish and Forshaw [41], the nanomagnets are arranged in an intersection of two lines, where the dipole coupling between the nanomagnets produces ferromagnetic ordering in the vertical line and antiferromagnetic ordering in the horizontal line. The magnet at the intersection of the two crossing lines switches according to the majority coupling of its three input neighbors, and thus performs majority-logic functionality. We have demonstrated the correct functioning of this logic gate for all combinations of inputs, and the details of these experiments can be found in the literature [40].

Our work demonstrates that logic functions can be realized in properly structured arrays of physically coupled nanomagnets. The technology for fabricating similar nanometer-scale magnets is currently under development by the magnetic data-storage industry. While the latter work focuses entirely on memory applications, and physical coupling between individual bits is undesirable, our work points out the possibility of also realizing logic functionality in such systems and points to the possibility of all-magnetic information processing systems that incorporate both memory and logic [42].

In addition to the above QCA implementations based on electrical and magnetic physical coupling, another possibility might be near-field optical coupling between plasmonic nanoparticles [43]. Recent work has shown that neighboring nanoparticles can be coupled through the strong fields surrounding them at plasmon resonances, and optic functionality has been shown in line segments of various geometries (such as corners and T-junctions) [44,45]. Based on these observations, QCA-like optical logic coupling appears to be feasible [46].

6.3 Single-Electron Circuits

The physics of single-electron tunneling is well understood [47,48], and several possible applications have been explored [49,50]. Single-electron transistors (SETs) can, in principle, be used in circuits similar to conventional silicon field-effect transistors (MOSFETs) [51,52], including complementary CMOS-type circuits [53]. In these applications, the state of each node in the circuit is characterized by a voltage, and one device communicates with other devices through the flow of current. The peculiar nonlinear nature of the SET I–V characteristic has led to proposals of SETs in synaptic neural-network circuits and for implementations of CNN [54–56].

There are, however, practical problems in using SETs as logic devices in conventional circuit architectures. One of the main problems is related to the presence of stray charges in the surrounding circuitry, which changes the SET characteristics in an uncontrollable way. Since the SET is sensitive to the charge of one electron, just a small fluctuation in the background potential is sufficient to change the Coulomb-blockade (CB) condition. Also, standard logic devices rely on the high gain of conventional metal-oxide semiconductor (MOS) transistors, which allows circuit design with fan-out. In contrast, the gain of SETs is rather small, which limits their usefulness in MOSFET-like circuit architectures.

An interesting SET logic family that does not rely on high gain has been proposed. This architecture is based on the binary decision diagram (BDD) [57,58]. The BDD consists of an array of current pathways connected by Coulomb-blockade switching nodes. These nodes do not need high gain, but only distinct on–off switching characteristics, which can be realized by switching between a blockaded state and a completely pinched-off state (which minimizes the influence of stray potentials). The functioning of such a CB BDD structure was demonstrated in an experiment that included the demonstration of the AND logic function [59,60].

In contrast to the above SET-based approaches, the quantized electronic charge can be used to directly encode digital information [61,62]. Korotkov and Likharev proposed the SET parametron [62], which is a wireless single-electron logic family. As schematically shown in Figure 6.5, the basic building block of this logic family consists of three conducting islands, where the middle island is slightly shifted off the line passing through the centers of the edge island. Electrons are allowed to tunnel through small gaps between the middle and edge islands but not directly between the edge islands (due to their larger spatial separation).

Let us assume that each cell is occupied by one additional electron and that a "clock" electric field is applied that initially pushes this electron onto the middle island (the direction of this clock field is perpendicular to the line connecting the edge islands). Now that the electron is located on the central island, the clock field is reduced, and eventually changes direction. At some point in time during this cycle, it will be energetically favorable for the electron to tunnel off the middle island and onto one of the edge islands. If both islands are identical, the choice of island will be random. However, this symmetry can be broken by a small switching field applied perpendicular to the clock field and along the line of the edge cells. This control over the left–right final position of the electron can be interpreted as one bit of binary information; the electron on the right island might mean logical "1" and on the left island, logical "0." (This encoding of logic information by the arrangement of electronic charge is similar to the QCA idea.) An interesting consequence of the asymmetric charge configuration in a switched cell is that the resulting electric dipole field can be used to switch neighboring cells during their decision-making moment.

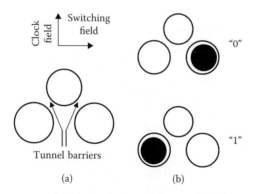

FIGURE 6.5 Schematic diagram of the SET parametron. (a) The basic cell consisting of three quantum dots, and occupied by a single electron. (b) The two basic electronic arrangements in the cell, which can be used to represent binary information. (From A.N. Korotkov and K.K. Likharev, *J Appl Phys*, 84, 11, 6114–6126 [1998]. With permission.)

Based on this near-neighbor interacting single-electron-parametron cell, a family of logic devices has been proposed [62]. A line of cells acts as an inverter chain and can be thought of as a shift register. Lines of cells can also be split into two, thus providing fan-out capability. In addition, SET parametron logic gates have been proposed that include NAND and OR gates. Recently, low-temperature prototypes of the single-electron parametron were experimentally demonstrated [63], and another group reported work on a similar device structure [64,65].

Another interesting possibility for single-electron circuits is the proposal by Kiehl and coworkers to encode information by the phase of phase-locked single-electron tunneling elements, which can be configured for bistable binary [66,67] or for ternary logic [68]. These phase-locked logic elements have also been arranged in cellular nonlinear network architectures [69].

6.4 Molecular Circuits

Chemical self-assembly processes look promising since they, in principle, allow vast amounts of devices to be fabricated very cheaply. But, there are key problems: (1) the need to create complex circuits for computers appears to be ill suited for chemical self-assembly, which yields mostly regular (periodic) structures; and (2) the need to deal with very large numbers of components and to arrange them into useful structures is a hard problem (NP-hard problem).

A review of the field of molecular electronics, starting from its early beginnings in the 1940s, is provided in Ref. [70], and recent developments are compiled in Ref. [71]. The latter volume contains a perspective on architectures for nanoprocessor systems integrated on the molecular scale [72].

One approach to molecular electronics is to build circuits in analogy to conventional silicon-based electronics. The idea is to find molecular analogs of electronic devices (such as wires, diodes, transistors, etc.) and then to assemble these into molecular circuits. This approach is reviewed and described in work by Ellenbogen and Love [73]. An electronically programmable memory device based on self-assembled molecular monolayers was reported by the groups of Reed and Tour [74–76].

The discovery of carbon nanotubes [77] provided a new building block for the construction of molecular-scale circuits. Dekker's group demonstrated a carbon nanotube SET operating at room temperature [78]. In subsequent work, the same group constructed logic circuits with FETs based on single-wall carbon nanotubes, which exhibited power gain (greater than 10) and large on–off ratios (greater than 10^5). A local-gate layout allowed for integration of multiple devices on a single chip, and one-, two-, and three-transistor circuits exhibiting digital logic operations such as an inverter, a logic NOR, and a static random-access memory cell were demonstrated [79]. In related work, Lieber's group has demonstrated logic circuits based on nanotube and semiconductor nanowires [80–82], and Avouris' group built an inverter from chemically doped nanotubes on a silicon substrate [83]. Appenzeller and coworkers developed and demonstrated carbon nanotube FETs with switching characteristics better than those of conventional MOSFETs, which are constrained by fundamental thermodynamic limits [84–86]. They also built a complete shift-register circuit based on a single-walled carbon nanotube, which is the first complex circuit implementation entirely done on a single molecule [87]. Chau and coworkers provided a recent benchmark study that showed that carbon nanotube FETs have device characteristics similar to industry-standard high-performance MOSFETs [88].

Another idea of a switch (and related circuitry) at the molecular level is the (mechanical) concept of an atom relay, which was proposed by Wada and coworkers [89,90]. The atom relay is a switching device based upon the controlled motion of a single atom. The basic configuration of an atom relay consists of a (conducting) atom wire, a switching atom, and a switching gate. The operation principle of the atom relay is that the switching atom is displaced from the atom wire due to an applied electric field on the switching gate ("off" state of the atom relay). Memory cell and logic gates (such as NAND and NOR functions) based on the atom relay configuration have been proposed and their operation was examined through simulation.

The above circuit approaches are patterned after conventional microelectronic circuit architectures, and they require the same level of device reliability and near-perfect fabrication yield. This is an area of concern since it is far from obvious that future molecular-electronics fabrication technologies will be able to rival the successes of the silicon-based microelectronics industry. There are several attempts to address these issues, and we will discuss the approach taken by the Hewlett-Packard and University of California research team [91,92]. This approach uses both chemistry (for the massively parallel construction of molecular components, albeit with unavoidable imperfections) and computer science (for a defect-tolerant reconfigurable architecture that allows one to download complex electronic designs).

This reconfigurable architecture is based on an experimental computer that was developed at Hewlett-Packard Laboratories in the mid-1990s [93]. Named "Teramac," it was first constructed using conventional silicon IC technology, in an attempt to develop a fault-tolerant architecture based on faulty components. Teramac was named for "Tera," 10^{12} operations per second (e.g., 10^6 logic elements operating at 10^6 Hz), and "mac" for "multiple architecture computer." It basically is a large, custom-configurable, highly parallel computer that consists of field-programmable gate arrays that can be programmed to reroute interconnections so as to avoid faulty components.

The Hewlett-Packard design is based on a cross-bar (Manhattan) architecture, in which two sets of overlapping nanowires are oriented perpendicular to each other. Each wire crossing becomes the location of a molecular switch, which is sandwiched between the top and bottom wires. Erbium disilicide wires, 2 nm in diameter and 9 nm apart, are used. The switches are realized by rotaxane molecules that preferentially attach to the wires and that are electrically addressable; i.e., electrical pulses lead to solid-state processes (analogous to electrochemical reduction or oxidation), which set or reset the switches by altering the resistance of the molecule. Attractive features of the crossbar architecture include its regular structure, which naturally lends itself to a chemical self-assembly of a large number of identical units and to a defect-tolerant strategy by having large numbers of potential "replacement parts" available.

Crossbars are natural for memory applications, but they can also be used for logic operations. They can be configured to compute Boolean logic expressions such as wired ANDs followed by wired ORs. A 6×6 diode crossbar can perform the function of a 2-bit adder [94].

The nanocell architecture pioneered by Tour and coworkers [95–97] is similar in the sense that functionality is programmed postfabrication, but it is not based on regular crossbar structures. The nanocell is contacted at the periphery by wires, and the interior (the so-called "molehole") consists of metal nanoparticles codeposited with molecular structures. The molecules provide connections between the particles, and chains of connected particles establish electrical connections between the external contacts. Each nanocell will have a different interior structure, and functionality is established by external probing and voltage pulses to program desired connectivity or destroy undesired ones. We refer to the literature for details [95–97].

The ability of a reconfigurable architecture to create a functional system in the presence of defective components represents a radical departure from today's microelectronics technology. Future nanoscale information processing systems may not have a central processing unit, but may instead contain extremely large configuration memories for specific tasks, which are controlled by a tutor that locates and avoids the defects in the system. The traditional paradigm for computation is to design the computer, build it perfectly, compile the program, and then run the algorithm. On the other hand, the Teramac and nanocell approaches are to build the computer (imperfectly), find the defects, configure the resources with software, compile the program, and then run it. This new paradigm moves tasks that are difficult to do in hardware into software tasks.

For completeness' sake, we would also like to mention two other molecular electronics approaches that are closer to existing CMOS technology and that significantly leverage existing fabrication technology. In the CMOL architecture, pioneered by Likharev and coworkers [98], a molecular layer consisting of crossed nanowires is to be fabricated on top of functional blocks in the CMOS layer. The molecular layer provides connectivity between the CMOS blocks, and programmable molecular switches at the cross-points of the nanowires provide programmability [99]. The approach pioneered by DeHon and coworkers [100,101] uses imprinted and self-assembled nanowires for both the interconnect and the device structures. On the architectural level, this work draws on well-established approaches for programmable logic, such as the programmable logic array [102,103].

6.5 Summary

If we are to continue to build complex systems of ever-smaller components, we must find new technologies, in conjunction with appropriate circuit architectures, that will allow massively parallel construction of electronic circuits at the atomic scale. In this chapter, we discussed several proposals for nanoelectronic circuit architectures, and we focused on those approaches that are different from the usual wired interconnection schemes employed in conventional silicon-based microelectronics technology. In particular, we discussed the QCA concept, which uses physical interactions between closely spaced nanostructures to provide local connectivity [104,105]. We also discussed several single-electron and molecular circuit architectures. In particular, we highlighted approaches to molecular electronics that are based on the (imperfect) chemical self-assembly of atomic-scale switches in regular crossbar arrays or irregular nanocells, and software solutions to provide defect-tolerant reconfiguration of the structure.

Acknowledgments

I would like to acknowledge many years of fruitful collaborations with my colleagues in Notre Dame's *Center for Nano Science and Technology*. Special thanks go to Professor Arpad Csurgay for many discussions that have strongly shaped my views of the field. This work was supported in part by grants from the Office of Naval Research and the W.M. Keck Foundation.

References

1. K. Goser, P. Glosekotter, J. Dienstuhl, *Nanoelectronics and Nanosystems; From Transistors to Molecular and Quantum Devices*, Springer Verlag, Berlin, 2004.
2. S.M. Goodnick, J. Bird, Quantum-effect and single-electron devices, *IEEE Trans Nanotech*, 2, 4, 368 (2003).
3. T. Ando, Y. Arakawa, K. Furuya, S. Komiyama, H. Nakashima, eds. *Mesoscopic Physics and Electronics*, NanoScience and Technology Series, Springer Verlag, 1998.
4. D. Goldhaber-Gordon, M.S. Montemerlo, J.C. Love, G.J. Opiteck, J.C. Ellenbogen, Overview of nanoelectronic devices, *Proc IEEE*, 85, 4, 541–557 (1997).
5. G.Y. Tseng, J.C. Ellenbogen, Towards nanocomputers, *Science*, 294, 1293–1294 (2001).
6. Editorial: Breakthrough of the year 2001, molecules get wired, *Science*, 294, 2429 and 2442–2443 (2001).

7. J.R. Barker, D.K. Ferry, Physics, synergetics, and prospects for self-organization in submicron semiconductor device structures, in *Proceedings of the 1979 International Conference on Cybernetics and Society*, IEEE Press, New York, 1979, p. 762.

8. J.R. Barker, D.K. Ferry, On the physics and modeling of small semiconductor devices – ii, *Solid State Electron*, 23, 531–544 (1980).

9. D.K. Ferry, W. Porod, Interconnections and architecture for ensembles of microstructures, *Superlatt Microstruct*, 2, 41 (1986).

10. C.S. Lent, P.D. Tougaw, W. Porod, G.H. Bernstein, Quantum cellular automata, *Nanotechnology*, 4, 49–57 (1993).

11. R.O. Grondin, W. Porod, C.M. Loeffler, D.K. Ferry, Cooperative effects in interconnected device arrays, in: Forrest L. Carter, ed., *Molecular Electronic Devices II*, Dekker, 1987, pp. 605–622.

12. V. Roychowdhuri, D.B. Janes, S. Bandyopadhyay, Nanoelectronic architecture for boolean logic, *Proc IEEE*, 85, 4, 574–588 (1997).

13. P. Bakshi, D. Broido, K. Kempa, Spontaneous polarization of electrons in quantum dashes, *J Appl Phys* 70, 5150 (1991).

14. L.O. Chua, L. Yang, Cellular neural networks: theory, *IEEE Trans Circuits and Systems* CAS-35, 1257–1272 (1988); and CNN: Applications, ibid. 1273–1290.

15. L.O. Chua ed., Special Issue on nonlinear waves, patterns and spatio-temporal chaos in dynamic arrays, *IEEE Trans Circuits Syst I Fundam Theory Appl*, 42, 10 (1995).

16. J.A. Nossek, T. Roska, Special Issue on cellular neural networks, *IEEE Trans Circuits Syst I Fundam Theory Appl*, 40, 3 (1993).

17. L.O. Chua, T. Roska, *Cellular Neural Networks and Visual Computing: Foundations and Applications*, Cambridge University Press, Cambridge, 2002.

18. C.S. Lent, P.D. Tougaw, W. Porod, Bistable saturation in coupled quantum dots for quantum cellular automata, *Appl Phys Lett*, 62, 714–716 (1993).

19. C.S. Lent, P.D. Tougaw, A device architecture for computing with quantum dots, *Proc IEEE*, 85, 4, 541–557 (1997).

20. W. Porod, Quantum-dot cellular automata devices and architectures, *Intern J High Speed Electron Sys*, 9, 1, 37–63 (1998).

21. W. Porod, C.S. Lent, G.H. Bernstein, A.O. Orlov, I. Amlani, G.L. Snider, J.L. Merz, Quantum-dot cellular automata: computing with coupled quantum dots, *Int J Electron*, 86, 5, 549–590 (1999).

22. G. Csaba, A.I. Csurgay, W. Porod, Computing architecture composed of next-neighbor-coupled optically-pumped nanodevices, *Int J Cir Theor Appl*, 29, 73–91 (2001).

23. G. Toth, C.S. Lent, P.D. Tougaw, Y. Brazhnik, W. Weng, W. Porod, R.-W. Liu, Y.-F. Huang, Quantum cellular neural networks, *Superlatt Microstruct*, 20, 473–477 (1996).

24. W. Porod, C.S. Lent, G. Toth, H. Luo, A. Csurgay, Y.-F. Huang, R.-W. Liu (Invited), Quantum-dot cellular nonlinear networks: computing with locally-connected quantum dot arrays, *Proceedings of the 1997 IEEE International Symposium on Circuits and Systems: Circuits and Systems in the Information Age*, 1997, pp. 745–748.

25. A.O. Orlov, I. Amlani, G.H. Bernstein, C.S. Lent, G.L. Snider, Realization of a functional cell for quantum-dot cellular automata, *Science*, 277, 928–30 (1997).

26. O. Orlov, I. Amlani, G. Toth, C.S. Lent, G.H. Bernstein, G.L. Snider, Experimental demonstration of a binary wire for quantum-dot cellular automata, *Appl Phys Lett*, 74, 19, 2875–2877 (1999).

27. I. Amlani, A.O. Orlov, G. Toth, G.H. Bernstein, C.S. Lent, G.L. Snider, Digital logic gate using quantum-dot cellular automata, *Science*, 284, 289–291 (1999).

28. R.K. Kummamuru et al., Operation of a quantum-dot cellular automata (QCA) shift register and analysis of errors, *IEEE Trans El Dev*, 50, 1906–1913 (2003).

29. G.L. Snider, A.O. Orlov, I. Amlani, X. Zuo, G.H. Bernstein, C.S. Lent, J.L. Merz, W. Porod, Quantum-dot cellular automata: review and recent experiments (invited), *J Appl Phys*, 85, 8, 4283–4285 (1999).

30. A.O. Orlov et al., Experimental studies of quantum-dot cellular automata devices, *Mesoscop Tunnel Dev,* 125–161 (2004).

31. M. Lieberman, S. Chellamma, B. Varughese, Y.L. Wang, C.S. Lent, G.H. Bernstein, G. Snider, F.C. Peiris, Quantum-dot cellular automata at a molecular scale, *Ann N Y Acad Sci,* 960, 225–239 (2002).

32. C.S. Lent, B. Isaksen, M. Lieberman, Molecular quantum-dot cellular automata, *J Am Chem Soc,* 125, 1056–1063 (2003).

33. C.S. Lent, Molecular electronics: bypassing the transistor paradigm, *Science,* 288, 1597–1599 (2000).

34. R.P. Cowburn, M.E. Welland, Room temperature magnetic quantum cellular automata, *Science,* 287, 1466–1468 (2000).

35. G. Csaba, W. Porod, Computing architectures for magnetic dot arrays, Presented at the *First International Conference and School on Spintronics and Quantum Information Technology,* Maui, Hawaii, May 2001.

36. G. Csaba, W. Porod, Simulation of field-coupled computing architectures based on magnetic dot arrays, *J Comput Electron,* 1, 87–91 (2002).

37. G. Csaba, A. Imre, G.H. Bernstein, W. Porod, V. Metlushko, Nanocomputing by field-coupled nanomagnets, *IEEE Trans Nanotech,* 1, 4, 209–213 (2002).

38. A. Imre, G. Csaba, A. Orlov, G.H. Bernstein, W. Porod, V. Metlushko, Investigation of shape-dependent switching of coupled nanomagnets, *Superlatt Microstruct,* 34, 513–518 (2003).

39. G. Csaba, W. Porod, A.I. Csurgay, A computing architecture composed of field-coupled single-domain nanomagnets clocked by magnetic fields, *Int J Circ Theor Appl,* 31, 67–82 (2003).

40. A. Imre, G. Csaba, L. Ji, A. Orlov, G.H. Bernstein, W. Porod, Majority logic gate for magnetic quantum-dot cellular automata, *Science,* 311, 205–208 (2006).

41. M.C.B. Parish, M. Forshaw, Physical constraints on magnetic quantum cellular automata, *Appl Phys Lett,* 83, 2046–2048 (2003).

42. R.P. Cowburn, Where have all the transistors gone? *Science,* 311, 183 (2006).

43. E. Ozbay, Plasmonics: merging photonics and electronics at nanoscale dimensions, *Science,* 311, 189 (2006).

44. S.A. Maier, P.G. Kik, H.A. Atwater, Observation of coupled plasmon-polariton modes in Au nanoparticle chain waveguides of different lengths: estimation of waveguide loss, *Appl Phys Lett,* 81, 1714 (2002).

45. M.L. Brongersma, J.W. Hartman, H.A. Atwater, Electromagnetic energy transfer and switching in nanoparticle chain arrays below the diffraction limit, *Phys Rev B,* 62, R16356 (2000).

46. A.I. Csurgay, W. Porod, Surface plasmon waves in nanoelectronic circuits, *Int J Circ Theor Appl,* 32, 339–361 (2004).

47. D.V. Averin, K.K. Likharev, in B.L. Altshuler et al., eds., *Mesoscopic Phenomena in Solids,* Elsevier, Amsterdam, 1991, p. 173.

48. H. Grabert, M.H. Devoret, eds., *Single Charge Tunneling,* Plenum Press, New York, 1992.

49. D.V. Averin, K.K. Likharev, Possible applications of single charge tunneling, in B.L. Altshuler et al., eds., *Mesoscopic Phenomena in Solids,* Elsevier, Amsterdam, 1991, pp. 311–332.

50. K. Likharev, SET: Coulomb blockade devices, *Nano Micro Technol,* 3, 1–2, 71–114 (2003).

51. K.K. Likharev, Single-electron transistors: electrostatic analogs to the DC SQUIDs, *IEEE Trans Magn* 23, 1142–1145 (1987).

52. R.H. Chen, A.N. Korotkov, K.K. Likharev, Single-electron transistor logic, *Appl Phys Lett,* 68, 1954–1956 (1996).

53. J.R. Tucker, Complementary digital logic based on the Coulomb blockade, *J Appl Phys,* 72, 4399–4413 (1992).

54. X. Wang, W. Porod, Single-electron transistor analytic I-V model for SPICE simulations, *Superlatt Microstruct,* 28, 5/6, 345–349 (2000).

55. C. Gerousis, S.M. Goodnick, W. Porod, Toward nanoelectronic cellular neural networks, *Int J Circ Theor Appl,* 28, 523–535 (2000).

56. C. Gerousis, S.M. Goodnick, W. Porod, Nanoelectronic single-electron transistor circuits and architectures, *Int J Circ Theor Appl*, 32, 323–338 (2004).

57. N. Asahi, M. Akazawa, Y. Amemiya, Binary-decision-diagram device, *IEEE Trans Electron Dev*, 42, 1999–2003 (1995).

58. N. Asahi, M. Akazawa, Y. Amemiya, Single-electron logic device based on the binary decision diagram, *IEEE Trans Electron Dev*, 44, 1109–1116 (1997).

59. K. Tsukagoshi, B.W. Alphenaar, K. Nakazato, Operation of logic function in a Coulomb blockade device, *Appl Phys Lett*, 73, 2515–2517 (1998).

60. S. Kasai, H. Hasegawa, A single electron binary-decision-diagram quantum logic circuit based on Schottky wrap gate control of a GaAs nanowire hexagon, *IEEE Electron Dev Lett*, 23, 446–448 (2002).

61. M.G. Ancona, Design of computationally useful single-electron digital circuits, *J Appl Phys*, 79, 526–539 (1996).

62. A.N. Korotkov, K.K. Likharev, Single-electron-parametron-based logic devices, *J Appl Phys*, 84, 11, 6114–6126 (1998).

63. D.V. Averin, A.N. Korotkov, *J Low Temp Phys*, 80, 173–185 (1991).

64. C. Kothandaramann, S.K. Iyer, S.S. Iyer, *IEEE Electron Dev Lett*, 23, 523–525 (2002).

65. E.G. Emiroglu, Z.A.K. Duranni, D.C. Hasko, D.A. Williams, *J Vac Sci Technol B*, 20, 2806–2809 (2002).

66. R.A. Kiehl, T. Oshima, Bistable locking of single-electron tunneling elements for digital circuitry, *Appl Phys Lett*, 67, 2494–2496 (1995).

67. T. Oshima, R.A. Kiehl, Operation of bistable phase-locked single-electron tunneling logic elements, *J Appl Phys*, 80, 912 (1996).

68. F.Y. Liu, F.-T. An, R.A. Kiehl, Ternary single electron tunneling phase logic, *Appl Phys Lett*, 74, 4040 (1999).

69. T. Yang, R.A. Kiehl, L.O. Chua, Tunneling phase logic cellular nonlinear networks, *Int J Bifurcat Chaos*, 11, 2895–2912 (2001).

70. N.S. Hush, An overview of the first half-century of molecular electronics, *Ann NY Acad Sci* 1006, 1–20 (2003).

71. G. Cuniberti, G. Fagas, K. Richter, eds., *Introducing Molecular Electronics*, Lecture Notes in Physics, Springer, New York, 2005, p. 680.

72. S. Das, G. Rose, M.M. Ziegler, C.A. Picconatto, J.C. Ellenbogen, Architectures and simulations for nanoprocessor systems integrated on the molecular scale, in: *Introducing Molecular Electronics*, Springer-Verlag, Heidelberg, 2005, pp. 479–514.

73. J.C. Ellenbogen, J.C. Love, Architectures for molecular electronic computers: 1. Logic structures and an adder built from molecular electronic diodes; and J.C. Ellenbogen, Architectures for molecular electronic computers: 2. Logic structures using molecular electronic FETs, MITRE Corp reports (1999), available at: http://www.mitre.org/technology/nanotech/.

74. J. Chen, M.A. Reed, A.M. Rawlett, J.M. Tour, Observation of a large on-off ratio and negative differential resistance in an electronic molecular switch, *Science*, 286, 1550–1552 (1999).

75. J. Chen, W. Wang, M.A. Reed, A.M. Rawlett, D.W. Price, J.M. Tour, Room-temperature negative differential resistance in nanoscale molecular junctions, *Appl Phys Lett*, 77, 1224–1226 (2000).

76. M.A. Reed, J. Chen, A.M. Rawlett, D.W. Price, J.M. Tour, Molecular random access memory cell, *Appl Phys Lett*, 78, 3735–3737 (2001).

77. S. Iijima, *Nature*, 354, 56 (1991).

78. H.W. Ch. Postma, T. Teepen, Z. Yao, M. Grifoni, C. Dekker, Carbon nanotube single-electron transistors at room temperature, *Science*, 293, 76–79 (2001).

79. A. Bachtold, P. Hadley, T. Nakanishi, C. Dekker, Logic circuits with carbon nanotube transistors, *Science*, 294, 1317–1320 (2001).

80. T. Rueckes, K. Kim, E. Joselevich, G. Tseng, C.-L. Cheung, C. Lieber, Carbon nanotube-based nonvolatile random access memory for molecular computing, *Science*, 289, 94–97 (2000).

81. Y. Cui, C. Lieber, Functional nanoscale electronic devices assembled using silicon nanowire building blocks, *Science*, 291, 851–853 (2001).

82. Y. Huang, X. Duan, Y. Cui, L.J. Lauhon, K.-H. Kim, C.M. Lieber, Logic gates and computation from assembled nanowire building blocks, *Science*, 294, 1313–1317 (2001).

83. V. Derycke, R. Martel, J. Appenzeller, P. Avouris, Carbon nanotube inter- and intramolecular logic gates, *Nano Lett*, 1, 453 (2001).

84. S.J. Wind, J. Appenzeller, R. Martel, V. Derycke, Ph. Avouris, Vertical scaling of carbon nanotube field-effect transistors using top gate electrodes, *Appl Phys Lett*, 80, 3817 (2002).

85. M. Radosavljevi, J. Appenzeller, Ph. Avouris, J. Knoch, High performance of potassium n-doped carbon nanotube field-effect transistors, *Appl Phys Lett*, 84, 3693 (2004).

86. J. Appenzeller, Y.-M. Lin, J. Knoch, Z. Chen, Ph. Avouris, Comparing carbon nanotube transistors — the ideal choice: a novel tunneling device design, *IEEE Trans Electron Dev*, 52, 12, 2568 (2005).

87. Z. Chen, J. Appenzeller et al., An integrated logic circuit assembled on a single carbon nanotube, *Science*, 311 1735 (2006).

88. Robert Chau et al., Benchmarking nanotechnology for high-performance and low-power logic transistor applications, *IEEE Trans Nanotechnol*, 4, 2 (2005).

89. Y. Wada, T. Uda, M. Lutwyche, S. Kondo, S. Heike, A proposal of nano-scale devices based on atom/molecule switching, *J Appl Phys*, 74, 7321–7328 (1993).

90. Y. Wada, Atom electronics, *Microelectron Eng*, 30, 375–382 (1996).

91. J.R. Heath, P.J. Kuekes, G.S. Snider, R.S. Williams, A defect tolerant computer architecture: opportunities for nanotechnology, *Science*, 280, 1716–1721 (1998).

92. C.P. Collier, E.W. Wong, M. Belohradsky, F.M. Raymo, J.F. Stoddart, P.J. Kuekes, R.S. Williams, J.R. Heath, Electronically configurable molecular-based logic gates, *Science*, 285 (1999).

93. R. Amerson, R.J. Carter, W.B. Culbertson, P. Kuekes, G. Snider, Teramac – configurable custom computing, *Proceedings of the 1995 IEEE Symposium on FPGA's for Custom Computing Machines*, pp. 32–38.

94. P. Kuekes, R.S. Williams, Molecular electronics, in *Proceedings of the European Conference on Circuit Theory and Design*, ECCTD'99.

95. J.M. Tour, W.L. Van Zandt, C.P. Husband, S.M. Husband, L.S. Wilson, P.D. Franzon, D.P. Nackashi, Nanocell logic gates for molecular computing, *IEEE Trans Nanotech* 1, 100–109 (2002).

96. C.P. Husband, S.M. Husband, J.S. Daniels, J.M. Tour, Logic and memory with nanocell circuits, *IEEE Trans Electron Dev* 50, 1865–1975 (2003).

97. J.M. Tour, L. Cheng, D.P. Nackashi, Y. Yao, A.K. Flatt, S.K. St. Angelo, T.E. Mallouk, P.D. Franzon, Nanocell electronic memories, *J Am Chem Soc*, 125, 13279–13283 (2003).

98. K.K. Likharev, D.B. Strukov, CMOL: devices, circuits, and architectures, in: G. Cuniberti, G. Fagas, K. Richter, eds., *Introducing Molecular Electronics*, Lecture Notes in Physics, Springer, New York, 2005, pp. 447–478.

99. D.B. Strukov, K.K. Likharev, CMOL FPGA: a reconfigurable architecture for hybrid digital circuits with two-terminal nano devices, *Nanotechnology*, 16, 888 (2005).

100. A. DeHon, Array-based architecture for FET-based, nanoscale electronics, *IEEE T-NANO*, 2, 23–32 (2003).

101. A. DeHon, P. Lincoln, J.E. Savage, Stochastic assembly of sublithographic nanoscale interfaces, *IEEE T-NANO*, 2, 165–174 (2003).

102. A. DeHon, M.J. Wilson, Nanowire-based sublithographic programmable logic arrays, in *Proc. ACM/SIGDA FPGA*, 123, ACM Press, Monterey, CA, 2004.

103. B. Gojman, R. Rubin, C. Pilotto, T. Tanamoto, A. DeHon, 3D nanowire-based programmable logic, in *Proceedings of the International Conference on Nano-Networks — Nanonets2006*, September 14–16, 2006.

104. A.I. Csurgay, W. Porod, C.S. Lent, Signal processing with near-neighbor-coupled time-varying quantum-dot arrays, *IEEE Trans Circ Sys* I, 47, 8, 1212–1223 (2000).

105. A.I. Csurgay, W. Porod, Equivalent circuit representation of arrays composed of Coulomb-coupled nanoscale devices: modeling, simulation and realizability, *Int J Circ Theor Appl*, 29, 3–35 (2001).

7

Molecular Computing and Processing Platforms

7.1 Introduction 7-2
7.2 Data and Signal Processing Platforms 7-4
7.3 Microelectronics and Nanoelectronics: Retrospect
and Prospect...7-8
7.4 Performance Estimates 7-11
Entropy and Its Application • Distribution Statistics • Energy
Levels • Device Switching Speed • Photon Absorption
and Transition Energetics • Processing Performance Estimates
7.5 Synthesis Taxonomy in Design of ^{M}ICs
and Processing Platforms 7-18
7.6 Biomolecular Processing and Fluidic Molecular
Electronics: Neurobiomimetics, Prototyping,
and Cognition 7-20
Neuroscience: Information Processing and Memory
Postulates • Applied Information Theory and Information
Estimates with Applications to Biomolecular Processing
and Communication • Fluidic Molecular Platforms
• Neuromorphological Reconfigurable Molecular Processing
Platforms
7.7 Design of Three-Dimensional Molecular Integrated
Circuits: Data Structures, Decision Diagrams,
and Hypercells..7-34
Molecular Electronics and Gates: Device and Circuits
Prospective • Decision Diagrams and Logic Design
of ^{M}ICs • ^{N}Hypercell Design
7.8 Three-Dimensional Molecular Signal/Data
Processing and Memory Platforms 7-43
7.9 Hierarchical Finite-State Machines and Their
Use in Hardware and Software Design 7-49
7.10 Adaptive Defect-Tolerant Molecular
Processing-and-Memory Platforms 7-52
7.11 Hardware–Software Design 7-55
7.12 Modeling and Analysis of Molecular
Electronic Devices...7-58
Introduction to Modeling Concepts • Heisenberg
Uncertainty Principle • Particle Velocity • Schrödinger
Equation • Quantum Mechanics and Molecular Electronic
Devices: Three-Dimensional Problem • Multiterminal
Quantum-Effect ^{ME}Devices

Sergey Edward Lyshevski
Rochester Institute of Technology

7.13 Conclusions 7-79
      Acknowledgments 7-79
      References 7-79

7.1 Introduction*

To devise and design molecular computing and processing platforms (MPPs), novel hardware solutions must be developed. Molecular electronics and new concepts should be utilized, including nanotechnology. Although progress in various applications of nanotechnology is being announced, many of those declarations have been largely acquired from well-known theories and accomplished technologies of material science, biology, chemistry, and other areas established in olden times and utilized for centuries. Atoms and atomic structures were envisioned by Leucippus of Miletus and Democritus around 440 BC, and the basic atomic theory was developed by John Dalton in 1803. The Periodic Table of Elements was established by Dmitri Mendeleev in 1869, and the electron was discovered by Joseph Thomson in 1897. The composition of atoms was discovered by Ernest Rutherford in 1910 using the experiments conducted under his direction by Ernest Marsden in the scattering of α-particles. The quantum theory was largely developed by Niels Bohr, Louis de Broglie, Werner Heisenberg, Max Planck, and other scientists at the beginning of the 20th century. Those developments were taken forward by Erwin Schrödinger in 1926. For many decades, comprehensive editions of chemistry and physics handbooks coherently reported thousands of organic and inorganic compounds, molecules, ring systems, purines, pyrimidines, nucleotides, oligonucleotides, organic magnets, organic polymers, atomic complexes, and molecules with dimensionality of the order of 1 nm. In the last 50 years, meaningful methods have been developed and commercially deployed to synthesize a great variety of nucleotides and oligonucleotides with various linkers and spacers, bioconjugated molecular aggregates, modified nucleosides, and other inorganic, organic, and biomolecules. The above-mentioned fundamental, applied, experimental, and technological accomplishments have provided essential foundations for many areas including biochemistry, chemistry, physics, electronics, etc.

Microelectronics has witnessed phenomenal accomplishments. For more than 50 years, the fields of microelectronic devices, integrated circuits (ICs), and high-yield technologies have matured and progressed, ensuring high-performance electronics. Many electronics-preceding processes and materials were improved and fully utilized. For example, crystal growth, etching, thin-film deposition, coating, photolithography, and other processes have been known and used for centuries. Etching was developed and performed by Daniel Hopfer from 1493 to 1536. Modern electroplating (electrodeposition) was invented by Luigi Brugnatelli in 1805. Photolithography was invented by Joseph Nicéphore Niépce in 1822, and he made the first photograph in 1826. In 1837, Moritz Hermann von Jacobi introduced and demonstrated silver, copper, nickel, and chrome electroplating. In 1839, John Wright, George Elkington, and Henry Elkington discovered that potassium cyanide can be used as an electrolyte for gold and silver electroplating. They patented this process, receiving the British Patent 8447 in 1840. The technologies used in the fabrication of various art and jewelry products, as well as Christmas ornaments, have been in existence for centuries.

In microfabrication technology, feature sizes have significantly decreased. The structural features of solid-state semiconductor devices have been scaled down to tens of nanometers, and the thickness of deposited thin films can be less than 1 nm. The epitaxy fabrication process, invented in 1960 by J.J. Kleimack, H.H. Loar, I.M. Ross, and H.C. Theuerer, led to the growing of layer after layer of silicon films identical in structure with the silicon wafer itself. Technological developments in epitaxy continued, resulting in the possibility of depositing uniform multilayered semiconductors and insulators with precise thicknesses in order to improve IC performance. Molecular beam epitaxy is the deposition of one or more pure materials on a single crystal wafer, one layer of atoms at a time, under high vacuum, forming

---

*This chapter is a modified and revised version of the chapter by S.E. Lyshevski, *Three-Dimensional Molecular Electronics and Integrated Circuits for Signal and Information Processing Platforms,* in *Handbook on Nano and Molecular Electronics,* Ed. S.E. Lyshevski, CRC Press, Boca Raton, FL, pp. 6-1–6-100, 2007.

a single-crystal epitaxial layer. Molecular beam epitaxy was originally developed in 1969 by J.R. Arthur and A.Y. Cho. The thickness of the insulator layer (formed by silicon dioxide, silicon nitride, aluminum oxide, zirconium oxide, or other high-k dielectrics) in field-effect transistors (FETs) was gradually reduced from tens of nanometers to less than 1 nm.

The aforementioned, as well as other meaningful fundamental and technological developments, were not referred to as nanoscience, nanoengineering, and nanotechnology until recent years. The recent trend of using the prefix *nano* in many cases is an excessive attempt to associate products, areas, technologies, and even theories with *nano*. Studies primarily focusing on atomic structures, examining atoms, researching subatomic particles, and studying molecules, biology, chemistry, physics, and other disciplines have been using the term *microscopic* even though they have dealt with the atomic theory of matter using pico- and femtometer atomic/subatomic dimensions, employing quantum physics, etc.

De Broglie's postulate provides a foundation of the Schrödinger theory, which describes the behavior of *microscopic* particles within the *microscopic* structure of matter composed of atoms. Atoms are composed of nuclei and electrons, and a nucleus consists of neutrons and protons. The *microscopic* theory has been used to examine *microscopic* systems (atoms and elementary particles) such as baryons, leptons, muons, mesons, partons, photons, quarks, etc. The electron and π-meson (pion) have masses 9.1×10^{-31} and 2×10^{-28} kg, while their radii are 2.8×10^{-15} and 2×10^{-15} m, respectively. For these subatomic particles, the term *microscopic* has been used. The femtoscale dimensionality of subatomic particles has not been a justification to define them to be "*femtoscopic*" particles or to classify these *microscopic* systems to be "*femtoscopic*."

Molecular electronics centers on developed science and engineering fundamentals, while progress in chemistry and biotechnology can be utilized to accomplish *bottom-up* fabrication. The attempts to invent appealing terminology for well-established theories and technologies has sometimes led to a broad spectrum of newly originated terms and revised definitions. For example, well-established molecular, polymeric, supramolecular, and other motifs sometimes have been renamed to be the directed nanostructured self-assembly, controlled biomolecular nanoassembling, etc. Controlling by the designer self-replication, though performed in biosystems through complex and not fully comprehended mechanisms, has been an ambitious target for many decades. This task can potentially be accomplished utilizing biochemistry and biotechnology. Many recently announced appealing declarations (molecular building blocks, molecular assembler, nanostructured synthesis, and others) are quite similar to those covered in *Aromatic Compounds*, *Chemistry of Coordination Compounds*, *Modern Materials*, and other chapters reported in undergraduate biology, biochemistry, and chemistry textbooks published many decades ago. In those chapters, different organic compounds, ceramics, polymers, crystals, composites, and other materials, as well as distinct molecules, are covered, with the corresponding synthesis processes known for decades or even centuries.

With the concentration on electronics, one may be interested in analyzing the major trends [1–4] as well as in defining microelectronics and nanoelectronics. Microelectronics is a well established and mature field, with more than a 150-billion-dollar market per year. As with microelectronics [1], the definition of nanoelectronics should stress on the underlined premises. The focus, objective, and major themes of nanoelectronics are defined as [4]: fundamental/applied/experimental research and technology developments in devising and the implementation of novel high-performance enhanced-functionality atomic/molecular devices, modules, and platforms (systems), as well as high-yield bottom-up fabrication. The nano (molecular) electronics centers on:

1. Discovery of novel devices that are based upon a new device physics
2. Utilization of exhibited unique phenomena and capabilities
3. Devising of enabling organizations and architectures
4. Bottom-up fabrication

Other features at the device, module, and system levels are emerging as subproducts of these four major themes. Compared with solid-state semiconductor (microelectronic) devices, molecular devices (^{M}devices) exhibit new phenomena and offer unique capabilities that should be utilized at the module and system levels. In order to avoid discussions in terminology and definitions, the term molecular, and not the prefix *nano*, is mostly used in this chapter.

At the device level, IBM, Intel, Hewlett-Packard, and other leading companies have been successfully conducting pioneering research and pursuing technological developments in *solid*, molecular electronic devices ([ME]devices), molecular wires, molecular interconnect, etc. Basic, applied, and experimental developments in *solid* molecular electronics are reported in Refs. [5–10]. Unfortunately, it seems that limited progress has been accomplished in molecular electronics, *bottom-up* fabrication, and technology developments. These revolutionary, high-risk, high-payoff areas have recently emerged, and they require time, readiness, commitment, acceptance, investment, infrastructure, innovations, and market needs. Among the most promising directions that will lead to revolutionary advances are the devising and designing of:

- Molecular signal/data processing platforms
- Molecular memory platforms
- Integrated molecular processing-and-memory platforms
- Molecular information processing platforms

Our ultimate objective is to contribute to the developments of a viable molecular architectronics ([M]architectronics) paradigm in order to radically increase the performance of processing (computing) and memory platforms. Molecular electronics holds the promise of guaranteeing information processing preeminence, computing superiority, and memory supremacy.

We present a unified synthesis taxonomy in the design of 3-D molecular integrated circuits ([M]ICs), which are envisioned to be utilized in processing and memory platforms for a new generation of arrays, processors, computers, etc. The design of [M]ICs is accomplished by using a novel technology-centric concept based on the use of [N]hypercells consisting of molecular gates ([M]gates). These [M]gates comprise interconnected multi-terminal [M]devices. Some promising [M]devices have been examined in sufficient detail. Innovative approaches in design of [M]PPs, formed from [M]ICs, are documented. Our major motivation is to further develop and apply a sound fundamental theory coherently supported by enabling solutions and technologies. We expand the basic and applied research toward technology-centric computer-aided design (CAD)-supported [M]ICs design theory and practice. The advancements and progress are ensured by using new sound solutions, and a need for a super-large-scale integration (SLSI) is emphasized. The fabrication aspects are covered. The results reported further expand the horizon of the molecular electronics theory and practice, information technology, and design of processing/memory platforms, as well as molecular technologies (nanotechnology).

7.2   Data and Signal Processing Platforms

We face a wide spectrum of challenges and problems. It seems that devising [M]devices, *bottom-up* fabrication, [M]ICs design, and technology-centric CAD developments are among the most complex issues. Before being engaged in [M]architectronics and its application, let us turn attention to the retrospect, and then focus on the prospect and opportunities. The history of data retrieval and processing tools is traced back thousands of years ago. To enter the data, retain it, and perform calculations, people used a mechanical "tool" called an *abacus*. The early *abacus*, known as a counting board, was a piece of wood, stone, or metal with carved grooves or painted lines between which movable beads, pebbles, or wood/bone/stone/metal disks were arranged. These beads were moved around, according to the "programming rules" memorized by the user, to solve some recording and arithmetic problems. The *abacus* was used for counting, tracking data, and recording facts even before the concept of numbers was invented. The oldest counting board, found in 1899 on the island of Salamis, was used by the Babylonians around 300 BC. As shown in Figure 7.1a, the Salamis' *abacus* is a slab of marble marked with two sets of 11 vertical lines (10 columns), a blank space between them, a horizontal line crossing each set of lines, and Greek symbols along the top and bottom. Another important invention around the same time was the astrolabe for navigation.

In 1623, Wilhelm Schickard built his "calculating clock" which is a 6-digit machine that can add, subtract, and indicate overflow by ringing a bell. Blaise Pascal is usually credited for building the first digital calculating machine. He made it in 1642 to assist his father who was a tax collector. This machine was able to add

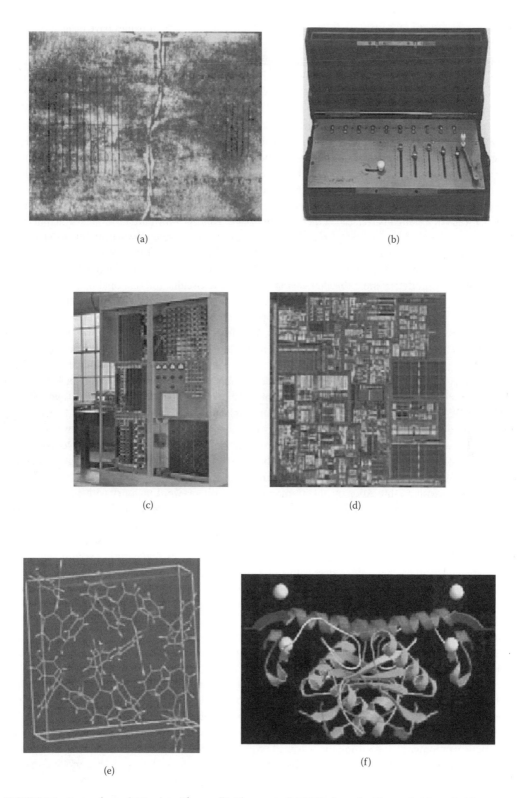

FIGURE 7.1 From *abacus* (300 BC) to Thomas "Arithmometer" (1820), from the Electronic Numerical Integrator and Computer (1946) to 1.5 × 1.5 cm 478-pin Intel® Pentium® 4 processor with 42 million transistors (2002), http://www.intel.com/, and toward 3-D *solid* and *fluidic* molecular electronics and processing.

numbers entered with dials. Pascal also designed and built a "Pascaline" machine in 1644. These 5- and 8-digit machines used a different concept compared with Schickard's "calculating clock." In particular, rising and falling weights instead of a gear drive were used. The Pascaline machine can be extended for more digits, but it cannot subtract. Pascal sold more than 10 machines, and several of them still exist. In 1674, Gottfried Wilhelm von Leibniz introduced a "Stepped Reckoner" using a movable carriage to perform multiplications. Charles Xavier Thomas applied Leibniz's ideas and, in 1820, made a mechanical calculator (Figure 7.1b). In 1822, Charles Babbage built a 6-digit calculator, which performed mathematical operations using gears. For many years, from 1834 to 1871, Babbage carried out the "Analytical Engine" project. His design integrated the stored-program (memory) concept, envisioning the idea that memory may hold more than 100 numbers. The proposed machine had a read-only memory in the form of punch cards. These cards were chained, and the motion of each chain could be reversed. Thus, the machine was able to perform the conditional manipulations and integrated coding features. The instructions depended on the positioning of metal studs in a slotted barrel, called the "control barrel." Babbage only partially implemented his ideas in designing a proof-of-concept programmable calculator because his innovative initiatives were far ahead of the technological capabilities and theoretical foundations. But the ideas and goals were set.

In 1926, Vannevar Bush proposed the "product integraph," which is a semiautomatic machine for solving problems in determining the characteristics of electric circuits. International Business Machines introduced in 1935 the "IBM 601" and made more than 1500 of them. This was a punch card machine with an arithmetic unit based on relays, which performed a multiplication in 1 sec. In 1937, George Stibitz constructed a 1-bit binary adder using relays. Alan Turing published a paper reporting "computable numbers" in 1937. In this paper, he solved mathematical problems and proposed a mathematical model of computing known as the *Turing machine*. The idea of the electronic computer is traced back to the late 1920s. However, the major breakthroughs appear later. In 1937, Claude Shannon in his master's thesis outlined the application of relays. He proposed an "electric adder to the base of two." George Stibitz, in 1937, developed a binary circuit based on Boolean algebra. He built and tested the proposed adding device in 1940. John Atanasoff completed a prototype of a 16-bit adder using diode vacuum tubes in 1939. The same year, Zuse and Schreyer examined the application of relay logic. Schreyer completed a prototype of the 10-bit adder using vacuum tubes in 1940, and he built memory using neon lamps. Zuse demonstrated the first operational programmable calculator in 1940. The calculator had floating point numbers with a 7-bit exponent, 14-bit mantissa, sign bit, 64-word memory with 1400 relays, and arithmetic and control units consisting of 1200 relays. Howard Aiken proposed a calculating machine that solved some problems of relativistic physics. He built an "Automatic Sequence Controlled Calculator Mark I." This project was finished in 1944, and "Mark I" was used to calculate ballistics problems. This electromechanical machine was 15 m long, weighed 5 tons, and had 750,000 parts (72 accumulators with arithmetic units and mechanical registers with a capacity of 23 digits +). The arithmetics was fixed-point, with a plug-board determining the number of decimal places. The input–output unit included card readers, a card puncher, paper tape readers, and typewriters. There were 60 sets of rotary switches, each of which could be used as a constant register, e.g., a mechanical read-only memory. The program was read from a paper tape, and data could be read from the other tapes, card readers, or constant registers. In 1943, the U.S. government contracted John Mauchly and Presper Eckert to design the Electronic Numerical Integrator and Computer, which likely was the first electronic digital computer built. The Electronic Numerical Integrator and Computer was completed in 1946 (Figure 7.1c). This machine performed 5000 additions or 400 multiplications per second, showing enormous capabilities for that time. The Electronic Numerical Integrator and Computer weighed 30 tons, consumed 150 kW, and had 18,000 diode vacuum tubes. John von Neumann with colleagues built the Electronic Discrete Variable Automatic Computer in 1945 using the so-called "von Neumann computer architecture."

Combinational and memory circuits comprised microelectronic devices, logic gates, and modules. Micro-electronics textbooks coherently document the developments starting from the discoveries of semiconductor devices to the design of ICs. The major developments are reported below. Ferdinand Braun invented the solid-state rectifier in 1874. Silicon diode was created and demonstrated by Pickard in 1906. The field-effect devices were patented by von Julius Lilienfeld and Oskar Heil in 1926 and 1935, respectively. The functional solid-state bipolar junction transistor (BJT) was built and tested on December 23, 1947 by John Bardeen and Walter Brattain. Gordon Teal made the first silicon transistor in 1948, and William Shockley invented the unipolar field-effect transistor in 1952. The first ICs were designed by Kilby and Moore in 1958.

Microelectronics has been utilized in signal processing and computing platforms. First, second, third, and fourth generations of computers emerged, and a tremendous progress was achieved. The Intel® Pentium® 4 processor, illustrated in Figure 7.1d, and Core™ Duo processor families were built using advanced Intel® microarchitectures. These high-performance processors are fabricated using 90 and 65 nm complimentary metal-oxide-semiconductor (CMOS) technology nodes. The CMOS technology was matured to fabricate high-yield high-performance ICs with trillions of transistors on a single die. The fifth generation of computers will utilize further scaled-down microelectronic devices and enhanced architectures. However, further progress and developments are needed. New solutions and novel enabling technologies are emerging.

The suggestion to utilize molecules as a molecular diode, which can be considered the simplest two-terminal *solid* ^{ME}device, was introduced by Ratner and Aviram in 1974 [11]. This visionary idea has been further expanded through meaningful theoretical, applied, and experimental developments [5–10]. Three-dimensional (3-D) molecular electronics and ^{M}ICs, designed within a 3-D organization, were proposed in Ref. [7]. These ^{M}ICs are designed as aggregated ^{N}hypercells comprising ^{M}gates engineered utilizing 3-D-topology multiterminal *solid* ^{ME}devices (Figure 7.1e).

The U.S. Patent 6,430,511 "Molecular Computer" was issued in 2002 to J.M. Tour, M.A. Reed, J.M. Seminario, D.L. Allara, and P.S. Weiss. The inventors envisioned a molecular computer as formed by establishing arrays of input and output pins, "injecting moleware," and "allowing the moleware to bridge the input and output pins." The proposed "moleware includes molecular alligator clip-bearing 2-, 3-, and molecular 4-, or multi-terminal wires, carbon nanotube wires, molecular resonant tunneling diodes, molecular switches, molecular controllers that can be modulated via external electrical or magnetic fields, massive interconnect stations based on single nanometer-sized particles, and dynamic and static random access memory (SRAM) components composed of molecular controller/nanoparticle or fullerene hybrids." Overall, one may find a great deal of motivating conceptual ideas while expecting the fundamental soundness and technological feasibility.

Questions regarding the feasibility of molecular electronics and ^{M}PPs arise. There does not exist conclusive evidence on the overall soundness of *solid* ^{M}ICs, as there was no analog for the solid-state microelectronics and ICs in the past. In contrast, biomolecular processing platforms (^{BM}PPs) exist in nature. We briefly focus our attention on the most primitive biosystems. Prokaryotic cells (bacteria) lack extensive intracellular organization and do not have cytoplasmic organelles, while eukaryotic cells have well-defined nuclear membranes as well as a variety of intracellular structures and organelles. However, even a 2 μm long single-cell *Escherichia coli*, *Salmonella typhimurium*, *Helicobacter pylori*, and other bacteria possess ^{BM}PPs, exhibiting superb information and signal/data processing. These bacteria also have molecular sensors, $\sim 50 \times 50 \times 50$ nm motors, as well as other numerous biomolecular devices and systems made from proteins. Though the bacterial motors (largest devices) have been studied for decades, baseline operating mechanisms are still unknown [12]. Biomolecular processing and memory mechanisms also have not been comprehended at the device and system levels. The fundamentals of biomolecular processing, memories, and device physics are not well understood even for single-cell bacteria. The information processing, memory storage, and memory retrieval are likely performed utilizing biophysical mechanisms involving ion (~ 0.2 nm)–biomolecule (~ 1 nm)–protein (~ 10 nm) electrochemomechanical interactions and transitions in response to stimuli. The *fluidic* molecular processing and ^{M}PPs, which mimic ^{BM}PPs, were first proposed in Ref. [7]. Figure 7.1f schematically illustrates the ion–biomolecule–protein complex. The electrochemomechanical interactions and transitions establish a possible device physics of a biomolecular device having feasibility and soundness of *synthetic* and *fluidic* molecular electronics.

Having emphasized the device levels, it should be stressed again that superb biomolecular 3-D organizations and architectures are not comprehended. Assume that in prokaryotic cells and neurons, processing and memory storage are performed by transitions in biomolecules such as folding transformations, induced potential, charge variations, bonding changes, etc. These electrochemomechanical changes are accomplished due to binding/unbinding of ions and/or biomolecules, enzymatic activities, etc. The experimental and analytic results show that protein folding is accomplished within nanoseconds and requires $\sim 1 \times 10^{-19}$ to 1×10^{-18} J of energy. Real-time 3-D image processing is ordinarily accomplished even by primitive insects and vertebrates that have less than 1 million neurons. To perform these and other immense processing tasks, less than 1 μW is consumed. However, real-time 3-D image processing cannot be performed by even envisioned processors with trillions of transistors, device switching speed ~ 1 THz, circuit speed ~ 10 GHz, device switching energy $\sim 1 \times 10^{-16}$ J, writing energy $\sim 1 \times 10^{-16}$ J/bit, read time ~ 10 nsec, etc. This is undisputable evidence of superb biomolecular processing that cannot be surpassed by any envisioned microelectronics enhancements and innovations.

7.3 Microelectronics and Nanoelectronics: Retrospect and Prospect

To design and fabricate planar CMOS ICs, which consist of FETs and BJTs as major microelectronic devices, processes and design rules have been defined. Taking note of the topological layout, the physical dimensions and area requirements can be estimated using the design rules which are centered on: (1) minimal feature size and minimum allowable separation in terms of absolute dimensional constraints; (2) lambda rule (defined using the *length unit* λ), which specifies the layout constraints taking note of nonlinear scaling, geometrical constraints, and minimum allowable dimensions, e.g., width, spacing, separation, extension, overlap, width/length ratio, etc. In general, λ is a function of exposure wavelength, image resolution, depth of focus, processes, materials, device physics, topology, etc. For different technology nodes, λ varies from $\sim 1/2$ to 1 of minimal feature size. For the current front-edge 65 nm technology node, introduced in 2005 and deployed by some high-technology companies in 2006, the minimal feature size is 65 nm. It is expected that the feature size could decrease to 18 nm by 2018. For n-channel metal-oxide-semiconductor FETs (MOSFETs) (physical cell size is $\sim 10\lambda \times 10\lambda$) and BJTs (physical cell size is $\sim 50\lambda \times 50\lambda$), the *effective* cell areas are in the range of hundreds and thousands of λ^2, respectively. For MOSFETs, the gate length is the distance between the active source and drain regions underneath the gate. This implies that if the channel length is 30 nm, it does not mean that the gate width or λ is 30 nm. For FETs, the ratio between the *effective* cell size and minimum feature size will remain ~ 20.

One cannot define and classify electronic, optical, electrochemomechanical, and other devices, or ICs, only by taking note of their dimensions (length, area, or volume) or minimal feature size. The device dimensionality is an important feature primarily from the fabrication viewpoint. To classify devices and systems, one examines the device physics, system organization/architecture, and fabrication technologies, assessing distinctive features, capabilities, and phenomena utilized. Even if the dimensions of CMOS transistors are scaled down to achieve 100×100 nm *effective* cell size for FETs by late 2020, these solid-state semiconductor devices may not be viewed as nanoelectronic devices because conventional phenomena and evolved technologies are utilized. The fundamental limits on microelectronics and solid-state semiconductor devices were known and reported for many years [1]. Although significant technology progress has been accomplished, ensuring high-yield fabrication of ICs, the basic physics of semiconductor devices has remained virtually unchanged for decades. Three editions (1969, 1981, and 2007) of the classic textbook, *Physics of Semiconductor Devices* [13–15], coherently cover the device physics. The evolutionary technological developments will continue beyond the current 65 nm technology node. The 45 nm CMOS technology node is expected to emerge in 2007. Assume that by 2018, 18 nm technology nodes will be deployed with the expected $\lambda = \sim 18$ nm and ~ 7 to 8 nm effective channel length for FETs. This will lead to the estimated footprint area of the interconnected FET to be in the range of tens of thousands of nm^2 because the *effective* cell area is at least $\sim 10\lambda \times 10\lambda$. Sometimes, a questionable, size-centered definition of nanotechnology surfaces, involving the 100 nm dimensionality criterion. It is unclear which dimensionality should be used. Also, it is unclear why 100 nm is declared, and not 1 or 999 nm? On the other hand, why not use a volumetric measure of 100 nm^3?

An electric current is a flow of charged particles. The current in conductors, semiconductors, and insulators is due to the movement of electrons. In aqueous solutions, the current is due to the movement of charged particles, e.g., ions, molecules, etc. These devices are classified using the dimension of the charged carriers (electrons, ions, or molecules). However, one may compare the device dimensionality with the size of the particle that causes the current flow or transitions. For example, considering a protein as a core component of a biomolecular device, and an ion as a charge carrier that affects the protein transitions, the device/carrier dimensionality ratio would be ~100. The *classical* electron radius r_0, called the Compton radius, is found by equating the electrostatic potential energy of a sphere with the charge e and radius r_0 to the relativistic rest energy of the electron, which is $m_e c^2$. We have $e^2/(4\pi\varepsilon_0 r_0) = m_e c^2$, where e is the charge on the electron, $e = 1.6022 \times 10^{-19}$ C; ε_0 is the permittivity of free space, $\varepsilon_0 = 8.8542 \times 10^{-12}$ F/m; m_e is the mass of electron, $m_e = 9.1095 \times 10^{-31}$ kg; c is the speed of light, and in the vacuum $c = 299,792,458$ m/sec. Thus, $r_0 = e^2/(4\pi\varepsilon_0 m_e c^2) = 2.81794 \times 10^{-15}$ m. With the achievable volumetric dimensionality of *solid* ^{ME}device in the order of $1 \times 1 \times 1$ nm, one finds that the device is much larger than the carrier. Up to 1×10^{18} devices can be placed in 1 mm^3. This upper-limit device density may not be achieved due to the synthesis constraints, technological challenges, expected inconsistency, aggregation/interconnect complexity, and other problems. The *effective* volumetric dimensionality of interconnected *solid* ^{ME}devices in ^{M}ICs is expected to be ~10 × 10 × 10 nm. For *solid* ^{ME}devices, quantum physics needs to be applied in order to examine the processes, functionality, performance, characteristics, etc. The device physics of *fluidic* and *solid* ^{M}devices are profoundly different. To emphasize the major premises, nanoelectronics implies the use of:

1. Novel high-performance devices, as devised using new device physics, which exhibit unique phenomena and capabilities to be exclusively utilized at the gate and system levels
2. Enabling 3-D organizations and advanced architectures, which ensure superb performance and superior capabilities. These developments rely on the device-level solutions, technology-centric SLSI design, etc.
3. Bottom-up fabrication

To design ^{M}ICs-comprised processing and memory platforms, one must apply novel paradigms and pioneering developments, utilizing 3-D-topology ^{M}devices, enabling organizations/architectures, sound *bottom-up* fabrication, etc. Tremendous progress has been made within the last 60 years in microelectronics, e.g., from inventions and demonstration of functional solid-state transistors to fabrication of processors that comprise trillions of transistors on a single die. Current high-yield 65 nm CMOS technology nodes ensure minimal features ~65 nm, and FETs were scaled down to achieve a channel length below 30 nm. Using this technology for the SRAM cells, a ~500,000 nm^2 footprint area was achieved by Intel. There are optimistic predictions that within 15 years, the minimal feature of planar (2-D) solid-state CMOS-technology transistors may approach ~10 nm, leading to an *effective* cell size for FET ~$20\lambda \times 20\lambda = 200 \times 200$ nm. However, the projected scaling trends are based on a number of assumptions and predicted enhancements [1]. Although the FET cell dimension can reach 200 nm, the overall prospects in microelectronics (technology enhancements, device physics, device/circuits performance, design complexity, cost, and other features) are troubling [1–4]. The near-absolute limits of the CMOS-centered microelectronics can be reached by the next decade. The general trends, prospects, and projections are reported in the *International Technology Roadmap for Semiconductors* [1].

The device size- and switching energy-centered version of Moore's first conjecture for high-yield room-temperature mass-produced microelectronics is reported in Figure 7.2 for past, current (90 and 65 nm), and predicted (45 and 32 nm) CMOS technology nodes. For the switching energy, one uses eV or J, and 1 eV = $1.602176462 \times 10^{-19}$ J. Intel expects to introduce 45 nm CMOS technology nodes in 2007. The envisioned 32 nm technology node is expected to emerge in 2010. The expected progress in the baseline characteristics, key performance metrics, and scaling abilities has been already slowed down due to the encountered fundamental and technological challenges and limitations. Correspondingly, new solutions and technologies have been sought and assessed [1]. The performance and functionality at the device, module, and system levels can be significantly improved by utilizing novel phenomena, employing innovative topological/organizational/architectural solutions, enhancing device functionality, increasing density, improving utilization, increasing switching speed, etc. Molecular electronics (nanoelectronics) is expected to depart from Moore's conjectures. High-yield affordable nanoelectronics is expected to ensure superior performance. The existing superb biomolecular processing/memory platforms and progress in

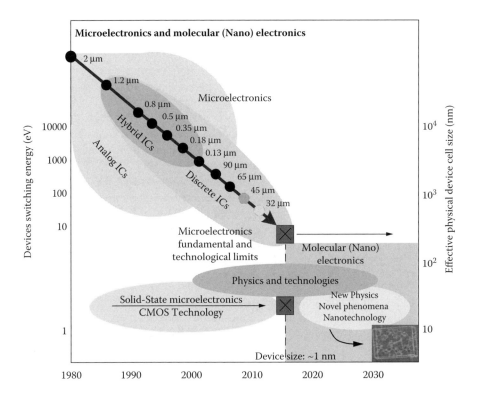

FIGURE 7.2 Envisioned molecular (nano) electronics advancements and microelectronics trends.

molecular electronics are assured evidence of the fundamental soundness and technological feasibility of molecular electronics and [M]PPs. Some data and expected developments, reported in Figure 7.2, are subject to adjustments because it is difficult to accurately foresee the fundamental developments and maturity of prospective technologies due to the impact of many factors. However, the overall trends are obvious and likely cannot be rejected. Having emphasized the emerging molecular (nano) electronics, it is obvious that solid-state microelectronics is a core 21st century technology. CMOS technology will remain a viable technology for many decades even as the limits will be reached and nanoelectronics will mature. It may be expected that, by 2030, core modules of super-high-performance processing (computing) platforms will be implemented using [M]ICs. However, microelectronics and molecular electronics will be complementary technologies, and [M]ICs will not diminish the use of ICs. Molecular electronics and [M]PPs are impetuous, revolutionary (not evolutionary) changes at the device, system, fundamental, and technological levels. The predicted revolutionary changes toward [M]devices are analogous to abrupt changes from the vacuum tube to a solid-state transistor.

The dominating premises of molecular (nano) electronics and [M]PPs have a solid bioassociation. There exist a great number of superb biomolecular systems and platforms. The device-level biophysics and system-level fundamentals of biomolecular processing are not fully comprehended, but, they are *fluidic* and molecule centered. For molecular electronics, theory, engineering practice, and technology are revolutionary advances compared with microelectronic theory and CMOS technology. From a 3-D-centered topology/organization/architecture standpoint, *solid* and *fluidic* molecular electronics evolution mimics superb [BM]PPs. Information processing, memory storage, and other relevant tasks, performed by biosystems, are a sound proving ground for the proposed developments. Molecular electronics will lead to novel [M]PPs. Compared with the most advanced CMOS processors, molecular platforms will greatly enhance functionality and processing capabilities, radically decrease latency, power and execution time, as well as drastically increase device density, utilization, and memory capacity. Many difficult problems at the device and system levels must be addressed, researched, and solved. For example, the following tasks should be carried out: design, analysis, optimization, aggregation, routing, reconfiguration, verification, evaluation, etc. Many of the above-mentioned problems have not been addressed yet. Due to significant challenges, much effort must be focused to solve these problems. We address and propose solutions to some of the aforementioned fundamental and applied problems, establishing an ^{M}architectronics paradigm. A number of baseline problems are examined, progressing from the system level to the module/device level and vice versa. Taking note of the diversity and magnitude of tasks under consideration, one cannot formulate, examine, and solve all challenging problems. A gradual step-by-step approach is pursued rather than attempting to solve abstract problems with a minimal chance of succeeding. There is a need to stimulate further developments and foster advanced research focusing on well-defined existing fundamentals and future perspectives emphasizing the near-, medium- and long-term prospects, visions, problems, solutions, and technologies.

7.4 Performance Estimates

The combinational and memory ^{M}ICs can be designed as aggregated ^{N}hypercells comprised of ^{M}gates and molecular memory cells [16]. At the device level, one examines functionality, studies characteristics, and estimates performance of 3-D-topology ^{M}devices. The device- and system-level performance measures are of great interest. The experimental results indicate that protein folding is performed within 1×10^{-6} to 1×10^{-12} sec and requires $\sim 1 \times 10^{-19}$ to 1×10^{-18} J of energy. These transition time and energy estimates can be used for some *fluidic* and *synthetic* ^{M}devices. To analyze protein-folding energetics, examine the switching energy in solid-state microelectronic devices, estimate *solid* ^{ME}devices energetics, and perform other studies, distinct concepts have been applied.

For solid-state microelectronic devices, the logic signal energy is expected to reduce to $\sim 1 \times 10^{-16}$ J, and the energy dissipated is $E = Pt = IVt = I^2Rt = Q^2R/t$, where P is the power dissipation; I and V are the current and voltage along the discharge path, respectively; R and Q are the resistance and charge, respectively.

The term $k_B T$ has been used in the solution of distinct problems. Here, k_B is the Boltzmann constant, $k_B = 1.3806 \times 10^{-23}$ J/K $= 8.6174 \times 10^{-5}$ eV/K. For example, expression $\gamma k_B T$ ($\gamma > 0$) has been used to perform energy estimates, and $k_B T \ln(2)$ was used in the attempt to assess the lowest energy bound for a binary switching. The applicability of distinct equations must be thoroughly examined and sound concepts must be applied. Statistical mechanics and entropy analysis coherently utilize the term $k_B T$ within a specific context as reported below, while, for some other applications and problems, the use of $k_B T$ may be impractical.

7.4.1 Entropy and Its Application

For an ideal gas, the kinetic-molecular Newtonian model provides the average translational kinetic energy of a gas molecule. In particular, $\frac{1}{2}m(v^2)_{av} = \frac{3}{2}k_B T$. One concludes that the average translational kinetic energy per gas molecule depends only on the temperature. The most notable equation of statistical thermodynamics is the Boltzmann formula for the entropy as a function of only the system state, e.g., $S = k_B \ln w$, where w is the number of possible arrangements of atoms or molecules in the system. Unlike energy, entropy is a quantitative measure of the system disorder in any specific state, and S is not related to each individual atom or particle. At any temperature above absolute zero, the atoms acquire energy, more arrangements become possible, and because $w > 1$, one has $S > 0$. The entropy and energy are very different quantities. When the interaction between the system and environment involves only reversible processes, the total entropy is constant, and $\Delta S = 0$. When there is any irreversible process, the total entropy increases, and $\Delta S > 0$. One may derive the entropy *difference* between two distinct states in a system that undergoes a thermodynamic process that takes the system from an initial *macroscopic* state 1 with w_1 possible *microscopic* states to a final *macroscopic* state 2 with w_2 associated *microscopic* states. The change in entropy is found as $\Delta S = S_2 - S_1 = k_B \ln w_2 - k_B \ln w_1 = k_B \ln(w_2/w_1)$. Thus, the entropy *difference* between two *macroscopic* states depends on the ratio of the number of possible *microscopic* states. The entropy change for any reversible isothermal process is given using an infinitesimal quantity of heat ΔQ. For initial and final states 1 and 2, one has

$$\Delta S = \int_1^2 \frac{dQ}{T}.$$

Example 7.1

To heat 1 ykg (1×10^{-24} kg) of silicon from 0 to 100°C, using the constant specific heat capacity $c = 702$ J/kg·K over the temperature range, the change of entropy is

$$\Delta S = S_2 - S_1 = \int_1^2 \frac{dQ}{T} = \int_{T_1}^{T_2} mc\frac{dT}{T} = mc\ln\frac{T_2}{T_1} = 1\times10^{-24} \ kg \times 702 \ \tfrac{J}{kg\cdot K} \times \ln\tfrac{373.15 \ K}{273.15 \ K} = 2.19\times10^{-22} \ J/K.$$

From $\Delta S = k_B\ln(w_2/w_1)$, one finds the ratio between *microscopic* states w_2/w_1. For the problem under consideration, $w_2/w_1 = 7.7078 \times 10^6$. If $w_2/w_1 = 1$, the total entropy is constant, and $\Delta S = 0$. The energy that must be supplied to heat 1×10^{-24} kg of silicon for $\Delta T = 100°$C is $Q = mc\Delta T = 7.02 \times 10^{-20}$ J. To heat 1 g of silicon from 0 to 100°C, one finds $\Delta S = S_2 - S_1 = mc\ln(T_2/T_1) = 0.219$ J/K and $Q = mc\Delta T = 70.2$ J. Taking note of equation $\Delta S = k_B \ln(w_2/w_1)$, it is impossible to derive the numerical value for w_2/w_1.

For a silicon atom, the covalent, atomic, and van der Waals radii are 117, 117, and 200 pm, respectively. The Si–Si and Si–O covalent bonds are 232 and 151 pm, respectively. One can examine the thermodynamics using the enthalpy, Gibbs function, entropy, and heat capacity of silicon in its solid and gas states. The atomic weight of a silicon atom is 28.0855 amu, where amu stands for the atomic mass unit, 1 amu = 1.66054×10^{-27} kg. Hence, the mass of a single Si atom is 28.0855 amu × 1.66054×10^{-27} kg/amu = 4.6637×10^{-26} kg. Therefore, the number of silicon atoms in 1×10^{-24} kg of silicon is $1 \times 10^{-24}/4.6637 \times 10^{-26} = 21.44$. Consider two silicon atoms to be heated from 0 to 100°C. For m = 9.3274×10^{-26} kg, we have $\Delta S = S_2 - S_1 = mc\ln(T_2/T_1) = 2.04 \times 10^{-23}$ J/K. One obtains an obscure result $w_2/w_1 = 4.39$. It should be emphasized again that the entropy and macroscopic/microscopic states analysis are performed for an ideal gas assuming the accuracy of the kinetic-molecular Newtonian model. In general, to examine the particle and molecule energetics, quantum physics must be applied.

For particular problems, using the results reported, one may carry out similar analyses for other atomic complexes. For example, while carbon has not been widely used in microelectronics, the organic molecular electronics is carbon-centered. Therefore, some useful information is reported. For a carbon atom, the covalent, atomic, and van der Waals radii are 77, 77 and 185 pm, respectively. Carbon can be in the solid (graphite or diamond) and gas states. Using the atomic weight of a carbon atom, which is 12.0107 amu, the mass of a single carbon atom is 12.0107 amu × 1.66054×10^{-27} kg/amu = 1.9944×10^{-26} kg.

Example 7.2

Letting $w = 2$, the entropy is found to be $S = k_B\ln2 = 9.57 \times 10^{-24}$ J/K = 5.97×10^{-5} eV/K. Having derived S, one cannot conclude that the minimal energy required to ensure the transition (*switching*) between two *microscopic* states or to erase a bit of information (energy dissipation) is $k_BT\ln2$, which for $T = 300$ K gives $k_BT\ln2 = 2.87 \times 10^{-21}$ J = 0.0179 eV. In fact, under this reasoning, one assumes the validity of the *averaging* kinetic-molecular Newtonian model and applies the assumptions of distribution statistics while at the same time allowing only two distinct *microscopic* system states. The energy estimates should be performed utilizing the quantum mechanics.

7.4.2   Distribution Statistics

Statistical analysis is applicable only to systems with a large number of particles and energy states. The fundamental assumption of statistical mechanics is that in thermal equilibrium, every distinct state with the same total energy is equally probable. Random thermal motions constantly change energy from one particle to another and from one form of energy to another (kinetic, rotational, vibrational, etc.) obeying the conservation of energy principle. The absolute temperature T has been used as a measure of the total energy of a system in thermal equilibrium. In semiconductor devices, enormous numbers of particles (electrons) are considered using the electrochemical potential $\mu(T)$. The Fermi–Dirac distribution function

$$f(E) = \frac{1}{1 + e[(E - \mu(T))/k_B T]}$$

gives the average (probable) number of electrons of a system (device) in equilibrium at temperature T to be found in a quantum state of energy E. The electrochemical potential at absolute zero is the Fermi energy E_F, and $\mu(0) = E_F$. The occupation probability that a particle would have the specific energy is not related to quantum indeterminacy. Electrons in solids obey Fermi–Dirac statistics. The distribution of electrons, leptons, and baryons (*identical fermions*) over a range of allowable energy levels at thermal equilibrium is expressed as

$$f(E) = \frac{1}{1 + e[(E - E_F)/k_B T]},$$

where T is the equilibrium temperature of the system. Hence, the Fermi–Dirac distribution function $f(E)$ gives the probability that an allowable energy state at energy E will be occupied by an electron at temperature T.

For *distinguishable* particles, one applies the Maxwell–Boltzmann statistics with a distribution function

$$f(E) = e\left(-\frac{E - E_F}{k_B T}\right).$$

The Bose–Einstein statistics are applied to *identical bosons* (photons, mesons, etc.). The Bose–Einstein distribution function is given as

$$f(E) = \frac{1}{e[(E - E_F)/k_B T] - 1}.$$

As was emphasized, the distribution statistics are applicable to electronic devices that consist of a great number of constituents, where particle interactions can be simplified by deducing the system behavior from the statistical consideration. Depending on the device physics, one must coherently apply the appropriate baseline theories and concepts.

Example 7.3

For $T = 100$ and 300 K, and $E_F = 5$ eV, the Fermi–Dirac distribution functions are reported in Figure 7.3a. Figure 7.3b documents the Maxwell–Boltzmann distribution functions $f(E)$.

7.4.3 Energy Levels

In ^{M}devices, one can calculate the energy required to excite the electron, and the allowed energy levels are quantized. In contrast, solids are characterized by energy band structures that define electric characteristics. In semiconductors, the relatively small band gaps allow excitation of electrons from the valance band to conduction band by thermal or optical energy. The application of quantum mechanics allows one to derive the expression for the quantized energy. For a hydrogen atom one has

$$E_n = -\frac{m_e e^4}{32\pi^2 \varepsilon_0^2 \hbar^2 n^2},$$

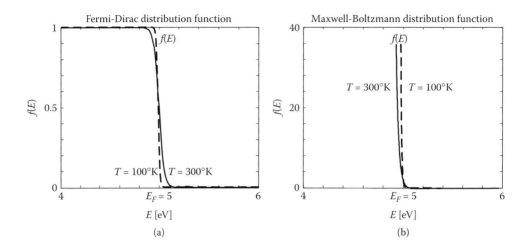

FIGURE 7.3 The Fermi–Dirac and Maxwell–Boltzmann distribution functions for $T = 100$K and $T = 300$K if $E_F = 5$ eV.

where $\hbar$ is the modified Planck's constant, $\hbar = h/2\pi = 1.055 \times 10^{-34}$ J·sec $= 6.582 \times 10^{-16}$ eV·sec. The energy levels depend on the quantum number n. As n increases, the total energy of the quantum state becomes less negative, and $E_n \to 0$ if $n \to \infty$. The state of lowest total energy is the most stable state for the electron, and the normal state of the electron for a hydrogen (one-electron atom) is at $n = 1$.

Thus, for the hydrogen atom, in the absence of a magnetic field **B**, the energy

$$E_n = -\frac{m_e e^4}{32\pi^2 \varepsilon_0^2 \hbar^2 n^2}$$

depends only on the principle quantum number n. The conversion 1 eV = $1.602176462 \times 10^{-19}$ J is commonly used, and $E_{n=1} = -2.17 \times 10^{-18}$ J = -13.6 eV. For $n = 2$, $n = 3$, and $n = 4$, we have $E_{n=2} = -5.45 \times 10^{-19}$ J, $E_{n=3} = -2.42 \times 10^{-19}$ J and $E_{n=4} = -1.36 \times 10^{-19}$ J. When the electron and nucleus are separated by an infinite distance ($n \to \infty$), one has $E_n \to 0$. The energy difference between the quantum states n_1 and n_2 is $\Delta E = E_{n=1} - E_{n=2}$, and

$$\Delta E = E_{n=1} - E_{n=2} = \frac{m_e e^4}{32\pi^2 \varepsilon_0^2 \hbar^2}\left(\frac{1}{n_2^2} - \frac{1}{n_1^2}\right),$$

where $m_e e^4 / 32\pi^2 \varepsilon_0^2 \hbar^2 = 2.17 \times 10^{-18}$ J = 13.6 eV.

The excitation energy of an exited state n is the energy above the ground state; e.g., for the hydrogen atom one has $(E_n - E_{n=1})$. The first exited state ($n = 2$) has the excitation energy $E_{n=2} - E_{n=1} = -3.4 + 13.6 = 10.2$ eV. In atoms, orbits are characterized by quantum numbers.

The De Broglie conjecture relates the angular frequency ν and energy E. In particular, $\nu = E/h$, where h is the Planck's constant, $h = 6.626 \times 10^{-34}$ J·sec $= 4.136 \times 10^{-15}$ eV·sec. The frequency of a photon electromagnetic radiation is found as $\nu = \Delta E/h$.

> **Remark.** The energy difference between the quantum states ΔE is not the energy uncertainty in the measurement of E, which is commonly denoted in the literature as ΔE. In this chapter, reporting the Heisenberg uncertainty principle, to ensure consistency, we use the notation $\Delta \hat{E}$. In particular, Section 7.12.2 reports the energy–time uncertainty principle as $\sigma_E \sigma_t \geq \frac{1}{2}\hbar$ or $\Delta \hat{E}\Delta t \geq \frac{1}{2}\hbar$, where σ_E and σ_t are the standard deviations, and notations $\Delta \hat{E}$ and Δt are used to define the standard deviations as uncertainties, $\Delta \hat{E} = \sqrt{\langle \hat{E}^2 \rangle - \langle \hat{E} \rangle^2}$.

For many-electron atoms, an atom in its normal (electrically neutral) state has Z electrons and Z protons. Here, Z is the atomic number, and for boron, carbon, and nitrogen, $Z = 5$, 6, and 7, respectively. The total electric charge of atoms is zero because the neutron has no charge while the proton and electron charges have the same magnitude but opposite sign. For the hydrogen atom, denoting the distance that separates the electron and proton by r, the Coulomb potential is $\Pi(r) = e^2/(4\pi\varepsilon_0 r)$. The radial attractive Coulomb potential felt by the single electron due to the nucleus having a charge Ze is $\Pi(r) = Z(r)e^2/(4\pi\varepsilon_0 r)$, where $Z(r) \to Z$ as $r \to 0$ and $Z(r) \to 1$ as $r \to \infty$. By evaluating the average value for the radius of the shell, the effective nuclear charge Z_{eff} is found. The common approximation to calculate the total energy of an electron in the outermost populated shell is

$$E_n = -\frac{m_e Z_{\text{eff}}^2 e^4}{32\pi^2 \varepsilon_0^2 \hbar^2 n^2},$$

and

$$E_n = -2.17 \times 10^{-18} \frac{Z_{\text{eff}}^2}{n^2} \, (J).$$

The effective nuclear charge Z_{eff} is derived using the electron configuration. For boron, carbon, nitrogen, silicon, and phosphorus, three commonly used Slater, Clementi, and Froese-Fischer Z_{eff} are: 2.6, 2.42, and 2.27 (for B), 3.25, 3.14, and 2.87 (for C), 3.9, 3.83, and 3.46 (for N), 4.13, 4.29, and 4.48 (for Si), 4.8, 4.89, and 5.28 (for P). Taking note of the electron configurations for the above-mentioned atoms, one concludes that ΔE could be of the order of $\sim 1 \times 10^{-19}$ to 1×10^{-18} J. If one supplies the energy greater than E_n to the electron, the energy excess will appear as kinetic energy of the free electron. The transition energy should be adequate to excite electrons. For different atoms and molecules with different exited states, as prospective *solid* ^{ME}devices, the transition (switching) energy can be estimated to be $\sim 1 \times 10^{-19}$ to 1×10^{-18} J. This energy estimate is valid for biomolecular and *fluidic* ^{M}devices.

The quantization of the orbital angular momentum of the electron leads to a quantization of the electron total energy. The space quantization permits only quantized values of the angular momentum component in a specific direction. The magnitude L_μ of the angular momentum of an electron in its orbital motion around the center of an atom and the z component L_z are $L_\mu = \sqrt{l(l+1)}\hbar$ and $L_z = m_l\hbar$, respectively, where l is the orbital quantum number; m_l is the magnetic quantum number which is restricted to integer values $-l, -l + 1, \ldots, l - 1, l$, e.g., $|m_l| \leq l$. If a magnetic field is applied, the energy of the atom will depend on the alignment of its magnetic moment with the external magnetic field.

In the presence of a magnetic field $\mathbf{B}$, the energy levels of the hydrogen atom are

$$E_n = -\frac{m_e e^4}{32\pi^2 \varepsilon_0^2 \hbar^2 n^2} - \mu_L \cdot \mathbf{B},$$

where μ_L is the orbital magnetic dipole moment, $\mu_L = -(e/2m_e)\mathbf{L}$, $\mathbf{L} = \mathbf{r} \times \mathbf{p}$. Let $\mathbf{B} = B_z \mathbf{z}$. One finds

$$E_n = -\frac{m_e e^4}{32\pi^2 \varepsilon_0^2 \hbar^2 n^2} + \frac{e}{2m_e} \mathbf{L} \cdot \mathbf{B} = -\frac{m_e e^4}{32\pi^2 \varepsilon_0^2 \hbar^2 n^2} + \frac{e}{2m_e} B_z L_z = -\frac{m_e e^4}{32\pi^2 \varepsilon_0^2 \hbar^2 n^2} + \frac{e}{2m_e} B_z m_l \hbar.$$

If the electron is in an $l = 1$ orbit, the orbital magnetic dipole moment is $\mu_L = (e\hbar/2m_e) = 9.3 \times 10^{-24}$ J/T $= 5.8 \times 10^{-5}$ eV/T. Hence, if the magnetic field is changed by 1 T, an atomic energy level changes by $\sim 1 \times 10^{-4}$ eV. The *switching* energy required to ensure the transitions between distinct *microscopic* states is straightforwardly derived using the wave function and allowed discrete energies.

7.4.4   Device Switching Speed

The transition (switching) speed of ^{M}devices largely depends on the device physics, phenomena utilized, and other factors. One examines dynamic evolutions and transitions by applying molecular dynamics theory, Schrödinger's equation, time-dependent perturbation theory, numerical methods, and other concepts. The analysis of state transitions and interactions allows one to coherently study the controlled device behavior, evolution, and dynamics. The simplified steady-state analysis is also applied to obtain estimates. Considering the electron transport, one may assess the device's features using the number of electrons. For example, for 1 nA current, the number of electrons that cross the molecule per second is $1 \times 10^{-9}/1.6022 \times 10^{-19} = 6.24 \times 10^9$ which is related to the device state transitions. The maximum carrier velocity places an upper limit on the frequency response of semiconductor and molecular devices. The state transitions can be accomplished by a single photon or electron. Using the Bohr postulates, the average velocity of an optically exited electron is $v = Ze^2 / 4\pi\varepsilon_0 \hbar n$. Taking into account that for all atoms $Z/n \approx 1$, one finds the orbital velocity of an optically excited electron to be $v = 2.2 \times 10^6$ m/sec, and $v/c \approx 0.01$. Considering an electron as a not relativistic particle, taking note of $E = mv^2/2$, we obtain the particle velocity as a function of energy as $v(E) = \sqrt{2E/m}$. Letting $E = 0.1$ eV $= 0.16 \times 10^{-19}$ J, one finds $v = 1.88 \times 10^5$ m/sec. Assuming 1 nm path length, the traversal (*transit*) time is $\tau = L/v = 5.33 \times 10^{-15}$ sec. Hence, ^{M}devices can operate at a high switching frequency. However, one may not conclude that the device switching frequency to be utilized is $f = 1/(2\pi\tau)$ due to device physics features (number of electrons, heating, interference, potential, energy, noise, etc.), system-level functionality, circuit specifications, etc. Having estimated the $v(E)$ for ^{M}devices, the comparison to microelectronics devices is of interest. In silicon, the electron and hole velocities reach up to 1×10^5 m/sec at a very high electric field with the intensity 1×10^5 V/cm. The reported estimates indicate that particle velocities in ^{M}devices exceed the carriers' saturated drift velocity in semiconductors.

7.4.5   Photon Absorption and Transition Energetics

Consider a rhodopsin, which is a highly specialized protein-coupled receptor, that detects photons in the rod photoreceptor cell. The first event in the monochrome vision process, after the photon (light) hits the rod cell, is the isomerization of the chromophore 11-*cis*-retinal to all-*trans*-retinal. When an atom or molecule absorbs a photon, its electron can move to the higher-energy orbital, and the atom or molecule makes a transition to a higher-energy state. In retinal absorption, a photon promotes a π electron to a higher-energy orbital, e.g., there is a $\pi - \pi^*$ excitation. This excitation breaks the π component of the double bond allowing free rotation about the bond between carbon 11 and carbon 12. This isomerization, which corresponds to switching, occurs in a picosecond range. The energy of a single photon is found as $E = hc/\lambda$, where λ is the wavelength. The maximum absorbance for rhodopsin is 498 nm. For this wavelength, one finds $E = 4 \times 10^{-19}$ J. This energy is sufficient to ensure transitions and functionality. It is important to emphasize that the photochemical reaction changes the shape of the retina, causing a conformational change in the opsin protein, which consists of 348 amino acids, covalently linked together to form a single chain. The sensitivity of the eye photoreceptor is one photon, and the energy of a single photon, which is $E = 4 \times 10^{-19}$ J, ensures the functionality of a molecular complex of ~5000 atoms that constitute 348 amino acids. We derived the excitation energy (signal energy) that is sufficient to ensure state transitions and processing. This provides conclusive evidence that ~1×10^{-19} to 1×10^{-18} J of energy is required to guarantee the state transitions for complex molecular aggregates.

7.4.6   Processing Performance Estimates

Reporting the performance estimates, we focus on molecular electronics, basic physics, and envisioned solutions. The 3-D-centered topology/organization of envisioned *solid* and *fluidic* devices and systems are analogous to the topology/organization of ^{BM}PPs. Aggregated brain neurons perform superb information processing, perception, learning, robust reconfigurable networking, memory storage, and other functions. The number of neurons in the human brain is estimated to be ~100 billion; mice and rats have ~100 million neurons, while honeybees and ants have ~1 million neurons. Bats use echolocation sensors for navigation, obstacle avoidance, and hunting. By processing the sensory data, bats can detect 0.1% frequency shifts caused by the Doppler effect. They distinguish echoes received ~100 μsec apart. To accomplish these, as well as to perform shift compensation and transmitter/receiver isolation, real-time signal/data processing should be accomplished within at least microseconds. Flies accomplish a real-time precisely coordinated motion due to remarkable actuation and an incredible visual system that maps the relative motion using the retinal photodetector arrays. The information from the visual system and sensors is transmitted and processed within the nanoseconds range requiring μW of power. The dimension of the brain neuron is ~10 μm, and the density of neurons is ~100,000 neurons/mm³. The review of electrical excitability of neurons is reported in Ref. [17]. The biophysics and mechanisms of biomolecular information and signal/data processing are not fully comprehended. The biomolecular state transitions are accomplished with a different rate. The electrochemomechanical biomolecular transformations (propagation of biomolecules and ions through the synaptic cleft and membrane channels, protein folding, binding/unbinding, etc.) could require microseconds. In contrast, photon- and electron-induced transitions can be performed within femtoseconds. The energy estimates were documented obtaining the transition energy $\sim 1 \times 10^{-19}$ to 1×10^{-18} J.

Performing enormous information processing tasks with immense performance that are far beyond predicted capabilities of envisioned parallel vector processors (which perform signal/data processing), the human brain consumes only ~20 W. Only some of this power is required to accomplish information and signal/data processing. This contradicts some postulates of slow processing, immense delays, high energy/power requirements, low switching speed, and other hypotheses reported in Refs. [18–21]. The human retina has 125 million rod cells and 6 million cone cells, and an enormous amount of data, among other tasks, is processed in real-time. Real-time 3-D image processing, ordinarily accomplished even by primitive vertebrates and insects that consume less than 1 μW to perform information processing, cannot be performed by envisioned processors with trillions of transistors, a device switching speed of 1 THz, circuit speed of 10 GHz, device switching energy 1×10^{-16} J, writing energy 1×10^{-16} J/bit, read time 10 nsec, etc. Molecular devices can operate with the estimated transition energy $\sim 1 \times 10^{-19}$ to 1×10^{-18} J, discrete energy levels (ensuring multiple-valued logics), and femtosecond transition dynamics, guaranteeing exceptional device performance. These 3-D-topology ᴹdevices result in the ability to design super-high-performance processing and memory platforms within 3-D organizations and enabling architectures ensuring unprecedented capabilities including massive parallelism, robustness, reconfigurability, etc.

Distinct performance measures, estimates, and indexes are used. For profoundly different paradigms (microelectronics vs. molecular electronics that are distinguished by distinct topologies, organizations, and architectures), Figure 7.4 reports some baseline performance estimates, e.g., transition (switching) energy, delay time, dimension, and number of modules/gates. It was emphasized that the device physics and system organization/architecture are dominating features as compared to the dimensionality or number of devices. Due to limited basic/applied/experimental results, as well as the attempts to use four performance variables, reported in Figure 7.4, some performance measures and projected estimates are expected to be refined. Molecular electronics and ᴹICs can utilize diverse molecular primitives and devices that: (1) operate due to different physics, such as electron transport, electrostatic transitions, photon emission, conformational changes, etc.; (2) exhibit distinct phenomena and effects. Therefore, biomolecular, *fluidic* and *solid* ᴹdevices and systems will exhibit distinct performance. As demonstrated in Figure 7.4, advancements are envisioned toward 3-D *solid* molecular electronics departing from ᴮᴹPPs by utilizing a familiar solid-state microelectronics solution. In Figure 7.4, a 3-D-topology neuron is represented as a biomolecular information processing/memory module that may consist of ᴹdevices.

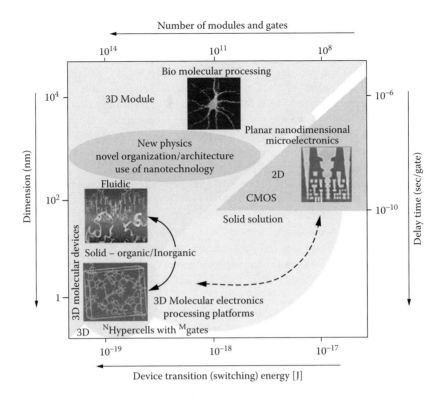

FIGURE 7.4 Towards molecular electronics and processing/memory platforms. Revolutionary advancements: from 2D microelectronics to 3-D molecular electronics; Evolutionary developments: from [BM]PPs to *solid* and *fluidic* molecular electronics and processing.

7.5 Synthesis Taxonomy in Design of [M]ICs and Processing Platforms

Molecular architectronics is a paradigm in the devising and designing of preeminent [M]ICs and [M]PPs. This paradigm is based on:

1. Discovery of novel topological/organizational/architectural solutions, as well as utilization of new phenomena and capabilities of 3-D molecular electronics at the system and device levels
2. Development and implementation of sound methods, technology-centric CAD, and SLSI design concurrently associated with bottom-up fabrication

Various design tasks for 3-D [M]ICs are not analogous to the CMOS-centered design, planar layout, placement, routing, interconnect, and other tasks, which were successfully solved. Conventional VLSI/ULSI design flow is based on the well-established system specifications, functional design, conventional architecture, verification (functional, logic, circuit and layout), as well as CMOS fabrication technology. The CMOS technology utilizes 2-D topology of conventional gates with FETs and BJTs. For [M]ICs, device- and system-level technology-centric design must be performed using novel methods. Figure 7.4 illustrates the proposed 3-D molecular electronics departing from 2-D multilayer CMOS-centered microelectronics. To synthesize [M]ICs, we propose to utilize a unified top-down (system level) and bottom-up (device/gate level) synthesis taxonomy within an *x*-domain flow map as reported in Figure 7.5. The core 3-D design themes are integrated within four domains:

- Devising with validation
- Analysis evaluation

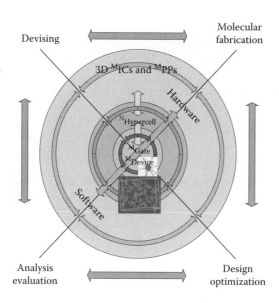

Devising

Molecular
fabrication

Analysis
evaluation

Design
optimization

FIGURE 7.5 Top-down and bottom-up synthesis taxonomy within an *x*-domain flow map.

- Design optimization
- Molecular fabrication

As reported in Figure 7.5, the synthesis and design of 3-D ^{M}ICs and ^{MP}Ps should be performed by utilizing a bidirectional flow map. Novel design, analysis, and evaluation methods must be developed. Design in 3-D space is radically different compared with VLSI/ULSI due to novel 3-D topology/organization, enabling architectures, new phenomena utilized, enhanced functionality, enabling capabilities, complexity, technology-dependence, etc. The unified top-down/bottom-up synthesis taxonomy should be coherently supported by developing innovative solutions to carry out a number of major tasks such as:

1. Devising and design of ^{M}devices, ^{M}gates, $^\aleph$hypercells, and networked $^\aleph$hypercells aggregates which form ^{M}ICs
2. Development of new methods in design and verification of ^{M}ICs
3. Analysis and evaluation of performance characteristics
4. Development of technology-centric CAD to concurrently support design at the system and device/gate levels

The reported unified synthesis taxonomy integrates:

1. *Top-down synthesis*: Devise super-high-performance molecular processing and memory platforms implemented by designed ^{M}ICs within 3-D organizations and enabling architectures. These 3-D ^{M}ICs are implemented as aggregated $^\aleph$hypercells composed of ^{M}gates that are engineered from ^{M}devices (Figures 7.6a and 7.6b)
2. *Bottom-up synthesis*: Engineer functional 3-D-topology ^{M}devices that compose ^{M}gates in order to form $^\aleph$hypercells (for example, multiterminal *solid* ^{ME}devices are engineered as molecules arranged from atoms)

The proposed synthesis taxonomy utilizes a number of innovations at the system and device levels. In particular, (1) innovative architecture, organization, topology, aggregation, and networking in 3-D; (2) novel enhanced-functionality ^{M}devices which form ^{M}gates, $^\aleph$hypercells, and ^{M}ICs; (3) unique phenomena, effects, and solutions (tunneling, parallelism, etc.); (4) *bottom-up* fabrication; (5) CAD-supported technology-centric SLSI design.

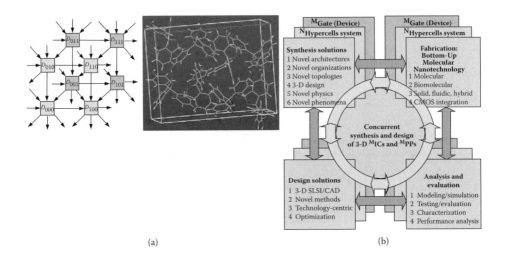

FIGURE 7.6 (a) Three-dimensional molecular electronics: aggregated $^{\aleph}$hypercells D_{ijk} composed of ^{M}gates that integrate multiterminal *solid* ^{ME}devices engineered from atomic complexes. (b) Concurrent synthesis and design at system, module, and gate (device) levels.

Super-high-performance molecular processing and memory platforms can be synthesized using $^{\aleph}$hypercells D_{ijk} within 3-D topology/organization, which are analogous to 3-D topology/organization of biomolecules and their aggregates. A vertebrate brain is of the most interest. However, not only vertebrates, but also single-cell bacteria possess superb 3-D ^{BM}PPs. We focus the major efforts on the *solid* molecular electronics due to a limited knowledge of the baseline processes, effects, mechanisms, and functionality of ^{BM}PPs. Insufficient knowledge makes it virtually impossible to comprehend and prototype biomolecular devices that operate utilizing different phenomena and concepts as compared to *solid* ^{ME}devices. Performance and baseline characteristics of *solid* ^{ME}devices are drastically affected by the molecular structures, aggregation, bonds, atomic orbitals, electron affinity, ionization potential, arrangement, sequence, assembly, folding, side groups, and other features. Molecular devices and ^{M}gates must ensure desired transitions, switching, logics, electronic characteristics, performance, etc. Enhanced functionality, high switching frequency, superior density, expanded utilization, low power, low voltage, desired $I–V$ characteristics, noise immunity, robustness, integration, and other characteristics can be ensured through a coherent design. In ^{M}devices, performance and characteristics can be changed and optimized by utilizing and controlling distinct transitions, states, and parameters. For *solid* ^{ME}devices, the number of quantum wells/barriers, their width, energy profile, tunneling length, dielectric constant, and other key features can be adjusted and optimized by engineering molecules with specific atomic sequences, bonds, side groups, etc. The goal is to ensure optimal achievable performance at the device, module, and system levels. The performance should be assessed by using the quantitative and qualitative performance measures, indexes, and metrics. The reported interactive synthesis taxonomy is coherently integrated within all tasks, including devising of ^{M}devices, discovering 3-D organization, synthesizing enabling architectures, designing ^{M}ICs, etc.

7.6 Biomolecular Processing and Fluidic Molecular Electronics: Neurobiomimetics, Prototyping, and Cognition

7.6.1 Neuroscience: Information Processing and Memory Postulates

Biosystems detect various stimuli, and the information is processed through complex electrochemomechanical phenomena and mechanisms at the molecular and cellular levels. Biosystems accomplish cognition, learning, perception, knowledge generation, storing, computing, coding, transmission, communication, adaptation, and other tasks related to the information processing. Appreciating neuroscience, neurophysiology, cellular biology, and other disciplines, this section addresses open-ended problems from engineering and technology standpoints reflecting some author's inclinations. Due to a lack of conclusive evidence, there does not exist an agreement regarding baseline mechanisms and phenomena (electrochemical, optochemical electromechanical, thermodynamic, and others), which ultimately result in signal/data and information processing in biosystems.

The human brain is a complex network of $\sim 1 \times 10^{11}$ aggregated neurons with more than 1×10^{14} synapses. Action potentials, and likely other information-containing signals, are transmitted to other neurons by means of very complex and not fully comprehended *axo-dendritic, dendro-axonic, axo-axonic*, and *dendro-dendritic* interactions utilizing axonic and dendritic structures. It is the author's belief that a neuron, as a complex system, performs information processing, memory storage, and other tasks utilizing ionic–biomolecular interactions and transitions. For example, biomolecules (neurotrans- mitters and enzymes) and ions propagate in the synaptic cleft, membrane channels, and cytoplasm. This controlled propagation of *information carriers* results in charge distribution, interaction, release, binding, unbinding, bonding, switching, folding, and other state transitions and events. The electroch- emomechanical transitions and interactions of *information carriers* under electrostatic, magnetic, hydro- dynamic, thermal, and other fields (forces) were examined in Ref. [22]. There are ongoing debates concerning system- and device-level considerations and neuronal aggregation, as well as fundamental phenomena observed, utilized, embedded, and exhibited by neurons and their organelles. There is no agreement on whether or not a neuron is a device (according to a conventional neuroscience postulate) or a system, or on how the information is processed, encoded, controlled, transmitted, routed, etc. The information processing and storage are far more complicated problems compared to data transmission, routing, communication, etc. Under these uncertainties, new theories, paradigms, and concepts have emerged.

By applying the possessed knowledge, there is a question whether it is possible to accomplish a coherent biomimetics (bioprototyping) and devise (discover and design) man-made bio-identical or bio-centered processing and memory platforms. Unfortunately, even for signal/data processing platforms, it seems unlikely that those objectives may be achieved in the near future. There are a great number of unsolved fundamental, applied, and technological problems. To some extent, a number of problems can be approached by examining and utilizing different biomolecular-centered processing postulates, concepts, and solutions. There is a need to develop general and application-centric foundations that will not rely on hypotheses, postulates, assumptions, and exclusive solutions, which depend on specific technologies, hardware, and fundamentals. Achievable technology-centric solid and fluidic molecular electronics are prioritized in this chapter due to noncomprehended cellular phenomena and mechanisms in [BM]PPs. Some postulates, concepts, and new solutions are reported.

The anatomist Heinrich Wilhelm Gottfried Waldeyer-Hartz found that the nervous system consists of nerve cells in which there are no mechanical joints in between. In 1891, he used the word *neuron*. The cell body of a typical vertebrate neuron consists of the nucleus (soma) and other cellular organelles. Neuron-branched projections (axons and dendrites) are packed with ~ 25 nm diameter microtubules, which may play a significant role in signal/data transmission, communication, processing, and storage. The cylindrical wall of each microtubule is formed by 13 longitudinal protofilaments of tubuline mole- cules, e.g., altering α and β heterodimers. The cross-sectional representation of a microtubule is a ring of 13 distinct subunits. Numerous and extensively branched dendrite structures are believed to transmit information toward the cell body. The information is transmitted from the cell body through axon structures. The axon originates from the cell body and terminates in numerous terminal branches. Each axon terminal branch may have thousands of synaptic axon terminals. These presynaptic axon terminals and postsynaptic dendrites establish the biomolecular-centered interface between neurons or between a neuron and target cells. Specifically, various neurotransmitters are released into the synaptic cleft and propagate to the postsynaptic membrane. It should also be emphasized that within a complex microtubule network, there are nucleus-associated microtubules.

Neurotransmitter molecules are: (1) synthesized (reprocessed) and stored into vesicles in the presynaptic cell; (2) released from the presynaptic cell, propagate, and bind to receptors on one or more postsynaptic cells; (3) removed and/or degraded. There are more than 100 known neurotransmitters, and the total number of neurotransmitters is not known. Neurotransmitters are classified as small-molecule neurotransmitters and neuropeptides (composed of 3 to 36 amino acids). It is reported that small-molecule neurotransmitters mediate rapid synaptic actions, while neuropeptides tend to modulate slower ongoing synaptic functions.

The conventional neuroscience theory postulates that in neurons, the information is transmitted by action potentials, which result due to ionic fluxes that are controlled by complex cellular mechanisms. The ionic channels are opened and closed by binding and unbinding of neurotransmitters that are released from the synaptic vesicles (located at the presynaptic axon sites). Neurotransmitters propagate through the synaptic cleft to the receptors at the postsynaptic dendrite (Figure 7.7). According to conventional theory, binding/unbinding of neurotransmitters in multiple synaptic terminals results in selective opening/closing of membrane ionic channels, and the flux of ions causes the action potential that is believed to contain and carry out information. At the cellular level, a wide spectrum of phenomena and mechanisms are not sufficiently studied or remain unknown. For example, the production, activation, reprocessing, binding, unbinding, and propagation of neurotransmitters, though they have been studied for decades, are not adequately comprehended. There are debates on the role of microtubules and microtubule-associated proteins (MAPs). With limited knowledge on signal transmission and communication in neurons, in addition to the action potential, other stimuli of different origin may exist and should be examined. Unfortunately, there do not exist a sound explanation, justification, and validation of information processing, memory storage, and other related tasks.

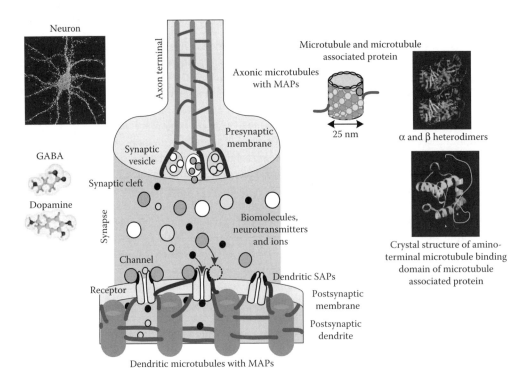

FIGURE 7.7 Schematic representation of the *axo-dendritic* organelles with AZ and PSD protein assemblies: (1) Binding and unbinding of the *information* carriers (biomolecules, neurotransmitters, and ions) result in the state transitions leading to information processing and memory storage; (2) 3-D-topology lattice of SAPs and microtubules with MAPs ensures reconfigurable 3-D organization utilizing *routing* carriers.

Binding and unbinding of neurotransmitters and ions cause electrochemomechanically induced transitions at the molecular and cellular levels due to charge variation, force generation, moment transformation, potential change, orbital overlap variation, vibration, resonance, folding, and other effects. For neurons and envisioned *synthetic fluidic* devices/modules, these transitions ultimately can result in information processing (with other directly related tasks) and memory storage. For example, a biomolecule (protein) can be used as a *biomolecular electrochemomechanical switch* utilizing the conformational changes, or as a *biomolecular electronic switch* using the charge changes that affect the electron transport. *Axo-dendritic* organelles with microtubules and MAPs, as well as the propagating ions and neurotransmitters in synapse, are schematically depicted in Figure 7.7. There are axonic and dendritic microtubules, MAPs, synapse-associated proteins (SAPs), endocytic proteins, etc. Distinct pre- and postsynaptic SAPs have been identified and examined. Large multidomain scaffold proteins, including SAP and MAP families, form the framework of the presynaptic active zones (AZ), postsynaptic density (PSD), endocytic zone (EnZ), and exocytic zone (ExZ) assemblies. There are numerous protein interactions between AZ, PSD, EnX, and ExZ proteins. With a high degree of confidence, one may conclude that there are the processing- and memory-associated state transitions in 3-D extracellular and intracellular protein assemblies.

In a microtubule, each tubulin dimer ($\sim 8 \times 4 \times 4$ nm) consists of positively and negatively charged α-tubulin and β-tubulin (Figure 7.7). Each heterodimer consists of ~ 450 amino acids, and each amino acid contains ~ 15 to 20 atoms. Tubulin molecules exhibit different geometrical conformations (states). The tubulin dimer subunits are arranged in a hexagonal lattice with different chirality. The interacting negatively charged C termini extend outward from each monomer (protrude perpendicularly to the microtubule surface), attracting positive ions from the cytoplasm. The intra-tubulin dielectric constant is $\varepsilon_r = 2$, while outside the microtubule $\varepsilon_r = 80$. The MAPs are proteins that interact with the microtubules of the cellular cytoskeleton. A large variety of MAPs have been identified. These MAPs accomplish different functions such as stabilization/destabilization of microtubules, guiding microtubules toward specific cellular locations, interconnect of microtubules and proteins, etc. Microtubule-associated proteins bind directly to the tubulin monomers. Usually, the carboxyl-terminus -COOH (C-terminal domain) of the MAP interacts with tubulin, while the amine-terminus -NH$_2$ (N-terminal domain) binds to organelles, intermediate filaments, and other microtubules. Microtubule-MAPs binding is regulated by phosphorylation. This is accomplished through the function of the microtubule-affinity-regulating-kinase protein. Phosphorylation of the MAP by the microtubule-affinity-regulating-kinase protein causes the MAP to detach from any bound microtubules. MAP1a and MAP1b, found in axons and dendrites, bind to microtubules differently from other MAPs, utilizing the charge-induced interactions. While the C terminals of MAPs bind the microtubules, the N terminals bind other parts of the cytoskeleton or the plasma membrane. MAP2 is found mostly in dendrites, while tau-MAP is located in the axon. These MAPs have a C-terminal microtubule-binding domain and variable N-terminal domains projecting outwards, interacting with other proteins. In addition to MAPs, there are many other proteins that affect microtubule behavior. These proteins are not considered to be MAPs because they do not bind directly to tubulin monomers, but affect the functionality of microtubules and MAPs. The mechanism of the so-called synaptic plasticity and the role of proteins, neurotransmitters and ions, which likely affect learning and memory, are not comprehended.

Innovative hypotheses of the microtubule-assisted quantum information processing are reported in Ref. [23]. The authors consider microtubules as assemblies of oriented dipoles and postulate that [23]: (1) conformational states of individual tubulins within neuronal microtubules are determined by mechanical London forces within the tubulin interiors which can induce a conformational quantum superposition; (2) in superposition, tubulins communicate/compute with entangled tubulins in the same microtubule, with other microtubules in the same neuron, with microtubules in neighboring neurons, and through macroscopic regions of brain by tunneling through gap junctions; (3) quantum states of tubulins/microtubules are isolated from environmental decoherence by biological mechanisms, such as quantum isolation, ordered water, Debye layering, coherent pumping, and quantum error correction; (4) microtubule quantum computations/superpositions are tuned by MAPs during a classical liquid phase that alternates with a quantum solid-state phase of actin gelation; (5) following periods of preconscious quantum computation, tubulin superpositions reduce or collapse by Penrose quantum gravity *objective reduction*; (6) the output states, which result from the *objective reduction* process, are nonalgorithmic (noncomputable) and govern neural events of binding of MAPs, regulating synapses, and membrane functions; (7) the reduction or self-collapse in the *orchestrated objective reduction* model is a *conscious moment*, related to Penrose's quantum gravity mechanism which relates the process to fundamental space–time geometry. The results reported in Ref. [23] suggest that tubulins can exist in quantum superposition of two or more possible states until the threshold for quantum state reduction (quantum gravity mediated by *objective reduction*) is reached. A double-well potential, according to Ref. [23], enables the inter-well quantum tunneling of a single electron and spin states because the energy is greater than the thermal fluctuations. The debates continue on the soundness of this concept examining the feasibility of utilization of quantum effects in tubulin dimers, relatively high width of the well (the separation is ~1.5 nm), decoherence, noise, etc.

In neurons, biomolecules (neurotransmitters and enzymes) and ions can be the *information* (processing) and *routing* carriers. Publications [22,24] suggest that signal and data processing (computing, logics, coding, and other tasks), memory storage, memory retrieval, and information processing (potentially) may be accomplished by using neurotransmitters and ions as the *information carriers*. There are distinct *information* carriers, e.g., *activating*, *regulating*, and *executing*.

Control of released specific neurotransmitters (*information carriers*) in a particular synapse and their binding to the receptors results in state transitions ensuring a cellular-level signal/data/information processing and memory mechanisms. The processing and memory may be robustly reconfigured utilizing *routing* carriers that potentially ensure networking. We originate the following major postulates:

1. Certain biomolecules and ions are the *activating*, *regulating*, and *executing information* carriers that interact with SAPs, MAPs, and other cellular proteins. Controlled binding/unbinding of *information* carriers lead to biomolecular-assisted electrochemomechanical state transitions (folding, bonding, etc.) affecting the processing- and memory-associated transitions in protein assemblies. This ultimately results in processing and memory storage. As the typifying examples: (i) binding/unbinding of *information* carriers ensures a combinational logics equivalent to *on* and *off* switching analogous to the AND- and OR-centered logics; (ii) charge change is analogous to the functionality of the molecular storage capacitor.

2. Specific biomolecules and ions are the *routing* carriers that interact with SAPs, MAPs, and other proteins. Binding and unbinding of *routing* carriers results in biomolecular-assisted state transitions ensuring robust reconfiguration, networking, adaptation, and interconnect.

3. Information processing and memories may be accomplished on a high radix by means of electromechanically induced events in specific neuronal protein complexes.

4. Presynaptic AZ and PSD (comprised of SAPs, MAPs, and other proteins), as well as microtubules, form a biomolecular 3-D-assembly (organization) within a reconfigurable processing-and-memory neuronal architecture.

A biomolecular processing includes various tasks, such as communication, signaling, routing, reconfiguration, coding, etc. Consider biomolecular processing between neurons using the *axo-dendritic* inputs and *dendro-axonic* outputs. We do not specify the information-containing signals (action potential, polarization vector, phase shifting, folding modulation, vibration, switching, etc.) with possible corresponding cellular mechanisms, which are due to complex biomolecular interactions and phenomena. The reported transitions can be examined using the *axo-dendritic* input vectors $\mathbf{x}_i$ (Figure 7.8). For example, the inputs to neuron $\mathbf{N}_0$ are $x_{0,1}, \ldots, x_{0,m}$, and $\mathbf{x}_0 = [x_{0,1}, \ldots, x_{0,m}]$. The first neuron $\mathbf{N}_0$ has m inputs (vector $\mathbf{x}_0$) and z outputs (vector $\mathbf{y}_0$). Spatially-distributed $\mathbf{y}_0 = [y_{0,1}, \ldots, y_{0,z}]$ furnish the inputs to neurons $\mathbf{N}_1, \mathbf{N}_2, \ldots, \mathbf{N}_{n-1}, \mathbf{N}_n$. The aggregated neurons $\mathbf{N}_0, \mathbf{N}_1, \ldots, \mathbf{N}_{n-1}, \mathbf{N}_n$ *process* the information by cellular transitions and mechanisms. The output vector $\mathbf{y}$ is $\mathbf{y} = f(\mathbf{x})$, where f is the nonlinear function, and, for example, in the logic design of ICs, f is the *switching* function. To ensure robustness, reconfigurability, and adaptiveness, we consider the feedback vector $\mathbf{u}$. Hence, the output of the neuron $\mathbf{N}_0$ is a nonlinear function of the input vector $\mathbf{x}_0$ and feedback vector $\mathbf{u} = [\mathbf{u}_0, \mathbf{u}_1, \ldots, \mathbf{u}_{n-1}, \mathbf{u}_n]$, e.g., $\mathbf{y}_0 = f(\mathbf{x}_0, \mathbf{u})$. As the information is processed by $\mathbf{N}_0$, it is fed to a neuronal aggregate $\mathbf{N}_1, \mathbf{N}_2, \ldots, \mathbf{N}_{n-1}, \mathbf{N}_n$. The neurotransmitter

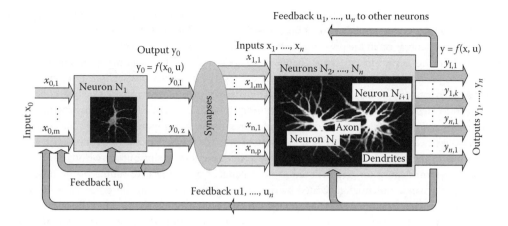

FIGURE 7.8 Input–output representation of $(n + 1)$ aggregated neurons with *axo-dendritic* inputs and *dendro-axonic* outputs.

release, performed by all neurons, is the *dendro-axonic* output y_i. As was emphasized, neurons have a branched dendritic tree with ending *axo-dendritic* synapses. The *fan-out* per neuron reaches 10,000. Figure 7.8 illustrates the 3-D aggregation of $(n + 1)$ neurons with the resulting input–output maps $y_i = f(\mathbf{x}_i, \mathbf{u})$. Dendrites may form *dendro-dendritic* interconnects, while in *axo-axonic* connects, one axon may terminate on the terminal of another axon and modify its neurotransmitter release.

7.6.2 Applied Information Theory and Information Estimates with Applications to Biomolecular Processing and Communication

Considering a neuron as a switching device, which could be an oversimplified hypothesis, the interconnected neurons are postulated to be excited only by the action potentials I_i. Neurons are modeled as a spatio-temporal lattice of aggregated processing elements (neurons) by the second-order linear differential equation [19,20]

$$\frac{1}{ab}\left(\frac{d^2 x_i}{dt^2} + (a+b)\frac{dx_i}{dt} + abx_i\right) = \sum_{\substack{j=1 \\ j \neq i}}^{N} [w_{1ij}Q(x_j, q_j) + w_{2ij}f_j(t, Q(x_j, q_j))] + I_i(t), \quad i = 1, 2, \ldots, N-1, N,$$

$$Q(x, q) = \begin{cases} q\left(1 - e\left(-\dfrac{e^x - 1}{q}\right)\right) & if x > \ln[1 - q\ln(1 + q^{-1})] \\ -1 & if x < \ln[1 - q\ln(1 + q^{-1})] \end{cases},$$

where a, b, and q are the constants; w_1 and w_2 are the topological maps. This model, according to Refs. [19,20], is an extension of the results reported in Refs. [26,27] by taking into consideration the independent dynamics of the dendrites' wave density and the pulse density for the parallel axons action.

Examining action potentials, synaptic transmission has been researched by studying the activity of the pre- and postsynaptic neurons [28–30] with the attempts to study communication, learning, cognition, perception, knowledge generation, etc. Ref. [31] proposes the learning equation for a synaptic adaptive weight $z(t)$ associated with a long-term memory as

$$\frac{dz}{dt} = f(x)[-Az + g(y)],$$

where x is the activity of a presynaptic (postsynaptic) cell; y is the activity of a postsynaptic (presynaptic) cell; $f(x)$ and $g(y)$ are the nonlinear functions; A is the matrix. Refs. [28–30] suggest that matching the action potential generation in the pre- and postsynaptic neurons equivalent to the condition of associative (Hebbian) learning results in a dynamic change in synaptic efficacy. The excitatory postsynaptic potential results due to presynaptic action potentials. After matching, the excitatory postsynaptic potential changes. Neurons fire irregularly at distinct frequencies. The changes in the dynamics of synaptic connections, resulting from Hebbian-type pairing, lead to significant modifications of the temporal structure of excitatory postsynaptic potentials generated by irregular presynaptic action potentials [25]. The changes that occur in synaptic efficacy due to the Hebbian pairing of pre- and postsynaptic activity substantially change the dynamics of the synaptic connection. The long-term changes in synaptic efficacy (long-term potentiation or long-term depression) are believed to be dependent on the relative timing of the onset of the excitatory postsynaptic potential generated by the pre- and postsynaptic action potentials [28–30]. The above reported, as well as other numerous concepts, have caused a lot of debates. The cellular mechanisms that are responsible for the induction of long-term potentiation or long-term depression are not known.

Analysis of distinct cellular mechanisms and even unverified hypotheses that exhibit sound merits have a direct application to molecular electronics, envisioned bioinspired processing, etc. For example, the design of processing and memory platforms may be performed by examining and comprehending baseline fundamentals at the device and system levels by making use of or prototyping/mimicking cellular organization, phenomena, and mechanisms. Based upon the inherent phenomena and mechanisms, distinct networking and interconnect of the *fluidic* and *solid* electronics can be envisioned. This interconnect, however, cannot likely be based on the semiconductor-centered interfacing reported in Ref. [32]. Biomolecular vs. envisioned *solid/fluidic* ^{ME}PPs can be profoundly different from the device and system-level standpoints.

Intelligent biosystems exhibit goal-driven behavior, evolutionary intelligence, learning, perception, and knowledge generation functioning in a non-Gaussian, nonstationary rapidly changing dynamic environment. There does not exist a generally accepted concept for a great number of key open problems such as bio-centered processing, memory, coding, etc. Attempts have been pursued to perform bioinspired symbolic, analog, digital (discrete-state and discrete-time), and hybrid processing by applying stochastic and deterministic concepts. To date, those attempts have not culminated in feasible and sound solutions. At the device/module level, utilizing biomolecules as the *information carriers*, novel devices and modules have been proposed for the envisioned *fluidic* molecular electronics [22,24]. The results were applied to control *information carriers* (intra- and outer-cellular ions and biomolecules) in cytoplasm, synaptic cleft, membrane channels, etc. The information processing platforms should be capable of mapping stimuli and capturing the goal-relevant information into the cognitive information processing, perception, learning, and knowledge generation [33]. For example, in bioinspired *fluidic* devices, to ensure processing, one should control propagation, production, activation, and binding/unbinding of biomolecules, which could be in *active*, *available*, *reprocessing*, and other states. Unfortunately, there is a significant gap between basic, applied, and experimental research as well as consequent engineering practice and technologies. Due to technological and fundamental challenges and limits, this gap may not be overcome in the nearest future.

Neurons in the brain, among various information processing and memory tasks, code and generate signals (stimuli), which are transmitted to other neurons through axon–synapse–dendrite *channels*. Unfortunately, we may not be able to coherently answer fundamental questions including how neurons process (compute, store, code, extract, filter, execute, retrieve, exchange, etc.) information. Even the communication in neurons is a disputed topic. The central assumption is that the information is transmitted and possibly processed by means of action potential (spikes mechanism). Unsolved problems exist in other critical areas, including information theory. Consider a series connection of processing elements (^{ME}device, biomolecule, or protein). The input signal is denoted as x, while the outputs of the first and second processing elements are y_1 and y_2. Even simplifying the data processing to a Markov chain $x \rightarrow y_1(x) \rightarrow y_2(y_1(x))$, the information measures used in communication theory can be applied only to a very limited class of problems. One may not be able to explicitly, quantitatively, and qualitatively examine the information-theoretic measures beyond communication and coding problems. Furthermore, the information-theoretic estimates in neurons and molecular aggregates, shown in Figure 7.8, can be applied to the communication-centered analysis only by assuming the availability of a great number of relevant data. Performing the communication and coding analysis, one examines the entropies of the variable x_i and y, denoted as $H(x_i)$ and $H(y)$. The probability distribution functions, conditional entropies $H(y|x_i)$ and $H(x_i|y)$, relative information $I(y|x_i)$ and $I(x_i|y)$, mutual information $I(y,x_i)$, as well as joint entropy $H(y,x_i)$ could be of interest.

In a neuron and its intracellular structures and organelles, baseline processes, mechanisms, and phenomena are not explicitly comprehended. The lack of ability to soundly examine and coherently explain the basic phenomena and processes has resulted in numerous hypotheses and postulates. From the signal/data processing standpoints, neurons are commonly studied as switching devices, while networked neuron ensembles have been considered assuming *stimulus-induced, connection-induced, adaptive,* and other *correlations.* Conventional neuroscience postulates that networked neurons transmit data, perform information processing, accomplish communication as well as perform other functions by means of sequences of spikes that propagate time-varying action potentials. Consider communication and coding in networked neurons, assuming the validity of conventional hypotheses. Each neuron usually receives inputs from many neurons. Depending on whether the input produces a spike (excitatory or inhibitory) and on how the neuron processes inputs determine the neuron's functionality. Excitatory inputs cause spikes, while inhibitory inputs suppress them. The rate at which spikes occur is believed to be changing, due to stimulus variations. Though the spike waveforms (magnitude, width, and profile) vary, these changes are usually considered to be irrelevant. In addition, the probability distribution function of the interspike intervals varies. Thus, input stimuli, as processed through a sequence of complex processes, result in outputs that are encoded as the pattern of action potentials (spikes). The spike duration is ~1 msec, and spike rate varies from one to thousands of spikes per second. The premise that the spike occurrence, timing, frequency, and its probability distribution encode the information has been extensively studied. It is found that the same stimulus does not result in the same pattern, and debates continue, with an alarming number of recently proposed hypotheses.

Let us discuss the relevant issues applying the information-theoretic approach. In general, one cannot determine if a signal (neuronal spike, voltage pulse in ICs, electromagnetic wave, etc.) is carrying information or not. There are no coherent information measures and concepts beyond communication- and coding-centered analysis. One of the open problems is to qualitatively and quantitatively define what the information is. It is not fully comprehended how neurons perform signal/data processing, not to mention information processing, but it is obvious that networked neurons are not analogous to combinational and memory ICs. Most importantly, by examining any signal, it is impossible to determine if it is carrying information or not as well as to coherently assess the signal/data processing, information processing, coding, or communication features. It is evident that there exists a need to further develop the information theory. Those meaningful developments, if successful, can be applied in the analysis of neurophysiological signal/data and information processing.

The entropy, which is the Shannon quantity of information, measures the complexity of the set, e.g., sets having larger entropies require more bits to represent them. For M objects (symbols) X_i that have probability distribution functions $p(X_i)$, the entropy is given as

$$H(X) = -\sum_{i=1}^{M} p(X_i)\log_2 p(X_i), \quad i = 1, 2, ..., M - 1, M.$$

Here, $H \geq 0$, and, hence, the number of bits required by the Source Coding Theorem is positive. Examining analog action potentials and considering spike trains, a *differential entropy* can be applied. For a continuous-time random variable X, the *differential entropy* is $H(X) = -\int p_X(x)\log_2 p_X(x)dx$, where $p_X(x)$ is a one-dimensional (1-D) probability distribution function of x, $\int p_X(x)dx = 1$. However, the *differential entropy* can be negative. For example, the *differential entropy* of a Gaussian random variable is $H(X) = 0.5\ln(2\pi e\sigma^2)$, and $H(X)$ can be positive, negative or zero depending on the variance. Furthermore, *differential entropy* depends on scaling. For example, if $Z = kX$, one has $H(Z) = H(X) + \log_2|k|$, where k is the scaling constant. To avoid the aforementioned problems, from the entropy analysis standpoints, continuous signals are discretized. Let X_n denote a discretized continuous random variable with a binwidth ΔT. We have, $\lim_{\Delta T \to 0} H(X_n) + \log_2 \Delta T = H(X)$. The problem though is to identify the information carrying signals for which ΔT should be obtained. One may use the a-order Renyi entropy measure as given by Ref. [34]:

$$R^a(X) = \frac{1}{1-a} \log_2 \int p_X^a(x)dx,$$

where a is the integer, $a \geq 1$. The first-order Renyi information ($a = 1$) leads to the Shannon quantity of information. However, Shannon's and Renyi's quantities measure the complexity of the set, and, even for this specific problem, the unknown probability distribution function should be obtained. The Fisher information

$$I_Y = \int \frac{(dp(x)/dx)^2}{p(x)} dx$$

is a metric for the estimations and measurements. In particular, I_F measures an adequate change in knowledge about the parameter of interest.

The entropy does not measure the complexity of a random variable, which could be voltage pulses in ICs, neuron inputs or outputs (response) such as spikes, or any other signals. The entropy can be used to determine whether random variables are statistically independent or not. Having a set of random variables denoted by $\mathbf{X} = \{X_1, X_2, ..., X_{M-1}, X_M\}$, the entropy of their joint probability function equals the sum of their individual entropies $H(\mathbf{X}) = \sum_{i=1}^{M} H(X_i)$, only if they are statistically independent.

One may examine the mutual information between the stimulus and the response in order to measure how similar the input and output are. We have

$$I(X,Y) = H(X) + H(Y) - H(X,Y),$$

$$I(X,Y) = \int p_{X,Y}(x,y) \log_2 \frac{p_{X,Y}(x,y)}{p_X(x)p_Y(y)} dxdy = \int p_{Y|X}(y|x)p_X(x) \log_2 \frac{p_{Y|X}(y|x)}{p_Y(y)} dxdy.$$

Thus, $I(X,Y) = 0$ when $p_{X,Y}(x,y) = p_X(x)p_Y(y)$ or $p_{Y|X}(y|x) = p_Y(y)$, i.e., when the input and output are statistically independent random variables of each other. When the output depends on the input, one has $I(X,Y) > 0$. The more the output reflects the input, the greater the mutual information. The maximum (infinity) occurs when $Y = X$. From a communications viewpoint, the mutual information expresses how much the output resembles the input. Taking into account that for discrete random variables $I(X,Y) = H(X) + H(Y) - H(X,Y)$ or $I(X,Y) = H(Y) - H(Y|X)$, one may utilize the conditional entropy $H(Y|X) = -\sum_{x,y} p_{X,Y}(x,y) \log_2 p_{Y|X}(y|x)$. Here, $H(Y|X)$ measures how random the *conditional* probability distribution of the output is, on the average, given a specific input. The more random it is, the larger the entropy, reducing the mutual information, and $I(X,Y) \leq H(X)$ because $H(Y|X) \geq 0$. The less random it is, the smaller the entropy until it equals zero when $Y = X$. The maximum value of mutual information is the entropy of the input (stimulus).

The channel capacity is found by maximizing the mutual information subject to the input probabilities, e.g. $C = \max_{p_X(i)} I(X,Y)$ [bit/symbol]. Thus, the analysis of mutual information results in the estimation of the channel capacity C which depends on $p_{Y|X}(y|x)$ that defines how the output changes with the input. In general, it is very difficult to obtain or estimate the probability distribution functions. Using conventional neuroscience hypotheses, the neuronal communication, to some extent is equivalent to the communication in the *point process channel* [35]. The *instantaneous* rate at which spikes occur cannot be lower than r_{min} and greater than r_{max} which are related to the discharge rate. Let the average sustainable spike rate be r_0. For a Poisson process, the channel capacity of the point processes, if $r_{min} \leq r \leq r_{max}$, is derived in Ref. [35] as

$$C = r_{min} \left[e^{-1} \left(1 + \frac{r_{max} - r_{min}}{r_{min}} \right)^{((1+r_{min})/(r_{max}-r_{min}))} - \left(1 + \frac{r_{min}}{r_{max} - r_{min}} \right) \ln \left(1 + \frac{r_{max} - r_{min}}{r_{min}} \right) \right],$$

which can be expressed in the following form [36]:

$$C = \begin{cases} \frac{r_{min}}{\ln 2} \left(e^{-1} \left(\frac{r_{max}}{r_{min}} \right)^{(r_{max}/(r_{max}-r_{min}))} - \ln \left(\frac{r_{max}}{r_{min}} \right)^{(r_{max}/(r_{max}-r_{min}))} \right), r_0 > e^{-1} r_{min} \left(\frac{r_{max}}{r_{min}} \right)^{(r_{max}/(r_{max}-r_{min}))} \\ \frac{1}{\ln 2} \left((r_0 - r_{min}) \ln \left(\frac{r_{max}}{r_{min}} \right)^{(r_{max}/(r_{max}-r_{min}))} - r_0 \ln \left(\frac{r_0}{r_{min}} \right) \right), r_0 < e^{-1} r_{min} \left(\frac{r_{max}}{r_{min}} \right)^{(r_{max}/(r_{max}-r_{min}))} \end{cases}$$

Let the minimum rate be zero. For $r_{min} = 0$, the expression for a channel capacity is simplified to

$$C = \begin{cases} \frac{r_{max}}{e \ln 2}, r_0 > \frac{r_{max}}{e} \\ \frac{r_0}{\ln 2} \ln \left(\frac{r_{max}}{r_0} \right), r_0 < \frac{r_{max}}{e} \end{cases}.$$

Example 7.4

Assume that the maximum rate varies from 300 to 1000 pulses/sec (or spikes/sec), and the average rate changes from 1 to 100 pulses/sec. Taking note of $r_{max}/e = 0.3679 r_{max}$, one obtains $r_0 < r_{max}/e$, and the channel capacity is given as

$$C = \frac{r_0}{\ln 2} \ln \left(\frac{r_{max}}{r_0} \right).$$

The channel capacitance $C(r_0, r_{max})$ is documented in Figure 7.9. For $r_0 = 100$ and $r_{max} = 1000$, one finds $C = 332.2$ bits or $C = 3.32$ bits/pulse.

The entropy is a function of the window size T and the time binwidth ΔT. For $\Delta T = 3 \times 10^{-3}$ sec and $18 \times 10^{-3} < T < 60 \times 10^{-3}$ sec, the entropy limit is reported to be 157 ± 3 bits/sec [37]. For the spike rate $r_0 = 40$ spikes/sec, $\Delta T = 3 \times 10^{-3}$ sec and $T = 0.1$ sec, the entropy is 17.8 bits [38]. These data agree with the above-made calculations for the capacity of the point process channel (Figure 7.9).

For $r_0 = 100$ and $r_{max} = 1000$, one finds that $C = 332.2$ bits ($C = 3.32$ bits/pulse). However, this does not mean that each pulse (spike) represents 3.32 bits or any other number of bits of information. In fact, the capacity is derived for digital communication. In particular, for a Poisson process, using r_{min}, r_{max} and r_0, we found specific rates with which digital signals (data) can be sent by a point process channel without incurring massive transmission errors.

For analog channels, the channel capacity is

$$C = \lim_{T \to \infty} \frac{1}{T} \max_{p_X(\cdot)} I(X,Y) \quad [bits/sec],$$

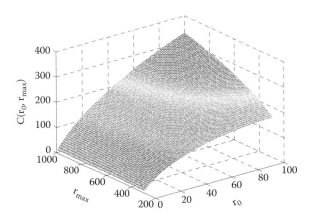

FIGURE 7.9 Channel capacity.

where T is the time interval during which communication occurs. In general, analog communication cannot be achieved through a noisy channel without incurring error. Furthermore, the probability distribution function as well as the distortion function must be known to perform the analysis. Probability distributions and distortion functions are not available, and processes are non-Poisson. Correspondingly, only some estimates may be made using a great number of assumptions. The focus can be directed rather on the application of biomimetics using sound fundamentals and technologies gained.

Another common critical assumption in the attempt to analyze bioprocessing features is a binary-centered hypothesis. Binary logics has a radix of two, meaning that it has two logic levels, e.g., 0 and 1. The radix r can be increased by utilizing r states (logic levels). Three- and four-valued logics are called ternary and quaternary [39]. The number of unique permutations of the truth table for r-valued logic is r^{r^2}. Hence for two-, three- and four-valued logic, we have 2^4 (16), 3^9 (19,683), and 4^{16} (4,294,967,296) unique permutations, respectively. The use of multiple-valued logic significantly reduces circuitry complexity, device number, power dissipation and improves interconnect, efficiency, speed, latency, packaging, and other features. However, sensitivity, robustness, noise immunity, and other challenging problems arise. An r-valued system has r possible outputs for r possible input values, and one obtains r^r outputs of a single r-valued variable [39]. For the radix $r = 2$ (binary logic), the number of possible output functions is $2^2 = 4$ for a single variable x. In particular, for $x = 0$ or $x = 1$, the output f can be 0 or 1, e.g., the output can be the same as the input (identity function), reversed (complement) or constant (either 0 or 1). With a radix of $r = 4$ for quaternary logic, the number of output functions is $4^4 = 256$. The number of functions of two r-valued variables is r^{r^2}, and for the two-valued case $2^{2^2} = 16$. The larger the radix, the smaller number of digits is necessary to express a given quantity. The radix (base) number can be derived from optimization standpoints. For example, mechanical calculators, including Babbage's calculator, mainly utilize 10-valued design. Though the design of multiple-valued memories is similar to the binary systems, multistate elements are used. A T-gate can be viewed as a universal primitive. It has $(r + 1)$ inputs, one of which is an r-valued control input whose value determines which of the other r (r-valued) inputs is selected for output. Due to quantum phenomena in *solid* ^ME^devices, or controlled release-and-binding/unbinding of specific *information carriers* in the *fluidic* ^M^devices, it is possible to employ enabling multiple-valued logics and memories.

7.6.3 Fluidic Molecular Platforms

The activity of brain neurons has been extensively studied using single microelectrodes as well as microelectrode arrays to probe and attempt to influence the activity of a single neuron or assembly of neurons in brain and neural culture. The integration of neurons and microelectronics has been studied in Refs. [32,40–42]. Motivated by a biological-centered hypothesis that a neuron is a processing module (system) that processes and stores the information, we propose a *fluidic* molecular processing device/module. This module emulates a brain neuron [22], and cultured neurons can be potentially utilized in implementation of 3-D processing and memory platforms. Signal/data processing and memory storage can be accomplished through release, propagation, and binding/unbinding of molecules. Binding of molecules and ions result in the state transitions to be utilized. Due to fundamental complexity and technological limits, one may not coherently mimic and prototype bioinformation processing. Therefore, we propose to emulate 3-D topologies and organizations of biosystems and utilize distinct molecules thereby ensuring a multiple-valued hardware solution. These innovations imply novel synthesis, design, aggregation, utilization, functionalization, and other features. Using molecules and ions as *information* and *routing* carriers, we propose a novel solution to solve signal, and potentially, information processing problems. We utilize 3-D topology/organization inherently exhibited by biomolecular platforms. The proposed *fluidic* molecular platforms can be designed within a processing-and-memory architecture. The *information* carriers are used as logic and memory inputs that lead to the state transitions. Utilizing *routing* carriers, persistent and robust morphology reconfiguration and reconfigurable networking are achieved. One may use distinct membranes and membrane lattices with highly selective channels, and different carriers can be employed. Computing, processing, and memory storage can be performed on the high radix. This ensures multiple-valued logics and memory.

Multiple *routing* carriers are steered in the fluidic cavity to the binding sites, resulting in the binding/unbinding of *routers* to the stationary molecules. The binding/unbinding events lead to a reconfigurable networking. Independent control of *information* and *routing* carriers cannot be accomplished through preassigned steady-state conditional logics, synchronization, timing protocols, and other conventional concepts. The motion and dynamics of the carrier release, propagation, binding/unbinding, and other events should be examined.

A 3-D-topology *synthetic fluidic* device/module is illustrated in Figure 7.10. The silicon inner enclosure can be made of proteins, porous silicon, or polymers to form membranes with fluidic channels that should ensure the selectivity. The *information* and *routing* carriers are encapsulated in the outer enclosure. The release and steering are controlled by the control apparatus.

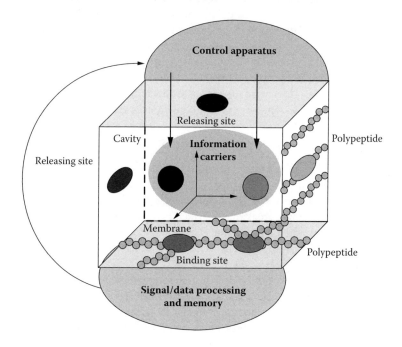

FIGURE 7.10 *Synthetic fluidic* molecular processing module.

The proposed device/module prototypes a neuron with synapses, membranes, channels, cytoplasm, and other components. Specific ions, molecules, and enzymes can pass through the porous membranes. These passed molecules (*information* and *routing* carriers) bind to the specific receptor sites, while enzymes free molecules from binding sites. Binding and unbinding of molecules result in the state transitions. The carriers that pass through selective fluidic channels and propagate through the cavity are controlled by changing the electrostatic potential or thermal gradient [22]. The goal is to achieve a controlled Brownian motion of carriers. Distinct control mechanisms (electrostatic, electromagnetic, thermal, hydrodynamic, etc.) allow one to uniquely utilize selective control, ensuring super-high performance and enabling functionality.

The controlled Brownian dynamics of molecules and ions in the fluidic cavity and channels was examined in Ref. [22]. Nonlinear stochastic dynamics of Brownian particles is of particular importance in cellular transport, molecular assembling, etc. It is feasible to control the propagation (motion) of carriers by changing the force $F_n(t,\mathbf{r},\mathbf{u})$ or varying the asymmetric potential $V_k(\mathbf{r},\mathbf{u})$. The high-fidelity mathematical model is given as

$$m_i \frac{d^2\mathbf{r}_i}{dt^2} = -F_{v_i}\left(\frac{d\mathbf{r}_i}{dt}\right) + \sum_{i,j,n} F_n(t,\mathbf{r}_{ij},\mathbf{u}) + \sum_{i,k} q_i \frac{\partial V_k(\mathbf{r}_i,\mathbf{u})}{\partial \mathbf{r}_i} + \sum_{i,j,k} \frac{\partial V_k(\mathbf{r}_{ij},\mathbf{u})}{\partial \mathbf{r}_{ij}} + f_r(t,\mathbf{r},\mathbf{q}) + \xi_{ri}, \quad i = 1,2,\ldots,N-1,N,$$

$$\frac{d\mathbf{q}_i}{dt} = f_q(t,\mathbf{r},\mathbf{q}) + \xi_{qi},$$

where $\mathbf{r}_i$ and $\mathbf{q}_i$ are the displacement and extended state vectors; $\mathbf{u}$ is the control vector; $\xi_r(t)$ and $\xi_q(t)$ are the Gaussian white noise vectors; F_v is the viscous friction force; m_i and q_i are the mass and charge; $f_r(t,\mathbf{r},\mathbf{q})$ and $f_q(t,\mathbf{r},\mathbf{q})$ are the nonlinear maps.

The Brownian particle velocity vector $\mathbf{v}$ is $\mathbf{v} = d\mathbf{r}/dt$. The Lorenz force on a Brownian particle possessing the charge q is $\mathbf{F} = q(\mathbf{E} + \mathbf{v} \times \mathbf{B})$, while using the surface charge density ρ_v, one obtains $\mathbf{F} = \rho_v(\mathbf{E} + \mathbf{v} \times \mathbf{B})$. The released carriers propagate in the fluidic cavity and are controlled by a control apparatus varying $F_n(t,\mathbf{r},\mathbf{u})$ and $V_k(\mathbf{r},\mathbf{u})$ [22]. This apparatus is comprised of polypeptide or molecular circuits which change the temperature gradient or the electric field intensity. The state transitions occur in the anchored processing polypeptide as *information* and *routing* carriers bind and unbind. For example, conformational *switching*, charge changes, electron transport, and other phenomena can be utilized. The settling time of electronic, photoelectric, and electrochemomechanical state transitions is from pico to microseconds. In general, it is possible to design, and potentially synthesize, aggregated 3-D networks of high-performance reconfigurable *fluidic* modules. These modules can be characterized in terms of input/output activity. The reported *fluidic* module, which emulates neurons, guarantees superior codesign features.

7.6.4 Neuromorphological Reconfigurable Molecular Processing Platforms

Consider a gate with binary inputs A and B. Taking note of the outputs to be generated by the universal logic gate, one has the following 16 functions: 0, 1, A, B, $\bar{A}$, $\bar{B}$, A+B, A+$\bar{B}$, $\bar{A}$+B, $\bar{A}$+$\bar{B}$, AB, A$\bar{B}$, $\bar{A}$B, $\bar{A}\bar{B}$, A$\bar{B}$+$\bar{A}$B, and AB+$\bar{A}\bar{B}$. The standard logic primitives (AND, NAND, NOT, OR, and others) can be implemented using a Fredkin gate, which performs conditional permutations. Consider a gate with a *switched* input A and a *control* input B. As illustrated in Figure 7.11, the input A is routed to one of two outputs, conditional on the state of B. The routing events change the output switching function which is AB or A$\bar{B}$.

Utilizing the proposed *fluidic* molecular processing paradigm, *routable* molecular universal logic gates (^{M}ULG) can be designed and implemented. We define an ^{M}ULG as a reconfigurable combinational gate that can be reconfigured to realize specified functions of its input variables. The use of specific multi-input ^{M}ULGs is defined by the technological soundness, requirements, and achievable performance. These ^{M}ULGs can realize logic functions using multi-input variables with the same delay as a two-input ^{M}gate. Logic functions can be efficiently factored and decomposed using ^{M}ULGs.

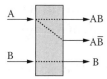

FIGURE 7.11 Gate schematic.

Figure 7.12 schematically depicts the proposed routing concepts for a reconfigurable logics. The typified 3-D-topologically reconfigurable routing is accomplished through the binding/unbinding of *routing* carriers to the stationary molecules. For illustrative purposes, Figure 7.12 documents reconfiguration of five ᴹgates depicting a reconfigurable networking-and-processing in 3-D. The *information carriers* are represented as the signals x_1, x_2, x_3, x_4, x_5, and x_6. The *routing* carriers ensure a reconfigurable routing and networking of ᴹgates and hypercells uniquely enhancing and complementing the ᴺhypercell design. In general, one may not be able to route any output of any gate/hypercell/module to any input of any other gate/hypercell/module. There are synthesis constraints, selectivity limits, complexity to control spatial motion of *routers* and other limits that should be integrated in the design.

It was documented that the proposed *fluidic* module can perform computation, implement complex logics, ensure memory storage, guarantee memory retrieval, etc. Sequences of conditional aggregation, carriers steering, 3-D directed routing, and spatial networking events form the basis of the logic gates and memory retrieval in the proposed neuromorphological reconfigurable *fluidic* ᴹPPs. In Section 7.6.3, we documented how to integrate the Brownian dynamics in the performance analysis and design. The *transit* time of *information* and *routing* carriers depends on the steering mechanism, control apparatus, particles used, sizing features, etc. From the design perspective, one applies the state-space paradigm using the *processing* and *routing* transition functions F_p and F_r that map previous states to the resulting new states in $[t, t_+], t_+ > t$. The output evolution is $\mathbf{y}(t_+) = F_i[t, \mathbf{x}(t), \mathbf{y}(t), \mathbf{u}(t)]$, where $\mathbf{x}$ and $\mathbf{u}$ are the state and control vectors. For example, $\mathbf{u}$ leads to the release and steering of the *routing* carriers with the resulting networking transitions. The reconfigurable system is modeled as $P \subset X \times Y \times U$, where X, Y and U are the inputs, output, and control sets, respectively.

The proposed neuromorphological reconfigurable *fluidic* ᴹPPs, which to some degree prototype ᴮᴹPPs, can emulate any existing ICs, surpassing the overall performance, functionality, and capabilities of envisioned microelectronic solutions. However, the theoretical and technological foundations of neuromorphological reconfigurable 3-D networking-processing-and-memory ᴹPPs remain to be developed and implemented.

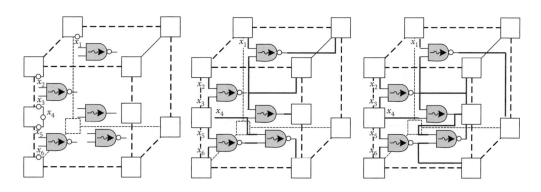

FIGURE 7.12 Reconfigurable routing and networking.

7.7 Design of Three-Dimensional Molecular Integrated Circuits: Data Structures, Decision Diagrams, and Hypercells

7.7.1 Molecular Electronics and Gates: Device and Circuits Prospective

Distinct ^{M}gates and ^{N}hypercells can be used to perform logic functions. To store the data, the memory cells are used. A systematic arrangement of memory cells and peripheral ^{M}ICs (to address and write the data into the cells as well as to delete data stored in the cells) constitute the memory. The ^{M}devices can be used to implement static and dynamic random access memory (RAM) as well as programmable and alterable read-only memory (ROM). Here, RAM is the read–write memory in which each individual molecular primitive can be addressed at any time, while ROM is commonly used to store instructions of an operating system. The static RAM may consist of a basic flip-flop ^{M}device with stable states (for example, 0 and 1). In contrast, the dynamic RAM, which can be implemented using one ^{M}device and storage capacitor, stores one bit of information charging the capacitor. As an example, the dynamic RAM cell is documented in Figure 7.13. The binary information is stored as the charge on the molecular storage capacitor ^MC_s (logic 0 or 1). This RAM cell is addressed by switching *on* the access ^{ME}device via the worldline signal, resulting in the charge being transferred into and out of ^MC_s on the dataline. The capacitor ^MC_s is isolated from the rest of the circuitry when the ^{ME}device is *off*. However, the leakage current through the ^{ME}device may require RAM cell refreshment to restore the original signal. Dynamic shift registers can be implemented using transmission ^{M}gates and ^{M}inverters, flip-flops can be synthesized by cross-coupling NOR ^{M}gates, while delay flip-flops can be built using transmission ^{M}gates and feedback ^{M}inverters.

   Among the specific characteristics under consideration are the read/write speed, memory density, power dissipation, volatility (data should be maintained in the memory array when the power is off), etc. The address, data, and control lines are connected to the memory array. The control lines define the function to be performed or the status of the memory system. The address and datalines ensure data manipulation and provide address into or out of the memory array. The address lines are connected to an address row decoder that selects a row of cells from an array of memory cells. A RAM organization, as documented in Figure 7.13, consists of an array (matrix) of storage cells arranged in an array of 2^n columns (bitlines) and 2^m rows (wordlines). To read the data stored in the array, a row address is supplied to the row decoder that selects a specific wordline. All cells along this wordline are activated and the contents of each of these cells are placed onto each of their corresponding bitlines. The storage cells can store

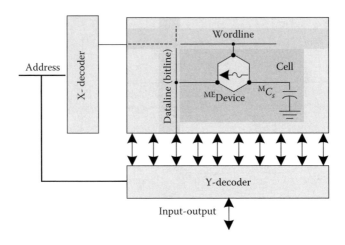

FIGURE 7.13 Dynamic RAM cell with ^{ME}device and storage molecular capacitor ^MC_s.

1 (or more) bits of information. The signal available on the bitlines is directed to a decoder. As reported in Figure 7.13, a binary (or high-radix) cell stores binary information utilizing an ^{ME}device at the intersection of the wordline and bitline. The ROM cell can be implemented as a: (1) parallel molecular NOR (^{M}NOR) array of cells; (2) series molecular NAND (^{M}NAND) array of cells requiring a single ^{M}device per storage cell. The ROM cell is programmed by either connecting or disconnecting the ^{M}device output (drain for FETs) from the bitline. Though a parallel ^{M}NOR array is faster, a series ^{M}NAND array ensures compacts and implementation feasibility.

In Figure 7.13, the multiterminal ^{ME}device is denoted as ⬡▶. There is a need to design ^{M}devices whose robustly controllable dynamics result in a sequence of quantum, quantum-induced, or not quantum state transitions that correspond to a sequence of computational, logic, or memory states. This is guaranteed even for quantum ^{M}devices, because quantum dynamics is deterministic, and nondeterminism of quantum mechanics arises when a device interacts with an uncontrolled outside environment or leaks information to an environment. In ^{M}devices, the *global* state evolutions (state transitions) should be deterministic, predictable, and controllable. The bounds posed by the Heisenberg uncertainty principle restrict the observability and do not impose limits on the device physics and device performance.

The logic device physics defines the mechanism of physical encoding of the logical states in the device. Quantum computing concepts emerged that proposed utilizing quantum spins of electrons or atoms to store information. In fact, a spin is a discrete two-state composition allowing a bit encoding. One can encode information using electromagnetic waves and cavity oscillations in optical devices. The information is encoded by DNA. The feasibility of different state encoding concepts depends on the ability to maintain the logical state for a required period. The stored information must be reliable, e.g., the probability of spontaneous changing of the stored logical state to other values should be small. One can utilize energy barriers and wells in the controllable energy space for a set of physical states encoding a given logical state. In order for the device to change the logical state, it must pass the energy barrier. To prevent this, the quantum tunneling can be suppressed by using high and wide potential barriers, minimizing excitation and noise, etc. To change the logical state, one varies the energy barrier as illustrated in Figure 7.14. If we examine the logical transition processes, we find that the logical states can be retained reliably by potential energy barriers that separate the physical states. The logical state is changed by varying the energy surface barriers as illustrated in Figure 7.14 for a 1-D case. The adiabatic transitions between logical states that are located at stable or meta-stable local energy minima results.

In VLSI design, resistor-transistor logic (RTL), diode-transistor logic (DTL), transistor-transistor logic (TTL), emitter-coupled logic (ECL), integrated-injection logic (IIL), merged-transistor logic (MTL), and other logic families have been used. All logic families and subfamilies (within TTL, there are Schottky, low-power Schottky, advanced Schottky, and others) have advantages and drawbacks. Molecular electronics offer unprecedented capabilities compared with microelectronics. Correspondingly, some logic families that ensure marginal performance using solid-state devices, provide superior performance as ^{M}devices are utilized. The ^{M}NOR gate, realized using the molecular resistor-transistor logic (^{M}RTL), is documented in Figure 7.15a. In electronics, NAND is one of the most important gates. The ^{M}NAND gate, designed by applying the molecular diode-transistor logic (^{M}DTL), is shown in Figure 7.15b. In Figures 7.15a and 7.15b, we use different symbols to designate molecular resistors ⊣⊓⊓⊢ ($^M r$), molecular diodes ▶⊢ ($^M d$) and molecular transistors ⊣ᵀ ($^M T$). It should be noted that the term $^M T$ should be

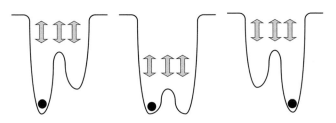

FIGURE 7.14 Logical states and energy barriers.

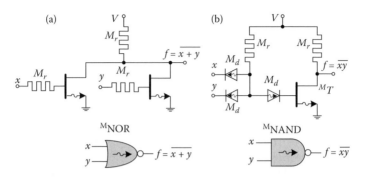

FIGURE 7.15 Circuit schematics: (a) Two-input ^{M}NOR gate. (b) Two-input ^{M}NAND gates.

used with great caution due to the distinct device physics of molecular and semiconductor devices. In order to introduce the subject, we use for a moment this incoherent terminology because ^MT may ensure characteristics similar to FETs and BJTs. However, the device physics of conventional three-, four- and many-terminal FETs and BJTs is entirely different when compared with even *solid* ^{ME}devices. Therefore, we depart from conventional terminology. Even a three-terminal *solid* ^{ME}device with the controlled I–V characteristics may not be referenced as a transistor. New terminology can be developed in the observable future that reflects the device physics of ^{M}device.

The ^{M}NAND gate, as implemented within an ^{M}DTL logic family, is illustrated within the $^{\aleph}$hypercell primitive schematics in Figure 7.16a. We emphasized the need for developing new symbols for molecular electronic devices. Quantum phenomena (quantum interaction, interference, tunneling, resonance, etc.) can be uniquely utilized. In Figure 7.16b, a multiterminal ^{ME}device (^{ME}D) is illustrated as ⬡⤳. Using the proposed ^{ME}D schematics, the illustrated ^{ME}D may have six *input*, *control*, and *output* terminals (ports), with the corresponding molecular bonds for the interconnect. As an illustration, a 3-D $^{\aleph}$hypercell primitive to implement a logic function $f(x_1,x_2,x_3)$ is shown in Figure 7.16b. Two-terminal molecular devices (^Md ⤲ and ^Mr ⟲) are shown. The input signals (x_1, x_2, and x_3) and output switching function f are documented in Figure 7.16b.

Molecular gates (^{M}AND and ^{M}NAND), designed within the molecular multiterminal ^{ME}D–^{ME}D logic family, are documented in Figure 7.17. Here, three-terminal cyclic molecules are utilized as ^{ME}devices, the physics of which is based on the quantum interaction and controlled electron tunneling. The inputs signals V_A and V_B are supplied to the *input* terminals, while the output signal is V_{out}. These ^{M}gates are designed using cyclic molecules within the carbon interconnecting framework as shown in Figure 7.17.

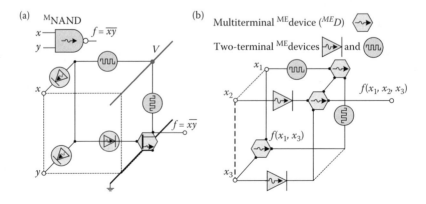

FIGURE 7.16 (a) Implementation of ^{M}NAND mapped by a $^{\aleph}$hypercell primitive. (b) $^{\aleph}$Hypercell primitive with two- and multiterminal ^{ME}devices.

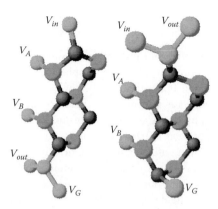

FIGURE 7.17 ^{M}AND and ^{M}NAND gates designed within the molecular ^{ME}D–^{ME}D logic family.

The details of synthesis, device physics, and phenomena utilized are reported in Ref. [4]. A coherent design should be performed in order to ensure the desired performance, functionality, characteristics, aggregability, topology, and other features. Complex ^{M}gates can be synthesized implementing $^{\aleph}$hypercells which form ^{M}ICs. The ^{M}AND and ^{M}NAND gates are documented in Figure 7.17, and Section 7.12.6 reports the device physics of the multiterminal ^{ME}devices.

7.7.2. Decision Diagrams and Logic Design of ^{M}ICs

Innovative solutions to perform the system-level logic design for 3-D ^{M}ICs should be examined. One needs to depart from 2-D logic design (VLSI, ULSI, and post-ULSI) as well as from planar ICs topologies and organizations. We propose SLSI design of ^{M}ICs, which mimics hierarchical 3-D bioprocessing platforms prototyping topologies and organizations observed in nature. This sound solution complies with the envisioned device-level outlook and fabrication technologies. In particular, using ^{M}devices, one may implement $^{\aleph}$hypercells that form ^{M}ICs. The use of $^{\aleph}$hypercells, as baseline primitives in the design of ^{M}ICs and processing/memory platforms, results in a technology-centric solution.

For 2-D CMOS ICs, a decision diagram (unique canonical structure) is derived as a reduced decision tree by using topological operators. In contrast, for 3-D ICs, a new class of decision diagrams and design methods must be developed to handle the complexity and 3-D features. A concept of design of linear decision diagrams, mapped by 3-D $^{\aleph}$hypercells, was proposed in Ref. [43]. In general, hypercell (cube, pyramid, hexagonal, or other 3-D topological aggregates) is a unique canonical structure, which is a reduced decision tree. Hypercells are synthesized by using topological operators (deleting and splitting nodes). Optimal and suboptimal technology-centric topology mappings of complex switching functions can be accomplished and analyzed. The major optimization criteria are: (1) minimization of decision diagram nodes and circuit terminals; (2) simplification of topological structures (linear arithmetic leads to simple synthesis and straightforward embedding of linear decision diagrams into 3-D topologies); (3) minimization of path length in decision diagrams; (4) routing simplification; (5) verification and evaluation. The optimal topology mapping results in power dissipation reduction, evaluation simplicity, testability enhancement, and other important features. For example, the switching power is not only a function of devices/gates/switches, but also a function of circuit topology, organization, design methods, routing, dynamics, switching activities, and other factors that can be optimized. In general, a novel CAD-supported SLSI should be developed to perform optimal technology-centric design of high-performance molecular platforms. Through a concurrent design, the designer should be able to perform the following major tasks:

1. Logic design of ^{M}ICs utilizing novel representations of data structures
2. Design and aggregation of $^{\aleph}$hypercells in functional ^{M}ICs

3. Design of multiple-valued and binary decision diagrams
4. CAD developments to concurrently support design tasks

SLSI utilizes a coherent top-down/bottom-up synthesis taxonomy as an important part of an ^{M}*archi-tectronics* paradigm. The design complexity should be emphasized. Current CAD-supported post-ULSI design does not allow one to design ICs with a number of gates more than 1,000,000. For ^{M}ICs, the design complexity significantly increases and novel methods are sought. The binary decision diagrams (BDDs) for representing Boolean functions are the state-of-the-art techniques in high-level logic design [43]. The reduced-order and optimized BDDs ensure large-scale data manipulations used to perform the logic design and circuitry mapping utilizing hardware description languages. The design scheme is: Function (Circuit) $\leftrightarrow$ BDD Model $\leftrightarrow$ Optimization $\leftrightarrow$ Mapping $\leftrightarrow$ Realization.

The dimension of a decision diagram (number of nodes) is a function of the number of variables and the variables' ordering. In general, the design complexity is $O(n^3)$. This enormous design complexity significantly limits the designer's abilities to design complex ICs without partitioning and decomposition. The commonly used word-level decision diagrams further increase the complexity due to processing of data in word-level format. Therefore, novel, sound software-supported design approaches are needed. Innovative methods in data structure representation and data structure manipulation are developed and applied to ensure the design specifications and objectives. We synthesize 3-D ^{M}ICs utilizing the linear word-level decision diagrams (LWDDs) that allow one to perform the compact representation of logic circuits using linear arithmetical polynomials (LP) [43,44]. The design complexity becomes $O(n)$. The proposed concept ensures compact representation of circuits compared with other formats and methods. The following design algorithm guarantees a compact circuit representation: Function (Circuit) $\leftrightarrow$ BDD Model $\leftrightarrow$ LWDD Model $\leftrightarrow$ Realization.

The LWDD is embedded in 3-D ^{M}hypercells that represent circuits in a 3-D space. The polynomial representation of logical functions ensures the description of multi-output functions in a word-level format. The expression of a Boolean function f of n variables $(x_1, x_2, \ldots, x_{n-1}, x_n)$ is

$$LP = a_0 + a_1 x_1 + a_2 x_2 + \cdots + a_{n-1} x_{n-1} + a_n x_n = a_0 + \sum_{j=1}^{n} a_j x_j.$$

To perform a design in 3-D, the mapping $LWDD(a_0, a_1, a_2, \ldots, a_{n-1}, a_n) \leftrightarrow LP$ is used. The nodes of LP correspond to a Davio expansion. The LWDD is used to represent any m-level circuit with levels L_i, $i = 1, 2, \ldots, m-1, m$ with elements of the molecular primitive library. Two data structures are defined in the algebraic form by a set of LPs as

$$L = \begin{cases} L_1 : \text{inputs } x_j; \text{ outputs } y_{1k} \\ L_2 : \text{inputs } y_{1k}; \text{ outputs } y_{2l} \\ \cdots\cdots\cdots\cdots\cdots\cdots \\ L_{m-1} : \text{inputs } y_{m-2,t}; \text{ outputs } y_{m-1,w} \\ L_m : \text{inputs } y_{m-1,w}; \text{ outputs } y_{m,n} \end{cases}$$

that corresponds to

$$LP_1 = a_0^1 + \sum_{j=1}^{n_1} a_j^1 x_j, \ldots, LP_m = a_0^n + \sum_{j=1}^{n_m} a_j^n y_{m-1,j}$$

or in the graphic form by a set of LWDDs as $LWDD_1(a_0^1, \ldots, a_{n_1}^1) \leftrightarrow LP_1, \ldots, LWDD_m(a_0^n, \ldots, a_{n_m}^n) \leftrightarrow LP_m$.

The use of LWDDs is a departure from the existing logic design tools. This concept is compatible with the existing software, algorithms, and circuit representation formats. Circuit transformation, format transformation, modular organization/architecture, library functions over primitives, and other features can be accomplished. All combinational circuits can be represented by LWDDs. The format transformation can be performed for circuits defined in Electronic Data Interchange Format (EDIF), Berkeley Logic Interchange Format (BLIF), International Symposium on Circuits and Systems Format (ISCAS), Verilog, etc. The library functions may have a library of LWDDs for multi-input gates, as well as libraries of ^{M}devices and ^{M}gates. The important feature is that these primitives are realized (through logic design) and synthesized as primitive aggregates within $^\aleph$hypercells. The reported LWDD simplifies analysis, verification, evaluation, and other tasks.

Arithmetic expressions underlying the design of LWDDs are canonical representations of logic functions. They are alternatives of the sum-of-product, product-of-sum, Reed-Muller, and other forms of representation of Boolean functions. Linear word-level decision diagrams are obtained by mapping LPs, where the nodes correspond to the Davio expansion and functionalizing vertices to the coefficients of the LPs. The design algorithms are given as: Function (Circuit) $\leftrightarrow$ LP Model $\leftrightarrow$ LWDD Model $\leftrightarrow$ Realization.

Any m-level logic circuits with a fixed order of elements are uniquely represented by a system of m LWDDs. The proposed concept is verified by designing 3-D ICs representing Boolean functions by hypercells. The CAD tools for logic design must be based on the principles of 3-D realization of logic functions with a library of primitives. Linear word-level decision diagrams are extended by embedding the decision tree into the hypercell structure. For two graphs $G = (V,E)$ and $H = (W,F)$, we embed the graph G into the graph H. The information in the resulting $^\aleph$hypercells is subdivided according to the new structural properties of the cell and the type of the embedded tree. The embedding of a guest graph G into a host graph H is a one-to-one mapping $M_{GV}:V(G)\rightarrow V(H)$, along with the mapping M that maps an edge $(u;v)\in E(G)$ to a path between $M_{GV}(u)$ and $M_{GV}(v)$ in H. Thus, the embedding of G into H is a one-to-one mapping of the nodes in G to the nodes in H.

In SLSI design, decision diagrams and decision trees are used. The information estimates can be evaluated [43]. Decision trees are designed using the Shannon and Davio expansions. There is a need to find the best variable and expansion for any node of the decision tree in terms of information estimates in order to optimize the design and synthesize optimal ^{M}ICs. The optimization algorithm should generate the *optimal paths* in a decision tree with respect to the design criteria. The decision tree is designed by arbitrarily choosing variables using either Shannon (S), positive Davio (pD), or negative Davio (nD) expansions for each node. The decision tree design process is a recursive decomposition of a switching function. This recursive decomposition corresponds to the expansion of switching function f with respect to the variable x. The variable x carries information that influences f. The initial and final state of the expansion $\sigma\in\{S,pD,nD\}$ can be characterized by the performance estimates. The information-centered optimization of ^{M}ICs design is performed in order to design optimal decision diagrams. A path in the decision tree starts from a node and finishes in a terminal node. Each path corresponds to a term in the final expression for f. For the benchmark c17 circuit, implemented using 3-D NAND ^{M}gates (^{M}NAND) as reported in Figure 7.18, Davio expansions ensure optimal design as compared with the Shannon expansion [43].

The software-supported logic design of proof-of-concept 3-D ^{M}ICs is successfully accomplished for complex benchmarking ICs in order to verify and examine the method proposed [43]. The size of LWDDs is compared with the best results received by other decision diagram packages developed for 2-D VLSI design. The method reported and software algorithms were tested and validated. The number of nodes, number of levels, and CPU time (in seconds) required to design decision diagrams for 3-D ^{M}ICs are examined. In addition, volumetric size, topological parameters, and other performance variables are analyzed. We assume: (1) feed-forward neural networked topology with no feedback; (2) threshold ^{M}gates as the processing primitives; (3) aggregated $^\aleph$hypercells comprised of ^{M}gates; (4) multilevel combinational circuits over the library of NAND, NOR, and EXOR ^{M}gates implemented using three-terminal ^{ME}devices. Experiments were conducted for a variety of ICs, and some results are reported in Table 7.1 [43]. The space size is given by X, Y, and Z that result in the volumetric quantity $V = X \times Y \times Z$. The topological

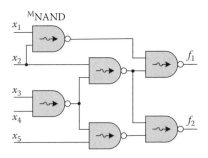

FIGURE 7.18   C17 with ^{M}NAND gates.

characteristics are analyzed using the total number of terminal (N_T) and intermediate (N_I) nodes. For example, c880 is an 8-bit arithmetic logic unit (ALU). The core of this circuit is in the form of the 8-bit 74,283 adder, which has 60 inputs and 26 outputs. A planar design leads to 383 gates. In contrast, 3-D design results in 294 ^{M}gates. A 3-D 9-bit ALU (c5315) with 178 inputs and 123 outputs is implemented using 1413 ^{M}gates, while a c6288 multiplier (32 inputs and 32 outputs) has 2327 ^{M}gates. Molecular gates are aggregated, networked, and grouped in 3-D within $^\aleph$hypercell aggregates. The number of incompletely specified $^\aleph$hypercells was minimized. The $^\aleph$hypercells in the ith layer were connected to the corresponding $^\aleph$hypercells in $(i-1)$th and $(i+1)$th layers. The number of terminal nodes and intermediate nodes are 3750 and 2813 for a 9-bit ALU, while, for a multiplier, we have 9248 and 6916 nodes. To combine all layers, more than 10,000 connections were generated. The design in 3-D was performed within 0.36 seconds for a 9-bit ALU. The studied 9-bit ALU performs arithmetic and logic operations simultaneously on two 9-bit input data words as well as computes the parity of the results. Conventional 2-D logic design for c5315 with 178 inputs and 123 outputs results in 2406 gates. In contrast, the proposed design, as performed using a proof-of-concept SLSI software, leads to 1413 ^{M}gates that are networked and aggregated in 3-D. In addition to conventional parameters (diameter, dilation cost, expansion, load, etc.), we use the number of variables in the logic function described by $^\aleph$hypercells, number of links, fan-out of the intermediate nodes, statistics, and others to perform the evaluation. To ensure the similarity to 2D design, binary three-terminal ^{ME}devices were used. The use of multiple-valued multiterminal ^{ME}devices results in superior performance.

The representative proof-of-concept CAD tools and software solutions were developed in order to demonstrate the 3-D design feasibility for combinational ^{M}ICs. The compatibility with hardware description languages is important. Three netlist formats (EDIF, ISCAS, and BLIF) are used and embedded in a proof-of-concept SLSI software that features [43]:

- a new design concept for 3-D ^{M}ICs
- synthesis and partitioning linear decision diagrams for given functions or circuits
- spectral representation of logic functions
- circuit testability and verification

TABLE 7.1   Design Results for 3-D ^{M}ICs

Circuit	I/O	Space Size				Nodes and Connections		CPU time (sec)
		G	X	Y	Z	N_T	N_I	
c432	36/7	126	66	64	66	2022	1896	<0.032
8-bit ALU c880	60/26	294	70	72	70	612	482	<0.047
9-bit ALU c5315	178/123	1413	138	132	126	3750	2813	<0.36
16×16 Multiplier c6288	32/32	2327	248	248	244	9246	6916	<0.47

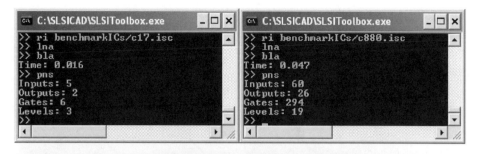

FIGURE 7.19 Design of 3-D ^{M}ICs using proof-of-concept SLSI software.

- compact format ensuring robustness and rapid-prototyping
- compressed optimal representation of complex ^{M}ICs

For 3-D ^{M}ICs, the results of design are reported in Figure 7.19 displaying the data in the Command Window. In particular, the design of a c17 circuit and 8-bit ALU (c880) are displayed.

7.7.3 ^{x}Hypercell Design

The binary tree is a networked description that carries information about dual connections of each node. The binary tree also carries information about functionality of the logic circuit and its topology. The nodes of the binary tree are associated with the Shannon and Davio expansions with respect to each variable and coordinate in 3-D. A node in the binary decision tree realizes the Shannon decomposition $f = x_i f_0 \oplus x_i f_1$, where $f_0 = f\,|_{x_i=0}$ and $f_1 = f\,|_{x_i=1}$ for all variables in f. Thus, each node realizes the Shannon expansion, and the nodes are distributed over levels. The classical hypercube contains 2^n nodes, while the ^{x}hypercell has $2^n + \sum_{i=0}^n 2^{n-1} C_i^m$ nodes in order to ensure a technology-centric design of ^{M}ICs. The ^{x}hypercell consists of terminal nodes, intermediate nodes, and roots. This ensures a straightforward ^{x}hypercell implementation, for example, by using the molecular multiplexer. The design steps are:

Step 1: Connect the terminal node with the intermediate nodes

Step 2: Connect the root with two intermediate nodes located symmetrically on the opposite faces

Step 3: Pattern the terminal and intermediate nodes on the opposite faces and connect them through the root

Figure 7.20a reports a 3-D ^{x}hypercell implemented using two-to-one molecular multiplexers.

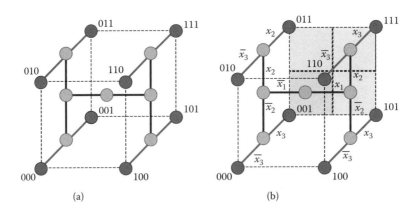

FIGURE 7.20 Multiplexer-based $^{\aleph}$hypercells: (a) $^{\aleph}$Hypercell with molecular multiplexers; (b) Implementation of a switching function $f = \bar{x}_1 x_2 \vee x_1 \bar{x}_2 \vee x_1 x_2 x_3$.

There are several methods for representing logic functions, and a hypercell solution is utilized. In general, $^{\aleph}$hypercell is a homogeneous aggregated assembly for massive super-high-performance parallel computing. We apply the enhanced switching theory integrated with a novel logic design concept. In the design, the graph-based data structures and 3-D topology are utilized. The $^{\aleph}$hypercell is a topological representation of a switching function by an n-dimensional graph. In particular, the switching function f is given as

$$\underset{\text{Switching Function}}{f} \Rightarrow \overset{\overset{\text{Coefficient}}{\Downarrow}}{\underset{\underset{\underset{\text{Operation}}{\Uparrow}}{i=0}}{\overset{2^n-1}{\mathbf{L}}} \mathbf{K}_i \left(x_1^{i_1}, \ldots, x_n^{i_n} \right)} \Rightarrow \overset{\text{Form of Switching Function}}{f_F} .$$

The data structure is described in matrix form using the truth vector $\mathbf{F}$ of a given switching function f as well as the vector of coefficients $\mathbf{K}$. The logic operations are represented by $\mathbf{L}$. $^{\aleph}$Hypercells compute f, and, for example, Figure 7.20b reports a $^{\aleph}$hypercell to implement $f = \bar{x}_1 x_2 \vee x_1 \bar{x}_2 \vee x_1 x_2 x_3$. From the technology-centric viewpoints, we propose a concept that employs ^{M}gates, coherently mapping the device/module/system-level and data structure solutions by using $^{\aleph}$hypercell. Aggregated $^{\aleph}$hypercells can implement switching functions f of arbitrary complexity. The logic design in spatial dimensions is based on the advanced methods and enhanced data structures to satisfy the requirements of 3-D topology. The appropriate data structure of logic functions and methods of embedding this structure into $^{\aleph}$hypercells are developed. The algorithm in a logic function's manipulation in order to change the carrier of information from the algebraic form (logic equation) to the hypercell structure consists of three steps, e.g.:

Step 1: The logic function is transformed to the appropriate algebraic form (Reed–Muller, arithmetic, or word-level in a matrix or algebraic representation)

Step 2: The derived algebraic form is converted to the graphical form (decision tree or decision diagram)

Step 3: The obtained graphical form is embedded and technologically implemented by $^{\aleph}$hypercells; the $^{\aleph}$hypercell aggregates form ^{M}ICs

The design is expressed as

$$\underset{\text{Step 1}}{\textit{Logic Function}} \Leftrightarrow \underset{\text{Step 2}}{\textit{Graph}} \Leftrightarrow \underset{\text{Step 3}}{\textit{Hypercell} / {}^{M}\textit{ICs}}.$$

The proposed procedure results in:

- Algebraic representations and robust manipulations of complex switching logic functions
- Matrix representations and manipulations providing consistency of logic relationships for variables and functions from the spectral theory viewpoint
- Graph-based representations using decision trees
- Direct mapping of decision diagrams into logical networks, as demonstrated for multiplexer-based $^\aleph$hypercells
- Robust embedding of data structures into $^\aleph$hypercells

From the synthesis viewpoint, the complexity of the molecular interconnect corresponds to the complexity of ^{ME}devices. We introduce a 3-D directly interconnected molecular electronics (^{3D}DIME) concept in order to reduce the synthesis complexity, minimize delays, ensure robustness, enhance reliability, etc. This solution minimizes the interconnect utilizing a direct atomic bonding of *input*, *control*, and *output* ^{M}device terminals (ports) by means of direct device-to-device aggregation. We have documented that ^{ME}devices and ^{M}gates are engineered and implemented using cyclic molecules within a carbon framework. For example, the *output* terminal of the ^{ME}device is directly connected to the *input* terminal of an other ^{ME}device. This ensures synthesis feasibility, compact implementation of $^\aleph$hypercells, applicability of ^{M}primitives, etc.

7.8   Three-Dimensional Molecular Signal/Data Processing and Memory Platforms

Advanced computer architectures (beyond von Neumann architecture) [43,45,46] can be devised and implemented to guarantee superior processing, communication, reconfigurability, robustness, networking, etc. In von Neumann computer architecture, the central processing unit (CPU) executes sequences of instructions and operands, which are fetched by the program control unit (PCU), executed by the data processing unit (DPU), and then, placed in memory. In particular, caches (high speed memory where data is copied when it is retrieved from the RAM improving the overall performance by reducing the average memory access time) are used. The instructions and data form instruction and data streams that flow to and from the processor. The CPU may have more than one processor and coprocessors with various execution units, multilevel instruction, and data caches. These processors can share or have their own caches. The *datapath* contains ICs to perform arithmetic and logical operations on words such as fixed- or floating-point numbers. The CPU design involves the trade-off between the hardware, speed, and affordability. The CPU is usually partitioned on the control and *datapath* units. The control unit selects and sequences the data-processing operations. The core interface unit is a switch that can be implemented as autonomous cache controllers operating concurrently and feeding the specified number (64 or 128) of bytes of data per cycle. This core interface unit connects all controllers to the data or instruction caches of processors. Additionally, the core interface unit accepts and sequences information from the processors. A control unit is responsible for controlling data flow between controllers that regulate the *in* and *out* information flows. There is the interface to input/output devices. On-chip debugging, error detection, sequencing logic, self-test, monitoring, and other units must be integrated to control a pipelined computer. The computer performance depends on the architecture and hardware components, and Figure 7.21 illustrates a conventional computer architecture.

Consider signal/data and information processing between nerve cells. The key to understanding processing, memory, learning, intelligence, adaptation, control, hierarchy, and other system-level basics lies in the ability to comprehend phenomena exhibited, organization utilized and architecture possessed by the central nervous system, neurons, and their organelles. Unfortunately, many problems have not been resolved. Each neuron in the brain, which performs processing and memory storage, has thousands of synapses with binding sites, membrane channels, MAPs, SAPs, etc. The *information carriers* accomplish transitions performing and carrying out various information processing, memory, communication, and

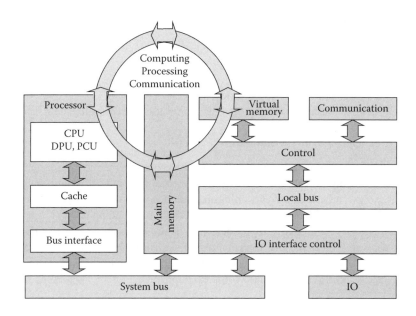

FIGURE 7.21 Computer architecture.

other tasks. The information processing and memories are reconfigurable and constantly adapt. Neurons function within a 3-D hierarchically distributed, robust, adaptive, parallel, and networked organization. Making use of the existing knowledge, Figure 7.22 documents a 3-D ^{M}PP. The processor executes sequences of instructions and operands, which are fetched (by the control unit) and placed in memory. The instructions and data form *instruction* and *data streams* that flow to and from the processor. The processor may have subprocessors with shared caches. The core interface unit concurrently controls operations and data retrieval. This interface unit interfaces all controllers to the data or processor instruction caches. The interface unit accepts and sequences information from the processors. A control unit is responsible for controlling data flow regulating the *in* and *out* information flows. The integrated processor-and-memory architecture that accomplishes the processing tasks is reported in Figure 7.22.

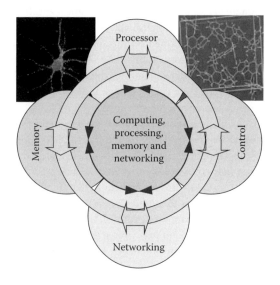

FIGURE 7.22 Molecular processing-and-memory platform.

The envisioned implementation of $^\mathrm{M}$PPs primarily depends on the progress in device physics, system organization/architecture, molecular hardware, and SLSI design. The critical problems in the design are the development, optimization, and utilization of hardware and software. The current status of fundamental and technology developments suggests that the $^\mathrm{M}$PPs likely will be designed utilizing a digital paradigm. Numbers in binary digital processors and memories are represented as a string of zeros and ones, and circuits perform Boolean operations. Arithmetic operations are performed based on a hierarchy of operations that are built upon simple operations. The methods to compute and algorithms used are different. Therefore, speed, robustness, accuracy, and other performance characteristics vary. The information is represented as a string of bits (zeros and ones). The number of bits depends on the length of the word (quantity of bits on which hardware is capable to operate). The operations are performed over the string of bits. There are rules that associate a numerical value X with the corresponding bit string $x = \{x_0, x_1, ..., x_{n-2}, x_{n-1}\}$, $x_i \in 0, 1$. The associated word (string of bits) is n bits long. If for every value X, there exists one, and only one, corresponding bit string x, the number system is nonredundant. If there can exist more than one x that represents the same value X, the number system is redundant. A *weighted* number system is used, and a numerical value is associated with the bit string x as $x = \sum_{i=0}^{n-1} x_i w_i$, $w_0 = 1, ..., w_i = (w_i - 1)(r_i - 1)$, where r_i is the *radix* integer. By making use of the multiplicity of instruction and data streams, the following classification can be applied:

> Single instruction stream/single data stream — conventional word-sequential architecture including pipelined computing platforms with parallel ALU
> Single instruction stream/multiple data stream — multiple ALU architectures, e.g., parallel-array processor (ALU can be either bit-serial or bit-parallel)
> Multiple instruction stream/single data stream
> Multiple instruction stream/multiple data stream — the multiprocessor system with multiple control units

In biosystems, multiple instruction streams/multiple data streams are observed. There is no evidence that technology will provide the abilities to synthesize biomolecular processors, not mentioning biocomputers, in the near future. Therefore, we concentrate efforts on computing platforms designed using *solid molecular electronics* that ensure soundness and technological feasibility. The performance estimates are reported in this chapter. Three-dimensional topologies and organizations significantly improve the performance of computing platforms guaranteeing, for example, massive parallelism and optimal utilization. Using the number of instructions executed (N), number of cycles per instruction (C_{PI}) and clock frequency (f_{clock}), the program execution time is $T_{ex} = NC_{PI}/f_{clock}$. In general, circuit hardware determines the clock frequency f_{clock}, software affects the number of instructions executed N, while architecture defines the number of cycles per instruction C_{PI}. Computing platforms integrate functional controlled hardware units and systems that perform processing, storage, execution, etc. The $^\mathrm{M}$PP accepts digital or analog input information, processes and manipulates it according to a list of internally stored machine instructions, stores the information, and produces the resulting output. The list of instructions is called a program, and internal storage is called memory. A memory unit integrates different memories. The processor accesses (reads or loads) the data from the memory systems, performs computations, and stores (writes) the data back to memory. The memory system is a collection of storage locations. Each storage location (memory word) has an address. A collection of storage locations forms an address space. Figure 7.23 documents the data flow and its control representing how a processor is connected to a memory system via address, control, and data interfaces. High-performance memory systems should be capable of serving multiple requests simultaneously, particularly for vector processors.

When a processor attempts to load or read the data from the memory location, the request is issued, and the processor stalls while the request returns. While $^\mathrm{M}$PPs can operate with overlapping memory requests, the data cannot be optimally manipulated if there are long memory delays. Therefore, a key performance parameter in the design is the effective memory speed. The following limitations are imposed on any memory system: the memory cannot be infinitely large, cannot contain an arbitrarily large amount of information, and cannot operate infinitely fast. Hence, the major characteristics are

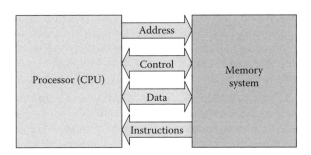

FIGURE 7.23 Memory–processor interface.

speed and capacity. The memory system performance is characterized by the latency (τ_l) and bandwidth (B_w). The memory latency is the delay as the processor first requests a word from memory until that word arrives and is available for use by the processor. The bandwidth is the rate at which information can be transferred from the memory system. Taking note of the number of requests that the memory can service concurrently $N_{request}$, we have $B_w = N_{request}/\tau_l$. Using 3-D ^{M}ICs, it becomes feasible to design and build superior memory systems with superior capacity, low latency, and high bandwidth approaching physical and technological limits. Furthermore, it becomes possible to match the memory and processor performance characteristics and capabilities.

Memory hierarchies ensure decreased latency and reduced bandwidth requirements, whereas parallel memories provide higher bandwidth. The ^{M}PP architectures can utilize a 3-D-organization with a fast memory located in front of a large but relatively slow memory. This significantly improves speed and enhances memory capacity. However, this solution results in the application of registers in the processor unit, and, most commonly accessed variables should be allocated at registers. A variety of techniques, employing either hardware, software, or a combination of hardware and software, must be employed to ensure that most references to memory are fed by the faster memory. The locality principle is based on the fact that some memory locations are referenced more often than others. The implementation of spatial locality, due to the sequential access, provides one with the property that an access to a given memory location increases the probability that neighboring locations will soon be accessed. Making use of the frequency of program looping behavior, temporal locality ensures access to a given memory location, increasing the probability that the same location will be accessed again soon. If a variable was not referenced for a while, it is unlikely that this variable will be needed soon. The performance parameter, which can be used to quantitatively examine different memory systems, is the effective latency τ_{ef}. We have $\tau_{ef} = \tau_{hit} R_{hit} + \tau_{miss}(1 - R_{hit})$, where τ_{hit} and τ_{miss} are the hit and miss latencies; R_{hit} is the hit ratio, $R_{hit} < 1$. If the needed word is found in a level of the hierarchy, it is called a hit. Correspondingly, if a request must be sent to the next lower level, the request is said to be a miss. The miss ratio is given as $R_{miss} = (1 - R_{hit})$. These R_{hit} and R_{miss} are affected by the program being executed and influenced by the high/low-level memory capacity ratio. The access efficiency E_{ef} of multiple-level memory ($i - 1$ and i) is found using the access time, hit, and miss ratios. In particular,

$$E_{ef} = \left(\frac{t_{access\ time\ i\text{-}1}}{t_{access\ time\ i}} R_{miss} + R_{hit} \right)^{-1}.$$

The hardware can dynamically allocate parts of the cache memory for addresses likely to be accessed soon. The cache contains only redundant copies of the address space. The cache memory can be associative or content-addressable. In an associative memory, the address of a memory location is stored along with its content. Rather than reading data directly from a memory location, the cache is given an address and responds by providing data that might or might not be the data requested. When a cache miss occurs, the memory access is then performed from main memory, and the cache is updated to include the new data. The cache should hold the most active portions of the memory, and the hardware dynamically selects portions of the main memory to store in the cache. When the cache is full, some data must be transferred to the main memory or deleted. A strategy for cache memory management is needed. These cache management strategies are based on the locality principle. In particular, spatial (selection of what is brought into the cache) and temporal (selection of what must be removed) localities are embedded. When a cache miss occurs, hardware copies a contiguous block of memory into the cache, which includes the word requested. This fixed-size memory block can be small, medium, or large. Caches can require all fixed-size memory blocks to be aligned. When a fixed-size memory block is brought into the cache, it is likely that another fixed-size memory block must be removed. The selection of the removed fixed-size memory block is based on effort to capture temporal locality.

The cache can integrate the data memory and the tag memory. The address of each cache line contained in the data memory is stored in the tag memory. The state can also track which cache line is modified. Each line contained in the data memory is allocated by a corresponding entry in the tag memory to indicate the full address of the cache line. The requirement that the cache memory be associative (content-addressable) complicates the design because addressing data by content is more complex than by its address (all tags must be compared concurrently). The cache can be simplified by embedding a mapping of memory locations to cache cells. This mapping limits the number of possible cells in which a particular line may reside. Each memory location can be mapped to a single location in the cache through direct mapping. There is no choice of where the line resides and which line must be replaced; however, poor utilization results. In contrast, a two-way set-associative cache maps each memory location into either of two locations in the cache. Hence, this mapping can be viewed as two identical directly mapped caches. In fact, both caches must be searched at each memory access, and the appropriate data selected and multiplexed on a tag match-hit and on a miss. Then, a choice must be made between two possible cache lines as to which is to be replaced. A single least-recently used bit can be saved for each such pair of lines to remember which line has been accessed more recently. This bit must be toggled to the current state each time. To this end, an M-way associative cache maps each memory location into M memory locations in the cache. Therefore, this cache map can be constructed from M identical direct-mapped caches. The problem of maintaining the least-recently used ordering of M cache lines is primarily due to the fact that there are $M!$ possible orderings. In fact, it takes at least $\log_2 M!$ bits to store the ordering. In general, a multi-associative cache may be implemented.

Multiple memory *banks*, formed by [M]ICs, can be integrated together to form a parallel main memory system. Since each *bank* can service a request, a parallel main memory system with N_{mb} *banks* can service N_{mb} requests simultaneously, increasing the bandwidth of the memory system by N_{mb} times the bandwidth of a single *bank*. The number of *banks* is a power of two, e.g., $N_{mb} = 2^p$. An n-bit memory word address is partitioned into two parts: a p-bit *bank* number and an m-bit address of a word within a *bank*. The p bits used to select a *bank* number could be any p bits of the n-bit word address. Let us use the low-order p address bits to select the *bank* number. The higher order $m = (n - p)$ bits of the word address is used to access a word in the selected *bank*. Multiple memory *banks* can be connected using *simple paralleling* and *complex paralleling*. Figure 7.24 shows the structure of a simple parallel memory system where m address bits are simultaneously supplied to all memory *banks*. All *banks* are connected to the same read/ write control line. For a read operation, the *banks* perform the read operation and accumulate the data in the latches. Data can then be read from the latches one by one by setting the switch appropriately. The *banks* can be accessed again to carry out another read or write operation. For a write operation, the latches are loaded one by one. When all latches have been written, their contents can be written into the memory *banks* by supplying m bits of address. In a simple parallel memory, all *banks* are cycled at the same time. Each *bank* starts and completes its individual operations at the same time as every other *bank*, and a new memory cycle starts for all *banks* once the previous cycle is complete. A complex parallel memory system is documented in Figure 7.24. Each *bank* is set to operate on its own, independent of the operation of the other *banks*. For example, the ith *bank* performs a read operation on a particular memory address, while the $(i + 1)$th *bank* performs a write operation on a different and unrelated memory

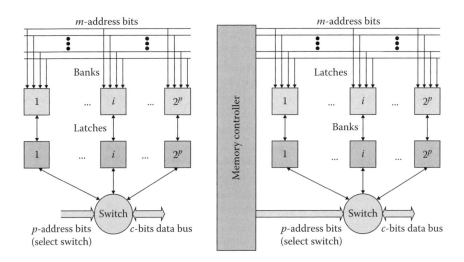

FIGURE 7.24 Simple and complex parallel main memory systems.

address. Complex paralleling is achieved using the address latch and a read/write command line for each *bank*. The *memory controller* handles the operation of the complex parallel memory. The processing unit submits the memory request to the memory controller, which determines which *bank* needs to be accessed. The controller then determines if the *bank* is busy by monitoring a busy line for each *bank*. The controller holds the request if the *bank* is busy, submitting it when the *bank* becomes available to accept the request. When the *bank* responds to a read request, the switch is set by the controller to accept the request from the *bank* and forward it to the processing unit. It can be seen that complex parallel main memory systems will be implemented as molecular vector processors. If consecutive elements of a vector are present in different memory *banks*, then the memory system can sustain a bandwidth of one element per clock cycle. Memory systems in ^MPPs can have thousands of *banks* with multiple memory controllers that allow multiple independent memory requests at every clock cycle.

Pipelining is a technique to increase the processor throughput with limited hardware in order to implement complex *datapath* (data processing) units (multipliers, floating-point adders, etc.). In general, a pipeline processor integrates a sequence of i data-processing molecular primitives, which cooperatively perform a single operation on a stream of data operands passing through them. Design of pipelining ^MICs involves deriving multistage balanced sequential algorithms to perform the given function. Fast buffer registers are placed between the primitives to ensure the transfer of data between them without interfering with one another. These buffers should be clocked at the maximum rate that guarantees the reliable data transfer between primitives. As illustrated in Figure 7.25, ^MPPs must be designed to guarantee the robust execution of overlapped instructions using pipelining. Four basic steps (fetch F_i — decode

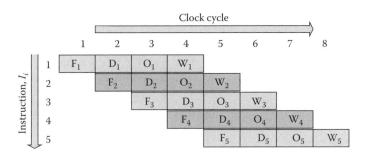

FIGURE 7.25 Pipelining of instruction execution.

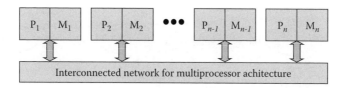

FIGURE 7.26 Multiprocessor architecture.

D_i — operate O_i — and — write W_i) and specific hardware units are needed to achieve these tasks. The execution of instructions can be overlapped. When the execution of some instruction I_i depends on the results of a previous instruction I_{i-1} that is not yet completed, instruction I_i must be delayed. The pipeline is said to be stalled, waiting for the execution of instruction I_{i-1} to be completed. While it is not possible to eliminate such situations, it is important to minimize the probability of their occurrence. This is a key consideration in the design of the instruction set and the design of the compilers that translate high-level language programs into machine language.

The parallel execution capability, called superscalar processing, when added to pipelining of the individual instructions, means that more than one instruction can be executed per basic step. Thus, the execution rate can be increased. The rate R_T of performing basic steps in the processor depends on the processor clock rate. The use of multiprocessors speeds up the execution of large programs by executing subtasks in parallel. The main difficulty in achieving this is decomposition of a given task into its parallel subtasks and ordering these subtasks to the individual processors in such a way that communication among the subtasks will be performed efficiently and robustly. Figure 7.26 documents a block diagram of a multiprocessor system with the interconnection network needed for data sharing among the processors P_i. Parallel paths are needed in this network in order to parallel activity to proceed in the processors as they access the global memory space as represented by the multiple memory units M_i. This is performed utilizing 3-D organization.

7.9 Hierarchical Finite-State Machines and Their Use in Hardware and Software Design

Simple register-level subsystems perform single data-processing operations, e.g., summation $X:=x_1 + x_2$, subtraction $X:=x_1 - x_2$, etc. To do complex data processing operations, multifunctional register-level subsystems should be designed. These register-level subsystems are partitioned as a data-processing unit (*datapath*) and a controlling unit (control unit). The control unit is responsible for collecting and controlling the data-processing operations (actions) of the *datapath*. To design the register-level sub-systems, one studies a set of operations to be executed and then designs ^{M}ICs using a set of register-level components that implement the desired functions. The ultimate goal is to achieve optimal achievable performance under the constraints. It is difficult to impose meaningful mathematical structures on register-level behavior using Boolean algebra and conventional gate-level design. Due to these difficulties, the heuristic synthesis is commonly accomplished as the following sequential algorithm:

> Define the desired behavior as a set of sequences of register-transfer operations (each operation can be implemented using the available components) comprising the algorithm to be executed
> Examine the algorithm to determine the types of components and their number to ensure the required *datapath*
> Design a complete block diagram for the *datapath* using the components chosen
> Examine the algorithm and *datapath* in order to derive the control signals with the ultimate goal to synthesize the control unit for the found *datapath* that meets the algorithm's requirements
> Test, verify, and evaluate the design performing analysis and simulation

Let us perform the design of virtual control units that ensures extensibility, flexibility, adaptability, robustness, and reusability. The design will be performed using the hierarchic graphs (HGs). A most

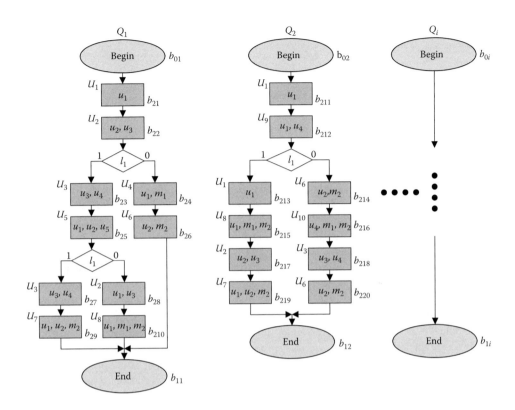

FIGURE 7.27   Control algorithm represented by HGs Q_1, Q_2, ..., Q_{i-1}, Q_i.

important problem is to develop straightforward algorithms that ensure implementation (nonrecursive and recursive calls) and utilize hierarchical specifications. We will examine the behavior, perform logic design, and implement reusable control units modeled as hierarchical finite-state machines with virtual states. The goal is to attain the top-down sequential well-defined decomposition in order to develop complex robust control algorithm step-by-step. We consider *datapath* and control units. The *datapath* unit consists of memory and combinational units. A control unit performs a set of instructions by generating the appropriate sequence of micro instructions that depend on intermediate logic conditions or on intermediate states of the *datapath* unit. To describe the evolution of a control unit, behavioral models are developed. We use the direct-connected HGs containing nodes. Each HG has an entry (*Begin*) and an output (*End*). Rectangular nodes contain micro instructions, macro instructions, or both.

A micro instruction set U_i includes a subset of micro operations from the set $U = \{u_1, u_2, ..., u_{u-1}, u_u\}$. Micro operations $\{u_1, u_2, ..., u_{u-1}, u_u\}$ control the specific actions in the *datapath* as shown in Figure 7.27. For example, one can specify that u_1 sends the data in the local stack, u_2 sends the data in the output stack, u_3 forms the address, u_4 calculates the address, u_5 forwards the data from the local stack, u_6 stores the data from the local stack in the register, u_7 forwards the data from the output stack to external output, etc. A micro operation is the output causing an action in the *datapath*. Any macro instruction incorporates macro operations from the set $M = \{m_1, m_2, ..., m_{m-1}, m_m\}$. Each macro operation is described by another lower level HG. Assume that each macro instruction includes one macro operation. Each rhomboidal node contains one element from the set $L \cup G$. Here, $L = \{l_1, l_2, ..., l_{l-1}, l_l\}$ is the set of logic conditions, while $G = \{g_1, g_2, ..., g_{g-1}, g_g\}$ is the set of logic functions. Using logic conditions as inputs, logic functions are derived by examining predefined set of sequential steps that are described by a lower level HG. Directed lines connect the inputs and outputs of the nodes. Consider a set $E = M \cup G$, $E = \{e_1, e_2, ..., e_{e-1}, e_e\}$. All elements $e_i \in E$ have HGs, and each e_i has the corresponding HG Q_i which specifies either an algorithm for performing e_i (if $e_i \in M$) or an algorithm for calculating e_i (if $e_i \in G$). Assume that $M(Q_i)$ is the subset of

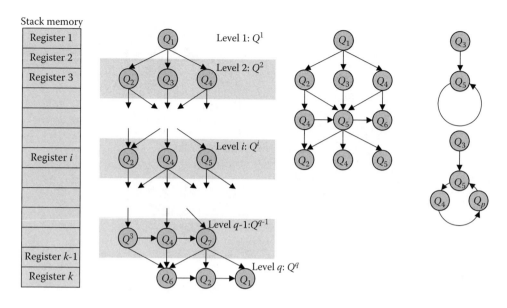

FIGURE 7.28 Stack memory with multiple-level sequential HGs with illustration of recursive call.

macro operations and $G(Q_i)$ is the subset of logic functions that belong to the HG Q_i. If $M(Q_i) \cup G(Q_i) = \varnothing$, the well-known scheme results [45]. The application of HGs enables one to gradually and sequentially synthesize complex control algorithms concentrating the efforts at each stage on a specified level of abstraction because specific elements of the set E are used. Each component of the set E is simple and can be checked and debugged independently. Figure 7.27 reports HGs $Q_1, Q_2, \ldots, Q_i$ which describe the control algorithm.

The execution of HGs is examined by studying complex operations $e_i = m_j \in M$ and $e_i = g_j \in G$. Each complex operation e_i that is described by an HG Q_i must be replaced with a new subsequence of operators that produces the result executing Q_i. In the illustrative example, shown in Figure 7.28, Q_1 is the first HG at the first level Q^1, the second level Q^2 is formed by Q_2, Q_3, and Q_4, etc. We consider the following hierarchical sequence of HGs: $Q^1_{\text{(level 1)}} \Rightarrow Q^2_{\text{(level 2)}} \Rightarrow \cdots \Rightarrow Q^{q-1}_{\text{(level q-1)}} \Rightarrow Q^q_{\text{(level q)}}$. All $Q_{i\text{ (level }i)}$ have the corresponding HGs. For example, Q^2 is a subset of the HGs that are used to describe elements from the set $M(Q_1) \cup G(Q_1) = \varnothing$, while Q^3 is a subset of the HGs that are used to map elements from the sets $\cup_{q \in Q^2} M(q)$ and $\cup_{q \in Q^2} G(q)$. In Figure 7.28, $Q^1 = \{Q_1\}$, $Q^2 = \{Q_2, Q_3, Q_4\}$, $Q^3 = \{Q_2, Q_4, Q_5\}$, etc.

Micro operations u^+ and u^- are used to increment and to decrement the stack pointer. The problem of switching to various levels can be solved using a stack memory (Figure 7.28). Consider an algorithm for $e_i \in M(Q_1) \cup G(Q_1) = \varnothing$. The stack pointer is incremented by the micro operation u^+, and a new register of the stack memory is set as the current register. The previous register stores the state when it was interrupted. New Q_i becomes responsible for the control until terminated. After termination of Q_i, the micro operation u^- is generated to return to the interrupted state. As a result, control is passed to the state in which Q_f is called. The design algorithm is formulated as: for a given control algorithm A, described by the set of HGs, construct the finite-state machine that implements A. In general, the design includes the following steps: (1) transformation of the HGs to the state transition table; (2) state encoding; (3) combinational logic optimization; (4) final structure design.

The first step is divided into three tasks as: (t1) mark the HGs with labels b (Figure 7.27); (t2) record transitions between the labels in the extended state transition table; (t3) convert the extended table to ordinary form. The labels b_{01} and b_{11} are assigned to the nodes *Begin* and *End* of the Q_1. The label $b_{02}, \ldots, b_{0i}$ and $b_{12}, \ldots, b_{1i}$ are assigned to nodes *Begin* and *End* of $Q_2, \ldots, Q_i$, respectively. The labels $b_{21}, b_{22}, \ldots, b_{2j}$ are assigned to other nodes of HGs, inputs, and outputs of nodes with logic conditions, etc.

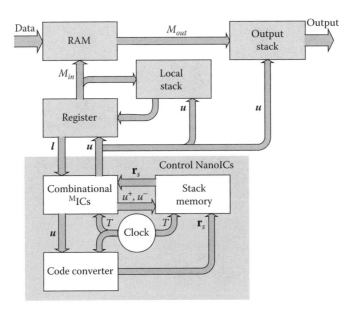

FIGURE 7.29 Hardware schematics.

Repeating labels are not allowed. The labels are considered as the states. The extended state transition table is designed using the state evolutions due to inputs (logic conditions) and logic functions which cause the transitions from $x(t)$ to $x(t + 1)$. All evolutions of the state vector $x(t)$ are recorded, and the state $x_k(t)$ has the label k. It should be emphasized that the table can be converted from the extended to the ordinary form. To program the Code Converter, as shown in Figure 7.29, one records the transition from the state x_1 assigned to the *Begin* node of the HG Q_1, e.g., $x_{01} \Rightarrow x_{21}(Q_1)$. The transitions between different HGs are recorded as $x_{ij} \Rightarrow x_{nm}(Q_j)$. For all transitions, the data-transfer instructions are derived. The hardware schematics is illustrated in Figure 7.29. Robust control algorithms are derived using the HGs employing the hierarchical behavior specifications and top-down decomposition. The reported method guarantees exceptional adaptation and reusability features through reconfigurable hardware and reprogrammable software for complex ICs and 3-D ^{M}ICs.

7.10 Adaptive Defect-Tolerant Molecular Processing-and-Memory Platforms

Some molecular fabrication processes, such as organic synthesis, self-assembly, and others, have been shown to be quite promising [5–10,47,48]. However, it is unlikely that near-future technologies will guarantee reasonable repeatable characteristics, affordable high-quality high-yield, satisfactory uniformity, desired failure tolerance, needed testability, and other important specifications imposed on ^{M}devices and ^{M}ICs. Therefore, design of robust defect-tolerant adaptive (reconfigurable) architectures (hardware) and software to accommodate failures, inconsistence, variations, nonuniformity, and defects is critical.

For conventional ICs, programmable gate arrays (PGAs) have been developed and utilized. These PGAs lead one to the on-chip reconfigurable circuits. The reconfigurable logics can be utilized as a functional unit in the *datapath* of the processor, having access to the processor register file and to on-chip memory ports. Another approach is to integrate the reconfigurable part of the processor as a coprocessor. For this solution, the reconfigurable logic operates concurrently with the processor. Optimal design and memory port assignments can guarantee the coprocessor reconfigurability and concurrency. In general, the reconfigurable architecture synthesis emphasizes a high-level design, rapid prototyping, and reconfigurability in order to reduce time and cost improving performance. The goal is to design and fabricate affordable

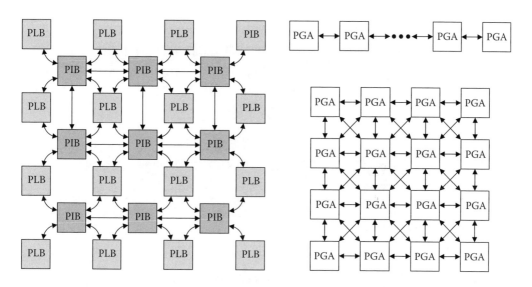

FIGURE 7.30 Programmable gate arrays and multiple PGAs organization.

high-performance high-yield [M]ICs and application-specific [M]ICs. These [M]ICs should be testable to detect the defects and faults. The design of the application-specific [M]ICs involves mapping application requirements into specifications implemented by [M]ICs. The specifications are represented at every level of abstraction including the system, behavior, structure, physical, and process domains. The designer should be able to differently utilize [M]ICs to meet the application requirements.

Reconfigurable [M]PPs should use reprogrammable logic units, such as PGAs, to implement a specialized instruction set and arithmetic units to optimize the performance. Ideally, reconfigurable [M]PPs should be reconfigured in real-time (runtime), enabling the existing hardware to be reused depending on its interaction with external units, data dependencies, algorithm requirements, faults, etc. The basic PGA architecture is built using the programmable logic blocks (PLBs) and programmable interconnect blocks (PIBs) (Figure 7.30). The PLBs and PIBs will hold the current configuration setting until adaptation is accomplished. The PGA is programmed by downloading the information in the file through a serial or parallel logic connection. The time required to configure a PGA is called the configuration time, and PGAs can be configured in series or in parallel. Figure 7.30 illustrates the basic architectures from which multiple PGA architectures can be derived. For example, pipelined interfaced PGA architecture fits for functions that have streaming data at specific intervals, while arrayed PGA architecture is appropriate for functions that require a systolic array. A hierarchy of configurability is different for the different PGA architectures, and the specifics of [M]ICs impose emphasized constraints on the technology-centric SLSI.

The goal is to design reconfigurable [M]PP architectures with corresponding software to cope with less-than-perfect, entirely or partially defective and faulty [M]devices, [M]gates, and [M]ICs used in arithmetic, logic, control, input–output, memory, and other units. To achieve our objectives, the redundant concept can be applied. The redundancy level is determined by the [M]ICs quality and software capabilities. Hardware and software evolutionary learning, adaptability, and reconfigurability can be achieved through decision-making, diagnostics, analysis, and optimization of software, as well as reconfiguring, pipelining, rerouting, switching, matching, controlling, and networking of hardware. Thus, one needs to design, optimize, build, test, and configure [M]PPs. The overall objective can be achieved by guaranteeing the evolution (behavior) matching between the ideal (CI) and fabricated (CF) molecular platform, its subsystems or components. The molecular compensator (CF1) can be designed and implemented for a fabricated CF2 such that the response of the CF will match the evolution of the CI (Figure 7.31). Both CF1 and CF2 represent [M]ICs hardware. The CI gives the reference ideal evolving model which provides the ideal input–output behavior, and the compensator CF1 should modify the evolution of CF2 such that CF, described by $C_F = C_{F1} \circ C_{F2}$ (series architecture), matches the CI behavior and functionality. Figure 7.31

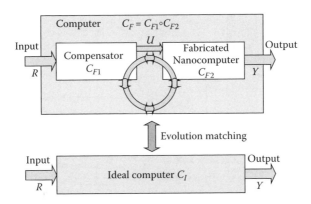

FIGURE 7.31 Molecular platform and evolution matching.

illustrates the concept. The necessary and sufficient conditions for strong and weak evolution matching based on CI and CF2 must be derived.

To address analysis, control, diagnostics, optimization, and design problems, the explicit mathematical models of a molecular platform or its units (subsystems) must be developed and applied. There are different levels of abstraction in modeling, simulation, and analysis. High-level models can accept streams of instruction descriptions and memory references, while the low-level (device/gate-level) modeling can be performed by making use of streams of input and output signals examining nonlinear transient behavior and steady-state characteristics of devices. The subsystem/unit-level modeling (medium-level) also can be formulated and performed. A subsystem can contain billions of ^{M}devices, and may not be modeled as queuing networks, difference equations, Boolean models, polynomials, information-theoretic models, etc. Different mathematical modeling concepts exist and have been developed for each level. In this section we concentrate on the high-, medium- and low-level systems modeling using the finite-state machine concept.

Molecular processors and memories accept the input information, process it according to the stored instructions, and produce the output. Any mathematical model is the mathematical idealization based upon the abstractions, simplifications, and hypotheses made. It is virtually impossible to develop and apply the complete mathematical model due to complexity and uncertainties. It is possible to concurrently model a molecular platform by the six-tuple $C = \{X, E, R, Y, F, X_0\}$, where X is the finite set of states with initial and final states $x_0 \in X$ and $x_f \subseteq X$; E is the finite set of events (concatenation of events forms a string of events); R and Y are the finite sets of the input and output symbols (alphabets); F are the transition functions mapping from $X \times E \times R \times Y$ to X (denoted as F_X), to E (denoted as F_E), or to Y (denoted as F_Y), $F \subseteq X \times E \times R \times Y$ (we assume that $F = F_X$, e.g., the transition function defines a new state to each quadruple of states, events, references and outputs, and F can be represented by a table listing the transitions or by a state diagram).

The evolution of a molecular platform is due to inputs, events, state evolutions, parameter variations, etc. A vocabulary (or an alphabet) A is a finite nonempty set of symbols (elements). A world (or a sentence) over A is a string of finite length of elements of A. The empty (null) string is the string which does not contain symbols. The set of all words over A is denoted as A_w. A language over A is a subset of A_w. A finite-state machine with output $C_{FS} = \{X, A_R, A_Y, F_R, F_Y, X_0\}$ consists of a finite set of states S, a finite input alphabet A_R, a finite output alphabet A_Y, a transition function F_Y that assigns a new state to each state and input pair, an output function F_Y that assigns an output to each state and input pair, and initial state X_0. Using the input–output map, the evolution of C can be expressed as $E_C \subseteq R \times Y$. That is, if C in state $x \in X$ receives an input $r \in R$, it moves to the next state $f(x,r)$, and produces the output $y(x,r)$. One can represent the molecular platform using state tables, which describe the state and output functions. In addition, the state transition diagram (direct graph whose vertices correspond to the states and edges correspond to the state transitions, and each edge is labeled with the input and output associated with the transition) is frequently used.

Quantum Computing — The quantum molecular platform is described by the seven-tuple $QC = \{X, E, R, Y, H, U, X_0\}$, where H is the Hilbert space; U is the unitary operator in the Hilbert space that satisfies the specific conditions.

The parameter set P should be used. Designing reconfigurable fault-tolerant architectures, sets P and P_0 are integrated, and $C = \{X, E, R, Y, P, F, X_0, P_0\}$. It is evident that the evolution of the C depends on P and P_0. The optimal performance can be achieved through adaptive synthesis, reconfiguration, and diagnostics. For example, one can vary F and variable parameters P_v to attain the best possible performance. The evolution of states, events, outputs, and parameters is expressed as

$$\underset{evolution\ 1}{(x_0, e_0, y_0, p_0)} \Rightarrow \underset{evolution\ 2}{(x_1, e_1, y_1, p_1)} \Rightarrow \cdots \Rightarrow \underset{evolution\ j-1}{(x_{j-1}, e_{j-1}, y_{j-1}, p_{j-1})} \Rightarrow \underset{evolution\ j}{(x_j, e_j, y_j, p_j)}.$$

The input, states, outputs, events, and parameter sequences are aggregated within the model as given by $C = \{X, E, R, Y, P, F, X_0, P_0\}$. The concept reported allows us to find and apply the minimal, but, complete functional description of molecular processing and memory platforms. The minimal subset of state, event, output, and parameter evolutions (transitions) can be used. That is, the partial description $C_{partial} \subset C$ results, and every essential quadruple (x_i, e_i, y_i, p_i) can be mapped by $(x_i, e_i, y_i, p_i)_{partial}$. This significantly reduces the complexity of modeling, simulation, analysis, and design problems.

Let the function F maps from $X \times E \times R \times Y \times P$ to X, e.g., $F : X \times E \times R \times Y \times P \to X$, $F \subseteq X \times E \times R \times Y \times P$. Thus, the transfer function F defines a next state $x(t+1) \in X$ based upon the current state $x(t) \in X$, event $e(t) \in E$, reference $r(t) \in R$, output $y(t) \in Y$, and parameter $p(t) \in P$. Hence, $x(t+1) = F(x(t), e(t), r(t), y(t), p(t))$ for $x_0(t) \in X_0$ and $p_0(t) \in P_0$.

The robust adaptive algorithms must be developed. The control vector $u(t) \in U$ is integrated into the model. We have $C = \{X, E, R, Y, P, U, F, X_0, P_0\}$, and the problem is to design the compensator. The strong evolutionary matching $C_F = C_{F1} \circ C_{F2} =_B C_I$ for given C_I and C_F is guaranteed if $E_{C_F} = E_{C_I}$. Here, $C_F =_B C_I$ means that the behaviors (evolution) of C_I and C_F are equivalent. The weak evolutionary matching $C_F = C_{F1} \circ C_{F2} \subseteq_B C_I$ for given C_I and C_F is guaranteed if $E_{C_F} \subseteq E_{C_I}$. Here, $C_F \subseteq_B C_I$ means that the evolution of C_F is contained in the behavior C_I. The problem is to derive a compensator $C_{F1} = \{X_{F1}, E_{F1}, R_{F1}, Y_{F1}, F_{F1}, X_{F10}\}$ such that for a given $C_I = \{X_I, E_I, R_I, Y_I, F_I, X_{I0}\}$ and $C_{F2} = \{X_{F2}, E_{F2}, R_{F2}, Y_{F2}, F_{F2}, X_{F20}\}$ the following conditions: $C_F = C_{F1} \circ C_{F2} =_B C_I$ (strong behavior matching) or $C_F = C_{F1} \circ C_{F2} \subseteq_B C_I$ (weak behavior matching) are satisfied. We assume that: (i) output sequences generated by C_I can be generated by C_{F2}; (ii) the C_I inputs match the C_{F1} inputs.

The output sequences means the state, event, output, and/or parameter vectors, e.g., we have (x, e, y, p). If there exists the state-modeling representation $\gamma \subseteq X_I \times X_F$ such that $C_I^{-1} \propto {}^{\gamma}_B C_{F2}^{-1}$ (if $C_I^{-1} \propto {}^{\gamma}_B C_{F2}^{-1}$, then $CI \propto {}^{\gamma}_B C_{F2}$), then the evolution matching problem is solvable. The compensator C_{F1} solves the strong matching problem $C_F = C_{F1} \circ C_{F2} =_B C_I$ if there exist the state-modeling representations $\beta \subseteq X_I \times X_{F2}$, $(X_{I0}, X_{F20}) \in \beta$ and $\alpha \subseteq X_{F1} \times \beta$, $(X_{F10}, (X_{I0}, X_{F20})) \in \alpha$ such that $C_{F1} = {}^{\alpha}_B C_I^{\beta}$ for $\beta \in \Gamma = \{\gamma | C_I^{-1} \propto {}^{\gamma}_B C_{F2}^{-1}\}$. The strong matching problem is tractable if there exist C_I^{-1} and C_{F2}^{-1}. The C can be decomposed using algebraic decomposition theory, which is based on the closed partition lattice. For example, consider the fabricated C_{F2} represented as $C_{F2} = \{X_{F2}, E_{F2}, R_{F2}, Y_{F2}, F_{F2}, X_{F20}\}$. A partition on the state set for C_{F2} is a set $\{C_{F2\ 1}, C_{F2\ 2}, \ldots, C_{F2\ i}, \ldots, C_{F2\ k-1}, C_{F2\ k}\}$ of disjoint subsets of the state set X_{F2} whose union is X_{F2}, e.g., $\bigcup_{i=1}^{k} C_{F2i} = X_{F2}$ and $C_{F2i} \cap C_{F2j} = \varnothing$ for $i \neq j$. Hence, one designs and implements the compensators $C_{F1\ i}$ for given $C_{F2\ i}$.

7.11 Hardware–Software Design

Significant research activities have been focused on the synthesis of novel processing and memory platforms. The aforementioned activities must be supported by a broad spectrum of hardware–software codesign including technology-centric CAD developments. The *Marchitectronics* paradigm can serve as

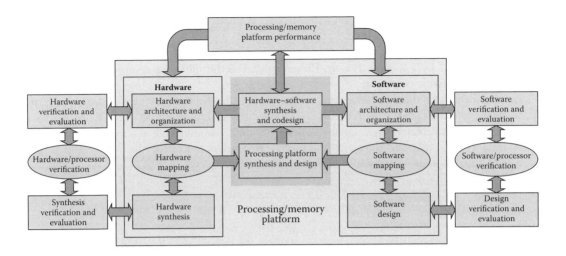

FIGURE 7.32 Hardware–software codesign for ᴹPPs.

the basis for the design and analysis of novel efficient, robust, homogeneous, and redundant ᴹPPs. Hardware and software codesign, integration, and verification are important problems to be addressed. The synthesis of concurrent architectures and their organization (collection of functional hardware components, modules, subsystems, and systems that can be software programmable and adaptively reconfigurable) are among the most important issues. It is evident that software depends on hardware and vice versa. The concurrency may indicate hardware and software compliance and matching. It is impractical to fabricate high-yield ideal (perfect) complex ᴹICs. Furthermore, it is unlikely that the software can be developed for not strictly defined configurations that must be adapted, reconfigured, and optimized. The not-perfect devices lead to the importance of diagnostics, reconfiguration, evaluation, testing, and other tasks to be implemented through robust software. The systematic synthesis, analysis, optimization, and verification of hardware and software, as illustrated in Figure 7.32, are applied to advance the design and synthesis.

The performance analysis, verification, evaluation, characterization, and other tasks can be formulated and examined only as the molecular processing/memory platforms are devised, synthesized, and designed. It is important to start the design process from a high-level, but explicitly defined abstraction domain which should:

Coherently capture the functionality and performance at all levels
Examine and verify the correctness of functionality, behavior, and operation of devices, modules, subsystems, and systems
Depict the specification of different organizations and architectures examining their adaptability, reconfigurability, optimality, etc.

System-level models describe processing and memory platforms as a hierarchical collection of modules, subsystems, and systems. For example, steady-state and dynamics of gates and modules are studied examining how these components perform and interact. The evolution of states, events, outputs, and parameters are of the designer's interest. Different discrete events, process networks, Petri nets, and other methods have been applied to model computers. Models based on synchronous and asynchronous finite-state machine paradigms with some refinements ensure meaningful features and map the essential behavior in different abstraction domains. Mixed control, data flow, data processing (encryption, filtering, and coding), and computing processes can be modeled.

A program is a set of instructions that one writes to define what the computer should do. For example, if the ICs consists of *on* and *off* logic switches, one can assign so that the first and second switches are *off*, while the third to eighth switches are *on* in order to receive the eight-bit signal 00111111. The program commands millions of switches, and a program should be written in circuitry-level language. For ICs, software developments have progressed to the development of high-level programming languages. A high-level programming language allows one to use a vocabulary of terms, e.g., read, write, or do instead of creating the sequences of *on–off* switching, which implements these functions. All high-level languages have their syntax, provide a specific vocabulary, and give explicitly defined sets of rules for using their vocabulary. A compiler is used to translate (interpret) the high-level language statements into machine code. The compiler issues the error messages if the programmer uses the programming language incorrectly. This allows one to correct the error and perform other translation by compiling the program again. The programming logic is an important issue because it involves executing various statements and procedures in the correct order to produce the desired results. One must use the syntax correctly and execute a logically constructed workable program. Two commonly used approaches to write computer programs are procedural and object-oriented programming. Through procedural programming, one defines and executes computer memory locations (variables) to hold values and writes sequential steps to manipulate these values. Object-oriented programming is the extension of procedural programming because it involves creating objects (program components) and creating applications that use these objects. Objects are made up of states, and these states describe the characteristics of an object. For 3-D ᴹICs, there is a need to develop novel software environments that may be organizationally/architecturally neutral or specific. It is unlikely that a single software toolbox can be used or will be functional for all classes of ᴹICs that utilize different hardware solutions, exhibit distinct phenomena, etc., for example, analog vs. digital, binary vs. multiple-valued, etc.

Specific hardware and software solutions must be developed and implemented. For example, ICs are designed by making use of hardware description languages (HDLs), for example, Very High Speed Integrated Circuit Hardware Description Language (VHDL) and Verilog. The design starts by interpreting the application requirements into architectural specifications. As the application requirements are examined, the designer translates the architectural specifications into behavior and structure domains. Behavior representation means the functionality required as well as the ordering of operations and completion of tasks in specified times. A structural description consists of a set of ᴹdevices and their interconnection. Behavior and structure can be specified and studied using HDLs. These languages efficiently manage quite complex hierarchies, which can include millions of logic ᴹgates. Another important feature is that HDLs are translated into netlists of library components using synthesis software.

The structural or behavioral representations are meaningful ways of describing a model. In general, HDLs can be used for design, verification, simulation, analysis, optimization, documentation, etc. For conventional ICs, VHDL and Verilog are among the standard design tools. In VHDL, a design is typically partitioned into blocks. These blocks are then integrated to form a complete design using the schematic capture approach. This is performed using a block diagram editor or hierarchical drawings to represent block diagrams. In VHDL, every portion of a VHDL design is considered as a block. Each block has an analog to an off-the-shelf IC, and is called an entity. The entity describes the interface to the block, schematics, and operation. The interface description is similar to a pin description and specifies the inputs and outputs to the block. A complete design is a collection of interconnected blocks. Consider a simple example of an entity declaration in VHDL. The first line indicates a definition of a new entity. The last line marks the end of the definition. The lines between, called the port clause, describe the interface to the design. The port clause provides a list of interface declarations. Each interface declaration defines one or more signals that are inputs or outputs to the design. Each interface declaration contains a list of names, mode, and type. As the interface declaration is accomplished, the architecture declaration is studied. As the basic building blocks using entities and their associated architectures are defined, one can combine them together to form other designs. The structural description of a design is a textual description of a schematic. A list of components and their connections is called a netlist. In the data flow domain, ICs are described by indicating how the inputs and outputs of built-in primitive components or pure combinational blocks are connected together. Thus, one describes how signals (data) flow through ICs. The architecture part describes the internal operation of the design. In the data flow domain, one specifies how data flows from the inputs to the outputs. In VHDL this is accomplished with the signal assignment statement. The evaluation of the expression is performed substituting the values of the signals in the expression and computing the result of each operator in the expression. The scheme used to model a VHDL design is called discrete event time simulation. When the value of a signal changes, this means that an event has occurred on that signal. The values of signals are only updated when discrete events occur. Since one event causes another, simulation proceeds in rounds. The simulator maintains a list of events that need to be processed. In each round, all events in a list are processed, and any new events that are produced are placed in a separate list (scheduled) for processing in a later round. Each signal assignment is evaluated once, when simulation begins to determine the initial value of each signal, to design [ME]ICs. In general, one needs to develop new technology-centric HDLs coherently integrating 3-D topologies/organization, enabling architectures, device physics, bottom-up fabrication, and other distinctive features of molecular electronics.

7.12 Modeling and Analysis of Molecular Electronic Devices

7.12.1 Introduction to Modeling Concepts

A great variety of molecules have been synthesized and examined for applications other than electronics. This section is devoted to the analysis of electron transport in [ME]devices that should ensure functionality, desired characteristics, and specified performance. These [ME]devices, composed of atomic aggregates ensuring chemical synthesis soundness, exhibit quantum phenomena that should be utilized. Molecular electronics devices should be examined by applying quantum mechanics. Coherent high-fidelity mathematical models are needed to carry out data-intensive analysis and examine electron transport in molecular complexes. Mathematical models should accurately describe the basic phenomena, be computationally tractable, and suit heterogeneous simulations as applied to carry out data-intensive analysis. Modeling and analysis of electronic devices are based on the Schrödinger equation, Green function, and other methods [7,56–58]. The kinetic energy, potentials, Fermi energy E_F, energy level broadening E_B, charge density, and other quantities, variables and parameters are used. Figure 7.33a schematically illustrates a 3-D-topology multiterminal and two-terminal [ME]devices.

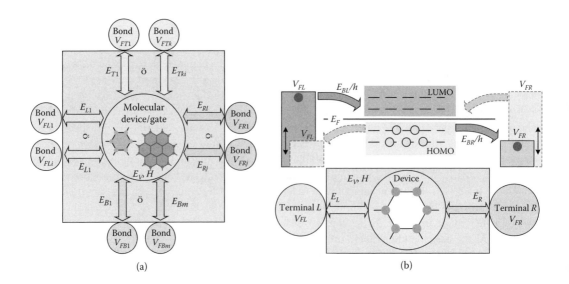

(a)   (b)

FIGURE 7.33 Molecular electronic devices. (a) Multiterminal [ME]device with the left (L), right (R), top (T) and bottom (B) bonds forming *input*, *control*, and *output* terminals; (b) two-terminal [ME]device with Hamiltonian H, single energy potential E_V, and varying left/right potentials V_{FL} and V_{FR}.

It will be reported that by using quantum mechanics, one can derive the dimensionless transmission probability of electron tunneling $T(E)$ which is a function of energy E, and $0 \le T(E) \le 1$. The conductance of molecular wires and some two-terminal ^{ME}devices was examined in Refs. [7,56–58]. A linear conductance that neglects thermal relaxation and other effects can be estimated by applying the so-called Landauer [59] or Landauer–Büttiker [60] expression $g(E) = (e^2/\pi\hbar)T(E)$. Here, the total transmission coefficient $T(E)$ is evaluated at the energy E which is equal to the Fermi energy E_F at zero voltage bias. The so-called quantum conductance is defined to be $g_0 = e^2/\pi\hbar = 7.75 \times 10^{-5} \, \Omega^{-1}$. The constant $e^2/3\pi^2\hbar^2$ in defining the expression for conductance was originally reported in Ref. [59], where the electron transport was studied in the electric field. By making use of the acceleration of electrons $d\mathbf{k}/dt = -e\mathbf{E}/\hbar$, the expression for conductivity was provided. In particular, assuming the equilibrium condition, Ref. [59] states: "For our isotropic band structure and isotropic background scattering the conductivity is given by $\sigma_B = (\tau_B/3\pi^2)(e^2/\hbar^2)k^2(dU/dk)$." Here, τ_B is the relaxation time; and k is the wave number.

Assuming the applicability of the Fermi–Dirac distribution, the current–voltage (I–V) characteristics for two-terminal ^{ME}devices (see Figure 7.33b) are commonly found by applying the following equation [58]

$$I(E) = \frac{2e}{h} \int_{-\infty}^{+\infty} T(E)[f(E_V, V_{FL}) - f(E_V, V_{FR})]dE,$$

where $f(E_V, V_{FL})$ and $f(E_V, V_{FR})$ are the Fermi–Dirac distribution functions, $f(E_V, V_{FL}) = (1 + e^{(E_V - V_{FL})/kT})^{-1}$ and $f(E_V, V_{FR}) = (1 + e^{(E_V - V_{FR})/kT})^{-1}$; E_V is the single energy potential that depends on the charge density $\rho(E)$ or the number of electrons N, $E_V = E_{V0} + V_{SC}$; V_{SC} is the self-consistent potential to be determined by solving the Poisson equation using the charge density, $V_{SC} = f_\rho(\rho)$ or $V_{SC} = V(N - N_0)$; N is the electron concentration; N_0 is the number of electrons at the equilibrium, $N_0 = 2f(E_{V0}, E_F) = 2(1 + e^{(E_{V0} - E_F)/kT})^{-1}$; V_{FL} and V_{FR} are the left and right electrochemical potentials related to the Fermi levels. The electrochemical potentials V_{FL} and V_{FR} vary, and there is no electron transport if $V_{FL} = V_{FR}$. The HOMO and LUMO orbitals, as well as the Fermi level, are documented in Figure 7.33b. Depending on the HOMO and LUMO levels, as well as E_F, the electron transport takes place through particular orbitals. The electron transport rates E_{BL}/h and E_{BR}/h are functions of the broadening energies E_{BL} and E_{BR}. One estimates the number of electrons and current as [58]

$$N = 2\frac{E_{BL}f(E_V, V_{FL}) + E_{BR}f(E_V, V_{FR})}{E_{BL} + E_{BR}}$$

and

$$I = \frac{eNE_{BR}}{h} = \frac{2eE_{BL}E_{BR}[f(E_V, V_{FL}) - f(E_V, V_{FR})]}{h(E_{BL} + E_{BR})}.$$

The approach reported above is well-suited for semiconductor microelectronic devices. For ^{ME}devices, many assumptions and postulates made may not apply. Correspondingly, other methods have been applied as was reported in Section 7.4. The application of quantum theory will be reported to examine the performance and baseline characteristics of ^{ME}devices. The wave function $\Psi(t,r)$, allowed energies, potentials, and other quantities must be studied to qualitatively and quantitatively examine time and spatial evolution of quantum system (^{M}device) states. This ensures a coherent analysis of behavior and phenomena including electron transport. For example, the transmission coefficient, expectation values of system variables, and other quantities are derived using the wave function, which is obtained by solving the Schrödinger equation.

7.12.2 Heisenberg Uncertainty Principle

We apply the quantum theory and perform some analyses from the experimental prospective by employing the Heisenberg uncertainty principle The Heisenberg uncertainty principle specifies that no experiments can be performed to furnish uncertainties below the limits defined by the uncertainty relationship. For a perturbed particle, using complementary observable variables A and B, the generalized uncertainty principle is given as

$$\sigma_A^2 \sigma_B^2 \ge \left(\frac{1}{2i} \left\langle [\hat{A}, \hat{B}] \right\rangle \right)^2,$$

where σ_A and σ_B are the standard deviations; $[\hat{A}, \hat{B}]$ is the commutator of two Hermitian operators $\hat{A}$ and $\hat{B}$, $[\hat{A}, \hat{B}] = \hat{A}\hat{B} - \hat{B}\hat{A}$.

We conclude that it is impossible to measure simultaneously two complementary observable variables with arbitrary accuracy. One may use the observable position x, for which $\hat{A} = x$, and the momentum p with the corresponding operator $\hat{B} = -i\hbar(\partial / \partial x)$. By taking note of the canonical commutation relation $[\hat{x}, \hat{p}] = i\hbar$, we obtain the position–momentum uncertainty principle as

$$\sigma_x^2 \sigma_p^2 \ge \left(\frac{1}{2i} i\hbar \right)^2 = \left(\tfrac{1}{2}\hbar \right)^2 \quad or \quad \sigma_x \sigma_p \ge \tfrac{1}{2}\hbar.$$

The energy–time uncertainty principle is

$$\sigma_E \sigma_t \ge \tfrac{1}{2}\hbar.$$

Notations Δx, Δp, ΔE, and Δt are frequently used to define the standard deviations as uncertainties. In Section 7.4 and quantum mechanics books, ΔE gives the energy difference between the quantum states. Hence, covering the Heisenberg uncertainty principle, we use the notation $\Delta\hat{E}$ which is not ΔE.

One defines the uncertainties ΔA and ΔB in the measurement of A and B by their dispersion, e.g.,

$$(\Delta A)^2 = \left\langle \left(\hat{A} - \left\langle \hat{A} \right\rangle \right)^2 \right\rangle = \left\langle \hat{A}^2 \right\rangle - \left\langle \hat{A} \right\rangle^2 \quad and \quad (\Delta B)^2 = \left\langle \left(\hat{B} - \left\langle \hat{B} \right\rangle \right)^2 \right\rangle = \left\langle \hat{B}^2 \right\rangle - \left\langle \hat{B} \right\rangle^2$$

or

$$\Delta A = \sqrt{\left\langle \hat{A}^2 \right\rangle - \left\langle \hat{A} \right\rangle^2} \quad and \quad \Delta B = \sqrt{\left\langle \hat{B}^2 \right\rangle - \left\langle \hat{B} \right\rangle^2}$$

The uncertainty relation is

$$\Delta A \Delta B \ge \tfrac{1}{2} \left| \left\langle [\hat{A}, \hat{B}] \right\rangle \right|.$$

The position–momentum and energy–time uncertainty principles are rewritten as

$$\Delta x \Delta p_x \ge \tfrac{1}{2}\hbar, \; \Delta y \Delta p_y \ge \tfrac{1}{2}\hbar, \; \Delta z \Delta p_z \ge \tfrac{1}{2}\hbar, \quad and \quad \Delta\hat{E}\Delta t \ge \tfrac{1}{2}\hbar$$

Example 7.5

Consider in detail the position–momentum uncertainty relation $\Delta x \Delta p_x \ge \tfrac{1}{2}\hbar$. The subscript x is used for the momentum p_x to indicate that $\Delta x \Delta p_x \ge \tfrac{1}{2}\hbar$ applies to motion of particle in a given direction and relates the uncertainties in position x and momentum p_x in that direction only. The relationship $\Delta x \Delta p_x \ge \tfrac{1}{2}\hbar$ gives an estimate (one cannot do better) of the minimum uncertainty that can result from any experiment, and measurement of the position and momentum of a particle will give uncertainties Δx and Δp_x. Hence, the Heisenberg uncertainty principle indicates: if the x-component of the momentum of a particle is measured with uncertainty Δp_x, then its x-position cannot be measured more accurately than $\Delta x \geq \hbar / 2\Delta p_x$. Thus, it is impossible to simultaneously measure two observable variables with an arbitrary accuracy.

Hence, there is a limit on the accuracy. One cannot perform experiments better than imposed by $\Delta x \Delta p_x \geq \frac{1}{2}\hbar$, $\Delta y \Delta p_y \geq \frac{1}{2}\hbar$, $\Delta z \Delta p_z \geq \frac{1}{2}\hbar$ and $\Delta \hat{E} \Delta t \geq \frac{1}{2}\hbar$, no matter which measuring hardware is used. It must be emphasized that the particle position, momentum, and energy are dynamic variables (measurable characteristics of the system or device) at any given time. In contrast, time is the independent variable of which the dynamic quantities are functions. That is, in $\Delta \hat{E} \Delta t \geq \frac{1}{2}\hbar$, Δt is the time it takes the system to change substantially. For example, Δt represents the amount of time it takes the expectation value of E to change by 1 standard deviation in order to ensure the observability of E.

The reported results impose constraints and limits on testing, evaluation and characterization of quantum systems, including ^{M}devices. The ability to conduct measurements for particular devices depends on the device physics, functionality, phenomena, carriers (photon, electron, or ion), etc. The uncertainty principle does not define or imply the dimensionality, switching time, power dissipation, switching energy, and other device characteristics. Those quantities must be found coherently applying other concepts reported in this section.

Example 7.6

For a single photon of energy E, the momentum is $p = E/c$. The de Broglie formula relates the momentum and the wavelength λ as $p = h/\lambda$. The rest energy of electron $m_e c^2$ is 5.1×10^5 eV. For the electron with the kinetic energy Γ, if $\Gamma << m_e c^2$, one may use nonrelativistic formalism to find momentum as $p = \sqrt{2m_e \Gamma}$ λ. Letting $\lambda = 1$ eV, we have $p = 5.4 \times 10^{-25}$ kg m/sec, which gives $\lambda = 1.2$ nm. The frequency of radiation is $v = c / \lambda$.

Example 7.7

Derive the position uncertainties Δx for a 9.1×10^{-31} kg electron (microscopic particles) and a 9.1×10^{-3} kg bullet (macroscopic particles). Let their speed $= 1000$ m/sec be measured with uncertainty 0.001%. Using $p = mv$, one finds $\Delta p = m\Delta v$. Hence, from $\Delta x \geq \hbar / 2\Delta p_x$, for an electron, one obtains $\Delta x \geq 0.00577$ m, while for a bullet we have $\Delta x \geq 5.77 \times 10^{-31}$ m. For the electron, taking note of the atomic radius of the silicon atom, which is 117 pm, one concludes that the position uncertainty Δx is 2.47×10^7 larger than the diameter of Si atom, while the dimension of a 1 cm bullet is 1.73×10^7 times larger than Δx, guaranteeing no restrictions on measurements for a bullet.

7.12.3 Particle Velocity

For ^{ME}devices, it is important to examine how wave packets evolve in time and space providing an answer on motion of quantum particles in space. The velocity of the group of matter waves is equal to the particle velocity whose motion they are governing. For the wave packets propagating in the x-direction, in order to examine the time evolution, we apply the following equation:

$$\Psi(t,x) = \frac{1}{\sqrt{2\pi}} \int_{-\infty}^{\infty} \phi(k) e^{i(kx - \omega t)} dk,$$

where $\phi(k)$ is the magnitude of the wave packet, k is the wave number, and ω is the angular frequency. Examining the time evolution of the wave packet, the group and phase velocities are given as

$$v_g = \frac{d\omega(k)}{dk} = v_{ph} + k\frac{dv_{ph}}{dk} = v_{ph} + p\frac{dv_{ph}}{dp}$$

and

$$v_{ph} = \frac{\omega(k)}{k}.$$

The group velocity represents the velocity of motion of the group of propagating waves that compose the wave packet. The phase velocity is the velocity of propagation of the phase of a single mth harmonic wave $e^{ik_m(\bar{x}-v_{ph}t)}$. The wave packet travels with the group velocity. Taking note of $E = \hbar\omega$ and $p = \hbar k$, one obtains $v_g = dE(p)/dp$ and $v_{ph} = E(p)/p$.

From $E = (p^2/2m) + \Pi$, assuming that $\Pi = $ const, we have

$$v_g = dE(p)/dp = p/m = v \text{ and } v_{ph} = E(p)/p = p/2m + \Pi/p.$$

Thus, the group velocity of the wave packet is equal to the particle velocity v.

For a free electron, the energy is

$$E = \frac{p^2}{2m} = \frac{\hbar^2 k^2}{2m} = \hbar\omega.$$

One finds

$$v_g = \frac{d\omega}{dk} = \frac{\hbar k}{m} = \frac{p}{m} = v.$$

Consider a free electron in the electric field with the intensity E_E. We have

$$dE = eE_E\, dx = eE_E \frac{dx}{dt} dt = eE_E v\, dt$$

and

$$dE = \hbar\, d\omega = \hbar \frac{d\omega}{dk} dk = \hbar v\, dk.$$

Thus, one finds

$$qE_E = \hbar \frac{dk}{dt}.$$

The time derivative of the electron velocity

$$v = \frac{d\omega}{dk} = \frac{1}{\hbar} \frac{dE}{dk}$$

gives the acceleration of the electron, and

$$a = \frac{dv}{dt} = \frac{1}{\hbar} \frac{d^2E}{dk\, dt} = \frac{1}{\hbar} \frac{d^2E}{dk^2} \frac{dk}{dt} = \frac{1}{\hbar^2} \frac{d^2E}{dk^2} eE_E.$$

The force acting on the electron is

$$F = \frac{dp}{dt} = \hbar \frac{dk}{dt},$$

or $F = eE_E$. Hence,

$$a = \frac{1}{\hbar^2} \frac{d^2E}{dk^2} F.$$

The expression

$$F = \hbar^2 \left(\frac{d^2E}{dk^2} \right)^{-1} \frac{dv}{dt}$$

is used in solid-state semiconductor devices to introduce the so-called *effective* mass of the electron which is

$$m_{eff} = \hbar^2 \left(\frac{d^2E}{dk^2} \right)^{-1}.$$

In solid ^{ME}devices, the device physics and 3-D topology must be coherently integrated. The derived expressions for the particle velocity can be used to obtain the *I–V* and *G–V* characteristics, estimate propagation delays, analyze switching speed, and examine other characteristics of ^{ME}devices.

Example 7.8

Consider a wave packet corresponding to a relativistic particle. The energy and momentum are

$$E = mc^2 = \frac{m_0 c^2}{\sqrt{1 - v^2/c^2}}$$

and

$$p = mv = \frac{m_0 v}{\sqrt{1 - v^2/c^2}},$$

where m_0 is the rest mass of the particle. From $E = c\sqrt{p^2 + m_0^2 c^2}$, one obtains

$$v_g = \frac{dE}{dp} = \frac{d\left(c\sqrt{p^2 + m_0^2 c^2} \right)}{dp} = \frac{pc}{\sqrt{p^2 + m_0^2 c^2}} = v$$

and

$$v_{ph} = \frac{E}{p} = \frac{c^2}{v}.$$

Example 7.9

Considering an electron as a not relativistic particle. From $E = mv^2/2$, one has $v = \sqrt{2E/m}$. Let $E = 0.1$ eV $= 0.1602176462 \times 10^{-19}$ J. For a not relativistic electron, we find $v = 1.88 \times 10^5$ m/sec. The time it takes the electron to travel 1 nm distance is $t = L/v = 5.33 \times 10^{-15}$ sec.

The particle (electron) traversal time is of interest to analyze the device performance [7,62]. For a 1-D case, for a particle with an energy E in $\Pi(x)$, one has

$$\tau(E) = \int_{x_0}^{x_f} \sqrt{\frac{m}{2[\Pi(x)-E]}}\,dx.$$

For a 1-D rectangular barrier with Π_0 and width L,

$$\tau(E) = \sqrt{\frac{m}{2(\Pi_0 - E)}}\,L.$$

By using the transmission probabilities of two particle states $T_1(E)$ and $T_2(E)$, we have [63]

$$\tau(E) = \lim_{\lambda \to 0}\left(\frac{\hbar}{|\lambda|}\sqrt{\frac{T_2(E)}{T_1(E)}}\right).$$

Example 7.10

Let $(\Pi_0 - E) = 0.1$ eV $= 0.16 \times 10^{-19}$ J and $L = 1$ nm. One finds $\tau = 5.33 \times 10^{-15}$ sec. The estimated τ agrees with the results reported for $\tau(E)$ in Ref. [63], where the transmission probabilities are used. As will be documented in Section 7.12.4, using the wave function, one may derive the expected value for the momentum to obtain $\tau(E)$.

7.12.4 Schrödinger Equation

The time-invariant (time-independent) Schrödinger equation for a particle in the Cartesian coordinate system is given as

$$-\frac{\hbar^2}{2m}\nabla^2\Psi(x,y,z) + \Pi(x,y,z)\Psi(x,y,z) = E(x,y,z)\Psi(x,y,z),$$

where ∇^2 is the Laplacian, $\nabla^2 = (\partial^2/\partial x^2) + (\partial^2/\partial y^2) + (\partial^2/\partial z^2)$; $\Pi(x,y,z)$ is the potential energy function; $E(x,y,z)$ is the total energy.

The Hamiltonian is $H = -(\hbar^2/2m)\nabla^2 + \Pi$.

Hence, $H(x,y,z)\Psi(x,y,z) = E(x,y,z)\Psi(x,y,z)$ or $H(\mathbf{r})\Psi(\mathbf{r}) = E(\mathbf{r})\Psi(\mathbf{r})$.

The time-dependent Schrödinger equation is

$$-\frac{\hbar^2}{2m}\nabla^2\Psi(t,x,y,z) + \Pi(t,x,y,z)\Psi(t,x,y,z) = i\hbar\frac{\partial\Psi(t,x,y,z)}{\partial t},$$

or

$$-\frac{\hbar^2}{2m}\nabla^2\Psi(t,\mathbf{r}) + \Pi(t,\mathbf{r})\Psi(t,\mathbf{r}) = i\hbar\frac{\partial\Psi(t,\mathbf{r})}{\partial t}.$$

The Schrödinger equation is: (1) consistent with the de Broglie–Einstein postulates $p = h/\lambda$ and $v = E/h$; (2) consistent with total, kinetic and potential energies, e.g., $E = p^2/2m + \Pi$; and (3) linear in $\Psi(t,\mathbf{r})$.

The Schrödinger equation should be solved using normalizing, boundary, and continuity conditions in order to find the wave function. In general, $\Psi(t,\mathbf{r})$ is a nonlinear function of energy, mass, etc. The probability of finding a particle within a volume V is $\int_V \Psi^*(t,\mathbf{r})\Psi(t,\mathbf{r})\,dV$, where $\Psi^*(t,\mathbf{r})$ is the complex conjugate of $\Psi(t,\mathbf{r})$. The wave function is normalized as $\int_{-\infty}^{\infty}\Psi^*(t,\mathbf{r})\Psi(t,\mathbf{r})\,dV = 1$, where in the Cartesian coordinate system $dV = dx\,dy\,dz$.

The time evolution of the system's states is defined by the wave function. The basic connection between the properties of $\Psi(t,\mathbf{r})$ and the behavior of the associated particle is expressed by the probability density $P(t,\mathbf{r})$. For example, the quantity $P(t,x)$ specifies the probability, per unit length, of finding the particle near x at time t. Thus, $P(t,x) = \Psi^*(t,x)\Psi(t,x)$. For a physical observable C that has an associated operator $\hat{C}$, the average expectation value of the observable is $\langle C \rangle = \int \Psi^*(t,\mathbf{r})\hat{C}\Psi(t,\mathbf{r})\,dV$. The following momentum and energy operators $p \leftrightarrow -i\hbar(\partial/\partial x)$ and $E \leftrightarrow i\hbar(\partial/\partial t)$ are applied. In general, for a momentum one has $p \leftrightarrow -i\hbar\nabla$.

For a given probability density $P(t,x)$, the expected values of any function of x can be derived. In particular,

$$\langle f(x) \rangle = \int_{-\infty}^{\infty} f(x)P(t,x)\,dx = \int_{-\infty}^{\infty} \Psi^*(t,x)f(x)\Psi(t,x)\,dx \ .$$

For example, the expectation values of x and x^2 are

$$7 \quad \langle x \rangle = \int_{-\infty}^{\infty} xP(t,x)dx = \int_{-\infty}^{\infty} \Psi^*(t,x)x\Psi(t,x)\,dx$$

and

$$\langle x^2 \rangle = \int_{-\infty}^{\infty} x^2 P(t,x)dx = \int_{-\infty}^{\infty} \Psi^*(t,x)x^2\,\Psi(t,x)\,dx.$$

For a 1-D case the expectation values of the momentum and total energy are

$$\langle p \rangle = \int_{-\infty}^{\infty} \Psi^*(t,x)\left(-i\hbar\frac{\partial}{\partial x}\right)\Psi(t,x)\,dx = -i\hbar\int_{-\infty}^{\infty}\Psi^*(t,x)\frac{\partial\Psi(t,x)}{\partial x}\,dx$$

and

$$\langle E \rangle = \int_{-\infty}^{\infty}\Psi^*(t,x)\left(i\hbar\frac{\partial}{\partial t}\right)\Psi(t,x)\,dx = i\hbar\int_{-\infty}^{\infty}\Psi^*(t,x)\frac{\partial\Psi(t,x)}{\partial t}\,dx = \int_{-\infty}^{\infty}\Psi^*(t,x)\left(-\frac{\hbar^2}{2m}\frac{\partial^2}{\partial x^2}+\Pi(t,x)\right)\Psi(t,x)\,dx.$$

For $f(p)$, we have

$$\langle f(p) \rangle = \int_{-\infty}^{\infty}\Psi^*(t,x)f\left(-i\hbar\frac{\partial}{\partial x}\right)\Psi(t,x)\,dx.$$

For example, one finds

$$\langle p^2 \rangle = \int_{-\infty}^{\infty}\Psi^*(t,x)\left(-i\hbar\frac{\partial}{\partial x}\right)^2\Psi(t,x)\,dx = -\hbar^2\int_{-\infty}^{\infty}\Psi^*(t,x)\frac{\partial^2\Psi(t,x)}{\partial x^2}dx.$$

For any dynamic quantity which is a function of x and p, e.g., $f(t,x,p)$, the expectation value is

$$\langle f(t,x,p) \rangle = \int_{-\infty}^{\infty}\Psi^*(t,x)f\left(t,x,-i\hbar\frac{\partial}{\partial x}\right)\Psi(t,x)\,dx\ .$$

As an illustration, for a potential $\Pi(t,x)$, we have

$$\langle \Pi(t,x) \rangle = \int_{-\infty}^{\infty} \Psi^*(t,x)\Pi(t,x)\Psi(t,x)\, dx .$$

Example 7.11

Let the wave function for the lowest energy state for a free particle be

$$\Psi(t,x) = \begin{cases} A\cos\dfrac{\pi x}{L} e^{-(iE/\hbar)t} & \text{for } -\tfrac{1}{2}L < x < \tfrac{1}{2}L \\ 0 & \text{for } x \leq -\tfrac{1}{2}L, x \geq \tfrac{1}{2}L \end{cases} .$$

As will be documented later, we consider a particle in a 1-D potential well with $\Pi(x) = 0$ in $-L/2 < x < L/2$, and $\Pi(x) = \infty$ otherwise.

One finds the total energy E by using the Schrödinger equation, which is

$$-\frac{\hbar^2}{2m}\frac{\partial^2 \Psi}{\partial x^2} = i\hbar\frac{\partial \Psi}{\partial t} \quad \text{for } -L/2 < x < L/2.$$

The expressions for the spatial and time derivatives are

$$\frac{\partial \Psi}{\partial x} = -\frac{\pi}{L}A\sin\frac{\pi x}{L}e^{-(iE/\hbar)t},$$

$$\frac{\partial^2 \Psi}{\partial x^2} = -\frac{\pi^2}{L^2}A\cos\frac{\pi x}{L}e^{-(iE/\hbar)t} = -\frac{\pi^2}{L^2}\Psi$$

and

$$\frac{\partial \Psi}{\partial t} = -\frac{iE}{\hbar}A\cos\frac{\pi x}{L}e^{-(iE/\hbar)t} = -\frac{iE}{\hbar}\Psi.$$

Thus, the Schrödinger equation gives

$$\frac{\hbar^2}{2m}\frac{\pi^2}{L^2}\Psi = -i\hbar\frac{iE}{\hbar}\Psi.$$

Therefore,

$$E = \frac{\pi^2\hbar^2}{2mL^2}.$$

The expectation values of x and x^2 are found by making use of

$$\langle x \rangle = \int_{-\infty}^{\infty} xP(t,x)\, dx = \int_{-\infty}^{\infty} \Psi^*(t,x)x\,\Psi(t,x)\, dx$$

and

$$\left\langle x^2 \right\rangle = \int_{-\infty}^{\infty} x^2 P(t,x)\, dx = \int_{-\infty}^{\infty} \Psi^*(t,x) x^2 \Psi(t,x)\, dx\,.$$

Taking note of $\Psi(t,x)$, we have

$$\left\langle x \right\rangle = \int_{-\frac{1}{2}L}^{\frac{1}{2}L} A\cos\frac{\pi x}{L} e^{(iE/\hbar)t} x A\cos\frac{\pi x}{L} e^{-(iE/\hbar)t}\, dx = A^2 \int_{-\frac{1}{2}L}^{\frac{1}{2}L} x\cos^2\frac{\pi x}{L}\, dx = 0$$

and

$$\left\langle x^2 \right\rangle = \int_{-\frac{1}{2}L}^{\frac{1}{2}L} A\cos\frac{\pi x}{L} e^{(iE/\hbar)t} x^2 A\cos\frac{\pi x}{L} e^{-(iE/\hbar)t}\, dx = A^2 \int_{-\frac{1}{2}L}^{\frac{1}{2}L} x^2\cos^2\frac{\pi x}{L}\, dx = 2A^2 \int_{0}^{\frac{1}{2}L} x^2\cos^2\frac{\pi x}{L}\, dx$$

$$= 2A^2 \frac{L^3}{\pi^3} \int_{-\frac{1}{2}L}^{\frac{1}{2}\pi} \left(\frac{\pi x}{L}\right)^2 \cos^2\frac{\pi x}{L}\, d\frac{\pi x}{L} = A^2 \frac{L^3}{24\pi^2}(\pi^2 - 6).$$

The wave function should be normalized, and the amplitude A can be found. One has

$$\int_{-\infty}^{\infty} \Psi^*(t,x)\Psi(t,x)\, dx = A^2 \int_{-\frac{1}{2}L}^{\frac{1}{2}L} \cos^2\frac{\pi x}{L}\, dx = 2A^2 \frac{L}{\pi} \int_{0}^{\frac{1}{2}\pi} \cos^2\frac{\pi x}{L}\, d\frac{\pi x}{L} = 2A^2 \frac{L}{\pi}\frac{\pi}{4}.$$

By normalizing the wave function as $\int_{-\infty}^{\infty} \Psi^*(t,x)\Psi(t,x)\, dx = 1$, we obtain $A = \sqrt{2/L}$. Hence,

$$\left\langle x^2 \right\rangle = \frac{2}{L}\frac{L^3}{24\pi^2}(\pi^2 - 6) = \frac{L^2}{12\pi^2}(\pi^2 - 6),$$

which gives the fluctuations of particles about the average, and the root-mean-square value is $\sqrt{\left\langle x^2 \right\rangle}$.

From

$$\left\langle p^2 \right\rangle = -\hbar^2 \int_{-\infty}^{\infty} \Psi^*(t,x)\frac{\partial^2 \Psi(t,x)}{\partial x^2}\, dx\,,$$

one has

$$\left\langle p^2 \right\rangle = \hbar^2 \frac{\pi^2}{L^2} \int_{-\infty}^{\infty} \Psi^*(t,x)\Psi(t,x)\, dx = \frac{\hbar^2 \pi^2}{L^2}.$$

Thus, the root-mean-square momentum is $\sqrt{\left\langle p^2 \right\rangle} = \pi\hbar/L$, and $\sqrt{\left\langle p^2 \right\rangle}$ represents the average momentum fluctuations about the average $\left\langle p \right\rangle = 0$. By making use of $E = \pi^2\hbar^2/2mL^2$, from $p = \pm\sqrt{2mE}$, one concludes that the magnitude of momentum is $\pi\hbar/L$.

For a 1-D problem, the probability current density $J(t,x)$ is given as

$$J(t,x) = \frac{i\hbar}{2m}\left(\Psi(t,x)\frac{\partial \Psi^*(t,x)}{\partial x} - \Psi^*(t,x)\frac{\partial \Psi(t,x)}{\partial x}\right).$$

The probability of finding a particle in the region $a < x < vb$ at time t is

$$P_{ab}(t) = \int_a^b \Psi^*(t,x)\Psi(t,x)dx,$$

and

$$\frac{dP_{ab}}{dt} = J(t,a) - J(t,b)$$

For the probability density $P(t,x) = \Psi^*(t,x)\Psi(t,x)$ one finds

$$\frac{\partial \rho(t,x)}{\partial t} + \frac{\partial J(t,x)}{\partial x} = 0.$$

Let the solution of the Schrödinger equation be $\Psi(t,x) = e^{-i(E/\hbar)t}\Psi(x)$. The probability density does not depend on time, $dP_{ab}/dt = 0$, and $J(t,x) = $ const. For example, if $\Psi(x) = A\,e^{ikx}$, we have $P_{ab} = |A|^2(b-a)$ and $P = |A|^2$. Hence,

$$J = \frac{\hbar k}{m}|A|^2 = \frac{\hbar k}{m}P.$$

For a 3-D problem, we have

$$\frac{\partial P(t,\mathbf{r})}{\partial t} + \nabla \cdot \mathbf{J}(t,\mathbf{r}) = 0.$$

Here, the probability density and probability current density are $P(t,\mathbf{r}) = \Psi^*(t,\mathbf{r})\Psi(t,\mathbf{r})$ and

$$\mathbf{J}(t,\mathbf{r}) = \frac{i\hbar}{2m}[\Psi(t,\mathbf{r})\nabla\Psi^*(t,\mathbf{r}) - \Psi^*(t,\mathbf{r})\nabla\Psi(t,\mathbf{r})].$$

Probability Current Density and Current Density. It must be emphasized that the probability current density $\mathbf{J}(t,\mathbf{r})$ and the current density $\mathbf{j}$ are entirely different variables. In semiconductor devices, one of the basic equations is $\mathbf{j} = Q\mathbf{v}$, where Q is the charge density; $\mathbf{v}$ is the velocity of the charge carrier (electron or hole) which is found by making use of the applied potential, electric field, and other quantities. Taking note of the volume charge density ρ_v, one has $\mathbf{j} = \rho_v\mathbf{v}$. Electric charges in motion constitute a current. As charged particles move from one region to another within a *conducting* path, electric potential energy is transformed. The current through the closed surface is $I = \oint_s \mathbf{j}\cdot d\mathbf{s}$ and $I = dQ/dt$. The current density in electronic devices is the number of electrons crossing a unit area per unit time $N_s\bar{v}_x$ (the unit for N_s is [electrons/cm^2]) multiplied by the electron charge. For a 1-D case $j_x = -eN\bar{v}_x$ or $j_x = -e\sum_i\bar{v}_{xi}$. Here, the average net velocity is found using the average momentum per electron, $\bar{v}_x = \bar{p}_x/m$. In contrast, in quantum mechanics, $\mathbf{J}(t,\mathbf{r})$ represents the rate of probability changes, allowing one to estimate $\langle p \rangle$ which is found using $\Psi(t,\mathbf{r})$.

Remark. Textbook [15] thoroughly reports the device physics and application of the basic laws to straightforwardly obtain and examine the steady-state and dynamic characteristics of FETs, BJTs, and other solid-state electronic devices. The deviations are straightforward, and some well-known basics are briefly reported below. For FETs, one may find the total charge in the channel Q and the *transit* time t which gives the time that it takes an electron to pass between source and drain. Thus, the drain-to-source current is $I_{DS} = Q/t$. The electron velocity is $\mathbf{v} = -\mu_n\mathbf{E}_E$, where μ_n is the electron mobility; $\mathbf{E}_E$ is the electric field intensity. One also has $\mathbf{v} = \mu_p \mathbf{E}_E$, where μ_p is the hole mobility. At room temperature for intrinsic silicon μ_n and μ_p reach ~1400 and ~450 cm²/V·s, respectively. It should be emphasized that μ_n and μ_p are functions of the field intensity, voltages and other quantities, therefore the *effective* μ_{neff} and μ_{peff} are used. Using the x component of the electric field, we have $E_{Ex} = -V_{DS}/L$, where L is the channel length. Thus, $v_x = -\mu_n E_{Ex}$, and $t = L/v_x = L^2/\mu_n V_{DS}$. The channel and the gate form a parallel capacitor with plates separated by an insulator (gate oxide). From $Q = CV$, taking note that the charge appears when the voltage between the gate and the channel V_{GC} exceeds the n-channel threshold voltage V_t, one has $Q = C(V_{GC} - V_t)$. Using the equation for a parallel-plate capacitor with length L, width W and plate separation equal to the gate-oxide thickness T_{ox}, the gate capacitance is $C = WL\varepsilon_{ox}/T_{ox}$, where ε_{ox} is the gate-oxide dielectric permittivity, and for silicon dioxide SiO_2, ε_{ox} is ~3.5 × 10⁻¹¹ F/m. We briefly reported the baseline equations in deriving the size-dependent quantities, such as current, capacitance, velocity, *transit* time, etc. Furthermore, the analytic equations for the $I-V$ characteristics for FETs and BJTs are straightforwardly obtained and reported in Ref. [15]. The derived expressions for the so-called Level 1 model of nFETs in the *linear* and *saturation* regions are

$$I_D = \mu_n \frac{\varepsilon_{ox}}{T_{ox}} \frac{W_c}{L_c - 2L_{GD}} \left[(V_{GS} - V_t)V_{DS} - \tfrac{1}{2}V_{DS}^2 \right] (1 + \lambda V_{DS}) \quad \text{for} \quad V_{GS} \geq V_t, \ V_{DS} < V_{GS} - V_t$$

and

$$I_D = \tfrac{1}{2}\mu_n \frac{\varepsilon_{ox}}{T_{ox}} \frac{W_c}{L_c - 2L_{GD}} (V_{GS} - V_t)^2 (1 + \lambda V_{DS}) \quad \text{for } V_{GS} \geq V_t, \ V_{DS} \geq V_{GS} - V_t.$$

where I_D is the drain current; V_{GS}, V_{DS} are the gate source and drain source voltages; L_c and W_c are the channel length and width; L_{GD} is the gate-drain overlap; λ is the channel length modulation coefficient. For pFETs, in the equations for I_D one uses μ_p. The coefficients and parameters used to calculate the characteristics of nFETs and pFETs are different. Due to distinct device physics, phenomena exhibited, and effects utilized, the foundations of semiconductor devices are not applicable to ^{ME}devices. For example, the electron velocity and $I-V$ characteristics can be found using $\Psi(t,\mathbf{r})$ which depends on 3-D $\mathbf{E}(\mathbf{r})$ as documented in this section.

For the potential barriers documented in Figures 7.34a and 7.34b, one studies a tunneling problem examining the incident and reflected wave function amplitudes. As shown, for $\Psi(x) = A\,e^{ikx} + B\,e^{-ikx}$, one has $J = (\hbar k/m)(|A|^2 - |B|^2)$, which can be defined as a difference between incident and reflected probability current densities, e.g., $J = J_I - J_R$. The reflection coefficient is $R = J_R/J_I = |B|^2/|A|^2$. One may find the velocity and probability density of incoming, injected, and backward electrons. The potential can vary as a function of the applied voltage (voltage bias is $\Delta V = V_L - V_R$), electric field, transitions, and other factors. Using the potential difference $\Delta\Pi$, the variation of a piecewise continuous energy potential barrier $\Pi(x)$ is shown in Figure 7.34a. The analysis of wave function and current (if $E < \Pi$ or $E > \Pi$) is of a specific interest. One may examine electrons that move from the region of negative values of coordinate x to the region of positive values of x. At x_{Lj} and x_{Rj} electrons encounter intermediate finite potentials Π_{0j} with width L_j (Figures 7.34a and 7.34b). At the left and right (x_{L1} and x_{RN}), the finite potentials are denoted as Π_{0L} and Π_{0R}. There is a finite probability for transmission and reflection. The electrons on the left side that occupy the energy levels E_n can tunnel through the barrier to occupy empty energy levels E_n on the right side. The currents have contribution from all electrons.

Consider the Schrödinger equation:

$$-\frac{\hbar^2}{2m}\frac{d^2\Psi(x)}{dx^2} + \Pi(x)\Psi(x) = E\Psi(x)$$

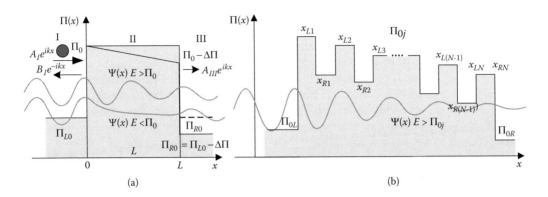

FIGURE 7.34 Electron tunneling through finite potential barriers: (a) single potential barrier; (b) multiple potential barriers.

which is given as a second-order differential equation

$$\frac{d^2\Psi(x)}{dx^2} = -k^2(x)\Psi(x)$$

to be numerically solved. Here,

$$k^2(x) = \frac{2m}{\hbar^2}[E - \Pi(x)].$$

The Euler approximation is used to represent the first spatial derivative as a first difference, e.g.,

$$\frac{d\Psi(x)}{dx} \approx \frac{\Psi_{n+1} - \Psi_n}{\Delta_h},$$

where Δ_h is the spatial discretization spacing. Thus, the Schrödinger equation can be numerically solved through discretization applying high-performance software. For example, MATLAB provides one with distinct application-specific differential equation solvers. Various discretization formulas and methods can be utilized. The Numerov three-point-difference expression is

$$\frac{d^2\Psi(x)}{dx^2} \approx \frac{\Psi_{n+1} - 2\Psi_n + \Psi_{n-1}}{\Delta_h^2}.$$

From

$$\frac{d^2\Psi(x)}{dx^2} = -k^2(x)\Psi(x),$$

one obtains a simple recursive equation

$$\Psi_{n+1} = \frac{2\left(1 - \frac{5}{12}k_n^2\Delta_h^2\right)\Psi_n - \left(1 + \frac{1}{12}k_{n-1}^2\Delta_h^2\right)\Psi_{n-1}}{1 + \frac{1}{12}k_{n+1}^2\Delta_h^2}.$$

Assigning initial values for Ψ_{n-1} and Ψ_n (for example, Ψ_0 and Ψ_1), the value of Ψ_{n+1} is derived. The *forward* or *backward* calculations of Ψ_i are performed with the accuracy $0(\Delta_h^6)$. The initial values of Ψ_{n-1} and Ψ_n can be assigned using the boundary conditions. One assigns and refines a *trial* energy E_n guaranteeing stability and convergence of the solution.

Using the Numerov three-point-difference expression, the Schrödinger equation is discretized as

$$\frac{\hbar^2}{2m}\left(\frac{(\Psi_{n+1}-\Psi_n)-(\Psi_n-\Psi_{n-1})}{\Delta_h^2}\right)-\Pi_n\Psi_n+E_n\Psi_n=0.$$

Using the Hamiltonian matrix $\mathbf{H}\ \mathbb{R}^{(N+2)\times(N+2)}$, vector $\boldsymbol{\Psi}\in\mathbb{R}^{N+2}$ that contains Ψ_i, and the source vector $\mathbf{Q}\in\mathbb{R}^{N+2}$, the following matrix equation $(E\mathbf{I}-\mathbf{H})\boldsymbol{\Psi}=\mathbf{Q}$ should be solved. Here, $\mathbf{I}\in\mathbb{R}^{(N+2)\times(N+2)}$ is the identity matrix. For a two-terminal ^{ME}device, the entities of the diagonal matrix $\mathbf{H}$ are

$$H_{n,n}=-\frac{\hbar^2}{2m\Delta_h^2}+\Pi_n$$

except $H_{0,0}$ and $H_{(N+1)(N+1)}$ which depend on the self-energies that account for the interconnect interactions. By taking note of notations used for the incoming wave function $\Psi(x)=Ae^{ik_Lx}+Be^{-ik_Lx}$, which leads to $\Psi_{-1}=Ae^{-ik_L\Delta_h}+Be^{ik_L\Delta_h}=Ae^{-ik_L\Delta_h}+(\Psi_0-A)e^{ik_L\Delta_h}$ and $\Psi_{N+2}=\Psi_{N+1}e^{ik_R\Delta_h}$ one has

$$H_{0,0}=-\frac{\hbar^2}{m\Delta_h^2}\left(1+\tfrac{1}{2}e^{ik_L\Delta_h}\right)+\Pi_0$$

and

$$H_{(N+1),(N+1)}=-\frac{\hbar^2}{m\Delta_h^2}\left(1+\tfrac{1}{2}e^{ik_R\Delta_h}\right)+\Pi_{N+1}.$$

Hence, the solution of the Schrödinger equation is reduced to the solution of linear algebraic equation. The probability current density is

$$J=\frac{i\hbar}{2m}\left(\Psi_n\frac{\Psi_{n+1}^*-\Psi_n^*}{\Delta_h}-\Psi_n^*\frac{\Psi_{n+1}-\Psi_n}{\Delta_h}\right).$$

7.12.5 Quantum Mechanics and Molecular Electronic Devices: Three-Dimensional Problem

The electron transport in ^{ME}devices must be examined in 3-D applying quantum mechanics. The time-independent Schrödinger equation

$$-\frac{\hbar^2}{2m}\nabla^2\Psi(\mathbf{r})+\Pi(\mathbf{r})\Psi(\mathbf{r})=E(\mathbf{r})\Psi(\mathbf{r})$$

can be solved in different coordinate systems depending on the problem under consideration. In the Cartesian system we have

$$\nabla^2\Psi(\mathbf{r})=\nabla^2\Psi(x,y,z)=\frac{\partial^2\Psi}{\partial x^2}+\frac{\partial^2\Psi}{\partial y^2}+\frac{\partial^2\Psi}{\partial z^2},$$

while in the cylindrical and spherical systems one solves

$$\nabla^2 \Psi(\mathbf{r}) = \nabla^2 \Psi(r,\phi,z) = \frac{1}{r}\frac{\partial}{\partial r}\left(r\frac{\partial \Psi}{\partial r}\right) + \frac{1}{r^2}\frac{\partial^2 \Psi}{\partial \phi^2} + \frac{\partial^2 \Psi}{\partial z^2}$$

and

$$\nabla^2 \Psi(\mathbf{r}) = \nabla^2 \Psi(r,\theta,\phi) = \frac{1}{r^2}\frac{\partial}{\partial r}\left(r^2\frac{\partial \Psi}{\partial r}\right) + \frac{1}{r^2 \sin\theta}\frac{\partial}{\partial \theta}\left(\sin\theta\frac{\partial \Psi}{\partial \theta}\right) + \frac{1}{r^2 \sin^2\theta}\frac{\partial^2 \Psi}{\partial \phi^2}.$$

The solution of the Schrödinger equation is obtained by using different analytical and numerical methods. The analytical solution can be found by using the separation of variables. For example, if the potential is $\Pi(x,y,z) = \Pi_x(x) + \Pi_y(y) + \Pi_z(z)$, one has $[H_x(x) + H_y(y) + H_z(z)]\Psi(x,y,z) = E\Psi(x,y,z)$, where the Hamiltonians are

$$H_x(x) = -\frac{\hbar^2}{2m}\frac{\partial^2}{\partial x^2} + \Pi_x(x), \quad H_y(y) = -\frac{\hbar^2}{2m}\frac{\partial^2}{\partial y^2} + \Pi_y(y) \quad and \quad H_z(z) = -\frac{\hbar^2}{2m}\frac{\partial^2}{\partial z^2} + \Pi_z(z).$$

The wave function is given as a product of three functions $\Psi(x,y,z) = X(x)Y(y)Z(z)$. This results in

$$\left[-\frac{\hbar^2}{2m}\frac{1}{X(x)}\frac{d^2 X(x)}{dx^2} + \Pi_x(x)\right] + \left[-\frac{\hbar^2}{2m}\frac{1}{Y(y)}\frac{d^2 Y(y)}{dy^2} + \Pi_y(y)\right] + \left[-\frac{\hbar^2}{2m}\frac{1}{Z(z)}\frac{d^2 Z(z)}{dz^2} + \Pi_z(z)\right] = E,$$

where the constant total energy is $E = E_x + E_y + E_z$. The separation of variables technique results in reduction of the 3-D Schrödinger equation to three independent 1-D equations, e.g.,

$$\left[-\frac{\hbar^2}{2m}\frac{d^2}{dx^2} + \Pi_x(x)\right]X(x) = E_x X(x),$$

$$\left[-\frac{\hbar^2}{2m}\frac{d^2}{dy^2} + \Pi_y(y)\right]Y(y) = E_y Y(y),$$

and

$$\left[-\frac{\hbar^2}{2m}\frac{d^2}{dz^2} + \Pi_z(z)\right]Z(z) = E_z Z(z).$$

The cylindrical and spherical systems can be effectively used to reduce the complexity and make the problem tractable. In the spherical system, one uses $\Psi(r,\theta,\phi) = R(r)Y(\theta,\phi)$. The Schrödinger partial differential equation is solved using the continuity and boundary conditions, and the wave function is normalized as $\int_V \Psi^*(\mathbf{r})\Psi(\mathbf{r})\,dV = 1$.

Example 7.12

Consider a microscopic particle in an infinite spherical potential well, let

$$\Pi(r) = \begin{cases} 0 & \text{for } r \le a \\ \infty & \text{for } r > a \end{cases}.$$

The Schrödinger equation is

$$-\frac{\hbar^2}{2m}\left[\frac{1}{r^2}\frac{\partial}{\partial r}\left(r^2\frac{\partial\Psi}{\partial r}\right)+\frac{1}{r^2\sin\theta}\frac{\partial}{\partial\theta}\left(\sin\theta\frac{\partial\Psi}{\partial\theta}\right)+\frac{1}{r^2\sin^2\theta}\frac{\partial^2\Psi}{\partial\phi^2}\right]+\Pi(r,\theta,\phi)\Psi(r,\theta,\phi)=E\Psi(r,\theta,\phi).$$

We apply the separation of variables concept. The wave function is given as $\Psi(r,\theta,\phi)=R(r)Y(\theta,\phi)$. Outside the well, when $r>a$, the wave function is zero. The stationary states are labeled using three quantum numbers n, l, and m_l. Our goal is to derive the expression for $\Psi_{nlm_l}(r,\theta,\phi)$. The energy depends only on n and l, e.g., E_{nl}. In general, $\Psi_{nlm_l}(r,\theta,\phi)=A_{nl}S_{Bl}(s_{nl}r/a)Y_l^{m_l}(\theta,\phi)$, where A_{nl} is the constant which must be found through the normalization of wave function; S_{Bl} is the spherical Bessel function of order l,

$$S_{Bl}(x)=(-x)^l\left(\frac{1}{x}\frac{d}{dx}\right)^l\frac{\sin x}{x},$$

and for $l=0$ and $l=1$, we have $S_{B0}=\sin x/x$ and $S_{B1}=\sin x/x^2-\cos x/x$; s_{nl} is the nth zero of the lth spherical Bessel function.

Inside the well, the radial equation is

$$\frac{d^2u}{dr^2}=\left(\frac{l(l+1)}{r^2}-k^2\right)u,\quad k^2=\frac{2mE}{\hbar^2}.$$

The general solution of this equation for an arbitrary integer l is $u(r)=ArS_{Bl}(kr)+BrS_{Nl}(kr)$, where S_N is the spherical Neumann function of order l,

$$S_{Nl}(x)=-(-x)^l\left(\frac{1}{x}\frac{d}{dx}\right)^l\frac{\cos x}{x},$$

and for

$$l=0\quad\text{and}\quad l=1,\text{ one finds }S_{N0}=-\cos x/x\quad\text{and}\quad S_{N1}=-\cos x/x^2-\sin x/x.$$

The radial wave function is $R(r)=u(r)/r$. We use the boundary condition $u(a)=0$. For $l=0$, from $d^2u/dr^2=-k^2u$, we have $u(r)=A\sin kr+B\cos kr$, where $B=0$. Taking note of the boundary condition, from $\sin ka=0$, one obtains $ka=n\pi$. The normalization of $u(r)$ gives $A=\sqrt{2/a}$.

The angular equation is

$$\sin\theta\frac{\partial}{\partial\theta}\left(\sin\theta\frac{\partial Y}{\partial\theta}\right)+\frac{\partial^2 Y}{\partial\phi^2}=-l(l+1)\sin^2\theta Y.$$

By applying $Y(\theta,\phi)=\Theta(\theta)\Phi(\phi)$, the normalized angular wave function (spherical harmonics) is known to be

$$Y_l^{m_l}(\theta,\phi)=\gamma\sqrt{\frac{2l+1}{4\pi}\frac{(l-|m_l|)!}{(l+|m_l|)!}}e^{im_l\phi}L_l^{m_l}(\cos\theta),$$

where $\gamma=(-1)^{m_l}$ for $m_l\geq0$ and $\gamma=1$ for $m_l\leq0$; $L_l^{m_l}(x)$ is the Legendre function,

$$L_l^{m_l}(x)=(1-x^2)^{\frac{1}{2}|m_l|}\left(\frac{d}{dx}\right)^{|m_l|}L_l(x);$$

$L_l(x)$ is the *l*th Legendre polynomial,

$$L_l(x) = \frac{1}{2^l l!} \left(\frac{d}{dx} \right)^l (x^2 - 1)^l.$$

Thus, the angular component of the wave function for $l = 0$ and $m_l = 0$ is

$$Y_0^0(\theta, \phi) = \frac{1}{\sqrt{4\pi}}.$$

Hence,

$$\Psi_{n00} = \frac{1}{\sqrt{2\pi a}} \frac{1}{r} \sin \frac{n\pi r}{a},$$

and the allowed energies are $E_{n0} = (\pi^2 \hbar^2 / 2ma^2)n^2$, $n = 1, 2, 3, \dots$. Using the *n*th order of the *l*th spherical Bessel function S_{Bnl}, the allowed energies are $E_{nl} = (\pi^2 \hbar^2 / 2ma^2)S_{Bnl}^2$.

The Schrödinger differential equation is numerically solved in all regions for the specified potentials, energies, potential widths, boundaries, etc. For 3-D-topology ^{ME}devices, using potentials, tunneling paths, interatomic bond lengths and other data, having found $\Psi(t,\mathbf{r})$, one obtains, $P(t,\mathbf{r})$, $T(t,E)$, expected values of variables, and other quantities of interest. For example, having found the velocity (or momentum) of a charged particle as a function of control variables (time-varying external electric or magnetic field) and parameters (mass, interatomic lengths, permittivity, etc.), the electric current is derived. As documented, the particle momentum, velocity, transmission coefficient, traversal time, and other variables change as functions of the time-varying external electromagnetic field. Therefore, depending on the device physics varying, for example, $\mathbf{E}(t\mathbf{r})$ or $\mathbf{B}(t\mathbf{r})$, one controls the electron transport. Different dynamic and steady-state characteristics are examined. For example, the steady-state experimental *I–V* and *G–V* characteristics are derived using theoretical fundamentals reported.

For the planar solid-state semiconductor devices, to derive the transmission coefficient $T(E)$, the Green function $G(E)$ has been used. In particular, we have $T(E) = tr[E_{BL}G(E)E_{BR}G^*(E)]$. To obtain the *I–V* characteristics, one self-consistently solves the coupled transport and Poisson's equations [7,58]. The Poisson equation $\nabla \cdot (\varepsilon(\mathbf{r})\nabla V(\mathbf{r})) = -\rho(\mathbf{r})$ is solved to find the electric field intensity and electrostatic potential. Here, $\rho(\mathbf{r})$ is the charge density, which is not a probability current density $\rho(t,\mathbf{r})$; $\varepsilon(\mathbf{r})$ is the dielectric tensor. For example, letting $\rho_x = \rho_0 \mathrm{sech}(x/L)\tanh(x/L)$ we solve $\nabla^2 V_x = -\rho_x/\varepsilon$ obtaining the following expressions

$$E_x = -\frac{\rho_0}{\varepsilon} L \mathrm{sech} \frac{x}{L}$$

and

$$V_x = 2\frac{\rho_0}{\varepsilon} L^2 \left(\tan^{-1} e^{x/L} - \frac{1}{4}\pi \right).$$

For 3-D-topology ^{ME}devices, the Poisson equation is of great importance to carry out the self-consistent solution. The Schrödinger and Poisson equations are solved utilizing robust numerical methods using the difference expressions for the Laplacian, integration–differentiation concepts, etc. It is possible to solve differential equations in 3-D using a finite difference method that gives lattices. Generalizing the results reported for the 1-D problem, for the Laplace equation one has

$$\frac{\partial^2 V(i,j,k)}{\partial^2 r} = \frac{V(i+1,j,k) - 2V(i,j,k) + V(i-1,j,k)}{\Delta_h^2},$$

where (i, j, k) gives a grid point; Δ_h is the spatial discretization spacing in the x, y, or z directions. For Poisson's equation, we have

$$\nabla \cdot (\varepsilon(\mathbf{r}) \nabla V(\mathbf{r})) = \frac{C_{i,j,k}^{i+1,j,k}(V_{i+1,j,k} - V_{i,j,k}) - C_{i,j,k}^{i,j,k}(V_{i,j,k} - V_{i-1,j,k})}{\Delta_x^2} + \frac{C_{i,j,k}^{i,j+1,k}(V_{i,j+1,k} - V_{i,j,k}) - C_{i,j,k}^{i,j,k}(V_{i,j,k} - V_{i,j-1,k})}{\Delta_y^2}$$

$$+ \frac{C_{i,j,k}^{i,j,k+1}(V_{i,j,k+1} - V_{i,j,k}) - C_{i,j,k}^{i,j,k}(V_{i,j,k} - V_{i,j,k-1})}{\Delta_z^2}, \quad C_{l,m,n}^{i,j,k} = \frac{2\varepsilon_{i,j,k}\varepsilon_{l,m,n}}{\varepsilon_{i,j,k} + \varepsilon_{l,m,n}}.$$

Thus, using the number of grid points, equation $\nabla \cdot (\varepsilon(\mathbf{r}) \nabla V(\mathbf{r})) = -\rho(\mathbf{r})$ is represented and solved as $\mathbf{AV} = \mathbf{B}$, where $\mathbf{A} \in \mathbb{R}^{N \times N}$ is the matrix; $\mathbf{B} \in \mathbb{R}^N$ is the vector of the boundary conditions. The self-consistent problem that integrates the solution of the Schrödinger (gives wave function, energy, etc.) and Poisson (provides the potential) equations is solved updating the potentials and other variables obtained through iterations. The convergence is enforced and specified accuracy is guaranteed by applying robust numerical methods.

7.12.6 Multiterminal Quantum-Effect ᴹᴱDevices

Quantum-well resonant tunneling diodes and FETs, Schottky-gated resonant tunneling, heterojunction bipolar, resonant tunneling bipolar, and other transistors have been introduced to enhance the microelectronic device performance. The tunneling barriers are formed using AlAs, AlGaAs, AlInAs, AlSb, GaAs, GaSb, GaAsSb, GaInAs, InP, InAs, InGaP, and other composites and spacers with the thickness in the range from 1 nm to tens of nm. The CMOS-technology high-speed double-heterojunction bipolar transistors ensure the cutoff frequency ~300 GHz, breakdown voltage ~5 V, and current density ~1 × 10^5 A/cm². The 1-D potential energy profile, shown in Figure 7.35, schematically depicts the first barrier (L_1, L_2), the well region (L_2, L_3), and the second barrier (L_3, L_4) with the quasi-Fermi levels E_{F1}, E_{F23}, and E_{F2}, respectively. The device physics of these transistors is reported [15], and the electron transport in double-barrier single-quantum-well is straightforwardly examined by applying a self-consistent approach and numerically solving the 1-D or 2-D Schrödinger and Poisson equations.

The ᴹAND and ᴹNAND gates were documented in Figure 7.17 utilizing multiterminal ᴹᴱdevices to form ᴹgates. Figure 7.36 illustrates the overlapping molecular orbitals for cyclic molecules used to implement these ᴹgates.

In Ref. [4], we reported 3-D-topology multiterminal ᴹᴱdevices formed using cyclic molecules with a carbon interconnecting framework. In this section, consider a three-terminal ᴹᴱdevice with the *input*, *control*, and *output* terminals as shown in Figure 7.37. The device physics of the proposed ᴹᴱdevice is based on the quantum interaction and controlled electron tunneling. The applied $V_{control}(t)$ changes the charge distribution $\rho(t,\mathbf{r})$ and $E_E(t,\mathbf{r})$ affecting the electron transport. This ᴹᴱdevice operates in the controlled electron-exchangeable environment due to quantum interactions. The controlled super-fast potential-assisted tunneling is achieved. The electron-exchangeable environment interactions qualitatively and quantitatively modify the device behavior and its characteristics. Consider the electron transport in the time- and spatial-varying metastable potentials $\Pi(t,\mathbf{r})$. From the quantum theory viewpoint, it is evident that the changes in the Hamiltonian result in: (1) changes of tunneling $T(E)$; (2) quantum

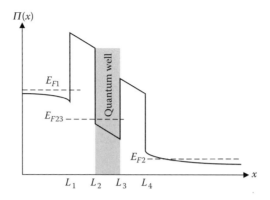

FIGURE 7.35 One-dimensional potential energy profile and quasi-Fermi levels in the double-barrier single-well heterojunction transistors.

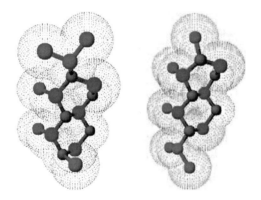

FIGURE 7.36 ^{M}NAND and ^{M}AND gates comprised from cyclic molecules.

FIGURE 7.37 Three-terminal ^{ME}device comprised from a cyclic molecule with a carbon interconnecting framework.

interactions due to variations of $\rho(t,\mathbf{r})$, $E_E(t,\mathbf{r})$, and $\Pi(t,\mathbf{r})$. The device controllability is ensured by varying $V_{control}(t)$ that affects the device switching, $I–V$ and other characteristics.

We solve high-fidelity modeling and data-intensive analysis problems for the studied ^{ME}device. For heterojunction microelectronic devices, one usually solves the 1-D Schrödinger and Poisson equations applying the Fermi–Dirac distribution function. In contrast, for the devised ^{ME}devices, a 3-D problem arises which cannot be simplified. Furthermore, the distribution functions and statistical mechanics postulates may not be straightforwardly applied.

For the studied cyclic molecule which forms an interconnected ^{ME}device, we consider 9 atoms with motionless protons with charges q_i. The radial Coulomb potentials are

$$\Pi_i(r) = -\frac{Z_{eff\,i}\,q_i^2}{4\pi\varepsilon_0 r}.$$

For example, for carbon $Z_{eff\,C} = 3.14$. Using the spherical coordinate system, the Schrödinger equation

$$-\frac{\hbar^2}{2m}\left[\frac{1}{r^2}\frac{\partial}{\partial r}\left(r^2\frac{\partial\Psi}{\partial r}\right) + \frac{1}{r^2\sin\theta}\frac{\partial}{\partial\theta}\left(\sin\theta\frac{\partial\Psi}{\partial\theta}\right) + \frac{1}{r^2\sin^2\theta}\frac{\partial^2\Psi}{\partial\phi^2}\right] + \Pi(r,\theta,\phi)\Psi(r,\theta,\phi) = E\Psi(r,\theta,\phi)$$

should be solved. For the problem under our consideration, it is virtually impractical to find the analytic solution as obtained in Example 7.12 by using the separation of variables concept. We represented the wave function as $\Psi(r,\theta,\phi) = R(r)Y(\theta,\phi)$ in order to derive and solve the radial and angular equations. In contrast, we discretize the Schrödinger and Poisson equations, as reported in this section, with the ultimate objective to numerically solve these differential equations. The magnitude of the time-varying potential applied to the control terminal is bounded due to the thermal stability of the molecule, e.g., $|V_{control}| \le V_{control\ max}$. In particular, we let $|V_{control}| \le 0.25$ V. The charge distribution is of our particular interest. Figure 7.38 documents a 3-D charge distribution in the molecule if $V_{control} = 0.1$ V and $V_{control} = 0.2$ V. The total molecular charge distribution is found by summing the individual orbital densities.

The Schrödinger and Poisson equations are solved using a self-consistent algorithm in order to verify the device physics soundness and examine the baseline performance characteristics. To obtain the current density $\mathbf{j}$ and current in the ^{ME}device, the velocity and momentum of the electrons are obtained by making use of

$$\langle p \rangle = \int_{-\infty}^{\infty} \Psi^*(t,\mathbf{r})\left(-i\hbar\frac{\partial}{\partial\mathbf{r}}\right)\Psi(t,\mathbf{r})\,d\mathbf{r}.$$

The wave function $\Psi(t,\mathbf{r})$ is derived for distinct values of $V_{control}$. The $I–V$ characteristics of the studied ^{ME}device for two different control currents (0.1 and 0.2 nA) are reported in Figure 7.39. The results

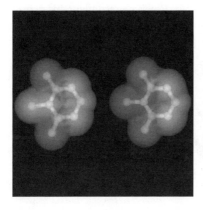
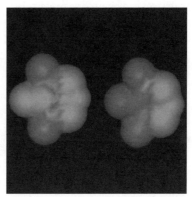

FIGURE 7.38 Charge distribution $\rho(\mathbf{r})$.

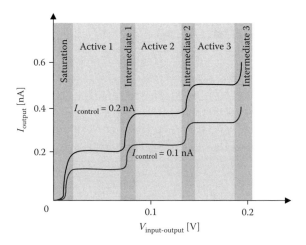

FIGURE 7.39 Multiple-valued *I–V* characteristics.

documented imply that the proposed ^{ME}device may be effectively used as a multiple-valued or symbolic ^{M}primitive in order to design enabling multiple-valued or symbolic logics and memories.

The traversal time of electron tunneling is derived from the expression

$$\tau(E) = \int_{r_0}^{r_f} \sqrt{\frac{m}{2[\Pi(\mathbf{r}) - E]}} \, d\mathbf{r}.$$

It is found that τ varies from 2.4×10^{-15} to 5×10^{-15} sec. Hence, the proposed ^{ME}device ensures superfast switching.

The reported monocyclic molecule can be used as a six-terminal ^{ME}device as illustrated in Figure 7.40. The proposed carbon-centered molecular hardware solution, in general,

- Ensures a sound *bottom-up* synthesis at the device, gate and module levels
- Guarantees aggregability to form complex ^{M}ICs
- Results in the experimentally characterizable ^{ME}devices and ^{M}gates

The use of the side groups R$_i$, shown in Figure 7.40, ensures the variations of the energy barriers and wells potential surfaces $\Pi(t,\mathbf{r})$. This results in the controlled electron transport and varying quantum interactions. As reported, the studied ^{ME}devices can be utilized in combinational and memory ^{M}ICs. In addition, those devices can be used as routers. Hence, one achieves a reconfigurable networking-processing-and-memory as covered in Section 7.6.4 for *fluidic* platforms. We conclude that neuromorphological reconfigurable *solid* ^{M}PPs can be designed using the proposed device-level solution.

FIGURE 7.40 Six-terminal ^{ME}devices.

7.13 Conclusions

Innovative developments in devising and design of novel ^{M}PPs were reported. A new hardware solution was proposed utilizing *solid* and *fluidic* molecular electronics. This leads to the application of novel 3-D-topology ^{M}devices. By using molecular technology (nanotechnology), novel 3-D organizations and enabling architectures were proposed and studied. For ^{M}PPs, enabling 3-D organizations and novel architectures were utilized at the module and system levels. A wide spectrum of fundamental, applied, and experimental issues, related to the device physics, phenomena, functionality, performance and capabilities, were researched. We advanced an ^{M}architectronics paradigm toward ^{M}PPs. Innovative concepts in SLSI design of ^{M}ICs and ^{M}PPs were reported and examined. The proposed ^{M}PPs ensure superior performance. Biomolecular processing platforms were examined in order to derive sound and feasible solutions for ^{M}PPs. Though a great number of problems remain to be solved, it was demonstrated that the proposed 3-D-centered topologies/organizations and novel architectures guarantee overall supremacy. Neuromorphological reconfigurable *solid* and *fluidic* ^{M}PPs were devised utilizing a $^{3\text{-}D}$networking-and-processing paradigm.

Acknowledgments

The author sincerely acknowledges partial support from *Microsystems and Nanotechnologies* under the U.S. Department of Defense, Department of the Air Force (Air Force Research Laboratory) contracts 8750024 and 8750058. *Disclaimer* — Any opinion, findings, and conclusions, or recommendations expressed in this chapter are those of the author and do not necessarily reflect U.S. Department of Defense, Department of the Air Force views. The software and CAD developments for a 3-D logic design, performed by Drs. S. Yanushkevich and V. Shmerko [43], are deeply appreciated.

References

1. Semiconductor Industry Association, *International Technology Roadmap for Semiconductors*, 2005 Edition, Semiconductor Industry Association, Austin, TX, 2006.
2. Brewer, J.E., Zhirnov, V.V. and Hutchby, J.A., Memory technology for the post CMOS era, *IEEE Circuits Devices Mag.*, 21(2), 13–20, 2005.
3. Ferry, D.K., Akis, R., Cummings, A., Gilbert, M.J. and Ramey, S.M., Semiconductor device scaling: physics, transport, and the role of nanowires, *Proceedings of IEEE Conference on Nanotechnology*, Cincinnati, OH, 2006.
4. Lyshevski, S.E. (Ed.), Three-dimensional molecular electronics and integrated circuits for signal and information processing platforms, in *Handbook on Nano and Molecular Electronics*, CRC Press, Boca Raton, FL, 2007.
5. Ellenbogen, J.C. and Love, J.C., Architectures for molecular electronic computers: logic structures and an adder designed from molecular electronic diodes, *Proc. IEEE*, 88(3), 386–426, 2000.
6. Heath, J.R. and Ratner, M.A., Molecular electronics, *Phys. Today*, 56(5), May, 43–49, 2003.
7. Lyshevski, S.E., *NEMS and MEMS: Fundamentals of Nano- and Microengineering*, CRC Press, Boca Raton, FL, 2005, 2nd ed.
8. Chen, J., Lee, T., Su, J., Wang, W., Reed, M.A., Rawlett, A.M., Kozaki, M., Yao, Y., Jagessar, R.C., Dirk, S.M., Price, D.W., Tour, J.M., Grubisha, D.S. and Bennett, D.W., Molecular electronic devices, in *Handbook of Molecular Nanoelectronics*, Reed, M.A. and Lee, L., Eds., American Science Publishers, 2003.
9. Tour, J.M. and James, D.K., Molecular electronic computing architectures, In *Handbook of Nanoscience, Engineering and Technology*, Goddard, W.A., Brenner, D.W., Lyshevski, S.E. and Lafrate, G.J., Eds., CRC Press, Boca Raton, FL, 2003, 4.1–4.28.
10. Wang, W., Lee, T., Kretzschmar, I. and Reed, M.A., Inelastic electron tunneling spectroscopy of an alkanedithiol self-assembled monolayer, *NanoLetters*, 4(4), 643–646, 2004.

11. Aviram, A. and Ratner, M.A., Molecular rectifiers, *Chem. Phys. Lett.*, 29, 277–283, 1974.

12. Berg, H.C., The rotary motor of bacterial flagella, *J. Annu. Rev. Biochem.*, 72, 19–54, 2003.

13. Sze, S.M., *Physics of Semiconductor Devices*, Wiley, New York, 1969.

14. Sze, S.M., *Physics of Semiconductor Devices*, Wiley, New York, 1981.

15. Sze, S.M. and Ng, K.K., *Physics of Semiconductor Devices*, Wiley, New York, 2007.

16. Lyshevski, S.E., Design of three-dimensional molecular integrated circuits and molecular architectronics, *Proceedings of IEEE Conference on Nanotechnology*, Cincinnati, OH, 2006.

17. Kaupp, U.B. and Baumann, A., Neurons — the molecular basis of their electrical excitability, in *Handbook of Nanoelectronics and Information Technology*, Waser, R., Ed., Wiley-VCH, Darmstadt, Germany, 2005, 147–164.

18. Churchland, P.S. and Sejnowski, T.J., *The Computational Brain*, MIT Press, Cambridge, MA, 1992.

19. Freeman, W., *Mass Action in the Nervous System*, Academic Press, New York, 1975.

20. Freeman, W., Tutorial on neurobiology from single neurons to brain chaos, *Int. J. Bifurcation Chaos*, 2(3), 451–482, 1992.

21. Laughlin, S., van Stevenink, R. and Anderson, J.C., The metabolic cost of neural computation, *Nat. Neurosci.* 1(1), 36–41, 1998.

22. Lyshevski, M.A. and Lyshevski, S.E., Fluidic nanoelectronics and Brownian dynamics, *Proceedings of NSTI Nanotechnology Conference*, Boston, MA, 2006, 3, 43–46.

23. Hameroff, S.R. and Tuszynski, J., Search for quantum and classical modes of information processing in microtubules: implications for "The Living State," in *Handbook on Bioenergetic Organization in Living Systems*, Musumeci, F. and Ho, M.-W., Eds., World Scientific, Singapore, 2003.

24. Lyshevski, M.A., Fluidic molecular electronics, *Proceedings of IEEE Conference on Nanotechnology*, Cincinnati, OH, 2006.

25. Dayan, P. and Abbott, L.F., *Theoretical Neuroscience: Computational and Mathematical Modeling of Neural Systems*, MIT Press, Cambridge, MA, 2001.

26. Grossberg, S., On the production and release of chemical transmitters and related topics in cellular control, *J. Theor. Biol.*, 22, 325–364, 1969.

27. Grossberg, S., *Studies of Mind and Brain*, Reidel, Amsterdam, 1982.

28. Abbott, L.F. and Regehr, W.G., Synaptic computation, *Nature*, 431, 796–803, 2004.

29. Markram, H., Lubke, J., Frotscher, M. and Sakmann, B., Regulation of synaptic efficacy by coincidence of postsynaptic APs and EPSPs, *Science*, 275, 213–215, 1997.

30. Rumsey, C.C. and Abbott, L.F., Equalization of synaptic efficacy by activity- and timing-dependent synaptic plasticity, *J. Neurophysiol.*, 91(5), 2004.

31. Grossberg, S., Birth of a learning law, *Neural Networks*, 11(1), 1968.

32. Frantherz, F., Neuroelectronics interfacing: semiconductor chips with ion channels, nerve cells, and brain, in *Handbook of Nanoelectronics and Information Technology*, Waser, R., Ed., Wiley-VCH, Darmstadt, Germany, 2005, 781–810.

33. Lyshevski, S.E., Molecular cognitive information-processing and computing platforms, *Proceedings of IEEE Conference on Nanotechnology*, Cincinnati, OH, 2006.

34. Renyi, A., On measure of entropy and information, *Proc. Berkeley Symp. Math. Statist. Prob.*, 1, 547–561, 1961.

35. Kabanov, Yu. M., The capacity of a channel of the Poisson type, *Theory Prob. Appl.*, 23, 143–147, 1978.

36. Johnson, D., Point process models of single-neuron discharges, *J. Comp. Neurosci.*, 3, 275–299, 1996.

37. Strong, S.P., Koberle, R., de Ruyter van de Steveninck, R.R. and Bialek, W., Entropy and information in neuronal spike trains, *Phys. Rev. Lett.*, 80(1), 1998.

38. Rieke, F., Warland, D., de Ruyter van Stevenink, R.R. and Bialek, W., *Spikes: Exploring the Neural Code*, MIT Press, Cambridge, MA, 1997.

39. Smith, K.C., Multiple-valued logic: a tutorial and appreciation, *Computer*, 21(4), 17–27, 1998.

40. Buitenweg, J.R., Rutten, W.L.C. and Marani, E., Modeled channel distributions explain extracellular recordings from cultured neurons sealed to microelectrodes, *IEEE Trans. Biomed. Eng.*, 49(11), 1580–1590, 2002.

41. Sigworth, F.J. and Klemic, K.G., Microchip technology in ion-channel research, *IEEE Trans. Nanobiosci.*, 4(1), 121–127, 2005.

42. Suzuki, H., Kato-Yamada, Y., Noji, H. and Takeuchi, S., Planar lipid membrane array for membrane protein chip, *Proc. Conf. MEMS*, 272–275, 2004.

43. Yanushkevich, S., Shmerko, V. and Lyshevski, S.E., *Logic Design of NanoICs*, CRC Press, Boca Raton, FL, 2005.

44. Malyugin, V.D., Realization of corteges of Boolean functions by linear arithmetical polynomials, *Automica Telemekhica*, (2), 114–121, 1984.

45. Lyshevski, S.E., Nanocomputers and Nanoarchitectronics, in *Handbook of Nanoscience, Engineering and Technology*, Goddard, W., Brenner, D., Lyshevski, S. and Iafrate, G., Eds., CRC Press, Boca Raton, FL, 2002, 6.1–6.39.

46. Porod, W., Nanoelectronic circuit architectures, in *Handbook of Nanoscience, Engineering and Technology*, Goddard, W., Brenner, D., Lyshevski, S. and Iafrate, G., Eds. CRC Press, Boca Raton, FL, 2003, 5.1–5.12.

47. Williams, S.R. and Kuekes, P.J., Molecular nanoelectronics, *Proceedings of the International Symposium on Circuits and Systems*, Geneva, Switzerland, 2000, 1, 5–7.

48. Kamins, T.I., Williams, R.S., Basile, D.P., Hesjedal, T. and Harris, J.S., Ti-catalyzed Si nanowires by chemical vapor deposition: microscopy and growth mechanism, *J. Appl. Phys.*, 89, 1008–1016, 2001.

49. Reichert, J., Ochs, R., Beckmann, D., Weber, H.B., Mayor, M. and Lohneysen, H.V., Driving current through single organic molecules, *Phys. Rev. Lett.*, 88(17), 2002.

50. Basch, H. and Ratner, M.A., Binding at molecule/gold transport interfaces. V. Comparison of different metals and molecular bridges, J. Chem. Phys., 119(22), 11926–11942, 2003.

51. Lee, K., Choi, J. and Janes, D.B., Measurement of *I–V* characteristic of organic molecules using step junction, *Proceedings of IEEE Conference on Nanotechnology*, Munich, Germany, 2004, 125–127.

52. Mahapatro, A.K., Ghosh, S. and Janes, D.B., Nanometer scale electrode separation (nanogap) using electromigration at room temperature, *Proc. IEEE Trans. Nanotechnol.*, 5(3), 232–236, 2006.

53. Carbone, A. and Seeman, N.C., Circuits and programmable self-assembling DNA structures, *Proc. Natl. Acad. Sci. USA*, 99(20), 12577–12582, 2002.

54. Porath, D., Cuniberti, G. and Di Felice, R., Charge transport in DNA-based devices, *Top. Curr. Chem.*, 237, 183–227, 2004.

55. Mahapatro, A.K., Janes, D.B., Jeong, K.J. and Lee, G.U., Electrical behavior of nano-scale junctions with well engineered double stranded DNA molecules, *Proceedings of IEEE Conference on Nanotechnology*, Cincinnati, OH, 2006.

56. Galperin, M. and Nitzan, A., NEGF-HF method in molecular junction property calculations, *Ann. NY Acad. Sci.*, 1006, 48–67, 2003.

57. Galperin, M., Nitzan, A. and Ratner, M.A., Resonant inelastic tunneling in molecular junctions, *Phys. Rev. B*, 73, 045314, 2006.

58. Paulsson, M., Zahid, F. and Datta, S., Resistance of a molecule, in *Handbook of Nanoscience, Engineering and Technology*, Goddard, W., Brenner, D., Lyshevski, S. and Iafrate, G., Eds., CRC Press, Boca Raton, FL, 2002, 12.1–12.25.

59. Landauer, R., Spatial variation of current and fields due to localized scatterers in metallic conduction, *IBM J.*, 1(3), 223–231, 1957. Reprinted in *IBM J. Res. Dev.*, 44(1/2), 251–259, 2000.

60. Büttiker, M., Quantized transmission of a saddle-point constriction, *Phys. Rev. B*, 41(11), 7906–7909, 1990.

61. Büttiker, M. and Landauer, R., Escape-energy distribution for particles in an extremely underdamped potential well, *Phys. Rev. B*, 30(3), 1551–1553, 1984.

62. Büttiker, M. and Landauer, R., Traversal time for tunneling, *Phys. Rev. Lett.*, 49(23), 1739–1742, 1982.

63. Galperin, M. and Nitzan, A., Traversal time for electron tunneling in water, *J. Chem. Phys.*, 114(21), 9205–9208, 2001.

64. Pariser, R. and Parr, R.G., A semiempirical theory of the electronic spectra and electronic structure of complex unsaturated molecules I, *J. Chem. Phys.*, 21, 466–471, 1953.

65. Pariser, R. and Parr, R.G., A semiempirical theory of the electronic spectra and electronic structure of complex unsaturated molecules II, *J. Chem. Phys.*, 21, 767–776, 1953.

66. Pople, J.A., Electron interaction in unsaturated hydrocarbons, *Trans. Faraday Soc.*, 49, 1375–1385, 1953.

8

Spin Field Effect Transistors

Supriyo Bandyopadhyay
Virginia Commonwealth University

Marc Cahay
University of Cincinnati

8.1 Introduction: The Spin Field Effect Transistor 8-1
8.2 Another SPINFET...8-3
8.3 Non-Idealities..8-4
8.4 Other Types of SPINFETs 8-4
8.5 What Motivates the Research in SPINFETs? 8-8
8.6 Can "Spin" Ever Produce a Power Advantage
Over "Charge"?..8-10
8.7 Conclusion 8-11
References 8-11

8.1 Introduction: The Spin Field Effect Transistor

In 1990, Datta and Das proposed a seminal concept that is the forbearer of all spin field effect transistors (SPINFETs)[1] [1]. They examined and analyzed a device structure identical to that of a MOSFET, except that the source and drain contacts are ferromagnetic. Figure 8.1 shows this structure. For simplicity, we will assume that the channel is a quantum wire, with only the lowest transverse subband occupied by carriers.

The two ferromagnetic contacts are magnetized along the direction of the channel and their magnetizations are parallel. One of them (the "source" contact) injects electrons into the channel with spins aligned along the direction of the source's magnetization, which, in this case, is the $+x$-direction. The spin injection efficiency is assumed to be 100%, so that every carrier has its spin aligned in the $+x$-direction. If there is no spin precession in the channel because of spin–orbit interaction, and no spin–flip interactions, then the injected carriers arrive at the "drain" contact with their spins still aligned in the original ($+x$) direction. The drain is a spin-selective transmitter, since it is also ferromagnetic. We will assume that it is a 100% efficient spin filter that passes only those carriers whose spins are aligned parallel to its magnetization (i.e., the $+x$-direction) and completely blocks carriers whose spins are antiparallel (i.e., pointing in the $-x$-direction). Since the arriving carriers have their spins aligned parallel to the drain's magnetization, the drain transmits all of them and current flows between the source and drain contacts.

---

[1] Datta and Das never termed their device a "transistor." With remarkable insight, they called their device an electronic analog of the electro-optic modulator, rather than a transistor. Although this device is now popularly known as a spin field effect transistor or the Datta–Das transistor, the original proponents still refrain from calling it a "transistor." This device (and its clones) may not perform as effective transistors because the ratio of the on-to-off conductance may be too low for mainstream applications, as we show in this chapter.

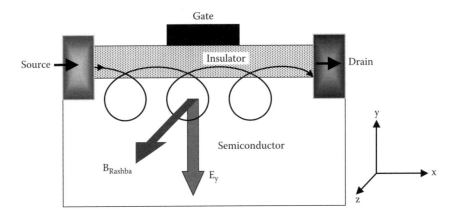

FIGURE 8.1 Structure of the spin field effect transistor. The helical path shows the spin orientation as a carrier travels from source to drain.

When an electrostatic potential is applied to the "gate" terminal, it induces an electric field transverse to the channel (in the y-direction). This electric field induces a Rashba spin–orbit interaction [2], which acts like an effective magnetic field that is oriented in a direction mutually perpendicular to the direction of current flow and the gate-induced electric field. Since we have a strictly 1-D channel, current flows only in the x-direction. Therefore, the effective magnetic field is directed along the z-direction. Because of the one-dimensionality of the channel, the axis of this magnetic field is fixed, and is always along the z-axis.

The strength of this magnetic field depends on the carrier's velocity[2] and is given by

$$B_{Rashba}(v) = \frac{2(m^*)^2 a_{46}}{e\hbar^2} E_y v \tag{8.1}$$

where v is the carrier velocity, E_y is the gate-induced electric field causing the Rashba interaction, m^* is the carrier effective mass, a_{46} is a material constant, and e is the electronic charge.

The spins of the injected carriers precess about this effective magnetic field (just like Larmor precession) with a frequency Ω given by the Larmor frequency

$$\Omega(v) = \frac{eB_{Rashba}(v)}{m^*} = \frac{2m^*}{\hbar^2} a_{46} E_y v \tag{8.2}$$

This precession takes place on the x–y plane since the axis of the magnetic field is along the z-direction. The rate at which the spin precesses in *space* can be obtained from the Larmor frequency

$$\Omega(v) = \frac{d\phi}{dt} = \frac{d\phi}{dx}\frac{dx}{dt} = \frac{d\phi}{dx} v = \frac{2m^*}{\hbar^2} a_{46} E_y v$$

$$\frac{d\phi}{dx} = \frac{2m^*}{\hbar^2} a_{46} E_y \tag{8.3}$$

where ϕ is the angle of spin precession.

---

[2] Some authors have assumed that this field is proportional to the electron's wavevector. This is not exactly correct since in the presence of spin orbit interaction, the velocity and wavevector are not proportional to each other.

Note that the spatial rate of spin precession $d\phi/dx$ is *independent* of the carrier velocity. Therefore, every electron, regardless of its injection velocity and any momentum-randomizing collision that it suffers in the channel[3], precesses by exactly the *same* angle as it traverses the distance between the source and the drain. This angle is given by

$$\Phi_{Rashba} = \frac{2m^*}{\hbar^2} a_{46} E_y L \tag{8.4}$$

where L is the source to drain separation (or the channel length).

If the electric field E_y is such that $\Phi_{Rashba} = (2n + 1)\pi$, where n is an integer, then the carriers arriving at the drain have their spins antiparallel to the drain's magnetization. These carriers are blocked and ideally no current flows. Thus, by changing E_y with a gate potential, one can change Φ_{Rashba} and modulate the source-to-drain current. This realizes field effect transistor action.

Note that this transistor can operate at elevated temperatures. High temperature will induce a thermal spread in the electron velocity and perhaps increase the rate of collisions that change an electron's velocity randomly, but it does not matter. Since Φ_{Rashba} is independent of electron velocity, thermal averaging has no effect on Φ_{Rashba} and therefore a high temperature does not degrade the performance of the SPINFET, as long as it has a 1-D channel.

8.2 Another SPINFET

In addition to the Rashba interaction, there can be other types of spin–orbit interaction in a semiconductor channel. An example is the Dresselhaus spin–orbit interaction [3], which exists in any material that lacks crystallographic inversion symmetry. This interaction also results in an effective magnetic field, just like the Rashba interaction. Assume that the channel of the transistor is in the [100] crystallographic direction. In that case, the effective magnetic field due to the Dresselhaus interaction will be directed along the x-axis and its strength will be given by

$$B_{Dresselhaus}(v) = \frac{2(m^*)^2 a_{42}}{e\hbar^2} \left[\left(\frac{\pi}{W_z} \right)^2 - \left(\frac{\pi}{W_y} \right)^2 \right] v \tag{8.5}$$

where W_z and W_y are the transverse dimensions of the quantum wire channel (assumed to be of rectangular cross section) and a_{42} is another material constant. Fortunately, this effective magnetic field is also proportional to the carrier velocity v.

The reader can easily understand that if we inject spins that are initially polarized along either the y- or z-axis[4], then these spins will precess about the effective magnetic field due to the Dresselhaus interaction, which is oriented along the x-axis. The precession takes place in the y–z plane. The angle by which the spin precesses in traveling between the source and the drain will be given by (compare with Equation [8.4])

$$\Phi_{Dresselhaus} = \frac{2m^* a_{42}}{\hbar^2} \left[\left(\frac{\pi}{W_z} \right)^2 - \left(\frac{\pi}{W_y} \right)^2 \right] L \tag{8.6}$$

This angle is independent of the carrier velocity. We can change $\Phi_{Dresselhaus}$ by varying W_z with a split gate potential. That will also realize transistor action and was the basis of another SPINFET proposed by us [4]. This device has all the advantages of the original SPINFET, namely that since $\Phi_{Dresselhaus}$ is independent

---

[3] Such collisions would change the electron's velocity randomly.
[4] This will require magnetizing the source and drain contacts along the y- or z-axis.

of carrier velocity, thermal averaging has no deleterious effect. Accordingly, this transistor is also able to operate at elevated temperatures, without any serious degradation in performance.

8.3   Non-Idealities

If both Rashba and Dresselhaus interactions are present, then the total effective magnetic field is the vector sum of the individual fields:

$$\mathbf{B}_{total} = B_{Dresselhaus}\hat{\mathbf{x}} + B_{Rashba}\hat{\mathbf{z}} \tag{8.7}$$

where $\hat{\mathbf{x}}$ and $\hat{\mathbf{z}}$ are the unit vectors along the x- and z-axes. This resultant field lies in the x–z plane and subtends an angle θ with the x-axis (channel axis) given by

$$\tan\theta = \frac{B_{Rashba}}{B_{Dresselhaus}} = \frac{a_{46}}{a_{42}} \frac{E_y}{\left[\left(\dfrac{\pi}{W_z}\right)^2 - \left(\dfrac{\pi}{W_y}\right)^2\right]} \tag{8.8}$$

Note that this angle is also independent of carrier velocity. Hence the axis of the effective magnetic field is again the same for every electron, at any fixed values of E_y, W_z, and W_y. If we inject spins with polarization normal to this axis, they will precess about this axis (on a plane normal to this axis) as they travel from the source to the drain. The precession angle will be given by (compare with Equation [8.4] and Equation [8.6])

$$\Phi = \frac{2m^*L}{\hbar^2}\sqrt{\left(a_{46}E_y\right)^2 + a_{42}^2\left(\left[\frac{\pi}{W_z}\right]^2 - \left[\frac{\pi}{W_y}\right]^2\right)^2} \tag{8.9}$$

We can change Φ by changing E_y with a gate potential, but doing that also changes the angle θ (see Equation [8.8]), and, therefore, the axis of the effective magnetic field will change. Thus, the precession plane will change if we change the gate voltage, unlike in the previous two cases. This is a complicated effect. The reader will understand that if we have the source and drain contacts magnetized in the same direction, then the current is never completely blocked at any gate voltage, and a large leakage will flow during the off state. Therefore, the simultaneous presence of both Rashba and Dresselhaus interactions is not desirable.

The ferromagnetic contacts can also induce a real magnetic field in the channel. This field is not proportional to the carrier velocity (it is a constant), and therefore the angle by which a spin precesses about it, as the carrier travels from the source to the drain, will depend on the carrier velocity. In that case, different electrons will precess by different angles, so that thermal averaging at high temperatures will reduce the current modulation significantly. In other words, both the on-to-off conductance ratio and the transconductance of the transistor will decrease. The magnetic field can also promote spin–flip scattering in the presence of spin–orbit interaction. These damaging effects have been discussed in Ref. [5].

8.4   Other Types of SPINFETs

A number of other SPINFETs have also been proposed. One of them — the so-called "nonballistic SPINFET" — works on the following principle:

Consider a 2-D semiconductor channel in the x–z plane. The Hamiltonian describing this system is

$$H = \frac{p_x^2 + p_y^2 + p_z^2}{2m^*} + V(y) - \frac{\eta}{\hbar}[p_x\sigma_z - p_z\sigma_x] - \frac{v}{\hbar}[p_x\sigma_x - p_z\sigma_z] \tag{8.10}$$

where η and v are the strengths of the Rashba and Dresselhaus interactions, respectively, the σ-s are the Pauli spin matrices, and $V(y)$ is the confining potential in the y-direction.

Since the Hamiltonian is invariant in the coordinates x and y, the wavevectors k_x and k_y are good quantum numbers and we can write the spatial part of the wavefunction as

$$\psi(x,y,z) = e^{ik_x x} e^{ik_z z} \lambda_n(y) \tag{8.11}$$

Using this wavefunction, we calculate the spatial average of the Hamiltonian as

$$\langle H \rangle = \varepsilon_n + \frac{\hbar^2}{2m^*}\left(k_x^2 + k_z^2\right) - \eta[k_x \sigma_z - k_z \sigma_x] - v[k_x \sigma_x - k_z \sigma_z] \tag{8.12}$$

where ε_n is the subband level in the quantum well.

Writing this Hamiltonian explicitly in terms of the Pauli matrices, we get

$$\langle H \rangle = \varepsilon_n + \frac{\hbar^2}{2m^*}\left(k_x^2 + k_z^2\right) - \eta\begin{bmatrix} k_x & 0 \\ 0 & -k_x \end{bmatrix} + \eta\begin{bmatrix} 0 & k_z \\ k_z & 0 \end{bmatrix} - v\begin{bmatrix} 0 & k_x \\ k_x & 0 \end{bmatrix} + v\begin{bmatrix} k_z & 0 \\ 0 & -k_z \end{bmatrix}$$

$$= \begin{bmatrix} E_n - \eta k_x + v k_z & \eta k_z - v k_x \\ \eta k_z - v k_x & E_n + \eta k_x - v k_z \end{bmatrix} \tag{8.13}$$

where $E_n = \varepsilon_n + \dfrac{\hbar^2}{2m^*}\left(k_x^2 + k_z^2\right)$.

Diagonalizing this Hamiltonian yields the dispersion relations of the eigenenergies and the eigenspinors:

$$E_{\pm}(n,k_x,k_y) = E_n \pm \sqrt{(\eta k_x - v k_z)^2 + (v k_x - \eta k_z)^2} \tag{8.14}$$

Thus, there are two spin–split subbands with eigenspinors

$$\varphi_+(k_x,k_y) = \begin{bmatrix} -\sin(\phi_k) \\ \cos(\phi_k) \end{bmatrix}$$

$$\varphi_-(k_x,k_y) = \begin{bmatrix} \cos(\phi_k) \\ \sin(\phi_k) \end{bmatrix} \tag{8.15}$$

where

$$\phi_k = \frac{1}{2}\arctan\left[\frac{-\eta k_z + v k_x}{\eta k_x - v k_z}\right] \tag{8.16}$$

Note that since ϕ_k is wavevector dependent, the eigenspinors in Equation (8.15) are also wavevector dependent. Therefore, the spin–split subbands whose dispersion relations are given in Equation (8.14) do not have a fixed spin quantization axis. The spin orientation depends on the wavevector. This situation is depicted in Figure 8.2(a).

But now consider the special case when $\eta = v$. In this case, $\phi_k = \pi/8$ so that the eigenspinors become wavevector independent. In that case, each spin–split subband will have a fixed spin quantization axis

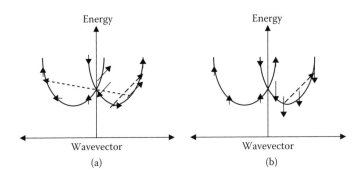

FIGURE 8.2 Energy dispersion relations of spin–split subbands showing the spin orientations at various wavevector states: (a) when $\eta \neq v$ and (b) when $\eta = v$.

and these axes will be antiparallel in the two subbands since the eigenspinors are orthogonal. This situation is shown in Figure 8.2(b).

When $\eta \neq v$, the eigenspinors in both subbands are wavevector dependent. Then, it is always possible to find two states in the two subbands at different wavevector values whose eigenspinors are not orthogonal, meaning that the spin orientations in these two states are not antiparallel. In that case, a momentum relaxing collision, caused by, say, a nonmagnetic impurity or a phonon, can induce a transition between these two states and thereby change the spin polarization since the spin orientations of the initial and final states are different. Therefore, any momentum-relaxing scattering will also relax spin. However, when $\eta = v$, the eigenspinors in the two subbands are perfectly orthogonal *at any wavevector state*. Therefore, it is impossible to have any intersubband scattering since the matrix element (and therefore the probability) for such a transition is exactly zero. It is of course possible to have intrasubband transition since the initial and final states have parallel spins, but such a transition does not relax spin at all because the spin orientations of the initial and final states are parallel. Therefore, phonons and nonmagnetic impurities cannot relax spin in the special situation when $\eta = v$.

Note that when $\eta = v$, the eigenspinors are

$$\varphi_{+} = \begin{bmatrix} -\sin(\pi/8) \\ \cos(\pi/8) \end{bmatrix}$$

$$\varphi_{-} = \begin{bmatrix} \cos(\pi/8) \\ \sin(\pi/8) \end{bmatrix}$$

The components of the eigenspin along the x-, y- and z-axis are found by

$$S_x = \begin{bmatrix} \varphi_{\pm} \end{bmatrix} \begin{bmatrix} \sigma_x \end{bmatrix} \begin{bmatrix} \varphi_{\pm} \end{bmatrix} = \pm \sin(\pi/4) = \pm \frac{1}{\sqrt{2}}$$

$$S_y = \begin{bmatrix} \varphi_{\pm} \end{bmatrix} \begin{bmatrix} \sigma_y \end{bmatrix} \begin{bmatrix} \varphi_{\pm} \end{bmatrix} = 0 \qquad\qquad (8.17)$$

$$S_z = \begin{bmatrix} \varphi_{\pm} \end{bmatrix} \begin{bmatrix} \sigma_x \end{bmatrix} \begin{bmatrix} \varphi_{\pm} \end{bmatrix} = \pm \cos(\pi/4) = \pm \frac{1}{\sqrt{2}}$$

Therefore, the eigenspin's polarization vector lies in the x–z plane and subtends an angle of 45˚ with the x- or z-axis.

The so-called nonballistic SPINFET proposed by two groups independently [6,7] works as follows. The device has exactly the same structure as the one in Figure 8.1, except that the source and drain contacts are magnetized (parallel to each other) in the x–z plane in a direction that subtends an angle of 45° with the x- or z-axis. This source injects spins into the channel in an eigenstate if the gate voltage is tuned to make $\eta = v$. Note that v is independent of the gate voltage, but η depends on it since $\eta = a_{46}E_y$. We assume that the spin injection efficiency is 100%, so that *every* spin, without exception, is injected in an eigenstate. The injected electrons do not suffer spin–flip interactions by nonmagnetic impurities or phonons, as already explained, and therefore arrive at the drain with their spins aligned along the drain's magnetization. These electrons are all transmitted by the drain and the current is a maximum. This is the "on" state of the transistor.

To turn the device off, the gate voltage is detuned to make $\eta \neq v$. Then, nonmagnetic impurities and phonons can induce spin flip. As a result, many of the electrons reaching the drain will have their spins flipped. They will be completely blocked by the drain (assuming that the drain is a 100% efficient spin filter) and the current will drop. This is interpreted as the "off" state.

A little bit of reflection will convince the reader that the maximum ratio of the on-to-off conductance is only 2. That is because, when spins are flipped randomly in the channel, what can happen at best is that 50% of the spins arriving at the drain will have their spins antiparallel to the drain's magnetization (blocked) while the remaining 50% will have their spins parallel (transmitted). Thus, the off current is no less than 50% of the on current, so that the maximum conductance ratio is 2. An actual simulation carried out by Safir et al. [8] found that the conductance ratio in realistic scenarios is not even 2, but only ~1.2. Transistors require a conductance modulation of about 10^5 for mainstream applications [9]. Therefore, this "transistor" is not suitable for any such application.

One can obviously improve the conductance ratio by making a minor alteration to the design. Let us consider the situation when the source and drain are magnetized in the *antiparallel* configuration, instead of the parallel configuration. In that case, when $\eta = v$, the drain blocks every electron and the current is ideally zero. When the gate voltage is changed to make $\eta \neq v$, spins flip and the drain now transmits the flipped spins. Then, the transmitted current will be nonzero. With the antiparallel arrangement, the conductance ratio is *ideally* infinite, which would have been excellent, but in reality it is far from that. The real ratio is quite small since it is impossible to inject or filter spins with 100% efficiency. We show this below.

Let us assume that the spin injection efficiency at the source contact is ξ. Then, by definition

$$\xi = \frac{I_{maj} - I_{min}}{I_{total}}$$

where I_{maj} is the current due to majority spins injected at the source, I_{min} is the current due to minority spins, and $I_{total} = I_{maj} + I_{min}$. Therefore, $I_{min} = [(1 - \xi)/2] I_{total}$. Assuming (generously) that the drain is a 100% efficient spin filter, I_{min} will be the leakage current (I_{off}) that flows through the transistor when it is "off." When the transistor is "on," the current that flows is due to flipped spins. Since, at best, 50% of the spins flip, the maximum possible value of the on current is $I_{on} = (1/2)I_{total}$. Consequently, the *maximum* on-to-off current ratio is

$$\frac{I_{on}}{I_{off}} = \frac{1}{1-\xi}$$
(8.18)

In order for this ratio to equal 10^5, which is needed for today's transistors, we must have $\xi = 99.999\%$. Furthermore, if we consider the fact that the drain cannot be a perfect spin filter either, then the required spin injection efficiency ξ will be even higher.

It is unlikely that we can achieve spin injection efficiency this high at room temperature. Not only is this impossible in the near term, it may be *forever* impossible, since there are fundamental barriers to ~100% spin injection efficiency, particularly at room temperature.

There are two known routes to achieving high spin injection efficiency: (a) using highly spin-polarized half metals as the ferromagnetic spin injector and (b) using spin-selective barriers that inject spins of a particular polarization only [10]. Unfortunately, there can be *no* half metals with 100% spin polarization at any temperature above absolute zero. Ref. [11] has shown that all half metals lose their high degree of spin polarization at temperature $T > 0$ K because of magnons and phonons. Even at $T = 0$ K, there are no ideal half metals with 100% spin polarization because of surfaces and inhomogeneities [11]. Therefore, half metals will not achieve $\sim$100% spin injection efficiency, even at 0 K, let alone room temperature. Consequently, half metals are not a viable route.

Spin-selective barriers can at best transmit one kind of spin at one specific injection energy. The best spin-selective barriers use resonant tunneling [10]. At 0 K, the transmission energy bandwidth can approach zero, so that nearly 100% spin injection efficiency is possible in principle, but at any nonzero temperature, thermal broadening of the carrier energy will ensure that the spin injection efficiency is far less than 100%. Therefore, this route will not work either at room temperature.

The highest spin injection efficiency demonstrated at or near room temperature is only about 70% [12] and at very low temperatures, a spin injection efficiency of $\sim$90% has been shown to be possible [13]. If we use these two values for ξ in Equation (8.18), then the maximum on-to-off ratio of the conductance is only 3.3 and 10, which is still a far cry from the 10^5 required. Therefore, these spin transistors are not viable as "transistors."

The reader will understand that the same problem afflicts the device of Ref. [1]. In fact, this is a generic problem that afflicts *all* SPINFETs that we are aware of.

Neither Ref. [1] nor Ref. [6,7] ever made the claim that their devices are competitive with the silicon MOSFET, which is the workhorse of electronics. The authors of Ref. [1], in particular, carefully avoided calling their device a "transistor." However, claims have been recently made by a group [14,15] that its SPINFET will be superior to a MOSFET. This group has proposed a device very similar (almost identical) to the device of Ref. [6,7] with the sole difference being that the source and drain contacts are magnetized in the antiparallel configuration, rather than the parallel configuration, in an effort to increase the conductance ratio. The way this device works is as follows. The source supposedly injects spins polarized parallel to the source's magnetization. When the gate voltage is zero, the spin orbit interaction in the channel is small and the injected spins do not flip. The drain therefore blocks them and the current is low. This is the off state. When the gate voltage is turned up, the spin orbit interaction increases and the spins flip more frequently. Then, a significant fraction of the spins (the flipped spins) arriving at the drain is transmitted and the current rises. This is the mode of switching the transistor from the off to the on state.

It should be obvious to the reader that the maximum conductance ratio of this device is given by Equation (8.18). If we generously assume a spin injection efficiency of even 90% at room temperature, then the maximum conductance ratio of this device is only 10, which immediately makes it noncompetitive with the silicon MOSFET by a long shot. We pointed this out in Ref. [16] to which the authors of Ref. [15] responded by saying that even though $\sim$ 100% spin injection efficiencies are not possible today, they may become possible in the future [17]. We find no basis for such optimism particularly in view of Ref. [11], which has shown that no $\sim$ 100% spin-polarized sources can even exist at any nonzero temperature. Consequently, $\sim$ 100% spin injection efficiency seems like an unachievable goal, not just today, but also in the long term. We pointed this out in Ref. [18].

8.5 What Motivates the Research in SPINFETs?

Research in SPINFETs, or for that matter any nontraditional transistor, is motivated by a desire to overcome the fundamental limitations of MOSFETs. The MOSFET has fundamental limits on speed and power. It is switched from the on state to the off state (or vice versa) by moving carriers into and out of the channel with a gate voltage. This physical motion of charges consumes a lot of energy and that leads to high levels of power dissipation in the device. There is no way to eliminate this dissipation. There is also a limit on the switching speed. No matter what we do, the shortest switching delay in the MOSFET

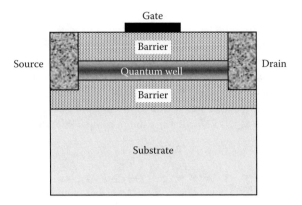

FIGURE 8.3 The structure of the VMT.

is the time it takes to move charges in and out, which is the transit time of carriers through the channel. That is limited by the minimum channel length achievable with the technology of the time and the maximum carrier velocity.

There is a great desire to overcome these limitations by adopting alternate strategies to modulate current. In the early 1980s, an idea took hold that came to be known as the "velocity modulation transistor" (VMT) [19,20]. It is based on the following concept.

The conductance of the channel in a transistor is

$$G = en_s\mu \tag{8.19}$$

where n_s is the surface carrier concentration and μ is the carrier mobility. Normally, the MOSFET is switched by changing n_s with the gate voltage. But can we change μ instead? In that case, we will not be moving charges around, thereby saving energy. The VMT does precisely that. The structure of the VMT is shown in Figure 8.3.

The electrons reside in the quantum well, which is a low bandgap material. The barriers are made of high bandgap materials. When a gate voltage is applied, the wavefunction of an electron in the quantum well is skewed towards one of the interfaces with the barrier layer because of electrostatic attraction or repulsion. The interface is rough and when electrons are close to it, they suffer increased elastic scattering that decreases the carrier mobility μ. Thus, the gate voltage can modulate the carrier mobility and alter the channel conductance without changing the carrier concentration. This will modulate the source-to-drain current. The hope was that since there is no change in n_s, one can effect switching in a time shorter than the transit time through the channel, and expend less power to switch.

The mobility of carriers changes in a time equal to the momentum relaxation time, which will be the switching time of a VMT. In the 1980s the transit time through the channel (which was relatively long at that time) was probably a lot longer than the momentum relaxation time, so that the VMT idea appeared attractive. Today, with 10 nm gate length MOSFETs in research production [21] (and perhaps soon to be in commercial production), the transit time is comparable to, or even less than, the momentum relaxation time. Therefore, the VMT is no longer particularly attractive from speed considerations. As far as power considerations go, the voltage that must be applied to the gate to change the mobility appreciably is not much smaller, if at all, than the voltage required to change the carrier concentration by several orders of magnitude. Thus, there is no power advantage either. But these are not the reasons why the VMT fell by the wayside. The real reason why this device never took hold is simply that while n_s can be modulated by several orders of magnitude with a gate voltage, μ can be modulated by only one order of magnitude at best. In Ref. [19], μ could be modulated by only 56%, or a factor of 1.56. Therefore, the conductance ratio of the VMT would have been no more than 10. This is of the same order that we expect for SPINFETs.

It should be obvious now that SPINFETs have an intrinsic similarity with the VMT. The channel conductance in a SPINFET is not modulated by changing the carrier concentration, but by changing the spin polarization. Does this take less energy? The answer is unfortunately *no*, as we have shown in Ref. [22]. The problem is that the gate voltage in a spin transistor changes the spin polarization of carriers by affecting *spin orbit interaction*. The gate voltage dependence of spin orbit interaction in the conduction band of most semiconductors is very weak, so that a lot of gate voltage is required to induce sufficient change in the spin polarization to turn a transistor from on to off, or vice versa. The ratio of the gate voltages required to switch the SPINFET of Ref. [1] and a comparable MOSFET was shown to be [22]

$$\frac{V_{SPINFET}}{V_{MOSFET}} \approx \frac{\hbar^2 \pi e}{2m^* \gamma E_F L} \tag{8.20}$$

where E_F is the Fermi energy in the channel and γ is the rate of change of spin orbit interaction strength with gate voltage ($\gamma = \partial \eta / \partial V_G$ where V_G is the gate voltage).

For realistic values of E_F and γ, the above ratio was shown to be smaller than unity only if the channel length of the transistor is several micrometers long [22]. Thus, no submicron SPINFET, of the type in Ref. [1], has any advantage over a comparable MOSFET in terms of power dissipation.

The same conclusion holds for the SPINFET of Ref. [15]. We have shown that if identical structures are used either as a SPINFET or as a ballistic silicon MOSFET, then the MOSFET would require less voltage to turn on and off than the SPINFET. Ref. [17] has disputed this analysis by claiming that our comparison assumes low temperature, while the SPINFET in Ref. [15] can supposedly work at room temperature. We disagree. The device of Ref. [15] requires greater than 99.999% spin injection efficiency to possess the claimed conductance ratio of 10^5, and that is nearly impossible to achieve even at 0 K, let alone room temperature. Thus, in any case, the low temperature comparison is the only one that is realistic and meaningful.

When all is said and done, SPINFETs, generally speaking, do not appear to have any power advantage over MOSFETs.

What about the switching delay of the spin transistor? In the device of Ref. [1], the minimum switching delay is either the transit time, or the time it takes to precess the spin by π radians, whichever is longer. In the device of Ref. [15], the minimum switching delay is the transit time or the spin relaxation time, whichever is longer. We can always make the transit time very small by exploiting velocity overshoot effects and also making the channel short. Therefore, the bottleneck will be the spin relaxation time or the spin precession time, both of which depend on the strength of the spin orbit interaction. Thus, once again, spin transistors require a very strong spin orbit interaction to be fast.

The switching delay of the MOSFET is limited by the transit time of carriers through the channel. Since the switching delay of the SPINFET is either the spin precession/relaxation time or the transit time, *whichever is longer*, therefore, the switching delay of the SPINFET is always equal to or longer than the switching delay of a comparable MOSFET. Thus, there is no speed advantage with SPINFETs. Of course, in a real circuit, the switching delay is most likely going to be determined by the gate charging time (RC time constant), which is a lot longer than the transit time or spin precession/relaxation time. The gate charging time is mostly determined by the peripherals (parasitic elements, interconnects, etc.) and has little relevance to the device itself.

8.6 Can "Spin" Ever Produce a Power Advantage Over "Charge"?

We have discussed the energy dissipated in switching a transistor from off to on, or vice versa. That is typically the energy dissipated in charging the gate, which is $(1/2)C_G V_G^2$, where C_G is the gate capacitance and V_G is the gate voltage required to turn the transistor on or off. In addition to this energy cost, there is also energy dissipated in the channel when the transistor is on. This energy is $I_{on}^2 R_C$ where I_{on} is the channel current and R_C is the channel resistance. Both SPINFETs and MOSFETs dissipate this energy since both sustain a channel current during the on state.

In order to truly reduce dissipation, one should eliminate current flow altogether. This requires a departure from the traditional transistor paradigm and embracing alternate ways to perform signal processing. "Spin" unlike "charge" is a pseudo vector and has both a magnitude (fixed) and a polarization (variable). One can encode logic bits 0 and 1 in the up- and down-spin polarizations, provided spin polarization can be made bistable. This can be achieved by placing an electron in a magnetic field, so that its spin can point either parallel or antiparallel to the field. That was the basis of the single spin logic idea [23], which represents binary digital information with spin polarization, and implements logic circuits by engineering the exchange interaction between spins. We have recently shown that these circuits are capable of extremely low power dissipation [24]. Therefore, "spin" can provide a significant power advantage, but probably not within the traditional transistor paradigm. Some "thinking outside the box" will be required to realize truly low-power spin devices.

8.7 Conclusion

In this article, we have presented the tempered view of SPINFETs. We have shown that the two major problems that these transistors countenance are low spin injection efficiency at a ferromagnet/semiconductor interface and weak spin orbit interaction in semiconductors. The spin orbit interaction may be stronger in the valence band of semiconductors than in the conduction band, so that p-channel SPINFETs may have an advantage over n-channel SPINFETs. However, if these two problems, which are both fundamental materials issues, can be overcome, then SPINFETs may become serious contenders against MOSFETs.

References

1. Datta, S. and Das, B., *Appl. Phys. Lett.*, 56, 665 (1990).
2. Rashba, E.I., *Sov. Phys. Semicond.*, 2, 1109 (1960); Bychkov, Y.A. and Rashba, E.I., *J. Phys. C.*, 17, 6039 (1984).
3. Dresselhaus, G., *Phys. Rev.*, 100, 580 (1955).
4. Bandyopadhyay, S. and Cahay, M., *Appl. Phys. Lett.*, 85, 1814 (2004).
5. Cahay, M. and Bandyopadhyay, S., *Phys. Rev. B*, 69, 045303 (2004); Cahay, M. and Bandyopadhyay, S., *Phys. Rev. B*, 68, 115316 (2003); Bandyopadhyay, S. and Cahay, M., *Physica E*, 25, 399 (2005).
6. Schliemann, J., Egues, J.C., and Loss, D., *Phys. Rev. Lett.*, 90, 146801 (2003).
7. Cartoixa, X., Ting, D.Z-Y, and Chang, Y.C., *Appl. Phys. Lett.*, 83, 1462 (2003).
8. Ehud Safir, Min Shen, and Semion Siakin, *Phys. Rev. B*, 70, 241302(R) (2004).
9. The International Technology Roadmap for Semiconductors, published by the Semiconductor Industry Association.
10. Koga, T., Nitta, J., Takayanagi, H., and Datta, S., *Phys. Rev. Lett.*, 88, 126601 (2002).
11. Dowben, P.A. and Skomski, R., *J. Appl. Phys.*, 95, 7453 (2004).
12. Salis, G., Wang, R., Jiang, X., Shelby, R.M., Parkin, S.S.P., Bank, S.R., and Harris, J.S., *Appl. Phys. Lett.*, 87, 26503 (2005).
13. Fiederling, R., Keim, M., Reuscher, G., Ossau, W., Schmidt, G., Waag, A., and Molenkamp, L.W., *Nature* (London), 402, 787 (1999).
14. Hall, K.C., Gundogdu, K., Hicks, J.L., Kocbay, A.N., Flatte, M.E., Boggess, T.F., Holabird, K., Hunter, A., Chow, D.H., and Zink, J.J., *Appl. Phys. Lett.*, 86, 202114 (2005).
15. Hall, K.C. and Flatte, M.E., *Appl. Phys. Lett.*, 88, 162503 (2006).
16. Bandyopadhyay, S., and Cahay, M., www.arXiv.org/cond-mat/0604532.
17. Flatte, M.E. and Hall, K.C., www.arXiv.org/cond-mat/0607432.
18. Bandyopadhyay, S. and Cahay, M., www.arXiv.org/cond-mat/0607659.

19. Sakaki, H., *Jpn. J. Appl. Phys.*, 21, L381 (1982); Hirakawa, K., Sakaki, H., and Yoshino, J., *Phys. Rev. Lett.*, 54, 1279 (1985).

20. Hamaguchi, C., Miyatsuji, K., and Hihara, H., *Jpn. J. Appl. Phys.*, Pt. 1, 23, 212 (1984).

21. Suman Datta, Intel Corporation, private communication.

22. Bandyopadhyay, S., and Cahay, M., *Appl. Phys. Lett.*, 85, 1433 (2004).

23. Bandyopadhyay, S., Das, B., and Miller, A.E., *Nanotechnology*, 5, 113 (1994).

24. Bandyopadhyay, S., *J. Nanosci. Nanotech.* (in press).

9

Electron Charge and Spin Transport in Organic and Semiconductor Nanodevices: Moletronics and Spintronics

9.1   Introduction 9-1
9.2   Electron Transport in Molecular Devices: Molecular
      Electronics 9-2
      Role of Molecule–Electrode Contact: Extrinsic Molecular
      Switching Due to Molecule Tilting • Molecular Quantum Dot
      Rectifiers • Molecular Switches • Role of Defects in Molecular
      Transport
9.3   Spin Injection: Processes and Devices 9-20
      Tunnel Magnetoresistance • Spin-Torque Domain Wall
      Switching in Nanomagnets • Spin Injection/Extraction into
      (from) Semiconductors • Spin Accumulation and Extraction
      • Conditions for Efficient Spin Injection and Extraction
      • High-Frequency Spin-Valve Effect • Spintronic Devices
      • Spin Injection into Organic Materials
9.4   Conclusions 9-36
References 9-37

A.M. Bratkovsky
Hewlett-Packard Laboratories

9.1   Introduction

Current interest in molecular electronics [1,2] and spintronics [3,4] is largely driven by expectations that these emerging technologies can be used to overcome bottlenecks of standard silicon technology, most importantly, scaling in terms of speed and power dissipation. In particular, it has been frequently envisaged in the past that molecules can be used as nanoelectronics components able to complement/replace standard silicon CMOS technology [1,2] on the way down to $\leq$10 nm circuit components. The first speculations about molecular electronic devices (diodes and rectifiers) were apparently made in the mid-1970s [5]. That original suggestion of a molecular rectifier has generated a large interest in the field and a flurry of suggestions of various molecular electronics components, especially coupled with premature estimates that silicon-based technology cannot scale to below 1 μm feature size. The Aviram–Ratner Donor–insulator–Acceptor construct TTF—σ—TCNQ (D^+—σ—A^-, see details below), where carriers were supposed to tunnel asymmetrically in two directions through an insulating saturated molecular "bridge," has never materialized, in spite of extensive experimental effort over a few decades [6]. The end result in some cases appears to be a slightly electrically anisotropic "insulator," rather than a diode, unsuitable as a replacement for silicon devices. This comes about because in order to assemble a reasonable quality monolayer of these molecules in a Langmuir–Blodgett trough (avoiding defects that will short the device after electrode deposition), one needs to attach a long "tail" molecule C18 [$\equiv(CH_2)18$] that can produce enough of van der Waals force to keep molecules together, but C18 is a wide-band insulator with a bandgap $Eg \approx 9$ to 10 eV. The outcome of these studies may have been anticipated, but if one were able to assemble the Aviram–Ratner molecules without the tail, they could not rectify anyway. Indeed, a recent ab-initio study [7] of $D^+\sigma A^-$ prospective molecule showed no appreciable asymmetry of its I–V curve. The molecule was envisaged by Ellenbogen and Love [8] as a 4-phenyl ring Tour wire with a dimethylene-insulating bridge in the middle directly connected to Au electrodes via thiol groups. Donor–acceptor asymmetry was produced by side NH_2^+ and NO_2^- moieties, which is a frequent motif in molecular devices using the Tour wires. The reason for poor rectification is simple: the bridge is too short; it is a transparent piece of one-dimensional insulator, whereas the applied field is three dimensional and it cannot be screened efficiently with an appreciable voltage drop on the insulating group in this geometry. Although there is only 0.7 eV energy separation between levels on the D and A groups, one needs about a 4 eV bias to align them and get a relatively small current because total resonant transparency is practically impossible to achieve. Remember that the model calculation implied an ideal coupling to electrodes, which is impossible in reality and which is known to dramatically change the current through the molecule (see below). We shall discuss below some possible alternatives to this approach.

Spintronics, on the other hand, is a less disruptive technology in that the goal is to use semiconductors for spin transport since the spin-diffusion length therein may be very large even at room temperature. The major difficulty here is that it requires efficient spin sources that can be integrated with semiconductors, and when ferromagnetic metals are used, it requires strongly modifying the interface to make the Schottky barrier transparent for electrons. Another popular choice is to avoid the problem with the Schottky barrier by using ferromagnetic semiconductors (FMSs) as a spin source. Those are, unfortunately, rather esoteric materials with weak ferromagnetism and a low Curie point, usually below room temperature. Therefore, FMS-based spintronic devices face very steep challenges. In the second part of this review, we shall, therefore, discuss exclusively the ferromagnetic metals–based spintronic devices.

9.2 Electron Transport in Molecular Devices: Molecular Electronics

It is worth noting that studies of energy and electron transport in molecular crystals [9] started in the early 1960s. In the mid-1960s it was established in what circumstances charge transport in biological molecules involves electron tunneling [10]. It was realized in the mid-1970s that, since the organic molecules are "soft," energy transport along linear biological molecules, proteins, etc., may proceed by low-energy nonlinear collective excitations, like Davydov solitons [11] (see review [12]).

To take over from current silicon CMOS technology, the molecular electronics should provide smaller, more reliable, functional components that can be produced and assembled concurrently and are compatible with CMOS for integration. The small size of units that molecules may hopefully provide is quite obvious. However, meeting other requirements seems to be a very long shot. To beat alternative technologies, for e.g., dense (and cheap) memories, one should aim at a few TB/in² ($>10^{12}$ to 10^{13} bit/cm²), which corresponds to linear bit (footprint) sizes of 3 to 10 nm, and an operation lifetime of ~10 years. The latter requirement is very difficult to meet with organic molecules that tend to oxidize and decompose, especially under conditions of a very high applied electric field (given the operational voltage bias of ~1 V for molecules integrated with CMOS and their small sizes in the order of a few nanometers). In terms of areal density, one should compare this with rapidly developing technologies like ferroelectric random access memories (FERAM) [13] or phase-change memories

(PCM) [14]. The current smallest commercial nanoferroelectrics are about 400×400 nm^2 and 20 to 150 nm thick [15], and the 128×128 arrays of switching ferroelectric pixel bits have been already demonstrated with a bit size ≤ 50 nm (with density ~TB/in^2) [16]. The PCM based on chalcolgenides GeSbTe seem to scale even better than the ferroelectrics. As we see, the mainstream technology for random-access memory approaches molecular size very rapidly. For instance, the so-called "nanopore" molecular devices by M.A. Reed et al. have comparable sizes and are yet to demonstrate a repeatable behavior (see below).

In terms of parallel fabrication of molecular devices, one is looking at "self-assembly" techniques (see, e.g., [17,18] and references therein). Frequently, the Langmuir–Blodgett technique is used for self-assembly of molecules on water, where molecules are prepared to have hydrophilic "head" and hydrophobic "tail" to make the assembly possible, see e.g., Refs. [19,20]. The allowances for a corresponding assembly, especially of hybrid structures (molecules integrated on silicon CMOS), are in the order of a fraction of an angstrom, so actually picotechnology is required [2]. Since it is problematic to reach such precision any time soon, the all-in-one molecule approach was advocated, meaning that a fully functional computing unit should be synthesized as a single supermolecular unit [2]. The hope is that perhaps directed self-assembly will help to accomplish building such a unit, but self-assembly on a large scale is impossible without defects [17,18], since the entropic factors work against it. Above some small defect concentration ("percolation") threshold, mapping of even a simple algorithm on such a self-assembled network becomes impossible [21].

There is also a big question about electron transport in such a device consisting of large organic molecules. Even in high-quality pentacene crystals, perhaps the best materials for thin film transistors, the mobility is a mere 1 to 2 cm^2/V·sec (see e.g., [22]), as a result of carrier trapping by interaction with a lattice. The situation with carrier transport through long molecules (>2 to 3 nm) is, of course, substantially different from the transport through short rigid molecules that have been envisaged as possible electronics components. Indeed, in short molecules the dominant mode of electron transport would be resonant tunneling through electrically active molecular orbital(s) (MOs) [23], which, depending on the work function of the electrode, affinity of the molecule, and symmetry of coupling between molecule and electrode, may be one of the lowest unoccupied molecular orbitals (LUMOs) or highest occupied molecular orbitals (HOMOs) [24,25]. Indeed, it is well known that in longer wires containing more than about 40 atomic sites, the tunneling time is comparable to or larger than the characteristic phonon times, so that the polaron (and/or bipolaron) can be formed inside the molecular wire [26]. There is a wide range of molecular bulk conductors with (bi)polaronic carriers. The formation of polarons (and charged solitons) in polyacetylene (PA) was discussed a long time ago theoretically in Refs. [27] and formation of bipolarons (bound states of two polarons) in Ref. [28]. Polarons in PA were detected optically in Ref. [29] and since then studied in great detail. There is an exceeding amount of evidence of the polaron and bipolaron formation in conjugated polymers such as polyphenylene, polypyrrole, polythiophene, polyphenylene sulfide [30], Cs-doped biphenyl [31], n-doped bithiophene [32], and polyphenylenevinylene-based light-emitting diodes [33], and in other molecular systems. Given the above problems with electron transport through large molecules, one should look at the short- to medium-size molecules first.

The latest wave of interest in molecular electronics is mostly related to recent studies of carrier transport in synthesized linear conjugated molecular wires (Tour wires [1]) with apparent nonlinear I–V characteristics (NDR) and "memory" effects [34–36], and of various molecules with a mobile "microcycle" that is able to move back and forth between metastable conformations in solution (molecular shuttles) [37] and demonstrate some sort of "switching" between relatively stable resistive states when sandwiched between electrodes in a solid-state device [38] (see also [39]). There are also various photochromic molecules that may change conformation ("switch") upon absorption of light [40], which may be of interest to some photonics applications but not for general purpose electronics. One of the most serious problems with using this kind of molecule is "power dissipation." Indeed, the studied organic molecules are, as a rule, very resistive (in the range of ~1 MΩ to 1 GΩ or more). Since usually the switching bias exceeds 0.5 V, the dissipated power density would be in excess of 10 kW/cm^2, which is many orders of magnitude higher than the presently manageable level. One can drop the density of switching devices, but this would undermine a main advantage of using molecular size elements. This is a common problem that CMOS faces too, but organic molecules do not seem to offer a tangible advantage yet. There are other outstanding problems, like understanding an actual switching mechanism, which seems to be rather molecule independent [39], stability, scaling, etc. It is not likely, therefore, that molecules will displace silicon technology, or become a large part of a hybrid technology in the foreseeable future.

First major moletronic applications would most likely come in the area of chemical and biological sensors. One of the current solutions in this area is to use the functionalized nanowires. When a target analyte molecule attaches from the environment to such a nanowire, it changes the electrostatic potential "seen" by the carriers in the nanowire. Since the conductance of the nanowire device is small, even one chemisorbed molecule could make a detectable change of a conductance [41]. Semiconducting nanowires can be grown from seed metal nanoparticles [17], or they can be carbon nanotubes, which are studied extensively due to their relatively simple structure and some unique properties like very high conductance [42].

In this chapter, we shall address various generic problems related to electron transport through molecular devices, and describe some specific molecular systems that may be of interest for applications as rectifiers and switches, and for some pertaining to physical problems. We shall first consider systems where an elastic tunneling is dominant, and interaction with vibrational excitations on the molecules only renormalizes some parameters describing tunneling. We shall also describe a situation where the coupling of carriers to molecular vibrons is strong. In this case, the tunneling is substantially inelastic and, moreover, it may result in current hysteresis when the electron–vibron interaction is so strong that it overcomes Coulomb repulsion of carriers on a central narrow-band/conjugated unit of the molecule separated from electrodes by wide-band gap saturated molecular groups like $(CH_2)_n$, which we shall call a molQD. Another very important problem is to understand the nature and the role of imperfections in organic thin films. It is addressed in the last section of the paper.

9.2.1 Role of Molecule–Electrode Contact: Extrinsic Molecular Switching Due to Molecule Tilting

We have predicted some time ago that there should be a strong dependence of the current through conjugated molecules (like the Tour wires [1]) on the geometry of molecule–electrode contact [24,25]. The apparent "telegraph" switching observed in scanning tunneling microscopy (STM) single-molecule probes of the three-ring Tour molecules, inserted into a self-assembled monolayer (SAM) of nonconducting shorter alkanes, has been attributed to this effect [35]. The theory predicts very strong dependence of the current through the molecule on the tilting angle between a backbone of a molecule and a normal to the electrode surface. Other explanations, like rotation of the middle ring, charging of the molecule, or effects of the moieties on the middle ring, do not hold. In particular, switching of the molecules "without" any NO_2 or NH_2 moieties has been practically the same as with them.

The simple argument in favor of the "tilting" mechanism of the conductance lies in a large anisotropy of the molecule–electrode coupling through π-conjugated MOs. In general, we expect the overlap and the full conductance to be maximal when the lobes of the p-orbital of the end atom at the molecule are oriented perpendicular to the surface and smaller otherwise, as dictated by the symmetry. The overlap integrals of a p-orbital with orbitals of other types differ by a factor about 3 to 4 for the two orientations. Since the conductance is proportional to the square of the matrix element, which contains a product of two metal-molecule hopping integrals, the total conductance variation with overall geometry may therefore reach two orders of magnitude, and in special cases be even larger.

In order to illustrate the geometric effect on current we have considered a simple two-site model with p-orbitals on both sites, coupled to electrodes with s-orbitals [25]. For nonzero bias the transmission probability has the resonant form (5) with line widths for hopping to the left (right) lead $\Gamma_L(\Gamma_R)$. The current has the following approximate form (with $\Gamma = \Gamma_L + \Gamma_R$):

$$I \approx \begin{cases} \left| \dfrac{q^2}{h} \dfrac{\Gamma}{t^2} V \propto \sin^4 \theta, \quad qV \ll E_{LUMO} - E_{HOMO}, \right. \\ \left| \dfrac{8\pi q}{h} \dfrac{\Gamma_L^\pi \Gamma_R}{\Gamma_L + \Gamma_R} \propto \sin^2 \theta, \quad qV > E_{LUMO} - E_{HOMO}, \right. \end{cases} \tag{9.1}$$

where θ is the tilting angle, t is the integral for π electrons, and q is the elementary charge [24,25] (Figure 9.1).

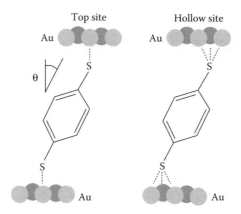

FIGURE 9.1 Schematic representation of the benzene-dithiolate molecule on top and hollow sites. End sulfur atoms are bonded to one and three surface gold atoms, respectively, θ is the tilting angle.

The tilting angle has a large effect on the I–V curves of benzene-dithiolate molecules, especially when the molecule is anchored to the Au electrode in the top position (Figure 9.2). By changing θ from 5° to just 15°, one drives the I–V characteristic from the one with a gap of about 2 V to an ohmic one with a large relative change of conductance. Even changing θ from 10° to 15° changes the conductance by about an order of magnitude. The I–V curve for the hollow site remains ohmic for tilting angles up to 75° with moderate changes of conductance. Therefore, if the molecule being measured snaps from the top to the hollow position and back, it will lead to an apparent switching [35]. It has recently been realized that the geometry of a contact strongly affects coherent spin transfer between molecularly bridged quantum dots [43]. It is worth noting that another frequently observed "extrinsic" mechanism of "switching" in organic layers is due to electrode material diffusing into the layer and forming metallic "filaments" (see below).

9.2.2 Molecular Quantum Dot Rectifiers

Aviram and Ratner speculated about a rectifying molecule containing donor (D) and acceptor (A) groups separated by a saturated σ-bridge (insulator) group, where the (inelastic) electron transfer will be more favorable from A to D [5]. The molecular rectifiers actually synthesized, $C_{16}H_{33}-\gamma$ Q3CNQ, were of

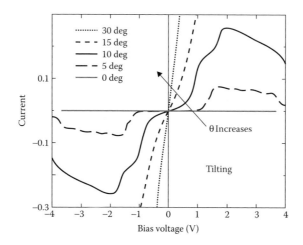

FIGURE 9.2 Effect of tilting on the I–V curve of the BDT molecule, Figure 9.1. Current is in units of $I_0 = 77.5\ \mu A$, θ is the tilting angle.

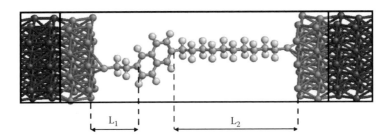

FIGURE 9.3 Stick figure representing the naphthalene conjugated central unit separated from the left(right) electrode by saturated (wide-band gap) alkane groups with length $L_{1(2)}$.

somewhat different $D - \pi - A$ type, i.e., the "bridge" group was conjugated [6]. Although the molecule did show rectification (with considerable hysteresis), it performed rather like an anisotropic insulator with tiny currents in the order of 10^{-17} A/molecule, because of the large alkane "tail" needed for Langmuir–Blodgett (LB) assembly. It was recently realized that in this molecule the resonance does not come from the alignment of the HOMO and LUMO, since they cannot be decoupled through the conjugated π–bridge, but rather due to an asymmetric voltage drop across the molecule [44]. Rectifying behavior in other classes of molecules is likely due to asymmetric contact with the electrodes [45,46], or an asymmetry of the molecule itself [47]. To make rectifiers, one should avoid using molecules with long insulating groups, and we have suggested using relatively short molecules with "anchor" end groups for their self-assembly on metallic electrodes, with a phenyl ring as a central conjugated part [48]. This idea has been tested in Ref. [49] with a phenyl and thiophene rings attached to a $(CH_2)_{15}$ tail by a CO group. The observed rectification ratio was $\lesssim 10$, with some samples showing ratios of about 37.

We have recently studied a more promising rectifier like $-S-(CH_2)_2-Naph-(CH_2)_{10}-S-$ with a theoretical rectification $\lesssim 100$ [50] (Figure 9.3). This system has been synthesized and studied experimentally [51]. To obtain an accurate description of transport in this case, we employ an ab-initio nonequilibrium Green's function (NEGF) method [52]. The present calculation takes into account only elastic tunneling processes. Inelastic processes may substantially modify the results in the case of strong interaction of the electrons with molecular vibrations, see Ref. [53] and below. There are indications in the literature that the carrier might be trapped in a polaron state in saturated molecules somewhat longer than those we consider in the present paper [54]. One of the barriers in the present rectifiers is short and relatively transparent, so there will be no appreciable Coulomb blockade effects. The structure of the present molecular rectifier is shown in Figure 9.3. The molecule consists of a central conjugated part (naphthalene) isolated from the electrodes by two insulating aliphatic chains $(CH_2)_n$ with lengths $L_1(L_2)$ for the left (right) chain.

The principle of molecular rectification by a molQD is illustrated in Figure 9.4, where the electrically "active" MO, localized on the middle conjugated part, is the LUMO, which lies at an energy Δ above the electrode Fermi level at zero bias. The position of the LUMO is determined by the work function of the metal $q\phi$ and the affinity of the molecule $q\chi$, $\Delta = \Delta_{LUMO} = q(\phi - \chi)$. The position of the HOMO is given by $\Delta_{HOMO} = \Delta_{LUMO} - E_g$, where E_g is the HOMO–LUMO gap. If this orbital is considerably closer to the electrode Fermi level E_F, then it will be brought into resonance with E_F prior to other orbitals. It is easy to estimate the forward and reverse bias voltages, assuming that the voltage mainly drops on the insulating parts of the molecule,

$$V_F = \frac{\Delta}{q}(1+\xi), \quad V_R = \frac{\Delta}{q}\left(1+\frac{1}{\xi}\right), \qquad (9.2)$$

$$V_F/V_R = \xi \equiv L_1/L_2, \qquad (9.3)$$

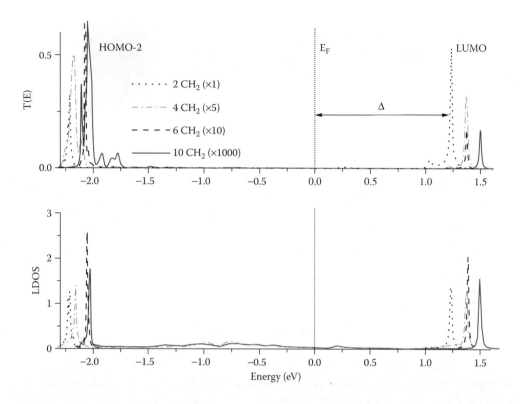

FIGURE 9.4 Transmission coefficient versus energy E for rectifiers $-S-(CH_2)_2-C_{10}H_6-(CH_2)_n-S-$; $n = 2, 4, 6, 10$. Δ indicates the distance of the closest MO to the electrode Fermi energy ($E_F = 0$).

where q is the elementary charge. A significant difference between forward and reverse currents should be observed in the voltage range $V_F < |V| < V_R$. The current is obtained from the Landauer formula

$$I = \frac{2q^2}{h} \int dE \, [f(E) - f(E + qV)] \, g(E,V) \tag{9.4}$$

We can make qualitative estimates in the resonant tunneling model, with the conductance $g(E,V) \equiv T(E,V)/q$, where $T(E,V)$ is the transmission given by the Breit–Wigner formula

$$T(E,V) = \frac{\Gamma_L \Gamma_R}{(E - E_{MO})^2 + (\Gamma_L + \Gamma_R)^2/4}, \tag{9.5}$$

E_{MO} is the energy of the MO. The width $\Gamma_{L(R)} \sim t^2/D = \Gamma_0 e^{-2\kappa L_{1(2)}}$, where t is the overlap integral between the MO and the electrode, D is the electron band width in the electrodes, and κ is the inverse decay length of the resonant MO into the barrier. The current above the resonant threshold is given as follows:

$$I \approx \frac{2q}{h} \Gamma_0 e^{-2\kappa L_2} \tag{9.6}$$

We see that increasing the spatial asymmetry of the molecule (L_2/L_1) changes the operating voltage range linearly, but it also brings about an "exponential" decrease in current [48]. This severely limits the ability to optimize the rectification ratio while simultaneously keeping the resistance at a reasonable value.

To calculate the I–V curves, we use an ab-initio approach that combines the Keldysh NEGF with pseudopotential-based real space density functional theory (DFT) [52]. The main advantages of our approach are (i) a proper treatment of the open boundary condition; (ii) a fully atomistic treatment of the electrodes and (iii) a self-consistent calculation of the nonequilibrium charge density using NEGF. The transport Green's function is found from the Dyson equation.

$$(G^R)^{-1} = (G_0^R)^{-1} - V, \tag{9.7}$$

where the unperturbed retarded Green's function is defined in operator form as $(G_0^R)^{-1} = (E + i0)\, \hat{S} - \hat{H}$, H is the Hamiltonian matrix for the scatterer (molecule plus screening part of the electrodes), S is the "overlap" matrix, $S_{i,j} = \langle \chi_i | \chi_j \rangle$ for nonorthogonal basis set orbitals χ_i, and the coupling of the scatterer to the leads is given by the Hamiltonian matrix $V = diag[\Sigma_{l,l}, 0, \Sigma_{r,r}]$, where $l(r)$ stands for the left (right) electrode. The self-energy part $\Sigma^<$, which is used to construct the nonequilibrium electron density in the scattering region, is found from $\Sigma^< = -2iIm[f(E)\Sigma_{l,l} + f(E+qV)\Sigma_{(r,r)}]$, where $\Sigma_{l,l(r,r)}$ is the self-energy of the left (right) electrode, calculated for the semi-infinite leads using an iterative technique [52]. $\Sigma^<$ accounts for the steady charge "flowing in" from the electrodes. The transmission probability is given by the following equation.

$$T(E, V) = 4Tr[(Im\Sigma_{l,l})G_{l,r}^R(Im\Sigma_{r,r})G_{r,l}^A], \tag{9.8}$$

where $G^{R(A)}$ is the retarded (advanced) Green's function, and Σ the self-energy part connecting left (l) and right (r) electrodes [52], and the current is obtained from Eq. (9.4). The calculated transmission coefficient $T(E)$ is shown for a series of rectifiers –S–$(CH_2)_m$–$C_{10}H_6$–$(CH_2)_n$–S– for $m = 2$ and $n = 2, 4$, 6, 10 at zero bias voltage in Figure 9.4. We see that the LUMO is the MO transparent to electron transport, which lies above E_F by an amount $\Delta = 1.2$ to 1.5 eV. The transmission through the HOMO and HOMO-1 states, localized on the terminating sulfur atoms, is negligible, but the HOMO-2 state conducts very well. The HOMO-2 defines the threshold reverse voltage V_R, thus limiting the operating voltage range. Our assumption, that the voltage drop is proportional to the lengths of the alkane groups on both sides, is quantified by the calculated potential ramp. It is close to a linear slope along the $(CH_2)_n$ chains [50]. The forward voltage corresponds to the crossing of the LUMO(V) and $\mu_R(V)$, which happens at about 2 V. Although LUMO defines the forward threshold voltages in all molecules studied here, the reverse voltage is defined by the HOMO-2 for "right" barriers $(CH_2)_n$ with $n = 6, 10$. The I–V curves are plotted in Figure 9.5. We see that the rectification ratio for current in the operation window $I_+ \neq I_-$ reaches a maximum value of 35 for the "2 to 10" molecule ($m = 2, n = 10$). Series of molecules with a central "single phenyl" ring [48] do not show any significant rectification. One can manipulate the system in order to increase the energy asymmetry of the conducting orbitals (reduce Δ). To shift the LUMO toward E_F, one can attach an electron withdrawing group, like –C≡N [50] to the middle conjugated group (naphthalene). The molecular rectification ratio is not great by any means, but one should bear in mind that this is a device necessarily operating in a ballistic quantum-mechanical regime because of the small size. This is very different from present Si devices with carriers diffusing through the system. As silicon devices become smaller, however, the same effects will eventually take over, and tend to diminish the rectification ratio, in addition to effects of finite temperature and disorder in the system.

9.2.3   Molecular Switches

There are various molecular systems that exhibit some kind of current "switching" behavior [35,37–39], "NDR" [34], and "memory" [36]. The switching systems are basically driven between two states with considerably different resistances. This behavior is not really sensitive to a particular molecular structure, since this type of bistability is observed in complex rotaxane-like molecules as well as in very simple alkane chains $(CH_2)_n$ assembled into LB films [39] and is not even exclusive to the organic films.

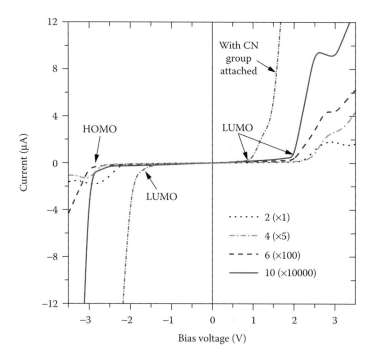

FIGURE 9.5   I–V curves for naphthalene rectifiers $-S-(CH_2)_2-C_{10}H_6-(CH_2)_n-S-$; $n = 2, 4, 6, 10$. The short-dash-dot curve corresponds to a cyano-doped (added group $-C\equiv N$) $n = 10$ rectifier.

The data strongly indicate that the switching has an extrinsic origin, and is related either to bistability of molecule–electrode orientation [24,25,35], or transport assisted by defects in the film [55,56].

9.2.3.1   Extrinsic Switching in Organic Molecular Films: Role of Defects and Molecular Reconfigurations

Evidently, large defects can be formed in organic thin films as a result of electromigration in very strong fields, as was observed long ago [57]. It was concluded some decades ago that the conduction through absorbed [55] and Langmuir–Blodgett [56] monolayers of fatty acids $(CH_2)_n$, which we denote as Cn, is associated with "defects." In particular, Polymeropoulos and Sagiv studied a variety of absorbed mono-layers from C7 to C23 on Al/Al$_2$O$_3$ substrates and found that the exponential dependence on the length of the molecular chains is only observed below the liquid nitrogen temperature of 77 K, and no discernible length dependence was observed at higher temperatures [55]. The temperature dependence of current was strong, and was attributed to transport assisted by some defects. The current also varied strongly with the temperature in Ref. [56] for LB films on Al/Al$_2$O$_3$ substrates in He atmosphere, which is not compatible with elastic tunneling. Since the He atmosphere was believed to hinder the Al$_2$O$_3$ growth, and yet the resistance of the films increased about 100-fold over 45 days, the conclusion was made that the defects somehow anneal out with time. Two types of switching have been observed in 3 to 30 μm thick films of polydimethylsiloxane, one as a standard dielectric breakdown with electrode material "jet evaporation" into the film with subsequent Joule melting of metallic filament under bias of about 100 V, and a low-voltage (<1 V) "ultraswitching" that has a clear "telegraph" character and resulted in inter-mittent switching into a much more conductive state [58]. The exact nature of this switching also remains unclear, but there is a strong expectation that the formation of metallic filaments that may even be in a ballistic regime of transport may be relevant to the phenomenon.

Recently, direct evidence was obtained of the formation of "hot spots" in the LB films that may be related to the filament growth through the film imaged with the use of AFM current mapping [59]. The system investigated in this work has been a Pt/stearic acid (C18)/Ti (Pt/C18/Ti) crossbar molecular structure,

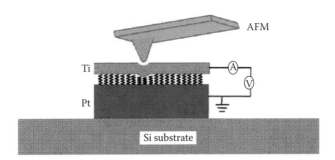

FIGURE 9.6 Experimental setup for mapping local conductance. AFM produces local deformation of the top electrode and underlying organic film. The total conductance of the device is measured and mapped. (From Lau, C.N., Stewart, D., Williams, R.S., and Bockrath, D., *Nano Lett.*, 4, 569, 2004.)

consisting of planar Pt and Ti electrodes sandwiching a monolayer of 2.6-nm-long stearic acid ($C_{18}H_{36}OH$) molecules with typical zero-bias resistance in excess of 10^5 ω. The devices have been switched reversibly and repeatedly to higher ("on") or lower ("off") conductance states by applying sufficiently a large bias voltage V_b to the top Ti electrode with regards to the Pt counterelectrode (Figure 9.6).

Interestingly, reversible switching was not observed in symmetric Pt/C18/Pt devices. The local conductance maps of the Pt/C18/Ti structure have been constructed by using an AFM tip and simultaneously measuring the current through the molecular junction biased to $V_b = 0.1$ V (the AFM tip was not used as an electrode, but only to apply local pressure at the surface). The study revealed that the film showed pronounced switching between electrically very distinct states, with zero-bias conductances 0.17 μsec ("off" state) and 1.45 μsec ("on" state) (Figure 9.7).

At every switching "on" there appeared a local conductance peak on the map with a typical diameter ~40 nm, which then disappeared upon switching "off" (Figure 9.7, top inset). The switching has been

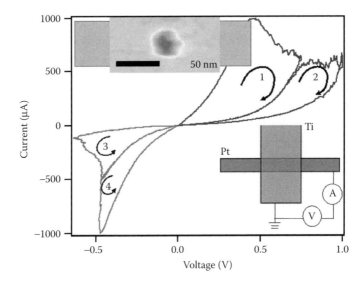

FIGURE 9.7 I–V characteristics showing the reversible switching cycle of the device (*bottom inset*) with organic film. The arrows indicate sweep direction. A negative bias switches the device to a high conductance state, while the positive one switches it to a low conductance state. The mapping, according to the schematic in Figure 9.6, shows the appearance of "hot spots" after switching (*top inset*). (From Lau, C.N., Stewart, D., Williams, R.S., and Bockrath, D., *Nano Lett.*, 4, 569, 2004.)

attributed to local conducting filament formation due to electromigration processes. It remains unclear exactly how the filaments dissolve under opposite bias voltage, why they tend to appear in new places after each switching, and why conductance in some cases strongly depends on temperature. It is clear, however, that switching in such a simple molecule without any redox centers, mobile groups, or charge reception centers should be "extrinsic." Interestingly, very similar switching between two resistive states has also been observed for tunneling through thin "inorganic" perovskite oxide films [60].

There have been plenty of reports on nonlinear I–V characteristics like NDR and random switching recently for molecules assembled on metal electrodes (gold) and silicon. Reports on NDR for molecules with metal contacts (Au and Hg) have been made in Refs. [34,46,61]. It became very clear, though, that most of these observations are related to molecular reconfigurations and bond breaking and making, rather than to any intrinsic mechanism, like redox states, speculated about in the original Ref. [34]. Thus, the NDR in Tour wires was related to molecular reconfiguration with respect to metallic electrodes [24,35]; NDR in ferrocene-tethered alkyl monolayers [61] was found to be related to oxygen damage at high voltage [62]. Structural changes and bond breaking have been found to result in NDR in experiments with STM [63–65] and mercury droplet contacts [66].

Several molecules, like styrene, have been studied on a degenerate Si surface and showed an NDR behavior [67]. However, those results have been carefully checked later and it was found that the styrene molecules do not exhibit NDR, but rather sporadically switch between states with different currents while held at the same bias voltage (the blinking effect) [68].

The STM map of the styrene molecules (indicated by arrows) on the Silicon (100) surface shows that the molecules are blinking (see Figure 9.8). The blinking is absent at clean Si areas, dark (D) and bright (C) defects. This may indicate a dynamic process occurring during the imaging. Comparing the panels (a) and (b), one may see that some molecules are actually decomposing. The height versus voltage spectra over particular points are shown in Figure 9.8c. The featureless curve 1 was taken over a clean silicon dimer.

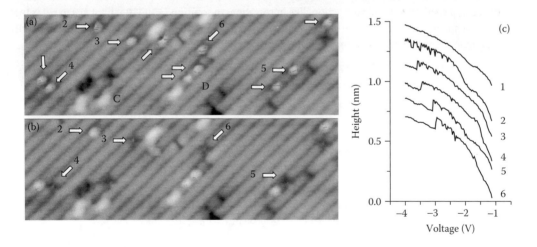

FIGURE 9.8 STM images or styrene molecules on clean Si (100) before and after spectroscopy over the area 75 × 240 Å. (*a*) Bias -2 V, current 0.7 nA. Only the styrene molecules (indicated by arrows) are blinking during imaging. The clean Si surface, bright defects (marked C), and dark sports (marked D) do not experience blinking. (*b*) Bias -2 V, current 0.3 nA. STM image of individual styrene molecules indicated with numbers 2 to 6. Styrene molecules 3 to 6 have decomposed. Decomposition involves the changing of the styrene molecule from a bright feature to a dark depression and also involves the reaction with an adjacent dimer. Styrene molecule 2 does not decompose and images as usual with no change of position. (*c*) Height–voltage spectra taken over clean silicon (1) and styrene molecules (2 to 6). The spectra taken over molecules show several spikes in height related to blinking in the images. In spectra 3 to 6, an abrupt and permanent change in height is recorded and is correlated with decomposition, as seen in the bottom left image. Spectrum 2 has no permanent height change, and the molecule does not decompose. (From Pitters, J.L. and Wolkow, R.A., *Nano Lett.*, 6, 390, 2006.)

The other spectra were recorded over individual molecules. Each of these spectra have many sudden decreases and increases in current as if the molecules are changing between different states during the measurement causing a change in current and a response of the feedback control, resulting in a change in height so there exist one or more configurations that lead to measurement of a different height. Evidently, these changes have the same origin as the blinking of molecules in STM images. Figure 9.8b reveals clear structural changes associated with those particular spectroscopic changes. In each case where a dramatic change in spectroscopy occurred, the molecules in the image have changed from a bright feature to a dark spot. This is interpreted as a decomposition of the molecules. A detailed look at each decomposed styrene molecule, at locations 3, 4, 5, and 6, shows that the dark spot is not in precise registry with the original bright feature, indicating that the decomposition product involves reaction with an adjacent dimer [68].

The fact that the structural changes and related NDR behavior are not associated with any resonant tunneling through the molecular levels or redox processes, but are perhaps related to inelastic electron scattering or other extrinsic processes, becomes evident from current versus time records shown in Figure 9.9, Ref. [68]. The records show either no change of the current with time, or one or a few random jumps between certain current states (telegraph noise). The observed changes in current at a fixed voltage obviously cannot be explained by shifting and aligning of molecular levels, as was suggested in Ref. [69], they must be related to adsorbate molecule structural changes with time. Therefore, the explanation by Datta et al. that the resonant level alignment is responsible for NDR does not apply [69]. As mentioned above, similar telegraph switching and NDR has been observed in Tour wires [35] and other molecules. Therefore, the observed NDR apparently has a similar origin in disparate molecules adsorbed on different substrates and has to do with molecular reconformation/reconfiguration on the surface. In fact, to the best of our knowledge, there is no convincing data about intrinsic switching of a resistance state of the molecule subject to the external electric field. Various molecules do show conformational switching, but it is limited to either photochromic molecules subject to illumination by photons with certain energy, or induced as a result of a chemical reaction, like in rotaxanes in solution [40]. What happens to rotaxanes in solid-state junctions is not known, but it is likely to be extrinsic, since rotaxanes behave there in exactly the same manner as benign insulating molecular alkane chains [39].

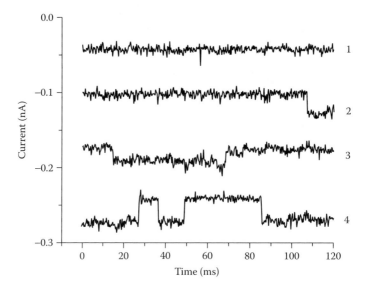

FIGURE 9.9 Variation of current through styrene molecules on Si (100) with time. Tunneling conditions were set at −3 V and 0.05 nA. Abrupt increases and decreases in current relate to changes of the molecule during the spectroscopy. Some experiments show no changes in current (*curve 1*), others show various kinds of telegraph switching. (From Pitters, J.L. and Wolkow, R.A., *Nano Lett.*, 6, 390, 2006.)

9.2.3.2 Intrinsic Polarization and Extrinsic Conductance Switching in Molecular Ferroelectric PVDF

The only well established, to the best of our knowledge, intrinsic molecular switching (of polarization, not current) under bias voltage was observed in molecular ferroelectric block copolymer polyvinylidene [70]. Ferroelectric polymer films have been prepared with the 70% vinylidene fluoride copolymer, P(VDF–TrFE 70:30), formed by horizontal LB deposition on aluminum-coated glass substrates with evaporated aluminum top electrodes. The polymer chains contain a random sequence of $(CH_2)_n(CF_2)_m$ blocks, the fluorine site carries a strong negative charge, and in the ferroelectric phase most of the carbon-fluorine bonds point in one direction. The fluorine groups can be rotated and aligned in a very strong electric field, ~5 MV/cm. As a result, the whole molecular chain orders, and in this way the macroscopic polarization can be switched between the opposite states. The switching process is extremely slow, however, and takes 1 to 10 sec (!) [71,72]. This is not surprising, given a strong Coulomb interaction between the charged groups and the metal electrodes, pinning by surface roughness, and steric hindrance to rotation. This behavior should be suggestive of other switching systems based on one of a few monolayers of molecules, and other nontrivial behavior involved [72].

The switching of current was also observed in films of PVDF 30 monolayers thick. The conductance of the film was following the observed hysteresis loop for polarization, ranging from $~1 \times 10^{-9}$ to $2 \times 10^{-6} \Omega^{-1}$ [73]. The phenomenon of conductance switching has these important features: (i) It is connected with bulk polarization switching; (ii) there is a large ~1000:1 contrast between the "on" and "off" states; (iii) the on state is obtained only when bulk polarization is switched in the positive direction; and (iv) conductance switching is much faster than bulk polarization switching. The conductance switches on only after 6 sec delay, after bulk polarization switching is nearly complete, presumably when the last layer switches into alignment with the others, while the conductance switches off without a noticeable delay after the application of reverse bias as even one layer reverses (this may create a barrier to charge transfer). The slow ~2 sec time constant for polarization switching is probably nucleation limited, as has been observed in high-quality bulk films with low nucleation site densities [74]. The duration of conductance switching transition ~2 msec may be limited only by the much faster switching time of individual layers.

The origin of conductance switching by three orders of magnitude is not clear. It may indeed be related to a changing amount of disorder for tunneling/hopping electrons. It is conceivable that the carriers are strongly trapped in polaron states inside PVDF and find the optimal path for hopping in the material, which is incompletely switched. This is an interesting topic that certainly is in need of further experimental and theoretical study.

9.2.3.3 Electrically Addressable Molecules

For many applications one needs an "intrinsic" molecular "switch," i.e., a bistable voltage-addressable molecular system with very different resistances in the two states that can be accessed very quickly. There is a trade-off between the stability of a molecular state and the ability to switch the molecule between two states with an external perturbation (we discuss an electric field; switching involving absorbed photons is impractical at a nanoscale). Indeed, the applied electric field, in the order of a typical breakdown field $E_b \lesssim 10^7$ V/cm, is much smaller than a typical atomic field ~10^9 V/cm, characteristic of the energy barriers. A small barrier would be a subject for sporadic thermal switching, whereas a larger barrier ~1 to 2 eV would be impossible to overcome with the applied field. One may only change the relative energy of the minima by external field and, therefore, redistribute the molecules statistically between the two states. An intrinsic disadvantage of the conformational mechanism, involving motion of an ionic group, exceeding the electron mass by many orders of magnitude, is a slow switching speed (~kHz). In case of supramolecular complexes like rotaxanes and catenanes [37], there are two entangled parts which can change mutual positions as a result of redox reactions (in solution). Thus, for the rotaxane-based memory devices a slow switching speed of ~10^{-2} sec was reported.

We have considered a bistable molecule with a $–CONH_2$ dipole group [75]. The barrier height is $E_b = 0.18$ eV. Interaction with an external electric field changes the energy of the minima, but the estimated switching field is huge, ~0.5 V/A. At nonzero temperatures, temperature fluctuations might result in

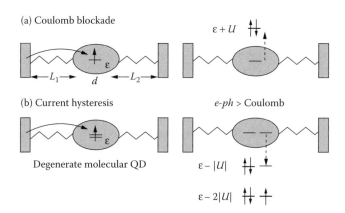

FIGURE 9.10   Schematic of the molQD with a central conjugated unit separated from the electrodes by wide-band insulating molecular groups. First, the electron tunnels into the dot and occupies an empty (degenerate) state there. If the interaction between the first and second incoming electrons is repulsive, $U > 0$, then the dot will be in a Coulomb blockade regime (a). If the electrons on the dot effectively attract each other, $U < 0$, the system will show current hysteresis (b).

statistical dipole flipping at lower fields. The I–V curve shows hysteresis in the 3 to 4 V window for two possible conformations. One can estimate the thermal stability of the state as 58 psec at room temperature, and 33 msec at 77 K.

9.2.3.4   Molecular Quantum Dot Switching

The molQD, as we define it, consists of a central "conjugated" unit (containing half-occupied, and, therefore, extended π–orbitals (Figure 9.10). Frequently, those are formed from the p-states on carbon atoms that are not "saturated" (i.e., they do not share electrons with other atoms forming strong σ–bonds, with a typical bonding–antibonding energy difference of about 1 Ry). Since the π–orbitals are half-occupied, they form the HOMO–LUMO states. The size of the HOMO–LUMO gap is then directly related to the size of the conjugated region d (Figure 9.10) by a standard estimate $E_{\text{HOMO–LUMO}} \sim \hbar^2/md^2 \sim 2$ to 5 eV. It is worth noting that in conjugated linear polymers like PA $(-C=C)n$ the spread of the π–electron would be $d = \infty$ and the expected $E_{\text{HOMO–LUMO}} = 0$. However, such a one-dimensional metal is impossible, Peierls distortion (C=C bond length dimerization) sets in and opens up a gap of about ~1.5 eV at the Fermi level [19]. In a molQD the central conjugated part is separated from electrodes by insulating groups with saturated σ-bonds, like, e.g., the alkane chains (Figure 9.3). Now, there are two main possibilities of carrier transport through the molQD. If the length of at least one of the insulating groups $L_{1(2)}$ is not very large (a conductance $G_{1(2)}$ is not much smaller than the conductance quantum $G_0 = 2e^2/h$), then the transport through the molQD will proceed by resonant tunneling processes. If, on the other hand, both groups are such that the tunnel conductance $G_{1(2)} \ll G_0$, the charge on the dot will be quantized. Then we will have another two possibilities: (i) the interaction of the extra carriers on the dot is "repulsive" $U > 0$, and we have a Coulomb blockade [76] or (ii) the effective interaction is "attractive," $U < 0$; then we would obtain the current "hysteresis" (see below). Coulomb blockade in molQDs has been demonstrated in Ref. [77]. In these works, and in Ref. [78], three-terminal active molecular devices have been fabricated and successfully tested.

Although the correlated electron transport through mesoscopic systems with repulsive electron–electron interactions received considerable attention in the past and continues to be the focus of current studies, much less has been known about a role of electron–phonon correlations in "molecular quantum dots" (molQD). Some time ago we proposed a negative-U Hubbard model of a d-fold degenerate quantum dot [79] and a polaron model of resonant tunneling through a molecule with degenerate level [53]. We found that the "attractive" electron correlations caused by any interaction within the molecule could lead to a molecular "switching effect" where I–V characteristics have two branches with high and low current at the same bias voltage. This prediction has been confirmed and extended further in the theory of "correlated" transport through degenerate molQDs with a full account of both the Coulomb repulsion and realistic electron–phonon (e–ph) interactions [53]. We have shown that while the phonon side-bands significantly modify the shape of hysteretic I–V curves in comparison with the negative-U Hubbard model, switching remains robust. It shows up when the effective interaction of polarons is attractive and the state of the dot is d-fold degenerate, $d > 2$. Different from the nondegenerate and two-fold degenerate dots, the rate equation for a multidegenerate dot, $d > 2$, weakly coupled to the leads, has multiple physical roots in a certain voltage range showing hysteretic behavior due to "correlations" between different electronic states of molQD [53]. Our conclusions are important for searching the current-controlled polaronic molecular switches. Incidentally, C_{60} molecules have the degeneracy $d = 6$ of the lowest unoccupied level, which makes them one of the most promising candidate systems, if weak-coupling with leads is secured.

9.2.3.5 What Molecular Quantum Dots Could Switch?

Note that switching required a degenerate MQD ($d > 2$) and the weak coupling to the electrodes, $\Gamma \ll \omega_0$. The case of strong coupling of nondegenerate level ($d = 1$) with $\Gamma > \omega_0$, where the electron transport is adiabatic, has been considered in Refs. [82,83]. Obviously, there is "no switching" in this case and this is exactly what these authors have found. The current hysteresis does not occur in their model, the current remains a single-valued function of bias with superimposed noise. Later, however, Galperin et al. [84] claimed that a strongly coupled ($\Gamma > \omega_0$, $\Gamma \approx 0.1$ to 0.3 eV) molecular bridge, which is nondegenerate ($d = 1$) does exhibit switching. This result is an artifact of their adiabatic approximation, as discussed in detail in Ref. [80]. Indeed, Galperin et al. [84] have obtained the current hysteresis in MQD as a result of illegitimate replacement of the occupation number operator $\hat{n}$ in the e–ph interaction by the average population n_0 (Eq. (2) of Ref. [84]) and found the average steady-state vibronic displacement $\langle d + d^\dagger \rangle$ proportional to n_0 (this is an explicit "neglect" of all quantum fluctuations on the dot accounted for in the exact solution). Then, by replacing the displacement operator $d + d^\dagger$ in the bare Hamiltonian, Eq. (11) of Ref. [84], by its average, Ref. [84], they obtained a new molecular level, $\tilde{\varepsilon}_0 = \varepsilon_0 - 2\varepsilon_{reorg} n_0$ shifted linearly with the average population of the level. The MFA spectral function turned out to be highly nonlinear as a function of the population, e.g., for the weak-coupling with the leads $\rho(\omega) = \delta(\omega - \varepsilon_0 - 2\varepsilon_{reorg} n_0)$, see Eq. (17) in Ref. [84]. As a result, the authors of Ref. 84 have found multiple ghost solutions for the steady-state population, Eq. (15) and Figure 1 of Ref. [84], and switching, Figure 4 of Ref. 84.

Note that the mean-field solution by Galperin et al. [84] applies at any ratio Γ / ω_0, including the limit of interest to us, $\Gamma \ll \omega_0$, where their transition between the states with $n_0 = 0$ and 1 only sharpens, but none of the results change. Therefore, MFA predicts a current bistability in the system where it does not exist at $d = 1$. The results in Ref. [84] are plotted for $\Gamma \geq \omega_0$, $\Gamma \approx 0.1$ to 0.3 eV, which corresponds to molecular bridges with a resistance of about a few 100 k ω. Such model "molecules" are rather "metallic" in their conductance and could hardly show any bistability at all because carriers do not have time to interact with vibrons on the molecule. Indeed, taking into account the coupling with the leads beyond the second order and the coupling between the molecular and both phonons could hardly provide any nonlinearity because these couplings do not depend on the electron population. This rather obvious conclusion for molecules strongly coupled to the electrodes can be reached in many ways; see, e.g., rather involved derivations in Refs. [82,83]. While Refs. [82,83] do talk about telegraph current noise in the model, there is no hysteresis in the adiabatic regime, $\Gamma \gg \omega_0$, either. This result certainly has nothing to do with our mechanism of switching [53] that applies to molQDs ($\Gamma \ll \omega_0$) with $d > 2$.

In reality, most of the molecules are very resistive, so the actual molQDs are in the regime we study, see Ref. [86]. For example, the resistance of fully conjugated three-phenyl ring Tour–Reed molecules chemically bonded to metallic Au electrodes [34] exceeds 1 GΩ. Therefore, most of the molecules of interest to us are in the regime that we discussed, not in that of Refs. [82,83].

9.2.4   Role of Defects in Molecular Transport

Interesting behavior of electron transport in molecular systems, as described above, refers to ideal systems without imperfection in ordering and composition. In reality, one expects that there will be considerable disorder and defects in organic molecular films. As mentioned above, the conduction through absorbed [55] and Langmuir–Blodgett [56] monolayers of fatty acids $(CH_2)_n$ was associated with defects. An absence of tunneling through self-assembled monolayers of C12 to C18 (inferred from an absence of thickness dependence at room temperature) has been reported by Boulas et al. [54]. On the other hand, the tunneling in alkanethiol SAMs was reported in [87,88], with an exponential dependence of monolayer resistance on the chain length L, $R_\sigma \propto \exp(\beta_\sigma L)$, and no temperature dependence of the conductance in C8 to C16 molecules was observed over the temperatures $T = 80$ to 300 K [88].

The electrons in alkane molecules are tightly bound to the C atoms by σ-bonds, and the band gap (between the HOMO and LUMO) is large, ~9 to 10 eV [54]. In conjugated systems with π-electrons the MOs are extended, and the HOMO–LUMO gap is correspondingly smaller, as in, e.g., polythiophenes, where the resistance was also found to scale exponentially with the length of the chain, $R_\pi \propto \exp(\beta_\pi L)$, with $\beta_\pi = 0.35$ Å^{-1} instead of $\beta_\sigma = 1.08$ Å^{-1} [87]. In stark contrast with the temperature-independent tunneling results for SAMs [88], recent extensive studies of electron transport through 2.8 nm thick eicosanoic acid (C20) LB monolayers at temperatures 2 to 300 K have established that the current is practically temperature independent below $T < 60$ K, but very strongly temperature dependent at higher temperatures $T = 60$ to 300 K [89].

A large amount of effort went into characterizing the organic thin films and possible defects there [20,90,91]. It has been found that the electrode material, like gold, gets into the body of the film, leading to the possibility of metal ions existing in the film as single impurities and clusters. Electronic states on these impurity ions are available for the resonant tunneling of carriers in very thin films (or hopping in thicker films, a crossover between the regimes depending on the thickness). Depending on the density of the impurity states, with increasing film thickness the tunneling will be assisted by impurity "chains," with an increasing number of equidistant impurities [92]. One-impurity channels produce steps on the I–V curve but no temperature dependence, whereas the inelastic tunneling through pairs of impurities at low temperatures defines the temperature dependence of the film conductance, $G(T) \propto T^{4/3}$, and the voltage dependence of current $I(V) \propto V^{7/3}$ [93]. This behavior has been predicted theoretically and observed experimentally for tunneling through amorphous Si [94] and Al_2O_3 [95]. Due to the inevitable disorder in a "soft" matrix, the resonant states on different impurities within a "channel" will be randomly moving in and out of resonance, creating mesoscopic fluctuations of the I–V curve. The tunneling may be accompanied by interaction with vibrons on the molecule, causing step-like features on the I–V curve [53,78].

During processing, especially top electrode deposition, small clusters of the electrode material may form in the organic film, causing Coulomb blockade, which also can show up as steps on the I–V curve. It has long been known that a strong applied field can cause localized damage to thin films, presumably due to electromigration and the formation of conducting filaments [57]. The damaged area was about 30 nm in diameter in 40 to 160 monolayer thick LB films [57](a) and 5 to 10 μm in diameter in films 500 to 5000 Å thick, and showed switching behavior under external bias voltage cycling [57](b). As discussed above, recent spatial mapping of a conductance in LB monolayers of fatty acids with the use of conducting AFM has revealed damage areas 30 to 100 nm in diameter, frequently appearing in samples after a "soft" electrical breakdown, which is sometimes accompanied by a strong temperature dependence of the conductance through the film [59].

A crossover from tunneling at low temperatures to an activation-like dependence at higher temperatures is expected for electron transport through organic molecular films. There are recent reports about such a crossover in individual molecules like the 2 nm long Tour wire with a small activation energy $E_a \approx 130$ meV [96]. Very small activation energies in the order of 10 to 100 meV have been observed in polythiophene monolayers [97]. Our present results suggest that this may be a result of interplay between the drastic renormalization of the electronic structure of the molecule in contact with electrodes and disorder in the film (Figure 9.12, right inset). We report the ab-initio calculations of point defect–assisted

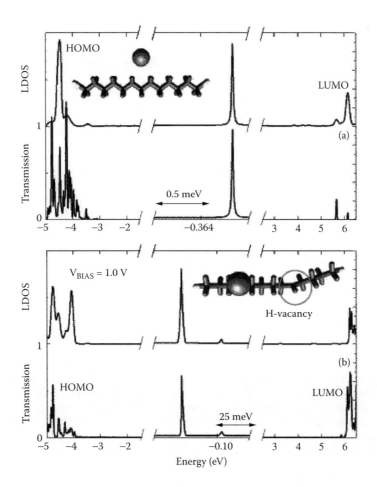

FIGURE 9.11 Local density of states and transmission as a function of energy for (a) C13 with Au impurity and (b) C13 with Au impurity and H-vacancy (*dangling bond*). Middle sections show closeups of the resonant peaks due to deep defect levels with respect to the HOMO and LUMO molecular states. The HOMO–LUMO gap is about 10 eV.

tunneling through alkanedithiols $S(CH_2)_nS$ and thiophene T3 (three rings SC_4) self-assembled on gold electrodes. The length of the alkane chain was in the range $n = 9$ to 15.

We have studied single and double defects in the film: (i) single Au impurity (Figure 9.10a and Figure 9.11a), (ii) Au impurity and H-vacancy (dangling bond) on the chain (Figure 9.12b and Figure 9.12c), (iii) a pair of Au impurities (Figure 9.11b), and (iv) Au and a "kink" on the chain (one C=C bond instead of a C–C bond). Single defect states result in steps on the associated I–V curve, whereas molecules in the presence of two defects generally exhibit an NDR. Both types of behavior are generic and may be relevant to some observed unusual transport characteristics of SAMs and LB films [55,56,59,89,96,97]. We have used an ab-initio approach that combines the Keldysh NEGF method with self-consistent pseudopotential-based real space DFT for systems with open boundary conditions provided by semi-infinite electrodes under external bias voltage [50,52]. All present structures have been relaxed with the Gaussian98 code prior to transport calculations [98]. The conductance of the system at a given energy is found from Eq. (9.8) and the current from Eq. (9.4).

The equilibrium position of an Au impurity is about 3 Å away from the alkane chain, which is a typical van-der-Waals distance. As the density maps show (Figure 9.12), there is an appreciable hybridization between the s- and d-states of Au and the sp-states of the carbohydrate chain. Furthermore, the Au^+ ion produces a Coulomb center trapping of a 6s electron state at an energy $\varepsilon_i = -0.35$ eV with respect to the

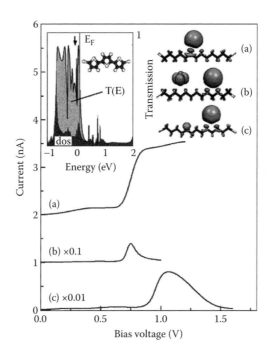

FIGURE 9.12 Current-voltage characteristics of an alkane chain C13 with (a) single Au impurity (6s-state), (b) two Au impurities (5d and 6s-states on left and right ions, respectively), and (c) Au impurity and H-vacancy (*dangling bond*). Double defects produce the NDR peaks (b) and (c). Inset shows the density of states, transmission, and stick model for polythiophene T3. There is significant transmission at the Fermi level, suggesting an ohmic I–V characteristic for T3 connected to gold electrodes. Disorder in the film may localize states close to the Fermi level (schematically marked by an arrow), which may assist in hole hopping transport with an apparently very low activation energy (0.01 to 0.1 eV), as is observed.

Fermi level, almost in the middle of the HOMO–LUMO ~10 eV gap in Cn. The tunneling evanescent resonant state is a superposition of the HOMOs and LUMOs. Those orbitals have a very complex spatial structure, reflected in an asymmetric line shape for the transmission. Since the impurity levels are very deep, they may be understood within the model of "short-range impurity potential" [99]. Indeed, the impurity wave function outside of the narrow well can be fairly approximated as follows:

$$\varphi(r) = \sqrt{\frac{2\pi}{\kappa}}\,\frac{e^{-\kappa r}}{r}, \tag{9.9}$$

where κ is the inverse radius of the state, $\hbar^2\kappa^2/2m^* = E_i$, where $E_i = \Delta - \varepsilon_i$ is the depth of the impurity level with respect to the LUMO, and Δ = LUMO – F is the distance between the LUMO and the Fermi level F of gold and, consequently, the radius of the impurity state $1/\kappa$ is small. The energy distance $\Delta \approx$ 4.8 eV in alkane chains $(CH_2)_n$ [54] ($\approx$ 5 eV from DFT calculations), and $m^* \sim 0.4$ the effective tunneling mass in alkanes [88]. For one impurity in a rectangular tunnel barrier [99] we obtain the Breit–Wigner form of transmission $T(E,V)$, as before, Eq. (9.5). Using the model with the impurity state wavefunction, we may estimate for an Au impurity in C13 ($L = 10.9$ Å) the width $\Gamma_L = \Gamma_R = 1.2 \times 10^{-6}$, which is within an order of magnitude compared with the calculated value 1.85×10^{-5} eV. The transmission is maximal and equals unity when $E = \varepsilon_i$ and $\Gamma_L = \Gamma_R$, which corresponds to a symmetrical position for the impurity with respect to the electrodes.

The electronic structure of the alkane backbone, through which the electron tunnels to an electrode, shows up in the asymmetric lineshape, which is substantially non-Lorentzian (Figure 9.11). The current remains small until the bias has aligned the impurity level with the Fermi level of the electrodes, resulting in a step in the current, $I_1 \approx \frac{2q}{\hbar} \Gamma_0 e^{-\kappa L}$ (Figure 9.11a). This step can be observed only when the impurity level is not very far from the Fermi level F, such that biasing the contact can produce alignment before a breakdown of the device may occur. The most interesting situations that we have found relate to the "pairs" of point defects in the film. If the concentration of defects is $c \ll 1$, the relative number of configurations with pairs of impurities will be very small, $\propto c^2$. However, they give an exponentially larger contribution to the current. Indeed, the optimal position of two impurities is symmetrical, a distance $L/2$ apart, with current $I_2 \propto e^{-\kappa L/2}$. The conductance of a two-impurity chain is [99]

$$g_{12}(E) = \frac{4q^2}{\pi\hbar} \frac{\Gamma_L \Gamma_R t_{12}^2}{\left| (E - \varepsilon_1 + i\Gamma_L)(E - \varepsilon_2 + i\Gamma_R) - t_{12}^2 \right|^2} \qquad (9.10)$$

For a pair of impurities with slightly differing energies $t_{12} = 2(E_1 + E_2)e^{-\kappa r_{12}}/\kappa r_{12}$, where r_{12} is the distance between them. The interpretation of the two-impurity channel conductance (10) is fairly straightforward: if there were no coupling to the electrodes, i.e., $\Gamma_L = \Gamma_R = 0$, the poles of g_{12} would coincide with the bonding and antibonding levels of the two-impurity "molecule." The coupling to the electrodes gives them a finite width and produces, generally, two peaks in conductance, whose relative positions in energy change with the bias. The same consideration is valid for longer chains too, and gives an intuitive picture of the formation of the impurity "band" of states. The maximal conductivity $g_{12} = q^2/\pi\hbar$ occurs when $\varepsilon_1 = \varepsilon_2$, $\Gamma_L \Gamma_R = t_{12}^2 = \Gamma_2^2$, where Γ_2 is the width of the two-impurity resonance, and it corresponds to the symmetrical position of the impurities along the normal to the contacts separated by a distance equal to half of the molecule length, $r_{12} = L/2$. The important property of the two-impurity case is that it produces NDR. Indeed, under external bias voltage the impurity levels shift as follows:

$$\varepsilon_i = \varepsilon_{i0} + qVz_i/L \qquad (9.11)$$

where z_i are the positions of the impurity atoms counted from the center of the molecule. Due to disorder in the film, under bias voltage, the levels will be moving in and out of resonance, thus producing NDR peaks on the I–V curve. The most pronounced NDR is presented by a gold impurity next to a Cn chain with an H-vacancy on one site (Figure 9.11b) (the defect corresponds to a dangling bond). The defects result in two resonant peaks in transmission. Surprisingly, the H-vacancy (dangling bond) has an energy very close to the electrode Fermi level F, with $\varepsilon_i = -0.1$ eV (Figure 9.11b, right peak). The relative positions of the resonant peaks move with an external bias and cross at 1.2 V producing a pronounced NDR peak in the I–V curve (Figure 9.12c). No NDR peak is seen in the case of an Au impurity and a kink C=C on the chain because the energy of the kink level is far from that of the Au 6s impurity level. The calculated values of the peak current through the molecules were large: $I_p \approx 90$ nA/molecule for an Au impurity with H-vacancy, and ≈ 5 nA/molecule for double Au impurities.

We have observed a new mechanism for the NDR peak in a situation with two Au impurities in the film. Namely, Au ions produce two sets of deep impurity levels in Cn films, one stemming from the 6s orbital, another from the 5d shell, as clearly seen in Figure 9.12b (inset). The 5d-states are separated in energy from 6s, so that now the tunneling through s–d pairs of states is allowed in addition to s–s tunneling. Since the 5d-states are at a lower energy than the s-state, the d- and s-states on different Au ions will be aligned at a certain bias. Due to the different angular character of those orbitals, the tunneling between the s-state on the first impurity and a d-state on another impurity will be described by the hopping integral analogous to the Slater–Koster $sd\sigma$ integral. The peak current in that case is smaller than for the pair Au–H-vacancy, where the overlap is of $ss\sigma$ type (see Figure 9.12b and Figure 9.12c).

Thiophene molecules behave very differently since the π-states there are conjugated and, consequently, the HOMO–LUMO gap is much narrower, just below 2 eV. The tail of the HOMO state in the T3 molecule (with three rings) has a significant presence near the electrode Fermi level, resulting in a practically "metallic" density of states and hence an ohmic I–V characteristic. This behavior is quite robust and is in apparent disagreement with experiments, where tunneling has been observed [87]. However, in actual thiophene devices the contact between the molecule and electrodes is obviously very poor, and it may lead to unusual current paths and temperature dependence [97].

We have presented the first parameter-free DFT calculations of a class of organic molecular chains incorporating single or double point defects. The results suggest that the present generic defects produce deep impurity levels in the film and cause a resonant tunneling of electrons through the film, strongly dependent on the type of defects. Thus, a missing hydrogen produces a level (dangling bond) with energy very close to the Fermi level of the gold electrodes F. In the case of a single impurity, it produces steps on the I–V curve when one electrode's Fermi level aligns with the impurity level under a certain bias voltage. The two-defect case is much richer, since in this case we generally see a formation of the NDR peaks. We found that the Au atom together with the hydrogen vacancy (dangling bond) produces the most pronounced NDR peak at a bias of 1.2 V in C13. Other pairs of defects do not produce such spectacular NDR peaks. A short-range impurity potential model reproduces the data very well, although the actual line shape is different.

There is a question remaining of what may cause the strong temperature dependence of conductance in "simple" organic films like $[CH_2]_n$. The activation-like conductance $\propto \exp(-E_a/T)$ has been reported with a small activation energy $E_a \sim 100$ to 200 meV in alkanes [89,96] and even smaller, 10 to 100 meV, in polythiophenes [97]. This is much smaller than the value calculated here for alkanes and expected from electrical and optical measurements on Cn molecules, $E_a \sim \Delta \sim 4$ eV [54], which correspond well to the present results. In conjugated systems, however, there may be rather natural explanation of small activation energies. Indeed, the HOMO in T3 polythiophene on gold is dramatically broadened, shifted to higher energies, and has a considerable weight at the Fermi level. The upward shift of the HOMO is just a consequence of the work function difference between gold and the molecule. In the presence of (inevitable) disorder in the film some of the electronic states on the molecules will be localized in the vicinity of E_F. Those states will assist the thermally activated hopping of holes within a range of small activation energies $\lesssim 0.1$ eV. Similar behavior is expected for Tour wires [96], where E_F - HOMO ~ 1 eV [52](c), if the electrode–molecule contact is poor, as is usually the case.

With regards to carrier hopping in monolayers of saturated molecules, one may reasonably expect that in many studied cases the organic films are riddled with metallic protrusions (filaments), emerging due to electromigration in a very strong electric field, and/or metallic, hydroxyl, etc., inclusions [57,59]. It may result in a much smaller tunneling distance d for the carriers and the image charge lowering of the barrier. The image charge lowering of the barrier in a gap of width d is $\Delta U = q^2 \ln 4 / (\varepsilon d)$, meaning that a decrease of about 3.5 eV may only happen in an unrealistically narrow gap $d = 2$ to 3 Å in a film with a dielectric constant $\varepsilon = 2.5$, but it will add to the barrier lowering. More detailed characterization and theoretical studies along these lines may help to resolve this very unusual behavior. We note that such a mechanism cannot explain the crossover with temperature from tunneling to hopping reported for single molecular measurements, which has to be a property of the device, but not a single molecule [96].

9.3 Spin Injection: Processes and Devices

Spin transport in metal-, metal-insulator, and semiconductor heterostructures holds promise for the next generation of high-speed low-power electronic devices [3,4,100–107]. Amongst important spintronic effects already used in practice one can indicate a giant magnetoresistance (MR) in magnetic multilayers [108] and tunnel magnetoresistance (TMR) in FM–insulator–FM (FM–I–FM) structures [109–115]. Another promising effect is domain wall switching in giant magnetoresistance (GMR) nanopillars directly by the flow of current through them [116–120]. Injection of spin-polarized electrons into semiconductors is of particular interest because of relatively large spin relaxation time (~1 nsec in semiconductors, ~1 msec in organics) [4] during which the electron can travel over macroscopic distances without losing polarization, or stay in a quantum dot/well. This also opens up possibilities, albeit speculative ones, for quantum information processing using spins in semiconductors.

The potential of spintronic devices is illustrated easiest with the simplest spin-dependent transport process, which is a TMR in FM–I–FM structure, described in the next section. The effect is a simple consequence of the golden rule that dictates a dependence of the tunnel current on the density of initial and final states for the tunneling electron. Most of the results for tunnel spin junctions would be reused later in describing the spin injection from FMs into semiconductors (or vice versa) in the later sections.

It is worth indicating right from the beginning that there are two major characteristics of the spin transport processes that will define the outcome of a particular measurement that are called a spin polarization and spin injection efficiency. They may be very different from each other, and this may lead (and frequently does) to confusion among researchers. The spin polarization measures the imbalance in the density of electrons with opposite spins (spin accumulation/depletion),

$$P = \frac{n_\uparrow - n_\downarrow}{n_\uparrow + n_\downarrow}, \tag{9.12}$$

whereas the injection efficiency is the polarization of injected "current" J

$$\Gamma = \frac{J_\uparrow - J_\downarrow}{J_\uparrow + J_\downarrow}, \tag{9.13}$$

where $\uparrow$ ($\downarrow$) refers to the electron spin projection on a quantization axis. In case of ferromagnetic materials the axis is antiparallel (AP) to the magnetization moment $\vec{M}$. Generally, $P \neq \Gamma$, but in some cases they can be close. Since in the FMs the spin density is constant, one can make a reasonable assumption that the current is carried independently by two electron fluids with opposite spins (Mott's two-fluid model [121]). Then, in the FM bulk the parameter

$$\Gamma = P_{FM} \equiv (\sigma_\uparrow - \sigma_\downarrow)/\sigma, \tag{9.14}$$

where $\sigma_{\uparrow(\downarrow)}$ are the conductivities of up-, down-spin electrons in FM, $\sigma = \sigma_\uparrow + \sigma_\downarrow$.

Looking at spin tunneling, we find that Γ characterizes the value of MR in magnetic tunnel junctions (MTJs), which is quite obvious since there one measures the difference between currents in two configurations: with parallel (P) and AP moments on electrodes. The tunnel current is small, hence the injected spin density is minute compared to metallic carrier densities. At the same time, in experiments where one injects spin (creates nonequilibrium spin population) in a quantum well, where it results in emission of polarized light (spinLED [123]) its measured intensity is, obviously, proportional to the spin polarization P.

We shall briefly outline the major spin-transport effects here in the following three sections. We shall start with an analysis of TMR, then GMR, naturally followed by the spin-torque (ST) switching in magnetic nanopillars. We then proceed with the main part, discussing an efficient spin injection in ferromagnet-semiconductor junctions through modified Schottky barriers, spin injection and extraction, ultrafast spin valve mechanism, and a few possible spintronic devices for low-power high-speed operation.

9.3.1 Tunnel Magnetoresistance

TMR is observed in metal–insulator–metal MTJs, usually with Ni–Fe, Co–Fe electrodes and (amorphous) Al_2O_3 tunnel barrier where one now routinely observes upward of a 40% change in conductance as a result of changing relative orientation of magnetic moments on electrodes. A considerably larger effect, about 200% TMR is found in Fe–MgO–Fe junctions with an epitaxial barrier that may be related to surface states and/or peculiarities of the band structure of the materials. As we shall see shortly, TMR is basically a simple effect of a difference between densities of spin-up and -down (initial and final tunneling) states. TMR is intimately related to GMR [108], i.e., a giant change in conductance of magnetic multilayers with relative orientation of magnetic moments in the stack.

Let us estimate the TMR using the golden rule that says that the tunnel current at small bias voltage V is $J_\sigma = G_\sigma V$, $G_\sigma \propto |M|^2 g_{i\sigma} g_{f\sigma}$, where $g_{i(f)\sigma}$ is the density of initial (final) tunneling states with a spin projection, and M is the tunneling matrix element. Consider the case of electrodes made with the same material. Denoting $D = g_\uparrow$ and $d = g_\downarrow$, we can write down the following expression for P and AP moments on the electrodes:

$$G_P \propto D^2 + d^2, \ G_{AP} \propto 2Dd, \tag{9.15}$$

and arrive at the expression for TMR first derived by Jullieres

$$TMR \equiv \frac{G_P - G_{AP}}{G_{AP}} = \frac{(D - d)^2}{2Dd} = \frac{2P^2}{1 - P^2}, \tag{9.16}$$

where we have introduced a definition of polarization

$$P = \frac{D - d}{D + d} \equiv \frac{g_\uparrow - g_\downarrow}{g_\uparrow + g_\downarrow} \tag{9.17}$$

It is worth noting that this definition of polarization is one of many. Indeed, we shall see immediately that obviously the "polarization" entering an expression for a particular process depends on particular physics and also the nature of the electronic states involved. Thus, the definition (Eq. 9.17) may lead one to believe that the "tunnel" polarization in elemental Ni should be negative, since at the Fermi level one has a sharp peak in the minority carrier density of states. The data, however, unambiguously suggest that the tunnel polarization in Ni is positive, $P > 0$ [111]. This finds a simple explanation in the model by Stearns who pointed out an existence of highly polarized d-states with small mass, close to one of the free electrons [139].

9.3.2   Spin-Torque Domain Wall Switching in Nanomagnets

Magnetic memory based on TMR is nonvolatile, may be rather fast (~1 nsec) and can be scaled down considerably toward a paramagnetic limit observed in nanomagnets. Switching, however, requires MTJ to be placed at a crosspoint of bit and word wires carrying current that produces a sufficient magnetic field for switching domain orientation in a "free" (unpinned) MTJ ferromagnetic electrode. The undesirable side effect is crosstalk between cells, rather complex layout and power budget.

Alternatively, one may take a GMR multilayer with AP orientation of magnetic moments and run the current perpendicular to layers (CPP geometry). In this case there will be a spin accumulation in the drain layer, i.e., the accumulation of minority spins. This means a transfer of spin (angular) moment across the GMR spacer layer. Injection of angular momentum means that there is a torque on the magnetic moment in the drain layer that may cause its switching. This is indeed what has been suggested by Berger and Slonczewski in 1996 [116,117], and confirmed experimentally in experiments on nanopillar multilayers by Tsoi et al. in 1998 [118].

The spin accumulation (and torque) is proportional to current density through the drain magnet or magnetic particle subject to polarized current (see below). It has been estimated by Slonczewski in a very simple effective mass model as

$$\vec{T}_I = \frac{gI}{qS}[\vec{n}_I \times \vec{n}], \tag{9.18}$$

$$g = A \sin\theta / (1 + B \cos\theta) \tag{9.19}$$

where $g(\theta)$ is the phenomenological ST function with $\cos\theta = \vec{n}_I \cdot \vec{n}$ [119,120], S is the total spin of the magnetic particle pointing along unit vector $\vec{n}$, I the total current through the particle with mean injected polarization pointing along $\vec{n}_I$, and q the elementary charge. The torque results in revolution of the moment on the particle, and it is described by the Landau–Lifshitz equation with the spin torque (18) added to its right-hand side. Time-resolved measurements of current-induced reversal of a free magnetic layer in Permalloy–Cu–Permalloy elliptical nanopillars at temperatures $T = 4.2$ to 160 K [120]. The values of A and B are in fair numerical agreement with those calculated from the two-channel model using the measured MR values of the nanopillar spin valve. There was, however, considerable device-to-device variation in the ST asymmetry parameter B. This is attributed to the presence of an antiferromagnetic oxide layer around the perimeter of the Permalloy free layer. Obviously, controlling this layer would be very important for the viability of the whole approach for applications. There are reports on the activation character of switching that may be related to the (de)pinning of the domain walls at the side walls of the pillar.

9.3.3 Spin Injection/Extraction into (from) Semiconductors

The principal difficulty of the spin injection is that the materials in the FM–S junction usually have very different electron affinity and, therefore, a high potential Schottky barrier forms at the interface [136] (Figure 9.13, curve 1). For GaAs and Si, the barrier height $\Delta \simeq 0.5$ to 0.8 eV with practically all metals, including Fe, Ni, and Co, [123,136] and the barrier width is large, $l \gtrsim 100$ nm for doping concentration $N_d = 10^{17}$ cm^{-3}. The spin injection corresponds to a reverse current in the Schottky contact, which is saturated and usually negligible due to such large l and Δ [136]. Therefore, a thin heavily doped n$^+$–S layer between FM metal and S is used to increase the reverse current [136] determining the spin injection [103,105,123,130]. This layer sharply reduces the thickness of the barrier, and increases its tunneling transparency [103,136]. Thus, a substantial spin injection has been observed in FM–S junctions with a thin n$^+$-layer [123].

A usually overlooked formal paradox of spin injection is that a current through Schottky junctions in prior theories depended solely on parameters of a semiconductor [136] and cannot formally be spin-polarized. Some authors even emphasize that in Schottky junctions "spin-dependent effects do not occur" [128]. In earlier works [125–134], spin transport through the FM–S junction, its spin-selective properties, and nonlinear I–V characteristics have not been actually calculated. They were described by various, often contradictory, boundary conditions at the FM–S interface. For example, Aronov and Pikus assumed that a spin polarization of current at the FM–S interface is a constant equal to that in the FM, $\Gamma = P_{FM}$ (Eq. 9.14), with $\sigma_{\uparrow(\downarrow)}$, the conductivities of up- and down-spin electrons in FM, $\sigma = \sigma_\uparrow + \sigma_\downarrow$, and then studied nonlinear spin accumulation in S considering spin diffusion and drift in the electric field [125]. The authors of Refs. [126–130] assumed a continuity of both the currents and the electrochemical potentials for both spins and found that a spin polarization of injected electrons depends on a ratio of conductivities of a FM and S (the so-called "conductivity mismatch" problem). At the same time, it has been asserted in Refs. [131–134] that the spin injection becomes appreciable when the electrochemical potentials have a substantial discontinuity (produced by, e.g., a tunnel barrier [132]). The effect, however, was described by the unknown spin-selective interface conductances $G_{i\sigma}$, which cannot be found with those theories.

We have developed a microscopic theory of the spin transport through FM–S junctions, which include an ultrathin heavily doped semiconductor layer (δ-doped layer) between FM and S [103,105]. We have studied nonlinear effects of spin accumulation in S near reverse-biased modified FM–S junctions with the δ-doped layer [103] and spin extraction from S near the modified forward-biased FM–S junctions [105]. We found conditions for the most efficient spin injection, which are opposite to the results of previous phenomenological theories. We show that (i) the current of the FM–S junction does depend on spin parameters of the ferromagnetic metal but not its conductivity, so, contrary to the results in Refs. [126–130,132–134], the "conductivity mismatch" problem does not arise for the Schottky FM–S junctions; we find also that (ii) a spin injection efficiency (polarization of current) Γ of the FM–S junction strongly depends on the current, contrary to the assumptions in Refs. [125–130,132–134]; (iii) the highest spin polarization of both the injected electrons P_n and spin injection efficiency can be realized at room temperatures and relatively small currents in high-resistance semiconductors, contrary to claims in Ref. [129], which are of most interest for spin injection devices [100,101,103]; we show that (iv) tunneling resistance of the FM–S junction has to be relatively small, which is "opposite" to the condition obtained in linear approximation in Ref. [132], and (v) the spin-selective interface conductances $G_{i\sigma}$ are not constants, as was assumed in Refs. [131–134], but vary with a current J in a strongly nonlinear fashion. We have suggested a new class of spin devices on the basis of the present theory.

In the following sections, we describe a general theory of spin current, spin injection and extraction (Section 9.3.4), followed by the discussion of the conditions of an efficient spin injection and extraction (Section 9.3.5). Further, we turn to the discussion of high-frequency spin valve effects in a system with two δ-doped Schottky junctions. This lays down the groundwork for the discussion (Section 9.3.6) of a new class of spin devices that include field detector, spin transistor, and square-law detector. The efficient spin injection and extraction may be a basis for efficient sources of (modulated) polarized radiation, and we discuss such a device (Section 9.3.7).

9.3.4 Spin Accumulation and Extraction

The modified FM–S junction with transparent Schottky barrier is produced by δ-doping the interface by sequential donor and acceptor doping. The Schottky barrier is made very thin by using large donor doping N_d^+ in a thin layer of thickness l. For reasons to become clear shortly, we would like to have a narrow spike followed by the narrow potential well with the width w and the depth $\sim rT$, where T is the temperature in units of $k_B = 1$ and r $\sim$ 2 to 3, produced by an acceptor doping N_a^+ a of the layer w (Figure 9.13). When donor and acceptor concentrations, N_d^+ and N_a^+, and the corresponding thicknesses of the doping profile, l and w, satisfy the conditions:

$$N_d^+ l^2 q^2 \simeq 2\varepsilon\varepsilon_0(\Delta - \Delta_0 - rT),$$

$$N_a^+ w^2 q^2 \simeq 2\varepsilon\varepsilon_0 rT, \tag{9.20}$$

and $l \lesssim l_0$, where $l_0 = \sqrt{\hbar^2 / [2m_*(\Delta - \Delta_0)]}$ ($l_0 \lesssim 2$ nm), the remaining low (and wide) barrier will have the height $\Delta_0 = (E_{c0} - F) > 0$, where E_{c0}, the bottom of the conduction band, is S in equilibrium, q is the

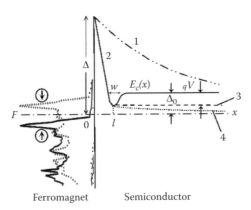

Ferromagnet   Semiconductor

FIGURE 9.13   Energy diagrams of a FM–S heterostructure with δ-doped layer (F is the Fermi level; Δ the height and l the thickness of an interface potential barrier; Δ_0 the height of the thermionic barrier in n-semiconductor). The standard Schottky barrier (*curve 1*); $E_c(x)$ the bottom of the conduction band in an n-semiconductor in equilibrium (*curve 2*), under small (*curve 3*), and large (*curve 4*) bias voltage. The spin polarized density of states in Ni is shown at $x < 0$.

elementary charge, and $\varepsilon (\varepsilon_0)$ is the dielectric permittivity of S (vacuum). A value of Δ_0 can be set by choosing a donor concentration in S,

$$N_d = N_c \exp[(F^S - E_{c0})/T] = N_c \exp(-\Delta_0 / T) = n \tag{9.21}$$

where F^S is the Fermi level in the semiconductor bulk, $N_c = 2M_c(2\pi m_* T)^{3/2} h^{-3}$ is the effective density of states and M_c is the number of effective minima of the semiconductor conduction band, and n and m_* is the concentration and effective mass of electrons in S [136]. Owing to small barrier thickness l, the electrons can rather easily tunnel through the δ-spike but only those with an energy $E \geq E_c$ can overcome the wide barrier Δ_0 due to thermionic emission, where $E_c = E_{c0} + qV$. We assume here the standard convention that the bias voltage $V < 0$ and current $J < 0$ in the reverse-biased FM–S junction and $V > 0$ ($J > 0$) in the forward-biased FM–S junction [136]. At positive bias voltage $V > 0$ we assume that the bottom of the conduction band shifts upward to $E_c = E_{c0} + qV$ with respect to the Fermi level of the metal.

Presence of the mini-well allows for keeping the thickness of the δ-spike barrier equal to $l \leq l_0$ and its transparency high at voltages $qV \lesssim rT$ (see below).

We assume elastic coherent tunneling, so that the energy E, spin σ, and $\vec{k}_{\parallel}$ (the component of the wave vector $\vec{k}$ parallel to the interface) are conserved. The exact current density of electrons with spin $\sigma = \uparrow, \downarrow$ through the FM–S junction containing the δ-doped layer (at the point $x = l$, Figure 9.13) can be written as [115,137]

$$J_{\sigma 0} = \frac{q}{h} \int dE [f(E - F_{\sigma 0}^S) - f(E - F_{\sigma 0}^{FM})] \int \frac{d^2 k_{\parallel}}{(2\pi)^2} T_{k\sigma} \tag{9.22}$$

where $T_{k\sigma}$ is the transmission probability, $f(E - F)$ the Fermi function, $F_{\sigma 0}^S$ ($F_{\sigma 0}^{FM}$) are the spin quasi-Fermi levels in the semiconductor (FM) near the FM–S interface, and the integration includes a summation with respect to a band index. Note that here we study a strong spin accumulation in the semiconductor. Therefore, we use "nonequilibrium" Fermi levels, $F_{\sigma 0}^{FM}$ and $F_{\sigma 0}^S$, describing distributions of electrons with spin $\sigma = \uparrow, \downarrow$ in the FM and the S, respectively, which is especially important for the semiconductor. This approach is valid when the spin relaxation time τ_s is much larger than the relaxation time of electron energy, τ_E, which is met in practically all semiconductors in a wide range of temperatures, including the room temperature. In particular, the electron density with spin σ in the S at the FM–S junction is given by

$$n_{\sigma 0} = (1/2)N_c \exp[(F_{\sigma 0}^S - E_c)/T], \tag{9.23}$$

where $F_{\sigma 0}$ is a quasi-Fermi level at a point $x = l$. One can see from (Eq. 9.22) that the current $J_{\sigma 0} = 0$ if we take $F_{\sigma 0}^{FM} = F_{\sigma 0}^S$, i.e., if we were to use the assumption of Refs. [126–130]. In reality, due to very high electron density in FM metal in comparison with electron density in S, $F_{\sigma 0}^{FM}$ differs negligibly from the equilibrium Fermi level F for currents under consideration, therefore, we can assume that $F_{\sigma 0}^{FM} = F$, as in Refs. [115,137] (see discussion below).

The current (Eq. 9.22) should generally be evaluated numerically for a complex band structure $E_{k\sigma}$ [138]. The analytical expressions for $T_{\sigma}(E, k_{\parallel})$ can be obtained in an effective mass approximation, $\hbar k_{\sigma} = m_{\sigma} v_{\sigma}$, where $v_{\sigma} = |\nabla E_{k\sigma}|/\hbar$ is the band velocity in the metal. This applies to "fast" free-like d-electrons in elemental FMs [115,139]. The present Schottky barrier has a "pedestal" with a height $(E_c - F) = \Delta_0 + qV$ which is opaque at energies $E < E_c$. For $E > E_c$ we approximate the δ-barrier by a triangular shape and one can use an analytical expression for $T_{\sigma}(E, k_{\parallel})$ [102] and find the spin current at the bias $0 < -qV \lesssim rT$, including at room temperature,

$$J_{\sigma 0} = j_{0d\sigma} \left[\frac{2n_{\sigma 0}(V)}{n} - \exp\left(-\frac{qV}{T}\right) \right] \tag{9.24}$$

$$j_0 = \alpha_0 nq v_T \exp(-\eta \kappa_0 l) \tag{9.25}$$

with the most important spin factor

$$d_\sigma = \frac{v_T v_{\sigma 0}}{v_{t0}^2 + v_{\sigma 0}^2}. \tag{9.26}$$

Here $\alpha_0 = 1.2(\kappa_0 l)^{1/3}$, $\kappa_0 \equiv 1/l_0 = (2m_*/\hbar^2)^{1/2}(\Delta - \Delta_0 - qV)^{1/2}$, $v_{t0} = \sqrt{2(\Delta - \Delta_0 - qV)/m_*}$ is the characteristic "tunnel" velocity, $v_\sigma = v_\sigma(E_c)$ the velocity of polarized electrons in FM with energy $E = E_c$, $v_T = \sqrt{3T/m_*}$ the thermal velocity. At larger reverse bias the miniwell on the right from the spike in Figure 9.13 disappears and the current practically saturates. Note that in the present case the carriers are subject to Boltzmann distribution $f(E-F) \approx \exp\frac{F-E}{T}$ in Eq. (22) for energies of interest, $E-F > rT$, and only small parallel momenta $k_\parallel \lesssim \sqrt{m_*T}/\hbar$ contribute to the integral [102]. Quite obviously, the tunneling electrons incident almost normally at the interface contribute most of the current, so that the peripheral areas of spin up and down Fermi surfaces do not matter much (more careful sampling can be done in special cases like, e.g., resonant tunneling levels in the barrier [115], or numerically when more quantitative results for d-states with complex Fermi surfaces are desired [138]).

One can see from Eq. (9.24) that the total current $J = J_{\uparrow 0} + J_{\downarrow 0}$ and its spin components $J_{\sigma 0}$ depend on a conductivity of a semiconductor but not that of an FM, as in usual Schottky junction theories [136]. On the other hand, $J_{\sigma 0}$ is proportional to the spin factor d_σ and the coefficient $j_0 d_\sigma \propto v_T^2 \propto T$, but not the usual Richardson's factor T^2 [136]. Expression (9.24) for current in a FM–S structure is valid for any sign of the bias voltage V. Note that at $V > 0$ (forward bias) it determines the spin current from S into FM. Hence, it describes spin extraction from S [105].

Consider "spin injection" that occurs at reverse bias, $V < 0$. For $-qV \approx rT \approx (2-3)T$ the value of $\exp(-qV/T) \gg 2n_{\sigma 0}/n \sim 1$ and, according to Eq. (9.24), the spin polarization of the current, $P_F = \Gamma$, and the spin current at the FM–S junction are equal, respectively,

$$P_F = \frac{J_{\uparrow 0} - J_{\downarrow 0}}{J_{\uparrow 0} + J_{\downarrow 0}} = \frac{d_\uparrow - d_\downarrow}{d_\uparrow + d_\downarrow} = \frac{(v_{\uparrow 0} - v_{\downarrow 0})(v_{t0}^2 - v_{\uparrow 0} v_{\downarrow 0})}{(v_{\uparrow 0} + v_{\downarrow 0})(v_{t0}^2 + v_{\uparrow 0} v_{\downarrow 0})} \tag{9.27}$$

$$J_{\uparrow 0} = (1 + P_F)J/2, \tag{9.28}$$

where $v_{\sigma 0} = v_\sigma(E_c)$ with $E_c = E_{c0} + qV$.

It is worth noting that, generally, $P_F \neq P_{FM}$ since this P_F is the polarization of the tunneling current from the FM through the modified Schottky barrier and is renormalized by tunneling, as it is in TMR. At the same time P_{FM} is the polarization of the current in the bulk FM, see Eq. (9.14). Moreover, P_F depends on a bias voltage V and differs from that in usual tunneling MIM structures [115], since in the present structure P_F refers to the electron states in FM "above" the Fermi level at energy $E = E_c > F$. This corresponds to high-energy equilibrium electrons, which may be highly polarized (see below). Following the pioneering work by Aronov and Pikus [125], one customarily assumes a boundary condition $J_{\uparrow 0} = (1 + P_{FM})J/2$, Eq. (9.14). Since there is a spin accumulation in S near the FM–S boundary, the density of electrons with spin σ in the semiconductor is $n_{\sigma 0} = n/2 + \delta n_{\sigma 0}$, where $\delta n_{\sigma 0}$ is a nonlinear function of the current J, and $\delta n_{\sigma 0} \propto J$ at small currents [125] (see also below). Therefore, the larger the J the higher the $\delta n_{\sigma 0}$ and the smaller the current $J_{\sigma 0}$ [see Eq. (9.24)]. In other words, there is a situation where a kind of a negative feedback is realized, which decreases the spin injection efficiency (polarization of current) Γ and makes it a nonlinear function of J, as we show below. We show that the spin injection efficiency, Γ_0, and the polarization, $P_{n0} = [n_\uparrow(0) - n_\downarrow(0)]/n$ in the semiconductor near FM–S junctions essentially differ and both are small at small bias voltage V (and current J) but increase with the current up to P_F. Moreover, P_F can essentially differ from P_{FM}, and may ideally approach 100%.

The current in a spin channel σ is given by the standard drift-diffusion approximation [125,134]

$$J_\sigma = q\mu n_\sigma E + qD\nabla n_\sigma \tag{9.29}$$

where E is the electric field, D and μ are the diffusion constant and mobility of the electrons, respectively. D and μ do not depend on the electron spin σ in the nondegenerate semiconductors. From current continuity and electroneutrality conditions

$$J(x) = \sum_\sigma J_\sigma = const, \qquad n(x) = \sum_\sigma n_\sigma = const \tag{9.30}$$

we find

$$E(x) = J / q\mu n = const, \qquad \delta n_\downarrow(x) = -\delta n_\uparrow(x) \tag{9.31}$$

Since the "injection" of spin polarized electrons from FM into S corresponds to a reverse current in the Schottky FM–S junction, one has $J < 0$ and $E < 0$ (Figure 9.13). The spatial distribution of density of electrons with spin σ in the semiconductor is determined by the continuity equation [125,129]

$$\nabla J_\sigma = \frac{q\delta n_\sigma}{\tau_s}, \tag{9.32}$$

where in the present one-dimensional case $\nabla = d/dx$. With the use of Eqs. (9.29) and (9.31), we obtain the equation for $\delta n_\uparrow(x) = -\delta n_\downarrow(x)$ [125,134]. Its solution, satisfying a boundary condition $\delta n_\uparrow \to 0$ at $x \to \infty$, is

$$\delta n_\uparrow(x) = C \frac{n}{2} \exp\left(-\frac{x}{L}\right) \equiv P_{n0} \exp\left(-\frac{x}{L}\right), \tag{9.33}$$

$$L_{inject(extract)} = \frac{1}{2}\left[\sqrt{L_E^2 + 4L_s^2} + (-)L_E\right] = \frac{L_s}{2}\left(\sqrt{\frac{J^2}{J_S^2} + 4} - \frac{J}{J_S}\right), \tag{9.34}$$

where the plus (minus) sign refers to forward (reverse) bias on the junction, $L_s = \sqrt{D\tau_s}$ is the usual spin-diffusion length, $L_E = \mu|E|\tau_s = L_s|J|/J_s$ the spin-drift length. Here we have introduced the characteristic current density

$$J_S \equiv qDn / L_s \tag{9.35}$$

and the plus and minus signs in the expression for the spin penetration depth L (Eq. 9.34) refer to the spin "injection" at a reverse bias voltage, $J < 0$, and spin "extraction" at a forward bias voltage, $J > 0$, respectively. Note that $L_{inject} > L_{extract}$, and the spin penetration depth or injection increases with current, at large currents, $|J| \gg J_s$, $L_{inject} = L_s|J|/J_s \gg L_s$, whereas $L_{extract} = L_s J_s / J \ll L_s$.

The degree of spin polarization of nonequilibrium electrons, i.e., a spin "accumulation" in the semiconductor near the interface is simply given by the parameter C in Eq. (9.33):

$$C = \frac{n_\uparrow(0) - n_\downarrow(0)}{n} = P_n(0) \equiv P_{n0} \tag{9.36}$$

By substituting Eq. (9.33) into Eqs. (9.29) and (9.24), we find

$$J_{\uparrow 0} = \frac{J}{2}\left(1 + P_{n0}\frac{L}{L_E}\right) = \frac{J}{2}\frac{(1 + P_F)(\gamma - P_{n0})}{\gamma - P_{n0}P_F} \tag{9.37}$$

where $\gamma = exp(-qV/T)-1$. From Eq. (9.37), one obtains a quadratic equation for $P_n(0)$ with a physical solution that can be written fairly accurately as

$$P_{n0} = \frac{P_F \gamma L_E}{\gamma L + L_E}.$$ (9.38)

By substituting (Eq. 9.38) into (Eq. 9.24), we find for the total current $J = J_{\uparrow 0} + J_{\downarrow 0}$:

$$J = -J_m \gamma = -J_m \left(e^{-qV/T} - 1\right),$$ (9.39)

$$J_m = \alpha_0 n q v_T (1 - P_F^2)(d_{\uparrow 0} + d_{\downarrow 0})e^{-\eta \kappa_0 l}$$ (9.40)

for the bias range $|qV| \lesssim rT$. The sign of the Boltzmann exponent is unusual because we consider the tunneling thermionic emission current in a modified barrier. Obviously, we have $J > 0 \ (< 0)$ when $V > 0 \ (< 0)$ for forward (reverse) bias.

We notice that at a reverse bias voltage $-|qV| \approx rT$ the shallow potential miniwell vanishes and $E_c(x)$ takes the shape shown in Figure 9.13 (curve 3). For $-|qV| > rT$, a wide potential barrier at $x > l$ (in S behind the spike) remains flat (characteristic length scale $\gtrsim 100$ nm at $N_d \lesssim 10^{17} cm^{-3}$), as in usual Schottky contacts [136]. Therefore, the current becomes weakly dependent on V, since the barrier is opaque for electrons with energies $E < E_c - rT$ (curve 4). Thus, Eq. (9.39) is valid only at $-|qV| \lesssim rT$ and the reverse current at $-|qV| \gtrsim rT$ practically saturates at the value

$$J_{sat} = q n \alpha_0 v_T (d_{\uparrow 0} + d_{\downarrow 0})(1 - P_F^2)\exp(r - \eta \kappa_0 l).$$ (9.41)

With the use of Eqs. (9.39) and (9.34), we obtain from Eq. (9.38) the spin polarization of electrons near the FM–S interface,

$$P_{n0} = -P_F \frac{2J}{2J_m + \sqrt{J^2 + 4J_S^2} - J}.$$ (9.42)

The spin injection efficiency at the FM–S interface is, using Eqs. (9.24), (9.37), (9.34), and (9.42),

$$\Gamma_0 \equiv \frac{J_{\uparrow 0} - J_{\downarrow 0}}{J_{\uparrow 0} + J_{\downarrow 0}} = P_n(0)\frac{L}{L_E} = -P_F \frac{\sqrt{4J_S^2 + J^2} - J}{2J_m + \sqrt{J^2 + 4J_S^2} - J}.$$ (9.43)

One can see that Γ_0 strongly differs from P_{n0} at small currents. As expected, $P_n \approx P_F |J|/J_m \rightarrow 0$ vanishes with the current (Figure 9.14), and the prefactor differs from those obtained in Refs. [125,129,131,133,134].

These expressions should be compared with the results for the case of a degenerate semiconductor for the polarization [107]

$$P_{n0} = -P_F \frac{6J}{3\left(\sqrt{J^2 + 4J_S^2} - J\right) + 10J_m}.$$ (9.44)

and the spin injection efficiency

$$\Gamma_0 = -P_F \frac{\sqrt{4J_S^2 + J^2} - J}{2J_m + \sqrt{J^2 + 4J_S^2} - J}.$$ (9.45)

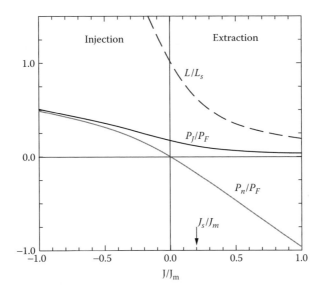

FIGURE 9.14 The spin accumulation $P_n = (n\uparrow - n\downarrow)/n$, the spin polarization of a current $P_J = (J\uparrow - J\downarrow)/J$, and the relative spin penetration depth L/L_s (*broken line*) in the semiconductor as the functions of the relative current density J/J_s for spin injection ($J < 0$) and spin extraction ($J > 0$) regimes. P_F is the spin polarization in the FM, the ratio $J_s/J_m = 0.2 = 1/5$, L_s is the usual spin-diffusion depth. The spin penetration depth considerably exceeds L_s for the injection and is smaller than L_s for the extraction.

In spite of very different statistics of carriers in a degenerate and nondegenerate semiconductor, the accumulated polarization as a function of current behaves similarly in both cases. An important difference comes from an obvious fact that the efficient spin accumulation in degenerate semiconductors may proceed at and below room temperature, whereas in present design an efficient spin accumulation in FM–S junctions with nondegenerate S can be achieved at around room temperature only.

In the reverse-biased FM–S junctions the current $J < 0$ and, according to Eqs. (9.42) and (9.43), $\text{sign}(\delta n_{\uparrow c}) = \text{sign}(P_F)$. In some realistic situations, like elemental Ni, the polarization at energies $E \approx F + \Delta_0$ would be negative, $P_F < 0$ and, therefore, electrons with spin $\sigma = \downarrow$ will be accumulated near the interface. For large currents $|J| \gg J_S$ the spin penetration depth L (Eq. 9.34) increases with current J and the spin polarization (of electron density) approaches the maximum value P_F. Unlike the spin accumulation P_{n0}, the spin injection efficiency (polarization of current) Γ_0 does not vanish at small currents, but approaches the value $\Gamma_0^0 = P_F J_S / (J_S + J_m) \ll P_F$ in the present system with a transparent tunnel δ-barrier. There is an important difference with the MTJs, where the tunnel barrier is relatively opaque and the injection efficiency (polarization of current) is high, $\Gamma \approx P_F$ [115]. However, the polarization of carriers P_{n0}, measured in, e.g., spin-LED devices [123], would be minute (see below). Both P_{n0} and Γ_0 approach the maximum P_F only when $|J| \gg J_S$, (Figure 9.15). The condition $|J| \gg J_S$ is fulfilled at $qV \approx rT \gtrsim 2T$ when $J_m \gtrsim J_S$.

Thus, we have shown that (i) an efficient spin injection in the reverse biased FM–S junctions at room temperature occurs in FM–S junctions when an ultrathin heavily n⁺-doped semiconductor layer (δ-doped layer) satisfying conditions of Eq. (9.20) is formed between the FM and nondegenerate n-type semiconductor; (ii) the reverse current of such modified Schottky junctions, which determines the spin injection from FMs into semiconductors, is due to tunneling and thermionic emission of spin polarized electrons; (iii) spin injection depends on parameters of both a semiconductor and an FM, in particular, on velocity of electrons with spin σ and energy $E \approx E_c$ and a conductivity of a semiconductor (but "not" an FM); (iv) spin injection efficiency (polarization of current), Γ_0, and polarization of carrier density, P_{n0}, in the semiconductor are two different quantities and both are small at low current; they increase with the total current and reach the maximal possible value $|P_F| \approx 1$ only at a relatively large current J_m, when the spin

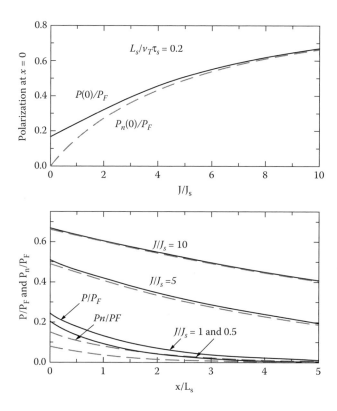

FIGURE 9.15   Spin polarization of a current $P = (J\uparrow - J\downarrow)/J$ (*solid line*) and spin accumulation $P_n = (n\uparrow - n\downarrow)/n$ (*broken line*) in the semiconductor as the functions of the relative current density J/J_s (*top panel*) and their spatial distribution for different densities of total current J/J_s (*bottom panel*) at $L_s / v_T \tau_s = 0.2$, where $J_s = qnL_s \, \tau_s$, P_F is the spin polarization in the FM (see text).

penetration depth L_E is much larger than the spin-diffusion length L_s; and (v) the smaller the semiconductor conductivity the lower the threshold current J_m for achieving an efficient spin injection. Below we also show that the most efficient spin injection can occur when (vi) the conduction band bottom of the semiconductor, $E_c = E_{c0} \simeq E_{c0} + rT$ is close to a peak in the density of minority electron states of elemental FMs Ni, Co, and Fe above the Fermi level.

Another situation is realized "in the forward-biased FM–S junctions" when $J > 0$. Indeed, according to Eqs. (9.42) and (9.43) at $J > 0$ the electron density distribution is such that $\text{sign}(\delta n_{\uparrow 0}) = -\text{sign}(P_F)$. If a system like elemental Ni is considered (Figure 9.13), then $P_F(F + \Delta_0) < 0$ and $\delta n_{\uparrow 0} > 0$, i.e., the electrons with spin $\sigma = \uparrow$ would be accumulated in a nonmagnetic semiconductor (NS) whereas electrons with spin $\sigma = \downarrow$ would be extracted from NS (the opposite situation would take place for $P_F(F + \Delta_0) > 0$). One can see from Eq. (9.42) that $|P_{n0}|$ can reach a maximum P_F only when $J \gg J_S$. According to Eq. (9.39), the condition $J \gg J_S$ can only be fulfilled when $J_m \gg J_S$. In this case Eq. (9.42) reduces to

$$P_{n0} = -P_F J/J_m = -P_F(1 - e^{-qV/T}). \tag{9.46}$$

Therefore, absolute magnitude of a spin polarization approaches its maximal value $|P_{n0}| \simeq P_F$ at $qV \ge 2T$ linearly with current (Figure 9.14). The maximum is reached when J approaches the value J_m, which depends weakly on bias V (see below). In this case, $\delta n_{\uparrow(\downarrow)}(0) \approx \mp P_F n/2$ at $P_F > 0$ $(d_\uparrow > d_\downarrow)$ so that the electrons with spin $\sigma = \uparrow$ are *extracted*, $n_\uparrow(0) \approx (1 - P_F)n/2$, from a semiconductor, while the electrons with spin $\sigma = \downarrow$ are "accumulated" in a semiconductor, $n_\uparrow(0) \approx (1 + P_F)n / 2$, near the FM–S interface. The penetration length of the accumulated spin (Eq. 9.34) at $J \gg J_s$ is

$$L = L_s^2 / L_E = L_s J_s / J \ll L_s \quad \text{at } J \gg J_s \tag{9.47}$$

i.e., it decreases as $L \propto 1/J$ (Figure 9.14). We see from Eq. (9.43) that at $J \gg J_s$

$$\Gamma_0 = P_F J_s^2 / J_m J \to 0. \tag{9.48}$$

Hence, the behavior of the spin injection efficiency at forward bias (extraction) is very different from a spin injection regime, which occurs at a reverse bias voltage: here the spin injection efficiency Γ_0 remains $\ll P_F$ and vanishes at large currents as $\Gamma_0 \propto J_s / J$. Therefore, we come to an unexpected conclusion that "the spin polarization of electrons," accumulated in an NS near a forward-biased FM–S junction can be relatively large for the parameters of the structure when the spin injection efficiency is actually very "small" [105]. Similar, albeit much weaker, phenomena are possible in systems with "wide opaque" Schottky barriers [140] and have been probably observed [141]. Spin extraction may be observed at a low temperature in FMS–S contacts as well [142]. The proximity effect leading to polarization accumulation in FM–S contacts [143] may be related to the same mechanism.

9.3.5 Conditions for Efficient Spin Injection and Extraction

According to Eqs. (9.40) and (9.35), the condition for maximal polarization of electrons P_n can be written as

$$J_m \gtrsim J_S, \tag{9.49}$$

or, equivalently, as a condition

$$\beta \equiv \alpha_0 v_T (d_{\uparrow 0} + d_{\downarrow 0})(1 - P_F^2) e^{-\eta l / l_0} \tau_s / L_s \gtrsim 1 \tag{9.50}$$

Note that when $l \lesssim l_0$, the spin injection efficiency at a small current is $\Gamma_0^0 = P_F / (1 + \beta) \ll P_F$, since in this case the value $\beta \approx (d_{\uparrow 0} + d_{\downarrow 0})\alpha_0 v_T \tau_s / L_s \gg 1$ for real semiconductor parameters. The condition $\beta \gg 1$ can be simplified and rewritten as a requirement for the spin-relaxation time

$$\tau_s \gg D \left(\frac{\Delta - \Delta_0}{2\alpha_0 v_{\sigma 0}^2 T} \right)^2 \exp \frac{2\eta l}{l_0} \tag{9.51}$$

It can be met only when the δ-doped layer is very thin, $l \lesssim l_0 \equiv \kappa_0^{-1}$. With typical semiconductor parameters at $T \simeq 300\,K$ ($D \approx 25$ cm^2/sec, $(\Delta - \Delta_0) \simeq 0.5$ eV, $v_{\sigma 0} \simeq 10^8$ cm/sec [136]) the condition (Eq. 9.51) is satisfied at $l \lesssim l_0$, when the spin-coherence time $\tau_s \gg 10^{-12}$ sec. It is worth noting that it can certainly be met: for instance, τ_s can be as large as ~ 1 nsec even at $T \simeq 300\,K$ (e.g., in ZnSe [144]).

Note that the higher the semiconductor conductivity, $\sigma_S = q\mu n \propto n$, the larger the threshold current $J > J_m \propto n$ (Eq. 9.40) is for achieving the maximal spin injection. In other words, the polarization P_{n0} reaches the maximum value P_F at a "smaller" current in "high-resistance" lightly doped semiconductors compared to heavily doped semiconductors. Therefore, the "conductivity mismatch" [128,132,133] is actually irrelevant for achieving an efficient spin injection.

The necessary condition $|J| \gg J_s$ can be rewritten at small voltages, $|qV| \ll T$, as

$$r_c \ll L_s / \sigma_s \tag{9.52}$$

where $r_c = (dJ/dV)^{-1}$ is the tunneling contact resistance. Here we have used the Einstein relation $D/\mu = T/q$ for nondegenerate semiconductors. We emphasize that Eq. (9.52) is "opposite" to the condition found by Rashba in Ref. [132] for small currents. Indeed, at small currents the "spin injection efficiency" may be large, $\Gamma_0 = P_F/(1 + \beta) \approx P_F$ only when $\beta \ll 1$, i.e., $r_c \gg L_s/\sigma_s$ (cf. Ref. [132]). This is exactly the situation with MTJs, where the current through the structure varies a great deal depending on mutual orientation of moments on ferromagnetic electrodes [115]. However, at such a large tunneling contact resistance r_c (near an opaque tunnel barrier) the saturation current J_{sat} of the FM–S junction is "much smaller" than J_s. Therefore, the degree of spin "accumulation" in the semiconductor is very small, $P_{n0} \ll 1$, but this P_{n0} is exactly the characteristic that determines the main spin effects [3,100–103]. Note that the conditions (Eq. 9.50) and (Eq. 9.52) do not depend on the electron concentration in the semiconductor, n, and are valid also for heavily doped degenerate semiconductors. We notice that the quasi-Fermi level $F_{\sigma 0}^{FM}$ for electrons with spin σ in FM differs very little from equilibrium Fermi level F. It is easy to see that $|F_{\sigma 0}^{FM} - F| \ll |F_{\sigma 0}^{S} - F|$ at current $J \lesssim J_{sat}$ since $n/n_{FM} \ll 1$, where n_{FM} is the electron density in FM metal. Thus, the assumption used above that $F_{\sigma 0}^{FM} = F$ indeed holds.

The spin factor $d_\sigma \propto v_{\sigma 0}^{-1}$ in the effective mass approximation since usually $v_{\sigma 0} \geq v_{t0}$. In a metal $v_{\sigma 0}^{-1} \propto g_{\sigma 0} = g_\sigma(E_c)$ [145], so that $d_\sigma \propto g_\sigma(E_c)$ where $g_{\sigma 0} = g_\sigma(E_c)$ is the density of states of the d-electrons with spin σ and energy $E = E_c$ in the FM. Thus, taking $m_\sigma = m$ we find from Eq. (9.27) $P_F \approx (g_{\uparrow 0} - g_{\downarrow 0})/(g_{\uparrow 0} + g_{\downarrow 0})$. One assumes that the same proportionality between the polarization and the density of states approximately holds in the general case of more complex band structures. Note that the polarization of d-electrons in elemental FMs Ni, Co, and Fe is reduced by the current of unpolarized s-electrons rJ_s, where $r < 1$ is a factor (roughly the ratio of the number of s-bands to the number of d-bands crossing the Fermi level). Together with the contribution of s-electrons the total polarization is approximately

$$P_F = \frac{J_{\uparrow 0} - J_{\downarrow 0}}{J_{\uparrow 0} + J_{\downarrow 0} + J_{s0}} \simeq \frac{g_{\uparrow 0} - g_{\downarrow 0}}{g_{\uparrow 0} + g_{\downarrow 0} + 2rg_{s0}} \tag{9.53}$$

Such a relation for P_F can be obtained from a usual "golden-rule"-type approximation for tunneling current (cf. Refs. [136,146,147]). The density of states $g_\downarrow$ for minority d-electrons in Fe, Co, and Ni has a peak at $E = E_F + \Delta_\downarrow$ ($\Delta_\downarrow \simeq 0.1$ eV) which is much larger than $g_\uparrow$ for the majority of d-electrons and g_s for s-electrons [148] (see Figure 9.13). The FM–S junction can be tailored to adjust the cutoff energy $E_c \simeq E_F + \Delta_\downarrow$ to the peak in the density of states of minority electrons. Thus, if one selects $\Delta_0 = \Delta_\downarrow + qV \simeq \Delta_\downarrow + rT$, then $g_{\downarrow 0} \gg g_{\uparrow 0} \gg g_{s0}$, and, according to Eq. (9.53), the polarization P_F may be close to 100% (note, that in the present case the polarization P_F is negative, $P_F \simeq -1$). We emphasize that the spin injection in structures considered in the literature [101,123–134] has been dominated by electrons at the Fermi level and, according to calculation [148], $g_\downarrow(F)$ and $g_\uparrow(F)$ are such that $P_F \lesssim 40\%$. We also notice that the condition Eq. (9.50) for parameters of the Fe/AlGaAs heterostructure studied in Ref. [123] ($l \simeq 3$ nm, $l_0 \simeq 1$ nm and $\Delta_0 = 0.46$ eV) is satisfied when $\tau_s \gtrsim 5 \times 10^{-10}$ sec and can be fulfilled only at low temperatures. Moreover, for the concentration $n = 10^{19}$ cm^{-3} E_c lies below F, so that the electrons with energies $E \simeq F$ are involved in tunneling, but for these states the polarization is $P_F \lesssim 40\%$. Therefore, the authors of Ref. [123] were indeed able to estimate the observed spin polarization as being $\approx 32\%$ at low temperatures.

Better control of the injection can be realized in heterostructures where a δ-layer between the FM and the n-semiconductor layer is made of a very thin heavily doped n$^+$-semiconductor with larger electron affinity than the n-semiconductor. For instance, FM–n$^+$–GaAs–n–Ga$_{1-x}$Al$_x$As, FM–n$^+$–Ge$_x$Si$_{1-x}$–n–Si or FM–n$^+$–Zn$_{1-x}$Cd$_x$Se–n–ZnSe heterostructures can be used for this purpose. The GaAs, Ge$_x$Si$_{1-x}$, or Zn$_{1-x}$Cd$_x$Se n$^+$-layer must have the width $l < 1$ nm and the donor concentration $N_d^+ > 10^{20}$ cm^{-3}. In this case, the ultrathin barrier forming near the FM–S interface is transparent for electron tunneling. The barrier height Δ_0 at Ge$_x$Si$_{1-x}$–Si, GaAs–Ga$_{1-x}$Al$_x$As or Zn$_{1-x}$Cd$_x$Se–ZnSe interface is controlled by the composition x and can be selected as $\Delta_0 = 0.05$ to 0.15 eV. When the donor concentration in Si, Ga$_{1-x}$Al$_x$As,

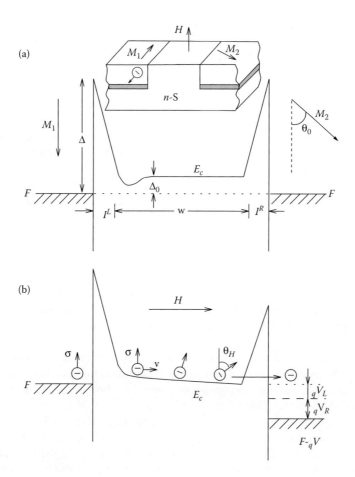

FIGURE 9.16 Energy diagram of the FM–S–FM heterostructure with δ-doped layers in equilibrium (a) and at a bias voltage V (b), with V_L (V_R) the fraction of the total drop across the left (right) δ-layer. F marks the Fermi level, Δ the height, $l^{L(R)}$ the thickness of the left (right) δ-doped layer, Δ_0 the height of the barrier in the n-type semiconductor (n-S), E_c the bottom of the conduction band in the n-S, ω the width of the n-S part. The magnetic moments on the FM electrodes M_1 and M_2 are at some angle θ_0 with respect to each other. The spins, injected from the left, drift in the semiconductor layer and rotate by the angle θ_H in the external magnetic field H. Inset: schematic of the device, with an oxide layer separating the ferromagnetic films from the bottom semiconductor layer.

or the ZnSe layer is $N_d < 10^{17}$ cm^{-3}, the injected electrons cannot penetrate a relatively low and wide barrier Δ_0 when its width $l_0 > 10$ nm.

9.3.6 High-Frequency Spin-Valve Effect

Here we describe a new high-frequency spin valve effect that can be observed in a FM–S–FM device with two back-to-back modified Schottky contacts (Figure 9.16). We find the dependence of current on a magnetic configuration in FM electrodes and an external magnetic field. The spatial distribution of spin-polarized electrons is determined by the continuity equation (9.32) and the current in the spin channel is given by Eq. (9.29). Note that $J < 0$, thus $E < 0$ in a spin injection regime. With the use of the kinetic equation and (9.29), we obtain the equation for $\delta n_\uparrow(x)$, Eq. (9.32) [125]. Its general solution is

$$\delta n \uparrow (x) = \frac{n}{2}\left(c_1 e^{-x/L_1} + c_2 e^{-(w-x)/L_2} \right) \qquad (9.54)$$

where $L_{1(2)} = (1/2)[\sqrt{L_E^2 + 4L_s^2} + (-)L_E]$ is the same as found earlier in Eq. (9.34). Substituting Eq. (9.54) into (9.29), we obtain

$$J_\uparrow(x) = (J/2)\left[1 + b_1 c_1 e^{-x/L_1} + b_2 c_2 e^{-(w-x)/L_2}\right] \qquad (9.55)$$

where $b_{1(2)} = L_1/L_E(-L_2/L_E)$.

Consider the case when $w \ll L_1$ and the transit time $t_{tr} \simeq \omega^2/(D + \mu|E|\omega)$ of the electrons through the n-semiconductor layer is shorter than τ_s. In this case a spin ballistic transport takes place, i.e., the spin of the electrons injected from the FM$_1$ layer is conserved in the semiconductor layer, $\sigma' = \sigma$. Probabilities of the electron spin $\sigma = \uparrow$ to have the projections along $\pm M_2$ are $\cos^2(\theta/2)$ and $\sin^2(\theta/2)$, respectively, where θ is the angle between vectors $\sigma = \uparrow$ and $\overline{M_2}$. Accounting for this, we find that the resulting current through the structure saturates at bias voltage $-qV > T$ at the value

$$J = J_0 \frac{1 - P_R^2 \cos^2\theta}{1 - P_L P_R \cos\theta} \qquad (9.56)$$

where J_0 is the prefactor similar to Eq. (9.25). For the "opposite bias" the total current J is given by Eq. (9.56) with the replacement $P_L \leftrightarrow P_R$. The current J is minimal for AP moments $\overline{M_1}$ and $\overline{M_2}$ in the electrodes when $\theta = \pi$ and near maximal for P magnetic moments $\overline{M_1}$ and $\overline{M_2}$. The ratio $\frac{J_{max}(P)}{J_{min}(AP)} = \frac{1 + P_L P_R}{1 - P_L P_R}$ is the same as for the tunneling FM–I–FM structure [114,115], hence, the structure also may be used as a memory cell.

The present heterostructure has an additional degree of freedom, compared to tunneling FM–I–FM structures that can be used for "magnetic sensing." Indeed, spins of the injected electrons can precess in an external magnetic field H during the transit time t_{tr} of the electrons through the semiconductor layer $(t_{tr} < \tau_s)$. The angle between the electron spin and the magnetization $\overline{M_2}$ in the FM$_2$ layer in Eq. (9.56) is in general $\theta = \theta_0 + \theta_H$, where θ_0 is the angle between the magnetizations M_1 and M_2, and θ_H is the spin rotation angle. The spin precesses with a frequency $\Omega = \gamma H$, where H is the magnetic field normal to the spin direction, and $\gamma = qg/(m_*c)$ is the gyromagnetic ratio, g is the g-factor. Therefore, $\theta_H = \gamma_0 g H t_{tr}(m_0/m_*)$ where m_0 is the mass of a free electron, $\gamma_0 g = 1.76 \times 10^7 \, Oe^{-1}s^{-1}$ for $g = 2$ (in some magnetic semiconductors $g \gg 1$). According to Eq. (9.56), with increasing H the current "oscillates" with an amplitude $(1 + P_L P_R)/(1 - P_L P_R)$ and period $\Delta H = (2\pi m_*)(\gamma_0 g m_0 t_{tr})^{-1}$ (Figure 9.17, top panel). Study of the current oscillations at various bias voltages allows finding P_L and P_R.

For magnetic sensing one may choose $\theta_0 = \pi/2 (\overline{M_1} \perp \overline{M_2})$. Then, it follows from Eq. (9.56) that for $\theta_H \ll 1$

$$J = J_0[1 + P_L P_R \gamma_0 g H t_{tr}(m_0/m_*)] = J_0 + J_H \qquad (9.57)$$

$$K_H = dJ/dH = J_0 P_L P_R \gamma_0 g t_{tr}(m_0/m_*), \qquad (9.58)$$

where K_H is the magneto-sensitivity coefficient. For example, $K_H \simeq 2 \times 10^{-3} J_0 P_L P_R$ A/Oe for $m_0/m_* = 14$ (GaAs) and $g = 2$, $t_{tr} \sim 10^{-11}$ sec, and the angle $\theta_H = \pi$ at $H \simeq 1$ kOe. Thus, $J_H \simeq 1$ mA at $J_0 = 25$ mA, $P_L P_R \simeq 0.2$, and $H \simeq 100$ Oe. The maximum operating speed of the field sensor is very high, since redistribution of nonequilibrium injected electrons in the semiconductor layer occurs over the transit time $t_{tr} = w/\mu|E| = J_s \omega \tau_s/(JL_s)$, $t_{tr} \lesssim 10^{-11}$ sec for $\omega < 200$ nm, $\tau_s = 3 \times 10^{-10}$ sec, and $J/J_s \gtrsim 10$ ($D \approx 25$ cm^2/sec at $T \simeq 300$ K [136]). Therefore, the operating frequency $f = 1/t_{tr} \gtrsim 100$ GHz ($\omega = 2\pi/t_{tr} \simeq 1$ THz) may be achievable at room temperature.

We see that (i) the present heterostructure can be used as a sensor for an ultrafast nanoscale reading of an inhomogeneous magnetic field profile, (ii) it includes two FM–S junctions and can be used for measuring spin polarizations of these junctions, and (iii) it is a multifunctional device where current depends on mutual orientation of the magnetizations in the ferromagnetic layers, an external magnetic

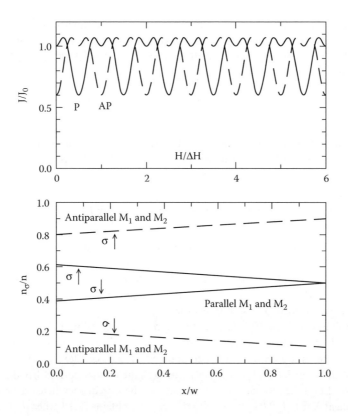

FIGURE 9.17 Oscillatory dependence of the current J through the structure on the magnetic field H (*top panel*) for parallel (P) and antiparallel (AP) moments M_1 and M_2 on the electrodes (Figure 9.1) and $P_L = P_R = 0.5$. Spatial distribution of the spin polarized electrons $n_{\uparrow(\downarrow)}/n$ in the structure for different configurations of the magnetic moments M_1 and M_2 in the limit of saturated current density J, $\omega = 60$ nm, $L_2 = 100$ nm (*bottom panel*).

field, and a (small) bias voltage, thus it can be used as a logic element, a magnetic memory cell, or an ultrafast read head.

9.3.7 Spintronic Devices

The high-frequency spin-valve effect, described above, can be used for designing a new class of ultrafast spin-injection devices like an amplifier, a frequency multiplier, and a square-law detector [103], a source of polarized radiation [106]. Their operation is based on injection of spin-polarized electrons from one FM to another through a semiconductor layer and spin precession of the electrons in the semiconductor layer in a magnetic field induced by a (base) current in an adjacent nanowire. The base current can control the emitter current between the magnetic layers with frequencies up to several 100 GHz. Here we shall describe a spintronic mechanism of ultrafast amplification and frequency conversion, which can be realized in heterostructures comprising a metallic ferromagnetic nanowire surrounded by a semiconductor (S) and ferromagnetic thin shells. Practical devices may have various layouts, with two examples shown in Figure 1 in Ref. 103, they contain two back-to-back modified Schottky barriers and an adjacent nanowire that carries the base current J_b. The calculation gives a current J_e (Eq. 9.56) through the structure as a function of the angle θ between the magnetization vectors $\overrightarrow{M_1}$ and $\overrightarrow{M_2}$ in the ferromagnetic layers. At small angles θ or $P_1 = P_L$ or $P_2 = P_R$ Eq. (9.56) reduces to

$$J_e = J_{0e}(1 + P_L P_R \cos\theta), \tag{9.59}$$

where $\theta = \theta_0 + \theta_H$, θ_0 is the angle between $\overrightarrow{M_1}$ and $\overrightarrow{M_2}$, and θ_H is the angle of the spin precesses with the frequency $\Omega = \gamma H_\perp$, where $H_\perp$ is the magnetic field component normal to the spin and γ is the gyromagnetic ratio. The angle of the spin rotation is equal to $\theta_H = \gamma H_b t_{tr} = \gamma t_{tr} J_b / 2\pi \rho_S$, where ρ_S is the characteristic radius of the S-layer. Then, according to Eq. (9.59),

$$J_e = J_{e0}[1 + P_1 P_2 \, cos(\theta_0 + k_j J_b)] \tag{9.60}$$

where $k_j = \gamma t_{tr} / 2\pi \rho_S = \gamma / \omega \rho_S$ and $\omega = 2\pi / t_{tr}$ is the frequency of a variation of the base current, $J_b = J_s cos(\omega t)$. Equation (9.60) shows that, when the magnetization M_1 is perpendicular to M_2, $\theta_0 = \pi/2$, and $\theta_H \ll \pi$,

$$J_e = J_{e0}(1 + k_j P_1 P_2 J_b), \; G = dJ_e / dJ_b = k_j P_1 P_2. \tag{9.61}$$

Hence, the "amplification" of the base current occurs with the gain G, which can be relatively high even for $\omega \gtrsim 100$ GHz. Indeed, $\gamma = q / (m_* c) \approx 2.2(m_0 / m_*) \times 10^5$ m/(A·s), where m_0 is the free electron mass, m_* the effective mass of electrons in the semiconductor, and c the velocity of light. Thus, the factor $k_j \approx 10^3$ A^{-1} when $\rho_S \approx 30$ nm, $m_0 = m_* = 14$ (GaAs) and $\omega = 100$ GHz, so that $G > 1$ at $J_{e0} > 0.1$mA/$(P_1 P_2)$. This device will perform as a spin transistor.

When M_1 is collinear with M_2 ($\theta_0 = 0, \pi$) and $\theta_H \ll \pi$, then, according to Eq. (9.60), the variation of emitter current is $\delta J_e(t) \propto J_b^2(t)$, and the device operates as a square-law detector. When $J_b(t) = J_{b0} cos(\omega_0 t)$, the emitter current has a component $\delta J_e(t) \propto cos(2\omega_0 t)$, and the device works as a frequency multiplier. When $J_b(t) = J_h cos(\omega_h t) + J_s cos(\omega_s t)$, the emitter current has the components proportional to $cos(\omega_h \pm \omega_s)t$, i.e., the device can operate as a high-frequency heterodyne detector with the conversion coefficient $K = J_{e0} J_h P_1 P_2 k_j^2 / 4$. For $k_j = 10^3$ A^{-1} one obtains $K > 1$ when $J_{e0} J_h > 4(mA)^2 / (P_1 P_2)$.

9.3.8 Spin Injection into Organic Materials

The main reason for long spin-diffusion length compared to the usual mean free path is a long spin relaxation time τ_s in semiconductors. This is a result of relatively small SO coupling that causes spin relaxation. Since the SO coupling scales as a fourth power of the atomic charge Z, it is very tempting to try carbon-based ($Z = 6$) organic materials. There the spin relaxation time is huge, $\tau_s \sim 10^{-3}$ sec. Unfortunately, the mobility is usually very bad, ~ 1 cm^2/V sec in the best case like penthacene, so the spin-diffusion length L_s is limited. Still, it has been found that in organic materials $L_s \sim 100$ nm [149], so this research is promising. Spin valves with organic materials have also been demonstrated [150]. There may be interesting developments in the organic spintronics.

9.4 Conclusions

Studying molecules as possible building blocks for ultradense electronic circuits is a fascinating quest that spans over 30 years. It was inspired decades ago by the notion that silicon technology is approaching an end of the roadmap estimated to be about 1 µm around 1985 [151]. More than 30 years later and with FET gate lengths getting below 10 nm [152], the same notion that silicon needs to be replaced at some point by other technologies floats up again. We do not know whether alternatives will continue to be steamrolled by silicon technology, which is a leading nanotechnology at the moment, but the mounting resistance to the famed Moore's law requires looking hard at other solutions for power dissipation, leakage current, crosstalk, speed, and other very serious problems. There are very interesting developments in studying electronic transport through molecular films but the mechanisms of some observed conductance "switching" and/or nonlinear electric behavior remain elusive, and this interesting behavior remains intermittent and not very reproducible. Most of the currently observed switching is extrinsic in nature. For instance, we have discussed the effect of molecule–electrode contact: the tilting of the angle at which the conjugated molecule attaches to the electrode may dramatically change its conductance, and that probably explains extrinsic "telegraph" switching observed in Tour wires [25,35] and molecule reconfigurations may lead to similar phenomena in other systems [68]. Defects in molecular films have also been discussed and may result in spurious peaks in I–V curves. We have outlined some designs of the molecules that may demonstrate rectifying behavior, which we call molecular "quantum dots." We have shown that at least in some special cases molQDs may exhibit fast (~THz) intrinsic switching.

Spintronics attempts to find ways of using spin degrees of freedom and/or long spin coherence time to find interesting new effects and applications. We have mentioned a variety of heterostructures where spin degree of freedom can be used to efficiently control the current: MTJs, metallic magnetic multilayers exhibiting GMR, ST effects in magnetic nanopillars. We described the method of facilitating an efficient spin injection/accumulation in semiconductors from standard ferromagnetic metals at room temperature. The main idea is to engineer the band structure near the FM–S interface by fabricating a δ-doped layer there, thus making the Schottky barrier very thin and transparent for tunneling. Long spin lifetime in a semiconductor allows us to suggest a few interesting new devices like high-frequency spin valve, field detectors, spin transistors, square law detectors, and sources of the polarized light. Organic materials with extremely long spin-relaxation time also may present an interesting opportunity. These developments may open up new opportunities in electronics applications that can potentially substantially extend current silicon technology toward low power high speed applications.

The author is grateful to Jeanie Lau for her Figures 9.6 and 9.7 and Jason Pitters and Robert Wolkow for Figures 9.8 and 9.9. The work has been partly supported by DARPA.

References

1. Tour, J.M., *Acc. Chem. Res.*, 33, 791, 2000.
2. Joachim, C., *Nanotechnology*, 13, R1, 2002.
3. Wolf, S.A., Awschalom, D.D., Buhrman, R.A., Daughton, J.M., von Molnar, S., Roukes, M.L., Chtchelkanova, A.Y., Treger, D.M., *Science*, 294, 1488, 2001; *Semiconductor Spintronics and Quantum Computation*, Awschalom, D.D., Loss, D., and Samarth, N., eds., Springer, Berlin, 2002.
4. Zutic, I., Fabian, J., and Das Sarma, S., *Rev. Mod. Phys.*, 76, 323, 2004.
5. Aviram, A., and Ratner, M.A., *Chem. Phys. Lett.*, 29, 277, 1974.
6. Martin, A.S., Sambles, J.R., and, Ashwell, G.J., *Phys. Rev. Lett.*, 70, 218, 1993; Metzger, R.M., Chen, B., Hopfner, U., Lakshmikantham, M.V., Vuillaume, D., Kawai, T., Wu, X., Tachibana, H., Hughes, T.V., Sakurai, H., Baldwin, J.W., Hosh, C., Cava, M.P., Brehmer, L., and Ashwell, C.J., *J. Am. Chem. Soc.*, 119, 10455, 1997.
7. Stokbro, K., Taylor, J., and Brandbyge, M., *J. Am. Chem. Soc.*, 125, 3674, 2003.
8. Ellenbogen, J.C., and Love, J., *IEEE Proc.*, 70, 218, 1993.
9. Davydov, A.S., *Theory of Molecular Excitons*, McGraw-Hill, New York, 1962.
10. Gutmann, F., *Nature*, 219, 1359, 1968.
11. Davydov, A.S., *J. Theor. Biol.*, 66, 379, 1977; Davydov, A.S., and, Kislukha, N.I., *Phys. Status Solid. (b)*, 75, 735, 1976.
12. Scott, A., *Phys. Reports*, 217, 1, 1992.
13. Scott, J.F., *Ferroelectric Memories*, Springer, Berlin, 2000.
14. Atwood, G., and Bez, R., Presentation at ISIF11, Honolulu, 21–27 April, 2006; Cho, W.Y., et al., *ISSCC Dig. Tech. Papers*, 2004; Wicker, G., *SPIE*, 3891, 2, 1999; Ovshinsky, S.R., *Phys. Rev. Lett.*, 21, 1450, 1968.
15. Jung, D.J., Kim, K., and Scott, J.F., *J. Phys. Condens. Mat.*, 17, 4843, 2005.
16. Hiranaga, Y., and Cho, Y., 11th Intl. Mtg. Ferroel., IMF 11, Iguassu Falls, Brazil, Sep 5–9, 2005.
17. Kamins, T.I., *Interface*, 14, 46, 2005; Self-assembled semiconductor nanowires, in *The Nano-Micro Interface*, Fecht, H.J. and Werner M., eds., Wiley-VCH, 2004, p. 195.
18. Rabani, E., Reichman, D.R., Geissler, P.L., and, Brus, L.E., *Nature*, 426, 271, 2003.

19. Petty, M.C., *Langmuir-Blodgett Films*, Cambridge University Press, Cambridge, 1996.

20. Ulman, A., *Characterization of Organic Thin Films*, Butterworth-Heinemann, Boston, 1995.

21. Snider, G., Kuekes, P., and Williams, R.S., *Nanotechnology*, 15, 881, 2004.

22. Stadlober, B., Zirkl, M., Beutl, M., Leising, G., Bauer-Gogonea, S., and Bauer, S., *Appl. Phys. Lett.*, 86, 242902, 2005; Shaw, J.M., and Seidler, P.F., *IBM J. Res. Dev.*, 45, 3, 2001.

23. Datta, S. et al., *Phys. Rev. Lett.*, 79, 2530, 1997.

24. Bratkovsky, A.M. and Kornilovitch, P.E., *Phys. Rev. B*, 67, 115307, 2003.

25. Kornilovitch, P.E. and Bratkovsky, A.M., *Phys. Rev. B*, 64, 195413, 2001.

26. Ness, N., Shevlin, S.A., and Fisher, A.J., *Phys. Rev. B*, 63, 125422, 2001.

27. Su, W.P. and Schrieffer, J.R., *Proc. Natl. Acad. Sci.*, 77, 5626, 1980; Brazovskii, S.A., *Sov. Phys.-JETP*, 51, 342, 1980.

28. Brazovskii, S.A. and Kirova, N.N., *Zh. Eksp. Teor. Fiz. Pis'ma Red.*, 33, 6, 1981; *JETP Lett.*, 33, 4, 1981.

29. Feldblum, A. et al., *Phys. Rev. B*, 26, 815, 1982.

30. Chance, R.R., Bredas, J.L., and Silbey, R., *Phys. Rev. B*, 29, 4491, 1984.

31. Ramsey, M.G. et al., *Phys. Rev. B*, 42, 5902, 1990.

32. Steinmuller, D., Ramsey, M.G., and Netzer, F.P., *Phys. Rev. B*, 47, 13323, 1993.

33. Swanson, L.S. et al., *Synth. Metals*, 55, 241, 1993.

34. Chen, J., Reed, M.A., Rawlett, A.M., and Tour, J.M., *Science*, 286, 1550, 1999.

35. Donhauser, Z.J., Mantooth, B.A., Kelly, K.F., Bumm, L.A., Monnell, J.D., Stapleton, J.J., Price, D.W., Jr., Rawlett, A.M., Allara, D.L., Tour, J.M., and Weiss, P.S., *Science*, 292, 2303, 2001; Donhauser, Z.J., Mantooth, B.A., Pearl, T.P., Kelly, K.F., Nanayakkara, S.U., and Weiss, P.S., *Jpn. J. Appl. Phys.*, 41, 4871, 2002.

36. Li, C., Zhang, D., Liu, X., Han, S., Tang, T., Zhou, C., Fan, W., Koehne, J., Han, J., Meyyappan, M., Rawlett, A.M., Price, D.W., and Tour, J.M., *Appl. Phys. Lett.*, 82, 645, 2003; Reed, M.A., Chen, J., Rawlett, A.M., Price, D.W., and Tour, J.M., *Appl. Phys. Lett.*, 78, 3735, 2001.

37. Collier, C.P., Wong, E.W., Belohradsky, M., Raymo, F.M., Stoddart, J.F., Kuekes, P.J., Williams, R.S., and Heath, J.R., *Science*, 285, 391, 1999; Collier, C.P., Mattersteig, G., Wong, E.W., Luo, Y., Beverly, K., Sampaio, J., Raymo, F.M., Stoddart, J.F., Heath, J.R., *Science*, 289, 1172, 2000.

38. Chen, Y., Ohlberg, D.A.A., Li, X., Stewart, D.R., and Williams, R.S., Jeppesen, J.O., Nielsen, K.A., Stoddart, J.F., Olynick, D.L., and Anderson, E., *Appl. Phys. Lett.*, 82, 1610, 2003; Chen, Y., Jung, G.Y., Ohlberg, D.A.A., Li, X., Stewart, D.R., Nielsen, K.A., Stoddart, J.F., and Williams, R.S., *Nanotechnology*, 14, 462, 2003.

39. Stewart, D.R., Ohlberg, D.A.A., Beck, P.A., Chen, Y., Williams, R.S., Jeppesen, J.O., Nielsen, K.A., and Stoddart, J.F., *Nanoletters*, 4, 133, 2004.

40. *Molecular Switches*, Feringa, B.L., ed., Wiley-VCH, 2001.

41. Zheng, G., Patolsky, F., Cui, Yi., Wang, W.U., Lieber, C.M., *Nature Biotechnol.*, 23, 1294, 2005.

42. Bachtold, A., Hadley, P., Nakanishi, T., and Dekker, C., *Science*, 294, 1317, 2001; Collins, P.G., Arnold, M.S., and Avouris, P., *Science*, 292, 706, 2001; Rueckes, T., Kim, K., Joselevich, E., Tseng, G.Y., Cheung, C.-L., and Lieber, C.M., *Science*, 289, 94, 2000.

43. Ouyang, M. and Awschalom, D.D., *Science*, 301, 1074, 2003.

44. Krzeminski, C., Delerue, C., Allan, G., Vuillaume, D., and Metzger, R.M., *Phys. Rev. B*, 64, 085405, 2001.

45. Zhou, C., Deshpande, M.R., Reed, M.A., Jones, L., II, and Tour, J.M., *Appl. Phys. Lett.*, 71, 611, 1997.

46. Xue, Y., Datta, S., Hong, S., Reifenberger, R., Henderson, J.I., Kubiak, C.P., *Phys. Rev. B*, 59, 7852(R), 1999.

47. Reichert, J., Ochs, R., Beckmann, D., Weber, H.B., Mayor, M., and Lohneysen, H.V., *Phys. Rev. Lett.*, 88, 176804, 2002.

48. Kornilovitch, P.E., Bratkovsky, A.M., and Williams, R.S., *Phys. Rev. B*, 66, 165436, 2002.

49. Lenfant, S., Krzeminski, C., Delerue, C., Allan, G., Vuillaume, D., *Nanoletters*, 3, 741, 2003.

50. Larade, B. and Bratkovsky, A.M., *Phys. Rev. B*, 68, 235305, 2003.

51. Chang, S., Li, Z., Lau, C.N., Larade, B., and Williams, R.S., *Appl. Phys. Lett.*, 83, 3198, 2003.

52. Taylor, J., Guo, H., and Wang, J., *Phys. Rev. B*, 63, R121104, 2001; ibid. 63, 245407, 2001; Jauho, A.P., Wingreen, N.S. and Meir, Y., *Phys. Rev. B*, 50, 5528, 1994.

53. Alexandrov, A.S. and Bratkovsky, A.M., *Phys. Rev. B*, 67, 235312, 2003.

54. Boulas, C., Davidovits, J.V., Rondelez, F., and Vuillaume, D., *Phys. Rev. Lett.*, 76, 4797, 1996.

55. Polymeropoulos, E.E. and Sagiv, J., *J. Chem. Phys.*, 69, 1836, 1978.

56. Tredgold, R.H. and Winter, C.S., *J. Phys. D*, 14, L185, 1981.

57. Carchano, H., Lacoste, R., and Segui, Y., *Appl. Phys. Lett.*, 19, 414, 1971; Couch, N.R., Movaghar, B., and Girling, I.R., *Sol. St. Commun.*, 59, 7, 1986.

58. Shlimak, I. and Martchenkov, V., *Sol. State Commun.* 107, 443, 1998.

59. Lau, C.N., Stewart, D., Williams, R.S., and Bockrath, D., *Nano Lett.*, 4, 569, 2004.

60. Rodriguez Contreras, J., Kohlstedt, H., Poppe, U., Waser, R., Buchal, C., and Pertsev, N.A., *Appl. Phys. Lett.*, 83, 4595, 2003.

61. Wassel, R.A., Credo, G.M., Fuierer, R.R., Feldheim, D.L., Gorman, C.B., *J. Am. Chem. Soc.*, 126, 295, 2004.

62. He, J. and Lindsay, S.M., *J. Am. Chem. Soc.*, 127, 11932, 2005.

63. Gaudioso, J., Lauhon, L.J., and Ho, W., *Phys. Rev. Lett.*, 85, 1918, 2000.

64. Hla, S.-W., Meryer, G., and Rieder, K.-H., *Chem. Phys. Lett.*, 370, 431, 2003.

65. Yang, G. and Liu, G., *J. Phys. Chem. B*, 107, 8746, 2003.

66. Salomon, A., Arad-Yellin, R., Shanzer, A., Karton, A., and Cahen, D.J., *Am. Chem. Soc.*, 126, 11648, 2004.

67. Guisinger, N.P., Greene, M.E., Basu, R., Baluch, A.S., Hersam, M.C., *Nano Lett.*, 4, 55, 2004.

68. Pitters, J.L. and Wolkow, R.A., *Nano Lett.*, 6, 390, 2006.

69. Rakshit, T., Liang, G.C., Ghosh, A.W., and Datta, S., *Nano Lett.*, 4, 1803, 2004.

70. Bune, A.V., Fridkin, V.M., Ducharme, S., Blinov, L.M., Palto, S.P., Sorokin, A., Yudin, S.G., and Zlatkin, A., *Nature*, 391, 874, 1998.

71. Ducharme, S., Fridkin, V.M., Bune, A.V., Palto, S.P., Blinov, L.M., Petukhova, N.N., and Yudin, S.G., *Phys. Rev. Lett.*, 84, 175, 2000.

72. Bratkovsky, A.M. and Levanyuk, A.P., *Phys. Rev. Lett.*, 87, 019701, 2001.

73. Bune, A., Ducharme, S., Fridkin, V., Blinov, L., Palto, S., Petukhova, N., and Yudin, S., *Appl. Phys. Lett.*, 67 3975, 1995.

74. Furukawa, T., Date, M., Ohuchi, M., and Chiba, A., *J. Appl. Phys.*, 56, 1481, 1984.

75. Kornilovitch, P.E., Bratkovsky, A.M., and Williams, R.S., *Phys. Rev. B*, 66, 245413, 2002.

76. Averin, D.V. and Likharev, K.K., in *Mesocopic Phenomena in Solids*, Altshuler, B.L. et al., eds., North-Holland, Amsterdam, 1991.

77. Park, H., Park, J., Lim, A.K.L., Anderson, E.H., Alivisatos, A.P., and McEuen, P.L., *Nature*, 407, 57, 2000; Park, J., Pasupathy, A.N., Goldsmith, J.I., Chang, C., Yaish, Y., Petta, J.R., Rinkoski, M., Sethna Abruna, J.P., McEuen, P.L., and Ralph, D.C., ibid. 417, 722, 2002; Liang, W., Shores, M.P., Bockrath, M., Long, J.R., and Park, H., ibid. 417, 725, 2002.

78. Zhitenev, N.B., Meng, H., and Bao, Z., *Phys. Rev. Lett.*, 88, 226801, 2002.

79. Alexandrov, A.S., Bratkovsky, A.M., and Williams, R.S., *Phys. Rev. B*, 67, 075301, 2003.

80. Alexandrov, A.S. and Bratkovsky, A.M., *Cond-Mat*/0603467.

81. Meir, Y. and Wingreen, N.S., *Phys. Rev. Lett.*, 68, 2512, 1992.

82. Mitra, A., Aleiner, I., and Millis, A., *Phys. Rev. Lett.*, 94, 076404, 2006.

83. Mozyrsky, D., Hastings, M.B., and Martin, I., *Phys. Rev. B*, 73, 035104, 2006.

84. Galperin, M., Ratner, M.A., and Nitzan, A., *Nano Lett.*, 5, 125, 2005.

85. Mitra, A., Aleiner, I., and Millis, A., *Phys. Rev. B*, 69, 245302, 2004.

86. Park, H., Park, J., Lim, A.K.L., Anderson, E.H., Alivisatos, A.P., and McEuen, P.L., *Nature*, 407, 57, 2000; Park, J., Pasupathy, A.N., Goldsmith, J.I., Chang, C., Yaish, Y., Retta, J.R., Rinkoski, M., Sethna, J.P., Abruna, H.D., McEuen, P.L., and Ralph, D.C., *Nature* (London), 417, 722, 2000;. Liang, W. et al., *Nature* (London), 417, 725, 2002.

87. Sakaguchi, H., Hirai, A., Iwata, F., Sasaki, A., and Nagamura, T., *Appl. Phys. Lett.*, 79, 3708, 2001; Cui, X.D., Zarate, X., Tomfohr, J., Sankey, O.F., Primak, A., Moore, A.L., Moore, T.A., Gust, D., Harris, G., and Lindsay, S.M., *Nanotechnology*, 13, 5, 2002.

88. Wang, W., Lee, T., and Reed, M.A., *Phys. Rev. B*, 68, 035416, 2003.

89. Stewart, D.R., Ohlberg, D.A.A., Beck, P.A., Lau, C.N., and Williams, R.S., *Appl. Phys. A*, 80, 1379, 2005.

90. Fisher, G.L., Hooper, A.E., Opila, R.L., Allara, D.L., and Winograd, N., *J. Phys. Chem. B*, 104, 3267, 2000; Seshadri, K., Wilson, A.M., Guiseppi-Elie A, Allara, D.L., *Langmuir*, 15, 742, 1999.

91. Walker, A.V., Tighe, T.B., Cabarcos, O., Reinard, M.D., Uppili, S., Haynie, B.C., Winograd, N., and Allara, D.L., *J. Am. Chem. Soc.*, 126, 3954, 2004.

92. Pollak, M. and Hauser, J.J., *Phys. Rev. Lett.*, 31, 1304, 1973; Lifshitz, I.M. and Kirpichenkov, V.Ya., *Zh. Eksp. Teor. Fiz.*, 77, 989, 1979.

93. Glazman, L.I. and Matveev, K.A., *Sov. Phys. JETP*, 67, 1276, 1988.

94. Xu, Y., Ephron, D., and Beasley, M.R., *Phys. Rev. B*, 52, 2843, 1995.

95. Shang, C.H., Nowak, J., Jansen, R., and Moodera, J.S., *Phys. Rev. B*, 58, 2917, 1998.

96. Selzer, Y., Cabassi, M.A., Mayer, T.S., Allara, D.L., *J. Am. Chem. Soc.*, 126, 4052, 2004.

97. Zhitenev, N.B., Erbe, A., and Bao, Z., *Phys. Rev. Lett.*, 92, 186805, 2004.

98. Frisch, M.J., GAUSSIAN98, Revision A.9, Gaussian, Inc., Pittsburgh, PA, 1998.

99. Larkin, A.I. and Matveev, K.A., *Zh. Eksp. Teor. Fiz.*, 93, 1030, 1987.

100. Datta, S. and Das, B., *Appl. Phys. Lett.*, 56, 665, 1990; Gardelis, S., Smith, C.G., Barnes, C.H.W., Linfield, E.H., and Ritchie, D.A., *Phys. Rev. B*, 60, 7764, 1999.

101. Sato, R., and Mizushima, K., *Appl. Phys. Lett.*, 79, 1157, 2001; Jiang, X., Wang, R., van Dijken, S., Shelby, R., Macfarlane, R., Solomon, G.S., Harris, J., and Parkin, S.S.P., *Phys. Rev. Lett.*, 90, 256603, 2003.

102. Bratkovsky, A.M. and Osipov, V.V., *Phys. Rev. Lett.*, 92, 098302, 2004.

103. Osipov, V.V. and Bratkovsky, A.M., *Appl. Phys. Lett.*, 84, 2118, 2004.

104. Osipov, V.V. and Bratkovsky, A.M., *Phys. Rev. B*, 70, 235302, 2004.

105. Bratkovsky, A.M. and Osipov, V.V., *J. Appl. Phys.*, 96, 4525, 2004.

106. Bratkovsky, A.M. and Osipov, V.V., *Appl. Phys. Lett.*, 86, 071120, 2005.

107. Osipov, V.V. and Bratkovsky, A.M., *Phys. Rev. B*, 72, 115322, 2005.

108. Baibich, M.N., Broto, J.M., Fert, A., Nguyen Van Dau, F., Petroff, F., Etienne, P., Creuzet, G., Friederich, A., and Chazelas, J., *Phys. Rev. Lett.*, 61, 2472, 1988; Berkowitz, A.E., Mitchell, J.R., Carey, M.J., Young, A.P., Zhang, S., Spada, F.E., Parker, F.T., Hutten, A., and Thomas, G., ibid. 68, 3745, 1992.

109. Julliere, M., *Phys. Lett.* 54A, 225, 1975.

110. Maekawa, S. and Gafvert, U., *IEEE Trans. Magn.*, 18, 707, 1982.

111. Meservey, R. and Tedrow, P.M., *Phys. Reports*, 238, 173, 1994.

112. Moodera, J.S., Kinder, L.R., Wong, T.M., and Meservey, R., *Phys. Rev. Lett.* 74, 3273, 1995.

113. Yuasa, S., Nagahama, T., Fukushima, A., Suzuki, Y., and Ando, K., *Nature Mater.*, 3, 858, 2004; Parkin, S.S.P., Kaiser, C., Panchula, A., Rice, P.M., Hughes, B., Samant, M., and Yang, S.-H., *Nature Mater.*, 3, 862, 2004.

114. Slonczewski, J.C., *Phys. Rev.*, B. 39, 6995, 1989.

115. Bratkovsky, A.M., *Phys. Rev.*, B 56, 2344, 1997.

116. Berger, L., *Phys. Rev.*, B 54, 9353, 1996.

117. Slonczewski, J., *J. Magn. Magn. Mater.*, 159, L1, 1996.

118. Tsoi, M.V. et al., *Phys. Rev. Lett.*, 80, 4281, 1998.

119. Slonczewski, J., *J. Magn. Magn. Mater.*, 247, 324, 2002.

120. Emley, N.C. et al., *Phys. Rev. Lett.*, 96, 247204, 2006.

121. Mott, N.F., *Proc. R. Soc. London Ser. A*, 153, 699, 1936.

122. Osipov, V.V., Viglin, N.A., and Samokhvalov, A.A., *Phys. Lett. A*, 247, 353, 1998; Ohno, Y., Young, D.K., Beschoten, B., Matsukura, F., Ohno, H., and Awschalom, D.D., *Nature*, 402, 790, 1999; Fiederling, R., Keim, M., Reuscher, G., Ossau, W., Schmidt, G., Waag, A., and Molenkamp, L.W., ibid. 402, 787, 1999.

123. Hanbicki, A.T., Jonker, B.T., Itskos, G., Kioseoglou, G., and Petrou, A., *Appl. Phys. Lett.* 80, 1240, 2002; Hanbicki, A.T., van't Erve, O.M.J., Magno, R., Kioseoglou, G., Li, C.H., and Jonker, B.T., ibid. 82, 4092, 2003; Adelmann, C., Lou, X., Strand, J., Palmstrom, C.J., and Crowell, P.A., *Phys. Rev. B*, 71, 121301, 2005.

124. Hammar, P.R., Bennett, B.R., Yang, M.J., and Johnson, M., *Phys. Rev. Lett.*, 83, 203, 1999; Zhu, H.J., Ramsteiner, M., Kostial, H., Wassermeier, M., Schonherr, H.-P., and Ploog, K.H., ibid. 87, 016601, 2001; Lee, W.Y., Gardelis, S., Choi, B.-C., Xu, Y.B., Smith, C.G., Barnes, C.H.W., Ritchie, D.A., Linfield, E.H., and Bland, J.A.C., *J. Appl. Phys.* 85, 6682, 1999; Manago, T. and Akinaga, H., *Appl. Phys. Lett.*, 81, 694, 2002; Motsnyi, A.F., De Boeck, J., Das, J., Van Roy, W., Borghs, G., Goovaerts, E., and Safarov, V.I., ibid. 81, 265, 2002; Ohno, H., Yoh, K., Sueoka, K., Mukasa, K., Kawaharazuka, A., and Ramsteiner, M.E., *Jpn. J. Appl. Phys.*, 42, L1, 2003.

125. Aronov, A.G. and Pikus, G.E., *Fiz. Tekh. Poluprovodn.*, 10, 1177, 1976 [*Sov. Phys. Semicond.*, 10, 698, 1976]

126. Johnson, M. and Silsbee, R.H., *Phys. Rev.*, B 35, 4959, 1987; Johnson, M. and Byers, J., ibid. 67, 125112, 2003.

127. van Son, P.C., van Kempen, H., and Wyder, P., *Phys. Rev. Lett.*, 58, 2271, 1987; Schmidt, G., Richter, G., Grabs, P., Gould, C., Ferrand, D., and Molenkamp, L.W., ibid. 87, 227203, 2001.

128. Schmidt, G., Ferrand, D., Molenkamp, L.W., Filip, A.T. and van Wees, B.J., *Phys. Rev.*, B 62, R4790, 2000.

129. Yu, Z.G., and Flatte, M.E., *Phys. Rev. B*, 66, R201202, 2002.

130. Albrecht, J.D. and Smith, D.L., *Phys. Rev. B*, 66, 113303, 2002.

131. Hershfield, S. and Zhao, H.L., *Phys. Rev. B*, 56, 3296, 1997.

132. Rashba, E.I., *Phys. Rev. B*, 62, R16267, 2000.

133. Fert, A. and Jaffres, H., ibid. 64, 184420, 2001.

134. Yu, Z.G. and Flatte, M.E., *Phys. Rev. B*, 66, 235302, 2002.

135. Shen, M., Saikin, S., Cheng, M.-C., *IEEE Trans. Nanotechnology*, 4, 40, 2005; *J. Appl. Phys.*, 96, 4319, 2004.

136. Sze, S.M., *Physics of Semiconductor Devices*, Wiley, New York, 1981; Monch, W., *Semiconductor Surfaces and Interfaces*, Springer, Berlin, 1995; Tung, R.T., *Phys. Rev. B*, 45, 13509, 1992.

137. Duke, C.B., *Tunneling in Solids*, Academic, New York, 1969.

138. Sanvito, S., Lambert, C.J., Jefferson, J.H., and Bratkovsky, A.M., *Phys. Rev. B*, 59, 11936, 1999; Wunnicke, O., Mavropoulos, Ph., Zeller, R., Dederichs, P.H., and Grundler, D., *Phys. Rev. B*, 65, 241306, 2002.

139. Stearns, M.B., *J. Magn. Magn. Mater.* 5, 167, 1977.

140. Ciuti, C., McGuire, J.P., and Sham, L.J., *Appl. Phys. Lett.*, 81, 4781, 2002; Ciuti, C., McGuire, J.P., and Sham, L.J., *Phys. Rev. Lett.*, 89, 156601, 2002.

141. Stephens, J., Berezovsky, J., Kawakami, R.K., Gossard, A.C., Awschalom, D.D., *cond-mat*/0404244, 2004.

142. Zutic, I., Fabian, J., and Das Sarma, S., *Phys. Rev. Lett.*, 88, 066603, 2002.

143. Epstein, R.J., Malajovich, I., Kawakami, R.K., Chye, Y., Hanson, M., Petroff, P.M., Gossard, A.C., and Awschalom, D.D., *Phys. Rev. B*, 65, 121202, 2002.

144. Kikkawa, J.M., Smorchkova, I.P., Samarth, N., Awschalom, D.D., *Science*, 277, 1284, 1997; Kikkawa, J.M. and Awschalom, D.D., *Nature*, 397, 139, 1999; Malajovich, I., Berry, J.J., Samarth, N., Awschalom, D.D., *Nature*, 411, 770, 2001; Hagele, D., Oestreich, M., and Ruhle, W.W., Nestle, N., and Eberl, K., *Appl. Phys. Lett.*, 73, 1580, 1998.

145. Ziman, J.M., *Principles of the Theory of Solids*, Cambridge Univ. Press, Cambridge, 1972.

146. Simmons, J.G., *J. Appl. Phys.*, 34, 1793, 1963; Simmons, J.G., *J. Phys. D*, 4, 613, 1971; Stratton, R., *Tunneling Phenomena in Solids*, Burstein, E. and Lundqvist, S., eds., Plenum, New York, 1969; Tung, R.T., *Phys. Rev. B*, 45, 13509, 1992.

147. Esaki, L., *Phys. Rev.*, 109, 603, 1958; Julliere, M., *Phys. Lett.*, 54A, 225, 1975.

148. Mazin, I.I., *Phys. Rev. Lett.*, 83, 1427, 1999; Moruzzi, V.L., Janak, J.F., and Williams, A.R., *Calculated Electronic Properties of Metals*, Pergamon, New York, 1978.
149. Dediu, V. et al., *Sol. State Commun.*, 122, 181, 2002; Arisi, E. et al., *J. Appl. Phys.*, 93, 7682, 2003.
150. Xiong, Z.H., Di Wu, Vardeny, Z.V., and Shi, J., *Nature*, 427, 821, 2004.
151. Rambidi, N.G. and Zamalin, V.M., *Molecular Microelectronics: Origins and Outlook*, Znanie, Moscow, 1985.
152. Ieong, M., Doris, B., Kedzierski, J., Rim, K., and Yang, M., *Science*, 306, 2057, 2004.

10

Nanoarchitectonics: Advances in Nanoelectronics

Kang Wang
Kosmas Galatsis
Roman Ostroumov
University of California, Los Angeles

Mihri Ozkan
University of California, Riverside

Kostantin Likharev
Stony Brook University

Youssry Botros
Intel Corporation

10.1 Introduction..**10**-1
10.2 The Nanoelectronics Landscape.....................................**10**-2
    Top-Down vs. Bottom-Up • Current CMOS Landscape
    • Beyond CMOS Technology Sequence
10.3 Fabrication of Nanostructures **10**-6
    Progress in Lithography Process • Future Nanoelectronic
    Self-Assembly
10.4 Nanodevices **10**-10
    Power Dissipation • Examples of Emerging Nanodevices
10.5 Nanoarchitectures **10**-13
10.6 Conclusions..**10**-18
Acknowledgments **10**-18
References **10**-19

10.1 Introduction

The Semiconductor Industry Association forecasts that worldwide sales of microchips will reach $309 billion in 2008 — an increase of 45% from the $213 billion record level of 2004 [1]. This includes sales of discrete components, optoelectronics, analogs, MOS logic devices, microprocessors, microcontrollers, digital signal processors, DRAM, and FLASH microchips. Such growth rates and market magnitude give the illusion of a bright future for the industry. However, an evaluation of the technical semiconductor roadmap beyond 10 years reveals various technical challenges for the industry's existing CMOS-based roadmap. Roadmaps [2,3], review papers [4–7], and even calls for proposals [8] highlight the expected challenges ahead. Power consumption, interconnects, scalability, and equipment costs are threatening the resilience of silicon CMOS technology, which has comforted mankind for over four decades. In response, new solutions beyond CMOS are being compiled and aligned to outmaneuver these haunting showstoppers. Possible solutions are being tried by various academic and industry-based research programs, funded by both government and the semiconductor community. Examples of such organized research focusing on nanoelectronics beyond CMOS include the Semiconductor Research Corporation (SRC), Microelectronics Advanced Research Corporation (MARCO), Nanoelectronics Research Corporation, and the National Science Foundation, which supports the National Nanotechnology Infrastructure Network and Nanoscale Science and Engineering Centers.

This chapter begins by analyzing the current landscape of semiconductors and explaining the current status of semiconductors from an industry viewpoint. We will review the anticipated CMOS roadmap and new nanostructure assembly techniques, and will provide examples of alternative nanodevices and nanoarchitectures beyond CMOS information processing.

10.2   The Nanoelectronics Landscape

After receiving the Nobel Prize in Physics for 2000, Herbert Kroemer, Zhores I. Alferov, and Jack S. Kilby were asked to comment on what will follow the microchip as we know it. Herbert Kroemer replied, "I am convinced things will not come to a stop, I just don't know in which direction they will go ... science drives applications, applications inspire science." There is no question that our hunger for faster information processing and larger storage capacity is inspiring science to scale CMOS to its limit. In addition, new scientific breakthroughs are creating the excitement that they will eventually unfold in a generation of new applications. At this time in history, nanoelectronics is the hope for providing solutions beyond CMOS' run of Moore's law. New functional materials, self-assembly techniques, and breakthroughs in quantum mechanical phenomena are bringing us even closer to complementary/alternative information processing. However, manufacturing beyond CMOS devices/systems is another major obstacle. In the interview, Jack S. Kilby highlighted the practical aspects of nanoelectronics, "... a big factor is a decrease in cost," as economic factors will have the last and final say. This section will attempt to provide a landscape of nanoelectronics and its future.

10.2.1   Top-Down vs. Bottom-Up

Manufacturing microchips is no different from baking your favorite chocolate chip cookies; conceptually, they are very much similar. For example, they are both made from a starting material, patterned, baked, and packaged. That said, cookies are far different from naturally occurring items such as apples and oranges. These "naturally" occurring items are made from a completely different approach, the bottom-up approach. All of Mother Nature's living products are derived in a bottom-up fashion. This fundamental difference is important as it provides new possibilities to the semiconductor industry, going beyond its well-defined top-down manufacturing capabilities. The interesting aspect is the progress that will occur within the next 10 to 20 years. Figure 10.1 illustrates the convergence of the microelectronics

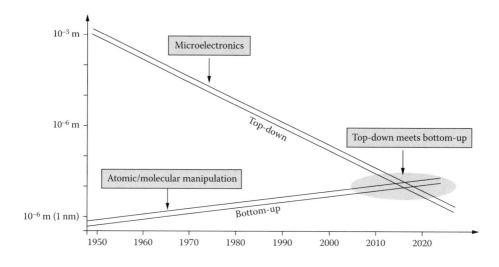

FIGURE 10.1 Possible convergence between top-down and bottom-up methods in patterning. As microelectronics top-down methods continue to be scaled down in feature size, while control and bottom-up methods are refined, a unique convergence is expected where features of both regimes could be integrated to produce alternative information processing systems.

top-down fabrication merging with bottom-up methods derived by chemistry, physics, and biology. It is anticipated to give rise to new opportunities in new information processing systems. Today, nanoelectronics research is routinely employing the predictability of top-down microstructures with the atomic/molecular density of bottom-up assembly approaches [9,10]. As presented in Figure 10.1, it may not be long before a "hybrid" approach is adopted for various semiconductor manufacturing processes. Before we delve into various self-assembly methods, let us first understand the current status and future of the top-down approach currently employed by the semiconductor industry.

10.2.2 Current CMOS Landscape

Over the past half century, the amount of information that computers are capable of processing (functional throughput) has followed the well-known Moore's law or doubled every 18 months [9]. This year, the most advanced silicon chips are based on the 65 nm node with a physical gate length about 35 nm. Such scaling will continue over the next decade and, according to the ITRS 2004 Roadmap, CMOS will reach the 18 nm technology node or 7 nm physical gate length by 2018 [9]. It is anticipated that beyond this point, CMOS scaling will likely become very difficult [11,12] and novel devices and new information processing architectures will be required.

There are many technological challenges arising while industry approaches the 18 nm node. Aggressive scaling of the gate length makes device parameter optimization quite difficult. While gate length is reducing, CMOS scaling rules require source and drain junction depths, gate thickness oxide, and a driving voltage that needs to be decreased while channel doping is to be increased [13]. This leads to various difficulties. Ultra-shallow junction formation cannot be achieved without incurring a significant increase in parasitic resistance. Doping requires very precise profile designs and process controls, whereas increasing channel-doping concentration degrades carrier mobility, lowering the drain current. Moreover, statistical fluctuation of channel dopants causes increasing variations in the threshold voltage, device characteristics, and variability, which poses difficulty in circuit design while scaling the supply voltage. Scaling of the gate oxide, on the other hand, leads to excessive gate leakage current, necessitating the introduction of new high-κ materials in the near future [9].

Another big challenge also exists in the methods of fine feature patterning. While optical lithography is consistently being pushed towards smaller feature sizes, it is expected that the most advanced immersion technology will extend optical lithography down to the 45 nm node. Lithography methods for nodes beyond that are unclear. There are several candidates to surpass optical lithography, which are extreme ultraviolet lithography (EUV) [14], electron projection lithography [5,15], mask-less lithography [5,16], and imprint technology [17]. This topic is further addressed in Section 10.3.1.

The situation with DRAM is even more challenging. Since a DRAM capacitor decreases physical dimensions with scaling the effective oxide thickness, it must be scaled aggressively to maintain appropriate storage capacitance. According to ITRS, conventional DRAMs can scale down to around the 32 nm node using metal insulator metal structures, with the insulator having a dielectric constant greater than 100. Accordingly, novel memory devices with nodes beyond 32 nm should be developed. A similar situation exists for nonvolatile memory (NVM). There are several kinds of NVM, with technological challenges for each kind, which differs, depending on the nature of the memory element. An example would be the charge in the floating gate for FLASH. However, the basic problems with scaling are the compromise of retention time, cell isolation, and reading sensitivity of the stored bit. The ITRS projects that current NVMs can be scaled down to the 45 nm node, while nodes beyond this would require radically new solutions.

The most worrisome issue among the various technological challenges would be device power dissipation. CPU power density has experienced an exponential growth rate, with a feature size miniaturization and increased density of integration, as shown in Figure 10.2. Increasing power dissipation is governed by technology miniaturization, which allows for higher integration density and faster switching speed, even though the power dissipation for a single device continues to decrease. If the trend continues, by some estimates future high performance processors may reach an unmanageable number of 5 to

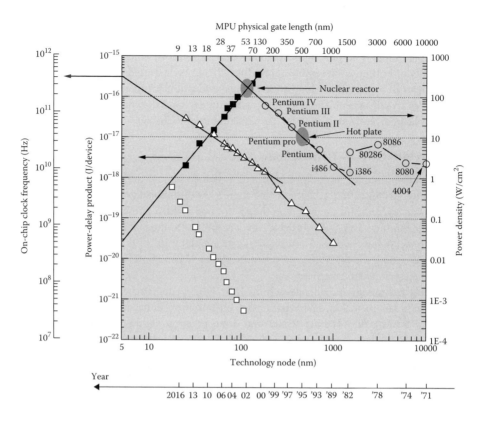

FIGURE 10.2 Power dissipation as a function of generation node. Power density on the surface of current and past microprocessor generations (*open circles*), on-chip clock frequency (*open triangles*), power-delay products for a single device (*solid squares*), estimated power density for the basic switch with linear driving function (*open squares*). (From Strukov, D.B. and K.K. Likharev, *Nanotechnology*, 2005, 16(1):137–148; Coey, J.M.D. and M. Venkatesan, *J Appl Phys*, 2002, 91(10):8345–8350; Unsal, O.S. and I. Koren, *Proc IEEE*, 2003, 91(7):1055–1069. With permission.)

10 MW/cm² [18], while ITRS estimates the 22 nm node may have a power dissipation of 93 W/cm², which is similar to the power density of a lightbulb filament of 100 W/cm². This issue may hinder further scaling and new devices and novel computer architectures may be required.

An example of the expected CMOS device roadmap is presented in Figure 10.3. The figure shows Intel's roadmap through CMOS scaling, where changes in device geometry and material advances is hoped will enable further scaling, past the 10 nm node. Through silicon innovations such as strained-Si channels [19–22], high-K/metal–gate gate stacks [22–25], and the nonplanar "Tri-gate" CMOS transistor architectures [22–24,26,27], CMOS transistor scaling and Moore's Law will continue at least into the early part of the next decade [27]. It is shown that conventional planar bulk MOSFET channel length scaling, which has driven the industry for the last 40 years, is slowing [23]. To continue Moore's law, new materials and structures are required [24]. As presented in Figure 10.3, new materials such as SiGe are incorporated within the 90 nm technology generation, resulting in a strain on the Si channel, for a 20 to 50% increase in mobility enhancement. For the next several logic technologies, MOSFETs will improve through higher levels of uniaxial process stress. After that, new materials that address MOSFET poly-Si gate depletion, gate thickness scaling, and alternate device structures (FinFET, tri-gate, or carbon nanotube) are the next possible technology directions [19–24,26,28–34]. Selecting which of these to implement depends on the magnitude of the performance benefit (and application) vs. manufacturing complexity and cost. Finally, for future material changes targeted toward enhanced transistor performance, there are three key points: 1) performance enhancement options need to be scalable to future technology nodes;

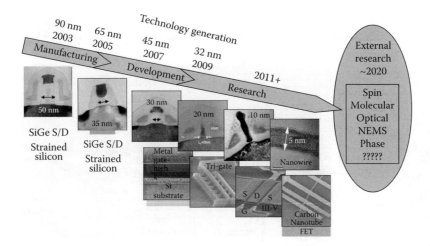

FIGURE 10.3 A prediction of the future beyond CMOS. Intel device roadmap showing continued CMOS scaling and transition into alternative devices such as those based on III–V and carbon nanotube FETs. (Courtesy of the Intel Corporation.)

2) new transistor features or structures that are not additive with current enhancement concepts may not be viable; 3) improving external resistance appears equally important as new channel materials (like carbon nanotubes) since the ratio of external to channel resistance is approaching one in nanoscale planar MOSFETs [27].

10.2.3 Beyond CMOS Technology Sequence

As already mentioned, guessing beyond CMOS would be like playing roulette at your favorite casino. Figure 10.3 shows a possible roadmap that includes alternative materials such as III–V semiconductors and new nanostructures such as carbon nanotubes. However, organized work has begun in surveying the landscape beyond CMOS. Some of the great minds in this area are collaborating with the SRC as part of a task force working on emerging research devices. They have composed a taxonomy of emerging technology sequence published in the ITRS [2], which has been referenced and elaborated in various publications [5,18,35–40]. Figure 10.4 has been reproduced from the ITRS, which gives a clear snapshot of alternatives beyond CMOS with various degrees that includes new devices, architectures, and state variables.

Devices: The device is the lowest level physical entity in an information processing system. It has the ability to switch a state variable by an external stimulus or by intrinsic/extrinsic rules, conditions, and characteristics. Our common, three-terminal field effect transistor (FET) relies on gate-to-control electron flow. Other devices such as two-terminal molecular switches [41] make use of bistable states, and two-terminal resonant tunneling diodes make use of nonlinear characteristics [42,43]. Other emerging devices include spintronic, CNT-based FETs, mechanical, molecular, phase change, and quantum properties.

Architectures: The Von Neumann architecture is employed in today's microprocessors and controllers that represent information in Boolean form. Address, data, and control are the three most important elements required for this architecture. Other architectures that support emerging devices have also been proposed, such as Crossbar, Neuromorphic Networks, Cellular Nonlinear Networks, Cellular Automata, and Quantum Cellular Automata.

State Variables: State variables refer to information representation. Today, electron charge is the primary state variable used (it exhibits voltage, current, and capacitance); other information processing regimes may take advantage of electron spin, molecular state, photon intensity or polarization, quantum state, phase state, or mechanical state.

Traditional solid-state electronics may soon be taken over by these new foreseeable alternatives beyond CMOS. Quantum mechanics, nanoscale forces, new materials, new assembly and fabrication techniques

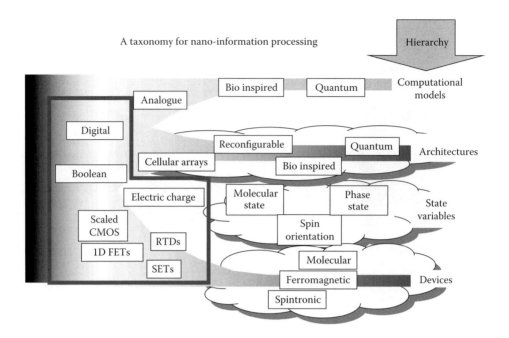

FIGURE 10.4 The ITRS 2004 Emerging Technology Sequence showing alternatives in architectures, logic devices, memory devices, and nonclassical CMOS designs. (From ITRS 2005 ed. With permission.)

and a new understanding of the meaning of information will be required to progress into a new paradigm that would offer advantages beyond CMOS. However, "there's plenty of room at the bottom"[44] and one thing is clear: smaller is better. Smaller enables the aforementioned state variables to be manipulated and allow for greater functionality. Given the case, the next question is: *How does one build these new nanostructures, nanodevices, and nanoarchitectures?*

10.3   Fabrication of Nanostructures

Other than conceptualizing a new device based on some new state variable, device and circuit fabrication is key in ensuring economic and scaling feasibility. This section will review the current lithography process and explore alternatives for nanoelectronics.

10.3.1   Progress in Lithography Process

Lithography was the key technology that propelled CMOS scaling for generations. Each new generation of lithography tools become more and more sophisticated and exponentially costly. One of the core elements of lithography is the wavelength of the light exposure. A shorter wavelength allows for smaller feature sizes; however, doing so challenges the performance of the light source and optics and further increases cost. The most advanced systems in development today are termed "extreme ultraviolet" (EUV). EUV operates at a wavelength of 193 nm, capable of producing structures down to 13.5 nm. This is great for achieving high resolution, although there are many technical difficulties. EUV is absorbed by lens material such as glass, so instead of lenses, mirrors must be used. Similarly, EUV will not pass through a glass mask, so a reflective mask is required (one that reflects the EUV in certain regions, to transfer the pattern onto the wafer). The reflective mask consists of a mirror consisting of 80 alternating layers of silicon and molybdenum, to maximize reflectivity. Production of EUV is not used for mass manufacture today. Tools such as Intel's EUV microexposure tool (MET), shown in Figure 10.5, are currently being

FIGURE 10.5 Lithography at its extreme. The EUV MET is used by Intel to develop its EUV technology for future device nodes. (Courtesy of the Intel Corporation.)

used to debug the technology for a new generation of steppers. The MET currently only prints 600×600 μm (less than 1 mm on a side), whereas a full die might be over 20 mm on a side.

Another key advance in lithography that is being pursued by equipment and optics manufacturers such as Applied Materials, Nikon, and Canon is water immersion lithography. This is a new approach for optical patterning that injects a liquid (with an index of reflection, n) between an exposure tool's projection lens and a wafer to achieve a better depth of focus and resolution over conventional lithography. In immersion lithography, a liquid is interposed between an exposure tool's projection lens and a wafer. Immersion technology offers better resolution enhancement over conventional projection lithography because the lens can be designed with a numerical aperture (NA) greater than one. This increases the ability to produce smaller images as shown by:

$$CD \text{ (resolution)} = k1 \text{ (process factor)} \times \lambda \text{(wavelength)/NA}$$

and,

$$NA = n \text{ (refraction index)} \times \sin \theta \text{ (the maximum incident angle of the exposure light)}$$

If the refraction index is greater than one (the value for air), such as ultrapure water ($n = 1.44$), light at the same angle will have an NA 1.44 times that of air. Normally, the semiconductor exposure equipment's projection lens and wafer are separated by air. However, if the gap is filled with ultrapure water to perform "immersed exposure," the equivalent wavelength will be 134 nm. This occurs even when using an excimer laser with a wavelength of 193 nm, which is expected to support printing down to 45 nm [45].

In addition to EUV and water immersion, other techniques such as phase-shift masking, modified illumination, optical proximity correction, and pupil filtering will assist in further driving optical lithography below the 65 nm node.

10.3.2 Future Nanoelectronic Self-Assembly

Alternative assembly techniques for nanoelectronics, other than top-down lithography, are being aggressively pursued that embrace the intricacies of the nanoworld. At the nanometer scale, the volume of

TABLE 10.1 List of Forces Acting at the Nanometer Scale

Type of Force	Strength (kJ/mol)	Example
Covalent	>210	C–C bond
Electrostatic	>190	Li$^+$–F$^-$
Dipole–dipole	5–40	H$^+$–Cl$^-$–H$^+$–Cl$^-$
π–π interaction	10–20	CNT–CNT[a]
Hydrogen bonding	5–40	ssDNA–ssDNA[b]
Dispersion	<5	H$^+$–O$^=$–H$^+$–Cl$^-$ Cl$^+$
Hydrophobic	5–40	H$_2$O—metal
Dative	20–380	S—Au
Ionic bonding	20–30	Na$^+$Cl$^-$ crystal
van der Waals	0.1–40	H$_2$O–H$_2$O

[a]Carbon nanotube.
[b]Single-stranded deoxyribonucleic acid.

objects is less dominant, as opposed to the total surface area of the same objects. Hence, the type of forces acting are different at this scale. For example, at the micron scale, gravity may still be an important force to consider in experimental conditions. However, gravity and inertia are negligible at the nanoscale. Table 10.1 lists different types of forces and their interaction strength. These unique forces at the nanoscale provide scientists with a broader toolset for manipulating properties at the nanoscale, while a much more complex world at the nanoscale exists in designing these nanosystems.

Examples of self-assembly can be found at different levels, ranging from atomic and molecular scales in organic or inorganic form. A high-level figure is presented in Figure 10.6. Assembly between carbon

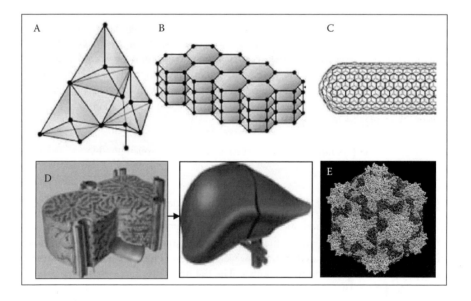

FIGURE 10.6 Self-assembly at different scales. (A) Atomic assembly showing a diamond has a face-centered cubic structure. (B) Graphite sheets assemble with weak van der Waals forces. (C) Single-walled carbon nanotubes can be in different forms, armchair or zig zag, based on the assembly of carbon atoms. (D) Hepatocytes interact with other hepatocytes, a fenestrated endothelium, stellate cells, extracellular matrix, and the bloodstream to assemble into liver. (E) Nanoscale assembly and cow pea mosaic virus (CPMV) capsid organization. A CPMV capsid comprises 60 repeating asymmetric subunits, each formed from large (L) and small (S) coat proteins, arranged in icosahedral symmetry. Each subunit is formed by two β-barrel domains from the L protein that assembles around the threefold axes, while each S protein contains one β-barrel domain, which assembles around the fivefold axes.

atoms in different arrangements can result in diverse materials. For example, diamonds, carbon nanotubes, and graphite are all formed by carbon atoms, although they all have different material properties such as mechanical strength, and electrical and optical properties. Similarly, at the molecular scale, arrangement of molecules can be found in different forms and thus in different structural and physical properties. A well-known example of molecular assembly is water. When water molecules are loosely arranged vs. being closely and more orderly arranged, liquid form will change into solid form. Similarly, a different assembly will yield different structural and physical properties.

How about assembly of organic yet biological systems? The best example of biological assembly is the human being. First, proteins, amino acids, and lipids form cells, and cells assemble in the form of tissue. Tissues assemble in the form of organs and they assemble in the form of a human body. More fascinating is the fact that after assembly of the proteins in a highly specific way, different functions can be activated.

Another assembly example is the formation of icosahedral viruses. Coat proteins form each subunit, forming pentamer structures around a fivefold axis 60 times to produce the entire cow pea mosaic virus (CPMV) capsid. Interaction of coat proteins or subunits among themselves is usually through a number of different relations including hydrogen bonding, steric hindrance, and hydrophobic/hydrophilic and van der Waals forces. Figure 10.6 summarizes atomic and biological assembly of different objects. By understanding nature's methods of using different forces at the nanoscale, one can nanoengineer highly functional structures with better electrical and optical properties. By combining the clever self-assembly ability of the organic and bioworld, it may be possible to use this property with inorganic-based microelectronic/nanoelectronic systems.

Some examples of organized academic research in self-assembly are presented by the Center of Functional Engineered Architectonics, which is funded by the SIA and Department of Defense. One focus area is alternative fabrication methods. For the predicted size, speed, and power advantages to be realized, new methods for creating complex patterns and assembly techniques with a feature size of 1 to 5 nm are being investigated. DNA, peptide nucleic acids (PNAs), viruses, tobacco mosaic virus, CPMV, proteins, microtubule (MT), and bacteriaphage (M13) are utilized as templates for the growth of nanowires or assembly of nanoparticles, including carbon nanotubes, zinc oxide nanowires, and quantum structures, into functional building blocks [46–57]. Metallic nanowires (Au, PT, Ni, etc.) are synthesized in varying lengths (nm-micron), diameters (4 nm to 100 μm), resulting in varying electrical resistances. Using DNA and PNA as linkers between inorganic nanoparticles, new device structures are designed, where DNA and PNA are simultaneously used as part of a diode or a transistor at the nanoscale [48–50,54–57]. In addition, inorganic and organic nanotemplates are used for the assembly of organometallic nanowires [53].

Figure 10.7 depicts images of some of the assembly projects. Here, MTs are used as templates to make Au nanowires while PNA and DNA are used to synthesize Pt and Au nanowires. In addition, both PNA and DNA are used as linkers to assemble single-walled carbon nanotubes, as shown in Figure 10.7.

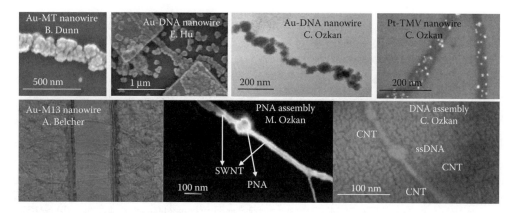

FIGURE 10.7 Samples of biotemplated synthesis of nanowires and assembly of nanoparticles within Theme-2 in the FENA center (www.fena.org). (*Top left to right*) MT, DNA, TMV-templated nanowires. (*Bottom left to right*) M13-templated nanowire, PNA, and DNA assembly of single-walled carbon nanotubes.

10.4   Nanodevices

The term "nanodevice" is indeed a broad term. Here, we will refer to devices at the nanoscale as, using ITRS' terminology, emerging research devices.

10.4.1   Power Dissipation

Among the many technological challenges, the major issue in information processing using millions of charge-based nanodevices is power dissipation. CPU power density has experienced an exponential growth rate, with a feature size reduction and increased density. Increasing power dissipation is governed by technology miniaturization, which allows for higher integration density and a faster switching speed, even though the power dissipation per device continues to decrease, as depicted in Figure 10.2. CPU power density already exceeds the power density of the hot plate at about 10 W/cm². If the trend continues, it could reach an unmanageable number of 5 to 10 MW/cm² [18]. In order to address these issues and assess emerging new technologies for performance improvement, one has to look at the fundamental limits of power dissipation from the viewpoint of information processing.

Stimulated by the desire to continue information throughput increase, research in new technologies and new information processing paradigms, such as those based on cellular automata, cellular nonlinear networks, cross net molecular computing, quantum computing, etc., have recently emerged [5,37,58–65]. These new computer architectures are aiming to enhance information throughput over the conventional Turing machine [66]. For example, by using nearest neighbor interactions in the cellular automata, one can define special rules for elements in a cell. By these means, parallel processing of the input is achieved. In quantum computing, one can bring the quantum system into superposition of all its possible states using such quantum phenomena as superposition and entanglement of the states, in addition to using special algorithms that provide the correct answer, with a greater probability upon measurement.

However, these new computing paradigms are able to outperform the conventional computer in some specific tasks only. For example, cellular automata is unbeatable in image processing. With quantum computing, one can factor large numbers in polynomial rather than exponential time or search immense databases in a time proportional to the square root of the number of elements. This is why conventional informational processing will still be of great importance and will need to continuously increase its performance. Devices that enable such architectures will be discussed below.

The question of physical limits on computation and minimal heat dissipation during information processing has been intensively studied during the past decades [67–72]. Computation can be performed without energy dissipation only by the physically reversible devices using the informationally reversible gates, such as a Toffoli gate, shown in Figure 10.8(a) [73,74]. One of the examples of reversible computing

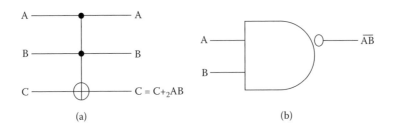

(a)   (b)

FIGURE 10.8 Informationally reversible and irreversible gates. (a) Toffoli gate — universal three-bit gate in reversible computing. This gate flips in the output C if A and B are both 1, and leaves C alone if they are not. A and B themselves remain unchanged. The Toffoli gate is symmetrical and can compute in either direction. (b) NAND gate — universal two-bit gate in informationally irreversible computing. This gate gives an output 0 only if A and B are both 1 and returns 0 in any other case. This gate is obviously asymmetric between inputs and outputs and can compute only in one direction.

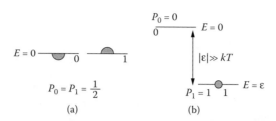

FIGURE 10.9 A two-level system for information processing. (a) Initially both levels are at zero energy and the system occupies each level with an equal probability of 1/2. (b) By applying an external field, we move level '1' down on the energy scale in such a way that it is always in thermal equilibrium with the bath — isothermal process. When $\varepsilon \gg kT$, the system is definitely in the '1' state.

is a quantum computer, where reversibility appears naturally due to the unitarity of quantum mechanics. However, if there is a reset of the information during the computational process (or information irreversible computation, which is the case in current computation technology, see Figure 10.8[b]), information bits would be processed with an energy dissipation of at least $kT \ln 2$ [75].

The minimum energy dissipation of $kT \ln 2$ is achieved for infinitesimally slow switching (timescale is much longer than energy relaxation time). It is also possible to obtain a simple relation for the energy dissipation during finite switching speed. One can consider any two-level system as a simple switch, as shown in Figure 10.9. For example, a scaled MOS transistor performing as a simple electronic switch can be represented as a single electron in two quantum wells. In this case, one can show that energy is related to the energy relaxation time (T_1) and switching speed (α) by the following expression [76]:

$$E_{dissip} = \frac{1}{2}\alpha T_1 + kT \ln 2 \tag{10.1}$$

In Figure 10.2, we plot this dependence (open squares) for the 1 ps energy relaxation time and ITRS 2004 projections for the switching speed and integration density. As one can see, fundamentally, power dissipation can be a few orders of magnitude lower than the interpolated CPU power density. This provides room for improvement compared to the current power dissipation of CMOS-based systems, where power is lost via technology imperfections such as leakage current, tunneling, and imperfect electrostatic control.

10.4.2 Examples of Emerging Nanodevices

Few emerging research devices identified in the ITRS and in Figure 10.4 will come close to fulfilling the stringent device criteria that CMOS has fulfilled for generations. New state variables that eliminate charge transport (and can offer the potential for minimal power dissipation) such as spin are receiving great attention. Spin is also identified as a potential state variable for new nanodevices. It has been observed that when current flows from a ferromagnetic metal into an ordinary metal, electrons retain their spin alignment. Hence, the spins aligned along a magnetic field can be transported just like charges [77]. This concept has resulted in several attempts to fabricate spin transistors that exploit spin-dependent transport of charge carriers. This is in order to yield a device with a high spin current gain and high magnetic sensitivity.

Research into spin transistors began in 1988 and has rapidly evolved into the field of spintronics over the past decade, and it promises the possibility of integrating memory and logic functions into a single device [78–82]. The first such attempt was Johnson's all-metal three-terminal device [83], which had a third terminal added to the middle paramagnetic layer of a GMR multilayer structure. The electrical characteristics of this purely ohmic device are magnetically tunable, but due to its all-metal construction, its operation yields only a small change of output voltage, with no power gain. The second variant, based on the proposal by Datta and Das [84], has been attempted. This transistor is a modification of a FET in which an applied electric field changes the width of the depletion region and hence the output current

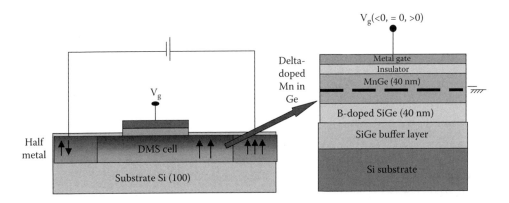

FIGURE 10.10 Schematic spin gain FET structure and schematic of the DMS cell structure with a MnGe/SiGe quantum well.

magnitude. Both the Johnson and the Datta/Das devices utilize a dependence of spin on the carrier transport (including conduction, injection efficiency, and relaxation time) to control the current flow through a device. The spin state is controlled by the magnetization of a ferromagnetic contact or by an external electromagnetic field. Nikonov and Bourianoff [85] postulate that a small base current with a small spin polarization can, like an electronic transistor with gain, stimulate a large spin polarization in the collector output current. Furthermore, it is expected that the polarization of the control current will be reflected in the output current. The spin gain is achieved by first biasing the base to produce a sufficiently high hole concentration for inducing a paramagnetic to ferromagnetic transition. It is then injected into a small spin-polarized control current, which acts as a stimulator to break the random orientation of the spins and to induce spontaneous magnetization in the diluted magnetic semiconductor (DMS) base. Structures such as those shown in Figure 10.10 aim to produce spin-gain based devices. The structure of the spin gain FET can be made up of DMS materials that include Mn-doped Ge or Mn-doped GaAs and AlGaAs and other group IV, II–VI, or III–V DMS materials. The source and drain contact may be constructed from ferromagnetic materials such as CrNi alloys or half metals such as CrO_2 [86]. For example, spin current can be injected from the half metal source (or from spin-polarized ferromagnetic contacts). The half metal source injection efficiency can reach 100% [87]. The gate voltage will control the surface potential and thus vary the hole concentration in the DMS cell. The transition of paramagnetism to ferromagnetism is anticipated to occur when the hole concentration exceeds a critical value. The spin gain will be obtained when polarized spins are injected from the source and when there is a high hole concentration present.

Other than logic devices, numerous emerging memory devices based on nanomaterials and nanostructures have also been identified by the ITRS [2], SRC Memory Task Force [8], and European Roadmap for Nanoelectronics [3]. In an effort to collate possible memory device alternatives, the ITRS roadmap identifies six emerging research memory areas beyond traditional RAM and flash devices. These include phase change memory, floating body DRAM, nano-floating gate memory, single-electron memory, insulator resistance change memory, and molecular memory, which are also positively reviewed in an article by Brewer et al. [36]. In addition, there have been various review papers [88–95] comparing and investigating alternative memory devices that include ferroelectric RAM, magnetic RAM, organic RAM, CNT electromechanical NVM, programmable metallization cell memory, magnetic tunneling junctions RAM, and thyristor RAM.

An example of a new nanodevice for a memory application is the multilevel memory device by Zhou [96]. This memory device explores multilevel memory states based on redox-active molecules containing multiple redox centers. A challenge in developing multilevel memory devices is the need for including additional charge-sensing circuitry to detect the various memory levels. The device employs functionalized nanowires that offer a high surface area to allow self-assembled packing of iron–terpyridine redox molecules. This provides for high stability and maximum storage density. Figure 10.11 shows a simple memory device that is able to be programmed into eight levels. The nanowire-based memory has

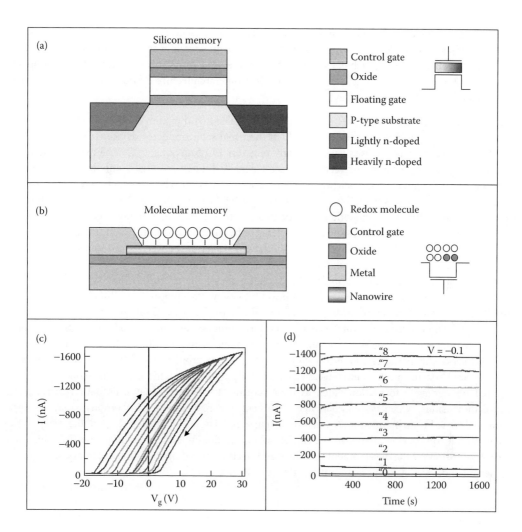

FIGURE 10.11 Schematic diagrams illustrating design and operation for a conventional silicon flash memory and the proposed molecular memory for multilevel nonvolatile data storage. (a) Schematic of a silicon flash memory; (b) schematic of the proposed molecular memory; (c) I–V_g hysteresis loops obtained by sweeping gate voltage from $-n*2.5$ V to $10 + n*2.5$ V and then back to the starting value; n is the index of levels from "2" (*the innermost curve*) to "8" (*the outermost curve*); (d) current recorded over time after the device was written into states "0" (*the bottom curve*) to "8" (*the top curve*). Little degradation in the stored signal was observed over 1500 sec, with a source-drain bias of $V = -0.1$ V. (Courtesy of C. Zhou at USC.)

been demonstrated with on/off ratios exceeding 10,000 and retention times greater than 600 h. The advantage of multilevel memory is the ability to achieve high bits/cm². Our memory utilizes an active area of 2 μm (channel length) by 10 nm (nanowire width), indicating a storage density of 40 Gbits/cm², assuming that each cell is programmed into eight levels.

10.5 Nanoarchitectures

From the standpoint of device structure, nanodevices may be divided into two major groups for architecture implementation:

1. Two-terminal devices, fabricated by self-assembly methods that have so far shown low yield. For these devices, low-cost VLSI fabrication may be rationally envisioned.

2. Three-terminal (and more complex) devices. These devices would require a major breakthrough in self-assembly. (Numerous suggestions to use nanowire crossings as FETs ignore the fundamental fact that below 10 nm, such devices operate as tunnel transistors and are inherently irreproducible [12,13,97]. Nevertheless, a few recent nanoelectronic architecture efforts have focused on using circuits with two-terminal devices.

These are two clearly distinguishable directions possible. The first direction could be developing circuits using mostly two-terminal nanodevices, in which the usual semiconductor devices (e.g., silicon MOS-FETs) are not used at all, or used very sparingly (e.g., for I/O functions, etc.). The main attraction of this approach is the possibility of very high circuit density, limited essentially only by the nanodevice footprint or by the pitch of nanowiring used for device connection. This approach may be represented by SET parametron circuits [98,99] (also known as "clocked quantum-dot cellular automata,") [100,101], nanocell circuits [102,103], and others. Unfortunately, to the best of our knowledge, the necessary combination of high functionality and defect tolerance has not been demonstrated (via reliable simulation) for either of these concepts. Moreover, the lack of isolation between input and output of two-terminal devices [97] makes the design of highly functional circuits difficult.

This fact serves as the motivation for the development of the second, hybrid type of circuit that incorporates two-terminal nanodevices and connecting nanowires along with a CMOS-based subsystem [104,105]. The clear advantage of this approach is that the FETs of the CMOS subsystem may readily provide signal amplification/restoration, which is hard to perform with two-terminal nanodevices. Other functions that can be implemented in CMOS are long-range communications and input/output operations.

Nevertheless, the hybrid circuit approach has its own set of problems, such as the interface between the CMOS subsystem and the nanodevice/nanowire subsystem. Indeed, the advanced techniques that are required for nanowire formation (such as nanoimprint [17,106,107]) may eventually reach sub-nm critical dimensions, which would challenge alignment and registration. This problem may be significantly alleviated if the nanodevice/nanowire system is regular and uniform, e.g., periodic, since in this case its shift by one period relative to the CMOS subsystem does not affect circuit properties. This is the prime motivation for the development of hybrid architectures based on nanowire crossbars, with similar two-terminal devices at each crosspoint, as shown in Figure 10.12(a) [108,109].

In the simplest case, the crosspoint devices are just "programmable diodes" (Figure 10.12[b]). At low voltages, such a device behaves just as a usual diode, but the application of higher voltages switches it between low-resistive (ON) and high-resistive (OFF) states. The sharp switching threshold allows the crosspoint device to be switched between the two internal states by applying voltages $\pm V_W$ (with $V_t < V_W < 2V_t$) to the two nanowires that it connects, so that voltage $V = \pm 2V_W$ applied to the selected nanodevice exceeds the threshold, while half-selected devices (with $V = \pm V_W$) are not disturbed. Such "programmable diode" (a.k.a. "latching switch") functionality may be achieved in several ways; for example, by switching between two atomic configurations of a molecule [41,110–112]. However, nanosecond-scale operations require fast electronic switches, e.g., single-electron latches (Figure 10.12[c]) [113]. Low-temperature prototypes of single-electron devices have already been demonstrated [114], and the molecular implementation of its main component, the single-electron transistor, has been reported by several groups [115–119]. We will show below that even with such simple devices, hybrid CMOS/nanodevice circuits may possess high performance and fault tolerance characteristics.

Interfacing even such a regular nanosystem as a crossbar to much coarser CMOS wiring is challenging. One solution for fabricating a demultiplexer-type interface is to use semiconductor wires with random doping; however, it requires very complex fabrication and assembly techniques and allows addressing only one nanowire at a time. Another approach is suggested in Ref. [120] (see also Ref. [105]). It is based on a cut of the ends of nanowires along a line that forms a small angle,

$$\alpha = \arctan(F_{nano}/\beta F_{CMOS}) \tag{10.2}$$

with the wire direction, where F_{nano} is the half-pitch of the nanowire crossbar and F_{CMOS} is that of the CMOS wires. (Factor β may be larger than 1 to accommodate proper layout of the CMOS subsystem.)

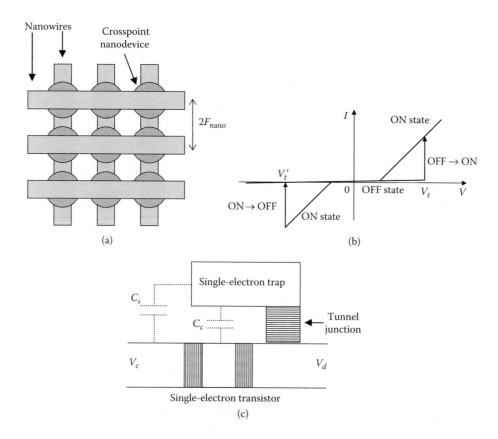

FIGURE 10.12 (a) Nanowire crossbar and (b,c) the simplest crosspoint device ("programmable diode" or "latching switch"); (b) *I–V* curve (schematically), and (c) possible single-electron implementation.

As a result of the cut, the ends of the adjacent nanowires stick out by distances (along the wire direction) differing by βF_{CMOS} and may be contacted individually by the similarly cut CMOS wires. Unfortunately, the latter (CMOS) cut has to be precisely aligned with the former (nanowire) one, and the existing nanoscale patterning technologies cannot provide such an alignment.

Based on our earlier work on neuromorphic circuits [121,122], we have shown [123,124] that the CMOS/nano-interfacing problem may be solved using the CMOS/crossbar interface distributed over all the crossbar area of the chip (Figure 10.13[a]). In these "CMOL" circuits, the interface is provided by pins with sharp, nanometer-scale tips. (The technology necessary for the fabrication of such pins has been already developed in the context of field-emission arrays [125].) As Figure 10.13(c) shows, pins of each type (reaching to either the lower or the upper nanowire level) are arranged into a square array with sides of $2\beta F_{CMOS}$. The nanowire crossbar is turned by an angle slightly different from the one given by Equation (10.1),

$$\alpha = \arcsin(F_{nano}/F_{CMOS}) \qquad (10.3)$$

relative to the CMOS pin array. By activating two pairs of perpendicular CMOS lines, two pins (and the two nanowires they contact) may be connected to CMOS data lines (Figure 10.12[b]). As Figure 10.12(c) illustrates, this approach allows a unique access to any nanodevice, even if $F_{nano} \ll F_{CMOS}$. For programmable diodes with sharp thresholds (such as Figure 10.11[c]), such access enables each device to be turned ON or OFF.

Of all other hybrid CMOL/nanocircuit ideas, the CMOL circuit concept is least component demanding. Still, like all bottom-up approaches, CMOL technology requires defect-tolerant circuit architectures, since

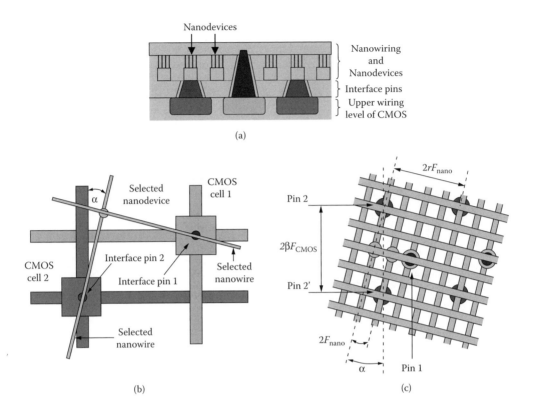

FIGURE 10.13 Low-level structure of the generic CMOL circuit: (a) schematic side view; (b) the idea of addressing a particular nanodevice; and (c) zoom-in on several adjacent interface pins to show that any nanodevice may be addressed via the appropriate pin pair (e.g., pins 1 and 2 for the left-most of the two shown devices, and interface pins 1 and 2′ for the right-most device). On (b), only the activated CMOS lines and nanowires are shown, while (c) shows only two devices.

the fabrication yield of nanodevices will hardly ever approach 100% as closely as that achieved for semiconductor transistors.

In the long term, the most important application of CMOL technology may be in mixed-signal neuromorphic networks, which may provide unparalleled performance for advanced information processing [117,118]. However, in the near term, digital CMOL circuits may provide a larger practical impact. The most straightforward application of such circuits are embedded memories and stand-alone memory chips, with their simple matrix structure. In such memories, each crosspoint nanodevice would play the role of a single-bit memory cell, while the CMOS subsystem may be used for coding, decoding, line driving, sensing, and input/output functions. We have carried out [126,127] a detailed analysis of CMOL memories with a global and quasi-local ("dash") structure of matrix blocks, including the synergy of two major techniques for increasing their defect tolerance: the memory matrix reconfiguration (the replacement of several rows and columns, with the largest number of bad memory cells, for spare lines), and advanced error correction codes. Figure 10.14 shows the final result of that analysis: the optimized total chip area per useful bit, as a function of the nanodevice yield.

The results show that with the natural requirement of very high density (exceeding their semiconductor counterparts by an order of magnitude), CMOL memories may be rather defect tolerant: a 90% chip yield may be achieved with 8 to 12% of bad nanodevices (depending on the required access time). In absolute terms, the density of such memories will be extremely impressive. For example, the normalized cell area $a \equiv A/N(F_{CMOS})^2 = 0.4$ at $F_{CMOS} = 32$ nm means that a memory chip of a reasonable size (2×2 cm^2) can store about 1 terabit of data — crudely, 100 *Encyclopaedia Britannicas*.

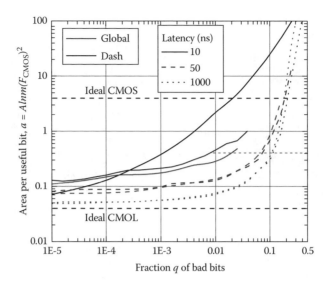

FIGURE 10.14 The graph presents the optimized area per useful bit as a function of single nanodevice yield, for two different CMOL memories (with "dash" and global block structure) and for several values of access time. (From Strukov, D.B. and K.K. Likharev, *Nanotechnology*, 2005, 16(1):137–148. With permission.)

Much less obvious has been the efficiency of CMOL applications for digital (Boolean) logic, especially in light of the necessary defect tolerance. In the usual custom logic circuits, the location of a defective gate from outside is hardly possible, while spreading around additional logic gates (e.g., providing von Neumann's majority multiplexing [128]) for error detection and correction becomes very inefficient for fairly low fractions q of defective devices. This is why the most significant previously published proposals for the implementation of logic circuits using CMOL-like hybrid structures had been based on a different approach: reconfigurable regular structures, which are to some extent similar to the usual CMOS-based field-programmable gate arrays (FPGAs). Before our recent work, two FPGA varieties had been analyzed, one based on look-up tables and another one using programmable-logic arrays. Unfortunately, all these approaches run into substantial problems [129].

An alternative approach [59] to Boolean logic circuits based on the CMOL concept can be considered that is close to the so-called cell-based FPGA. In this approach, an elementary CMOS cell includes two pass transistors and an inverter, and is connected to the nanowire/nanodevice subsystem via two pins (Figure 10.15[a]). Disabling the CMOS inverters allows for carrying out the circuit reconfiguration via the cell's pass transistors. On the other hand, enabling the inverter turns the cell into a NOR gate (Figure 10.15[b]), generally with an almost arbitrary fan-in.

First results for CMOL FPGA have been obtained using a simple, two-step approach to reconfiguration, in which the desired circuit is first mapped on the apparently perfect (defect-free) CMOL fabric, and then is reconfigured around defective components using a simple algorithm [59]. The Monte Carlo simulation (so far only for the "stack-on-open"-type defects which are expected to dominate in CMOL circuits) has shown that even this simple configuration procedure may ensure very high defect tolerance. For example, the reconfiguration of a 32-bit Kogge–Stone adder, mapped on the CMOL fabric with realistic values of parameters, may allow for achievment of the 99% circuit yield (sufficient for a ~90% yield of properly organized VLSI chips), with as much as 22% of defective devices, while the defect tolerance of another key circuit, a fully connected 64-bit crossbar switch is about 25%.

Our most striking result was that such high defect tolerance may coexist with high density and performance, at acceptable power consumption. For example, calculations have shown [59] that for the total power of 200 W/cm² (planned by the ITRS for the long-term CMOS technology nodes), an optimization may bring the logic delay of the 32-bit Kogge–Stone adder down to just 1.9 ns, at the total

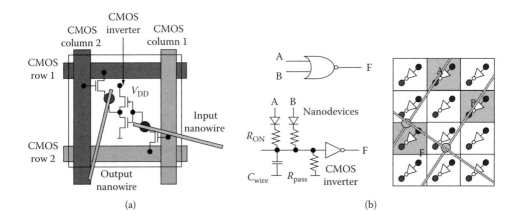

FIGURE 10.15 An illustration of the CMOL FPGA concept showing (a) logic cell schematics and (b) implementation of a fan-in-two NOR gate, where the left diagram shows the nanodevices used within the array.

area of 110 μm²; i.e., provide an area-delay product of 150 ns-μm², for realistic values of F_{CMOS} = 32 nm and F_{nano} = 8 nm. A minor error in that work (which was found and corrected later [130]) led to the underestimation of the actual delay by a factor close to 2.5. Still, the area-delay product compares very favorably with the estimated 70,000 ns-μm² (with 1.7 ns delay and 39,000 μm² area) for a fully CMOS FPGA implementation of the same circuit (with the same F_{CMOS}).

The advantage of CMOL FPGA circuits over custom CMOS circuits (with the same F_{CMOS} and chip power) still has to be calculated, but probably will be less spectacular, though still significant (perhaps a factor of three to four in the area-delay product). This advantage may be substantially increased by using programmable diodes with negative differential resistance, which would allow for carrying out signal restoration on the nanodevice crossbar level.

To summarize, there is a good chance for the development, within perhaps the next 10 to 15 years and maybe substantially earlier, of hybrid "CMOL" integrated circuits that will allow the extension of Moore's law to the few-nm range. Estimates show that such circuits could be used for several important applications, including terabit-scale memories and reconfigurable digital circuits with multi-teraflops-scale performance. There are also good prospects for mixed-signal neuromorphic networks, which may, for the first time, compete with biological neural systems in a real density, far exceeding them in speed, at an acceptable power dissipation [121,122].

10.6   Conclusions

A key driver of nanoelectronics presented in this chapter comes from the shortcomings of the current CMOS-based information processing technology. The focus is on nanotechnology and nanoelectronics to provide answers to nanostructure assembly, development of smarter materials such as functional materials, less power-dependant devices using new state variables such as spin, and more efficient, higher-throughput, and fault-tolerant architectures (e.g., neuromorphic architectures such as CMOL). The nanoscale provides us with a great array of tools, phenomena, effects, and intricacies that, if manipulated correctly, are likely to fulfill and complete the roadmap beyond CMOS.

Acknowledgments

The authors would like to acknowledge the contributions by Angela Belcher (MIT), Evelyn Hu (UCSB), Chongwu Zhou (USC), Cengiz Ozkan (UCR), Bruce Dunn (UCLA), and George Bourianoff (Intel). In addition, they acknowledge the assistance by J.H. Lee, J. Li, X. Ma, N. Simonian, and D. Strukov for the

CMOL work and P. Adams, J. Barhen, W. Chen, S. Das, A. DeHon, J. Ellenbogen, D. Hammerstrom, X. Liu, J. Lukens, A. Mayr, V. Protopopescu, M. Reed, M. Stan, and Ö. Türel for fruitful discussions and collaborations. We would like to thank the MARCO FENA Center for financially supporting the afore-mentioned research areas, and, most importantly, thank the students and researchers who participated in the work presented within this chapter.

References

1. Greenagel, J., SIA forecast: chip sales will surpass $300 billion. *SIA Media Release*, San Jose, 2005.
2. *The International Technology Roadmap for Semiconductors*. Semiconductor Industry Associations, 2004.
3. Compano, R., *Technology Roadmap for Nanoelectronics*. European Commission IST Programme, Future and Emerging Technologies, 2000.
4. Meindl, J.D., Beyond Moore's Law: the interconnect era. *Comput Sci Eng*, 2003, 5(1):20–24.
5. Risch, L., The end of the CMOS roadmap — new landscape beyond. *Mat Sci Eng C Biomimet Supramol Syst*, 2002, 19(1/2):363–368.
6. Wang, K.L., Issues of nanoelectronics: a possible roadmap. *J Nanosci Nanotech*, 2002, 2(3/4):235–266.
7. Deleonibus, S., Devices architectures and materials for nanoCMOS at the end of the roadmap and beyond. *Mat Sci Semiconduct Proc*, 2004, 7(4–6):167–174.
8. SRC, *Research Needs for Advanced Memory for 32nm Technology Node and Beyond*. SRC: Research Triangle Park NC, 2005.
9. Cheng, J.C., A.M. Mayes, and C.A. Boss, Nanostructure engineering by templated self-assembly of block copolymers. *Nat Mater*, 2004, 3(11):823–828.
10. Stoykovich, M.P. et al., Directed assembly of block copolymer blends into nonregular device-oriented structures. *Science*, 2005, 308(5727):1442–1446.
11. Walls, T.J., V.A. Sverdlov, K.K. Likharev. Quantum mechanical modeling of advanced sub-10 nm MOSFETs. In: *Third IEEE Conference on Nanotechnology. IEEE-NANO 2003, Proceedings* (Cat. No.03TH8700), NJ, vol. 2, 2003, pp. 28–31.
12. Sverdlov, V.A., T.J. Walls, K.K. Likharev, Nanoscale silicon MOSFETs: a theoretical study. *IEEE Trans Electron Dev*, 2003, 50(9):1926–1933.
13. Frank, D.J. et al., Device scaling limits of Si MOSFETs and their application dependencies. *Proc IEEE*, 2001, 89(3):259–288.
14. Silverman, P.J., Extreme ultraviolet lithography: overview and development status. *J Microlith, Microfab, Microsyst*, 2005, 4(1):11006-1-5.
15. Yamabe, M., Status and issues of electron projection lithography. *J Microlith, Microfab, Microsyst*, 2005, 4(1):11005-1-10.
16. Brandstatter, C. et al., Projection maskless lithography. In: *SPIE-Int. Soc. Opt. Eng. Proc.*, 2004, 5374(1):601–609.
17. Zankovych, S., et al. Nanoimprint lithography: challenges and prospects. *Nanotechnology*, 2001, 12(2):91–95.
18. Zhirnov, V.V. et al., Limits to binary logic switch scaling — a gedanken model. *Proc IEEE*, 2003, 91(11):1934–1939.
19. Thompson, S. et al., A 90 nm logic technology featuring 50 nm strained silicon channel transistors, 7 layers of Cu interconnects, low k ILD, and 1 mu m/sup 2 SRAM cell. *International Electron Devices Meeting. Technical Digest* (Cat. No.02CH37358), 2002, pp. 61–64.
20. Thompson, S.E. et al., A 90-nm logic technology featuring strained-silicon. *IEEE Trans Electron Dev*, 2004, 51(11):1790–1797.
21. Thompson, S.E. et al., A logic nanotechnology featuring strained-silicon. *IEEE Electron Dev Lett*, 2004, 25(4):191–193.
22. Chau, R. et al., Advanced CMOS transistors in the nanotechnology era for high-performance, low-power logic applications. In: *Seventh International Conference on Solid-State and Integrated Circuits Technology Proceedings, 2004* (IEEE Cat. No.04EX862), vol. 1, 2005, pp. 26–30.

23. Chau, R. et al., Silicon nano-transistors for logic applications. *Phys E Low-Dimension Syst Nano-struct*, 2003, 19(1/2):1–5.

24. Chau, R. et al., Silicon nano-transistors and breaking the 10 nm physical gate length barrier. *61st Device Research Conference. Conference Digest* (Cat. No.03TH8663), 2003, pp. 123–126.

25. Chau, R. et al., High-kappa/metal-gate stack and its MOSFET characteristics. *IEEE Electron Dev Lett*, 2004, 25(6):408–410.

26. Chau, R., Benchmarking nanotechnology for high-performance and low-power logic transistor applications. In: *Fourth IEEE Conference on Nanotechnology, 2004* (IEEE Cat. No.04TH8757), 2004, pp. 3–6.

27. Thompson, S.E. et al., In search of "Forever," continued transistor scaling one new material at a time. *IEEE Trans Semiconduct Manufact*, 2005, 18(1):26–36.

28. Thompson, S.E., Strained Si and the future direction of CMOS. *Proceedings of the Fifth International Workshop on System-on-Chip for Real-Time Applications*, 2005, pp. 14–16.

29. Chau, R. et al., Application of high-kappa gate dielectrics and metal gate electrodes to enable silicon and nonsilicon logic nanotechnology. *Microelectron Eng*, 2005, 80:1–6.

30. Chau, R. et al., Emerging silicon and nonsilicon nanoelectronic devices: opportunities and challenges for future high-performance and low-power computational applications. In: *2005 IEEE VLSI-TSA. International Symposium on VLSI Technology* (VLSI-TSA-TECH) (IEEE Cat. No. 05TH8802), 2005, pp. 13–16.

31. Chau, R. et al., Benchmarking nanotechnology for high-performance and low-power logic transistor applications. *IEEE Trans Nanotech*, 2005, 4(2):153–158.

32. Datta, S. et al., Advanced Si and SiGe strained channel NMOS and PMOS transistors with high-k/metal-gate stack. *Proceedings of the 2004 Bipolar/BiCMOS Circuits and Technology Meeting* (IEEE Cat. No.04CH37593), 2004, pp. 194–197.

33. Datta, S. and R. Chau, Silicon and III–V nanoelectronics. *Proceedings of the International Conference on Indium Phosphate and Related Materials, 2005* (IEEE Cat. No. 05CH37633), 2005, pp. 7–8.

34. Mohta, N. and S.E. Thompson, Mobility enhancement. *IEEE Circ Dev Mag*, 2005, 21(5):18–23.

35. Bourianoff, G., The future of nanocomputing. *Computer*, 2003, 36(8):44–53.

36. Brewer, J.E., V.V. Zhirnov, and J.A. Hutchby, Memory technology for the post CMOS era. *IEEE Circ Dev Mag*, 2005, 21(2):13–20.

37. Hutchby, J.A. et al., Extending the road beyond CMOS. *IEEE Circ Dev Mag*, 2002, 18(2):28–41.

38. Zhirnov, V.V. and D.J.C. Herr, New frontiers: self-assembly and nanoelectronics. *Computer*, 2001, 34(1):34–43.

39. Zhirnov, V.V. et al., Emerging research memory and logic technologies. *IEEE Circ Dev Mag*, 2005, 21(3):47–51.

40. Zhirnov, V.V. et al., Emerging research logic devices. *IEEE Circ Dev Mag*, 2005, 21(3):37–46.

41. Collier, C.P. et al., A 2-catenane-based solid state electronically reconfigurable switch. *Science*, 2000, 289(5482):1172–1175.

42. Fay, P. et al., Fabrication of monolithically-integrated InAlAs/InGaAs/InP HEMTs and InAs/AlSb/GaSb resonant interband tunneling diodes. *IEEE Trans Electron Dev*, 2001, 48(6):1282–1284.

43. Duschl, R. and K. Eberl, Physics and applications of Si/SiGe/Si resonant interband tunneling diodes. *Thin Solid Films*, 2000, 380(1/2):151–153.

44. Feynman, R.P., There's plenty of room at the bottom. *Eng Sci*, 1960:22–36.

45. Nikon, Purified water extends the limits of ArF steppers. *Immers Lithogr Technol*, 2006.

46. Tsai, C., C.S.O. directed self-assembly of virus-based hybrid nanostructures. In: *Technical Proceedings of the 2005 Nanotechnology Conference and Trade Show*, Anaheim.

47. Zhou, J., T.-J.M.L., Y. Gao, M. Xue, T. Hamasaki, E. Hu, K. Wang, B. Dunn, Nanoscale assembly of nanowires templated by microtubules. In: *MRS Fall Meeting*, 2005, Boston, MA.

48. Singh, K., R.P., X. Wang, R. Lake, C. Ozkan, K. Wang, M. Ozkan, SWNT-PNA-SWNT conjugates: synthesis, characterization and modeling. In press.

49. Singh, K., K.A., X. Wang, A. Balandin, C. Ozkan, M. Ozkan. Bio-assembly of nanoparticles for device applications. In: *Nanotech 2005*, Los Angeles.

50. Singh, K.V., X.W., R.R. Pandey, R. Lake, C.S. Ozkan, M. Ozkan. Functionally engineered carbon nanotubes-peptide nucleic acid nanocomponents. *Mater Res Soc Symp Proc,* 2005.

51. Mao, C. et al., Virus-based toolkit for the directed synthesis of magnetic and semiconducting nanowires. *Science,* 2004, 303(5655):213–217.

52. Portney, N.G. et al., Organic and inorganic nanoparticle hybrids. *Langmuir,* 2005, 21(6):2098–2103.

53. Ravindran, S., C.T., K.V. Singh, S. Andavan, G.T., Y. Gao, M. Ozkan, E. Hu, C.S. Ozkan, Nanopatterned liquid metal electrode for the synthesis of novel prussian blue nanotubes and nanowires. *Nanotechnology,* 2005.

54. Ravindran, S., S.A.G.T., C.S. Ozkan, Selective and controlled self assembly of zinc oxide hollow spheres on single walled carbon nanotube templates. *Nanotechnology,* 2005.

55. Wang, X., R.P., K.V. Singh, C. Tsai, R. Lake, M. Ozkan, C.S. Ozkan, Synthesis and characterization of peptide nucleic acid-platinum complexes. *Nanotechnology,* 2005.

56. Wang, X., K.S., C. Tsai, R. Lake, A. Balandin, M. Ozkan, C. Ozkan, Oligonucleotide metallization for conductive bio-inorganic interfaces in self assembled nanoelectronics and nanosystems. In: *Micro- and Nanosystems-Materials and Devices, Mater. Res. Soc. Symp. Proc. 2005,* Warrendale, PA.

57. Wang, X., R.R.P., R. Lake, C.S. Ozkan, Hybrid nanomaterials for active electronics and bio-nanotechnology. In: *Technical Proceedings of the 2005 Nanotechnology Conference and Trade Show,* 2005, Anaheim.

58. Wang, K.L., Issues of nanoelectronics: a possible roadmap. *J Nanosci Nanotech,* 2002, 2(3/4):235–266.

59. Strukov, D.B. and K.K. Likharev, CMOL FPGA: a reconfigurable architecture for hybrid digital circuits with two-terminal nanodevices. *Nanotechnology,* 2005, 16(6):888–900.

60. Turel, O., et al., Neuromorphic architectures for nanoelectronic circuits. *Inter J Circ Theor Appl,* 2004, 32(5):277–302.

61. Khitun, A. and K.L. Wang, Cellular nonlinear network based on semiconductor tunneling structure with a self-assembled quantum dot layer. In: *2004 Fourth IEEE Conference on Nanotechnology* (IEEE Cat. No.04TH8757), 2004, pp. 161–163.

62. Li-Ju, L. et al., The quantum-dot large-neighborhood cellular nonlinear network (QLN-CNN) in nanotechnology. In: *Proceedings of the 2001 first IEEE Conference on Nanotechnology. IEEE-NANO 2001* (Cat. No.01EX516), 2001, pp. 331–334.

63. Blair, E.P. and C.S. Lent, Quantum-dot cellular automata: an architecture for molecular computing. In: *IEEE International Conference on Simulation of Semiconductor Processes and Devices, 2003* (Cat. No.03TH8679), 2003, pp. 14–18.

64. Orlov, A.O. et al., Experimental demonstration of a latch in clocked quantum-dot cellular automata. *Appl Phys Lett,* 2001, 78(11):1625–1627.

65. Chuang, M.A.N.a.I.L., *Quantum Computation and Quantum Information.* Cambridge University Press, 2000.

66. Turing, A.M., On computable numbers, with an application to the Entscheidungs problem. *Proc Lond Math Soc,* 1937, 42(2):230–265; correction, *Proc Lond Math Soc,* 1937, 43:544–546.

67. Keyes, R.W. and R. Landauer, Minimal energy dissipation in logic. *IBM J Res Dev,* 1970, 14(2):152–157.

68. Keyes, R.W., Fundamental limits in digital information processing. *Proc IEEE,* 1981, 69(2):267–278.

69. Keyes, R.W., Fundamental limits of silicon technology. *Proc IEEE,* 2001, 89(3):227–239.

70. Landauer, R., Irreversibility and heat generation in the computing process. *IBM J Res Dev,* 1961, 5:183.

71. Landauer, R., Computation: a fundamental physical view. *Physica Scripta,* 1987, 35(1):88–95.

72. Bate, R.T., Quantum-mechanical limitations on device performance. In: N.G. Einspruch, ed., *VLSI Electronics: Microstructure Science,* Academic Press, New York, 1982, pp. 359–386.

73. Bennett, C.H., Logical reversibility of computation. *IBM J Res Dev,* 1973, 17(6):525–532.

74. Fredkin, E. and T. Toffoli. Conservative logic. *Int J Theoret Phys,* 1982, 21(3/4):219–253.

75. This is a well known result from the works of Shannon and Landauer. In Zhirnov's work [4] this minimal energy dissipation was used to estimate the ultimate limits on scaling and switching speed.

76. Ostroumov, R.P. and K.L. Wang, Fundamental power dissipation in scaled CMOS and beyond. Unpublished, 2005.

77. Itoh, K. et al. Topology and spin alignment in model compounds of organic ferro- and ferrimagnets. *J Mol Electron*, 1988, 4(3):181–186.

78. Kato, Y., J. Berezovsky, and D.D. Awschalom, Spintronics: semiconductors, molecules and quantum information. In: *2004 International Electron Devices Meeting* (IEEE Cat. No. 04CH37602), 2005, pp. 537–538.

79. Awschalom, D.D., M.E. Flatte, and N. Samarth, Spintronics. *Sci Am*, 2003, 288(6):52–59.

80. Awschalom, D.D., M.E. Flatte, and N. Samarth, Spintronics. *Sci Am*, 2002, 286(6):52–59.

81. Wolf, S.A. et al., Spintronics: a spin-based electronics vision for the future. *Science*, 2001, 294(5546):1488–1495.

82. Malajovich, I. et al., Persistent sourcing of coherent spins for multifunctional semiconductor spintronics. *Nature*, 2001, 411(6839):770–772.

83. Johnson, M., Bipolar spin switch. *Science*, 1993, 260(5106):320–323.

84. Datta, S. and B. Das, Electronic analog of the electro-optic modulator. *Appl Phys Lett*, 1990, 56(7):665–667.

85. Nikonov, D.E. and G.I. Bourianoff, Spin gain transistor in ferromagnetic semiconductors — the semiconductor Bloch-equations approach. *IEEE Trans Nanotech*, 2005, 4(2):206–214.

86. Coey, J.M.D. and M. Venkatesan, Half-metallic ferromagnetism: example of CrO/sub 2. *J Appl Phys*, 2002, 91(10):8345–8350.

87. Park, Y.D. et al., A group-IV ferromagnetic semiconductor: Mn/sub x/Ge/sub 1-x. *Science*, 2002, 295(5555):651–654.

88. Goronkin, H. and Y. Yang, High-performance emerging solid-state memory technologies. *MRS Bull*, 2004, 29(11):805–813.

89. Muller, G. et al. Status and outlook of emerging nonvolatile memory technologies. In: *2004 International Electron Devices Meeting* (IEEE Cat. No.04CH37602), 2005, pp. 567–570.

90. Natarajan, S. and A. Alvandpour, Emerging memory technologies — mainstream or hearsay? In: *2005 IEEE VLSI-TSA International Symposium on VLSI Design, Automation & Test* (VLSI-TSA-DAT) (Cat. No. 05TH8803), 2005, pp. 222–228.

91. Natarajan, S., Emerging memory technologies, in *SPIE-Int. Soc. Opt. Eng. Proc.* 2003, 5274(1):7–13.

92. Natori, K., Emerging memory devices. *Oyo Buturi*, 2001, 70(2):192–200.

93. Suizu, K., T. Ogawa, and K. Fujishima, Emerging memory solutions for graphics applications. *IEICE Trans Electron*, 1995, E78-C(7):773–781.

94. Tran, L.C., Challenges of DRAM and flash scaling — potentials in advanced emerging memory devices. In: *2004 Seventh International Conference on Solid-State and Integrated Circuits Technology Proceedings* (IEEE Cat. No. 04EX862), vol. 1, 2005, pp. 668–672.

95. Tran, L.C., Beyond nanoscale DRAM and flash challenges and opportunities for research in emerging memory devices. In: *2004 IEEE Workshop on Microelectronics and Electron Devices* (IEEE Cat. No.04EX810), 2004, pp. 35–38.

96. Zhou, L. et al., Multilevel memory based on molecular devices. *Appl Phys Lett*, 2004, 84(11):1949–1951.

97. Likharev, K.K., Electronics below 10 nm. In: *Nano and Giga Challenges in Microelectronics*, 2003, Elsevier, Amsterdam, pp. 27–68.

98. Likharev, K.K. and A.N. Korotkov, Single-electron parametron: reversible computation in a discrete-state system. *Science*, 1996, 273(5276):763–765.

99. Korotkov, A.N. and K.K. Likharev, Single-electron-parametron-based logic devices. *J Appl Phys*, 1998, 84(11):6114–6126.

100. Lent, C.S. and P.D. Tougaw, A device architecture for computing with quantum dots. *Proc IEEE*, 1997, 85(4):541–557.

101. Niemier, M.T. and P.M. Kogge, Problems in designing with QCAs: layout = timing. *Int J Circ Theor Appl*, 2001, 29(1):49–62.

102. Tour, J.M. et al., Nanocell logic gates for molecular computing. *IEEE Trans Nanotech*, 2002, 1(2):100–109.

103. Husband, C.P. et al., Logic and memory with nanocell circuits. *IEEE Trans Electron Dev*, 2003, 50(9):1865–1875.

104. Das, S., G. Rose, M.M. Ziegler, C.A. Picconatto, and J.E. Ellenbogen, Architectures and simulations for nanoprocessor systems integrated on the molecular scale. In *Introducing Molecular Electronics*, Springer, Berlin, 2005, pp. 448–481.

105. Stan, M.R. et al., Molecular electronics: from devices and interconnect to circuits and architecture. *Proc IEEE*, 2003, 91(11):1940–1957.

106. Resnick, D.J. et al., Imprint lithography for integrated circuit fabrication. *J Vac Sci Tech B*, 2003, 21(6):2624–2631.

107. Choi, J., K. Nordquist, A. Cherala, L. Casoose, K. Gehoski, W.J. Dauksher, S.V. Sreenivasan, and D.R. Resnick, Distortion and overlay accuracy of UV step and repeat imprint lithography. *Microelectron Eng*, 2005, 78–79(Special Issue):633–640.

108. Kuekes, P.J., G.S. Snider, and R.S. Williams, Crossbar nanocomputers. *Sci Am*, 2005, 293(5):72.

109. DeHon, A.a.K.L., Hybrid CMOS/nanoelectronic digital circuits: devices, architectures, and design automation. *Proc ICCAD*, 2005.

110. Collier, C.P. et al., Electronically configurable molecular-based logic gates. *Science*, 1999, 285(5426):391–394.

111. Yong, C. et al., Nanoscale molecular-switch crossbar circuits. *Nanotechnology*, 2003, 14(4):462–468.

112. Wu, W. et al., One-kilobit cross-bar molecular memory circuits at 30-nm half-pitch fabricated by nanoimprint lithography. *Appl Phys (Mater Sci Proc)*, 2005, A80(6):1173–1178.

113. Fölling, S., Ö. Türel, and K.K. Likharev. Single-electron latching switches as nanoscale synapses. *Proc IJCNN*, 2001.

114. Dresselhaus, P.D. et al., Measurement of single electron lifetimes in a multijunction trap. *Phys Rev Lett*, 1994, 72(20):3226–3229.

115. Hongkun, P. et al., Nanomechanical oscillations in a single-C/sub 60/transistor. *Nature*, 2000, 407(6800):57–60.

116. Gubin, S.P. et al., Molecular clusters as building blocks for nanoelectronics: the first demonstration of a cluster single-electron tunnelling transistor at room temperature. *Nanotechnology*, 2002, 13(2):185–194.

117. Zhitenev, N.B., H. Meng, and Z. Bao, Conductance of small molecular junctions. *Phys Rev Lett*, 2002, 88(22):226801/1–4.

118. Jiwoong, P. et al., Coulomb blockade and the Kondo effect in single-atom transistors. *Nature*, 2002, 417(6890):722–725.

119. Kubatkin, S. et al., Single-electron transistor of a single organic molecule with access to several redox states. *Nature*, 2003, 425(6959):698–701.

120. Ziegler, M.M. and M.R. Stan, CMOS/nano co-design for crossbar-based molecular electronic systems. *IEEE Trans Nanotech*, 2003, 2(4):217–230.

121. Turel, O. and K. Likharev, CrossNets: possible neuromorphic networks based on nanoscale components. *Int J Cir Theor Appl*, 2003, 31(1):37–53.

122. Likharev, K.K., A. Mayr, I. Muckra, and Ö. Türel, CrossNets: high-performance neuromorphic architectures for CMOL circuits. *Ann N Y Acad Sci*, 2003, 1006:146–156.

123. Likharev, K.K., CMOL: A new concept for nanoelectronics. In: *12th International Symposium on Nanostructures Physics and Technology*, St. Petersburg, Russia, 2004.

124. Likharev, K.K., CMOL: A silicon-based bottom-up approach to nanoelectronics. *Interface*, 2005, 14(1):43–45.

125. Jensen, K.L., Field emitter arrays for plasma and microwave source applications. *Phys Plasma*, 1999, 6(5):2241–2253.

126. Strukov, D.B. and K.K. Likharev, Prospects for terabit-scale nanoelectronic memories. *Nanotechnology*, 2005, 16(1):137–148.

127. Strukov, D.B.a.K.K.L., Architectures for defect-tolerant CMOL memories. Paper in preparation.

128. von Neumann, J., Probabilistic logics and the synthesis of reliable organisms from unreliable components. In *Automata Studies*, Princeton University Press, Princeton, NJ, 1956, pp. 329–378.

129. Likharev, K.K.a.D.B.S., CMOL: devices, circuits, and architectures. In: G.C.e. al., ed., *Introducing Molecular Electronics*, 2005, pp. 447–477.

130. Strukov, D.B. and K.K. Likharev, A reconfigurable architecture for hybrid CMOS/nanodevice circuits. Paper accepted for presentation at FPGA'06, 2006, Monterey, CA.

131. Unsal, O.S. and I. Koren, System-level power-aware design techniques in real-time systems. *Proc IEEE*, 2003, 91(7):1055–1069.

11

Molecular Machines

11.1 Introduction 11-1
Overview of Molecular Machinery • Enabling Technologies
• The Controlled Molecular Motion Quandary • Models
from Nature and Bottom-Up Assembly • Molecular
Machine Taxonomy

11.2 Transferring Molecular Movement to the Solid State 11-6
Comparison of Molecular Environments • The $CBPQT^{4+}$
Recognition System — A Case Study • Bistable [2]Rotaxanes
• LB Studies • Computational Modeling • Electronic
Transport • Observing Condensed Phase Movement

11.3 Molecular Muscle Systems 11-16
The Potential of Artificial Muscles • Transition Metal
Controlled Actuators • Redox-Controlled Annulene-Based
Muscles • Photo-Controlled Polymeric Muscles
• Flexing Cantilevers with Palindromic [3]Rotaxanes

11.4 Molecular Nanovalves 11-21
Typical Valve Systems • Irreversible Thin Film Regulators
• Light-Regulated Azobenzene and Coumarin Valves
• Supramolecular Gatekeepers • Molecular Gatekeepers
• Polymeric Valves • Biological Nanovalves

11.5 Molecular Rotors 11-26
Rotation as a Fundamental Molecular Motion • Rotation by
Design • Molecular Compasses and Gyroscopes
• Light-Driven Unidirectional Rotation • Surface–Rotor
Interactions

11.6 Surfaces with Controllable Wettability 11-35
Water — Love to Hate It, Hate to Love It • Energetic
Contributors to Wettability • Light-Responsive Surfaces
• Electrostatic Control of Surface Properties • Nanofluidics

11.7 Conclusions 11-40
Acknowledgments 11-40
References 11-41

Brian H. Northrop
Adam B. Braunschweig
Paula M. Mendes
William R. Dichtel
J. Fraser Stoddart
University of California, Los Angeles

11.1 Introduction

11.1.1 Overview of Molecular Machinery

Modern synthetic chemistry has equipped researchers with the ability to construct molecules of ever increasing structural complexity. In some ways, synthetic chemists can be likened to architects or engineers working in a molecular world. This analogy is more appropriate then with respect to the current endeavor to design and synthesize molecular machines since chemists, like architects, must consider design, construction, and interactions with surroundings. This chapter will explore the realm of synthetic molecular machines by highlighting the design, construction, and function of a number of examples in the fields of rotors, muscles, switches, etc., where the common theme is a change in material properties as a result of some stereochemical rearrangement at the molecular level.

Macroscopic machines rely upon the coordinated function of multiple constituent parts in order to perform a specific task. Machines consume fuel (input) in order to accomplish their specified function (output). Molecular machines[1–5] can be envisioned as functioning in a similar fashion. A molecular machine may be thought of in terms of molecules that can, with an appropriate stimulus, be temporarily lifted out of equilibrium, and, upon their returning to equilibrium, inducing a change in the observable macroscopic properties of the system. However, there is considerable debate currently as to the exact constitution and properties of a "molecular machine," always assuming that the term "machine" can even be applied to a molecular system, and how one separates a molecular machine from its biological counterparts. In this chapter, however, we do not aim to set limits on what does and does not constitute a molecular machine, simply because of the present lack of consensus regarding its definition. By setting such limits, we would most likely render this article obsolete from the start. Rather, the aim is to focus attention on how the controlled motions of synthetic molecular systems have been harnessed to cause observable macroscopic changes in bulk systems as a result of stereochemical rearrangements at the molecular level. The design and evaluation of such molecular systems has matured as a result of con-comitant advances in three important scientific areas:

1. Organic synthesis, which has progressed such that almost any molecular architecture can be synthesized
2. Powerful computational techniques, which are capable of bringing understanding to complex intra- and intermolecular interactions in molecular systems
3. The advent of a collection of powerful single-molecule analytical tools

11.1.2 Enabling Technologies

The construction of molecular machines has been led by progress in major areas of chemistry, including (1) the introduction of new synthetic methodologies; (2) the development of supramolecular chemistry; and (3) the improvement of spectroscopic techniques used to characterize the structures and dynamics of increasingly complex organic molecules. With advancements in synthetic methods, chemists are able to perform transformations with greater selectivity, sensitivity, and stereocontrol. Living free radical polymerizations,[6,7] asymmetric catalysis,[8] metal-catalyzed cross-coupling reactions,[9] and metathesis[10] are but only a few examples of the new methods that aid and abet the construction of diverse molecules and macromolecules. Major developments in the field of supramolecular chemistry[11–17] have enabled the design of molecular components to interact favorably with each other in such a way that they can self-organize and self-assemble into larger, well-defined architectures through noncovalent interactions prior to the template-directed synthesis of mechanically interlocked molecular compounds. Analytical techniques, such as high-resolution mass spectrometry, multidimensional nuclear magnetic resonance (NMR) spectroscopy, and spectroelectrochemical and spectrophotometric techniques, have provided chemists with an increasing number of tools with which to investigate the structures and functions of synthetically made compounds in solution. With these three powerful contemporary tools — advanced synthetic methods, supramolecular chemistry, and new analytical techniques — many novel molecular compounds have been constructed and characterized.

The use of computational techniques in developing appropriate structural and superstructural targets and calculating the noncovalent interactions within infinitely complex systems has become standard practice in the development, design, and understanding of nanoscopic systems. The development of massively parallel computer clusters, as well as more efficient and accurate computational algorithms, coupled with increases in processing speeds, has enabled larger and larger molecular and supramolecular systems to be investigated with increasing accuracy. The development of density functional theory and quantum mechanical methods, for which Pople[18] and Kohn[19] were awarded the 1998 Nobel Prize in chemistry, allows researchers to compute accurately many physical and electronic properties that are of particular relevance to molecular machines and electronics. Electron transfer rates, orbital energies, solvent effects, excited states, and many other electronic and magnetic properties can be quantified. In addition, molecular dynamics simulations[20] provide valuable insight into the structures and superstructures of large molecules or ensembles of molecules, how they interact, and how they evolve over time — all of which are especially relevant for determining interactions between molecules and surfaces. Hybrid computational methods, such as ONIOM[21] and Car-Parrinello molecular dynamics,[22] combine the accuracy of quantum mechanics with the speed of molecular mechanics.

Advances in nanotechnology have led to the creation of tools for the controlled deposition of molecular systems onto a variety of substrates. Self-assembled monolayers (SAMs)[23] of thiols on gold[24,25] and triethoxysilanes on silicon dioxide[26] are examples of two widely used techniques to attach organic compounds onto metallic and inorganic substrates. Langmuir–Blodgett (LB) techniques[27,28] have enabled researchers to transfer monolayers of amphiphilic molecules onto almost any substrate. More recently, many techniques have emerged for patterning molecules on surfaces. For example, Whitesides[29] has developed soft-lithographic techniques for the patterning of thiols on gold using poly(dimethylsiloxane) stamps. Dip-pen nanolithography, which relies upon the tip of an atomic force microscope (AFM) and can be used to pattern molecular "inks" on a variety of surfaces with 10 to 15 nm resolution, has been developed by Mirkin.[30] Linford[31,32] has developed a number of chemo-mechanical techniques to pattern various functional groups on silicon simply by etching the surface with a diamond tip, a tungsten carbide ball, or an AFM in the presence of the desired molecules. Electrografting, a technique introduced by Buriak,[33] uses a conducting AFM to pattern alkynes onto silicon. These are just a few of the recent advances in surface attachment and patterning techniques that provide researchers with the tools to link, precisely and selectively, functional organic molecules to surfaces in anticipation of developing coordinated and coherent surface-bound machines.

The variety of methods available to study these functionalized surfaces with single-molecule precision has grown in step with the ability to create sophisticated, functionalized surfaces. Probe microscopies, such as atomic force microscopy[34] and scanning-tunneling microscopy[35] (STM), for which Ruska, Binnig, and Heinrich were awarded the 1986 Nobel Prize in physics, are vital analytical tools that are capable of imaging surfaces down to the nanometer scale. Surface-sensitive spectroscopic techniques such as x-ray photoelectron spectroscopy[36,37] (XPS), ellipsometry,[38] x-ray reflectometry[39] (XR), and contact angle measurements[40] are some examples of analytical techniques that are able to evaluate the composition, orientation, and homogeneity of surfaces in order to optimize and analyze the orientations and stabilities of molecules at surfaces.

11.1.3 The Controlled Molecular Motion Quandary

Even with the powerful tools made available through chemical, computational, and analytical techniques, creating molecular actuators that function analogously to their macroscopic counterparts remains a considerable challenge. The opportunities of harnessing the actuation of molecules to cause macroscopically detectable changes makes overcoming these challenges worthwhile. Many of these challenges lie in the fundamental differences between the macroscopic and molecular worlds. Macroscopic actuators are hardly affected by the random thermal motions (e.g., vibrations and rotations) of molecules whereas these Brownian forces play a major part in the transfer of energy, heat, and motion at the molecular level. Additionally, the law of microscopic reversibility states that passage rates over barriers of equal height will be the same in either direction.[41] Thus biasing motion in one direction using thermal energy is a monumental task, the challenge of which can be demonstrated by considering (Figure 11.1) the Feynman ratchet.[41] In a Feynman ratchet, a paddle is connected to a saw-toothed gear, which is held in place by a spring-loaded pawl. The saw-tooth geometry of the gear induces an asymmetry to the system that enables the gear to rotate in one direction but not in the other. Feynman considered whether the collective thermal fluctuations of many molecules striking the paddle would be capable of driving rotation in only one direction as a result of the asymmetric saw-toothed gearing, resulting in perpetual motion.

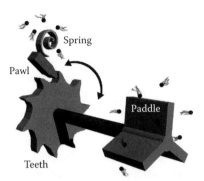

FIGURE 11.1 Schematic of a Feynman ratchet. Given sufficient time, there is a statistical possibility that enough of the collective random motions of enough of the molecules striking the paddle in unison may cause the ratchet to turn. The asymmetry of the teeth would ensure that rotation occurs only in one direction, resulting in perpetual unidirectional motion. However, this attractive possibility is thwarted by the fact that it is equally likely for enough molecules to strike and compress the spring, allowing the pawl to disengage, and leaving the ratchet more likely to rotate in the opposite direction. Ultimately, Feynman demonstrated how the second law of thermodynamics prevails to ensure that random Brownian motion cannot be used to achieve perpetual unidirectional rotation.

However, unidirectional rotation is not possible because it is equally probable that thermal fluctuations will compress the spring, releasing the pawl, and allow the paddle to slip backward. As this example demonstrates, the second law of thermodynamics and the law of microscopic reversibility tend to counteract any attempts to impart a directional bias into molecular structures for the purpose of performing useful work. However, these limitations are overcome with appropriately, and often times cleverly, designed wholly synthetic molecules. The inspiration for systems able to overcome the realities posed by the laws of microscopic reversibility and thermodynamics to perform useful work comes both from natural examples of molecular actuators, including muscles and proton pumps, and from the many potential benefits that can be gained from precise structural control and the resulting properties of atomically well-defined systems.

11.1.4 Models from Nature and Bottom-Up Assembly

Blueprints for designing molecular machinery are found commonly in nature (Figure 11.2), which is rich with examples of molecules that are capable of performing complex functions.[42,43] Kinesins, a class of motor proteins that shuttle various cargos along microtubules, achieve actuation by what has been referred to as a "walking" mechanism. The movement of myosin along actin filaments, driven by the hydrolysis of ATP, causes muscle contraction and expansion. Whip-like and sinusoidal movements of cilia and flagella, respectively, enable them to transport cells throughout the body. The rotary motion of F_1-ATPase, which is also powered by ATP, is capable of pumping protons across a membrane. In all these examples, the concerted movements of molecules or groups of molecules are used to perform vital biological functions. By drawing analogies between these natural motor proteins and macroscopic machines, many research groups have set out to design molecular machines capable of performing a veritable collection of functions.

In addition to gaining inspiration from nature, the drive to synthesize molecular actuators stems from the advantages provided by the bottom-up assembly[1–5,44] of molecular components. The top-down approach to constructing functional devices and actuators reaches a number of fundamental limitations as device features approach the molecular scale. Lithographic techniques for fabricating devices become both less precise and increasingly costly as features become smaller.[44,45] Building up from the molecular scale, however, has distinct advantages over the top-down approach. With the assistance of supramolecular chemistry, molecular recognition, self-assembly, and template-directed synthesis, specific molecular architectures of varying complexity can be created. This approach benefits from the fact that by virtue

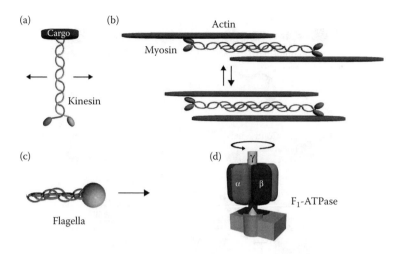

FIGURE 11.2 (a) A schematic representation of the motor protein kinesin, which is able to transport various cargos along microtubules. (b) The directional movements of myosin along actin filaments are responsible for muscle contraction and expansion. (c) Flagella and cilia (not shown) are able to propel materials across the cell surface as a result of their sinusoidal and whip-like movements, respectively. (d) F_1-ATPase undergoes unidirectional rotation during the hydrolysis of ATP to ADP.

of molecular recognition and self-assembly, individual components do not have to be sequentially pieced together. In addition, template-directed synthesis can be particularly modular and the physical properties of constituent parts can be optimized to perform various desired functions. Moreover, the molecular nature of molecular machines makes them amenable to a wide variety of inputs, including pH changes, oxidation, reduction, ion addition, solvent polarity, and light. The characteristics of this chemical, bottom-up approach to the design of molecular-scale machinery with specific architectures, properties, and functions are some of the greatest motivators for research in this area.

11.1.5 Molecular Machine Taxonomy

Among the many designs for externally controllable molecular machines are molecular shuttles, switches, muscles, nanovalves, rotors, and surfaces with controlled wettability (Figure 11.3). Molecular shuttles are molecules wherein a ring component and a dumbbell component are mechanically interlocked and, thus, are considered [2]rotaxanes. The presence of two identical recognition sites that are capable of becoming complexed by the ring component through noncovalent interactions located on the dumbbell component results in what is referred to as a degenerate molecular shuttle. The thermally activated movement of the ring component between the identical sites along the dumbbell exchanges the two coconformations that the molecule is capable of adopting and is referred to as shuttling. The incorporation of two different recognition sites into the dumbbell gives rise to a molecular switch. With the appropriate external stimulus, a molecular switch is capable of being selectively stimulated to be in one of two identifiably different coconformations. Molecular muscles are compounds that can, when triggered by an external stimulus, be induced to expand or contract reversibly. This expansion and contraction is reminiscent of the same process that occurs in muscle fibers. Harnessing the ability to change molecular structures in such a way that they can cover or uncover porous materials, and thus either trap or release their content(s), forms the basis of a molecular nanovalve. Molecular rotors are molecules composed of two moieties that are able to rotate relative to each other in a controlled manner. Using structural changes in surface-bound molecules to induce changes in the hydrophobicity or hydrophilicity of a surface allows for selective control of surface wettability and has potential applications in microfluidic devices.

This chapter will highlight some of the recent accomplishments made in the design, synthesis, and modeling of synthetic organic molecular machines. Particular attention will be paid to those examples

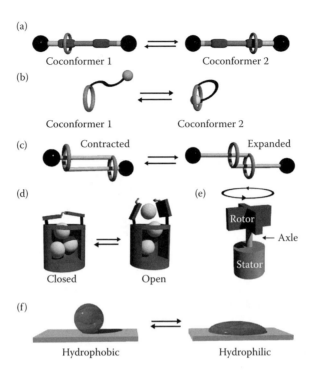

FIGURE 11.3 (a) A molecular shuttle consisting of a ring component that is mechanically interlocked onto a dumbbell-shaped component and is able to shuttle between two recognition sites as a result of thermal activation. Here coconformers 1 and 2 are degenerate. (b) Molecular switches are compounds that can be externally stimulated to exist in either of two observably different states or conformers. (c) A molecular muscle is able to expand and contract reversibly upon external stimulation. (d) Molecular valves can be used to trap and release other molecules as the result of controlled molecular motions. (e) Molecular rotors undergo controlled rotational motion of a rotor unit relative to a stator, which are connected via an axle. (f) Surfaces with controlled wettability can be stimulated to be hydrophobic or hydrophilic.

that involve molecules in condensed phases or attached to surfaces, as they provide perhaps the greatest insight into the development of functional devices. The first section will focus on molecular shuttles and switches based on mechanically interlocking rotaxanes and noncovalently associated pseudorotaxanes based on the π-electron-rich cyclophane, cyclobis(paraquat-*p*-phenylene)vinylene[46] (CBPQT[4+]), simply because they have been thoroughly studied in solution as well as in condensed phases and on solid substrates. The idea behind this agenda is twofold — (1) to provide specific examples of a class of molecular machines and (2) to introduce how such machines can be transferred from the solution phase onto surfaces in such a way that they retain their desirable physical and dynamic properties. From there on, the next three sections of this chapter will be devoted to recent advancements in designing organic molecules to function as molecular muscles, nanovalves, and rotors. The final section will feature some current research that takes advantage of the stimuli-responsive motions of surface-bound organic molecules to control surface wettability.

11.2 Transferring Molecular Movement to the Solid State

11.2.1 Comparison of Molecular Environments

The solution phase is often an ideal medium in which to study the structure, function, and dynamic properties of molecular machines, given the wealth of spectroscopic and analytical techniques that are available for solution-phase investigations. However, designing and fabricating useful devices based on

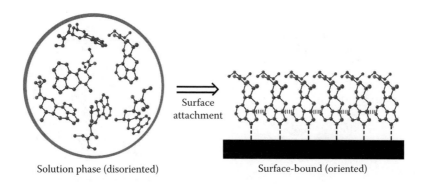

Solution phase (disoriented)   Surface-bound (oriented)

FIGURE 11.4 A comparison of the disorder found in the solution phase (*left*) with the order that may be imparted to molecules when they are assembled or attached to surfaces. The ordered configurations of surface-bound molecules can be attributed to the surface–substrate and noncovalent substrate–substrate interactions.

the functional properties of molecular machines requires that they operate in a coordinated and cooperative manner. In this regard, the solution phase is not an ideal environment for functional nanomechanical devices. Molecular conformations and motions are in disordered and disorientated states when molecules are in solution (Figure 11.4). While a specifically designed molecular machine may be capable of externally controlled expansion and contraction in solution, there is no means of harnessing such motion unless the molecule is ordered within a condensed phase or attached to a solid substrate. The attachment of molecular machines to surfaces provides a means of organizing and orienting them such that their motions can be coordinated so that they may operate in a collective manner. However, benefits gained by attaching molecular machines to a surface can be complicated by the inherent difficulties of studying organic molecules in the condensed phase for which spectroscopic techniques such as ^{1}H NMR and UV/Vis spectroscopies are not well suited. Furthermore, the molecular machines must be designed such that the surface–substrate interactions occur in a manner that enables rather than inhibits the desired molecular function.

Nonetheless, a thorough understanding of the solution-phase properties of molecular machines is a prerequisite to understanding their function in condensed phases. In addition, intermolecular and surface–substrate interactions can have a profound effect upon the packing, orientation, and ultimately, function of molecules in condensed phases and on surfaces. For these reasons, it is important to investigate the effects of structure on the deposition of molecules onto surfaces, the conformations they adopt, and how they pack. Through iterative structure–function feedback loops, which take advantage of the knowledge gained in solution and condensed phase studies to redesign molecular systems, the properties of surface-bound molecular machines can be optimized.

11.2.2 The CBPQT⁴⁺ Recognition System — A Case Study

Mechanically interlocked [2]rotaxanes[11,47–49] and noncovalently associated [2]pseudorotaxanes[50,51] based (Figure 11.5) upon the recognition between the π-electron-poor cyclophane CBPQT^{4+} and a variety of π-electron-rich guests — classes of molecules and supermolecules that have been developed into a number of functional surface-bound nanodevices[52] — have been studied extensively in solution, in a variety of condensed phases, and when attached to a number of different surfaces. Information gained from these investigations across a range of environments provides an example of how the structure–function relationships of molecular machines can be used to aid in the design of operational, surface-bound nanodevices. Molecular recognition occurs when π-electron-rich guests, such as 1,5-dioxynaphthalene (DNP) and tetrathiafulvalene (TTF), are threaded within the cavity of the CBPQT^{4+} host as a result of stabilizing [C–H⋯O], [C–H⋯ π], and [π⋯π] interactions. If the threaded complex is not interlocked, meaning the guest can escape reversibly from and return to the host in a dynamic equilibrium, the complex is a

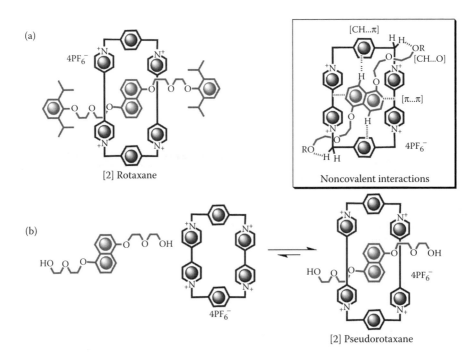

FIGURE 11.5 (a) A [2]rotaxane consists of a ring-shaped component that is confined to a dumbbell-shaped component by the presence of large "stoppering" units at both ends of the dumbbell. (b) In a [2]pseudorotaxane, the ring-shaped component is able to move on and off of a thread-like component via processes referred to as threading and dethreading. The ratio of free and complexed species is determined by the free energy of binding in the host–guest complex. (*Inset*) Binding observed in (a) and (b) is favorable as a result of stabilizing noncovalent [C–H$\cdots\pi$], [C–H$\cdots$O], and [$\pi\cdots\pi$] interactions.

[2]pseudorotaxane. This threaded complex can be mechanically interlocked, creating a [2]rotaxane with the incorporation of large stoppering units at each end of the guest such that the CBPQT^{4+} host is unable to disassociate from the guest dumbbell, i.e., a [2]rotaxane is a molecule. Stoppering can be achieved through template-directed[53–58] slipping,[59,60] clipping,[61,62] or stoppering[63] to form [2]rotaxanes. Alternatively, these same synthetic procedures may be used to create [2]catenanes, molecules consisting of two mutually interlocking rings. By varying the structure of the guests, the spacers, the stoppers, and even the ring, a molecular toolkit[52] has been developed by which a large number of molecular machines can be designed to perform a wide variety of tasks.

11.2.3 Bistable [2]Rotaxanes

One of the most interesting variations of the [2]rotaxane structure is the bistable form (Figure 11.6), which undergoes redox-controlled mechanical actuation, either chemically or electrochemically, causing the ring component to move along the dumbbell between two different molecular recognition units.[64–71] In such a nondegenerate [2]rotaxane, two different coconformations are said to exist and there is an equilibrium between having the ring positioned at one recognition site or the other. The ratio of the two different coconformations (N) depends on the difference in free energy[72–74] ($\Delta\Delta G°$) with which either recognition site is bound by the ring. In the case of a bistable [2]rotaxane composed of a CBPQT^{4+} ring and a dumbbell containing the π-electron-rich recognition sites DNP and TTF, the CBPQT^{4+} ring, at equilibrium, interacts preferentially with the TTF site, resulting in the ground state coconformation (GSCC). However, because this is a system in dynamic equilibrium, in some of the molecules, the CBPQT^{4+} ring will be interacting with the DNP site in what is known as the metastable state coconformation (MSCC). Under

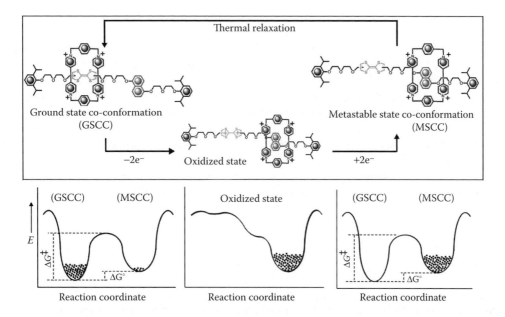

FIGURE 11.6 Molecular structures and representative energy diagrams for the redox-controlled switching process of a bistable [2]rotaxane. In the GSCC, the CBPQT^{4+} ring binds preferentially to the TTF recognition unit. A two-electron oxidation causes the ring to move to the DNP recognition unit as a result of charge–charge repulsion between the tetracationic cyclophane and the dicationic TTF^{2+}. After reduction of TTF^{2+} to its neutral state, the CBPQT^{4+} ring continues to encircle the DNP unit in what is referred to as the MSCC. Thermal relaxation allows translation of the CBPQT^{4+} ring back to the GSCC.

ambient conditions in acetonitrile (MeCN), the ring prefers the TTF site to the DNP one by 1.6 kcal mol^{-1}, and so the bistable [2]rotaxane prefers to adopt the GSCC to the MSCC in >90% of the molecules. The equilibrium ratio of GSCC to MSCC (K_{eq}) depends on differences in free energy of binding ($\Delta\Delta G°$) of the ring between the two sites and is quantified by the relation

$$K_{eq} = e^{-(\Delta\Delta G°)/RT} \tag{11.1}$$

where R is the universal gas constant and T is the temperature in degrees Kelvin. Energetically, this bistability can be thought of as a two-well potential surface, with the deeper well corresponding to the GSCC and the shallower one corresponding to the MSCC. Chemical or electrochemical oxidation of the TTF to its dication (TTF^{2+}) causes[75] the disappearance of the energy minimum, corresponding to the GSCC, because of Coulombic repulsion between the tetracationic CBPQT^{4+} ring and the TTF^{2+} dication. As a result, the CBPQT^{4+} ring interacts preferentially with the DNP site, leading to the movement of the ring to this site. Thus chemical or electrochemical oxidation is the basis of the redox-controlled mechanical actuation of this system. Reduction of the TTF^{2+} back to the ground state TTF reestablishes the well on the potential energy surface corresponding to the GSCC. However, the activation barrier that exists between the two wells on the energy surface delays the return of the molecule from the metastable state to the ground state. Eventually, thermal relaxation causes the molecule to overcome this barrier and return to the GSCC, where the CBPQT^{4+} again encircles TTF, thus completing a full switching cycle.

The range of tetracationic hosts and π-electron-rich guests that exist in the molecular toolkit has led to a wide array of different bistable [2]rotaxanes. Because the ratio of MSCC to GSCC (N_{MSCC}/N_{GSCC}) at equilibrium and the relaxation rate from MSCC to GSCC both result from the differences in thermodynamic parameters between the recognition sites and the CBPQT^{4+} ring, varying these sites will also vary their association with the CBPQT^{4+} ring and consequently the switching properties of the bistable molecules. The difference in free energies, $\Delta\Delta G°$, between two different recognition sites can be estimated from the binding (K_a) of individual guests excised from the bistable [2]rotaxane with the CBPQT^{4+} host. These data can be obtained by studying the binding of [2]pseudorotaxanes composed of the CBPQT^{4+} host and the relevant guests. The thermodynamic binding parameters (ΔG^o, ΔH^o, ΔS^o) for [2]pseudo-rotaxanes formed between CBPQT^{4+} as the host and a variety of guests, such as TTF with diethylene glycol chains (TTF-DEG) and DNP with DEG chains (DNP-DEG), have been determined by (1) isothermal titration calorimetry; (2) the single-point ^{1}H NMR spectroscopic method; (3) UV/Vis titration. The differences in the thermodynamic binding parameters ($\Delta\Delta G^o$, $\Delta\Delta H^o$, $\Delta\Delta S^o$) between any two of the guests can be deduced by simple subtraction. Thus, the behavior of a bistable [2]rotaxane containing different recognition sites can be estimated from the experimental differences in thermodynamic parameters of the individual components.

The equilibrium constant, K_{eq}, between the two different coconformations is determined using Equation (11.1) and changes that occur as a result of temperature can be estimated using Equation (11.2):

$$\Delta\Delta H^o/T - \Delta\Delta S^o = -R\ln K_{eq} \tag{11.2}$$

Of significant interest in the design of bistable [2]rotaxanes is the fact that the large changes in K_{eq}, which occur when T is varied, are the result of big differences in $\Delta\Delta H^o$, whereas if $\Delta\Delta H^o$ is small, then almost no change in N_{MSCC}/N_{GSCC} occurs as T changes. This theoretical model has been validated experimentally through subtle structural modifications of the dumbbell components in bistable [2]rotaxanes. For example, exchanging TTF for bispyrrolotetrathiafulvalene (BPTTF) affects significantly the temperature dependence of N_{MSCC}/N_{GSCC} because of the differences in enthalpy of binding parameters between TTF and BPTTF with the CBPQT^{4+} ring. This same phenomenon — the $\Delta\Delta H^o$ control of the temperature-dependent switching that is seen in solution — has also been observed in a polymer matrix and in SAMs. Because this thermodynamically controlled switching mechanism is so consistent across environments, the properties of bistable molecules on surfaces can be predicted by studying the complexation of model [2]pseudorotaxanes in solution.

In contrast to the effects of different recognition sites on the ground state equilibrium position of the CBPQT^{4+} ring, the rate at which bistable [2]rotaxanes return to equilibrium upon reduction of the TTF^{2+} dication to its neutral state is more strongly controlled by environmental factors[72–74] such as viscosity and temperature. All bistable [2]rotaxanes investigated so far show a decrease in the rate of return to the GSCC from the MSCC upon reduction of the TTF^{2+} dication as temperature is lowered, an observation that reinforces the fact that the relaxation rates from the MSCC to the GSCC depend upon an activated process. This temperature dependence is observed because less thermal energy is available to overcome the activation barrier ($\Delta G^{\ddagger}$) between the GSCC and the MSCC. Increased viscosity also slows down the movement of the ring and, consequently, the switching speed. Viscosity effects have been tested by measuring differences in thermal relaxation rates both in MeCN solution and in a condensed polymer matrix, where viscosity increases from 3.5 (MeCN) to 50,000 cP (polymer matrix).[72,73] In the case[74] of a bistable, amphiphilic [2]rotaxane, where the dumbbell component contains DNP and BPTTF sites, this change in viscosity causes the relaxation from the MSCC to the GSCC to slow down from 1.26 to 10.2 sec at 298 K. Also, when bistable [2]rotaxanes are compressed into a closely packed monolayer, switching slows down to 624 sec at 298 K.

11.2.4 LB Studies

Slowing down of the switching speed, in addition to other changes that result from altering the molecular environment, happen because of new inter- and intramolecular and surface interactions that, as a result of solvation and dilution, do not exist in solution. In an effort to gage the effects of the new interactions that appear in condensed phases, amphiphilic bistable [2]rotaxanes were investigated by Langmuir–Blodgett (LB) studies[76,77] in SAMs. Since the properties of SAMs at the air–water interface provide insight into the structure and packing of these [2]rotaxanes on a variety of surfaces, by rationally and methodically varying the molecular structure of the molecules studied in LB films, specific monolayer properties

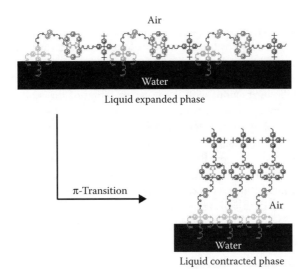

Liquid expanded phase

π-Transition

Liquid contracted phase

FIGURE 11.7 Molecular structures of amphiphilic bistable [2]rotaxanes at the air–water interface of a LB film. In the liquid-expanded phase, the bistable [2]rotaxanes adopt an elongated conformation in which the hydrophilic CBPQT^{4+} ring of each [2]rotaxane molecule is able to come in contact with the water. Increasing the surface pressure results in a π-transition to the liquid-condensed phase where the conformations of [2]rotaxanes are more upright and elongated.

can be obtained. In this manner, the structure–property relationships of these amphiphilic [2]rotaxanes can be established. A series of LB monolayers with various dumbbells and bistable [2]rotaxanes incorporating the CBPQT^{4+} ring and TTF and DNP recognition sites were prepared, and the effects of structural modifications on monolayer stability, packing area, conformation, and thickness were determined. The consequences of changing hydrophilic and hydrophobic tail lengths of the stoppers, the relative positions of the TTF and DNP recognition sites along the dumbbell, and the presence or absence of the CBPQT^{4+} ring were all investigated in some detail. Experimental parameters, such as subphase temperature, monolayer compression rates, spreading solvents, and salt concentrations were also investigated. The CBPQT^{4+} ring, DEG, tetraethylene glycol (TEG), and TTF are all hydrophilic, while the extended aromatic surfaces of DNP and the tetraarylmethane stoppers render these positions of the amphiphilic dumbbells and bistable [2]rotaxanes hydrophobic. They are expected to have a folded conformation (Figure 11.7) at the air–water interface as a result of the alternating hydrophobic and hydrophilic nature of their constituent parts. The inter- and intramolecular forces caused by the CBPQT^{4+} ring led to the most pronounced changes in the pressure-area isotherm. Films of dumbbells (i.e., without the CBPQT^{4+} ring) had an average collapse pressure of ~45 mN m^{-1}, while their respective bistable [2]rotaxanes (i.e., with the CBPQT^{4+} ring in place) had an average collapse pressure of ~63 mN m^{-1}. The added structural integrity of the bistable [2]rotaxane films most likely results from intermolecular repulsion between adjacent tetracationic moieties, as well as hydrophilic anchoring of the CBPQT^{4+} ring as it interacts with the air–water interface in the liquid expanded phase (Figure 11.7). The interactions of the hydrophilic stoppers with both the air–water interface and adjacent molecules also affect monolayer quality. Altering the lengths of the polyether chains attached to the hydrophilic stopper from DEG to TEG increases compressibility and decreases viscosity at high surface pressures. The TEG tails appear to be more spread out over the air–water interface, and, as a result, interact more strongly with the TEG tails of adjacent molecules than the DEG tails, which only interact under conditions of sufficiently high compression. It is not unlikely that this phenomenon occurs because the CBPQT^{4+} ring is approximately the same size as the DEG tails. Therefore, when the smaller stopper with DEG tails is used, the interactions between the CBPQT^{4+} ring and adjacent molecules are dominant. When the larger TEG stopper is used, the effects of one stopper interacting with the stoppers of adjacent molecules dominate the behavior of the monolayer. The fact that a more upright conformation of the molecules is observed when the shorter DEG tails are used also explains the smaller mean molecular area, stronger intermolecular interactions, decreased compressibility, and greater viscosity when compared to bistable [2]rotaxanes with the longer, more flexible TEG tails. Film thicknesses (21–39 Å), as measured by ellipsometry, were significantly less than the calculated lengths (65–75 Å) of the fully elongated molecules, further suggesting that the molecules adopt a folded conformation as a result of either hydrophilic interactions with the water or tilting of the molecules with respect to the surface. The changes that resulted from variations in experimental parameters — e.g., compression rate, subphase temperature, spreading solvent, and subphase salt concentration — only affected film stability and did not influence molecular conformation, viscosity, or area. This data indicates that interactions resulting from structural effects — more than any other parameter — have the greatest influence on the organization of molecules in the resulting films. The LB studies show very clearly that a variety of inter- and intramolecular, and surface interactions, which do not occur in solution, dominate the structure of the molecular switches at surfaces. This knowledge that has been gained from LB studies is useful when considering the rational design of functional surface-bound molecular systems.

11.2.5　Computational Modeling

In addition to the structural information gained from organic thin films on surfaces by LB studies, computational modeling[78,79] has also provided insight into the packing, orientation and energy changes that occur in monolayers as molecules are switched between the MSCC and the GSCC. Fully atomistic molecular dynamics simulations were carried out on simulated Au (111) surfaces at 300 K on cells of 16 bistable [2]rotaxane molecules — which incorporate a DNP recognition site and a monopyrrolotetrathiafulvalene recognition site within its dumbbell unit and contain a disulfide tether at one terminus and a hydrophobic stopper at the other — ordered into a 4×4 grid. The dimensions were varied to represent the under-packed, liquid-expanded phase (353 Å^2 molecule^{-1}) as well as the over-packed, liquid-condensed phase (65 Å^2 molecule^{-1}). Both the MSCC and the GSCC of the molecules were calculated to determine the surface tension, stress distributions, and relative stabilities of each conformation. Molecular dynamics simulations determined an optimized packing area of 115 Å^2 molecule^{-1}, which is comparable to the area of 120 to 180 Å^2 molecule^{-1} determined through LB studies. The surface tension was calculated to increase from 45 to 65 dyn cm^{-1} as the molecules were switched from the GSCC to the MSCC, the lower surface tension of the GSCC resulting from the bistable [2]rotaxanes being in the more energetically favorable conformation. This insight explains why the surface of bistable [2]rotaxane films in the GSCC are more hydrophobic, and thus have a higher contact angle, than films of bistable [2]rotaxanes in the MSCC. In fact, for all cell dimensions calculated, the GSCC was shown to be more stable than the MSCC by 14 to 16 kcal mol^{-1} for the entire 16 molecule ensemble (approximately a 0.9 Kcal mol^{-1} contribution per molecule). Also confirmed by the calculations is the strong interaction between hydrophilic portions of the molecules and the Au surface. In simulations with large cell dimensions, representing the liquid-expanded phase, the molecules adopt a folded conformation where the CBPQT^{4+} ring lies parallel to the surface. As the cell dimensions are decreased, representing an increase in surface pressure, the tilt angle of the ring and the dumbbell relative to the surface increases. These computations predict, for the most stable 4×4 grid, tilt angles of 39 and 61° for the GSCC and the MSCC, respectively, which leads to an optimal thickness of 40.5 Å for the GSCC and 40.0 Å for the MSCC. Computational results are, therefore, in reasonable agreement with experimental values of 21 to 39 Å. These results show that computational modeling, carried out in tandem with experimental results, provides the feedback necessary to optimize structural details of functional molecular machines by accounting for intermolecular, intramolecular, and surface interactions.

11.2.6　Electronic Transport

Besides strongly affecting the structure of molecules in the condensed phase, surface composition also influences the electronic properties of molecular machines. In fact, one of the goals of the field of molecular electronics[80–82] is to develop solid-state devices with useful transport signatures.[83] Traditional models[84–86] for correlating transport to structure are limited, however, to the solution phase, whereas any theory attempting to describe a system consisting of a molecule and an electrode must treat the two as a single, inseparable unit.[87–90] Understanding the interactions between metal electrodes and bistable [2]rotaxanes containing the CBPQT[4+] ring along with DNP and TTF recognition sites provides insight that can be applied to the design of molecular electronic devices.[91] To this end, experiments were carried out[70] by sandwiching bistable [2]rotaxanes into a three-terminal device Pt break junction[91] that uses the gates to correlate the molecular energy levels of a single molecule with the Fermi energies of the electrodes. These studies show that device characteristics are extremely sensitive to the structure of the molecule. The primary structural difference between the two molecules investigated was that, while one of the two molecules was pretty much symmetric, with disulfide units at each end, the other was decidedly unsymmetric and had a disulfide tether at one end and a large hydrophobic stopper at the other. The key difference between the disulfide tether and hydrophobic stopper end groups is that the disulfide tether chemisorbs onto the Pt electrodes whereas the hydrophobic stopper weakly physisorbs onto the Pt electrodes. Three main differences were observed between the "symmetric" and unsymmetric [2]rotaxanes when the current (I), measured as gate voltage (V_G), was swept at 1.7 K: (1) the conductance spectrum of the [2]rotaxane with both disulfide tethers was symmetric even though the molecule is not completely symmetric, while the conductance spectrum of the unsymmetrically ended [2]rotaxane was asymmetric; (2) the junction resistance of the molecule with the two disulfides was two orders of magnitude lower than the [2]rotaxane with the hydrophobic stopper; and (3) I disappeared for the molecule with the hydrophobic stopper when T was raised above 20 K, whereas the symmetrically terminated molecule retained its current characteristics up to about 100 K. This last observation lends credence to the fact that differences in conduction are a result of the nature of the molecule–electrode bond because it is assumed that the hydrophobic stopper, which is weakly physisorbed to the Pt electrode at 1.7 K, lifts off the electrode as the temperature is raised, and thus no longer bridges the source and drain. This fundamental study of the interactions between molecular structure and conductance further confirms that, in solid-state, molecular electronic devices, the interactions between the surface and the molecules is a key feature of device performance that must be thoroughly investigated to ensure proper device design.

11.2.7 Observing Condensed Phase Movement

Once the interactions that control the behavior of a molecular switch in the solid state have been considered, demonstrating actuation in a condensed phase device must still be carried out in convincing fashion. Designing a system in which the different intermolecular, intramolecular, and surface interactions work in unison to achieve molecular actuation in an oriented, stimulus-controlled fashion remains the primary goal in the field of molecular machines. In the aforementioned LB studies,[92] the molecules were switched while (1) on a water surface in (2) a Langmuir trough. In the computational studies, the molecules were modeled in each conformation of interest, and then the film properties were calculated and compared. Neither of these investigations, however, constitutes a direct experimental verification of molecular actuation in a condensed phase. Recently, the movement of the CBPQT[4+] ring between two stations in a bistable [2]rotaxane in thin films was confirmed by XPS,[92] surface-sensitive XR,[93] impedance spectroscopy[94] measurements, and in a molecular switch tunnel junction[74] (MSTJ).

XPS experiments directly demonstrated[92] the mechanical movement of the CBPQT[4+] ring between DNP and TTF recognition sites in tightly packed LB films of two constitutionally isomeric, amphiphilic, bistable [2]rotaxanes deposited on Si wafers. The amphiphilic, bistable [2]rotaxanes investigated contain DNP and TTF recognition sites as well as a hydrophobic, aromatic stopper at one end and a hydrophilic tris-DEG substituted stopper at the other. The arrangement of the recognition sites relative to the different stoppers distinguishes the two constitutional isomers. Evidence for the redox-controlled mechanical switching was based on the fact that the photoemission intensity of electrons ejected from an atom is exponentially related to the depth of the atom within a film and, therefore, the depths of different atoms within a film can be differentiated. Quantitative analysis demonstrated that the up and down movement of the CBPQT^{4+} ring from one recognition site to another resulted from the introduction of the oxidizing agent Fe(ClO$_4$)$_3$ into the aqueous subphase of the Langmuir trough. The switching process was observed statically by measuring relative differences in the intensity of N1s nuclei which are only present in the CBPQT^{4+} ring and by comparing the photoemission of LB films prepared with Fe(ClO$_4$)$_3$ to films without it in the subphase. The changes in intensity correspond to a difference in height of 14 Å between the position of the ring in the two different systems, directly supporting the hypothesis that the ring switches between upper and lower recognition sites on the dumbbell in a closely packed monolayer as a result of the introduction of a chemical oxidant.

A similar study[93] demonstrating the switching of LB monolayers was carried out with XR, a technique that measures the electron density profile within a monolayer and provides direct structural information about the isomeric structures associated with the location of the ring in the constitutionally isometric bistable [2]rotaxanes. Using a synchrotron x-ray source, XR functions by simultaneously varying incident and grazing angles while recording the intensity pattern that results from the interference of the x-rays reflected from different depths of the sample. This data yields the out-of-plane electron density across the water/monolayer/air interface. As in the XPS experiment, the monolayer was assembled and the XR spectrum was measured both before and after the addition of the Fe(ClO$_4$)$_3$ oxidant to the aqueous subphase. Both constitutional isomers of the bistable [2]rotaxane, having either the TTF recognition unit closer to the hydrophilic stopper and DNP recognition site closer to the hydrophobic stopper or with the DNP recognition site closer to the hydrophilic stopper and the TTF recognition site closer to the hydrophobic stopper, were subjected to the XR experiment. Upon oxidation, a change in electron density corresponding to a 12 Å shift of the ring in the direction of the DNP recognition site was observed, a value which is in good agreement with the value of 14 Å determined by XPS. This value is well short of the calculated value of 34 Å calculated for the fully extended [2]rotaxanes and further confirms that the molecules adopt folded conformations as a result of hydrophilic interactions at the air–water interface. However, neither the XPS nor XR experiments track, in real time, the movement of the ring because the monolayers must first be switched by introduction of an oxidant, and then measured; more recent work[94] has attempted to measure the movement of the CBPQT^{4+} ring *in situ*.

The rate of electron transfer to the CBPQT^{4+} ring has been measured and its subsequent movement has been demonstrated in a close-packed, viscous SAM on an Au surface by impedance spectroscopy.[94] A [2]rotaxane monolayer consisting of the CBPQT^{4+} ring and an π-electron-rich diaminobenzene recognition site was self-assembled on a gold surface (Figure 11.8). Ring movement was demonstrated by first reducing the CBPQT^{4+} ring to its biradical dication followed by subsequent reoxidation to its tetracation, and measuring differences in the rate of electron transfer for each process. The rate of electron-transfer reduction to the biradical was measured to be 80 sec^{-1}, whereas back electron-transfer to form the tetracation was observed to be significantly greater, namely 1100 sec^{-1}. This data can be explained by the fact that reduction of the tetracationic cyclophane to its biradical dication decreases its strength as an electron-acceptor, which, in turn, results in the destruction of the favorable noncovalent interactions that exist between the cyclophane and the π-electron-rich diaminobenzene. Given that the electrode is negatively charged and the biradical dication is positively charged, the cyclophane presumably moves towards the electrode surface. The resulting proximity of the cyclophane to the electrode surface results in the increased electron transfer rate during oxidation. In this manner, impedance spectroscopy was successfully used to measure the movement of the cyclophane in a closely packed monolayer.

The switching behavior of amphiphilic, bistable [2]rotaxanes has also been confirmed by studying[74] the redox-controlled switching behavior in solution, polymer gels, SAMs, and MSTJs. The switching cycles can be detected by a number of experimental observations in the cyclic voltammogram (CV) studied, providing information which can be used to understand switching behavior across different environments. The first oxidation potential of the GSCC corresponds to the one-electron oxidation of TTF to TTF$^{+\cdot}$ and occurs at $^+$490 mV, while the same oxidation of TTF to TTF$^{+\cdot}$ in the MSCC is +310 mV (potentials referenced to an Ag/AgCl electrode).[72,73] Knowing that relaxation of the MSCC back to the GSCC is temperature-dependant, thermal activation parameters may be quantified by time- and temperature-dependent CV measurements. The ratio of MSCC to GSCC is dependent on the structural

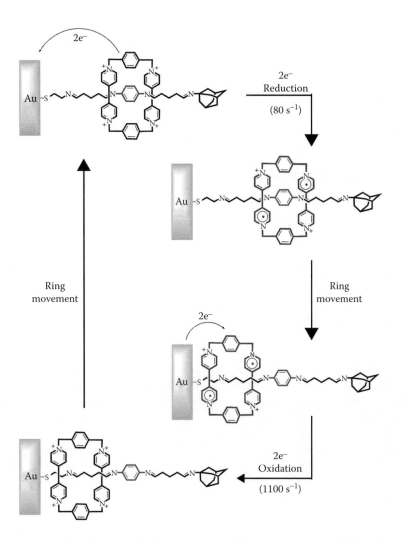

FIGURE 11.8 Demonstration of the mechanical movements of the tetracationic cyclophane CBPQT⁴⁺ along a diaminobenzene-containing thread that is tethered to a gold surface. A two-electron reduction of the CBPQT⁴⁺ ring erases the favorable binding interactions that exist between the ring host and diaminobenzene guest with an electron transfer rate of 80 Hz. The resulting dicationic CBPQT²⁺⁻ is attracted to and moves toward the gold surface. A two-electron oxidation of the CBPQT²⁺⁻ returns the cyclophane host back to its tetracationic state and restores the favorable noncovalent interactions between the CBPQT⁴⁺ host and diaminobenzene guest. The rate of oxidative electron transfer is 1100 Hz, which is indicative of the close proximity of the reduced ring to the gold surface and is demonstrative of the reversible redox-controllable mechanical motions of the interlocked molecule on a gold surface.

features of the particular bistable molecules, whereas changes in kinetic parameters (lifetime and decay) are more sensitive to physical environment. Thus CV was used to quantify the ratios of different coconformations by measuring the integration of oxidation peaks corresponding to MSCC and GSCC, in solution, polymer gels, and SAMs. While at equilibrium, the N_{MSCC}/N_{GSCC} ratios for the different amphiphilic, bistable molecules were very similar in all three environments, relaxation rates from the MSCC to the GSCC slowed down and lifetimes increased, as the viscosity increased from solution to polymer gel to SAMs.

Using the knowledge that the switching mechanism appears to be universal across the three environments, this analysis was extended to a functioning nanodevice incorporating the amphiphilic bistable [2]rotaxane molecular switches in MSTJs.[74] The detailed procedures regarding MSTJ manufacture, design and operation have been reviewed extensively[70,82,95] and will not be discussed in great detail here. A brief description, however, of those aspects relevant to this discussion will be made. The MSTJs investigated contained a Langmuir deposited layer of amphiphilic bistable [2]rotaxanes — which contain TTF and DNP recognition sites along their dumbbells along with a hydrophobic stopper at one end and a hydrophilic stopper at the other — sandwiched between an n-type poly(silicon) bottom electrode and a metallic top electrode. These MSTJs have been used in molecular logic devices — such as a 64 kbit RAM — and are able to function because they can be switched reversibly between low conductance states and hysteretic, high-conductance states. Several critical pieces of evidence indicate that the MSCC is the high-conductance state: (1) high conductance occurs subsequent to an oxidizing bias being passed between the electrodes which, in other environments, would lead to the MSCC; (2) the lower first-oxidizing potential of the MSCC in CV experiments suggests that it is of higher conductance than the GSCC; (3) modeling[78] has shown that switching to the MSCC causes a narrowing of the HOMO-LUMO gap with respect to the GSCC, which, in turn, would lead to a higher conductance state. However, the switching could not be tested directly by CV in MSTJs, so other methods were used to determine that the switching mechanism in MSTJs was consistent with the switching mechanism in other environments. The equilibrium thermodynamics could be inferred within the MSTJs by assuming that the high and low conductance states correspond to the MSCC and GSCC, respectively. Thus, the temperature-dependent switching characteristics, normalized against the transport characteristics of an MSTJ, provide insight into the thermodynamics of the molecules within the junctions. Conductance changes in MSTJs are measured as remnant molecular signatures, hysteretic switching loops that represent the normalized current response to a voltage pulse as the gate voltage is swept. Because the current stays on after the bias has been removed, the high conductance is most likely a result of the bistable [2]rotaxane remaining in the MSCC before thermal relaxation to the GSCC, which reduces the conductance to the normalized baseline. The remnant molecular signatures, along with the decay rate to the baseline current, can be used to determine both N_{MSCC}/N_{GSCC} and provide thermodynamic data, the decay rate, and ultimately provide kinetic data. Using this hypothesis, it has been shown that, even in a full device setting, amphiphilic bistable [2]rotaxanes demonstrate switching thermodynamics that are controlled by structure while switching kinetics in these systems are still predominantly controlled by environmental factors.

The observation that molecular structure plays a dominant role in the thermodynamic properties of bistable [2]rotaxanes and environmental factors govern their kinetics is vital to the development of surface-bound organic molecular devices based upon these particular molecules. Systematic investigations of the interplay between structure, function, and environment, such as those outlined in this section, enable researchers to fine-tune the dynamic properties of organic molecular machines as they are moved from the solution environment to more condensed phases and onto surfaces. Surface–substrate interactions, molecular packing, and molecular conformation at the organic–solid interface all influence the dynamic properties of functional molecules and affect the ultimate performance of surface-bound molecular machines. With these important considerations in mind, a number of research groups have recently been able to develop molecular machines that operate in condensed phases or on solid supports. The following sections will highlight advancements that have been made in the design, synthesis, fabrication, and modeling of four types of synthetic organic molecular machines — (1) molecular muscle systems, (2) nanovalves, (3) rotors, and (4) surfaces with controllable wettability.

11.3 Molecular Muscle Systems

11.3.1 The Potential of Artificial Muscles

Despite the significant progress that has been made recently, emulating the high strength, high displacement, and low weight of natural muscle tissue (Figure 11.2[b]) with synthetic molecular systems, the challenge is one that remains at the forefront of the field of molecular machinery. Such artificial systems promise to impact a wide range of technologies, from nanoelectromechanical systems to exotic applications such as futuristic body armor. Investigations into actuating systems which switch in response to

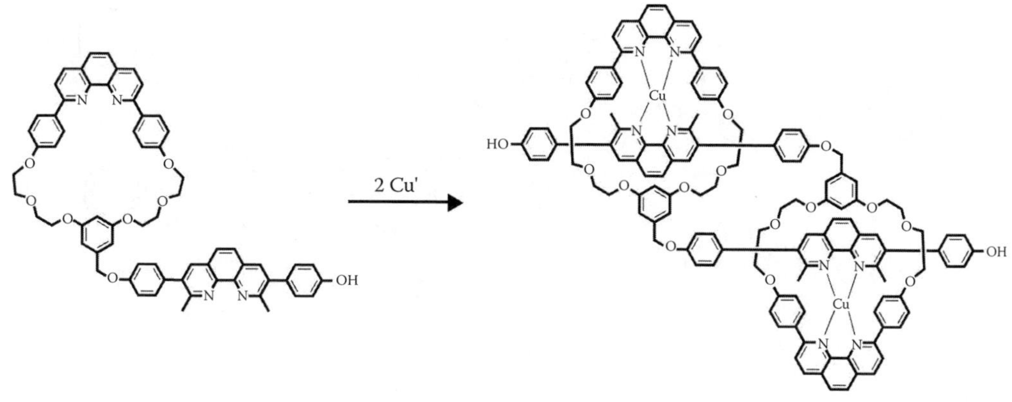

FIGURE 11.9 Cu(I)-templated synthesis of an interlocked dimer.

chemical, photochemical, or electrochemical stimuli have run a wide gauntlet of approaches, including hydrogel swelling,[96] osmotic expansion of conjugated polymers,[97] motions within supramolecular polymer systems, ion intercalation in nanoparticle films,[98] and the bending of surfaces in response to DNA hybridization.[99] Most of these approaches to actuation rely on bulk material responses, rather than controlled linear motion at the single-molecule level.

No synthetic system reported to date approaches the beauty and efficiency of the sarcomere, the cellular unit responsible for skeletal muscle contraction.[100] Muscle contraction occurs as a result of an ATP hydrolysis-driven power stroke that causes myosin and actin filaments to slide over one another. This process is suggestive of a biomimetic approach in which two or more interlocked molecular components may be induced to slide through one another in response to some external stimulus, resulting in the contraction and expansion of the single-molecule system. The challenge is to turn this singular event into a coherent one involving many molecules in a materials setting.

11.3.2 Transition Metal Controlled Actuators

Pioneering work by Sauvage and coworkers resulted in the development of a transition metal templated synthesis of a mechanically interlocked dimer (Figure 11.9).[101] The dimer's components consist of a macrocycle containing a bidentate ligand (1,10-phenanthroline) and a complementary ligand on the thread portion. Addition of a stoichiometric amount of Cu(I) ions resulted in the formation of the metal-templated dimer in quantitative yield.[102] It is important to note that the formation of dimer structures from self-complementary rotaxane components is a general phenomenon observed with other binding motifs — the dimer structure produces favorable binding interactions while, at the same time, maximizing the entropy of the system relative to alternative oligomeric and polymeric structures.[103–105] The generality of this approach bodes well for future actuating systems based on dimers with interlocked topologies.

Elaboration of the rotaxane dimer to a linear actuator was accomplished by incorporating a second binding site onto the thread portion of the dimer. The system now exhibits two potential isomers, an extended coconformation in which metal coordination occurs with the macrocycle and the inner ligand, and a contracted coconformation in which the rings coordinate metal ions at the outer ligand set. Stimuli capable of switching the coordination preference from one site to the other should expand or contract the molecule in a manner reminiscent of myosin and actin motion in natural muscles.

These transition-metal controlled molecular actuators can be switched (Figure 11.10) by means of a sequential metalation/demetalation procedure. The interlocked dicopper complex exists exclusively in the extended coconformation as a result of the preference of Cu(I) for a tetracoordinate-binding geometry. The molecule loses its strong coconformational preference following demetalation using an excess of potassium cyanide; the bulky stoppers, however, at both ends of the threads prevent dissociation of the dimer. The addition of Zn(II) ions to the demetalated dimer produces the contracted isomer in quantitative

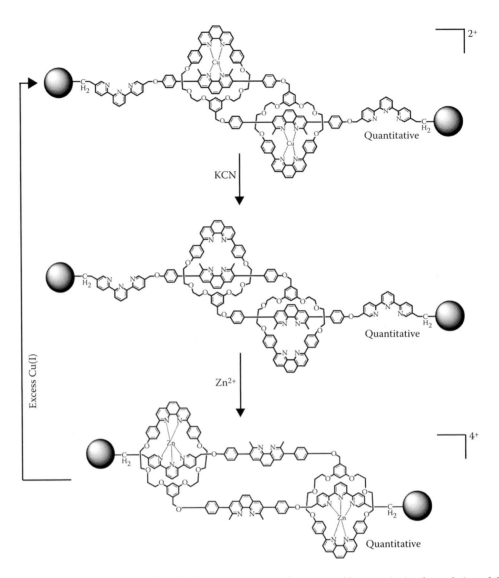

FIGURE 11.10 Contraction of the bistable dimer structure was demonstrated by quantitative demetalation of the Cu(I) ions using KCN, followed by treatment with Zn(NO$_3$)$_2$ to give the contracted structure. The extended form was recovered after treatment with excess [Cu(MeCN)$_4$]PF$_6$.

yield. The extended isomer may be recovered by treating the *bis*-Zn complex with an excess of Cu(I). As a result of the switching process, it has been estimated that the end-to-end length of the molecule changes from 83 to 65 Å during the switching process, giving rise to a contraction ratio (27%) similar to that of natural muscle.

Though natural muscles are also contracted via a chemical means, artificial systems are more likely to find practical application if they can be switched as a result of electrochemical or photochemical inputs. In the system discussed in Figure 11.10, a dimer based on the binding of two Cu(II) ions should exist in the contracted coconformation. However, electrochemical oxidation of the *bis*-Cu(I) extended coconformation resulted in a *bis*-Cu(II) complex in the extended coconformation that was far too kinetically stable to undergo efficient contraction. The Strasbourg group has also demonstrated photochemical shuttling in Ru (II)-based rotaxane systems. Through modification of the molecular design, it may prove possible to extend these more desirable switching mechanisms to linear molecular actuators of this type.

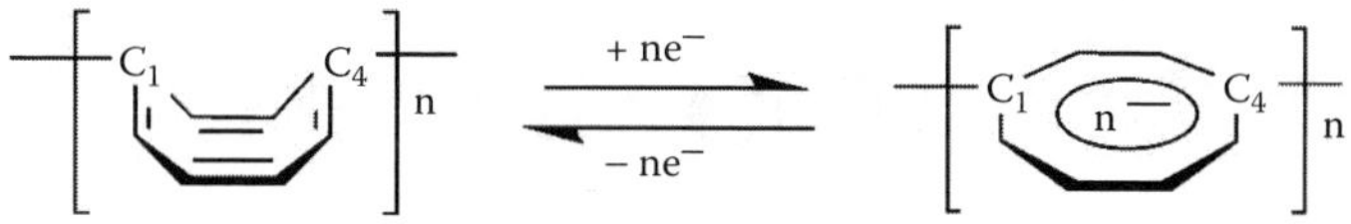

FIGURE 11.11 Expansion of poly(cyclooctatetraene) upon electrochemical reduction.

11.3.3 Redox-Controlled Annulene-Based Muscles

Marsella and coworkers[106,107] reported a strategy for electrochemical actuation based on the well-known tub-to-planar conformational change upon one or two electron reduction of [8]annulenes. Poly(cyclooctatetraene) itself could theoretically undergo this actuation process at each repeat unit, amplifying (Figure 11.11) this molecular reorganization into macroscopic motion. However, the group chose to explore thiophene-fused annulene derivatives, such as tetra(2,3-thienyelene), hoping to take advantage of the desirable synthetic convenience and relatively stable redox chemistry of thiophene derivatives, while retaining the fundamental redox-driven conformational changes of the parent annulene system.

The dibutyl tetra(2,3-thienylene) parent system was synthesized and characterized electrochemically.[108] This small-molecule derivative exhibited a reversible one-electron oxidation. In contrast to the parent cyclooctatetraene, however, a second oxidation is not observed. The molecule's apparent disinterest in complying with Huckel's $4n + 2$ aromaticity rule was explained by the inability of the system to attain planarity as a result of steric iterations between the β-hydrogen atoms. Despite this observation, the single-electron redox cycling of the parent tetra(2,3-thienylene) still results in a 6.7% dimensional change, close to that of the best bulk-actuating systems.

Successful halogenation and polymerization of the tetra(2,3-thienylene) under standard Ni-mediated cross-coupling polymerization conditions has resulted (Figure 11.12) in a model polymeric actuator. The solid-state electrochemistry of the poly[tetra(2,3-thienylene)] derivative is promising though somewhat inconclusive. The polymer undergoes reversible oxidation to an average charge of +0.6/monomer unit. Attempts at further oxidation, however, led to irreversible behavior of unknown origin. The authors have commented on some difficulties in determining whether actuation observed in the solid-state polymer films is a result of bulk effects — such as swelling arising from intercalation of counterions into the polymer film — or of molecular processes. They have also reported a second generation thiophene-fused didehydro[12]annulene system, which was shown — following DFT calculations — to exhibit an 18% maximum dimensional change and stability over a broader range of oxidation states.[109]

11.3.4 Photo-Controlled Polymeric Muscles

The challenge raised by the poly[tetra(2,3-thieylene)] system of operating actuators in the solid state should not be overlooked: mechanical work done by artificial muscles in solution is typically lost. Thus, the challenge of solid-state operation must be realized in order for these systems to reach their full potential. Recently, Gaub et al.[110] elegantly measured the force exerted on an AFM cantilever upon photoisomerization of a single azobenzene polypeptide molecule (Figure 11.13). Briefly, a single polypeptide molecule is covalently attached to an amine-functionalized glass substrate via standard carbodiimide-coupling chemistry. A thiol group, deprotected *in situ*, served as the attachment point to the gold-coated

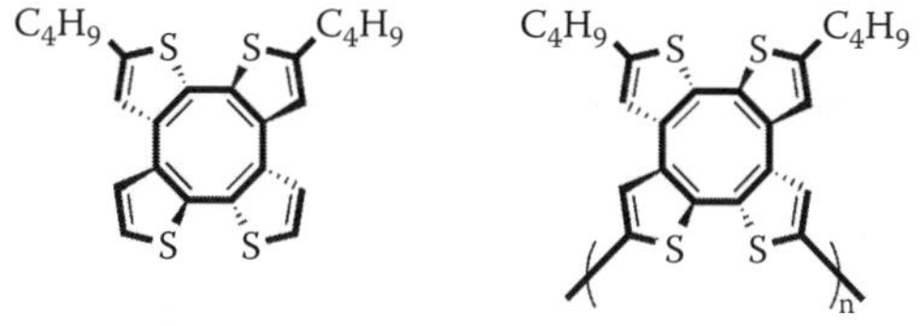

FIGURE 11.12 Tetra(2,3-thienylene) and poly[tetra(2,3-thienylene)] reported by Marsella et al.

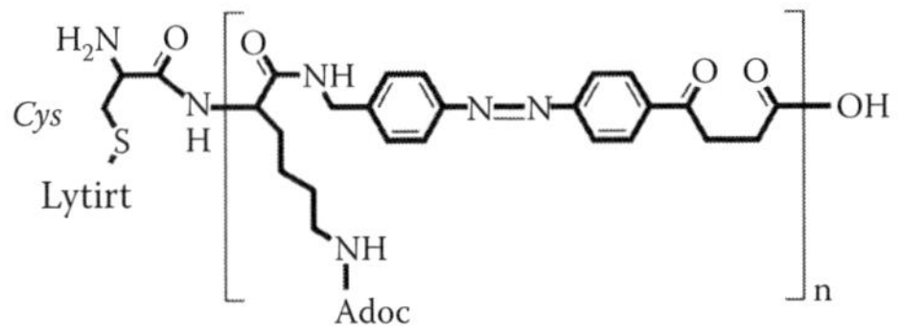

FIGURE 11.13 Azobenzene polypeptide utilized for light-driven single-molecule force measurements.

AFM cantilever tip. Force–extension curves of the polypeptide molecules when the molecule is switched between the *trans*-dominated photostationary state at 420 nm and the *cis*-dominated photostationary state at 365 nm are in accordance with the expected shortening of the polymer chain.

The mechanical work done by the photoisomerization of the polypeptide was measured through a full work cycle. First a force–extension curve of the predominately *trans* isomer was measured over the range of 80 to 200 pN. The polymer chain was then irradiated at 365 nm, resulting in contraction of the polymer to the predominately *cis*-form against the 200 pN applied force. A new force–extension curve was measured by reducing the restoring force to 85 pN. The cycle could then be repeated by isomerizing the chain to the preferential *trans*-state. The mechanical work performed by a single molecule was calculated to be 4.5×10^{-20} J, corresponding to an operating efficiency (W_{out}/W_{in}) of 10^{-18}! Though this efficiency is prohibitively low to be of any practical use, it reflects limitations of the experimental setup and is not indicative of an inherent limit on molecule performance. The precise measurement of the mechanical work of the photoisomerization of single polymer chains represents a significant advance in harnessing molecular actuators for practical use.

11.3.5 Flexing Cantilevers with Palindromic [3]Rotaxanes

Of course, practical solid-state devices must harness the work of large assemblies of molecules that switch quickly and coherently. Bistable rotaxanes developed in our own laboratories at UCLA are particularly well suited for this challenge as a result of their ability to undergo electrochemically controlled mechanical movements. A general discussion of the structure and function of bistable [2]rotaxanes was presented in Section 11.2.3. Employing the knowledge gained from the design and use of bistable [2]rotaxanes in molecular electronic devices,[82,111–114] a symmetrical doubly bistable [3]rotaxane was designed for electromechanical applications.[115] The design incorporates two TTF and two DNP recognition sites distributed symmetrically about the rigid, triphenylene center of the palindromic dumbbell component. The two CBPQT[4+] rings that encircle the dumbbell position themselves predominantly on the TTF recognition sites at equilibrium (Figure 11.14). In such a coconformation, the molecule is fully extended, and the distance between the two CBPQT[4+] rings is roughly 4.2 nm. A four-electron oxidation induces the two CBPQT[4+] rings to switch to the more central DNP recognition sites, where they are only about 1.4 nm apart. The process is reversed upon a four-electron reduction of the TTF[2+] dications. This reversible electrochemically controlled switching, and resulting contraction from 4.2 to 1.4 nm, results in a mechanical strain of 67%. Disulfide tethers were incorporated onto each of the two CBPQT[4+] rings such that the mobile portions of the molecule can be anchored onto gold-coated silicon microcantilever beams. Even without alignment of the doubly bistable [3]rotaxane molecules relative to one another, the mutual inward motion of the two CBPQT[4+] rings induces a sheer force on the substrate, causing an upward bending motion.

Chemical oxidation of the TTF units of SAMs of the doubly bistable [3]rotaxane formed on the gold-coated cantilevers resulted in a 35 nm upward deflection (Figure 11.14) of the cantilever tips, corresponding to a force of 10.2 pN/molecule. Addition of a reducing agent caused the cantilevers to return to their original position. Although the bending behavior was cycled 25 times, a decrease in the magnitude of the response was observed, possibly as a result of either degradation of the molecule or rearrangement of the dynamic S–Au attachment sites. Nevertheless, this experiment demonstrates that many linear

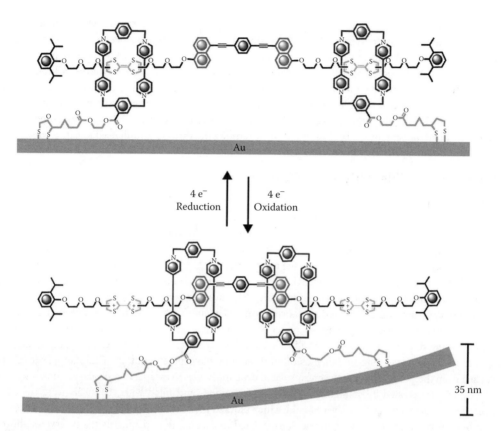

FIGURE 11.14 Structure and graphical representation of a palindromic [3]rotaxane capable of binding to Au-coated cantilevers. Oxidation of the TTF units causes the CBPQT^{4+} rings (blue) to encircle the I, S-DNP stations, resulting in upward deflection in the cantilever beams.

molecular muscles (ca. 1 billion molecules) can work cooperatively in the solid state to perform mechanical work upon objects that are five orders of magnitude larger than themselves.

Artificial molecular muscles will continue to be an active area of research, especially as control fabrication at the nanoscale level continues to develop. The ability to align the molecular actuators, potentially in ordered liquid crystalline phases, phase-separated diblock copolymers, or on other ordered substrates might result in systems with the molecules pulling in a single direction. At the same time, new molecular actuators with coherent, rapid, and stable switching will continue to attract synthetic interest.

11.4 Molecular Nanovalves

11.4.1 Typical Valve Systems

A macroscopic valve is a device with a mobile control element that regulates the flow of gases, liquids, or solid particles by opening and closing their passageways. These valves are essential regulators for devices that can be natural, such as those that control blood flow in the heart, or artificial, such as the inlet and exhaust valves of a combustion engine. Inspired by the molecular machinery found in nature, significant advances in the design and synthesis of analogous molecular nanovalve devices, be they irreversible or reversible, have been made using nanoscale components. One of the driving forces for work in this field is the controlled release of drug molecules *in vivo*. Many therapeutic treatments require that the active agent is not released until the target is reached, but in most systems, release begins immediately upon the introduction of the drug orally, subcutaneously, or intravenously into the body.

Nanovalves that can control the release of therapeutic agents from a carrier in which the agents are encapsulated are ideal systems for solving the problem of targeted delivery because the release can be initiated by an external stimulus once the target has been reached. Also, the size of molecular nanovalves is compatible with biological systems. The growth in the field of molecular nanovalves has followed an evolutionary pathway in which irreversible systems that first showed molecular control of gating evolved into more sophisticated, reversible valves that resulted directly from solution-based molecular machines and polymers. Among the many molecular-based valve systems, light-induced *cis/trans* isomerization, redox-controlled movement, and pH modulation are among the various stimuli used to gate a passageway.

11.4.2 Irreversible Thin Film Regulators

A precursor to using molecular machines as reversible gating mechanisms involves the usage of thin anode membranes or removable nanoparticle surfaces, upon whose removal from the covering of microreservoirs filled with chemicals, a passageway is opened through which the chemicals may diffuse. Using standard lithographic techniques, reservoirs can be made and filled with an analyte before the anode membrane is deposited on top to lock the analyte within the reservoirs. Both CdS nanoparticles and gold films have been used to function as caps for the controlled release of chemical compounds from the pores, albeit in an irreversible fashion. Mesoporous silica nanospheres with average pore diameters of 2.3 nm were filled[116] with various drug molecules — including vancomycin and adenosine triphosphate — and then derivatized with CdS nanospheres with an average diameter of 200 nm. The particles were tethered to the opening of the pores using disulfide bonds, as a result of which the pores were effectively sealed. Upon addition of disulfide-reducing release triggers, such as dithiothreitol or mercaptoethanol, the CdS nanoparticles were detached from the pores, allowing the encapsulated drug molecules to escape. In a similar system, microreservoirs that were created by standard lithographic techniques and filled with chemicals were coated by chemical vapor deposition with a layer of gold, which blocked the escape of the chemicals from the reservoir. Using both fluorescein and $CaCO_3$ as the encapsulated molecules, the opening of the reservoirs and the subsequent release of these molecules was detected upon dissolution of the gold by the application of an oxidizing bias. Because of the biocompatibility of gold, this system was tested *in vivo* to determine whether an implanted device consisting of the reservoirs with gold coatings could be used to affect the controlled release of drug molecules. By specifically addressing a single reservoir with the application of an oxidizing voltage, the gold was released from only a single reservoir while the other reservoirs remained sealed. This mechanism allows for specific dosing as well as slow release over time. Although these functional valve systems have the potential to deliver therapeutic agents *in vivo*, they lack the essential reversibility that may be advantageous for various applications. Also, devices manufactured by lithographic techniques are larger and more cumbersome than systems that consist wholly of small molecules.[116,117–119]

11.4.3 Light-Regulated Azobenzene and Coumarin Valves

The systems discussed in the previous section only allow for one cycle of storage and release of guest molecules. By exploiting the reversible *cis→trans* isomerization of the N=N bond in azobenzene, however, a fully reversible nanovalve has been constructed (Figure 11.15) to control mass transport in a nanoporous film. This system takes advantage of the change in azobenzene molecular dimensions, which are decreased by approximately 3.4 Å upon isomerization from the *trans* to *cis* conformation, by placing azobenzene molecules on the surface of a membrane composed of monodisperse pores, synthesized by surfactant-directed self-assembly. Irradiation with UV light causes the azobenzene molecules on the pore surfaces to occupy less volume in the *cis* form, effectively opening the pores, and upon exposure to visible light or heat, the molecules revert to the *trans* form, thereby closing the pores. The passage of ferrocene dimethanol (FDM) through a nanoporous film decorated with azobenzene molecules has been used to show that the pores do in fact open and close as a result of this isomerization, using chronoamperometric measurements. Porous membranes with azobenzene valves have been assembled onto an ITO film with FDM in a supernatant solution. Upon exposure of the film to UV light, an increase in oxidative current

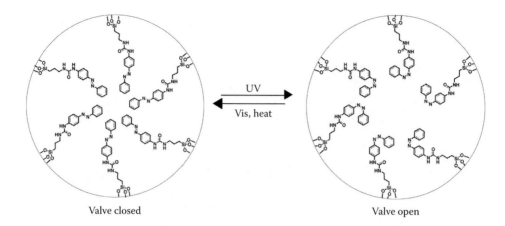

FIGURE 11.15 An azobenzene-operated molecular nanovalve. The *trans–cis* isomerization of azobenzene derivatives tethered to a nanoporous silica substrate results in the reversible opening and closing of the silica pores.

is observed as a result of the transport of the FDM from the supernatant, through the porous membrane, to the ITO layer. When the surface is exposed to heat or visible light, the oxidative current returned to pre-UV exposure levels by a mechanism that is the result of the closing of the pores halting the flow of FDM. This measurement was carried out three successive times to show the reproducibility of the opening and closing of the pores that the azobenzene allows. A similar photocontrolled nanovalve system (Figure 11.16) was made using coumarin-modified mesoporous silica to control the mass transport of guests encapsulated within the silica matrix.[120,121] Here, the dimers act as a net to control the access of guest molecules (cholestane, pyrene and phenanthrene) to and from the pores of derivatized hexagonal mesoporous silica upon photoactivation and subsequent de-dimerization. When the coumarin moiety is dimerized by irradiation with UV light of a wavelength longer than 310 nm, the pore entrances are

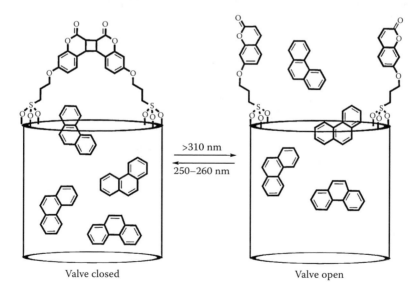

FIGURE 11.16 The photoinduced [2 + 2] dimerization of coumarin derivatives forms the basis of a molecular nanovalve. When dimerized, the tethered coumarin derivatives sterically block the openings of silica chambers. A [2 + 2] retrocyclization opens the pores and allows pyrene guests to escape.

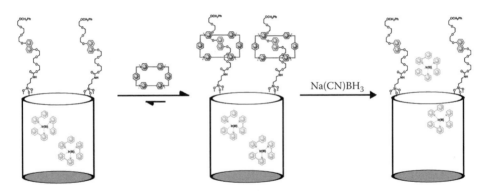

FIGURE 11.17 Example of an operational supramolecular nanovalve wherein the complexation of CBPQT⁴⁺ by DEG–DNP-covered SiO$_2$ nanopores results in the formation of [2]pseudorotaxanes that trap Ir(PPy)$_3$ dye molecules inside. Reduction of the CBPQT⁴⁺ rings induces dethreading of the [2]pseudorotaxanes, releasing the dye molecules.

blocked sterically by the cyclobutane coumarin dimers, and guest molecules, which have been previously loaded into the pores, are trapped inside. However, irradiation of the dimerized-coumarin-modified silica nanoparticles with shorter wavelength UV light ($\lambda \approx 250$ nm) regenerates the coumarin monomer derivative by inducing a [2 + 2] retrocyclization, resulting in the photocleavage of the cyclobutane rings and release of the guest molecules, which are included inside, from the pore void. These investigations demonstrate effectively that photocontrolled molecular actuation control reverses mass transport through porous silica, thereby demonstrating the potential of these systems as molecular-scale valves.

11.4.4 Supramolecular Gatekeepers

Another means by which the orifices of mesoporous silica have been gated is through the use (Figure 11.17) of [2]pseudorotaxanes as gateposts. Redox-active supramolecular systems have been applied[122–124] to control the uptake, storage, and release of dye molecules in mesoporous silica. Initially, a supramolecular nanovalve was developed[123] by the tethering of [2]pseudorotaxanes as gatekeepers at the entrances of cylindrical pores (2 nm in diameter) in mesostructured thin silica films. In the [2]pseudorotaxane, a 1,5-dioxynaphthalene-containing derivative (DNPD) acts as the gatepost, and CBPQT⁴⁺, which recognizes the DNP units on account of a cooperative array of [π···π], [CH···π], and [CH···O] noncovalent interactions, serves as the gate that controls access into and out of the nanopores. The DNPD molecules were first of all covalently tethered to the orifices of the nanopores, which were subsequently diffusion-filled with a luminescent metal complex, tris(2,2′-phenylpyridyl)iridium(III), or Ir(PPy)$_3$, which is 1 nm in diameter. Because the diameter of the dye is large with respect to the pore size, any blockage of the pore opening will result in the encapsulation of the dye within the pore. The DNPD-containing threads were then complexed with the CBPQT⁴⁺ rings (0.8 × 1.0 nm in size), thereby sterically sealing off the nanopores and trapping the Ir(PPy)$_3$ dye molecules inside the nanopores. Reduction of the CBPQT⁴⁺ ring using an external reducing reagent (NaCNBH$_3$) induces the dethreading of the [2]pseudorotaxane, a situation which results in the unblocking of the pore orifice and subsequent release of the dye.

11.4.5 Molecular Gatekeepers

A more robust and reusable nanovalve (Figure 11.18) design has been developed[124] wherein the [2]pseudorotaxanes were replaced by bistable [2]rotaxane as the gatekeepers, presiding over the opening and closing of the nanoscale passageways of spherical mesoporous silica particles. The moving part of the molecular valve is the CBPQT⁴⁺ ring, which switches, under redox control, between the two different recognition sites — one TTF and the other DNP — on the dumbbell component of the [2]rotaxane. The bistable [2]rotaxane that is attached to silica particles was designed such that the DNP unit was

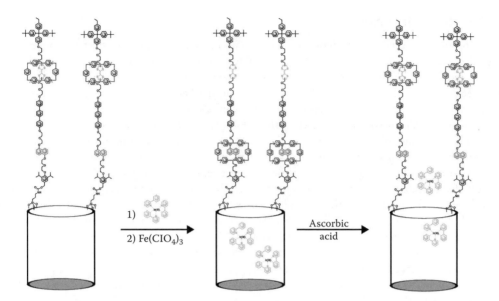

FIGURE 11.18 [2]Rotaxanes tethered to the surface of nanoporous silica form the basis of reversible nanovalves. Dye molecules can be loaded into the nanoporous silica while the [2]rotaxanes are in their GSCC. Chemical oxidation traps the dye molecules within the nanoporous SiO_2 as a result of the $CBPQT^{4+}$ ring sterically blocking the pore openings. Reduction with ascorbic acid restores the GSCC, releasing the dye molecules.

closer to the nanopores, and the TTF unit was further from the nanopores. In its GSCC, the $CBPQT^{4+}$ ring prefers to encircle the TTF unit, rather than the DNP one on the dumbbell component. The porous silica can be loaded with $Ir(PPy)_3$ molecules when the surface-bound [2]rotaxanes are in their GSCC, since the $CBPQT^{4+}$ ring is sufficiently far away from the nanopore openings to allow the dye molecules to diffuse inside. In the presence of an oxidant, $Fe(ClO_4)_3$, however, the TTF unit is oxidized to its dication, and the $CBPQT^{4+}$ ring moves to the DNP unit, sitting significantly close to the nanopores so that the dye cannot escape because of steric blockage of the nanopore entrances by the $CBPQT^{4+}$ ring. Upon reduction of the TTF dication to its neutral state by ascorbic acid, the $CBPQT^{4+}$ ring undergoes a mechanical movement away from the openings of the nanopores and the dye is released. In contrast to the [2]pseudorotaxane-based nanovalve, the nanopores controlled by [2]rotaxane-based gatekeepers can be opened and closed reversibly without the re-addition of $CBPQT^{4+}$, a structural modification which marks an improvement over the previous design.

11.4.6 Polymeric Valves

Polymeric materials have also been used to gate the pores of nanoporous materials by the manipulation of either temperature or ionic strength. The sequestration and release of proteins by charge selection in mesoporous silica materials have been reported.[125] Protonation of NH_2 groups of silica pores functionalized with (3-aminopropyl)-triethoxysilane induces the electrostatic encapsulation inside the pores of anionic proteins, which are fully released upon increasing the ionic strength of the solution. Temperature control of nanopores was also achieved by using the changes in volume that accompany the swelling and deswelling of temperature-responsive polymers as the source of hindrance at the pores' orifices.[126,127] Poly(N-isopropyl acrylamide) is a polymer that is hydrated and extended at low temperatures, inhibiting the passage of molecules from silica nanopores. Upon lowering the temperature, the polymer collapses, which allows for the passage of molecules through the pores. Furthermore, cross-linked poly((N-isopropylacrylamide)-co-N-ethylacrylamide), for example, has been shown[126] to behave as a reversible valve in microfluidic chips in much the same manner.

11.4.7   Biological Nanovalves

The incorporation of molecular nanovalves into proteins has led to a system in which the movement of molecules across a lipid membrane could be reversibly controlled by exposure to light. In a melding of biology and chemistry, Feringa et al.[128] have created a membrane-bound nanovalve with an addressable photosensitive molecular switch to mediate a charge-gating mechanism in a protein whose pore normally opens only in response to increased tension. The valve consists of a bacterial homopentamer protein, the mechanosensitive channel of large conductance from *Escherichia coli*, modified by the selective attachment of one photochromic spiropyran moiety to each of the protein's five subunits at the 22nd amino acid position. Upon irradiation with UV light ($\lambda = 366$ nm), the neutral spiropyran group isomerizes to the zwitterionic merocyanine moiety via an electrocyclic ring opening. The presence of the zwitterions within the pore channel causes the opening of the pore through charge repulsion to a diameter of approximately 3 nm. The opening of the pore was determined by measuring the ionic current flowing through the modified channel using a patch clamp experiment. As a result of irradiation with 366 nm light, current increased, demonstrating that the pore had opened. Visible light ($\lambda > 460$ nm) reverses the reaction, neutralizing the localized charge and leading to valve closure, in a mechanism that was demonstrated by the reduction of ionic current across the membrane. This experiment showed how a conformational change results in the modification of a membrane protein from a tension-gated nanovalve to a photo-controlled biological nanovalve.

Molecular scale nanovalve systems have the potential to play many important roles as regulatory devices in functional nanoscale devices. The applications that exist for molecular scale nanovalves include sensors, photomemories, light-driven displays, microfluidics, and controlled drug release. Reversible nanovalves may be used as control elements in micro- and nanofluidics by regulating the rate and direction of fluid flow in lab-on-a-chip devices. Irreversible nanovalves have potential in drug delivery systems, provided they are biocompatible. In this regard, it is especially desirable to develop noninvasive procedures, such as light, to control the operation of nanovalves. The recent advances in the fabrication of nanovalves can be attributed to evolution of the chemical and material sciences which, together, provide researchers with the tools to tailor functional molecules capable of performing specific tasks and also to link these functional molecules to solid substrates in order to exploit their capabilities in a device setting.

11.5   Molecular Rotors

11.5.1   Rotation as a Fundamental Molecular Motion

One of the ubiquitous processes in molecules is rotational motion. With respect to the design of molecular machinery,[3] rotational control is one of the most desirable functions for controlling through an external stimulus. Whether it be the controlled rotation about a single bond, the concerted movements of geared aromatic units, or the continuous revolution of a dipole under an applied electric field, the analogies between specifically designed organic rotors and macroscopic rotary motors have piqued the curiosity of those interested in the construction of molecular scale machines. Having the ability to design and synthesize molecular rotors,[129] i.e., molecules whose rotation can be selectively and repeatedly controlled, is one of the first and most important steps toward building multicomponent molecular machinery with the capacity to do useful work.

A molecular rotor can be thought of as any molecule containing two or more distinct moieties that rotate relative to each other either spontaneously (as induced by thermal energy) or through driven rotation (such as with light, an oscillating electric field, or gaseous flow). A rotor consists of three main parts: a rotator, an axle, and a stator. The rotator is the portion of the molecule that, as implied by its name, undergoes rotational motion. It is connected through an axle, about which it rotates, to the stationary portion of the molecule, the stator. In the macroscopic world, the distinction between rotor and stator is easy to make. Helicopters, windmills, and ceiling fans, for example, all have rotors (propellers) that are easily distinguished from their stators. At the molecular level, however, distinguishing the rotor from the stator is difficult, especially for solution phase rotors, because of the similarity in sizes between

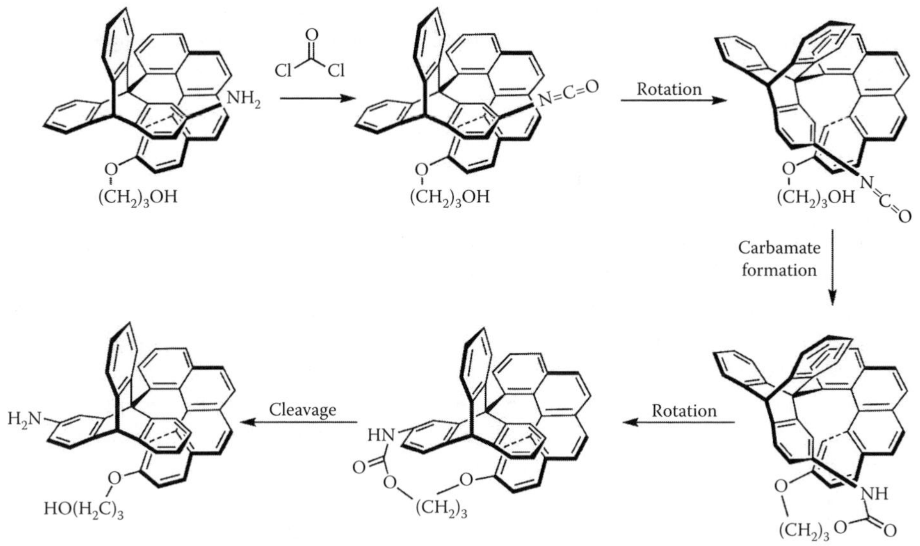

FIGURE 11.19 Using phosgene as a chemical fuel, Kelly and coworkers were able to achieve controlled unidirectional rotation of a tryptycene rotor unit relative to a helacine stator that is connected by a C–C single bond axle.

rotor and stator. A convention may be adopted in which the stator has a greater inertial mass than the rotor and, hence, is the more stationary molecular component. By contrast, for crystalline rotors and surface-mounted rotors, there is little or no ambiguity as the stator can be thought of as the crystal lattice in the former case or the surface or molecular component attached to the surface in the latter.

11.5.2 Rotation by Design

The number of molecular rotors synthesized to date and the vast literature that has been published describing their function is beyond the scope of this relatively brief overview of the topic. Readers interested in a more thorough exploration of the topic are directed to an extensive review of artificial molecular rotors that has recently been published by Michl.[129] This section will highlight the major principles in the design and synthesis of solid-state, liquid crystalline, and surface-mounted molecular rotors. It can be argued that condensed phase and surface-mounted molecular machines have some of the greatest potential to be developed into functional devices on account of greater control over alignment, coordination, and coherence in condensed phases and in the solid state. However, only through extensive and pioneering solution phase studies[68,130–141] of molecular rotors have these recent advances been made possible. Solution phase rotational properties of molecular propeller compounds,[130–133] gears,[130–133] porphyrin rotors,[134] rotations about triple bonds,[135–137] and in supramolecular systems[68,138–141] have all been investigated. Two solution phase examples of molecular rotors have been designed in the laboratories of Kelly[142] and Leigh.[143,144] In Kelly's system (Figure 11.19), an amino-triptycyl rotor was induced to undergo unidirectional rotation relative to an alcohol-substituted [4]helicene stator using phosgene as a chemical fuel. Leigh's design involved first a [3]catenane[143] and, shortly thereafter, a [2]catenane[144] capable of unidirectional rotation when driven with the appropriate sequence of chemical and photochemical transformations. Here, a large macrocyclic ring acts as a stator upon which two, in the case of the [3]catenane, or one, in the case of the [2]catenane, smaller macrocycles are promoted to pirouette around it. While these two solution phase molecular rotors highlight some of the tremendous progress that has been made in the control of molecular motions, the solution phase is inherently chaotic and lacks the coordination, cooperativity, orientation, and coherence that may be gained by operating in the solid-state, in liquid crystals, or onto surfaces.

Before exploring some of the recent condensed phase molecular rotors, it is important to understand the factors that influence rotor performance and the various means by which they operate. The rotation of a 1-D rotor system is described by the Langevin[145] equation:

$$I\frac{d^2\theta}{dt^2} = \frac{-\partial V_{net}}{\partial \theta} - \eta\frac{d\theta}{dt} + \xi(T,t)$$

the components of which describe the moment of inertia (I), torsional potential (θ), applied potential (V_{net}), constant of friction (η), and the effects of stochastic thermal fluctuations (ξ, which is a function of temperature, T, and time, t). The moment of inertia of a molecular rotor relates to how difficult, or easy as the case may be, it is to start and then stop rotation. At the molecular as well as macroscopic levels, inertia is usually dominated by frictional forces, η. In molecules, the constant of friction most often takes the form of rotational energy lost to other modes in the system and surroundings — e.g., the axle, stator, solvent, surface, or lattice — through molecular vibrations. Random thermal energy contributions, $\xi(T,t)$, can be the driving force behind molecular rotation, but this stochastic torque is more commonly associated with returning a system to equilibrium through transfer of rotational energy from the rotor to the thermal bath until a steady state is reached. In order to induce controlled rotational motion, an external force, V_{net}, must be applied that is sufficient to drive the rotor against the random thermal and frictional forces of all other components of the system. Examples of driving potentials include light, electric fields, chemical reactions, and heat, each of which is used to modify or interact with the torsional potential of a rotor in such a way as to temporarily drive the system out of equilibrium — i.e., its resting state — in order to achieve driven motion upon its return to the previous or a new equilibrium state. When the driving force takes the rotor out of equilibrium and imparts an asymmetry to the system, it is possible to achieve unidirectional rotation.[146] If an asymmetry is not introduced into the system, the law of microscopic reversibility ensures that rotation is equally likely in either the forward or reverse direction and unidirectional rotation cannot be achieved. However, as in the case of the F_1-ATPase rotary motor, as well as synthetic, organic examples that will be discussed in this section, rotation in one direction can be accomplished. Unidirectional rotation has the potential to impart greater uniformity, cooperativity, and coherence and so is especially desirable.

11.5.3　Molecular Compasses and Gyroscopes

Garcia-Garibay[147–149] has been particularly active in the design, synthesis, and characterization of molecules that display internal rotational motion in the solid state, with the goal of creating molecular compasses and gyroscopes possibly possessing photonic applications. The design of the molecular gyroscope incorporates an internal dipolar rotor connected to triply bridged triphenylene stators via acetylene axles with near barrierless C–C bond rotation (Figure 11.20). Internal rotation of the dipolar rotor will thus be shielded from any steric or otherwise frictional influences by the bridged stators, and the dipolar nature of the rotor allows it to be oriented and reoriented by electric, magnetic, or optical stimuli. This protection of the inner core is important, for it opens up the possibility of communication between rotors via isolated dipole–dipole interactions without steric interference. A similarly designed organometallic molecular gyroscope has recently been reported by the Gladysz group[150] and studied in solution; it lacks, however, a dipolar rotor unit and is yet to be studied in condensed phases.

Preliminary work toward the realization of a molecular gyroscope by Garcia-Garibay et al. has led to the development of a myriad of molecular compasses based on the rotation of a very similar system they use to make molecular gyroscopes.[147–149] Although the molecular compass design incorporates most of the characteristics of the molecular gyroscope design, it lacks the bridging units between triphenylene stators. For example, crystals of 1,4-bis(3,3,3,-triphenylpropynyl)benzene (TPPB) have been grown and the rotation of the central *p*-phenylene unit has been investigated.[147,149] Crystallization of TPPB from benzene resulted in clathrates, which could be desolvated upon heating.[149] Variable temperature ^{13}C cross-polarization and magic-angle spinning (CPMAS) NMR spectroscopy was used to observe the coalescence of magnetically

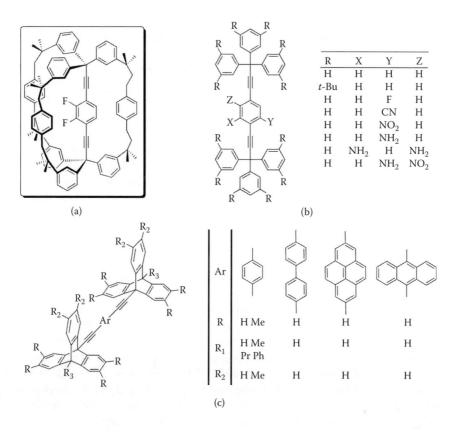

R	X	Y	Z
H	H	H	H
t-Bu	H	H	H
H	H	F	H
H	H	CN	H
H	H	NO_2	H
H	H	NH_2	H
H	NH_2	H	NH_2
H	H	NH_2	NO_2

(a)   (b)

	Ar				
R	H Me	H	H	H
R_1	H Me Pr Ph	H	H	H
R_2	H Me	H	H	H

(c)

FIGURE 11.20 (a) Molecular structure of a proposed molecular gyroscope with an environmentally isolated central difluorophenylene rotor. Terphenylene (b) and tryptycene-based (c) molecular rotors containing a variety of nonpolar as well as polar rotators that have been investigated by Garcia-Garibay and coworkers.

and crystographically nonequivalent *p*-phenylene protons that can be exchanged by rotation. CPMAS NMR showed a barrier to rotation of 12.8 kcal/mol for the clathrate.[147,149] These experiments have recently been repeated with isotopically labeled stators in order to provide better resolution of peaks exchanged upon *p*-phenylene rotation.[151] Arrhenius parameters were derived that indicated a barrier of 11.3 kcal/mol with a preexponential factor of 2.9×10^{11} sec^{-1}. The difference in calculated rotational barriers may be the result of different experimental temperature ranges or the change in vibrational amplitudes and effective sizes of the isotopically labeled compound. Interestingly, benzene molecules of the clathrate showed a sharp CPMAS signal, indicating these trapped solvent molecules rotate rapidly in the solid state.[149] While the two-fold rotation of the *p*-phenylene rotors was shown to be in the KHz range, the six-fold rotation of benzene occurs in the 100 MHz range, indicating that the motion of the rotors and incorporated solvent are not coordinated or "geared." Crystals of TPPB are packed in such a way that triphenylene stators of neighboring molecules are interdigitated, i.e., the stator units of one molecule are aligned between stator units and proximal to the rotors of adjacent molecules.[147,149] The 11 to 12 kcal/mol barrier to *p*-phenylene rotation was attributed to these steric interactions. Desolvated crystals of TPPB were also investigated. Desolvation was accompanied by a 23.2% decrease in volume,[149] resulting in slightly closer packing of adjacent molecules. Quadrupolar echo ^{2}H NMR line-shape analysis between 300 and 480 K revealed a barrier of 14.6 kcal/mol for *p*-phenylene rotation. The slightly higher barrier could be the result of closer packing and greater steric interactions. Modifications to design were made in order to provide greater shielding from intermolecular steric interactions and to lower the barrier to rotation: bulky 3,5-ditertbutyl groups were incorporated onto the stator units of molecular compasses.[152] X-ray crystal structures revealed greater isolation of internal rotor units by the bulkier stators. Although *p*-phenylene rotation was observed to increase to the GHz range, quantitative rotational rates could not be fitted to a single rate constant. This situation implies that there is some loss of homogeneity in the crystal and the presence of bulky *t*-butyl groups results in a more amorphous solid.

A series of molecular compasses was also prepared with triptycene units functioning as the stators[148,153] and with benzene, anthracene, pyrene, and biphenyl rotors to find the optimal design by testing a variety of structural parameters. AM1 level calculations revealed that rotations around the acetylene axles of the phenyl, pyrene, and biphenyl molecular compasses were, within error, barrierless.[148] Hydrogen atoms at the 1, 4, 1', and 4' positions of anthracene, however, undergo steric interactions with the endo hydrogen atoms of triptycene stators, resulting in a barrier of about 4 kcal•mol^{-1}. Only compasses with phenyl rotors gave x-ray quality crystals, both with unsubstituted and with methyl-substituted triptycene stators. In the solid state, the bulkier triptycene stators provided a greater amount of free volume to the *p*-phenylene rotors. By investigating atomic displacement parameters from x-ray crystal structure data, a barrier of 3.3 kcal/mol at 100 K for *p*-phenylene rotation was calculated.[153] This value is only 0.4 kcal/mol greater than the internal rotational barrier of ethane[154] and indicates how efficient molecular rotors can be when properly designed, even in the solid state.

Molecular compasses containing rotors with dipoles have also been synthesized to make molecules that can be subjected to the influence of electromagnetic fields.[155] These dipolar compasses are all based upon the initial TPPB design with the addition of F, CN, NO$_2$, NH$_2$, *o*-diNH$_2$, and *p*-NH$_2$NO$_2$ substituents on the central *p*-phenylene rotor. AM1 calculations predict dipoles varying from 0.74 to 7.30 Debye for the series of compounds studied. Crystal structures were obtained for all but the diamino and aminonitro derivatives. Interestingly, x-ray crystal structure data revealed disorder in the crystals that corresponded to a center of inversion. A dynamic process could only account for this disorder through inversion of the two triphenylene stators, followed by a 180° rotation of the dipolar rotor, or vice versa. Such a process is not possible in the solid state as it would completely destroy the crystal lattice. Therefore, it must be concluded that the disorder is static and molecular rotors can adopt one of two centrosymmetric conformations in the crystal. A qualitative understanding of the rotational dynamics of dipolar rotors in the crystal can be gained from this static disorder. Any rotations must involve either a complete 360° rotation or two sequential 180° jumps. The authors argue[155] that complete 360° rotation implies a singular energetic minimum in the rotational potential while two sequential 180° jumps would imply an asymmetric double-welled potential where the dipolar rotor, originally in the most favorable conformation, rotates 180°, spends very little time in the less favorable conformation, and rotates 180° back to the global minimum. Signal averaging and peak overlap in the ^{13}C CPMAS NMR spectrum precluded quantitative results regarding rates of rotation. The response of a fluorinated molecular rotor to an applied electric field has been investigated in collaboration with Price[156] using a combination of dielectric spectroscopy and ^{2}H NMR. These results suggest a two-fold (13.7 kcal/mol) barrier with an asymmetry of about 1.5 kcal/mol for clathrates and 1.9 kcal/mol for desolvated crystals. About one-third of the rotors were observed to have no response to the applied field, most likely a result of intermolecular steric ones. It is therefore believed that steric interactions are, in fact, more dominant than dipolar rotor–rotor interactions. However, a more polar substituent than fluorine may cause the response to an electric field to dominate over steric interactions. Alternatively, steric interactions could be overcome and the correlation of rotor–rotor dipoles could be investigated with the successful synthesis of an enclosed dipolar molecular gyroscope. This structure remains one of the central goals of rotor design, and the extensive studies of both nonpolar and polar solid-state rotors by Garcia-Garibay have provided a firm foundation on which to build.

11.5.4 Light-Driven Unidirectional Rotation

Considerable progress has also been made in the development of unidirectional molecular rotors in the Feringa laboratories — first in solution,[157–161] then in liquid crystals,[162] and recently on surfaces.[163] Early work on biphenanthrylidene chiroptic molecular switches,[164] helically chiral compounds capable of undergoing *cis–trans* isomerizations and helical inversions with the appropriate wavelength of light, provided the basis for the development of synthetic unidirectional molecular rotors. By inducing a second

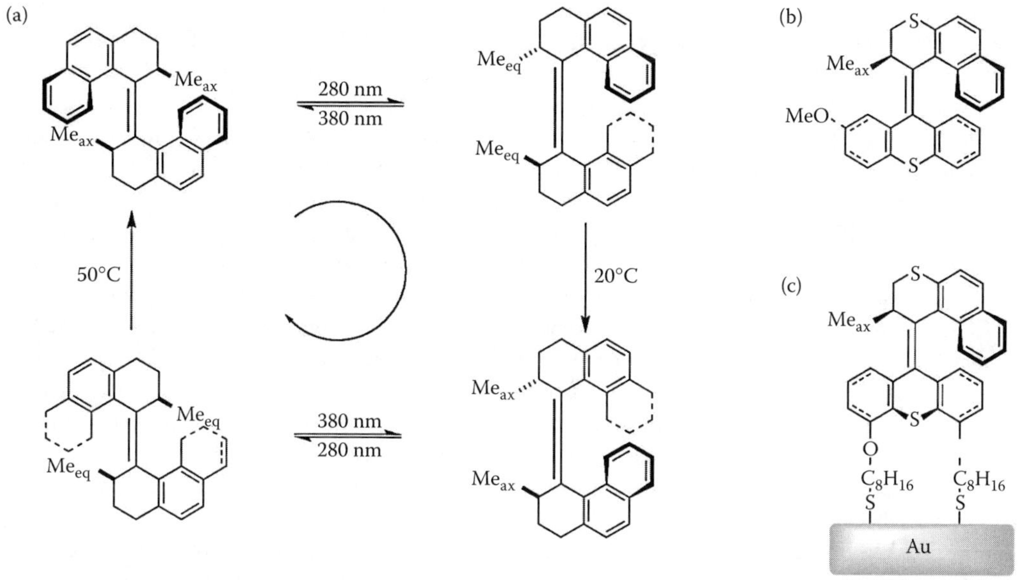

FIGURE 11.21 (a) Molecular structure of a first-generation unidirectional molecular rotor where photoinduced *cis–trans* isomerization steps followed by thermal helical inversions ensure unidirectional rotation about the central alkene axle. (b) A second-generation unidirectional molecular rotor. (c) The attachment of alkanethiol "legs" to the stator of a second-generation unidirectional molecular rotor allows the rotor to be attached to a gold surface.

element of chirality,[157] in the form of a pseudo-axial methyl substituent (Figure 11.21) to both the upper and lower helical "propellers" connected by a central C–C double bond, unidirectional rotation can be achieved through a four-step process: two energetically uphill *cis–trans* photoisomerizations each followed by a thermally promoted energetically downhill helical inversion. The unidirectionality of rotation was controlled through two key structural elements. The pseudo-axial orientation of stereogenic methyl groups ensured that irradiation of the (*P,P*)-*trans* isomer (Figure 11.21[a]) at >280 nm gave an exclusively (*M,M,*)-*cis* isomer, where the chiral methyl substituents now adopt an energetically unfavorable pseudo-equatorial orientation. The conformational flexibility of upper and lower portions of the molecular rotors in the vicinity of the central olefin axle allows for a thermally promoted helical inversion step that is irreversible, exothermic, and results in the formation of a (*P,P*)-*cis* isomer where methyl groups again adopt a pseudo-axial orientation. Another stereochemically controlled photoisomerization, followed by a thermally controlled helical inversion, completes the cycle. The direction of rotation has been shown to be completely dependent upon the configurations of the stereogenic centers, which provide the asymmetry necessary for unidirectional motion.[158] Stepwise unidirectional rotation could be monitored by circular dichroism (CD) spectroscopy and chiral HPLC showed no racemization during photochemical or thermal steps. The desymmetrization of upper and lower paddles, incorporation of heteroatoms, and addition of a series of substituents produced a number of second-generation[158–161] molecular rotors and resulted in speeding up rotation by lowering thermal inversion barriers and improving the photoequilibrium of photostationary states. Further research in the solution phase on light driven, unidirectional molecular rotors has led to rotors that undergo continuous rotation at 60°C with 365 nm light.[160,161] Recently, Feringa has also demonstrated unidirectional rotation in a solution phase chemically driven molecular motor.[165]

Feringa has shown that, when incorporated into a nematic liquid crystal film, the controlled motions of sterically overcrowded molecular rotors can lead to macroscopic changes in the film.[162] It has been previously demonstrated[166] that doping a liquid crystal with a chiral dopant can greatly influence the organization of the liquid crystalline matrix. More specifically, the wavelength of light reflected from a cholesteric liquid crystal is dependent upon the pitch, angle of incident light, and average refractive index of the film.[162] Furthermore, the pitch is inversely proportional to the concentration, enantiomeric excess, and helical twisting power (β) of a chiral dopant. By doping the nematic liquid crystal E7 with the (*P,P*)-*trans*, (*P,P*)-*cis*, and (*M,M*)-*trans* isomers of the molecular rotor shown in Figure 11.21(a), Feringa was able to measure helical twisting powers of +69 (right-handed cholesteric), +12 (right-handed cholesteric), and −5 (left-handed cholesteric) μm^{-1}, respectively. The fast helical inversion of the thermally unstable (*M,M*)-*cis* isomer prohibited measurement of its helical twisting power. When a spin-coated liquid crystalline film of E7 doped with 6.16 wt% of the first generation unidirectional rotor (Figure 11.21[a]) was irradiated with >280 nm light, a bathochromic shift in the wavelength of reflected light was observed. Over a period of 80 sec, the color of the film changed continuously over the entire visible spectrum from violet to red on account of the photoinduced increase in concentration of both (*P,P*)-*cis* and (*M,M*)-*trans* isomers and the concomitant lowering of β-values and pitch of the film. Upon heating the film to 60°C, helical inversion of the (*M,M*)-*trans* isomer to the (*P,P*)-*trans* isomer resulted in a hypsochromic shift back to violet as a result of the strongly positive β-value and smaller pitch of the (*P,P*)-*trans* isomer. These results demonstrate the potential of using the rotational motions of a synthetic molecular rotor to change selectively the macroscopic properties of a liquid crystalline film. By simply changing the irradiation time, intensity, or wavelength, any color of the visible light spectrum can be produced.[162] Such molecular control over the macroscopic properties has potential uses in photoinduced information storage with the added benefit of nondestructive readout capabilities.

More recently, Feringa has attached modified second-generation[158,160] molecular rotors successfully to a solid support and observed their unidirectional rotation.[163] Demonstrating that the second-generation molecular rotors are capable of rotating unidirectionally on a surface in the same manner as they do in solution is an important step toward functional devices. In solution, the random orientations of such rotors ensure that no useful work can be extracted from the system. Because surface attachment allows all the stators and rotators to be aligned in the same direction, by replacing the two methoxy groups of a second-generation molecular rotor (Figure 11.21[c]) with octanethiols, the rotors could be self-assembled onto gold nanoparticles. Attaching the lower portion of the rotor to Au nanoparticles via two tethered thiols rather than one results in unambiguous assignment of the rotor and stator. Octanethiol legs were chosen over shorter alkanethiols in order to minimize electronic interactions between the chromophores and the surface as well as to provide enough separation between the two for rotational motion to occur relatively unobstructed. Transmission electron microscopy data was combined with CD spectroscopy and the known density of gold in order to derive the average chemical formula for one nanoparticle, revealing a ratio of about 26 molecular rotors per 251 Au atoms. Irradiation of the nanoparticle-bound molecular rotors with 365 nm light resulted[163] in an inversion of the molecules' CD spectrum, indicating that the same photoisomerization that is well studied in solution[159] is also operative when rotors are attached to a solid support. Heating the nanoparticles to 50°C resulted in helical inversion, also verified by CD. The observation that ^{1}H NMR signals for surface-bound molecular rotors remained broad after heating at 70°C for 2 h or irradiation with 365 nm light for 3 h and that no signals for free rotors were ever observed, was taken as evidence that rotation occurs while the molecules are attached to the surface. Despite the relative liability of the Au–S bond, it is unlikely that molecules undergo successive cycles of desorption, rotation, and resorption as the equilibrium with free rotors would be observed by ^{1}H NMR spectroscopy. An energy of activation of 22.9 kcal/mol was determined from variable-temperature CD spectroscopy. This value is only 0.5 kcal/mol greater than that of the same rotor in solution, indicating that surface attachment has little effect on the dynamics of rotation.[163] Additional proof of unidirectional rotation on gold was provided by isotopically labeling one of the legs of the molecular rotors. Self-assembly onto gold followed by irradiation at 365 nm resulted in the formation of a 1.2:1 ratio of unstable-*trans* to stable-*cis* isomers of the surface-bound rotors. This sample was then divided into two. In the first sample, rotors were removed from the nanoparticles by treatment with KCN. Thermal isomerization at 70°C resulted in the formation of a 1.2:1 ratio of stable-*trans* to stable-*cis* isomers, with none of the unstable-*trans* isomers reported. The second sample was thermally isomerized while still attached to the surface. Subsequent detachment with KCN revealed the same ratio of isomers as was found for the unbound sample, indicating that unidirectional rotation is achieved for surface-bound molecular rotors in exactly the same manner as in solution. The achievements of Feringa et al. in moving from solution phase chiroptic switches,[164] to unidirectional

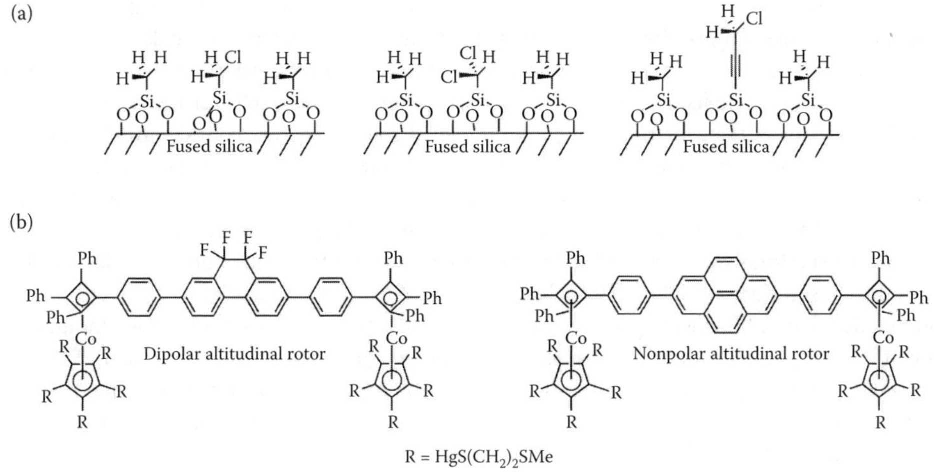

FIGURE 11.22 (a) Molecular structure of chloromethyl, dichloromethyl, and 1-chloropropyl-2-yne azimuthal rotors mounted on a fused silica surface. (b) Structures of dipolar and nonpolar altitudinal molecular rotors that have been investigated by Michl and coworkers.

rotors,[157–161] to the incorporation of these rotors into liquid crystals,[162] and now onto solid substrates[163] represents the evolution of structures necessary for the development of molecular machinery. The goal of using addressable molecular functions to induce changes in the macroscopic properties of materials in a controlled manner has been achieved with doped liquid crystals. The development of functional, surface-bound unidirectional rotors is an initial step toward studying how the motions of larger ensembles of rotors might be oriented, coordinated, and cooperative. These systems may have future applications in the design and fabrication of ever more complex molecular machinery.

11.5.5 Surface–Rotor Interactions

Michl and coworkers[167] have studied extensively a number of surface-bound molecular rotors, both experimentally and theoretically. These studies differ from those of Garcia-Garibay and Feringa in that they focus largely on the interactions between synthetic rotors and the surfaces to which they are attached, how these interactions are affected by structural modifications to the rotors, and to what extent the surface–rotor interactions govern the dynamic properties of the rotors. Initial experiments[167] focused on chlorinated rotors chemisorbed (Figure 11.22) onto silicon substrates. Mixtures of methyltrichlorosilane and either chloromethyltrichlorosilane or dichloromethyltrichlorosilane were vapor deposited onto fused silica substrates. In this manner, azimuthal chloromethyl or dichloromethyl rotors were embedded in a 3 to 5 Å thick monolayer of $(-O)_3Si-Me$. Capacitance and dissipation factor measurements indicated that dipole relaxation occurred through a thermally activated hopping mechanism. Data obtained for rates of relaxation were used to calculate rotational barriers for surface-mounted chloromethyl rotors and a large distribution of barriers was observed, with 75% of the barriers falling between 1.5 and 3.0 kcal/mol. Molecular modeling of surface-mounted rotors was performed using the universal force field[20] in order to gain insight into the factors contributing to such a wide distribution of rotational barriers. It was concluded from the combination of experimental data and molecular simulations that the broad distribution was a direct result of disorder on fused silica surfaces. In the case of surface-mounted chloromethyl rotors, the axle is simply the Si–C between the chloromethyl rotor and fused silica stator. Such a short axle puts the rotor and stator in very close proximity to each other. Any slight variations in the fused silica surface can greatly affect the ability of a bulky chlorine substituent to be able to rotate around its Si–C axle because of nonbonding surface–rotor interactions. Therefore, variations in the fused silica surface can be observed in the wide distribution of rotational barriers.[167]

A longer axle was used to separate the rotor from the surface and, therefore, lessen the contribution of such nonbonding interactions, thereby extending understanding of surface–molecule effects. Michl and Horinek[168] have studied the rotational dynamics of surface mounted 1-chloropropyl-2-yne rotors on fused silica through molecular modeling. Barriers were shown to range between 0.65 and 3.1 kcal/mol.[129,168] While these barriers are lower than for the smaller chloromethyl rotors, they are considerably higher than the barriers for acetylene rotation as calculated[148] with semi-empirical AM1 methods by Garcia-Garibay (~0.05 kcal/mol); computational error, especially when comparing two different computational methods, can account for the 0.6 kcal/mol difference between the near barrierless rotation calculated by Garcia-Gariaby and 0.65 kcal/mol barriers calculated by Michl. However, the large distribution in calculated barriers and the fact that some are as high as 3.0 kcal/mol required further investigation.

Computational modeling that focused on the important structural parameters of surface-mounted rotors provided insight into the relationship between their structure, orientation, and function. Molecular modeling simulations showed that the rotational barriers for 1-chloropropyl-2-yne rotors were highly dependent upon the polar angle formed between the rotor and the surface.[168] Rotors that were positioned nearly perpendicular to the surface showed the lowest rotational barriers and also had torsional potentials with just one minimum and one maximum. The minima and maxima were directly related to the van der Waals attraction between the surface and the chlorine substituent of the rotor. As the polar angle was decreased and the rotor was brought closer and closer to the surface, rotational barriers increased and torsional potentials became more complex. The increase in rotational barriers was simply the result of increasing van der Waals interactions. More complex torsional potentials with two or three minima and maxima resulted from van der Waals interactions between the surface and methylene hydrogen atoms of the molecular rotors. The experimental and theoretical results pertaining to chloromethyl and 1-chloropropyl-2-yne rotors provide insight into the subtle surface-rotor interactions that greatly influence the dynamics of molecular rotors.

More elaborate altitudinal molecular rotors, both nonpolar and dipolar, have also been synthesized[169] and investigated[169,170] by Michl et al. using tunneling barrier height imaging. The two rotors are shown in Figure 11.22(b). Rotation of the central tetrahydropyrene and tetrafluoro[3]helicene cores were too fast to be observed by NMR spectroscopy, though the barrier was estimated to be around 3.0 kcal·mol^{-1}, given known rotational barriers of their constituent parts in solution.[169] Altitudinal rotors were adsorbed onto an Au(111) surface at both monolayer and submonolayer concentrations. Strong binding was observed, even after exposure to air for several days, on account of the mercury and sulfur atoms present in the legs of the stator units. Surfaces were characterized by XPS, ellipsometry, and STM. Tunneling barrier height imaging was used to investigate whether dipolar molecular rotors were capable of rotation while assembled onto surfaces. Molecular modeling showed that dipolar molecular rotors were capable of adopting conformations with their rotors either sterically locked parallel to the surface or free to flip rapidly between perpendicular orientations.[169,170] The difference in locked and free conformations resulted from the positions of the leg moieties of the stators, which are capable of blocking rotation if proximal to the rotor units. The local work function of a gold surface is dependent upon the presence of a dipole[171]; hence, rotors that are free to rotate can have an effect upon the local work function of the surface depending on the orientation of its fluorine substituents. The electric field potential of an STM tip is capable of locking the freely rotating dipolar rotors into orientations aligned either with or against the electric field. Changing the direction of the STM electric field changes the orientation of dipolar rotors and, thus, the work function of the metal surface. These differences were measured with tunneling barrier height measurements. Bright spots could be seen in barrier height images where differences in the work function of the surface were measured. About two-thirds of dipolar molecular rotors were shown to rotate in response to the electric field; the remaining third were presumed to be in the sterically locked conformation and therefore dark. Repeated scans showed "blinking" in some of the bright spots,[169,170] indicative of a certain amount of mobility in the stator legs, allowing rotors to switch between locked (dark image) and freely rotating (bright image) conformations. Similar barrier height images were taken of the nonpolar motors and no differences in work function were observed. Molecular dynamics simulations have demonstrated[169,170] the possibility of unidirectional rotation of dipolar motors under the influence of an alternating electric field.

Molecular rotors[129] continue to be an active area of research in many laboratories around the world. These examples of solid state, liquid crystalline, and surface-bound molecular rotors are a sampling of the recent achievements of a few laboratories in moving out of the solution phase and into and onto more condensed phases in hopes of developing functional molecular devices. Studies performed by Garcia-Garibay, Feringa, Michl, and their coworkers concerning the influences of molecular orientation, intra-, and intermolecular steric interactions, response to electric fields, and incorporation of electric dipoles have provided a great deal of insight into the factors that govern condensed phase molecular rotors and will continue to aid in the development of more advanced rotors. Investigations into the concerted actions of large ensembles of molecular rotors are greatly anticipated.

11.6 Surfaces with Controllable Wettability

11.6.1 Water — Love to Hate It, Hate to Love It

In the previous section, much attention was paid to amphiphilic [2]rotaxanes, whose hydrophobic and hydrophilic regions allowed for their self-assembly into thin films by LB techniques. Similarly, surfaces may be classified as hydrophobic or hydrophilic in the same manner as molecules or sections of molecules merit these classifications. While the hydrophobic or hydrophilic nature of molecules dictates their behavior in different solvents and on air–liquid interfaces, the hydrophobicity or hydrophilicity of a surface is most commonly associated with the surfaces' wettability; that is, the extent to which a drop of liquid will spread across — or wet — a surface. For example, fluorinated surfaces are particularly hydrophobic, and a drop of water on a fluorinated surface will adopt a near-spherical shape. In contrast, a surface terminated by hydroxy groups is hydrophilic, and a drop of water will take the shape of a truncated hemisphere as it wets the surface, and these qualitative observations are quantified through contact angle measurements. Having the ability to manipulate the hydrophobicity and hydrophilicity of a surface dynamically allows for the shapes and movements of liquids on such surfaces to be controlled. In such a way, the design and fabrication of surfaces that have externally controllable wettability properties as a result of molecular actuation are a form of molecular machinery. Interest in the ability to control surface wettability has increased[172,173] dramatically, and manipulation of surface wettability properties using a readily controlled external stimulus holds promise as a means of developing biocompatible, self-cleaning, contaminant-free[174,175] materials. Our aim in this section is to describe the current progress in developing smart surfaces for controlling wettability, with special emphasis on their switching mechanism and reversibility.

11.6.2 Energetic Contributors to Wettability

The wetting behavior of a surface is the result of the surface free energies of the three interfaces: the surface–vapor interface, the surface–liquid interface, and the liquid–vapor interface. At equilibrium, these free energies are balanced and a characteristic contact angle (θ), defined as the angle created between the liquid–vapor interface and the surface–liquid interface, is observed. The balance of this equilibrium, and resulting contact angle, is controlled by chemical composition[176,177] and surface geometry.[175,178,179] Changes in any of the interfacial surface free energies will be reflected in changes in the contact angle of a drop on the surface; however, it is the manipulation of the surface–liquid interfacial free energy that is most commonly used to influence surface wettability. When the surface–liquid interfacial free energy is increased or decreased, a droplet will spread or contract on the surface until the minimum interfacial free energies, as determined by the cohesive forces in the liquid and the adhesion between the solid–liquid interface, are reached. By taking advantage of processes that allow for the functionalization of various surfaces — from metallic gold[24,180] to insulating SiO_2/Si[26–28,181] — with organic molecules or polymers, smart surfaces that exhibit dynamic changes in interfacial energies in response to photonic, [179,182–197] electrical potential,[94,198–202] solvent[203–206] temperature,[207–209] and pH[210] stimuli have been achieved.

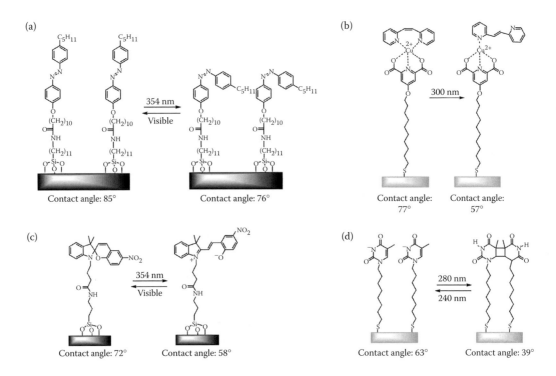

FIGURE 11.23 Selected examples of chemically modified surfaces wherein the photoinduced structural changes in the azobenzene (a), 2,2'-dipyridylethylene (b), spyropyran (c), and pyrimidine (d) moieties of these surface-bound organic molecules give rise to changes in the wettability of the surface.

11.6.3 Light-Responsive Surfaces

Chemical systems that cause changes in wettability upon irradiation with light include azobenzene,[184–190] spiropyran,[179,191] phenylazoacrylanilide,[192] pyrimidine,[193] 2,2'-dipyridylethylene[194] and 2-hydroxyphenyldiphenylmethanol[195] moieties (Figure 11.23). Photoresponsive organic molecules have been assembled onto surfaces by tethering them to triethoxysilanes, thiols, disulfides, or polymer backbones and then chemisorbing or physisorbing the tethered moieties to surfaces such as gold, silicon, or silicon dioxide. Once assembled, these systems take advantage of the actuation of surface-bound synthetic organic molecules to affect changes in the wettability of a surface. A significant number of investigations of photoresponsive surfaces have relied[184–190] on the *trans–cis* isomerization of azobenzene derivatives. The azo chromophore isomerizes by illumination with UV light ($\lambda = 300$ to 400 nm) from the stable *trans* form to the *cis* state, while reverse isomerization can be triggered by irradiation with visible light ($\lambda = 425$ to 500 nm). This *trans–cis* photoisomerization[184,211] is reversible, with little degradation occurring after many switching cycles. Isomerization of azobenzene is accompanied by an appreciable shape change as the *trans* isomer adopts a more linear conformation than the *cis* isomer. When derivatized with hydrophobic or hydrophilic functional groups, this change in molecular conformation selectively exposes the hydrophobic or hydrophilic functional groups to the surface–liquid interface. In addition, the *trans* and *cis* isomers of azobenzene exhibit different dipole moments,[186] which in turn has an impact on the corresponding wetting behavior. The photoinduced isomerization of surface-bound *p*-phenylazoacrylanilides from *trans* to *cis* results in a large dipole change and concomitant increase in hydrophilicity ($\Delta\theta \sim 10°$). Recently, Jiang et al.[187] observed superhydrophobicity in a self-assembled azobenzene monolayer with the introduction of patterned square pillars etched into a flat silicon substrate by photolithography. With a spacing of 40 μm between pillars, a larger change in wettability ($\Delta\theta \sim 70°$) was observed than had been observed in the absence of the patterned pillars ($\Delta\theta \sim 2°$).

Similarly, the photoinduced isomerization of spiropyrans can change surface–liquid interfacial free energies. Garcia et al.[179,191] tethered spiropyrans to a glass surface via a 3-aminopropyltriethoxysilyl tether. Spiropyran-coated surfaces exhibited contact angles of roughly 73°. Irradiation of the spiropyrans with UV light induces isomerization from the more hydrophobic spiro conformation to the polar, hydrophilic zwitterionic merocyanine conformation. This isomerization is accompanied by a decrease in the contact angle of 14° as a result of the more polar nature of the merocyanine zwitterion. Irradiation with visible light reverses this process and results in a subsequent increase in contact angle. Pyrimidine-coated surfaces have also been used to influence surface wettability. Pyrimidine moieties were tethered to a gold surface via an alkane thiol. Photodimerization of the monolayer with 280 nm light was shown to decrease surface charge and, therefore, render the surface more hydrophobic. This process could be reversed upon irradiation with 240 nm light, thereby causing reversible contact angle changes as large as 25°. More recently, Ralston and coworkers[193] have demonstrated how the DNA base thymine and other uracil derivatives, when alkylated with a hydrocarbon chain and assembled at a gold interface, change surface hydrophobicities and hydrophilicities upon their dimerization[196] when subjected to UV irradiation. The process is accompanied by a reversible decrease in wettability.

Photochromic molecules have also been incorporated[194] into noncovalent multilayer films to provide a means of switching surface wettability. These multilayer films consist of (1) an SAM of 4-[(10-mercaptodecyl)oxy]pyridine-2,6-dicarboxylic acid attached to a gold surface via a decanethiol tether, (2) a layer of Cu(II) ions that complex with the pyridine-2,6-dicarboxylic acid headgroups of the SAM, and (3) a layer of photoactive *cis*-2,2′-dipyridylethylene groups. In the *cis* isomer, the bidentate Cu(II) ions form a symmetric pentavalent complex with the hydrophobic *cis* ethylene exposed to the surface (contact angle of 77°). Irradiation with UV light ($\lambda = 300$ nm) induces a *cis*–*trans* photoisomerization of the 2,2′-dipyridylethylene moiety. In the *trans* isomer, only one of the two pyridyl groups of the 2,2′-dipyridylethylene are able to complex the Cu(II) ion while the other is exposed to the surface. This conformation — a tetravalent Cu(II) complex and a surface-exposed pyridyl nitrogen — results in a more hydrophilic surface ($\theta = 57°$). Knowing that irradiation of 2-hydroxyphenyldiphenylmethanol derivatives with >270 nm light converts the hydrophilic diol to a hydrophobic carbonyl with the concomitant liberation of a water molecule, Irie[193] was able to induce contact angle changes of ~14° by copolymerizing 2-hydroxyphenyldiphenylmethanol with butyl methacrylate and casting the resulting polymer onto a Teflon plate. Conversion of the hydrophobic carbonyl back to the more hydrophilic 2-hydroxyphenyldiphenylmethanol was achieved simply by placing the surface in the dark.

11.6.4 Electrostatic Control of Surface Properties

The use of organic molecules to control surface wettability is not limited to surface-bound photoactive compounds. In pioneering work, Langer et al.[200] developed a novel switching surface design that enables surface wettability to reversibly switch between the hydrophilic state and the hydrophobic state in response to an electrical potential, without altering the chemical identity of the surface (Figure 11.24). Langer's design relies upon surface-bound molecules that are capable of undergoing externally controlled conformational changes that expose either hydrophobic or hydrophilic moieties at the surface–liquid interface. To achieve such conformational changes, monolayers of the addressable organic compounds must have sufficiently low packing density to cause homogenous structural changes across the entire surface without hindrance from intermolecular steric interactions. Toward this aim, a gold surface was functionalized with a SAM of (16-mercapto)hexadecanoic acid (MHA) at a low packing density (~0.65 nm²/ molecules) to ensure[212] sufficient spatial freedom for synergistic molecular reorientation of the surface-bound molecules. In order to achieve the desired low packing density, a derivative of MHA with a bulky globular end group ((2-chlorophenyl)diphenylmethyl ester) was self-assembled on gold and subsequent cleavage of the bulky end group resulted in a low-density SAM of MHA. Additionally, cleavage of the (2-chlorophenyl)diphenylmethyl ester exposed terminal carboxylate anions on the gold surface, rendering the surface hydrophilic. Upon application of an electrical potential, the negatively charged carboxylate groups experienced an attractive force to the gold surface, leading to a conformational reorientation of

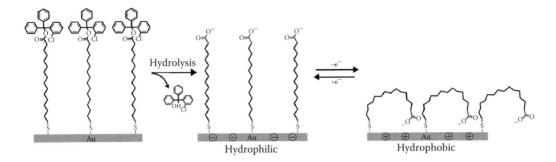

FIGURE 11.24 The self-assembly of (16-mercapto)hexadecanoic acid with sterically bulky (2-chlorophenyl)diphenylmethyl ester end groups onto a gold surface. Hydrolysis of the end groups exposes terminal anionic carboxylate moieties that adopt an extended, hydrophilic conformation with the application of a negative bias to the surface or a bent, hydrophobic, conformation when the bias is switched to being positive.

the MHA molecules from a hydrophilic (straight chains with carboxylate anions exposed at the surface) to a hydrophobic (bent chains with greasy alkyl chains exposed at the surface) state. Reversible conformational transitions were confirmed at a molecular level with the use of sum-frequency generation spectroscopy and at a macroscopic level with the use of contact angle measurements. An alternative method for preparing loosely packed ω-carboxyalkyl monolayers for potential-controlled surface wettability was later reported,[202] which involved the assembly of a preformed inclusion complex — a cyclodextrin (CD)-wrapped alkanethiolate — on gold, followed by the release of the CD space-filling group from the anchored pseudorotaxane. Removal of the noncovalently bound spacing group was a means by which a low density, regular monolayer could be formed. This system was then used to show the switchable adsorption of proteins under redox control, an elegant demonstration of the potential of surfaces with switchable surface properties.

In an alternative approach, Willner et al.[201] reported the potential-controlled bending of an alkanethiol-tethered bipyridinium monolayer on a gold electrode. As in the studies mentioned above, surface coverage of dicationic bipyridinium groups was purposely kept low (~11%) in order to allow for unimpeded potential-induced bending and stretching of the monolayer constituents. The bipyridinium monolayer was prepared by the covalent linkage of N-methyl-N'-carboxydecyl-4,4'-bipyridinium to a mercapto-ethanol monolayer self-assembled onto a gold electrode. In this manner, a hydrophilic bipyridinium moiety was tethered to a gold surface by a hydrophobic hydrocarbon spacer. Upon application of a positive potential to the gold electrode, the dicationic bipyridinium moieties were repelled electrostatically from the electrode surface, resulting in a hydrophilic interface ($\theta = 70°$). Biasing the potential of the electrode to −0.7 V induced the reduction of the bipyridium moieties to the radical-cations, which were subsequently attracted to the negatively charged electrode surface. The conformational rearrangement of the bipyridinium moieties resulted in the exposure of the hydrocarbon spacer chains to the interface, yielding a hydrophobic surface ($\theta = 79°$). The same research team further demonstrated[94] the modulation of interfacial properties caused by electrochemically responsive surfaces based on switchable bipyridinium molecular shuttles. For this purpose, a thiol-tethered rotaxane, consisting of the hydrophilic cyclophane CBPQT[4+] threaded on a hydrophobic diiminobenzene unit and stoppered by an adamantane unit, was self-assembled onto a gold electrode. Prior to reduction, the cyclophane was localized on the rotaxane via the π-donor–acceptor complex with the diiminobenzene unit, yielding a hydrophilic interface. Subsequent electrochemical reduction of the cyclophane to the corresponding biradical dication resulted in its dissociation from the π-donor diiminobenzene site and its molecular mechanical translocation along the rotaxane towards the electrode. This mechanical movement exposed the hydrophobic diiminobenzene unit and, together with the less hydrophilic cyclophane, resulted in a more hydrophobic interface upon reduction of the cyclophane. Finally, oxidation of the biradical dications gave the reverse result as the tetracationic cyclophane returned to the diiminobenzene π-donor sites. The contact angle of the system

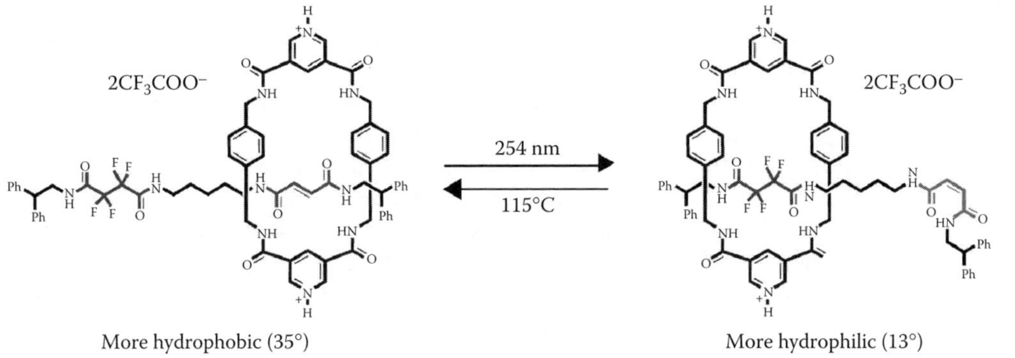

More hydrophobic (35°)   More hydrophilic (13°)

FIGURE 11.25 Molecular structure of a [2]rotaxane that can be switched between a coconformation with the more hydrophobic tetrafluorosuccinamide recognition unit or the more hydrophilic fumaramide recognition unit exposed. Leigh and coworkers have shown that a SAM of the switchable [2]rotaxanes allows for the controlled directional movement of fluids along both flat as well as up inclined (12°) surfaces.

changed reversibly from 55°, when the cyclophane was in its oxidized state, to 105° for the reduced cyclophane.

11.6.5 Nanofluidics

Beyond the possibility of developing reconfigurable hydrophobic–hydrophilic surfaces on substrates, the possibility of a guided motion of droplets across surfaces (Figure 11.25) may have important implications on the development of micro- and nanofluidic systems. Designing a surface tension gradient on a substrate enables control of droplet positions as the random movements of droplets are biased toward the more wettable side of the surface.[182,183,197,213–215] Such a surface tension gradient can be achieved by altering the wettability of specific regions of a surface selectively while leaving other regions unchanged. Those droplets that are positioned on or proximate to the altered regions will strive to minimize interfacial surface free energies by either wetting or dewetting the surface. This asymmetric wetting or dewetting can be used to induce movement of the droplet as it reorganized to reach a new equilibrium. Surface energy heterogeneity brought about by photoinduced wetting/dewetting has been used to move droplets of liquid across surfaces.

Ichimura et al.[182,183] used the asymmetric irradiation — i.e., irradiation of one edge of a droplet and not the other — of photoisomerizable SAMs composed of a calix[4]resorcinarene derivative having photochromic azobenzene units to produce gradients in the free energy of the surface. Selectively wetting the front edge of a macroscopic oil droplet while dewetting its back edge, causing the drop to move forward, was achieved in a two-stage process: (1) the azobenzene-terminated surface was photoisomerized at the leading edge of an oil droplet, resulting in the hydrophilic *cis* conformation and wetting of the surface by that portion of the microscopic oil droplet in the irradiated region, and (2) selectively irradiating at the rear of the macroscopic oil droplet with a second wavelength, inducing a *cis*-to-*trans* isomerization and selective dewetting of the back edge of the droplet. The overall effect was movement of the oil droplet in the direction of the *cis* (hydrophilic) region and away from the *trans* (hydrophobic) region. The direction and velocity of the droplets' motion were tunable by varying the direction and steepness of the light intensity gradient.

More recently, Leigh et al.[197] reported the macroscopic directional transport of a droplet on a photoresponsive surface resulting from the mechanical movements of light-switchable [2]rotaxane molecular shuttles capable of exposing or concealing fluoroalkane residues, thereby modifying the interfacial properties. The light switchable [2]rotaxane incorporates fumaramide and tetrafluorosuccinamide recognition sites in its dumbbell component and a benzylic amide ring component that predominantly encircles the fumaramide site. In this coconformation, the hydrophobic tetrafluorosuccinamide moiety is exposed, rendering the surface more hydrophobic. Upon irradiation at 240 to 400 nm, the fumaramide unit isomerizes to the corresponding *cis*-maleamide, causing the macrocycle to shuttle toward the tetrafluo-rosuccinamide portion of the dumbbell. As a consequence, the fluorine-rich recognition site is shielded, making the surface more polar. The [2]rotaxane was bound[216] onto a SAM of 11-mercaptoundecanoic acid on gold as a result of hydrogen bonding between the pyridine moieties of the macrocycle and the acid functions of the SAM. The rotaxane molecules were oriented with the thread parallel to the SAM surface and plane of the macrocycle perpendicular to the surface. A drop of diiodomethane was placed on the [2]rotaxane-functionalized SAM and the surface was irradiated with a perpendicular beam of 240 to 400 nm light focused on one side of the drop and the adjacent surface to produce a gradient in the surface free energy across the length of the drop. As the contact angle on the irradiated edge decreases, that part of the surface becomes the leading edge of a creeping drop. Within 5 min, the leading edge creeps so far forward that the following edge contracts and the drop moves toward the area of higher wettability until the isomerization of the [2]rotaxane reaches a photostationary state at the illuminated area. The partial photoisomerization of the [2]rotaxane SAM performed macroscopic work against the force of gravity by driving a diiodomethane drop up a 12° incline.

These studies represent some of the recent advancements in controlling surface hydrophobicity and hydrophilicity. Changes in the wettability of functionalized surfaces have been achieved by taking advantage of the stimulus-driven molecular motions of specifically designed and synthesized molecules as well as surface functionalization techniques that enable the proper patterning of or spacing between these functional molecules when they are attached to a surface. In addition to demonstrating photochemical,[179,182–197] electrical,[94,198–202] solvent,[203–206] temperature,[207–209] and pH[210] control over surface wettability, recent work[182,183,197] has shown that the movements of functional, surface-bound molecules can be used to drive droplets of various liquids across these surfaces in a controlled manner. Further development of these molecular machines may lead to a number of applications in the fields of micro- and nanofluidics.

11.7 Conclusions

As a science, synthetic chemistry has evolved to the point where even the most complex molecular structures are within reach. In addition, advanced analytical and computational techniques endow researchers with the ability to characterize and model the structures, interactions, and dynamics of molecular compounds to a degree that was not possible only decades ago. Such capabilities have led to numerous advances in the design and fabrication of molecular machines. By the virtues of molecular recognition, self-assembly, and bottom-up construction, functional molecular machinery has the potential to provide the next generation of advanced materials and devices where control over structure and function at the molecular level results in tremendous gains in device performance. This chapter has highlighted selected examples of molecular machines — i.e., molecular shuttles, switches, muscles, valves, rotors, and molecules capable of controlling surface wettability — that are operative in condensed phases, the solid state, or on surfaces where their dynamics are more coordinated than in a solution-phase environment. These examples, as advanced as they are, will likely provide the backbone for even more advanced materials wherein the operative, functional elements are individual molecular machines working in a cooperative manner to perform specific tasks.

Acknowledgments

We are grateful to the National Science Foundation for financial support through an ACS Division of Organic Chemistry for a graduate fellowship, sponsored by the Nelson J. Leonard ACS DOC Fellowship, sponsored by Organic Synthesis, Inc. to B.H.N., and a GK-12 program grant DGE 02-31988 fellowship to A.B.B.

References

1. Balzani, V., Gomez-Lopez, M. and Stoddart, J.F., Molecular machines, *Acc. Chem. Res.* 31, 405, 1998.
2. Balzani, V. et al., Artificial molecular machines, *Angew. Chem. Int. Ed.* 39, 3349, 2000.
3. Stoddart, J.F., Molecular machines, *Acc. Chem. Res.* 34, 410, 2001.
4. Balzani, V., Credi, A. and Venturi, M., *Molecular Devices and Machines: A Journey into the Nanoworld,* Wiley-VCH, Weinheim, 2003.
5. Flood, A.H. et al., Meccano on the nanoscale - a blueprint for making some of the world's tiniest machines, *Aust. J. Chem.* 57, 301, 2004.
6. Hawker, C.J., Living free radical polymerization: a unique technique for the preparation of controlled macromolecular architectures, *Acc. Chem. Res.* 30, 373, 1997.
7. Hawker, C.J. and Wooley, K.L., The convergence of synthetic organic and polymer chemistries, *Science* 309, 1200, 2005.
8. Dalko, P.I. and Moisan, L., Enantioselective organocatalysis, *Angew. Chem. Int. Ed.* 40, 3726, 2001.
9. Hassan, J. et al., Aryl-aryl bond formation one century after the discovery of the Ullmann reaction, *Chem. Rev.* 102, 1359, 2002.
10. Trnka, T.M. and Grubbs, R.H., The development of L2X2Ru:CHR olefin metathesis catalysts: an organometallic success story, *Acc. Chem. Res.* 34, 18, 2001.
11. Schill, G., *Catenanes, Rotaxanes and Knots,* Academic Press, New York, 1971.
12. Walba, D.M., Topological stereochemistry, *Tetrahedron* 41, 3161, 1985.
13. Cram, D.J., The design of molecular hosts, guests, and their complexes, *Angew. Chem. Int. Ed. Engl.* 27, 1009, 1988.
14. Lehn, J.M., Supramolecular chemistry — scope and perspectives molecules, supermolecules, and molecular devices, *Angew. Chem. Int. Ed. Engl.* 27, 89, 1988.
15. Pedersen, C.J., The discovery of crown ethers, *Angew. Chem. Int. Ed. Engl.* 27, 1021, 1988.
16. Busch, D.H. and Stephenson, N.A., Molecular-organization, portal to supramolecular chemistry - structural-analysis of the factors associated with molecular-organization in coordination and inclusion chemistry, including the coordination template effect, *Coord. Chem. Rev.* 100, 119, 1990.
17. Lehn, J.M., Perspectives in supramolecular chemistry - from molecular recognition towards molecular information-processing and self-organization, *Angew. Chem. Int. Ed. Engl.* 29, 1304, 1990.
18. Pople, J.A., Quantum chemical models, *Angew. Chem. Int. Ed.* 38, 1894, 1999.
19. Kohn, W., Electronic structure of matter-wave functions and density functionals, *Rev. Mod. Phys.* 71, 1253, 1999.
20. Rappe, A.K. et al., Uff, a full periodic-table force-field for molecular mechanics and molecular-dynamics simulations, *J. Am. Chem. Soc.* 114, 10024, 1992.
21. Svensson, M. et al., ONIOM: a multilayered integrated MO + MM method for geometry optimizations and single point energy predictions. A test for Diels-Alder reactions and Pt(P(t-Bu)3)2 + H2 oxidative addition, *J. Phys. Chem.* 100, 19357, 1996.
22. Car, R. and Parrinello, M., Unified approach for molecular-dynamics and density-functional theory, *Phys. Rev. Lett.* 55, 2471, 1985.
23. Ulman, A., Formation and structure of self-assembled monolayers, *Chem. Rev.* 96, 1533, 1996.
24. Bain, C.D. et al., Formation of monolayer films by the spontaneous assembly of organic thiols from solution onto gold, *J. Am. Chem. Soc.* 111, 321, 1989.
25. Love, J.C. et al., Self-assembled monolayers of thiolates on metals as a form of nanotechnology, *Chem. Rev.* 105, 1103, 2005.
26. Haller, I., Covalently attached organic monolayers on semiconductor surfaces, *J. Am. Chem. Soc.* 100, 8050, 1978.
27. Sagiv, J., Organized monolayers by adsorption. 1. Formation and structure of oleophobic mixed monolayers on solid-surfaces, *J. Am. Chem. Soc.* 102, 92, 1980.

28. Maoz, R. and Sagiv, J., On the formation and structure of self-assembling monolayers. 1. A comparative ATR-wettability study of Langmuir-Blodgett and adsorbed films on flat substrates and glass microbeads, *J. Colloid Interface Sci.* 100, 465, 1984.

29. McDonald, J.C. and Whitesides, G.M., Poly(dimethylsiloxane) as a material for fabricating microfluidic devices, *Acc. Chem. Res.* 35, 491, 2002.

30. Piner, R.D. et al., Dip-pen nanolithography, *Science* 283, 661, 1999.

31. Lua, Y.Y. et al., Chemomechanical production of submicron edge width, functionalized, ~20 μm features on silicon, *Langmuir* 19, 985, 2003.

32. Wacaser, B.A. et al., Chemomechanical surface patterning and functionalization of silicon surfaces using an atomic force microscope, *Appl. Phys. Lett.* 82, 808, 2003.

33. Hurley, P.T., Ribbe, A.E. and Buriak, J.M., Nanopatterning of alkynes on hydrogen-terminated silicon surfaces by scanning probe-induced cathodic electrografting, *J. Am. Chem. Soc.* 125, 11334, 2003.

34. Binnig, G., Quate, C.F. and Gerber, C., Atomic force microscope, *Phys. Rev. Lett.* 56, 930, 1986.

35. Binnig, G. et al., Surface studies by scanning tunneling microscopy, *Phys. Rev. Lett.* 49, 57, 1982.

36. Fadley, C.S. et al., Surface analysis and angular-distributions in x-ray photoelectron-spectroscopy, *J. Electron Spectrosc. Relat. Phenom.* 4, 93, 1974.

37. Auger, *Surface Analysis and X-Ray Photoelectron Spectroscopy,* SurfaceSpectra Ltd and IM Publications, Manchester, UK, 2003.

38. McCrackin, F.L. et al., Measurement of thickness and refractive index of very thin films and optical properties of surfaces by ellipsometry, *J. Res. Natl. Stand. Technol. A.* A 67, 363, 1963.

39. Alsnielsen, J. et al., Principles and applications of grazing-incidence x-ray and neutron-scattering from ordered molecular monolayers at the air-water-interface, *Phys. Rep. Rev. Sec. Phys. Lett.* 246, 252, 1994.

40. Good, R.J., Contact-angle, wetting, and adhesion - a critical review, *J. Adhes. Sci. Technol.* 6, 1269, 1992.

41. Atkins, P.W., *The Second Law,* Scientific American Books, New York, 1984.

42. *Guidebook to the Cytoskeletal and Motor Proteins,* 2nd ed., Oxford University Press, Oxford, UK, 1999.

43. Howard, J., *Mechanics of Motor Proteins and the Cytoskeleton,* Sinauer Associates, Sunderland, MA, 2001.

44. Feynman, R.P., There's plenty of room at the bottom, *Eng. Sci.* 23, 22, 1960.

45. Moore, G., Cramming more components into integrated circuits, *Electronics* 1965.

46. Odell, B. et al., Cyclobis(paraquat-para-phenylene) - a tetracationic multipurpose receptor, *Angew. Chem. Int. Ed. Engl.* 27, 1547, 1988.

47. Schill, G. et al., Studies on the statistical synthesis of rotaxanes, *Chemische Berichte-Recueil* 119, 2647, 1986.

48. Dietrich-Buchecker, C.O. and Sauvage, J.P., Interlocking of molecular threads - from the statistical approach to the templated synthesis of catenands, *Chem. Rev.* 87, 795, 1987.

49. Sauvage, J.P. and Dietrich-Buchecker, C.O., *Molecular Catenanes, Rotaxanes and Knots,* Wiley-VCH, Weinheim, 1999.

50. Anelli, P.L. et al., Self-assembling [2]pseudorotaxanes, *Angew. Chem. Int. Ed. Engl.* 30, 1036, 1991.

51. Loeb, S.J. and Wisner, J.A., A new motif for the self-assembly of [2]pseudorotaxanes; 1,2-bis(pyridinium)ethane axles and [24]crown-8 ether wheels, *Angew. Chem. Int. Ed.* 37, 2838, 1998.

52. Braunschweig, A.B., Northrop, B.H. and Stoddart, J.F., Structural control at the organic-solid interface, *J. Mater. Chem.* 16, 32, 2006.

53. Anderson, S., Anderson, H.L. and Sanders, J.K.M., Expanding roles for templates in synthesis, *Acc. Chem. Res.* 26, 469, 1993.

54. Cacciapaglia, R. and Mandolini, L., Catalysis by metal-ions in reactions of crown-ether substrates, *Chem. Soc. Rev.* 22, 221, 1993.

55. Hoss, R. and Vogtle, F., Template syntheses, *Angew. Chem. Int. Ed. Engl.* 33, 375, 1994.

56. Schneider, J.P. and Kelly, J.W., Templates that induce alpha-helical, beta-sheet, and loop conformations, *Chem. Rev.* 95, 2169, 1995.

57. Diederich, F. and Stang, P.J., *Templated Organic Synthesis*, Wiley-VCH, Weinheim, 1999.

58. Stoddart, J.F. and Tseng, H.R., Chemical synthesis gets a fillip from molecular recognition and self-assembly processes, *Proc. Natl. Acad. Sci. U.S.A.* 99, 4797, 2002.

59. Ashton, P.R. et al., Slippage - an alternative method for assembling [2]rotaxanes, *J. Chem. Soc. Chem. Commun.* 1269, 1993.

60. Raymo, F.M., Houk, K.N. and Stoddart, J.F., The mechanism of the slippage approach to rotaxanes. Origin of the all-or-nothing substituent effect, *J. Am. Chem. Soc.* 120, 9318, 1998.

61. Ashton, P.R. et al., Isostructural, alternately-charged receptor stacks - the inclusion complexes of hydroquinone and catechol dimethyl ethers with cyclobis(paraquat-para-phenylene), *Angew. Chem. Int. Ed. Engl.* 27, 1550, 1988.

62. Doddi, G. et al., Template effects in the self-assembly of a [2]rotaxane and a [2]pseudorotaxane with the same binding sites in the linear component, *J. Org. Chem.* 66, 4950, 2001.

63. Rowan, S.J., Cantrill, S.J. and Stoddart, J.F., Triphenylphosphonium stoppered [2]rotaxanes, *Org. Lett.* 1, 129, 1999.

64. Asakawa, M. et al., Molecular and supramolecular synthesis with dibenzofuran-containing systems, *Chem. Eur. J.* 3, 1136, 1997.

65. Ashton, P.R. et al., Simple mechanical molecular and supramolecular machines: photochemical and electrochemical control of switching processes, *Chem. Eur. J.* 3, 152, 1997.

66. Anelli, P.L. et al., Toward controllable molecular shuttles, *Chem. Eur. J.* 3, 1113, 1997.

67. Asakawa, M. et al., A chemically and electrochemically switchable [2]catenane incorporating a tetrathiafulvalene unit, *Angew. Chem. Int. Ed.* 37, 333, 1998.

68. Asakawa, M. et al., Molecular meccano, 49 - pseudorotaxanes and catenanes containing a redox-active unit derived from tetrathiafulvalene, *Eur. J. Org. Chem.*, 985, 1999.

69. Ashton, P.R. et al., A three-pole supramolecular switch, *J. Am. Chem. Soc.* 121, 3951, 1999.

70. Pease, A.R. et al., Switching devices based on interlocked molecules, *Acc. Chem. Res.* 34, 433, 2001.

71. Jeppesen, J.O. et al., Honing up a genre of amphiphilic bistable [2]rotaxanes for device settings, *Eur. J. Org. Chem.*, 196, 2005.

72. Flood, A.H. et al., The role of physical environment on molecular electromechanical switching, *Chem. Eur. J.* 10, 6558, 2004.

73. Steuerman, D.W. et al., Molecular-mechanical switch-based solid-state electrochromic devices, *Angew. Chem. Int. Ed.* 43, 6486, 2004.

74. Choi, J.W. et al., Ground-state equilibrium thermodynamics and switching kinetics of bistable [2]rotaxanes switched in solution, polymer gels, and molecular electronic devices, *Chem. Eur. J.* 12, 261, 2006.

75. Nielsen, M.B. et al., Binding studies between tetrathiafulvalene derivatives and cyclobis(paraquat-p-phenylene), *J. Org. Chem.* 66, 3559, 2001.

76. Tseng, H.R. et al., Redox-controllable amphiphilic [2]rotaxanes, *Chem. Eur. J.* 10, 155, 2004.

77. Lee, I.C. and Frank, C.W., Langmuir and Langmuir-Blodgett films of amphiphilic bistable rotaxanes, *Langmuir* 20, 5809, 2004.

78. Jang, S.S. et al., Structures and properties of self-assembled monolayers of bistable [2]rotaxanes on Au(111) surfaces from molecular dynamics simulations validated with experiment, *J. Am. Chem. Soc.* 127, 1563, 2005.

79. Jang, Y.H., Jang, S.S. and Goddard, W.A., Molecular dynamics simulation study on a monolayer of half [2]rotaxane self-assembled on Au(111), *J. Am. Chem. Soc.* 127, 4959, 2005.

80. Aviram, A. and Ratner, M.A., Molecular rectifiers, *Chem. Phys. Lett.* 29, 277, 1974.

81. Heath, J.R. and Ratner, M.A., Molecular electronics, *Phys. Today* 56, 43, 2003.

82. Flood, A.H. et al., Whence molecular electronics?, *Science* 306, 2055, 2004.

83. Jortner, J. et al., Superexchange mediated charge hopping in DNA, *J. Phys. Chem. A* 106, 7599, 2002.

84. Yaliraki, S.N., Kemp, M. and Ratner, M.A., Conductance of molecular wires: influence of molecule-electrode binding, *J. Am. Chem. Soc.* 121, 3428, 1999.

85. Hipps, K.W., Molecular electronics - it's all about contacts, *Science* 294, 536, 2001.

86. Cui, X.D. et al., Reproducible measurement of single-molecule conductivity, *Science* 294, 571, 2001.
87. Reed, M.A. et al., Conductance of a molecular junction, *Science* 278, 252, 1997.
88. Park, H. et al., Nanomechanical oscillations in a single-C-60 transistor, *Nature* 407, 57, 2000.
89. Park, J. et al., Coulomb blockade and the Kondo effect in single-atom transistors, *Nature* 417, 722, 2002.
90. Liang, W.J. et al., Kondo resonance in a single-molecule transistor, *Nature* 417, 725, 2002.
91. Yu, H.B. et al., The molecule-electrode interface in single-molecule transistors, *Angew. Chem. Int. Ed.* 42, 5706, 2003.
92. Huang, T.J. et al., Mechanical shuttling of linear motor-molecules in condensed phases on solid substrates, *Nano Lett.* 4, 2065, 2004.
93. Norgaard, K. et al., Structural evidence of mechanical shuttling in condensed monolayers of bistable rotaxane molecules, *Angew. Chem. Int. Ed.* 44, 7035, 2005.
94. Katz, E., Lioubashevsky, O. and Willner, I., Electromechanics of a redox-active rotaxane in a monolayer assembly on an electrode, *J. Am. Chem. Soc.* 126, 15520, 2004.
95. Luo, Y. et al., Two-dimensional molecular electronics circuits, *ChemPhysChem* 3, 519, 2002.
96. Juodkazis, S. et al., Reversible phase transitions in polymer gels induced by radiation forces, *Nature* 408, 178, 2000.
97. Bay, L. et al., A conducting polymer artificial muscle with 12% linear strain, *Adv. Mater.* 15, 310, 2003.
98. Raguse, B., Muller, K.H. and Wieczorek, L., Nanoparticle actuators, *Adv. Mater.* 15, 922, 2003.
99. Fritz, J. et al., Translating biomolecular recognition into nanomechanics, *Science* 288, 316, 2000.
100. Spudich, J.A. and Rock, R.S., A crossbridge too far, *Nature Cell Biol.* 4, E8, 2002.
101. Collin, J.P. et al., Shuttles and muscles: linear molecular machines based on transition metals, *Acc. Chem. Res.* 34, 477, 2001.
102. Jimenez, M.C., Dietrich-Buchecker, C. and Sauvage, J.P., Towards synthetic molecular muscles: contraction and stretching of a linear rotaxane dimer, *Angew. Chem. Int. Ed.* 39, 3284, 2000.
103. Chiu, S.H. et al., An hermaphroditic [2]daisy chain, *Chem. Comm.* 2948, 2002.
104. Cantrill, S.J. et al., Supramolecular daisy chains, *J. Org. Chem.* 66, 6857, 2001.
105. Rowan, S.J. et al., Toward daisy chain polymers: "Wittig exchange" of stoppers in 2 rotaxane monomers, *Org. Lett.* 2, 759, 2000.
106. Marsella, M.J. and Reid, R.J., Toward molecular muscles: design and synthesis of an electrically conducting poly[cyclooctatetrathiophene], *Macromolecules* 32, 5982, 1999.
107. Marsella, M.J., Classic annulenes, nonclassical applications, *Acc. Chem. Res.* 35, 944, 2002.
108. Marsella, M.J. et al., Tetra[2,3-thienylene]: a building block for single-molecule electromechanical actuators, *J. Am. Chem. Soc.* 124, 12507, 2002.
109. Marsella, M.J., Piao, G. and Tham, F.S., Expanding tetra[2,3-thienylene]-based molecular muscles to larger [4n]annulenes, *Synthesis*, 1133, 2002.
110. Hugel, T. et al., Single-molecule optomechanical cycle, *Science* 296, 1103, 2002.
111. Collier, C.P. et al., Molecular-based electronically switchable tunnel junction devices, *J. Am. Chem. Soc.* 123, 12632, 2001.
112. Luo, Y. et al., Two-dimensional molecular electronics circuits, *Chemphyschem* 3, 519, 2002.
113. Yu, H.B. et al., The molecule-electrode interface in single-molecule transistors, *Angew. Chem. Int. Ed.* 42, 5706, 2003.
114. Heath, J.R., Stoddart, J.F. and Williams, R.S., More on molecular electronics, *Science* 303, 1136, 2004.
115. Liu, Y. et al., Linear artificial molecular muscles, *J. Am. Chem. Soc.* 127, 9745, 2005.
116. Lai, C.Y. et al., A mesoporous silica nanosphere-based carrier system with chemically removable CdS nanoparticle caps for stimuli-responsive controlled release of neurotransmitters and drug molecules, *J. Am. Chem. Soc.* 125, 4451, 2003.
117. Santini, J.T., Cima, M.J. and Langer, R., A controlled-release microchip, *Nature* 397, 335, 1999.
118. Li, Y.W. et al., In vivo release from a drug delivery MEMS device, *J. Control Release* 100, 211, 2004.

119. Shawgo, R.S. et al., Repeated in vivo electrochemical activation and the biological effects of micro-electromechanical systems drug delivery device, *J. Biomed. Mater. Res.* 71A, 559, 2004.

120. Mal, N.K., Fujiwara, M. and Tanaka, Y., Photocontrolled reversible release of guest molecules from coumarin-modified mesoporous silica, *Nature* 421, 350, 2003.

121. Mal, N.K. et al., Photo-switched storage and release of guest molecules in the pore void of coumarin-modified MCM-41, *Chem. Mater.* 15, 3385, 2003.

122. Chia, S.Y. et al., Working supramolecular machines trapped in glass and mounted on a film surface, *Angew. Chem. Int. Ed.* 40, 2447, 2001.

123. Hernandez, R. et al., An operational supramolecular nanovalve, *J. Am. Chem. Soc.* 126, 3370, 2004.

124. Nguyen, T.D. et al., A reversible molecular valve, *Proc. Natl. Acad. Sci. U.S.A.* 102, 10029, 2005.

125. Han, Y.J., Stucky, G.D. and Butler, A., Mesoporous silicate sequestration and release of proteins, *J. Am. Chem. Soc.* 121, 9897, 1999.

126. Luo, Q.Z. et al., Monolithic valves for microfluidic chips based on thermoresponsive polymer gels, *Electrophoresis* 24, 3694, 2003.

127. Fu, Q. et al., Control of molecular transport through stimuli-responsive ordered mesoporous materials, *Adv. Mater.* 15, 1262, 2003.

128. Kocer, A. et al., A light-actuated nanovalve derived from a channel protein, *Science* 309, 755, 2005.

129. Kottas, G.S. et al., Artificial molecular rotors, *Chem. Rev.* 105, 1281, 2005.

130. Mislow, K., Stereochemical consequences of correlated rotation in molecular propellers, *Acc. Chem. Res.* 9, 26, 1976.

131. Berg, U. et al., Steric interplay between alkyl-groups bonded to planar frameworks, *Acc. Chem. Res.* 18, 80, 1985.

132. Iwamura, H. and Mislow, K., Stereochemical consequences of dynamic gearing, *Acc. Chem. Res.* 21, 175, 1988.

133. Oki, M., *The Chemistry of Rotational Isomers,* Springer-Verlag, Berlin, 1993.

134. *The Porphyrin Handbook,* Academic Press, New York, 2000.

135. Glass, T.E., Cooperative chemical sensing with bis-tritylacetylenes: pinwheel receptors with metal ion recognition properties, *J. Am. Chem. Soc.* 122, 4522, 2000.

136. Raker, J. and Glass, T.E., General synthetic methods for the preparation of pinwheel receptors, *Tetrahedron* 57, 10233, 2001.

137. Raker, J. and Glass, T.E., Cooperative ratiometric chemosensors: pinwheel receptors with an integrated fluorescence system, *J. Org. Chem.* 66, 6505, 2001.

138. Cram, D.J. et al., Host-guest complexation. 59. Two chiral [1.1.1]orthocyclophane units bridged by three biacetylene units as a host which binds medium-sized organic guests, *J. Am. Chem. Soc.* 113, 8909, 1991.

139. Simanek, E.E. et al., Observation of diastereomers of the hydrogen-bonded aggregate Hub(M)3center·3CA using 1H nuclear magnetic resonance spectroscopy when CA is an optically-active isocyanuric acid, *J. Org. Chem.* 62, 2619, 1997.

140. Prins, L.J. et al., Complete asymmetric induction of supramolecular chirality in a hydrogen-bonded assembly, *Nature* 398, 498, 1999.

141. Lutzen, A. et al., Encapsulation of ion-molecule complexes: second-sphere supramolecular chemistry, *J. Am. Chem. Soc.* 121, 7455, 1999.

142. Kelly, T.R., De Silva, H. and Silva, R.A., Unidirectional rotary motion in a molecular system, *Nature* 401, 150, 1999.

143. Leigh, D.A. et al., Unidirectional rotation in a mechanically interlocked molecular rotor, *Nature* 424, 174, 2003.

144. Hernandez, J.V., Kay, E.R. and Leigh, D.A., A reversible synthetic rotary molecular motor, *Science* 306, 1532, 2004.

145. Joachim, C. and Gimzewski, J.K., Single molecular rotor at the nanoscale, in: *Molecular Machines and Motors,* Springer-Verlag, Berlin, 2001, p. 1.

146. Mandl, C.P. and Konig, B., Chemistry in motion - unidirectional rotating molecular motors, *Angew. Chem. Int. Ed.* 43, 1622, 2004.

147. Dominguez, Z. et al., Molecular compasses and gyroscopes. I. Expedient synthesis and solid state dynamics of an open rotor with a bis(triarylmethyl) frame, *J. Am. Chem. Soc.* 124, 2398, 2002.

148. Godinez, C.E., Zepeda, G. and Garcia-Garibay, M.A., Molecular compasses and gyroscopes. II. Synthesis and characterization of molecular rotors with axially substituted bis 2-(9-triptycyl)ethynyl arenes, *J. Am. Chem. Soc.* 124, 4701, 2002.

149. Dominguez, Z. et al., Molecular compasses and gyroscopes. - III. Dynamics of a phenylene rotor and clathrated benzene in a slipping-gear crystal lattice, *J. Am. Chem. Soc.* 124, 7719, 2002.

150. Shima, T., Hampel, F. and Gladysz, J.A., Molecular gyroscopes: (Fe(CO)(3){ and {Fe(CO)(2)(NO)}(+) rotators encased in three-spoke stators; facile assembly by alkene metatheses, *Angew. Chem. Int. Ed.* 43, 5537, 2004.

151. Karlen, S.D. and Garcia-Garibay, M.A., Highlighting gyroscopic motion in crystals in 13C CPMAS spectra by specific isotopic substitution and restricted cross polarization, *Chem. Commun.* 189, 2005.

152. Khuong, T.A.V. et al., Molecular compasses and gyroscopes: engineering molecular crystals with fast internal rotation, *Cryst. Growth Des.* 4, 15, 2004.

153. Godinez, C.E. et al., Molecular crystals with moving parts: synthesis, characterization, and crystal packing of molecular gyroscopes with methyl-substituted triptycyl frames, *J. Org. Chem.* 69, 1652, 2004.

154. Hirota, E., Saito, S. and Endo, Y., Barrier to internal-rotation in ethane from the microwave-spectrum of CH3CHD2, *J. Chem. Phys.* 71, 1183, 1979.

155. Dominguez, Z. et al., Molecular compasses and gyroscopes with polar rotors: synthesis and characterization of crystalline forms, *J. Am. Chem. Soc.* 125, 8827, 2003.

156. Horansky, R.D. et al., Dielectric response of a dipolar molecular rotor crystal, *Phys. Rev. B* 72, 2005.

157. Koumura, N. et al., Light-driven monodirectional molecular rotor, *Nature* 401, 152, 1999.

158. Koumura, N. et al., Light-driven molecular rotor: unidirectional rotation controlled by a single stereogenic center, *J. Am. Chem. Soc.* 122, 12005, 2000.

159. Feringa, B.L., In control of motion: from molecular switches to molecular motors, *Acc. Chem. Res.* 34, 504, 2001.

160. Koumura, N. et al., Second generation light-driven molecular motors. unidirectional rotation controlled by a single stereogenic center with near-perfect photoequilibria and acceleration of the speed of rotation by structural modification, *J. Am. Chem. Soc.* 124, 5037, 2002.

161. ter Wiel, M.K.J. et al., Increased speed of rotation for the smallest light-driven molecular motor, *J. Am. Chem. Soc.* 125, 15076, 2003.

162. van Delden, R.A. et al., Unidirectional rotary motion in a liquid crystalline environment: color tuning by a molecular motor, *Proc. Natl. Acad. Sci. U.S.A.* 99, 4945, 2002.

163. van Delden, R.A. et al., Unidirectional molecular motor on a gold surface, *Nature* 437, 1337, 2005.

164. Feringa, B.L. et al., Chiroptical molecular switches, *Chem. Rev.* 100, 1789, 2000.

165. Fletcher, S.P. et al., A reversible, unidirectional molecular rotary motor driven by chemical energy, *Science* 310, 80, 2005.

166. Solladie, G. and Zimmermann, R.G., Liquid-crystals — a tool for studies on chirality, *Angew. Chem. Int. Ed. Engl.* 23, 348, 1984.

167. Clarke, L.I. et al., The dielectric response of chloromethylsilyl and dichloromethylsilyl dipolar rotors on fused silica surfaces, *Nanotechnology* 13, 533, 2002.

168. Horinek, D. and Michl, J., Molecular dynamics simulation of an electric field driven dipolar molecular rotor attached to a quartz glass surface, *J. Am. Chem. Soc.* 125, 11900, 2003.

169. Zheng, X.L. et al., Dipolar and nonpolar altitudinal molecular rotors mounted on an Au(111) surface, *J. Am. Chem. Soc.* 126, 4540, 2004.

170. Horinek, D. and Michl, J., Surface-mounted altitudinal molecular rotors in alternating electric field: single-molecule parametric oscillator molecular dynamics, *Proc. Natl. Acad. Sci. U.S.A.* 102, 14175, 2005.

171. Akiyama, R., Matsumoto, T. and Kawai, T., Capacitance of a molecular overlayer on the silicon surface measured by scanning tunneling microscopy, *Phys. Rev. B* 62, 2034, 2000.

172. Lahann, J. and Langer, R., Smart materials with dynamically controllable surfaces, *MRS Bulletin* 30, 185, 2005.

173. Liu, Y. et al., Controlled switchable surface, *Chem. Eur. J.* 11, 2622, 2005.

174. Wang, R. et al., Light-induced amphiphilic surfaces, *Nature* 388, 431, 1997.

175. Blossey, R., Self-cleaning surfaces - virtual realities, *Nature Mater.* 2, 301, 2003.

176. Chaudhury, M.K. and Whitesides, G.M., Correlation between surface free-energy and surface constitution, *Science* 255, 1230, 1992.

177. Abbott, N.L., Folkers, J.P. and Whitesides, G.M., Manipulation of the wettability of surfaces on the 0.1-micrometer to 1-micrometer scale through micromachining and molecular self-assembly, *Science* 257, 1380, 1992.

178. Zhang, J.L. et al., Reversible superhydrophobicity to superhydrophilicity transition by extending and unloading an elastic polyamide, *Macromol. Rapid Commun.* 26, 477, 2005.

179. Rosario, R. et al., Lotus effect amplifies light-induced contact angle switching, *J. Phys. Chem. B* 108, 12640, 2004.

180. Nuzzo, R.G. and Allara, D.L., Adsorption of bifunctional organic disulfides on gold surfaces, *J. Am. Chem. Soc.* 105, 4481, 1983.

181. Lee, S.W. and Laibinis, P.E., Protein-resistant coatings for glass and metal oxide surfaces derived from oligo(ethylene glycol)-terminated alkyltrichlorosilanes, *Biomaterials* 19, 1669, 1998.

182. Ichimura, K., Oh, S.K. and Nakagawa, M., Light-driven motion of liquids on a photoresponsive surface, *Science* 288, 1624, 2000.

183. Oh, S.K., Nakagawa, M. and Ichimura, K., Photocontrol of liquid motion on an azobenzene monolayer, *J. Mater. Chem.* 12, 2262, 2002.

184. Feng, C.L. et al., Reversible wettability of photoresponsive fluorine-containing azobenzene polymer in Langmuir-Blodgett films, *Langmuir* 17, 4593, 2001.

185. Feng, C.L. et al., Reversible light-induced wettability of fluorine-containing azobenzene-derived Langmuir-Blodgett films, *Surf Inter Anal* 32, 121, 2001.

186. Raduge, C. et al., Controlling wettability by light: illuminating the molecular mechanism, *Euro. Phys. J. E* 10, 103, 2003.

187. Jiang, W.H. et al., Photo-switched wettability on an electrostatic self-assembly azobenzene monolayer, *Chem. Comm.* 3550, 2005.

188. Delorme, N. et al., Azobenzene-containing monolayer with photoswitchable wettability, *Langmuir* 21, 12278, 2005.

189. Hamelmann, F. et al., Light-stimulated switching of azobenzene-containing self-assembled monolayers, *Appl. Surface Sci.* 222, 1, 2004.

190. Siewierski, L.M. et al., Photoresponsive monolayers containing in-chain azobenzene, *Langmuir* 12, 5838, 1996.

191. Rosario, R. et al., Photon-modulated wettability changes on spiropyran-coated surfaces, *Langmuir* 18, 8062, 2002.

192. Ishihara, K. et al., Photoinduced change in wettability and binding ability of azoaromatic polymers, *J. Appl. Polym. Sci.* 27, 239, 1982.

193. Abbott, S. et al., Reversible wettability of photoresponsive pyrimidine-coated surfaces, *Langmuir* 15, 8923, 1999.

194. Cooper, C.G.F. et al., Non-covalent assembly of a photoswitchable surface, *J. Am. Chem. Soc.* 126, 1032, 2004.

195. Irie, M. and Iga, R., Photoresponsive polymers - reversible wettability change of poly butyl methacrylate-co-alpha-(2-hydroxyphenyl)-alpha-(4-vinylphenyl)benzyl alcohol, *Macromol. Rapid Commun.* 8, 569, 1987.

196. Lake, N., Ralston, J. and Reynolds, G., Light-induced surface wettability of a tethered DNA base, *Langmuir* 21, 11922, 2005.

197. Berna, J. et al., Macroscopic transport by synthetic molecular machines, *Nature Mater.* 4, 704, 2005.
198. Sondaghuethorst, J.A.M. and Fokkink, L.G.J., Potential-dependent wetting of electroactive ferrocene-terminated alkanethiolate monolayers on gold, *Langmuir* 10, 4380, 1994.
199. Abbott, N.L. and Whitesides, G.M., Potential-dependent wetting of aqueous-solutions on self-assembled monolayers formed from 15-(ferrocenylcarbonyl)pentadecanethiol on gold, *Langmuir* 10, 1493, 1994.
200. Lahann, J. et al., A reversibly switching surface, *Science* 299, 371, 2003.
201. Wang, X.M. et al., Potential-controlled molecular machinery of bipyridinium monolayer-functionalized surfaces: an electrochemical and contact angle analysis, *Chem. Comm.* 1542, 2003.
202. Liu, Y. et al., Controlled protein assembly on a switchable surface, *Chem. Comm.* 1194, 2004.
203. Julthongpiput, D. et al., Y-shaped amphiphilic brushes with switchable micellar surface structures, *J. Am. Chem. Soc.* 125, 15912, 2003.
204. Julthongpiput, D. et al., Y-shaped polymer brushes: nanoscale switchable surfaces, *Langmuir* 19, 7832, 2003.
205. Zhao, B. et al., Nanopattern formation from tethered PS-b-PMMA brushes upon treatment with selective solvents, *J. Am. Chem. Soc.* 122, 2407, 2000.
206. Motornov, M. et al., Reversible tuning of wetting behavior of polymer surface with responsive polymer brushes, *Langmuir* 19, 8077, 2003.
207. Takei, Y.G. et al., Dynamic contact-angle measurement of temperature-responsive surface-properties for poly(N-isopropylacrylamide) grafted surfaces, *Macromolecules* 27, 6163, 1994.
208. Crevoisier, G.B. et al., Switchable tackiness and wettability of a liquid crystalline polymer, *Science* 285, 1246, 1999.
209. Sun, T.L. et al., Reversible switching between superhydrophilicity and superhydrophobicity, *Angew. Chem. Int. Ed.* 43, 357, 2004.
210. Wilson, M.D. and Whitesides, G.M., The anthranilate amide of polyethylene carboxylic-acid shows an exceptionally large change with pH in its wettability by water, *J. Am. Chem. Soc.* 110, 8718, 1988.
211. Rau, H., Photochemistry and photophysics, in: *Photoisomerization of Azobenzenes*, Rabek, J.F., ed. CRC Press, Boca Raton, FL, 1990, p. 119.
212. Pei, Y. and Ma, J., Electric field induced switching behaviors of monolayer-modified silicon surfaces: surface designs and molecular dynamics simulations, *J. Am. Chem. Soc.* 127, 6802, 2005.
213. Daniel, S., Chaudhury, M.K. and Chen, J.C., Past drop movements resulting from the phase change on a gradient surface, *Science* 291, 633, 2001.
214. Grunze, M., Surface science - driven liquids, *Science* 283, 41, 1999.
215. Wasan, D.T., Nikolov, A.D. and Brenner, H., Fluid dynamics - droplets speeding on surfaces, *Science* 291, 605, 2001.
216. Cecchet, F. et al., Structural, electrochemical, and photophysical properties of a molecular shuttle attached to an acid-terminated self-assembled monolayer, *J. Phys. Chem. B* 108, 15192, 2004.

3

Molecular Electronics Devices

12   Molecular Conductance Junctions: A Theory and Modeling Progress Report
Vladimiro Mujica, Mark A. Ratner 12-1
Introduction • Experimental Techniques for Molecular Junction Transport
• Coherent Transport: The Generalized Landauer Formula • Gating and Control of
Junctions: Diodes and Triodes • The Onset of Inelasticity • Molecular Junction
Conductance and Nonadiabatic Electron Transfer • Onset of Incoherence and
Hopping Transport • Advanced Theoretical Challenges • Remarks

13   Modeling Electronics at the Nanoscale *Narayan R. Aluru, Jean-Pierre Leburton,*
William McMahon, Umberto Ravaioli, Slava V. Rotkin, Martin Staedele,
Trudy van der Straaten, Blair R. Tuttle, Karl Hess 13-1
Introduction • Nanostructure Studies of the Si-SiO$_2$ Interface • Modeling of
Quantum Dots and Artificial Atoms • Multiscale Theory and Modeling of
Carbon Nanotube Nanoelectromechanical Systems • Simulation of
Ionic Channels • Conclusions

14   Resistance of a Molecule *Magnus Paulsson, Ferdows Zahid, Supriyo Datta* 14-1
Introduction • Qualitative Discussion • Coulomb Blockade?
• Nonequilibrium Green's Function (NEGF) Formalism • An Example: Quantum
Point Contact (QPC) • Concluding Remarks • MATLAB® Codes

12

Molecular Conductance Junctions: A Theory and Modeling Progress Report

12.1 Introduction **12**-1
12.2 Experimental Techniques for Molecular
      Junction Transport **12**-3
12.3 Coherent Transport: The Generalized
      Landauer Formula..**12**-4
      Length Scales in Mesoscopic Systems • Landauer
      Transmission Model of Conductance and Its Extensions
      to Molecular Wires
12.4 Gating and Control of Junctions: Diodes
      and Triodes..**12**-10
12.5 The Onset of Inelasticity **12**-11
12.6 Molecular Junction Conductance and Nonadiabatic
      Electron Transfer **12**-13
12.7 Onset of Incoherence and Hopping Transport...........**12**-14
12.8 Advanced Theoretical Challenges **12**-17
      Electrostatic Potentials and Image Charges • Geometry
      • Reactions • Chirality Effects • Photo-assisted Transitions
      • Dynamical Analysis
12.9 Remarks **12**-22
Acknowledgments...**12**-23
References..**12**-23

Vladimiro Mujica
Universidad Central de Venezuela

Mark A. Ratner
Northwestern University

12.1 Introduction

The disciplinary area of molecular electronics is concerned with electronic phenomena that are controlled by the molecular organization of matter. The two simplest thematic parts of this research world are molecular wire junctions (in which an electrical current between two electrodes is modulated by a single molecule or a few molecules that form a junction) and molecular optoelectronics (in which a molecule or complex multimolecular structure acts either as a light-emitting diode producing light from electrical input or as a photo-conversion device producing electrical output from incident photons). Several of the fundamental problems and attributes of these systems are the same: they depend on electrical transduction between a continuum electrode and a discrete molecular structure. This fundamental theoretical problem in molecular electronics will be our entire focus in this chapter.

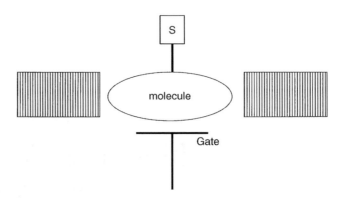

FIGURE 12.1 Schematic of a molecular junction consisting of a single molecule suspended between source and drain macroscopic electrodes. The transport through the molecule can be modulated by tuning the electrostatic potential through the gate or by bringing up another molecule or ionic species, denoted as S, that interacts covalently with the molecule.

Because the molecular structures are characterized by discrete energy levels and well-defined wave functions, the problem of transport in molecular wire junctions is inherently quantum mechanical. Therefore, in this chapter we will discuss the simplest quantal formulations of conductance properties in molecular wire junctions, how calculations of such conductance might be completed, and how they compare with the available experimental data.

Molecular electronics[1–10] deals with two kinds of materials. In the discrete molecular wires themselves, the characterization is most simply done in terms of a molecular wave function; in independent particle models such as Huckel, Kohn–Sham, or Hartree–Fock, these can be characterized in terms of single delocalized molecular orbitals. The electrodes, on the other hand, are continuum structures. Simple metallic electrodes can be characterized by their band structures, and semiconductor electrodes involve multiple band structures and band bending. All of these phenomena are crucial in determining the conductance, and all must therefore be included in a simple conductance formulation.

To construct a molecular wire junction such as that schematically shown in Figure 12.1, it is necessary to make the molecules, to assemble the junction, and to measure the transport. The issues of assembly and synthesis are crucial to the whole field of molecular electronics, but we will not deal with them in this chapter. Our aim here is a conceptual review of how to calculate and understand molecular conductance in wire junctions and some of the mechanisms that underlie it. As Figure 12.1 makes clear, there are several problems involved in moving charge between the macroscopic anode and macroscopic cathode. These processes include charge injection at the interface, charge transport through the molecule, modulation of charge by external fields, and the nature of the electrostatic potential acting at finite voltages.

Molecular electronics is a very young field, and neither the experimental generalities nor the appropriate computational/theoretical formulation is entirely agreed upon. The viewpoint taken here is that the role of theory and modeling is to aid the interpretation and prediction of experiments. Mechanistic insight is crucial and will be our focus. Nevertheless, the development of accurate quantitative methodology to compare with accurate quantitative experimental data must be one of the aims of the entire endeavor of molecular electronics. Only under those conditions can this subject become a well-grounded science as opposed to a series of beautiful episodic measurements and interpretations.

These topics are sufficiently important to have been surveyed several times. References 1–17 are some of the overviews and surveys that have appeared in the recent literature; other articles in this book provide descriptions of the experimental situation, device applications, and preparation methodologies.

The current chapter is organized as follows: Section 12.2 recaps very briefly some of the relevant experiments in simple molecular wire junctions. Section 12.3 is devoted to the simplest behavior — that of coherent conductance as formulated by Landauer — and its generalization to molecular wire junctions.

Section 12.4 briefly discusses electronic control and gating processes. Section 12.5 deals with the onset of inelastic behavior and how that modulates the current in tunneling junctions. Section 12.6 is devoted to the relationship between intramolecular nonadiabatic electron transfer and molecular wire junction conductance. The transition from coherent to incoherent behavior and the appearance of hopping mechanisms are discussed in Section 12.7. Section 12.8 lists some issues that have been so far unaddressed in the understanding of molecular wire junctions. Finally, in Section 12.9 we discuss very briefly the possible development of a reliable structure/function interpretation of molecular transport junctions.

12.2   Experimental Techniques for Molecular Junction Transport

Because this entire book is devoted to subjects involving molecular and nanoscale electronics, mostly with an experimental focus, we limit ourselves here to a very brief discussion of the measurements that have been made and are being made, and the challenges that they pose to computation.

Molecular electronics really began with the pioneering work of Kuhn[18] and collaborators, who measured current through molecular adlayer films. This work raised very significant questions concerning transport in molecular junctions, but the methodology both for their preparation and for their measurement was quite crude by contemporary standards.

The two great experimental advances that heralded the growth of molecular electronics as a science were the development of self-assembly methodologies[2,19] for preparation of molecular nanostructures and of scanning probe methodologies both for their preparation and measurement.[20] These are reviewed extensively elsewhere; suffice it to say here that without such techniques, the subject would never have developed. This is not to say that the techniques are ideal: self-assembled monolayers give monolayers on surfaces, not isolated molecules. It is entirely clear (changing work functions of the metal, interaction among wires, mutual polarization of the surface, long-range electrostatic interactions among charges on different wires) that transport through a single molecule within a molecular adlayer structure is not the same as transport through a single isolated molecular junction. This issue has not been very well attacked either experimentally or computationally, and it is one to which the field will have to return if the understanding of junctions is to be complete.

Similarly, the use of scanning probe methodologies for measuring transport is not entirely straightforward. If either a scanning atomic force microscope (AFM) tip or a scanning tunneling microscope (STM) tip directly contacts the film, the junction is very poorly defined geometrically; and such geometric effects may completely overwhelm the intrinsic transport that one wishes to measure. Perturbation in the structure caused by this scanning probe tip can also be important. The issue of geometries in scanning probe junctions is one that has not been effectively addressed and will be dealt with briefly in Sections 12.3 and 12.8.

It is also important to distinguish soft molecular structures from hard molecular structures. In hard molecular transport junctions, particularly those based on nanowires and carbon nanotubes,[21–23] the problems are intrinsically different from those in soft junctions using traditional organic molecules. Table 12.1 sketches some of the differences. Using nanotubes, tremendous progress has been made in addressing important issues such as the limit of quantized conductance, multiple channels, the sensitivity of the measurements to the nature of the contacts, field effects, modulations, and a panoply of other

TABLE 12.1 Junctions in Different Limits

	Soft (Molecular) Junction	Hard (Semiconductor) Junction
Typical species	p-benzene dithiol, alkene thiols, DNA	Carbon nanotubes, InP nanowires
Atomic count	Accurate to one atom	Accurate to 10^2–10^4 atoms
Interfacial binding	Coordinate covalent	Metallic or physisorptive
Orbital structure	Discrete molecular orbitals	Dense orbital structure
Transport mechanism	Nonresonant tunneling, resonant tunneling, hopping	Ballistic transport, quantized conductance

fascinating topics.[21–24] The current chapter will not discuss these structures; its scope will be limited to soft molecular junctions.

The simplest issue, then, is that of transport through a junction containing one molecule, as schematically illustrated in Figures 12.1 and 12.3. Several measurements, notably molecular break junctions,[25,26] may well measure current through single molecules. Such measurements will be easiest to interpret, and indeed some of the early results of Reed's work using molecular break junctions have generated the most computational effort in the single molecule transport situation — the primary focus in this chapter.[27–29]

More interesting and complex behavior certainly will arise in molecular junctions. Otherwise, they would be interesting only as rather poor interconnects that are dominated by their surface properties and, as such, much less interesting from the point of view of applications and technology. Molecular junctions can exhibit fascinating properties. These include applications as diodes (molecular rectifiers)[30–33] and triodes (molecular field-effect transistors).[34–36] There are also significant applications of molecular wires as sensors, as switches, and even as logic structures.[37–38] Molecules have been critical for all development of organic light-emitting diodes, and the general area of optoelectronics and photo conversion is clearly a significant one involving molecular wire junctions.[38–40]

To keep this review at a reasonable length, we have chosen to focus directly on simple single wire interconnect structures. These have been studied using a variety of techniques including STM,[41] AFM,[42] nanopore,[32] break junction,[32] crossed wires,[43] ESDIAD,[44] and direct nanoscale measurements. While all of these measurements raise some issues of interpretation, they are the raw data with which it is best to compare computations of molecular junction transport.

One important remark should be made here and is amplified a bit in Section 12.6: the fundamental process of molecular junction transport consists of electron transfer. It is therefore closely related both to measurements and models of the intramolecular nonadiabatic electron transfer phenomena in molecules and to measurements on microwave conductivity in isolated molecular structures. The huge difference between those measurements, which are made directly on molecules, and measurements on molecules in junctions is that in the junction structures the electrodes are crucial to the structure, the performance, and the conductance spectrum. This is a major focus of this chapter, and the most important lesson from this work is a straightforward one: transport in molecular tunnel junctions is not a measurement of the conductance of a molecular wire but rather a measurement of the conductance of the molecular wire junction that includes the electrodes. This is most obviously clear from the formal analysis involving the overall system, given in the next section.

12.3 Coherent Transport: The Generalized Landauer Formula

The simplest mesoscopic circuit involving a molecular wire consists of a molecule coupled to two or more nanoelectrodes, or contacts, which in turn are connected to an external voltage source. This basic circuit, where charge and energy transport occur, has been the subject of a very intensive research effort in the last 10 years. It can be used as a model for molecular imaging via the STM, for break junctions, and for molecular conductance.[45–52]

A remarkable feature of the research effort in the mesoscopic domain is that it has brought up the existence of a common theoretical framework for transport in very different systems: quantum dots, tunneling junctions, electrostatic pores, STM and AFM imaging, and break junctions. Concepts such as conductance, once ascribed only to bulk matter, can now be extended to a single molecular junction, revealing a deep connection between the models used in intramolecular electron transfer (ET) and those for transport.

12.3.1 Length Scales in Mesoscopic Systems

The word *mesoscopic* was coined by van Kampen [53] to refer to a system whose size is intermediate between microscopic and macroscopic. Microscopic systems are studied using either quantum mechanics or semi-classical approximations to it. Usually a system approaches macroscopic behavior once its size is much larger than all relevant correlation lengths ξ characterizing quantum phenomena.[54] Important systems for electronics are usually *hybrid* — that is, they involve both microscopic and macroscopic parts. This property is, to a large extent, responsible for some of the most interesting and challenging aspects of both fabrication and understanding of such devices.

In its original meaning, mesoscopic systems were studied to understand the macroscopic limit and how it is achieved by building up larger and larger clusters to go from the molecule to the bulk. In current research, many novel phenomena exist that are intrinsic to mesoscopic systems. An example is provided by very small conducting systems that show coherent quantum transport; that is, an electron can propagate across the whole system without inelastic scattering, thereby preserving phase memory.[54]

Roughly speaking, for a system of size L, there are four important lengths, in increasing magnitude, that characterize the system and define the type of transport: the Fermi wavelength λ_F, the electronic mean free path I_e, the coherence length ξ, and the localization length L_ϕ. For L between λ_F and L_ϕ the system is considered to be mesoscopic and the transport mechanism goes from ballistic to diffusive and then localized, depending on whether $\lambda_F \leq L \leq I_e$, $I_e \leq L \leq \xi$, or $\xi \leq L \leq L_\phi$, respectively. In all these regimes, transport must be described quantum mechanically. For $L > L_\phi$, the system is considered to be macroscopic; and a Boltzmann equation is a good approximation to the transport equation.[54]

We will concentrate on electron/hole transport in mesoscopic systems. For hybrid systems, both the contacts and the molecular intervening medium must be taken into account within a single framework. Contacts can be either metallic or semiconductors, and they bring in the continuum of electronic states that is necessary to have conduction. Molecular systems by themselves show gaps distributed in the whole energy spectrum that would, in principle, preclude transport.

From size considerations only, it is clear that transport through a mesoscopic system will involve the three transport mechanisms mentioned above. But a molecule has an electronic structure of its own that must be taken into account for the description of the current. This means that considering electron transport through an individual molecular junction brings in a much richer structure because of the discrete nature of the energy spectrum and the fact that, upon charging, a molecule virtually becomes a new chemical species.[50–51,55–57]

12.3.2 Landauer Transmission Model of Conductance and Its Extensions to Molecular Wires

Landauer[54,58–61] provided the most influential work in the study of charge transport in mesoscopic systems. The basic idea in Landauer's approach is to associate conductance with transmission through the inter-electrode region. This approach is especially suitable for mesoscopic transport and requires the use of the full quantum mechanical machinery, because now the carriers can have a coherent history within the sample. The physics of Landauer's model is essentially a generalization of models for quantum tunneling along several paths. These paths interfere according to the rules of quantum mechanics; and in orbital models of a molecule, they arise from molecular orbitals.

Conductance, g, is defined by the simple relation:

$$g = \frac{j}{V},$$ (12.1)

where j is the net current flow and V is the external voltage. For an ideal conducting channel, with no irregularities or scattering mechanisms along its length, and with the additional assumption that the tube is narrow enough so that only the lowest of the transverse eigenstates in the channel has its energy below the Fermi level of the contacts, the resulting conductance is given by the quantum of conductance:

$$g = \frac{e^2}{\pi\hbar}.$$ (12.2)

This is the conductance of an ideal ballistic channel with transmission equal to unity. The potential drop, associated with the resistance $r = g^{-1}$, occurs at the connections to the contacts. Therefore, Equation (12.2) specifies a limiting value of the contact resistance, equal to 12.8 kΩ, that is known as the *Sharvin limit*.[62] This alone is a most remarkable result that corresponds to the impossibility of totally short-circuiting a device operating through quantum channels — an event that for a classical circuit would correspond to zero resistance. Equation (12.2) also expresses the fact that conductance is quantized, a result that has been confirmed experimentally.

If a scattering obstacle, for instance a molecule, is inserted into the channel, the transmission probability T generally becomes smaller than one and the conductance is reduced accordingly to

$$g = \frac{e^2 T}{\pi \hbar}. \tag{12.3}$$

The Landauer formalism, in the simple version presented here, assumes that both the temperature and the voltage are very small and there are no incoherent processes involved in the transport. These constraints can be released in generalized versions of the theory.[60–67]

The resistance associated to Equation (12.3) can be written as[68]:

$$r = \frac{\pi \hbar}{e^2}\left[1 + \frac{1-T}{T}\right] = \frac{\pi \hbar}{e^2}\left[1 + \frac{R}{T}\right], \tag{12.4}$$

where R is the reflection coefficient. Equation (12.4) makes it explicit that, in the simple case considered here, the total resistance associated to a channel consists of a contact and a molecular resistance. Under some conditions the molecular resistance can be brought to zero — a situation that is entirely equivalent to a transparent barrier in tunneling phenomena, leaving only the contact resistance associated to the molecule contact interface.

A very convenient way to adapt Landauer's formalism to molecular circuits is the use of stationary scattering theory to compute the transmission coefficient associated with each channel.[45,46] This provides a description of the stationary current driven by an external voltage V through a molecular circuit. The relevant quantity is the total transition probability per unit time that an electron undergoes a transition from one electrode to the other as time evolves from the remote past to the remote future. This quantity, integrated over the whole energy range accessible for the tunneling electrons and multiplied by the electron charge, is the stationary current in the circuit[33]:

$$I(V) = \frac{2e}{h}\int_{\mu_L}^{\mu_R} d\varepsilon\, T(\varepsilon, \phi(V))\,. \tag{12.5}$$

where $T(\varepsilon, \phi(V))$ is a dimensionless transmission function; μ_L and μ_R are the chemical potential on the left and right electrodes corresponding to a voltage difference V, i.e., $\mu_L - \mu_R = eV$; and $\phi(V)$ is a function that describes the spatial profile of the voltage drop across the interface, which must be determined self-consistently for the calculation of the current.

The transmission function is given explicitly by[33]

$$\text{Tr}[G(\varepsilon, \phi(V))\Delta_L(\varepsilon, \phi(V))G^T(\varepsilon, \phi(V))\Delta_R(\varepsilon, \phi(V))], \tag{12.6}$$

with $\Delta_{L(R)}$ being the spectral density corresponding to each contact, and $G(\varepsilon, V)$, the Green function matrix, defined by

$$G(z) = (z\mathbf{S} - \mathbf{H}_M - \Sigma)^{-1} \tag{12.7}$$

where **S** and Σ are the overlap and self-energy matrices, respectively, and $\mathbf{H}_M$ is the molecular Hamiltonian. The self-energy matrix is defined by

$$\Sigma(z) = \mathbf{V}_{ME}(z - \mathbf{H}_E)^{-1}\mathbf{V}_{EM}, \tag{12.8}$$

where $\mathbf{V}_{ME}$ is the matrix specifying the coupling between the molecule and the electrodes, whereas $\mathbf{H}_E$ is the Hamiltonian matrix or the electrodes. Δ_L is related to the self-energy by

$$\Delta_L = -\frac{1}{\pi}\mathrm{Im}(\Sigma_L), \tag{12.9}$$

and a similar expression holds for the right electrode.

Using this formalism we can derive an expression for the differential conductance, $g = \partial I/\partial V$, that can be used in both the linear and nonlinear voltage regimes and that includes the effect of the reservoirs in a more complete fashion. Equation (12.5) has been used to compute the conductance of real molecules, to study nonlinear effects, and to include the local variation of the external field through the use of a self-consistent procedure to solve Poisson and Schrödinger equations simultaneously[70] or via density functional theory.[29] Green's function techniques of the type associated with Equation (12.5) constitute a powerful tool to describe an open system where the total Hamiltonian has been partitioned so that the reservoirs modify the molecular Hamiltonian through the self-energy term that represents the influence of the contacts. This approach takes its most general form in the Keldysh formalism.[68,71–72]

The use of the scattering description permits a natural interpretation of molecular orbitals, or delocalized molecular states, as the analogue of the transverse channels in electrostatic junctions. The conductance associated with a molecular channel can be much lower than the corresponding one for a ballistic channel because it depends both on the specific nature of the chemical bond between the wire and the metal surface and on the delocalization in the molecular orbital. This corresponds to the nonresonant regime.[46,57] The resonant regime, on the other hand, is entirely analogous to the ballistic behavior, and the wire-electrode conductance becomes equal to the Sharvin conductance multiplied by the number of degenerate molecular channels. These ideas are fundamental to the understanding of the I–V curves involving molecular wires.[46,57,73–84]

Some of the concepts discussed in terms of the Landauer model are illustrated in Figures 12.2, 12.3, 12.4, and 12.5. Figure 12.2 demonstrates the simplest form of the Landauer formula — the only contribution to the transport comes from scattered waves that proceed through the molecular junction constriction, while back-scattered waves do not contribute. This is precisely in accord with the Landauer formulation of Equation (12.3), where the transmission probability T describes the probability for a wave to be scattered forward through the junction.

Figure 12.3 schematically shows transport through a molecular junction. Figure 12.3A interprets the results of Equation (12.6): the spectral density Δ describes the interface between the electrode and the molecule, while the Green's function G describes transport through the molecule between the interfaces. Generally, one cannot limit oneself only to the molecular component — the so-called *extended molecule*

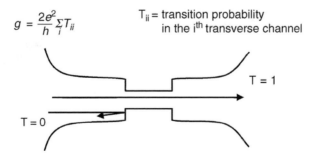

FIGURE 12.2 Illustration of the Landauer coherent conductance picture. Two waves are incident upon a molecular constriction — the wave that passes through has a transition probability of 1, and the wave that is back-scattered has a transition probability of zero. The overall conductance is the atomic unit of conductance times the sum of the transition probabilities.

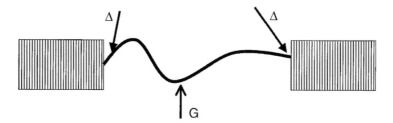

FIGURE 12.3A Physical situation underlying the generalized coherent Landauer transport for a molecular wire junction. The molecule itself, indicated by the solid line, interacts with the electrodes at the left and the right. The spectral densities denoted Δ describe interaction at the interfaces, while the delocalization over the molecule is described by Green's function G.

(Figure 12.3B) structure[85] keeps a few metallic atoms in its complete orbital representation to describe the binding of the molecule to the metallic interface.

Figure 12.4 shows characteristic conductances as a function of voltage for the parabenzene dithiol bridge between gold electrodes. The two graphs show the measurement reported by Reed's group using molecular break junctions[25] and the current/voltage characteristics calculated using self-consistent density functional theory by Xue.[86] Note that the voltage corresponding to the peak in the conductance is roughly the same in the calculated and experimental data, but that the conductance as measured is roughly a factor of 600 smaller than the conductance as calculated. We believe that this has to do with the nature of the electrode interfaces in Reed's break junction measurements. Work by Schön and collaborators[87] indeed suggests that with optimal interfacial transport, conductance values quite similar to those calculated here are observed. Note also the symmetric nature of this computed and measured transport: this is because the junction is at least putatively totally symmetric. The first peak corresponds to transport through the occupied (hole-type super exchange) levels in the molecular structure.

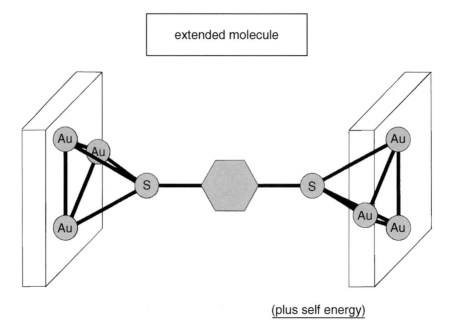

FIGURE 12.3B Schematic of the extended molecule used in the calculation of coherent transport through the benzene dithiol structure. The six gold atoms are treated as valence species along with the benzene dithiol. The bulk electrodes appear as a self-energy term.

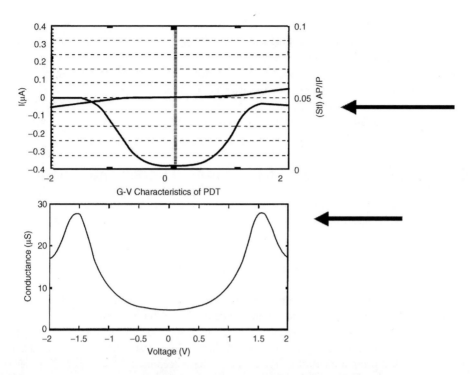

FIGURE 12.4 Comparison of the measured conductance through the benzene dithiol structure of Figure 12.3B as measured by Reed and collaborators (see Figure 12.9 caption) using break junctions (*upper curve*) with that calculated by Xue (A. Xue and M.A. Ratner, to be published) using the extended molecule scheme in a nonequilibrium Green's function formulation with full density functional treatment. The maxima occur at roughly the same voltage, near 1.5 volts, but the measured reported conductance is nearly a factor of 600 smaller than the calculated one.

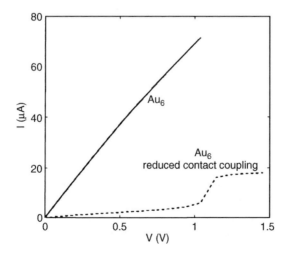

FIGURE 12.5 Calculated current/voltage characteristics for a wire consisting of six gold atoms. The upper curve shows near-perfect conductance quantization, with good contacts (covalent distances) between the gold wire and the bulk gold. The dotted line corresponds to artificially weakened interfacial mixing and shows highly nonlinear current/voltage characteristics due to the interfacial mixing. (From P.S. Damle, A.W. Ghosh and S. Datta, *Phys. Rev. B*, 64, 201403, 2001. With permission.)

In Figure 12.5,[88] results are presented for the calculation of a gold wire consisting of discrete gold atoms between gold interfaces. The two lines show the transport corresponding to gold atoms within a covalent radius of the interface and precisely the same junction with the gold atoms extended away from the interface. With poor interfacial mixing, the conductance is reduced from the quantum conductance — even in a situation where the electrode and metal are made of the same material — so that no Schottky (polarization) barriers exist at the interface.

The calculations and measurements discussed in this section are characteristic for the Landauer (coherent transport) regime in molecular wires.

12.4   Gating and Control of Junctions: Diodes and Triodes

Some of the earliest suggestions of molecular electronic behavior involved molecular rectifiers[30] — that is, the possibility for molecules to act as actual electronic active devices, rather than simply to be used only as wires or interconnects (previously examined by Kuhn[18]). Accordingly, much of the recent activity has indeed focused on active molecular structures. We have already referred to work on light-emitting diodes and photovoltaics. Other activities include molecular magnetic structures, ferroelectrics, actuators, and other behaviors.[89]

Keeping with our limited focus, it is worth mentioning the modification of current flow by molecular binding or by electrostatic potential interactions that can change the transport. This is schematically indicated in Figure 12.1, where the molecule is shown with two possible functionalization sites. The schematic label S involves the binding of a substrate, while the electrostatic gate is similar to what is done in ordinary field-effect transistors. Both of these schemes have been used for sensing capabilities as well as switches; a large gain can be exhibited by either FET or CHEMFET structures, while binding of molecules has also been demonstrated to change the current substantially either because of wave function mixing or (more generally and more easily) by modification of structure.[90–91]

The simplest way to understand this behavior is in terms of Equations (12.5) and (12.6). Writing the Green's function (that corresponds to the molecular component) as:

$$G_{\ell r} = \sum_s \frac{\langle l|s\rangle\langle s|r\rangle}{E_s - \eta - \Sigma^{(s)}_{(\ell r)}(\eta)} \tag{12.10}$$

we can easily understand how the current is modified. We recall that in Equation (12.10), the state $|s>$ is an eigenstate of an extended molecule, with energy E_s. When the electrostatic potential is changed (FET behavior), then the state $|s>$ will be changed by the electrostatic potential. This in turn will modify both the overlap integrals in the numerator of Equation (12.10) and the energies of the denominator. Figure 12.6 shows some very early calculations by Mao,[92] using an extended Huckel model for the molecule and a simple Newns–Anderson model for the electrode. We see that the current at zero voltage and its shape with changing injection energy are substantially modified by the presence of what is effectively an external potential — in this case resulting from the reduction of a quinone species to its di-anion. Because the charging is local, the effects can in fact be very large.

The general scheme of utilizing nearby molecular structures to change the electrostatic potential indeed underlies both the historical work on CHEMFET structures[34] involving binding at macroscopic surfaces, and more recent work on both molecular and nanowire logic.[37] In all these cases, the simplest theoretical analysis is simply based on modification of the wave function in the numerator of Equation (12.10) and the energy in the denominator.

Switching that involves actual geometric change is also of major interest. In particular, work by the UCLA–Hewlett Packard collaboration[93] has been based on substantial geometric change in rotaxane structures. Here the modifications are much larger, because actual chemical entities (in this case cyclic aromatic compounds) have changed position. The switching is relatively slow because it involves major molecular motion, but it also has several advantages. Indeed, the work in California has been very successful in producing real molecular electronic switches and architectures based on this motif.

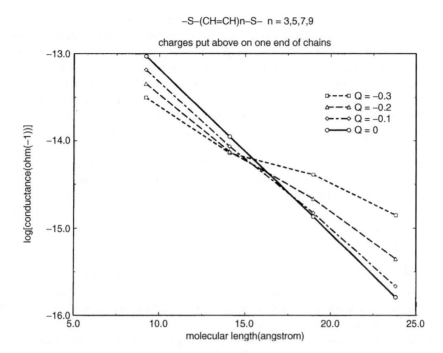

FIGURE 12.6 Results of an early calculation on molecular-type field-effect transistors. A quinone molecule is placed above one end of the conjugated dithiol structure, and the conductance was calculated using a simple Huckel-type representation and the Landauer conductance formula of Section 12.3. The charge Q is simply charge localized on the quinone moiety suspended above one end of the chain. Notice the substantial modification in current due to the external potential arising from charged quinone species. (From Y. Mao, Ph.D. thesis, Northwestern University. 2000.)

Computationally, what is required to understand the switching is again quite similar to Equation (12.10); but now the states in the numerator and energy denominator will be changed, not because of an external field but because of quite different chemical binding.

In extended molecular wire structures, the coherent formulation will fail, and the analysis in terms of Equation (12.10) is no longer adequate. It is then necessary to employ electron transfer hopping models that describe the motion from one site along the wire to the next.[15] This is almost certainly the appropriate limit to describe experiments on extended conductive polymer molecular structures[94] that can be gated by binding or by electrostatic potentials. In this case, the change in current arises because of change in the local hopping rates. Because these are understood using generalized Marcus/Hush/Jortner theory,[95] the role of the gating will appear in the calculation of the three major factors that enter into that model (reorganization energies, tunneling barriers, and exergicities). Some initial attempts in this area have been completed.[96]

While there have been few theoretical efforts to describe with quantitative accuracy gating and switching processes, it seems quite clear that the overall formulation of the rate processes for both the coherent and incoherent limits will permit, quite directly, analysis of these phenomena. These should be one of the more exciting and important areas in the molecular electronics of the next 5 years.

12.5   The Onset of Inelasticity

Inelastic transport arises in mesoscopic systems due to a number of processes where the energy of the scattered electron is not preserved. Chief among these are electron–electron and electron–phonon scattering. The former involves many-particle correlation, and the latter is related to the coupling between vibrational and electronic degrees of freedom. They may have important consequences in modifying the transport properties of molecular devices and also in inducing charging, desorption, and chemical reactions.

Inelasticity is also related to the loss of coherence. This is clearly the case for electron–electron interaction, whereas electron–phonon coupling can induce both coherent and incoherent inelastic transport. Here we comment on the description of inelastic processes caused by electron–phonon interaction.

Electron–phonon coupling can be discussed as a pure quantum problem or in the context of semi-classical models. In either description, the most relevant physical parameter in assessing the influence of electron–phonon coupling in the problem of molecular conductance is the Buttiker–Landauer tunneling time. As will be discussed in Section 12.7, this time is longer in the resonant regime and scales with the length of the bridge.

Electron–phonon coupling is responsible for a number of collective excitations that influence the transport properties of many materials.[97] In metals, it accounts for the resistivity behavior and metal–insulator transitions; in semiconductors and other polarizable media, it is related to the formation of polarons; in conducting polymers, it induces the formation of solitons, polarons, and bipolarons. Ness et al.[98] have presented a quantum description of coherent electron–phonon coupling and polaron-like transport in molecular wires. They show that when the electron–lattice interaction is taken into account, transport in the wire is due to polaron-like propagation. The electron transmission can be strongly enhanced in comparison with the case of elastic scattering through the undistorted molecular wire. A static model of this type of coupling had been previously presented in Reference 99.

Burin et al.[100] have developed a semi-classical theory of off-resonance tunneling electrons interacting with a single vibrational mode of the medium. Depending on the difference between the tunneling time and the vibrational period, two different regimes are found: one that corresponds to coherent superexchange and a second involving polaron-like vibronic transport. Both results are qualitatively in agreement with the quantum model despite the fact that the semi-classical treatment is inherently incoherent because the quantum–classical coupling introduces random dephasing of the electronic wave function.

Figure 12.7 shows one of the major results of the semi-classical study. The energy exchange between the tunneling electron and the bridge vibrations is strongest for moderately slow tunneling. In this regime, the

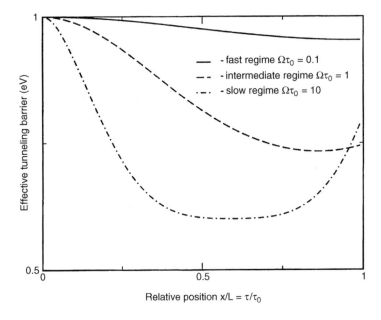

FIGURE 12.7 The effective tunneling barrier for an electron passing through a rectangular barrier in the potential. The modifications in height are due to interactions between the electron and polarizing environment. The tunnel barrier and the applied voltage are each taken to be two volts; the nearly straight line corresponds to fast tunneling, while the two lower curves correspond to slower tunneling in which the electron has more opportunity to polarize the environment. Notice on the far right that the barrier is higher at the end of the reaction than at the beginning; this is because electrons, tunneling through the barrier, have deposited energy in the vibrational medium. (From A.L. Burin, Y.A. Berlin, and M.A. Ratner, *J. Phys. Chem. A*, 105, 2652. 2001. With permission.)

electron tunneling process is inelastic, and (in the case shown here) energy is lost to the vibrations of the medium.

Incoherent processes leading to dephasing can be introduced via a density matrix approach to transport. In a series of papers, Nitzan and coworkers[15,101–103] have studied the effect of dephasing on electron transfer and conduction. The main result of these studies is that in addition to the coherent superexchange mechanism, there appears an incoherent path that is closely related to activated hopping.[104–105] This topic is the subject of Section 12.7.

12.6 Molecular Junction Conductance and Nonadiabatic Electron Transfer

Electron transfer is one of the most important areas of chemical kinetics, biochemical function, materials chemistry, and chemical science in general.[106–107] In electron transfer reactions, electronic charge is moved either from one molecular center to another or from one section of a molecule to another section of that same molecule. The latter processes are called *intramolecular electron transfer* and are advantageous for study because one does not have to worry about the work terms involved in assembling the molecular pair.

Intramolecular electron transfer reactions pervade both natural and synthetic systems. A common structure is the so-called *triad structure*, consisting of a donor, a bridge, and an acceptor. One can then study electron transfer either in the ground state (by first reducing the electron donor, and then watching the kinetics as the electron moves to the acceptor) or photochemically (by photo-exciting the donor to form D^*BA, which then decays by intramolecular photo-excited electron transfer to D^+BA^-, which can then decay by back-transfer to the original reactant, DBA).

In both molecular junctions and electron transfer reactions, the fundamental process is that of an electron moving from one part of space into another. Not surprisingly, then, the two processes can be closely related to one another. In particular, we consider the so-called nonadiabatic electron transfer situation. This is the condition that is obtained when the electronic mixing between donor and acceptor through the bridge is relatively small compared with, for example, typical vibrational frequencies or characteristic temperatures. Under these conditions, perturbation theory shows that the rate for moving from the donor to the acceptor site depends both on the electronic mixing and on the vibronic coupling that permits energy to be exchanged between the electronic and the nuclear motion systems. Under these conditions, the nonadiabatic electron transfer rate, k_{et}, is given by[106–109]:

$$k_{ET} = \frac{2\pi}{\hbar} V_{DA}^2 (DWFC) \tag{12.11}$$

This form follows from the golden rule of perturbation theory, expressing the electron transfer (ET) rate in terms of the electronic mixing V_{DA} between the donor and the acceptor sites and a density of states-weighted Franck–Condon factor (DWFC) that describes vibrational overlap between the initial and final states, weighted by the number of states. For thermal electron transfer reactions, the Marcus/Jortner/Hush form (Equation 12.11) is broadly applicable and very useful.

Because both nonadiabatic electron transfer and molecular junction tunneling transport from the metal occur by means of electron tunneling, they can be compared in a useful fashion. Table 12.2 shows

TABLE 12.2

Molecular Junction Transport	Nonadiabatic Electron Transfer
Electron tunneling	Electron tunneling
Electrode continuum	Vibronic continuum
Conductance	Rate constant
Landauer formula	Marcus/Jortner/Hush expression

the similarities and differences. The great similarity is that electron tunneling is the process by which charge is actually transferred. Everything else differs: in particular, the continuum of states (that causes relaxation and chemical kinetics to occur) arises not from the multiple electronic states of the electrode in the junction transport problem, but from the vibrations and solvent motions in the nonadiabatic electron transfer situation. The observables also differ: conductance of the junction and rate constants in electron transfer systems. The appropriate analytical expressions are also quite different: the Marcus/ Jortner/Hush formula of Equation 12.11 derives from second-order perturbation theory and is valid when the electron mixing is small. The Landauer form of Section 12.3 derives from an identity between scattering and transport and makes no particular assumptions about the smallness of any given term.

Nitzan has developed the similarity substantially farther.[15] Indeed, he has extended it beyond the coherent tunneling regime discussed in Section 12.3 to include the inelasticity discussed in Section 12.5, and even the hopping limits that we will discuss in the next section. Nitzan finds that the electron transfer rate and the transmission can be usefully compared.[15,110] In particular, he finds that:

$$g = \frac{e^2}{\pi\hbar} \frac{k_{ET}}{(\text{DWFC})} \frac{8\hbar}{\pi\Delta_L\Delta_R} \tag{12.12}$$

Here the relationship is between the conductance g and the electron transfer rate constant k_{ET}. The tunneling terms are common to both and therefore disappeared from this expression. What appears instead is the ratio of the bath densities: the spectral density terms $\Delta_L\Delta_R$ describe the mixing of the molecular bridge with the electrodes on the left and right, whereas DWFC describes the mixing of nonadiabatic electron transfer tunneling with the vibrations of the molecular environment.

The assumptions underlying Equation 12.12 are not general, and Nitzan remarks on conditions under which they will and will not hold. Still, it is an extremely useful rule of thumb. Moreover, because semi-classical expressions are available for DWFC, one can choose typical values for the parameters to get a relationship between the conductance and the electron transfer rate. Nitzan derives the form:

$$g(\Omega^{-1}) \sim 10^{-17} k_{ET}(S^{-1}) \tag{12.13}$$

This means that for typical fast nonadiabatic electron transfers of the order of 10–^{12}/s, one would expect a conductance of the order of $10^{-5}/\Omega$.

The fundamental observation that electron tunneling is a common process between junction transport and nonadiabatic electron transfer permits not only the semi-quantitative relationships of Equations (12.12) and (12.13) but also comparison of the mechanisms. If the two processes were fundamentally controlled by state densities, dephasing processes, and electron tunneling, then the mechanisms that should inhere in one might also be important in the other. This has led to a very fruitful set of comparisons between the mechanistic behaviors corresponding to electron transfer and molecular junction conductance. Because electron transfer has been studied for so much longer, because the experiments have been so much better developed, and because of the signal importance of electron transfer as a field in chemistry, materials, physics, biology and engineering, it is possible to learn a substantial amount about the mechanisms that one expects in molecular junction conductance from the behaviors involved in electron transfer. This fundamental identification is significant and is the topic of the next section.

12.7 Onset of Incoherence and Hopping Transport

The Landauer approach of Section 12.3 assumes that all scattering is elastic. Section 12.5 dealt with the onset of inelasticity and with simple dynamic polarization of the environment by the tunneling electron. More generally, one expects that sufficiently strong interaction between the tunneling electron and its environment can lead to loss of coherence and to a different mechanism for wire transport.

A significant qualitative insight can be gained by estimating the so-called *contact time*; this is not the inverse of the rate constant for motion but rather an estimate of the actual time for the tunneling process to take place — that is, the time for the tunneling electron to be in contact with the medium with which it will interact. Landauer and Buttiker originally developed this concept[111] and analyzed tunnel processes for heterostructures. In molecular wires, the idea of a tunneling barrier is probably less appropriate than a tight binding picture, effectively because molecules are better described in terms of Huckel-type models than simple barrier structures.

Utilizing the Huckel model for the wire and continuum damping pictures for the electrodes, the general development of Landauer and Buttiker can be modified[112] to give a very simple form for the contact time. This simple form is:

$$\tau_{LB} \approx N\hbar/\Delta E_G \tag{12.14}$$

Here the Landauer Buttiker time, τ_{LB}, is roughly equal to the uncertainty product times the length of the wire. More precisely, ΔE_G is the gap energy separating the electrostatic potential of the electrode with the appropriate frontier orbital energy on the molecule, and N is the number of subunits within the molecule. A more general form, one that goes smoothly to the classical Landauer Buttiker limit, can be derived by making corrections in the denominator.[112] The important point is that for large gaps and short wires, the Landauer Buttiker time will be substantially shorter than any characteristic vibrational or orientational period of the molecular medium. Under these conditions one expects simple tunneling processes, and the coherent limit of Section 12.3 should be relevant. As the gap becomes smaller and the length longer, the LB time will increase. One then expects that when the LB time becomes comparable to the timescale of some of the motions of the molecular medium in which the electron finds itself, inelastic events should become more important, and the coherent picture needs to be generalized.

The two dominant interaction processes between the molecular bridge and the molecular environment both arise from polarization of the environment by the electronic charge. (A third polarization interaction, image effects, has not been properly dealt with and is mentioned in Section 12.8.) These polarizations include whatever solvent may be present and polarization of chemical bond multipoles by electronic charge. These result in both relaxation processes (energy transfer between the environment and the electronic system) and in dephasing processes (energy fluctuations in the electronic levels). The most straightforward way to treat such structures is through the use of density matrices.[113–116] The density matrix describes the evolution of a system that may or may not consist of a pure state (that is, a single wave function). It is ideal for discussion of systems in which environments that are not of primary interest interact strongly with electronic systems that are of primary interest. They are familiar, for example, in the treatment of NMR problems, where the longitudinal and transverse relaxation times (T_1 and T_2) arise because the nuclear spin interacts with its environment.[117] The diagonal elements of the density matrix correspond to populations in electronic levels; and the off-diagonal elements of the density matrix describe so-called coherences, which are effectively bond orders between electronic sites. In the presence of interactions between the system and the environment, it is straightforward to write the time evolution of the density matrix as:

$$\dot{\rho} = \frac{i}{\hbar}[H,\rho] + \dot{\rho}_{diss} \tag{12.15}$$

Here the first term on the right is the causal one arising from behavior of the system Hamiltonian, while the second term refers to dissipative interactions between the system of interest and its environment. While there are many approaches for analyzing the effects of the dissipative term (and none of them is exact[118]), the simplest and most frequently used approach is that taken by Bloch for magnetic resonance.[117] Here one differentiates the relaxation times for pure dephasing, called T_2^*, and for energy relaxation, called T_1. One then assumes that the populations relax at a rate inversely proportional to T_1, and the coherences dephase at a rate inversely proportion to T_2^* (here δ_{ij} is the Kronecker δ, which vanishes unless i and j are the same, in which case it equals unity):

$$\rho_{ij})_{dis} = (1-\delta_{ij})\rho_{ij}/T_2^* + \delta_{ij}\rho_{ii}/T_1 \qquad (12.16)$$

In the absence of dissipation terms, the solutions to Equation (12.15) are multiply oscillatory, so equilibrium is never approached. When the dissipative terms are added, the systems indeed approach equilibrium; and a rate is determined both by the frequency values due to the Hamiltonian and by the relaxation processes T_1 and T_2^*.

For an N-level system, the equation system of (12.15) involves N^2 equations and is generally complex to solve. In some limits, however, the solution can be obtained analytically.[15,101] It has been shown that, for a very simple model of bridge-assisted transfer in which only dephasing occurs and in the limit where the dephasing is relatively weak, the overall rate constant for electron transfer reaction between donor and acceptor actually breaks into a sum of two independent contributions:

$$k_{ET} = k_{coh} + k_{incoh} \qquad (12.17)$$

Here the coherent term corresponds to electron tunneling through the bridge sites and will generally behave or decay exponentially with distance as expected by the McConnell relationship.[119]

$$k_{coh} \sim e^{-\beta R} \qquad (12.18)$$

Here R is the length between donor and acceptor (or in the molecular junction between the molecule/metal interfaces), and β is a characteristic fall-off parameter. The incoherent term arises from the dephasings and can generally be written as:[15,103]

$$k_{incoh} \sim \frac{1}{(b+cR+dR^2)} F(T) \qquad (12.19)$$

The constants b, c, and d depend on the nature of the experiment and the relative magnitudes of the Hamiltonian parameters. Very commonly, the first term of the denominator will dominate, so that there is effectively no decrease of the incoherent transfer rate with length. The temperature dependence is determined by the function $F(T)$, which in the simple activated case becomes

$$F(T) = \exp(-\Delta G^{\ddagger}/k_B T) \qquad (12.20)$$

Equation (12.17) therefore predicts that one should see two different rates, one arising from coherent tunneling and the other from incoherent dephased motion, that can become thermally activated hopping when equilibrium is obtained (that is, when the longitudinal relaxation time T_1 is large).

The expectation of Equations (12.17) to (12.20) is shown in Figure 12.8, for the case where the first term of the denominator of Equation (12.19) indeed dominates. Based on the arguments of Equation (12.14), we would expect the relative magnitude of the incoherent and coherent term to depend on the parameters: for long wires, small gaps, and high temperatures, the incoherent term should be favored. At very low temperatures and with large gaps, one would expect coherent tunneling as is generally seen for donor/bridge/acceptor electron transfer for short bridges.

Because the onset of incoherence really corresponds to vibronic coupling, it is possible to successfully derive more detailed predictions, utilizing an actual coupling model of the spin/boson type to describe the interaction between the tunneling electron and its environment.[104,105] This sort of analysis has been reported by Segal and Nitzan,[103] who also make important remarks on the relationships between subsystems and on the nature of the energy flow that occurs when transport undergoes the mechanism modification from tunneling to hopping.

In polymeric structures, the hopping mechanism has long been known to dominate. *Solitons, polarons,* and *bipolarons* are terms describing particular polarization characteristics, in which charge polarizes the

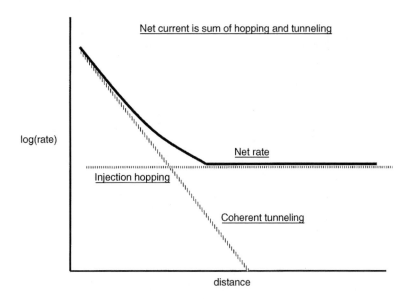

FIGURE 12.8   Schematic of the rate behavior through a molecular wire bridge in the presence of interactions between the system and its environment. The two limiting cases of injection hopping (fully incoherent) and tunneling (fully coherent) are indicated by the dashed lines. The net rate is the sum of those two contributors.

vibrational environment, causing trapping and localization.[94,120] Charge motion then corresponds to the motion of these quasi particles that consist of the electronic carrier plus its accompanying polarization cloud. For molecular transport junctions, this same sort of behavior should be seen for long distances, and the situation has been treated in a simple model. An important set of observations by Reed and his collaborators[32] in extended molecular wires in nanopore junctions indicates that, indeed, one can observe both hopping and tunneling processes, depending on the nature of the energy barriers and the applied potential (Figure 12.9).

The transitions between coherent and incoherent motion have been observed extensively in donor/bridge/acceptor systems, although much less work has to date been reported in tunnel junctions. One particularly marked case is DNA, where Figures 12.10 and 12.11 show clear limiting cases for the dependence of the intramolecular electron transfer rates on distance. Figure 12.10[121] shows photo-excited electron transfer in DNA hairpin structures, where both the charge separation and charge recombination rates are indeed exponential, with characteristic values for the β of Equation (12.18). Figure 12.11[122] shows a comparison with a model very much like that of Equations (12.15) to (12.17) with results for differential cleavage reactions reported by Giese, Michel–Beyerle, and collaborators.[123] The tight binding model parameters corresponding to the energy gap and the local (Huckel) electron tunneling amplitude have been calculated by the Munich and Tel Aviv groups.[124] The fit is quite good, and the two different regimes (tunneling at short distances and hopping at long distances) appear quite straightforward.

12.8   Advanced Theoretical Challenges

While there has been extensive and (in some cases) excellent theory devoted to transport in molecular wire junctions, if molecules simply act as interconnects, their role in molecular electronics is finite. There are far more sophisticated and challenging possibilities that have not yet been extensively addressed in the theoretical community. These processes go beyond conductance in simple wire circuits and still make up major challenges for the community. Some of these issues include:

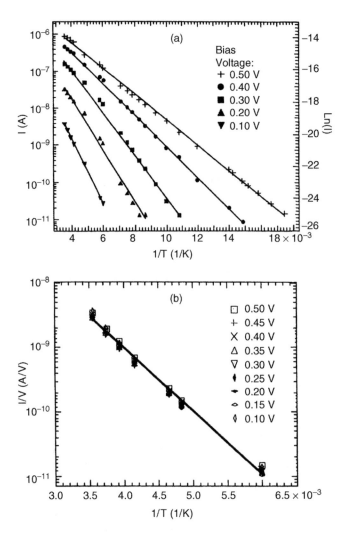

FIGURE 12.9 Temperature dependence of the current at forward (a) and backward (b) biases for a nanopore junction containing the p-biphenylthiol molecule sandwiched between titanium and gold electrodes, with the sulfur next to the gold. In (a), the forward bias is fitted to the thermionic emission through an image potential, while in (b), a simple hopping model is used. (From M.A. Reed et al., *Ann. N.Y. Acad. Sci.*, 852, 133, 1998. With permission.)

12.8.1 Electrostatic Potentials and Image Charges

When a molecule is placed between electrodes and voltage is applied, there is clearly a changing chemical potential between the two ends of the molecular junction. A very important issue then becomes exactly how this charge is distributed across the molecular interface. The three simplest approximations are to assume that all the charge drops at the interfaces,[125–126] that the charge drops smoothly across the molecular junction in a linear fashion,[127] and that the electrostatic potential changes across the molecular junction in a fashion determined self-consistently with the total charge density.[28,86,88,128–130] The third of these seems the most elegant and has been used in self-consistent density functional theory calculations, with the usual Green's function approximation used for the electrodes. It is then generally found (Figure 12.12 shows one example) that the electrostatic potential is complex, a substantial drop at the interfaces with substantial structure within the molecule itself. The suggestion that most of the voltage drops at the interfaces arises from the relatively poor contact between molecule and interface, compared with delocalization along well-conjugated molecules. It was used in some of the early investigations[125] in the field and probably represents

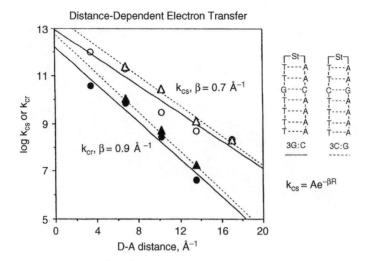

FIGURE 12.10   Rate behavior as a function of distance for hairpin DNA compounds. Initially, the bridging stilbene (noted St) was photo-excited; the results are given for the charge separation (CS) and charge recombination (CR). Both are fully coherent and display an exponential decay with length. (From F.D. Lewis et al., *Science* 227, 673, 1997. With permission.)

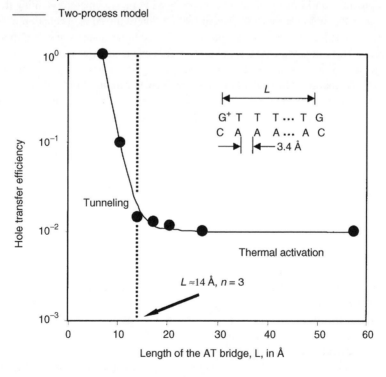

FIGURE 12.11   Comparison of calculated hole transfer efficiency (proportional to rate constant) for extended DNA structures. The tunneling regime at short distances and the thermal hopping at long distances are indicated. The solid dots refer to the experimental work, and the line through them corresponds to theoretical treatment utilizing the two processes of Figure 12.10. All parameters were evaluated theoretically. (Experiment from B. Giese et al., *Nature,* 412, 318, 2001; modeling from Y.A. Berlin et al, *Chem. Phys.,* 275, 61, 2002. With permission.)

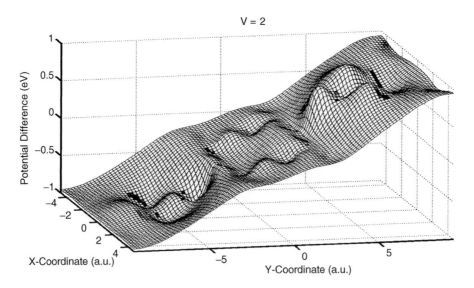

FIGURE 12.12 Calculated electrostatic potential, from a full self-consistent density functional theory calculation on the benzene dithiol bridge between gold electrodes at a voltage of two volts. Notice the substantial flatness near the interface, and the highly structured potential, arising from a combination of electrostatic forces and nuclear trapping. (From A. Xue, M.A. Ratner, and S. Datta, to be published. With permission.)

a reasonable approximation in many situations, especially STM measurement. Assuming that the voltage drops linearly across the molecular junction, which is a solution to the Poisson equation in the absence of all charge, is the simplest way to treat voltage and has been used in the analysis of localization phenomena in charged wires.[127] This seems too bold a simplification for any quantitative work.

Image charges occur, in classical electrodynamics, because of the polarization of the charge within the metal by external charges. Most calculations on molecular wire junctions either ignore the image charge altogether or hope that the image charge is taken into account either by the metallic particles included in the extended molecule treatment or by the self-consistent solution to the density functional. It is not at all clear that this is correct — image charge stabilization is a substantial contribution to the energy, and in areas such as molecular optoelectronics, the image effects can lead to substantial barriers at the interface,[131] which either have to be tunneled through or hopped over. One expects similar behavior in molecular wire junctions,[132] and no explicit treatment of image charge has, to our knowledge, yet been given.

12.8.2 Geometry

Two major geometric questions are involved in trying to understand what happens in molecular wire junctions. For the zero-field conductance, one normally assumes that the molecule in the junction is of the same geometry as the free molecule. This is generally unjustified and is assumed for simplicity. Even if this molecular geometry is unchanged, there is the issue of binding to the electrode interfaces. In thiol/gold (the most investigated situation), it has been normally assumed that the sulfur is either within a three-fold site on the surface or directly over a metal atom. These give very different calculated conductances.[27]

Recent experimental data by Weber and colleagues[26] suggest that, even in the gold/thiol interaction, the geometries are not necessarily regular and may evolve dynamically in time. *Ab-initio* calculations[133] also suggest that the geometry of the gold/thiol interface is highly fluid and not at all regular and that the potential energy surface for this motion is fairly flat. Therefore, the assumptions on the binding that have been used to compute self-energies are not necessarily highly accurate.

Once the current is applied, the situation can be even worse. It is known that with enough current, bonds can break. Molecules in very high electromagnetic fields such as those that occur at the molecular junction interface (exceeding 10^6 volts/cm) may well have geometries that are substantially different from those in the unperturbed molecule. Simple calculations of Stark changes in the geometry suggest that this is definitely true for molecules with large dipole moments. Therefore, it is really necessary to understand the geometry of the molecule within the junction before junction conductance can be appropriately calculated.

12.8.3   Reactions

The behavior of molecules with large amounts of current passing through may well be a new kind of spectroscopy, and even a new kind of catalytic chemistry. It is known that reactions very often occur when currents are passing through the molecules: the entire idea of positioning using scanning tunneling microscopes is predicated on the fact that one can break the surface bonding, move the molecule or atom to the tip, and then replace it.[134] This obviously means that either the electrostatic field or the current is causing bond breaking. Such bond-breaking kinetics has not really been discussed in connection with molecular junctions but might be one of the fascinating applications of them.

In nanotubes, Avouris and his colleagues[135] have developed a general cleaning technique in which currents are passed through the nanotubes and particular adlayer structures are excised by bond breaking. Similarly, Wolkow and collaborators[136] have demonstrated that under substantial current conditions, alkane species remain conductive, while unsaturated species can be thermally destroyed. Some excellent theoretical modeling has been done on these problems by Seideman.[137]

Actual chemical reactions in scanning probe environments have been reported by Ho[138] (Figure 12.13) and by other groups. The bond cleavage here is due to the current passing through the molecule, and development of appropriate models is still in its infancy. Similarly, Sagiv and Maoz[139] have shown that in the presence of a conductive AFM current, alkane adlayers can be oxidized by oxygen and water to form carboxylic acid structures. This sort of electrochemistry in a tip-field has not been effectively modeled but might represent both a very important application of molecular junction behavior and a challenging theoretical problem.

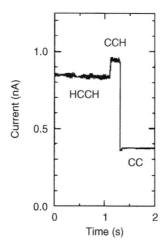

FIGURE 12.13   The current as a function of time for a molecular junction consisting of a single acetylene molecule. After roughly one second, enough energy has been deposited in the molecule to break the bond, giving the CCH radical. After roughly half a second more, the second hydrogen is removed and the bridge consists only of two carbons. (From L.J. Lauhon et al, *Phys. Rev. Lett.*, 84, 1527, 2000. With permission.)

12.8.4   Chirality Effects

Single electron currents might be very weakly affected by local chirality because of broken symmetry structures. One expects such relativistic interactions to be weak, but Naaman and collaborators[140] have shown substantial effects of chiral adlayers on the transport of polarized spins. This problem is, once again, experimentally challenging and theoretically largely unaddressed.

Closely related to chirality is the issue of magnetic effects on transport. More sophisticated theory[141] and experiments are needed to understand the possible effects of magnetic interactions on currents through molecular junctions.

12.8.5   Photo-assisted Transitions

In semiconductor electrode structures, photo-assisted chemistry is very important — for example, a so-called Graetzel cell,[142] in which photo-excited dye molecules transfer electrons into the conduction band of nanoscale titania, is of major interest both intrinsically and for applications. The more general problem of photo-assisted behavior in molecular junctions is beginning to be addressed[143] theoretically using several techniques; but once again, without comparison with direct experiments, these are incomplete.

It is clear that despite the high interest and beautiful work in the general area of molecular electronics, some of the major theoretical problems have not yet been properly approached.

12.8.6   Dynamical Analysis

Like spectroscopy,[109] current flow can be evaluated dynamically. This has several advantages, including (in particular) increasing understanding of how currents actually flow in molecule-conductor junctions. The first such calculations[144] are starting to appear.

12.9   Remarks

Transport in molecular wire junctions and the fundamental scientific and technological issues that it raises have produced extensive scientific excitement and some very impressive work. In the area of carbon nanotubes, a great deal is now understood: effects of interfacial mixing at the electrode can dominate transport, the behavior can be ballistic in well-formed single-walled structures, both semiconducting and metallic tubes are known and understood, and issues such as defects, adsorbates, branches, and crossings can now be discussed and calculated. Therefore, in the nanotube area, computational capabilities have meshed with excellent experimental work to produce a situation in which we understand quite well what the expected accuracies and errors of any given computational approach might be.

The field of transport through more traditional molecular junctions is much more problematic. In discussing the comparison of theory with experiment for simple break junction measurements on straightforward organic molecules, we saw that disagreements in the conductance by factors of 100 or more between theory and experiment are common and that neither the theoretical approaches nor the experimental ones are necessarily converged. Even the simple suggestion that under favorable conditions, transport through organic molecules can show quantized conductance has never been reported experimentally until 2001,[145] although theory straightforwardly predicts such behaviors in the absence of strong vibronic coupling when contact resistances are sufficiently small.[12,46,146,147]

More sophisticated problems, such as those discussed in the last section, have been approached using model calculations. Significant insight into mechanistic behaviors has been gleaned, but these calculations have not been normed — there is really no independent verification of their validity because the appropriate comparison of quantitative calculations with quantitative measurements has not yet been made.

One major challenge for the community over the next few years, then, will be the development of well-defined model systems in which such comparisons can be made. The notion of defining a model science[148] — by comparison of a well-defined set of accurate experimental data with a series of calculations involving different approximations to the wave function, the mixing, the geometrics, the level of theory, and the method used — is necessary if quantitative estimates and predictions of transport behavior are to be obtained. A number of experimental groups[26,32,42,43] are now pursuing this issue, using reliable geometries and methodologies that can, perhaps, yield accurate and reliable data for a series of molecular wire structures.

In addressing the challenge of understanding molecular wire junctions, then, the community has made great progress. The next step will entail systematic comparison with good data, refinement of methodologies, and extension of the simple schemes described in this chapter toward the understanding of more complex, and more fascinating, molecular junction behavior.

Since completion of this overview, several important advances in molecular conductance junctions have been reported. Probably the most significant has been the development,[A1,A2] by groups at Harvard and Cornell, of methods to measure transport in individual molecules. The technology is based on use of a break junction that is prepared by rapid heating of a gold thin wire. The break occurs with characteristic gap sizes of roughly 10 nM; subsequently, individual molecules can be positioned at the break, and their conductance properties measured. These measurements can be controlled by a third, gate electrode that is positioned on a layer below the gold wire.

These measurements allow for solving the band lineup problem, because the gate electrode can be used to bring the molecular levels into resonance with the Fermi levels of the source and drain. Quantized conductance has indeed been observed, as discussed above in the theory section. Additionally, Kondo transport (conductance maxima at zero applied source-drain voltage due to pinning of the unpaired occupancy orbitals at the Fermi level) and spin effects on the molecule have been observed.

Work by Seideman, Guo and collaborators[A3] has employed perturbation methods to calculate both the elastic and inelastic responses in junctions containing buckyball and small-molecule junctions. This is significant for the development of approaches to the issues concerning inelastic scattering and chemical reactivity referred to in Section 12.8.

Acknowledgments

We are grateful to the Northwestern Molecular Electronics group, especially Abe Nitzan, Alex Burin, Alessandro Troisi, and Alex Xue for useful discussions. This work was sponsored by the Chemistry Divisions of the NSF and ONR, by the DARPA Moletronics program, and by DoD/MURI collaboration.

References

1. M.C. Petty, M.R. Bryce, and D. Bloor, *Introduction to Molecular Electronics* (Oxford University Press, New York, 1995).
2. C.A. Mirkin and M.A. Ratner, *Ann. Revs. Phys. Chem.* 43, 719 (1992).
3. A. Aviram (Ed.), *Molecular Electronics — Science and Technology* (AIP, New York, 1992).
4. A. Aviram (Ed.), *Molecular Electronics* (Engineering Foundation, New York, 1989).
5. A. Aviram and M.A. Ratner (Eds.), Molecular electronics: science and technology, *Ann. N.Y. Acad. Sci.* 852, (1998).
6. A. Aviram, M.A. Ratner, and V. Mujica (Eds.), Molecular electronics, *Ann. N.Y. Acad. Sci.* 960 (2002).
7. J. Jortner and M.A. Ratner (Eds.), *Molecular Electronics* (Blackwell, Oxford, 1997).
8. S. Datta, *Superlattice Microst.* 28, 253 (2000).
9. M.D. Ward, *J. Chem. Educ.* 78, 1021 (2001).
10. C. Joachim, J.K. Gimzewski, and A. Aviram, *Nature*, 408, 541 (2000).
11. M.A. Ratner and M.A. Reed, *Encyclopedia of Science and Technology,* Third ed. 10, 123 (2002).
12. A. Troisi and M.A. Ratner, to be published.
13. V. Mujica, *Rev. Mex. Phys.* 47, 59 Suppl. (2001).
14. F. Zahid, M. Paulsson, and S. Datta, H. Morkoc (Ed.), in *Advanced Semiconductors and Organic Nano-Techniques*, (Academic Press, New York, in press).

15. A. Nitzan, *Ann. Rev. Phys. Chem.* 52, 681 (2001).
16. *Chemical Physics*, Special Issue, to be published.
17. J.C. Ellenbogen and J.C. Love, *Proc. IEEE* 88, 386 (2000).
18. B. Mann and H. Kuhn, *J. Appl. Phys.* 42, 4398 (1971).
19. Y.N. Xia, J.A. Rogers, K.E. Paul, and G.M. Whitesides, *Chem. Rev.* 99, 1823 (1999).
20. H.-J. Güntherodt and R. Wiesendanger (Eds.), *Scanning Tunneling Microscopy I*, (Springer-Verlag, Berlin, 1992).
21. C. Dekker, *Phys. Today* 52, 22 (1999).
22. M.S. Dresselhaus, G. Dresselhaus, and P. Avouris (Eds.), *Carbon Nanotubes: Synthesis, Structure, Properties, and Applications* (Springer, Berlin, 2001).
23. T.W. Odom, J.H. Hafner, and C.M. Lieber, *Top. Appl. Phys.* 80, 173 (2001).
24. R. Martel, V. Derycke, C. Lavoie, J. Appenzeller, K.K. Chan, J. Tersoff, and P. Avouris, *Phys. Rev. Lett.* 87, 256805 (2001).
25. M.A. Reed, C. Zhou, C.J. Muller, T.P. Burgin, and J.M. Tour, *Science* 278, 252 (1997).
26. J. Reichert, R. Ochs, D. Beckmann, H.B. Weber, M. Mayor, and H. von Lohneysem, *Phys. Rev. Lett.* 88, 176804 (2002).
27. M. Di Ventra, S.T. Pantelides, and N.D. Lang, *Phys. Rev. Lett.* 84, 979 (2000).
28. C.K. Wang, Y. Fu, and Y. Luo, *Phys. Chem. Chem. Phys.* 3, 5017 (2001).
29. Y. Xue and M.A. Ratner, *J. Chem. Phys.*, submitted.
30. A. Aviram and M.A. Ratner, *Chem. Phys. Lett.* 29, 277 (1974).
31. R.M. Metzger et al., *J. Phys. Chem.* B105, 7280 (2001).
32. M.A. Reed, C. Zhou, M.R. Deshpande, and C.J. Muller, in A. Aviram and M.A. Ratner (Eds.), *Molecular Electronics: Science and Technology, Ann. N.Y. Acad. Sci.* 852 (1998).
33. V. Mujica, A. Nitzan, and M.A. Ratner. *Chem. Phys.* in press.
34. D.G. Wu, D. Cahen, P. Graf, R. Naaman, A.Nitzan, and D. Shvarts, *Chem. Eur. J.* 7, 1743 (2001).
35. A. Bachtold, P. Hadley, T. Nakanishi, and C. Dekker, *Science* 294, 1317 (2001); M. Ouyang, J.L. Huang, and C.M. Lieber, *Phys. Rev. Lett.* 88, 066804 (2002).
36. J.H. Schon and Z. Bao, *Appl. Phys. Lett.* 80, 332 (2002).
37. G.Y. Tseng and J.C. Ellenbogen, *Science* 294, 1293 (2001).
38. M.A. Ratner, *Mater. Today* 5, 20 (2002).
39. U. Mitschke and P. Bauerle, *J. Mater. Chem.* 10, 1471 (2000).
40. D.B. Mitzi, K. Chondroudis, and C.R. Kagan, *IBM J. Res. Dev.* 45, 29 (2001).
41. M. Dorogi, J. Gomez, R. Osifchin, and R.P.A. Andres, *Phys. Rev. B* 53, 9071 (1995).
42. X.D. Cui, A. Primak, X. Zarate, J. Tomfohr, O.F. Sankey, A.L. Moore, T.A. Moore, D. Gust, G. Harris, and S.M. Lindsay, *Science* 294, 571 (2001).
43. R. Shashidhar et. al., submitted.
44. J.G. Lee, J. Ahner, and J.T. Yates, Jr., *J. Chem. Phys.* 114, 1414 (2001).
45. V. Mujica, M. Kemp, and M.A. Ratner, *J. Chem. Phys.* 101, 6849 (1994).
46. V. Mujica, M. Kemp, and M.A. Ratner, *J. Chem. Phys.* 101, 6856 (1994).
47. M. Sumetskii, *Phys. Rev. B* 48, 4586 (1993).
48. M. Sumetskii, *J. Phys.: Condens. Matter* 3, 2651 (1991).
49. C. Joachim and J.F. Vinuesa, *Europhys. Lett.* 33, 635 (1996).
50. M.P. Samanta, W. Tian, S. Datta, J.I. Henderson, and C.P. Kubiak, *Phys. Rev. B* 53, R7626 (1996).
51. S. Datta, W. Tian, S.Hong, R. Reifenberger, J.J. Henderson, and C.P. Kubiak, *Phys. Rev. Lett.* 79, 2350 (1997).
52. C. Joachim, J.K. Gimzewski, R.R. Schlittler, and C. Chavy, *Phys. Rev. Lett.* 74, 2102 (1995).
53. N.G. van Kampen, *Stochastic Processes in Physics and Chemistry* (Elsevier Science, New York, 1983).
54. Y. Imry, *Introduction to Mesoscopic Physics* (Oxford University Press, New York 1997).
55. V. Mujica, A. Nitzan, Y. Mao, W. Davis, M. Kemp, A. Roitberg, and M.A. Ratner, *Adv. Chem. Phys. Series*, 107, 403 (1999).

56. E.G. Emberly and G. Kirczenow, *Phys. Rev. B* 58, 10911 (1998).
57. C. Kergueris, J.P. Bourgoin, S. Palacin, D. Esteve, C. Urbina, M. Magoga, and C. Joachim, *Phys. Rev. B* 59, 12505 (1999).
58. R. Landauer, *Philos. Mag.* 218, 863 (1970).
59. R. Landauer, *IBM J. Res. Dev.* 32, 306 (1988).
60. R. Landauer, *Physica Scripta* T42, 110 (1992).
61. Y. Imry and R. Landauer, *Rev. Mod. Phys.* 71, S306 (1999).
62. Y.V. Sharvin, *JETP* 48, 984 (1965).
63. J.L. D'Amato and H. Pastawski, *Phys. Rev. B* 41, 7411 (1990).
64. H. Pastawski, *Phys. Rev. B* 44, 6329 (1991).
65. H. Pastawski, *Phys. Rev. B* 46, 4053 (1992).
66. Y. Meir and N.S. Wingreen, *Phys. Rev. Lett.* 68, 2512 (1992).
67. N.S. Wingreen, A.P. Jauho, Y. Meir, *Phys. Rev. B* 48, 8487 (1993).
68. S. Datta, *Electronic Transport in Mesoscopic Systems* (Cambridge University Press, New York, 1995).
69. S.N.Yaliraki, A.E. Roitberg, C. González, V. Mujica, and M.A. Ratner, *J. Chem. Phys.* 111, 6997 (1999).
70. V. Mujica, A. Roitberg, and M.A. Ratner, *J. Chem. Phys.* 112, 6834 (2000).
71. L.V. Keldysh, *Soviet Phys. JETP* 20, 1018 (1965).
72. G.D. Mahan, *Many-Particle Physics*, (Plenum Press, New York 1990).
73. L.E. Hall, J.R. Reimers, N.S. Hush, and K. Silverbrook, *J. Chem. Phys.* 112, 1510 (2000).
74. W. Tian, S. Datta, S. Hong, R. Reifenberger, J.I. Henderson, and C.P. Kubiak, *J. Chem. Phys.* 109, 2874 (1998).
75. W. Häusler, B. Kramer, and J. Masek, *Z. Phys. B* 85, 435 (1991).
76. N.D. Lang, *Phys. Rev. B* 55, 4113 (1997).
77. A. Nakamura, M. Brandbyge, L.B. Hansen, and K.W. Jacobsen, *Phys. Rev. Lett.* 82, 1538 (1999).
78. H. Nakatsuji and K. Yasuda, *Phys. Rev. Lett.* 76, 1039 (1996).
79. P.A. Serena and N. García (Eds.), *Nanowires,* (Kluwer Academic Publishers, Dordrecht, 1997).
80. G. Treboux, P. Lapstun, and K. Silverbrook, *J. Phys. Chem. B* 102, 8978 (1998).
81. G. Treboux, P. Lapstun, Z. Wu, and K. Silverbrook, *Chem. Phys. Lett.* 301, 493 (1999).
82. M. Dorogi, J. Gomez, R.P. Andres, and R. Reifenberger, *Phys. Rev. B* 52, 9071 (1995).
83. P. Andres, S. Datta, D.B. Janes, C.P. Kubiak, and R. Reifenberger, in H.S. Nalwa (Ed.), *The Handbook of Nanostructured Materials and Technology* (Academic Press, New York, 1998).
84. C. Joachim and S. Roth (Eds.), *Atomic and Molecular Wires* (Kluwer Academic, Dordrecht, 1997).
85. M. Magoga and C. Joachim, *Phys. Rev. B* 56, 4722 (1997).
86. Y. Xue and M.A. Ratner, *Chem. Phys.* in press.
87. J.H. Schon, H. Meng, and Z. Bao, *Nature* 413, 713 (2001).
88. P.S. Damle, A.W. Ghosh, and S. Datta, *Phys. Rev. B* 64, 1403 (2001).
89. A.P. Alivisatos et al,. *Adv. Mat.* 10, 1297 (1998).
90. C.P. Kubiak et al., *J. Phys. Chem. B.* in press.
91. V. Mujica, S. Datta, A. Nitzan, C.P. Kubiak, and M.A. Ratner, *J. Phys. Chem.,* submitted.
92. Y. Mao, Ph.D. Thesis, Northwestern University (2000).
93. C.P. Collier, J.O. Jeppesen, Y. Luo et al., *J. Am. Chem. Soc.* 123, 12632 (2001); C.P. Collier, E.W. Wong et al., *Science* 285, 391 (1999).
94. H.S. Nalwa (Ed.), *Handbook of Organic Conductive Molecules and Polymers* (Wiley, New York, 1997).
95. M. Bixon, and J. Jortner, *Adv. Chem. Phys.* 106, 35 (1999).
96. G. Hutchinson, Y. Berlin, M.A. Ratner, and J. Michl, *J. Phys. Chem.,* submitted.
97. O. Madelung, *Introduction to Solid-State Theory* (Springer, Berlin, 1978).
98. H. Ness, S.A. Shevlin, and A.J. Fisher, *Phys. Rev. B* 63, 125422 (2001).
99. M. Olson, Y. Mao, T. Windus, et al., *J. Phys. Chem. B* 102, 941 (1998).
100. A.L. Burin, Y.A. Berlin, and M.A. Ratner, *J. Phys. Chem. A* 105, 2652 (2001).

101. D. Segal, A. Nitzan, W.B. Davis, M.R. Wasielewski, and M.A. Ratner, *J. Phys. Chem. B* 104, 3817 (2000).
102. D. Segal, A. Nitzan, M. Ratner, and W.B. Davis, *J. Phys. Chem. B* 104, 2709 (2000).
103. D. Segal, A. Nitzan, W.B. Davis et al., *J. Phys. Chem. B* 104, 3817 (2000).
104. S.S. Skourtis and S. Mukamel, *Chem. Phys.* 197, 367 (1995).
105. A.K. Felts et al., *J. Phys. Chem.* 99, 2929 (1995).
106. M. Bixon, and J. Jortner (Eds.), Special number on electron transfer, *Adv. Chem. Phys.* 106 (1999).
107. P.F. Barbara, T.J. Meyer, and M.A. Ratner, *J. Phys. Chem.* 100, 13148 (1996).
108. J. Jortner, *J. Chem. Phys.* 64, 4860 (1976).
109. G.C. Schatz and M.A. Ratner, *Quantum Mechanics in Chemistry* (Dover, New York, 2002), Chapter 10.
110. A. Nitzan, *J. Phys. Chem. A* 105, 2677 (2001).
111. M. Buttiker, and R. Landauer, *Phys. Scripta* 32, 429 (1985).
112. A. Nitzan, J. Jortner, J. Wilkie, A.L. Burin, and M.A. Ratner, *J. Phys. Chem. B* 104, 3817 (2000).
113. K. Blum, *Density Matrix Theory and Applications*, (Plenum Press, New York, 1981).
114. G.C. Schatz and M.A. Ratner, *Quantum Mechanics in Chemistry* (Dover, New York, 2002), Chapter 11.
115. S. Mukamel, *Principles of Nonlinear Optical Spectroscopy* (Oxford University Press, New York, 1995).
116. M.D. Fayer, *Elements of Quantum Mechanics* (Oxford University Press, New York, 2001).
117. C.P. Slichter, *Principles of Magnetic Resonance* (Springer-Verlag, New York, 1990).
118. D. Kohen, C.C. Marston, and D.J. Tannor, *J. Chem. Phys.* 107, 5236 (1997).
119. H.M. McConnell, *J. Chem. Phys.* 35, 508 (1961).
120. A.J. Heeger, *Rev. Mod. Phys.* 73, 681 (2001).
121. F.D. Lewis et al., *Science* 227, 673 (1997).
122. Y.A. Berlin, A.L. Burin, and M.A. Ratner, *Chem. Phys.* 275, 61 (2002).
123. B. Giese, *Accounts Chem. Res.* 33, 631 (2000); F.D. Lewis, *Accounts Chem. Res.* 34, 159 (2001); G.B. Schuster, *Accounts Chem. Res.* 33, 253 (2000).
124. A.A. Voityuk et al., *J. Phys. Chem. B* 104, 9740 (2000).
125. S. Datta, *Superlattice Microst.* 28, 253 (2000).
126. R.A. Marcus, *J. Chem. Soc. Faraday Trans.* 92, 3905 (1996).
127. V. Mujica, M. Kemp, A. Roitberg, and M.A. Ratner, *J. Chem. Phys.* 104, 7296 (1996).
128. C. Gonzalez, V. Mujica, and M.A. Ratner, *Ann. N.Y. Acad. Sci.* 960, 163–176 (2002).
129. S.T. Pantelides, M. Di Ventra, and N.D. Lang, *Physica B* 296, 72 (2001).
130. T.W. Kelley and C.D. Frisbie, *J. Vac. Sci. Technol. B* 18, 632 (2000).
131. G. Ingold et al., *J. Chem. Phys.*, submitted.
132. Y.Q. Xue, S. Datta, and M.A. Ratner, *J. Chem. Phys.* 115, 4292 (2001).
133. H. Basch and M.A. Ratner, *J. Chem. Phys.*, submitted.
134. M.F. Crommie et al., *Physica D* 83, 98 (1995).
135. P.C. Collins, M.S. Arnold, and P. Avouris, *Science* 292, 706 (2001).
136. R.A. Wolkow, *Annu. Rev. Phys. Chem.* 50, 413 (1999).
137. S. Alavi and T. Seideman, *J. Chem. Phys.* 115, 1882 (2001).
138. L.J. Lauhon and W. Ho, *J. Phys. Chem. B* 105, 3987 (2001).
139. R. Maoz and J. Sagiv, *Langmuir* 3, 1034 (1987).
140. R. Naaman, private communication.
141. E.G. Petrov, I.S. Tolokh, and V. May, *J. Chem. Phys.* 108, 4386 (1998).
142. M. Gratzel, *Nature* 414, 338 (2001).
143. J.T. York, R.D. Coalson, and Y. Dahnovsky, *Phys. Rev. B* 65, 235321 (2002); A. Tikhonov, R.D. Coalson, and Y. Dahnovsky, *J. Chem. Phys.* 117, 567–580 (2002).
144. R. Baer and R. Gould, *J. Chem. Phys.* 114, 3385 (2001).
145. J.H. Schon and Z. Bao, *Appl. Phys. Lett.* 80, 847 (2002).

146. Y.V. Sharvin, *Zh. Eksp. Teor. Fiz.* 48, 984 (1965).
147. L.E. Hall et al., *J. Chem. Phys.* 112, 1510 (2000).
148. J.A. Pople, *Rev. Mod. Phys.* 71, 1267 (1999).
A1. W.J. Liang, M.P. Shores, M. Bockrath, J.R. Long, and H. Park, *Nature* 417, 725–729 (2002).
A2. J. Park, A.N. Pasupathy, J.L. Goldsmith, C. Chang, Y. Yaish, J.R. Petta, M. Rinkoski, J.P. Sethna, H.D. Abruna, P.L. McEuen, and D.C. Ralph, *Nature* 417, 722–725 (2002).
A3. S. Alavi, R. Rousseau, G.P. Lopinski, R.A. Wolkow, and T. Seideman, *Faraday Discussions* 117, 213–229 (2000); S. Alavi, B. Larade, J. Taylor, H. Guo, and T. Seideman, *Chem. Phys.* in press.

13

Modeling Electronics at the Nanoscale

13.1 Introduction 13-1
References 13-2
13.2 Nanostructure Studies of the Si-SiO$_2$ Interface 13-2
Si-H Bonds at the Si-SiO$_2$ Interface • Reliability
Considerations at the Nanoscale • Tunneling in Ultra
Thin Oxides
References 13-7
13.3 Modeling of Quantum Dots and Artificial Atoms 13-8
Introduction • The Many-Body Hamiltonian of Artificial
Atoms • Full Scale Simulation of Quantum Dot Devices
• Quantum Modeling of Artificial Molecules and Exchange
Engineering
Acknowledgments 13-19
References 13-19
13.4 Multiscale Theory and Modeling of Carbon
Nanotube Nanoelectromechanical Systems 13-20
Operation of NEM Switches • Nanotube Mechanics
• Electrostatics • Analytical Consideration of the Pull-In
• Outlook
Acknowledgments 13-30
References 13-30
13.5 Simulation of Ionic Channels 13-32
Hierarchical Approach to Modeling Ion Channels
• Drift-Diffusion Models • Monte Carlo Simulations
13.6 Conclusions 13-37
Acknowledgments 13-38
References 13-38

Narayan R. Aluru
Jean-Pierre Leburton
William McMahon
Umberto Ravaioli
Slava V. Rotkin
University of Illinois

Martin Staedele
Infineon Technologies

Trudy van der Straaten
University of Illinois

Blair R. Tuttle
Penn State University

Karl Hess
University of Illinois

13.1 Introduction

Karl Hess

Nanostructure research is defined by a scale — the nanometer length scale. Simulation of nanostructures, however, must be multiscale in its very nature. It is not the nanostructures themselves that open the horizon to new opportunities and applications in all walks of life; it is the integration of nanostructures into large systems that offers the possibility to perform complex electrical, mechanical, optical, and chemical tasks.

Conventional electronics approaches to nanometer dimensions and simulation techniques must increasingly use atomistic methods to compute, for example, tunneling and size quantization effects as well as the features of the electronic structure of the solids that define the nanometer-sized device. The atomistic properties need then to be linked to macroscopic electromagnetic fields and to the equations of Maxwell and, ultimately, to systems performance and reliability. The transition from the quantum and atomistic scale to the classical macroscopic scale is of great importance for the accuracy of the simulation. It can be described by the Landauer–Buettiker formalism, by Bardeen's transfer Hamiltonian method, or by more demanding methods such as the Schroedinger Equation Monte Carlo approach.[1] To encompass all of these scales and transitions, a hierarchy of methods (sets of equations) that supply each other with parameters is needed even for conventional silicon technology. Similar hierarchical approaches will be needed for future devices and their integration in electronics as well as electromechanics. One can already anticipate the demand for simulation methods that merge electronics, mechanics, and optics as well as the highly developed methods of chemistry.

A theoretical tool of ever-increasing use and usefulness is density functional theory (DFT). DFT describes, for example, the electrical and optical properties of a quantum dot (the prototype for future electron devices) and is also widely used in chemistry. The simulation methods become altogether more fundamental and powerful as simulation of nanostructure technology, both present and more futuristic, progresses. For example, the same simulation methods that have been developed in the last decade for silicon technology can also be applied to some biological systems, the carbon-based devices of nature, and the newly emerging field of carbon nanotubes. In turn, the methods developed in biochemistry become increasingly useful to answer questions in electronics and electromechanics at the nanoscale.

It is currently not possible to give an overview of all these opportunities in the limited space of this chapter. We present therefore only four vignettes that demonstrate the wide range of knowledge that is needed in nanostructure simulations and what can be anticipated in the future for simulations ranging from silicon-based electronics and nanoelectromechanics to biological systems such as protein-based ion channels.

References

1. Hess, K., *Advanced Theory of Semiconductor Devices*, IEEE Press, 2000.

13.2   Nanostructure Studies of the Si-SiO$_2$ Interface

William McMahon, Martin Staedele, Blair R. Tuttle, and Karl Hess

In this section we discuss modeling of the Si-SiO$_2$ interface, mostly in the context of Metal-Oxide-Semiconductor Field-Effect Transistors (MOSFETs). New insights are gained by explicitly calculating material properties using nanostructure and atomic-level techniques. This section thus offers an example of how nanostructure simulation is already necessary for conventional silicon technology as encountered in the highly integrated chips of today.

13.2.1   Si-H Bonds at the Si-SiO$_2$ Interface

Hydrogen has long been used in the processing of MOSFETs in order to passivate electrically active defects that occur, for instance, at the Si-SiO$_2$ interface. The Si-H binding energy was commonly assumed to be the threshold energy for H-related degradation in MOSFETs.[1] We have used density functional calculations to investigate the energetics of the hydrogen dissociation process itself.[2–7] These calculations show that there are several mechanisms by which hydrogen can desorb through processes that involve much lower energies than the Si-H binding energy of ~3.6 eV. These results explain continued hot-electron degradation in MOSFETs even as operating voltages have been scaled to below 3.6 eV.[4] Moreover, a distribution of dissociation energies due to disorder at the interface is expected. Such a distribution indicates that the probability of degradation will increase dramatically as MOSFETs are scaled to sub-100 nanometer channel lengths.[8,9]

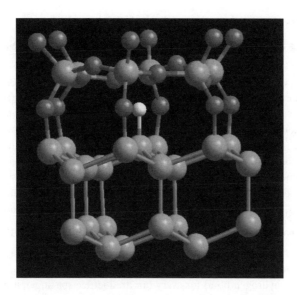

FIGURE 13.1   Atomic ball-and-stick model of an isolated Si-H bond at the Si-SiO$_2$ interface. Smaller balls represent oxygen atoms, and larger balls represent silicon.

13.2.1.1 Density Functional Calculations

Density functional theory (DFT) has become the leading theoretical tool for understanding nanoscale phenomena in physics and chemistry. This is because DFT allows an accurate determination of electronic structure and also efficiently scales with the number of atoms in a calculation. We have performed a comprehensive DFT study of the mechanisms of Si-H bond breaking at the Si-SiO$_2$ interface. We have used several atomic models of the interface including the cluster model shown in Figure 13.1. These studies demonstrate how DFT can be used to model electronics on the nanoscale.

Our main results for Si-H at the Si-SiO$_2$ interface are as follows. The energy needed to dissociate an isolated silicon–hydrogen bond (placing the hydrogen in a vacuum state at infinity) is found to be ~3.6 eV. For an Si-H bond at the Si-SiO$_2$ interface, if the dissociated hydrogen atom enters bulk SiO$_2$, then the dissociation or dissociation energy is also 3.6 eV because atomic hydrogen interacts only weakly with the rather open, insulating oxide. However, the Si-H dissociation energy can be significantly reduced for Si-H bonds at the Si-SiO$_2$ interface because hydrogen can desorb by first entering bulk silicon. The energy needed to place a neutral hydrogen atom, arising from the silicon dangling bond site, into bulk silicon far from any defects is ~2.5 eV. As hydrogen diffuses to a surface or interface, it can passivate other defects or combine with another hydrogen atom to form H$_2$. At a surface or an open interface such as the Si-SiO$_2$ interface, H$_2$ molecules can easily diffuse away, leaving behind the silicon dangling bonds. Experimentally, the thermally activated dissociation of hydrogen from the (111)Si-SiO$_2$ interface is measured at 2.56 eV. This is consistent with our calculated mechanism with H entering bulk silicon before leaving the system as H$_2$.

In addition to the above considerations, the threshold energy for hot-electron degradation can be greatly reduced if dissociation occurs by multiple vibrational excitations. For low voltages, Si-H dissociation involving multiple vibrational excitations by the transport electrons becomes relatively more likely. Because hydrogen is very light, the hydrogen in an Si-H bond is a quantum oscillator. Hot electrons can excite the hydrogen quantum oscillator from the ground state into an excited state. Because the Si-H vibrational modes are well above the silicon phonon modes, the excited state will be long-lived, allowing for multiple vibrational excitation. In this case, the Si-H dissociation can take place at channel electron energies lower than 2.5 eV and perhaps as low as 0.1 eV, the vibrational energy of the Si-H bending mode.[10]

13.2.2 Reliability Considerations at the Nanoscale

13.2.2.1 Increasing Effect of Defect Precursor Distribution at the Nanoscale

For micron-sized devices, many Si-H bonds at the Si-SiO$_2$ interface must be broken before the device has significantly degraded. For nanoscale devices, a much smaller number of defects (possibly on the order of 10s or lower) could cause a device to fail. Because of the smaller number of defects required, there is an increasing probability with decreasing device size of having a significant percentage of defect precursors with lifetimes in the short-lifetime tail of the Si-H dissociation energy distribution mentioned above. This results in an increasing number of short-time failures for smaller devices. In order to quantify this result, the shape of the distribution of dissociation energies must be known. Fortunately, this shape can be determined from the time dependence of trap generation under hot-electron stress, and from this the effect of deviations from this distribution on the reliability can be calculated. To understand how the reliability can be understood, we specifically look at the example of interface trap generation in nMOSFETs.

13.2.2.2 Hot-Electron Interface Trap Generation for Submicron nMOSFETs

A sublinear power law of defect generation with time is observed for the generation of interface traps at the Si-SiO$_2$ interface. Because the hydrogen is relatively diffusely spread throughout the interface (with only around one silicon–hydrogen bond for every hundred lattice spacings), it is clear that any process which breaks these bonds will be first order in the number of Si-H bonds. That is, the rate equation for this process can be written

$$\frac{dN(E_b)}{dt} = \frac{N(E_b)}{\tau(E_b)}$$

where $N(E_b)$ is the number of silicon–hydrogen bonds (which must be a function of E_b, the bond energy), and τ is some lifetime that, for a hot-electron-driven process, would involve an integration of the electron distribution with the cross section for defect creation, also a function of the bond energy. In order to get the true number of defects as a function of time for an average device, the solution to this rate equation must be integrated over the distribution of bond energies. This gives

$$N_{tot}(t) = \int_0^\infty f(E_b) N_0 \exp\left(-\frac{t}{\tau(E_b)}\right) dE_b$$

This integral is what produces the sublinearity of the time dependence of the generation of interface traps. The importance of this integral lies in the fact that it relates the sublinearity of the time dependence of the generation of interface traps with the average distribution of defect energies, which can be related to the distribution of defect generation lifetimes. This distribution can be used to determine the failure function for the failure mode involving this type of defect. One can extract the defect activation energy distribution from this integral once one knows the sublinear power law for defect generation with time.[11]

13.2.2.3 Reliability from Defect Precursor Distribution

Again utilizing the assumption of independent defects, the failure function of a device (defined as the probability of having a sufficient number of defects that will fail before some time t) will be a binomial or, approximately, a Poisson distribution. One of the characteristics of this failure function is an exponential increase in the probability of failure as the number of defects required to cause failure gets small. This is demonstrated in Figure 13.2, where we compare the reciprocal of the number of devices on a chip (which gives an idea of how much the reliability of a single device on that chip must increase) with the variation in the Poisson distribution with the number of interface traps required for the failure of a device. This is done for four gate lengths, with the gate lengths and number of devices at a given gate length taken from the semiconductor roadmap. The number of defects required for device failure comes from ISE-TCAD simulations. Notice the number of failures increases exponentially as the gate length is reduced below 100 nanometers.

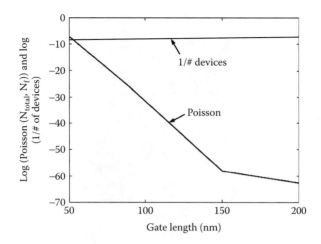

FIGURE 13.2 Example of failure function for interface trap generation.

Using knowledge of the type of defect involved in a degradation process, one can analytically derive the expected failure function for that type of degradation. This is not restricted to interface trap generation by Si-H dissociation, as the assumptions that went into the model are very few: first-order kinetics and a distribution of dissociation energies.

13.2.3 Tunneling in Ultra Thin Oxides

The thickness of gate oxides in MOSFETs is approaching 1–2 nm, i.e., only a few Si-O bond lengths. Consequently, gate leakage currents have become a major design consideration. For such ultra thin oxides, it is increasingly important to understand the *influence* of microscopic structure and composition of the oxide and its interface with silicon *on* the magnitude of oxide transmission probabilities and tunneling currents.

To fully explore the microscopic nature of gate leakage currents, an atomic-orbital formalism for calculating the transmission probabilities for electrons incident on microscopic models of Si-SiO$_2$-Si heterojunction barriers was implemented.[12–14] Subsequently the magnitude of leakage currents in *real* MOSFETs was calculated by incorporating the incident electron density from device simulations.[15,16] Such an approach allows one to examine the influence of atomic structure on tunneling. Significant results include assessing the validity of the bulk band structure picture of tunneling, determining the energy dependence of tunneling effective mass, and quantifying the nature of resonant tunneling through defects. Below, we will briefly discuss the most important details and results.

The microscopic supercell models of Si[100]-SiO$_2$-Si[100] heterojunctions that have been used were constructed by sandwiching unit cells of (initially) tridymite or *beta*-quartz polytype of SiO$_2$ between two Si[100] surfaces. The models are periodic in the plane perpendicular to the interface with periodic lengths of 0.5–1.5 nm. As more detail is desired, e.g., to examine the effects of interfacial morphology, the lateral periodic length scale can be increased with added computational costs. As an example, Figure 13.3 shows a ball-and-stick skeleton of a tridymite-based cell.

Reflection and transmission coefficients of the supercells described above were calculated using a transfer-matrix-type scheme embedded in a tight-binding framework. We solve the Schrödinger equation with open boundary conditions for the whole junction at a fixed energy E (measured relative to the silicon conduction band minimum on the channel side of the oxide) and in-plane momentum $k_\parallel$ (that is a good quantum number due to the lateral periodicity) in a *layer-orbital basis*. An empirical sp^3 tight-binding basis with second-nearest neighbor interactions for both silicon and the oxide were used. The tight-binding parameters were chosen to yield experimental bulk band gaps and to reproduce density functional calculations of the effective masses of the lowest conduction bands. An electron state propagating toward the oxide from the channel side of the junction, characterized by E, $k_\parallel$ and its wavevector

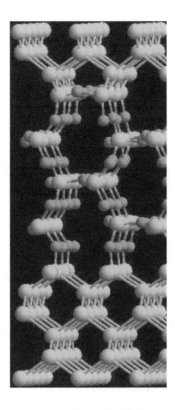

FIGURE 13.3 Ball-and-stick model of an $Si[100]$-SiO_2-$Si[100]$ model heterojunction based on the 1.3 nm thin gate oxide based on the tridymite polytype of SiO_2. (Dark = oxygen and light = silicon.)

component normal to the interface ($k_{perp,in}$), is scattered into sets of reflected and transmitted states (characterized by wavevector components $k_{perp,out}$). From the scattering wavefunctions, transmission amplitudes and dimensionless transmission coefficients are obtained.

The present microscopic models allow one to predict the intrinsic decay properties of the wavefunctions into the gate oxide. Because of the local nature of bonding in the oxide, a bulk picture of tunneling persists qualitatively even for the thinnest oxide barriers. We have analyzed the complex bands of the present bulk oxide models and find that (1) only one single complex band is relevant for electron tunneling; (2) several different bands are involved in hole tunneling; and (3) all complex oxide bands are highly nonparabolic. Because of the mismatch in the Brillouin zones for the oxide on top of the silicon, the bulk silicon $k_{\parallel}$ is not conserved and different states have differing decay constants. The energy dependence of the integrated transmission is shown for oxide thicknesses between 0.7 and 4.6 nm in Figure 13.4, which also includes effective–mass-based results with a constant (EM) and the energy-dependent (EM*) electron mass, which was fitted to our tight-binding complex band structures. The parabolic effective mass approximation overestimates the transmission for oxides thinner than ~1 nm. As oxide thicknesses increase, the tight-binding transmission is underestimated at low energies and overestimated at higher energies. The higher slope of the transmission obtained in the parabolic effective mass approximation is consistent with the findings for the tunneling masses and explains previous errors in oxide thicknesses derived from tunneling experiments and a constant parabolic effective mass model.[12] Using the correct tight-binding dispersion of the imaginary bands in an effective mass calculation (i.e., the EM* results in Figure 13.4) leads to qualitatively correct slopes for transmission; however, the absolute values are typically overestimated by one to two orders of magnitude. A possible reason for much of this discrepancy may be that the effective-mass-based transmission calculation underestimates the full band structure mismatch of silicon and its oxide.

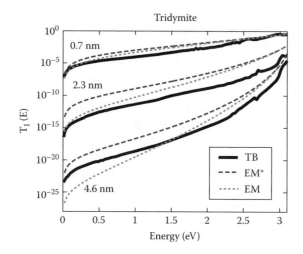

FIGURE 13.4 Integrated transmission (T_I) vs. the energy of the incident electron for tridymite-based oxides with thicknesses of 0.7, 2.3, and 4.6 nm. Results are for calculations with our atomic-level tight-binding method (TB, *solid line*), effective mass approaches with constant (EM, *dotted line*), and energy dependent (EM*, *dashed line*) effective masses.

The transmission coefficients were combined with electron densities and the corresponding distribution functions at the Si-SiO$_2$ interface of prototypical MOSFETs with channels of 50 nm and 90 nm. These quantities were obtained from full-band Monte Carlo simulations and were used to calculate the absolute magnitudes for gate leakage currents which, for oxide thicknesses smaller than ~4 nm, are dominated by tunneling of cold electrons in the source and drain contacts for defect-free oxides. As a consequence, the tunneling current densities (integrated over the entire gate length) decrease upon applying a drain-source voltage. The elastic gate leakage currents were recalculated including oxygen vacancies for a given energy E_{vac} in the oxide band gap from 0 to 3 eV above the silicon conduction band edge. The leakage currents at an arbitrary vacancy density were calculated using an interpolation formula.[15,16] Interestingly, we find that for all possible combinations of vacancy energy and density, the gate currents are still dominated by cold electrons originating in the contact regions. We have calculated the direct gate current densities from the source contact for the 50 nm transistor with a 1.3 nm oxide and the 90 nm transistor with a 2.9 nm oxide for defect densities in the range of 10^{10}–10^{13} cm^{-2} and a homogeneous as well as various Gaussian distributions of E_{vac} in energy space. The magnitude of the defect-induced current increase is very sensitive to the density and the energy distribution of the defects. For defect densities greater than 10^{12} cm^{-2}, the enhancement can be as high as 2–3 orders of magnitude. Also, the resonant effects are somewhat less pronounced for the thinner oxide.

We regard this work as the first steps toward the full understanding of oxide tunneling from a microscopic point of view. The theoretical approach presented here[12,15,16] could certainly be applied to other systems; and there are other methods to calculate electron transport at the atomic scale, which are of general interest for those interested in modeling nanoelectronic devices.[17–19]

References

1. Hu, C., Tam, S.C., Hsu, H., Ko, P., Chan, T., and Terrill, K.W., *IEEE Trans. Electron. Devices* ED-32, 375, 1985.
2. Tuttle, B.R., Hydrogen and PB defects at the Si(111)-SiO2 interface: an *ab initio* cluster study, *Phys. Rev. B* 60, 2631, 1999.
3. Tuttle, B.R. and Van de Walle, C., Structure, energetics and vibrational properties of Si-H bond dissociation in silicon, *Phys. Rev. B* 59, 12884, 1999.

4. Hess, K., Tuttle, B.R., Register, L.F., and Ferry, D., Magnitude of the threshold energy for hot-electron damage in metal oxide semiconductor field-effect transistors by hydrogen desorption, *Appl. Phys. Lett.* 75, 3147, 1999.

5. Tuttle, B.R., Register, L.F., and Hess, K., Hydrogen related defect creation at the Si-SiO2-Si interface of metal-oxide-semiconductor field-effect transistors during hot electron stress, *Superlattices Microstruct.* 27, 441, 2000.

6. Tuttle, B.R., McMahon, W., and Hess, K., Hydrogen and hot electron defect creation at the Si(100)-SiO2 interface of metal-oxide-semiconductor field effect transistors, *Superlattices Microstruct.* 27, 229, 2000.

7. Tuttle, B.R., Energetics and diffusion of hydrogen in SiO2, *Phys. Rev. B* 61, 4417, 2000.

8. Hess, K., Register, L.F., McMahon, W., Tuttle, B.R., Ajtas, O., Ravaioli, U., Lyding, J., and Kizilyalli, I.C., Channel hot carrier degradation in MOSFETs, *Physica B* 272, 527, 1999.

9. Hess, K., Haggag, A., McMahon, W., Cheng, K., Lee, J., and Lyding, J., The physics of determining chip reliability, *IEEE Circuits Device* 17, 33–38, 2001.

10. Van de Walle, C. and Tuttle, B.R., Microscopic theory of hydrogen in silicon devices, *IEEE Trans. Electron. Devices* 47, 1779, 2000.

11. Haggag, A., McMahon, W., Hess, K., Cheng, K., Lee, J., and Lyding, J., *IEEE Intl. Rel. Phys. Symp. Proc.*, 271, 2001.

12. Staedele, M., Tuttle, B.R., and Hess, K., Tunneling through ultrathin SiO_2 gate oxide from microscopic models, *J. Appl. Phys.* 89, 348, 2002.

13. Staedele, M., Fischer, B., Tuttle, B.R., and Hess, K., Influence of defects on elastic gate tunneling currents through ultrathin SiO2 gate oxides: predictions from microscopic models, *Superlattices Microstruct.* 28, 517, 2000.

14. Staedele, M., Tuttle, B.R., and Hess, K., Tight-binding investigation of tunneling in thin oxides, *Superlattices Microstruct.* 27, 405, 2000.

15. Staedele, M., Tuttle, B.R., Fischer, B., and Hess, K., Tunneling through ultrathin thin oxides — new insights from microscopic calculations, *Intl. J. Comp. Electr.*, 2002.

16. Staedele, M., Fischer, B., Tuttle, B.R., and Hess, K., Resonant electron tunneling through defects in ultrathin SiO2 gate oxides in MOSFETs, *Solid State Electronics*, 46, 1027–1032 (2001).

17. Klimeck, G., Bowen, R.C., and Boykin, T., Off-zone-center or indirect band-gap-like hole transport in heterostructures, *Phys. Rev. B* 63, 195310, 2001.

18. DiVentra, M. and Pantelides, S., Hellmann–Feynman theorem and the definition of forces in quantum time-dependent and transport problems, *Phys. Rev. B* 61, 16207, 2000.

19. Damle, P.S., Ghosh, A.W., and Datta, S., Unified description of molecular conduction: from molecules to metallic wires, *Phys. Rev. B* 64, 16207, 2001.

13.3 Modeling of Quantum Dots and Artificial Atoms

Jean-Pierre Leburton

13.3.1 Introduction

In the last 10 years, the physics of quantum dots has experienced a considerable development because of the manifestation of the discreteness of the electron charge in single-electron charging devices, as well as the analogy between three-dimensionally (3-D) quantum-confined systems and atoms.[1] Early studies were motivated by the observation of single-electron charging in granular metallic islands containing a "small" number of conduction electrons ($N \sim 100$ to 1000) surrounded by an insulator characterized by a small capacitance C.[2,3] In metallic dots, however, quantum confinement is relatively weak and the large effective mass of conduction electrons makes the energy spectrum a quasi-continuum, with negligible separation between electron states even at low temperature, $\Delta E << k_B T$. Hence, the addition of an electron to the island requires the charging energy $e^2/2C$ from a supply voltage source to overcome the

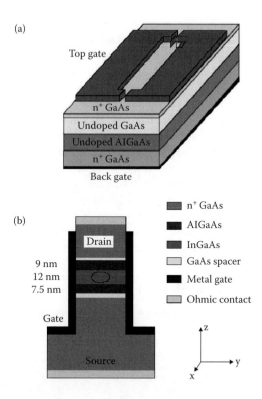

FIGURE 13.5 (a) Schematic representation of a planar quantum dot structure with the layered materials; the dark areas represent the confining metallic gates on the top surface. (b) Vertical quantum dot structure with the different constituting materials; the vertical dark areas on the side represent the controlling metallic gate.

electrostatic repulsion or Coulomb blockade from the electrons present in the dot, with negligible influence of the energy quantization in the system.[4]

Advances in patterning and nanofabrication techniques have made possible the realization of semiconductor quantum dots with precise geometries and characteristic sizes comparable to the de Broglie wavelength of charge carriers.[5] These quantum dots are realized in various configurations by combining heterostructures and electrostatic confinement resulting from biased metal electrodes patterned on the semiconductor surfaces. In 3-D confined III–V compound semiconductors, the small effective mass of conduction electrons results in an energy spectrum of discrete bound states with energy separation comparable to, or even larger than, the charging energy $e^2/2C$. The ability to vary the electrostatic potential over large voltage ranges allows for fine tuning of the quantum dot charge of just a few electrons ($N \sim$ 1 to 10).[6] Early experiments on single electron charging were made with layered (alternate) AlGaAs/GaAs structures by patterning several Schottky metal gates on top of a two-dimensional (2-D) electron gas to achieve lateral confinement. A back gate controls the number of electrons in the 2-D gas and the dot (Figure 13.5[a]).[5] In these planar structures, the current flows parallel to the layers, and the tunneling barriers between the dot and the 2-D gas are electrostatically modulated by the top gates. In vertical quantum dots, the charging electrons are sandwiched vertically between two tunneling heterobarriers, while the lateral confinement results from a vertical Schottky barrier achieved by the deep mesa etching of the multilayer structure (Figure 13.5[b]).[7,8] In this case, the current flows perpendicularly to the 2-D gas between the two heterobarriers, which are usually high and thin because they are made of different semiconductor materials; e.g., InGaAs or AlGaAs. In general, planar dots have a poor control of the exact number of electrons while vertical dots lack the barrier tunability of lateral structures.

In semiconductor quantum dots, discrete energy levels with Coulomb interaction amongst electrons for achieving the lowest many-body state of the system are reminiscent of atomic structures. In cylindrical quantum dots, shell structures in the energy spectrum and Hund's rule for spin alignment with partial shell filling of electrons have recently been observed.[8] One of the peculiarities of these nanostructures is the ability to control not only the shape of the dot, but also the number of electrons through gate electrodes.[9] Hence, "artificial atoms" can be designed to depart strongly from the 3-D spherical symmetry of the central Coulomb potential and its nucleus charge. In this context the physics of a few electrons, quantum dots offers new opportunities to investigate fundamental concepts such as the interaction between charge carriers in arbitrary 3-D confining potentials and their elementary excitation from the equilibrium. Moreover, since Hund's rule is the manifestation of spin effects with shell filling in quantum dots, the electron spin can, in principle, be also controlled by the electric field of a transistor gate.[10] The idea of controlling spin polarization independently of the number of electrons in quantum dots has practical consequences since it provides the physical ingredients for processing quantum information and making quantum computation possible.[11] In addition, spin degrees of freedom can be utilized for storing information in new forms of memory devices. Aside from the investigation of basic quantum phenomena, "artificial atoms" are also promising for applications in high-functionality nanoscale electronic and photonic devices such as ultrasmall memories or high-performance lasers.[2,12,13]

13.3.2   The Many-Body Hamiltonian of Artificial Atoms

The electronic spectrum of N-electron quantum dots are computed by considering the many-body Hamiltonian:

$$\hat{H} = \sum_i \hat{H}_{0i} + \sum_{i,j} \hat{H}_{ij} \tag{13.1}$$

where $\hat{H}_{0i}$ is the single-particle Hamiltonian of the ith electron and

$$H_{ij} = \frac{e^2}{\varepsilon \left| \vec{r}_i - \vec{r}_j \right|} \tag{13.2}$$

is the interaction Hamiltonian describing the Coulomb interaction between carriers. Here ε is the dielectric constant of the material. In the second term of Equation (13.1), the sum is carried out on $i \neq j$, avoiding the interaction of carriers with themselves. Quite generally, the Hamiltonian Eq. (13.1) is used for solving the Schrödinger equation for the many-particle energies and wave functions,

$$E = E_N(1,2,3,...,N) \tag{13.3}$$

$$\Psi = \Psi_N(\vec{r}_1, \vec{r}_2, \vec{r}_3, ..., \vec{r}_N) \tag{13.4}$$

which, given the two-body interaction Eq. (13.2), can only be solved exactly for $N = 2$. In this section, we will describe a natural approach toward the solution of this problem for a general number N of electrons by considering successive approximations.

13.3.2.1   Single-Particle Hamiltonian and Shell Structures

We start by considering a system of independent and 3-D confined electrons in the conduction band. By neglecting the interaction $\hat{H}_{ij}$, the Hamiltonian Eq. (13.1) is reduced to a sum of single particle Hamiltonians, each of the same form

$$\hat{H}_{0i} = \hat{H}_i = \frac{\hat{p}_{xi}^2 + \hat{p}_{yi}^2 + \hat{p}_{zi}^2}{2m^*} + \hat{V}(\vec{r}_i) \tag{13.5}$$

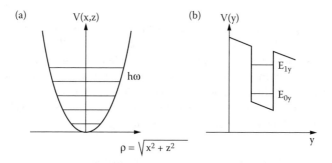

FIGURE 13.6 Schematic representation of (a) a 2-D parabolic potential with cylindrical symmetry in the x–z plane showing equally spaced energy levels and (b) the square potential with the first two quantized levels in the y-direction with $E_{1y} - E_{0y} >> \hbar\omega$.

Here we assume the electrons can be described with an effective mass m^*; $\hat{p}_{xi}, \hat{p}_{yi}, \hat{p}_{zi}$ are the components of the ith electron momentum and $\hat{V}(\vec{r}_i)$ is the external potential, which contains several contributions according to the confinement achieved in the quantum dot. We will assume that the quantum dot is realized by the confinement of the electrons in a heterostructure quantum well along the y-direction and electrostatic confinement in the x–z plane (Figure 13.6[a]). The latter confinement results usually from dopant atoms in neighboring semiconductor layers and from the fringing field of the metal electrodes on the semiconductor surface. This configuration is most commonly achieved in planar quantum dots and vertical quantum dots and results in a first approximation in a 2-D parabolic potential in the x–z plane (Figure 13.6[b]). Confinement at the heterostructure along the y-direction is generally strong ($\sim$10 nm), with energy separation of the order of 50 to 100 meV, while the x–z planar confinement is much weaker, with energy separation of the order of 1 meV over a larger distance ($>\sim$100 nm). In that case, the external potential is separable in a first approximation,

$$\hat{V}(\vec{r}) = \hat{V}_1(x,z) + \hat{V}_2(y) \tag{13.6}$$

which results in the energy spectrum $E_{v,n_x,n_z} = E_v + E_{n_x,n_z}$ with corresponding wave functions $\psi_v(y)\psi_{n_x,n_z}(x,z)$ where $E_v(E_{n_x,n_z})$ is the spectrum resulting from the y-potential (x–z potential). Hence, each value of the v-quantum number gives a series of x–z energy levels. At low temperature, given the large separation between the E_v energy states, only the first levels of the lowest series $v = 0$ are occupied by electrons. If one further assumes that the $\hat{V}_1(x,z)$ potential is cylindrically symmetric, the $v = 0$ energy spectrum is written as[14]

$$E_{0,n_x,n_z} = E_{0,m,l} = E_0 + (2m + |l| + 1)\hbar\omega \tag{13.7}$$

where ω is the frequency of the cylindrical parabolic potential. Here each m-level is $2m$-times degenerate, with the factor 2 accounting for the spin degeneracy. The number m ($=1,2,3,\ldots$) is the radial quantum number and the number l ($=0, +1, +2,\ldots$) is the angular momentum quantum number. Hence the 2-D cylindrical parabolic potential results into 2-D s,p,d,f,...-like orbitals supporting 2, 4, 6, 8, ... electrons, which give rise to shell structures filled with 2, 6, 12, 20, ... particles, thereby creating a sequence of numbers that can be regarded as the 2-D analogs of "magic numbers" in atomic physics.[8,15]

In the absence of cylindrical or square symmetry, the parabolic potential is characterized by two different frequencies, ω_x and ω_z, which lifts the azimuthal degeneracy on the l-number of the 2-D artificial atoms. Therefore electronic states are spin-degenerate only and determine a sequence of shell filling numbers 2, 4, 6, 8, ... of period or increment 2. Only when the ratio ω_x/ω_z is commensurable does the sequence of filling numbers deviate from the period 2, and provide a new sequence of numbers for particular combinations of the n_x and n_z quantum numbers in the case of accidental degeneracy.[16]

Another important class of 3-D confined systems includes quantum dots obtained by self-assembled or self-organized Stranski–Krastanov (SK) epitaxial growth of lattice-mismatched semiconductors, which results in the formation of strained induced nanoscale islands of materials. InAs and InGaAs islands on GaAs have been obtained with this technique in well-controlled size and density.[13,17,18] For these materials, shapes vary between semi-spherical and pyramidal forms, and the size is so small that these quantum dots only contain one 3-D fully quantized level for conduction electrons.

13.3.2.2 Hartree–Fock Approximation and Hund's Rules

The natural extension of the atomic model for independent 3-D confined electrons is the consideration of the Coulomb interaction between particles in the Hartree–Fock (HF) approximation. The HF scheme has the advantage of conserving the single particle picture for the many-body state of the system by representing the total wavefunction as a product of single-particle wavefunctions in a Slater determinant that obeys Fermi statistics. The main consequence of the HF approximation for the Coulomb interaction amongst particles is a correction of two-terms to the single particle energies derived from the H_0 Hamiltonian[19]

$$\left(\hat{H}_{0i} + \frac{e^2}{\varepsilon} \sum_j \int d\vec{r}' \left| \psi_j(\vec{r}') \right|^2 \frac{1}{\left| \vec{r} - \vec{r}' \right|} - \frac{e^2}{\varepsilon} \sum_j \int d\vec{r}' \frac{1}{\left| \vec{r} - \vec{r}' \right|} \psi_j^*(\vec{r}') \psi_i(\vec{r}') \psi_j(\vec{r}) \delta_{S_i S_j} \right) \psi_i(\vec{r}) = E_i \psi_i(\vec{r}) \qquad (13.8)$$

where the first term in the Hamiltonian is the single-particle Hamiltonian. The second term is the Hartree potential and the sum is carried on all occupied *j*-states, irrespectively of their spins, and accounts for the classical repulsion between electrons. The third term accounts for the attractive exchange interaction that occurs amongst carriers with parallel spins, i.e., $S_i = S_j$. In this scheme, the wavefunctions $\psi_i(\vec{r}_i)$ satisfy the HF integro–differential equation where the Coulomb interaction term depends upon all the single-particle wavefunctions of the occupied states. The HF equation is therefore nonlinear and must be solved self-consistently for all wavefunctions of occupied states.

One of the important consequences of the HF approximation for inter-electron interaction (Equation [13.2]) is the prediction of spin effects in the shell filling of artificial atoms similar to Hund's rules in atomic physics.[8] These effects are illustrated in the charging energy of a few electron quantum dots with cylindrical parabolic potential achieved in planar or vertical quantum structures.[6] In Figure 13.7(a), we show schematically the Coulomb staircase resulting from charging a quantum dot with a few electrons as a function of the charging energy or voltage between the metal electrode and the semiconductor substrate. The relative step sizes of the staircase represent the amount of energy needed to put an additional electron in the dot. The arrows on each step represent the spin of each individual electron on the successive orbitals during the charging process. The filling of the first shell (s-orbital with two electrons) consists of one electron with spin up followed by an electron with spin down. The step size of the spin-up electron measures the charging energy needed to overcome the Coulomb repulsion against the spin-down electron, which is just the Hartree energy between the two particles. The larger step size of the second (spin-down) electron is due to the fact that the charging of the third electron requires the charging energy augmented by the energy to access the next quantized level, which is the first p-orbital. The latter process starts the second shell filling with the third electron on either one of the degenerate $l = \pm 1$ orbitals of either spin (here we choose the $l = -1$ and the spin up). At this stage, the configuration with the fourth electron on the $l = 1$ orbital with a parallel spin becomes more favorable because it minimizes the Hartree energy between orbitals of different quantum numbers and results in an attractive exchange energy between the two electrons (Figure 13.7[c]). This is the reason why the third step is smaller than the first and second steps, requiring less energy and demonstrating Hund's rule in the electron filling of the 2-D artificial atom. The fourth step is long because the addition of the fifth

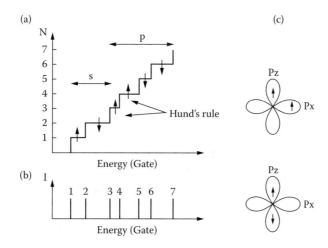

FIGURE 13.7 (a) Coulomb staircase as a function of the charging energy with the spin states of each electron. N is the number of electrons, and the horizontal two head arrows indicate the occupation of the s- and p-orbitals in the dot. (b) Electron current through the dot vs. the charging energy. (c) 2-D p-orbitals illustrating the two possible occupations of two electrons with parallel (*top diagram*) and antiparallel spins (*bottom diagram*).

electron on either of the p-orbitals with $l = \pm 1$ must correspond to a spin-down electron that undergoes a repulsion from the two other p-electrons without benefiting from the exchange with them since its spin is antiparallel. Figure 13.7(b) shows the current peaks resulting from the single-electron charging of the quantum dot, which is obtained by differentiating the Coulomb staircase. Current characteristics with similar structure have recently been observed in gated, double-barrier GaAs/AlGaAs/InGaAs/AlGaAs/GaAs vertical quantum dot tunneling devices that revealed the shell structure for a cylindrical parabolic potential as well as spin effects obeying Hund's rule in the charging of the dot.[8]

The HF approximation provides a reasonable picture of the contribution of electron–electron interaction and spin effects in the spectrum of quantum dots. However, it is well known from atomic physics and theoretical condensed matter physics that this approximation suffers from two important drawbacks: the neglect of electron correlation and the fact that it overestimates the exchange energy.[20] Moreover, it leads to a tedious solution of the self-consistent problem when a large number of electrons are involved.

13.3.3 Full Scale Simulation of Quantum Dot Devices

Advances in computer simulation combine the sophistication of realistic device modeling with the accuracy of computational physics of materials based on the density functional theory (DFT).[21–25] These powerful methods provide theoretical tools for analyzing the fine details of many-body interactions in nanostructures in a 3-D environment made of heterostructures and doping, with realistic boundary conditions. *Microscopic* changes in the quantum states are described in terms of the variation of *macroscopic* parameters such as voltages, structure size, and the physical shape of the dots without *a priori* assumption of the confinement profile. Consequently, engineering the exchange interaction among electrons for achieving controllable spin effects in quantum devices becomes possible.

The implementation of a spin-dependent scheme for the electronic structure of artificial molecules involves the solution of the Kohn–Sham equation for each of the spins: i.e., up ($\uparrow$) and down ($\downarrow$). Under the local spin density approximation within the DFT, the Hamiltonian H $^{\uparrow(\downarrow)}$ for the spin $\uparrow$ ($\downarrow$) electrons in the presence of an external magnetic field reads[20,26,39]

$$\hat{H}^{\uparrow(\downarrow)} = \frac{1}{2}\left(-i\hbar\nabla + \frac{e}{c}\vec{A}\right)\frac{1}{m^*(\vec{r})}\left(-i\hbar\nabla + \frac{e}{c}\vec{A}\right) + E_c(\vec{r}) + \mu_{xc}^{\uparrow(\downarrow)}[n] \tag{13.9}$$

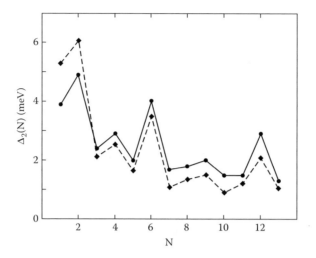

FIGURE 13.8 Addition energy of a vertical single quantum dot. *Dashed line*: experimental data from Tarucha et al.; *solid line*: theory. (From S. Tarucha, D.G. Austing, T. Honda, R.J. van der Hage and L.P. Kouwenhoven, *Phys. Rev. Lett.* 77, 3613 (1996); P. Matagne, J.P. Leburton, D.G. Austing and S. Tarucha, *Phys. Rev. B* 65, 085325 (2002). With permission.)

where $m^\star(r)$ is the position-dependent effective mass of the electron in the different materials, $A = 1/2$ $(By, -Bx, 0)$ is the corresponding vector potential, $E_c(r) = e\phi(r) + \Delta E_{os}$, is the effective conduction band edge, and $\phi(r)$ is the electrostatic potential, which contains the Coulomb interaction between electrons, and ΔE_{os} is the conduction band offset between GaAs and AlGaAs. The respective Hamiltonians are identical in all respects, except for the exchange-correlation potential, which is given by

$$\mu_{xc}^{\uparrow(\downarrow)} = \frac{d(n\varepsilon_{xc}[n])}{dn^{\uparrow(\downarrow)}} \tag{13.10}$$

where ε_{xc} is the exchange-correlation energy as a function of the total electron density $n(r) = n^\uparrow(r) + n^\downarrow(r)$ and the fractional spin polarization $\xi = (n^\uparrow - n^\downarrow)/n$, as parameterized by Ceperley and Alder.[27] While it is known that the DFT underestimates the exchange interaction between electrons, which leads to incorrect energy gaps in semiconductors, it provides a realistic description of spin–spin interactions in quantum nanostructures, as shown in the prediction of the addition energy of vertical quantum dots (see Figure 13.8).

The 3-D Poisson equation for the electrostatic potential $\phi(r)$ reads

$$\vec{\nabla} \cdot [\varepsilon(\vec{r})\vec{\nabla}\phi(\vec{r})] = -\rho(\vec{r}) \tag{13.11}$$

Here $\varepsilon(r)$ is the permittivity of the material, and the charge density ρ is comprised of the electron and hole concentrations as well as the ionized donor and acceptor concentrations present in the respective regions of the device. The dot region itself is undoped or very slightly p-doped. At equilibrium, the electron concentrations for each spin in the dots are computed from the wavefunctions obtained from the respective Kohn–Sham equations, i.e., $\rho(r) = en(r)$ with $n^{\uparrow(\downarrow)} = \Sigma_i |\psi_i^{\uparrow(\downarrow)}(r)|^2$. In the region outside the dots, a Thomas–Fermi distribution is used, so that the electron density outside the dot is a simple local function of the position of the conduction band edge with respect to the Fermi level, ε_F. The various gate voltages — V_{back}, V_t, and those on the metallic pads and stubs — determine the boundary conditions on the potential $\phi(r)$ in the Poisson equation. For the lateral surfaces in the x–z plane in Figure 13.5, vanishing electric fields are assumed.

Self-consistent solution of the Kohn–Sham and Poisson equations proceeds by solving the former for both spins, calculating the respective electron densities and exchange correlation potentials, solving the Poisson equation to determine the potential $\phi(r)$ and repeating the sequence until the convergence criterion is satisfied.[24] Typically, this criterion is such that variations in the energy levels and electrostatic potential between successive iterations are below 10^{-6} eV and 10^{-6} V, respectively.

The determination of N_{eq}, the number of electrons in the dots at equilibrium for each value of the gate and tuning voltages, is achieved by using Slater's transition rule:[28]

$$E_T(N+1) - E_T(N) = \int_0^1 \varepsilon_{LAO}(n) dn \cong \varepsilon_{LAO}\left(\frac{1}{2}\right) - \varepsilon_F \tag{13.12}$$

where $E_T(N)$ is the total energy of the dot for N electrons and $\varepsilon_{LAO}(1/2)$ is the eigenvalue of the lowest-available-orbital when it is occupied by 0.5 electrons. From the latter equation, it is seen that if the right-hand side is positive $N_{eq} = N$, otherwise $N_{eq} = N + 1$. Thus the $N \rightarrow N + 1$ transition points are obtained by populating the system with $N + 0.5$ electrons and varying V_{back} until $\varepsilon_{LAO}(1/2) - \varepsilon_F$ becomes negative. It should be noted that the approximation made in the latter equation is valid only if ε_{LAO} varies linearly with N. This approach has been very successful in the analysis of the electronic spectra and charging characteristics of vertically confined quantum dots.[29] Figure 13.8 shows the addition energy spectrum of a single vertical quantum dot as a function of the number N of electrons in the dot. The addition energy measures the energy required to add a new electron in the dot given the presence of other electrons already in the dot, and the restriction imposed by the Pauli principle on the electron energy spectrum. The peaks at $N = 2, 6$, and 12 are the signature of the existence of a 2-D shell structure in the dot, while the secondary peaks at $N = 4$ and 9 reflect the existence of Hund's rule at half-filled shells. The agreement between theory and experimental data is excellent for the position of the peaks as well as for their magnitude.

The 3-D self-consistent technique provides the direct extraction of the exchange interaction, $J = E_T - E_S$ defined as the energy difference between the triplet and singlet state for two-electron systems as a function of magnetic fields; the results are shown in Figure 13.9. Aside from the fact that in both structures, J changes sign over the range of investigated magnetic fields, resulting in a spin state transition where the triplet becomes the ground state, one can see that the behavior of J in the circular and rectangular mesa structures is radically different. First, the value of J for the rectangular QD is about three times smaller

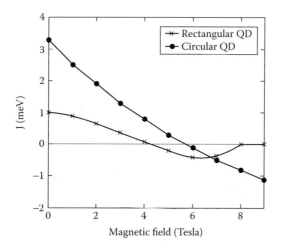

FIGURE 13.9 Calculated singlet triplet energy separation J in the two electron system as a function of the magnetic field. The aspect ratio for rectangular QD is 1.444.

than for the circular QD at zero magnetic field. Also, in the limit of a large magnetic field, J vanishes in the rectangular mesa QD, while it becomes more negative without showing any saturation in the circular QD. This difference in the magnetic field dependence of J for the two structures is due to symmetry breaking in the rectangular QD, where, with increasing magnetic fields, the two electrons localize on opposite sides of the dot, resulting in a vanishing J; no such effect is found in circular QDs, up to $B = 9$ T.[39]

The technique is also useful in designing double quantum dots with variable inter-dot barriers for controlling electron–electron interactions [30]. The devices have a planar geometry made of GaAs/AlGaAs heterostructure that contains a 2-D electron gas (2DEG). The dots are defined by a system of gate pads and stubs that are negatively biased to deplete the 2DEG, leaving two pools of electrons that form two quantum dots connected in series (Figure 13.9).[26] In these "artificial diatomic molecules," electron states can couple to form covalent states that are delocalized over the two dots, with electrons tunneling between them without being localized to either.[31] These *bonding states* have lower energies than the constituent dot states by an amount that is equivalent to the binding energy of the molecule. In our case, the dimensions are such that the electron–electron interaction energy is comparable to the single-particle energy level spacing. The number of electrons in the dot, N, is restricted to low values in a situation comparable to a light diatomic molecule such as H–H or B–B. The coupling between dots can be adjusted by varying the voltage on the tuning gates, V_t, to change the height of the barrier between the two dots. The number of electrons N in the double dot is varied as the 2DEG density, with the back gate voltage V_{back} for a fixed bias on the top gates. Hence, controllable exchange interaction that gives rise to spin polarization can be engineered with this configuration by varying N and the system spin, independently.[32]

13.3.4 Quantum Modeling of Artificial Molecules and Exchange Engineering

In order to simulate these effects, we consider a structure that consists of a 22.5 nm layer of undoped $Al_{0.3}Ga_{0.7}As$ followed by a 125 nm layer of undoped GaAs, and finally an 18 nm GaAs cap layer (Figure 13.10[a]). The latter is uniformly doped to 5×10^{18} cm^{-3} so that the conduction band is just above the Fermi level at the boundary between the GaAs cap layer and the undoped GaAs. The inverted heterostructure is grown on the GaAs substrate. The lateral dimension of the gates and spacing are shown in Figure 13.10(b). Figure 13.10(c) shows the schematic of the lowest four states with their wavefunctions in the double dot for two different tuning voltages.[26] For both values of V_t, the ground state in the individual dots is s-like and forms a degenerate pair. Here, we borrow the terminology of atomic physics to label the quantum dot states. The first excited states, which are p_x- and p_z-like, are degenerate for weak inter-dot coupling, whereas for strong coupling, the p_z-like states mix to form symmetric (*bonding*) and antisymmetric (*antibonding*) states that are lower in energy than the p_x-like states as seen in Figure 13.10 (c). This reordering of the states has an important bearing on the spin-polarization of the double-dot system, as shall be fully explained below.

In the present double dot structure, we focus on the spin states of the electron system and allow N to vary from zero to eight for two values of V_t: $V_t = -0.67$ V defined as the weak coupling regime, and $V_t = -0.60$ V defined as the strong coupling regime. Electron spin states that are relevant in this analysis are designated by $s_1^{\uparrow(\downarrow)}$, $s_2^{\uparrow(\downarrow)}$ (lower energy s-states in dots 1 and 2, and $p_{x1}^{\uparrow(\downarrow)}$, $p_{x2}^{\uparrow(\downarrow)}$, $p_{z1}^{\uparrow(\downarrow)}$, and $p_{z2}^{\uparrow(\downarrow)}$; higher energy p-states where the x and z indices indicate the orientation of the wavefunctions). For $N = 0$, in the weak coupling regime, the computer model shows that s-states in dots 1 and 2 have negligible overlap because of the relatively high and wide barrier. Indeed, the bonding–antibonding energy separation resulting from the coupling between these states is orders of magnitude smaller than the Coulomb charging energy so that s-electrons are practically localized in each dot. A similar situation arises for the p_{x}- and p_{z}- states, which although experiencing slight overlap because they are higher in energy, see a lower and thinner barrier, and are quasi-degenerate within each dot. In fact, in the weak coupling limit, p_{z1}- and p_{z2}- states that are oriented along the coupling direction between dots, experience a bigger overlap than the corresponding p_x-states, and consequently lie slightly lower in energy than the

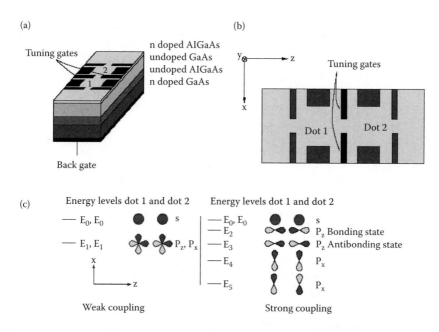

FIGURE 13.10 Schematic representation of the planar coupled quantum dot device. (a) Layer structure with top and back gates; (b) top view of the metal gate arrangement with sizes and orientations; (c) schematic representation of the six lowest orbitals in the weak (*left hand side*) and strong (*right hand side*) coupling regimes. In both cases, the s-states are strongly localized in their respective dots. Left: p_x- and p_z-like orbitals are degenerated within each dot and decoupled from the corresponding state in the other dot. Right: increasing coupling lifts the p-orbitals, degeneracy with a reordering of the states. The gray and dark orbitals indicate positive and negative parts of the wavefunctions, respectively.

latter. Hence, as far as the lower s- and p-states are concerned, the double dot system behaves as two quasi-independent dots (Figure 13.10[c], left). In addition, because of the large distance separating the two lower s-states for $V_t = -0.67$ V, Coulomb interaction between electrons in dots 1 and 2 is negligibly small, and both dots can be charged simultaneously through double charging[26] to completely fill the s_{1-} and s_{2-} states. Therefore, for $N = 4$, there is no net spin polarization in the double dot, since both contain an equal number of spin ↑ and spin ↓ electrons.

When the double dot is charged with the fifth electron, the latter occupies either the p_{z1-} or p_{z2}-state (e.g., ↑ spin, i.e., $p_{z1}{}^\uparrow$ or $p_{z2}{}^\uparrow$) that has the lowest available energy. The sixth electron takes advantage of the nonzero p-state overlap and occupies the other $p_z{}^\uparrow$-state with a parallel ↑ spin. The seventh and eighth electrons find it energetically favorable to occupy successively $p_{x1}{}^\uparrow$ and $p_{x2}{}^\uparrow$, but not any of the spin ↓ states, because of the attractive nature of the exchange-correlation energy among the spin ↑ electrons that results from the nonzero p-state overlap, and lowers the energy of the double dot. This particular high spin polarization configuration among p-orbitals in the "artificial" diatomic molecule deserves special attention because it appears to violate one of Zener's principles on the onset of magnetism in transition elements; this principle forbids spin alignment for electrons on similar orbitals in adjacent atoms.[33] Therefore, it could be argued that the high spin polarization configuration obtained in the calculation is the consequence of a DFT artifact. Recently, however, Wensauer et al. confirmed the DFT results based on a Heitler–London approach.[34] Similar conclusions have also been obtained by an "exact" diagonalization technique on vertically coupled quantum dots for $N = 6$ electrons.[35,36] Let us point out that Zener's principle is purely empirical, as it is based on the observation of the magnetic properties of natural elements that lack the tunability of "artificial" systems. Therefore, the total spin of the double dot can possibly steadily increase by $1/2\hbar$ for each electron added after the fourth electron to $2\hbar$ for $N = 8$, and there is no contradiction with Zener's principle applied to natural elements. After all, high spin configurations have been shown to compete for the ground state of light diatomic molecules such as B_2.[37]

TABLE 13.1 Spin of the Double Dot for Various Occupation Numbers in the Two Coupling Regimes

$\sqrt[V]{N}$		1	2	3	4	5	6	7	8	9	10	11	12
Spin	−0.67 V	1/2	?	1/2	0	1/2	1	3/2	2	3/2	1	1/2	0
($\hbar$)	−0.67 V	1/2	0	1/2	0	1/2	0	1/2	0	1/2	0	1/2	0

The question mark at N = 2 in the weak coupling regime indicates that the spins are uncorrelated.

The variation of the total spin S in the double dot with N is shown in Table 13.1. It is also seen that as N increases above eight electrons, the spin $\downarrow$ states start to be occupied, thereby decreasing S by $1/2\hbar$ for each additional electron, forming antiparallel pairs to complete the shell until $N = 12$ when S is reduced to zero. The sequence of level filling with the occupation of degenerate states by electrons of parallel spins is observed in atoms, and is governed by Hund's rules; it is therefore impressive that similar rules successfully govern level filling in the double dot in the weak coupling regime. Let us point out that even though $S = 2\hbar$ is the most favored state of the double dot, energetically it is not significantly lower than other competing states for $N = 8$. For instance, the excited states with $S < 2\hbar$ for $N = 8$ are only about 0.1 meV higher in energy. Consequently, for this particular double dot structure, any attempt to observe experimentally the parallel alignment of the spins of unpaired electrons is restricted to low temperatures for which $k_BT \ll 0.1$ meV, or any kinds of electrostatic fluctuations smaller than this value. However, it must be noted that the structure is not optimized and that the evidence of spin polarization among p-states in the double dot system may be achieved in smaller dots with stronger exchange interaction. The key issue here is the fact that the quantum mechanical coupling between the two dots in this bias regime is not strong enough to lift the spatial quasi-degeneracy among p_x- and p_z-states which, for our particular configuration, were separated by no more than a few microelectron volts. Stronger coupling between the quantum dots eliminates this effect. Accordingly, if V_t increases to −0.60 V, also referred to as the strong coupling regime, the p-state spatial degeneracy is completely lifted, while deeper s-states also couple, although to a slighter extent, to lead to the spectrum of Figure 13.10(c, right). Therefore, the spin sequence as a function of N is alternatively $S = 1/2\hbar$ for odd N when the last occupying electron is unpaired, and $S = 0$ for even N, when it pairs up with an electron of the opposite spin (Table 13.1).

The variation of inter-dot coupling by varying V_t provides a control of direct exchange interaction between p-like electrons in the two dots, which may be more robust than for s-electrons. Hence, a lowering of the inter-dot barrier results in a reordering of the single-particle levels, thereby transforming the double-dot (for $N = 8$) from a spin polarized $S = 2\hbar$ to an unpolarized state $S = 0$. An important result from Table 13.1 is that the Loss–DiVincenzo scheme for quantum computing with double dots could also be achieved for $N = 6$ electrons where the control of qubit entanglement for a quantum control-not (XOR) gate operation would be realized with the $S = 1/2\hbar$ spin states of two p-electrons instead of two single electrons ($N = 2$) in the original scenario.[38]

The electrostatic nature of the confinement potential, specifically the coupling barrier, is central to the occurrence of the effects mentioned above. Indeed, the barrier is not uniform, but wider (and higher) for the lower quantum dot s-states than for the higher p-states (Figure 13.10[c]). This situation is similar to the electronic properties of natural diatomic molecules where the strongly localized s-states correspond to atomic core states and the delocalized p-states to covalent bonding states. It is therefore possible to engineer exchange interaction in the "artificial molecule" by suitably tailoring the coupling barrier between quantum dots. This is achievable by proper device designs, e.g., by adjusting gate size and shape, the doping profiles, the distance between the GaAs/AlGaAs heterojunction and the control gates, and possibly choosing other III–V semiconductor systems to optimize the energy separation between singlet and multiplet states.

Although the DFT calculation provides a reasonably good description of the exchange interaction J between two electrons in single quantum dot systems, it leads to spurious results in coupled quantum dot systems (Figure 13.10). Specifically, the presence of the fictitious "self-interaction" term in the

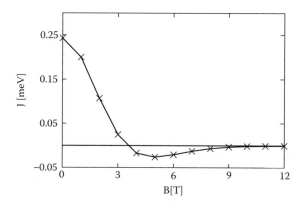

FIGURE 13.11 Exchange interaction as a function of magnetic field in a model coupled dot system. $V_1 = V_r = 25$ meV and $d = 50$ nm.

Kohn–Sham Hamiltonian has more dramatic effects than in single QDs, yielding unphysical results for J. In particular, we found that the ground state energy with one electron in a coupled quantum dot is lower than that with zero electrons, which leads to failure to determine the charging point bias using Slater's rule. In order to avoid this issue, we use the method of exact diagonalization to obtain the exchange interaction J in a double dot system.[40] In Figure 13.11, we show the magnetic field dependence of J for a 2-D Gaussian model potential

$$V(x,y) = -V_l e^{-[(x+d/2)^2 + y^2]} - V_r e^{-[(x-d/2)^2 + y^2]}$$ (13.13)

It is shown that the singlet–triplet transition occurs at a relatively small field ($B \sim 3$ T) and that J saturates at large magnetic fields ($B > 8$ T), which is in good qualitative agreement with model calculations using analytic approaches such as Heiter–London and Hund–Mulliken [41]. We point out that for realistic studies of many-body effects, the potential obtained from the 3-D self-consistent scheme can, in principle, be coupled to the exact diagonalization method to obtain quantitative assessment of the exchange interaction in coupled QD systems.

Acknowledgments

JPL is indebted to Drs. D.G. Austing, R.M. Martin, S. Nagaraja, L.-X. Zhang, and S. Tarucha for fruitful discussions, and to Dr. P. Matagne for technical assistance. This work was supported by NSF grants DESCARTES ECS-98-02730, the Materials Computational Center (DMR-99-76550), and the DARPA-QUIST program (DAAD19-01-1-0659).

References

1. M.A. Kastner, *Physics Today* 46, 24 (1993).
2. D.V. Averin and K.K. Likharev, in *Mesoscopic Phenomena in Solids*, eds. B.L. Altshuler, P.A. Lee and R.A. Webb, Elsevier, Amsterdam pp.173–271 (1991).
3. M.H. Devoret and H. Grabert, in *Single Charge Tunneling: Coulomb Blockade Phenomena in Nanostructures*, NATO ASI B294, eds. M.H. Devoret and H. Grabert, Plenum, New York, pp. 1–19 (1991).
4. K.K. Likharev, *Proc. IEEE*, 87, 606 (1999).
5. V. Meirav and E.B. Foxman, *Semicond. Sci. Technol.* 10, 255 (1995).
6. R.C. Ashoori, H.L Stoermer, J.S. Weiner, L.N. Pfeiffer, S.J. Pearton, K.W. Bladwin and K.W. West, *Phys. Rev. Lett.* 68, 3088 (1992).
7. R. Ashoori, *Nature*, 379, 413 (1996).

8. S. Tarucha, D.G. Austing, T. Honda, R.J. van der Hage and L.P. Kouwenhoven, *Phys. Rev. Lett.* 77, 3613 (1996).
9. D.G. Austing, T. Honda and S. Tarucha, *Jpn. J. Appl. Phys.* 36, 4151 (1997).
10. S.M. Sze, *Physics of Semiconductor Devices,* 2nd ed. John Wiley, New York (1981).
11. D.P. Di Vincenzo, *Nature* 393, 113 (1998).
12. Y. Arakawa and A. Yariv, *IEEE J. Quant. Electron.* 22, 1887 (1986).
13. D. Bimberg, M. Grundmann and N.N. Ledentsov, *Quantum Dot Heterostructures*, Wiley, London (1998).
14. N.F. Johnson, *J. Phys. Condens. Matter* 7, 965 (1995).
15. M. Maccucci, K. Hess and G.J. Iafrate, *J. Appl. Phys.* 77, 3267 (1995).
16. S. Nagaraja, P. Matagne, V.Y. Thean, J.P. Leburton, Y-H. Kim and R.M Martin, *Phys. Rev. B*, 56 15752 (1997).
17. D. Leonard, K. Pond and P.M. Petroff, *Phys. Rev. B* 50, 11687 (1994).
18. M.S. Miller, J.O. Malm, M.E. Pistol, S. Jeppesen, B. Kowalski, K. Georgsson and L. Samuelson, *Appl. Phys. Lett.* 80, 3360 (1996).
19. O. Madelung, *Introduction to Solid State Physics*, Springer Series in Solid-Sate Science, 2. Springer Verlag, Berlin (1978).
20. R.O Jones. and Gunnarsonn, *Rev. Mod. Phys.* 61, 689 (1989).
21. A. Kumar, S.E. Laux and F. Stern, *Phys. Rev. B* 42, 5166 (1990).
22. M. Stopa, *Phys. Rev. B* 54, 13767 (1996).
23. M. Koskinen, M. Manninen and S.M Rieman, *Phys. Rev. Lett.* 79, 1389 (1997).
24. D. Jovanovic and J. P. Leburton, *Phys. Rev. B* 49, 7474. (1994).
25. I.H. Lee, V. Rao, R.M. Martin and J.P. Leburton, *Phys. Rev. B* 57, 9035 (1998).
26. S. Nagaraja, J.P. Leburton and R.M. Martin, *Phys. Rev. B* 60, 8759 (1999).
27. J.P. Perdew and A. Zunger, *Phys. Rev. B* 23, 5048 (1981).
28. J.C. Slater, *Quantum Theory of Molecules and Solids*, McGraw-Hill, New York, (1963).
29. P. Matagne, J.P. Leburton, D.G. Austing and S. Tarucha, *Phys. Rev. B* 65, 085325 (2002).
30. R.H. Blick, R.J. Haug, J. Weis, D. Pfannkuche, K.V. Klitzing and K. Eberl, *Phys. Rev. B* 53, 7899 (1996).
31. W.G. van der Wiel et al., *Rev. Mod. Phys.* 75, 1 (2003).
32. This is not the case in shell filling of single quantum dots since the total spin of the electronic systems is directly related to the number *N* of electrons in the dot.
33. C. Zener, *Phys. Rev.* 81, 440 (1951).
34. A. Wensauer, O. Steffens, M. Suhrke and U. Roessler, *Phys. Rev. B* 62, 2605 (2000).
35. H. Imamaura, P.A. Maksym and H. Aoki, *Phys. Rev. B* 59, 5817 (1999).
36. M. Rotani, F. Rossi, F. Manghi and E. Molinari, *Solid State Commun.* 112, 151 (1999).
37. C.F. Bender and E.R Davidson. *J. Chem. Phys.* 46, 3313 (1967).
38. D. Loss and D.P. Di Vincenzo, *Phys. Rev. A.* 57, 120 (1998).
39. D. V. Melnikov et al., *Phys. Rev. B* 72, 085331 (2005).
40. L.-X. Zhang, D.-V. Melnikov and J. P. Leburton, to be published.
41. G. Burkard, D. Loss and D.P. DiVincenzo, *Phys. Rev. B* 59, 2070 (1999).

13.4 Multiscale Theory and Modeling of Carbon Nanotube Nanoelectromechanical Systems

Slava V. Rotkin, Narayan R. Aluru, and Karl Hess

The device aspects of carbon nanotubes represent an interesting new area of nanoscience and nanotechnology.[1] Various (carbon, nitride, and chalcogenide) nanotubes are promising for applications because of their unusual mechanical and electronic properties, stability, and functionality.[2]

The lattice structure of single-wall carbon nanotubes follows the lattice structure of graphene (mono-layer of graphite): a hexagonal pattern is repeated with translational symmetry along the tube axis and with axial (chiral) symmetry along the tube circumference. Nanotubes are labeled using two numbers [n,m]. These are components of the vector that generates the tube circumference after scrolling, in terms of basic vectors of the graphene lattice (see Figure 13.12a). It is easy to find that only two types of single-wall nanotubes (SWNTs) have a pure axial symmetry: the so-called armchair (A) and zig-zag (Z) nanotubes. The graphene rectangle shown in Figure 13.12a gives an armchair (A) nanotube when wrapped from top to bottom (Figure 13.12b) and a zig-zag (Z) nanotube when wrapped from left to right. Any other type of nanotube is chiral, which means that it belongs to a screw-axis symmetry group.[2]

Graphite-like systems and materials such as fullerenes, nanotubes, nanographites, and organic mac-romolecules are well known to have valence/conduction band systems generated by pi and sigma valence electrons.[3] The latter ones are localized and, normally, contribute only to the mechanical properties of the graphitic material. In contrast, pi-electrons are mobile and highly polarizable and define transport, electrical, and electromechanical properties. The pi-electronic structure of a monolayer of graphite (graphene)

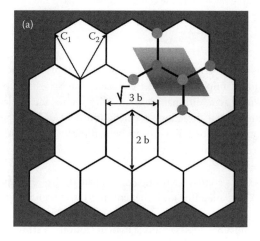

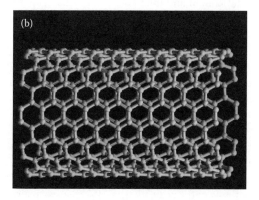

FIGURE 13.12 (a) Honeycomb lattice structure of graphene having a rhombic unit cell with two carbon atoms. Translated along basal vectors, c1 and c2, it forms two interconnected sublattices. The carbon–carbon bond length, b, is ~0.14 nm. The edge direction, in basal vectors, is denoted by two integers (shown fragment has left/right edge of type [2,2]). (b) Lattice structure of [10,10] armchair SWNT. Wrapping the honeycomb lattice along some chosen axis will form a nanotube.

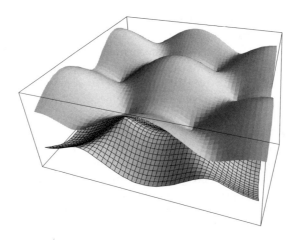

FIGURE 13.13 TBA electronic structure of the valence bands of a monolayer of graphite. The three-fold symmetry of the lattice results in six Fermi points where the conduction band meets the valence band. Only two of these six Fermi points are not equivalent.

is shown in Figure 13.13. It has six Fermi points that separate an empty conduction band from an occupied (symmetrical) valence band. A simple but correct picture of the electronic structure of a SWNT follows from a band-folding argument: an additional space quantization for the pi-electrons appears due to their confinement in the circumferential direction. It can be thought of as a mere cross-sectioning of the electronic structure of graphene along the nanotube symmetry direction. Depending on the lattice symmetry of the tube, three different situations can be realized:

1. The armchair SWNT has a cross-section passing through the Fermi point (Figure 13.14, top). In this case the SWNT is metallic and the conduction band merges with the valence band (Figure 13.15, left).
2. The zig-zag/chiral nanotube cross-section is distant from the Fermi point (Figure 13.14, bottom). This tube has a nonzero gap as shown in Figure 13.15(right), and it is a semiconductor tube.
3. One third of zig-zag and chiral nanotubes have a very small gap, which follows from arguments other than simple band-folding. In our simplified picture, these SWNTs will have a zero band gap and be metals.

The band gap of a semiconductor nanotube depends solely on the tube radius, R. A simple rule follows from the band-folding scheme: $E_g = tb/R$, where $t \sim 2.7$ eV is the hopping integral for pi-electrons. This gap dependence was experimentally measured by scanning tunneling spectroscopy,[4] optical spectroscopy,[2,5–7] and resonance Raman spectroscopy.[2,8]

Nanotubes with conductivities ranging from metallic to semiconductor were indeed synthesized. The temperature dependence of the conductivity of the nanotube indicates reliably the metallic or semiconductor character. The field effect is also very useful in distinguishing between two types. To measure this effect, the nanotube is placed between two electrodes on top of a backgate contact that is covered with an insulating layer. After synthesis and purification, a SWNT is normally p-type; i.e., the majority of carries are holes. A typical density of $\sim 10^7$ cm^{-1} holes defines the conductivity in the ON state of a SWNT when operated as a field-effect transistor (FET) at zero gate voltage. External positive voltage, applied to the backgate, can deplete the holes and switch the SWNT-FET to the OFF state. Experiments have shown a drop of five orders of magnitude of the source–drain current when the gate voltage was changed by 3 V for very thin insulator layers (thickness less than 2 nm).[9] In the case of a metallic nanotube bridging two electrodes, only a weak dependence, if any, of the conductivity on the gate voltage is seen.

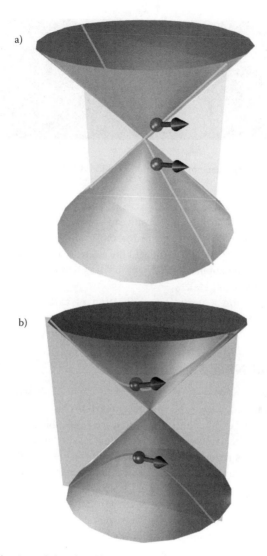

FIGURE 13.14 Lowest conduction sub-band and highest valence sub-band of a metallic tube (*top*) compared with semiconductor nanotube (*bottom*). A nanotube quantization condition makes a cut from the cone-shaped bands of graphite. In the case of a metallic nanotube, this cross-section passes through the Fermi point and no gap develops between the sub-bands. The electron dispersion is linear in the longitudinal wave vector. In the case of a semiconductor nanotube, the cross-section is shifted away from the Fermi point. The carrier dispersion is a hyperbola.

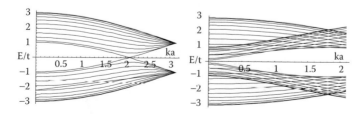

FIGURE 13.15 Electronic structure of a metallic armchair [10,10] nanotube (*left*) and a semiconductor zig-zag [17,0] nanotube (*right*). The pi-electron energy is plotted in units of hopping integral, $t \sim 2.7$ eV vs. the dimensionless product of the longitudinal wave vector, k, and the bond length, $a \sim 0.14$ nm (half of the Brillouin zone is shown).

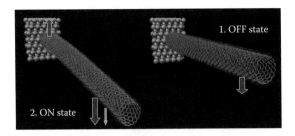

FIGURE 13.16 ON and OFF states of a nanotube electromechanical switch. Arrows show applied forces: electrostatic, van der Waals, and elastic.

The electronic structure of SWNT is highly sensitive to external fields,[10–12] and lattice distortions cause changes in the electronic structure. A lattice distortion moves the Fermi point of graphite and results in the closing/opening of an energy gap,[13,14] a change in the electron density, and charging of the tube. This opens many possibilities for the application of nanotubes as nanobiosensors, mesoscopic devices, and nanoelectromechanical systems (NEMs).

In this chapter, we focus on a particular application of carbon nanotubes — NEM switches (Figure 13.16) and nanotweezers (Figure 13.17). The three basic energy domains that describe the physical behavior of NEM switches — mechanics, electrostatics, and van der Waals — are described below.

13.4.1 Operation of NEM Switches

Shown in Figure 13.18 is the NEM operation of a carbon nanotube-based cantilever switch. Since this theoretical study was conducted, a few experimental realizations have been demonstrated by researchers all over the world.[15-22] The key components of a switch are a movable structure, which can be a SWNT or a multiwall carbon nanotube, nanowire, nanocrystalline needle, or other 1-D conducting object, and a fixed ground plane, which is modeled by a graphite bulk (in Ref. 23). When a potential difference is created between the movable structure and the ground plane, electrostatic charges are induced on both the movable structure and the ground plane. The electrostatic charges give rise to electrostatic forces, which deflect the movable tube. In addition to electrostatic forces, depending on the gap between the movable tube and the ground plane, the van der Waals forces (see below) also act on the tube and deflect it. The directions of the electrostatic and van der Waals forces are shown in Figure 13.18. Counteracting the electrostatic and van der Waals forces are elastic forces, which try to restore the tube to its original straight position. For an applied voltage, an equilibrium position of the tube is defined by the balance of the elastic, electrostatic, and van der Waals forces. As the tube deflects, all forces are subject to change, and a self-consistent analysis is necessary to compute the equilibrium position of the tube.

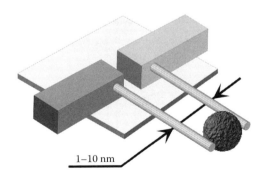

1–10 nm

FIGURE 13.17 Nanotube nanotweezers device.

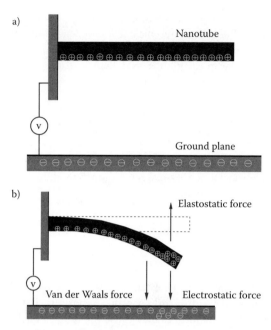

FIGURE 13.18 Force balance for a nanotube over a ground plane: (*top*) position of the tube when $V = 0$; (*bottom*) deformed position of the tube when $V \neq 0$.

When the potential difference between the tube and the ground plane exceeds a critical value, the deflection of the tube becomes unstable and the tube collapses onto the ground plane. The potential that causes the tube to collapse is defined as the pull-in voltage or the collapse voltage. When the pull-in voltage is applied, the tube comes into contact with the ground plane, and the device is said to be in the ON state (Figure 13.8 top). When the potential is released and the tube and the ground plane are separated, the device is said to be in the OFF state (Figure 13.8 bottom).

When compared with microelectromechanical switches, the operation of NEM switches is different because of the importance of the van der Waals forces, which can be neglected at the micrometer scale. The sticking of NEM devices becomes an increasing problem at the nanoscale and can limit the range of operability of NEMs. If the gap between the cantilever tube and the ground plane is very small, even without an applied voltage, the tube can collapse onto the ground plane because of the van der Waals forces. In addition, the separation of the tube from the ground plane after the contact becomes an issue as the van der Waals forces will tend to keep the tube and the ground plane together.

13.4.2 Nanotube Mechanics

The mechanical and structural properties of nanoscale systems have been studied both theoretically and experimentally over the last decade.[24] The strong correlation between the structure and electronic properties of a nanosystem requires a proper understanding of the nanomechanical and NEM behavior of nanotubes. Such studies can lead to new design tools for microscopy and characterization studies as well as the development of highly sensitive detectors. The mechanical behavior of a small structure differs from that of a bulk structure. New phenomena such as super-low friction,[25] super-high stiffness,[26] and high cohesion at small distances[27] are encountered.

The mechanical behavior of nanotubes can be modeled either by simple continuum approaches or by more complex atomistic approaches based on molecular dynamics simulations. The elastic properties of pure single-wall and multiwall nanotubes were studied by, for example, Sanchez–Portal et al.[28] and Yakobson and Avouris.[29] Atomistic approaches have the advantage of capturing the mechanical behavior accurately; however, they require large computational resources. Continuum theories, when properly parameterized and calibrated,[30] can be more efficient for understanding the mechanical behavior of nanotubes. A simple continuum approach for modeling the mechanical behavior of NEM switches is based on the beam theory. The beam equation is given by:

$$EI \frac{\partial^4 r}{\partial x^4} = q$$

where r is the gap between the conductor and the ground plane, x is the position along the tube, q is the force per unit length acting normal to the beam, E is the Young's modulus, and I is the moment of inertia, which, for nanotubes, can be estimated as:

$$I = \frac{\pi}{4} \left(R_{ext}^4 - R_{int}^4 \right)$$

where R_{int} is the interior radius and R_{ext} is the exterior radius of the nanotube.

The beam theory can, however, suffer from several limitations. For very large loads, the stress concentration at the edges of the nanotubes may cause the tube to buckle and form kinks. In such cases, the deflection deviates from the beam theory locally. The buckling happens at a certain strain, depending on the device geometry, the nanotube symmetry, and the load. If buckling is to be simulated, one can try advanced continuum theories such as a shell theory or a full elasticity theory.[31]

Many-body corrections to van der Waals interactions from semiclassical Casimir forces were calculated and applied in the continuum modeling of nanotube mechanics.[32] The basic analysis of the role of van der Waals terms in electromechanical systems has demonstrated its significance at the subnanometer scale.[23,33] A recent theory[34] of van der Waals interaction for shells of pure carbon was based on the universal principles formulated in the 1930s.[35] The new approach is based on the quantum electrodynamical description of the van der Waals/Casimir forces. A simple and effective model has been developed to estimate the many-body contribution due to collective modes (plasmons).[32] This contribution is believed to be a major portion of the total van der Waals energy because of the high oscillator strength of the plasmons. The theory reveals many-body terms that are specific for various low-dimensional graphite nanostructures and are not taken into account by standard one-body calculations within the dispersionless model by Lennard-Jones.[35] We have demonstrated the use of the model for several systems (shown in Figure 13.19): a double-walled nanotube (a), a nanotube on the surface (b), and a pair of single-wall tubes (c). A significant difference has been shown for the dependence of the van der Waals energy on distance, which is a consequence of our quantum correction.[32] Further studies have been published on the van der Waals forces in NEMs since the publication of the first edition of this handbook.[36,37]

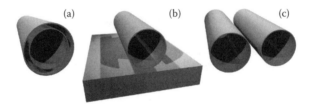

FIGURE 13.19 Geometry of nanotube systems for which a quantum correction to van der Waals forces has been calculated: (a) double-wall nanotube; (b) single-wall nanotube on a surface; (c) two single-wall nanotubes.

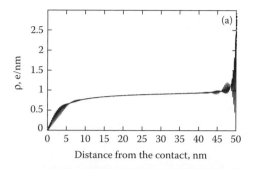

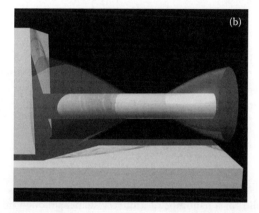

FIGURE 13.20 (a) Self-consistent charge density of a [10,10] armchair nanotube at 5 V applied between side and backgate contacts. (b) Sketch of the simulated device geometry. The distance between the tube center and the backgate is 5 nm, and the tube radius and length are 0.6 and 60 nm, respectively.

13.4.3 Electrostatics

The 3-D character of the electromagnetic eigenmodes and 1-D charge density distribution of a SWNT system result in a weak screening of the Coulomb interaction and the external field. We present a quantum mechanical calculation of the polarizability of the metallic [10,10] tube. The nanotube polarizability is not defined solely by the intrinsic properties of the tube.[38] It depends also on the geometry of the nanotube and closest gates/contacts.[2,39] Hence, the charge distribution has to be treated self-consistently. Local perturbations of the electronic density will influence the entire system, unlike in common semiconductor structures. For example, a point charge placed near the tube surface will generate an induced-charge density along the tube length, which decays very slowly with the distance from the external charge.

Figure 13.20 (after Ref. 38) is a sketch of the depolarization of the tube potential (induced-charge density) by the side electrode and the backgate (part b of Figure 13.20 shows the geometry of the device simulated). The continuous line is the statistical approximation (Boltzmann–Poisson equations), which coincides well with the quantum mechanical result (dotted line) except for the quantum beating oscillations at the tube end. The depolarization manifests itself as a significant nonuniformity of the charge along the tube length. This effect is described by the self-consistent compact modeling, which is outlined below.

The potential ϕ_{act} that is induced by a charge density, ρ_{ind}, in 1-D systems is proportional to the charge density. Thus, for a degenerate electronic structure of a metallic nanotube in the low-temperature limit, the Poisson equation is effectively reduced to[38,40]

$$\rho_{ind}(z) = -e^2 \nu_M \, \phi_{act}(z).$$

Here nM stands for the nanotube density of states, which is constant in a studied voltage range. We have demonstrated that $e2nM$ acts as an atomistic quantum capacitance of a SWNT

$$C_Q^{-1} = \frac{1}{e^2 v_M}$$

(a similar quantity for a 2-D electron gas system has been introduced by Luryi[41]), and the geometric capacitance

$$C_g^{-1} = 2\log\left(\frac{2h}{R}\right)$$

is a function of distance to the backgate and SWNT radius. In case of the straight SWNT (as in Figure 13.20), the geometric capacitance is a logarithmic function of the distance between the tube and the gate. In equilibrium, we have the following relation between the equilibrium charge density and external potential (gate voltage), which comprises both the atomistic and the geometric capacitances:

$$\rho_\infty = -\frac{\phi_{ext}}{C_g^{-1} + C_Q^{-1}} \simeq -\phi_{ext} C_9 \left(1 - \frac{C_g}{C_Q}\right)$$

This equation is still valid for a nanotube of arbitrary shape, although no simple expression for the geometric capacitance can be written.

13.4.4 Analytical Consideration of the Pull-In

We finish this section with an analytical model that can be used for a quick estimation of pull-in voltages of the nanotube system within continuum modeling. Assuming that the elastic energy of the NEM device is given by

$$T = k(h - x)^2/2$$

and the external (electrostatic) force is the gradient of the energy component given by

$$V = C\phi^2/2$$

we can calculate elastic and electrostatic forces. Then we include the van der Waals energy term:

$$W \simeq \varepsilon x^{-\alpha}$$

and write analytically the pull-in voltage and pull-in gap as functions of the device stiffness, k, the device capacitance, C, and van der Waals energy, W:

$$x_o = hA_1 \frac{1}{2}\left(1 + \sqrt{1 + A_2 \frac{W(x_o)}{kh^2}}\right)$$

$$V_o = B_1 \frac{\sqrt{2kh}}{\sqrt{C(x_o)}} \sqrt{\frac{1}{2} - B_2 \frac{W(x_o)}{kh^2} + \frac{1}{2}\sqrt{1 + A_2 \frac{W(x_o)}{kh^2}}}$$

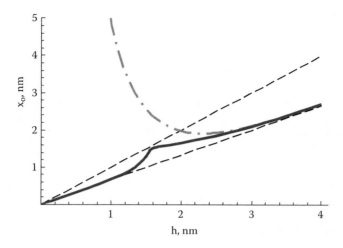

FIGURE 13.21 The pull-in gap as a function of the initial device gap. The solid curve represents the self-consistent analytical result. The dash–dotted curve shows the dependence when neglecting the van der Waals correction.

Here four constants $A1$, $B1$, $A2$, and $B2$ describe the specific dependence of C and W on x, the dynamic gap or the internal coordinate of the NEM device. In case of a planar switch and the Lennard-Jones potential, these constants are $3/2$, $\sqrt{2}/3$, 36, and 36, respectively.

As a result of the van der Waals attraction to the gate, the NEM device cannot operate at very small gaps, h. The critical gap, h_c (at which $xo = 0$) is about 2 nm for the switch with $k \sim W/1$ nm^2, and $C \simeq 2k^{1/2}/(3$ V/nm). Next, Figure 13.21 and Figure 13.22 show that, by neglecting the van der Waals correction to the pull-in gap, xo, one underestimates the critical pull-in voltage by 15%.

The self-consistent solution for the pull-in gap is plotted in Figure 13.22. Again, neglecting the van der Waals terms results in an unphysical divergence of the pull-in gap when approaching the critical distance h_c.

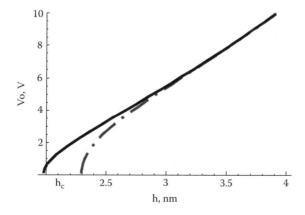

FIGURE 13.22 The pull-in voltage as a function of the gap. The solid curve represents the analytical result explained in the text. The dash–dotted curve shows the dependence when neglecting the van der Waals correction for the pull-in gap.

13.4.5 Outlook

Development of fast and precise approaches for 3-D device modeling of nanotube systems becomes clearly important in view of the existence of prototypes for nanotube electronics made at the IBM and Delft groups[9,42] and also after the recent successes of NEMS fabrication.[15–22] The physics of carbon nanotube devices is rather distinct from the physics of standard semiconductor devices, and it is unlikely that semiconductor device modeling tools can be simply transferred to nanotube device modeling.[2,37]

Development of device modeling tools for nanotubes can be very complicated because of the breakdown of continuum theories. Molecular mechanics (MM) and molecular dynamics (MD) can be used reliably when continuum theories break down. However, both MM and MD can be very computer time consuming. A good compromise is to develop a multiscale approach where continuum theories are combined with atomistic approaches. Multiscale methods can be accurate and more efficient compared with atomistic approaches. The highest level in the multiscale hierarchy is represented by the quantum mechanical result for the single-tube polarizability, which is the atomistic analog of the bulk dielectric function. It contains the complete information for the electronic structure and charge distribution and gives the means for calculating the screened Coulomb and the van der Waals/Casimir forces.[32] The main difficulty here is the requirement to solve the problem for device structures. The electronic structure and the polarizability change during device operation, and this requires a self-consistent treatment.[2] At the intermediate level, classical MD provides a detailed knowledge for geometry and material parameters of the system. This is a prerequisite for calculating the mechanical response of the system. It also supplies proper boundary conditions for electrostatic calculations through the actual device geometry.[39] At the lowest level of the simulation hierarchy, the only level that can be used to simulate and understand larger systems of devices, continuum theories can and must be applied. The parameters of the continuum models will, of course, need to be derived from the higher level simulations.

Modeling and simulation of large-scale nanoscale circuits where carbon nanotubes are interconnected with other nanoelectronic, nanomechanical, chemical, and biological molecules are beyond the capability of currently existing supercomputers. Development of compact models for nanotubes and other nanodevices can enable the design of large-scale nanocircuits for breakthrough engineering applications.

Acknowledgments

N.R. Aluru is thankful to M. Dequesnes for technical assistance and for allowing the use of some of his results. S.V. Rotkin is indebted to Dr. L. Rotkina and Dr. I. Zharov for fruitful discussions and to K.A. Bulashevich for technical support. N.R. Aluru and S.V. Rotkin acknowledge the support of the CRI grant of UIUC. S.V. Rotkin acknowledges support of DoE grant DE-FG02-01ER45932 and the Beckman Fellowship from the Arnold and Mabel Beckman Foundation. K. Hess acknowledges support by the Office of Naval Research (NO0014-98-1-0604).

References

1. Hess, K., *Advanced Theory of Semiconductor Devices*, IEEE Press, 2000.
2. Rotkin, S.V. and Subramoney, S., eds., *Applied Physics of Nanotubes: Fundamentals of Theory, Optics and Transport Devices*, Ph. Avouris, Ser. Ed., Nanoscience and Nanotechnology Series, Springer, 2005.
3. Dresselhaus, M.S., Dresselhaus, G., and Eklund, P.C., *Science of Fullerenes and Carbon Nanotubes*, Academic Press, 1996.
4. Wildoer, J.W.G., Venema, L.C., Rinzler, A.G., Smalley, R.E., and Dekker, C., Electronic structure of atomically resolved carbon nanotubes, *Nature* 391(6662), 59–62, 1998.
5. O'Connell, M., Bachilo, S.M., Huffman, C.B., Moore, V., Strano, M.S., Haroz, E., Rialon, K., Boul, P.J., Noon, W.H., Kittrell, C., Ma, J., Hauge, R.H., Weisman, R.B., and Smalley, R.E., *Science* 297, 593, 2002.

6. Hagen, A., Moos, G., Talalaev, V., and Hertel, T., *Appl. Phys. A* 78, 1137, 2004.
7. Lefebvre, J., Homma, Y., and Finnie, P., *Phys. Rev. Lett.* 90, 217401/1, 2003.
8. Kuzmany, H., Plank, W., Hulman, M., Kramberger, C., Gruneis, A., Pichler, T., Peterlik, H., Kataura, H., and Achiba, Y., Determination of SWCNT diameters from the Raman response of the radial breathing mode, *Eur. Phys. J. B* 22(3), 307–320, 2001.
9. Bachtold, A., Hadley, P., Nakanishi, T., and Dekker, C., Logic circuits with carbon nanotube transistors, *Science* 294(5545), 1317–1320, 2001.
10. Li, Y., Rotkin, S.V., and Ravaioli, U., Electronic response and bandstructure modulation of carbon nanotubes in a transverse electrical field, *Nano Lett.* 3(2), 183–187, 2003, Li, Y., Rotkin, S.V., Ravaioli, U., Metal-semiconductor transition in armchair carbon nanotubes by symmetry breaking, *Appl. Phys. Lett.* 85(18), 4178–4180, 2004, Li, Y., Rotkin, S.V., and Ravaioli, U., Metal-semiconductor transition and Fermi velocity renormalization in metallic carbon nanotubes, *Phys. Rev. B* 73(3), 035415–10, 2006.
11. Park, C.J., Kim, Y.H., and Chang, K.J., *Phys. Rev. B* 60, 10656, 1999, Tien, L.-G., Tsai, C.-H., Li, F.-Y., and Lee, M.-H., Band-gap modification of defective carbon nanotubes under a transverse electric field, *Phys. Rev. B* 72, 245417–245426, 2005.
12. Son, Y.-W., Ihm, J., Cohen, M.L., Louie, S.G., and Choi, H.J., Electrical switching in metallic carbon nanotubes, *Phys. Rev. Lett.* 95, 216602–216604, 2005.
13. Kleiner, A. and Eggert, S., Band gaps of primary metallic carbon nanotubes, *Phys. Rev. B* 63, 073408, 2001.
14. Yang, L. and Han, J., Electronic structure of deformed carbon nanotubes, *Phys. Rev. Lett.* 85, 154–157, 2000.
15. Jang, J.E., Cha, S.N., Choi, Y., Amaratunga, G.A.J., Kang, D.J., Hasko, D.G., Jung, J.E., and Kim, J.M., Nanoelectromechanical switches with vertically aligned carbon nanotubes, *Appl. Phys. Lett.* 87, 163114–1631203, 2005; Cha, S.N., Jang, J.E., Choi, Y., Amaratunga, G.A.J., Kang, D.J., Hasko, D.G., Jung, J.E., and Kim, J.M., Fabrication of a nanoelectromechanical switch using a suspended carbon nanotube, *Appl. Phys. Lett.* 86, 3105–3107, 2005.
16. Dujardin, E., Derycke, V., Goffman, M.F., Lefevre, R., and Bourgoin, J.P., Self-assembled switches based on electroactuated multiwalled nanotubes, *Appl. Phys. Lett.* 87, 193107–193113, 2005.
17. Ke, C.H., Pugno, N., Peng, B., and Espinosa, H.D., Experiments and modeling of carbon-nanotube based NEMS devices, *J. Mech. Phys. Solids* 53, 1314–1333, 2005.
18. Sapmaz, S., Blanter, Y.M., Gurevich, L., and van der Zant, H.S.J., Carbon nanotubes as nanoelectromechanical systems, *Phys. Rev. B* 67, 235414–235417, 2003.
19. Kinaret, J.M., Nord, T., and Viefers, S., A carbon-nanotube-based nanorelay, *Appl. Phys. Lett.*, 82, 1287–1289, 2003; Jonsson, L.M., Axelsson, S., Nord, T., Viefers, S., and Kinaret, J.M., High frequency properties of a CNT-based nanorelay, *Nanotechnology* 15, 1497–1502, 2004; Axelsson, S., Campbell, E.E.B., Jonsson, L.M., Kinaret, J., Lee, S.W., Park, Y.W., and Sveningsson, M., Theoretical and experimental investigations of three-terminal carbon nanotube relays, *New J. Phys.* 245, 2005.
20. Ziegler, K.J., Lyons, D.M., Holmes, J.D., Erts, D., Polyakov, B., Olin, H., Svensson, K., and Olsson, E., Bistable nanoelectromechanical devices, *Appl. Phys. Lett.* 84, 4074–4076, 2004.
21. Postma, H.W.C., Kozinsky, I., Husain, A., and Roukes, M.L., Dynamic range of nanotube- and nanowire-based electromechanical systems, *Appl. Phys. Lett.* 86, 223105–223113, 2005.
22. Lee, J. and Kim, S., Manufacture of a nanotweezer using a length controlled CNT arm, *Sensors and Actuators A — Physical*, 120, 193–198, 2005.
23. Dequesnes, M., Rotkin, S.V., and Aluru, N.R., Calculation of pull-in voltages for carbon nanotube-based nanoelectromechanical switches, *Nanotechnology* 13, 120–131, 2002.
24. Ebbesen, T.W., Potential applications of nanotubes, in *Carbon Nanotubes*, CRC Press, Boca Raton, FL, 1997, p. 296.
25. Falvo, M.R., Steele, J., Taylor, R.M., and Superfine, R., Gearlike rolling motion mediated by commensurate contact: carbon nanotubes on HOPG, *Phys. Rev. B* 62(16), R10665–R10667, 2000.

26. Yu, M.F., Files, B.S., Arepalli, S., and Ruoff, R.S., Tensile loading of ropes of single wall carbon nanotubes and their mechanical properties, *Phys. Rev. Lett.* 84(24), 5552–5555, 2000.

27. Hertel, T., Walkup, R.E., and Avouris, P., Deformation of carbon nanotubes by surface van der Waals forces, *Phys. Rev. B Condens. Matt.* 58(20), 13870–13873, 1998.

28. Sanchez–Portal, D., Artacho, E., Solar, J.M., Rubio, A., and Ordejon, P., Ab initio structural, elastic, and vibrational properties of carbon nanotubes, *Phys. Rev. B Condens. Matt.* 59(19), 12678–12688, 1999.

29. Yakobson, B.I. and Avouris, P., Mechanical properties of carbon nanotubes, in *Carbon Nanotubes: Synthesis, Structure, Properties, Applications,* Dresselhaus, M. et al., eds., 80, 287–327, 2001.

30. Dequesnes, M., Rotkin, S.V., and Aluru, N.R., Parameterization of continuum theories for single wall carbon nanotube switches by molecular dynamics simulations, *J. Comput. Electron.* 1(3), 313–316, 2002.

31. Yakobson, B.I., Brabec, C.J., and Bernholc, J., Nanomechanics of carbon tubes — instabilities beyond linear response, *Phys. Rev. Lett.* 76(14), 2511–2514, 1996.

32. Rotkin, S.V. and Hess, K., Many-body terms in van der Waals cohesion energy of nanotubules, *J. Comp. Electr.* 1, 294–297, 2002.

33. Rotkin, S.V., Analytical calculations for nanoscale electromechanical systems, in Hesketh, P.J., Ang, S.S., Davidson, J.L., Hughes, H.G., and Misra, D., eds., *Microfabricated Systems and MEMS — VI,* vol. PV 2002–2006, Electrochemical Society Symposium Proceedings, ECS Inc., Pennington, NJ, 2002, pp. 90–97.

34. Girifalco, L.A., Hodak, M., and Lee, R.S., Carbon nanotubes, buckyballs, ropes, and a universal graphitic potential, *Phys. Rev. B Condens. Matt.* 62(19), 13104–13110, 2000.

35. Lennard-Jones, J.E., Perturbation problems in quantum mechanics, *Proc. R. Soc. London A* 129, 598–615, 1930.

36. Palasantzas, G. and DeHosson, J.T.M., Phase maps of microelectromechanical switches in the presence of electrostatic and Casimir forces, *Phys. Rev. B* 72, 121409–121414, 2005, Palasantzas, G. and De Hosson, J.T.M., Pull-in characteristics of electromechanical switches in the presence of Casimir forces: influence of self-affine surface roughness, *Phys. Rev. B* 72, 115426–115435, 2005.

37. Dequesnes, M., Tang, Z., and Aluru, N.R., Static and dynamic analysis of carbon nanotube-based switches, *J. Eng. Mater. Technol. Trans. Asme.* 126, 230–237, 2004.

38. Rotkin, S.V., Bulashevich, K.A., and Aluru, N.R., *Atomistic Models for Nanotube Device Electrostatics,* ECS Centennial Meet, ECS, Philadelphia, 2002, pp. V5–V1164.

39. Rotkin, S.V., Formalism of dielectric function and depolarization in SWNT: application to nanooptical switches and probes, in Lakhtakia, A., Maksimenko, S.A., eds., *Nanomodeling,* 5509, 145–159, 2004.

40. Odintsov, A.A. and Tokura, Y., Contact phenomena and Mott transition in carbon nanotubes, *J Low Temp. Phys.* 118(5/6), 509–518, 2000.

41. Luryi, S., Quantum capacitance devices, *Appl. Phys. Lett.* 52(6), 501–503, 1988.

42. Derycke, V., Martel, R., Appenzeller, J., and Avouris, P., Carbon nanotube inter- and intramolecular logic gates, *Nano Lett.* 1(9), 453–456, 2001.

13.5   Simulation of Ionic Channels

Umberto Ravaioli, Trudy van der Straaten, and Karl Hess

Nature has created many forms of nanostructures. Ion channels are of particular importance and have become accessible to the simulation methods that are widely used in computational electronics and for the nanostructures that have been described above. We therefore add this section to emphasize the importance of merging the understanding of biological (i.e., carbon-based) and silicon-based nano-structures.

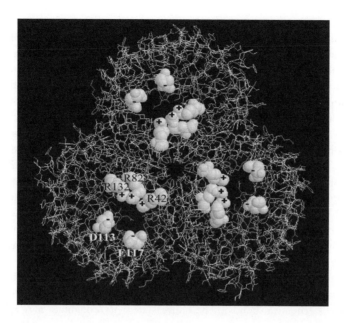

FIGURE 13.23 Molecular structure of *ompF,* a porin channel found in the outer membrane of the *E. coli* bacterium. This projection along the length of the channel shows the threefold symmetry of the trimer. Several ionized amino acids in the constriction region of each pore are highlighted.

Found in all life forms, ion channels are in a class of proteins that forms nanoscopic aqueous tunnels in the otherwise almost impermeable membranes of biological cells. An example of an ion channel, *ompF* porin, which resides in the outer membrane of the *E. coli* bacterium, is illustrated in Figure 13.23. Every ion channel consists of a chain of amino acids carrying a strong and rapidly varying permanent electric charge. By regulating the passive transport of ions across the cell membrane, ion channels maintain the correct internal ion composition that is crucial to cell survival and function. Ion channels directly control electrical signaling in the nervous system, muscle contraction, and the delivery of many clinical drugs.[1] Most channels have the ability to selectively transmit or block a particular ion species, and many exhibit switching properties similar to electronic devices. From a device point of view, ion channels can be viewed as transistors with unusual properties: exquisite sensitivity to specific environment factors, ability to self-assemble, and desirable properties for large-scale integration such as the infinite ON/OFF current ratio. By replacing or deleting one or more of the amino acids, many channels can be mutated, altering the charge distribution along the channel.[2] Engineering channels with specific conductances and selectivities are thus conceivable, as well as incorporating ion channels in the design of novel bio-devices.

Experimentally, the electrical and physiological properties of ion channels can be measured by inserting the channel into a lipid bilayer (membrane) and solvating the channel/membrane in an electrolyte solution. An electrochemical gradient is established across the membrane by immersing electrodes and using different concentrations of salt in the baths on either side of the membrane.

13.5.1 Hierarchical Approach to Modeling Ion Channels

Detailed simulation of ion transport in protein channels is very challenging because of the disparate spatial and temporal scales involved. A suitable model hierarchy is desirable to address different simulation needs. Continuum models, based on the drift-diffusion equations for charge flow, are the fastest approach; but they require large grids and extensive memory to resolve the three-dimensional channel geometry. Ion traversal of the channel is a very rare event on the usual timescale of devices, and the flow is actually a granular process. Continuum models, therefore, are useful mainly to probe the steady state of the system. We suppose the system to be ergodic. At a given point in the simulation domain, the steady-state ion concentration represents the probability of ion occupation at that position, averaged over very long times or, equivalently, averaged over many identical channels at any given instant. Despite some limitations, continuum models can be parameterized to match current–voltage characteristics by specifying a suitable space and/or energy-dependent diffusion coefficient, which accounts for the ions' interactions with the local environment.

A step above in the hierarchy we find particle models, where the trajectories of individual ions are computed. The simpler model is based on a Brownian dynamics description of ion flow, in which ion trajectories evolve according to the Langevin equation. Ions move in the local electric field, calculated from all the charges in the system as well as any externally applied fields. The energy dissipated via ion-water scattering is modeled by including a simple frictional term in the equation of motion, while the randomizing effect of the scattering is accounted for by including a zero-mean Gaussian noise term.[3] Ionic core repulsion can also be included by adding a suitable repulsive term (e.g., Lennard–Jones) to the total force acting on the ion. When the latter is neglected, the simulation is equivalent to a discrete version of the drift-diffusion model.[4] If the ion motion is assumed to be strongly overdamped, relatively long time steps can be used (e.g., picoseconds), making this a very practical approach.

At the next level in the hierarchy are particle models, where the ion flow is resolved with a self-consistent transient, following Monte Carlo or MD approaches, as they are known in semiconductor device simulation. MD simulations resolve the motion and forces among all particles, both free (ions and water molecules) and bound (e.g., protein atoms) in the system. Bound particles are modeled as charged balls connected by springs (chemical bonds). The entire system is brought to a simulated experimental temperature and then equilibrated by allowing the system to evolve according to Newtonian mechanics.[5] While this methodology is the most complete, due to the extreme computational costs involved, it can only be applied today to very small systems on very short time scales of simulation. Monte Carlo methods, originally developed for semiconductor device simulation, provide a more practical compromise. Water and protein are treated as a background dielectric medium, as is done with Brownian dynamics, and only the individual ion trajectories are resolved. The key difference between Brownian dynamics and Monte Carlo techniques lies in the way the ion dynamics are handled. In Monte Carlo models the ion trajectories evolve according to Newtonian mechanics; but individual ion–water collision interactions are replaced with an appropriate scattering model, which is resolved on the natural timescales of the problem.[6] In the limit of high friction, both approaches should give the same result.

13.5.2 Drift-Diffusion Models

Drift-diffusion models are useful for studying ion transport in open-channel systems over timescales that cannot be resolved practically by detailed particle models. Water, protein, and membrane are treated as uniform background media with specific dielectric constants; and the macroscopic ion current in the water is resolved by assigning an appropriate space or energy-dependent mobility and diffusion coefficient to each ionic species. The solution of Poisson's equation over the entire domain provides a simple way to include external boundary conditions and image force effects at dielectric discontinuities. Complete three-dimensional models of flow in ionic channels can be implemented with the established tools of semiconductor device simulation.

In order to define the various regions of the computational domain, the molecular structure of the protein must be mapped onto a grid. Protein structures are known with atomic resolution for a number of important channels, but considerable processing is still necessary to determine the charge and the dielectric permittivity distribution corresponding to the individual molecular components. With this information, one can assemble a grid defining the boundaries between water and protein, as illustrated in Figure 13.24. The current density $j_\pm$ arising from the flow of ions down the electrochemical gradient in the aqueous region of the domain is given by the drift-diffusion equation:

$$\vec{j}_\pm = -(\mu_\pm \rho_\pm \vec{\nabla}\varphi - D_\pm \vec{\nabla}\rho_\pm)$$

where $\rho_\pm$ are the ionic charge densities and $\mu_\pm$ and $D_\pm$ are, respectively, the mobilities and diffusion coefficients of each ionic species. For the purposes of this discussion, we restrict ourselves to systems

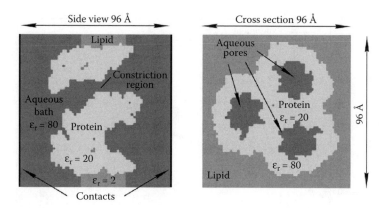

FIGURE 13.24 Mesh representation of the *ompF* trimer *in situ* in a membrane, immersed in a solution of potassium chloride — longitudinal and cross-sectional slices through the three-dimensional computational domain generated on a uniform rectilinear grid (1.5Å spacing). Electrodes immersed in the baths maintain a fixed bias across the channel/membrane system.

with only two ionic species of opposite charge, but the same treatment can be extended to allow for multiple ionic species by including a drift-diffusion equation for each additional species. Conservation of charge is enforced by a continuity equation for each species, given by

$$\vec{\nabla} \cdot \vec{j}_{\pm} + \frac{\partial \rho_{\pm}}{\partial t} = S_{\pm}$$

The term $S_{\pm}$ is set to zero for simple transport simulation, but it can be set to any functional form to describe higher order effects, such as the details of ion binding and other chemical phenomena that populate or deplete the ion densities. The electrostatic potential φ is described by Poisson's equation:

$$\vec{\nabla} \cdot (\varepsilon \vec{\nabla} \varphi) = -(\rho_{fixed} + \rho_{+} + \rho_{-})$$

where ρ_{fixed} represents the density of fixed charge residing within and on the surface of the protein. When solved simultaneously, this system of coupled equations provides a self-consistent description of ion flow in the channel. The equations are discretized on the grid and solved iteratively for steady-state conditions, subject to specific boundary conditions for applied potential and for ionic solution concentrations in the baths at the ends of the channel. In a typical semiconductor device, the mobile charge in the contacts is originated by fixed ionized dopants. In an ion-channel system, the salt concentration in the electrolyte far from the protein determines the density of mobile ionic charges, which, from an electrical point of view, behave similarly to the intrinsic electron/hole concentrations in an undoped semiconductor at a given temperature.

13.5.2.1 Application of the Drift-Diffusion Model to Real Ion Channels

Complete three-dimensional drift-diffusion models have been implemented using the computational platform PROPHET[7] and used to study transport in ion channels like gramicidin and porin, for which detailed structure and conductivity measurements are available. Porin in particular presents a very challenging problem because, as shown in Figure 13.23, the channel is a *trimer* consisting of three identical parallel channels, connected through a common anti-chamber region. Memory requirements for continuum simulations of porin are currently at the limit of available workstation resources; however, simulations are now performed routinely on distributed shared memory machines. Figure 13.25 compares the current–voltage curves computed with a three-dimensional drift-diffusion simulation with those measured experimentally.[8] These results were generated in approximately 8 hours on an SGI origin 2000.

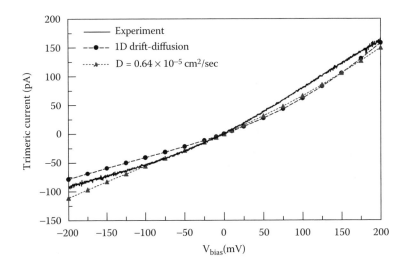

FIGURE 13.25 Comparison of measured and computed current–voltage curves for *ompF* in 100 mM potassium chloride, assuming a spatially dependent diffusion coefficient and a spatially uniform diffusion coefficient.

13.5.3 Monte Carlo Simulations

The Monte Carlo simulation technique, as it is known in the tradition of semiconductor device simulation, can be coupled with a particle-mesh model to provide a self-consistent, time-resolved picture of ion dynamics in a channel system.[6,9] The starting point is the grid, which defines the regions accessible to ions as well as the dielectric topography of the system. In reality the boundaries between aqueous, protein, and membrane regions are not static but move over atomic length scales due to the thermal fluctuations of the atoms of the protein. Such fluctuations, which are resolved in MD simulations, are ignored in Monte Carlo simulations (although in principle they could be included).

13.5.3.1 Resolving Single-Ion Dynamics

Ions are distributed throughout the aqueous region according to a given initial concentration profile. The charge of each mobile ion, and of each static charge within the protein, is interpolated to the grid using a prescribed weighting scheme to construct a charge density at the discrete grid points. The electrostatic field due to the charge density distribution, as well as any externally applied field, is found by solving Poisson's equation on the grid. The field at the grid points is interpolated back to the ion positions and used to move the ions forward in time by integrating Newton's second law over small time steps. At the end of the time step, the new ion positions are used to recalculate a new charge density distribution and hence a new field to advance the ions over the next time step. This cycle is iterated either until a steady-state is reached or until quantities of interest (e.g., diffusion coefficient) have been calculated. The effects of ion volume can also be incorporated by including an ionic core repulsive term in the force, acting on each ion as is done in Brownian dynamics.

13.5.3.2 Modeling Ion–Water Interactions

Ion motion is treated as a sequence of free flights interrupted by collisions with water molecules, which are modeled by assuming a particular ion-water scattering rate $\nu(E_{ion}(t))$, generally a function of ion energy. The scattering rate represents the average number of collisions per unit time that an ion would experience if it maintained a constant energy. The probability for an ion to travel for a time t without scattering is given by

$$P(t) = \exp\left(-\int_0^t \nu(E_{ion}(t'))dt'\right)$$

The probability density function (probability per unit time) for a flight to have duration t is given by

$$p(t) = \nu(E_{ion}(t))P(t)$$

Ion flight times can be randomly selected from the probability density function by integrating the latter over the (unknown) flight time T_f and equating the integral to a uniformly distributed random number r on the unit interval. Thus,

$$-\log(r) = \int_0^{T_f} \nu(E_{ion}(t))dt$$

The integral on the right-hand side is trivial only for constant scattering rates, but in general it cannot be performed analytically. A number of methods have been introduced to solve the integral; an extended discussion is given at the Internet location given in Ref. 6.

13.6 Conclusions

The combination of a three-dimensional drift-diffusion and three-dimensional Monte Carlo approach provides the essential hierarchy for looking at biological systems from the point of view of device-like applications. There are, however, significant differences between solid-state devices and biological systems, which require different choices in the definition of a Monte Carlo simulation strategy. In a typical device, the ensemble must include many thousands of particles; but a reasonable steady state is reached after several picoseconds of simulation (on the order of ten to twenty thousand time steps). In a practical simulation domain for a biological channel, only a very small number of ions is present in the system; but because the ion traversal of the channel is a rare event, measurable current levels can only be established by extending the simulation to the millisecond range. Because the number of time steps required to resolve ion dynamics is typically on the order of tens of femtoseconds, this would require a number of iteration steps on the order of 10^{12}, which is still extremely expensive. For a fully self-consistent simulation, the solution of Poisson's equation in three-dimensions presents the real bottleneck, while the computational cost of resolving the few particle trajectories is minimal. Alternative schemes for evaluating the electrostatic potential self-consistently include precalculating the potential for various ion pair configurations, storing the results in look-up tables, and employing the superposition principle to reconstruct the potential at the desired point by interpolating between table entries.[10] A prototype Monte Carlo simulation of sodium chloride transport in the gramicidin channel has been successfully implemented,[11] adapting the grid developed for the continuum simulations, as shown in Figure 13.26. For this simulation Poisson's equation is solved approximately every 10 time steps using an accurate conjugate gradient method.

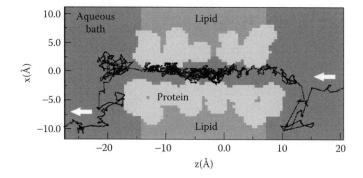

FIGURE 13.26 Geometric representation of the gramicidin channel as used in the three-dimensional Monte Carlo transport model. The successful trajectory of a single sodium ion traversing the channel is shown.

Acknowledgments

N. R. Aluru is thankful to M. Dequesnes for technical assistance and allowing the use of some of his results. J.-P. Leburton is indebted to G. Austing, R. M. Martin, S. Nagaraja, and S. Tarucha for fruitful discussions, and to P. Matagne for technical assistance. U. Ravaioli and T. van der Straaten are indebted to Bob Eisenberg (Rush Medical College) and Eric Jakobsson (University of Illinois) for introducing them to the subject of ion channels and for many useful discussions; to R. W. Dutton, Z. Yu, and D. Yergeau (Stanford University) for assistance with PROPHET; and T. Schirmer (University of Basel) for providing molecular structures for the porin channel. S. V. Rotkin is indebted to L. Rotkina and I. Zharov for fruitful discussions and to K. A. Bulashevich for technical support. N. R. Aluru, J.-P. Leburton, and S. V. Rotkin acknowledge the support of the CRI grant of UIUC. The work of J.-P. Leburton was supported by NSF grants DESCARTES ECS-98-02730, the Materials Computational Center (DMR-99–76550) and the DARPA-QUIST program (DAAD19–01–1–0659). The work of U. Ravaioli and T. van der Straaten was funded in part by NSF Distributed Center for Advanced Electronics Simulation (DesCArtES) grant ECS 98–02730, DARPA contract F30 602-012-0513 (B.E.), and a NSF KDI grant to the University of Illinois. S. V. Rotkin acknowledges the support of DoE grant DE-FG02-01ER45932 and the Beckman Fellowship from the Arnold and Mabel Beckman Foundation. For his calculations, B. R. Tuttle primarily utilized the SGI-ORIGIN2000 machines at the National Center for Supercomputing Applications in Urbana, IL. K. Hess acknowledges the Army Research Office (DAAG55-98-1-03306) and the Office of Naval Research (NO0014-98-1-0604) and ONR–MURI.

References

1. Hille, B., *Ionic Channels of Excitable Membranes*, Sinauer Associates, Massachusetts, 1992.
2. Phale, P.S., Philippsen, A., Widmer, C., Phale, V.P., Rosenbusch, J.P., and Schirmer, T., Role of charged residues at the ompF porin channel constriction probed by mutagenesis and simulation, *Biochemistry* 40, 6319–6325, 2001.
3. Reif, F., *Fundamentals of Statistical and Thermal Physics*, McGraw-Hill, Singapore, 1987.
4. Schuss, Z., Nadler, B., and Eisenberg, R.S., Derivation of Poisson and Nernst-Planck equations in a bath and channel from a molecular model, *Phys. Rev. E* 64, 036116-(1–14), 2001.
5. Allen, M.P. and Tildesley, D.J., *Computer Simulation of Liquids*, Clarendon, Oxford, 1987.
6. http://www.ncce.ce.g.uiuc.edu/ncce.htm.
7. http://www.tcad.stanford.edu/.
8. van der Straaten, T., Varma, S., Chiu, S.-W., Tang, J., Aluru, N., Eisenberg, R., Ravaioli, U., and Jakobsson, E., Combining computational chemistry and computational electronics to understand protein ion channels, in *The 2002 Intl. Conf. Computational Nanosci. Nanotechnol.*, 2002.
9. Hockney, R.W. and Eastwood, J.W., *Computer Simulation Using Particles*, McGraw-Hill, 1981.
10. Chung, S.-H., Allen, T.W., Hoyles, M., and Kuyucak, S., Permeation of ions across the potassium channel: Brownian dynamics studies, *Biophys. J.* 77, 2517–2533, 1999.
11. van der Straaten, T. and Ravaioli, U., Self-consistent Monte-Carlo/P3M simulation of ion transport in the gramicidin ion channel, unpublished.

14

Resistance of a Molecule

14.1   Introduction **14-1**
14.2   Qualitative Discussion **14-3**
        Where Is the Fermi Energy? • Current Flow as a Balancing Act
14.3   Coulomb Blockade? **14-8**
        Charging Effects • Unrestricted Model • Broadening
14.4   Nonequilibrium Green's Function (NEGF)
        Formalism **14-12**
14.5   An Example: Quantum Point Contact (QPC) **14-15**
14.6   Concluding Remarks **14-19**
Acknowledgments..**14-20**
14.7.A MATLAB®Codes..**14-20**
        Discrete One-Level Model • Discrete Two-Level Model
        • Broadened One-Level Model • Unrestricted Discrete
        One-Level Model • Unrestricted Broadened One-Level Model
References **14-23**

Magnus Paulsson
Ferdows Zahid
Supriyo Datta
Purdue University

14.1   Introduction

In recent years, several experimental groups have reported measurements of the current–voltage (I-V) characteristics of individual or small numbers of molecules. Even three-terminal measurements showing evidence of transistor action have been reported using carbon nanotubes[1,2] as well as self-assembled monolayers of conjugated polymers.[3] These developments have attracted much attention from the semiconductor industry, and there is great interest from an applied point of view to model and understand the capabilities of molecular conductors. At the same time, this is also a topic of great interest from the point of view of basic physics. A molecule represents a quantum dot, at least an order of magnitude smaller than semiconductor quantum dots, which allows us to study many of the same mesoscopic and/or many-body effects at far higher temperatures.

So what is the resistance of a molecule? More specifically, what do we see when we connect a short molecule between two metallic contacts as shown in Figure 14.1 and measure the current (I) as a function of the voltage (V)? Most commonly we get I-V characteristics of the type sketched in Figure 14.2. This has been observed using many different approaches including breakjunctions,[4–8] scanning probes,[9–12] nanopores,[13] and a host of other methods (see, for example, Reference 14). A number of theoretical models have been developed for calculating the I-V characteristics of molecular wires using semi-empirical[12,15–18] as well as first-principles[19–25] theory.

Our purpose in this chapter is to provide an intuitive explanation for the observed I-V characteristics using simple models to illustrate the basic physics. However, it should be noted that molecular electronics is a rapidly developing field, and much of the excitement arises from the possibility of discovering novel physics beyond the paradigms discussed here. To cite a simple example, very few experiments to date[3] incorporate the gate electrode shown in Figure 14.1, and we will largely ignore the gate in this chapter.

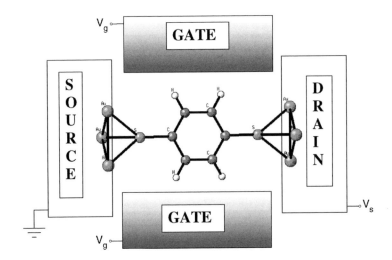

FIGURE 14.1 Conceptual picture of a "molecular transistor" showing a short molecule (Phenyl dithiol, PDT) sandwiched between source and drain contacts. Most experiments so far lack good contacts and do not incorporate the gate electrodes.

However, the gate electrode can play a significant role in shaping the I-V characteristics and deserves more attention. This is easily appreciated by looking at the applied potential profile U_{app} generated by the electrodes in the absence of the molecule. This potential profile satisfies the Laplace equation without any net charge anywhere and is obtained by solving:

$$\nabla \cdot (\varepsilon \nabla U_{app}) = 0 \tag{14.1}$$

subject to the appropriate boundary values on the electrodes (Figure 14.3). It is apparent that the electrode geometry has a significant influence on the potential profile that it imposes on the molecular species, and this could obviously affect the I-V characteristics in a significant way. After all, it is well known that a three-terminal metal/oxide/semiconductor field-effect transistor (MOSFET) with a gate electrode has a very different I-V characteristic compared with a two terminal n-i-n diode. The current in a MOSFET saturates under increasing bias, but the current in an n-i-n diode keeps increasing indefinitely. In contrast to the MOSFET, whose I-V is largely dominated by classical electrostatics, the I-V characteristics of molecules are determined by a more interesting interplay between nineteenth-century physics (electrostatics) and twentieth-century physics (quantum transport); and it is important to do justice to both aspects.

We will start in Section 14.2 with a qualitative discussion of the main factors affecting the I-V characteristics of molecular conductors, using a simple toy model to illustrate their role. However, this toy model misses two important factors: (1) shift in the energy level due to *charging effects* as the molecule

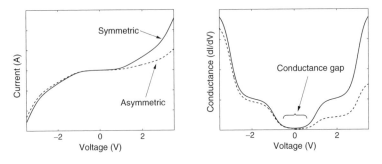

FIGURE 14.2 Schematic picture, showing general properties of measured current–voltage (I-V) and conductance (G-V) characteristics for molecular wires. Solid line: symmetrical I-V. Dashed line: asymmetrical I-V.

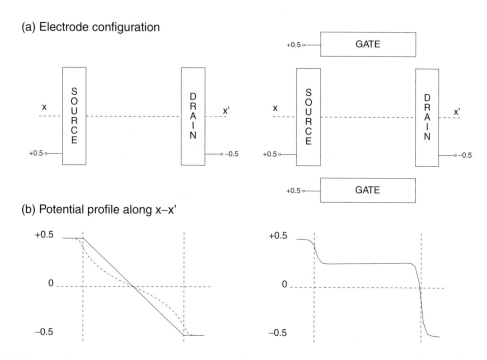

(a) Electrode configuration

(b) Potential profile along x–x'

FIGURE 14.3 Schematic picture, showing potential profile for two geometries, without gate (*a*) and with gate (*b*).

loses or gains electrons and (2) *broadening* of the energy levels due to their finite lifetime arising from the coupling (Γ_1 and Γ_2) to the two contacts. Once we incorporate these effects (Section 14.3) we obtain more realistic I-V plots, even though the toy model assumes that conduction takes place independently through individual molecular levels. In general, however, multiple energy levels are simultaneously involved in the conduction process. In Section 14.4 we will describe the nonequilibrium Green's function (NEGF) formalism, which can be viewed as a generalized version of the one-level model to include multiple levels or conduction channels. This formalism provides a convenient framework for describing quantum transport[26] and can be used in conjunction with *ab initio* or semi-empirical Hamiltonians as described in a set of related articles.[27,28] Then in Section 14.5 we will illustrate the NEGF formalism with a simple semi-empirical model for a gold wire, *n* atoms long and one atom in cross-section. We could call this an Au$_n$ molecule, though that is not how one normally thinks of a gold wire. However, this example is particularly instructive because it shows the lowest possible *resistance of a molecule* per channel, which is $\pi\hbar/e^2 = 12.9\ k\Omega$.[29]

14.2 Qualitative Discussion

14.2.1 Where Is the Fermi Energy?

Energy-Level Diagram: The first step in understanding the current (I) vs. voltage (V) curve for a molecular conductor is to draw an energy-level diagram and locate the Fermi energy. Consider first a molecule sandwiched between two metallic contacts but with very weak electronic coupling. We could then line up the energy levels as shown in Figure 14.4 using the metallic work function (WF), the electronic affinity (*EA*), and ionization potential (*IP*) of the molecule. For example, a (111) gold surface has a work function of ~ 5.3 eV, while the electron affinity and ionization potential, EA_0 and IP_0, for isolated phenyl dithiol (Figure 14.1) in the gas phase have been reported to be ~ 2.4 eV and 8.3 eV, respectively.[30] These values are associated with electron emission and injection to and from a vacuum and may need some modification

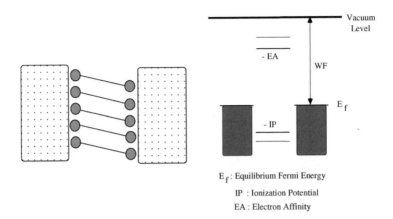

E_f : Equilibrium Fermi Energy

IP : Ionization Potential

EA : Electron Affinity

FIGURE 14.4 Equilibrium energy level diagram for a metal–molecule–metal sandwich for a weakly coupled molecule.

to account for the metallic contacts. For example, the actual *EA, IP* will possibly be modified from EA_0, IP_0 due to the image potential W_{im} associated with the metallic contacts[31]:

$$EA = EA_0 + W_{im} \qquad (14.2)$$

$$IP = IP_0 - W_{im} \qquad (14.3)$$

The probability of the molecule losing an electron to form a positive ion is equal to $e^{(WF-IP)/k_BT}$, while the probability of the molecule gaining an electron to form a negative ion is equal to $e^{(EA-WF)/k_BT}$. We thus expect the molecule to remain neutral as long as both $(IP - WF)$ and $(WF - EA)$ are much larger than k_BT, a condition that is usually satisfied for most metal–molecule combinations. Because it costs too much energy to transfer one electron into or out of the molecule, it prefers to remain neutral in equilibrium.

The picture changes qualitatively if the molecule is chemisorbed directly on the metallic contact (Figure 14.5). The molecular energy levels are now broadened significantly by the strong hybridization with the delocalized metallic wave functions, making it possible to transfer fractional amounts of charge to or from the molecule. Indeed there is a charge transfer, which causes a change in the electrostatic potential inside the molecule; and the energy levels of the molecule are shifted by a contact potential (CP), as shown.

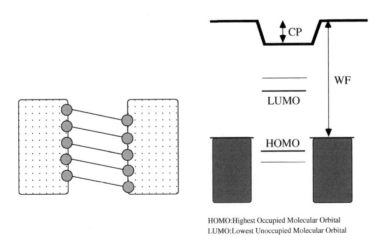

HOMO:Highest Occupied Molecular Orbital

LUMO:Lowest Unoccupied Molecular Orbital

WF : Metal Work Function

CP : Contact Potential

FIGURE 14.5 Equilibrium energy level diagram for a metal–molecule–metal sandwich for a molecule strongly coupled to the contacts.

It is now more appropriate to describe transport in terms of the HOMO–LUMO levels associated with incremental charge transfer[32] rather than the affinity and ionization levels associated with integer charge transfer. Whether the molecule–metal coupling is strong enough for this to occur depends on the relative magnitudes of the single electron charging energy (U) and energy level broadening (Γ). As a rule of thumb, if $U \gg \Gamma$, we can expect the structure to be in the Coulomb Blockade (CB) regime characterized by integer charge transfer; otherwise it is in the self-consistent field (SCF) regime characterized by fractional charge transfer. This is basically the same criterion that one uses for the Mott transition in periodic structures, with Γ playing the role of the hopping matrix element. It is important to note that, for a structure to be in the CB regime, both contacts must be weakly coupled, as the total broadening Γ is the sum of the individual broadening due to the two contacts. Even if only one of the contacts is coupled strongly, we can expect $\Gamma \sim U$, thus putting the structure in the SCF regime. Figure 14.14 in Section 14.3.2 illustrates the I-V characteristics in the CB regime using a toy model. However, a moderate amount of broadening destroys this effect (see Figure 14.17), and in this chapter we will generally assume that the conduction is in the SCF regime.

Location of the Fermi energy: The location of the Fermi energy relative to the HOMO and LUMO levels is probably the most important factor in determining the current (I) vs. voltage (V) characteristics of molecular conductors. Usually it lies somewhere inside the HOMO–LUMO gap. To see this, we first note that E_f is located by the requirement that the number of states below the Fermi energy must be equal to the number of electrons in the molecule. But this number need not be equal to the integer number we expect for a neutral molecule. A molecule does not remain exactly neutral when connected to the contacts. It can and does pick up a fractional charge depending on the work function of the metal. However, the charge transferred (δn) for most metal–molecule combinations is usually much less than one. If δn were equal to +1, the Fermi energy would lie on the LUMO; while if δn were −1, it would lie on the HOMO. Clearly for values in between, it should lie somewhere in the HOMO–LUMO gap.

A number of authors have performed detailed calculations to locate the Fermi energy with respect to the molecular levels for a phenyl dithiol molecule sandwiched between gold contacts, but there is considerable disagreement. Different theoretical groups have placed it close to the LUMO[16,21] or to the HOMO.[12,19] The density of states inside the HOMO–LUMO gap is quite small, making the precise location of the Fermi energy very sensitive to small amounts of electron transfer — a fact that could have a significant effect on both theory and experiment. As such it seems justifiable to treat E_f as a fitting parameter within reasonable limits when trying to explain experimental I-V curves.

Broadening by the contacts: Common sense suggests that the strength of coupling of the molecule to the contacts is important in determining the current flow — the stronger the coupling, the larger the current. A useful quantitative measure of the coupling is the resulting broadening Γ of the molecular energy levels, see Figure 14.6. This broadening Γ can also be related to the time τ it takes for an electron placed in that level to escape into the contact: $\Gamma = \hbar/\tau$. In general, the broadening Γ could be different for different energy levels. Also it is convenient to define two quantities Γ_1 and Γ_2, one for each contact, with the total broadening $\Gamma = \Gamma_1 + \Gamma_2$.

One subtle point: Suppose an energy level is located well below the Fermi energy in the contact, so that the electrons are prevented from escaping by the exclusion principle. Would Γ be zero? No, the broadening would still be Γ, independent of the degree of filling of the contact as discussed in Reference 26. This observation is implicit in the NEGF formalism, though we do not invoke it explicitly.

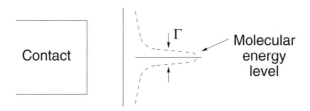

FIGURE 14.6 Energy level broadening.

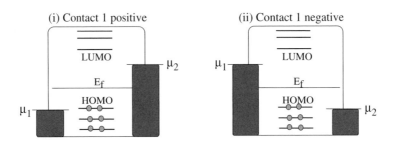

FIGURE 14.7 Schematic energy level diagram of metal–molecule–metal structure when contact 1 is (i) positively biased and when contact 1 is (ii) negatively biased with respect to contact 2.

14.2.2 Current Flow as a Balancing Act

Once we have drawn an equilibrium energy-level diagram, we can understand the process of current flow, which involves a nonequilibrium situation where the different reservoirs (e.g., the source and the drain) have different electrochemical potentials μ (Figure 14.7). For example, if a positive voltage V is applied externally to the drain with respect to the source, then the drain has an electrochemical potential lower than that of the source by eV : $\mu_2 = \mu_1 - eV$. The source and drain contacts thus have different Fermi functions, and each seeks to bring the active device into equilibrium with itself. The source keeps pumping electrons into it, hoping to establish equilibrium. But equilibrium is never achieved as the drain keeps pulling electrons out in its bid to establish equilibrium with itself. The device is thus forced into a balancing act between two reservoirs with different agendas that send it into a nonequilibrium state intermediate between what the source and drain would like to see. To describe this balancing process we need a kinetic equation that keeps track of the in- and outflow of electrons from each of the reservoirs.

Kinetic equation: This balancing act is easy to see if we consider a simple one-level system, biased such that the energy ε lies in between the electrochemical potentials of the two contacts (Figure 14.8). An electron in this level can escape into contacts 1 and 2 at a rate of $\Gamma_1/\hbar$ and $\Gamma_2/\hbar$, respectively. If the level were in equilibrium with contact 1, then the number of electrons occupying the level would be given by

$$N_1 = 2(\text{for spin}) \, f(\varepsilon, \mu_1) \tag{14.4}$$

where

$$f(\varepsilon, \mu) = \frac{1}{1 + e^{\frac{\varepsilon - \mu}{k_B T}}} \tag{14.5}$$

is the Fermi function. Similarly, if the level were in equilibrium with contact 2, the number would be:

$$N_2 = 2(\text{for spin}) \, f(\varepsilon, \mu_2) \tag{14.6}$$

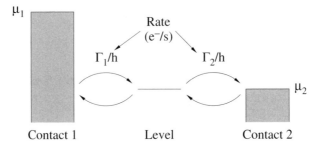

FIGURE 14.8 Illustration of the kinetic equation.

Under nonequilibrium conditions, the number of electrons N will be somewhere in between N_1 and N_2. To determine this number we write a steady-state *kinetic equation* that equates the net current at the left junction:

$$I_L = \frac{e\Gamma_1}{\hbar}(N_1 - N) \tag{14.7}$$

to the net current at the right junction:

$$I_R = \frac{e\Gamma_2}{\hbar}(N - N_2) \tag{14.8}$$

Steady state requires $I_L = I_R$, from which we obtain

$$N = 2\frac{\Gamma_1 f(\varepsilon, \mu) + \Gamma_2 f(\varepsilon, \mu_2)}{\Gamma_1 + \Gamma_2} \tag{14.9}$$

so that from Equation (14.7) or Equation (14.8) we obtain the current:

$$I = \frac{2e\Gamma_1\Gamma_2}{\hbar\Gamma_1 + \Gamma_2}(f(\varepsilon, \mu_1) - f(\varepsilon, \mu_2)) \tag{14.10}$$

Equation (14.10) follows very simply from an elementary model, but it serves to illustrate a basic fact about the process of current flow. No current will flow if $f(\varepsilon, \mu_1) = f(\varepsilon, \mu_2)$. A level that is way below both electrochemical potentials μ_1 and μ_2 will have $f(\varepsilon, \mu_1) = f(\varepsilon, \mu_2) = 1$ and will not contribute to the current, just like a level that is way above both potentials μ_1 and μ_2 and has $f(\varepsilon, \mu_1) = f(\varepsilon, \mu_2) = 0$. It is only when the level lies between μ_1 and μ_2 (or within a few $k_B T$ of μ_1 and μ_2) that we have $f(\varepsilon, \mu_1) \neq f(\varepsilon, \mu_2)$, and a current flows. Current flow is thus the result of the *difference in opinion* between the contacts. One contact would like to see more electrons (than N) occupy the level and keeps pumping them in, while the other would like to see fewer than N electrons and keeps pulling them out. The net effect is a continuous transfer of electrons from one contact to another.

Figure 14.9 shows a typical *I* vs. *V* calculated from Equation (14.10), using the parameters indicated in the caption. At first the current is zero because both μ_1 and μ_2 are above the energy level.

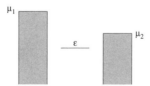

Once μ_2 drops below the energy level, the current increases to I_{max}, which is the maximum current that can flow through one level and is obtained from Equation (14.10) by setting $f(\varepsilon, \mu_1) = 1$ and $f(\varepsilon, \mu_2) = 0$:

$$I_{max} = \frac{2e}{\hbar}\Gamma_{eff} = \frac{2e\Gamma_1\Gamma_2}{\hbar\Gamma_1 + \Gamma_2} \tag{14.11}$$

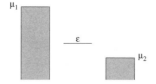

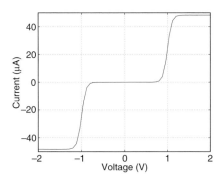

FIGURE 14.9 The current–voltage (I-V) characteristics for our toy model with $\mu_1 = E_f - eV/2$, $\mu_2 = E_f + eV/2$, $E_f = -5.0$ eV, $\varepsilon_0 = -5.5$ eV and $\Gamma_1 = \Gamma_2 = 0.2$ eV. MATLAB code in Section 14.A.1 ($U = 0$).

Note that in Figure 14.9 we have set $\mu_1 = E_f - eV/2$ and $\mu_2 = E_f + eV/2$. We could, of course, just as well have set $\mu_1 = E_f - eV$ and $\mu_1 = E_f$. But the average potential in the molecule would be $-V/2$ and we would need to shift ε appropriately. It is more convenient to choose our reference such that the average molecular potential is zero, and there is no need to shift ε.

Note that the current is proportional to Γ_{eff}, which is the parallel combination of Γ_1 and Γ_2. This seems quite reasonable if we recognize that Γ_1 and Γ_2 represent the strength of the coupling to the two contacts and as such are like two *conductances in series*. For long conductors we would expect a third conductance in series representing the actual conductor. This is what we usually have in mind when we speak of conductance. But short conductors have virtually zero resistance, and what we measure is essentially the contact or interface resistance.[*] This is an important conceptual issue that caused much argument and controversy in the 1980s. It was finally resolved when experimentalists measured the conductance of very short conductors and found it approximately equal to $2e^2/h$, which is a fundamental constant equal to 77.8 µA/V. The inverse of this conductance $h/2e^2 = 12.9$ kΩ is now believed to represent the minimum contact resistance that can be achieved for a one-channel conductor. Even a copper wire with a one-atom cross-section will have a resistance at least this large. Our simple one-level model (Figure 14.9) does not predict this result because we have treated the level as discrete, but the more complete treatment in later sections will show it.

14.3 Coulomb Blockade?

As we mentioned in Section 14.2, a basic question we need to answer is whether the process of conduction through the molecule belongs to the Coulomb Blockade (CB) or the Self-Consistent Field (SCF) regime. In this section, we will first discuss a simple model for charging effects (Section 14.3.1) and then look at the distinction between the simple SCF regime and the CB regime (Section 14.3.2). Finally, in Section 14.3.3 we show how moderate amounts of level broadening often destroy CB effects, making a simple SCF treatment quite accurate.

14.3.1 Charging Effects

Given the level (ε), broadening (Γ_1, Γ_2), and the electrochemical potentials μ_1 and μ_2 of the two contacts, we can solve Equation (14.10) for the current I. But we want to include charging effects in the calculations. Therefore, we add a potential U_{SCF} due to the change in the number of electrons from the equilibrium value ($f_0 = f(\varepsilon_0, E_f)$):

$$U_{SCF} = U(N - 2f_0) \tag{14.12}$$

---

[*] Four-terminal measurements have been used to separate the contact from the device resistance (see, for example, Reference 33).

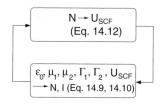

FIGURE 14.10 Illustration of the SCF problem.

similar to a Hubbard model. We then let the level ε float up or down by this potential:

$$\varepsilon = \varepsilon_0 + U_{SCF} \tag{14.13}$$

Because the potential depends on the number of electrons, we need to calculate the potential using the self-consistent procedure shown in Figure 14.10.

Once the converged solution is obtained, the current is calculated from Equation (14.10). This very simple model captures much of the observed physics of molecular conduction. For example, the results obtained by setting $E_f = -5.0$ eV, $\varepsilon_0 = -5.5$ eV, $\Gamma_1 = 0.2$ eV, $\Gamma_2 = 0.2$ eV are shown in Figure 14.11 with $(U = 1.0$ eV) and without $(U = 0$ eV) charging effects. The finite width of the conductance peak (with $U = 0$) is due to the temperature used in the calculations $(k_B T = 0.025$ eV). Note how the inclusion of charging tends to broaden the sharp peaks in conductance, even though we have not included any extra level broadening in this calculation. The size of the conductance gap is directly related to the energy difference between the molecular energy level and the Fermi energy. The current starts to increase when the voltage reaches 1 V, which is exactly $2|E_f - \varepsilon_0|$, as would be expected even from a theory with no charging. Charging enters the picture only at higher voltages, when a chemical potential tries to cross the level. The energy level shifts in energy (Equation [14.13]) if the charging energy is nonzero. Thus, for a small charging energy, the chemical potential easily crosses the level, giving a sharp increase of the current. If the charging energy is large, the current increases gradually because the energy level follows the chemical potential due to the charging.

What determines the conductance gap? The above discussion shows that the conductance gap is equal to $4(|E_f - \varepsilon_0| - \Delta)$, where Δ is equal to $\sim 4k_B T$ (plus $\Gamma_1 + \Gamma_2$ if broadening is included, see Section 14.3.3); and ε_0 is the HOMO or LUMO level, whichever is closest to the Fermi energy, as pointed out in Reference 34.

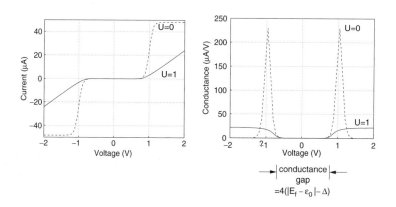

FIGURE 14.11 The current–voltage (I-V) characteristics (*left*) and conductance–voltage (G-V) (*right*) for our toy model with $E_f = -5.0$ eV, $\varepsilon_0 = -5.5$ eV and $\Gamma_1 = \Gamma_2 = 0.2$ eV. Solid lines: charging effects included ($U = 1.0$ eV), Dashed line: no charging ($U = 0$). MATLAB code in section 14.A.1.

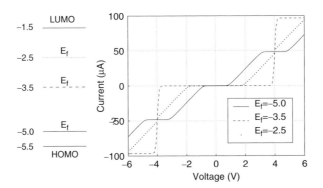

FIGURE 14.12 Right, the current–voltage (I-V) characteristics for the two level toy model for three different values of the Fermi energy (E_f). Left, the two energy levels (LUMO = –1.5 eV; HOMO = –5.5 eV) and the three different Fermi energies (–2.5, –3.5, –5.0) used in the calculations. (Other parameters used: U = 1.0 eV; $\Gamma_1 = \Gamma_2 = 0.2$ eV). MATLAB code in Section 14.A.1.

This is unappreciated by many who associate the conductance gap with the HOMO–LUMO gap. However, we believe that what conductance measurements show is the gap between the Fermi energy and the nearest molecular level.* Figure 14.12 shows the I-V characteristics calculated using a two-level model (obtained by a straightforward extension of the one-level model) with the Fermi energy located differently within the HOMO–LUMO gap giving different conductance gaps corresponding to the different values of $|E_f - \varepsilon_0|$. Note that with the Fermi energy located halfway in between, the conductance gap is twice the HOMO–LUMO gap; and the I-V shows no evidence of charging effects because the depletion of the HOMO is neutralized by the charging of the LUMO. This perfect compensation is unlikely in practice, because the two levels will not couple identically to the contacts as assumed in the model.

A very interesting effect that can be observed is the asymmetry of the I-V characteristics of $\Gamma_1 \neq \Gamma_2$ as shown in Figure 14.13. This may explain several experimental results which show asymmetric I-V,[6,12] as discussed by Ghosh et al.[35] Assuming that the current is conducted through the HOMO level ($E_f > \varepsilon_0$), the current is less when a positive voltage is applied to the strongly coupled contact (Figure 14.13a). This is due to the effects of charging as has been discussed in more detail in Reference 35. Ghosh et al. also show that this result will reverse if the conduction is through the LUMO level. We can simulate this

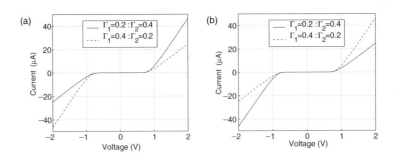

FIGURE 14.13 The current–voltage (I-V) characteristics for our toy model (E_f = –5.0 eV and U = 1.0 eV). (a) Conduction through LUMO ($E_f < \varepsilon_0 = -4.5$ eV). Solid lines: $\Gamma_1 = 0.2$ eV $< \Gamma_2 = 0.4$ eV. Dashed lines: $\Gamma_1 = 0.4$ eV $> \Gamma_2 = 0.2$ eV. MATLAB code in Section 14.A.1. Here positive voltage is defined as a voltage that lowers the chemical potential of contact 1.

---

*With very asymmetric contacts, the conductance gap could be equal to the HOMO–LUMO gap as commonly assumed in interpreting STM spectra. However, we believe that the picture presented here is more accurate unless the contact is so strongly coupled that there is a significant density of metal-induced gap states (MIGS).[28]

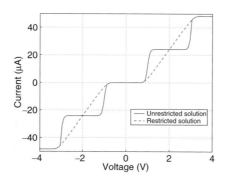

FIGURE 14.14 The current–voltage (I-V) characteristics for restricted (*dashed line*) and unrestricted solutions (*solid line*). $E_f = -5.0$ eV, $\Gamma_1 = \Gamma_2 = 0.2$ eV and $U = 1.0$ eV. MATLAB code in Sections 14.A.1 and 14.A.4.

situation by setting ε_0 equal to -4.5 eV, 0.5 eV above the equilibrium Fermi energy E_f. The sense of asymmetry is now reversed as shown in Figure 14.13b. The current is larger when a positive voltage is applied to the strongly coupled contact. Comparison with STM measurements seems to favor the first case, i.e., conduction through the HOMO.[35]

14.3.2 Unrestricted Model

In the previous examples (Figures 14.11 and 14.13) we have used values of $\Gamma_{1,2}$ that are smaller than the charging energy U. However, under these conditions one can expect CB effects, which are not captured by a *restricted solution*, which assumes that both spin orbitals see the same self-consistent field. However, an unrestricted solution, which allows the spin degeneracy to be lifted, will show these effects.[*] For example, if we replace Equation (14.13) with ($f_0 = f(\varepsilon_0, E_f)$):

$$\varepsilon_\uparrow = \varepsilon_0 + U(N_\downarrow - f_0) \tag{14.14}$$

$$\varepsilon_\downarrow = \varepsilon_0 + U(N_\uparrow - f_0) \tag{14.15}$$

where the up-spin level feels a potential due to the down-spin electrons and vice versa, then we obtain I-V curves as shown in Figure 14.14.

If the SCF iteration is started with a spin-degenerate solution, the same restricted solution as before is obtained. However, if the iteration is started with a spin-nondegenerate solution, a different I-V is obtained. The electrons only interact with the electron of the opposite spin. Therefore, the chemical potential of one contact can cross one energy level of the molecule because the charging of that level only affects the opposite spin level. Thus, the I-V contains two separate steps separated by U instead of a single step broadened by U.

For a molecule chemically bonded to a metallic surface, e.g., a PDT molecule bonded by a thiol group to a gold surface, the broadening Γ is expected to be of the same magnitude or larger than U. This washes out CB effects as shown in Figure 14.17. Therefore, the CB is not expected in this case. However, if the coupling to both contacts is weak, we should keep the possibility of CB and the importance of unrestricted solutions in mind.

---

[*] The unrestricted one-particle picture discussed here provides at least a reasonable qualitative picture of CB effects, though a complete description requires a more advanced many particle picture.[36] The one-particle picture leads to one of many possible states of the device depending on our initial guess, while a full many-particle picture would include all states.

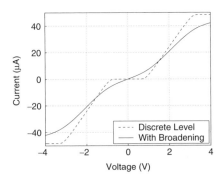

FIGURE 14.15 The current–voltage (I-V) characteristics: solid line, includes broadening of the level by the contacts; dashed line, no broadening, same as the solid line in Figure 14.11. MATLAB code in Sections 14.A.3 and 14.A.1 ($E_f =$ −5.0 eV, $\varepsilon_0 =$ −5.5, $U = 1$, and $\Gamma_1 = \Gamma_2 = 0.2$ eV).

14.3.3 Broadening

So far we have treated the level ε as discrete, ignoring the broadening $\Gamma = \Gamma_1 + \Gamma_2$ that accompanies the coupling to the contacts. To take this into account we need to replace the discrete level with a Lorentzian density of states $D(E)$:

$$D(E) = \frac{1}{2\pi} \frac{\Gamma}{(E-\varepsilon)^2 + (\Gamma/2)^2} \tag{14.16}$$

As we will see later, Γ is in general *energy-dependent* so that $D(E)$ can deviate significantly from a Lorentzian shape. We modify Equations (14.9) and (14.10) for N and I to include an integration over energy:

$$N = 2\int_{-\infty}^{\infty} dED(E) \frac{\Gamma_1 f(E,\mu_1) + \Gamma_2 f(E,\mu_2)}{\Gamma_1 + \Gamma_2} \tag{14.17}$$

$$I = \frac{2e}{\hbar} \int_{-\infty}^{\infty} dED(E) \frac{\Gamma_1 \Gamma_2}{\Gamma_1 + \Gamma_2} (f(E,\mu_1) - f(E,\mu_2)) \tag{14.18}$$

The charging effect is included as before by letting the center ε, of the molecular density of states, float up or down according to Equations (14.12) and (14.13) for the restricted model or Equations (14.14) and (14.15) for the unrestricted model.

For the restricted model, the only effect of broadening is to smear out the I-V characteristics as evident from Figure 14.15. The same is true for the unrestricted model as long as the broadening is much smaller than the charging energy (Figure 14.16). But moderate amounts of broadening can destroy the Coulomb Blockade effects completely and make the I-V characteristics look identical to the restricted model (Figure 14.17). With this in mind, we will use the restricted model in the remainder of this chapter.

14.4 Nonequilibrium Green's Function (NEGF) Formalism

The one-level toy model described in the last section includes the three basic factors that influence molecular conduction, namely, $E_f - \varepsilon_0$, $\Gamma_{1,2}$, and U. However, real molecules typically have multiple levels that often broaden and overlap in energy. Note that the two-level model (Figure 14.12) in the last section treated the two levels as *independent*, and such models can be used only if the levels do not overlap. In general, we need a formalism that can do justice to multiple levels with arbitrary broadening and overlap. The nonequilibrium Green's function (NEGF) formalism described in this section does just that.

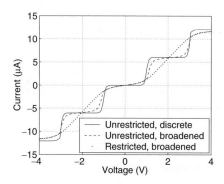

FIGURE 14.16 Current–voltage (I-V) characteristics showing the Coulomb blockade: discrete unrestricted model (*solid line*, MATLAB code in Section 14.A.4) and the broadened unrestricted mode (*dashed line*, 14.A.5). The dotted line shows the broadened restricted model without Coulomb blockade (14.A.3). For all curves the following parameters were used: $E_f = -5.0$, $\varepsilon_0 = -5.5$, $U = 1$, and $\Gamma_1 = \Gamma_2 = 0.05$ eV.

In the last section we obtained equations for the number of electrons N and the current I for a one-level model with broadening. It is useful to rewrite these equations in terms of the Green's function $G(E)$, defined as follows:

$$G(E) = \left(E - \varepsilon + i\frac{\Gamma_1 + \Gamma_2}{2} \right)^{-1} \tag{14.19}$$

The density of states $D(E)$ is proportional to the spectral function $A(E)$ defined as:

$$A(E) = -2\text{Im}\{G(E)\} \tag{14.20}$$

$$D(E) = \frac{A(E)}{2\pi} \tag{14.21}$$

while the number of electrons N and the current I can be written as:

$$N = \frac{2}{2\pi} \int_{-\infty}^{\infty} dE(|G(E)|^2 \, \Gamma_1 f(E,\mu_2) + |G(E)|^2 \, \Gamma_2 f(E,\mu_2)) \tag{14.22}$$

$$I = \frac{2e}{h} \int_{-\infty}^{\infty} dE \Gamma_1 \Gamma_2 |G(E)|^2 \, (f(E,\mu_1) - f(E,\mu_2)) \tag{14.23}$$

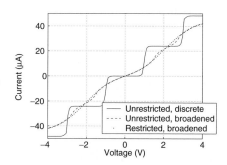

FIGURE 14.17 Current–voltage (I-V) characteristics showing the suppression of the Coulomb blockade by broadening: discrete unrestricted model (*solid line*) and the broadened unrestricted model (*dashed line*). The dotted line shows the broadened restricted model, $\Gamma_1 = \Gamma_2 = 0.2$ eV.

In the NEGF formalism the single energy level ε is replaced by a Hamiltonian matrix $[H]$, while the broadening $\Gamma_{1,2}$ is replaced by a complex energy-dependent self-energy matrix $[\Sigma_{1,2}(E)]$ so that the Green's function becomes a matrix given by

$$G(E) = (ES - H - \Sigma_1 - \Sigma_2)^{-1} \qquad (14.24)$$

where S is the identity matrix of the same size as the other matrices, and the broadening matrices $\Gamma_{1,2}$ are defined as the imaginary (more correctly as the anti-Hermitian) parts of $\Sigma_{1,2}$:

$$\Gamma_{1,2} = i(\Sigma_{1,2} - \Sigma_{1,2}^\dagger) \qquad (14.25)$$

The spectral function is the anti-Hermitian part of the Green's function:

$$A(E) = i(G(E) - G^\dagger(E)) \qquad (14.26)$$

from which the density of states $D(E)$ can be calculated by taking the trace:

$$D(E) = \frac{Tr(AS)}{2\pi} \qquad (14.27)$$

The density matrix $[\rho]$ is given by (c.f., Equation [14.22]):

$$\rho = \frac{1}{2\pi} \int_{-\infty}^{\infty} [f(E,\mu_1)G\Gamma_1 G^\dagger + (E,\mu_2)G\Gamma_2 G^\dagger] dE \qquad (14.28)$$

from which the total number of electrons N can be calculated by taking a trace:

$$N = Tr(\rho S) \qquad (14.29)$$

The current is given by (c.f., Equation [14.23]):

$$I = \frac{2e}{h} \int_{-\infty}^{\infty} [Tr(\Gamma_1 G\Gamma_2 G^\dagger)(f(E,\mu_1) - f(E,\mu_2))] dE \qquad (14.30)$$

Equations (14.24) through (14.30) constitute the basic equations of the NEGF formalism, which have to be solved self-consistently with a suitable scheme to calculate the self-consistent potential matrix $[U_{SCF}]$ (c.f., Equation [14.13]):

$$H = H_0 + U_{SCF} \qquad (14.31)$$

where H_0 is the bare Hamiltonian (like ε_0 in the toy model) and U_{SCF} is an appropriate functional of the density matrix ρ:

$$U_{SCF} = F(\rho) \qquad (14.32)$$

This self-consistent procedure is essentially the same as in Figure 14.10 for the one-level toy model, except that scalar quantities have been replaced by matrices:

$$\varepsilon_0 \rightarrow [H_0] \qquad (14.33)$$

$$\Gamma \rightarrow [\Gamma], [\Sigma] \qquad (14.34)$$

$$N \rightarrow [\rho] \qquad (14.35)$$

$$U_{SCF} \rightarrow [U_{SCF}] \qquad (14.36)$$

The size of all these matrices is $(n \times n)$, n being the number of basis functions used to describe the *molecule*. Even the self-energy matrices $\Sigma_{1,2}$ are of this size although they represent the effect of infinitely large contacts. In the remainder of this section and the next section, we will describe the procedure used to evaluate the Hamiltonian matrix H, the self-energy matrices $\Sigma_{1,2}$, and the functional F used to evaluate the self-consistent potential U_{SCF} (see Equation [14.32]). But the point to note is that once we know how to evaluate these matrices, Equations (14.24) through (14.32) can be used straightforwardly to calculate the current.

Nonorthogonal basis: The matrices appearing above depend on the basis functions that we use. Many of the formulations in quantum chemistry use nonorthogonal basis functions; and the matrix Equations (14.24) through (14.32) are still valid as is, except that the elements of the matrix $[S]$ in Equation (14.24) represent the overlap of the basis function $\phi_{mn} = (\overline{r})$:

$$S_{mn} = \int d^3 r \phi_m^*(\overline{r}) \phi_n(\overline{r}) \tag{14.37}$$

for orthogonal bases, $S_{mn} = \delta_{mn}$ so that S is the identity matrix as stated earlier. The fact that the matrix Equations (14.24) through (14.32) are valid even in a nonorthogonal representation is not self-evident and is discussed in Reference 28.

Incoherent Scattering: One last comment about the general formalism: the formalism as described above neglects all incoherent scattering processes inside the molecule. In this form it is essentially equivalent to the Landauer formalism.[37] Indeed, our expression for the current (Equation [14.30]) is exactly the same as in the transmission formalism, with the transmission T given by $Tr(\Gamma_1 G \Gamma_2 G^\dagger)$. But it should be noted that the real power of the NEGF formalism lies in its ability to provide a first-principles description of incoherent scattering processes — something we do not address in this chapter and leave for future work.

A practical consideration: Both Equations (14.28) and (14.30) require an integral over all energy. This is not a problem in Equation (14.30) because the integrand is nonzero only over a limited range, where $f(E, \mu_1)$ differs significantly from $f(E, \mu_2)$. But in Equation (14.28), the integrand is nonzero over a large energy range and often has sharp structures, making it numerically challenging to evaluate the integral. One way to address this problem is to write

$$\rho = \rho_{eq} + \Delta\rho \tag{14.38}$$

where ρ_{eq} is the equilibrium density matrix given by

$$\rho_{eq} = \frac{1}{2\pi} \int_{-\infty}^{\infty} f(E, \mu)[G\Gamma_1 G^\dagger + G\Gamma_2 G^\dagger] dE \tag{14.39}$$

and $\Delta\rho$ is the change in the density matrix under bias:

$$\Delta\rho = \frac{1}{2\pi} \int_{-\infty}^{\infty} G\Gamma_1 G^\dagger [f(E, \mu_1) - f(E, \mu)] + G\Gamma_2 G^\dagger [f(E, \mu_2) - f(E, \mu)] dE \tag{14.40}$$

The integrand in Equation (14.40) for $\Delta\rho$ is nonzero only over a limited range (like Equation [14.30] for I) and is evaluated relatively easily. The evaluation of ρ_{eq} (Equation [14.39]), however, still has the same problem; but this integral (unlike the original Equation [14.28]) can be tackled by taking advantage of the method of contour integration as described in References 38 and 39.

14.5   An Example: Quantum Point Contact (QPC)

Consider, for example, a gold wire stretched between two gold surfaces as shown in Figure 14.18. One of the seminal results of mesoscopic physics is that such a wire has a quantized conductance equal to $e^2/\pi\hbar \sim 77.5\ \mu A/V \sim (12.9\ k\Omega)^{-1}$. This was first established using semiconductor structures[29,40,41] at 4 K, but

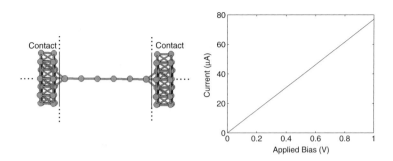

FIGURE 14.18 Left: wire consisting of six gold atoms forming a quantum point contact (QPC). Right: quantized conductance ($I = (e^2/\pi\hbar)V$).

recent experiments on gold contacts have demonstrated it at room temperature.[42] How can a wire have a resistance that is independent of its length? The answer is that this resistance is really associated with the interfaces between the narrow wire and the wide contacts. If there was scattering inside the wire, it would give rise to an additional resistance in series with this fundamental interface resistance. The fact that a short wire has a resistance of 12.9 kΩ is a nonobvious result that was not known before 1988. This is a problem for which we do not really need a quantum transport formalism; a semiclassical treatment would suffice. The results we obtain here are not new or surprising. What is new is that we treat the gold wire as an Au_6 molecule and obtain well-known results commonly obtained from a continuum treatment.

In order to apply the NEGF formalism from the last section to this problem, we need the Hamiltonian matrix $[H]$, the self-energy matrices $\Sigma_{1,2}$, and the self-consistent field $U_{SCF} = F([\rho])$. Let us look at these one by one.

Hamiltonian: We will use a simple semi-empirical Hamiltonian which uses one s-orbital centered at each gold atom as the basis functions, with the elements of the Hamiltonian matrix given by

$$H_{ij} = \varepsilon_o \quad \text{if } i = j$$
$$= -t \quad \text{if } i, j \text{ are nearest neighbors} \tag{14.41}$$

where $\varepsilon_0 = -10.92$ eV and $t = 2.653$ eV. The orbitals are assumed to be orthogonal, so that the overlap matrix S is the identity matrix.

Self Energy: Once we have a Hamiltonian for the entire molecule-contact system, the next step is to partition the device from the contacts and obtain the self-energy matrices $\Sigma_{1,2}$ describing the effects of the contacts on the device. The contact will be assumed to be essentially unperturbed relative to the surface of a bulk metal so that the full Green's function (G_T) can be written as (the energy E is assumed to have an infinitesimal imaginary part $i0^+$):

$$G_T = \begin{pmatrix} ES - H & ES_{dc} - H_{dc} \\ ES_{cd} - H_{cd} & ES_c - H_c \end{pmatrix}^{-1} = \begin{pmatrix} G & G_{dc} \\ G_{cd} & G_c \end{pmatrix} \tag{14.42}$$

where c denotes one of the contacts (the effect of the other contact can be obtained separately). We can use straightforward matrix algebra to show that:

$$G = (ES - H - \Sigma)^{-1} \tag{14.43}$$

$$\Sigma = (ES_{dc} - H_{dc})(ES_c - H_c)^{-1}(ES_{cd} - H_{cd}) \tag{14.44}$$

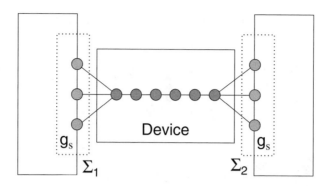

FIGURE 14.19 Device, showing surface Green's function (g_s) and self-energies (Σ).

The matrices S_{dc}, S_c, H_{dc}, and H_c are all infinitely large because the contact is infinite. But the element of S_{dc}, H_{dc} is nonzero only for a small number of contact atoms whose wavefunctions significantly overlap the device. Thus, we can write:

$$\Sigma = \tau g_s \tau^\dagger \tag{14.45}$$

where τ is the nonzero part of $ES_{dc}-H_{dc}$, having dimensions ($d \times s$) where d is the size of the device matrix, and s is the number of surface atoms of the contact having a nonzero overlap with the device. g_s is a matrix of size $s \times s$ which is a subset of the full infinite-sized contact Green's function $(ES_c - H_c)^{-1}$. This surface Green's function can be computed exactly by making use of the periodicity of the semi-infinite contact, using techniques that are standard in surface physics.[31] For a one-dimensional lead, with a Hamiltonian given by Equation (14.41), the result is easily derived[29]:

$$g_s(E) = \frac{e^{ika}}{t} \tag{14.46}$$

where ika is related to the energy through the dispersion relation:

$$E = \varepsilon_0 - 2t \cos(ka) \tag{14.47}$$

The results presented below were obtained using the more complicated surface Green's function for an FCC (111) gold surface as described in Reference 28. However, using the surface Green's function in Equation (14.46) gives almost identical results.

Electrostatic potential: Finally we need to identify the electrostatic potential across the device (Figure 14.20) by solving the Poisson equation:

$$-\nabla^2 U_{tot} = \frac{e^2 n}{\varepsilon_o} \tag{14.48}$$

with the boundary conditions given by the potential difference V_{app} between the metallic contacts (here ε_0 is the dielectric constant). To simplify the calculations we divide the solution into an applied and self-consistent potential ($U_{tot} = U_{app} + U_{SCF}$), where U_{app} solves the Laplace equation with the known potential difference between the metallic contacts:

$$-\nabla^2 U_{app} = 0 \qquad U_{app} = -eV_n \qquad \text{on electrode n} \tag{14.49}$$

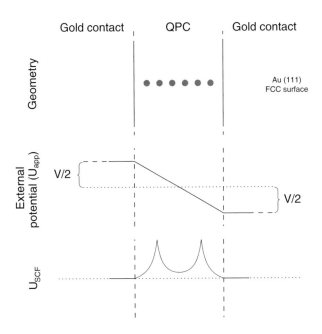

FIGURE 14.20 The electrostatic potentials divided into the applied (U_{app}) and the self-consistent field (U_{SCF}) potentials. The boundary conditions can clearly be seen in the figure; U_{SCF} is zero at the boundary and $U_{app} = \pm V_{app}/2$.

Thus, U_{tot} solves Equation (14.48) if U_{SCF} solves Equation (14.48) with zero potential at the boundary:

$$-\nabla^2 U_{SCF} = \frac{e^2 n}{\varepsilon_o} \qquad U_{SCF} = 0 \quad \text{on all electrodes} \tag{14.50}$$

In the treatment of the electrostatic we assume the two contacts to be semi-infinite classical metals separated by a distance (W). This gives simple solutions to both U_{app} and U_{SCF}. The applied potential is given by (capacitor)

$$U_{app} = \frac{V}{W} x \tag{14.51}$$

where x is the position relative to the midpoint between the contacts. The self-consistent potential is easily calculated with the method of images where the potential is given by a sum over the point charges and all their images. However, to avoid the infinities associated with point charges, we adopt the Pariser–Parr–Pople (PPP) method[43,17] in the Hartree approximation. The PPP functional describing the electron–electron interactions is

$$H_{ij}^{e-e} = \delta_{ij} \sum_k \left(\rho_{kk} - \rho_{kk}^{eq} \right) \gamma_{ik} \tag{14.52}$$

where ρ is the charge density matrix, ρ^{eq} the equilibrium charge density (in this case ρ_{ii}^{eq} as we are modeling the s-electrons of gold), and the one-center two-electron integral γ_{ij}. The diagonal elements γ_{ii} are obtained from experimental data, and the off-diagonal elements (γ_{ij}) are parameterized to describe a potential that decreases as the inverse of the distance ($1/R_{ij}$):

$$\gamma_{ij} = \frac{e^2}{4\pi\varepsilon_o R_{ij} + \dfrac{2e^2}{\gamma_{ii} + \gamma_{jj}}} \tag{14.53}$$

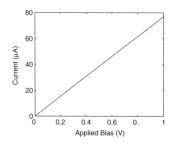 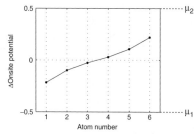

FIGURE 14.21 I-V (*left*) and potential drop for an applied voltage of 1 V (*right*) for a six-atom OPC connected to two contacts. The potential plotted is the difference in onsite potential from the equilibrium case.

Calculations on the QPC: The results for the I-V and potential for a QPC are shown in Figure 14.21. The geometry used was a linear chain of six gold atoms connected to the FCC (111) surface of the contacts in the center of a surface triangle. The Fermi energy of the isolated contacts was calculated to be $E_f = -8.67$ eV, by requiring that there is one electron per unit cell.

As evident from the figure, the I-V characteristics are linear and the slope gives a conductance of 77.3 μA/V close to the quantized value of $e^2/\pi\hbar \sim 77.5$ μA/V as previously mentioned. What makes the QPC distinct from typical molecules is the strong coupling to the contacts, which broadens all levels into a continuous density of states; and any evidence of a conductance gap (Figure 14.2) is completely lost. Examining the potential drop over the QPC shows a linear drop over the center of the QPC with a slightly larger drop at the end atoms. This may seem surprising because transport is assumed to be *ballistic*, and one expects no voltage drop across the chain of gold atoms. This can be shown to arise because the chain is very narrow (one atom in cross-section) compared with the screening length.[19]

We can easily imagine an experimental situation where the device (QPC or molecule) is attached asymmetrically to the two contacts with one strong and one weak side. To model this situation we artificially decreased the interaction between the right contact and the QPC by a factor of 0.2. The results of this calculation are shown in Figure 14.22. A weaker coupling gives a smaller conductance, as compared with the previous figure. More interesting is that the potential drop over the QPC is asymmetric. Also in line with our classical intuition, the largest part of the voltage drop occurs at the weakly coupled contact with smaller drops over the QPC and at the strongly coupled contact. The consequences of asymmetric voltage drop over molecules has been discussed by Ghosh et al.[35]

14.6 Concluding Remarks

In this chapter we have presented an intuitive description of the current–voltage (I-V) characteristics of molecules using simple toy models to illustrate the basic physics (Sections 14.1–14.3). These toy models were also used to motivate the rigorous nonequilibrium Green's Function (NEGF) theory (Section 14.4). A simple example was then used in Section 14.5 to illustrate the application of the NEGF formalism.

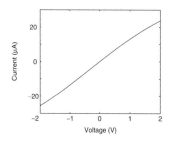 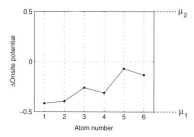

FIGURE 14.22 I-V and potential drop for an applied voltage of 1 V for the QPC asymmetrically connected to the gold contacts. The coupling to the right contact used is $-0.2t$.

The same basic approach can be used in conjunction with a more elaborate Hückel Hamiltonian or even an *ab initio* Hamiltonian. But for these advanced treatments we refer the reader to References 27 and 28.

Some of these models are publicly available through the Purdue Simulation Hub (www.nanohub.purdue.edu) and can be run without any need for installation. In addition to the models discussed here, there is a Hückel model which is an improved version of the earlier model made available in 1999. Further improvements may be needed to take into account the role of inelastic scattering or polaronic effects, especially in longer molecules such as DNA chains.

Acknowledgments

Sections 14.2 and 14.3 are based on material from a book by Supriyo Datta entitled *Quantum Phenomena: From Atoms to Transistors*. It is a pleasure to acknowledge helpful feedback from Mark Ratner, Phil Bagwell, and Mark Lundstrom. The authors are grateful to Prashant Damle and Avik Ghosh for helpful discussions regarding the *ab initio* models. This work was supported by the NSF under Grant No. 0085516–EEC.

14.7.A   MATLAB® Codes

The MATLAB codes for the toy models can also be obtained at www.nanohub.purdue.edu.

14.7.A.1   Discrete One-Level Model

```
% Toy model, one level
% Inputs (all in eV)
E0 = -5.5;Ef = -5;gam1 = 0.2;gam2 = 0.2;U = 1;
%Constants (all MKS, except energy which is in eV)
hbar = 1.06e-34;q = 1.6e-19;IE = (2*q*q)/hbar;kT = .025;
% Bias (calculate 101 voltage points in [-4 4] range)
nV = 101;VV = linspace(-4,4,nV);dV = VV(2)-VV(1);
N0 = 2/(1+exp((E0-Ef)/kT));
for iV = 1:nV% Voltage loop
     UU = 0;dU = 1;
     V = VV(iV);mu1 = Ef-(V/2);mu2 = Ef+(V/2);
     while dU>1e-6%SCF
          E = E0+UU;
          f1 = 1/(1+exp((E-mu1)/kT));f2 = 1/(1+exp((E-mu2)/kT));
          NN = 2*((gam1*f1)+(gam2*f2))/(gam1+gam2);% Charge
          Uold = UU;UU = Uold+(.05*((U*(NN-N0))-Uold));
          dU = abs(UU-Uold);[V UU dU];
     end
     curr = IE*gam1*gam2*(f2-f1)/(gam1+gam2);
     II(iV) = curr;N(iV) = NN;[V NN];
end
G = diff(II)/dV;GG = [G(1) G];% Conductance
h = plot(VV,II*10^6,'k');% Plot I-V
```

14.7.A.2   Discrete Two-Level Model

```
% Toy model, two levels
% Inputs (all in eV)
Ef = -5;E0 = [-5.5 -1.5];gam1 = [.2 .2];gam2 = [.2 .2];U = 1*[1 1;1 1];
```

```
% Constants (all MKS, except energy which is in eV)
hbar = 1.06e-34;q = 1.6e-19;IE = (2*q*q)/hbar;kT = .025;
n0 = 2./(1+exp((E0-Ef)./kT));
nV = 101;VV = linspace(-6,6,nV);dV = VV(2)-VV(1);Usc = 0;
for iV = 1:nV
     dU = 1;
     V = VV(iV);mu1 = Ef+(V/2);mu2 = Ef-(V/2);
     while dU>1e-6
          E = E0+Usc;
          f1 = 1./(1+exp((E-mu1)./kT));f2 = 1./(1+exp((E-mu2)./kT));
          n = 2*(((gam1.*f1)+(gam2.*f2))./(gam1+gam2));
          curr = IE*gam1.*gam2.*(f1-f2)./(gam1+gam2);
          Uold = Usc;Usc = Uold+(.1*(((n-n0)*U')-Uold));
          dU = abs(Usc-Uold);[V Usc dU];
     end
     II(iV) = sum(curr);N(iV,:) = n;
end
G = diff(II)/dV;GG = [G(1) G];
h = plot(VV,II);% Plot I-V
```

14.7.A.3   Broadened One-Level Model

```
% Toy model, restricted solution with broadening
% Inputs (all in eV)
E0 = -5.5;Ef = -5;gam1 = 0.2;gam2 = 0.2;U = 1.0;
% Constants (all MKS, except energy which is in eV)
hbar = 1.06e-34;q = 1.6e-19;IE = (2*q*q)/hbar;kT = .025;
% Bias (calculate 101 voltage points in [-4 4] range)
nV = 101;VV = linspace(-4,4,nV);dV = VV(2)-VV(1);
N0 = 2/(1+exp((E0-Ef)/kT));
for iV = 1:nV% Voltage loop
     UU = 0;dU = 1;
     V = VV(iV);mu1 = Ef-(V/2);mu2 = Ef+(V/2);
     nE = 400;% Numerical integration over 200 points
     id = diag(eye(nE))';
     EE = linspace(-10,0,nE);dE = EE(2)-EE(1);
     f1 = 1./(1+exp((EE-id*mu1)/kT));
     f2 = 1./(1+exp((EE-id*mu2)/kT));
     while dU>1e-4% SCF
          E = E0+UU;
          g = 1./(EE-id*(E+i/2*(gam1+gam2)));
          NN = 2*sum(g.*conj(g).*(gam1*f1+gam2*f2))/(2*pi)*dE;
          Uold = UU;UU = Uold+(.2*((U*(NN-N0))-Uold));
          dU = abs(UU-Uold);[V UU dU];
     end
     curr = IE*gam1*gam2*sum((f2-f1).*g.*conj(g))/(2*pi)*dE;
     II(iV) = real(curr);N(iV) = NN;[V NN curr E mu1 mu2];
end
G = diff(II)/dV;GG = [G(1) G];% Conductance
h = plot(VV,II,'.'); Plot I-V
```

14.7.A.4   Unrestricted Discrete One-Level Model

```
% Toy model unrestricted solution
% Inputs (all in eV)
E0 = -5.5;Ef = -5;gam1 = 0.2;gam2 = 0.2;U = 1;
% Constants (all MKS, except energy which is in eV)
hbar = 1.06e-34;q = 1.6e-19;IE = (q*q)/hbar;kT = .025;
% Bias (calculate 101 voltage points in [-4 4] range)
nV = 101;VV = linspace(-4,4,nV);dV = VV(2)-VV(1);
N0 = 1/(1+exp((E0-Ef)/kT));
for iV = 1:nV% Voltage loop
     U1 = 0;U2 = 1e-5;dU1 = 1;dU2 = 1;
     V = VV(iV);mu1 = Ef-(V/2);mu2 = Ef+(V/2);
     while (dU1+dU2)>1e-6% SCF
          E1 = E0+U1;E2 = E0+U2;
          f11 = 1/(1+exp((E1-mu1)/kT));f21 = 1/(1+exp((E1-mu2)/kT));
          f12 = 1/(1+exp((E2-mu1)/kT));f22 = 1/(1+exp((E2-mu2)/kT));
          NN1 = ((gam1*f12)+(gam2*f22))/(gam1+gam2);
          NN2 = ((gam1*f11)+(gam2*f21))/(gam1+gam2);
          Uold1 = U1;Uold2 = U2;
          U1 = Uold1+(.05*((2*U*(NN1-N0))-Uold1));
          U2 = Uold2+(.05*((2*U*(NN2-N0))-Uold2));
          dU1 = abs(U1-Uold1);dU2 = abs(U2-Uold2);
     end
     curr1 = IE*gam1*gam2*(f21-f11)/(gam1+gam2);
     curr2 = IE*gam1*gam2*(f22-f12)/(gam1+gam2);
     (iV) = curr1;I2(iV) = curr2;
     N1(iV) = NN1;N2(iV) = NN2;[V NN1 NN2];
end
G = diff(I1+I2)/dV;GG = [G(1) G];% Conductance
h = plot(VV,I1+I2,'-'); Plot I-V
```

14.7.A.5   Unrestricted Broadened One-Level Model

```
% Toy model, unrestricted solution with broadening
% Inputs (all in eV)
E0 = -5.5;Ef = -5;gam1 = 0.2;gam2 = 0.2;U = 1;
% Constants (all MKS, except energy which is in eV)
hbar = 1.06e-34;q = 1.6e-19;IE = (q*q)/hbar;kT = .025;
% Bias (calculate 101 voltage points in [-4 4] range)
nV = 101;VV = linspace(-4,4,nV);dV = VV(2)-VV(1);
N0 = 1/(1+exp((E0-Ef)/kT));
nE = 200;% Numerical integration over 200 points
id = diag(eye(nE))';
EE = linspace(-9,-1,nE);dE = EE(2)-EE(1);
for iV = 1:nV% Voltage loop
     U1 = 0;U2 = 1;dU1 = 1;dU2 = 1;
     V = VV(iV);mu1 = Ef-(V/2);mu2 = Ef+(V/2);
     f1 = 1./(1+exp((EE-id*mu1)/kT));
     f2 = 1./(1+exp((EE-id*mu2)/kT));
     while (dU1+dU2)>1e-3% SCF
```

```
        E1 = E0+U1;E2 = E0+U2;
        g1 = 1./(EE-id*(E1+i/2*(gam1+gam2)));
        g2 = 1./(EE-id*(E2+i/2*(gam1+gam2)));
        NN1 = sum(g1.*conj(g1).*(gam1*f1+gam2*f2))/(2*pi)*dE;
        NN2 = sum(g2.*conj(g2).*(gam1*f1+gam2*f2))/(2*pi)*dE;
        Uold1 = U1;Uold2 = U2;
        U1 = Uold1+(.2*((2*U*(NN2-N0))-Uold1));
        U2 = Uold2+(.2*((2*U*(NN1-N0))-Uold2));
        dU1 = abs(U1-2*U*(NN2-N0));dU2 = abs(U2-2*U*(NN1-N0));
    end
    curr = IE*gam1*gam2*sum((f2-f1).*(g1.*conj(g1)+...
    g2.*conj(g2)))/(2*pi)*dE;
    II(iV) = real(curr);N(iV) = NN1+NN2;
    [V NN1 NN2 curr*1e6 E1 E2 mu1 mu2];
end
G = diff(II)/dV;GG = [G(1) G];% Conductance
h = plot(VV,II,' - ');% Plot I-V
```

References

1. P. Avouris, P.G. Collins, and M.S. Arnold. Engineering carbon nanotubes and nanotube circuits using electrical breakdown. *Science*, 2001.
2. A. Bachtold, P. Hadley, T. Nakanishi, and C. Dekker. Logic circuits with carbon nanotube transistors. *Science*, 294:1317, 2001.
3. H. Schön, H. Meng, and Z. Bao. Self-assembled monolayer organic field-effect transistors. *Nature*, 413:713, 2001.
4. R.P. Andres, T. Bein, M. Dorogi, S. Feng, J.I. Henderson, C.P. Kubiak, W. Mahoney, R.G. Osifchin, and R. Reifenberger. "Coulomb staircase" at room temperature in a self-assembled molecular nanostructure. *Science*, 272:1323, 1996.
5. M.A. Reed, C. Zhou, C.J. Muller, T.P. Burgin, and J.M. Tour. Conductance of a molecular junction. *Science*, 278:252, 1997.
6. C. Kergueris, J.-P. Bourgoin, D. Esteve, C. Urbina, M. Magoga, and C. Joachim. Electron transport through a metal–molecule–metal junction. *Phys. Rev. B*, 59(19):12505, 1999.
7. C. Kergueris, J.P. Bourgoin, and S. Palacin. Experimental investigations of the electrical transport properties of dodecanethiol and α, ω bisthiolterthiophene molecules embedded in metal–molecule–metal junctions. *Nanotechnology*, 10:8, 1999.
8. J. Reichert, R. Ochs, H.B. Weber, M. Mayor, and H.V. Löhneysen. Driving current through single organic molecules. *cond-mat/0106219*, June 2001.
9. S. Hong, R. Reifenberger, W. Tian, S. Datta, J. Henderson, and C.P. Kubiak. Molecular conductance spectroscopy of conjugated, phenyl-based molecules on au(111): the effect of end groups on molecular conduction. *Superlattices Microstruct.*, 28:289, 2000.
10. J.J.W.M. Rosink, M.A. Blauw, L.J. Geerligs, E. van der Drift, and S. Radelaar. Tunneling spectroscopy study and modeling of electron transport in small conjugated azomethine molecules. *Phys. Rev. B*, 62(15):10459, 2000.
11. C. Joachim and J.K. Gimzewski. An electromechanical amplifier using a single molecule. *Chem. Phys. Lett.*, 265:353, 1997.
12. W. Tian, S. Datta, S. Hong, R. Reifenberger, J.I. Henderson, and P. Kubiak. Conductance spectra of molecular wires. *J. Chem. Phys.*, 109(7):2874, 1998.
13. J. Chen, W. Wang, M.A. Reed, A.M. Rawlett, D.W. Price, and J.M. Tour. Room-temperature negative differential resistance in nanoscale molecular junctions. *Appl. Phys. Lett.*, 77(8):1224, 2000.

14. D. Porath, A. Bezryadin, S. de Vries, and C. Dekker. Direct measurement of electrical transport through DNA molecules. *Nature*, 403:635, 2000.

15. M. Magoga and C. Joachim. Conductance of molecular wires connected or bonded in parallel. *Phys. Rev. B*, 59(24):16011, 1999.

16. L.E. Hall, J.R. Reimers, N.S. Hush, and K. Silverbrook. Formalism, analytical model, and *a priori* Green's function-based calculations of the current–voltage characteristics of molecular wires. *J. Chem. Phys.*, 112:1510, 2000.

17. M. Paulsson and S. Stafström. Self-consistent field study of conduction through conjugated molecules. *Phys. Rev. B*, 64:035416, 2001.

18. E.G. Emberly and G. Kirczenow. Multiterminal molecular wire systems: a self-consistent theory and computer simulations of charging and transport. *Phys. Rev. B*, 62(15):10451, 2000.

19. P.S. Damle, A.W. Ghosh, and S. Datta. Unified description of molecular conduction: from molecules to metallic wires. *Phys. Rev. B*, 64:201403(r), 2001.

20. J. Taylor, H. Gou, and J. Wang. *Ab initio* modeling of quantum transport properties of molecular electronic devices. *Phys. Rev. B*, 63:245407, 2001.

21. M. Di Ventra, S.T. Pantelides, and N.D. Lang. First-principles calculation of transport properties of a molecular device. *Phys. Rev. Lett.*, 84:979, 2000.

22. P. Damle, A.W. Ghosh, and S. Datta. First-principles analysis of molecular conduction using quantum chemistry software. *Chem. Phys.*, 281, 171, 2001.

23. Y.Q. Xue, S. Datta, and M.A. Ratner. Charge transfer and "band lineup" in molecular electronic devices: a chemical and numerical interpretation. *J. Chem. Phys.*, 115:4292, 2001.

24. J.J. Palacios, A.J. Pérez–Jiménez, E. Louis, and J.A. Vergés. Fullerene-based molecular nanobridges: a first-principles study. *Phys. Rev. B*, 64(11):115411, 2001.

25. J.M. Seminario, A.G. Zacarias, and J.M. Tour. Molecular current–voltage characteristics. *J. Phys. Chem. A*, 1999.

26. S. Datta. Nanoscale device modeling: the Green's function method. *Superlattices Microstruct.*, 28:253, 2000.

27. P.S. Damle, A.W. Ghosh, and S. Datta. Molecular nanoelectronics, in M. Reed (Ed.), *Theory of Nanoscale Device Modeling*. (To be published in 2002; for a preprint, e-mail: datta@purdue.edu.)

28. F. Zahid, M. Paulsson, and S. Datta. Advanced semiconductors and organic nano-techniques, in H. Markoc (Ed.), *Electrical Conduction through Molecules*. Academic Press. (To be published in 2002; for a preprint, e-mail: datta@purdue.edu.)

29. S. Datta. *Electronic Transport in Mesoscopic Systems*. Cambridge University Press, Cambridge, UK, 1997.

30. S.G. Lias et al. Gas-phase ion and neutral thermochemistry, *J. Phys. Chem.* Reference Data. American Chemical Society and Americal Institute of Physics, 1988.

31. C. Desjoqueres and D. Spanjaard. *Concepts in Surface Physics*. 2nd ed. Springer-Verlag, Berlin, 1996.

32. R.G. Parr and W. Yang. *Density Functional Theory of Atoms and Molecules*, Oxford University Press, 1989, p.99

33. de Picciotto, H.L. Stormer, L.N. Pfeiffer, K.W. Baldwin, and K.W. West. Four-terminal resistance of a ballistic quantum wire. *Nature*, 411:51, 2001.

34. S. Datta, W. Tian, S. Hong, R. Reifenberger, J.I. Henderson, and C.P. Kubiak. Current–voltage characteristics of self-assembled monolayers by scanning tunneling microscopy. *Phys. Rev. Lett.*, 79:2530, 1997.

35. A.W. Ghosh, F. Zahid, P.S. Damle, and S. Datta, Insights from I-V asymmetry in molecular conductors. Preprint. *cond-mat/0202519.*

36. Special issue on single-charge tunneling, *Z. Phys. B.*, 85, 1991.

37. P.F. Bagwell and T.P. Orlando. Landauer's conductance formula and its generalization to finite voltages. *Phys. Rev. B*, 40(3):1456, 1989.

38. M. Brandbyge, J. Taylor, K. Stokbro, J.-L. Mozos, and P. Ordejon. Density functional method for nonequilibrium electron transport. *Phys. Rev. B*, 65(16):165401, 2002.

39. R. Zeller, J. Deutz, and P. Dederichs. *Solid State Commun.*, 44:993, 1982.

40. Y. Imry. *Introduction to Mesoscopic Physics.* Oxford University Press, Oxford, UK, 1997.

41. D.K. Ferry and S.M. Goodnick. *Transport in Nanostructures.* Cambridge University Press, London, 1997.

42. K. Hansen, E. Laegsgaard, I. Stensgaard, and F. Besenbacher. Quantized conductance in relays. *Phys. Rev. B*, 56:1022, 1997.

43. J.N. Murrell and A.J. Harget. *Semi-Empirical SCF MO Theory of Molecules.* John Wiley & Sons, London, 1972.

4

Manipulation and Assembly

15 Magnetic Manipulation for the Biomedical Sciences *J.K. Fisher, L. Vicci, K. Bloom, E. Timothy O'Brien, C.W. Davis, R.M. Taylor, II, R. Superfine* 15-1
Introduction • Background • Description of Magnetic System • Description of the Magnetic Systems • Computer Control and Data Acquisition • Conclusion • Acknowledgments

16 Nanoparticle Manipulation by Electrostatic Forces *Michael Pycraft Hughes* 16-1
Introduction • Theoretical Aspects of AC Electrokinetics • Applications of Dielectrophoresis at the Nanoscale • Biomolecular Applications of Dielectrophoresis • Particle Separation • Conclusion

17 Biological- and Chemical-Mediated Self-Assembly of Artificial Micro- and Nanostructures *S.W. Lee, R. Bashir* 17-1
Introduction • Bio-inspired Self-Assembly • The Forces and Interactions of Self-Assembly • Biological Linkers • State of the Art in Bio-inspired Self-Assembly • Future Directions • Conclusions

18 Nanostructural Architectures from Molecular Building Blocks *Damian G. Allis, James T. Spencer* 18-1
Introduction • Bonding and Connectivity • Molecular Building Block Approaches

19 Building Block Approaches to Nonlinear and Linear Macromolecules *Stephen A. Habay, Christian E. Schafmeister* 19-1
Introduction • Building Block Approaches to Nonlinear Macromolecules • Building Block Approaches to Linear Macromolecules • Conclusion

20 Introduction to Nanomanufacturing *Ahmed Busnaina* 20-1
Introduction • Nanomanufacturing Challenges • Top-down Approach • Bottom-up Approach • Combined Top-down and Bottom-up Nanomanufacturing Approaches • Registration and Alignment • Reliability and Defect Control

21 Textile Nanotechnologies *B. Pourdeyhimi, N. Fedorova, William Dondero, Russell E. Gorga, Stephen Michielsen, Tushar Ghosh, Saurabh Chhaparwal, Carola Barrera, Carlos Rinaldi, Melinda Satcher, Juan P. Hinestroza* 21-1
Introduction • Nanofibers • Electrospun Nanofibers with Magnetic Domains for Smart Tagging of Textile Products • Carbon Nanotubes • Surface Activation • Nanoadditives in Textiles

15

Magnetic Manipulation for the Biomedical Sciences

J.K. Fisher

L. Vicci

K. Bloom

E. Timothy O'Brien

C.W. Davis

R.M. Taylor, II

R. Superfine

University of North Carolina

15.1 Introduction 15-1

15.2 Background 15-2

15.3 Description of Magnetic System 15-3
Magnetic Forces

15.4 Description of the Magnetic Systems 15-4
Pole Materials and Methods • Position Detection • Implementation

15.5 Computer Control and Data Acquisition 15-9
System Performance

15.6 Conclusion 15-19

Acknowledgments 15-19

References 15-19

15.1 Introduction

Force generation is ubiquitous in biological systems. At the cellular level, adhesion, motility, alteration of cell morphology, and cell division are produced by the forces generated by thousands of molecular motors and the assembly of protein polymers.[1-5] Inside the cell, these forces produce active transport of vesicles and organelles, the movement of chromosomes during mitosis, and even replication, transcription, and translation of genetic information into proteins.[6-12] Moreover, cells both respond to external forces and impose forces, for example, to produce macroscopic fluid flow in the airway.[13] Quantification of the forces generated during these cellular processes, the role of force as a feedback mechanism, and the determination of the mechanical properties of involved structures has been enabled by recently developed nanoscale manipulation devices.

The majority of the existing microbiological force application methods can be divided into two categories: those that can apply relatively high forces through the use of a physical connection to a probe and those that apply smaller forces with a detached probe. Existing magnetic manipulators utilizing high fields and high-field gradients have been able to reduce this gap in the maximum applicable force, but generally with compromises between force direction, maximum force, and integration with high numerical aperture (NA) microscopy.

Here we present two different magnetic manipulator systems. Our first-generation system, a tetrapole design, uses relatively large coils with a high number of ampere-turns to generate forces in the nN range on 4.5 μm magnetic beads. This sizeable magnetic system is limited to microscopes with low NA (long working distance) objectives, but it requires no special machining processes to construct. The second-generation system, a high-bandwidth hexapole design compatible with high NA microscopy, requires more sophisticated fabrication techniques due to the dimensions and materials used. We refer to these magnetic manipulators and the accompanying tracking and computer control systems as 3-D force microscopes (3DFMs).[14,15]

15.2   Background

Measurement and application of forces at the nanoscale can be accomplished using multiple techniques. Among the mechanical probe techniques, glass fibers or microneedles[16–18] have been used to measure the effects of forces on the movement of chromosomes and the force exerted by myosin on actin.[19] Recently, atomic force microscopy (AFM) has emerged as a suitable method for obtaining subnanometer spatial resolution[20] with picoNewton force sensitivity[21] using techniques relying on the deformation of a cantilever spring element.[22] In addition, Evans[23] has developed a method using a deformable vesicle attached to a pipette to measure the forces between membrane-bound molecules and target specimens such as other vesicles or flat substrates. Although they provide important insights within their domains, these methods suffer from the invasiveness of the attached cantilever, pipette, or fiber, as well as an inherent limitation in the sensitivity of the measurement (typically 10 pN for AFM and 1 pN for micropipette[24]) and the directionality of applied forces.

To address these shortcomings, methods have been developed that use a refractive microbead, often below 1 μm in diameter, as a mechanical probe. The bead can be free to move throughout the accessible volume within a specimen or can be functionalized to be attached to specific molecular groups or proteins. Optical tweezers use the optical power gradient of a focused laser beam to attract refractive materials toward the waist of the focused beam.[25,26] The position of the beam may be adjusted using steering optics, and the force generated on microbeads by the optical trap can be varied by changing the intensity of the laser. This force can be accurately calibrated using multiple methods,[27] including analyses of the bead's position power spectrum and the effect of drag forces on the trapped bead. Laser tweezers have been applied to a wide variety of biological problems, from measurements of the forces generated during DNA transcription[28] to the properties of neuronal membranes[29] and to the forces generated by the molecular motors dynein, kinesin, and myosin.[30–34] Additionally, laser tweezers have been used to study the mechanical properties of single molecules such as actin, titin, and DNA, and to investigate the forces required for nucleosome disruption in chromatin.[35–40] This method offers increased sensitivity over the mechanical probe methods discussed above, with sensitivity down to approximately 0.01 pN.[41] Its limitations are in the achievable force (generally less than 200 pN), specimen heating at higher forces[42,43] (approximately 10°C/watt of laser power at 1064 nm laser wavelength in water), and the nonspecificity of forces that act on all refracting particles and macromolecules within the range of the optical trap. Although laser heating is minimized with a trap using an 850 nm laser wavelength,[44] such lasers are not widely available at significant powers.

Using a magnetically permeable microbead and magnetic field gradients, it is possible to perform manipulations similar to those by optical tweezers without the generation of specimen-damaging heat. In addition, since typical biological materials are at most weakly magnetically active, this method is more specific than optical tweezers. Typically, magnetic systems can measure forces down to approximately 0.01 pN[45] and are limited by remnant magnetization of the magnetic materials.

Beginning with Crick's[46] *in vitro* studies of the viscoelastic properties of cytoplasm in 1949, magnetic forces have been used to investigate a wide range of biophysical properties. Many of the systems that have been reported have applied forces in a single direction, often with one pole tip.[47–50] Among these single-tip systems is a device developed by Bausch that is capable of applying up to 10 nN on a 4.5-μm paramagnetic bead.[50] In addition to the single-tip systems, various multipole systems have been constructed. These multipole devices include Valberg's magnetic system designed for applying torques,[51] a multipole geometry designed by Strick to apply forces upward while applying a torque to a ferromagnetic bead,[45] an eight-pole instrument constructed by Amblard[52] to apply torques as well as forces within the specimen plane, Gosse's six-pole design with poles above the specimen plane with no magnetic forces available in the downward direction,[53] and systems designed by Haber[54] and Huang[55] to deliver a uniform gradient and allow for the use of a high NA objective. Huang's full octapole system utilized a backiron to complete the magnetic flux circuit, resulting in increased field efficiency.

The wide range of applications for the systems discussed above motivated the development of our first-generation magnetic manipulator. Our desire to perform 3-D manipulations, while maintaining the ability to apply high forces, led to our construction of a closed-frame tetrapole system with sharp pole tips. For our second-generation system, we shrunk the size of our device, making it compatible with commercially available microscopes utilizing high NA objectives and condensers. We also added the versatility of removable pole pieces, allowing the user to design appropriate pole pieces for their specific applications. This system can be configured with pole tips in a hexapole geometry that provides nearly uniform 3-D force directionality over the full $4\pi sr$ of a solid angle. Finally, through our choice of materials, we were able to increase the manipulation bandwidth of the second-generation system into the kilohertz range.

15.3   Description of Magnetic System

15.3.1   Magnetic Forces

Force on a magnetic bead is caused by an interaction between its magnetic dipole moment m and the gradient ∇B of an incident magnetic field. For a soft, magnetically permeable bead, m is entirely induced by the incident field. Subject to saturation properties of the magnetic material in the bead,

$$m = \frac{\pi d^3}{2\mu_0}\left(\frac{\mu_r - 1}{\mu_r - 2}\right)B \tag{15.1}$$

where μ_0 is the permeability of free space in SI units, μ_r is the relative permeability of the bead, and d is the diameter of the bead. The magnetic force is

$$\mathbf{F} = \frac{\pi d^3}{4\mu_0}\left(\frac{\mu_r - 1}{\mu_r + 2}\right)\nabla B^2 \tag{15.2}$$

The field is produced by multiple electromagnet pole tips arranged in space to provide the necessary directional capability. Except for very near a pole tip, the field's behavior can be modeled by a monopole. According to this model, the magnitude of B from a singly excited magnetic pole is proportional to B_p/r^2, where B_p is the pole strength and r is the distance from the pole. Correspondingly, $\nabla(B^2) = -4B_p^2/r^5$, directed toward the pole. Clearly, the distance between a pole tip and the bead is of primary importance in optimizing bead force.

The standard design for generating magnetic fields at a specimen is to couple the flux from a current-carrying coil to the sample region through a permeable core that narrows at the specimen. The coil is typically wound around the core, with the end closest to the specimen tapered to concentrate the flux so that a large field and field gradient (when high forces are desired) is created. The analogy between electric circuits and magnetic circuits (Figure 15.6) provides an immediate insight into magnetic system design. In a series electrical circuit, it is obvious that, for a fixed voltage, the highest electric current will be produced when the circuit resistance is minimized. Using the magnetic–electrical circuit analogy,[56] for a fixed magnetomotive force (MMF) as generated by the current in the coils, the highest magnetic field will be produced when the circuit reluctance is minimized. This implies that the system should minimize air gaps and attempt to provide a high-permeability path for the flux through a return loop.

15.3.1.1 Magnetic Circuit Topology

Schematic diagrams of electrical analogs to the magnetic circuits are shown in Figure 15.6. MMFs are represented by voltage sources, and the reluctances of the various pieces of the magnetic path are represented by resistors. For simplicity in comparison, the number of coils for both system analogs has been made the same.

Figure 15.6(a) shows the first-generation as-built topology. Due to spatial constraints, the four coils were moved away from the pole pieces and instead encircle four return path pieces that connect the back ends of the pole cores in a ring. These are shown as MMFs M_i in series with their respective return path circuit reluctances R_f. Pole reluctances R_p connect the junctions between the R_f to edges of the specimen space, and specimen reluctances R_a connect their respective pole tips to the bead.

The topology of the second-generation design with two drive coils omitted is shown in Figure 15.6(b). A coil encircling each pole provides an MMF M_i in series with pole reluctance R_p and specimen reluctance R_a due to the air/glass/liquid gap between the pole tip and the location of the bead. This is conceptually straightforward, where the B_p of each pole is proportional to its coil current, with due regard to magnetic saturation of the pole.

For a given current, the coil dissipates heat proportional to its resistance, which is inversely proportional to its volume. Thus, large coils are needed to obtain high B_p without excessive heating. B_p is also inversely proportional to magnetic path reluctance R, which is the sum of the reluctances R_i of the constituent pieces of the total magnetic path, each of which can be approximated as $R_i \approx l_i/\mu_r^i A_i$ where l_i is the length and A_i the cross-sectional area of piece i, and μ_r^i is its relative magnetic permeability. A simplified model of the magnetic path for each pole consists of the pole itself, a gap through the specimen space, and, in the case of the tetrapole system, the parallel combination of the other three poles and the return path between the backsides of all four poles. Magnetic path materials are commonly available, having $\mu_r > 5000$. Consequently, the path reluctance is dominated by the gaps in the circuit. To provide a good magnetic return path, a high μ_r magnetic shell or complete drive ring can be added to the configuration to connect the backsides of the poles.[55]

It should be appreciated that both of the magnetic circuit analogs presented in Figure 15.6 are simplified models, and that parasitic reluctances actually exist between every pair of circuit nodes, some of them being significant. They are omitted here for clarity in understanding the behavior of the as-built topology.

15.4 Description of the Magnetic Systems

For our first-generation tetrapole system, we used a 0.7 NA Mitutoyo APO 100 lens (Mitutoyo America Corporation, Aurora, IL), due to its generous 6-mm working distance. The lens bodies are wider than the optical path itself and fill an angle of 110°, which is very close to the 109° included angle between tetrahedral pole axes. The finite diameter of the magnetic poles precludes arranging their axes in equilateral tetrahedral angles, unless the lens bodies or the poles themselves are modified with relief cuts. A simpler solution was to modify the pole axis angles to 135° (or from a 35.3 to 22.5° inclination from horizontal), while preserving the tetrahedral pole tip positions. A pole tip taper angle of 9.5° leaves a 7.75° angular clearance from the lens body. However, this clearance is diminished as the size of the tetrahedral working volume is increased to accommodate a usable specimen space.

The pole tips may be regarded as spherical. For high-permeability core material, the modeled magnetic monopoles are located approximately at the centers of these spheres. Thus, for a specimen chamber consisting of two horizontal coverslips enclosing a liquid cell, the vertical separation between the upper and lower monopole locations is the sum of the sphere radii (≤ 50 μm), the specimen chamber thickness ($\cong 400$ μm), and 100 μm working clearance between the pole tips and the specimen chamber. This comes to 600 μm of vertical separation. The total distance from each monopole to the center of the specimen chamber is therefore $(600/2)/\cos(109/2) = 517$ μm. For this specimen chamber size, 517 μm is the closest we can reasonably expect to get the magnetic poles in this design.

Our desire to use a magnetic shell to minimize return path reluctance, along with the need for a large number of ampere-turns in a limited conductor cross-section, motivated the positioning of the coils in

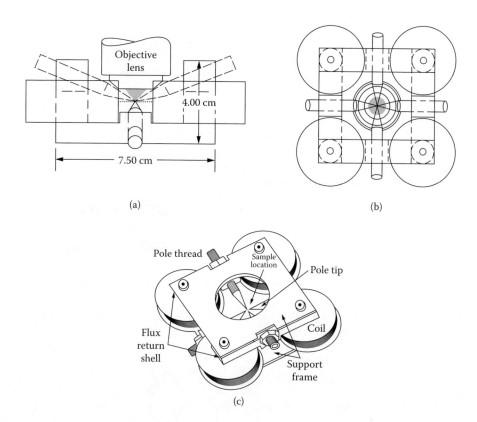

FIGURE 15.1 Engineering drawing for the first-generation system showing the four magnetic poles and four coils: side view, bottom-up view, and perspective view, respectively. The upper objective lens is shown in the first two views.

locations where they are sufficiently removed from the optical path. To do this, the return path geometry was configured as a frame, with coils encircling parts of the return path rather than the poles themselves, as shown in Figure 15.1. To allow for a large number of ampere-turns, the coils were wound on cylindrical bobbins with a generous cross-sectional area, inner diameter of 1.7 cm, outer diameter of 4.25 cm, and length of 2.0 cm. They consisted of 340 turns of #22AWG Belden #8077 (Belden Division, Richmond, IN) magnet wire. The temperature rise of the coils driven at 3 A for 3 min was 22.7°C. This provides coils capable of generating an MMF greater than 1000 A turns.

For our second-generation system, we again turned to the circuit analogy to gain design insights. As seen in Figure 15.6, the magnetic system necessarily includes an air gap between pole tips at the specimen region. The reluctance of this gap is in series with other magnetic circuit reluctances, such as other gaps where the magnetic permeability is low. If the reluctance of these other gaps is significantly below that of the specimen air gap, then the total circuit reluctance and, hence, magnetic performance will not suffer. With this device, we have taken advantage of this freedom in design by separating the pole tips from the current carrying coils and the flux return path. This provides flexibility in the implementation of a wide range of field geometries with facile exchange of pole tips.

The fixed drivers consist of a pair of symmetrically opposed magnetic drive rings, respectively, above and below a specimen chamber. The use of a drive ring assures a completed magnetic flux circuit, which is essential for efficient field use.[55] Each drive ring is a castellated annular magnetic core with a coil wound around each of its six castellations as shown in Figure 15.3(c), thereby forming six drive poles. For the flux return path through the drive ring, we chose corrosion-resistant Metglas alloy 2417A (Honeywell, International Inc., Morriston, NJ) tape-wound toroidal cores with a relative permeability of over 30,000

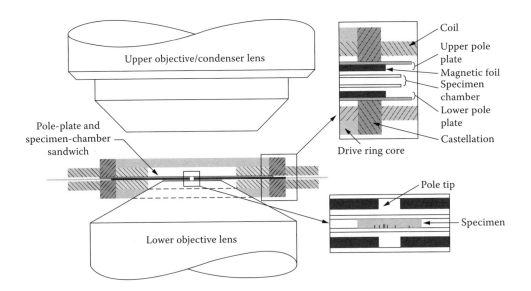

FIGURE 15.2 Second-generation system sketch. Vertically symmetric magnetic driver assembly consisting of drive-ring cores and coils closely coupled to thin-foil poles in the pole plate and specimen chamber sandwich. The high-NA lower objective lens places tight geometric constraints on both the driver and the sandwich. A long working distance condenser leaves space for other (e.g., microfluidic) subsystems.

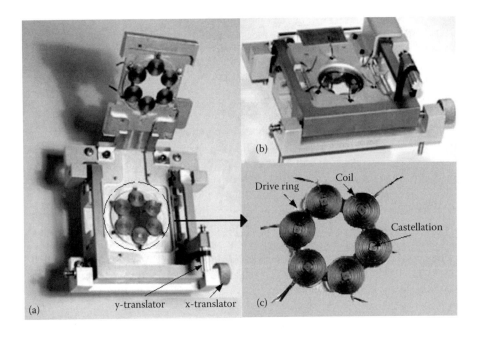

FIGURE 15.3 (a) Second-generation system in the open position showing double hinges, kinematic dowel pins, magnetic drivers, and an xy-adjustable specimen slide holder. (b) Stage in closed position showing specimen chamber x and y adjustment knobs (no specimen present). (c) A magnetic driver assembly comprising a drive ring and six coils. The drive ring is a laminated Metglas alloy 2714A (Honeywell, International Inc. Morriston, NJ) with high permeability, low magnetostriction, and remanance, and excellent high-frequency performance. The coils are each 25 turns of 6×40 mil flat magnet wire for a high fill factor. The design current is 2.5 A/coil.

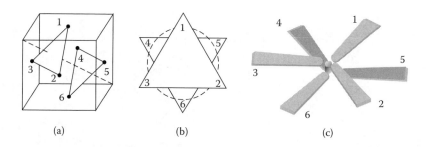

(a)   (b)   (c)

FIGURE 15.4 Implementation of a symmetric FCC pole tip placement with thin-foil poles in two closely spaced parallel planes. (a) Optical axis (*dashed line*) is perpendicular to two planes, each containing three FCC points. (b) Two plates form parallel equilateral triangles with a cylindrical working volume between them. (c) Magnetic flux is conducted by thin-foil poles in two parallel planes to tips having centers at the FCC locations.

up to 30 kHz for field strengths above 0.01 T, a saturation induction of 0.57 T, and near zero magnetostriction. After machining the six castellations in this material, coils were wound with 6 × 40 mil flat magnet wire, providing a high conductor fill ratio in the available space. In operation, the upper and lower drive-ring poles are precisely aligned with each other and with the pole plates. Corresponding upper and lower drive coils are connected in series to receive the same electrical current, such that their magnetic polarities are the same. This provides six magnetomotive excitations to drive magnetic flux into the pole plates.

The second-generation system relies on high-field gradients (∇B) and reduced magnetic path reluctance to produce large forces. To maximize the gradient, pole tips need to be sharp and are located close to the specimen. Tip manufacturing processes used in the second-generation system allow for sharper tips and will be discussed in more detail. To decrease the magnetic path reluctance, the reluctance of the specimen space has been minimized by reducing the distance by which the pole tips are separated. In addition to the reduced tip separation, the pole tips are necessarily thin to fit into the tightly constrained space limited by the close working distance of a high NA microscope lens (Figure 15.2). A specimen can be placed directly on a pole plate or in a separate coverslip sandwich that sits below one pole plate or in between two pole plates. The total thickness of the specimen and pole plate space can range from 150 to 500 μm. The upper-drive ring is on a hinged mechanism allowing it to be lowered onto the specimen chamber, such that all the magnetic components are properly aligned. This provides easy removal of experiment-specific specimen chambers without the need to change magnetic drivers. Specimen chambers may be selected from a standard library of configurations, or custom pole shapes and chambers can be fabricated for special purposes.

15.4.1 Pole Materials and Methods

The removable poles in the second-generation system have been fabricated using two different methods. In the first method, pulsed electrodeposition[57–59] was used to deposit magnetic material on the surface of a coverslip in a pattern described by a photolithography process. This method has the benefit of being able to develop complex pole geometries at the expense of processing complexity. The second method, laser machining of thin permalloy foils (Laserod Inc., Gardena, CA), has been used to fabricate poles from commercially available magnetic foils. This process is capable of cutting materials with thicknesses up to approximately 400 μm, and offers 10 μm lateral resolution. Additionally, the use of commercially available materials ensures that the properties of the materials are well characterized. Figure 15.5(a) shows a 175-μm thick laser-machined three-pole pole plate. This is one half of the hexapole geometry used in experiments where the directionality of applied force is critical. Figure 15.5(b) is an example of the tip-flat geometry used to create high gradients for experiments where it is sufficient to pull in only one direction.

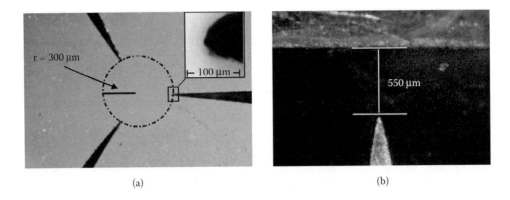

(a)   (b)

FIGURE 15.5 (a) Three-pole design using laser-machined pole pieces cut from .007″ thick low permeability, high saturation foil (Mushield). (b) Tip-flat geometry used in maximum force experiments.

15.4.2 Position Detection

To determine the effect of the force applied via our magnetic systems, we must monitor the position of the object(s) being manipulated. Low bandwidth 2-D position information for an entire field of view may be obtained using video tracking methods that rely on postprocessing of individual video frames to determine the location of selected features. This method is limited by image type (brightfield, fluorescence, etc.) frame rate of the camera (typically 30 to 120 frames/sec), pixel size of the camera, and the algorithm used to compute the object's location. Alternatively, 3-D laser tracking may be used to collect position information for a single object with high temporal and spatial resolution. This method is discussed in more detail below.

The physical principle used in laser tracking is interferometry, which requires a sufficiently coherent light source, for which we use a near-infrared laser diode. The tracker uses the interference between a reference beam and a measurement beam to produce a diffraction pattern that represents the magnitude and phase of the measurement beam in a detection plane.

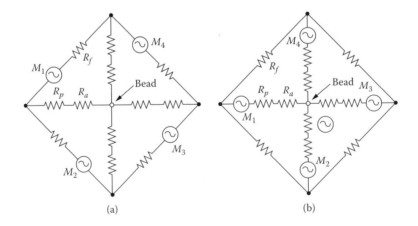

(a)   (b)

FIGURE 15.6 Electric Circuit Analog. Electric circuit analogs to the magnetic paths. The reluctances R represent the return path reluctance R_f, the pole tip reluctance R_p, and the reluctance R_a for the gap between the pole tip and the bead. The MMFs M_i are generated by currents through coils wound around their respective legs. Labels of symmetrically defined reluctances have been omitted for the sake of clarity. (a) Electric circuit analog for the first-generation design with coils wrapped around the flux returns path segments. (b) Electric circuit for the second-generation system with two coils omitted to ease comparison.

In the technique we have implemented,[60–62] a laser beam is focused by a condenser lens to a diffraction limited beam waist. An objective lens is placed on the opposite side and focused on the waist to capture the exiting beam (see Figure 15.13 for optical system diagram). This reference beam is in the shape of two reentrant cones between the confocally arranged lenses. In the far field, the reference beam behaves as if it were emitted from a point source located at the beam waist. If a small sphere (bead) with index of refraction different from its surrounding medium is placed within the reference beam, it will scatter some fraction of the beam. The scattering pattern can be quite complicated, but for beads less than half a wavelength in diameter, only the principal forward scattered lobe lies within the solid angle of the reference beam. The scattered light from the bead can be approximately regarded as originating from an isotropic point source located at the center of the bead. Larger beads also behave as nearly point sources, but with nonisotropic radiation patterns, a detail we shall not discuss here. In the far field, the interference of these two beams carries 3-D position information in the form of a diffraction pattern.

A property of an ideal lens is that the light fields at the front and back focal planes of the lens are Fourier transforms of each other. Accordingly, the diffraction pattern at the back focus of the objective lens is a Fourier transform containing bead displacement information. A relay lens is used to reimage the back focal plane of the objective lens onto a quadrant photodiode (QPD) to capture the information in this pattern. A bead exactly at the beam waist will scatter a measurement beam in some fixed phase relative to the reference beam. If the bead is displaced in x, a fringe pattern with an x dependence will occur; similarly for displacements in y.

Measuring z displacement relies on a phenomenon near the beam waist known as the Gouy phase shift in which light in the near-field about the beam waist propagates at apparently superluminal speed. Intuitively, this has to happen for the concave spherical wavefronts entering the beam waist zone to deform to planar at the waist, then to convex as they leave the zone. If the bead is displaced from the beam waist in z, it will scatter a measurement beam in some fixed phase relative to the Gouy-shift-dependent phase of the incident beam. This z-dependent phase variation is propagated to the far field where it interferes with the fixed phase of the reference beam. To be sure, a distance-dependent "bull's eye" diffraction pattern occurs anyway, but it is too subtle to be of much use by itself. The Gouy phase shift causes a local but marked variation in the intensity of the center of the bull's eye with z displacement of the bead.

15.4.3 Implementation

The available analytical models describing the mapping of QPD signals into XYZ position relative to the beam waist[60,63] put stringent constraints on the shape, size, and composition of the tracked bead. For biological experiments that involve application of force, beads of larger size are preferable to achieve higher magnetic pull. Thus, we require more flexibility in probe characteristics than offered by an analytical model. As a result, we have developed a novel technique where instead of relying on an *a priori* model of the light scatter, we estimate the mapping function before (and potentially during) each experiment.

We use a standard system identification technique to determine the relationship between the QPD signals and bead position in three dimensions. Before each experiment, small noise signals are injected into the three-axis piezo-driven stage (model Nano-LP 100; Mad City Labs Inc., Madison, WI), causing the bead position to change in a calibrated manner. These small perturbations in the bead position result in small changes in the QPD signals. Correlations of the injected noise with corresponding photodiode signals are then analyzed to estimate the mapping function. Tracking software uses this newly estimated mapping function to improve its performance. A publication providing detailed information about the algorithm is in preparation.

15.5 Computer Control and Data Acquisition

The 3DFM system is controlled by five PCs: one for the tracking subsystem; one for the magnet subsystem; one for high-resolution high-speed video capture; one for low-resolution, low-speed video capture; and one for the user interface to the entire system. We elected to use several computers for aggregate processing

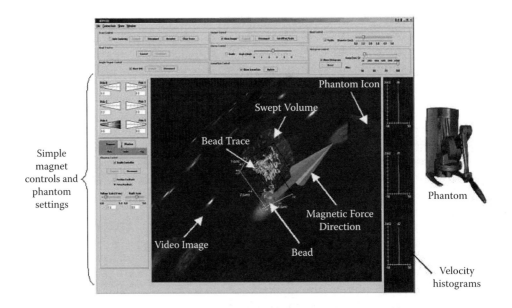

FIGURE 15.7 3DFM user interface: The 3-D display (shown in a monoscopic view but also displayable in stereo) displays the current location of a tracked bead as a green wireframe sphere. This sphere is centered in the live video display of 2-D microscope data (the gray plane with white spots surrounding the sphere in the image). This plane of video moves with the bead through the volume and shows objects in the specimen that lie on or near the focal plane passing through the bead. The sphere leaves a yellow line as a trail, showing where the bead has moved during an experiment. A transparent shell is drawn around the volume that has been "carved out" by the bead as it has moved along the trace; this shows the boundary of the region explored by the bead. During an experiment, the scientist will control the forces on (and position of) the bead using the Phantom force-feedback controller (SensAble Technologies Inc., Woburn, MA) or via force-driven functions. This control can be fully manual, where either force-clamp or position-clamp control can be used with the magnetic drive system. The control can be fully automatic, where the user specifies a force vs. time relationship that is carried out by the computer.

power, convenience, and flexibility. For those that desire an inexpensive alternative, it would be possible to run both the magnet subsystem and a video capture device from the same computer. Both the tracking and the high-resolution/high-speed video capture computers are based on workstation-class computers with dual 3 GHz Pentium Xeon processors, 1 GB of main memory, and 140 GB of RAID0 disk. The tracking computer uses an analog output board (model PCI-6733; National Instruments, Austin, TX) for stage positioning and a multifunction I/O board for stage, QPD, and laser-intensity sensing (model PCI-6052E; National Instruments, Austin, TX). The high-resolution/high-speed video capture computer controls a CoolSNAP HQ camera (Photometrics, Tucson, AZ) via a supplied PCI card. The camera has a maximum resolution of 1392 by 1040 12-bit pixels digitized at 20 MHz. The magnet computer is desktop-class with a 3 GHz Pentium 4 processor and 512 MB of main memory. A National Instruments analog output board (model PCI-6713) is used to drive the magnetics' electronics. Low-resolution/low-speed video capture is accomplished using a desktop-class computer with a 3 GHz Pentium 4 processor and 512 MB of main memory. The user-interface computer is a workstation-class computer with dual 2.2 GHz Pentium Xeon processors, 1 GB of main memory, and a Quadro4 (Nvidia, Santa Clara, CA) graphics card. The dual processors, large memory, and the high-end graphics card are useful for computationally and display-intensive visualization tasks.

The 3DFM user interface (Figure 15.7) brings together several data streams from the microscope: bead trajectory, magnetic drive force, and 2-D fluorescent microscopy, and it enables control over instrument parameters while displaying results from the instrument's subsystems.[64] The interface allows a scientist to control currents sent to the magnetic system with a Phantom force-feedback controller (allowing the user to "feel" the force he or she is applying) or via predefined functions while observing the specimen's

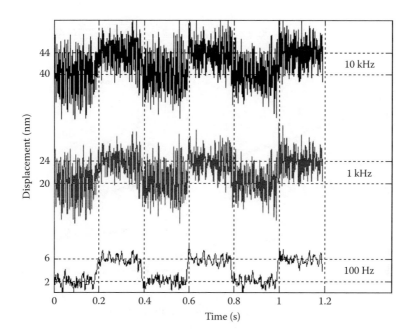

FIGURE 15.8 Tracking system response to 4 nm bead motion. The QPD signals for three bandwidths of the measurement, i.e., 10 kHz, 1 kHz, and 100 Hz, are shown. DC offsets were added to the 1 and 10 kHz data sets in the figure for clarity.

response to the applied force. The Phantom controller provides flexible, immediate control for exploration, while the predefined functions are used for experiments with defined parameter spaces that are investigated with repetition.

Real-time position information from the laser tracking system is displayed in three manners: a bead-step histogram, a strip chart, and a 3-D bead trace. The bead-step histogram shows binned counts of observed bead velocities in a fixed X, Y, and Z coordinate system and may be used to generate information about the viscous properties of the specimen. The strip chart graphs the x, y, and z position information vs. time. For the 3-D bead trace, a polyline is used to show the path of the bead over time and a wireframe sphere scaled to the size of the tracked bead displays the particles' current location. The user interface allows the scientist to rotate and zoom into the 3-D bead trace, enabling real-time inspection of nanometer-scale position fluctuations or observations of the bead's movement with respect to its environment.

15.5.1 System Performance

15.5.1.1 Tracking Resolution Characterization

The performance of the interferometric tracking subsystem has been tested using 0.957-μm polystyrene beads (Polysciences, Inc., Warrington, PA) immobilized in agarose. To test the system, a single bead was placed into the beam waist of the tracking laser. The bead was then moved by 4-nm square pulses in the positive x direction using the three-axis piezo-driven stage. In Figure 15.8, the QPD signal of the 4 nm displacement is shown for three different bandwidths of the measurement, i.e., 10 kHz, 1 kHz, and 100 Hz. From these experiments we have determined that the lateral resolution of the system is 2.4 nm at 10 kHz. A similar experiment for the axial resolution of the system resulted in a value of 4.4 nm at 10 kHz.

15.5.1.2 Magnetic Force Calibration

The magnetic forces were determined by measuring the velocity of the beads in fluids of known viscosity, including a neutrally buoyant solution of corn syrup. The viscosity of the corn syrup was measured as 1.31 Pa s at 25°C using a commercial viscometer (Cannon Fenske, Viscometer Type No.513, State College, PA).

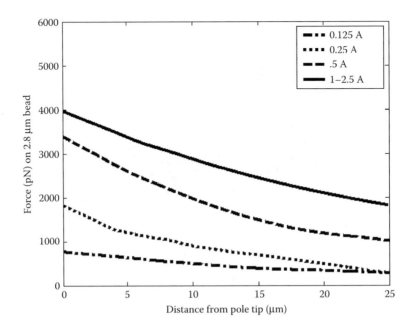

FIGURE 15.9 Forces generated on a 2.8 µm bead using the first-generation system. The force did not increase at higher currents due to saturation of both the pole tip and the bead.

The magnetic forces were calculated from Stokes formula, $F = 6\pi\eta a_b v$, where η is the fluid viscosity, a_b is the bead radius, and v the bead velocity. In maximum force experiments, video data taken at 120 frames/second was tracked to determine particle velocities for beads close to a pole tip. For these experiments, symmetry was sacrificed, and the pole was allowed to enter the specimen chamber. In the first-generation system, close to a pole tip, we measured forces as large as 200 pN on 0.94-µm beads, 4 nN on 2.8-µm beads, and 9 nN on 4.4-µm beads. Forces measured in the center of the specimen chamber were considerably smaller, but directionally far more symmetric. Figure 15.9 shows the relation between force and bead position for multiple currents. At currents above 1 A the force did not increase, indicating that both the pole tip and the bead had been saturated. For our second-generation system, maximum force values of 700 pN and 13 nN were determined for the 1- and 4.5-µm beads, respectively, using a point-flat geometry with a 550-µm gap (see Figure 15.5b) made from 350-µm thick material with a saturation of approximately 20,000 Gauss and a permeability of 300 (MuShield, Manchester, NH). This geometry was chosen for its simplicity and high-field gradient near the pole tip. Force vs. position data for 1- and 4.5-µm beads are displayed in Figure 15.10. The force bandwidth for this second-generation system was also determined by oscillating a 1 µm bead in between opposite poles in a planar, six-pole geometry. This analysis revealed that the -3 dB roll off for the bead's response is greater than 3 kHz.

15.5.1.3 Magnetic Force Directionality

To establish the available force directions for the first-generation (tetrahedral) and second-generation (hexapole) systems, simulations using monopole approximations of the pole tips for both geometries were used to model the field and field gradient generated by a given pole tip excitation.[15] For the first-generation system, these simulations showed that forces can be generated in approximately triangular patches about the poles and on lines between poles but not in directions opposite to the poles for the tetrapole geometry. For the second-generation system with a hexapole geometry, simulations showed that forces can be generated over the full $4\pi\ sr$ of a solid angle in the hexapole geometry.

A plot of the simulation for the first-generation system is shown in Figure 15.11(a) where the brightness is proportional to the average of the forces that occurred in that direction. Within the monopole model

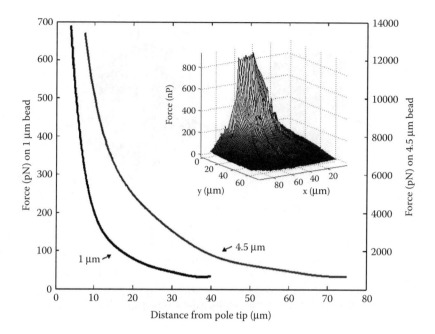

FIGURE 15.10 Thin-foil (second-generation) 3DFM magnetic force vs. distance. Maximum forces obtained for 1 and 4.5 μm superparamagnetic beads in the second-generation system. The Y axis on the left-hand side of the figure corresponds to force on the 1-μm bead. Forces on the 4.5-μm bead are on the right-hand Y axis. Insert image shows forces vs. position on a 1-μm bead for an *xy* plane.

of pole tip field generation, the direct problem of calculating the magnitude and direction of the B field from a given set of pole tip strengths (coil currents) is straightforward. However, the inverse problem of obtaining the set of coil currents for a given B field has no simple solution. We therefore chose to explore the B field space through the generation of 10,000 random sets of coil currents to experimentally confirm the simulation (Figure 15.11[a]). To approximate access to a full 3-D force space in the tetrapole system, we have implemented a control program that cyclically pulls the bead toward different poles with strengths and durations such that the time average of the bead force is the magnitude and direction desired. This method produces a trajectory that is a zig-zag path with step directions determined by the pole locations. The step size would be determined by the viscosity of the specimen and the forces obtained at the sample position, with temporal widths determined by the bandwidth of the magnetics system. As an example of this method, we estimate the motion of a 4.5-μm bead seeking the largest "effective" force in an otherwise forbidden direction. For this instrument, in the specimen location of the cilia experiment (described later), with the sample presumed to be the viscosity of water, we would obtain step sizes of approximately 130 nm, with temporal widths of about 0.1 sec. This is limited by our current magnetics bandwidth of about 10 Hz. As a result, the time multiplexing approach is not generally applicable to high sensitivity experiments but is useful for bead positioning.

For the second-generation system, simulation data established that our hexapole design with a face-centered-cubic (FCC) pole tip placement around the specimen chamber would provide nearly uniform 3-D force directionality over the full 4π sr of solid angle. To demonstrate this ability, we performed two separate experiments; the first to show large-scale magnetic symmetry and the second to demonstrate small-scale bead control over one octant of the sphere representing all possible excitation directions.

Large-scale magnetic symmetry was demonstrated by pulling a 2.8-μm superparamagnetic bead (M-280; Dynal Biotech, Oslo, Norway) toward each magnetic axis of symmetry (Figure 15.12[a]). This required a total of 26 different excitations toward each of the six pole tips individually, between two adjacent poles, and between each set of three adjacent poles. In this experiment, movement in the expected

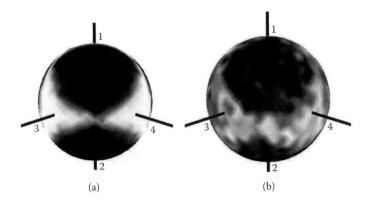

(a)   (b)

FIGURE 15.11 Spherical plots for directionality of forces for the first-generation system. Plot (a) is of a magnetic monopole model, while plot (b) is of experimental data. The four solid lines projecting from the spheres show directions to the magnetic poles. In both cases, a uniform random distribution of normalized pole excitations was applied and the resulting forces averaged over sampling bins. The averages of the bins are plotted as brightness. Dark regions indicate that no set of coil currents generated force in that direction. Both model and experimental data show that forces can be generated in the directions of the poles and along the lines joining the poles but not in directions opposite the poles.

direction is seen, but it is off from the expected location by 6 to 12° (depending on the axis of rotation). The deviation of the lab coordinate system from its theoretical location is most likely the source of this difference. Additionally, the measured maximum percent difference for the average force generated by the one, two, and three pole excitations was 31%, with an estimated average force of approximately 1.5 pN. This average force could easily be increased to approximately 15 pN for a 2.8-μm bead at the center of the geometry by reducing the distance from the center of the FCC geometry to each pole tip and increasing the coil currents.

Small-scale, fine control of bead position is demonstrated in Figure 15.12(b). Here, force vectors were generated to sample the space between three poles, filling one octant of the surface of the sphere. Forces were applied in each direction for 3 sec, with the bead being returned back to the origin after each excitation via a force in the opposite direction. The small-scale bead control (filling of the octant) shown in

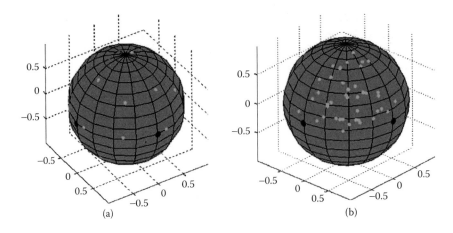

(a)   (b)

FIGURE 15.12 Experimentally obtained force directionality data for the second-generation system. Large black dots represent pole locations and light gray dots indicate force directions (directions in which the bead was pulled). Axis of symmetry data (a) where forces were applied toward each pole, in between two poles and in between three poles. Force directions for one octant of pole excitations (b).

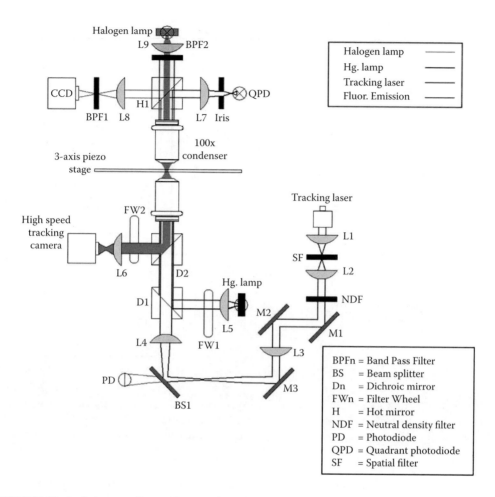

FIGURE 15.13 Optical system diagram for second-generation 3DFM.

Figure 15.12(b), combined with the symmetry data of the first experiment (Figure 15.12[a]), indicated that we would be able to fill all eight octants on the surface of the sphere, and thus, pull the bead in all directions.

15.5.1.4 Magnetic Manipulation of Biological Specimen

The magnetic manipulator's ability to noninvasively apply relatively large forces to biological specimen, while avoiding localized heating, is the cornerstone of this manipulation technique. Force directionality, magnitude, and bandwidth requirements must be taken into account when designing a magnetic manipulator for a specific set of experiments, as well as imaging requirements that may place constraints on the optics and thus the dimensions of the magnetic manipulator. Here we present two separate biological experiments; the first, a cilia manipulation experiment, performed using our first-generation 3DFM. The second, a chromatin manipulation experiment, performed using our second-generation 3DFM.

15.5.1.4.1 Application: Cilia Manipulation

Cilia are ubiquitous actuating structures in biology and are present in unicellular, simple multicellular, and complex multicellular organisms including vertebrates.[65] The rapid oscillation or "beating" of cilia is used for locomotion and translocation of food particles by unicellular ciliates such as paramecia. It is also used for transporting food particles and fluids by stationary multicellular organisms such as tunicates.[65] Of the many functions served by cilia in humans, we are studying their role in the clearance of infectious agents and particulates from the lungs.

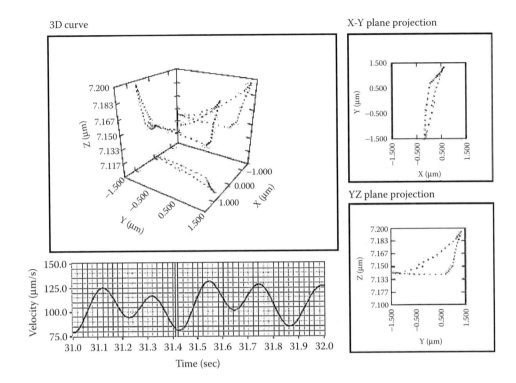

FIGURE 15.14 Cilia tracking with a first-generation 3DFM system. A bead is attached to the end(s) of one or more beating cilia. The spatial plots show tracking of the cilia during one complete stroke. The motion extends over 3 μm and appears to lie nearly in the *y–z* plane. The velocity of the waveform to the left of the cursor in the temporal plot corresponds to the spatial plots.

The alveolar surface of the human lung is bathed in a thin aqueous layer covered by a thicker and more complex layer of mucus. Mucus is a nonuniform viscoelastic fluid composed of mucin proteins, glycosaminoglycans, and cellular remnants such as actin and DNA. The mucus layer sticks to and traps particulates, bacteria, and other infectious agents. The beating cilia propel this layer out through the airways toward the throat,[66] where the mucus and its entrained detritus are swallowed and disposed of safely in the gut. A basic understanding of this process and how it can fail is fundamental to several important research areas: environmental factors affecting lung function, developing new drug delivery methods, studying the underlying causes of diseases such as cystic fibrosis,[67] and developing new treatments for them.

The pattern of the ciliary beat and the force applied by the forward or "power" stroke of the cilia tips as they engage the mucus are essential to understanding how the cilia propel mucus. Previous experiments studying the beat pattern have employed optical microscopy and high-speed video imaging,[68–71] both inherently 2-D instruments. The measurements of forces exerted by cilia have heretofore been limited to the compound cilium of single-cell organisms[72] and to the flagella of bull sperm.[17] In this study, we used the 3DFM to track the motion of cilia in living human lung cell cultures and to explore their response to applied forces.

15.5.1.4.2 *Results*

A real-time 3-D trajectory of a magnetic bead attached to the cilia was measured, the plot of which is shown in Figure 15.14. The bead size of 1 μm and the spacing between cilia being less than 300 nm precludes our being able to definitively rule out multiple attachments at this time. A stroke length of 3 μm was measured for the trajectory of the cilia beat. Our tracking system is capable of measuring the beat cycle in full 3-D, significantly including the *z* motion of the cilia tip. This vertical component of

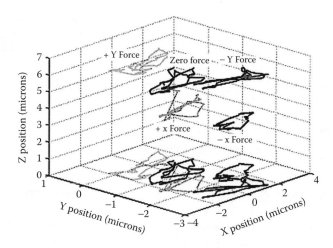

FIGURE 15.15 Cilia manipulation. The trajectories of approximately two cilia beats during the application of force in each of the tetrahedral directions of our pole tips. Projections of the trajectories onto the *X–Y* plane are shown at the bottom. The directions in which the cilia were pulled correspond to the tetrahedral directions of the pole tips. The poles designated as "−Y" and "+Y" are above the specimen chamber, whereas those designated as "−X" and "+X" are below the chamber.

motion is understood to be a critical factor in the operation of the cilia–mucus system as it allows the cilia to retract into the less viscous lower layer during the retraction stroke. During the power stroke, the cilia are at their greatest extension, allowing them to couple more effectively to the more viscous overlying mucus. This *z* motion is observed in our 3-D traces.

We applied forces to the magnetic beads attached to cilia in an initial attempt to explore the response of the cilia to forces. Figure 15.15 shows that sequential excitation of individual poles results in the average position of the orbit of the cilia being shifted in the direction of the energized pole. Since the poles lie at the corners of a tetrahedron, an energized pole will, in general, pull a bead both laterally and vertically at the same time. This was observed. The applied magnetic force, estimated to be about 2 pN, appears to be enough to shift the beat trajectory significantly without stalling the cilia. This is consistent with the measurements performed on sperm flagella where a stall force of about 250 pN was measured.[72]

15.5.1.4.3   Application: Chromatin Manipulation

In addition to the cilia force response experiments carried out with our first-generation system, we have used our second-generation system to investigate the interaction of single molecules with proteins. In these experiments, chromatin fibers are extended with the goal of investigating the strength of the DNA–protein interactions that maintain the higher-order chromatin structure. The system is challenged to maintain nanometer-scale position tracking sensitivity, while the bead moves over several microns. This is achieved using our stage tracking routine that moves the nanometric specimen stage to keep the bead within the linear range of the laser tracking.

Chromatin, the condensed form of DNA, is made up of DNA and histone proteins. The association of DNA with these histones forms the nucleosome, a structure that condenses the DNA by wrapping it 1.65 times around a histone octomer. The histone octomer is made up of two copies of each histone H2A, H2B, H3, and H4 and is known as the nucleosome core particle.[73] Further compaction of the DNA is accomplished through interactions between core histone N-terminal domains and linker histones. The structure and conformational changes that may take place throughout the cell cycle are important to gene regulation and understanding the mechanisms behind transcription, replication, and repair.

15.5.1.4.4   Results

Chromatin fibers were manipulated, and extension (change in bead position) was monitored using the 3DFM and a "ramp and hold" manipulation method (Figure 15.16[b]). Fiber extension was monitored,

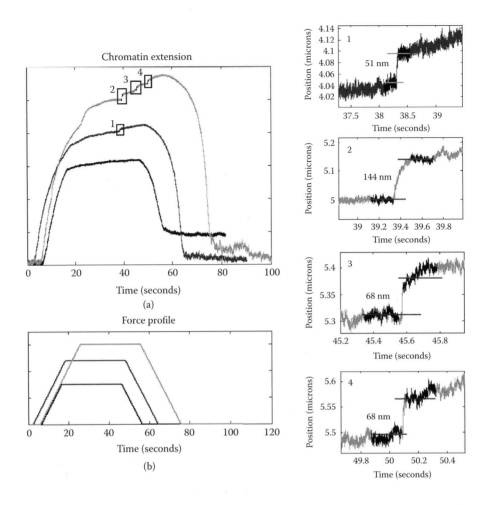

FIGURE 15.16 Chromatin manipulation. (a) Extension profile for consecutive manipulations of the same chromatin fiber. For extension with a maximum applied force of approximately 15 pN, no nucleosome disruption events were observed. In the subsequent traces (24 and 30 pN) the increased maximum force resulted in a total of four nucleosome disruption events. [(c), 1–4], with possible multiple disruptions taking place in event 2. (b) Force profile corresponding to each extension in (a)

with specific attention paid to sudden increases in overall extension (an indication of a possible nucleosome disruption event).[37,38,74–77] Three consecutive extensions of the same fiber are shown in Figure 15.16(a). For the initial application of force, where the maximum applied force was approximately 15 pN, the tension on the nucleosome (histone–DNA complex) was not large enough to cause a disruption event. The overall extension of the chromatin fiber was significantly less (<50%) than what would be expected for b-form DNA alone, indicating that the nucleosome organization of the fiber remained intact. In the second extension of the fiber, with a maximum force of approximately 24 pN, one nucleosome disruption event was observed during the "hold" interval (Figure 15.16[c, 1]). The amplitude of the observed disruption events has been determined by taking the difference of the average of 1000 data points immediately before and 1000 data points immediately after the event. Using this method, the amplitude of this disruption event was determined to be 51 nm. For the third extension of the fiber (maximum force approximately 30 pN), three nucleosome disruption events were observed, with amplitudes of 144, 68, and 68 nm for the first, second, and third events, respectively (Figure 15.16[c, 2–4). Overall, the amplitude of the nucleosome disruption events shown in Figure 15.16 are in reasonable agreement with published results for "full"[40] nucleosome disruptions, with the second disruption event most likely being the result of the removal of two nucleosomes.

It is important to note that the amplitude of the disruption events as observed by the 3DFM may be slightly larger than those viewed using traditional laser tweezers due to the inherent differences in these force application methods. Traditional laser tweezers operate in a "position clamp" mode where restorative forces act to maintain an object's position at the center of the laser trap. When the DNA that is wrapped around a nucleosome is released, it causes a decrease in the force necessary to return the bead's position to the center of the laser trap, thus reducing the overall tension on the chromatin fiber. Manipulation techniques based on magnetics, such as the 3DFM, inherently operate in a "force clamp" mode where force remains relatively constant as position changes. For these chromatin experiments, the constant force delivered by the 3DFM will extend the DNA released during a nucleosome disruption event slightly more than the reduced force of the laser trap. In addition, the constant force will allow for investigations into the cooperativity exhibited by nucleosome–nucleosome interactions, a property that would be indicated by increases in the number of multiple nucleosome disruption events.

15.6 Conclusion

Magnetic manipulators working on the cellular and subcellular levels provide a means for the investigation of the biomechanical properties essential for organisms to function. These systems offer force sensitivities on a par with those of the most sensitive probe-based manipulators, while offering a broad range of applicable forces and the ability to manipulate multiple beads in parallel. The magnetic system's unique ability to perform manipulations noninvasively, without the use of an attached cantilever or specimen-damaging temperature elevations, offers potential for biological experimentation ranging from single molecule to cell culture studies. The two systems presented here, while examples of different levels of complexity, demonstrate the utility of this manipulation technique for biological investigations. The first-generation system with its fixed pole geometry, high ampere-turn, and low-bandwidth magnetic system, combined with low NA imaging, may not be as versatile as our second-generation system, but the ease and relative low cost with which such a system may be constructed makes for a simple yet powerful tool. Our second generation, thin-foil magnetic force system provides additional capabilities such as compatibility with high NA optics, interchangeable pole geometries, and high-frequency manipulation. The increased bandwidth provides means for the investigation of the dynamic properties of viscoelastic materials, cells, and proteins.

Acknowledgments

The authors would like to thank the members of the Computer Integrated Systems for Microscopy & Manipulation (CISMM) group at UNC for their assistance, the National Institute of Biomedical Imaging and Bioengineering (grant numbers P41-EB002025-23A1 and R01-EB00761), the National Heart, Lung, and Blood Institute (grant number R01-HL077546-01A2), and NASA (grant number NCC-1-02037) for their support.

References

1. T. P. Stossel, *Science* 260, 1086 (1993).
2. C. L. Rieder and E. D. Salmon, *J Cell Biol* 124, 223 (1994).
3. L. P. Cramer, T. J. Mitchison, and J. A. Theriot, *Curr Opin Cell Biol* 6, 82 (1994).
4. V. Mermall, P. L. Post, and M. S. Mooseker, *Science* 279, 527 (1998).
5. J. M. Scholey, I. Brust-Mascher, and A. Mogilner, *Nature* 422, 746 (2003).
6. S. A. Endow, *Bioessays* 25, 1212 (2003).
7. P. Huitorel, *Biol Cell* 63, 249 (1988).
8. M. J. Davey, D. Jeruzalmi, J. Kuriyan, et al., *Nat Rev Mol Cell Biol* 3, 826 (2002).

9. V. Ellison and B. Stillman, *Cell* 106, 655 (2001).
10. K. S. Wilson and H. F. Noller, *Cell* 92, 337 (1998).
11. D. Signor and J.M. Scholey, *Essays Biochem* 35, 89 (2000).
12. W. Wintermeyer, M.V. Rodnina, *Essays Biochem* 35, 117 (2000).
13. H. Matsui, S. H. Randell, S. W. Peretti, et al., *J Clin Invest* 102, 1125 (1998).
14. J. K. Fisher, J. Cribb, K. V. Desai, et al., *Rev Sci Instrum* 77 (2006).
15. J. K. Fisher, J. R. Cummings, K. V. Desai, et al., *Rev Sci Instrum* 76 (2005).
16. R. B. Nicklas, *J Cell Biol* 97, 542 (1983).
17. K. A. Schmitz, D. L. Holcomb-Wygle, D. J. Oberski, C. B. Lindemann, *Biophys J* 79, 468 (2000).
18. T. Yanagida, M. Nakase, K. Nishiyama, et al., *Nature* 307, 58 (1984).
19. R. V. Skibbens and E. D. Salmon, *Exp Cell Res* 235, 314 (1997).
20. A. A. Baker, W. Helbert, J. Sugiyama, et al., *Biophys. J* 79, 1139 (2000).
21. E.-L. Florin, A. Pralle, E. H. K. Stelzer, et al., *Appl Phys A* 66, S75 (1998).
22. P. C. Braga and D. Ricci, in *Methods in Molecular Biology* (Humana Press, Totowa, NJ, 2003), Vol. 242.
23. E. Evans, *Annu Rev Biophys Biomol Struct* 30, 105 (2001).
24. R. S. Conroy and C. Danilowicz, *Contemp Phys* 45, 277 (2004).
25. A. Ashkin, J. M. Dziedzic, J. E. Bjorkholm, et al., *Opt. Lett* 11, 288 (1986).
26. S. M. Block, D. F. Blair, and H. C. Berg, *Nature* 338, 514 (1989).
27. M. J. Lang and S. M. Block, *Am J Phys* 71, 201 (2003).
28. H. Yin, M. D. Wang, K. Svoboda, et al., *Science* 270, 1653 (1995).
29. J. W. Dai and M. P. Sheetz, *Biophysl J* 68, 988 (1995).
30. E. Hirakawa, H. Higuchi, and Y. Y. Toyoshima, Proceedings of the National Academy of Sciences of the United States of America 97, 2533 (2000).
31. S. M. Block, L. S. Goldstein, and B. J. Schnapp, *Nature* 348, 348 (1990).
32. S. C. Kuo and M. P. Sheetz, *Science* 269, 232 (1993).
33. K. Visscher, M. J. Schnitzer, and S. M. Block, *Nature* 400, 184 (1999).
34. M. Rief, R. S. Rock, A. D. Mehta, et al., *Proc Natl Acad Sci U S A* 97, 9482 (2000).
35. B. D. Brower-Toland, C. L. Smith, R. C. Yeh, et al., *Proc Natl Acad Sci U S A* 99, 1960 (2002).
36. C. Claudet, D. Angelov, P. Bouvet, et al., *J Biol Chem* 280, 19958 (2005).
37. B. D. D. Brower-Toland, R. C. Yeh, C. L. Smith, et al., *Biophys J* 82, 907 (2002).
38. M. L. Bennink, S. H. Leuba, G. H. Leno, et al., *Nat Struct Biol* 8, 606 (2001).
39. Y. Cui and C. Bustamante, *Proc Natl Acad Sci U S A* 97, 127 (2000).
40. L. H. Pope, M. L. Bennink, K. A. van Leijenhorst-Groener, et al., *Biophys J* 88, 3572 (2005).
41. J. C. Meiners and S. R. Quake, *Phys Rev Lett* 84, 5014 (2000).
42. E. Peterman, F. Gittes, and C. F. Schmidt, *Biophys J* 84, 1308 (2003).
43. P. N. Prasad, *Optics Lett* 28, 2288 (2003).
44. A. Schonle and S. W. Hell, *Optics Lett* 23, 325 (1998).
45. T. R. Strick, J. F. Allemand, D. Bensimon, et al., *Science* 271, 1835 (1996).
46. F. H. C. Crick and A. F. W. Hughes, *Exp Cell Res* 1, 36 (1949).
47. F. Ziemann, J. Radler, and E. Sackmann, *Biophys J* 66, 2210 (1994).
48. F. Assi, R. Jenks, J. Yang, et al., *J Appl Phys* 92, 5584 (2002).
49. M. Barbic, J. J. Mock, A. P. Gray, et al., *Appl Phys Lett* 79, 1897 (2001).
50. A. R. Bausch, F. Ziemann, A. A. Boulbitch, et al., *Biophys J* 75, 2038 (1998).
51. P. A. Valberg and D. F. Albertini, *J Cell Biol* 101, 130 (1985).
52. F. Amblard, B. Yurke, A. Pargellis, et al., *Rev Sci Instrum* 67, 818 (1996).
53. C. Gosse and V. Croquette, *Biophys J* 82, 3314 (2002).
54. C. Haber and D. Wirtz, *Rev Sci Instrum* 71, 4561 (2000).
55. H. Huang, C. Y. Dong, H. S. Kwon, et al., *Biophys J* 82, 2211 (2002).
56. D. C. P. Lorrain, F. Lorrain, *Electromagnetic Fields and Waves* (W. H. Freeman, New York, 1988).
57. B. N. Popov, K. M. Yin, and R. E. White, *J Electrochem Soc* 140, 1321 (1993).

58. P. T. Tang, *Electrochimica Acta* 47, 61 (2001).

59. T. Osaka, M. Takai, K. Hayashi, et al., *Nature* 392, 796 (1998).

60. A. Pralle, M. Prummer, E. L. Florin, et al., *Microsc Res Tech* 44, 378 (1999).

61. I. M. Peters, B. G. d. Grooth, J. M. Schins, et al., *Rev Sci Instrum* 69, 2762 (1998).

62. A. Rohrbach and E. H. K. Stelzer, *J Appl Phys* 91, 5474 (2002).

63. A. Rohrbach and E. H. K. Stelzer, *Appl Opt* 41, 2494 (2002).

64. D. Marshburn, C. Weigle, B. G. Wilde, et al., in *IEEE Visualization 2005* (IEEE Computer Society, Minneapolis, MN, 2005).

65. R. A. Wallace, J. L. King, G. P Sanders, *Biology: The Science of Life* (Foresman and Co., Glenview, IL, 1981).

66. B. K. Rubin, *Respiratory Care* 23, 761 (2002).

67. R. C. Boucher, *Eur Respir J* 23, 146 (2004).

68. M. J. Sanderson and M. A. Sleigh, *J Cell Sci* 47, 331 (1981).

69. M. Rautiainen, S. Matsune, S. Shima, et al., *Acta Oto-Laryngologica* 112, 845 (1992).

70. P. F. M. Teunis and H. Machemer, *Biophys J* 67, 381 (1994).

71. M. A. Chilvers and C. O'Callaghan, *Thorax* 55, 314 (2000).

72. M. J. Moritz, K. A. Schmitz, and C. B. Lindemann, *Cell Motil Cytoskeleton* 49, 33 (2001).

73. K. E. van Holde, *Chromatin* (Springer-Verlag, New York, 1988).

74. B. D. Brower-Toland, C. L. Smith, R. C. Yeh, et al., *Proc Natl Acad Sci U S A* 99, 1960 (2002).

75. L. H. Pope, M. L. Bennink, M. P. Arends, et al., *Biophys J* 82, 2476 (2002).

76. B. Brower-Toland, D. A. Wacker, R. M. Fulbright, et al., *J Mol Biol* 346, 135 (2005).

77. S. H. Leuba, M. A. Karymov, M. Tomschik, et al., *Proc Natl Acad Sci U S A* 100, 495 (2003).

16

Nanoparticle Manipulation by Electrostatic Forces

16.1 Introduction **16**-1
16.2 Theoretical Aspects of AC Electrokinetics **16**-2
Dielectrophoresis • Dielectrophoretic Behavior of Solid
Particles • Double Layer Effects • Modeling of Complex
Spheroids: The Multishell Model • Modeling Nonspherical
Ellipsoids • Phase-Related Effects: Electrorotation and
Traveling Wave Dielectrophoresis • Analysis of Key Spectrum
Events • Limitations on Minimum Particle Trapping Size:
Brownian Motion, Conduction, and Convection
16.3 Applications of Dielectrophoresis at the Nanoscale **16**-23
Trapping Single Nanoparticles
16.4 Biomolecular Applications of Dielectrophoresis **16**-26
16.5 Particle Separation **16**-26
16.6 Conclusion **16**-28
References **16**-28

Michael Pycraft Hughes
University of Surrey

16.1 Introduction

Methods for the manipulation of nanometer-scale objects can be grouped into two philosophies [1]: self-assembly of molecular components, a process widely seen in nature, is described as the "bottom-up approach" (small things make larger things); the opposite is the "top-down approach," where large machines (such as the tools of modern semiconductor manufacture) are used to make much smaller products. While the former is considered to be ultimately preferable, modern technology has only been able to manipulate molecular-scale objects at will using the top-down approach, such as writing using xenon atoms as ink and a scanning tunneling microscope as a pen [2].

However, with the use of very small electrodes (usually formed by top-down methods), it is possible to generate electric fields with such complex local geometry that the manipulation of single molecules in solution becomes possible — and these fields can be manipulated so that the particles can be steered across the electrodes, and used to assemble miniature electric circuits [3]. The means by which this is achieved is through the interaction between time-variant, nonuniform electric fields, which can be described as a number of subphenomena collectively referred to as electrokinetics.

Alternating current (AC) electrokinetic techniques such as dielectrophoresis and electrorotation [4,5] have been used for many years for the manipulation, separation, and analysis of objects with lengths of the order of 1 μm to 1 mm in solution. The induction of a force (or torque) occurs due to the interaction of an induced dipole with the imposed electric field. This can be used to cause the object to exhibit a variety of motions including attraction, repulsion, and rotation, by changing the nature of the dynamic field — in many ways, these forces may be viewed as an electrostatic equivalent to optical tweezers [6] and optical spanners [7] in that they exert translational and rotational forces on a body due to the interaction between a body and an imposed field gradient.

Recent advances in semiconductor manufacturing technology have enabled researchers to develop electrodes for manipulating proteins [8,9], to concentrate 14 nm beads from solution [10], and to trap single viruses and 93-nm-diameter latex spheres in contactless potential energy cages [11]. AC electro-kinetic techniques are simple and cheap, and require no moving parts, relying entirely on the electrostatic interactions between the particle and the dynamic electric field. In this chapter, we will consider the ways in which this form of electrostatic manipulation can benefit nanotechnology, and the constraints on the technique due to factors such as Brownian motion, heating of the medium, and electrode dimensions as the electrode array is miniaturized to the nanometer scale.

16.2 Theoretical Aspects of AC Electrokinetics

16.2.1 Dielectrophoresis

If a dielectric particle is suspended in an electric field, it will *polarize* (positive and negative charges will accumulate at the ends of the particle nearest to the electrode of opposite sign), or to use an alternate phrase, a *dipole* is *induced* in the body. If that electric field is spatially uniform, then the Coulombic attraction induced at either side of the particle is equal and there is no net movement. However, if the electric field strength is higher on one end of the particle than another, the net force is nonzero and the particle will move toward the region of highest electric field; that is, it will move *up the field gradient*. This is *dielectrophoresis* [12], and is shown schematically in Figure 16.1. Note that the force is related to

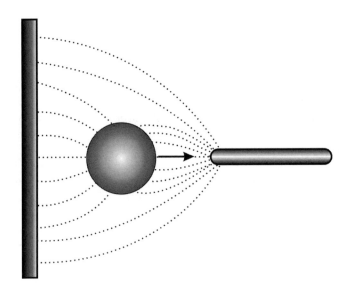

FIGURE 16.1 A schematic of a polarizable particle suspended within a point–plane electrode system. When the particle polarizes, the interaction between the dipolar charges and the local electric field produces a force. Due to the inhomogeneous nature of the electric field, the force is greater on the side facing the point than that on the side facing the plane, and there is net motion toward the point electrode. This effect is called positive dielectrophoresis. If the particle is less polarizable than the surrounding medium, the dipole will align counter to the field, and the particle will be repelled from the high-field regions, which is called negative dielectrophoresis.

the direction of the field gradient and is independent of the direction of the electric field; consequently, we can use AC as well as DC fields.

If the particle is more polarizable than the medium around it (as shown in the figure), the dipole aligns with the field and the force acts up the field gradient toward the region of highest electric field. If the particle is less polarizable than the medium, the dipole aligns against the field and the particle is repelled from regions of high electric field. The polarizability is related to the ability of the charges to realign with the changing field; the dipole is related to differences in the conduction — either resistive or capacitive — in the particle and medium, and depends on frequency through the change in capacitive reactance of these materials. This effect is called a *dielectric dispersion* and its frequency is a characteristic of the particle and its environment. Since the alignment of the field is irrelevant, using AC fields has the advantage of reducing any *electrophoretic* force (due to any net particle charge) to zero.

The dielectrophoretic force, $\mathbf{F}_{DEP}$, acting on a spherical body is given by:

$$\mathbf{F}_{DEP} = 2\pi r^3 \varepsilon_m Re[K(\omega)]\nabla E^2 \tag{16.1}$$

where r is the particle radius, ε_m is the permittivity of the suspending medium, ∇ is the Del vector (gradient) operator, E is the *rms* electric field, and $Re[K(\omega)]$ is the real part of the Clausius–Mossotti factor, given by:

$$K(\omega) = \frac{\varepsilon_p^* - \varepsilon_m^*}{\varepsilon_p^* + 2\varepsilon_m^*} \tag{16.2}$$

where ε_m^* and ε_p^* are the complex permittivities of the medium and particle respectively, and $\varepsilon^* = \varepsilon - j\sigma/\omega$, with σ being the conductivity, ε the permittivity, and ω the angular frequency of the applied electric field. The limiting (DC) case of Equation (16.2) is

$$K(\omega = 0) = \frac{\sigma_p - \sigma_m}{\sigma_p + 2\sigma_m} \tag{16.3}$$

The frequency dependence of $Re[K(\omega)]$ indicates that the force acting on the particle varies with the frequency. The magnitude of $Re[K(\omega)]$ varies depending on whether the particle is more or less polarizable than the medium. If $Re[K(\omega)]$ is positive, then particles move to regions of highest field strength (positive dielectrophoresis); the converse is negative dielectrophoresis, where particles are repelled from these regions. By careful construction of the electrode geometry, which creates the electric field, it is possible to create electric field morphologies so that potential energy minima are bounded by regions of increasing electric field strengths. An example of such an electrode (often called a *quadrupolar* or *polynomial* electrode array) is shown in Figure 16.2. In such electrodes, particles experiencing positive dielectrophoresis are attracted to the regions of highest electric field (typically the electrode edges, particularly where adjacent electrodes are close), while particles experiencing negative dielectrophoresis are trapped in an isolated field minimum at the center of the array.

16.2.2 Dielectrophoretic Behavior of Solid Particles

Dielectrophoretic force can be induced in a wide range of submicrometer particles from molecules to viruses. However, before we study how dielectrophoresis might be applied to manipulating and studying these complex particles, it is wise to consider a simple case to see how the basic principles of dielectrophoresis work at the submicrometer range. This is important, because when the diameter becomes significantly smaller than 1 μm, a number of factors that have relatively little effect on the dielectrophoretic response of larger particles, such as cells, increase in importance and begin to dominate the response.

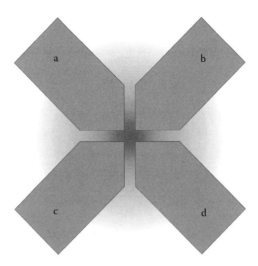

FIGURE 16.2 A schematic of a geometry of electrodes typically used in nanoscale dielectrophoresis work, known as a quadrupolar or polynomial electrode array. The gap between opposing electrodes in the center of the array is typically of the order of 1 to 10 μm for nanometer-scale work, but arrays as small as 500 nm have been used. To induce dielectrophoretic motion in particles suspended near the electrode array, electrodes would be energized such that *a* and *c* are of the same phase, and *b* and *d* are in antiphase to them. Changing the phase relationship allows other effects such as electrorotation to be observed.

In order to understand these effects fully, we shall begin by examining a simple particle, consisting of one material only. By examining this, we can develop a model that can later be adapted for more complex shapes and structures.

One of the most useful tools to understanding the fundamental mechanisms underlying the dielectrophoresis of particles on the nanometer scale is the homogeneous sphere or bead, typically made from polymers such as latex [13], or occasionally from metals such as palladium [3]. Most common in dielectrophoresis research are latex spheres. These are (as their name suggests) spherical blobs of latex that have been impregnated with fluorescent molecules, enabling the observation of very small particles (sizes as small as 14 nm diameter are available) with a fluorescence microscope. The primary advantage of using latex spheres is that they are very much a known quantity. They are solid and homogeneous (that is, they consist of one material and are consistent throughout). The internal conductivity and permittivity are known, as are the surface properties. Since they are spherical, their dielectrophoretic behavior is easy to model. Furthermore, there are straightforward chemical methods for changing those surface properties. Finally, they are readily available in a wide variety of sizes and colors.

Consider a typical experiment in which a sample of latex beads in a solution of known conductivity (such as ultra-pure water with a small quantity of potassium chloride [KCl]) is placed on an electrode array written onto a microscope slide, and covered with a coverslip. The electrode slide is then placed onto a fluorescence microscope (required in order to see particles this small) and the electrodes are connected, via attached wires, to a power source (typically a benchtop signal generator, providing perhaps 5 $V_{pk\text{-}pk}$ at a frequency between 10 kHz and 10 MHz or more). When the voltage is applied, the particles are observed to move quickly to the electrodes. Within a few seconds, collections such as those shown in Figure 16.3 are observed; the particles collect in the inter-electrode "arms" (Figure 16.3[a]) or in the center of the array (Figure 16.3[b]) depending on the frequency of the applied voltage being low or high, respectively. At one specific frequency, the force appears to vanish and the particles float freely. Varying the voltage also changes the force, making the particles travel more quickly or slowly to the trap. If the particles are small enough, then the magnitude of Brownian motion is sufficient to require a large voltage to be applied to ensure the particles remain trapped.

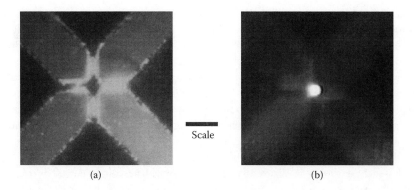

(a)   (b)

FIGURE 16.3 Fluorescence photograph of 200 nm latex beads collecting in an electrode array, approximately 5 sec after the application of a 20 V_{pk-pk} electric field. (a) A 1 MHz signal produces positive dielectrophoresis; (b) a 10 MHz signal gives negative dielectrophoresis. Scale bar in both pictures: 20 μm.

The frequency dependence of ε^* and hence $Re[K(\omega)]$ implies that the force on the particle also varies with frequency. Under certain conditions, and at a specific value of frequency, the force on the particle goes to zero when $Re[K(\omega)] = 0$. Above that frequency, a homogeneous particle will experience negative dielectrophoresis; below that, positive dielectrophoresis. The frequency at which the polarizability of the particle (and hence the force exerted on it) goes to zero is commonly referred to as the *crossover frequency*. Crossover frequencies are a product of dielectric dispersions, which cause the polarizability of the particle to change sign. It is possible to monitor the effects of changing the medium conductivity on the crossover frequency in order to estimate the properties of the particle.

Consider the following example. The polarizability of a homogeneous spherical particle (conductivity 10 mSm⁻¹, relative permittivity 2.55ε0, radius 2 μm) will exhibit a single dielectric dispersion, such as the one shown in Figure 16.4 if suspended in an aqueous medium of conductivity 1 mSm⁻¹, as calculated

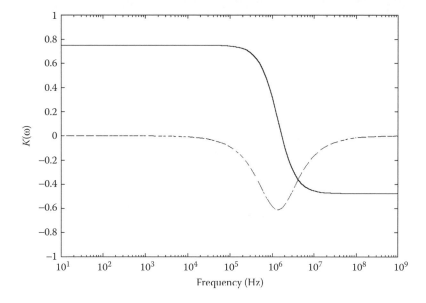

FIGURE 16.4 A plot of the real (*solid line*) and imaginary (*dotted line*) parts of the Clausius–Mossotti factor calculated for a 216 nm latex bead in a 1 mSm⁻¹ solution, neglecting surface charge effects. The magnitude and signs of the real and imaginary parts govern the magnitude and direction of the dielectrophoretic force and electrorotational torque, respectively.

using the Clausius–Mossotti factor (Equation [16.2]). The crossover frequency is found to rise as the medium conductivity is increased. Eventually, a threshold is observed in the medium conductivity above which the crossover frequency is observed to drop rapidly. However, at medium conductivities above the threshold, positive dielectrophoresis is still observed (in contradiction to expected behavior) at much lower frequencies than before. As the medium conductivity is increased, the polarizability of the particle compared with the medium drops, resulting in the predispersion (positive) side of the curve having a lower value. Eventually the low-frequency polarizability becomes so low that it is below zero at all frequencies; that is, the particle always experiences negative dielectrophoresis. This can be seen in Figure 16.5, where the polarizability is plotted for a range of suspending medium conductivities.

If we plot the polarizability as a function of both frequency and conductivity of the suspending medium, we find a plot such as shown in Figure 16.6. Ideally, it would be convenient to directly measure the polarizability as a mechanism for determining the dielectric properties of the particle by curve-fitting data to Equation (16.2), a method often used for the measurements of cells by determining the rate at which particles collect under positive dielectrophoresis for different frequencies. However, this is not easy in the case of submicrometer particles, where electrohydrodynamic and Brownian motions can easily disrupt the stable collection of particles. While successful attempts have been made to use a modified collection rate technique to study both the herpes viruses and latex beads, a far more convenient method of determining dielectric properties is to examine the intercept on the X–Y plane in Figure 16.6 — the plot of frequency where the value of $Re[K(\omega)]$ is zero, against conductivity — and infer the dielectric properties from that graph. This technique has been used to study latex spheres [13–15], viruses [16–18], and proteins [19], as well as larger particles such as cells [20–22]. This method is convenient for the measurement of colloids because the zero force frequency can always be seen quite clearly, even in the presence of disruptive fluid flow or Brownian motion. Data are collected at a range of conductivities (typically at five conductivities per decade), and a best-fit line is used to determine the most likely data set for the experimental data.

For micrometer-scale homogeneous particles in media whose conductivity and permittivity are known, there is a solution for Equation (16.2) that matches the crossover spectrum. However, as the diameter

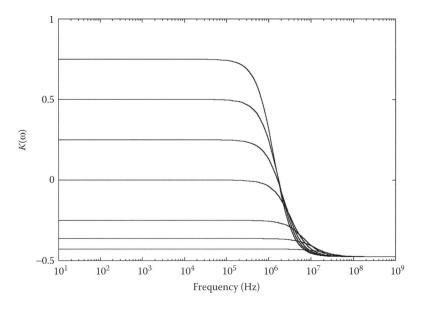

FIGURE 16.5 The real part of the Clausius–Mossotti factor as a function of frequency for a 200 nm diameter latex bead, for different values of suspending medium conductivity. The conductivity varies from 0.1 mSm⁻¹ (*top line*) to 500 mSm⁻¹ (*bottom line*). At conductivities above 20 mSm⁻¹, $Re[K(\omega)]$ is always negative; that is, the particles always experience negative dielectrophoresis. At lower conductivities, particles cross from positive to negative dielectrophoresis at about 3 MHz.

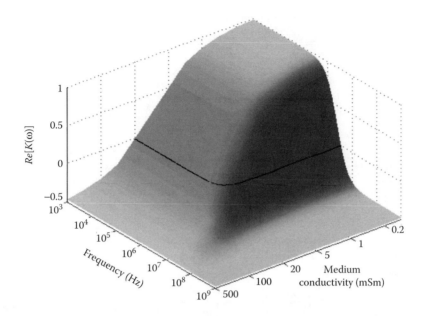

FIGURE 16.6 The data presented in Figure 16.3 plotted with conductivity on a third axis. The combinations of frequency and conductivity where $Re[K(\omega)] = 0$ form a distinct shape, indicated by the black line.

of the particle under study is reduced below 1 µm, this model becomes increasingly inaccurate. The crossover is found to rise with increasing medium conductivity, and a point above the threshold is reached where the crossover drops rapidly and only negative dielectrophoresis should be seen, and the particle still exhibits a crossover but at a much lower frequency. The reason for this change in behavior is due to the increasing effect of the surface charge, and more specifically, the electrical double layer.

16.2.3 Double Layer Effects

While simple models of dielectrophoretic behavior using only the Clausius–Mossotti factor will suffice for particles on the micrometer scale and larger, they are relatively poor at predicting the response of smaller particles. This is due to the influence of *surface charge* effects. While these effects can be observed in both micrometer- and nanometer-scale dielectrophoresis, they are much more pronounced in nanoscale particles. This is because the particles are much nearer in size to that of the ion cloud that surrounds them — the *electrical double layer* — and are therefore far more influenced by it.

The electrical double layer is the name given to the cloud of ions that are attracted to any charged surface in solution. The charges on the surface attract ions in the solution that have opposite charge — the *counterions* — while repelling charges of the same sign, or *coions*. At the surface itself, ions and water molecules will be adsorbed onto the surface, forming the *stagnant layer* or *Stern layer*. Between the outside face of the Stern layer (sometimes called the *slip plane*) and the bulk solution (where the ion distribution is unaffected by the presence of the charged surface) lies the *diffuse layer*, where the ion concentrations vary due to the effect of the surface, but they remain in motion in the solution. The thickness of the diffuse double layer is called the Debye length, and is given the symbol $1/\kappa$. We cannot directly measure the electrostatic potential at the surface, but we can measure the electrostatic potential at the slip plane, which is termed the ζ (zeta) potential.

The actual experimental crossover spectrum for 216 nm latex beads is shown in Figure 16.7. There are a number of significant differences from our original model. The response is not constant over the lower range of conductivities and does not exhibit only negative dielectrophoresis at higher conductivities; the crossover frequency exhibits a rise with increasing conductivity, and when it reaches the threshold and the crossover frequency drops, it only does so by about one order of magnitude.

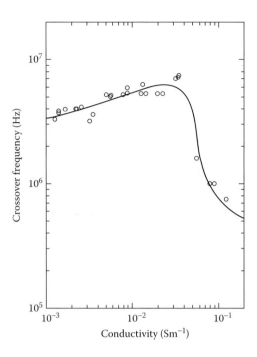

FIGURE 16.7 Experimental data for crossover frequencies of 216 nm latex beads in KCl solutions (*squares*) with a best-fit line according to the models described in the text.

In order to fit data such as those shown in Figure 16.7, a number of changes must be made to the model. If the only model used were the Clausius–Mossotti factor, then the conductivity of the particle in the model must be much greater than that which we know latex to possess. Experiments with latex beads of different sizes show that these effects become increasingly prominent as the size of the particle is decreased. Although there are different causes for these effects, they are all related to the movement of charges in the electrical double layer around the bead; specifically, the movement of charge around the Stern and diffuse layers and the dielectric dispersion experienced by the charges in the double layer (the alpha relaxation). We will examine these separately to examine how they each affect the dielectric response of the particle.

16.2.3.1 Charge Movement in the Stern Layer

It has been known for some years that the surface charge affects the dielectric response of particles. Studies by Arnold and Zimmermann [23] demonstrated that the electrorotation of latex spheres produced anomalously high values of internal conductivity — which in latex spheres should be near-zero — which was later attributed to the movement of charge *around* the particle. This charge is attracted to the charges on the surface of the particle; when placed in an electric field, the charges move in a laminar (i.e., planar) fashion around it. Arnold and Zimmermann determined that the component of aggregate particle conductivity σ_p which could be attributed to surface charge movement could be determined using the following equation:

$$\sigma_p = \sigma_{pbulk} + \frac{2K_s}{r} \tag{16.4}$$

where σ_{pbulk} is the conductivity of the particle interior, K_s is the surface conductance of the particle, and r is the radius of the particle. This formula was used, for example, in the determination of the surface charge of cells infected by malarial parasites by Gascoyne et al. [24]. In our study of nanometer-scale particles, this is very useful as it explains why our models of the behavior of our particles indicate a significantly higher conductivity than we know latex to possess. This effect becomes increasingly significant as particle radius is decreased, due to the inverse relationship between radius and the additional conductivity term due to surface conductivity.

For latex spheres, the bulk conductivity is negligible, so that that the effective conductivity of the particle is dominated by the surface conductance, K_s, where typical values of K_s are of the order of 1 nS. We can extend this further; according to Lyklema [25], the surface conductance can be calculated directly from the surface charge density, provided the mobility of the ions in the Stern layer is known and the mobilities of the counterions and coions are approximately equal, using the formula:

$$K_s = u^i \mu^i \tag{16.5}$$

where u^i represents the charge density that exists on the surface of the particle, and μ^i is the mobility of the counterion in the Stern layer, which is usually slightly lower than the value of mobility in the bulk solution. It is possible to measure this by determining the surface charge density by some other means, such as by the use of a Coulter counter [14], and using the value of K_s determined by dielectrophoretic means to establish the Stern layer mobility.

If the ionic mobilities or valences are not equal, then the equation must be adapted. It was first demonstrated by Green and Morgan [14] that latex beads exhibited different behavior in solutions of KCl and KPO_4, despite the fact that the counterion was K^+ in both cases. Further investigation by Hughes and Green [26] suggested that the conductivity of the Stern layer is influenced by the mobility of the counterions *and* coions in the bulk solution, by the equation:

$$K_s^i = \frac{u \mu_s^i \sigma_m}{2 z^i F c^i \mu_m^i} \tag{16.6}$$

where μ_s^i and μ_m^i are the mobilities of the ion species in the Stern layer and bulk medium respectively, c is the electrolyte concentration (mol m^{-3}), z is the valency of the ion, and F is the Faraday constant. If the electrolyte is symmetrical, it is possible to replace the conductivity term and concentration c^i with molar conductivity (S m^2 mol^{-1}):

$$K_s^i = \frac{u \mu_s^i \Lambda}{2 z^i F \mu_m^i} \tag{16.7}$$

Values of Λ are constant for given electrolytes (a table of values is given by Bockris and Reddy [27]). Equation (16.7) reduces to Equation (16.5) only if the values of the mobilities of the coion and counterion are equal, in which case the value of $\Lambda / 2 z^i F \mu_m^i$ goes to 1, as is the case for solutions such as KCl; in this instance, Equation (16.5) holds. Note that the value for ion mobility in the Stern layer is not equal to that found in the bulk medium, being somewhat lower [25].

16.2.3.2 Charge Movement in the Diffuse Double Layer

The movement of charge though the Stern layer is an important factor in the contribution of the double layer to the net conductivity of the particle. However, there is a second layer of charge movement in the *diffuse* double layer. This is different to, and distinct from, charge movement in the Stern layer; where the Stern layer charge is bound to the surface of the particle and moves in a laminar manner, charge distributed in the diffuse layer forms an amorphous ionic cloud around the particle. Significantly, the size of this cloud has an inverse relationship with the conductivity of the suspending medium — the greater the ionic strength of the medium, the thinner the diffuse double layer.

Analysis of the surface conductivity of a particle by Hughes et al. [15] led to a model with two separate components, instead of a single surface conductance K_s, containing terms due to both the charge movement in the Stern layer *and* charge movement in the diffuse part of the double layer [25]. The total surface conductance can then be written as:

$$K_s = K_s^i + K_s^d \tag{16.8}$$

where K_s^i and K_s^d are the Stern layer and the diffuse layer conductances respectively; this then corresponds to a net particle conductivity given by the expression

$$\sigma_p = \sigma_{pbulk} + \frac{2K_s^i}{r} + \frac{2K_s^d}{r} \tag{16.9}$$

Unlike the processes within the Stern layer, charge movement in the diffuse part of the double layer is related to *electro-osmotic transport* rather than straightforward conduction. Electro-osmosis is a process of fluid movement due to an applied potential across a nearby charged surface; the counter-charge accumulates near the surface, and then moves in the electric field due to Coulombic attraction. The presence of the surface creates a viscous drag, which impedes the motion of the charges. This effect is widely studied and is a common method of propelling analytes in capillary electrophoresis, where the counterions to the charged glass capillary are attracted to the electrode at the end of the capillary tube, "dragging" the analyte with them by viscous forces.

Lyklema [25] gives the following expression for the effective conductance of the diffuse layer:

$$K_s^d = \frac{(4F^2 cz^2 D^d (1 + 3m/z^2))}{RT\kappa} \left(\cosh\left[\frac{zq\zeta}{2kT} \right] - 1 \right) \tag{16.10}$$

where D^d is the ion diffusion coefficient, z the valence of the counterion, F the Faraday constant, k the Boltzmann's constant, R the gas constant, q the charge on the electron, and T the temperature; κ is the inverse Debye length given by

$$\kappa = \sqrt{\left(\frac{2czF^2}{\varepsilon RT} \right)} \tag{16.11}$$

where c is the electrolyte concentration (mol m^{-3}) and ζ is the ζ-potential, and the dimensionless parameter m is given by:

$$m = \left(\frac{RT}{F} \right)^2 \frac{2\varepsilon_m}{3\eta D^d} \tag{16.12}$$

where η is the viscosity. A key factor in this expression is the relationship between the surface conductance and the concentration of ions in the bulk medium, which appears twice in this expression. There is a c in the expression itself, and a $c^{1/2}$ in the expression for κ. This gives a net contribution of $c^{1/2}$ to the total diffuse layer conductance. Since the concentration defines the medium conductivity, this expression indicates that as the conductivity of the medium is increased, so the conductivity of the particle will increase but by a lesser degree. This is what we see when the crossover frequency of the particle rises when the medium conductivity is increased; the effective conductivity of the particle is *also* increased. The remaining values in the equations are more or less constants; the principal unknown variable is the

ζ-potential. This is known to vary slightly as a function of medium ionic strength but the variation is small, and its mechanism is not fully understood; however, as the concentration of ions is known, determining the diffuse layer conductance allows the direct measurement of the ζ-potential.

16.2.3.3 The Alpha Relaxation

The above formulae describe the way in which the electrical properties of the particle, as represented in the Clausius–Mossotti equation, are augmented by the movement of charge around the particle. However, in order to describe the low-frequency dispersion visible in high conductivity media (where no positive dielectrophoresis is expected), it is necessary to add an *additional* dispersion. Such additional polarizations follow the Debye model of the form $1/(1 + j\omega\tau_e)$, where τ_e is the relaxation frequency of the additional dispersion [14]. One possibility is that this additional term derives from the dielectric dispersion, by surface conduction, of the charge in the Stern layer. Unlike the diffuse layer, the Stern layer is of fixed size and charge, these being dictated by the surface charge density of the particle. Hence, the frequency of the dielectric dispersion would be expected to be stable over a range of medium conductivities, but vary proportionally to the particle radius and the surface conductance, as has been observed [13–17]. A good fit to the published data is given when the dispersion has a relaxation time

$$\tau_e = \frac{\varepsilon_o \varepsilon_s a}{K_s^i} \qquad (16.13)$$

where ε_s is the relative permittivity of the Stern layer. The best fit is provided when ε_s is approximately $14\varepsilon_0$. Since the Stern layer consists of bound ions and water molecules held in specific orientations by electrostatic interactions with the charged particle surface, this may be reasonable.

If we consider all the above factors, the Clausius–Mossotti factor, surface conduction in the double layer, and Stern layer relaxation, we can find best-fit lines that correspond well to our data. For example, we can determine the net effects of all these factors on the 216 nm latex beads shown in Figure 16.7. Superimposed on the data is a best-fit line derived using the above equations and values $K_s^i = 0.9$ nS, $\zeta = -100$ mV, and $\varepsilon_s = 14\varepsilon_0$. The relative permittivities of the particle and medium are 2.55 and 78, respectively; the internal conductivity of the beads was considered to be negligible. As can be seen, the model accurately predicts the behavior both below and above the decade transition in a crossover frequency at 40 to 50 mSm^{-1}.

16.2.4 Modeling of Complex Spheroids: The Multishell Model

Thus far, we have examined the dielectrophoretic response of solid, homogeneous spheres. However, many nanometer-scale particles such as viruses are not solid, homogeneous spheres, consisting of a number of "shells" surrounding a central "core." In the simplest case, a virus might consist of a protein case enclosing a central space wherein the viral DNA lies; a more complex virus such as herpes simplex encloses that protein shell in a thick protein gel, which is in turn surrounded by a lipid membrane similar to that which encloses a cell. Some viruses are not at all spherical, but are long and cylindrical; we will deal with them later in the chapter.

It is possible to extend the model of dielectric behavior to account for more complex particle structures. Developed by Irimajiri et al. [28], it works by considering each layer as a homogeneous particle suspended in a medium, where that medium is in fact the layer surrounding it. So, starting from the core, we can determine the dispersion at the interface between the core and the layer surrounding it, which we will call shell 1. This combined dielectric response is then treated as a particle suspended in shell 2, and a second dispersion due to that interface is determined. Then a third, due to the interface between shells 2 and 3, and so on. In this way, the dielectric properties of all the shells combine to give the total dielectric response for the entire particle. This is illustrated schematically in Figure 16.8.

In order to examine this mathematically, let us consider a spherical particle with N shells surrounding a central core. To each layer we assign an outer radius a_i, with a_1 being the radius of the core and a_{N+1}

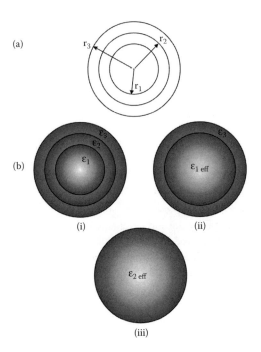

FIGURE 16.8 (a) A sphere comprises a core and inner and outer shells, with radii r_1, r_2, and r_3. (b) These three layers have complex permittivity ε_1^*, ε_2^*, and ε_3^*. We can find the total polarizability of the particle by successively combining the two innermost layers to find the effective combined complex permittivity.

being the radius of the outer shell (and therefore the radius of the entire particle). Similarly each layer has its own complex permittivity given by

$$\varepsilon_i^* = \varepsilon_i - j\frac{\sigma_i}{\omega} \tag{16.14}$$

where i has values from 1 to $N+1$. In order to determine the effective properties of the whole particle, we first replace the core and the first shell surrounding it with a single, homogeneous "core." This new core has a radius a_2 and a complex permittivity given by

$$\varepsilon_{1\mathit{eff}}^* = \varepsilon_2^* \frac{\left(\frac{a_2}{a_1}\right)^3 + 2\frac{\varepsilon_1^* - \varepsilon_2^*}{\varepsilon_1^* + 2\varepsilon_2^*}}{\left(\frac{a_2}{a_1}\right)^3 - \frac{\varepsilon_1^* - \varepsilon_2^*}{\varepsilon_1^* + 2\varepsilon_2^*}} \tag{16.15}$$

We now have a "core" surrounded by $N-1$ shells. We then proceed by repeating the above calculation, but combining the "new" core with the second shell, thus:

$$\varepsilon_{2\mathit{eff}}^* = \varepsilon_3^* \frac{\left(\frac{a_3}{a_2}\right)^3 + 2\frac{\varepsilon_{1\mathit{eff}}^* - \varepsilon_3^*}{\varepsilon_{1\mathit{eff}}^* + 2\varepsilon_3^*}}{\left(\frac{a_3}{a_2}\right)^3 - \frac{\varepsilon_{1\mathit{eff}}^* - \varepsilon_3^*}{\varepsilon_{1\mathit{eff}}^* + 2\varepsilon_3^*}} \tag{16.16}$$

If this procedure is repeated a further $N - 2$ times, the final step will replace the final shell and the particle will be replaced by a single homogeneous particle with effective complex permittivity ε^*_{Peff} given by

$$\varepsilon^*_{Peff} = \varepsilon^*_{Neff} \frac{\left(\frac{a_{N+1}}{a_N}\right)^3 + 2\frac{\varepsilon^*_{N-1eff} - \varepsilon^*_{N+1}}{\varepsilon^*_{N-1eff} + 2\varepsilon^*_{N+1}}}{\left(\frac{a_{N+1}}{a_N}\right)^3 - \frac{\varepsilon^*_{N-1eff} - \varepsilon^*_{N+1}}{\varepsilon^*_{N-1eff} + 2\varepsilon^*_{N+1}}} \tag{16.17}$$

This value provides an expression for the combined complex permittivity of the particle at any given frequency ω. It can also be combined with the complex permittivity of the medium to calculate the Clausius–Mossotti factor, as demonstrated by Huang et al. [29] for yeast cells.

16.2.5 Modeling Nonspherical Ellipsoids

A second special case of the formula for dielectrophoretic force that is often required is that of a long cylinder. This is needed in order to model the dielectric response of common nanotechnological components such as nanotubes and nanowires, as well as complex biological particles such as certain viruses. It is possible in these cases to adapt our model to compensate for the change in shape by deriving a general expression for the Clausius–Mossotti factor for elliptical particles, of which the spherical model is a special case; the cylinder can then be approximated to a long thin ellipsoid, with reasonable accuracy [18,30,31].

When an elliptical particle polarizes, the magnitude of the dipole moment is different along each axis; for example, a prolate (football or rugby ball-shaped) ellipsoid will have a dispersion along its long axis of a relaxation frequency different from the dispersion across its short (but equal) axes. The dispersion frequency of the dipole formed along the long axis will be of a lower frequency than that formed across the shorter axes (because the charges have further to travel from end to end in an alternating field), but the magnitude of the dipole formed will be greater due to the greater separation between the charges.

Consider an elliptical particle such as that shown in cross-section in Figure 16.9. It consists of two axes in projection, x and y, plus a third axis projecting from the page, z. The radii of the object along these axes are a, b, and c respectively. It can be demonstrated [4,31] that the particle will undergo three dispersions at different frequencies according to the thickness of the ellipsoid along each axis. However, in addition to the dielectrophoretic force experienced by the particle, it will also experience a torque acting so as to align the longest nondispersed axis with the field. This phenomenon, often observed in practical dielectrophoresis, is *electro-orientation* [4]. When a nonspherical object is suspended in an electric field (for example, but not solely, when experiencing dielectrophoresis), it rotates such that the dipole along the longest nondispersed axis aligns with the field. Since each axis has a different dispersion, the particle

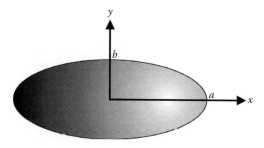

FIGURE 16.9 A schematic diagram of an elliptical particle, showing axes x and y, along which the particle extends by distances a and b. The particle extends along the z axis (out of the page) by a length c. If $c = b$, the particle is prolate; if $c = a$, the particle is oblate.

orientation will vary according to the applied frequency. For example, at lower frequencies, a rod-shaped particle experiencing positive dielectrophoresis will align with its longest axis along the direction of the electric field; the distribution of charges along this axis has the greatest moment and therefore exerts the greatest torque on the particle to force it into alignment with the applied field. As the frequency is increased, the dipole along this axis reaches dispersion, but the dipole formed *across* the rod does not and the particle will rotate 90° and align perpendicular to the field. This smaller axis has a shorter distance for charges to travel between cycles and so the dispersion frequency will be higher; however, the shorter distance means the dipole moment is smaller. This will result in the force experienced by the particle being smaller in this mode of behavior.

When aligned with one axis parallel to the applied field, a prolate ellipsoid experiences a force given by the equation:

$$F_{DEP} = \frac{2\pi abc}{3}\varepsilon_m Re[X(\omega)]\nabla E^2 \tag{16.18}$$

where

$$X(\omega) = \frac{\varepsilon_p^* - \varepsilon_m^*}{(\varepsilon_p^* - \varepsilon_m^*)A_\alpha + \varepsilon_m^*} \tag{16.19}$$

where α represents either the x, y, or z axis and A is the *depolarization factor*. This factor represents the different degrees of polarization along each axis, such that:

$$A_x = \frac{abc}{2}\int_0^\infty \frac{ds}{(s+a^2)\sqrt{(s+a^2)(s+b^2)(s+c^2)}}$$

$$A_y = \frac{abc}{2}\int_0^\infty \frac{ds}{(s+b^2)\sqrt{(s+a^2)(s+b^2)(s+c^2)}} \tag{16.20}$$

$$A_z = \frac{abc}{2}\int_0^\infty \frac{ds}{(s+c^2)\sqrt{(s+a^2)(s+b^2)(s+c^2)}}$$

where s is the variable of integration. The polarization factors are interrelated such that $A_x + A_y + A_z = 1$ [4,32].

The most useful version of these expressions is the simplified one for the case of prolate ellipsoids ($a > b$, $b = c$). This expression is useful because many nanoparticles can be approximated to prolate ellipsoids. In that case, Equation (16.18) may be rewritten as:

$$F_{DEP} = \frac{2\pi abc}{3}\varepsilon_m Re\left[\frac{\varepsilon_p^* - \varepsilon_m^*}{1+\left(\frac{\varepsilon_p^* - \varepsilon_m^*}{\varepsilon_m^*}\right)A}\right]\nabla E^2 \tag{16.21}$$

where A is given by the expansion

$$A = \frac{1}{3\gamma^{-2}}\left[1+\frac{3}{5}(1-\gamma^{-2})+\frac{3}{7}(1-\gamma^{-2})^2+\cdots\right] \tag{16.22}$$

and where $\gamma = a/b$. For a spherical particle, $\gamma = 1$ and $A = 1/3$, and Equation (16.18) can be rearranged to the expression for the force on a sphere as shown in Equation (16.1).

As with spherical particles, multishell prolate ellipsoids may also be modeled provided their dimensions are known. The procedure is as demonstrated earlier for spheroids, but with the expression for the equivalent permittivity of the ellipsoid being replaced by Equation (16.23), thus:

$$\varepsilon_{(i)\mathit{eff}}^{*} = \varepsilon_i^{*} \frac{\varepsilon_i^{*} + (\varepsilon_{(i+1)\mathit{eff}}^{*} - \varepsilon_i^{*})[A_{i\alpha} + v_i(1 - A_{(i-1)\alpha})]}{\varepsilon_i^{*} + (\varepsilon_{(i+1)\mathit{eff}}^{*} - \varepsilon_i^{*})(A_{i\alpha} + v_i A_{(i-1)\alpha})} \tag{16.23}$$

for $i = 1$ to $N - 2$ as before, and where

$$v_i = \frac{a_i b_i c_i}{a_{i-1} a_{i-1} a_{i-1}} \tag{16.24}$$

For a more complete exploration of the mathematics underlying the dielectrophoresis of elliptical particles, readers are again referred toward the excellent book by Jones [4].

16.2.6 Phase-Related Effects: Electrorotation and Traveling Wave Dielectrophoresis

Thus far we have explored only the effects of interactions between a dipole that acts in phase with the applied electric field. However, there is a second class of AC electrokinetic phenomena that depend on the interactions between an *out-of-phase* dipole with a spatially moving electric field. Since an induced dipole may experience a force with both an in-phase and an out-of-phase component simultaneously, the induced forces due to these components will be experienced at the same time, with the respective induced forces superimposed.

If a polarizable particle is suspended in a rotating electric field, the induced dipole will form across the particle and should rotate in synchrony with the field. However, if the angular velocity of the field is sufficiently large, the time taken for the dipole to form (the *relaxation time* of the dipole) becomes significant and the dipole will lag behind the field. This results in a nonzero angle between the field and the dipole, which induces a torque in the body and causes it to rotate asynchronously with the field; the rotation can be with or against the direction of rotation of the field, depending on whether the lag is less or more than 180°. This phenomenon was called *electrorotation* by Arnold and Zimmermann [33], and is shown schematically in Figure 16.10. The general equation for time-averaged torque Γ experienced by a spherical polarizable particle of radius r suspended in a rotating electric field E is given by

$$\Gamma = -4\pi\varepsilon_m r^3 \, \mathrm{Im}[K(\omega)] E^2 \tag{16.25}$$

where $\mathrm{Im}[K(\omega)]$ represents the imaginary component of the Clausius–Mossotti factor shown in Figure 16.4. The minus sign indicates that the dipole moment lags behind the electric field. When viscous drag is accounted for, the rotation rate $R(\omega)$ of the particle is given by [23]

$$R(\omega) = -\frac{\varepsilon_m \, \mathrm{Im}[K(\omega)] E^2}{2\eta} \tag{16.26}$$

where η is the viscosity of the medium. Note that unlike the dielectrophoretic force equation (Equation [16.1]), the relationship with the electric field is as a function of the square of the electric field rather than of the *gradient* of the square of the electric field. Furthermore, the torque depends on the *imaginary* rather than the real part of the Clausius–Mossotti factor. A particle may experience both dielectrophoresis

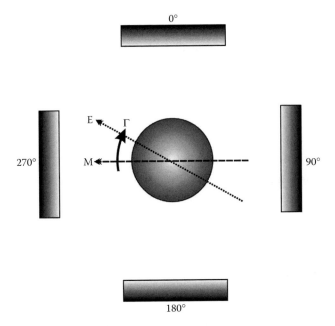

FIGURE 16.10 A schematic of a polarizable particle suspended in a rotating electric field generated by four electrodes with a 90° advancing phase. If the electric field **E** rotates sufficiently quickly, the induced dipole **M** will lag behind the electric field by an angle related to the time taken for the dipole to form (the relaxation time). The interaction between the electric field and the lagging dipole induces a torque Γ in the particle, causing the particle to rotate. This effect is known as *electrorotation.*

and electrorotation simultaneously, and the magnitudes and directions of both are related to the interaction between the dielectric properties of the particle and the medium; the relative magnitudes of force and torque are proportional to the real and imaginary parts of the Clausius–Mossotti factor.

Electrorotation was first examined scientifically in the early 1960s [34] but was most fully explored by Arnold and Zimmermann [23,33], who described the processes causing the observed motion. For cellular-scale objects, the principal attraction of electrorotation has been the ability to measure the rotation rate of single cells at different frequencies and thereby gain an accurate value of polarizability, which is not available through crossover measurements, where ensembles of particles are more easily observed. However, we may speculate that electrorotation may provide a means for the actuation of nanometer-scale electric motors. Berry and Berg [35] have used electrorotation to drive the molecular motor of *E. coli* bacteria backwards at speeds of up to 2000 Hz. Since the electrodes involved in that work were applying 10 V across inter-electrode gaps of 50 μm or more, considerably higher fields (and hence induced torques) could be applied with electrodes, with inter-electrode gaps of the order of 1 μm. Electric motors using electrorotation have been extensively studied by Hagedorn et al. [36], who demonstrated that such motors are capable of providing similar power outputs to synchronous dielectric motors for fluid pumping applications. It has been speculated that such nanomotors might be actuated by rotation-mode lasers; there are a number of advantages that favor electrorotation over its optical equivalent, most notably that electrorotation-induced torque is easily controlled by altering the frequency of the electric field, and that there does not need to be a direct optical path to the part to be manipulated.

There is a linear analog to electrorotation, which induces a translational force rather than torque in particles, an effect known as *traveling wave* dielectrophoresis. This shares with dielectrophoresis the phenomenon of translational induced motion, but rather than acting toward a specific point (that of the highest electric field strength), the force acts to move particles *along* an electrode array in the manner of an electrostatic "conveyor belt."

Consider a particle in a sinusoidal electric field that travels — that is, rather than merely changing magnitude, the field maxima and minima move through space, like waves on the surface of water [37]. These waves move across a particle, and a dipole is induced by the field. If the speed at which the field crosses the particle is great enough, then there will be a time lag between the induced dipole and the electric field in much the same way as there is an angular lag in a rotating field that causes electrorotation. This physical lag between dipole and field induces a force on the particle, resulting in induced motion; the degree of lag, related to the velocity (and hence the frequency) of the wave, will dictate the speed and direction of any motion induced in the particle. This is shown schematically in Figure 16.11. The underlying principle is closely related to electrorotation; it could be argued that the name "traveling wave dielectrophoresis" is misleading because the origin of the effect is not dielectrophoretic; that is, it does not involve the interaction of dipole and field *gradient*. Instead, the technique is a linear analog of electrorotation, in a manner similar to the relationship between rotary electric motors and the linear electric motors used to power magnetically levitated trains. As with the rotation of particles, the movement is asynchronous with the moving field, with rates of movement of 100 μm/sec being reported. The value of the force $\mathbf{F}_{TWD}$ is given by [29]:

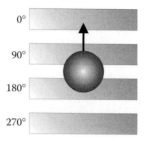

0°

90°

180°

270°

FIGURE 16.11 A schematic showing a polarizable particle suspended in a traveling electric field generated by electrodes on which the applied potential is 90° phase-advanced with respect to the electrode above. If the electric field moves sufficiently quickly, the induced dipole will lag behind the electric field, inducing a force in the particle. This causes the particle to move along the electrodes, a phenomenon known as *traveling wave dielectrophoresis*.

$$\mathbf{F}_{TWD} = \frac{-4\pi\varepsilon_m r^3 \operatorname{Im}[K(\omega)]\mathbf{E}^2}{\lambda} \quad (16.27)$$

where λ is the wavelength of the traveling wave, and is usually equal to the distance between electrodes, which have signals of the same phase applied.

Traveling wave dielectrophoresis was first observed by Batchelder [38], and subsequently by Masuda et al. [39]. However, it was not until the work of Fuhr et al. [40] that the phenomenon was fully explored and its origins known. Subsequent work by Huang [37] demonstrated both the equations outlined above and demonstrated the potential for separation by traveling wave dielectrophoresis. A large corpus of work now exists, including theoretical studies [37,41], devices for electrostatic pumping [42], and large-scale cell separators [43].

The application of traveling wave dielectrophoresis is largely used as a means of transporting particles. While the majority of work on traveling wave dielectrophoresis has concerned micrometer-sized objects such as blood cells [43], some work has been performed on the concentration of nanoparticles on a surface using the so-called "meander" electrodes [44]. These structures use four electrodes in a series of interlocking spirals to generate a traveling wave; at the center of the spiral, the electrodes form a quadrupole-type electrode array. It has been demonstrated that by careful manipulation of the amplitudes and phases of the potentials on these electrode structures, it is possible to "steer" the motion of particles across the array. Tools such as these could be used as the basis for "conveyor belts" for *factories on a chip*, wherein different chemical processes may be carried out on the same chip, with operations performed by electrostatic or chemical means and the resultant output transferred to a new process by AC electro-kinetic means [45].

16.2.7 Analysis of Key Spectrum Events

For a given shelled sphere, where the shell is of relatively low conductivity compared to the inner compartment and outer medium, we anticipate a polarizability spectrum of the type shown in Figure 16.12. This displays two characteristic dispersions, one rising at lower frequency and one falling at higher frequency. The frequency where the polarizability crosses from negative to positive for a homogenous sphere allows the direct determination of the properties of the sphere from Equation (16.2) by equating it to zero; however, for a shelled sphere, the expression is much more complicated to determine such factors. Some work has been performed in deriving expressions for the lower crossover frequency, but the upper frequency has largely been ignored because of limitations with signal generation equipment.

Benguigui and Lin [46] demonstrated that it is possible to determine an expression for the low-frequency crossover of a homogeneous sphere of conductivity and permittivity ε_p and σ_p, respectively; this was extended by Huang et al. [21,29] to consider the low-frequency crossover of a shelled sphere by considering it to be equivalent to a homogeneous sphere of effective conductivity and permittivity of

$$\left.\begin{aligned} \varepsilon_p &= \varepsilon_2\left(\tfrac{r2}{r2-r1}\right) \\ \sigma_p &= \sigma_2\left(\tfrac{r2}{r2-r1}\right) \end{aligned}\right\} \tag{16.28}$$

Using this approximation, Huang and coworkers produced an expression for the low-frequency crossover F_{x1}:

$$F_{x1} = \frac{1}{2\pi}\sqrt{\frac{2\sigma_3^2 - A\sigma_2\sigma_3 - A^2\sigma_{23}^2}{A^2\varepsilon_2^2 - A\varepsilon_2\varepsilon_3 - 2\varepsilon_3^2}} \tag{16.29}$$

where $A = r_2/(r_2 - r_1)$; that is, the ratio of the radius of the cell to the thickness of the membrane.

The high-frequency crossover has thus far not been considered, as no similar approximation can be made, and the solution to the Clausius–Mossotti factor for a shelled sphere is long and complicated.

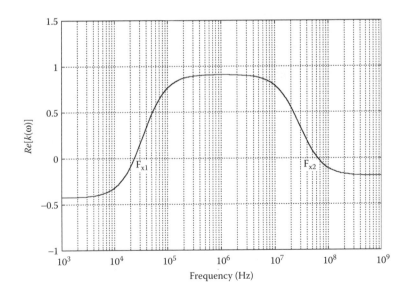

FIGURE 16.12 A spectrum of the Clausius–Mossotti factor, with the crossover frequencies noted.

It is known that the thickness of the cell membrane is considerably smaller than the radius of the cell; that is, the ratio $a = (r_2/r_1)^3$ has a value very near 1. If we make the approximation $a = 1$, we find that the high-frequency crossover frequency F_{x2} is given by the following:

$$F_{x2} = \frac{1}{2\pi} \sqrt{\frac{\sigma_1^2 - \sigma_1\sigma_3 - 2\sigma_3^2}{2\varepsilon_3^2 - \varepsilon_1\varepsilon_3 - \varepsilon_1^2}} \tag{16.30}$$

Numerical studies of the output of the full expression indicate that this approximation holds for the upper-crossover frequency for a wide range of electrical values provided $a < 1.15$ (approx.), corresponding to a ratio of cell radius to membrane thickness of 20:1. Within this limit, which represents all cases of biological cells, Equation (16.30) holds. The only case where the equation does not work is where no upper crossover exists, which happens when $\varepsilon_1 > \varepsilon_3$ (meaning the real part of the Clausius–Mossotti factor remains positive for all frequencies above F_{x1}) or $\sigma_1 < \sigma_3$ (meaning the real part of the Clausius–Mossotti factor never has a value greater than zero). The latter can be controlled by the use of low-conductivity media; where the former condition occurs, the final value of $K(\omega)$ is greater than 1 and no crossover occurs. Provided these conditions are not met, ε_1 and σ_3 have very little effect on F_{x2}. Since the upper-crossover frequency is independent of size, largely independent (in biological cells) of cytoplasmic permittivity or membrane properties, and independent of the remaining variables, those relating to the medium can be precisely defined; F_{x2} provides a direct measurement of the cytoplasmic conductivity.

Where no crossover occurs, due to the dispersion ending with a value of $K(\omega)$ greater than zero, we can still infer information from the behavior of the collection spectrum by observing fragments of the dispersion. Although the analytical expression for $\mathrm{Re}[K(\omega)]$ is highly complex, the behavior of the dielectrophoretic collection spectrum is ultimately governed by dielectric dispersions of the form quoted by Benguigui and Lin [46] for a homogeneous sphere:

$$\left. \begin{aligned} \mathrm{Re}[k(\omega)] &= \frac{\varepsilon_2 - \varepsilon_1}{\varepsilon_2 + 2\varepsilon_1} + \frac{3(\sigma_2\varepsilon_1 - \sigma_1\varepsilon_2)}{\tau_{MW}(\sigma_2 + 2\sigma_1)^2 \left(1 + \omega^2\tau_{MW}^2\right)} \\[2ex] \tau_{MW} &= \frac{\varepsilon_2 + \varepsilon_1}{\sigma_2 + 2\sigma_1} \end{aligned} \right\} \tag{16.31}$$

The spectrum changes between plateaux during interfacial dispersions in a manner which varies in terms of the start- and end-values, but not in terms of the frequency range over which this occurs. The dependence of the spectrum on the ω^2 term means that over a range of one decade centered on the midpoint frequency (F_m) of the dispersion, the value of $\mathrm{Re}[K(\omega)]$ sweeps across 82% of the change in polarizability due to the dispersion. The predictable nature of the dispersion with respect to frequency means that at higher frequencies (beyond those generated by conventional benchtop function generators), we need only observe the onset of a dispersion (say, from the plateau region to the midpoint of the dispersion, or even less) in order to predict the remainder of the dispersion behavior. Analysis of the first derivative yields the frequency of the midpoints of the two transitions, given by

$$\left. \begin{aligned} F_{m1} &= \frac{1}{2\pi} \sqrt{\frac{a\sigma_2^2 + 4a(a-1)\sigma_2\sigma_3 + 4(a-1)^2\sigma_3^2}{\varepsilon_3^2 - 2\varepsilon_1\varepsilon_3 - 2\varepsilon_1^2}} \\[2ex] F_{m2} &= \frac{1}{2\pi} \sqrt{\frac{\sigma_1^2 + 4\sigma_1\sigma_3 + 4\sigma_3^2}{\varepsilon_1^2 + 4\varepsilon_1\varepsilon_3 + 4\varepsilon_3^2}} \end{aligned} \right\} \tag{16.32}$$

where $a = (r_2/r_1)^3$ as before. Specifically, F_{m1} and F_{m2} are related as roots in $a - 1$ and a, though as $a \approx 1$ (as before), it does not appear in the final expression for F_{m2}.

Although the actual crossover frequency is often too high to be observed with conventional function generators, the behavior of the dielectric dispersion — taking approximately one decade to transit between stable plateaux of $\text{Re}[K(\omega)]$ — means that the frequency at which $\text{Re}[K(\omega)]$ begins to decline is *also* dependent only on σ_1. This is useful as this measurement is well within the capabilities of benchtop signal generators, while the actual value for intact, viable cells is often in the range of hundreds of megahertz. If the experimental data are sufficiently accurate, then knowing both the frequency at which the collection changes by 10% and the collection frequency is 3.16× higher means that the initial and terminal values (those tending to zero or infinite frequency) of the polarizability can in theory be estimated without being observed directly.

16.2.8 Limitations on Minimum Particle Trapping Size: Brownian Motion, Conduction, and Convection

When attempting to exert a force on nanoparticles in suspension, in order to direct them with great precision in solution, there are a number of factors related to the presence of the suspending medium that act to disrupt the controlled movement. While many of these effects also affect micrometer-scale particles, their influence is far more significant when dealing with nanoparticles. There are two reasons for this: the fact that the particles are nearer the size of the molecules of the suspending medium, and the fact that the electric field gradient (and therefore the electric field strength) is considerable enough to impart significant forces in particles with small volumes.

The first of these — the relative size of molecule and object — increases the effect of Brownian motion, the movement of colloidal particles due to the impacts of moving water molecules colliding with the surface of the particle. Brownian motion has been widely studied [47], and within the field of dielectrophoresis, it has been examined both as a problem to be overcome [8,11,12] and as a means to provide propulsion [48–57]. In some cases, statistical analysis is required to discriminate between dielectrophoretic and Brownian motion; any force applied to a particle will ultimately result in what Ramos et al. [58] describe as "observable deterministic motion," but if the force is small, the time taken to observe it may be far longer than the duration of the experiment. However, the forces used in the experiments described in the literature are sufficiently large for the applied force to observably overcome Brownian motion and be of the order of a second or (often significantly) less.

The second major effect that disrupts dielectrophoretic measurement on the nanoscale is the motion of the suspending medium due to the interaction between the matter molecules and the electric field, a concept known as electrohydrodynamics (EHD). This large and complex subject is beyond the scope of this chapter, and the interested reader is pointed toward excellent in-depth reviews by Ramos et al. [58] and Green et al. [59]. The two major EHD forces that are significant in driving fluid flow around electrodes are electrothermal and electro-osmotic in origin. The former force is due to localized heating of the medium, causing discontinuities of medium conductivity and permittivity, which was found to be insignificant due to the small volumes these discontinuities occupy (as the electrodes are so small). The latter, and far more significant, force is due to the interaction of the tangential electric field with the diffuse double layer above the electrode surfaces themselves. This creates a fluid pumping action, the magnitude of which can be several orders of magnitude larger than the dielectrophoretic force — suspending medium above quadrupole electrodes, for example, is pumped into the center of the chamber along the electrode surfaces, whereupon it forms a spout which forces the particles up. Under other conditions, this fluid flow can be reversed, drawing material to the electrodes. Careful electrode design can allow the EHD forces to be used as an aid to particle trapping and separation, as discussed by Green and Morgan [60].

To stably trap submicrometer particles, the dielectrophoretic force acting to move the particle into the center of the trap must exceed the action of Brownian motion on the particle, which, if large enough, will cause a particle to escape. In the case of positive dielectrophoresis, the applied force acts toward a single point (that of greatest field strength), and the force attracting it to that point must exceed the action of Brownian motion. For particles trapped in planar electrode arrays by negative dielectrophoresis, the particle is held in an dielectrophoretic force field "funnel."

The trapping of particles using positive dielectrophoresis is the simplest case to analyze; particles are attracted to the point of highest electric field strength, rising up the field gradient. Once at that point, the particle remains unless Brownian motion displaces it a sufficient distance that the field is unable to bring it back. The nature of positive dielectrophoresis is such that, given a long enough period of time (be it seconds or years), any particle polarized such that it experiences positive dielectrophoresis will ultimately fall into the trap [58]. As the electric field gradient extends to infinity, there will be an underlying average motion that will over time cause a displacement toward the high-field trap.

The concentration of particles from solution by negative dielectrophoresis is slightly different in concept from the above case. Particles are trapped in regions where the electric field strength is very low, and are prevented from escaping by a surrounding force-field "wall" that encloses the particles (although it may be in the form of an open-topped funnel). However, since in both the above cases, the particle must achieve the same effect — the overcoming of a dielectrophoretic energy barrier that forces the particle into the trap — the mathematical treatment of both cases is similar.

Given this approach to the trapping of particles by overcoming the action of Brownian motion through the application of a quantifiable force, it is possible to determine what the relationship is between the magnitude of the electric field applied by the electrodes and the smallest particle that may be trapped by such electrodes. This approach was pioneered by Smith et al. [61] for determining the smallest particle that may be trapped by laser tweezers, a technique similar to positive dielectrophoresis. However, it is equally applied to both positive and negative dielectrophoresis.

Consider the force on a particle of radius r suspended in a nonuniform electric field, experiencing a trapping F_{DEP} from Stokes' law; the particle's terminal velocity v is given by:

$$v = \frac{F_{DEP}}{6\pi\eta r} \tag{16.33}$$

Considering a small region of thickness Δd over which the force is constant, the time t_{DEP} taken for the particle to traverse this region is given by

$$t_{DEP} = \frac{6\pi\eta r \Delta d}{F_{DEP}} \tag{16.34}$$

Brownian motion acts to displace the particle from its position. From Einstein's equation (Equation 16.47), the mean time $\langle \tau_B \rangle$ taken for a particle to move a distance Δd in one dimension is given by

$$\langle \tau_B \rangle = \frac{3\pi\eta r (\Delta d)^2}{kT} \tag{16.35}$$

For stable trapping to occur, the time for the particle to move along the field gradient by dielectrophoresis should be significantly less than the time taken for the particle to move out from it by Brownian motion, so that any displacement from the trap is immediately countered by dielectrophoresis; a factor of ×10 was suggested by Smith et al. [61], though this is arbitrary. From Equation (16.34) and Equation (16.35), the conditions are

$$\frac{6\pi\eta r \Delta d}{F_{DEP}} < \frac{1}{10}\left(\frac{3\pi\eta r (\Delta d)^2}{kT} \right) \tag{16.36}$$

and

$$\Delta d > \frac{20kT}{F_{DEP}} \tag{16.37}$$

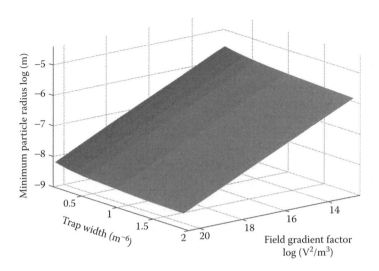

FIGURE 16.13 A graph showing the variation in the minimum radius of particles that could be trapped according to the expression in Equation (16.38).

The smallest particle radius that meets this criterion is given by

$$r > \sqrt[3]{\frac{10kT}{\pi \varepsilon_m \Delta d Re[K(\omega)] \nabla \mathrm{E}^2}}$$ (16.38)

The variation r can be calculated from Equation (16.38) as a function of field gradient ∇E^2 and trap width Δd. At a temperature of 300 K, and with $\varepsilon_m = 78\varepsilon_0$ and $Re[K(\omega)] = 1$, this variation is shown in Figure 16.13. In cases where ∇E^2 varies as a function of distance, the trapping efficiency is given as the maximum value of the function $\Delta d \nabla E^2$ for the particular trap.

In order to determine what this might mean in terms of a given electrode geometry, it is necessary to simulate the electric field gradient around that geometry. The most accurate method of deriving the trapping force is to integrate the force across all given paths from the center of the trap to infinity (or at least to the edges of the trap), thereby determining the value of $\Delta d \nabla E^2$ for all possible escape paths. To do this, numerical models must be employed to determine the nature of the electric field around the electrodes, which is rarely determinable by analytical methods. For example, a numerical model based on the Moments method [62] was used to calculate ∇E^2 around the polynomial electrode array. Figure 16.14 shows a 3-D plot of both the *rms* electric field strength and the magnitude of ∇E^2 across the center of the electrode at a height of 7 µm above the electrode array shown in Figure 16.2, where particles are observed to be trapped by negative dielectrophoresis. The simulation was performed with an applied voltage of 5 V_{pk-pk}. The trap efficiency is governed by the smallest distance that a trapped particle has to travel in order to escape from the trap. In the case of a particle trapped by positive dielectrophoresis, it is principally governed by the magnitude of the electric field, since the particle is trapped at the point of the highest field, which diminishes with increasing distance from that point. For particles trapped by negative dielectrophoresis, Δd is governed by the magnitude of the field barrier that *encloses* the particle. For example, as can be seen in Figure 16.14, the distribution of the electric field strength (as one might expect) has high field regions in the inter-electrode gaps and a field null at the center. However, the force pattern is more complex, with a force barrier surrounding the field null. Repulsion by this barrier prevents particles trapped by negative dielectrophoresis from escaping. In order to trap large numbers of particles, the dielectrophoretic force must overcome diffusion, but not necessarily by a factor as great as 10. If the condition is merely $\mathbf{F}_{\mathrm{DEP}} > \mathbf{F}_{\mathrm{BROWNIAN}}$, then the dielectrophoretic force on

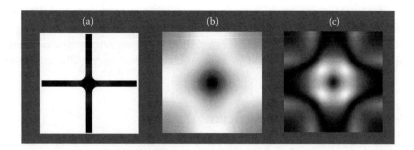

FIGURE 16.14 A simulation of the electric fields and forces generated by quadrupolar electrodes such as those shown in Figure 16.2. (a) A schematic of the electrodes covered by the simulation, and the spatial variations of (b) the electric field magnitude, and (c) dielectrophoretic force magnitude across the electrode array. The electrode array was modeled as having gaps of 2 μm between adjacent electrodes and 5 μm across the center of the array, and the simulation shows the field 7 μm above the electrodes. As can be seen, the field strength is greatest above the inter-electrode "arms" of the array; there is a field null at the center of the array, enclosed on all sides by a region of high electric field.

particles will be on average greater than Brownian motion and there will be a net force on the particle mass toward the trap, even if individual particles occasionally escape from the trap.

16.3 Applications of Dielectrophoresis at the Nanoscale

Although it is not the first application of dielectrophoresis to be demonstrated at the nanoscale, recent research into applications of this phenomenon has been dominated by the manipulation of nanoscale filaments such as nanowires and, more ubiquitously, carbon nanotubes. These are molecular constructs typically a few nanometers in diameter, but several hundred nm (or even microns) long. These particles have remarkable electronic properties, which lend their application to many novel electronic devices. However, there are two primary problems with the use of nanotubes and nanowires in electronic applications. The first is that unlike conventional electronic device manufacture (where layers of metal and other materials are deposited from vapor or plasma through well-defined "masks"), nanowire devices require the placement of the wires or tubes in specific locations, in order to make contact both with energizing electrodes and commonly with each other. Secondly, carbon nanotubes are produced in large numbers of mixed semiconducting and metallic tubes, with a requirement that they be separated before the appropriate nanotubes' chirality can be placed within an electronic device. Dielectrophoresis offers a method to circumvent both problems.

Consider first the assembly of electronic devices from suspended populations of nanotubes. The first example of such self-assembled nanoconstruction was the manipulation of large nanowires into functional electronic devices in the early 2000s [63–66]. This was achieved by activating electrodes that pointed toward each other across a central chamber — similar to the quadrupole electrodes' array but with more "pointed" electrode structures — and applying a solution containing only p-doped semiconducting nanowires. Only one opposing pair of electrodes was energized in order to trap a single nanowire across the center of the chamber. The solution was removed and a second applied, containing only n-doped nanowires. By then, energizing the other electrodes, a second nanowire was attracted, crossing the first at 90°. When the solutions were removed, the device consisting of the crossed nanowires — one p-type and one n-type — remained and was electrically characterized, and demonstrated strong optical characteristics, indicating a possible application as a nanoscale light-emitting diode for nanooptical and nanoelectronic applications.

The first experiments reporting the dielectrophoresis of carbon nanotubes were conducted in the late 1990s [67]. However, the manipulation of carbon nanotubes gained massive momentum when Ralph Krupke and colleagues demonstrated for the first time that dielectrophoresis could be used to separate metallic and semiconducting carbon nanotubes [68]. Other workers have used dielectrophoresis for a number of assembly applications using carbon nanotubes. For example, it has been used to fabricate thermal sensors [70], field effect transistors [71], gas sensors to detect NO_2 [72] and NH_3 [73], and AFM tips [74]. In terms of electronics applications, one of the most important parts of the nanotube circuit is the interconnect between nanotubes and electrode, which has been extensively characterized [75,76]. Much subsequent work has been directed at large-scale separation of metallic and semiconducting nanotubes for industrial applications [77,78].

16.3.1 Trapping Single Nanoparticles

One of the many advantages of manipulating particles on the nanometer scale is the possibility of manipulating single particles. Such a technology could potentially open up new fields in the study of single-molecule chemistry and molecular biology, and is presently being pursued by a number of workers [11,79]. The majority of this research is performed using optical trapping — the so-called "laser tweezers" — in which focused laser beams are used to exert pressure on particles [80]. However, dielectrophoresis offers many advantages over laser tweezers, including the fact that the technique allows trapping from solution followed by contact on a sensor of some sort, allowing an analyte to be studied.

The trapping of single particles is somewhat different from and somewhat more difficult to achieve than trapping a larger population of particles. In the latter case, particles need only to tend to move toward the trapping region; if some particles leave the trap, it is not considered a problem, provided greater numbers of particles are moving into the trap. Furthermore, as can be seen in Figure 16.3, dielectrophoretic traps — even those constructed on the micrometer scale — tend to collect large numbers of particles. A spherical trapping volume 1 μm across could contain over 500 particles of diameter 100 nm.

There are two ways to increase the selectivity of the trapping mechanism. One method is to reduce the size of the electrodes to the order of the size of the particle to be trapped. The other mechanism is to use a larger trap, but alternate between a regime wherein a single particle is attracted to the electrodes and a second regime to prevent other particles approaching the trap. This can be achieved by either applying negative dielectrophoresis to keep extraneous particles away from an isolated energy well or removing the field to prevent further particles being attracted. Both methods have advantages and disadvantages, and applications to which they best present themselves.

Different electrode geometries are required for each trapping method. The basis of one electrode design for single-particle applications is the dual need to both attract a single particle and repel all others. Unlike bulk nanoparticle trapping where the aim is merely to attract particles to a region, it is necessary to both attract a particle to a point and trap it, while excluding all other particles from that trap. There is a second design strategy that may be used — that the particle experiences no force at all, either by removing the electric field or by retaining the particle at a point where a field null exists. Either of these two methods may be used to prevent particles nearing the trapping point.

Single-particle trapping by positive dielectrophoresis was first demonstrated by Bezryadin and coworkers in 1997 [3]. The electrode geometry consisted of two needle-type platinum electrodes facing one another, suspended in free space by etching the silicon substrate beneath the point where the electrodes met. The distance between opposing electrode tips was 4 nm. The potential was applied through a high-value (100 MΩ) resistor; a 4.5 V, DC field was used. As we have seen, AC fields are far more common for dielectrophoresis, but they are not a prerequisite.

Colloidal palladium particles with sizes down to 5 nm diameter were introduced in solution. These particles became polarized and were attracted up the field gradient to the electrode tips by positive dielectrophoresis. However, as soon as the first palladium sphere reached the center point between two opposing electrodes, a circuit between the electrodes was made. This resulted in current flowing in the circuit, in which the majority of the supply voltage was dropped across the resistor; with virtually no voltage dropped across the electrodes, the magnitude of the electric field generated in the inter-electrode gap was diminished, preventing other colloidal particles from reaching the electrode tips. Once in place, the trapped particle was sufficiently attached to the electrodes for the solution to be removed and the assembly to be observed using a scanning electron microscope, as shown in Figure 16.15. A similar approach has been used to study the trapping of single fluorescent protein molecules in suspension [81].

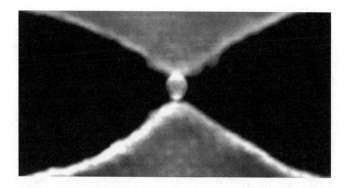

FIGURE 16.15 A separation of herpes simplex virions (in the electrode arms) from Herpes Simplex capsids (in the ball at the center of the array). The ball is levitated approximately 10 μm above the electrode array. (Courtesy of Dr. H. Morgan, University of Southampton.)

Single-particle trapping is also possible by negative dielectrophoresis. It is more complex than the positive trapping technique described above; however, particles of biological origin such as viruses are nonconducting, and therefore the use of negative dielectrophoresis is more appropriate. Furthermore, since the size of the trapping volume at the center of the arrays is defined not by the electrodes but by the geometry of the generated field, the technology required to construct the electrodes is far more readily available. Negative dielectrophoretic trapping of single particles has been achieved by this method using viruses, latex spheres, viral substructures, and macromolecules such as DNA, as well as larger structures such as cells [11,82,83].

There are a number of drawbacks to this form of trapping, the principal one being related to the observation of the particle. Unlike positive trapping where the particle may be detected electrically, particles trapped by negative dielectrophoresis are suspended in the medium at an indeterminate height above the electrode structure. Ultimately, the only means by which such particles can be observed is by fluorescent staining. This is a general problem in the field of single-nanoparticle detection, and other methods such as laser tweezers also require the use of fluorescent staining.

Electrodes used to trap particles are generally quadrupolar arrays such as the design shown in Figure 16.2. This array geometry has the advantage of a well-defined, enclosed field minimum surrounded by regions of high field strength. Ideally, the potential energy minimum would be small enough to contain only one particle. Where this is not the case, if particles are first attracted to the electrode tips by positive dielectrophoresis and the field frequency is then switched to induce negative dielectrophoresis, only those particles on the inward-facing tips of the electrodes will fall into the trap; the others will be repelled into the bulk. It is possible to trap single particles this way, though occasionally two or three particles may fall into the potential energy minimum at the center of the trap. Under these conditions, 97 nm diameter latex beads have been retained in the center of the electrode array while other particles were forced away. Single herpes virus particles could also be held in the same electrode array.

Single particles trapped by this method move within the confines of the electric field cage under the influence of Brownian motion. During trapping of an object with density greater than that of water, such as a single herpes virus, the particle is levitated in a stable vertical position above the electrodes; Brownian motion is balanced against the weight of the particle. However, particles such as the latex spheres (which have a density approximately equal to that of water) are not constrained in this way because Brownian motion causes constant random movement in the z-direction. Such particles may eventually diffuse out of the top of a "funnel"-type or "open" trap, in which the field gradient is generated by one set of planar electrodes "beneath" the trapped particle, though particles have been held for 30 min or more. In order to ensure that a particle remains within the trap, a second layer of electrodes may be introduced above the first, so that a closed field "cage" such as those employed by Fuhr and coworkers [83] is created. These have the additional advantage of allowing a degree of 3-D positional control by varying the intensity of the field strength at the various positional electrodes, as has been demonstrated using a 1 μm diameter latex bead. Alternatively, a planar (2-D) electrode array can be sufficient provided that any coverslip used to contain the solution above the electrode is sufficiently close to the electrode plane, so that any field trap constraining the particle in the x–y plane extends to the full height of the solution, creating a force field of cylindrical aspect.

16.4 Biomolecular Applications of Dielectrophoresis

It was suggested for many years [12] that due to the action of electrohydrodynamic forces, it would be impossible to successfully manipulate particles on the nanometer scale. It is then ironic that the first research to significantly break the suggested lower particle limit of 1 μm should do so with the manipulation of proteins — macromolecules of the order of a few nanometers across [8]. The manipulation of molecules is somewhat more complex than the manipulation of latex beads, since they contain not only an induced dipole due to the electrical double layer, but also a permanent dipole due to the position of fixed charges across the protein molecule. Washizu and coworkers demonstrated not only that dielectrophoresis of proteins was feasible, but also that the technique could be used for the separation of proteins of different molecular weights on an electrode array. Another example of manipulation on this scale is the coating of microstructures of micrometer-scale devices using nanopowdered diamond to form cold cathode devices [84].

Perhaps the greatest macromolecule of interest today is DNA, and many researchers have explored the possible applications of using dielectrophoresis for genomics applications. As with protein studies, the pioneering work was performed by Washizu and coworkers [85], who demonstrated trapping of DNA molecules in 1990. Subsequent work showed that not only could DNA be trapped, but by careful manipulation of the field, strands of DNA could be stretched across an inter-electrode gap, and then cut to fragments of specific sizes using a laser [86,87]. Another demonstration of the application of dielectrophoresis to DNA research was the trapping of single DNA fragments 10 nm long by Bezryadin and coworkers [88]. This group subsequently used the technique to study the electrical properties of single fragments of DNA, which demonstrated interesting electrical properties that may have potential applications in nanometer-scale electronics. As electrodes and understanding of the underlying mathematics have improved, the applications have become more finessed; for example, dielectrophoresis has been used to measure the mechanical properties of DNA [89], to measure the electrical properties (including the effects of humidity on the molecule) [90], separate DNA strands by size [91], and organize molecules into regular patterns on a chip [92].

Similar approaches have been taken with proteins. The first report of the dielectrophoretic manipulation of proteins by positive dielectrophoresis was by Washizu and colleagues in 1990 [8], with the first report of negative dielectrophoresis and protein characterization coming over a decade later [19]. Since then it has been used as a tool for trapping proteins in a nanopipette for subsequent analysis [93]. Similar approaches have been used to manipulate the suspended droplets in emulsions [94]. Finally, recent work has used a combination of dielectrophoresis and electrohydrodynamic forces to concentrate a range of particles, from cell-sized to proteins, 20 nm latex beads, short DNA strands, and dissolved proteins [95–97].

16.5 Particle Separation

Ever since Gascoyne et al. [98] demonstrated the separation of healthy and leukemic mouse blood cells using microelectrodes early in the 1990s, dielectrophoresis has increasingly been used as a tool for the separation of heterogeneous mixtures of particles into homogeneous populations in different parts of an electrode array. The method underlying the technique is simple; since polarizable particles demonstrate a crossover frequency that is dependent on those particles' dielectric properties, particles with different properties may under specific conditions exhibit different crossover frequencies. As those particles experience positive dielectrophoresis below the crossover frequency and negative dielectrophoresis above, it follows that at a frequency between the two crossover frequencies of the two particle types, one will experience positive dielectrophoresis while the other experiences negative dielectrophoresis. This will result in one group being attracted to regions of high field strength, with the other group being repelled; hence, the two populations are separated. Such separations are typically carried out using electrode arrays with well-defined regions of high and low electric field strengths.

Consider the separation of a mixture of two populations of latex beads, identical except for having different radii. Since the effective conductivity (and hence the polarizability), as expressed in Equation (16.2) of a latex sphere is dependent on the double of surface conductances divided by the particle radius (Equation [16.9]), it follows that the value of Re[$K(w)$] will be strongly affected by particle radius. This is indeed the case, with larger particles exhibiting lower crossover frequencies than smaller (but otherwise identical) particles.

Secondly, particles of identical size and internal composition can be separated according to their surface properties. This was first demonstrated by Green and Morgan in 1997 [99], who reported the separation of 93 nm latex spheres. By using a castellated electrode array with 4 μm feature sizes, the researchers demonstrated that the particles exhibited a narrow range of surface conductances rather than each having an identical value of surface conductance. This caused the population of particles to have crossover frequencies dispersed across a narrow frequency window, and by applying a frequency in the middle of that range, it was demonstrated that the particles could in fact be separated.

This effect was expanded upon by Hughes et al. [15], by actively modifying the surfaces of latex particles to improve the separation and to identify possible biotechnological applications of the technique. The surfaces of some of the beads were chemically modified using 1-ethyl-3-(3-dimethylaminopropyl) carbodiimide (EDAC), a reagent used for the chemical coupling of protein to the carboxyl surface of the beads. This caused a significant reduction in the crossover frequency, which was found to equate to a similar reduction in surface conductance from 1.1 to 0.55 nS. The EDAC-activated beads were then mixed with antibodies and the crossover behavior was measured again. As the surfaces of the beads were covered by the antibodies, the crossover spectrum exhibited a further drop in frequency, equating to a further drop in K_s and ζ-potential. However, the crossover frequencies of the IgG-labeled beads varied by up to a factor of 2 between different beads as a result of different amounts of antibody coupling between beads, allowing the separation of beads containing different amounts of protein on their surfaces.

There are a number of potential applications of such a system. First, since the crossover frequency is directly related to the amount of protein attached to the bead surface, it allows the rapid assaying of the amount of protein attached to a sphere, which in turn relates to the amount of protein in the environment, making a single sphere a potential biosensor. If the system is calibrated such that the crossover frequency in a particular medium that corresponds to a specific protein coverage is known, then observing the frequencies at which a single bead, or an ensemble, changes dielectrophoretic behavior allows the measurement of the protein content in the medium. This could be used for a number of different proteins or other compounds by mixing fluorescent beads of different colors, each with a different surface functionality. By constructing electrodes over a suitable photosensor, systems such as this may form the basis of "lab on a chip" systems.

A second application concerns the fact that very small latex spheres have a large surface area-to-volume ratio, so that a small volume of beads has a potentially huge surface area. For example, a 1 ml sample containing 1% (by volume) of 200 nm diameter beads (as used by Hughes et al. [15]) has a total surface area of 300 m^2 — which in biosensor terms makes such a sensor exceptionally sensitive. This can, for example, be used to detect very low quantities of target molecules; if a large number of small, activated beads are held near their crossover frequency, then a single molecule attaching to the surface of one bead may change the surface charge of the bead enough to cause that bead to pass the crossover frequency and be detected. A similar system, on the micrometer scale, has been developed at the University of Wales at Bangor as a means of detecting water-borne bacteria [100].

Just as latex beads of different sizes, properties, or surface functionalities can be separated into sub-populations on an electrode array, so also can we separate bioparticles with different properties. An example of this is shown in Figure 16.16, which illustrates the separation of herpes simplex virus particles, which have collected in the "arms" of the electrodes and here appear pale, from herpes simplex capsids,

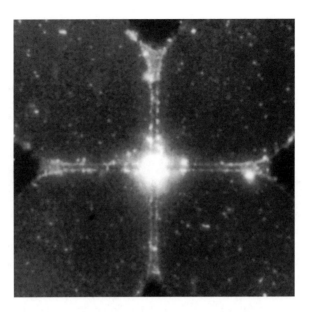

FIGURE 16.16 An electron micrograph of a single 17 nm palladium sphere trapped between two electrodes by positive dielectrophoresis. The image measures 200 nm wide and 100 nm high. (From Bezryadin, A., Dekker, C., and Schmid, G., *Appl. Phys. Lett.*, 71, 1273, 1997. With permission.)

which appear here as a bright ball at the center of the array. Similar separations have been performed for herpes simplex virions and Tobacco Mosaic virions [101]. The procedure for such separations usually follows a categorization of the dielectric response of the two particle species in order to find the optimum frequency and suspending medium conductivity. Another method of separation was demonstrated by Washizu et al. [8], who fractioned a mixture of protein and DNA molecules by a combination of field flow and dielectrophoresis. Separation and identification of biological material is, perhaps, the most important application of dielectrophoresis to nanomedicine — allowing, for example, the point-of-care analysis of blood samples to determine the cause of an infection without the need for lengthy analyses at remote laboratories.

16.6 Conclusion

The techniques of AC electrokinetics, and particularly that of dielectrophoresis, have much to offer the expanding science of nanotechnology. Whereas dielectrophoretic methods have been applied to the manipulation of objects in the microscale in the past, there can be no doubt of their tremendous potential for nanoparticle manipulation. This opens a wide range of potential applications for AC electrokinetics to the development of mainstream nanotechnology. The fundamental challenge in the advancement of nanotechnology is the development of precision tools for large-scale self-assembly of nanometer-scale components. The techniques described here go some way to addressing this by providing tools for the trapping, manipulation, and separation of molecules and other nanoparticles using tools (electrodes) on the micrometer scale.

References

1. Drexler, K.E., *Nanosystems; Molecular Machinery, Manufacturing and Computation*, Wiley, New York, 1992.
2. Eigler, D.M. and Schweizer, E.K., Positioning single atoms with a scanning tunnelling microscope, *Nature*, 344, 524, 1990.

3. Bezryadin, A., Dekker, C., and Schmid, G., Electrostatic trapping of single conducting nanoparticles between nanoelectrodes, *Appl. Phys. Lett.*, 71, 1273, 1997.

4. Jones, T.B., *Electromechanics of Particles*, Massachusetts Institute of Technology, Cambridge, 1995.

5. Zimmermann, U. and Neil, G.A., *Electromanipulation of Cells*, CRC Press, Boca Raton, FL, 1996.

6. Ashkin, A., Dziedzic, J.M., Bjorkholm, J.E., and Chu, S., Observation of a single-beam gradient force optical trap for dielectric particles, *Opt. Lett.*, 11, 288, 1986.

7. Simpson, N.B., Dholakia, K., Allen, L., and Padgett, M.J., Mechanical equivalence of spin and orbital angular momentum of light: an optical spanner, *Opt. Lett.*, 22, 52, 1997.

8. Washizu, M., Suzuki, S., Kurosawa, O., Nishizaka, T., and Shinohara, T., Molecular dielectrophoresis of biopolymers, *IEEE Trans. Ind. Appl.*, 30, 835, 1994.

9. Bakewell, D.J.G., Hughes, M.P., Milner, J.J., and Morgan, H., Dielectrophoretic manipulation of avidin and DNA, in *Proceedings 20th Annual International Conference of the IEEE Engineering in Medicine and Biology Society*, 1998.

10. Müller, T., Gerardino, A., Schnelle, T., Shirley, S.G., Bordoni, F., DeGasperis, G., Leoni, R., and Fuhr, G., Trapping of micrometre and sub-micrometre particles by high-frequency electric fields and hydrodynamic forces, *J. Phys. D*, 29, 340, 1996.

11. Hughes, M.P. and Morgan, H., Dielectrophoretic manipulation of single sub-micron scale bioparticles, *J. Phys. D*, 31, 2205, 1998.

12. Pohl, H.A., *Dielectrophoresis*, Cambridge University Press, Cambridge, 1978.

13. Green, N.G. and Morgan, H., Dielectrophoretic investigations of sub-micrometre latex spheres, *J. Phys. D*, 30, 2626, 1997.

14. Green, N.G. and Morgan, H., Dielectrophoresis of submicrometre latex spheres. 1. Experimental results, *J. Phys. Chem.*, 103, 41, 1999.

15. Hughes, M.P., Morgan, H., and Flynn, M.F., Surface conductance in the diffuse double-layer observed by dielectrophoresis of latex nanospheres, *J. Coll. Int. Sci.*, 220, 454, 1999.

16. Hughes, M.P., Morgan, H., and Rixon, F.J., Dielectrophoretic manipulation and characterisation of herpes simplex virus-1 capsids, *Eur. Biophys. J.*, 30, 268, 2001.

17. Hughes, M.P., Morgan, H., and Rixon, F.J., Measurements of the properties of herpes simplex virus type 1 virions with dielectrophoresis, *Biochim. Biophys. Acta*, 1571, 1, 2002.

18. Morgan, H. and Green, N.G., Dielectrophoretic manipulation of rod-shaped viral particles, *J. Electrostat.*, 42, 279, 1997.

19. Hughes, M.P., *Nanoelectromechanics in Engineering and Biology*, CRC Press, Boca Raton, FL, 2002.

20. Gascoyne, P.R.C., Pethig, R., Burt, J.P.H., and Becker, F.F., Membrane changes accompanying the induced differentiation of Friend murine erythroleukemia cells studied by dielectrophoresis, *Biochim. Biophys. Acta*, 1149, 119, 1993.

21. Huang, Y., Wang, X.-B., Becker, F.F., and Gascoyne, P.R.C., Membrane changes associated with the temperature-sensitive P85$^{gag-mos}$-dependent transformation of rat kidney cells as determined by dielectrophoresis and electrorotation, *Biochim. Biophys. Acta*, 1282, 76, 1996.

22. Gascoyne, P.R.C., Noshari, J., Becker, F.F., and Pethig, R., Use of dielectrophoretic collection spectra for characterizing differences between normal and cancerous cells, *IEEE Trans. Ind. Appl.*, 30, 829, 1994.

23. Arnold, W.M. and Zimmermann, U., Electro-rotation — development of a technique for dielectric measurements on individual cells and particles, *J. Electrostat.*, 21, 151, 1988.

24. Gascoyne, P.R.C., Pethig, R., Satayavivad, J., Becker, F.F., and Ruchirawat, M., Dielectrophoretic detection of changes in erythrocyte membranes following malarial infection, *Biochim. Biophys. Acta*, 1323, 240, 1997.

25. Lyklema, J., *Fundamentals of Interface and Colloid Science*, Academic Press, London, 1995.

26. Hughes, M.P. and Green, N.G., The influence of Stern layer conductance on the dielectrophoretic behaviour of latex nanospheres, *J. Coll. Int. Sci.*, 250, 266, 2002.

27. Bockris, J.O'M. and Reddy, A.K.N., *Modern Electrochemistry*, Plenum Press, New York, 1973.

28. Irimajiri, A., Hanai, T., and Inouye, V., A dielectric theory of "multi-stratified shell" model with its application to lymphoma cell, *J. Theor. Biol.*, 78, 251, 1979.

29. Huang, Y., Holzel, R., Pethig, R., and Wang, X.-B., Differences in the AC electrodynamics of viable and nonviable yeast-cells determined through combined dielectrophoresis and electrorotation studies, *Phys. Med. Biol.*, 37, 1499, 1992.

30. Lipowicz, P.J. and Yeh, H.C., Fiber dielectrophoresis, *Aerosol. Sci. Technol.*, 11, 206, 1989.

31. Kakutani, T., Shibatani, S., and Sugai, M., Electrorotation of nonspherical cells: theory for ellipsoidal cells with an arbitrary number of shells, *Biochem. Bioenerg.*, 31, 131, 1993.

32. Hasted, J.B., *Aqueous Dielectrics*, Chapman and Hall, London, 1973.

33. Arnold, W.M. and Zimmermann, U., Rotating-field-induced rotation and measurement of the membrane capacitance of single mesophyll cells of *Avena sativa*, *Z. Naturforsch.*, 37c, 908, 1982.

34. Teixeira-Pinto, A.A., Nejelski, L.L., Cutler, J.L., and Heller, J.H., The behaviour of unicellular organisms in an electromagnetic field, *J. Exp. Cell Res.*, 20, 548, 1960.

35. Berry, R.M. and Berg, H.C., Torque generated by the flagellar motor of *Escherichia coli* while driven backward, *Biophys. J.*, 76, 580, 1999.

36. Hagedorn, R., Fuhr, G., Müller, T., Schnelle, T., Schnakenberg, U., and Wagner, B., Design of asynchronous dielectric micromotors, *J. Electrostat.*, 33, 159, 1994.

37. Huang, Y., Wang, X.-B., Tame, J., and Pethig, R., Electrokinetic behaviour of colloidal particles in travelling electric fields: studies using yeast cells, *J. Phys. D*, 26, 312, 1993.

38. Batchelder, J.S., Dielectrophoretic manipulator, *Rev. Sci. Instrum.*, 54, 300, 1983.

39. Masuda, S., Washizu, M., and Iwadare, M., Separation of small particles suspended in liquid by nonuniform traveling field, *IEEE Trans. Ind. Appl.*, 23, 474, 1987.

40. Fuhr, G., Hagedorn, R., Müller, T., Benecke, W., Wagner, B., and Gimsa, J., Asynchronous traveling-wave induced linear motion of living cells, *Studia Biophys.*, 140, 79, 1991.

41. Hughes, M.P., Pethig, R., and Wang, X.-B., Forces on particles in travelling electric fields: computer-aided simulations, *J. Phys. D*, 29, 474, 1996.

42. Fuhr, G., Schnelle, T., and Wagner, B., Travelling-wave driven microfabricated electrohydrodynamic pumps for liquids, *J. Micromech. Microeng.*, 4, 217, 1994.

43. Morgan, H., Green, N.G., Hughes, M.P., Monaghan, W., and Tan, T.C., Large-area travelling-wave dielectrophoresis particle separator, *J. Micromech. Microeng.*, 7, 65, 1997.

44. Fuhr, G., Fiedler, S., Müller, T., Schnelle, T., Glasser, H., Lisec, T., and Wagner, B., Particle micromanipulator consisting of two orthogonal channels with travelling-wave electrode structures, *Sens. Actua. A*, 41, 230, 1994.

45. Ward, M., Devilish tricks with tiny chips, *New Scientist*, 1st March, 22, 1997.

46. Benguigui, L. and Lin, I.J., More about the dielectrophoretic force, *J. Appl. Phys.*, 53, 1141, 1982.

47. Einstein, A., *Investigations on the Theory of Brownian Movement*, Dover, New York, 1956.

48. Ajdari, A. and Prost, J., Drift induced by a spatially periodic potential of low symmetry: pulsed dielectrophoresis, *C.R. Acad. Sci. Paris*, 315, 1635, 1992.

49. Chauwin, J.-F., Ajdari, A., and Prost, J., Mouvement sans force, in *Proc. SFP 4émes Journées de la Matire Condensèe*, Rennes, 1994.

59. Magnasco, M.O., Forced thermal ratchets, *Phys. Rev. Lett.*, 71, 1477, 1993.

51. Astumian, R.D. and Bier, M., Fluctuation driven ratchets: molecular motors, *Phys. Rev. Lett.*, 72, 1766, 1994.

52. Rousselet, J., Salome, L., Ajdari, A., and Prost, J., Directional motion of Brownian particles induced by a periodic asymmetric potential, *Nature*, 370, 446, 1994.

53. Doering, C.R., Horsthemke, W., and Riordan, J., Nonequilibrium fluctuation-induced transport, *Phys. Rev. Lett.*, 72, 2984, 1994.

54. Prost, J., Chauwin, J.-F., Peliti, L., and Ajdari, A., Asymmetric pumping of particles, *Phys. Rev. Lett.*, 72, 2652, 1994.

55. Faucheux, L.P. and Libchaer, A., Selection of Brownian particles, *J. Chem. Soc. Faraday Trans.*, 91, 3163, 1995.

56. Gorre-Talini, L., Spatz, J.P., and Silberzan, P., Dielectrophoretic ratchets, *Chaos*, 8, 650, 1998.

57. Chauwin, J.F., Ajdari, A., and Prost, J., Force-free motion in asymmetric structures — a mechanism without diffusive steps, *Europhys. Lett.*, 27, 421, 1994.

58. Ramos, A., Morgan, H., Green, N.G., and Castellanos, A., AC electrokinetics: a review of forces in microelectrode structures, *J. Phys. D*, 31, 2338, 1998.

59. Green, N.G., Ramos, A., Morgan, H., and Castellanos, A., Sub-micrometre AC electrokinetics: particle dynamics under the influence of dielectrophoresis and electrohydrodynamics, *Inst. Phys. Conf. Ser.*, 163, 89, 1999.

60. Green, N.G. and Morgan, H., Separation of submicrometre particles using a combination of dielectrophoretic and electrohydrodynamic forces, *J. Phys. D*, 31, L25, 1998.

61. Smith, P.W., Ashkin, A., and Tomlinson, W.J., Four-wave mixing in an artificial Kerr medium, *Opt. Lett.*, 6, 284, 1981.

62. Birtles, A.B., Mayo, B.J., and Bennett, A.W., Computer technique for solving 3-dimensional electron-optics and capacitance problems, *Proc. IEEE*, 120, 213, 1973.

63. Duan, X., Huang, Y., Cui, Y., Wang, J., and Lieber, C.M., Indium phosphoide nanowires as building blocks for nanoscale electronic and optoelectronic devices, *Nature*, 409, 66, 2001.

64. Cui, Y. and Leiber, C.M., Functional nanoscale electronic devices assembled using silicon nanowires building blocks, *Science*, 291, 851, 2001.

65. Huang, Y., Duan, X.F., Wie, Q.Q., and Leiber, C.M., Directed assembly of one-dimensional nano-structures into functional networks, *Science*, 291, 630, 2001.

66. Smith, P.A., Nordquist, C.D., Jackson, T.N., Mayer, T.S., Martin, B.R., Mbindyo, J., and Mallouk, T.E., Electric-field assisted assembly and alignment of metallic nanowires, *Appl. Phys. Lett.*, 77, 1399, 2000.

67. Yamamoto, K., Akita, S., and Nakayama, Y., Orientation and purification of carbon nanotubes using AC electrophoresis, *J. Phys. D*, 31, L34, 1998.

68. Krupke, R., Hennrich, F., Löhneysen, H.V., and Kappes, M.M., Separation of metallic from semi-conducting single-walled carbon nanotubes, *Science*, 301, 344, 2003.

69. Krupke, W., Hennrich, F., Kappes, M.M., and Löhneysen, H.V., Surface conductance induced dielectrophoresis of semiconducting single-walled carbon nanotubes, *Nano Lett.*, 4, 1395, 2004.

70. Chan, R.H.M., Fung, C.K.M., and Li, W.J., Rapid assembly of carbon nanotubes for nanosensing by dielectrophoretic force, *Nanotechnology*, 15, S672, 2004.

71. Li, J., Zhang, Q., Yang, D., and Tian, J., Fabrication of carbon nanotubes field effect transistors by a dielectrophoresis method, *Carbon*, 42, 2263, 2004.

72. Suehiro, J., Zhou, G., Imakiire, H., Ding, W., and Hara, M., Controlled fabrication of carbon nanotubes NO_2 gas sensor using dielectrophoretic impedance measurement, *Sens. Actua. B*, 108, 398, 2005.

73. Lucci, M., Regoliosi, P., Reale, A., Di Carlo, A., Orlanducci, S., Tamburri, E., Terranova, M.L., Lugli, P., DiNatale, C., D'Amico, A., and Paolesse, R., Gas sensing using single wall carbon nanotubes ordered with dielectrophoresis, *Sens. Actua. B*, 111–112, 181, 2005.

74. Tang, J., Yang, G., Zhang, Q., Pharat, A., Maynor, B., Liu, J., Qin, L.-C., and Zhou, O., Rapid and reproducible fabrication of carbon nanotubes AFM probes by dielectrophoresis, *Nano Lett.*, 5, 11, 2005.

75. Chen, Z., Yang, Y., Chen, F., Qing, Q., Wu, Z., and Liu, Z., Controllable interconnection of single-walled carbon nanotubes under AC electric field, *J. Phys. Chem. B*, 109, 11420, 2005.

76. Dong, L., Bush, J., Chirayos, V., Solanki, R., and Jiao, J., Dielectrophoretically controlled fabrication of single-crystal nickel silicide nanowire interconnects, *Nano Lett.*, 5, 2112, 2005.

77. Lutz, T. and Donovan, K.J., Macroscopic scale separation of metallic and semiconducting nano-tubes by dielectrophoresis, *Carbon*, 43, 2508, 2005.

78. Maeda, Y., Kimura, S.I., Kanda, M., Hirashima, Y., Hasegawa, T., Wakahara, T., Lian, Y., Nakahodo, T., Tsuchiya, T., Akasaka, T., Lu, J., Zhang, X., Gao, Z., Yu, Y., Nagase, S., Kazaoui, S., Minami, N., Shimizu, T., Tokumoto, H., and Saito, R., Large-scale separation of metallic and semiconducting single-walled carbon nanotubes, *J. Am. Chem. Soc.*, 127, 10287, 2005.

79. Watarai, H., Sakamoto, T., and Tsukahara, S., In situ measurement of dielectrophoretic mobility of single polystyrene microspheres, *Langmuir*, 13, 2417, 1997.

80. Chiu, D.T. and Zare, R.N., Optical detection and manipulation of single molecules in room-temperature solutions, *Chemistry*, 3, 335, 1997.

81. Hölzel, R., Calander, N., Chiragwandi, Z., Willander, M., and Bier, F.F., Trapping single molecules by dielectrophoresis, *Phys. Rev. Lett.*, 95, 128102, 2005.

82. Schnelle, T., Hagedorn, R., Fuhr, G., Fiedler, S., and Müller, T., Three-dimensional electric field traps for manipulation of cells — calculation and experimental verification, *Biochim. Biophys. Acta*, 1157, 127, 1993.

83. Schnelle, T., Müller, T., and Fuhr, G., Trapping in AC octode field cages, *J. Electrostat.*, 50, 17, 2000.

84. Alimova, A.N., Chubun, N.N., Belobrov, P.I., Ya Detkov, P., and Zhirnov, V.V., Electrophoresis of nanodiamond powder for cold cathode fabrication, *J. Vac. Sci. Technol.*, 17, 715, 1999.

85. Washizu, M. and Kurosawa, O., Electrostatic manipulation of DNA in microfabricated structures, *IEEE Trans. Ind. Appl.*, 26, 1165, 1990.

86. Washizu, M., Kurosawa, O., Arai, I., Suzuki, S., and Shimamato, N., Applications of electrostatic stretch and positioning of DNA, *IEEE Trans. Ind. Appl.*, 31, 447, 1995.

87. Yamamoto, T., Kurosawa, O., Kabata, H., Shimamato, N., and Washizu, M., Molecular surgery of DNA based on electrostatic manipulation, *IEEE Trans. Ind. Appl.*, 36, 1010, 2000.

88. Porath, D., Bezryadin, A., de Vries, S., and Dekker, C., Direct measurement of electrical transport through DNA molecules, *Nature*, 403, 635, 2000.

89. Germishulzen, W.A., Tosch, P., Middelberg, A.P.J., Wälti, C., Davies, A.G., Wirtz, R., and Pepper, M., Influence of alternating current electrokinetic forces and torque on the elongation of immobilized DNA, *J. Appl. Phys.*, 97, 014702, 2005.

90. Tuukkanen, S., Kuzyk, A., Toppari, J.J., Hytönen, V.P., Ihalainen, T., and Törmä, P., Dielectrophoresis of nanoscale double-stranded DNA and humidity effects on its electrical conductivity, *Appl. Phys. Lett.*, 87, 183102, 2005.

91. Nedelcu, S. and Watson, J.H.P., Size separation of DNA molecules by pulsed electric field dielectrophoresis, *J. Phys. D*, 37, 2197, 2005.

92. Zheng, L., Brody, J.P., and Burke, P.J., Electronic manipulation of DNA, proteins, and nanoparticles for potential circuit assembly, *Biosens. Bioelectron.*, 20, 606, 2004.

93. Clarke, R.W., White, S.S., Zhou, D., Ying, l., and Klenerman, D., Trapping of proteins under physiological conditions in a nanopipette, *Angew. Chem.*, 44, 3747, 2005.

94. Flores-Rodrigues, N., Bryning, Z., and Markx, G.H., Dielectrophoresis of reverse phase emulsions, *IEEE Proc. Nanobiotechnol.*, 152, 137, 2005.

95. Hoettges, K.F., Cotton, A., Hopkins, N.A.E., McDonnell, M.B., and Hughes, M.P, Combined dielectrophoretic/electrohydrodynamic/evanescent-light-scattering biosensors for the detection of pathogenic organisms, *Eng. Med. Biol.*, 22, 68, 2003.

96. Hoettges, K.F., McDonnell, M.B., and Hughes, M.P., Use of combined dielectrophoretic/electro-hydrodynamic forces for biosensor enhancement, *J. Phys. D*, 36, L101, 2003.

97. Hübner, Y., Hoettges, K.F., McDonnell, M.P., Carter, M.J., and Hughes, M.P., Applications of dielectrophoretic/electro-hydrodynamic "zipper" electrodes for detection of biological nanoparticles, *Int. J. Nanomed.* (in press).

98. Gascoyne, P.R.C., Huang, Y., Pethig, R., Vykoukal, J., and Becker, F.F., Dielectrophoretic separation of mammalian cells studied by computerized image analysis, *Meas. Sci. Technol.*, 3, 439, 1992.

99. Green, N.G. and Morgan, H., Dielectrophoretic separation of nanoparticles, *J. Phys. D*, 30, L41, 1997.

100. Burt, J.P.H., Pethig, R., and Talary, M.S., Microelectrode devices for manipulating and analysing bioparticles, *Trans. Inst. Meas. Control*, 20, 82, 1998.

101. Morgan, H., Hughes, M.P., and Green, N.G., Separation of submicron bioparticles by dielectrophoresis, *Biophys. J.*, 77, 516, 1999.

17

Biological- and Chemical-Mediated Self-Assembly of Artificial Micro- and Nanostructures

17.1 Introduction..17-1
    On Size and Scale • Miniaturization
    • Engineering ←→ Biology

17.2 Bio-inspired Self-Assembly...17-3
    Self-Assembly Defined • Self-Assembly Categorized
    • Motivation for Bio-inspired Self-Assembly

17.3 The Forces and Interactions of Self-Assembly 17-5
    van der Waals–London Interactions • Hydrogen Bonding
    • Ionic Bonds • Hydrophobic Interactions

17.4 Biological Linkers...17-6
    DNA as Biolinkers • Protein Complexes as Biolinkers
    • Strategies for Assembly with Biolinkers

17.5 State of the Art in Bio-inspired Self-Assembly 17-12
    Biological Entity-Mediated Self-Assembly • Chemically
    Treated Surfaces and Self-Assembly • Electrically Mediated
    • Fluidics Mediated

17.6 Future Directions..17-30
    Bio-inspired Active Device Assembly • New Biolinkers

17.7 Conclusions..17-31
Acknowledgments 17-32
References 17-32

S.W. Lee
Yonsei University

R. Bashir
Purdue University

17.1 Introduction

17.1.1 On Size and Scale

Recent advances in the field of nanotechnology and nanobiotechnology have been fueled by the advancement in fabrication technologies that allow construction of artificial structures that are of the same size or smaller than many biological entities. Figure 17.1 shows the size and scale of many biological and artificial structures. It is interesting to note that the minimum feature in modern day integrated circuits,

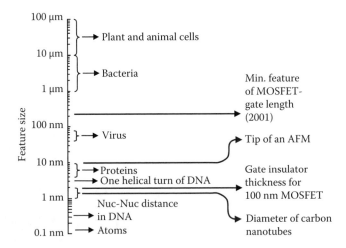

FIGURE 17.1 Size and scale of naturally occurring structures as compared to human-made structures.

which is 0.05 μm in 2005, is an order of magnitude smaller than cells and bacteria. The tip of an atomic force microscope, a key tool in advancing the field of nanotechnology, is smaller than most viruses. The gate insulator thickness of a modern day metal-oxide-semiconductor (MOS) transistor is thinner than one helical turn of a DNA. Thus, it is clear that the top-down fabrication technologies have progressed enough to allow the fabrication of micro- and nanostructures that can be used to interface, interrogate, and integrate biological structures with artificial structures.

17.1.2 Miniaturization

Since the invention of the junction transistor in 1947 and the subsequent invention of the integrated circuit, the complexity of microelectronic integrated circuits and devices has increased exponentially. Figure 17.2 shows the trends in miniaturization and complexity using silicon complementary

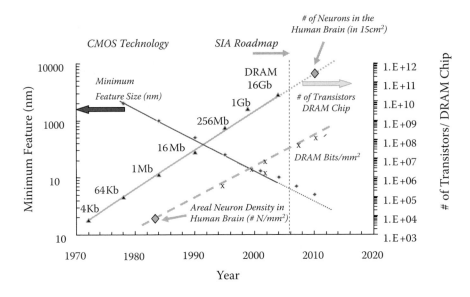

FIGURE 17.2 Trends in miniaturization of integrated circuits in the last 25 years.

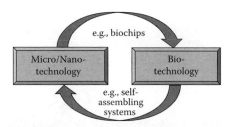

FIGURE 17.3 Micro- and nanotechnology can be used to solve problems in life sciences, or life sciences can be used to impact micro- and nanotechnology.

metal-oxide-semiconductor (CMOS) technology. The minimum feature size has decreased from 2 μm in 1980 to 0.05 μm in 2005 in volume production [1]. In research labs, minimum features sizes, which are a factor of 5 to 10 smaller, have been demonstrated. The Semiconductor Industry Association roadmap projects that these trends will continue for another 15 to 20 years, but it is becoming increasingly difficult to continue to downscale due to real physical limitations including size of atoms, wavelengths of radiation used for lithography, interconnect schemes, etc. No known solutions currently exist for many of these problems [2,3].

As the construction of artificial computational systems, i.e., integrated circuits, continues to become insurmountably difficult, more and more engineers and scientists are turning toward nature for answers and solutions. A variety of extremely sophisticated and complicated molecular systems occur in nature that vary in density, sense, and relay information, perform complex computational tasks, and self-assemble into complex shapes and structures. Two examples can be considered, i.e., that of the human brain and that of genomic DNA in the nucleus of the cell. There are about 10^{11} neurons in the human brain in a volume of about 15 cm^3 [4]. The total number of transistors on a 2-D chip will actually reach the number of neurons in the human brain by about year 2010. The area density of the neurons was actually surpassed in the mid-1980s but as is known, the 3-D nature and parallel interconnectivity of the neurons is what makes the exquisite functions of the brain possible. So even though humans have achieved or will soon achieve a similar density of basic computational elements to that of the brain, the replication of brain functions is far from reality. Similarly, the case of DNA is also far-reaching and intriguing. The human DNA is about 6 mm long, has about 2×10^8 nucleotides, and is tightly packed in a volume of 500 um^3 [4]. If a set of three nucleotides can be assumed to be analogous to a byte (since a three codon set from mRNA is used to produce an amino acid), then these numbers represent about 1 Kb/μm (linear density) or about 1.2 Mb/μm^3 (volume density). These numbers are not truly quantitative but can give an appreciation of how densely stored information is in the DNA molecules. Certainly, a memory chip based on DNA as the active elements could have extremely high density!

17.1.3 Engineering ←→ Biology

As the dimensions of artificially machined structures become smaller than biological entities, numerous very exciting possibilities arise to solve important and complex problems. One way to categorize the types of possible research is shown in Figure 17.3. Micro- and nanotechnology can be used to solve important problems in life sciences. Examples of this category would be biochips, which even now are being commercialized. On the other hand, knowledge from life sciences can be used to solve important problems in engineering, and examples of this would include bio-inspired self-assembly and biomimitic devices. All biological organisms are essentially self-assembled, beginning from one cell, resulting in entire organisms.

17.2 Bio-inspired Self-Assembly

17.2.1 Self-Assembly Defined

Self-assembly can be defined as the process of self-organization of one or more entities as the total energy of the system is minimized to result in a more stable state. This process of self-assembly inherently implies the following: (1) some mechanism where movement of entities takes place using diffusion, electric fields,

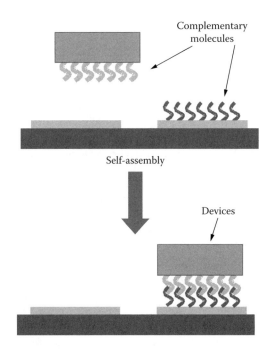

FIGURE 17.4 A basic concept schematic showing the process of self-assembly.

etc.; (2) the concept of recognition between different elements, or biolinkers, that result in self-assembly; and (3) where the recognition results in binding of the elements dictated by forces (electrical, covalent, ionic, hydrogen, van der Waals, etc.), such that the resulting physical placement of the entities results in the state of lowest energy.

The basic idea behind the concept of biologically mediated self-assembly is shown in Figure 17.4. An object is coated with molecules (biolinkers) that recognize and are complementary to other molecules on another object. The two objects are somehow brought in proximity and under appropriate conditions; the two linkers would bind together and as a result form an ordered arrangement. This is how cells are captured in an ELISA plate and single strands of DNA hybridize, for example, and how devices can possibly be self-assembled on heterogeneous substrates. The forces that bind the biolinkers could include any of the forces described in more detail in subsequent sections. For example, the molecules could be single strands of DNA that are complementary to each other and help to bind the two objects together. The molecules on one or both objects could also be charged and electrostatic forces could be used to bring and bind the two objects.

17.2.2 Self-Assembly Categorized

There is a significant amount of literature available on self-assembly. Broadly speaking, the research on self-assembly can be divided into three categories:

1. Formation of objects themselves. These objects could include clusters, molecules and macromolecules, quantum dots, nanowires and nanotubes, etc. Crystal growth could be included in this category.
2. Formation of 2-D/3-D arrays and networks. These could include self-assembled monolayers (SAMs), close packed arrays of particles, clusters, and shells, and 3-D close packed assemblies, SK strain-induced material self-assembly, etc.
3. Directed selective assembly of objects at specific locations due to variety of forces such as biological, chemical, or electrical forces.

The discussion in this chapter has focused mostly on the third category where micro- and nanoscale objects are assembled at specific sites due to biologically inspired forces and phenomena.

17.2.3 Motivation for Bio-inspired Self-Assembly

Self-assembly processes not only are interesting from a scientific point of view but can also have a wide variety of applications. These applications can include any case where micro- or nanoscale objects of one type need to be placed or assembled at specific sites on another substrate. As will be evident from the Section 17.5 below, the applications can be in the area of (1) detection and diagnostics; (2) fabrication of novel electronic/optoelectronic systems; and (3) in the area of new material synthesis. For example, in the case of detection of DNA oligonucleotides, avidin-coated gold or polystyrene beads are assembled onto a biotinylated target DNA to indicate complementary binding. Proteins and DNA when attached to carbon nanotubes or silicon nanowire or devices can be used to assemble these devices on substrates for ultradense electronics or flexible displays. Heterogeneous integration of materials can be achieved using such biologically mediated assembly of components. Since these assembly techniques provide micro- and nanoscale placement of objects, the assembly can be repeated multiple times to result in novel 3-D material synthesis.

17.3 The Forces and Interactions of Self-Assembly

Molecular and macromolecular recognition processes are key to bio-mediated self-assembly and various forces comprise the molecular recognition processes. We will now review these forces and the resulting chemical bonds briefly [4]. These bonds can be categorized as covalent and noncovalent. Covalent bonds, as is well known, are strong, and a single C–C bond has energy of about 90 kcal/mol and is responsible for the bonding in between the subunits of the macromolecules. The much weaker noncovalent bonds in biological molecules are due to (1) van der Waals interactions (~0.1 kcal/mole/atom); (2) hydrogen bonding (~1 kcal/mol); (3) ionic bonding (~3 kcal/mol in water, 80 kcal/mole in vacuum); and (4) hydrophobic interactions. These interactions (or a combination thereof) determine the specific recognition and binding between specific molecules and macromolecules that are used in the self-assembly processes described in this chapter.

17.3.1 van der Waals–London Interactions

This type of force is generated between two identical inert atoms that are separated from each other with a distance that is large in comparison with the radii of the atom. Since the charge distributions are not rigid, each atom causes the other to slightly polarize and induce a dipole moment. These resulting dipole moments cause an attractive interaction between the two atoms. This attractive interaction varies as the minus sixth power of the separation of the two atoms, $\Delta U = -\frac{A}{R^6}$, where A is a constant. Van der Waals–London interaction is a quantum effect and will result between any two charged bodies where the charge distributions are not rigid so that they can be perturbed in space to induce a dipole moment. Two atoms will be attracted to each other till the distance between them equals the sum of their van der Waals radii. Brought closer than that, the two atoms will repel each other. Individually, these forces are very weak but can play a very important role in determining the binding of two macromolecular surfaces.

17.3.2 Hydrogen Bonding

Another very important interaction that binds different molecules together is that of hydrogen bonding. This is how the nucleotides of a DNA (described below) complementarily bind to each other. Hydrogen bonding is largely ionic in nature and can result when an hydrogen atom, which is covalently bonded to a small electronegative atom (e.g., F, N, and O), develops a positive-induced dipole charge. This positive charge can then interact with the negative end of a neighboring dipole resulting in a hydrogen bond. This interaction typically varies as $1/R^3$ where R is the distance between the dipoles. Hydrogen bonding is an important part of the interaction between water molecules.

17.3.3 Ionic Bonds

Ionic interaction can also take place in partially charged groups or fully charged groups (ionic bonds). The force of attraction is given by Coulomb's law as $F = K(q^+q^-/R^2)$, where R is the distance of separation. When present among water and counterions in biological mediums, ionic bonds become weak since the charges are partially shielded by the presence of counterions. Still, these ionic interactions between groups are very important in determining the recognition between different macromolecules.

17.3.4 Hydrophobic Interactions

Hydrophobic interactions between two hydrophobic groups are produced when these groups are placed in water. The water molecules will tend to move these groups close in such a way as to keep these hydrophobic regions close to each other. Hence this coming together of hydrophobic entities in water can be termed as a hydrophobic bond. Again, these interactions tend to be weak but are important, for example, when two hydrophobic groups on the surface of two macromolecules can come together when placed in water and thus result in binding of the two surfaces. The surface energy at the interface of a hydrophobic surface and water is high since water will tend to be moved away from the surface. Hence two hydrophobic surfaces will join to minimize the total exposed surface area and hence the total energy. The phenomenon plays a very important part in the protein folding (Section 17.4.2.1) and also to assemble objects in fluids using capillary forces (Section 17.5.2).

17.4 Biological Linkers

Biologically mediated self-assembly utilizes complementary molecules, which can be termed "biolinkers." These molecules act like a lock and key and bind to each other under appropriate conditions. The binding of these linker molecules (actually macromolecules) is a result of the noncovalent interactions described above. Two possible biolinkers include complementary nucleic acid molecules and protein complexes, and these are described below in more detail.

17.4.1 DNA as Biolinkers

17.4.1.1 DNA Fundamentals

DNA is the basic building block of life. Hereditary information is encoded in the chemical language of DNA and reproduced in all cells of living organisms. The double-stranded helical structure of the DNA is key to its use in self-assembly applications. Each strand of the DNA is about 2 nm wide and composed of a linear chain of four possible bases (adenine, cytosine, guanine, and thymine) on a backbone of alternating sugar molecules and phosphate ions (Figure 17.5). Each unit of a phosphate, a sugar molecule, and base is called a nucleotide and each nucleotide is about 0.34 nm long. The specific binding through hydrogen bonds between adenine and thymine and cytosine and guanine as shown in Figure 17.5(b) can result in the joining of two complementary single-stranded (ss) DNA to form a double-stranded (ds) DNA. The phosphate ion carries a negative charge in the DNA molecule, a property used in the drift of the molecule under an electric field, e.g., in electrophoresis applications. The negative charges result in electrostatic repulsion of the two strands and hence to keep the two strands together, positive ions need to be present in the ambient to keep the negative charges neutralized. The joining of two ssDNA through hydrogen bonding (described earlier) to form a dsDNA is called hybridization. Hence two single strands of DNA can be designed to have complementary sequences and made to join under appropriate conditions. If a dsDNA is heated above a certain temperature, called the melting temperature T_m, the two strands will separate into single strands. The melting temperature is a function of temperature, ion concentration of the ambient, and the G–C content in the sequence. When the temperature is reduced,

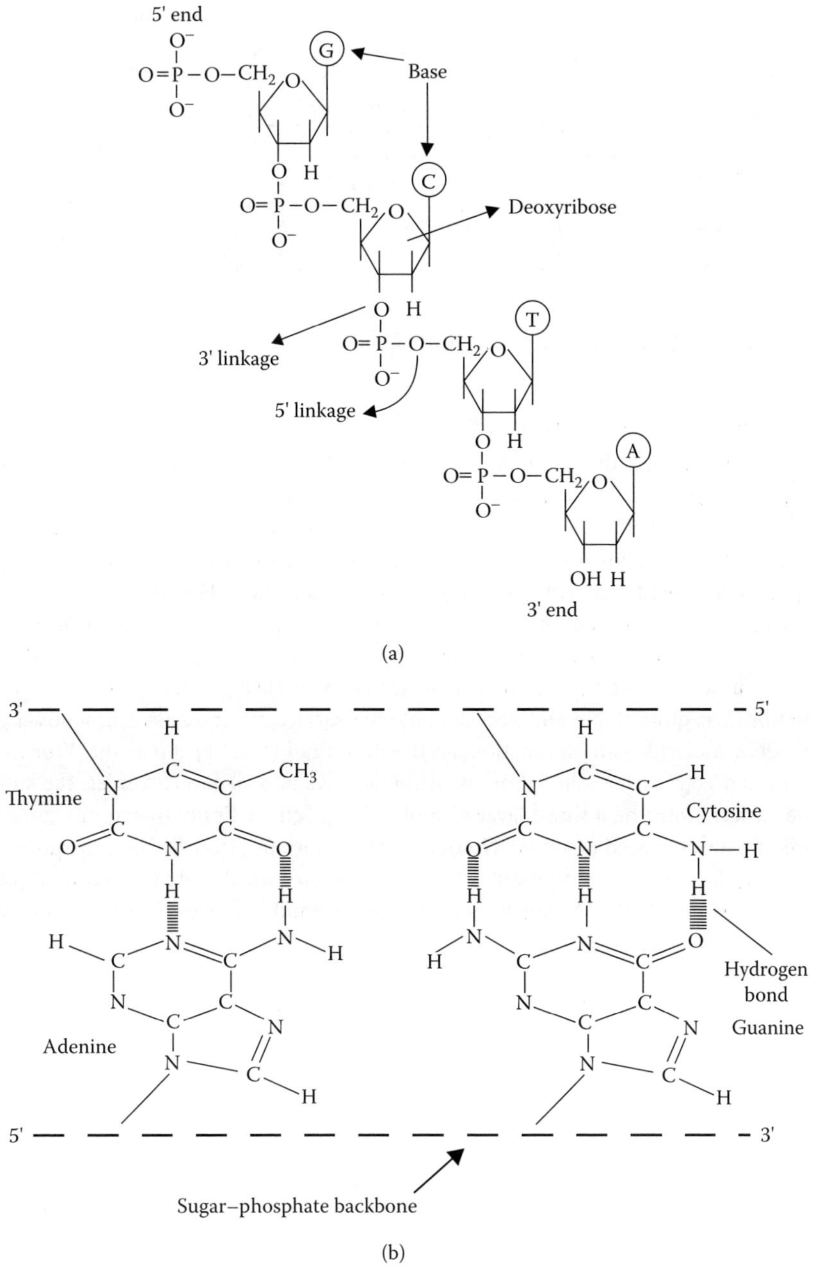

FIGURE 17.5 (a) Sugar–phosphate backbone of a DNA. (b) The four bases of a DNA showing their complementary binding properties. (From Alberts, B., Bray, D., Lewis, J., Raff, M., Roberts, K., and Watson, J.D., *Molecular Biology of the Cell*, 3rd ed., Garland Publishing, New York. Reprinted with permission.)

the two strands will eventually come together by diffusion and rehybridize or renature to form the double-stranded structure as shown Figure 17.6. These properties of the DNA can be utilized in the ordering and assembly of artificial structures if these structures can be attached to ssDNA. It should also be pointed out that the sequence of the DNA can be chosen and the molecules can now be obtained from a variety of commercial sources [5].

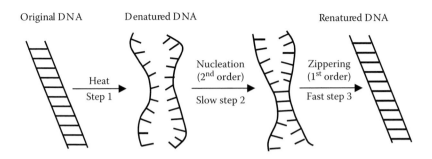

FIGURE 17.6 Schematic showing denaturing and hybridization of DNA.

17.4.1.2 Attachment of DNA to Gold Surfaces

It is also important to review the methodology to attach DNA molecules to surfaces. The most widely used attachment scheme utilizes the covalent bond between sulfur and gold [6–19]. The formation of long-chain ω-substituted dialkyldisulfide molecules on a gold substrate was first reported in 1983 [6]. Films of better quality were formed and reported by the adsorption of alkyl thiols [7–13]. Bain and Whitesides presented a model system consisting of long-chain thiols, $HS(CH_2)_nX$ (where X is the end group) that adsorb from a solution onto gold and form densely packed, oriented monolayers [7,8]. The schematic of the Au–S bond is shown in Figure 17.7 [9]. The bonding of the sulfur head group to the gold substrate is in the form of a metal thiolate, which is a very strong bond (~44 kcal/mol), and hence the resulting films are quite stable and very suitable for surface attachment of functional groups. For example, the DNA molecule can be functionalized with a thiol (S-H) group at the 3′ or 5′ end. Upon immersion of clean gold surfaces in solutions of thiol-derivatized oligonucleotides, the sulfur adsorbs onto the gold surfaces forming a single layer of molecules as schematically shown in Figure 17.7, where the hydrocarbon is now replaced with a ssDNA or a dsDNA molecule [16–19]. Chemisorption of thiolated ssDNA leads to surface coverages of about 10^{13} molecules/cm², which corresponds to about 1 strand/10nm² [18]. Hickman et al. also demonstrated the selective and orthogonal self-assembly of disulfide

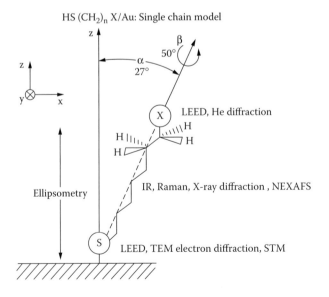

FIGURE 17.7 Schematic of a long-chain thiol molecule on a gold surface. (From Dubois, L. and Nuzzo, R.G., *Ann. Rev. Phys. Chem.*, 43, 437, 1992. Copyright, Annual Reviews, www.AnnualReviews.org. Reprinted with permission.)

with gold and isocyanide with platinum [10]. This work can be important in the orientation-dependent self-assembly of structures that have both platinum and gold surface exposed for functionalization. Hence, the thiol-based chemistry has served as the fundamental attachment scheme for DNA and oligonucleotides for the self-assembly of artificial nanostructures.

17.4.2 Protein Complexes as Biolinkers

17.4.2.1 Protein Fundamentals

Proteins are made of amino acids. Amino acids are small chemical compounds containing an amino (NH_2) group and a carboxylic group (COOH) connected to a central carbon atom. There are different side-chains connected to this central carbon atom (called α-carbon), hence resulting in different amino acids, as shown in Figure 17.8(a) and Figure 17.8(b). There are many amino acids, but 20 of them are most commonly used in the formation of proteins. Proteins are formed when chains of amino acids are connected by peptide bonds between the amine group of one amino acid and the carboxylic group of another amino acid. Once these chains are placed in water, they fold upon themselves into complex 3-D globular structures through covalent and noncovalent interactions (described in Section 17.4). The hydrophobic side-chains of the amino acids tend to cluster in the interior. The overall structure of the proteins is stabilized by noncovalent interactions between the various parts of the amino acid chain. Antibodies, ligands, etc. are all proteins that have regions on these macromolecules that exhibit specificity to other proteins, receptors, or molecules. This specificity comes about due to a combination of the noncovalent interaction forces described above and can be used for self-assembly. Ligand and receptors can also be used for self-assembly, and the best-known and most widely used ligand/receptor system is that of avidin and biotin. Avidin (a receptor with four binding sites to biotin) is a protein (molecular mass = 68 kDa) found in egg white. Biotin (a vitamin with a molecular mass of 244 Da) is the ligand that binds to avidin with a very high affinity ($K_a = 10^{15}$ M^{-1}).

17.4.2.2 Protein Attachment to Surfaces

The attachment of proteins on microfabricated surfaces will be vital to the success of self-assembly that is protein mediated. These attachment strategies are also useful for other applications such as protein chips and attachment and capture of cells on microfabricated surfaces. Certainly a lot needs to be learned from the prior work of adsorption of proteins at metal surfaces and electrodes [20,21]. The attachment of proteins on surfaces is a lot more complex when compared to attachment of DNA to surfaces. The proteins have to be attached in such a way that their structure and functionality should be retained.

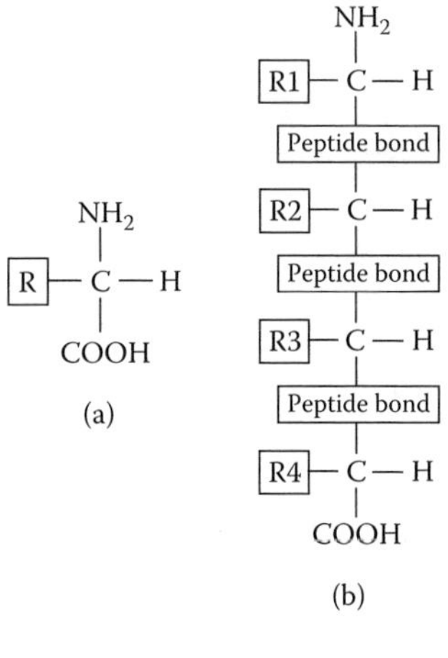

FIGURE 17.8 (a) An amino acid with a side-chain [R]; (b) a chain of amino acids linked by peptide bonds to form a protein.

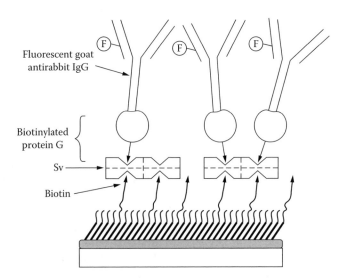

FIGURE 17.9 Schematic of antibody attachment using microconstant printing of a SAM. (From Lahiri, J., Ostuni, E., and Whitesides, G.M., *Langmuir*, 15, 2055, 1999. Copyright, American Chemical Society. Reprinted with permission.)

The attachment of antibodies and proteins has been demonstrated on microfabricated surfaces using functional groups such as silane [22,23], amine [24], carboxyl [25], and thiols [26]. Attachment of avidin on microfabricated surfaces using bovine serum albumin layers has also been demonstrated in such a way that the avidin retains its binding ability to biotin and hence any biotinylated protein [27]. Patterning of ligands has also been demonstrated using alkenethiolate SAMs, which were produced on Au layers, and then ligands such as biotin were printed on the SAMs using microcontact printing [28]. As shown in Figure 17.9, subsequent binding of streptavidin, biotinylated protein G, and fluorescently labeled goat antirabbit IgG protein within 5 μm squares demonstrated the patterning of these proteins. Protein microarrays have been demonstrated, where proteins were immobilized by covalently attaching them on glass surfaces that were treated with aldehyde-containing silane reagents [29]. These aldehydes react with the primary amines on the proteins such that the proteins still stay active and interact with other proteins and small molecules. 1600 spots were produced on a square centimeter using robotic nanoliter dispensing where each site was about 150 to 200 μm in diameter. All the above approaches take a protein and devise a technique to attach it to a microfabricated surface by functionalizing one end of the protein with chemical groups that have affinity to that particular surface. One of the most exciting techniques, that is truly bio-inspired, has been demonstrated by Belcher and coworkers who used natural selection and evolutionary principles with combinatorial phage display libraries to evolve peptides that bind to semi-conductor surfaces with high specificity [30]. They screened millions of peptides with unknown binding properties and evolved these peptides by rereacting them with semiconductor surfaces under increasingly stringent conditions. 12-mer peptides were developed that showed selective binding affinity to GaAs but not to $Al_{0.98}Ga_{0.02}As$ and Si and to 100 GaAs but not to 111B GaAs. These proteins can possibly be designed with multiple recognition sites and hence be used to design nanoparticle heterostructures in two and three dimensions.

17.4.3 Strategies for Assembly with Biolinkers

Different strategies can be employed when using DNA or protein complexes as the biolinker for the self-assembly. These strategies need techniques to attach the molecules to surfaces, as summarized above. Figure 17.10 shows four possible cases of assembling structures using DNA or protein complexes (more strategies are, of course, possible). The first case consists of thiolated DNA, with complementary strands

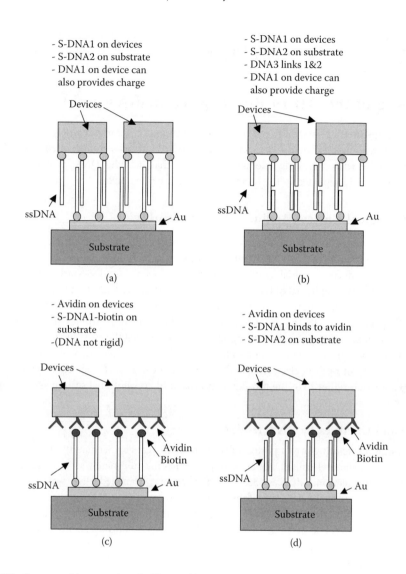

FIGURE 17.10 Four possible strategies of self-assembly using DNA and ligand/receptor complexes.

attached to the device and the substrate. Direct hybridization can be used to place and bind the device to the substrate. The second case is that of indirect hybridization where a third linking strand, each half being complementary to the two strands, is used to bind the device to the substrate. The third case shows a ssDNA with a thiol at one end providing the attachment to a gold layer and a biotin at the other end. The devices or objects to be assembled are coated with avidin and hence can be captured by the biotin on the DNA strand. The single strand is expected to not be rigid and lay on the Au layer in such a way that the biotin is not physically accessible to avidin for binding. The fourth case shows a single strand attached to the gold layer, which is then hybridized with a complementary strand functionalized with a biotin molecule. The device, which is coated with avidin, can now be captured by the biotin, and the more rigid double strand provides better binding capability between the avidin–biotin complex. For the first two cases, the DNA attached to the devices can also provide charges (due to the phosphate groups in the backbone), so that the devices can also be brought closer to the binding site using electrostatics. Of course additional scenarios can also be envisioned that utilize the long-range electrostatic, magnetic, or capillary forces to bring the structures/devices close to the binding sites and then use the short-range chemical and biological forces to result in intimate binding at the desired site. The use of DNA provides the attractive possibility of making selective addressable sites where different types of devices can be assembled at different sites simultaneously, bringing nano- and microstructure fabrication in a beaker closer to reality.

17.5   State of the Art in Bio-inspired Self-Assembly

The work in bio-inspired self-assembly can be grouped into four categories, namely (1) biological entity mediated; (2) chemically mediated; (iii) electrically mediated; and (iv) fluidics mediated. Much of the work has been done with chemically mediated assembly that includes DNA, proteins, and SAMs. A summary of these works reported in the literature will be provided below.

17.5.1   Biological Entity-Mediated Self-Assembly

There has been a tremendous interest in recent years to develop concepts and approaches for self-assembled systems for electronic and optical applications. Material self-assembly has been demonstrated in a variety of semiconductor materials (GaAs, InSb, SiGe, etc.) using Stranksi-Krastanov strain-dependent growth of lattice mismatch epitaxial films [31–34]. Periodic structures with useful optical properties have been demonstrated, but no reports of actual electronic functions can be found in literature using this lattice strain-dependent growth and assembly. While significant work continues along that direction, it has also been recognized by engineers, chemists, and life scientists that the exquisite molecular recognition of various natural biological materials can also be used to form a complex network of potentially useful particles for a variety of optical, electronic, and sensing applications. This approach can be considered a bottom-up approach rather than the top-down approach of conventional scaling and much work has been reported toward this front.

17.5.1.1   Nanostructures by DNA Itself

Pioneering research extending over a period of more than 15 years by Seeman has laid a foundation for the construction of structures using DNA as scaffolds, which may ultimately serve as frameworks for the construction of nanoelectronic devices [35–39]. In that work, branched DNA was used to form stick figures by properly choosing the sequence of the complementary strands. Macrocycles, DNA quadrilateral, DNA knots, Holliday junctions, and other structures were designed. Figure 17.11(a) shows a stable branched DNA junction made by DNA molecules. The hydrogen bonding is indicated by dots between the nucleotides. It is also possible to take this structure and devise a 2-D lattice as shown in Figure 17.11(b) if hybridization regions (sticky ends) are provided in region B. It was also pointed out that it was easier to synthesize these structures but more difficult to validate the synthesis. The same group also reported on the design and observation via atomic microscope (AFM) of 2-D crystalline forms of DNA double cross-over molecules that are programmed to self-assemble by the complementary binding of the sticky ends of the DNA molecules [39]. Single-domain crystal sizes, which were as large as 2×8 μm, were shown by AFM images. Recently, Rothemund et al. [40] demonstrated programmable DNA nanotubes made from the DNA double cross-over molecules (DAE-E tiles), which had 7 to 20 nm diameter. Moreover, these tubes were assembled with each other to form tens of microns in length at room temperature and were disassembled by heating, with an energy barrier $\sim 180k_bT$ [41]. These lattices and tubes can also serve as scaffolding material for other biological materials. It should be noted that in this work, the 2-nm wide stiff DNA molecules themselves are used to form the 2- and 3-D structures.

17.5.1.2   DNA-Mediated Assembly of Nanostructures

Among roles envisioned for nucleic acids in nanoelectronic devices, the self-assembly of DNA-conjugated nanoparticles has received the most attention in recent literature. Mirkin et al. [42] and Alivisatos et al. [43] were the first to describe self-assembly of gold nanoclusters into periodic structures using DNA. A method of assembling colloidal gold nanoparticles into macroscopic aggregates using DNA as linking elements was described [42]. The method involved attaching noncomplementary DNA oligonucleotides

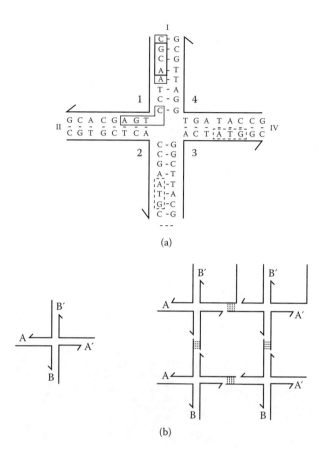

(a)

(b)

FIGURE 17.11 (a) A four-armed stable branched junction made from DNA molecules; (b) use of the branched junction to form periodic crystals. (From Seeman, N.C., *Nanotechnology*, 149, 1991. Copyright, IOP Publishing Limited. Reprinted with permission.)

to the surfaces of two batches of 13-nm gold particles capped with thiol groups, which bind to gold. When another oligonucleotide duplex with ends that are complementary to the grafted sequence is introduced, the nanoparticles self-assemble into aggregates. The process flow is shown in Figure 17.12, and this process could also be reversed when the temperature was increased due to the denaturation of the DNA oligonucleotides. Closed packed assemblies of aggregates with uniform particle separations of about 60 Å were demonstrated in this study as shown in Figure 17.13(a). In the same journal issue, techniques were also reported where discrete numbers of gold nanocrystals are organized into spatially defined structures based on DNA base-pair matching [43]. Gold particles, 1.4 nm in size, were attached to either the 3′ or 5′ of 19 nucleotide long ssDNA codon molecules through the well-known thiol attachment scheme. Then, 37 nucleotide long, ssDNA template molecules were added to the solution containing the gold nanoparticles functionalized with ssDNA. The authors showed that the nanocrystals could be assembled into dimers (parallel and antiparallel) and trimers upon hybridization of the codon molecules with that of the template molecule. Due to the ability to choose the number of nucleotides, the gold particles can be placed at defined positions from each other as schematically shown in Figure 17.14. TEM results showed that the distance between the parallel and antiparallel dimers was 2.9 to 10 nm and 2.0 to 6.3 nm, respectively. These structures could potentially be used for applications such as chemical sensors, spectroscopic enhancers, nanostructure fabrication, etc. These techniques have been used to devise sensitive colorimetric schemes for the detection of polynucleotides based on distance-dependent optical properties of aggregated gold particles in solutions [44].

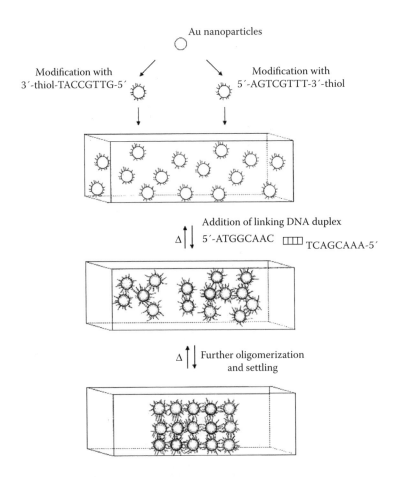

FIGURE 17.12 Fabrication process for the aggregated assembly of DNA-conjugated gold nanoparticles. (From Mirkin, C.A., Letsinger, R.L., Mucic, R.C., and Storhoff, J.J., *Nature*, 382, 607, 1996. Copyright, Macmillan Magazine Limited. Reprinted with permission.)

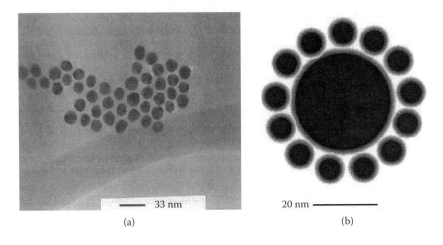

FIGURE 17.13 (a) TEM image of the aggregated DNA/Au colloidal particles. (b) TEM image of a nanoparticle satellite constructed via DNA-mediated docking of 9-nm gold particles onto a 31-nm gold particle. (Redrawn from Mucic, R.C., Storhoff, J.J., Mirkin, C.A., and Letsinger, R.L., *J. Am. Chem. Soc.*, 120, 12674, 1998. Copyright, American Chemical Society. Reprinted with permission.)

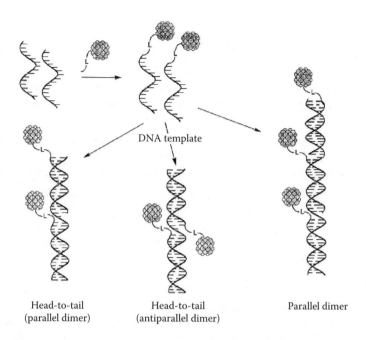

Head-to-tail
(parallel dimer)

Head-to-tail
(antiparallel dimer)

Parallel dimer

FIGURE 17.14 Assembly of nanocrystals to form dimers and trimers based on DNA hybridization. (From Alivisatos, A.P., Johnsson, K.P., Peng, X., Wilson, T.E., Loweth, C.J., Bruchez, M.P., and Schultz, P.G., *Nature*, 382, 609, 1996; Loweth, C.J., Caldwell, W.B., Peng, X., Alivisatos, A.P., and Schultz, P.G., *Agnew. Chem. Int. Ed.*, 38(12), 1808, 1999. Copyright, Macmillan Magazine Limited. Reprinted with permission.)

Mucic et al. have also described the construction of binary nanoparticle networks composed of 9 and 31 nm particles, both composed of citrate-stabilized colloidal gold [45]. These 9(±1) and 31(±3) nm particles were coated with different 12-mer oligonucleotides via a thiol bond. When a third DNA sequence (24-mer), which was complementary to the oligonucleotides on both particles, was added, hybridization led to the association of particles. When a ratio of 9 to 31 nm particles was large, the assembly illustrated in Figure 17.13(b) was formed (redrawn from [45]). Loweth et al. have presented further details of the formation of the heterodimeric and heterotrimeric nonperiodic nanocluster molecules based upon earlier work of Alivisatos [46]. The authors showed exquisite control of the placement of 5- and 10-nm gold nanoclusters that were derivatized with ssDNA. Various schemes of heterodimers and heterotrimers were designed and demonstrated using TEM images. This nanoparticle DNA-mediated hybridization also forms the basis of genomic detection using colorometric analysis [47]. Hybridization of the target with the probes results in the formation of a nanoparticle/DNA aggregate, which causes a red to purple color change in solution, due to the red-shift in the surface plasmon resonance of the Au nanoparticles. The networks show a very sharp melting transition curve that allows for single-base mismatches, deletions, or insertions to be detected. The same approach, which can also be taken on a surface combined with reduction of silver at the site of nanoparticle capture, allows the use of a conventional flatbed scanner as a reader. Sensitivities that are 100Xgreater than that of conventional fluorescence-based assays have been described [48]. Reviews on these topics have also been published and can provide more information than what is presented here [49–51].

Csaki showed the use of gold nanoparticles (with mean diameter 15, 30, and 60 nm) as a means for labeling DNA and characterizing DNA hybridization on a surface [52,53]. Single strands of DNA were attached to unpatterned gold substrates. Colloidal gold particles were labeled with thiolated complementary DNA strands, which were then captured by the strands on surface. Niemeyer et al. [54] have also showed site-specific immobilization of 40 nm gold nanoparticles that were citrate passivated and then modified with 5′ thiol derivatized 24-mer DNA oligomers. The capture DNA was placed at specific sites using nanoliter dispensing of the solution. 1-D gold nanoparticle arrays [55] were created using rolling-circle polymerization and DNA-encoded self-assembly. The authors attached a gold nanoparticle (5 nm) at the end of a DNA1 with a 5′ thiol group. Then, templates DNA (T_{king}), which were complementary to the DNA1, combined with rolling-circle polymerization. The DNA (T_{king}) on the rolling-circle polymerization hybridized with the DNA1, and, as a result, long 1-D gold nanoparticle arrays were formed. Using a DNA tile described in Figure 17.11(a), 2-D and periodical gold nanoparticle arrays were also formed [56]. The method to demonstrate the arrays was as follows:

1. Gold nanoparticles of about 5-nm were attached into complementary DNAs of the DNA tiles.
2. The complementary DNAs with the gold nanoparticles hybridized with the DNA tiles.
3. The DNA tiles with gold nanoparticles formed 2-D periodic arrays as described in Figure 17.11(b).

Mirkin and coworkers have also demonstrated the formation of supramolecular nanoparticle structures where up to four layers of gold nanoparticles have been produced. The scheme is shown in Figure 17.15 where a linking strand can bind the DNA strand, which is immobilized on the surface, and a DNA-derivatized nanoparticle [57]. The linking strand can then be used to bind another layer of nanoparticle, and a multilayered network structures can be produced.

17.5.1.3 DNA-Directed Micro- and Nanowires

The concepts of DNA-mediated self-assembly of gold nanostructures have been extended to metallic nanowires/rods [58,59]. The concept, though feasible, has not been completely demonstrated yet. The basic idea behind this work is to fabricate gold and/or platinum metal wires, functionalize these wires with ssDNA, and assemble them on substrates, which have the complementary ssDNA molecules, attached at specific sites. Thus, self-assembly of interconnects and wires can be made possible. The metallic wires are formed by electroplating in porous alumina membranes with pores sizes of about 200 nm [59]. The processes for the formation of alumina films with nanohole arrays have been developed and demonstrated by many authors [61,62]. Metallic rods, ranging from 1 to 6 μm in length, were produced, depending on the electroplating conditions. The goal of the work would be to form Pt rods with Au at the ends or vice versa. The same authors showed attachment and quantification of ssDNA on the Au ends of the Au/Pt/Au rods [63]. DNA strands were also attached to gold substrates, and it was shown that the rods attached to the substrates only when the DNA strands were indeed complementary. It should be noted that the attachment of the DNA strands was not patterned, and the complementary binding of the rods was not shown to be site specific; however, it is the next logical step for that work.

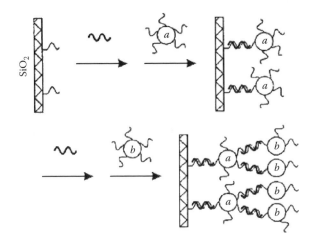

FIGURE 17.15 Synthesis of 3-D assembly using DNA and nanoparticles. (From Taton, T.A., Mucic, R.C., Mirkin, C.A., and Letsinger, R.L., *J. Am. Chem. Soc.*, 122, 6305, 2000. Copyright, American Chemical Society. Reprinted with permission.)

The use of DNA as a template for the fabrication of nanowires has been demonstrated through a very interesting process by Braun et.al. [64–66]. The authors formed a DNA bridge between two gold electrodes, again using thiol attachment. Once a DNA bridge is formed between the 12 to 16 μm spacing of the electrodes, a chemical deposition process is used to vectorially deposit silver ions along the DNA through Ag^+/Na^+ ion exchange and formation of complexes between the gold and the DNA bases (Figure 17.16). The result is a silver nanowire that is formed using the DNA as a template or skeleton. Current-voltage characteristics were measured to demonstrate the possible use of these nanowires. The authors also reported the formation of luminescent self-assembled poly(p-phenylene vinylene) wires for possible optical applications [65]. The work has a lot of potential and much room for further research exists to control the wire width, the contact resistances between the gold electrode and the silver wires, and use of other metals and materials. DNA-templated gold wires have also been developed by Keren et al. [67]. First, ssDNA and RecA monomer were polymerized, resulting in a nucleoprotein filament, as shown Figure 17.17(i). Consequently, the filament bound to the aldehyde-derived dsDNA substrate, as shown in Figure 17.17(ii). The substrate was immersed into $AgNO_3$ solution and was incubated. As a

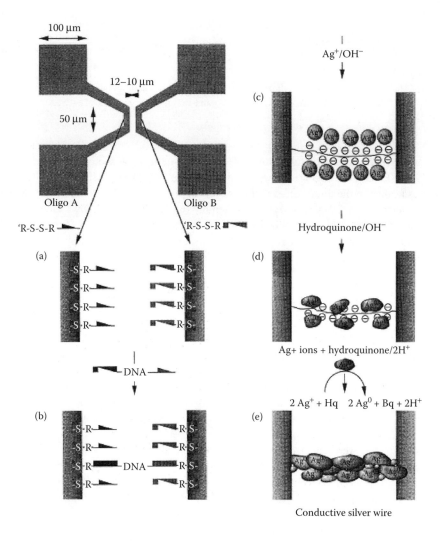

FIGURE 17.16 The process flow for the formation of DNA-directed silver nanowires. (From Braun, E., Eichen, Y., Sivan, U., and Yoseph, G.J., *Nature*, 391(19), 775, 1998. Copyright, Macmillan Magazine Limited. Reprinted with permission.)

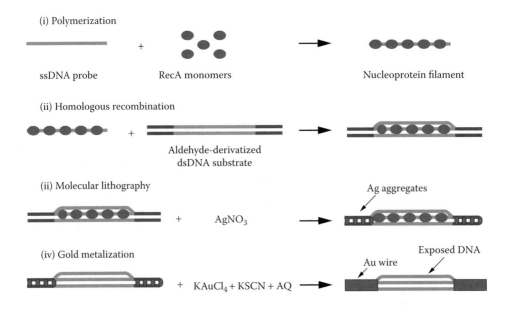

FIGURE 17.17 Schematic diagram of the homologous recombination reaction and molecular lithography. (From Keren, K., Krueger, M., Gilad, R., Ben-Yoseph, G., Sivan, U., and Braun, E., *Science*, 297, 72–75, 2002. Reprinted with permission.)

result, Ag aggregated along the substrate molecule except the area protected RecA, as shown in Figure 17.17(iii). Finally, the aggregated Ag served as catalysts for gold deposition and then conductive gold wire was formed in the unprotected region, as shown in Figure 17.17(iv). A 2-D nanowire network was also demonstrated by DNA-templated self-assembly [68,69]. The 4 × 4 DNA tile strand structure containing nine oligonucleotides, which had a four-arm junction oriented in each direction (N, S, E, and W) and each junction had a T_4 loop that allowed the arms to point to four direction connecting adjacent junctions, was constructed. Then, each tile with the sticky end of the tiles was connected, resulting in 2-D nanogrids and nanoribbons. To DNA metallization, the ribbons were seeded with silver using the glutaraldehyde method and silver was deposited [70]. As a result, the nanoribborns with 43 nm diameter nanoribbons [68] or 15 nm diameter nanoribbons [69] were formed.

17.5.1.4 DNA Inspired Self-Assembly of Electronic Devices

Many researchers have investigated the development of materials applied for molecular electronics, photo-detector, and full cell or field-effect-transistors by combining DNA with inorganic materials such as copper ions, acridine orange, tin ions, or carbon nanotubes. Kentaro Tanaka et al. [71] have demonstrated an artificial DNA by copper ion-mediated base pairing. A series of artificial oligonucleotides, $d(5'-GH_nC-3')$, with hydroxypyridone nucleobases (H) as flat bidentate ligands, was synthesized. Subsequently, copper ions were immersed into the artificial oligonucleotides and then double helices of the oligonucleotides, $nCu^+-d(5'-GH_nC-3')$, were formed through copper ion base pairing ($H-Cu^+-H$), as shown in Figure 17.18. A dye–DNA network [72] was incubated through mixing a DNA solution and an acridine orange solution. The electrical conductivity of the network, measured by a conducting AFM, was enhanced while the sample was exposed to visible light. DNA composite membranes as applied to fuel cell applications were investigated by Won et al. [73]. Nanoporous polycarbonate membranes were modified with tin (Sn) ion after immersing the membranes into a methanol/water solution that contains $SnCl_2$ and CF_3COOH for 45 min. Then, the membranes were transferred into a DNA solution. After 1 day, the membranes absorbed by DNA were irradiated with UV light in order to make cross-linking of DNA film. As a result, the membranes with DNA film have high proton conductivity.

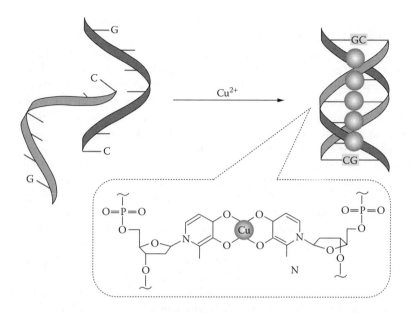

FIGURE 17.18 Schematic diagram of Cu^{2+}-mediated duplex formation between two artificial DNA strands. (From Yan, H., Park, S.H., Finkelstein, G., Relif, J.H., and LaBean, T.H., *Material and methods, http://www.sciencemag.org/cgi/content/full/301/5641/1882/DC1*. Reprinted with permission.)

Implementation of carbon nanotube electronic devices using the DNA-inspired self-assembled technique has also been explored by many researchers. Hazani et al. [74] have reported the assembly of single-walled carbon nanotubes (SWNTs) on gold electrodes using hybridization of DNA. The strategy of this approach is shown in Figure 17.10(a). They incubated the self-assembled DNA monolayers on gold electrode using 3′-thiolated oligonucleotides. Meanwhile, oxidized SWNTs with 3′-amino-modified oligonucleotides were prepared. Then, the solution containing DNA-modified oxidized SWNTs was introduced into the electrode in order to achieve the hybridization of DNA. Multicomponent structure using carbon nanotubes and gold nanoparticles has been also described by Li et al. [75] using hybridization of DNA. The approach mentioned earlier consisting of DNA-templated assembly in combination with RecA protein as reported by Keren et al. [76] has also been used to form transistors. In this approach, the precise localization of an SWNT operating as a field effect transistor was demonstrated. DNA-inspired self-assembly of active devices (complementary strands of DNA on a device and a substrate) has been proposed by Heller and coworkers [77] for assembling optical and optoelectronic components on a host substrate, but the basic concept has not been demonstrated yet.

17.5.1.5 Protein Complex-Mediated Assembly of Nano- and Microstructures

Semiconductor nanoscale quantum dots have gained a lot of attention in recent years due to the improvement in the synthesis techniques, for example, CdSe or ZnSe quantum dots of predetermined sizes can be synthesized. The optical properties of these structures have been studied in great detail [78,79]. These quantum dots have many advantages over conventional fluorophores such as narrow, tunable emission spectrum, and photochemical stability. The programmed assembly of such quantum dots using DNA has also been reported [80]. Typically these quantum dots are soluble only in nonpolar solvents, which makes it difficult to functionalize them with DNA by a direct reaction. The authors successfully demonstrated the use of 3-mercaptopropionic acid to initially passivate the QD surface and act as a pH trigger for controlling water solubility. The use of DNA-functionalized quantum dots, thus, allows the synthesis of hybrid assemblies with a different type of optical nanoscale building block.

The biological tagging of the quantum dots briefly described above is an interesting example of bio-inspired self-assembly. If these optical emitters can be tagged with specific proteins and antibodies, then site-specific markers can be developed. Examples of such applications have been demonstrated in the literature where CdSe/CdS core/shell quantum dots with a silica shell were used for biological staining of mouse fibroblast cells [81]. The surface of green-colored nanocrystals was modified with trimethoxysilylpropyl urea and acetate groups so that they were found to selectively go inside the cell nucleus. The authors also incubated fibroblasts with phalloidin-biotin and streptavidin and labeled the actin filaments of these cells with biotinylated red nanocrystals. Hence, penetration of the green nanoprobes inside the nucleus and the red staining of the actin fibers were simultaneously demonstrated. Covalently coupled protein-labeled nanocrystals for use in biological detection have also been demonstrated [82]. In this case, the ZnS-capped CdSe quantum dot was covalently coupled to a protein by mercaptoacetic acid. The mercapto group binds to the Zn atom. The carboxyl group can be used for covalent coupling of various biological molecules through amine group cross-linking as shown in Figure 17.19. It was postulated that for steric reasons maybe 2 to 5 protein molecules (100 kD) or so can be attached to a 5 nm quantum dot. The authors demonstrated receptor-mediated endocytosis and that the quantum dots were transferred inside the cells. These quantum dots can also be made as nanoshells consisting of a dielectric core with a metallic shell of nanometer thickness [83]. By varying the relative dimensions of the core and shell, the optical resonance of these nanoparticles can be varied over hundreds of nanometers in wavelength, across the visible and into the infrared region of the spectrum. When proteins and antibodies are attached to these nanoshells, these can be used as optical markers for biological diagnostic applications.

The assembly of microbeads using DNA and the avidin/biotin complex has also been demonstrated. This scheme is the same as that shown in Figure 17.10(d) where a single strand of captured DNA is first attached to a substrate. Then a second biotinylated target strand is brought in and, if complementary, will hybridize to the first strand exposing the biotin. Next, beads coated with avidin (or related receptors) are exposed to the surface and if the target strand did hybridize then the beads will be captured due to the avidin–biotin interaction. Thus, the presence of beads signals the presence of the complementary strands. Figure 17.20 shows an optical picture of a chip surface where avidin-coated polystyrene beads were captured when the DNA strands were indeed complementary [84]. The same scheme has been used to develop detectors for biological warfare agents by attaching the DNA on thin films exhibiting giant

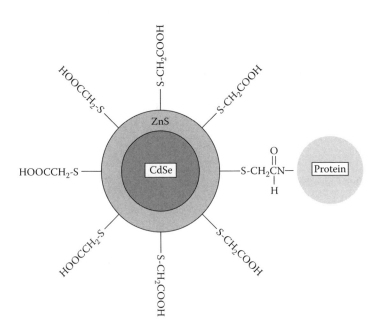

FIGURE 17.19 A CdSe/ZnS core/shell quantum dot with protein attachment scheme. (From Chan, W.C.W. and Nie, S., *Science*, 281, 2016, 1998. Copyright, AAAS. Reprinted with permission.)

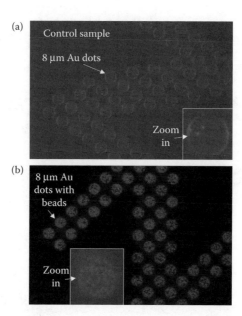

FIGURE 17.20 Optical micrographs of 0.8-μm avidin-coated polystyrene beads collected on biotinylated DNA on a surface: (a) a control sample with no DNA attached, hence beads are not collected; (b) sample with complementary strands of DNA, hence beads are collected on the pads. (From McNally, H., Pingle, M., Lee, S.W., Guo, D., Bergstrom, D., Bashir, R., *Appl. Surf. Sci.*, 214, 109–119, 2003. With permission.)

magnetoresistive (GMR) effect. Capture of target DNA and the presence of 1-μm sized magnetic beads can be electrically detected using the GMR sensors [85]. The intensity and location of the signal can indicate the concentration and identity of pathogens present in the sample.

Genetically engineered polypeptides, which are sequences of amino acids for selective binding to an inorganic surface, can be used for the assembly of functional nanostructures as shown below:

1. Inorganic-binding polypeptides are selected through display methods.
2. These selected polypeptides are modified by molecular biology to be used as linkers to bind nanoparticles, functional polymer, or other nanostructures on molecular templates.
3. Through self-assembly using modified polypeptides, ordered and multifunctional nanostructures can be implemented [86,87].

For the first step, phage display is a very powerful technique for selecting the proper polypeptide for binding to the inorganic materials. Randomized oligonucleotides are inserted into certain genes encoded on phage genomes or on bacterial plasmids, and libraries are created (Step 1 in Figure 17.21). As a result, the randomized polypeptide sequence is incorporated within a protein residing on the surface of the organism (Step 2 of Figure 17.21). Eventually, each phase or cell produces and displays a different, but random, peptide. (Step 3 in Figure 17.21). Then, these phases or cells are interacted on an inorganic substrate (Step 4 in Figure 17.21). After the washing steps, nonbound phases or cells are eliminated (Step 5 in Figure 17.21). Bound phases or cells are removed from the substrate (Step 6 in Figure 17.21). Then, the phases or cells are replicated and grown (Step 7 in Figure 17.21). Finally, DNA fragments for binding to the target substrate are extracted from the phases or cells. Using these display techniques, many researchers have developed binding sequences and implemented nanostructures. Brown et al. have demonstrated short Pt-and-Pd-binding sequences using cell surface display [88], and the 12 amino acid Au-binding sequences were developed by Naik et al. using phase display (PD) [89]. Moreover, metal-oxide-binding sequences for ZnO [90] and Zeilites [91], semiconductor-binding sequences for GaAs [92] and ZnS [93], and ionic crystal-binding sequences for $CaCO_2$ [94] and Cr_2O_3 [95] were reported using the PD method.

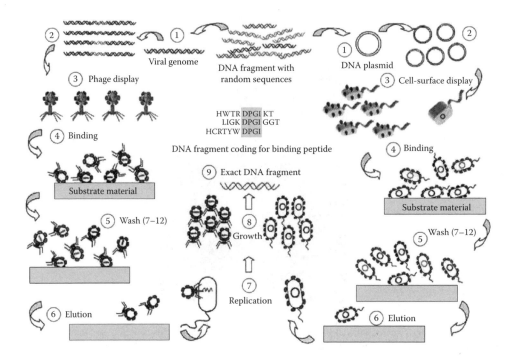

FIGURE 17.21 PD and cell-surface display. (From Sarikaya, M., Tamerler, C., Jen, A.K. –Y., Schulten, K., and Baneyx, F., *Nature Mater.*, 2, 577–585, 2003. Reprinted with permission.)

17.5.2 Chemically Treated Surfaces and Self-Assembly

The assembly of structures can also be performed using specific surface treatment with chemicals that make two surfaces bind to each other through small molecules. As an example, the assembly of arrays of nanoscale gold clusters and the measurement of electronic properties through the Au/molecule wires have been demonstrated by Andres, Datta, and coworkers [49,96]. They produced 2 to 5 nm Au clusters, which were initially encapsulated with dodecanethiol. The clusters can be precipitated and cross-linked into an ordered linked cluster network using molecules with a thiol at both ends, which links the adjacent clusters. The molecules used were from the aryl dithiol family of molecules. This work demonstrated guided self-assembly, high-quality ordering at the nanoscale, and low-resistance coupling to a semiconductor surface through the docking molecule [97].

A powerful approach to chemical self-assembly is the formation of SAMs to make surfaces that are hydrophobic and hydrophilic such that these surface forces can be utilized to place structures at specific sites [98], as shown in Figure 17.22. The authors used microcontact printing to pattern a gold surface into a grid of hydrophobic and hydrophilic SAMs of alkanethiolates. SAMs that were CH_3 terminated were hydrophobic while those that were COOH terminated were hydrophilic. Aqueous solutions with $CuSO_4$ or KNO_3 crystals would only be attached to the hydrophilic areas, and evaporation of the fluid medium would leave the crystals within those hydrophilic regions. Assembly of crystals and particles with a size down to 150 nm on a size within a 2×2 μm^2 hydrophilic region was demonstrated.

Whitesides and coworkers have also demonstrated procedures that use capillary forces to form millimeter-scale plastic objects into aggregates and 3-D objects [99–103]. Capillary forces scale with the length of a solid–liquid interface, whereas pressure and body forces (gravity) scale with area and volume. Hence capillary forces become dominant when compared to these other forces. The reported techniques rely on coating selective surfaces of an object with lubricants such as alkanes and photocurable methacrylates to create surfaces of high interfacial energies. These surfaces are then agitated in a second liquid medium in which the lubricant is not soluble. Since the interfacial energy of the liquid medium and the

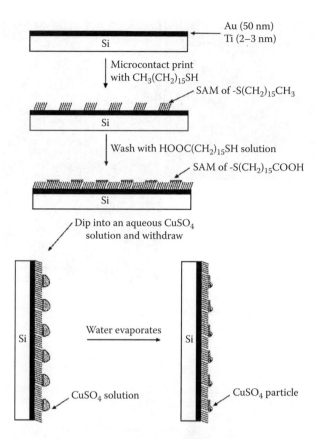

FIGURE 17.22 Process flow for formation of hydrophobic and hydrophilic surfaces using SAMs and microcontact printing. (From Qin, D., Xia, Y., Xu, B., Yang, H., Zhu, C., and Whitesides, G.M., *Adv. Mater.*, 11(17), 1433, 1999. Copyright, Wiley Interscience. Reprinted with permission.)

lubricant is high, it is energetically favorable for the surfaces to combine and align to minimize the exposed surface area. Micrometer- and millimeter-scale objects were shown to assemble using these hydrophobic effect- and shape-mediated complementarities. Using the same concept, Jacobs et al. [104] fabricated a cylindrical display where GaAs/GaAlAs LEDs (280 × 280 × 200 μm) were assembled on copper squares on a polymide substrate so that a 2-D array of LEDs was generated, as shown in Figure 17.23. The same technique was also applied to package semiconductor dies onto substrate [105,106]. The assembly of silicon microstructures on hydrophobic substrates with the use of capillary forces has also been reported by Srinivasan et al. [107,108]. A hydrophobic adhesive layer was formed on binding sites on patterned substrates. When the hydrophobic pattern on the micromachined silicon parts ($150 \times 150 \times 15$ μm^3 to $400 \times 400 \times 50$ μm^3) came in contact with the adhesive-coated substrate-binding sites, shape matching self-assembly occurred due to minimization of surface energies. Optical micromirrors for MEMS applications [108] were assembled on silicon and glass substrates, and alignment precisions of less than 0.2 μm and rotational misalignment of 0.3° were demonstrated.

One of the most exciting developments in the field of nanotechnology has been the discovery and development of carbon nanotubes. These tubes are very attractive and promising due to their unique electronic properties and can possibly be used to make nanoscale transistors and switches [109]. Some work toward site-specific assembly of individual nanotubes has been demonstrated [110]. The authors report the adsorption of these carbon nanotubes onto amino-functionalized surfaces. Trimethylsilyl SAM layers formed on an oxide substrate were patterned using AFM or e-beam lithography and a second SAM was formed with –NH2 functionality (Figure 17.24). Individual nanotubes were shown to be selectively

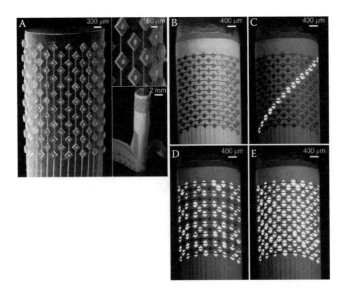

FIGURE 17.23 Scanning electron micrograph of a cylindrical display. (From Jacobs, H.O., Tao, A.R., Schwartz, A. Gracias, D.H., and Whitesides, G.M. *Science*, 296, 323–325, 2002. Reprinted with permission.)

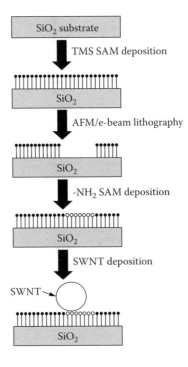

FIGURE 17.24 Chemical self-assembly of a SWNT. (From Liu, J., Casavant, J., Cox, M., Walters, D.A., Boul, P., Lu, W., Rimberg, A.J., Smith, K.A., Colbert, D.T., and Smalley, R.E., *Chem. Phys. Lett.*, 303, 125, 1999. Copyright, Elsevier Science. Reprinted with permission.)

placed between electrodes on these amino sites. These methods were expanded to implement an array of carbon nanotubes by Rao et al. [111]. Their group applied polar chemical groups such as amino ($-NH_2$) and nonpolar groups such as methyl ($-CH_3$) into a substrate so that two distinct regions were created on the substrate. After the substrate was immersed into a solution containing SWNTs, the nanotubes were attracted toward the polar regions and formed an aligned array. Another demonstration of a SWNT array was also reported by Huang et al. [112]. However, there still remains a large gap between these nanotube array structures and realization of functional circuits using these components. How the gap will be narrowed is one of the most important issues. More work is needed in order to solve these issues.

17.5.3 Electrically Mediated

A variety of electrically mediated assembly techniques can be found in the literature. Basically, these can be divided into assembly due to charged molecules only and assembly due to electrokinetic effects (e.g., electrophoresis, electroosmotic, and dielectrophoresis).

Electrostatic forces due to charged molecules can be used to provide site-specific assembly on microfabricated surfaces. The basic idea is simple and is that the object to be assembled be charged, and the site where the device is to be assembled provides the opposite charge. Electrostatic forces, along with external forces such as simple agitation, will then bring the device to the assembly sites. The force acting between the charges is simply given as $F = qE$, where q is the charge and E is the resulting electric field due to the presence of the charges. Figure 17.25 shows the basic concept of the electric field-mediated assembly. Tien et al. [113] demonstrated the use of electrostatic interactions to direct the placement and patterning of 10-μm diameter gold disks on substrates functionalized with charged molecules. The disks were produced by electrodeposition of gold in a photoresist pattern and then released using sonication in a charged thiol solution. Meanwhile, substrates with patterned (negative and positive) charged surfaces were produced by microcontact printing and attachment of thiolated molecules. Many solutions such as $HS(CH_2)_{11}NH_3^+Cl^-$, $HS(CH_2)_{11}NMe_2$, $HS(CH_2)_{11}NMe_3^+Br^-$, etc. were used to provide a positive charge, while $HS(CH_2)_{15}COOH$ and $HS(CH_2)_{11}PO_3H_2$ yielded negative charges. Patterned assemblies over 1 cm^2 areas were demonstrated. However, it was noted that in-plane ordering was still lacking presumably because the objects cannot move laterally once they are attached to a site. The required lateral mobility can come from capillary forces at the fluid–fluid interface [96,114] or by simply reducing the size of the charged pattern on the substrate and increasing the concentration of the objects in the solution above the pattern.

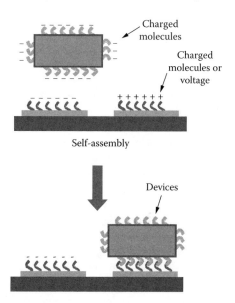

FIGURE 17.25 Conceptual schematic of charged devices assembling on oppositely charged regions on a substrate.

The charge on the devices can be provided by molecules while the charges on the substrate can also be provided by applying a voltage potential, all within a fluid medium. This is the well-known principle of electrophoresis used to separate charged molecules and macromolecules based on differences in their sizes. The fixed charges on the devices and objects are generally neutralized by the presence of counterions in the fluid, but upon application of a DC voltage, the molecule is polarized and the charged object would move to one electrode, whereas the counterions would move to the other electrode. Perhaps the most well-known example is that by Heller and coworkers [115–118], who have shown that electrophoretic placement of DNA captures strands at specific sites on biochips to realize DNA arrays. Subsequent hybridization of fluorescently labeled target probes at specific sites provides insight into the sequence of the target probes. These active microelectronic array devices allow electrophoretic fields to be used to carry out accelerated DNA hybridization reactions and to improve selectivity for single nucleotide polymorphism, short tandem repeat, and point mutation analysis [119]. The ability to generate electric fields at the microscale allows charged molecules (DNA, RNA, proteins, enzymes, antibodies, nanobeads, and even micron-scale semiconductor devices) to be electrophoretically transported to or from any microscale location on the planar surface of the device [120]. Edman et al. [121] fabricated 20-μm diameter InGaAs LEDs and demonstrated a process for releasing them from the host substrate and an electrophoretic process for assembling them on silicon substrates. No details were given on the charging of devices or the mechanisms behind the device transport. Most likely the mechanism consisted of some variant of electrokinetic transport, where movement of ions in the fluid medium would force the movement of objects along the flow contours. The process will require a higher current density as compared to a purely electrophoretic transport. The devices can generally develop a charge on their surfaces when immersed in a liquid; for example, a clean oxide surface with a native oxide can have Si–OH groups at the surface and develop a negative charge at pH 7 in water [113]. Recently, the use of silicon-on-insulator (SOI) wafers to fabricate trapezoidal-shaped silicon islands, which were 4×4 μm at the top and about 8×8 μm at the base and had a thin gold layer on one side, has been demonstrated [122,123]. The Au surface was functionalized with 4-mer DNA or a charged molecule (2-mercaptoethane sulfonic acid sodium salt) to provide negative charges on the islands [124]. The islands were released from the substrate into a fluid medium over an electrode array and then manipulated at the microscale using voltages applied at the electrodes, as schematically shown in Figure 17.26. The molecules provided negative charges on the devices, hence allowing the electrophoretic transport of these devices under electric fields to specific sites. O'Riordan et al. [125] have demonstrated programmed self-assembly using electric fields. They built a 4×4 array circular receptor with 100 μm diameter and 250 μm pitch electrode sites on a silicon substrate. A GaAs-based light-emitting diode with 50 μm diameter was transported, positioned, and localized on the selected binding site using an electric field, as shown in Figure 17.27.

A third type of effect named dielectrophoresis is worth mentioning, which can also be used to manipulate particles at the micro- and nanoscale. Dielectrophoretic forces are developed on a particle of a dielectric constant different than the medium that it is in while it is exposed to a nonuniform AC electric field. Both positive and negative dielectrophoretic forces can be produced and since the dielectric constant can be a function of the applied signal frequency, the direction of the force on an object can be changed as a function of frequency [127,128]. This effect has been used to assemble and align metallic nanowires within patterns of interdigitated electrode arrays, in an effort to develop a molecular-scale interconnect technology [129]. The self-assembly using silicon islands and resistors [130,131] has also been demonstrated. As the difference of conductivity of the silicon devices become obvious, the devices were moved toward the position having minimum electric field gradient (negative dielectrophoresis) or maximum electric field gradient (positive dielectrophoresis). Dielectrophoretic force or dielectrophoretic force in combination of 1,9-nonanedithiol was used for the implementation of a three-terminal single-crystal silicon MOS field effect transistor self-assembly [132]. The devices with 2 μm width, 15 μm length, and 1.3 μm thickness were fabricated on an SOI wafer and successfully released from an original substrate into DI water as shown in Figure 17.28. Then, the devices suspended in DI water were assembled on a right binding position using dielectrophoretic force and the force combining dielectrophoretic force

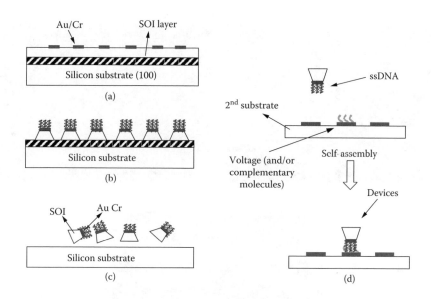

FIGURE 17.26 Process flow for active device fabrication, release, and self-assembly. (a) Define patterns of Au/Cr contacts on an SOI wafer; (b) etch the silicon down to the buried oxide and attach molecules (DNA, etc.) to the top layer; (c) release the islands (devices) from the substrate and collect/concentrate; (d) assemble the devices on another surface with voltage (or the complementary ssDNA). (Adapted from Edman, C,F., Swint, R.B., Furtner, C., Formosa, R.E., Roh, S.D., Lee, K.E., Swanson, P.D., Ackley, D.E., Coleman, J.J., and Heller, J.J., *IEEE Photonics Technol. Lett.*, 12(9), 1198, 2000. With permission.)

and molecular binding as shown in Figure 17.29. The electrical characteristics such as transfer (V_{ds} vs. I_{ds}) and output (V_{ds} vs. I_{ds}) of the assembled device were also measured and analyzed. This technique can be applied for the realization of 3-D integrated circuits, as shown in Figure 17.30.

17.5.4 Fluidics Mediated

Fluidic self-assembly (FSA) is a process for assembling devices and objects using fluidic shear forces along with other forces such as gravity [133–137] or chemical modification [138,139]. Shape-mediated FSA,

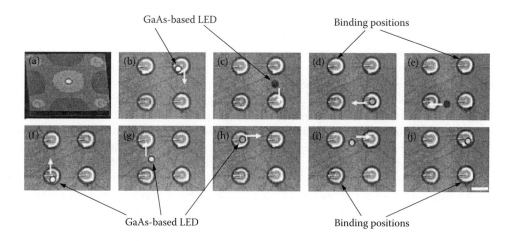

FIGURE 17.27 Microscopic images described to figure configured assembly. (From O'Riordan, A., Delaney, P., and Redmond, R., *NanoLetters*, 4, 761–765, 2004. Reprinted with permission.)

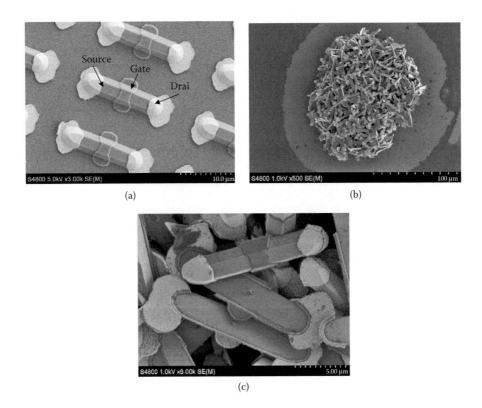

FIGURE 17.28 The field emission scanning electron microscope (FESEM) images. (a) The completed silicon MOSFETs on a BESOI wafer. (b) FESEM pictures of an aggregate of devices after drying. (c) Close up FESEM of the released device before assembly. (From Lee, S.W. and Bashir, R., *Adv. Mater.*, 17, 2671–2677, 2005. With permission.)

which uses gravity along with fluidics, starts with the fabrication of the devices to be self-assembled and target sites on substrates, which have matching shapes. The technique can also be termed "shape-mediated assembly." The objects are transported in a fluid and distributed over the target surface, which then are deposited in the holes in the substrate. Since Brownian motion and diffusion are negligible for large devices (larger than 10 μm or so), fluid motion is used to bring the device close to the assembly site. Gravity and van der Waals forces then hold the device in place. Figure 17.31 shows the basic concept of FSA, which can be applicable to different device types and can also be used for heterogeneous integration of materials.

Smith and coworkers pioneered the FSA method by demonstrating a quasimonolithic integration of GaAs LEDs on silicon substrates using this fluidic transport and shape mediation for proper orientation and placement [133,136]. LEDs were designed to be 10 × 10 μm at the base and 18 × 18 μm at the top and were assembled in trapezoidal wells etched in silicon by anisotropic etching. The authors also extended their work to demonstrate silicon device regions (30 to 1000 μm on a side) that were from silicon wafers. These silicon blocks, with prebuilt and tested transistors, then have been assembled into wells etched in plastic substrates. The resulting planer surfaces are then metalized to form interconnects at the top side. This technology is now being commercialized for producing flexible displays [137], where the silicon devices are being assembled on plastics substrates within ± 1 μm precision to form one side of a liquid crystal display. Flexible and rollable electronics can also be made possible using these approaches.

Lieber and coworkers have demonstrated the assembly of 1-D nanowires and nanostructures using a combination of fluidic alignment and surface patterning techniques [138]. They fabricated the nanowires

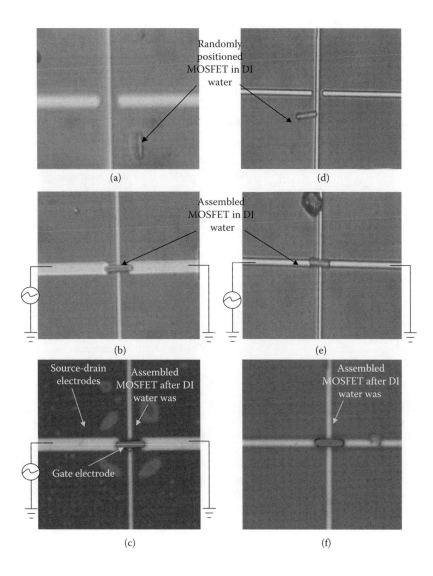

FIGURE 17.29 Nonfunctionalized electrode. (a) Randomly distributed silicon MOSFET in DI water (0.05% Tween-20) before AC signal is on. (b) Assembled silicon MOSFET in DI water after AC signal is on. (c) Assembled silicon MOSFET after DI water is completely dried, where the AC signal was on during the drying process. Functionalized electrode. (d) Randomly distributed silicon MOSFET in DI water (0.05% Tween-20) prior to application of AC signal. (e) Assembled silicon MOSFET in DI water following application of AC signal. (f) Assembled silicon MOSFET showing a good alignment after complete evaporation of the DI water, where the AC signal was off during the drying process. (From Lee, S.W. and Bashir, R., *Adv. Mater.*, 17, 2671–2677, 2005. With permission.)

of silicon, gallium phosphide, or indium phosphide using laser-assisted catalytic growth. The released nanowires were aligned along the direction of fluid flow as shown in Figure 17.32, and it was also shown that the nanowires exhibited more selectivity to the NH_2-terminated monolayers that have a partial positive charge. The authors went on to show functional electronic elements such as transistors and inverter circuits using these silicon quantum wires [139]. These results are very exciting since they demonstrate the fabrication, assembly, and synthesis of functional circuit blocks using these 1-D semiconductor wires.

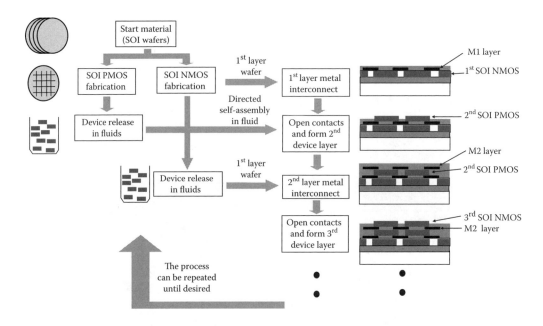

FIGURE 17.30 Proposed schematic process flow for the realization of 3-D integrated circuits using directed FSA. (From Lee, S.W. and Bashir, R., *Adv. Mater.*, 17, 2671–2677, 2005. With permission.)

17.6 Future Directions

17.6.1 Bio-inspired Active Device Assembly

A lot has been accomplished in the last 10 years toward the biologically mediated assembly of artificial nano- and microstructures. Nevertheless, a great deal remains to be done to bridge the gap between the electronic devices that will be constructed on a 20 to 50-nm scale within the next 10 years and molecules of a few nanometers or less in size. Much work has been done on assembly of passive electronic components and devices (gold clusters, metal rods, etc.). Work has been done on optical devices (quantum dots), etc. and, lately, there are more reports of assembly of carbon nanotubes and quantum wires, which can be

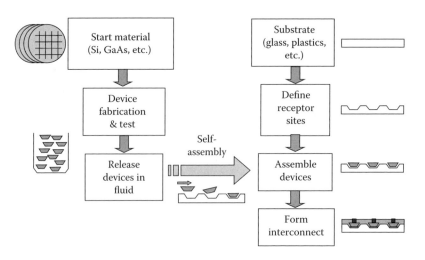

FIGURE 17.31 The process of FSA of silicon regions in plastic substrates. Adapted from www.alientechnology.com.

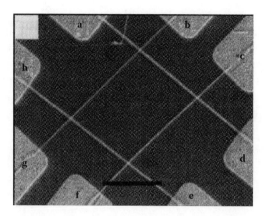

FIGURE 17.32 Assembly of silicon nanowires using fluidics and a surface treatment. (From Huang, Y. Duan, X. Wei, Q, Lieber, C.M., *Science*, 291, 630, 2001. Copyright AAAS. Reprinted with permission.)

used as active devices. The processes for self-assembly of active devices such as carbon nanotubes [109,110], silicon quantum wires [138,139], and silicon MOSFETs [132] need to be pursued more actively. As an example, the entire footprint of a current silicon CMOS transistor in production, with $W/L = 0.8/0.15$ μm, is less than 1.1 μm on a side. Can such active devices of such sizes be assembled in two and three dimensions using bio-inspired assembly using DNA, proteins, etc.? Doing so, we can potentially reduce the cost of fabrication of microelectronic systems and circuits. The active devices should include scaled-down silicon transistors, silicon nanowires and related materials, and carbon nanotubes, due to their likely potential promise (see Figure 17.32). Can silicon nanowires and or carbon nanoubes be assembled into regular arrays for memory and logic applications? Currently, these 1-D devices are either placed one at a time at desired sites by manipulation using AFM tips or grown at specific sites. If hundreds of thousands of these devices are to be used in a circuit, self-assembly using linker molecules could be an attractive way to assemble these devices. It is also important to note that for use in future scaled nanoelectronic systems will need devices as well as interconnects. This direction of research will also result in heterogenous integration of material at the micro- and nanoscale in dimensions smaller than what is already underway, as described in previous sections. For example, can GaAs/InGaAs HBTs be fabricated for RF applications and then assembled on silicon CMOS chips?

17.6.2 New Biolinkers

Two types of biolinkers that were discussed in this chapter were DNA and protein complexes. There is a need for the development of additional binding strategies for attaching linkers to specific metals (more than just Au and Pt). There is also a need for developing and identifying new complementary biolinkers (molecular glues) that have specific recognition and binding properties that are also electrically conductive and are able to carry useful amounts of current densities. Good ohmic contacts to the devices through these linkers are also needed. This way, additional contacts will not be needed and the linkers themselves could possibly provide good contacts to other devices after self-assembly. It would also be quite interesting and useful if such linkers were reversible to possibly realize reconfigurable circuits by causing physical movement of the devices or by reprogramming a link.

17.7 Conclusions

The field of nanotechnology has emerged as one of the most important areas of research for the future, and biologically and chemically mediated self-assembly has the potential to profoundly impact this field. This chapter presented the motivations and fundamentals behind these assembly concepts, with a focus on biologically mediated assembly. Biolinker fundamentals and attachment to surfaces were reviewed.

The state of the art in biologically mediated assembly of artificial nano- and microstructures was presented. Examples of chemically, electrically, and fluidics-mediated self-assembly were also presented, all with a focus on assembly of nano- and microscale objects at specific sites and locations. Work needs to continue in this area to eventually realize bottom-up fabrication to produce fully self-assembling, self-healing, and self-repairing systems that are capable of sensing, computing, and reacting to specific stimuli.

Acknowledgments

The author would like to acknowledge many valuable discussions with Prof. D. Janes, Prof. D. Bergstrom (Co-PI with the author on a NSF funded active device self-assembly project), Prof. M. Lundstrom, and Prof. S. Datta. The author would like to thank, Dr. H. Mcnally, D. Guo, and Dr. M. Pingle for results, assistance with references, and manuscript review, and the NSF, State of Indiana, and Purdue University for funding.

References

1. Samsung Electronics, Seoul Korea, www.samsung.com/PressCenter/PressRelease, Sep 12, 2005.
2. *The National Technology Roadmap for Semiconductors (NTRS)*, SIA (Semiconductor Industry Association) 1997.
3. *The International Technology Roadmap for Semiconductors (ITRS)*, SIA (Semiconductor Industry Association) 1999.
4. Alberts, B., Bray, D., Lewis, J., Raff, M., Roberts, K., and Watson, J.D., *Molecular Biology of the Cell*, 3rd edi., Garland Publishing, N.Y.
5. http://alphadna.com/, http://www.biosyn.com/, http://www.genemedsyn.com/, etc.
6. Nuzzo, R.G. and Allara, D.L., Adsorption of bifunctional organic disulfides on gold surfaces, *J. Am. Chem. Soc.*, 105, 4481, 1983.
7. Bain, C.D. and Whitesides, G.M., Modeling organic-surfaces with self-assembled monolayers, *Agnew. Chem. Int. Ed. Engl.*, 28(4), 506, 1989.
8. Bain, C.D., Troughton, E.B., Tao, Y.T., Evall, J., Whitesides, G.M., and Nuzzo, R.G., Combining spontaneous molecular assembly with microfabrication to pattern surfaces - selective binding of isonitriles to platinum microwires and characterization by electrochemistry and surface spectroscopy, *J. Am. Chem. Soc.*, 111, 321, 1989.
9. Dubois, L. and Nuzzo, R.G., Synthesis, structure, and properties of model organic-surfaces, *Ann. Rev. Phys. Chem.*, 43, 437, 1992.
10. Hickman, J.J., Laibinis, P.E., Auerbach, D.I., Zou, C., Gardner, T.J., Whitesides, G.M., and Wrighton, M.S., Toward orthogonal self-assembly of redox active molecules on Pt and Au-selective reaction of disulfide with Au and isocyanide with Pt, *Langmuir*, 8, 357, 1992.
11. Tour, J.M., Jones, L., Pearson, D.L., Lamba, J.J.S., Burgin, T.P., Whitesides, G.M., Allara, D.L., Parikh, A.N., and Atre, S.V., Self-assembled monolayers and multilayers of conjugated thiols, alpha,omega-dithiols, and thioacetyl-containing adsorbates — understanding attachments between potential molecular wires and gold surfaces, *J. Am. Chem. Society*, 117, 9529, 1995.
12. Weisbecker, C.S., Merritt, M.V., and Whitesides, G.M., Molecular self-assembly of aliphatic thiols on gold colloids, *Langmuir*, 12, 3763, 1996.
13. Nelles, G.H., Schonherr, H., Jaschke, M., Wolf, H., Schaur, M., Kuther, J., Tremel, W., Bamberg, E., Ringsdorf, H., and Butt, H.J., Two-dimensional structure of disulfides and thiols on gold(111), *Langmuir*, 14(4), 808, 1998.
14. Jung, C., Dannenberger, O., Xu, Y., Buch, M., and Grunze, M., Self-assembled monolayers from organosulfur compounds: a comparison between sulfides, disulfides, and thiols, *Langmuir*, 14(5), 1103, 1998.
15. Tarlov, J.M. and Newman, J.G., Static secondary ion mass-spectrometry of self-assembled alkanethiol monolayers on gold, *Langmuir*, 8, 1398, 1992.

16. Peterlintz, K.A., Georgiadis, R.M., Herne, T.M., and Tarlov, M.J., Observation of hybridization and dehybridization of thiol-tethered DNA with two color surface plasmon resonance, *J. Am. Chem. Soc.*, 119, 3401, 1997.

17. Herne, T.M. and Tarlov, M.J., Characterization of DNA probes immobilized on gold surfaces, *J. Am. Chem. Soc.*, 119(38), 8916, 1997.

18. Steel, A.B., Herne, T.M., and Tarlov, M.J., Electrochemical quantitation of DNA immobilized on gold, *Anal. Chem.*, 70, 4670, 1998.

19. Yang, M., Yau, H.C.M, and Chan, H.L., Adsorption kinetics and ligand-binding properties of thiol-modified double-stranded DNA on a gold surface, *Langmuir*, 14, 6121, 1998.

20. Fukuzaki, S., Urano, H., and Nagata, K.N., Adsorption of bovine serum albumin onto metal oxide surfaces, *J. Ferment. Bioeng.*, 81(2), 163, 1996.

21. Roscoe, S.G. and Fuller, K.L., Interfacial behavior of globular proteins at platinum electrode, *J. Colloid Interfacial Sci.*, 152(2), 429, 1992.

22. Britland, S., Arnaud, E.P., Clark, P., McGinn, B., Connolly, P., and Moores, G., Micropatterning proteins and synthetic peptides on solid supports: a novel application for microelectronic fabrication technology, *Biotechnol. Prog.* 8, 155, 1992.

23. Mooney, J.F., Hunt, A.J., McIntosh, J.R., Liberko, C.A., Walba, D.M., Rogers, C.T., Patterning of functional antibodies and other proteins by photolithography of silane monolayers, *Proc. Natl. Acad. Sci.*, 93(22), 12287, 1996.

24. Nicolau, D.V., Taguchi, T., Taniguchi, H., and Yoshikawa, S., Micron-sized protein patterning on diazonaphthoquinone/novolak thin polymeric films, *Langmuir*, 14(7), 1927, 1998.

25. Williams, R.A. and Blanch, H.W., Covalent immobilization of protein monolayers for biosensors applications, *Biosens. Bioelectron.*, 9, 159, 1994.

26. Lahiri, J., Isaacs, L., Tien, J., and Whitesides, G.M., A strategy for the generation of surfaces presenting ligands for studies of binding based on an active ester as a common reactive intermediate: a surface plasmon resonance study, *Anal. Chem.*, 71, 777, 1999.

27. Bashir, R., Gomez, R., Sarikaya, A., Ladisch, M., Sturgis, J., and Robinson, J.P., Adsorption of avidin on micro-fabricated surfaces for protein biochip applications, *Biotech. Bioeng.*, 73(4), 324, 2001.

28. Lahiri, J., Ostuni, E., and Whitesides, G.M., Patterning ligands on reactive SAMs by microcontact printing, *Langmuir*, 15, 2055, 1999.

29. MacBeath, G. and Schreiber, S.L., Printing proteins as microarrays for high-throughput function determination, *Science*, 289, 1760, 2000.

30. Whaley, S.R., English, D.S., Hu, E.L., Barbara, P.F., and Belcher, A.M., Selection of peptides with semi-conductor binding specificity for directed nanocrystal assembly, *Nature*, 405(8), 665, 2000.

31. Madhukar, A., Xie, Q., Cheng, P., and Knocker, A., Nature of strained INAS 3-dimensional island formation and distribution on GaAs (100), *Appl. Phys. Lett.*, 64(20), 2727, 1994.

32. Moison, J.M., Houzay, F., Barthe, F. and Leprince, L., Self-organized growth of regular nanometer-scale INAS dots on GaAs, *Appl. Phys. Lett.*, 64(2), 197, 1994.

33. Kamins, T.I., Carr, E.C., Williams, R.S., and Rosner, S.J., Deposition of three-dimensional GE islands on Si(001) by chemical vapor deposition at atmospheric and reduced pressures, *J. Appl. Phys.*, 81(1), 211, 1997.

34. Bashir, R., Kabir, A.E., and Chao, K., Formation of self-assembled si1-xgex islands using reduced pressure chemical vapor deposition and subsequent thermal annealing of thin germanium-rich films, *Appl. Surf. Sci.*, 152, 99, 1999.

35. Seeman, N.C., Nucleic acid junctions and lattices, *J. Theor. Biol.*, 99, 237, 1982.

36. Seeman, N.C., The use of branched DNA for nanoscale fabrication, *Nanotechnology*, 149, 1991.

37. Seeman, N.C., Zhang, Y., and Chen, J., DNA nanoconstructions, *J. Vac. Sci. Technol. A*, 12, 1895, 1994.

38. Winfree, E., Liu, F., Wenzler, L., and Seeman, N.C., Design and self-assembly of two-dimensional DNA crystals, *Nature*, 394, 539, 1998.

39. Seeman, N.C., DNA nanotechnology: novel DNA constructions, *Ann. Rev. Biophys. Biomol. Struc.*, 27, 225, 1998.

40. Rothemund, P.W.K., Ekani-Nkodo, A., Papadakis, N., Kumar, A., Fygenson, D.K., and Winfree, E., Design and characterization of programmable DNA nanotubes, *J. Am. Chem. Soc.*, 126, 16344, 2004.

41. Ekani-Nkodo, A., Papadakis, N., Kumar, A., and Fygenson, D.K., Joining and scission in the self-assembly of nanotubes from DNA tiles, *Phys. Rev. Lett.*, 93, 268301–268311, 2004.

42. Mirkin, C.A., Letsinger, R.L., Mucic, R.C., and Storhoff, J.J., A DNA-based method for rationally assembling nanoparticles into macroscopic materials, *Nature*, 382, 607, 1996.

43. Alivisatos, A.P., Johnsson, K.P., Peng, X., Wilson, T.E., Loweth, C.J., Bruchez, M.P., and Schultz, P.G., Organization of 'nanocrystal molecules' using DNA, *Nature*, 382, 609, 1996.

44. Elghanian, R., Storhoff, J.J., Mucic, R.C., Letsinger, R.L., and Mirkin, C.A., Selective colorimetric detection of polynucleotides based on the distance-dependent optical properties of gold nanoparticles, *Science*, 277, 1078, 1997.

45. Mucic, R.C., Storhoff, J.J., Mirkin, C.A., and Letsinger, R.L., DNA-directed synthesis of binary nanoparticle network materials, *J. Am. Chem. Soc.*, 120, 12674, 1998.

46. Loweth, C.J., Caldwell, W.B., Peng, X., Alivisatos, A.P., and Schultz, P.G., DNA-based assembly of gold nanocrystals, *Agnew. Chem. Int. Ed.*, 38(12), 1808, 1999.

47. Storhoff, J.J., Elghanian, R., Mucic, R.C., Mirkin, C.A., and Letsinger, R.L., One pot colorimetric differentiation of polnucleotides with single base imperfections using gold nanoparticle probes, *J. Am. Chem. Soc.*, 120, 1959, 1998.

48. Taton, T.A., Mirkin, C.A., and Letsinger, R.L. Scanometric DNA array detection with nanoparticle probes, *Science*, 289, 1757, 2000.

49. Andres, R.P., Datta, S., Janes, D.B., Kubiak, C.P. and Reifenberger, R., A good review of self-assembled Au clusters and S-H attachment to Au, in: *The Handbook of Nanostructured Materials and Nano-technology*, Academic Press, 1998.

50. Niemeyer, C.M., Progress in engineering up nanotechnology devices utilizing DNA as a construction material, *Appl. Phys. A.*, 69, 119, 1999.

51. Storhoff, J.J. and Mirkin, C.A., Programmed materials synthesis with DNA, *Chem. Rev.*, 99, 1849, 1999.

52. Csaki, A., Moller, R., Straube, W., Kohler, J.M., and Fritzsche, W., DNA monolayer on gold substrates characterized by nanoparticle labeling and scanning force microscopy, *Nucleic Acid Res.*, 29(16), 81, 2001.

53. Reichert, J., Csaki, A., Kohler, J.M., and Fritzsche, W., Chip-based optical detection of DNA hybridization by means of nanobead labeling, *Anal. Chem.*, 72, 6025, 2000.

54. Niemeyer, C.M., Ceyhan, B., Gao, S., Chi, L., Peschel, S., and Simon, U., Site-selective immobilization of gold nanoparticles functionalized with DNA oligomers, *Coll. Polym. Sci.*, 279, 68, 2001.

55. Deng, Z., Tian, Y., Lee, S.H., Ribbe, A.E., and Mao, C., DNA-encoded self-assembly of gold nanoparticles into one-dimensional arrays, *Angew. Chem. Int. Ed.*, 44, 3582–3585, 2005.

56. Sharma, J., Chhabra, R., Liu, Y., Ke, Y., and Yan H., DNA-templated self-assembly of two-dimensional and periodical gold nanoparticle arrays, *Angew. Chem. Int. Ed.* 45, 730–735, 2006.

57. Taton, T.A., Mucic, R.C., Mirkin, C.A., and Letsinger, R.L., The DNA-mediated formation of supramolecular mono- and multilayered nanoparticle structures, *J. Am. Chem. Soc.*, 122, 6305, 2000.

58. Huang, S., Martin, B., Dermody, D., Mallouk, T.E., Jackson, T.N., and Mayer, T.S., *41st Electronic Materials Conference Digest*, 41, 1999.

59. Martin, B.R., Dermody, D.J., Reiss, B.D., Fang, M., Lyon, L.A., Natan, M.J., and Mallouk, T.E., Orthogonal self-assembly on colloidal gold-platinum nanorods. *Adv. Mater.*, 11(12), 1021, 1999.

60. Mayer, T.S., Jackson, T.N., Natan, M.J., and Mallouk, T.E., *1999 Materials Research Society Fall Meeting Digest*, 157, 1999.

61. Masuda, H. and Fukuda, K., Ordered metal nanohole arrays made by a 2-step replication of honeycomb structures of anodic alumina, *Science*, 268, 1466, 1995.

62. Li, A.P., Muller, F., Birner, A., Nielsch, K., and Gosele, U., Hexagonal pore arrays with a 50-420 nm interpore distance formed by self-organization in anodic alumina, *J. Appl. Phys.*, 84(11), 6023, 1998.

63. Mbindyo, J.K.N., Reiss, B.D., Martin, B.J., Keating, C.D., Natan, M.J., and Mallouk, T.E., DNA-directed assembly of gold nanowires on complementary surfaces, *Adv. Mater.*, 13(4), 249, 2001.

64. Braun, E., Eichen, Y., Sivan, U., and Yoseph, G.J., DNA-templated assembly and electrode attachment of a conducting silver wire, *Nature*, 391(19), 775, 1998.

65. Eichen, Y., Braun, E., Sivan, U., and Yoseph, G.B., Self-assembly of nanoelectronic components and circuits using biological templates, *Acta Polym.*, 49, 663, 1998.

66. Braun, E., Sivan, U., Eichen, Y., and Yoseph, G.B., Self-assembly of nanometer scale electronics by biotechnology, *24ᵗʰ International Conf. on the Physics of Semiconductors*, 269, 1998.

67. Keren, K., Krueger, M., Gilad, R., Ben-Yoseph, G., Sivan, U., and Braun, E., Sequence-specific molecular lithography on single DNA molecules, *Science*, 297, 72–75, 2002.

68. Yan, H., Park, S.H., Finkelstein, G., Relif, J.H., and LaBean, T.H., DNA-templated self-assembly of protein arrays highly conductive nanowires, *Science*, 301, 1882–1884, 2003.

69. Park, S.H., Yan, H., Relif, J.H., Labean, T.H., and Finkelstein, G., Electronic nanostrutures templated on self-assembled DNA scaffolds, *Nanotechnology*, 15, S525–S527, 2004.

70. Yan, H., Park, S.H., Finkelstein, G., Relif, J.H., and LaBean, T.H., Material and methods, http://www.sciencemag.org/cgi/content/full/301/5641/1882/DC1.

71. Tanaka, K., Tengeiji, S., Kato, T., Toyama, N., and Shionoya, M., A discrete self-assembled metal array in artificial DNA, *Science*, 299, 1212, 2003.

72. Gu, J., Tanaka, S., Otsuka, Y., Tabata, H., and Kawai, T., Self-assembled dye-DNA network and its photoinduced electrical conductivity, *Appl. Phys. Lett.*, 80, 688, 2002.

73. Won, J., Chae, S.K., Kim, J.H., Park, H.H., Kang, Y.S., and Kim, H.S., Self-assembled DNA composite membranes, *J. Memb. Sci.*, 245, 113, 2005.

74. Hazani, M., Hennrich, F., Kappes, M., Naaman, R., Peled, D., Sidorov, V., and Shvarts, D., DNA-mediated self-assembly of carbon nanotube-based electronic devices, *Chem. Phys. Lett.*, 391, 389–392, 2004.

75. Li, S., He, P., Dong, J., Guo, Z., and Dai, L., DNA-directed self-assembling of carbon nanotubes, *J. Am. Chem. Soc.*, 127, 14–15, 2004.

76. Keren, K., Bermen, R.S., Buchstab, E., Sivan, U., and Braun, E., DNA-templated carbon nanotube field-effect transistor, *Science*, 302, 1380–1382, 2003.

77. Ackley, D.E., Heller, M.J., and Edman, C.F, DNA technology for optoelectronics, *Proc. - Lasers and Electro-Optics Society, Annual Meeting-LEOS*, 1, 85, 1998.

78. Alivasatos, A.P., Perspectives on the physical chemistry of semiconductor nanocrystals, *J. Phys. Chem.*, 10, 13226, 1996.

79. Klein, D.L., Roth, R., Kim, A.K.L., Alivisatos, A.P., and McEuen, P.L., A single-electron transistor made from a cadmium selenide nanocrystal, *Nature*, 699, 1997.

80. Mitchell, G.P., Mirkin, C.A., and Letsinger, R.L., Programmed assembly of DNA functionalized quantum dots, *J. Am. Chem. Soc.*, 121, 8122, 1999.

81. Bruchez, M., Moronne, M., Gin, P., Weiss, S., and Alivisatos, A.P., Semiconductor nanocrystals as fluorescent biological labels, *Science*, 281, 2013, 1998.

82. Chan, W.C.W. and Nie, S., Quantum dot bioconjugates for ultrasensitive nonisotopic detection, *Science*, 281, 2016, 1998.

83. Oldenburg, S.J., Averitt, R.D., Westcott, S.L., Halas, N.J., Nanoengineering of optical resonances, *Chem. Phys. Lett.*, 288 (2–4), 243–277, 1998, .

84. McNally, H., Pingle, M., Lee, S.W., Guo, D., Bergstrom, D., Bashir, R., Self-assembly of micro and nano-scale particles using bio-inspired events, *Appl. Surf. Sci.*, 214, 109–119, 2003.

85. Edelstein, R.L., Tamanaha, C.R., Sheehan, P.E., Miller, M.M., Baselt, D.R., Whitman L.J., and Colton, R.J., BARC biosensor applied to the detection of biological warfare agents, *Biosens. Bioelectron.*, 14 (10), 2000.

86. Sarikaya, M., Tamerler, C., Jen, A.K. –Y., Schulten, K., and Baneyx, F., Molecular biomimetics: nanotechnology through biology, *Nature Mater.*, 2, 577–585, 2003.

87. Sarikaya, M., Tamerler, C., Schwartz, D.T., and Baneyx, F., Materials assembly formation using engineered polypeptides, *Annu. Rev. Mater. Res.*, 34, 373–408, 2004.

88. Brown, S., Metal-recognition by repeating polypeptides, *Nature Biotechnol.*, 15, 269–272, 1997.
89. Naik, R.R., Stringer, S.J., Agarwal, G., Johns, S.E., and Stone, M.O., Biomimetic synthesis and patterning of silver nanoparticles, *Nature Mater.*, 1, 169–172, 2002.
90. Kiargaard, K., Sorensen, J.K., Schembri, M.A., and Klemm, P., Sequestration of zinc oxide by fimbrial designer chelators, *App. Env. Microbiol.*, 66, 10–14, 2002.
91. Nygaard, S., Wendelbo, R., and Brown, S., Surface-specific zeolite-binding proteins, *Adv. Mater.*, 24, 1853–1856, 2002.
92. Whaley, S.R., English, D.S., Hu, E.L., Barabara, P.H., and Belcher, A.M., Selection of peptides with semiconducting binding specificity for directed nanocrystal assembly, *Nature*, 2, 1–6, 2002.
93. Lee, S.W., Mao, C., Flynn, C.E., and Belcher, A.M., Ordering of quantum dots using genetically engineered viruses, *Science*, 296, 892–895, 2002.
94. Gaskin, D.J.H., Starck, K., and Vulfson, E.N., Identification of inorganic crystal specific sequences using phase display combinatorial library of short peptides: a feasibility study, *Biotechnol. Lett.*, 22, 1211–1216, 2000.
95. Scembri, M.A., Kjaergaard, K., Klemm, P., Bioaccumulation of heavy metals by fimbrial designer adhesions, *FEMS Microbial. Lett.*, 170, 363–371, 1999.
96. Andres, R.P., Bielefeld, J.D., Henderseon, J.I., Janes, D.B., Kolangunta, V.R., Kubiak, C.P., Mahoney, W.J., and Osifchin, R.G., Self-assembly of a two-dimensional superlattice of molecularly linked metal clusters, *Science*, 273, 1690–1693, 1996.
97. Liu, J., Lee, T., Janes, D.B., Walsh, B.L., Melloch, M.R., Woodall, J.M., Riefenberger, R., and Andres, R.P., Guided self-assembly of Au nanocluster arrays electronically coupled to semiconductor device layers, *Appl. Phys. Lett.*, 77(3), 373–375, 2000.
98. Qin, D., Xia, Y., Xu, B., Yang, H., Zhu, C., and Whitesides, G.M., Fabrication of ordered two-dimensional arrays of micro- and nanoparticles using patterned self-assembled monolayers as templates, *Adv. Mater.*, 11(17), 1433, 1999.
99. Terfort, A., Bowden, N., and Whitesides, G.M., Three-dimensional self-assembly of millimeter scale objects, *Nature*, 386, 162, 1997.
100. Tien, J., Breen, T.L., and Whitesides, G.M., Crystallization of millimeter scale objects using capillary forces, *J. Am. Chem. Soc.*, 120, 12670, 1998.
101. Breen, T.L., Tien, J., Oliver, S.R.J., Hadzic, T., and Whitesides, G.M., Design and self-assembly of open regular 3-D mesostructures, *Science*, 284, 948, 1999.
102. Terfort, A., and Whitesides, G.M., Self-assembly of an operating electrical circuit based on shape complementarity and the hydrophobic effect, *Adv. Mat.*, 10(6), 470, 1998.
103. Gracias, D.H., Tien, J., Breen, T.L., Hsu, C., and Whitesides G.M., Forming electrical networks in three dimensions by self-assembly, *Science*, 289, 1170–1172, 2000.
104. Jacobs, H.O., Tao, A.R., Schwartz, A., Gracias, D.H., and Whitesides, G.M. Fabrication of a cylindrical display by patterned assembly, *Science*, 296, 323–325, 2002.
105. Zheng, W. and Jacobs, H.O., Shape-and-solder-directed self-assembly to package semiconductor device segments, *Appl. Phys. Lett.*, 85, 3635–3637, 2004.
106. Zheng, W., Buhlmann, P., and Jacobs, H.O., Sequential shape-and-solder-directed self-assembly of functional microsystems, *Proc. Natl. Acad. Sci. USA*, 101, 12814–12817, 2004.
107. Srinivasan, U., Liepmann, D., and Howe, R.T., Microstructure to substrate self-assembly using capillary forces, *J. Microelectromechanic. Syst.*, 10(1), 17, 2001.
108. Srinivasan, U., Helmbrecht, M.A., Rembe, C., Muller, R.S., and Howe, R.T., Fluidic self-assembly of micromirrors onto microactuators using capillary forces, *IEEE J. Select. Topics Quantum Electron.*, 8, 4–11, 2002.
109. Derycke, V., Martel, R., Appenzeller, J., and Avouris, Ph. Carbon nanotube inter- and intramolecular logic gates, *NanoLetters*, 1(9), 453, 2001.
110. Liu, J., Casavant, J., Cox, M., Walters, D.A., Boul, P., Lu, W., Rimberg, A.J., Smith, K.A., Colbert, D.T., and Smalley, R.E., Controlled deposition of individual single walled carbon nanotubes on chemically functionalized templates, *Chem. Phys. Lett.*, 303, 125, 1999.

111. Rao. S.G., Huang, L., Setyawan, W. Hong, S., Large-scale assembly of carbon nanotubes, *Nature*, 425, 36–37, 2003.
112. Huang L., Cui, X., Dukovic, G., and O'Brien, S.P., Self-organizing high-density single-walled carbon nanotube arrays form surfactant suspensions, *Nanotechnology*, 15, 1450–1544, 2004.
113. Tien, J., Terfort, A., and Whitesides, G.M., Microfabrication through electrostatic self-assembly, *Langmuir*, 13, 5349, 1997.
114. Bowden, N., Terfort, A., Carbeck, J., and Whitesides, G.M., Self-assembly of mesoscale objects into ordered two-dimensional arrays, *Science*, 276, 233, 1997.
115. Heller, M.J., An active microelectronics device for multiplex DNA analysis, *IEEE Eng. Med. Biol. Mag.*, 15(2), 100, 1996.
116. Sosnowski, R.G., Tu, E., Butler, W.F., O'Connell, J.P., and Heller, J.J., Rapid determination of single base mismatch mutations in DNA hybrids by direct electric field control, *Proc. of the National Academy of Sciences of the United States of America.*, 94(4), 1119, 1997.
117. Huang, Y., Ewalt, K.L., Tirado, M., Haigis, R., Forster, A., Ackley, D., Heller, M.J., O'Connell, J.P., and Krihak, M., Electric manipulation of bioparticles and macromolecules on microfabricated electrodes, *Anal. Chem.*, 73(7), 1549, 2001.
118. www.nanogen.com.
119. Heller, M.J., Forster, A.H., Tu, E., Active microelectronic chip devices which utilize controlled electrophoretic fields for multiplex DNA hybridization and other genomic applications, *Electrophoresis*, 21(1), 157, 2000.
120. Fan, C., Shih, D.W., Hansen, M.W., Hartmann, D., Van Blerkom, D., Esener, S.C., and Heller, M., Heterogeneous integration of optoelectronic components, *SPIE-Int. Soc. Opt. Eng. Proceedings of SPIE — the International Society for Optical Engineering, USA*, 3290, 2, 1997.
121. Edman, C.F., Swint, R.B., Furtner, C., Formosa, R.E., Roh, S.D., Lee, K.E., Swanson, P.D., Ackley, D.E., Coleman, J.J., and Heller, J.J., Electric field directed assembly of an InGaAs LED onto silicon circuitry, *IEEE Photonics Technol. Lett.*, 12(9), 1198, 2000.
122. Bashir, R., Lee, S.W., Guo, D., Pingle, M., Bergstrom, D., McNally, H.A., and Janes, D., *Proc. of the MRS Fall Meeting*, Boston, MA., 2000.
123. Bashir, R., DNA-mediated artificial nanobiostructures, *Superlattice Microstruct.*, 29, 1, 2001.
124. Lee, S., McNally, H., Guo, D., Pingle, M., Bergstrom, D., and Bashir, R., Electric-field-mediated assembly of silicon islands coated with charged molecules, *Langmuir*, 18, 3383–3386, 2002.
125. O'Riordan, A., Delaney, P., and Redmond, R., Field configured assembly: programmed manipulation and self-assembly at the mesocale, *NanoLetters*, 4, 761–765, 2004.
126. Wang, X., Huang, Y., Gascoyne, P.R.C., and Becker, F.F., Dielectrophoretic manipulation of particles, *IEEE Trans. Industry Appl.*, 33(3), 660, 1997.
127. Pethig, R. and Markx, G.H., Application of dielectrophoresis in biotechnology, *TIBTECH*, (15), 426, 1997.
128. Ramos, A., Morgan, H., Green, N.G., and Castellanos, A., AC electrokinetics: a review of forces in microelectrode structures, *J. Phys. D: Appl. Phys.*, 31, 2338, 1998.
129. Smith, P., Nordquist, C.D., Jackson, T.N., Mayer, T.S., Martin, B.R., Mbindyo, J., and Mallouk, T.E., Electric field assisted assembly and alignment of metallic nanowires, *App. Phys. Lett.*, 77(9), 1399, 2000.
130. Lee, S.W., McNally, H.A., and Bashir, R., Electric Field and Charged Molecules Mediated Self-Assembly for Electronic Devices, *Materials Research Society Symposium-Proceedings*, 735, 49–53, 2002.
131. Lee, S.W. and Bashir, R., Dielectrophoresis and electrohydrodynamics mediated fluidic assembly of silicon resistors, *Appl. Phys. Lett.*, 83, 3833–3855, 2003.
132. Lee, S.W. and Bashir, R., Dielectrophoresis and chemically mediated directed self-assembly of micron scale 3-terminal MOSFETs, *Adv. Mater.*, 17, 2671–2677, 2005.
133. Yeh, H.J. and Smith, J.S., Fluidic self-assembly for the integration of GaAs light-emitting diodes on Si substrates, *IEEE Photonics Technol. Lett.*, 6(6), 706, 1994.

134. Tu, J.K., Talghader, J.J., Hadley, M.A., and Smith, J.S., Fluidic self-assembly of InGaAs vertical cavity surface emitting lasers onto silicon, *Electron. Lett.*, 31, 1448, 1995.

135. Talghader, J.J., Tu, J.K., and Smith, J.S., Integration of fluidically self-assembled optoelectronic devices using a silicon based process, *IEEE Photonics Technol. Lett.*, 7, 1321, 1995.

136. Smith, J.S., High density, low parasitic direct integration by fluidic self-assembly (FSA), *Proceedings of the 2000 IEDM*, 201, 2000.

137. *Alien Technology*, Morgan Hill, CA. www.alientechnology.com.

138. Huang, Y., Duan, X., Wei, Q., Lieber, C.M., Directed assembly of on-dimensional nanostructures into functional networks, *Science*, 291, 630, 2001.

139. Cui, Y. and Lieber, C.M., Functional nanoscale electronic devices assembled using silicon nanowire building blocks, *Science*, 291, 851, 2001.

18

Nanostructural Architectures from Molecular Building Blocks

18.1 Introduction 18-1
18.2 Bonding and Connectivity 18-3
Covalent Bonding • Coordination Complexes • Dative Bonds
• π-Interactions • Hydrogen Bonds
18.3 Molecular Building Block Approaches 18-13
Supramolecular Chemistry • Covalent Architectures
and the Molecular Tinkertoy Approach • Transition Metals
and Coordination Complexes • Biomimetic Structures
• Dendrimers
References 18-60

Damian G. Allis
James T. Spencer
Syracuse University

18.1 Introduction

The concept of a *molecular building block* (MBB) has been used prominently in describing a particular application of small molecules in the design of macromolecules, such as biomolecules, supramolecular structures, molecular crystal lattices, and some forms of polymeric materials. It is also common to refer to MBBs as "molecular subunits, modular building blocks,"[1,2] or synthons, which have been defined as "structural units within supermolecules which can be formed and/or assembled by known or conceivable synthetic operations involving intermolecular interactions."[3] MBBs are, therefore, the structural intermediates between atoms, the most basic of all building units, and macromolecules or extended arrays, of which the MBBs are the common structural element. While many MBB approaches are not directed toward the design of nanostructures or nanoscale materials, all share the same design considerations and are consistent with the criteria used to distinguish the MBB approaches considered here from other nanoscale fabrication techniques.

The fabrication of any structure or material from building blocks requires that the design strategy meet specific criteria. First, relying on a building block as the basis of a fabrication process indicates that this starting material is not the smallest possible component from which the manufacturing process can proceed, but it is itself pre-assembled from more fundamental materials for the purpose of simplifying the building process. It is assumed that the subunit, as a prefabricated structure, has been engineered with an important function in the assembly process of a larger, more complex structure. Second, it is assumed that a means to subunit interconnectivity has been considered in the design process. The method of connectivity between subunits may be either intrinsic to the subunit, such as a direct bonding connection between them, or available externally, such as a stabilizing electrostatic force between subunits. Third, it is assumed that the subunit is capable of being positioned correctly and precisely in the fabrication process. Fourth, and perhaps most important from a design perspective, is that the subunit provides an intermediate degree of control in the properties of the larger structure. A fabrication process based upon the manipulation of designed subunits may not provide the ultimate in stability, customizability, or structural detail when compared with the design of a system from the most basic materials, but it certainly offers enough control and flexibility for useful applications.

The defining feature of the MBB approach is the use of a molecular subunit that has incorporated within its covalent framework the means for a directed connectivity between subunits. As the MBB is itself a molecule, its synthesis can be considered among the preparative steps in the overall fabrication process and not necessarily an integral part of the actual supramolecular assembly. If "supermolecules are to molecules and the intermolecular bond what molecules are to atoms and the covalent bond,"[4] then the individual molecule forms the fundamental component in the design of MBB-based nanostructures. The starting point for the final product is the MBB, and the means to assembling the final product is through manipulation of the MBB. The assembly of a nanostructure can be predicted based upon the covalent framework of the MBB and its assembly-forming features. Because this intermolecular connectivity is an integral part of the design process, the means for controlling subunit–subunit interactions needs to be incorporated early in the design of the nanostructure. The self-assembly or self-directing interactions between subunits are based upon the properties of the MBB. The ability to customize the stability and functionality of the resulting materials is, therefore, based upon MBB modification.

The merits and limitations of building block approaches transcend scale. In all cases, the selection of suitable building materials is dictated by their ability to fit together in a precise and controllable manner. Limitations to a particular design or application are imposed partly by the properties of the subunit and partly by the design itself. While all building block designs suffer from one or more limitations, designs can often be successfully employed for a specific application or in a specific environment. For instance, bricks are ideal building materials for the construction of permanent structures, blocks of ice are appropriate for use in below-freezing conditions, and canvas is ideal for structures that require mobility. One would not select ice as a building material in temperate climates, canvas for arctic conditions, or brickwork for temporary residences. Given a set of environmental conditions and the properties of available materials, certain combinations will invariably make more sense than others. In nanostructure design, the important concerns often include solubility, thermal stability, means to assembly, defect tolerance, error correction capabilities, functionality, and chemical reactivity. Chemical environments and ambient conditions limit the feasibility of certain nanostructures just as they limit the choice of molecular subunits. These same issues are key to synthetic chemistry, where factors such as temperature, solvent, reaction duration, and choice of chemical functionalities will always play key roles in the design of chemical pathways and molecular fabrication processes.

All MBB approaches benefit from the ability to accurately predict intermolecular interactions from conceptual and theoretical treatments. Additionally, a vast synthetic background exists from which to make and modify subunits. Experimental precedent for the basic preparative methodology in a number of naturally occurring and man-made systems form a firm foundation for MBB pathways. Not only are the means to nanostructure fabrication facilitated through theoretical investigations, cognizant design strategies, and even Edisonian efforts, but many examples of macromolecular formation exist currently that provide the means for understanding how molecules can be used to construct supermolecular arrays. Concurrent with the design of new nanostructures from MBB approaches is the continued growth of the field of supramolecular chemistry and an enhanced understanding of molecular phenomena "beyond the molecule."[5]

The emphasis on design in molecular nanotechnology from MBBs connotes a certain deliberation in the choice of materials and the means to assembly. It is, therefore, important to stress efforts to engineer macromolecular assemblies from known molecular systems. This chapter begins with a discussion of the chemical and electrostatic interactions important in macromolecular formation. The discussion of these interactions as applied to nanostructure formation begins with two limiting cases in MBB design, covalent and electrostatic connectivity. With the formal groundwork of connectivity and some useful boundaries established to focus the discussion, a few important areas of MBB-based nanostructure formation are presented to demonstrate the application of the approach and related issues. This chapter is not meant to be rigorously complete, but instead provides a broad overview of current techniques involving the use of molecules as building components in larger systems.

18.2 Bonding and Connectivity

A structure is of limited value for an application without a means of maintaining its strength and functional integrity over the duration of its anticipated lifetime. At the macroscale, stabilization may come in the form of interlocking parts, mechanical or adhesive fixtures, fusing or melting at connection points between materials, or, in much larger structures, gravity. At the nanoscale, the role of gravity becomes unimportant in the formation of supramolecular assemblies,[6] and nearly all stability comes from electronic interactions. These interactions take forms ranging from strong covalent bonds to weak intermolecular (noncovalent) interactions. All molecular-based nanostructures incorporate various combinations of these interactions to maintain shape and impart function. It is therefore important to understand the range and form of the stabilization energies associated with these interactions and their relationship to the structures that incorporate them.

18.2.1 Covalent Bonding

Of singular importance in synthetic chemistry is the manipulation of the covalent bond. The design of any nanoscale architecture from simpler molecules must first address the design of the covalent framework of the MBB itself. The means to any macromolecular stabilization is a result of the inclusion of chemical functionalities onto this stable framework. The role of covalent bonds in the MBB approach is then twofold. First, these bonds are required within the subunit to provide the structural integrity necessary for the prediction and synthesis of nanostructures from MBB components. Second, covalent bonds may be employed as one of the means for fastening MBBs together into larger structures.

Covalent bonds are formed by the sharing of pairs of electrons between atoms.[7] The most familiar examples of covalent bonding are the connections between carbons in organic molecules. The importance of organic chemistry as a field underscores our desire to understand and modify the covalent framework of carbon-containing molecules for many important applications. Typical covalent bond energies range from 100 to 500 kJ/mol.[8] In the case of multiple bonds between atoms, the total energy may exceed 1000 kJ/mol. While this is a very large range of energies, even the covalent bonds at the low end of the spectrum are rather strong interactions, especially when compared with the noncovalent energies frequently responsible for macromolecular stabilization (*vide infra*). It is because of these large covalent bond strengths that the subunits involved in MBB approaches provide significant internal structural stability and predictability.

Covalent bonding includes a variety of useful motifs in the structural customization of a subunit. The strong σ-bonds, in which a pair of electrons is shared directly along the interatomic axis of two atoms, provide for low-energy rotation in straight-chain molecules and low-energy twisting in closed-ring systems (Figures 18.1A and 18.1B). In organic molecules, σ-bonding plays the initial role of defining the connectivity and general shape of the structure. The formation of π-bonds in molecules involves electrons in atomic orbitals that are not involved in the σ-bonding framework. In such instances, main group atoms involved in the molecular backbone are either sp^2- or sp-hybridized, leaving either one or two p-orbitals through which π-bonding can occur (Figure 18.1C). The π-bonded portion of the molecule is then held planar to maximize p-orbital overlap between atoms. These π-bonds may be delocalized over the entire length of the available π-orbital framework, making them well suited to molecular electronic applications that require both structural stability and electron mobility.[9] Structurally, the π-bonds remove the low-energy rotational freedom from the underlying σ-bond framework. In cases where two π-bonds

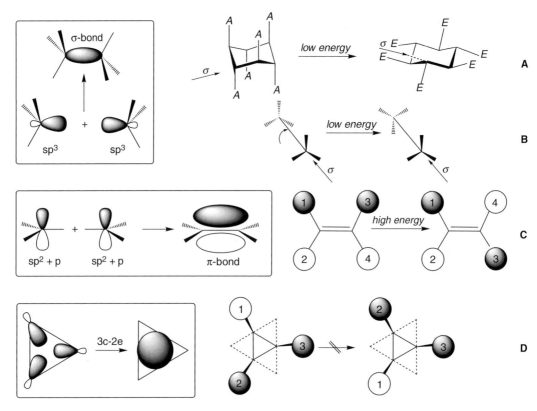

FIGURE 18.1 Rotation in covalent bonds. (A) Ring twisting about a σ-bond (indicated by arrow) with a change in orientation of one set of the substituents from axial (A, left) to equatorial (E, right). (B) Free rotation about the σ-bonds in linear chains. (C) High-energy bond breaking is required for rotation about a π-bond. (D) Reorientation of substituents in 3c-2e bonds is highly restricted (extension of cluster framework indicated by dashed lines).

are formed between two adjacent atoms, the resulting π-electron density around the σ-bond is cylindrical, and the molecular fragment behaves as a rigid linear rod.[10] A third motif involves what is often referred to as either *electron deficient* or *three-center-two-electron bonding*. Structural flexibility is fully restricted in three-center-two-electron (3c-2e) bonds. These bonds, observed in main group polyhedra and in many metal clusters, involve three adjacent atoms sharing a single pair of electrons (Figure 18.1D). Molecules employing this mode of bonding are generally three-dimensional, meaning that overall structural flexibility within the molecule is lost due to the cage-like interconnections between atoms. These clusters share some electronic properties with π-bonds, although their delocalized nature is largely limited to their internal skeletal frameworks.[11] As a result, radial bonds from these structures behave very much like typical σ-bonds.

Neglecting the covalent framework of the MBBs and focusing only on the connectivity between subunits, a number of advantages are derived from the application of covalent bonds in nanostructural design. First, intermolecular covalent bonding leads to extremely stable nanostructures. Whereas weaker electrostatic interactions are greatly affected by factors such as temperature and choice of solvent, covalent bonds retain their connectivity until concerted efforts are made to break them. Covalent bonds are, then, structurally dependable, prohibiting the reorganization often observed in the continuous breaking and reforming of the other types of intermolecular interactions. It is this feature that similarly allows the covalent architecture of the subunit to be held constant within the context of the larger nanostructure. Finally, an extensive synthetic precedent also exists for connecting almost any molecular fragment or functional group to another. Where specific types of connections have not been previously addressed, their formation is generally possible by a modification of some other known reaction.

The strength and chemistry of covalent bonding also has some important limitations. First, the strength of these bonds frequently limits the flexibility of larger molecules.[6,10] Weaker electrostatic interactions must be used if motion and structural rearrangement are required. With this greater flexibility in the weaker electrostatic interactions comes a higher degree of error tolerance. An unplanned covalent bond between two subunits in a molecular architecture is difficult to correct, requiring far more intensive efforts than simple thermodynamic manipulation. Structural designs based on covalent bonding must, therefore, be well conceived initially to avoid subsequent problems in the fabrication process. Finally, the use of the covalent bond in nanoscale assembly requires direct chemical manipulation. Consequently, two subunits may be self-directing in the formation of their bond by the choice of functionalities, but they are typically not self-assembling. A chemical workup is generally required to form a covalent bond and, as necessary, isolate a product from a reaction mixture.

18.2.2 Coordination Complexes

Lying between the strong covalent bonds of the smaller main group elements and the variety of noncovalent interactions are the coordination bonds of metal–ligand complexes. The initial descriptions of metal–ligand compounds as *complexes* stems from the ability of metals to coordinate small, electron-donating molecules (ligands) beyond the typical maximum of four-point substitutions possible with many main group elements.[8] Metal complexes are known to exist with the metal coordinated to anywhere from one to 12 ligands, although the vast majority of coordination compounds exist in the four-coordinate to eight-coordinate regime (Figure 18.2). The interest in the properties and applications of metals in discrete molecules has enriched such diverse fields as molecular orbital theory, crystallography, catalysis, molecular electronics, supramolecular chemistry, and medicinal chemistry.[8] The availability of d-orbitals in the transition metals and f-orbitals in the lanthanides and actinides results in an extension of the geometric and structural variety available with main group elements. A well-developed synthetic precedent also provides the means to exploiting this rich structural variety within a single molecule context.[12]

Metal–ligand bonds form either through the covalent association of ligands to pair single electrons in metal orbitals or, most often, through the coordination of paired electrons from ligands to fill the valence shell of the metal. Examples include single-ligand lone-pair/metal bonds (the metal analogue

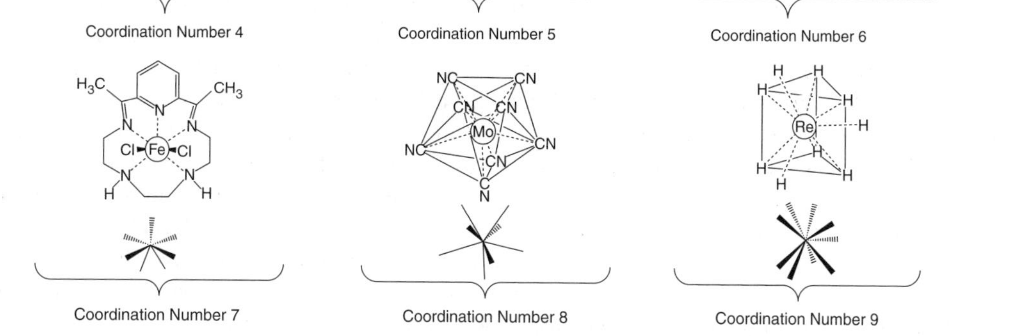

FIGURE 18.2 Examples of coordination geometries among a number of metal complexes.

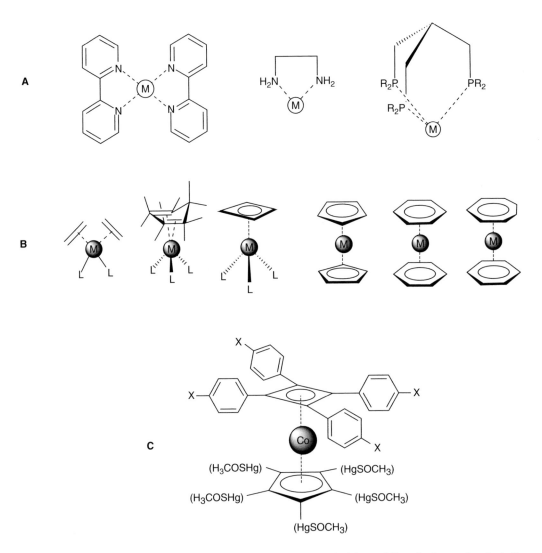

FIGURE 18.3   Metal-ligand bonding. (A) A selection of chelating ligands. (B) Metal–ligand π-interactions including metallocenes. (C) A surface-mounted molecular rotor design.

of a main group σ-bond), chelating ligand bonds (where the ligand is coordinated to the metal by more than one pair of electrons (Figure 18.3A), and metal–ligand π bonds (Figure 18.3B). The low-energy *sharing* of pairs of electrons arises from the coordination sphere of the metal, which can readily accommodate the available electron pairs. Metal–ligand bonds are usually far stronger than other electrostatic interactions because they involve the sharing of pairs of electrons through direct orbital interactions, yet they are generally weaker than the covalent bonds found in organic compounds. To specifically address issues of connectivity, the extensive use of lone-pair coordination to saturate the valence shells of many of the metals adds electron density well in excess of the nuclear charge, pushing the limits of the ability of some metal nuclei to fully accommodate all of the required electrons. Also, the majority of coordinating ligands are stable molecules, and any intermolecular destabilization is typically directed first to the weaker metal–ligand bond. Finally, the molecular volume of the ligand can have a significant effect on the stability of the metal–ligand bond in cases where the metal has a high coordination number, requiring many lone pairs to saturate its valence shell. This last feature of steric saturation is of primary importance in rare earth complexes. In these compounds, orbital interactions between the metal and ligand are significantly attenuated; and stabilization arises primarily from charge balance and steric saturation of the metal center.

A series of metal–ligand coordination complexes is shown in Figure 18.2 to demonstrate some of the structural variety available from metal coordination. Ligand lone-pair coordination is essentially σ-bonding; and the properties of these bonds are consistent with σ-bonding in organic frameworks, including low barriers to rotation and geometric predictability. One special subset of these lone-pair ligands is the chelating ligands, which coordinate to a single metal center through two or more lone-pair donors on the same ligand (Figure 18.3). This class of ligands, driven to higher metal coordination numbers through entropic effects[8] has a significant role in the design of nanostructures from coordination-based approaches (*vide infra*).

An important case of metal–ligand π-coordination occurs in the metallocenes, where the entire π-system of an organic ring can be coordinated to the metal center[13] (Figure 18.3B). The most familiar of these systems is the neutral ferrocene, which saturates an iron(II) center by the coordination of two five-member aromatic cyclopentadienyl rings ($[C_5H_5]^-$). In the design of some of the smallest functional nanostructures, such π-coordinated molecules have distinct advantages, including (1) high stability, (2) incorporation of organic frameworks with the potential to substitute onto the framework, and (3) very low barriers to rotation about the axis of the metal and ring center. Small, surface-mounted metal-ring compounds have already been demonstrated as potential systems for molecular rotors[14] (Figure 18.3C).

Metal–ligand bonds, as the intermediary between main group covalent bonding and weaker electro-static interactions, are well suited to the fabrication of many types of macromolecules and nanoscale arrays. First among their advantages are the higher coordination numbers of these atoms. While a single nonmetallic main group atom generally provides the structural flexibility required to link together from one to four substituents, main group molecules are required to achieve higher connectivity. Instead of designing a six-coordinate center from the smallest molecular octahedron, *closo*-$[B_6H_6]^{-2}$, single-metal atoms readily perform the same task (Figure 18.2). A second advantage of metal-based structures is the number of available metals from which to choose, both for structural complexity and functionality. With this large selection of metal atoms also comes an extensive synthetic precedent,[8,12] allowing the selection of a particular coordination geometry for its known structural features, stability, and chemical accessibility. In instances where lone-pair coordination is used to saturate the valence shell of a metal, the required chemical manipulation is typically too mild to affect the covalent structure of the ligand. Furthermore, because ligands coordinate through weaker bonds, they are also often thermally and photochemically labile under moderate conditions. This ability to form stable structures by thermodynamic or photolytic methods, however, also carries with it the disadvantage of having to control the environment carefully in order to maintain the structural integrity of the final products.

18.2.3 Dative Bonds

A dative bond is an intermolecular interaction between a lone pair of electrons on one atom and a vacant, atom-centered orbital on another. These bonds behave as covalent σ-bonds in many respects, making them close analogs to metal–ligand coordinate bonds (the distinction is made here by limiting dative bonding to main group–main group or metal–metal interactions).[15] While a lone pair of electrons and two atom centers are involved, these interactions are relatively weak when compared with the covalent bonding of the main group elements. The molecules involved in these bonds are themselves independently stable species. The strength of the dative bond is determined by several factors, all of which provide their own means to customization depending on the application.

Dative bonds are most common among pairs of molecules incorporating Group III[(13)] and Group V[(15)] atoms.[16] In such cases, the formation of a dative bond requires the presence of the Group III atom, where an empty orbital remains after the σ-bonds are formed from the three available valence electrons (Figure 18.4). Elements including and beyond Group V usually have at least one lone pair available for

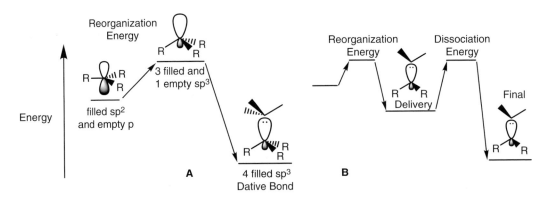

FIGURE 18.4 Dative bonding in main group elements. (A) General pathway for dative bond formation. (B) Energetic considerations of dative bond acceptor "delivery" pathway.

donation in bond formation. Dative structures are classic examples of Lewis acids and bases,[17] in which the lone-pair donor is the Lewis base and the lone-pair acceptor is the Lewis acid. Among the strongest and most studied dative bonds are those between boron (Group III) and nitrogen (Group V) in cases where the boron is treated as an electron precise (2c-2e) atom.[16] The formation of the three electron precise σ-bonds to boron results in the molecule adopting a trigonal planar conformation, leaving an unoccupied p-orbital to act as an electron pair acceptor (Figure 18.4A). The coupling of an atom with a lone pair of electrons to boron results in a reorganization of the boron center,[18] causing it to change shape and hybridization from trigonal planar (sp^2) to tetrahedral (sp^3). The stability of a dative bond is then dependent upon (1) the choice of lone-pair donor and acceptor, (2) the substituents on the donor and acceptor, and (3) the reorganization energy. These bonds typically range from 50 to 85 kJ/mol, although some have been shown to have bond strengths of 100 kJ/mol.[16] In small systems, such as $H_3B:NH_3$ and $F_3B:NH_3$, the stabilizing energy is large because there is very little steric congestion from the substituents. Among the systems with significant steric congestion, boraadamantane forms uniquely stable dative structures (Figure 18.5).[18] In boraadamantane, the adamantyl framework forces the boron to be sp^3-hybridized regardless of the presence of a lone-pair donor. The reorganization energy is effectively included in the synthesis of the boraadamantane Lewis acid, leaving the entirety of the lone-pair interaction to form a particularly stable dative bond.[18] Dative bonds provide the directionality of covalent bonds with the lower stabilization energy of electrostatic interactions, giving them useful features for nanoscale design. Dative-based molecular assemblies require the selection of building blocks that limit the lone pairs and vacant orbitals to structurally important sites.[19] The design of connectivity is then a matter of limiting dative bonding everywhere else in the subunits.

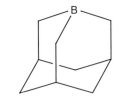

FIGURE 18.5 Boraadamantane.

In organic molecules, incorporating the dative components into the "correct" structural sites on the molecule and including only C-H bonds everywhere else effectively accomplishes this. While the chemistry of organoboron compounds might not be as well developed as that of organonitrogen compounds, a considerable synthetic precedent exists for both. The inclusion of active centers for dative design is possible through the addition of many known organic components. The strength of dative bonds can also be tailored by either changing the donor and acceptor substituents or by changing the initial hybridization of the electron pair acceptor.[16]

The limitations of the dative bond approach stem primarily from the lone-pair acceptor. While the lone-pair donor is often unreactive, lone-pair acceptors, such as the many organoboron compounds, are highly electrophilic and will coordinate with any available electron pairs. Part of the design of these systems must include potential problems with *delivery* to the donor (Figure 18.4B). When a *delivery molecule* is required initially to coordinate to the acceptor prior to assembly, this molecule must be chosen to be more stable than any possible lone-pair donors in solution, yet weakly coordinating enough such that the delivery molecule is easily displaced from the system during assembly. In some instances, the selection of a good delivery molecule can be nontrivial, since an effective choice involves the subtle interplay between the strength of the delivery–acceptor and the final donor–acceptor bond strengths.

18.2.4 π-Interactions

The variety of electrostatic interactions involving the π-systems of aromatic molecules have been shown to play important roles in such diverse areas as the packing of molecules in molecular crystals, the base stacking (as opposed to base pairing) interactions in DNA, polymer chemistry, the structure and reactivity of many organometallic complexes, and the formation, shapes, and function of proteins.[20] The accessible and highly delocalized pool of electrons above and below an aromatic molecular plane is well suited to forming electrostatic interactions with cations, neutral molecular pairs with complementary electron density differences, and other aromatic π-molecular systems.

Because π-systems may be thought of as regions of approachable electron density, noncovalent interactions with aromatic rings occur when a system with a net-positive region is brought within proximity of the aromatic molecule. Three of the most familiar types of π-interactions are (1) aromatic ring/electrophile interactions, where the electrophile is highly positive, (2) phenyl/perfluorophenyl interactions, where the electronegativities of the σ-bond periphery have an overall effect on the charge distribution of the molecule, and (3) lower energy quadrupolar interactions, where weak π-interactions occur based on electron density differences across the molecular plane (Figure 18.6).

Cation-π interactions have long been known to play important roles in molecular recognition, biochemical processes, and catalysis.[21] The most familiar cations used to study these interactions are Group I[(1)] elements and small protonated Lewis bases (such as NR_4^+). The binding energies of these pairs can be quite large, with the strongest interactions approaching the strengths of weak covalent bonds.[21] Important to the nature of these interactions is the ability of the aromatic rings to compete successfully with polar solvents for the cation. Remarkably, the π-system of the nonpolar benzene molecule has been shown to bind K^+ ions more strongly than the oxygen lone pairs in water.[22] The customization of the cation-π binding energy can be controlled by the choice of Group I[(1)] cation or the substitutions on the molecular cations, where bulky substituents tend to lower the stabilization by forcing the cation further from the π-system.

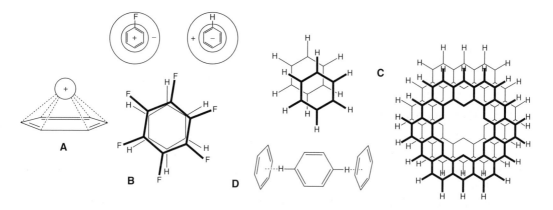

FIGURE 18.6 A selection of π-interactions. (A) π-cation interactions. (B) Ideal π-stacking arrangement of benzene/perfluorobenzene. (C) Staggered π-stacking in benzene (left) and kekulene (right). (D) Preferred herringbone π-stacking configuration of benzene, with hydrogen atoms centered on the π-system of adjacent benzene rings.

Benzene/hexafluorobenzene stacking is a specific example of the general type of π-stabilization that occurs with the pairing of molecules that have large quadrupole moments of the opposite sign.[23] In benzene, the regions of highest electron density are the π-system and σ-system of the carbons, leaving the peripheral hydrogen atoms net-positive from inductive effects (Figure 18.6B). In hexafluorobenzene, the charge density is reversed, with the peripheral fluorine atoms containing the highest electron density. Together, the molecular pair is ideally suited for stacking due to the complementary arrangement of its electron densities inside the rings and along the outer periphery of the two molecules. In a classic study of this form of π-stacking, one equivalent of benzene (m.p. 5.5°C) was combined with one equivalent of hexafluorobenzene (m.p. 4°C) to form a mixture with a melting point of 24°C.[24] The actual stacking of these rings was subsequently confirmed by a variety of spectroscopic methods.[25] This same stabilization has been used successfully in the formation of other π-stacking species[26] to align various arenes in molecular crystals, providing a facile means for molecular alignment of thermal and photochemical polymerization reactions.[27]

There are many other important examples and structural motifs in π-stabilizing interactions. These include the stacking of DNA base pairs, the stabilization of tertiary structures in proteins, the aggregation of large porphyrins, and the formation of molecular crystals incorporating aromatic moieties.[20] In benzene and kekulene, for example, stacked structures are most stable when slightly offset, maximizing the overlap of the net-positive periphery and the electron-rich π-system (Figure 18.6C).[20] The offset stacking of the purine and pyrimidine base pairs in DNA plays an important function in stabilizing the double helix. Benzene and many other aromatic systems crystallize as herringbone-shaped structures, with the peripheral hydrogen atom on one ring placed along the central axis of the π-system of a perpendicular ring (Figure 18.6D).[23]

The variety of π-interaction types and structural motifs available among the aromatic rings leads to a number of important features for nanoscale design. First, the π-orbitals exist above and below the molecular plane. An interaction with a π-system is, therefore, often just as likely to occur above the molecular plane as below, allowing these stacking interactions to occur over long distances with many repeating units. For example, crystals of benzene/hexafluorobenzene and the extended π-stacking arrangement in base pairs in a single strand of DNA provide considerable electrostatic stability and alignment. Second, the stability of a π-interaction can be directly controlled by the chemical substituents attached to the ring. The significant change in the properties of benzene/hexafluorobenzene solutions attest to this chemical flexibility.[24] Aromatic heterocycles, such as the purines and pyrimidines in DNA nucleotides, demonstrate the ability to customize these interactions based on directly changing the π-system through hetero-atom substitution. Third, depending upon the surroundings, the π-interactions can be modified by solvent effects. This is demonstrated in the base stacking of DNA, where the stability from heterocycle π-stacking interactions is in addition to the stability gained from minimizing the surface area of the rings exposed to the aqueous environment. Similar arguments have been used to describe the formation of tertiary structure and aggregation of proteins.[20,28] The use of these types of interactions for designing nanostructures is limited, however, by the relatively unpredictable stacking arrangements observed and the sizes of these complex aromatic rings. Stability from aromatic π-stacking requires the use of rings which, when compared to the more direct hydrogen bond or metal–ligand coordination bond, need a larger space and more flexibility to allow for the optimized stacking arrangement to occur.

18.2.5   Hydrogen Bonds

Hydrogen bonding is "the most reliable directional interaction in supramolecular chemistry,"[3] and its role in numerous macromolecular phenomena has been well studied. As a frequently employed electrostatic interaction with vast synthetic and theoretical precedent, a rigorous analysis of this interaction in its many forms is beyond the scope of this discussion on nanoscale design and is left to significantly more detailed treatments in many excellent reviews.[29,30] Important to understanding this type of interaction from a nanoscale design perspective, however, is the nature of the bond, the functional groups responsible for its occurrence, and the relative stabilities that come with different functional groups. Appropriately, these topics are covered here in general with specific examples used to highlight the discussion.

A hydrogen bond is formed when the hydrogen in a polar bond approaches the lone pair of electrons on an ion or atom.[8] A polar bond to hydrogen occurs when the hydrogen is attached to an atom of high electronegativity, such as nitrogen, oxygen, or fluorine. Because hydrogen atoms have no inner core of electrons, the pull of electron density from them exposes a significant positive nuclear charge to interact electrostatically with nearby electron density. This is further strengthened by the very small size of the hydrogen atom. The electronegativity difference between carbon and hydrogen is small enough that a significant dipole is not produced, resulting in very weak hydrogen bonds involving C-H bonds. The strength of the hydrogen bond is determined by the polarity of the bond in which the hydrogen is covalently bound and the electronegativity of the atom to which the hydrogen is electrostatically attracted. Hydrogen bonds can be divided into *strong* (20–40 kJ/mol) and *weak* (2–20 kJ/mol) interactions,[3] each of which is important to certain types of supramolecular assembly.

Hydrogen bonds can be used to stabilize structures ranging from small dimers to extended arrays of massive molecules. The most commonly encountered strong molecular hydrogen bonds tend to favor the use of oxygen or nitrogen, a result of their large electronegativity differences with hydrogen. Also important for MBB assembly is the ability of oxygen and nitrogen to covalently bond to more than one atom, allowing them to be incorporated into larger molecular frameworks. This is in contrast with fluorine, which can only be used to terminate a covalent framework, making its role in typical hydrogen-bonded nanostructures rather limited. There are numerous combinations of hydrogen bonding interactions that can be incorporated into a covalent framework from the available organic precedent for the manipulation of functional groups such as O-H, C = O, N-H, C = N, COOH, NH_2, and NOO^- (Figure 18.7A). Weak hydrogen bonds have also been shown to play important roles in the shapes and stabilities of

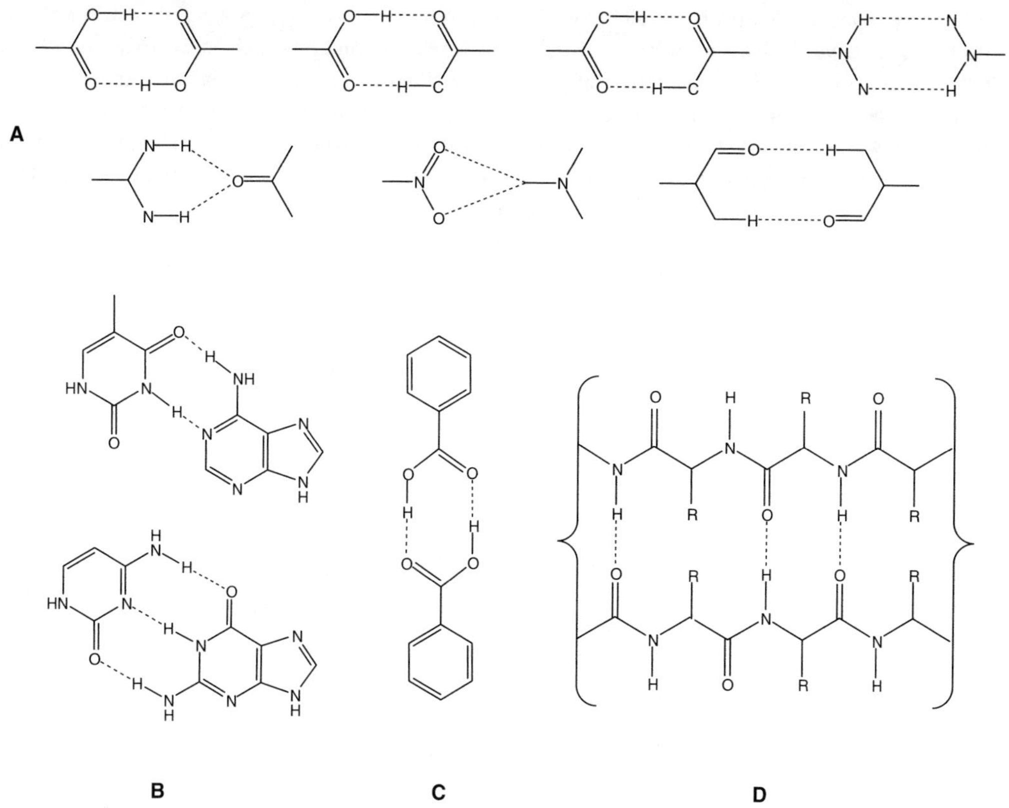

FIGURE 18.7 Hydrogen bonded structures. (A) A selection of hydrogen-bonded structures. (B) Thymine–adenine (top) and cytosine–guanine (bottom) base pairing. (C) Hydrogen-bonded carboxylate dimer. (D) Portion of hydrogen bonding network in peptide β-sheets.

macromolecular assemblies and crystals that do not include functional groups capable of strong hydrogen bonds.[31,32] These weaker bonds include interactions such as OH···π and NH···π.

A small selection of relevant hydrogen-bonded complexes is provided in Figures 18.7B through 18.7D. The most familiar hydrogen-bonding interaction, outside of ice crystals, occurs in the nucleotide base pairs of DNA, where strongly bonding functional groups are incorporated into small, aromatic heterocycles. The bonds form so as to stabilize particular pairs (thymine/adenine and cytosine/guanine) in the formation of the double-helical structure. Strong hydrogen bonding also occurs between the C=O and N-H groups of amino acids in the formation of the secondary structure of proteins (i.e., α-helices and β-sheets). Artificial superstructures employing hydrogen bonding include simple dimers, linear arrays, two-dimensional networks, and, with the correct covalent framework, three-dimensional structures.

There are many advantages to using hydrogen bonding in the formation of macromolecules and extended arrays. First, these interactions are both self-assembling and self-directing. Stable structures based solely on electrostatic interactions are free to form and dissociate with relatively little energy required. Unlike covalent bonds, which require specific reaction conditions, hydrogen bonds (and other electrostatic interactions) require only the appropriate medium through which to form stable structures. The spontaneity of protein secondary structure formation in aqueous media is, perhaps, the most remarkable example of this phenomenon. Second, there are many functional groups that can act as either hydrogen donors (X-H bond) or acceptors (lone pair). This availability comes from both an extensive synthetic precedent and a large number of different donors and acceptors that can be employed to customize the strengths of hydrogen bonds. Third, hydrogen bonds are typically directed interactions with small steric requirements. Whereas π-stacking requires both a large surface area and very specific electronic distributions in the aromatic rings, hydrogen bonds can form with molecules as small as hydrogen fluoride. Fourth, directional interactions such as hydrogen bonds are relatively easy to incorporate into larger molecules, provided the attached covalent frameworks are shaped correctly to allow the interactions to occur. The pairing of nucleotides in DNA are specific examples of where the selected covalent frameworks determine the optimum orientations of the hydrogen bonding interactions. In crystal engineering, many molecular architectures are based on the inclusion of known pairs of hydrogen-bonding functionalities into organic frameworks.[3] Another advantage that stems from the small size and unidirectional nature of the hydrogen bond is the ability to incorporate multiple interactions within a very small space. Again, base pairing in DNA is an example of where either two (A with T) or three (C with G) hydrogen bonds occur in small heterocytes (Figure 18.7B). The ability to incorporate multiple hydrogen bonds into a single framework also allows orientational specificity to be designed into a structure. Not only do nucleotides pair specifically according to the number of hydrogen bonds (A with T and C with G), but they form stable interactions in only one dimeric conformation.

The greatest limitation in hydrogen bonding comes from the relative stabilities of these bonds and the potential for such bonding throughout an ensemble of molecules. While certain interactions can be predicted to be most stable based on their conformation and functional groups, there are usually many other interactions that form the macromolecular equivalent of metastable structures in solution; and the directing of a single, preferential hydrogen-bonded framework can be difficult to predict or control. In polar solvents, such as water, this predictability becomes even more difficult. The local hydrogen bonds that form with aqueous solvation approach the strengths of many other hydrogen-bonding interactions. Although the formation of the DNA double helix in aqueous media is driven by entropy, the relative stability of nucleotide–water interactions is significant, providing local instabilities in the DNA double helix.[33] This same dynamic equilibrium in DNA between water–nucleotide and nucleotide–nucleotide interactions, however, is also partially responsible for its biological activity, as a DNA helix unable to be destabilized and "unzip" is poorly suited to providing genetic information. As with all of the bonding motifs discussed, the merits and limitations of hydrogen bonding in nanostructural design and formation are sometimes subjective; and the specifics of a system and its surroundings play important roles in determining the best choice of macromolecular stabilization.

18.3 Molecular Building Block Approaches

The overriding goal of the MBB approach is the assembly of nanostructures or nanoscale materials through the manipulation of a subunit by chemical methods or electrostatic interactions. The MBB is selected or designed with this manipulation in mind. The MBB is, ideally, divisible into one or more chemically or electrostatically active regions and a covalent framework, the purpose of which is simply to support the active regions of the subunit. With the division between covalent architectures and lower energy electrostatic systems in mind, the range of MBB designs can be bounded by those systems fabricated through only covalent bonds between subunits and those including only weak interactions between otherwise covalently isolated subunits. Appropriately, these two cases will be considered first. With the definition of the boundaries of what can be done with MBBs in the limit of structural inter-connectivity requirements, intermediate systems that balance relative degrees of covalent and electrostatic character, including familiar biological systems, coordination nanostructures, and dendritic systems, are then considered.

18.3.1 Supramolecular Chemistry

Supramolecular chemistry is the science of electrostatic interactions at the molecular level. Direct correlations of structure and function exist between molecular chemistry and supramolecular chemistry, and many parallels can be drawn between the two that highlight the utility and importance of chemical design from noncovalent interactions. The range of covalent bonding and chemical functionalities within a molecular framework gives rise to a range of noncovalent interactions that can be used to form stable structures composed of many molecules. The chemistry of the covalent bond also allows for the engineering of electrostatic interactions. Just as a molecular chemist would employ reaction conditions and various functionalities to direct a particular chemical synthesis, the supramolecular chemist employs the surroundings and the entire molecule to tailor stabilizing interactions into a macromolecular framework. The energies of the interactions between molecules in supramolecular design are far weaker than those interactions within the molecular framework. Consequently, in supramolecular chemistry, the entirety of the covalent framework of the molecular subunit is treated as a whole; and the assembly of the supramolecular array progresses from the MBB just as the synthesis of a molecule is treated as an assembly of discrete atoms.

Supramolecular chemistry is, however, unique in many respects. The formation of new structures in both molecular chemistry and supramolecular chemistry is based upon understanding and predicting chemical interactions. In the case of molecular chemistry, structure formation is based on reaction centers with the covalent framework of the molecule altered to form a new structure. In supramolecular chemistry, superstructure formation is based on interaction centers in which the covalent framework of the molecule as a whole remains unaffected by the stabilizing interactions that occur beyond it. In molecular chemistry, the covalent frameworks of the precursor molecules must be altered through energy-intensive chemical manipulation. Reactions may be self-directing based on the positions of functional groups and reaction conditions; but the actual formation of a molecule requires some form of external manipulation, such as a naturally occurring enzyme or catalyst, or a particular reaction pathway to facilitate the breaking and formation of chemical bonds. A self-assembling molecular reaction is then a fortuitous occurrence of both the correct molecules and the correct chemical environment. In supramolecular chemistry, interactions between molecules are self-directing and spontaneous in solution. Because significant changes to the covalent framework of the subunits are not part of the superstructure formation process, stabilization from noncovalent interactions is based only on localized chemical environments. Provided that the stabilizing interactions between subunits are sufficiently large, molecules will spontaneously form into larger structures. The goal of supramolecular chemistry is the application of this spontaneity in the rational design of larger structures. The total stabilization energy for a supramolecular array from its component molecules is smaller than the total covalent energy between a molecule and its component atoms. Consequently, the formation and degradation of a supramolecular array is far less energy-intensive than the formation and breakage of covalent bonds. In many instances, stabilization in supramolecular designs benefits from the similarities in energy between MBB interactions and the energy of the surroundings, including the stability gained from the interactions between subunits and solvent molecules. The dynamics of proteins in aqueous media are excellent examples of where a macromolecular structure and the environment can be used in concert to create both stability and function in chemically massive molecules.

Supramolecular chemistry broadly encompasses the use of any electrostatic interaction in the formation of larger molecule-based structures. As such, any system that is based on interactions *beyond the molecule* falls under the supramolecular heading. Supramolecular chemistry, as it is then loosely defined, is an outgrowth of many related disciplines which serve to study phenomena beyond the molecular boundary, including biochemistry, crystal engineering, and significant portions of inorganic chemistry. Much of our initial understanding of molecular interactions comes from the study of naturally occurring structures in these well-established fields. To study the secondary structure of proteins and DNA is to study specific examples of the supramolecular aspects of biochemistry. The functions of these macromolecules in the intracellular matrix are based on noncovalent interactions, including the enzymatic activity of proteins on a substrate, the binding of cations to a protein, the dynamics of DNA duplication, and protein folding. The periodic lattices of many molecular crystals provide examples of how electrostatic interactions direct the alignment of molecules in the solid state. For instance, the unique properties of ice crystals relative to liquid water demonstrate how intermolecular interactions can be just as important as intramolecular interactions in defining structure and properties.

As a unique discipline, supramolecular chemistry emphasizes the design of novel molecular architectures based on the rational incorporation of electrostatic interactions into molecular frameworks. The discussion of supramolecular chemistry here will emphasize the design of macromolecules using only electrostatic interactions. Specifically, supramolecular structures formed from hydrogen bonding and π-interactions are detailed. Dative-based designs, while offering a number of attractive properties for noncovalent stabilization, have seen limited application for the design of nanoscale architectures. The division between entirely electrostatic assemblies and mixed covalent/electrostatic assemblies is stressed when possible to examine how specific noncovalent interactions can be used as the primary means to define the shape of supermolecular structures. Specific instances of nanostructure formation employing both covalent and noncovalent bonding are addressed subsequently in two sections, where the importance of both structure and function can be considered in context. The interactions between metal centers and organic ligands for the formation of coordination nanostructures is also treated as separate from this general discussion in order to provide emphasis on this particularly well-defined segment of supramolecular chemistry.

18.3.1.1 Hydrogen Bonding in Supramolecular Design

Hydrogen bonding is used extensively in supramolecular chemistry to provide strength, structural selectivity, and orientational control in the formation of molecular lattices and isolated macromolecules. The advantages inherent to hydrogen bonding interactions are universal among the different areas of supramolecular chemistry, whether the application is in the stabilization of base pairs in DNA or the alignment of synthons in infinite crystal lattices. The functional groups most familiar in hydrogen bonding have significant precedent in organic chemistry and are, therefore, readily incorporated into other molecules through chemical methods.[3] The complementary components of a hydrogen bond can be incorporated into molecules with very different chemical and electronic properties. In benzoic acid, for example, a polar carboxylate group is covalently linked to a nonpolar benzene ring to form a molecule with two distinct electrostatic regions (Figure 18.8A). The formation of benzoic acid dimers in solution is strongly directed by the isolation of polar and nonpolar regions in the individual molecules and the stability that comes with forming hydrogen bonds between the highly directing donor/acceptor groups.[34] The predictability of hydrogen bond formation in solution and the directional control that comes with donor/acceptor pairing allows for MBBs incorporating these functionalities to be divided into distinct structural regions based on their abilities to form strong hydrogen bonding interactions. This simplifies

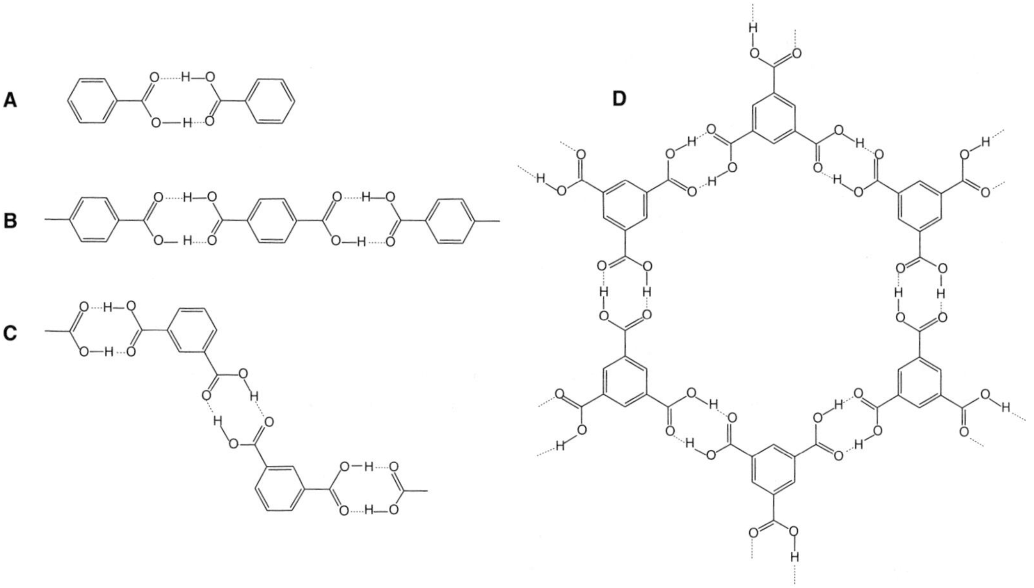

FIGURE 18.8 Hydrogen-bonded aromatic/carboxylic acid assemblies. (A) Benzoic acid dimers. (B) Linear chains of terephthalic acid. (C) Chains of isophthalic acid. (D) Hexagonal arrays of trimesic acid.

the design process in molecules that are tailored to form stable interactions only in specific regions, allowing for the identification of structural patterns in macromolecular formation.[35] The general shape of the nonpolar backbone in benzene, for instance, creates a geometric template from which it becomes possible to predict the shapes of the larger macromolecular structures that result from hydrogen bond formation. To illustrate this template approach with molecular hexagons, a series of examples of both arrays and isolated nanostructures are considered below that use only hydrogen bonding and the shapes of the subunits to direct superstructure formation.

18.3.1.1.1 Crystal Engineering

The hydrogen bond has been used extensively in the design of simple molecular crystals. Crystal engineering has been defined as "the understanding of intermolecular interactions in the context of crystal packing and in the utilization of such understanding in the design of new solids with desirable physical and chemical properties."[36] Many researchers in the field of crystal engineering have been guided by the very predictable and directional interactions that come with hydrogen bonding in its various forms. The cognizant design of extended arrays of hydrogen-bonded structures in molecular crystals is made possible by the broad understanding of these interactions in other systems, especially from the formation of biomolecules and small guest–host complexes. Among those examples that best demonstrate the rational design of molecular crystals from simple subunits and well-understood interactions are the aromatic/carboxylate structures (Figure 18.8). From the very predictable dimerization of benzoic acid in solution comes a number of similar structures whose geometries are singly dependent on the shape of the hexagonal benzene core. Isophthalic acid[37] and terephthalic acid[38] are simple extensions of the benzoic acid motif that form hydrogen-bonded chains (Figure 18.8B, C). The hexagonal trimesic acid structure[39] stems directly from the placement of strong hydrogen bonding groups on the benzene frame, yielding two-dimensional arrays of hexagonal cavities in the solid state (Figure 18.8D). The same chemical design has also been considered with the amide linkages, in which a higher connectivity is possible through four hydrogen bonding positions (Figure 18.9). Linear chains of benzamide[40] form from each amide linkage, forming four strong hydrogen bonds to three adjacent benzamide molecules. The repeating subunit of these chains is a dimer very similar to that of the benzoic acid dimer, with additional hydrogen bonding groups extending perpendicularly from each dimer to facilitate linear connectivity to other pairs (Figure 18.9A).

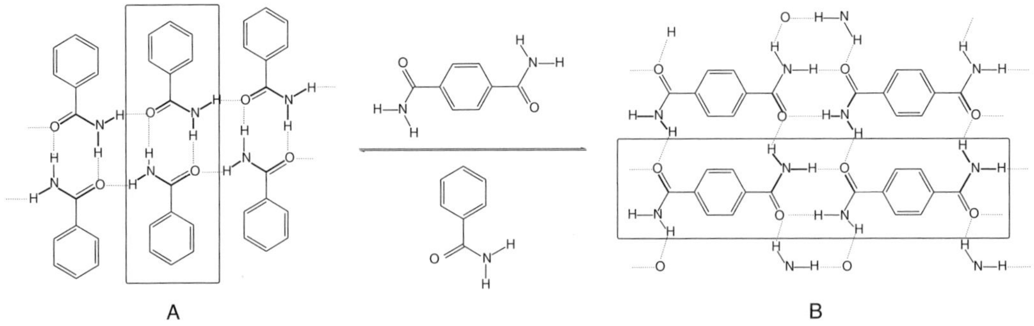

FIGURE 18.9 Hydrogen-bonded structures from amide linkages. (A) Benzamide dimers form linear chains. (B) Terephthalamide forms highly connected sheets. The corresponding aromatic/carboxylate motifs are enclosed in boxes.

Planar sheets of terephthalamide[41] form from the same extended linear chain motif found in *para*-substituted terephthalic acid (Figure 18.9B). Again, the perpendicular hydrogen bonding groups direct the connectivity of these linear chains into two-dimensional sheets. The commonality among all of these benzene-based MBBs is the division between the rigid alignment of the functionalities on a covalent framework and the positions of the interaction centers beyond the molecular frame.

18.3.1.1.2 Supramolecular Structures

A number of isolated supramolecular structures are known that use only hydrogen bonding to direct their formation. In some instances, this has been accomplished through modifying the substituents on array-forming MBBs to promote the formation of isolated systems. While unsubstituted isophthalic acid in solution was found to form linear ribbons in the solid state, the addition of bulky substituents at the *meta*-positions of the two carboxylic acid moieties resulted in the formation of isolated molecular hexagons — structures that mimic exactly the hexagonal cavities formed through hydrogen bonding in the trimesic acid arrays[42] (Figure 18.10). For greater control in the formation of complex supermolecules, the engineering of highly directional hydrogen bonding regions is often required. The customization of interactions between MBBs is performed by either attaching more than two hydrogen bonding pairs onto the same framework (to prohibit free rotation when single σ-bonds are used to connect the donor/acceptor assemblies) or by embedding two or more functionalities directly into a covalent framework (Figure 18.11). In both routes, the resulting structures are no longer limited to stable designs based solely on single donor/acceptor pairs or sets of hydrogen bonding fragments isolated to σ-bound molecular fragments.

By fixing the positions of the donor and acceptor groups in a framework, the connectivity of subunits must occur with orientational specificity, creating what are commonly known as *molecular recognition* sites. In hydrogen-bonded systems, each interaction region of the molecular recognition site is clearly identified by the arrangement of the donor/acceptor groups, such as shown in Figure 18.11B. For crystal engineering and nanostructure formation, where stability and the fitting of subunits to one another define the shape of the entire system, both the hydrogen bonding arrangement and the shapes of the molecules are important to the success of a molecular recognition site (Figure 18.11C).

Two specific MBB designs have been extensively used together to illustrate the roles of structure and orientation in the formation of hydrogen-bonded nanostructures. Cyanuric acid and melamine are two highly symmetric molecules with complementary hydrogen bonding regions along each molecular face (Figure 18.12). In solution, 1:1 mixtures of these molecules form insoluble complexes of extended hexagonal cavities[43] (Figure 18.13C). By the removal of a hydrogen bonding interaction from each molecule, two different assemblies have been shown to form. In both instances, cyanuric acid is converted into a barbituric acid-based molecule by the removal of one N-H fragment from the central ring, while the melamine structure is altered by the removal of one nitrogen atom from its central ring (Figures 18.13D and 18.13E). The formation of linear chains has been shown to be favored in the native structures and when the substituents on the MBBs are kept small[44] (Figure 18.13F). The addition

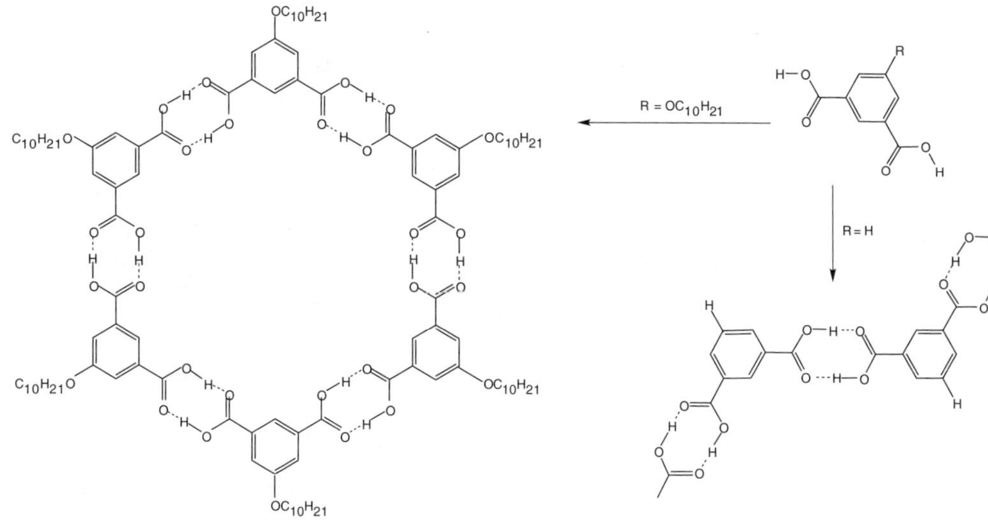

FIGURE 18.10 Isophthalic acid derivatives direct the formation of different hydrogen-bonded networks.

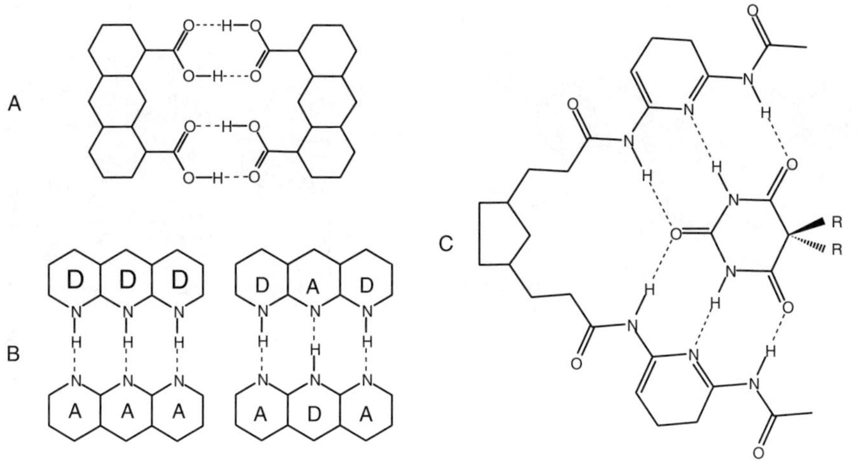

FIGURE 18.11 Engineering orientational specificity into hydrogen-bonded structures. (A) Multiple interaction zones fix the orientation of guest–host complexes by prohibiting rotation. (B) Donor (D) and acceptor (A) interactions between hydrogen-bonded fragments embedded within molecular frameworks. (C) Size and orientation direct the binding of barbituric acid within a molecular recognition zone.

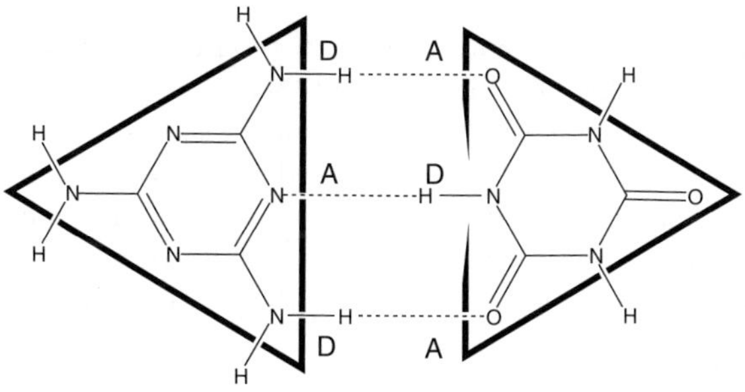

FIGURE 18.12 Complementary DAD:ADA hydrogen bonding in melamine (*left*) and cyanuric acid (*right*).

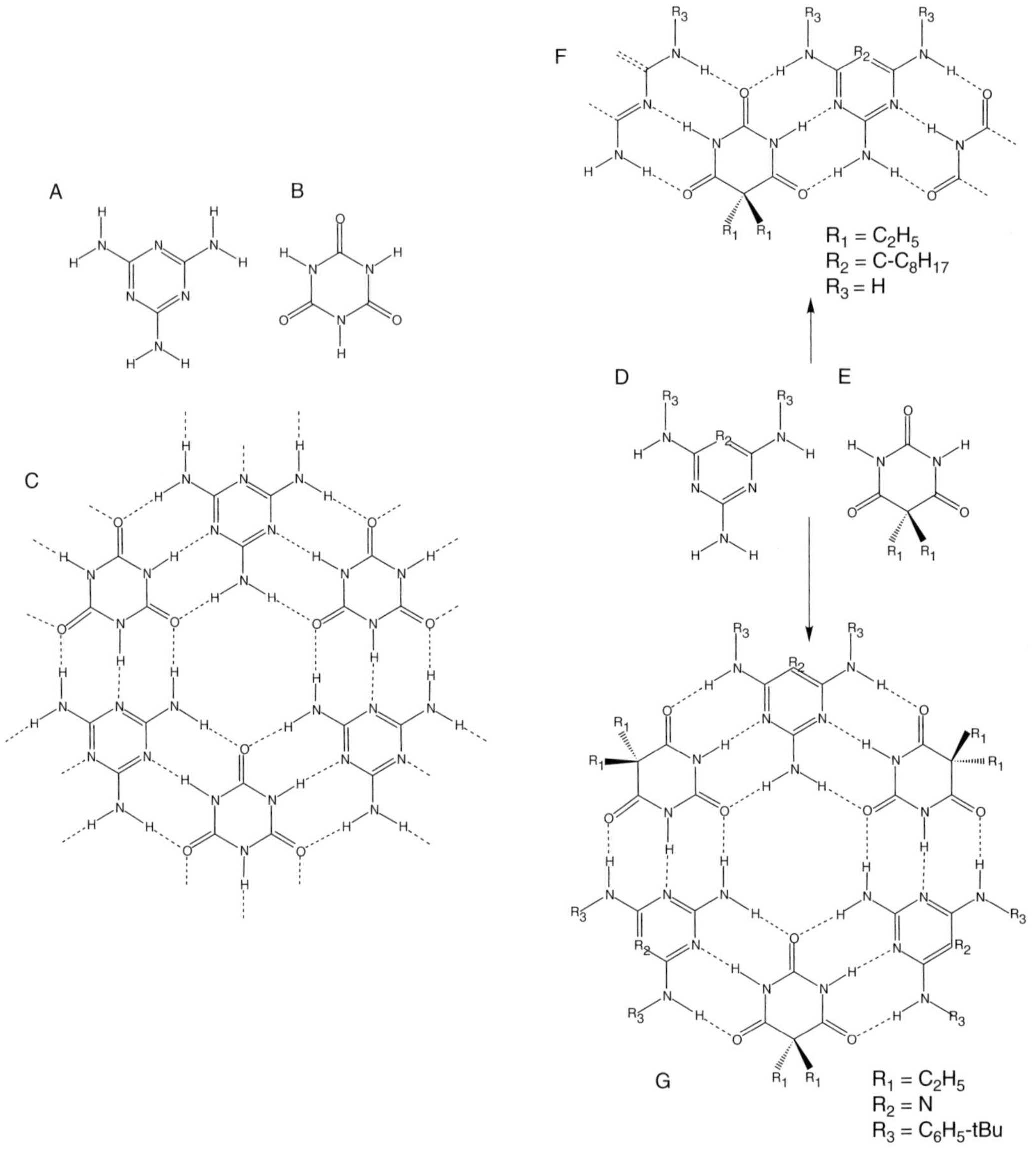

FIGURE 18.13 1:1 mixtures of A and B form extended arrays. Structures utilizing D and E form either linear chains (F) or supramolecular hexagons (G) depending on the choice of R groups.

of bulky substituents to the subunits (similar to the method used to form hexagons of isophthalic acid) directs the hexagonal species shown in Figure 18.13G to self-assemble in solution.[45]

These supramolecular designs are easily rationalized from the shapes of the hexagonal frames to which hydrogen bonding fragments are attached. Hydrogen bonding has been used frequently in the design of smaller guest–host interactions and molecular recognition sites, with much of this work derived from extensive biochemical precedent. Subsequent sections on biomimetic designs and dendrimers illustrate a few of these specific instances of isolated hydrogen bonding interaction in specific MBB designs.

18.3.1.2 π-Interactions

The use of π-interactions has been shown to be important for a number of biological and molecular assembly applications. In biological structures, π-stabilization and the hydrophobicity of the aromatic

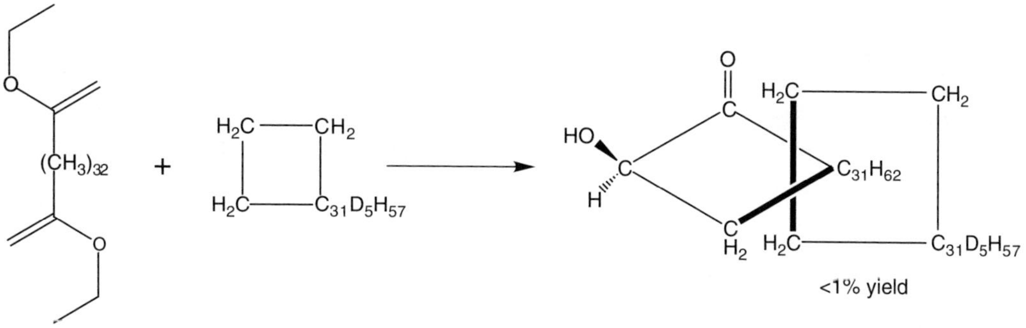

FIGURE 18.14 The first reported catenane.

rings both contribute to the formation of secondary and tertiary (aggregate) structure in DNA and proteins. The stability gained from stacking π-systems with complementary electron densities has been used as a driving force for a number of crystal engineering-based structures. The herringbone stacking pattern of aromatic π-systems with peripheral substituents is a very familiar motif in crystal engineering and has been shown to be responsible for the observed packing of many molecular crystals.[3] For the formation of supramolecular assemblies, however, the role of the π-interaction as a singular driving force is rather limited. Interactions with the π-systems of small aromatic groups are difficult to utilize because the energies of the different orientations can be very similar. Consequently, many supramolecular structures employing π-stacking interactions either use π-stacking in conjunction with other interaction types or use π-π-interactions between highly polarized species to direct the formation of supramolecular structures. Three specific examples are discussed below to illustrate how π-interactions can be employed in the electrostatic-based supramolecular formation of nonbiological structures.

18.3.1.2.1 Catenanes

Catenanes are a unique class of supramolecular structures formed by the interpenetration of two or more macrocycles to form what is often referred to as a *topological bond*. The assembly of interlocked rings has been demonstrated both by statistical and directed techniques. The statistical method used to form the first isolated catenane[46] is shown in Figure 18.14 and gave very poor yields, demonstrating the limitations of self-assembly without direction from electrostatic interactions. The other types of catenanes have been synthesized with far greater success by relying on local stabilization from π-interactions in aromatic rings in conjunction with other electrostatic interaction types.

The formation of two coordination-based catenanes has been proposed to arise from guest–host interactions between π-systems (Figure 18.15). These two structures, identical except for the choice of metal center (either palladium(II) or platinum(II)), form initially as single-ring systems from 1,4-bis (4-pyridylmethyl)benzene. In the palladium(II) complexes,[47] concentration was found to play a key role in determining the relative populations of rings (low concentrations) and catenanes (high concentrations) at ambient temperatures. The equivalent platinum(II) catenane[48] was found to form irreversibly as a function of temperature. Here, raising the temperature of the system to break the strong platinum–nitrogen coordination bond opens the ring systems for monocycle insertion. In both cases, catenane formation is promoted by π-interactions between the ring systems that stabilize the molecular interlocks long enough to allow for the formation of the metal–ligand topological bond.

Perhaps the most familiar catenanes are those composed of paraquat–crown complexes[49] (Figure 18.16). In these systems, the interlocking of a neutral crown ether and a paraquat ring is directed and stabilized by two strong electrostatic interactions. First, strong hydrogen bonding between the crown oxygens and the acidic hydrogens on the aromatic rings of the paraquat serve to fix part of the paraquat within the crown ring. Second, strong π-π-interactions between the crown aromatic rings and the positively charged aromatic rings of the paraquat serve to direct the insertion of the crown ring into the open paraquat assembly prior to its covalent ring closure.

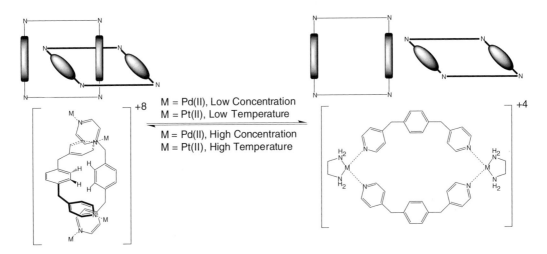

FIGURE 18.15 The effects of metal, concentration, and temperature on the formation of coordination catenanes of palladium(II) and platinum(II).

18.3.1.2.2 Molecular Zippers

One of the most interesting pairings of edge-to-face π-π-interactions and hydrogen bonding comes in the form of molecular *zipper* structures formed from amide oligomers[50] (Figure 18.17A). The formation of double strands of the amide oligomers is rationalized based on ^{1}H NMR titration studies in the nonpolar solvent chloroform and the known structural features of oligomer chain pairs used in the dimerization study. In the general design scheme, the oligomer chains A and B have complementary binding regions capable of forming stable A:A, B:B, or A:B dimers. Based on the chain lengths of the two monomers, however, A:A and B:B dimers are found to not maximize the total possible number of π-π-interactions and amide hydrogen bonds (Figure 18.17B). The A:B dimer maximizes the total number of possible interactions along the entire length of the dimer complex, thereby promoting its formation in solution from equal mixtures of both A and B. Among the number of dimer systems examined, the commonalities to all are the increase in stability with increased oligomer lengths (providing more interactions between dimers) and the decrease in stability in polar solvents, such as methanol, which competitively bind to the polar amide functionalities and weaken the zipper structure.

18.3.1.2.3 Aedemers

The preferential face-to-face stacking of aromatic molecules with complementary ring charge densities has been demonstrated in many instances. The application of this phenomenon to nanoscale design

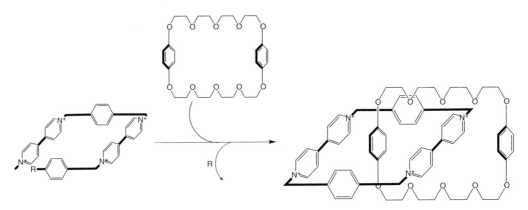

FIGURE 18.16 General crown ether/paraquat catenane and assembly mechanism.

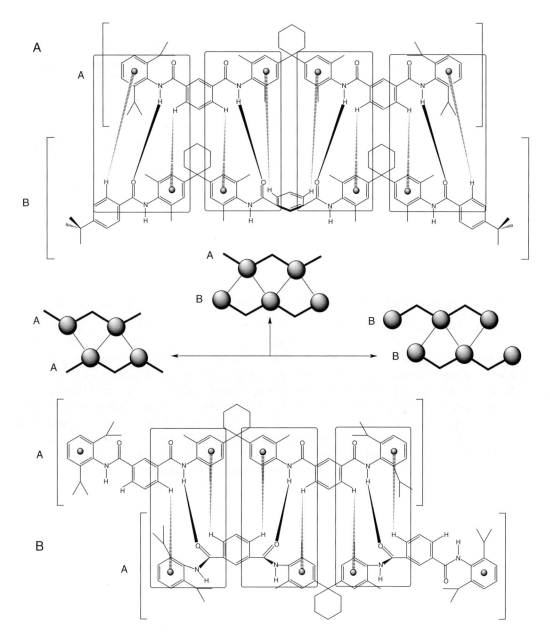

FIGURE 18.17 Molecular zippers from amide oligomers. The number of π-stacking interactions (dashed lines) and hydrogen bonding interactions (bold lines) is maximized with A:B dimers (top).

beyond the alignment of molecules in extended arrays has not, however, been exploited far beyond biological designs. The stabilizing interactions of two complementary π-stacking pairs have been shown to direct the formation of secondary structure in at least one other type of covalently linked macromolecule. The aedemers[51] are synthetic oligomers incorporating π-system donors and acceptors attached by long-chain tethers (Figure 18.18). In aqueous media, the strong π-π-interactions between the donor/acceptor pairs are enhanced by the respective hydrophobicity of the rings and the polar carboxylate groups attached to the tether. In water, the π-systems are found to self-assemble into single stacks of either two or three discrete donor/acceptor pairs.[51]

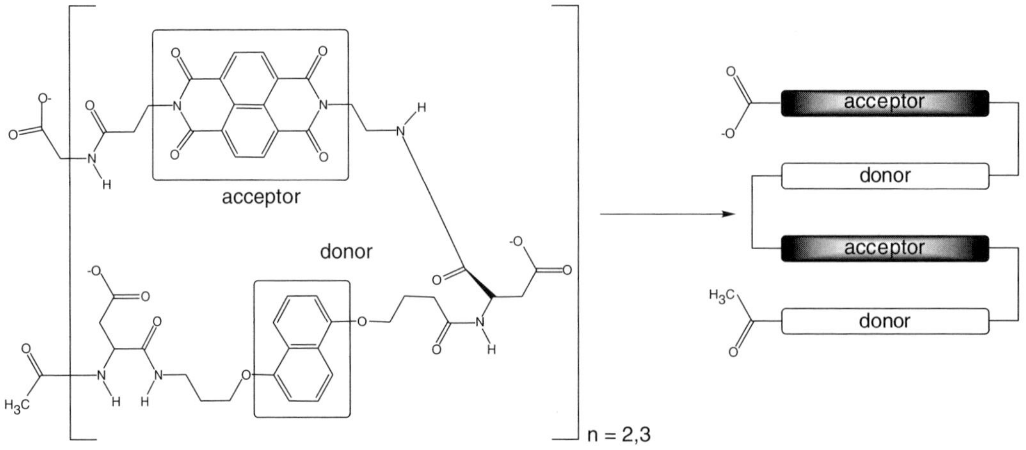

FIGURE 18.18 Aedemer molecules (left) and their directed stacking in water (right).

18.3.2 Covalent Architectures and the Molecular Tinkertoy Approach

The covalent bond is central to all MBB designs. The strength and directionality of these bonds define the shape of the subunits, thereby directing the formation of all larger structures stabilized by covalent bonding or noncovalent interactions. Covalent bonds are typically insensitive to environmental variables, such as the choice of solvent or the ambient temperature. The electrostatic interactions used to stabilize multimolecular structures, in contrast, are often strongly affected by these environmental factors. Covalent bonds offer far greater positional specificity and structural invariance than their noncovalent analogues. The chemical reactions used to form covalent bonds occur preferentially at specific positions on a molecular framework through the placement of suitable functional groups and the control of reaction conditions. Furthermore, covalent bond formation, in contrast with noncovalent interactions, is typically irreversible without concerted efforts to break them. Beyond the formation of the strong connections, the predictability of covalent architectures also allows for control of structure with great accuracy.

The fabrication of larger structures from covalently linked MBBs is based upon the use of individual subunits as rigid building blocks to incrementally build highly stable structures. Covalently linked nanostructures and covalent molecular scaffolding offer the same advantages that stable support structures provide at the macroscale. The shapes of rigidly bound structures are usually reliable over long periods of time. Covalent bond energies for familiar organic structures are an order of magnitude stronger than many of the electrostatic interactions currently employed for the formation of many supramolecular lattices. The continual breaking and reforming of these electrostatic interactions in supramolecular systems, while providing these structures with fault-tolerance and energy-driven self-maintenance, make their interconnectivity very sensitive to their surroundings. Covalently linked structures are themselves structurally stable under similar conditions, and any structural variance comes in the form of deformations instead of bond breaking and reforming. The chemistry involved in forming nanostructures from covalent bonds can be well defined and unidirectional with the correct choices of functional groups and reaction conditions. While the self-assembly methods of supramolecular chemistry provide a means to forming stable structures through the engineering of specific interactions into subunits, covalent connectivity can be directed with great positional control through the rational use of reaction pathways.

The formation of covalently bound nanoscale structures from molecular subunits is common in chemistry and materials science. The most common examples come from polymer chemistry, where small lengths of randomly oriented monomers become long chains of highly interwoven materials as the scale of the system is increased from Angstroms to nanometers and beyond. The formation of highly ordered, covalently bound nanoscale architectures and macromolecules is far less common in chemistry, as the controlled formation of nanoscale structures from covalent bonding is problematic in both of the routes currently proposed. In the engineering-based, top-down approaches, the positional specificity required for fabricating macromolecules from covalent bonds is simply not available, as the MBBs used for their formation are too small to be controlled and placed with any specificity. One might consider the assembly methodology of these approaches to be "too precise" for the selection of MBBs, as the desired level of control places severe restrictions on the design process and the choices of MBBs. In the bottom-up approaches of solvent-based chemistry and atomic manipulation, the reliability of positional accuracy becomes suspect in assemblies formed from rigid, highly stable connections. Errors in the placement of atoms or MBBs within a given framework, because they are irreversible without a level of chemical manipulation that also jeopardizes the structural integrity of the remaining covalent bonds, can potentially render a fabricated assembly useless with a single misplaced bond. Here, the idealized assembly process of solution-based methods may be considered as "too statistical."

To overcome the limits of both approaches, a fabrication process must successfully address positional control, connectivity, and the chemical manipulation of the reaction centers. The basis of supramolecular chemistry is the formation of a macromolecular assembly from weaker, noncovalent interactions; a wealth of examples demonstrates the validity of the approach.[36,52] The means to covalent supramolecular chemistry need not be dissimilar from this already proven approach to macromolecular formation. A covalent-based approach must, however, rigorously control the reaction conditions and the assembly progress of the larger structures to prohibit the unwanted interactions that are, in supramolecular chemistry, easily removed through the control of the ambient conditions. The scope of synthetic chemistry is narrowed considerably when the discussion is limited to the formation of nanoscale architectures from covalent bonding between MBBs instead of only the manipulation of covalent bonds within a single molecule. To illustrate the considerations and limitations of covalent-based nanostructure design from MBBs, one of the most well-developed chemical approaches is detailed below.

The "molecular Tinkertoy" approach[53] to nanoscale scaffolding is based upon the treatment of molecules as simple, rigid construction components or *modules*. The features of the modules that are considered most important in this approach are those required for the construction of the assembly, such as the module length and the availability of suitable bonding positions on the module for connectivity to other subunits. Within the Tinkertoy paradigm, all of the required components and critical fabrication issues are based upon only covalent bonding. The engineering kit of the modular chemist consists of (1) rigid rod molecules of variable lengths, (2) connectors to act as corners or intersections for the scaffolding, and (3) a chemical means to control the assembly of the rods and connectors[53] (Figure 18.19). Such a kit at the macroscale is already familiar to any student of organic chemistry in the form of

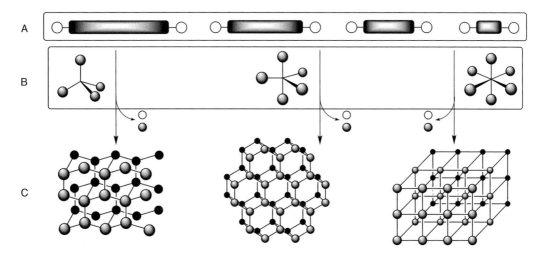

FIGURE 18.19 The engineering kit of the Tinkertoy chemist. (A) Rigid rods of various lengths. (B) Connectors and junctions. (C) A means to covalent assembly to create nanoscale scaffolding.

molecular models, although the construction of scaffolding from the molecular kit is far more challenging than the fitting together of pieces of plastic. Within the context of covalent bonding, each of these three aspects of Tinkertoy design can be treated independently. The fabrication of rigid rods, for instance, can take inspiration from any chemical designs that result in linear structures, regardless of the choice of connectors or the development of the chemical pathways to assemble the nanostructures. The shape of a molecular scaffolding is defined by the connectors; and the engineering of a repeating structural motif, be it a simple cube or a diamondoid-based tetrahedral motif, is accessible based on the choice of the appropriate connector from among the available molecules that allow for the specific connectivity (Figure 18.19). The issue of chemical control becomes the most difficult of the three to handle, as the ordered assembly of extended arrays from simple rods and connectors cannot be controlled from the highly orchestrated procedures used for macroscale scaffolding construction, although the required chemistry is easily applied to the individual connector–junction reactions.

The concepts of the Tinkertoy approach are applicable to all structural features, including the formation of junctions and the assembly of the larger structures in solution. The most exhaustive treatment of the approach thus far has been for the linear, rigid rods used to define the dimensions of the scaffolding. While the number of molecules capable of acting as subunits for linear rods is large, the initial series of proposed subunits has been limited to a select set of twenty-four. The scope of this discussion is limited to the manipulation of these different modules for both the formation of linear rods and the design of molecular junctions. The chemistry of the twenty-four modules has been extensively developed and reviewed in the interest of firmly establishing the precedent for the first components of the engineering kit.[10] These twenty-four linear modules, shown in Figure 18.20, share a number of important characteristics that are briefly described below.

1. Stability

   The most important features to consider with respect to the environment of a nanostructure are the stability and reactivity of its components. Unless chemical functionality is required for an application, the best choices of MBBs are those that will react only during the formation of the covalent architecture. The subunit should, therefore, be inert with respect its chemical environment after assembly.

   The most common structures from among the initial MBBs that provide this level of chemical predictability are the saturated hydrocarbons (Figure 18.20A). These molecules rely exclusively on the use of strong σ-bonding between carbons and hydrogens to form rigid structures and are ideal for rigid rod fabrication. Their interconnected frameworks limit their flexibility while at the same time providing a molecular axis through which linear dimers, trimers, etc., can be formed via single σ-connections. The remaining saturated hydrocarbons (Figure 18.20D) differ from the cage structures by the inclusion of two bonding sites per pair of axial carbons. With these modules, either several σ-bonds can be used to form rigid structures, or both σ- and π-bonding can be used to create single connection points with restricted rotation (Figure 18.20D). The carboranes, a second class of molecules, display extreme stability and unique connectivity within a very small space (Figure 18.20B). The deltahedral framework of the cluster skeleton prohibits appreciable flexibility within the subunit, while the radial bonds of the apical carbons in $C_2B_{10}H_{12}$ and $C_2B_8H_{10}$ provide rigorously linear external linkages. Furthermore, these clusters have been shown in many instances to be remarkably stable compounds under very harsh conditions.[54] The remaining modules contain one or more π-electron systems. While π-systems are more susceptible to chemical reactions than saturated systems, much of this reactivity can be limited through the proper control of the surroundings. The molecules containing π-electrons are the only systems from the original series of subunits that provide a means to form stabilizing electrostatic interactions in solution (e.g., π-interactions, hydrogen bonding, or dative bonds).

2. Size

   Greater control of the size of a nanoscale assembly is possible by using many smaller subunits rather than few larger subunits. The twenty-four initial modules are among the smallest rigid

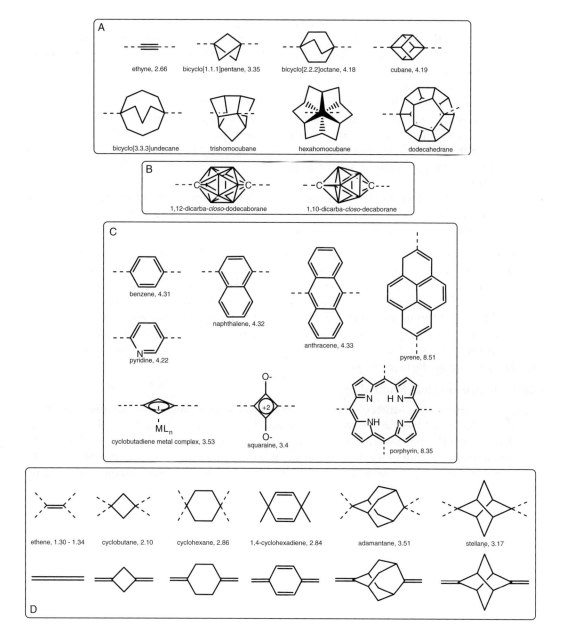

FIGURE 18.20 The 24 original modules. (A) Saturated hydrocarbons. (B) Carboranes. (C) Linear-connecting π-systems. (D) σ/σ and σ/π connectors. As applicable, names and incremental lengths (in Angstroms) are provided.

molecules known, and no single module crosses the nanometer threshold. Ethyne, for instance, is the smallest organic subunit available that provides linear connectivity on both ends through σ-bonding. A linear rod in a nanoscale scaffold can be fabricated from the available modules to "add up" to some required length. The rigid bonding within each module results in the structure having some fixed distance between the axial connection points which, when added to a typical single C-C bond length to account for the extra-module σ-linkage, defines a distance termed an *incremental length* (Figure 18.20). In order to construct a rod of some predetermined length, the only feature that needs to be considered from among the available modules is the incremental length between axial connection points. Having determined which modules are required to fabricate a rod of some predefined length, a chemical pathway can be employed based upon the known reaction chemistry of each subunit. As necessary, the general approach may be applied to any other molecules or combinations of molecular subunits for the fabrication of rods of an absolute length.

3. Chemical Precedent

The design of linear rods from the available modules is both flexible and straightforward. With few exceptions, chemical precedent exists for the syntheses and linking of all twenty-four modules.[10] Furthermore, the chemistry required for linking together different modules has also been demonstrated. Co-oligomers, chains of subunits composed of two or more different modules, are important both for customizing the lengths of the linear rods and for altering the solubility properties of the larger structures. Of particular importance in the linear rod treatment is the ethyne bridge. Ethynyl linkages are ideal for improving the solubility of molecular rods while minimizing the increase in chain length. A great deal of chemical precedent also exists for their inclusion into a number of modular structures.

Many linear molecular rods have been synthesized from the collection of modules. Beyond the formation of the rods is their connection to either two-dimensional junctions to form planar molecular grids or three-dimensional junctions to form molecular scaffolding. While covalent junctions have not been fully addressed, a number of the original twenty-four modules offer both structural flexibility and chemical precedent beyond their useful axial bonding. Specifically, the symmetry and connectivity of certain modules are appropriate for the formation of diamondoid, honeycomb (hexagonal), and cubic molecular lattices through familiar chemical manipulation. These lattices and the modules appropriate for their juncture are discussed below.

18.3.2.1 Diamondoid Scaffolding

Diamondoid structures are networks of tetrahedra in a molecular or macromolecular lattice (Figure 18.21). Within the lattice are two basic structural features. The first and most fundamental feature is the tetrahedral center (Figure 18.21A), to which four adjacent tetrahedra are attached. The smallest tetrahedral-based structural motif in the diamondoid lattice is the adamantanoid framework (Figure 18.21B). In the actual diamond framework, the tetrahedral centers are sp³-hybridized carbon atoms, and the

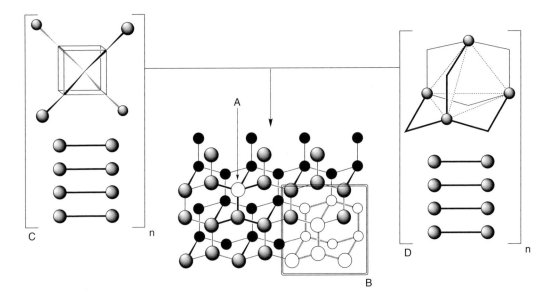

FIGURE 18.21 Diamondoid nanoscaffolding. (A) Tetrahedral module. (B) Adamantyl subunit. (C) Cubane assembly. (D) Adamantane assembly.

repeating motif is the adamantyl frame. The strength of diamond at the macroscale stems from the strength of the carbon–carbon σ-bonds and the extensive connectivity of the carbon atoms within the diamond network. The formation of MBB-based diamondoid frameworks has been explored in a number of coordination and supramolecular designs.[55–57] The noncovalent interactions within these diamondoid lattices offer reasonable strengths, the same high connectivities, and the spontaneous self-assembly of the subunits into rigid lattices. For the fabrication of extended arrays of diamondoid lattices, this self-assembly feature is particularly attractive, because the synthesis of molecular diamond has been limited to small molecules based more on incremental growth of adamantane frames[58] than the actual formation of rigid, covalent arrays.

Covalent diamondoid structures offer structural rigidity and controllable assembly intermediate between molecular diamond and the noncovalent MBB designs. The Tinkertoy approach offers a plausible means to the formation of such covalent diamondoid arrays. To construct these arrays with linear rods, the required molecular junctions must have tetrahedral symmetry elements that can connect through σ-bonds at the tetrahedral centers. Adamantane and cubane provide both the required tetrahedral symmetry elements for the placement of the linear rods and the synthetic precedent for their covalent attachment. Among the modules bicyclo[2.2.2]octane, bicyclo[1.1.1]pentane, bicyclo[3.3.3]undecane, trishomocubane, hexahomocubane, and dodecahedrane, structures with either tetrahedral centers or quasi-tetrahedral bonding positions (threefold rotation axes exist that include the axial connection points for the linear rods), either the chemistry has not been developed for tetrahedral assembly or the structures are too flexible to adequately control the diamondoid assembly. The control of functional group placement at the tetrahedral corners of both adamantane and cubane has been well developed, with many of these same functional groups employed for the syntheses of linear rods from these two modules. The control of tetrahedral adamantane functionalization has already been exploited for the formation of supramolecular building blocks in diamondoid lattice formation. In these MBBs, carboxylate groups are used to form strong hydrogen bonding interactions with neighboring adamantane frames, effectively extending the connectivity of the trimesic acid complex into a third dimension.[55] The covalent attachment of linear rod modules has also been demonstrated by way of a tetraphenyl adamantane derivative (Figure 18.22) that has been used as an MBB for subsequent macromolecular syntheses.[59]

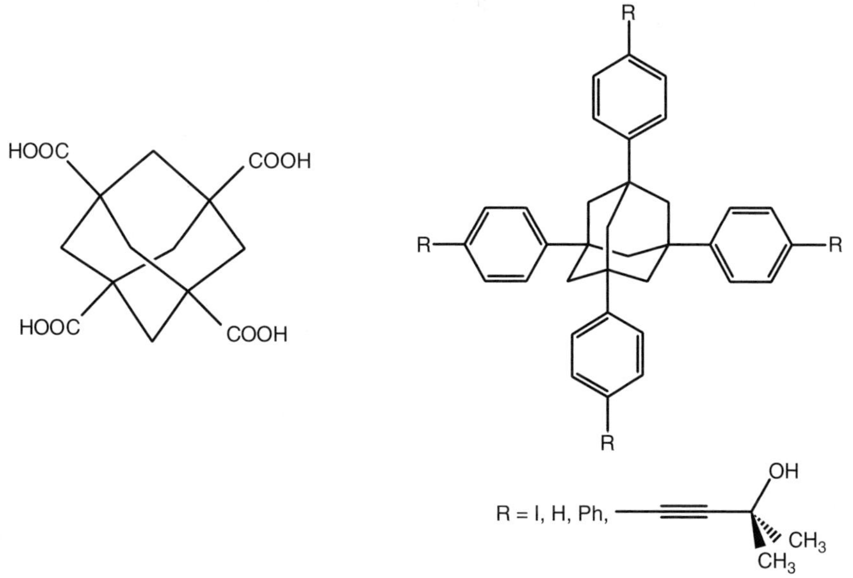

FIGURE 18.22 Adamantane-based MBBs for supramolecular design. (A) Adamantane-1,3,5,7-tetracarboxylic acid for supramolecular designs from hydrogen bonding. (B) Adamantane-based fragment with module linkages and known substituents.

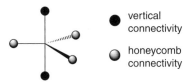

FIGURE 18.23 Honeycomb/vertical stacking connectivity in D$_3$-symmetric modules.

18.3.2.2 Honeycomb Lattices

Macromolecular honeycombs require two different modes of connection (Figure 18.23). The hexagonal planar array is formed by the connection of linear molecules to triangular junctions. With the hexagonal plane formed, the vertical stacking of these structures is performed by attachment of the triangular junctions through chemical bonds perpendicular to the hexagonal plane. The ideal junctions for honeycomb designs are then molecules with trigonal bipyramidal symmetry, providing the ideal connectivity for linear rod structures in all directions. Such junctions are readily available from familiar coordination compounds. These structures, however, do not provide the structural stability of covalently bound junction/rod linkages. Although no single module addresses all of the design issues entirely, three are available that individually account for specific aspects of the honeycomb design.

Planar hexagonal scaffolding has already been addressed in the structure and chemistry of benzene. The placement of functional groups at the 1,3,5-positions of the benzene ring (Figure 18.24) yields the required triangular connectivity for the junctions, while an extensive chemical precedent for benzene functionalization makes the ring ideal for such applications. The propensity of trimesic acid to form

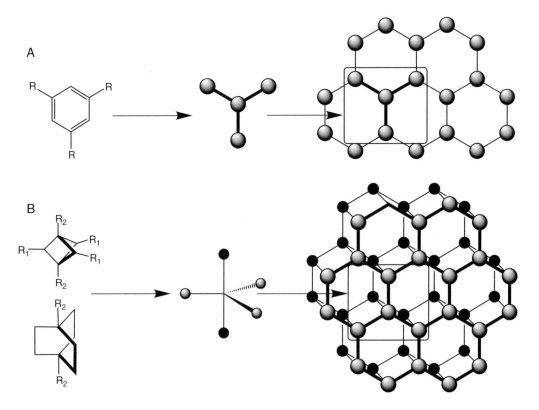

FIGURE 18.24 Examples of honeycomb scaffolding. (A) Planar structures formed from 1,3,5-substituted benzene rings. (B) bicyclo[1.1.1]pentane and bicyclo[2.2.2]octane modules as potential subunits for three-dimensional honeycomb structures.

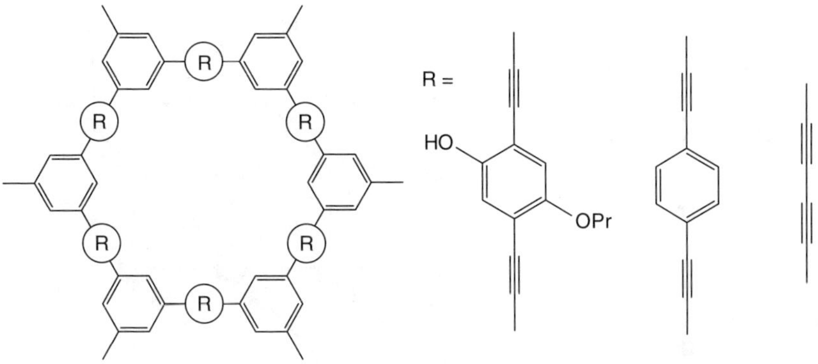

FIGURE 18.25 Isolated hexagonal macromolecules from benzene junctions.

extended arrays of hexagonal cavities from carboxylate-based hydrogen-bonding interactions clearly demonstrates the importance of the geometry of the junction in directing the formation of the larger structures in solution (Figure 18.8). This same chemical design can be and has been employed successfully in a number of isolated benzene-based systems employing linear rods (Figure 18.25). Among the many known hexagonal macromolecules employing benzene junctions, many incorporate linear structures similar or identical to rod designs from the selected modules.[60–62]

The limitation of the benzene ring for scaffolding design is its planarity. While π-stacking interactions might be employed to form vertical honeycomb scaffolding, the covalent connection of hexagonal arrays into the third dimension is impossible with the benzene ring alone. From a structural standpoint, however, it is important to note that the only function of the benzene junction is to provide a triangular framework. Any other modules that incorporate equilateral triangles within their covalent frames will perform the same task. From among the remaining modules, the bicyclo[1.1.1]pentane and bicyclo[2.2.2]octane cages provide the correct symmetry and structural elements for the formation of planar arrays and vertical stacking through covalent bonds (Figure 18.25). The bicyclo[1.1.1]pentane is the better choice for designing such systems, as the structure is less flexible than the octane cage, and the carbons used for forming the hexagonal array from the linear rods have their available σ-bonds oriented in the hexagonal plane. The current limitation with the pentane cage for hexagonal designs is the synthetic precedent for the functionalization of the equatorial carbons, although these issues have recently received significant attention.[63]

18.3.2.3 Cubic Scaffolding

Idealized cubic lattices from the molecular Tinkertoy approach share a number of similarities with both the diamondoid and honeycomb designs. Structural connectivity in cubic lattices begins with octahedral junctions (Figure 18.26). Provided the junctions have ideal octahedral symmetry, the cubic lattices appear uniform with respect to all perpendicular sets of axes. The high symmetry of the idealized junction, as was found in diamondoid structures, permits the outward growth of the lattice from a single point by the addition of quantities of junction and linear rod without orientational preference. This simplifies the required control of the growth process relative to honeycomb structures, which have two different types of covalent connectivity that must be considered. Unlike the diamondoid structures, however, lattices formed from octahedral junctions have a very well-defined layering scheme along each axis. Therefore, a plausible growth process for the entire cubic lattice can mimic the same processes used for honeycomb growth, where a single layer is formed from two orthogonal sets of connections, while a third set perpendicular to the growing lattice plane remains unused until vertical connectivity is required. In instances where the growth process is selected to mimic the honeycomb methodology, the idealized octahedral junction can be separated into a square planar component and a perpendicular axial component. The selection of planar or vertical connectivity can be controlled during the growth process by chemical manipulation of the two distinct growth directions (Figure 18.26B).

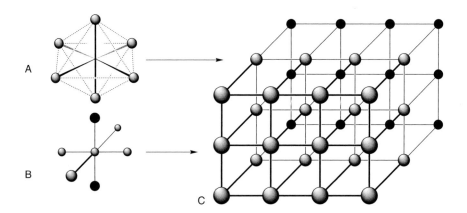

FIGURE 18.26 Cubic scaffolding. (A) Octahedral subunits for uniform structure growth. (B) Square-planar connectivity and vertical stacking connections for deformed cubic lattices. (C) A cubic lattice.

No single module provides the idealized octahedral connectivity required for uniform lattice growth in all directions. The design of two-dimensional square planar lattices can be readily designed from single σ-bond connectivity using porphyrins and cyclobutadiene metal complexes or double σ-bond/mixed σ–π connectivity using cyclobutane rings, stellanes, or adamantanes. Beyond the initial designs, however, the limited chemical precedent of a number of these modules prohibits their current usability. From these initial five modules, the porphyrins have been successfully employed in a number of rectangular and square planar arrays because of their extensive synthetic precedent and the availability of subsequent vertical connectivity through slight structural modification[64–65] (Figure 18.27). A number of linear rods have been used to connect porphyrins together, including ethynyl chains,[66] benzene chains,[67,68] chelating ligands,[69–72] and other porphyrins.[73–75] While the square planar framework has also been demonstrated

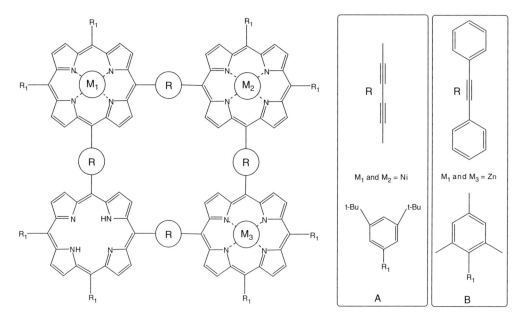

FIGURE 18.27 Porphyrin squares, connecting linear rods, and peripheral substitutions. (Set A from Sugiura, K., Fujimoto, Y., and Sakata, Y., A porphyrin square: synthesis of a square-shaped π-conjugated porphyrin tetramer connected by diacetylene linkages, *J. Chem. Soc., Chem. Commun.*, 1105, 2000; Set B from Wagner, R.W., Seth, J., Yang, S.I., Kim., D., Bocian, D.F., Holten, D., and Lindsey, J.S., Synthesis and excited-state photodynamics of a molecular square containing four mutually coplanar porphyrins, *J. Org. Chem.*, 63, 5042, 1998. With permission.)

with the cyclobutadiene metal complexes, the extension of these arrays into the third dimension is prohibited by the use of metal complexation to stabilize the highly reactive four-member ring.

Because no single module provides a chemically feasible route to vertical stacking after the formation of the square planar array, alternative stacking interactions must be employed for the formation of quasi-octahedral complexes. The porphyrins provide this added functionality by way of metal complexation within the central core. The coordination center within the porphyrin core then requires the use of metal–ligand complexation to form the vertical stacking interactions. The same directionality provided by covalent σ-bonding is still available from metal–ligand coordination, however, and the relative strengths of these stabilizing interactions can be controlled by the choice of metal. While a module-derived dipyridine structure is plausible based on the axial positions of the nitrogen lone pairs, the known vertical stacking motif has been performed with 1,4-diazabicyclo[2.2.2]octane,[76] the axial coordination analogue of bicyclo[2.2.2]octane (Figure 18.28).

The exclusive reliance on covalency for the fabrication of a nanostructure is not without important limitations. One limitation stems from the essentially irreversible formation of covalent bonds. In the formation of larger systems, extreme care must be taken to make chemical reactions as predictable and unidirectional as possible. The thermodynamically driven self-correction mechanisms of biological systems and supramolecular crystals cannot be used to repair an "incorrect" covalent bond without jeopardizing the structural integrity of the remaining structure. When an unwanted covalent bond forms, the means to correcting the error often involves harsh chemical manipulation. Thus, when a chemical route is chosen to correct some structural error, the pathway must be tailored to avoid reacting with any other part of the molecular superstructure. Also, because a chemical reaction is required to form a covalent bond, any structures employing a covalent bond are not strictly self-assembling. In a hydrogen-bonding network, for instance, the lattice forms due to electrostatic interactions between donors and acceptors. The stability that comes with these weak interactions may be small, but the formation of the larger network provides significant stabilization and the structure spontaneously forms. The formation of covalent architectures typically requires control of environmental conditions and subsequent purification of the desired product from the remainder of the reaction mixture.

A variety of chemical considerations associated with the synthesis and characterization of these structures has also been considered within the context of the Tinkertoy approach.[10] First among these considerations is the solubility of the progressively larger structures. The growth of larger structures is often limited by the ability to keep the assembly in solution. The chemical methods most likely to keep a larger structure in solution, such as the addition of side chains or the use of charged species, often have their own drawbacks. For instance, the application of these solvation techniques can affect the function of the nanostructure in unpredictable and undesired ways. With issues of solubility come problems of separation and purification. Such issues are familiar to biochemists, however, and many of the same techniques that have permitted the separation of biomolecules can also be applied to nanostructures.

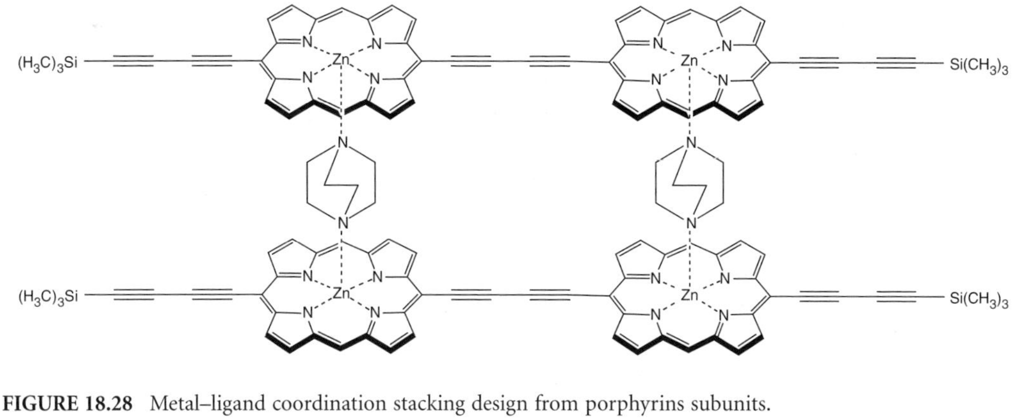

FIGURE 18.28 Metal–ligand coordination stacking design from porphyrins subunits.

18.3.3 Transition Metals and Coordination Complexes

One of the great advances in macromolecular design has been in the development of a variety of metal complexation motifs for the formation of two- and three-dimensional nanostructures. The design features here are based on the chemistry of small metal–ligand compounds, where the coordination requirements of the metal direct the attachment and orientation of ligands. The formation of larger geometric structures from metal–ligand compounds typically comes through the use of ligands with two or more separate metal-coordinating regions (Figure 18.29). In two-dimensional designs, the ligands typically constitute the sides of the structure while metal complexes define the corners. In three-dimensional designs, the ligands delineate the faces of the structures with the metals occupying the vertices. The chemistry involved in the formation of these nanostructures is often straightforward. The nanoscale assembly of coordination complexes is typically accomplished by the removal of labile ligands from some coordinately saturated metal complex in solution, a process greatly simplified by the relatively weak strengths of many metal–ligand bonds.[77] Coordination-based methods not only allow for the formation of symmetric molecular nanostructures but also provide for the formation of molecular cavities through ligand encapsulation pathways[78–80] (Figure 18.30).

The vast majority of coordination nanostructures have been based upon the use of chelating organic ligands, with either nitrogen atoms as the lone-pair donors or cyclic ligands with hydroxyl (-OH) groups used to provide metal connectivity through relatively weak covalent metal–oxygen bonds. Nitrogen-based ligands have been used far more often in coordination-based nanostructure design and are preferred among other ligand types for a number of structural reasons. The nitrogen atom is a close structural analogue and is isoelectronic with a covalent C-H fragment, making it quite versatile in the modification of organic ligands for metal complex formation (Figure 18.31). Whether incorporated into a saturated

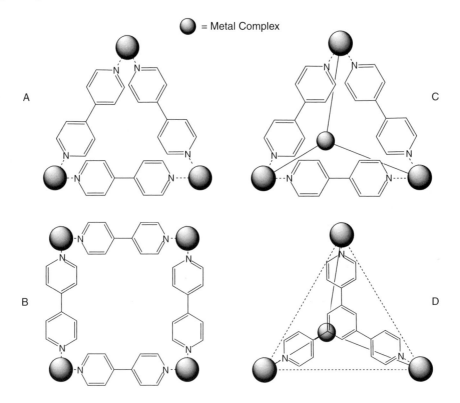

FIGURE 18.29 Metal–ligand structural motifs. A (triangle) and B (square) are two-dimensional structures with the ligands defining the sides. C and D (tetrahedrons) are three-dimensional structures with the ligands defining either the sides (C) or the faces (D).

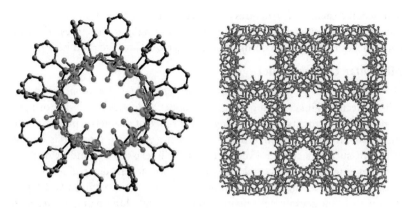

FIGURE 18.30 Structures with metal–ligand coordination cavities and channels.

aliphatic framework or directly into an aromatic ring to form a heterocycle, a nitrogen atom and a C-H fragment are nearly identical in terms of hybridization and geometry except that a hydrogen atom is required to form a complete electronic octet for carbon. The radial orientation of the lone pair from the backbone of many ligands provides accessibility to coordinating metals, while the σ-bond quality of the lone pair provides predictable, unidirectional coordination based on the geometry of the nitrogen atom in the ligand. These organic ligands can be designed such that the nitrogen lone pair is the only site on the ligand available for coordination to the metal center in nanostructure formation. The nitrogen lone pair, in the absence of a Lewis acid, becomes the reaction center for complexation only when the metal center becomes coordinatively unsaturated, typically by chemical methods too gentle to affect the ligand framework. This predictability in nitrogen–metal coordination comes from a vast synthetic precedent, ranging from the simple coordination of NH₃ to the complexation of multidentate ligands that serve to singly saturate the metal coordination sphere. Furthermore, the dissociation of the nitrogen-based ligand from the metal center has little effect on the stability of the ligand itself, providing a thermodynamic means for controlling the formation and self-maintenance of these systems. The chemical modification of these ligands also has significant structural implications for the resulting assemblies. Simple modification to the organic framework of these ligands can target macromolecular structures to within a few Angstroms of some specified size (Figure 18.31). Similar to the molecular Tinkertoy approach, ligands can be modified either step-wise through the addition of linear linkers (such as acetylene) or more subtly using nonlinear linkers, such as either flexible ring systems or saturated organic chains.

Similar to the study of structure and function in biomolecules (*vide infra*), much of the initial work in metal complexation involved the modification of known structures to create new structures. As the

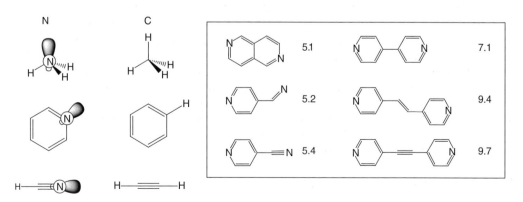

FIGURE 18.31 The inclusion of nitrogen (left) lone-pair donors into simple organic frameworks and a selection of N-N distances (in Angstroms) in common organic ligands.

field has progressed, the catalog of structures and reactions has increased to the point where trends and designs have been focused into general strategies for fabricating new structures. The two most actively investigated approaches to designing coordination architectures are discussed below.

18.3.3.1 Molecular Library Model

The molecular library model,[81] also known as the *directional-bond approach*,[77] is the metal–ligand analogue to the molecular Tinkertoy approach. The model addresses the design of nanostructures by using a set of molecular fragments encompassing a wide range of geometric patterns for the fabrication of two- and three-dimensional structures. In this approach, a *geometric fragment* is simply some subunit of a larger structure, such as a corner, a vertex, or a side. To classify a ligand or metal complex into a particular fragment category, a structural analysis is performed to determine the angles among all available coordination sites in the molecular framework. The choice of ligands is typically limited to rigid molecules with monodentate coordination modes (single lone pairs) in order to improve the predictability of the method for nanoscale design.[82] The number of candidate ligands is very large, however, and the restriction to molecules with limited degrees of freedom does not significantly affect the flexibility of the method. The rigidity of both the ligands and the metals is used only to restrict the choices of geometric fragments for particular designs, and a small amount of flexibility in the ligands and metal coordination sphere is expected in the assembly process. An important aspect of this approach is that both ligands and metals can be used as the fragments to form a structural feature. A nonlinear or multi-branched ligand, for instance, can be used as a corner or a vertex just as a metal with axial coordination sites can be used as a side. It is the higher coordination of ligands to a metal center that sets the metal apart from organic systems, however, and the metal is most frequently employed as the more complicated geometric fragment.

The range of available ligands and metal complexes has been divided into two libraries based on the dimensionality of the desired structure[81] (Figure 18.32). For the design of two-dimensional nano-structures, such as regular polygons or polycyclic assemblies, the classification of doubly connecting, or *ditopic*, geometric fragments requires only three points. In the ligand, these points are composed of two lone-pairs and the center of the covalent framework of the ligand (Figure 18.33). In the metal complex, these three points are the two coordination sites for the connected ligands and the metal atom (Figure 18.33). The internal angles of the desired nanostructure then determine which fragments can be used for its fabrication. For the fabrication of cyclic polygons with three to six sides, the internal angles and combinations of geometric fragments required are summarized in Figure 18.32 (A–I). It is important to note that these ditopic classifications define only individual sets of binding angles within a molecule. Within a molecule used as a geometric fragment in a larger structure, it is possible to have independent sets of binding angles. Consequently, structures with multiple planar rings are possible (Figure 18.33).

Three-dimensional nanostructures are fabricated from combinations of tritopic and ditopic geometric fragments. Symmetric three-dimensional structures resulting from various combinations of tritopic and ditopic fragments are shown in Figure 18.32 (J–M). The design strategy for new nanostructures is the same in both two- and three-dimensional systems, except the additional level of complication of three-dimensional structures requires more elaborate geometric fragments for the assembly. In both library sets, the linear linkage serves the important roles of length extender and coupler for identical fragments. It should be noted that length is not a factor considered in the classification process. Modifying the length of a structure is a matter of either modifying the molecular bridge between coordinating regions of a ligand or using linear subunits of the appropriate length with metal complexes at the corners (two dimensions) or vertices (three dimensions).

18.3.3.2 Symmetry Interaction Model

The symmetry interaction model[80,83] is founded in the understanding that many highly symmetric, naturally occurring structures are formed as a consequence of incommensurate lock-and-key interactions between the subunits.[84] The method, as applied to metallocycles, is then retrosynthetic in principle, using

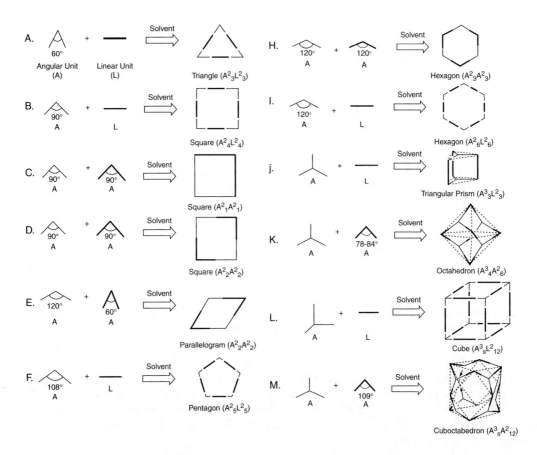

FIGURE 18.32 Metal complex and ligand classifications for two- and three-dimensional coordination nanostructures in the molecular library approach. (From Stang, P.J. and Olenyuk, B., Self-assembly, symmetry, and molecular architecture: coordination as the motif in the rational design of supramolecular metallacyclic polygons and polyhedra, *Acc. Chem. Soc.*, 30, 502, 1997. With permission.)

the known geometric features of highly symmetric polyhedra to direct the formation of new metal–ligand assemblies. This model differs from the molecular library model in two important respects. First, there is a definite division between the role of the metal and the role of the ligand in the symmetry interaction model. This is in contrast to the molecular library, where both ligands and metal complexes can be used

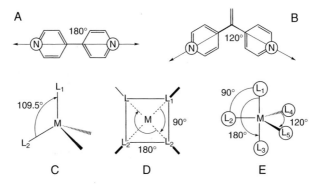

FIGURE 18.33 Binding angles in a selection of ligands (A, B) and metal complexes (C–E). Note that two unique angles are available in square planar structures (D), and three unique angles are available in trigonal bipyramidal structures (E).

anywhere within the skeleton of a nanostructure to create corners, vertices, or sides. Second, through the selection of geometric fragments in the molecular library approach, the binding angles within each metal are determined by the orientation of the leaving groups on the metal. The remainder of the metal coordination sphere is saturated with other ligands that retain their coordination positions during the nanostructure fabrication process. The symmetry interaction model relies on the strong binding of chelating ligands to saturate the entire coordination sphere of the metal ion.[82] The coordination sphere of the metal, then, is responsible for defining the orientation of the ligands in the final structure, while the ligands are responsible for forming the sides (between metal–metal pairs) or faces (binding three metals) of the structure (Figure 18.29). The use of chelating ligands in the symmetry interaction model has the benefit of increased stability in the final structures through the formation of multiple coordination bonds per ligand and the inherent kinetic stability that comes from the chelate effect.[8] The important components in the symmetry interaction model are the orientation of the lone pairs of the chelating ligands within the organic framework and the geometry of the coordination sphere of the main group or transition metal atoms.

The development of a rational design strategy for the symmetry interaction model is more complex than for the molecular library model. In the symmetry interaction model, a library of angles and interactions based solely on the choice of metal or ligand is not employed. Instead, the design of nanostructures from this approach requires an understanding of the chelating ligands and the coordination sphere of the metal ion. Among the coordination nanostructures, most of the designs applicable to the symmetry interaction approach are based on the use of tetrahedral (4-coordinate), square planar (4-coordinate), and octahedral (6-coordinate) structures (Figure 18.34). As the ligands themselves are not responsible for imparting dimensionality to these designs, polyhedral coordination nanostructures based on the symmetry interaction approach employ metal ions with octahedral coordination spheres. This limitation does simplify the design process because it is possible to classify the available metal ions according to their coordination numbers. Because this methodology requires that the metal be stripped of ligands prior to nanostructure formation, it is also possible to select metal–ligand starting materials based on the lability of the metal complex ligands under certain reaction conditions.

A means has been developed to understand the spatial relationships between the metal coordination sphere and the attached ligands by defining common geometric features and determining their importance in the fabrication process.[80,84] Because the method relies heavily on the use of symmetry to define the geometric features of both the interactions and the assemblies themselves, highly symmetric structures, such as Platonic solids, can be fully analyzed by considering their vertices. The coordination sphere of the metal, where all of the connectivity and structural determination occurs, is divided into a *coordinate vector*, a *chelate plane*, and an *approach angle* (Figure 18.35). The coordinate vector is defined as the vector between the coordinating atom(s) of the ligand and the metal. In chelating ligands, the bisection point of the lone pairs and the metal atom forms this vector. In monodentate ligands, this vector is simply along the lone pair–metal bond. The coordinate vectors and the rotation axis of the metal that would

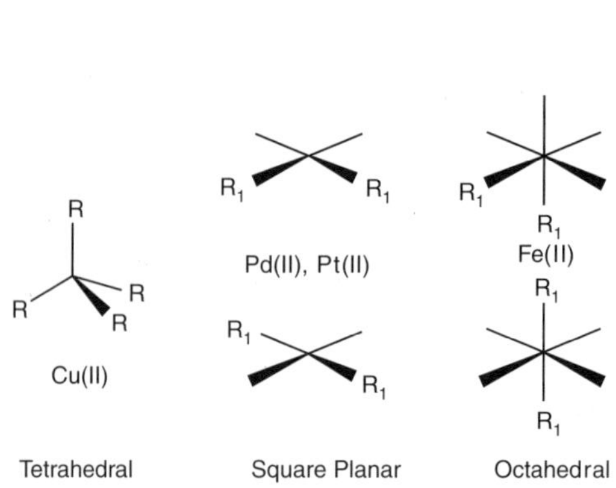

FIGURE 18.34 Coordination geometries for common metal ions.

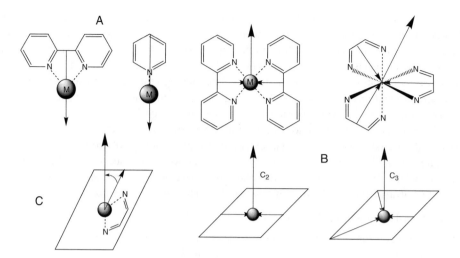

FIGURE 18.35 (A) Coordinate vectors. (B) Chelate plane. (C) Approach angle.

transform one ligand into an adjacent ligand then define the chelate plane. In metals with three ligands, only the three coordinate vectors of the ligands are required to define the chelate plane. In metals with two ligands, defining a third axis perpendicular to the coordinate vectors and the major rotation axis of the metal–ligand complex designates the chelate plane. The approach angle is defined as the angle between the major axis of the metal center and the plane of the chelating ligands in the final structure.

A demonstration of the geometric features in assembled macromolecules is provided here for two structures (Figure 18.36). In a helical, D_3-symmetry structure composed of two metals and three chelating ligands (simplified to M_2L_3), the orientation of the ligands in the coordination sphere of the two metals requires that the two chelate planes be parallel to one another. The C_3-rotation axes and the C_2-rotation axes that bisect each shared ligand of the two metals are then automatically aligned. Within any such helical or rod-like structure of D_{3h} symmetry, the local features of the metal coordination sphere are consistent. The selection of metal–ligand sets can be directed by the known geometric requirements of the structure. The formation of a molecular tetrahedron from chelating ligands and metals can be completed by either an M_4L_6 combination, where six ligands are required to form each of the skeletal components, or an M_4L_4 combination, where each ligand consists of the three chelating sites related by a C_3-rotation axis through the plane of the ligand (Figure 18.36). Each metal is bound to three chelating

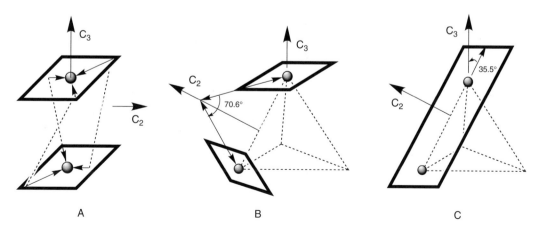

FIGURE 18.36 Structural examples of the symmetry interaction approach. (A) D_3-symmetry helix. (B) Tetrahedron from coordinate vector description. (C) Tetrahedron from approach angle description.

ligands, with the chelate plane defined by the coordinate vectors at each vertex. In the M_4L_6 case, the C_2-rotation axes bisect the ligands along the tetrahedral skeleton. The C_2-rotation axes in the M_4L_4 case are at the same positions in the tetrahedral frame, although the skeletal framework of the tetrahedron is only inferred because all ligands are now facial. The idealized tetrahedron is defined from the coordinate vectors of two coordination centers along a single C_2-axis, making a theoretical angle of 70.6° with respect to the relative orientations of the chelate planes (Figure 18.36). In cases where the chelating ligands are held planar to one another and are orientated antiparallel, the approach angle can be used as the defining feature of the metal–ligand interaction. In a tetrahedron, the use of these ligands requires that the approach angle be 35.3°. Given a coordination nanostructure, all of the isolated metal–ligand complexes can be treated by similar symmetry descriptions. From the geometric relationships of the metal–ligand interactions and the structural features of the isolated metal–ligand complexes, an understanding of many macropolyhedral structures made from metal–ligand interactions becomes possible.

18.3.3.3 Two-Dimensional Structures

Two-dimensional polygons fabricated from metal–ligand interactions have extensive synthetic precedent, and there is virtually no limit to the possible structural combinations that can be made from very simple synthetic modifications.[82] As systematized in the molecular library approach,[81] the specific structural features (side lengths and internal angles) of many regular polygons are chemically accessible by employing bis-monodentate ligands (although bis-bidentate ligands are also appropriate with certain metal centers) and coordinatively unsaturated metal centers. While the control of structural features may not be absolute compared to proposed atomistic[6] or molecular Tinkertoy[53] methods, the approach and the many available structures provide the means for controlling size and shape well below the nanometer threshold in a highly predictable manner. This control over structural features is incorporated both within the extensive synthetic precedent for the organic ligands and the selectivity for coordination number and ligand type among the available metal complexes. With the structural variety of geometric fragments in the molecular library also comes the ability to select for chemical reactivity.

The flexibility of the metal–ligand bond in polygon formation is highlighted here by a selection of palladium- and copper-based systems below (Figures 18.37 and 18.38). The commonly employed

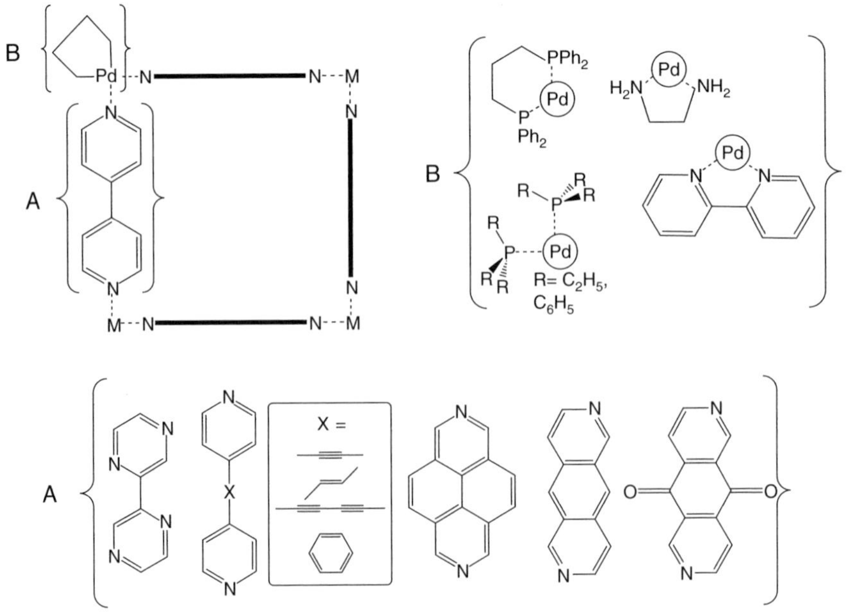

FIGURE 18.37 A selection of known square planar palladium(II) coordination structures. (A) Coordinating ligands. (B) Metal centers with auxiliary ligands.

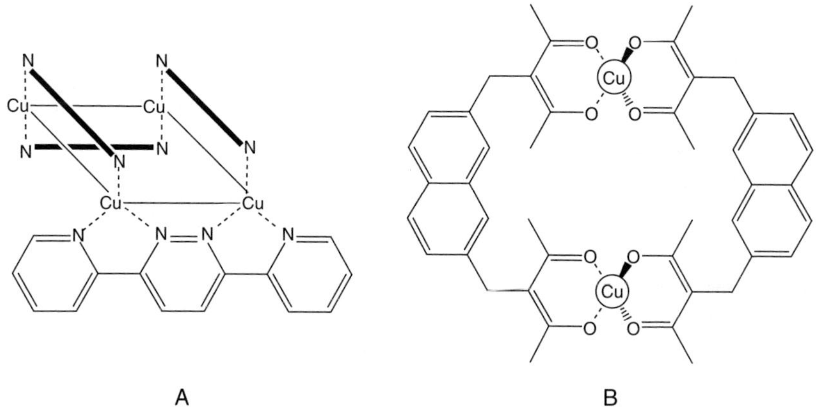

A   B

FIGURE 18.38 Tetrahedral metal–ligand centers for copper(I) complexation structures.

palladium(II) ion (as well as platinum(II)) is square planar, providing a perpendicular pair of coordination sites for monodentate ligands. Consequently, molecular squares are a familiar result of their use. The copper(I) ion, as a tetrahedral coordination center, is useful for connecting bidentate ligands perpendicular to one another.[85–89] Besides the obvious differences in ligand placement and orientation come the differences in ligand mobility. In the palladium-based nanostructures, single metal–ligand σ-bonds provide free rotation about the coordination site and greater structural flexibility. With bidentate coordination in the copper complexes, structural flexibility is greatly limited.

A selection of known palladium–nitrogen coordination combinations is shown in Figure 18.37. Although polygons with more sides are possible, structures usually contain from three to six sides.[82] In the structures where the palladium is used as a corner, the metal is delivered to the reaction mixture with one bidentate ligand (typically diamine [H_2N-R-R-NH_2] or diphosphine [Ph_2P-R-R-PPh_2]) and two labile ligands. Many nanostructures based on the molecular library model incorporate both strongly binding and weakly binding ligands in the same metal complex to allow for greater control over the ligand coordination position.[82] In tetrahedral metal complexes, where the two labile ligands are always next to one another, the difference in metal–ligand bond strengths serves to control which ligands are removed during nanostructure formation (Figure 18.34). In the square planar and octahedral cases, different isomers place the labile ligands at nonadjacent positions. Both the bond strengths and the ligand positions must be accounted for in the selection of the metal complex. When labile ligands are oriented 180° to one another in the palladium systems, these complexes become linear linkages suitable for use as the sides of polygons.

The formation of the palladium(II) nanostructure begins with the removal of the labile ligands. Two common labile ligands in palladium(II) complexes are triflate (OTf⁻) and nitrate (NO_3^-) anions. Their removal leaves both an open coordination site and a positive charge on the metal. The oxidation state of the metal changes as a result of the loss of an unpaired electron to a highly electronegative atom. In OTf⁻ and NO_3^-, the two electrons are lost to form the palladium(II) ion due to the electron-withdrawing oxygens on each ligand. Coordination of bis-monodentate ligands then leads to the formation of the polygon sides. The process is repeated until each metal has lost its labile ligands and coordinated an equal number of nitrogen ligands.

The square planar geometry of the palladium(II) does not limit its applicability to polygons with more or fewer than four sides. The otherwise disfavored formation of strained complexes, such as molecular triangles from square planar palladium(II) cations, can be forced to occur in a system by steric[90] concentration (enthalpy/entropy arguments),[91,92] or guest–complexation effects.[92] For instance, the replacement of the small bidentate ethylenediamine ligand with 2,2′-bipyridine results in the formation of both squares (the preferred structure with the smaller ligand) and triangles in solution from steric effects[92] (Figure 18.39). Both concentration–dependence and guest–complexation were found to play important

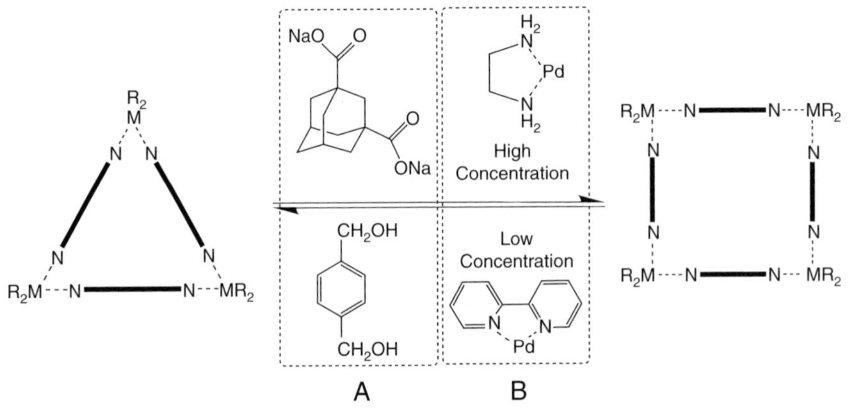

FIGURE 18.39 The direction of triangle or square formation in solution. (A) Steric bulk of polar groups within the nanostructure. (B) Steric bulk of auxiliary metal–ligands or concentration of the coordination nanostructures in solution.

roles in controlling the equilibrium of one palladium(II)-based assembly[47,48] (Figure 18.39). By varying the concentration of palladium(II) complexes (salts of ethylenediaminepalladium with either triflate or nitrate) and bidentate ligands (trans-1,2-bis(4-pyridyl)ethylene), the formation of triangles or squares could be directed.[92] At low concentrations (0.1 mM), triangles were favored due to entropic effects. At higher concentrations (10 mM), the more stable molecular squares were favored. Guest–complexation was found to affect the concentrations of trimer and tetramer in solution by directing either the triangle or the square to form with the addition of p-dimethoxybenzene or a disodium salt of 1,3-adamantanedicarboxylic acid, respectively[92] (Figure 18.39). Alternately, the incorporation of flexible bis-monodentate ligands can be used to form triangles from the square planar palladium(II) ion.[93] In instances where the labile ligands of the palladium(II) complex were oriented 180° from one another, coordination polygons with various numbers of sides were fabricated by altering the binding angle of the ligands. This is the method employed in one instance for forming molecular hexagons and pentagons in solution.[82,94]

Copper(II) has been used as the coupling element for a number of both two- and three-dimensional nanostructures.[85–89] The formation of small molecular squares using four copper(I) ions was made possible by the use of 3,6-bis(2′-pyridyl)pyridazine and copper(I) triflate.[88] In this design, the tetrahedral coordination center of the copper(I) ion fixes two pairs of bidentate ligands perpendicular to one another on opposite sides of the coordination plane of the four metal centers (Figure 18.38A). The characterization of this molecule indicated that the close proximity of the ligand rings to one another allows for a favorable π-stacking interaction, increasing the overall stability of the entire molecule.[88] Two copper(I) ions can also be used with ligands containing flexible bidentate regions to form molecular squares.[89] The free rotation of the bidentate branches in a bis-dione allow for one such dinuclear copper(I) complex[90] (Figure 18.38B). As will be discussed below, the copper(I) ion is very well suited to using the same types of bidentate coordination to form three-dimensional structures.

18.3.3.4 Three-Dimensional Structures

The spontaneous formation of three-dimensional architectures from noncovalent self-assembly is a common occurrence in biological systems. In proteins, the spontaneous formation of structure and function occurs at the most basic level, with hydrogen bonding along the polypeptide chain to form the secondary structure, and also among the largest of the aggregate protein interactions, such as in many viral and bacterial capsids. The tetrahedral bonding in carbon places significant restrictions on the geometric flexibility of the designs. This requires organic systems to use larger molecular subunits, such as amino acids and nucleotides, in order to gain enough structural flexibility to create complex structures. In contrast, the use of metals in the design of nanostructures, especially in smaller nanosystems, offers far greater flexibility for the formation of structurally complex macromolecules from controllable chemical and electrostatic interactions. The same advantages for creating two-dimensional structures from metal–ligand interactions are also realized in the third dimension, and both mono-dentate and bidentate (chelating) ligands have been used prominently in the formation of three-dimensional nanostructures.

The synthetic precedent for three-dimensional architectures can be divided into two broadly defined categories. The first of these categories includes linear coordination complexes such as ladders, racks, rods, and helices. Such systems are based on the vertical stacking of identical ligand–metal coordination regions and are extendable by modifying the lengths of the subunits that define their walls (Figure 18.40). The second category encompasses the polyhedral macromolecules. These systems confine the coordination regions to vertices instead of linear arrays, resulting in unimolecular architectures with dimensional customizability confined to modification of the ligands that define the sides (bis-chelating ligands) or faces (tri-chelating ligands) (Figure 18.41).

Ladders and rods are fabricated using tetrahedral-coordinating metals that act as *spiro*-centers between two different ligands to lock them in place and perpendicular to one another (Figure 18.40). Both ladders and rods utilize the same coordination center and a repeating sequence of covalently bound bidentate ligands to act as the vertical stabilizers (walls). Two such ladder structures employ a tetraphenyl derivative of the tetradentate bipyrimidine as the horizontal ligands, or rungs, and copper(I) as the metal centers to coordinate the rungs to 2,2'-bipyridine chains[95] (Figure 18.40). Three known rods were fabricated similarly, utilizing tetrahedral metal centers (copper(I) or silver(I)) and the same ligand chains of 2,2'-bipyridine as the vertical supports. These rod structures, however, employ hexaphenyl derivatives of hexaazatriphenylene as the tridentate ligands for the horizontal supports.[86] Molecular racks are based on the same basic ideas. These systems, however, employ tridentate ligands and six-coordinate metal centers to form rigid arrays. Because only one extended vertical chain is required to form these structures, racks can be formed from isomers of the same repeating tridentate motif.[96] A variety of *syn*- and *trans*-isomers of ruthenium(II) racks have been synthesized through thermodynamic self-assembly using vertical chains with both two- and three-tridentate subunits[97,98] (Figure 18.40).

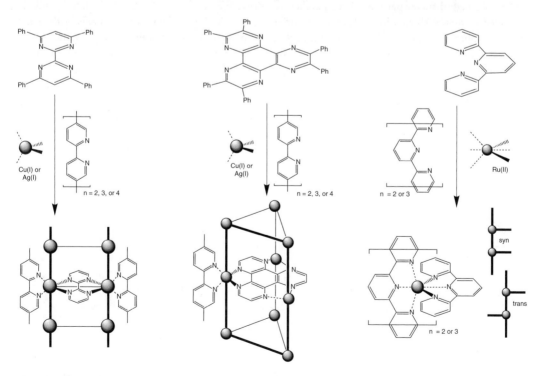

FIGURE 18.40 Ladders (left), rods (middle), and racks (right) formed from metal–ligand coordination.

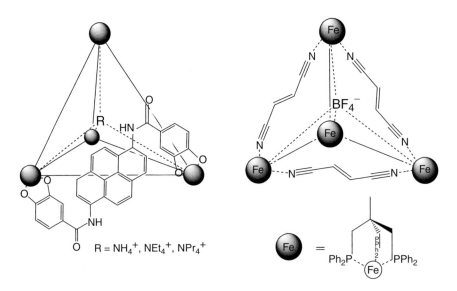

FIGURE 18.41 Coordination tetrahedra and encapsulation of guest molecules. (A) Symmetry interaction-based coordination tetrahedra and a series of encapsulated cations. (B) Molecular library-based tetrahedra and encapsulated boron tetrafluoride ion.

The symmetry interaction approach has been used extensively in the design and study of homodimetallic helicates.[77] The majority of these structures share the same design features, including the utilization of two octahedral coordination centers and three ligands composed of two bidentate regions and various organic bridges that provide unrestricted rotation about the bridge–bidentate bond. The chelate planes of the octahedral metal pairs are held parallel, requiring that each set of three bidentate ligands be provided with enough rotational flexibility to orient themselves along the C_3 rotation axis of the coordination spheres (Figure 18.42). The customization of the helicate shape is limited to modification of the ligand lengths. From among a set of common bidentate motifs, including those shown in Figure 18.42, any of a number of organic structures have been employed as bridges to vary the helicate length.[80]

It is remarkable that often the only requirements for the formation of macromolecular polyhedra are highly coordinating metal centers and multi-branching ligands. Many of the resulting macromolecular polyhedra, because their formation and stability are based only on metal–ligand coordination along the periphery, are skeletal structures with hollow cavities. Because the exteriors of these hollow polyhedra often can be deformed or broken through thermodynamic manipulation, it becomes possible to incorporate

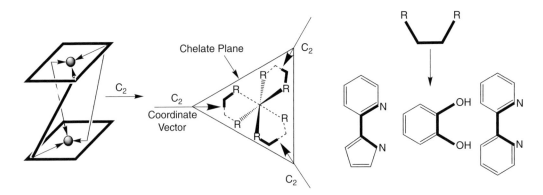

FIGURE 18.42 Helicate formation from the symmetry interaction model. At right is a selection of employed chelating ligand fragments.

smaller molecules into the polyhedral cavity in solution. Consequently, these macromolecular coordination polyhedra can act as large host molecules for the incorporation of single-guest molecules or collections of molecules for isolated chemical or structural studies.

Coordination tetrahedra are the smallest of the polyhedral structures to be formed both through the use of monodentate ligands and chelating ligands. The chelate-based systems are ideally suited to formation based on the principles of the symmetry interaction model, and their formation and structural features have been extensively studied.[80] Two basic motifs in coordination tetrahedra exist: those utilizing four metal centers and six ligands to form the edges of the structure (M_4L_6, Figure 18.A1.C) and those using four metal centers and four tridentate ligands to form the faces (M_4L_4, Figure 18.A1.D). It has been shown that small tetrahedral coordination structures can be used as a way of isolating small molecules in solution[99] (Figure 18.41). This work demonstrated two important features of coordination polyhedra. First, it showed that these coordination polyhedra are dynamic, with their metal–ligand bonds continually being broken and reformed in solution in order to establish an equilibrium with the guest molecules. Second, it showed that guest molecules can be preferentially selected and encapsulated within a tetrahedron (in the order $NEt_4^+ > NPr_4^+ > NMe_4^+$). In an example of the molecular library approach to three-dimensional nanostructure formation, the linear bidentate molecule fumaronitrile was used as the linking ligand to form the sides of a tetrahedron employing iron(II) vertices[100] (Figure 18.41). The remainder of the iron(II) coordination sphere was saturated using a tridentate phosphine ligand. Again, the tetrahedron was shown to encapsulate a counterion guest (BF_4^-). In this system, however, the tetrahedral symmetry elements are aligned in conjunction with the symmetry elements of the macrostructure. It is believed that the anion may be acting as a template over which the assembly of the cluster proceeds.[100]

18.3.4 Biomimetic Structures

The most versatile and, arguably, most important use of the MBB approach for the formation of nanostructures occurs in biochemistry, where intricate and highly specialized molecular "machinery" controls the manipulation of simple molecules to create functional structures. Biochemistry, as applied to the synthesis of nanostructures, is a special case of supramolecular chemistry. In these biomolecules, the covalent and electrostatic interactions of individual MBBs are used in concert with their aqueous surroundings to impose a sequential order and preferred orientation in the self-assembly of complex structures. The mechanisms and the raw materials of biochemical nanotechnology are not only self-sustaining, where the means for synthesizing and modifying the subunits are internally available to the system, but also self-regulating, where enzymatic activity controls such features as the availability, degradation, and reconstitution of materials into new macromolecules. The MBB approach, when considered from a biomimetic or biochemically inspired perspective, provides both an extensive background from which to understand design- and preparation-related issues at the nanoscale and a wealth of elegant examples from which to conceive novel structures. By studying the dynamics of biomolecular interactions, the role of the subunit in the formation of larger systems and the effects of environment on the formation and operation of these nanostructures may be better understood within a very important context.

Apart from the structural beauty of biomolecular systems, the greatest advantage of relying on biomimetic approaches to form nanostructures is that entire classes of functional structures already exist for study and modification. Nature has provided both a conceptual scaffolding from which to study structure/property relationships on chemically massive structures and a wealth of example systems that are often easily obtained. The biomimetic approach also has the unusual quality of being based upon a "finished product." The goals of biomimetic design are then achieved through retro-analysis, working from a known model to construct a new system based in biochemical precedent through chemical derivitization of the known subunits, environmental manipulation, or the application of biomimetic principles to other non-biological subunits.

The foundations of biochemical design are well understood from an MBB perspective. A great deal of knowledge of the structure and function of the subunits and a detailed understanding of the electrostatic interactions responsible for imparting secondary structure is available for these systems. The literature

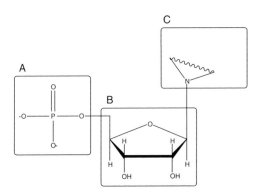

FIGURE 18.43 Structural components of DNA nucleotide bases. (A) Phosphodiester linkage. (B) 2′-ribose sugar. (C) Purine or pyrimidine nitrogen base.

on this subject is vast, and more detailed discussions are presented elsewhere.[28,101] A great amount of biochemical detail has been omitted from this discussion in order to focus on the actual MBB aspects of these structures and how the biomimetic approach can be readily applied to new systems. While the intricacies of protein folding, enzymatic activity, and tertiary structure are all important aspects of biochemistry, the fundamental understanding of molecular interactions at the macromolecular level are available from even small segments of DNA or small peptide chains in a protein.

18.3.4.1 DNA

Each nucleic acid molecule, the MBB of DNA, can be divided into three parts, with each portion of the molecule contributing significantly to the structure and electrostatic properties of the resulting DNA double helix (Figure 18.43). The covalent architecture of each helix, the primary structure responsible for maintaining the order of the nucleic acids, is composed of a phosphodiester and a 2′-deoxyribose residue in each subunit. The primary structure of the helix is formed via a condensation reaction between a phosphodiester and the 2′-hydroxyl group of a deoxyribose sugar, resulting in the elimination of one water molecule for each nucleotide linkage. Attached to each deoxyribose is either a monocyclic pyrimidine or dicyclic purine nitrogen base. Base pairs are then stabilized through hydrogen bonding interactions between a purine (adenine or guanine) and a pyrimidine (cytosine or thymine) on different (complementary) helices (Figure 18.44). For the purposes of encoding genetic information, two different purine/pyrimidine pairs (A with T and C with G) occur naturally. It is, however, sufficient to simply define the pyrimidine/purine pairing sequence in order to form the double helical structure of DNA. The complete secondary structure of DNA is a product of two types of electrostatic interactions. First, the formation of the double helix results from the correct hydrogen-bonded pairing of complementary bases between helices. Second, a π-stacking interaction, largely isolated within each helix between adjacent bases, further stabilizes the structure. This stacking is not completely isolated within a single helix, however, as the twisting of the double helix creates a slight overlap between bases on opposite strands.[102]

18.3.4.2 Proteins

Amino acids, the building blocks of proteins, are also composed of three structurally and electrostatically important parts. This division of structure begins with the covalent framework of the amino acid sequence, which is limited to the repeating peptide linkage (N-C-C) formed through rotationally unrestricted σ-interactions (Figure 18.45A). Directly attached to the N-C-C backbone are alternating donor (N-H) and acceptor (O = C) pairs for the formation of hydrogen bonds (Figure 18.45B). Any of a number of possible pendant (R) groups may be incorporated into the structure (Figure 18.45C). These functional substitutions on each amino acid are responsible for some of the secondary structure stabilization and enzymatic activity of the protein. The naturally occurring side-groups fall into four major categories based on their behavior in aqueous media.[101] These categories are (1) hydrophobic, (2) polar, (3) positively

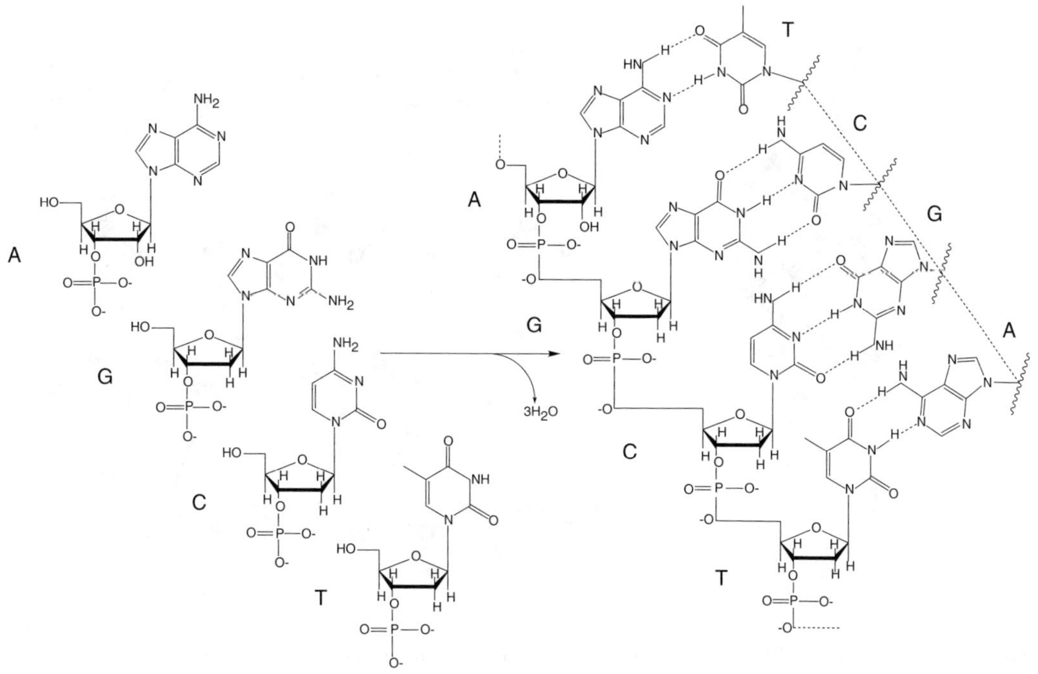

FIGURE 18.44 Connectivity and hydrogen bonding among nucleotide sequences in DNA.

charged, and (4) negatively charged. Since these are all neutral molecules in their isolated forms, their charge is a function of the pH of the intracellular environment.[101] For the general discussion, the R groups can be temporarily neglected, although their importance in imparting function to these structures cannot go unnoticed. As with DNA, the protein backbone is formed through a condensation reaction. The formation of secondary structure then occurs within small sequences of the polypeptide chain through intrachain hydrogen bonds between a N-H hydrogen and a C=O oxygen (α-helices) or between pairs of longer polypeptide chains through intrachain hydrogen bonding between the N-H hydrogens of one chain and the C=O oxygens of another (β-sheets) (Figure 18.46). The twists and bends responsible for the overall three-dimensional structure of proteins are a result of local breaks in the α-helices and β-sheets. The sequences responsible for these local breaks typically extend over many fewer amino acids than do the more regular helices and sheets.[28]

Nucleotides and amino acids share important similarities in subunit design. The formation of polypeptide chains and single helices occur through the removal of water, by far the most prevalent molecule in the intracellular matrix. The availability of subunits for macrostructure formation is regulated by either direct synthesis or modification of externally acquired subunits. In both nucleotides and amino acids, the subunits contain a covalent framework through which to interact with adjacent subunits and a highly directed noncovalent framework capable of stabilizing arrangements of subunits through electrostatic interactions. The majority of all superstructure formation occurs through hydrogen bonding, electrostatic

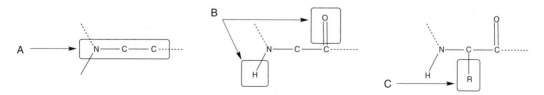

FIGURE 18.45 Structural features of amino acids. (A) N-C-C backbone. (B) Hydrogen-bonding regions. (C) Pendant group.

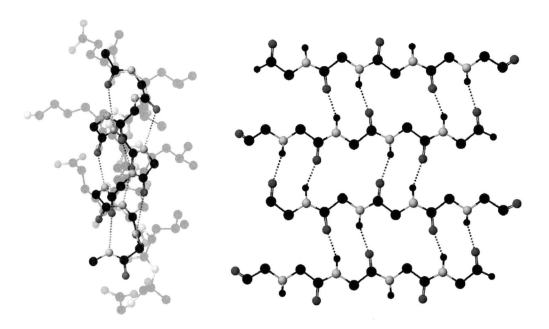

FIGURE 18.46 α-Helix (left) and β-sheet (right) secondary structures from amino acid sequences.

interactions that are easily broken in aqueous solutions. Hydrogen bonding promotes added functionality by allowing for the low-energy error correction of structural mismatches. Both π-π-interactions and the hydrophobic environment promote the π-stacking of the nucleotides, where this stacking serves to minimize the total surface area of the rings in contact with the polar aqueous surroundings. The anionic nature of the phosphodiester backbones at typical *in vivo* pH promotes the solvation of the exterior of the helices and increases stabilization in solution. The aqueous environment also has a destabilizing effect, since broken base pairs can hydrogen-bond to nearby water molecules. The similar strengths of hydrogen bonds between either amino acids or nucleotides and water means that rearrangements of the structures can and do occur dynamically, driving these structures to their energetic minima during their formation and allowing these structures to readily change shape or to be disassembled. While hydrogen bonding with the solvent can occur in the unpaired bases, their correct base pairing pattern provides greater stability, both through entropic affects and the proper alignment of π-stacking pairs. The effect of solvent on structure and biological function is best demonstrated by considering the folding and enzymatic activity of proteins in other solvents. In such nonaqueous instances, significant structural deformations from the aqueous structure and limited function are often observed, which result from improper folding.

This biochemical precision in design and function provides a very complete model from which to design similar macromolecules. Approaches to nanoscale design based on biomimicry begin with the realization that the possible variations of structure and function are enormous. The structures and functions of many of the naturally occurring biomolecules are still being investigated, and much more work still needs to be done to understand how these molecules interact with one another in the intracellular matrix. The extension of the biomimetic approach beyond biochemistry provides researchers with both a synthetic framework and a familiar nanoscale motif from which to design new structures. Two very broad methodologies based on the current understanding of structure and function in biochemistry have emerged in recent years. In one methodology, the known chemistry of nucleotides and amino acids are being exploited to develop novel, nonbiological nanostructures. In this approach, the *biochemical properties* of the subunits are being applied in new ways to form structures based in biochemistry but without any direct biological relevance. In the other methodology, the fundamental principles of biomolecular formation are being applied to new synthetic subunits. The emphasis on biomimetic design leads to the use of molecular subunits that are designed to behave like nucleotides and amino acids based on covalency and electrostatic stabilization features. In these designs, the choice of subunit can include anything from nucleotides that have been slightly altered from their naturally occurring forms to completely novel molecules applied in a biomimetic fashion. It is important to note here that the design approach from the synthetic subunits is directed specifically toward biomolecule mimicry, even when the subunits are ideal for other designs. These two approaches are closely related within the biomimetic context, as both approaches are founded directly from the guiding principles of biochemistry.

18.3.4.3 New Designs from Old Subunits

Nucleotide sequences and polypeptide chains are simply large molecules made up of a series of connected, structurally similar subunits. A particular order of nucleotides in DNA leads to the complementary pairing of bases and the storage of genetic information for the formation of specific polypeptide sequences. A specific order of amino acids in polypeptide chains is responsible for directing the spontaneous formation of a secondary structure by way of hydrogen bond-directed folding. From the final product of this protein formation comes a macromolecule with biological activity. In both DNA and proteins, a limited number of different combinations of nucleotides and amino acids control every biochemical process that occurs in an organism. In all other possible combinations of these MBBs, the potential exists to form a macromolecule with some unique nonbiological structure or function. In instances where a new sequence is nearly identical to a natural sequence, one might expect the structures and functions of both to be very similar. This is often the case, although examples exist where the substitution of one key subunit by another leads to the complete loss of biological activity. As more deviations from a natural sequence are incorporated into a synthetic sequence, the new structure loses these similarities. As a new structure, however, its properties may prove ideal to some other function. As general MBBs, there is essentially no limit on their application to the creation of other macromolecules or nanoscale materials.

Great structural variety and chemical function are available from different combinations of amino acids. The current limits on our ability to understand their interactions, however, prohibit the design of very complex structures. In contrast, the interactions responsible for the formation of DNA double helices are well understood because the separation of covalent backbone and electrostatic moieties is pronounced in the nucleotides. Our understanding of the noncovalent interactions of nucleotide bases with one another are specifically relevant in this respect. Consequently, the cognizant design of new structures from naturally occurring nucleotides has proven to be far more manageable than similar efforts from amino acids. Among the many nanostructural designs employing DNA as a key structural element, the most intriguing of these designs uses DNA as a construction element in the same way that molecular Tinkertoy approaches use linear molecules as components in skeletal frameworks. Many complex supermolecular structures from simple DNA fragments have been synthesized by relying on the strength of the double helix and the very predictable interactions of nucleotide bases. In these approaches, the DNA strands are divisible into rigid sections of stable base pairs and sticky sections of unpaired bases (Figure 18.47). In the fabrication of materials, rigid sections are responsible for defining the sides of structures while the manipulation of the sticky ends is responsible for forming and stabilizing corners.

One of the first structural applications of rigid/sticky nanoscale assembly was in the formation of tetravalent DNA junctions[103] (Figure 18.48). Each junction is composed of three regions which facilitate the self-assembly of two-dimensional lattices in solution. The formation of one base-paired arm exposes

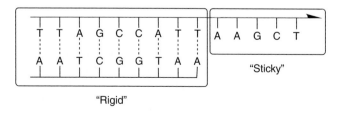

FIGURE 18.47 *Rigid* (paired) and *sticky* (unpaired) regions of DNA building blocks.

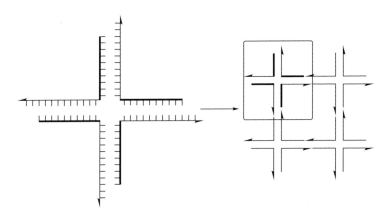

FIGURE 18.48 Two-dimensional DNA junctions from *rigid* and *sticky* engineering.

a sticky region and aligns two single sequences that will become a perpendicular set of arms. The perpendicular arms have identical base sequences and are unable to pair with each other. An arm/unpaired fragment with complementary bases to the unpaired arms of another fragment then pair to form the remainder of the rigid portion of the junction. Self-assembly of these junctions into a two-dimensional lattice occurs with the pairing of the extended sticky ends at each junction corner.

This same DNA design strategy of engineering strongly binding regions within junctions and incorporating unpaired strands to the ends of these junctions has been used for the formation of corners or vertices in a number of complex geometric structures, including isolated polygons and a number of polyhedral nanostructures.[104,105] All cases thus far demonstrate the importance of a rational design approach to the formation of DNA-based structures, because the extension of base pairing beyond two dimensions requires that base-pair complementarity be precisely controlled in order to direct structural formation beyond simple linear sequences. Most recently, a nanomechanical rotary device has been shown to operate by way of conformational changes between the device DNA strands and a second set of strongly binding DNA fragments.[106] The strong noncovalent binding of trigger fragments to the device strands causes conformational changes in regions that find new energetic minima through rotation. In effect, a DNA device has been created which is powered by a very site-specific kind of DNA "fuel."[106] Among other applications of DNA for nanostructural formation are those that rely solely on the complementary binding of strands to direct and stabilize other structures. For instance, complementary binding has been used as the noncovalent stabilizer to direct the formation of simple polygons from oligonucleotide/organic hybrids[105] (Figure 18.49).

18.3.4.4 Old Designs from New Subunits

The reproduction of biomolecular structures by synthetic subunits provides chemists with both an interesting challenge in supramolecular chemistry and a well-established set of guiding principles. The emphasis on designing subunits for the sole purpose of reproducing bioarchitectures is founded in our increased understanding of structural interactions within DNA and proteins. In both DNA and proteins, the vast majority of this structural precedent is based at the subunit level. Much of the work has been based on the use of subunit modification for the purpose of understanding the formation of secondary structures.

At one end of the biomimetic design regime is the use of synthetic nucleotides and amino acids to alter the properties of familiar biomolecules and to make novel structures based on the known interactions of these subunits.[107–111] The modification of amino acids in peptide sequences has been extensively used as a means to study protein folding, the enzymatic processes of these structures, and novel molecular scaffolding designs based on common supramolecular protein motifs, such as artificial α-helices and β-sheets. The design advantages responsible for the proliferation of nanostructures based on nucleotide interactions have also been responsible for the extensive modification of nucleotides as a direct means

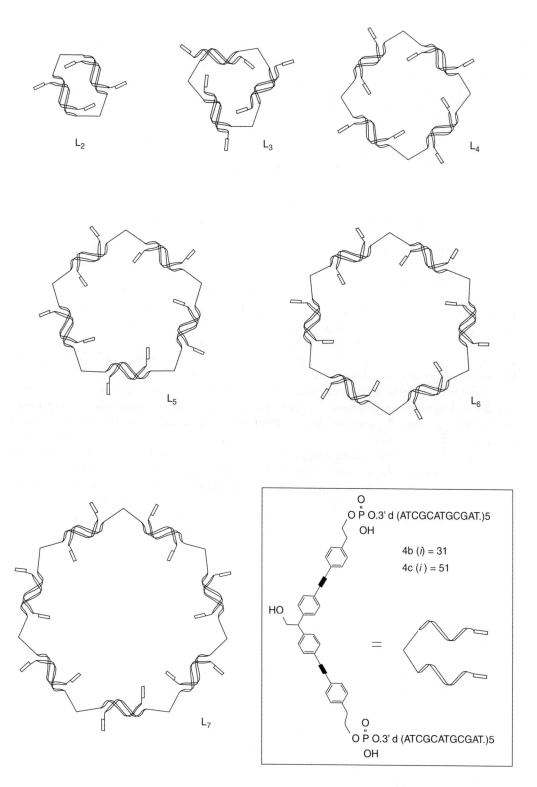

FIGURE 18.49 Polygons from DNA/organic hybrids. (From Shi, J. and Bergstrom, D.E., Assembly of novel DNA cycles with rigid tetrahedral linkers, *Angew. Chem. Intl. Ed. Engl.*, 36, 111, 1997. With permission.)

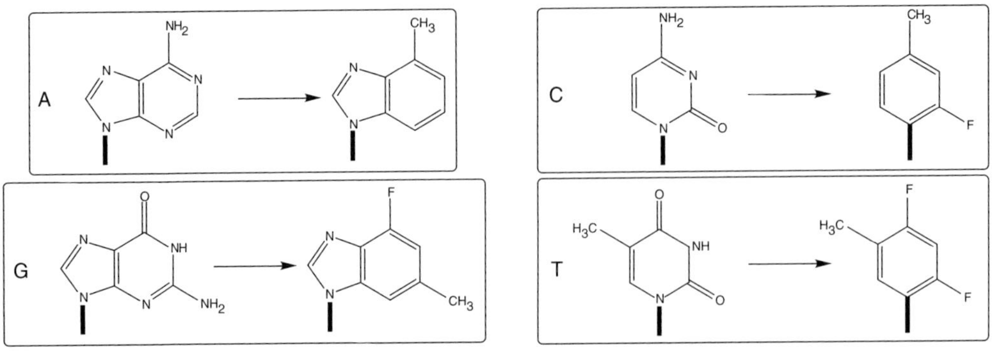

FIGURE 18.50 Nucleotide mimic structures without hydrogen bonding groups (right) from native structures (left).

to study the structure and function of DNA. A selection of synthetic nucleotides and their corresponding natural nucleotides is provided in Figure 18.50. Among these particular designs, the modifications have involved the removal of hydrogen bonding from the nitrogen bases, and they were used specifically to demonstrate the importance of aromatic stacking in the stabilization of the DNA double helix and to provide key insights into the importance of hydrogen bond stacking stabilization in the formation of DNA double helices and the molecular recognition events of DNA replication.[107]

Much of this work, which has emphasized altering the interactions between individual base pairs while causing minimal deformations in the double-helical structure, is also directly applicable to the novel DNA-based design strategies described above, as the modifications are typically rather subtle and the integrity of the nucleotide architecture remains intact. With the structural benefits of artificial MBBs in biomimetic design come many potential biomedical applications, as these synthetic subunits are generally not degraded by enzymatic processes, making them interesting candidates for the synthesis of novel therapeutics and biomaterials.[107]

At the other end of this biomimetic design regime is the reliance on only the biomimetic design strategy for the creation of biomolecular architectures. Such structures follow directly from the implementation of the structure–property relationships found in nucleotides and amino acids as the guide for the synthesis of new MBBs. These new subunits then share many of the same important design features as nucleotides and amino acids but have marginal structural similarities to the native subunits. The design features most important in the biomimetic design of novel subunits include consideration of primary and secondary structural features.

1. Primary Structure
    The covalent backbones of DNA and proteins define the order of the subunits while also providing some degree of structural flexibility to allow the noncovalent assembly of the larger structures. Within each subunit and in the subunit–subunit connections in both structures, this flexibility is incorporated by way of σ-bonding. Positional control is a function of rotation at specific points in the nucleotide/amino acid framework. As one limiting case, the covalent framework of a subunit can be designed to have no structural flexibility except for freedom of rotation at the subunit–subunit connection points and at the point of attachment for the fragment responsible for electrostatic stabilization (Figure 18.51A). This is similar to the freedom of movement in DNA, as the deoxyribose ring does strongly limit the positional freedom of the attached nitrogen base and the phosphodiester linkage. As a second limiting case, only freedom of movement at the subunit–subunit connections is allowed; and the remainder of the structure is rigid with respect to reorientation about the subunit–subunit bond (Figure 18.51B).

2. Secondary Structure
    Secondary structure is determined by the electrostatic stabilization introduced in the subunits and the positional freedom of these subunits as defined by the subunit–subunit connectivity. In DNA, not only are the nitrogen bases connected through a rotationally unrestricted bond to

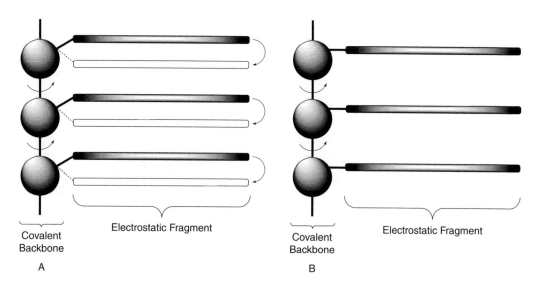

FIGURE 18.51 Limiting cases of MBB flexibility in biomimetic design. (A) Rotational freedom in both the covalent backbone and electrostatic regions. (B) Rotational freedom only in the covalent backbone.

the deoxyribose ring, but a number of pivot points are available for further orientational control. This flexibility is limited, however, by the use of ring structures in both the electrostatic component and the covalent framework. In amino acids, considerable rotational flexibility is available within the covalent backbone, allowing for different structural motifs to form from the subunits (α-helices, β-sheets).

The choice of electrostatic stabilization is also a factor to be considered. In both DNA and proteins, hydrogen bonding predominates. From an engineering perspective, the use of hydrogen bonds is ideal for both the degree of stability required of these structures and the environment in which these structures must function. The interactions of the subunits with the aqueous surroundings are the fundamental means by which all secondary structure formation occurs. Water plays the role of the medium, as it is the solvation of the larger structures that allows for electrostatic interactions to form and reform on the way to a stable minimum. Water, as a small molecule capable of forming stable hydrogen bonds with the noncovalent framework of DNA and proteins, is also responsible for the local destabilization of the larger structures. This local instability is responsible for the structural dynamics of proteins and DNA in solution and can be viewed as an integral part of the function of enzymes in all intracellular processes.

The design of a structural analogue to proteins or DNA from these guidelines must begin with the design of subunits that embody the same fundamental properties as nucleotides or amino acids. While the predictability of protein folding is still difficult, structure and enzymatic activity can be rationalized based on the final structure. DNA, however, has been found to be very amenable to structural manipulation. The predictability of the double helix from naturally occurring sequences has become familiar enough that DNA has been used to build artificial scaffolding and simple devices. Based on the ability to rationally design structures from the familiar structure–property relationships of the DNA nucleotides, alternative helical and double-helical structures based on novel subunits should also offer a certain degree of macromolecular predictability. One example of this approach is provided below.

The design of a new structural subunit employing the limiting cases in bonding and interactions is shown in Figure 18.52. Here, the covalent framework that defines the macromolecular backbone is based on rigid carboranes that are held together through a structurally inflexible five-member ring. By fixing the two inflexible carboranes to one another through the small ring system, structural flexibility within the subunit is greatly diminished. Connectivity between subunits is made by way of either a direct subunit–subunit linkage or through the use of some small, flexible spacer. As a consequence of the design, large-scale flexibility

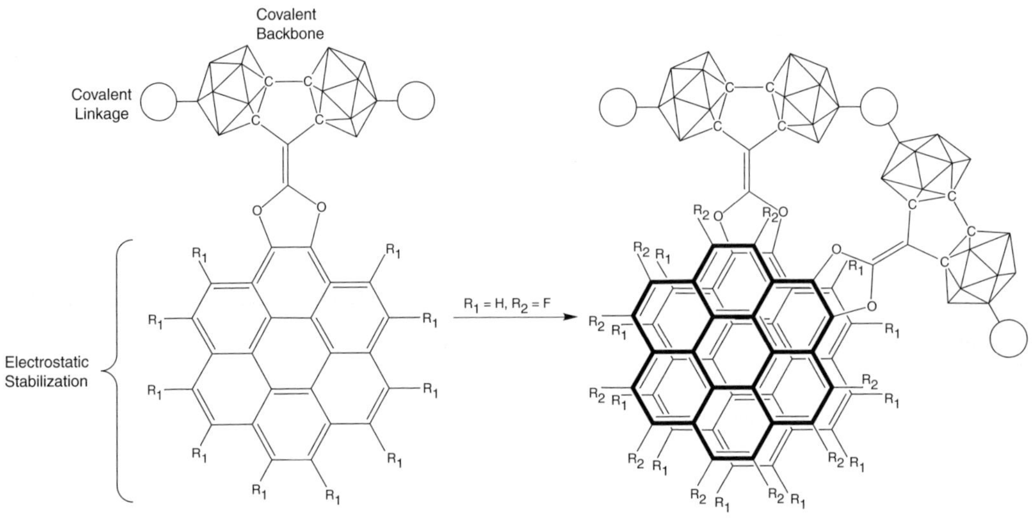

FIGURE 18.52 Synthetic bis-*ortho*-carborane MBB for biomimetic design. Cage boron–hydrogen bonds, oriented in the eclipsed conformation, are shown as unlabeled vertices.

is limited to rotation at a single point in the covalent backbone. All secondary structure formation, therefore, must occur through the rotational reorientation of subunits with respect to one another.

The means to secondary structure stabilization occurs through the interaction of functional groups pendant on the subunit frame. In these structures, the functional groups are placed at the noncarborane-substituted position of the five-member ring. The removal of rotational flexibility in this structure is by way of σ- and π-bonding between the covalent framework and the functional group. With both the interior of the subunit and the functional group held fixed through covalent bonding, interactions between subunits can only occur through rotational interactions.

The reliance on direct interactions, like hydrogen bonding, requires additional degrees of orientational flexibility within the subunit framework in order to form the most stabile interactions when the positions of the subunits themselves are not ideally arranged spatially. The reliance on rigorously directional interactions in solution can be removed by the selection of functional groups that do not interact through directed interactions. This route requires the removal of polar interactions as the means to forming stable interactions. The use of π-stacking interactions in the DNA double helix provides both significant stability and direction for the formation of a helical network with unfavorable interactions with the aqueous surroundings. As π-stacking can be engineered to be most favorable with actual stacking of the π-electrons between rings, the use of this type of interaction for the formation of helical structures should be possible. This helical stacking can be accomplished by limiting the positional flexibility of the π-systems to motions that align them in a vertical manner with limited opportunity to form other stable π-stacking arrangements. The rotational limitations of the carborane subunits allow such limited flexibility. Within the subunit formed from the linking of carborane-based MBBs to one another through bonds that only provide rotation and stable π-stacking interactions, the helical structure is both controllable and favored (Figure 18.53A). This preferential formation can be enhanced by the inclusion of polar functionalities on the exterior of the carborane subunits, forcing the π-stacking alignment within the helices by hydrophilic/hydrophobic interactions. Furthermore, the formation of double helices from the same stacking arrangement can be enhanced by the use of π-stacking pairs with alternating ring-periphery electron densities (Figure 18.53B). From the stability shown for benzene–perfluorobenzene pairs, similar MBB designs based on the same covalent subunit framework and π-system containing modified substituents becomes an interesting possibility for directing the formation of such designs. With the exclusive use of π-stacking for the formation of secondary structure, however, a larger space must be employed between stacking moieties.

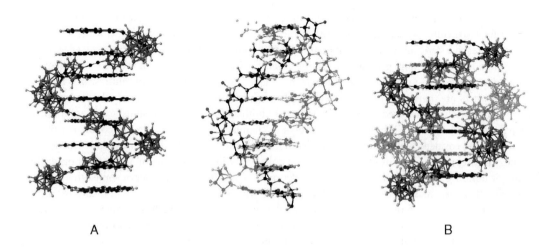

FIGURE 18.53 Helical (A) and double-helical (B) designs from synthetic MBBs. DNA provided at center (all structures to the same scale).

In carborane-based subunit designs, a double-helical structure can be designed by alternating the covalent backbone of each helix with subunits containing π-stacking functionalities.

18.3.5 Dendrimers

Dendrimers, also commonly referred to as *starburst polymers, cascade polymers*, or *arborols*, compose a special subset of supramolecular chemistry[112–115] that employs an MBB methodology in their formation. Dendrimers can be defined as "highly ordered, regularly branched, globular macromolecules prepared by a stepwise iterative fashion."[116] While the growth process of these structures is based in polymer chemistry, dendrimers offer exceptional control of structural and chemical properties within a predictable, unimolecular architecture. Further, the control of chemical functionality is available both within and along the periphery of dendrimers at any step in the growth process. Consequently, dendrimers can be either synthesized for a specific function or can be designed to behave as a nanoscale chemical environment itself for a number of applications (*vide infra*). With increasing interest in the use of dendrimers in materials science, biomedical applications, and in nanoscale laboratory applications,[114–117] the rapid progress in their development has emphasized both the basic methods for their fabrication and selective methods for the incorporation of function.

A number of structural and synthetic features separate dendritic polymers from the two remaining classes in polymer chemistry: hyperbranched polymers and linear polymers.[116] First, the dimensionality of a dendrimer is controllable from the very beginning of its growth. Linear polymers, while their random assembly in solution is three-dimensional, are formed through one-dimensional bonds. Because their orientation is statistical during this linear assembly process, there is little control over their secondary structure. The dimensionality of a dendrimer is determined from the shape of its structural core, which then directs the polymerization process over a length (one-dimensional), an area (two-dimensional), or a volume (three-dimensional). Because the growth of a dendrimer occurs radially from the inner core, the initial branching of the structure must take on the dimensionality of the inner core. Second, dendrimers are formed through a controllable, iterative process. Both linear and hyperbranched polymers, in contrast, are formed through chaotic, noniterative reactions, limiting both the control of their shapes and the degree of their polymerization. A dendrimer can be grown with no polydispersity, yielding a single, uniform structure of chemically massive unimolecular proportions. The largest of these unimolecular dendrimers have been shown to grow to sizes of up to 100 nm and molecular weights of 10^3 kDa.[117] Third, and perhaps most useful for nanoscale fabrication, is that the growth of a dendrimer can potentially be designed to be self-limiting regardless of the availability of monomer or reaction conditions. This is possible because the exponential addition of monomers to the dendrimer periphery rapidly surpasses the increase in the volume of the final structure, which only increases as the cube of the radius. Consequently, a dendrimer will eventually reach a steric limit past which monomer addition is impossible, a condition known as *De Gennes dense packing*[116] or the more general term *starburst limit*.[118] This steric limitation is based on a theoretical limit, however, and the understanding of dendrimer shape is still an area of significant research interest.

Dendrimers are, perhaps, the most controllable of the covalently bound supermolecular structures because the reactions involved in their formation are both self-directing and statistical in solution. The preparation of dendrimers is based in linear polymer chemistry, where a simple A/B copolymer motif is used to create covalent bonds between complementary reaction pairs. In dendrimers, this reaction pair strategy utilizes both a small molecule from which polymerization begins and an A monomer onto which multiple bonding sites for B monomers are incorporated. The initial A monomer or some other template molecule then becomes the seed, or *focal point*, from which *n* (typically 2 or 3) branches extend. By defining a dendritic focal point, it is not required that the point from which the growth process occurs be the absolute center of the dendrimer. In fact, dendrimers can be formed with the focal point on almost any type of molecule at almost any position, and a number of structures have been synthesized using aspects of the dendritic growth for purely functional purposes.

The structure of a dendrimer may be divided into a focal point and branched generations (Figure 18.54). A *generation* is simply a shell of B monomers around either the focal point (then referred to as the *inner core*) or a previous growth generation. Uniform dendritic growth then requires the addition of a stoichiometric quantity of B monomers for the number of A regions available along the dendrimer periphery. Uniform dendrimer growth is then most directly limited by the availability of monomer, steric constraints, and solubility.

While few alternative routes are known, the vast majority of all dendrimer syntheses is based on either *divergent* or *convergent* strategies. In the divergent approach,[119–121] the site from which dendritic growth begins becomes the focal point of the entire dendrimer framework (Figure 18.54). Each additional generation of monomer adds such that *n* of these monomers covalently bonds to the dendritic periphery at the tail of each previous generation (which then becomes a local focal point in the growth process). The uniformity of each generation is controllable by the inclusion of chemical functionalities onto the ends of each monomer, rendering the newly added generation incapable of undirected growth. The growth process is then continued by the removal of these chemical functionalities. This control of the periphery during the growth process results in divergent-based dendrimers having limited polydispersity in the final structures. Uniform dendritic growth is halted with the depletion of available monomers or the steric congestion of branches along the periphery. It is important to note that only uniform dendritic growth is stopped due to steric congestion. Irregularities in the peripheral branches result from continued addition of

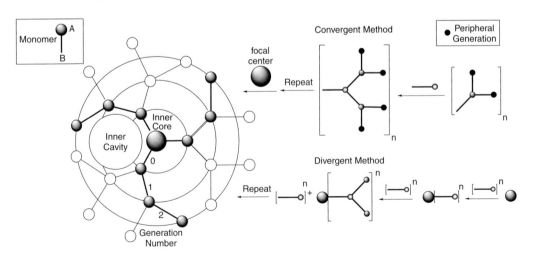

FIGURE 18.54 Dendrimer framework and convergent and divergent synthetic methods.

monomers beyond the starburst limit. Consequently, the control of the absolute size and packing in these structures is difficult to predict with great accuracy beyond a certain generation. It is because of the number of defect structures possible with the radial growth mechanism beyond a certain peripheral steric bulk that the divergent methods have some uncontrollable degree of polydispersity in the final structures.[116]

The convergent approach[116,122,123] to dendrimer formation begins with the peripheral generation and builds inward to a focal point by the coupling of progressively larger branches (Figure 18.54). This reversal from the divergent approach has the effect of switching the important advantages and limitations of the two methods. The fabrication of larger dendrimers is possible by divergent methods, as smaller monomers are added to the periphery of an otherwise sterically congested structure. In convergent methods, large branches are combined with one another, making the proximity of the reaction centers a critical factor in controlling the synthesis of larger structures. Convergent methods lead to greater uniformity of macromolecules, however, as the physical separation of defect structures is a far easier task.[116]

By the divergent method, two dendrimers might have identical molecular weights but great variability in branch lengths due to misdirected polymerization in larger structures. In the convergent methods, large branches either connect together to form a much larger branch or remain unconnected. The resulting increase in mass of bound branches then provides a direct means for separating structures. While the ultimate connection of these branches to the focal center may be a difficult task due to steric crowding, the completed structures are far more massive than any other components left in the reaction mixture and are therefore easier to isolate. The coupling of progressively larger branches, however, does ultimately limit the size of the dendrimers possible by the convergent method.[117]

The design of dendrimers and dendritic structures has begun to move beyond the polymerization chemistry of the branches and into the regime of structure- and application-specific modifications. The design of these functional dendrimers begins with the choice of the inner core. Among the synthesized dendrimers with functionalized cores, some of the most useful interiors for nanoscale applications include those with guest–host binding sites,[124,125] "dendritic probe" potential,[126,127] catalytic activity,[128–130] redox activity,[131–134] and those which employ dendrons, or larger dendritic branches, to act as stoppers for molecular assemblies.[135] A number of these applications are discussed below.

In the design of a dendrimer with an application-specific focal core, the method of dendrimer formation must be chosen carefully. Because dendrimer growth begins at the focal point in divergent methods, the application of a divergent growth scheme requires that the active portion of the core be chemically inert to the polymerization process and that this inertness continue over subsequent polymerization cycles. The convergent method, however, directs the growth of uniform branches until the focal core is ultimately added to the system. As the final formation of the dendrimer in the convergent method requires a chemical step that need not be a polymerization reaction, it is possible to add functional cores with far greater control. Consequently, a number of the discussed functional structures have been synthesized based on convergent approaches.[116]

Beyond the core, the customizability of both the monomers and the periphery has been used to engineer large-scale structural features and functionality into dendrimers. Between the core and the periphery, the inner-branching structure of dendrimers has been found to be highly customizable both for the formation of microenvironments within the cavities formed during the dendrimer growth process and for the inclusion of a number of host–receptors for the selective binding of guest molecules. The chemical modification of the periphery has proven to be a critical feature in the application of dendrimers. The exponential increase in dendrimer growth results in the rapid increase of peripherally bound substituents. As the dendrimer grows, the interactions between the periphery and the environment become the principle features governing dendrimer solubility and morphology. A number of studies have demonstrated that dendrimers incorporating either polar or nonpolar moieties along their periphery have significantly different solubility properties.[136,137] Furthermore, it has been shown that incorporating both highly polar and nonpolar regions into the dendrimer framework gives these structures unique yet controllable molecular encapsulation behavior.[137,138]

Combinatorial strategies have also been used in the dendrimer polymerization process as a means to alter the properties of both their interiors and periphery. A combinatorial approach to dendrimer

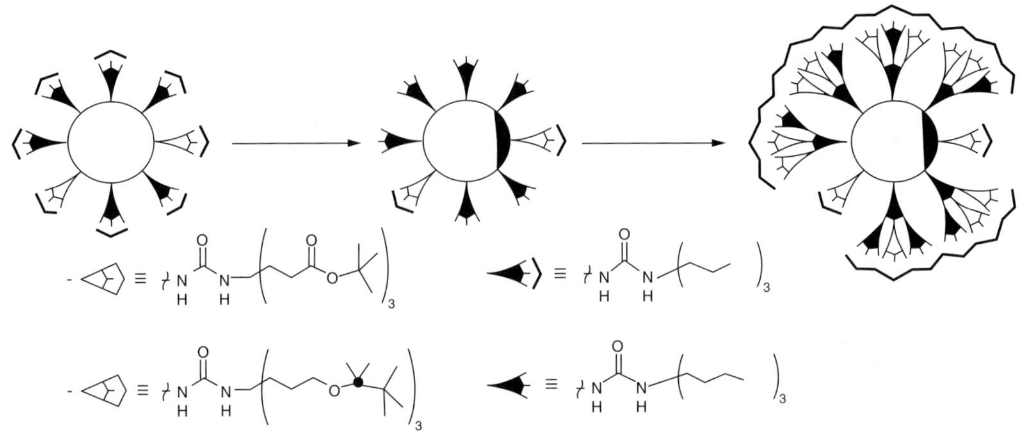

FIGURE 18.55 Dendrimer polycells from the inclusion of heterogeneous monomers. (From Newkome, G.R., Supra-supermolecular chemistry: the chemistry within the dendrimer, *Pure App. Chem.*, 70, 2337, 1998. With permission.)

synthesis is one in which different monomers are made available during polymerization at various steps in the growth process. In this process, the incorporation of different chemical branches during dendrimer growth can be accomplished either from the very beginning of the dendrimer formation, where the entire dendrimer is then made up of structurally unique branches, or after some number of identical generations have been added. Both approaches result in different local environments within the dendrimer, because the internal cavities typically span multiple generations. These heterogeneous structures, formed by altering the concentrations of different monomers during the growth process, have been termed *polycells*. The first instances of polycells employed a selection of isocyanate-based monomers with either reactive or chemically inert ends[139] (Figure 18.55). Not only was it shown that different monomers were readily incorporated into the same dendritic framework, but the combination of reactive and unreactive monomers demonstrated the ability to form dendritic branches with different generation numbers and chemical functionalities.[140] By this method, both the internal cavities and the dendritic periphery form molecular-sized pockets within which encapsulation, trapping, or noncovalent binding can occur.

As a class of supermolecules, dendrimers share similarities in MBB design methodology with both the biomimetic and molecular Tinkertoy approaches. A repeating subunit is connected covalently to other subunits to define a stable, although flexible and highly branched, skeleton. The shape of the final structure is then determined by the interactions of the subunits as constrained by the covalent framework. The reliance on covalency as the principle means of structure formation and the application of covalency within the context of a controlled-growth approach is what gives dendrimers a molecular Tinkertoy quality. Also, the many finger-like projections of the branches that give dendrimers their random, dynamic morphology are still anchored at structurally well-defined focal centers, as in the skeletal framework of rigid architectures. Finally, it is possible to impart structural rigidity to both sets of structures beyond any local stability that comes with noncovalent interactions, although this rigidity in dendrimer design must come at a cost of significant steric congestion, which can make a predictable, uniform growth process difficult.

The MBB similarities between dendritic methods and the biomimetic approach come from the use of subunit properties and interactions to define the secondary structure of the macromolecule, including the customization of both classes of macromolecules to control such features as solubility and aggregate interactions (tertiary structure). Dendrimers, because they are made from simple subunits in solution, can be grown specifically for particular environments. Similar to biomolecules, the electrostatic properties of the monomer can give rise to local environments within the dendrimer itself, as has been demonstrated in many instances by the incorporation of nonpolar/polar functionalities into polar/nonpolar monomers. For example, water-soluble, unimolecular micelles and other large dendritic structures have been synthesized with nonpolar centers by incorporating charged functional groups, such as carboxylate anions, into the periphery.[141] Biomimicry is taken further in dendrimers with the use of redox-active porphyrin focal centers and dendritic outgrowth to model the enzymatic behavior of some proteins.[131,142] By engineering hydrophobic/hydrophilic regions into a macromolecule to direct the formation of secondary structure in solution, this approach is similar to the chemical design of DNA and proteins.

The interactions between subunits that define the final structure in dendrimers are not necessarily based upon the formation of a directed secondary structure (biomimetic approaches) or by fixing the subunits within a larger covalent framework (Tinkertoy approaches). There are no intramolecular features governing the absolute size of the DNA double helix. This holds for proteins to a lesser extent, as it is the intramolecular interactions between the larger subunits (α-helices and β-sheets) in the protein that direct the formation of a localized, three-dimensional structure. Uniform dendrimer growth will, however, eventually succumb to steric crowding along the periphery. Also, the study of dendrimer formation for specific structural applications beyond the radial growth mechanism is still in its infancy. The formation of dendritic superstructures, including monolayer and multilayer formations on surfaces, has been demonstrated as a function of aggregate interactions and general molecular packing. The applicability of these designs, however, is currently limited to "bulk material" uses, such as chemical sensors,[143] catalysis, and chromatographic applications.[144,145]

Dendrimers are not just an interesting class of macromolecular structures. They can be synthesized to include the properties and functions of many customizable monomeric subunits and focal centers. Furthermore, this functionality can be wholly incorporated into a growth generation *via* stoichiometric control of the monomers, introduced statistically by the addition of dissimilar monomers, or performed by post-synthetic modification. Both the *microenvironment* and *functionalization* possibilities of dendrimers have been studied with great success. A brief discussion of two of the applications is provided below.

18.3.5.1 Guest–Host Interactions

One of the functional similarities between dendrimers and nanostructures employing electrostatic interactions, such as molecular crystals and biochemical structures, is the ability to integrate guest–host regions into the covalent skeleton through direct modification of the MBB subunits. A monomer generation can have incorporated into it a region customized to bind a specific molecule or type of chemical functionality. One benefit to introducing chemical functionality by way of monomer-based methods is that the tailoring of noncovalent interactions can be accomplished prior to the incorporation of the monomer into the dendritic framework. Furthermore, as has been demonstrated in the design of dendrimers with polar/nonpolar regions, it is possible to selectively exclude intramolecular or aggregate interactions between the guest–host binding regions from the remainder of the macromolecule simply by the exclusion of certain chemical functionalities from the remaining monomer generations. A dendrimer synthesized with a host interaction designed from strong hydrogen bonds, for instance, can be grown to include large pockets of nonpolar regions (such as long-chain alkanes) in subsequent generations.

The ability to bind molecules in solution by these engineered guest–host interactions depends upon the size of the dendrimer, the amount of branching, and the generation to which the guest–host region is added. In dendrimers with regions of limited steric congestion, it becomes possible to form stable guest–host interactions with many types of molecules. The applications here range from the trapping of molecules in solution by guest–host interactions to the formation of dendrimers themselves by noncovalent means. In both instances, the orientation of the host-binding region with respect to the focal point provides a means for controlling the exact orientation of the bound guest. In the case of dendrimer formation by guest–host interactions, the relative orientation of one branch to another can be controlled. In both cases, however, it is important to note that the size of the dendrimer is a critical feature, as the formation of stable guest–host interactions requires that the guest bind in the absence of steric strain. A number of studies have shown how the steric bulk of the dendrimer can affect both the orientational specificity of the guest binding region and the strength of the host–guest interaction.[146–148]

The use of hydrogen bonding within a dendrimer framework for forming stable guest–host interactions with small molecules has been demonstrated.[140] In a series of dendritic motifs, a hydrogen-bonding region composed of diacylaminopyridine was introduced early in the growth process (Figure 18.56). The binding pocket of the dendrimer with the inclusion of diacylaminopyridine is then donor–acceptor–donor (DAD) in nature, which can be used to bind selectively to guest molecules with a complementary acceptor–donor–acceptor (ADA) arrangement (Figure 18.56). The molecule selected for studying the guest–host binding interaction in these dendrimers was barbituric acid, which contains two such ADA structures. NMR (^{1}H) titration methods were used to show that pairs of dendrimer arms were able to form stable interactions with a barbituric acid molecule. The assembly of large dendrimers from noncovalent interactions has also been elegantly demonstrated.[149] The focal centers of dendritic branches were engineered with two isophthalic acid fragments incorporated into a small aromatic spacer, providing four hydrogen bonding regions (or eight possible hydrogen bond pairs) per core fragment. Hexameric dendrimers were found to form preferentially in solution by way of strong hydrogen bonding between donor–acceptor pairs at the focal centers of each branch (Figure 18.57).

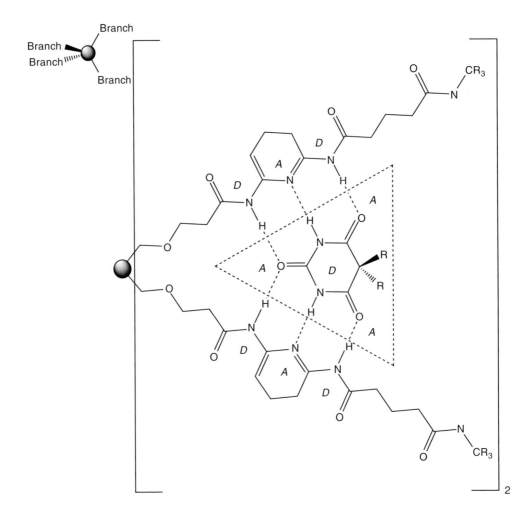

FIGURE 18.56 Guest–host assembly from dendrimer engineering (barbituric acid at center).

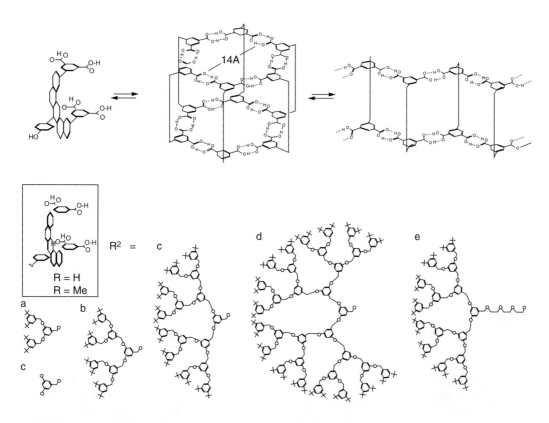

FIGURE 18.57 Dendrimer formation from hydrogen-bonding interactions. (From Zeng, F. and Zimmerman, S.C., Dendrimers in supramolecular chemistry: from molecular recognition to self-assembly, *Chem. Rev.*, 97, 1681, 1997. With permission.)

18.3.5.2 Microenvironments

In much the same way that transition metal nanostructures have been shown to encapsulate small molecules, the cavities formed by the overgrowth of generations along the periphery of large dendrimers have been shown to create microenvironments within which molecules can become trapped and bound. The isolation of single molecules or small ensembles of molecules within macromolecular enclosures has obvious utility in nanoscale laboratory applications, a field of chemistry just beginning to develop as an outgrowth of supramolecular chemistry. Molecular cavities within larger dendritic structures benefit from the variety of available monomers, the reproducibility of the cavities using dendritic growth methods, and the wide variety of polar and nonpolar solvents by which to promote solubility and encapsulation. For instance, a macromolecule can be synthesized with multiple regions that behave very differently in different solvents. In dendrimers large enough to encapsulate molecules, the properties of the cavity interior can be very different from the environment at the dendrimer periphery. One notable example of how molecules can be preferentially separated from solution based on polar/nonpolar interactions is provided in the encapsulation of Bengal Rose or 4-nitrobenzoic acid within the nonpolar cavities of a unimolecular micelle composed of long-chain alkane interiors and hydrophilic aliphatic acid exteriors[150,151] (Figure 18.58). Aqueous environments promote the encapsulation of the molecules in the nonpolar interior, while nonpolar solvents, such as toluene, were found to promote their release. Differences in local hydrophilicity/hydrophobicity are easily

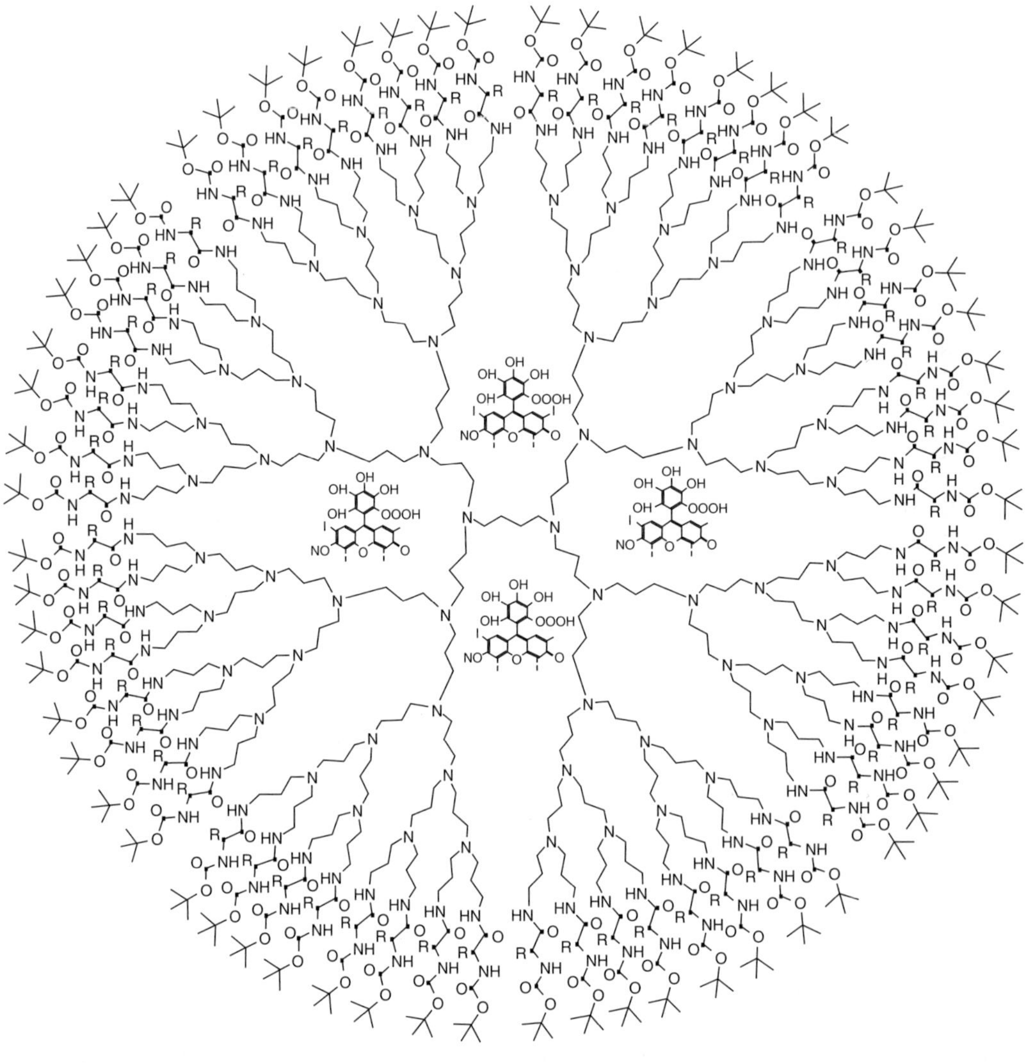

FIGURE 18.58 Bengal Rose encapsulation in dendrimer center. (From Zeng, F. and Zimmerman, S.C., Dendrimers in supramolecular chemistry: from molecular recognition to self-assembly, *Chem. Rev.*, 97, 1681, 1997. With permission.)

controlled in dendrimers by either the choice of the initial monomer or the post-synthetic functionalization of the dendrimer periphery. The differences in the spectroscopic properties of many molecules that come with different solvent shells have been the key to studying many encapsulated molecule/dendrimer systems.[117]

References

1. Mullen, K. and Rabe, J.P., Macromolecular and supramolecular architectures for molecular electronics, *Ann. N. Y. Acad. Sci.*, 852, 205, 1998.
2. Lindsey, J.S., Prathapan, S., Johnson, T.E., and Wagner, R.W., Porphyrin building blocks for modular construction of bioorganic model systems, *Tetrahedron*, 50, 8941, 1994.
3. Desiraju, G.R., Supramolecular synthons in crystal engineering – a new organic synthesis, *Angew. Chem. Intl. Ed. Engl.*, 34, 2311, 1995.

4. Lehn, J.-M., Perspectives in supramolecular chemistry: from molecular recognition to molecular information processing and self-organization, *Angew. Chem. Intl. Ed. Eng.*, 29, 1304, 1990.

5. Lehn, J.-M., Perspectives in supramolecular chemistry — from molecular recognition towards self-organization, *Pure Appl. Chem.*, 66, 1961, 1994.

6. Drexler, K.E., *Nanosystems*, Wiley-Interscience, New York, 1992.

7. Ebbing, D.D., *General Chemistry*, 5th ed., Houghton Mifflin, Boston, 1996.

8. Huheey, J.E., Keiter, E.A., and Keiter, R.L., Appendix E of *Inorganic Chemistry*, 4th ed., Harper Collins, New York, 1993.

9. Ellenbogen, J.C. and Love, J.C., Architectures for Molecular Electronic Computers: 1. Logic Structures and an Adder Built from Molecular Electronic Diodes, The Mitre Corp., 1999.

10. Schwab, P.F.H., Levin, M.D., and Michl, J., Molecular rods. 1. simple axial rods, *Chem. Rev.*, 99, 1863, 1999.

11. Allis, D.G. and Spencer, J.T., Polyhedral-based nonlinear optical materials. Part 1. Theoretical investigation of some new high nonlinear optical response compounds involving carboranes and charged aromatic donors and acceptors, *J. Organomet. Chem.*, 614–615, 309, 2000.

12. Collman, J.P., Hegedus, L.S., Norton, J.R., and Finke, R.G., *Principles and Applications of Organotransition Metal Chemistry*, University Science Books, Mill Valley, CA, 1987.

13. Long, N.J., *Metallocenes: An Introduction to Sandwich Complexes*, Blackwell Science, Oxford, 1998.

14. Vacek, J. and Michl, J., Molecular dynamics of a grid-mounted molecular dipolar rotor in a rotating electric field, *Proc. Natl. Acad. Sci.*, 98, 5481, 2001.

15. Purcell, K.F. and Kotz, J.C., *Inorganic Chemistry*, W.B. Saunders, London, 1977.

16. Stone, F.G.A., Stability relationships among analogous molecular addition compounds of group III elements, *Chem. Rev.*, 58, 101, 1958.

17. Lewis, G.N., *Valence and the Structure of Atoms and Molecules*, Chemical Catalogue, New York, 1923.

18. Mikhailov, B.M., The chemistry of 1-boraadamantane, *Pure Appl. Chem.*, 55, 1439, 1983.

19. Merkle, R.C., Molecular building blocks and development strategies for molecular nanotechnology, *Nanotechnology*, 11, 89, 2000.

20. Hunter, C.A. and Sanders, J.K.M., The nature of π-π interactions, *J. Am. Chem. Soc.*, 112, 5525, 1990.

21. Ma, J.C. and Dougherty, D.A., The cation-π interaction, *Chem. Rev.*, 97, 1303, 1997.

22. Sunner, J., Nishizawa, K., and Kebarle, P., Ion-solvent molecule interactions in the gas phase. The potassium ion and benzene, *J. Phys. Chem.*, 85, 1814, 1981.

23. Müller-Dethlefs, K. and Hobza, P., Noncovalent interactions: a challenge for experiment and theory, *Chem. Rev.*, 100, 143, 2000.

24. Patrick, C.R. and Prosser, G.S., A molecular complex of benzene and hexafluorobenzene, *Nature*, 187, 1021, 1960.

25. Williams, J.H., Cockcroft, J.K., and Fitch, A.N., Structure of the lowest temperature phase of the solid benzene-hexafluorobenzene adduct, *Angew. Chem. Intl. Ed. Engl.*, 31, 1655, 1992.

26. Collings, J.C., Roscoe, K.P., Thomas, R.L., Batsanov, A.S., Stimson, L.M., Howard, J.A.K., and Marder, T.B., Arene-perfluoroarene interactions in crystal engineering. Part 3. Single-crystal structure of 1:1 complexes of octafluoronaphthalene with fused-ring polyaromatic hydrocarbons, *New J. Chem.*, 25, 1410, 2001.

27. Coates, G.W., Dunn, A.R., Henling, L.M., Dougherty, D.A., and Grubbs, E.H., Phenyl-perfluorophenyl stacking interactions: a new strategy for supermolecule construction, *Angew. Chem. Intl. Ed. Engl.*, 36, 248, 1997.

28. Stryer, L., *Biochemistry*, 4th ed., W.H. Freeman, New York, 1995.

29. Scheiner, S., *Hydrogen Bonding: A Theoretical Perspective*, Oxford University Press, New York, 1997.

30. Joesten, M.D., *Hydrogen Bonding*, Marcel Dekker, New York, 1974.

31. Desiraju, G.R., The C-H⋯O hydrogen bond in crystals: what is it? *Acc. Chem. Res.*, 24, 290, 1991.

32. Steiner, T. and Saenger, W., Geometry of carbon-hydrogen⋯oxygen hydrogen bonds in carbohydrate crystal structures, analysis of neutron diffraction data, *J. Am. Chem. Soc.*, 114, 10146, 1992.

33. Kool, E.T., Morales, J.C., and Guckian, K.M., Mimicking the structure and function of DNA: insights into DNA stability and replication, *Angew. Chem. Intl. Ed. Engl.*, 39, 990, 2000.

34. Bruno, G. and Randaccio L., A refinement of the benzoic acid structure at room temperature, *Acta Crystallogr.*, B36, 1711.

35. Aakeroy, C.B. and Leinen, D.S., Hydrogen-bond assisted assembly of organic and organic-inorganic solids, in Braga, D., Grepioni, F., and Orpen, A.G. (Eds.), *Crystal Engineering: From Molecules and Crystals to Materials*, NATO Science Series, Kluwer Academic Publishers, London, 1999.

36. Desiraju, G.R., *Crystal Engineering: The Design of Organic Solids*, Elsevier, Amsterdam, 1989.

37. Derissen, J.L., Isophthalic acid, *Acta Crystallogr.*, B30, 2764, 1974.

38. Bailey, M. and Brown, C.J., The crystal structure of terephthalic acid, *Acta. Crystallogr.*, 22, 387, 1967.

39. Duchamp, D.J. and Marsh, R.E., The crystal structure of trimesic acid, *Acta Crystallogr.*, B25, 5, 1969.

40. Penfold, B.R. and White, J.C.B., The crystal and molecular structure of benzamide, *Acta Crystallogr.*, 12, 130, 1959.

41. Cobbledick, R.E. and Small, R.W.H., The crystal structure of terephthalamide, *Acta Crystallogr.*, B28, 2893, 1972.

42. Yang, J., Marendaz, J.-L., Geib, S.J., and Hamilton, A.D., Hydrogen bonding control of self-assembly: simple isophthalic acid derivatives form cyclic hexameric aggregates, *Tetrahedron Lett.*, 35, 3665, 1994.

43. Wang, Y., Wei, B., and Wang, Q., Crystal structure of melamine cyanuric acid complex (1:1) trihydrochloride, MCA.3HCl, *J. Crystallogr. Spectrosc. Res.*, 20, 79, 1990.

44. Lehn, J.-M., Mascal, M., DeCian, A., and Fischer, J., Molecular ribbons from molecular recognition directed self-assembly of self-complementary molecular component, *J. Chem. Soc., Perkin Trans. 2*, 461, 1992.

45. Zerkowski, J.A., Seto, C.T., and Whitesides, G.M., Solid-state structures of "rosette" and "crinkled tape" motifs derived from cyanuric acid–melamine lattice, *J. Am. Chem. Soc.*, 114, 5473, 1992.

46. Wasserman, E., The preparation of interlocking rings: a catenane, *J. Am. Chem. Soc.*, 82, 4433, 1960.

47. Fujita, M., Ibukuro, F., Seki, H., Kamo, O., Imanari, M., and Ogura, K., Catenane formation from two molecular rings through very rapid slippage, a Mobius strip mechanism, *J. Am. Chem. Soc.*, 118, 899, 1996.

48. Fujita, M., Ibukuro, F., Hagihara, H., and Ogura, K., Quantitative self-assembly of a [2]catenane from two preformed molecular rings, *Nature*, 367, 721, 1994.

49. Raymo, F.M. and Stoddart, J.F., Interlocked macromolecules, *Chem. Rev.*, 99, 1643, 1999.

50. Bisson, A.P., Carver, F.J., Eggleston, D.S., Haltiwanger, R.C., Hunter, C.A., Livingstone, D.L., McCabe, J. F., Rotger, C., and Rowan, A.E., Synthesis and recognition properties of aromatic amide oligomers: molecular zippers, *J. Am. Chem. Soc.*, 122, 8856, 2000.

51. Lokey, R.S. and Iverson, B.L., Synthetic molecules that fold into a pleated secondary structure in solution, *Nature*, 375, 303, 1995.

52. Lehn, J.-M., *Supramolecular Chemistry*, VCH, Weinheim, 1995.

53. Michl, J., The "molecular Tinkertoy" approach to materials, in Harrod, J.F. and Laine, R.M. (Eds.), *Applications of Organometallic Chemistry in the Preparation and Processing of Advanced Materials*, Kluwer Academic, Netherlands, 1995, p. 243.

54. Grimes, R.N., *Carboranes*, Academic Press, New York, 1970.

55. Ermer, O., Fivefold-diamond structure of adamantane-1,3,5,7-tetracarboxylic acid, *J. Am. Chem. Soc.*, 110, 3747, 1988.

56. Zaworotko, M.J., Crystal engineering of diamondoid networks, *Chem. Soc. Rev.*, 283, 1994.

57. Reddy, D.S., Craig, D.C., and Desiraju, G.R., Supramolecular synthons in crystal engineering. 4. Structure simplification and synthon interchangeability in some organic diamondoid solids, *J. Am. Chem. Soc.*, 118, 4090, 1996.

58. McKervey, M.A., Synthetic approaches to large diamondoid hydrocarbons, *Tetrahedron*, 36, 971, 1980.

59. Mathias, L.J., Reichert, V.R., and Muir, A.V.G., Synthesis of rigid tetrahedral tetrafunctional molecules from 1,3,5,7-tetrakis(4-iodophenyl)adamantane, *Chem. Mater.*, 5, 4, 1993.

60. Tobe, Y., Utsumi, N., Nagano, A., and Naemura, K., Synthesis and association behavior of [4.4.4.4.4.4]metacyclophanedodecayne derivatives with interior binding groups, *Angew. Chem. Intl. Ed. Engl.*, 37, 1285, 1998.

61. Höger, S. and Enkelmann, V., Synthesis and x-ray structure of a shape-persistent macrocyclic amphiphile, *Angew. Chem. Intl. Ed. Engl.*, 34, 2713, 1995.

62. Zhang, J., Pesak, D.J., Ludwick, J.L., and Moore, J.S., Geometrically-controlled and site-specifically functionalized phenylacetylene macrocycles, *J. Am. Chem. Soc.*, 116, 4227, 1994.

63. Levin, M.D., Kaszynski, P., and Michl, J., Bicyclo[1.1.1]pentanes, [n]staffanes, [1.1.1]propellanes, and tricyclo[2.1.0.02,5]pentanes, *Chem. Rev.* 100, 169, 2000.

64. Sugiura, K., Fujimoto, Y., and Sakata, Y., A porphyrin square: synthesis of a square-shaped π-conjugated porphyrin tetramer connected by diacetylene linkages, *J. Chem. Soc., Chem. Commun.*, 1105, 2000.

65. Wagner, R.W., Seth, J., Yang, S.I., Kim., D., Bocian, D.F., Holten, D., and Lindsey, J.S., Synthesis and excited-state photodynamics of a molecular square containing four mutually coplanar porphyrins, *J. Org. Chem.*, 63, 5042, 1998.

66. Arnold, D.P. and James, D.A., Dimers and model monomers of nickel(II) octaethylporphyrin substituted by conjugated groups comprising combinations of triple bonds with double bonds and arenes. 1. Synthesis and electronic spectra., *J. Org. Chem.*, 62, 3460, 1997.

67. Hammel, D., Erk, P., Schuler, B., Heinze, J., and Müllen, K., Synthesis and reduction of 1,4-phenylene-bridged oligoporphyrins, *Adv. Mater.*, 4, 737, 1992.

68. Kawabata, S., Tanabe, N., and Osuka, A., A convenient synthesis of polyyne-bridged porphyrin dimers, *Chem. Lett.*, 1797, 1994.

69. Collin, J.-P., Dalbavie, J.-O., Heitz, V., Sauvage, J.-P., Flamigni, L., Armaroli, N., Balzani, V., Barigelletti, F., and Montanari, I., A transition-metal-assembled dyad containing a porphyrin module and an electro-deficient ruthenium complex, *Bull. Soc. Chim. Fr.*, 133, 749, 1996.

70. Collin, J.-P., Harriman, A., Heitz, V., Obodel, F., and Sauvage, J.-P., Photoinduced electron- and energy-transfer processes occurring within porphyrin-metal-bisterpyridyl conjugates, *J. Am. Chem. Soc.*, 116, 5679, 1994.

71. Odobel, F. and Sauvage, J.-P., A new assembling strategy for constructing porphyrin-based electro- and photoactive multicomponent systems, *New J. Chem.*, 18, 1139, 1994.

72. Harriman, A., Obodel, F., and Sauvage, J.-P., Multistep electron transfer between porphyrin modules assembled around a ruthenium center, *J. Am. Chem. Soc.*, 117, 9461, 1995.

73. Osuka, A. and Shimidzu, H., Meso,meso-linked porphyrin arrays, *Angew. Chem. Intl. Ed. Engl.*, 36, 135, 1997.

74. Yoshida, N., Shimidzu, H., and Osuka, A., Meso-meso linked diporphyrins from 5,10,15-trisubstituted porphyrins, *Chem. Lett.*, 55, 1998.

75. Ogawa, T., Nishimoto, Y., Yoshida, N., Ono, N., and Osuka, A., One-pot electrochemical formation of meso,meso-linked porphyrin arrays, *J. Chem. Soc., Chem. Commun.*, 337, 1998.

76. Anderson, H.L., Conjugated porphyrin ladders, *Inorg. Chem.*, 33, 972, 1994.

77. Holliday, B.J. and Mirkin, C.A., Strategies for the construction of supramolecular compounds through coordination chemistry, *Angew. Chem. Intl. Ed. Engl.*, 40, 2022, 2001.

78. Bonavia, G., Haushalter, R.C., O'Connor, C.J., Sangregorio, C., and Zubieta, J., Hydrothermal synthesis and structural characterization of a tubular oxovanadium organophosphonate, $(H_3O)[V_3O_4](H_2O)(PhPO_3)_3]\cdot xH_2O(x = 2.33)$, *J. Chem. Soc., Chem. Commun.*, 1998, 2187.

79. Khan, M.I., Meyer, L.M., Haushalter, R.C., Schewitzer, A.L., Zubieta, J., and Dye, J.L., Giant voids in the hydrothermally synthesized microporous square pyramidal-tetrahedral framework vanadium phosphates $[HN(CH_2CH_2)_3NH]K_{1.84}-[V_5O_9(PO_4)_2]\cdot xH_2O$ and $Cs_3[V_5O_9(PO_4)_2]\cdot xH_2O$, *Chem. Mater.*, 8, 43, 1996.

80. Caulder, D.L. and Raymond, K.N., The rational design of high symmetry coordination clusters, *J. Chem. Soc., Dalton Trans.*, 1185, 1999.

81. Stang, P.J. and Olenyuk, B., Self-assembly, symmetry, and molecular architecture: coordination as the motif in the rational design of supramolecular metallacyclic polygons and polyhedra, *Acc. Chem. Soc.*, 30, 502, 1997.

82. Leininger, S., Olenyuk, B., and Stang, P.J., Self-assembly of discrete cyclic nanostructures mediated by transition metals, *Chem. Rev.*, 100, 853, 2000.

83. Albrecht, M., Dicatechol ligands: novel building blocks for metallo-supramolecular chemistry, *Chem. Soc. Rev.*, 27, 281, 1998.

84. Beissel, T., Powers, R.E., and Raymond, K.N., Coordination number incommensurate cluster formation. Part 1. Symmetry-based metal complex cluster formation, *Angew. Chem. Intl. Ed. Engl.*, 35, 1084, 1996.

85. Baxter, P.N.W., Lehn, J.-M., Baum, G., and Fenske, D., The design and generation of inorganic cylindrical cage architectures by metal-ion-directed multicomponent self-assembly, *Chem. Eur. J.*, 5, 102, 1999.

86. Baxter, P.N.W., Lehn, J.-M., Kneisel, B.O., Baum, G., and Fenske, D., The designed self-assembly of multicomponent and multicompartmental cylindrical nanoarchitectures, *Chem. Eur. J.*, 5, 113, 1999.

87. Berl, V., Huc, I., Lehn, J.-M., DeCian, A., and Fischer, J., Induced fit selection of a barbiturate receptor from a dynamic structural and conformational/configurational library, *Eur. J. Org. Chem.*, 11, 3089, 1999.

88. Youinou, M.-T., Rahmouri, N., Fischer, J., and Osborn, J.A., Self-organization of a tetranuclear complex with a planar arrangement of copper(I) ions: synthesis, structure, and electrochemical properties, *Angew. Chem. Intl. Ed. Engl.*, 31, 733, 1992.

89. Maverick, A.W., Ivie, M.L., Waggenspack, J.W., and Fronzek, F.R., Intramolecular binding of nitrogen bases to a cofacial binuclear copper(II) complex, *Inorg. Chem.*, 29, 2403, 1990.

90. Fujita, M., Sasaki, O., Mitsuhashi, T., Fujita, T., Yazaki, J., Yamaguchi, K., and Ogura, K.J., On the structure of transition-metal-linked molecular squares, *J. Chem. Soc., Chem. Commun.*, 1535, 1996.

91. Fujita, M., Supramolecular self-assembly of finite and infinite frameworks through coordination, *Synth. Org. Chem. Jpn.*, 54, 953, 1996.

92. Lee, S.B., Hwang, S.G., Chung, D.S., Yun, H., and Hong, J.-I., Guest-induced reorganization of a self-assembled Pd(II) complex, *Tetrahedron Lett.*, 39, 873, 1998.

93. Schnebeck, R.-D., Randaccio, L., Zangrando, E., and Lippert, B., Molecular triangle from en-Pt(II) and 2,2'-bipyrazine, *Angew. Chem. Intl. Ed. Engl.*, 37, 119, 1998.

94. Stang, P.J., Persky, N., and Manna, J., Molecular architecture via coordination: self-assembly of nanoscale platinum containing molecular hexagons, *J. Am. Chem. Soc.*, 119, 4777, 1997.

95. Baxter, P.N.W., Hanan, G.S., and Lehn, J.-M., Inorganic arrays via multicomponent self-assembly: the spontaneous generation of ladder architectures, *Chem. Commun.*, 2019, 1996.

96. Swiegers, G.F. and Malefetse, T.J., New self-assembled structural motifs in coordination chemistry, *Chem. Rev.*, 100, 3483, 2000.

97. Hanan, G.S., Arana, C.R., Lehn, J.-M., Baum, G., and Fenske, D., Coordination arrays: synthesis and characterization of rack-type dinuclear complexes, *Chem. Eur. J.*, 2, 1292, 1996.

98. Hanan, G.S., Arana, C.R., Lehn, J.-M., Baum, G., and Fenske, D., Synthesis, structure, and properties of dinuclear and trinuclear rack-type Ru(II) complexes, *Angew. Chem. Intl. Ed. Engl.*, 34, 1122, 1995.

99. Johnson, D.W. and Raymond, K.N., The self-assembly of a $[Ga_4L_6]^{+12-}$ tetrahedral cluster thermodynamically driven by host-guest interactions, *Inorg. Chem.*, 40, 5157, 2001.

100. Mann, S., Huttner, G., Zsolnia, L., and Heinze, K., Supramolecular host–guest compounds with tripod-metal templates as building blocks at the corners, *Angew. Chem. Intl. Ed. Engl.*, 35, 2808, 1997.

101. Lehninger, A.L., *Biochemistry*, Worth Publishers, New York, 1970.

102. Kool, E.T., Preorganization of DNA: design principles for improving nucleic acid recognition by synthetic oligonucleotides, *Chem. Rev.*, 97, 1473, 1997.

103. Seeman, N.C., Nucleic acid junctions and lattices, *J. Theor. Biol.*, 99, 237, 1982.

104. Seeman, N.C., Nucleic acid nanostructures and topology, *Angew. Chem. Intl. Ed. Engl.*, 37, 3220, 1998.

105. Shi, J. and Bergstrom, D.E., Assembly of novel DNA cycles with rigid tetrahedral linkers, *Angew. Chem. Intl. Ed. Engl.*, 36, 111, 1997.

106. Yan, H., Zhang, X., Shen, Z., and Seeman, N.C., A robust DNA mechanical device controlled by hybridization topology, *Nature*, 415, 62, 2002.

107. Kool, E.T., Morales, J.C., and Guckian, K.M., Mimicking the structure and function of DNA: insights into DNA stability and replication, *Angew. Chem. Intl. Ed. Engl.*, 39, 990, 2000.

108. Leumann, C.J., Design and evaluation of oligonucleotide analogues, *Chimia*, 55, 295, 2001.

109. Wu, C.W., Sanborn, T.J., Zuckermann, R.N., and Barron, A.E., Peptoid oligomers with α-chiral, aromatic side chains: effects of chain length on secondary structure, *J. Am. Chem. Soc.*, 123, 2958, 2001.

110. Wu, C.W., Sanborn, T.J., Huang, K., Zuckermann, R.N., and Barron, A.E., Peptoid oligomers with α-chiral, aromatic side chains: sequence requirements for the formation of stable peptoid helices, *J. Am. Chem. Soc.*, 123, 6778, 2001.

111. Offord, R.E., *Semisynthetic Proteins*, Wiley Interscience, New York, 1980.

112. Newkome, G.R., Moorefield, C.N., and Vögtle, F., *Dendritic Macromolecules: Concepts, Syntheses, Perspectives*, VCH, Weinheim, Germany, 1996.

113. Matthews, O.A., Shipway, A.N., and Stoddart, J.F., Dendrimers — branching out from curiosities into new technologies, *Prog. Polym. Sci.*, 23, 1, 1998.

114. Voit, B.I., Dendritic polymers — from aesthetic macromolecules to commercially interesting materials, *Acta. Polym.*, 46, 87, 1995.

115. Tomalia, D.A., Naylor, A.M., and Goddard, W.A., III., Starburst dendrimers: control of size, shape, surface chemistry, topology and flexibility in the conversion of atoms to macroscopic materials, *Angew. Chem. Intl. Ed. Engl.*, 29, 138, 1990.

116. Grayson, S.M. and Fréchet, J.M.J., Convergent dendrons and dendrimers: from synthesis to applications, *Chem. Rev.*, 101, 3819, 2001.

117. Zeng, F. and Zimmerman, S.C., Dendrimers in supramolecular chemistry: from molecular recognition to self-assembly, *Chem. Rev.*, 97, 1681, 1997.

118. De Gennes, P.G. and Hervet, H., Statistics of "starburst" polymers, *J. Phys. Lett.*, 44, 351, 1983.

119. Tomalia, D.A., Starburst/cascade dendrimers: fundamental building blocks for a new nanoscopic chemistry set, *Aldrichimica Acta*, 26, 91, 1993.

120. Newkome, G.R., Gupta, V.k., Baker, G.R., and Yao, Z.-Q., Cascade molecules: a new approach to micelles. A [27]-Arborol, *J. Org. Chem.*, 50, 2003, 1985.

121. de Brabander-van de Berg, E.M.M. and Meijer, E.W., Poly-(propylene imine) dendrimers — large-scale synthesis by heterogeneously catalyzed hydrogenations, *Angew. Chem. Intl. Ed. Engl.*, 32, 1308, 1993.

122. Hawker, C.J. and Fréchet, J. M.J., Preparation of polymers with controlled molecular architectures. A new convergent approach to dendritic macromolecules, *J. Am. Chem. Soc.*, 112, 7638, 1990.

123. Xu, Z.F. and Moore, J. S., Stiff dendritic macromolecules. 3. Rapid construction of large-size phenylacetylene dendrimers up to 12.5 nanometers in molecular diameter, *Angew. Chem. Intl. Ed. Engl.*, 32, 1354, 1993.

124. Numata, M., Ikeda, A., Fukuhara, C., and Shinkai, S., Dendrimers can act as a host for [60]fullerene, *Tetrahedron Lett.*, 40, 6945, 1999.

125. Zimmerman, S.C., Wang, Y., Bharathi, P., and Moore, J.S., Analysis of amidinium guest complexation by comparison of two classes of dendrimer hosts containing a hydrogen bonding unit at the core, *J. Am. Chem. Soc.*, 120, 2172, 1998.

126. Hawker, C.J., Wooley, K.L. and Fréchet, J.M.J., Unsymmetrical three-dimensional macromolecules: preparation and characterization of strongly dipolar dendritic macromolecules, *J. Am. Chem. Soc.*, 115, 4375, 1993.

127. Smith, D.K. and Müller, L., Dendritic biomimicry: microenvironmental effects on tryptophan fluorescence, *J. Chem. Soc., Chem. Commun.*, 1915, 1999.

128. Rheiner, P.B. and Seebach, D., Dendritic TADDOLs: synthesis, characterization and use in the catalytic enantioselective addition of Et$_2$Zn to benzaldehyde, *Chem. Eur. J.*, 5, 3221, 1999.

129. Yamago, S., Furukawa, M., Azumaa, A., and Yoshida, J., Synthesis of optically active dendritic binaphthols and their metal complexes for asymmetric catalysis, *Tetrahedron Lett.*, 39, 3783, 1998.

130. Bhyrappa, P., Young, J. K., Moore, J. S., and Suslick, K.S., Dendrimer-metalloporphyrins: synthesis and catalysis, *J. Am. Chem. Soc.*, 118, 5708, 1996.

131. Dandliker, P.J., Deiderich, F., Gross, M., Knobler, C.B., Louati, A., and Sanford, E.M., Dendritic porphyrins: modulation of the redox potential of the electroactive chromophore by peripheral multifunctionality, *Angew. Chem. Intl. Ed. Engl.*, 33, 1739, 1994.

132. Newkome, G.R., Güther, R., Moorefield, C.N., Cardullo, F., Echegoeyen, L., Pérez-Cordero, E., and Luftmann, H., Chemistry of micelles, routes to dendritic networks: bis-dendrimers by coupling of cascade macromolecules through metal centers, *Ang. Chem. Int. Ed. Engl.*, 34(18): 2023–2026, 1995.

133. Avent, A.G., Birkett, P.R., Paolucci, F., Roffia, S., Taylor, R., and Wachter, N.K., Synthesis and electrochemical behavior of [60]fullerene possessing poly(arylacetylene) dendrimer addends, *J. Chem. Soc., Perkin Trans. 2*, 1409, 2000.

134. Gorman, C.B. and Smith, J.C., Structure–property relationships in dendritic encapsulation, *Acc. Chem. Res.*, 34, 60, 2001.

135. Amabilino, D.B., Ashton, P.R., Balzani, V., Brown, C.L., Credi, A., Frechet, J.M.J., Leon, J.W., Raymo, F.M., Spencer, N., Stoddart, J.F., and Venturi, M., Self-assembly of [n]rotaxanes bearing dendritic stoppers, *J. Am. Chem. Soc.*, 118, 12012, 1996.

136. Wooley, K.L., Hawker, C.J., and Fréchet, J.M.J., Unsymmetrical three-dimensional macromolecules: preparation and characterization of strongly dipolar dendritic macromolecules, *J. Am. Chem. Soc.*, 115, 11496, 1993.

137. Hawker, C.J., Wooley, K.L., and Fréchet, J.M.J., Unimolecular micelles and globular amphiphiles — dendritic macromolecules as novel recyclable solubilization agents, *J. Chem. Soc., Perkin Trans 1.*, 21, 1287, 1993.

138. Jansen, J.F.G.A., Meijer, E.W., and de Brabander-van den Berg, E.M.M., The dendritic box: shape-selective liberation of encapsulated guests, *J. Am. Chem. Soc.*, 117, 4417, 1995.

139. Newkome, G.R., Suprasupermolecular chemistry: the chemistry within the dendrimer, *Pure App. Chem.*, 70, 2337, 1998.

140. Newkome, G.R., Woosley, B.D., He, E., Moorefield, C.N., Guther, R., Baker, G.R., Escamilla, G.H., Merrill, J., and Luftmann, H., Supramolecular chemistry of flexible, dendritic-based structures employing molecular recognition, *J. Chem. Soc., Chem. Commun.*, 2737, 1996.

141. Stevelmans, S., Van Hest, J.C.M., Jansen, J.F.G.A., Van Boxtel, D.A.F.J., de Brabander-van den Berg, E.M.M., and Meijer, E.W., Synthesis, characterization, and guest–host properties of inverted uni-molecular dendritic micelles, *J. Am. Chem. Soc.*, 118, 7398, 1996.

142. Chow, H.-F., Chan, I.Y.-K., Chan, D.T.W., and Kwok, R.W.M., Dendritic models of redox proteins: x-ray photoelectron spectroscopy and cyclic voltammetry studies of dendritic bis(terpyridine) iron(II) complexes, *Chem. Eur. J.*, 2, 1085, 1996.

143. Castagnola, M., Cassiano, L., Lupi, A., Messana, I., Patamia, M., Rabino, R., Rossetti, D.V., and Giardina, B., Ion-exchange electrokinetic capillary chromatography with starburst (PAM–AM) dendrimers — a route towards high-performance electrokinetic capillary chromatography, *J. Chromatogr.*, 694, 463, 1995.

144. Muijselaar, P.G.H.M., Claessens, H.A., Cramers, C.A., Jansen, J.F.G.A., Meijers, E.W., de Brabander-Van den Berg, E.M.M., and Vanderwal, S., Dendrimers as pseudo-stationary phases in electrokinetic chromatography, *HRC J. High. Res. Chromat.*, 18, 121, 1995.

145. Newkome, G.R., Weis, C.D., Moorefield, C.N., Baker, G.R., Childs, B.J. and Epperson, J., Isocyanate-based dendritic building blocks: combinatorial tier construction and macromolecular-property modification, *Angew. Chem. Intl. Ed. Engl.*, 37, 307, 1998.

146. Smith, D.K. and Diederich, F., Dendritic hydrogen bonding receptors: enantiomerically pure dendroclefts for the selective recognition of monosaccharides, *J. Chem Soc., Chem. Commun.*, 22, 2501, 1998.

147. Smith, D.K. Zingg, A., and Diederich, F., Dendroclefts. Optically active dendritic receptors for the selective recognition and chirooptical sensing of monosaccharide guests, *Helv. Chim. Acta.*, 82, 1225, 1999.

148. Cardona, C.M., Alvarez, J., Kaifer, A.E., McCarley, T.D., Pandey, S., Baker, G.A., Bonzagni, N.J., and Bright, F.V. Dendrimers functionalized with a single fluorescent dansyl group attached "off center", synthesis and photophysical studies, *J. Am. Chem. Soc.*, 122, 6139, 2000.

149. Zimmerman, S.C., Zeng, F.W., Reichert, D.E.C., and Kolotuchin, S.V., Self-assembling dendrimers, *Science*, 271, 1095, 1996.

150. Jansen, J.F.G.A., de Brabander-van der Berg, E.M.M., and Meijer, E.W., Encapsulation of guest molecules into a dendritic box, *Science*, 266, 1226, 1994.

151. Jansen, J.F.G.A., Meijer, E.W., and de Brabander-van der Berg, E.M.M., Bengal rose-at-dendritic box, *Macromol. Symp.*, 102, 27, 1996.

19

Building Block Approaches to Nonlinear and Linear Macromolecules

19.1 Introduction 19-1
19.2 Building Block Approaches to Nonlinear
Macromolecules...19-3
Nanocars • Catenanes • Rotaxanes • Molecular Containers
from Building Blocks
19.3 Building Block Approaches to Linear
Macromolecules...19-17
m-Phenylene Ethynylene Foldamers • Functional Achiral
Foldamers That Bind Heparin • Aedomers • β-Peptides
• BLOCK Approach to Large Structures • b*is*-Amino Acids
19.4 Conclusion..19-32
References..19-32

Stephen A. Habay
University of California–Irvine

Christian E. Schafmeister
University of Pittsburgh

19.1 Introduction

Within biology, the class of molecules that carry out most of the functions of life are proteins. Proteins act as catalysts, sensors, membrane channels, pumps, motors, and computational elements. Proteins are chains of amino acids that vary in length from 30 to 50 amino acids to those containing more than 1500 amino acids. Although proteins carry out an enormous array of functions, they are constructed from just 20 molecular building blocks. What distinguishes one protein from another is its precise sequence of amino acids. In order to carry out their functions, proteins must first fold into a well-defined, 3-D structure. The broad outlines of how and why proteins fold are understood; however, due to the complexity of the folding process, the details of how individual proteins fold still eludes us.[1]

Proteins have several properties that enable them to carry out their diverse functions. First, proteins are large: they are macromolecules with molecular weights on the order of 10,000 to 500,000 Da, and many proteins have dimensions on the order of two to several dozen nanometers. This allows them to be large enough to contain cavities that can bind small molecules and channels through which small molecules and ions can pass. Secondly, proteins can position and hold chemically reactive groups in precise positions relative to one another. This allows them to carry position chemical groups that can stabilize transition states and abstract and donate protons to perform fast and efficient catalysis, and complementary groups for molecular recognition with high selectivity. Thirdly, proteins are dynamic;

they change shape in controlled ways in response to binding and releasing small molecules, which enables them to convert chemical energy into mechanical motion and to store information.

It is a goal of many synthetic chemists to develop the ability to construct macromolecules that are as capable and as complex as biological proteins. This is important, because with this capability, we could create macromolecules with new functions that are not found in nature and macromolecules that are stable under wider environmental conditions than are biological proteins. Among other things, we could create new catalysts, new medicines, and new molecular devices for converting solar energy into electricity and fuels. Another reason for developing this capability is to better understand the relationship between molecular structure and function. Once we can construct functional macromolecules that operate as proteins do, we will have a much better understanding of how structure determines function in molecules such as proteins. If we are to truly understand how functional macromolecules such as proteins operate, we need to construct functional macromolecules that operate as proteins do. This is best expressed by a quote by Lord Kelvin:

"I never satisfy myself until I can make a mechanical model of a thing. If I can make a mechanical model I can understand it. As long as I cannot make a mechanical model all the way through I cannot understand…."[2]

Proteins are the best example of a molecular building block approach to nanoscale architecture. Several groups are developing *de novo* designs of new proteins as a route to functional macromolecules.[3] The sophistication and power of recombinant DNA technology coupled with heterologous protein expression makes it straightforward to create artificial proteins with designed amino acid sequences. However, the complexity of the protein-folding problem makes it very challenging to create new proteins that fold into well-defined 3-D structures and even more challenging to create new proteins with designated functions.

Chemists have been developing approaches to functional nanoscale molecules for decades. Chemists use sequences of reactions to add and remove fragments from their working molecule until they achieve their target molecule. Synthesis of complex molecules that approach the size of small proteins (1000 to 10,000 Da) is challenging because almost every chemical step of a multistep synthesis results in the loss of the working molecules. To illustrate, a synthesis involving 20 steps in which each reaction has a 90% yield of the intended product will result in a final yield of the target molecule of only 12.2% (0.90^{20}). A synthesis involving 50 steps in which each reaction has a 90% yield of the intended product will result

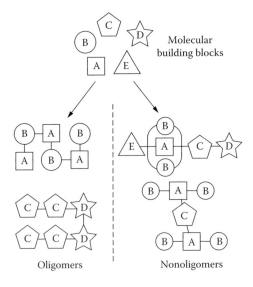

Oligomers   Nonoligomers

FIGURE 19.1 Molecular building blocks are stable, storable compounds that are activated and coupled to other building blocks to synthesize macromolecules with structures and properties that emerge from the specific arrangement of the component building blocks. Oligomers are linear chains of monomers and include proteins, DNA, and most artificial foldamers. Nonoligomers can have topologies that are limited only by imagination and include rotaxanes, catenanes, calixarenes, and star-shaped molecules.

in a final yield of target molecule of only 0.5% (0.90^{50}). Synthesis of large molecules requires many steps, and this makes it challenging to synthesize macromolecules.

One strategy for synthesizing large molecular structures with complex architectures is to synthesize them from a collection of smaller component building blocks. These monomers can be assembled, through highly reliable coupling reactions, into larger rigid systems of interest. The building blocks must be stable and storable and they must be capable of activation to form reactive compounds that undergo a controlled reaction with other monomers. The building blocks may be commercially available, or they may be synthesized themselves using synthetic chemistry. Building blocks may be achiral or they may contain one or more stereocenters. The coupling reactions used to join the blocks can be any reaction within the synthetic toolbox of organic chemistry, but they are usually chosen from those reactions that give reproducibly high yields with minimal side reactions such as amide bond formation between amines and activated carboxylic acids. One advantage that the chemist has over nature is that macromolecules with almost any topology imaginable can be constructed using synthetic chemistry. Organic chemists have assembled building blocks using ring topologies, branching topologies, and chain-like linear topologies such as those of proteins and DNA (Figure 19.1). The shapes of the macromolecules and functional group positioning are incorporated into the design aspects of the synthesis and should be general enough to allow flexibility in assembly.

19.2 Building Block Approaches to Nonlinear Macromolecules

19.2.1 Nanocars

There is currently a great deal of interest in nanometer-scale machines with controlled mechanical motion.[4] Tour and coworkers have designed small molecular machines, dubbed nanocars, that exhibit translational movement via a fullerene-based wheel-like rolling motion.[5] This movement is evidenced to be a rolling motion specifically, and not stick-slip or sliding, by both direct and indirect manipulation of the nanocars and examination of similar molecular systems.

The nanocar molecules are studied by scanning tunneling microscopy (STM) on a gold surface at varying temperatures. Synthesis of the nanocars is accomplished through Pd-catalyzed coupling of phenylene ethynylene building blocks to form the chassis and axles of the molecules, followed by attachment of four C_{60}-fullerene wheels. The nanocars are stable and stationary on the gold surface at room temperature. This is attributed to the strong adhesion force between the fullerenes and underlying gold. The fullerene wheels are brightly imaged during STM, which allow tracking of motion across the metal surface (Figure 19.2).

Motion of the nanocars is induced indirectly by heating of the gold surface above approximately 200°C (Figure 19.2[c]). The molecules are moved in two dimensions by both translation and pivoting and can be followed through a series of 1 min images. Translational motion occurs perpendicular to the axles, demonstrating a directional preference based upon molecular structure. Above 225°C, the rate of motion of the nanocars is too rapid to be followed by the slow-acquiring STM imaging.

The nanocars can also be moved directly by pulling with the STM tip (Figure 19.2[d]). Attempts at pushing the molecules along the surface with the STM tip do not facilitate translational motion; only pushing aside or pivoting of the molecules is observed. These studies corroborate the ability to synthesize nanoscale molecular machines through a building block approach and control their molecular motion on surfaces. Future studies will hopefully lead to electric field-induced motion of a wider variety of nanoscale molecular machines.

19.2.2 Catenanes

One interesting target of nonlinear building block synthesis is the catenane. A catenane comprises two or more mechanically interlocked molecular rings. Catenanes are interesting for both their unique topologies and their potential for use in nanoscale devices. Many diverse mechanical templating procedures have been developed to thread one building block through the other, followed by "clipping" of a second ring that interlocks the two pieces.[6] Transition metals such as ruthenium, palladium, and iron

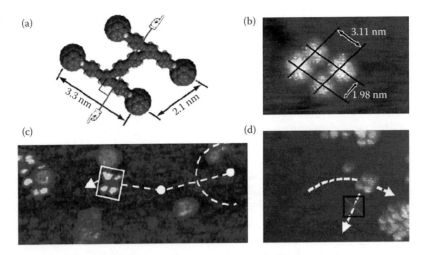

FIGURE 19.2 (a) Modeled distances at 3.3 nm across the width and 2.1 nm across the length, or the axle and chassis directions, respectively, as shown in this alkyl-free space-filling model. The fingers indicate the expected direction of rolling perpendicular to the axles. (b) High-resolution STM image (V_b = 0.4 V, I_t = 60 pA). The orientation of the molecules can be determined by the fullerene peak-to-peak distances. (c) Nanocar rolling on an Au surface at ~200°C with small translations intermittent with small-angle pivots. (d) Direct STM manipulation of the nanocar pulled with the tip perpendicular to its axles (30 pA, 0.1 V imaging conditions; 3.5 nA, 0.1 V manipulation conditions). The tip was lowered in front of the molecule in the direction of motion and pulled along the dotted arrow. The same technique failed to move the nanocar when the tip was lowered to the side and dragged away at 90° to its previous motion.

have been utilized to form a mechanically threaded coordination complex between the two monomers, allowing for increased yields during the interlocking process.[7] Cations and anions have also been used to facilitate threading as well as hydrogen bonding and π-donor/acceptor interactions.[8]

The final step in catenane production is the closure of a second ring to interlock the two building blocks. Typically this reaction has been the result of a kinetically controlled covalent bond formation, but in recent years, thermodynamically controlled processes have been explored as well. Specifically, dynamic covalent chemical processes[9] have proven extremely useful in assembling nanoscale structures. Kinetically controlled processes can form large amounts of noninterlocked by-products during the interlocking reaction. A thermodynamically controlled process allows the product distribution to reflect the stability of the intermediates at equilibrium, which contributes to increased yields of catenane.

Stoddart and coworkers provide a nice example of reversible catenane formation with the olefin metathesis reaction.[10] When cyclic polyether building block **1** is mixed with dibenzylammonium macrocycle **2**, in the presence of Grubbs' ruthenium catalyst, a catenation reaction occurs where two individual macrocycles "magically" interlock to form catenane **3** in 75% yield (Scheme 19.1).

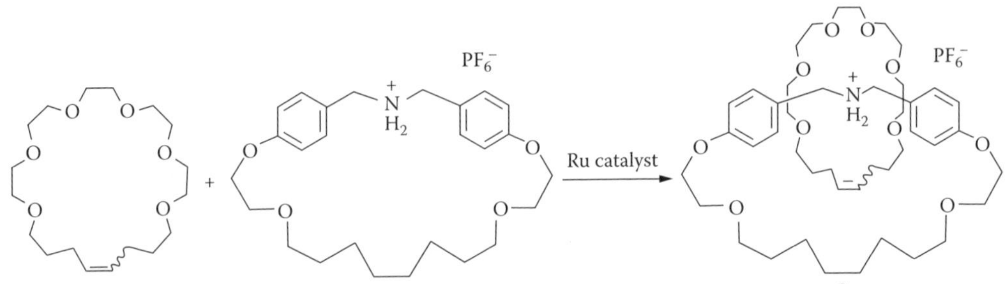

SCHEME 19.1 Magic ring catenation.

The catenation reaction is confirmed using nuclear magnetic resonance (NMR) spectroscopy and takes place by ring-opening metathesis of **1** followed by threading of the linear polyether through **2** and then ring-closing metathesis to interlock the two macrocycles. This process is comparable to the biological function of topoisomerase enzymes on circular double-stranded DNA, or, more generally, the conjuror's classic ring-linking magic trick.

The ring-closing metathesis strategy has been demonstrated in concert with other threading strategies such as metal coordination.[7] Sauvage and coworkers have synthesized a macrocycle that is interlocked with a handcuff-like compound (Scheme 19.2).[11] The "handcuff" monomer **5** was synthesized in 70 to 80% yield from a homocondensation of macrocyclic dione **4** in melted ammonium acetate (180°C). Two equivalents of the dibutenylic monomer **6** were threaded using metal templation with Cu (I). The tetrahedral complexation of Cu (I) allows concatenation via ring-closing metathesis to form the interlocked compound **8** in 80% after 10 days as a mixture of E–E, E–Z, and Z–Z isomers, the structures of which are confirmed unambiguously by ROESY NMR spectroscopy. Other metals such as Ru (II) and Fe (II) are used to template catenane formation with terpyridine building blocks; although a mixture of catenane and "figure-eight" products are formed during the metathesis reaction.[12] These examples demonstrate some of the rather interesting topologies that can be constructed using this method.

A final example of ring-closing metathesis-based catenane formation comes via anion-templated assembly (Figure 19.3). Beer and coworkers synthesized a [2]- and [3]catenane (**10** and **11**, respectively) using chloride in a tight ion pair with a pyridinium cation.[13] This pairing templates the formation of a pseudorotaxane **9** that can be clipped together through a ring-closing metathesis reaction. The molecular

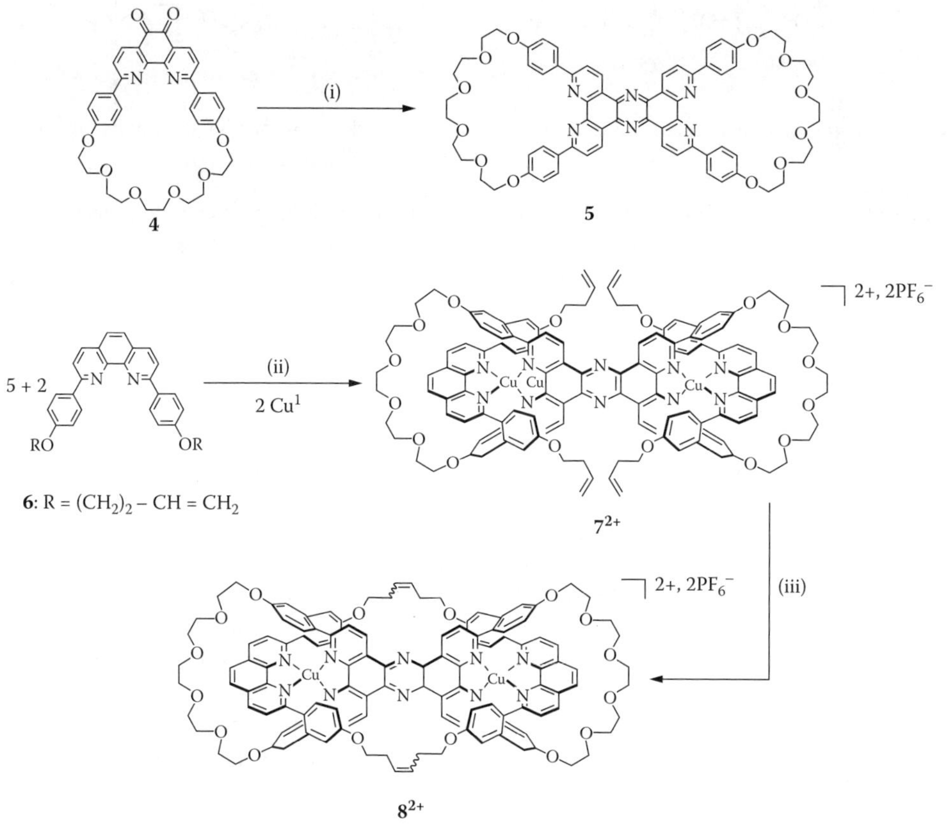

SCHEME 19.2 Metal-templated catenane formation. Conditions: (i) NH_4OAc, 180°C, 2 h; (ii) $Cu(MeCN)_4(PF_6)$, $CH_2Cl_2/MeCN$ 2:1, 5 d; and (iii) $Cl_2(PCy_3)_2Ru = CHPh$, CH_2Cl_2, 10 d.

(a)

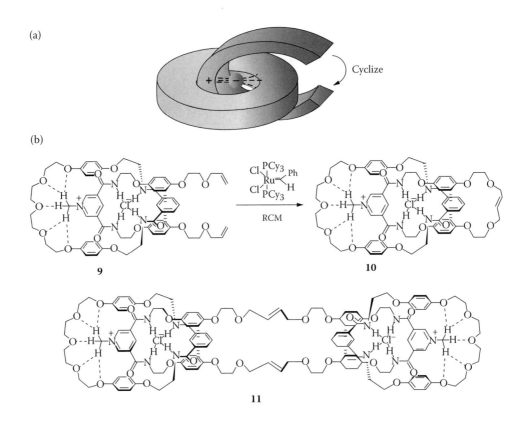

(b)

9

10

11

FIGURE 19.3 (a) Ion-pair-templated catenane formation. (b) Ring-closing metathesis strategy for clipping of the pseudorotaxane.

building blocks are a pyridinium thread component and a hydroquinone-containing macrocycle. The major product is the [2]catenane **10** (45%) with only a small amount of [3]catenane **11** produced (5%). Within compound **10**, secondary hydrogen bonding and π–π stacking interactions are designed to complement the binding sites of the macrocyclic component and stabilize anion templation. Anion titrations reveal that the catenane **10** is highly selective for chloride, demonstrating the unique and constrained binding pocket of the catenane.

Other dynamic covalent processes can produce macrocycles and catenanes. Recently dipeptide hydrazone building blocks (e.g., Figure 19.4, compound **12**) were assembled, at thermodynamic equilibrium, into a [2]catenane **18** consisting of six building blocks.[14] The catenane self-assembles around an acetylcholine ligand template through reversible hydrazone linkages to form two interlocked 42-membered rings, on a preparative scale, in 67% yield after many days.

Figure 19.4 depicts the various intermediates formed before amplification of one of two possible diastereomeric catenane products. Dimers (**13**), trimers (**14**), and larger macrocycles up to hexamers (**17**) were observed in the HPLC trace. The formation of a [2]catenane at such high yield from such a large mixture of intermediates is typically highly unfavorable, which demonstrates the high selectivity of the catenane for the acetylcholine ligand. This is confirmed by a measured binding constant of 1.4×10^7 M^{-1}, several orders of magnitude higher than acetylcholine binding to the trimer (**14**) and tetramer (**15**) species. The power of dynamic covalent chemistry to select diverse and complex structures from a library of simple building blocks is remarkable.

Complex interlocked architectures can be synthesized, using kinetically controlled processes as well, examples are pretzelanes, such as **23**, and cyclic *bis*[2]catenanes, such as **24**. Stoddart and coworkers synthesized these nanoscale structures[15] with building blocks that thread by electron-rich naphthyl/electron-poor bipyridinium donor–acceptor interactions and interlock by intramolecular nucleophilic

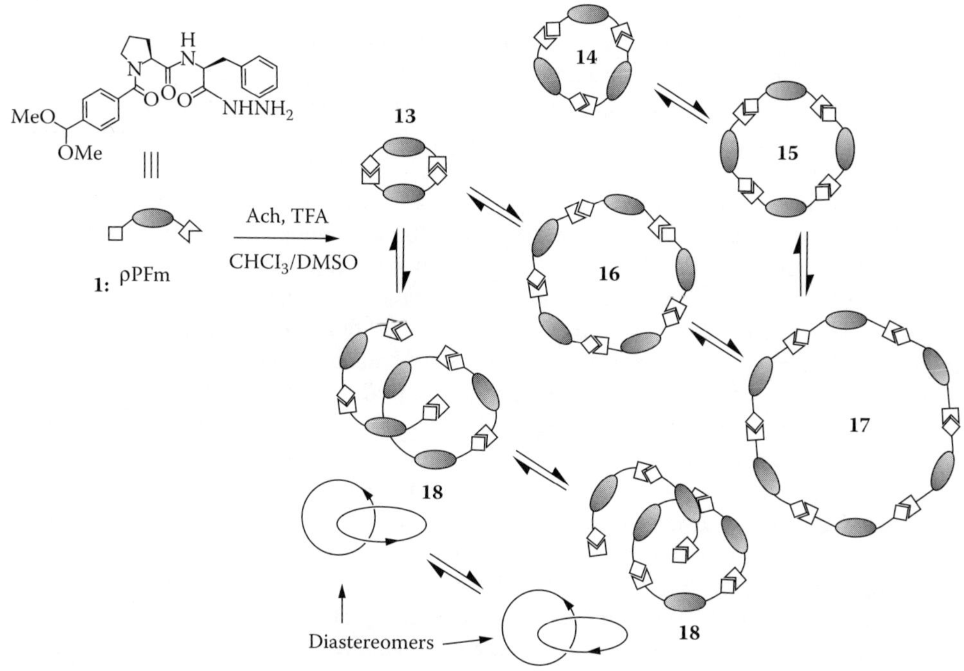

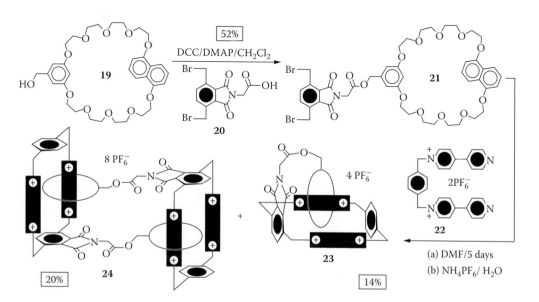

FIGURE 19.4 Acetylcholine-templated catenane formation.

displacement of bromide ions (Scheme 19.3). The *para*-xylylene dibromide building block **20** is tethered to a macrocyclic, napthyl-containing crown ether **19**. The resulting compound **21** is treated with a dicationic cyclobis(paraquat-*p*-phenylene) (CBPQT⁴⁺) monomer **22** that threads through the crown ether and can be trapped by dibromide displacement. The length of the tether dictates whether the pretzelane **23** or cyclic *bis*[2]catenane **24** is produced, the more flexible tether favoring pretzelane formation.

SCHEME 19.3 Pretzelane **23** and cyclic bis[2]catenane **24** formation.

The pretzelane structure **23** is characterized by ROESY NMR and ESI mass spectrometry and exists as two enantiomeric pairs of diastereomers, resulting from two elements of chirality. Helical chirality results from the asymmetry of the tetracationic cyclophane, and planar chirality is associated with the 1,5-dioxynapthylene (DNP) ring system. These chirality elements, along with a high barrier to enantiomerization (17.5 kcal/mol), could potentially lead to electrochemically switchable isomerism among bistable pretzelanes.

Functional catenane devices have also been synthesized. Recently a tristable [2]catenane system was developed[16] as an electrochemically color-switchable RGB dye containing three donor building blocks in one macrocycle with an interlocked acceptor CBPQT[4+] monomer (Figure 19.5). The acceptor ring orients itself initially around a tetrathiafulvalene (TTF) unit when all three stations are neutral (green). The application of a particular voltage oxidizes the donor TTF building block to a doubly charged species, which causes the migration of the CBPQT[4+] acceptor to a second donor benzidene (BZD) station (blue). A second oxidation step allows the acceptor to migrate to a DNP donor (red). When the voltage is reset back to zero, the CBPQT[4+] returns to the original donor TTF building block (green). In this instance, building block selection is crucial for obtaining the proper color for each station of the tristable catenane. Optimization of the oxidation potentials for each donor monomer adjusts the HOMO(donor)/LUMO (acceptor) band gap to provide the correct absorption wavelength. This optimization is done by testing various arrays of fluorine-substituted analogs of the BZD unit, of which the disubstituted unit has an adsorption of ~600 nm, giving the desired blue color. This type of dye could be embedded in a polymer matrix that would provide a cheap, simple RGB cell used in paper-like electronic displays.

A second functional catenane device was also recently engineered to act as a reversible rotary motor (Figure 19.6).[17] The clockwise or counterclockwise circumrotation is driven by the Brownian motion

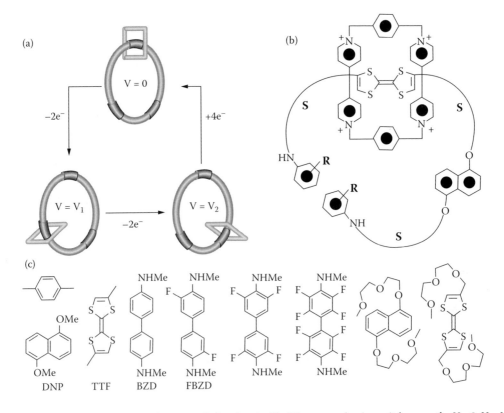

FIGURE 19.5   (a) As the external voltage is applied to the tristable [2]catenane, the ring switches over the $V = 0$, $V = V_1$ and $V = V_2$ stations. (b) The layout of the RGB tristable [2]catenane. The units refer to the donor building blocks TTF, R-BZD, and DNP. (c) Candidate donor units that were screened.

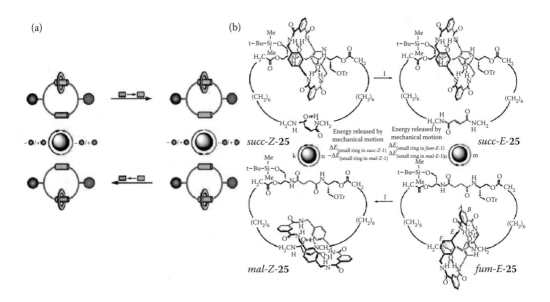

FIGURE 19.6 (a) A schematic representation of the molecular motor. The ring preferentially resides on one or another of the two binding sites (stations), represented by colored cylinders. The colored spheres are bulky groups, each of which sterically blocks one of the two tracks preventing the ring from moving between the stations. The blue-to-green and green-to-blue transformations represent chemical reactions that change the affinity of a station for the small ring, providing a driving force for the ring to redistribute itself between the stations. Removal of a bulky group (red or purple sphere) allows the ring to move between stations by a particular route. Reattachment of the sphere prevents the ring from falling back to the previous station, ratcheting the net transported quantity of rings. (b) Synthetic operation of the molecular motor: j. hv 254 nm, 5 min, 50%; k. TBAF, 20 min then cool to −78°C and add 2,4,6-collidine, TBDMSOTf, 1 h, overall 61%; l. piperidine, 1 h, ~100%; m. $Me_2S \cdot BCl_3$, −10°C, 10 min, and then TrCl, Bu_4NClO_4, 2,4,6-collidine, 16 h, overall 74%; n. $Me_2S \cdot BCl_3$, −10°C, 15 min, then cool to −78°C and add 2,4,6-collidine, TrOTf, 5 h, overall 63%; o. TBAF, 20 min, then cool to −10°C and add 2,4,6-collidine, TBDMSOTf, 40 min, overall 76%.

inherent in the molecular system. The smaller ring, consisting of isophthaloyl and *p*-xylylene diamine building blocks, rests originally on the *E*-fumaramide monomer station (*fum-E*-**25**) of the larger ring. Irradiation of this compound at 254 nm isomerizes the fumaramide olefin to the *Z* configuration (*mal-Z*-**25**), weakening the binding affinity of the smaller ring at this station. Selective deprotection/protection of either the TBS or the trityl ether dictates the direction of circumrotation about the larger ring. The smaller ring migrates to the station containing the succinamide monomer (*succ-Z*-**25**). Treatment with piperidine isomerizes the fumaramide moiety back to the *E* isomer (*succ-E*-**25**), and another round of deprotection/protection allows the smaller ring component to return to the original station. What is interesting, from a design perspective, is that the molecular motor is rationally constructed such that each building block and protecting group is chemically orthogonal to the others. This allows complete control over the directionality of motion and properties of the device.

19.2.3 Rotaxanes

Rotaxanes are mechanically interlocked molecules consisting of a macrocyclic wheel wrapped around a dumbbell-like axle containing bulky end caps. Rotaxanes have been constructed from many diverse molecular components such as calixarenes,[18] cyclodextrins,[19] and cucurbiturils,[20] for which there are excellent recent reviews. Diverse building block approaches to the syntheses of rotaxanes have been developed and the structural modification of existing rotaxane structures through mild chemical transformations[21,22] are now becoming more commonplace.

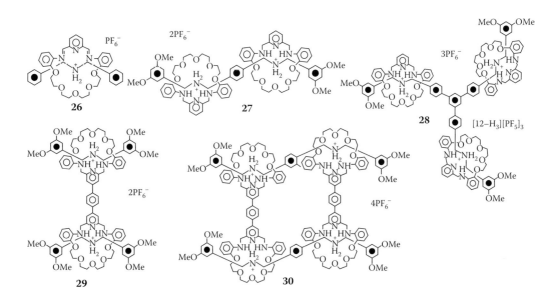

FIGURE 19.7 Reversible rotaxane construction through imine bond formation. Various complex architectures are synthesized.

An excellent example of the utility of building blocks for the construction of complex nanostructures comes from the synthesis of increasingly intricate rotaxane molecules by Stoddart and coworkers.[23] They demonstrate that thermodynamically controlled imine bond formation can be used to stitch together a wide variety of interlocked rotaxanes (Figure 19.7). Dialdehyde and diamine building blocks assemble around a templating axle (typically a dialkylammonium cation) and reversibly react to form the interlocked wheel component of the rotaxane, which can be kinetically trapped by reduction of the imine products with BH$_3$·THF. The robust chemistry yields a [2]rotaxane **26**, [3]rotaxane **27**, branched [4]rotaxane **28**, bis[2]rotaxane **29**, and a cyclic [4]rotaxane **30** in very good yield. This method has also been used to generate the equally impressive interlocked rotaxane dendrimers.[24]

Another templating reaction, different from threading of alkyl ammonium salts, used in the syntheses of rotaxanes, is Leigh's diamide hydrogen bonding motif (Scheme 19.4).[25] In this case, two separate hydrogen bond donor amide moieties of the axle building block template allow the formation of the macrocyclic wheel component around the axle. This strategy has been employed in the synthesis of a chemically reversible molecular catenane motor[17] and a molecular rotaxane shuttle.[26]

Onagi and Rebek have used this templating effect of diamide hydrogen bonding groups to create a rotaxane in which they could monitor the shuttling of the ring over the axle using fluorescence resonance energy transfer (FRET).[27] FRET is a very useful phenomenon in which energy is transferred from a donor fluorescent group to an acceptor group when they are within a few nanometers of each other. When light energy of a particular wavelength excites the donor fluorophore it can then transfer some of that energy to the acceptor. In the absence of an acceptor, the donor would emit light at its normal emission wavelength, but when an acceptor is nearby, the fluorescence emission of the donor is diminished because of loss of energy to the acceptor. In addition, the acceptor emission is enhanced. FRET is an excellent measurement tool because the net effect dies off with the sixth power of the distance between the two fluorophores. Thus, the donor and acceptor must be within the proximity of one another (20–100 Å) to measure the effect. FRET is especially useful because it allows the measurement in dynamic systems. Onagi and Rebek developed the rotaxane **35** that has two donor groups attached to a gly–gly station on one end of an axle and an acceptor group attached to the ring (Figure 19.8). In Onagi and Rebek's system, the donor moiety is a coumarin 2 attached to one end of the axle and the acceptor is a coumarin 343 connected to the interlocked macrocycle. The rotaxane contains both characteristic absorptions of the donor and acceptor. Efficient FRET

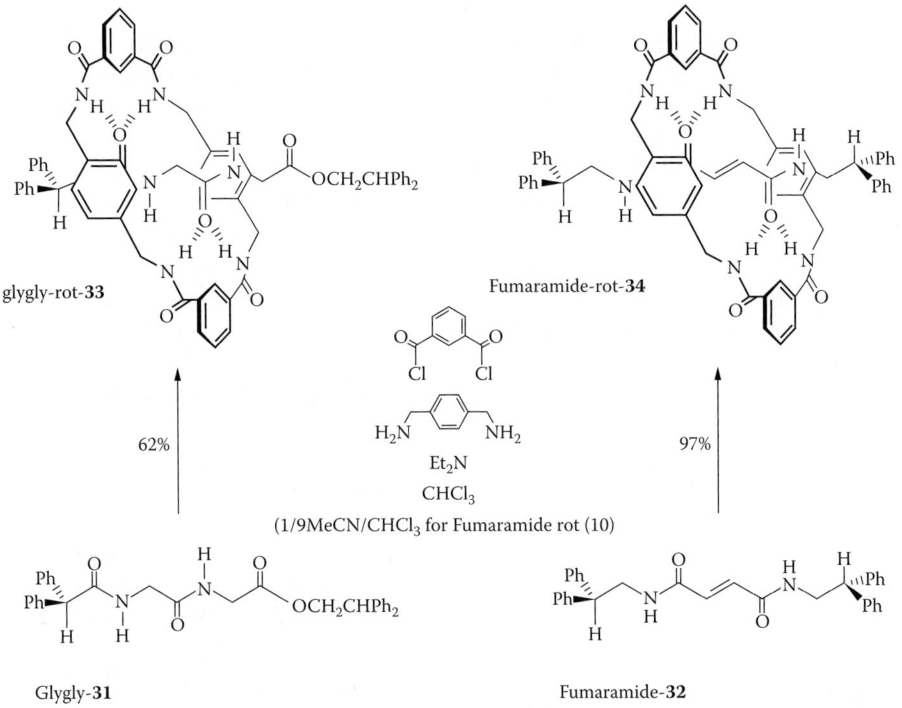

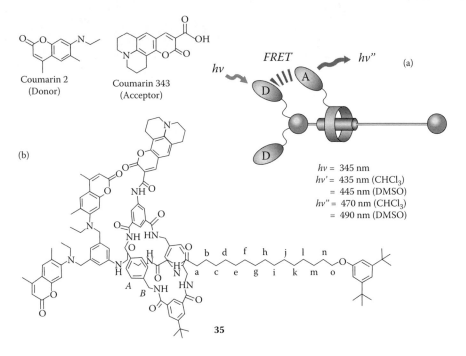

SCHEME 19.4 Leigh's hydrogen-bonding templation of rotaxanes. Two axle molecules, glygly-**31** and fumaramide-**32** template the formation of rotaxanes glygly-rot-**33** and fumaramide-rot-**34**, respectively.

FIGURE 19.8 A schematic representation of the rotaxane **35** showing that a macrocycle was mechanically interlocked by a dumbbell composed of an axle (black) with a hydrogen bonding site terminated by the bulky end group. Interlocked components were tagged by the donor D and acceptor A.

is measured when the acceptor is excited at 345 nm. The macrocyclic ring is stationed, in nonpolar environments, around the diamide moiety of the axle, which was confirmed by ^{1}H and ROESY NMR experiments. In polar solvents, the hydrogen-bonding interactions within the rotaxane are perturbed enough that the ring can move away from the diamide station, causing a decrease in the FRET intensity. FRET has also been used within Rebek's group to measure the self-assembly of molecular capsules.[28]

Yet another useful mechanism for rotaxane formation was accomplished by Asakawa and coworkers.[29] The method, known as "threading-followed-by-shrinking," is an alternative to the other known methods of rotaxane formation, such as "clipping," "threading-followed-by-stoppering," and "slipping." This method requires a macrocycle building block with an incorporated salophen [N,N'-o-phenylenebis-(salicylidene-iminato) dianion] moiety and a transition metal. Salophens are tetradentate ligands that form square planar complexes with metals and are currently of interest for catalysis, molecular recognition, and electron transfer (see references within[29]).

The dibenzyl ammonium axle monomer is threaded, as usual, through the large macrocyclic crown ether wheel followed by the addition of palladium (II) acetate (Figure 19.9). The metal forms a square planar complex with the salophen moiety, effectively shrinking the size of the macrocyclic cavity. As a result, the axle component becomes trapped inside the compressed wheel, forming the interlocked rotaxane. The rotaxane is fully characterized by NMR, mass spectral, and x-ray analyses. This new method of rotaxane formation may eventually be useful for generating catalytically active rotaxane species.

An interesting application of rotaxane construction is the encapsulation of fluorescent dyes. Smith and coworkers have used a Leigh-type amide rotaxane to sterically shield a squaraine-based near-infrared

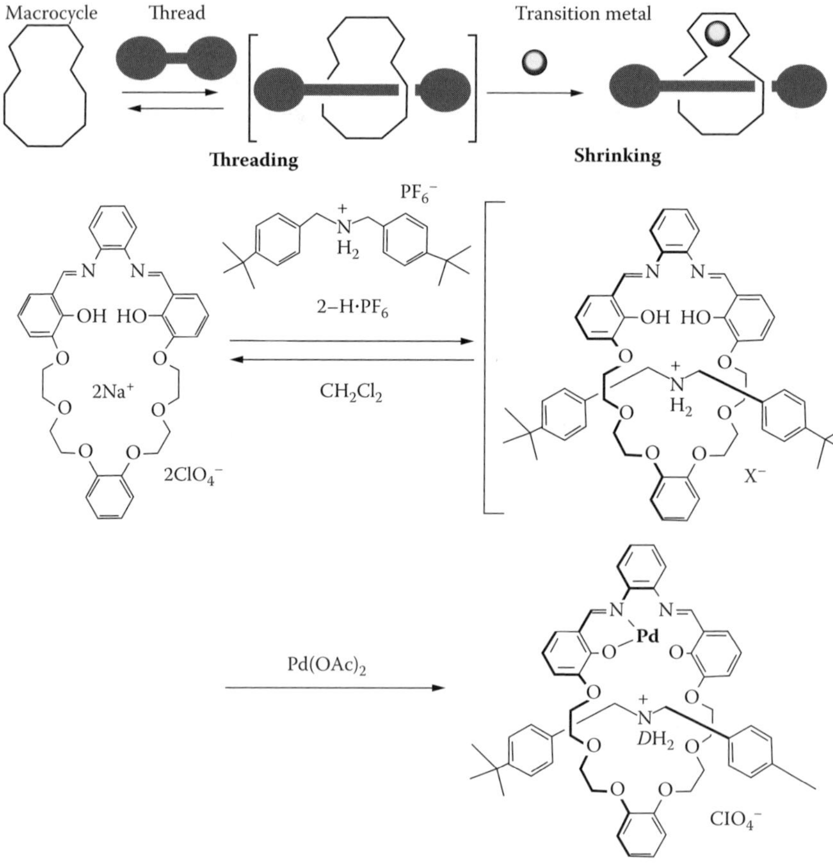

FIGURE 19.9 Schematic of the "threading-followed-by-shrinking" mechanism and synthetic procedure for the formation of the interlocked rotaxane.

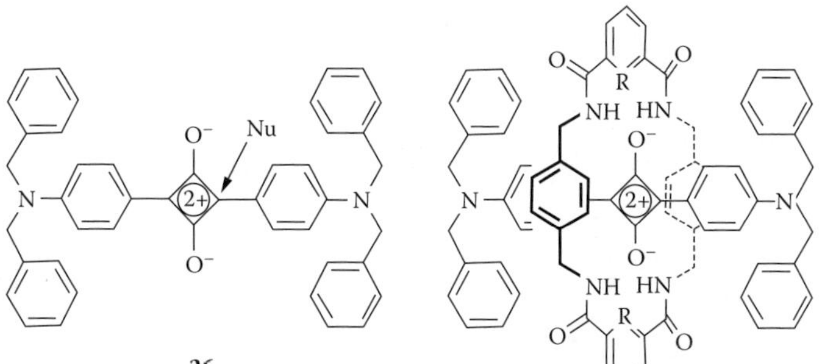

36

37 R = CH
38 R = N

FIGURE 19.10 Near-IR squaraine dye and its encapsulation into rotaxanes.

(IR) dye.[30] Dyes that absorb near-IR radiation are finding uses in nanotechnology, materials science, and biomedical science. Some problems of squaraine-containing dyes, such as susceptibility to nucleophilic addition and aggregation, stand to limit the use of these valuable tools. One method of addressing these concerns is to surround the dye with a protective molecular shield (Figure 19.10). The strategy chosen by Smith and coworkers involves the synthesis of a dumbbell-shaped squaraine molecule **36**. The squaraine contains two oxyanions attached to a cyclobutene dication with bulky end caps. Typically, the cyclobutene ring is susceptible to addition by nucleophiles, causing the loss of its blue color. Using a hydrogen bond-templated clipping reaction, Smith and coworkers enclosed a macrocyclic ring around the squaraine, thereby protecting it from outside agents. The rotaxane exists in a conformation where the two oxyanions can hydrogen bond with the amide hydrogens of the macrocycle. In the case of the pyridyl-containing macrocycle **38**, internal hydrogen bonding occurs between the pyridyl nitrogen and the adjacent NH moieties. As a result, **38** wraps more tightly around the squaraine core than the isophthalamide-containing macrocycle **37**.

By encapsulating the squaraine dye inside a rotaxane core, the chemical stability is dramatically increased. Reaction of **36** with cysteine has a half-life of approximately 5 min, **37** has a substantially longer half-life, and **38** is chemically inert. Additionally, the rate of hydrolytic decomposition of the squaraine **36** (in 4:1 THF-H$_2$O) occurs within 48 h, leaving the solution colorless. The rotaxane **37** becomes colorless within a week, and **38** maintains a blue color for months.

Finally, squaraines are known to aggregate in DMSO–water solutions, leading to broadening of the absorption spectrum. Indeed, a 1:1 solution of **36** shows significant blue-shifted and red-shifted bands, attributed to aggregation. However, even in a 9:1 water–DMSO solution it shows a relatively sharp absorption band at 637 nm. These outstanding results should make possible the synthesis of many more useful and interesting squaraine-derived rotaxanes.

19.2.4   Molecular Containers from Building Blocks

Molecular containers encompass an enormous subfield of nanotechnology.[31] Their applications include, but are not limited to, sensors, catalysts, encapsulated dyes, separation tools, memory storage devices, solar cells, radiation therapy, drug encapsulation, and more. As a result of their extensive utility, rapid and modular approaches to their synthesis are needed. There are many diverse building blocks that comprise molecular containers such as calixarenes, resorcarenes, cryptands, carcerands, crown ethers, spherands, and cyclodextrins. Some very interesting, recent nonlinear approaches to a few of these building blocks have been published.

For instance, a new convergent strategy for the synthesis of calixarenes from Gopalsamuthiram and Wulff[32] entails the reaction of a *bis*-carbene complex **39** with a *bis*-propargyl benzene **40** (Figure 19.11[b]). This reaction results in the formation of two of the phenol rings and the macrocycle of the calixarene

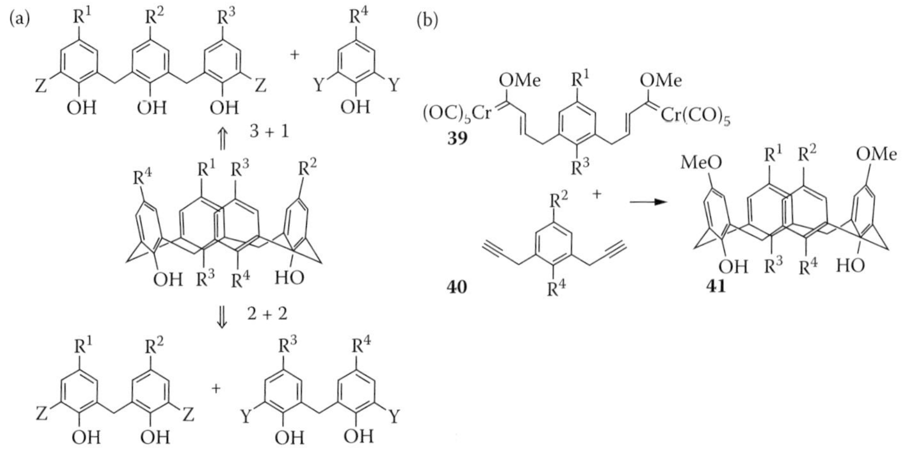

FIGURE 19.11 (a) Two separate, established and convergent strategies (3+1 and 2+2) for the synthesis of calixarenes. These routes are necessarily longer but have been the routes of choice for the construction of calixarenes designed for specific applications. (b) A new method for the synthesis of calixarenes from the reaction of a *bis*-carbene complex with *bis*-propargyl benzene. In this reaction, the phenol groups are assembled along with the macrocycle in one step. Yields for this transformation are good compared to previously studied methods.

41 in one step. Formation of the reactive *bis*-carbene complex can be accomplished in two steps in moderate yield; however, the reaction with the *bis*-propargyl benzene component provides calixarenes in excellent yields (30–40%) compared to previously published 2 + 2 and 3 + 1 strategies (Figure 19.11[a]).

Bohmer and coworkers have shown that urea-containing calixarenes can be used as building blocks to template the formation of complex multicatenanes.[33] In aprotic solvents open chain tetraureas **42** and **43a** (Figure 19.12 and Figure 19.13) form well-defined homodimers. In a 1:1 solution of **45** and **42** or

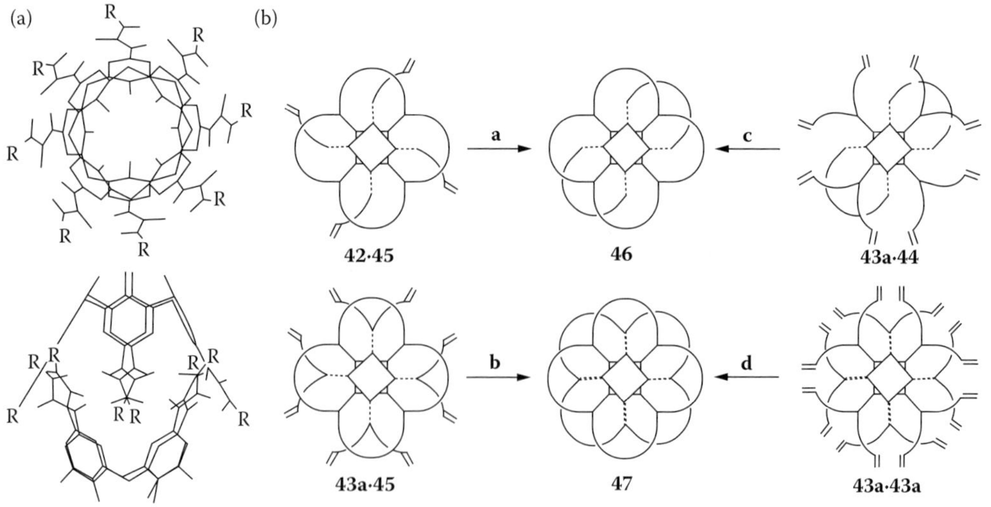

FIGURE 19.12 (a) Hydrogen-bonded dimer of a tetraurea calix[4]arene **42** or **43**, showing the mutual orientation of the urea residues R. Ether groups are omitted for clarity. (b) Schematic representation of the synthesis of multi-catenanes **46** and **47** by metathesis reaction of selectively formed heterodimers **42·45** and **43a·45** followed by hydro-genation (reactions **a** and **b**). While reaction **d** led to a complicated mixture of products, reaction **c** (which has not yet been checked) seems at least an alternative to **a**, although wrong connections between double bonds are possible for **c** in contrast to **a**.

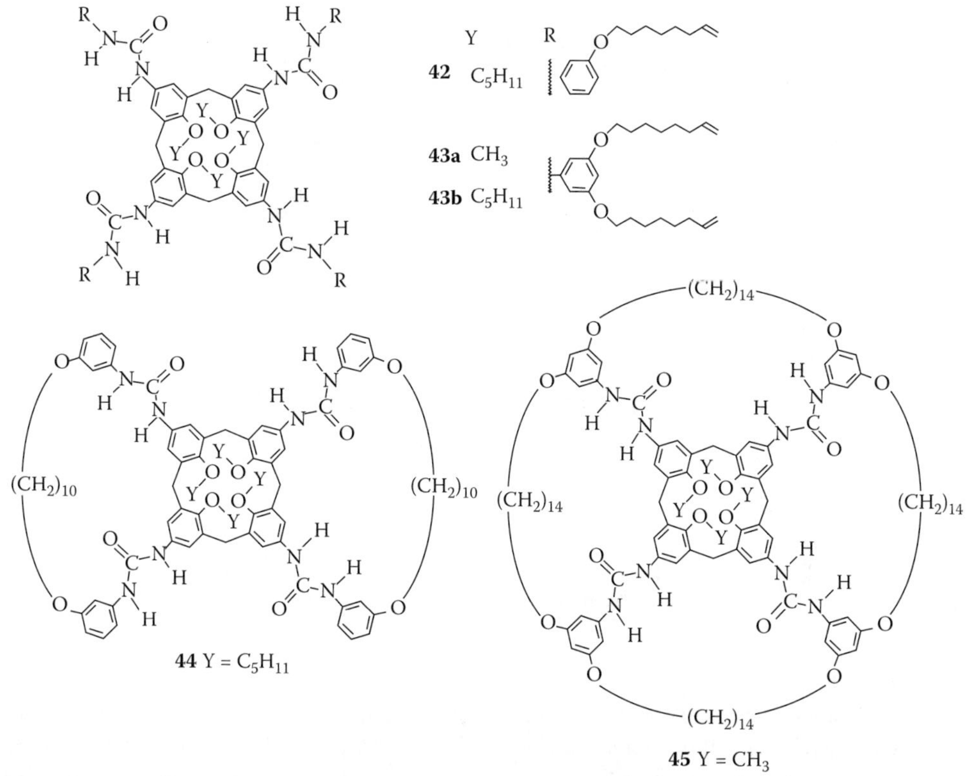

FIGURE 19.13 Structures of the calixarene components **42**, **43**, **44**, and **45**.

43a heterodimers form exclusively (as evidenced by [^1]H NMR). When the pseudorotaxane heterodimers are exposed to metathesis reaction conditions, amazingly, only single products **46** and **47** are formed in greater than 50% yield! The unprecedented structures are unambiguously determined by ESI-MS, [^1]H-, and [^13]C-NMR. The [8]catenane structure **47** was also determined by x-ray analysis and consists of two conically shaped rings of four annulated loops attached to the wide rims of two calix[4]arenes that are interwoven such that each loop of one of the rings penetrates two adjacent loops of the other ring and vice versa.

A chiral calixarene/cryptand hybrid has been shown to bind cations and small neutral organic molecules within its well-defined cavity. Jabin and coworkers synthesized the first known enantiopure calix[6]aza-cryptand[34] and found that it adopts a rigid cone conformation ideal for host–guest chemistry applications. Typically calix[6]arenes suffer from a large degree of inherent flexibility and propensity to undergo cone–cone interconversion that prevents them from acting as good host molecules. In this case, the calix[6]arene is constrained by the tripodal aza cap, thereby locking it into a well-defined conformation (Figure 19.14).

The chiral calix[6]aza-cryptand **48** is synthesized in five steps in a straightforward manner from (*S*)-2-aminopropan-1-ol, *N*-dinosylated trisamine and 1,3,5-*tris*-tosylated calix[6]arene building blocks. The lone stereogenic center of the aza cap translates its chirality to the cavity of the host molecule, which can be seen in the dissymmetrical nature of its [^1]H NMR spectrum. The ability of the host molecule **48** to bind small guests is evaluated by NMR by first protonating all the amines of the cap with four equivalents of trifluoroacetic acid, leading to the corresponding tetra-ammonium salt **49**. Upon addition of polar neutral molecules such as imidazolidin-2-one (IMI), DMSO, EtCONH$_2$, or EtOH, new dissymmetrical high-field NMR patterns are observed. NOESY experiments indicate that these high-field resonances belong to one equivalent of a guest molecule inside the deep calixarene cavity of the host.

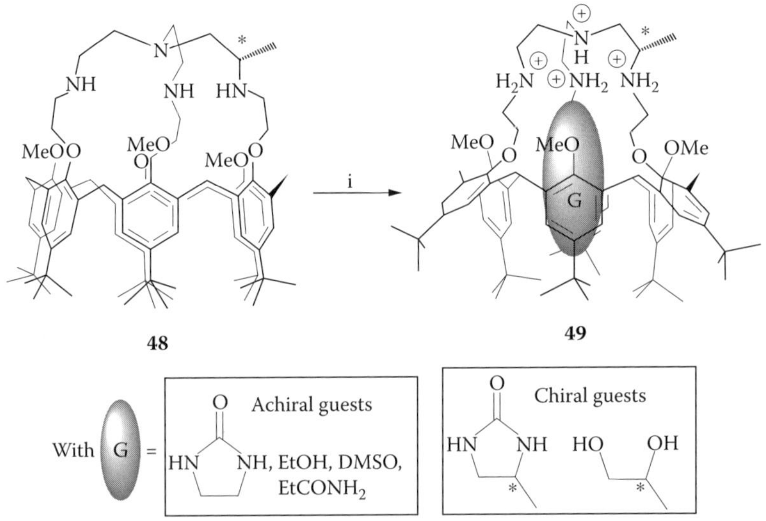

48   **49**

With **G** = | Achiral guests: HN—NH, EtOH, DMSO, EtCONH₂ | Chiral guests

FIGURE 19.14 Binding of small organic molecules into a calix[6]aza-cryptand cavity. Discrimination between enantiomeric pairs of racemic mixtures is seen in NMR studies. (i) 4 equivalents trifluoroacetic acid, **G** (guest).

Another set of experiments demonstrates the ability of the host molecule **49** to bind the chiral racemic guests (±)-propane-1,2-diol (PPD) and (±)-4-methylimidazolidin-2-one (MIMI). In both cases, a 2:1 mixture of diastereomeric endocomplexes is formed. In the case of PPD, a subsequent addition of (R)-(−)-PPD leads to a reversal of the diastereomeric mixture, allowing the discrimination between each enantiomeric guest. This work represents a solid starting point for future exploration of enantioselective molecular receptors and catalysts from calix[6]arenes.

An example of selective anion binding by amidocryptands is presented by Bowman-James and coworkers.[35] Bicyclic cryptands offer advantages over monocyclic analogs due to their cage-like structure that can capture and sequester anions. In monocyclic analogs, quaternization of tertiary amines improves selectivity for anions considerably by adding positive charge. A similar effect is expected in cryptands. Compound **50** (Figure 19.15) is synthesized from tris(3-aminopropyl)amine (trpn) and 2,6-pyridinedi-carbonyl dichloride building blocks, and the quaternized **51** is made by the treatment of **50** with methyl iodide. The anion–cryptand complexes are crystallized, revealing interesting structural features. **50** exists as a cage-like structure with two loops pointing in the same direction and the third loop in the opposite direction. This shape may be partially stabilized by π-stacking interactions between two of the pyridine spacer units. In the chloride complex, two chlorides are bridged by a "cascading" water molecule, typical of certain cryptand complexes. The sulfate complex contains one sulfate anion bridged by two water molecules.

In the case of the quaternized **51**, an entirely different "bowl-shaped" conformation is observed, with all three loops pointing in the same direction. Water again plays an important role, with multiple water molecules lying inside the bowl with the anions suspended on top (no encapsulation of anions observed in these complexes). In the chloride complex, one anion is centered between the two quaternized amines at the top of the bowl and linked by four hydrogen bonds to water molecules and amide hydrogens. In the oxalate complex, the oxalate anion also lies between the two positively charged quaternary ammonium sites, bound by five hydrogen bonds. The fact that both anions lie directly between the two positively charged "poles" demonstrates well balanced charge distribution that may be a driving force for the formation of a bowl-shaped conformation. Preliminary binding studies with the quaternized **51** indicate that it has a high selectivity for $H_2PO_4^-$, whereas **50** shows a strong affinity for fluoride. These amido-cryptands may provide another structural motif of importance to the development of selective capsules and containers for anion chemistry.

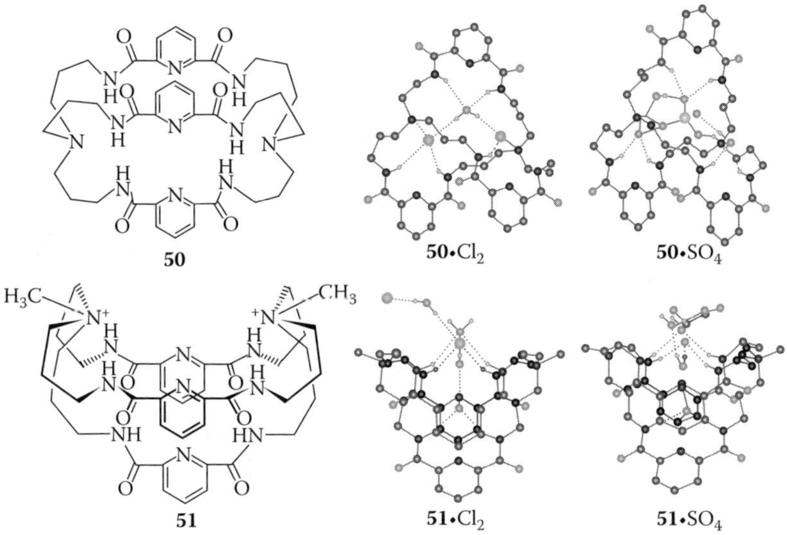

FIGURE 19.15 Amidocryptands and their anion inclusion complexes. For **50**, the complexes with chloride and sulfate are shown and for **51** the complexes with chloride and oxalate are shown.

Typically interlocked molecular structures are synthesized from a linear guest threaded through a macrocyclic host. However, novel nonlinear threaded structures are being realized due to their unique topologies and potential uses as nanoscale devices. Gibson and coworkers have devised slow exchange C_3-symmetric inclusion complexes based on a new cryptand/trispyridinium recognition motif, in which 1,3,5-tris(N-pyridiniummethyl)benzene salts act as guests.[36] Cryptand **52** and tripod **53** (Figure 19.16) are the requisite building blocks for the synthesis of the C_3-symmetric complex. The x-ray structure of **53a** shows that all three legs of the tripod are on the same side, which could envelop the phenylene ring of **52**. NMR spectra reveal that a 1:1 mixture of **52** and **53** form a slow exchange, completely C_3-symmetric complex (on the NMR timescale). Only one set of complexed signals is observed, with no isomeric signals present, which indicates that the tripodal arms of the complex are rapidly flipping up and down. The complex is held together most likely by a combination of hydrogen-bonding interactions between the α-pyridinium protons of the tripod and the oxygen atoms of the cryptand and π-stacking interactions between the core of the tripod and the phenylene rings of the cryptand. Interestingly, the pyridinium ring nitrogen atoms are sterically shielded, in the complex, from pairing with the PF_6^- anions. Finally, the complex could be potentially interlocked by choosing the appropriate end caps for the tripod and being reduced to a neutral interlocked complex.

19.3 Building Block Approaches to Linear Macromolecules

Oligomeric molecules are molecules that are composed of linear sequences of building blocks; examples are proteins, DNA, and RNA. Linear molecules are attractive platforms for the development of functional molecules because they are relatively easy to synthesize and can be very large, complex, and asymmetric. Linear molecules are typically synthesized by assembling building blocks that are coupled together using one highly optimized chemical reaction. In some coupling reactions, such as amide bond formation, coupling yields of 99% and higher are routinely achievable, allowing the synthesis of very large oligomers containing dozens of building blocks. The building blocks used to construct oligomeric molecules and the chemistry used to couple them together are limited only by the chemist's imagination. Oligomeric molecules can be synthesized in either a convergent fashion in solution, where pairs of shorter oligomers

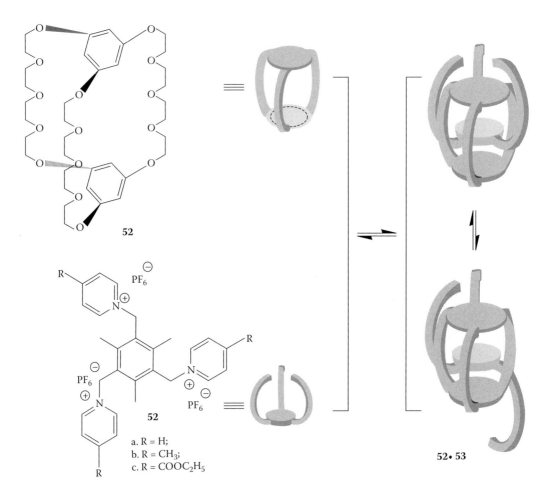

FIGURE 19.16 Cryptand **52** and tripod **53** form a slow exchange complex (on the NMR timescale) where the tripodal component is threaded through the cryptand. The arms of the tripod can rapidly flip up and down, yielding an entirely C_3-symmetric system.

are coupled to form longer oligomers, or in a linear fashion on solid support. The advantage of convergent synthesis is that large molecules can be synthesized with few linear steps. A drawback of this approach is that the product must be isolated at the end of each coupling reaction. As products get larger and larger, their solubility can change in unexpected ways, complicating isolation. Oligomeric molecules can also be synthesized in a linear fashion on solid support. The advantages of solid-phase assembly are that the growing product is attached to a plastic resin and can be isolated at the end of each reaction by simple filtration. Solid-phase synthesis also allows the use of a large excess of activated building blocks to drive coupling reactions to very high yields, although this can be a disadvantage if the building blocks are expensive. At the end of a solid-phase synthesis, the product is typically cleaved from the resin and purified using chromatography.

Oligomeric molecules that have a tendency to fold into well-defined 3-D structures have been termed "foldamers" by Gellman and have been the subject of several excellent reviews.[37–39] In the past few years, remarkable progress has been made by chemists in the development of foldamers that have some of the properties of biological proteins. Artificial oligomers have been developed that can reliably fold into well-defined 3-D architectures and display a variety of functions such as selective binding of small molecules. We provide a few examples of linear oligomers that display nanoscale architecture.

19.3.1 *m*-Phenylene Ethynylene Foldamers

An active area of research in the past decade has been the development of foldamers that, analogous to proteins, create cavities and bind small guest molecules within them. Among the most highly developed foldamers are the *m*-phenylene ethynylene oligomers. *m*-Phenylene ethynylene oligomers are a class of foldamers that fold into a compact helical conformation in polar solvents such as acetonitrile and water (Figure 19.17[a]).[40] Moore and coworkers have developed the synthesis of *m*-phenylene ethynylene oligomers using both convergent[40] and solid-phase approaches.[41] These oligomers form unfolded, disordered chains in apolar solvents such as chloroform and in experiments where the solvent composition is systematically changed from pure chloroform to pure acetonitrile, the oligomers containing more than

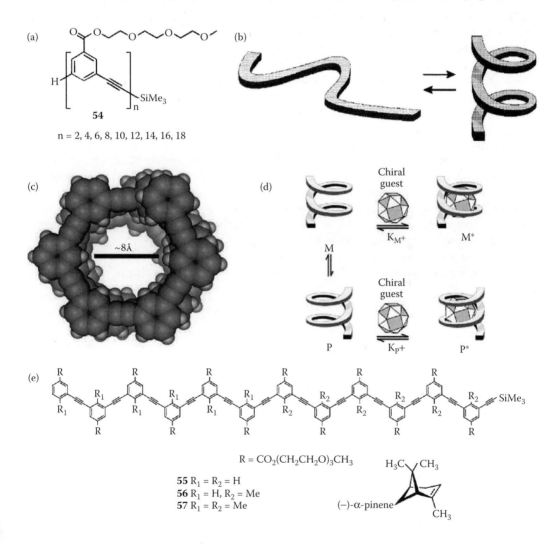

FIGURE 19.17 (a) The structure of *m*-phenylene ethynylene (PE) oligomers **54**. (b) PE oligomers are in equilibrium between an unfolded state (*left*) and a folded state with a helical conformation. (c) A top view of a PE oligomer. PE oligomers have a tubular cavity with an internal diameter of about 8Å. (d) A folded PE oligomer is in equilibrium between two equal energy enantiomeric helical states M and P. When they bind a chiral guest, they adopt two diastereomeric states M* and P* that have different relative energies. The population of the M* and P* state depends on their relative energy. (e) The structure of PE oligomers **55**, **56**, and **57** and the guest (−)-α-pinene. (From Matsuda, K., Stone, M.T., and Moore, J.S., *J Am Chem Soc* 124 (40), 11836–11837, 2002; Prince, R.B., Barnes, S.A., and Moore, J.S., *J Am Chem Soc* 122 (12), 2758–2762, 2000. With permission.)

eight monomers undergo a cooperative transition from the unfolded state to a folded helical state (Figure 19.17[b]).[40] By synthesizing a series of double spin-labeled oligomers containing 4, 5, and 6 phenylene ethynylene monomers between two spin-labeled monomers and analyzing their electron spin resonance spectra in the folded state, Moore and coworkers demonstrated that there are six monomers in each turn of the helix.[42] Molecular modeling suggests that the helices contain a tubular cavity with an internal diameter of about 8 Å (Figure 19.17[c]). While *m*-phenylene ethynylene oligomers contain no chiral centers, individual oligomers adopt left-handed and right-handed helical conformations of identical energy.

Moore and coworkers have demonstrated that *m*-phenylene ethynylene (PE) oligomers bind a variety of chiral small molecules including (−)-α-pinene in 1:1 complexes. When the *m*-phenylene ethynylene dodecamer **55** binds a chiral guest molecule, it form two diastereomeric host–guest complexes in a ratio that depends on the difference in energy between the two diastereomeric complexes (Figure 19.17[d]).[43] The binding for different guests was somewhat selective, with binding constants ranging from 6830 M^{-1} for (−)-α-pinene to 1790 M^{-1} for [(1R)-*endo*]-(+)-fenchyl alcohol. Modified dodecamers **56** and **57** were synthesized that projected methyl groups into the tubular cavity within the helices, reducing the space available to guests. These modified oligomers **56** and **57** still bound (−)-α-pinene but with association constants that were 24- and 171-fold lower (respectively) than that of **55**. In more recent work, Moore and coworkers have demonstrated that PE oligomers bind rod-like guests with affinity that depends on the oligomers' chain length (Figure 19.18).[44]

Moore and coworkers have demonstrated that PE oligomers can contain a variety of different building blocks and still form stable helices in polar solvents.[45–49] Increasing the number of ethylene glycol units on the side chain of each PE monomer allows the synthesis of oligomers such as **59** that are soluble in water (Figure 19.19[a]).[47] The helical folding of PE oligomers is very robust. The folding stability of PE oligomers incorporating a wide range of single-site modifications (Figure 19.19[b]) was probed using chloroform/acetonitrile titrations; all of the oligomers formed stable helices with folding free energies that ranged between −4.3 and −6.0 kCal/mol.[45]

Several of the groups incorporated previously as single-site modifications were basic *para*-substituted pyridinium groups (Figure 19.19[c]), which range in basicity, with pKa values ranging from 5.4 to 14.0.

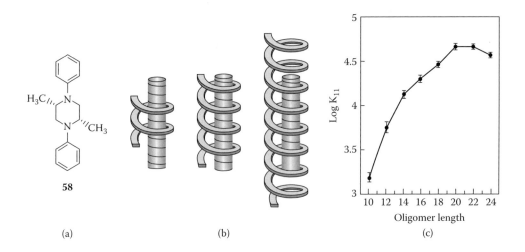

(a)   (b)   (c)

FIGURE 19.18   (a) The structure of **58**, the rod-like guest (2S,5S)-2,5-dimethylpiperazine that binds PE oligomers. (b) A cartoon illustrating how PE oligomers of different lengths bind a single rod-like host. Oligomers that are too short leave the host exposed to solvent (*left*), oligomers that form helices of the ideal length bury the guest (*center*), and oligomers that are too long bury the host but leave part of the tubular cavity exposed to solvent (*right*). (c) The relationship between association constant K_{11} and PE oligomers' length in the binding of **58** shows a maximum association constant for PE oligomers containing 20 and 22 monomers. (From Tanatani, A., Mio, M.J., and Moore, J.S., *J Am Chem Soc* 123 (8), 1792–1793, 2001. With permission.)

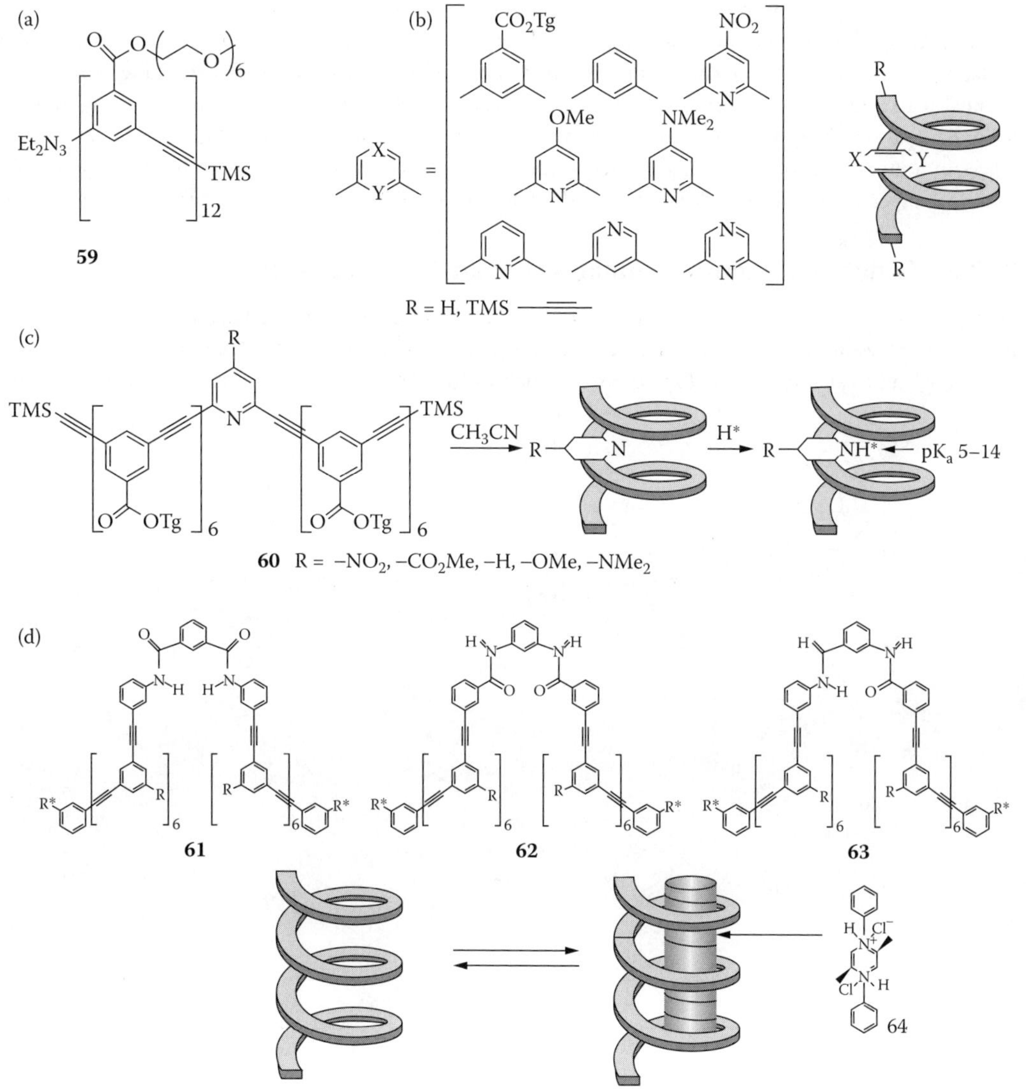

FIGURE 19.19 (a) The structure of the water-soluble PE oligomer **59**. (b) PE oligomers were synthesized incorporating single-site modifications consisting of the eight groups within the box and two different types of end modifications. All of these modified oligomers formed stable helices in acetonitrile. (c) A series of PE oligomers were synthesized, incorporating modified pyridine bases with pKas that range from 5 to 14. (d) Three PE oligomers (**60**, **61**, and **62**) were synthesized with backbone modifications that presented hydrogen-bond donor and acceptor groups into the tubular cavity. Only **60** bound the HCl salt of a rod-like guest. (From Stone, M.T. and Moore, J.S., *Org Lett* 6 (4), 469–472, 2004; Goto, H., Heemstra, J.M., Hill, D.J., and Moore, J.S., *Org Lett* 6 (6), 889–892, 2004. With permission.)

Moore and coworkers examined the basicity of these pyridinium groups in acetonitrile solvent within short PE trimers that are incapable of forming helices and longer tridecamers **60** that form stable helices.[49] They demonstrated that the formation of helices only perturbed the pKas of two of the groups and by no more than 0.4 pKa units. This demonstrates that the binding cavity of PE oligomers can be functionalized with basic pyridinium groups, opening the door to the use of these foldamers as pyridinium-based nucleophilic catalysts for alkyl and acyl transfer reactions.

In another study, Moore and coworkers demonstrated that backbone modifications of PE oligomers lead to sequence-specific binding of a piperazinium dichloride guest.[46] Three PE oligomers **61**, **62**, and **63** containing pairs of amide bonds on the central monomer were constructed. These amide bonds were oriented with both amide NH groups pointing into the cavity **61**, both carbonyl groups pointing into the cavity **62**, and one NH and one carbonyl pointing into the cavity **63**. All three oligomers formed stable helices in acetonitrile. None of the oligomers bound the rod-like guest **58**. However, the PE oligomer **61** bound the piperazinium dichloride **64**, the dihydrochloride salt of **58**, while the other two oligomers did not.

19.3.2 Functional Achiral Foldamers That Bind Heparin

Choi et al.[50] have developed a series of 1,3-substituted arylamide foldamers that are able to bind heparin, an anionic, polysulfonated, polysaccharide based anticoagulant (Figure 19.20). Heparin is used clinically to prevent and treat thrombotic diseases and in dialysis to prevent clotting as the blood passes through

Compound	R^4	R^5	IC_{50} [μm]	K_B [μm][a]	HC_{50} [μm][b]
66	H	NH_2	256	6.7	>1540
67	H	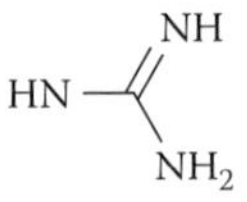 (guanidine group)	77.9	3.2	>1363
68	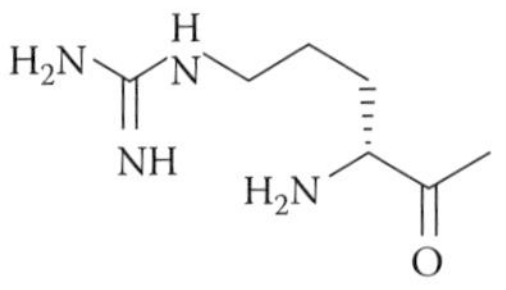 (arginine-derived group)	NH_2	22.5	1.8	>1087
69	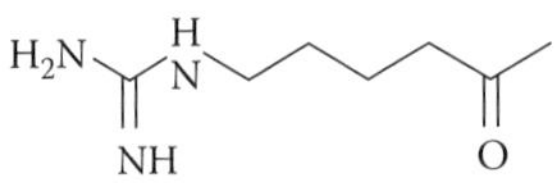 (agmatine-derived group)	NH_2	28.1	2.0	927

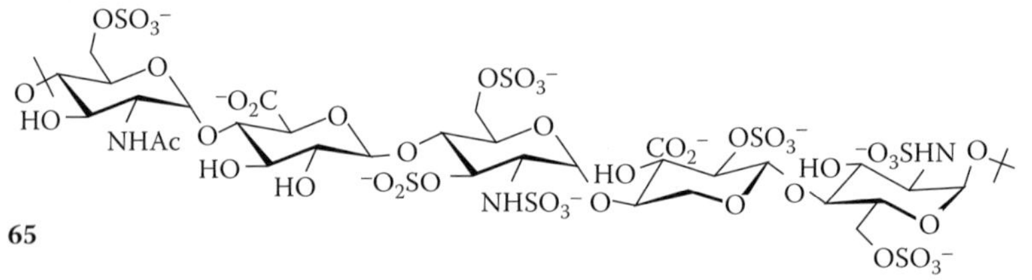

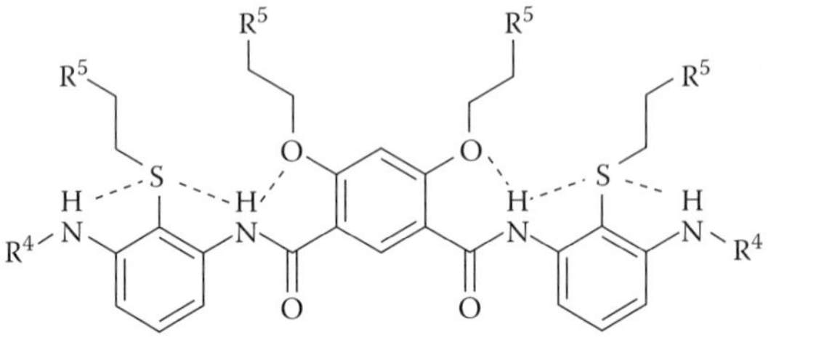

[a] The dissociation constant (K_B) was measured by Schild plot analysis by using the anti-factor Xa assay.
[b] The HC_{50} value (measurement of hemolytic activity) was obtained by measuring 50% lysis of human erythrocytes.

FIGURE 19.20   (a) The antithrombin III-binding pentasaccharide of heparin. (b) The structure of foldamers **66–69** and their biological activity.

a dialysis machine. There are major adverse effects associated with excessive heparin therapy, including hemorrhage and heparin-induced thrombocytopenia. To prevent these adverse effects, protamine is used clinically as a heparin antidote[51] to neutralize the anticoagulation function of excess heparin. Heparin exerts its anticoagulative effects by binding to antithrombin III, a plasma serine proteinase inhibitor to form a complex that inhibits coagulation. Heparin binds antithrombin through its antithrombin III-binding pentasaccharide **65**. Protamine acts as a heparin antidote by binding heparin and preventing its interaction with antithrombin. Antithrombin, protamine, and other heparin-binding proteins share a common motif responsible for their ability to bind heparin: they display several basic residues (lysine and arginine) with a spacing that allows them to bind the anionic charges of heparin. Choi et al.[50] synthesized small molecule foldamers **66–69** that display four or six cationic groups with a spacing that allows them to bind heparin. These foldamers are based on a 1,3-substituted arylamide backbone. Hydrogen bonding between the NH groups of the backbone with the side chain thio-ether and ether groups of the oligomer (Figure 19.20[b]) helps to increase the conformational rigidity of the foldamer. The lead compound **66** displayed four cationic amines and inhibited the ability of low-molecular-weight (LMW) heparin to bind antithrombin with an IC_{50} of 256 μM and dissociation constant with LMW heparin of 6.7 μM. To create a more potent foldamer, Choi and coworkers replaced the primary amines of **66** with guanidinyl groups to form compound **67**, resulting in a threefold improvement in IC_{50} and a reduction of the K_d by a factor of two. To create even more potent foldamers, Choi and coworkers attached to both ends of the oligomer two additional guanidinyl groups in the form of 5-(amidinoamino) pentanoic acid to create **68** and arginine to create **69**. Both **68** and **69** were approximately three times more potent than **67** according to their IC_{50} values and almost twice as potent according to their K_d values. Despite these compounds being considerably smaller than protamine, the most active compounds **68** and **69** were only two- to threefold less potent than protamine (average Mw = 5100 Da). As a test of toxicity, Choi and coworkers measured the ability of these compounds to lyse human erythrocytes. None of the compounds lysed human red blood cells at concentrations up to 1000 μM.

19.3.3 Aedomers

Aedomers are a class of highly developed foldamers developed by Iverson and coworkers[52] These foldamers are assembled from achiral amino acid monomers incorporating 1,5-dialkoxynaphthalene (Dan) groups and 1,4,5,8-napthalene-tetracarboxylic diimide (Ndi) groups, which are aromatic donors and acceptors, respectively. They are synthesized using sequential solid-phase synthesis. Within oligomers, the Dan and Ndi groups are interspersed with chiral amino acids such as aspartic acid to increase solubility of the oligomers. In an early work, Iverson and coworkers demonstrated that foldamers created from alternating Dan and Ndi groups fold with a pleated structure in aqueous solution (Figure 19.21[b])[52] The electron-rich aromatic donors and electron-poor acceptors stack on top of each other in an alternating fashion stabilized by electrostatic and hydrophobic interactions.

In a more recent work, Iverson and coworkers have created homo-oligomers of Ndi **73–76** and homo-oligomers of Dan **77–80** and combined them to form heteroduplexes in aqueous solution (Figure 19.22). The combinations of shorter sequences **73:77** and **74:78** display 1:1 binding at room temperature using NMR titration and monitoring the proton chemical shifts of the Ndi groups. For the longer sequences **75:79** and **76:80,** Iverson and coworkers turned to isothermal titration calorimetry (ITC) to study the association, because the NMR spectra of these longer sequences show many overlapped peaks. At room temperature the ITC data for **75:79** and **76:80** did not show saturation and did not fit any binding models well. However when the ITC data was collected at T = 318 K, all of the duplex pairs demonstrated well-behaved 1:1 binding. This suggests that at ambient temperatures, the Ndi and Dan oligomers are associating as 1:2 or 2:1 complexes or that they are aggregating. At higher temperatures this more complex behavior is disfavored.

Even more recently, Iverson and coworkers have demonstrated a sophisticated level of control; they have created aedomers with controlled folding topologies.[53] Iverson and coworkers synthesized four aedomers **81–84**. The first aedomer **81** has the sequence Dan-Ndi-Dan and its ultraviolet–visible (UV–vis)

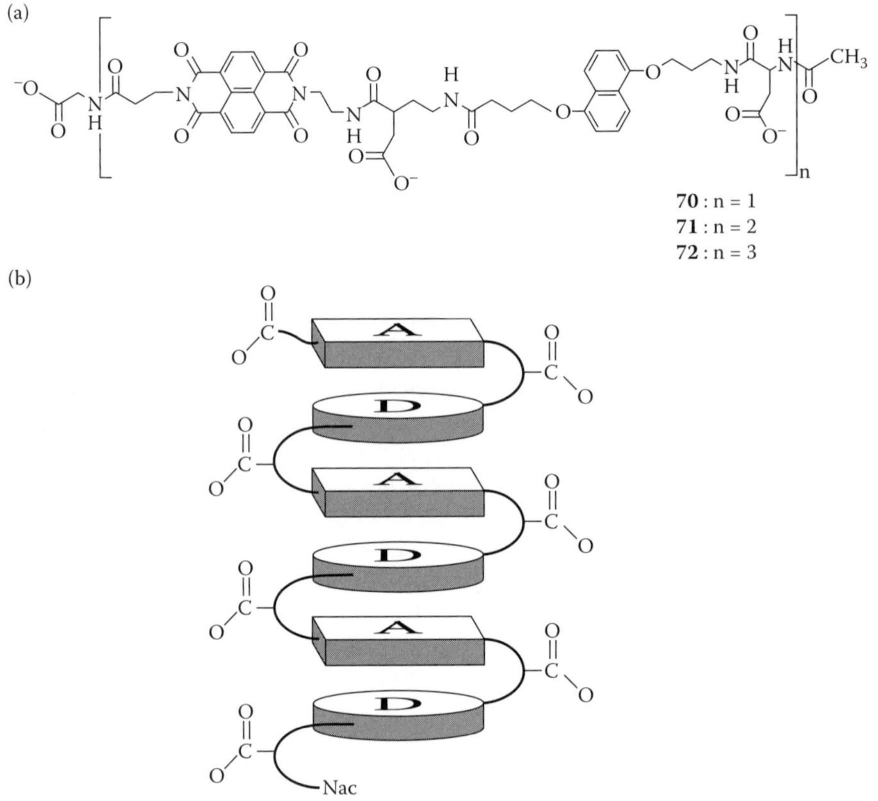

(a)

70 : n = 1
71 : n = 2
72 : n = 3

(b)

FIGURE 19.21 (a) The chemical structure of aedomers **70, 71,** and **72.** (b) A cartoon illustrating how compound **72** folds into a compact structure.

spectrum and 2-D NMR spectroscopy is consistent with a pleated folding topology similar to that of earlier compounds. The aedomer **82** was synthesized with a Dan-Dan-Ndi sequence and it was designed to adopt an intercalative folding topology where the terminal Ndi monomer intercalates between the first and second Dan monomers (Figure 19.23). Compound **83** has the Dan–Dan–Ndi sequence of **82**

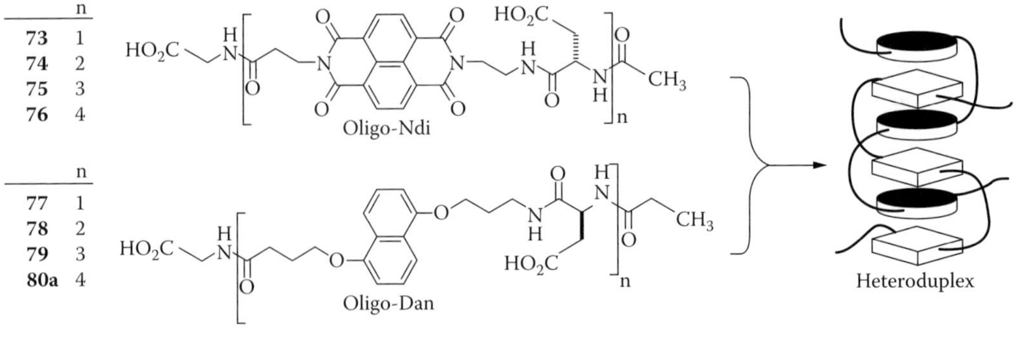

	n
73	1
74	2
75	3
76	4

Oligo-Ndi

	n
77	1
78	2
79	3
80a	4

Oligo-Dan

Heteroduplex

FIGURE 19.22 Homo-oligomers of Ndi **73–76** were combined with homo-oligomers of Dan **77–80** to create heteroduplexes. The black disks represent Dan units and the white squares represent Ndi units.

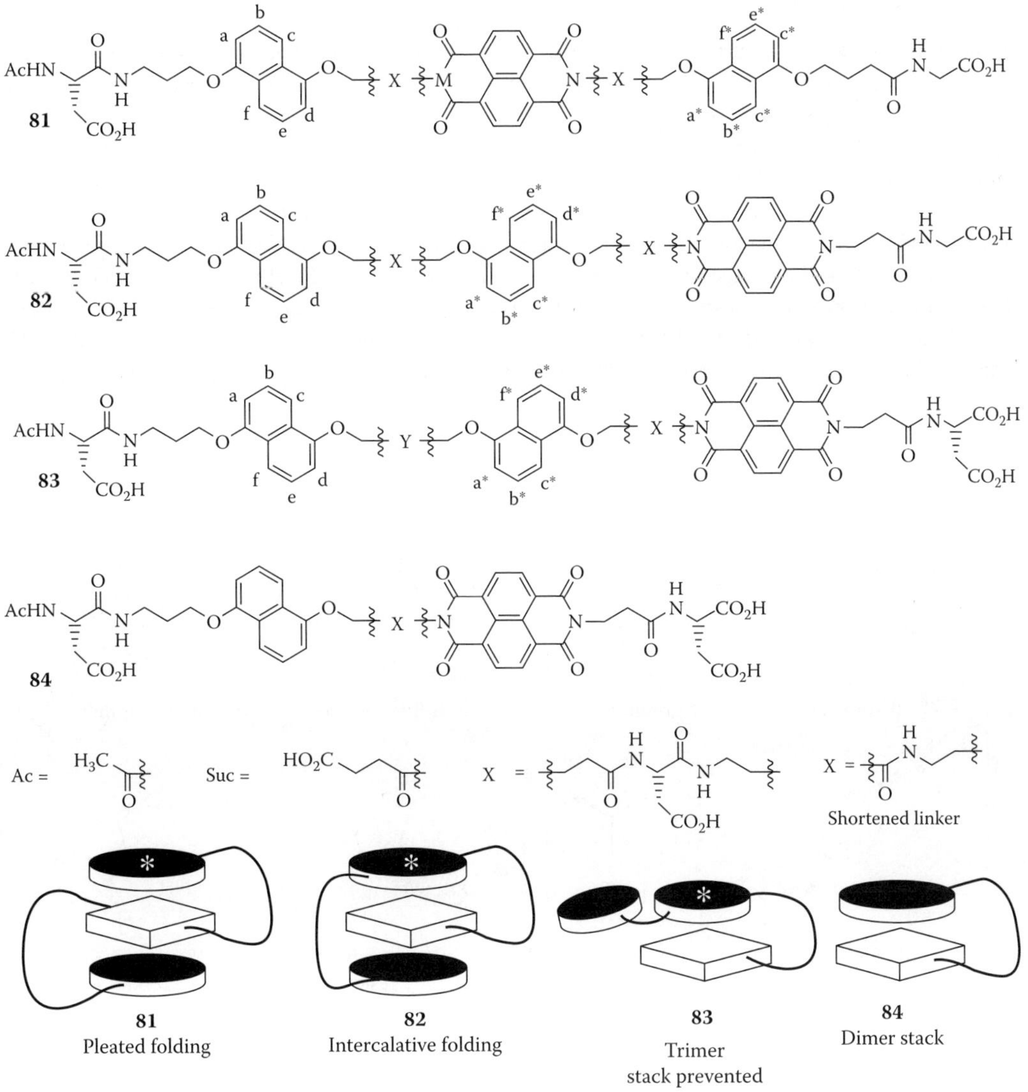

FIGURE 19.23 The structures of Aedomers **81–84** and cartoons representing the target-folding topology for each of them. Black disks represent Dan or Dan* units.

but with a shorter linker between the first and second Dan monomers designed to prevent the Ndi monomer from being able to intercolate between the two Dan monomers. Compound **84** was synthesized as a control. The aedomers **81** and **82** have almost identical UV–vis spectra, suggesting that they are both folded to the same extent and their aromatic units are stacked on top of each other. On the other hand, compound **83** has a UV–vis spectrum that suggests it is not as well folded and does not have the stacking geometry of **81** and **82**. A 2-D NOESY spectrum of **82** reveals close contacts between the two Dan aromatic rings and the Ndi ring (resonances are seen between the Ndi aromatic protons and protons f*, e*, c*, and a* of the Dan* unit as well as between the Ndi aromatic protons and the c proton of the Dan unit). All of this is consistent with **82** adopting an intercalative folding topology.

19.3.4 β-Peptides

The molecular building blocks that are the closest unnatural analogs to the natural α-amino acids are the β-amino acids. Many researchers have assembled β-amino acids into β-peptides with diverse structures and functions and several excellent reviews have been written.[37–39,54] One of the most interesting features of β-peptides is that they form remarkably stable helical structures with sequences as short as six residues, in contrast to α-peptides which fold only when they contain at least 20 amino acids.[38,54] Because they contain one extra carbon in their backbone, β-amino acids are more diverse than α-amino acids, with multiple substitution patterns and stereochemical configurations (Figure 19.24). β-Peptides have been demonstrated to adopt a wide variety of stable secondary structures (Figure 19.24 and Figure 19.25). In an early report, Gellman and coworkers demonstrated that β-peptides constructed from sequences of four and five (*R,R*)-*trans*-ACHC residues formed 14-helix structures in the solid state and that the hexamer **85** formed a well-defined helical structure in solution.[55] At the same time, Seebach and coworkers demonstrated that the

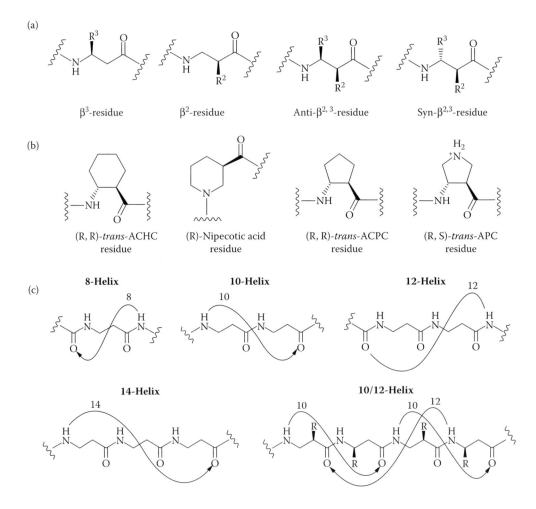

FIGURE 19.24 (a) Structures of the most common acyclic β-amino acid residues. (b) Structures of some of the most common cyclic β-amino acid residues. (c) The nomenclature used to describe the most common helices formed by β-peptides. The arrows indicate the hydrogen-bonding pattern formed in the folded helix. The numbers represent the number of atoms in each hydrogen-bonded ring within the helix.

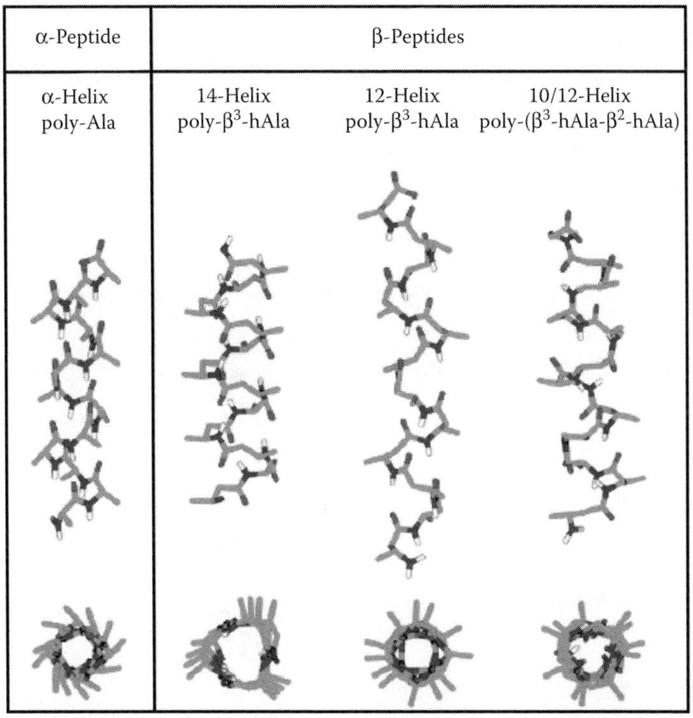

α-Peptide	β-Peptides		
α-Helix poly-Ala	14-Helix poly-β³-hAla	12-Helix poly-β³-hAla	10/12-Helix poly-(β³-hAla-β²-hAla)

FIGURE 19.25 Models of the 3-D structures of the α-helix, and the main helices formed by β-peptides.

β-peptide **86** constructed from acyclic β-amino acids formed a 14-helix (Seebach nomenclature (M) 3_1 helix) in solution.[56] Since that time, there have been many reports of β-peptides that form stable 14-helices in both organic solvents and in water.[57–60] In general, 14-helices are favored in β-peptides constructed from β-amino acids that stabilize a gauche conformation between the C^2 to C^3 bond.[38] These include β^3-mono-substituted residues, β^2-monosubstituted residues, and conformationally constrained *trans*-ACHC residues. Gellman and coworkers have demonstrated that β-peptides constructed from cyclopentane containing *trans*-ACPC residues favor a 12-helix structure.[61] The water-soluble β-peptide **87**, which forms a 12-helix, was also demonstrated through the incorporation of the (*R,S*)-*trans*-3-aminopyrrolidine-4-carboxylic acid building block into the oligomers (Figure 19.26).[62]

The organization of peptide helices into bundles is a common feature in natural proteins. This organization has been mimicked by complex β-peptides developed by Gellman and coworkers.[63] In this work, Gellman and coworkers assembled several β-peptides that displayed nucleobase recognition units on every third β-amino acid (Figure 19.27). Two of the nucleobase-bearing β-peptides, **89** and **90**, are shown (Figure 19.27). These β-peptides were designed to fold into 14-helix forms and associate with each other through Watson–Crick base pairing as shown in Figure 19.27(b). By monitoring the UV spectrum as a function of temperature, Gellman and coworkers were able to demonstrate that a 1:1 mixture of **89** and **90** formed complexes that were significantly more stable than solutions of **89** by itself and **90** by itself. The 1:1 mixture of **89** and **90** had a melting temperature of 44°C while the melting temperature of a solution of **90** had a melting temperature of only 28°C while the melting temperature of **89** was less than 0°C.

19.3.5 BLOCK Approach to Large Structures

A very interesting synthetic approach to macromolecules with nanoscale architecture is the BLOCK approach of Warrener et al.[64] They have synthesized a small collection of A-type and B-type building

85

86

87

88

FIGURE 19.26 The chemical structures of four β-peptides.

(a)

89

H-(β-HLys-β-HalA-ACHC-β-HLys-β-HalT-ACHC-β-HLys-β-HalC-ACHC-β-HLys-β-HalA-ACHC-β-HGly)-NH$_2$

90

H-(β-HLys-β-HalT-ACHC-β-HLys-β-HalG-ACHC-β-HLys-β-HalA-ACHC-β-HLys-β-HalT-ACHC-β-HGly)-NH$_2$

(b)   (c)

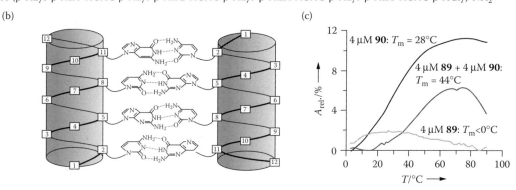

FIGURE 19.27 (a) The chemical structure of β-peptides **89** and **90**. (b) A model of the expected β-peptide helix association through antiparallel Watson–Crick base pairing. (c) The observed UV melting curves for 4 μM **89**, 4 μM **90**, and **89** and **90** combined at a 1:1 ratio with each at 4 μM concentration. The melting curve of the 1:1 mixture of **89:90** suggests that the two β-peptides are forming a complex.

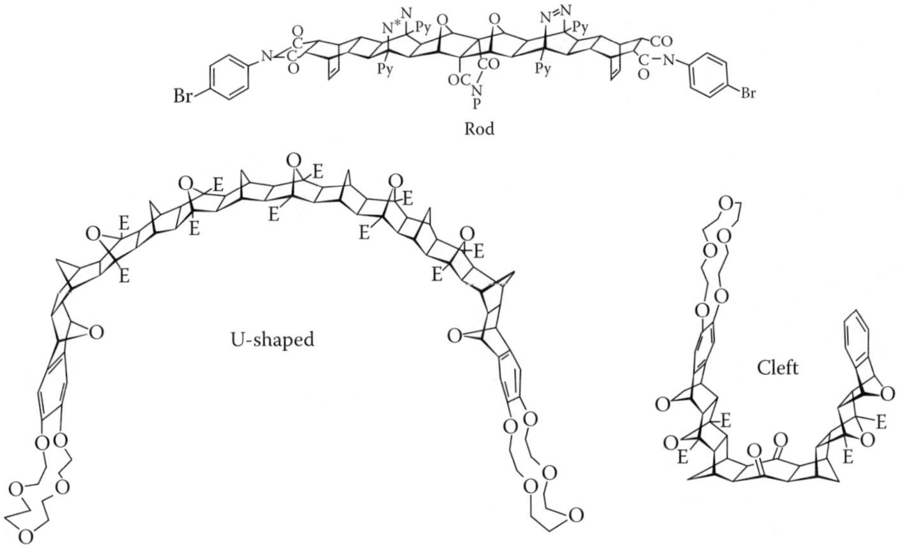

FIGURE 19.28 Examples of molecular architectures synthesized by the BLOCK approach.

blocks in which members of the A group can react with members of the B group. The resulting structure of the products is a function of the types of building blocks used. Rods, spacers, U-shaped cavity structures, and platform molecules, in which functional groups are separated and oriented precisely on rigid frameworks, have been achieved (Figure 19.28).

In one particular example from Warrener et al.,[65] A-type BLOCKS consist of a functionalized nor-bornene compound (**91** or **92**), and B-type BLOCKS are fused cyclobutene epoxides (**93–96**) that open to 1,3-dipolar species upon heating, which can be trapped stereospecifically by the norbornene BLOCKS to form a number of crown ether-, ligand- , and redox center-containing molecules (Figure 19.29). While an enormous number of combinations of A + B are possible, only a representative portion are shown here. For example, heating the dimethoxynaphthaline-containing **91** with **95** yields only the *exo, exo*-coupled product **97**. Similarly, **98** can be synthesized from a combination of either **92** and **93** or **91** and **94**. Additionally, large polynorbornanes containing two of the same functionality, such as **99**, are synthesized by incorporating dual BLOCK **96** with two equivalents of **91**. Finally, a versatile route to differentially functionalized structures is illustrated by the reaction of spacer **91** with A-type BLOCK **100** to yield **102**. This intermediate can be transformed into latent epoxide **103** and treated with **91** to give coupled product **104**. This general method is useful for the production of large, functional polyalicyclic structures that may lead to new substrates for supramolecular applications.

19.3.6 b*is*-Amino Acids

In our laboratory, we have developed a new approach to macromolecules with designed shapes. Our approach is to develop building blocks that connect through pairs of bonds and to assemble them into ladder-oligomers with shapes that are defined by the structure of the monomers and the sequence in which they are assembled (Figure 19.30[a]). This is similar to the BLOCK approach to macromolecules; however, in our approach we use chiral building blocks and sequential amide bond-forming reactions to assemble our oligomers. Our goal is to develop a small set of monomers that hold their partners in different relative orientations. We seek to synthesize a small collection of rod and turn monomers and then assemble them to construct chains of monomers that follow any designed path through 3-D space desired. Rather than relying exclusively on weak intramolecular interactions to fold, our molecules create nanoscale architecture using strong covalent bonds, the rigidity of rings and stereochemistry.

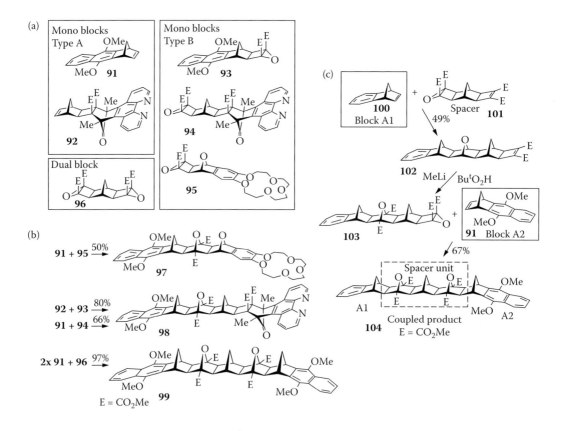

FIGURE 19.29 (a) Representative examples of A-type and B-type building blocks. (b) Large functionalized polynorbornanes made from various combinations of building blocks. (c) Synthesis of a larger spacer compound **104** via transformation of intermediate **102** into a B-type monomer followed by coupling to an A-type monomer **91**.

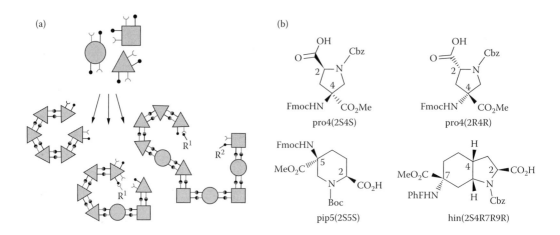

FIGURE 19.30 (a) A cartoon illustrating the *bis*-amino acid building-block approach to macromolecules. Different sequences of conformationally restrained building blocks coupled through pairs of bonds will create different conformationally restrained 3-D structures. (b) Four *bis*-amino acid building blocks.

Our monomers are *bis*-amino acids in which two suitably protected amino acid groups are displayed on a cyclic structure (Figure 19.30[a]). We assemble these monomers on a solid support using solid-phase peptide synthesis and then close a second set of amide bonds between each adjacent pair of monomers to create conformationally restrained spiroladder oligomers (Figure 19.31). Using this approach, we have synthesized a variety of nanoscale molecules with different shapes. A very favorable property of these molecules is that most structures we have synthesized are water soluble, and this will allow us to apply them to biological problems.[66] We have assembled two pentamer sequences using the first two monomers pro4(2S4S) and pro4(2R4R), one of which forms a molecular rod **105** and the other forms a curved structure **106**. These structures were characterized using FRET to measure the distance across the napthyl donor and dansyl acceptor displayed on the ends of the molecules (Figure 19.31[a]).[67] We developed the synthesis of the third monomer called hin(2S4R7R9R) from natural L-tyrosine.[68] We assembled this monomer in a heterosequence of pro4(2S4S)-hin(2S4R7R9R)-pro4(2S4S)-(S)-tyrosine and determined its structure in water using 2-D NMR. We demonstrated that the hin monomer causes the oligomer to form a very tight turn. In very recent work, we have demonstrated the synthesis of the pip5(2S5S) monomer. With this monomer we synthesized a trimer of pip5(2S5S)–pip5(2S5S)–pip5(2S5S) and demonstrated that it adopts a left-handed helical structure.[69]

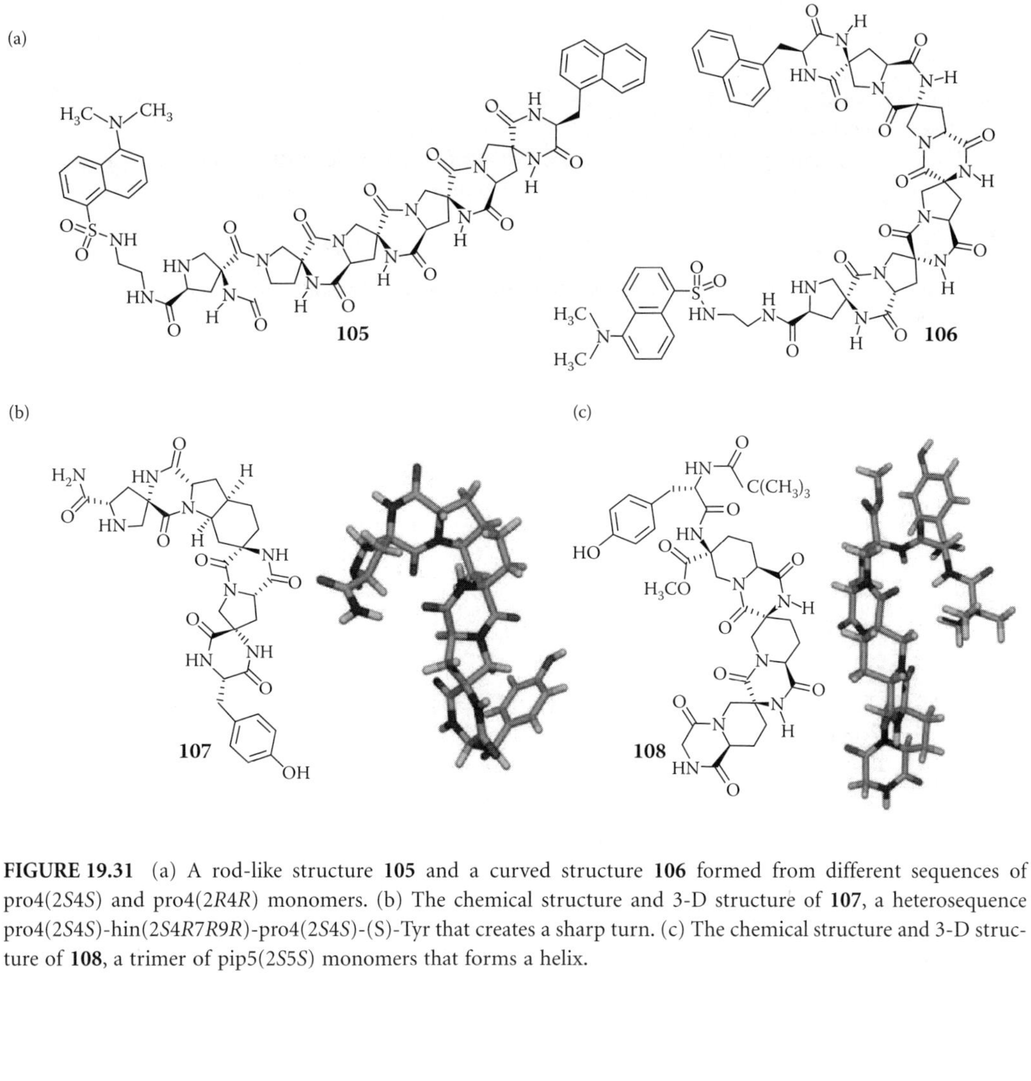

FIGURE 19.31 (a) A rod-like structure **105** and a curved structure **106** formed from different sequences of pro4(2S4S) and pro4(2R4R) monomers. (b) The chemical structure and 3-D structure of **107**, a heterosequence pro4(2S4S)-hin(2S4R7R9R)-pro4(2S4S)-(S)-Tyr that creates a sharp turn. (c) The chemical structure and 3-D structure of **108**, a trimer of pip5(2S5S) monomers that forms a helix.

19.4   Conclusion

In the past few years, a tremendous amount of progress has been made by chemists in using molecular building block approaches to synthesize molecules with nanoscale architectures. What makes this area so exciting is that there is still so much to learn and so much to invent as we work toward the capability of creating functional macromolecules that approach and perhaps 1 day surpass the power of biological proteins. The amazing capabilities of natural proteins show us the power of controlled nanoscale architecture, therein lies the ultimate ability to control matter and to convert energy from one form into another with the ultimate in efficiency. Just as we have built airplanes that fly higher and faster than any bird, automobiles that move faster than any running animal, and computers that perform calculations faster than any nervous system, one day we will develop functional macromolecules that outperform natural proteins and that will more directly service human needs.

References

1. Daggett, V. and Fersht, A.R., Is there a unifying mechanism for protein folding? *Trends Biochem Sci* 28 (1), 18–25, 2003.
2. Kelvin, L., Baltimore Lectures on Molecular Dynamics and the Wave Theory of Light, 1904.
3. Cubberley, M.S. and Iverson, B.L., Models of higher-order structure: foldamers and beyond, *Curr Opin Chem Biol* 5 (6), 650–653, 2001.
4. Kelly, T.R. and Sestelo, J.P., Rotary motion in single-molecule machines, *Struct Bond* 99, 19–53, 2001.
5. Shirai, Y., Osgood, A.J., Zhao, Y., Kelly, K.F., and Tour, J.M., Directional control in thermally driven single-molecule nanocars, *Nano Lett* 5 (11), 2330–2334, 2005.
6. Raehm, L., Hamilton, D.G., and Sanders, J.K.M., From kinetic to thermodynamic assembly of catenanes: error checking, supramolecular protection and oligocatenanes, *Synlett* 11, 1743–1761, 2002.
7. Chambron, J.-C., Collin, J.-P., Heitz, V., Jouvenot, D., Kern, J.-M., Mobian, P., Pomeranc, D., and Sauvage, J.P., Rotaxanes and catenanes built around octahedral transition metals, *Eur J Org Chem* 8, 1627–38, 2004.
8. Hubin, T.J., Kolchinski, A.G., Vance, A.L., and Busche, D.H., Template control of supramolecular architecture, *Adv Supramol Chem* 5, 237–357, 1999.
9. Rowan, S.J., Cantrill, S.J., Cousins, G.R.L., Sanders, J.K.M., and Stoddart, J.F., Dynamic covalent chemistry, *Angewandte Chemie International Edition* 41 (6), 898–952, 2002.
10. Guidry, E.N., Cantrill, S.J., Stoddart, J.F., and Grubbs, R.H., Magic ring catenation by olefin metathesis, *Org Lett* 7 (11), 2129–2132, 2005.
11. Frey, J., Kraus, T., Heitz, V., and Sauvage, J.P., A catenane consisting of a large ring threaded through both cyclic units of a handcuff-like compound, *Chem Commun* (42), 5310–5312, 2005.
12. Loren, J.C., Gantzel, P., Linden, A., and Siegel, J.S., Synthesis of achiral and racemic catenanes based on terpyridine and a directionalized terpyridine mimic, pyridyl-phenanthroline, *Org Biomol Chem* 3 (17), 3105–3116, 2005.
13. Sambrook, M.R., Beer, P.D., Wisner, J.A., Paul, R.L., and Cowley, A.R., Anion-templated assembly of a [2]catenane, *J Am Chem Soc* 126 (47), 15364–15365, 2004.
14. Lam, R.T.S., Belenguer, A., Roberts, S.L., Naumann, C., Jarrosson, T., Otto, S., and Sanders, J.K.M., Amplification of acetylcholine-binding catenanes from dynamic combinatorial libraries, *Science* 308 (5722), 667–669, 2005.
15. Liu, Y., Bonvallet, P.A., Vignon, S.A., Khan, S.I., and Stoddart, J.F., Donor-acceptor pretzelanes and a cyclic bis[2]catenane homologue, *Angewandte Chemie International Edition* 44 (20), 3050–3055, 2005.
16. Deng, W.Q., Flood, A.H., Stoddart, J.F., and GoddardIii, W.A., An electrochemical color-switchable RGB dye: tristable [2]catenane, *J Am Chem Soc* 127 (46), 15994–15995, 2005.

17. Hernandez, J.V., Kay, E.R., and Leigh, D.A., A reversible synthetic rotary molecular motor, *Science* 306 (5701), 1532–1537, 2004.
18. Arduini, A., Ciesa, F., Fragassi, M., Pochini, A., and Secchi, A., Selective synthesis of two constitutionally isomeric oriented calix[6]arene-based rotaxanes, *Angewandte Chemie International Edition* 44 (2), 278–281, 2004.
19. Arunkumar, E., Forbes, C.C., and Smith, B.D., Improving the properties of organic dyes by molecular encapsulation, *Eur J Org Chem* 2005 (19), 4051–4059, 2005.
20. Kim, K., Mechanically interlocked molecules incorporating cucurbituril and their supramolecular assemblies, *Chem Soc Rev* 31 (2), 96–107, 2002.
21. Kihara, N., Motoda, S., Yokozawa, T., and Takata, T., End-cap exchange of rotaxane by the tsuji-trost allylation reaction, *Org Lett* 7 (7), 1199–1202, 2005.
22. Sasabe, H., Kihara, N., Mizuno, K., Ogawa, A., and Takata, T., Efficient synthesis of [2]- and higher order rotaxanes via the transition metal-catalyzed hydrosilylation of alkyne, *Tetrahed Lett* 46 (22), 3851–3853, 2005.
23. Aricó, F., Chang, T., Cantrill, S.J., Khan, S.I., and Stoddart, J.F., Template-directed synthesis of multiply mechanically interlocked molecules under thermodynamic control, *Chemistry* 11 (16), 4655–4666, 2005.
24. Leung, K.C.F., Arico, F., Cantrill, S.J., and Stoddart, J.F., Template-directed dynamic synthesis of mechanically interlocked dendrimers, *J Am Chem Soc* 127 (16), 5808–5810, 2005.
25. Leigh, D.A., Venturini, A., Wilson, A.J., Wong, J.K.Y., and Zerbetto, F., The mechanism of formation of amide-based interlocked compounds: prediction of a new rotaxane-forming motif, *Chemistry* 10 (20), 4960–4969, 2004.
26. Perez, E.M., Dryden, D.T.F., Leigh, D.A., Teobaldi, G., and Zerbetto, F., A generic basis for some simple light-operated mechanical molecular machines, *J Am Chem Soc* 126 (39), 12210–12211, 2004.
27. Onagi, H. and Rebek, J., Fluorescence resonance energy transfer across a mechanical bond of a rotaxane, *Chem Commun* (36), 4604–4606, 2005.
28. Castellano, R.K., Craig, S.L., Nuckolls, C., and Rebek, J., Detection and mechanistic studies of multicomponent assembly by fluorescence resonance energy transfer, *J Am Chem Soc* 122 (33), 7876–7882, 2000.
29. Yoon, I., Narita, M., Shimizu, T., and Asakawa, M., Threading-followed-by-shrinking protocol for the synthesis of a [2]rotaxane incorporating a Pd(ii)-salophen moiety, *J Am Chem Soc* 126 (51), 16740–16741, 2004.
30. Arunkumar, E., Forbes, C.C., Noll, B.C., and Smith, B.D., Squaraine-derived rotaxanes: sterically protected fluorescent near-IR dyes, *J Am Chem Soc* 127 (10), 3288–3289, 2005.
31. Turner, D.R., Pastor, A., Alajarin, M., and Steed, J.W., Molecular containers: design approaches and applications, *Struct Bond* 108, 97–168, 2004.
32. Gopalsamuthiram, V. and Wulff, W.D., A new convergent strategy for the synthesis of calixarenes via a triple annulation of Fischer carbene complexes, *J Am Chem Soc* 126, 13936–13937, 2004.
33. Wang, L., Vysotsky, M.O., Bogdan, A., Bolte, M., and Bohmer, V., Multiple catenanes derived from calix[4]arenes, *Science* 304, 1312–1314, 2004.
34. Garrier, E., Le Gac, S., and Jabin, I., First enantiopure calix[6]aza-cryptand: synthesis and chiral recognition properties towards neutral molecules, *Tetrahed Asymm* 16 (23), 3767–3771, 2005.
35. Kang, S.O., Powell, D., and Bowman-James, K., Anion binding motifs: topicity and charge in amidocryptands, *J Am Chem Soc* 127, 13478–13479, 2005.
36. Huang, F., Fronczek, F.R., Ashraf-Khorassani, M., and Gibson, H.W., Slow-exchange C3-symmetric cryptand/trispyridinium inclusion complexes containing non-linear guests: a new type of threaded structure, *Tetrahed Lett* 49 (39), 6765–6769, 2005.
37. Gellman, S.H., Foldamers: a manifesto, *Acc Chem Res* 31 (4), 173–180, 1998.
38. Cheng, R.P., Gellman, S.H., and DeGrado, W.F., Beta-peptides: from structure to function, *Chem Rev* 101 (10), 3219–3232, 2001.

39. Hill, D.J., Mio, M.J., Prince, R.B., Hughes, T.S., and Moore, J.S., A field guide to foldamers, *Chem Rev* 101 (12), 3893–4011, 2001.

40. Prince, R.B., Saven, J.G., Wolynes, P.G., and Moore, J.S., Cooperative conformational transitions in phenylene ethynylene oligomers: chain-length dependence, *J Am Chem Soc* 121 (13), 3114–3121, 1999.

41. Nelson, J.C., Young, J.K., and Moore, J.S., Solid-phase synthesis of phenylacetylene oligomers utilizing a novel 3-propyl-3-(benzyl-supported) triazene linkage, *J Org Chem* 61 (23), 8160–8168, 1996.

42. Matsuda, K., Stone, M.T., and Moore, J.S., Helical pitch of *m*-phenylene ethynylene foldamers by double spin labeling, *J Am Chem Soc* 124 (40), 11836–11837, 2002.

43. Prince, R.B., Barnes, S.A., and Moore, J.S., Foldamer-based molecular recognition, *J Am Chem Soc* 122 (12), 2758–2762, 2000.

44. Tanatani, A., Mio, M.J., and Moore, J.S., Chain length-dependent affinity of helical foldamers for a rodlike guest, *J Am Chem Soc* 123 (8), 1792–1793, 2001.

45. Goto, H., Heemstra, J.M., Hill, D.J., and Moore, J.S., Single-site modifications and their effect on the folding stability of *m*-phenylene ethynylene oligomers, *Org Lett* 6 (6), 889–892, 2004.

46. Goto, K. and Moore, J.S., Sequence-specific binding of *m*-phenylene ethynylene foldamers to a piperazinium dihydrochloride salt, *Org Lett* 7 (9), 1683–1686, 2005.

47. Stone, M.T. and Moore, J.S., A water-soluble *m*-phenylene ethynylene foldamer, *Org Lett* 6 (4), 469–472, 2004.

48. Cary, J.M. and Moore, J.S., Hydrogen bond-stabilized helix formation of a *m*-phenylene ethynylene oligomer, *Org Lett* 4 (26), 4663–4666, 2002.

49. Heemstra, J.M. and Moore, J.S., Pyridine-containing m-phenylene ethynylene oligomers having tunable basicities, *Org Lett* 6 (5), 659–662, 2004.

50. Choi, S., Clements, D.J., Pophristic, V., Ivanov, I., Vemparala, S., Bennett, J.S., Klein, M.L., Winkler, J.D., and DeGrado, W.F., The design and evaluation of heparin-binding foldamers, *Angewandte Chemie International Edition* 44 (41), 6685–6689, 2005.

51. Gabriel, G.J. and Iverson, B.L., Aromatic oligomers that form hetero duplexes in aqueous solution, *J Am Chem Soc* 124 (51), 15174–15175, 2002.

52. Lokey, R.S. and Iverson, B.L., Synthetic molecules that fold into a pleated secondary structure in solution, *Nature* 375 (6529), 303–305, 1995.

53. Gabriel, G.J., Sorey, S., and Iverson, B.L., Altering the folding patterns of naphthyl trimers, *J Am Chem Soc* 127 (8), 2637–2640, 2005.

54. Seebach, D. and Matthews, J.L., Beta-peptides: a surprise at every turn, *Chem Commun* (21), 2015–2022, 1997.

55. Appella, D.H., Christianson, L.A., Karle, I.L., Powell, D.R., and Gellman, S.H., Beta-peptide foldamers: robust helix formation in a new family of beta-amino acid oligomers, *J Am Chem Soc* 118 (51), 13071–13072, 1996.

56. Seebach, D., Overhand, M., Kuhnle, F.N.M., Martinoni, B., Oberer, L., Hommel, U., and Widmer, H., Beta-peptides: synthesis by Arndt-Eistert homologation with concomitant peptide coupling. Structure determination by NMR and CD spectroscopy and by x-ray crystallography. Helical secondary structure of a beta-hexapeptide in solution and its stability towards pepsin, *Helv Chim Acta* 79 (4), 913–941, 1996.

57. Appella, D.H., Barchi, J.J., Durell, S.R., and Gellman, S.H., Formation of short, stable helices in aqueous solution by beta-amino acid hexamers, *J Am Chem Soc* 121 (10), 2309–2310, 1999.

58. Seebach, D., Ciceri, P.E., Overhand, M., Jaun, B., Rigo, D., Oberer, L., Hommel, U., Amstutz, R., and Widmer, H., Probing the helical secondary structure of short-chain beta-peptides, *Helv Chim Acta* 79 (8), 2043–2066, 1996.

59. Seebach, D., Abele, S., Gademann, K., Guichard, G., Hintermann, T., Jaun, B., Matthews, J.L., and Schreiber, J.V., Beta(2)- and beta(3)-peptides with proteinaceous side chains: synthesis and solution structures of constitutional isomers, a novel helical secondary structure and the influence of solvation and hydrophobic interactions on folding, *Helv Chim Acta* 81 (5), 932–982, 1998.

60. Seebach, D., Mathad, R.I., Kimmerlin, T., Mahajan, Y.R., Bindschadler, P., Rueping, M., Jaun, B., Hilty, C., and Etezady-Esfarjani, T., NMR-solution structures in methanol of an alpha-heptapeptide, of a beta(3)/beta(2)-nonapeptide, and of an all-beta(3)-icosapeptide carrying the 20 proteinogenic side chains, *Helv Chim Acta* 88 (7), 1969–1982, 2005.

61. Appella, D.H., Christianson, L.A., Klein, D.A., Powell, D.R., Huang, X.L., Barchi, J.J., and Gellman, S.H., Residue-based control of helix shape in beta-peptide oligomers, *Nature* 387 (6631), 381–384, 1997.

62. Wang, X.F., Espinosa, J.F., and Gellman, S.H., 12-helix formation in aqueous solution with short beta-peptides containing pyrrolidine-based residues, *J Am Chem Soc* 122 (19), 4821–4822, 2000.

63. Bruckner, A.M., Chakraborty, P., Gellman Samuel, H., and Diederichsen, U., Molecular architecture with functionalized beta-peptide helices, *Angewandte Chemie* (International ed. in English) 42 (36), 4395–4399, 2003.

64. Warrener, R.N., Butler, D.N., and Russell, R.A., Fundamental principles of block design and assembly in the production of large, rigid molecules with functional units (effectors) precisely located on a carbocyclic framework, *Synlett* 566–573, 1998.

65. Warrener, R.N., Schultz, A.C., Butler, D.N., Wang, S., Mahadevan, I.B., and Russell, R.A., A new building block technique based on cycloaddition chemistry for the regiospecific linking of alicyclic sub-units as a route to large, custom-functionalised structures, *Chem Commun*, 1023–1024, 1997.

66. Levins, C.G. and Schafmeister, C.E., The synthesis of functionalized nanoscale molecular rods of defined length, *J Am Chem Soc* 125 (16), 4702–4703, 2003.

67. Levins, C.G. and Schafmeister, C.E., The synthesis of curved and linear structures from a minimal set of monomers, *J Org Chem* 70 (22), 9002–9008, 2005.

68. Habay, S.A. and Schafmeister, C.E., Synthesis of a bis-amino acid that creates a sharp turn, *Org Lett* 6 (19), 3369–3371, 2004.

69. Gupta, S., Das, B.C., and Schafmeister, C.E., Synthesis of a pipecolic acid-based bis-amino acid and its assembly into a spiro ladder oligomer, *Org Lett* 7 (14), 2861–2864, 2005.

20

Introduction to Nanomanufacturing

20.1 Introduction **20**-1
20.2 Nanomanufacturing Challenges **20**-2
20.3 Top-down Approach **20**-3
    Nanoimprint Lithography for Nanoscale Devices
20.4 Bottom-up Approach **20**-4
20.5 Combined Top-down and Bottom-up
    Nanomanufacturing Approaches **20**-5
    Nanoscale Patterning • Possible Approaches to Directed
    Self-Assembly of Nanoelements • Directed Self-Assembly of
    Nanoelements Using Nanotemplates • Nanoscale Patterning
    Using Block Copolymers • Directed Self-Assembly of
    Conductive Polymers Using Nanoscale Templates
20.6 Registration and Alignment **20**-11
20.7 Reliability and Defect Control **20**-11
    Reliability and Characterization Tools • Removal of Defects
    Due to Micro- and Nanoscale Contamination
Acknowledgments **20**-14
References **20**-14

Ahmed Busnaina
Northeastern University

20.1 Introduction

In January 1996, scientists at IBM's Zurich Research Laboratory showed for the first time that individual molecules can be moved and precisely positioned at room temperature using a scanning tunneling microscope (STM). In 1989, scientists at IBM's Almaden Research Center in California became the first to show that it is possible to position individual atoms, when they wrote the letters "IBM" with 35 xenon atoms. However, this was possible only at a very low temperature ($-270°C$). This pioneering research proved that it is possible to directly manipulate molecules and atoms, and opened the door to the possibility of nanoscale-directed assembly. Since the advent of the STM and AFM, scientific breakthroughs in nanoscience have come at a rapid rate. The transfer of nanoscience accomplishments into technology, however, is severely hindered by a lack of understanding of the barriers to manufacturing in the nanoscale dimension. For example, while shrinking dimensions hold the promise of exponential increases in data-storage densities, realistic commercial products cannot be realized without first answering the question of how one can assemble and connect billions of nanoscale elements, or how one can prevent failures and avoid defects in such an assembly. Most nanotechnology research focuses on manipulating several to several hundred nanoelements (nanoparticles, nanotubes, molecules, etc.) to assemble into a specific structure or pattern. There is a need to conduct fast, massive, directed assemblies of nanoscale elements at high rates and over large areas. To move scientific discoveries from the laboratory to commercial products, a different set of fundamental research issues must be addressed such as scale-up of assembly process to production volumes, process robustness and reliability, and integration of nanoscale structures and devices into micro-, meso-, and macroscale products. The field of nanomanufacturing is very broad and highly interdisciplinary. This chapter gives an overview of nanomanufacturing challenges, top-down and bottom-up approaches, and combined top-down and bottom-up approaches, as well as a short discussion on nanoscale registration and alignment and reliability and defect control.

Many workshops have been organized by the Nanoscale Science, Engineering, and Technology Subcommittee of the National Science and Technology Council's Committee on Technology in the last few years to address challenges facing the National Nanotechnology Initiative (NNI). The NNI's goal is to accelerate the research, development, and deployment of nanotechnology to address national needs, enhance the economy, and improve the quality of life in the U.S. and around the world. NNI seeks to do this through the coordination of activities and programs across the federal government. The workshops also help to identify funding priorities and long-term goals toward commercializing nanotechnology. The "grand challenges" identified by the NNI are directly related to applications of nanotechnology and have the potential to make a significant economic and societal impact. The nine grand challenge areas are as follows:

1. Nanostructured materials by design
2. Manufacturing at the nanoscale
3. Chemical–biological–radiological–explosive detection and protection
4. Nanoscale instrumentation and metrology
5. Nanoelectronics, nanophotonics, and nanomagnetics
6. Health care, therapeutics, and diagnostics
7. Nanoscience research for energy needs
8. Microcraft and robotics
9. Nanoscale processes for environmental improvement

20.2 Nanomanufacturing Challenges

The NNI grand challenges and the NSF Workshop on 3-D Nanomanufacturing [1,2] held in Birmingham, AL, in January 2003, identified three critical and fundamental technical barriers to nanomanufacturing:

1. How can we control the assembly of 3-D heterogeneous systems, including the alignment, registration, and interconnection at three dimensions and with multiple functionalities?
2. How can we handle and process nanoscale structures in a high-rate/high-volume manner, without compromising the beneficial nanoscale properties?
3. How can we test the long-term reliability of nanocomponents, and detect, remove, or prevent defects and contamination?

In addition, the first and second joint workshops sponsored by the NNI and the Semiconductor Research Corp. ("Silicon Nanoelectronics and Beyond: Challenges and Research Directions" held in December 2004 and 2005) identified the need for research into new noncharge-based switches but also stressed the need to develop nanomanufacturing technologies. Among these were: fabricating nanobuilding blocks and nanostructures to assemble nanodevices with precise orientation, location, and size and shape control; new structures to enable ballistic transport; and contacts and contact engineering, interconnects, and structures to manage heat removal. In addition, the workshops also indicated that research is needed into nanoscale materials by design; self-assembly for functionality; nanoscale materials' characterization and metrology; and properties of materials at the nanoscale, as well as biomimetic concepts, predictive modeling of directed self-assembly, and assembly of components at a variety of scales by self-assembly. Some of the research gaps that need to be addressed are:

a current assessment of emerging devices in terms of functionality and performance followed by reliability and eventual manufacturability

maintaining initial focus on hybridization with CMOS along with parallel options that may not involve CMOS

new research directions needed to meet the NNI goals in addition to current research efforts going on in many industries, universities, and government research centers and laboratories

heterogeneous process integration such as combination of hierarchical-directed assembly techniques with other processing techniques

nanoscale metrology tools such as in line or *in situ* monitoring and feedback

high-throughput hierarchical-directed assemblies

nanoscale components and interconnect reliability

nanoscale defect mitigation and removal, and defect tolerant materials, structures, and processes such as self-healing

probabilistic design for manufacturing that addresses variability and noise at the atomic scale.

20.3 Top-down Approach

Top-down approaches using many relatively new techniques such as FIB, EUV lithography, e-beam lithography, AFM (dip pen or AFM field evaporation) lithography, plasmonic imaging lithography, nanoimprint lithography, and many others have been pursued for many years. The work published in this area includes the lithography work done in semiconductor manufacturing, which is covered in thousands of articles. This topic is too broad and diverse to be covered here. The development work has been incremental, and no significant breakthrough has been reported in the last year. One notable development, HP and UCLA (Professor Yong Chen) have made progress in making molds for nanoimprint lithography that have nanoscale features smaller than 10 nm, using thin film deposition techniques to produce a mold.

20.3.1 Nanoimprint Lithography for Nanoscale Devices

Nanoimprint lithography (sometimes called soft lithography) is a promising economic nanoscale-patterning technique (Figure 20.1) that made much progress in the past few years on tool designs and

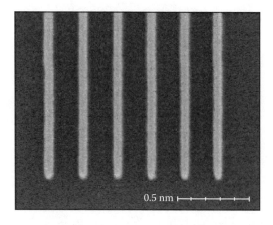

0.5 nm

FIGURE 20.1 30-nm lines on semi-isolated pitch made using a UV–NIL process (step and flash imprint lithography variant).

processing techniques [3]. Recent research and development efforts have focused on developing new materials for specific nanoimprint applications. Materials proposed for nanoimprinting include imprintable dielectrics, conducting polymers, biocompatible materials, and materials for microfluidic devices.

Enabling UV–NIL for nanoscale device manufacturing will require the development of new photocurable precursors. Photocuring adds another constraint on materials design, but offers the advantages of using a low-viscosity imprint resist especially in high-throughput and multilevel device fabrication. The development of a photocurable interlayer dielectric may have a significant impact on the semiconductor industry by simplifying the fabrication processes.

20.4   Bottom-up Approach

Patterning, templating, and surface functionalization are commonly used for directed assembly. Geometrical shaping and structuring processes at the nanoscale are used in many applications to produce functional devices, templates, or integrated multielement systems. For example, many lithography techniques could be combined with focused ion beam, two-photon lithography, or probe-based methods including AFM, STM, and near-field optical and mechanical tip scribing, as well as soft lithography techniques. These could also be extended to 3-D patterning by processes such as stereolithographic layering. These patterned substrates made using the processes mentioned above could be used as templates to enable the precise assembly of various nanoelements. However, in order to extend these tools to a true nanomanufacturing process, the assembly needs to be conducted in a continuous or high-rate/high-volume process (for example multistep or reel-to-reel process), with repeatability, scalability, and control. This way, nanobuilding blocks (such as nanotubes or block copolymers) can be guided to assemble in prescribed patterns (2-D or 3-D) over large areas in high-rate, scalable, and commercially relevant processes such as injection molding or extrusion.

Figure 20.2 shows how a large-scale-directed assembly process could work. Nanotemplates that could be electrostatically (or chemically) addressable can control the placement and positioning of carbon nanotubes (CNTs) (or other nanoelements) [4]. The nanotubes (or nanotube bundles) align on the charged wires of the nanotemplate (step b); the assembled (patterned) nanoelements can then be transferred onto another substrate, as shown in step c.

Biologically inspired assembly/molecular manufacturing, a new approach, is one of the most challenging nanomanufacturing techniques. It is ideal to think of utilizing the many directed self-assembly

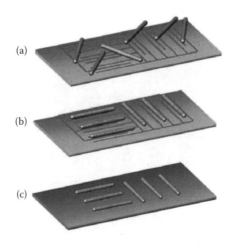

FIGURE 20.2 Nanotubes deposited (a) and assembled (b) on a nanotemplate, and then transferred to a second substrate (c). (Courtesy of the NSF Nanoscale Science and Engineering Center for High-Rate Nanomanufacturing, Northeastern University, Boston, MA. With permission.)

techniques inherent in nature to make a wide range of hierarchical structures. There are many barriers to mimicking nature, including precision synthesis or the ability to obtain the same building blocks repeatedly and reliably (sequence, composition, block, and chain lengths, etc.). To go beyond self-assembly (uniform structures) and have the ability to fabricate super molecular structures, we need to utilize the same interaction potentials (e.g., shape, electrostatics, hydrophobicity, metal coordination, and controlled arrangement of functional sites) [5]. The modification of viruses and proteins to serve as assemblers of newly designed materials [6] has shown a promising potential for using them in nanomanufacturing. Most of the directed bottom-up approaches use templating. This is especially true if the desired patterns are nonuniform. The next section discusses different approaches to using templates in assembling nanoelements such as nanoparticles, nanotubes, and polymers.

20.5 Combined Top-down and Bottom-up Nanomanufacturing Approaches

Commercial scale-up will not be realized unless high-rate/high-volume assemblies of nanoelements can be performed economically and using environmentally benign processes. High-rate/high-volume directed self-assembly will accelerate the creation of highly anticipated commercial products and enable the creation of an entirely new generation of novel applications, with scalability and integration. This includes understanding what is essential for a rapid multistep or reel-to-reel process, as well as for accelerated-life testing of nanoelements and defect tolerance. For example, a fundamental understanding of the interfacial behavior and forces required for assembling, detaching, and transferring nanoelements is needed.

20.5.1 Nanoscale Patterning

Nanoscale patterns can be created using e-beam, dip pen, or nanoimprint lithography (Figure 20.3). In many ways, dip pen nanolithography (DPN) represents a bridge between the top-down and bottom-up approaches. It is a tool for the direct or manual deposition of organic molecules onto solid substrates [7–9]. Because the organic molecules interact with the substrate as well as other organic molecules, DPN has a self-assembly component. The e-beam and DPN lithography are not suitable for high-rate manufacturing, but they are suitable for making the above nanotemplates.

For smaller patterns, self-ordering growth of nanoarrays on strained interfaces is an attractive option for preparing highly ordered nanotemplates with specific feature sizes and densities [10–12]. Reconstructed surfaces, e.g., Au(111) or Pt(111), and monolayer thick strained films, e.g., Ag or Cu on Ru(0001) and $Si_{0.25}Ge_{0.75}$ on Si(001), exhibit well-ordered networks of misfit dislocations that can be engineered to create nanotemplates with specific feature sizes, density, and structure [13–20].

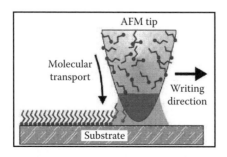

FIGURE 20.3 Diagram depicting the basic concepts of DPN. (From Piner, R.D., Zhu, J., Xu, F., Hong, S., and Mirkin, C.A., *Science* 1999, 283, 661–663; Hong, S., Zhu, J., and Mirkin, C.A., *Science* 1999, 286, 523–525; Hong, S. and Mirkin, C.A., *Science* 2000, 288, 1808. With permission.)

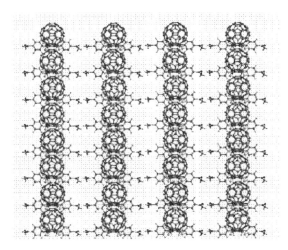

FIGURE 20.4 Fullerene nanowires with 1 to 10 nm spacing. (Courtesy of the NSF Nanoscale Science and Engineering Center for High-Rate Nanomanufacturing, Northeastern University, Boston, MA. With permission.)

Using self-assembly, these patterns can be even smaller and more compact, such as the case when nanowires are formed using functionalized fullerene [21] as shown in Figure 20.4. The fields of supramolecular chemistry and self-assembled monolayers are well established; however, using supramolecular chemistry to pattern surfaces (bottom-up self-assembly) is new. The challenges are to use functionalized fullerenes to pattern surfaces and synthesize suitably functionalized fullerenes for self-assembly on substrate surfaces. Fullerene molecules are ideal building blocks for the bottom-up self-assembly of nanotemplates because they are soluble in a host of solvents, can be functionalized using selective chemistries, have a relatively high cohesive energy [22], and bind well to a variety of substrates [23]. Functionalized fullerenes with multiple supramolecular synthons can spontaneously self-organize into fullerene nanowires [60] with spacings of 1 to 10 nm, controlled by functional groups, as shown in the figure. These nanowires could be used for high-resolution patterns on nanotemplates that can be utilized for directed self-assembly.

20.5.2 Possible Approaches to Directed Self-Assembly of Nanoelements

Assembly techniques such as microchannels [24,25] and electric fields [26] have been explored for local assembly of CNTs in interconnects and electromechanical probes [27,28]. These techniques, however, do not provide precise large-scale assembly at high rates and high volumes. The electrostatically addressable nanotemplate offers a simple means of controlling the placement and positioning of nanoelements for transfer using conductive nanowires. Gold nanowires have been used initially, and other conductors will be developed for use in templates.

20.5.2.1 Directed Assembly

To demonstrate how a large-scale assembly process will work, steps for conducting an electrostatically addressable nanotemplate for the placement and positioning of CNTs, nanoparticles, or other nanoelements are shown in Figure 20.5. The nanotubes align on the charged wires of the nanotemplate. The nanotemplate and nanoelements (step 2) can form a device or can function as a template to transfer patterned arrays of nanoelements onto another substrate as shown in steps 3 and 4. When the nanotemplates are moved with nanoprecision accuracy and alignment, they can be used to assemble a wide variety of nanoelements into very closely packed columns or rows with a very narrow pitch. Figure 20.6 shows red fluorescent negatively charged polystyrene latex (PSL) particles assembled on positively charged wires only.

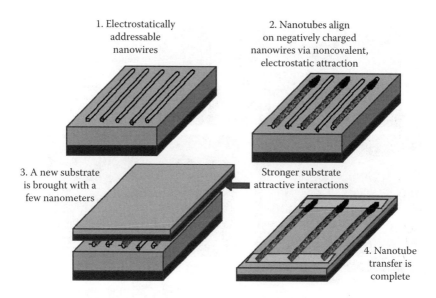

1. Electrostatically addressable nanowires

2. Nanotubes align on negatively charged nanowires via noncovalent, electrostatic attraction

3. A new substrate is brought with a few nanometers

Stronger substrate attractive interactions

4. Nanotube transfer is complete

FIGURE 20.5 Steps of 2-D molecular assembly. (Courtesy of the NSF Nanoscale Science and Engineering Center for High-Rate Nanomanufacturing, Northeastern University, Boston, MA. With permission.)

20.5.3 Directed Self-Assembly of Nanoelements Using Nanotemplates

Nanotemplates can be used to enable the precise assembly and orientation of various nanoelements such as nanoparticles and nanotubes. The directed assembly of colloidal nanoparticles into nonuniform 2-D nanoscale features has been demonstrated via template-assisted electrophoretic deposition. The assembly process is controlled by adjusting the applied voltage, assembly time, particle charge and concentration, and the geometric design of the templates. The assembly of PSL particles in trenches is shown in Figure 20.7. The figure shows that assembly processes can be controlled to produce monolayers or multilayers as well as a full or partial assembly of nanoparticles. PSL and silica nanoparticles as small as 10 nm were used and assembled into nanoscale features [29,30]. This approach offers a simple, fast, and scalable means of nanoscale-directed self-assembly of nanoparticles and other nanoelements.

Electrostatically addressable nanotemplates could also be used to directly assemble CNTs. Aligned nanotubes in nanoscale trenches (80 to 300 nm wide and 100,000 nm long) are shown in Figure 20.8. The figure shows directed assembly, using electrophoresis, through a conductor at the bottom of the trench.

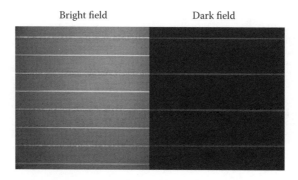

Bright field   Dark field

FIGURE 20.6 Assembly of 300 nm PSL particles (negatively charged) on positively charged Au microwires. (Courtesy of the NSF Nanoscale Science and Engineering Center for High-Rate Nanomanufacturing, Northeastern University, Boston, MA. With permission.)

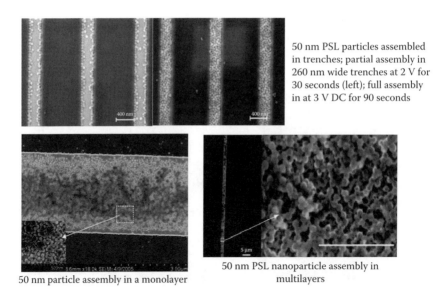

50 nm PSL particles assembled in trenches; partial assembly in 260 nm wide trenches at 2 V for 30 seconds (left); full assembly in at 3 V DC for 90 seconds

50 nm particle assembly in a monolayer

50 nm PSL nanoparticle assembly in multilayers

FIGURE 20.7 Assembly of nanoparticles in nanoscale trenches. (Courtesy of the NSF Nanoscale Science and Engineering Center for High-Rate Nanomanufacturing, Northeastern University, Boston, MA. With permission.)

The voltage used was varied from 3 to 5 V. The density of single-walled nanotubes (SWNTs) assembled inside the trenches was dependent on the trench size, the voltage applied, the concentration of the nanotube in the used suspension, and nanotube functionalization. In all cases, the nanotubes assembled inside the trenches oriented along the direction of the PMMA trench [31]. When the PMMA was dissolved, the nanotubes remained at the location of assembly. At voltages lower than 5 V, no assembly was observed inside trenches of widths less than 100 nm. At a higher voltage (5 V), the SWNTs assembled inside trenches of widths less than 100 nm.

20.5.3.1 Nanotemplates for Guided Self-Assembly of Polymer Melts

Block copolymers are of considerable interest because of their ability to self-assemble into a variety of interesting and useful morphologies [32]. These morphologies can be used as flexible templates for the assembly of nanodevices [33], etc., that are appropriately modified to "mate" with the block copolymer [34,35]. They have already been used to prepare ordered structures [36] incorporating nanorods [37]

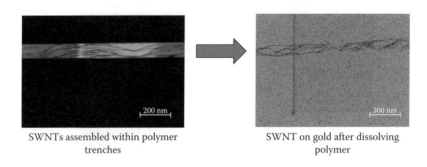

SWNTs assembled within polymer trenches

SWNT on gold after dissolving polymer

FIGURE 20.8 Assembly of aligned nanotubes in nanotrench-based templates. (Courtesy of the NSF Nanoscale Science and Engineering Center for High-Rate Nanomanufacturing, Northeastern University, Boston, MA. With permission.)

and nanoparticles [38–41], and also as nanoreactors [42]. Unguided, the type of morphology depends on polymer type, composition, and processing conditions, and results in structures that are not defect free over large areas. Approaches for morphology control of block copolymers include nanopatterned surfaces [43,44] and electric fields [45,46]. Recently, Kim et al. [47] used a chemically modified surface to prepare defect-free nanopatterns over large areas. Nanotemplate use for the control of nanoscale morphology in high-rate/high-volume manufacturing methods would open the door to commercial production of nanoscale morphology in polymeric materials and would also allow for the manufacturing of 3-D structures with controlled surface morphology via injection molding.

20.5.4 Nanoscale Patterning Using Block Copolymers

Nanoscale patterning using block copolymers involves combining "bottom-up" and "top-down" processes. The block copolymers are two polymer chains that are covalently linked at one end. Immiscible block copolymers in a thin film self-assemble into highly ordered morphologies, where the size scale of the features is only limited by the size of the polymer chains. For advanced nanoelectronics, self-assembly is insufficient, and there is a need for directed self-assembly processes to produce complex patterns. This may require the synthesis of polymers that have well-defined characteristics to enable fine control over the morphology and interfacial properties. Among nano-scale features, sharp angles present severe curvature constraints on the copolymer such that the microdomains of the copolymer cannot follow these features. One way to overcome this is to use templates to pattern block copolymers, and a homopolymer toward this purpose was introduced by Nealey and coworkers [48].

They used small amounts of homopolymer added to the copolymer, as shown in Figure 20.9. The homopolymer segregated to the areas of high curvature, alleviating the strain on the copolymer, and, as a consequence, the template features could be reproduced with high fidelity. This suggests that the copolymer microdomains can correct small defects in the patterning from the lithographic step and possibly improve the aspect ratio of the features for subsequent etching processes.

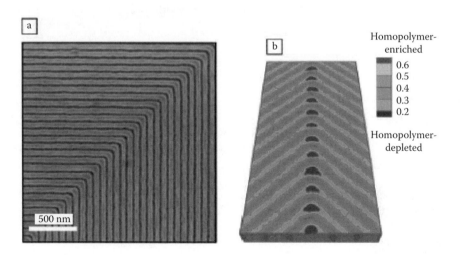

FIGURE 20.9 (a) FESEM image of a spin-coated thin film of a blend of PS-b-PMMA, having a lamellar microdomain morphology, with PS and PMMA. The mixture is seen to replicate the underlying pattern with high fidelity. (b) Redistribution of homopolymers facilitates assembly: concentration map of the homopolymers on the surface shows that the homopolymers are concentrated at the sharp edges to alleviate the curvature constraints caused by the patterning. (From Stokovich, M.P., Muller, M., Kim, S.O., Solak, H.H., Edwards, E.W., de Pablo, J.J., and Nealey, P.F., *Science*, 2005, 308, 1442–1446. With permission.)

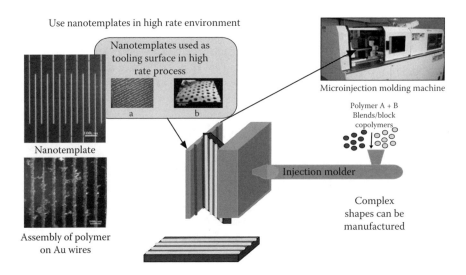

FIGURE 20.10   Guided self-assembly of polymer melts at high rates. (Courtesy of the NSF Nanoscale Science and Engineering Center for High-Rate Nanomanufacturing, Northeastern University, Boston, MA. With permission.)

20.5.5   Directed Self-Assembly of Conductive Polymers Using Nanoscale Templates

The approach presented here utilizes the "rigid" nanotemplate as the assembly or the mold surface, as in an injection molding process (or as a die in an extrusion process). This approach would allow the preparation of unique patterns and the ability to pattern much smaller feature sizes. Nanotemplates, suitably patterned, are used to control the block copolymer or blend morphology under high-rate processes from the melt [49], as shown in Figure 20.10. The assembly of conductive polymers (polyaniline [PANI]) using a template with microfeatures (1 to 2 μm line width) is shown in Figure 20.11. The figure

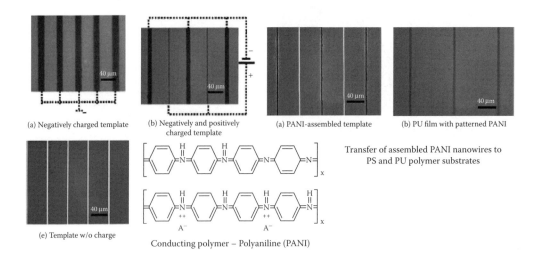

FIGURE 20.11   Assembly of polymers using electrostatically addressable templates. (Courtesy of the NSF Nanoscale Science and Engineering Center for High-Rate Nanomanufacturing, Northeastern University, Boston, MA. With permission.)

shows a successful assembly of PANI on the template. It also shows successful transfer of the assembled PANI to polystyrene and polyurethane substrates [50].

20.6   Registration and Alignment

The high-rate transfer of nanoelements from a nanotemplate to another substrate is needed for directed assembly. However, several technologies are needed to enable uniform contact or a very small uniform gap for the purpose of nanoelement transfer and nanoscale registration over the length scale of the substrate for multiple layers. Although no method currently exists to meet these requirements, techniques used in bump bonding of chips to circuit boards and in wafer bonding offer suggestions for a feasible approach. One approach could be that the two complementary surfaces are brought into contact at one edge using a prealignment procedure. Next, chemical forces (through chemical functionalization of patterned guides) take over and very slightly distort the substrate and template as the contact area propagates forward to bring matching features together and completes the physical registration. One of the manufacturing challenges will be to understand the interaction between the forces involved in the alignment process and the interface between the substrate and the template.

20.7   Reliability and Defect Control

As the assembled devices are manufactured, there is a need to address reliability and failure. Since the functionality of manufactured devices becomes dependent on nanoscale structures, reliability becomes a critical issue. Establishing a robust process and system can be broken down into three distinct, but interrelated, functions:

1. Prevention through a better understanding of failure mechanisms
2. Removal of defects
3. Development of fault tolerance and self-repair

20.7.1   Reliability and Characterization Tools

A critical barrier to the design of nanostructures and devices is the lack of available data on the reliability and properties of nanoscale materials to feed into the modeling efforts to predict nanoscale material behavior and reliability. One approach is to use MEMS-based material testing devices to conduct a range of tests of nanoscale structures such as nanowires, nanotubes, and nanofibers. There are relatively few characterization methods for the mechanical properties of the individual nanofibers available [51,52]. The MEMs devices could consist of three classes. One class contains structures for electrical characterization. The second class of devices has moving or suspended parts [53] and will permit rapid cycling (10^3 to 10^5 Hz) of temperature, strain, and current flow in, e.g., deposited nanowires in order to accelerate the generation of defects. The third class of devices consists of suspended nanostructrues. For example, microscopic defect generation can be tracked (Figure 20.12) during the measurement of the tensile stress–strain curve and yielding of nanowires. Similarly, the reliability of the connection between wires and interconnects can also be investigated using the same approach since these devices allow *in situ* electrical measurements during the application of cyclic and other stresses.

20.7.2   Removal of Defects Due to Micro- and Nanoscale Contamination

It is expected that controlling contamination and the detection and removal of defects will be critical for nanomanufacturing. Surfaces prepared for nanoscale applications such as deposition of monolayers or self-assembly of nanoelements need to be free of particulates and other contamination on the order of a nanometer or less. This does not necessarily mean a cumbersome and expensive clean room environment. Substrates could be isolated from the manufacturing environment and contaminants can be kept at bay. Contamination and defects, typically process generated, are responsible for as much as 75% or

Moving structure

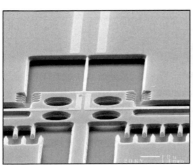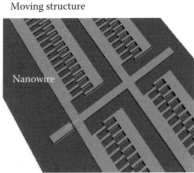

Nanowire

FIGURE 20.12 MEMs comb drive designed to perform pull tests to investigate nanoscale material properties. FESEM image (*left*) of the fabricated MEMs structure. (Courtesy of the NSF Nanoscale Science and Engineering Center for High-Rate Nanomanufacturing, Northeastern University, Boston, MA. With permission.)

more of the yield loss in integrated circuit fabrication [54] and other microfabrication processes. Currently in the semiconductor industry, the state of the art only offers nonselective removal of contaminants, although it is relatively fast and is applied over a large area. The removal-of-defects processes amount to about 20 to 25% of the total number of manufacturing processes. The upcoming challenges in nanomanufacturing will be even greater. There will be a need for selective removal of defects and impurities (e.g., oxygen in CNTs or particles attached to or residing next to assembled nanotubes). Chemistry will play a much larger role than it does now. There will be a greater need to understand the adhesion of surfaces, particles, and nanoelements in a variety of conditions and situations. In addition, the removal of defects will have to be accomplished without disturbing or destroying assembled nanoelements and nanostructures, which is not a large concern now in the semiconductor industry, with the exception of the nanoscale polysilicon structure, which is more susceptible to damage.

It has been shown that the dependence of the adhesion force on time and environmental conditions, especially in nanoscale polymer particles, is significant [55]. Large increases in the adhesion force makes defect removal extremely difficult. Nanoscale contamination removal from nanostructures will thus be studied both experimentally and theoretically using numerical modeling. One promising approach is the use of high-frequency pulsating flow (acoustic streaming) to enhance transport from small cavities (e.g., deep trenches and vias) by enhanced convective mixing [56–58]. This method has already been used to completely remove nanoparticles down to 50 nm from large substrates [59] as well as particles (300 to 800 nm) from trenches as deep as 500 μm [60].

20.7.2.1 Selective Removal in Assembly Application

A gold surface has a higher Hamaker constant than a silicon surface and induces higher adhesion force on PSL particles. Particles deposited anywhere else can be removed by cleaning control (high-frequency flow rinse). Figure 20.13 shows the forces and moments acting on particles as a result of drag forces applied through a moving fluid on the surface and those assembled on the gold wires. It also shows the equation for the moment ratio (ratio of the removal moment to the adhesion moment), which is used as an indicator for the removal of particles. A moment ratio of 1 or more indicates that the particles will be removed from the surface.

Experimental results for selective removal are shown in Figure 20.14. Figure 20.14(a) shows that fluorescent nanoparticles assembly without cleaning control leaves many undesirable particles deposited between the wires, which could interfere with the transfer process. Figure 20.14(b) shows that when cleaning control (selective removal) is applied, most of the particles are removed. This can be accomplished by understanding the adhesion and removal force of these nanoparticles on different surfaces and controlling the applied fluid velocity (used for cleaning) on the particle such that only the undesirable particles are removed.

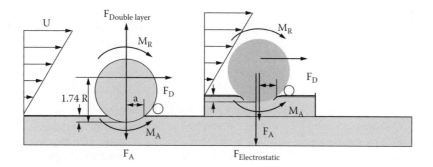

MR is the moment ratio, F_D is the drag force,

$$MR = \frac{Removal\ moment}{Adhesion\ resisting\ moment} = \frac{M_R}{M_a}$$

Moment ratio for the particle on the surface

Moment ratio for the particle on the gold surface

$$MR = \frac{F_D \cdot (1.74R - \delta) + F_{Double\ layer} \cdot \alpha}{F_\alpha \cdot \alpha}$$

$$MR = \frac{F_D(1.74R - \delta)}{(F_\alpha + F_{Electrostatic}) \cdot \alpha}$$

A stronger removal force is required to remove the particle on the gold wire compared to the particle on the surface. There is a certain flow velocity which will selectively remove the particle from the surface and not the one located on the gold wire.

FIGURE 20.13 Particle removal forces.

20.7.2.2 Selective Removal of Particles from CNT Films

CNT synthesis produces many carbon particles and other contaminants along with CNTs. Chemical purification does not ensure complete removal of these particles. A standard semiconductor cleaning procedure (megasonic cleaning using SC1 chemistry) was employed for the removal of carbon particles from CNT films on substrates (CNTs were spun on 6″ wafers). However, the process completely removed the CNT film. Upon further analysis of the removal forces on the particle and the SWNTs and by

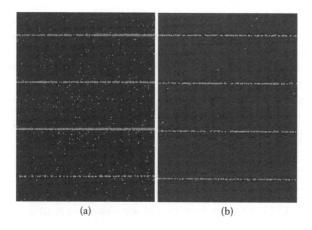

(a)   (b)

FIGURE 20.14 Assembly of fluorescent nanoparticles assembly (a) without cleaning control and (b) with cleaning control.

controlling the pH, ionic strength, and the acoustic streaming velocity, the wafers were cleaned using megasonic cleaning in a solution with a pH slightly less than 11. Experimental results showed that the carbon particles were efficiently removed without damaging or removing the CNT film.

Acknowledgments

Some of the directed assembly and reliability work reported here is supported by the National Science Foundation Nanoscale Science and Engineering Center (NSEC) for High-Rate Nanomanufacturing (NSF grant - 0425826). The author would also like to thank the executive committee members of the NSF Nanoscale Science and Engineering Center for High-Rate Nanomanufacturing especially professors Joey Mead, Carol Barry, Nick McGruer, and Glen Miller for their valuable contributions to this chapter.

References

1. Busnaina, A., Barry, C., Mead, J., and Miller, G., *Proceedings, MANCEF-COMS 2003, NSF Workshop on 3-D Nanomanufacturing: Partnering with Industry; Conclusions and Report*, Amsterdam, September 8–11, 2003, pp. 263–268.
2. www.nano.neu.edu/nsf_workshop_agendaII.html.
3. Stewart, M.D. and Willson, C.G., Imprint materials for nanoscale devices, *MRS Bull* 2005, 30, 947–951.
4. Courtesy of the NSF Nanoscale Science and Engineering Center for High-Rate Nanomanufacturing, Northeastern University, Boston, MA.
5. Tirrell, M., New molecular systems (in research): directed self-assembling, *NNI Workshop on Manufacturing at the Nanoscale*, Washington, D.C., March 31, 2004.
6. Belcher, A., New molecular systems (in research): biologically inspired assembly, *NNI Workshop on Manufacturing at the Nanoscale*, Washington, D.C., March 31, 2004.
7. Piner, R.D., Zhu, J., Xu, F., Hong, S., and Mirkin, C.A., Dip-pen nanolithography, *Science* 1999, 283, 661–663.
8. Hong, S., Zhu, J., and Mirkin C.A., Multiple ink nanolithography: toward a multiple-pen nano-plotter, *Science* 1999, 286, 523–525.
9. Hong, S. and Mirkin, C.A., A nanoplotter with both parallel and serial writing capabilities, *Science* 2000, 288, 1808.
10. Brune, H. et al., Self-organized growth of nanostructure arrays, *Nature* 1998, 394, 451.
11. Yokoyama, T. et al., Selective assembly on a surface of supramolecular aggregates with controlled size and shape, *Nature* 2001, 413, 619.
12. Rosei, F. et al., Organic molecules acting as templates on metal surfaces, *Science* 2002, 296, 328.
13. Sandy, A.R. et al., Structure and phases of the Au(111) surface: x-ray-scattering measurements, *Phys Rev B* 1991, 43, 4667.
14. Bott, M. et al., Pt(111) reconstruction induced by enhanced Pt gas-phase chemical potential, *Phys Rev Lett* 1993, 70, 1489–1492.
15. Hwang, R.Q. et al., Near-surface buckling in strained metal overlayer systems, *Phys Rev Lett* 1995, 75, 4242.
16. Tersoff, J. et al., Self-organization in growth of quantum dot superlattices, *Phys Rev Lett* 1996, 76, 1675.
17. Pohl, K. et al., Identifying the forces responsible for self-organization of nanostructures at crystal surface, *Nature* 1999, 397, 238.
18. Hrbek, J. et al., A prelude to surface chemical reaction: imaging the induction period of sulfur interaction with a strained Cu layer, *J Phys Chem B* 1999, 103, 10557.
19. Alerhand, O.L. et al., Spontaneous formation of stress domains on crystal surfaces, *Phys Rev Lett* 1988, 61, 1973.
20. Pohl, K. et al., Thermal vibrations of a two-dimensional vacancy island crystal in a strained metal film, *Surf Sci* 1999, 433–435, 506.

21. Miller, G.P., University of New Hampshire, Private communication.

22. Girifalco, L.A., and Hodak, M., Van der Waals binding energies in graphitic structures, *Phys Rev B* 2002, 65, 125404.

23. Girard, C., Lambin, P., Dereux, A., and Lucas, A.A., Van der Waals attraction between two C_{60} fullerene molecules and physical adsorption of C_{60} on graphite and other substrates, *Phys Rev B* 1994, 49, 425.

24. Messer, B., Song, J.H., and Yang, P., Microchannel networks for nanowire patterning, *J Am Chem Soc* 2000, 122, 10232.

25. Huang, Y., Duan, X., Wei, Q., and Lieber, C.M., Directed assembly of one dimensional nanostructures into functional networks, *Science* 2001, 291, 630.

26. Duan, X., Huang, Y., Cui, Y., Wang, J., and Lieber, C.M., Indium phosphide nanowires as building blocks for nanoscale electronic and optoelectronic devices, *Nature* 2001, 409, 66.

27. Guillorn, M.A., McKnight, T.E., Melechko, V.I., Britt, P.F., Austin, D.W., Lowndes, D.H., and Simpson, M.L., Individually addressable vertically aligned carbon nanofiber-based electrochemical probes, *J Appl Phys* 2002, 91(6), 3824.

28. Li, J., Ye, Q., Cassell, A., H.T., Ng, R., Stevens, J., Han, J., and Meyyappan, M., Bottom-up approach for carbon nanotube interconnects, *Appl Phys Lett* 2003, 82(15), 2491.

29. Xiong, X., Makaram, P., Bakhatri, K., Busnaina, A., Small, J., Somu, S., Miller, G., and Park, J., Directed assembly of nanoelements using electrostatically addressable templates, *Proceedings of MRS Fall 2005 Meeting*, Boston, MA, November 28–December 2, 2005, p. 418.

30. Xiong, X., Makaram, P., and Busnaina, A., Large scale directed-assembly of nanoparticles using nanotrench templates, *Appl Phys Lett*, in press.

31. Makaram, P., Xiong, X., Bakhatri, K., Busnaina, A., and Miller, G., SWNT directed assembly using nanotemplates, *Proceedings of MRS Fall 2005 Meeting*, Boston, MA, November 28–December 2, 2005, p. 476.

32. Hadjichristidis, N., Pispas, S., and Floudas, G., *Block Copolymers: Synthetic Strategies, Physical Properties, and Applications*, John Wiley & Sons, Hoboken, NJ, 2003.

33. McClelland, G.M., Hart, M.W., Rettner, C.T., Best, M.E., Carter, K.R., and Terris, B.D., Nanoscale patterning of magnetic islands by imprint lithography using a flexible mold, *Appl Phys Lett* 2002, 81, 1483.

34. Kim, D.H., Lin, Z., Kim, H.-C., Jeong, U., and Russell, T.P., On the replication of block copolymer templates by poly(dimethylsiloxane) elastomers, *Adv Mater* 2003, 15, 811.

35. Kim, Y.S., Lee, H.H., and Hammond, P.T., High density nanostructure transfer in soft molding using polyurethane acrylate molds and polyelectrolyte multilayers, *Nanotechnology* 2003, 14, 1140.

36. Maldovan, M., Carter, W.C., and Thomas, E.L., Three-dimensional dielectric network structures with large photonic band gaps, *Appl Phys Lett* 2003, 83, 5172.

37. Chen, K. and Ma, Y., Ordering stripe structures of nanoscale rods in diblock copolymer scaffolds, *J Chem Phys* 2002, 116(18), 7783.

38. Tokuhisa, H. and Hammond, P.T., Nonlithographic micro- and nanopatterning of TiO2 using polymer stamped molecular templates, *Langmuir* 2004, 20, 1436.

39. Ali, H.A., Iliadis, A.A., Mulligan, R.F., Cresce, A.V.W., Kofinas, P., and Lee, U., Properties of self-assembled ZnO nanostructures, *Solid-State Electron* 2002, 46, 1639.

40. Clay, R.T. and Cohen, R.E., Synthesis of metal nanoclusters within microphase-separated diblock copolymers: ICP-AES analysis of metal ion uptake, *Supramol Sci* 1997, 4, 113.

41. Sohn, B.H. and Cohen, R.E., Electrical properties of block copolymers containing silver nanoclusters within oriented lamellar microdomains, *J Appl Polym Sci* 1997, 65, 723.

42. Liu, T., Burger, C., and Chu, B., Nanofabrication in polymer matrices, *Prog Polym Sci* 2003, 28, 5.

43. Rockford, L., Mochrie, S.G.J., and Russell, T.P., Propagation of nanopatterned substrate templated ordering of block copolymers in thick films, *Macromolecules* 2001, 34, 1487.

44. Yang, X.M., Peters, R.D., Kim, T.K., Nealy, P.F., Brandow, S.L., Chen, M.-S., Shirey, L.M., and Dressick, W.J., Proximity x-ray lithography using self-assembled alkylsilixone films: resolution and pattern transfer, *Langmuir* 2001, 17, 228.

45. Schaffer, E.T., Thurn-Albrecht, E.T., Russell, P., and Steiner, U., Electrically induced structure formation and pattern transfer, *Nature* 2000, 403(6772), 874.

46. Thurn-Albrecht, T., DeRouchy, J., Russell, T.P., Jaeger, H.M., Overcoming interfacial interactions with electric fields, *Macromolecules* 2000, 33, 3250.

47. Kim, S.O., Solak, H.H., Stoykovich, M.P., Ferrier, N.J., DePablo, J.J., and Nealy, P.F., Epitaxial self-assembly of block copolymers on lithographically defined nanopatterned substrates, *Nature* 2003, 424, 411.

48. Stokovich, M.P., Muller, M., Kim, S.O., Solak, H.H., Edwards, E.W., de Pablo, J.J., and Nealey, P.F., Directed assembly of block copolymer blends into non-regular device oriented structures, *Science*, 2005, 308, 1442–1446.

49. Mead, J., University of Massachusetts, Lowell, Private communication.

50. Wei, M., Tao, Z., Xiong, X., Kim, M., Lee, J., Somu, S., Sengupta, S., Busnaina, A., Barry, C., and Mead, J., Directed assembly of conducting polymers using electrostatically addressable templates and pattern transfer to polymeric substrate, *Macromol Rapid Communic* (in press).

51. Buer, A., Ugbolue, S.C., and Warner, S.B., Electrospinning and properties of some nanofibers, *Textile Res J* 2001, 41(4), 323.

52. Ko, F.K., Khan, S., Ali, A., Gogotsi, Y., Naguib, N., Yang, G., Li, C., Shimoda, H., Zhou, O., Bronikowski, M., Smalley, R.E., and Willis, P.A., Structure and properties of carbon nanotube reinforced nanocomposites, *Am Institut Aeronaut Astronaut* 2002, 1426.

53. Van Arsdell, W.W. and Brown, S. B., Subcritical crack growth in silicon, *IEEE J MEMS* 1999, 8(3), 319–327.

54. Hattori, T., Contamination control: problems and prospects, *Solid State Technol* 1990, 33(7), s1.

55. Krishnan, S., Busnaina, A.A., Rimai, D.S, and DeMejo, D.P., The adhesion-induced deformation and the removal of submicrometer particles, *J Adhesion Sci Technol* 1994, 8(11), 1357.

56. Busnaina, A.A., Lin, H., and Suni, I.I., Cleaning of high aspect ratio submicron trenches, *Proceedings of IEEE/SEMI Advanced Semiconductor Manufacturing Conference*, Boston, MA, April 30–May 2, 2002, pp. 304–308.

57. Lin, H., Busnaina, A.A., and Suni, I.I., Physical modeling of rinsing and cleaning of submicron trenches, *Proceedings of the IEEE 2000 International Interconnect Technology Conference*, IEEE Electron Devices Society, San Francisco, CA, June 5–7, 2000, p. 49.

58. Lin, H., Busnaina, A.A., and Suni, I.I., Modeling of rinsing and cleaning of trenches, *International Sematech Wafer Cleaning and Surface Preparation Workshop 2000*, Austin, TX, April 11–12, 2000 (CD).

59. Busnaina, A., Bakhtari, K., Guldiken, O., Makaram, P., and Park, J., Nanoparticle removal using acoustic streaming, *J Eletrochem Soc* (in press).

60. Busnaina, A., Bakhtari, K., Guldiken, O., and Park, J., Experimental and analytical study of sub-micron particle removal from deep trenches, *J Eletrochem Soc* (in press).

21

Textile Nanotechnologies

21.1 Introduction 21-1
21.2 Nanofibers 21-2
  Electrospinning • The Meltblowing Process • "Splittable"
  Bicomponent Fibers • Partially "Soluble" Bicomponent Fibers
  • Summary
References 21-13
21.3 Electrospun Nanofibers with Magnetic Domains
  for Smart Tagging of Textile Products 21-17
  Electrospinning of Polymeric Nanofibers • PEO Nanofibers
  • Composite Nanofibers via Electrospinning • Magnetic
  Nanoparticles: Synthesis and Functionalization • Magnetic
  Characterization of Electrospun Nanofibers • Conclusions
  and Technological Implications
References 21-25
21.4 Carbon Nanotubes...21-27
  Structure • Production Methods • Properties • From Nanoscale
  to Macroscale • Polymer/Nanotube Composites
  • Polypropylene/Nanotube Composites • Conclusion
References 21-34
21.5 Surface Activation 21-38
  Small Molecules (Molecular Weight < 1000 G/mol = 1 kD)
  • Polymers or Large Molecules (Molecular Weight > 1 kD)
References 21-41
21.6 Nanoadditives in Textiles...21-42
  Classification • Equiaxed Nanomaterials • Carbonaceous
  Materials • Thermal Stability and FR • Barrier Properties
References 21-61

B. Pourdeyhimi
N. Fedorova
William Dondero
Russell E. Gorga
Stephen Michielsen
Tushar Ghosh
Saurabh Chhaparwal
North Carolina State University

Carola Barrera
Carlos Rinaldi
University of Puerto Rico–Mayaguez

Melinda Satcher
Juan P. Hinestroza
Cornell University

21.1 Introduction

B. Pourdeyhimi

The *Dictionary of Fibers and Textile Technology* commonly defines "nano" as "the precise manipulation of individual atoms and molecules to create layer structures" [1]. This definition is very inclusive and fitting for this chapter in that it not only includes nanofibers, but can also include nanocoatings, nanoparticles, and nanolayer assemblies. Textiles provide the link to all these various nanotechnologies. That is, textile substrates can be composed of nanofibers and can also be functionalized as by other nanotechnologies. This forms the topic of this chapter. For clarity, the chapter is divided into sections, with appropriate references for each.

21.2 Nanofibers

N. Fedorova and B. Pourdeyhimi

Fabrics composed of nanofibers offer small pore size and large surface area, and they are expected to bring value to applications where the properties, such as sound and temperature insulation, fluid holding capacity, softness, durability, luster, barrier property enhancement, and filtration performance are needed. In particular, liquid and aerosol filtration could benefit greatly from the introduction of nanofibers, since these fibers will improve their performance significantly. The use of nanofibers in composite materials for protective gear, such as face masks, medical gowns and drapes, and protective clothing applications is also being aggressively explored [2]. Other employments of nanofibers include barrier fabrics, wipes, personal care, and pharmaceutical applications. Enormous applications of these fibers are expected in nanocatalysis, tissue scaffolds, and optical engineering [3]. Hollow and core-sheath nanofibers have great potential for optical and microelectronics uses [4]. Currently nanofibers are widely applied in ultrasuede and other synthetic leather products as well as in commercial air filtration applications [5–9]. In today's environment, it is not surprising at all to witness an explosion of renewed interest in nanofibers and nanomaterials.

In the fiber industry, there is no commonly accepted definition of nanofibers. Some authors refer to them as materials with the diameter ranging from 100 to 500 nm [3], others consider filaments with the diameter less than 1 μm as nanofibers [5,10] and some hold the opinion that nanofibers are materials with diameters below 100 nm [11]. Here, we will define nanofibers as fibers whose diameters measure 500 nm or less. Filaments with diameters ranging from 0.5 to 10 μm (about 1 denier) will be considered as microfibers. Manufacturing techniques associated with the production of polymeric micro- and nanofibers are electrospinning, meltblowing and the use of "splittable" segmented pie and/or "soluble" islands-in-the-sea (I/S) bicomponent fibers. In electrospinning, a fiber is drawn from a polymer solution or melt by electrostatic forces [12]. This process is able to produce filaments with diameters in the range from 40 to 2000 nm [13]. In meltblowing, melted polymers are extruded from dies, attenuated by heated, high-velocity air streams and spun into micro-sized fibers with diameters in the range of 0.5 to 10 μm [7,14–16]. Even though sub-micron filaments can be obtained via meltblowing, most commercially available media fall short of that and are generally about 2 μm and above. The third technique allows the production of fibers with a diameter from 100 nm to 5 μm [8,17]. This method includes spinning of the bicomponent fibers via conventional meltspinning processes, such as spunbonding or meltblowing, after which the conjugate fibers are split into smaller fibers by mechanical means (hydroentangling, carding, twisting, drawing, etc.) or by dissolving one of the components.

21.2.1 Electrospinning

The very first available records indicate that electrospinning may be the earliest method for the production of nanofibers. The origin of this process could be tracked back to the early 1930s when in 1934 Formhals patented his invention for production of artificial filaments using electric charges [18]. He electrospun cellulose acetate fibers using acetone as the solvent. In the 1960s, fundamental studies on the jet-forming process during electrospinning of fibers were initiated by Taylor and the importance of the conical shape of the jet was stated [19,20]. It was established that the conical shape of the jet is important because it defines the onset of the extensional velocity gradients in the fiber-forming process. The electrospinning of acrylic fibers, whose diameters ranged from 500 to 1100 nm, was described by Baumgarten [21]. He determined the spinnability limits of polyacrylonitrile/dimethylformamide solution and observed the dependence of fiber diameter on the viscosity of the solution. Larrondo and Mandley produced polyethylene (PE) and polypropylene (PP) fibers from the melts [22–24]. They found that meltspun fibers had relatively larger diameters than solvent-spun fibers. However, until the mid-1990s there was little interest in electrospun nanofibers. Research on these fibers was triggered by the work of Doshi and Reneker who studied the characteristics of the polyethylene oxide nanofibers by varying the solution concentration and applied electric potential [25]. Before this time, the process of forming fine fibers using electrical charges was commonly referred to as electrostatic spinning. Since then, the process attracted a rapidly growing interest triggered by the potential applications of nanofibers in fields other than filtration and the term "electrospinning" was coined and now is widely used in the literature. More recently, three major breakthroughs were made: the development of methods of uniaxial alignment of electrospun nanofibers; the method of producing continuous ceramic nanofibers; and the method of coaxial electrospinning for production of hollow and sheath-core nanofibers [4,10,26–32].

Note, however, that electrospinning does not result in nanofibers, but rather a nonwoven web composed of nanofibers, perhaps limiting its general applicability in many textile uses. The main application of electrospun fibers continues to be in filtration and medical products [24,33–35]. One of the major producers of electrospun products in Europe and the U.S. is Freudenberg Nonwovens of Weinheim (Germany), which has been practicing electrospinning for over 20 years, producing electrospun filter media from a continuous web feed for ultra-high efficiency filtration markets [33]. Donaldson Company Inc. also has been using nanofiber web (Ultra-Web), consisting of fibers with sub-half-micron diameters for air filtration in commercial, industrial, and defense applications since 1981 [5,8]. Smaller companies are now beginning to electrospin nanofibers, including eSpin Technologies in Chattanooga, TN and Foster Miller, Inc. in Waltham, MA [7]. The potential application of electrospun web is as a protective membrane layer for providing protection from extreme weather conditions, enhancing fabric breathability, increasing wind resistance, and improving the chemical resistance of clothing to toxic chemical exposure [8,36]. Other potential uses of electrospun nanofibers are nanoreinforcement, tissue engineering, implants, drug delivery, as supports for enzymes and catalysts, for fabrication of high-performance lithium batteries, nanoscale electronic and optoelectronic devices, and for use as nanofluidic channels [37,38].

Because of the continuing interest in this technology, a separate section is devoted to this topic, and the rest of this discussion is focused on the development of nanofibers as well as webs composed of nano- or near-nano fibers using other methodologies.

21.2.2   The Meltblowing Process

The concept of meltblowing was first demonstrated in 1954 by Van A. Wente, of the Naval Research Laboratories, who was interested in developing fine fibers to collect radioactive particles in the upper atmosphere to monitor worldwide testing of nuclear weapons. In this process, an extruder forced a molten polymer through a row of fine orifices directly into two converging high-velocity streams of heated air or other gas. It was claimed that fibers as small as 0.1 to 1 μm can be formed by this method [39,40]. In the late 1960s and early 1970s, Exxon Research launched the first semiworks line, licensing the technology and providing the name for a new process: "meltblowing" Thus, Exxon became the first to demonstrate, patent, publicize, and license the use of Wente's concept as a very practical one-step process to produce unique types of nonwoven webs. Early successful licensees included Kimberly-Clark, Johnson & Johnson, James River, Web Dynamics, and Ergon Nonwovens, followed by many other companies [41].

Among the meltblowing equipment developers, Accurate Products Co. was the first to successfully build a 40-in meltblowing die [42], and Reifenhauser was the first to significantly improve the meltblowing die design [43] 3M built equipment to produce high-temperature stable nonwoven webs based on multilayer blown microfibers [44]. Biax FiberFilm has designed meltblowing equipment that uses multiple rows of orifices to provide higher productivity [45,46], and Kimberly-Clark has patented a slot die for meltblowing to minimize orifice plugging [47]. Chisso Corporation developed the equipment to produce conjugate meltblown I/S [48] and side-by-side [49] web types. Kimberly-Clark patented the process to produce an in-line perturbation of the flow of the fluid (air) to make crimped or uncrimped fibers at reduced energy costs [50], and Mitsui Petrochemical Industries obtained patents on the use of capillary meltblowing dies [51,52]. Fiber Web North America Inc. patented a process to produce sub-denier and micro-denier fibers utilizing multicomponent dies to produce continuous, easily splittable hollow fibers of low orientation [53]. This became an alternative process to make microfibers. The use of modular dies to produce a mixture of fibers in a range 0.5 to 1 μm was patented in 2000 by Fabbricante et al. [16].

Considerable efforts have been made in the last 30 years on the process study. For example, the study of the influence of the airflow rate on the fiber diameter showed that significantly smaller fiber diameter is observed at higher flow rates under the same polymer throughput [15,54]. It was also shown that the fiber attenuation occurred mostly in the first 5 cm from the die [15]. The model of a steady-state meltblowing process, which showed a rapid decrease of the filament temperature and rapid increase of filament elongational viscosity within the first 5 cm from the die, was developed as well [55]. However, the most advanced development is a new bicomponent meltblown technology [14,15,48,49,56,57].

The initial application of the meltblown web was as a battery separator [58–61]. Other uses include face masks, respirators, and filter media. The replacement of glass fibers with meltblown fibers in face masks and respirators was initiated by Johnson & Johnson in the early 1970s, and it is now essentially complete [41]. One of the earliest filtration markets targeted for the meltblown webs was cigarette filters, and a sheath-core bicomponent web was developed for such use [62]. However, despite these early activities, commercial success of the meltblown webs for cigarette filters has not been achieved to date [41]. Meltblowing technology is successfully used for air and liquid filtration applications [63–67]. FiberWeb North America Inc. patented the production of fine web as filtration media for disposable medical products [68]. Kimberly-Clark was one of the first companies to meltblow elastomeric materials for potential use in medical products [41]. More recently, the meltblowing process has been used to form meltblown adhesive filaments for bonding substrates in the production of a variety of bodily fluid absorbing hygienic articles, such as disposable diapers and incontinence pads, sanitary napkins, patient underlays, wipes, and surgical dressings [69]. Meltblown web is used for artificial leather application as well [41].

Meltblowing is an extrusion technology that produces fiber webs directly from a polymer. The schematic of the meltblowing process is presented in Figure 21.1.

A thermoplastic fiber-forming polymer is extruded through a linear die containing closely arranged small orifices. The extruded filaments are attenuated by two convergent streams of high-velocity hot air to form fine fibers. After the polymer threads are attenuated by hot air, the resulting fibers are expanded into the free air of room temperature. Due to the mixture of high-speed hot air and fibers with ambient

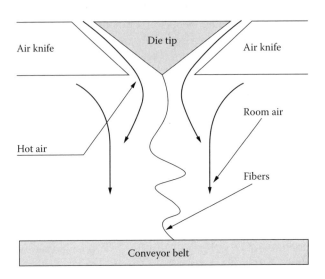

FIGURE 21.1 Schematic representation of the meltblowing process.

air, the fiber bundles start their movements forward and backward (Figure 21.1). These movements help to stretch the filaments even more due to the so-called "form drag." This form drag appears with every change in fiber direction and leads to a variability in the meltblown fiber diameter.

Generally, fiber attenuation is achieved by three different forces: aerodynamic drag near the die, aerodynamic drag near the collector, and the fiber elongation due to fiber vibration movements along spinline. However most of the attenuation is appearing near the die, as it has been reported by Bresee and Ko [70]. An additional function of hot air streams is to transport the fibers to a collector (conveyor belt), where they self-bond at the contact points [14,70].

Filaments produced by the meltblowing process have generally low or no molecular orientation. Also fibers do not often crystallize until reaching the collector, as it has been reported by Bresee and Ko [70]. The processing conditions influencing the final properties of the meltblown fibers and webs include: the melt temperature, the throughput, the die geometry, the airflow rate and its temperature, the die-to-collector distance, and collector speed [15,54]. By varying any of these input parameters final properties of fibers, such as the cross-sectional shape, the diameter, morphology, and the web structure can be changed.

There are three distinct regimes for the meltblowing process [41]. The most common is a very-high airflow rate regime that allows production of fibers in the range from 2 to 5 μm. This regime is in current commercial use. An ultra-high flow rate regime allows production of ultrafine fibers with a diameter of less than 1 μm. Although fibers as small as 0.1 μm can be produced via this regime, it is still under development [71]. A low airflow rate regime produces 1 denier (approximately 10 μm) and larger fibers [41]. The effort to produce sub-micron fibers by using splittable cross-sectional fiber morphology in the meltblowing process was made; however, the smallest achieved fiber diameter was generally in the range of 1 to 2 μm [53,72].

The fine fibers of the conventional meltblowing process result in a soft, self-bonded fabric having excellent covering power and opacity. Because of the fineness and tremendous number of fibers comprising these webs, the meltblown nonwovens can develop significant bonding strength through fiber entanglements. Also, meltblown fiberwebs are characterized by their high surface area per unit weight and fine porosity [41]. Nevertheless, meltblowing has few drawbacks. Only low viscosity materials could be spun into meltblown webs to avoid excessive polymer swelling upon the exit of the spinneret. It is estimated that over 90% of all meltblown nonwovens are made of PP, with the melt flow rate ranging from 1000 to 1500 g/10 min [56]. The inability of using different polymers limits many potential applications of the meltblown webs. Another disadvantage is low strength of the meltblown fibers, caused by little or no fiber molecular orientation and low molecular weight of the used polymers [70]. Like electrospun nanofibers, meltblown fibers typically need a supporting structure and are generally employed in a composite structure [7]. This allows for the meltblown web to optimize its filtration properties; it, however, becomes an expensive way to meet the customer's needs and adds complexity to the manufacturing process. The brittleness of meltblown nonwovens causes difficulties with their downstream processing as well. These fabrics are difficult to dye and incorporate into other nonwoven filter media structures, such as carded, air-laid, needle-punched, or wet-laid composites. Finally, meltblown webs typically exhibit broad fiber diameter distributions that can be inappropriate for some applications [54].

Even though fibers with diameters less than 1 μm can be made through the meltblowing process, the mean diameters of meltblown fibers are much larger than those of electrospun fibers. They are typically in the range of from 2 to 10 μm [7,15,73]. There have been various attempts to reduce the diameter of the meltblown fibers. One example of such attempts includes reducing the polymer throughput or the orifice diameter. However, this direct controlling approach limits the productivity of the process extremely. A few years ago, another method of production of polymeric fine fibers was introduced. In this technique, fibers are created by meltblowing with a modular die [16]. This approach allows the production of a mixture of fibers in the range of 0.5 to 1 μm. Nevertheless, from the abovementioned, it follows that the meltblowing process is applicable for the production of microfibers rather than nanofibers.

21.2.3 "Splittable" Bicomponent Fibers

An alternative way to produce ultrafine fibers is splitting bicomponent (conjugate) fibers, which consist of two polymers having different chemical or physical properties. The splitting fiber technology was first developed commercially in Japan in the mid- to late 1960s for synthetic suede fabrics [74]. The production of microfibers or even nanofibers involves spinning and processing of the bicomponent fibers in the range from 2 to 4 denier per filament (20 to 40 μm), after which the fibers are split into smaller filaments with denier of 0.1 (approximately 1 μm) or even less [17]. The particular process employed for the production of ultrafine filaments from the bicomponent fibers depends on the specific combination of components comprising the fiber and their configuration. One common method involves mechanically working the fiber by drawing, needle-punching, beating, twisting, carding, or hydroentangling [53,72,75,76]. Hydroentanglement is a mechanical bonding process in which fibers are interlocked by a series of very fine, parallel, high-pressure water streams (jets). Hydroentangling energy is a major factor influencing the degree to which the web is bonded or split. This energy is proportionally dependent on water pressure and inversely dependent on the fabric basis weight and processing speed.

When mechanical action is used to separate the bicomponent fibers, the fiber components must bond poorly with each other to facilitate subsequent separation. In other words, the polymer constituents must differ from each other significantly to ensure minimal inter-filamentary bonding. For this reason, polymers having different chemistries (incompatible polymers) are usually chosen for the production of "splittable" fibers. The term "incompatible polymers" is used to indicate that the chosen polymers will not form a miscible blend when the melt is blended. The examples of particularly desirable pairs of incompatible polymers useful for the "splittable" conjugate fibers include polyolefin–polyamide, polyolefin–polyester, and polyamide–polyester.

The production of mechanically "splittable" bicomponent fibers presents challenges that are not encountered in the making of other types of composite fibers. In particular, when mechanical action is used to separate these fibers, the fiber components must be selected carefully to provide an adequate balance between adhesive and dissociative properties, because poor bonding is known to facilitate the separation process. Conversely, the components should remain bonded during at least a portion of the downstream processing incurred in fabric formation. To add to this difficulty, many conventional textile processes impart a significant stress to a fiber, thus promoting its untimely splitting. Premature splitting is highly undesirable because conventional textile equipment is frequently not designed to process extremely fine filaments, and quickly becomes fouled by them. In addition to their adhesive properties, the melt rheologies of the polymers comprising the bicomponent "splittable" fiber strongly influence the splitting process. For example, the melt rheologies of the two components must be such that one component does not totally encapsulate the other during meltspinning, thus complicating later splitting [77].

Many different types of the "splittable" bicomponent fibers, such as segmented ribbon, tipped, solid, and hollow segmented pie were used to obtain ultrafine fibers. The examples of tipped trilobal and segmented ribbon fibers are depicted in Figure 21.2. The most readily "splittable" bicomponent fiber is a segmented ribbon (Figure 21.2[a]). However, this fiber is one of the most difficult to process. Tipped fibers are materials in which one polymer is placed on the tip of a trilobal or delta cross-section (Figure 21.2[b]). After spinning, mechanical action is applied to the tipped fibers, causing the polymer on the tips of the fiber to break apart into microfibers having a denier of 0.2 (about 1 to 1.2 μm) and spiral around core polymer [17].

The typical cross-sections of the solid and hollow segmented pie (pie wedge) fibers are depicted in Figure 21.3. The main difference between them is that the use of the hollow segmented pie fibers can simplify the fiber-splitting process [78]. Hollow pie wedge fibers can be split into smaller filaments upon exiting a spinneret, by drawing and stretching, or attenuating the filaments in a pressurized gaseous stream, including air steam, or by developing a triboelectric charge in at least one of the components through application of an external electrical field, or by a combination of some or all of these. The downside of using the hollow segmented pie fibers is that they are typically difficult to spin [78]. Splitting of the solid segmented pie fibers can be achieved by mechanical action, such as drawing, carding,

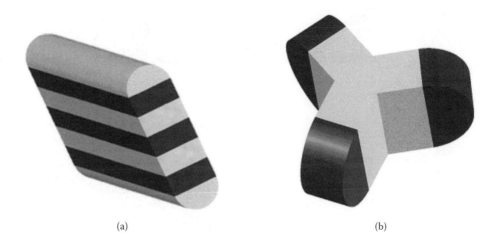

(a)   (b)

FIGURE 21.2 Schematic representation of bicomponent fibers: (a) tipped trilobal and (b) segmented ribbon fiber.

beating, hydroentangling, ultrasonics, etc., or by dissolving one of the components after nonwoven fabric processing [53,76,79–83]. The hydroentanglement process is the most widely used for splitting the wedge pie fibers [77,84]. It simultaneously splits and entangles the fibers to form a bonded nonwoven web.

The segmented pie fibers can be produced either through meltblowing or spunbonding. Spunbonding involves the extrusion of polymer melts through dies, with subsequent cooling and attenuation of fibers by high-velocity air jet stream. The filaments are then deposited on a moving belt and bonded mechanically, thermally, or otherwise. Bicomponent fiber spinning and fabric formation are similar to single component fiber spinning and fabric formation. The only difference between them is that instead of one polymer stream, two separate polymer streams are extruded. These streams pass through filters, a spin beam, and a spinpack distribution system separately, and form a conjugate stream entering the spinneret hole. After exiting the spinneret orifices, the bicomponent fibers are quenched, attenuated, and deposited on the moving belt. A typical spunbonding process consists of the following elements: polymer feed, extruder, metering pump, die assembly, filament spinning, drawing and deposition system, collecting belt, bonding zone, and winding unit. The schematic of an open spunbond process for bicomponent fiber spinning is depicted in Figure 21.4.

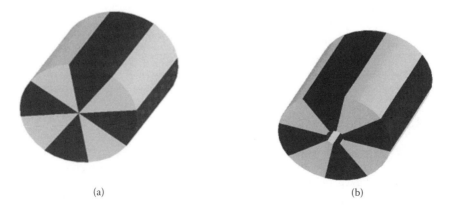

(a)   (b)

FIGURE 21.3 Typical bicomponent segmented pie fiber: (a) solid and (b) hollow.

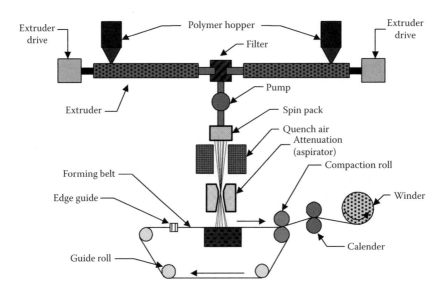

FIGURE 21.4 Schematic diagram of the open bicomponent spunbond process with belt collector.

When the segmented pie fiber webs are made via the meltblowing process, they consist of very fine, breakable fibers and contain inter-fiber bonds that restrict fiber movements. Thus, the hydroentangling process or any other mechanical action is difficult to use for splitting these fibers without their breakage [72]. One possible way to produce ultrafine fibers from the meltblown webs is to dissolve one of the polymers with a solvent, preferably water [79–81]. For this purpose, one fiber component should be soluble and the other should not be soluble in the applied solvent. Although this method could produce sub-denier or low-denier fibers, it causes issues such as wastewater treatment, cost, etc. To improve the fiber and fabric performance and reduce environmental challenges, the "splittable" fibers can be spun via the spunbonding process and split by hydroentanglement. For example, 16, 18, and 32 spunbonded segmented pie fibers were used for quite some time in the Far East to obtain microfibers [17,85]. An example of successful spunbonded microfiber fabric produced from "splittable" fibers is Evolon, introduced to the U.S. market by Freudenberg Nonwovens a few years ago [78]. It is made from hollow and solid segmented pie fibers split by hydroentanglement. The most common polymer combination used in Evolon fabric is PET/PA 6.6 in the ratio 65/35. The typical fiber configuration is a 16-segmented pie. Evolon applications are in sports and leisure wear, work wear, automotive wear, shoe lining, and many others.

Many types of polymer combinations were spun into the segmented pie fibers including PP/PMP (poly(4-methyl-1-penten) [83], PET/PP, PA/PP [81], ((poly)lactic acid) (PLA)/PET [82], polyolefin/ (poly(vinyl) alcohol) (PVA) [79], PET/PA [84], and others. Overall, filaments as small as 0.1 to 0.2 denier (approximately 0.5 to 1 µm) were produced via the "splittable" fiber approach [17,85].

The fiber size after splitting the pie wedge fibers depends on the number of segments and the initial diameter of the meltspun fibers influenced by the applied spinning process. The higher the count of segments, the smaller the fiber diameter is, as can be seen from Figure 21.5.

Additionally, the smaller the initial bicomponent fiber diameter, the smaller the fiber after splitting. Since the spunbonding process typically produces larger fibers than the meltblowing process, to obtain smaller filaments via spunbonding, the number of segments in the pie wedge fiber should be increased. For instance, filament sub-micron dimensions can be reached by splitting a spunbonded 64 segmented pie or by splitting meltblown 16 segmented pie fibers [86,87]. An increase in segment count makes the process of the fiber formation more challenging, because of the difficulty of controlling polymer elongational viscosities during spinning. As a result, a rise in the number of segments leads to the formation of structures in which one polymer completely encapsulates another (Figure 21.6). This may create problems with subsequent splitting of the resulting bicomponent fibers.

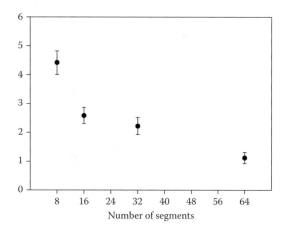

FIGURE 21.5 Spunbonded fiber diameter as a function of the number of segments.

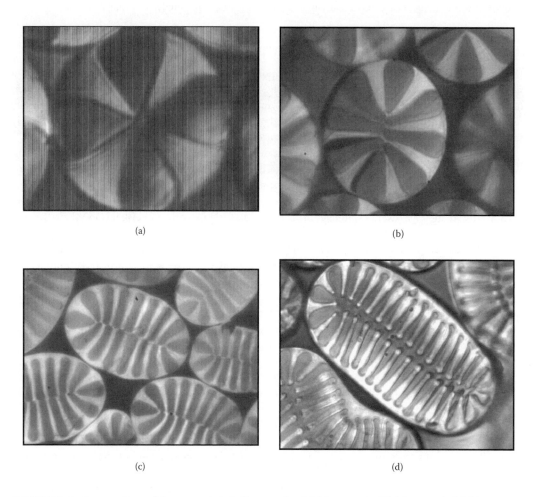

FIGURE 21.6 Cross-sections of the segmented pie fibers made of: (a) 8 segments; (b) 16 segments; (c) 32 segments; and (d) 64 segments.

FIGURE 21.7 Typical bicomponent islands-in-the-sea fiber.

The applications of microfiber nonwovens obtained by splitting segmented pie fibers include air and liquid filtration, medical care, technical wipes, synthetic leather and suede, and many others.

21.2.4 Partially "Soluble" Bicomponent Fibers

Ultrafine fibers also can be prepared using bicomponent fibers comprised of a desired polymer and a soluble polymer. The soluble polymer is then dissolved out of the composite fiber, leaving micro- or even nanofilaments of the other remaining insoluble polymer [14,79–81]. Islands-in-the-sea fibers are typically used in such approaches. These fibers are composed of numerous much smaller filaments, all completely separated from each other by the sea (matrix) of another polymer (Figure 21.7).

Several methods of producing I/S bicomponent fibers are described in various U.S. patents [88–90]. For example, Moriki and Ogasawara employed a technique wherein a number of core polymer streams were injected into a matrix or sheath stream via small tubes, one tube for each core stream [88]. The combined sheath–core streams converge inside the spinneret hole and form one I/S conjugate stream. Each of Moriki and Ogasawara's spinneret orifices produces a fiber with seven islands in the matrix of a sea polymer. Such a spinneret is suitable for the extrusion of continuous fibers with 126 filaments of perhaps 0.3 denier per filament (approximately 3 μm), when the sheath polymer is dissolved away. However, the Moriki and Ogasawara technique is not suitable for the extrusion of large numbers of multicomponent fibers from each spinneret, as is necessary for economical production of fibers via meltspinning. Another problem is the cleaning of the tubes that inject the polymer streams. Because of the extreme fineness of the tube, damaging them during cleaning of the spinpack parts is hard to avoid. Another method was described by Kiriyama et al. [89] that involved mixing different polymer streams with the static mixer in the spinning process. The static mixer divides and redivides a multicomponent stream, forming the stream with hundreds, or thousands, of core streams within the matrix stream. When the matrix is dissolved away from the resulting fiber, a bundle of extremely fine fibers is produced. The disadvantages of this method include nonuniform distribution of the islands in the matrix of the sea, which can lead to clustering and variability in the island fiber sizes and formation of discontinuous island filaments. Another well-known method is due to Hills [90], in which the spinning and fabric formation are similar to the spinning of standard homopolymer fibers. This method allows for the melt and solution processing of plural component fibers. Moreover, Hills' spinpack offers some attractive features, such as high spinneret hole density for complex bicomponent cross-sections, flexibility in selection of polymer types, and bicomponent cross-sections. The Hills patent describes how two separate polymer streams after passing through the filters, metering plate, and distribution plates, form a conjugate stream entering the spinneret hole and then pass into the spinneret orifice, forming I/S fibers. In the production of the I/S filaments, the spinneret design and the distribution plates are crucial because the fiber diameter, the cross-sectional area, and the number of islands depend on the diameter and the shape of the spinneret orifice and the polymer distribution in the distribution plates.

One of the approaches to producing I/S fibers is the spunbonding process. The structure and properties of spunbonded fibers and fabrics are determined by processing conditions such as extrusion temperature, melt throughput, quench air temperature, and drawing conditions, and by the material properties of the polymers used. The melting temperature, melting viscosity, initial structure, glass transition temperature, solidification point, and many other parameters of fiber-forming polymers play an important role in the final fiber structure and properties. When I/S fibers are produced via the spunbonding, the number of islands, polymer composition, and materials used as the islands and sea also play an important role in the final fiber properties. At the stage of the selection of polymers for islands or sea, their spinnability in the I/S bico-configuration has to be accounted for. The most profound effect on the spinnability of the two polymers in the bico-configuration is their elongational viscosities [91–95]. The difference in the viscosity of polymers can cause serious migration and deformation of an interface between polymers, resulting from unequal concentration of stresses in the spinline. It is believed that during conjugated spinning of polymers having significantly different elongational viscosities, the spinline tension is mainly concentrated on the component having the higher viscosity and affecting the component little. Higher spinline tension results in a higher molecular orientation and crystallinity of one component, but in a lower molecular orientation and crystallinity of the other component as a result of orientation relaxation [92]. The cases of fracturing of one component having very high melt viscosities are also known [96]. This could result from the fast solidification of the high-viscosity polymer upstream in the spinline, while the low-viscosity polymer is still in a molten state and can be deformed. Even though the high-viscosity polymer has completed its thinning, the low-viscosity polymer had not yet reached the spinning speed and could be considered to continue flowing until reaching the solidification temperature itself, trying to make the high-viscosity polymer accelerate. However, because the high-viscosity polymer has already been solidified, it cannot deform anymore. This could lead to an appearance of cohesive fracture of this polymer, and cases are known where, instead of sheath–core fibers, hollow fibers were obtained [96]. Thus, the search for processing conditions under which viscosities of the two components of the bicomponent fibers are identical or close is critical for conjugated spinning. The additional requirement for the sea polymer is its ease of removal, preferably through dispersion or dissolution in water. The ability of this polymer to be recycled is also an important factor. Typical resins usable as the island are polyolefins, such as PE, PP, etc., polyamides, such as nylon 6, nylon 66, etc., thermoplastic polyesters, such as PET, PBT, etc., and polyurethane (TPU) [9,97–99]. The examples of resins usable as the sea are those that are removable without having a bad effect upon the island, such as partially saponified PVA [48,97,98], copoly(ethylene-terephthalate-5-sodiumsulfoisophthalate) hydrolyzable with alkaline, polystyrene (PS) [9,99], PLA [100], etc.

For some time now, the 16, 36, 24, and 32 spunbonded I/S fibers have been used for making ultrasuede and artificial leather. Higher island counts (1000 or more) are now possible to spin. It has been stated that spinning of 16, 37, and 1000 I/S fibers allows the production of filaments with fineness of 0.15, 0.05, and 0.001 to 0.01 denier (approximately 1, 0.3, and 0.1 to 0.5 μm, respectively), after the sea removal [8]. Hills reported the spinning of fibers containing 1120 islands with 50% of PP, PET, and PA6 as the island polymers and 50% of EVOH as the sea polymer. Fibers as small as 300 nm were obtained after EVOH removal.

The method of the sea polymer removal usually involves its dissolving in solvents, such as hot water, hot water with caustic soda, alkali solution, carbon tetrachloride, trichloroethylene, etc. [9,97–100]. From the environmental point of view, hot water is preferable to any other solvent. However, nanofibers released by dissolving the sea from the I/S fibers are costly to produce because of wasting of the sea polymers and the need of wastewater treatments. This cost can be reduced in

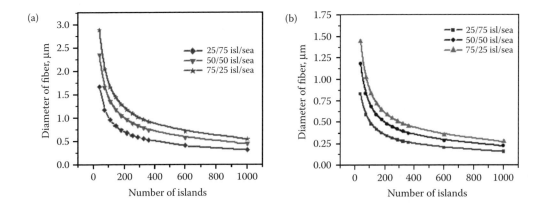

FIGURE 21.8   The effect of the number of islands and polymer composition on the spunbonded final fiber diameter for initial fiber diameter: (a) 20 μm and (b) 10 μm.

several ways: by decreasing the ratio of the sea polymer in a fiber, by selecting the sea component without the usual requirements of its usability in the product since it will be ultimately removed, or by selecting a water-soluble sea polymer that could be recycled. Splitting spunbond I/S fibers mechanically is ecologically friendlier. This method of obtaining ultrafine filaments from I/S fibers was introduced by Kawano [97]. He reported the meltspinning of side-by-side and sheath–core conjugate fibers in which one side (sheath) was the I/S configuration and another side (core) was purely made of PP. Kawano proposed the method of splitting the I/S part of the fiber, made of PP (island) and PVA (sea), by water needle-punching (or by removing PVA in hot water). He claimed that spinning several hundred islands and splitting them by water jets (or hot water treatment) resulted in fibers ranging from 0.0002 to 0.1 denier (approximately 0.01 to 0.5 μm). Okamoto [9] also suggested that for better splittability by hydroentanglement, the ratio of sea polymer should be less than 5% of the total weight of a fiber.

In general, the size of the filaments after sea polymer removal depends on several factors, such as the initial diameter of the fiber, the number of islands, and polymer composition. An increase in the number of islands, or sea polymer ratio, or decrease in the initial diameter of the fiber leads to reduction in the filament size after the sea removal, as can be seen from Figure 21.8.

To obtain even smaller filaments than in the spunbonding process, meltblowing could be potentially applied for the spinning of the I/S fibers. This technique of production of the bicomponent fibers and then dissolving the sea component from the meltblown fiberwebs was patented in 1994 [48]. The patent claimed that PP as the island and PVA as the sea could be meltspun and then PVA could be removed by treatment in hot water, releasing the islands as less than 0.01 denier (approximately 0.5 μm). Generally, it was estimated that islands with diameters as low as 50 nm can be obtained by meltblowing of 600 I/S [101], while by spinning the same number of islands via spunbonding, filaments with diameters of about 300 nm may be obtained. The use of the meltblowing process for the spinning of the I/S fibers has few drawbacks. First of all, there is a limitation on the polymer selection. Only low-molecular-weight polymers can be used here. Second, because of the poor strength of meltblown fibers, the dissolving of the sea from the ultrafine meltblown webs can result in the disintegration of the web structure, especially if agitation of the fabric is required for better polymer solubility. To avoid damaging the fabric, the meltblown web has to be laid over another substrate first and then subjected to sea removal treatments. This, in turn, may complicate sea removal and increase manufacturing cost. To date, the spinning of I/S fibers via meltblowing has not been practiced commercially.

The major application of fibers obtained via the I/S technique is in synthetic leather and suedes [9]. In the case of synthetic leathers, a subsequent step introduces coagulated polyurethane into the fabric, and may also include a top coating. Another use includes technical wipes, ultrahigh filtration media, etc.

21.2.5 Summary

Nanofibers, measuring 500 nm or less, can be produced via different methods, such as electrospinning or meltblowing, or by using "splittable" or "soluble" bicomponent fibers. Filaments made by meltblowing or electrospinning have lower strength and higher variability in the fiber diameters than fibers produced via the segmented pie or the I/S technique. Webs made of electrospun or meltblown fibers should be laid over a suitable substrate that has to provide appropriate mechanical properties and complementary functionality to the web. Meltblowing produces microfibers rather than nanofibers; moreover, it can process only a limited number of polymers. Electrospinning, on the other hand, is able to make nanowebs with substantially more and smaller micropores than meltblown or spunbonded web; however, it has very low productivity. The segmented pie and I/S fibers produce highly developed and strong ultrafine and nanofibers. However, despite a lot of similarities between these techniques, the I/S method has fewer limitations on the polymer choice and more options for the fiber cross-sections, and it allows the production of smaller fibers than the segmented pie approach. Moreover, I/S fibers are less challenging to spin than the wedge pie fibers, especially when an appropriate spinpack design is chosen. Based on the abovementioned, the I/S approach, which involves spinning of the bicomponent fibers via the spunbonding process and splitting them by hydroentanglement or by hot water treatment, becomes a promising method for the production of nanofibers.

References

1. *Dictionary of Fibers and Textile Technology*, Hoechst Celanese, 1990.
2. Graham, K., Gogins, M. and Schreuder-Gibson, H., Incorporation of electrospun nano fibers into functional structures, *INJ*, 13(2), 21–27, 2004.
3. Subbiah, T., Bhat, G.S., Tock, R.W., Parameswaran, S. and Ramkumar, S.S., Electrospinning of nano fibers, *J. Appl. Polym. Sci.*, 96, 557–569, 2005.
4. Sun, Z., Zussman, E., Yarin, A.Y., Wendorff, J.H. and Greiner, A., Compound core-shell polymer nano fibers by co-electrospinning, *Adv. Mater.*, 15(22),1929–1932, 2003.
5. Grafe, T., Gogins, M., Barris, M., Schaefer, J. and Canepa R., Nano fibers in filtration applications in transportation, Filtration 2001 International Conference and Exposition of the INDA, Chicago, IL, December 3–5, 2001.
6. Baker, B., Bicomponent fibers: a personal perspective, *IFJ*, 13(3), 26–42, 1998.
7. Grafe, T. and Graham, K., Polymeric nano fibers and nanofiber webs: a new class of nonwovens, *INJ*, 12(2), 51–55, 2003.
8. Cheng, K.-K., This artificial leather beats the hide off the real thing, *IFJ*, October, 40–41, 1998.
9. Okamoto, M., Multi-core composite filaments and process for producing same, U.S. Patent 4,127,696, November 28, 1978.
10. McCann, J.T., Li, D. and Xia, Y., Electrospinning of nano fibers with core-sheath, hollow, or porous structures, *J. Mater. Chem.*, 15, 735–738, 2005.
11. MacDiarmid, A.G., Jones, W.E., Jr., Norris, I.D., Gao, J., Johnson, A.T., Jr., Pinto, N.J., Hone, J., Han, B., Ko, F.K. and Okuzaki, H., Electrostatically-generated nano fibers of electronic polymers, *Synth. Metals*, 119, 27–30, 2001.
12. Kessick, R. and Tepper G., Microscale electrospinning of polymer nanofiber interconnections, *Appl. Phys. Lett.*, 83(3), 557–559, 2003.
13. Reneker, D.H. and Chun, I., Nanometer diameter fibres of polymer, produced by electrospinning, *Nanotechnology*, 7, 216–223, 1996.

14. Zhang, D., Sun, C. and Song, H., An investigation of fiber splitting of bicomponent meltblown/ microfiber nonwovens by water treatment, *J. Appl. Polym. Sci.*, 94, 1218–1226, 2004.

15. Zhao, R. and Wadsworth, L.C., Attenuating PP/PET bicomponent melt blown microfibers, *Polym. Eng. Sci.*, 43(2), 463–469, 2003.

16. Fabbricante, A., Ward, G. and Fabbricante, T., Micro-denier nonwovens materials made using modular die units, U.S. Patent 6,114,017, September 5, 2000.

17. Hagewood, J., Ultra microfibers: beyond evolution, *IFJ*, October, 47–48, 1998.

18. Formhals, A., Process and apparatus for preparing artificial threads, U.S. Patent 1,975,504, 1934.

19. Taylor, G.I., Disintegration of water drops in an electric field, *Proc. R. Soc. Lond. A Math. Phys. Sci.*, 280(1382), 383–397, 1964.

20. Taylor, G.I., Electrically driven jets, *Proc. R. Soc. Lond. A Math. Phys. Sci.*, 313(1515), 453–475, 1969.

21. Baumgarten, P.K., Electrostatic spinning of acrylic microfibres, *J. Colloid Interface Sci.*, 36, 71–79, 1971.

22. Larrondo, L. and Mandley, R.St.J., Electrostatic fibre spinning from polymer melts. I. Experimental observations on fibre formation and properties, *J. Polym. Sci. Polym. Phys.*, 19, 909–919, 1981.

23. Larrondo, L. and Mandley, R.St.J., Electrostatic fibre spinning from polymer melts. II. Examination of flow field in an electrically driven jet, *J. Polym. Sci. Polym. Phys.*, 19, 921–932, 1981.

24. Larrondo, L. and Mandley, R.St.J., Electrostatic fibre spinning from polymer melts. III. Electrostatic deformation of a pendant drop of polymer melt, *J. Polym. Sci. Polym. Phys.*, 19, 933–940, 1981.

25. Doshi, J. and Reneker, D.H., Electrospinning process and applications of electrospun fibers, *J. Electrostatics*, 35(2–3), 151–160, 1995.

26. Theron, A., Zussman, E. and Yarin, A.L., Electrostatic field-assisted alignment of electrospun nanofibres, *Nanotechnology*, 12(3), 384–390, 2001.

27. Desch, R., Liu, T., Schaper, A.K., Greiner, A. and Wendorff, J.H., Electrospun nanofibers: internal structure and intrinsic orientation, *J. Polym. Sci. A Polym. Chem.*, 41, 545–553, 2003.

28. Li, D., Wang, Y. and Xia, Y., Electrospinning of polymeric and ceramic nanofibers as uniaxially aligned arrays, *Nanoletters*, 3(8), 1167–1171, 2003.

29. Deitzel, J.M., Kleinmeyer, J.D., Hirvonen, J.K. and Beck Tan, N.C., Controlled deposition of electrospun poly(ethylene oxide) fibers, *Polymer*, 42(19), 8163–8170, 2001.

30. Kameoka, J. and Craighead, H.G., Fabrication of oriented polymeric nanofibers on planar surfaces by electrospinning, *Appl. Phys. Lett.*, 83 (2), 371–373, 2003.

31. Dai, H., Gong, J., Kim, H. and Lee, D., A novel method for preparing ultra-fine alumina-borate oxide fibres via an electrospinning technique, *Nanotechnology*, 13(5), 674–677, 2002.

32. Li, D. and Xia, Y., Direct fabrication of composite and ceramic hollow nanofibers by electrospinning, *NanoLetters*, 4(5), 933–938, 2004.

33. Hayati, I., Bailey, A.I. and Tadros, T.F., Investigations into the mechanisms of electrohydrodynamic spraying of liquids. I. Effect of electric field and the environment on pendant drops and factors affecting the formation of stable jets and atomization, *J. Colloid Interface Sci.*, 117(1), 205–221, 1987.

34. Deitzel, J.M., Kleinmeyer, J., Harris, D. and Beck Tan, N.C., The effect of processing variables on the morphology of electrospun nanofibers and textiles, *Polymer*, 42, 261–272, 2001.

35. Hohman, M.M., Shin, M., Rutledge, G. and Brenner, M.P., Electrospinning and electrically forced jets. I. Stability theory, *Phys. Fluids*, 13, 2201–2220, 2001.

36. Hohman, M.M., Shin, M., Rutledge, G. and Brenner, M.P., Electrospinning and electrically forced jets. II. Applications, *Phys. Fluids*, 13, 2221–2236, 2001.

37. Shin, Y.M., Hohman, M.M., Brenner, M.P. and Rutledge, G.C., Experimental characterization of electrospinning: the electrically forced jet and instabilities, *Polymer*, 42(25), 9955–9967, 2001.

38. Shin, Y.M., Hohman, M.M., Brenner, M.P. and Rutledge, G.C., Electrospinning: a whipping fluid jet generates submicron polymer fibers, *Appl. Phys. Lett.*, 78(8), 1149–1151, 2001.

39. Wente, V.A., Manufacture of superfine organic fibers, U.S. Department of Commerce, Office of Technical Services Report No. PBI 11437, Naval Research Laboratory, Report 4364, 1954.
40. Wente, V.A., Superfine thermoplastic fibers, *Ind. Eng. Chem.*, 48, 1342–1346, 1956.
41. McCulloch, J.G., The history of the development of melt blowing technology, *INJ*, 8(1), 66–72, 1999.
42. Buehning, P.G., Melt blowing die, U.S. Patent 4,986,743, January 22, 1991.
43. Rubhausen, A. and Roock, D., Apparatus for blow-extruding filaments for making a fleece, U.S. Patent 5,248,247, September 28, 1993.
44. Joseph, E.G., High temperature stable nonwoven webs based on multi-layer blown microfibers, U.S. Patent 5,232,770, August 3, 1993.
45. Schwartz, E., Apparatus and process for uniformly melt-blowing a fiberforming thermoplastic polymer in a spinnerette assembly of multiple rows of spinning orifices, U.S. Patent 5476616, December 19, 1995.
46. Schwartz, E., Apparatus and process for melt-blowing a fiberforming thermoplastic polymer and product produced thereby, U.S. Patent 4,380,570, April 19, 1983.
47. Appel, D.W., Drost, A.D. and Lau, J.C., Slotted melt-blown die head, U.S. Patent 4,720,252, January 19, 1988.
48. Nishioi, H., Ogata, S. and Tsujiyama, Y., Microfibers-generating fibers and a woven or non-woven fabric of microfibers, U.S. Patent 5,290,626, March 1, 1994.
49. Terakawa, T. and Nakajima, S., Spinneret device for conjugate melt-blow spinning, U.S. Patent 5,511,960, April 30, 1996.
50. Lau, J.C. and Haynes, B.D., Apparatus for the production of fibers and materials having enhanced characteristics, U.S. Patent 5,711,970, January 27, 1998.
51. Mende, T. and Sakai, T., Melt-blowing die, U.S. Patent 5017112, May 21, 1991.
52. Mende, T. and Sakai, T., Melt-blowing method having notches on the capillary tips, U.S. Patent 5,171,512, December 15, 1992.
53. Gillespie, J.D., Christopher, D.B., Thomas, H.E., Phillips, J.H., Gessner, S.L., Trimble, L.E. and Austin, J.A., Meltspun multicomponent thermoplastic continuous filaments, products made therefrom, and methods therefore, U.S. Patent 5,783,503, July 21, 1998.
54. Bresee, R.R., Qureshi, U.A. and Pelham, M.C., Influence of processing conditions on melt blown web structure: Part 2 — primary airflow rate, *INJ*, 14(2), 11–18, 2005.
55. Zhao, R. and Wadsworth, L.C., Study of polypropylene/poly(ethylene terephthalate) bicomponent melt-blowing process: the fiber temperature and elongational viscosity profiles of the spinline, *J. Appl. Polym. Sci.*, 89, 1145–1150, 2003.
56. Zhao, R., Wadsworth, L.C., Sun, C. and Zhang, D., Properties of PP/PET bicomponent melt blown microfiber nonwovens after heat-treatment, *Polym. Int.*, 52, 133–137, 2003.
57. Zhang, D., Sun, C., Beard, J., Brown, H., Carson, I. and Hwo, C., Development and characterization of poly(trimethylene terephthalate)-based bicomponent meltblown nonwovens, *J. Appl. Polym. Sci.*, 83, 1280–1287, 2002.
58. Prentice, J.S., Laminated non-woven sheet, U.S. Patent 4,078,124, March 7, 1978.
59. Komatsu, M., Narukawa, K. and Yamamoto, N., Process for producing hydrophilic polyolefin nonwoven fabric, U.S. Patent 4,743,494, May 10, 1988.
60. Kanno, T., Matsushima, Y. and Suzuki, M., High-strength non-woven fabric, method of producing same and battery separator constituted thereby, U.S. Patent 5,089,360, February 18, 1992.
61. Howard, R.E. and Young, J., Nonwoven webs of microporous fibers and filaments, U.S. Patent 5,230,949, July 27, 1993.
62. Berger, R.M., Bicomponent fibers and tobacco smoke filters formed therefrom, U.S. Patent 5,509,430, April 23, 1996.
63. Shipp, P.W. and Vogt, C.M., Melt-blown material with depth fiber size gradient, U.S. Patent 4,714,647, December 22, 1987.
64. Pall, D.B., Melt-blown fibrous web, U.S. Patent 5,582,907, December 10, 1996.

65. Allen, M.A., Melt blowing of tubular filters, U.S. Patent 5,409,642, April 25, 1995.
66. Mozelack, B., Schmitt, R.J., Barboza, S.D., Jana, P., Nguyen, S.N., Gschwandtner, R.R., Connor, R.D. and Yingling, T.W., Apparatus for making melt-blown filter cartridges, U.S. Patent 6,662,842, December 16, 2003.
67. Midkiff, D.G., Filtration media and articles incorporating the same, U.S. Patent 6,322,604, November 27, 2001.
68. Watt, J.M. and Lickfield, D.K., Meltblown barrier webs and processes of making same, U.S. Patent 5,645,057, July 8, 1997.
69. Kwok, K.-C., Bolyard, E.W., Jr. and Riggan, L.E., Jr., Meltblowing method and system, U.S. Patent 5,904,298, May 18, 1999.
70. Bresee, R.R. and Ko, W.-C., Fiber formation during melt blowing, *INJ*, 12(2), 21–28, 2003.
71. Farer, R., Ghosh, T.K., Seyam, A.M., Grant, E. and Batra, S.K., Study of meltblown structures formed by robotic and meltblowing integrated system: impact of process parameters on fiber orientation, *INJ*, 11(4),14–21, 2002.
72. Sun, C., Zhang, D., Liu, Y. and Xiao, R., Preliminary study on fiber splitting of bicomponent meltblown fibers, *J. Appl. Polym. Sci.*, 93, 2090–2094, 2004.
73. Tsai, P., Chen, W.W. and Roth, J.R., Investigation of the fiber, bulk, and surface properties of meltblown and electrospun polymeric fabrics, *INJ*, 13(3), 17–23, 2004.
74. Dugan, J. and Homonoff, E., Synthetic Split Microfiber Technology for Filtration, INDA Filtration, Philadelphia, PA, November 28–30, 2000.
75. Zapletalova, T., Introduction to bicomponent fibers, *IFJ*, June, 20–24, 1998.
76. Perez, M.A., Swan, M.D.T. and Louks, J.W., Microfibers and method of making, U.S. Patent 6,110,588, August 29, 2000.
77. Dugan, J., Splittable multicomponent fibers containing a polyacrylonitrile polymer component, U.S. Patent 6,444,312, September 3, 2002.
78. Groitzsch, D., Ultrafine microfiber spunbond for hygiene and medical application, EDANA Non-wovens Symposium, 2000.
79. Pike, R.D., Super fine microfiber nonwoven web, U.S. Patent 5,935,883, August 10, 1999.
80. Pike, R.D., Microfiber nonwoven web laminates, U.S. Patent 6,624,100, September 23, 2003.
81. Pike, R.D., Sasse, P.A., White, E.J. and Stokes, T.J., Fine denier fibers and fabrics made therefrom, U.S. Patent 5,759,926, June 2, 1998.
82. Dugan, J. and Harris, F.O., Splittable multicomponent polyester fibers, U.S. Patent 6,780,357, August 24, 2004.
83. Dugan, J., Splittable multicomponent polyolefin fibers, U.S. Patent 6,461,729, October 8, 2002.
84. Wagner, R. and Groten, R., Method for the production of a synthetic leather, U.S. Patent 6,838,043, January 4, 2005.
85. Schreuder-Gibson, H., Gibson, P. and Hsiesh, Y.-L., Transport properties of electrospun nonwoven membranes, *INJ*, 11(2), 21–27, 2002.
86. Hagewood, J., Properties study of new bico spunbond and meltblown barrier fabrics, Proceedings of INTC, September 21–24, 2004.
87. Ward, D., Bicomponent technology: an alternative route to microfiber fabrics, *IFJ*, December, 20, 1997.
88. Moriki, Y. and Ogasawara, M., Spinneret for production of composite filaments, U.S. Patent 4,445,833, May 1, 1984.
89. Kiriyama, T., Norota, S., Segawa, Y., Emi, S., Imoto, T. and Azumi, T., Novel assembly of composite fibers, U.S. Patent 4,414,276, November 8, 1983.
90. Hills, W.H., Method of making plural component fibers, U.S. Patent 5,162,074, November 10, 1992.
91. Choi, Y.B. and Kim, S.Y., Effects of interface on the dynamic mechanical properties of PET/nylon 6 bicomponent fibers, *J. Appl. Polym. Sci.*, 74, 2083–2093, 1999.
92. Kikutani, T., Radhakrishnan, J., Arikawa, S., Takaku, A., Okui, N., Jin, X., Niwa, F. and Kudo, Y., High-speed melt spinning of bicomponent fibers: mechanism of fiber structure development in poly(ethylene terephthalate)/polypropylene system, *J. Appl. Polym. Sci.*, 62, 1913–1924, 1996.

93. El-Salmawy, A., Miyamoto, M. and Kimura, Y., Preparing a core-sheath bicomponent fiber of poly(butylene terephthalate). Poly(butylenes succinate-co-L-lactide), *Tex. Res. J.*, 70(11), 1011–1018, 2000.

94. Cho, H.H., Kim, K.H., Kang, Y.A., Ito, H. and Kikutani, T., Fine structure and physical properties of polyethylene/poly(ethylene terephthalate) bicomponent fibers in high-speed spinning. I. Poly-ethylene sheath /poly(ethylene terephthalate) core fibers, *J. Appl. Polym. Sci.*, 77, 2254–2266, 2000.

95. Rwei, S.P., Jue, Z.F. and Chen, F.L., PBT/PET conjugated fibers: melt spinning, fiber properties and thermal bonding, *Polym. Eng. Sci.*, 44(2), 331–344, 2004.

96. Yoshimura, M., Iohara, K., Nagai, H., Takahashi, T. and Koyama, K., Structure formation of blend and sheath/core conjugated fibers in high-speed spinning of PET, including a small amount of PMMA, *J. Macr. Sci. B Phys.*, B42(2), 325–339, 2003.

97. Kawano, M., Micro-fibers-generating conjugate fibers and woven or non-woven fabric thereof, U.S. Patent 4,966,808, October 30, 1990.

98. Kawano, M., Hot-melt-adhesive, micro-fiber-generating conjugate fibers and a woven or non-woven fabric using the same, U.S. Patent 5,124,194, June 23, 1992.

99. Okamoto, M., Multi-component composite filament, U.S. Patent 4381335, April 26, 1983.

100. Dugan, V., Novel properties of PLA fibers, *INJ*, 10(3), 29–33, 2001.

101. Hagewood, J., Polymeric nanofibers — fantasy or future. http://www.hillsinc.net.

21.3 Electrospun Nanofibers with Magnetic Domains for Smart Tagging of Textile Products

Carola Barrera, Carlos Rinaldi, Melinda Satcher, and Juan P. Hinestroza

This section discusses the use of the electrospinning process to generate polymeric nanofibers containing discrete magnetic domains. The AC magnetic response of these nanofibers has been found to be unique and fully traceable to their precursor polymeric solution, making them good candidates for anticounter-feiting devices as well as for positive identification and smart tagging of textile products.

21.3.1 Electrospinning of Polymeric Nanofibers

Electrospinning is a simple, well-known technique capable of producing fibers with diameters as small as 100 nm [1–5]. This process was first used in the mid-1930s and has now been rediscovered as an alternative for producing polymer nanofibers [6]. During this process, an electric field induces a charge on the surface of a polymer solution held by its surface tension at the end of a capillary tube. The induced charges are distributed across the surface of the drop in such a way that mutual charges repel each other. A force is created directly opposite to that due to surface tension, which will end up forming the well-known Taylor cone [7]. When charges in the fluid reach a critical value, a jet will erupt from the tip of the capillary, which, upon stretching and bending, will generate fibers. Hohman et al. [1] developed a theory that predicts two types of instabilities a charged jet may experience when subjected to an external electric field. The first instability is called a varicose instability, where the charged jet remains straight but its radius is modulated. The second instability is called a whipping instability, because the jet radius remains constant while the centerline is modulated. This whipping instability results in a rapid elongation of the jet with concomitant solvent evaporation and polymer nanofiber deposition on a collector [8–10]. Operating diagrams are commonly used to show the states of instability for a specific solution at a certain flow rate and electric field [1,5,8]. The electrospinning operating conditions can be highly influenced by parameters such as the solution viscosity, conductivity, and surface tension, with viscosity being one of the most relevant [11]. As the viscosity of the solution increases, higher flow rates and electric fields are required, until a maximum value is achieved where electrospinning is no longer possible [12].

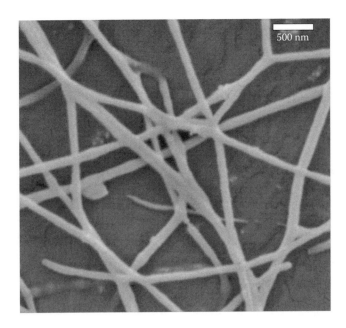

FIGURE 21.9   PEO nanofibers electrospun from a water-based solution containing 0.01 volume fraction of magnetite nanoparticles.

Typical electrospinning setups consist of a high-voltage source used to create an electric field between two electrodes, and a syringe pump used to supply the polymer solution through a capillary tube. A wide range of polymers have been successfully electrospun, with poly(ethylene oxide) (PEO) being one of the most studied (see Figure 21.9).

21.3.2   PEO Nanofibers

PEO is a homopolymer of ethylene oxide, with a structural unit represented as $-[CH_2CH_2O]_n-$. It can be synthesized by cationic polymerization of oxirane, using a protonic or Lewis acid system to initiate the polymerization. The same structural unit $(-[CH_2CH_2O]_n-)$ can be obtained when the monomer ethylene glycol $(HOCH_2CH_2OH)$ undergoes intermolecular dehydration, but in this case, the polymer is referred to as poly(ethylene glycol) (PEG). The molecular weight of PEG polymers is relatively low (below 25,000 g/mol), because during dehydration, other side reactions compete with polymerization.

PEO nanofibers have been primarily used to study the electrospinning process, and a few specific applications have been reported. Fong and Reneker [12] used electrospun PEO fibers to study the effect of solution viscosity, surface tension, and net charge density on the formation of beaded nanofibers. In a recent work, Deitzel [13] and Kameoka and Craighead [14] reported the construction of a novel electrospinning apparatus that allows greater control over PEO nanofiber deposition and collection. The electrospinning processing conditions have also been studied using PEO solutions, suggesting that properties like viscosity, conductivity, surface tension, hydrostatic pressure on the capillary, electric field, and distance between plates all have an effect on fiber morphology and size [1,5,11,15].

21.3.3   Composite Nanofibers via Electrospinning

Composite nanofibers are attracting considerable attention in areas such as wound dressings, medical prostheses, drug delivery, protective clothing, and catalysis, through the incorporation of functional nanoparticles into the polymer matrix during the electrospinning process. For example, electrospun

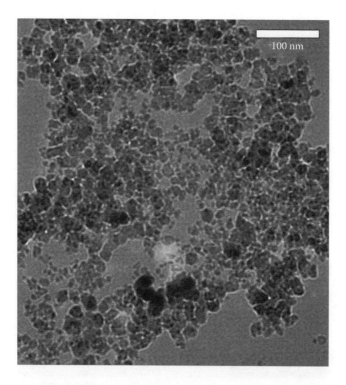

FIGURE 21.10 TEM image of magnetic nanoparticles from a water-based commercial ferrofluid.

titania nanofibers loaded with gold nanoparticles and carbon nanotubes for use in photocatalysis and chemical-sensing applications [16,17] have been reported. Studies on electrospun polycaprolactone nanofibers have also been reported, where calcium carbonate nanoparticles have been embedded in fibers for use in bone regeneration therapies [18]. Silver has also been successfully incorporated into nanofibers for use in wound-care treatment [19]. In 2003, Li et al. [20] reported for the first time the use of electrospinning to form magnetic nanofibers made of nickel ferrite ($NiFe_2O_3$). Their intention was to produce *in situ* linear chains of $NiFe_2O_3$ nanoparticles that will exhibit high-coercivity values, using two metal alkoxides of nickel and iron dissolved in poly(vinyl pyrrolidone). They compared the *in situ* synthesis of $NiFe_2O_3$ using electrospinning with the conventional sol–gel method. X-ray diffraction and TEM showed that after 2 h of hydrolysis and subsequent calcinations, the electrospun nanofibers contained $NiFe_2O_3$ nanoparticles with smaller and more uniform diameters than the conventional sol–gel method. In 2004, Wang et al. [21] reported field-responsive superparamagnetic nanofibers of PEO and poly(vinyl alcohol) (PVA) loaded with magnetite nanoparticles formed by electrospinning. In this case, the magnetic nanoparticles were synthesized in the presence of a copolymer of PVA and Jeffamine ($PEO/PVA–NH_2$), which attaches to the particle surface and confers steric stabilization on the nanoparticle when suspended in the electrospinning polymer solution. Fibers produced through this method were heavily loaded with magnetic nanoparticles, and no further studies were made regarding particle distribution across the fiber. The most recent work in this area was reported by Tan et al. [22], where nanofibers composed of magnetite dispersed in poly(hydroxyethyl methacrylate) and poly-L-lactide were formed by electrospinning, for application in drug delivery and as drug carriers.

We recently produced PEO nanofibers containing embedded magnetic nanoparticles at several loadings of magnetic nanoparticles. Figure 21.11 is a TEM image of a PEO nanofiber electrospun from a water-based solution containing a magnetite volume fraction of 0.10 v/v. The TEM image indicates a uniform distribution of the magnetic nanoparticles.

FIGURE 21.11 TEM images of PEO nanofibers electrospun from a solution containing 0.1 v/v magnetite nanoparticles.

The uniformity of the loading of the magnetic nanoparticles inside individual nanofibers can also be analyzed using electron diffraction patterns, such as the one shown in Figure 21.12. While electron diffraction patterns are most frequently used to probe the crystal structure of solids, our group is pioneering its use in determining the position and distribution of the magnetic nanoparticles embedded in a single nanofiber as well as in determining short-range order in the amorphous PEO matrix. This technique is very powerful as it is able to resolve subangstrom features.

21.3.4   Magnetic Nanoparticles: Synthesis and Functionalization

Metal oxide nanoparticles can be synthesized through well-known coprecipitation techniques [23]. During this process, an aqueous solution of metal salts is made to react with an alkaline solution, resulting in the formation of a black magnetic precipitate. The stability of these particles in a liquid is of utmost importance when preparing magnetic fluids, or ferrofluids as they are commonly known. Magnetic fluids are colloidal suspensions of permanently magnetized nanoparticles in a carrier liquid such as water or oil. The nature of these suspensions gives them a unique combination of fluidity and the capability to interact with external magnetic fields, which is highly dependent on particle size and medium interactions. To avoid particle agglomeration, a stabilizer such as a surfactant or polymer is usually added during particle synthesis and adsorbed on the particle surface. A wide range of synthetic and natural polymers have been reported in the literature as stabilizers of magnetic nanoparticles, by the use of ligands that attach to the particle surface [24].

Of special interest is the use of PEO or PEG as a stabilizer of magnetic nanoparticles. Most of the work reported in this area involves the use of an engineered copolymer of PEO with a backbone that contains carboxyl groups available for attachment to the particle surface, and grafted PEO chains that confer particle stability in the dispersing medium. Moeser et al. [25] reported the synthesis of a copolymer of

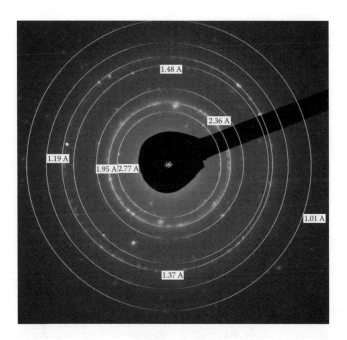

FIGURE 21.12 TEM electron diffraction pattern of a PEO nanofiber electrospun from a solution containing 0.1 v/v magnetite nanoparticles from a water-based ferrofluid.

poly(acrylic acid) grafted with PEO and poly(propylene oxide) (PPO) as side chains, with most of its carboxylic groups left unreacted. In 2005, Jain et al. [26] reported the synthesis of oleic acid and pluronic-coated nanoparticles, which consisted of an alternating copolymer of PEO and PPO, with carboxyl groups available for surface attachment. PEO has been used not only in copolymers but also as a modified PEG with functional groups like amines and silane that will serve as ligands for particle attachment [27–29]. Another type of ligands that is becoming very attractive in biological applications is the thiol group. Only a few reports have appeared on the synthesis of thiol-terminated PEOs [30], and most of the presented techniques do not allow one-step thiolation or addition of more than one thiol group to the employed monomer. So far, this thiol-ended PEO has only been used in self-assembled monolayers on gold surfaces for biomedical applications [31,32].

A common synthesis route for magnetic nanoparticles of magnetite and cobalt ferrite follows the coprecipitation method according to the following reactions:

$$Fe^{+2} + 2Fe^{+3} + 8OH^- \rightarrow Fe_3O_4 + 4H_2O \tag{21.1}$$

$$Co^{+2} + 2Fe^{+3} + 8OH^- \rightarrow CoFe_2O_4 + 4H_2O \tag{21.2}$$

X-ray diffraction is commonly used to determine the particle's crystallinity and average size, using patterns reported in the literature [36,37] and the Scherrer equation [38]:

$$D = \frac{0.9\lambda}{\beta\cos\theta} \tag{21.3}$$

where D is the mean particle diameter, λ is the wavelength, β is the width of the most intense peak at half height, and θ is the angle.

Particle size can be also be determined by direct reading of transmission electron microscopy (TEM) images, such as the one shown in Figure 21.10.

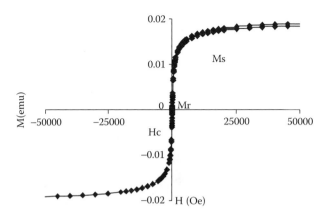

FIGURE 21.13 Room temperature equilibrium magnetization of a bundle of PEO nanofibers electrospun from a water-based PEO solution containing a 0.01 volume fraction of magnetite.

21.3.5 Magnetic Characterization of Electrospun Nanofibers

The magnetic response of magnetic nanoparticles to an oscillating magnetic field provides unique information about the particles' magnetization dynamics (see Figure 21.13). This response can be modeled mathematically using the relaxation equation:

$$\frac{dM}{dt} = -\frac{1}{\tau}(M - \chi_0 H) \tag{21.4}$$

where M is the magnetization of the suspension, χ_0, the initial susceptibility, H, the applied magnetic field, and τ, the magnetic relaxation time. If we consider an oscillating magnetic field H defined as $H = H_0 \cos(\Omega t)$, the magnetization response becomes $M = \chi' H_0 \cos(\Omega t) + \chi'' H_0 \sin(\Omega t)$, where Ω is the field frequency. χ' and χ'' are known as the real and imaginary components of the complex or dynamic susceptibility, and are defined by

$$\chi' = \frac{\chi_0}{1 + \Omega^2 \tau^2}, \qquad \chi'' = \frac{\chi_0 \Omega \tau}{1 + \Omega^2 \tau^2} \tag{21.5}$$

Our group has measured the magnetic response of PEO-magnetite nanofibers using a Quantum Design MPMS-XL7 SQUID magnetometer. The equilibrium magnetization of a nanofiber electrospun from a solution containing 0.01 volume of magnetic nanoparticles is shown in Figure 21.12. It can be observed that the magnetization curve does not show magnetic hysteresis, indicating superparamagnetic behavior typical of single-domain magnetite nanoparticles.

The magnetic relaxation time, τ, is dependent on the mechanism by which particles respond to a change in an external magnetic field. Two possibilities are rotational Brownian motion and Néel relaxation. The first mechanism is related to particle rotation along with its magnetic moment and the second is due to the rotation of the magnetic vector within the particles [33]. Brownian relaxation corresponds to a rotational diffusion time, τ_B, given by

$$\tau_B = \frac{3\eta_0 V_h}{kT} \tag{21.6}$$

where V_h is the particle hydrodynamic volume, η_0 is the viscosity of the carrier liquid, and kT is thermal energy. The Néel relaxation is given by

$$\tau_N = \frac{1}{f_0}\exp\left(\frac{KV}{kT}\right) \tag{21.7}$$

where f_0 is an attempt frequency having an approximate value of 10^9 Hz and K is the magnetocrystalline anisotropy constant. Because both mechanisms occur in parallel, the effective relaxation time is given by

$$\tau = \frac{\tau_B \tau_N}{\tau_B + \tau_N} \tag{21.8}$$

The Langevin equation is commonly used to describe the behavior of monodisperse suspensions of magnetic nanoparticles with negligible particle–particle interactions:

$$\frac{M}{\phi M_d} = \coth\alpha - \frac{1}{\alpha} \equiv L(\alpha) \tag{21.9}$$

$$\alpha = \frac{\pi}{6}\frac{\mu_0 M_d H d^3}{kT} \tag{21.10}$$

where ϕ is the volume fraction of the magnetic cores, M_d is the domain magnetization of the material (446 kA/m for magnetite), $\mu_0 = 4\pi \times 10^{-7}$ Henries/m is the magnetic permeability of free space, d is the diameter of the magnetic particle core, $k = 1.38 \times 10^{-23}$ J/K is Boltzmann's constant, and T is the absolute temperature.

A Taylor series expansion for small α shows that

$$L(\alpha) = \frac{\alpha}{3}, \quad \text{therefore, } M = \chi_i H \tag{21.11}$$

where χ_i is the initial susceptibility of the suspension, which is related to the properties of the nanoparticles by

$$\chi_i = \frac{\pi}{18}\phi\mu_0 \frac{M_d^2 d^3}{kT} \tag{21.12}$$

At high magnetic fields (large α), the magnetic moment approaches saturation. In this case, Equation (21.10) has the limit,

$$M = \phi M_d\left(1 - \frac{6}{\pi}\frac{kT}{\mu_0 M_d H d^3}\right) \tag{21.13}$$

Because ferrofluids are typically polydisperse, it is important to distinguish between the particle diameters determined from Equation (21.10) and Equation (21.13). At low fields, Brownian motion dominates the behavior of the smaller particles; hence it is the larger particles that contribute primarily to M_d. At higher fields, the larger particles have already saturated and it is the smaller particles that contribute primarily to M_d. A more detailed consideration of the effect of particle size polydispersity on the equilibrium magnetic response of ferrofluids is given by Chantrell et al. [34]. In this case, a median particle diameter and a standard deviation can be calculated for a ferrofluid with a log–normal distribution using magnetization curves. In this analysis, the magnetization of the ferrofluid is given by the sum of the contributions from each particle diameter weighted by a volume fraction distribution function defined as the total magnetic volume having a diameter between D_p and $D_p + dD_p$. For the asymptotic cases of the Langevine function, the equilibrium magnetization at high fields and low fields are now given by Equation (21.6) to Equation (21.14), and Equation (21.7) to Equation (21.15), respectively:

$$M = \phi M_d \left[1 - \frac{\pi}{6} \frac{kT \exp(9\ln^2 \sigma_g / 2)}{\mu_0 M_d H D_{pgv}^3} \right] \tag{21.14}$$

$$\chi_i = \frac{\pi}{18} \frac{\phi \mu_0 M_d^2}{kT} D_{pgv}^3 \exp\left(\frac{-9\ln^2(\sigma_g)}{2} \right) \tag{21.15}$$

where D_{pgv} is the volume median diameter and σ_g is the shape factor or the geometric standard deviation.

The out-of-phase component of the AC susceptibility signal for a bundle of electrospun nanofibers and their precursor solution is shown in Figure 21.14. Characteristic signals at 40 and 100 Hz are noted in both traces. The signal from the electrospun nanofibers appears to closely map the signal from the precursor solution. These characteristic signals are functions of the size and nature of the magnetic nanoparticles.

Because these nanofibers are not visible to the human eye but their presence can be noted via their magnetic signals, it is expected that novel anticounterfeiting devices could be constructed through judicious selection of the size of the nanoparticles as well as their chemical composition.

In addition to PEO, our group has electrospun nanofibers containing controllable magnetic domains out of nylon, polyester, and cellulose. Due to the versatility of the electrospinning process, these nanofibers can be directly spun either onto textile raw materials such as yarns and bundles or over final products such as fabrics or apparel, allowing their use as tracers throughout the textile supply chain.

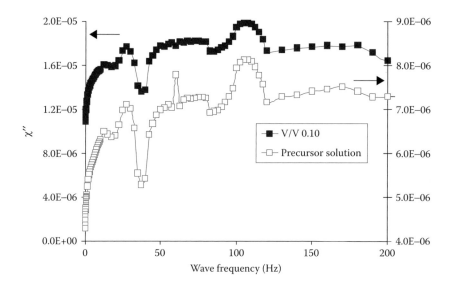

FIGURE 21.14 Out-of-phase component of the AC susceptibility as a function of frequency for a bundle of PEO nanofibers containing magnetite nanoparticles as well as for their precursor solution.

21.3.6 Conclusions and Technological Implications

Counterfeiting of high-value goods is no longer a fake problem. U.S. apparel import preference programs require the use of U.S.-made fabrics and yarns. To enforce these programs, the origin of the fabrics and yarns in finished textiles must be determined. Due to the importance of the U.S. market, increased counterfeiting and piracy have long affected the American textile industry, with well-known brands being major victims. Due to globalization trends in the industry, the problem has resulted in major economic losses for American textile and apparel companies, as design theft and forgeries arise all over the world. At the core of the problem are the facts that anticounterfeiting measures are usually applied to the manufactured product rather than to the raw materials, so labels and holograms can be easily stolen from these products or simply falsified. According to the UN Organization for Economic Cooperation and Development, counterfeiting cost the U.S. industry $450 billion in 2002 and accounted for approximately 7 to 9% of world trade in general. It is estimated that these actions resulted in the loss of about 17,000 jobs/year. Among the U.S. Customs' seizures of counterfeited goods, the amount of textile and clothing has doubled since 2001 to 9.2 million pieces, a rise of 10.9% year-on-year.

For a long time, counterfeiting and piracy have been considered as matters of more direct concern to private interests. However, during recent years, government agencies have shown increasing interest in the problem as public safety may be jeopardized with the importation of large amounts of counterfeited technical and medical textiles such as surgery gowns, fire-resistance fabrics, bandages, prostheses, etc.

The use of embedded magnetic nanoparticles in nanofibers and the ability to control the geometric and magnetic distributions of these particles in an individual nanofiber opens up a great number of possibilities for smart textile products. For example, a textile product may change shape rapidly to accommodate a sudden change in environmental conditions such as those experienced by fighter pilots during high *G* maneuvers, or a glove may be engineered to release a medicine in a controlled way for people with arthritis by simply activating an external magnetic field. Other potential smart clothing applications include the use of dyes and pigments that are responsive to magnetic fields for camouflage purposes, and the use of nanofibers with a unique AC susceptibility signature as positive identification mechanisms for official uniforms of first responders and military personnel.

References

1. Hohman, M.M., Shin, M., Rutledge, G., and Brenner, M.P. Electrospinning of electrically forced jets. II. Applications, *Phys Fluids*, vol. 13, pp. 2221–2236, 2001.
2. Huang, Z.M., Zhang, Y.Z., Kotaki, M., and Ramakrishna, S. A review on polymer nanofibers by electrospinning and their applications in nanocomposites, *Compos Sci Technol*, vol. 63, 2003.
3. Jayaraman, K., Kotaki, M., Zhang, Y.Z., Mo, X.M., and Ramakrishma, S. Recent advances in polymer nanofibers, *J Nanosci Nanotechnol*, vol. 4, pp. 52–65, 2004.
4. Reneker, D.H. and Chun, I. Nanometre diameter fibres of polymer, produced by electrospinning, *Nanotechnology*, vol. 7, pp. 216–223, 1996.
5. Shin, Y.M., Hohman, M.M., Brenner, M.P., and Rutledge, G.C. Electrospinning: a whipping fluid jet generates submicron polymer fibers, *Appl Phys Lett*, vol. 78, pp. 1149–1151, 2001.
6. Formhals, A. Apparatus for producing artificial filaments from material such as cellulose acetate, in *Us: Schreiber-Gastell*, Richard, 1934.
7. Taylor, G. Electrically driven jets, *Proc Roy Soc Lond*, vol. 313, pp. 453–475, 1969.
8. Hohman, M.M., Shin, M., Rutledge, G., and Brenner, M. Electrospinning and electrically forced jets: stability theory, *Phys Fluid*, vol. 13, pp. 2201–2220, 2001.
9. Russell, T.P., Lin, Z.Q., Schaffer, E., and Steiner, U. Aspects of electrohydrodynamics instabilities at polymer interfaces, *Fibers Polym*, vol. 4, pp. 1–7, 2003.
10. Shin, Y.M., Hohman, M.M., Brenner, M.P., and Rutledge, G.C. Experimental characterization of electrospinning: the electrically forced jets and instabilities, *Polymer*, vol. 19, pp. 9955–9967, 2001.

11. Theron, S.A., Zussman, E., and Yarin, A.L. Experimental investigation of the governing parameters in the electrospinning of polymer solutions, *Polymer*, vol. 45, pp. 2017–2030, 2004.

12. Fong H. and Reneker, D.H. Beaded nanofibers formed during electrospinning, *Polymer*, vol. 40, pp. 4585–4592, 1999.

13. Deitzel, J.M., Kleinmeyer, J.D., Hirvonen, J.K., and Tan, N.C.B. Controlled deposition of electrospun poly(ethylene oxide) fibers, *Polymer*, vol. 42, pp. 8163–8170, 2001.

14. Kameoka, J. and Craighead, H.G. Fabrication of oriented polymeric nanofibers on planar surfaces by electrospinning, *Appl Phys Lett*, vol. 83, pp. 371–373, 2002.

15. Doshi, J. and Reneker, D.H. Electrospinning process and applications of electrospun fibers, *J Electrostat*, vol. 35, pp. 151–160, 1995.

16. Kedem, S., Schmidt, J., Paz, Y., and Cohen, Y. Composite polymer nanofibers with carbon nanotubes and titanium dioxide particles, *Langmiur*, vol. 21, pp. 5600–5604, 2005.

17. Li, D., McCann, J.T., Gratt, M., and Xia, Y. Photocatalytic deposition of gold nanoparticles on electrospun nanofibers of titania, *Chem Phys Lett*, vol. 394, pp. 387–391, 2004.

18. Fujihara, K., Kotaki, M., and Ramakrishna, S. Guided bone regeneration membrane made of polycaprolactone/calcium carbonate composite nanofibers, *Biomaterials*, vol. 26, pp. 4139–4147, 2005.

19. Melaiye, A., Sun, Z., Hindi, K., Milsted, A., and Ely, D. Silver (I)-imidazole cyclophane gem-diol complexes encapsulated by electrospun tecophilic nanofibers: formation of nanosilver particles and antimicrobial activity, *J Am Chem Soc*, vol. 127, pp. 2285–2291, 2005.

20. Li, D., Herricks, T., and Xia, Y. Magnetic nanofibers of nickel ferrite prepared by electrospinning, *Appl Phys Lett*, vol. 83, pp. 4586–4588, 2003.

21. Wang, M., Singh, H., Hatton, T.A., and Rutledge, G.C. Field-responsive superparamagnetic composite nanofibers by electrospinning, *Polymer*, vol. 45, pp. 5505–5514, 2004.

22. Tan, S.T., Wendorff, J.H., Pietzonka, C., Jia, Z.H., and Wang, G.Q. Biocompatible and biodegradable polymer nanofibers displaying superparamagnetic properties, *Chem Phys Chem*, vol. 6, pp. 1–6, 2005.

23. Reimers, G.W. and Khalafalla, S.E. Preparing magnetic fluids by a peptizing method, Technical Progress Report, U.S. Dept. of the Interior, 1972.

24. Gupta, A.K. and Gupta, M. Synthesis and surface engineering of iron oxide nanoparticles for biomedical applications, *Biomaterials*, vol. 25, pp. 3995–4021, 2005.

25. Moeser, G.D., Roach, K.A., Green, W.H., Laibinis, P.E., and Hatton, T.A. Water-based magnetic fluids as extractant for synthetic organic compounds, *Ind Eng Chem*, vol. 41, pp. 4739–4749, 2002.

26. Jain, T.K., Morales, M.A., Sahoo, S.K., Leslie-Pelecky, D.L., and Labhasetwar, V. Iron oxide nanoparticles for sustained delivery of anticancer agents, *Mol Pharmaceut*, vol. 2, pp. 194–205, 2005.

27. Kim, D.K., Mikhaylova, M., Zhang, Y., and Muhammed, M. Protective coating of superparamagnetic iron oxide nanoparticles, *Chem Mater*, vol. 15, pp. 1617–1627, 2003.

28. Illum, L., Church, A.E., Butterworth, M.D., Arien, A., Whetstone, J., and Davis, S.S. Development system for targeting the regional lymph nodes for diagnostic imaging: in vivo behaviour of colloidal PEG-coated magnetite nanospheres in the rat following interstitial administration, *Pharmaceut Res*, vol. 18, pp. 640–645, 2001.

29. Zhang, Y., Kohler, N., and Zhang, M. Surface modification of superparamagnetic magnetite nanoparticles and their intracellular uptake, *Biomaterials*, vol. 23, pp. 1553–1561, 2002.

30. Du, Y.J. and Brash, J.L. Synthesis and characterization of thiol-terminated poly(ethylene oxide) for chemisorption to gold surface, *J Appl Polym Sci*, vol. 90, pp. 594–607, 2003.

31. Wuelfing, W.P., Gross, S.M., Miles, D.T., and Murray, R.W. Nanometer gold clusters protected by surface-bound monolayers of thiolated poly(ethylene oxide) polymer electrolyte, *J Am Chem Soc*, vol. 120, pp. 12696–12697, 1998.

32. Loo, C.H., Lee, M., Hirsch, L.R., and West, J.L. Nanoshell bioconjugate for integrated imaging and therapy of cancer, *Proc SPIE*, vol. 5327, pp. 1–4, 2004.

33. Rosensweig, R.E. *Ferrohydrodynamics*, Cambridge University Press, 1985.

34. Satoh, A., Coverdale, D., Chantrell, R., Stokesian dynamics simulations of ferromagnetic colloidal dispersions subjected to a sinusoidol shear flow, *J Colloid Interface Sci* 231 (2): 238–246.

21.4   Carbon Nanotubes

William Dondero and Russell E. Gorga

Carbon nanotubes are graphitic sheets rolled into seamless tubes (i.e., arrangements of carbon hexagons into tube-like fullerenes) and have diameters ranging from about a nanometer to tens of nanometers with lengths up to centimeters. Nanotubes have received much attention due to their interesting properties (high modulus and electrical/thermal conductivity) since their discovery by Iijima in 1991 [1,2]. Since then, significant effort has gone into incorporating nanotubes into conventional materials (such as polymers) for improved strength and conductivity [3–12].

21.4.1   Structure

Nanotubes can be synthesized in two structural forms, single-wall and multiwall (as shown in Figure 21.15). The first tubules Iijima discovered exhibited the multiwall structure of concentric nanotubes forming one tube, defining a multiwall nanotube (MWNT) [1]. Later, he observed a single-shell structure believed to be the precursor to the MWNTs [2]. This single graphitic sheet rolled into a tube with a cap at either end and diameter around 1 nm is defined as a SWNT. Additionally, nanotubes are described using one of three morphologies: armchair, zig zag, and chiral (as shown in Figure 21.16). The packing of the carbon hexagons in the graphitic sheets defines a chiral vector and angle. The indices of the vector determine the morphology of the nanotube. Variations in the nanotube morphology can lead to changes in the properties of the nanotube. For instance, the electronic properties of an armchair nanotube are metallic; however, the electronic properties of zig zag and chiral nanotubes are semiconducting [13]. The behavior is determined based on a mathematical model developed using the chiral vector indices [14].

21.4.2   Production Methods

The three main methods of manufacturing nanotubes include direct-current arc discharge, laser ablation, and chemical vapor deposition (CVD). A thorough discussion on each of these production methods can be found in Ref. [16]. Direct-arc discharge and laser ablation were the first techniques used to produce gram quantities of SWNTs. In both methods, the evaporation of solid carbon is used to condense carbon gas. The products of such methods are normally tangled and poorly oriented. CVD produces nanotubes from the decomposition of a continuously supplied carbon-containing gas onto a substrate. Due to the continuous supply of the gas, high-purity nanotubes can be produced on a larger (or industrial) volume scale. Producing the nanotubes in an ordered array with controlled length and diameter can also be achieved via CVD methods. Furthermore, plasma-enhanced CVD results in further nanotube uniformity within the array [17]. More recently, a team led by Schlittler has developed a self-assembly method for making an ordered array of nanotubes with identical geometry and high purity [18].

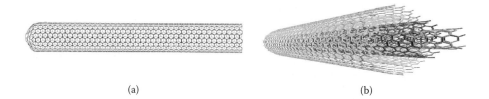

(a)   (b)

FIGURE 21.15   Schematic of (a) a SWNT and (b) a MWNT.

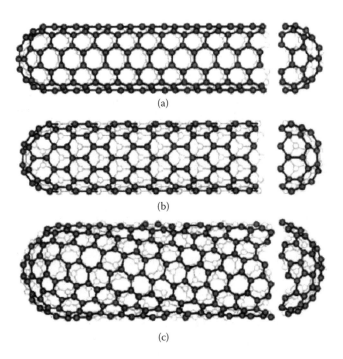

FIGURE 21.16 Schematic of nanotube morphologies: (a) armchair, (b) zig zag, (c) chiral. (From Harris, P.J.F., *Int Mat Rev*, 2004, 49(1):31–43. With permission.)

21.4.3 Properties

Based on their very high aspect ratio (>1000), carbon nanotubes possess an interesting combination of physical properties. Attempts to determine the physical properties of carbon-nanotubes have been conducted both experimentally and theoretically. Originally, researchers used the vibration of the nanotubes as a function of temperature to calculate a Young's modulus of 1 TPa [19]. Such high strength as compared to other materials, as shown in Figure 21.17, led to a pronounced interest in carbon nanotube-enabled materials. Common methods to measure the elastic properties of individual nanotubes include micro-Raman spectroscopy [20], thermal oscillations by transmission electron microscopy [19,21], and application of a force to a nanotube rope suspended across a pit using an atomic force microscope cantilever [22,23]. Other groups measured the properties of a rope and obtained an average value for each tube based on the number of nanotubes in the rope [24,25]. The experimental values measured ranged from significantly below theoretical values to values in agreement with theory. These methods have produced tensile modulus and strength values for single-wall nanotubes (SWNTs) and MWNTs ranging from 270 GPa to 1 TPa and 11 to 200 GPa, respectively [14].

Modeling such as molecular dynamics, empirical potentials and first-principles total energy, continuum shell, and empirical lattice have been used to describe the elastic properties of nanotubes [26–28]. The empirical lattice models, previously used to calculate the elastic properties of graphite, led to tensile modulus values in the range of 1 TPa for MWNTs and SWNTs [26]. These values compare well with the diamond structure and outperform conventional carbon fibers.

In addition to the mechanical properties, researchers have investigated the electrical and thermal conductivity of nanotubes. In a similar fashion, modeling has been used to determine the conductivity via the structure of the nanotubes as compared with graphite. Much of the theoretical work found that conductivity is greatly dependent on the small structural variations in the nanotubes [29]. For example Berber et al. found unusually high thermal conductance of 6600 W/m K at room temperature for a

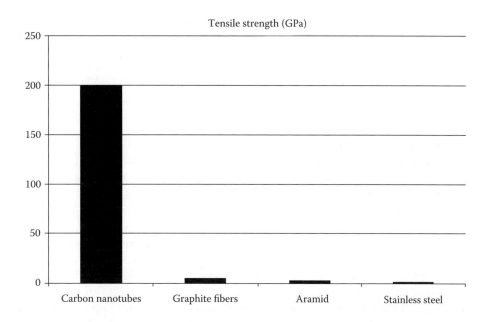

FIGURE 21.17 Tensile strength comparison of common engineering materials. (From Lau, K.T. and D. Hui, *Composites Part B. Eng*, 2002, 33(4):263–277. With permission.)

particular nanotube structure using nonequilibrium and equilibrium molecular dynamics [30]. In addition, the static electrical conductive and superconductive nature of nanotubes was modeled based on the conductivity of the graphite sheet structure [31]. Experimentally, Hone et al. reported slightly lower values in the range of 1750 to 5800 W/m K for the room temperature of a single nanotube rope by comparing the temperature drop across a constantan rod of known thermal conductance to a nanotube mat sample in series with the rod [32]. For electrical conductivity, experimental four-probe measurements of individual nanotubes showed resistivities in the range of 5.1×10^{-6} – 5.8 Ω cm (or conductivities ranging from 0.17 to 196,078 S/cm) [29].

21.4.4 From Nanoscale to Macroscale

Currently many methods of producing nanotube-enabled materials are under investigation. Some methods attempt to form fibers from nanotubes alone, whereas other procedures use a matrix to support the nanotubes. Without a support medium, research teams expect to achieve property values closer to the values of the individual nanotubes. However, the use of a matrix or binder often makes production of macroscale materials more feasible. Forming fibers directly from nanotube production methods, such as arc discharge [33] and CVD [34], has been investigated. Furthermore, research has shown the ability to create nanotube yarns by twisting them together as they are pulled out of a nanotube forest [35]. Other studies have proposed using super-acid solutions as a medium to support the fiber through a conventional spinning process followed by the removal of the medium [36,37]. However, the most common method has been to incorporate them in a polymer matrix as a reinforcing material. The combined polymer/nanotube structure is often referred to as a nanocomposite. Research has proposed that the nanotubes will provide load transfer in the same way noncontinuous fibers do in conventional composite systems. Experimentally, property improvement in nanotube-enabled materials has been marginal. Mechanical properties have varied from no change [5,38] to moderate increases (25 to 50% modulus increase; 80 to 150% toughness increase) [39,40], electrical properties in the range of 2 S/m (with a percolation threshold of 0.0025 wt%) [39], and 125% increase in thermal properties at room temperature [15]. However, optimal property improvements have not been achieved due to deficiencies in nanotube dispersion and alignment.

The remainder of this section will focus on polymer/nanotube composites, since composite research has been the primary route to realize nanotube properties on the macroscale.

21.4.5 Polymer/Nanotube Composites

21.4.5.1 Nanotube Dispersion and Orientation

Dispersion of SWNTs and higher concentrations of MWNTs into a polymer matrix has been one of the largest challenges to date [5], due to aggregation of the nanotubes as a result of the van der Waals interactions between individual tubes.

Two essential components for optimal reinforcement in particle-reinforced composite systems are filler dispersion and orientation. Consistent dispersion of the reinforcing material throughout the matrix leads to consistent load transfer from matrix to particle. Moreover, it can assist with the realization of a network for conductivity of electrical and thermal energy. Similarly, orientation in the direction of applied forces allows for greater load transfer. If the particle is oriented in a direction other than the direction of the applied force, the full potential of the particle cannot be realized. In addition, having all the particles oriented in the same direction allows for easier transfer of energy (electrical or thermal). Achieving consistent dispersion and orientation will allow for optimal property improvements.

Successfully dispersing these 1-D nanoparticles in a polymer matrix continues to be of interest. Researchers have used many different techniques to attempt to disperse nanotubes in polymer matrices including solution chemistry to functionalize the nanotube surface [41–46], the use of polymers to coat the nanotube surface [47], *in situ* polymerization of the nanocomposite [48,49], ultrasonic dispersion in solution [11,50], melt processing [7,51–55], the use of surfactants [46,56], electrospinning [57], gelation/crystallization [58], and electrode chemistry [59].

In gelation/crystallization experiments, the nanotubes were dissolved in a solvent, polymer solution was added, a gel was formed, the gel was formed into a film, and finally the solvent was evaporated [58]. Nanotube surface modifications used plasma treatment or chemical oxidation to attach functional groups. These groups allowed the nanotubes to bond better to the matrix and overcome the van der Waals interactions between nanotubes [45]. Melt compounding involves creating a preblend by dry mixing polymer powder with the nanotube powder. The preblend is fed into an extruder allowing for control of shear, temperature, and residence time. After the residence time, the material is extruded in the film or fiber form [52].

Good dispersion alone has shown moderate property improvements, but nanotube alignment, or orientation, has led to further improvements. Using melt compounding followed by melt drawing has shown a significant increase in mechanical properties [40,52]. Transmission and scanning electron microscopy have shown good dispersion and orientation using such methodology [40,51,52]. These improvements have been shown to increase until an optimal loading level and then decrease above this concentration [52,60]. In addition to mechanical drawing, the inherent conductive nature of carbon nanotubes has been utilized to induce alignment. For example, the application of a magnetic field to a nanocomposite sample during or after processing has shown nanotube alignment [61,62]. Again, the optimal properties, as predicted by such models as Halpin–Tsai, have not been achieved due to poor interfacial bonding, variations in nanotube structure, and nanotube properties variation.

21.4.5.2 SWNTs vs. MWNTs

Carbon nanotubes tend to agglomerate due to the strong attractive forces. SWNTs agglomerate more easily than MWNTs due to their size difference (i.e., greater surface area) and can form ropes or aligned bundles of SWNTs. SWNTs often require more specialization to produce than MWNTs. Therefore, the cost of purified SWNTs tends to be greater than that of MWNTs. MWNTs, on the other hand, have been found to demonstrate lower mechanical, electrical, and thermal properties due to the ability of the concentric nanotubes to slide past each other. Due to the inherent tube-within-a-tube structure, the MWNTs tend to have a larger diameter (10 nm) as compared with SWNTs (~1 nm). However, improvements in nanotube fabrication have led to MWNTs with more precise, smaller diameters. This may lead to nanotubes with improved properties over larger-diameter MWNTs with less agglomeration than SWNTs.

Recent research has shown that the tensile toughness of 1 wt% MWNTs in PMMA exhibited the same tensile properties as composites fabricated with 1 wt% SWNTs in PMMA [63] using the dispersion process outlined by Sabba et al. [64,65]. Both samples were drawn to a 12:1 draw ratio. In each case, the dispersion of the nanotubes was excellent (as shown by electron microscopy); however, no special dispersion treatment (other than melt mixing) was needed for the MWNT composite, making the MWNTs the preferable toughening particle for PMMA.

21.4.5.3 Matrix Material

Most researchers have incorporated nanotubes into thermoplastic polymers due to their ease of processibility. This allows fibers or films to be made and other objects to be easily melt-processed into molded parts. Research has been performed using both amorphous and semicrystalline polymers with varying degrees of success. The nanotubes have been shown to toughen and stiffen amorphous brittle materials, such as poly(methyl methacrylate) [7,52,66] as well as in semicrystalline polymers such as polypropylene [40]. In addition, the toughening and stiffening effects and morphological changes are the subject of discussion for semicrystalline polymers such as polypropylene. Research has suggested that nanotubes act as nucleating agents in these materials, leading to new crystallographic morphologies and ultimately increased strength [51,67–69].

Although most polymer/nanotube composite research has utilized thermoplastic matrices, research has also been conducted with thermosetting materials [12,70–78]. The thermosets are converted from liquid resin to hard rigid solids by chemical cross-linking, often at an elevated cure temperature [79]. The change in viscosity as a function of cross-linking can be problematic for optimizing dispersion and orientation. Research has been done to study mechanical [12,70,76–78], thermal [76], and electrical [71,72,74,78] properties of thermosetting polymer/nanotube composites, with various results. As with thermoplastics, nanotube dispersion [72,75] and alignment [73] are important factors in achieving property improvements. Of the various thermosetting polymers, epoxies have been the most commonly used.

Due to the large amount of work that has focused on the incorporation of nanotubes into polypropylene matrices, the last part of the nanotube section will focus on polypropylene/nanotube composites.

21.4.6 Polypropylene/Nanotube Composites

21.4.6.1 Polypropylene

Extensive structure–property relationships have been developed for polypropylene [80–82]. For example, polypropylene demonstrates polymorphism with α and β crystal forms (as well as a highly aligned amorphous region, called the mesophase [82]). Furthermore, methods of processing polypropylene into fibers, films, or other forms have been extensively studied. Melt processing polypropylene into fibers for applications such as textile fibers is the most common method studied. The availability, ease of use, and ease of manufacture has led to wide use of polypropylene for many industrial applications.

21.4.6.2 Crystallization

Studies have proposed that nanotubes could nucleate crystallization, increase crystallization rates, change the crystal form, and alter the overall percent crystallinity. Such changes have been proposed as possible mechanisms for property improvements, such as modulus. In such studies, both SWNTs [51,68,83,84] and MWNTs [67,69,85] have been used to study such crystallization behavior.

For SWNTs, the crystallization temperature, T_c, increases with increased loading levels of 1 to 20% SWNTs [51,68,83,84]. The studies proposed that the increase in T_c suggests that the nanotubes nucleate crystal growth [51,83,84]. Furthermore, optical microscopy shows large, well-defined grains in pure polypropylene but smaller grains in polypropylene/nanotube composites [51,83,84]. These smaller grains coupled with shifts observed by Raman and infrared spectroscopy further indicate nucleation via nanotube addition [83,84]. Nucleating agents can lead to the formation of β crystals in polypropylene [51]. However, wide angle x-ray diffraction showed only the common α polypropylene crystals [51]. Further crystallization kinetics experiments showed no changes in the Avarami exponent at a constant loading level [51]. However, changing the loading level changes the Avarami exponent [68]. Further crystallization experiments showed that the crystallization rate increased with the addition of nanotubes but did not increase with increased loading levels [51,68]. The addition of SWNTs to polypropylene increases T_c, nucleates α crystal growth, changes the Avarami exponent with increasing loading, and increases the rate of crystallization.

Similar to the case of SWNTs, MWNTs demonstrated an increase in the crystallization temperature, T_c [67,69,85]. Additionally, no changes in the overall percent crystallinity were observed with the MWNT-reinforced composites [67,69]. The combination of increased T_c and no change in the overall percent crystallinity led to the conclusion that the nanotubes are nucleating smaller polypropylene crystals at higher temperatures [67,69,85]. Bright field optical microscopy images showed further evidence of nanotubes acting as nucleation sites [69]. These images showed crystal lamella growing off the nanotubes perpendicular to the main polymer chain axis [69]. Crystallization kinetics experiments also showed an increase in the rate of crystallization using Avarami's equation [69,85]. In contrast to SWNTs, the addition of MWNTs led to a small decrease in the Avarami exponent [69]. Furthermore, Ozama's m also decreased with the addition of MWNTs, suggesting a change in the crystal dimensionality [67]. Similar to SWNTs, MWNTs increased the crystallization temperature, increased the crystallization rate, acted as nucleation sites, and did not increase the overall crystallinity. However, in contrast, MWNTs demonstrated a change in the dimensionality of the crystal growth.

21.4.6.3 Morphology

Closely related to the effects nanotubes have on the crystallinity of a semicrystalline polymer is the morphology of the composite. Much work has been done to determine how the addition of nanotubes could affect polymer morphology. In the case of polypropylene, the neat polymer exhibits a spherulitic morphology [67]. Scanning electron microscopy depicts the morphology as large circular grains growing and impinging on each other [67]. Using electron microscopy, researchers have studied the effects nanotubes have on this morphology. Different effects are reported for SWNTs [51,84] and MWNTs [67,69,86], similar to those seen when looking at crystallization.

Research has shown that the addition of SWNTs maintains the general spherulitic morphology at loading levels of 1 to 20% SWNTs [51,84]. Optical micrographs show large spherulites impinging on each other with neat polypropylene [51,84]. In contrast, a large quantity of smaller spherulites are seen in the micrographs of the nanocomposite samples [51,84]. However, as the loading level of nanotubes increases to 15% and greater, aggregation of the nanotubes begins to occur [84]. The aggregates show up as self-assembled bundles in the optical micrographs. Furthermore, Raman spectroscopy of 0, 5, 10, 15, and 20% samples showed trends in the D-band graphitic structure peak. The relative area under the peak increased significantly until it was 15% nanotubes at which point the increase slowed. Additionally, the peak width increased until 15% and stopped. Finally, the position shifted until 15% and then returned. From these results, the researchers concluded that until 15%, the polymer was intercalated into the nanotubes [84]. Above this point, the nanotubes became saturated and did not allow more polymer to penetrate into the nanotubes. At low loading levels, polymer intercalates SWNTs, leading to crystal nucleation and the large quantity of smaller grains seen in optical micrographs. Above the optimal level (15%), the nanotubes do not intercalate and instead aggregate into self-assembled bundles. As discussed above, this is supported by the constant Avarami exponent denoting 3-D crystal growth at low concentrations.

In contrast, MWNTs were shown to have a lower value for Ozama's *m,* denoting change from 3-D to 2-D crystal growth [67]. Optical micrographs of polypropylene/MWNTs exhibited fibrillar morphology [67]. The normal spherulitic morphology of polypropylene has 3-D growth, whereas fibrillar morphology exhibits 2-D growth. Although the morphology changed, the sample still appeared homogenous throughout, revealing good dispersion [67]. In another study using highly oriented nanocomposite films and fibers, a different morphology was observed [69]. Due to the orientation of the molecular chains imparted by drawing the sample, the MWNTs were also oriented in the direction of the drawing [69]. However, the crystal lamella were seen growing perpendicular to the molecular axis and the oriented nanotubes [69]. Finally, another study looked closely at the interface between chlorinated polypropylene and MWNTs [86]. Close to the surface of the nanotubes, a crystalline layer was observed and measured. The study suggested that this layer increased polymer/nanotube adhesion and ultimately led to property improvements, such as increased toughness and modulus. As with crystallization, morphological changes have been proposed as possible factors in the improved mechanical properties of a polymer/nanotube composite system.

21.4.6.4 Mechanical Properties

Research has shown mechanical property improvements with the addition of both SWNTs and MWNTs to polypropylene. As discussed earlier, dispersion of the nanotubes plays a large role in the property improvements. One study showed that the addition of 1 wt% SWNTs via melt blending did not significantly improve mechanical properties [51]. In the study, scanning electron microscopy images showed micrometer-scale aggregates and nanotube ropes within the sample. Some research has suggested that the aggregates could act as stress concentrators inhibiting property improvement. However, from the slight increases, the aggregates either were not big enough to create stress concentrations, or the strength of the ropes was greater than the losses from the stress concentrations. The research concluded that better nanotube dispersion would lead to more rope, less aggregates, and ultimately significant mechanical property improvements.

In another study, significant property improvements were observed when adding SWNTs to polypropylene [8]. The nanotubes were first dispersed in the polymer using solution processing. After removing the solvent, the fibers were meltspun and postdrawn for improved dispersion and orientation. The addition of 0.5 and 1 wt% demonstrated increasing tensile strength and modulus values. At 1 wt%, the tensile strength and modulus were 40 and 55% higher than the neat polypropylene fibers. However, additions of 1.5 and 2 wt% made spinning the fibers difficult and led to lower mechanical properties.

A similar maximum in mechanical properties as a function of nanotube loading was also noted in another study in both conventional tensile experiments and dynamic mechanical analysis [87]. The study looked at the properties of both as-spun fiber and postdrawn polypropylene/SWNT fibers. Additionally, the studies investigated any differences in low meltflow rate and high meltflow rate polypropylene. The basic tensile qualitative behavior of neat polypropylene (high initial modulus, sharp yield, and stretching to very long elongations) was observed in the as-spun nanocomposites. However, the nanocomposites exhibited lower elongation to break values. Increased tensile strength was noted for the low meltflow rate samples and the opposite for the high meltflow rate samples. Furthermore, the postdrawn low flow rate samples increased in tensile strength up to 1% and then decreased. The higher concentrations caused the fibers to become brittle and difficult to draw. All the high meltflow rate samples exhibited decreasing ultimate tensile strength. The dynamic mechanical analysis produced results similar to the conventional tensile tests. Within temperature limits, the storage modulus, E', increased with increasing nanotube additions. At the higher temperatures, E' of the postdrawn samples began to decrease. The concentration of 1% was found to be the optimal value for E'. Moreover, as evidenced by the tensile test, the postdrawing process created a far stiffer material. In contrast, the loss modulus, E'', show little to no discernable trend as related to polypropylene molecular weight or nanotube concentration.

Research into polypropylene/MWNTs has been performed to a lesser extent than SWNTs. In one study, functionalized nanotubes were added to chlorinated polypropylene to create composite films [86]. The study noted an upward shift in the yield point on the stress–strain curve and consequently increased yield strength. Additionally, the Young's modulus, ultimate tensile strength, and toughness increased with increasing nano-tube volume fraction by factors of 3.1, 3.9, and 4.4, respectively. From the data and previously derived composite models, mathematical models based on volume fraction were developed for the Young's modulus and composite strength. In another study, polypropylene/MWNT powder was created using a pan-milling process [85]. The pan milling of the neat polypropylene reduced the elongation to break significantly. However, the addition of nanotubes increased the elongation to break (over pan-milled polypropylene) through crystal structure changes and induced polypropylene orientation. Again an optimal 1% nanotube loading exhibited increased Young's modulus and yield strength. The pan milling allows for better dispersion, reduces the number of defects, curvature, and entanglements, and enhances the polymer/nanotube adhesion.

Most recently, Dondero and Gorga [40] showed that 0.25 wt% MWNTs in drawn polypropylene (with a 12:1 draw ratio) increased tensile toughness by 32% and modulus by 138% over drawn polypropylene. Differential scanning calorimetry supported the conclusion of modulus increases via load transfer rather than increased crystallinity. However, wide angle x-ray diffraction indicated a crystal structure change from α and mesophase to α phase only with the addition of nanotubes in oriented samples. The activation energy, calculated from dynamic mechanical experiments, also revealed an increasing trend to a maximum as a function of nanotube loading. In drawn samples, transmission electron microscopy and scanning electron microscopy showed well-dispersed and highly oriented nanotubes.

21.4.7   Conclusion

The mechanical, electrical, and thermal properties of individual nanotubes have been shown to be exceptionally high in both theory and experiment. Therefore, there exists a tremendous potential to realize the nanoscale properties of nanotubes at the macroscale. To create such macroscale materials, many issues surrounding the incorporation of nanotubes into a matrix, strategies for property improvement, and the mechanisms responsible for those property improvements still remain. Since only moderate success has been achieved over the last 15 years, researchers continue to investigate strategies to optimize the fabrication of nanotube-enabled materials to achieve both improved mechanical and improved transport properties.

References

1. Iijima, S., Helical microtubules of graphitic carbon. *Nature*, 1991, 354(6348):56–58.
2. Iijima, S. and T. Ichihashi, Single-shell carbon nanotubes of 1-nm diameter. *Nature*, 1993, 363(6430):603–605.
3. Bower, C., et al., Deformation of carbon nanotubes in nanotube-polymer composites. *Appl Phys Lett*, 1999, 74(22):3317–3319.
4. Cooper, C.A., et al., Distribution and alignment of carbon nanotubes and nanofibrils in a polymer matrix. *Comp Sci Tech*, 2002, 62(7,8):1105–1112.
5. Haggenmueller, R., et al., Aligned single-wall carbon nanotubes in composites by melt processing methods. *Chem Phys Lett*, 2000, 330(3,4):219–225.
6. Jin, L., C. Bower, and O. Zhou, Alignment of carbon nanotubes in a polymer matrix by mechanical stretching. *Appl Phys Lett*, 1998, 73(9):1197–1199.
7. Jin, Z., et al., Dynamic mechanical behavior of melt-processed multi-walled carbon nanotube/poly(methyl methacrylate) composites. *Chem Phys Lett*, 2001, 337(1–3):43–47.
8. Kearns, J.C. and R.L. Shambaugh, Polypropylene fibers reinforced with carbon nanotubes. *J Appl Polym Sci*, 2002, 86(8):2079–2084.
9. Lozano, K. and E.V. Barrera, Nanofiber-reinforced thermoplastic composites. I. Thermoanalytical and mechanical analyses. *J Appl Polym Sci*, 2001, 79(1):125–133.

10. Potschke, P., T.D. Fornes, and D.R. Paul, Rheological behavior of multiwalled carbon nanotube/polycarbonate composites. *Polymer*, 2002, 43(11):3247–3255.
11. Safadi, B., R. Andrews, and E.A. Grulke, Multiwalled carbon nanotube polymer composites: synthesis and characterization of thin films. *J Appl Polym Sci*, 2002, 84(14):2660–2669.
12. Schadler, L.S., S.C. Giannaris, and P.M. Ajayan, Load transfer in carbon nanotube epoxy composites. *Appl Phys Lett*, 1998, 73(26):3842–3844.
13. Dresselhaus, M.S., et al., Electronic, thermal and mechanical properties of carbon nanotubes. *Phil Trans Roy Soc London A*, 2004, 362(1098):2065–2098.
14. Lau, K.T. and D. Hui, The revolutionary creation of new advanced materials — carbon nanotube composites. *Composites. Part B. Eng*, 2002, 33(4):263–277.
15. Harris, P.J.F., Carbon nanotube composites. *Int Mat Rev*, 2004, 49(1):31–43.
16. Liu, J., S. Fan, and H. Dai, Recent advances in methods of forming carbon nanotubes. *MRS Bull*, 2004:244–250.
17. Thostenson, E.T., Z. Ren, and T.W. Chou, Advances in the science and technology of carbon nanotubes and their composites: a review. *Comp Sci Tech*, 2001, 61(13):1899–1912.
18. Schlittler, R.R., et al., Single crystals of single-walled carbon nanotubes formed by self-assembly. *Science*, 2001, 292(5519):1136–1139.
19. Treacy, M.M.J., T.W. Ebbesen, and J.M. Gibson, Exceptionally high Young's modulus observed for individual carbon nanotubes. *Nature*, 1996, 381(6584):678–680.
20. Lourie, O., D.M. Cox, and H.D. Wagner, Buckling and collapse of embedded carbon nanotubes. *Phys Rev Lett*, 1998, 81(8):1638–1641.
21. Krishnan, A., et al., Young's modulus of single-walled nanotubes. *Phys Rev B*, 1998, 58(20):14013–14019.
22. Salvetat, J.P., et al., Elastic and shear moduli of single-walled carbon nanotube ropes. *Phys Rev Lett*, 1999, 82(5):944–947.
23. Walters, D.A., et al., Elastic strain of freely suspended single-wall carbon nanotube ropes. *Appl Phys Lett*, 1999, 74(25):3803–3805.
24. Li, F., et al., Tensile strength of single-walled carbon nanotubes directly measured from their macroscopic ropes. *Appl Phys Lett*, 2000, 77(20):3161–3163.
25. Pan, Z.W., et al., Tensile tests of ropes of very long aligned multiwall carbon nanotubes. *Appl Phys Lett*, 1999, 74(21):3152–3154.
26. Lu, J.P., Elastic properties of single and multilayered nanotubes. *J Phys Chem Solids*, 1997, 58(11):1649–1652.
27. Robertson, D.H., D.W. Brenner, and J.W. Mintmire, Energetics of nanoscale graphitic tubules. *Phys Rev B*, 1992, 45(21):12592–12595.
28. Yakobson, B.I., C.J. Brabec, and J. Bernholc, Nanomechanics of carbon tubes: instabilities beyond linear response. *Phys Rev Lett*, 1996, 76(14):2511–2514.
29. Ebbesen, T.W., et al., Electrical conductivity of individual carbon nanotubes. *Nature*, 1996, 382(6586):54–56.
30. Berber, S., Y.K. Kwon, and D. Tomanek, Unusually high thermal conductivity of carbon nanotubes. *Phys Rev Lett*, 2000, 84(20):4613–4616.
31. Benedict, L.X., et al., Static conductivity and superconductivity of carbon nanotubes — relations between tubes and sheets. *Phys Rev B*, 1995, 52(20):14935–14940.
32. Hone, J., et al., Thermal conductivity of single-walled carbon nanotubes. *Phys Rev B*, 1999, 59(4):R2514–R2516.
33. Li, H.J., et al., Direct synthesis of high purity single-walled carbon nanotube fibers by arc discharge. *J Phys Chem B*, 2004, 108(15):4573–4575.
34. Li, Y.L., I.A. Kinloch, and A.H. Windle, Direct spinning of carbon nanotube fibers from chemical vapor deposition synthesis. *Science*, 2004, 304(5668):276–278.
35. Zhang, M., K.R. Atkinson, and R.H. Baughman, Multifunctional carbon nanotube yarns by downsizing an ancient technology. *Science*, 2004, 306(5700):1358–1361.

36. Ericson, L.M., et al., Macroscopic, neat, single-walled carbon nanotube fibers. *Science*, 2004, 305(5689):1447–1450.

37. Zhou, W., et al., Single wall carbon nanotube fibers extruded from super-acid suspensions: preferred orientation, electrical, and thermal transport. *J Appl Phys*, 2004, 95(2):649–655.

38. Sennett, M., et al., Dispersion and alignment of carbon nanotubes in polycarbonate. *Appl Phys A: Mat Sci Process*, 2003, 76(1):111–113.

39. Andrews, R. and M.C. Weisenberger, Carbon nanotube polymer composites. *Current Opin Solid State Mater Sci*, 2004, 8(1):31–37.

40. Dondero, W.E. and R.E. Gorga, Morphological and mechanical properties of carbon nanotube polymer composites via melt compounding, *J Pol Sci B–Pol Phys*, 44(5): 864–878, 2006.

41. Chen, J., et al., Solution properties of single-walled carbon nanotubes. *Science*, 1998, 282(5386):95–98.

42. Mitchell, C.A., et al., Dispersion of functionalized carbon nanotubes in polystyrene. *Macromolecules*, 2002, 35(23):8825–8830.

43. Bubert, H., et al., Characterization of the uppermost layer of plasma-treated carbon nanotubes. *Diamond Rel Mater*, 2003, 12(3–7):811–815.

44. Eitan, A., et al., Surface modification of multiwalled carbon nanotubes: toward the tailoring of the interface in polymer composites. *Chem Mater*, 2003, 15(16):3198–3201.

45. Jang, J., J. Bae, and S.H. Yoon, A study on the effect of surface treatment of carbon nanotubes for liquid crystalline epoxide-carbon nanotube composites. *J Mater Chem*, 2003, 13(4):676–681.

46. Shaffer, M.S.P., X. Fan, and A.H. Windle, Dispersion and packing of carbon nanotubes. *Carbon*, 1998, 36(11):1603–1612.

47. Star, A., et al., Preparation and properties of polymer-wrapped single-walled carbon nanotubes. *Angewandte Chemie, Int. Ed.*, 2001, 40(9):1721–1725.

48. Jia, Z., et al., Study on poly(methyl methacrylate)/carbon nanotube composites. *Mater Sci Eng A*, 1999, A271:395–400.

49. Deng, J., et al., Carbon nanotube-polyaniline hybrid materials. *Eur Polym J*, 2002, 38(12):2497–2501.

50. Qian, D., et al., Load transfer and deformation mechanisms in carbon nanotube-polystyrene composites. *Appl Phys Lett*, 2000, 76(20):2868–2870.

51. Bhattacharyya, A.R., et al., Crystallization and orientation studies in polypropylene/single wall carbon nanotube composite. *Polymer*, 2003, 44(8):2373–2377.

52. Gorga, R.E. and R.E. Cohen, Toughness enhancements in poly(methyl methacrylate) by addition of oriented multiwall carbon nanotubes. *J Polym Sci Part B-Poly Phys*, 2004, 42(14):2690–2702.

53. Potschke, P., A.R. Bhattacharyya, and A. Janke, Melt mixing of polycarbonate with multiwalled carbon nanotubes: microscopic studies on the state of dispersion. *Eur Polym J*, 2004, 40(1):137–148.

54. Siochi, E.J., et al., Melt processing of SWCNT-polyimide nanocomposite fibers. *Composites Part B. Eng*, 2004, 35(5):439–446.

55. Tang, W.Z., M.H. Santare, and S.G. Advani, Melt processing and mechanical property characterization of multi-walled carbon nanotube/high density polyethylene (MWNT/HDPE) composite films. *Carbon*, 2003, 41(14):2779–2785.

56. Gong, X., et al., Surfactant-assisted processing of carbon nanotube/polymer composites. *Chem Mater*, 2000, 12(4):1049–1052.

57. Dror, Y., et al., Carbon nanotubes embedded in oriented polymer nanofibers by electrospinning. *Langmuir*, 2003, 19(17):7012–7020.

58. Bin, Y.Z., et al., Development of highly oriented polyethylene filled with aligned carbon nanotubes by gelation/crystallization from solutions. *Macromolecules*, 2003, 36(16):6213–6219.

59. Chen, G.Z., et al., Carbon nanotube and polypyrrole composites: coating and doping. *Adv Mater*, 2000, 12(7):522–526.

60. Wilbrink, M.W.L., et al., Toughenability of nylon-6 with $CaCO_3$ filler particles: new findings and general principles. *Polymer*, 2001, 42:10155–10180.

61. Fischer, J.E., et al., Magnetically aligned single wall carbon nanotube films: preferred orientation and anisotropic transport properties. *J Appl Phys*, 2003, 93(4):2157–2163.

62. Kimura, T., et al., Polymer composites of carbon nanotubes aligned by a magnetic field. *Adv Mater*, 2002, 14(19):1380–1383.

63. Gorga, R.E., et al., Mechanical properties of single-wall and multi-wall carbon nanotube composites. Submitted 2007.

64. Sabba, I. and E.L. Thomas, High-concentration dispersion of single-wall carbon nanotubes. *Macromolecules*, 2004, 37(13):4815–4820.

65. Sabba, Y. and E.L. Thomas, High-concentration dispersion of single-wall carbon nanotubes. *Macromolecules*, 2004, 37(13):4815–4820.

66. Zeng, J.J., et al., Processing and properties of poly(methyl methacrylate)/carbon nano fiber composites. *Composites Part B. Eng*, 2004, 35(2):173–178.

67. Assouline, E., et al., Nucleation ability of multiwall carbon nanotubes in polypropylene composites. *J Polym Sci Part B-Polym Phys*, 2003, 41(5):520–527.

68. Grady, B.P., et al., Nucleation of polypropylene crystallization by single-walled carbon nanotubes. *J Phys Chem B*, 2002, 106(23):5852–5858.

69. Sandler, J., et al., Crystallization of carbon nanotube and nanofiber polypropylene composites. *J Macromol Sci Phys*, 2003, B42(3–4):479–488.

70. Xiao, K.Q. and L.C. Zhang, The stress transfer efficiency of a single-walled carbon nanotube in epoxy matrix. *J Mater Sci*, 2004, 39(14):4481–4486.

71. Valentini, L., et al., Dielectric behavior of epoxy matrix/single-walled carbon nanotube composites. *Comp Sci Tech*, 2004, 64(1):23–33.

72. Sandler, J., et al., Development of a dispersion process for carbon nanotubes in an epoxy matrix and the resulting electrical properties. *Polymer*, 1999, 40(21):5967–5971.

73. Martin, C.A., et al., Electric field-induced aligned multi-wall carbon nanotube networks in epoxy composites. *Polymer*, 2005, 46(3):877–886.

74. Martin, C.A., et al., Formation of percolating networks in multi-wall carbon-nanotube-epoxy composites. *Comp Sci Tech*, 2004, 64(15):2309–2316.

75. Liao, Y.H., et al., Investigation of the dispersion process of SWNTs/SC-15 epoxy resin nanocomposites. *Mater Sci Eng*, 2004, 385(1–2):175–181.

76. Lau, K.T., et al., Thermal and mechanical properties of single-walled carbon nanotube bundle-reinforced epoxy nanocomposites: the role of solvent for nanotube dispersion. *Comp Sci Tech*, 2005, 65(5):719–725.

77. Gojny, F.H., et al., Carbon nanotube-reinforced epoxy-compo sites: enhanced stiffness and fracture toughness at low nanotube content. *Comp Sci Tech*, 2004, 64(15):2363–2371.

78. Allaoui, A., et al., Mechanical and electrical properties of a MWNT/epoxy composite. *Comp Sci Tech*, 2002, 62(15):1993–1998.

79. Hull, D. and T.W. Clyne, An Introduction to Composite Materials. 2nd ed. Cambridge Solid State Science Series, S. Suresh, D.R. Clarke, I.M. Ward, Eds., Cambridge University Press, Cambridge, UK, 1996.

80. Morawetz, H., ed. *Polymer Monographs*. Vol. 2. Gordon and Breach Science Publishers, New York, 1968.

81. Karger-Kocsis, J., ed. *Polypropylene: An A-Z Reference*. Kluwer Academic Publishers, Boston, 1999.

82. Broda, J., Polymorphism in polypropylene fibers. *J Appl Polym Sci*, 2003, 89(12):3364–3370.

83. Valentini, L., et al., Effects of single-walled carbon nanotubes on the crystallization behavior of polypropylene. *J Appl Polym Sci*, 2003, 87(4):708–713.

84. Valentini, L., et al., Morphological characterization of single-walled carbon nanotubes-PP composites. *Comp Sci Tech*, 2003, 63(8):1149–1153.

85. Xia, H.S., et al., Preparation of polypropylene/carbon nanotube composite powder with a solid-state mechanochemical pulverization process. *J Appl Polym Sci*, 2004, 93(1):378–386.

86. Coleman, J.N., et al., High-performance nanotube-reinforced plastics: understanding the mechanism of strength increase. *Adv Funct Mater*, 2004, 14(8):791–798.

87. Moore, E.M., et al., Enhancing the strength of polypropylene fibers with carbon nanotubes. *J Appl Polym Sci*, 2004, 93(6):2926–2933.

21.5   Surface Activation

Stephen Michielsen

Often it is difficult to make nanofibers that have both mechanical properties *and* the desired surface properties in a one-step process. This can easily be remedied by making fibers that have the desired mechanical properties and modifying their surfaces to have the desired surface properties. However, in using this approach, one must be cognizant of the thickness of the surface-modified layer. Many techniques are routinely practiced in which the surface modification is several micrometers thick. On small diameter fibers, these "thick" coatings dominate the mechanical properties and increase the size of the nanofibers into the micrometer range. In addition, these coatings can block pores or channels in nanofiber webs, greatly altering the barrier and filtration properties.

It is also worth noting that many of the desired surface properties depend only on the material in the outer 0.5 to 5 nm. For example, surface tension depends on 0.1 to 1.0 nm.[1–3] Friction can be controlled by the outer 1 to 5 nm.[4] Static charge generation or discharge is primarily controlled by the outer 1 to 3 nm.[5] We would expect adsorption of proteins and other biological materials to depend on this same thickness. For surface properties that depend only on this outer 1 to 5 nm, any additional material applied to the surface does not provide additional benefit, but does increase the fiber diameter, the fiber stiffness, weight, and cost while reducing pore sizes.

In the remainder of this section, we will discuss several approaches for modifying the surface. There are several methods that can be used to arrange these techniques. We have elected to separate them according to the size of the molecular units that provide this modification.

21.5.1   Small Molecules (Molecular Weight < 1000 G/mol = 1 kD)

There are several methods that can be used to apply small molecules to the surface. These include topical coatings, self-assembled monolayers (SAM), low-temperature plasmas, corona treatment, and ultraviolet (UV) grafting.

Topical coatings have been applied to fibers since the first synthetic fibers were made. Typical surface treatments include the application of antistatic agents, lubricants, and antimicrobial agents. In these treatments, the small molecules are applied as neat oils, dissolved in suitable solvents and applied, or converted to water emulsions and applied. With careful control of concentration and application rate, monolayers of these materials can be applied. A coating can remain as a uniform coating if the surface-modifying material wets the surface or it can withdraw into droplets on the surface if it does not wet the surface well. Although this process is fast and inexpensive, the material can easily be washed off the surface. Thus this approach is best for applications where the treatment is only desired for short periods of time or where the fibers are never exposed to water or solvents.

Self-assembling monolayers are applied to surfaces to form dense, uniform, 1 to 2 nm thick coatings. There are two major approaches using alkyl-trichlorosilanes on silicon/silicon dioxide and alkanethiols or dialkyl disulfides chemisorbed on gold.[1] In this approach, the surface active agents are allowed to adsorb and react with the surface. In a typical application, the molecules are all the same length, and, since they adsorb alongside of each other, they form a dense coating in which all the chains line up. The thickness of the layer is directly controlled by the length of the alkyl chains when fully stretched. This approach suffers from several shortcomings. First, the surface has to be coated with silicon or gold (or other suitable metal.) Then, only certain materials can pack tightly enough to form SAM. This restricts the range of surface treatments that can readily be attached. However, this approach has been used as a substrate for further modification. It will not be discussed further in this chapter since it is currently impractical for most fiber applications.

Plasma, corona, UV treatments all involve the creation of highly reactive species that can react with the surface. In cold plasma treatment, a radio frequency electric field is used to strip electrons from the molecules in the vapor phase. Corona treatment uses high electric fields to do the same while UV uses high-energy light to excite a UV-absorbing molecule, which then transfers the energy to another molecule that is designed to break at a weak point, creating two highly reactive radicals. In each of these processes, a portion of the resulting ions and radicals collide with the fiber surface and react with it. Thus, the small molecules form covalent bonds with the surface, modifying its properties. Since single molecules bond to the surface, very thin layers can be attached. However, the reactive species are so reactive that they can react with each other as well as the fiber surface. This can result in a complex, thick layer that is difficult to characterize. Nonetheless, when properly controlled, very thin layers can be deposited.

Both plasma treatment and corona treatment can be performed at atmospheric pressure or under vacuum. Atmospheric plasma reduces cost while vacuum plasma increases specificity. When atmospheric pressure is used, the treatment gas is usually air and the surface is typically modified by oxygen or a water-derived species. Thus the surface usually consists of the original material, ketones, alcohols, and carboxylic acids. If the surface was initially hydrophobic, these treatments can readily make them hydrophilic. By flooding the surface with a gas other than air, it is possible to attach other materials to the surface. In this process, the substrate is simultaneously exposed to a moderate vacuum and atmospheric pressure of the desired treatment gas. This increases the likelihood of treatment with the desired gas, but it is very difficult to remove all of the trapped air and moisture from a fibrous structure. Thus the resulting surfaces usually are a combination of the treatment gas derivatives and air-based derivatives. These approaches have been used to apply fluorochemicals, carboxylic acids, and hydroxyl groups. The use of this approach to deposit polymers is described in a subsequent section.

UV-photoinitiated chemical modification is nearly always done in a vacuum to enhance yield. The photo-absorber is often benzophenone. After absorbing a UV-photon, the photo-absorber extracts a hydrogen atom from the surface or from other molecules in the gas phase. This creates two radicals that can react further, and each fragment can graft to the surface. Again, the reactive species is so reactive that multiple layers can deposit if great care is not taken.

These processes can be used for a wide variety of surface modifications. However, except in the simplest cases, it can be difficult to deposit uniform, thin (<100 nm) coatings. For very thin coatings, it is nearly impossible to obtain uniform coverage. This can be due to different functional groups on the substrate that react differently with the ions and radicals produced above or due to the highly reactive species that react with the first suitable material they collide with. Thus the surface is often mottled with regions consisting of the desired treatment mixed with regions that are untreated. Only when the coatings are relatively thick do the coatings become uniform.

21.5.2 Polymers or Large Molecules (Molecular Weight > 1 kD)

By attaching large molecules capable of covering a large area on the surface, it is possible to get relatively thin coatings *and* uniform surface treatments. There are three major categories of modifying the surface with polymers: topical application, growing the polymer from the surface by polymerizing monomers from an initiator that is attached to the surface (grafting-from), and grafting a preformed polymer to the surface (grafting-to).

Topical coatings can again be used to deposit thin polymeric coatings. These suffer from the same difficulties as described previously. However, to stabilize the structure, polymers can be crosslinked after deposition. This provides an inexpensive method for applying a relatively uniform coating. This approach has been used to attach antimicrobial agents.[6] One shortcoming is that, since the coating may not be covalently attached to the surface but merely surrounding it, any defect in the coating can lead to delamination and loss of the coating.

Graft-from — In the *graft-from* process, the surface is first activated using a cold plasma, corona treatment, or UV light. Next, the surface is exposed to the monomer, and polymerization occurs. Typically, dense, brush-like polymer structures grow from the surface. There are three main characteristics of the polymers grown in this process. First, the density of the graft sites is much larger than the size of the growing polymer. Second, the growing polymer avoids adjacent growing polymers. Third, a considerable amount of homopolymerization (ungrafted polymer) occurs. The first two effects result in the grafted polymer chains growing in an elongated and nonentangled form, while the third adds weight and bulk. Typical thicknesses of "graft-from" coatings are greater than 40 nm.[3] They have been used to alter the surface tension, protein adsorption, friction, zeta potential, and antistatic behavior.

Although the graft-from approach is very versatile, it is not compatible with many fiber-based processes. Often the graft-from techniques require specific gaseous atmospheres that differ substantially from air, or they may be dipped into highly reactive liquids. They often require several minutes to hours to carry out. However, synthetic fiber spinning is practiced commercially at speeds of 1000 to 8000 m/min. In these cases, the treatment should be completed in a fraction of a second. Although fabric processing occurs much more slowly, the fibers and yarns contained in the fabric trap a large amount of air, which often must be removed before a suitable graft-from polymerization can occur. These problems can be reduced if the surface-modifying polymer is synthesized separately and then grafted to the surface.

Graft-to — In this approach, preformed polymers are grafted to reactive sites on the surface of the substrate. The three main requirements are that the polymer to be grafted contains suitable reactive sites, that the substrate has a sufficient number of complementary reactive sites, and that the grafting reaction occurs fast enough and in a suitable environment to be compatible with textile processes. The thickness of these layers depends on the molecular weight of the "graft-to" polymer, the density of reactive surface sites, and the strength of the adsorption of the "graft-to" polymer to the surface. If the strength of the adsorption is too high, the surface-modifying polymer adsorbs irreversibly and it cannot spread on the surface. In this case, the coverage of the surface is limited by the "jamming limit" to approximately 54.7%.[7] Similarly, if the graft-site density is too high and if the reaction occurs quickly, the coverage is again limited by the jamming limit. On the other hand, if the adsorption energy is low, the polymer can adsorb, desorb, and spread on the surface. With a suitable balance between these factors and a delayed grafting reaction, nearly complete surface coverage can be obtained.[8]

Recently, in studies in which they grafted poly(acrylic acid) (PAA), onto nylon-6,6 films, Thompson and Michielsen[9] showed that if adsorption and grafting occurred simultaneously, coverage of the surface was limited to 55%, in agreement with the jamming limit. However, if they were allowed to adsorb first and grafting was delayed for several days, they could achieve surface coverage of ~72%. On the other hand, if the adsorption was weak and the number of reactive sites in the grafted polymer was low, surface coverage of ~75% could be attained. Michielsen used this later approach to permanently attach ~5 nm thick lubricant layer to the surface of fibers.[4] Thompson and Michielsen[9] also found that, even when adsorption was strong, grafting of nonadsorbing or weakly adsorbing small molecules to PAA could alter the balance between adsorption and desorption. If PAA was tethered to the surface at only a few points, the grafted PAA could spread across the entire surface (100% coverage) by reducing the number of strongly adsorbing groups after adsorption and tethering to the surface.

Once a polymer has been grafted to the surface, if there are residual reactive sites, they can be used to further modify the surface. By choosing a "graft-to" polymer with many reactive sites, one can attach a large number of materials to the surface. In the PAA case above, a single reactive amino site on nylon was converted to more than 1000 reactive carboxylic acid sites. Thus, the grafted polymer can act as a scaffold for further modification *and* a reactive site amplifier. Michielsen and coworkers[10–12] have used this approach to graft amine-containing small molecules to PAA that had previously been grafted to nylon. They were able to attach $-SO_3^-$, $-NH_2$, $-C_nF_{2n+1}$, and $-C_2H_{2n+1}$ groups. They also used this procedure to attach antimicrobial agents.[13] Analyses of the surface of these materials indicated that the coating thickness was less than 12 nm.[8] The surfaces could be provided with a wide range of properties — hydrophobic, hydrophilic, acidic, basic, and neutral. They could attract dyes or repel them.[12] A careful choice of grafted-to polymer and subsequent modification by small molecules allows a near limitless variety of modifications that can be performed in 1 to 10 nm thick coatings that, except for the smallest fibers, will not contribute to their mechanical performance. In addition, they can be applied so that they do not occlude pores or significantly alter the flow of fluids through small channels in nanofiber webs.

Several groups have grafted polymers to various surfaces to alter the boundary friction.[14] In these cases, the lubricant adsorbs weakly on the substrate to reduce the shear strength, but is grafted to the surface to prevent the lubricant from coming off and leaving unprotected regions. Michielsen[4] grafted polydimethyl siloxane (PDMS) and polyethylene (PE) to the surface of nylon films and fibers. The PDMS molecules contained carboxylic acid or epoxy groups while the PE contained anhydride groups. All three groups reacted readily with the amino end-groups of nylon-6,6. At elevated temperatures, grafting occurred in less than 10 msec. The PDMS lubricant could even be grafted to the surface of nylon fibers during the meltspinning process. The boundary friction coefficient could be varied from 0.3 for untreated nylon to 0.05 for a PDMS that exhibited 75% coverage. The friction coefficient varied linearly with the surface area fraction:

$$\mu = \mu_n(1 - \phi_{PDMS}) \tag{21.16}$$

where μ is the friction coefficient for the PDMS-modified fiber, μ_n is the friction coefficient for the unmodified nylon fiber, and ϕ_{PDMS} is the fraction of the surface covered by grafted PDMS molecules. ϕ_{PDMS} was found to depend on the radius of gyration, R_G, of the PDMS molecule and on the density of amino groups on the surface of nylon:

$$\phi_{PDMS} = \pi R_G^2 / A \tag{21.17}$$

where A is the average area per amino group, i.e., 1/amino group density. In these experiments, PDMS grafted to the surface was believed to be in the mushroom regime where the individual molecules take a hemispherical shape and do not interpenetrate. Since there is no interpenetration and no strong adhesion forces, the shear strength of the interface between two fibers treated in this manner is low and hence the friction is low. The PDMS- and PE-grafted layers were found to be only 1 to 5 nm thick, which was comparable to their R_Gs.

In other studies, Bakhshaee[12] and Michielsen[11] attached PAA to the surface of nylon-6,6 and subsequently converted them to $-SO_3^-$ and $-NH_3^+$ by grafting taurine or ethylene diamine to the residual $-COOH$ groups. They found that the $-NH_3^+$ moieties increased acid dye uptake by ~10% as they acted as additional dye sites while the $-SO_3^-$ moieties reduced dye uptake by 20% as they repelled acid dyes. Again, these groups were located in less than 10 nm thick surface grafted layers. These are substantial changes for such a thin coating.

Using a similar approach, Thompson and Michielsen[10] were able to convert PAA that had been grafted to nylon into a fluorochemical-based surface. The surface was segregated into several layers, a fluorochemical-rich outer layer, a PAA-rich intermediate layer, and the nylon substrate. The overall thickness was less than 12 nm.[8] This treatment imparted the wetting behavior of fluorochemicals, a soft feel to the fibers, reduced dye uptake, and improved antistain behavior.

In another set of experiments, Sherrill et al.[13] attached protoporphyrin IX and zinc protoporphyrin IX to the surface of nylon fibers by grafting them to PAA that had been previously grafted to nylon and acted as a scaffold for attachment of the porphyrins. Bozja et al.[15] showed that these fibers exhibited a light-activated antimicrobial effect that efficiently killed *S. aureus*.

Considerable work remains to be done to understand how to control surface coverage and obtain the desired surface properties. Nonetheless, surface modification of fibers with treatments that are only a few nanometers thick promises to greatly enhance the performance of the fibers.

References

1. Whitesides, G.M., Laibinis, P.E., *Langmuir* 1990, 6, 87–96.
2. Horr, T.J., Ralston, J., Smart, R., *Coll Surf A: Physicochem Eng Aspect* 1995, 97, 183–196.
3. Kato, K., Uchida, E., Kang, E.-T., Uyama, Y., and Ikada, Y., *Prog Polym Sci* 2003, 28, 209–259.
4. Michielsen, S., *J Appl Polym Sci* 1999, *73*, 129–136.
5. Brennan, W.J., Lowell, J., O'Neill, M.C., Wilson, M.P.W., *J Phys D: Appl Phys* 1992, 25, 1513–1517.
6. Vandendaele, I.P., Langerock, I.A., White, W.C., Krueger, J., *Proceedings of the 2003 Medical Textile Conference in Bolton, U.K.*, 2003.
7. Feder, J., *J Theor Biol* 1980, 87, 237–254.

8. Thompson, K., *School of Polymer, Textile and Fiber Engineering*, Georgia Institute of Technology, Atlanta, GA, 2005, p. 172.

9. Thompson, K. and Michielsen, S., personal communication, 2004.

10. Thompson, K., Michielsen, S., personal communication, 2005.

11. Michielsen, S., *Abstr Pap Am Chem S* 2003, 225, U558–U558.

12. Bakhshaee, M., in: *School of Textile and Fiber Engineering*, Georgia Institute of Technology, Atlanta, GA, 2002, p. 91.

13. Sherrill, J., Michielsen, S., and Stojiljkovic, I., *J Polym Sci Pol Chem* 2003, 41, 41–47.

14. Tsukruk, V.V., *Adv Mater* 2001, 13, 95–108.

15. Bozja, J., Sherrill, J., Michielsen, S., and Stojiljkovic, I. *J Polym Sci Pol Chem* 2003, 41, 2297–2303

21.6   Nanoadditives in Textiles

Tushar Ghosh and Saurabh Chhaparwal

Bulletproof vests, golf clubs, skies, tennis and badminton rackets, vehicle tires, spacecraft, and numerous other products are all around us as excellent examples of strong but lightweight composite structures. Sandwich structures such as laminates, reinforced polymers, and fiber-reinforced composites (fiberglass and carbon fiber [CF]) are examples of macrocomposites and microcomposites, one comes across in daily life.[1] In recent times, progress in the composite material development has shifted to a new paradigm with the development of nanomaterials and their incorporation in other polymers to form nanocomposites.

Nanomaterials refers to the materials having at least one of their dimensions in nanometer scale, and their properties vary significantly from their bulky counterparts. While "nanocomposite" refers to a special class of materials made from combination(s) of these two or more nanomaterials, distinctly different in phase and mixed at nanometer level. The two phases may be inorganic–organic, inorganic–inorganic, or organic–organic and the resultant material may be amorphous, crystalline, or even semicrystalline.[2] Recent developments in manufacture and processability of these materials with novel properties have stimulated research to develop multifunctional materials at macroscopic scale by engineering structures at nanometer scale. Nanotechnology is also known as a "bottom-up" technology as the material with the desired property would be engineered bottom-up from atoms to a higher-level hierarchical structure with a fundamentally new molecular organization. This can be illustrated by electron micrographs in Figure 21.18 where carbon nanotubes (CNTs) (10 to 20 nm) have been deposited on the surface of a CF (7 μm in diameter) in yarn bundle (in millimeters) woven into a fabric (in meters).

The basis of nanomaterials lies in the fact that the properties of the material changes dramatically when its size is changed and reduced, in this case, to nanometer scale. As described in the report published by the National Nanotechnology Initiative,[6] "Nanotechnology is concerned with materials and systems whose structures and components exhibit novel and significantly improved physical, chemical, and biological properties, phenomena, and processes due to their nanoscale size." The unique properties may be due to the fact that the behavior of atoms and molecules is governed by quantum mechanics while that of the bulk materials is by classical mechanics. Between these two domains, nanometer range is the threshold for transition in the material's behavior. For example, ceramic though brittle can be made deformable by reducing the grain size to nanometer range. A gold particle of 1 nm across shows red color.[7] CNTs of 1 to 30 nm diameter have tensile strength of >100 Gpa,[8] and each of these materials can be successfully incorporated in or on textile and polymer matrices/surfaces without changing much of their processability.

Nanomaterials used for fillers or reinforcement can be classified broadly as nanoparticles or equiaxed nanoparticulate material (Figure 21.19[a]), nanofibers (or rods) (Figure 21.19[b]), and nanoplatelets (Figure 21.19[c]). These materials could be further classified into subcategories discussed in detail in later sections. These materials incorporated in a polymer matrix, as fillers, are collectively known as polymer nanocomposites (PNCs). These PNCs, due to these fillers, have unique mechanical, thermal, physical and

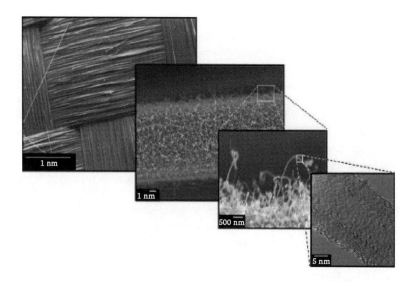

FIGURE 21.18 Variation in reinforcement scales from millimeters to nanometers: (*from left*) from woven fabric of yarn bundles, to a single CF with entangled CNTs grown on the surface, to the nanometer diameter and wall structure of the CNT. (From Thostenson ET, Chunyu, Chou, *Comp. Sci. Technol.* 65, 2005, 491–516; Thostenson ET et. al., CNT-reinforced composites: processing, characterization and modeling, Ph.D. Dissertation, University of Delaware, 2004; Thostenson ET, Chou TW, *J. Phys. D* 36(5), 2003, 573–582. With permission.)

chemical barriers, optical, electrical and magnetic properties, or even combinations of these. Additionally, PNC materials can be easily extruded or molded to near-final shapes, thus simplifying their manufacturing. Reducing the size of particulates to the nanometer range results in high volume and surface area for a given weight of the particulate, thus, with the uniform distribution of the particulates or fillers, the active number of sites for interaction with the polymer matrix increases substantially and, hence, the properties. As a high degree of property enhancement is achieved at relatively low levels of particulate additions, these nanocomposites are much lighter than conventional composites and could result in a significantly reduced impact on the environment along with other benefits. For example, it has been estimated that large-scale use of PNCs in U.S. vehicle manufacturing could save 1.5 billion liters of gasoline over the life of one year's production and reduce related carbon dioxide emissions by more than 10 billon lbs.[9]

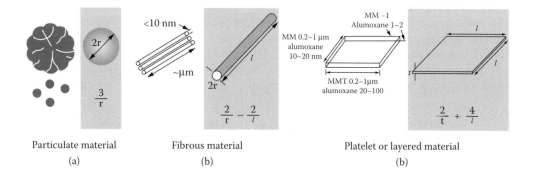

FIGURE 21.19 Surface area/volume ratios for various reinforcement filler geometries. (a) Spherical shape particles such as ceramics, metals, and their oxides; (b) rod-like CFs; (c) platelet-like structure for clays such as MMT and nanographite. (From ATP Project, Nanocomposites New Low-Cost, High Strength Materials for Automotive Parts: National Institutes of Technology, 97–02-0047, 1997; Lee et al., 2005, In Press. With permission.)

Major parameters that govern the PNC performance are matrix polymer and nanomaterials characteristics and arrangement of these constituents within the composite. Since the nanomaterials have one of the dimensions in nanometer scale, 10 to 100 times the order of molecular chains, their presence limits the mobility of chains and conformations that polymer molecules can adopt, and the free energy of the chains and molecules in this interfacial region are different than the chains and molecules far from there, i.e., bulk.[11] Thus, polymer matrix can be considered as a nanoscopically confined polymer, and, as a result, the thermal transition behavior of the polymer is altered. The fillers might also affect the degree of ordering and packing perfection and, thus, crystallinity and crystal size in the polymer. The size of the nanomaterials also greatly affects and governs the PNC characteristics. For example, nanosized titanium dioxide (TiO_2) can enhance the dielectric strength of polypropylene (PP) without affecting the transparency of it, which is not the case with microsized TiO_2, which is used as a delustrant in many polymers.

Finally, the arrangement of these nanomaterials in the polymer matrix critically determines their properties and can be as important a factor as particle–particle association, clustering, percolation, and distribution of nanomaterials in polymer matrix.[12,13]

21.6.1 Classification

"Nanoparticle" or nanoparticulate refers to materials having isotropic geometry and most of their dimensions in nanometer scale (1 to 100 nm). The existence of these types of tiny particulates was known long ago in the form of *aurum potabile* (potable gold) accounted by Paracelsus around 1570 and *luna potabile* (potable silver) in 1677.[14]

Metal colloids have also been used to dye glass and as paints for enamel. Today not only metals but also a large number of other inorganic and organic nanoparticles are used to reinforce polymers and ceramics, which could be classified as inorganic particulates in organic matrix or organic particulates in inorganic matrix (ceramics). In this chapter, metal and inorganic oxide particulates in organic matrix, as relevant to textiles (Table 21.1), are considered in detail.

21.6.2 Equiaxed Nanomaterials

Nanostructured materials vary from 0-D atom clusters to 3-D equiaxed grain structure. By definition, equiaxed particles possess approximately equal dimensions in all directions.

21.6.2.1 Metals

Metallic nanoparticulates possess a wide range of novel chemical and physical properties such as electrical, optical, magnetic, application in catalysis, etc. due to their intermediate structures between the atomic state and the bulk. Noble metal nanoparticles/organic polymer composite films have their potential applications for photonics and electro-optics.[16–18] These nanoparticles when dispersed in polymer matrices do not coalescence and have long-term stability. Thus, nanoparticulate-reinforced composites have the specific properties of nanoparticulates and the processability of the polymer (e.g., elasticity, transparency, ease of synthesis, and processability). Among various nanoparticulates being developed for textile application, silver is one of the most used in textile and biomedical applications as it has been medically proven to kill over 650

TABLE 21.1   List of Various Nanoparticulates Useful in Textile Applications

	Compounds		
Metals	Organic	Inorganic	Polymers
Pd/Pt	Vitamins	TiO_2	Block-copolymers
Ag	DNA	ZnO	Dispersion, etc.
Fe	Hydroxylapatite	Fe_2O_3	
Ni	Color pigments	Al_2O_3	
Cu		SiO_2	

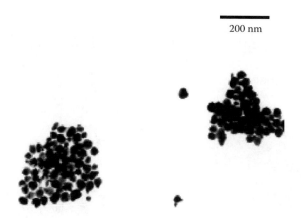

200 nm

FIGURE 21.20 TEM images of Ag nanoparticles synthesized with potassium bitartrate as a reductant with TDPC-Pd nanoparticles obtained at a TDPC/Ag molar ratio of 1. (From Rong MZ, et al., *J. Mat. Sci. Lett.* 20, 2001, 1473–1476 With permission.)

disease-causing organisms in the body and is also safer as an antibacterial agent than many of the organic compounds.[19–21] Silver nanoparticles (Figure 21.20) are also widely used as photosensitive components,[22,23] as catalysts, or in photocatalysis[24,25] and chemical analysis.[26] If incorporated in embolic materials, silves has an advantage of curing and detectability and improved material properties at the same time.[27]

The other nanoparticles available are Fe, Pd/Pt, Au, Cu, etc. The usual synthetic technique for making such nanoparticles involves chemical or electrochemical reduction of metal ions in the presence of a stabilizer (protective agent) such as linear polymers and ligands, which prevent the nanoparticles from aggregation and allow isolation of the nanoparticles. The other methods are metal vapor deposition, sonication, chemistry, etc.

21.6.2.2 Oxides

Another class of nanosize particles that are gaining ground in textile applications is oxides especially of metals such as TiO_2, Al_2O_3, ZnO, and MgO. They possess photocatalytic ability, electrical conductivity, and ultraviolet (UV) absorption, and antibacterial, antifungal, antiodor, self-decontaminating, UV blocking, and photo-oxidizing capacity against chemical and biological species. Intensive research is going on for incorporating nano-oxides for antimicrobial and UV blocking functions for military protection gear and civil health products. It has been shown that nylon fibers filled with ZnO nanoparticles can provide UV shielding and also have improved static behavior. A composite fiber with TiO_2/MgO can provide self-sterilization.[28]

Silica (SiO_2) nanoparticles having a diameter of 7 to 27 nm and surface area ranging from 100 to 380 m^2g^{-1} have been used for deep dyeing polyester fibers. The process included polymerization of SiO_2 nanoparticles into polyester (1 to 2 wt%), followed by drawing and heat treatment in caustic soda to dissolve the SiO_2 from the surface resulting in a porous surface that is receptive to the dye. The process also claims to have imparted a soft, cotton-like feel to the fabric. SiO_2 in ethylene glycol has been used to improve abrasion resistance and antiblocking of polyester films, with minimal effect on the optical and flatness of the film.[29]

Titania (TiO_2) also has plenty of practical application owing to its optical, photocatalytic, and high dielectric properties. It acts as a disinfectant by absorbing UV radiation (for example, from sunlight), which kills microbes upon contact. A thin layer of TiO_2 has been used to engender a "lotus leaf" effect on substrates including fabrics. The coating results in nanoscale roughness, which, when combined with a hydrophobic substance, reduces the wetting ability of water.[30] When irradiated by UV radiation it results in superhydrophilic surfaces that have application as self-cleaning glasses[31] and can be applied to textiles. Due to its high refractive index and bright white color, TiO_2 acts as an effective opacifier and thus is used as an antiluster agent also.

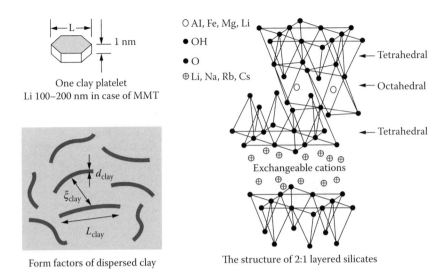

FIGURE 21.21 Schematic illustration of a layered silicates structure. (From Sinha Ray S, Okamoto K, Okamoto M, *Macromolecules*, 36, 2003, 2355–2367 With permission.)

Condensation from the gas phase (aerosol method) by flame hydrolysis is a commercial way to make some of these nano-oxides. A large range of metal oxides including SiO_2, TiO_2, Al_2O_3, CuO, CeO_2, ZnO, ZrO_2, Fe_2O_3, etc. can be made by this process.[32,34] Laser ablation and reduction of precursors are also widely used methods to synthesize the oxides.[33]

21.6.2.3 Platelet Nanomaterials

Layered clay is the most widely used platelet nanomaterial having a crystal structure consisting of 2-D layers formed by fusing two SiO_2 tetrahedrals with an edge-shared octahedral sheet of alumina, magnesia, iron, etc. Lateral dimension may be in micrometers, but the thickness is around 1 nm. (Figure 21.21). [35]

PNCs using these clays were first reported by Blumstein[37] early in 1961 for improved thermal stability as polymethyl meth acrylate (PMMA) and montmorillonite (MMT) clay nanocomposite. A report from a Toyota research group, about improvement of thermal and mechanical properties of nylon-6 for a small addition of MMT[38] and the possibility of melt mixing clay in polymer matrix without the use of organic solvents stimulated scientific and industrial interest in these materials.[39] Thereafter a large number of layered silicates such as hectorite, saponite, fluoromica, fluorohectorite, vermiculite, kaolinite, magadiite, etc. have been developed for various uses such as higher heat distortion temperatures and flame retardancy (FR) combined with higher stiffness, strength, and improved barrier properties such as lower gas and liquid permeability.[40] For example, nylon-layered silicate nanocomposites with 2 vol% fillers have been shown to double their tensile modulus and strength along with a 63% reduction in heat release at a heat flux of 50 kW/m² without increasing carbon mono-oxide or soot release.[41,42] These nanocomposites can also be used as a UV blocker.[7]

Fan et al.[43] have reported enhanced dyeability of PP by incorporating MMT clay. Zhang and coworkers[44] have incorporated organomontmorillanite in PP matrix spun into fiber form. It was found that the crystallinity was improved, but orientation was lower than that of the pure PP for the same draw ratios. Pavlikova al.[45] have reported improvement in tensile property for PP filled with organophilic-layered silicates for 5 wt% loadings. As shown in Figure 21.22(a) and Figure 21.22(b), draw ratios of the fibers influence the level of exfoliation of clay. Figure 21.22(a) with a 1% loading and higher drawing ratio (λ = 3) has better distribution and exfoliation than 5% loading at λ = 2 (Figure 21.22[b]).

Properties and performance of the clay-reinforced polymer matrix are mainly governed by the distribution or morphology of the clay in the matrix, which is of three types: (1) miscible or phase separated; (2) intercalated; and (3) exfoliated (Figure 21.23). Published literature supports the fact that exfoliated

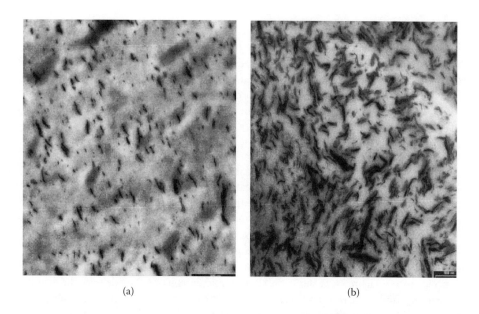

(a)   (b)

FIGURE 21.22 TEM micrographs of PP nanocomposite fiber nanoparticles of SOMASIF MEC16. (a) 1% loading at $\lambda = 3$; (b) 5% loading at $\lambda = 2$. (From Pavlikova S, Thomann R, Reichert P, Mulhaupt R, Marcincin A, Borsig E, *J. Appl. Polym. Sci.* 89, 2003, 604–611. With permission.)

structures have slightly better mechanical properties, particularly a higher modulus, than intercalated nanocomposites.[45–47] Masenelli-Varlot et al.[46] studied the physical properties of polyamide-6 with intercalated and exfoliated MMT. Exfoliated samples showed higher stress-to-failure, while intercalated samples showed higher strain-to-failure. Intercalated samples were slightly weaker due to weaker mechanical coupling.

There are several ways to make these platelet nanomaterials including melt compounding, solution blending, *in situ* polymerization, template synthesis, chemical/thermal conversion, etc.[1,48]

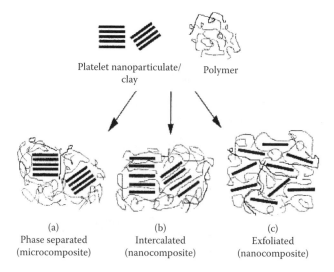

FIGURE 21.23 Morphologies of polymer/clay composites. (a) Conventional miscible or phase separated, (b) intercalated, and (c) exfoliated. (From Masenelli-Varlot K, Reynaud E, Vigier G, Varler J, *J. Poly. Sci. B: Polym. Phys.* 40, 2002, 272–283. With permission.)

21.6.3 Carbonaceous Materials

Carbon in various forms such as carbon black (CB), graphite powder, graphite platelets, CFs, carbon nanofibers (CNFs), and CNT have been used mainly for enhancing mechanical as well as electrical properties. Carbonaceous materials are revolutionizing various fields in material science. They have unexplored potential in many areas of applications such as the aerospace, automobile, energy, or chemical industries in which they can be used as absorbents, templates, actuators and sensors, reinforcements, catalysts, etc.[8]

21.6.3.1 Carbon Black

CB has been used for centuries in pigment inks and mural paints and later on as reinforcement for rubber and in the automotive industry. Today, it is used extensively as a reinforcement and conductive filler in rubber, for protection of plastics from UV degradation, and in paints and ink coatings.[49] The structure of CB is determined from its size and shape and the number of particles/aggregate. It is categorized as *low structure* and *high structure* depending on the size of the primary aggregate, with few prime particles as *low structure* and substantial branching and chaining as *high structure*.[50]

In general, the size of the CB particle may vary from 10 nm to more than 10 µm depending on the manufacturing process. Figure 21.24 shows the electron micrograph of CB made from the furnace process. CB is characterized by its structure, particle size (surface area) and size distribution, and surface chemistry, especially when used as conductive fillers.[51–53] Also, CB surface is chemically active as it consists of hydrogen, oxygen, and sulfur as surface atoms combined with carbon and thus is used as filler for filter applications.[54]

21.6.3.2 Graphite

Graphite can be incorporated in polymer matrices in various forms such as granules, natural graphite flakes, and expanded graphite.[56–59] Denser than CB, graphite has a layered type hexagonal structure with 3-D ordering of the layers having strong covalent carbon–carbon bonds (Figure 21.25).

Natural graphite flakes have good electrical conductivity (10^6 S/m at ambient temperature). They have been used with many polymer systems such as PMMA,[60] nylon,[61] polystyrene (PS),[62] epoxy,[63] etc. for enhancing thermal and electrical properties of the composite.

21.6.3.3 Carbon Fibers

CFs are essentially hexagonal carbon layers in the form of ribbons or sheets. Figure 21.26 shows the schematic representation of a 2-D section through the structure parallel to the fiber axis. There are two

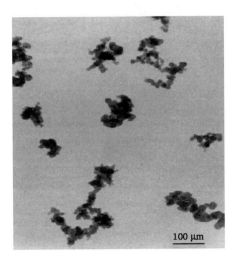

100 µm

FIGURE 21.24 Electron micrograph of typical furnace process carbon black. (From Kraus G, *Die Angewandte Makromolekulare Chemie* 13/61 (1 977) 215–248 (Nr. 843). With permission.)

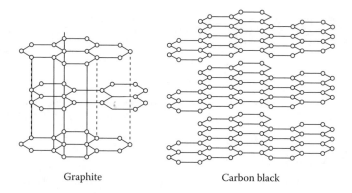

Graphite   Carbon black

FIGURE 21.25 Crystallographic arrangements of carbon atoms for graphite and CB. (From Accorsi J, Romero E, *Plastics Eng.* 29, 1995. With permission.)

types of CFs commonly in use: polyacrylonitrile (PAN)-based and pitch-based fibers. CFs based on cellulose fibers (rayon) are produced for very specific use. CFs when treated above 2000°C are termed as "graphite fibers." Having high aspect ratios, CFs are mainly used for enhancing mechanical and electrical properties such as electromagnetic interference shielding (EMI), antistatic coating, electrically conductive CF–polymer composites, resistance heating elements, and CFs with conductive and superconducting coatings. Intercalated CFs have many potential uses (batteries, catalysts, selective absorbents, etc.).[64–66]

21.6.3.4 Carbon Nanotubes

In 1991, Japanese microscopist, Sumio Iijima observed graphitic carbon needles, ranging from 4 to 30 nm in diameter and up to 1 mm in length, as by-products of an arc-discharge evaporation of carbon in an argon environment.[67] Later, single-sheet tubules with diameters of 0.7 to 1.6 nm were developed as single-wall nanotubes (SWNTs).[68] Multiwalled carbon nanotubes (MWNT) consist of several tens of graphite shells with adjacent shell separation of ~0.34 nm.[8] These SWNTs and MWNTs have interesting properties such as Young's modulus of 1TPa[70], tensile strength possibly greater than 100 Gpa,[8] high flexibility,[71] bending fully reversible up to a 110° critical angle for SWNT,[72] and thermal conductivity twice that of diamond,[73] and they can carry current density as high as 109 A/cm^2.[74] Thus, they have potential to be used in areas that have never been explored before by any of the carbon forms discussed above. Figure 21.27 shows SWNT and MWNT morphologies for various configurations.

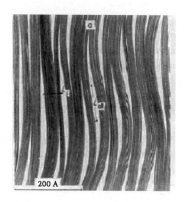

FIGURE 21.26 Schematic presentation of a 2-D section through the structure parallel to the fiber axis. (From Wilham R, *Adv. Mater.* 2(11), 1990, 528–536. With permission.)

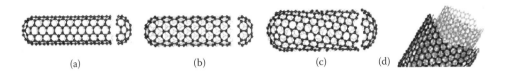

(a)   (b)   (c)   (d)

FIGURE 21.27 Schematic of nanotube morphologies. (a) armchair, (b) zig zag, (c) chiral), (d) MWNT. (From Harris PJF, *Int. Mater. Rev.* 49(1), 2004, 31–43.; Breuer O, Sundararaj U, *Polym. Compos.* 25(6), 2004. With permission.)

21.6.3.5 Carbon Nanofibers

CNFs (diameter range, 3 to 100 nm; length range, 0.1 to 1000 mm or 50 to 200 nm in aggregated form) have been known since 1889[77] as emerged during catalytic conversion of carbon-containing gases. The recent explosion of interest in these materials originates from their potential for unique applications as well as their chemical similarity to fullerenes and CNTs. These are found in a wide range of morphologies, from disordered bamboo-like structures (Figure 21.28[a])[78] to highly graphitized "cup-stacked" structures (Figure 21.28[b]) and (Figure 21.28[c])[79,80]. CNFs have a unique high surface area (~200 m²/gm) and excellent thermal and electrical properties. CNTs have been blended with epoxy, polyimide, PMMA, polycarbonate (PC), Polyethylene (PE), PP, Poly(ethylene terepthalate) (PET), etc. to enhance their electrical and thermal behavior.[80–85]

21.6.3.6 Properties and Areas of Applications

Due to the several advantages discussed earlier, nanomaterials and nanocomposite fibers and films are gaining importance. They have been used in textiles for improving performance or creating unprecedented functionalities, ranging from enhanced dyeability to superhydrophilic finishing. Nanomaterials have also been used in fibers for applications such as electromagnetic/infrared shielding. Most of the research is focused on using nanomaterials for making nanocomposite fibers as well as finishing textile fabrics.

21.6.3.7 Mechanical Reinforcement

Micromaterials (5 to 50 μm) have been used for years as fillers for reinforcement and enhancement of tensile and impact strength of thermoplastics. However, these materials did not make significant inroads in fibers because of dimensional incompatibility. But with the advent of nanomaterials, they have been used increasingly for fiber and film reinforcements and many other applications. CNTs having theoretical

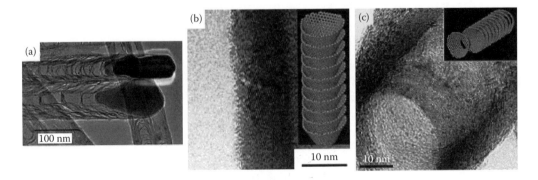

FIGURE 21.28 TEM micrographs of CNF representing (a) disordered bamboo-like structures, (b) highly graphitized sidewall of a cup-stacked (molecular models inset) nanofibers showing the shell tilt angle, and (c) a nesting of the stacked layers. (From Endo M, Kim YA, Ezaka M, Osada K, Yanagisawa T, Hayashi T, et al. *Nano. Lett.* 3, 2003, 723; Merkulov VI, Lowndes DH, Wei YY, Eres G, Voelkl E, *Appl. Phys. Lett.* 76, 2000, 3555. With permission.)

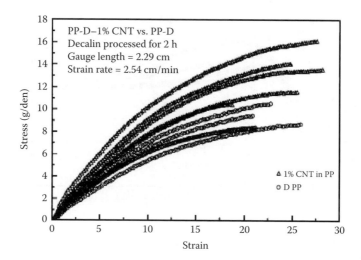

FIGURE 21.29 Replicate stress–strain behavior of PP-D fibers and PP-D-1% CNT fibers postdrawn at 125% to a diameter of 15 µm and the maximum possible draw ratio (8.3 for PP-D-1% CNT; 9.3 for PP-D). (From Vaccarini L, De´sarmot G, Almairac R, Tahir S, Goze C, Bernier P., *Electr. Prop. Nov. Mater-Mol. Nano.* 2000, 521–525. With permission.)

strength of 200 Gpa[86] have been widely used in polymer composites in several types of matrices: PMMA,[87–89] epoxy resin,[86,90,91] PP (Figure 21.29),[92,93] PS,[94,95] etc. The mechanical behavior of such composites depends on factors such as type of nanotube used (single walled or multiwalled), their geometric characteristics, nature of polymer used,[96] and the mixing process. Some of the processes used for these composite's synthesis are *in situ* polymerization,[87] solvent method,[94] and melt mixing.[97] Haggenmuller and coworkers[98] investigated a PMMA/SWNT composite and reported a 54% increase in tensile strength and 94% increase in modulus with 8% SWNT loading. Weisenberger et al.[99] also reported similar results for dry-jet wet spun PAN/dimethylacetamide with 5% MWNTs. They found a 31% increase in breaking strength, 36% increase in initial modulus, 46% increase in yield strength, and 83% increase in energy to break when compared to control samples having no MWNTs.

Other polymer–nanoparticulate systems with enhanced mechanical performance reported are $CaCO_3$ (44 nm) with PP, (Figure 21.30),[100] poly(vinyl chloride) (PVC),[101] PE,[102] alumina with PMMA,[103] and organoclay with isotactic PP.[104] A large number of clay-reinforced nanocomposites with different polymers including PC, PAN, and PP have been examined in film form. Shelley et al.[105] examined a polyamide-6 system with clay platelets ($1 \times 10 \times 10$ nm) at 2 and 5% weight fraction. Increased elastic modulus was reported for both the loadings with a 40% increase for 2% loading compared to a pure polymer system and 175% for a 5% loading. Yield stress along with failure strain was also improved. Further increases in loading led to deterioration of properties.

Pavlikova et al.[45] studied the effect of organoclay on tensile properties of fibers and have reported that the properties improved up to two times especcially for higher draw ratios. The increase in reinforcement efficiency is attributed to: (1) efficient exfoliation of the clay due to increased drawing and (2) a considerable increase in self-exfoliated structures in fibers.

It is important to note that dispersion of nanoparticulates in the polymer matrix is often the key to improved behavior, and thus compatibalizers may be used for improving the dispersion of nanoparticulates in the polymer matrix. Kaemfer et al.[106] examined the use of silicate modification and addition of maleic anhydride-grafted isotactic PP as compatibilizers for organophilic-layered silicates in syndiotactic PP and reported to have very significant improvement in the mechanical properties. Wu et al.[101] have reported similar effects of chlorinated PE on $CaCO_3$ and a PVC system. But Pavlikova et al.[45]

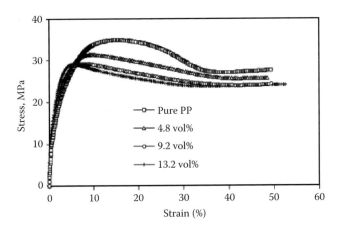

FIGURE 21.30 Stress–strain behavior of the nanocomposites and PP. (From Weisenberger MC, Grulke EA, Jacques D, Rantell AT, Andrewsa R, *J. Nanosci. Nanotechnol.* 3(6), 2003, 535–539(5). With permission.)

found a decrease in mechanical strength of PP fibers with the addition of compatibilizers (10% PP-grafted maleic anhydride) at low drawing ratios ($\lambda = 2$).

21.6.4 Thermal Stability and FR

Use of nanoclays for thermal stability of polymers and FR was first demonstrated in 1965 for MMT and PMMA composites. The degradation temperature increased about 50°C for about 10 wt% additives.[107]

Chiu Chih-Wei[108] has reported a significant increase in temperature of decomposition of PP fibers with the addition of small quantities (less than 3%) of TiO_2 (Figure 21.31).

Degradation temperature of polydimethylsiloxane was reported to increase by about 140°C by the addition of 10% MMT for delaminated composites.[109] The prevalent hypothesis on the MMT FR activity is based on gasification and precipitation of MMT. It is considered that polymer is burned away while precipitating the MMT remains on the surface, thereby acting as a barrier to oxygen diffusion into polymer and also to volatiles from bulk. It is also suggested that the MMT in the bulk migrates to surface above the glass transition temperature. Gilman and coworkers,[110] on the other hand, suggest that the FR activity is

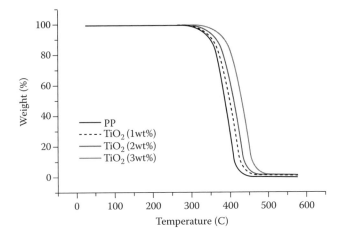

FIGURE 21.31 *TGA* analysis of the PP–TiO_2 nanocomposite fibers at 1%, 2%, and 3% loading by weight. (From Blumstein A, *J. Polym. Sci., Part A: Poly. Chem.* 3, 1965, 2665. With permission.)

TABLE 21.2   Cone Calorimeter Data from Selected PNCs

Sample	Residue Yield %	Peak HRR (kw m^{-2})
Nylon	1	1010
Nylon 6-silicate 2%, delaminated	3	686
Nylon 6-silicate 5%, delaminated	6	378
PP	0	1525
PP-silicate 2%, intercalated	5	450
PP-silicate 2%, intercalated	—	305

Source: Burnside SD, Giannelis EP, *Chem. Mater.* 7, 1995, 1597; Gilman JW, Kashiwagi T, Lichtenhan J. Recent advances in flame retardant polymer nanocomposites. In: Proceedings of Conference on Fire and Materials, Jan 22–24. London: Interscience Communications, 2001, 273–284.

due to a difference in condensation phase decomposition processes and not to a gas phase effect. Gilman et al.[111] supported their theory by Cone calorimeter data for nylon-6 and PP-silicate nanocomposites (Table 21.2).

Other nanomaterials that can be used to enhance the FR of polymers, in general, are TiO$_2$, Sb$_2$O$_3$, and borosiloxanes.[112,113]

Addition of nanoclay to the polymer matrices leads to higher heat deflection temperature (HDT). For example, addition of MMT to the PP matrices enhances their HDT by about 40˚C compared to neat PP for 6 wt% loading. The improvement in HDT is due to better mechanical stability of the composite structure. Addition of clays also increases ablation resistance of the polymer. Unlike traditional fillers, which require up to 30 wt% loading, making filament production very difficult, 2 to 5 wt% nanoclay-filled nylon 6 composites showed excellent ablation properties, thus enhancing the potential application of textiles for aerospace applications.

Yano et al.[114] have reported a reduction in the thermal coefficient of a polyimide–clay system for an increase in clay wt% implying improved dimensional stability even for high temperature (Figure 21.32).

21.6.5 Barrier Properties

Nanomaterials-filled composites have reduced gas and liquid permeability, thus, making them suitable for use as barrier membranes or nanocomposite films for common uses such as packaging as well as for

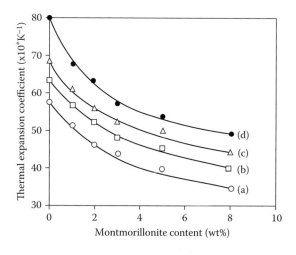

FIGURE 21.32 Dependence of the thermal coefficient on MMT content in a polyimide clay system at (a) 150°C, (b) 200°C, (c) 250°C, and (d) 295°C. (From Yano K, Ususki A, Okada A, *J. Polym. Sci. A: Polym. Chem.* 35 1997, 2289–2294. With permission.)

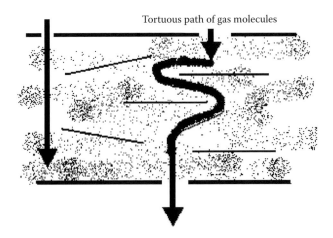

Tortuous path of gas molecules

FIGURE 21.33 Illustration of Neilson's tortuous path model for barrier enhancement of nanocomposites.

military applications such as protective suits and tents, etc. The high aspect ratios of nanoclay and their dispersion in the fully exfoliated state make them ideal for such applications. Neilson[115] developed a tortuous path model stating that the increased tortuosity resulting in increase in effective path length for diffusion of gas or liquid molecules enhances barrier properties (Figure 21.33). The permeability ($P_{nanocomposite}$) of the nanocomposite is predicted for a given aspect ratio (α) of platelet-like nanomaterials and volume fraction (ϕ) of platelet present in the polymer matrix by Equation (21.18).

$$P_{nanocomposite} = \frac{(1-\phi)P_{matrix}}{1+\alpha\phi/2} \tag{21.18}$$

where P_{matrix} is the permeability of the polymer matrix.

Yano et al.[115] have compared data for clay-filled polyimide nanocomposites with the Neilson model (Equation [21.18]) and found a good fit. Addition of 2% exfoliated clay results in large reduction in permeability of the composites for oxygen, helium, and water vapor, (Figure 21.34[a]), (Figure 21.34[b]), and (Figure 21.34[c]). Similar results have been reported for PET,[120] nylon,[116] and polycaprolactone–clay composites.[117]

Barrier property is dependent on the degree of dispersion and alignment of platelet nanomaterials. Also the size of the platelet affects the permeability coefficient (Figure 21.35).[114]

Water barrier property is important for applications such as aerospace, electronics, and electrical engineering. Nylon showed increased water resistance from 2% absorption to 1% at 5 wt% loading of clay.[116] Messersmith and Giannelis[118] showed a significant reduction in water vapor permeability for a low addition of 4.8 vol% of clay to poly(ε-caprolactum). Many commercial nanocomposites for barrier applications are also available, such as vermiculite clay–PET composite with oxygen permeation values of $\leq$1 cc m^{-2} d^{-2}. Ube industries produce nylon–clay nanocomposite having half the permeability of pure nylon for 2 wt% loading.[119]

21.6.5.1 Optical Properties

Nanomaterials have their dimensions less than the wavelength of light; thus they do not have any Rayleigh scattering. Thus, the composite structure may have the improved properties without affecting the optical appearance. Yano et al.[115] have showed that addition of 2% MMT to polyimide imparts excellent gas barrier behavior without changing the transparency of the polymer.

Poly(vinyl acetate), poly(vinyl alcohol) (PVA), and PVA/MMT nanocomposites containing 4 and 10 wt% MMT showed UV resistance in spite of being transparent even for a high loading of 10%. (Figure 21.36) Similar were the results obtained for PP–MMT and PP-g-MA-MMT (i.e., MMT treated by maleic

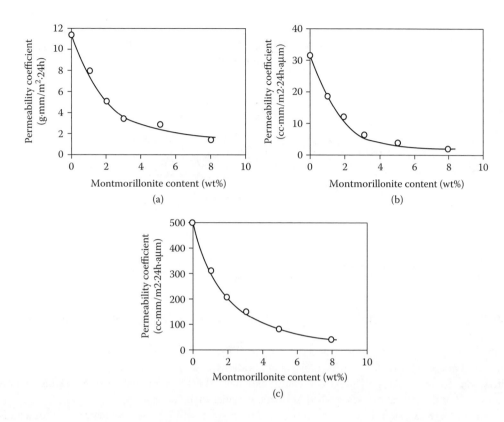

FIGURE 21.34 Permeability coefficient of (a) water vapor, (b) O_2, (c) He in polyimide–clay hybrid films. (From Yano K, Ususki A, Okada A, *J. Polym. Sci. A: Polym. Chem.* 35, 1997, 2289–2294. With permission.)

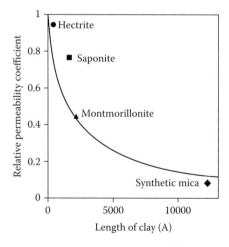

FIGURE 21.35 Clay length dependence on the relative permeability coefficient. (From Zhang S, Horrocks AR, *Prog. Polym. Sci.* 28, 2003, 1517–1538. With permission.)

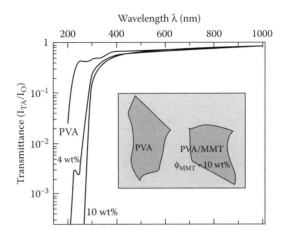

FIGURE 21.36 UV–VIS transmittance spectra of PVA and PVA/MMT nanocomposites containing 4 and 10 wt% MMT. (From *Modern Plastics*, February 1998. With permission)

anhydride). But at higher loading percentages (such as >20%) particles agglomerate, thus causing haziness in the films.[121]

Polymer-based nanocomposites have been used to fabricate photonic devices to control the physical properties of optical beams for a wide range of applications such as optical fibers, lenses, glasses, optical filters, etc. For example, polymer–CNT composites have effectively filtered intense 532-nm optical pulses,[122] but the saturation was reached for loading higher than 3.8 wt%.[123]

Some of these nanoadditives have been reported to interact with UV light, which can lead to deterioration in mechanical properties and discoloration of dyed textile materials. Thus, stabilizers have been commonly used with the polymers. Other than organic compounds, some of the inorganic oxides such as TiO_2,[124,125] ZnO,[125,126] and SiO_2 (UV reflectivity of 99.9% SiO_2 is 83%)[127] absorb UV radiation due to their wide band gap. Leodidou et al.[125] prepared a film of colloidal TiO_2 or ZnO in a copolymer of ethylene and vinyl acetate by *in situ* hydrolysis of incorporated titanium tetrachloride or diethyl zinc. TiO_2 (particle size: 70 nm) and ZnO (average diameter: 15 nm) showed UV absorption, but sheets made by incorporating TiO_2 had slight opacity, which was not the case with ZnO. ZnO[128,129] and TiO_2[130] have also been incorporated in functional monocomponent as well as bicomponent fibers.

Nanomaterials have been used as coloring agents for polymers. It has been known for ages that gold nanoparticles showed different colors for different particle sizes (Figure 21.37).

Nanomaterials also have interesting application as clear nanocomposites having refractive indices over the entire range of <1 to >3.9. Zimmerman et al.[132] have made PbS–gelatin nanocomposite films having a refractive index ranging from 1.5 to 2.5 by varying PbS volume fraction from 0 to 55%. Addition of nanoscale FeS to PE increased the films, refractive index to 2.5 and 2.8.[133]

21.6.5.2 Surface Modification

Surface properties play a very important role for many textile applications and nanomaterials are giving a new dimension to them by having multifunctionality. Carefree fabrics with superior liquid, oil, dirt, and stain repellency,[134] with wrinkle resistance without any change in feel or breathability throughout the life of the garment, based on nanotechnology, are now becoming widely popular. The case is similar for other functionalities such as UV-resistance, electromagnetic shielding[135] and antibacterial, antiodor, moisture control, etc. made from natural fibers such as cotton, wool, silk, or any synthetic fiber. Ideally, discrete nanomaterials can be arranged in a controlled fashion on textiles by thermodynamic, spray,[28] microemulsification, electrostatic, or many other newly developed technologies for specific finish systems and applications. Multifunctional self-assembled layers, by electrostatic attraction, having a self-healing capability is a promising potential technology for nanocoating but still in the embryo stage.[136,7]

FIGURE 21.37 Bright colors of gold nanoparticles: red suspension (*extreme left*) contains the smallest particles while blue suspension (*extreme right*) contains the largest particles. The origin of the color is easily understood from the absorption spectra of each suspension (*above each suspension*). The relative position of the visible range has been superimposed at the appropriate location. The sample at the extreme right absorbs the least in the blue region and hence it appears blue, similar reasoning is valid for the rest. Inset scale bar 10 nm TEM image of the "red"gold colloids, particles are <10 nm. (From Hartley SM, Axtel Holly, The Next Generation of Chemical and Biological Protective Material Utilizing Reactive Nanoparticles, Gentex Corporation, Carbondale, PA, 18407. With permission.)

One of the interesting applications of nanocoatings is water and soil repellency by a "lotus effect" on textiles. Surface roughness at the nanolevel with low surface energy causes water droplets to roll and not to stick to the surface, picking up dirt, etc. while rolling, (Figure 21.38[a–d]). Igor[137] has made a highly hydrophobic PET fabric by a combination of PS (low surface energy component) and silver nanoparticles (roughness initiation component) (Figure 21.41). They have also used PS and triblock copolymer PS-b-(ethylene-co-butylene)-b-styrene simultaneously on the PET substrate and then created a rough surface by dissolving PS and not the triblock copolymer. They have also demonstrated that a PET surface can be made hydrophilic or hydrophobic by treating the grafted chains of PS and poly(2-vinylpyridine) by

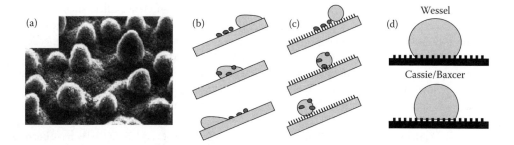

FIGURE 21.38 (a) Scanning electron micrograph of the lotus leaf surface. Schematic representation of the self-cleaning mechanism: moving of water droplet along contaminated smooth (b) and rough (c) surfaces. (d) Effect of roughness on the contact angle. (Igor Luzinov, Ultrahydrophobic Fibers: Lotus Approach, National Textile Center Annual Report: November 2004 NTC Project: C04-CL06. With permission.)

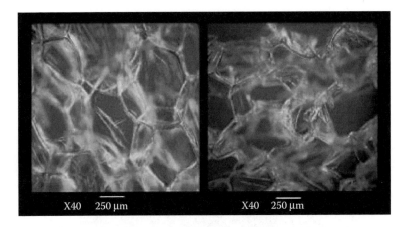

FIGURE 21.39 Optical micrograph of (a) pure polyurethane and (b) nanosilver-coated polyurethane. (From Walid AD, John HX, Xiaoming T, *J. Am. Ceram. Soc.* 87(9), 2004, 1782–1784. With permission.)

ethylene or toluene treatment respectively (Figure 21.42). Walid et al.[138] had prepared similar transparent and durable superhydrophobic SiO_2-coating films on cotton substrates at low temperatures.

Jain et. al.[140] have studied the method of coating silver nanoparticles on the poly(urethane) (PU) foams by solution coating (Figure 21.39). Lee et al.[141] have studied the antibacterial efficacy of nanosized silver colloidal solution on the cellulosic and synthetic fabrics.

Sheng Hung Industrial of Taiwan has launched nonwoven fabric with a nanocoating, which can be used for all color ink-jet printers, plotters, and lithographic presses.[142] Suzutora of Gamagori in Japan has developed a patented technology for nanometal coating on fibers, wovens, knits, and other sheet-like materials having high UV, infrared, and electromagnetic wave barrier properties.[135] The technology is versatile and can be used to impart properties such as photocatalysis, superhydrophilicity, transparent conductivity, coloration design features, anticorrosion, antibacterial, deoderization, etc. by different choices of metals in the film. Eric Devaux et al.[143] have reported the use of MMT and polyhedral oligomeric silsesquioxanes for imparting FR along with improved mechanical behavior, water repellency, and air permeability for PU resins and cotton. Yet another strategically important application is high-performance body armors. Kevlar bulletproof vests have been dipped into shear-thickening colloidal SiO_2 particles that cluster and jam due to stresses on the impact of a bullet but do not affect wearability in the absence of impact.[144]

21.6.5.3 Electrical and Thermal Properties

Some polymers are intrinsically conductive, such as PS, polyethylene oxide, PMMA, PVC, nylon, etc, while others such as PP, PET, etc. are made conductive by adding conductive fillers such as silver,[145] gold, Ni,[146] Cu, carbon particles, and CNTs. When in nanodimensions they have an advantage of a lower percolation threshold than the conventional particles along with enhanced mechanical properties and optical clarity. The properties are different than their bulk counterparts as, first, quantum effects begin to become important[33] and, secondly, the interparticle distances are decreased tremendously, leading to percolation at low loadings, as low as 0.0025 wt% of CNT (Figure 21.40).[147] CNTs have excellent properties and can be incorporated in almost all kinds of polymer matrices as such or by CNT surface modification.[148,149]

Other than CNTs and graphite,[150,151] Ag[158] metal oxides such as ZnO and Fe_3O_4[152,153] have been used to alter conductivity and dielectric constants for embedded capacitor application.[154] Presence of these materials also improves the thermal conductivity of the polymer matrices (Figure 21.41)[76,155] and therefore can be used in printed circuit boards, connectors, heat sinks, lids, etc. and thermal management materials from satellite structures to packaging. These PNCs also have potential magnetic and electro-chemical applications such as EMI, wave absorption,[156] electronic device packaging,[157] etc.

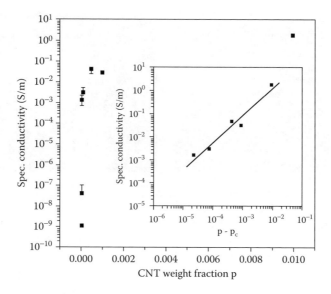

FIGURE. 21.40 Semilog plot of the specific composite conductivity as a function of CNT weight fraction p. The insert shows a log–log plot of the conductivity as a function of p-p_c with an exponent t of 1.2; the critical weight concentration is $p_c = 0:0025$ wt%. (From McCluskey P, Nagvanski M, Nanocomposite materials offer higher conductivity and flexibility, Adhesive Joining and Coating Technology in Electronics Manufacturing, 1998. Proceedings of 3rd International Conference on 28–30 Sept. 1998, 282–286. With permission.)

21.6.5.4 Biodegradability and Bio-Related Properties

Within the last decade or so significant efforts have been undertaken to develop bio-based or natural polymer-based products and innovative processes. A new class of biopolymers having multifunctionality, biodegradability, and biocompatibility based on renewable resources such as cellulosic plastics (plastic made

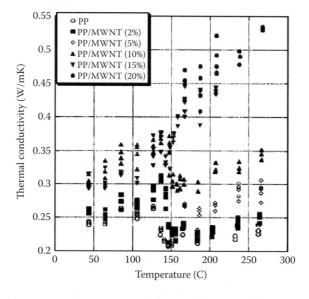

FIGURE 21.41 Comparison of thermal conductivity between PP and PP/MWNT nanocomposite. (From Pothuku-chi S, Li Y, Wong CP, *J. Appl. Polym. Sci.* 93(4), 2004, 1531–1538. With permission.)

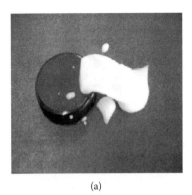 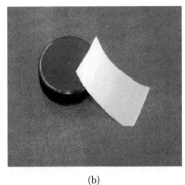

(a)   (b)

FIGURE 21.42   Casted films after they had been stored for 8 h in water. (a) Pure thermoplastic starch-casted film. (b) Starch/clay nanocomposite casted film.(From Chung DDL, Composite Materials: Functional Materials for Modern Technologies, Springer-Verlag, London, 2003. With permission.)

from wood), corn derivative plastics, and polyhydroxyalkanoates (plastic made from bacterial sources) in combination with nanoclay reinforcement has been developed for a variety of applications such as packaging, gas barrier or absorption, etc.[159] But until today most of these bio-based polymeric materials, i.e., polysaccharides (cellulose and starch), proteins (collagens and gelatin), and polyesters have had high production costs and also have limitations such as: (1) they are insoluble or sparingly soluble in water but have high water uptake leading to swelling (hence, low-dimensional stability); (2) they are rapid and undergo uncontrollable degradation by bacteria; and (3) they have low mechanical and processing properties. Thus, only starch and chitosan, when transformed as thermoplastic materials, are used along with nanoclay systems for bioplastic applications. Fisher[159] investigated starch–clay nanocomposites having high water resistivity, (Figure 21.42[a]) and (Figure 21.42[b]), and reported that they could be blow molded.

Some of the other biodegradable polymers such as poly(ε-caprolactone), PVA, poly(lactide) (PLA), poly(butylene succinate), unsaturated polyester,[158] poly(hydroxy butyrate), aliphatic polyester, and thermoplastic starch[155,156] have also been used to make nanocomposites.[157] Sinha et al.[158,159] reported the biodegradability of neat PLA and trimethyloctadecyl ammonium-modified MMT (C3C18-MMT) nanocomposites tested at $580 \pm 20°C$ in actual composite made from food waste (Figure 21.43[a]) and (Figure 21.43[b]).

21.6.5.5   Future of Nanomaterials

Nanotechnology has come a long way with increased efforts from governments and research communities all over the world not only in electronics but also in biotechnology, material sciences, textiles, medicines, and many other fields. The increasing effort is evident from increased spending of over 2.2 billion dollars from 430 million in 1997 as well as over a 10-fold increase in citations to over 20,000 since 1990 worldwide.[160] In textiles, most research efforts are directed toward improving existing functionalities and performances of conventional products. However, considerable research is underway in the development of so-called "smart textiles" having novel functionalities never imagined before. Much of the research in the U.S. is driven by the needs of the armed forces for increasing protection from conventional efforts in many cases including detection capabilities as well as protection. Soldier System Center in Natick, MA, is aimed at making combat gear for soldiers, which would be more than 20% lighter, would protect them from chemical or biological hazards yet be more breathable, bulletproof yet flexible. Further, it would be able to sense an oncoming attack (such as sensing harmful gas or electromagnetic rays), change chameleon-like to blend with surroundings, make temperature adjustments, monitor health and have self-healing functions, provide backup power for up to several hours, have ferrofluid-filled hollow

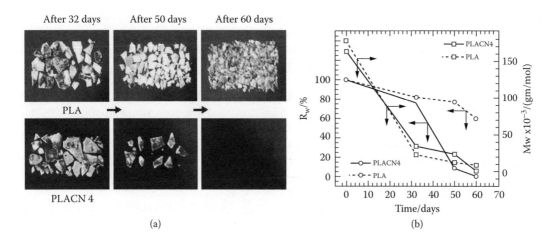

FIGURE 21.43 (a) Actual picture of biodegradability of neat PLA and PLACN4 recovered from compost with time. The initial shape of the crystallized samples was $3 \times 10 \times 0\%1$ cm³. (b) Time dependence of residual weight, Rw and of matrix, Mw, of PLA and PLACN4 under compost at 58 ± 2°C. (From Yahamoto T, Kutoba E, Taniguchi A, Dev S, Tanaka K, Osakada K, Sumita M, *Chem. Mater.* 4, 570–576, 1992; Chung DDL, *Composite Materials: Functional Materials for Modern Technologies*, Springer-Verlag, London, 2003. With permission.)

fibers that can turn into a rigid shell or splint for a broken bone, and have self-cleaning, self-repairing, and many other functions that would be similar to those of soldiers seen in science fiction movies.[162] Another potential development is inserting "nanomuscle fibers" that can simulate muscles giving soldiers mammoth strength. Commercial applications of nanotechnology and incorporation of nanomaterials for daily life products with better functionalities are increasing. One can find its application in products ranging from Toyota cars to Nike trainer shoes, but still it is in its insipient stage. The development of nanofibers, nanomaterial-based functional finishes, and smart textiles has endless possibilities with millions of dollars of impact along with great social impact that could change the present way of living.[161]

References

1. Oriakhi C, Polymer nanocomposition approach to advanced materials, *J. Chem. Edu.* 77(9), 2000, 1138–1146.
2. Oriakhi C. *Chem. Br.* 1998, 11, 59.
3. Thostenson, ET, Chunyu, Chou, TW Nanocomposite in context, *Comp. Sci. Technol.* 65, 2005, 491–516.
4. Thostenson, ET et. al., Carbon nanotube-reinforced composites: processing, characterization and modeling, Ph.D. Dissertation, University of Delaware, 2004.
5. Thostenson ET, Chou TW, On the elastic properties of carbon nanotube-based composites: modeling and characterization, *J. Phys. D* 36(5), 2003, 573–582.
6. Siegel RW, Hu E, Roco MC, Nanostructure Science and Technology-A Worldwide Study, September 1999.
7. Qian Lei, Hinestroza, Application of nanotechnology for high performance textiles, JTATM 4(1), 2004.
8. Rupesh Khare, Suryasarathi Bose, Carbon nanotube based composites — a review, *J. Min. Mat. Charact. Eng.* 4(1), 2005, 31–46.
9. ATP Project, Nanocomposites New Low-Cost, High Strength Materials for Automotive Parts: National Institutes of Technology, 97–02-0047, 1997.

10. Lee et al., Polymer nanocomposite foams composite science and technology, 2005, In Press.
11. Naganuma T, Kagawa Y, Effect of particle size on the optically transparent nano meter-order glass particle-dispersed epoxy matrix composites, *Compos. Sci. Technol.* 62(9), 2002, 1187–1189.
12. Vaia R, Krishnamoorti R, Polymer nanocomposites: synthesis, characterization and modeling, *Am. Chem. Soc.* 2001, 1–6.
13. Vaia R, Wagner D, Framework for nanocomposites, *Mater. Today*, 2004, 32–37.
14. Yang P. Nanocompoistes, The Chemistry of Nanostructured Materials, World Scientific, 359–386.
15. http://www.thebritishmuseum.ac.uk/science/lycurguscup/sr-lycugus-p1.html
16. Halperin WP, *Rev. Modern Phys.* 58, 1986, 533.
17. Henglein A, *Chem. Rev.* 89, 1989, 1861.
18. Kimura K, Phase transitions, 493, 1990, 24–26.
19. Kim HC, Kong YD, *Poly. Sci. Technol.* 7, 1996, 59.
20. Wasino W, Functional fibers: Trends in Technology and Product Development in Japan, Toray Research Center Inc, Japan 1993.
21. Oya, *J. Antibac. Antifungal Agents* (Jpn), 24, 1996, 429.
22. Hailstone RK, *J. Phys. Chem.* 99, 1995, 4414.
23. Sebestein J, et al. SU8-silver photosensitive nanocomposite, *Adv. Eng. Mat.* 6(9), 719–724, 2004.
24. Sun T, Seff K, *Chem. Rev.* 94, 1994, 857.
25. Tada H, Teranishi K, Inubushi Y, Ito S, *Langmuir* 16, 2000.
26. Pal T, *J. Chem. Educ.* 71, 1994, 679.
27. Rong MZ et al., Interfacial interaction in Ag/polymer nanocomposite films, *J. Mat. Sci. Lett.* 20, 2001, 1473–1476.
28. Hartley SM, Axtell H, The Next Generation of Chemical and Biological Protective Material Utilizing Reactive Nanoparticles, Gentex Corporation, Carbondale, PA 18407.
29. Catone DL, Nano-particle additives for PET, *Chem. Fiber Int.* 54, 2004, 25,26.
30. Sugunan A, Dutta J, Nanoparticles for Nanotechnology, PSI Jilid 4 2004, 1,2, 50–57.
31. Wang R, Hashimoto K, Fujishima A, Light induced amphilic surfaces, *Nature*, 388, 1997, 431–432.
32. Nanophase Technologies Corp., http://www.nanophase.com
33. Ajayan PM, Schadler, Buran PV. Nanocomposite Science and Technology, Wiley–VCH, 2003, 77–144.
34. Iskandar, Lenggoro, Xia, Okuyama, Spherical particle derived from colloidal nanoparticles, *J. Nanoparticle Res.* 2001.
35. Mai YW, Yu Z, Xie X, Zhang Q, Ma J, Polymer nanocomposite and their application, *Transaction* 10(4), 67–72.
36. Sinha Ray S, Okamoto K, Okamoto M, Structure–property relationship in biodegradable poly(butylene succinate)/layered silicate nanocomposites, *Macromolecules* 36, 2003, 2355–2367.
37. Blumstein A, *Bull. Chim. Soc.* 899, 1961.
38. Okada A, Kawasumi M, Usuki A, Kojima Y, Kurauchi T, Kamigaito O. Synthesis and properties of nylon-6/clay hybrids. In: Schaefer DW, Mark JE, eds. Polymer Based Molecular Composites. MRS Symposium Proceedings, Pittsburgh, 171, 1990, 45–50.
39. Vaia RA, Ishii H, Giannelis EP, Synthesis and properties of two-dimensional nanostructures by direct intercalation of polymer melts in layered silicates, *Chem. Mater.* 5, 1993, 1694–1696.
40. Kaemfer D, Thomann R, Mulhaupt R, Melt compounding of syndiotactic polypropylene nano-composites containing organophilic layered silicates and in situ formed core/shell nanoparticles, *Polymer* 43, 2002, 2909–2916
41. Giannelis EP, Polymer-layered silicate nanocomposites: synthesis, properties and applications, *Appl. Organomettal. Chem.* 12, 1998, 675–680.
42. Thostenson ET, Li WZ, Wang DZ, Ren ZF, Chou TW, Carbon nanotube/carbon fiber hybrid multiscale composites, *J. Appl. Phys.* 919, 1998, 6034–6037.
43. Fan Q, Samuel C, Alton R, Wilson, et al., Nanoclay modified polypropylene dyeable with acid and disperse dyes, *AATCC Rev.*, 2003, 25.

44. Zeng R, et al., Nanostructured Silver/Polystyrene Composite Film: Preparation and Ultrafast Time Optical Nonlinearity.

45. Pavlikova S, Thomann R, Reichert P, Mulhaupt R, Marcincin A, Borsig E, Fiber spinning from poly(propylene)-organoclay nanocomposite, *J. Appl. Polym. Sci.* 89, 2003, 604–611.

46. Masenelli-Varlot K, Reynaud E, Vigier G, Varler J, *J. Poly. Sci. B: Polym. Phys.* 40, 2002, 272–283.

47. Gopakumar TG, Lee JA, Kontopoulou M, Parent JS, Influence of clay exfoliation on the physical properties of montmorillonite/polyethylene composites, *Polymer* 43, 2002, 5483–5491.

48. Alexander M, Dubiois P. Polymer-layered silicate nanocomposites: preparation, properties and use of a new class of materials, *Mater. Sci. Eng.* 28, 2000, 1–63.

49. Donnet JB, Bansal RC, Wang MJ, Carbon Black–Science and Technology, 2nd ed., Marcel Decker, 1993, 1–64.

50. Accorsi J, Romero E, *Plastics Eng.* 29, 1995.

51. Sichel EK, Carbon Black-Polymer Composites, Marcel Dekker, New York, 1982.

52. Lyon F, Encyclopedia of Polymer Science and Engineering, 2nd ed., Wiley, New York, 2, 1985, 623.

53. Donnet J.-B, *Carbon* 32, 1994, 1305.

54. Rothon R, *Particulate-Filled Polymer Composites*, Longman Scientific and Technical, 1995.

55. Kraus G, Reinforcement of elastomers by carbon black, *Die Angewandte Makromolekulare Chemie* 13/61 (1 977) 215–248 (Nr. 843).

56. Thongruang W, Spontak RJ, Balik CM, *Polymer* 43, 2002, 2279–2286.

57. Pan YX, Yu ZZ, Ou YC, Hu GH, *J. Polym. Sci. B: Polym. Phys.* 38, 2000, 1626–1633.

58. Nagatta K, Iwabuki H, Nigo H, *Compos. Inter.* 6, 1999, 1223.

59. Zheng W, Lu X, Wong S-C, *J. Appl. Polym. Sci.* 91, 2004, 2781.

60. Zheng W, Wong SC, *Compos. Sci. Technol.* 63, 2003, 225–235.

61. Chen GH, Wu DJ, Weng WG, Yan WL, *J. Appl. Polym. Sci.* 82, 2001, 2506–2513.

62. Pan YX, Yu ZZ, Ou YC, Hu GH, *J. Polym. Sci. B: Polym. Phys.* 38, 2000, 1626–1633.

63. Celzard A, Furdin G, Mareche JF, McRae E, *Solid State Comm.* 92, 1994, 377–383.

64. Feller JF, Linossier I, Grohens Y, *Mater. Lett.* 57, 2002, 64.

65. Vilcakova J, Saha P, Kresalek V, Quadrat O, *Synth. Met.* 113, 2000, 83.

66. Agari Y, Ueda A, Nagai S, *J. Appl. Polym. Sci.* 52, 1994, 1223.

67. Wilham R, Carbon fibers, *Adv. Mater.* 2(11), 1990, 528–536.

68. Iijima S, *Nature* 354(6348), 1991, 56–58.

69. Iijima S, Ichihashi T, *Nature* 363(6430), 1993, 603–605.

70. Treacy MMJ, Ebbesen TW, Gibson JM, *Nature*, 381(6584), 1996, 678–680.

71. Despres JF, Daguerre E, Lafdi K, Flexibility of graphene layers in carbon nanotubes, *Carbon* 33, 1995, 87–89.

72. Iijima S, Brabec Ch, Maiti A, Bernholc J, Structural flexibility of carbon nanotubes, *J. Phys. Chem.*104, 1996, 2089–2092.

73. Hone J, *Appl. Phys.* 273, 2001.

74. Sander J, ARM, Verschueren, C Deeker, *Nature* 393, 1998, 49.

75. Harris PJF, *Int. Mater. Rev.* 49(1), 2004, 31–43.

76. Breuer O, Sundararaj U, Big returns from small fibers: a review of polymer/carbon nanotube composites, *Polym. Compos.* 25(6), 2004.

77. Hughes TV, Chambers CR, U.S. Patent 405,480 1889.

78. Endo M, Kim YA, Hayashi T, Fukai Y, Oshida K, Terrones M, et al., *Appl. Phys. Lett.* 80, 2002, 1267.

79. Endo M, Kim YA, Ezaka M, Osada K, Yanagisawa T, Hayashi T, et al. *Nano. Lett.* 3, 2003, 723.

80. Merkulov VI, Lowndes DH, Wei YY, Eres G, Voelkl E, *Appl. Phys. Lett.* 76, 2000, 3555.

81. Sandler JKW, Kirk JE, Kinloch IA, Shaffer MSP, Windle AH, *Polymer* 44, 2003, 5893.

82. Park C, Ounaies Z, Watson KA, Pawlowski K, Lowther SE, Connell JW, Siochi EJ, Harrison JS, Clair TL. St., ICASE NASA Langley Research Center Hampton, Virginia, October 2002.

83. Pötschke P, Fornes TD, Paul DR, *Polym.* 43, 2002, 3247.

84. Du FM, Fischer JE, Winey KI, *J. Polym. Sci. B* 41, 2003, 3333.

85. Mierczynska A, Friedrich J, Maneck HE, Boiteux G, Jeszka JK, *Cent. Eur. J. Chem.* 2, 2004, 363.

86. Andrews R, Jacques D, Minot M, Rantell T, *Macromol. Mater. Eng.* 287, 2002, 395.

87. Jia Z, Wang Z, Xu C, Liang J, Wei B, Wu D, Zhu S, *Mater. Sci. Eng. A*, 271, 1999, 395–400.

88. Stéphan C, Nguyen TP, Lamy de la Chapelle M, Lefrant S, Journet C, Bernier P. *Synth. Met.*, 108, 2000, 139–149.

89. Cooper CA, Ravich D, Lips D, Mayer J, Wagner HD, *Compos. Sci. Technol.* 62, 2002, 1105–1112.

90. Gong X, Liu J, Baskaran S, Voise RD, Young JS. *Chem. Mater.* 12, 2000, 1049–1052.

91. Schadler LS, Giannaris SC, Ajayan PM, *Appl. Phys. Lett.* 73, 1998, 3842–3844.

92. Vaccarini L, De´sarmot G, Almairac R, Tahir S, Goze C, Bernier P, Proceedings of the XIV International Winterschool, Kirchberg, Tyral, 2000, Kuzmany, H, Fink, J, Mehring, M, Roth, S, eds. (AIP Conference Proceedings 544, Woodbury, NY). *Electr. Prop. Nov. Mater-Mol. Nano.* 2000, 521–525.

93. Kearns JC, Shambaugh RL, Polypropylene fibers reinforced with carbon nanotubes, *J. Appl. Polym. Sci.* 86, 2002, 2079–2084.

94. Dondero W, Morphological and Mechanical Properties of Carbon Nanotube/Polymer Composite via Melt Compounding. Ph.D thesis, NC State University, 2005.

95. Safadi B, Andrews R, Grulke EA, *J. Appl. Polym. Sci.* 84, 2001, 2660–2669.

96. Thostenson ET, Chou TW, *J. Phys. D: Appl. Phys.* 35, 2002, L77–L80.

97. Moore EM, Ortiz DL, Marla VT, Shambaugh RL, Grady BP, Enhancing the strength of polypropylene fibers with carbon nanotubes, *J. Appl. Polym. Sci.* 93, 2004, 2926–2933.

98. Pötschke P, Fornes TD, Paul DR, *Polymer* 43, 2002, 3247–3255.

99. Haggenmuller R, Gommans H, Rinzler AG, Fischer JE, Winey KI, *Chem. Phys. Lett.* 330, 2000, 219.

100. Weisenberger MC, Grulke EA, Jacques D, Rantell AT, Andrewsa R, Enhanced mechanical properties of polyacrylonitrile/multiwall carbon nanotube composite fibers, *J. Nanosci. Nanotechnol.* 3(6), 2003, 535–539(5).

101. Chang CM, Wu J, Li JX, Cheung YK, Polypropylene/calcium carbonate nanocomposites, *Polymer* 4, 2002, 2981-2992.

102. Wu D, Wang X, Song Y, Jin R, Nanocomposites of poly(vinyl chloride) and nanometric calcium carbonate particles: effect of chlorinated polyethylene on mechanical properties, morphology, and rheology, *J. Appl. Polym. Sci.* 92, 2004, 2714–2723.

103. Suwanprateeb J, Calcium carbonate filled polyethylene: correlation of hardness and yield stress, *Composites Part A: Appl. Sci. and Manufact.* (Incorporating Composites and Composites Manufacturing) 31(4), 2000, 353–359(7).

104. Ash BJ, Stone J, Rogers DF, Schandler LS, Siegel BC, Benicewicz T, *Mater. Res. Soc. Symp. Proc.* 661, 2000.

105. Mlynaŕckova Z, Borsig E, Lege´n J, Marcinc A, Alexy P, Influence of the composition of polypropylene/organoclay nanocomposite fibers on their tensile strength. *J. Macromol. Sci., Part A: Pure and Appl. Chem.* 42, 2005, 543–554.

106. Shelley JS, Mather PT, DeVries KL, *Polymer* 42, 2001, 5849–5858.

107. Kaempfer D, Thomann R, Mulhaupt R, Melt compounding of syndiotactic polypropylene nanocomposites containing organophilic layered silicates and in situ formed core/shell nanoparticles, *Polymer* 43, 2002, 2096–2916.

108. Blumstein A, *J. Polym. Sci., Part A: Poly. Chem.* 3, 1965, 2665.

109. Chiu Chih-Wei, Spinning and Elongational Flow Behavior of Nanoscale TiO2/pp Fibers, e-thesis Feng Chia University, June 2004, 66.

110. Burnside SD, Giannelis EP, *Chem. Mater.* 7, 1995, 1597.

111. Gilman JW, Kashiwagi T, Lichtenhan J. Recent advances in flame retardant polymer nanocomposites. In: Proceedings of Conference on Fire and Materials, Jan 22–24. London: Interscience Communications, 2001, 273–284.

112. Gilman JW. Flammability and thermal stability studies of polymer layered-silicate (clay) nanocomposites. *Appl. Clay. Sci.* 15(1–2), 1999, 31–49.

113. Mai YW, Yu Z, Xie X, Zhang Q, Ma J, Polymer nanocomposites and their application, *Transaction*, 10(4), 67–72.
114. Zhang S, Horrocks AR, A review of flame retardant polypropylene fibers, *Prog. Polym. Sci.* 28, 2003, 1517–1538.
115. Yano K, Ususki A, Okada A, Synthesis and properties of polyimide-clay hybrid films, *J. Polym. Sci. A: Polym. Chem.* 35 1997, 2289–2294.
116. Yano K, Ususki A, Okada A, Synthesis and properties of polyimide-clay hybrid films, *J. Polym. Sci. A: Polym. Chem.* 31, 1993, 2493–2498.
117. Neilson LE, *J. Macromol. Sci. (Chem.)* A1(5), 1967, 929–942.
118. Kojimoto Y, Usuki A, Kawasumi M, Okada A, Kuraucki T, Kamigiato O, *J. Appl. Polym. Sci.* 49, 1993, 1259.
119. Messersmith PB, Giannelis EP, *Chem. Mater.* 6, 1994, 1719.
120. Barrier Messersmith, Giannelis EP, *J. Poly. Sci.* 33, 1995, 1047–1057.
121. *Modern Plastics*, February 1998
122. *Modern Plastics*, February 1999
123. Strawhecker KE, Manias E. Structure and properties of poly(vinyl alcohol)/Na -montmorillonite nanocomposites. *Chem. Mater.* 12, 2000, 2943–2949.
124. Tang BZ, Xu H, *Macromolecule* 32, 1999, 2569–2576.
125. O'Flaherty SM, Murphy R, Hold SV, Cadak M, Coleman JN, Blau WJ, *J. Phy. Chem. B.* 107, 2003, 958–964.
126. Beringer J, Hofer D, Nanotechnology and its Application, Milliand English, September 2004, E107.
127. Leodidou TK, Margraf P, Caseri W, Suter UW, Walther P, Polymer sheets with a thin nanocomposite layer acting as a UV filter, *Polym. Adv. Technol.* 8, 8, 505–512.
128. Xiong M, Gu G, You B, Wu L, Preparation and characterization of poly(styrene butylacrylate) latex/nano-ZnO nanocomposite, *J. Appl. Polym. Sci.* 90, 2005, 1923–1931.
129. Zheng Y, Zheng Y, Ning R, Effects of nanoparticles SiO2 on the performance of nanocomposites, *Mater. Lett.* 57, 2003, 2940–2944.
130. Liu CH, Lai PL, The development of Nano nO polyester fiber, *J. Chine Textile Institute*, 14(3), 2004.
131. Hartley SM, Axtel Holly, The Next Generation of Chemical and Biological Protective Material Utilizing Reactive Nanoparticles, Gentex Corporation, Carbondale, PA, 18407.
132. Chiu Chih-Wei, Spinning and Elongational Flow Behavior of Nanoscale TiO2/pp Fibers, Feng Chia University e-thesis, June 2004, 66.
133. Sugunan A, Dutta J, Nanoparticles for nanotechnology, *Jilid PSI* 4, 2004, 1–2.
134. Zimmermann L, Weibel M, Caseri W, and Suter U, High refractive index films of polymer nano-composites, *J. Mater. Res.* 8(7), 1742.
135. Leodidou KT, Althaus HJ, Wyser Y, Vetter D, Büchler, Caseri W, Suter U, High refractive index materials of iron sulfides and poly(ethylene oxide), *J. Mater. Res.* 12(8), 1997.
136. Hegemann D, Stain repellent finishing on fabrics, *Adv. Eng. Mater.* 7(5), 401–404
137. New Metal Coating, Nonwovens Report International, August 2005, 28.
138. Rubner MF, Multi layer thin films, 2003, 133–154.
139. Igor Luzinov, Ultrahydrophobic Fibers: Lotus Approach, National Textile Center Annual Report: November 2004 NTC Project: C04-CL06.
140. Walid AD, John HX, Xiaoming T, Superhydrophobic silica nanocomposite coating by a low-temperature process, *J. Am. Ceram. Soc.* 87(9), 2004, 1782–1784.
141. www.biologie.uni-hamburg.de.
142. Jain P, Pradeep T, Potential of silver nanoparticle-coated polyurethane foam as an antibacterial water filter, *Biotechnol. Bioeng.* 90(1), 2005.
143. Lee HJ, Yeo SY, Jeong SH, Antibacterial effect of nanosized silver colloidal solution on textile fabrics, *J. Mater. Sci.* 38, 2003, 2199–2204.
144. New products — Alternative Canvas, Nonwoven report International, February 2004, 19.

145. Devaux E, Rochery M, Bourbigot S. Polyurethane/clay and polyurethane/POSS nanocomposite as flame retardant coating for polyester and cotton fabrics, *Fire Mater.* 26, 4–5, 149–154.

146. Ian H, Nanotechnolgy: small is beautiful, *Int. Dyer*, 190, 2005, July 6, 5–6.

147. Mc Cluskey P, Nagvanski M, Nanocomposite materials offer higher conductivity and flexibility, Adhesive Joining and Coating Technology in Electronics Manufacturing, 1998. Proceedings of 3rd International Conference on 28–30 Sept. 1998, 282–286.

148. Vysotskii VV, Pryamova TD, Roldugin VI, Shamurina MV, Percolation transition and conductivity mechanisms in metal-filled polymer, *Kolloidnyi Zhurnal*, Sept–Oct, 1995, 649–654.

149. Sandler JKW, Kirk JE, Kinloch IA, Shaffer MSP, Windle AH, Ultra-low electrical percolation threshold in carbon-nanotube-epoxy composites, *Polymer* 44, 2003, 5893–5899.

150. Smith Jr., JG, Delozier DM, Connell JW, Watson KA, *Polymer*, 45, 2004, 6133.

151. Burton DJ, Glasgow DG, Lake ML, Kwag C, Finegan JC, 46th International SAMPE Symposium and Exhibition: Materials and Processes Odyssey, Long Beach, CA, May 2001.

152. Zheng W, Lu X,, Wong SC, Electrical and mechanical properties of expanded graphite-reinforcd high density polyethylene, *J. Appl. Polym. Sci.* 91(5), 2004, 2781–2788.

153. Zheng W, Wong, SC, Electrical conductivity and dielectric properties of PMMA/expanded graphite composites, *Composite Sci. Technol.* 63(2), 2003, 225–235.

154. Thompson CM, Preparation and characterization of metal oxide/polyimide nanocomposites, *Composites Sci. Technol.* 63(11), 2003, 1591–1598.

155. Pothukuchi S, Li Y, Wong CP, Development of novel polymer-metal nanocomposite obtained through the route of in-situ reduction for integral capacitor application, *J. Appl. Polym. Sci.* 93(4), 2004, 1531–1538.

156. Kashiwagi T, Grulke E, Hilding J, Groth, H, Harris, R, Butler, K, Shields J, Kharchenko S, Douglas J, Thermal and flammability properties of polypropylene/carbon nanotube nanocomposites, *Polymer* 4228 45, 2004, 4227–4239.

157. Yi Li, Suresh Pothukuchi, CP Wong, Formation and Dielectric Properties of a Novel Polymer-Metal Nanocomposite, 9th International Symposium on Advanced Packaging Materials, 175–181.

158. Yahamoto T, Kutoba E, Taniguchi A, Dev S, Tanaka K, Osakada K, Sumita M, *Chem. Mater.* 4, 570–576, 1992.

159. Chung DDL, Composite Materials: Functional Materials for Modern Technologies, Springer-Verlag, London, 2003.

160. Deng J, Ding X, Zhang W, Png Y. Wang J, Long X, Li P. Chan ASC, *Polymer* 43, 2002, 2179–2184.

161. Fisher S, Nanoparticle reinforced Natural Plastics, Book Natural Fibers, Plastics and Composites, Wallenberger F, Weston N, eds. Kluwere Academic Publishers, 345–365.

162. Kornmann LA, Berglund JS, Giannelis EP, *Polym. Eng. Sci.* 38, 1998, 1351.

163. Okamoto M, Biodegradable polymer/layered silicate nanocomposites: a review, Handbook of Biodegradable Polymeric Materials and Their Applications, Surya Mallapragada, Balaji Narasimhan, eds. Vol. 1, 1–45.

164. Park HM, Li X, Jin CZ, Park CY, Cho WJ, Ha CS. *Macromol. Mater. Eng.* 8, 2002, 553.

165. Park HM, Li X, Jin CZ, Park CY, Cho WJ, Ha CS, *J. Mater. Sci.* 38, 2003, 909.

166. Sinha Ray S, Okamoto M, Yamada K, Ueda K, *Nano. Lett.* 2, 2002, 1093.

167. Sinha Ray, S Yamada K, Okamoto M, Ueda K, *Polymer* 44, 2003, 857.

168. Jhala PB. Nanotechnology in textiles: the rising Wave, *Indian Textile* J., 2004 13–18.

169. Murday S, The coming revolution — Science and Technology of Nanoscale structures, AMPTIAC quarterly, 6:1 5–10.

170. http://www.defenselink.mil/news/Jul2004/n07272004_2004072705.html July 27 2004.

5

Functional Structures

22 Carbon Nanotubes *M. Meyyappan, Deepak Srivastava* 22-1

Introduction • Structure and Properties of CNTs • Computational Modeling and Simulation • Nanotube Growth • Material Development • Application Development • Concluding Remarks

23 Mechanics of Carbon Nanotubes *Dong Qian, Gregory J. Wagner, Wing Kam Liu, Min-Feng Yu, Rodney S. Ruoff* 23-1

Introduction • Mechanical Properties of Nanotubes • Experimental Techniques • Simulation Methods • Mechanical Applications of Nanotubes • Conclusions

24 Dendrimers — an Enabling Synthetic Science to Controlled Organic Nanostructures *D.A. Tomalia, S.A. Henderson, M.S. Diallo* 24-1

Introduction • The Dendritic State • Unique Dendrimer Properties • Structural and Physical Properties of Dendrimers in Solution • Dendrimers as Nanopharmaceuticals and Nanomedical Devices • Dendrimers as Functional Nanomaterials • Dendrimers as Reactive Modules for the Synthesis of More Complex Nanoscale Architectures • Core–Shell Patterns Influencing the Modular Reactivity of Dendrimers • Conclusions

25 Design and Applications of Photonic Crystals *Dennis W. Prather, Ahmed S. Sharkawy, Shouyuan Shi, Caihua Chen* 25-1

Introduction • Photonic Crystals? How Do They Work? • Analogy between Photonic And Semiconductor Crystals • Analyzing Photonic Bandgap Structures • Electromagnetic Localization in Photonic Crystals • Doping of Photonic Crystals • Microcavities in Photonic Crystals • Photonic Bandgap Applications • Applications Utilizing Irregular Dispersive Phenomena in Photonic Crystals • Future Applications and Concluding Remarks

26 Progress in Nanofluidics for Cell Biology *Todd Thorsen, Joshua S. Marcus* 26-1

Introduction • General Manipulation and Immobilization of Cells in Microdevices • Cell Sorting in Microdevices • Devices to Monitor Single-cell Physiology • Cell Culture in Microdevices • Conclusions

27 Carbon Nanostructures and Nanocomposites *Yanhong Hu, Zushou Hu, Clifford W. Padgett, Donald W. Brenner, Olga A. Shenderova* 27-1

Introduction to Composites • Modern History of Carbon Nanostructures • The Extended Carbon Family • Structure, Notation, and Synthesis of the Common Carbon Nanostructures • Nanocomposites Using Fullerenes • Nanocomposites Using Nanotubes • Nanocomposites Using Diamond-Like Structures • Remaining Challenges and Opportunities

28 **Contributions of Molecular Modeling to Nanometer-Scale Science
and Technology** *Donald W. Brenner, Olga A. Shenderova, J.D. Schall,
D.A. Areshkin, S. Adiga, J.A. Harrison, S.J. Stuart* **28**-1

Opening Remarks • Molecular Simulations • First-Principles Approaches: Forces
on the Fly • Applications • Concluding Remarks

29 **Accelerated Design Tools for Nanophotonic Devices and Applications**
*James P. Durbano, Ahmed S. Sharkawy, Shouyuan Shi, Fernando E. Ortiz,
Petersen F. Curt, Dennis W. Prather*..**29**-1

Introduction • The Celerity™ Hardware Accelerator • Mie Theory
Comparison • Coupled-Resonator Optical Waveguides • Analysis of Negative Index
(Left-Handed) Materials • Subwavelength Optical Systems • Summary and Conclusion

30 **Nanoparticles for Drug Delivery** *Meredith L. Hans, Anthony M. Lowman* **30**-1

Introduction • Synthesis of Solid Nanoparticles • Processing
Parameters • Characterization • Nanoparticulate Delivery Systems • Targeted Drug
Delivery Using Nanoparticles • Drug Release

22

Carbon Nanotubes

22.1	Introduction	22-1
22.2	Structure and Properties of CNTs	22-2
22.3	Computational Modeling and Simulation	22-3
	Nanomechanics of Nanotubes and Composites • Vibrational Characteristics and Thermal Transport in Nanotubes • Molecular Electronics with Nanotube Junctions • Sensors and Actuators	
22.4	Nanotube Growth	22-14
	Arc Process and Laser Ablation • Chemical Vapor Deposition • Catalyst Preparation • Continuous, High-Throughput Processes	
22.5	Material Development	22-19
	Purification • Characterization • Functionalization	
22.6	Application Development	22-20
	CNTs in Microscopy • CNT-Based Nanoelectronics • Sensors • Field Emission • Nanotube–Polymer Composites • Other Applications	
22.7	Concluding Remarks	22-27
	Acknowledgments	22-27
	References	22-27

M. Meyyappan
Deepak Srivastava
NASA Ames Research Center

22.1 Introduction

Carbon nanotubes (CNTs) were discovered by Iijima [1] as elongated fullerenes in 1991. Since then, research on growth, characterization, and application development has exploded due to the unique electronic and extraordinary mechanical properties of CNTs. The CNT can be metallic or semiconducting, and thus it offers possibilities to create semiconductor–semiconductor and semiconductor–metal junctions useful in electronic devices. The high-tensile strength, Young's modulus, and other mechanical properties hold promise for high-strength composites for structural applications. Researchers have been exploring the potential of CNTs and making progress in the past decade in a wide range of applications: nanoelectronics, sensors, field-emission-based displays, batteries, polymer matrix composites, reinforcement material, and electrodes, to name a few. In this chapter, an overview of this rapidly emerging field is provided. First, the structure of the nanotube and properties are explained. Unlike many other fields in science and engineering, the evolution of CNTs to its current level owes significantly to the contributions from modeling and simulation. Computational nanotechnology has played an early and major role in predicting as well as explaining the interesting properties of CNTs. So, a section is devoted to modeling and simulation after the description of the properties. Then, nanotube growth is covered in detail followed by material development functions such as purification, characterization, etc. Finally, a review of the current status of various applications is provided. For a detailed discussion on all these aspects, the reader is referred to a recent textbook [2].

22.2   Structure and Properties of CNTs

A single-wall CNT (SWCNT) is best described as a rolled-up tubular shell of graphene sheet (Figure 22.1[a]), which is made of benzene-type hexagonal rings of carbon atoms [2]. The body of the tubular shell is thus mainly made of hexagonal rings (in a sheet) of carbon atoms, whereas the ends are capped by half-dome-shaped half-fullerene molecules. The natural curvature in the sidewalls is due to the rolling of the sheet into the tubular structure, whereas the curvature in the end caps is due to the presence of topological (pentagonal rings) defects in the otherwise hexagonal structure of the underlying lattice. The role of a pentagonal ring defect is to give a positive (convex) curvature to the surface, which helps in closing the tube at the two ends. A multiwall carbon nanotube (MWCNT) is a rolled-up stack of graphene sheets into concentric cylinders, with the ends again either capped by half-fullerenes or kept open. A nomenclature (n,m), used to identify each SWCNT, refers to integer indices of two graphene unit lattice vectors corresponding to the chiral vector of a nanotube [2]. Chiral vectors determine the directions along which the graphene sheets are rolled to form tubular shell structures perpendicular to the tube axis vectors, as explained in Ref. [2]. The nanotubes of type (n,n), as shown in Figure 22.1(b), are commonly called armchair nanotubes because of the $\backslash\!\!\!\!\_\!\!\!/\backslash\!\!\!\!\_\!\!\!/$ shape, perpendicular to the tube axis, and have a symmetry along the axis with a short unit cell (0.25 nm) that can be repeated to make the

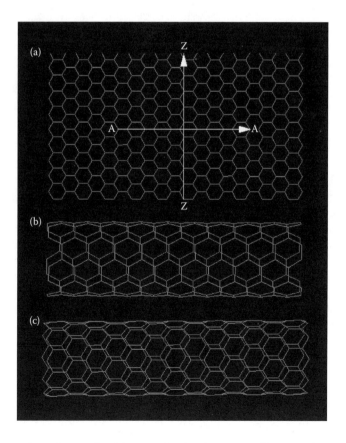

FIGURE 22.1   (a) A graphene sheet made of C atoms placed at the corners of hexagons forming the lattice; arrows A–A and Z–Z denote the rolling direction of the sheet to make the nanotube; (b) a (5,5) armchair nanotube; (c) a (10,0) zig-zag nanotube.

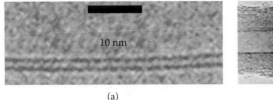 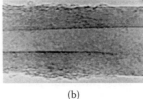

(a)   (b)

FIGURE 22.2 TEM images of (a) SWCNT and (b) MWCNT.

entire section of a long nanotube. Another type of nanotube $(n,0)$ is known as the zig-zag nanotube (Figure 22.1[c]) because of the /\/\ shape perpendicular to the axis; it also has a short unit cell (0.43 nm) along the axis [3]. All the remaining nanotubes are known as chiral or helical nanotubes and have longer unit cell sizes along the tube axis. Transmission electron microscopy (TEM) images of a SWCNT and a MWCNT are shown in Figure 22.2. The symmetry properties of the nanotubes of different chiralities are explained in detail in Ref. [2].

The SWCNTs and MWCNTs are interesting nanoscale materials for the following reasons.

A SWCNT can be either metallic or semiconducting, depending on its chiral vector (n,m), where n and m are two integers. A metallic nanotube is obtained when the difference $n - m$ is a multiple of three. If the difference is not a multiple of three, a semiconducting nanotube is obtained. In addition, it is also possible to connect nanotubes with different chiralities, creating nanotube heterojunctions, which can form a variety of nanoscale molecular electronic device components.

Single- and multiwall nanotubes have very good elastic and mechanical properties because the 2-D arrangement of carbon atoms in a graphene sheet allows large out-of-plane distortions, while the strength of carbon–carbon in-plane bonds keeps the graphene sheet exceptionally strong against any in-plane shear distortion or fracture. These structural and materials characteristics of nanotubes lead to their possible use in making lightweight but highly elastic and very strong composite materials.

Nanotubes are high aspect-ratio structures with good electrical and mechanical properties. Consequently, the application of nanotubes in field emission displays, scanning probe microscopic tips for metrology, and other areas have started to materialize in the commercial sector, as will be discussed later in this chapter.

Since nanotubes are hollow, tubular, and caged molecules, they have been proposed as lightweight, large-surface-area packing material for gas storage and hydrocarbon fuel storage devices, gas or liquid filtration devices, nanoscale containers for molecular drug delivery, and casting structures for making nanowires and nanocapsulates.

It is also worth noting that carbon-based materials are ideally suitable as molecular-level building blocks for nanoscale system design, fabrication, and applications. From a structural or functional materials perspective, carbon is the only element that exists in a variety of shapes and forms with varying physical and chemical properties. For example, diamond and layered graphite forms of carbon are well known, but the same carbon also exists in planar sheet, rolled-up tubular, helical spring, rectangular hollow box, and nanoconical forms. All basic shapes and forms needed to build any complex molecular-scale architectures are thus readily available with carbon. Additionally, by coating any carbon-based nanoscale devices with biological lipid layers and protein molecules, it also may be possible to extend into the rapidly expanding area of bionanotechnology.

22.3 Computational Modeling and Simulation

The structural, electronic, mechanical, and thermal properties of interacting, bulk condensed matter systems were studied in the earlier days with analytical methods for infinite systems. Numerical simulations of the finite size systems have become more common recently due to the availability of powerful computers. Molecular dynamics (MD) refers to an approach where the motion of atoms or molecules is treated in approximate finite difference equations of Newtonian mechanics. The use of classical mechanics is well justified, except when dealing with very light atoms and very low temperatures. The dynamics of complex condensed phase systems such as metals and semiconductors is described with explicit or implicit many-body force field functions using embedded atom method type potentials for metals [4], and Stillinger-Weber, [5] and Tersoff-Brenner (T-B) [6,7] type potentials for semiconductors [8]. The T-B type potentials are parameterized and particularly suited for carbon-based systems (such as CNTs) and have been used in a wide variety of scenarios, yielding results in agreement with experimental observations. However, currently, there is no universal force field function that works for all materials and in all scenarios. Consequently, one needs to be careful where true chemical changes (involving electronic rearrangements with forming and breakup of bonds) with large atomic displacements are expected to occur.

In its global structure, a general MD code typically implements an algorithm to find a numerical solution for a set of coupled first-order ordinary differential equations given by the Hamiltonian formulation of Newton's second law. The equations of motion are numerically integrated forward in finite time steps using a predictor–corrector method. A major distinguishing feature of the T-B potential [6,7] is that short-range bonded interactions are reactive so that chemical bonds can form and break during the course of a simulation. Therefore, compared to other MD codes, the neighbor list describing the environment of each atom includes only a few atoms and needs to be updated more frequently. The computational cost of the many-body bonded interactions is relatively high compared to the cost of similar methods with nonreactive interactions with simpler functional forms. As a result, the overall computational costs of both short-range interactions and long-range, nonbonding van der Waals (VDW) (Lennard Jones [6–12]) interactions are roughly comparable.

For large-scale atomistic modeling (10^5 to 10^8 atoms), multiple processors are used for MD simulations, and the MD code needs to be parallelized. A route to the parallelization of a standard MD code involves decoupling the neighbor list construction from the computation of the atomic forces, and parallelizing each part in the most efficient way possible. Parallelization of the MD code using T-B potentials for carbon atom interactions was attempted and achieved recently. An example of the parallel implementation of this classical MD code is described in detail in Ref. [9]. This parallelized MD code has been utilized in simulations of mechanical properties of the nanotubes, nanotube–polymer composites, mechanical strain-driven chemistry of CNTs, and molecular gears and motors powered by laser fields [3].

In recent years, several more accurate quantum MD schemes have been developed in which the forces between atoms are computed at each time step via quantum mechanical calculations within the Born–Oppenheimer approximation. The dynamic motion for ionic positions is still governed by Newtonian or Hamiltonian mechanics and described by MD. The most widely known and accurate scheme is the Car–Parrinello (CP) molecular dynamic method [10], where the electronic states and atomic forces are described using the ab initio density functional method (usually within the local density approximation [LDA]). In the intermediate regimes, the tight-binding MD (TBMD) [11] approach for up to a few thousand atoms provides an important bridge between the ab initio quantum MD and classical MD methods. The computational efficiency of the tight-binding method derives from the fact that the quantum Hamiltonian of the system can be parameterized. Furthermore, the electronic structure information can be easily extracted from the tight-binding Hamiltonian, which, in addition, also contains the effects of angular forces in a natural way. In a generalized nonorthogonal TBMD scheme, Menon and Subbaswami have used a minimal number of adjustable parameters to develop a transferable scheme applicable to clusters as well as bulk systems containing Si, C, B, N, and H [12,13]. The main advantage of this approach is that it can be used to find an energy-minimized structure of a nanoscale system under consideration without symmetry constraints. The parallelization of the TBMD code involves parallelization of the direct diagonalization part (of the electronic Hamiltonian matrix) as well as that of the MD part. The parallelization of a sparse symmetric matrix giving many eigenvalues and eigenvectors is a complex step in the simulation of large intermediate range systems and needs development of new algorithms.

The ab initio or first principles method is an approach to solve complex quantum many-body Schrödinger equations using numerical algorithms [14]. The ab initio method provides a more accurate description of the quantum mechanical behavior of materials properties even though the system size is currently limited to only a few hundred atoms. Current ab initio simulation methods are based on a rigorous mathematical foundation of the density functional theory (DFT) [15,16]. This is derived from the fact that the total electronic energy in the ground state is a function of the density of the system. For practical applications, the DFT–LDA method has been implemented with a pseudo-potential approximation and a plane-wave-basis expansion of single-electron wave functions [14]. These approximations reduce the electronic structure problem to a self-consistent matrix diagonalization problem. A popular DFT simulation program is the Vienna Ab Initio Simulation Package, which is available through a license agreement [17].

In computational nanotechnology research, these three simulation methods can be used in a complementary manner to improve computational accuracy and efficiency. Based on experimental observations or theoretical dynamic and structure simulations, the atomic structure of a nanosystem can first be investigated. After the nanoscale system configurations are finalized, the electronic behaviors of the system are investigated through static ab initio electronic energy minimization schemes [14] or through studies of the quantum conductance [18] behavior of the system. The structural, mechanistic, thermal transport, and chemical reaction characteristics on the other hand are generally simulated first with large-scale atomistic MD methods, and smaller local regions involving chemical rearrangement of atoms are then validated with more accurate quantum tight-binding and ab initio DFT methods. This strategy has been covered in detail in a recent review article focusing on computational nanotechnology of CNTs [3,19].

In the following sections, we describe several representative examples where computational nanotechnology has clearly played an important role in either explaining some recent experimental observations or predicting structures (or properties) that have been later fabricated (or measured) in experiments.

22.3.1 Nanomechanics of Nanotubes and Composites

SWCNTs and MWCNTs have been shown to have exceptionally strong and stiff mechanical characteristics along the axis of the tube and very flexible characteristics along the normal to the axis of the tube [19–22]. For axial deformations, the Young's modulus of the SWCNTs can reach beyond 1 TPa, and the yield strength can be as large as 120 GPa. The initial investigations, using classical MD simulations withT-B potential, showed that the tubes are extremely stiff under axial compression, and that the system remains within the elastic limit even for very large deformations (up to 15% strain) [9,20]. Nonlinear elastic instabilities, with the appearance of "fin"-like structures, are observed during these deformations, but the system remains within the elastic limit and returns to the original unstrained state as soon as the external constraining forces are removed. As shown in Figure 22.3, when compressed beyond elastic limits, the SWCNT and MWCNT undergo sideways bucklings, and plastic deformations occur mainly through extreme bending situations in the sideways-buckled tubes. A significantly different deformation mode, however, was also predicted where the nanotube essentially remains straight, but the structure locally collapses at the location of the deformation [21] (Figure 22.4). This generally occurs for thin tubes and for tube lengths shorter than the Euler buckling limit. The local collapse is driven by graphitic to a diamond-like bonding transition at the location of the collapse. Both the simulated collapsing mechanisms have been observed in experiments [23]. The tensile strain on the other hand causes the formation of Stone-Wallace (SW) type topological defects in the tube (Figure 22.5[a]), which leads to thinning and collapse of the tube when stretched further [24,25].

It turns out that the yielding of a CNT under tension, via the formation of SW defects, is inherently a kinetic phenomenon, and a more relevant quantity is the activation energy or barrier to the formation of the SW-type defects in the CNTs under tension. The static in-plane activation barriers are found to be about 10 eV in nanotubes with no strain, and the value drops to about 5 to 6 eV for tensile strain between 5 and 10% [24,25]. Using simple transition state theory (TST), the breaking strain of SWCNTs under tensile stretch is predicted to be about 17% [26], which is smaller than the earlier predictions of about a 30% breaking strain simulated by the same group [20] but still much larger than the breaking strain of 6 to 12% observed in experiments on ropes or bundles of SWCNTs and on MWCNTs [27,28].

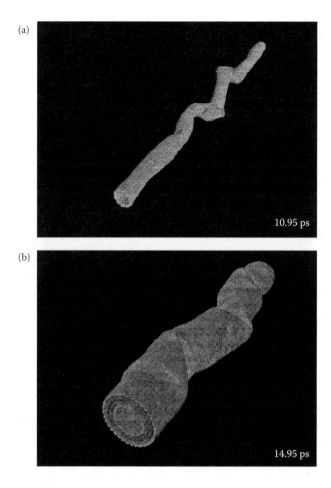

FIGURE 22.3 (a) An axially compressed SWCNT within elastic limit shows sideways buckling and accumulation of strain at the tip of the sideways-buckled structure; (b) same as in (a) except that it is a MWCNT with four walls.

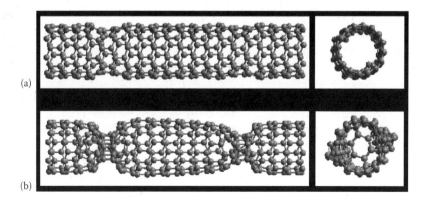

FIGURE 22.4 A 12% axially compressed (8,0) nanotube at the (a) beginning and (b) end of a spontaneous local plastic collapse of the tube, which is driven by diamond-like bonding transitions at the location of the collapse. See cross-sectional view in (b).

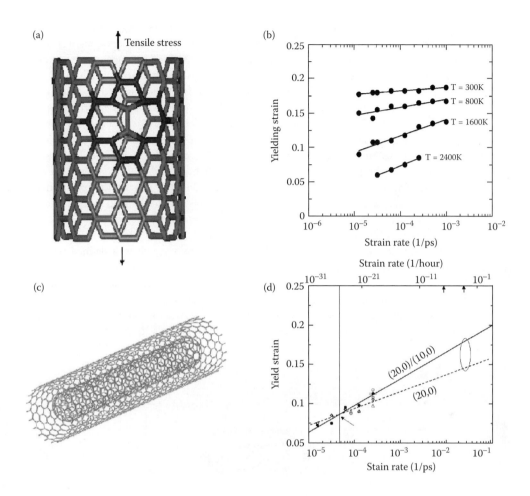

FIGURE 22.5 (a) The Stone-Wallace bond rotation defect on a zig zag CNT; (b) the yielding strain of a SWCNT as a function of temperature and strain rate; (c) a double-wall CNT as a model for yielding of MWCNTs; (d) the tensile yielding of the double-wall CNT as a function of strain rate, with arrows showing the strain rates that can be measured in experiments. (From C. Wei, D. Srivastava, and K. Cho, *Phys. Rev. B* 67, 115407 (2003); C. Wei, K. Cho, and D. Srivastava, *Appl. Phys. Lett.* 82, 2512 (2003). With permission.)

Yielding strains are found to be sensitive to both the strain rate and the kinetic temperature during the simulation. At the strain rate of 10^{-5}/ps, the yielding strain can vary between 15% at low temperature and 5% at high temperature [29]. Very extensive MD simulations over large variations in temperature and strain rate changes by three orders of magnitude reveal a complex dependence of the yielding strain on these parameters [29]. For example, the yielding strain of a 6 nm long (10,0) SWCNT at several temperatures and strain rates varying between 10^{-3}/ps and 10^{-5}/ps are shown in Figure 22.5(b). A strain rate- and temperature-dependent model of yielding for SWCNTs, within the TST-based framework, has been derived, which gives the yielding strain as

$$\varepsilon_Y = \frac{\bar{E}_v}{VK} + \frac{k_B T}{VK} \ln\left(\frac{N\bar{\varepsilon}}{n_{site}\bar{\varepsilon}_0}\right)$$

Here the yielding strain ε_Y depends on the activation volume V, force constant K, the temperature and time-averaged dynamic activation energy $\bar{E}_v$, the temperature T, the number of activation sites n_{site}, the intrinsic strain rate $\bar{\varepsilon}_0$, and the strain rate $\bar{\varepsilon}$. Some of the intrinsic parameters such as dynamic activation energy, activation volume, and intrinsic strain rates can be fitted from the MD simulation data as shown in Figure 22.5(b), while the remaining parameters such as the temperature, number of possible activation sites, and strain rates can be chosen to reflect the experimental reality. The room-temperature yielding strain of a 1-μm long (10,10) CNT comes out to be about 9% for a realistic and experimentally feasible strain rate of 1%/h [29]. This compares well with the yielding strain of 6 to 12% observed in experiments on bundles of SWCNTs and MWCNTs. The model has been extended to MWCNTs as well, and finds that the double-wall MWCNTs (Figure 22.5[c]) are stronger than the equivalent SWCNTs by a couple of percentage points of yielding strain at experimentally feasible strain rates and temperatures as shown in Figure 22.5(d) [30].

One of the major reasons for studying the mechanical and thermal properties of individual SWCNTs or MWCNTs is to explore the possibility of using them for lightweight, very strong, multifunctional composite materials. The structural strength or thermal characteristics of such composite materials depend on the transfer characteristics of such properties from the fiber to the matrix and the coupling between the two. In some cases, the coupling is through chemical interfacial bonds, which can be covalent or noncovalent in nature, while in other cases, the coupling could be purely physical in nature through nonbonded VDW interactions. Additionally, the aspect ratio of the fiber, which is defined as L/D (L is the length of the fiber and D is the diameter), is also an important parameter for the efficiency of mechanical load transfers, because the large surface area of the fiber is better for large load transfer.

Several recent experiments on the preparation and mechanical characterization of CNT–polymer composites have been reported [31–35]. The MD simulations of the structural and mechanical properties of polymer composites with embedded nanotubes (Figure 22.6[a]) have also been reported recently [36].

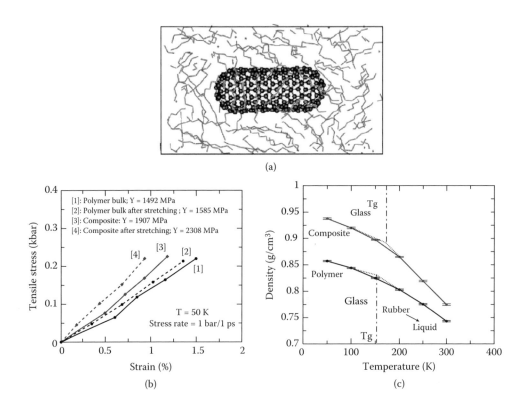

(a)

(b)

(c)

FIGURE 22.6 (a) The MD simulation cell of a CNT-PE composite, showing increase in (b) the Young's modulus, and (c) the glass transition temperature and thermal expansion coefficient (slope of the curves) of the composite over the pure PE system above the glass transition temperature. (From C. Wei, D. Srivastava, and K. Cho, *Nano. Lett.* 2, 647 (2002). With permission.)

The coupling at the interface was through nonbonded VDW interactions. Shown in Figure 22.6(b) is the strain–stress curve for both the composite system and the pure polyethylene (PE) matrix system. The Young's modulus of the composite is found to be 1900 MPa, which is about 30% larger than that of the pure polymer matrix system. A further enhancement of the Young's modulus of the same sample can be achieved by carrying the system through repeated cycles of loading–unloading of the tensile strain on the composite matrix. In agreement with the experimental observation, this tends to align the polymer molecules with the nanotube fibers, causing a better load transfer between the two. Frankland et al. [37] studied the load transfer between polymer matrix and SWCNTs and found that there is no permanent stress transfer for 100-nm long (10,10) CNTs within PE only if VDW interaction is present. They estimated that the interfacial stress could be 70 MPa with chemical bonding between SWCNT and the polymer matrix, while it is only 5 MP for the nonbonding case. Additionally, the density of the CNT–polyethylyne composite for short (10 monomers) and long (100 monomers) polymer chains has been simulated as a function of temperature [36]. The results (Figure 22.6[c]) show that, due to mixing of 8% by volume of SWCNT in PE, the glass transition temperature increases by about 20% and, more significantly, the thermal expansion coefficient above the glass transition temperature increases by as much as 142%. The enhanced thermal expansion coefficient is attributed mostly to an equivalent increase in the excluded volume of the embedded CNT as a function of temperature [36]. Since both excited vibrational phonon modes and Brownian motion contribute to the dynamic excluded volume of the embedded CNT, as the temperature is increased, their contributions toward the excluded volume increase significantly.

Recent MD simulations of CNTs in PE matrices have revealed nonbonded VDW interaction-induced structural ordering in the surrounding PE matrix molecules [38]. To minimize the interface energy between the CNT and the surrounding PE molecules, the PE molecules are found to form discrete absorption layers as a function of radial distance from the axis of the nanotube (Figure 22.7[a] and

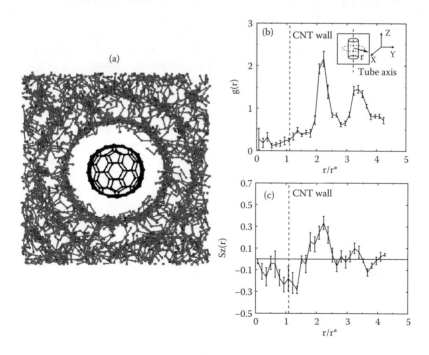

FIGURE 22.7 (a) Formation of PE layers around the CNT as an indication of the observed structural order in the MD-simulated sample with (b) radial distribution function with respect to tube axis as a function of distance from the axis, and (c) the orientation order parameter representing alignment of PE molecules along the axis of the tube in each layer. (From C. Wei, D. Srivastava, and K. Cho, *Nano. Lett.* 4, 1949 (2004). With permission.)

Figure 22.7[b]), and within each layer, the PE molecules prefer to align parallel to the axis of the tube and not wrap circumferentially around the nanotube (Figure 22.7[c]). An increase in the related structural order parameter is found to contribute toward the mechanical modulus of the composite, and the above-described orientation has been found to be inelastic in nature [38]. It is expected that such structural ordering will affect the thermal and electrical transport characteristics of the composite as well.

22.3.2 Vibrational Characteristics and Thermal Transport in Nanotubes

The simulated vibrational or phonon spectra of the nanotubes were initially used for characterization of the structure by identification and assignment of the observed peaks in the resonance Raman experiments [39]. The phonon spectrum is obtained by constructing and diagonalizing the dynamical matrix from position-dependent interatomic force constants. The eigenvalues of the dynamical matrix give the frequency of vibrations, and the corresponding eigenvectors give the nature of the corresponding modes of the vibration for atoms within a unit cell. The accuracy of the computed spectra at 0 K depends on the accuracy of the interatomic forces used in constructing the dynamical matrix; many studies on the zero temperature phonon spectra of CNTs using the ab initio density functional [40], tight-binding [41,42], and T-B interatomic potentials [43] have been conducted recently. Agreement between the experimental peaks and zero temperature vibrational peaks is generally good because only the configurations at or very close to equilibrium structures are investigated.

Thermal transport through CNTs, on the other hand, is sensitive to the choice of good atomic interaction potential, including the anharmonic part, because the atomic displacements far from equilibrium positions are also sampled. The room temperature phonon spectra or density of states and vibrational amplitudes were computed recently [44]. The spectra were simulated through Fourier transforms of the temperature-dependent velocity autocorrelation functions computed from MD trajectories using the T-B potential for C–C interactions. Good agreement with zero temperature phonon spectra, computed with higher-accuracy ab initio DFT and tight-binding methods, was obtained [40–43]. The simulations of thermal transport through CNTs under equilibrium and shock pulse conditions are rather few and relatively recent. The thermal conductivity of SWCNTs has been computed using direct MD simulation methods [45–47]. The typical room-temperature thermal conductivity of SWCNTs of 1 to 3 nm diameters is found to be around 2500 W/mK, as shown in Figure 22.8. The thermal conductivity shows a peaking behavior as a function of temperature. As the temperature is raised from very low values, more and more phonons are excited and contribute toward the heat flow in the system. However, at higher temperatures, phonon–phonon scattering starts to dominate and streamlined heat flow in the CNTs decreases, causing a peak in thermal conductivity in the intermediate temperature range. For all the simulated CNTs (of 1–2 nm diameter) as a function of the tube diameter and chirality, it was found that the peak position is around room temperature and is sensitive to the radius of the CNTs and not the chirality or the helicity [47]. This means that the thermal transport in SWCNTs is mostly through excitations of a low-frequency radial phonon mode and the coupling of the radial mode with the axial or longitudinal phonons in the low-frequency region [47,48]. The convergence of the thermal conductivity as a function of the length of the CNT, however, is still an issue in all the direct MD simulation results reported so far. A significant variation in the room temperature thermal conductivity of (10,10) CNT has been reported, partly because some of the studies have used very small CNT lengths in the simulations [45]. Preliminary investigations of the convergence with respect to the temperature gradient (dT/dX) across the CNT length have shown an inverse power law dependence of the thermal conductivity on the thermal gradient. Rigorous, long-time, and length-scale simulations are required to investigate the convergence behavior as a function of CNT length. It is noted, however, that only the absolute value of thermal conductivity changes but not the qualitative behavior of the relative peak heights and position as a function of temperature.

Another feasible type of heat transport involves subjecting CNTs to intense heat pulses [49]. A heat pulse propagating in a crystalline material at low temperatures is expected to generate the propagation of a second sound wave that is not attenuated by dissipative scattering processes. The MD simulation

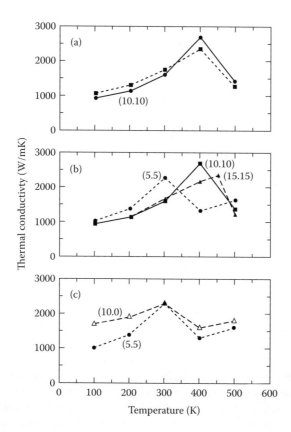

FIGURE 22.8 Thermal conductivity as a function of temperature for (a) armchair and zig-zag CNTs, (b) armchair nanotube of increasing radius, and (c) armchair and zig-zag nanotubes of the same radius. (From M.A. Osman and D. Srivastava, *Nanotechnology* 12, 21 (2001). With permission.)

technique is ideally suited for the investigation of such phenomena because the temperature, shape, and duration of heat pulses can be simulated in a controlled way in a CNT. The heat pulse duration was taken to be 1 psec, during which the temperature was ramped up and down according to a desired shape. The first sound, or the leading waves, move at higher speed, but the shape of the wave changes and intensity also decays. The leading wave is followed by a second sound wave that maintains its shape and intensity for the entire duration of the simulation. The propagation of both the first and the second sound waves was investigated. In zig-zag CNTs, the leading wave is found to move at the sound velocity of longitudinal acoustic phonons, whereas in the armchair CNTs, the leading wave moves with the sound velocity of transverse acoustic phonons. The main conclusion is that the leading stress waves under heat pulse conditions travel slower in armchair CNTs compared to those in zig-zag CNTs [49], which is consistent with the earlier thermal conductivity simulations [47] that showed a higher thermal conductivity for a zig-zag CNT than for an equivalent-size armchair CNT. The current status of vibrational phonon spectra and modeling and direct simulations of thermal transport under steady state and pulsed conditions is described in a recent review [50].

22.3.3 Molecular Electronics with Nanotube Junctions

The possibility of using carbon in place of silicon in the field of nanoelectronics has generated considerable enthusiasm. Metallic and semiconducting behavior and electronic transport through individual SWCNTs have been extensively investigated. The main thrust has been to see if the individual (or bundles of) nanotube(s) could be used as quantum molecular wires for interconnects in future computer systems (the experimental development is discussed under Section 22.6). Ballistic electron transport through individual nanotubes has been supported by many independent studies, and it is considered to be one of the reasons that nanotubes exhibit high current density compared to other materials at similar nanoscales [51].

Inspired by the above, possibilities of connecting nanotubes of different diameters and chirality in nanotube heterojunctions as molecular electronic devices or switching components have also been investigated [52]. The simplest way is to introduce pairs of heptagons and pentagons in an otherwise perfect hexagonal lattice structure of the material. The resulting junction can act like a rectifying diode (Figure 22.9). Such two-terminal rectifying diodes were first postulated theoretically [52], and recently, they have been observed in experiments [53]. There are two ways to create heterojunctions with more than two terminals: first, connecting different nanotubes through topological defect-mediated junctions [54]; second, laying down crossed nanotubes over each other and simply forming physically contacted or touching junctions [55]. The differences in the two approaches are the nature and characteristics of the junctions forming the device. In the first case, the nanotubes are chemically connected through bonding networks, forming a stable junction that could possibly give rise to a variety of switching, logic, and transistor applications [56]. In the second case, the junction is merely through a physical contact and will be amenable to changes in the nature of the contact. The main applications in the second category will be in electromechanical bistable switches and sensors [55,57]. The bistable switches can act as bits in a CNT-based computing architecture.

Novel structures of CNT "T-junctions" and "Y-junctions," as models of three-terminal nanoscale monomolecular electronic devices, have been proposed [54,56]. The T-junctions can be considered as a specific case of a family of Y-junctions in which the two connecting nanotubes are perpendicular to each other (see Figure 22.10). The pentagon–heptagon defect pair rule was found to be not applicable in the formation of the Y-junctions [56]. Experimentalists have also succeeded in producing MWCNT junctions using template-based chemical vapor deposition (CVD) [58] and pyrolysis of an organometallic precursor with nickelocene and thiophene [59]. The template-based method reports junctions consisting of large-diameter stems with two smaller branches with an acute angle between them resembling "tuning forks." The pyrolysis method reports multiple Y-junctions along a continuous MWCNT. The electrical conductance measurements on these Y-junctions show intrinsic nonlinear and asymmetric I–V behavior with

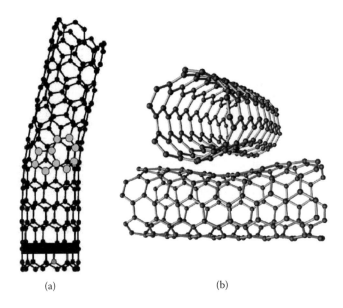

(a)   (b)

FIGURE 22.9 Example of (a) topological defect-mediated chemically connected two-terminal junction, and (b) physical cross-bar junctions of CNTs. (From D. Srivastava, Computational nanotechnology of carbon nanotubes; D. Srivastava, M. Menon, and P.M. Ajayan, *J. Nanoparticle Res.* 5, 395 (2003). With permission.)

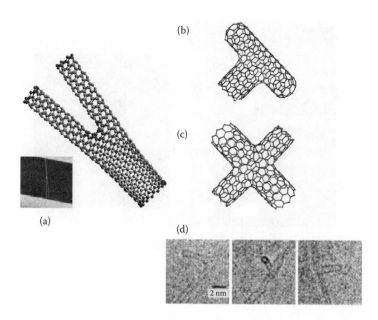

FIGURE 22.10 Examples of three-terminal CNT junctions: (a) Y-, (b) T-, (c) X-junctions of CNTs shown with formation of T-junctions using nanotube welding shown in (d). (From D. Srivastava, Computational nanotechnology of carbon nanotubes; D. Srivastava, M. Menon, and P.M. Ajayan, *J. Nanoparticle Res.* 5, 395 (2003). With permission.)

rectification at room temperature. The quantum conductivity of a variety of SWCNT Y-junctions shows current rectification under changes in the bias voltage [60,61]. The degree of rectification is found to depend on the type and nature of Y-junctions. Some junctions show good rectification while others show small "leakage" currents. Moreover, simulations also show that the molecular switches thus produced can easily function as three-terminal bistable switches that are controlled by a control or "gate" voltage applied at a branch terminal [60,61]. Further, under certain biasing conditions, nanotube Y-junctions are shown to work as "OR" or "XOR" gates as well. The possible reasons for rectification in SWCNT Y-junctions include constructive or destructive interference of the electronic wave functions through two different channels at the location of the junction; hence the rectification is strongly influenced by the structural asymmetry across the two branches in a junction [62]. The T-, X-, and Y-junctions of SWCNTs have also been fabricated by direct welding of nanotubes using localized heating by controlled exposure to electron beams [63]. Progress in the synthesis and electronic transport analysis of CNT junctions is described in a recent review [64].

In considering the architecture for future computing systems, we need not constrain ourselves to the specifications of the silicon-based devices, circuitry, and architecture. For example, a possible alternative architecture could be based on the structure and functioning of dendritic neurons in biological neural logic and computing systems. The tree shown in Figure 22.11 has a four-level branching structure and is made of 14 CNT "Y-junctions." Such a structure is conceptually amenable to fabrication via the template-based CVD method, which is used for growing individual Y-junctions, and provides a first model of a biomimetic neural network made of SWCNTs or MWCNTs [3]. In principle, such a "tree" could be trained to perform complex computing and switching applications in a single pass. A generic synthetic approach to rationally designing and fabricating such a hierarchy of multiple-branched structures made of CNTs has been proposed and demonstrated in template-based growth of nanotubes and nanowires using the CVD method. The number and frequency of branching, dimensions, and the overall architecture for such "tree-like" structures are controlled by controlling the nanopore design in the underlying anodic aluminum oxide template. Several levels of multiple-branched structures of CNT trees have been fabricated with this approach [65].

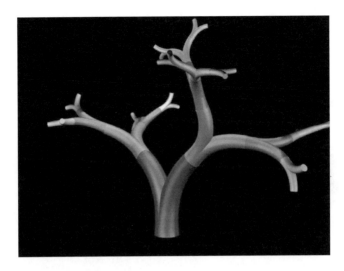

FIGURE 22.11 An illustration of a four-level dendritic neural tree made of 14 symmetric Y-junctions of the type in Figure 22.10. (From D. Srivastava, M. Menon, and K. Cho, *Computing in Engineering and Sciences* (CISE), an IEEE and APS publication, 42 (2001). With permission.)

22.3.4 Sensors and Actuators

As mentioned earlier, CNTs have different electronic properties depending on their chiral vector ranging from metals to semiconductors (1 eV bandgap). Semiconducting SWCNTs are very promising candidates for novel sensing applications, since the surface modifications, due to chemical adsorption or mechanical deformation of the nanotubes, can directly modify the electronic conductance of nanotubes. Experimental and theoretical works have proved that SWCNTs are extremely sensitive to gas molecules [66,67], and both chemical reactivity and electronic properties are strongly dependent on the mechanical deformation of nanotubes [68,69]. These characteristics have given rise to possibilities for their use in chemical, vibrational, and pressure-sensor applications. The role of computational investigations in this case has been to precisely define how chemical reactivity [68] and electronic properties [69] change in a mechanically strained tube, and to determine the effect of gas phase chemisorption [69] on the electronic characteristics of the system. It has been suggested that the mechanically tunable chemistry of nanotubes [68] will have possible applications in chemisorption-induced electronics as well [70]. Experiments have shown that nanotube sensors can detect ppm level gas molecules at room temperature, and this opens the possibility of developing nanotube-based chemical sensors operating at room temperature.

For mechanical and vibration-sensor applications, the full range of electronic-bandgap changes have been computed as a function of axial compression, tensile stretch, torsion, and bending strain [69]. Additionally, as the cross-section of (8,0) SWCNT is flattened up to 40%, the bandgap of the nanotube decreases from 0.57 eV and disappears at 25% deformation. As the deformation further increases to 40%, the bandgap reopens and reaches 0.45 eV. This strong dependence of the SWCNT band structure on mechanical deformation can be applied to develop nanoscale mechanical sensors [71].

22.4 Nanotube Growth

The earliest approach to producing nanotubes was an arc process [72] pioneered by Iijima [1]. This was shortly followed by a laser ablation technique developed at Rice University [73]. CVD and related techniques have become common since then [74–80]. All these processes are described below. The figure-of-merit for an ideal growth process depends on the application. For the development of composites and other structural applications, the expected metric is the ability to produce "tons a day." In contrast, the ability to achieve controlled growth (of specified thickness) on patterns is important for applications in nanoelectronics, field-emission-based displays, and sensors and similar applications. Even in these cases, such needs for patterned growth — undoubtedly originating from the microelectronics fabrication technology and knowledge base — may become irrelevant if circuits could be assembled from nanotubes in solution. Regardless of the applications and growth approach, the ability to control the diameter and chirality of the nanotubes is critical to realizing the promise of CNTs, but this has been elusive to date.

22.4.1 Arc Process and Laser Ablation

The arc process involves striking a DC arc discharge in an inert gas (such as argon or helium) between a set of graphite electrodes [1,2,72]. The electric arc vaporizes a hollow graphite anode packed with a mixture of a transition metal (such as Fe, Co, or Ni) and graphite powder. The inert gas flow is maintained at 50 to 600 torr. Nominal conditions involve 2000 to 3000°C, 100 A, and 20 V. This produces SWCNTs in a mixture of MWCNTs and soot. The gas pressure, flow rate, and metal concentration can be varied to change the yield of nanotubes, but these parameters do not seem to change the diameter distribution. The typical diameter distribution of SWCNTs by this process appears to be 0.7 to 2 nm.

In laser ablation, a target consisting of graphite mixed with a small amount of transition-metal particles as catalysts is placed at the end of a quartz tube enclosed in a furnace [2,73]. The target is exposed to an argon ion laser beam that vaporizes graphite and nucleates CNTs in the shock wave just in front of the target. Argon flow through the reactor, heated to about 1200°C by the furnace, carries the vapor and the nucleated nanotubes, which continue to grow. The nanotubes are deposited on the cooler walls of the quartz tube downstream from the furnace. This produces a high percentage of SWCNTs (~70%), with the rest being catalyst particles and soot.

22.4.2 Chemical Vapor Deposition

CVD is a popular technique in the silicon integrated circuit (IC) manufacturing industry to grow a variety of metallic, semiconducting, and insulating thin films. Typical CVD relies on thermal generation of active radicals from a precursor gas that leads to the deposition of the desired elemental or compound film on a substrate. Sometimes, the same film can be grown at a much lower temperature by dissociating the precursor with the aid of highly energetic electrons in a glow discharge. In either case, catalysts are almost never required. In contrast, a transition metal catalyst is necessary to grow CNTs from some form of hydrocarbon feedstock (CH_4, C_2H_2, C_2H_4,...) or CO, and in principle, extensive dissociation of the feedstock in the vapor phase is not necessary, as the feedstock can dissociate on the catalyst surface. Nevertheless, this approach is still called CVD in the nanotube literature.

A thermal CVD reactor is simple and inexpensive to construct in the laboratory, and consists of a quartz tube enclosed in a furnace. Typical laboratory reactors use a 1 or 2″ quartz tube, capable of holding small substrates. The substrate material may be Si, mica, quartz, or alumina. The setup needs a few mass-flow controllers to meter the gases and a pressure transducer to measure the pressure. The growth may be carried out at atmospheric pressure or slightly reduced pressures using a hydrocarbon or CO feedstock. The growth temperature is in the range of 700 to 900°C. A theoretical study of CNT formation suggests that a high kinetic energy (and thus a high temperature, ≥900°C) and limited, low supply of carbon are necessary to form SWCNTs [81]. Not surprisingly, CO and CH_4 are the two gases that have been reported mostly to give SWCNTs. MWCNTs are grown using CO and CH_4, as well as other higher hydrocarbons at lower temperatures of 600 to 750°C. As mentioned earlier, CNT growth requires a transition-metal catalyst. The type of catalyst, particle size, and the catalyst preparation techniques dictate the yield and quality of CNTs, and this will be covered in more detail shortly.

As in silicon IC manufacturing, the CNT community has also looked to low-temperature plasma processing to grow nanotubes at low temperatures. The conventional wisdom in choosing plasma processing is that the precursor is dissociated by highly energetic electrons, and as a result, the substrate temperature can be substantially lower than in thermal CVD. However, this does not exactly apply here since the surface-catalyzed nanotube growth needs its own minimum activation temperature.

This temperature depends on the nature of the precursors and catalyst, but this is not well established now. In any case, there are no reliable reports on low-temperature growth relative to thermal CVD. Nevertheless, several plasma-based growth techniques have been reported at temperatures similar to thermal CVD [80,82–84], and, in general, the plasma-grown nanotubes appear to be more vertically oriented than is possible by thermal CVD. This feature is attributed to the electric field in the plasma normal to the substrate. Since the plasma is very efficient in tearing apart the precursors and creating radicals, it is also hard to control and keep the supply of carbon low to the catalyst particles and, hence, plasma-based growth always results in MWCNTs and filaments.

The plasma reactor consists of a high-vacuum chamber to hold the substrate, mass-flow controllers, a mechanical (roughing) pump and, if necessary, a turbopump since plasma reactors almost always run at reduced pressures (0.1 to 50 torr), pressure gauges, and a discharge source. The latter is the heart of the plasma-processing apparatus, and in CNT growth, using a microwave source is one of the widely used approaches, probably because of the popularity this source enjoyed in the diamond community. The microwave system consists of a power supply at 2.45 GHz and waveguides coupled to the growth chamber. Although the plasma can provide intense heating of the substrate, it is normal to have an independent heater for the substrate holder. Other plasma sources include an inductive source and a hot-filament DC discharge. The latter uses a tungsten, tungsten carbide, or similar filament heated to about 2500 K, with the gas flow maintained at 1 to 20 torr. Simple DC or RF discharges can also be used in nanotube growth, and, indeed, the CNT literature has numerous reports on the use of DC discharges. The DC- and RF-based reactors consist of two parallel plate electrodes, with one grounded and the other connected to either a DC or a 13.56 MHz power supply.

22.4.3 Catalyst Preparation

Several catalyst preparation techniques reported in the literature consist of some form of solution-based catalysts. One recipe for growing MWCNTs is as follows [76]: First, 0.5 g (0.09 mmol) of Pluronic P-123 triblock copolymer is dissolved in 15 cc of a 2:1 mixture of ethanol and methanol. Next, $SiCl_4$ (0.85 cc, 7.5 mmol) is slowly added, using a syringe, into the triblock copolymer/alcohol solution and stirred for 30 min at room temperature. Stock solutions of $AlCl \cdot 6H_2O$, $CoCl_2 \cdot 6H_2O$, and $Fe(NO_3)_3 \cdot 6H_2O$ are prepared at the concentration of the structure-directing agent (SDA) and inorganic salts. The catalyst solutions are filtered through 0.45-μm polytetrafluoroethylene (PTFE) membranes before applying them to the substrate. The substrate with the catalyst formulation is loaded into a furnace and heated at 700°C for 4 h in air to render the catalyst active by the decomposition of the inorganic salts and removal of the SDA. Figure 22.12 shows SEM images of MWCNTs grown using this formulation in a thermal CVD reactor. A nanotube tower with millions of multiwalled tubes supporting each other by van der Waals force is seen. If the catalyst solution forms a ring during annealing, then a hollow tower results.

Several variations of solution-based techniques have been reported in the literature, and all of them have done well in growing CNTs. A common problem with the above approaches is that it is difficult to confine the catalyst from solutions within small patterns. Another problem is the excessive time required to prepare the catalyst; a typical solution-based technique for catalyst preparation involves several steps lasting hours. In contrast, physical processes such as sputtering and e-beam deposition not only can deal with very small patterns but are also quick and simple in practice [79,82]. Delzeit et al. reported catalyst preparation using ion-beam sputtering wherein an underlayer of Al (~10 nm) is deposited first, followed by 1 nm of an Fe active catalyst layer [79]. Figure 22.13 shows SWCNTs grown by thermal CVD on a 400 mesh TEM grid used to pattern the substrate. Methane feedstock at 900°C was used to produce these nanotubes. The same but thicker underlayer and catalyst films at 750°C with ethylene as the source gas yielded MWCNT towers (not shown here).

One of the most successful approaches to obtaining oriented arrays of nanotubes uses a nanochannel alumina template for catalyst patterning [85]. First, aluminum is anodized on a substrate such as Si or quartz, which provides ordered, vertical pores. Anodizing conditions are varied to tailor the pore diameter, height, and spacing between pores. This is followed by electrochemical deposition of a cobalt catalyst at

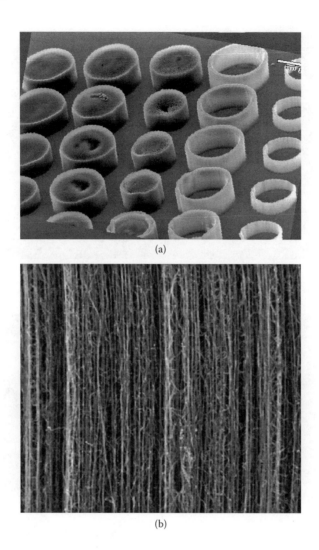

FIGURE 22.12 MWCNTs grown by thermal CVD. (a) Different catalyst solution concentrations result in towers and ring-like structures. (b) Close-up view of one of these structures in (a) showing a forest of nanotubes supporting each other by VDW force, thus resulting in a vertical structure.

the bottom of the pores. The catalyst is activated by reduction at 600°C for 4 to 5 h. Figure 22.14 shows an ordered array of MWCNTs (mean diameter 47 nm) grown by CVD from 10% acetylene in nitrogen. The use of a template not only provides uniformity but also provides vertically oriented nanotubes.

Another approach to obtaining vertical structures is of course to use a plasma process that does not require a template. Figure 22.15 shows vertically aligned nanotubes obtained using a DC discharge. These structures exhibit a bamboo-like morphology instead of ideal parallel walls of a MWCNT; for that reason, these structures are more commonly called carbon nanofibers (CNFs) or multiwalled CNFs. These CNFs, in general, are more vertical (than the ideal MWCNTs), individual, and freestanding structures.

22.4.4 Continuous, High-Throughput Processes

The nanotube growth on substrates is driven by the desire to develop devices. Nanotubes have also been produced using catalysts floating in the gas phase, instead of supported catalysts. This approach is

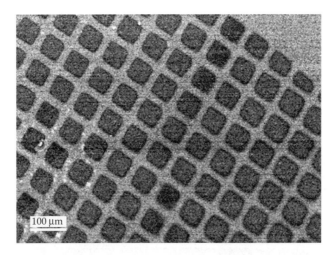

FIGURE 22.13 SWCNTs grown by thermal CVD on a 400 mesh TEM grid.

FIGURE 22.14 An ordered array of MWCNTs grown using an alumina template. (From J. Li, C. Papadopoulos, J.M. Xu, M. Moskovits, *Appl. Phys. Lett.* 75, 367 (1999). With permission.)

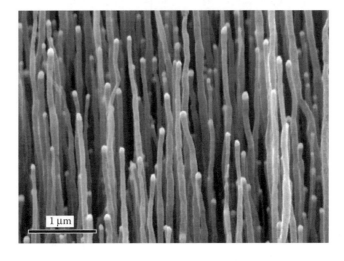

FIGURE 22.15 Vertically aligned CNFs from a DC plasma-enhanced CVD process. (Image courtesy of Alan Cassell.)

designed for producing large quantities of nanotubes in a continuous process. The earliest report [86] of such a process involved pyrolysis of a mixture containing benzene and a metallocene (such as ferrocene, cobaltocene, or nickelocene). In the absence of metallocene, only nanospheres of carbon were seen, but a small amount of ferrocene yielded large quantities of nanotubes. The growth system uses a two-stage furnace wherein a carrier gas picks up the metallocene vapor at around 200°C in the first stage, and the decomposition of the metallocene as well as pyrolysis of the hydrocarbon and catalytic reactions occur in the second stage at elevated temperatures (>900°C). Acetylene with ferrocene or iron pentacarbonyl at 1100°C has been shown to yield SWCNTs in a continuous process in the same two-stage system. Andrews et al. also reported a continuous process for multiwall tubes using a ferrocyne–xylene mixture at temperatures as low as 650°C [87]. A high-pressure process using CO (HiPCO) has been developed for SWCNTs [88]. In this process, the catalyst is generated *in situ* by thermal decomposition of iron pentacarbonyl. The products of this decomposition include iron clusters in the gas phase that act as nuclei for the growth of SWCNTs, using CO. The process is operated at pressures of 10 to 50 atm and temperatures of 800 to 1200°C. SWCNTs as small as 0.7 nm in diameter are produced by this process, and the process seems to be amenable for scale-up for commercial production.

22.5 Material Development

22.5.1 Purification

The as-grown material typically contains a mixture of SWCNTs, MWCNTs, amorphous carbon, and catalyst metal particles, and the ratio of the constituents varies from process to process and depends on the growth conditions for a given process. Purification processes have been developed to remove all the unwanted constituents [89]. The purpose is to remove all the unwanted material and obtain the highest yield of nanotubes with no damage to the tubes. A typical process, used in Ref. [90], to purify HiPCO-derived SWCNTs [88] is as follows: First, a sample of 50 mg is transferred to a 50 ml flask with 25 ml of concentrated HCl and 10 ml of concentrated HNO_3. The solution is heated for 3 h and constantly stirred with a magnetic stirrer in a reflux apparatus equipped with a water-cooled condenser. This is done to remove iron and graphite nanocrystallites. The resulting suspension is transferred into centrifuge tubes and spun at 3220 g for 30 min. After pouring off the supernatant, the solid is resuspended and spun (30 min) thrice in deionized water. Next, the solid is treated with NaOH (0.01 M) and centrifuged for 30 min. This process yields nanotube bundles with tube ends capped by half-fullerenes. Finally, the sample is dried overnight in a vacuum oven at 60°C.

22.5.2 Characterization

High-resolution TEM is used (as in Figure 22.2) to obtain valuable information about the nanotube structure such as diameter, open vs. closed ends, presence of amorphous material and defects, and nanotube quality. In MWCNTs, the spacing between layers is 0.34 nm and, therefore, a count of the number of walls readily provides the diameter of the structure. Since the nature of catalyst and particle size distribution appear to be critical in determining the nanotube-growth characteristics, researchers have used TEM, EDX, and AFM to characterize the catalyst surface prior to loading in the growth reactor. These techniques together provide information on the particle size and chemical composition of the surface [79].

Besides TEM, Raman spectroscopy is perhaps the most widely used characterization technique to study nanotubes [91,92]. Raman spectroscopy of the SWCNTs is a resonant process associated with optical transition between spikes in the 1-D electronic density of states. Both the diameter and the metallic vs. semiconducting nature of the SWCNTs dictate the energy of the allowed optical transitions and, hence, this characteristic is used to determine the diameter and nature of the nanotubes. An example from Ref. [79] is presented in Figure 22.16. The spectra was obtained in the backscattering configuration using a 2 mW laser power (633 nm excitation) on the sample with a 1 μm focus spot. The nanotube G band zone is formed through graphite Brilliouin zone folding. The spectrum in Figure 22.16 shows the

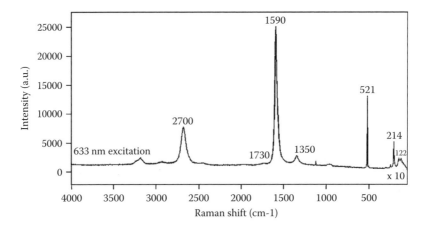

FIGURE 22.16 Raman spectra of CVD-grown SWCNT sample.

characteristic narrow G band at 1590 cm^{-1} and the signature band at 1730 cm^{-1} for SWCNTs. A strong enhancement in the low frequency is also observed in the low-frequency region for the radial breathing mode (RBM). As shown in Ref. [91], a 1-D density of state of metallic nanotubes near the Fermi level has the first singularly bandgap between 1.7 and 2.2 eV, which will be resonant with the 633 nm (1.96 eV) excitation line to the RBM. Using $\alpha = 248$ cm^{-1} in $\Omega_{RBM} = \alpha/d$, a diameter distribution of 1.14 to 2 nm is computed for the sample in Figure 22.16, with the dominant distribution around 1.16 nm. The strong enhancement in the 633 nm excitation [92] indicates that this sample largely consists of metallic nanotubes.

22.5.3 Functionalization

Functionalization of nanotubes with other chemical groups on the sidewall may help modify the properties required for an application in hand. For example, chemical modification of the sidewalls may improve the adhesion characteristics of nanotubes in a host polymer matrix to make functional composites. Functionalization of the nanotube ends can lead to useful chemical sensors and biosensors.

Chen et al. [93] reported that reaction of soluble SWCNTs with dichlorocarbene led to the functionalization of nanotubes with Cl on the sidewalls. The saturation of 2% of the carbon atoms in SWCNTs with C–Cl is sufficient to result in dramatic changes in the electronic band structure. Michelson et al. [94] reported fluorination of SWCNTs with F_2 gas flow at temperatures of 250 to 600°C for 5 h. The fluorine is shown to attach covalently to the sidewall of the nanotubes. Two-point probe measurements showed that the resistance of the fluorinated sample increased to ~20 MΩ from ~15 Ω for the untreated material. While most functionalization experiments use a wet chemical or exposure to high-temperature vapors or gases, Khare et al. reported [90,95] a low-temperature cold-plasma-based approach to attach functional groups to the nanotubes. They have been able to show functionalization with atomic hydrogen from a H_2 discharge, as evidenced by the C–H stretching modes observed in the FTIR analysis of the samples. They have also demonstrated attaching of F, NH_x, and N atoms or groups to the sidewall using appropriate source gases (such as CF_4, NH_3, or N_2) to strike the glow discharge [96,97].

22.6 Application Development

22.6.1 CNTs in Microscopy

Atomic force microscopes (AFM) are widely used now by the research community to image and characterize various surfaces and also by the IC industry as a metrology tool. A typical AFM probe consists of a silicon or silicon nitride cantilever with a pyramidal tip. This tip now can be made as

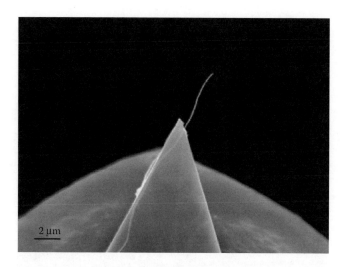

FIGURE 22.17 SWCNT tip grown by CVD at the end of an AFM cantilever.

small as 20 nm, with reasonable resolution. However, the large cone angle of this tip (30 to 35°) makes it difficult for probing narrow and deep features such as trenches in IC manufacturing. Another serious drawback is that the tip is brittle, thus limiting its use in applications; either the tip breaks after only a limited use or becomes blunt during continued use. For these reasons, a CNT probe has become an attractive alternative. Although early CNT probes [98] were manually attached to the tip of the AFM cantilever, currently it is possible to directly grow a SWCNT or MWCNT probe by CVD [99]. The CNT probe not only offers extraordinary nanometer scale resolution but is also robust, due to its high strength and the ability to retain structural integrity even after deformation within the elastic limit. Figure 22.17 shows a SWCNT probe prepared by CVD, with a diameter of ~2 nm. In a well-characterized CVD process, it may be possible to control the length of the probe by selecting the growth time. In the absence of such growth vs. time knowledge, it is possible to shorten the tip to the desired length by applying an electric field to etch away the nanotube. Figure 22.18 shows a photoresist pattern produced by interferometric lithography. The image, if obtained using a conventional silicon pyramidal tip, would show sloping sides for the photoresist, an artifact due to the pyramidal shape of the tip. The image in Figure 22.18, obtained using a MWCNT probe, shows no such artifact, but shows the vertical walls for the photoresist lines as confirmed by SEM.

FIGURE 22.18 AFM profile of a 280-nm line/space photoresist pattern.

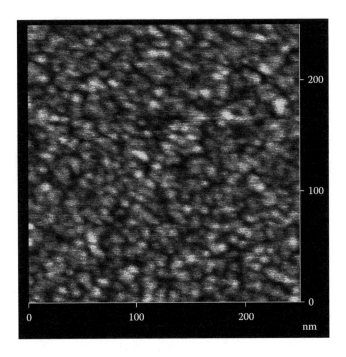

FIGURE 22.19   AFM image of a 2 nm Ir surface collected using a SWCNT tip.

In addition to profilometry, CNT tips are also useful in imaging thin films in semiconductor metrology. For example, Figure 22.19 shows an AFM image of a 2 nm film of Ir on a mica surface collected with a SWCNT probe (~2 nm tip diameter). This affords a very high lateral resolution with grain sizes as small as 4 to 7 nm. Interestingly, even after continuous scanning for hours, the tip exhibits no detectable degradation in the lateral resolution of the grains. In contrast, the resolution capability becomes worse with time in the case of the silicon probe. As characterization tools in IC industry, CNT probes would require less frequent changing of probes, thus offering higher throughput. More recently, Nguyen et al. [100] sharpened MWCNT probes to the dimensions of a SWCNT, which provided imaging as good as in Figure 22.19 but retained the probe robustness of the MWCNT structures. Researchers have also successfully used CNT probes to image biological samples; although early samples were dried DNA or proteins, Stevens et al. [101] showed that biological samples in their native aqueous environment can be imaged if the CNT probe is rendered hydrophilic by coating it with ethylene diamine.

22.6.2   CNT-Based Nanoelectronics

The unique electronic properties of CNTs and possible devices from theoretical perspectives were discussed in detail in Section 22.2 and Section 22.3, respectively. Room temperature demonstration of conventional switching mechanisms such as in field effect transistors (FETs) first appeared in 1998 [102,103]. As shown in Figure 22.20, a SWCNT is placed to bridge a pair of metal electrodes serving as source and drain. The electrodes are defined using lithography on a layer of SiO_2 in a silicon wafer, which acts as the back gate. The variation of drain current with gate voltage at various source–drain biases for a 1.6 nm nanotube [102,103] clearly demonstrated that the gate can strongly control the current flow through the nanotube. In this device, the holes are the majority carriers of current as evidenced by an increase in current at negative gate voltages. Ref. [104] provides a theoretical analysis of how the nanotube FETs work and shows that the electrode–nanotube contact influences the subthreshold channel conductance vs. gate voltage.

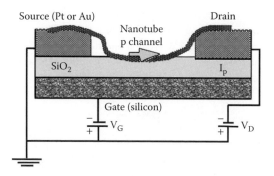

FIGURE 22.20 A CNT FET.

Since the pioneering demonstration of nanotube FETs, CNT logic circuits have been successfully fabricated. Figure 22.21 shows a CNT-based field effect inverter consisting of n-type and p-type transistors [105]. The nanotube is grown using CVD between the source and the drain and on top of an SiO_2 layer with a silicon back gate. The as-grown nanotube exhibits p-type doping. Masking a portion of the nanotube and doping the remaining part with potassium vapor creates an n-type transistor. By combining the p- and n-type transistors, complementary inverters are created. Applying a 2.9 V bias to the V_{DD} terminal in the circuit and sweeping the gate voltage from 0 to 2.5 V yields an output voltage going from 2.5 V to 0. The inverter in Figure 22.21 shows a gain of 1.7 but does not exhibit an ideal behavior, since in this early demonstration, the PMOS is suspected to be leaky. Bachtold et al. [106] reported logic circuits with CNT transistors that have high gain (greater than 10), a large on–off ratio (greater than 10^5), and room temperature operation. Their demonstration included operations such as an inverter, a logic NOR, a static random-access memory cell, and an AC ring oscillator.

Fabrication of nanotube-based three-terminal devices involves horizontally placing nanotubes between the metal electrodes. While earlier demonstrations carefully transplanted a simple nanotube from bulk material, recent works have used CVD to bridge the electrodes with a nanotube [105]. This is indeed anything but routine since growth naturally occurs vertically from a surface. In this regard, successful fabrication of CNT–Y junctions has been reported, and the I–V measurements on these CVD-grown

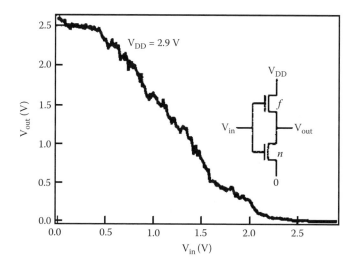

FIGURE 22.21 A CNT-based inverter.

multiwall Y tubes show reproducible rectification at room temperature, as discussed in Section 22.3. Alternatively, a concept proposed by Rueckes et al. [55] plans to use directed assembly of nanotubes from solution. This approach involves a suspended, crossed nanotube geometry that leads to bistable, electrostatically switchable on–off states. Construction of nonvolatile random access memory and logic functions using this approach is expected to reach a level of 10^{12} elements/cm^2.

22.6.3 Sensors

Significant research is in progress to develop CNT-based chemical, biological, and physical sensors. These efforts can be broadly classified into two categories: one that utilizes certain properties of the nanotube (such as change in conductivity with gas adsorption) and the second that relies on the ability to functionalize the nanotube (tip and sidewall) with molecular groups that serve as sensing elements.

Gas sensors for sensing NO_2, NH_3, etc. have previously used semiconducting metal oxides (SnO_2, for example) and conducting polymers. Most oxide-based sensors work at temperatures above 200°C but exhibit good sensitivity. Kong et al. [66] reported that the room temperature conductivity of CNTs changes significantly with exposure to NH_3 and NO_2, a property useful for developing sensors. Current–voltage characteristics of a 1.8 nm SWCNT placed between a pair of Ni/Au electrodes and a back gate (as in Figure 22.20) were studied before and after exposure to various doses of NH_3 and NO_2. The SWCNT samples used in the experiments were hole-doped semiconductors. Ammonia shifts the valence band of the CNT from the Fermi level and thus reduces the conductance due to hole depletion. The conductance was observed to decrease by a factor of 100 upon exposure to 1% NH_3. The response time, which may be defined as the time needed to see one order of magnitude change in conductance, was 1 to 2 min. The sensitivity, defined as the ratio of resistances before and after gas exposure, was 10 to 100. These results at room temperature compare favorably with metal oxides operating at 300 to 500°C. Similar large conductance changes upon exposure to oxygen have also been reported [107]. One current drawback in deploying CNTs in sensors is the slow recovery of the CNTs to the initial state. In the above experiments, it took up to 12 h for the conductivity of the CNTs to return to the original value after the source gas was withdrawn. Heating the sensor to 200°C reduced this period to about 1 h. Alternatively, exposing the CNTs to UV light allowed a speedier recovery or desorption process.

Although the above results from a transistor configuration sensor are interesting, this type of sensor is unlikely to become practical. First, a 3-D device is relatively expensive to fabricate; note that the extent of cost reduction normally achieved in CMOS downscaling is not possible, because the sensor market is a lot smaller than that for computer chips. Second, bridging a source and drain by a single SWCNT using CVD is a low-probability process and the yield would be very poor. To overcome these limitations, Li et al. [108,109] fabricated a simpler, interdigitated electrode structure, using simple microfabrication techniques. A film of SWCNTs was laid across the electrodes by a solution-casting process. This involves drop-depositing a few drops of a solution of SWCNTs in dimethyl formamide. When the solvent evaporates rapidly, it leaves a film behind on the electrodes that consists of a statistically meaningful number of nanotubes. This simple structure has been shown to be sensitive to NH_3, NO_2, nitrotoulene, benzene, acetone, etc. [2,108]. Interestingly, CNT sensors based on a change in conductivity normally operate on the principle of charge transfer between SWCNT and the molecules. But this is true only with small molecules such as NO_2 and NH_3. However, the sensor response to large molecules in the case of the interdigitated structure could be due to intertube modulation of the SWCNT network [108]. The sensitivity reported for all the gases investigated ranged from low ppm to ppb levels.

CNTs are also ideal for developing biosensors. The role of CNT itself is to serve as a nanoelectrode. With macro- and microelectrodes, the size difference between the electrode and the molecules of interest can lead to signal-to-noise ratio (SNR) problems. When the same area is covered with nanoelectrodes, not only are the SNR problems diminished due to the comparable size of the electrodes and the bio-objects but also the signal is amplified due to the number of nanoelectrodes in the electrode array. Li and coworkers [110,111] demonstrated this by using an array of vertically aligned CNFs. Their electrode array was created by growing CNFs by using a plasma CVD process and isolating each CNF or electrode from its neighbor through intercalation of CNT gaps with SiO_2 deposited by thermal CVD. A mechanical polishing step after this provides a flat top surface of the nanoelectrode array, with only tiny ends of the nanotubes sticking out. Li et al. functionalized these ends with DNA and brought complementary strands to demonstrate hybridization. The biosensor is currently capable of sensing less than 1000 target molecules, as shown in Ref. [111].

22.6.4 Field Emission

The geometric properties of nanotubes such as the high aspect ratio and small tip radius of curvature, coupled with the extraordinary mechanical strength and chemical stability, make them an ideal candidate for electron field emitters. CNT field emitters have several industrial and research applications: flat panel displays (FPD), outdoor displays, traffic signals, electron microscopy, and any other applications that need a source of electrons. De Heer et al. [112] demonstrated the earliest high-intensity electron gun based on field emission from a film of nanotubes. A current density of 0.1 mA/cm^2 was observed for voltages as low as 200 V. For comparison, most conventional field emitter displays operate at 300 to 5000 V, whereas cathode ray tubes use 30,000 V. Since this early result, several groups have studied the emission characteristics of SWCNTs and MWCNTs [113–117].

A typical field emission test apparatus consists of a cathode and anode enclosed in an evacuated cell at a vacuum of 10^{-9} to 10^{-8} torr. The cathode consists of a glass or PTFE substrate with metal-patterned lines where a film of nanotubes can be transplanted after the bulk-produced material is purified. Instead, nanotubes can be directly grown on the cathode in the CVD or plasma CVD chamber described earlier. The anode operating at positive potentials is placed at a distance of 20 to 500 μm from the cathode. The turn-on field, arbitrarily defined as the electric field required for generating 1 nA, can be as small as 1.5 V/μm. The threshold field — the electric field needed to yield a current density of 10 mA/cm^2 — is in the range of 5 to 8 V/μm. At low emission levels, the emission behavior follows the Fowler-Nordheim (F-N) relation; i.e., the plot of $\ln(I/V^2)$ vs. $\ln(1/V)$ is linear. The emission current significantly deviates from the F-N behavior in the high-field region, and, indeed, the emission current typically saturates. While most works report current densities of 0.1 to 100 mA/cm^2, very high current densities up to 4 A/cm^2 have been reported by Zhu et al. [116].

Working full-color FPD and CRT-lighting elements have been demonstrated by groups from Japan and Korea. In the case of FPD, the anode structure consists of a glass substrate with phosphor-coated indium tin oxide stripes. The anode and cathode are positioned perpendicular to each other to form pixels at their intersections. Appropriate phosphors such as Y_2O_2S:Eu, ZnS:Cu, Al, and ZnS:Ag, Cl are used at the anode for red, green, and blue colors respectively. Full color displays as large as 40″ have been demonstrated. Manufacturing issues in terms of uniformity, lifetime, robust behavior, etc., would determine if this technology can compete with plasma and LED displays. For lighting elements, the phosphor screen is printed on the inner surface of the glass and backed by a thin Al film (~100 nm) to give electrical conductivity. A lifetime test of the lighting element suggests a lifetime of over 10,000 h [114].

22.6.5 Nanotube–Polymer Composites

Using nanotubes as reinforcing fibers in composite materials is still a developing field from both theoretical and experimental perspectives. A detailed discussion and review are given by Berrara [34]. Several experiments regarding the mechanical properties of nanotube–polymer composite materials with MWCNTs have been reported [31–33,118]. Wagner et al. [32] experimentally studied the fragmentation of MWCNTs within thin polymeric films (urethane/diacrylate oligomer EBECRYL 4858) under compressive and tensile strain. They found that the nanotube–polymer interfacial shear stress τ is of the order of 500 MPa, which is much larger than that of conventional fibers with a polymer matrix. This has suggested the possibility of chemical bonding between the nanotubes and the polymer in their composites, but the nature of the bonding is not clearly known. Lourie et al. [23] have studied the fragmentation of SWCNT within the epoxy resin under tensile stress. Their experiment also suggests a good bonding between the nanotube and the polymer in the sample. Schadler et al. [31] have studied the mechanical properties of 5 wt% MWCNTs within an epoxy matrix by measuring the Raman peak shift when the composite is under compression and under tension. The tensile modulus of the composites in this experiment is found to enhance much less than the enhancement of the same composite under compression. This difference has been attributed to the sliding of inner shells of the MWCNTs when a tensile stress was applied. In cases of SWCNT–polymer composites, the possible sliding of individual tubes in the SWCNT rope may also reduce the efficiency of load transfer. It is suggested that for the SWCNT rope case, interlocking using polymer molecules might bind the SWCNT rope more strongly. Andrews et al. [118] have also studied the composites of 5 wt% of SWCNT embedded in petroleum pitch matrix, and their measurements show an enhancement of the Young's modulus of the composite under tensile stress. Measurements by Qian et al. [33] of a 1 wt% MWCNT–polystyrene composite under tensile stress also showed a 36% increase of the Young's modulus compared with the pure polymer system. The possible sliding of inner shells in MWCNT and individual tubes in a SWCNT rope is not discussed in the above two studies. There are currently no clean experiments available on SWCNT–polymer composites, perhaps because SWCNTs are not available in large quantities for experimentation on bulk composite materials.

22.6.6 Other Applications

In addition to the fields discussed above, CNTs are being investigated for several other applications. The possibility of storing hydrogen for fuel cell development has received much attention [119]. A H_2 storage capability of 8% by weight appears to be the target for use in automobiles. Though interesting basic science of hydrogen storage has been emerging, no technological breakthrough has been reported yet. Most reports appear to provide a storage capacity of 1 to 2%. Higher capacities including those from well-cited references suggesting 6 to 8% appear to be unreproducible, unreliable, and totally wrong, to this date. Filling nanotubes with a variety of metals has also been attempted [120–123]. The most noteworthy of these attempts involves lithium storage for battery applications [122,123].

The thermal conductivity of SWCNTs is reported to be about 2500 W/mK (in Section 22.3) in the axial direction, which is second only to epitaxial diamond. This property can be exploited for cooling semiconductor chips and heat pipes. However, this application has poor prospects, since SWCNTs produced by any of the methods discussed in Section 22.4 appear mostly like spaghetti on a plate. With lack of orientation, the SWCNT films exhibit thermal conductivities far lower than commonly used materials in such applications. Ngo et al. [124] instead used vertically aligned CNFs grown by PECVD in chip-cooling applications. Although these CNFs are far inferior to SWCNTs in terms of thermal conductivity and other properties, they are at least amenable to processing and integration with today's electronic-device fabrication steps. Intercalating an array of CNFs with copper using electrodeposition, Ngo et al. showed that the CNT–Cu complex can serve as a thermal interface material (replacing thermal grease and other polymeric interface materials) to cool future computer chips.

CNTs also exhibit properties suitable for their use as next generation interconnects. Copper, the current interconnect material, begins to suffer from electromigration at current densities exceeding 10^6 A/cm^2. In contrast, CNTs can support current densities of 10^6 to 10^9 A/cm^2 without suffering any problems. As in the case of chip cooling, the problem again is in developing processes that enable integration with current IC manufacturing. Axial conduction being dominant, it is important to obtain vertically aligned structures, but this is impossible with SWCNTs. Again Li et al. [125,126] used vertically aligned CNFs — though inferior to SWCNTs but amenable to processing in a manner compatible with present IC manufacturing steps — to demonstrate interconnects for DRAM via filling applications.

SWCNTs and MWCNTs are useful as high surface area materials in a variety of applications. CNTs are being studied for gas adsorption, for separation, and as supports for catalysts. Cinke et al. [127] measured the surface area of purified HiPCO SWCNTs to be 1587 m^2/g, which is the highest value reported to date. Such large surface areas and pore volumes are ideal for gas adsorption applications. Other interesting applications include nanoscale reactors, ion channels, and drug delivery systems.

22.7 Concluding Remarks

CNTs show a remarkable potential in a wide variety of applications such as future nanoelectronics devices, field emission devices, high-strength composites, sensors, and many related fields. This potential has propelled concentrated research activities across the world in growth, characterization, modeling, and application development. Though significant progress has been made in the last five years, numerous challenges remain and there is still a great deal of work to be done before CNT-based products become ubiquitous. The biggest challenge now is to have control over the chirality and diameter during the growth process; in other words, the ability to specify a priori and obtain the desired chiral nanotubes is important in developing electronics and related applications. Issues related to contacts, novel architectures, and development of inexpensive manufacturing processes are additional areas warranting serious consideration in electronics applications. The major roadblock now to developing structural applications is the lack of raw materials in large quantities. With the current bulk production rate of SWCNTs hovering around a few kilograms a day, large-scale composite development efforts are nonexistent at present. This scenario will change with breakthroughs in large-scale production of CNTs bringing the cost per pound to reasonable levels for structural applications. Sensor development efforts are showing promise and progress, and numerous areas such as functionalization, signal processing and integrity, system integration, etc., require significant further developments. In all the above areas, modeling and simulation is expected to continue to play a critical role as it has done since the discovery of CNTs.

Acknowledgments

The authors thank the members of the experimental group at NASA Ames Research Center for Nanotechnology for their contributions to the material presented in this chapter. DS is with the University Affiliated Research Center at NASA Ames, operated by the University of California, Santa Cruz.

References

1. S. Iijima, *Nature* 354, 56 (1991).
2. M. Meyyappan, ed., *Carbon Nanotubes: Science and Applications*, CRC Press, Boca Raton, FL, 2004.
3. D. Srivastava, M. Menon, and K. Cho, *Computing in Engineering and Sciences (CISE)*, an IEEE and APS publication, 42 (2001).
4. M.S. Daw and M.I. Baskes, *Phys. Rev. Lett.* 50, 1285 (1983); S.M. Foiles, M.I. Baskes, and M.S. Daw, *Phys. Rev. B* 33, 7983 (1986).
5. F.H. Stillinger and T.A. Weber, *Phys. Rev. B* 31, 5262 (1985).
6. J. Tersoff, *Phys. Rev. B* 38, 9902 (1988).
7. D.W. Brenner, *Phys. Rev. B* 42, 9458 (1990).
8. B.J. Garrison and D. Srivastava, *Ann. Rev. Phys. Chem.* 46, 373 (1995).
9. D. Srivastava and S. Barnard, in *Proceedings of IEEE Supercomputing '97* (SC '97) (1997).
10. R. Car and M. Parrinello, *Phys. Rev. Lett.* 55, 2471 (1985).
11. W.A. Harrison, *Electronic Structure and the Properties of Solids*, Freeman, San Francisco, 1980.
12. M. Menon and K.R. Subbaswamy, *Phys. Rev. B* 55, 9231 (1997).
13. M. Menon, *J. Chem. Phys.* 114, 7731 (2000).
14. Payne et al., *Rev. Mod. Phys.* 68, 1045 (1992).
15. P. Hohenberg and W. Kohn, *Phys. Rev.* 136, 864B (1964).
16. W. Kohn and L.J. Sham, *Phys. Rev.* 140, 1133A (1965).
17. Details available at http://cms.mpi.univie.ac.at/vasp/.
18. S. Datta, *Electronic Transport in Mesoscopic Systems (Methodology and References)*, Cambridge University Press, Cambridge, 1995.
19. D. Srivastava, Computational nanotechnology of carbon nanotubes, in M. Meyyappan, ed., *Carbon Nanotubes: Science and Applications*, CRC Press, Boca Raton, FL, 2004.

20. B.I. Yakobson, C.J. Brabec, and J. Bernholc, *Phys. Rev. Lett.* 76, 2511 (1996).
21. D. Srivastava, M. Menon, and K. Cho, *Phys. Rev. Lett.* 83, 2973 (1999).
22. B.I. Yakobson and Ph. Avouris, Mechanical properties of carbon nanotubes, in M.S. Dresselhaus and Ph. Avouris, eds., *Carbon Nanotubes*, Springer Verlag, Berlin-Heidelberg, 2001, p. 293.
23. O. Lourie, D.M. Cox, and H.D. Wagner, *Phys. Rev. Lett.* 81, 1638 (1998).
24. M.B. Nardelli, B.I. Yakobson, and J. Bernholc, *Phys. Rev. Lett.* 81, 4656 (1998).
25. P. Zhang, and V. Crespi, *Phys. Rev. Lett.* 81, 5346 (1998).
26. G.G. Samsonidze and B.I. Yakobson, *Phys. Rev. Lett.* 88, 065501 (2002).
27. M. Yu, O. Lourie, M. Dyer, K. Moloni, T. Kelly, and R. Ruoff, *Science* 287, 637 (2000).
28. B. Vigolo, A. Pénicaud, C. Coulon, C. Sauder, R. Pailler, C. Journet, P. Bernier, and P. Poulin, *Science* 290, 1331 (2000).
29. C. Wei, D. Srivastava, and K. Cho, *Phys. Rev. B* 67, 115407 (2003).
30. C. Wei, K. Cho and D. Srivastava, *Appl. Phys. Lett.* 82, 2512 (2003).
31. L.S. Schadler, S.C. Giannaris and P.M. Ajayan, *Appl. Phys. Lett.* 73, 3842 (1998).
32. H.D. Wagner, O. Lourie, Y. Feldman, and R. Tenne, *Appl. Phys. Lett.* 72, 188 (1998).
33. Q. Qian, E.C. Dickey, R. Andrews, and T. Rantell, *Appl. Phys. Lett.* 76, 2868 (2000).
34. E.V. Barrera, M.L. Shofner, and E.L. Corral, Applications: composites, in M. Meyyappan, ed., *Carbon Nanotubes: Science and Applications*, CRC Press, Boca Raton, FL, 2004.
35. E.T. Thostenson, C. Li, and T-W. Chou, *Comp. Sci. Technol.* 65, 491 (2005).
36. C. Wei, D. Srivastava, and K. Cho, *Nano Lett.* 2, 647 (2002).
37. S.V.J. Frankland et al., *MRS Proceedings* 593, A14.17 (2000).
38. C. Wei, D. Srivastava, and K. Cho, *Nano Lett.* 4, 1949 (2004).
39. A.M. Rao, E. Richter, S. Bandow, B. Chase, P.C. Eklund, K.A. Williams, S. Fang, K.R. Subbaswamy, M. Menon, A. Thess, R.E. Smalley, G. Dresselhaus, and M.S. Dresselhaus, *Science* 275, 187 (1997).
40. J. Yu, R.K. Kalia, and P. Vashishta, *J. Chem. Phys.*, 103, 6697 (1995).
41. R. Ernst, and K.R. Subbaswami, *Phys. Rev. Lett.* 78, 2738 (1997).
42. R. Saito, T. Takeya, T. Kimura, G. Dresselhaus, and M.S. Dresselhaus, *Phys. Rev. B* 59, 2388 (1999).
43. Y.W. Kwon, C. Manthena, J.J. Oh, and D. Srivastava, *J. Nanosci. Nanotechnol.* 5, 703 (2005).
44. V.P. Sokhan, D. Nicholson, and N. Quirke, *J. Chem. Phys.* 113, 2007 (2000).
45. S. Berber, Y.K. Kwon and D. Tomanek, *Phys. Rev. Lett.* 84, 4613 (2000).
46. J. Che, T. Cagin, and W.A. Goddard, *Nanotechnology* 11, 65 (2001).
47. M.A. Osman and D. Srivastava, *Nanotechnology* 12, 21 (2001).
48. E.G. Noya, D. Srivastava, L. Chernozetonskii, and M. Menon, *Phys. Rev. B* 70, 115416 (2004).
49. M. Osman and D. Srivastava, Molecular dynamics simulations of heat pulse propagation in single-wall carbon nanotubes, *Phys. Rev. B* 72, in press.
50. M.A. Osman, A. Cummings, and D. Srivastava, Thermal transport in carbon nanotubes, in G.A. Manossri, T.F. George, G.P. Zhang, L. Assoufid, eds., *Molecular Building Blocks for Nanotechnology: from Diamondoids to Nanoscale Materials and Applications*, Springer-Verlag, 2005.
51. P. Collins and P. Avouris, *Scientific American*, 62 (2000).
52. L. Chico, V.H. Crespi, L.X. Benedict, S.G. Louie, and M.L. Cohen, *Phys. Rev. Lett.* 76, 971 (1996).
53. Z. Yao, H.W.C. Postma, L. Balants, and C. Dekker, *Nature* 402, 273 (1999).
54. M. Menon and D. Srivastava, *Phys. Rev. Lett.* 79, 4453 (1997).
55. T. Rueckes, K. Kim, E. Joselevich, G.Y. Tseng, C-L Cheung, and C.M. Lieber, *Science* 289, 94 (2000).
56. M. Menon and D. Srivastava, *J. Mat. Res.* 13, 2357 (1998).
57. C. Joachim, J.K. Gimzewski, and A. Aviram, *Nature* 408, 541 (2000).
58. C. Papadopoulos, A. Rakitin, J. Li, A.S. Vedeneev, and J.M. Xu, *Phys. Rev. Lett.* 85, 3476 (2000).
59. C. Satishkumar, P.J. Thomas, A. Govindraj, and C.N.R. Rao, *Appl. Phys. Lett.* 77, 2530 (2000).
60. A. Antonis, M. Menon, D. Srivastava, G. Froudakis, and L.A. Chernozatonskii, *Phys. Rev. Lett.* 87, 66802 (2001).
61. A. Andriotis, M. Menon, D. Srivastava, and L. Chernozatonski, *Phys. Rev. B* 65, 165416 (2002).
62. A. Andriotis, D. Srivastava, and M. Menon, *Appl. Phys. Lett.* 82/83, 2512 (2003).

63. M. Terrones, F. Banhart, N. Grobert, J.-C. Charlier, H. Terrones, and P.M. Ajayan, *Phys. Rev. Lett.* 89, 75505 (2002).
64. D. Srivastava, M. Menon, and P.M. Ajayan, *J. Nanoparticle Res.* 5, 395 (2003).
65. G. Meng, Y.J. Jung, A. Cao, R. Vajtai, and P.M. Ajayan, *PNAS* 102, 7074 (2005).
66. J. Kong, N.R. Franklin, C. Zhou, M.G. Chapline, S. Peng, K.J. Cho, and H. Dai, *Science* 287, 622 (2000).
67. S. Peng and K.J. Cho, *Nanotechnology* 11, 57 (2000).
68. D. Srivastava, D.W. Brenner, J.D. Schall, K.D. Ausman, M. Yu, and R.S. Ruoff, *J. Phys. Chem. B* 103, 4330 (1999).
69. L. Yang and J. Han, *Phys. Rev. Lett.* 85, 154 (2000).
70. O. Gulseran, T. Yildirium, and S. Ciraci, *Phys. Rev. Lett.* 87, 116802 (2001).
71. S. Peng and K. Cho, *J. Appl. Mech.-Trans. ASME*, 69, 451 (2002).
72. C.H. Kiang, W.A. Goddard, R. Beyers, and D.S. Bethune, *Carbon* 33, 903 (1995).
73. T. Guo, P. Nikolaev, A. Thess, D.T. Colbert, and R.E. Smalley, *Chem. Phys. Lett.* 243, 49 (1995).
74. H. Dai, A.G. Rinzler, P. Nikolaev, A. Thess, D.T. Colbert, R.E. Smalley, *Chem. Phys. Lett.* 260, 471 (1996).
75. J. Kong, H.T. Soh, A.M. Cassell, C.F. Quate, and H. Dai, *Nature* 395, 878 (1998).
76. A.M. Cassell, S. Verma, J. Han, and M. Meyyappan, *Langmuir* 17, 260 (2001).
77. H. Kind, J.M. Bonard, L. Forro, K. Kern, K. Hernadi, L. Nilsson, L. Schlapbach, *Langmuir* 16, 6877 (2000).
78. M. Su, B. Zheng, J. Liu, *Chem. Phys. Lett.* 322, 321 (2000).
79. L. Delzeit, B. Chen, A. Cassell, R. Stevens, C. Nguyen, and M. Meyyappan, *Chem. Phys. Lett.* 348, 368 (2001).
80. M. Meyyappan, L. Delzeit, A. Cassell, and D. Hash, *Plasma Sources Sci. Technol.* 12, 205 (2003).
81. H. Karzow and A. Ding, *Phys. Rev. B* 60, 11180 (1999).
82. Y.Y. Wei, G. Eres, V.I. Merkulov, D.H. Lowndes, *Appl. Phys. Lett.* 78, 1394 (2001).
83. C. Bower, W. Zhu, S. Jin, and O. Zhou, *Appl. Phys. Lett.* 77, 830 (2000).
84. L. Delzeit, I. McAninch, B.A. Cruden, D. Hash, B. Chen, J. Han, and M. Meyyappan, *J. Appl. Phys.* 91, 6027 (2002).
85. J. Li, C. Papadopoulos, J.M. Xu, M. Moskovits, *Appl. Phys. Lett.* 75, 367 (1999).
86. B.C. Satishkumar, A. Govindraj, R. Sen, C.N.R. Rao, *Chem. Phys. Lett.* 293, 47 (1998).
87. R. Andrews, D. Jacques, A.M. Rao, F. Derbyshire, D. Qian, X. Fan, E.C. Dickey, and J. Chen, *Chem. Phys. Lett.* 303, 467 (1999).
88. P. Nikolaev, M.J. Bronikowski, R.K. Bradley, F. Rohmund, D.T. Colbert, K.A. Smith, and R.E. Smalley, *Chem. Phys. Lett.* 313, 91 (1999).
89. J. Liu, A.G. Rinzler, H. Dai, J.H. Hafner, R.K. Bradley, P.J. Boul, A. Lu, T. Iverson, K. Shelimov, C.B. Huffman, F. Rodriguez-Macias, Y.-S. Shon, T.R. Lee, D.T. Colbert, and R.E. Smalley, *Science* 280, 1253 (1998).
90. B. Khare, M. Meyyappan, A.M. Cassell, C.V. Nguyen, and J. Han, *Nano Lett.* 2, 73 (2002).
91. K. McGuire and A.M. Rao, in *Carbon Nanotubes: Science and Applications*, CRC Press, Boca Raton, FL, 2004.
92. M.A. Pimenta, A. Marucci, S.A. Empedocles, M.G. Bawendi, E.B. Hanlon, A.M. Rao, P.C. Eklund, R.E. Smalley, G. Dresselhaus, and M.S. Dresselhaus, *Phys. Rev. B* 58, 16016 (1998).
93. J. Chen, M.A. Hamon, H. Hu, Y. Chen, A.M. Rao, P.C. Eklund, and R.C. Haddon, *Science* 282, 95 (1998).
94. E.T. Mickelson, C.B. Huffman, A.G. Rinzler, R.E. Smalley, R.H. Hauge, and J.L. Margrave, *Chem. Phys. Lett.* 296, 188 (1998).
95. B.N. Kharc, M. Meyyappan, J. Kralj, P. Wilhite, M. Sisay, H. Imanaka, J. Koehne, and C.W. Bauschlicher, *Appl. Phys. Lett.* 81, 5237 (2002).
96. B.N. Khare, P. Wilhite, R.C. Quinn, B. Chen, R.H. Schingler, B. Tran, H. Imanaka, C.R. So, C.W. Bauschlicher Jr., and M. Meyyappan, *J. Phys. Chem. B* 108, 8166 (2004).

97. B.N. Khare, P. Wilhite, and M. Meyyappan, *Nanotechnology* 15, 1650 (2004).

98. H. Dai, N. Franklin, and J. Han, *Appl. Phys. Lett.* 73, 1508 (1998).

99. C.V. Nguyen, K.J. Cho, R.M.D. Stevens, L. Delzeit, A. Cassell, J. Han, and M. Meyyappan, *Nanotechnology* 12, 363 (2001).

100. C.V. Nguyen, C. So, R.M. Stevens, Y. Li, L. Delziet, P. Sarrazin, M. Meyyappan, *J. Phys. Chem. B* 108, 2816 (2004).

101. R.M. Stevens, C.V. Nguyen, and M. Meyyappan, *IEEE Trans. Nanobioscience* 3, 56 (2004).

102. S.J. Tans, A.R.M. Verschueren, and C. Dekker, *Nature* 393, 49 (1998).

103. R. Martel, T. Schmidt, H.R. Shea, T. Hertel, and Ph. Avouris, *Appl. Phys. Lett.* 73, 2447 (1998).

104. T. Yamada, *Appl. Phys. Lett.* 76, 628 (2000).

105. X. Liu, C. Lee, C. Zhou, and J. Han, *Appl. Phys. Lett.* 79, 3329 (2001).

106. A. Bachtold, P. Hadley, T. Nakanishi, and C. Dekker, *Science* 294, 1317 (2001).

107. P.G. Collins, K. Bradley, M. Ishigami, and A. Zettl, *Science* 287, 1801 (2000).

108. J. Li, Y. Lu, Q. Ye, M. Cinke, J. Han, and M. Meyyappan, *Nano Lett.* 3, 929 (2003).

109. Y. Lu, J. Li, J. Han, H.-T. Ng, C. Binder, C. Partridge, and M. Meyyappan, *Chem. Phys. Lett.* 391, 344 (2004).

110. J. Koehne, H. Chen, J. Li, A.M. Cassell, Q. Ye, H.T. Ng, J. Han, and M. Meyyappan, *Nanotechnology* 14, 1239 (2003).

111. J.E. Koehne, H. Chen, A.M. Cassell, Q. Ye, J. Han, M. Meyyappan, and J. Li, *Clin. Chem.* 50, 1886 (2004).

112. W.A. de Heer, A. Chatelain, D. Ugarte, *Science* 270, 1179 (1995).

113. P.G. Collins and A. Zettl, *Appl. Phys. Lett.* 69, 1969 (1996).

114. Y. Saito, S. Uemura, and K. Hamaguchi, *Jpn. J. Appl. Phys. Part 2*, 37, L 346 (1998).

115. X. Xu and G.R. Brandes, *Appl. Phys. Lett.* 74, 2549 (1999).

116. W. Zhu, C. Bower, O. Zhou, G. Kochanski, and S. Jin, *Appl. Phys. Lett.* 75, 873 (1999).

117. W.B. Choi, Y.H. Lee, D.S. Chung, N.S. Lee, and J.M. Kim, in D. Tomanek and R. Enbody, eds., *Science and Application of Nanotubes*, Kluwer Academic/Plenum Publishers, New York, 2000, p. 355.

118. R. Andrews, D. Jacques, A.M. Rao, T. Rantell, F. Derbyshire, Y. Chen, J. Chen, and R.C. Haddon, *Appl. Phys. Lett.* 75, 1329 (1999).

119. Y. Ye et al., *Appl. Phys. Lett.* 304, 207 (1999).

120. M. Terrones et al., *MRS Bull.* 43 (1999).

121. C.-H. Kiang, J.-S. Choi, T.T. Tran, and A.D. Bacher, *J. Phys. Chem. B* 103, 7449 (1999).

122. I. Mukhopadhyay, N. Hoshino, S. Kawasaki, F. Okino, W.K. Hsu, and H. Touhara, *J. Electrochem. Soc.* 149, A39 (2002).

123. J.S. Sakamoto and B. Dunn, *J. Electrochem. Soc.* 149, A26 (2002).

124. Q. Ngo, B.A. Cruden, A.M. Cassell, G. Sims, M. Meyyappan, J. Li, and C.Y. Yang, *Nano Lett.* 4, 2403 (2004).

125. J. Li, Q. Ye, H.T. Ng, A. Cassell, R. Stevens, J. Han, and M. Meyyappan, *Appl. Phys. Lett.* 82, 2491 (2003).

126. Q. Ngo, D. Petranovic, S. Krishnan, A.M. Cassell, Q. Ye, J. Li, M. Meyyappan, and C.Y. Yang, *IEEE Trans. Nanotech.* 3, 311 (2004).

127. M. Cinke, J. Li, B. Chen, A. Cassell, L. Delzeit, J. Han, and M. Meyyappan, *Chem. Phys. Lett.* 365, 69 (2002).

23

Mechanics of Carbon Nanotubes[1]

23.1 Introduction **23**-1

23.2 Mechanical Properties of Nanotubes.............................**23**-2
Molecular Structure of CNTs • Modeling of Nanotubes as
Elastic Materials • Strength of Nanotubes • Elastic Properties
Based on Crystal Elasticity

23.3 Experimental Techniques..**23**-20
Instruments for the Mechanical Study of Carbon Nanotubes
• Methods and Tools for Mechanical Measurement
• Challenges and New Directions

23.4 Simulation Methods..**23**-29
Ab Initio and Tight-Binding Methods • Classical Molecular
Dynamics • Continuum and Multiscale Models

23.5 Mechanical Applications of Nanotubes.........................**23**-39
Nanoropes • Filling the Nanotubes • Nanoelectromechanical
Systems (NEMS) • Nanotube-Reinforced Polymer

23.6 Conclusions **23**-48

Acknowledgments..**23**-49

References **23**-49

Dong Qian
Gregory J. Wagner
Wing Kam Liu
Northwestern University

Min-Feng Yu
University of Illinois

Rodney S. Ruoff
Northwestern University

23.1 Introduction

The discovery of multi-walled carbon nanotubes (MWCNTs) in 1991[1] has stimulated ever-broader research activities in science and engineering devoted entirely to carbon nanostructures and their applications. This is due in large part to the combination of their expected structural perfection, small size, low density, high stiffness, high strength (the tensile strength of the outermost shell of an MWCNT is approximately 100 times greater than that of aluminum), and excellent electronic properties. As a result, carbon nanotubes (CNTs) may find use in a wide range of applications in material reinforcement, field-emission panel display, chemical sensing, drug delivery, and nanoelectronics.

Indeed, NASA is developing materials using nanotubes for space applications, where weight-driven cost is the major concern, by taking advantage of their tremendous stiffness and strength. Composites based on nanotubes could offer strength-to-weight ratios beyond any materials currently available. Companies such as Samsung and NEC have invested tremendously and demonstrated product quality devices utilizing carbon nanotubes for field-emission display.[2,3] Such devices have shown superior

---

[1] This chapter appeared previously in *Applied Mechanics Reviews*, 56, 6, 2002. It is reprinted here with permission of ASME International.

qualities such as low turn-on electric field, high-emission current density, and high stability. With the advance of materials synthesis and device processing capabilities, the importance of developing and understanding nanoscale engineering devices has dramatically increased over the past decade. Compared with other nanoscale materials, single-walled carbon nanotubes (SWCNTs) possess particularly outstanding physical and chemical properties. SWCNTs are remarkably stiff and strong, conduct electricity, and are projected to conduct heat even better than diamond, which suggests their eventual use in nanoelectronics. Steady progress has been made recently in developing SWCNT nanodevices and nanocircuits,[4,5] showing remarkable logic and amplification functions. SWCNTs are also under intensive study as efficient storage devices, both for alkali ions for nanoscale power sources and for hydrogen for fuel cell applications.

On other fronts, CNTs also show great potential for biomedical applications due to their biocompatibility and high strength. The current generation of composites used for replacement of bone and teeth are crude admixtures of filler particles (often glass) that have highly inadequate mechanical properties compared with skeletal tissue. Fiber-based composites have been investigated in the past but have not worked well because of eventual degradation of the filler/matrix interface, usually due to attack by water. Thus, a major issue is interfacial stability under physiological conditions. Graphitic materials are known to resist degradation in the types of chemical environments present in the body. Recent demonstration of CNT artificial muscle[6] is one step along this direction and promises a dramatic increase in work density output and force generation over known technologies, along with the ability to operate at low voltage. CNTs are also being considered for drug delivery: they could be implanted without trauma at the sites where a drug is needed, slowly releasing a drug over time. They are also of considerable promise in cellular experiments, where they can be used as nanopipettes for the distribution of extremely small volumes of liquid or gas into living cells or onto surfaces. It is also conceivable that they could serve as a medium for implantation of diagnostic devices.

The rapid pace of research development as well as industry application has made it necessary to summarize the current status about what we know and what we do not know about this particularly interesting nanostructure. A number of excellent reviews on the general properties of CNTs can be found in References 7–9. In this paper we have made this effort with emphasis on the mechanical aspects.

The chapter is organized as follows: Section 23.2 focuses on the mechanical properties of CNTs. This includes the basic molecular structure, elastic properties, strength, and crystal elasticity treatment. Section 23.3 summarizes the current experimental techniques that were used in the measurement of the mechanical properties of CNTs. A brief discussion on the challenges in the experiment is also presented. Various simulation methods are discussed in Section 23.4. The length scale spans from quantum level to continuum level, which highlights the multiscale and multiphysics features of the nanoscale problem. Section 23.5 considers a few important mechanical applications of CNTs. Finally, conclusions are made in Section 23.6.

23.2 Mechanical Properties of Nanotubes

23.2.1 Molecular Structure of CNTs

23.2.1.1 Bonding Mechanisms

The mechanical properties of CNTs are closely related to the nature of the bonds between carbon atoms. The bonding mechanism in a carbon nanotube system is similar to that of graphite, as a CNT can be thought of as a rolled-up graphene sheet. The atomic number for carbon is 6, and the atom electronic structure is $1s^2 2s^2 2p^2$ in atomic physics notation. For a detailed description of the notation and the structure, readers may refer to basic textbooks on general chemistry or physics.[10] When carbon atoms combine to form graphite, sp^2 hybridization[10] occurs. In this process, one s-orbital and two p-orbitals combine to form three hybrid sp^2-orbitals at 120° to each other within a plane (shown in Figure 23.1). This in-plane bond is referred to as a σ-bond (*sigma*-bond). This is a strong covalent bond that binds

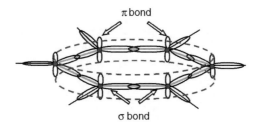

FIGURE 23.1 Basic hexagonal bonding structure for one graphite layer (the *graphene sheet*). Carbon nuclei shown as filled circle, out-of-plane π-bonds represented as delocalized (dotted line), and σ-bonds connect the C nuclei in-plane.

the atoms in the plane and results in the high stiffness and high strength of a CNT. The remaining *p*-orbital is perpendicular to the plane of the σ-bonds. It contributes mainly to the interlayer interaction and is called the π-bond (*pi*-bond). These out-of-plane, delocalized π-bonds interact with the π-bonds on the neighboring layer. This interlayer interaction of atom pairs on neighboring layers is much weaker than a σ-bond. For instance, in the experimental study of *shell-sliding*,[11] it was found that the shear strength between the outermost shell and the neighboring inner shell was 0.08 MPa and 0.3 MPa according to two separate measurements on two different MWCNTs. The bond structure of a graphene sheet is shown in Figure 23.1.

23.2.1.2 From Graphene Sheet to Single-Walled Nanotube

There are various ways of defining a unique structure for each carbon nanotube. One way is to think of each CNT as a result of rolling a graphene sheet, by specifying the direction of rolling and the circumference of the cross-section. Shown in Figure 23.2 is a graphene sheet with defined roll-up vector *r*. After rolling to form an NT, the two end nodes coincide. The notation we use here is adapted from References 8, 12, and 13. Note that *r* (bold solid line in Figure 23.2) can be expressed as a linear combination of base vectors *a* and *b* (dashed line in Figure 23.2) of the hexagon, i.e.,

$$r = na + mb$$

with *n* and *m* being integers. Different types of NT are thus uniquely defined by the values of *n* and *m*, and the ends are closed with caps for certain types of fullerenes (Figure 23.3).

Three major categories of NT can also be defined based on the chiral angle θ (Figure 23.2) as follows:

$$\theta = 0 \qquad \textit{Zigzag}$$
$$0 < \theta < 30 \qquad \textit{Chiral}$$
$$\theta = 30 \qquad \textit{Arm Chair}$$

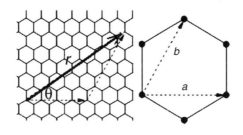

FIGURE 23.2 Definition of a roll-up vector as linear combinations of base vectors *a* and *b*.

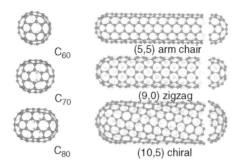

FIGURE 23.3 Examples of zig zag, chiral, and armchair nanotubes and their caps corresponding to different types of fullerenes. (From Dresselhaus MS, Dresselhaus G, and Saito R, (1995), *Carbon*. 33(7): 883–891. With permission from Elsevier Science.)

Based on simple geometry, the diameter d and the chiral angle θ of the NT can be given as:

$$d = 0.783\sqrt{n^2 + nm + m^2}\,\text{Å}$$

$$\theta = \sin^{-1}\left[\frac{\sqrt{3}m}{2(n^2 + nm + m^2)}\right]$$

Most CNTs to date have been synthesized with closed ends. Fujita et al.[14] and Dresselhaus et al.[8,15] have shown that NTs which are larger than (5,5) and (9,0) tubes can be capped. Based on Euler's theorem of polyhedra,[16] which relates the numbers of the edges, faces, and vertices, along with additional knowledge of the minimum energy structure of fullerenes, they conclude that any cap must contain six pentagons that are isolated from each other. For NTs with large radii, there are different possibilities of forming caps that satisfy this requirement. The experimental results of Iijima et al.[17] and Dravid[18] indicate a number of ways that regular-shaped caps can be formed for large-diameter tubes. *Bill*-like[19] and semi-toroidal[17] types of termination have also been reported. Experimental observation of CNTs with open ends can be found in Reference 17.

23.2.1.3 Multi-Walled Carbon Nanotubes and Scroll-Like Structures

The first carbon nanotubes discovered[1] were multi-walled carbon nanotubes (shown in Figures 23.4 and 23.5). Transmission electron microscopy studies on MWCNTs suggest a Russian doll-like structure (nested shells) and give interlayer spacing of approximately ~0.34 nm,[20,21] close to the interlayer separation of graphite, 0.335 nm. However, Kiang et al.[22] have shown that the interlayer spacing for MWCNTs can range from 0.342 to 0.375 nm, depending on the diameter and number of nested shells in the MWCNT. The increase in intershell spacing with decreased nanotube diameter is attributed to the increased repulsive force as a result of the high curvature. The experiments by Zhou et al.,[21] Amelinckx et al.,[23] and Lavin et al.[24] suggested an alternative *scroll* structure for some MWCNTs, like a cinnamon roll. In fact, both forms might be present along a given MWCNT and separated by certain types of defects. The energetics analysis by Lavin et al.[24] suggests the formation of a scroll, which may then convert into a stable multi-wall structure composed of nested cylinders.

23.2.2 Modeling of Nanotubes as Elastic Materials

23.2.2.1 Elastic Properties: Young's Modulus, Elastic Constants, and Strain Energy

Experimental fitting to mechanical measurements of the Young's modulus and elastic constants of nanotubes have been mostly made by assuming the CNTs to be elastic beams. An extensive summary is given

FIGURE 23.4 Top left: High-resolution transmission electronic microscopy (HRTEM) image of an individual MWCNT. The parallel fringes have ~0.34 nm separation between them and correspond to individual layers of the coaxial cylindrical geometry. Bottom left: HRTEM image showing isolated SWCNT as well as bundles of such tubes covered with amorphous carbon. The isolated tubes shown are approximately 1.2 nm in diameter. Top right: HRTEM image showing the tip structure of a closed MWCNT. The fringe (layer) separation is again 0.34 nm. Bottom right: The image of a MWCNT showing the geometric changes due to the presence of five and seven membered rings (position indicated in the image by P for pentagon and H for heptagon) in the lattice. Note that the defects in all the neighboring shells are conformal. (From Ajayan PM and Ebbesen TW, (1997), *Rep. Prog. Phys.* 60(10): 1025–1062. With permission.)

in Section 23.2.2.2. Aside from the use of the beam assumption, there are also experimental measurements that were made by monitoring the response of a CNT under axial load. Lourie and Wagner[27] used micro-Raman spectroscopy to measure the compressive deformation of a nanotube embedded in an epoxy matrix. For SWCNT, they obtained a Young's modulus of 2.8–3.6 TPa, while for MWCNT, they measured 1.7–2.4 TPa.

Yu et al.[28] presented results of 15 SWCNT bundles under tensile load and found Young's modulus values in the range from 320 to 1470 GPa (mean: 1002 GPa). The stress vs. strain curves are shown

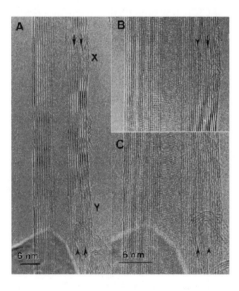

FIGURE 23.5 HRTEM image of a MWCNT. Note the presence of anomalously large interfringe spacings indicated by arrows. (From Amelinckx S, Lucas A, and Lambin P, (1999), *Rep. Prog. Phys.* 62(11): 1471–1524. With permission.)

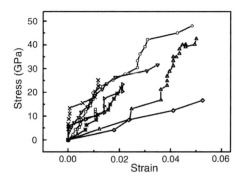

FIGURE 23.6 Eight stress vs. strain curves obtained from the tensile-loading experiments on individual SWCNT bundles. The values of the nominal stress are calculated using the cross-sectional area of the perimeter SWCNTs assuming a thickness of 0.34 nm. (From Yu MF, Files BS, Arepalli S, and Ruoff RS, (2000), *Phys. Rev. Lett.* 84(24): 5552–5555. With permission.)

in Figure 23.6. Another experiment performed by Yu et al.[29] on the tensile loading of MWCNTs yielded a Young's modulus from 270 to 950 GPa (Figure 23.7). It should be pointed out that concepts such as Young's modulus and elastic constants belong to the framework of continuum elasticity; an estimate of these material parameters for nanotubes implies the continuum assumption. Because each individual SWCNT involves only a single layer of rolled graphene sheet, the thickness t will not make any sense until it is given based on the continuum assumption. In the above-mentioned experiments, it is assumed that the thickness of the nanotube is close to the interlayer distance in graphite, i.e., 0.34 nm.

In the following, we focus on the theoretical prediction of these parameters. Unless explicitly given, a thickness of 0.34 nm for the nanotube is assumed. The earliest attempt to predict Young's modulus theoretically seems to have been made by Overney et al.[30] Using an empirical Keating Hamiltonian with parameters determined from first principles, the structural rigidity of (5,5) nanotubes consisting of 100, 200, and 400 atoms was studied. Although the values are not explicitly given, it was later pointed out by Treacy et al.[31] that the results from Overney et al. implied a Young's modulus in the range of 1.5 to 5.0 TPa. The earliest energetics analysis of CNTs was presented by Tibbetts[32] using elastic theory. He pointed out that the strain energy of the tube is proportional to $1/R^2$ (where R is the radius of the CNT). Ruoff and Lorents[33] suggested the use of elastic moduli of graphite by neglecting the change in the atomic structure when a piece of graphene sheet is rolled into a nanotube. Because the mechanical behavior of single-crystal graphite is well understood, it would be a good approximation to use the in-plane modulus

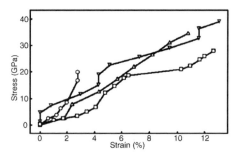

FIGURE 23.7 Plot of stress vs. strain curves for five individual MWCNTs. (Reprinted from Yu MF, Lourie O, Dyer MJ, Moloni K, Kelly TF, and Ruoff RS, (2000), *Science.* 287(5453): 637–640. © 2000 American Association for the Advancement of Science. With permission.)

(1.06 TPa of graphite). However, whether such an approximation is good for SWCNTs with small radius was not known.

Robertson et al.[34] examined the energetics and elastic properties of SWCNTs with radii less than 0.9 nm using both Brenner's potential[35] and first-principles total energy methods. Their results showed a consistent linear proportionality to $1/R^2$ of the strain energy, which implies that small deformation beam theory is still valid even for the small radius limit. An elastic constant (C_{11}) close to that of graphite was predicted using the second set of parameters from Brenner's potential[35]; however, it was also shown that the first set of parameters from the same potential results in excess stiffness. Gao et al.[36] carried out a similar study on SWCNTs of larger radius (up to 17 nm) with a potential that is derived from quantum mechanics. A similar linear relation of the strain energy to $1/R^2$ was found. By computing the second derivative of the potential energy, values of Young's modulus from 640.30 GPa to 673.49 GPa were obtained from the MD simulation for closest-packed SWCNTs.

Yakobson et al.[37] compared particular molecular dynamics (MD) simulation results to the continuum shell model[38] and thereby fitted both a value for Young's modulus (~5.5 TPa) and for the effective thickness of the CNTs ($t = 0.066$ nm). Lu[39] derived elastic properties of SWCNTs and MWCNTs using an empirical model in his MD simulation. A Young's modulus of ~1TPa and a shear modulus of ~0.5 TPa were reported based on a simulated tensile test. Lu also found from his simulation that factors such as chirality, radius, and the number of walls have little effect on the value of Young's modulus. Yao et al.[40] used a similar approach but with a different potential model and obtained a Young's modulus of approximately 1 TPa. In addition, they treated the dependence of Young's modulus on both the radius and chirality of the tube. They employed an MD model that included bending, stretching, and torsion terms, and their results showed that the strain in the tube was dominated by the torsional terms in their model. An alternate method is to derive Young's modulus based on the energy-per-surface area rather than per-volume. This was used in the study by Hernandez et al.[41] Using a nonorthogonal tight-binding scheme, they reported a *surface* Young's modulus of 0.42 TPa-nm, which, when converted to Young's modulus assuming the thickness of 0.34 nm, resulted in a value of 1.2 TPa. This value is slightly higher than that obtained by Lu.[39]

Zhou et al.[42] estimated strain energy and Young's modulus based on electronic band theory. They computed the total energy by taking account of all occupied band electrons. The total energy was then decomposed into the rolling energy, the compressing or stretching energy, and the bending energy. By fitting these three values with estimates based on the continuum elasticity theory, they obtained a Young's modulus of 5.1 TPa for SWCNT having an effective wall thickness of 0.71 Å. Note that this is close to the estimate by Yakobson et al.[37] because the rolling energy and stretching energy terms were also included in the shell theory that Yakobson et al. used. In addition, the accuracy of the continuum estimate was validated by the derivation of a similar linear relation of the strain energy to $1/R^2$.

Although no agreement has been reached among these publications regarding the value of the Young's modulus at this moment, it should be pointed out that a single value of Young's modulus cannot be uniquely used to describe both tension/compression *and* bending behavior. The reason is that tension and compression are mainly governed by the in-plane σ-bond, while pure bending is affected mainly by the out-of-plane π-bond. It may be expected that different values of elastic modulus should be obtained from these two different cases unless different definitions of the thickness are adopted, and that is one reason that accounts for the discrepancies described above. To overcome this difficulty, a consistent continuum treatment in the framework of crystal elasticity is needed; and we present this in Section 23.2.4.

23.2.2.2 Elastic Models: Beams and Shells

Although CNTs can have diameters only several times larger than the length of a bond between carbon atoms, continuum models have been found to describe their mechanical behavior very well under many circumstances.[43] Indeed, their small size and presumed small number of defects make CNTs ideal systems for the study of the links between atomic motion and continuum mechanical properties such as Young's modulus and yield and fracture strengths. Simplified continuum models of CNTs have taken one of two forms: simple beam theory for small deformation and shell theory for larger and more complicated distortions.

Assuming small deformations, the equation of motion for a beam is:

$$\rho A \frac{\partial^2 u}{\partial t^2} + EI \frac{\partial^4 u}{\partial x^4} = q(x) \tag{23.1}$$

where u is the displacement, ρ is the density, A the cross-sectional area, E Young's modulus, I the moment of inertia, and $q(x)$ a distributed applied load. This equation is derived assuming that displacements are small and that sections of the beam normal to the central axis in the unloaded state remain normal during bending; these assumptions are usually valid for small deformations of long, thin beams, although deviations from this linear theory are probable for many applications of CNTs.[44,45] The natural frequency of the ith mode of vibration is then given by:

$$\omega_i = \frac{\beta_i^2}{L^2} \sqrt{\frac{EI}{\rho A}} \tag{23.2}$$

where β_i is the root of an equation that is dictated by the boundary conditions. For a beam clamped at one end (zero displacement and zero slope) and free at the other (zero reaction forces), this equation is:

$$\cos \beta_i \cosh \beta_i + 1 = 0 \tag{23.3}$$

Thus, the frequencies of the first three modes of vibration of a clamped-free beam can be computed from Equation (23.3) with $\beta_1 \approx 1.875$, $\beta_2 \approx 4.694$, and $\beta_3 \approx 7.855$.

Measurements on vibrating CNTs can therefore be used to estimate Young's modulus. The first experimental measurements of Young's modulus in MWCNTs were made by Treacy et al.,[31] who used a vibrating beam model of a MWCNT to estimate a modulus of about 1.8 TPa. The authors observed TEM images of MWCNTs that appeared to be undergoing thermal vibration, with a mean-square vibration amplitude that was found to be proportional to temperature. Assuming equipartition of the thermal energy among vibrational modes and a hollow cylinder geometry of the tube, this allowed Young's modulus to be estimated based on the measured amplitude of vibration at the tip of the tube. The spread in the experimental data is quite large, with modulus values for 11 tubes tested ranging from 0.40 to 4.15 TPa with a mean value of 1.8 TPa; the uncertainty was ±1.4 TPa. A similar study by Krishnan et al.[46] of SWCNTs found an average modulus of about 1.3–0.4/+0.6 TPa for 27 SWCNTs.

Rather than relying on estimates of thermal vibrations, Poncharal et al.[44] used electromechanical excitation as a method to probe the resonant frequencies of MWCNTs. For tubes of small diameter (less than about 12 nm), they found frequencies consistent with a Young's modulus in the range of 1 TPa. However, for larger diameters, the bending stiffness was found to decrease by up to an order of magnitude, prompting the authors to distinguish their measured *bending modulus* (in the range 0.1 to 1 TPa) from the true Young's modulus. One hypothesis put forth for the decrease in effective modulus is the appearance of a mode of deformation, in which a wavelike distortion or *ripple* forms on the inner arc of the bent MWCNT. This mode of deformation is not accounted for by simple beam theory. Liu et al.[45] used a combination of finite element analysis (using the elastic constants of graphite) and nonlinear vibration analysis to show that nonlinearity can cause a large reduction in the effective bending modulus. However, Poncharal et al. reported that there was no evidence of nonlinearity (such as a shift of frequency with varying applied force) in their experimental results, which causes some doubt as to whether a large nonlinear effect such as the rippling mode can explain the decrease in the bending modulus with increasing

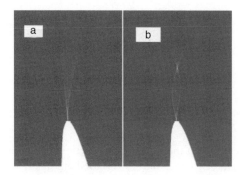

FIGURE 23.8 Scanning electron microscope (SEM) images of electric field-induced resonance of an individual MWCNT at its fundamental resonance frequency (a) and at its second-order harmonic(b).

tube diameter. Similar resonance excitation of MWCNTs has recently been realized by Yu et al.[47] inside a scanning electron microscope (SEM) (Figure 23.8).

Static models of beam bending can also be used to measure mechanical properties. Wong et al.[48] measured the bending force of MWCNTs using atomic force microscopy (AFM). Assuming the end displacement of an end-loaded cantilevered beam is given by PL3/3EI, where P is the applied force, they fit a Young's modulus of 1.28 ± 0.59 TPa. Salvetat et al.[49,50] measured the vertical deflection vs. the applied force dependence of MWCNT and SWCNT ropes spanning one of the pores in a well-polished alumina nanopore membrane using AFM. They fit values of about 1 TPa for MWCNTs grown by arc discharge, whereas those grown by the catalytic decomposition of hydrocarbons had a modulus 1–2 orders of magnitude smaller.

Govinjee and Sackman[51] studied the validity of modeling MWCNTs with Euler beam theory. They showed the size dependency of the material properties at the nanoscale, which does not appear in the classical continuum mechanics. The beam assumption was further explored by Harik.[52,53] From scaling analysis, he proposed three nondimensional parameters to check the validity of the continuum assumption. The relation between these parameters and MD simulation was discussed. Ru[54–59] has used a shell model to examine the effects of interlayer forces on the buckling and bending of CNTs. It is found that, for MWCNTs, the critical axial strain is decreased from that of a SWCNT of the same outside diameter,[54–56] in essence because the van der Waals forces between layers always cause an inward force on some of the tubes. Note that although the critical axial strain is reduced, the critical axial force may be increased due to the increased cross-sectional area. The phenomenon is also seen when the CNT is embedded in an elastic matrix.[57,58] Ru uses a similar analysis to treat the buckling of columns of SWCNTs arranged in a honeycomb pattern.[59]

23.2.2.3 Elastic Buckling and Local Deformation of NT

Experiments have shown a few cases of exceptional tensile strength of carbon nanotubes (see Section 23.2.3 "Strength of Nanotubes"). In addition, experiment and theory have addressed structural instability for tubes under compression, bending, or torsion. Buckling can occur in both the axial and transversal direction. Buckling can also occur in the whole structure or locally. Analyses based on continuum theory and the roles of interlayer potential are discussed in Section 23.4.2.2.

Yakobson et al.[37,43] modeled buckling of CNTs under axial compression and used the Brenner potential in their MD simulation. Their simulation also showed four *snap-throughs* during the load process, resulting from instability. The first buckling pattern starts at a nominal compressive strain of 0.05. Buckling due to bending and torsion was demonstrated in References 37, 60, 61, and 62. In the case of bending, the pattern is characterized by the collapse of the cross-section in the middle of the tube, which confirms the experimental observations by Iijima et al.[63] and Ruoff et al.[64,65] using HRTEM

(a)   (b)

FIGURE 23.9   (a) Buckling of SWCNT under bending load. (b) Buckling of SWCNT under torsional load.

and by Wong et al.[48] using AFM. When the tube is under torsion, flattening of the tube, or equivalently a collapse of the cross-section, can occur due to the torsional load. Shown in Figure 23.9 are the simulation results of buckling patterns using a molecular dynamics simulation code developed by the authors. Falvo et al.[66] bent MWCNTs by using the tip of an AFM. They showed that MWCNTs could be bent repeatedly through large angles without causing any apparent fracture in the tube. Similar methods were used by Hertel et al.[67] to buckle MWCNTs due to large bending. Lourie et al.[68] captured the buckling of SWCNTs under compression and bending by embedding them into a polymeric film. Unlike the single kink seen by Iijima et al.,[63] the buckling pattern under bending was characterized by a set of local rippling modes.

The radial deformability for tubes has also been studied. Ruoff et al.[69] first studied radial deformation between adjacent nanotubes (Figure 23.10). Partial flattening due to van der Waals forces was observed in TEM images of two adjacent and aligned MWCNTs along the contact region. This was the first observation that CNTs are not necessarily perfectly cylindrical. Indeed, in an anisotropic physical environment, all CNTs are likely to be, at least to some degree, not perfectly cylindrical due to mechanical deformation. Tersoff and Ruoff[70] studied the deformation pattern of SWCNTs in a closest-packed crystal and concluded that rigid tubes with diameters smaller than 1 nm are less affected by the van der Waals attraction and hardly deform. But for diameters over 2.5 nm, the tubes flatten against each other and form a honeycomb-like structure. This flattening of larger diameter SWCNTs could have a profound effect on factors such as storage of molecular hydrogen in SWCNT crystals composed of such larger diameter tubes, if they can be made, because the interstitial void space is dramatically altered. Lopez et al.[70a] have reported the observation, with HRTEM, of polygonized SWCNTs in contact.

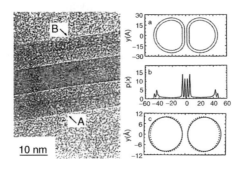

FIGURE 23.10   Left: HRTEM image of two adjacent MWCNTs *a* and *b*. Nanotube *a* has 10 fringes, and nanotube *b* has 22. The average interlayer spacings for inner layers and outer layers belonging to MWCNT *a* are 0.338 nm and 0.345 nm, respectively. For MWCNT *b*, these are 0.343 nm and 0.351 nm, respectively. The 0.07 and 0.08 nm differences, respectively, are due to the compressive force acting in the contact region, and the deformation from perfectly cylindrical shells occurring in both inner and outer portions of each MWCNT. Right from top to bottom: (a) Calculated deformation resulting from van der Waals forces between two double-layered nanotubes. (b) Projected atom density from (a). The projected atom density is clearly higher in the contact region, in agreement with the experimental observation of the much darker fringes in the contact region as compared to the outer portions of MWCNTs *a* and *b*. (c) Calculated deformation for adjacent single-layer nanotubes. (From Ruoff RS, Tersoff J, Lorents DC, Subramoney S, and Chan B, (1993), *Nature*, 364(6437): 514–516. With permission.)

Chopra et al.[71] observed fully collapsed MWCNTs with TEM and showed that the collapsed state can be energetically favorable for certain types of CNTs having a certain critical radius and overall wall thickness. Benedicts et al.[72] proposed the use of the ratio of mean curvature modulus to the interwall attraction of graphite to predict whether the cross-section will collapse and applied their model for a collapsed MWCNT observed in TEM by the same authors. In addition, their experiment also provided the first microscopic measurement of the intensity of the inter-shell attraction.

Hertel et al.[73] and Avouris et al.[74] studied the van der Waals interaction between MWCNTs and a substrate by both experiment (using AFM) and simulation. They found that radial and axial deformations of the tube may lead to a high energy binding state, depending on the types of CNTs present. Fully or partially collapsed MWCNTs on surfaces have also been reported by Yu et al.,[75,76] and their work included a careful analysis of the mechanics of tubes when in contact with surfaces. Lordi and Yao[77] simulated the radial deformation of CNTs due to a local contact and compared the model structures with experimental images of CNTs in contact with nanoparticles. With simulation of both SWCNT and MWCNT up to the radius of a (20, 20) nanotube (13.6 Å), no collapse due to local contact was reported. Instead, the radial deformation was found to be reversible and elastic. Based on these results, they suggested that mechanically cutting a nanotube should be rather difficult, if not impossible.

Gao et al.[36] carried out an energetics analysis on a wide range of nanotubes. They showed the dependence of geometry on the radius of the tube. To be more specific, a circular cross-section is the stable configuration for a radius below 1 nm, in line with the prior results obtained by Tersoff and Ruoff[70] mentioned above. If the radius is between 1 and 2 nm, both circular and collapsed forms are possible. Beyond 3 nm the SWCNT tends to take the collapsed form. However, we should note that the treatment by Gao et al. is of isolated SWCNTs, and their results would be modified by contact with a surface (which might accelerate collapse. Yu et al.,[75,76] on the other hand, as already demonstrated by Tersoff and Ruoff,[70] showed that completely surrounding a SWCNT by similar sized SWCNTs might stabilize against complete collapse.

It is interesting to speculate whether larger diameter SWCNTs would *ever* remain completely cylindrical in any environment. A nearly isotropic environment would be a liquid comprised of small molecules, or perhaps a homogeneous polymer comprised of relatively small monomers. However, because the CNTs have such small diameters, fluctuations are present in the environment on this length scale, locally destroying the time-averaged isotropicity present at longer length scales, may trigger collapse. Alternatively, if full collapse does not occur, the time-averaged state of such a CNT in a (time-averaged) isotropic environment might be perfectly cylindrical. Perhaps for this reason, CNTs might be capable of acting as probes of minute fluctuations in their surrounding (molecular) environment.

Shen et al.[78] conducted a radial indentation test of ~10 nm diameter MWCNTs with scanning probe microscopy. They observed deformability (up to 46%) of the tube and resilience to a significant compressive load (20 μN). The radial compressive elastic modulus was found to be a function of compressive ratios and ranged from 9.0 to 80.0 GPa.

Yu et al.[79] performed a nanoindentation study by applying compressive force on individual MWCNTs with the tip of an AFM cantilever in tapping-mode (Figure 23.11) and demonstrated a deformability similar to that observed by Shen et al. They estimated the effective elastic modulus of a range of indented MWCNTs to be from 0.3 GPa to 4 GPa by using the Hertzian contact model. The reader should note that the difference between this effective elastic modulus and the elastic modulus discussed in Section 23.2.2.1 is that the effective modulus refers to the elastic response to deformation of an anisotropic indentation load applied in the radial direction. We thus also distinguish this type of load from the isotropic load that could, for example, be applied by high pressure for CNTs suspended in a liquid pressure medium, which is more appropriately referred to as *isotropic radial compressive loading*. Yu et al.[76] provided further energetics analysis on MWCNTs that are in configurations of partial or full collapse, or collapsed combined with a twist. They showed that interlayer van der Waals interactions play an important role in maintaining the collapsed configuration.

Chesnokov et al.[80] observed remarkable reversible volume compression of SWCNTs under quasi-hydrostatic pressure up to 3 GPa and obtained a volume compressibility of 0.0277 GPa^{-1}, which suggests the use of CNT as energy-absorbing material. The volume compressibility of SWCNTs having a diameter

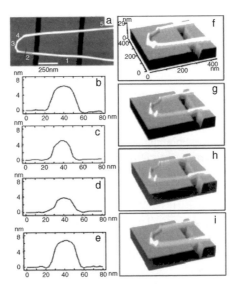

FIGURE 23.11 Deformability of an MWCNT deposited on a patterned silicon wafer as visualized with tapping-mode AFM operated far below mechanical resonance of a cantilever at different set points. The height in this and all subsequent images was coded in grayscale, with darker tones corresponding to lower features. (a) Large-area view of a MWCNT bent upon deposition into a hairpin shape. (b)–(e) Height profiles taken along the thin marked line in (a) from images acquired at different set-point (S/S_0) values: (b) 1.0; (c) 0.7; (d) 0.5; (e) 1.0. (f)–(i) Three-dimensional images of the curved region of the MWCNT acquired at the corresponding set-point values as in (b)–(e). (From Yu MF, Kowalewski T, and Ruoff RS, (2000), *Phys. Rev. Lett.* 85(7): 1456–1459. With permission.)

of 1.4 nm under hydrostatic pressure was also studied by Tang et al.[81] by *in situ* synchrotron X-ray diffraction. The studied SWCNT sample, which consisted primarily of SWCNT bundles and thus not individual or separated SWCNTs, showed linear elasticity under hydrostatic pressure up to 1.5 GPa at room temperature with a compressibility value of $0.024 GPa^{-1}$, which is smaller than that of graphite ($0.028 GPa^{-1}$). However, the lattice structure of the SWCNT bundles became unstable for pressure beyond ~4 GPa and was destroyed upon further increasing the pressure to 5 GPa.

A very subtle point that comes into the picture is the effect of interlayer interactions when CNTs collapse. This effect can produce some results that cannot be described by traditional mechanics. For instance, Yu et al.[82] observed fully collapsed MWCNT ribbons in the twisted configuration with TEM. One such cantilevered MWCNT ribbon had a twist present in the freestanding segment (Figure 23.12a). Such a configuration cannot be accounted for by elastic theory because no external load (torque) is present to hold the MWCNT ribbon in place and in the twisted form. However, it is known that the difference of approximately 0.012 eV/atom in the interlayer binding energy of the *AA* and the *AB* stacking configurations (Figure 23.12d, c) exists for two rigid graphitic layers spaced 0.344 nm apart. The analysis by Yu et al.[82] suggests that the elastic energy cost for the twist formation of this particular CNT ribbon can be partially compensated for by achieving more favorable atomic registry. The observation of the existence of this freestanding twist in the ribbon thus suggests that an energy barrier exists to keep the twisted ribbon from untwisting. The mechanics analysis performed suggests that this twisted structure is metastable. More details on the modeling of interlayer interaction with the account of this registry effect are presented in Section 23.4.2.2.

As a brief summary, compared with their high rigidity in the axial direction, CNTs are observed to be much more compliant in the radial direction. Thus a CNT readily takes the form of a partially or fully collapsed nanoribbon when the radius and wall thickness are in particular ranges, when either isolated in free space, or in contact with a surface. A CNT may also locally flatten when surrounded, as occurs, for example, in the SWCNT bundles.

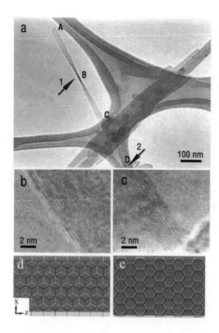

FIGURE 23.12   A free-standing twisted MWCNT ribbon. (a) A TEM image of this ribbon anchored on one end by a carbon support film on a lacy carbon grid. Arrows point to the twists in the ribbon. (b) – (c) Eight resolved fringes along both edges of the ribbon imaged near the anchor point. (d) A schematic depicting the AB stacking between armchair CNT shells (the two layers are the layer having brighter background and black lattices vs. the layer having darker background and white lattices). The AB stacking can be achieved by shifting the layer positions along the x direction that is perpendicular to the long axis of the MWCNT. (e) A schematic depicting the lattice alignment between the zig–zag CNT shells by allowing the relative shifting of the layers along the x direction. The AB stacking is not possible; only AA stacking or other stacking (as shown in the schematic) is possible. (From Yu MF, Dyer MJ, Chen J, Qian D, Liu WK, and Ruoff RS, (2001), *Phys. Rev. B, Rapid Commn.* 64: 241403R. With permission.)

23.2.3 Strength of Nanotubes

The strength of a CNT will likely depend largely on the distribution of defects and geometric factors. In the case of geometric factors, buckling due to compression, bending, and torsion has been discussed in Section 23.2.2.3. Note that even in these loading cases, it is still possible for plastic yielding to take place due to highly concentrated compressive force. Another geometric factor is the interlayer interaction in the case of MWCNTs and bundles of CNTs.

23.2.3.1   Strength Due to Bond Breaking or Plastic Yielding

Unlike bulk materials, the density of the defects in nanotubes is presumably less; and therefore the strength is presumably significantly higher at the nanoscale. The strength of the CNT could approach the theoretical limit depending on the synthesis process. There are several major categories of synthesis method: carbon evaporation by arc current discharge,[1,83,84] laser ablation,[85,86] or chemical vapor deposition (CVD).[87–89] MWCNTs produced by the carbon plasma vapor processes typically possess higher quality in terms of defects than those produced by the shorter time, lower temperature CVD processes. However, it is not known at this time whether the SWCNTs produced by the laser ablation method, for example, are better than those produced by CVD growth from preformed metal catalyst particles present on surfaces. It is thus worth mentioning the essential role of various catalysts in the production of SWCNTs.[90]

There have been few experimental reports on testing the tensile strength of nanotubes. The idea is simple but seems rather difficult to implement at this stage. Yu et al.[28] were able to apply a tensile load on 15 separate SWCNT bundles and measure the mechanical response (Figure 23.13). The maximum tensile strain they obtained was 5.3%, which is close to the theoretical prediction made by Nardelli et al.[91]

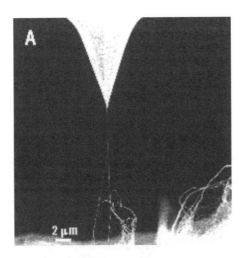

FIGURE 23.13 SEM image of a tensile-loaded SWCNT bundle between an AFM tip and a SWCNT buckytube paper sample. (From Yu MF, Files BS, Arepalli S, and Ruoff RS, (2000), *Phys. Rev. Lett.* 84(24): 5552–5555. With permission.)

The *average* SWCNT tensile strength for each bundle ranged from 13 to 52 GPa, calculated by assuming the load was applied primarily on the SWCNTs present at the perimeter of each bundle.

Tensile loading of 19 individual MWCNTs was reported by Yu et al.,[29] (Figure 23.14) and it was found that the MWCNTs broke in the outermost layer by a "sword-in-sheath" breaking mechanism with tensile strain at break of up to 12%. The tensile strength of this outermost layer (equivalent to a large-diameter SWCNT) ranged from 11 to 63 GPa.

By laterally stretching suspended SWCNT bundles fixed at both ends using an AFM operated in lateral force mode, Walters et al.[92] were able to determine the maximum elongation of the bundle and thus the breaking strain. The tensile strength was thus estimated to be 45±7 GPa by assuming a Young's modulus of 1.25TPa for SWCNT. The tensile strength of very long (~2mm) ropes of CVD-grown aligned MWCNTs was measured to be 1.72±0.64 GPa by Pan et al.[93] using a modified tensile testing apparatus, by constantly monitoring the resistance change of the ropes while applying the tensile load. The much lower value they obtained is perhaps to be expected for CVD-grown MWCNTs, according to the authors. But another factor may be the much longer lengths tested, in that a much longer CNT may be more likely to have a critical concentration of defects that could lead to failure present somewhere along their length. The dependence of fracture strength on length has not yet been experimentally addressed.

Another way of applying tensile load is to use the load transferred by embedding the CNTs in a matrix material. Wagner et al.[94] observed fragmentation in SWCNTs using this method and reported a tensile

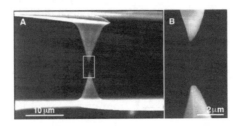

FIGURE 23.14 Tensile loading of individual MWCNTs. (a) A SEM image of an MWCNT attached between two AFM tips. (b) Large magnification image of the indicated region in (a) showing the MWCNT between the AFM tips. (Reprinted from Yu MF, Lourie O, Dyer MJ, Moloni K, Kelly TF, and Ruoff RS, (2000), *Science.* 287(5453): 637–640. ©2000 American Association for the Advancement of Science. With permission.)

strength of 55 GPa. Tensile strength measurements on a resin-based SWCNT composite were performed by Li et al.[95] Through a treatment that includes modeling the interfacial load transfer in the SWCNT–polymer composite, the tensile strength of the SWCNTs was fit to an average value of ~22 GPa. A compression test of the CNT has been performed by Lourie et al.[68] in which, in addition to the buckling mode that is mentioned in Section 23.2.2.3, they also observed plastic collapse and fracture of thin MWCNTs. The compressive strength and strain corresponding to these cases were estimated to be approximately 100–150 GPa and > 5%, respectively.

Theoretical prediction of CNT strength has emphasized the roles of defects, loading rate, and temperature using MD simulation. Yakobson et al.[37,43,61,96] performed a set of MD simulations on the tensile loading of nanotubes. Even with very high strain rate, nanotubes did not break completely in half; and the two separated parts were instead connected by a chain of atoms. The strain and strength from these simulations are reported to be 30% and 150 GPa, respectively. The fracture behavior of CNT has also been studied by Belytschko et al.[97] using MD simulations. Their results show moderate dependence of fracture strength on chirality (ranges from 93.5 GPa to 112 GPa), and fracture strain between 15.8% and 18.7% were reported. In these simulations, the fracture behavior is found to be almost independent of the separation energy and to depend primarily on the inflection point in the interatomic potential. The values of fracture strains compare well with experimental results by Yu et al.[29]

A central theme that has been uncovered by quantum molecular dynamics simulations of plastic yielding in CNTs is the effect of pentagon/heptagon (or 5/7) defects (Figure 23.15, where the bond rotation leading to the 5–7–7–5 defect initially formed. This is referred to as the *Stone–Wales bond-rotation*[8]) which, for certain types of SWCNTs and at sufficiently high temperature, can lead to plastic yielding.[96] Nardelli et al.[99] studied the mechanism of strain release under tensile loading using both classical and quantum molecular dynamics (MD and QMD). They found that in the case of tension, topological defects such as the 5–7–7–5 defect tend to form when strain is greater than 5% in order to achieve the relaxation of the structure. At high temperature (which for these carbon systems means temperatures around 2000 K) the 5–7–7–5 defect can, for a subset of the SWCNT types, separate into two 5,7 pairs that can glide with respect to each other. Nardelli et al.[91] observed the ductile–to–brittle failure transition as a function of both temperature and strain in their MD simulation. Generally, high strain (15%) and low temperature (1300K) lead to brittle behavior (crack extension or separation), while low strain (3%) and high temperature (3000K) make nanotubes more ductile (dislocation motion without cracking). The reader is reminded of the short timescale of these simulations.

Srivastava et al.[100] demonstrated that a local reconstruction leading to formation of sp^3 bonds can occur under compressive load, and obtained a compressive strength of approximately 153 GPa.

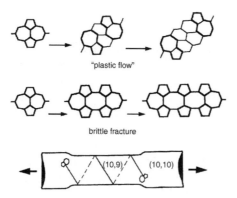

FIGURE 23.15 The 5–7–7–5 dislocation evolves either as a crack (brittle cleavage) or as a couple of dislocations gliding away along the spiral slip plane (plastic yield). In the latter case, the change of the nanotube chirality is reflected by a step-wise change of diameter and by corresponding variations of electrical properties. (From Yakobson BI, 1997, in *Recent Advances in the Chemistry and Physics of Fullerenes and Related Materials*, Kadish KM (Ed.), Electrochemical Society, Inc., Pennington, NJ, 549. With permission of Elsevier Science.)

Wei et al.[101] and Srivastava et al.[102] demonstrated that finite temperature (starting from 300 K) can help the nanotube to overcome a certain energy barrier to achieve plastic deformation. They conducted a similar study of the temperature effect using MD with compressive loading at 12% strain and observed both the formation of sp^3 bonds (at 300K) and 5/7 defects (1600K) for an (8,0) tube. In addition, it was reported[102] that a slower strain rate tends to trigger plastic yield. The conclusion is very similar to the one obtained by Nardelli et al.,[99] but a subtle question is whether the ductile behavior is due to strain or strain rate.

A detailed analysis on the mechanism of defects and dislocation is presented by Yakobson.[61,103] It was shown that the glide direction of the 5/7 defects is dependent on the chirality of the nanotube and, consequently, an irreversible change in the electronic structure also takes place in the vicinity of the chirality change (an armchair tube changes from metallic to semiconducting). Similar conclusions about the effect of strain release on the electronic structure of the tube have been given by Zhang et al.[104] Zhang et al. also observed the dependence of the elastic limit (the onset of plastic yielding) on the chirality of the tube: for the same radius, an (n,0), thus zig-zag, tube has nearly twice the limit of an (n,n), thus armchair, tube. This is due mostly to the different alignment of the defects with the principle shear direction. The compressive strength they obtained ranged from 100 to 170 GPa. Another mechanism of defect nucleation is described by Zhang and Crespi,[105] where plastic flow can also occur due to the spontaneous opening of double-layered graphitic patches.

As a brief summary, experimental and simulation efforts have recently been undertaken to assess the strength of CNT. A major limitation for simulation currently is the small timescale that current MD methods can address, much shorter than are actually implemented in experiment. Implicit methods and bridging scale methods show promise in alleviating this difficulty.

23.2.3.2 Strength Due to Interlayer Sliding

Most synthesized nanotubes are either randomly agglomerated MWCNTs or bundles of closest-packed SWCNTs.[85,90] The SWCNTs have a narrow diameter distribution as synthesized and are consequently typically tightly and efficiently packed in bundles.[86,106] The tensile response of SWCNT nanoropes has been measured by Yu et al.[28]; tensile loading of individual MWCNTs was presented by Yu et al.[29] They demonstrated that MWCNTs that are mounted by attachment to the outermost shell and then loaded in tension break in the outermost shell, thus indicating negligible load transfer to the inner shells. This limits the potential of MWCNTs for structural applications — in terms of exhibiting high stiffness and strength in tension, they are a victim of their own high perfection of bonding, as there is evidently no covalent bonding between the nested shells and thus little or no load transfer from the outer shell to the inner shells.

The weak intershell interaction has been measured by Yu et al.[11] (Figure 23.16) and estimated by Cumings and Zettl.[107] In the study of *shell-sliding* of two MWCNTs by Yu et al.,[11] direct measurement of the dependence of the pulling force on the contact length between the shells was performed, and the shear strength between the outermost shell and the neighboring inner shell for two MWCNTs was found to be 0.08 and 0.3 Mpa, respectively. These values are on the low end of the experimental values of the shear strength in graphite samples,[108] thus emphasizing the weak intershell interactions in high-quality, highly crystalline MWCNTs, similar to the type originally discovered by Iijima.[1] Cumings et al.[107] demonstrated inside a TEM that a *telescope process* can be repeated on the MWCNTs that they tested up to 20 times without causing any apparent damage. (The reader should note that assessment of damage at the nanoscale is quite challenging. Here, the authors simply note that there was no apparent change as imaged by high-resolution TEM.) These two studies motivate contemplation of the possibility of a nanoscale-bearing system of exceptional quality. The shear strengths corresponding to static friction and dynamic friction were estimated to be 0.66 and 0.43 MPa, respectively.[107] These values are on the same order as what Yu et al. obtained by experimental measurement.[11]

For SWCNT bundles, there have been theoretical estimates that, in order to achieve load transfer so that the full bundle cross-section would be participating in bearing load up to the intrinsic SWCNT breaking strength (when loading has, for example, been applied only to those SWCNTs on the perimeter), the SWCNT contact length must be on the order of 10 to 120 microns.[109,110] This is much longer than the typical length of individual SWCNTs in such bundles,[111,112] where mean length values on the order

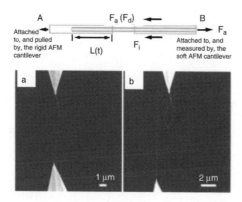

FIGURE 23.16 The forces involved in the shell-sliding experiment can be described by $F_a = F_s + F_i = \pi d\tau L(t) + F_i$, where F_a is the applied pulling force as a function of time, τ is the shear strength, L is the contact length, d the shell diameter, and F_i is a diameter-dependent force originating from both surface tension and edge effects. SEM images showing the sword-in-sheath breaking mechanism of MWCNTs. (a) A MWCNT attached between AFM tips under no tensile load. (b) The same MWCNT after being tensile loaded to break. Notice the apparent overall length change of the MWCNT fragments after break compared to the initial length and the curling of the top MWCNT fragment in (b). (From Yu MF, Yakobson BI, and Ruoff RS, (2000), *J. Phys. Chem. B.* 104(37): 8764–8767. With permission.)

of a few hundred nanometers have been obtained. This means that load transfer in a parallel bundle containing such relatively short SWCNTs is very likely to be very small and thus ineffective. Quantitative analysis of actual dynamics by modeling using MD requires an interlayer potential that precisely accounts for the interlayer interaction as a function of relative slide position as well as correct incorporation of thermal effects. A general discussion of interlayer potentials is presented in Section 23.3.2.2.

23.2.4 Elastic Properties Based on Crystal Elasticity

It now can be seen that all the theoretical studies presented in Section 23.2.2.1 have either failed to distinguish between the case of infinitesimal strain and finite strain or have directly applied the result at infinitesimal strain to the case of finite strain. In this section, the elastic properties of CNTs at finite strain will be discussed. The significance of this study is obvious from the various deformation patterns due to the load. Of particular interest is the derivation of the elastic constants based on a known atomic model, from either an empirical expression or *ab initio* calculation.

Early attempts to derive elastic constants based on the potential energy of a crystal system were made by Born and Huang.[113] In their treatment, the potential energy was expressed as a function of the displacements of atoms. Because the formulation does not generally guarantee the rotational invariance, the so-called Born–Huang conditions were proposed. Keating[114–117] showed the inconsistencies in the Born–Huang conditions and stated that the potential energy of the crystal system can be alternatively expressed in terms of the variables that intrinsically preserve the invariance property. Based on Brugger's thermodynamic definition[118] of elastic constants, Martin[119–121] derived elastic constants for a crystal system in which the energy density is a sum of the contributions from many-body interactions. Martin's approach is in fact a hyperelastic approach because the energy density was considered as a function of the Lagrangian strain. Based on the embedded-atom method (EAM),[122,123] Tadmor et al.[124,125] derived the corresponding elasticity tensor at finite deformation in their quasicontinuum analysis. Friesecke and James[126] proposed another approach of bridging between continuum and atomic structure, with emphasis on nanostructures in which the size of one dimension is much larger than the other.

The essence of hyperelasticity has been discussed in detail in References 127 through 129. The advantage of the hyperelastic formulation is that it is inherently material frame-invariant; therefore, no special treatment is needed in the large deformation computation.

As a brief summary, if the energy density W of the material is known, the relation between the nominal stress P and the deformation gradient F is given as[127]:

$$P = \frac{\partial W}{\partial F^T} \quad \text{or} \quad P_{ij} = \frac{\partial W}{\partial F_{ji}} \tag{23.4}$$

In Equation (23.4), $F_{ij} = \partial x_i / \partial X_j$, x and X are, respectively, the spatial and material coordinates. The lowercase subscript denotes the dimension. An equivalent form is to express in terms of the second Piola–Kirchhoff stress S (referred to as 2nd PK stress) and Lagrangian strain E, i.e.,

$$S = \frac{\partial W}{\partial E} \quad \text{or} \quad S_{ij} = \frac{\partial W}{\partial E_{ij}} \tag{23.5}$$

Correspondingly, there are two sets of elasticity tensors:

$$C^{PF} = \frac{\partial^2 W}{\partial F^T \partial F^T} \quad \text{or} \quad C_{ijkl}^{PF} = \frac{\partial^2 W}{\partial F_{ji} \partial F_{lk}} \tag{23.6}$$

and

$$C^{SE} = \frac{\partial^2 W}{\partial E \partial E} \quad \text{or} \quad C_{ijkl}^{SE} = \frac{\partial^2 W}{\partial E_{ij} \partial E_{kl}} \tag{23.7}$$

C^{PF} and C^{SE} are generally referred to as the *first elasticity tensor* and the *second elasticity tensor*, respectively. Their difference is mainly governed by different stress and strain measures that are involved in their definition. It can be shown that the two elasticity tensors are related by

$$C_{ijkl}^{PF} = C_{imnk}^{SE} F_{jm} F_{ln} + S_{ik} \delta_{lj} \tag{23.8}$$

where δ_{ij} is the Kronecker delta.

In addition, there are also the *third elasticity tensor* $C_{ijkl}^{(3)}$ and *fourth elasticity tensor* $C_{ijkl}^{(4)}$. The *third elasticity tensor* relates the velocity gradient $L_{ij} = \partial v_i / \partial x_j$ to the push-forward of the rate of nominal stress P, i.e.,

$$F_{ir} \dot{P}_{rj} = C_{ijkl}^{(3)} L_{kl}^T \tag{23.9}$$

The fourth elasticity tensor is the spatial form of the second elasticity tensor and is defined by

$$C_{ijkl}^{(4)} = F_{im} F_{jn} F_{kp} F_{lq} C_{mnpq}^{SE} \tag{23.10}$$

It can be shown that the fourth elasticity tensor is essentially the tangent moduli in the spatial form and it relates the convective rate of the Kirchhoff stress to the rate of deformation:

$$D_{ij} = \frac{1}{2}(L_{ij} + L_{ji}) \tag{23.11}$$

The convected rate of Kirchhoff stress corresponds to the mathematical concept of Lie derivatives, which consistently define the time derivatives of tensors. Therefore, the fourth elasticity tensor plays an important role in maintaining objectivity during stress update and material stability analysis. More detailed descriptions can be found in Chapters 5 and 6 in the book by Belytschko, Liu, and Moran.[127]

To apply hyperelasticity to the crystal system, the Cauchy–Born rule[124,130,131] must be imposed. This rule assumes that the local crystal structure deforms homogeneously and that the mapping is characterized by the deformation gradient *F*. With this assumption, we can apply Equations (23.4) to (23.7) to specific atomic models of the nanotube system. We consider the undeformed state of the CNT to be the same as that of a graphene sheet and adopt the classical Tersoff–Brenner model.[35,132–135] The empirical equation is given as follows:

$$\Phi_{ij} = \Phi_R(r_{ij}) - \bar{B}_{ij}\Phi_A(r_{ij}) \tag{23.12}$$

in which *i*, *j* are the indices for carbon atoms, Φ_R and Φ_A represent the repulsive and attractive parts of the potential, respectively. The effect of bonding angle is considered in the term $\bar{B}_{ij}$. The detailed expression for each individual term looks a little tedious due to the consideration of many-body effects. Readers may consult the original paper[35] for parameters and functional forms.

The basic element of the graphite structure is a single hexagon, with each side of the hexagon a result of the covalent bond due to sp^2 hybridization described in Section 23.2.1.1. Because each bond is also shared by the other neighboring hexagon, the total bonding energy Φ for one hexagon can be given as half of the summation of the bonding energy from all six covalent bonds. This summation can be further reduced if the symmetry is considered. The energy density, due to the fact that the single layer of graphite is only a result of repeating the hexagon structure, is given as:

$$W = \frac{\Phi}{A_0} \tag{23.13}$$

with $\Phi = \sum_{l=1}^{3}\Phi_l$. In Equation (23.13), A_0 is the area of the undeformed hexagon and Φ_l is the bond energy for the *l*th bond. According to Brenner's potential:

$$\Phi_l = \Phi_l(r^l, r^{l1}, r^{l2}, \cos\theta^{l1}, \cos\theta^{l2}) \tag{23.14}$$

in which *l*1 and *l*2 refer to the neighboring bonds of bond *l* and Φ_l is a function of bond lengths and bonding angles. Note that *W* is in fact the surface energy density of the system, and the thickness term *t* is not needed because we are dealing with a single layer of graphite. As a result of this, the units for the stress and elasticity tensors are different from those used in the conventional procedures. According to Section 23.2.1.2, the deformation gradient is determined by the roll-up vector, which is composed of the direction and length of rolling operation and the subsequent mechanical relaxation after rolling. These effects are embedded in the deformation gradient F_{ij}. The relaxed structure can be obtained from MD simulation for a specific type of CNT. We have evaluated elastic constants and Young's modulus for various types of nanotubes. For the purpose of comparison, we have converted the elastic constants and Young's modulus by assuming an artificial thickness *t* = 0.34 nm. With this assumption, the second elasticity tensor in the hyper-elastic formulation is given as:

$$C_{ijkl}^{SE} = \frac{\partial S_{ij}}{\partial E_{kl}} \tag{23.15a}$$

with

$$S_{ij} = \frac{1}{A_0 t}\sum_{l=1}^{3}\left(\sum_{N=1,l1,l2}\frac{\partial \Phi}{\partial r^N}\frac{\partial r^N}{\partial E_{ij}} + \sum_{N=1,2}\frac{\partial \Phi_l}{\partial \cos\theta_{lN}}\frac{\partial \cos\theta_{lN}}{\partial E_{ij}}\right) \tag{23.15b}$$

We view the concept of Young's modulus as a tangent modulus that is defined in the deformed configuration (the CNT as a result of rolling the graphene sheet.). Therefore it can be calculated from both the above equation and the expression for the fourth elasticity tensor (Equation (23.10)). The standard definition of Young's modulus is the ratio of the uniaxial stress exerted on a thin rod (a nanotube, in our case) to the resulting normal strain in the same direction. If we define 1 to be the axial direction and 2 and 3 to be the other two orthogonal directions, then it can be shown that the relation between Young's modulus Y and the elastic constants is given as:

$$Y = \frac{C_{11}C_{23}^2 + C_{12}C_{13}^2 + C_{33}C_{12}^2 - C_{11}C_{22}C_{33} - 2C_{12}C_{13}C_{23}}{C_{23}^2 - C_{22}C_{33}} \tag{23.16a}$$

$$C_{ijkl} = F_{im}F_{jn}F_{kp}F_{lq}\left(\frac{\partial S_{mn}}{\partial E_{pq}}\right) \tag{23.16b}$$

Note that the Voigt notation has been used in Equation (23.16a), and elastic constants correspond to the components of the fourth elasticity tensor defined in Equation (23.16b). By plugging the model parameters and the deformation gradient, the Young's modulus for a (10,10) and (100,100) nanotube are obtained as 0.7 and 1TPa, respectively. It is observed that the Young's modulus from the case of (10,10) is quite different from that of graphite due to the effect of rolling. Such effect becomes small as the radius of CNT increases. Note that this value of Young's modulus is determined *consistently* based on crystal elasticity *combined* with the use of MD simulation. In contrast with some of the approaches described in Section 23.2.2.1, which use purely molecular dynamics to determine the Young's modulus, the approach is semi-analytical and serves as a link between the continuum and atomistic scale. In addition, as we extend to the finite deformation case, the issue of anisotropy naturally arises as a result of this formulation, which can only be qualitatively reproduced by MD. It is emphasized that the thickness assumption is only used in the comparison with the experiments or theoretical predictions that have taken the interlayer separation as the thickness. Clearly this thickness t is not needed in our formulation.

The procedure described above reveals certain limitations of the standard hyperelastic approach and Cauchy–Born rule, which can be described as follows:

- The deformation is in fact not homogenous as the graphene sheet is rolled into CNT. Correspondingly, the energy of the CNT not only depends on the deformation gradient but also on higher order derivatives of *F*. In such a case, a set of high-order elastic constants that belongs to the framework of *multipolar* theory[136] needs to be determined.

- Another aspect that has been missing in the hyperelastic theory is the dependence of the energy on the so-called inner displacement of the lattice, which can be defined as the relative displacement between two overlapping Bravais lattices. Note that the inner displacement can occur without violating the Cauchy–Born rule. According to Cousins,[137] the consideration of these variables results in the so-called *inner elastic constants*.

In general, the factors mentioned above are difficult to evaluate through purely analytical methods. A continuum treatment, which accounts for the effect of inner displacement, has been proposed recently in References 138 through 140. For computational implementations, see Section 23.4.3.

23.3 Experimental Techniques

The extremely small dimensions of CNTs — diameters of a few tens of nanometers for MWCNTs and about 1 *nm* for SWCNTs, and length of a few microns — impose a tremendous challenge for experimental study of mechanical properties. The general requirements for such study include (1) the challenge of CNT placement in an appropriate testing configuration, such as of picking and placing, and in certain cases the fabrication of clamps; (2) the achievement of desired loading; and (3) characterizing and measuring the mechanical deformation at the nanometer length scale. Various types of high-resolution microscopes are indispensable instruments for the characterization of nanomaterials, and recent innovative developments in the new area of *nanomanipulation* based on inserting new tools into such instruments has enhanced our ability to probe nanoscale objects. We discuss several of these below.

23.3.1 Instruments for the Mechanical Study of Carbon Nanotubes

Electron microscopy (EM, SEM, and TEM) and scanning probe microscopy (SPM) have been the most widely used methods for resolving and characterizing nanoscale objects. Electron microscopy uses high-energy electron beams (several keV up to several hundred keV) for scattering and diffraction, which allows the achievement of high resolving power, including down to subnanometer resolution because of the extremely short wavelength (a fraction of an Angstrom) of electrons at high kinetic energy.

We and others have primarily used transmission electron microscopy (TEM) and scanning electron microscopy (SEM) to study nanotube mechanics. In TEM, an accelerated electron beam from a thermal or a field emitter is used. The beam transmits through the sample and passes several stages of electro-magnetic lenses, projecting the image of the studied sample region to a phosphor screen or other image recording media. In SEM, a focused electron beam (nanometers in spot size) is rastered across the sample surface; and the amplified image of the sample surface is formed by recording the secondary electron signal or the back scattering signal generated from the sample.

Sample requirements typically differ between TEM and SEM. In TEM, a thin (normally several tens of nanometers or less in thickness) and small (no more than 50mm^2) sample is a requirement as a small sample chamber is available; and a dedicated holder is typically used in these expensive instruments (which are thus typically time-shared among many users) for sample transfer. In SEM there is no strict limit on sample size in principle, and normally a large sample chamber is available so that samples can be surveyed over large areas. As to the difference in the ultimate resolution, TEM is limited by such factors as the spread in energy of the electron beam and the quality of the ion optics; and SEM is limited by the scattering volume of the electrons interacting with the sample material. TEM normally has a resolution on the order of 0.2 nm, while SEM is capable of achieving a resolution up to 1 nm.

An exciting new development in electron microscopy is the addition of aberration correction, also referred to as *corrective ion optics*. We provide no extensive review of this topic here, but note for the reader that a coming revolution in electron microscopy will allow for image resolution of approximately 0.04 nm with TEM and of about 0.1 nm for SEM, and perhaps better. There are two types of corrective ion optics: one corrects for spherical aberration (to correct for aberration in the lenses) and one for achromatic aberration (meant to correct for the spread in the wavelengths of the electrons emanating from the emitter). As newer instruments are installed in the next few years, such as at national laboratories, improvement in the image resolution for mechanics studies of nanosized specimens is an enticing goal (see, for example, http://www.ornl.gov/reporter/no22/-dec00.htm and http://www.nion.com/).

The scanning or atomic force microscope (SFM, also referred to as AFM) has also been particularly useful for mechanical studies of CNTs. Since the invention of the STM in 1986, it has been quickly accepted as a standard tool for many applications related to surface characterization. High-resolution (nanometer up to atomic resolution) mapping of surface morphology on almost any type of either conductive or nonconductive material can be achieved with a SFM.

The principle of the microscope is relatively simple. A probe, having a force-sensitive cantilever with a sharp tip, is used as a sensor to physically scan, in close proximity to, the sample surface. The probe is driven by a piezoelectric tube capable of nanometer-resolution translations in the *x*, *y* and *z* directions; and the tip normally has a radius of curvature on the order of 10 nm. The force interaction between tip and sample results in deflection of the cantilever. While scanning the sample surface in the *x* and *y* directions, the deflection of the cantilever is constantly monitored by a simple optical method or other approaches. A feedback electronic circuit that reads the deflection signal and controls the piezoelectric tube is responsible for keeping a constant force between the tip and the sample surface, and a surface profile of the sample can thus be obtained. Depending on the type of interaction force involved for sensing, an SPM instrument can include a host of methods, such as AFM, friction-force microscopy (FFM), magnetic-force microscopy (MFM), electric-force microscopy (EFM), and so on. Depending on the mechanism used for measuring the force interaction, scanning probe microscopy also includes many modes of operation, such as contact mode, tapping mode, and force modulation mode. For more information on scanning force microscopy, see, for example, References 141 and 142 as well as http://-www.thermomicro.com/spmguide/contents.htm.

23.3.2 Methods and Tools for Mechanical Measurement

Following is a brief summary of the methods used for measuring the mechanical properties of CNTs, and especially isolated individual CNTs; and new tools specifically designed for such tasks will be introduced.

23.3.2.1 Mechanical Resonance Method

Treacy and coworkers[31] deduced values for Young's modulus for a set of MWCNTs by measuring the amplitude from recorded TEM images of the thermal vibration of the free ends of each when naturally cantilevered. Krishnan et al.[46] succeeded in measuring the Young's modulus of SWCNTs (Figure 23.17) using a similar method. The amplitude was measured from the blurred spread of tip positions of the free end of the cantilevered CNT compared with the clamped end. The amplitude can be modeled by considering the excitation of mechanical resonance of a cantilever:

$$\sigma^2 = \frac{16L^3 kT}{\pi Y(a^4 - b^4)} \sum_n \beta_n^{-4} \approx 0.4243 \frac{L^3 kT}{Y(a^4 - b^4)} \tag{23.17}$$

where σ is the amplitude at the free end, L is the length of the cantilevered beam, k is the Boltzmann constant, T is the temperature, Y is the Young's modulus, a is the outer radius and b is the inner radius of CNT, and β_n is a constant for free vibration mode n. The tip blurring originates from thermal activation of vibrations (the CNT behaves classically because of the low frequency modes that are populated by the expected kT of thermal vibrational energy), and the amplitude can be modeled by considering the excitation of mechanical resonance of a cantilever. The image is blurred simply because the frequency is

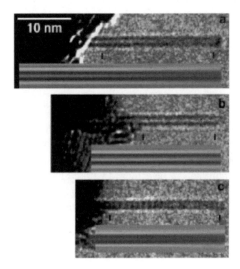

FIGURE 23.17 TEM images of vibrating single-walled nanotubes. Inserted with each micrograph is the simulated image corresponding to the best least-square fit for the adjusted length L and tip vibration amplitude σ. The tick marks in each micrograph indicate the section of the nanotube shank that was fitted. The nanotube length, diameter W, tip amplitude, and the estimated Young's modulus E are (a) $L = 36.8$ nm, $\sigma = 0.33$ nm, $W = 1.50$ nm, $E = 1.33 \pm 0.2$ TPa; (b) L = 24.3 nm, $\sigma = 0.18$ nm, $W = 1.52$ nm, $E = 1.20 \pm 0.2$ TPa; and (c) $L = 23.4$ nm, $\sigma = 50.30$ nm, $W = 1.12$ nm, $E = 1.02 \pm 0.3$ TPa. (From Krishnan A, Dujardin E, Ebbesen TW, Yianilos PN, and Treacy MMJ, (1998), *Phys. Rev. B.* 58(20): 14013–14019. With permission.)

high relative to the several-second integration time needed for generating the TEM image. Because the resonance is a function of the cantilever stiffness, and the geometry is directly determined by the TEM imaging, the Young's modulus values could be fit. The advantage of this method is that it is simple to implement without the need for additional instrument modification or development — only a variable-temperature TEM holder is needed. The principle can also be applied for the study of other nanowire-type materials as long as the tip blurring effect is obvious. The drawback is that a model fit is needed to determine the real cantilever length, and human error is inevitable in determining the exact amplitude of the blurred tip. As pointed out by the authors, the error for the Young's modulus estimation using such a method is around ±60%.

Poncharal et al.[44] introduced an electric field excitation method (Figure 23.18) for the study of mechanical resonance of cantilevered MWCNTs and measured the bending modulus. In the experiment, a specially designed TEM holder was developed that incorporated a piezo-driven translation stage and a mechanical-driven translation stage. The translation stages allowed the accurate positioning of the MWCNT material inside the TEM, in this case relative to a counter electrode. Electrical connections were made to the counter electrode and the electrode attaching the MWCNT materials, so that DC bias as well as AC sinusoidal voltage could be applied between the counter electrode and the MWCNT. The generated AC electric field interacts with induced charges on the MWCNT, which produces a periodic driving force. When the frequency of the input AC signal matched the mechanical resonance frequency of the MWCNT, obvious oscillation corresponding to the resonance mode of the cantilever MWCNT was observed (Figure 23.18) and the resonance frequency of the MWCNT thereby determined. Using continuum beam mechanics, the bending modulus of the MWCNT was calculated from the measured resonance frequency and CNT geometry according to Equation (23.2). The benefits of such an approach are the efficient method for driving mechanical resonances and, because the whole experiment is done inside a TEM, the ability to analyze the high-aspect ratio nanostructures in detail.

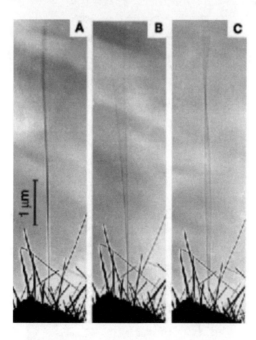

FIGURE 23.18   Electric field-driven resonance of MWCNT. (A) In the absence of a potential, the nanotube tip $(L = 6.25$ mm, $D = 14.5$ nm) vibrated slightly because of thermal effects. (B) Resonant excitation of the fundamental mode of vibration $(f_1 = 530$ kHz). (C) Resonant excitation of the second harmonic $(f_2 = 3.01$ MHz). For this nanotube, a value of $E_b = 0.21$ TPa was fit to the standard continuum beam mechanics formula. (Reprinted from Poncharal P, Wang ZL, Ugarte D, and de Heer WA, (1999), *Science*. 283(5407): 1513–1516. ©1999 American Association for the Advancement of Science. With permission.)

23.3.2.2 Scanning-Force Microscopy Method

The atomic force microscope operated in lateral-force mode, contact mode, or tapping mode, has been the main tool in studying the mechanical response of individual CNTs under static load and when in contact with surfaces. Falvo et al.[66] used a nanomanipulator and contact mode AFM to manually manipulate and bend MWCNTs deposited on a substrate surface. The strong surface force between the MWCNTs and the substrate allowed such an operation to be performed. By intentionally creating large curvature bends in MWCNTs, buckles and periodic ripples were observed. These authors estimated that, based on the local curvature of the bend found, some MWCNTs could sustain up to a 16% strain without obvious structural or mechanical failure (Figure 23.19)

Wong et al.[48] measured the bending modulus of individual MWCNTs using an AFM operated in lateral-force mode (Figure 23.20). The MWCNTs, deposited on a low-friction MoS2 surface, were pinned down at one end by overlaying SiO pads using lithography. AFM was then used to locate and measure the dimension of the MWCNT, and lateral force was applied at the different contact points along the length of the MWCNT (Figure 23.20). By laterally pushing the MWCNT, lateral force vs. deflection data were recorded. The data were then analyzed using a beam mechanics model that accounted for the friction force, the concentrated lateral force, and the rigidity of the beam. The bending modulus value for the MWCNT was obtained by fitting the measured force vs. deflection curve. Such a method also allowed the bending strength to be determined by deflecting the beam past the critical buckling point.

Salvetat et al. used another approach to deflect under load MWCNT[49] and SWCNT ropes[50] by depositing them onto a membrane having 200 nm pores (Figure 23.21). By positioning the AFM tip directly on the midpoint of the CNT spanning the pore and applying an indentation force (Figure 23.21), force vs. deflection curves were obtained and compared with theoretical modeling based on beam mechanics. Elastic moduli

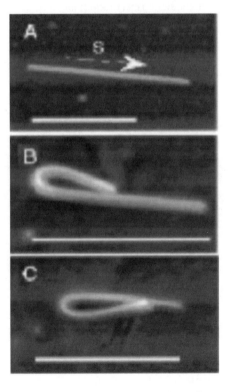

FIGURE 23.19 Bending and buckling of MWCNTs. (A) An original straight MWCNT. (B) The MWCNT is bent upwards all the way back onto itself. (C) The same MWCNT is bent all the way back onto itself in the other direction. (Reprinted from Falvo MR, Clary GJ, Taylor RM, Chi V, Brooks FP, Washburn S, and Superfine R, (1997), *Nature.* 389(6651): 582–584. ©1997 Macmillan Publishers. With permission.)

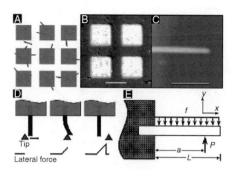

FIGURE 23.20 Overview of one approach used to probe mechanical properties of nanorods and nanotubes. (A) SiC nanorods or carbon nanotubes were deposited on a cleaved MoS_2 substrate, and then pinned by deposition of a grid of square SiO pads. (B) Optical micrograph of a sample showing the SiO pads and the MoS_2 substrate. The scale bar is 8 mm. (C) An AFM image of a 35.3 nm diameter SiC nanorod protruding from an SiO pad. The scale bar is 500 nm. (D) Schematic of beam bending with an AFM tip. The tip (triangle) moves in the direction of the arrow, and the lateral force is indicated by the trace at the bottom. (E) Schematic of a pinned beam with a free end. The beam of length L is subjected to a point load P at $x = a$ and to a distributed friction force f. (Reprinted from Wong EW, Sheehan PE, and Lieber CM, (1997), *Science.* 277(5334): 1971–1975. ©1997 American Association for the Advancement of Science. With permission.)

for individual MWCNTs and separately for SWCNT bundles were deduced. In principle, such a measurement requires a well-controlled and stable environment to eliminate the errors induced by unexpected tip–surface interactions and instrument instability as well as a very sharp AFM tip for the experiment.

The radial deformability of individual MWCNTs was studied by Shen et al.[78] using an indentation method and Yu et al.[79] using a tapping mode method in AFM. Load was applied along the radial direction (perpendicular to the axial direction that is defined as along the longaxis of the CNT) of MWCNTs, and the applied force vs. indentation depth curve was measured. Using the classic Hertz theory, the deformability of the MWCNT perpendicular to the long axis direction was obtained. The reader should note that this is not the radial compressibility, because the force is not symmetrically applied in the radial direction (not isotropic). Thus, one might think of this as *squashing* the MWCNT locally by indentation.

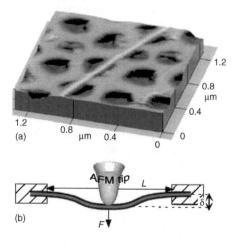

FIGURE 23.21 (a) AFM image of a SWCNT bundle adhered to the polished alumina ultrafiltration membrane, with a portion bridging a pore of the membrane. (b) Schematic of the measurement: the AFM is used to apply a load to the nanobeam and to determine directly the resulting deflection. A closed-loop feedback ensured an accurate scanner positioning. Si_3N_4 cantilevers with force constants of 0.05 and 0.1 N/m were used as tips in the contact mode. (From Salvetat JP, Briggs GAD, Bonard JM, Bacsa RR, Kulik AJ, Stockli T, Burnham NA, and Forro L, (1999), *Phys. Rev. Lett.* 82(5): 944–947. With permission.)

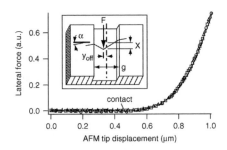

FIGURE 23.22   Lateral force on a SWCNT bundle as a function of AFM tip position. The four symbols represent data from four consecutive lateral force curves on the same rope, showing that this rope is straining elastically with no plastic deformation. Inset: The AFM tip moves along the trench, in the plane of the surface, and displaces the rope as shown. (From Walters DA, Ericson LM, Casavant MJ, Liu J, Colbert DT, Smith KA, and Smalley RE, (1999), *Appl. Phys. Lett.* 74(25): 3803–3805. With permission.)

These two approaches are technically different. In the indentation method, the image scan is stopped and the AFM tip is held steady to apply a vertical force on a single point on the MWCNT through the extension and retraction of the piezoelectric tube along the z direction. A force vs. indentation depth curve is obtained by monitoring the AFM cantilever deflection under the extension or retraction. In the tapping mode method, an off-resonance tapping technique is used so that the tapping force can be quantitatively controlled by adjusting the free cantilever amplitude and the set point. The set point is a control parameter in tapping mode AFM for keeping a constant cantilever amplitude (thus a constant distance between the AFM tip and the sample surface) in imaging scan mode. AFM images of each MWCNT are acquired using different set points, and force vs. indentation depth curves are obtained by plotting the curve of the set point vs. MWCNT height. The advantage of using the tapping mode method is that the squashing deformability of the MWCNT along its whole length can be obtained through several image acquisitions, though care must be taken to choose the appropriate tapping mode imaging parameters for such an experiment.

Walters et al.[92] studied the elastic strain of SWCNT nanobundles by creating a suspended SWCNT bundle that was clamped at both ends by metal pads over a trench created with standard lithographic methods. Using an AFM operated in lateral-force mode, they were able to repeatedly stretch and relax the nanoropes elastically as shown in Figure 23.22, including finally stretching to the breaking point to determine the maximum strain. The absolute force used to stretch the SWCNT bundle was not measured, and the breaking strength was estimated by assuming the theoretical value of ~1 TPa for Young's modulus for the SWCNTs in the rope.

23.3.2.3   Measurement Based on Nanomanipulation

The response to axial tensile loading of individual MWCNTs was realized by Yu et al.[29] using a new testing stage based on a nanomanipulation tool operating inside an SEM. The nanomanipulation stage allowed for the three-dimensional manipulation — picking, positioning, and clamping — of individual MWCNTs. The individual MWCNTs were attached to AFM probes having sharp tips by a localized electron beam-induced deposition (EBID) of carbonaceous material inside the SEM. An MWCNT so clamped between two AFM probes was then tensile loaded by displacement of the rigid AFM probe (Figure 23.23), and the applied force was measured at the other end by the cantilever deflection of the other, compliant AFM probe. The measured force vs. elongation data were converted, by SEM measurement of the MWCNT geometry, to a stress vs. strain curve; and the breaking strength of each MWCNT was obtained by measuring the maximum tensile loading force at break.

Yu et al.[28] applied a similar approach for the tensile strength measurement of small bundles of SWCNTs. The entangled and web-like agglomeration of SWCNTs in raw samples made it difficult to find an individual SWCNT and resolve it by SEM or to pick out individual SWCNT nanobundles; so a modified

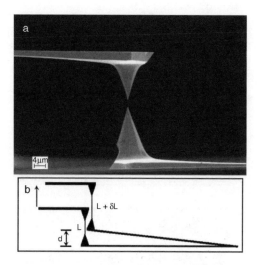

FIGURE 23.23 (a) Individual MWCNT is clamped in place and stretched by two opposing AFM tips. (b) Schematic of the tensile loading experiment. (Reprinted from Yu MF, Lourie O, Dyer MJ, Moloni K, Kelly TF, and Ruoff RS, (2000), *Science.* 287(5453): 637–640. ©2000 American Association for the Advancement of Science. With permission.)

approach was used for the experiment. SWCNT bundles having a strong attachment at one end to the sample surface were selected as candidates for the measurement. The free end of the SWCNT bundle was then approached and attached to an AFM tip by the same EBID method outlined above. The AFM tip was used to stretch the SWCNT bundle to the breaking point, and the same AFM tip also served as the force sensor to measure the applied force (Figure 23.24). Stress vs. strain curves for SWCNT bundles were obtained as well as the breaking strength; these stress vs. strain curves were generated by assuming a model in which only the perimeter SWCNTs in a bundle actually carried the load. The reader is referred to Reference 26 for the full explanation of this model.

The shear strength between the shells of an MWCNT is also an interesting subject for experimental study. Yu et al.[11] were able to directly measure the friction force between the neighboring layers while pulling the inner shells out of the outer shells of a MWCNT, using the same apparatus for measuring the tensile strength of individual MWCNTs. The possibility of such measurement was based on the discovery that tensile-loaded MWCNTs normally broke with a sword-in-sheath breaking mechanism.[29] The separated outer shell can still be in contact with the underlying inner shell in certain cases (in other cases, the *snap back* of the loading and force-sensing cantilevers leads to two separated fragments). The consecutive measurement of force and contact length (the overlap length between the outer shell and its neighbor) provided the necessary data for obtaining the dynamic and static shear strength between the shells.

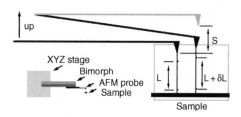

FIGURE 23.24 Schematic showing the principle of the experiment for the measurement of the tensile strength of SWCNT bundles. The gray cantilever indicates where the cantilever would be if no rope were attached on the AFM tip after its displacement upward to achieve tensile loading.

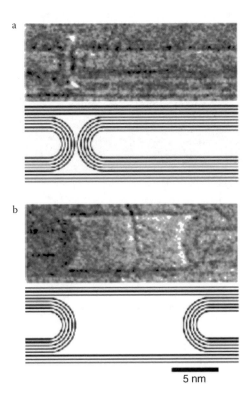

FIGURE 23.25 (a) An as-grown bamboo section. (b) The same area after the core tubes on the right have been telescoped outward. The line drawings beneath the images are schematic representations to guide the eye. (Reprinted from Cumings J and Zettl A, (2000), *Science.* 289(5479): 602–604. ©2000 American Association for the Advancement of Science. With permission.)

A similar experiment done in a TEM rather than a SEM was that of Cummings et al.,[107] who used a TEM holder having a piezoelectric-driven translation stage for approaching and opening the end of a MWCNT. The MWCNT cap was opened by *eroding* it away, using an electric discharge method inside the TEM.[143] The end of the exposed core part was then spot-welded to the moving probe using a short electrical pulse, and the MWCNT was telescoped by drawing out the core part from the outer shell housing. It was then possible to disengage the core part from the welding spot and observe the retraction of the core part back into its housing by the surface-driven forces (Figure 23.25). By analyzing the surface force and the friction forces involved in such a retraction using published parameters and modeling, the upper limit values for the dynamic and static friction force between the shells were estimated but not experimentally measured

23.3.3 Challenges and New Directions

The new developments in the area of nanoscale manipulation and measurement as reflected in the studies presented in the last section have certainly helped our understanding of the mechanics of CNTs. Because CNTs possess unique one-dimensional structures that maintain their conformations while being manipulated (in contrast, e.g., to biomolecules or certain polymer systems), they represent a "nanotinkertoy" for manipulation on the nanoscale. Therefore, such types of approaches also provide a window on current capabilities for exploring and exploiting the nanoworld and an avenue for future advancement in methods and tools useful in nanotechnology. In general, mechanical characterization takes a totally different approach than does electrical characterization. Mechanical characterization normally requires dynamic physical interaction with the object — for example, in the case for CNT, stretching, bending, compressing, and twisting with nanometer positioning accuracy, while performing measurement with nano-Newton force resolution and nanometer dimensional change resolution. Thus, in order for a successful and reliable mechanical study on nanoscale structures, extreme care must be taken to evaluate the three-dimensional stability and accuracy of the instrument and the effect of any significant external factors such as surface contact, mechanical attachment, stray electric charge, and so on. The experiments described in the last section represent the current state of the art in the manipulation and mechanical characterization of CNTs and should provide useful references for further instrument development, which can then be applied to many other low-dimensional nanostructures for mechanical studies.

But what has the community not yet achieved? We have not yet measured the tensile-loading response of an individual SWCNT, nor have we applied a known torque and controllably introduced a twist or series of twists along a CNT. The challenges here include: (1) the visualization, manipulation, precise placement, and fixation of a *flexible* one-dimensional nanostructure, the SWCNT, onto a device having displacement and force-sensing capabilities for the tensile measurement; and (2) a technical breakthrough in generating repeatable coaxial rotation with subnanometer runoff and with sufficient force output for applying torque to CNT. The influence of environment on NT mechanics — such as effects of temperature, chemical environment, or loading rate — has not yet been explored in any detail; nor do we have a clear and detailed picture of the initial defect distribution, or the nucleation, propagation, and ultimate failure resulting from defects. From the experimental perspective, such advances will come with new approaches and automated tools with subnanometer or atomic scale resolving power and stability generated by innovative thinking. It is clear that to attain further advances in nanoscale mechanics, focused effort is necessary in developing new measurement tools that can be integrated into high spatial resolution imaging instruments and that incorporate micro- or nanoelectromechanical system designs.

23.4 Simulation Methods

There are two major categories of molecular simulation methods for NT systems: classical molecular dynamics (MD) and *ab initio* methods. In general, *ab initio* methods give more accurate results than MD, but they are also much more computationally intensive. A hybrid method, tight-binding molecular dynamics (TBMD), is a blend of certain features from both MD and *ab initio* methods. In addition to these methods, continuum and multiscale approaches have also been proposed.

23.4.1 *Ab Initio* and Tight-Binding Methods

The central theme of *ab initio* methods is to obtain accurate solutions to the Schrödinger equation. A comprehensive description of these methods can be found in the book by Ohno et al.[144] Some notation is also adapted in this section. For general background in quantum mechanics, the reader is referred to any of the standard textbooks, such as References 145 through 149. In general, the state of a particle is defined by a wavefunction ψ based on the well-known wave–particle duality. The Schrödinger equation is

$$H\psi = E\psi \tag{23.18}$$

where H is the Hamiltonian operator of the quantum mechanical system, and ψ is the energy eigenfunction corresponding to the energy eigenvalue E.

Although the phrase *ab initio* is used, analytical or exact solutions are available only for a very limited class of problems. In general, assumptions and approximations need to be made. One of the most commonly used approximations is the Born–Oppenheimer approximation. In this approximation, it is assumed that the electrons are always in a steady state derived from their averaged motion because their positions change rapidly compared to the nuclear motion. Therefore, the motion of the electrons can be considered separately from the motion of the nuclei — as if the nuclei were stationary.

For an N-electron system, the Hamiltonian operator for each electron can be expressed as:

$$H_i = -\frac{1}{2}\nabla_i^2 + \sum_{j>i}^{N}\frac{1}{|r_i - r_j|} + v(r_i \tag{23.19}$$

The Hamiltonian operator in the above equation is composed of three parts. The first term in Equation (23.19) gives the kinetic energy when operating on the electron wave function; the second term gives the electron–electron Coulomb interaction; and the last term comes from the Coulomb potential from the nuclei. The total Hamiltonian operator of the N-electron system is then:

$$H = \sum_{i=1}^{N} H_i \tag{23.20}$$

and the electron state is solved from the following eigenvalue equation:

$$H\Psi_{\lambda_1,\lambda_2,...,\lambda_N} = E_{\lambda_1,\lambda_2,...,\lambda_N}\Psi_{\lambda_1,\lambda_2,...,\lambda_N} \tag{23.21a}$$

in which λ_i denotes an eigenstate that corresponds to the one-electron eigenvalue equation:

$$H_i\Psi_{\lambda_i} = E_{\lambda_i}\Psi_{\lambda_i} \tag{23.21b}$$

Because obtaining exact solutions to Equation (23.21b) is generally very difficult, approximation methods have been developed. In the following, we will introduce two of the most commonly used approaches.

23.4.1.1 The Hartree–Fock Approximation

In the Hartree–Fock approximation,[150–152] the ground state of the Hamiltonian H is obtained by applying the variational principle with a normalized set of wavefunctions ψ_i. The methodology is identical to the Ritz method, i.e., to seek the solution by minimizing the expectation value of H with a trial function:

$$\langle\Psi|H|\Psi\rangle = \sum_{S_1}\sum_{S_2}\sum_{S_3}\int\Psi^* H\Psi\, dr_1 dr_2 \ldots dr_N \tag{23.22}$$

One possible choice of the trial function is the Slater determinant of the single-particle wavefunctions, i.e.,

$$\Psi = \frac{1}{\sqrt{N!}}\begin{vmatrix} \Psi_1(1) & \Psi_2(1) & \cdots & \psi_N(1) \\ \Psi_1(2) & \Psi_2(2) & & \Psi_N(2) \\ . & . & & . \\ . & . & & . \\ \Psi_1(N) & \Psi_2(N) & & \Psi_N(N) \end{vmatrix} \tag{23.23}$$

In the above equation, the number in the bracket indicates the particle coordinate, which is composed of the spatial coordinate r and internal spin degree of freedom. The subscript denotes the energy level of the wavefunction, and $\psi_\lambda(i)$ forms an orthonormal set.

The Hamiltonian, on the other hand, is decomposed into a one-electron contribution H_0 and two-body electron–electron Coulomb interaction U as follows:

$$H = \sum_i H_0(i) + \frac{1}{2}\sum_{i,j} U(i,j) \tag{23.24}$$

The one-electron contribution H_0 consists of the kinetic energy and nuclear Coulomb potential:

$$H_0(i) = -\frac{1}{2}\nabla_i^2 + v(r_i) \tag{23.25}$$

in which

$$v(r_i) = -\sum_j \frac{Z_j}{|r_i - R_j|} \tag{23.26}$$

with Z_j being the nuclear charge of the jth atom. The two-body electron–electron interaction is given as the Coulomb interaction:

$$U(i,j) = \frac{1}{|r_i - r_j|} \tag{23.27}$$

With the trial function and decomposition of H, the expectation value of the Hamiltonian can then be rewritten as:

$$\langle\Psi|H|\Psi\rangle = \sum_{\lambda=1}^{N}\langle\Psi_\lambda|H_0|\Psi_\lambda\rangle + \frac{1}{2}\sum_{\lambda,\nu}\langle\Psi_\lambda\Psi_\nu|U|\Psi_\lambda\Psi_\nu\rangle - \frac{1}{2}\sum_{\lambda,\nu}\langle\Psi_\lambda\Psi_\nu|U|\Psi_\nu\Psi_\lambda\rangle \tag{23.28}$$

Applying the variational principle to Equation (23.28), it can be shown that solving the electron state of the system can now be approximated by solving the following equation for the one-electron wavefunction:

$$H_0\Psi_\lambda(i) + \left[\sum_{\nu=1}^{N}\sum_{S_j}\int \Psi_\nu^*(j)U(i,j)\Psi_\nu(j)dr_j\right]\Psi_\lambda(i)$$

$$-\left[\sum_{\nu=1}^{N}\sum_{S_j}\int \Psi_\nu^*(j)\,U(i,j)\Psi_\lambda(j)dr_j\right]\Psi_\nu(i) = \varepsilon_\lambda\Psi_\lambda(i) \tag{23.29}$$

in which ε_j is the Lagrangian multiplier used to enforce the orthonormal condition for the eigenfunction. The Hartree–Fock approximation has been used in many *ab initio* simulations. A more detailed description and survey of this method can be found in Reference 153.

23.4.1.2 Density Functional Theory

The density functional theory was originally proposed in a paper by Hohenberg and Kohn.[154] In this paper, they showed that the ground-state electronic energy is a unique functional of the electronic density. In most cases, the potential due to the external field comes mainly from the nuclei. The electronic energy can then be expressed as:

$$E = T[\rho(r)] + \int \frac{[\rho(r)\rho(r')]}{(|r-r'|)}drdr' + \int V_N(r)dr + E_{XC}[\rho(r)] \tag{23.30}$$

In Equation (23.30), $T[\rho(r)]$ is the kinetic energy and is a function of the electron density; the second term represents the electrostatic potential; the third term denotes the contribution from the nuclei; and the last term is the exchange–correlation functional. Kohn and Sham[155] presented a procedure to calculate the electronic state corresponding to the ground state using this theory, and their method is generally referred to as the *local-density approximation,* or LDA. LDA is another type of widely used *ab initio* method. The term *local density* comes from the assumption that the exchange–correlation function corresponding to the homogeneous electron gas is used. This assumption is only valid locally when the inhomogeneity due to the presence of the nuclei is small.

The essence of the LDA method is to obtain the ground state by introducing the variational principle to the density functional. This leads to a one-electron Schrödinger equation (also called the Kohn–Sham equation) for the Kohn–Sham wavefunction ψ_λ

$$\left\{ -\frac{1}{2}\nabla^2 - v(r) + \int \frac{\rho(r')}{(r-r')}dr' + \mu_{XC}[\rho](r) \right\} \Psi_\lambda(r) = \varepsilon_\lambda \Psi_\lambda(r) \tag{23.31}$$

Note that the term $\mu_{XC}[\rho](r)$ is the derivative of the exchange–correlation functional with respect to the electron density. Different functional forms for the exchange–correlation energy have been proposed.[156–159] The problem is reduced to obtaining the solutions to systems of one-electron equations. Once Ψ_λ and ε_λ are solved, the total energy can be obtained from Equation (23.30). The major advantage of using LDA is that the error in the electron energy is second-order between any given electron density and ground-state density.

The solution procedure requires an iterative diagonalization process, which in general involves $O(N^3)$ order of computation. A single electron wavefunction with a plane-wave basis and pseudo-potential have been used in the application of the LDA method.[160] Major improvements have been made using the Car–Parrinello MD method[161] and conjugate gradient (CG) method.[160] The Car–Parrinello method reduces the order from $O(N^3)$ to $O(N^2)$. As shown in Reference 160, the CG method can even be more efficient.

23.4.1.3 The Tight-Binding Method

The tight-binding theory was originated by Slater and Koster.[162] The advantage of the tight-binding method is that it can handle a much larger system than the *ab initio* method while maintaining better accuracy than MD simulation. A survey of the method can be found in Reference 163. In the tight-binding method, a linear combination of atomic orbitals (referred to as LCAO) is adopted in the wavefunction. Although the exact forms of the basis are not known, the Hamiltonian matrix can be parameterized; and the total energy and electronic eigenvalues can be deduced from the Hamiltonian matrix. The interatomic forces are evaluated in a straightforward way based on the Hellmann–Feynman theorem, and the rest of the procedure is almost identical to the MD simulation. For this reason sometimes the tight-binding method is also referred to as the *tight-binding MD method* or simply *TBMD*.

As shown by Foulkes and Haydock,[164] the total energy can be expressed as the sum of the eigenvalues of a set of occupied nonself-consistent one-electron molecular eigenfunctions in addition to certain analytical functions. The analytical function is usually assumed to take the form of a pair–additive sum. For example, the total energy can be given as

$$E = \sum_i \sum_{j>i} E_{ij} + \sum_k \varepsilon_k \tag{23.32}$$

The first term on the right side is the interatomic interaction, and the second term is the sum of the energies of occupied orbitals. A simple scheme in constructing the second term is to expand the wavefunction in a localized orthonormal minimal basis with parameterized two-center Hamiltonian matrix elements. The parameterization process can be performed by fitting to results from the *ab initio* methods[165,166] or computing the matrix exactly based on the localized basis.[167–169] A major problem with the TBMD method is the way that the parameterization of the Hamiltonian limits its applicability, or *transferability* as referred to in the computational physics community. As a simple example, when one switches from diamond structure to graphite, the nature of the nearest neighbor changes. In the early development of TBMD, Harrison[163] attempted to use a set of universal parameters; this approach turns out to be neither transferable nor accurate. The solution is then to add in modifications[170,171] or to use a completely different basis.[172]

23.4.2 Classical Molecular Dynamics

Many reviews are available on the subject of classical molecular dynamics.[173–175] MD is essentially a particle method[176,177] because the objective is to solve the governing equations of particle dynamics based on Newton's second law, i.e.,

$$m_j \frac{d^2 \mathbf{r}_i}{dt^2} = -\nabla V \tag{23.33}$$

in which m_i and $\mathbf{r}_i$ are the mass and spatial coordinates of the ith atom, respectively; V is the empirical potential for the system; and ∇ denotes the spatial gradient. Due to the small timescale involved, explicit integration algorithms such as the Verlet method[178,179] and other high-order methods are commonly used to ensure high-order accuracy.

An alternate but equivalent approach is to solve the Hamiltonian system of ordinary differential equations:

$$dp_i/dt = -\partial H/\partial q_i \tag{23.34}$$

$$dq_i/dt = -\partial H/\partial p_i \tag{23.35}$$

in which (q_i, p_i) are the set of canonically conjugate coordinates and momenta, respectively. H is the Hamiltonian function given as:

$$H = \sum_{i=1}^{N} \frac{p_i^2}{2m_i} + V \tag{23.36}$$

Symplectic integrators[180] have been developed to solve the above Hamiltonian equations of motion. The major advantages of this class of methods are that certain invariant properties of the Hamiltonian system can be preserved,[180] and it is easy to implement in large-scale computations.

23.4.2.1 Bonding Potentials

The basic formulation of MD requires that the spatial gradient of the potential function V be evaluated. Different empirical potentials for an NT system have been developed to satisfy this requirement. Allinger and coworkers developed[181,182] a molecular mechanics force field #2 (MM2) and an improved version known as the MM3 force field. Their model has been applied in the analysis of a variety of organic and inorganic systems. It should be noted that MM2/MM3 is designed for a broad class of problems. It is expected that the model may not work well under certain conditions. For example, the model is known to yield unrealistic results when interatomic distance is in the region of highly repulsive interactions. Mayo et al.[183] presented a generic force field based on simple hybridization considerations. The proposed form of the potential is a combination of bond length (two-body), angle bend (three-body), and torsion (four-body) terms. This empirical model has been used by Guo et al.[184] in the analysis of crystal structures of C_{60} and C_{70}, and by Tuzun et al.[185,186] in the analysis of carbon NT filled with fluid and inert gas atoms. Like the MM2/MM3 model, this force field covers a wide range of nonmetallic main group elements.

Another class of empirical potentials for CNT is characterized by the quantum-mechanical concept of bond order formalism originally introduced by Abell.[187] An alternate interpretation of this formalism can also be found in Reference 188. Using a Morse-type potential, Abell showed that the degree of bonding universality can be well maintained in molecular modeling for similar elements. Tersoff[132–135] introduced this important concept for the modeling of Group IV elements such as C, Si, and Ge, and reasonably accurate results were reported. In Tersoff–Abell bond order formalism, the energy of the system is a sum of the energy on each bond. The energy of each bond is composed of a repulsive part and attractive part. A bond order function is embedded in the formulation. The bond order depends on the local atomic environment such as angular dependency due to the bond angles. Nordlund et al.[189] modified the Tersoff potential such that the interlayer interaction is also considered. Brenner[35] made further improvements to the Tersoff potential by introducing additional terms into the bond order function. The main purpose of these extra terms is to correct the overbinding of radicals. Compared with the Tersoff potential, Brenner's potential shows robustness in the treatment of conjugacy; and it allows for forming and breaking of the bond with the correct representation of bond order. Brenner's potential has enjoyed success in the analysis of formation of fullerenes and their properties,[34,190–192] surface patterning,[193] indentation and friction at nanoscale,[194–203] and energetics of nanotubes.[34] An improved version of Brenner's potential has recently been proposed.[188,204] Based on the approximation of the many-atom expansion for the bond order within the two-center, orthogonal tight-binding (TB) model. Pettifor and Oleinik[205,206] have derived analytical forms that handle structural differentiation and radical formation. The model can be thought of as semiempirical as it is partly derived from TB. Application of this model to hydrocarbon systems can be found in Reference 205. Depending on the range of applicability, a careful selection of a potential model for a specific problem is needed.

23.4.2.2 Interlayer Potentials

Another important aspect of modeling in the analysis of CNT systems is the interlayer interaction. There are two major functional forms used in the empirical model: the inverse power model and the Morse function model. A very widely used inverse power model, the Lennard–Jones (LJ) potential, was introduced by Lennard–Jones[207,208] for atomic interactions:

$$\mu(R_{ij}) = 4\varepsilon \left[\left(\frac{R}{R_{ij}} \right)^{12} - \left(\frac{R}{R_{ij}} \right)^6 \right] \tag{23.37}$$

where R_{ij} denotes the interatomic distance between atoms i and j, σ is the collision diameter (the interatomic distance at which $\phi(R)$ is zero), and ε is the energy at the minimum in $\phi(R_{ij})$. This relationship is shown in Figure 23.26. In the figure, the energy u, interatomic distance R, and interatomic force F are all normalized as $u^* = u/\varepsilon$, $r^* = R_{ij}/R$, and $F^* = FR/\varepsilon$. The corresponding force between the two atoms as a function of interatomic distance is also shown in Figure 23.26 and can be expressed as:

$$F_{ij} = -\frac{\partial u(R_{ij})}{\partial R_{ij}} = 24 \frac{\varepsilon}{R} \left[2 \left(\frac{R}{R_{ij}} \right)^{13} - \left(\frac{R}{R_{ij}} \right)^7 \right] \tag{23.38}$$

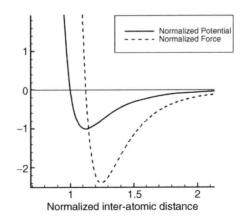

FIGURE 23.26 The pair potential and inter-atomic force in a two-atom system.

TABLE 23.1 Model Parameters for LJ Potential

Parameter Source	A	Σ	y_0
LJ1 [209]	24.3 × 10⁻⁷⁹ J·m⁶	1.42 A	2.7
LJ2 [210]	32 × 10⁻⁷⁹ J·m⁶	1.42 A	2.742

For the carbon–carbon system, the LJ potential energy has been treated by Girifalco and Lad[209,210] and is given as:

$$\phi_i = \frac{A}{\sigma^6}\left[\frac{1}{2}y_0^6\left(\frac{1}{\left(\frac{r_i}{\sigma}\right)^{12}} - \frac{1}{\left(\frac{r_i}{\sigma}\right)^6}\right)\right] \tag{23.39}$$

In Equation (23.39), σ is the bond length, y_0 is a dimensionless constant, and r_i is the distance between the *i*th atom pair. Two sets of parameters have been used: one for a graphite system[209] and the second for an fcc crystal composed of C_{60} molecules.[210] The converted parameters from the original data are given in Table 23.1.

Wang et al.[211] derived the following Morse-type potential for carbon systems based on local density approximations (LDA):

$$U(r) = D_e[(1 - e^{-\beta(r-r_e)})^2 - 1] + E_r e^{-\beta' r} \tag{23.40}$$

where $D_e = 6.50 \times 10^{-3}$ eV, is the equilibrium binding energy, $E_r = 6.94 \times 10^{-3}$ eV, is the hard-core repulsion energy, $r_e = 4.05$ Å, is the equilibrium distance between two carbon atoms, $\beta = 1.00/Å$, and $\beta' = 4.00/Å$. In a comparison study by Qian et al.,[212] it was found that the two LJ potentials yield much higher atomic forces in the repulsive region than the LDA potential; while in the attractive region, the LDA potential gives a much lower value of the binding energy than the two LJ potentials.

Further verification of this LDA potential is obtained by computing the equation of state (EOS) for a graphite system by assuming no relaxation within the graphene plane. The results are compared with published experimental data by Zhao and Spain[213] (referred to as EXP1) and by Hanfland et al. (referred to as EXP2),[214] and with the *ab initio* treatment (which included in-plane relaxation and treated an infinite crystal) by Boetgger.[215] The LDA model fits this experimental data and Boetgger's model reasonably well. The LJ potential, in contrast, deviates strongly from the experimental high-pressure data for graphite (and Boettger's high-level computational treatment) where the relative volume is smaller (Figure 23.27). Based on this comparison, Qian et al.[212] proposed to use LDA for interatomic distances less than 3.3 Å

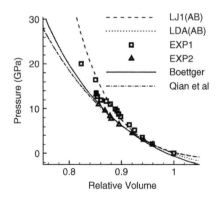

FIGURE 23.27 Comparison of EOS for graphite using different models with experimental data. (From Qian D, Liu WK, and Ruoff RS, (2001), *J. Phys. Chem. B.* 105: 10753–10758. With permission of Elsevier Science.)

and LJ for interatomic distance greater than 3.4 Å. The transition region is handled by curve fitting to ensure the continuity. A comparison of this model with the rest is also shown in Figure 23.27. Girifalco et al.[216] replaced the discrete sum of the atom pair potentials with a continuous surface integral. In their approach, different model parameters were derived for the cases of two parallel nanotubes and between C_{60} and nanotubes, although the LJ functional form is unchanged.

One of the major disadvantages of the potentials mentioned above is that the difference in interlayer binding energy of the AA and the AB stacking configurations for two rigid graphitic layers spaced at ~0.34 nm is probably not well represented; in short, the *corrugation* in the interlayer energy, due to the *pi* bonds projecting orthogonal to the plane of the layers, is not well captured. In addition, this registration effect also leads to unique nanoscale tribological features. Particularly, the corrugation between neighboring layers will play a central role in the friction present in such a nano-bearing system. The effects of interlayer registration between CNT and the surface it is sliding on have been studied recently in a number of interesting experiments by Falvo et al.[217–220] In these experiments, atomic force microscopy is used to manipulate the MWCNTs on the surface of graphite. A transition of slip-to-roll motion was reported as the MWCNT was moved to a certain particular position. This phenomenon uncovers the effects of commensurance at the nanoscale. Both MD[221] and quasi-static simulation[222] have been performed to verify the experiments. Kolmogorov and Crespi[223] developed a new *registry-dependent graphitic potential* which accounts for the exponential atomic-core repulsion and the interlayer delocalization of π orbitals in addition to the normal two-body van der Waals attraction. This model derives an approximate 12 meV/atom difference between the AB stacking and the AA stacking. As discussed in Section 23.2.2.3, Yu et al.[82] applied this model to treat the mechanics of a free-standing twisted MWCNT observed experimentally. In more general settings than, for example, perfectly nested perfect cylinders or two perfectly parallel graphene sheets, the formulation of appropriate models of interlayer interactions remains an important challenge for modeling. Indeed at the nanoscale, the importance of surface interactions cannot be underestimated; and we envision further work by theoreticians and experimentalists to treat problems where some useful discussion between the two camps can be achieved.

23.4.3   Continuum and Multiscale Models

Despite constant increases in available computational power and improvement in numerical algorithms, even classical molecular dynamics computations are still limited to simulating on the order of 10^6–10^8 atoms for a few nanoseconds. The simulation of larger systems or longer times must currently be left to continuum methods. From the crystal elasticity approach in Section 23.2.4, one might immediately see the possibility of applying the finite element method (FEM) to the computational mechanics of nanotubes because the continuum concept of stress can be extracted from a molecular model. However, the fundamental assumption of the continuum approximation — that quantities vary slowly over lengths on the order of the atomic scales — breaks down in many of the most interesting cases of nanomechanics. Thus it would be very useful to have in hand a method that allows the use of a molecular dynamics-like method in localized regions, where quantities vary quickly on the atomic length scale, seamlessly blended with a continuum description of the surrounding material in which, presumably, small scale variations are unimportant or can be treated in an averaged sense.

Several promising methods have been developed toward this goal. The quasicontinuum method, introduced by Tadmor et al.[124,125] and extended and applied to several different problems over the last few years,[224–232] gives a theory for bridging the atomistic and continuum scales in quasistatic problems. In this method, a set of atoms L making up a Bravais lattice has selected from it a subset L_h. A triangulation of this subset allows the introduction of finite element-like shape functions $\varphi_h(\mathbf{l}_h)$ at lattice points $\mathbf{l}_h \in L_h$, allowing the interpolation of quantities at intermediate points in the lattice. For example, the deformed coordinates $\mathbf{q}$ at a lattice point $\mathbf{l}$ can be interpolated:

$$q_h(\mathbf{l}) = \sum_{l_h \in L_h} \varphi_h(\mathbf{l}) q_h(\mathbf{l}_h) \qquad (23.41)$$

In this way, the problem of the minimization of energy to find equilibrium configurations can be written in terms of a reduced set of variables. The equilibrium equations then take the form, at each reduced lattice point $\mathbf{l}_h$:

$$f_h(\mathbf{1}_h) = \sum_{l \in L} f(1)\varphi_h(1) = 0 \qquad (23.42)$$

The method is made practical by approximating summations over all atoms, as implied by the above equation, by using summation rules analogous to numerical quadrature. These rules rely on the smoothness of the quantities over the size of the triangulation to ensure accuracy. The final piece of the method is therefore the prescription of adaptivity rules, allowing the reselection of representative lattice points in order to tailor the computational mesh to the structure of the deformation field. The criteria for adaptivity are designed to allow full atomic resolution in regions of large local strain — for example, very close to a dislocation in the lattice.

The quasicontinuum method has been applied to the simulation of dislocations,[124,125,229,232] grain boundary interactions,[224,227] nanoindentation,[125,227 228] and fracture.[225,226] An extension of the method to finite temperatures has been proposed by Shenoy et al.[233] We have successfully extended the quasicontinuum method to the analysis of the nanotube (CNT) system, although careful treatment is needed. The major challenge in the simulation of CNTs using the quasicontinuum method is the fact that CNT is composed of atomic layers. The thickness of each layer is the size of one carbon atom. In this case, it can be shown that the direct application of the Cauchy–Born rule results in inconsistency in the mapping. Arroyo and Belytshko[234] corrected this inconsistent mapping by introducing the concept of exponential mapping from differential geometry. An alternate approach is to start with the variational principle and develop a method without the Cauchy–Born rule. We have developed the framework of this method with the introduction of a mesh-free approximation.[235–241] For a survey of mesh-free and particle methods and their applications, please see Reference 177; an online version of this paper can be found at http://-www.tam.nwu.edu/wkl/liu.html. In addition, two special journal issues[242,243] have been devoted to this topic. Odegard et al.[244] have proposed modeling a CNT as a continuum by equating the potential energy with that of the representative volume element (RVE). This assumption seems to be the same as that in the quasicontinuum method; however, the Born rule is not used. The method has been recently applied in the constitutive modeling of CNT-reinforced polymer composite systems.[245]

The mesh-free method has been directly applied by Qian et al.[212] in the modeling of a CNT interacting with C_{60}. The C_{60} molecule is, on the other hand, modeled with a molecular potential. The interaction between the C_{60} and the continuum is treated based on the conservation of momentum. The interaction forces on the CNT are obtained through the consistent treatment of the weak formulation.

In this approach, a continuous deformation mapping will be constructed through the mesh-free mapping function ϕ. The final form of the discrete equation can be shown to be:

$$f_l^{ext} + f_l^{int} + f_l^{inert} = 0 \qquad (23.43)$$

The internal force term that is related to the internal energy is expressed as:

$$f_l^{int} = \sum_{l \in I} w_l f_i(\varphi(r,\theta)) \qquad (23.44)$$

Another approach to the coupling of length scales is the FE/MD/TB model of Abraham et al.[246,247] In this method, three simulations are run simultaneously using finite element (FE), molecular dynamics (MD) and semiempirical tight-binding (TB). Each simulation is performed on a different region of the domain, with a coupling imposed in *handshake* regions where the different simulations overlap. The method is designed for implementation on supercomputers via parallel algorithms, allowing the solution of large problems. One example of such a problem is the propagation of a crack in a brittle material.[246] Here, the

TB method is used to simulate bond breaking at the crack tip, MD is used near the crack surface, and the surrounding medium is treated with FE. This method has also been used by Nakano et al. in large-scale simulations of fracture.[248] Rafii-Tabar et al.[249,250] have presented a related method combining FE and MD for the simulation of crack propagation.

A related method, coarse-grained molecular dynamics (CGMD),[251,252] has been introduced as a replacement for finite elements in the FE/MD/TB simulations. In this approach, the continuum-level (or coarse-grained) energy is given by an ensemble average over the atomic motions in which the atomic positions are constrained to give the proper coarse-scale field. In this way, the fine-scale quantities that are not included in the coarse-scale motion are not neglected completely, as their thermodynamic average effect is retained.

We are currently developing a method for the coupling of continuum and MD simulations at finite temperature which, rather than enforcing coupling though boundary conditions in a hand-shake region, allows the continuum and MD representations to coexist in areas of interest in the computational domain. The methodology of the multiscale method can be traced back to the original paper by Liu et al.[253,254] and has been successfully applied in the multiple-scale problems involving strain localization,[255] boundary layers,[256] and coupling of finite elements with mesh-free shape functions.[257] In the current problem setting, this is done by writing a multiple-scale decomposition of the atomic displacements u_α in terms of finite element node displacements d_I and MD displacements d_α; the total scale is given by the usual finite element interpolation, plus the MD displacements, minus the projection of the MD displacements onto the finite element basis. Designating this projection operator as P:

$$u_\alpha = \sum_I N_I(x_\alpha)d_I + d_\alpha - Pd_\alpha \qquad (23.45)$$

where $N_I(x_\alpha)$ is the finite element shape function for node I evaluated at atom α. The key to the method is the subtraction of the projection of d_α, which we term the *bridging scale* and which allows for a unique decomposition into coarse and fine scales. With this decomposition, the Lagrangian (kinetic minus potential energies) can be written

$$L(d_I, \dot{d}_I, d_\alpha, \dot{d}_\alpha) = \sum_{I,J} \frac{1}{2}\dot{d}_I M_{IJ}\dot{d}_J + \sum \frac{1}{2}\dot{d}_\alpha M_{\alpha\beta}\dot{d}_\beta - U(d_I, d_\alpha) \qquad (23.46)$$

where M_{IJ} is the usual finite element mass matrix, $M_{\alpha\beta}$ is a fine-scale mass matrix, and U is the potential energy function. The equations of motion that can be derived from this Lagrangian are the usual finite element equation (plus a contribution to the internal force due to d_α), and the standard MD equations of motion (plus a driving term due to the continuum scale), i.e.,

$$\sum_J M_{IJ}\ddot{d}_J = -\frac{\partial U(d_I, d_\alpha)}{\partial d_I} \qquad (23.47a)$$

$$\sum_\beta M_{\alpha\beta}\ddot{d}_\beta = \frac{\partial U(d_I, d_\alpha)}{\partial d_\alpha} \qquad (23.47b)$$

These equations can be solved using existing FE and MD codes along with suitable methods for exchanging information about internal forces and boundary conditions. An energy equation can also be derived by considering the continuum-level temperature to be described by the fluctuations in the fine-scale motions. Shown in Figure 23.28 is an example of implementing the multiscale method. It can be seen that the mesh-free discretization and MD coexist and are coupled in the fine-scale region. For the coarse-scale region, a mesh-free approximation is used. For the case shown in Figure 23.28, the error in the bending energy is less than 1% when compared with a purely MD approach. A systematic description of this method is to be described in an upcoming paper.[258]

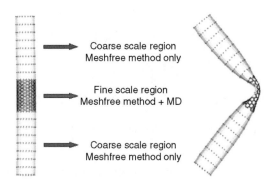

FIGURE 23.28 Multiscale analysis of a carbon nanotube.

23.5 Mechanical Applications of Nanotubes

23.5.1 Nanoropes

As discussed in Section 23.2.3.2, the primary product of current methods of SWCNT synthesis contains not individual, separated SWCNTs, but rather bundles of closest-packed SWCNTs.[85,90] Load transfer to the individual SWCNTs in these bundles is of paramount importance for applications involving tensile load bearing. It is estimated that to achieve load transfer so that the full bundle cross-section would be participating in load bearing up to the intrinsic SWCNT breaking strength, the SWCNT contact length must be on the order of 10 to 120 microns. There is strong evidence, however, that the typical length of individual SWCNTs in such bundles is only about 300 nm.[111,112]

From continuum mechanics analysis of other types of wire or fiber forms,[259,260] it is found that twisting the wires or *weaving* the fibers can lead to a cable or rope that has a much better load transfer mechanism in tension than a straight bundle would have. Compared with parallel wires, the major difference in terms of the mechanics of a wire rope is that wire ropes have a radial force (in the direction of vector **N** and **N'** as shown in Figure 23.29) that presses the surrounding wires to the core.

Correspondingly, the advantages of having wires in the form of a rope are

- A rope provides better load transfer and structural reliability. For example, when one wire component breaks, the broken sections of that particular wire can still bear load transferred from the other wires through the strong friction force that is a consequence of the radial compression.
- Wire rope has a smaller bending stiffness, therefore it is desirable for applications in which the rope has to be bent frequently. The fatigue life is significantly longer.
- The radial force component gives the rope structure more stability than wires in parallel.

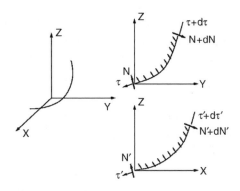

FIGURE 23.29 A three-dimensional section of a wire and force components in the *x-z* and *x-y* planes.

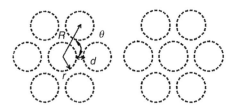

FIGURE 23.30 (*Left*) geometric parameters used in relaxed nanotube bundles. (*Right*) Configuration of bundled nanotubes after relaxation.

Based on preliminary modeling we have done of a seven-element twisted SWCNT bundle, it is believed that better load transfer in tension can be achieved by making nanoscale ropes and textiles of SWCNTs. We briefly describe some of these new mechanisms and issues below; a detailed analysis is given in Reference 261. We have used MD based on the empirical Tersoff–Brenner potential to analyze a bundle of SWCNTs under twist. A single strand composed of six (10,10) SWCNTs with a length of 612 Å surrounding a core (10,10) tube is studied. Both ends of the core tube are fixed, and a twist around the center of the core on the six neighboring SWCNTs is applied at an angular velocity of 20π/ns (see Figure 23.30 for the cross-section before (a) and after (b) relaxation). The geometric parameters corresponding to the initial configuration are: $r = 6.78$ Å, $d = 3.44$ Å, $R = 2r + d = 17$ Å, $\theta = \pi/3$.

Shown in Figure 23.31 are three snapshots of the deformation of the bundle after twisting is introduced. The corresponding change in the cross-section is shown in Figure 23.32 at the midpoint of the bundle. Clearly, radial deformation strongly depends on the twist angle. Further calculation[262] is performed on a bundle of SWCNTs with the same cross-sectional configuration but with a length of 153.72 Å. The total number of atoms is 17,500. Both ends of the core tube are fixed, and a twist around the center of the core on the six neighboring SWCNTs is applied to achieve a desired angle of twist. After this the whole twisted structure is relaxed to obtain the equilibrium configuration. A constant incremental displacement is then imposed on the core tube while holding the surrounding nanotubes fixed. Plotted in Figure 23.33 is the axial load transferred to the inner tube as a function of the twist angle. We use the transferred axial load as an index for the effectiveness of the load transfer mechanism. As can be seen in Figure 23.33, small twist has very little effect. For the case of 0 degrees (no twist), a force of only ~0.048 eV/Å is transferred to the center tube, an indication of a very smooth inter-tube contact condition; however, for a twist angle of 120 degrees, the transferred load increases to 1.63 eV/Å, about 34 times higher. Our calculation also indicates that too much twist results in unstable structures, i.e., the inner tube is being "squeezed out" when the twisting angle is 180 degrees or higher. This calculation clearly indicates that a great enhancement in the load transfer mechanism can be achieved by making the nanorope. More study on quantitatively determining the effects of various factors is under way.[261] For these significantly collapsed SWCNTs in highly twisted SWCNT bundles, the effective contact area is significantly increased; and this may contribute to better load transfer. In such twisted SWCNT bundles, the load transfer is likely enhanced by an expected increase in shear modulus due to the decrease in the interlayer separation. Recently, Pipes and Hubert[263] applied both textile mechanics and anisotropic elasticity to analyze polymer matrix composites consisting of discontinuous CNTs assembled in helical geometry. The effective elastic properties were predicted, and their study showed the strong dependence of the mechanical behavior on the helical angle of the assembly.

FIGURE 23.31 Snapshots of twisting of the SWCNT bundle. The twisting angles are 30, 60, 90, 120, 150, and 180 degrees, respectively.

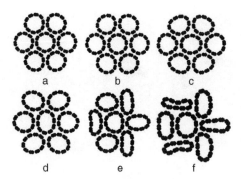

FIGURE 23.32 Change in cross-section at the midpoint of the SWNT bundle as a function of twist angle. From *a* to *f*, the twisting angles are 30, 60, 90, 120, 150, and 180 degrees, respectively.

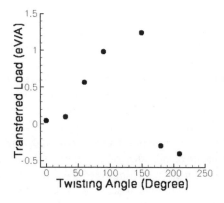

FIGURE 23.33 Transferred load as a function of twisting angle.

In the experiment (Figure 23.34), we attach the bundles using our previously developed nanoclamping methods (Figure 23.35)[11,29,264] and our new method of deposition of low electrical resistance W clamps. One goal involves measuring whether twisting enhances load transfer in terms of increased stiffness and strength. We measure the bundle stiffness without twisting, and then as a function of twisting, in the low-strain regime to assess the effective modulus of the bundle as a function of twist. To study the influence of twisting on strength, we will measure the load at break of similar diameter and length bundles with and without twisting, as a function of diameter, length, and number of twists. The boundary conditions also are assessed. We expect that the carbonaceous deposit made by electron beam-induced decomposition of residual hydrocarbons in the SEM, or the W deposit made, will (largely) be in contact with the outermost (perimeter) SWCNTs in the bundles. In untwisted perfect crystal SWCNT bundles having, for example, little load transfer to core SWCNTs from the perimeter SWCNTs, and with only perimeter SWCNTs clamped (from the method of deposition of our nanoclamp), we should be able to measure if load bearing and breakage occur only at the perimeter SWCNTs. We should be able to differentiate this from the load transfer we hypothesize will take

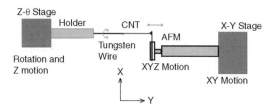

FIGURE 23.34 Proposed experimental stage for twisting the nanoropes.

FIGURE 23.35 Home-built nanomanipulation testing stage, which fits in the palm of the hand and is used in high-resolution scanning electron microscopy. Further details are available in Yu et al., (2000), *J. Phys. Chem. B.* 104(37): 8764–8767; Yu et al., (2000), *Phys. Rev. Lett.* 84(24): 5552–5555; Yu et al., (2000), *Science.* 287(5453): 637–640.

place upon twisting. One method of monitoring such effects would be to simultaneously monitor the electrical conductivity of the bundle during mechanical loading.

We have recently been developing methods for simultaneous measurement of the electrical conductivity of nanofilaments spanning the opposing AFM tip cantilevers by using conductive AFM tip cantilevers and the W nanoclamp. We propose that such measurements will help to elucidate the dynamics of the nanobundle. The video recording allows for a time resolution of only ~1/30 of a second, the time between video frames. However, electrical conductivity can be measured on a much finer timescale; doing so may allow us to infer when individual SWCNTs have broken prior to the whole bundle breaking, to study interlayer interactions between SWCNTs in the bundle as a function of twisting, and to study changes in conductivity that occur simply due to mechanical deformation of individual tubes in the bundle. Mapping out the electromechanical response of SWCNT bundles as a function of compressive, tensile, and torsional, loading will provide a database useful for assessing their application as actuators, sensors, and NEMS components. Measuring the electromechanical response in real applications such as cabling in a suspension bridge could also be a useful method of monitoring the reliability of nanorope components in everyday use.

23.5.2 Filling the Nanotubes

The mechanical benefits of filling nanotubes with various types of atoms are the following:

- Filling provides reinforcements for the hollow tube in the radial direction, thus preventing it from buckling, which, from the discussion in Section 23.2.2.3, is known to take place easily.
- Filling provides the smoothest and smallest nano-bearing system.
- Filling also provides an efficient storage system.

Currently there are two major experimental techniques that fill the CNT with foreign materials: arc evaporation, in which foreign materials are put in the anode for their incorporation into the CNTs formed from the carbon plasma, and opening the CNT by chemical agents followed by subsequent filling from either solution-based transport or by vapor transport. Because the chemical and physical environment in the solution-based filling method is less intensive, fragile materials such as biomaterials can also be put into the CNTs using that method.

Early attempts to make filled CNTs using arc evaporation have also resulted in filled carbon nanoparticles rather than filled CNTs. The foreign fillers include most metallic elements,[265–276] magnetic materials,[277–281] and radioactive materials,[282,283] surrounded by a carbon shell. One of the most well-known commercial applications of such filled nanoparticles is the invention of the so-called Technegas,[284,285] which is essentially radioactive material coated with carbon and used as an imaging agent in the detection of lung cancer.

A major breakthrough in filling CNTs was made by Ajayan et al.,[286,287] in which CNTs were opened by an oxidation process and foreign materials (molten lead in their case) were drawn into the CNT due to capillary action. This is the first experiment that showed the possibility of opening the CNT, although

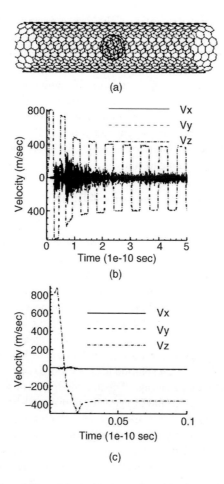

FIGURE 23.36 Computational modeling of C_{60} inside nanotubes. (a) The configuration for the problem. (b) The three velocity components history of the C_{60} as it shuttles through the (10,10) nanotube 20 times (V_z corresponds to the axial direction with an initial value of 0). (c) Same as (b) but for the case of an (8,8) nanotube. (From Wang Y, Tomanek D, and Bertsch GF, (1991), *Phys. Rev. B.* 44(12): 6562–6565. With permission of Elsevier Science.)

the method used is not universally applicable. Numerous efforts have been undertaken along this direction, including combining oxidation with heating[287–289] and the use of liquid-based approaches, such as nitric acid,[290] hydrochloric acid,[291] or other oxidants.[292,293] Using these processes and the arc evaporation method, different types of materials were also incorporated into CNTs, e.g., compounds of metals or their carbides[268,290,294,295] and biological molecules,[296,297] hydrogen,[298] and argon.[299] Filling of SWCNTs was first reported by Sloan et al.[291,300] Experimental observation of fullerenes inside CNTs has also been reported by Smith et al.[301–303] and Burteaux et al.[304] using pulsed laser evaporation (PLV) of a graphite target containing a catalyst, and by Sloan et al.[305] and Zhang et al.[306] using arc vaporization of carbon with a mixed Ni/Y catalyst.

Filled CNTs have recently been treated by theory and modeling. Pederson et al.[307] discussed the capillary effect at the nanoscale. A major topic is the effect of the change in the dimensionality (from three-dimensional to one-dimensional) and scale (from macroscale to nanoscale) on the resulting physical properties. The effect of the size of the CNT as a geometric constraint on the crystallization process is discussed by Prasad et al.[308] Tuzun et al. studied flow of helium and argon[185] and mixed flow of helium and C_{60}[186] inside CNTs using MD. Berber et al.[309] studied various configurations of putting C_{60} inside a CNT using an *ab initio* method. Stan et al. presented analysis of the hydrogen storage problem.[310] The diffusion properties of molecular flow of methane, ethane, and ethylene inside CNTs were studied using MD by Mao et al.[311]

A general analysis of the statistical properties of the quasi-one-dimensional structure inside a filled CNT is treated by Stan et al.[312] The mechanics of C_{60} inside CNTs was studied by Qian, Liu, and Ruoff et al.[212] using MD simulation. An interesting phenomenon is observed in which C_{60} is sucked into the (10,10) or (9,9) SWCNTs by the sharp surface tension force present in the front of the open end, following which it then oscillates between the two open ends of the nanotube, never escaping. Moreover, the oscillation shows little decay after stabilizing after a few cycles. Both the C_{60} molecule and the nanotube show small deformations as a function of time and position. Furthermore, C_{60}, even when fired on axis with an initial speed up to 1600 m/sec, cannot penetrate into any of the (8,8), (7,7), (6,6) or (5,5) NTs. The simulation results suggest the possibility of using C_{60} in CNTs for making nanodevices such as nanobearings or nanopistons. We continue to study this nanocomposite of fullerenes in CNTs.

23.5.3   Nanoelectromechanical Systems (NEMS)

23.5.3.1   Fabrication of NEMS

Nanoelectromechanical systems (NEMS) are evolving, with new scientific studies and technical applications emerging. NEMS are characterized by small dimensions, where the dimensions are relevant for the function of the devices. Critical feature sizes may be from hundreds to a few nanometers. New physical properties, resulting from the small dimensions, may dominate the operation of the devices. Mechanical devices are shrinking in size to reduce mass, decrease response time, and increase sensitivity.

NEMS systems defined by photolithography processes are approaching the dimensions of carbon nanotubes. NEMS can be fabricated with various materials and integrated with multiphysics systems such as electronic, optical, and biological systems to create devices with new or improved functions. The new class of NEMS devices may provide a revolution in applications such as sensors, medical diagnostics, functional molecules, displays, and data storage. The initial research in science and technology related to nanomechanical systems is taking place now in a growing number of laboratories throughout the world.

The fabrication processes combine various micro/nanomachining techniques including micro/nanostructure fabrication and surface chemistry modification of such fabricated structures to allow site-specific placement of nanofilaments (NF) for actuation and sensing. Different kinds of NFs such as SWCNTs and MWCNTs, nanowhiskers (NWs), and etched nanostructures require different processing steps.

Controlled deposition of individual SWCNTs on chemically functionalized templates was first introduced by Liu et al.[313] Reliable deposition of well-separated individual SWCNTs on chemically functionalized nanolithographic patterns was demonstrated. The approach offers promise in making structures and electronic circuits with nanotubes in predesigned patterns.

SWCNTs generally exist as bundles when purchased, so they have to be dispersed into individual SWCNTs in a solution before being used for device experimentation. One controllable procedure is done by diluting the SWCNT suspension into N, N-dimethylformamide (DMF) at a ratio of 1:100. The vial is then sonicated in a small ultrasonic bath for 4–8 hours. Such steps are repeated at least 3 times until the clear suspension contains at least 50% of the SWCNTs have completely separated as individual SWCNTs (Figure 23.37).

Recently, Chung, Ruoff, and Lee[314] have developed new techniques for nanoscale gap fabrication (50–500nm) and the consecutive integration of CNTs. These techniques are essential for the batch assembly of CNTs (diameter ~ 10nm) for various applications.[29,264,315–319] Figure 23.38 shows an ideal configuration for chemical sensing by electromechanical transduction involving only a single CNT. So far, they have successfully deposited CNT bundles across the circular gap based on microlithography (Figure 23.39). The fabrication steps do not use serial and time-consuming processes such as e-beam lithography, showing the potential for batch production. It is also noted that these deposition steps were performed at standard environment conditions, thus providing more process freedom. Details of this new fabrication technique can be found in Reference 314.

23.5.3.2   Use of Nanotubes as Sensors in NEMS

The small size and unique properties of carbon nanotubes suggest that they can be used in sensor devices with unprecedented sensitivity. One route to making such sensors is to utilize the electrical properties of CNTs.

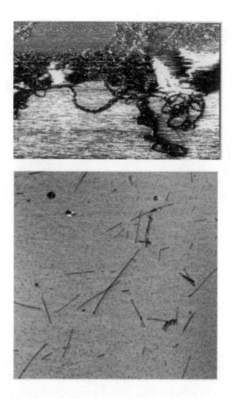

FIGURE 23.37 Before (*upper*) and after (*lower*) untangling of the nanotubes in the suspension.

Experiment[320] and computation[321,322] have shown that the conductivity of CNTs can change by several orders of magnitude when deformed by the tip of an atomic force microscope. CNT chemical sensors have also been demonstrated: it has been found experimentally that the electrical resistance of a semi-conducting SWCNT changes dramatically upon adsorption of certain gaseous molecules, such as NO_2, NH_3,[318] H_2,[323] and O_2.[317] This phenomenon has been modeled numerically by Peng and Cho[324] using a DFT–LDA technique.

The high sensitivity of the Raman spectra of CNTs to their environment also makes them useful as mechanical sensors. Raman spectra are known from experiment to give shifted peaks when CNTs undergo stress or strain.[325–327] This phenomenon has been used to detect phase transitions and to measure stress fields in polymers with embedded nanotubes.[328,329]

Yet another interesting application of nanotubes as sensors is through the use of the mechanical resonance frequency shift to detect adsorbed molecules or groups of molecules. When the mass of a

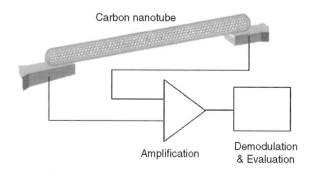

FIGURE 23.38 Carbon nanotube-based sensor.

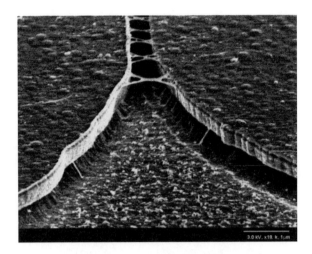

FIGURE 23.39   CNT deposition in a round gap by AC electrophoresis.

vibrating cantilever changes, as by the addition of an adsorbed body, the fundamental frequency of the cantilever as described by Equation (23.2) decreases. Ilic et al.[330] demonstrated micromachined silicon cantilevers with typical dimensions on the order of 20 $\mu m \times$ 320 nm $\times$ 100 μm to detect *E. coli* cells; the device was shown to be sensitive enough to detect as few as 16 cells under atmospheric conditions. Because the sensitivity of the device is hampered by the mechanical quality factor *Q* of the oscillating cantilever, even greater sensitivity might be expected in vacuum where the lack of air damping enables higher *Q*; of course, operation in vacuum limits the utility of the device. Shrinking the size of the design by using a carbon nanotube rather than a micromachined cantilever may give unprecedented sensitivity.

Poncharal et al.[44] were able to use the resonant frequency of a vibrating cantilevered nanotube (Figure 23.18) to estimate the mass of a ~30 fg carbon particle attached to the end of the tube. Our own calculations predict that detection of a 1% change in the resonant frequency of a 100-nm-long (10,10) CNT allows the measurement of an end mass of around 800 amu, approximately the mass of a single C_{60} molecule. The frequency shift of a vibrating cantilevered CNT due to an adsorbed mass at the free end can be estimated using beam theory (Section 23.2.2.2). To the first order approximation, the change in resonance frequency of mode *i* is

$$\Delta f_i = -\frac{2 M f_i}{\rho A L} \tag{23.48}$$

where *M* is the mass of the adsorbed body and ρ, *A*, and *L* are, respectively, the density, cross-sectional area, and length of the CNT. Sensitivity may be even further enhanced by taking advantage of critical points in the amplitude–frequency behavior of the vibrating CNT. For example, it is likely that the nonlinear force–deflection curve of the nanotube under certain conditions can result in a bistable response, as has been seen for other vibrating structures at this scale.[331,331a] Recently we have measured four parametric resonances of the fundamental mode of boron nanowires, the first observation of parametric resonance in a nanostructure.[332] It is likely that extraordinarily sensitive molecular sensors can be based on exploiting the parametric resonance of nanostructures (such as nanotubes, nanowires, or nanoplatelets).

23.5.4 Nanotube-Reinforced Polymer

The extremely high modulus and tunable electrical and thermal properties of carbon nanotubes offer an appealing mechanism to dramatically improve both strength and stiffness characteristics, as well as add multifunctionality to polymer-based composite systems. Experimental results to date, however, demonstrate only modest improvements in important material properties amidst sometimes contradictory data.[333–337] Much of the discrepancy in the published results can be attributed to nonuniformity of material samples. In order to obtain optimal property enhancement, key issues to be resolved include improved dispersion of nanotubes, alignment of nanotubes, functionalization of the nanotubes to enhance matrix bonding/load transfer, and efficient use of the different types of nanotube reinforcements (single-wall vs. nanorope vs. multi-wall).

Ongoing investigations are focusing on improved processing and design of nanocomposites with emphasis on controlled nanotube geometric arrangement. Techniques to obtain homogeneous dispersion and significant alignment of the nanotubes include application of electric field during polymerization, extrusion, and deformation methods.[338–340] In addition to dispersion and alignment, recent work[341,342] has demonstrated that the waviness of the nanotubes (see Figure 23.40) in the polymer decreases the potential reinforcing factor by an additional 50% beyond a twofold decrease due to random orientation.

Among the encouraging results is work by Qian[335] for a MWCNT-reinforced polymer, where good dispersion and matrix bonding was achieved. In this case, using only 0.5 vol% nanotube reinforcement with no alignment and moderate NT waviness, elastic stiffness was improved 40% over that of the neat matrix material; and strength values improved nearly 25%. Figure 23.40 shows nanotubes bridging a

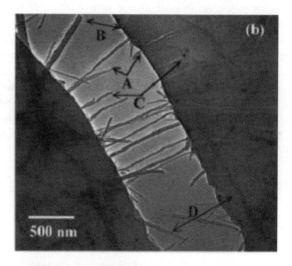

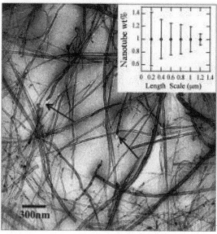

FIGURE 23.40 Upper: *In situ* TEM image of a crack in polystyrene film with nanotubes bridging the crack. Lower: arrangement of nanotubes in polymer, good dispersion, random orientation, and moderate waviness. (From Qian D, Dickey EC, Andrews R, and Rantell T, (2000), *Appl. Phys. Lett.* 76(20): 2868–2870. With permission.)

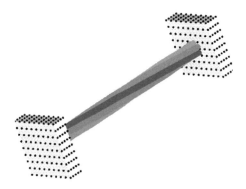

FIGURE 23.41   Multiscale analysis of nanorope-reinforced materials.

matrix crack and demonstrates excellent bonding between nanotubes and matrix material. Viscoelastic properties have also been investigated, with some evidence that well-dispersed nanotubes impact the mobility of the polymer chains themselves, causing changes in glass transition temperature and relaxation characteristics,[343,344] a feature not observed in polymers with a micron-sized reinforcing phase. Limited work on electrical properties shows that percolation can be reached with nanotube volume fractions of less than 1%,[333] leading to dramatic changes in electrical response of the polymer. This enables applications such as polymer coatings with electrostatic discharge capability.[345]

Understanding the mechanisms involved and the degree of property improvement possible for nanotube-reinforced polymers remains a goal of intensive research. Current work emphasizes surface modification of CNTs; control of matrix–NT adhesion; processing methods to control nanotube geometry; hybridization of nanoscale and microscale reinforcements; multifunctional capabilities; and integration of atomistic, micromechanics, and continuum modeling for predictive capability and understanding.

The multiscale method described in Section 23.4.3 shows great potential in the analysis and modeling of nanotube-reinforced material. Shown in Figure 23.41 is an illustration of the approach, combined with the nanorope application discussed in Section 23.5.1. In this method, the domain of the problem is decomposed into three regions: (1) the coarse-scale region, (2) the coupling-scale region, and (3) the fine-scale region. We will solve the coupled equation (Equations (23.47a and 23.47b)) based on the area of interest. A unique advantage of this method is the multiscale decomposition of the displacement field. The governing equations for continuum scale and molecular scale are unified and no special treatment is needed in handling the transition. More details of the modeling approach are included in Reference 346.

23.6   Conclusions

The development of successful methods for the synthesis of high-quality CNTs[85] has led to a worldwide R&D effort in the field of nanotubes. Part of the focus in this paper has been to address the particularly promising mechanical attributes of CNTs as indicated in numerous experiments, with emphasis on the inherent high stiffness and strength of such nanostructures. As stated in Section 23.3.3, much work still remains on the experimental measurement of the mechanical properties of CNTs. Progress made on the theoretical and computational fronts has also been briefly summarized.

It is amazing to see the applicability of continuum elasticity theory even down to the nanometer scale, although this fact should not lead to overconfidence in its use or to any ignorance of the physics that lies beneath. There are also a lot of unresolved issues in theoretical analysis and simulation due to the highly cross-disciplinary nature of the problem. In fact, numerous research efforts are under way in developing multiscale multiphysics simulation schemes. These efforts depend largely on gaining a basic understanding of the phenomena from the quantum level to the continuum level. Therefore, the terms *nanomechanics* or *nanoscale mechanics* indeed go beyond the nanoscale and serve as a manifestation of the link between fundamental science and important engineering applications. This is a major challenge for the engineering community as well as the science community in the sense that the traditional boundary of each field has to be redefined to establish the new forefront of nanotechnology. A number of applications that are described in Section 23.5 serve as the best examples. Yet there are still numerous other likely applications of CNTs that we would like to mention; for instance, there has also been interesting research on the use of CNTs as ferroelectric devices, nanofluidic devices, and energy storage devices; and there exists the possibility of their use in biosurgical instruments. We have not been able to address these topics due to time and space limitations.

As we try to give an up-to-date status of the work that has been done, the field is rapidly evolving, and a tremendous amount of information can be found from various media. It is conceivable that the discovery of carbon nanotubes and other multifunctional nanostructures will parallel the importance of the transistor in technological impact. The size and the mechanical, electrical, thermal, and chemical properties of CNTs — as well as the fact that they are hollow and that matter can thus be located inside them and transported through them — have suggested an astonishingly wide array of potential applications, many of which are under testing now. Statements[347] are being made about factories being set up with ton-level production capabilities. We believe the unique structure and properties of CNTs and related nanomaterials will bring a fundamental change to technology.

Acknowledgments

R.S. Ruoff appreciates support from the NASA Langley Research Center Computational Materials: Nanotechnology Modeling and Simulation Program, and support to G.J. Wagner, R.S. Ruoff, and W.K. Liu from the ONR Miniaturized Intelligent Sensors Program (MIS) and from the grant: Nanorope Mechanics, NSF (Oscar Dillon and Ken Chong, Program Managers). The work of D. Qian and G.J. Wagner has also been supported by the Tull Family Endowment. R.S. Ruoff and W.K. Liu have, for some of the work described here, been supported by several grants from NSF. D. Qian also acknowledges the support of the Dissertation Year Fellowship from Northwestern University. The contributions from Junghoon Lee, Jae Chung, and Lucy Zhang (also supported by ONR MIS) on NEMS and from Cate Brinson (also supported by the NASA Langley Research Center Computational Materials: Nanotechnology Modeling and Simulation Program) on nanotube reinforced polymers are gratefully acknowledged.

References

1. Iijima S, (1991), Helical microtubules of graphitic carbon, *Nature.* 354(6348): 56–58.
2. Normile D, (1999), Technology — nanotubes generate full-color displays, *Science.* 286(5447): 2056–2057.
3. Choi WB, Chung DS, Kang JH, Kim HY, Jin YW, Han IT, Lee YH, Jung JE, Lee NS, Park GS, and Kim JM, (1999), Fully sealed, high-brightness carbon-nanotube field-emission display, *Appl. Phys. Lett.* 75(20): 3129–3131.
4. Bachtold A, Hadley P, Nakanishi T, and Dekker C, (2001), Logic circuits with carbon nanotube transistors, *Science.* 294(5545): 1317–1320.
5. Derycke V, Martel R, Appenzeller J, and Avouris P, (2001), Carbon nanotube inter- and intramolecular logic gates, *Nano Lett.* 10.1021/nl015606f.
6. Baughman RH, Cui CX, Zakhidov AA, Iqbal Z, Barisci JN, Spinks GM, Wallace GG, Mazzoldi A, De Rossi D, Rinzler AG, Jaschinski O, Roth S, and Kertesz M, (1999), Carbon nanotube actuators, *Science.* 284(5418): 1340–1344.
7. Harris PJF, (1999), *Carbon Nanotube and Related Structures: New Materials for the 21st Century.* Cambridge University Press. Cambridge, UK.
8. Dresselhaus MS, Dresselhaus G, Eklund PC, (1996), *Science of Fullerenes and Carbon Nanotubes.* Academic Press. San Diego.
9. Dresselhaus MS and Avouris P, (2001), Introduction to carbon materials research, *Carbon Nanotubes*, 1–9.

10. Brown TLL, Bursten BE, and Lemay HE, (1999), *Chemistry: The Central Science*. 8th ed. Prentice Hall. PTR.

11. Yu MF, Yakobson BI, and Ruoff RS, (2000), Controlled sliding and pullout of nested shells in individual multi-walled carbon nanotubes, *J. Phys. Chem. B*. 104(37): 8764–8767.

12. Saito R, Fujita M, Dresselhaus G, and Dresselhaus MS, (1992), Electronic-structure of chiral graphene tubules, *Appl. Phys. Lett.* 60(18): 2204–2206.

13. Dresselhaus MS, Dresselhaus G, and Saito R, (1995), Physics of carbon nanotubes, *Carbon*. 33(7): 883–891.

14. Fujita M, Saito R, Dresselhaus G, and Dresselhaus MS, (1992), Formation of general fullerenes by their projection on a honeycomb lattice, *Phys. Rev. B*. 45(23): 13834–13836.

15. Dresselhaus MS, Dresselhaus G, Eklund PC, (1993), Fullerenes, *J. Mater. Res.* 8: 2054.

16. Yuklyosi K, Ed. (1977), *Encyclopedic Dictionary of Mathematics*. The MIT Press. Cambridge.

17. Iijima S, (1993), Growth of carbon nanotubes, *Mater. Sci. Eng. B Solid State Mater. Adv. Technol.* 19(1–2): 172–180.

18. Dravid VP, Lin X, Wang Y, Wang XK, Yee A, Ketterson JB, and Chang RPH, (1993), Buckytubes and derivatives — their growth and implications for buckyball formation, *Science*. 259(5101): 1601–1604.

19. Iijima S, Ichihashi T, and Ando Y, (1992), Pentagons, heptagons and negative curvature in graphite microtubule growth, *Nature*. 356(6372): 776–778.

20. Saito Y, Yoshikawa T, Bandow S, Tomita M, and Hayashi T, (1993), Interlayer spacings in carbon nanotubes, *Phys. Rev. B*. 48(3): 1907–1909.

21. Zhou O, Fleming RM, Murphy DW, Chen CH, Haddon RC, Ramirez AP, and Glarum SH, (1994), Defects in carbon nanostructures, *Science*. 263(5154): 1744–1747.

22. Kiang CH, Endo M, Ajayan PM, Dresselhaus G, and Dresselhaus MS, (1998), Size effects in carbon nanotubes, *Phys. Rev. Lett.* 81(9): 1869–1872.

23. Amelinckx S, Bernaerts D, Zhang XB, Vantendeloo G, and Vanlanduyt J, (1995), A structure model and growth-mechanism for multishell carbon nanotubes, *Science*. 267(5202): 1334–1338.

24. Lavin JG, Subramoney S, Ruoff RS, Berber S, and Tomanek D, (2001), Scrolls and nested tubes in multiwall carbon tubes, unpublished.

25. Ajayan PM and Ebbesen TW, (1997), Nanometre-size tubes of carbon, *Rep. Prog. Phys.* 60(10): 1025–1062.

26. Amelinckx S, Lucas A, and Lambin P, (1999), Electron diffraction and microscopy of nanotubes, *Rep. Prog. Phys.* 62(11): 1471–1524.

27. Lourie O and Wagner HD, (1998), Evaluation of Young's modulus of carbon nanotubes by micro-Raman spectroscopy, *J. Mater. Res.* 13(9): 2418–2422.

28. Yu MF, Files BS, Arepalli S, and Ruoff RS, (2000), Tensile loading of ropes of single wall carbon nanotubes and their mechanical properties, *Phys. Rev. Lett.* 84(24): 5552–5555.

29. Yu MF, Lourie O, Dyer MJ, Moloni K, Kelly TF, and Ruoff RS, (2000), Strength and breaking mechanism of multi-walled carbon nanotubes under tensile load, *Science*. 287(5453): 637–640.

30. Overney G, Zhong W, and Tomanek D, (1993), Structural rigidity and low-frequency vibrational-modes of long carbon tubules, *Zeitschr. Phys. D-Atoms Molecules Clusters*. 27(1): 93–96.

31. Treacy MMJ, Ebbesen TW, and Gibson JM, (1996), Exceptionally high Young's modulus observed for individual carbon nanotubes, *Nature*. 381(6584): 678–680.

32. Tibbetts GG, (1984), Why are carbon filaments tubular? *J. Crystal Growth*. 66(3): 632–638.

33. Ruoff RS and Lorents DC, (1995), Mechanical and thermal properties of carbon nanotubes, *Carbon*. 33(7): 925–930.

34. Robertson DH, Brenner DW, and Mintmire JW, (1992), Energetics of nanoscale graphitic tubules, *Phys. Rev. B*. 45(21): 12592–12595.

35. Brenner DW, (1990), Empirical potential for hydrocarbons for use in simulating the chemical vapor deposition of diamond films, *Phys. Rev. B*. 42(15): 9458–9471.

36. Gao GH, Cagin T, and Goddard WA, (1998), Energetics, structure, mechanical and vibrational properties of single-walled carbon nanotubes, *Nanotechnology*. 9(3): 184–191.

37. Yakobson BI, Brabec CJ, and Bernholc J, (1996), Nanomechanics of carbon tubes: instabilities beyond linear response, *Phys. Rev. Lett.* 76(14): 2511–2514.

38. Timoshenko S and Gere J, (1988), *Theory of Elastic Stability*. McGraw-Hill. New York.

39. Lu JP, (1997), Elastic properties of carbon nanotubes and nanoropes, *Phys. Rev. Lett.* 79(7): 1297–1300.

40. Yao N and Lordi V, (1998), Young's modulus of single-walled carbon nanotubes, *J. Appl. Phys.* 84(4): 1939–1943.

41. Hernandez E, Goze C, Bernier P, and Rubio A, (1998), Elastic properties of c and bxcynz composite nanotubes, *Phys. Rev. Lett.* 80(20): 4502–4505.

42. Zhou X, Zhou JJ, and Ou-Yang ZC, (2000), Strain energy and Young's modulus of single-wall carbon nanotubes calculated from electronic energy-band theory, *Phys. Rev. B.* 62(20): 13692–13696.

43. Yakobson BI and Smalley RE, (1997), Fullerene nanotubes: C-1000000 and beyond, *Am. Sci.* 85(4): 324–337.

44. Poncharal P, Wang ZL, Ugarte D, and de Heer WA, (1999), Electrostatic deflections and electro-mechanical resonances of carbon nanotubes, *Science*. 283(5407): 1513–1516.

45. Liu JZ, Zheng Q, and Jiang Q, (2001), Effect of a rippling mode on resonances of carbon nanotubes, *Phys. Rev. Lett.* 86(21): 4843–4846.

46. Krishnan A, Dujardin E, Ebbesen TW, Yianilos PN, and Treacy MMJ, (1998), Young's modulus of single-walled nanotubes, *Phys. Rev. B.* 58(20): 14013–14019.

47. Yu MF, Dyer MJ, Chen J, and Bray K (2001), Multiprobe nanomanipulation and functional assembly of nanomaterials inside a scanning electron microscope. *Intl. Conf. IEEE-NANO2001*, Maui.

48. Wong EW, Sheehan PE, and Lieber CM, (1997), Nanobeam mechanics: elasticity, strength, and toughness of nanorods and nanotubes, *Science*. 277(5334): 1971–1975.

49. Salvetat JP, Kulik AJ, Bonard JM, Briggs GAD, Stockli T, Metenier K, Bonnamy S, Beguin F, Burnham NA, and Forro L, (1999), Elastic modulus of ordered and disordered multi-walled carbon nanotubes, *Adv. Mater.* 11(2): 161–165.

50. Salvetat JP, Briggs GAD, Bonard JM, Bacsa RR, Kulik AJ, Stockli T, Burnham NA, and Forro L, (1999), Elastic and shear moduli of single-walled carbon nanotube ropes, *Phys. Rev. Lett.* 82(5): 944–947.

51. Govindjee S and Sackman JL, (1999), On the use of continuum mechanics to estimate the properties of nanotubes, *Solid State Comm.* 110(4): 227–230.

52. Harik VM, (2001), Ranges of applicability for the continuum-beam model in the mechanics of carbon-nanotubes and nanorods, *Solid State Comm.* 120(331–335).

53. Harik VM, (2001), Ranges of applicability for the continuum-beam model in the constitutive analysis of carbon-nanotubes: nanotubes or nano-beams? in NASA/CR-2001–211013, also in *Computational Material Science* (Submitted).

54. Ru CQ, (2000), Effect of van der Waals forces on axial buckling of a double-walled carbon nanotube, *J. Appl. Phys.* 87(10): 7227–7231.

55. Ru CQ, (2000), Effective bending stiffness of carbon nanotubes, *Phys. Rev. B.* 62(15): 9973–9976.

56. Ru CQ, (2000), Column buckling of multi-walled carbon nanotubes with interlayer radial displacements, *Phys. Rev. B.* 62(24): 16962–16967.

57. Ru CQ, (2001), Degraded axial buckling strain of multi-walled carbon nanotubes due to interlayer slips, *J. Appl. Phys.* 89(6): 3426–3433.

58. Ru CQ, (2001), Axially compressed buckling of a double-walled carbon nanotube embedded in an elastic medium, *J. Mech. Phys. Solids*. 49(6): 1265–1279.

59. Ru CQ, (2000), Elastic buckling of single-walled carbon nanotube ropes under high pressure, *Phys. Rev. B.* 62(15): 10405–10408.

60. Bernholc J, Brabec C, Nardelli MB, Maiti A, Roland C, and Yakobson BI, (1998), Theory of growth and mechanical properties of nanotubes, *App. Phys. A — Mater. Sci. Proc.* 67(1): 39–46.

61. Yakobson BI and Avouris P, (2001), Mechanical properties of carbon nanotubes, in *Carbon Nanotubes*, 287–327.

62. Qian D, Liu WK, and Ruoff RS, (2002), Bent and kinked multi-shell carbon nanotubes — treating the interlayer potential more realistically. *43rd AIAA/ASME/ASCE/AHS Struct., Struct. Dynamics, Mater. Conf.* Denver, CO.

63. Iijima S, Brabec C, Maiti A, and Bernholc J, (1996), Structural flexibility of carbon nanotubes, *J. Chem. Phys.* 104(5): 2089–2092.

64. Ruoff RS, Lorents DC, Laduca R, Awadalla S, Weathersby S, Parvin K, and Subramoney S, (1995), *Proc. Electrochem. Soc.* 95–10: 557–562.

65. Subramoney S, Ruoff RS, Laduca R, Awadalla S, and Parvin K, (1995), *Proc. Electrochem. Soc.* 95–10: 563–569.

66. Falvo MR, Clary GJ, Taylor RM, Chi V, Brooks FP, Washburn S, and Superfine R, (1997), Bending and buckling of carbon nanotubes under large strain, *Nature.* 389(6651): 582–584.

67. Hertel T, Martel R, and Avouris P, (1998), Manipulation of individual carbon nanotubes and their interaction with surfaces, *J. Phys. Chem. B.* 102(6): 910–915.

68. Lourie O, Cox DM, and Wagner HD, (1998), Buckling and collapse of embedded carbon nanotubes, *Phys. Rev. Lett.* 81(8): 1638–1641.

69. Ruoff RS, Tersoff J, Lorents DC, Subramoney S, and Chan B, (1993), Radial deformation of carbon nanotubes by van der Waals forces, *Nature.* 364(6437): 514–516.

70. Tersoff J and Ruoff RS, (1994), Structural properties of a carbon-nanotube crystal, *Phys. Rev. Lett.* 73(5): 676–679.

70a. Lopez, MJ, Rubio A, Alonso JA, Qin LC, and Iijima S, (2001) Novel polygonized single-wall carbon nanotube bundles, *Phys. Rev. Lett.* 86(14): 3056–3059.

71. Chopra NG, Benedict LX, Crespi VH, Cohen ML, Louie SG, and Zettl A, (1995), Fully collapsed carbon nanotubes, *Nature.* 377(6545): 135–138.

72. Benedict LX, Chopra NG, Cohen ML, Zettl A, Louie SG, and Crespi VH, (1998), Microscopic determination of the interlayer binding energy in graphite, *Chem. Phys. Lett.* 286(5–6): 490–496.

73. Hertel T, Walkup RE, and Avouris P, (1998), Deformation of carbon nanotubes by surface van der Waals forces, *Phys. Rev. B.* 58(20): 13870–13873.

74. Avouris P, Hertel T, Martel R, Schmidt T, Shea HR, and Walkup RE, (1999), Carbon nanotubes: nanomechanics, manipulation, and electronic devices, *Appl. Surf. Sci.* 141(3–4): 201–209.

75. Yu MF, Dyer MJ, and Ruoff RS, (2001), Structure and mechanical flexibility of carbon nanotube ribbons: an atomic-force microscopy study, *J. Appl. Phys.* 89(8): 4554–4557.

76. Yu MF, Kowalewski T, and Ruoff RS, (2001), Structural analysis of collapsed, and twisted and collapsed, multi-walled carbon nanotubes by atomic force microscopy, *Phys. Rev. Lett.* 86(1): 87–90.

77. Lordi V and Yao N, (1998), Radial compression and controlled cutting of carbon nanotubes, *J. Chem. Phys.* 109(6): 2509–2512.

78. Shen WD, Jiang B, Han BS, and Xie SS, (2000), Investigation of the radial compression of carbon nanotubes with a scanning probe microscope, *Phys. Rev. Lett.* 84(16): 3634–3637.

79. Yu MF, Kowalewski T, and Ruoff RS, (2000), Investigation of the radial deformability of individual carbon nanotubes under controlled indentation force, *Phys. Rev. Lett.* 85(7): 1456–1459.

80. Chesnokov SA, Nalimova VA, Rinzler AG, Smalley RE, and Fischer JE, (1999), Mechanical energy storage in carbon nanotube springs, *Phys. Rev. Lett.* 82(2): 343–346.

81. Tang J, Qin LC, Sasaki T, Yudasaka M, Matsushita A, and Iijima S, (2000), Compressibility and polygonization of single-walled carbon nanotubes under hydrostatic pressure, *Phys. Rev. Lett.* 85(9): 1887–1889.

82. Yu MF, Dyer MJ, Chen J, Qian D, Liu WK, and Ruoff RS, (2001), Locked twist in multi-walled carbon nanotube ribbons, *Phys. Rev. B, Rapid Commn.* 64: 241403R.

83. Ebbesen TW and Ajayan PM, (1992), Large-scale synthesis of carbon nanotubes, *Nature.* 358(6383): 220–222.

84. Iijima S, Ajayan PM, and Ichihashi T, (1992), Growth model for carbon nanotubes, *Phys. Rev. Lett.* 69(21): 3100–3103.

85. Thess A, Lee R, Nikolaev P, Dai HJ, Petit P, Robert J, Xu CH, Lee YH, Kim SG, Rinzler AG, Colbert DT, Scuseria GE, Tomanek D, Fischer JE, and Smalley RE, (1996), Crystalline ropes of metallic carbon nanotubes, *Science.* 273(5274): 483–487.

86. Guo T, Nikolaev P, Thess A, Colbert DT, and Smalley RE, (1995), Catalytic growth of single-walled nanotubes by laser vaporization, *Chem. Phys. Lett.* 243(1–2): 49–54.

87. Kong J, Soh HT, Cassell AM, Quate CF, and Dai HJ, (1998), Synthesis of individual single-walled carbon nanotubes on patterned silicon wafers, *Nature.* 395(6705): 878–881.

88. Cassell AM, Raymakers JA, Kong J, and Dai HJ, (1999), Large-scale CVD synthesis of single-walled carbon nanotubes, *J. Phys. Chem. B.* 103(31): 6484–6492.

89. Li WZ, Xie SS, Qian LX, Chang BH, Zou BS, Zhou WY, Zhao RA, and Wang G, (1996), Large-scale synthesis of aligned carbon nanotubes, *Science.* 274(5293): 1701–1703.

90. Dal HJ, Rinzler AG, Nikolaev P, Thess A, Colbert DT, and Smalley RE, (1996), Single-wall nanotubes produced by metal-catalyzed disproportionation of carbon monoxide, *Chem. Phys. Lett.* 260(3–4): 471–475.

91. Nardelli MB, Yakobson BI, and Bernholc J, (1998), Brittle and ductile behavior in carbon nanotubes, *Phys. Rev. Lett.* 81(21): 4656–4659.

92. Walters DA, Ericson LM, Casavant MJ, Liu J, Colbert DT, Smith KA, and Smalley RE, (1999), Elastic strain of freely suspended single-wall carbon nanotube ropes, *Appl. Phys. Lett.* 74(25): 3803–3805.

93. Pan ZW, Xie SS, Lu L, Chang BH, Sun LF, Zhou WY, Wang G, and Zhang DL, (1999), Tensile tests of ropes of very long aligned multi-wall carbon nanotubes, *Appl. Phys. Lett.* 74(21): 3152–3154.

94. Wagner HD, Lourie O, Feldman Y, and Tenne R, (1998), Stress-induced fragmentation of multi-wall carbon nanotubes in a polymer matrix, *Appl. Phys. Lett.* 72(2): 188–190.

95. Li F, Cheng HM, Bai S, Su G, and Dresselhaus MS, (2000), Tensile strength of single-walled carbon nanotubes directly measured from their macroscopic ropes, *Appl. Phys. Lett.* 77(20): 3161–3163.

96. Yakobson BI, Campbell MP, Brabec CJ, and Bernholc J, (1997), High strain rate fracture and c-chain unraveling in carbon nanotubes, *Computational Mater. Sci.* 8(4): 341–348.

97. Belytschko T, Xiao SP, Schatz GC, and Ruoff RS, (2001), Simulation of the fracture of nanotubes, *Phys. Rev. B*, accepted.

98. Yakobson BI, (1997), in *Recent Advances in the Chemistry and Physics of Fullerenes and Related Materials*, Kadish KM (Ed.), Electrochemical Society, Inc., Pennington, NJ, 549.

99. Nardelli MB, Yakobson BI, and Bernholc J, (1998), Mechanism of strain release in carbon nanotubes, *Phys. Rev. B.* 57(8): R4277-R4280.

100. Srivastava D, Menon M, and Cho KJ, (1999), Nanoplasticity of single-wall carbon nanotubes under uniaxial compression, *Phys. Rev. Lett.* 83(15): 2973–2976.

101. Wei CY, Srivastava D, and Cho KJ, (2001), Molecular dynamics study of temperature-dependent plastic collapse of carbon nanotubes under axial compression, *Comp. Modeling Eng. Sci.* 3, 255.

102. Srivastava D, Wei CY, and Cho KJ, (2002) Computational nanomechanics of carbon nanotubes and composites, ASME, *Applied Mechanics Reviews* (special issue on nanotechnology), in press.

103. Yakobson BI, (1998), Mechanical relaxation and intramolecular plasticity in carbon nanotubes, *Appl. Phys. Lett.* 72(8): 918–920.

104. Zhang PH, Lammert PE, and Crespi VH, (1998), Plastic deformations of carbon nanotubes, *Phys. Rev. Lett.* 81(24): 5346–5349.

105. Zhang PH and Crespi VH, (1999), Nucleation of carbon nanotubes without pentagonal rings, *Phys. Rev. Lett.* 83(9): 1791–1794.

106. Bockrath M, Cobden DH, McEuen PL, Chopra NG, Zettl A, Thess A, and Smalley RE, (1997), Single-electron transport in ropes of carbon nanotubes, *Science.* 275(5308): 1922–1925.

107. Cumings J and Zettl A, (2000), Low-friction nanoscale linear bearing realized from multi-wall carbon nanotubes, *Science.* 289(5479): 602–604.

108. Kelly BT, (1981), *Physics of Graphite.* Applied Science. London.

109. Ausman KD and Ruoff RS, (2001), Unpublished.

110. Yakobson BI, (2001), Unpublished.

111. Geohegan DB, Schittenhelm H, Fan X, Pennycook SJ, Puretzky AA, Guillorn MA, Blom DA, and Joy DC, (2001), Condensed phase growth of single-wall carbon nanotubes from laser annealed nanoparticulates, *Appl. Phys. Lett.* 78(21): 3307–3309.

112. Piner RD and Ruoff RS, (2001), Unpublished.

113. Born M and Huang K, (1954), *Dynamical Theory of Crystal Lattices.* Oxford University Press. Oxford, UK.

114. Keating PN, (1966), Theory of third-order elastic constants of diamond-like crystals, *Phys. Rev.* 149(2): 674.

115. Keating PN, (1966), Effect of invariance requirements on elastic strain energy of crystals with application to diamond structure, *Phys. Rev.* 145(2): 637.

116. Keating PN, (1967), On sufficiency of Born–Huang relations, *Phys. Lett. A.* A 25(7): 496.

117. Keating PN, (1968), Relationship between macroscopic and microscopic theory of crystal elasticity. 2. Nonprimitive crystals, *Phys. Rev.* 169(3): 758.

118. Brugger K, (1964), Thermodynamic definition of higher order elastic coefficients, *Phys. Rev.* 133(6A): A1611.

119. Martin JW, (1975), Many-body forces in metals and Brugger elastic-constants, *J. Phys. C — Solid State Phys.* 8(18): 2837–2857.

120. Martin JW, (1975), Many-body forces in solids and Brugger elastic-constants. 2. INNER elastic-constants, *J. Phys. C — Solid State Phys.* 8(18): 2858–2868.

121. Martin JW, (1975), Many-body forces in solids — elastic-constants of diamond-type crystals, *J. Phys. C — Solid State Phys.* 8(18): 2869–2888.

122. Daw MS and Baskes MI, (1984), Embedded-atom method — derivation and application to impurities, surfaces, and other defects in metals, *Phys. Rev. B.* 29(12): 6443–6453.

123. Daw MS, Foiles SM, and Baskes MI, (1993), The embedded-atom method — a review of theory and applications, *Mater. Sci. Rep.* 9(7–8): 251–310.

124. Tadmor EB, Ortiz M, and Phillips R, (1996), Quasicontinuum analysis of defects in solids, *Philos. Mag. A — Phys. Cond. Matter Struct. Defects Mech. Prop.* 73(6): 1529–1563.

125. Tadmor EB, Phillips R, and Ortiz M, (1996), Mixed atomistic and continuum models of deformation in solids, *Langmuir.* 12(19): 4529–4534.

126. Friesecke G and James RD, (2000), A scheme for the passage from atomic to continuum theory for thin films, nanotubes and nanorods, *J. Mech. Phys. Solids.* 48(6–7): 1519–1540.

127. Belytschko T, Liu WK, and Moran B, (2000), *Nonlinear Finite Elements for Continua and Structures.* John Wiley & Sons, LTD.

128. Marsden JE and Hughes TJR, (1983), *Mathematical Foundations of Elasticity.* Prentice-Hall. Englewood Cliffs, NJ.

129. Malvern LE, (1969), *Introduction to the Mechanics of a Continuous Medium.* Prentice-Hall. Englewood Cliffs, NJ.

130. Milstein F, (1982), Crystal elasticity, in *Mechanics of Solids*, Sewell MJ (Ed.), Pergamon Press, Oxford, UK.

131. Ericksen JL, (1984), in *Phase Transformations and Material Instabilities in Solids*, Gurtin M (Ed.), Academic Press, New York.

132. Tersoff J, (1986), New empirical model for the structural properties of silicon, *Phys. Rev. Lett.* 56(6): 632–635.

133. Tersoff J, (1988), New empirical approach for the structure and energy of covalent systems, *Phys. Rev. B.* 37(12): 6991–7000.

134. Tersoff J, (1988), Empirical interatomic potential for carbon, with applications to amorphous-carbon, *Phys. Rev. Lett.* 61(25): 2879–2882.

135. Tersoff J, (1989), Modeling solid-state chemistry — interatomic potentials for multicomponent systems, *Phys. Rev. B.* 39(8): 5566–5568.

136. Green AE and Rivlin RS, (1964), Multipolar continuum mechanics, *Arch. Rat. Mech. An.* 17: 113–147.

137. Cousins CSG, (1978), Inner elasticity, *J. Phys. C — Solid State Phys.* 11(24): 4867–4879.

138. Zhang P, Huang Y, Geubelle PH, and Hwang KC, (2002), On the continuum modeling of carbon nanotubes. *Acta Mechanica Sinica.* (In press).

139. Zhang P, Huang Y, Geubelle PH, Klein P, and Hwang KC, (2002), The elastic modulus of single-wall carbon nanotubes: a continuum analysis incorporating interatomic potentials, *Intl. J. Solids Struct.* In press.

140. Zhang P, Huang Y, Gao H, and Hwang KC, (2002), Fracture nucleation in single-wall carbon nanotubes under tension: a continuum analysis incorporating interatomic potentials. *J. Appl. Mech.* (In press).

141. Wiesendanger R, (1994), *Scanning Probe Microscopy and Spectroscopy: Methods and Applications.* Cambridge University Press. Oxford, UK.

142. Binnig G and Quate CF, (1986), Atomic force microscope, *Phys. Rev. Lett.* 56: 930–933.

143. Cumings J, Collins PG, and Zettl A, (2000), Materials — peeling and sharpening multi-wall nanotubes, *Nature.* 406(6796): 586–586.

144. Ohno K, Esfarjani K, and Kawazoe Y, (1999), *Computational Material Science: From Ab Initio to Monte Carlo Methods.* Solid State Sciences, Cardona M, et al. (Ed.) Springer. Berlin.

145. Dirac PAM, (1958), *The Principles of Quantum Mechanics.* Oxford University Press. London.

146. Landau LD and Lifshitz EM, (1965), *Quantum Mechanics; Non-Relativistic Theory.* Pergamon. Oxford, UK.

147. Merzbacher E, (1998), *Quantum Mechanics.* Wiley. New York.

148. Messiah A, (1961), *Quantum Mechanics.* North-Holland. Amsterdam.

149. Schiff LI, (1968), *Quantum Mechanics.* McGraw-Hill. New York.

150. Fock V, (1930), Naherungsmethode zur losung des quantenmechanis-chen mehrkorperproblems, *Z. Physik.* 61: 126.

151. Hartree DR, (1928), The wave mechanics of an atom with a non-Coulomb central field, part I, theory and methods, *Proc. Cambridge Phil. Soc.* 24: 89.

152. Hartree DR, (1932–1933), A practical method for the numerical solution of differential equations, *Mem. and Proc. Manchester Lit. Phil. Soc.* 77: 91.

153. Clementi E, (2000), *Ab initio* computations in atoms and molecules (reprinted from *IBM Journal of Research and Development* 9, 1965), *IBM J. Res. Dev.* 44(1–2): 228–245.

154. Hohenberg P and Kohn W, (1964), Inhomogeneous electron gas, *Phys. Rev. B.* 136: 864.

155. Kohn W and Sham LJ, (1965), Self-consistent equations including exchange and correlation effects, *Phys. Rev.* 140(4A): 1133.

156. Perdew JP, McMullen ER, and Zunger A, (1981), Density-functional theory of the correlation-energy in atoms and ions — a simple analytic model and a challenge, *Phys. Rev. A.* 23(6): 2785–2789.

157. Perdew JP and Zunger A, (1981), Self-interaction correction to density-functional approximations for many-electron systems, *Phys. Rev. B.* 23(10): 5048–5079.

158. Slater JC, Wilson TM, and Wood JH, (1969), Comparison of several exchange potentials for electrons in cu+ ion, *Phys. Rev.* 179(1): 28.

159. Moruzzi VJ and Sommers CB, (1995), *Calculated Electronic Properties of Ordered Alloys: A Handbook.* World Scientific. Singapore.

160. Payne MC, Teter MP, Allan DC, Arias TA, and Joannopoulos JD, (1992), Iterative minimization techniques for *ab initio* total-energy calculations — molecular-dynamics and conjugate gradients, *Rev. Mod. Phys.* 64(4): 1045–1097.

161. Car R and Parrinello M, (1985), Unified approach for molecular-dynamics and density-functional theory, *Phys. Rev. Lett.* 55(22): 2471–2474.

162. Slater JC and Koster GF, (1954), Wave functions for impurity levels, *Phys. Rev.* 94(1498).

163. Harrison WA, (1989), *Electronic Structure and the Properties of Solids: The Physics of the Chemical Bond.* Dover. New York.

164. Matthew W, Foulkes C, and Haydock R, (1989), Tight-binding models and density-functional theory, *Phys. Rev. B.* 39(17): 12520–12536.

165. Xu CH, Wang CZ, Chan CT, and Ho KM, (1992), A transferable tight-binding potential for carbon, *J. Phys. Condensed Matter.* 4(28): 6047–6054.

166. Mehl MJ and Papaconstantopoulos DA, (1996), Applications of a tight-binding total-energy method for transition and noble metals: elastic constants, vacancies, and surfaces of monatomic metals, *Phys. Rev. B.* 54(7): 4519–4530.

167. Liu F, (1995), Self-consistent tight-binding method, *Phys. Rev. B.* 52(15): 10677–10680.

168. Porezag D, Frauenheim T, Kohler T, Seifert G, and Kaschner R, (1995), Construction of tight-binding-like potentials on the basis of density-functional theory — application to carbon, *Phys. Rev. B.* 51(19): 12947–12957.

169. Taneda A, Esfarjani K, Li ZQ, and Kawazoe Y, (1998), Tight-binding parameterization of transition metal elements from LCAO *ab initio* Hamiltonians, *Computational Mater. Sci.* 9(3–4): 343–347.

170. Menon M and Subbaswamy KR, (1991), Universal parameter tight-binding molecular-dynamics — application to c-60, *Phys. Rev. Lett.* 67(25): 3487–3490.

171. Sutton AP, Finnis MW, Pettifor DG, and Ohta Y, (1988), The tight-binding bond model, *J. Phys. C — Solid State Phys.* 21(1): 35–66.

172. Menon M and Subbaswamy KR, (1997), Nonorthogonal tight-binding molecular-dynamics scheme for silicon with improved transferability, *Phys. Rev. B.* 55(15): 9231–9234.

173. Haile JM, (1992), *Molecular Dynamics Simulation.* Wiley Interscience, New York.

174. Rapaport DC, (1995), *The Art of Molecular Dynamics Simulation.* Cambridge University Press, London.

175. Frenkel D and Smit, B., (1996), *Understanding Molecular Simulation: From Algorithms to Applications.* Academic Press, New York.

176. Hockney RW and Eastwood, J.W., (1989), *Computer Simulation Using Particles.* IOP Publishing Ltd. New York.

177. Li. SF and Liu WK, (2002), Mesh-free and particle methods, *Appl. Mech. Rev.* 55(1): 1–34.

178. Berendsen HJC and van Gunsteren W.F., (1986), *Dynamics Simulation of Statistical Mechanical Systems,* Ciccotti GPF, Hoover, W.G. (Eds.) Vol. 63. North Holland. Amsterdam. 493.

179. Verlet L, (1967), Computer experiments on classical fluids. I. Thermodynamical properties of Lennard-Jones molecules, *Phys. Rev.* 159(1): 98.

180. Gray SK, Noid DW, and Sumpter BG, (1994), Symplectic integrators for large-scale molecular-dynamics simulations — a comparison of several explicit methods, *J. Chem. Phys.* 101(5): 4062–4072.

181. Allinger NL, (1977), Conformational-analysis.130. MM2 — hydrocarbon force-field utilizing V1 and V2 torsional terms, *J. Am. Chem. Soc.* 99(25): 8127–8134.

182. Allinger NL, Yuh YH, and Lii JH, (1989), Molecular mechanics — the MM3 force-field for hydrocarbons. 1, *J. Am. Chem. Soc.* 111(23): 8551–8566.

183. Mayo SL, Olafson BD, and Goddard WA, (1990), Dreiding — a generic force-field for molecular simulations, *J. Phys. Chem.* 94(26): 8897–8909.

184. Guo YJ, Karasawa N, and Goddard WA, (1991), Prediction of fullerene packing in c60 and c70 crystals, *Nature.* 351(6326): 464–467.

185. Tuzun RE, Noid DW, Sumpter BG, and Merkle RC, (1996), Dynamics of fluid flow inside carbon nanotubes, *Nanotechnology.* 7(3): 241–246.

186. Tuzun RE, Noid DW, Sumpter BG, and Merkle RC, (1997), Dynamics of he/c-60 flow inside carbon nanotubes, *Nanotechnology.* 8(3): 112–118.

187. Abell GC, (1985), Empirical chemical pseudopotential theory of molecular and metallic bonding, *Phys. Rev. B.* 31(10): 6184–6196.

188. Brenner DW, (2000), The art and science of an analytic potential, *Physica Status Solidi B — Basic Res.* 217(1): 23–40.

189. Nordlund K, Keinonen J, and Mattila T, (1996), Formation of ion irradiation induced small-scale defects on graphite surfaces, *Phys. Rev. Lett.* 77(4): 699–702.

190. Brenner DW, Harrison JA, White CT, and Colton RJ, (1991), Molecular-dynamics simulations of the nanometer-scale mechanical properties of compressed buckminsterfullerene, *Thin Solid Films.* 206(1–2): 220–223.

191. Robertson DH, Brenner DW, and White CT, (1992), On the way to fullerenes — molecular-dynamics study of the curling and closure of graphitic ribbons, *J. Phys. Chem.* 96(15): 6133–6135.

192. Robertson DH, Brenner DW, and White CT, (1995), Temperature-dependent fusion of colliding c-60 fullerenes from molecular-dynamics simulations, *J. Phys. Chem.* 99(43): 15721–15724.

193. Sinnott SB, Colton RJ, White CT, and Brenner DW, (1994), Surface patterning by atomically-controlled chemical forces — molecular-dynamics simulations, *Surf. Sci.* 316(1–2): L1055-L1060.

194. Harrison JA, White CT, Colton RJ, and Brenner DW, (1992), Nanoscale investigation of indentation, adhesion and fracture of diamond (111) surfaces, *Surf. Sci.* 271(1–2): 57–67.

195. Harrison JA, White CT, Colton RJ, and Brenner DW, (1992), Molecular-dynamics simulations of atomic-scale friction of diamond surfaces, *Phys. Rev. B.* 46(15): 9700–9708.

196. Harrison JA, Colton RJ, White CT, and Brenner DW, (1993), Effect of atomic-scale surface-roughness on friction — a molecular-dynamics study of diamond surfaces, *Wear.* 168(1–2): 127–133.

197. Harrison JA, White CT, Colton RJ, and Brenner DW, (1993), Effects of chemically-bound, flexible hydrocarbon species on the frictional properties of diamond surfaces, *J. Phys. Chem.* 97(25): 6573–6576.

198. Harrison JA, White CT, Colton RJ, and Brenner DW, (1993), Atomistic simulations of friction at sliding diamond interfaces, *MRS Bull.* 18(5): 50–53.

199. Harrison JA and Brenner DW, (1994), Simulated tribochemistry — an atomic-scale view of the wear of diamond, *J. Am. Chem. Soc.* 116(23): 10399–10402.

200. Harrison JA, White CT, Colton RJ, and Brenner DW, (1995), Investigation of the atomic-scale friction and energy — dissipation in diamond using molecular-dynamics, *Thin Solid Films.* 260(2): 205–211.

201. Tupper KJ and Brenner DW, (1993), Atomistic simulations of frictional wear in self-assembled monolayers, *Abstr. Papers Am. Chem. Soc.* 206: 172.

202. Tupper KJ and Brenner DW, (1993), Molecular-dynamics simulations of interfacial dynamics in self-assembled monolayers, *Abstr. Papers Am. Chem. Soc.* 206: 72.

203. Tupper KJ and Brenner DW, (1994), Molecular-dynamics simulations of friction in self-assembled monolayers, *Thin Solid Films.* 253(1–2): 185–189.

204. Brenner DW, (2001), Unpublished.

205. Pettifor DG and Oleinik II, (1999), Analytic bond-order potentials beyond Tersoff-Brenner. Ii. Application to the hydrocarbons, *Phys. Rev. B.* 59(13): 8500.

206. Pettifor DG and Oleinik II, (2000), Bounded analytic bond-order potentials for sigma and pi bonds, *Phys. Rev. Lett.* 84(18): 4124–4127.

207. Jones JE, (1924), On the determination of molecular fields-I. From the variation of the viscosity of a gas with temperature, *Proc. R. Soc.* 106: 441.

208. Jones JE, (1924), On the determination of molecular fields-II. From the equation of state of a gas, *Proc. R. Soc.* 106: 463.

209. Girifalco LA and Lad RA, (1956), Energy of cohesion, compressibility and the potential energy functions of the graphite system, *J. Chem. Phys.* 25(4): 693–697.

210. Girifalco LA, (1992), Molecular-properties of c-60 in the gas and solid-phases, *J. Phys. Chem.* 96(2): 858–861.

211. Wang Y, Tomanek D, and Bertsch GF, (1991), Stiffness of a solid composed of c60 clusters, *Phys. Rev. B.* 44(12): 6562–6565.

212. Qian D, Liu WK, and Ruoff RS, (2001), Mechanics of c60 in nanotubes, *J. Phys. Chem. B.* 105: 10753–10758.

213. Zhao YX and Spain IL, (1989), X-ray-diffraction data for graphite to 20 gpa, *Phys. Rev. B.* 40(2): 993–997.

214. Hanfland M, Beister H, and Syassen K, (1989), Graphite under pressure — equation of state and first-order Raman modes, *Phys. Rev. B.* 39(17): 12598–12603.

215. Boettger JC, (1997), All-electron full-potential calculation of the electronic band structure, elastic constants, and equation of state for graphite, *Phys. Rev. B.* 55(17): 11202–11211.

216. Girifalco LA, Hodak M, and Lee RS, (2000), Carbon nanotubes, buckyballs, ropes, and a universal graphitic potential, *Phys. Rev. B.* 62(19): 13104–13110.

217. Falvo MR, Clary G, Helser A, Paulson S, Taylor RM, Chi V, Brooks FP, Washburn S, and Superfine R, (1998), Nanomanipulation experiments exploring frictional and mechanical properties of carbon nanotubes, *Microsc. Microanal.* 4(5): 504–512.

218. Falvo MR, Taylor RM, Helser A, Chi V, Brooks FP, Washburn S, and Superfine R, (1999), Nanometre-scale rolling and sliding of carbon nanotubes, *Nature.* 397(6716): 236–238.

219. Falvo MR, Steele J, Taylor RM, and Superfine R, (2000), Evidence of commensurate contact and rolling motion: AFM manipulation studies of carbon nanotubes on hopg, *Tribol, Lett.* 9(1–2): 73–76.

220. Falvo MR, Steele J, Taylor RM, and Superfine R, (2000), Gearlike rolling motion mediated by commensurate contact: carbon nanotubes on hopg, *Phys. Rev. B.* 62(16): R10665-R10667.

221. Schall JD and Brenner DW, (2000), Molecular dynamics simulations of carbon nanotube rolling and sliding on graphite, *Molecular Simulation.* 25(1–2): 73–79.

222. Buldum A and Lu JP, (1999), Atomic scale sliding and rolling of carbon nanotubes, *Phys. Rev. Lett.* 83(24): 5050–5053.

223. Kolmogorov AN and Crespi VH, (2000), Smoothest bearings: Interlayer sliding in multi-walled carbon nanotubes, *Phys. Rev. Lett.* 85(22): 4727–4730.

224. Shenoy VB, Miller R, Tadmor EB, Phillips R, and Ortiz M, (1998), Quasicontinuum models of interfacial structure and deformation, *Phys. Rev. Lett.* 80(4): 742–745.

225. Miller R, Ortiz M, Phillips R, Shenoy V, and Tadmor EB, (1998), Quasicontinuum models of fracture and plasticity, *Eng, Fracture Mech.* 61(3–4): 427–444.

226. Miller R, Tadmor EB, Phillips R, and Ortiz M, (1998), Quasicontinuum simulation of fracture at the atomic scale, *Modelling Sim. Mater. Sci. Eng.* 6(5): 607–638.

227. Shenoy VB, Miller R, Tadmor EB, Rodney D, Phillips R, and Ortiz M, (1999), An adaptive finite element approach to atomic-scale mechanics — the quasicontinuum method, *J. Mech. Phys. Solids.* 47(3): 611–642.

228. Tadmor EB, Miller R, Phillips R, and Ortiz M, (1999), Nanoindentation and incipient plasticity, *J. Mater. Res.* 14(6): 2233–2250.

229. Rodney D and Phillips R, (1999), Structure and strength of dislocation junctions: an atomic level analysis, *Phys. Rev. Lett.* 82(8): 1704–1707.

230. Smith GS, Tadmor EB, and Kaxiras E, (2000), Multiscale simulation of loading and electrical resistance in silicon nanoindentation, *Phys. Rev. Lett.* 84(6): 1260–1263.

231. Knap J and Ortiz M, (2001), An analysis of the quasicontinuum method, *J. Mech Phys. Solids.* 49(9): 1899–1923.

232. Shin CS, Fivel MC, Rodney D, Phillips R, Shenoy VB, and Dupuy L, (2001), Formation and strength of dislocation junctions in fcc metals: a study by dislocation dynamics and atomistic simulations, *J. Physique Iv.* 11(PR5): 19–26.

233. Shenoy V, Shenoy V, and Phillips R, (1999), Finite temperature quasicontinuum methods, in *Multiscale Modelling of Materials*, Ghoniem N (Ed.), Materials Research Society, Warrendale, PA, 465–471.

234. Arroyo M and Belytschko T, (2002), An atomistic-based membrane for crystalline films one atom thick *J. Mech. Phys. Solids.* 50: 1941–1977.

235. Liu WK and Chen YJ, (1995), Wavelet and multiple scale reproducing kernel methods, *Intl. J. Numerical Meth. Fluids.* 21(10): 901–931.

236. Liu WK, Jun S, and Zhang YF, (1995), Reproducing kernel particle methods, *Intl. J. Numerical Meth. Fluids.* 20(8–9): 1081–1106.

237. Liu WK, Jun S, Li SF, Adee J, and Belytschko T, (1995), Reproducing kernel particle methods for structural dynamics, *Intl. J. Numerical Meth. Eng.* 38(10): 1655–1679.

238. Liu WK, Chen YJ, Uras RA, and Chang CT, (1996), Generalized multiple scale reproducing kernel particle methods, *Computer Meth. Appl. Mech. Eng.* 139(1–4): 91–157.

239. Liu WK, Chen Y, Chang CT, and Belytschko T, (1996), Advances in multiple scale kernel particle methods, *Computational Mech.* 18(2): 73–111.

240. Liu WK, Jun S, Sihling DT, Chen YJ, and Hao W, (1997), Multiresolution reproducing kernel particle method for computational fluid dynamics, *Intl. J. Numerical Meth. Fluids.* 24(12): 1391–1415.

241. Liu WK, Li SF, and Belytschko T, (1997), Moving least-square reproducing kernel methods. 1. Methodology and convergence, *Computer Meth. Appl. Mech. Eng.* 143(1–2): 113–154.

242. Liu WK, Belytscho T, and Oden JT. (Eds.), (1996), *Computer Meth. Appl. Mech. Eng.* Vol. 139.

243. Chen JS and Liu WK (Eds.), (2000), *Computational Mech.* Vol. 25.

244. Odegard GM, Gates TS, Nicholson LM, and Wise KE, (2001), *Equivalent-Continuum Modeling of Nano-Structured Materials.* NASA Langley Research Center, NASA-2001-TM 210863.

245. Odegard GM, Harik VM, Wise KE, and Gates TS, (2001), *Constitutive Modeling of Nanotube-Reinforced Polymer Composite Systems.* NASA Langley Research Center, NASA-2001-TM 211044.

246. Abraham FF, Broughton JQ, Bernstein N, and Kaxiras E, (1998), Spanning the continuum to quantum length scales in a dynamic simulation of brittle fracture, *Europhys. Lett.* 44(6): 783–787.

247. Broughton JQ, Abraham FF, Bernstein N, and Kaxiras E, (1999), Concurrent coupling of length scales: methodology and application, *Phys. Rev. B.* 60(4): 2391–2403.

248. Nakano A, Bachlechner ME, Kalia RK, Lidorikis E, Vashishta P, Voyiadjis GZ, Campbell TJ, Ogata S, and Shimojo F, (2001), Multiscale simulation of nanosystems, *Computing Sci. Eng.* 3(4): 56–66.

249. Rafii-Tabar H, Hua L, and Cross M, (1998), Multiscale numerical modelling of crack propagation in two-dimensional metal plate, *Mater. Sci. Tech.* 14(6): 544–548.

250. Rafii-Tabar H, Hua L, and Cross M, (1998), A multi-scale atomistic-continuum modelling of crack propagation in a two-dimensional macroscopic plate, *J. Phys. Condensed Matter.* 10(11): 2375–2387.

251. Rudd RE and Broughton JQ, (1998), Coarse-grained molecular dynamics and the atomic limit of finite elements, *Phys. Rev. B.* 58(10): R5893-R5896.

252. Rudd RE and Broughton JQ, (2000), Concurrent coupling of length scales in solid state systems, *Physica Status Solidi B-Basic Research.* 217(1): 251–291.

253. Liu WK, Zhang Y, and Ramirez MR, (1991), Multiple scale finite-element methods, *Intl. J. Numerical Meth. Eng.* 32(5): 969–990.

254. Liu WK, Uras RA, and Chen Y, (1997), Enrichment of the finite element method with the reproducing kernel particle method, *J. Appl. Mech. Trans. ASME.* 64(4): 861–870.

255. Hao S, Liu WK, and Qian D, (2000), Localization-induced band and cohesive model, *J. Appl. Mech. Trans. ASME.* 67(4): 803–812.

256. Wagner GJ, Moes N, Liu WK, and Belytschko T, (2001), The extended finite element method for rigid particles in stokes flow, *Intl. J. Numerical Meth. Eng.* 51(3): 293–313.

257. Wagner GJ and Liu WK, (2001), Hierarchical enrichment for bridging scales and mesh-free boundary conditions, *Intl. J. Numerical Meth. Eng.* 50(3): 507–524.

258. Wagner GJ and Liu WK, (2002), Coupling of atomistic and continuum simulations (in preparation).

259. Costello GA, (1978), Analytical investigation of wire rope, *Appl. Mech. Rev. ASME.* 31: 897–900.

260. Costello GA, (1997), *Theory of Wire Rope*. 2nd ed. Springer. New York.

261. Qian D, Liu WK, and Ruoff RS, (2002), Load Transfer Mechanism in Nano-ropes, Computational Mechanics Laboratory Research Report (02-03) Dept. of Mech. Eng., Northwestern University.

262. Ruoff RS, Qian D, Liu WK, Ding WQ, Chen XQ, and Dikin D (Eds.), (2002), What kind of carbon nanofiber is ideal for structural applications? *43rd AIAA/ASME/ASCE/AHS Struct. Struct. Dynamics Mater. Conf.* Denver, CO.

263. Pipes BR and Hubert P, (2001), Helical carbon nanotube arrays: mechanical properties (submitted).

264. Yu MF, Dyer MJ, Skidmore GD, Rohrs HW, Lu XK, Ausman KD, Von Ehr JR, and Ruoff RS, (1999), Three-dimensional manipulation of carbon nanotubes under a scanning electron microscope, *Nanotechnology*. 10(3): 244–252.

265. Ruoff RS, Lorents DC, Chan B, Malhotra R, and Subramoney S, (1993), Single-crystal metals encapsulated in carbon nanoparticles, *Science*. 259(5093): 346–348.

266. Tomita M, Saito Y, and Hayashi T, (1993), Lac2 encapsulated in graphite nano-particle, *Jpn. J. App. Phys. Part 2–Lett.* 32(2B): L280-L282.

267. Seraphin S, Zhou D, Jiao J, Withers JC, and Loutfy R, (1993), Selective encapsulation of the carbides of yttrium and titanium into carbon nanoclusters, *Appl. Phys. Lett.* 63(15): 2073–2075.

268. Seraphin S, Zhou D, Jiao J, Withers JC, and Loutfy R, (1993), Yttrium carbide in nanotubes, *Nature*. 362(6420): 503–503.

269. Seraphin S, Zhou D, and Jiao J, (1996), Filling the carbon nanocages, *J. Appl. Phys.* 80(4): 2097–2104.

270. Saito Y, Yoshikawa T, Okuda M, Ohkohchi M, Ando Y, Kasuya A, and Nishina Y, (1993), Synthesis and electron-beam incision of carbon nanocapsules encaging yc2, *Chem. Phys. Lett.* 209(1–2): 72–76.

271. Saito Y, Yoshikawa T, Okuda M, Fujimoto N, Sumiyama K, Suzuki K, Kasuya A, and Nishina Y, (1993), Carbon nanocapsules encaging metals and carbides, *J. Phys. Chem. Solids*. 54(12): 1849–1860.

272. Saito Y and Yoshikawa T, (1993), Bamboo-shaped carbon tube filled partially with nickel, *J. Crystal Growth*. 134(1–2): 154–156.

273. Saito Y, Okuda M, and Koyama T, (1996), Carbon nanocapsules and single-wall nanotubes formed by arc evaporation, *Surf. Rev. Lett.* 3(1): 863–867.

274. Saito Y, Nishikubo K, Kawabata K, and Matsumoto T, (1996), Carbon nanocapsules and single-layered nanotubes produced with platinum-group metals (Ru, Rh, Pd, Os, Ir, Pt) by arc discharge, *J. Appl. Phys.* 80(5): 3062–3067.

275. Saito Y, (1996), Carbon cages with nanospace inside: fullerenes to nanocapsules, *Surf. Rev. Lett.* 3(1): 819–825.

276. Saito Y, (1995), Nanoparticles and filled nanocapsules, *Carbon*. 33(7): 979–988.

277. McHenry ME, Majetich SA, Artman JO, Degraef M, and Staley SW, (1994), Superparamagnetism in carbon-coated Co particles produced by the Kratschmer carbon-arc process, *Phys. Rev. B*. 49(16): 11358–11363.

278. Majetich SA, Artman JO, McHenry ME, Nuhfer NT, and Staley SW, (1993), Preparation and properties of carbon-coated magnetic nanocrystallites, *Phys. Rev. B*. 48(22): 16845–16848.

279. Jiao J, Seraphin S, Wang XK, and Withers JC, (1996), Preparation and properties of ferromagnetic carbon-coated Fe, Co, and Ni nanoparticles, *J. Appl. Phys.* 80(1): 103–108.

280. Diggs B, Zhou A, Silva C, Kirkpatrick S, Nuhfer NT, McHenry ME, Petasis D, Majetich SA, Brunett B, Artman JO, and Staley SW, (1994), Magnetic properties of carbon-coated rare-earth carbide nanocrystallites produced by a carbon-arc method, *J. Appl. Phys.* 75(10): 5879–5881.

281. Brunsman EM, Sutton R, Bortz E, Kirkpatrick S, Midelfort K, Williams J, Smith P, McHenry ME, Majetich SA, Artman JO, Degraef M, and Staley SW, (1994), Magnetic-properties of carbon-coated, ferromagnetic nanoparticles produced by a carbon-arc method, *J. Appl. Phys.* 75(10): 5882–5884.

282. Funasaka H, Sugiyama K, Yamamoto K, and Takahashi T, (1995), Synthesis of actinide carbides encapsulated within carbon nanoparticles, *J. Appl. Phys.* 78(9): 5320–5324.

283. Kikuchi K, Kobayashi K, Sueki K, Suzuki S, Nakahara H, Achiba Y, Tomura K, and Katada M, (1994), Encapsulation of radioactive gd-159 and tb-161 atoms in fullerene cages, *J. Am. Chem. Soc.* 116(21): 9775–9776.

284. Burch WM, Sullivan PJ, and McLaren CJ, (1986), Technegas – a new ventilation agent for lung-scanning, *Nucl. Med. Commn.* 7(12): 865.

285. Senden TJ, Moock KH, Gerald JF, Burch WM, Browitt RJ, Ling CD, and Heath GA, (1997), The physical and chemical nature of technegas, *J. Nucl. Med.* 38(8): 1327–1333.

286. Ajayan PM and Iijima S, (1993), Capillarity-induced filling of carbon nanotubes, *Nature.* 361(6410): 333–334.

287. Ajayan PM, Ebbesen TW, Ichihashi T, Iijima S, Tanigaki K, and Hiura H, (1993), Opening carbon nanotubes with oxygen and implications for filling, *Nature.* 362(6420): 522–525.

288. Tsang SC, Harris PJF, and Green MLH, (1993), Thinning and opening of carbon nanotubes by oxidation using carbon-dioxide, *Nature.* 362(6420): 520–522.

289. Xu CG, Sloan J, Brown G, Bailey S, Williams VC, Friedrichs S, Coleman KS, Flahaut E, Hutchison JL, Dunin-Borkowski RE, and Green MLH, (2000), 1d lanthanide halide crystals inserted into single-walled carbon nanotubes, *Chem. Commn.* (24): 2427–2428.

290. Tsang SC, Chen YK, Harris PJF, and Green MLH, (1994), A simple chemical method of opening and filling carbon nanotubes, *Nature.* 372(6502): 159–162.

291. Sloan J, Hammer J, Zwiefka-Sibley M, and Green MLH, (1998), The opening and filling of single walled carbon nanotubes (SWTS), *Chem. Commn.* (3): 347–348.

292. Hiura H, Ebbesen TW, and Tanigaki K, (1995), Opening and purification of carbon nanotubes in high yields, *Adv. Mater.* 7(3): 275–276.

293. Hwang KC, (1995), Efficient cleavage of carbon graphene layers by oxidants, *J. Chem. Soc. - Chem. Commn.* (2): 173–174.

294. Ajayan PM, Colliex C, Lambert JM, Bernier P, Barbedette L, Tence M, and Stephan O, (1994), Growth of manganese filled carbon nanofibers in the vapor-phase, *Phys. Rev. Lett.* 72(11): 1722–1725.

295. Subramoney S, Ruoff RS, Lorents DC, Chan B, Malhotra R, Dyer MJ, and Parvin K, (1994), Magnetic separation of gdc2 encapsulated in carbon nanoparticles, *Carbon.* 32(3): 507–513.

296. Tsang SC, Davis JJ, Green MLH, Allen H, Hill O, Leung YC, and Sadler PJ, (1995), Immobilization of small proteins in carbon nanotubes – high-resolution transmission electron-microscopy study and catalytic activity, *J. Chem. Soc. Chem. Commn.* (17): 1803–1804.

297. Tsang SC, Guo ZJ, Chen YK, Green MLH, Hill HAO, Hambley TW, and Sadler PJ, (1997), Immobilization of platinated and iodinated oligonucleotides on carbon nanotubes, *Angewandte Chemie.* 36(20): 2198–2200. International ed. in English.

298. Dillon AC, Jones KM, Bekkedahl TA, Kiang CH, Bethune DS, and Heben MJ, (1997), Storage of hydrogen in single-walled carbon nanotubes, *Nature.* 386(6623): 377–379.

299. Gadd GE, Blackford M, Moricca S, Webb N, Evans PJ, Smith AN, Jacobsen G, Leung S, Day A, and Hua Q, (1997), The world's smallest gas cylinders? *Science.* 277(5328): 933–936.

300. Sloan J, Wright DM, Woo HG, Bailey S, Brown G, York APE, Coleman KS, Hutchison JL, and Green MLH, (1999), Capillarity and silver nanowire formation observed in single walled carbon nanotubes, *Chem. Commn.* (8): 699–700.

301. Smith BW and Luzzi DE, (2000), Formation mechanism of fullerene peapods and coaxial tubes: a path to large scale synthesis, *Chem. Phys. Lett.* 321(1–2): 169–174.

302. Smith BW, Monthioux M, and Luzzi DE, (1998), Encapsulated c-60 in carbon nanotubes, *Nature.* 396(6709): 323–324.

303. Smith BW, Monthioux M, and Luzzi DE, (1999), Carbon nanotube encapsulated fullerenes: a unique class of hybrid materials, *Chem. Phys. Lett.* 315(1–2): 31–36.

304. Burteaux B, Claye A, Smith BW, Monthioux M, Luzzi DE, and Fischer JE, (1999), Abundance of encapsulated c-60 in single-wall carbon nanotubes, *Chem. Phys. Lett.* 310(1–2): 21–24.

305. Sloan J, Dunin-Borkowski RE, Hutchison JL, Coleman KS, Williams VC, Claridge JB, York APE, Xu CG, Bailey SR, Brown G, Friedrichs S, and Green MLH, (2000), The size distribution, imaging and obstructing properties of c-60 and higher fullerenes formed within arc-grown single walled carbon nanotubes, *Chem. Phys. Lett.* 316(3–4): 191–198.

306. Zhang Y, Iijima S, Shi Z, and Gu Z, (1999), Defects in arc-discharge-produced single-walled carbon nanotubes, *Philos. Mag. Lett.* 79(7): 473–479.

307. Pederson MR and Broughton JQ, (1992), Nanocapillarity in fullerene tubules, *Phys. Rev. Lett.* 69(18): 2689–2692.

308. Prasad R and Lele S, (1994), Stabilization of the amorphous phase inside carbon nanotubes — solidification in a constrained geometry, *Philos. Mag. Lett.* 70(6): 357–361.

309. Berber S, Kwon YK, and Tomanek D, (2001), Unpublished.

310. Stan G and Cole MW, (1998), Hydrogen adsorption in nanotubes, *J. Low Temp. Phys.* 110(1–2): 539–544.

311. Mao ZG, Garg A, and Sinnott SB, (1999), Molecular dynamics simulations of the filling and decorating of carbon nanotubules, *Nanotechnology.* 10(3): 273–277.

312. Stan G, Gatica SM, Boninsegni M, Curtarolo S, and Cole MW, (1999), Atoms in nanotubes: small dimensions and variable dimensionality, *Am. J. Phys.* 67(12): 1170–1176.

313. Liu J, Casavant MJ, Cox M, Walters DA, Boul P, Lu W, Rimberg AJ, Smith KA, Colbert DT, and Smalley RE, (1999), Controlled deposition of individual single-walled carbon nanotubes on chemically functionalized templates, *Chem. Phys. Lett.* 303(1–2): 125–129.

314. Chung J, Lee JH, Ruoff RS, and Liu WK, (2001), Nanoscale gap fabrication and integration of carbon nanotubes by micromachining (in preparation).

315. Ren Y and Price DL, (2001), Neutron scattering study of h-2 adsorption in single-walled carbon nanotubes, *Appl. Phys. Lett.* 79(22): 3684–3686.

316. Zhang YG, Chang AL, Cao J, Wang Q, Kim W, Li YM, Morris N, Yenilmez E, Kong J, and Dai HJ, (2001), Electric-field-directed growth of aligned single-walled carbon nanotubes, *Appl. Phys. Lett.* 79(19): 3155–3157.

317. Collins PG, Bradley K, Ishigami M, and Zettl A, (2000), Extreme oxygen sensitivity of electronic properties of carbon nanotubes, *Science.* 287(5459): 1801–1804.

318. Kong J, Franklin NR, Zhou CW, Chapline MG, Peng S, Cho KJ, and Dai HJ, (2000), Nanotube molecular wires as chemical sensors, *Science.* 287(5453): 622–625.

319. Yamamoto K, Akita S, and Nakayama Y, (1998), Orientation and purification of carbon nanotubes using AC electrophoresis, *J. Phys. D Appl. Phys.* 31(8): L34-L36.

320. Tombler TW, Zhou CW, Alexseyev L, Kong J, Dai HJ, Lei L, Jayanthi CS, Tang MJ, and Wu SY, (2000), Reversible electromechanical characteristics of carbon nanotubes under local-probe manipulation, *Nature.* 405(6788): 769–772.

321. Maiti A, (2001), Application of carbon nanotubes as electromechanical sensors – results from first-principles simulations, *Physica Status Solidi B-Basic Research.* 226(1): 87–93.

322. Maiti A, Andzelm J, Tanpipat N, and von Allmen P, (2001), Carbon nanotubes as field emission device and electromechanical sensor: results from first-principles dft simulations, *Abstr. Papers Am. Chem. Soc.* 222: 204.

323. Kong J, Chapline MG, and Dai HJ, (2001), Functionalized carbon nanotubes for molecular hydrogen sensors, *Adv. Mater.* 13(18): 1384–1386.

324. Peng S and Cho KJ, (2000), Chemical control of nanotube electronics, *Nanotechnology.* 11(2): 57–60.

325. Wood JR, Frogley MD, Meurs ER, Prins AD, Peijs T, Dunstan DJ, and Wagner HD, (1999), Mechanical response of carbon nanotubes under molecular and macroscopic pressures, *J. Phys. Chem. B.* 103(47): 10388–10392.

326. Wood JR and Wagner HD, (2000), Single-wall carbon nanotubes as molecular pressure sensors, *Appl. Phys. Lett.* 76(20): 2883–2885.

327. Wood JR, Zhao Q, Frogley MD, Meurs ER, Prins AD, Peijs T, Dunstan DJ, and Wagner HD, (2000), Carbon nanotubes: from molecular to macroscopic sensors, *Phys. Rev. B.* 62(11): 7571–7575.

328. Zhao Q, Wood JR, and Wagner HD, (2001), Using carbon nanotubes to detect polymer transitions, *J. Polymer Sci. Part B-Polymer Phys.* 39(13): 1492–1495.

329. Zhao Q, Wood JR, and Wagner HD, (2001), Stress fields around defects and fibers in a polymer using carbon nanotubes as sensors, *Appl. Phys. Lett.* 78(12): 1748–1750.

330. Ilic B, Czaplewski D, Craighead HG, Neuzil P, Campagnolo C, and Batt C, (2000), Mechanical resonant immunospecific biological detector, *Appl. Phys. Lett.* 77(3): 450–452.

331. Carr DW, Evoy S, Sekaric L, Craighead HG, and Parpia JM, (1999), Measurement of mechanical resonance and losses in nanometer scale silicon wires, *Appl. Phys. Lett.* 75(7): 920–922.

331a. Yu MF, Wagner GJ, Ruoff RS, and Dyer MJ, (2002), Realization of parametric resonances in a nanowire mechanical system with nanomanipulation inside a scanning electron microscope, *Phys. Rev. B*, in press.

332. Turner KL, Miller SA, Hartwell PG, MacDonald NC, Strogatz SH, and Adams SG, (1998), Five parametric resonances in a microelectromechanical system, *Nature.* 396(6707): 149–152.

333. Sandler J, Shaffer MSP, Prasse T, Bauhofer W, Schulte K, and Windle AH, (1999), Development of a dispersion process for carbon nanotubes in an epoxy matrix and the resulting electrical properties, *Polymer.* 40(21): 5967–5971.

334. Schadler LS, Giannaris SC, and Ajayan PM, (1998), Load transfer in carbon nanotube epoxy composites, *Appl. Phys. Lett.* 73(26): 3842–3844.

335. Qian D, Dickey EC, Andrews R, and Rantell T, (2000), Load transfer and deformation mechanisms in carbon nanotube-polystyrene composites, *Appl. Phys. Lett.* 76(20): 2868–2870.

336. Ajayan PM, Schadler LS, Giannaris C, and Rubio A, (2000), Single-walled carbon nanotube-polymer composites: strength and weakness, *Adv. Mater.* 12(10): 750–753.

337. Thostenson ET, Ren ZF, and Chou TW, (2001), Advances in the science and technology of carbon nanotubes and their composites: a review, *Composites Sci. Technol.* 61(13): 1899–1912.

338. Jin L, Bower C, and Zhou O, (1998), Alignment of carbon nanotubes in a polymer matrix by mechanical stretching, *Appl. Phys. Lett.* 73(9): 1197–1199.

339. Haggenmueller R, Gommans HH, Rinzler AG, Fischer JE, and Winey KI, (2000), Aligned single-wall carbon nanotubes in composites by melt processing methods, *Chem. Phys. Lett.* 330(3–4): 219–225.

340. Barrera EV, (2000), Key methods for developing single-wall nanotube composites, *Jom-J. Minerals Metals Mater. Soc.* 52(11): A38-A42.

341. Fischer JE, (2002) Nanomechanics and the viscoelastic behavior of carbon nanotube-reinforced polymers. Ph.D. thesis. Northwestern University, Evanston, IL.

342. Fisher FT, Bradshaw RD, and Brinson LC, (2001), Effects of nanotube waviness on the mechanical properties of nanoreinforced polymers, *Appl. Phys. Lett.* 80(24): 4647–4649.

343. Shaffer MSP and Windle AH, (1999), Fabrication and characterization of carbon nanotube/poly(vinyl alcohol) composites, *Adv. Mater.* 11(11): 937–941.

344. Gong XY, Liu J, Baskaran S, Voise RD, and Young JS, (2000), Surfactant-assisted processing of carbon nanotube/polymer composites, *Chem. Mater.* 12(4): 1049–1052.

345. Lozano K, Bonilla-Rios J, and Barrera EV, (2001), A study on nanofiber-reinforced thermoplastic composites (II): Investigation of the mixing rheology and conduction properties, *J. Appl. Polymer Sci.* 80(8): 1162–1172.

346. Qian D and Liu WK, (2002), In preparation.

347. Mitsubishi chemical to mass-produce nanotech substance, (2001), in *Kyodo News.*

24

Dendrimers — an Enabling Synthetic Science to Controlled Organic Nanostructures

24.1 Introduction **24**-2
Civilizations, Technology Periods, and Historical Revolutions
• Importance of Controlled Organic Nanostructures in
Biology • The "Wet and Dry World" of Nanotechnology
• Potential Bottom-up Synthesis Strategies for Organic
Nanostructures

24.2 The Dendritic State **24**-6
A Comparison of Organic Chemistry and Traditional Polymer
Chemistry with Dendritic Macromolecular Chemistry
• Dendritic Polymers — A Fourth Major New Architectural
Class • Dendrons/Dendrimers

24.3 Unique Dendrimer Properties **24**-19
Nanoscale Monodispersity • Nanoscale Container/Scaffolding
Properties • Amplification and Functionalization
of Dendrimer Surface Groups • Nanoscale Dimensions
and Shapes-Mimicking Proteins

24.4 Structural and Physical Properties of Dendrimers
in Solution **24**-20
Size and Structure of Dendrimers in Solutions • Shape,
Conformation, and Solvation of Dendrimers • Dynamics
of Dendrimers in Solutions

24.5 Dendrimers as Nanopharmaceuticals
and Nanomedical Devices **24**-24
Dendrimers as Genetic Material Transfer Agents
• Dendrimer–Carbohydrate Conjugates for Polyvalent
Recognition • Dendrimers as Targeted Drug-Delivery Agents
• Dendrimers as Magnetic Resonance Imaging Contrast Agents
• Dendrimers as Antiviral Agents • Dendrimers as
Angiogenesis Inhibitors

24.6 Dendrimers as Functional Nanomaterials **24**-27
Dendrimers as Chelating Agents for Cations • Dendrimers
as Ligands for Anions • Dendrimers as Unimolecular Micelle
Mimics • Dendrimers as Catalysts and Redox Active
Nanoparticles • Applications to Chemical and Environmental
Process Engineering

D.A. Tomalia
*Dendritic Nanotechnologies, Inc.,
and Central Michigan University*

S.A. Henderson
Starpharma Pty Limited

M.S. Diallo
*California Institute of Technology
and Howard University*

24.7 Dendrimers as Reactive Modules for the Synthesis
       of More Complex Nanoscale Architectures **24**-30
       Nanostructure Control beyond the Dendrimer • Dendrimers
       as Reactive Modules for the Synthesis of More Complex
       Nanoscale Architectures (Megamers)
24.8 Core–Shell Patterns Influencing the Modular
       Reactivity of Dendrimers **24**-34
24.9 Conclusions...**24**-35
Acknowledgments **24**-35
References **24**-35

24.1 Introduction

24.1.1 Civilizations, Technology Periods, and Historical Revolutions

Throughout the last 36 centuries of human history, a mere handful of civilizations have emerged to dominate both social order and define the human condition in the world (Figure 24.1). These unique cultures generally emerged as a result of certain converging societal parameters that included evolving politics, religion, social order, or major military/scientific paradigm shifts. Historical patterns will show, however, that the ultimate magnitude and duration of the civilizations were inextricably connected to their investment in certain key emerging technologies. These broad technologies not only underpinned these cultures, but also defined the "cutting edge" of human knowledge at that time in history and are referred to as "technology ages or periods" (Figure 24.1).

The first 118 centuries (i.e., 10,000 B.C. to 1800) were distinguished by only three major technology periods (i.e., the stone, bronze, and iron ages). Each period was based on materials (*building blocks*)

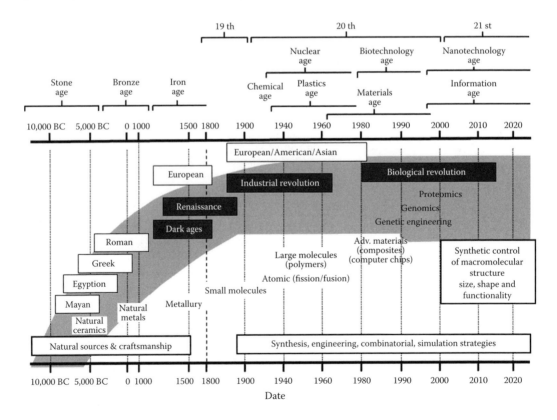

FIGURE 24.1 Civilizations, technology periods (ages), and historical revolutions as a function of time.

derived from natural sources and involved empirical knowledge gained through craftsmanship. Developments in these technology periods provided the resources and intellectual forces for dominance by these earlier civilizations. These successful reigns were followed by regression into the "Dark Ages." Emergence of the European influence concurrent with the "Renaissance" not only led to an advancement of the "Iron Age," but also to the initiation of the "chemical age" and ultimately aligned critical forces leading to the great Industrial Revolution. During the past two centuries, there has been a dramatic proliferation of technologies, with at least four major technology ages emerging in the last seven decades: nuclear, plastics, materials, and biotechnology. These technical advancements have been aligned primarily with Euro-American societies; however, more recently, there has been substantial Asian/Pacific Basin influence. Such technical advancements are very dependent upon certain critical enabling sciences including *synthesis*, *engineering*, and, more recently, *combinatorial/simulation* strategies.

Generally speaking, significant new paradigm shifts have initiated each of these technology periods. Typical of each period has been the systematic characterization and exploitation of novel structural/materials properties. Based on property patterns observed within the complexity boundaries of the specific technology, new scientific rules and principles evolve and become defined. Contemporary society has benefited substantially from both the knowledge and the materials created by these technology periods.

The general trend for succession of these technology periods has involved progression from simpler materials to more complex forms. In this fashion, earlier precursor technologies become the platforms upon which subsequent more complex structures (i.e., symmetries) are based. As described by Anderson [1], emerging new properties, principles, and rules are exhibited as a function of *symmetry breaking*. This occurs as one progresses from more basic structures to higher orders of complexity. Such a pattern is noted as one follows the hierarchical/development of complexity that has been observed in biological systems over the past several billion years of evolution. Simply stated "the whole becomes not only more than, but different from the sum of its parts"[1]. It is this premise that has driven the human quest for understanding new and higher forms of complexity. It is from this pattern of inquiry that the great Industrial Revolution was spawned with the convergence of knowledge, and new materials created by just five technology areas (Figure 24.1) providing the critical environment and synergy for this historical revolution. Economic benefits and enhancements to the human condition in areas such as transportation, shelter, clothing, energy, and agriculture are now recognized to be immeasurable. It is from this perspective that we now consider the emergence of new technologies such as *genomics*, *proteomics*, and, particularly, *nanotechnology* as we examine the prospects for the next revolution. The biological revolution which began with the elucidation of the DNA structure by Crick and Watson is currently predicting life-enhancing strategies that sound almost like science fiction. Such predictions include the total elimination of human disease, human life expectancy beyond 100 years, and, perhaps, extraterrestrial emigration in the next century.

24.1.2   Importance of Controlled Organic Nanostructures in Biology

Critical to the successful creation of all biological structures required for life has been the evolutionary development of strategies to produce controlled organic nanostructures. It is speculated that the evolutionary development of biological complexity occurred in two significant phases and involved *bottom-up synthesis*. Clearly, critical parameters such as mass and dimensions had to increase in size to define the appropriate building modules. The first phase was abiotic and involved molecular evolution from atoms to small molecules. This began approximately 13 billion years ago and progressed for nearly 8 to 9 billion years. Development of this complexity was necessary to create appropriate building blocks for the subsequent evolution of life, namely those referred to as macromolecules and including DNA, RNA, and proteins. This latter phase is believed to have begun 3.5 billion years ago. These modules were generally collections of precisely bonded atoms that occupied space with dimensions ranging from 1 to 100 nm. Such structures required the controlled assembly of as many as 10^3 to 10^9 atoms and possessed molecular weights ranging from 10^4 to 10^{10} Da. These assemblies required the rigorous control of *size*, *shape*, *surface chemistries*, *scaffolding*, *container properties*, and reside in the generally accepted range of nanoscale dimensions as described in Figure 24.2.

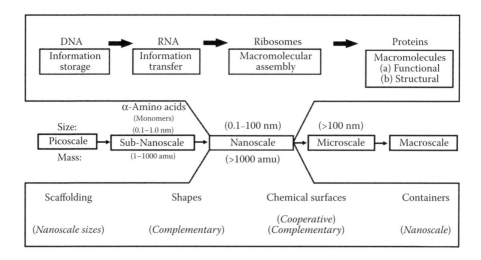

FIGURE 24.2 Biological structure control strategy, leading to nanoscale scaffolding, shapes, chemical surfaces, and containers, found in proteins.

24.1.3 The "Wet and Dry World" of Nanotechnology

According to Nobel Laureate Smalley [2], the world of nanotechnology can be subdivided into two major areas: the *wet* and *dry* sides. The former, of course, includes the biological domain, wherein the water-based science of living entities is dependent upon *hydrophilic* nanostructures and devices that may function within biological cells. Dendritic nanostructures, especially *dendrons/dendrimers* [3], fulfill many applications in the wet world of nanotechnology (Figure 24.3). In contrast, the dry side is expected to include those applications focused on more *hydrophobic* architectures and strategies. Progress in this second area may be expected to enhance the tensile strength of materials, increase the conductivity of electrons, or allow the size reduction of computer chips to levels unattainable with traditional bulk materials.

Although substantial progress has been made concerning the use of fullerenes and carbon nanotubes for dry nanotech applications, their use in biological applications has been hindered by the fact that they are highly hydrophobic and available in only several specific sizes (i.e., usually approximately 1 nm). However, recent advances involving the functionalization of fullerenes may offer future promise for these materials in certain biological applications [3,4].

24.1.4 Potential Bottom-up Synthesis Strategies for Organic Nanostructures

At least three major strategies are presently available for covalent synthesis of organic nanostructures, namely: (A) *traditional organic chemistry*, (B) *traditional polymer chemistry*, and more recently (C) *dendritic macromolecular chemistry* (Figure 24.4).

Broadly speaking, traditional organic chemistry produces higher complexity by involving the formation of relatively few covalent bonds between small heterogeneous aggregates of atoms or reagents to give well-defined small molecules (i.e., <1000 Da). On the other hand, polymerization strategies such as (B) and (C) involve the formation of large multiples of covalent bonds between homogenous small molecule monomers to produce macromolecules (i.e., >1000 Da) or infinite networks with a broad range of structure control [5,6].

By comparing these three covalent synthesis strategies (i.e., [A], [B], and [C]), it is now apparent that strategy (C) (i.e., especially using dendrons/dendrimers) offers a very broad and versatile approach to quantized nanoscale building blocks [7–10]. In fact, dendrons/dendrimers may be viewed as nanoscale monomers suitable for the synthesis of more complex precise nanostructures referred to as *megamers* that will be discussed later (Figure 24.5).

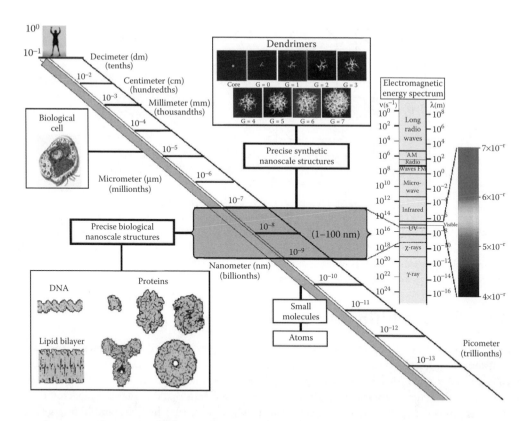

FIGURE 24.3 Nanoscale dimensional comparison of poly(amidoamine) dendrimers [nh$_3$ core] (gen = 0–7) with biological building blocks (i.e., cell, proteins, DNA, lipid bilayer), small molecules, atoms, and electromagnetic energy spectrum.

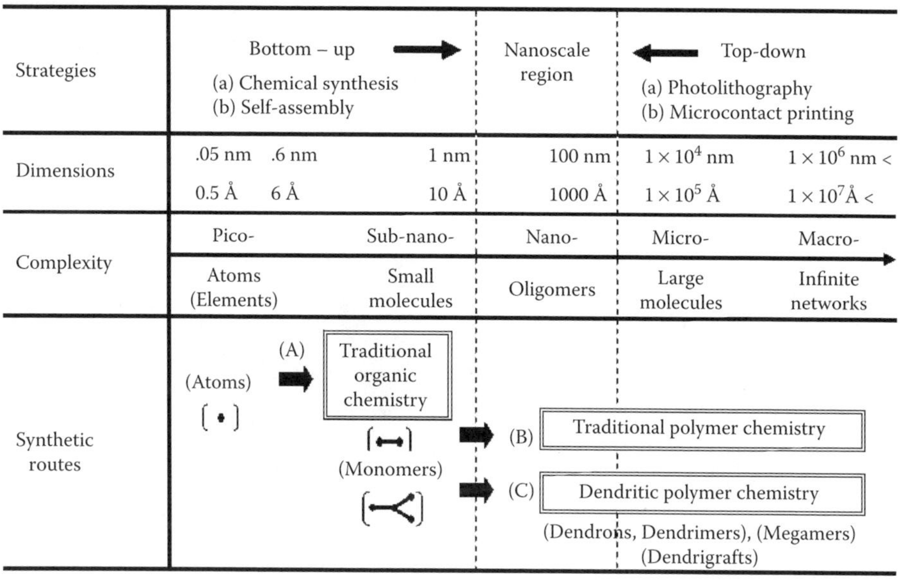

Strategies	Bottom – up ➡ (a) Chemical synthesis (b) Self-assembly			Nanoscale region	⬅ Top-down (a) Photolithography (b) Microcontact printing	
Dimensions	.05 nm 0.5 Å	.6 nm 6 Å	1 nm 10 Å	100 nm 1000 Å	1×10^4 nm 1×10^5 Å	1×10^6 nm < 1×10^7 Å <
Complexity	Pico- Atoms (Elements)	Sub-nano- Small molecules		Nano- Oligomers	Micro- Large molecules	Macro- ➡ Infinite networks

Synthetic routes

(Atoms) [•] (A) ➡ Traditional organic chemistry

[↔] (Monomers) ➡ (B) Traditional polymer chemistry

[◁] ➡ (C) Dendritic polymer chemistry

(Dendrons, Dendrimers), (Megamers) (Dendrigrafts)

FIGURE 24.4 Molecular complexity as a function of covalent synthesis strategies and molecular dimensions.

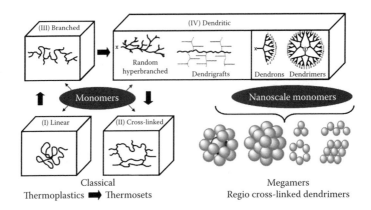

FIGURE 24.5 A comparison of traditional polymer architectures (i.e., (I) linear, (II) cross-linked, (III) branched with (IV) dendritic) and their relationship in the transition from the thermoplastic to the thermoset state. The derivation of all architectural classes from classical monomers is noted, whereas, dendrons/dendrimers may function as nanoscale monomers to megamers.

24.2 The Dendritic State

Dendritic architecture is one of the most pervasive topologies observed at the micro- and macro-dimensional length scales (i.e., μm-m). At the nanoscale (molecular) level, relatively few examples of this architecture are known. Most notable are glycogen and amylopectin hyperbranched structures that nature uses for biological energy storage. Presumably, the many-chain ends that decorate these macromolecules facilitate enzymatic access to the glucose components when needed for high demand bioenergy events [11]. Another nanoscale example of dendritic architecture in biological systems is found in proteoglycans. These macromolecules appear to provide energy-absorbing, cushioning properties and determine the viscoelastic properties of connective tissue (Figure 24.6).

In the past two decades, versatile chemistry strategies have been developed that allow the design, synthesis, and functionalization of a wide variety of such dendritic structures.

24.2.1 A Comparison of Organic Chemistry and Traditional Polymer Chemistry with Dendritic Macromolecular Chemistry

Beginning with Wöhler in 1828, organic chemistry has led to the synthesis of literally millions of small molecules. Organic synthesis involves formation of molecular structure by using the three well-known

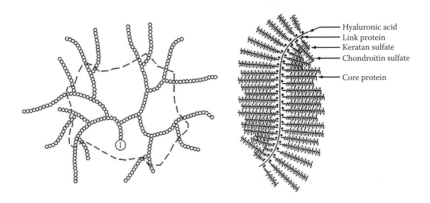

FIGURE 24.6 Topologies for (a) amylopectin and (b) proteoglycans.

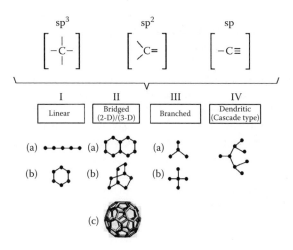

FIGURE 24.7 Four major small molecular architectures derived from the hybridization states of carbon.

hybridization states of carbon, combined with specific heteroatoms to produce key hydrocarbon building blocks (modules) or functional groups (connectors). These two construction parameters have been used to assemble literally millions of more complex structures by either *divergent* or *convergent* strategies involving a limited number of stepwise, covalent bond-forming events [12]. Relatively small molecules (i.e., <1 nm) are produced, allowing the precise control of shape, mass, flexibility, and functional group placement; wherein, product isolation is involved at each stage. The divergent and convergent strategies are recognized as the essence of traditional organic synthesis.

Based on the various hybridization states of carbon (Figure 24.7), at least four major carboskeletal architectures are recognized [12–14] and include: (I) *linear*, (II) *bridged* (two-dimensional/three-dimensional), (III) *branched*, and (IV) *dendritic* (i.e., cascade type) [12,13]. It should be noted that buckminsterfullerenes, a subset of Class (II), bridged (three-dimensional) structures and cascade molecules, a low-molecular-weight subset of Class (IV) dendritic architectures, are relatively recent examples of organic small molecule topologies. The former recognized as *bucky balls*, and *carbon nanotubes* have enjoyed considerable attention as precise, hydrophobic nanomodules suitable for applications in the dry nanotechnology world. The low-molecular-weight dendritic architectures [14] have been used as precursors to true dendrimer nanostructures [15,16] as will be described later.

Over the past 70 years, traditional polymer chemistry has evolved based on the catenation of reactive small molecular modules or monomers. Broadly speaking, these propagations involve the use of reactive (AB-type monomers) that may be covalently bonded to produce a variety of large, nanoscale molecules with polydispersed masses. Such multiple bond-formation strategies may be driven by a variety of mechanisms including: (1) *chain growth*, (2) *ring opening*, (3) *step-growth condensation*, or (4) *enzyme-catalyzed processes*. Staudinger [17–19] first introduced this paradigm in the 1920s by demonstrating that reactive monomers could be used to produce a statistical distribution of one-dimensional (linear) molecules with very high molecular weights (i.e., >10^6 Da). As many as 10,000 or more covalent bonds may be formed in a single-chain reaction of monomers. Although macro/mega-molecules with nanoscale dimensions may be attained, structure control of critical macromolecular design parameters such as *size*, *molecular shape*, *positioning of atoms*, or covalent connectivity — other than those affording linear or cross-linked topologies — is difficult. However, recent progress has been made using "living polymerization" techniques which afford better control over molecular weight and some structural elements as described elsewhere [20,21].

$$n[AB] \, (\text{monomers}) \rightarrow [AB]_n$$

Traditional polymerizations usually involve AB-type monomers based on substituted ethylenes or strained small ring compounds. These monomers may be propagated by using chain reactions initiated by free radical, anionic, or cationic initiators [22]; or alternatively, AB-type monomers may be used in polycondensation reactions (Figure 24.5) [6].

These polymerization strategies generally produce linear architectures, however, branched topologies may be formed either by chain transfer processes, or intentionally introduced by grafting techniques. In any case, the linear and branched architectural classes have traditionally defined the broad area of *thermoplastics*. Of equal importance is the major architectural class that is formed by the introduction of covalent bridging bonds between linear or branched polymeric topologies. These cross-linked (bridged) topologies were studied by Flory in the early 1940s and constitute the second major area of traditional polymer chemistry, namely *thermosets*.

Therefore, approximately 50 years after the introduction of the "macromolecular hypothesis" by Staudinger [19], the entire field of polymer science was simply viewed as consisting of only two major architectural classes: (I) *linear topologies* as found in thermoplastics and (II) *cross-linked architectures* as found in thermosets [5,22]. The major focus of polymer science between the 1920s and the 1970s was on architecturally driven properties manifested by either linear or cross-linked topologies. Based on unique properties exhibited by these two architectural types, many natural polymers critical to success in World War II were replaced with synthetic polymers [5]. In the 1960s and 1970s, pioneering investigation into long chain branching in polyolefins and other related branching systems began to emerge [23,24]. More recently, intense commercial interest has been focused on new polyolefin architectures based on *random long branched* and *dendritic topologies* [25]. These architectures are produced by *metallocene* and *Brookhart-type* catalysts. In summary, by the end of the 1970s, only three major architectural classes of polymers were recognized and referred to as classical polymers (Figure 24.5).

In fact, it is now recognized that these three open-assembly topologies (i.e., linear, branched, dendritic) represent a graduated continuum of architectural intermediacy between thermoplastic and thermoset behavior [26]. In the past several years, the emergence of *dendritic polymers* as a fourth major class of polymer architecture has become widely recognized (Figure 24.5) [15].

In summary, traditional organic chemistry offers exquisite structure control over a wide variety of compositions up to, but *not including higher nanoscale structural dimensions* (e.g., buckminsterfullerene diameters $\cong 1\,nm$). Furthermore, such all-carbon nanostructures are limited to only one hydrophobic compositional form and are hampered by the difficulties of functionalizing their surfaces to produce hydrophilic structures [3].

Classical polymer chemistry offers relatively little structural control, but facile access to statistical distributions of polydispersed nanoscale structures. Living polymerizations provide slightly better, but still imperfect control over product size and mass distribution or polydispersity [21]. In contrast, as will be seen below, dendritic macromolecular chemistry provides essentially all the features required for unparalleled control over *topology, composition, size, mass, shape*, and *functional group placement* [20]. These are features that truly distinguish many successful nanostructures found in nature [27].

The quest for nanostructures and devices based on the biomimetic premise of architectural and functional precision is intense and remains an ultimate challenge. As illustrated in Figure 24.8, the fourth major class of polymer architecture, namely dendritic topologies (i.e., dendrons/dendrimers), is converging with nanotechnology to produce unique new "dendritic effects" and "nanoscale effects" that are emerging in new properties, products, therapies, and diagnostics (Figure 24.8). The rest of this chapter will attempt to provide an overview of key features of the dendritic state that address issues of interest to the nanoscientist.

24.2.2 Dendritic Polymers — A Fourth Major New Architectural Class

Dendritic topologies are recognized as a fourth major class of macromolecular architecture [28–32]. The signature for such a distinction is the unique repertoire of new properties exhibited by this class of polymers. These new architecturally driven properties are referred to as *dendritic effects* [10]. Many of these unique dendritic effect properties are illustrated in Figure 24.9 as they are compared to "linear

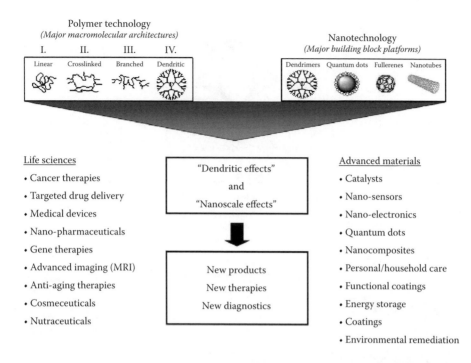

FIGURE 24.8 Convergence of polymer technology with nanotechnology to produce "dendritic effects" and "nano-scale effects."

polymer" properties. Innumerable new synthetic strategies have been reported for the preparation of these materials, thus providing access to a broad range of dendritic structures. Presently, this architectural class consists of three dendritic subclasses, namely: (IVa) *random hyperbranched polymers*, (IVb) *dendrigraft polymers*, and (IVc) *dendrons/dendrimers* (Figure 24.5 and Figure 24.10). This subset order (IVa to IVc) reflects the relative degree of structural control present in each of these dendritic architectures (Figure 24.10).

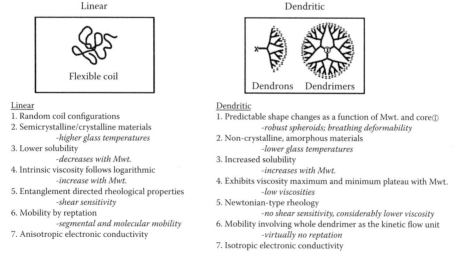

Linear
1. Random coil configurations
2. Semicrystalline/crystalline materials
    -higher glass temperatures
3. Lower solubility
    -decreases with Mwt.
4. Intrinsic viscosity follows logarithmic
    -increase with Mwt.
5. Entanglement directed rheological properties
    -shear sensitivity
6. Mobility by reptation
    -segmental and molecular mobility
7. Anisotropic electronic conductivity

Dendritic
1. Predictable shape changes as a function of Mwt. and core①
    -robust spheroids; breathing deformability
2. Non-crystalline, amorphous materials
    -lower glass temperatures
3. Increased solubility
    -increases with Mwt.
4. Exhibits viscosity maximum and minimum plateau with Mwt.
    -low viscosities
5. Newtonian-type rheology
    -no shear sensitivity, considerably lower viscosity
6. Mobility involving whole dendrimer as the kinetic flow unit
    -virtually no reptation
7. Isotropic electronic conductivity

FIGURE 24.9 Comparison of unique dendritic architecture effects (properties) with traditional linear polymer properties.

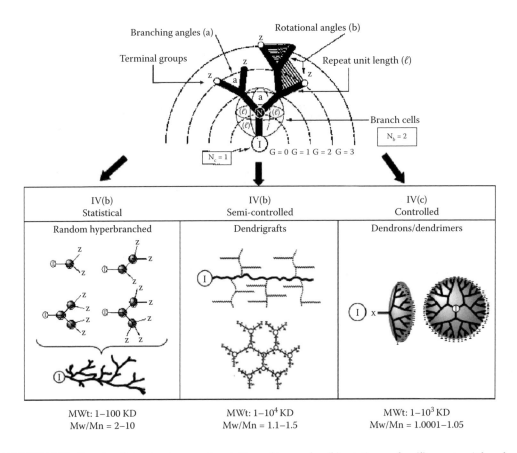

FIGURE 24.10 Branch cell structural parameters: (a) branching angles, (b) rotation angles, (*l*) repeat unit lengths, (Z) terminal groups and dendritic subclasses derived from branches, (IVa) random hyperbranched, (IVb) dendrigrafts, and (IVc) dendrons/dendrimers.

All dendritic polymers are open, covalent nano-assemblies of branch cells (BCs). They may be organized as very symmetrical, monodispersed arrays, as is the case for dendrimers, or as irregular polydispersed assemblies that typically define random hyperbranched polymers. As such, the respective subclasses and the level of structure control are defined by the propagation methodology, as well as by the BC construction parameters used to produce these assemblies. These BC parameters are determined by the composition of the BC monomers, as well as the nature of the "excluded volume" defined by the BC. The excluded volume of the BC is determined by the length of the arms, the symmetry, rigidity/flexibility, as well as by the branching and rotation angles involved within each of the BC domains [33]. As shown in Figure 24.10, these dendritic arrays of BCs usually manifest covalent connectivity relative to some molecular reference marker (I) or core. As such, these BC arrays may be very nonideal/polydispersed (e.g., $M_w/M_n \cong 2$ to 10), as observed for random hyperbranched polymers (IVa), or very ideally organized into highly controlled, core–shell-type structures as noted for dendrons/dendrimers (IVc): $M_w/M_n \cong 1.01$ to 1.0001. Dendrigraft (arborescent) polymers reside between these two extremes of structure control, frequently manifesting rather narrow polydispersities of M_w/M_n 1.1 to 1.5 depending on their mode of preparation.

24.2.3 Dendrons/Dendrimers

Dendrons and dendrimers are the most intensely investigated subset of dendritic polymers. In the past decade, over 8000 literature references have appeared on this unique class of structure-controlled polymers. The word "dendrimer" is derived from the Greek words *dendri* (branch tree-like) and *meros* (part of).

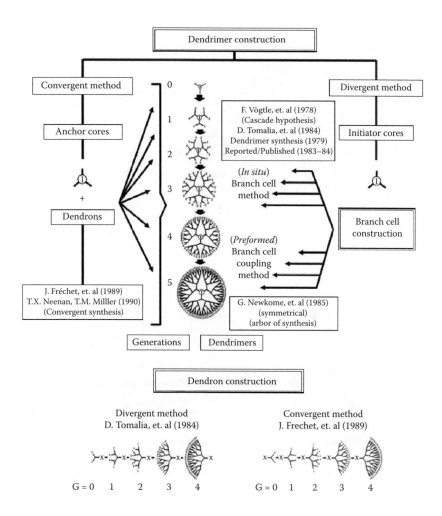

FIGURE 24.11 Overview of synthetic strategies for (a) branch cell construction, (b) dendron construction, and (c) dendrimer construction annotated with discovery scientists.

The term was coined by Tomalia et al. [15,16,34] over 20 years ago in the first full paper describing Tomalia-type poly(amidoamine) (PAMAM) dendrimers. PAMAM dendrimers constitute the first dendrimer family to be commercialized and represents the most extensively characterized and best understood series at this time. In view of the vast amount of literature in this field, the remaining overview will focus on PAMAM dendrimers with a particular focus on their use as (i) functional nanomaterials in biomedical, chemical, and environmental applications and (ii) reactive modules for the synthesis of more complex nanoscale architectures (e.g., megamers, core–shell tecto(dendrimers), etc.).

24.2.3.1 Synthesis — Divergent and Convergent Methods

In contrast to traditional polymers, dendrimers are unique core–shell structures possessing three basic architectural components, namely: (I) *a core*, (II) *an interior of shells* (*generation*) consisting of repetitive BC units, and (III) *terminal functional groups* (i.e., the outer shell or periphery) as illustrated in Figure 24.11 and Figure 24.12.

In general, synthesis of dendrimers involves hierarchical assembly strategies that require the following construction components:

Cores : Monomers → Branch cells → Dendrons → Dendrimers

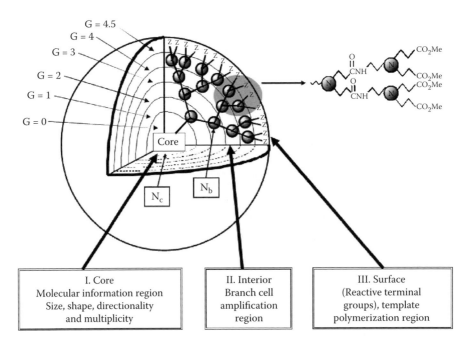

I. Core	II. Interior	III. Surface
Molecular information region Size, shape, directionality and multiplicity	Branch cell amplification region	(Reactive terminal groups), template polymerization region

FIGURE 24.12 Three-dimensional projection of dendrimer core-shell architecture for G = 4.5 poly(amidoamine) (PAMAM) dendrimer with principal architectural components (I) core, (II) interior, and (III) surface.

Many methods for assembling these components have been reported; however, they can be broadly categorized as either "divergent" or "convergent" strategies. Within each of these major approaches, there may be variations in methodology for "BC construction" (i.e., *in situ* vs. *preformed*) or dendron construction (i.e., *divergent* vs. *convergent*), as overviewed in Figure 24.11.

Historically, early developments in the field were based on the "divergent methods." Vögtle and coworkers [14] first reported the synthesis of several low-molecular-weight (<900 Da; G = 0 to 2) cascade structures using the divergent, "in situ" BC method. This synthesis was based on a combination of acrylonitrile and reduction chemistry. As Moors and Vögtle [35] reported later, higher-generation cascade structures and indeed dendrimers could not be obtained by this process due to synthetic and analytical difficulties. Nearly simultaneously, a completely characterized series of high-molecular-weight (i.e., >58,000 Da; G = 0 to 7) Tomalia-type PAMAM dendrimers were synthesized [15,16,34]. Success with their approach was based on "click chemistry" as now defined by Sharpless et al. [36,37]. It involved a two-step iterative reaction sequence consisting of acrylate Michael addition and amidation chemistry [34,38–42]. This methodology provided the first commercial route to dendrimers, as well as the first opportunity to observe unique dendrimer property development that occurs *only* at higher generations (i.e., G = 4 or higher). Many of these observations were described in a publication that appeared in 1985 [34,40] and later reviewed extensively in 1990 [41,42].

The first published use of "preformed BC" methodology was reported in a communication by Newkome et al. [43]. This approach involved the coupling of preformed BC reagents around a core to produce low-molecular-weight (i.e., <2000 Da, G = 3) "arborol structures." This approach has been used to synthesize many other dendrimer families including *dendri-poly(ethers)* [44], *dendri-poly(thioethers)* [45], and others [20,46]. Each of these methods involved the systematic divergent growth of BCs around a core that defined shells within the "dendrons." The multiplicity and directionality of the initiator sites (N_c) on the core determine the number of dendrons and the ultimate shape of the dendrimer. In essence, dendrimers propagated by this method constitute groups of molecular trees (i.e., two or more dendrons) that are propagated outwardly from their roots (cores). This occurs in stages (generations); wherein, the functional leaves of these trees become reactive precursor templates (scaffolding) upon which to assemble the next generation of branches. This methodology can be used to produce multiples of trees (dendrimers) or single trees (dendrons) as shown in Figure 24.11.

Using a totally novel approach, Hawker and Fréchet [47] followed by Miller and Neenan [48] reported the "convergent" construction of such molecular trees by first starting with the leaves or surface BC reagents. By amplifying with these reagents in stages (generations), one produces a dendron possessing a single reactive group at the root or focal point of the structure. Subsequent coupling of these reactive dendrons through their focal point to a common "anchoring core" yields the corresponding dendrimers. Because of the availability of orthogonal functional groups at the focal point and periphery of the dendrons, the convergent synthesis is particularly useful for the preparation of more complex macromolecular architectures [49] such as linear dendritic hybrids, block copolymeric dendrimers, or dendronized polymers. Another significant difference is that the divergent approach requires an exponential increase in the number of coupling steps for generation growth, whereas the convergent method involves only a constant number of reactions (typically two to three) at each stage of the synthesis. Today, several hundred reports utilize the original poly(ether), Fréchet-type dendron [47] method, making this the best understood and structurally the most precise family of convergent dendrimers.

Overall, each of these dendrimer construction strategies offer their respective advantages and disadvantages. Some of these issues, together with experimental laboratory procedures, are reviewed elsewhere [20].

24.2.3.2 Dendrimer Features of Interest to Nanoscientists

Dendrimers may be viewed as unique, information processing, nanoscale devices. Each architectural component manifests a specific function, while at the same time defining properties for these nanostructures as they are grown generation by generation (Figure 24.12). For example, the *core* may be thought of as the molecular information center from which *size*, *shape*, *directionality*, and *multiplicity* are expressed via the covalent connectivity to the outer shells. Within the *interior*, one finds the *BC amplification region*, which defines the type and amount of interior void space that may be enclosed by the terminal groups as the dendrimer is grown. BC multiplicity (N_b) determines the density and degree of amplification as an exponential function of generation (G). The interior composition and amount of solvent filled void space determines the extent and nature of guest–host (endo-receptor) properties that are possible within a particular dendrimer family and generation. Finally, the surface consists of reactive or passive terminal groups that may perform several functions. With appropriate function, they serve as a *template polymerization region* as each generation is amplified and covalently attached to the precursor generation. Secondly, the surface groups may function as passive or reactive gates controlling entry or departure of guest molecules from the dendrimer interior. These three architectural components determine the physical/chemical properties, as well as the overall sizes, shapes, and flexibility of dendrimers. It is important to note that dendrimer diameters increase linearly as a function of shells or generations added, whereas the terminal functional groups increase exponentially as a function of generation. This dilemma enhances "tethered congestion" of the anchored dendrons, as a function of generation, due to the steric crowding of the end groups. As a consequence, lower generations are generally open, floppy structures, whereas higher generations become robust, less deformable spheroids, ellipsoids, or cylinders depending on the shape and directionality of the core.

Tomalia-type PAMAM dendrimers are synthesized by the divergent approach. This methodology involves *in situ* BC construction in stepwise, iterative stages (i.e., generation = 1, 2, 3, ...) around a desired core to produce mathematically defined nanoscale *core–shell* structures. Typically, ethylenediamine ($N_c = 4$) or ammonia ($N_c = 3$) is used as nucleophilic cores and allowed to undergo reiterative two-step reaction sequences involving: (a) exhaustive alkylation of primary amines (Michael addition) with methyl acrylate and (b) amidation of amplified ester groups with a large excess of ethylenediamine to produce primary amine terminal groups as illustrated in Scheme 24.1.

This first reaction sequence on the exposed dendron (Figure 24.12) creates $G = 0$ (i.e., the core BC), wherein the number of arms (i.e., dendrons) anchored to the core is determined by N_c. Iteration of the alkylation/amidation sequence produces an amplification of terminal groups from 1 to 2 with the *in situ*

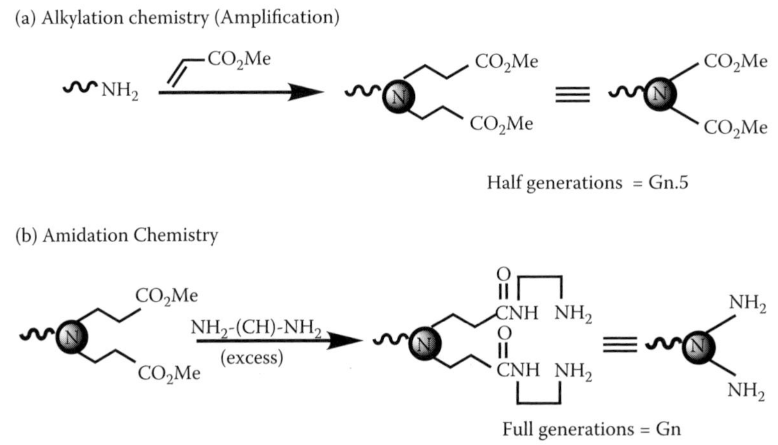

(a) Alkylation chemistry (Amplification)

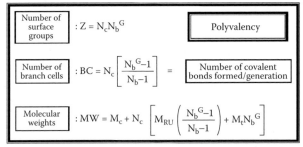

Half generations = Gn.5

(b) Amidation Chemistry

Full generations = Gn

SCHEME 24.1

creation of a BC at the anchoring site of the dendron that constitutes $G = 1$. Repeating these iterative sequences (Scheme 24.1) produces additional shells (generations) of BCs that amplify mass and terminal groups according to the mathematical expressions described in Figure 24.13.

As early as 2001, Nobel Laureate Prof. B. Sharpless popularized a modular approach to organic synthesis that he referred to as click chemistry [36,37]. This strategy was defined in the context of four major organic reaction categories that included the following:

1. Addition of nucleophiles to activated double bonds (i.e., Michael addition chemistry)
2. "Nonaldol"-type carbonyl chemistry (i.e., formation of amides, hydazones, etc.)

Number of surface groups	$: Z = N_c N_b{}^G$		Polyvalency	
Number of branch cells	$: BC = N_c \left[\dfrac{N_b{}^G - 1}{N_b - 1} \right]$	$=$	Number of covalent bonds formed/generation	
Molecular weights	$: MW = M_c + N_c \left[M_{RU} \left(\dfrac{N_b{}^G - 1}{N_b - 1} \right) + M_t N_b{}^G \right]$			

Generation	Surface groups (Z)	Molecular formula	MW	Diameter (nm)
0	4	$C_{22}H_{48}N_{10}O_4$	517	1.4
1	8	$C_{62}H_{128}N_{26}O_{12}$	1,430	1.9
2	16	$C_{142}H_{288}N_{50}O_{28}$	3,256	2.6
3	32	$C_{302}H_{608}N_{122}O_{60}$	6,909	3.6
4	64	$C_{622}H_{1248}N_{250}O_{124}$	14,215	4.4
5	128	$C_{1262}H_{2528}N_{506}O_{252}$	28,826	5.7
6	256	$C_{2542}H_{5088}N_{1018}O_{508}$	58,048	7.2
7	512	$C_{5102}H_{10208}N_{2042}O_{1020}$	116,493	8.8
8	1,024	$C_{10222}H_{20448}N_{4090}O_{2044}$	233,383	9.8
9	2,048	$C_{20462}H_{40928}N_{8186}O_{4092}$	467,162	11.4
10	4,096	$C_{40942}H_{81888}N_{16378}O_{8188}$	934,720	~13.0

FIGURE 24.13 Dentritic branching mathematics for predicting number of dendrimer surface groups, number of branch cells, and molecular weights. Calculated values for [ethylenediamine core]; *dendri-poly*(amidoamine) series with nanoscale diameters (nm).

3. Nucleophilic ring opening of strained heterocyclic electrophiles (i.e., aziridines, epoxides, etc.)
4. Huisgen-type 1,3-dipolar cycloaddition of azides to alkynes.

It should be noted that the first three reaction categories of click chemistry have been used by Tomalia [34,41] and Vögtle [14] as preferred iterative synthetic routes to the first reported examples of dendrimers and low-molecular-weight cascade molecules, respectively.

In 1968, Huisgen [50] reported the facile, high-yield, chemoselective cycloaddition of organic azides with alkynes to form covalent 1,4-disubstituted 1,2,3-triazole linkages. More recently, Sharpless et al. [37,51] have shown that terminal alkynes may be catalyzed by Cu^{1+} salts in an orthogonal fashion to form the corresponding triazoles in very high yields. Because of the high chemoselectivity of these reactions, they may be selectively performed in the presence of a wide variety of competing or parallel reactions/functionalities without interference. These features make this approach very attractive for dendrimer syntheses. Click chemistry based on these copper-catalyzed Huisgen reactions has been used recently to synthesize dendrimers [51–53], dendronized linear polymers [54], and other dendritic architecture [55].

It is apparent that both the core multiplicity (N_c) and BC multiplicity (N_b) determine the precise number of terminal groups (Z) and mass amplification as a function of generation (G). One may view those generation sequences as quantized polymerization events. The assembly of reactive monomers [38,41], BCs [20,41,46], or dendrons [20,47,56] around atomic or molecular cores to produce dendrimers according to divergent/convergent dendritic branching principles has been well demonstrated. Such systematic filling of space around cores with BCs, as a function of generational growth stages (BC shells), to give discrete, quantized bundles of mass has been shown to be mathematically predictable (Figure 24.13) [45,57]. Predicted molecular weights have been confirmed by mass spectroscopy [58–60] and other analytical methods [41,47,61–63]. Predicted numbers of BCs, terminal groups (Z), and molecular weights as a function of generation for an ethylenediamine core ($N_c = 4$) PAMAM dendrimer are shown in Figure 24.13. It should be noted that the molecular weights approximately double as one progresses to the next generation. The surface groups (Z) and BCs amplify mathematically according to a power function, thus producing discrete, monodispersed structures with precise molecular weights and nanoscale diameter enhancement, as described in Figure 24.13. These predicted values are routinely verified by mass spectroscopy for the earlier generations (i.e., $G = 0$ to 5); however, with divergent dendrimers, minor mass defects are often observed for higher generations as congestion induced *de Gennes dense packing* begins to take affect (Figure 24.14).

24.2.3.3 Dendrimer Shape Changes — A Nanoscale Molecular Morphogenesis

As illustrated in Figure 24.14, dendrimers undergo *congestion-induced* molecular shape changes from flat, floppy conformations to robust spheroids as first predicted by Goddard and coworkers [64]. Shape change transitions were subsequently confirmed by extensive photophysical measurements, pioneered by Turro et al. [65–68] and solvatochromic measurements by Hawker et al. [69]. Depending upon the accumulative core and BC multiplicities of the dendrimer family under consideration, these transitions were found to occur between G = 3 and 5. Ammonia core PAMAM dendrimers ($N_c = 3$, $N_b = 2$) exhibited a molecular morphogenesis break at G = 4.5, whereas the ethylenediamine (EDA) PAMAM dendrimer family ($N_c = 4$, $N_b = 2$) manifested a shape change break around $G = 3$ to 4 [64] and the Fréchet-type convergent dendrons ($N_b = 2$) around $G = 4$ [69]. It is readily apparent that increasing the core multiplicity from $N_c = 3$ to 4 accelerates congestion and forces a shape change at least one generation earlier. Beyond these generational transitions, one can visualize these dendrimeric shapes as nearly spheroidal or slightly ellipsoidal *core–shell-type architectures*.

24.2.3.4 de Gennes Dense Packing — A Nanoscale Steric Phenomenon

As a consequence of the excluded volume associated with the core, interior, and surface BCs, steric congestion is expected to occur due to tethered connectivity to the core. Furthermore, the number of dendrimer surface groups, Z, amplifies with each subsequent generation (G). This occurs according to

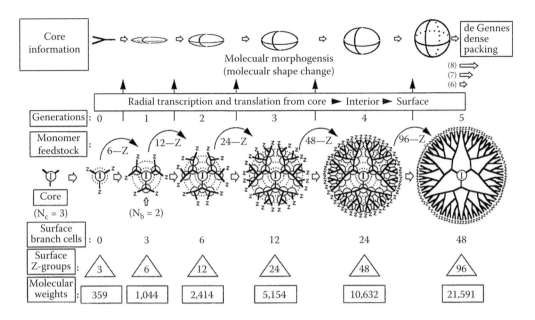

FIGURE 24.14 Comparison of molecular shape change, two-dimensional branch cell amplification surface branch cells, surface groups (Z) and molecular weights as function of generation: G = 0-6.

geometric branching laws, which are related to core multiplicity (N_c) and BC multiplicity (N_b). These values are defined by the following equation:

$$Z = N_c N_b{}^G$$

Since the radii of the dendrimers increase in a linear manner as a function of G, whereas the surface cells amplify according to $N_c N_b{}^G$, it is implicit from this equation that generational reiteration of BCs ultimately will lead to a so-called "dense-packed state."

As early as 1983, de Gennes and Hervet [15,70] proposed a simple equation derived from fundamental principles, to predict the dense-packed generation, for PAMAM dendrimers. It was predicted that at this generation, ideal branching can no longer occur since the available surface space becomes too limited for the mathematically predicted number of surface branch cells to occupy. This produces a "closed geometric structure." The surface is "crowded" with exterior groups, which although potentially chemically reactive are sterically prohibited from participating in ideal dendrimer growth.

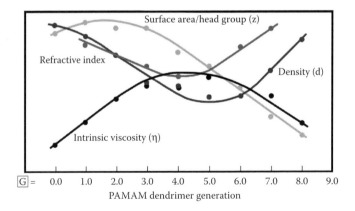

FIGURE 24.15 Comparison of surface area/head group (Z), refractive index, density (d), and intrinsic viscosity (η) as a function of generation: G = 1-9.

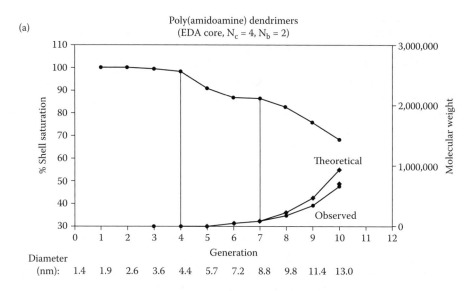

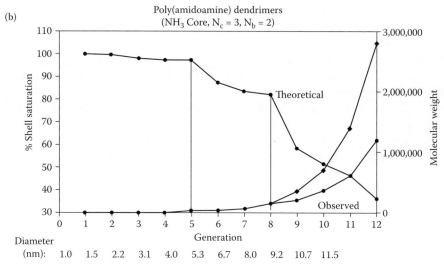

FIGURE 24.16 (a) Comparison of theoretical/observed molecular weights and % shell filling for EDA core poly(amidoamine) (PAMAM) dendrimers as a function of generation: G = 1-10. (b) Comparison of theoretical/observed molecular weights and % shell filling for NH₃ core poly(amidoamine) (PAMAM) dendrimers as a function of generation: G = 1-12.

This "critical packing state" does not preclude further dendrimer growth beyond this point in the genealogical history of the dendrimer preparation. On the contrary, although continuation of dendrimer step-growth beyond the dense-packed state cannot yield structurally ideal, next-generation dendrimers, it can nevertheless occur, as indicated by further increases in the molecular weight of the resulting products. Predictions by de Gennes and Hervet [70] suggested that the PAMAM dendrimer series should reach a critical packing state at generations 9 to 10. Experimentally, we observed a moderate molecular weight deviation from predicted ideal values beginning at generations 4 to 7 (Figure 24.16). This digression became very significant at generations 7 to 8 as dendrimer growth was continued to generation 12 [28]. The products thus obtained are of "imperfect" structure because of the inability of all surface groups to undergo further reaction. Presumably a fraction of these surface groups remain trapped or are sterically encumbered under the surface of the newly formed dendrimer shell, yielding a unique architecture possessing two types of terminal groups. This new surface group population will consist of both those groups that are accessible to subsequent reiteration reagents and those that will be sterically screened. The total number of these groups will not, however, correspond to the predictions of the mathematical branching law, but will fall between that value which was mathematically predicted for the next generations (i.e., $G + 1$) and that expected for the precursor generation (G). Thus, a mass-defective dendrimer "generation" is formed.

Dendrimer surface congestion can be appraised mathematically as a function of generation, from the following simple relationship:

$$A_z = \frac{A_D}{N_Z} \alpha \frac{r^2}{N_c N_b^G}$$

where A_Z is the surface area per terminal group Z, A_D the dendrimer surface area, and N_Z the number of surface groups Z per generation. This relationship predicts that at higher generations G, the surface area per Z group becomes increasingly smaller and experimentally approaches the cross-sectional area or van der Waals dimension of the surface groups Z. The generation G thus reached is referred to as the "*de Gennes" dense-packed generation* [20,41,45]. Ideal dendritic growth without branch defects is possible only for those generations preceding this dense-packed state. This critical dendrimer property gives rise to self-limiting dendrimer dimensions, which are a function of the BC segment length (l), the core multiplicity N_c, the BC juncture multiplicity N_b, and the steric dimensions of the terminal group Z (Figure 24.10). Although the dendrimer radius r in the above expression is dependent on the BC segment lengths l, large l values delay this congestion. In contrast, larger N_c, N_b values, and larger Z dimensions dramatically hasten it.

Additional physical evidence supporting the development of congestion as a function of generation is shown in the composite comparison of dendrimer property changes as illustrated in Figure 24.15. Plots of intrinsic viscosity [η] [41,71], density z, surface area per Z group (A_Z), and refractive index n as a function of generation clearly show maxima or minima at generations 3 to 5, paralleling computer-assisted molecular-simulation predictions [64,72], as well as extensive photochemical probe experiments reported by Turro et al. [65–68].

The intrinsic viscosities [η] increase in a very classical fashion as a function of molar mass (generation), but decline beyond a certain generation because of change from an extended to a globular shape [41,64,72]. In effect, once this critical generation is reached, the dendrimer begins to act more like an Einstein spheroid. The intrinsic viscosity is a physical property that is expressed in dL/g (i.e., the ratio of volume to mass). As the generation number increases and transition to a spherical shape takes place, the volume of the spherical dendrimer roughly increases in cubic fashion, whereas its mass increases exponentially, hence the value of [η] must decrease once a certain generation is reached. This prediction has now been confirmed experimentally [41,71].

The dendrimer density z (atomic mass units per unit volume) clearly minimizes between generations 4 and 5. It then begins to increase as a function of generation due to the increasingly larger, exponential accumulation of surface groups. Since refractive indices are directly related to density parameters, their values minimize and parallel the above density relationship.

Clearly, this de Gennes dense-packed congestion would be expected to contribute to (a) sterically inhibited reaction rates and (b) sterically induced stoichiometry [41]. Each of these effects was observed experimentally at higher generations. The latter would be expected to induce dendrimer mass defects at higher generations which we have used as a diagnostic signature for appraising the *de Gennes dense-packing* effect.

Theoretical dendrimer mass values were compared to experimental values by performing electrospray and MALDI-TOF mass spectral analysis on the respective PAMAM families (i.e., $N_c = 3$ and 4) [59]. Note that there is essentially complete shell filling for the first five generations of the (NH_3) core ($N_c = 3$, $N_b = 2$) PAMAM series (Figure 24.16b). A gradual digression from theoretical masses occurs for $G = 5$ to 8, followed by a substantial break (i.e., $\Delta = 23\%$) between $G = 8$ and 9. *This discontinuity in shell saturation is interpreted as a signature for de Gennes dense packing.* It should be noted that shell saturation values continue to decline monotonically beyond this breakpoint to a value of 35.7% of theoretical at $G = 12$. A similar trend is noted for the EDA core, PAMAM series ($N_c = 4$, $N_b = 2$), however, the shell saturation inflection point occurs at least one generation earlier (i.e., $G = 4$ to 7, see Figure 24.16a). This suggests that the onset of *de Gennes dense packing* may be occurring between $G = 7$ and 8.

Unique features offered by the "dendritic state" that have no equivalency in classical polymer topologies are found almost exclusively in the dendron/dendrimer subset or to a slightly lesser degree in the dendrigrafts. They include:

1. Nearly complete nanoscale monodispersity
2. The ability to control nanoscale container/scaffolding properties
3. Exponential amplification and functionalization of dendrimer surfaces
4. Nanoscale dimension and shape mimicry of proteins

These features are captured to some degree with *dendrigraft* polymers, but are either absent or present to a vanishing small extent for random *hyperbranched* polymers.

24.3   Unique Dendrimer Properties

24.3.1   Nanoscale Monodispersity

The monodispersed nature of dendrimers has been verified extensively by mass spectroscopy, size exclusion chromatography, gel electrophoresis, and transmission electron microscopy [20,73]. As is always the case, the level of monodispersity is determined by the skill of the synthetic chemist, as well as the isolation/purification methods utilized.

In general, convergent methods produce the most nearly isomolecular dendrimers. This is because the convergent growth process allows purification at each step of the synthesis and eliminates cumulative effects due to failed couplings [47,49]. Appropriately purified, convergent dendrimers are probably the most precise synthetic macromolecules that exist today.

As discussed earlier, mass spectroscopy has shown that PAMAM dendrimers (Figure 24.16a and b) produced by the "divergent method" are very monodisperse and have masses consistent with predicted values for the earlier generations (i.e., $G = 0$ to 5). Even at higher generations, as one enters the de Gennes dense-packed region, the molecular weight distributions remain very narrow (i.e., 1.05) and consistent, in spite of the fact that experimental masses deviate substantially from predicted theoretical values. Presumably, de Gennes dense packing produces a very regular and dependable effect that is manifested by the observed narrow molecular weight distribution.

24.3.2   Nanoscale Container/Scaffolding Properties

Unimolecular container/scaffolding behavior appears to be a periodic property that is specific to each dendrimer family or series. These properties are determined by the size, shape, and multiplicity of the construction components that are used for the core, interior, and surface of the dendrimer (Figure 24.10). Higher multiplicity components and those that contribute to "tethered congestion" will hasten the development of "container properties" or rigid surface scaffolding as a function of generation. Within the PAMAM dendrimer family, these periodic properties are generally manifested in three phases as shown in Figure 24.17.

The earlier generations (i.e., $G = 0$ to 3) exhibit no well-defined interior characteristics, whereas interior development related to geometric closure is observed for the intermediate generations (i.e., $G = 4$ to 6/7). Accessibility and departure from the interior is determined by the *size and gating properties* of the surface groups. At higher generations (i.e., $G = >7$) where de Gennes dense packing is severe, rigid scaffolding properties are observed, allowing relatively little access to the interior except for very small guest molecules. The site-isolation and encapsulation properties of dendrimers have been reviewed recently by Esfand and Tomalia [74] Hecht and Fréchet [75], and Meijer et al. [76].

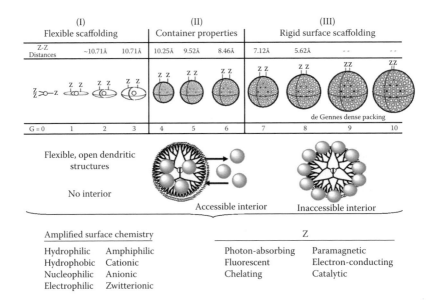

FIGURE 24.17 Periodic properties for poly(amidoamine) (PAMAM) dendrimers as a function of generation G = 0-10. (I) Flexible scaffolding (G = 0-3), (II) container properties (G = 4-6), and (III) rigid surface scaffolding (G = 7-10). Various chemo/physical dendrimer surfaces amplified according to: $Z - N_cN_b G$ where: N_c = core multiplicity, N_b = branch cell multiplicity, G = generation.

24.3.3 Amplification and Functionalization of Dendrimer Surface Groups

Dendrimers within a generational series can be expected to present their terminal groups in at least three different modes, namely: as *flexible, semi-flexible,* or *rigid functionalized scaffolding.* Based on mathematically defined dendritic branching rules (i.e., $Z = N_cN_b{}^G$), the various surface presentations become more congested and rigid as a function of increasing generation level. It is implicit that this surface amplification can be designed to control gating properties associated with unimolecular container development. Furthermore, dendrimers may be viewed as versatile, nanosized objects that can be surface functionalized with a vast array of chemical and application features (Figure 24.14). The ability to control and engineer these parameters provides an endless list of possibilities for utilizing dendrimers as modules for nanodevice designs [28,57,77,78]. Recent reviews have begun to focus on this area [41,75,78–80].

24.3.4 Nanoscale Dimensions and Shape-Mimicking Proteins

In view of the extraordinary structure control and nanoscale dimensions observed for dendrimers, it is not surprising to find extensive interest in their use as globular protein mimics. Based on their systematic, dimensional length scaling properties (Figure 24.18) and electrophoretic/hydrodynamic [61,63] behavior, they are referred to as *artificial proteins* [28,74]. Substantial effort has been focused recently on the use of dendrimers for "site isolation" mimicry of proteins [41], enzyme-like catalysis [81], as well as other biomimetic application [28,82], drug delivery [74,83], surface engineering [84], and light harvesting [85,86]. These fundamental properties have in fact led to their commercial use as globular protein replacements for gene therapy, immunodiagnostics [87,88], and a variety of other biological applications described below.

24.4 Structural and Physical Properties of Dendrimers in Solution

Although dendrimers such as PAMAM dendrimers are being evaluated for a variety of applications ranging from drug delivery to catalysis and gene therapy, their structures and dynamics in solutions are

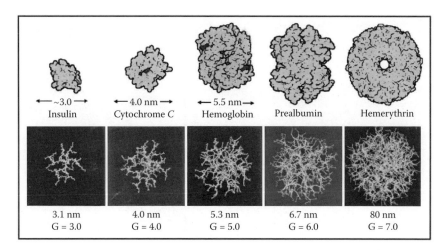

FIGURE 24.18 Comparison of selected proteins showing the close dimensional size/scaling (nm) to respective generations [-ammonia core]; *dendri*-poly(amidoamine) dendrimer.

still the subject of considerable interest in the literature. Key structural and physical properties of dendrimers include size, shape, conformation, solvent penetration, and internal dynamics.

24.4.1 Size and Structure of Dendrimers in Solutions

During the last 10 years, a broad range of experiment probes (e.g., small-angle neutron and x-ray scattering) and computational methods (e.g., molecular dynamics and Monte Carlo simulations) have been used to characterize the size and shape of PAMAM dendrimers [70,72,89–99]. For example, small-angle neutron scattering (SANS) experiments by Topp et al. [90] show that the radius of gyration of PAMAM dendrimers in the series of deuterated solvents [$D(CD_2)OD$ with $m = 0, 1, 2$ and 4] decrease by ~10% with decreasing solvent quality. Monte Carlo simulation studies by Welch and Muthukumar [92] also suggest a twofold increase in dendrimer size as solution pH and ionic strength decrease. Goddard and coworkers [72,98] recently have performed fully atomistic molecular dynamic (MD) simulations to investigate systematically the behavior of EDA core Gx-NH_2 PAMAM dendrimers in the gas phase, as well as in water at various protonation levels in explicit solvent. The estimated radii of gyration of the dendrimers are given in Table 24.1. Note that the radii of gyration of the dendrimers in water agree well with the results of SANS and small-angle x-ray scattering (SAXS) measurements in methanol.

Solution pH has a significant impact on the structures of PAMAM dendrimers. Note that the radii of gyration of the dendrimers increase as solution pH decreases. The estimated solvent accessible surface area and solvent excluded volumes are given in available internal surface area, and volumes also increase as the solution pH increases (Table 24.1). At high pH (i.e., no protonated amine groups), the MD simulations show the presence of smaller pores and cavities inside the PAMAM dendrimers [98]. Conversely, at neutral pH (i.e., protonated primary amine groups) and low pH (i.e., protonated primary and tertiary amine groups), some of these channels open up and connect to the outer surface (Figure 24.19).

24.4.2 Shape, Conformation, and Solvation of Dendrimers

Since the first publication on the de Gennes dense-packing phenomenon in 1983 [70], the conformation of dendrimers such as PAMAM have been the subject of ongoing debate in the literature. Small-angle scattering experiments and computer simulations can give key data on the shape and conformation of dendrimers including (i) fractal dimension, (ii) shape tensor, (iii) monomer radial density profiles, and (iv) location of dendrimer terminal groups. MD simulations [72,94] and SAXS experiments [97] yield fractal dimensions of ~3 for PAMAM dendrimers ($G > 2$). This suggests that these dendrimers have compact and globular structures. Estimates of the three principal moments of inertia (I_1, I_2, and I_3) of

TABLE 24.1 Radii of Gyration (R_g) (Å) for EDA Core Gx-NH$_2$ PAMAM Dendrimers

| | No Solvent | | Water | | | | | | Experiments | | |
| | | | High pH | | Neutral pH | | Low pH | | | | |
G	R_g	R_N	R_g	R_N	R_g	R_N	R_g	R_N	SAXS [2]	SAXS [9]	SANS [3]
4	14.50 ± 0.28	16.81 ± 0.32	16.78 ± 0.15	18.49 ± 0.22	17.01 ± 0.1	18.84 ± 0.12	19.01 ± 0.08	21.20 ± 0.15	17.10	18.60	
5	18.34 ± 0.37	20.26 ± 0.68	20.67 ± 0.09	22.71 ± 0.47	22.19 ± 0.14	24.43 ± 0.07	24.76 ± 0.14	27.38 ± 0.18	24.10	23.07	22.10

Note: R_N is the radius of gyration considering only the location of the primary amine groups [98].

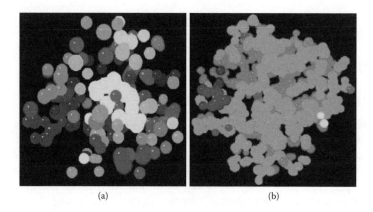

(a)   (b)

FIGURE 24.19   Inner channels and cavities for G5-NH$_2$ PAMAM dendrimer as determined from MD simulations [97]. In the absence of solvent (a), there exists only small pockets. At low pH (b), we see the existence of a single continuous channel.

EDA core Gx-NH$_2$ PAMAM dendrimers show that the dendrimers are more "spherical" in aqueous solutions [98]. Most computer simulations suggest that the terminal groups of PAMAM dendrimers are distributed throughout the macromolecules. This suggests significant folding back of the poly(amidoam-ine) branches of PAMAM dendrimers in sharp contrast to the model by de Gennes and Hervet [70], which suggests that dendrimer terminal groups are preferentially located at the periphery.

Since dendrimers such as PAMAM are being evaluated as *protein mimics*, it is critical to evaluate water penetration into the interior of these macromolecules. Experimental characterization of water behavior inside a dendrimer and near its surface is very difficult and challenging. MD simulations, however, can provide key insights and data on water penetration within a dendrimer. Maiti et al. [98] have used MD simulations to estimate the extent of binding (EOB) of water molecules inside Gx-NH$_2$ EDA core PAMAM dendrimers. For G5-NH$_2$ PAMAM dendrimers, they reported EOB values of 378, 524, and 754 water molecules at high pH (~10), neutral pH (~7.0), and low pH (~4.0). Lin et al. [99] also used MD simulations to characterize water penetration within a G5-NH$_2$ EDA core PAMAM dendrimer at high, neutral, and low pH. They were able to distinguish between three environmentally defined types of water molecules: (a) water buried inside the dendrimer, (b) water residing at the dendrimer–water interface, and (c) bulk water residing outside the dendrimer (Figure 24.20). Lin et al. [99] found that interior and surface water exhibit two differentiated relaxation times: a fast relaxation (~1 ps) assigned to the liberation of water molecules within each domain and a slower relaxation (~20 ps) assigned to the diffusion of water molecules between the two domains. Conversely, the bulk water molecules exhibit slow and fast relaxation times, respectively, of ~0.4 and ~14 ps. Lin et al. [99] were able to estimate the enthalpies and

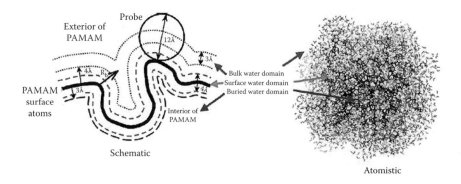

FIGURE 24.20   Water domains in G5-NH$_2$ EDA Core PAMAM dendrimers as characterized by MD simulations [99].

entropies of three classes of water molecules. They found that water molecules located at the surface and inside the G5-NH$_2$ PAMAM dendrimers have excess free energies of 0.1 to 0.2 and 0.4 to 1.3 kcal/mol, respectively. This excess free energy should provide additional driving force for guest uptake by PAMAM dendrimers in aqueous solutions.

24.4.3 Dynamics of Dendrimers in Solutions

The characterization of dendrimer dynamics in solutions is a challenging undertaking. One of the most extensive investigations of the dynamics of this behavior was reported by Rathgeber et al. [96]. They combined neutron spin echo (NSE) spectroscopy with SAXS to assess the relationships between the structures of EDA core Gx-NH$_2$ PAMAM dendrimers ($G = 0$ to 8) in dilute methanol solutions and their dynamics. To interpret these data, Rathgeber et al. [96] formulated a model that describes dendrimers as "colloidal" particles with covalently bonded polymeric chains. This allowed them to probe both the dynamics of the dendrimer particle (e.g., diffusion) and the internal dynamics of its polymeric chains (e.g., self-avoiding random walk). The overall results of the SAXS and NSE data analyses suggest that dendrimer dynamics is dominated by particle center of mass diffusion in dilute solutions of a good solvent such as methanol. For G5-G8 PAMAM dendrimers, the estimated diffusion coefficients from the NSE experiments range from $9.4 \pm 0.4 \times 10^{-7}$ to $5.7 \pm 0.2 \times 10^{-7}$ cm^2/s. These values are close to those calculated values from the Stokes–Einstein relation using SAXS estimated particle diameters.

24.5 Dendrimers as Nanopharmaceuticals and Nanomedical Devices

Many promising biomedical applications for dendrimers are emerging with several currently under development or entering clinical trials. Several major reviews have appeared recently on this area [100,101].

As described earlier, the nature of dendrimer synthesis allows complete control over the molecular weight, chemical composition, and the ability to tailor physical properties (e.g., including pharmacokinetics for a desired therapeutic/biomedical application). Dendrimers offer the possibility of cleaner, monodispersed products; wherein, small molecule criteria as applied in the traditional pharmaceutical world may be applied to large molecules. This property of dendrimers may avoid the purity/side effect issues generally found with traditional polymer therapeutics [102].

24.5.1 Dendrimers as Genetic Material Transfer Agents

Gene regulation has many potential applications in treating disease states such as cancer, tissue graft survival, and multiple drug resistance as well as other conditions where it would be beneficial to reduce or eliminate messages produced from genes causing the adverse effect. A major drawback preventing further exploitation of genetic therapies is the delivery of therapeutic nucleic material to the cell and, in particular, to the cell nucleus. A number of reviews on the subject have been published [103,104].

The application of PAMAM dendrimers as transfection agents for antisense oligonucleotides and plasmids was first described by Tomalia and coworkers in 1996. They found that stable DNA–dendrimer complexes formed and could successfully suppress luciferase expression in an *in vitro* cell culture system [105,106]. Further work [107–109] provided additional evidence that the stability of these DNA–dendrimer complexes was dependent on electrostatic charge of the PAMAM dendrimer. These complexes were also stable to a range of pH values and restriction endonucleases.

Most work on dendrimers for gene delivery focuses on the cationic Tomalia-type PAMAM dendrimers which are commercially available from Dendritic Nanotechnologies, Inc., Mt. Pleasant, MI. The complexation of PAMAM dendrimers with nucleic acids is based on the electrostatic interaction between cationic primary amines of the dendrimer and the anionic phosphates of nucleic acid and lacks any sequence specificity [110,111]. Maximum transfection efficiency occurs when the complexes exhibit a net positive charge (i.e. an excess of primary amines with respect to the nucleic acid phosphates).

Thermally fractured PAMAM dendrimers (i.e., SuperFect®*) [112] (*Registered trademark of Qiagen, Valencia, CA) is a widely applicable transfection agent which is becoming a standard *in vitro* transfection vector for many cell and molecular biologists. However, the *in vivo* application of such dendrimer gene-delivery systems has so far been met with limited success [104] with the exception that Schätzlein *et al.* [113] have allegedly demonstrated that poly(propyleneimine) (PPI) dendrimers are suitable for *in vivo* administration.

Florence and coworkers [114,115] have reported lipidic dendrimers for *in vitro* gene delivery of DNA (double- and single-stranded) and RNA. These dendrimers are composed of cationic lysine components for interaction with nucleic acids and a neutral lipidic core that facilitates membrane lipid-bilayer transport.

24.5.2 Dendrimer–Carbohydrate Conjugates for Polyvalent Recognition

Carbohydrate–protein interactions play a vital role in many biological functions such as cell adhesion, receptor-mediated events, cellular recognition processes, and microbial–host cell interactions. These individual sugar-receptor complexes are often weak and nonselective [116]. Pioneering work by Roy et al., Stoddart et al., Lindhorst et al., and Okada et al. lead to the development of carbohydrate dendrimers (see review by Turnbull and Stoddart [117]). These carbohydrate-linked dendrimers possess several sugar groups attached to the outside of the dendrimer shell. This leads to a greatly enhanced affinity and selectivity of the sugar-receptor reaction, which is believed to be mediated via multivalent interactions [118]. Such enhanced affinity has been demonstrated for carbohydrate-linked polylysine [119], PAMAM [120], and PPI [121] dendrimers. Many potential biomedical applications exist for glycodendrimers [122] as antivirals [123] to prevent tumor metastases [124], for lectin targeting [125–127], bacterial [128], and anticell adhesion agents in general [129].

24.5.3 Dendrimers as Targeted Drug-Delivery Agents

Therapeutic drug delivery by dendrimers may be achieved in several ways. The dendrimer may act as a nanocontainer to encapsulate therapeutic guest molecules within its structure, or alternatively drugs can be covalently coupled to the surface of the dendrimer. Such unimolecular encapsulation of guest molecules within dendrimers was first described by Tomalia et al. [41,64,74,78] and subsequently by Meijer et al. [130–132]. Many potential applications of dendrimers as drug-delivery vehicles have been recently described [74,82,133]. Duncan and coworkers [134] demonstrated that an anionic terminated PAMAM dendrimer conjugated to cisplatin (producing a dendrimer-platinate) was highly water soluble and released platinum slowly *in vitro*. When tested *in vivo*, the dendrimer-platinate displayed greater antitumor activity than naked cisplatin and exhibited lower toxicity. More recent studies have shown that these PAMAM cisplatin conjugates involve encapsulation within the dendrimer interior rather than exterior attachment [135].

Additionally, the surface of the dendrimer may present covalently attached recognition functionality which is suitable for targeting the dendrimer–guest complexes to a specific biological destination [136]. An early example of this application was shown by Moroder and coworkers [137] wherein antibody-labeled dendrimers containing radioactive boron isotopes were targeted to tumor cells.

A number of groups have used folate-modified dendrimers to selectively target cancer cells demonstrating the potential for these dendrimers to deliver cytotoxic compounds [138–142].

"Bow-tie" dendrimer structures formed from two covalently attached dendrons, one with multiple attachment points for drug molecules and the other for the attachment of solubilizing poly(ethylene oxide) chains, have been studied by Gillies and Fréchet [143].

Baker and coworkers [144] have used RGD-4C peptide to target PAMAM dendrimers to human umbilical vein endothelial cells as a model for targeting the angiogenic vasculature of tumors.

Backer et al. [145] have covalently attached a boronated PAMAM dendrimer to vascular endothelial growth factor (VEGF) as a way of targeting boron neutron capture therapy to the neovasculature of tumors.

24.5.4　Dendrimers as Magnetic Resonance Imaging Contrast Agents

Since the first studies of dendrimer-based metal chelates as magnetic resonance imaging (MRI) contrast agents [146,147], continued work with dendrimer-based gadolinium chelates has been reported extensively by Kobayashi and Brechbiel [148,149]. Contrast agents with a diameter of 3 to 6 nm were observed to excrete through the kidney, offering potential as functional renal contrast agents [150]. In contrast, larger (7 to 12 nm) complexes were retained in circulation and performed as blood pool contrast agents [151]. Larger hydrophilic agents are useful for lymphatic imaging [152], and hydrophobic agents were useful as liver contrast agents [153]. Dendrimer-gadolinium chelates conjugated to monoclonal antibodies were found to function as tumor-specific contrast agents [154]. An example of tumor targeting was first reported by Wiener et al. [155] who uses folate conjugation to PAMAM dendrimer-gadolinium chelates to target tumor cells. These folate-conjugated chelates targeted the complex to folate-binding proteins on the tumor cell surfaces. Such high-affinity folate-binding sites are upregulated on the cell surface of many human cancers of epithelial origin [156].

Margerum et al. [157] have also shown that PAMAM–gadolinium complexes exhibit reduced liver uptake and increased bioavailability when grafted with polyethylene glycol side chains.

Schering AG has developed the first dendritic gadolinium-based contrast agent for MRI to enter clinical trials [158]. The structure of this agent, also referred to as Gadomer-17, consists of 24 macrocyclic cyclen-based gadolinium chelates linked to the 24 amino groups of 12 L-lysine residues at the surface of the dendritic skeleton. As a result of its high molecular weight (17.5 kDa), this compound has excellent blood pool properties — it remains in the intravascular space for a prolonged period of time after i.v. administration, without relevant extravasation. This longer residence time allows "steady-state" measurements in MR angiographic applications (e.g., coronary angiography) to be performed. The quantitative renal elimination, high acute tolerance, and favorable pharmacokinetics [159,160] make Gadomer-17 a promising blood pool agent.

24.5.5　Dendrimers as Antiviral Agents

Dendrimers possessing anionic surface groups have been studied as antiviral agents since the mid-1990s and have been subject of several reviews [83,160].

Reuter et al. [123] first reported the use of PAMAM dendrimers with sialic acid functional groups as effective inhibitors of influenza virus binding to host cell receptors and noted the potential for these compounds to be used therapeutically. The compounds have been tested *in vitro* and *in vivo* using a murine influenza pneumonitis model [161]. Whitesides and coworkers [162] had previously shown that sialic acid containing polyacrylamide inhibitors prevented the attachment of influenza virus to mammalian erythrocytes. They proposed that the underlying mechanism for this inhibition was due to polyvalent interactions coupled with steric stabilization by the dendrimers that inhibit the interaction between the virus and the erythrocyte.

Poly(lysine) dendrimers with sulfonated naphthyl surface groups have been shown to exhibit antiviral activity against enveloped viruses such as herpes simplex virus (HSV) 1 and 2 [163]. The prophylactic efficacy of these dendrimers *in vitro* suggested that they have the potential as topical microbicides — products intended for application to vaginal or rectal mucosa to protect against sexually transmitted infections. Subsequent evaluation of these dendrimers in a mouse model against genital HSV-2 infection led to highly active dendrimer candidates that provide significant protection when applied up to 30 min prior to viral challenge. This is the first report of microbicidal activity by dendrimers *in vivo* [163]. The sulfonated dendrimers inhibit both the early stage of HSV adhesion events by blocking virus entry into cells as well as late-stage HSV replication [164,165].

Witvrouw et al. [166,167] reported the synthesis and evaluation of polyanionic dendrimers for their antiviral effects against human immunodeficiency virus (HIV). Utilizing "time of addition" studies to determine the mode of action, they found that specific PAMAM dendrimers inhibit both viral attachment to cells, as well as the action of HIV viral reverse transcriptase and virally encoded integrase which occur intracellularly during the virus replicative cycle. This is consistent with other polyanionic compounds such as dextran sulfate, heparin, and suramin that inhibit HIV viral attachment by blocking the binding of the viral envelope glycoprotein gp120 to the cellular CD4 receptor of T lymphocytes. Thus, these polyanionic compounds can inhibit virus binding to cells, virus-induced syncytium formation, and, consequently, viral transmission [168,169]. As a result of this work and further lead optimization at Starpharma Pty Ltd, Melbourne, Australia, a poly(lysine) dendrimer with sulfonated naphthyl surface groups (SPL7013) was selected for development as a topical microbicide. In pivotal nonhuman primate studies, SPL7013 gel provided significant protection from infection following vaginal challenge with a chimeric Simian-human immunodeficiency virus (SHIV) [170]. Following these and other preclinical studies [171,172], Starpharma submitted an investigational new drug application (IND) for SPL7013 gel (VivaGel™) (Starpharma Pty Ltd., Melbourne, Australia) with the U.S. Food and Drug Administration (FDA) in June 2003. The first clinical trials were completed in 2004 [173] and represent the first time a dendrimer-based nanopharmaceutical has been tested in humans. A thorough review of the clinical data revealed no evidence of irritation, inflammation, absorption into the blood, or detrimental effect on microflora [174].

In another example of dendrimers with anti-HIV activity, Kensinger et al. [175] have reported a polysulfated galactose-derivatized DAB dendrimer (PS Gal 64mer) that inhibits infection of cultured indicator cells by HIV-1. The surface group of the dendrimer was chosen by using an enzyme-linked immunosorbent assay to determine which of a number of glycosphingolipids was the best ligand for gp120. The level of activity in the viral inhibition assays was comparable to dextran sulfate (50 kDa) with EC_{50} values in the nanomolar range.

24.5.6 Dendrimers as Angiogenesis Inhibitors

Polyanionic dendrimers are claimed as angiogenesis inhibitors in a patent by Holan and Matthews [176]. It was postulated that these dendrimers mimic heparin or heparan sulfate, sequestering fibroblast growth factor, and possibly other growth factors, thereby disrupting the formation of new blood vessels. These large amino-acid-containing structures are nonantigenic when compared with other peptide-based macromolecules [177].

Shaunak et al. [178] report that the application of PAMAM dendrimer conjugates of glucosamine and glucosamine-6-sulfate exhibits immunomodulatory and antiangiogenic properties that prevent scar tissue formation during glaucoma filtration surgery.

Poly(lysine) dendrimers with arginine surface groups, designed to mimic the surface structure of endostatin, have shown antiangiogenic activity in the chicken embryo chorioallantoic membrane assay [179]. Endostatin is an endogenous inhibitor of angiogenesis, the binding of which to heparin or heparan sulfate proteoglycan results in its antiangiogenic activity [180].

Marano et al. [181] used lipid-lysine dendrimers to deliver a sense oligonucleotide (ODN-1) that possesses antihuman VEGF activity. These positively charged dendrimers were shown to deliver ODN-1 and mediate a reduction in VEGF concentration both *in vitro* and *in vivo*. This study also demonstrated that the dendrimers protected ODN-1 from nuclease degradation.

24.6 Dendrimers as Functional Nanomaterials

Dendrimers are perhaps the most versatile category of nanomaterials available to date. These "soft" nanoparticles can be used as hosts for cations, anions, and organic/inorganic solutes. Furthermore, dendrimers may also be used as scaffolds or nanocontainer templates for the preparation of metal-bearing nanoparticles with tunable electronic, optical, and catalytic properties [182,183].

24.6.1 Dendrimers as Chelating Agents for Cations

Chelating agents are widely used in a variety of industrial, environmental, and biomedical processes including selective extractants in hydrometallurgy, high-capacity polymeric ligands for water treatment, contrast agent carriers for magnetic resonance imaging, and templates for the synthesis of metal-bearing nanostructures [184].

Metal ion complexation is an acid–base reaction that depends on several parameters including metal ion size acidity, ligand basicity, and molecular architecture and solution physical–chemical conditions. Three major milestones in coordination chemistry include the discoveries of the hard and soft acids and bases principle, the *chelate effect*, and the *macrocyclic effect* [184]. The emergence of dendrimers may be viewed as a recent significant milestone in ligand architecture and coordination chemistry.

PAMAM dendrimers possess functional nitrogen and amide groups arranged in regular "branched-upon-branched" patterns which are displayed in geometrically progressive numbers as a function of generation level. This high density of nitrogen ligands in concert with the attachment of various functional groups such as carboxyl, hydroxyl, etc. to PAMAM dendrimers make them particularly attractive as high-capacity chelating agents for metal ions including: transition metal ions, lanthanides, and actinides such as Cu(II), Ni(II), Pd(II), Pt(II), Hg(II), Co(II), Ag(I), Au(I), Gd(III), or U(VI) [182,185–199]. Several investigators have used Cu(II) as a probe to characterize the fundamental mechanisms of metal ion uptake by PAMAM dendrimers in aqueous solutions. A broad range of experimental tools have been employed to characterize Cu(II) binding to PAMAM dendrimers including atomic absorption spectroscopy [185–187], potentiometric titrations [192], UV–visible spectroscopy [182,192], MALDI-TOF mass spectrometry [189], EPR spectroscopy [188], and EXAFS spectroscopy [186,190]. DFT calculations [200,201] and thermodynamic modeling [186] have also been used to characterize Cu(II) binding to PAMAM dendrimers. These investigations suggest that the uptake of metal ions such as Cu(II) in dilute aqueous solutions by PAMAM dendrimers involves several processes, including: (1) specific binding (e.g., ionic and/or covalent bonding) to the dendrimers through coordination with their primary/tertiary amine and amide internal groups and (2) nonspecific binding through coordination with water molecules trapped inside the dendrimers.

24.6.2   Dendrimers as Ligands for Anions

Anions are ubiquitous in biological systems; wherein, DNA and enzymes such as carboxypeptidase A are good examples [202]. Anions such as perchlorate (ClO_4^-), pertechnetate (TcO_4^-), chromate (CrO_4^{2-}), arsenate (AsO_4^{3-}), phosphate (HPO_4^{2-}), and nitrate (NO_3^-) have emerged as major water contaminants throughout the world. Although significant research efforts have been devoted to the design and synthesis of selective chelating agents for cation separations [184], anion separations have received relatively limited attention [202]. The design of selective ligands for anions is a very challenging undertaking. Unlike cations, anions have filled orbitals and thus cannot covalently bind to ligands [202,203]. Anions have a variety of geometries (e.g., spherical for Cl^- and tetrahedral for ClO_4^-) which are sensitive to solution pH in many cases. Thus, shape-selective and pH-responsive receptors may be required to effectively target anions. The charge-to-radius ratios of anions are also lower than those of cations [202,203]. Thus, anion binding to ligands through electrostatic interactions tends to be weaker than cation binding. Anion binding and selectivity also depend on anion hydrophobicity and solvent polarity [202,203].

Dendrimers provide unprecedented opportunities for developing new classes of high capacity and selective ligands for anions. Tomalia and coworkers [64] were the first investigators to establish that dendrimers such as PAMAM can encapsulate anions. They used ^{13}C NMR spin-relaxation measurements (T_1) to show that G4 and G5 PAMAM dendrimers with methyl ester terminal groups can encapsulate organic anions such as aspirin and 2,4-dichlorophenoxyacetic acid in chloroform. Subsequently, Twyman et al. [204] showed that PAMAM dendrimers possessing methyl ester terminal groups can bind organic anions such as benzoic acid. Birnbaum et al. [205] have found that PAMAM dendrimers provide ideal building blocks for the development of water-soluble ligands for anions such as AsO_4^{3-}, CrO_4^{2-}, and HPO_4^{2-}. Astruc and coworkers [206,207] also reported the development of polyamidoferrocene dendrimers that can bind anions such as $H_2PO_4^-$, HSO_4^-, and adenosine 5′-triphosphate (ATP) in dichloromethane.

24.6.3   Dendrimers as Unimolecular Micelle Mimics

Since traditional micelles are known to provide a compatible nanoenvironment for the partitioning of organic solutes, aqueous solutions of surfactants above their critical micelle concentration can significantly enhance organic solubility within their interiors. Micellization involves free energies of the order of 10 RT; thus, micelles tend to be dynamic and flexible structures with finite lifetimes [208]. Conversely, dendrimers can be designed and synthesized as stable covalently fixed micelle mimics that can encapsulate and bind organic solutes in aqueous or nonaqueous solutions. The uptake of organic solutes by dendrimers may involve several mechanisms including: (a) partitioning into the dendrimer core/shell, (b) hydrogen bonding to the macromolecule interior or terminal groups, and (c) shape-specific interactions with the macromolecule internal and terminal groups. During the last decade, several investigators have probed the uptake of organic solutes by dendrimers. Pistolis et al. [209] combined UV–VIS absorption and fluorescence spectroscopy to investigate pyrene solubilization in aqueous solutions of G0, G1, and G2 PAMAM dendrimers possessing terminal amino groups. They found that the amount of pyrene solubilized increases linearly with dendrimer generation. Watkins et al. [210] have evaluated the interactions of Red Nile dye (a probe that fluoresces intensely in hydrophobic lipids and organic solvents) with a series of modified PAMAM dendrimers; wherein, their EDA core was replaced with diaminoalkanes derived from various hydrocarbon lengths. Measurements of the probe fluorescence spectra in dilute aqueous solutions of the modified PAMAM dendrimers showed significant emission for PAMAM dendrimer G3(C12) (i.e., generation 3 G3 with a C_{12} diaminoalkane core). Newkome et al. [211] have synthesized a dendrimer with an alkane core and 36 carboxyl terminal groups that can bind hydrophobic probes such as phenol blue, 7-chlorotetracycline, and diphenylhexatriene in aqueous solutions. Hawker et al. [212] also reported the synthesis of a dendrimer with a polyaromatic ether core and 32 carboxyl terminal groups that can solubilize hydrophobic organic compounds such as pyrene and 2,3,6,7-tetranitrofluorenone in aqueous solutions. Kleinman et al. [213] have recently shown that 2-naphthol binds preferentially to the tertiary amine groups within the dendrimer interior. More recently, Caminade and Majoral [214] have described the preparation of water-soluble phosphorous dendrimers that can bind organic solutes.

24.6.4 Dendrimers as Catalysts and Redox Active Nanoparticles

Because of their well-defined size, shape, and molecular composition, dendrimers provide unprecedented opportunities for developing a new category of homogeneous/heterogeneous redox active nanoparticles and catalysts. Vassilev and Ford [215] reported that complexes of Cu(II), Zn(II), and Co(III) with PPI dendrimers catalyze the hydrolysis of *p*-nitrophenyl diphenyl phosphate in zwitterionic-buffered aqueous solutions. During the last decade, a number of redox active dendritic catalysts have been synthesized and characterized including dendrimers with ferrocene terminal groups that can oxidize glucose or reduce nitrates [216–218]. A number of dendritic catalytic systems have also been successfully implemented in continuous membrane reactors [219–222].

The use of dendrimers as templates for the synthesis of metal-bearing nanoparticles with catalytic properties is a very active area of research [182,196–199,223]. These nanoparticles, commonly referred to as dendrimer-encapsulated nanoparticles (DENs), can be efficiently prepared by reactive encapsulation. This involves the complexation of guest metal ions followed by their reduction and immobilization inside a dendrimer host and/or at its surface. The metal ion-dendrimer load and size (i.e., generation) of the dendrimer can be optimized to produce metal-bearing nanoparticles with tunable catalytic/redox activity. DENs can also functionalize with surface groups that provide appropriate solubility in various media (e.g., water and organic solvents) or bind to appropriate surfaces (e.g., silica and titania). Crooks and co-workers [223] have used PAMAM and PPI dendrimers to prepare a variety of metal-bearing nanoparticles (~3 nm) including Pd DENs, Pt DENs, and bimetallic Pd/Pt DENs. They found that these DENs can serve as homogeneous and heterogeneous catalysts for hydrogenation, Heck coupling, or Suzuki coupling reactions.

24.6.5 Applications to Chemical and Environmental Process Engineering

Dendrimers exhibit a number of critical physicochemical properties that make them attractive as separation/reaction media for chemical and environmental process engineering. They can function as high capacity/selectivity and recyclable ligands for toxic metal ions, radionuclides, organic and inorganic solutes/anions in aqueous and nonaqueous solutions. Dendrimers are also providing unprecedented

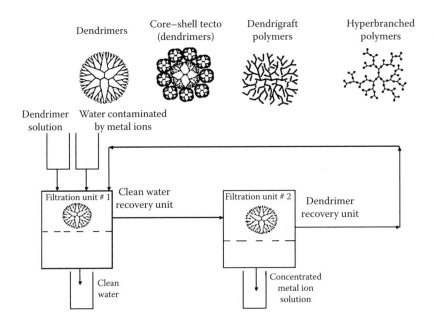

FIGURE 24.21 Recovery of metal ions from aqueous solutions by dendrimer enhanced filtration [188,225].

opportunities to develop more efficient catalysts and redox active media. The combination of functionalized dendrimers with the well-established technology of nanofiltration (NF) and ultrafiltration (UF) is expected to lead to the development of more efficient chemical and environmental processes. In the area of chemical processing, several research groups are exploring the use of dendrimer catalysts to develop a new generation of membrane reactors [220–222]. Work in this area is expected to lead to the development of more efficient and cost-effective and environmentally acceptable processes for the synthesis of commercial products. Dendrimers have exhibited great potential as functional materials for water purification. Diallo has developed a dendrimer-enhanced filtration (DEF) process (Figure 24.21) for recovering metal ions from aqueous solutions [4,40]. Other applications of DEF to water purification are discussed by Diallo [222,224].

24.7   Dendrimers as Reactive Modules for the Synthesis of More Complex Nanoscale Architectures

24.7.1   Nanostructure Control beyond the Dendrimer

Dendrimer synthesis strategies now provide virtual control of macromolecular nanostructures as a function of size [225,226], shape [227,228], and surface or interior functional groups [41]. These strategies involve the covalent assembly of hierarchical components such as reactive monomers (A) [38], BCs (B) [20,45,46], and dendrons (C) [56] around atomic or molecular cores according to divergent or convergent dendritic branching principles (Figure 24.22) [20,45,229].

Systematic filling of space around a core with shells (layers) of BCs (i.e., generations) produces discrete core–shell dendrimer structures. Dendrimers are quantized bundles of mass that possess amplified surface functionality and are mathematically predictable [57]. Predicted molecular weights and surface stoichiometry have been confirmed experimentally by mass spectrometry [58,59], gel electrophoresis [61,63], and other analytical methods [225,230]. It is now recognized that empirical structures (i.e., modules) such as B, C, and D may be used to define these hierarchical constructions. Such synthetic strategies have produced traditional dendrimers with dimensions that extend well into the lower nanoscale region (i.e., 1 to 20 nm) [231]. The precise structure control and unique new properties exhibited by these dendrimeric

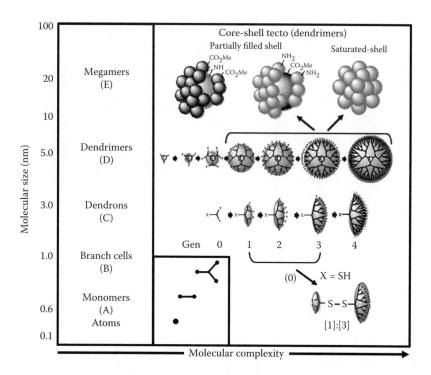

FIGURE 24.22 Approximate nanoscale dimensions as a function of atoms, monomers, branch cells, dendrimers, and megamers.

architectures have yielded many interesting advanced material properties [75,130,232]. Nanoscale dendrimeric containers [130,182,233] and scaffoldings [41,234] have been used to template zero-valent metal nanodomains [79,182], nanoscale magnets [235–237], electron-conducting matrices [238,239], as well as provide a variety of novel optoelectronic properties [240,241]. However, the use of such traditional strategies for the synthesis of precise nanostructures (i.e., dendrons [C] and dendrimers [D]) larger than 15 to 20 nm has several serious disadvantages. First, it is hampered by the large number of reiterative synthetic steps required to attain these higher dimensions (e.g., Gen 9; PAMAM dendrimer, diameter ~10 nm, requires 18 reaction steps). Secondly, these constructions are limited by the de Gennes dense-packing phenomenon, which precludes ideal dendritic construction beyond certain limiting generations [20,70]. For these reasons, our attention has turned to the use of dendrimers as reactive modules for the rapid construction of control nanoarchitectures possessing a higher complexity and dimensions beyond the dendrimer. We refer to these generic poly(dendrimers) as *megamers* [242]. Both randomly assembled megamers [242] and structure-controlled megamers [226,242,243] have been demonstrated. Recently, new mathematically defined megamers (i.e., dendrimer clusters) or core–shell tecto(dendrimers) have been reported [226,232,243,244]. The principles of these structure-controlled megamer syntheses mimic those used for the traditional core–shell construction of dendrimers. First, a megamer-core reagent (usually a spheroid) is selected. Next, a limited amount of this reactive core reagent is combined with an excess of a megamer-shell reagent. The objective is to completely saturate the spheroid target core surface with covalently bonded spheroidal megamer-shell reagent. Since the diameters of the megamer-core and shell reagents are very well defined, it is possible to mathematically predict the number of megamer-shell molecules required to saturate a targeted core dendrimer [245].

These core–shell relationships have been analyzed mathematically as a function of the ratio of core (r_1) and shell (r_2) radii [245]. At low r_1/r_2 values (i.e., 0.1 to 1.2), very important symmetry properties emerge as shown in Figure 24.23. It can be seen that, when the core reagent is small and the shell reagent is larger, only a very limited number of shell-type dendrimers can be attached to the core dendrimer

(a)

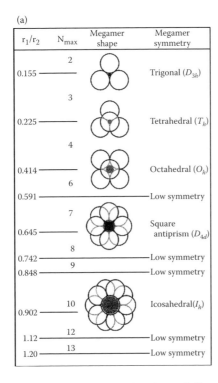

r_1/r_2	N_{max}	Megamer shape	Megamer symmetry
0.155	2		Trigonal (D_{3h})
0.225	3		Tetrahedral (T_h)
0.414	4		Octahedral (O_h)
0.591	6		Low symmetry
0.645	7		Square antiprism (D_{4d})
0.742	8		Low symmetry
0.848	9		Low symmetry
0.902	10		Icosahedral(I_h)
1.12	12		Low symmetry
1.20	13		Low symmetry

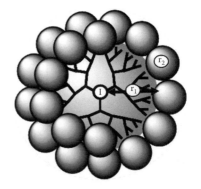

(b) r_1 = radius of core dendrimer
r_2 = radius of shell dendrimer

N_{max} = total theoretical number of shell-like spheroids with radius r_2 that can be ideally parked around a core spheroid with radius r_1.

(c)

Mansfield-Tomalia-Rakesh equation

$$N_{max} = 2\pi/\sqrt{3}\,(r_1/r_2 + 1)^2$$

When $r_1/r_2 > 1.20$

FIGURE 24.23 (a) Symmetry properties of core-shell structure, where $r_1/r_2 <1.20$. (b) Sterically induced stoichiometry (sis) based on the respective radii of core and shell dendrimers. (c) Mansfield–Tomalia–Rakesh equation for calculating the maximum shell filling when $r_1/r_2 > 1.20$.

based on available space. However, when $r_1/r_2 \geq 1.2$, the available space around the core allows attachment of many more spheroidal shell reagents up to a discrete saturation level. The saturation number (N_{max}) is well defined and can be predicted from a general expression that is described by the Mansfield–Tomalia–Rakesh equation (Figure 24.23).

24.7.2 Dendrimers as Reactive Modules for the Synthesis of More Complex Nanoscale Architectures (Megamers)

24.7.2.1 Saturated-Shell-Architecture Approach

The general chemistry used in this approach involves the combination of a limited amount of an amine-terminated, dendrimer core reagent (e.g., $G = 5$ to 7; NH$_2$-terminated PAMAM dendrimer) with an excess of a carboxylic acid terminated (e.g., PAMAM) dendrimeric shell reagent [243]. These two charge-differentiated species are allowed to self-assemble into the electrostatically driven, supramolecular, core–shell tecto(dendrimer) architectures. After equilibration, covalent bond formation at these charge-neutralized, dendrimer contact sites is induced with carbodiimide reagents (Figure 24.24a) [243,244].

The carboxylic-acid-terminated shell-reagent dendrimers (e.g., $G = 3$ or 5) were synthesized by ring opening of succinic anhydride with the appropriate amine-terminated PAMAM dendrimers. All reactions leading to core–shell tecto(dendrimers) were performed in the presence of LiCl at room temperature as dilute solutions (~0.5 wt.%) in water. Equilibration times of 16 to 20 h were required to complete the charge-neutralized self-assembly of excess shell reagent around the limited core dendrimer reagent. Following this self-assembly and equilibration, a linking reagent, 1-(3-dimethylaminopropyl)-3-ethylcarbodiimide hydrochloride, was added to covalently bond the assembly of dendrimeric shell reagents to a single dendrimeric core reagent at the amine–carboxylic acid interaction sites. These sites are presumed to reside primarily at the exterior of the core dendrimer reagent [243,244].

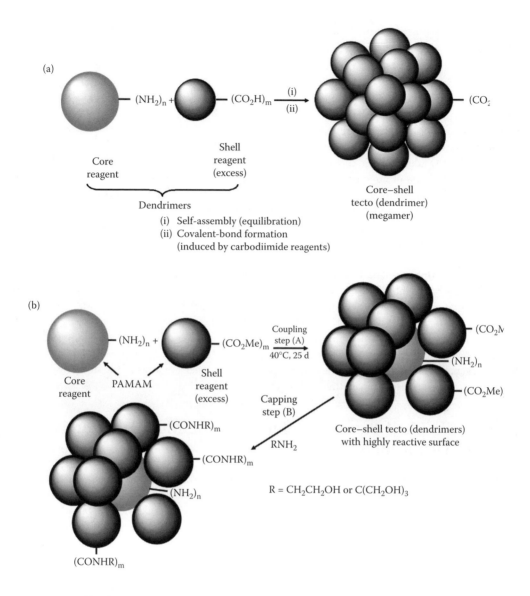

FIGURE 24.24 (a) The saturated-shell architecture approach to megamer synthesis. All surface dendrimers are carboxylic acid terminated. (b) Step a: the unsaturated-shell architecture approach to megamer synthesis. Step b describes surface-capping reactions.

Remarkably monodispersed products were obtained by performing the core–shell self-assembly reactions in the presence of LiCl. In the absence of LiCl, these reactions yielded bimodal or trimodal product-mass distributions (as observed by size-exclusion chromatography [SEC]). Core–shell products formed in the absence of LiCl are multimodal and are presumed to be due to clustering of the amine-terminated core reagent into various domain sizes. Such clustering of amine-terminated PAMAM dendrimers has been noted in earlier work [225]. Attempts to subsequently charge neutralize these polydispersed domains with anionic dendrimeric shell reagent produced a broad product distribution. Reversing the terminal functional groups on the core and shell reagents, respectively (i.e., using a carboxylic-acid-terminated PAMAM dendrimer as the core and an excess amine-terminated PAMAM dendrimer as the shell reagent) under identical reaction conditions, did not yield the desired product. The reason for this is not evident from our studies so far.

24.7.2.2 Unsaturated-Shell-Architecture Approach

The second method, the direct covalent-bond-formation method, produces semi-controlled, partially filled shell structures [28,242]. It involves the reaction of a limited amount of a nucleophilic dendrimeric core reagent with an excess of electrophilic dendrimeric shell reagent as illustrated in Figure 24.24b [244]. This route involves the random parking of the reactive shell reagent on a core-substrate surface. As a consequence, partially filled shell products are obtained, which possess relatively narrow, but not precise molecular weight distributions as noted for saturated-shell architectures [243]. These distributions are determined by the core–shell parking efficiency prior to covalent bond formation.

Various PAMAM dendrimeric core reagents (either amine- or ester-functionalized) were each allowed to react with an excess of an appropriate PAMAM dendrimeric shell reagent. The reactions were performed at 40°C in methanol and monitored by FT-IR, ^{13}C NMR, SEC, and gel electrophoresis. Conversions in Step A (see Figure 24.24b) were monitored by SEC and confirmed by observing the formation of shorter-retention-time products, consistent with higher-molecular-weight structures. Additional evidence was gained by observing the loss of the migratory band associated with the dendrimeric core reagent present in the initial reaction mixture, accompanied by the formation of a higher-molecular-weight product, which displayed a much shorter migratory band position on the electrophoretic gel. In fact, the molecular weights of the resulting core–shell tecto(dendrimers) could be estimated by comparing the migratory time of the core–shell products with the migration distances of the PAMAM dendrimer reagents (e.g., $G = 2$ to 10) used for their construction [61,63].

It was important to perform capping reactions on the surface of the resulting unsaturated, ester-terminated core–shell products, in order to pacify the highly reactive amine cleft surfaces against further reaction. Preferred capping reagents for pacifying the ester domains of the surface were either 2-aminoethanol or tris(hydroxymethyl)aminomethane [28].

24.8 Core–Shell Patterns Influencing the Modular Reactivity of Dendrimers

Dendritic species, possessing an unsaturated outer monomer shell consisting of ester and amine domains, exhibited autoreactive behavior. This autoreactive behavior was observed whenever a completely saturated state of either ester or amine groups was not attained. These species, which included missing-branch structures, led to the formation of monodendrimers containing macrocyclic terminal groups as well as moderate amounts of megamers (i.e., dimeric, trimeric, etc. species). Ideal dendrimer structures (i.e., saturated-outer-monomer-shell products) could, however, be separated from these side products by silica gel column chromatography and preparative TLC isolation techniques. Ideal dendrimer structures that exhibited mathematically predictable masses, as well as unsaturated-monomer-shell products exhibiting mass defects, were readily characterized by electrospray (ESI) and MALDI-TOF mass spectrometry [58–60,246].

Recently, we have reported work that offers additional evidence that unfilled-outer-monomer-shell species are autoreactive intermediates and do indeed lead to megamer formation. In general, saturated-shell PAMAM dendrimers (i.e., all-amine- or all-ester-group-saturated surfaces) are very robust species (i.e., are analogous to inert gas configurations observed at the atomic level). In this regard, *they do not exhibit autoreactive characteristics*. Such samples may be stored for months or years without change. In contrast, PAMAM dendrimer samples possessing unfilled monomer shells (i.e., amine and ester group domains on the dendrimer surface) are notorious for exhibiting autoreactive properties leading to terminal looping (i.e., macrocycle formation) and megamer formation [57,247].

Remarkably, these autoreactivity patterns are also observed for the dimensionally larger core–shell tecto(dendrimer) architectures. For example, saturated-shell, core–shell tecto(dendrimer) architectures *exhibit no autoreactivity*, whereas partially filled shell, core–shell tecto(dendrimers) *exhibit profound autoreactivity*, unless pacified by reagents possessing orthogonally reactive functionalities. This behavior is comparable to the modular reactivity patterns of atoms and basic dendrimers as illustrated in Figure 24.25 [28,244,247].

	Atoms	Dendrimers	Core–shell tecto (dendrimers)
Dimensions	0.05–0.6 nm	1–15 nm	5.0 ≥ 100 nm
Valency (reactivity)	Unfilled outer electron shell	Unfilled outer branch cell shell	Unfilled outside dendrimer shell
(Core–shell) Architecture-induced reactivity (unfilled shells)	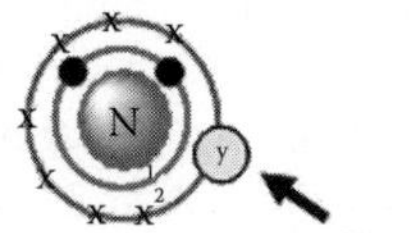 (e.g., fluorine) Unfilled shell (x)	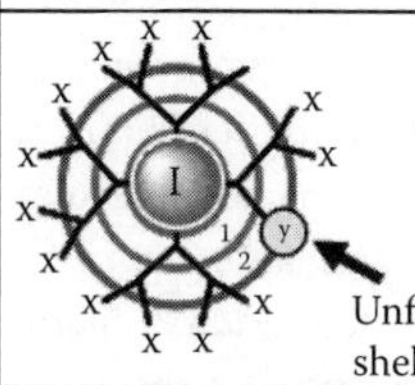 Unfilled shell (x)	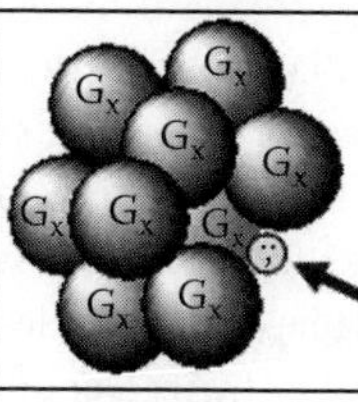 Unfilled shell (x)
Functional components directing valency	Missing one electron (y) in outer shell (x) penultimate to saturated noble gas configuration	Missing one terminal branch cell in outer shell (x) exposing functionality (y)	Missing one dendrimer shell reagent exposing functionality (y)

FIGURE 24.25 Quantized module (building block) reactivity patterns at the subnanoscale (atoms), lower nanoscale (dendrimers), and higher nanoscale (core-shell tecto [dendrimers]) levels involving unsaturated electron, monomer, or dendrimer shells.

24.9   Conclusions

In summary, using strictly abiotic methods, it has been widely demonstrated over the past decade that dendrimers [20] can be routinely constructed with control that rivals the structural regulation found in biological systems. The close scaling of size [74,248], shape, and quasi-equivalency of surfaces [249–251] observed between nanoscale biostructures and various dendrimer families/generational levels are both striking and provocative [74,75,248–254]. These remarkable similarities suggest a broad strategy based on rational biomimicry as a means for creating a repertoire of structure-controlled, size, and shape-variable dendrimer assemblies. Successful demonstration of such a biomimetic approach has proved to be a versatile and powerful synthetic strategy for systematically accessing virtually any desired combination of size, shape, and surface in the nanoscale region. Future extensions will involve combinational variation of dendrimer module parameters such as (*i*) families (interior compositions), (*ii*) surfaces, (*iii*) generational levels, or (*iv*) architectural shapes (i.e., spheroids, rods, etc.).

Acknowledgments

We express sincere appreciation to Linda S. Nixon for graphics and manuscript preparation. Diallo thanks the National Science Foundation (NSF Grants CTS-0086727, CTS-0329436, and CTS-0506951) and the U.S. Environmental Protection Agency (NCER STAR Grant R829626) for funding his research on the use of dendritic polymers as functional materials for water purification. Partial funding for this research was also provided by the Department of Energy (Cooperative Agreement EW15254), the W. M. Keck Foundation, the National Water Research Institute (Research Project Agreement No. 05-TT-004) and the NSF-sponsored Cornell University Nanobiotechnology Center. This center is funded by the STC Program of the National Science Foundation under Agreement No. ECS-9876771.

References

1. Anderson, P.W., More is different, *Science,* 177, 393, 1972.
2. Smalley, R., Testimony to U.S. Science and Technology Office, 1999.

3. Richardson, C.F., Schuster, D.I., and Wilson, S.R., Synthesis and characterization of water-soluble amino fullerene derivatives, *Org. Lett.*, 2, 1011, 2000.

4. Jensen, A.W., Maru, B.S., Zhang, X., Mohanty, D., Fahlman, B.D., Swanson, D.R., and Tomalia, D.A., Preparation of fullerene-shell dendrimer-core nanoconjugates, *Nano Lett.*, 5, 1171, 2005.

5. Morawetz, H., *Polymers: The Origin and Growth of a Science*, Wiley, New York, 1985.

6. Elias, H.-G., *An Introduction to Polymer Science*, VCH, Weinheim, 1997.

7. Tomalia, D.A., Birth of a new macromolecular architecture: dendrimers as quantized building blocks for nanoscale synthetic polymer chemistry, *Prog. Polym. Sci.*, 30, 294, 2005.

8. Tomalia, D.A., Birth of a new macromolecular architecture: dendrimers as quantized building blocks for nanoscale synthetic organic chemistry, *Aldrichim. Acta*, 37, 39, 2004.

9. Tomalia, D.A., The dendritic state, *Mater. Today*, 34, 34–36, March 2005.

10. Tomalia, D.A., Dendrons/dendrimers: the convergence of quantized dendritic building blocks/architectures for applications in nanotechnology, *Chem. Today*, 23, 52, 2005.

11. Sunder, A., Heinemann, J., and Frey, H., Controlling the growth of polymer trees: concepts and perspectives for hyperbranched polymers, *Chem. Eur. J.*, 6, 2499, 2000.

12. Corey, E.J. and Cheng, X.-M., *The Logic of Chemical Synthesis*, Wiley, New York, 1989.

13. Berzelius, J., *J. Fortsch. Phys. Wissensch.*, 11, 44, 1832.

14. Buhleier, E., Wehner, W., and Vögtle, F., Cascade- and nonskid-chain-like syntheses of molecular cavity topologies, *Synthesis*, 405, 155, 1978.

15. Tomalia, D.A. and Fréchet, J.M.J., Discovery of dendrimers and dendritic polymers: a brief historical perspective, *J. Polym. Sci. A1*, 40, 2719, 2002.

16. Tomalia, D.A., Dewald, J.R., Hall, M.J., Martin, S.J., and Smith, P.B., Preprints of the 1st SPSJ International Polymeric Conference, Society of Polymer Science, Kyoto, Japan, August 1984, pp. 65.

17. Staudinger, H., *Schweiz. Chem. Z.*, 105, 1919.

18. Staudinger, H., *Ber., Deutsch Chem. Ges.*, 53, 1073, 1920.

19. Staudinger, H., *From Organic Chemistry to Macromolecules: A Scientific Autobiography*, Wiley, New York, 1961.

20. Fréchet, J.M.J. and Tomalia, D.A., *Dendrimers and Other Dendritic Polymers*, Wiley, Chichester, 2001.

21. Matyjaszewski, K. and Spanswick, J., Controlled/living radical polymerization, *Mater. Today*, 26, 2005.

22. Elias, H.-G., *Mega Molecules*, Springer, Berlin, 1987.

23. Roovers, J., *Advances in Polymer Science, Branched Polymers I*, Vol. 142, Springer, Berlin, 1999.

24. Roovers, J., *Advances in Polymer Science, Branched Polymers II*, Vol. 143, Springer, Berlin, 2000.

25. Scheirs, J. and Kaminsky, W., *Metallocene-Based Polyolefins*, Vols. 1 and 2, Wiley, Brisbane, 2000.

26. Dusek, K. and Duskova-Smrckova, M., Formation, structure and properties and the crosslinked state relative to precursor architecture, in *Dendrimers and Dendritic Polymers*, Wiley, West Sussex, 2001, p. 111.

27. Goodsell, D.S., Biomolecules and nanotechnology, *Am. Sci.*, 88, 230, 2000.

28. Tomalia, D.A., Brothers, H.M., II, Piehler, L.T., Durst, H.D., and Swanson, D.R., Partial shell-filled core–shell tecto(dendrimers): a strategy to surface differentiated nano-clefts and cusps, *Proc. Natl. Acad. Sci. USA*, 99 (8), 5081, 2002.

29. Tomalia, D.A., Starburst dendrimers — nanoscopic supermolecules according to dendritic rules and principles, *Macromol. Symp.*, 101, 243, 1996.

30. Tomalia, D.A., Brothers, H.M., II, Piehler, L., and Hsu, Y., Dendritic macromolecules: a fourth major class of macromolecular architecture, *Proc. Am. Chem. Soc. Div. PMSE*, 73, 75, 1995.

31. Naj, A.K., Persistent inventor markets a molecule, *Wall Street Journal*, LXXVII, 92, Feb. 26, New York, 1996, p. B1.

32. Tomalia, D.A., SPSJ award for outstanding achievement in polymer science and technology for the invention and development of dendritic polymers, in 52nd Society of Polymer Science Japan (SPSJ) Annual Meeting, Nagoya, Japan, May 29, 2003.

33. Tomalia, D.A., Hall, M., and Hedstrand, D.M., Starburst dendrimers. III. The importance of branch junction symmetry in the development of topological shell molecules, *J. Am. Chem. Soc.*, 109, 1601, 1987.

34. Tomalia, D.A., Baker, H., Dewald, J., Hall, M., Kallos, G., Martin, S., Roeck, J., Ryder, J., and Smith, P., A new class of polymers: starburst dendritic macromolecules, *Polym. J.* (*Tokyo*), 17, 117, 1985.

35. Moors, R. and Vögtle, F., Dendrimere polyamine, *Chem. Ber.*, 126, 2133, 1993.

36. Kolb, H.C., Finn, M.G., and Sharpless, K.B., Click chemistry: diverse chemical function from a few good reactions, *Angew. Chem. Int. Ed.*, 40, 2004, 2001.

37. Kolb, H.C. and Sharpless, K.B., The growing impact of click chemistry on drug discovery, *Drug Discovers Today*, 8, 1128, 2003.

38. Tomalia, D.A., Dendrimer molecules, *Sci. Am.*, 272, 42, 1995.

39. Esfand, R. and Tomalia, D.A., Laboratory synthesis of poly(amidoamine) (PAMAM) dendrimers, in *Dendrimers and Other Dendritic Polymers*, Fréchet, J.M.J. and Tomalia, D.A., Eds., Wiley, Chichester, , 2001, p. 587.

40. Tomalia, D.A., Baker, H., Dewald, J., Hall, M., Kallos, G., Martin, S., Roeck, J., Ryder, J., and Smith, P., Dendritic macromolecules: synthesis of starburst dendrimers, *Macromolecules*, 19, 2466, 1986.

41. Tomalia, D.A., Naylor, A.M., and Goddard, W.A., III, Starburst dendrimers: molecular level control of size, shape, surface chemistry, topology and flexibility from atoms to macroscopic matter, *Angew. Chem. Int. Ed. Engl.*, 29, 138, 1990.

42. Tomalia, D.A., Hedstrand, D.M., and Wilson, L.R., Dendritic polymers, in *Encyclopedia of Polymer Science and Engineering*, 2nd ed., Index vol., Wiley, New York, 1990, p. 46–92.

43. Newkome, G.R., Yao, Z.-Q., Baker, G.R., Jr., and Gupta, V.K., Cascade molecules: a new approach to micelles, *J. Org. Chem.*, 50, 2003, 1985.

44. Padias, A.B., Hall, H.K., Jr., Tomalia, D.A., and McConnell, J.R., Starburst polyether dendrimers, *J. Org. Chem.*, 52, 5305, 1987.

45. Lothian-Tomalia, M.K., Hedstrand, D.M., and Tomalia, D.A., A contemporary survey of covalent connectivity and complexity: the divergent synthesis of poly(thioether) dendrimers. Amplified, genealogical directed synthesis leading to the de Gennes dense packed state, *Tetrahedron*, 53, 15495, 1997.

46. Newkome, G.R., Moorfield, C.N., and Vögtle, F., *Dendritic Molecules*, VCH, Weinheim, 1996.

47. Hawker, C.J. and Fréchet, J.M.J., Preparation of polymers with controlled molecular architecture: a new convergent approach to dendritic macromolecules, *J. Am. Chem. Soc.*, 112, 7638, 1990.

48. Miller, T.M. and Neenan, T.X., Convergent synthesis of monodisperse dendrimers based upon 1,3,5-trisubstituted benzenes, *Chem. Mater.*, 2, 346, 1990.

49. Fréchet, J.M.J., Functional polymers and dendrimers: reactivity, molecular architectures, and interfacial energy, *Science*, 263, 1710, 1994.

50. Huisgen, R., Cycloadditions — definition, classification, and characterization, *Angew. Chem. Int. Ed.*, 7, 321, 1968.

51. Wu, P., Feldman, A.K., Nugent, A.K., Hawker, C.J., Scheel, A., Voit, B., Pyun, J., Frechet, J.M., Sharpless, K.B., and Fokin, V.V., Efficiency and fidelity in a click-chemistry route to triazole dendrimers by the copper(I)-catalyzed ligation of azides and alkynes, *Angew. Chem. Int. Ed.*, 43, 3928, 2004.

52. Joralemon, M.J., O'Reilly, R.K., Matson, J.B., Nugent, A.K., Hawker, C.J., and Wooley, K.L., Dendrimers clicked together divergently, *Macromolecules*, 38, 5436, 2005.

53. Wu, P., Malkoch, M., Hunt, J.N., Vestberg, R., Kaltgrad, E., Finn, M.G., Fokin, V.V., Sharpless, K.B., and Hawker, C.J., Multivalent, bifunctional dendrimers prepared by click chemistry, *Chem. Commun.*, 5775, 2005.

54. Helms, B., Mynar, J.L., Hawker, C.J., and Frechet, J.M., Dendronized linear polymers via "click chemistry," *J. Am. Chem. Soc.*, 126, 15020, 2004.

55. Joralemon, M.J., O'Reilly, R.K., Hawker, C.J., and Wooley, K.L., Shell click-crosslinked (SCC) nanoparticles: a new methodology for synthesis and orthogonal functionalization, *J. Am. Chem. Soc.*, 127, 16892, 2005.

56. Zeng, F. and Zimmerman, S.C., Dendrimers in supramolecular chemistry: from molecular recognition to self-assembly, *Chem. Rev.*, 97, 1681, 1997.

57. Tomalia, D.A., Starburst/cascade dendrimers: fundamental building blocks for a new nanoscopic chemistry set, *Adv. Mater.*, 6, 529, 1994.

58. Kallos, G.J., Tomalia, D.A., Hedstrand, D.M., Lewis, S., and Zhou, J., Molecular weight determination of a polyamidoamine starburst polymer by electrospray ionization mass spectrometry, *Rapid Commun. Mass Spectrom.*, 5, 383, 1991.

59. Dvornic, P.R. and Tomalia, D.A., Genealogically directed syntheses (polymerizations): direct evidence by electrospray mass spectroscopy, *Macromol. Symp.*, 98, 403, 1995.

60. Hummelen, J.C., van Dongen, J.L.J., and Meijer, E.W., Electrospray mass spectrometry of poly(propylene imine) dendrimers — the issue of dendritic purity of polydispersity, *Chem. Eur. J.*, 3, 1489, 1997.

61. Brothers, H.M., II, Piehler, L.T., and Tomalia, D.A., Slab-gel and capillary electrophoretic characterization of polyamidoamine dendrimers, *J. Chromatogr. A*, 814, 233, 1998.

62. Tomalia, D.A. and Dewald, J.R., U.S. Patent, 4,587,329, 1986.

63. Zhang, C. and Tomalia, D.A., Gel electrophoresis characterization of dendritic polymers, in *Dendrimers and Other Dendritic Polymers*, Fréchet, J.M.J. and Tomalia, D.A., Eds., Wiley, Chichester, 2001, p. 239.

64. Naylor, A.M., Goddard, W.A., III, Keifer, G.E., and Tomalia, D.A., Starburst dendrimers. 5. Molecular shape control, *J. Am. Chem. Soc.*, 111, 2339, 1989.

65. Turro, N.J., Barton, J.K., and Tomalia, D.A., Molecular recognition and chemistry in restricted reaction spaces, *Acc. Chem. Res.*, 24 (11), 332, 1991.

66. Gopidas, K.R., Leheny, A.R., Caminati, G., Turro, N.J., and Tomalia, D.A., Photophysical investigation of similarities between starburst dendrimer and anionic micelles, *J. Am. Chem. Soc.*, 113, 7335, 1991.

67. Ottaviani, M.F., Turro, N.J., Jockusch, S., and Tomalia, D.A., Characterization of starburst dendrimers by EPR. 3. Aggregational processes of a positively charged nitroxide surfactant, *J. Phys. Chem.*, 100, 13675, 1996.

68. Jockusch, J., Ramirez, J., Sanghvi, K., Nociti, R., Turro, N.J., and Tomalia, D.A., Comparison of nitrogen core and ethylenediamine core starburst dendrimers through photochemical and spectroscopic probes, *Macromolecules*, 32, 4419, 1999.

69. Hawker, C.J., Wooley, K.L., and Fréchet, J.M.J., Solvatochromism as a probe of the microenvironment in dendritic polyethers: transition from an extended to a globular structure, *J. Am. Chem. Soc.*, 115, 4375, 1993.

70. de Gennes, P.G. and Hervet, H.J., Statistics of starburst polymers, *J. Phys. Lett.* (*Paris*), 44, 351, 1983.

71. Mourey, T.H., Turner, S.R., Rubinstein, M., Frechet, J.M.J., Hawer, C.J., and Wooley, K.L., Unique behavior of dendritic macromolecules: intrinsic viscosity of polyether dendrimers, *Macromolecules*, 25, 2401, 1992.

72. Maiti, P.K., Cagin, T., Wang, G., and Goddard, W.A., III, Structure of PAMAM dendrimers: generation 1 through 11, *Macromolecules*, 37, 6236, 2004.

73. Bauer, B.J. and Amis, E.J., Characterization of dendritically branched polymers by SANS, small angle x-ray scattering (SAXS), and transmission electron microscopy (TEM), in *Dendrimers and Other Dendritic Polymers*, Fréchet, J.M.J. and Tomalia, D.A., Eds., Wiley, Chichester, 2001, p. 255.

74. Esfand, R. and Tomalia, D.A., Poly(amidoamine) (PAMAM) dendrimers: from biomimicry to drug delivery and biomedical applications, *Drug Discov. Today*, 6 (8), 427, 2001.

75. Hecht, S. and Fréchet, J.M.J., Dendritic encapsulation of function: applying nature's site isolation principle from biomimetics to materials science, *Angew. Chem. Int. Ed.*, 40 (1), 74, 2001.

76. Weener, J.-W., Baars, M.W.P.L., and Meijer, E.W., Some unique features of dendrimers based upon self-assembly and host–guest properties, in *Dendrimers and Other Dendritic Polymers*, Fréchet, J.M.J. and Tomalia, D.A., Eds., Wiley, Chichester, 2001, p. 387.

77. de A.A. Soler-Illia, G.J., Rozes, L., Boggiano, M.K., Sanchez, C., Turrin, C.-O., Caminade, A.-M., and Majoral, J.-P., New mesotextured hybrid materials made from assemblies of dendrimers and titanium (IV)-oxo-organo clusters, *Angew. Chem. Int. Ed.*, 39, 4250, 2000.

78. Tomalia, D.A., Dendrimeric supramolecular and supramacromolecular assemblies, in *Supramolecular Polymers*, Ciferri, A., Ed., CRC Press, Boca Raton, FL, 2005, p. 187.

79. Crooks, R.M., Lemon, B., III, Sun, L., Yeung, L.K., and Zhao, M., Dendrimer-encapsulated metals and semiconductors: synthesis, characterization, and applications, in *Topics in Current Chemistry*, Vol. 212, Springer, Berlin, 2001.

80. Freeman, A.W., Koene, S.C., Malenfant, P.R.L., Thompson, M.E., and Frechet, J.M.J., Dendrimer-containing light-emitting diodes: toward site-isolation of chromophores, *J. Am. Chem. Soc.*, 122, 12385, 2000.

81. Piotti, M.E., Rivera, F., Bond, R., Hawker, C.J., and Frèchet, J.M.J., Synthesis and catalytic activity of unimolecular dendritic reverse micelles with "internal" functional groups, *J. Am. Chem. Soc.*, 121, 9471, 1999.

82. Bieniarz, C., Dendrimers: applications to pharmaceutical and medicinal chemistry, in *Encyclopedia of Pharmaceutical Technology*, Vol. 18, Marcel Dekker, New York, 1998, p. 55.

83. Svenson, S. and Tomalia, D.A., Dendrimers in biomedical applications — reflections on the field, *Adv. Drug Deliv. Rev.*, 57, 2106, 2005.

84. Tully, D.C. and Fréchet, J.M.J., Dendrimers at surfaces and interfaces: chemistry and applications, *Chem. Commun.*, 1229, 2001.

85. Jiang, D.-L. and Aida, T., Dendritic polymers: optical and photochemical properties, in *Dendrimers and Other Dendritic Polymers*, Fréchet, J.M.J. and Tomalia, D.A., Eds., Wiley, West Sussex, 2001, p. 425.

86. Adronov, A. and Fréchet, J.M.J., Light-harvesting dendrimers, *Chem. Commun.*, 1701, 2000.

87. Singh, P., Moll, F., III, Lin, S.H., and Ferzli, C., Starburst dendrimers: a novel matrix for multi-functional reagents in immunoassays, *Clin. Chem.*, 42 (9), 1567, 1996.

88. Singh, P., Dendrimer-based biological reagents: preparation and applications in diagnostics, in *Dendrimers and Dendritic Polymers*, Fréchet, J.M.J. and Tomalia, D.A., Eds., Wiley, Chichester, 2001, p. 463.

89. Prosa, T.J., Bauer, B.J., Amis, E.J., Tomalia, D.A., and Scherrenberg, R., A SAXS study of the internal structure of dendritic polymer systems, *J. Polym. Sci. B*, 35, 2913, 1997.

90. Topp, A., Bauer, B.J., Tomalia, D.A., and Amis, E.J., Effect of solvent quality on the molecular dimensions of PAMAM dendrimers, *Macromolecules*, 32, 7232, 1999.

91. Topp, A., Bauer, B.J., Klimash, J.W., Spindler, R., and Tomalia, D.A., Probing the location of the terminal groups of dendrimers in dilute solution, *Macromolecules*, 32, 7226, 1999.

92. Welch, P. and Muthukumar, M., Tuning the density profile of dendritic polyelectrolytes, *Macromolecules*, 31, 5892, 1998.

93. Nisato, G., Ivkov, R., and Amis, E.J., Size invariance of polyelectrolyte dendrimer, *Macromolecules*, 33, 4172, 2000.

94. Murat, M. and Grest, G.S., Molecular dynamics study of dendrimer molecules in solvents of varying quality, *Macromolecules*, 29, 1278, 1996.

95. Lyulin, A.V., Davies, G.R., and Adolf, D.B., Location of terminal groups of dendrimers: Brownian dynamics simulation, *Macromolecules*, 33, 6899, 2000.

96. Rathgeber, S., Monkenbusch, M., Kreitschmann, M., Urban, V., and Brulet, A., Dynamics of starburst dendrimers in solution in relation to their structural properties, *J. Chem. Phys.*, 117, 4047, 2002.

97. Rathgeber, S., Pakula, T., and Urban, V., Structure of star-burst dendrimers: a comparison between small angle x-ray scattering and computer simulation results, *J. Chem. Phys.*, 121, 3840, 2004.

98. Maiti, P.K., Cagin, T., Lin, S.T., and Goddard, W.A., III, Effect of solvent and pH on the structure of PAMAM dendrimers, *Macromolecules*, 38, 979, 2005.

99. Lin, S.T., Maiti, P.K., Lin, S.T., and Goddard, W.A., III, Dynamics and thermodynamics of water in PAMAM dendrimers at subnanosecond time scales, *J. Phys. Chem. B*, 109, 8663, 2005.

100. Florence, A.T., Dendrimers: a versatile targeting platform, *Adv. Drug Deliv. Rev.*, 57, 2104, 2005.

101. Lee, C.A., MacKay, J.A., Frechet, J.M.J., and Szoka, F.C., Designing dendrimers for biological applications, *Nat. Biotechnol.*, 23, 1517, 2005.

102. Duncan, R., The dawning era of polymer therapeuties, *Nat. Rev. Drug. Dev,* 2, 347, 2003.

103. Kubasiak, L.A. and Tomalia, D.A., Cationic dendrimers as gene transfection vectors: *dendri*-poly(amidoamines) and *dendri*-poly(propylenimines), in *Polymeric Gene Delivery Principles and Applications*, Amiji, M.M., Ed., CRC Press, Boca Raton, FL, 2005, p. 133.

104. Dufes, C., Uchegbu, I.F., and Schätzlein, A.G., Dendrimers in gene delivery, *Adv. Drug Deliv. Rev.*, 57, 2177, 2005.

105. Bielinska, A., Kukowska-Latallo, J.F., Johnson, J., Tomalia, D.A., and Baker, J.R., Jr., Regulation of *in vitro* gene expression using antisense oligonucleotides or antisense expression plasmids transfected using starburst PAMAM dendrimers, *Nucleic Acids Res.*, 24, 2176, 1996.

106. Tomalia, D.A. et al., Bioactive and/or targeted dendrimer conjugates, U.S. Pat. 5714166, 1998.

107. Kukowska-Latallo, J.F., Bielinska, A.U., Johnson, J., Spindler, R., Tomalia, D.A., and Baker, J.R., Jr., Efficient transfer of genetic material into mammalian cells using starburst polyamidoamine dendrimers, *Proc. Natl. Acad. Sci. USA*, 93, 4897, 1996.

108. Eichman, J.D., Bielinska, A.U., Kukowska-Latallo, J.F., Donovan, B.W., and Baker, J.R., Jr., Bioapplications of PAMAM dendrimers, in *Dendrimers and Other Dendritic Polymers*, Fréchet, J.M.J. and Tomalia, D.A., Eds., Wiley, West Sussex, 2001, p. 441.

109. DeLong, R., Stephenson, K., Loftus, T., Fisher, M., Alahari, S., Nolting, A., and Juliano, R.L., Characterization of complexes of oligonucleotides with polyamidoamine starburst dendrimers and effects on intracellular delivery, *J. Pharm. Sci.*, 86, 762, 1997.

110. Tang, M.X. and Szoka, F.C., The influence of polymer structure on the interactions of cationic polymers with DNA and morphology of the resulting complexes, *Gene Ther.*, 4, 823, 1997.

111. Chen, W., Turro, N.J., and Tomalia, D.A., Using ethidium bromide to probe the interactions between DNA and dendrimers, *Langmuir*, 16, 15, 2000.

112. Tang, M.X., Redemann, C.T., and Szoka, F.C., *In vitro* gene delivery by degraded polyamidoamine dendrimers, *Bioconjug. Chem.*, 7, 703, 1996.

113. Schätzlein, A.G., Zinselmeyer, B.H., Elouzi, A., Dufes, C., Chim, Y.-T. A., Roberts, C.J., Davis, M.C., Munro, A., Gray, A.I., and Uchegbu, I.F., Preferential liver gene expression with polypropylenimine dendrimers, *J. Control Release*, 101, 247, 2005.

114. Toth, I., Sakthivel, T., O'Donnell, M., Pasi, K.J., Wilderspin, A.F., Lee, C.A., Toth, I., and Florence, A.T., Novel cationic lipidic peptide dendrimer vectors *in vitro* gene delivery, *STP Pharma Sci.*, 9, 93, 1999.

115. Bayele, H.K., Sakthivel, T., O'Donnell, M., Pasi, K.J., Wilderspin, A.F., Lee, C.A., Toth, I., and Florence, A.T., Versatile peptide dendrimers for nucleic acid delivery, *J. Pharm. Sci.*, 94, 446, 2005.

116. Kiessling, L.L. and Pohl, N.L., Strength in numbers: non-natural polyvalent carbohydrate derivatives, *Chem. Biol.*, 3, 71, 1996.

117. Turnbull, W.B. and Stoddart, J.F., Design and synthesis of glycodendrimers, *Rev. Mol. Biotechnol.*, 90, 231, 2002.

118. Mammen, M., Choi, S.-K., and Whitesides, G.M., Polyvalent interactions in biological systems: implications for design and use of multivalent ligands and inhibitors, *Angew. Chem. Int. Ed.*, 37, 2754, 1998.

119. Roy, R., Zanini, D., Meunier, S.J., and Romanowska, A., Solid-phase synthesis of dendritic sialoside inhibitors of influenza A virus haemagglutinin, *J. Chem. Soc. Commun.*, 24, 1869, 1993.

120. Zanini, D. and Roy, R., Synthesis of new α-thiosialodendrimers and their binding properties to the sialic acid specific lectin from *Limax flavus*, *J. Am. Chem. Soc.*, 119, 2088, 1997.

121. Peerlings, H.W.I., Nepogodiev, S.A., Stoddard, J.F., and Meijer, E.W., Synthesis of spacer-armed glucodendrimers based on the modification of poly(propylene imine) dendrimers, *Eur. J. Org. Chem.*, 1879, 1998.

122. Roy, R., A decade of glycodendrimer chemistry, *Trends Glycosci. Glycotech.*, 15, 291, 2003.

123. Reuter, J.D., Myc, A., Hayes, M.M., Gan, Z., Roy, R., Qin, D., Yin, R., Piehler, L.T., Esfand, R., Tomalia, D.A., and Baker, J.R., Jr., Inhibition of viral adhesion and infection by sialic-acid-conjugated dendritic polymers, *Bioconjug. Chem.*, 10, 271, 1999.

124. Roy, R. and Baek, M.-G., Glycodendrimers: novel glycotope isosteres unmasking sugar coding. Case study with T-antigen markers from breast cancer MUC1 glycoprotein, *Rev. Mol. Biotechnol.*, 90, 291, 2002.

125. Andre, S., Ortega, J.C., Perez, M.A., Roy, R., and Gabius, H.-J., Lactose-containing starburst dendrimers: influence of dendrimer generation and binding-site orientation of receptors (plant/animal lectins and immunoglobulins) on binding properties, *Glycobiology*, 9, 1253, 1999.

126. Page, D., Zanini, D., and Roy, R., Macromolecular recognition: effect of multivalency in the inhibition of binding of yeast mannan to concanavalin A and pea lectins by mannosylated dendrimers, *Bioorg. Med. Chem.*, 4, 1949, 1996.

127. Woller, E.K. and Cloninger, M.C., The lectin-binding properties of six generations of mannose-functionalized dendrimers, *Org. Lett.*, 4, 7, 2002.

128. Hansen, H.C., Haataja, S., Finne, J. and Magnusson, G., Di-, tri-, and tetravalent dendritic galabiosides that inhibit hemagglutination by *Streptococcus suis* at nanomolar concentration, *J. Am. Chem. Soc.*, 119, 6974, 1997.

129. Roy, R., Dendritic and hyperbranched glycoconjugates as biomedical anti-adhesion agents, in *Dendrimer and Dendritic Polymers*, Fréchet, J.M.J. and Tomalia, D.A., Eds., Wiley, Chichester, NY, 2001, p. 361.

130. Jansen, J.F.G.A., de Brabander-van den Berg, E.M.M., and Meijer, E.W., Encapsulation of guest molecules into a dendritic box, *Science*, 266, 1226, 1994.

131. Jansen, J.F.G.A., Janssen, R.A.J., de Brabander-van den Berg, E.M.M., and Meijer, E.W., Triplet radical pairs of 3-carboxyproxyl encapsulated in a dendritic box, *Adv. Mater.*, 7, 561, 1995.

132. Jansen, J.F.G.A., Meijer, E.W., and de Brabander-van den Berg, E.M.M., The dendritic box: shape-selective liberation of encapsulated guests, *J. Am. Chem. Soc.*, 117, 4417, 1995.

133. Behr, J.-P., Synthetic gene-transfer vectors, *Acc. Chem. Res.*, 26, 274, 1993.

134. Malik, N., Evagorou, E.G., and Duncan, R., Dendrimer-platinate: a novel approach to cancer chemotherapy, *Anticancer Drugs*, 10, 767, 1999.

135. (a) Malik, N. and Duncan, R., Dendritic-platinate drug delivery system, U.S. Pat. 6,585,956, July 1, 2003; (b) Malik, N. and Duncan, R., Method of treating concerous tumors with a dendritic-platinate drug delivery systme, U.S., Pat. 6,790,437, September 14, 2004; (c) Malik, N., Duncan, R., Tomalia, D.A. and Esfand, R., Dendritic-antineoplastic drug delivery system, U.S. Pat. 7,005,124, February 28, 2006.

136. Baker, J.R., Jr., Quintana, A., Piehler, L., Banazak-Holl, M., Tomalia, D., and Raczka, E., The synthesis and testing of anti-cancer therapeutic nanodevices, *Biomed. Microdevices*, 3, 61, 2001.

137. Qualmann, B., Kessels, M.M., Musiol, H.-J., Sierralta, W.D., Jungblut, P.W., and Moroder, L., Synthesis of boron-rich lysine dendrimers as protein labels in electron microscopy, *Angew. Chem. Int. Ed. Engl.*, 35, 909, 1996.

138. Kono, K., Liu, M., and Frechet, J.M.J., Design of dendritic macromolecules containing folate or methotrexate residues, *Bioconjug. Chem.*, 10, 1115, 1999.

139. Quintana, A., Raczka, E., Piehler, L., Lee, I., Myc, A., Majoros, I., Patri, A.K., Thomas, T., Mule, J., and Baker, J.R., Jr., Design and function of a dendrimer-based therapeutic nanodevice targeted to tumor cells through the folate receptor, *Pharm. Res.*, 19, 1310, 2002.

140. Shukla, S., Wu, G., Chatterjee, M., Yang, W., Sekido, M., Diop, L.A., Muller, R., Sudimack, J.J., Lee, R.J., Barth, R.F., and Tjarks, W., Synthesis and biological evaluation of folate receptor-targeted boronated PAMAM dendrimers as potential agents for neutron capture therapy, *Bioconjug. Chem.*, 14, 158, 2003.

141. Majoros, I.J., Thomas, T.P., Mehta, C.B., and Baker, J.R., Jr., Poly(amidoamine) dendrimer-based multifunctional engineered nanodevice for cancer therapy, *J. Med. Chem.*, 48, 5892, 2005.

142. Kukowska-Latallo, J., Candido, K.A., Cao, Z., Nigavekar, S.S., Majoros, I.J., Thomas, T.P., Balogh, L.P., Khan, M.K., and Baker, J.R., Jr., Nanoparticle targeting of anticancer drug improves therapeutic response in animal model of epithelial cancer, *Cancer Res.*, 65, 5317, 2005.

143. Gillies, E.R. and Fréchet, J.M.J., Designing macromolecules for therapeutic applications: polyester dendrimer–poly(ethylene oxide) "bow tie" hybrids with tunable molecular weight and architectures, *J. Am. Chem. Soc.*, 124, 14137, 2002.

144. Shukla, R., Thomas, T.P., Peters, J., Kotlyar, A., Myc, A., and Baker, J.R., Jr., Tumor angiogenic vasculature targeting with PAMAM dendrimer-RGD conjugates, *Chem. Commun.*, 46, 5739, 2005.

145. Backer, M.V., Gaynutdinov, T.I., Patel, V., Bandyopadhyaya, A.K., Thirmumamagal, B.T.S., Tjarks, W., Barth, R.F., Claffey, K., and Backer, J.M., Vascular endothelial growth factor selectively targets boronated dendrimers to tumor vasculature, *Mol. Cancer Ther.*, 4, 1423, 2005.

146. Wiener, E.C., Brechbiel, M.W., Brothers, H., Magin, R.L., Gansow, O.A., Tomalia, D.A., and Lauterbur, P.C., Dendrimer-based metal chelates: a new class of magnetic resonance imaging contrast agents, *Magn. Reson. Med.*, 31, 1, 1994.

147. Wiener, E.C., Tomalia, D.A., and Lauterbur, P.C., Relaxivity and stabilities of metal complexes of starburst dendrimers: a new class of MRI contrast agents, in *Works in Progress*, Society of Magnetic Resonance in Medicine, 9th Annual Meeting and Exhibition, New York, 1990, p. 1106.

148. Kobayashi, H. and Brechbiel, M.W., Nano-sized MRI contrast agents with dendrimer cores, *Adv. Drug Deliv. Rev.*, 57, 2271, 2005.

149. Kobayashi, H. and Brechbiel, M.W., Dendrimer-based macromolecular MRI contrast agents: characteristics and application, *Mol. Imaging*, 2, 1, 2003.

150. Kobayashi, H., Jo, S.-K., Kawamoto, S., Yasuda, H., Hu, X., Knopp, M.V., Brechbiel, M.W., Choyke, P.L., and Star, R.A., Polyamine dendrimer-based MRI contrast agents for functional kidney imaging to diagnose acute renal failure, *J. Magn. Reson. Imaging*, 20, 512, 2004.

151. Sato, N., Kobayashi, H., Hiraga, A., Saga, T., Togashi, K., Konishi, J., and Brechbiel, M.W., Pharmacokinetics and enhancement of macromolecular MR contrast agents with various sizes of polyamidoamine dendrimer cores, *Magn. Reson. Med.*, 46, 1169, 2001.

152. Kobayashi, H., Kawamoto, S., Star, R.A., Waldmann, T.A., Tagaya, Y., and Brechbiel, M.W., Micromagnetic resonance lymphangiography in mice using a novel dendrimer-based magnetic resonance imaging contrast agent, *Cancer Res.*, 63, 271, 2003.

153. Kobayashi, H., Kawamoto, S., Saga, T., Sato, N., Hiraga, A., Ishimori, T., Akita, Y., Mamede, M.H., Konishi, J., Togashi, K., and Brechbiel, M.W., Novel liver macromolecular MR contrast agent with a polypropylenimine diaminobutyl dendrimer core: comparison to the vascular MR contrast agent with the polyamidoamine dendrimer core, *Magn. Reson. Med. Med.*, 46, 795, 2001.

154. Kobayashi, H., Sato, N., Saga, T., Nakamoto, Y., Ishimori, T., Toyama, S., Togashi, K., Konishi, J., and Brechbiel, Monoclonal antibody–dendrimer conjugates enable radiolabeling of antibody with markedly high specific activity with minimal loss of immunoreactivity, *Eur. J. Nucl. Med.*, 27, 1334, 2000.

155. Wiener, E.C., Konda, S., Shadron, A., Brechbiel, M., and Gansow, O., Targeting dendrimerchelates to tumors and tumor cells expressing the high-affinity folate receptor, *Invest. Radiol.*, 32, 748, 1997.

156. Gareis, M., Harrer, P., and Bertling, W.M., Homologous recombination of exogenous DNA fragments with genomic DNA in somatic cells of mice, *Cell. Mol. Biol.*, 37, 191, 1991.

157. Margerum, L.D., Campion, B.K., Koo, M., Shargill, N., Lai, J.-J., Marumoto, A., and Sontum, P.C., Gadolinium(III) DO3A macrocycles and polyethylene glycol coupled to dendrimers: effect of molecular weight on physical and biological properties of macromolecular magnetic resonance imaging contrast agents, *J. Alloys Compounds*, 249, 185, 1997.

158. Platzek, J. and Schmitt-Willich, H., Synthesis and development of gadomer, a dendritic MR contrast agent, *ACS Symp. Ser.*, 903, 192, 2005.

159. Misselwitz, B., Schmitt-Willich, H., Ebert, W., Frenzel, T., and Wienmann, H.-J., *MAGMA*, 12, 128, 2001.

160. Boas, U. and Heegaard, P.M.H., Dendrimers in drug research, *Chem. Soc. Rev.*, 33, 43, 2004.

161. Landers, J.J., Cao, Z., Lee, I., Piehler, L.T., Myc, P.P., Myc, A., Hamouda, T., Galecki, A.T., and Baker, J.R., Jr., Prevention of influenza pneumonitis by sialic acid conjugated dendritic polymers, *J. Infect. Dis.*, 186, 1222, 2002.

162. Mammen, M., Dahmann, G., and Whitesides, G.M., Effective inhibitors of haemagglutination by influenza virus synthesized from polymers having active ester groups: insight into mechanism of inhibition, *J. Med. Chem.*, 38, 4179, 1995.

163. Bourne, N., Stanberry, L.R., Kern, E.R., Holan, G., Matthews, B., and Bernstein, D.I., Dendrimers, a new class of candidate topical microbicides with activity against herpes simplex virus infection, *Antimicrob. Agents Chemother.*, 44, 2471, 2000.

164. Gong, E., Matthews, B., McCarthy, T., Chu, J., Holan, G., Raff, J., and Sacks, S., Evaluation of dendrimer SPL7013, a lead microbicide candidate against herpes simplex viruses, *Antiviral Res.*, 68, 139, 2005.

165. Gong, Y., Matthews, B., Cheung, D., Tam, T., Gadawski, I., Leung, D., Holan, G., Raff, J., and Sacks, S., Evidence of dual sites of action of dendrimers: SPL-2999 inhibits both virus entry and late stages of herpes simplex virus replication, *Antiviral Res.*, 55, 319, 2002.

166. Witvrouw, M., Pannecouque, C., Matthews, B., Schols, D., Andrei, G., Snoeck, R., Neyts, J., Leyssen, P., Desmyter, J., Raff, J., DeClerq, E., and Holan, G., Dendrimers inhibit the replication of human immunodeficiency virus by a dual mechanism of action, *Antiviral Res.*, 41, A25, 1999.

167. Witvrouw, M., Fikkert, V., Pluymers, W., Matthews, B., Mardel, K., Schols, D., Raff, J., Debyser, Z., De Clerq, E., Holan, G., and Pannecouque, C., Polyanionic (i.e., polysulfonate) dendrimers can inhibit the replication of human immunodeficiency virus by interfering with both virus adsorption and later steps (reverse transcriptase/integrase) in the virus replicative cycle, *Mol. Pharmacol.*, 58, 1100, 2000.

168. de Clercq, E., Anti-HIV activity of sulfated polysaccharides, in *Carbohydrates and Carbohydrate Polymers, Analysis, Biotechnology, Modification, Antiviral, Medical and Other Applications*, Yalpani, M., Ed., ATL Press, Mt. Prospect, IL, 1993, p. 87.

169. de Clercq, E., Toward improved anti-HIV chemotherapy: therapeutic strategies for intervention with HIV infections, *J. Med. Chem.*, 38, 2491, 1995.

170. Jiang, Y.H., Emau, P., Cairns, J.S., Flanary, L., Morton, W.R., McCarthy, T.D., and Tsai, C.C., SPL7013 gel as a topical microbicide for prevention of vaginal transmission of SHIV89.6P in macaques, *AIDS Res. Hum. Retroviruses*, 21, 207, 2005.

171. Bernstein, D.I., Stanberry, L.R., Sacks, S., Ayisi, N.K., Gong, Y.H., Ireland, J., Mumper, R.J., Holan, G., Matthews, B., McCarthy, T., and Bourne, N., Evaluations of unformulated and formulated dendrimer-based microbicide candidates in mouse and guinea pig models of genital herpes, *Antimicrob. Agents, Chemother.*, 47, 3784, 2003.

172. Dezzutti, C.S., James, V.N., Ramos, A., Sullivan, S.T., Siddig, A., Bush, T.J., Grohskopf, L.A., Paxton, L., Subbarao, S., and Hart, C.E., *In vitro* comparison of topical microbicides for prevention of human immunodeficiency virus type 1 transmission, *Antimicrob. Agents, Chemother.*, 48, 3834, 2004.

173. McCarthy, T.D., Karellas, P., Henderson, S.A., Giannis, M., O'Keefe, D.F., Heery, G., Paull, J.R.A., Matthews, B.R., and Holan, G., Dendrimers as drugs: discovery and preclinical and clinical development of dendrimer-based microbicides for HIV and STI prevention, *Mol. Pharmacol.*, 2, 312, 2005.

174. Product focus: VivaGel, Starpharma Limited, Melbourne, Australia. See: www.starpharma.com.

175. Kensinger, R.D., Catalone, B.J., Krebs, F.C., Wigdahl, B., and Schengrund, C.-L., Novel polysulfated galactose-derivatized dendrimers as binding antagonists of human immunodeficiency virus type 1 infection, *Antimicrob. Agents Chemother.*, 48, 1614, 2004.

176. Holan, G. and Matthews, B., Angiogenic inhibitory compounds, Australian Patent PCT/AU97/ 00447, 1997.

177. Holan, G., personal communication.

178. Shaunak, S., Thomas, S., Gianasi, E., Godwin, A., Jones, E., Teo, I., Mireskandari, K., Luthert, P., Duncan, R., Patterson, S., Khaw, P., and Brocchini, S., Polyvalent dendrimer glucosamine conjugates prevent scar tissue formation, *Nat. Biotechnol.*, 22 (8), 977, 2004.

179. Kasai, S., Nagasawa, H., Shimamura, M., Uto, Y., and Hori, H., Design and synthesis of antiangiogenic/heparin-binding arginine dendrimer mimicking the surface of endostatin, *Bioorg. Med. Chem. Lett.*, 12, 951, 2002.

180. Sasaki, T., Larsson, H., Kreuger, J., Salmivirta, M., Claesson-Welsh, L., Lindahl, U., Hohenester, E., and Timpl, R., *EMBO J.*, 18, 6240, 1999.

181. Marano, R.J., Wimmer, N., Kearns, P.S., Thomas, B.G., Toth, I., Brankov, M., and Rakoczy, P.E., Inhibition of *in vitro* VEGF expression and choroidal neovascularization by synthetic dendrimer peptide mediated delivery of a sense oligonucleotide, *Exp. Eye Res.*, 79, 525, 2004.

182. Balogh, L. and Tomalia, D.A., Poly(amidoamine) dendrimer-templated nanocomposites. 1. Synthesis of zero valent copper nanoclusters, *J. Am. Chem. Soc.*, 120, 7355, 1998.

183. Tomalia, D.A. et al., U.S. Patent 6,664,315 B2, 2003.

184. Martell, A.E. and Hancock, R.D., *Metal Complexes in Aqueous Solutions*, Plenum Press, New York, 1996.

185. Diallo, M.S., Balogh, L., Shafagati, A., Johnson, J.H., Jr., Goddard, W.A., III, and Tomalia, D.A., Poly(amidoamine) dendrimers: a new class of high capacity chelating agents for Cu(II) ions, *Environ. Sci. Technol.*, 33, 820, 1999.

186. Diallo, M.S., Christie, S., Swaminathan, P., Balogh, L., Shi, X., Um, W., Papelis, C., Goddard, W.A., III, and Johnson, J.H., Jr. Dendritic chelating agents. 1. Cu(II) binding to ethylene diamine core poly(amidoamine) dendrimers in aqueous solutions, *Langmuir*, 20, 2640, 2004.

187. Diallo, M.S., Christie, S., Swaminathan, P., Johnson, J.H., Jr., and Goddard, W.A., III, Dendrimer enhanced ultrafiltration. 1. Recovery of Cu(II) from aqueous solutions using Gx-NH2 PAMAM dendrimers with ethylene diamine with ethylene diamine core, *Environ. Sci. Technol.*, 39, 1366, 2005.

188. Ottaviani, M.F., Montalti, F., Turro, N.J., and Tomalia, D.A., Characterization of starburst dendrimers by the EPR technique: copper(II) ions binding full-generation dendrimers, *J. Phys. Chem. B*, 101, 158, 1997.

189. Zhou, L., Russell, D.H., Zhao, M., and Crooks, R.M., Characterization of poly(amidoamine) dendrimers and their complexes with Cu^{2+} by matrix-assisted laser desorption ionization mass spectrometry, *Macromolecules*, 34, 3567, 2001.

190. Tran, M.L., Gahan, L.R., and Gentle, I.R., Structural studies of copper(II)-amine terminated dendrimer complexes by EXAFS, *J. Phys. Chem. B*, 108, 20130, 2004.

191. Krot, K.A., de Namor, A.F.D., Aguilar-Cornejo, A., and Nolan, K.B., Speciation, stability constants and structures of complexes of copper(II), nickel(II), silver(I) and mercury(II) with PAMAM dendrimer and related tetraamide ligands, *Inorg. Chim. Acta*, 358, 3497, 2005.

192. Ottaviani, M.F., Favuzza, P., Bigazzi, M., Turro, N.J., Jockusch, S., and Tomalia, D.A., A TEM and EPR investigation of the competitive binding of uranyl ions to starburst dendrimers and liposomes: potential use of dendrimers as uranyl ion sponges, *Langmuir*, 16, 7368, 2000.

193. Zhang, Z.S., Yu, X.M., Fong, L.K., and Margerum, L.D., Ligand effects on the phosphoesterase activity of Co(II) Schiff base complexes built on PAMAM dendrimers, *Chim. Acta*, 317, 72, 2001.

194. Kobayashi, H., Kawamoto, S., Saga, T., Sato, N., Hiraga, A., Konishi, J., Togashi, K., and Brechbiel, M.W., Micro-MR angiography of normal and intratumoral vessels in mice using dedicated intravascular MR contrast agents with high generation of polyamidoamine dendrimer core: reference to pharmacokinetic properties of dendrimer-based MR contrast agents, *J. Magn. Reson. Imaging*, 14, 705, 2001.

195. Kobayashi, H., Kawamoto, S., Jo, S.-K., Brechbiel, M.W., and Star, R.A., Macromolecular MRI contrast agents with small dendrimers: pharmacokinetic differences between sizes and cores, *Bioconjug. Chem.*, 14, 388, 2003.

196. Esumi, K., Hosoya, T., Suzuki, A., and Torigoe, K., Spontaneous formation of gold nanoparticles in aqueous solution of sugar-persubstituted poly(amidoamine) dendrimers, *Langmuir*, 16, 2978, 2000.

197. Balogh, L., Swanson, D.R., Tomalia, D.A., Hagnauer, G.L., and McManus, A.T., Dendrimer — silver complexes and nanocomposites as antimicrobial agents, *Nano Lett.*, 1 (1), 18, 2001.

198. Crooks, R.M., Zhao, M., Sun, L., Chechik, V., and Yeung, L.K., Dendrimer-encapsulated metal nanoparticles: synthesis, characterization and application to catalysis, *Acc. Chem. Res.*, 34, 181, 2001.

199. Wilson, O.M., Scott, R.W.J., Garcia-Martinez, J.C., and Crooks, R.M., Separation of dendrimer-encapsulated Au and Ag nanoparticles by selective extraction, *Chem. Mater.*, 16, 4202, 2004.

200. Soto-Castro, D. and Guadarrama, P., Macrocyclic vs. dendrimeric effect: a DFT study, *J. Comp. Chem.*, 25, 1215, 2004.

201. Tarazona-Vasquez, F. and Balbuena, P.B., Complexation of Cu(II) ions with the lowest generation poly(amido-amine)-OH dendrimers: a molecular simulation study, *J. Phys. Chem. B*, 109, 12480, 2005.

202. Beer, P.D. and Gale, P.A., Anion recognition and sensing: the state of the art and future perspectives, *Angew. Chem. Int. Ed. Engl.*, 40, 487, 2001.

203. Gloe, K., Stephan, H., and Grotjahn, M., Where is the anion extraction going? *Chem. Eng. Technol.*, 26, 1107, 2003.

204. Twyman, L.J., Beezer, A.E., Esfand, R., Hardy, M.J., and Mitchell, J.C., The synthesis of water soluble dendrimers and their application of possible drug delivery systems, *Tetrahedron Lett.*, 40, 1743, 1999.

205. Birnbaum, E.R., Rau, K.C., and Sauer, N.N., Selective anion binding from water using soluble polymer, *Sep. Sci. Technol.*, 38, 389, 2003.

206. Daniel, M.-C., Ruiz, J., Blais, J.-C., Daro, N., and Astruc, D., Synthesis of five generations of redox-stable pentamethylamidoferrocenyl dendrimers and comparison of amidoferrocenyl-and pentamethylamidoferrocenyl dendrimers as electrochemical exoreceptors for the selective recognition of $H_2PO_4^-$, HSO_4^-, and adenosine 5'-triphosphate (ATP) anions: stereoelectronic and hydrophobic roles of cyclopentadienyl permethylation, *Chem. Eur. J.*, 9, 4371, 2003.

207. Valerio, C., Fillaut, J.L., Ruiz, J., Guittard, J., Blais, J.-C., and Astruc, D., The dendritic effect in molecular recognition: ferrocene dendrimers and their use as supramolecular redox sensors for the recognition of small inorganic anions, *J. Am. Chem. Soc.*, 119, 2588, 1997.

208. Israelachvili, J.N., *Intermolecular and Surface Forces*, Academic Press, San Diego, CA, 1992.

209. Pistolis, G., Malliaris, A., Paleos, C.M., and Tsiourvas, D., Study of poly(amidoamine) starburst dendrimers by fluorescence probing, *Langmuir*, 13, 5870, 1997.

210. Watkins, D.M., Sayed-Sweeet, Y., Klimash, J.W., Turro, N.J., and Tomalia, D.A., Dendrimers with hydrophobic cores and the formation of supramolecular dendrimer-surfactant assemblies, *Langmuir*, 13, 3136, 1997.

211. Newkome, G.R., Moorfield, C.N., Baker, G.R., Jr., Saunders, M.J., and Grossman, S.H., Unimolecular micelles, *Angew. Chem. Int. Ed.*, 30 (9), 1178, 1991.

212. Hawker, C.J., Wooley, K.L., and Fréchet, J.M.J., Unimolecular micelles and globular amphiphiles: dendritic macromolecules as novel recyclable solubilizing agents, *J. Chem. Soc., Perkin Trans. I*, 1287, 1993.

213. Kleinman, M.H., Flory, J.H., Tomalia, D.A., and Turro, N.J., Effect of protonation and PAMAM dendrimer size on the complexation and dynamic mobility of 2-naphthol, *J. Phys. Chem. B*, 104, 11472, 2000.

214. Caminade, A.-M. and Majoral, J.P., Water-soluble phosphorous-containing dendrimers, *Prog. Polym. Sci.*, 30, 491, 2005.

215. Vassilev, K. and Ford, W.T.J., Poly(propyleneimine) dendrimer complexes of Cu(II), Zn(II), and Co(III) as catalysts of hydrolysis of *p*-nitrophenyl diphenyl phosphate, *J. Polym. Sci. A*, 37, 2727, 1999.

216. Astruc, D. and Chardac, F., Dendritic catalysts and dendrimers in catalysis, *Chem. Rev.*, 101, 2991, 2001.

217. Ooe, M., Murata, M., Mizugaki, T., Ebitani, K., and Kaneda, K., Supramolecular catalysts by encapsulating palladium complexes within dendrimers, *J. Am. Chem. Soc.*, 126, 604, 2004.

218. Knapen, J.W.J., van der Made, A.W., de Wilde, J.C., van Leeuwen, P.W.N.M., Wijkens, P., Grove, D.M., and Van Koten, G., Homogeneous catalysts based on silane dendrimers functionalized with arylnickel(II) complexes, *Nature*, 372, 659, 1994.

219. van de coevering, R., Gebbink, J.M.K., and Van Koten, G., Soluble organic supports for the non-covalent immobilization of homogeneous catalysts: modular approaches towards sustainable catalysts, *Prog. Polym. Sci.*, 30, 474, 2005.

220. Brinkman, N., Giebel, D., Lohmer, M., Reetz, M.T., and Kragi, U., Allylic substitution with dendritic palladium catalysts in a continuously operating membrane reactor, *J. Catal.*, 183, 163, 1999.

221. van Heerbeek, R., Kamer, P.C.J., van Leeuwen, P.W.N.M., and Reek, J.N.H., Dendrimers as support for recoverable catalysts and reagents, *Chem. Rev.*, 102, 3717, 2002.

222. DeGroot, D., Reek, J.N.H., Kamer, P.C.J., and van Leeuwen, P.W.N.M., Palladium complexes of phosphane-functionalised carbosilane dendrimers as catalysts in a continuous-flow membrane reactor, *Eur. J. Org. Chem.*, 2002, 1085, 2002.

223. Scott, R.W.J., Wilson, O.M., and Crooks, R.M., Synthesis, characterization, and applications of dendrimer-encapsulated nanoparticles, *J. Phys. Chem. B*, 109, 692, 2005.

224. Diallo, M.S., Water treatment by dendrimer enhanced filtration, U.S. Patent, pending.

225. Jackson, J.L., Chanzy, H.D., Booy, F.P., Drake, B.J., Tomalia, D.A., Bauer, B.J., and Amis, E.J., Visualization of dendrimer molecules by transmission electron (TEM): staining methods and cryo-TEM of vitrified solutions, *Macromolecules*, 31, 6259, 1998.

226. Li, J., Swanson, D.R., Qin, D., Brothers, H.M., II, Piehler, L.T., Tomalia, D.A., and Meier, D.J., Characterization of core–shell tecto(dendrimers) molecules by tapping mode atomic force microscopy, *Langmuir*, 15, 7347, 1999.

227. Tomalia, D.A., Huang, B., Swanson, D.R., Brothers, H.M., II, and Klimash, J.W., Structure control within poly(amidoamine) dendrimers: size, shape and regio-chemical mimicry of globular proteins, *Tetrahedron*, 59, 3799, 2003.

228. Yin, R., Zhu, Y., and Tomalia, D.A., Architectural copolymers: rod-shaped, cylindrical dendrimers, *J. Am. Chem. Soc.*, 120, 2678, 1998.

229. Matthews, O.A., Shipway, A.N., and Stoddart, J.F., Dendrimers — branching out from curiosities into new technologies, *Prog. Polym. Sci.*, 23, 1, 1998.

230. Li, J., Piehler, L.T., Qin, D., Baker, J.R., Jr., and Tomalia, D.A., Visualization and characterization of poly(amidoamine) dendrimers by atomic force microscopy, *Langmuir*, 16, 5613, 2000.

231. Majoral, J.-P. and Caminade, A.-M., Dendrimers containing heteroatoms (Si, P, B, Ge, or Bi), *Chem. Rev.*, 99, 845, 1999.

232. Freemantle, M., Blossoming dendrimers, *Chem. Eng. News*, 77 (44), 27, 1999.

233. Tomalia, D.A. and Esfand, R., Dendrons, dendrimers and dendrigrafts, *Chem. Ind.*, 11, 416, 1997.

234. Astruc, D., Organometallic chemistry at the nanoscale: dendrimers for redox processes and catalysis, *Pure Appl. Chem.*, 75, 461, 2003.

235. Shull, R.D., Balogh, L., Swanson, D.R., and Tomalia, D.A., Magnetic dendrimers and other nanocomposites, in *Book of Abstracts*, 216th ACS National Meeting, Boston, August 23–27, 1998, MACR-069 Publisher, ACS, Washington.

236. Rajca, A. and Utampanya, S., Dendrimer-based metal chelates: a new class of magnetic resonance imaging contrast agents, *J. Am. Chem. Soc.*, 115, 10688, 1993.

237. Rajca, A., Wongsriratanakul, J., Rajca, S., and Cerny, R., A dendritic macrocyclic organic polyradical with a very high spin of S = 10, *Angew. Chem. Int. Ed.*, 37, 1229, 1998.

238. Tabakovic, I., Miller, L.L., Guan, R.G., Tully, D.C., and Tomalia, D.A., Dendrimers peripherally modified with anion radicals that form Pi-dimers and Pi-stacks, *Chem. Mater.*, 9, 736, 1997.

239. Miller, L.L., Duan, R.G., Tully, D.C., and Tomalia, D.A., Electrically conducting dendrimers, *J. Am. Chem. Soc.*, 119, 1005, 1997.

240. Kawa, M. and Fréchet, J.M.J., Self-assembled lanthanide-cored dendrimer complexes: enhancement of the luminescence properties of lanthanide ions through site-isolation and antenna effects, *Chem. Mater.*, 10, 286, 1998.

241. Sato, T., Jiang, D.-L., and Aida, T., A blue-luminescent dendritic rod: poly(phenyleneethynylene) within a light-harvesting dendritic envelope, *J. Am. Chem. Soc.*, 121, 10658, 1999.

242. Tomalia, D.A., Uppuluri, S., Swanson, D.R., and Li, J., Dendrimers as reactive modules for the synthesis of new structure controlled, higher complexity — megamers, *Pure Appl. Chem.*, 72, 2343, 2000.

243. Uppuluri, S., Piehler, L.T., Li, J., Swanson, D.R., Hagnauer, G.L., and Tomalia, D.A., Core–shell tecto(dendrimers). I. Synthesis and characterization of saturated shell models, *Adv. Mater.*, 12 (11), 796, 2000.

244. Tomalia, D.A. and Swanson, D.R., Laboratory synthesis and characterization of megamers: core–shell tecto(dendrimers), in *Dendrimers and Other Dendritic Polymers*, Fréchet, J.M.J. and Tomalia, D.A., Eds., Wiley, Chichester, NY, 2001, p. 617.

245. Mansfield, M.L., Rakesh, L., and Tomalia, D.A., The random parking of spheres on spheres, *J. Chem. Phys.*, 105, 3245, 1996.

246. Peterson, J., Allikmaa, V., Subbi, J., Pehk, T., and Lopp, M., Structural deviations in poly(amidoamine) dendrimers: a MALDI-TOF MS analysis, *Eur. Polym. J.*, 39, 33, 2003.

247. Tomalia, D.A., Starburst/cascade dendrimers: fundamental building blocks for a new nanoscopic chemistry set, *Aldrichim. Acta*, 26 (4), 91, 1993.

248. Ottaviani, M.F., Sacchi, B., Turro, N.J., Chen, W., Jockush, S., and Tomalia, D.A., An EPR study of the interactions between starburst dendrimers and polynucleotides, *Macromolecules*, 32, 2275, 1999.

249. Percec, V., Johansson, G., Ungar, G., and Zhou, J.P., Fluorophobic effect induces the self-assembly of semifluorinated tapered monodendrons containing crown ethers into supramolecular columnar dendrimers which exhibit a homeotropic hexagonal columnar liquid crystalline phase, *J. Am. Chem. Soc.*, 118, 9855, 1996.

250. Percec, V., Ahn, C.-H., Unger, G., Yeardly, D.J.P., and Moller, M., Controlling polymer shape through the self-assembly of dendritic side-groups, *Nature*, 391, 161, 1998.

251. Hudson, S.D., Jung, H.-T., Percec, V., Cho, W.-D., Johansson, G., Ungar, G., and Balagurusamy, V.S.K., Direct visualization of individual cylindrical and spherical supramolecular dendrimers, *Science*, 278, 449, 1997.

252. Goodson, T., III, Optical effects manifested by PAMAM dendrimer metal nano-composites, in *Dendrimers and Other Dendritic Polymers*, Fréchet, J.M.J. and Tomalia, D.A., Eds., Wiley, Chichester, NY, 2001, p. 515.

253. Bosman, A.W., Janssen, H.M., and Meijer, E.W., About dendrimers: structure, physical properties, and applications, *Chem. Rev.*, 99, 1665, 1999.

254. Sayed-Sweet, Y., Hedstrand, D.M., Spindler, R., and Tomalia, D.A., Hydrophobically modified poly(amidoamine) (PAMAM) dendrimers: their properties at the air–water interface and use as nanoscopic container molecules, *J. Mater. Chem.*, 7 (7), 1199, 1997.

25

Design and Applications of Photonic Crystals

25.1 Introduction...25-1
25.2 Photonic Crystals? How Do They Work? 25-3
25.3 Analogy between Photonic and Semiconductor
Crystals...25-4
25.4 Analyzing Photonic Bandgap Structures.......................25-5
25.5 Electromagnetic Localization in Photonic
Crystals...25-10
25.6 Doping of Photonic Crystals.......................................25-10
25.7 Microcavities in Photonic Crystals 25-11
25.8 Photonic Bandgap Applications..................................25-11
    Low-Loss Optical Waveguides • Waveguide Bends • High-Q
    Microcavities • High-Quality Filter • Channel Drop Filters in
    Photonic Crystals • Optical Limiter • Beam Splitter • Bragg
    Reflector • Zero Cross-Talk • Waveguiding through Localized
    Coupled Cavities • 2-D Distributed Feedback Laser Generator
    • Photonic Crystal Fiber • Second Harmonic Generation
    • Air-Bridge Microcavity • Control of Spontaneous Emission
    • Enhancing Patch Antenna Performance Using Photonic
    Crystals • Surface-Emitting Laser Diode Using Photonic
    Bandgap Crystal Cavity • Optical Spectrometer • Hybrid
    Photonic Crystal Structures • Tunable Photonic Crystals
    • Optical Switch Using PBG • Photonic Crystal Optical
    Networks • Fabrication Error Analysis
25.9 Applications Utilizing Irregular Dispersive
Phenomena in Photonic Crystals..................................25-29
    Super-Prism Phenomenon • Spot-Size Converter
    • Self-Collimation • Dispersion-Based Nonchannel
    Waveguiding • Dispersion-Based Routing Structure
    • Dispersion-Based Variable Beam Splitter
25.10 Future Applications and Concluding Remarks 25-33
References 25-33

Dennis W. Prather
Ahmed S. Sharkawy
Shouyuan Shi
Caihua Chen
University of Delaware

25.1 Introduction

Throughout the last two decades, there has been significant activity in the development of photonic devices that can confine, control, and route light on a scale comparable to modern electronic devices, namely the nanometer scale. A key motivation for this is to realize photonic circuits having a density approaching that of modern electronic circuits. However, in order for this to be done, such devices would need the ability to confine light on a subwavelength scale and exist in a material compatible with the microelectronics manufacturing infrastructure. Although the latter requirement was readily satisfied through a proper choice of materials, i.e., silicon, the former one was more elusive. The reason for this arose from the fact that reflective, or conducting, devices are very lossy at optical wavelengths and refractive, or total internally reflective, devices do not offer mode confinement on a small enough scale. For this reason, researchers turned to the field of photonic crystal (PhC) devices, and their associated photonic bandgap (PBG) devices, which offer both low loss and high confinement, and can be readily fabricated in silicon. However, before we begin discussing the various aspects of silicon-based PhCs, we first present a brief perspective on their current status.

To a large extent, the field of PhCs can be thought of as having its origins in the rather conventional 1-D thin-film stack, wherein a quarter wave thickness of alternating materials is arranged in a periodic fashion. In this device, it is well known that certain frequencies are transmitted whereas others are not. As the index contrast between the alternating layers is increased, the selectivity of the transmitted and reflected wavelengths is also increased.

When extended to two and three dimensions, this selectivity is what gives rise to high-mode confinement. While 1-D thin-film stacks have been known for over a century, their generalization to higher dimensions was not proposed until the 1970s by Bykov[1,2] as a possible way of inhibiting spontaneous emission. In essence, Bykov proposed the use of a periodic structure that served to inhibit certain electromagnetic frequencies, thereby disallowing spontaneous emission. These devices ultimately became known as photonic bandgap (PBG) structures, as they suppressed a band of frequencies from existing, the so-called "photonic bandgap." In a similar vein, Yablonovitch[3] proposed a structure where an electronic and a photonic gap overlapped, thereby making it possible to enhance the performance of lasers, heterojunction bipolar transistors, and solar cells. Subsequent to this work, John[4] proposed using such structures for the localization of light in strongly scattering dielectric structures. In each of these cases, the basic idea was to tailor the properties of photons in a PhC in a way directly analogous to how atomic crystals tailor the properties of electrons. That is to say, in the electronic case, the wavefunctions of electrons interact with the periodic potential of the atomic lattice, and for a certain range of energies (similar to frequencies for photons), electronic states cease to exist, thus giving rise to an electronic bandgap. For PhCs, the analogue of the electronic potential in an atomic crystal is the dielectric constants of the constituent materials of the PBG structure. And, due to the periodic interaction, certain PBGs appear wherein certain modes, or frequencies, are disallowed. In such a structure, one can then introduce a line or point defect, which amounts to the absence of the periodic lattice, wherein a mode is localized by virtue of being suppressed within the lattice. For this reason, these devices offer extreme mode confinement as well as the ability to control and route light very efficiently.

With these advances, the field of PhCs and in particular their realization in silicon-based materials has been a very active field of research over the last two decades. During this time, great success was achieved in identifying suitable periodic structures, dielectric materials, and both theoretical as well as experimental demonstration and characterization of 1-D, 2-D, and 3-D PhC and bandgap structures. However, developing suitable structures for novel device applications and systems that are economically feasible remains a challenge. This difficulty arises from the challenging aspects of their fabrication, which often requires high-resolution lithography and high-aspect-ratio etching. Although progress has been, and is being, made in these areas, significant market opportunities are yet to arise. Nonetheless, the field of PhCs offers significant potential and is poised to find its place in the realm of high-technology applications.

Thus, in accordance with this perspective, the remainder of this chapter presents a brief historical overview and attempts to highlight milestones in the development of functional PhC-based devices and applications.

More recently (1987), Yablonovitch[3] and John[4] proposed the idea that a periodic arrangement of metallic or dielectric objects can possess the property of a bandgap for certain regions in the frequency spectrum, depending on the material they are constructed from, whether it be metallic for the microwave regime or dielectric for the optical regime. The structure was called a "photonic crystal" and it can prohibit the propagation of light over a certain band of wavelengths, while allowing other bands to

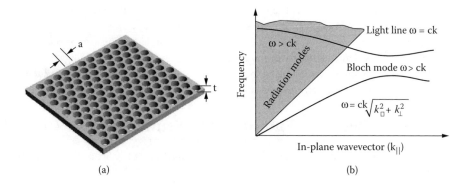

FIGURE 25.1 (a) PhC slab and (b) dispersion diagram for an in-plane wavevector of a periodic structure overlapped with a light line.

propagate. Such behavior gives rise to a PBG, analogous to the electronic bandgap in semiconductor materials. An example of a periodic structure that may exhibit the property of a bandgap is shown in Figure 25.1a.

2-D PhCs can be realized using either a periodic array of dielectric rods of any shape or geometry, or a perforated dielectric slab of air holes. Such structures can be further optimized to achieve a wider, or a narrower, bandgap based on the desired application. 2-D PhCs impose periodicity in two dimensions, whereas the third dimension is either infinitely long (PhC fiber) or has a finite height (PhC slabs). 3-D PhCs impose periodicity in all three dimensions.

As they are easier to fabricate and analyze, 2-D PhCs have attracted the attention of a large number of researchers and engineers. Planar PhC circuits such as splitters,[5–9] high Q-microcavities,[10–16] and channel drop/add filters[17–20] have been investigated both theoretically[21–23] and experimentally.[24–42]

Once a PhC has been designed, its properties can be engineered in a manner similar to that done in electronic crystals. For bandgap-related applications, this is attained through the process of doping. In a PhC, doping is achieved by either adding or removing dielectric material to a certain area. The area to which we have added, or from which we have removed, dielectric material then acts as a defect region that can be used to localize an electromagnetic wave. Doping a PhC opens a broad range of possibilities for optical device development through the localization of light. For nonbandgap-related applications, the dispersion properties of the host PhC structure can be engineered in a similar fashion, to implement various functionalities.

25.2 Photonic Crystals? How Do They Work?

The physical phenomenon that clearly describes the operation of a PhC is the localization of light, which is achieved from the scattering and interference produced by a coherent wave in a periodic structure. Upon an incident radiation, the periodic scatterers constructing a PhC could reflect an incident radiation at the same frequency in all directions. Then, wherever in space the radiation interferences occur constructively, sharp peaks would be observed. This portion of the radiation spectrum is then forbidden to propagate through the periodic structure and this band of frequencies is what was called a stop band, or a PBG. On the other hand, wherever in space an incident radiation destructively interferes with the periodic scatterers in a certain direction, this part of the radiation spectrum will propagate through the periodic structure with minimal attenuation, and this band of frequencies was called a pass band.

For an electromagnetic wave propagating within a dielectric material, scattering takes place on a scale much larger than the wavelength of light. The localization of light occurs when the scale of the coherent multiple scatterers is reduced to the wavelength itself. In this case, a photon located in a lossless dielectric media provides an ideal realization of a single excited state in a static medium, at room temperature. Unlike electron localization, which requires an electron–electron interaction and electron–phonon interactions, photon localization offers the unique possibility of studying the angular, spatial, and temporal dependence of wavefield intensities near localized transitions.

Light localization has fundamental consequences at the quantum level. This can be seen for a periodic array of high dielectrics that have dimensions comparable to the wavelength of light, by exhibiting a complete PBG in a certain range, analogous to the electronic energy bandgap in a semiconductor material. In a PhC, there are no allowed electromagnetic states in the forbidden frequency range.

25.3   Analogy between Photonic and Semiconductor Crystals

In a semiconductor crystal, electron localization can be described using the Schrödinger equation for an electron with an effective mass m^*

$$\left[\frac{-h^2}{8\pi^2 m^*} \nabla^2 + V(x) \right] \varphi(x) = E\varphi(x) \tag{25.1}$$

where h is Planck's constant, m^* the effective mass of electron, $V(x)$ the potential function, $\varphi(x)$ the wavefunction, and E the total energy. The probability of finding an electron at x is given by $|\varphi(x)^2|$. The electron can be trapped by a random potential $V(x)$ in deep local potential fluctuations if the energy E is sufficiently negative. As the energy increases, the probability for the trapped electron to tunnel to a nearby potential fluctuation also increases.

In the case of monochromatic electromagnetic waves of frequency ω propagating in an inhomogeneous, but nondissipative dielectric medium, Maxwell's equations are used to describe the wave propagation through space.

Starting with four macroscopic Maxwell equations:

$$\nabla \cdot \mathbf{B} = 0 \tag{25.2}$$

$$\nabla \cdot \mathbf{D} = \rho \tag{25.3}$$

$$\nabla \times \mathbf{E} + \frac{\partial \mathbf{B}}{\partial t} = 0 \tag{25.4}$$

$$\nabla \times \mathbf{H} - \frac{\partial \mathbf{D}}{\partial t} = \mathbf{J}, \tag{25.5}$$

and using the two constitutive equations

$$\mathbf{D(r)} = \varepsilon(\mathbf{r})\mathbf{E(r)} \tag{25.6}$$

$$\mathbf{B(r)} = \mu(\mathbf{r})\mathbf{H(r)}, \tag{25.7}$$

while keeping in mind that for a dielectric material

$$\mu(\mathbf{r}) = 1.0, \tag{25.8}$$

we can substitute Eq. (25.6) to Eq. (25.8) into Eq. (25.4) and Eq. (25.5) and write them in frequency (steady state) domain form. Doing so results in the following equations:

$$\nabla \times \mathbf{E} + j\omega\mu(\mathbf{r})\mathbf{H(r)} = 0 \tag{25.9}$$

$$\nabla \times \mathbf{H} - j\omega\varepsilon(\mathbf{r})E(\mathbf{r}) = 0. \tag{25.10}$$

In deriving Eq. (25.10), we have assumed that there are no sources of current ($\mathbf{J} = 0$).

Taking the curl of Eq. (25.9) and using Eq. (25.10) to eliminate $\mathbf{H}(\mathbf{r})$ we get

$$\nabla \times \left[\nabla \times \mathbf{E}(\mathbf{r}) \right] = \omega^2 \mu(\mathbf{r}) \varepsilon(\mathbf{r}) \mathbf{E}(\mathbf{r}) \tag{25.11}$$

$$\nabla \times \left[\nabla \times \mathbf{H}(\mathbf{r}) \right] = -\omega^2 \mu(\mathbf{r}) \varepsilon(\mathbf{r}) \mathbf{H}(\mathbf{r}). \tag{25.12}$$

The right-hand side of Eq. (25.11) can be further expanded using vector identities

$$-\nabla^2 \mathbf{E} + \nabla(\nabla \cdot \mathbf{E}) = \omega^2 \mu(\mathbf{r}) \varepsilon(\mathbf{r}) \mathbf{E}(\mathbf{r}). \tag{25.13}$$

The total dielectric constant $\varepsilon(\mathbf{r})$ can be separated into two parts as

$$\varepsilon(\mathbf{r}) = \varepsilon_0 + \varepsilon_{\text{spatial}}(\mathbf{r}), \tag{25.14}$$

where ε_0 is the average value of the dielectric function and $\varepsilon_{\text{spatial}}(\mathbf{r})$ the spatial component of the dielectric function, which is analogous to the potential $V(x)$ in the Schrödinger equation. Eq. (25.12) then can be written as

$$-\nabla^2 \mathbf{E} + \nabla\left(\nabla \cdot \mathbf{E}\right) = \omega^2 \left[\varepsilon_0 + \varepsilon_{\text{spatial}}(\mathbf{r}) \right] \mu(\mathbf{r}) \mathbf{E}(\mathbf{r}). \tag{25.15}$$

The quantity $\varepsilon_0 \omega^2$ is similar to the total energy E in the Schrödinger equation.

For an electronic system, lowering the electron energy usually enhances the electron localization. For a PhC, lowering the photon energy leads to a complete disappearance of the scattering mechanism itself, where at a high photon energy, geometric and ray optic theory becomes more valid, and interference corrections to optical transport become less and less effective.

Eq. (25.12) can be formulated in the form

$$-\nabla^2 \mathbf{E} + \nabla(\nabla \cdot \mathbf{E}) - \omega^2 [\varepsilon(\mathbf{r}) - 1] \mathbf{E}(\mathbf{r}) = \omega^2 \mathbf{E}(\mathbf{r}), \tag{25.16}$$

which is another form of the Schrödinger equation. By comparison, it can be seen that positive dielectric scatterers are analogous to regions of negative potential energy in a quantum system.

We can also see from the above equation that, because the increase in dielectric strength is analogous to an increase in the potential well depth of a quantum mechanical system, the overall effect is to lower the frequency of all modes of the system; hence the band edges will move downward in frequency with a general frequency dependence of $1/\sqrt{\varepsilon_r}$.

25.4 Analyzing Photonic Bandgap Structures

The propagation of electromagnetic waves through a structure having a periodic modulation of material properties on the scale of the wavelength is profoundly different from that of the homogeneous case. This is well known in solid-state physics where the periodicity of the atomic lattice is responsible for the formation of energy bands and corresponding energy gaps of electronic states in metal, semiconductor, and insulator crystals. In a similar fashion, an electromagnetic wave propagating through a periodic medium will also exhibit the formation of allowable states, or modes, and their corresponding gaps. Moreover, when combined with localized defects, such devices can be used to extract the spatial and temporal properties of photons. However, to realize this potential, the initial challenge for researchers within the community was to develop accurate and fast modeling and simulation tools that were capable of analyzing various globally and locally periodic structures with complex material properties.

During the early 1990s, most research efforts were focused on developing efficient and accurate algorithms to perform the calculations of the band structures. Among these algorithms were: the plane wave expansion method (PWM),[43–45] finite-difference time-domain (FDTD) method,[46,47] transfer matrix method,[48] and the finite element method (FEM).[49,50] In addition to these techniques, several other interesting approaches were developed.[51–53]

The method initially used for the theoretical analyses of PBG structures is the PWM, which makes use of an important principle: that normal modes in periodic structures can be expressed as a superposition of a set of plane waves, which is also known as Floquet's theorem.[54]

The PWM represents the periodic fields using a Fourier expansion in terms of harmonic functions defined by the reciprocal lattice vectors, wherein the application of the Fourier expansion turns Maxwell's equation into an eigenvalue problem. This simplicity, together with the development of powerful numerical procedures, has made the plane wave method one of the most widely used tools for finding Bloch modes and eigenfrequencies of an infinite periodic system of scattering objects. A most widely used version of the method is applying the preconditioned conjugate gradient minimization of the Rayleigh quotient for finding eigenstates and frequencies.[44] The minimization of the Rayleigh quotient allows us to handle thousands of plane waves. However, the PWM was limited to simulating infinitely periodic structures, which was constrained by multiple symmetries and the assumption that the structure was lossless. It also assumed the structure to be perfectly periodic and, hence, fabrication tolerances, which highly modulate the spatial and temporal responses of a periodic structure, could not be easily simulated in the PWM technique. In addition, it was not capable of calculating the transmission or reflection spectra. Nevertheless, PWM remains to be a useful platform for quickly determining whether a periodic structure does or does not have a bandgap for a specific polarization, and remains the platform for extracting the highly complex dispersive properties of such periodic structures. The photon dispersion relations inside a PhC have been calculated using PWM,[43,55–57] where Eq. (25.11) is solved as an eigenvalue problem with $\mathbf{E}(\mathbf{r})$ as its eigenfunctions, and ω^2 its eigenvalues. The solution over an irreducible Brillouin zone is plotted in the form of a dispersion diagram, as shown in Figure 25.1b. A dispersion diagram is a 2-D plot of different eigenmodes for different wavevectors, or propagation angles, within a PhC lattice. While a 2-D dispersion diagram is sufficient to show whether or not a bandgap may, or may not, exist for a certain PBG structure, it may not be sufficient for applications where nonlinear behavior of PhCs is being analyzed. An example is the negative refractive index phenomenon and its applications to the super prismatic effect[58–98] in PhCs. For these applications, a 3-D dispersion diagram, or a dispersion surface, will provide a more detailed view on a PhC spatial response for various bands of frequencies both inside and outside the bandgap. Even though the PWM produces an accurate solution for the dispersion properties of a PhC structure, it is still limited due to the fact that transmission spectra, field distribution, and back reflections cannot be easily extracted.

The FDTD method[99] is widely used to calculate transmission and reflection spectra for general computational electromagnetic problems, and it is generally considered to be one of the most applicable for the PhCs. In this case, a wave propagating through the PhC structure is found by a direct discretization of Maxwell's equations in point form, wherein the partial difference equations are discretized in both time and space on a staggered grid. In addition to discretization, the proper boundary conditions, i.e., absorbing and periodic boundary conditions, can be applied. If one defines the input signal as a continuous wave or pulse, the excitation can be propagated through the structure by time stepping through the entire grid repeatedly. Using this approach, several algorithms have been developed for calculating PhC band structures. However, the basic FDTD implementation on a single computer is extremely time-consuming, as computational requirements grow exponentially with problem size. Various approaches have been taken to tackle this problem. The initial solution was to parallelize the FDTD algorithm over a Beowulf cluster of tens or hundreds of PC nodes;[100,101] this, however, provided only a short-term resolution to the ongoing issue of computational time. The mean time to failure of the number of nodes constituting the cluster grew exponentially as did the maintenance cost and the physical space necessary to host such clusters. Recently, a different approach to solving the same problem was taken. This approach relies on implementing the FDTD algorithm over a hardware accelerator-based workstation,[102–106] where a dedicated FPGA-based chip is programmed to execute the FDTD algorithm at computational speeds equivalent to a 150 PC node cluster.

In a manner similar to the FDTD method, the transfer matrix method is implemented by discretizing Maxwell's equations. However, in this approach, the initial excitation is limited to a monochromatic wave. The structure under consideration is divided into a set of layers with the same number of grid nodes in each layer. Then, using the discretized form of Maxwell's equations, the field E_i in the nodes of one layer may be connected to the field E_{i+1} in the nodes of the neighboring layers via the transfer matrix $E_{i+1} = T_i E_i$. Thus, by integrating all layers, the output field is connected to the input field by the transfer matrix, which is a product of individual layer-to-layer transfer matrices. As in the case of the FDTD method, proper boundary conditions need to be used. Although the transfer matrix method is less universal due to numerical instabilities during the integration, it is generally more computationally efficient than the FDTD method.

The FEM is a frequency domain method used to solve Maxwell's equations. In fact, it is also based on a variational principle, as in the case of the plane wave method. However, instead of using a plane wave expansion basis, which is defined over the extent of an entire unit cell, FEM uses a subdomain basis to discretize within the computational unit cell. As such, FEM more efficiently takes into account material discontinuities in the dielectric structure, which helps overcome the slow convergence of the PWEM. To solve the resulting matrix eigenvalue problem, a preconditioned subspace iteration algorithm may be applied to find the most relevant set of eigenvalues within the large system of equations.

For silicon-based PhCs, the most promising class of PhC structures is the PhC slab, which has 2-D (in-plane) periodicity and a height that is comparable to the wavelength of light. The PhC slab is much easier to fabricate than a corresponding 3-D PhC structure and, consequently, more attractive for chip-level integration of different optical devices.[23,107,108] However, being finite in height requires another mechanism for light confinement in the third dimension, namely total internal reflection (TIR). Therefore, for these devices, it is a combination of these two phenomena that serves to localize the in-plane light within the slab.

For these devices, the boundary between the guided and radiation modes is described as the light cone, where the radiation modes are the states that extended to infinity in the clad region outside the slab, and the guided modes are those localized to the plane of the slab, as shown in Figure 25.1b. States that lie below the light line in the band diagram cannot couple with modes in the bulk background. Thus, the discrete bands below the light cone are confined. Mathematically, we express the wavevector $\mathbf{k}$ as $\mathbf{k} = \mathbf{k}_0 + \mathbf{k}_z$, where $\mathbf{k}_0$ is the in-plane wavevector and $\mathbf{k}_z$ is the out-plane wavevector. If the guided modes have an imaginary k_z component, then their modes decay in the cladding. However, if the radiation modes have a real k_z component, then they will leak to the cladding, or radiate to infinity.

PhC slab structures have many potential applications, and most of them rely on their corresponding band structures. To employ the plane wave method for the band structure calculations of a PhC slab, which has only 2-D periodicity, a 3-D periodicity was imposed by introducing a periodic sequence of slabs separated by a sufficient amount of background region to ensure electromagnetic isolation, which is commonly referred to as the supercell technique. In this way, the guided modes are localized within the slab so that the additional periodicity of periodic slabs having a large separation will not affect their eigenfrequencies. However, for the radiation (leaky) modes, which lie above the light cone, this technique is no longer appropriate due to the artificial periodicity in the out-of-plane direction.

To overcome this and to determine the leaky modes above the light cone requires the application of a perfectly matched layer (PML) in the z-direction to absorb the radiation from the slab. The PML-absorbing boundary condition,[109] which was introduced by Berenger as a means to truncate the computational region in the FDTD method, has fast become one of the most proficient ways to absorb waves for any frequency and angle of incidence. This technique is based on an anisotropic-material-based formulation that offers special advantages in that it does not require modification of Maxwell's equations.[110] As such, by using PMLs, the artificial periodicity in the z-direction can be used without affecting the accuracy, as any wave propagating out of the supercell will be absorbed and not give rise to an artificial resonance within it. Thus,

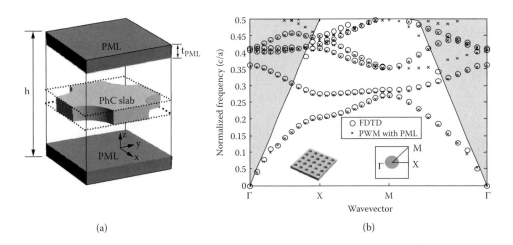

(a)   (b)

FIGURE 25.2 (a) Unit cell for the band structure calculations and (b) comparison of the dispersion diagram between the FDTD and the proposed methods for even modes.

combining the PMLs with the PWM to cast Maxwell's equations into a generalized complex eigenvalue problem[111] has been shown to be a very accurate tool for determining the band diagrams for PhCs.

Although the introduction of the PMLs into the PWM does sufficiently suppress spurious modes, it does give rise to the so-called PML modes, which are generated due to the periodic boundary conditions applied along the z-direction. Therefore, an additional tool is required to distinguish the guide modes, leaky modes, and PML modes. Along these lines, two concepts can be used to distinguish those modes: one is based on the evaluation of the Q-factor of complex resonance modes and the other is based on the fact that guided modes are characterized by a relatively high power concentration within the PhC slab.

To see this, consider a square lattice with air holes embedded in a slab with a high dielectric constant of 12.25 and a thickness of $0.6a$, as shown in Figure 25.2a. The air holes are of a circular cross-section with radius $r = 0.3a$. As shown in Figure 25.2b, there is a good agreement between the modified PWM with the PML method and a 3-D FDTD method.

For a PhC slab, another possible analysis tool is the effective index method, which can reduce a full 3-D problem to a 2-D analysis. For structures with a relatively low index contrast between the slab and cladding layers, the conventional effective index scheme offers a good approximation over a wide frequency range.[112] However, for those structures with a high index contrast, the effective index method is only valid within a very narrow frequency range because the effective index varies significantly over a typical frequency range of interest. For instance, for a perforated silicon slab with a permittivity of 12.25 and a slab thickness of $0.6a$, the effective index of the fundamental mode varies from 1 to 3.25 within the frequency range between 0 and $0.5c/a$. To obtain a good approximation of the effective index, a stepwise effective index method combined with the PWM has been proposed to determine the dispersion diagram.[113]

This revised PWM method starts with a 2-D PhC, for which the problem can be decoupled into two sets of polarizations, namely transverse electric (TE) and transverse magnetic (TM). The eigenproblem can be obtained by following a similar procedure as in the conventional PWM derivation, which for the case of TE modes is:[114]

$$\frac{1}{k_0}\begin{bmatrix} -k_0[G_y] & k_0^2 - (k_x + [G_x])[\kappa]^{-1}(k_x + [G_x]) \\ k_0^2[\kappa] & -k_0[G_y] \end{bmatrix}\begin{bmatrix} [e_x] \\ [h_z] \end{bmatrix} = k_y \begin{bmatrix} [e_x] \\ [h_z] \end{bmatrix} \qquad (25.17)$$

As the above equation shows, instead of solving for the eigenfrequencies as done in the conventional PWM, the revised PWM solves for the wavevectors as eigenvalues for a given frequency.

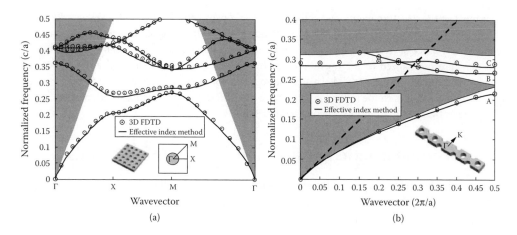

FIGURE 25.3 Comparison of the dispersion diagrams for the TE modes of a PhC slab between the FDTD and the revised PWM: (a) square lattice and (b) triangle line defect.

To understand this, consider a square lattice with air holes embedded in a slab with a high dielectric constant of 12.25 and a thickness of $0.6a$. The air holes are of a circular cross-section with radius $r = 0.3a$. Figure 25.3a shows agreement for the TE mode between the FDTD results and those obtained by the proposed method, within the normalized eigenfrequency range between 0 and 0.5. In addition, a comparison of the computational cost between the two algorithms is worth noting: it took 128 min to obtain the eigenfrequencies for each given wavevector using the FDTD method, whereas it took only 0.05 sec to calculate the eigenwave vectors for each given frequency using the proposed algorithm, which is a significant reduction in computational cost. This method was also used to examine a PhC line defect waveguide, in which the waveguide is formed by removing a row of air holes along the Γ–K direction of the triangular lattice. The suspended slab had a thickness of $0.6a$ and dielectric constant of 11.56. The radii of air holes were $0.3a$ and the results are shown in Figure 25.3b.

Over the past 15 years, numerous modeling and simulation tools to design and analyze complex PhC structures were introduced into the commercial market;[115–120] these tools reflect an ongoing effort within the research community to continue to develop and refine suitable and efficient algorithms and simulation tools for the future development of PhC devices and applications. With a range of such tools available, the community next moved towards realization of the various devices and applications in terms of fabrication and experimental demonstration. Therefore, in the following section, we present a detailed description of the efforts and techniques developed and optimized for this aspect of PhC development.

Eq. (25.11) defines the main design parameters associated with a PhC, such as the fill factor (defined as the ratio of the area of the lattice filled by dielectric to the total area of the whole crystal), the refractive index contrast between the dielectric material and the host material, the ratio of the lattice constant to the radius of the cylinders (for the case of a cylindrical rod), and the wavelength to lattice constant ratio. These parameters define the location, size of the bandgap, and whether a bandgap may, or may not, exist for a specific polarization, either TE field, electric field parallel to the plane, or TM field, magnetic field parallel to the plane. As an example, we present the analysis of a 2-D PhC structure built on a rectangular lattice or silicon rods ($\varepsilon_r = 11.56$), on air background ($\varepsilon_b = 1$), and for a dielectric rod to lattice constant ratio of ($r/a = 0.2$), using the FDTD method. Numerical calculations for transmission coefficients using FDTD, with PML-absorbing boundaries on two sides of the computational region and block periodic boundary conditions on the other two boundaries (to simulate an infinite crystal) were previously presented in Reference 121. As stated before, within-the-bandgap transmission is prohibited (highly attenuated) for which a certain PBG structure resembles an optical mirror with a high reflectivity as shown[121], where the horizontal axis represents the normalized frequency, and the vertical axis represents the amplitude transmission coefficient, in units of dB for six layers of dielectric rods built on a rectangular lattice. We can also see that for the periodic arrangement of the structure mentioned above, a bandgap opens between $a/l_{high} = 0.2452$ and $a/l_{low} = 0.4329$ with a center frequency at $a/l_{center} = 0.339$ and a bandgap size Df/f_{center} equal to 55.4%. This structure can operate for any range of frequencies ranging from microwave frequencies to optical frequencies because all calculations are normalized to the operating wavelength. For example, in the optical frequency regime, where $l = 1550$ nm the structure will have a lattice constant equal to $a = 525.5$ nm and the silicon rods will have a radius $r = 105$ nm. In the following section, we discuss electromagnetic localization in PhCs through analogy to electron localization in semiconductor crystals.

25.5 Electromagnetic Localization in Photonic Crystals

The analogy between electron and photon localization has led to the scaling theory of localization, where the localization critical point is defined by a condition described by

$$\rho(\omega)D(\omega)\ell \cong 1 \tag{25.18}$$

where $\rho(\omega)$ is the photon density of states (DOS), $D(\omega)$ the diffusion coefficient of light in a multiple scattering medium, and ℓ the transport mean free path.

If the scattering microstructures do not significantly alter the photon DOS from its free space value

$$\rho_{vac}(\omega) = \frac{1}{c}\left(\frac{\omega}{c}\right)^2 . \tag{25.19}$$

Then the localization criterion becomes

$$\left(\frac{\omega}{c}\ell\right)^2 \approx 1 \quad \text{assuming that} \quad D(\omega) \approx c\ell \tag{25.20}$$

In general, we can interpret the factor $4\pi(\omega/c)^2$ as the total phase space available for propagation of a photon of frequency ω. Thus, the localization criterion is given by

$$(phase\ space) \times \ell^2 = 4\pi, \tag{25.21}$$

The concepts of DOS, diffusion coefficient, mean free path, phase space, and their interrelationships provide the basis for further understanding PBGs.

The spatial localization of light in a PhC is achieved by introducing defects, which can take the form of a line defect, in which case the PhC resembles a waveguide. Photons lying within the bandgap from which they are not allowed to propagate through the crystal are confined to the defect region, as defined by the PhC walls. Another kind of defect is that of a point defect, in which case the PhC creates a cavity that confines a single or a multiple of closely separated modes to the spatial location of the point defect, centered within the cavity. In the following section, we discuss in more detail the introduction of a point defect into a PhC.

25.6 Doping of Photonic Crystals

In a PhC, there exist a dielectric band and an air band, analogous to the valance and the conduction bands, respectively, in semiconductor material. Between the dielectric and air bands is the PBG, within which no energy state exists and as a result propagation is prohibited.

Doping a semiconductor material can be achieved by either adding a donor or an acceptor atom. Both result in a change in the electrical properties of an atomic crystal by either having a p-type or

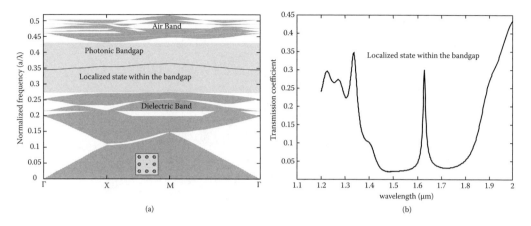

FIGURE 25.4 A photonic crystal with a point defect will allow a single or multiple localized mode to exist within the bandgap as shown in (a) band diagram (b) attenuation diagram.

an n-type material. In a similar fashion, the optical properties of a PhC can be changed by introducing point defects, i.e., either adding or removing a certain amount of dielectric material. When dielectric material is added to a unit cell, it behaves like a donor atom in an atomic crystal, which corresponds to a donor mode and has its origin at the bottom of the air band of the PhC. Alternatively, removing dielectric material from a unit cell makes it behave like an acceptor atom in an atomic crystal, which corresponds to an acceptor mode and has its origin at the top of the dielectric band of the PhC, as shown in Figure 25.4. Thus, acceptor modes are preferable for making single-mode laser microcavities, because they allow a single localized mode to oscillate in the cavity. By adding or removing a certain amount of dielectric material to the PhC, we are disrupting the symmetry of the photonic lattice. By doing so, we are allowing either a single state or a multiple of closely separated states to exist within the bandgap. This phenomenon of localizing states, by introducing point defects, can be useful in designing high-Q-value microcavities in PhCs.

25.7 Microcavities in Photonic Crystals

As stated above, introducing a point defect in a PhC can make a microcavity. As such, the defect can have any shape, size, or dielectric constant. By varying any one of these parameters, the number of modes and the center frequency of the localized mode, or modes, inside the cavity can be changed. If we consider the case of a square lattice of cylindrical rods with a difference in dielectric constant much greater than 2, between the host material and the lattice material, we can introduce a point defect by simply changing one of the parameters of a given rod within the crystal. For example, a point defect consisting of a rod with a radius smaller than those surrounding it will guarantee a single mode to be localized at the point defect. Alternatively, as we increase the radius of the defect, to be equal to or greater than those surrounding it, we will introduce a multiple of closely separated modes localized within the cavity.

The quality factor of the microcavity plays a major role in designing a high-density wavelength division multiplexing (WDM) system. The quality factor depends mainly on the size of the crystal. For high-Q values, the size of the crystal surrounding the cavity needs to be large. It was also shown that the spectral widths of the defect modes decrease rapidly with an increasing number of lattice layers, which is more favorable in WDM because it maximizes the selectivity of the available bandwidth.

25.8 Photonic Bandgap Applications

In this section, we review a few applications that have been introduced in the literature using PhCs.

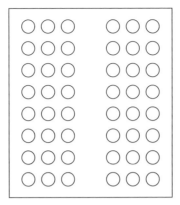

FIGURE 25.5 A waveguide created in a photonic crystal by introducing line defect by either removing or adding dielectric to a certain row or column.

25.8.1 Low-Loss Optical Waveguides

When line defects are introduced into a PhC lattice, an electromagnetic wave having a frequency within the bandgap of the structure can be guided through the crystal. In this case, the line defect resembles a waveguide, as shown in Figure 25.5. In this way, line defects can be formed by either adding or removing dielectric material to a certain row or column along one of the directions of the PhC. To this end, PhC waveguides can be used as an optical wire to guide an optical signal between different points, or devices, within an optical integrated circuit or an optical network.

To create such a channel, one can decrease the radius of a certain row to the point that it no longer exists. By doing so, we have created a waveguide that has a width of

$$W_{grect} = (\Omega + 1)a - 2r, \tag{25.22}$$

where Ω is the number of rows or columns, where the line defect will be created, and r the radius of the rods from which the PhC was created. The width of the line defect/waveguide is proportional to the number of guided modes for a certain wavevector.[56,122] Field patterns for every eigenmode, as well as energy flow, can also be calculated using the FDTD method on the 2-D structure presented above.[7]

Waveguides can also be created on a 2-D triangular lattice PhC of air holes in silicon (Si) background, for which TE modes can be guided through a line defect. In such a structure, the line defect is created in the crystal by increasing the dielectric constant of the line defect, as opposed to decreasing the dielectric constant of the line defect for the 2-D rectangular lattice case, which was used to guide TM waves. For the triangular lattice, the waveguide width can be calculated using

$$W_{gtri} = (\Omega + 1)\frac{\sqrt{3}}{2}a - 2r. \tag{25.23}$$

Note that for the case of a perforated dielectric slab, elimination of a single row or column will not be sufficient to have a single mode of propagation through the line defect, and further design considerations must be taken to achieve that goal.[122,123]

By removing a column, or a row, we can confine the optical beam to the waveguide in a fashion very similar to the TIR concept, which is used to confine the optical signal in optical fibers. However, in PhCs, the mechanism of in-plane optical confinement for a wave propagating through the defect is through multiple Bragg reflections, or distributed Bragg reflections. For finite height PhC structures (slabs), vertical confinement is achieved through TIR at the interface between the PhC slab and lower dielectric constant material, e.g., air.[124,125]

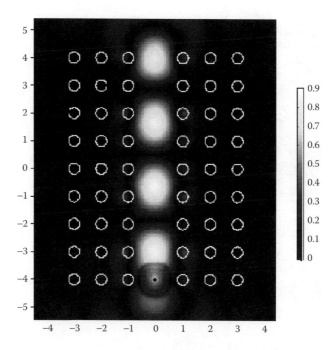

FIGURE 25.6 Steady state field solution of a TM pulse propagating through a PBG waveguide, simulations were performed using FDTD with PML-absorbing boundary conditions.

The main idea of operation for this kind of waveguide is that an incident beam with a frequency within the bandgap of the structure will not propagate through the structure, but will propagate through the waveguide with minimal field leakage. Using this approach, a throughput efficiency as high as 100% can be achieved through the waveguide.[126–135] A snapshot of a FDTD simulation of an optical pulse propagating through the above structure is shown in Figure 25.6.

From Figure 25.6, we can see that by using only three layers on each side of the PBG channel, we were able to achieve high lateral confinement.

25.8.2 Waveguide Bends

A low-loss waveguide that includes a sharp 90° bend in the 2-D PhCs has been reported.[7,9,18,129,136,137] Theoretically, it was shown by a simple scattering theory that 100% transmission is possible. Experimentally, over 80% transmission was demonstrated at a frequency of 100 GHz by using a square array of circular alumina rods having a dielectric constant of 8.9 and a radius of $0.2a$, where a is the lattice constant of the array. For $a = 1.27$ mm, the crystal had a large bandgap extending from 76 to 105 GHz. A line defect was created inside the crystal by removing a row of rods. The optical-guided mode produced by the defect had a large bandwidth, extending over the entire bandgap. A snapshot of FDTD simulation of the structure described above is shown in Figure 25.7.

25.8.3 High-Q Microcavities

A microcavity can be created in a PhC through doping, or the introduction of point defects to a unit lattice as explained previously. If 2-D periodicity is broken by a local defect, local defect modes can occur within the forbidden bandgap. The local defect can then be introduced by either adding or removing a certain amount of dielectric material. If extra dielectric material is added to one of the unit cells, the defect will behave like a donor atom in a semiconductor, and gives rise to donor modes. On the other hand, removing some dielectric materials will introduce defects similar to acceptor atoms in semiconductors

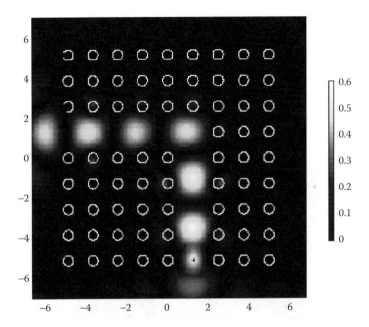

FIGURE 25.7 A sharp waveguide bend built on a PBG.

that produce acceptor modes. It was found that acceptor modes are particularly well suited for making laser microresonator cavities.[11,16,24,122,138]

High-Q microcavities could be constructed with 2-D PhCs if light scattering in the vertical direction, due to the finite depth of the crystal, is minimized. This was demonstrated by the FDTD calculations of the Q-factor for an optical microcavity defined by a three-layer slab waveguide and a 2-D PhC and mirrors. Studying the effect of the finite depth of the crystal on the cavity modes and the loss mechanisms within the cavity optimized the performance of the PhC mirror. It was shown that the Q-factor of the cavity mode depends strongly on the depth of the holes defining the PhC and the refractive index of the material surrounding the waveguide core.[13–15,139–141]

A snapshot of the FDTD simulation, where a point defect was introduced to a PhC for which it was used as a microcavity, is shown in Figure 25.8.

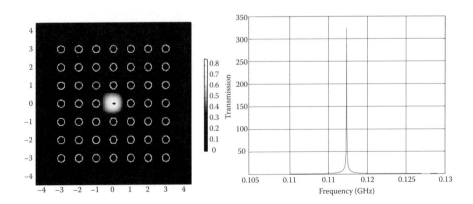

FIGURE 25.8 A microcavity built in a PBG by removing a single rod from the center of a rectangular unit lattice. FDTD simulation is on the left, and the spectral result for the field inside the cavity is on the right.

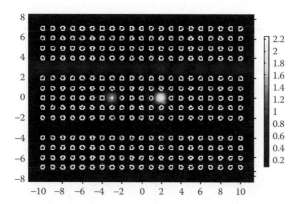

FIGURE 25.9 A channel drop filter realized using a channel, a bus, and two microcavities built on PBG.

25.8.4 High-Quality Filter

High-quality filters were realized by using a perturbed PhC.[142] The perturbed crystal was constructed by randomly repeated stacking of a number of identical unit cells. Each unit cell was an alternating array of M identical high-dielectric components and M identical low-dielectric components. The unit cells were separated by N spacers made of the same material as that of the low-dielectric components. It was shown that although most states are localized due to randomness, there existed states of certain wavelengths. This leads to high-quality resonant tunneling with a transmission peak much more narrow than that of the tunneling through a perfect crystal.

25.8.5 Channel Drop Filters in Photonic Crystals

A channel drop filter was realized in PhCs using two waveguides and an optical resonator system.[19,20,143–145] Maximum transfer between the two waveguides occurs by creating resonant states of different symmetry and by forcing an accidental degeneracy between them. For this case, the optical resonator system consists of two high-Q microcavities. A snapshot of the FDTD simulation for the structure explained above is shown in Figure 25.9, and a plot for the frequency response of the time-varying field stored at the detector on the dropped channel is shown in Figure 25.10. Examining the frequency response of the field in the channel reveals that even though the above system achieves high selectivity (3 mm line width), there exists a considerable amount of interference with the dropped frequency in the channel. This interference can be minimized by narrowing the incident pulse to the extent that the noise is outside the transmission band of the structure but that will be the cost of reducing the number of allowed channels.

25.8.6 Optical Limiter

An effective 2-D optical limiter operating at 514.5 nm for a pulse duration of 0.1 to 4 ms was investigated.[146] The PhC consisted of 180 to 230 nm spatial-period nanochannel glass containing a thermal nonlinear ethanol–toluene liquid. A dynamic range in excess of 130 was achieved in a single-element device with a threshold current of 200 mJ/cm². It was also shown that the spectral width could be broadened to 100 nm by increasing the index differential between the organic liquid and the glass matrix.

25.8.7 Beam Splitter

Waveguide branching,[7,8] and beam splitters[6] have been demonstrated in PhCs. Efficient waveguide branching was achieved by satisfying the rate-matching condition

$$\frac{1}{t_1} = \sum_{n=2}^{N} \frac{1}{t_n}, \quad n = 2, 3, \ldots N \tag{25.24}$$

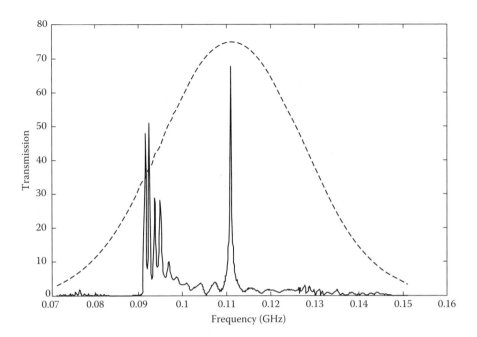

FIGURE 25.10 Frequency spectrum of the time varying electric field in the channel normalized to the incident field. Incident field is shown in the dashed line, while dropped frequency is shown in the solid line.

where $1/t_n$ is the decay rate in waveguide number n in n waveguide branches, and ideal splitting will occur when all $1/t_n$ are equal, which is achieved by placing extra point defects between the input and the output waveguides.

To split electromagnetic waves in 3-D PhCs, a coupled-cavity PhC waveguide was used,[6] where square-shaped alumina rods, with $e = 10.24$ were used. Sample dimensions were $0.32 \times 0.32 \times 15.25 \text{ cm}^3$ at the microwave frequencies. A center-to-center separation between the rods was 1.12 cm. The structure had a 3-D full PBG extending from 10.6 to 12.8 GHz.

25.8.8 Bragg Reflector

A circular Bragg reflector was made with disk-shaped microcavities, of approximately 10 μm² in area, in a GaAs/AlGaAs waveguide structure by etching deep vertical concentric trenches.[147] The trenches formed the Bragg reflector confining light in two lateral dimensions. From photoluminescence excited in the waveguide, the confinement of discrete disk modes was demonstrated. The wavevector of these discrete disk modes was mainly in the radial direction, in contrast with whispering gallery modes in the tangential direction. The high-Q factor of up to 650 indicated that in-plane reflectivities could approach 90%.

25.8.9 Zero Cross-Talk

Numerical simulations showed that a cross-talk reduction of up to eight orders of magnitude could be achieved with a 2-D PhC,[148] as compared to a conventional high-index-contrast waveguide crossing. It was proposed that the design principles could also apply to 3-D systems and are not restricted to PhCs. Tuning the intersection to reduce radiation losses or using a system of resonators to flatten the resonant peak could improve the performance of these devices.

25.8.10 Waveguiding through Localized Coupled Cavities

Coupled PhC cavities were used as a waveguiding mechanism in 3-D PBG structures.[149–152] Here it was shown that photons propagate through strongly localized cavities due to coupling between adjacent cavity modes. Transmission values as high as 100% were observed for various waveguide structures even for cavities placed along arbitrarily shaped paths.

25.8.11   2-D Distributed Feedback Laser Generator

A new type of laser resonator with 2-D distributed feedback from a PhC was reported.[153] The gain medium consisted of a thin film of organic material doped with Coumarin 490 and DCM. The thickness of the film was 150 nm and the period of the pattern was 400 nm. The planar waveguide consisting of the organic core layer and the air and Si cladding layers were supported by only the lowest-order TE and TM modes. With a 337-nm pulsed nitrogen laser as the pump source, laser action was observed in the wavelength range of 580 to 600 nm. The threshold pump power was ~50 kW/cm². Numerical calculations predicted two peaks of different polarization at the wavelengths, in close agreement with the experimental results. It was also predicted that it would be possible to achieve laser action from 2-D PhCs with a complete bandgap using organic media in conjunction with advanced Si microfabrication technology.

25.8.12   Photonic Crystal Fiber

PhC fiber was first reported in 1996.[123,154–203] The PhC fiber consisted of pure silica core with a higher refractive index surrounded by silica/air PhC material with hexagonal symmetry. It was shown that the fiber supported only a single low-loss guided mode over a very broad spectral range of at least 458 to 1550 nm. Further experimental investigation and theoretical analysis using the effective index model revealed that such a PhC fiber can be single mode for any wavelength, and its useful single-mode range within the transparency window of silica is bounded by a bending loss edge both at the short and long wavelengths.[157,204] The critical parameter is the ratio d/Γ, where d is the air hold diameter and Γ the spacing between adjacent holes.

25.8.13   Second Harmonic Generation

Second harmonic generation was observed experimentally in a centrosymmetric crystalline lattice of dielectric spheres.[205] The PhC was composed of polystyrene spherical particles of optical dimensions. The inversion symmetry of the centrosymmetric-face-centered cubic lattice is broken locally at the surface of each sphere in such a way that the scattered second harmonic light interferes constructively, leading to a nonvanishing macroscopic field. To enhance the second-order nonlinear interaction, a layer of strongly nonlinear molecules is adsorbed on the surface of each sphere of 115-nm diameter. It was also observed that phase matching of the fundamental and second harmonic waves is due to the long-range periodic distribution of dielectric material, which provides the bending of the photon dispersion curve at the edge of the Bragg reflection band of a given lattice plane. It was also pointed out that the flexibility in selecting the nonlinear molecules makes the nonlinear PhC very attractive for the study of surface chemical processes and for the improvement of nonlinear optoelectronic devices.

25.8.14   Air-Bridge Microcavity

A new type of high-Q microcavity consisting of a channel waveguide and a 1-D PhC was investigated.[99, 206–207] A bandgap for the guided modes is opened, and a sharp resonant state is created by adding a single defect in the periodic system. Numerical analyses of the eigenstates show that strong field confinement of the defect state can be achieved with a modal volume less than that of half a cubic half-wavelength. In the structures proposed, coplanar microcavities use index guiding to confine light along two dimensions and a 1-D PhC to confine light along the third. The microcavities are made of high-index channel waveguides in which a strong periodic variation of the refractive index is introduced by vertically etching a series of holes through the guide. The guided modes undergo multiple scattering by the periodic array of holes causing a gap to open between the first and the second guided mode bands. The size of the gap is determined by the dielectric constant of the waveguide and by the size of the holes relative to their central distances. By introducing a defect in the periodic array of holes, a sharp resonant mode can be introduced within the gap. Good confinement of the radiation in the microcavity was achieved by having a large contrast between the waveguide and the substrate, which also kept the mode from extending significantly into the substrate. It was also found that maximum confinement could be reached by completely surrounding the cavity by air.

25.8.15 Control of Spontaneous Emission

Enhancement and suppression of spontaneous emission in thin-film InGaAs/InP PhC have been studied.[3,208] Angular resolved photoluminescence measurements were used to experimentally determine the band structure of such a PhC and the overall enhancement of spontaneous emission. It was shown that emission into leaky conduction bands of the crystal has the same effect as cavity-enhanced spontaneous emission, provided these bands are flat enough relative to the emission band of the material. A MOCVD-grown $In_{0.47}Ga_{0.53}As/InP$ single quantum well double heterostructure was used for these experiments.

25.8.16 Enhancing Patch Antenna Performance Using Photonic Crystals

PhCs were used to enhance the performance of a patch antenna.[209] Traditionally, patch antennas have some limitations such as restricted bandwidth of operation, low gain, and potential decrease in radiation efficiency due to surface-wave losses. Using a PBG substrate for a patch antenna minimized the surface-wave effects compared to conventional patch antennas, thus improving the gain and far-field radiation pattern.

25.8.17 Surface-Emitting Laser Diode Using Photonic Bandgap Crystal Cavity

A surface-emitting laser diode consisting of a 3-D PBG crystal cavity was presented.[29] Spontaneous emission was controlled so as to radiate in the lasing direction with a narrow radiation angle by introducing a plane phase shift region into the cavity. The radiation pattern of the localized phase shift mode in the PhC is analyzed with a plane-wave method and using a 2-D model. It was also shown that the radiation angle of the spontaneous emission in the PhC cavity is as narrow as that of the stimulated emission of the conventional surface-emitting laser. The PBG crystal cavity laser operated as a light source without threshold and spatial emission noise; therefore, this approach is very attractive for use as a light source in spatially integrated optical circuits.

25.8.18 Optical Spectrometer

Both single-channel and multiple-channel optical spectrometers were reported.[17] For the theoretical analysis of such a device, the computational region consisted of a 2-D PhC having a square lattice with a lattice constant $a = 350$ nm. The lattice is made of dielectric rods with a dielectric constant of 11.56 (corresponding to Si) and radius $r = 70$ nm, on an air background. The transmission spectra for the above structure can be obtained using either finite-difference frequency-domain or FDTD with periodic boundary conditions. The above structure has a bandgap located between $\lambda = 0.833$ and 1.25 μm.

25.8.18.1 Single-Channel Optical Spectrometers

A single-channel optical spectrometer was implemented by combining a PhC waveguide evanescently coupled with a PhC microcavity, such that for an incident broadband pulse with a frequency content within the bandgap of the structure, a single frequency will be efficiently selected through the microcavity and further directed to its destination through another waveguide as shown in Figure 25.11.

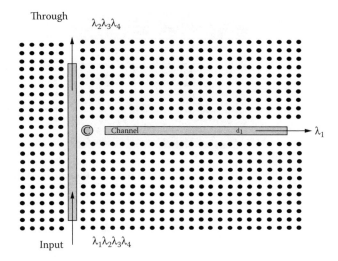

Through

$\lambda_2\lambda_3\lambda_4$

Channel d_1 $\rightarrow \lambda_1$

Input $\lambda_1\lambda_2\lambda_3\lambda_4$

FIGURE 25.11 A single channel optical spectrometer with a point defect (C) of diameter d = 52.5 nm and dielectric constant εr = 7. A broadband incident pulse evanescently coupled to the microcavity will drop a narrow wavelength through a channel toward a detector placed at d1.

To simulate the structure shown in Figure 25.11, the FDTD method with PML-absorbing boundary conditions was used. The computational space had a sampling rate of ($\lambda/40$), where λ is the wavelength of the light in vacuum. To this end, a pulse of center wavelength ($\lambda = 1$ μm) and of width $\Delta\lambda = 0.6$ μm was transmitted through the waveguide, which excited a single mode of oscillation inside the cavity. The field in the cavity was then coupled to the channel through an evanescent field. A detector was placed inside the channel to obtain the wavelength spectrum of the field in the channel, which is shown in Figure 25.12. The spectrum was obtained by taking the Fourier transform of the time-dependent field at the detector. From Figure 25.12, we can see that the quality factor of the cavity is about 2000 and the point defect of $r = 52.5$ nm corresponded to a center wavelength of $\lambda = 1.025$ μm and had a spectral line width of $\Delta\lambda = 2$ nm. This means that for an incident pulse width of $\Delta\lambda = 0.6$ μm, we can achieve nearly 300 different channels by fine-tuning the defect size of the center rod in the cavity while maintaining its dielectric constant at $\varepsilon_r = 7.0$.

25.8.18.2 Multiple-Channel Optical Spectrometers

A multiple-channel optical spectrometer can be achieved by cascading a number of single-channel spectrometers that are branched from a main waveguide channel. In this case, six cavities, each having a different defect size and its own guiding channel, were included. Each channel was branched from the main waveguide, as shown in Figure 25.13. Such a topology allows for better utilization of the structure by maximizing the density of the channels within the computational region. In addition to the pervious cavity, which had a point defect size of $r = 52.5$ nm, five more cavities were added with different point defects: $r = 8.75$ nm, $r = 17.5$ nm, $r = 26.25$ nm, $r = 35$ nm, and $r = 43.75$ nm, while maintaining the dielectric constant of all point defects constant at $\varepsilon_r = 7.0$. A separate analysis for each point defect was performed prior to this case, which corresponded to central wavelengths of $\lambda_2 = 0.875$ μm, $\lambda_3 = 0.895$ μm, $\lambda_4 = 0.925$ μm, $\lambda_5 = 0.94$ μm, and $\lambda_6 = 0.96$ μm, respectively. A spectral analysis of the above structure is shown in Figure 25.18. The spectrum was obtained by taking the Fourier transform of the time-dependent field at each detector. They were determined to have Lorentzian line shapes, as shown in Figure 25.14. Also shown in Figure 25.14 is that different point defect sizes corresponded to different localized modes, or different center frequencies. They were found to match previous calculations for the case of a single cavity. Also note that the central wavelength of each channel is directly proportional to the radius of the defect; in other words, as the radius of the defect increases, it is spanned through the available bandwidth

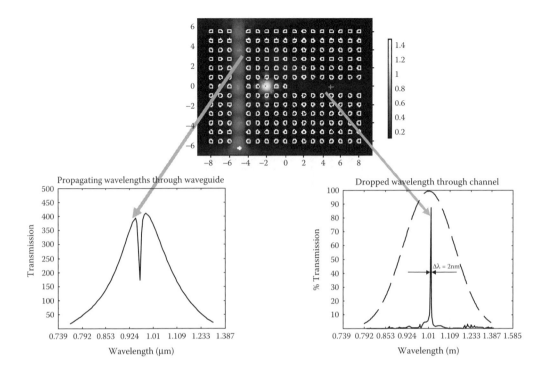

FIGURE 25.12 FDTD simulation results for the structure shown in Figure 25.11. Shown on the left is a selected wavelength from a propagating pulse, and on the right is the propagating pulse after a single wavelength has been detected (dropped) through the cavity.

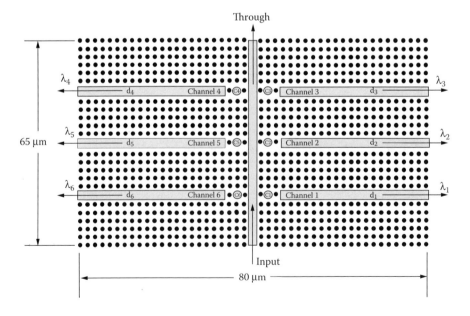

FIGURE 25.13 An optical spectrometer created in a PBG or rectangular lattice is six channel for simplicity but the device can be expanded to include N number of channels.

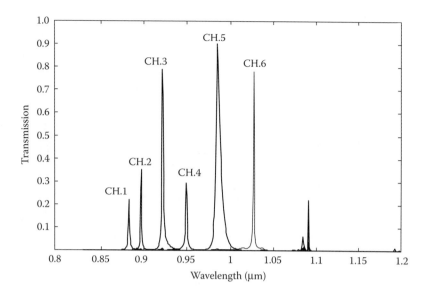

FIGURE 25.14 Spectral results for the optical spectrometer shown in Figure 25.13. Six spectrally separated wavelengths corresponding to six different cavities were detected.

of the incident pulse. Also, there is a limit to how much one can increase the radius of the defect while maintaining a single mode (acceptor mode) inside the cavity. Once the size of the defect starts getting close in size to that of the lattice rods, in which case more dielectric material is added instead of removing dielectric material, multiple modes begin to exist in the cavity. Lastly, the difference in the spectral line widths between different channels is due to the difference in Q-values of different cavities, which can be optimized for equally high Q-values.

The above new topology offers a more flexible design freedom and fewer constraints in contrast to its prior counterparts, in which achieving a high fan-out will require a much larger area.

25.8.19 Hybrid Photonic Crystal Structures

When a PhC of single-crystalline structure is formed, there is a natural matching of the crystal lattice, and a high-quality single crystal layer results. On the other hand, if a PhC is formed from multiple-crystalline structures (e.g., rectangular, triangular), the newly created structure is no longer a single-crystalline structure but a hybrid structure, which in turn contains the optical characteristics of both structures. Such a structure was called a heterostructure PhC.[210,211]

Heterostructure PhCs were used to further optimize the bandgap size obtained from a single PBG structure, such that a bandgap size of 94% was obtained from a heterostructure of a rectangular PhC lattice (bandgap size of 55%) and a triangular PhC lattice (bandgap size of 39 and 19%), as shown in Figure 25.15. Wide bandgap PhCs are advantageous for applications such as wide band optical mirrors, wide band optical matching elements, and wide band optical couplers.

When PhCs of different lattice structures are brought together, discontinuities in their respective energy bands is expected, as both structures will have different bandgaps, as shown in Figure 25.16. The discontinuities in the dielectric bands ΔE_{diel} and the air band ΔE_{air} accommodate the difference in bandgap between the two PhCs, ΔE_g, and the barrier

$$\Delta E_g = E_{g1} - E_{g2} \tag{25.25}$$

on either side of the wide bandgap lattice form what is in principle similar to an electronic quantum well in semiconductors, which in this case can be called a photonic quantum well. In PhCs, though, the

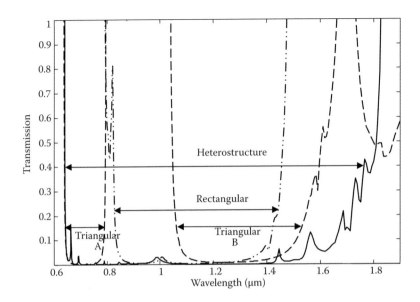

FIGURE 25.15 Band diagram of a rectangular PC lattice (–) with a bandgap size 54.87%, Band diagram of a triangular lattice (-..-.) with a short wavelength bandgap size 19.44% and a long wavelength bandgap size 39.39%, and band diagram of a heterostructure PC lattice (solid line) with a bandgap size 94.02%.

bands above and below the bandgap are generally unoccupied and a transition from one Bloch state to another state is only possible if the phase-matching condition is satisfied.[212]

By careful design of both rectangular and triangular lattices, respectively, the discontinuities in the dielectric band as well as the air band shown in Figure 25.16 can be minimized if not completely eliminated. This basically means that bandgap matching between two different lattices can be achieved, and heterostructure PhCs can be used as a matching element. An attenuation diagram for a triangular lattice matched with an attenuation diagram for a rectangular lattice is shown in Figure 25.17. Another application where heterostructure PhCs were used to improve the performance of a device was in an

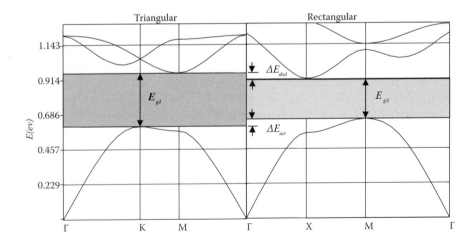

FIGURE 25.16 Band diagram of a triangular PC lattice with E_{g1} and a rectangular PC lattice with E_{g2}. As both lattices are brought together to form heterostructure, PC band edge discontinuities ΔE_{diel} and ΔE_{air} start to appear at the dielectric and air bands, respectively.

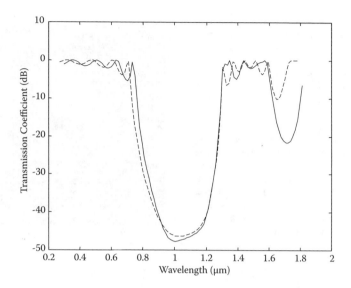

FIGURE 25.17 Transmission spectra for a rectangular lattice (solid line) with a ratio of r/a = 0.4 versus transmission spectra for a triangular lattice (dashed line) with a ratio of r/a = 0.43.

optical beam splitter. An optical beam splitter implemented in a single PhC structure (rectangular) is shown in Figure 25.18. Such a device achieved an overall throughput efficiency of 50% with most of the loss in transmission contributed by the back reflections at the splitting section.

Using a heterostructure PhC, the device performance increased from 50 to 90% as shown in Figure 25.19. Other examples utilizing heterostructure PhCs are a 1-to-4-beam splitter shown in Figure 25.20 and a beam splitter/combiner shown in Figure 25.21.

25.8.20 Tunable Photonic Crystals

PhCs are built using low-loss dielectric material (either rods on an air background or air holes on dielectric background). Once they are built with certain geometry and dimensions, they will behave in a way

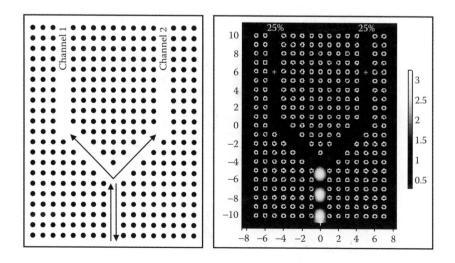

FIGURE 25.18 An optical beam splitter to demonstrate the idea of an optical matching element.

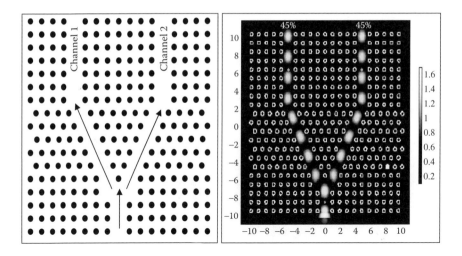

FIGURE 25.19 A schematic diagram of an optical beam splitter using heterostructure lattice.

consistent with those properties. Any desired alteration to their performance requires a new design. However, if one can dynamically vary some of the properties of an existing PhC, one can modify or control its performance. Such a change can be done by changing the index of refraction, which can be achieved by applying an electric field, which is called the electro-optic effect. In this case, applying an electric field to a PhC will change the dielectric constant of the crystal, which in turn will change the transmission properties of the PhC. This way one can tune a PhC without changing any of the geometrical properties.

Tuning the band structures of PhCs was first proposed using semiconductor-based PhCs.[213] A second method that was later proposed was the use of liquid crystal infiltration.[211,214] Semiconductor-based PhCs can be made tunable if the free-carrier density is sufficiently high, and the photonic band structure is strongly dependent on the temperature T and on the impurity concentration N. This was shown for a 2-D intrinsic InSb or extrinsic Ge PhCs. The disadvantage of this technique is that absorption cannot be avoided but can be minimized by careful selection of the materials.

Liquid crystals can also be used to tune a 2-D PhC using the temperature-dependent refractive index of a liquid crystal. Liquid crystal E7 was infiltrated into the air pores of a macroporous silicon PhC with a triangular

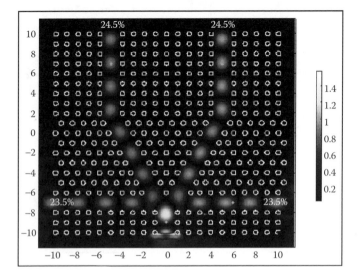

FIGURE 25.20 A 1-to-4 optical splitter using heterostructure PBG lattice.

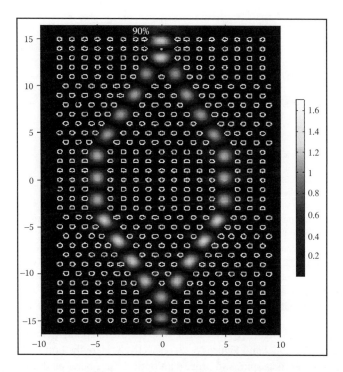

FIGURE 25.21 An optical beam splitter and combiner using heterostructure PBG lattice.

lattice pitch of 1.58 μm and a bandgap wavelength range of 3.3 to 5.7 μm. After infiltration, the bandgap for the TE mode shifted dramatically to 4.4 to 6.0 μm, whereas that of the TM mode collapsed. The sample was further heated to the nematic-isotropic phase transition temperature of the liquid crystal (59°C), the short-wavelength band edge of the TE gap shifted by 70 nm, whereas the long wavelength edge was constant.

25.8.21 Optical Switch Using PBG

Ultrafast all-optical switching in a silicon-based PhC has been investigated,[214] wherein the effect of two-photon absorption with Kerr nonlinearity on the optical properties of PhCs made with amorphous silicon and SiO_2 was studied. A stop band appearing near 1.5 μm is monitored with a peak probe beam and modulated by changes in the refractive index caused by a pump pulse at 1.71 μm with 18 GW/cm² peak intensity. Nonlinear optical characterization of the sample using Z-scan points to two-photon absorption as the main contributor to free-carrier excitation in silicon at that power level. Modulation in the transmittance near the band edge is found to be dominated by the optical Kerr effect within the pulse overlap (~400 fs), whereas free-carrier index changes are observed for 12 ps.

Other switching techniques in PhCs have also been presented both theoretically as well as experimentally.[214–220]

25.8.22 Photonic Crystal Optical Networks

High-density optical interconnects using PhCs were proposed as a technique for 3-D optical signal distribution and routing through multiple planes and in different directions.[221–224] These optical networks offer the ability to guide light analogously to electrical printed circuit boards, which transport electrons through electrical networks. Owing to their unique ability to confine and control light on the sub-wavelength scale, PhCs have led to a challenging prospect of miniaturization and large-scale integration of high-density optical interconnects.

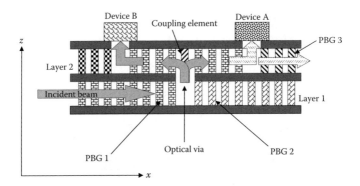

FIGURE 25.22 A cross-sectional view (xz plane) of an on chip optical network composed of two finite height planar photonic crystal layers. In layer 1 there exist two PBGs pre-designed to specific frequency bands, such that an optical beam coming from the right-hand side will propagate through PBG1 (through a waveguide) and will be prohibited from propagation through PBG2. Similar to PBG1 and 2, in layer 2 there exists either PBG1, 2 and/or 3 which may or may not be tuned to the same frequency. To efficiently couple layer 1 and 3 we have used an optical via combined with a coupling element to either enhance or diminish coupling. Lateral confinement is achieved via multiple Bragg reflections through PBGs, while vertical confinement is achieved through total internal reflections (TIR).

A schematic diagram showing a cross-sectional view (*xz*-plane and *xy*-plane) of a two-layer optical network is shown in Figure 25.22 and Figure 25.23, respectively. Here the bottom layer consists of a PhC slab, which is used to both guide, split, and process an optical wave between different points within that layer. On the top layer, another PhC slab of either the same dimension (diameter, pitch, and thickness)

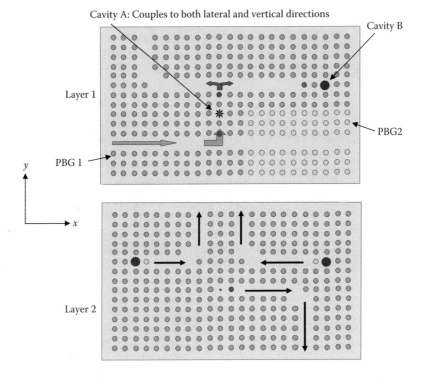

FIGURE 25.23 Cross-sectional view (xy plane) for layers 1 and 2 shown in Figure 25.22. In layer 1, cavity A was used to couple both in lateral dimension within layer 1 and in vertical direction to layer 2. While cavity B was used only for vertical coupling between layer 1 and layer 3.

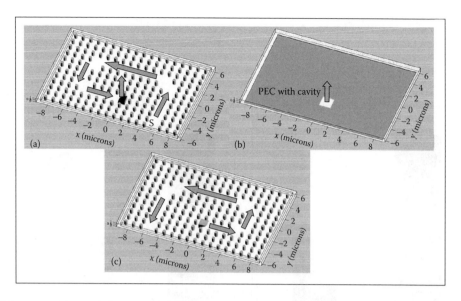

FIGURE 25.24 Dielectric deflector can be used as an optical via to couple light between either adjacent or non-adjacent layers. (a) Layer 1 contains a waveguide and a dielectric deflector. (b) Confining layer contains a cavity for vertical coupling. (c) Layer (2) contains another deflector, which will in turn refract an optical beam toward the waveguide in layer 2.

or structure as the one on the bottom layer is again used for different optical guiding and processing. Finally, optical signals of specific wavelengths are guided to their final destination at either a specific photodetector or an optoelectronic device, or even guided back through an optical fiber to another chip. In between the top and bottom layers, a confining layer is used to simulate the effect of TIR, where for the case of dielectric rods on an air background the confining layer was a PEC layer and for the case of a perforated dielectric slab we used a thin layer of SiO_2 as a confining layer. Also in between the top and bottom layers, an optical via was used to efficiently couple a propagating electromagnetic wave from one layer to the next.

One technique that can be used for broadband vertical coupling between adjacent and nonadjacent PBG layered networks is through the use of dielectric deflectors. In this case, we use two deflectors of the same dimensions to couple from one layer to the next higher layer. One deflector is located in the original layer and the other one is located in the next layer, such that a wave propagating in a direction perpendicular to the deflector surface in the original layer will be completely reflected vertically to the deflector in the destination layer, which in turn will reflect the wave towards a waveguide in the top layer. A schematic diagram showing an implementation deflector coupling is shown in Figure 25.24, where two deflectors of the same dimensions were used. Note that a PEC cavity is still needed to provide a point-to-point communication between the two deflectors. Deflector coupling can be used for either static coupling, where the angle is fixed, or dynamic coupling, where the angle can be varied, and hence coupling between different layers can be either enhanced or vanished based on the system requirements. Dynamic coupling can be achieved by either tuning the angle of the deflector to a certain direction, which can be achieved using MEMS technology, or by optically tuning the absorption properties of the deflector to adaptively control coupling to certain layers. Fabricating the deflector from two dielectric materials can again provide another way for dynamically controlling the coupling mechanism between two different layers in our optically interconnected network.

For the case of a perforated dielectric slab, dielectric deflectors will be replaced with air deflectors, and the dielectric air interface will again redirect an optically propagating wave to its destination layer.

Four consecutive snapshots of FDTD simulations for a two-layer network utilizing a deflector are shown in Figure 25.25, where two confining layers were used to confine an electromagnetic wave propagating through layer 2 (Figure 25.25a). The wave is then incident on the deflector, which was then coupled to layer 4, that was also sandwiched between two confining layers (Figure 25.25b–d).

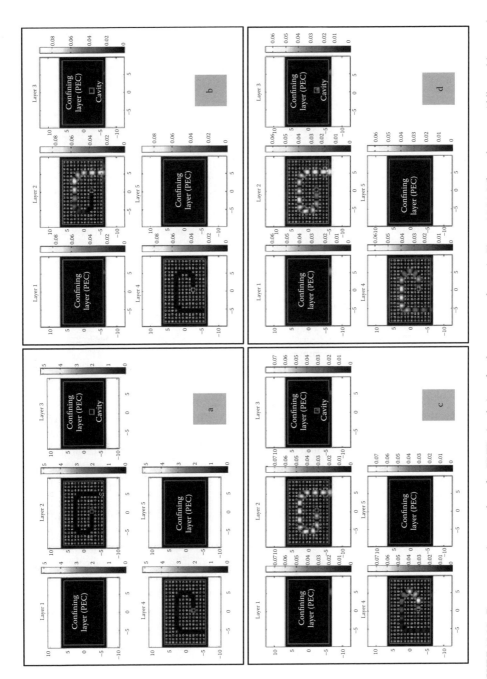

FIGURE 25.25 Four consecutive snapshots for FDTD simulations for the structure shown in Figure 25.24. Where the top, middle, and bottom layers (layers 1, 3, and 5) are perfect electric conductors (PEC) for vertical confinement and layers 2 and 4 are finite height photonic crystal waveguides. Source (S) is located at the bottom right corner.

25.8.23 Fabrication Error Analysis

Structural fluctuations on the PBG during fabrication for the case of PhC with a finite number of periods have been investigated.[224–226] The emphasis was on determining the effects of misalignment of basic structural elements and overall surface roughness, because of their general fabrication relevance. It was found that refractive index disorder affects the longer wavelength part of the first photonic band. Interestingly, positional disorder in the cylinder centers mainly affects the first gap and has little effect on the second, whereas thickness disorder has a smoothing effect on the band diagram. Even with a misalignment as much as 18%, the bandgap remained as large as 10%. Considering the disorder of all parameters at once shows that the refractive index and radius disorder have the most effect on the bandgap for both polarizations.

25.9 Applications Utilizing Irregular Dispersive Phenomena in Photonic Crystals

For applications utilizing irregular dispersive phenomena in PhCs, a bandgap need not exist at all but instead certain equi-frequency contours (EFC) may be utilized for specific applications depending on the contour shape, which can be controlled by engineering the fill factor and the geometrical lattice of the periodic structure.

Unlike in isotropic materials where only circular equi-frequency contours are used to describe the dispersive properties in homogenous media, in PhCs, noncircular contours can be found and hence various phenomena and applications may be found within a PhC structure. PhCs offer unique dispersion properties for various lattice geometries and various index contrasts. Such properties can be engineered to control the propagation of an electromagnetic light wave through these structures and hence allow various applications to emerge.

While 2-D dispersion diagrams are sufficient for applications utilizing strong confinement in PhC, they do not provide sufficient information for intra-bandgap applications exploring the dispersion properties of PhC, which are manifested by dispersion surfaces. Dispersion surfaces are more complex than dispersion diagrams and hence advanced analysis tools and theoretical models are necessary to extract PhC dispersion data at various frequency points. Once available, the dispersion data along with the PhC can be engineered to control the propagation of electromagnetic waves through the PhC lattice. In order to find the 3-D dispersion diagram, we have to solve a harmonic problem to find the various eigenmodes of a PhC structure. We also noticed the high spatial variation of the available eigenmodes through the complex 3-D dispersion diagrams obtained. At different frequency values, the intersection of the complex dispersion surfaces produces what we called in Chapter 5 an equi-frequency contour or EFC, where for a homogenous medium, an EFC always takes a circular shape such that if a light source is placed in the structure, electromagnetic waves emanating from it will propagate equally in all directions without any confinement in the lateral dimension. This is the reason why a planar dielectric waveguide needs a boundary to confine the electromagnetic wave in a homogenous structure.

In PhCs, various EFC shapes provide anomalous dispersion properties, which cannot be found in a homogenous medium, and hence interesting interaction between electromagnetic waves and a PhC structure will result in control of propagation of the electromagnetic waves in the lateral dimension. Various applications utilizing various EFC shapes have emerged in both microwave and optical frequency regimes. In the following section, we discuss a few of those applications.

25.9.1 Super-Prism Phenomenon

Conventional prisms made of a homogenous medium can deflect an electromagnetic light wave by about ±5°. The homogenous medium could be replaced by a PhC structure having an EFC with a star shape.[58,60] Such a contour provides a highly dispersive behavior, and hence a small variation in the incident electromagnetic wave will result in a much pronounced variation in the propagation angle. Such strong dispersive properties cannot be achieved in conventional dielectric prisms, and hence a PhC structure with such behavior was called a super-prism.[227]

Super-prism phenomena were demonstrated at optical frequencies with a 3-D periodic structure fabricated on a Si substrate.[60,227–235] The extraordinary angular sensitivity of light propagation was demonstrated by the following: with a ±12° change in the incident angle, the direction of the transmitted beam varied from −70° to +70°. This effect, together with wavelength sensitivity, is two orders of magnitude stronger than that of a conventional prism or grating. It was also shown by photonic band calculations that the angular dependence of negative refraction and multiple-beam branching was due to highly anisotropic dispersion surfaces. The application of this super-prism phenomenon enables the fabrication of photonic integrated circuits[63–65, 69–71,73,76] for WDM.[67,68]

25.9.2 Spot-Size Converter

A spot-size converter was demonstrated in PhCs with a conversion ratio of 10:1 for a 1.0 μm wavelength as discussed in Reference 236. The PhC was fabricated by depositing alternate layers of amorphous Si and SiO_2 on a patterned Si substrate having a hexagonal array of holes with a lattice constant of 0.3 μm. The replication of the surface holes causes the structure to be self-organized. A polarized light was incident on the crystal from the edge at an angle of 15° from the Γ–M direction, using a VCSEL with $\lambda = 0.956$ μm. For a 40-μm wide incident beam, the propagating beam inside the PhC reduced to 4 μm.

25.9.3 Self-Collimation

Self-collimation of an electromagnetic wave was achieved with a PhC fabricated on silicon.[232] It was shown that the divergence of the collimated beam is insensitive to that of the incident beam and much smaller than the divergence of conventional Gaussian collimators. The phenomenon was interpreted in terms of highly modulated dispersion surfaces with inflection points,[237–242] where the curvature changes from downward to upward corresponding to, respectively, a concave/convex lens.

25.9.4 Dispersion-Based Nonchannel Waveguiding

Waveguiding in periodic dielectric structures was only implemented by using the confinement properties of PBG structures. This was the only viable solution prior to exploiting the true potential of engineering the dispersion properties of such periodic structures to achieve nonchannel waveguiding. On the other hand, this was not the only choice available if the operational bandwidth was to expand beyond the forbidden band to the pass band of the periodic structure. In such a case, the interaction of the electromagnetic wave propagating through the periodic structure will govern the directions over which the wave is allowed to propagate. To illustrate this behavior, consider the structure introduced in Section 6.2.4 where a periodic array of air holes with $r = 133$ nm are arranged on a square lattice with lattice constant $a = 442$ nm and in high-index dielectric background such as silicon. If this structure was illuminated with a light source operating at $\lambda = 1480$ nm, the electromagnetic light wave will be self-guided by the square-like EFC of the PhC structure at that wavelength.

25.9.5 Dispersion-Based Routing Structure

Consider the structure used for nonchannel waveguiding in Section 25.9.6, where the dispersion properties of the periodic structure acted naturally as a guiding mechanism in the absence of an actual guiding structure. Such a structure can only guide an electromagnetic wave along two orthogonal directions, x or y. A question that might arise is: how can one redirect an electromagnetic wave propagating along either one of the orthogonal directions to propagate along the other direction? Remember that the structure was designed to operate outside the bandgap and did not include any actual guiding media.

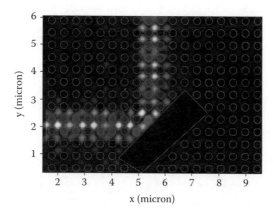

FIGURE 25.26 A horizontal cross-sectional view of the steady state amplitude of Hz component on the central plane of the photonic crystal slab when d = 0.107a. The bending efficiency is 82.9%.

The answer would be to include an element to obstruct the propagation direction of the electromagnetic wave and redirect it efficiently towards the other orthogonal direction. A possible implementation of such an element is a dielectric mirror positioned in such a way as to reflect an incident electromagnetic light wave towards one of the possible orthogonal directions.[233] Recall from Snell's law of reflection that the angle of incidence is equal to the reflection angle at the normal to the interface between two media. Hence if we position such a mirror at 45° obstructing the propagation path of an incident light beam, the reflected beam will also make a 45° angle with respect to the mirror surface and 90° with respect to the incident beam; an example of such a structure is shown in Figure 25.26.

An incoming light wave incident on the mirror is laterally confined or collimated by the existence of the PhC structure possessing a square EFC. At the mirror/PhC interface, the incident light beam is redirected in the orthogonal direction due to TIR at the silicon–air boundary.

A 3-D FDTD method was used to evaluate the performance of the routing structure shown in Figure 25.26. The steady-state amplitude of the magnetic field component H_z on the central plane of the slab is shown in Figure 25.26 when $d = 0.107a$, where d is defined as the distance between the nearest edge of air holes and the edge of the etched mirror surface.

25.9.6 Dispersion-Based Variable Beam Splitter

The dielectric mirror in Figure 25.26 was replaced with a PhC structure, which may take the same geometry as the host PhC structure or may be different. In either case, controlling the parameters of such a structure can be used to control the percentage transmission through such structure. By doing so, a hybrid PhC structure combines a dispersion-based and a confinement-based structure, as shown in Figure 25.27. Such a structure was used as a variable beam splitter. To this end, the optical beam splitter consists of two sections as shown in Figure 25.27: a dispersion guiding PhC structure and a beam splitting structure. These two kinds of PhC structures with different radii are both arranged on the same square lattice in a high index background of $n = 3.5$. The PhC guiding structure has an air hole with a ratio $r_g = 0.3a$, where a is the lattice constant. The 45° rotated splitting structure has an air hole of radius r_s, which varies from 0.3a to 0.435a. If the light is launched from port 0, then it will propagate through the dispersion guiding structure and arrive at the splitting structure. While passing through the splitting structure, the signal will be split between two orthogonal branches: one along the same direction as the incident wave, and exiting at port 1, and the other portion along a direction orthogonal to the incident wave, and exiting at port 2. The splitting percentage is proportional to the radius and number of layers of the splitting structure. In this section, we study the effect of varying the radii of the splitting section to achieve an arbitrary power ratio between output ports.

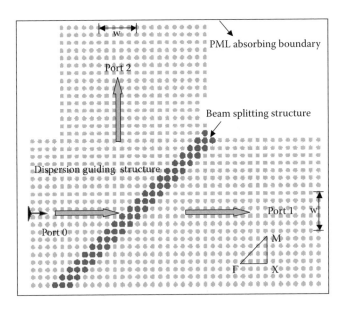

FIGURE 25.27 Dispersion based tunable beam splitter utilizing a hybrid PhC structure.

If the radius of air holes in the splitting region is the same as the one in the dispersion guiding structure, it is clear that the wave will completely output at port 1, as shown in Figure 25.28. This output energy, in fact, is used as a reference value for the calculation of splitting percentages. Figure 25.28c shows the steady-state result of magnetic field (Hz) for the radius of 0.42a in the splitting structure, most of the energy exits at port 2. At $r_s = 0.36a$, one obtains approximately 3 dB splitting, and the corresponding steady-state result of the magnetic field is shown in Figure 25.28b.

As the radius continually varies from 0.3a to 0.435a, the percentage of optical power vs. the normalized radius is plotted in Figure 25.29. At radius of 0.36a, the output ports have nearly the same power. Within the radius range between 0.3a and 0.345a, there is no stop band. As such, the dispersion contour at frequency 0.26c/a is still a square-like shape; hence, the light will propagate in the same direction as the incident wave and a small amount of light will propagate to port 2 due to the interface of two slightly different types of PhC structures. As the radius of air holes continues to increase, the frequency 0.26c/a falls into the bandgap. However, as we mentioned above, as the frequency is close to the edge of the bandgap, a few PhC layers are not sufficient to completely reflect the light, and a portion of the signal will output to port 1.

The beam splitting device shown in Figure 25.27 was experimentally validated by fabricating the respective structure in a 260-nm thick device layer of a silicon-on-insulator substrate. The fabricated

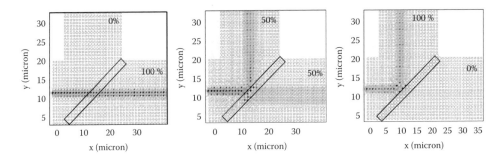

FIGURE 25.28 Steady state result of Hz field with radii of air holes of (a) 0.3a (b) 0.36a, (c) 0.42a for the splitting structure.

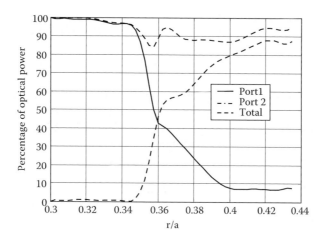

FIGURE 25.29 Percentages of output optical power of port 1 and port 2 vary with the radius of air holes of splitting structure.

device consisted of air holes with a lattice constant $a = 442$ nm and radii $r_g = 0.26a$ and $r_s = 0.35a$ for the guiding and splitting regions, respectively.[240,244]

25.10 Future Applications and Concluding Remarks

In this chapter, we have presented an overview of various computational methods for modeling and simulating PhCs, suitable fabrication processes, and past/present device developments in order to develop and optimize photonic integrated circuits (PICs) using PhC and PBG structures for a broad range of applications. The implication of this work is the ability to incorporate on-chip optical functionality, including confinement and routing, on a scale comparable to, or less than, the wavelength of light.

While attempting to present a brief historical overview, in the course of this chapter, it has also become clear that PhC technology is just beginning to come of age and many applications have yet to be realized. Optical components that can permit the miniaturization of an application-specific optical integrated circuit (ASPIC) on a subwavelength scale represent good candidates for next-generation high-density optical interconnects and integration. Admittedly, in recent years, there has been a growing effort in the realization of PhCs for a number of new devices and applications. To this end, we acknowledge that we have summarized but only a subset of those that have been proposed and regret having omitted any particular bodies of work.

Results of this study will be used to realize a new generation of optoelectronic/photonic systems that serve to satisfy the growing demand in terms of next-generation high-technology systems.

References

1. Bykov, V., *Z. Exp. Teoret. Fiz.*, 62, 1972.
2. Bykov, V., *Z. Tekhnich. Fiz.*, 48, 1978.
3. Yablonovitch, E., *Phys. Rev. Lett.*, 58, 2059–2062, 1987.
4. John, S., *Phys. Rev. Lett.*, 58, 2486–2489, 1987.
5. Parker, G. and Charlton, M., *Phys. World*, 13, 2000.
6. Bayindir, M., Temelkuran, B., and Ozbay, E., *Appl. Phys. Lett.*, 77, 3902–3904, 2000.
7. Sondergaard, T. and Dridi, K.H., *Phys. Rev. B*, 61, 15688–15696, 2000.
8. Fan, S.H., Johnson, S.G., Joannopoulos, J.D., Manolatou, C., and Haus, H.A., *J. Opt. Soc. Am. B*, 18, 162–165, 2001.

9. Ziolkowski, R.W., *Opt. Quant. Electron.*, 31, 843–855, 1999.

10. *Microcavities and Photonic Bandgaps: Physics and Applications*, Vol. 324, Kluwer Academic Publishers, Elounda, Crete, 1995.

11. D'Urso, B., Painter, O., O'Brien, J., Tombrello, T., Yariv, A., and Scherer, A., *J. Opt. Soc. Am. B*, 15, 1155–1159, 1998.

12. Sun, H.-B., Mizeikis, V., Xu, Y., Juodkazis, S., Ye, J.-Y., Matsuo, S., and Misawa, H., *Appl. Phys. Lett.*, 79, 1–3, 2001.

13. Fan, S., Villeneuve, P.R., and Joannopoulos, J.D., *Phys. Rev. B*, 54, 7837–7842, 1996.

14. Johnson, S.G., Shanhui, F., Mekis, A., and Joannopoulos, J.D., *Appl. Phys. Lett.*, 78, 3388–3390, 2001.

15. Meade, R.D., Devenyi, A., Joannopoulos, J.D., Alerhand, O.L., Smith, D.A., and Kash, K., *J. Appl. Phys.*, 75, 4753–4755, 1994.

16. Scherer, A., Painter, O., D'Urso, B., Lee, R., and Yariv, A., *J. Vac. Sci. Technol. B*, 16, 3906–3910, 1998.

17. Sharkawy, A., Shi, S., and Prather, D.W., *Appl. Opt.*, 40, 2247–2252, 2001.

18. Stoffer, R., Hoekstra, H.J.W.M., Ridder, R.M.D., Groesen, E.V., and Beckum, F.P.H.V., *Opt. Quant. Electron.*, 32, 947–961, 2000.

19. Fan, R., Villeneuve, R., Joannopoulos, J.D., and Haus, H.A., *Phys. Rev. Lett.*, 80, 960–963, 1998.

20. Haus, H.A., Fan, S., Villeneuve, P.R., and Joannopoulos, J.D., *Opt. Express*, 3, 4–11, 1998.

21. Johnson, S.G., Villeneuve, P.R., Fan, S.H., and Joannopoulos, J.D., *Phys. Rev. B*, 62, 8212–8222, 2000.

22. Adibi, A., Xu, Y., Lee, R.K., Yariv, A., and Scherer, A., *J. Lightwave Technol.*, 18, 1554–1564, 2000.

23. Johnson, S.G., Fan, S., Villeneuve, P.R., and Joannopoulos, J.D., *Phys. Rev. B*, 60, 5751–5758, 1999.

24. Painter, O., Vuckovic, J., and Scherer, A., *J. Opt. Soc. Am. B*, 16, 275–285, 1999.

25. Charlton, M.D.B., Zoorob, M.E., Parker, G.J., Netti, M.C., Baumberg, J.J., Cox, S.J., and Kemhadjian, H., *Mater. Sci. Eng. B*, 74, 17–24, 2000.

26. Gander, M., McBride, R., Jones, J., Mogilevtsev, D., Birks, T., Knight, J., and Russell, P., *Electron. Lett.*, 35, 63–64, 1999.

27. Smith, C.J.M., Rue, R.M.D.L., Rattier, M., Olivier, S., Benisty, H., Weisbuch, C., Krauss, P.R., Houdre, R., and Oesterle, U., *Appl. Phys. Lett.*, 78, 1487–1489, 2001.

28. Lin, S.-Y. and Fleming, J.G., *J. Lightwave Technol.*, 17, 1944–1947, 1999.

29. Hirayama, H., Hamano, T., and Aoyagi, Y., *Appl. Phys. Lett.*, 69, 791–793, 1996.

30. Xia, Y., Gates, B., and Park, S.H., *J. Lightwave Technol.*, 17, 1956–1962, 1999.

31. Rowson, S., Chelnokov, A., Cuisin, C., and Lourtioz, J.-M., *J. Opt. A. Pure Appl. Opt.*, 1, 483–489, 1999.

32. Rowson, S., Chelnokov, A., and Lourtioz, J.M., *J. Lightwave Technol.*, 17, 1989–1995, 1999.

33. Noda, S., Yamamoto, N., Imada, M., Kobayashi, H., and Okano, M., *J. Lightwave Technol.*, 17, 1948–1955, 1999.

34. Moosburger, J., H. Th., and Forchel, A., *J. Vac. Sci. Technol. B*, 18, 3501–3504, 2000.

35. Sabouroux, P., Tayeb, G., and Maystre, D., *Opt. Commun.*, 160, 33–36, 1999.

36. Cheng, C.C. and Scherer, A., *J. Vac. Sci. Technol. B*, 13, 2696–2700, 1995.

37. Cheng, C.C., Scherer, A., Tyan, R.-C., Fainman, Y., Witzgall, G., and Yablonovitch, E., *J. Vac. Sci. Technol. B*, 15, 2764–2767, 1997.

38. Edrington, A.C., Urbas, A.M., DeRege, P., Chen, C.X., Swager, T.M., Hadjichristidis, N., Xenidou, M., Fetters, L.J., Joannopoulos, J.D., Fink, Y., and Thomas, E.L., *Adv. Mater.*, 13, 421–425, 2001.

39. Norris, D.J. and Vlasov, Y., *Adv. Mater.*, 13, 371–376, 2001.

40. Jiang, P., Ostojic, G.N., Narat, R., Mittleman, D.M., and Colvin, V.L., *Adv. Mater.*, 13, 389–393, 2001.

41. Xia, Y., Gates, B., and Li, Z.-Y., *Adv. Mater.*, 13, 409–413, 2001.

42. Feiertag, G., Ehrfeld, W., Freimuth, H., Kolle, H., Lehr, H., Schmidt, M., Sigalas, M.M., Soukoulis, C.M., Kiriakidis, G., Pedersen, T., Kuhl, J., and Koenig, W., *Appl. Phys. Lett.*, 71, 1441–1443, 1997.

43. Ho, K.M., Chan, C.T., and Soukoulis, C.M., *Phys. Rev. Lett.*, 65, 3152–3155, 1990.

44. Johnson, S.G. and Joannopoulos, J.D., *Opt. Express*, 8, 173–180, 2001.

45. Johnson, S.G., Mekis, A., Fan, J., and Joannopoulos, J.D., *Comput. Sci. Eng.*, 3, 38–47, 2001.

46. Chan, C.T., Yu, Q.L., and Ho, K.M., *Phys. Rev. B*, 51, 16635–16642, 1995.
47. Xiao, S.S. and He, S.L., *Phys. Rev. B*, 324, 403–408, 2002.
48. Pendry, J.B., *J. Phys.: Condens. Matter*, 8, 1085–1108, 1996.
49. Axmann, W. and Kuchment, P., *J. Comput. Phys.*, 150, 468–481, 1999.
50. Koshiba, M., *IEICE Trans. Electron.*, E85C, 881–888, 2002.
51. Robertson, W.M., *J. Opt. Soc. Am. B*, 14, 1066–1073, 1997.
52. Dobson, D.C., Gopalakrishnan, J., and Pasciak, J.E., *J. Comput. Phys.*, 161, 668–679, 2000.
53. Botten, L.C., Nicorovici, N.A., McPhedran, R.C., de Sterke, C.M., and Asatryan, A.A., *Phys. Rev. E*, 6404, 6603+, 2001.
54. Ashcroft, N.W. and Mermin, N.D., *Solid State Physics*, Holt, Rinehart and Winston, 1976.
55. Leung, K.M. and Liu, Y.F., *Phys. Rev. B*, 41, 10188–10190, 1990.
56. Liu, L. and Liu, J.T., *Eur. Phys. J. B*, 9, 381–388, 1999.
57. Zhang, Z. and Satpathy, S., *Phys. Rev. Lett.*, 65, 2650–2653, 1990.
58. Kosaka, H., Kawashima, T., Tomita, A., Notomi, M., Tamamura, T., Sato, T., and Kawakami, S., *Phys. Rev. B*, 58, R10096–R10099, 1998.
59. Wu, L.J., Mazilu, M., Gallet, J.F., Krauss, T.F., Jugessur, A., and De La Rue, R.M., *Opt. Lett.*, 29, 1620–1622, 2004.
60. Kosaka, H., Kawashima, T., Tomita, A., Notomi, M., Tamamura, T., Sato, T., and Kawakami, S., *J. Lightwave Technol.*, 17, 2032–2034, 1999.
61. Baba, T. and Ohsaki, D., *Jpn. J. Appl. Phys., Part 1*, 40, 5920–5924, 2001.
62. Ochiai, T. and Sánchez-Dehesa, *J. Phys. Rev. B*, 64, 5113–5119, 2001.
63. Baba, T. and Nakamura, M., *IEEE J. Quant. Electron.*, 38, 909–914, 2002.
64. Baba, T. and Matsumoto, T., *Appl. Phys. Lett.*, 81, 2325–2327, 2002.
65. Baba, I. and M. Nakamura, *IEEE Journal of Quantum Electronics*, 38(7): 909–914, 2002.
66. Busch, K., *Comptes Rendus Physique*, 3, 53–66, 2002.
67. Chung, K.B. and Hong, S.W., *Appl. Phys. Lett.*, 81, 1549–1551, 2002.
68. Enoch, S. and Tayeb, G., *The Richness of the Dispersion Relation of Photonic Crystals: Application to Superprism Effect and Other Remarkable Effects*, Washington, 2002.
69. Lijun, W.U., Mazilu, M., Karle, T., and Krauss, T.F., *IEEE J. Quant. Electron.*, 38, 915–918, 2002.
70. Ma, H., Jen, A.K.Y., and Dalton, L.R., *Adv. Mater.*, 14, 1339–1365, 2002.
71. Park, W. and Summers, C.J., *Opt. Lett.*, 27, 1397–1399, 2002.
72. Wu, L., Mazilu, M., Karle, T., and Krauss, T.F., *IEEE J. Quant. Electron.*, 38, 915–918, 2002.
73. Bravo-Abad, J., Ochiai, T., and Sanchez-Dehesa, *J., Phys. Rev. B*, 67, 2003.
74. Gerken, M. and Miller, D.A.B., *Appl. Opt.*, 42, 1330–1345, 2003.
75. Kee, C.S., Kim, K., and Lim, H., *Phys. B: Condens. Matter*, 338, 153–158, 2003.
76. Krauss, T.F., *Phys. Stat. Sol. A*, 197, 688–702, 2003.
77. Li, B., Zhou, J., Li, L.T., Wang, X.J., Liu, X.H., and Zi, J., *Appl. Phys. Lett.*, 83, 4704–4706, 2003.
78. Momeni, B. and Adibi, A., *Appl. Phys. B*, 77, 555–560, 2003.
79. Panoiu, N.C., Bahl, M., and Osgood, R.M., *Opt. Lett.*, 28, 2503–2505, 2003.
80. Prasad, T., Colvin, V., and Mittleman, D., *Phys. Rev. B*, 67, 2003.
81. Pustai, D.M., Sharkawy, A., Shi, S.Y., Jin, G., Murakowski, J., and Prather, D.W., *J. Microlithogr. Microfab. Microsyst.*, 2, 292–299, 2003.
82. Qiu, M., Thylen, L., Swillo, M., and Jaskorzynska, B., *IEEE J. Sel. Top. Quant. Electron.*, 9, 106–110, 2003.
83. Scrymgeour, D., Malkova, N., Kim, S., and Gopalan, V., *Appl. Phys. Lett.*, 82, 3176–3178, 2003.
84. Wu, L., Mazilu, M., and Krauss, T.F., *J. Lightwave Technol.*, 21, 561–566, 2003.
85. Xiong, S. and Fukshima, H., *J. Appl. Phys.*, 94, 1286–1288, 2003.
86. Ye, Y.H., Jeong, D.Y., Mayer, T.S., and Zhang, Q.M., *Appl. Phys. Lett.*, 82, 2380–2382, 2003.
87. Zayats, A.V., Dickson, W., Smolyaninov II, and Davis, C.C., *Appl. Phys. Lett.*, 82, 4438–4440, 2003.
88. Cabuz, A.I., Centeno, E., and Cassagne, D., *Appl. Phys. Lett.*, 84, 2031–2033, 2004.

89. Chen, H.L., Lee, H.F., Chao, W.C., Hsieh, C.I., Ko, F.H., and Chu, T.C., *J. Vac. Sci. Technol. B*, 22, 3359–3362, 2004.

90. Di Falco, A., Conti, C., and Assanto, G., *J. Lightwave Technol.*, 22, 1748–1753, 2004.

91. Gerken, M. and Miller, D.A.B., *J. Lightwave Technol.*, 22, 612–618, 2004.

92. Li, Z.F., Chen, H.B., Song, Z.T., Yang, F.H., and Feng, S.L., *Appl. Phys. Lett.*, 85, 4834–4836, 2004.

93. Luo, C.Y., Soljacic, M., and Joannopoulos, J.D., *Opt. Lett.*, 29, 745–747, 2004.

94. Lupu, A., Cassan, E., Laval, S., El Melhaoui, L., Lyan, P., and Fedeli, J.M., *Opt. Express*, 12, 5690–5696, 2004.

95. Neal, R.T., Zoorob, M.E., Charlton, M.D., Parker, G.J., Finlayson, C.E., and Baumberg, J.J., *Appl. Phys. Lett.*, 84, 2415–2417, 2004.

96. Panoiu, N.C., Bahl, M., and Osgood, R.M., *J. Opt. Soc. Am. B*, 21, 1500–1508, 2004.

97. Qin, X.Y., Huang, B.Q., Chen, H.X., Yang, L.G., and Gu, P.F., *Acta Phys. Sin.*, 53, 3794–3799, 2004.

98. Witzens, J., Hochberg, M., Baehr-Jones, T., and Scherer, A., *Phys. Rev. E*, 69, 2004.

99. Taflove, A. and S.C. Hagness, *Computational Electrodynamics: The Finite Difference Time Domain Method, Second Edition*. 2000, Boston, MA: Artech House.

100. Ahuja, V. and Long, L.N., *J. Comput. Phys.*, 137, 299–320, 1997.

101. Guiffaut, C. and Mahdjoubi, K., *IEEE Antennas Propagation Mag.*, 43, 94–103, 2001.

102. Durbano, J. and Prather, D.W., US Patent and Trademark Office, Vol. UD02-16 (131*278), 2002.

103. Durbano, J., Curt, P.F., Humphrey, J.R., Ortiz, F.E., and Prather, D., *FPGA-based Acceleration of the Three-Dimensional Finite Difference Time Domain Method for Electromagnetic Calculations*, 2004.

104. Durbano, J., Curt, P.F., Humphrey, J.R., Ortiz, F.E., Prather, D., and Mirotznik, M.S., *Hardware Acceleration of the 3-D Finite-Difference Time-Domain Method*, 2004.

105. Durbano, J., Curt, P.F., Humphrey, J.R., Ortiz, F.E., Prather, D., and Mirotznik, M.S., *IEEE 3-D Antennas Wireless Propagation Lett.*, 2, 54–57, 2003.

106. Durbano, J., Curt, P.F., Humphrey, J.R., Ortiz, F.E., Prather, D., and Mirotznik, M.S., *Proceedings of the 11th Annual IEEE Symposium on Field Programmable Custom Computing Machines* (FCCM), 2003, pp. 269–270.

107. Baba, T., Motegi, A., Iwai, T., Fukaya, N., Watanabe, Y., and Sakai, A., *IEEE J. Quant. Electron.*, 38, 2002.

108. Scherer, A., Painter, O., Vuckovic, J., Loncar, M., and Yoshie, T., *IEEE Trans. Nanotechnol.* 1, 4, 2002.

109. Berenger, J.P., *J. Comp. Phys.*, 114, 185–200, 1994.

110. Sacks, Z.S., Kingsland, D.M., Lee, R., and Lee, J., *IEEE Trans. Antennas Propagat.*, AP-43, 1460–1463, 1995.

111. Shi, S., Chen, C., and Prather, D.W., *J. Opt. Soc. Am. A*, 21, 1769, 2004.

112. Qiu, M., *Appl. Phys. Lett.*, 81, 1163–1165, 2002.

113. Chen, C., Shi, S., and Prather, D., *Appl. Phys. Lett.*, 2004.

114. Shi, S., Chen, C., and Prather, D.W., *Appl. Phys. Lett.*, 86, 2005.

115. Photonics, A., *APSS*. 2004: Hamilton, Ontario.http://www.apollophoton.com/apollo.

116. Design, P., *Crystal Wave*. 2005 Oxford, UK. http//www.photond.com/

117. Rsoft, *BandSolve*. 2005: New York. http//www.rsoftdesign.com/

118. Photonics, E., *PBGI Lab*. 2005:Newark.http//www.emphotonics.com/pbglab.html

119. FEM Lab. *COMSOL*. 2005: MA

120. Optiwave, *OptiFDTD*. 2004. http://www.optiwave.com/

121. Prather, D.W., A. Sharkawy, and S. Shi, Design and Applications of Photonic Crystals, in *Handbook of Nanoscience, Engineering, and technology*, W.A. Goddard III, et al., Editors. 2002, CRC Press LLC: Boca Raton, FL, 211–232.

122. Yablonovitch, E. and Gmitter, T.J., *Phys. Rev. Lett.*, 67, 3380–3383, 1991.

123. Knight, J.C., Birks, T.A., Russell, P.S., and Atkin, D.M., *Opt. Lett.*, 21, 1547–1549, 1996.

124. Sondergaard, T., Arentoft, J., Bjarklev, A., Kristensen, M., Erland, J., Broeng, J., and Barkou, S.E., *Opt. Commun.*, 194, 341–351, 2001.

125. Sondergaard, T., Bjarklev, A., Kristensen, M., Erland, J., and Broeng, J., *Appl. Phys. Lett.*, 77, 785–787, 2000.

126. Fan, S.H., S.G. Johnson, J.D. Joannopoulos, C. Manolatou, and H.A. Haus, *J. Opt. Soc. of Am. B-Optical Physics*, 18(2): pp. 162–165, 2001.

127. Johnson, S.G., P.R. Villeneuve, S.H. Fan, and J.D. Joannopoulos, *Physical Review B*, 62(12): 8212–8222, 2000.

128. Charlton, M.D.B., M.E. Zoorob, G.J. Parker, M.C. Netti, J.J. Baumberg, S. Cox, and H. Kemhadjian, *Materials Science and Engineering B-Solid State Materials for Advanced Technology*, 74(1-3): 17–24, 2000.

129. Mekis, A., Chen, J.C., Kurland, I., Fan, S., Villeneuve, P.R., and Joannopoulos, J.D., *Phys. Rev. Lett.*, 77, 3787–3790, 1996.

130. Mekis, A. and Joannopoulos, J.D., *J. Lightwave Technol.*, 19, 861–865, 2001.

131. Chow, E., Lin, S.Y., Wendt, J.R., Johnson, S.G., and Joannopoulos, J.D., *Opt. Lett.*, 26, 286–288, 2001.

132. El-Kady, I., Sigalas, M.M., Biswas, R., and Ho, K.M., *J. Lightwave Technol.*, 17, 2042–2049, 1999.

133. Loncar, M., Doll, T., Vuckovic, J., and Scherer, A., *J. Lightwave Technol.*, 18, 1402–1411, 2000.

134. McGurn, A.R., *Phys. Lett. A*, 260, 314–321, 1999.

135. Smith, C.J.M., Benisty, H., Olivier, S., Rattier, M., Weisbuch, C., Krauss, T.F., Rue, R.M.D.L., Houdre, R., and Oseterle, U., *Appl. Phys. Lett.*, 77, 2813–2815, 2000.

136. Weisbuch, C., Benisty, H., Olivier, S., Rattier, M., Smith, C.J.M., and Krauss, T.F., *IEICE Trans. Electron.*, E84-C, 660–668, 2001.

137. Tokushima, M., Kosaka, H., Tomita, A., and Yamada, H., *Appl. Phys. Lett.*, 76, 952–954, 2000.

138. Pottier, P., Seassal, C., Letartre, X., Leclercq, J.L., Viktrorovitch, P., Cassagne, D., and Jouanin, C., *J. Lightwave Technol.*, 17, 2058–2062, 1999.

139. Meade, R.D., Brommer, K.D., Rappe, A.M., and Joannopoulos, J.D., *Phys. Rev. B*, 44, 13772–13774, 1991.

140. Meade, R.D., Rappe, A.M., Brommer, K.M., Joannopoulos, J.D., and Alerhand, O.L., *Phys. Rev. B*, 48, 8434–8437, 1993.

141. Painter, O., Husain, A., Scherer, A., O'Brien, J., Kim, I., and Dapkus, P.D., *J. Lightwave Technol.*, 17, 2082–2088, 1999.

142. Lei, X.-Y., Li, H., Ding, F., Zhang, W., and Ming, N.-B., *Appl. Phys. Lett.*, 71, 2889–2891, 1997.

143. Fan, S., Villeneuve, P.R., and Joannopoulos, J.D., *Phys. Rev. B*, 59, 15882–15892, 1999.

144. Jian, F. and Sai-Ling, H., *Chin. Phys. Lett.*, 17, 737–739, 2000.

145. Centeno, E., Guizal, B., and Felbacq, D., *J. Opt. A: Pure Appl. Opt.*, 1, L10–L13, 1999.

146. Lin, H.-B., Tonucci, R.J., and Campillo, A.J., *Opt. Lett.*, 23, 94–96, 1998.

147. Labilloy, D., Benisty, H., Weisbuch, C., Krauss, T.F., Smith, C.J.M., Houdre, R., and Oesterle, U., *Appl. Phys. Lett.*, 73, 1314–1316, 1998.

148. Johnson, S.G., Manolatou, C., Fan, S.H., Villeneuve, P.R., Joannopoulos, J.D., and Haus, H.A., *Opt. Lett.*, 23, 1855–1857, 1998.

149. Bayindir, M., Temmelkuran, B., and Ozbay, E., *Phys. Rev. B*, 61, R11855–R11858, 2000.

150. Zheltikov, A.M., Magnitskil, S.A., and Tarasishin, A.V., *JETP Lett.*, 70, 323–328, 1999.

151. Bayindir, M. and Ozbay, E., *Phys. Rev. B*, 62, R2247–R2250, 2000.

152. Bayindir, M., Temelkuran, B., and Ozbay, E., *Phys. Rev. Lett.*, 84, 2140–2143, 2000.

153. Meier, M., Mekis, A., Dodabalapur, A., Timko, A., Slusher, R.E., Joannopoulos, J.D., and Nalamasu, O., *Appl. Phys. Lett.*, 74, 7–9, 1999.

154. Knight, J.C., T.A. Birks, P.S. Russell, and D.M. Atkin, *Opt. Lett.*, 21(19): 1547–1549, 1996.

155. Birks, T.A., Knight, J.C., and Russell, P.S., *Opt. Lett.*, 22, 961–963, 1997.

156. Broeng, J., Barkou, S.E., Bjarklev, A., Knight, J.C., Birks, T.A., and Russell, P.S., *Opt. Commun.*, 156, 240–244, 1998.

157. Knight, J.C., Birks, T.A., Russell, P.S.J., and de Sandro, J.P., *J. Opt. Soc. Am. A*, 15, 748–752, 1998. (rep. of 122)

158. Knight, J.C., T.A. Birks, P.S.J. Russell, and J.P. de Sandro, *J. Opt. Soc. Am. A-Optics Image Science and Vision*, 15(3): 748–752, 1998.

159. Knight, J.C., Birks, T.A., Russell, P.S.J., and Rarity, J.G., *Appl. Opt.*, 37, 449–452, 1998.

160. Mogilevtsev, D., Birks, T.A., and Russell, P.S., *Opt. Lett.*, 23, 1662–1664, 1998.

161. Silvestre, E., Russell, P.S., Birks, T.A., and Knight, J.C., *J. Opt. Soc. Am. A*, 15, 3067–3075, 1998.

162. Barkou, S.E., Broeng, J., and Bjarklev, A., *Opt. Lett.*, 24, 46–48, 1999.

163. Birks, T.A., Mogilevtsev, D., Knight, J.C., and Russell, P.S., *IEEE Photon. Technol. Lett.*, 11, 674–676, 1999.

164. Broeng, J., Mogilevstev, D., Barkou, S.E., and Bjarklev, A., *Opt. Fiber Technol.*, 5, 305–330, 1999.

165. Cregan, R.F., Mangan, B.J., Knight, J.C., Birks, T.A., Russell, P.S., Roberts, P.J., and Allan, D.C., *Science*, 285, 1537–1539, 1999.

166. Cregan, R.F., Knight, J.C., Russell, P.S., and Roberts, P.J., *J. Lightwave Technol.*, 17, 2138–2141, 1999.

167. Eggleton, B.J., Westbrook, P.S., Windeler, R.S., Spalter, S., and Strasser, T.A., *Opt. Lett.*, 24, 1460–1462, 1999.

168. Ferrando, A., Silvestre, E., Miret, J.J., Andres, P., and Andres, M.V., *Opt. Lett.*, 24, 276–278, 1999.

169. Gander, M.J., McBride, R., Jones, J.D.C., Birks, T.A., Knight, J.C., Russell, P.S., Blanchard, P.M., Burnett, J.G., and Greenaway, A.H., *Opt. Lett.*, 24, 1017–1019, 1999.

170. Ghosh, R., Kumar, A., and Meunier, J.P., *Electron. Lett.*, 35, 1873–1875, 1999.

171. Monro, T.M., Richardson, D.J., Broderick, N.G.R., and Bennett, P.J., *J. Lightwave Technol.*, 17, 1093–1102, 1999.

172. Brechet, F., Marcou, J., Pagnoux, D., and Roy, P., *Opt. Fiber Technol.*, 6, 181–191, 2000.

173. Diez, A., Birks, T.A., Reeves, W.H., Mangan, B.J., and Russell, P.S., *Opt. Lett.*, 25, 1499–1501, 2000.

174. Eggleton, B.J., Westbrook, P.S., White, C.A., Kerbage, C., Windeler, R.S., and Burdge, G.L., *J. Lightwave Technol.*, 18, 1084–1100, 2000.

175. Ferrando, A., Silvestre, E., Miret, J.J., Andres, P., and Andres, M.V., *Opt. Lett.*, 25, 1328–1330, 2000.

176. Ferrando, A., Silvestre, E., Miret, J.J., Andres, P., and Andres, M.V., *J. Opt. Soc. Am. A*, 17, 1333–1340, 2000.

177. Ferrando, A., Silvestre, E., Miret, J.J., and Andres, P., *Opt. Lett.*, 25, 790–792, 2000.

178. Holzwarth, R., Udem, T., Hansch, T.W., Knight, J.C., Wadsworth, W.J., and Russell, P.S.J., *Phys. Rev. Lett.*, 85, 2264–2267, 2000.

179. Jones-Bey, H., *Laser Focus World*, 36, 15–16, 2000.

180. Kawanishi, T. and Izutsu, M., *Opt. Express*, 7, 10–22, 2000.

181. Knight, J.C., Arriaga, J., Birks, T.A., Ortigosa-Blanch, A., Wadsworth, W.J., and Russell, P.S.J., *IEEE Photon. Technol. Lett.*, 12, 807–809, 2000.

182. Mangan, B., Knight, J., Birks, T., Russell, P., and Greenaway, A., *Electron. Lett.*, 36, 1358–1359, 2000.

183. Midrio, M., Singh, M.P., and Someda, C.G., *J. Lightwave Technol.*, 18, 1031–1037, 2000.

184. Monro, T.M., Richardson, D.J., Broderick, N.G.R., and Bennett, P.J., *J. Lightwave Technol.*, 18, 50–56, 2000.

185. Ortigosa-Blanch, A., Knight, J.C., Wadsworth, W.J., Arriaga, J., Mangan, B.J., Birks, T.A., and Russell, P.S.J., *Opt. Lett.*, 25, 1325–1327, 2000.

186. Ranka, J.K., Windeler, R.S., and Stentz, A.J., *Opt. Lett.*, 25, 796–798, 2000.

187. Zheltikov, A.M., *Usp. Fizicheskikh Nauk*, 170, 1203–1215, 2000.

188. Bagayev, S.N., Dmitriyev, A.K., Chepurov, S.V., Dychkov, A.S., Klementyev, V.M., Kolker, D.B., Kuznetsov, S.A., Matyugin, Y.A., Okhapkin, M.V., Pivtsov, V.S., Skvortsov, M.N., Zakharyash, V.F., Birks, T.A., Wadsworth, W.J., Russell, P.S., and Zheltikov, A.M., *Laser Phys.*, 11, 1270–1282, 2001.

189. Coen, S., Chan, A.H.L., Leonhardt, R., Harvey, J.D., Knight, J.C., Wadsworth, W.J., and Russell, P.S.J., *Opt. Lett.*, 26, 1356–1358, 2001.

190. Eggleton, B.J., Ahuja, A.K., Feder, K.S., Headley, C., Kerbage, C., Mermelstein, M.D., Rogers, J.A., Steinvurzel, P., Westbrook, P.S., and Windeler, R.S., *IEEE J. Sel. Top. Quant. Electron.*, 7, 409–424, 2001.

191. Fedotov, A.B., Zheltikov, A.M., Tarasevitch, A.P., and von der Linde, D., *Appl. Phys. B*, 73, 181–184, 2001.
192. Ferrando, A., Silvestre, E., Andres, P., Miret, J.J., and Andres, M.V., *Opt. Express*, 9, 687–697, 2001.
193. Fitt, A.D., Furusawa, K., Monro, T.M., and Please, C.P., *J. Lightwave Technol.*, 19, 1924–1931, 2001.
194. Furusawa, K., Malinowski, A., Price, J.H.V., Monro, T.M., Sahu, J.K., Nilsson, J., and Richardson, D.J., *Opt. Express*, 9, 714–720, 2001.
195. Hansen, T., Broeng, J., Libori, S., Knudsen, E., Bjarklev, A., Jensen, J., and Simonsen, H., *IEEE Photon. Technol. Lett.*, 13, 588–590, 2001.
196. Holzwarth, R., Zimmermann, M., Udem, T., Hansch, T.W., Russbuldt, P., Gabel, K., Poprawe, R., Knight, J.C., Wadsworth, W.J., and Russell, P.S.J., *Opt. Lett.*, 26, 1376–1378, 2001.
197. Husakou, A.V. and Herrmann, J., *Phys. Rev. Lett.*, 8720, 3901+, 2001.
198. Knight, J.C., Birks, T.A., Mangan, B.J., and Russell, P.S.J., *MRS Bull.*, 26, 614–617, 2001.
199. Koch, K.W., *Opt. Express*, 9, 675–675, 2001.
200. Koshiba, M. and Saitoh, K., *IEEE Photon. Technol. Lett.*, 13, 1313–1315, 2001.
201. Mangan, B.J., Arriaga, J., Birks, T.A., Knight, J.C., and Russell, P.S., *Opt. Lett.*, 26, 1469–1471, 2001.
202. Monro, T.M., Pruneri, V., Broderick, N.G.R., Faccio, D., Kazansky, P.G., and Richardson, D.J., *IEEE Photon. Technol. Lett.*, 13, 981–983, 2001.
203. Muntele, I., *MRS Bull.*, 26, 659–660, 2001.
204. Birks, T., J. Knight, and P. Russell, *Opt. Lett.*, 1997. 22(13): pp. 961–963, 1997.
205. Marorell, J., Vilaseca, R., and Corbalan, R., *Appl. Phys. Lett.*, 70, 702–704, 1997.
206. Villeneuve, P.R., Fan, S., and Joannopoulos, J.D., *Appl. Phys. Lett.*, 67, 167–169, 1995.
207. Ripin, D.J., Lim, K.-Y., Petrich, G.S., Villeneuve, P.R., Fan, S., Thoen, E.R., Joannopoulos, J.D., Ippen, E.P., and Kolodziejski, L.A., *J. Lightwave Technol.*, 17, 2152–2160, 1999.
208. Boroditsky, M., Vrijen, R., Krauss, P.R., Coccioli, R., Bhat, R., and Yablonovitch, E., *J. Lightwave Technol.*, 17, 2096–2112, 1999.
209. Gonzalo, R., Maagt, P.D., and Sorolla, M., *IEEE Trans. Microwave Theory Tech.*, 47, 2131–2138, 1999.
210. Sharkawy, A., Shi, S., and Prather, D.W., *Appl. Opt.*, 41, 7245–7253, 2002.
211. Leonard, S.W., Mondia, J.P., van Driel, H.M., Toader, O., John, S., Busch, K., Birner, A., Gosele, U., and Lehmann, V., *Phys. Rev. B*, 61, R2389–R2392, 2000.
212. Winn, J.N., Fan, S., and Joannopoulos, J.D., *Condens. Matter Mater. Phys.*, 59, 1551–1554, 1999.
213. Halevi, P. and Mendieta, F.R., *Phys. Rev. B*, 85, 1875–1878, 2000.
214. Hache, A. and Bourgeois, M., *Appl. Phys. Lett.*, 77, 4089–4091, 2000.
215. Lan, S., Nishikawa, S., and Wada, O., *Appl. Phys. Lett.*, 78, 2101–2103, 2001.
216. Florescu, M. and John, S., *Phys. Rev. A*, 64, 338011–3380121, 2001.
217. Tran, P., *Opt. Lett.*, 21, 1138–1140, 1996.
218. Petrosyan, D. and Kurizki, G., *Phys. Rev. A*, 64, 238101–23806, 2001.
219. Lustrac, A.D., Gadot, F., Cabaret, S., Lourtioz, J.-M., Brillat, T., Priou, A., and Akmansoy, E., *Appl. Phys. Lett.*, 75, 1625–1627, 1999.
220. Lousses, V. and Vigeneron, J.P., *Phys. Rev. E*, 63, 2001.
221. Sharkawy, A., Shouyuan, S., and Prather, D.W., *Optical Networks on a Chip Using Photonic Bandgap Materials*, SPIE, San Diego, CA, 2001, pp. 179–190.
222. Fan, S.H., Villeneuve, P.R., and Joannopoulos, J.D., *J. Appl. Phys.*, 78, 1415–1418, 1995.
223. Chutinan, A. and Noda, S., *J. Opt. Soc. Am. B*, 16, 240–244, 1999.
224. Asatryan, A.A., Robinson, P.A., Botten, L.C., McPhedran, R.C., Nicorovici, N.A., and de Sterke, C.M., *Phys. Rev. E*, 62, 5711–5720, 2000.
225. Chutinan, A. and Noda, S., *J. Opt. Soc. Am. B*, 16, 1398–1402, 1999.
226. Fan, S., P.R. Villeneuve, and J.D. Joannopoulos, *J. Appl. Phy.*, 1995. 78(3): 1415–1418, 1995.
227. Kosaka, H., T. Kawashima, Akihisa Tomita, M. Notomi, T. Tamamura, T. Sato, and S. Kawakami, *Phys. Rev. B*, 1998. 58(16): R10096–R10099, 1998.
228. Shelby, R.A., Smith, D.R., and Schultz, S., *Science*, 292, 77–79, 2001.
229. Notomi, M., *Phys. Rev. B*, 62, 10696–10705, 2000.

230. Notomi, M. and Tamamura, T., *Phys. Rev. B*, 61, 7165–7168, 2000.
231. Kosaka, H., Kawashima, T., Tomita, A., Notomi, M., Tamamura, T., Sato, T., and Kawakami, S., *Appl. Phys. Lett.*, 74, 1370–1372, 1999.
232. Kosaka, H., Kawashima, T., Tomita, A., Notomi, M., Tamamura, T., Sato, T., and Kawakami, S., *Appl. Phys. Lett.*, 74, 1212–1214, 1999.
233. Kosaka, H., Tomita, A., Kawashima, T., Sato, T., and Kawakami, S., *Phys. Rev. B*, 62, 1477–1480, 2000.
234. Lin, S.-Y., Hietala, V.M., Wang, L., and Jones, E.D., *Opt. Lett.*, 21, 1771–1773, 1996.
235. Halevi, P., *Phys. Rev. Lett.*, 82, 719–722, 1999.
236. Kosaka, H., Kawashima, T., Tomita, A., Sato, T., and Kawakami, S., *Appl. Phys. Lett.*, 76, 268–270, 2000.
237. Witzens, J., Loncar, M., and Scherer, A., *IEEE J. Sel. Top. Quant. Electron.*, 8, 2002.
238. Hu, X.H. and Chan, C.T., *Appl. Phys. Lett.*, 85, 1520–1522, 2004.
239. Prather, D.W., Shi, S.Y., Pustai, D.M., Chen, C.H., Venkataraman, S., Sharkawy, A., Schneider, G.J., and Murakowski, J., *Opt. Lett.*, 29, 50–52, 2004.
240. Pustai, D.M., Shi, S.Y., Chen, C.H., Sharkawy, A., and Prather, D.W., *Opt. Express*, 12, 9, 1823–1831, 2004.
241. Miao, B., Chen, C., Shi, S., and Prather, D.W., *IEEE Photon. Technol. Lett.*, 17, 61–63, 2005.
242. Prather, D.W., Shi, S., Venkataraman, S., Lu, Z., Murakowski, J., and Schneider, G., *Proc. SPIE*, 5733, 84–93, 2005.
243. Chen, C., Sharkawy, A., Pustai, D., Shi, S., and Prather, D.W., *Opt. Express*, 11, 3153–3159, 2003.
244. Shi, S., Sharkawy, A., Chen, C., Pustai, D.M., and Prather, D.W., *Opt. Lett.*, 29, 617, 2004.

26

Progress in Nanofluidics for Cell Biology

26.1 Introduction...**26**-1
26.2 General Manipulation and Immobilization
      of Cells in Microdevices **26**-2
      Mechanical Cell Trapping • Biomolecular Cell Trapping
      • Electrokinetic Cell Trapping
26.3 Cell Sorting in Microdevices............................**26**-5
      Electroosmotic and Dielectrophoretic Cell Sorting
      • Pressure-Based Cell Sorting • Optical Sorting
26.4 Devices to Monitor Single-Cell Physiology **26**-7
      Single-Cell Electroporation • Cellular Nucleic Acid Isolation
      • Nucleic Acid Amplification/Detection • Nanofluidic
      Electrophoresis • Proteomics Applications • Quantifying
      Intracellular Protein and Small Molecule Detection
26.5 Cell Culture in Microdevices............................**26**-15
      Surface Chemistry and Cell Culture • Microchannel Geometry
      and Cell Culture • Flow Control in Microscale Culture Systems
26.6 Conclusions..**26**-20
Acknowledgments **26**-20
References **26**-20

Todd Thorsen
*Massachusetts Institute
of Technology*

Joshua S. Marcus
California Institute of Technology

26.1 Introduction

The need for faster and cheaper technologies to extract biological information, both at the molecular and cellular levels, has driven the trend to miniaturize laboratory techniques in the last two decades. Just as the integrated circuit revolutionized information technology, multiplexed microscale devices capable of manipulating and processing DNA, protein, and cell-based samples in nanoliters of fluid have the potential to have the same impact on biology and medicine. The majority of the devices reported to date, often referred to by the descriptor "lab-on-a-chip," have focused on the automation of molecular biology techniques, including the amplification of nucleic acids by polymerase chain reaction (PCR)[1–9] separating and sizing complex mixtures of DNA or protein samples,[10–20] and performing molecular hybridization assays.[21–23] While the earliest reported microscale devices consisted of channels etched in hard substrates such as silicon,[24,25] glass,[26] and plastic,[27] micro-electro-mechanical-system-based fabrication technologies have been increasingly applied to fabricate sophisticated devices from a variety of materials, including soft elastomers such as polydimethylsioxane (PDMS),[28] with hundreds of microchannels and integrated sensors to measure physiological parameters like pH and temperature.

The ability to precisely control parameters such as substrate, flow rate, buffer composition, and surface chemistry in these microscale devices makes them ideal for a broad spectrum of cell biology-based applications, ranging from the high-throughput screening (HTS) of single cells to 3-D scaffolds for tissue and organ culture. Microscale devices offer the possibility of solving system integration issues for cell biology, while minimizing the necessity for external control hardware. Many applications, such as enzymatic library screening in bacteria, are currently carried out as a series of multiple, labor-intensive steps required in the assay process, from growing single colony isolates to plate-based assays and recovery. While the industrial approach to complexity has been to develop elaborate mechanical HTS workstations, this technology comes at a price, requiring considerable expense, space, and labor in the form of operator training and maintenance. For small laboratories or research institutions, this technology is simply out of reach. Devices consisting of addressable microscale fluidic networks can dramatically simplify the screening process, providing an environment where bulk cultures can be injected and compartmentalized into subnanoliter aliquots for sensitive single-cell assays. At the other end of the spectrum, the ability to regulate fluid flow within microscale devices with subnanoliter precision has also generated interest in using them as tools for tissue and organ culture. Advances in substrate micropatterning techniques to mimic capillaries, the implementation of biocompatible and biodegradable substrates, and the ability to pattern surfaces with molecules to stimulate cell adhesion have provided researchers with excellent tools to engineer artificial organs.

In this chapter, we will examine how microscale devices are being applied to both the manipulation and the interrogation of single cells in nanoliter volumes. At the single-cell level, major sections are dedicated to cell trapping, cell sorting, and assays to monitor the physiological parameters of the cell such as enzymatic activity, and gene and protein expression. Transitioning from cellular assays to cell culture, the chapter concludes with a discussion of recent research in the engineering of surfaces, geometrical and scale considerations, and control mechanisms for cell culture in microfabricated systems.

26.2 General Manipulation and Immobilization of Cells in Microdevices

Recent advances in the "-omics"-based sciences, notably genomics and proteomics, have catalyzed a strong interest in the understanding of molecular processes at the cellular level, reaching out to both scientists and engineers alike interested in research areas like gene expression and cell biomechanics. The ability to manipulate single cells has been accomplished over the past few decades in bulk solutions using a variety of techniques, including mechanical tools (micropipettes and robotic micromanipulators),[29] optical tweezers,[30,31] and microelectrodes.[32,33] Adapting these general classes of techniques to nanofluidic systems, using microchannels or microfabricated structures incorporating nanoliters of fluid, offers an unprecedented level of control over the placement and patterning of single cells.

26.2.1 Mechanical Cell Trapping

The physical isolation of cells from bulk culture solutions using mechanical methods such as pipetting and serial dilution is a very tedious, time-consuming, and repetitive process for the laboratory technician. Potential problems with this technique include contamination with bacteria and mold, and shearing of the cells as they are pulled through the pipette nozzle, resulting in low viability. Additionally, the isolation and culture of the isolated cells, typically carried out in microwells containing microliters of media, can be very stressful on a single cell, as the microenvironment surrounding the cell is quite different from bulk cell culture, with growth being promoted by cytokines and other factors secreted into the media.

Several groups have recently reported on the development of nanofluidic systems to mechanically manipulate and isolate single or small groups of cells, using microscale conduits and culture chambers. Thorsen et al. used multilayer soft lithography (MSL) to fabricate integrated PDMS-based devices for programmable cell-based assays with thousands of integrated elastomeric valves (Figure 26.1).[34] MSL is a process used to create stacked 2-D microscale channel networks from elastomeric materials. Elastomeric

Row multiplexor   Horizontal compartmentalization valve   Sample input

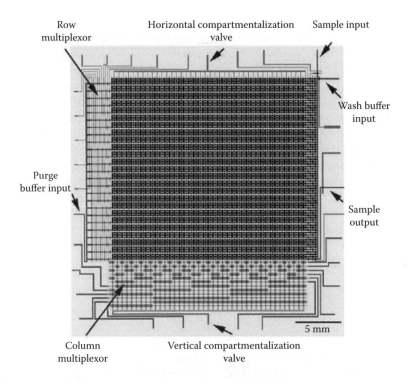

Wash buffer input

Purge buffer input

Sample output

Column multiplexor   Vertical compartmentalization valve

5 mm

FIGURE 26.1 Multilayer PDMS microfluidic array device channel. The device contains an array of 25×40 chambers with volumes of ~250 pL that can be individually addressed by the selective pneumatic switching of multiplexed integrated elastomeric valves. (From Thorsen, T., Maerkl, S.J. and Quake, S.R., *Science*, 298, 580, 2002. Copyright 2002, AAAS. With permission.)

valve structures are made by bonding a thick layer of microchannel-patterned PDMS, the "control" layer, to a thin PDMS layer spin-coated on a second microchannel-patterned silicon wafer, the "flow" layer. The spin-coating process provides a method for precisely controlling the thickness of the silicone layer over the microchannels, which function as valve structures in the assembled devices at the interface where the control and flow layer channels overlap. Pneumatic actuation of the valves is accomplished through interconnect holes punched through the thick layer prior to the secondary bonding step, connecting the control channels to a pressurized air supply controlled by miniature solenoid valves while interconnect holes on the bottom flow layer provide access for sample input and output. Within these devices, they isolated single *E. coli* bacteria in subnanoliter chambers created by closing down an array of valve structures and assayed them for cytochrome c peroxidase activity.

Khademhosseini et al. reported on the use of polyethylene glycol (PEG)-based microwells within microchannels to dock small groups of cells in predefined locations in an array format.[35,36] Photopolymerizable PEG microwells were fabricated using a positive-relief PDMS stamp patterned with cylinders (50 to 800 μm width, ~80 μm in height). To fabricate the microwells, the PDMS stamp was pressed into a thick film of PEG, which was subsequently cross-linked with ultraviolet (UV) light. After removal of the PDMS stamp, the microwells were reversibly covered with a parallel PDMS microchannel structure for multiphenotype cell loading, including fibroblasts, hepatocytes, and embryonic stem cells. Using a syringe pump for loading with nL sec^{-1} flow rates, cells docked in the wells were cultured under standard culture conditions (37°C/5% CO_2). The microwells provided a low shear environment conducive to cell growth and division, while the media flowing through the PDMS microchannel manifold kept the cultures hydrated.

Microscale fabricated structures have been used to filter and trap cells, consisting of groups of weirs and gratings in microchannels with feature-to-feature spacing smaller than the target cell type.[37–41]

Zhu and coworkers used a microchannel containing an array of weirs with 1 to 2 μm gaps to trap and concentrate microbial cells.[40] Two common water-borne pathogens larger in diameter than the weir gap, *Cryptosporidium parvum* and *Giardia lamblia*, were trapped in the weirs under constant flow and subsequently identified using fluorescent antibodies. Li et al. combined micromachined weirs with electroosmotic flow (EOF), using U-shaped microstructures in microchannels to capture and retain cells (yeast and Jurket T lymphocytes) in a flowing media.[41] The device was fabricated as an etched glass plate (Micralyne, Edmonton, Canada) with 15 μm deep U-shaped cell traps centered in a crosshair shaped set of microchannels. With the opening of the U-traps parallel to the default flow direction, cells were flowed through the device and past the trap openings electroosmotically. By switching the electrode off-on configuration, cells passing by the U-trap opening were selectively pulled into the trap for interrogation. The sidewalls of the U-traps consisted of etched weirs with 5 μm spacing, permitting continuous follow-through of media. The utility of this method to both capture and maintain cell viability was confirmed using fluorogenic substrates added to the media, which were internalized and hydrolyzed to fluorescent products in the healthy trapped cells.

26.2.2   Biomolecular Cell Trapping

The trapping of cells using biomolecules in nanofluidic systems has been demonstrated using antibodies and proteins with a high affinity for the target cell.[42–44] Chang et al. used square silicon micropillars in a channel coated with the target protein, an E-selectin-IgG chimera, to mimic the rolling and tethering behavior of leukocyte recruitment to blood vessel walls.[43] HL-60 promyelocytic leukemia cells flowed through the channels specifically bound to the E-selectin-coated regions via a ligand present on the cell surface. By altering the flow rate, the researchers were able to selectively capture and release the cells from the micropillars, resulting in a 100-fold enrichment of the cells in the flowing media upon release. Zhang et al. employed a multiple coating strategy to attach fibronectin, a cell adhesion promoter, to the sides of the PDMS microchannel.[44] An IgG antifibronectin antibody was covalently attached to the PDMS substrate via a silane linker, followed by the addition of fibronectin, which was firmly immobilized on the PDMS surface upon antibody capture. After coating, murine S180 cancer cells flowed through the device, forming a saturated monolayer on the device walls in the regions where fibronectin was present.

26.2.3   Electrokinetic Cell Trapping

Using electric fields to both induce flow and separate molecules are popular laboratory techniques that have been widely adapted to microscale devices for the separation of nucleic acids and proteins.[45–54] Extending this work to the general manipulation of cells, especially mammalian cells, is challenging due to low cell viability in the high voltage/high salt concentration buffer systems used when applying electrical fields to separate small molecules. pH, osmolarity and temperature (due to joule heating by electrodes) become significant parameters that must be also carefully monitored in the electrokinetic manipulation of cells.

Dielectrophoresis (DEP), in which a nonuniform alternating current is applied to separate cells on the basis of their polarizability, has been readily adapted to microscale devices for cell capture.[54–63] The intrinsic dielectric property of an individual cell coupled with that of the surrounding media determines if a cell is repelled by or attracted to the applied electric field. Voldman et al. developed a quadrapole negative DEP trap, in which HL-60 cells were repelled and focused between four AC-field driven electrodes[62] (Figure 26.2). The repulsion and subsequent trapping of the cells was accomplished by using a highly conductive buffer solution [Hanks buffered saline solution with 1% bovine serum albumin (1 S m^{-1})]. At all AC-driven frequencies, the cells were less polarizable than the buffer, repelling the cells into the trap at the center of the quadrapole. Gascoyne et al. also used DEP and a quadrapole electrode array patterned in a PDMS microchannel for the detection of malaria-infected cells in erythrocyte samples.[63] Using a low-conductivity suspension buffer (RPMI culture media [55 mS m^{-1}]), normal erythrocytes, with a higher polarizability than the surrounding buffer, were attracted to the electrodes while malaria-infected erythrocytes were repelled and remained in suspension. The low polarizability of

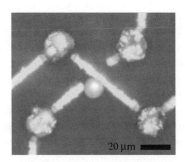

FIGURE 26.2 Microfabricated quadrapole DEP trap. Calcein AM-loaded HL-60 cell (*center*) is trapped between four trapezoidally arranged gold posts. (Figure courtesy of Dr. Joel Voldman, MIT, Cambridge, MA.)

the malaria-infected erythrocytes, a property exploited by the researchers for the device design, is attributed to the presence of cell membrane pores expressed by the parasites and altered lipid membrane composition in the cell membrane, increasing the ionic exchange with the external buffer.

Toriello et al. recently developed a direct current system for single-cell capture on gold electrodes.[64] Chinese hamster ovary (CHO) cells were initially treated with a thiol-containing peptide motif (RGD) that binds to the cell membrane. After labeling, the cells, suspended in 1× phosphate-buffered saline pH 7.4, were flowed into a PDMS microchannel containing the patterned electrodes, until the cell concentration was constant. Flow was then stopped and a 50 V cm⁻¹ was applied to the interdigitated electrodes for 10 min, creating a net positive charge on the gold surface, which electrophoretically attracted the cells in solution, which have an intrinsically negative surface charge.[65] The thiol groups on the attached peptides interacted with the gold and firmly attached the cells to the electrodes, with a single cell deposited on 63% of the electrodes. The ability to spatially pattern different cell populations was further demonstrated using two CHO cell populations, depositing the first population on the odd-numbered electrodes, and then reversing the field to deposit the second population on the even ones.

26.3 Cell Sorting in Microdevices

Although conventional commercial fluorescent-activated cell sorters (FACS) are extremely efficient, with a very high throughput (up to 3×10^4 cells/sec, MoFlo Cytometer [Cytomation]), they are susceptible to frequent microbial contamination and prohibitively expensive for small laboratories.[66] In an effort to miniaturize the cell-sorting process, microscale devices have been fabricated that use several strategies, including electrokinetic, pressure, and optical deflection-based methods.[67–75] While the first prototypes developed in the late 1990s were capable of tens of cells per second, state-of-the-art microscale FACS devices are quite impressive, comparable to commercial sorters.

26.3.1 Electroosmotic and Dielectrophoretic Cell Sorting

EOF, in which flow is induced by applying a voltage potential that induces flow through the migration of ions in solution, was one of the earliest technologies to be used in microscale flow cytometers. Fu et al. developed a simple PDMS T-microchannel FACS device in the late 1990s consisting of a single input and two collection outputs connected to electrodes, with the ability to sort cells by switching the applied voltage potential between the two outputs.[67] Using a blue argon ion laser (488 nm), focused slightly upstream of the T-channel junction as the excitation source to illuminate cells flowing through the channels and a photomultiplier tube (PMT) detector, *E. coli* expressing green fluorescent protein (GFP) were sorted from nonfluorescent control bacteria at a rate of 20 cells sec⁻¹ with a 30-fold enrichment rate in the positive output channel (GFP bacteria/control). Unlike conventional FACS machines, the ability to reinterrogate cells was demonstrated in the electroosmotically driven microscale devices simply by reversing the potential, flowing them back across the detection region of the devices. However, EOF has several disadvantages that proved to be detrimental for maintaining cell viability in microscale FACS devices, including ionic depletion, evaporation, and the use of high-voltage potentials on the order of hundreds of volts cm^{-1} that compromise cell membranes by electroporation.[67,68] In the devices developed by Fu et al., sorted bacteria viability was ~20%.

DEP has been applied to the development of a microfabricated device for the high throughput sorting of rare cell populations. While DEP generates high electric fields that, like EOF, can be detrimental to cell viability, exposing cells to short bursts of DEP (on the order of μs) is well tolerated by cells and can be applied as a cell-sorting tool.[69] Hu et al. developed a polyimide-based microfabricated device with an integrated quadrupole electrode for the marker-specific sorting of rare bacteria populations.[70] The device was designed to exploit the difference in dielectrophoretic response between a control *E. coli* population and a bead-tagged population expressing a rare surface marker. The rare population was tagged using a marker-specific antibody conjugated to a strepavidin polystyrene bead (Bangs Laboratories, Carmel, IN). After labeling, the rare cells were spiked in the control bacteria solution (1:5000) and a syringe pump was used to flow the cell suspension through the device at a rate of 10^4 sec^{-1}. Exploiting the equivalent dielectric properties of the cells and beads, a negative DEP was applied that was strong enough to deflect the bead-tagged cells into a output channel separate from the bulk population, with a 250-fold single pass enrichment factor. The total flow rate of 2 to 3 × 10^7 cells/h is comparable to conventional cell sorters.

26.3.2 Pressure-Based Cell Sorting

Pressure-based microfabricated cell sorters have been developed for the sorting of both prokaryotic and eukaryotic cells. Flow switching to the collection and waste outputs in these devices has been achieved using a number of techniques, including integrated elastomeric valves in multilayer PDMS-based devices and hydro-dynamic switching.

Quake's group fabricated PDMS-based multilayer devices for the pressure-based sorting of *E. coli* expressing GFP in a wild-type *E. coli* control population.[71] The device, arranged as a T-channel, contained pairs of integrated elastomeric sorting valves overlapping the two output channels and a set of valves configured as a peristaltic pump to control flow rate. Pneumatic actuation of the on-chip sorting valves was based on the laser-induced fluorescence (LIF) of the cells (argon ion laser excitation source, 488 nm) and a PMT detector. A computer equipped with a data acquisition card (National Instruments, Austin, TX) was used to process the PMT signal, triggering the actuation of either the waste or the collection port valves, which were controlled by pressurized air gated by external solenoid valves. Using the device, 83-fold enrichment rates for the the GFP-expressing bacteria were achieved at a flow rate of 1.3 × 10^5 bacteria/h.

Hydrodynamic focusing and switching provides an alternative to integrated valves in pressure-driven micro-fabricated cytometers, instead using pressured liquid introduced through microchannels connected to the main cell-sorting flow channel to manipulate the flow trajectory of the cells.[72,73] Focusing is achieved using two microchannels that symmetrically intersect with the main flow channel, to which a balanced load of pressurized liquid can be applied to force the cells into the center streamline. After focusing, cell sorting on the basis of LIF is achieved using a secondary set of intersecting symmetric microchannels. Using off-chip valves to selectively apply pressurized liquid through just one of the intersecting microchannels, the trajectory of the focused cells is biased toward the sides of the main flow channel for downstream passive sorting using a v-shaped intersection that splits the stream down the middle into two microchannels that lead to the collection and waste ports. Although one of the challenges in operating such devices is back pressure problems created by the impinging flow streams, recent prototypes have demonstrated impressive sorting speeds. Wolff et al. developed a microfabricated sorter using hydrodynamic switching, in which sorting speeds up to 12,000 sec^{-1} were reported for the sorting of fluorescent latex beads from chicken red blood cells.[73]

26.3.3 Optical Sorting

The use of photonic pressure to manipulate cells either by trapping or repulsion in microchannels is an appealing technology. Ozkan et al. used vertical cavity surface-emitting laser arrays to trap and manipulate individual polystyrene spheres and cells in PDMS microchannels.[74] While field-trapping strengths are characteristically weak for larger objects like mammalian cells, making high-throughput sorting by capture a tedious process, the use of photonic pressure to repel cells has been shown to be an effective tool for cell sorting in microfabricated devices. Wang et al. recently used a 20 W Ytterbium fiber laser (1070 nm) like a hydrodynamic switch to sort flow-focused eukaryotic HeLa cells (GFP-transformed vs. nontransformed control) in an etched glass microchannel.[75] Using an LIF detection platform (5 mW 488 nm semiconductor laser, PMT emission detector), the Ytterbium laser was applied to deflect the GFP population from the main streamline into a separate collection channel. In single-pass sorting runs, enrichment rates up to 71-fold were observed. While sorting rates were low (up to 103 cells sec^{-1}), the authors were able to demonstrate no loss in cell viability using the optical switch. With its inherent scalability, coupled with the rapid switching time of the optical components, faster, more efficient examples of optical sorting in microscale devices are likely to emerge in the near future.

26.4 Devices to Monitor Single-Cell Physiology

It has become increasingly clear that studying cell populations, regardless of what is being assayed (e.g., mRNA, protein, or small molecule levels) is only sufficient for obtaining average values over the particular population. Vast cellular heterogeneity is a common theme in all biological organisms, and the ability to interrogate cells in an individual manner could elicit the presence of subpopulations and provide insight into specific processes, such as stem cell differentiation events, which are masked at the population level. Although questions pertaining to what genes and proteins are expressed in single cells and to how individual cells communicate and respond to different stimuli are all active research areas, tools to answer these questions are generally lacking. The main reason for the lack of technologies to study the single cell can be contributed to the poor yields encountered when subjecting these precious samples to multistep processes.

The typical channel dimensions found in microscale devices (10 to 100 μm in *x*, *y*, and *z*) and the ability to manipulate nanoliters of reagents on-chip have made the devices encouraging platforms for the analysis of single cells.[76] Furthermore, the economy-of-scale benefits along with the ability to parallelize and automate processes are significant advantages not found with conventional biological assays. The focus of this section will be on micro/nanofluidic assays developed to interrogate the intracellular content of single cells, although cell population assays that are amenable to individual cells will also be considered. We will begin by discussing assays in which single cells can be biochemically modified, followed by assays to determine expression of genes, proteins, and other signaling molecules. It is not the intention to provide a comprehensive list of intracellular assays applied in nanofluidic devices, but rather to highlight techniques that have the most promise for single-cell analysis.

26.4.1 Single-Cell Electroporation

Electroporation, or the application of an electrical field to induce changes in membrane permeability, has been harnessed in batch assays to introduce polar and charged agents such as nucleic acids, proteins, dyes, and drugs into cells. Single-cell electroporation allows for intracellular biochemical manipulation as well as investigation of cell-to-cell variation in response to various substances. Recently, electroporation assays have been developed to perform the technique on single cells in microchannels.[77–80] These efforts utilized cell loading with a syringe in order to control flow rate and to apply back pressure to the microchannel for effective trapping. The earlier devices[77,80] were fabricated with standard silicon microfabrication techniques and trapped an individual cell in a micropore present between two electrodes. More recent work took advantage of soft lithography[81] to trap cells in tapered microchannels. Cross-sections of the devices that employ the two electroporation schemes are given in Figure 26.3. Each group effectively trapped a single cell between two electrodes and was able to show electroporation events by measuring either molecular uptake or molecular release by fluorescence. Current jumps due to membrane poration could also be studied by treating the cell as a resistor in the existing electrical circuit. Although cells were successfully trapped by either strategy, a significant difference between the two strategies is the effective voltage needed to electroporate cells. In the PDMS chip,[78] ≤1 V was needed to electroporate a

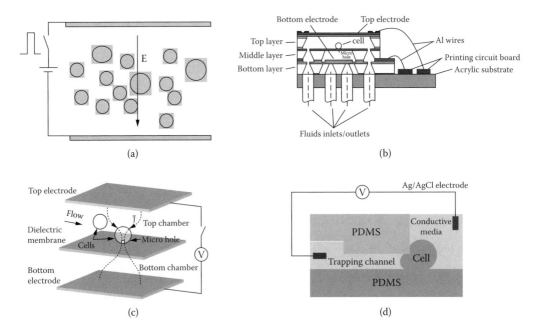

FIGURE 26.3 Electroporation schemes. (a) Bulk electroporation. Only a portion of the cells are electroporated because the electric field is not uniform with respect to the cells exposed to it. (b and c) Cross-sections of silicon single-cell electroporation devices. Cells are transported in a microchannel and captured in a pore by applying back pressure to the channel. (From Huang, Y. and Rubinsky, B., *Sens. Actua. A*, 89, 242–249, 2001; Huang, Y. and Rubinsky, B., *Sens. Actua. A*, 104, 205–212, 2003. With permission from Elsevier.) (d) Cross-section of a PDMS-based single-cell electroporation device. Cells are captured between electrodes by hardwiring the device to constrict channels to a height smaller than a cell's diameter. (From Khine, M. et al., *Lab Chip*, 5, 38, 2005. With permission from the Royal Society of Chemistry.)

single HeLa cell, while in the silicon devices,[77,80] 15 to 20 V was needed to reversibly electroporate various individual mammalian cells. In each case, however, the molecular transfer rate was close to 100%, a direct contrast to conventional schemes.[4] A main avenue of research in this area will be the integration of downstream on-chip steps after electroporation/molecular transfer events, as well as development of more efficient trapping strategies to increase sample throughput.

26.4.2 Cellular Nucleic Acid Isolation

The isolation of gDNA or mRNA from bacterial and/or mammalian single cells is a crucial step for many biological and medical applications. Construction of cDNA or gDNA libraries at the single-cell level could allow for gene discovery in rare cells and the elucidation of molecules important for various processes. The following section will focus on the isolation of nucleic acids on-chip from a single cell or a small number of cells. It is known that pipetting DNA or RNA results in its shearing,[82] and therefore the focus will be on reports that take advantage of the low Reynolds number flow inside nanofluidic channels to lyse cells.

Early work on cell lysis/gDNA isolation[83] focused on lysing single GFP expressing *E. coli* bacterial cells (treated off-chip with lysozyme to break down the peptidoglycan layer) by diffusional mixing with dI water at the junction of a T channel that was molded in PDMS and bonded to a glass coverslip by oxygen plasma treatment. Efficient lysis was demonstrated by the release of GFP into the microchannel. The lysate was then transported via DEP to a trapping region of the device, where the channel constricted and contained a series of 10 μm gaps to both maximize the electric field and trap the *E. coli* chromosomes. DNA was visualized with a nucleic acid stain, with cells being stained before loading on-chip. It was shown that manipulating the applied DC and AC currents allows for chromosome shuttling to downstream DEP traps.

Anderson et al.[84] developed a polycarbonate-machined device capable of automated microarray sample preparation and subsequent hybridization. A solid phase of cellulose is packed into a chamber, and homogenous lysate generated off-chip is drawn by vacuum through the extraction chamber. The nucleic acid is then eluted for downstream PCR. The nucleic acid isolation procedure yielded a sensitivity of 300 copies.

The first reported integrated mammalian cell capture, lysis, and nucleic acid isolation chip-based assays[85] utilized the technique of MSL[86,87] for device fabrication. In this report, both *E. coli* and murine NIH/3T3 cells were isolated in nanofluidic channels by active mechanical valves. Cells were lysed on-chip by diffusional mixing, followed by nucleic acid isolation by affinity capture. Columns of microbeads were built against a partially closed microvalve, with valve pressure and lysate flow rate having to be tuned precisely to carry out successful assays. Hong et al. demonstrated single mammalian cell lysis followed by mRNA capture/purification as well as bacterial cell lysis/gDNA isolation and purification from as little as 30 bacterial cells.[85] Recent advances by the Quake group have made column construction more robust and digital in nature, as opposed to the analog predecessor (Figure 26.4). Marcus et al. demonstrated single mammalian cell resolution with the new method.[88] In addition, first strand cDNA synthesis was integrated with the aforementioned processes by utilizing the oligo(dT)$_{25}$ derivatized beads, which comprised the affinity column, as both primers and a solid-phase support. The additional step

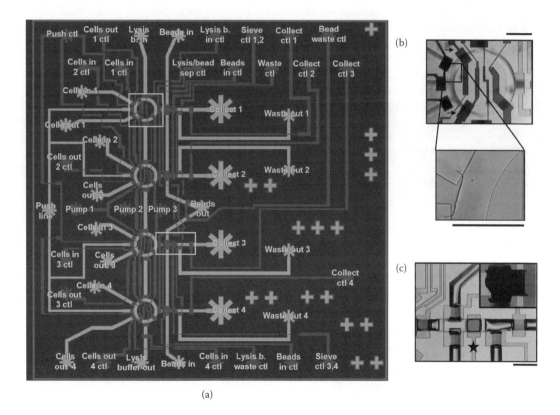

(a)

FIGURE 26.4 4plex mRNA isolation/1st strand synthesis device. (a) AutoCAD drawing of a device with inputs and outputs labeled according to function. Rounded flow channels are depicted in light gray and control channels are shown in dark gray. Unrounded (rectangular profile) flow channels for affinity column construction are shown in dark gray. Portions of the drawing in white boxes are shown in b and c, respectively. (b) Optical micrographs of the lysis ring and an NIH/3T3 cell captured in the ring. A cell is captured by opening the respective cell in-and-out valves and keeping pump 1 and pump 2 closed (marked with arrows in figure). Cells are then loaded under pneumatic pressure with a pipette tip. (c) Optical micrographs of the affinity column construction area and a stacked column against a sieve valve (marked with a star in figure). Scale bars are 400 μm. (From Marcus, J.S. Anderson, W.F. and Quake, S.R., *Anal. Chem.*, 78, 3084, 2006. With permission from the American Chemical Society.)

did not have adverse effects on sensitivity. By utilizing off-chip quantitative RT-PCR (RT-qPCR) on the bead:mRNA, or the bead:cDNA complexes, the authors demonstrated the capability to detect mRNA copy numbers spanning six orders of magnitude (10 to 10^6 copies).

26.4.3 Nucleic Acid Amplification/Detection

PCR has proven to be the most sensitive method for amplifying small amounts of nucleic acids for the purpose of global and specific gene identification, with single-cell studies now commonplace in the literature.[89–97] Nanofluidic channels offer a robust platform for PCR[9,84,98–104] for various reasons. Nanofluidic devices provide efficient thermal transfer due to high surface-to-volume ratios and offer diagnostic and forensic applications based on the ability for faster thermal cycling and on-site detection. The capacity to partition reagents into thousands of chambers[9,34] enables concentrated reactions, digital detection,[105] and therefore absolute copy number determination in single cells or otherwise.[103] Despite the overwhelming promise for applying nanofluidic PCR to single cells, little has been reported[84,99,106] on devices that integrate cell lysis and RNA or DNA purification with an amplification reaction. Integration of cell lysis is critical because of the aforementioned shearing issue and because it establishes the temporal resolution of the particular downstream measurement(s).[107] Furthermore, to realize true economy-of-scale and to maintain concentrated samples throughout the full purification and amplification processes, process integration is essential. On-chip PCR has been reviewed extensively elsewhere,[82,108] and therefore only one application of nanofluidic PCR will be discussed here. Other gene-expression microanalytical techniques will also be reviewed.

Unger et al.[103] harnessed MSL technology to design a PCR device capable of partitioning PCR reactions into tens of thousands of picoliter compartments (Figure 26.5). The proof-of-principle work isolated

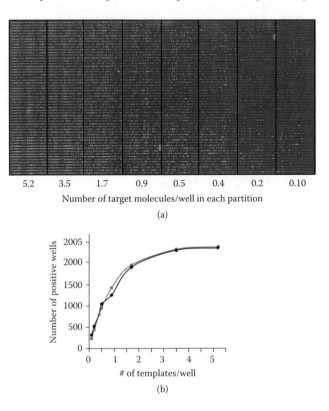

(a)

(b)

FIGURE 26.5 Digital PCR results from Unger et al. (a) Scan of a chip after 30 cycles of PCR. β-actin cDNA molecules were serially diluted and detected with Taqman™ chemistry and a flatbed scanner. (b) The positive reactors in (a) are nearly identical to what is calculated by Poisson statistics. The black line is the number of observed positive wells while the gray line is the number calculated by Poisson statistics. (From http://www.fluidigm.com/didIFC.htm.)

varying concentrations of β-actin cDNA molecules down to 0.1 copies/chamber and employed fluorescent hydrolysis probes specific to β-actin for readout of the method. The observed number of positive chambers matched almost identically with what is calculated by Poisson statistics. More recently, this technology has been applied by Quake's group to study multiplexed gene expression in single mammalian cells, as well as to determine phenotypes of unculturable bacteria found in the termite gut.[85] In the case of mammalian cells; single cells are seeded in 96 well plates or microcentrifuge tubes followed by chemical lysis off-chip in a modified RT-PCR buffer. The lysate is then loaded onto the device and partitioned via actuation of the valve array. The chip is then thermal cycled and fluorescence detected with a flatbed scanner. Besides fluorescence detection with oligonucleotide probes or DNA-binding dyes, the presence of on-chip PCR products has been elucidated by electrochemical[99,109–111] and electrophoretic methods.

Nucleic acid sequence-based amplification has been applied in nanoliter volumes inside microscale devices, although not to single cells.[112–114] The technique is gaining popularity in nanosystems because the process is isothermal (41°C) and therefore simplifies chip design; and obviates the problem of evaporation at hot start and denaturation temperatures (~95°C). Jayaraman's group[115,116] developed PDMS-based devices to study TNF-α-activated NF-κB gene expression dynamics with EGFP reporters inside living mammalian cells. The authors presented two similar devices, both of which allowed for cells to be seeded in chambers along a nanofluidic network, and subsequently grown to the desired confluency. Medium exchange every 12 h was accomplished with syringe pumps to replenish metabolites and to remove toxic wastes and dead cells. The devices differed in fluidic architecture so that eight[115] or four[116] different concentrations of activator (applied via syringe pumps) could be applied to cells in parallel, enabling dose–response curves to be generated for populations of cells as well as single cells (Figure 26.6). Future work based on these studies will surely incorporate the monitoring of multiplexed gene expression dynamics using multiple reporters, as well as dynamics of heterogeneous cells, grown together in culture,

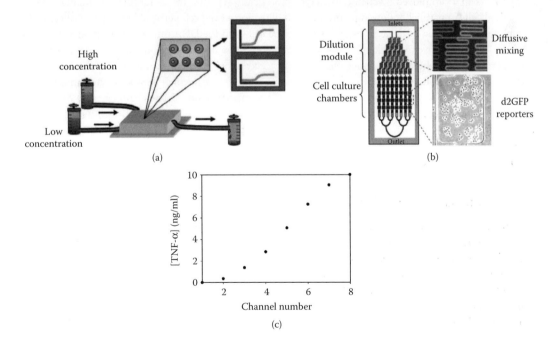

FIGURE 26.6 Living cell array. (a) Chip-loading scheme. Cells are seeded into the device with a syringe pump through one of the inlets. Chemicals and medium are also loaded with syringe pumps. (b) Medium and chemicals are diluted via diffusive mixing in the dilution module and are delivered to the cells. TNF-α (spiked with fluorescein) gradient is formed via diffusive mixing. Cells are visualized with d2GFP reporters. (c) TNF-α dose–response curve for chambers in the LCA. (From Thompson, D.M. et al., *Anal. Chem.*, 76, 4098, 2004. With permission from the American Chemical Society.)

by the utilization of distinct reporters in each cell type. Furthermore, intrinsic and extrinsic noise[117] in single-cell gene expression will be able to be resolved by these methodologies.

26.4.4   Nanofluidic Electrophoresis

An extensive body of work has been done to study intracellular contents of single cells by electrophoresis inside nanofluidic channels.[107,118–124] The majority of applications have focused on the detection of small molecules and metabolites,[118,120–122] as well as DNA content.[119] Electrophoretic detection of specific protein expression has been accomplished in nanofluidic devices in immunoassay-based formats[125] as well as by probing for aromatic amino acids, but is yet to be integrated on-chip with single-cell samples.[126] However, a device integrating cell lysis with detection of a GFP chimera has been disclosed.[127]

Fang and coworkers[120,122] published two reports in 2005, disclosing a method that utilized electrophoresis to detect reactive oxygen species (ROS) and glutathione (GSH) inside single human erythrocytes. These molecules are known to mediate numerous pathological processes, including cancer and brain trauma. The researchers utilized photolithographic and wet chemical etching to define a 12 μm deep and 48 μm wide T-channel in soda-lime glass (Figure 26.7). Single cells were transported into the lysis/separation portion of the channel by EOF by applying electrical potentials to the reagent reservoirs. Lysis of the single cell was accomplished either by subjecting the cell to a stronger electric field[122] or by subjecting it to the same electric field after a buffer exchange.[120] For detection, the authors took advantage of native nonfluorescent dyes rhodamine 123 (DHR 123) and 2,3-naphthalene-dicarboxaldehyde, which fluoresce when oxidized by ROS or GSH, respectively. Labeling of the cells was accomplished either before sample loading or directly after cell lysis, by diffusive mixing. The authors utilized LIF for detection, and were able to resolve ROS and GSH at sensitivities two orders of magnitude lower than conventional detection limits.

Munce et al.[121] employed a scheme to electrophoretically separate the intracellular contents of four individual cells in parallel, followed by multiplexed detection of the cell permeable molecule calcein AM (which fluoresces after cleavage by esterases inside the cell) and the DNA minor groove binder, Hoechst 3342, by LIF. Acute myeloid leukemia cells were transported via optical tweezing to a tapered portion of the microchannel, where cells underwent lysis by electromechanical shearing. The technique offers selective lysis of cells that contact the channel opening (Figure 26.8), which allows for the simultaneous injection of the cell's cytoplasmic contents into predetermined channels. The channels were defined in PMMA by laser ablation, and electrodes were formed by inserting stainless steel wires into the cathode and anode reservoirs in the nanofluidic device. The majority of future studies in this field will most likely harness soft materials, such as PDMS,[81,86] to achieve greater parallelization and process integration and separation via mechanical valves.

26.4.5   Proteomics Applications

In general, reported single-cell protein expression studies, whether specific or on the scale of the proteome, are far fewer than single-cell gene expression studies. Numerous factors contribute to this fact, including the inability to amplify proteins from a single cell and the variability of gene products expressed due to truncations, splice variants, and posttranslational modifications.[128] Tools that operate on the length scale of proteins, such as devices harnessing nanofluidic channels, could have far-reaching implications for single-cell protein expression experiments.

Benchtop-based immunoassays, such as ELISA, are the workhorses for determining intracellular protein expression in bulk samples. ELISA has been transferred to microscale formats,[126] but has not been utilized to detect analytes from cell lysates. The key consideration for these assays is how to immobilize the first layer of the immunostack inside a nanochannel. Two methods that immediately come to mind are surface derivatization and solid phase extraction with antibody-coated microbeads. Because of the high surface-to-volume ratio found in nanofluidic channels, surface derivatization will employ a greater capacity for capture than beads or conventional microtiter-plate-based assays. However, bead assays with the scheme employed by Marcus et al. (Figure 26.4) may provide a robust way to interact cellular lysate

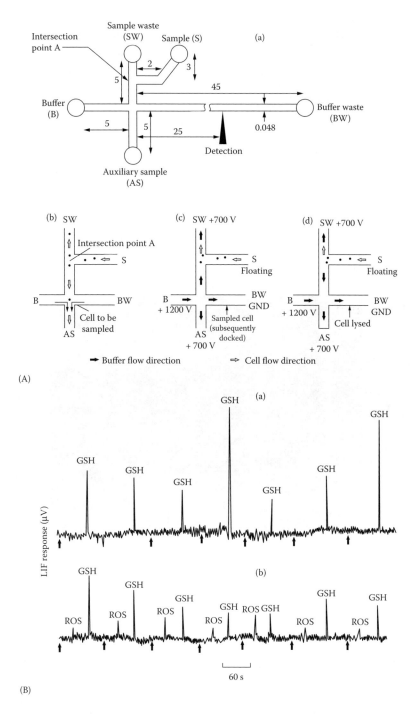

FIGURE 26.7 Determination of erythrocyte contents by capillary electrophoresis. (a) — (a) Schematic of the T-microchannel etched in soda-lime glass. Channels are 12 μm high and 48 μm wide. (b) A single cell is transported from S through intersection point A to AS and SW by hydrostatic pressure. (c) A set of electrical potentials is applied to inject a single cell from the sample channel into the separation channel. Voltages are turned off once the cell is docked in channel. (d) The docked cell is lysed by applying the same voltage used for cell transport under different buffer conditions. (B) A recorded electropherograms of seven cells injected consecutively. The separation distance was 25 mm and took place with a voltage of +240 V/cm. Arrows indicate the starting points for separations. (*Top*) Fresh cells; (*Bottom*) stimulated with H_2O_2 for 10 min. (From Ling, Y.Y., Yin, X.F. and Fang, Z.L., *Electrophoresis*, 26, 4759, 2005. With permission from John Wiley and Sons.)

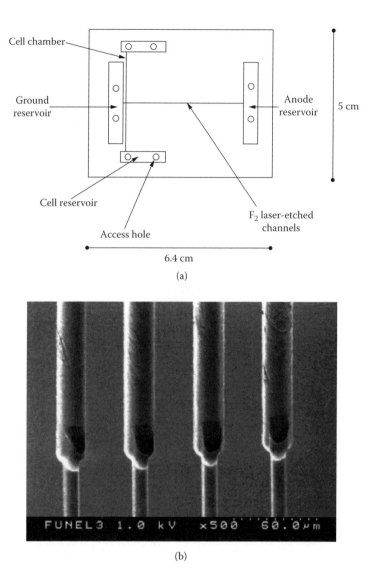

FIGURE 26.8 Parallel intracellular study by microfluidic lysis/electrophoresis. (a) Layout of the device utilized for parallel cellular electromechanical lysis/electrophoresis. One laser-etched channel is shown for clarity. (b) Scanning electron micrograph of the tapered intersection of the injection and separation modules of the channels. (From Munce, N.R. et al., *Anal. Chem.*, 76, 4983, 2004. With permission from the American Chemical Society.)

with antibodies for proteomic applications. Kartalov et al. derivatized epoxide floors of PDMS nanochannels with antibodies to cancer markers.[129] Because the device utilized active valving, specific channels could be individually addressed with antibodies of choice, enabling multiplexed detection from one sample. Once antibodies are bound to the surface, the antigen-containing sample can be pressure-driven over each section of the device, enabling all antibody parameter space to be sampled. Protein digests have also been detected in nanofluidic systems by 2-D electrophoresis, although not from cellular samples. Ramsey's group[17,130] has separated tryptic digests of model proteins by micellar electrokinetic chromatography followed by capillary electrophoresis. The second report demonstrated higher separation resolution because of the optimized channel geometry. The more recent work employed asymmetric turns and decreased the second dimension's injection plug length, therefore achieving faster sampling and an order of magnitude improvement in peptide peaks resolved.

Other proteomics tools being transferred to the nanoscale are devices that prepare samples for mass spectrometry (MS) and x-ray diffraction analysis. MS is a powerful tool in that it can elucidate global networks of proteins, and x-ray diffraction of macromolecular crystals is the most sensitive method for determining molecular structure. Microscale MS applications have been reviewed extensively elsewhere[126] and therefore will not be discussed further. Because structure implies function, elucidating protein structure is an extremely active pursuit in both biochemistry and cell biology laboratories. Hansen et al. took advantage of the unique fluid physics present on the microscale[131,132] to screen protein crystallization conditions by barrier-interface metering. The PDMS device fabricated by MSL is capable of screening 144 protein crystallization conditions with only 10 nL of protein for each condition, which is 2 orders of magnitude less than conventional methods such as hanging drop.

26.4.6 Quantifying Intracellular Protein and Small Molecule Detection

Detecting proteins and small molecules in single living cells has been accomplished via fluorescence assays inside nanofluidic devices. In the case of small molecules, most reported work[76,133–135] has been carried out by probing the intracellular Ca^{2+} concentration, $[Ca^{2+}]_i$, in mammalian cells, mainly due to the established commercial dyes available and the importance of Ca^{2+} as a universal second messenger. Protein expression has been examined by quantifying GFP variants transfected inside cells.[34,116,136,137]

In the $[Ca^{2+}]_i$ reports referred to above, either cells were stained with the respective calcium indicators off-chip,[76,133,134] or staining was implemented via diffusive mixing on-chip.[135] Mixing the fluorescent dye with cells on-chip was inefficient, taking nearly an hour to maximize signal to noise with respect to the cell and its surrounding channel. Although dye loading on-chip minimizes cell damage, more robust and efficient methods are needed to shorten the diffusion length scale, possibly by using valves[86] to enclose the reaction. Manipulation of cells on-chip also varied from researcher to researcher. Two groups[76,135] reported microfabricated structures inside fluidic channels to retain single cells, while another group took advantage of a hydrodynamic pressure difference between two parallel flow channels to dock cells in a hard-wired dam between the channels.[62] The final group[133] flowed cells through microchannels and numerous detection zones, to gain real-time monitoring of $[Ca^{2+}]_i$. The $[Ca^{2+}]_i$ assays implemented were similar in nature, employing diffusive mixing of calcium agonists with cells followed by detection of a fluorescent dye product. The next generation of calcium assays will certainly integrate the culture of mammalian cells along with real-time calcium monitoring. The ability to partition cell subpopulations in nanoliter compartments will allow for HTS applications as well as for the study of intrapopulation effects.

26.5 Cell Culture in Microdevices

The ability to confine single or small groups of cells in nanoliters of fluid presents new opportunities and challenges. As cells undergo gas exchange with the local environment, consume nutrients, and secrete waste products, confining them in micron-sized spaces can induce undesirable stress, particularly for extended culture periods. For some applications, such as cellular assays, where the retention time in the devices is relatively short (on the order of tens of minutes), optimizing culture conditions is less important. However, as interest grows for devices to study applications such as long-term cell behavior, toxicology, and cell–cell interactions, engineering microdevices for long-term, sustained cell-viability growth is critical.

26.5.1 Surface Chemistry and Cell Culture

The design of microfabricated devices for long-term cell culture can be challenging, as, even in bulk, culture conditions are quite specialized for different cell types. While many cell lines grow in suspension, others are adherent and need extracellular matrices of proteins and small molecules for proper attachment. In this section, we will present recent advances in surface chemistry treatment as they pertain to cell culture in microdevices. For a more comprehensive review of the topic, several excellent reviews are recommended.[138–140]

The patterning of molecules and polymers on surfaces to promote cell adhesion has been accomplished using several methods, including photolithography,[141–149] microcontact printing,[81,150–152] and selective molecular deposition in microchannels using laminar flow.[153,154] While many of the pioneering photolithography experiments focused on the direct attachment of proteins and peptides to glass surfaces,[141–145] photopolymerizable hydrogel matrices have recently emerged as a powerful tool for micropatterned cell culture, using materials like PEG[146–148] and polyacrylamide (PAM).[149] Langer and coworkers created a device with an integrated PDMS microchannel-based mixer to dynamically fabricate PEG-based hydrogels with controllable gradients of tethered molecules on a glass substrate.[147] A hydrogel gradient was set up containing PEG conjugated to a peptide ligand (RGDS) known to promote cell adhesion, and differential cell attachment across the gradient was demonstrated using human umbilical vein endothelial cells. Koh et al. used PEG hydrogel microstructures to create multiphenotype cell arrays.[148] RGD peptide-modified PEG hydrogels containing encapsulated cells (fibroblasts, macrophages, and hepatocytes) were injected into PDMS microchannels and polymerized using UV light (365 nm). After patterning, the cells exhibited excellent viability (in excess of 1 week in culture). While a significant amount of research has focused on the surface chemistry of photopolymerizable materials for cell culture, the importance of the mechanical interaction between hydrogels and cell culture has also been investigated.[149,155–157] Zaari et al. used a PDMS-based device to generate a PAM hydrogel matrix on glass, controlling the overall stiffness of the matrix using a gradient of different *bis*-acrylamide formulations. The elastic modulus of the patterned product was characterized using atomic force microscopy (3 to 40 kPa). Using bovine vascular smooth muscle cells, clear differences in cell attachment and proliferation were observed across the matrix, with better adhesion and growth found on the stiffest regions.

Microcontact printing for cell biology applications involves the direct transfer or stamping of small molecules onto a substrate using a positive-relief elastomeric mold with predetermined feature sizes, ranging from nanometers to microns.[81,150–152] To promote cell adhesion to specific surfaces, many groups have used SAM, in which alkanethiols are stamped on a gold surface.[152] By changing the functional group at the surface, SAMs can be used both to selectively promote cell adhesion and to provide inert, protein-repelling surfaces to confine cells to precise areas of a patterned substrate.[140,158] McClary et al. studied the adhesion and proliferation of fibroblasts on carboxyl- and methyl-terminated alkanethiol-on-gold substrates.[159] Strong cell proliferation was observed on the carboxyl-terminated SAMs relative to the methyl surface. Recently, Hammond's group reported a new patterning approach using polymer-on-polymer stamping.[160] A polyamine surface patterned onto a poly(acrylic acid)/poly(allylamine hydrochloride) cell-resistant substrate was used as a template for the deposition of colloidal particles. The colloidal particles, derivatized with the peptide RGD to promote cell adhesion, were deposited on the stamped regions in loose- and tight-packed configurations. Fibroblasts cultured on these self-assembled arrays showed good adhesion and proliferation on the loosely packed colloids, and were generally nonadherent to the closely packed configuration, opening up the potential for interesting future research on the effect of surface roughness on micropatterned cell culture.

The characteristic laminar flow profile of fluids in microchannels has also been exploited as a tool for the direct patterning of solid substrates with cells and biomolecules to promote cell adhesion.[153,154] Takayama et al. used a PDMS-based device containing multiple inputs to selectively copattern different cell types across a common microchannel.[153] With the inlets converging into a central microchannel, various cell lines (chicken erythrocytes, bovine endothelial cells, *E. coli* bacteria) were deposited as stripes parallel to the microchannel as they were swept toward the outlet under laminar flow (Re < 1). By changing the volumetric flow rate, the width of the individual stripes could be dynamically tuned from 5 to 100 μm. Takayama later applied this technique to the subcellular delivery of small molecules to different regions of a bovine endothelial capillary cell immobilized in a microchannel[154] (Figure 26.9). Using two streams of small molecules targeting the mitochondria in the cell conjugated to red and green fluorescent probes, a colaminar flow profile of the reagents was established over the cell, labeling one half of the cell red and the other half green. This ability to locally deliver reagents to cells confined in microchannels was exploited by Kaji et al. to stimulate single cardiomyocytes.[161] Single cells prepared by microcontact printing were subjected to multiple colaminar streams of stimulants and inhibitors to measure contractility.

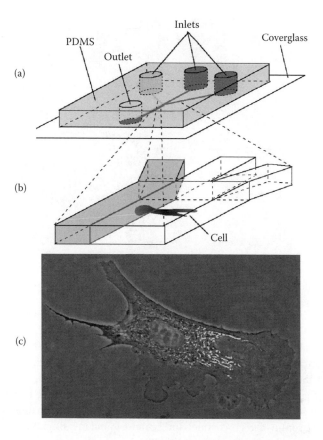

FIGURE 26.9 (a,b) Experimental set-up. (b) Close-up of the point at which the inlet channels combine into one main channel. (c) Fluorescence images of a single cell after treatment of its right pole with Mitotracker Green FM and its left pole with Mitotracker Red CM-H2XRos. The entire cell is treated with the DNA-binding dye Hoechst 33342. (From Takayama, S. et al., *Nature*, 411, 1016, 2001. With permission from Macmillan Publishers Ltd.)

Local delivery of compounds like 1-octanol inactivated parts of the myocytes' contraction patterns, while other regions of the cell maintained activity.

26.5.2 Microchannel Geometry and Cell Culture

Culturing cells *in vitro* is a challenging task, as they survive in a specialized environment *in vivo* that is difficult to replicate. Consequently, as researchers continue to develop microscale devices for cell culture, careful thought is being given to their geometry. Many devices are being developed for the growth and analysis of specialized cell lines, such as fibroblasts,[162] neurons,[163–166] and hepatocytes.[162,167–169]

As the interest in neural networks continues to grow, microscale devices have the potential to be used as an important culture tool for the selective patterning, growth, and analysis of neural cell cultures. Previous research using grooved substrates has shown that the morphology and orientation of individual neurites in culture is geometry specific.[166,170,171] When cultured on narrow, shallow grooves (1 to 4 μm wide, 14 nm to 1.1 μm deep), neurites orient both perpendicular and parallel to the grooves, depending on cell type.[166,170] Mahoney et al. used a polyimide-based device to study geometrical influences on a model neuronal cell line (PC12) grown in microchannels.[166] Using 10 μm (h) × 20 to 60 μm (w) microchannels, they studied the effect of microchannel geometry on the orientation of filamentatious outgrowths of neurites from the main cell body and the number of neurites projecting from each cell. In the narrower channels (20 to 30 μm), the neurites preferentially oriented parallel to the microchannels,

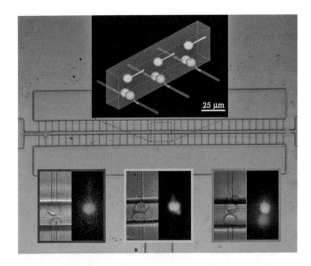

FIGURE 26.10 PDMS device for the analysis of pairwise cell–cell interactions. Cell pairs are trapped and confined at the junctions of smaller side channels perpendicular to main flow stream (*top, inset*), facilitating Calcein AM dye transfer between cells (*bottom, inset*). (Figure courtesy of Dr. Luke Lee, University of California, Berkeley, CA.)

while perpendicular outgrowths were observed in the wider channels. Interestingly, in comparison to control PC12 cells grown in a bulk liquid culture, the cells extended half as many neurites in the microchannels, and those that were observed were a factor of two longer than the controls. These differences may be due to mechanical interactions between the cells and the microchannel boundaries, gradients of local cell stimulants such as collagen that are amplified in the nanofluidic environment, or metabolite levels in the confined microchannels unlike those found in bulk culture. Beebe's group recently published a study on diffusion-dependent cell behavior in microenvironments that found a sharp difference between the proliferation of cells in bulk and cell cultures in microchannels.[172] Using armyworm ovarian cells (Sf9) cultured in a PDMS microchannel, a significant decrease in cell proliferation (vs. bulk culture) that could be directly correlated with the height of the microchannels was observed. Using 1 mm width channels, lower proliferation rates were seen in 250 µm (h) channels (2.4-fold) compared to 2 mm (h) channels (5.2-fold) and bulk culture (ninefold) over 6 days. As diffusion dominates over convection in microscale devices, microchannel geometry becomes a critical design factor in avoiding localized nutrient depletion and has a major impact on cell culture.

The confining geometry of microscale culture systems is ideal not only for investigating the growth characteristics of single cells, but also for exploring cell–cell interactions. The ability to both manipulate and precisely position cells within these platforms provides a unique tool to look at intercellular communication pathways, with the potential to scale up experiments for high-throughput analysis. Lee et al. developed a PDMS-based device for parallel pair-wise trapping of mouse fibroblast cells (NIH3T3), measuring cell–cell interaction by visualization on a fluorescent intracellular dye (Calcein AM) through the cell–cell junctions[173] (Figure 26.10). The main PDMS microchannel (20 µm [w] × 50 µm [h]) was used to load the cells, while negative pressure applied to regularly spaced 2 µm (w) × 2 µm (h) side channels intersecting the main channel was used to trap and position cells on both sides. The width of the main channel (20 µm) was designed such that two fibroblasts trapped across from each other (12 µm diameter each) form a snug fit, creating an ideal interface to study transport at the cell–cell interface.

26.5.3   Flow Control in Microscale Culture Systems

Establishing proper cell culture conditions in microscale platforms depends not only on channel geometry, but also on the ability to control the local environment around the cells. Flow rates need to

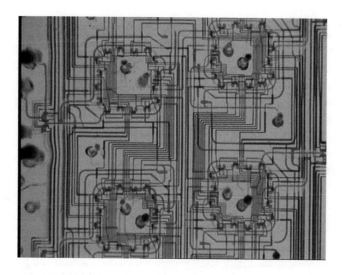

FIGURE 26.11 Microfabricated PDMS bacterial chemostat fabricated by MSL. Each of the four reactors shown (square regions), designed to culture *E. coli*, contains integrated elastomeric valves to carry out automated mixing, dilution, and waste removal. (Figure courtesy of Dr. Stephen Quake, Stanford University.)

be balanced to deliver nutrients to cells, remove waste products, and permit gas exchange with the local environment, while minimizing shear forces that can compromise both cell adhesion and viability.

Many platforms have been developed for automated flow control for long-term microscale cell culture platforms, using tools like integrated elastomeric valves,[174] Braille pins,[175] and syringe pumps[176,177] to regulate liquid flow. Balagadde et al. developed a microfabricated PDMS bacterial chemostat with fluid flow regulated by peristaltic pumps fabricated from integrated elastomeric valves[86,174] (Figure 26.11). The device, containing six 16 nL reactors, was used to culture *E. coli* carrying a synthetic "population control" that regulates cell density through a quorum-sensing mechanism. Each reactor functioned as a microchemostat, providing automated mixing, dilution, and waste removal (spent media). A population control circuit, under the control of a synthetic promoter in the bacteria, was used to dynamically regulate the cell population through the expression of a killer gene, *LacZα-ccdB*, which was turned on in response to high levels of a secreted hormone (acyl-homoserine lactone), whose levels correlate with cell density. Using the circuit, the researchers were able to visualize oscillations in cell culture population over test periods up to 500 h. Another active, on-chip pumping scheme was demonstrated by Gu et al., which used an addressable Braille pin array, with 320 pins, combined with a microscale PDMS device, creating a novel computer-controlled cell culture platform.[175] The molded PDMS microchannels, sealed with a 140 µm layer of PDMS that functioned as an interface between the channels and the Braille pins, were aligned with multiple pins under the channels to enable complex operations such as pumping and mixing. Using most of the pins on the array, long-term cell culture of mouse myocytes was demonstrated (3 weeks) using a device with multiple parallel perfusion channels to provide separate streams of fresh media to different areas of the culture. As a final example, Hung and coworkers developed a PDMS-based 10 × 10 array of 1 mm (dia.) microchambers surrounded by 2 µm (w) channels for continuous media perfusion.[177] Human carcinoma cells used as a model cell line were loaded in the chambers and cultured up to 14 days at 37°C. Good cell viability, monitored using an intracellular viability dye, Calcein AM, and the repassaging of cells grown in the devices were observed over the entire culture period.

Cell culture has also been demonstrated in microchannels using passive fluidic manipulation techniques, such as hydrostatic pressure.[178–180] Using bulk media reservoirs positioned above the microchannels, Tourvoskaia et al. created a device for the long-term differentiation of muscle cells (greater than 2 weeks).[180] A PDMS-based device was used, with microchannels sealed to a borosilicate substrate patterned with alternating regions of cell-repellent polymer (P[Aam-co-EG]) and cell-adhesion-promoting Matrigel (BD Biosciences,

San Jose, CA). After cell seeding, the growth media reservoir was positioned at a height to provide a constant flow rate of ~0.5 ml d^{-1}. Over the course of the experiment, cell differentiation was observed, with cells assembling from individual myoblasts into myotubes. Zhu et al. used a gravity-fed horizontally oriented array of reservoirs for myocyte cell culture in a PDMS microchannel.[179] The multiple-reservoirs configuration was exploited to set up a chemical gradient across the channel, demonstrating differential myocyte cell growth in the presence of different concentrations of the cytotoxic drug colchicine. These gravity-fed microculture systems represent a good, basic automated culture platform that greatly reduces the need for external equipment or power sources, with excellent potential for incorporation in field-based applications.

26.6   Conclusions

As nanofluidic techniques for cell biology continue to be developed, we are likely to see more totally integrated platforms, complete with real-time sensors for physiological parameters such as nutrient composition, pH, and media gas concentrations that will auto-regulate the local environment in the microdevices. Nanofluidic assays for cell interrogation and manipulation show great promise for robust single-cell analysis. Harnessing the length scales present within these devices will provide highly concentrated assays and the ability to analyze single cells in a massively parallel fashion. The fluid physics present in the nanoliter regime, as well as the ability to manipulate single cells in channels on the order of their diameter, may provide the answers in the field of cell biology that elude conventional culture methods.

Acknowledgments

The authors thank W. French Anderson for critical reading of the chapter. J.S.M. is a graduate student in Stephen R. Quake's group at the California Institute of Technology and is supported by National Institutes of Health (NIH) grant 1R01 HG00264401A1.

References

1. Shoffner, M. et al., Surface passivation of microfabricated silicon-glass chips for PCR, *Nucleic Acids Res.*, 24, 375, 1996.
2. Cheng, J. et al., Chip PCR. II. Investigation of different PCR amplification systems in microfabricated silicon-glass chips, *Nucleic Acids Res.*, 24, 380, 1996.
3. Taylor, T. et al., Optimization of the performance of the polymerase chain reaction in silicon-based micro-structures, *Nucleic Acids Res.*, 25, 3164, 1997.
4. Kopp, M., de Mello, A., Manz, A., Chemical amplification: continuous-flow PCR on a chip, *Science*, 280, 1046, 1998.
5. Giordano, B. et al., Polymerase chain reaction in polymeric microchips: DNA amplification in less than 240 seconds, *Anal. Biochem.*, 291, 124, 2001.
6. Wilding, P. et al., Integrated cell isolation and polymerase chain reaction analysis using silicon microfilter chambers, *Anal. Biochem.*, 257, 95, 1998.
7. Hong, J. et al., PDMS (polydimethylsiloxane)-glass hybrid microchip for gene amplification, in *Annu. IEEE EMBS Conf. Microtechnology Med. Biol.*, Dittmar, A. and Beebe, D., eds., 1st, Lyon, 2000, 407.
8. Obeid, P.J. and Christopoulos, T.K., Continuous-flow DNA and RNA amplification chip combined with laser-induced fluorescence detection, *Anal. Chim. Acta.*, 494, 1, 2003.
9. Liu, J., Hansen, C. and Quake, S.R., Solving the "world-to-chip" interface problem with a microfluidic matrix, *Anal. Chem.*, 75, 4718, 2003.
10. Manz, A., Graber, N. and Widmer, H.M., Miniaturized total chemical analysis systems: a novel concept for chemical sensing, *Sensors Actu. B*, 1, 244, 1990.
11. Harrison, D.J. et al., Micromachining a miniaturized capillary electrophoresis-based chemical-analysis system on a chip, *Science*, 261, 895, 1993.

12. Chou, H.P. et al., A microfabricated device for sizing and sorting DNA molecules, *Proc. Natl. Acad. Sci. U.S.A.*, 96, 1, 1999.

13. Effenhauser, C.S. et al., High-speed separation of antisense nucleotides on a micromachined capillary electrophoresis device, *Anal. Chem.*, 66, 2949, 1994.

14. Woolley, A.T. and Mathies, R.A., Ultra-high speed DNA fragment separation using microfabricated capillary array, *Proc. Natl. Acad. Sci. U.S.*, 91, 11348, 1994.

15. Duke, T.A.J. and Austin, R.H., Microfabricated sieve for the continuous sorting of macromolecules, *Phys. Rev. Lett.*, 80, 1552, 1998.

16. Yao, S. et al., SDS capillary gel electrophoresis of proteins in microfabricated channels, *Proc. Natl. Acad. Sci. U.S.*, 96, 5372, 1999.

17. Ramsey, J.D. et al., High-efficiency, two-dimensional separations of protein digests on microfluidic devices, *Anal. Chem.*, 75, 3758, 2003.

18. Herr, A.E. et al., On-chip coupling of isoelectric focusing and free-solution electrophoresis for multidimensional separations, *Anal. Chem.*, 75, 1180, 2003.

19. Wang, Y.-C., Choi, M.H. and Han, J., On-chip IEF peak manipulation for 2D protein separation and MS coupling, *Proc. MicroTAS*, 2003.

20. Chen, X.X. et al., A prototype two-dimensional capillary electrophoresis system fabricated in poly(dimethylsiloxane), *Anal. Chem.*, 74, 1772, 2002.

21. Selvaganapathy, P.R., Carlen, E.T. and Mastrangelo, C.H., Recent progress in microfluidic devices for nucleic acid and antibody assays, *P. IEEE*, 91, 954, 2003.

22. Fan, Z.H. et al., Dynamic DNA hybridization on a chip using paramagnetic beads, *Anal. Chem.*, 71, 4851, 1999.

23. Liu, J.Y. et al., DNA amplification and hybridization in integrated monolithic plastic microfluidic devices, *Anal. Chem.*, 74, 3063, 2002.

24. Wilding, P. et al., Manipulation and flow of biological fluids in straight channels micromachined in silicon, *Clin. Chem.*, 40, 43, 1994.

25. Manz, A. et al., Design of an open-tubular column liquid chromatograph using silicon chip technology, *Sens. Actuators B*, 1, 249, 1990.

26. Harrison, D.J. et al., Micromachining a miniaturized capillary-electrophoresis based chemical-analysis system on a chip, *Science*, 261, 895, 1993.

27. Martynova, L. et al., Fabrication of plastic microfluidic channels by imprinting methods, *Anal. Chem.*, 69, 4783, 1997.

28. Duffy, D.C. et al., Rapid prototyping of microfluidic systems in poly(dimethylsiloxane), *Anal. Chem.*, 70, 4974, 1998.

29. Matsuoka, H. et al., High throughput easy microinjection with a single-cell manipulation supporting robot, *J. Biotech.*, 116, 185, 2005.

30. Buican, T.N. et al., Automated single-cell manipulation and sorting by light trapping, *Appl. Opt.*, 26, 5311, 1987.

31. Ashkin, A. and Dziedzic, J.M., Optical trapping and manipulation of viruses and bacteria, *Science*, 235, 1517, 1987.

32. Schnelle, T. et al., 3-dimensional electrical field traps for manipulation of cells – calculation and experimental verification, *Biochim. Biophys. Acta*, 1157, 127, 1993.

33. Muller, T. et al., A 3-D microelectrode system for handling and caging single cells and particles, *Biosens. Bioelect.*, 14, 247, 1999.

34. Thorsen, T., Maerkl, S.J. and Quake, S.R., Microfluidic large-scale integration, *Science*, 298, 580, 2002.

35. Khademhosseini, A. et al., Molded polyethylene glycol microstructures for capturing cells within microchannnels, *Lab Chip*, 4, 425, 2004.

36. Khademhosseini, A. et al., Cell docking inside microwells with reversibly sealed microfluidic channels for fabricating multiphenotype cell arrays, *Lab Chip*, 5, 1380, 2005.

37. Carlson, R. et al., Self-sorting of white blood cells in a lattice, *Phys. Rev. Lett.*, 79, 2149, 1997.

38. Bakajin, O. et al., Sizing, fractionation and mixing of biological objects via microfabricated devices, *microTAS*, 193–198, 1998.

39. Wilding, P. et al., Integrated cell isolation and polymerase chain reaction analysis using silicon microfilter chambers, *Anal. Biochem.*, 257, 95, 1998.

40. Zhu, L. et al., Filter-based microfluidic device as a platform for immunofluorescent assay of microbial cells, *Lab Chip*, 4, 337, 2004.

41. Li, P.C.H. et al., Transport, retention, and fluorescence measurement of single biological cells studied in microfluidic chips, *Lab Chip*, 4, 174, 2004.

42. Andersson, H. and ver den Berg, A., Microfluidic devices for cellomics: a review, *Sens. Actua. B*, 92, 315, 2003.

43. Chang, W.C., Lee, L.P. and Liepmann, D., Biomimetic technique for adhesion-based collection and separation of cells in a microfluidic channel, *Lab Chip*, 5, 64, 2005.

44. Zhang, Z.L. et al., *In situ* biofunctionalization and cell adhesion in microfluidic devices, *Microelect. Eng.*, 78–79, 556, 2005.

45. Xu, N.Q. et al., A microfabricated dialysis device for sample cleanup in electrospray ionization mass spectrometry, *Anal. Chem.*, 70, 3553, 1998.

46. Lion, N. et al., On-chip protein sample desalting and preparation for direct coupling with electrospray ionization mass spectrophotometry, *J. Chrom. A.*, 1003, 11, 2003.

48. Oleschuk, R.D. et al., Trapping of bead-based reagents within microfluidic systems: on-chip solid-phase extraction and electrochromatography, *Anal. Chem.*, 72, 585, 2000.

49. Bergkvist, J. et al., Improved chip design for integrated solid-phase microextraction in on-line proteomic sample preparation, *Proteomics*, 2, 422, 2002.

50. Cohen, A.S. and Karger, B.L., High performance sodium dodecyl sulfate polyacrylamide gel electrophoresis of peptides and proteins, *J. Chrom.*, 397, 409, 1987.

51. Hofmann, O. et al., Adaptation of capillary isoelectric focusing to microchannels on a glass chip, *Anal. Chem.*, 71, 678, 1999.

52. Herr, A.E. et al., On-chip coupling of isoelectric focusing and free-solution electrophoresis for multidimensional separations, *Anal. Chem.*, 75, 1180, 2003.

53. Chen, X.X. et al., A prototype two-dimensional capillary electrophoresis system fabricated in poly(dimethylsiloxane), *Anal. Chem.*, 74, 1772, 2002.

54. Fiedler, S. et al., Dielectrophoretic sorting of particles and cells in a microsystem, *Anal. Chem.*, 70, 1909, 1998.

55. Markx, G. et al., Dielectrophoretic characterization and separation of microorganisms. *Microbiology*, 140, 585, 1994.

56. Cheng, J. et al., Electric field controlled preparation and hybridization analysis of DNA/RNA from *E. coli* on microfabricated bioelectronic chips, *Nat. Biotechnol.*, 16, 541, 1998.

57. Becker, F. et al., Separation of human breast cancer cells from blood by differential dielectric affinity, *Proc. Natl. Acad. Sci. U.S.*, 92, 860, 1995.

58. Wang, X.-B. et al., Cell separation by dielectrophoretic field-flow fractionation, *Anal. Chem.*, 72, 832, 2000.

59. Arai, F. et al., High-speed separation system of randomly suspended single living cells by laser trap and dielectrophoresis, *Electrophoresis*, 22, 283, 2001.

60. Xu, J. et al., Dielectrophoretic separation and transportation of cells and bioparticles on microfabricated chips, *MicroTAS*, 565–566, 2001.

61. Huang, Y. et al., Electric manipulation of bioparticles and macromolecules on microfabricated electrodes, *Anal. Chem.*, 273, 1549, 2001.

62. Voldman, J. et al., A microfabrication-based dynamic array cytometer, *Anal. Chem.*, 74, 3984, 2002.

63. Gascoyne, P. et al., Microsample preparation by dielectrophoresis: isolation of malaria, *Lab Chip*, 2, 70, 2002.

64. Toriello, N.M., Douglas, E.S. and Mathies, R.A., Microfluidic device of electric field-driven single-cell capture, *Anal. Chem.*, 77, 6935, 2005.

65. Mehrishi, J.N. and Buer, J., Electrophoresis of cells and the biological relevance of surface charge, *Electrophoresis*, 23, 1984, 2002.

66. Thorsen, T., Manipulation of biomolecules and reactions, in *Nanolithography and Patterning Techniques in Microelectronics*, Bucknall D, ed., Woodhead Publishing Ltd., UK, 2005.

67. Fu, A.Y. et al., A microfabricated fluorescence-activated cell sorter, *Nature Biotech.*, 17, 1109,1999.

68. McClain, M.A. et al., Flow cytometry of Escherichia coli on microfluidic devices, *Anal. Chem.*, 73, 5334, 2001.

69. Archer, S. et al., Cell reactions to dielectrophoretic manipulation, *Biochem. Biophys. Res. Commun.*, 257, 687, 1999.

70. Hu, X.Y. et al., Marker-specific sorting of rare cells using dielectrophoresis, *Proc. Natl. Acad. Sci. U.S.*, 102, 15757, 2005.

71. Fu, A. et al., An integrated microfabricated cell sorter, *Anal. Chem.*, 74, 2451, 2002.

72. Kruger, J. et al., Development of a microfluidic device for fluorescence activated cell sorting, *J. Micromech. Microeng.*, 12, 486, 2002.

73. Wolff, A. et al., Integrating advanced functionality in a microfabricated high-throughput fluorescent-activated cell sorter, *Lab Chip*, 3, 22, 2003.

74. Oskan, M. et al., Optical manipulation of objects and biological cells in microchannels, *Biomed. Microdev.*, 5, 61, 2003.

75. Wang, M.W. et al., Microfluidic sorting of mammalian cells by optical force switching. *Nature Biotech.*, 23, 83, 2005.

76. Wheeler, A.R. et al., Microfluidic device for single-cell analysis, *Anal. Chem.*, 75, 3581, 2003.

77. Huang, Y. and Rubinsky, B., Flow-through micro-electroporation chip for high efficiency single-cell genetic manipulation, *Sens. Actua. A*, 104, 205, 2003.

78. Khine, M. et al., A single cell electroporation chip, *Lab Chip*, 5, 38, 2005.

79. Olofsson, J. et al., Single-cell electroporation, *Curr. Opin. Biotechnol.*, 14, 29, 2003.

80. Huang, Y. and Rubinsky, B., Microfabricated electroporation chip for single cell membrane permeabilization, *Sens. Actua. A*, 89, 242, 2001.

81. Xia, Y. and Whitesides, G.M., Soft lithography, *Angew. Chem. Int. Ed.*, 37, 550, 1998.

82. Tegenfeldt, J.O. et al., Micro- and nanofluidics for DNA analysis, *Anal. Bioanal. Chem.*, 378, 1678, 2004.

83. Prinz, C. et al., Bacterial chromosome extraction and isolation, *Lab Chip*, 2, 207, 2002.

84. Anderson, R.C. et al., A miniature integrated device for automated multistep genetic assays, *Nucleic Acids Res.*, 28, E60, 2000.

85. Hong, J.W. et al., A nanoliter-scale nucleic acid processor with parallel architecture, *Nat. Biotechnol.*, 22, 435, 2004.

86. Unger, M.A. et al., Monolithic microfabricated valves and pumps by multilayer soft lithography, *Science*, 288, 113, 2000.

87. Studer, V. et al., Scaling properties of a low-actuation pressure microfluidic valve, *J. Appl. Phys.*, 95, 393, 2004.

88. Marcus, J.S., Anderson, W.F. and Quake, S.R., Microfluidic single cell mRNA isolation and analysis, *Anal. Chem.*, 78, 3084, 2006.

89. Chiang, M.K. and Melton, D.A., Single-cell transcript analysis of pancreas development, *Dev. Cell*, 4, 383, 2003.

90. Tietjen, I. et al., Single-cell transcriptional analysis of neuronal progenitors, *Neuron*, 38, 161, 2003.

91. Gentile, L. et al., Single-cell quantitative RT-PCR analysis of Cpt1b and Cpt2 gene expression in mouse antral oocytes and in preimplantation embryos, *Cytogenet. Genome Res.*, 105, 215, 2004.

92. Han, S.H. et al., Single-cell RT-PCR detects shifts in mRNA expression profiles of basal forebrain neurons during aging, *Brain Res. Mol. Brain Res.*, 98, 67, 2002.

93. Lindqvist, N., Vidal-Sanz, M. and Hallbook, F., Single cell RT-PCR analysis of tyrosine kinase receptor expression in adult rat retinal ganglion cells isolated by retinal sandwiching, *Brain Res. Brain Res. Protoc.*, 10, 75, 2002.

94. Liss, B., Improved quantitative real-time RT-PCR for expression profiling of individual cells, *Nucleic Acids Res.*, 30, e89, 2002.

95. Silbert, S.C., Quantitative single-cell RT-PCR for opioid receptors and housekeeping genes, *Methods Mol. Med.*, 84, 107, 2003.

96. Todd, R. and Margolin, D.H., Challenges of single-cell diagnostics: analysis of gene expression, *Trends Mol. Med.*, 8, 254, 2002.

97. Wagatsuma, A. et al., Determination of the exact copy numbers of particular mRNAs in a single cell by quantitative real-time RT-PCR, *J. Exp. Biol.*, 208, 2389, 2005.

98. Liu, J., Enzelberger, M. and Quake, S., A nanoliter rotary device for polymerase chain reaction, *Electrophoresis*, 23, 1531, 2002.

99. Liu, R.H. et al., Self-contained, fully integrated biochip for sample preparation, polymerase chain reaction amplification, and DNA microarray detection, *Anal. Chem.*, 76, 1824, 2004.

100. Obeid, P. and Christopoulos, T., Continuous-flow DNA and RNA amplification chip combined with laser-induced fluorescence detection, *Anal. Chim. Acta*, 494, 1, 2003.

101. Yuen, P.K. et al., Microchip module for blood sample preparation and nucleic acid amplification reactions, *Genome Res.*, 11, 405, 2001.

102. Zhou, X. et al., Determination of SARS-coronavirus by a microfluidic chip system, *Electrophoresis*, 25, 3032, 2004.

103. Unger, M. et al., European Patent. Vol. EP1463796, 2004.

104. Marcus, J.S., Anderson, W.F. and Quake, S.R., Parallel picoliter RT-PCR assays using microfluidics, *Anal. Chem.*, 78, 956, 2006.

105. Vogelstein, B. and Kinzler, K.W., Digital PCR, *Proc. Natl. Acad. Sci. U.S.*, 96, 9236, 1999.

106. Lee, C.-Y. et al., Integrated microfluidic systems for cell lysis, mixing/pumping and DNA amplification, *J. Micromech. Microeng.*, 15, 1215, 2005.

107. McClain, M.A. et al., Microfluidic devices for the high-throughput chemical analysis of cells, *Anal. Chem.*, 75, 5646, 2003.

108. Auroux, P.A. et al., Miniaturised nucleic acid analysis, *Lab Chip*, 4, 534, 2004.

109. Galloway, M. and Soper, S.A., Contact conductivity detection of polymerase chain reaction products analyzed by reverse-phase ion pair microcapillary electrochromatography, *Electrophoresis*, 23, 3760, 2002.

110. Galloway, M. et al., Contact conductivity detection in poly(methyl methacrylate)-based microfluidic devices for analysis of mono- and polyanionic molecules, *Anal. Chem.*, 74, 2407, 2002.

111. Shiddiky, M.J., Park, D.S. and Shim, Y.B., Detection of polymerase chain reaction fragments using a conducting polymer-modified screen-printed electrode in a microfluidic device, *Electrophoresis*, 26, 4656, 2005.

112. Foote, R.S. et al., Preconcentration of proteins on microfluidic devices using porous silica membranes, *Anal. Chem.*, 77, 57, 2005.

113. Gulliksen, A. et al., Real-time nucleic acid sequence-based amplification in nanoliter volumes, *Anal. Chem.*, 76, 9, 2004.

114. Gulliksen, A. et al., Parallel nanoliter detection of cancer markers using polymer microchips, *Lab Chip*, 5, 416, 2005.

115. Thompson, D.M. et al., Dynamic gene expression profiling using a microfabricated living cell array, *Anal. Chem.*, 76, 4098, 2004.

116. Wieder, K.J. et al., Optimization of reporter cells for expression profiling in a microfluidic device, *Biomed. Microdevices*, 7, 213, 2005.

117. Elowitz, M.B. et al., Stochastic gene expression in a single cell, *Science*, 297, 1183, 2002.

118. Gao, J., Yin, X.F. and Fang, Z.L., Integration of single cell injection, cell lysis, separation and detection of intracellular constituents on a microfluidic chip, *Lab Chip*, 4, 47, 2004.

119. Kleparnik, K. and Horky, M., Detection of DNA fragmentation in a single apoptotic cardiomyocyte by electrophoresis on a microfluidic device, *Electrophoresis*, 24, 3778, 2003.

120. Ling, Y.Y., Yin, X.F. and Fang, Z.L., Simultaneous determination of glutathione and reactive oxygen species in individual cells by microchip electrophoresis, *Electrophoresis*, in press, 2005.

121. Munce, N.R. et al., Microfabricated system for parallel single-cell capillary electrophoresis, *Anal. Chem.*, 76, 4983, 2004.

122. Sun, Y. et al., Determination of reactive oxygen species in single human erythrocytes using microfluidic chip electrophoresis, *Anal. Bioanal. Chem.*, 382, 1472, 2005.

123. Xia, F. et al., Single-cell analysis by electrochemical detection with a microfluidic device, *J. Chromatogr. A*, 1063, 227, 2005.

124. Roper, M.G. et al., Microfluidic chip for continuous monitoring of hormone secretion from live cells using an electrophoresis-based immunoassay, *Anal. Chem.*, 75, 4711, 2003

125. Cheng, S.B. et al., Development of a multichannel microfluidic analysis system employing affinity capillary electrophoresis for immunoassay, *Anal. Chem.*, 73, 1472, 2001.

126. Lion, N. et al., Microfluidic systems in proteomics, *Electrophoresis*, 24, 3533, 2003.

127. Hellmich, W. et al., Single cell manipulation, analytics, and label-free protein detection in microfluidic devices for systems nanobiology, *Electrophoresis*, 26, 3689, 2005.

128. Li, J. et al., Application of microfluidic devices to proteomics research: identification of trace-level protein digests and affinity capture of target peptides, *Mol. Cell Proteomics*, 1, 157, 2002.

129. Kartalov, E.P. et al., High-throughput multi-antigen microfluidic fluorescence immunoassays, *Biotechniques*, 40, 85, 2006.

130. Rocklin, R.D., Ramsey, R.S. and Ramsey, J.M., A microfabricated fluidic device for performing two-dimensional liquid-phase separations, *Anal. Chem.*, 72, 5244, 2000.

131. Hansen, C. and Quake, S.R., Microfluidics in structural biology: smaller, faster … better, *Curr. Opin. Struct. Biol.*, 13, 538, 2003.

132. Hansen, C.L. et al., A robust and scalable microfluidic metering method that allows protein crystal growth by free interface diffusion, *Proc. Natl. Acad. Sci. U.S.*, 99, 16531, 2002.

133. Tran, L. et al., Agonist-induced calcium response in single human platelets assayed in a microfluidic device, *Anal. Biochem.*, 341, 361, 2005.

134. Yang, M., Li, C.W. and Yang, J., Cell docking and on-chip monitoring of cellular reactions with a controlled concentration gradient on a microfluidic device, *Anal. Chem.*, 74, 3991, 2002.

135. Li, X. and Li, P.C., Microfluidic selection and retention of a single cardiac myocyte, on-chip dye loading, cell contraction by chemical stimulation, and quantitative fluorescent analysis of intracellular calcium, *Anal. Chem.*, 77, 4315, 2005.

136. Buhlmann, C. et al., A new tool for routine testing of cellular protein expression: integration of cell staining and analysis of protein expression on a microfluidic chip-based system, *J. Biomol. Tech.*, 14, 119, 2003.

137. Li, P.C. et al., Transport, retention and fluorescent measurement of single biological cells studied in microfluidic chips, *Lab Chip*, 4, 174, 2004.

138. Chen, C.S., Jiang, X. and Whitesides, G.M., Microengineering the environment of mammalian cells in culture, *MRS Bull.*, 30, 194, 2005.

139. Park, T.H. and Shuler, M.L., Integration of cell culture and microfabrication technology, *Biotechnol. Prog.*, 19, 243, 2003.

140. Weibel, D.B., Garstecki, P. and Whitesides, G.M., Combining microscience and neurobiology, *Curr. Opin. Neurobiol.*, 15, 560, 2005.

141. Britland, S., Clark, P. and Moores, G., Micropatterned substratum adhesiveness: a model for morphogenic cues controlling cell behavior, *Exp. Cell. Res.*, 198, 124, 1992.

142. Kleinfeld, D., Kahler, K.H. and Hockberger, P.E., Controlled outgrowth of dissociated neurons on patterned substrates, *J. Neurosci.*, 8, 4098, 1988.

143. Bhatia, S.N., Yarmush, M.L. and Toner, M., Controlling cell interactions by micropatterning in co-cultures: hepatocytes and 3T3 fibroblasts, *J. Biomed. Mater. Res.*, 34, 189, 1997.

144. Britland, S. et al., Micropatterning proteins and synthetic peptides on solid supports: a novel application for microelectronics fabrication technology, *Biotechnol. Prog.*, 8, 155, 1992.

145. Lom, B., Healy, K.E. and Hockberger, P.E., A versatile technique for patterning biomolecules onto glass coverslips. *J. Neurosci. Methods*, 50, 385, 1993.

146. Heo, J. and Crooks, R.M., Microfluidic biosensor based on an array of hydrogel-trapped enzymes, *Anal. Chem.*, 77, 6843, 2005.

147. Burdick, J.A., Khademhosseini, A. and Langer, R., Fabrication of gradient hydrogels using a microfluidic photopolymerization process, *Langmuir*, 20, 5153, 2004.

148. Koh, W.-G., Itle, L.J. and Pisho, M., Molding of hydrogel microstructures to create multiphenotype cell arrays, *Anal. Chem.*, 75, 5783, 2003.

149. Zaari, N. et al., Photopolymerization in microfluidic gradient generators: microscale control of substrate compliance to manipulate cell response, *Adv. Mater.*, 16, 2133, 2004.

150. Kane, R. et al., Patterning proteins and cells using soft lithography, *Biomaterials*, 20, 2363, 1999.

151. Kim, E., Xia, Y. and Whitesides, G.M., Polymer microstructures formed by moulding in capillaries, *Nature*, 376, 581, 1995.

152. Lopez, G.P. et al., Convenient methods for patterning the adhesion of mammalian cells to surfaces using self-assembled monolayers of alkanethiolates on gold, *J. Am. Chem. Soc.*, 115, 5877, 1993.

153. Takayama, S. et al., Patterning cells and their environments using multiple laminar fluid flows in capillary networks, *Proc. Natl. Acad. Sci. U.S.*, 96, 5545, 1999.

154. Takayama, S. et al., Subcellular positioning of small molecules, *Nature*, 411, 1016, 2001.

155. Pelham, R.J. and Wang, Y.L., High resolution detection of mechanical forces exerted by locomoting fibroblasts on the substrate, *Mol. Biol. Cell*, 10, 935, 1999.

156. Galbraith, C.G., Yamada, K.M. and Sheetz, M.P., The relationship between force and focal complex development, *J. Cell Biol.*, 159, 605, 2002.

157. Engler, A. et al., Substrate compliance vs. ligand density in cell on gel response, *Biophys. J.*, 86, 617, 2004.

158. Jiang, X.Y. et al., Palladium as a substrate for self-assembled monolayers used in biotechnology, *Anal. Chem.*, 76, 6116, 2004.

159. McClary, K.B., Ugarova, T. and Grainger, D.W., Modulating fibroblast adhesion, spreading and proliferation using self-assembled monolayer films of alkylthiolates on gold, *J. Biomed. Mat. Res.*, 50, 428, 2000.

160. Zheng, H.P. et al., Controlling cell attachment selectively onto biological polymer-colloid templates using polymer-on-polymer stamping, *Langmuir*, 20, 7215, 2004.

161. Kaji, H., Nishizawa, M. and Matsue, T., Localized chemical stimulation to micropatterned cells using multiple laminar flows, *Lab Chip*, 3, 208, 2003.

162. Prokop, A. et al., NanoLiterBioReactor: long-term mammalian cell culture at nanofabricated scale, *Biomed. Microdev.*, 6, 325, 2004.

163. Chung, B.G. et al., Human neural stem cell growth and differentiation in a gradient-generating microfluidic device, 5, 401, 2005.

164. Taylor, A.M. et al., A microfluidic culture platform for CNS axonal injury, regeneration, and transport, *Nat. Methods*, 2, 599, 2005.

165. Rhee, S.W. et al., Patterned cell culture inside microfluidic devices, *Lab Chip*, 5, 102, 2005.

166. Mahoney, M.J. et al., The influence of microchannels on neurite growth and architecture, *Biomaterials*, 26, 771, 2005.

167. Toh, Y.C. et al., A configurable three-dimensional microenvironment in a microfluidic channel for primary hepatocyte culture, *Assay Drug Dev. Technol.*, 3, 169, 2005.

168. Ostrovidov, S. et al., Membrane-based PDMS bioreactor for perfused 3D primary rat hepatocyte cultures, *Biomed. Microdev.*, 6, 279, 2004.

169. Leclerc, E., Sakai, Y. and Fujii, T., Perfusion culture of fetal human hepatocytes in microfluidic environments, *Biochem. Eng. J.* 20, 143, 2004.

170. Rajnicek, A.M., Britland, S., McCaig, C.D., Contact guidance of CNS neurites on grooved quartz: influence of groove dimensions, neuronal age and cell type, *J. Cell Sci.*, 110, 2905, 1997.

171. Clark, P. et al., Topographical control of cell behavior: II. Multiple grooved substrata, *Development*, 108, 635, 1990.

172. Yu, H. et al., Diffusion dependent cell behavior in microenvironments, *Lab Chip*, 5, 1089, 2005.

173. Lee, P.J. et al., Microfluidic application-specific integrated device for monitoring direct cell-cell communication via gap junctions between individual cell pairs, *Appl. Phys. Lett.*, 86, 223902, 2005.

174. Balagadde, F.K. et al., Long term monitoring of bacteria undergoing programmed population control in a microchemostat, *Science*, 309, 137, 2005.

175. Gu, W. et al., Computerized microfluidic cell culture using elastomeric channels and braille displays, *Proc. Natl. Acad. Sci. U.S.*, 101, 15861, 2004.

176. Prokop, A. et al., NanoLiterBioReactor: Long-term mammalian cell culture at nanofabricated scale, *Biomed. Microdev.*, 6, 325, 2004.

177. Hung, P.J. et al., Continuous perfusion microfluidic cell culture array for high-throughput cell-based assays, *Biotech. Bioeng.*, 89, 1, 2005.

178. Peng, X.Y. and Li, P.C.H., A three-dimensional flow control single-cell experiments on a microchip. 1. Cell selection, cell retention, cell culture, cell balancing, and cell scanning, *Anal. Chem.*, 76, 5273, 2004.

179. Zhu, H. et al., Arrays of horizontally-oriented mini-reservoirs generate steady microfluidic flows for continuous perfusion cell culture and gradient generation, *Analyst*, 129, 1026, 2004.

180. Tourovskaia, A., Figueroa-Masot, X. and Folch, A., Differentiation-on-a-chip: a microfluidic platform for long-term cell culture studies, *Lab Chip*, 5, 14, 2005.

27

Carbon Nanostructures and Nanocomposites

Yanhong Hu
Zushou Hu
Clifford W. Padgett
Donald W. Brenner
North Carolina State University

Olga A. Shenderova
International Technology Center

27.1 Introduction to Composites **27**-1
27.2 Modern History of Carbon Nanostructures **27**-2
27.3 The Extended Carbon Family **27**-3
27.4 Structure, Notation, and Synthesis
of the Common Carbon Nanostructures **27**-5
27.5 Nanocomposites Using Fullerenes **27**-9
27.6 Nanocomposites Using Nanotubes **27**-9
27.7 Nanocomposites Using Diamond-Like
Structures **27**-19
27.8 Remaining Challenges and Opportunities **27**-21
Acknowledgments **27**-21
References **27**-21

27.1 Introduction to Composites

Composites are materials made from two or more components that are constructed to take advantage of the combination of properties offered by the components. Conventional composite structures have a long history in the development of materials with useful and often unique properties. An important historical example is the mixture of clay and straw used to build adobe structures. The clay provides a high-volume solid matrix, whereas the straw reinforces the brick against fracture. Modern examples of composites include carbon fiber composites used in structures that range from advanced aircraft to tennis rackets, fiberglass that is used for relatively strong yet lightweight structures such as boat hulls, Portland concrete, which is composed of a combination of Portland cement, sand, rock, and water, "carbon black" particles in automobile tires which increase wear resistance, and reinforced concrete, where the metal rebar helps stabilize the structure against fracture under shear stress.

Conventional composites can be conveniently divided into three classes of structures. The first is composed of roughly spherical particles embedded into a matrix. An example is Portland concrete mentioned above. The second class is composed of fibers embedded within a matrix. An example is fiberglass, which is composed of glass fibers with a high tensile modulus that are embedded into an epoxy polymer matrix. The final class is composed of layered structures, examples of which include laminates and plywood. Several factors contribute to the thermo-mechanical properties of conventional composites. In fiber-reinforced composites, for example, the volume fraction, the diameter, and the degree of dispersion and alignment of the fibers all contribute to the mechanical properties of the system, as does the fiber yield strength and the strength of the fiber–matrix interaction. The first set of characteristics listed are, to a large degree, dependent on the synthesis and processing of the structure, whereas the yield strength and fiber–matrix interaction are largely intrinsic functions of the components (although in some cases the fibers can be preprocessed to enhance the fiber–matrix interaction).

In nanocomposites, at least one dimension of a component has a characteristic scale of less than 100 nm. Similar to conventional composites, the reinforcing structures in nanocomposites can be spherical (e.g., nanoclusters), fiber-like (e.g., carbon nanotubes), or plate-like (e.g., layered silicates and graphite sheets). Nanocomposites can exhibit unique properties compared to conventional composites (Ajayan et al. 2003). For example, it is possible to make nanocomposites with improved mechanical or electrical properties without losing transparency because nano-reinforcements do not significantly scatter light. The small size of nanoscale reinforcements also overcomes a common problem found in conventional composites where microscale elements lead to an increase in strength and stiffness but result in a decrease in toughness because of large local stress concentrations. The small size of nano-reinforcements can also facilitate composite processing, making it possible, e.g., to mold polymer nanocomposites into complex shapes because nano-reinforcements can be processed with the polymer using normal methods such as injection molding or extrusion (Calvert 1999).

There are also several disadvantages to nanocomposite structures. The small size of the reinforcing structures in polymer nanocomposites can cause a substantial increase in viscosity of the mixture at a high loading, which can be detrimental to processing and some applications (Han and Lem 1983). It can also be difficult to achieve a good dispersion of nanoelements in nanocomposites because of their large surface-to-volume ratio. As discussed in more detail below, in composites containing carbon nanotubes, the nanotubes typically form bundles or ropes (Ebbesen 1997), which reduces the aspect ratio and potentially limits the load-carrying efficiency of the nanocomposite. Because of the strong inter-nanotube interactions, breaking up these aggregates is nontrivial.

The creation of new classes of particles, fibers, and layered structures with nanometer-scale dimensions in bulk quantities has generated tremendous interest in creating advanced nanocomposites with new and potentially unique properties that cannot be achieved within conventional composite structures. Central to these nanotechnology efforts is the use of carbon nanostructures in both graphitic and diamond-like forms to create a wide range of unique materials. The history, properties, structure, and bonding of carbon nanostructures are discussed in the following sections.

27.2   Modern History of Carbon Nanostructures

As discussed in the chapters by Meyyappan and Srivastava and by Qian et al. in this handbook, carbon nanotubes (CNTs) form one of the central pillars of nanotechnology that cuts across nanoelectronics and nanomaterials applications. Their unique combination of ultrahigh tensile modulus and strength, ballistic electron and heat transport properties, and strong coupling between structure and electronic properties combined with the ability to produce and process them at the bulk scale have made them attractive for a wide variety of applications. Indeed, it would not be surprising if CNTs and their related structures became both the "steel" and "silicon" of the twenty first century. Remarkably, these possibilities have only been recognized over the last two decades, despite a long history of studying carbon structures in the physical and biological sciences. By the mid-1980s, the structure, bonding, and resulting technological applications of carbon appeared to be well established. With only six electrons per atom, highly accurate first principles calculations on carbon were possible that led to a detailed understanding of its bonding in molecules and in condensed phases (Fahy et al. 1987). At the macroscopic scale, the phase diagram of carbon appeared to be well known, and metastable phases such as the Wurtzite structure were well established. Much of the detailed physics and chemistry of soot formation was also beginning to be well understood (Frenklach et al. 1983). Applications of carbon-based materials had appeared in mature technologies such as lubrication, rubber processing, and energy storage via battery technology, and advanced lightweight and high-strength composites were being created using carbon fibers. Even polymeric structures, such as polyacetylene and related electrically conducting structures, that had until then intrigued scientists, were reasonably well understood from both an experimental and theoretical viewpoint (Su et al. 1979). Most of the remaining research in the mid-1980s on carbon structures focused on the deposition of diamond and diamond-like coatings for tribological applications (Devries 1987), and the growth and doping of single-crystal diamond for high-power microelectronics applications (Collins 1989).

At the same time, many academic and industrial researchers were moving away from carbon-based structures to materials such as the nitrides. Basic research done by Smalley, Kroto, Curl, and their co-workers at Rice University was leading to intriguing results that hinted that carbon was not as well understood as many in the scientific community believed (Kroto et al. 1985). The discovery of the fullerenes by this group, and the subsequent production of bulk quantities of these clusters by Kratschmer et al. (1990) created renewed significant interest in carbon. It was the discovery by Iijima (1991) of CNTs, however, that started the intense research on carbon nanostructures that continues today. As discussed elsewhere in this book, new technological applications of fullerenes and CNTs continue to be developed (Terrones 2003). These applications include reinforcing structures and heat pipes for nanocomposites, electron emitters for display technologies (Deheer et al. 1995), x-ray emitters for medical applications (Zhang et al. 2005), one-dimensional junctions and nanowires for nano- and microelectronics applications (Graham et al. 2005), and scanning probe tips for nanoscale metrology (Nguyen et al. 2005).

In work that predates that on fullerenes and CNTs, research primarily in Russia focused on the bulk production and the development of technology applications of diamond nanoparticles (Shenderova et al. 2002). Similarly, research on the synthesis of diamond rods was initiated in Russia over 30 years ago (Shenderova et al. 2005). Studies aimed at producing and processing these structures have continued to the present.

One of the central features of carbon research over the last two decades has been the strong relationship between theory and experiment. For example, theory first predicted the unique dependence of the bandgap on the helical structure of CNTs (Mintmire et al. 1992) that was subsequently confirmed by experiment (Wildoer et al. 1998). The reversible formation of kinks in bent CNTs was observed almost simultaneously in experiments and in simulations (Iijima et al. 1996). Calculations at Los Alamos in 1990 suggested that diamond nanoclusters become more stable than graphitic polyaromatics with an increasing hydrogen-to-carbon ratio, with the crossover in stability occurring at about the nanometer scale (Badziag et al. 1990). These calculations explained the experimental observation of diamond nanoparticles as significant byproducts of detonations (Greiner et al. 1988). More recently, theoretical predictions of the stability and the mechanical properties of diamond nanorods (DNRs) have prompted an experimental synthesis of these structures, which in turn has resulted in a new material, agglomerated DNRs, that reportedly has the smallest measured compressibility of any known material (Dubrovinskaia et al. 2005).

The general properties of CNTs, as well some examples of their application in various nanotechnologies, are covered in chapters elsewhere in this handbook. Rather than repeat this material, this chapter is focused specifically on nanocomposites. At the same time, however, exclusively discussing CNTs in the context of composites does not adequately describe the possibilities offered by competing structures such as DNRs for mechanical reinforcement and thermal management. Therefore, the content of this chapter includes not only CNTs, but also fullerenes, diamond nanoparticles, and DNRs as components of nanocomposites.

27.3 The Extended Carbon Family

For the sake of this chapter, carbon nanostructures are defined as carbon materials produced such that either their size or their structure is controlled at the nanometer scale (less than 100 nm). The entire range of dimensionalities is represented within the extended carbon family: zero-dimensional structures (fullerenes, diamond clusters), one-dimensional structures (CNTs, DNRs), two-dimensional structures (graphite sheets, diamond nanoplatelets), and three-dimensional structures (nanocrystalline diamond films, fullerite, CNT ropes). The existence of a wide variety of carbon structures and nanostructures originates from the unique ability of carbon to rehybridize to sp^2 and sp^3 electronic configurations with almost identical energies.

The nature of the chemical bonds in carbon structures leads to a classification scheme where each valence state corresponds to a certain form of a simple substance (Heimann et al. 1997). Elemental carbon exists in three bonding states corresponding to sp^3, sp^2, and sp hybridization of the atomic orbitals. The corresponding three carbon allotropes with an integer degree of carbon bond hybridization are diamond, graphite, and carbyne (Heimann et al. 1997), the last being a controversial allotrope, the structure of which is still debated. All other forms of carbon constitute "transitional" forms that can be divided into two large groups. The first group comprises mixed carbon forms with short-range order that have more or less arranged carbon atoms of different hybridization states (e.g., diamond-like carbon, vitreous carbon, soot, carbon blacks), as well as numerous hypothetical structures such as graphynes and "super-diamond." The second group includes intermediate carbon forms with a noninteger degree of carbon bond hybridization, sp^n. The subgroup with $1 < n < 2$ includes various monocyclic carbon structures. For $2 < n < 3$, the intermediate carbon forms comprise closed-shell carbon structures such as fullerenes, carbon anions, and CNTs. The fractional degree of hybridization in this group of carbon structures is due to the curvature of the framework.

The value of this scheme is that any form of carbon material is included in the classification. A classification scheme for carbon forms that is based on hybridization was also suggested by Inagaki (2000), who considered diamond, graphite, fullerenes, and carbyne as the four basic carbon forms. In addition, Inagaki's scheme demonstrates interrelations between organic/inorganic carbon substances at the molecular scale and emphasizes the interdisciplinary nature of carbon nanotechnology, which is based on both materials science and chemistry.

The two classification schemes outlined above have been combined to classify carbon nanostructures within a general hierarchy of carbon materials (Shenderova et al. 2002, 2005) (Figure 27.1). The scheme is based on two major characteristics: the type of carbon atom hybridization and the characteristic sizes of nanostructures. Starting with a description of the bonding nature of carbon atoms, the concept is to analyze how different classes of carbon networks are formed with the increasing characteristic size of a carbon structure. Starting with small organic molecules (inner circle), the hierarchy of carbon materials can be described as an extension of organic molecular species to bulk inorganic all-carbon materials

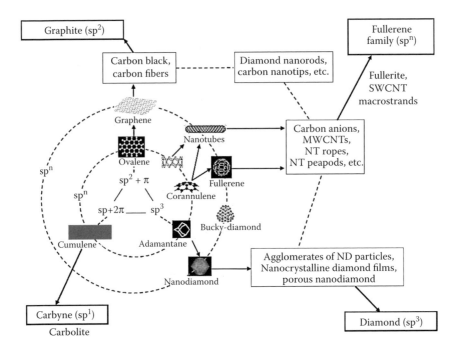

FIGURE 27.1 Family of carbon structures at the nanometer scale.

through a variety of carbon entities in the nanoscopic size range. If we consider fullerenes, CNTs, and graphene of finite size and the currently observed smallest nanodiamond clusters as basic structural units in the carbon family, prototype molecules can be assigned to these units at the scale of molecules (inner circle). It should be noted that this scheme does not mean that these molecules are involved in the synthesis of the units; rather, it emphasizes topological similarities between organic species and inorganic materials.

Although the sizes of representative members of sp^2 and sp^n ($2 < n < 3$) families experimentally identified change rather smoothly from the molecular scale (few nanometers) to the scale of nanostructures (tens of nanometers), for sp^3 carbon forms there is currently a gap in size between observed molecular forms of the highest diamondoids (Dahl et al. 2003) (~1 nm in size, containing up to 50 carbon atoms) and the smallest nanodiamond particles (~2 to 3 nm in size, few thousands of carbon atoms). Recently, a hybrid form of an entity with a diamond core and fullerene-like outer shells called bucky-diamond has been suggested and experimentally confirmed (Raty et al. 2003). The next structural level in Figure 27.1, with a corresponding increase of the characteristic sizes, consists of assemblies of structural units, ranging from simple forms, such as multi-walled CNTs or carbon anions to more complicated carbon architectures such as carbon black, schwarzites, and ultrananocrystalline diamond films. Finally, at the upper micro/macroscopic scale, there is diamond, graphite, carbolite, fullerite, and recently discovered single-walled CNT strands of macroscopic sizes (Zhu et al. 2002). Although the described scheme corresponds to a bottom-up approach of molecular synthesis, it is also necessary to consider nanostructures obtained by top-down approaches using different nanopatterning techniques such as, e.g., fabrication of DNRs by reactive ion etching of diamond films (Baik et al. 2000). Obviously, structural units from different families can be combined to form hybrid nanostructures (Shenderova et al. 2003; Gruen et al. 2005).

27.4 Structure, Notation, and Synthesis of the Common Carbon Nanostructures

Fullerenes refer to a class of structure in which carbon atoms are arranged in a closed cage. Observed sizes of fullerenes range from 30 atoms (C_{30}) (von Helden et al. 1993) to a structure with 960 atoms (C_{960}) (Yarris 1993). The most common member in the fullerene family is the 60-carbon atom structure C_{60}, also known as the "buckyball," in which the 60 carbon atoms form 20 hexagons and 12 pentagons (Kroto et al. 1985). The arrangement of bonds between these atoms resembles the stitches of a typical soccer ball. This unique hollow structure of fullerenes, together with the aromatic bonding, results in unusual physical, chemical, and biological properties (Dresselhaus et al. 1996). The chemical properties of C_{60} are similar to those of electron-deficient polyalkenes (Wudl 1992; Taylor and Walton 1993). This electrophilic property facilitates the modification of buckyballs by using organic chemistry and other synthetic techniques. Various atoms or molecular fragments have been bonded to the C_{60} molecule, while leaving its cage structure almost intact (Haddon et al. 1991; Fagan et al. 1992; Hawkins 1992; Olah et al. 1992; Wudl 1992; Taylor and Walton 1993). Pure C_{60} crystals show semiconducting behavior based on electronic structure investigations (Dresselhaus et al. 1996). There are no free charge carriers in a pristine C_{60} solid unless it is thermally or optically excited (Dresselhaus et al. 1996). However, when doped with other atoms or molecules, especially electron donor dopants such as the alkali metals, materials based on buckyballs can be conductors or even superconductors (Dresselhaus et al. 1996).

CNTs can be envisioned as cylinders composed of rolled up graphite planes. Depending on how they are produced and processed, they can be single-walled CNTs (SWCNTs), multi-walled CNTs (MWCNTs), or in the form of ropes composed of arrays of CNTs. The vector that wraps around the circumference of a SWCNT can be chosen to lie within a 30° wedge of the in-plane graphite lattice. A customary notation for a SWCNT is to describe this vector in terms of the two primitive in-plane lattice vectors. Within this notation, CNTs that are denoted by (n, n) and $(n, 0)$ both have reflection symmetry through the long axis of the cylinder. All of the other structures have only a helical symmetry.

In its bulk form, graphite is a semimetal, with the occupied and unoccupied electronic states meeting at the Brillouin zone boundary (zero electron momentum). For CNTs, the electronic states, and therefore the bandgap, are determined by their cylindrical structure. Theory predicts that for sufficiently small radii, CNTs for which $(2n + m)/3$ is an integer have a zero bandgap, whereas all CNTs for which this relation is not satisfied have a finite bandgap (Mintmire et al. 1992; White et al. 1993; Saito et al. 1998). This unique property of CNTs, together with characteristics such as ballistic electron transport (White and Todorov 1998), has made CNTs a target for nano- and microelectronics applications.

The tensile modulus of a CNT, which is largely independent of helical structure and radius (except at very small radii), is about 1 TPa depending on how the cross-sectional area of the structure is defined (Robertson et al. 1992). CNTs also have a large thermoconductivity, although the exact value appears to be uncertain as discussed below. Also as discussed in more detail below, bent CNTs and CNTs compressed along their axis can reversibly form kinks, and CNTs can be chemically functionalized, which can alter their electronic and thermomechanical properties.

There are many other structures predicted by modeling and/or seen experimentally that are closely related to CNTs. These structures include a variety of end caps, tapered rods, springs, filled structures (including "peapods" with fullerenes inside CNTs), collapsed "ribbon" structures, T-, X- and Y-junctions, and many others too numerous to list here.

The discovery of CNTs is credited to Iijima, as discussed elsewhere in this book, and he was indeed the first to fully document their structure and to pursue their properties. There are, however, indications in the research literature as to their existence prior to Iijima's initial studies (Dresselhaus and Avouris 2001). Shown in Figure 27.2, for example, is a scanning electron microscopy (SEM) image of a carbon surface that was generated in the mid-1970s as part of research involving G. Cuomo and associates at IBM (Cuomo). The image is a carbon surface after it had been exposed to ion bombardment in the presence of an electrostatic extraction field. The hair-like structures coming up from the surface may be CNTs, or they may be some form of diamond fiber. The purpose of this experiment was to create a controlled carbon ion source for deposition of hard coatings rather than to produce a nanostructure, and so refined synthesis and characterization of these structures was not further pursued.

A variety of techniques have been developed for synthesizing CNTs since the first report by Iijima (1991). Common methods include electric arc discharge (Iijima 1991; Ebbesen and Ajayan 1992; Ajayan et al. 1993; Bethune et al. 1993; Ebbesen et al. 1993; Iijima and Ichihashi 1993; Ebbesen 1994; Journet et al. 1997), laser ablation (Guo et al. 1995a, 1995b; Thess et al. 1996; Qin and Iijima 1997; Yudasaka et al.

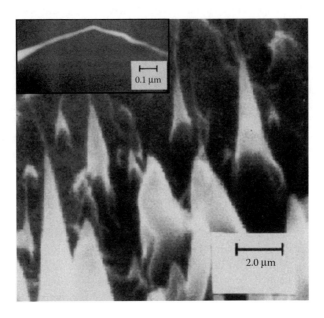

FIGURE 27.2 SEM image of carbon protrusions formed by sputtering and extraction *via* an electrostatic field.

1997, 1998, 1999; Liu et al. 1998; Rinzler et al. 1998; Zhang et al. 1998; Zhang and Iijima 1999), and chemical vapor deposition (CVD) (Li et al. 1996; Sen et al. 1997; Cheng et al. 1998; Huang et al. 1998; Ren et al. 1998; Cassell et al. 1999; Nikolaev et al. 1999; Bower et al. 2000; Su et al. 2000; Ho et al. 2001; Lee et al. 2001; Delzeit et al. 2002; Ducati et al. 2002; Franklin et al. 2002; Hata et al. 2004). High temperatures are reached in both the arc discharge and laser ablation methods, ensuring good annealing of defects in CNT growth (Moravsky et al. 2005). However, arc discharge and laser ablation are not suitable for large-scale production due to difficulties in the scale-up of the system (Sinnott and Andrews 2001; Thostenson et al. 2001; Meyyappan 2005). CVD and related techniques appear to have the greatest potential to generate large quantities of CNTs for bulk applications (Thostenson et al. 2001; Meyyappan 2005).

The atoms in CNT caps are more reactive than those on the sidewall due to the extra strain in the cap region (Srivastava et al. 1999). Chemical functionalization of CNT ends was first discovered during the purification of CNTs. Oxidants such as an acidic potassium permanganate solution (Hiura et al. 1995) or HNO_3/H_2SO_4 (Liu et al. 1998; Saito et al. 2002) open the ends and convert the capped CNTs into open fullerene pipes. The radicals at the ends are stabilized by bonding with carboxylic acid (–COOH) or hydroxylic (–OH) groups. These functional groups have rich chemistry and they can be used as precursors for further chemical reactions (Ebbesen 1996; Chen et al. 1998; Hamon et al. 1999; Saito et al. 2002). The modified CNTs can also have increased solubility in organic solvents (Chen et al. 1998; Liu et al. 1998; Hamon et al. 1999), which implies an improved dispersion in composite processing.

Successful functionalization of the sidewalls of CNTs was first carried out by exposing purified CNTs to a fluorine-containing gas at room temperature (Hamwi et al. 1997). In a subsequent study by Mickelson et al. (1998), it was reported that nondestructive fluorination of purified SWCNTs occurs at temperatures up to 325°C, and that the process is reversible, with anhydrous hydrazine easily removing the added fluorine. These fluoronanotubes have C–F bonds that are weaker than those in alkyl fluorides (Kelly et al. 1999; Stevens et al. 2003), and that provides substitutional sites for additional functionalization (Mickelson et al. 1998; An et al. 2002; Marcoux et al. 2002; Touhara et al. 2002; Plank et al. 2003). Successful replacements of the fluorine atoms by amino (Stevens et al. 2003) and hydroxyl groups (Zhang et al. 2004) have been achieved.

Modifications in CNT structures after atomic irradiation are common, including the creation of various atomic-scale defects, a decrease in CNT diameter, and merging of adjacent CNTs through the formation of cross-links (Kiang et al. 1996; Ajayan et al. 1998; Terrones et al. 2000; Banhart 2001; Kis et al. 2004). In addition to these structural changes, fluorine or hydrogen atoms can also be introduced into the CNT wall via techniques such as ion deposition, plasma irradiation, and proton bombardment (Chen et al. 2001; Ni et al. 2001; Khare et al. 2003). The formation of intertube bridging in CNT bundles (Ni and Sinnott 2000; Terrones et al. 2000; Kis et al. 2004) and the creation of cross-links between shells in MWCNTs (Ni et al. 2001; Huhtala et al. 2004) are of particular importance in making CNT composites. The cross-links can inhibit the intertube/intershell sliding and enhance efficient load transfer between adjacent tubes or the outer and inner shells in MWCNTs. Simulations carried out by several groups have shed light on the mechanisms of irradiation-induced structural modifications (Ajayan et al. 1998; Ni and Sinnott 2000; Krasheninnikov et al. 2001; Ni et al. 2001; Salonen et al. 2002; Huhtala et al. 2004; Jang et al. 2004; Kis et al. 2004; da Silva et al. 2005; Sammalkorpi et al. 2005), which have the advantage of greater control compared to wet chemical techniques (Stahl et al. 2000; Krasheninnikov et al. 2001).

As discussed above, nanosized diamond represents another important class of carbon nanostructure with potential applications in nanocomposites. Experimental observations of nanosized diamond go back more than two decades; these are summarized in Shenderova et al. (2002) and Shenderova and McGuire (2006). Methods for synthesizing nanodiamond particles involve processes such as gas-phase nucleation at ambient pressure, chlorination of carbide material at moderate temperatures, ion irradiation of graphite, electron irradiation of carbon anions, high-pressure–high-temperature graphite transformation within a shock wave, and carbon condensation during explosive detonation. Observations suggest that as much as 10 to 20% of the interstellar carbon is in the form of nanodiamonds.

Ultrananocrystalline diamond powder can be produced by detonation of carbon-containing explosives (Dolmatov 2001; Gruen et al. 2005; Shenderova and Gruen 2006), which yields bulk quantities of the so-called detonation nanodiamond (DND). The availability of bulk quantities of material facilitates its use in practical applications. DNDs are synthesized at high-pressure–high-temperature conditions within a shock wave during detonation of carbon-containing explosives with a negative oxygen balance. In this method, diamond clusters are formed from carbon atoms contained within the explosives molecules themselves, so only the explosive material is used as a precursor. A wide variety of explosive materials can be used. One example is a mixture of 2-methyl-1,3,5-trinitrobenzene (i.e., TNT) and hexahydro-1,3,5-trinitro-1,3,5-triazine (i.e., hexogen or RDX), which is composed of C, N, O, and H with a negative oxygen balance (i.e., with the oxygen content lower than the stoichiometric value required to react with all carbon of the explosives) so that "excess" carbon is present in the system. The explosion takes place in a nonoxidizing medium that acts as a coolant and is either a gas (e.g., N_2, CO_2, Ar, or other medium under pressure) or ice, the so-called "dry" or "wet" synthesis, respectively. Typically, the average primary particle size of DNDs is ~3 to 5 nm. The product obtained by detonation, called detonation soot, contains the diamond nanoparticles (up to 80 wt% depending on the coolant media) along with other carbon structures (amorphous carbon, nanographite, etc.). A variety of techniques have been used to separate the DND phase from soot, e.g., by oxidizing the nondiamond carbon. In the final product, nanodiamond primary particles form tightly and loosely bonded aggregates ranging in their largest dimension from several tens to several hundreds of nanometers. More details on synthesis and processing of this material can be found in Dolmatov (2001), Shenderova et al. (2002), Gruen et al. (2005), and Shenderova and Gruen (2006).

High-resolution transmission electron microscopic images of a single nanodiamond cluster on the surface of a molybdenum tip indicate the presence of facets at the particle surface, with the cluster resembling a polyhedral shape (Tyler et al. 2003). The shape of a diamond cluster is inherently connected to its stability, which in turn depends on the coordination of the surface atoms and defects present in the bulk of the particles (e.g., twins in particles with a pentagonal shape). The stability of nanodiamond particles also depends on the presence of specific functional groups on a particle surface or on the interface energy between a DND particle and the surrounding matrix if a particle is incorporated into the matrix. Formation energies and preferred morphologies of hydrogenated diamond clusters have been calculated using a bond-order interatomic potential for hydrocarbons (Brenner et al. 2002a). Four different morphologies of nanodiamond particles were considered: octahedral, cuboctahedral, spherical, and pentagonal. It was concluded that the most stable morphologies for hydrogenated nanodiamond are octahedral, followed by pentagonal and spherical.

DNRs, which are the diamond analogues of CNTs (Figure 27.3), hold promise for a number of applications. The relatively inert chemistry, the mechanical strength, and the high thermal conductivity of bulk diamond, for example, suggest that DNRs may find important applications as fillers in nanocomposites for structural and thermal management applications. Calculations have also shown that DNRs may be insulating, semimetallic, or semiconducting depending on the nanowire diameter, surface morphology, and degree of surface hydrogenation (Barnard 2005), suggesting possible applications in nanoelectronics and functional nanocomposites. DNRs and related structures have been produced experimentally by etching diamond substrates, by diamond deposition within a template, and by compression of fullerenes (Shenderova et al. 2005).

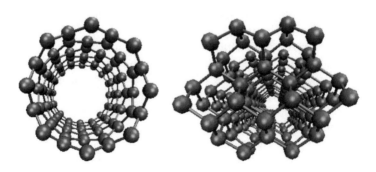

FIGURE 27.3 Illustrations of a CNT (*left*) and DNR (*right*).

The thermodynamic stability of DNRs and preferable morphologies have been investigated theoretically by Barnard (2005) and are summarized in a recent review. Barnard (2005) investigated several morphologies of DNRs with hydrogen-free surfaces using first-principles calculations. It was concluded that the presence of (111) facets leads to partial graphitization. It was also found that bare (001) surfaces undergo dimer pairing surface reconstructions to reduce the number of radical sites, whereas rods with (011) facets largely preserve the diamond structure. The latter DNR morphology also had the smallest heat of formation. The same classes of DNRs with hydrogen surface termination were found to preserve diamond surface morphologies (Barnard 2005).

Using molecular modeling, Shenderova et al. (2003) calculated the binding energy for four DNRs with directional orientations and hydrogen-terminated low-index surfaces that represent a subset of the most stable structures. The data were plotted against the carbon-to-hydrogen ratio for each structure and compared to (17,0) SWCNTs with finite lengths and hydrogen termination at their ends. The carbon-to-hydrogen ratio for the CNTs was varied by changing the length of the CNT. On the basis of these results, it was concluded that the binding energy of DNRs is comparable to that of SWCNTs, especially for large carbon-to-hydrogen ratios.

27.5 Nanocomposites Using Fullerenes

Although the use of fullerenes as nanocomposite components is not as well explored, e.g., CNTs, there has been some exciting work in this area. C_{60}, for example, has been mixed with various metals such as tin, copper, and aluminum (Barrera et al. 1994). The C_{60}/aluminum composite, which was prepared using ball milling, showed an enhanced hardness over the pure metal (Garibay-Febles et al. 2000). A C_{60}/polyethylene composite film also demonstrated an enhanced hardness with an increasing concentration of fullerenes (Calleja et al. 1996). A study of C_{60}/polystyrene composites has demonstrated that the introduction of fullerenes increases the packing density of polystyrene chains, resulting in a composite film with an improved gas permeation selectivity (Gladchenko et al. 2002). The presence of these C_{60} molecules also apparently lowers the thermal degradation temperature of polystyrene, which means a better thermal stability of the resultant material (Gladchenko et al. 2002). In addition, Banerjee et al. (2005) noticed that a C_{60}/glass composite exhibits a moderate value of third-order nonlinear susceptibility and optical limiting properties, which could be used in nonlinear optical devices.

Wang (1992) discovered that a composite film of polyvinylcarbazole (PVK) doped with a mixture of C_{60} and C_{70} exhibits a fast and complete photoinduced discharge at the fullerene–polymer interface. The same behavior was also noticed by Sariciftci et al. (1992) when they doped poly[2-methoxy,5-(2′-ethyl-hexyloxy)-p-phenylene vinylene] (MEH-PPV) with C_{60}. Using light-induced electron spin resonance experiments, Sariciftci et al. (1992) were able to show that photoinduced electron transfer occurs from the polymer to the C_{60}, thereby creating C_{60} anions and mobile holes in the polymer. This behavior can be explained by the extraordinary electron acceptability of the C_{60} molecules. In addition, based on the decay rate of the photoluminescence in a C_{60}/(MEH-PPV) composite, Sariciftci et al. (1992) estimated that charge transfer occurs on a picosecond timescale. This same "ultrafast photoinduced charge transfer" is found in composite materials made from other conjugating polymers and fullerene derivatives (Zakhidov et al. 1996; Miller et al. 1997; Brabec et al. 1998; Dyakonov et al. 1999; Yoshino et al. 1999; Pasimeni et al. 2001a, 2001b; Zerza et al. 2001; Lin et al. 2003; Marumoto et al. 2003; Sensfuss et al. 2003; Chirvase et al. 2004). This effect is potentially useful for photoconductors, rectifying diodes, photorefractive devices, and photovoltaic energy-conversion applications (Dresselhaus et al. 1996; Lee and Kim 2004).

27.6 Nanocomposites Using Nanotubes

Ongoing experimental research on CNT-reinforced nanocomposites has shown exciting results, although the full promise of composites using these structures has yet to be achieved. Polymers, ceramics, and metals have all been used as the matrix material, but CNT/polymer composites are the most studied class of system (Desai and Haque 2005). Both MWCNTs and SWCNTs have been investigated, but SWCNTs are usually favored over MWCNTs as structural reinforcements (Belytschko et al. 2002; Ding et al. 2003) because only the outer shell of MWCNTs is found to contribute to tensile loading.

Significant mechanical improvements have been obtained for polymer-based nanocomposites using CNTs. Cadek et al. (2002) noticed an 80% improvement in the tensile modulus for polyvinyl alcohol (PVA) by adding only 1 wt% CNTs. Ruan et al. (2003) reported an increase of about 140% in ductility for ultrahigh-molecular-weight polyethylene when 1 wt% of MWCNTs were introduced. By dispersing 2 wt% MWCNTs in nylon-6, Liu et al. (2004) showed that the elastic modulus and the yield strength of the composite were increased by 214 and 162%, respectively. Chang et al. (2005) studied the mechanical properties of SWCNT/polypropylene composites at different CNT concentrations. They reported a three-fold increase in the Young's modulus when 1 wt% SWCNTs was added, but the effect was attenuated with further addition of CNTs. More remarkably, Dalton et al. (2003) showed that SWCNT/PVA fibers with 60 wt% SWCNTs had an extraordinary combination of toughness and strength.

CNTs can have higher thermal and electrical conductivities than copper, resulting in multifunctional materials with alterations to the electrical and thermal properties of the polymer upon addition of the CNTs. For example, the introduction of conductive CNTs into insulating polymers can help avoid electrostatic charging of the polymer in addition to the mechanical improvement (Sandler et al. 1999), and nano-assemblies of polymers and CNTs with predictable electronic properties can be made when the polymer is able to coat around the SWCNTs (McCarthy et al. 2001). Thermal degradation of polymers limits their applications at high temperatures. By adding CNTs, the thermal stability of the resultant composites can be greatly improved (Shaffer and Windle 1999; Yang et al. 2004).

The adhesion between the reinforcements and the matrix materials in composites helps to determine the influence of the reinforcements on the thermomechanical properties of the system. The interfacial area in CNT-reinforced nanocomposites is extremely large due to the high aspect ratio of the CNT. The interfacial interaction in CNT-reinforced composites is therefore even more critical to the ultimate performance of the composite. Several studies have shown evidence of strong bonding between polymers and the CNT (Wagner et al. 1998; Lourie and Wagner 1999; Qian et al. 2000; McCarthy et al. 2002; Barber et al. 2003). In contrast, there are studies that suggest a poor interfacial interaction. Schadler et al. (1998) investigated the load transfer in CNT/epoxy composites in both tension and compression. By monitoring the shift of the Raman peak, they concluded that the load transfer in compression was effective, whereas it was poor in tension, which implies a weak interaction in the interfacial area. Song and Youn (2004) also noticed the poor interfacial bonding between the CNT and epoxy as evidenced by the CNT pullout in a field-emission SEM image.

Another important factor that controls the performance of CNT-reinforced composites is the dispersion of CNTs. CNTs tend to aggregate into bundles. Although a recent study showed that the interfacial sliding within CNT bundles results in very high mechanical damping (Suhr et al. 2005), slipping of CNTs when they are assembled in ropes adversely affects their load-carrying capability (Salvetat et al. 1999). As a result, breaking the bundles into individual CNTs is preferable to take full advantage of the reinforcements (Ajayan et al. 2000). This was confirmed by the investigations of Liu and Wagner (2005). In their studies, rubbery and glassy epoxy resins were used as the matrix materials. Strong covalent bonds between the CNT and the matrix were found in both systems. The Young's modulus increased by 28% in the CNT/rubbery epoxy system, whereas no improvement could be observed in the glassy epoxy-based composites. This is attributed to a better dispersion of the CNTs in the less viscous rubbery polymer than in the glassy resin.

One way to effectively improve the interfacial bonding and the CNT dispersion is to chemically modify the surfaces of the CNTs (Gojny et al. 2003, 2005). Studies carried out by Skakalova et al. (2005) have explicitly shown that when the SWCNT is pretreated by thionyl chloride, an improved interfacial bonding between the CNT and poly(methyl methacrylate) is obtained. This improved bonding results in remarkable increases in the Young's modulus, tensile strength, and toughness as compared to the composites made from pristine CNTs (Skakalova et al. 2005). Chemical functionalization of the CNT sidewall by attaching hydroxyl, amino, or carboxyl acid groups successfully provides multiple bonding sites to the surrounding polymer and improves the dispersion when CNTs are mixed with the matrix (Stevens et al. 2003; Zhu et al. 2003, 2004; Zhang et al. 2004; Ramanathan et al. 2005). Most recently, unconventional groups such as organometallic molecules (Viswanathan et al. 2003; Blake et al. 2004) and even proteins (Bhattacharyya et al. 2005) have been chemically attached to CNT sidewalls. The interfacial properties of the resultant composites were improved substantially, which led to dramatic enhancement in their mechanical properties.

The processing technique is also important to achieve a good dispersion. In addition to conventional techniques such as shear mixing (Stephan et al. 2000; Gojny et al. 2004, 2005; Potschke et al. 2004), melt compounding (Liu et al. 2004; Zhang et al. 2004), and surfactant-assisted processing (Gong et al. 2000; Regev et al. 2004), other methods such as ultrasonication (Safadi et al. 2002; Graff et al. 2005) and *in situ* polymerization (Tang and Xu 1999; Viswanathan et al. 2003; Zhang et al. 2004; Datsyuk et al. 2005; Uchida and Kumar 2005) have found successful applications in dispersing CNTs in various polymer matrices sometimes even without chemical pretreatments.

Alignment of the reinforcing CNTs can make nanocomposites with anisotropic mechanical and electrical properties. Alignment can be realized by shear (Ajayan et al. 1994; Haggenmueller et al. 2000) or uniaxial stretching of the composite at a high temperature (Jin et al. 1998). Applying an electric field during processing can also induce the formation of aligned CNT networks within the polymer (Martin et al. 2005). The method introduced recently by Raravikar et al. (2005) is unique. In their method, aligned arrays of CNTs were first grown using a CVD technique. *In situ* polymerization was initiated when monomers were introduced into the CNT array. As a result, a nanocomposite with both well-aligned and dispersed CNTs was produced.

An electrospinning process that is capable of forming nanoscale fibers electrostatically from polymer solutions or melts appears very promising to convert CNTs to macroscopic structures (Ko et al. 2002). The researchers were able to incorporate 10 wt% of SWCNTs into polyacrylonitrite (PAN). The coelectrospinning of SWCNT/PAN can produce continuous filaments with enhanced thermal and mechanical properties. These nanocomposite fibrils can be used as precursors for linear, planar, and three-dimensional fiber assemblies for macroscopic composites. A layer-by-layer (LbL) assembly technique that possesses great potential for large-scale fabrication of CNT-filled composites was also reported (Olek et al. 2004). In this technique, CNT/polyelectrolyte multilayer composite films are formed on glass slides or silicon wafers by sequential deposition of oppositely charged CNTs and polyelectrolytes such as polyethylene-imine. The LbL assembly technique can easily achieve a high CNT content (50±5 wt%) and maintain a homogeneous distribution of CNTs. The resultant composites can be conveniently removed from the substrate and cut into desired sizes and shapes for further processing.

The exceptional resilience of CNTs is particularly desirable to compensate for the brittleness of ceramics; therefore, using CNTs as a second phase to improve the performance of ceramic (Hwang and Hwang 2001; Zhan et al. 2003; Borrmann et al. 2004) and even metallic (Kuzumaki et al. 1998; Dong et al. 2001; Noguchi et al. 2004) materials has been attempted. To make CNT/metal-oxide nanocomposites, a two-step preparation scheme was demonstrated (Laurent et al. 1998; Peigney et al. 1998; Flahaut et al. 2000). In the first step, the nanocomposite powder is fabricated by growing CNTs *in situ*. The dense CNT/metal-oxide material is then prepared by hot-pressing the composite powder. However, the CNT-containing composite showed lower fracture strength than the carbon-free metal-oxide structure. An early study to make a CNT/aluminum composite by a hot-pressing and a hot-extrusion method also achieved no enhancement in the mechanical strength (Kuzumaki et al. 1998). It is believed that major causes of the disappointing results in these early studies are the weak interfacial interaction in the composite and the poor dispersion of the CNTs in the matrix.

A surfactant-assisted method apparently achieved a good dispersion of silica-coated CNTs in a ceramic matrix, thereby enhancing the mechanical strength of the resultant composite by ~100% (Hwang and Hwang 2001). Zhan et al. (2003) and Zhan and Mukherjee (2004) prepared CNT/alumina nanocomposites using the spark-plasma sintering (SPS) technique. The fracture toughness of the prepared composite was nearly three times that of pure alumina at a loading of 10 vol% SWCNT. This high performance resulted from an optimum dispersion of the CNTs in the matrix, a strong cohesion between the CNT and the alumina, and a lessened amount of damage to the CNT structure in the SPS technique than in the traditional hot-pressing method. As opposed to the *in situ* growth of CNTs in the metal-oxide powder by Laurent et al. (1998), Peigney et al. (1998), and Flahaut et al. (2000), Lupo et al. (2004) synthesized a CNT/zirconia composite by growing zirconia on CNTs via a hydrothermal process. A homogeneous mixture that is expected to be an ideal starting material for CNT/ceramic composites was produced. An et al. (2004) reported another technique to make CNT-reinforced ceramic composites with improved properties. In their method, the nanocomposites were prepared using polymer-derived ceramics instead of a ceramic powder. Because CNTs can be easily dispersed in liquid-phase polymer precursors prior to pyrosis, a homogeneous mixture of the CNT and the ceramic was conveniently obtained. Modifications of the CNT surface and/or the polymer precursor also provide the possibility to improve the CNT–ceramic interfacial properties.

Extensive computational studies have been carried out of CNT nanocomposites. In many cases, these studies have provided new and powerful insights into the structure, properties, and bonding of systems, including predictions of new systems with optimized and in some cases unique characteristics. The computational methods used to study CNTs and CNT-filled composites can be broadly divided into two classes: continuum and atomistic methods. There are several approaches to continuum modeling, ranging from solving purely analytical equations that represent some quantity, to finite element methods (FEMs) that divide the continuum into small deformable grains, to the multiscale methods that can bridge several of the continuum levels down to the atomistic.

Macroscale continuum models generally assume uniform materials within the boundaries of the system; however, with composite materials this assumption is problematic. Micromechanics methods assume that the system is composed of smaller components (or volume elements) that allow for non-uniformity in materials properties, while these smaller components are treated as continua with boundaries specified by neighboring components and local geometry. Thus, micromechanics models can treat composite systems as being composed of several different types of components and can account for interfaces between constituents, voids, discontinuities, and surfaces without losing accuracy. Micromechanics models provide a transition from macroscale models to the microscale.

Several groups have reported the use of analytical micromechanical models for predicting properties of CNT/polymer composites. Odegard et al. (2003) predicted the mechanical properties of a (6,6) CNT/polyimide composite with a poly(*m*-phenylene vinylene-*co*-2,5-dioctoxy-*p*-phenylene)-CNT/polyimide interface. They examined the effects of CNT volume fraction, orientation, and length on the mechanical properties of the composite. In a similar study, the effects of functionalization on the properties of a (10,10) CNT/polyethylene composite were characterized. It was found that chemical functionalization degraded most of the elastic properties of composites (Odegard et al. 2003). Lagoudas and Seidel (2004) examined the properties of CNT/epoxy composites and found that clusters of CNTs in the composite played a role in the reduction of composite properties; poor CNT–matrix bonding and cluster misalignment have a much greater effect on the drop in composite performance. Pipes and Hubert (2002, 2003a, 2003b) predicted properties of composites containing helical arrays of CNTs and later extended their model to predict thermal expansion properties and to predict large-scale properties for these and related systems. Xiao and Liao (2004) developed a model for nonlinear pullout of CNT-reinforced composites in which thermal residual stresses, Poisson's contraction, and nonlinear elastic behavior of the CNTs are considered. Their results suggested that the distribution of interfacial shear stress along the CNT length is sensitive to the CNT's elastic nonlinearity (Xiao and Liao 2004). Others have looked at MWCNTs. Thostenson and Chou (2003) predicted the elastic moduli of MWCNT/polystyrene composites and the effects of CNT structure, geometry, and constituents on the moduli. They discovered that composite elastic properties are very dependent on CNT diameter. Yoon et al. (2003) used a multiple-elastic-beam model to predict the vibrational modes and resonant frequencies of MWCNTs in an elastic medium. Zhang and Wang (2005) have modeled thermal effects on interfacial stress transfer in CNT/polymer composites. Their results showed that a mismatch of thermal expansion coefficients between CNT and polymer matrix was one of the most important factors governing interfacial stresses, whereas other factors such as temperature fields, volume faction, and CNT radius also affect the stress.

Several groups have used FEMs to characterize the mechanical and thermal properties of CNT/polymer composites. Fisher et al. (2003), for example, used a FEM to determine the influence of fiber waviness in CNT-reinforced polymer on their overall mechanical properties. In particular, they noted that CNT waviness may significantly limit the magnitude of the modulus enhancement of the composite. Bradshaw et al. (2003) used this model to determine the effective stiffness of a CNT/polymer composite containing wavy CNTs; again a reduction in mechanical properties was observed with increased CNT waviness. Li and Chou (2003) used an FEM to study stress distribution, with emphasis on interfacial shear stress between the CNT and polymer in the composite. Zhang et al. (2004) simulated heat conduction in CNT-filled composites. In their simulation, they treated the CNTs as a heat superconductor and only modeled heat flowing in the host polymers. The thermoelastic properties of CNT-reinforced polymers have been simulated by Lusti and Gusev (2004). They studied how CNT alignment effects thermoplastic properties of the composite and found that the effect was the greatest for fully aligned CNTs (in the axial direction); however, composites containing randomly aligned CNTs exhibited moderate enhancements.

Multiscale modeling is a combination of several modeling methods that are used to span multiple time and length scales. Generally information from an atomic-scale simulation is used in or coupled to a continuum-based method. Several groups have used this hybrid approach for modeling properties of CNT/polymer composite materials. Maiti et al. (2005) performed mesoscale simulations based on Flory–Huggins theory. They calculated the solubility parameters for CNTs as a function of tube radius and investigated the alignment of CNTs as a function of shear and dispersion through functionalization. Chandra et al. (2004) and Namilae and Chandra (2005) used a hierarchical modeling method to combine an explicit interatomic potential energy function with finite elements. They linked the MD simulation to the finite element through an atomically informed cohesive zone model with parameters representing the interfaces. They used their model to simulate fiber pullout, the CNT matrix interface, and the effects of interface strength on composite stiffness. Further details of multiscale modeling methods can be found in reviews by Gates et al. (2005) and Park and Liu (2004).

Atomic-scale modeling has been used to characterize a variety of properties related to the structure, bonding, mechanical, and thermal properties of CNT/polymer composites. For example, modeling studies have focused on calculating nonbonded adhesion strengths and critical shear stresses for CNTs interacting with polyethylene, polyaromatic polymers, epoxy resins, and other polymer matrices. The use and production of chemical cross-links between CNTs and polymer matrices to enhance load transfer has also been explored. In addition to mechanical properties, atomic modeling is being used to characterize the thermal properties of CNT/polymer composites, including thermal conductivity of pristine and functionalized CNTs and the influence of CNTs on the glass-transition temperature of polymers. All of these studies have been useful in both providing unique and critical insights into the atomic-level physical and chemical interactions that play an important role in the thermomechanical properties of CNT/polymer compositions, and in providing important data needed for larger-scale modeling of these systems that is not readily available from experiment.

From an atomic modeling viewpoint, the most straightforward system to study is a CNT embedded within a polyethylene matrix because only two types of atoms are involved (carbon and hydrogen) and the bonding types are well defined. Frankland et al. (2002, 2003) used two force field expressions, one that assumed a "united-atom" approximation and one that explicitly includes hydrogen atoms, to model crystalline and amorphous polyethylene matrices within which single-shelled (10,10) CNTs were embedded. Critical shear stresses needed to slide a nanotubule through the matrix were calculated via a simulated pull-through experiment in which an increasing load was applied to a CNT during an MD simulation. With this method, critical shear stresses were estimated from the applied load at which the CNT began to slide through the matrix and the estimated contact area between the CNT and matrix. The calculated critical shear stresses ranged from 0.7 MPa for an amorphous matrix modeled with united-atom potentials to 2.8 MPa for a crystalline matrix that included explicit hydrogen atoms. Assuming a tensile fiber strength of 50 GPa for the CNTs, these critical shear stress values imply that lengths exceeding ·-25 μm would be required for significant load transfer between the matrix and CNTs. MD simulations were also used to model loading strain for composites containing capped (10,10) CNTs of varying lengths.

The largest system modeled contained a 100-nm long CNT in a united-atom/amorphous polyethylene matrix. Consistent with the relatively small critical shear stresses determined from the pull-through simulations, after straining the system the CNTs released from the matrix starting at their ends and regained their initial length.

Similar calculations were reported by Griebel and Hamaekers (2004), where stress–strain relations were calculated using an applied external stress. As in the simulations by Frankland et al., the systems consisted of a CNT of infinite length via periodic boundaries and a 30-nm capped CNT, both of which were embedded in a polyethylene matrix. For the capped and infinite structures, enhancements in the longitudinal modulus of a factor of 2 and 30, respectively, were reported. These results are comparable to predictions using a macroscopic rule of mixtures that account for fiber length.

In a related molecular modeling study, Wei et al. (2004) noted that a CNT embedded into a polyethylene matrix induces discrete layers within the matrix that form a series of density "rings" along the axis of the CNT. Using an order parameter that characterizes the orientation of the bonds within the polyethylene with respect to the axis of the CNT, it was also shown that the chains within these rings have a tendency to align parallel to the CNT axis. This ordering of the polyethylene matrix was driven by an increase in CNT–matrix adhesion, although the resulting increase was not sufficient to induce significant load transfer from the matrix to the CNT. However, chain ordering within the matrix can increase the elastic modulus of the system along the ordering direction, leading to a mechanism for mechanical reinforcement within the composite that does not necessarily require high critical shear forces between the matrix and CNT.

The influence of the aspect ratio on the adhesion energy between a CNT and an amorphous polyethylene matrix was characterized by Al-Haik et al. (2005) using molecular modeling. In these calculations, the aspect ratio was changed through incremental changes in the chiral angle of CNTs with a fixed number of carbon atoms. They reported that lower chiral angles, which have higher aspect ratios in these simulations, have higher adhesion energy. This is consistent with classical elasticity theory where the adhesion energies are inversely proportional to the square of a fiber radius. They also reported that the CNT with the highest aspect ratio underwent significant distortion to enhance the CNT–matrix adhesion, whereas the CNT with the largest radius underwent only a moderate degree of deformation.

Using data from MD simulations of CNT fiber pull-through, Frankland and Harik (2003) developed an interfacial friction model for a polyethylene matrix that includes a viscous friction term. In these simulations, a load was applied incrementally to the CNT. Below a critical shear stress, the CNT displayed thermal vibrations but no net motion. At applied loads above the critical shear stress, the CNT underwent steady sliding motion with semiperiodic variations in the sliding speed that matched the periodic structure of the carbon rings in the CNT. Using the average velocity and value of the applied stress led to a viscous friction coefficient, which together with the critical shear stress defined a friction model for the entire pull-through process.

Owing to the aromatic structure of CNTs, conjugated polymers or polymers containing aromatic sidegroups may have CNT–matrix binding energies that are larger than polyethylene matrices and hence enhanced load transfer in a nanocomposite. This assumption is consistent with experimental studies that report for CNTs heavy coating and strong bonding with conjugated polymers (Ajayan et al. 1994; Jin et al. 1998). Liao and Li (2001) used molecular mechanics to characterize the adhesion between styrene oligomers and a single graphite sheet and the binding energy and fiber pullout shear stress for a single- and double-walled CNT in a polystyrene matrix. For the oligomer studies, the simulations predict an increase in binding energy on a per-monomer basis with increasing chain length up to a limiting value of about 0.22 eV for an 80-monomer chain. For their CNT/polystyrene studies, they noted that the polymer and CNT have different thermal expansion coefficients and that this difference in thermal properties can lead to a radial stress and deformation of the CNT with an associated increase in critical shear stress. Using a simulated pullout experiment, the calculations indicate similar pullout energy for the single- and double-walled CNTs, and a critical shear stress for each of about 160 MPa, a value that is well above that calculated by others for polyethylene matrices (see above). Although Liao and Li noted that the combination of chemical interactions, the difference in thermal properties between the matrix and CNT, and the radial deformation of the CNT all contributed to the critical shear stress, their modeling suggests that the latter two are the most important factors that contribute to their relatively large calculated critical shear stress.

Yang et al. (2005) modeled the interaction of CNTs with different radii interacting with a series of oligomer chains containing conjugated backbones and side groups. Their results indicate that polymers with aromatic rings in the backbones are more strongly bound to CNTs than are polymers of similar size with aromatic side groups, and that the binding energy becomes stronger as the CNT radius increases up to a limiting value that corresponds to interaction with a flat single graphite sheet. The difference in binding energy between the polymers with an aromatic backbone and those with aromatic side groups is a result of the geometry of these two structures relative to the CNT orientation. In both cases, the polymer backbone aligns with the CNT axis. For structures with an aromatic backbone, the polymer is able to align with the CNT such that the area between the backbone and CNT is maximized. In the case of aromatic side groups, both the side groups and backbone are not able to align with the CNT axis. Instead, the plane of the aromatic rings on the side groups tend to lie more perpendicular to the CNT, which maximizes the mutual interaction between these side groups.

Lordi and Yao (2000) performed an extensive set of atomic simulations on a series of polymers that included aromatic constituents in the chain backbone as well as aromatic side groups. The calculations were used to characterize preferred chain-CNT configurations, calculate adhesive energies, and determine frictional forces for CNT-chain sliding in a matrix. A key conclusion of this work is that polymers that tend to form hollow helical structures are best able to maximize contact between aromatic rings in the backbone and a CNT, leading to enhanced adhesive energies. In the case of cisoidal poly(phenyl-acetylene), the polymer backbone can wrap around a (10,10) CNT such that a helical structure of the CNT and polymer are commensurate. The authors conclude that the magnitude of the frictional forces at a sliding CNT–polymer chain interface do not depend on the degree of CNT–polymer adhesion, and hence strong adhesion may not lead directly to strong load transfer. In contrast, although the calculated frictional forces between the polymer matrices and the CNT are relatively weak, the authors note that they are greater than frictional forces for CNTs sliding past one another in bundles. Hence, although strong CNT–matrix adhesion may not lead directly to better load transfer, it can lead indirectly to enhanced mechanical properties by helping to disperse CNT ropes and therefore create more CNT–polymer interfaces with a relatively higher sliding friction.

Several modeling studies have been carried out of CNT–polymer interactions for polymers containing N, O, and B atoms. In the work of Lordi and Yao (2000) discussed above, the polymers poly(hydroxyl amino ether), poly(hydroxyl amide ether), and poly(methyl methacrylate) interacting with a CNT were studied. In addition to the conclusions discussed above, the calculations indicate that hydroxy side groups, and to a lesser extent phenyl side groups, are advantageous for creating strong CNT–matrix interfaces.

Gou et al. (2005) modeled a three CNT rope of capped (10,10) CNTs within a molecular model of a cured epoxy resin. In their simulations, the epoxy did not penetrate the region between the CNTs within a rope during system equilibration, but instead wrapped around the outside of the rope. Consistent with this structure, the simulations gave a stronger binding between the CNTs within a rope than between the CNT–epoxy resin. Pullout of either one CNT from the three CNT ropes in the epoxy or all three CNTs together was modeled. They report a shear stress of 61 and 36 MPa for pullout of one CNT and all three CNTs, respectively. Both values are lower than the roughly 88 MPa they estimate for pull-out of a CNT from the interior of a CNT rope with no surrounding matrix.

Using first-principles total energy methods, Simeoni et al. (2005) studied the interaction of two iso-electronic polymers, poly(para-phenylene) (PPP) and poly(para-borazylene) (PBZ) with a CNT. The intent of this study was not to characterize CNT–polymer binding for composite applications, but rather to characterize a CNT–polymer system for applications in optoelectronic devices. Nonetheless, this study does provide some useful insights into the influence of electronic effects on binding that are not available from molecular modeling studies. For example, CNTs can be metallic or semiconducting depending on their helical structure and radius, which can in principle change their interaction with a matrix depending on the polymer. The PPP is composed of a backbone of connected benzene rings with a bandgap of

1.8 eV (as given by density functional calculations within the local density approximation). PBZ has the same basic chain structure except that every pair of carbon atoms is replaced with a B–N atom pair. This structure has a calculated bandgap of 4.66 eV, which is larger than that of the PPP. All hydrogen atoms in the PPP are positively charged due to charge transfer to the carbon, whereas in the PBZ hydrogen atoms are positively or negatively charged depending on the atom type to which they are bonded. Binding to (12,0), (16,0), and (18,0) CNTs was studied. Both the first and last CNTs are metallic, whereas the intermediate structure is semiconducting. The authors report that the PBZ does not significantly alter the electronic properties of either the metallic or semiconducting CNTs, whereas the PPP alters the electronic properties of the two metallic CNTs but not the semiconducting one. Consistent with these results, the minimum energy distance between the three CNTs and the PBZ on the outside of the CNTs is constant at 0.33 nm, whereas that between the PPP and the CNTs varies from 0.32 to 0.335 nm. Similarly, the reported binding energies between the PBZ and the (12,0), (16,0), and (18,0) CNTs are 210, 210, and 160 meV, respectively, with corresponding values for the PPP of 179, 200, and 280 meV. Although the effect of the electronic structure of the polymer and the CNT is relatively small, which helps to validate the use of molecular modeling for studying these systems, it does illustrate how electronic effects can influence binding properties.

Several experimental studies have reported the formation of strong chemical bonds between CNTs and polymer matrices with an associated enhancement in mechanical reinforcement of the composite (Gojny et al. 2003, 2005; Zhu et al. 2004). To further explore this, molecular modeling has been used to characterize the influence of these cross-links on load transfer as well as on the mechanical properties of the CNTs. In particular, there is a potential for a trade-off between enhancing the load transfer between the matrix and the CNT, and the degradation of the mechanical properties of the CNTs due to atomic rehybridization associated with the functionalization.

Several molecular modeling studies have been carried out to probe how covalent bond formation to a CNT alters elastic and plastic properties of a CNT. Using MD simulations, Garg and Sinnott (1998) characterized the influence of chemical functionalization on the dynamics of the CNTs under uniaxial compression for a range of radii and helical structures. The CNTs were functionalized by chemical addition of $H_2C=C$ species to the center region of the structure, followed by compression of the ends of the CNT toward the center of the structure. For both functionalized and pristine structures, the CNTs form buckles as compression progresses, with the buckling dynamics corresponding to distinct modulations of the force vs. displacement data generated by the simulations. For each CNT studied, however, the force at which buckling first occurs was reduced by between 12.5 and 17% for the functionalized CNTs relative to the pristine structures, with little influence of helical structure on the buckling force. For the two smallest radii simulated, 0.339 and 0.391 nm, buckling resulted in loss of the chemisorbed species from the CNT, whereas the species remained bound to the structure for the larger radii studied (0.67 to 1.7 nm).

Using similar modeling techniques, Brenner et al. (2002b) characterized the stress–strain behavior under uniaxial tension of a (10,10) CNT onto which methyl radicals were chemisorbed. In contrast to the compression studies (Garg and Sinnott 1998), these simulations predict little change in the tensile modulus even up to 15% of the carbon atoms functionalized. The reason for this behavior can be traced to the characteristics of carbon bonding. Upon functionalization, the hybridization of the carbon atoms in the CNT goes from sp^2 (ignoring the curvature of the CNT) to sp^3. This rehybridization results in the carbon atom onto which the adduct is bonded being raised away from the axis of the CNT due to a change in bond angle to approximately 109.5° (Figure 27.4). This change in angle is accompanied by an increase in equilibrium bond length, which helps compensate for the change in bond angle. The net result is that elastic properties of the CNT are not as changed by chemisorption as might be expected. This result is encouraging for the use of chemical cross-links to enhance load transfer in composites between CNTs and a polymer matrix.

Namilae et al. (2004) performed an in-depth simulation study of the mechanical properties of different CNT–hydrocarbon functional group combinations under an applied tensile uniaxial strain. Several measures of local stress were used to analyze the simulations. They report that the local stiffness of CNTs

FIGURE 27.4 Illustrations of chemisorption to the sidewall of a CNT. Left: methyl radical chemisorbed to a CNT. Right: chemical cross-link formed between a CNT and a polyethylene matrix.

increases with chemical functionalization independent of the functional group, with the effect being larger for smaller radii CNTs. They also report local residual stresses that arise from the functionalization. Citing preliminary MD simulations of functionalized CNTs subject to tensile deformation at 3000K, they report that functionalization results in the formation of plastic deformation via Stone–Wales transformations at smaller strains in functionalized CNTs compared to pristine CNTs, and correspondingly CNT failure initiates near the functional attachments at lower strains relative to a pristine CNT. Taken together, these results suggest that functionalizing CNTs via covalent bond formation results in stiffer and more brittle structures in tension compared to pristine structures.

Frankland et al. (2002) characterized the change in critical stress determined from a simulated pull-through experiment for CNTs cross-linked into a polyethylene matrix compared to strictly nonbonded CNT–matrix interactions (Figure 27.4). Two critical stresses were reported, a stress at which a CNT first starts to move with respect to the matrix, and a larger stress at which motion of the CNT significantly loads the cross-links. They report that chemical cross-links between CNTs and matrices involving less than 1% of carbon atoms on an SWCNT can increase critical shear strengths by a factor of about 40. Assuming a fiber strength of 50 GPa for the CNTs, the critical lengths needed for good load transfer drop from ~25 to 0.6–2.0 μm with cross-linking.

Several groups have performed molecular simulations aimed at exploring possible mechanisms for cross-link formation between CNTs and a polymer matrix. Srivastava et al. (1999) characterized the reactivity of a kink formed in a bent CNT (Figure 27.5) using hydrogen-binding energies as a metric. The simulations predict a strong preference for chemical bonding at the apex sites of the kink, leading to a correlation between mechanical deformation that induces the kink and chemical reactivity. Analysis of the electronic states from a tight-binding calculation indicated formation of a radical state at the kink apex with an energy near the Fermi level that is responsible for the enhanced reactivity. CNTs will form

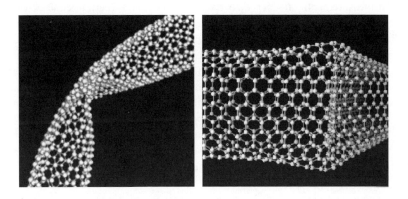

FIGURE 27.5 Illustrations of a kink formed in a bent CNT.

reversible kinks relatively easily, and presumably such structures can be formed during processing of a CNT/polymer composite. The reactivity of these sites could lead to cross-link formation.

Hu et al. (2003) and Hu and Sinnott (2004) performed MD simulations to examine the chemical modification of CNT/polystyrene composites through energetic ion beam depositions. The simulations showed that cross-links between otherwise unfunctionalized CNTs and polymer matrices can be produced, which would result in the enhancement of the load transfer capability across the CNT–polymer interface. The authors pointed out that the formation of cross-links not only was related to the deposition energy, but also depended on the composite geometry. They also noted that the modification capability of the energetic deposition technique might be limited to the near surface because of the rapid energy decay within the composite. Nevertheless, these studies suggest that ion deposition could be an effective method to modify the CNT–polymer interface *in situ* without first sacrificing the CNTs by strong acidic or other harsh chemical treatments.

Using first-principles methods, Mylvaganam and Zhang (2004) calculated the chemical binding energies between mono- and di-radical states on polyethylene and CNTs. Because of the computational demands of these calculations, the polymer was represented by short hydrocarbon chain segments, and the CNTs were modeled by polyaromatic molecules with hydrogen termination that were chosen to simulate sections of (5,0) and (17,0) CNTs. They concluded that bond formation between radicals in an otherwise saturated hydrocarbon and a CNT is energetically favorable, and that reactions may take place at multiple sites on a CNT corresponding to different initial radical sites on the hydrocarbon. In addition, they concluded that the energetics for bond formation is enhanced by the CNT curvature, and therefore binding is most favorable with small-diameter CNTs. On the basis of these calculations, the authors suggest that free radical initiators added to a composite may help induce cross-linking to the CNT and hence enhance load transfer between the matrix and CNT.

Compared to the simulations of mechanical properties, much less has been done in the area of atomic-level modeling of the thermal properties of CNT/polymer composites. Wei et al. (2002) used molecular modeling to characterize the change in density at constant pressure as a function of temperature for a system composed of a capped 2-nm long (10,10) CNT embedded into a polyethylene matrix. The simulations show a distinct change in slope in the density vs. temperature data that indicates a glass transition in the system. The glass transition temperature T_g reported for their pure polyethylene sample was 150K, which is in reasonable agreement with experiments. With the CNT embedded in the matrix, the simulations produce a T_g of 170K. The authors attribute this increase in T_g to a coupling between the CNT and polymer motion that tends to slow the motion of the polymer chains when the system is below T_g. They also noted that the volume thermal expansion, which is determined from the slope of the density vs. temperature data, is larger for the composite than for the pure polyethylene system both above and below T_g. This is attributed to an increase in the excluded volume associated with the CNT (which is large due to the hollow structure) as a function of temperature compared to the pure polymer system.

Experimental measurements on CNT mats have produced large thermal conductivities of 1750 to 5850 W/m K (Hone et al. 1999, 2000; Yi et al. 1999). For comparison, copper and silver, which have the highest conductivities of any metal, have thermal conductivities of around 400 W/m K at room temperature. The large apparent thermal conductivity of CNTs has raised interest in using them for thermal management in polymer composites. Biercuk et al. (2002), for example, report a significant increase in the thermal transport properties of an industrial epoxy with the introduction of 1 wt% CNTs. Many of the same issues that are important for mechanical reinforcement also arise in thermal management, e.g., whether well-dispersed CNTs or CNT ropes would be best for thermal management, the efficiency of is heat transfer from a matrix to a CNT, the issues that contribute to the percolation threshold for CNTs in polymer matrices (Shenogina et al. 2005), and whether cross-linking the CNT to the matrix could enhance heat transfer without compromising the thermal properties of the CNTs. Atomic modeling is just beginning to address these and related issues.

At present, there is considerable confusion and conflicting predictions for the thermal conductivities of pristine CNTs, with predicted values ranging from about 300 to 6600 W/m K (Berber et al. 2000;

Che et al. 2000; Osman and Srivastava 2001; Maruyama 2002; Padgett and Brenner 2004). It is not clear whether these differences are due to inherent properties of the various potential functions used in the simulations or the computational methods used to obtain the thermal conductivity values. Indeed, phonon lifetimes are very sensitive to anharmonic regions of a potential energy expression, and this region of the potential is generally less well established than the harmonic terms. In addition, it has been recently shown that classical dynamics can predict thermal conductivities for CNTs that are considerably above the upper limit set by quantum effects regardless of the potential energy expression or the method used to extract the thermal conductivities from the simulations (Mingo and Broido 2005). Given the current state of this field, relative thermal conductivities calculated from atomic simulations appear to be much more reliable at present than are absolute reported values.

To evaluate the influence of chemical functionalization on thermal transport in CNTs, Padgett and Brenner (2004) used molecular modeling to calculate thermal conductivities of isolated (10,10) CNTs on which phenyl groups are chemically attached to random positions on the CNT (Figure 4.7). In contrast to the tensile modulus, where up to 15% functionalization was predicted to have little influence on the tensile uniaxial stress–strain relation (Brenner et al. 2002b), these simulations predict that chemisorbing phenyl groups to as little as 0.25% of the carbon atoms in the CNT reduces the thermal conductivity by greater than a factor of 3. The vibrational power spectrum shows a slight broadening of the highest frequency optic modes due to chemisorption, but little change in the acoustic modes by which most of the heat is transferred. This result suggests that the decrease in thermal conductivity is likely not due to changes in the local vibrational properties of the system, but rather the functionalization introduces scattering centers that reduce the phonon lifetime. This is an unfortunate result with respect to using cross-links to enhance thermal management because it implies that the reduction in thermal conductivity is independent of the species chemisorbed to the CNT.

27.7 Nanocomposites Using Diamond-Like Structures

The addition of DND to polymers has been shown to result in an increase in mechanical strength, wear resistance, and heat-aging resistance (Dolmatov 2001, 2006). Highly effective coatings based on the incorporation of DND in fluoroelastomers and polysiloxanes have been developed, and the elastic strength of rubbers based on polyisoprene, butadiene–styrene, butadiene–nitrile, and natural rubbers have been considerably improved (Dolmatov 2001; Shenderova and McGuire 2006). For example, fluo-roelastomers filled with DND particles showed an increase in the tensile modulus at 100% elongation (from 280 to 480%) and in the conditional rupture strength (from 8.5 to 92 and from 15.7 to 173 MPa, respectively). An increase in cross-linking is one mechanism to explain the increased polymer composite strength resulting from DND particles (Dolmatov 2001). According to Dolmatov (2001), the addition of DND into the rubbers decreases attrition wear by an average of three to five times, and increases the rupture strength and breaking temperature by 30 and 15%, respectively. The specific utilization of DND or a diamond blend (a mixture containing DND and a significant percentage of other carbon-based products of detonation) is 1 to 5 kg per 1000 kg of rubber and 1 to 5 kg per 1000 m² of polymer coating or film. Experiments conducted on poly(methyl methacrylate), benzocyclobutene, and polyimide showed that the incorporation of 1 to 2% by mass of nanodiamond particles results in an increase in thermal stability of at least 30°C (Shenderova and McGuire, 2006).

The enhanced mechanical and thermal properties of composites with incorporated DNDs can be understood as follows. Within the current nanocomposite concept (Banerjee et al. 2001; Vaia and Giannelis 2001), there are three major characteristics defining the polymer nanocomposite performance: a nanoscopically confined polymer matrix, nanoscale inorganic reinforcing elements, and the nanoscale arrangement of these elements. The effect of the nanoelements on the surrounding matrix is related to a fundamental length scale of the adjacent matrix, which for polymers is on the order of their radii of gyration (5 to 10 nm). The distances between reinforcing elements in composites with a few volume percent of nanoelements are of the same order of magnitude as the radius of gyration. Therefore, it appears that a majority of polymer molecules reside near nanoelements. Because an interface limits the number of conformations of a polymer molecule, the entire matrix may be considered to be nanoscopically confined. These restrictions on chain conformations alter molecular mobility, relaxation behavior, free volume, and thermal transitions (Banerjee et al. 2001; Vaia and Giannelis 2001). Configurational confinement provided by nanoelements results in increased mechanical strength and thermal stability of the polymer composite. Thus, mechanically and thermally stable nanoelements well dispersed in the matrix provide beneficial effects. Formation of chemical bonds between polymer molecules and nanodiamond particles will also reduce the mobility of the chain molecules and improve the overall rigidity of the matrix resulting in enhanced thermal and mechanical properties. Importantly, nanodiamond-based nanoelements are expected to exert significant influence on polymer properties because the surface of nanodiamond elements contains a significant number of carbon sites for bonding with the matrix. Nanodiamond-based nanoelements are readily functionalized and it is possible to design DND surface functionalization that will result in cross-linking with polymer chains during nanocomposite curing.

MD simulations were performed to understand mechanisms responsible for improved mechanical properties of nanodiamond/polymer composites. Three systems were considered. The first was a pure polyethylene matrix. The second was a nanodiamond/polyethylene composite in which nanodiamonds were embedded into an amorphous polyethylene matrix. No chemical cross-links were created in the simulation between the polyethylene matrix and the nanodiamond particles. The third system was the same nanodiamond/polyethylene composite but with bonds between the nanodiamonds and polyethylene chains. After equilibration, the systems were uniaxially strained, and the stress was calculated, resulting in simulated stress–strain curves. Almost no enhancement relative to the pure polymer was observed in the simulated stress–strain curves for the systems containing the nanodiamonds when the strain was small (less than 2.5%). However, the stiffness was improved by adding the nanodiamonds into the polymer matrix when the strain was larger than ~2.5%. The composites with cross-links were slightly stiffer than the structures without cross-links when the strain was between ~2.5 and 13%. Interestingly, the trend reversed for strains larger than about 13%; the composite without the cross-links became slightly stiffer than that with cross-links at very large strains.

Mechanical and thermal properties of DNRs were predicted based on the results of atomistic modeling (Shenderova et al. 2003, 2005). Using properties for individual carbon–carbon bonds taken from *ab initio* calculations, Shenderova et al. (2003) estimated the elastic stiffness and tensile fracture force for DNRs with ⟨111⟩, ⟨011⟩, and ⟨001⟩ orientations along the DNR axis as a function of diameter. On the basis of these calculations, it was predicted that DNRs would have smaller loads for brittle fracture and smaller zero strain stiffness than SWCNTs for diameters less than 2 to 6 nm, depending on the DNR orientation. For larger diameters, however, both the fracture force and stiffness for DNRs was predicted to exceed those for CNTs. The result for small diameters is due to the high fracture force and stiffness of graphite carbon–carbon bonds compared to those in diamond. Above the diameter threshold, however, the larger number of bonds per unit cross-sectional area for the DNRs compared to the SWCNTs dominates the relative properties for the two structures. The weight-normalized ratio between the fracture forces of the DNRs and the SWCNTs is 0.68. At larger diameters, DNRs are stronger, but at the cost of a lower strength-to-weight ratio. In the case of MWCNTs, it was predicted that at all diameters the fracture force and stiffness for the CNTs is larger than that for the ⟨111⟩- and ⟨011⟩-oriented DNRs but smaller than that for a DNR with a ⟨001⟩ orientation.

Carbon nanostructures have potential for thermal management applications in polymer composites provided that heat can be effectively transferred between the polymer matrix and the carbon nanostructure. One possible way to accomplish this coupling is to cross-link the carbon nanostructure into the polymer matrix. In the case of the CNTs, however, MD simulations (discussed above) have suggested that phonon scattering from defects in the CNTs, created by the cross-links on the CNT walls, significantly reduces the thermal conductivity of the CNT (Padgett and Brenner 2004). Modeling suggests that enhancing heat coupling for this system through cross-linking sacrifices the thermal management properties of the CNT.

For characterization of the thermal transport properties of DNRs, a series of MD simulations of ⟨011⟩-(111)/(001) DNRs with and without chemisorbed species was carried out (Shenderova et al. 2005). A periodic boundary condition was applied along each rod axis, and the length of the supercell was varied to explore the dependence of thermal conductivity on DNR length. The DNRs were chemically functionalized by replacing surface hydrogen atoms with phenyl groups. For the degree of functionalization less than about 3.5%, and assuming a solid cylinder for the CNT, the thermal conductivity for the CNT is higher than that for the DNR. For higher degrees of surface functionalization, however, the DNR is predicted to have a higher thermal conductivity, e.g., the thermal transport in the DNR is less effected by functionalization than is transport in a CNT. Therefore, the modeling implies that DNRs may be more efficient at transporting heat out of a polymer matrix via chemical cross-linking than CNTs with a comparable volume fraction of the composite.

27.8 Remaining Challenges and Opportunities

There are a number of challenges to transitioning carbon-based nanocomposites from laboratory studies and limited applications to widespread use. One of these is the prohibitive cost of large quantities of CNTs. At current rates, e.g., the retail cost for purchasing low-quality SWCNTs is at least $50 per gram. This high cost is in large part due to the small yields of current synthesis methods. Similarly, processing and functionalizing large quantities of CNTs is an immature and hence expensive technology, and better processing methods are needed to produce well-dispersed and well-aligned nanotubes. These challenges, however, are largely of an engineering nature, with most of the enabling science now reasonably well understood. Because of this, the rapid development of new materials based on CNT composites is anticipated over the next decade.

Issues of affordability and processing also arise for the widespread use of composites containing nanodiamond particles and DNRs. Currently, there are commercial centers that produce bulk quantities of DND particles located in Russia, Ukraine, Belorussia, Germany, Japan, and China. Despite the commercial availability, however, the synthesis of monocrystalline nanodiamond particles of controllable size has yet to be achieved, and methods to avoid particle aggregation are still being optimized. Being carbon structures, however, much of the fundamental chemistry needed to functionalize these particles such that they can be well dispersed with strong particle–matrix bonding is known from decades of organic synthesis. Developing new composites from these structures is therefore also largely an engineering problem.

The scale-up production of DNRs with controllable properties is even further in its infancy, with cost-effective methods to produce bulk quantities of these structures largely lacking. As discussed above, however, initial experimental and modeling studies on these structures have yielded promising results, and as research in this area continues it is anticipated that reliable scale-up methods will be developed.

Acknowledgments

We thank S. Sinnott, D. Srivastava, D. Irving, and S.-J. Frankland for helpful discussions regarding several of the concepts and results discussed in this chapter. We also thank the National Science Foundation, the Department of Energy, NASA, and the Office of Naval Research for financial support.

References

Ajayan, P.M., J.M. Lambert, *et al.* (1993). Growth morphologies during cobalt-catalyzed single-shell carbon nanotube synthesis, *Chem. Phys. Lett.*, 215 (5), 509–517.

Ajayan, P.M., L.S. Schadler, *et al.* (2000). Single-walled carbon nanotube-polymer composites: strength and weakness, *Adv. Mater.*, 12 (10), 750–753.

Ajayan, P.M., L.S. Schadler, *et al.* (2003). *Nanocomposites Science and Technology*, Wiley-VCH, Weinheim.

Ajayan, P.M., O. Stephan, *et al.* (1994). Aligned carbon nanotube arrays formed by cutting a polymer resin–nanotube composite, *Science*, 265 (5176), 1212–1214.

Ajayan, P.M., V. Ravikumar, *et al.* (1998). Surface reconstructions and dimensional changes in single-walled carbon nanotubes, *Phys. Rev. Lett.*, 81 (7), 1437–1440.

Al-Haik, M., M.Y. Hussaini, *et al.* (2005). Adhesion energy in carbon nanotube–polyethylene composite: effect of chirality, *J. Appl. Phys.*, 97 (7).

An, K.H., J.G. Heo, *et al.* (2002). X-ray photoemission spectroscopy study of fluorinated single-walled carbon nanotubes, *Appl. Phys. Lett.*, 80 (22), 4235–4237.

An, L.N., W.X. Xu, *et al.* (2004). Carbon-nanotube-reinforced polymer-derived ceramic composites, *Adv. Mater.*, 16 (22), 2036+.

Badziag, P., W.S. Verwoerd, *et al.* (1990). Nanometre-sized diamonds are more stable than graphite, *Nature*, 343 (6255), 244–245.

Baik, E.S., Y.J. Baik, *et al.* (2000). Fabrication of diamond nano-whiskers, *Thin Solid Films*, 377, 295–298.

Banerjee, D., R. Sahoo, *et al.* (2005). Complex formation by bismuth and boron with fullerene (C_{60}): a reaction that opens up a novel route for synthesis of C_{60}-inorganic hybrid composites, *J. Mater. Res.*, 20 (5), 1113–1121.

Banerjee, K., S.J. Souri, *et al.* (2001). 3-D ICs: a novel chip design for improving deep-submicrometer interconnect performance and systems-on-chip integration, *Proc. IEEE*, 89 (5), 602–633.

Banhart, F. (2001). The formation of a connection between carbon nanotubes in an electron beam, *Nano Lett.*, 1 (6), 329–332.

Barber, A.H., S.R. Cohen, *et al.* (2003). Measurement of carbon nanotube–polymer interfacial strength, *Appl. Phys. Lett.*, 82 (23), 4140–4142.

Barnard, A.S. (2005). From nanodiamond to nanowires, in *Synthesis, Properties and Applications of Ultrananocrystalline Diamond*, D. Gruen, O. Shenderova, and A. Vul, Eds., Springer, Amsterdam, pp. 25–38.

Barrera, E.V., J. Sims, *et al.* (1994). Processing of fullerene-reinforced composites, *J. Mater. Res.*, 9 (10), 2662–2669.

Belytschko, T., S.P. Xiao, *et al.* (2002). Atomistic simulations of nanotube fracture, *Phys. Rev. B*, 65 (23), 235–430.

Berber, S., Y.K. Kwon, *et al.* (2000). Unusually high thermal conductivity of carbon nanotubes, *Phys. Rev. Lett.*, 84 (20), 4613–4616.

Bethune, D.S., C.H. Kiang, *et al.* (1993). Cobalt-catalyzed growth of carbon nanotubes with single-atomic layerwalls, *Nature*, 363 (6430), 605–607.

Bhattacharyya, S., C. Sinturel, *et al.* (2005). Protein-functionalized carbon nanotube–polymer composites, *Appl. Phys. Lett.*, 86 (11), 113104.

Biercuk, M.J., M.C. Llaguno, *et al.* (2002). Carbon nanotube composites for thermal management, *Appl. Phys. Lett.*, 80 (15), 2767–2769.

Blake, R., Y.K. Gun'ko, *et al.* (2004). A generic organometallic approach toward ultra-strong carbon nanotube polymer composites, *J. Am. Chem. Soc.*, 126 (33), 10226–10227.

Borrmann, T., K. Edgar, *et al.* (2004). Calcium silicate–carbon nanotube composites, *Curr. Appl. Phys.*, 4 (2–4), 359–361.

Bower, C., W. Zhu, *et al.* (2000). Plasma-induced alignment of carbon nanotubes, *Appl. Phys. Lett.*, 77 (6), 830–832.

Brabec, C.J., V. Dyakonov, *et al.* (1998). Investigation of photoexcitations of conjugated polymer/fullerene composites embedded in conventional polymers, *J. Chem. Phys.*, 109 (3), 1185–1195.

Bradshaw, R.D., F.T. Fisher, *et al.* (2003). Fiber waviness in nanotube-reinforced polymer composites. II. Modeling via numerical approximation of the dilute strain concentration tensor, *Compos. Sci. Technol.*, 63 (11), 1705–1722.

Brenner, D.W., O.A. Shenderova, *et al.* (2002a). Atomic modeling of carbon-based nanostructures as a tool for developing new materials and technologies, *Comput. Model. Eng. Sci.*, 3 (5), 643–673.

Brenner, D.W., O.A. Shenderova, *et al.* (2002b). A second-generation reactive empirical bond order (REBO) potential energy expression for hydrocarbons, *J. Phys.: Condens. Matter*, 14 (4), 783–802.

Cadek, M., J.N. Coleman, *et al.* (2002). Morphological and mechanical properties of carbon-nanotube-reinforced semicrystalline and amorphous polymer composites, *Appl. Phys. Lett.*, 81 (27), 5123–5125.

Calleja, F.J.B., L. Giri, *et al.* (1996). Structure and mechanical properties of polyethylene–fullerene composites, *J. Mater. Sci.*, 31 (19), 5153–5157.

Calvert, P. (1999). Nanotube composites — a recipe for strength, *Nature*, 399 (6733), 210–211.

Cassell, A.M., J.A. Raymakers, *et al.* (1999). Large scale CVD synthesis of single-walled carbon nanotubes, *J. Phys. Chem. B*, 103 (31), 6484–6492.

Chandra, N., S. Namilae, *et al.* (2004). Linking atomistic and continuum mechanics using multiscale models, in *Materials Processing and Design: Modeling, Simulation and Applications*, S. Ghosh, J.C. Castro, and J.K. Lee, Eds., American Institute of Physics, pp. 1571–1576.

Chang, T.E., L.R. Jensen, *et al.* (2005). Microscopic mechanism of reinforcement in single-wall carbon nanotube/polypropylene nanocomposite, *Polymer*, 46 (2), 439–444.

Che, J.W., T. Cagin, *et al.* (2000). Thermal conductivity of carbon nanotubes, *Nanotechnology*, 11 (2), 65–69.

Chen, J., M.A. Hamon, *et al.* (1998). Solution properties of single-walled carbon nanotubes, *Science*, 282 (5386), 95–98.

Chen, Q.D., L.M. Dai, *et al.* (2001). Plasma activation of carbon nanotubes for chemical modification, *J. Phys. Chem. B*, 105 (3), 618–622.

Cheng, H.M., F. Li, *et al.* (1998). Large-scale and low-cost synthesis of single-walled carbon nanotubes by the catalytic pyrolysis of hydrocarbons, *Appl. Phys. Lett.*, 72 (25), 3282–3284.

Chirvase, D., J. Parisi, *et al.* (2004). Influence of nanomorphology on the photovoltaic action of polymer–fullerene composites, *Nanotechnology*, 15 (9), 1317–1323.

Collins, A.T. (1989). Diamond electronic devices — a critical-appraisal, *Semicond. Sci. Technol.*, 4 (8), 605–611.

Cuomo, G.

da Silva, A.J.R., A. Fazzio, *et al.* (2005). Bundling up carbon nanotubes through Wigner defects, *Nano Lett.*, 5 (6), 1045–1049.

Dahl, J.E., S.G. Liu, *et al.* (2003). Isolation and structure of higher diamondoids, nanometer-sized diamond molecules, *Science*, 299 (5603), 96–99.

Dalton, A.B., S. Collins, *et al.* (2003). Super-tough carbon-nanotube fibres — these extraordinary composite fibres can be woven into electronic textiles, *Nature*, 423 (6941), 703.

Datsyuk, V., C. Guerret-Piecourt, *et al.* (2005). Double walled carbon nanotube/polymer composites via *in situ* nitroxide mediated polymerisation of amphiphilic block copolymers, *Carbon*, 43 (4), 873–876.

Deheer, W.A., A. Chatelain, *et al.* (1995). A carbon nanotube field-emission electron source, *Science*, 270 (5239), 1179–1180.

Delzeit, L., I. McAninch, *et al.* (2002). Growth of multiwall carbon nanotubes in an inductively coupled plasma reactor, *J. Appl. Phys.*, 91 (9), 6027–6033.

Desai, A.V. and M.A. Haque (2005). Mechanics of the interface for carbon nanotube–polymer composites, *Thin-Walled Struct.*, 43 (11), 1787–1803.

Devries, R.C. (1987). Synthesis of diamond under metastable conditions, *Annu. Rev. Mater. Sci.*, 17, 161–187.

Ding, W., A. Eitan, *et al.* (2003). Direct observation of polymer sheathing in carbon nanotube–polycarbonate composites, *Nano Lett.*, 3 (11), 1593–1597.

Dolmatov, V.Y. (2001). Detonation synthesis ultradispersed diamonds: properties and applications. *Russ. Chem. Rev.*, 70, 687–708.

Dolmatov, V.Y. (2006). Applications of UNCD particulate, in *Ultrananocrystalline Diamond: Synthesis, Properties and Applications*, O. Shenderova and D. Gruen, Eds., William-Andrew Publishing, Norwich, NY, Chap. 14.

Dong, S.R., J.P. Tu, *et al.* (2001). An investigation of the sliding wear behavior of Cu-matrix composite reinforced by carbon nanotubes, *Mater. Sci. Eng. A*, 313 (1–2), 83–87.

Dresselhaus, M.S. and P. Avouris (2001). Introduction to carbon materials research, in *Carbon Nanotubes: Synthesis, Properties and Applications*, M.S. Dresselhaus, G. Dresselhaus, and P. Avouris, Eds., Springer, Berlin, pp. 1–8.

Dresselhaus, M.S., G. Dresselhaus, *et al.* (1996). *Science of Fullerenes and Carbon Nanotubes*. Academic Press, San Diego.

Dubrovinskaia, N., L. Dubrovinsky, *et al.* (2005). Aggregated diamond nanorods, the densest and least compressible form of carbon, *Appl. Phys. Lett.*, 87 (8), 083106.

Ducati, C., I. Alexandrou, *et al.* (2002). Temperature selective growth of carbon nanotubes by chemical vapor deposition. *J. Appl. Phys.*, 92 (6), 3299–3303.

Dyakonov, V., G. Zoriniants, *et al.* (1999). Photoinduced charge carriers in conjugated polymer–fullerene composites studied with light-induced electron-spin resonance, *Phys. Rev. B*, 59 (12), 8019–8025.

Ebbesen, T.W. (1994). Carbon nanotubes, *Annu. Rev. Mater. Sci.*, 24, 235–264.

Ebbesen, T.W. (1996). Wetting, filling and decorating carbon nanotubes, *J. Phys. Chem. Sol.*, 57 (6–8), 951–955.

Ebbesen, T.W. (1997). Production and purification of carbon nanotubes, in *Carbon Nanotubes: Preparation and Properties*, T.W. Ebbesen, Eds., CRC Press, Boca Raton, FL, pp. 139–162.

Ebbesen, T.W. and P.M. Ajayan (1992). Large-scale synthesis of carbon nanotubes, *Nature*, 358 (6383), 220–222.

Ebbesen, T.W., H. Hiura, *et al.* (1993). Patterns in the bulk growth of carbon nanotubes, *Chem. Phys. Lett.*, 209 (1–2), 83–90.

Fagan, P.J., J.C. Calabrese, *et al.* (1992). Metal-complexes of buckminsterfullerene (C-60), *Acc. Chem. Res.*, 25 (3), 134–142.

Fahy, S., S.G. Louie, *et al.* (1987). Theoretical total-energy study of the transformation of graphite into hexagonal diamond. *Phys. Rev. B*, 35 (14), 7623–7626.

Fisher, F.T., R.D. Bradshaw, *et al.* (2003). Fiber waviness in nanotube-reinforced polymer composites-1: modulus predictions using effective nanotube properties, *Compos. Sci. Technol.*, 63 (11), 1689–1703.

Flahaut, E., A. Peigney, *et al.* (2000). Carbon nanotube-metal-oxide nanocomposites: microstructure, electrical conductivity and mechanical properties, *Acta Mater.*, 48 (14), 3803–3812.

Frankland, S.J.V., A. Caglar, *et al.* (2002). Molecular simulation of the influence of chemical cross-links on the shear strength of carbon nanotube–polymer interfaces, *J. Phys. Chem. B*, 106 (12), 3046–3048.

Frankland, S.J.V. and V.M. Harik (2003). Analysis of carbon nanotube pull-out from a polymer matrix, *Surf. Sci.*, 525 (1–3), L103–L108.

Frankland, S.J.V., V.M. Harik, *et al.* (2003). The stress–strain behavior of polymer–nanotube composites from molecular dynamics simulation, *Compos. Sci. Technol.*, 63 (11), 1655–1661.

Franklin, N.R., Q. Wang, *et al.* (2002). Integration of suspended carbon nanotube arrays into electronic devices and electromechanical systems, *Appl. Phys. Lett.*, 81 (5), 913–915.

Frenklach, M., S. Taki, *et al.* (1983). A conceptual-model for soot formation in pyrolysis of aromatic-hydrocarbons, *Combust. Flame*, 49 (1–3), 275–282.

Garg, A. and S.B. Sinnott (1998). Effect of chemical functionalization on the mechanical properties of carbon nanotubes, *Chem. Phys. Lett.*, 295 (4), 273–278.

Garibay-Febles, V., H.A. Calderon, *et al.* (2000). Production and characterization of (Al, Fe)-C (graphite or fullerene) composites prepared by mechanical alloying, *Mater. Manuf. Process.*, 15 (4), 547–567.

Gates, T.S., G.M. Odegard, *et al.* (2005). Computational materials: multi-scale modeling and simulation of nanostructured materials, *Compos. Sci. Technol.*, 65 (15–16), 2416–2434.

Gladchenko, S.V., G.A. Polotskaya, *et al.* (2002). The study of polystyrene–fullerene solid-phase composites, *Tech. Phys.*, 47 (1), 102–106.

Gojny, F.H., J. Nastalczyk, *et al.* (2003). Surface modified multi-walled carbon nanotubes in CNT/epoxy-composites, *Chem. Phys. Lett.*, 370 (5–6), 820–824.

Gojny, F.H., M.H.G. Wichmann, *et al.* (2004). Carbon nanotube-reinforced epoxy-composites: enhanced stiffness and fracture toughness at low nanotube content, *Compos. Sci. Technol.*, 64 (15), 2363–2371.

Gojny, F.H., M.H.G. Wichmann, *et al.* (2005). Influence of different carbon nanotubes on the mechanical properties of epoxy matrix composites — a comparative study, *Compos. Sci. Technol.*, 65 (15–16), 2300–2313.

Gong, X.Y., J. Liu, *et al.* (2000). Surfactant-assisted processing of carbon nanotube/polymer composites, *Chem. Mater.*, 12 (4), 1049–1052.

Gou, J.H., Z.Y. Liang, *et al.* (2005). Computational analysis of effect of single-walled carbon nanotube rope on molecular interaction and load transfer of nanocomposites, *Composites Part B*, 36 (6–7), 524–533.

Graff, R.A., J.P. Swanson, *et al.* (2005). Achieving individual-nanotube dispersion at high loading in single-walled carbon nanotube composites, *Adv. Mater.*, 17 (8), 980–984.

Graham, A.P., G.S. Duesberg, *et al.* (2005). How do carbon nanotubes fit into the semiconductor roadmap? *Appl. Phys A*, 80 (6), 1141–1151.

Greiner, N.R., D.S. Phillips, *et al.* (1988). Diamonds in detonation soot, *Nature*, 333 (6172), 440–442.

Griebel, M. and J. Hamaekers (2004). Molecular dynamics simulations of the elastic moduli of polymer–carbon nanotube composites, *Comput. Meth. Appl. Mech. Eng.*, 193 (17–20), 1773–1788.

Gruen, D., L. Curtiss, *et al.* (2005). Synthesis of ultrananocrystalline diamond/nanotube self-composites by direct insertion of carbon dimer molecules into carbon bonds, European Diamond Conference, Toulouse, France.

Gruen, D., O. Shenderova, *et al.* (2005). *Synthesis, Properties and Applications of Ultrananocrystalline Diamond*, Springer, Amsterdam.

Guo, T., P. Nikolaev, *et al.* (1995a). Self-assembly of tubular fullerenes, *J. Phys. Chem.*, 99 (27), 10694–10697.

Guo, T., P. Nikolaev, *et al.* (1995b). Catalytic growth of single-walled nanotubes by laser vaporization, *Chem. Phys. Lett.*, 243 (1–2), 49–54.

Haddon, R.C., A.F. Hebard, *et al.* (1991). Conducting films of C60 and C70 by alkali-metal doping, *Nature*, 350 (6316), 320–322.

Haggenmueller, R., H.H. Gommans, *et al.* (2000). Aligned single-wall carbon nanotubes in composites by melt processing methods, *Chem. Phys. Lett.*, 330 (3–4), 219–225.

Hamon, M.A., J. Chen, *et al.* (1999). Dissolution of single-walled carbon nanotubes, *Adv. Mater.*, 11 (10), 834–840.

Hamwi, A., H. Alvergnat, *et al.* (1997). Fluorination of carbon nanotubes, *Carbon*, 35 (6), 723–728.

Han, C.D. and K.W. Lem (1983). Rheology of unsaturated polyester resins. 1. Effects of filler and low-profile additive on the rheological behavior of unsaturated polyester resin, *J. Appl. Polym. Sci.*, 28 (2), 743–762.

Hata, K., D.N. Futaba, *et al.* (2004). Water-assisted highly efficient synthesis of impurity-free single-walled carbon nanotubes, *Science*, 306 (5700), 1362–1364.

Hawkins, J.M. (1992). Osmylation of C-60 — proof and characterization of the soccer-ball framework. *Acc. Chem. Res.*, 25 (3), 150–156.

Heimann, R.B., S.E. Evsyukov, *et al.* (1997). Carbon allotropes: a suggested classification scheme based on valence orbital hybridization, *Carbon*, 35 (10–11), 1654–1658.

Hiura, H., T.W. Ebbesen, *et al.* (1995). Opening and purification of carbon nanotubes in high yields, *Adv. Mater.*, 7 (3), 275–276.

Ho, G.W., A.T.S. Wee, *et al.* (2001). Synthesis of well-aligned multiwalled carbon nanotubes on Ni catalyst using radio frequency plasma-enhanced chemical vapor deposition, *Thin Solid Films*, 388 (1–2), 73–77.

Hone, J., M.C. Llaguno, *et al.* (2000). Electrical and thermal transport properties of magnetically aligned single wall carbon nanotube films, *Appl. Phys. Lett.*, 77 (5), 666–668.

Hone, J., M. Whitney, *et al.* (1999). Thermal conductivity of single-walled carbon nanotubes, *Phys. Rev. B*, 59 (4), R2514–R2516.

Hu, Y., I. Jang, *et al.* (2003). Modification of carbon nanotube–polystyrene matrix composites through polyatomic-ion beam deposition: predictions from molecular dynamics simulations, *Compos. Sci. Technol.*, 63 (11), 1663–1669.

Hu, Y.H. and S.B. Sinnott (2004). Molecular dynamics simulations of polyatomic-ion beam deposition-induced chemical modification of carbon nanotube/polymer composites, *J. Mater. Chem.*, 14 (4), 719–729.

Huang, Z.P., J.W. Wu, *et al.* (1998). Growth of highly oriented carbon nanotubes by plasma-enhanced hot filament chemical vapor deposition, *Appl. Phys. Lett.*, 73 (26), 3845–3847.

Huhtala, M., A.V. Krasheninnikov, *et al.* (2004). Improved mechanical load transfer between shells of multiwalled carbon nanotubes, *Phys. Rev. B*, 70 (4), 045404.

Hwang, G.L. and K.C. Hwang (2001). Carbon nanotube reinforced ceramics, *J. Mater. Chem.*, 11 (6), 1722–1725.

Iijima, S. (1991). Helical microtubules of graphitic carbon, *Nature*, 354 (6348), 56–58.

Iijima, S. and T. Ichihashi (1993). Single-shell carbon nanotubes of 1-nm diameter, *Nature*, 363 (6430), 603–605.

Iijima, S., C. Brabec, *et al.* (1996). Structural flexibility of carbon nanotubes, *J. Chem. Phys.*, 104 (5), 2089–2092.

Inagaki, M. (2000). *New Carbons*, Elsevier, Amsterdam.

Jang, I., S.B. Sinnott, *et al.* (2004). Molecular dynamics simulation study of carbon nanotube welding under electron beam irradiation, *Nano Lett.*, 4 (1), 109–114.

Jin, L., C. Bower, *et al.* (1998). Alignment of carbon nanotubes in a polymer matrix by mechanical stretching, *Appl. Phys. Lett.*, 73 (9), 1197–1199.

Journet, C., W.K. Maser, *et al.* (1997). Large-scale production of single-walled carbon nanotubes by the electric-arc technique, *Nature*, 388 (6644), 756–758.

Kelly, K.F., I.W. Chiang, *et al.* (1999). Insight into the mechanism of sidewall functionalization of single-walled nanotubes: an STM study, *Chem. Phys. Lett.*, 313 (3–4), 445–450.

Khare, B., M. Meyyappan, *et al.* (2003). Proton irradiation of carbon nanotubes, *Nano Lett.*, 3 (5), 643–646.

Kiang, C.H., W.A. Goddard, *et al.* (1996). Structural modification of single-layer carbon nanotubes with an electron beam, *J. Phys. Chem.*, 100 (9), 3749–3752.

Kis, A., G. Csanyi, *et al.* (2004). Reinforcement of single-walled carbon nanotube bundles by intertube bridging, *Nat. Mater.*, 3 (3), 153–157.

Ko, F.K., S. Khan, *et al.* (2002). Structure and properties of carbon nanotube reinforced nanocomposites, 43rd AIAA/ASME/ASCE/AHS Structures, Structural Dynamics, and Materials Conference, Denver, CO.

Krasheninnikov, A.V., K. Nordlund, *et al.* (2001). Formation of ion-irradiation-induced atomic-scale defects on walls of carbon nanotubes, *Phys. Rev. B*, 63 (24), 245–405.

Kratschmer, W., L.D. Lamb, *et al.* (1990). Solid C_{60} — a new form of carbon, *Nature*, 347 (6291), 354–358.

Kroto, H.W., J.R. Heath, *et al.* (1985). C_{60}: buckminsterfullerene, *Nature*, 318 (6042), 162–163.

Kuzumaki, T., K. Miyazawa, *et al.* (1998). Processing of carbon nanotube reinforced aluminum composite, *J. Mater. Res.*, 13 (9), 2445–2449.

Lagoudas, D.C. and G.D. Seidel (2004). Effective elastic properties of carbon nanotube reinforced composites, 45th AIAA/ASME/ASCE/AHS Structures, Structural Dynamics, and Materials Conference, Palm Springs, CA.

Laurent, C., A. Peigney, *et al.* (1998). Carbon nanotubes Fe alumina nanocomposites. Part II. Microstructure and mechanical properties of the hot-pressed composites, *J. Eur. Ceram. Soc.*, 18 (14), 2005–2013.

Lee, C.J., S.C. Lyu, *et al.* (2001). Diameter-controlled growth of carbon nanotubes using thermal chemical vapor deposition, *Chem. Phys. Lett.*, 341 (3–4), 245–249.

Lee, K. and H. Kim (2004). Polymer photovoltaic cells based on conjugated polymer–fullerene composites, *Curr. Appl. Phys.*, 4 (2–4), 323–326.

Li, C.Y. and T.W. Chou (2003). Multiscale modeling of carbon nanotube reinforced polymer composites, *J. Nanosci. Nanotechnol.*, 3 (5), 423–430.

Li, W.Z., S.S. Xie, *et al.* (1996). Large-scale synthesis of aligned carbon nanotubes, *Science*, 274 (5293), 1701–1703.

Liao, K. and S. Li (2001). Interfacial characteristics of a carbon nanotube–polystyrene composite system, *Appl. Phys. Lett.*, 79 (25), 4225–4227.

Lin, H.Z., H.M. Huang, *et al.* (2003). Photophysics and applications in plastic solar cells of conjugated polymer/fullerene composites, *Polym. Compos.*, 11 (8), 679–689.

Liu, J., A.G. Rinzler, *et al.* (1998). Fullerene pipes, *Science*, 280 (5367), 1253–1256.

Liu, L.Q. and H.D. Wagner (2005). Rubbery and glassy epoxy resins reinforced with carbon nanotubes, *Compos. Sci. Technol.*, 65 (11–12), 1861–1868.

Liu, T.X., I.Y. Phang, *et al.* (2004). Morphology and mechanical properties of multiwalled carbon nanotubes reinforced nylon-6 composites, *Macromolecules*, 37 (19), 7214–7222.

Lordi, V. and N. Yao (2000). Molecular mechanics of binding in carbon-nanotube–polymer composites, *J. Mater. Res.*, 15 (12), 2770–2779.

Lourie, O. and H.D. Wagner (1999). Evidence of stress transfer and formation of fracture clusters in carbon nanotube-based composites, *Compos. Sci. Technol.*, 59 (6), 975–977.

Lupo, F., R. Kamalakaran, *et al.* (2004). Microstructural investigations on zirconium oxide–carbon nanotube composites synthesized by hydrothermal crystallization, *Carbon*, 42 (10), 1995–1999.

Lusti, H.R. and A.A. Gusev (2004). Finite element predictions for the thermoelastic properties of nanotube reinforced polymers, *Modell. Simul. Mater. Sci. Eng.*, 12 (3), S107–S119.

Maiti, A., J. Wescott, *et al.* (2005). Nanotube–polymer composites: insights from Flory–Huggins theory and mesoscale simulations, *Mol. Simul.*, 31 (2–3), 143–149.

Marcoux, P.R., J. Schreiber, *et al.* (2002). A spectroscopic study of the fluorination and defluorination reactions on single-walled carbon nanotubes, *Phys. Chem. Chem. Phys.*, 4 (11), 2278–2285.

Martin, C.A., J.K.W. Sandler, *et al.* (2005). Electric field-induced aligned multi-wall carbon nanotube networks in epoxy composites, *Polymer*, 46 (3), 877–886.

Marumoto, K., Y. Muramatsu, *et al.* (2003). Light-induced ESR studies of polarons in regioregular poly(3-alkylthiophene)-fullerene composites, *Synth. Met.*, 135 (1–3), 433–434.

Maruyama, S. (2002). A molecular dynamics simulation of heat conduction in finite length SWNTs, *Physica B*, 323 (1–4), 193–195.

McCarthy, B., J.N. Coleman, *et al.* (2001). Complex nano-assemblies of polymers and carbon nanotubes, *Nanotechnology*, 12 (3), 187–190.

McCarthy, B., J.N. Coleman, *et al.* (2002). A microscopic and spectroscopic study of interactions between carbon nanotubes and a conjugated polymer, *J. Phys. Chem. B*, 106 (9), 2210–2216.

Meyyappan, M. (2005). Growth: CVD and PECVD, in *Carbon Nanotubes: Science and Applications*, M. Meyyappan, Ed., CRC Press, Boca Raton, FL, pp. 99–116.

Mickelson, E.T., C.B. Huffman, *et al.* (1998). Fluorination of single-wall carbon nanotubes, *Chem. Phys. Lett.*, 296 (1–2), 188–194.

Miller, E.K., K. Lee, *et al.* (1997). Observation of photoinduced charge transfer in conducting polymer fullerene composites using a high-bandgap polymer, *Synth. Met.*, 84 (1–3), 631–632.

Mingo, N. and D.A. Broido (2005). Carbon nanotube ballistic thermal conductance and its limits, *Phys. Rev. Lett.*, 95 (9).

Mintmire, J.W., B.I. Dunlap, *et al.* (1992). Are fullerene tubules metallic, *Phys. Rev. Lett.*, 68 (5), 631–634.

Moravsky, A.P., E.M. Wexler, *et al.* (2005). Growth of carbon nanotubes by arc discharge and laser ablation, in *Carbon Nanotubes: Science and Applications*, M. Meyyappan, Ed., CRC Press, Boca Raton, FL, pp. 65–97.

Mylvaganam, K. and L.C. Zhang (2004). Chemical bonding in polyethylene-nanotube composites: a quantum mechanics prediction, *J. Phys. Chem. B*, 108 (17), 5217–5220.

Namilae, S. and N. Chandra (2005). Multiscale model to study the effect of interfaces in carbon nanotube-based composites, *J. Eng. Mater. Technol.*, 127 (2), 222–232.

Namilae, S., N. Chandra, *et al.* (2004). Mechanical behavior of functionalized nanotubes, *Chem. Phys. Lett.*, 387 (4–6), 247–252.

Nguyen, C.V., Q. Ye, *et al.* (2005). Carbon nanotube tips for scanning probe microscopy: fabrication and high aspect ratio nanometrology, *Measure. Sci. Technol.*, 16 (11), 2138–2146.

Ni, B. and S.B. Sinnott (2000). Chemical functionalization of carbon nanotubes through energetic radical collisions, *Phys. Rev. B*, 61 (24), R16343–R16346.

Ni, B., R. Andrews, *et al.* (2001). A combined computational and experimental study of ion-beam modification of carbon nanotube bundles, *J. Phys. Chem. B*, 105 (51), 12719–12725.

Nikolaev, P., M.J. Bronikowski, *et al.* (1999). Gas-phase catalytic growth of single-walled carbon nanotubes from carbon monoxide, *Chem. Phys. Lett.*, 313 (1–2), 91–97.

Noguchi, T., A. Magario, *et al.* (2004). Carbon nanotube/aluminium composites with uniform dispersion, *Mater. Trans.*, 45 (2), 602–604.

Odegard, G.M., S.J.V. Frankland, *et al.* (2003). The effect of chemical functionalization on mechanical properties of nanotube/polymer composites, 44th AIAA/ASME/ASCE/AHS Structures, Structural Dynamics, and Materials Conference, Norfolk, VA.

Odegard, G.M., T.S. Gates, *et al.* (2003). Constitutive modeling of nanotube-reinforced polymer composites, *Compos. Sci. Technol.*, 63 (11), 1671–1687.

Olah, G.A., I. Bucsi, *et al.* (1992). Chemical-reactivity and functionalization of C_{60} and C_{70} fullerenes, *Carbon*, 30 (8), 1203–1211.

Olek, M., J. Ostrander, *et al.* (2004). Layer-by-layer assembled composites from multiwall carbon nanotubes with different morphologies, *Nano Lett.*, 4 (10), 1889–1895.

Osman, M.A. and D. Srivastava (2001). Temperature dependence of the thermal conductivity of single-wall carbon nanotubes, *Nanotechnology*, 12 (1), 21–24.

Padgett, C.W. and D.W. Brenner (2004). Influence of chemisorption on the thermal conductivity of single-wall carbon nanotubes, *Nano Lett.*, 4 (6), 1051–1053.

Park, H.S. and W.K. Liu (2004). An introduction and tutorial on multiple-scale analysis in solids, *Comput. Meth. Appl. Mech. Eng.*, 193 (17–20), 1733–1772.

Pasimeni, L., L. Franco, *et al.* (2001a). Evidence of high charge mobility in photoirradiated poly-thiophene–fullerene composites, *J. Mater. Chem.*, 11 (4), 981–983.

Pasimeni, L., M. Ruzzi, *et al.* (2001b). Spin correlated radical ion pairs generated by photoinduced electron transfer in composites of sexithiophene/fullerene derivatives: a transient EPR study, *Chem. Phys.*, 263 (1), 83–94.

Peigney, A., C. Laurent, *et al.* (1998). Carbon nanotubes Fe alumina nanocomposites. Part I. Influence of the Fe content on the synthesis of powders, *J. Eur. Ceram. Soc.*, 18 (14), 1995–2004.

Pipes, R.B. and P. Hubert (2002). Helical carbon nanotube arrays: mechanical properties, *Compos. Sci. Technol.*, 62 (3), 419–428.

Pipes, R.B. and P. Hubert (2003a). Helical carbon nanotube arrays: thermal expansion, *Compos. Sci. Technol.*, 63 (11), 1571–1579.

Pipes, R.B. and P. Hubert (2003b). Scale effects in carbon nanostructures: self-similar analysis, *Nano Lett.*, 3 (2), 239–243.

Plank, N.O.V., J. Liudi, *et al.* (2003). Fluorination of carbon nanotubes in CF_4 plasma, *Appl. Phys. Lett.*, 83 (12), 2426–2428.

Potschke, P., A.R. Bhattacharyya, *et al.* (2004). Melt mixing of polycarbonate with multiwalled carbon nanotubes: microscopic studies on the state of dispersion, *Eur. Polym. J.*, 40 (1), 137–148.

Qian, D., E.C. Dickey, *et al.* (2000). Load transfer and deformation mechanisms in carbon nanotube–polystyrene composites, *Appl. Phys. Lett.*, 76 (20), 2868–2870.

Qin, L.C. and S. Iijima (1997). Structure and formation of raft-like bundles of single-walled helical carbon nanotubes produced by laser evaporation, *Chem. Phys. Lett.*, 269 (1–2), 65–71.

Ramanathan, T., H. Liu, *et al.* (2005). Functionalized SWCNT/polymer nanocomposites for dramatic property improvement, *J. Polym. Sci. Part B*, 43 (17), 2269–2279.

Raravikar, N.R., L.S. Schadler, *et al.* (2005). Synthesis and characterization of thickness-aligned carbon nanotube–polymer composite films, *Chem. Mater.*, 17 (5), 974–983.

Raty, J.Y., G. Galli, *et al.* (2003). Quantum confinement and fullerene-like surface reconstructions in nanodiamonds, *Phys. Rev. Lett.*, 90 (3), 037401.

Regev, O., P.N.B. ElKati, *et al.* (2004). Preparation of conductive nanotube–polymer composites using latex technology, *Adv. Mater.*, 16 (3), 248–251.

Ren, Z.F., Z.P. Huang, *et al.* (1998). Synthesis of large arrays of well-aligned carbon nanotubes on glass, *Science*, 282 (5391), 1105–1107.

Rinzler, A.G., J. Liu, *et al.* (1998). Large-scale purification of single-wall carbon nanotubes: process, product, and characterization, *Appl. Phys. A*, 67 (1), 29–37.

Robertson, D.H., D.W. Brenner, *et al.* (1992). Energetics of nanoscale graphitic tubules, *Phys. Rev. B*, 45 (21), 12592–12595.

Ruan, S.L., P. Gao, *et al.* (2003). Toughening high performance ultrahigh molecular weight polyethylene using multiwalled carbon nanotubes, *Polymer*, 44 (19), 5643–5654.

Safadi, B., R. Andrews, *et al.* (2002). Multiwalled carbon nanotube polymer composites: synthesis and characterization of thin films, *J. Appl. Polym. Sci.*, 84 (14), 2660–2669.

Saito, R., G. Dresselhaus, *et al.* (1998). *Physical Properties of Carbon Nanotubes*, Imperial College Press, London.

Saito, T., K. Matsushige, *et al.* (2002). Chemical treatment and modification of multi-walled carbon nanotubes, *Physica B*, 323 (1–4), 280–283.

Salonen, E., A.V. Krasheninnikov, *et al.* (2002). Ion-irradiation-induced defects in bundles of carbon nanotubes, *Nucl. Instrum. Meth. Phys. Res. Sect. B*, 193, 603–608.

Salvetat, J.P., G.A.D. Briggs, *et al.* (1999). Elastic and shear moduli of single-walled carbon nanotube ropes, *Phys. Rev. Lett.*, 82 (5), 944–947.

Sammalkorpi, M., A.V. Krasheninnikov, *et al.* (2005). Irradiation-induced stiffening of carbon nanotube bundles. *Nucl. Instrum. Meth. Phys. Res. Sect. B*, 228, 142–145.

Sandler, J., M.S.P. Shaffer, *et al.* (1999). Development of a dispersion process for carbon nanotubes in an epoxy matrix and the resulting electrical properties, *Polymer*, 40 (21), 5967–5971.

Sariciftci, N.S., L. Smilowitz, *et al.* (1992). Photoinduced electron-transfer from a conducting polymer to buckminsterfullerene, *Science*, 258 (5087), 1474–1476.

Schadler, L.S., S.C. Giannaris, *et al.* (1998). Load transfer in carbon nanotube epoxy composites, *Appl. Phys. Lett.*, 73 (26), 3842–3844.

Sen, R., A. Govindaraj, *et al.* (1997). Carbon nanotubes by the metallocene route, *Chem. Phys. Lett.*, 267 (3–4), 276–280.

Sensfuss, S., A. Konkin, *et al.* (2003). Optical and ESR studies on poly(3-alkylthiophene)/fullerene composites for solar cells, *Synth. Met.*, 137 (1–3), 1433–1434.

Shaffer, M.S.P. and A.H. Windle (1999). Fabrication and characterization of carbon nanotube/poly(vinyl alcohol) composites, *Adv. Mater.*, 11 (11), 937–941.

Shenderova, O.A., D. Areshkin, *et al.* (2003). Bonding and stability of hybrid diamond/nanotube structures, *Mol. Simul.*, 29 (4), 259–268.

Shenderova, O., D. Brenner, *et al.* (2003). Would diamond nanorods be stronger than fullerene nanotubes? *Nano Lett.*, 3 (6), 805–809.

Shenderova, O. and D. Gruen (2006). *Ultrananocrystalline Diamond: Synthesis, Properties and Applications*, Willliam-Andrew Publishing, Norwich, NY.

Shenderova, O., Z. Hu, *et al.* (2005). Carbon family at the nanoscale, in *Synthesis, Properties and Applications of Ultrananocrystalline Diamond*, D. Gruen, O. Shenderova, and A. Vul, Eds., Springer, Amsterdam, pp. 1–14.

Shenderova, O. and G. McGuire.

Shenderova, O. and G. McGuire (2006). Nanocrystalline diamond, in *Handbook of Nanomaterials*, Y. Gogotsi, Ed., CRC Press, Boca Raton, FL, pp. 201–235.

Shenderova, O.A., C.W. Padgett, *et al.* (2005). Diamond nanorods, *J. Vac. Sci. Technol. B*, 23 (6), 2457–2464.

Shenderova, O.A., V.V. Zhirnov, *et al.* (2002). Carbon nanostructures, *Crit. Rev. Solid State Mater. Sci.*, 27 (3–4), 227–356.

Shenogina, N., S. Shenogin, *et al.* (2005). On the lack of thermal percolation in carbon nanotube composites, *Appl. Phys. Lett.*, 87 (13), 133106.

Simeoni, M., C. De Luca, *et al.* (2005). Interaction between zigzag single-wall carbon nanotubes and polymers: a density-functional study, *J. Chem. Phys.*, 122 (21), 214710.

Sinnott, S.B. and R. Andrews (2001). Carbon nanotubes: synthesis, properties, and applications, *Crit. Rev. Solid State Mater. Sci.*, 26 (3), 145–249.

Skakalova, V., U. Dettlaff-Weglikowska, *et al.* (2005). Electrical and mechanical properties of nanocomposites of single wall carbon nanotubes with PMMA, *Synth. Met.*, 152 (1–3), 349–352.

Song, Y.S. and J.R. Youn (2004). Properties of epoxy nanocomposites filled with carbon nanomaterials, *E-Polymers*, 080.

Srivastava, D., D.W. Brenner, *et al.* (1999). Predictions of enhanced chemical reactivity at regions of local conformational strain on carbon nanotubes: kinky chemistry, *J. Phys. Chem. B*, 103 (21), 4330–4337.

Stahl, H., J. Appenzeller, *et al.* (2000). Intertube coupling in ropes of single-wall carbon nanotubes, *Phys. Rev. Lett.*, 85 (24), 5186–5189.

Stephan, C., T.P. Nguyen, *et al.* (2000). Characterization of singlewalled carbon nanotubes-PMMA composites, *Synth. Met.*, 108 (2), 139–149.

Stevens, J.L., A.Y. Huang, *et al.* (2003). Sidewall amino-functionalization of single-walled carbon nanotubes through fluorination and subsequent reactions with terminal diamines, *Nano Lett.*, 3 (3), 331–336.

Su, M., B. Zheng, *et al.* (2000). A scalable CVD method for the synthesis of single-walled carbon nanotubes with high catalyst productivity, *Chem. Phys. Lett.*, 322 (5), 321–326.

Su, W.P., J.R. Schrieffer, *et al.* (1979). Solitons in polyacetylene, *Phys. Rev. Lett.*, 42 (25), 1698–1701.

Suhr, J., N. Koratkar, *et al.* (2005). Viscoelasticity in carbon nanotube composites, *Nat. Mater.*, 4 (2), 134–137.

Tang, B.Z. and H.Y. Xu (1999). Preparation, alignment, and optical properties of soluble poly(phenylacetylene)-wrapped carbon nanotubes, *Macromolecules*, 32 (8), 2569–2576.

Taylor, R. and D.R.M. Walton (1993). The chemistry of fullerenes, *Nature*, 363 (6431), 685–693.

Terrones, M. (2003). Science and technology of the twenty-first century: synthesis, properties and applications of carbon nanotubes, *Annu. Rev. Mater. Res.*, 33, 419–501.

Terrones, M., H. Terrones, *et al.* (2000). Coalescence of single-walled carbon nanotubes, *Science*, 288 (5469), 1226–1229.

Thess, A., R. Lee, *et al.* (1996). Crystalline ropes of metallic carbon nanotubes, *Science*, 273 (5274), 483–487.

Thostenson, E.T. and T.W. Chou (2003). On the elastic properties of carbon nanotube-based composites: modelling and characterization, *J. Phys. D*, 36 (5), 573–582.

Thostenson, E.T., Z.F. Ren, *et al.* (2001). Advances in the science and technology of carbon nanotubes and their composites: a review, *Compos. Sci. Technol.*, 61 (13), 1899–1912.

Touhara, H., J. Inahara, *et al.* (2002). Property control of new forms of carbon materials by fluorination, *J. Fluorine Chem.*, 114 (2), 181–188.

Tyler, T., V.V. Zhirnov, *et al.* (2003). Electron emission from diamond nanoparticles on metal tips, *Appl. Phys. Lett.*, 82 (17), 2904–2906.

Uchida, T. and S. Kumar (2005). Single wall carbon nanotube dispersion and exfoliation in polymers, *J. Appl. Polym. Sci.*, 98 (3), 985–989.

Vaia, R.A. and E.P. Giannelis (2001). Polymer nanocomposites: status and opportunities, *MRS Bull.*, 26 (5), 394–401.

Viswanathan, G., N. Chakrapani, *et al.* (2003). Single-step *in situ* synthesis of polymer-grafted single-wall nanotube composites, *J. Am. Chem. Soc.*, 125 (31), 9258–9259.

von Helden, G., M.T. Hsu, *et al.* (1993). Carbon cluster cations with up to 84 atoms — structures, formation mechanism, and reactivity, *J. Phys. Chem.*, 97 (31), 8182–8192.

Wagner, H.D., O. Lourie, *et al.* (1998). Stress-induced fragmentation of multiwall carbon nanotubes in a polymer matrix, *Appl. Phys. Lett.*, 72 (2), 188–190.

Wang, Y. (1992). Photoconductivity of fullerene-doped polymers, *Nature*, 356 (6370), 585–587.

Wei, C.Y., D. Srivastava, *et al.* (2002). Thermal expansion and diffusion coefficients of carbon nanotube–polymer composites, *Nano Lett.*, 2 (6), 647–650.

Wei, C.Y., D. Srivastava, *et al.* (2004). Structural ordering in nanotube polymer composites, *Nano Lett.*, 4 (10), 1949–1952.

White, C.T., D.H. Robertson, *et al.* (1993). Helical and rotational symmetries of nanoscale graphitic tubules, *Phys. Rev. B*, 47 (9), 5485–5488.

White, C.T. and T.N. Todorov (1998). Carbon nanotubes as long ballistic conductors, *Nature*, 393 (6682), 240–242.

Wildoer, J.W.G., L.C. Venema, *et al.* (1998). Electronic structure of atomically resolved carbon nanotubes, *Nature*, 391 (6662), 59–62.

Wudl, F. (1992). The chemical-properties of buckminsterfullerene (C_{60}) and the birth and infancy of fulleroids, *Acc. Chem. Res.*, 25 (3), 157–161.

Xiao, T. and K. Liao (2004). A nonlinear pullout model for unidirectional carbon nanotube-reinforced composites, *Composites Part B*, 35 (3), 211–217.

Yang, M.J., V. Koutsos, *et al.* (2005). Interactions between polymers and carbon nanotubes: a molecular dynamics study, *J. Phys. Chem. B*, 109 (20), 10009–10014.

Yang, S.Y., J.R. Castilleja, *et al.* (2004). Thermal analysis of an acrylonitrile-butadiene-styrene/SWNT composite, *Polym. Degrad. Stab.*, 83 (3), 383–388.

Yarris, L. (1993). *Carbon Cages: LBL Scientists Study Fullerenes.*

Yi, W., L. Lu, *et al.* (1999). Linear specific heat of carbon nanotubes, *Phys. Rev. B*, 59 (14), R9015–R9018.

Yoon, J., C.Q. Ru, *et al.* (2003). Vibration of an embedded multiwall carbon nanotube, *Compos. Sci. Technol.*, 63 (11), 1533–1542.

Yoshino, K., H. Kajii, *et al.* (1999). Electrical and optical properties of conducting polymer–fullerene and conducting polymer–carbon nanotube composites, *Fullerene Sci. Technol.*, 7 (4), 695–711.

Yudasaka, M., F. Kokai, *et al.* (1999). Formation of single-wall carbon nanotubes: comparison of CO_2 laser ablation and Nd:YAG laser ablation, *J. Phys. Chem. B*, 103 (18), 3576–3581.

Yudasaka, M., T. Ichihashi, *et al.* (1998). Roles of laser light and heat in formation of single-wall carbon nanotubes by pulsed laser ablation of $C_xNi_yCo_y$ targets at high temperature, *J. Phys. Chem. B*, 102 (50), 10201–10207.

Yudasaka, M., T. Komatsu, *et al.* (1997). Single-wall carbon nanotube formation by laser ablation using double-targets of carbon and metal, *Chem. Phys. Lett.*, 278 (1–3), 102–106.

Zakhidov, A.A., H. Araki, *et al.* (1996). Fullerene-conducting polymer composites: intrinsic charge transfer processes and doping effects, *Synth. Met.*, 77 (1–3), 127–137.

Zerza, G., C.J. Brabec, *et al.* (2001). Ultrafast charge transfer in conjugated polymer–fullerene composites, *Synth. Met.*, 119 (1–3), 637–638.

Zhan, G.D. and A.K. Mukherjee (2004). Carbon nanotube reinforced alumina-based ceramics with novel mechanical, electrical, and thermal properties, *Int. J. Appl. Ceram. Technol.*, 1 (2), 161–171.

Zhan, G.D., J.D. Kuntz, *et al.* (2003). Single-wall carbon nanotubes as attractive toughening agents in alumina-based nanocomposites, *Nat. Mater.*, 2 (1), 38–42.

Zhang, D.W., L. Shen, *et al.* (2004). Carbon nanotubes reinforced nylon-6 composite prepared by simple melt-compounding, *Macromolecules*, 37(2), 256–259.

Zhang, J., Y. Cheng, *et al.* (2005). A nanotube-based field emission x-ray source for microcomputed tomography, *Rev. Sci. Instrum.*, 76 (9), 094301.

Zhang, J.M., M. Tanaka, *et al.* (2004). A simplified approach for heat conduction analysis of CNT-based nano-composites, *Comput. Meth. Appl. Mech. Eng.*, 193 (52), 5597–5609.

Zhang, L., V.U. Kiny, *et al.* (2004). Sidewall functionalization of single-walled carbon nanotubes with hydroxyl group-terminated moieties, *Chem. Mater.*, 16 (11), 2055–2061.

Zhang, X.T., J. Zhang, *et al.* (2004). Surfactant-directed polypyrrole/CNT nanocables: synthesis, characterization, and enhanced electrical properties, *Chem. Phys. Chem.*, 5 (7), 998–1002.

Zhang, Y. and S. Iijima (1999). Formation of single-wall carbon nanotubes by laser ablation of fullerenes at low temperature, *Appl. Phys. Lett.*, 75 (20), 3087–3089.

Zhang, Y., H. Gu, *et al.* (1998). Single-wall carbon nanotubes synthesized by laser ablation in a nitrogen atmosphere, *Appl. Phys. Lett.*, 73 (26), 3827–3829.

Zhang, Y.C. and X. Wang (2005). Thermal effects on interfacial stress transfer characteristics of carbon nanotubes/polymer composites, *Int. J. Solids Struct.*, 42 (20), 5399–5412.

Zhu, H.W., C.L. Xu, *et al.* (2002). Direct synthesis of long single-walled carbon nanotube strands, *Science*, 296 (5569), 884–886.

Zhu, J., H.Q. Peng, *et al.* (2004). Reinforcing epoxy polymer composites through covalent integration of functionalized nanotubes, *Adv. Funct. Mater.*, 14 (7), 643–648.

Zhu, J., J.D. Kim, *et al.* (2003). Improving the dispersion and integration of single-walled carbon nanotubes in epoxy composites through functionalization, *Nano Lett.*, 3 (8), 1107–1113.

28

Contributions of Molecular Modeling to Nanometer-Scale Science and Technology

Donald W. Brenner
Olga A. Shenderova
J.D. Schall
D.A. Areshkin
S. Adiga
North Carolina State University

J.A. Harrison
United States Naval Academy

S.J. Stuart
Clemson University

Opening Remarks 28-1
28.1 Molecular Simulations 28-2
    Interatomic Potential Energy Functions and Forces
    • The Workhorse Potential Energy Functions • Quantum Basis
    of Some Analytic Potential Functions
28.2 First-Principles Approaches: Forces on the Fly 28-12
    Other Considerations in Molecular Dynamics Simulations
28.3 Applications 28-13
    Pumps, Gears, Motors, Valves, and Extraction Tools
    • Nanometer-Scale Properties of Confined Fluids
    • Nanometer-Scale Indentation • New Materials and Structures
28.4 Concluding Remarks 28-27
Acknowledgments 28-28
References 28-28

Opening Remarks

Molecular modeling has played a key role in the development of our present understanding of the details of many chemical processes and molecular structures. From atomic-scale modeling, for example, much of the details of drug interactions, energy transfer in chemical dynamics, frictional forces, and crack propagation dynamics are now known, to name just a few examples. Similarly, molecular modeling has played a central role in developing and evaluating new concepts related to nanometer-scale science and technology. Indeed, molecular modeling has a long history in nanotechnology of both leading the way in terms of what in principle is achievable in nanometer-scale devices and nanostructured materials, and in explaining experimental data in terms of fundamental processes and structures.

There are two goals for this chapter. The first is to educate scientists and engineers who are considering using molecular modeling in their research, especially as applied to nanometer-scale science and technology, beyond the black box approach sometimes facilitated by commercial modeling codes. Particular emphasis is placed on the physical basis of some of the more widely used analytic potential energy expressions. The choice of bonding expression is often the first choice to be made in instituting a molecular model, and a proper choice of expression can be crucial to obtaining meaningful results. The second goal of this chapter is to discuss some examples of systems that have been modeled and the type of information that has been obtained from these simulations. Because of the enormous breadth of molecular modeling studies related to nanometer-scale science and technology, we have not attempted a comprehensive literature survey. Instead, our intent is to stir the imagination of researchers by presenting a variety of examples that illustrate what molecular modeling has to offer.

28.1   Molecular Simulations

The term *molecular modeling* has several meanings, depending on the perspective of the user. In the chemistry community, this term often implies molecular statics or dynamics calculations that use as interatomic forces a valence-force-field expression plus nonbonded interactions (see below). An example of a bonding expression of this type is Allinger's molecular mechanics (Burkert, 1982; Bowen, 1991). The term molecular modeling is also sometimes used as a generic term for a wider array of atomic-level simulation methods (e.g., Monte Carlo modeling in addition to molecular statics/dynamics) and interatomic force expressions. In this chapter, the more generic definition is assumed, although the examples discussed below emphasize molecular dynamics simulations.

In molecular dynamics calculations, atoms are treated as discrete particles whose trajectories are followed by numerically integrating classical equations of motion subject to some given interatomic forces. The numerical integration is carried out in a stepwise fashion, with typical timesteps ranging from 0.1 to about 15 femtoseconds depending on the highest vibrational frequency of the system modeled. Molecular statics calculations are carried out in a similar fashion, except that minimum energy structures are determined using either integration of classical equations of motion with kinetic energies that are damped, by steepest descent methods, or some other equivalent numerical method.

In Monte Carlo modeling, *snapshots* of a molecular system are generated, and these configurations are used to determine the system's properties. Time-independent properties for equilibrium systems are usually generated either by weighting the contribution of a snapshot to a thermodynamic average by an appropriate Boltzmann factor or, more efficiently, by generating molecular configurations with a probability that is proportional to their Boltzmann factor. The Metropolis algorithm, which relies on Markov chain theory, is typically used for the latter. In kinetic Monte Carlo modeling, a list of possible dynamic events and the relative rate for each event is typically generated given a molecular structure. An event is then chosen from the list with a probability that is proportional to the inverse of the rate (i.e., faster rates have higher probabilities). The atomic positions are then appropriately updated according to the chosen event, the possible events and rates are updated, and the process is repeated to generate a time-dependent trajectory.

The interatomic interactions used in molecular modeling studies are calculated either from (1) a sum of nuclear repulsions combined with electronic interactions determined from some first principles or semi-empirical electronic structure technique, or (2) an expression that replaces the quantum mechanical electrons with an energy and interatomic forces that depend only on atomic positions. The approach for calculating atomic interactions in a simulation typically depends on the system size (i.e., the number of electrons and nuclei), available computing resources, the accuracy with which the forces need to be known to obtain useful information, and the availability of an appropriate potential energy function. While the assumption of classical dynamics in molecular dynamics simulations can be severe, especially for systems involving light atoms and other situations in which quantum effects (like tunneling) are important, molecular modeling has proven to be an extremely powerful and versatile computational technique.

For convenience, the development and applications of molecular dynamics simulations can be divided into four branches of science: chemistry, statistical mechanics, materials science, and molecular biology. Illustrated in Figure 28.1 are some of the highlights of molecular dynamics simulation as applied to problems in each of these fields. The first study using classical mechanics to model a chemical reaction was published by Eyring and coworkers in 1936, who used a classical trajectory (calculated by hand) to model the chemical reaction $H+H_2 -> H_2+H$ (Hirschfelder, 1936). Although the potential energy function was crude by current standards (it produced a stable H_3 molecule), the calculations

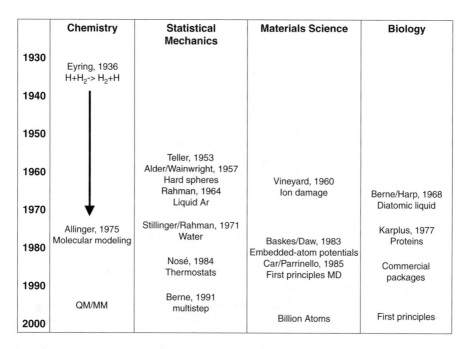

	Chemistry	Statistical Mechanics	Materials Science	Biology
1930				
	Eyring, 1936 $H+H_2 \rightarrow H_2+H$			
1940				
1950				
		Teller, 1953		
1960		Alder/Wainwright, 1957 Hard spheres	Vineyard, 1960 Ion damage	
		Rahman, 1964 Liquid Ar		Berne/Harp, 1968 Diatomic liquid
1970				
	Allinger, 1975 Molecular modeling	Stillinger/Rahman, 1971 Water		Karplus, 1977 Proteins
1980			Baskes/Daw, 1983 Embedded-atom potentials	
		Nosé, 1984 Thermostats	Car/Parrinello, 1985 First principles MD	Commercial packages
1990				
	QM/MM	Berne, 1991 multistep		
2000			Billion Atoms	First principles

FIGURE 28.1 Some highlights in the field of molecular dynamics simulations.

themselves set the standard for numerous other applications of classical trajectories to understanding the dynamics of chemical reactions. In the 1950s and early 1960s, the first dynamic simulations of condensed phase dynamics were carried out. These early calculations, which were performed primarily at National Laboratories due to the computational resources available at these facilities, were used mainly to test and develop statistical mechanical descriptions of correlated many-particle motion (Rahman, 1964). A major breakthrough in the application of molecular modeling to statistical mechanics occurred in the 1980s with the derivation of thermostats that not only maintain an average kinetic energy corresponding to a desired temperature but, more importantly, produce kinetic energy *fluctuations* that correctly reproduce those of a desired statistical mechanical ensemble (Nosé, 1984).

The first reported application of molecular modeling to materials science was by Vineyard and coworkers, who in 1960 reported simulations of ions impinging on a solid (Gibson, 1960). The development since then of many-body potential energy functions that capture many of the details of bonding in metals and covalent materials, together with the ability to describe forces using first-principles methods, has allowed molecular modeling to make seminal contributions to our understanding of the mechanical properties of materials.

The application of molecular modeling techniques to biological systems did not begin in earnest until the 1970s, when computers began to get sufficiently powerful to allow simulations of complex hetero-molecules. This research was facilitated by the availability of commercial modeling packages a decade later, which allowed research in areas such as drug design to move from the academic research laboratory to the drug companies. The search for new drugs with specific biological activities continues to be an extremely active area of application for molecular simulation (Balbes, 1994).

28.1.1 Interatomic Potential Energy Functions and Forces

Much of the success of an atomic simulation relies on the use of an appropriate model for the interatomic forces. For many cases relatively simple force expressions are adequate, while for simulations from which quantitative results are desired, very accurate forces may be needed. Because of the importance of the interatomic force model, several of the more widely used potential energy expressions and the derivation of some of these models from quantum mechanics are discussed in detail in the following two subsections.

28.1.2　The Workhorse Potential Energy Functions

There are available in the literature hundreds, if not thousands, of analytic interatomic potential energy functions. Out of this plethora of potential energy functions have emerged about a dozen *workhorse* potentials that are widely used by the modeling community. Most of these potential functions satisfy the so-called *Tersoff test*, which has been quoted by Garrison and Srivastava as being important to identifying particularly effective potential functions (Garrison, 1995). This test is (a) has the person who constructed the potential subsequently refined the potential based on initial simulations? and (b) has the potential been used by other researchers for simulations of phenomena for which the potential was not designed? A corollary to these tests is that either the functional form of the potential should be straightforward to code, or that an implementation of the potential be widely available to the modeling community. An example, which is discussed in more detail below, is the Embedded-Atom Method (EAM) potentials for metals that were originally developed by Baskes, Daw, and Foiles in the early 1980s (Daw, 1983). Although there were closely related potential energy functions developed at about the same time, most of which are still used by modelers, the EAM potentials remain the most widely used functions for metals, due at least in part to the developers' making their computer source code readily available to researchers.

The quantum mechanical basis of most of the potential energy functions discussed in this section is presented in more detail in the next section, and therefore some of the details of the various analytic forms are given in that section. Rather than providing a formal discussion of these potentials, the intent of this section is to introduce these expressions as practical solutions to describing interatomic forces for large-scale atomic simulations. The workhorse potential functions can be conveniently classified according to the types of bonding that they most effectively model, and therefore this section is organized according to bonding type.

28.1.2.1　Metals

A number of closely related potential energy expressions are widely used to model bonding in metals. These expressions are the EAM (Daw, 1983), effective medium theory (Jacobsen, 1987; Jacobsen, 1988), the glue model (Ercolessi 1988), the Finnis–Sinclair (Finnis, 1984), and Sutton–Chen potentials (Sutton, 1990). Although they all have very similar forms, the motivation for each is not the same. The first three models listed are based on the concept of *embedding* atoms within an electron gas. The central idea is that the energy of an atom depends on the density of the electron gas near the embedded atom and the mutually repulsive pairwise interaction between the atomic cores of the embedded atom and the other atoms in the system. In the EAM and glue models, the electron density is taken as a pairwise sum of contributions from surrounding atoms at the site of the atom whose energy is being calculated. There are thus three important components of these potentials: the contribution of electron density from a neighboring atom as a function of distance, the pairwise-additive interatomic repulsive forces, and the embedding function relating the electron density to the energy. Each of these can be considered adjustable functions that can be fit to various properties such as the lattice constant and crystal cohesive energy. In contrast, effective medium theory attempts to use the average of the electron density in the vicinity of the atoms whose energy is being calculated, and a less empirical relationship between this electron density and the energy (Jacobsen, 1988).

Although the functional forms for the Finnis–Sinclair and Sutton–Chen potentials are similar to other metal potentials, the physical motivation is different. In these cases, the functional form is based on the so-called *second moment approximation* that relates the binding energy of an atom to its local coordination through the spread in energies of the local density of electronic states due to chemical binding (see the next section for more details). Ultimately, though, the functional form is very similar to the electron-density-based potentials. The main difference is that the EAM and related potentials were originally developed for face-centered cubic metals, while the Finnis–Sinclair potentials originally focused on body-centered cubic metals.

28.1.2.2　Covalent Bonding

Molecular structure calculations typically used by chemists have relied heavily on expressions that describe intramolecular bonding as an expansion in bond lengths and angles (typically called valence-force fields) in combination with nonbonded interactions. The nonbonded interactions typically use pair-additive functions that mimic van der Waals forces (e.g., Lennard–Jones potentials) and Coulomb forces due to partial charges. Allinger and coworkers developed the most widely recognized version of this type of expression (Burkert, 1982; Bowen, 1991). There are now numerous variations on this approach, most of which are tuned to some particular application such as liquid structure or protein dynamics. While very accurate energies and structures can be obtained with these potentials, a drawback to this approach is that the valence-force expressions typically use a harmonic expansion about the minimum energy configuration that does not go to the noninteraction limit as bond distances get large.

A major advance in modeling interatomic interactions for covalent materials was made with the introduction of a potential energy expression for silicon by Stillinger and Weber (Stillinger, 1985). The basic form is similar to the valence-force expressions in that the potential energy is given as a sum of two-body bond-stretching and three-body angle-bending terms. The Stillinger–Weber potential, however, produces the correct dissociation limits in both the bend and stretch terms and reproduces a wide range of properties of both solid and molten silicon. It opened a wide range of silicon and liquid phenomena to molecular simulation and also demonstrated that a well-parameterized analytic potential can be useful for describing bond-breaking and bond-forming chemistry in the condensed phase. The Stillinger–Weber potential is one of the few breakthrough potentials that does not satisfy the first of the Tersoff rules mentioned above in that a better parameterized form was not introduced by the original developers for silicon. This is a testament to the careful testing of the potential function as it was being developed. It is also one of the few truly successful potentials whose form is not directly derivable from quantum mechanics as described in the next section.

Tersoff introduced another widely used potential for covalent materials (Tersoff, 1986; Tersoff, 1989). Initially introduced for silicon, and subsequently for carbon, germanium, and their alloys, the Tersoff potentials are based on a quantum mechanical analysis of bonding. Two key features of the Tersoff potential function are that the same form is used for structures with high and low atomic coordination numbers, and that the bond angle term comes from a fit to these structures and does not assume a particular orbital hybridization. This is significantly different from both the valence-force type expressions, for which an atomic hybridization must be assumed to define the potential parameters, and the Stillinger–Weber potential, which uses an expansion in the angle bend terms around the tetrahedral angle. Building on the success of the Tersoff form, Brenner introduced a similar expression for hydrogen and carbon that uses a single potential form to model both solid-state and molecular structures (Brenner, 1990). Although not as accurate in its predictions of atomic structures and energies as expressions such as Allinger's molecular mechanics, the ability of this potential form to model bond breaking and forming with appropriate changes in atomic hybridization in response to the local environment has led to a wide range of applications. Addition of nonbonded terms into the potential has widened the applicability of this potential form to situations for which intermolecular interactions are important, such as in simulations of molecular solids and liquids (Stuart, 2000).

To model covalent bonding, Baskes and coworkers extended the EAM formalism to include bond angle terms (Baskes, 1989). While the potential form is compelling, these modified EAM forms have not been as widely used as the other potential forms for covalent bonding described above.

Increases in computing power and the development of increasingly clever algorithms over the last decade have made simulations with forces taken from electronic structure calculations routine. Forces from semi-empirical tight-bonding models, for example, are possible with relatively modest computing resources and therefore are now widely used. Parameterizations by Sutton and coworkers, as well as by Wang, Ho, and their coworkers, have become de facto standards for simulations of covalent materials (Xu, 1992; Kwon, 1994; Sutton, 1996).

28.1.2.3 Ionic Bonding

Materials in which bonding is chiefly ionic have not to date played as large a role in nanometer-scale science and technology as have materials with largely covalent or metallic bonding. Therefore, fewer potentials for ionic materials have been developed and applied to model nanoscale systems. The standard for bulk ionic materials is the shell model, a functional form that includes polarization by allowing ionic charges of fixed magnitude to relax away from the nuclear position in response to a local electric field. This model is quite successful for purely ionic systems, but it does not allow charge transfer. Many ionic systems of interest in nanoscale material science involve at least partial covalent character and, thus, require a potential that models charge transfer in response to environmental conditions. One of the few such potentials was introduced by Streitz and Mintmire for aluminum oxide (Streitz, 1994). In this formalism, charge transfer and electrostatic interactions are calculated using an electronegativity equalization approach (Rappé, 1991) and added to EAM type forces. The development of similar formalisms for a wider range of materials is clearly needed, especially for things like piezoelectric systems that may have unique functionality at the nanometer scale.

28.1.3　Quantum Basis of Some Analytic Potential Functions

The functional form of the majority of the workhorse potential energy expressions used in molecular modeling studies of nanoscale systems can be traced to density functional theory (DFT). Two connections between the density functional equations and analytic potentials as well as specific approaches that have been derived from these analyses are presented in this section. The first connection is the Harris functional, from which tight-binding, Finnis–Sinclair, and empirical bond-order potential functions can be derived. The second connection is effective medium theory that leads to the EAM.

The fundamental principle behind DFT is that all ground-state properties of an interacting system of electrons with a nondegenerate energy in an external potential can be uniquely determined from the electron density. This is a departure from traditional quantum chemical methods that attempt to find a many-body wavefunction from which all properties (including electron density) can be obtained. In DFT the ground-state electronic energy is a unique functional of the electron density, which is minimized by the correct electron density (Hohenberg, 1964). This energy functional consists of contributions from the electronic kinetic, Coulombic, and exchange–correlation energy. Because the form of the exchange–correlation functional is not known, DFT does not lead to a direct solution of the many-body electron problem. From the viewpoint of the development of analytic potential functions for condensed phases, however, DFT is a powerful concept because the electron density is the central quantity rather that a wavefunction, and electron densities are much easier to approximate with an analytic function than are wavefunctions.

The variational principle of DFT leads to a system of one-electron equations of the form:

$$[T + V_H(\mathbf{r}) + V_N(\mathbf{r}) + V_{xc}(\mathbf{r})] \; \phi_i^{K-S} = \varepsilon_i \; \phi_i^{K-S} \tag{28.1}$$

that can be self-consistently solved (Kohn, 1965). In this expression T is a kinetic energy operator, $V_H(\mathbf{r})$ is the Hartree potential, $V_N(\mathbf{r})$ is the potential due to the nuclei, and ε_i and ϕ_i^{K-S} are the eigenenergies and eigenfunctions of the one-electron Kohn–Sham orbitals, respectively. The exchange–correlation potential, $V_{xc}(\mathbf{r})$, is the functional derivative of the exchange-correlation energy E_{xc}:

$$V_{xc}(\mathbf{r}) = \delta E_{xc}[\rho(\mathbf{r})]/\delta\rho(\mathbf{r}) \tag{28.2}$$

where $\rho(\mathbf{r})$ is the charge density obtained from the Kohn–Sham orbitals. When solved self-consistently, the electron densities obtained from Equation (28.1) can be used to obtain the electronic energy using the expression:

$$E^{KS}[\rho^{sc}(\mathbf{r})] = \sum_k \varepsilon_k - \int \rho^{sc}[V_H(\mathbf{r})/2 + V_{xc}(\mathbf{r})] \; d\mathbf{r} + E_{xc}[\rho^{sc}(\mathbf{r})] \tag{28.3}$$

where ρ^{sc} is the self-consistent electron density and ε_k is the eigenvalues of the one-electron orbitals. The integral on the right side of Equation (28.3) corrects for the fact that the eigenvalue sum includes the exchange–correlation energy and double-counts the electron–electron Coulomb interactions. These are sometimes referred to as the double-counting terms.

A strength of DFT is that the error in electronic energy is second-order in the difference between a given electron density and the true ground-state density. Working independently, Harris as well as Foulkes and Haydock showed that the electronic energy calculated from a single iteration of the energy functional,

$$E^{Harris}(\rho^{in}(\mathbf{r})) = \sum_k \varepsilon_k^{out} - \int \rho^{in}[V_H^{in}(\mathbf{r})/2 + V_{xc}^{in}(\mathbf{r})] \; d\mathbf{r} + E_{xc}[\rho^{in}(\mathbf{r})] \tag{28.4}$$

is also second-order in the error in charge density (Harris, 1985; Foulkes, 1989). This differs from the usual density functional equation in that while the Kohn–Sham orbital energies ε_k^{out} are still calculated, the double-counting terms involve only an input charge density and not the density given by these orbitals. Therefore this functional, generally referred to as the *Harris functional*, yields two significant computational benefits over a full density functional calculation — the input electron density can be chosen to simplify the calculation of the double-counting terms, and it does not require self-consistency. This property has made Harris functional calculations useful as a relatively less computationally intensive variation of DFT compared with fully self-consistent calculations (despite the fact that the Harris functional is not variational). As discussed by Foulkes and Haydock, the Harris functional can be used as a basis for deriving tight-binding potential energy expressions (Foulkes, 1989).

The tight-binding method refers to a non-self-consistent semi-empirical molecular orbital approximation that dates back to before the development of density functional theory (Slater, 1954). The tight-binding method can produce trends in (and in some cases quantitative values of) electronic properties; and if appropriately parameterized with auxiliary functions, it can produce binding energies, bond lengths, and vibrational properties that are relatively accurate and that are transferable within a wide range of structures, including bulk solids, surfaces, and clusters.

Typical tight-binding expressions used in molecular modeling studies give the total energy E_{tot} for a system of atoms as a sum of eigenvalues ε of a set of occupied non-self-consistent one-electron molecular orbitals plus some additional analytic function A of relative atomic distances:

$$E_{tot} = A + \sum_k \varepsilon_k \tag{28.5}$$

The idea underlying this expression is that quantum mechanical effects are roughly captured through the eigenvalue calculation, while the analytic function applies corrections to approximations in the electronic energy calculation needed to obtain reasonable total energies. The simplest and most widespread tight-binding expression obtains the eigenvalues of the electronic states from a wavefunction that is expanded in an orthonormal minimal basis of short-range atom-centered orbitals. One-electron molecular orbital coefficients and energies are calculated using the standard secular equation as done for traditional molecular orbital calculations. Rather than calculating multi-center Hamiltonian matrix elements, however, these matrix elements are usually taken as two-center terms that are fit to reproduce electronic properties such as band structures; or they are sometimes adjusted along with the analytic function to enhance the transferability of total energies between different types of structures. For calculations involving disordered structures and defects, the dependence on distance of the two-center terms must also be specified.

A common approximation is to assume a pairwise additive sum over atomic distances for the analytic function:

$$A = \sum_i \sum_j \theta(r_{ij}) \tag{28.6}$$

where r_{ij} is the scalar distance between atoms i and j. The function $\theta(r_{ij})$ models Coulomb repulsions between positive nuclei that are screened by core electrons, plus corrections to the approximate quantum mechanics. While a pairwise sum may be justified for the interatomic repulsion between nuclei and core electrons, there is little reason to assume that it can compensate for all of the approximations inherent in tight-binding theory. Nevertheless, the tight-binding approximation appears to work well for a range of covalent materials.

Several widely used potentials have attempted to improve upon the standard tight-binding approach. Rather than use a simple pairwise sum for the analytic potential, for example, Xu et al. assumed for carbon the multi-center expression:

$$A = P\left[\sum_i \sum_j \theta(r_{ij})\right] \tag{28.7}$$

where P represents a polynomial function and $\theta(r_{ij})$ is an exponential function splined to zero at a distance between the second and third nearest neighbors in a diamond lattice (Xu, 1992). Xu et al. fit the pairwise terms and the polynomial to different solid-state and molecular structures. The resulting potential produces binding energies that are transferable to a wide range of systems, including small clusters, fullerenes, diamond surfaces, and disordered carbon.

A number of researchers have suggested that for further enhancement of the transferability of tight-binding expressions, the basis functions should not be assumed to be orthogonal. This requires additional parameters describing the overlap integrals. Several methods have been introduced for determining these. Menon and Subbaswamy, for example, have used proportionality expressions between the overlap and Hamiltonian matrix elements from Hückel theory (Menon, 1993). Frauenheim and coworkers have calculated Hamiltonian and overlap matrix elements directly from density functional calculations within the local density approximation (Frauenheim, 1995). This approach is powerful because complications associated with an empirical fit are eliminated, yet the relative computational simplicity of a tight-binding expression is retained.

Foulkes and Haydock have used the Harris functional to justify the success of the approximations used in tight-binding theory (Foulkes, 1989). First, taking the molecular orbitals used in tight-binding expressions as corresponding to the Kohn–Sham orbitals created from an input charge density justifies the use of non-self-consistent energies in tight-binding theory. Second, if the input electron density is approximated with a sum of overlapping, atom-centered spherical orbitals, then the double-counting terms in the Harris functional are given by

$$\sum_a C_a + \frac{1}{2} \sum_i \sum_j U(r_{ij}) + U_{np} \tag{28.8}$$

where C_a is a constant intra-atomic energy, $U_{ij}(r_{ij})$ is a short-range pairwise additive energy that depends on the scalar distance between atoms i and j, and U_{np} is a non-pairwise additive contribution that comes from the exchange–correlation functional. Haydock and Foulkes showed that if the regions where overlap of electron densities from three or more atoms are small, the function U_{np} is well approximated by a pairwise sum that can be added to U_{ij}, justifying the assumption of pair-additivity for the function A in Equation (28.6). Finally, the use of spherical atomic orbitals leads to the simple form:

$$V_{xc}(\mathbf{r}) = \sum_i V_i(\mathbf{r}) + U(\mathbf{r}) \tag{28.9}$$

for the one-electron potential used to calculate the orbital energies in the Harris functional. The function $V_i(\mathbf{r})$ is an additive atomic term that includes contributions from core electrons as well as Hartree and exchange–correlation potentials, and $U(\mathbf{r})$ comes from nonlinearities in the exchange–correlation functional. Although not two-centered, the contribution of the latter term is relatively small. Thus, the use of strictly two-center matrix elements in the tight-binding Hamiltonian can also be justified.

Further approximations building on tight-binding theory, namely the second-moment approximation, can be used to arrive at some of the other analytic potential energy functions that are widely used in molecular modeling studies of nanoscale systems (Sutton, 1993). In a simplified quantum mechanical picture, the formation of chemical bonds is due to the splitting of atomic orbital energies as molecular orbitals are formed when atoms are brought together. For condensed phases, the energy and distribution of molecular orbitals among atoms can be conveniently described using a local density of states. The local density of states is defined for a given atom as the number of electronic states in the interval between energy e and e+δe weighted by the "amount" of the orbital on the atom. For molecular orbitals expanded in an orthonormal linear basis of atomic orbitals on each atom, the weight is the sum of the squares of the linear expansion coefficients for the atomic orbitals centered on the atom of interest. The electronic bond energy associated with an individual atom can be defined as twice (two electrons per orbital) the integral over energy of the local density of states multiplied by the orbital energies, where the upper limit of the integral is the Fermi energy. With this definition, the sum of the energies associated with each atom is equal to twice the sum of the energies of the occupied molecular orbitals.

An advantage of using a density of states rather than directly using orbital energies to obtain an electronic energy is that like any distribution, the properties of the local density of states can often be conveniently described using just a few moments of the distribution. For example, as stated above, binding energies are associated with a spread in molecular orbital energies; therefore the second moment of the local density of states can be related to the potential energy of an atom. Similarly, formation of a band gap in a half-filled energy band can be described by the fourth moment (the Kurtosis) of the density of states that quantifies the amount of a distribution in the middle compared with that in the wings of the distribution.

There is a very powerful theorem, called the *moments theorem*, that relates the moments of the local density of states to the bonding topology (Cyrot–Lackmann, 1968; Sutton, 1993). This theorem states that the *nth moment of the local density of states on an atom i is determined by the sum of all paths of n hops between neighboring atoms that start and end at atom i.* Because the paths that determine the second moment involve only two hops, one hop from the central atom to a neighbor atom and one hop back, according to this theorem the second moment of the local densities of states is determined by the number of nearest neighbors. As stated above, the bond energy of an atom can be related to the spread in energy of the molecular orbitals relative to the atomic orbitals out of which molecular orbitals are constructed. The spread in energy is given by the square root of the second moment, and therefore it is reasonable to assume that the electronic contribution to the bond energy E_i of a given atom i is proportional to the square root of the number of neighbors z:

$$E_i \propto z^{1/2} \tag{28.10}$$

This result is called the second-moment approximation.

The definition of neighboring atoms needs to be addressed to develop an analytic potential function from the second-moment approximation. Exponential functions of distance can be conveniently used to count neighbors (these functions mimic the decay of electronic densities with distance). Including a proportionality constant between the electronic bond energy and the square root of the number of neighbors, and adding pairwise repulsive interactions between atoms to balance the electronic energy, yields the Finnis–Sinclair analytic potential energy function for atom i:

$$E_i = \sum_j 1/2 A e^{-\alpha r_{ij}} - \left[\sum_j B e^{-\beta r_{ij}} \right]^{1/2} \tag{28.11}$$

where the total potential energy is the sum of the atomic energies (Finnis, 1984). This is a particularly simple expression that captures much of the essence of quantum mechanical bonding.

Another variation on the second-moment approximation is the Tersoff expression for describing covalent bonding (Tersoff, 1986; Tersoff, 1989). Rather than describe the original derivation due to Abell (Abell, 1985), which differs from that outlined above, one can start with the Finnis–Sinclair form for the electronic energy of a given atom i and arrive at a simplified Tersoff expression through the following algebra (Brenner, 1989):

$$E_i^{el} = -B \left(\sum_j e^{-\beta r_{ij}} \right)^{1/2} \tag{28.12a}$$

$$= -B \left(\sum_j e^{-\beta r_{ij}} \right)^{1/2} \times \left[\left(\sum_j e^{-\beta r_{ij}} \right)^{1/2} \Big/ \left(\sum_k e^{-\beta r_{ik}} \right)^{1/2} \right] \tag{28.12b}$$

$$= -B \left(\sum_j [e^{-\beta r_{ij}}] \right) \times \left(\sum_k e^{-\beta r_{ik}} \right)^{-1/2} \tag{28.12c}$$

$$= \sum_j \left[-B e^{-\beta r_{ij}} \times \left(\sum_k e^{-\beta r_{ik}} \right)^{-1/2} \right] \tag{28.12d}$$

$$= \sum_j \left[-B e^{-\beta r_{ij}} \times \left(e^{-\beta r_{ij}} + \sum_{k \neq i,j} e^{-\beta r_{ik}} \right)^{-1/2} \right] \tag{28.12e}$$

$$= \sum_{j} \left[-B\, e^{-\beta/2\, r_{ij}} \times \left(1 + \sum_{k \neq i,j} e^{-\beta\, (r_{ik} - r_{ij})} \right)^{-1/2} \right] \tag{28.12f}$$

This derivation leads to an attractive energy expressed by a two-center pair term of the form:

$$-B e^{-\beta/2\, r_{ij}} \tag{28.13}$$

that is modulated by an analytic bond order function of the form:

$$\left(1 + \sum_{k \neq i,j} e^{-\beta\, (r_{ik} - r_{ij})} \right)^{-1/2} \tag{28.14}$$

While mathematically the same form as the Finnis–Sinclair potential, the Tersoff form of the second-moment approximation lends itself to a slightly different physical interpretation. The value of the bond order function decreases as the number of neighbors of an atom increases. This in turn decreases the magnitude of the pair term and effectively mimics the limited number of valence electrons available for bonding. At the same time as the number of neighbors increases, the number of bonds modeled through the pair term increases, the structural preference for a given material — e.g., from a molecular solid with a few neighbors to a close-packed solid with up to 12 nearest neighbors — depends on a competition between increasing the number of bonds to neighboring atoms and bond energies that decrease with increasing coordination number.

In the Tersoff bond-order form, an angular function is included in the terms in the sum over k in Equation (28.14), along with a few other more subtle modifications. The net result is a set of robust potential functions for group IV materials and their alloys that can model a wide range of bonding configurations and associated orbital hybridizations relatively efficiently. Brenner and coworkers have extended the bond-order function by adding *ad hoc* terms that model radical energetics, rotation about double bonds, and conjugation in hydrocarbon systems (Brenner, 1990). Further modifications by Stuart, Harrison, and associates have included non-bonded interactions into this formalism (Stuart, 2000).

Considerable additional effort has gone into developing potential energy expressions that include angular interactions and higher moments since the introduction of the Finnis–Sinclair and Tersoff potentials (Carlsson, 1991; Foiles, 1993). Carlsson and coworkers, for example, have introduced a matrix form for the moments of the local density of states from which explicit environment-dependent angular interactions can be obtained (Carlsson, 1993). The role of the fourth moment, in particular, has been stressed for half-filled bands because, as mentioned above, it describes the tendency to introduce an energy gap. Pettifor and coworkers have introduced a particularly powerful formalism that produces analytic functions for the moments of a distribution that has recently been used in atomic simulations (Pettifor, 1999; Pettifor, 2000). While none of these potential energy expressions has achieved the workhorse status of the expressions discussed above, specific applications of these models, particularly the Pettifor model, have been very promising. It is expected that the use of this and related formalisms for modeling nanometer-scale systems will significantly increase in the next few years.

A different route through which analytic potential energy functions used in nanoscale simulations have been developed is effective medium theory (Jacobsen, 1987; Jacobsen, 1988). The basic idea behind this approach is to replace the relatively complex environment of an atom in a solid with that of a simplified host. The electronic energy of the true solid is then constructed from accurate energy calculations on the simplified medium. In a standard implementation of effective medium theory, the simplified host is a homogeneous electron gas with a compensating positive background (the so-called *jellium model*). Because of the change in electrostatic potential, an atom embedded in a jellium alters the initially homogeneous electron density. This difference in electron density with and without the embedded atom can be calculated within the local density approximation of DFT and expressed as a spherical function about the atom embedded in the jellium. This function depends on both the identity of the embedded atom and the initial density of the homogeneous electron gas. The overall electron density of the solid can then be approximated by a superposition of the perturbed electron densities. With this *ansatz*, however, the differences in electron density are not specified until the embedding electron density associated with each atom is known. For solids containing defects Norskov, Jacobsen, and coworkers have suggested using spheres centered at each site with radii chosen so that the electronic charge within each sphere cancels the charge of each atomic nucleus (Jacobsen, 1988). For most metals, Jacobsen has shown that to a good approximation the average embedding density is related exponentially to the sphere radius, and these relationships for 30 elements have been tabulated.

Within this set of assumptions, for an imperfect crystal the average embedding electron density at each atomic site is not defined until the total electron density is constructed; and the total electron density in turn depends on these local densities, leading to a self-consistent calculation. However, the perturbed electron densities need only be calculated once for a given electron density; so this method can be applied to systems much larger than can be treated by full density functional calculations without any input from experiment.

Using the variational principle of density functional theory, it has been shown that the binding energy E_B of a collection of atoms with the assumptions above can be given by the expression:

$$E_B = \sum_i E_i(\rho_i^{ave}) + \sum_i \sum_j E_{ov} + E_{1e} \tag{28.15}$$

where ρ_i^{ave} is the average electron density for atomic site i, $E_i(\rho_i^{ave})$ is the energy of an atom embedded in jellium with density ρ_i^{ave}, E_{ov} is the electrostatic repulsion between overlapping neutral spheres (which is summed over atoms i and j), and E_{1e} is a one-electron energy not accounted for by the spherical approximations (Jacobsen, 1988). For most close-packed metals, the overlap and one-electron terms are relatively small, and the embedding term dominates the energy. Reasonable estimates for shear constants, however, require that the overlap term not be neglected; and for non-close-packed systems, the one-electron term becomes important. Equation (28.15) also provides a different interpretation of the Xu et al. tight-binding expression Equation (28.7) (Xu, 1992). The analytic term A in Equation (28.7), which is taken as a polynomial function of a pair sum, can be interpreted as corresponding to the energy of embedding an atom into jellium. The one-electron tight-binding orbitals are then the one-electron terms typically ignored in Equation (28.15).

The EAM is essentially an empirical non-self-consistent variation of effective medium theory. The binding energy in the EAM is given by:

$$E^B = \sum_i F(\rho_i) + \sum_i \sum_j U(r_{ij}) \tag{28.16}$$

where ρ_i is the electron density associated with i, F is called the embedding function, and $U(r_{ij})$ is a pair-additive interaction that depends on the scalar distance r_{ij} between atoms i and j (Daw, 1983). The first term in Equation (28.16) corresponds to the energy of the atoms embedded in jellium, the second term represents overlap of neutral spheres, and the one-electron term of Equation (28.15) is ignored. In the embedded-atom method, however, the average electron density surrounding an atom within a neutral sphere used in effective medium theory is replaced with a sum of electron densities from neighboring atoms at the lattice site i:

$$\rho_i = \sum_j \phi(r_{ij}) \tag{28.17}$$

where $\phi(r_{ij})$ is the contribution to the electron density at site i from atom j. Furthermore, the embedding function and the pair terms are empirically fit to materials properties, and the pairwise–additive electron contributions $\phi^{ij}(r_{ij})$ are either taken from atomic electron densities or fit as empirical functions.

28.2 First-Principles Approaches: Forces on the Fly

The discussion in the previous section has focused on analytic potential energy functions and inter-atomic forces. These types of expressions are typically used in molecular modeling simulations when large systems and/or many timesteps are needed, approximate forces are adequate for the results desired, or computing resources are limited. Another approach is to use forces directly from a total energy calculation that explicitly includes electronic degrees of freedom. The first efficient scheme and still the most widespread technique for calculating *forces on the fly* is the Car–Parrinello method (Car, 1985). Introduced in the middle 1980s, the approach yielded both an efficient computational scheme and a new paradigm in how one calculates forces. The central concepts are to include in the equations of motion both the nuclear and electronic degrees of freedom, and to simultaneously integrate these coupled equations as the simulation progresses. In principle, at each step the energy associated with the electronic degrees of freedom should be *quenched* to the Born–Oppenheimer surface in order to obtain the appropriate interatomic forces. What the Car–Parrinello method allows one to do, however, is to integrate the equations of motion without having to explicitly quench the electronic degrees of freedom, effectively allowing a trajectory to progress *above*, and hopefully parallel to, the Born–Oppenheimer potential energy surface. Since the introduction of the Car–Parrinello method, there have been other schemes that have built upon this idea of treating electronic and nuclear degrees of freedom on an even footing. Clearly, though, the Car–Parrinello approach opened valuable new avenues for modeling — avenues that will dominate molecular simulation as computing resources continue to expand.

Worth noting is another approach, popular largely in the chemistry community, in which a region of interest, usually where a reaction takes place, is treated quantum mechanically, while the surrounding environment is treated with a classical potential energy expression. Often referred to as the *Quantum Mechanics/Molecular Mechanics* (QM/MM) method, this approach is a computationally efficient compromise between classical potentials and quantum chemistry calculations. The chief challenge to the QM/MM method appears to be how to adequately treat the boundary between the region whose forces are calculated quantum mechanically and the region where the analytic forces are used.

A similar approach to the QM/MM method has been used to model crack propagation. Broughton and coworkers, for example, have used a model for crack propagation in silicon in which electrons are included in the calculation of interatomic forces at the crack tip, while surrounding atoms are treated with an analytic potential (Selinger, 2000). To treat long-range stresses, the entire atomically resolved region is in turn embedded into a continuum treated with a finite element model. These types of models, in which atomic and continuum regions are coupled in the same simulation, are providing new methods for connecting atomic-scale simulations with macroscale properties.

28.2.1 Other Considerations in Molecular Dynamics Simulations

In addition to the force model, there are other considerations in carrying out a molecular dynamics simulation. For the sake of completeness, two of these are discussed. The first consideration has to do with the choice of an integration scheme for calculation of the dynamics. The equations of motion that govern the atomic trajectories are a set of coupled differential equations that depend on the mutual interaction of atoms. Because these equations cannot in general be solved analytically, numerical integration schemes are used to propagate the system forward in time. The first step in these schemes is to convert the differential equations to difference equations, and then solve these equations step-wise using some finite timestep. In general, the longer the timestep, the fewer steps are needed to reach some target total time and the more efficient the simulations. At the same time, however, the difference equations are better approximations of the differential equations at smaller timesteps. Therefore, long timesteps can introduce significant errors and numerical instabilities. A rule of thumb is that the timestep should be no longer than about 1/20 of the shortest vibrational period in the system. If all degrees of freedom are included, including bond vibrations, timestep sizes are typically on the order of one femtosecond. If vibrational degrees of freedom can be eliminated, typically using rigid bonds, timesteps an order of magnitude larger can often be used. In addition to large integration errors that can occur from step to step, it is also important to consider small errors that are cumulative over many timesteps, as these can adversely affect the results of simulations involving many timesteps. Fortunately, a great deal of effort has gone into evaluating both kinds of integration errors, and integrators have been developed for which small errors tend to cancel one another over many timesteps. In addition, many calculations rely on dynamic simulations purely as a means of sampling configuration space, so that long-time trajectory errors are relatively unimportant as long as the dynamics remain in the desired ensemble.

Another consideration in molecular dynamics simulation is the choice of thermostat. The simplest approach is to simply scale velocities such that an appropriate average kinetic energy is achieved. The drawback to this approach is that fluctuations in energy and temperature associated with a given thermodynamic ensemble are not reproduced. Other methods have historically been used in which frictional forces are added (e.g., Langevin models with random forces and compensating frictional terms), or constraints on the equations of motion are used. The Nosé–Hoover thermostat, which is essentially a hybrid frictional force/constrained dynamics scheme, is capable of producing both a desired average kinetic energy and the correct temperature fluctuations for a system coupled to a thermal bath (Nosé, 1984). Similar schemes exist for maintaining desired pressures or stress states that can be used for simulations of condensed phase systems.

28.3 Applications

Discussed in the following sections are some applications of molecular modeling to the development of our understanding of nanometer-scale chemical and physical phenomena. Rather than attempt to provide a comprehensive literature review, we have focused on a subset of studies that illustrate both how modeling is used to understand experimental results, and how modeling can be used to test the boundaries of what is possible in nanotechnology.

28.3.1 Pumps, Gears, Motors, Valves, and Extraction Tools

Molecular modeling has been used to study a wide range of systems whose functionality comes from atomic motion (as opposed to electronic properties). Some of these studies have been on highly idealized structures, with the intent of exploring the properties of these systems and not necessarily implying that the specific structure modeled will ever be created in the laboratory. Some studies, on the other hand, are intended to model experimentally realizable systems, with the goal of helping to guide the optimization of a given functionality.

Goddard and coworkers have used molecular dynamics simulations to model the performance of idealized planetary gears and neon pumps containing up to about 10,000 atoms (Cagin, 1998). The structures explored were based on models originally developed by Drexler and Merkle (Drexler, 1992), with the goal of the Goddard studies being to enhance the stability of the structures under desired operating conditions. The simulations used primarily generic valence force-field and non-bonded interatomic interactions. The primary issue in optimizing these systems was to produce gears and other mutually moving interfaces that met exacting specifications. At the macroscale these types of tolerances can be met with precision machining. As pointed out by Cagin et al., at the nanometer scale these tolerances are intimately tied to atomic structures; and specific parts must be carefully designed atom-by-atom to produce acceptable tolerances (Cagin, 1998). At the same time, the overall structure must be robust against both shear forces and relatively low-frequency (compared with atomic vibrations) system vibrations caused by sudden accelerations of the moving parts. Illustrated in Figure 28.2 is a planetary gear and a neon pump modeled by the Goddard group. The gear contains 4235 atoms, has a molecular weight of 72491.95 grams/mole, and contains eight moving parts, while the pump contains 6165 atoms. Molecular simulations in which the concentric gears in both systems are rotated have demonstrated that in designing such a system, a careful balance must be maintained between rotating surfaces being too far apart, leading to gear slip, and being too close together, which leads to lock-up of the interface.

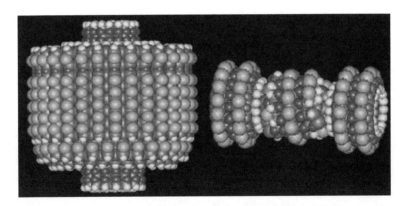

FIGURE 28.2 Illustrations of a molecular pump and gear. (Courtesy of T. Cagin and W.A. Goddard, from Cagin, T., Jaramillo–Botero, A., Gao, G., Goddard, W.A. (1998) Molecular mechanics and molecular dynamics analysis of Drexler–Merkle gears and neon pump, *Nanotechnology,* 9, 143–152. With permission.)

Complicating this analysis are vibrations of the system during operation, which can alternately cause slip and lock-up as the gears move. Different modes of driving these systems were modeled that included single impulses as well as time-dependent and constant angular velocities, torques, and accelerations. The simulations predict that significant and rapid heating of these systems can occur depending on how they are driven, and therefore thermal management within these systems is critical to their operation.

Several models for nanometer-scale bearings, gears, and motors based on carbon nanotubes have also been studied using molecular modeling. Sumpter and coworkers at Oak Ridge National Laboratory, for example, have modeled bearings and motors consisting of nested nanotubes (Tuzen, 1995). In their motor simulations, a charge dipole was created at the end of the inside nanotube by assigning positive and negative charges to two of the atoms at the end of the nanotube. The motion of the inside nanotube was then driven using an alternating laser field, while the outer shaft was held fixed. For single-laser operation, the direction of rotation of the shaft was not constant, and beat patterns in angular momentum and total energy occurred whose period and intensity depended on the field strength and placement of the charges. The simulations also predict that using two laser fields could decrease these beat oscillations. However, a thorough analysis of system parameters (field strength and frequency, temperature, sleeve and shaft sizes, and placement of charges) did not identify an ideal set of conditions under which continuous motion of the inner shaft was possible.

Han, Srivastava, and coworkers at NASA Ames have modeled gears created by chemisorbing benzynyl radicals to the outer walls of nanotubes and using the molecules on two nanotubes as interlocking gear teeth (Han, 1997; Srivastava, 1997). These simulations utilized the bond order potentials described above to describe the interatomic forces. Simulations in which one of the nanotubes is rotated such that it drives the other nanotube show some gear slippage due to distortion of the benzynyl radicals, but without bond breakage. In the simulations, the interlocking gears are able to operate at 50–100 GHz. In a simulation similar in spirit to the laser-driven motor gear modeled by Tuzun et al., Srivastava used both a phenomenological model and molecular simulation to characterize the NASA Ames gears with a dipole driven by a laser field (Srivastava, 1997). The simulations, which used a range of laser field strengths and frequencies, demonstrate that a molecular gear can be driven in one direction if the frequency of the laser field matches a natural frequency of the gear. When the gear and laser field frequency were not in resonance, the direction of the gear motion would change as was observed in the Oak Ridge simulations.

Researchers at the U.S. Naval Research Laboratory have also modeled nanotube gears using the same interatomic potentials as were used for the NASA gears (Robertson, 1994). In these structures, however, the gear shape was created by introducing curvature into a fullerene structure via five- and seven-membered rings (see Figure 28.3). The primary intent of these simulations was not necessarily to model a working gear, but rather to demonstrate the complexity in shape that can be introduced into fullerene structures. In the simulations the shafts of the two gears were allowed to rotate while full dynamics of

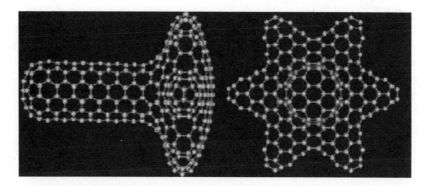

FIGURE 28.3 Illustration of the gears simulated by Robertson and associates.

the cams was modeled. The system was driven by placing the gears such that the gear teeth interlocked, and then rotating the shaft of one of the gears. Different driving conditions were studied, including the rate at which the driving gear was accelerated and its maximum velocity. For example, when the rotational speed of the driving gear was ramped from 0 to 0.1 revolutions per picosecond over 50 picoseconds, the driven gear was able to keep up. However, when the rotation of the driving gear was accelerated from 0 to 0.5 revolutions per picosecond over the same time period, significant slippage and distortion of the gear heads were observed that resulted in severe heating and destruction of the system.

Molecular modeling has recently been used to explore a possible nanoscale valve whose structure and operating concept is in sharp contrast to the machines discussed above (Adiga, 2002). The design is motivated by experiments by Park, Imanishi, and associates (Park, 1998; Ito, 2000). In one of these experiments, polymer brushes composed of polypeptides were self-assembled onto a gold-plated nanoporous membrane. Permeation of water through the membrane was controlled by a helix–coil transformation that was driven by solvent pH. In their coiled states, the polypeptide chains block water from passing through the pores. By changing pH, a folded configuration can be created, effectively opening the pores.

Rather than using polypeptides, the molecular modeling studies used a ball-and-spring model of polymer comb molecules chemisorbed to the inside of a slit pore (Adiga, 2002). Polymer comb molecules, also called molecular bottle brushes, consist of densely grafted side chains that extend away from a polymer backbone (Figure 28.4). In a good solvent (i.e., a solvent in which the side chains can be dissolved), the side chains are extended from the backbone, creating extensive excluded volume interactions. These interactions can create a very stiff structure with a rod configuration. In a poor solvent, the

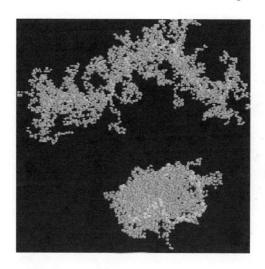

FIGURE 28.4 Illustration of a bottle brush molecule. *Top*: Rod-like structure. *Bottom*: Globular structure.

side chains collapse to the backbone, relieving the excluded volume interactions and allowing the total system to adapt a globular configuration. The concept behind the valve is to use this rod–globular transition in comb polymers assembled into the inside of a nanometer-scale pore to open or close the pore in response to solvent quality. The size of the pore can in principle be controlled to selectively pass molecular species according to their size, with the maximum species size that is allowed to pass controlled by solvent quality (with respect to the comb polymers).

Illustrated in Figure 28.5 are snapshots from a molecular modeling simulation of a pore of this type. The system modeled consists of comb polymer molecules 100 units long with sides of 30 units grafted to the backbone every four units. Periodic boundary conditions are applied in the two directions parallel to the grafting surface. The brushes are immersed in monomeric solvent molecules of the same size and mass as those of the beads of the comb molecules. The force-field model uses a harmonic spring between monomer units and shifted Lennard–Jones interactions for nonbonded, bead–bead, solvent–solvent, and bead–solvent forces. Purely repulsive nonbonded interactions are used for the brush–wall and solvent–wall interactions. The wall separation and grafting distances are about 2 times and one-half, respectively, of the end-to-end distance of a single comb molecule in its extended state, and about 8 and 2 times, respectively, of a comb molecule in its globule state. The snapshots in Figure 28.5 depict the pore structure at three different solvent conditions created by altering the ratio of the solvent–solvent to solvent–polymer interaction strength. In the simulations, pore opening due to a change in solvent quality required only about 0.5 nanoseconds. The simulations also showed that oligomer chain molecules with a radius smaller than the pore size can translate freely through the pore, while larger molecules become caught in the comb molecules, with motion likely requiring chain reptation, a fairly slow process compared with free translation. Both the timeframe of pore opening and the size selectivity for molecules passing through the pore indicate an effective nanoscale valve.

A molecular abstraction tool for patterning surfaces, which was first proposed by Drexler, has also been studied using molecular modeling (Sinnott, 1994; Brenner, 1996). The tool is composed of an ethynyl radical chemisorbed to the surface of a scanned-probe microscope tip, which in the case of the modeling studies was composed of diamond. When brought near a second surface, it was proposed that the radical species could abstract a hydrogen atom from the surface with atomic precision. A number of issues related to this tool were characterized using molecular modeling studies with the bond-order potential discussed above. These issues included the timescales needed for reaction and for the reaction energy to flow away from the reaction site, the effect of tip crashes, and the creation of a signal that abstraction had occurred. The simulation indicated that the rates of reaction and energy flow were very fast, effectively creating an irreversible abstraction reaction if the tip is left in the vicinity of the surface

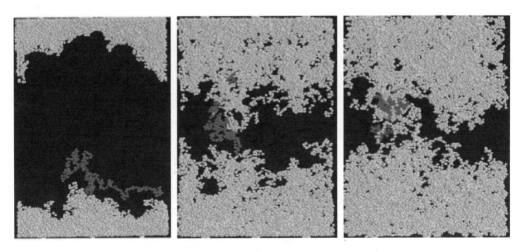

FIGURE 28.5 Illustration of an array of bottle brush molecules grafted to the inside of nanometer-scale slit pores at three different solvent qualities.

from which the hydrogen is abstracted. As expected, however, with the ethynyl radical exposed at the end of the tip, tip crashes effectively destroy the system. To study how this can be avoided, and to create a system from which a signal for abstraction could be detected, a structure was created in the modeling studies in which asperities on the tip surround the ethynyl radical. With this configuration, the simulations predicted that a load on the tip can be detected as the tip comes into contact with a surface, and that abstraction can still occur while the asperities protect the ethynyl from further damage if a larger load is applied to the tip. While this system has not been created experimentally, the modeling studies nonetheless represent a creative study into what is feasible at the nanometer scale.

28.3.2 Nanometer-Scale Properties of Confined Fluids

The study of friction and wear between sliding bodies, now referred to as the field of *tribology*, has a long and important history in the development of new technologies. For example, the ancient Egyptians used water to lubricate the path of sleds that transported heavy objects. The first scientific studies of friction were carried out in the sixteenth century by Leonardo da Vinci, who deduced that the weight, but not the shape, of an object influences the frictional force of sliding. Da Vinci also first introduced the idea of a friction coefficient as the ratio of frictional force to normal load. This and similar observations led, 200 years later, to the development of Amonton's law, which states that for macroscopic systems the frictional force between two objects is proportional to the normal load and independent of the apparent contact area. In the eighteenth century, Coulomb verified these observations and clarified the difference between static and dynamic friction.

Understanding and ultimately controlling friction, which traditionally has dealt with macroscale properties, is no less important for the development of new nanotechnologies involving moving interfaces. With the emergence of experimental techniques, such as the atomic-force microscope, the surface-force apparatus, and the quartz crystal microbalance, has come the ability to measure surface interactions under smooth or single-asperity contact conditions with nanometer-scale resolution. Interpreting the results of these experiments, however, is often problematic, as new and sometimes unexpected phenomena are discovered. It is in these situations that molecular modeling has often played a crucial role.

Both experiments and subsequent molecular modeling studies have been used to characterize behavior associated with fluid ordering near solid surfaces at the nanometer scale and the influence of this ordering on liquid lubrication at this scale. Experimental measurements have demonstrated that the properties of fluids confined between solid surfaces become drastically altered as the separation between the solid surfaces approaches the atomic scale (Horn, 1981; Chan, 1985; Gee, 1990). At separations of a few molecular diameters, for example, an increase in liquid viscosity by several orders of magnitude has been measured (Israelachvili, 1988; Van Alsten, 1988). While continuum hydrodynamic and elasto-hydrodynamic theories have been successful in describing lubrication by micron-thick films, these approaches start to break down when the liquid thickness approaches a few molecular diameters. Molecular simulations have been used to great advantage at length scales for which continuum approaches begin to fail. Systems that have been simulated include films of spherical molecules, straight-chain alkanes, and branched alkanes confined between solid parallel walls.

Using molecular simulations involving both molecular dynamics and Monte Carlo methods, several research groups have characterized the equilibrium properties of spherical molecules confined between solid walls (Schoen, 1987, 1989; Bitsanis, 1987; Thompson, 1990; Sokol, 1992; Diestler, 1993). These studies have suggested that when placed inside a pore, fluid layers become layered normal to the pore walls, independent of the atomic-scale roughness of the pore walls (Bitsanis, 1990). Simulations have also shown that structure in the walls of the pore can induce transverse order (parallel to the walls) in a confined atomic fluid (Schoen, 1987). For example, detailed analysis of the structure of a fluid within a layer, or epitaxial ordering, as a function of wall density and wall–fluid interaction strengths was undertaken by Thompson and Robbins (Thompson, 1990b). For small values of wall-to-liquid interaction strength, fluid atoms were more likely to sit over gaps in the adjacent solid layer. Self-diffusion within this layer, however, was roughly the same as in the bulk liquid. Increasing the strength of the wall–fluid interactions by a factor of 4.5 resulted in the epitaxial locking of the first liquid layer to the solid. While diffusion in the first layer was too small to measure, diffusion in the second layer was approximately half of its value in the bulk fluid. The second layer of liquid crystallized and became locked to the first *liquid* layer when the strength of the wall–liquid interaction was increased by approximately an order of magnitude over its original value. A third layer never crystallized.

Confinement by solid walls has been shown to have a number of effects on the equilibrium properties of static polymer films (Ribarsky, 1992; Thompson, 1992; Wang, 1993). For example, in simulations of linear chains by Thompson et al., film thickness was found to decrease as the normal pressure on the upper wall increases. At the same time, the simulations predict that the degree of layering and in-plane ordering increases, and the diffusion constant parallel to the walls decreases. In contrast to films of spherical molecules, where there is a sudden drop in the diffusion constant associated with a phase transition to an ordered structure, films of chain molecules are predicted to remain highly disordered, and the diffusion constant drops steadily as the pressure increases. This indicates the onset of a glassy phase at a pressure below the bulk transition pressure. This wall-induced glass phase explains dramatic increases in experimentally measured thin-film relaxation times and viscosities (Van Alsten, 1988; Gee, 1990).

In contrast to the situation for simple fluids and linear chain polymers, experimental studies (Granick, 1995) have not indicated oscillations in surface forces of confined highly-branched hydrocarbons such as squalane. To understand the reason for this experimental observation, Balasubramanian et al. used molecular modeling methods to examine the adsorption of linear and branched alkanes on a flat Au(111) surface (Balasubramanian, 1996). In particular, they examined the adsorption of films of n-hexadecane, three different hexadecane isomers, and squalane. The alkane molecules were modeled using a united atom model with a Lennard–Jones 12–6 potential used to model the interactions between the united atoms. The alkane–surface interactions were modeled using an external Lennard–Jones potential with parameters appropriate for a flat Au(111) substrate. The simulations yield density profiles for n-hexadecane and 6-pentylundecane that are nearly identical to experiments and previous simulations. In contrast, density profiles of the more highly branched alkanes such as heptamethylnonane and 7,8-dimethyltetradecane exhibit an additional peak. These peaks are due to methyl branches that cannot be accommodated in the first liquid layer next to the gold surface. For thicker films, oscillations in the density profiles for heptamethylnonane were out of phase with those for n-hexadecane, in agreement with the experimental observations.

The properties of confined spherical and chain molecular films under shear have been examined using molecular modeling methods. Work by Bitsanis et al., for example, examined the effect of shear on spherical, symmetric molecules, confined between planar, parallel walls that lacked atomic-scale roughness (Bitsanis, 1990). Both Couette (simple shear) and Poiseuille (pressure-driven) flows were examined. The density profiles in the presence of both types of flow were identical to those under equilibrium conditions for all pore widths. Velocity profiles, defined as the velocity of the liquid parallel to the wall as a function of distance from the center of the pore, should be linear and parabolic for Couette and Poiseuille flow, respectively, for a homogeneous liquid. The simulations yielded velocity profiles in the two monolayers nearest the solid surfaces that deviate from the flow shape expected for a homogeneous liquid and that indicate high viscosity. The different flow nature in molecularly thin films was further demonstrated by plotting the effective viscosity vs. pore width. For a bulk material the viscosity is independent of pore size. However, under both types of flow, the viscosity increases slightly as the pore size decreases. For ultrathin films, the simulations predict a dramatic increase in viscosity.

Thompson and Robbins also examined the flow of simple liquids confined between two solid walls (Thompson, 1990). In this case, the walls were composed of (001) planes of a face-center-cubic lattice. A number of wall and fluid properties, such as wall–fluid interaction strength, fluid density, and temperature, were varied. The geometry of the simulations closely resembled the configuration of a surface-force apparatus, where each wall atom was attached to a lattice site with a spring (an Einstein oscillator model) to maintain a well-defined solid structure with a minimum number of solid atoms. The thermal roughness and the response of the wall to the fluid was controlled by the spring constant, which was adjusted so that the atomic mean-square displacement about the lattice sites was less than the Lindemann criterion for melting. The interactions between the fluid atoms and between the wall and fluid atoms were modeled by different Lennard–Jones potentials. Moving the walls at a constant velocity in opposite directions simulated Couette flow, while the heat generated by the shearing of the liquid was dissipated using a Langevin thermostat. In most of the simulations, the fluid density and temperature were indicative of a compressed liquid about 30% above its melting temperature. A number of interesting phenomena were observed in these simulations. First, both normal and parallel ordering in the adjacent liquid was induced by the well-defined lattice structure of the solid walls. The liquid density oscillations also induced oscillations in other microscopic quantities normal to the walls, such as the fluid velocity in the flow direction and the in-plane microscopic stress tensor, that are contrary to the predictions of the continuum Navier–Stokes equations. However, averaging the quantities over length scales that are larger than the molecular lengths produced smoothed quantities that satisfied the Navier–Stokes equations.

Two-dimensional ordering of the liquid parallel to the walls affected the flow even more significantly than the ordering normal to the walls. The velocity profile of the fluid parallel to the wall was examined as a function of distance from the wall for a number of wall–fluid interaction strengths and wall densities. Analysis of velocity profiles demonstrated that flow near solid boundaries is strongly dependent on the strength of the wall–fluid interaction and on wall density. For instance, when the wall and fluid densities are equal and wall–fluid interaction strengths are small, the velocity profile is predicted to be linear with a no-slip boundary condition. As the wall–fluid interaction strength increases, the magnitude of the liquid velocity in the layers nearest the wall increases. Thus the velocity profiles become curved. Increasing the wall–fluid interaction strength further causes the first two liquid layers to become locked to the solid wall. For unequal wall and fluid densities, the flow boundary conditions changed dramatically. At the smallest wall–fluid interaction strengths examined, the velocity profile was linear; however, the no-slip boundary condition was not present. The magnitude of this slip decreases as the strength of the wall–fluid interaction increases. For an intermediate value of wall–fluid interaction strength, the first fluid layer was partially locked to the solid wall. Sufficiently large values of wall–fluid interaction strength led to the locking of the second fluid layer to the wall.

While simulations with spherical molecules are successful in explaining many experimental phenomena, they are unable to reproduce all features of the experimental data. For example, calculated relaxation times and viscosities remain near bulk fluid values until films crystallize. Experimentally, these quantities typically increase many orders of magnitude before a well-defined yield stress is observed (Israelachvili, 1988; Van Alsten, 1988). To characterize this discrepancy, Thompson et al. repeated earlier shearing simulations using freely jointed, linear-chain molecules instead of spherical molecules (Thompson, 1990, 1992). The behavior of the viscosity of the films as a function of shear rate was examined for films of different thickness. The response of films that were six to eight molecular diameters thick was approximately the same as for bulk systems. When the thickness of the film was reduced, the viscosity of the film increased dramatically, particularly at low shear rates, consistent with the experimental observations.

Based on experiments and simulations, it is clear that fluids confined to areas of atomic-scale dimensions do not necessarily behave like liquids on the macroscopic scale. In fact, depending on the conditions, they may often behave more like solids in terms of structure and flow. This presents a unique set of concerns for lubricating moving parts at the nanometer scale. However, with the aid of simulations such as the ones mentioned here, plus experimental studies using techniques such as the surface-force apparatus, general properties of nanometer-scale fluids are being characterized with considerable precision. This, in turn, is allowing scientists and engineers to design new materials and interfaces that have specific interactions with confined lubricants, effectively controlling friction (and wear) at the atomic scale.

28.3.3 Nanometer-Scale Indentation

Indentation is a well-established experimental technique for quantifying the macroscopic hardness of a material. In this technique, an indenter of known shape is loaded against a material and then released, and the resulting permanent impression is measured. The relation between the applied load, the indenter shape, and the profile of the impression is used to establish the hardness of the material on one of several possible engineering scales of hardness. Hardness values can in turn be used to estimate materials properties such as yield strength. Nanoindentation, a method in which both the tip–surface contact and the resulting impression have nanometer-scale dimensions, has become an important method for helping to establish properties of materials at the nanometer scale. The interpretation of nanoindentation data usually involves an analysis of loading and/or unloading force-vs.-displacement curves. Hertzian contact mechanics, which is based on continuum mechanics principles, is often sufficient to obtain properties such as elastic moduli from these curves, as long as the indentation conditions are elastic. In many situations, however, the applicability of Hertzian mechanics is either limited or altogether inappropriate, especially when plastic deformation occurs. It is these situations for which molecular modeling has become an essential tool for understanding nanoindentation data.

Landman was one of the first researchers to use molecular modeling to simulate the indentation of a metallic substrate with a metal tip (Landman, 1989, 1990, 1991, 1992, 1993). In an early simulation, a pyramidal nickel tip was used to indent a gold substrate. The EAM was used to generate the interatomic forces. By gradually lowering the tip into the substrate while simultaneously allowing classical motion of tip and surface atoms, a plot of force vs. tip-sample separation was generated, the features of which could be correlated with detailed atomic dynamics. The shape of this virtual loading/unloading curve showed a jump to contact, a maximum force before tip retraction, and a large loading–unloading hysteresis. Each of these features match qualitative features of experimental loading curves, albeit on different scales of tip–substrate separation and contact area. The computer-generated loading also exhibited fine detail that was not resolved in the experimental curve. Analysis of the dynamics in the simulated loading curves showed that the jump to contact, which results in a relatively large and abrupt tip–surface attraction, was due to the gold atoms bulging up to meet the tip and subsequently wetting the tip. Advancing the tip caused indentation of the gold substrate with a corresponding increase in force with decreasing tip–substrate separation. Detailed dynamics leading to the shape of the virtual loading curve in this region consisted of the flow of the gold atoms that resulted in the pile-up of gold around the edges of the nickel indenter. As the tip was retracted from the sample, a connective neck of atoms between the tip and the substrate formed that was largely composed of gold atoms. Further retraction of the tip caused adjacent layers of the connective neck to rearrange so that an additional row of atoms formed in the neck. These rearrangement events were the essence of the elongation process, and they were responsible for a fine structure (apparent as a series of maxima) present in the retraction portion of the force curve. These elongation and rearrangement steps were repeated until the connective neck of atoms was severed.

The initial instability in tip–surface contact behavior observed in Landman's simulations was also reported by Pethica and Sutton and by Smith et al (Pethica, 1988; Smith, 1989). In a subsequent simulation by Landman and associates using a gold tip and nickel substrate, the tip deformed toward the substrate during the jump to contact. Hence, the softer material appears to be displaced. The longer-range jump-to-contact typically observed in experiments can be due to longer-ranged surface adhesive forces, such as dispersion and possibly wetting of impurity layers, as well as compliance of the tip holder, that were not included in the initial computer simulations.

Interesting results have been obtained for other metallic tip–substrate systems. For example, Tomagnini et al. used molecular simulation to study the interaction of a pyramidal gold tip with a lead substrate using interatomic forces from the glue model mentioned above (Tomagnini, 1993). When the gold tip was brought into close proximity to the lead substrate at room temperature, a jump to contact was initiated by a few lead atoms wetting the tip. The connective neck of atoms between the tip and the surface was composed almost entirely of lead. The tip became deformed because the inner-tip atoms were pulled more toward the sample surface than toward atoms on the tip surface. Increasing the substrate temperature to 600K caused the formation of a liquid lead layer approximately four layers thick on the surface of the substrate. During indentation, the distance at which the jump to contact occurred increased by approximately 1.5 Å, and the contact area also increased due to the diffusion of the lead. The gold tip eventually dissolved in the liquid lead, resulting in a liquid-like connective neck of atoms that followed the tip upon retraction. As a result, the liquid–solid interface moved farther back into the bulk lead substrate, increasing the length of the connective neck. Similar elongation events have been observed experimentally. For example, scanning tunneling microscopy experiments on the same surface demonstrate that the neck can elongate approximately 2500 Å without breaking.

Nanoindentation of substrates covered by various overlayers have also been simulated via molecular modeling. Landman and associates, for example, simulated indentation of an n-hexadecane-covered gold substrate with a nickel tip (Landman, 1992). The forces governing the metal–metal interactions were derived from the EAM, while a variation on a valence-force potential was used to model the n-hexadecane film. Equilibration of the film on the gold surface resulted in a partially ordered film where molecules in the layer closest to the gold substrate were oriented parallel to the surface plane. When the nickel tip was lowered, the film swelled up to meet and partially wet the tip. Continued lowering of the tip toward the film caused the film to flatten and some of the alkane molecules to wet the sides of the tip. Lowering the tip farther caused drainage of the top layer of alkane molecules from underneath the tip and increased wetting of the sides of the tip, pinning of hexadecane molecules under the tip, and deformation of the gold substrate beneath the tip. Further lowering of the tip resulted in the drainage of the pinned alkane molecules, inward deformation of the substrate, and eventual formation of an intermetallic contact by surface gold atoms moving toward the nickel tip, which was concomitant with the force between the tip and the substrate becoming attractive.

Tupper and Brenner have used atomic simulations to model loading of a self-assembled thiol overlayer on a gold substrate (Tupper, 1994). Simulations of compression with a flat surface predicted a reversible structural transition involving a rearrangement of the sulfur head groups bonded to the gold substrate. Concomitant with the formation of the new overlayer structure was a change in slope of the loading curve that agreed qualitatively with experimental loading data. Simulations of loading using a surface containing a nanometer-scale asperity were also carried out. Penetration of the asperity through the self-assembled overlayer occurred without an appreciable loading force. This result suggests that scanning-tunneling microscope images of self-assembled thiol monolayers may reflect the structure of the head group rather than the end of the chains, even when an appreciable load on the tip is not measured.

Experimental data showing a large change in electrical resistivity during indentation of silicon has led to the suggestion of a load-induced phase transition below the indenter. Clarke et al., for example, report forming an Ohmic contact under load; and using transmission electron microscopy, they have observed an amorphous phase at the point of contact after indentation (Clarke, 1988). Based on this data, the authors suggest that one or more high-pressure electrically conducting phases are produced under the indenter, and that these phases transform to the amorphous structure upon rapid unloading. Further support for this conclusion was given by Pharr et al., although they caution that the large change in electrical resistivity may have other origins and that an abrupt change in force during unloading may be due to sample cracking rather than transformation of a high pressure phase (Pharr, 1992). Using micro-Raman microscopy, Kailer et al. identified a metallic β-Sn phase in silicon near the interface of a diamond indenter during hardness loading (Kailer, 1999). Furthermore, upon rapid unloading they detected amorphous silicon as in the Clarke et al. experiments, while slow unloading resulted in a mixture of high-pressure polymorphs near the indent point.

Using molecular dynamics simulations, Kallman et al. examined the microstructure of amorphous and crystalline silicon before, during, and after simulated indentation (Kallman, 1993). Interatomic forces governing the motion of the silicon atoms were derived from the Stillinger–Weber potential mentioned above. For an initially crystalline silicon substrate close to its melting point, the simulations indicated a tendency to transform to the amorphous phase near the indenter. However, an initially amorphous silicon substrate was not observed to crystallize upon indentation; and no evidence of a transformation to the β-Sn structure was found. In more recent simulations by Cheong and Zhang that used the Tersoff silicon potential, an indentation-induced transition to a body-centered tetragonal phase was observed, followed by transformation to an amorphous structure after unloading (Cheong, 2000). A transition back to the high-pressure phase upon reloading of the amorphous region was observed in the simulations, indicating that the transition between the high-pressure ordered phase and the amorphous structure is reversible.

Smith, Tadmore, and Kaxiras revisited the silicon nanoindentation issue using a quasi-continuum model that couples interatomic forces from the Stillinger–Weber model to a finite element grid (Smith, 2000). This treatment allows much larger systems than would be possible with an all-atom approach. They report good agreement between simulated loading curves and experiment, provided that the curves are scaled by the indenter size. Rather than the β-Sn structure, however, atomic displacements suggest formation of a metallic structure with fivefold coordination below the indenter upon loading and a residual simple cubic phase near the indentation site after the load is released rather than the mix of high-pressure phases characterized experimentally. Smith et al. attribute this discrepancy to shortcomings of the Stillinger–Weber potential in adequately describing high-pressure phases of silicon. They also used a simple model for changes in electrical resistivity with loading involving contributions from both a Schottky barrier and spreading resistance. Simulated resistance-vs.-loading curves agree well with experiments despite possible discrepancies between the high-pressure phases under the indenter, suggesting that the salient features of the experiment are not dependent on the details of the high-pressure phases produced.

Molecular simulations that probe the influence of nanometer-scale surface features on nanoindentation force–displacement curves and plastic deformation have recently been carried out. Simulations of shallow, elastic indentations, for example, have been used to help characterize the conditions under which nanoindentation could be used to map local residual surface stresses (Shenderova, 2000). Using the embedded-atom method to describe interatomic forces, Shenderova and associates performed simulations of shallow, elastic indentation of a gold substrate near surface features that included a trench and a dislocation intersecting the surface. The maximum load for a given indentation depth of less than one nanometer was found to correlate to residual stresses that arise from the surface features. This result points toward the application of nanoindentation for nondestructively characterizing stress distributions due to nanoscale surface features.

Zimmerman et al. have used simulations to characterize plastic deformation due to nanoindentation near a step on a gold substrate (Zimmerman, 2001). The simulations showed that the load needed to nucleate dislocations is lower near a step than on a terrace, although the effect is apparently less than that measured experimentally due to different contact areas.

Vashishta and coworkers have used large-scale, multi-million-atom simulations to model nanoindentation of Si_3N_4 films with a rigid indenter (Walsh, 2000). The simulations demonstrated formation of an amorphous region below the indenter that was terminated by pile-up of material around the indenter and crack formation at the indenter corners.

The utility of hemispherically capped single-wall carbon nanotubes for use as scanning probe microscope tips has been investigated using molecular dynamics simulation by Harrison and co-workers and by Sinnott and coworkers (Harrison, 1997; Garg, 1998a). In the work reported by Harrison et al., it was shown that (10,10) armchair nanotubes recover reversibly after interaction with hydrogen-terminated diamond (111) surfaces. The nanotube exhibits two mechanisms for releasing the stresses induced by indentation: a marked inversion of the capped end, from convex to concave, and finning along the tube's axis. The cap was shown to flatten at low loads and then to invert in two discrete steps. Compressive stresses at the vertex of the tip build up prior to the first cap-inversion event. These stresses are relieved by the rapid popping of the three layers of carbon atoms closest to the apex of the tip inside the tube. Continued application of load causes the remaining two rings of carbon atoms in the cap to be pushed inside the tube. Additional stresses on the nanotube caused by its interaction with the hard diamond substrate are relieved via a *finning* mechanism, or flattening, of the nanotube. That is, the nanotube collapses so that opposing walls are close together. These conformational changes in the tube are reversed upon pull-back of the tube from the diamond substrate. The tube recovers its initial shape, demonstrating the potential usefulness of nanotubes as scanning probe microscope tips.

The same capped (10,10) nanotube was also used to indent n-alkane hydrocarbon chains with 8, 13, and 22 carbon atoms chemically bound to diamond (111) surfaces (Tutein, 1999). Both flexible and rigid nanotubes were used to probe the n-alkane monolayers. The majority of the torsional bonds along the

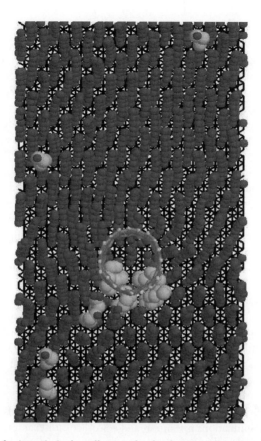

FIGURE 28.6 Illustration of a (10,10) single-wall nanotube that has partially indented a monolayer composed of C_{13} chains on a diamond substrate. Looking down along the tube, it is apparent that gauche defects (light gray, largest spheres) form under and adjacent to the nanotube. Hydrogen atoms on the chains and the tube cap atoms have been omitted from the picture for clarity.

carbon backbone of the chains were in their anti-conformation prior to indentation. Regardless of the nanotube used, indentation of the hydrocarbon monolayers caused a disruption in the ordering of the monolayer, pinning of hydrocarbon chains beneath the tube, and formation of gauche defects with the monolayer below and adjacent to the tube (see Figure 28.6). The flexible nanotube is distorted only slightly by its interaction with the softer monolayers because nanotubes are stiff along their axial direction. In contrast, interaction with the diamond substrate causes the tube to fin, as it does in the absence of the monolayer. Severe indents with a rigid nanotube tip result in rupture of chemical bonds with the hydrocarbon monolayer. This was the first reported instance of indentation-induced bond rupture in a monolayer system. Previous simulations by Harrison and coworkers demonstrated that the rupture of chemical bonds (or fracture) is also possible when a hydrogen-terminated diamond asperity is used to indent both hydrogen-terminated and hydrogen-free diamond (111) surfaces (Harrison, 1992).

28.3.4 New Materials and Structures

Molecular modeling has made important contributions to our understanding of the properties, processing, and applications of several classes of new materials and structures. Discussing all of these contributions is beyond the scope of this chapter (a thorough discussion would require several volumes). Instead, the intent of this section is to supplement the content of some of the more detailed chapters in this book by presenting examples that represent the types of systems and processes that can be examined by atomic simulation.

28.3.4.1 Fullerene Nanotubes

Molecular modeling has played a central role in developing our understanding of carbon-based structures, in particular molecular fullerenes and fullerene nanotubes. Early molecular modeling studies focused on structures, energies, formation processes, and simple mechanical properties of different types of fullerenes. As this field has matured, molecular modeling studies have focused on more complicated structures and phenomena such as nonlinear deformations of nanotubes, nanotube functionalization, nanotube filling, and hybrid systems involving nanotubes. Molecular modeling is also being used in conjunction with continuum models of nanotubes to obtain deeper insights into the mechanical properties of these systems.

Several molecular modeling studies of nanotubes with sidewall functionalization have been recently carried out. Using the bond order potential discussed earlier, Sinnott and coworkers modeled CH_3+ incident on bundles of single-walled and multi-walled nanotubes at energies ranging from 10 to 80 eV (Ni, 2001). The simulations showed chemical functionalization and defect formation on nanotube sidewalls, as well as the formation of cross-links connecting either neighboring nanotubes or between the walls of a single nanotube (Figure 28.7). These simulations were carried out in conjunction with experimental studies that provided evidence for sidewall functionalization using $CF3+$ ions deposited at comparable incident energies onto multi-walled carbon nanotubes.

Molecular modeling has also predicted that kinks formed during large deformations of nanotubes may act as reactive sites for chemically connecting species to nanotubes (Figure 28.8). In simulations by Srivastava et al., it was predicted that binding energies for chemically attaching hydrogen atoms to a nanotube can be enhanced by over 1.5 eV compared with chemical attachment to pristine nanotubes (Srivastava, 1999). This enhancement comes from mechanical deformation of carbon atoms around kinks and ridges that force bond angles toward the tetrahedral angle, leading to radical sites on which species can strongly bond.

Several modeling studies have also been carried out that have examined the mechanical properties of functionalized nanotubes. Simulations by Sinnott and coworkers, for example, have predicted that covalent chemical attachment of $H_2C = C$ species to single-walled nanotubes can decrease the maximum compressive force needed for buckling by about 15%, independent of tubule helical structure or radius (Garg, 1998b). In contrast, similar simulations predict that the tensile modulus of single-walled (10,10)

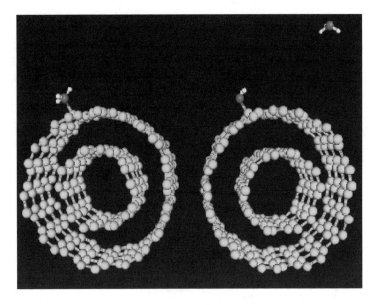

FIGURE 28.7 Illustration of fullerene nanotubes with functionalized sidewalls. (Courtesy of S.B. Sinnott, University of Florida.)

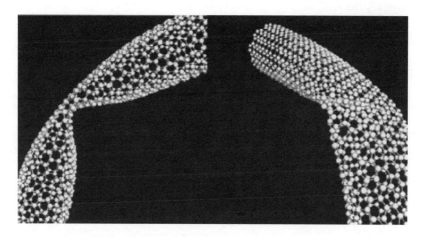

FIGURE 28.8 Illustrations of a kinked fullerene nanotube.

nanotubes is largely unchanged for configurations on which up to 15% of the nanotube carbon atoms (the largest degree modeled) are of the carbon atoms being covalently bonded to CH_3 groups. These simulations also predict a slight decrease in nanotube length due to rehybridization of the nanotube carbon atom valence orbitals from sp^2 to sp^3 (Brenner, 2002).

Several applications of sidewall functionalization via covalent bond formation have been suggested. For example, molecular simulations suggest that the shear needed to start pulling a nanotube on which 1% of carbon atoms are cross-linked to a model polymer matrix is about 15 times that needed to initiate motion of a nanotube that interacts with the matrix strictly via nonbonded forces (Frankland, 2002). This result, together with the prediction that the tensile strength of nanotubes is not compromised by functionalization, suggests that chemical functionalization leading to matrix–nanotube cross-linking may be an effective mode for enhancing load transfer in these systems without sacrificing the elastic moduli of nanotubes (Brenner, 2002). Other applications of functionalized nanotubes include a means for controlling the electronic properties of nanotubes (Brenner, 1998; Siefert, 2000) and a potential route to novel quantum dot structures (Orlikowski, 1999).

The structure and stability of several novel fullerene-based structures have also been calculated using molecular modeling. These structures include nanocones, tapers, and toroids, as well as hybrid diamond cluster-nanotube configurations (Figure 28.9) (Han, 1998; Meunier, 2001; Shenderova, 2001; Brenner, 2002). In many cases, novel electronic properties have also been predicted for these structures.

28.3.4.2 Dendrimers

As discussed in Chapter 24 by Tomalia et al., dendritic polymer structures are starting to play an important role in nanoscale science and technology. While techniques such as nuclear magnetic resonance and infrared spectroscopies have provided important experimental data regarding the structure and relaxation dynamics of dendrimers, similarities between progressive chain generations and complex internal structures make a thorough experimental understanding of their properties difficult. It is in these cases that molecular simulation can provide crucial data that is either difficult to glean from experimental studies or not accessible to experiment.

Molecular modeling has been used by several groups as a tool to understand properties of these species, including their stability, shape, and internal structure as a function of the number of generations and chain stiffness. Using a molecular force field, simulations by Gorman and Smith showed that the equilibrium shape and internal structure of dendrimers varies as the flexibility of the dendrimer repeat unit is changed (Gorman, 2000). They report that dendrimers with flexible repeat units show a somewhat globular shape, while structures formed from stiff chains are more disk-like. These simulations also showed that successive branching generations can fold back, leading to branches from a given generation that can permeate the entire structure.

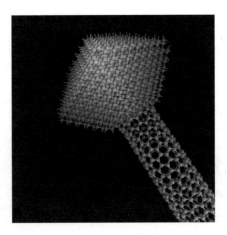

FIGURE 28.9   Illustration of a diamond-nanotube hybrid structure.

In related molecular modeling studies, Karatasos, Adolf, and Davies (Karatasos, 2001) as well as Scherrenberg et al. (Scherrenberg, 1998) studied the structure and dynamics of solvated dendrimers, while Zacharopoulos and Economou modeled a melt (Zacharopoulos, 2002). These studies indicate that dendritic structures become more spherical as the number of generations increases, and that the radius of gyration scales as the number of monomer units to the 1/3 power. Significant folding of the chains inside the structures was also observed (as was seen in the Gorman and Smith study).

28.3.4.3   Nanostructured Materials

Nanostructured materials can have unusual combinations of properties compared with materials with more conventional grain sizes and microstructures. Molecular simulations have contributed to our understanding of the origin of several of these properties, especially how they are related to deformation mechanisms of strained systems.

Using an effective medium potential, Jacobsen and coworkers simulated the deformation of strained nanocrystalline copper with grain sizes that average about 5 nm (Schiotz, 1998). These simulations showed a softening for small grain sizes, in agreement with experimental measurements. The simulations indicate that plastic deformation occurs mainly by grain boundary sliding, with a minimal influence of dislocation motion on the deformation.

Van Swygenhoven and coworkers performed a series of large-scale molecular dynamics simulations of the deformation of nanostructured nickel and copper with grain sizes ranging from 3.5 nm to 12 nm (Van Swygenhoven, 1999). The simulations used a second-moment-based potential as described above, with constant uniaxial stress applied to the systems. The simulations revealed different deformation mechanisms depending on grain size. For samples with grain sizes less than about 10 nm, deformation was found to occur primarily by grain boundary sliding, with the rate of deformation increasing with decreasing grain size. For the larger grain sizes simulated, a change in the deformation mechanism was reported in which a combination of dislocation motion and grain boundary sliding occurred. Characteristic of this apparent new deformation regime was that the strain rate was independent of grain size. In subsequent simulations, detailed mechanisms of strain accommodation were characterized that included both single-atom motion and correlated motion of several atoms, as well as stress-assisted free-volume migration (Van Swygenhoven, 2001).

Wolf and coworkers have also carried out detailed studies of the deformation of nanostructured metals. In studies of columnar structures of aluminum, for example, emission of partial dislocations that were formed at grain boundaries and triple junctions was observed during deformation (Yamakov, 2001). The simulations also showed that these structures can be reabsorbed upon removal of the applied stress, which the authors suggest may contribute to the fact that dislocations are not normally observed experimentally in systems of this type after external stresses are released.

Simulations have also been used to characterize the dynamics in nanostructured materials during ion bombardment and to understand the origin of apparently anomalous vibrational modes in nanostructured materials (Derlet, 2001; Samaras, 2002). In studies of the latter, for example, it was shown that enhancements in both the low- and high-vibrational frequencies for nanostructured nickel and copper arise from atoms at the grain boundaries, and that the vibrational frequencies of atoms in the grains are largely unaffected by the grain size.

Wolf and coworkers have recently simulated the dynamics of grain growth in nanocrystalline face-center-cubic metals (Haslam, 2001). Assuming columnar structure and grain sizes of about 15 nm, these simulations indicate that grain rotation can play a role in grain growth that is as equally important as grain boundary migration. The simulations predict that necessary changes in the grain shape during grain rotation in columnar polycrystalline structures can be accommodated by diffusion either through the grain boundaries or through the grain interior (Moldovan, 2001). Based on this result, the authors have suggested that both mechanisms, which can be coupled, should be accounted for in mesoscopic models of grain growth. Moreover, Moldovan et al. have recently reported the existence of a critical length scale in the system that enables the growth process to be characterized by two regimes. If the average grain size is smaller than the critical length, as in the case of nanocrystals, grain growth is dominated by the grain-rotation coalescence mechanism. For average grain sizes exceeding the critical size, the growth mechanism is due to grain boundary migration.

Large-scale simulations of the structure, fracture, and sintering of covalent materials have also been studied using large-scale atomic modeling (Vashishta, 2001). Vashishta and coworkers, for example, have simulated the sintering of nanocluster-assembled silicon nitride and nanophase silicon carbide. The simulations, which used many-body potentials to describe the bonding, revealed a disordered interface between nanograins. This is a common feature of grain boundaries in polycrystalline ceramics, as opposed to more ordered interfaces typical of metals. In the silicon nitride simulations, the amorphous region contained undercoordinated silicon atoms; and because this disordered region is less stiff than the crystalline region, the elastic modulus was observed to decrease in systems with small grain sizes within which more of the sample is disordered. In simulations of crack propagation in this system, the amorphous intergranular regions were found to deflect cracks, resulting in crack branching. This behavior allowed the simulated system to maintain a much higher strain than a fully crystalline system. In the silicon carbide simulations, onset of sintering was observed at 1500 K, in agreement with neutron-scattering experiments. This temperature is lower than that for polycrystalline silicon carbide with larger grain sizes, and therefore is apparently due to the nanocrystalline structure of the samples. The simulations also predict that bulk modulus, shear modulus, and Young's modulus all have a power–law dependence on density with similar exponents.

In related studies, Keblinski et al. have used atomic simulation to generate nanocrystalline samples of silicon and carbon (Keblinski, 1997; Keblinski, 1999). For silicon, which used the Stillinger–Weber potential, disordered layers with structures similar to bulk amorphous silicon between grains were predicted by the simulations. This result suggests that this structure is thermodynamically stable for nanocrystalline silicon, and that some mechanical properties can be understood by assuming a two-phase system. In the case of carbon, grains with the diamond cubic structure connected by disordered regions of sp^2-bonded carbon have been revealed by molecular simulation. These disordered regions may be less susceptible to brittle fracture than crystalline diamond.

28.4 Concluding Remarks

By analyzing trends in computing capabilities, Vashishta and coworkers have concluded that the number of atoms that can be simulated with analytic potentials and with first-principle methods is increasing exponentially over time (Nakano, 2001). For analytic potentials, this analysis suggests that the number of atoms that can be simulated has doubled every 19 months since the first liquid simulations using continuous potential by Rahman in the early 1960s (Rahman, 1964). For simulations using first-principles forces, the same analysis suggests that the number of atoms that can be simulated has doubled every 12 months since the Car–Parrinello method was introduced in the middle 1980s (Nakano, 2001). With these extrapolations, modeling a gold interconnect 0.1 μm on a side and 100 μm long is just about feasible now with potentials such as the EAM, while first-principles simulations will have to wait almost two decades (or longer) to attack a problem of this size. However, increases in computing are only one side of a convergence among modeling, experiment, and technology. Over the same period of time it will take for first-principles modeling to rise to the scale of current interest in electronic device technology, there will be a continuing shrinkage of device dimensions. This means that modeling and technological length scales will converge over the next decade. Indeed, a convergence of sorts is already apparent in nanotube electronic properties and in molecular electronics, as is apparent from the chapters on these subjects in this handbook. This is clearly an exciting time, with excellent prospects for modeling and theory in the next few years and beyond.

Acknowledgments

Helpful discussions with Kevin Ausman, Jerzy Bernholc, Rich Colton, Brett Dunlap, Mike Falvo, Dan Feldheim, Alix Gicquel, Al Globus, Chris Gorman, Jan Hoh, Richard Jaffe, Jackie Krim, J.-P. Lu, John Mintmire, Dorel Moldovan, A. Nakano, Airat Nazarov, Boris Ni, John Pazik, Mark Robbins, Daniel Robertson, Chris Roland, Rod Ruoff, Peter Schmidt, Susan Sinnott, Deepak Srivastava, Richard Superfine, Priya Vashishta, Kathy Wahl, Carter White, Sean Washburn, Victor Zhirnov, and Otto Zhou are gratefully acknowledged. The authors wish to acknowledge support for their research efforts from the Air Force Office of Scientific Research, the Army Research Office, the Department of Defense, the Department of Energy, the National Aeronautics and Space Administration, the Petroleum Research Fund, the National Science Foundation, the Office of Naval Research, and the Research Corporation.

References

Abell, G.C. (1985) Empirical chemical pseudopotential theory of molecular and metallic bonding, *Phys. Rev. B* 31, 6184–6196.

Adiga, S.P. and Brenner, D.W. (2002) Virtual molecular design of an environment-responsive nanoporous system, *Nanoletters*, 2, 567–572.

Balasubramanian, S., Klein, M., and Siepmann, J.I. (1996) Simulation studies of ultrathin films of linear and branched alkanes on a metal substrate, *J. Phys. Chem.* 100, 11960.

Balbes, L.M., Mascarella, S.W., and Boyd, D.B. (1994) Perspectives of modern methods in computer-aided drug design, in *Reviews in Computational Chemistry*, K.B. Lipkowitz and D.B. Boyd, Eds., VCH Publishers, New York, 1994, Vol. 5, pp. 337–379.

Baskes, M.I., Nelson, J.S., and Wright, A.F. (1989) Semiempirical modified embedded-atom potentials for silicon and germanium, *Phys. Rev. B* 40, 6085.

Bitsanis, I., Magda, J., Tirrell, M., and Davis, H. (1987) Molecular dynamics of flow in micropores, *J. Chem. Phys.* 87, 1733–1750.

Bitsanis, I., Somers, S.A., Davis, T., and Tirrell, M. (1990) Microscopic dynamics of flow in molecularly narrow pores, *J. Chem. Phys.* 93, 3427–3431.

Bowen, J.P. and Allinger, N.L. (1991) Molecular mechanics: the art and science of parameterization, in *Reviews in Computational Chemistry*, K.B. Lipkowitz and D.B. Boyd, Eds., VCH Publishers, New York, Vol. 2, pp. 81–97.

Brenner, D.W. (1989) Relationship between the embedded-atom method and Tersoff potentials, *Phys. Rev. Lett.* 63, 1022.

Brenner, D.W. (1990) Empirical potential for hydrocarbons for use in simulating the chemical vapor deposition of diamond films, *Phys. Rev. B* 42, 9458

Brenner, D.W., Schall, J.D., Mewkill, J.P., Shenderova, O.A., and Sinnott, S.B. (1998) Virtual design and analysis of nanometer-scale sensor and device components, *J. Brit. Interplanetary Soc.*, 51 137 (1998).

Brenner, D.W., Shenderova, O.A., Areshkin, D.A., Schall, J.D., and Frankland, S.-J.V. (2002) Atomic modeling of carbon-based nanostructures as a tool for developing new materials and technologies, in *Computer Modeling in Engineering and Science*, in press.

Brenner, D.W., Sinnott, S.B., Harrison, J.A., and Shenderova, O.A. (1996) Simulated engineering of nanostructures, *Nanotechnology*, 7, 161.

Burkert, U. and Allinger, N.L. (1982) *Molecular Mechanics*, ACS Monograph 177, American Chemical Society, Washington, D.C.

Cagin, T., Jaramillo–Botero, A., Gao, G., and Goddard, W.A. (1998) Molecular mechanics and molecular dynamics analysis of Drexler–Merkle gears and neon pump, *Nanotechnology*, 9, 143–152.

Car, R. and Parrinello, M. (1985) Unified approach for molecular dynamics and density-functional theory, *Phys. Rev. Lett.* 55, 2471.

Carlsson, A.E. (1991) Angular forces in group-VI transition metals: application to W(100), *Phys. Rev. B* 44, 6590.

Chan, D.Y.C. and Horn, R.G. (1985) The drainage of thin liquid films between solid surfaces, *J. Chem. Phys.* 83, 5311–5324.

Cheong, W.C.D. and Zhang, L.C. (2000) Molecular dynamics simulation of phase transformations in silicon monocrystals due to nano-indentation, *Nanotechnology* 11, 173.

Clarke, D.R., Kroll, M.C., Kirchner, P.D., Cook, R.F., and Hockey, B.J. (1988) Amorphization and conductivity of silicon and germanium induced by indentation, *Phys. Rev. Lett.* 60, 2156.

Cyrot-Lackmann, F. (1968) Sur le calcul de la cohesion et de la tension superficielle des metaux de transition par une methode de liaisons fortes, *J. Phys. Chem. Solids* 29, 1235.

Daw M.S. and Baskes, M.I. (1983) Semiempirical quantum mechanical calculation of hydrogen embrittlement in metals, *Phys. Rev. Lett.* 50, 1285.

Derlet, P.M., Meyer, R., Lewis, L.J., Stuhr, U., and Van Swygenhoven, H. (2001) Low-frequency vibrational properties of nanocrystalline materials, *Phys. Rev. Lett.* 87, 205501.

Diestler, D.J., Schoen, M., and Cushman, J.H. (1993), On the thermodynamic stability of confined thin films under shear, *Science* 262, 545–547.

Drexler, E. (1992) *Nanosystems: Molecular Machinery, Manufacturing and Computation*, Wiley, New York.

Ercolessi, F., Parrinello, M., and Tossatti, E. (1988) Simulation of gold in the glue model, *Philos. Mag. A* 58, 213.

Finnis M.W. and Sinclair, J.E. (1984) A simple empirical n-body potential for transition metals, *Philos. Mag. A* 50, 45.

Foiles, S.M. (1993) Interatomic interactions for Mo and W based on the low-order moments of the density of states, *Phys. Rev. B* 48, 4287.

Foulkes, W.M.C. and Haydock, R. (1989) Tight-binding models and density-functional theory, *Phys. Rev. B* 39, 12520.

Frankland, S.J.V., Caglar A., Brenner D.W., and Griebel M. (2002) Molecular simulation of the influence of chemical cross-links on the shear strength of carbon nanotube-polymer interfaces, *J. Phys. Chem. B* 106, 3046.

Frauenheim, Th., Weich, F., Kohler, Th., Uhlmann, S., Porezag, D., and Seifert, G. (1995) Density-functional-based construction of transferable nonorthogonal tight-binding potentials for Si and SiH, *Phys. Rev. B* 52, 11492.

Garg, A., Han, J., and Sinnott, S.B. (1998a) Interactions of carbon-nanotubule proximal probe tips with diamond and graphene, *Phys. Rev. Lett.* 81, 2260.

Garg, A. and Sinnott, S.B. (1998b) Effect of chemical functionalization on the mechanical properties of carbon nanotubes, *Chem. Phys. Lett.* 295, 273.

Garrison, B.J. Srivastava, D. (1995) Potential energy surface for chemical reactions at solid surfaces, *Annu. Rev. Phys. Chem.* 46, 373.

Gee, M.L., McGuiggan, P.M., Israelachvili, J.N., and Homola, A.M. (1990) Liquid to solidlike transitions of molecularly thin films under shear, *J. Chem. Phys.* 93, 1895.

Gibson, J.B., Goland, A.N., Milgram, M., and Vineyard, G.H. (1960) Dynamics of radiation damage, *Phys. Rev.* 120, 1229.

Gorman, C.B. and Smith, J.C. (2000) Effect of repeat unit flexibility on dendrimer conformation as studied by atomistic molecular dynamics simulations, *Polymer* 41, 675.

Granick, S., Damirel, A.L., Cai, L.L., and Peanasky, J. (1995) Soft matter in a tight spot: nanorheology of confined liquids and block copolymers, *Isr. J. Chem.* 35, 75–84.

Han, J. (1998) Energetics and structures of fullerene crop circles, *Chem. Phys. Lett.* 282, 187.

Han, J., Globus, A., Jaffe R., and Deardorff, G. (1997) Molecular dynamics simulations of carbon nanotube-based gears, *Nanotechnology* 8, 95.

Harris, J. (1985) Simplified method for calculating the energy of weakly interacting fragments, *Phys. Rev. B* 31, 1770.

Harrison, J.A., Stuart, S.J., Robertson, D.H., and White, C.T. (1997) Properties of capped nanotubes when used as SPM tips, *J. Phys. Chem. B.* 101 9682.

Harrison, J.A., White, C.T., Colton, R.J., and Brenner, D.W. (1992) Nanoscale investigation of indentation, adhesion, and fracture of diamond (111) surfaces, *Surf. Sci.* 271, 57.

Haslam, A.J., Phillpot, S.R., Wolf, D., Moldovan, D., and Gleiter, H. (2001) Mechanisms of grain growth in nanocrystalline fcc metals by molecular-dynamics simulation, *Mater. Sci. Eng.* A318, 293.

Hirschfelder, J., Eyring, H., and Topley, B. (1936) Reactions involving hydrogen molecules and atoms, *J. Chem. Phys* 4, 170.

Hohenberg P. and Kohn, W. (1964) Inhomogeneous electron gas, *Phys. Rev.* 136, B864.

Horn, R.G. and Israelachvili, J.N. (1981) Direct measurement of structural forces between two surfaces in a nonpolar liquid, *J. Chem. Phys.* 75, 1400–1411.

Israelachvili, J.N., McGuiggan, P.M., and Homola, A.M. (1988) Dynamic properties of molecularly thin liquid films, *Science* 240, 189–191.

Ito, Y., Park, Y.S., and Imanishi, Y. (2000) Nanometer-sized channel gating by a self-assembled polypeptide brush, *Langmuir* 16, 5376.

Jacobsen, K.W. (1988) Bonding in metallic systems: an effective medium approach, *Comments Cond. Matter Phys.* 14, 129.

Jacobsen, K.W., Norskov, J.K., and Puska, M.J. (1987) Interatomic interactions in the effective-medium theory, *Phys. Rev. B* 35, 7423.

Kailer, A., Nickel, K.G., and Gogotsi, Y.G. (1999) Raman microspectroscopy of nanocrystalline and amorphous phases in hardness indentations, *J. Raman Spectrosc.* 30, 939.

Kallman, J.S., Hoover, W.G., Hoover, C.G., De Groot, A.J., Lee, S.M., and Wooten, F. (1993) Molecular dynamics of silicon indentation, *Phys. Rev. B* 47, 7705.

Karatasos, K., Adolf, D.B., and Davies, G.R. (2001) Statics and dynamics of model dendrimers as studied by molecular dynamics simulations, *J. Chem. Phys.* 115, 5310.

Keblinski, P., Phillpot, S.R., Wolf, D., and Gleiter, H. (1997) On the thermodynamic stability of amorphous intergranular films in covalent materials, *J. Am. Ceramic Soc.* 80, 717.

Keblinski, P., Phillpot, S.R., Wolf, D., and Gleiter, H. (1999) On the nature of grain boundaries in nanocrystalline diamond, *Nanostruct. Mater.* 12, 339.

Kohn, W. and Sham, L.J. (1965) Self-consistent equations including exchange and correlation effects, *Phys. Rev. A* 140, 1133.

Kwon, I., Biswas, R., Wang, C.Z., Ho, K.M., and Soukoulis, C.M. (1994) Transferable tight-binding model for silicon, *Phys. Rev. B* 49, 7242.

Landman, U., Luedtke, W.D., and Ribarsky, M.W. (1989a) Structural and dynamical consequences of interactions in interfacial systems, *J. Vac. Sci. Technol.* A 7, 2829–2839.

Landman, U., Luedtke, W.D., and Ribarsky, M.W. (1989b) Dynamics of tip–substrate interactions in atomic force microscopy, *Surf. Sci. Lett.* 210, L117.

Landman, U., Luedtke, W.D., Burnham, N.A., and Colton, R.J. (1990) Atomistic mechanisms and dynamics of adhesion, nanoindentation, and fracture, *Science* 248, 454.

Landman, U. and Luedtke, W.D. (1991) Nanomechanics and dynamics of tip–substrate interactions, *J. Vac. Sci. Technol. B* 9, 414–423.

Landman, U., Luedtke, W.D., and Ringer, E.M. (1992) Atomistic mechanisms of adhesive contact formation and interfacial processes, *Wear* 153, 3.

Landman, U., Luedtke, W.D., Ouyang, J., and Xia, T.K. (1993) Nanotribology and the stability of nano-structures, *Jpn. J. App. Phys.* 32, 1444.

Menon, M. and Subbaswamy, K.R. (1993) Nonorthogonal tight-binding molecular-dynamics study of silicon clusters, *Phys. Rev. B* 47, 12754.

Meunier, V, Nardelli M.B., Roland, C., and Bernholc, J. (2001) Structural and electronic properties of carbon nanotube tapers, *Phys. Rev. B* 64, 195419.

Moldovan, D., Wolf, D., and Phillpot, S.R. (2001) Theory of diffusion-accommodated grain rotation in columnar polycrystalline microstructures, *Acta Mat.* 49, 3521.

Moldovan, D., Wolf, D., Phillpot, S.R., and Haslam, A.J. (2002) Role of grain rotation in grain growth by mesoscale simulation, *Acta Mat.* in press.

Nakano, A., Bachlechner, M.E., Kalia, R.K., Lidorikis, E., Vashishta, P., Voyiadjis, G.Z., Campbell, T.J., Ogata, S., and Shimojo, F. (2001) Multiscale simulation of nanosystems, *Computing Sci. Eng.*, 3, 56.

Ni, B., Andrews R., Jacques D., Qian D., Wijesundara M.B.J., Choi Y.S., Hanley L., and Sinnott, S.B. (2001) A combined computational and experimental study of ion-beam modification of carbon nanotube bundles, *J. Phys. Chem. B* 105, 12719.

Nosé, S. (1984a) A unified formulation of the constant-temperature molecular dynamics method, *J. Chem. Phys.* 81, 511–519.

Nosé, S. (1984b) A molecular dynamics method for simulations in the canonical ensemble, *Mol. Phys.* 52, 255–268.

Orlikowski, D., Nardelli, M.B., Bernholc, J., and Roland, C. (1999) Ad-dimers on strained carbon nanotubes: a new route for quantum dot formation? *Phys. Rev. Lett.* 83, 4132.

Park, Y.S., Toshihiro, I., and Imanishi, Y. (1998) Photocontrolled gating by polymer brushes grafted on porous glass filter, *Macromolecules* 31, 2606.

Pethica, J.B. and Sutton, A.P. (1988) On the stability of a tip and flat at very small separations, *J. Vac. Sci. Technol. A* 6, 2490.

Pettifor D.G. and Oleinik I.I. (1999) Analytic bond-order potentials beyond Tersoff–Brenner. I. Theory, *Phys. Rev. B* 59, 8487.

Pettifor D.G. and Oleinik I.I. (2000) Bounded analytic bond-order potentials for sigma and pi bonds, *Phys. Rev. Lett.* 84, 4124.

Pharr, G.M., Oliver, W.C., Cook, R.F., Kirchner, P.D., Kroll, M.C., Dinger, T.R., and Clarke, D.R. (1992) Electrical resistance of metallic contacts on silicon and germanium during indentation, *J. Mat. Res.* 7, 961.

Rahman, A. (1964) Correlations in the motion of liquid argon, *Phys. Rev.* 136A, 405.

Rappé, A.K. and Goddard, W.A. (1991) Charge equilibration for molecular dynamics simulations, *J. Phys. Chem.* 95, 3358.

Ribarsky, M.W. and Landman, U. (1992) Structure and dynamics of n-alkanes confined by solid surfaces. I. Stationary crystalline boundaries, *J. Chem. Phys.* 97, 1937–1949.

Robertson, D.H., Brenner, D.W., and White, C.T. (1994) Fullerene/tubule based hollow carbon nanogears, *Mat. Res. Soc. Symp. Proc.* 349, 283.

Samaras, M., Derlet, P.M., Van Swygenhoven, H., and Victoria, M. (2002) Computer simulation of displacement cascades in nanocrystalline Ni, *Phys. Rev. Lett.* 88, 125505.

Scherrenberg, R., Coussens, B., van Vliet, P., Edouard, G., Brackman, J., and de Brabander, E. (1998) The molecular characteristics of poly(propyleneimine) dendrimers as studied with small-angle neutron scattering, viscosimetry, and molecular dynamics, *Macromolecules* 31, 456.

Schiøtz, J., Di Tolla, F.D., and Jacobsen, K.W. (1998) Softening of nanocrystalline metals at very small grain sizes, *Nature* 391, 561.

Schoen, M., Rhykerd, C.L., Diestler, D.J., and Cushman, J.H. (1987) Fluids in micropores. I. Structure of a simple classical fluid in a slit-pore, *J. Chem. Phys.* 87, 5464–5476.

Schoen, M., Rhykerd, C.L., Diestler, D.J., and Cushman, J.H. (1989) Shear forces in molecularly thin films, *Science* 245, 1223.

Selinger, R.L.B., Farkas, D., Abraham, F., Beltz, G.E., Bernstein, N., Broughton, J.Q., Cannon, R.M., Corbett, J.M., Falk, M.L., Gumbsch, P., Hess, D., Langer, J.S., and Lipkin, D.M. (2000) Atomistic theory and simulation of fracture, *Mater. Res. Soc. Bull.* 25, 11.

Shenderova, O., Mewkill, J., and Brenner, D.W. (2000) Nanoindentation as a probe of nanoscale residual stresses: atomistic simulation results, *Molecular Simulation* 25, 81.

Shenderova, O.A., Lawson, B.L., Areshkin, D., and Brenner, D.W. (2001) Predicted structure and electronic properties of individual carbon nanocones and nanostructures assembled from nanocones, *Nanotechnology* 12, 291.

Siefert, G., Kohler, T., and Frauenheim, T. (2000) Molecular wires, solenoids, and capacitors and sidewall functionalization of carbon nanotubes, *App. Phys. Lett.* 77, 1313.

Sinnott, S.B., Colton, R.J., White, C.T., and Brenner, D.W. (1994) Surface patterning by atomically controlled chemical forces – molecular dynamics simulations, *Surf. Sci.* 316, L1055.

Slater, J.C. and Koster, G.F. (1954) Simplified LCAO method for the periodic potential problem, *Phys. Rev.* 94, 1498.

Smith, G.S., Tadmor, E.B., and Kaxiras, E. (2000) Multiscale simulation of loading and electrical resistance in silicon nanoindentation, *Phys. Rev. Lett.* 84, 1260.

Smith, J.R., Bozzolo, G., Banerjea, A., and Ferrante, J. (1989) Avalanche in adhesion, *Phys. Rev. Lett.* 63, 1269–1272.

Sokol, P.E., Ma, W.J., Herwig, K.W., Snow, W.M., Wang, Y., Koplik, J., and Banavar, J.R. (1992) Freezing in confined geometries, *Appl. Phys. Lett.* 61, 777–779.

Srivastava, D. (1997) A phenomenological model of the rotation dynamics of carbon nanotube gears with laser electric fields, *Nanotechnology* 8, 186.

Srivastava D., Brenner, D.W., Schall, J.D., Ausman, K.D., Yu, M.F., and Ruoff, R.S. (1999) Predictions of enhanced chemical reactivity at regions of local conformational strain on carbon nanotubes: kinky chemistry, *J. Phys. Chem. B* 103, 4330.

Stillinger, F. and Weber, T.A. (1985) Computer simulation of local order in condensed phases of silicon, *Phys. Rev. B* 31, 5262.

Streitz, F.H. and Mintmire, J.W. (1994) Electrostatic potentials for metal-oxide surfaces and interfaces, *Phys. Rev. B* 50, 11996–12003.

Stuart, S.J., Tutein, A.B., and Harrison, J.A. (2000) A reactive potential for hydrocarbons with intermolecular interactions, *J. Chem. Phys.* 112, 6472.

Sutton, A.P. and Chen, J. (1990) Long-range Finnis–Sinclair potentials, *Philos. Mag. Lett.* 61, 139.

Sutton, A.P. (1993) *Electronic Structure of Materials*, Clarendon Press, Oxford.

Sutton, A.P., Goodwin P.D., and Horsfield, A.P. (1996) Tight-binding theory and computational materials synthesis, *Mat. Res. Soc. Bull.* 21, 42.

Tomagnini, O., Ercolessi, F., and Tosatti, E. (1993) Microscopic interaction between a gold tip and a Pb(110) surface, *Surf. Sci.* 287/288, 1041–1045.

Tupper, K.J. and Brenner, D.W. (1994a) Compression-induced structural transition in a self-assembled monolayer, *Langmuir* 10, 2335–2338.

Tupper, K.J., Colton, R.J., and Brenner, D.W. (1994b), Simulations of self-assembled monolayers under compression: effect of surface asperities, *Langmuir* 10, 2041–2043.

Tutein, A.B., Stuart, S.J., and Harrison, J.A. (1999) Indentation analysis of linear-chain hydrocarbon monolayers anchored to diamond, *J. Phys. Chem. B*, 103, 11357.

Thompson, P.A. and Robbins, M.O. (1990a) Origin of stick-slip motion in boundary lubrication, *Science* 250, 792–794.

Thompson, P.A. and Robbins, M.O. (1990b) Shear flow near solids: epitaxial order and flow boundary conditions, *Phys. Rev. A* 41, 6830–6837.

Thompson, P.A., Grest, G.S., and Robbins, M.O. (1992) Phase transitions and universal dynamics in confined films, *Phys. Rev. Lett.* 68, 3448–3451.

Tersoff, J. (1986) New empirical model for the structural properties of silicon, *Phys. Rev. Lett.* 56, 632.

Tersoff, J. (1989) Modeling solid-state chemistry: interatomic potentials for multicomponent systems, *Phys. Rev. B* 39, 5566.

Tuzun, R., Noid, D.W., and Sumpter, B.G. (1995a) Dynamics of a laser-driven molecular motor, *Nanotechnology* 6, 52.

Tuzun, R., Noid, D.W., and Sumpter, B.G. (1995b) The dynamics of molecular bearings, *Nanotechnology* 6, 64.

Van Alsten, J. and Granick, S. (1988) Molecular tribometry of ultrathin liquid films, *Phys. Rev. Lett.* 61, 2570.

Van Swygenhoven H. and Derlet, P.M. (2001) Grain-boundary sliding in nanocrystalline fcc metals, *Phys. Rev. B* 64, 224105.

Van Swygenhoven, H., Spaczer, M., Caro, A., and Farkas, D. (1999) Competing plastic deformation mechanisms in nanophase metals, *Phys. Rev. B* 60, 22.

Vashishta, P., Bachlechner, M., Nakano, A, Campbell, T.J., Kalia, R.K., Kodiyalam, J., Ogata, S., Shimojo, F., and Walsh, P. (2001) Multimillion atom simulation of materials on parallel computers —nanopixel, interfacial fracture, nanoindentation and oxidation, *Appl. Surf. Sci.* 182, 258.

Walsh, P., Kalia, R.K., Nakano, A., Vashishta, P., and Saini, S. (2000) Amorphization and anisotropic fracture dynamics during nanoindentation of silicon nitride: a multimillion atom molecular dynamics study, *Appl. Phys. Lett.* 77, 4332.

Wang, Y., Hill, K., and Harris, J.G. (1993a) Thin films of n-octane confined between parallel solid surfaces. Structure and adhesive forces vs. film thickness from molecular dynamics simulations, *J. Phys. Chem.* 97, 9013.

Wang, Y., Hill, K., and Harris, J.G. (1993b) Comparison of branched and linear octanes in the surface force apparatus. A molecular dynamics study, *Langmuir* 9, 1983.

Xu, C.H., Wang, C.Z., Chan, C.T., and Ho, K.M. (1992) A transferable tight-binding potential for carbon, *J. Phys. Condens. Matt.* 4, 6047.

Yamakov, V., Wolf, D., Salazar, M., Phillpot, S.R., and Gleiter, H. (2001) Length-scale effects in the nucleation of extended dislocations in nanocrystalline Al by molecular dynamics simulation, *Acta Mater.* 49, 2713–2722.

Zacharopoulos, N. and Economou, I.G. (2002) Morphology and organization of poly(propylene imine) dendrimers in the melt from molecular dynamics simulation, *Macromolecules* 35, 1814.

Zimmerman, J.A., Kelchner, C.L., Klein, P.A., Hamilton, J.C., and Foiles, S.M. (2001) Surface step effects on nanoindentation, *Phys. Rev. Lett.* 87, 165507.

29

Accelerated Design Tools for Nanophotonic Devices and Applications

James P. Durbano
Ahmed S. Sharkawy
Shouyuan Shi
Fernando E. Ortiz
Petersen F. Curt
EM Photonics, Inc.

Dennis W. Prather
University of Delaware

29.1 Introduction 29-1
29.2 The Celerity™ Hardware Accelerator 29-2
29.3 Mie Theory Comparison 29-3
29.4 Coupled-Resonator Optical Waveguides 29-5
   Introduction • Numerical Simulations • Summary
   and Conclusion
29.5 Analysis of Negative Index (Left-Handed) Materials...29-11
   Introduction • LHM Modeling and Simulation • Numerical
   Demonstration of Negative Refractive Index • Conclusion
29.6 Subwavelength Optical Systems 29-15
   Introduction • Numerical Validation in Hardware • Asymmetrical
   Lens Simulation and Iterative Design • Conclusion
29.7 Summary and Conclusion...29-21
References 29-21

29.1 Introduction

The explosion of nanophotonic design over the past two decades has led to amazing breakthroughs in areas such as telecommunications, biological and physical sciences, integrated circuits, and computing. Behind this technology lies an integral, often overlooked factor in the design process: simulation. Before the devices of tomorrow are fabricated, they are first modeled and extensively simulated. It is here that new ideas are tested, modified, retested, and constantly manipulated to create the optimal device. Unfortunately, as nanophotonic designs have grown ever more complicated, researchers have taken the current generation of modeling and simulation tools to their limits. Despite the wealth of computational algorithms available to model such devices, current computer system technology remains unable to fully and accurately model many devices and systems. In fact, the lack of a powerful simulation platform has directly impeded progress in this field. For example, in their 2003 *Optics Express* paper on negative index materials (NIMs), Greegor et al. state that "the direct detailed simulation of a large number of unit cells is not possible due to computer memory and computational time requirements" [1]. Although supercomputers and clusters of computers can be used to shorten the computational time, these solutions can be prohibitively expensive and frequently impractical. As a result, an approach that increases simulation speeds in a relatively inexpensive and practical way is required.

To this end, EM Photonics, Inc., has developed a computational electromagnetic (CEM) accelerator, and special-purpose hardware that implements the finite-difference time-domain (FDTD) method, in order to perform rapid simulations of large problems. This hardware, known as Celerity™, rivals the performance of a 150-node PC cluster and is capable of solving problems in excess of 270 million nodes, roughly $(33 \lambda)^3$, far outperforming the capabilities of single PCs [2]. In this chapter, we show how this hardware platform can significantly accelerate the analysis of several problems of interest to the nanophotonics design community, including coupled-resonator optical waveguides (CROW), NIMs (also known as a left-handed materials [LHMs]), and subwavelength optical lenses. Each of these structures will be discussed and the simulation results produced by the hardware accelerator will be presented. Whenever possible, analytic or published experimental solutions associated with each model are included for comparison. The reader will find that CEM hardware accelerators are incredibly fast, accurate, and robust, and are viable platforms for both academic and industrial applications.

29.2 The Celerity™ Hardware Accelerator

The acceleration platform used to model and simulate each of the devices presented in this chapter consists of a host PC, a CAD interface, and a custom, FPGA-based PCI card [2,3]. We modeled some of the structures shown in our custom CAD environment and others with the OptiFDTD MAX™ platform by Optiwave Systems, Inc., which integrates the Celerity acceleration board with the OptiFDTD CAD front end [4]. The front-end CAD software then sends the appropriate data, such as the mesh size and the number of timesteps to execute, to the hardware via the PCI bus. The FDTD accelerator proceeds to update the fields, periodically sending the results back to the host computer for postprocessing and visualization. The accelerator board itself supports up to 16 Gb of DDR SDRAM, 36 Mb of DDR SRAM, a Xilinx Virtex-II 8000 FPGA, and a PLX 9656 external PCI controller (see Figure 29.1). This card fits into standard PCs and supports the PCI 64/66 bus for maximum PC accelerator throughput. These components provide a peak memory bandwidth of 25 Gb/s and a computational throughput of 40 GFLOPS.

In the next section, we discuss one of the most common benchmarking problems, comparison to Mie theory, and demonstrate the accuracy of the hardware solver through comparison with analytic results.

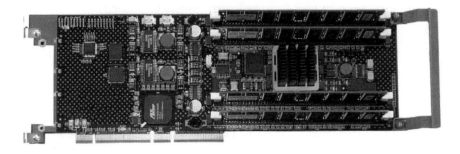

FIGURE 29.1 Hardware acceleration platform. This is the FPGA-based board that houses the acceleration architecture. It includes a Xilinx Virtex-II 8000 FPGA, 16 Gb DDR SDRAM, and 36 Mb DDR SRAM.

29.3 Mie Theory Comparison

In his 1908 paper, Gustav Mie described how spherical particles scatter electromagnetic radiation [5]. Specifically, Mie theory provides a mathematical representation of how an incident, uniform plane wave, which irradiates a homogenous, isotropic sphere, will scatter. Perhaps the most popular mathematical formulation of Mie theory was described in 1941 by Julius Stratton [6]. In Stratton's formulation, radiating spherical vector wavefunctions are used to describe the incident wave and scattering fields, while regular spherical vector wavefunctions are used to describe the internal fields. The expansion coefficients of the scattered fields are then computed by enforcing the boundary conditions on the surface [7]. The interested reader is directed to chapter 4 of Ref. [8] for an excellent discussion of this topic.

Because Mie theory comparisons are frequently used by developers of CEM simulation tools to verify accuracy, one of the first simulations run on the hardware accelerator was a dielectric sphere in free space, irradiated by a uniform plane wave. A z-directed, theta-polarized sinusoidal plane wave of frequency 2.4 GHz and unity amplitude was launched at a glass sphere. The solution space was $250 \times 250 \times 500$ mm ($2\lambda \times 2\lambda \times 4\lambda$) with uniform sampling every 1 mm (approximately 31 million nodes). The dielectric sphere (relative permittivity 2.7225) had a radius of 62.5 mm and was centered at (125 mm, 125 mm, 125 mm). Perfectly matched layer (PML) boundaries, 16 cells thick, surrounded the computational mesh. A steady-state plane detector was placed in the YZ plane at the $x = 125$ mm slice to sample the electric fields after 4000 timesteps. To verify accuracy against Mie theory, the steady-state scattered electric fields along a line scan of length 2λ located 2λ from the center of the sphere were compared (see Figure 29.2).

Figure 29.3 and Figure 29.4 illustrate the numerical differences between the hardware and analytic solutions. Pictured are the magnitude and phase for x-directed electric fields. The y- and z-directed fields

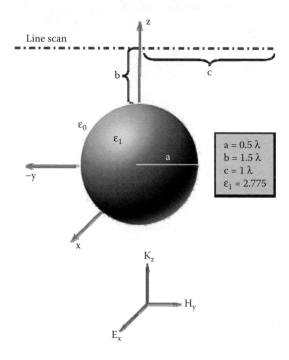

FIGURE 29.2 Scattering off a dielectric sphere. A plane wave was launched at a sphere of radius 0.5λ and the steady-state fields of both the hardware solver and the analytic solution were compared along a line two wavelengths from the center of the sphere.

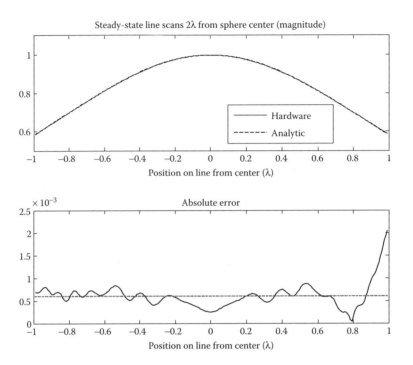

FIGURE 29.3 Comparison of hardware and analytic solvers — magnitude. (*Top*) Magnitude of steady-state E_x fields as computed by Mie series and hardware. Because the results are almost identical, only one line can be seen. (*Bottom*) Absolute difference between the two solutions. Note the average difference (*dashed line*) is on the order of 10^{-4}.

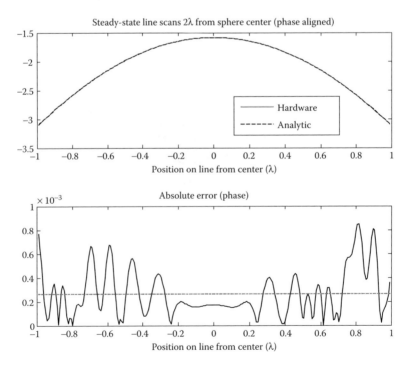

FIGURE 29.4 Comparison of hardware and analytic solvers — phase. (*Top*) Phase of steady-state E_x fields as computed by Mie series and hardware. Because the results are almost identical, only one line can be seen. (*Bottom*) Absolute difference between the two solutions. Note the average difference (*dashed line*) is on the order of 10^{-4}.

are not shown because they do not compose a significant fraction of the overall wave (E_y and E_z magnitudes are of the order of 10^{-17}). The average absolute error between the hardware and analytic results was on the order of 10^{-4}, with the peak error on the order of 10^{-3}. This is indicative of an incredibly accurate solver.

In this section, we compared the results of the hardware accelerator against the analytic results provided by Mie theory and demonstrated correspondence. Now that the accuracy of the platform has been demonstrated, we can begin discussing some of the applications that have been simulated with this platform, beginning with CROWs.

29.4 Coupled-Resonator Optical Waveguides

29.4.1 Introduction

The CROW is a new type of waveguide in which light propagates due to the coupling between adjacent resonators [9,10]. A CROW can be constructed from various resonator types, including ring resonators, microdisk resonators, and photonic crystal cavities [11–14]. Although these implementations differ in the fine details, such as confinement and coupling mechanisms, the general characteristics, such as the dispersion relation and band structure, are very similar and can be determined by free space range and the quality factor (Q) of each resonator, as well as the coupling between adjacent resonators. CROWs have attracted significant interest within the photonics community because of their ability to manipulate light and provide better control over the optical dispersion characteristics on a microscopic scale. Furthermore, the weak coupling between adjacent, high-Q resonator cavities can significantly reduce the group velocity of light in such structures and may potentially lead to applications in delaying, storing, and buffering of optical pulses, as well as laser systems [15–17].

Infinite CROW chains have been previously analyzed using a tight-binding formalism, in which the electric field of an eigenmode of the CROW can be expanded by a Bloch wave superposition of the individual resonator mode, whereas finite CROWs can be modeled using transfer matrix and time-domain analysis methods [9,10,12,18]. The transfer matrix method calculates transmission through a finite sequence of resonators and couplers by successively multiplying propagation matrices, which characterize the coupling between two adjacent ring resonators. The time-domain analysis method is based on temporal coupled mode theory, which provides a qualitative description by introducing a phenomenological coupling constant and is more favorable for the analysis of pulse propagation through the CROW. However, these models are unrealistic when designing physical systems, as they do not account for input/output coupling, loss, different resonator sizes, and variations in coupling strength [12].

Clearly, a quantitative and self-consistent study of light propagation in CROWs is needed to address various issues such as propagation losses, speed, and the efficiency of coupling light into and out of these devices. As such, a rigorous numerical electromagnetic algorithm is desirable to solve such problems. Further, such an analysis requires support for very large problem sizes, too large to be practically simulated using standard software tools, in order to support multiple rings in the structure. To this end, a hardware-based FDTD solver was used to simulate several CROW structures for an optical delay application. We also demonstrate how this tool was used to uncover previously unknown scalability relationships in CROW structures.

29.4.2 Numerical Simulations

In this section we will perform full 3-D FDTD simulations of several CROWs, including structures with 2, 5, 10, and 20 rings, to determine the delay from when an electromagnetic wave is launched at the input waveguide until it reaches the output waveguide. To determine this, two simulations are required. The first simulation will determine the resonant frequency of the CROW system and the second simulation will launch a pulse with this frequency. To validate our results, the 3-D hardware simulation of a two-ring CROW structure will be compared against the 2-D simulation results presented in [12]. It is important to note that while CROW structures can be simulated using a 2-D FDTD engine, as in [12], this requires utilizing the effective index approximation to account for the finite height of the structure. Thus, 2-D simulations will not provide any indication of the losses associated with the thickness of the structure and, therefore, cannot be used to provide accurate design models for physical CROW structures. Furthermore, 2-D simulations will produce inaccurate estimates of the resonant frequencies and the guided mode profiles within the CROW structure. Thus, a full 3-D simulation, such as the one performed by the hardware accelerator, is desired.

29.4.2.1 Two-Ring Simulation

The first simulation was performed with two coupled-ring resonators, as shown in Figure 29.5. The in- and out-coupling waveguides and rings (radius = 5 μm) are constructed as freestanding Si waveguides (refractive index 3.5) and each has the same width ($w = 0.3$ μm) and thickness ($h = 0.3$ μm). The gap between each waveguide and ring, as well as between adjacent rings, is $g = 0.3$ μm.

To find the optical resonance of this system, a broadband, Gaussian-modulated pulse (center frequency = 200 THz, bandwidth = 60 THz), was launched at the input port. The FDTD mesh size was chosen as $dx = dy = dz = 0.05$ μm with a timestep $dt = 9.526 \times 10^{-8}$ nsec. After meshing the design, the computational region contained 5.71 million FDTD nodes (i.e., Yee cells) [19]. Point detectors were placed at the input and output of the waveguides to measure transmission through the CROW structure. The simulation was run for 150,000 timesteps, requiring just over 5.5 h on the Celerity accelerator card, corresponding to roughly 42 million nodes per second (Mnps) of sustained computational throughput (see Refs. [3,20] for a detailed discussion of this performance measurement). 150,000 timesteps were required to ensure that high resolution would be achieved in the frequency domain, thus allowing accurate determination of the resonant frequency. After Fourier transforming the detector responses and normalizing the results against the input spectrum, the throughput and output responses were computed (Figure 29.6). From the figure, we determine the proper resonant frequency (i.e., the frequency that has the highest Q factor) to be 191.4 THz. Also note that, at lower frequencies, the signal is over-coupled from the input waveguide to the rings, while it is under-coupled at higher frequencies. This behavior is expected given the effective increase in gap length (g), otherwise known as the electrical length, between the waveguides, that arises with increasing the frequency.

After determining the proper resonant frequency of the structure, a pulse with center frequency 191.4 THz, bandwidth 0.4 THz, and a width of 1.2 psec was launched down the input waveguide. This second simulation was run for 200,000 timesteps and required 7.55 h of computation time on the Celerity platform. Figure 29.7 shows several snapshots of the field distribution at the center plane of CROW structure, clearly demonstrating the pulse propagating along the CROW structure. The temporal

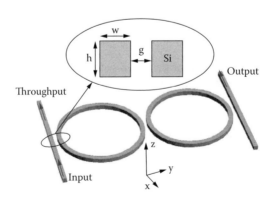

FIGURE 29.5 A CROW with two coupled-ring resonators. Because of its ability to store/delay optical signals, a CROW structure is a common building block in many optical communication systems. The input and output waveguides run parallel to one another, but are separated by ring resonators.

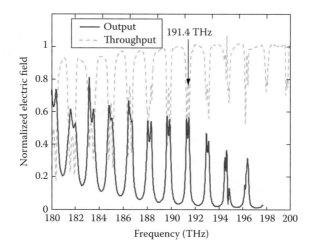

FIGURE 29.6 Transmission spectra at throughput and output ports. Here we see the input-normalized field values at the throughput and output ports of the CROW for various frequencies in the bandwidth. The split peaks of each resonance are attributed to the Fabry-Perot effect of two ring resonators. From this, we note the resonant frequency is at 191.4 THz, which is used as the center frequency in the next simulation.

responses of input and output detectors are plotted in Figure 29.8 and show how the input pulse is placed on the output after a delay of 1.88 psec, which is in agreement with the two-dimensional simulation results presented in [12]. The reader may also notice a "tail" on the output pulse. This can be attributed to back reflection of the wave as it propagates along the ring resonator and can be minimized by using a finer mesh.

29.4.2.2 Five-Ring Simulation

Following the same procedure, we consider a CROW structure with five rings. This particular structure required 13.02 million FDTD nodes to model. Again, a 150,000 timestep simulation was performed to determine the proper resonant frequency using a broadband Gaussian pulse, which required less than 13 h with the hardware accelerator. The transmission spectra at the throughput and output ports are shown in Figure 29.9. In comparing these results against those of the two-ring structure, note that the resonant frequency is the same, but the overall transmission efficiency has decreased, which is due to the increased propagation losses associated with additional rings, including bending loss, surface roughness loss, and coupling loss.

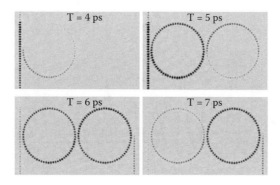

FIGURE 29.7 Propagation through a two-ring CROW structure. Here we see snapshots of the incident pulse (electric field magnitude) as it travels through the ring resonators and, ultimately, to the output waveguide. The delay from input to output directly corresponds to previously published results.

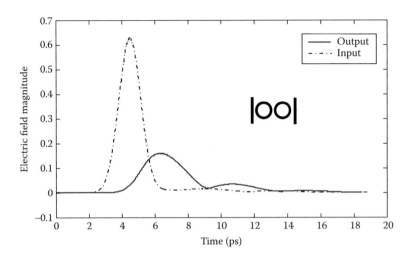

FIGURE 29.8 Pulse propagation through a two-ring CROW. Here we see the pulse as seen at the input and output ports. Note that the output pulse is delayed by 1.88 psec as compared to the input pulse, which corresponds to the published results in Ref. [12].

To investigate the performance of the delay line as the number of rings increase, we next launched the same Gaussian-modulated pulse as in the two-ring simulation. This time, however, the simulation was run for 300,000 timesteps and required 26 h. Once again, the input and output responses were sampled and are plotted in Figure 29.10. From this figure, we see that there is a 4.4 psec time delay from input to output using a five-ring structure as opposed to a 1.8 psec delay using only two rings. Figure 29.11 shows several transient results at the center plane of the CROW structure. Note the decrease in amplitude as the pulse propagates, which is a result of the propagation losses.

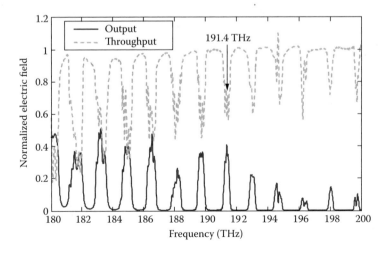

FIGURE 29.9 Transmission spectra at throughput and output ports. Here we see the input-normalized field values at the throughput and output ports of the CROW for various frequencies in the bandwidth. Once again, we note the resonant frequency is at 191.4 THz, which is used as the center frequency in the next simulation.

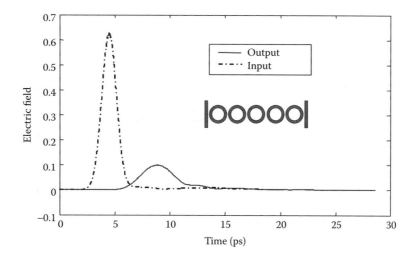

FIGURE 29.10 Pulse propagation through a five-ring CROW. Here we see the pulse as seen at the input and output ports. Note that the output pulse is delayed by 4.4 psec as compared to the input pulse.

29.4.2.3 10- and 20-Ring Simulations

Finally, we consider 10- and 20-ring resonator structures, with computational region sizes of 25.22 million and 49.61 million FDTD nodes, respectively. These simulations were performed for 300,000 and 450,000 timesteps and required 50 and 147.6 h of computation time, respectively. Previously, such a simulation would have been impossible in a standard desktop computing environment. The 10-ring CROW achieves a delay of 8.13 psec (Figure 29.12) while the 20-ring CROW achieved a 15.78 psec delay (Figure 29.13). As with the five-ring CROW, the amplitude of the pulse decays as it propagates along the waveguide, again demonstrating that the propagation loss, or the photon's lifetime (Q factor), is a critical parameter in the design of such devices.

29.4.3 Summary and Conclusion

In this section we presented results from several simulations of CROW structure delay devices using the Celerity hardware accelerator. Our initial simulations were compared against published data and demonstrated the accuracy of the platform, while the later simulations were used to model CROWs with

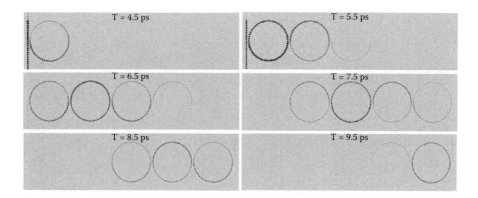

FIGURE 29.11 Propagation through a five-ring CROW structure. Here we see snapshots of the incident pulse (electric field magnitude) as it travels through the ring resonators and, ultimately, to the output waveguide. Note that the magnitude of the pulse decays over time because of propagation losses.

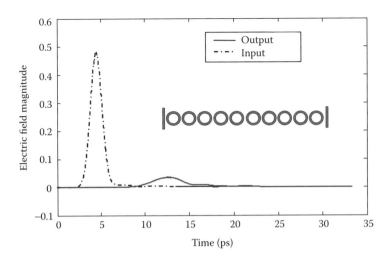

FIGURE 29.12 Pulse propagation through a 10-ring CROW. Here we see the pulse as seen at the input and output ports. Note that the output pulse is delayed by 8.13 psec as compared to the input pulse.

5-, 10-, and 20-ring resonators, whose simulation would have previously been impractical given the time and memory constraints of most simulation platforms.

To summarize our findings, the time delay and pulse amplitude as a function of the number of ring resonators is plotted in Figure 29.14. From this figure it is easy to see that time delay is a linear function of the number of rings. Using this information, it is noted that a detailed analysis of a CROW with numerous rings could be performed by using a small number of rings that are optimized for coupling efficiency, and extrapolating the results.

By harnessing the power of custom-hardware computing, full 3-D, rigorous electromagnetic analyses of realistic CROWs can be performed. This enables a complete simulation of the electromagnetic coupling and interactions of subdevices, which will provide a path to the optimal design of such systems.

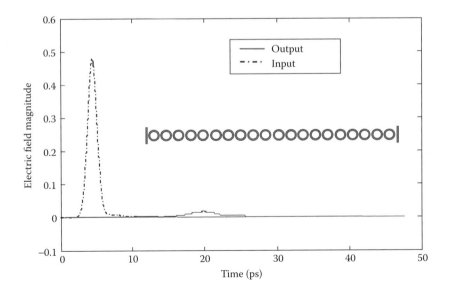

FIGURE 29.13 Pulse propagation through a 20-ring CROW. Here we see the pulse as seen at the input and output ports. Note that the output pulse is delayed by 15.78 psec as compared to the input pulse.

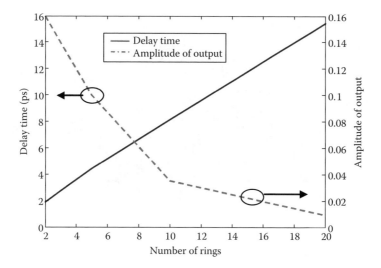

FIGURE 29.14 Delay time and amplitude as a function of the number of rings. Note that adding rings results in a linear increase in delay time, but also results in a decrease in amplitude as seen at the output, which is a result of increased propagation losses.

29.5 Analysis of Negative Index (Left-Handed) Materials

29.5.1 Introduction

Although materials exhibiting negative permeability and permittivity, referred to as metamaterials, have been postulated for many years, it is only recently that such materials have been physically realized [21–25]. The initial theory, proposed by V.G. Veselago in 1968, suggested that a slab of uniform, isotropic material with $\mu = \varepsilon = -1$ will refocus a point image perfectly by exploiting both the propagating and the evanescent waves emitted by the point object (source) [26]. Such materials are also called LHMs, because the electromagnetic triad, electric field, **E**, magnetic field, **H** and propagation wavevector, **k**, follow a left-handed relationship (i.e., $\mathbf{E} \times \mathbf{H} = -\mathbf{k}$). This unconventional electromagnetic behavior, which arises from the negative permittivity and permeability, leads to negative refraction, and, hence, an NIM may result. Veselago expected such structures to provide a reversed Doppler effect, reversed Cerenkov radiation, negative index of refraction, and superlensing. In addition, metamaterials have found application in antenna design and radar cross-section design, as well as lenses.

Recently, NIMs have been realized by forming a periodic arrangement of metallic wires, which are used to provide a negative permittivity for a specific frequency range, and an array of split-ring resonators (SRRs), which are used to provide a negative permeability (Figure 29.15) [23,24,27]. In previous attempts to model these structures, a material was embedded with a negative index of refraction as a black box and the effective index approximation was utilized [28]. While such attempts are sufficient for demonstrating the concepts of negative refraction, they do not utilize physically realizable elements to build LHMs and, hence, cannot account for various spectral and temporal material properties. Thus, in order to determine the frequency range over which such structures exhibit a negative refraction, a rigorous numerical modeling of the structure is necessary. To this end, it is necessary to model a large array of SRRs, along with the wires, to accurately extract the frequency response of the LHM structure. However, this is a nontrivial task, given the level of detail that is required in the model, which translates into massive computational size and time. In fact, Greegor et. al state that "the direct simulation of a large number of unit cells is not possible due to computer memory and computational time requirements" [1]. Although some researchers propose the use of single-cell or small-array simulations coupled with periodic boundary conditions to solve this problem, this approach causes an under-prediction of the negative

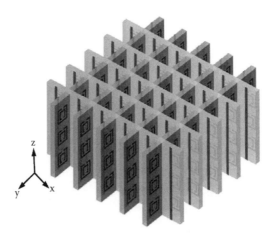

FIGURE 29.15 LHM structure. This is a model (scaled down) of the LHM structure simulated on the hardware platform. In this model, it is easy to see both the SRRs and the wires.

refractive index, resulting in an inaccurate design of the LHM material [29]. Hence, a numerical platform capable of handling such computationally intense problems, such as a hardware-based solver, is necessary. In this section, we utilize our hardware acceleration platform for the analysis of LHM structures. Specifically, the hardware accelerator is used to calculate the transmission spectra of the LHM in order to identify the frequency regions where the permittivity and permeability are negative.

29.5.2 LHM Modeling and Simulation

The LHM structure to be modeled was composed of a periodic array of unit-cells. Each unit-cell consisted of a metallic, SRR structure patterned on one side of a dielectric substrate ($\varepsilon_r = 3.4$) (Figure 29.16) and a metallic wire ($0.5 \times 1.0 \times 20$ mm) patterned on the other side of the substrate. The thickness of the SRR, the wire, and the substrate were chosen to be 0.5 mm. This unit-cell, which measured $20 \times 20 \times 20$ mm,

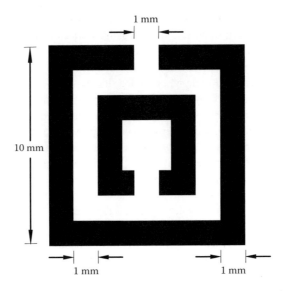

FIGURE 29.16 A metallic split-ring resonator. Here we see the SRR used in the LHM model. The length and width of each SRR was 10 mm, and the azimuthal and inter-ring gaps were 1 mm.

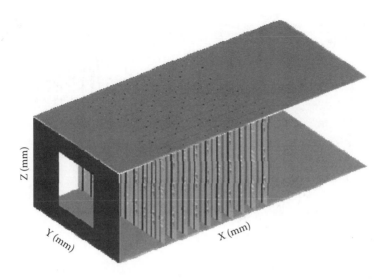

FIGURE 29.17 LHM numerical characterization setup. The LHM structure is placed inside a two-dimensional waveguide formed by parallel, perfectly conducting plates, in order to provide vertical confinement and more accurately reflect a guided mode source.

was then replicated 20 times in the *XZ* plane, 10 times in the *XY* plane, and three times in *YZ* plane to form an array of 600 SRR+wire pairs (lattice constant $a = 30$ mm) (Figure 29.15). This LHM structure was placed inside a two-dimensional waveguide formed by parallel, perfectly conducting plates, each 40.15 cm wide and 24 cm long, in order to provide vertical confinement and more accurately reflect a guided mode source (Figure 29.17).

Ultimately, the computational domain for this structure measured $24 \times 40.15 \times 5$ cm, or $481 \times 803 \times 101$ cells (~ 40 million nodes), not including the PML absorbing boundary region. For an FDTD discretization of 0.5 mm, the corresponding timestep is 9.63×10^{-4} nsec. The 20,000 timestep simulation required less than 2 GB of memory and approximately 5.5 h of computation time on the Celerity™ accelerator card. However, because the acceleration platform contains 16 GB of RAM, an LHM structure consisting of up to 4700 unit-cells could be simulated.

To determine frequencies at which negative permittivity and permeability might exist, the transmission spectra of the LHM structure is measured to identify stopband frequency regions. Once these frequency regions are identified, a continuous wave is used to examine the steady-state behavior of the LHM structure. For this particular LHM structure, a broadband, *z*-polarized windowed plane wave propagating along the *x*-direction was used. A point detector was placed at the far end of the LHM structure in order to measure the transmission spectra through the periodic array of SRRs. Specifically, the point detector recorded the time-varying electric and magnetic field amplitudes, which were then normalized to the source. The frequency response was then obtained by performing a fast Fourier transform of the normalized data. Two simulations were performed, one with the SRRs alone (without wires) and one with both SRRs and wires (Figure 29.18). From these results, we see that the SRR structure at resonance has a stopband 38 dB down at 2.75 GHz. For both wires and SRRs together, a small passband exists between 3.75 and 4.25 GHz, near the resonance of the rings.

Once an accurate measurement of the transmission spectra of the metamaterial structure has been identified, and regions of negative permittivity and permeability have been highlighted, the next step is to validate that the material does indeed posses a negative refractive index for the frequency range of interest based on the geometrical dimensions of the unit-cell containing the SRR and the backplane wires. In the next section, we further analyze this structure to demonstrate negative refraction in LHM materials.

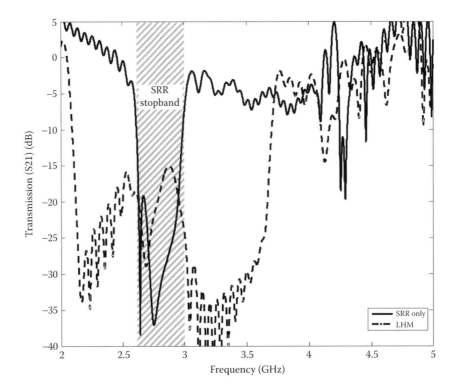

FIGURE 29.18 Transmission spectra of SRRs and SRRs + wires (LHM). From these results, we see that the SRR structure at resonance has a stopband 38 dB down at 2.75 GHz, with regions of positive and negative permeability on both sides. For both wires and SRRs together, a small passband exists between 3.75 and 4.25 GHz, near the resonance of the rings.

29.5.3 Numerical Demonstration of Negative Refractive Index

To numerically demonstrate that the composite structure formed from the combination of SRRs and wires possesses a negative index of refraction, we constructed a 26° prism and embedded the LHM structure within it (Figure 29.19). By measuring the direction of the power leaving the prism, it is possible to calculate the index of refraction using Snell's law.

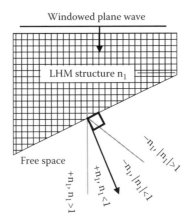

FIGURE 29.19 LHM/prism model used to demonstrate negative refractive index. By embedding the LHM structure in a prism and using Snell's law, it is possible to demonstrate the left-handed behavior of the structure.

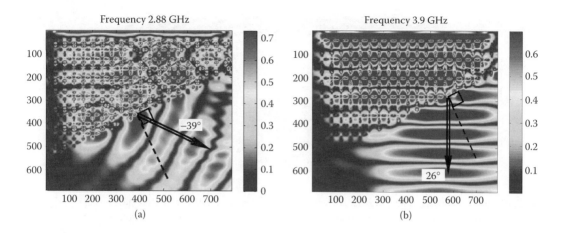

FIGURE 29.20 Demonstrated negative refractive index. Here we see the wave exiting the LHM/prism structure. (a) Note that the wave front exists to the left (negative angle) of the normal, indicating negative refraction. (b) Here, the wave front exits to the right (positive angle) of the normal, indicating positive refraction.

The prism/LHM structure was simulated using a windowed plane wave source. The frequency of the incident beam was varied according to the transmission spectra results previously obtained (Figure 29.18). Specifically, simulations were performed at frequencies of 2.88, 3.0, 3.8, and 3.9 GHz. The first frequency, 2.88 GHz, was chosen because it lies within the resonance band of the SRR structure. Simulations demonstrate that the index of refraction within the prism structure has been changed ($n = -1.4356$). Specifically, the wave front is now directed away from the surface of the prism, indicating a negative refraction index prism material (Figure 29.20a). Next, we shifted the frequency away from the resonance band of the SRRs to the edge of the stopband at 3.0 GHz. At this frequency, the overall LHM structure is highly dispersive and possesses a positive index of refraction (Figure 29.20b). Similarly, frequencies of 3.8 and 3.9 GHz were tested and both exhibited a positive index of refraction. Thus, we have shown, through numerical simulation, that synthetically engineered metamaterials do possess a negative index of refraction at a certain frequency range proportional to the geometrical dimensions of the SRR and wire unit-cells.

29.5.4 Conclusion

In conclusion, in order to accurately design and simulate LHM structures, it is necessary to examine a large number of unit-cells. Such analysis requires a numerical platform capable of handling computationally intense problems. To this end, a hardware-based solver was used to analyze an LHM structure composed of SRRs and wires. Specifically, the hardware accelerator was used to calculate the transmission spectra of the LHM in order to identify the frequency regions where the permittivity and permeability are negative. From these simulations, we confirmed a negative refractive index exists at specific frequencies for this structure. Thus, we have demonstrated that a hardware-based solver enables the analysis of LHM structures that would otherwise be impractical with standard software simulation suites.

29.6 Subwavelength Optical Systems

29.6.1 Introduction

Optical lenses and systems have historically been of interest to the design and simulation community due to their extensive use in a variety of applications, including telecommunications and imaging. Diffractive optical elements (DOEs) in particular have made a significant impact in both industry and

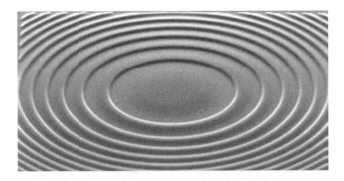

FIGURE 29.21　DOE lens with subwavelength features. This particular glass lens is used for steering light for optically interconnected systems. The image was captured using a scanning electron microscope (SEM) at 450× magnification.

academia with the advent of lithographic techniques that enable the creation of miniaturized optical systems, otherwise unachievable with conventional methods (Figure 29.21). Furthermore, the use of low-cost materials and simple fabrication techniques make these systems economically attractive for mass production applications. In addition to the economical benefits, DOEs enable reductions in the number of optical elements required for a given application, decreasing system weight and increasing durability and reliability.

In order to simulate these systems, it is often possible to take advantage of symmetry or apply physical approximations, such as ray tracing or beam propagation, in order to avoid a rigorous solution of Maxwell's equations [30]. However, in cases where these techniques are not applicable, such as when feature sizes become comparable to the wavelength of interest, full-wave electromagnetic simulations are required.

In the remainder of this section, we will demonstrate the advantages of hardware acceleration for the simulation of optical applications, DOEs particularly. Specifically, we will present a DOE lens simulation, comparing results with previously published analytic and experimental data. We then modify the basic lens arrangement to produce an optical beam splitter. Because such a device cannot be analytically described using symmetry, this example will demonstrate the unique advantages of powerful hardware solvers for the analysis of DOE applications. Finally, we discuss how hardware-based solvers are critical for iterative designs by means of a 1-to-3 beam splitter example.

29.6.2 Numerical Validation in Hardware

We begin by verifying the accuracy of the hardware accelerator by modeling a DOE lens that has been previously designed, simulated, and fabricated [31]. This particular problem is an excellent candidate for such evaluation because, in addition to the published experimental results, it is also possible to obtain an exact analytical solution by using body of revolution (BOR) symmetry simplifications [32]. These analytical results are exempt from the errors introduced by using finite differences to approximate derivatives and numerical artifacts that arise from the limited precision format in digital computers.

This particular glass lens (index of refraction 1.5) has a 5 μm focal length, a 10 × 10 μm aperture, and a minimum feature size of 1 mm (Figure 29.22). The lens was designed using a technique that combines scalar diffraction and effective medium theory and modeled using the CAD tool by EM Photonics, Inc., following the design parameters outlined in [31]. The 3-D design was sampled at $\lambda/20$ (requiring 20 million FDTD cells) and was excited for 5000 timesteps (approximately half a femtosecond) with a 300 THz plane wave in order to reach steady state. PML-absorbing boundaries, 16 layers deep, surrounded the computational space and were used to minimize back-reflections into the computational region.

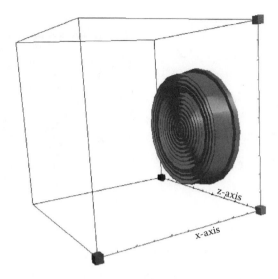

FIGURE 29.22 Subwavelength lens geometry. The lens was modeled using the CAD tool by EM Photonics, Inc., following the design parameters outlined in [31]. The design required 20 million FDTD cells and was excited for 5000 timesteps (approximately half a femtosecond) with a 300 THz plane wave.

The results presented in Figure 29.23 and Figure 29.24 are in agreement with both the analytical and experimental data published in [31]. Although a nonchallenging simulation for the hardware accelerator, this example enables the validation of hardware platform results against previously published data.

According to Ref. [31], this problem required 8 h of CPU time (albeit, in a 250 MHz workstation), but required only 15 min on the hardware acceleration platform. More importantly, this problem required only 10% of the available memory in the acceleration platform, providing significant resources for larger designs incorporating multiple components to be run in relatively short times.

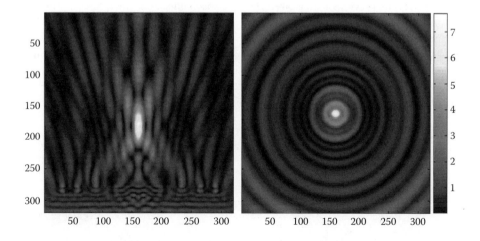

FIGURE 29.23 Focusing light. This figure shows *XY* (*left*) and *YZ* (*right*) slices of the computational space after reaching steady state. A strong electric field concentration can be seen around the focal point, qualitatively validating the focusing behavior of the lens.

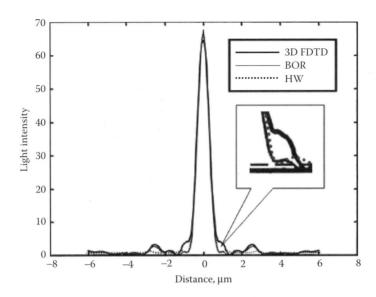

FIGURE 29.24 Result comparison. This plot compares intensity vs. distance data from three sources: analytical computations using BOR, a third-party 3-D FDTD software simulation, and the hardware (HW) accelerator. The differences between the software and hardware FDTD simulations are due to the improvements in the boundary conditions, which reflect in side-lobes that are closer to the analytical results. (From M.S. Mirotznik, D.W. Prather, J.N. Mait, W.A. Beck, S. Shi, and X. Gao, *Appl Opt*, 39, 2871–2880, 2000. With permission.)

29.6.3 Asymmetrical Lens Simulation and Iterative Design

After validating the results obtained with the hardware accelerator, we now present a problem that has important practical applications in fiber optics and optical data networks, but that cannot be easily solved using analytical methods: a DOE-based beam splitter. For this simulation, the goal was to create a lens that will split a single beam of light into four equal portions. This was achieved by dividing the lens created above into four sections and reassembling as shown in Figure 29.25. By modifying the lens into an asymmetrical structure, the BOR technique can no longer be applied and numerical techniques must be utilized. As with the previous lens, a 300 THz wave was launched at the splitter and was simulated for 5000 timesteps. In examining the simulation results (Figure 29.26), we note that intensity of the electromagnetic fields is highest (clearly focused) at the four corners of the lens, as expected. Using the hardware acceleration platform enabled the rapid, accurate simulation of a problem that would otherwise require many hours of computation time on a standard PC.

Finally, we discuss one of main advantages of hardware acceleration for computational simulations: iterative designs. For many applications, including lens design, it is necessary to run numerous simulations

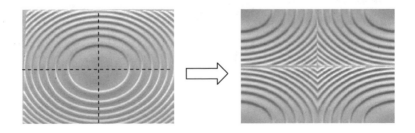

FIGURE 29.25 Beam splitter lens configuration. A beam splitter is easily constructed by starting from a focusing lens (*left*) and rearranging the quadrants to direct the energy in four different directions (*right*).

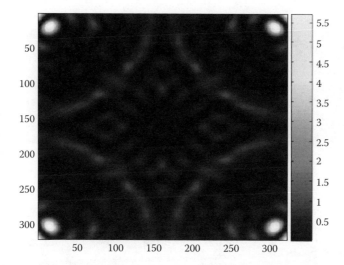

FIGURE 29.26 Beam splitter simulation results. Here we see the intensity of the simulated electric fields on the focal plane, showing that the splitter behaves as intended.

and modify the parameters after each run in order to optimize the system. Because of their rapid solution speed, hardware-accelerated platforms directly enable iterative designs. One such design, a 1-to-3 beam splitter, is shown in Figure 29.27. Here, an initial lens profile was entered into the hardware-accelerated FDTD tool and controlled by an optimization algorithm. After each iteration, the optimization algorithm modified the lens profile and a new profile was tested in the acceleration platform. This process continued until the desired efficiency was reached. Figure 29.28 and Figure 29.29 show the beam-splitter results for both the initial and optimized profiles. Note that, in the optimized profile, the beams are more intense within the focal plane. Without a hardware-accelerated solver, iterative design would be highly impractical for many designs because of the long runtimes associated with typical software simulations.

29.6.4 Conclusion

In this section, we used a popular application, DOEs, to validate the results from the hardware accelerator against published analytical and experimental data. Further, we showed how this technology could be applied to problems without simple analytical solutions. Finally, we discussed how the speed of hardware-accelerated simulations enables iterative design, which is useful for finding optimal design parameters numerically.

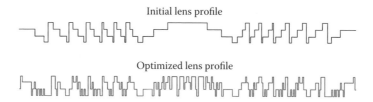

FIGURE 29.27 Optimized 1-to-3 beam splitter. This figure shows lens profiles for two 1-to-3 beam splitters, as discussed in [33]. An initial lens profile (*top*) was entered into the hardware-accelerated FDTD tool and controlled by an optimization algorithm. After numerous iterations, the second profile (*bottom*) was obtained, which is significantly more efficient. (From J.P. Durbano, Hardware implementation of a 1-dimensional FDTD algorithm for the analysis of electromagnetic propagation, University of Delaware, Newark, M.E.E. Thesis, 2002. With permission.)

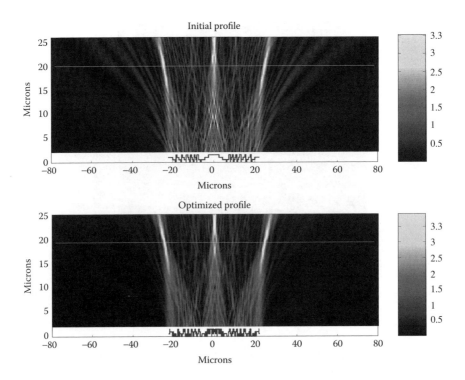

FIGURE 29.28 Amplitude propagation plots of the initial and optimized profile. Note that in the optimized profile, the beams are better focused as more energy is contained in the focal plane.

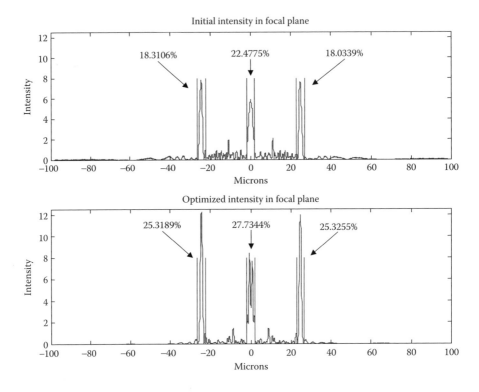

FIGURE 29.29 Initial and optimized intensity in focal plane. One sees that the optimized lens outperforms the initial design, as more energy is located in the focal plane.

29.7 Summary and Conclusion

In this chapter we demonstrated the advantages of hardware acceleration for nanophotonic device simulation. In particular, a FDTD hardware solver was used, which greatly increases the maximum problem size and dramatically decreases the computational time. Several nanophotonic applications were run on the hardware platform, with results that matched analytic and experimental solutions over a variety of problems, including a coupled-ring optical waveguide, an NIM, and a subwavelength optical lens. We also demonstrated that the hardware platform is capable of analyzing problems that had never been attempted because of their sheer size.

The development of such a remarkable tool will directly enable more advanced research in the nanophotonics field and beyond, as the acceleration hardware can accommodate larger problems than PCs and can solve them orders of magnitude faster. Previously, where only individual components could be modeled, now entire systems can be simulated. Similarly, problems which required 1 week of computation time are now finished after several hours. This system represents a new era in CEM simulations, enabling the design of next-generation devices today.

References

1. R.B. Greegor, C.G. Parazzoli, K. Li, B.E.C. Koltenbah, and M. Tanielian, Experimental determination and numerical simulation of the properties of negative index of refraction materials, *Opt Express*, 11, 688–695, 2003.
2. J.P. Durbano, F.E. Ortiz, J.R. Humphrey, and D.W. Prather, FPGA-based acceleration of the 3D finite-difference time-domain method, Presented at 12th Annual IEEE Symposium on Field-Programmable Custom Computing Machines (FCCM), 2004.
3. J.P. Durbano, J.R. Humphrey, F.E. Ortiz, P.F. Curt, D.W. Prather, and M.S. Mirotznik, Hardware acceleration of the 3D finite-difference time-domain method, Presented at IEEE AP-S International Symposium on Antennas and Propagation, 2004.
4. Optiwave Systems, Inc., www.optiwave.com.
5. G. Mie, Beiträge zur optik trüber medien, speziell kolloidaler metallösungen (Contributions on optical properties of turbid solutions, with special reference to colloid metallic solutions), *Ann Physik*, 25, 377–445, 1908.
6. J.A. Stratton, *Electromagnetic Theory*, New York: McGraw-Hill, 1941.
7. Wikipedia, http://en.wikipedia.org/wiki/Mie_theory, Mie Theory.
8. C.F. Bohren and D.R. Huffman, *Absorption and Scattering of Light by Small Particles*, New York: Wiley-Interscience, 1983.
9. A. Yariv, Y. Xu, R.K. Lee, and A. Scherer, Coupled-resonator optical waveguide: a proposal and analysis, *Opt Lett*, 24, 711–713, 1999.
10. S. Mookherjea and A. Yariv, Coupled resonator optical waveguides, *IEEE J Select Topics Quant Electron*, 8, 448–456, 2002.
11. S. Mookherjea, Dispersion characteristics of coupled-resonator optical waveguides, *Opt Lett*, 30, 2406–2408, 2005.
12. J.K.S. Poon, J. Scheuer, Y. Xu, and A. Yariv, Designing coupled-resonator optical waveguide delay lines, *J Opt Soc Am B*, 21, 1665–1673, 2004.
13. L.J. Feng, H.T. Jiang, H.Q. Li, Y.W. Zhang, and H. Chen, The dispersive characteristics of impurity bands in coupled-resonator optical waveguides of photonic crystals, *Acta Physica Sinica*, 54, 2102–2105, 2005.
14. M. Soltani, A. Adibi, Y. Xu, and R.K. Lee, Systematic design of single-mode coupled-resonator optical waveguides in photonic crystals, *Opt Lett*, 28, 1978–1980, 2003.
15. S. Mookherjea and A. Yariv, Optical pulse propagation and holographic storage in a coupled-resonator optical waveguide, *Phy Rev E*, 64, 2001.

16. J.K.S. Poon, J. Scheuer, S. Mookherjea, G.T. Paloczi, Y.Y. Huang, and A. Yariv, Matrix analysis of microring coupled-resonator optical waveguides, *Opt Express*, 12, 90–103, 2004.

17. Y. Xu, R.K. Lee, and A. Yariv, Propagation and second-harmonic generation of electromagnetic waves in a coupled-resonator optical waveguide, *J Opt Soc Am B*, 17, 387–400, 2000.

18. Y. Xu, Y. Li, R.K. Lee, and A. Yariv, Scattering theory analysis of waveguide-resonator coupling, *Phys Rev E*, 62, 7389–7404, 2000.

19. K.S. Yee, Numerical solution of initial boundary value problems involving Maxwell's equations in isotropic media, *IEEE Trans Antennas Propagation*, 14, 302–307, 1966.

20. J.P. Durbano, F.E. Ortiz, J.R. Humphrey, D.W. Prather, and M.S. Mirotznik, Hardware implementation of a three-dimensional finite-difference time-domain algorithm, *IEEE Antennas Wireless Propagation Lett*, 2, 54–57, 2003.

21. J.B. Pendry, A chiral route to negative refraction, *Science*, 306, 1353–1355, 2004.

22. A.A. Houck, J.B. Brock, and I.L. Chuang, Experimental observations of a left-handed material that obeys Snell's Law, *Phys Rev Lett*, 90, 1–4, 2003.

23. J.B. Pendry, Negative refraction makes a perfect lens, *Phys Rev Lett*, 85, 3966–3969, 2000.

24. R.A. Shelby, D.R. Smith, and S. Schultz, Experimental verification of a negative index of refraction, *Science*, 292, 77–79, 2001.

25. A. Grbic and G.V. Eleftheriades, Experimental verification of backward-wave radiation from a negative refractive index metamaterial, *J Appl Phys*, 92, 5930–5935, 2002.

26. V.G. Veselago, The electromagnetic substances with simultaneously negative values of permittivity and permeability, *Soviet Physics Uspekhi*, 10, 509–514, 1968.

27. D.R. Smith, W.J. Padilla, D.C. Vier, S.C. Nemat-Nasser, and S. Schultz, Composite medium with simultaneously negative permeability and permittivity, *Phys Rev Lett*, 84, 4184–4187, 2000.

28. P. Kolinko and D.R. Smith, Numerical study of electromagnetic waves interacting with negative index materials, *Opt Express*, 11, 640–648, 2003.

29. C.D. Moss, T.M. Grzegorczyk, Y. Zhang, and J.A. Kong, Numerical studies of left handed metamaterials, *PIER*, 35, 315–334, 2002.

30. C.A. Balanis, *Advanced Engineering Electromagnetics*, New York: John Wiley & Sons, 1989.

31. M.S. Mirotznik, D.W. Prather, J.N. Mait, W.A. Beck, S. Shi, and X. Gao, Three-dimensional analysis of subwavelength diffractive optical elements with the finite-different time-domain method, *Appl Opt*, 39, 2871–2880, 2000.

32. D.B. Davidson and R.W. Ziolkowski, Body-of-revolution finite difference time-domain modeling of space-time focusing by a three-dimensional lens, *J Opt Soc Am A*, 11, 1471–1490, 1994.

33. J.P. Durbano, Hardware implementation of a 1-dimensional finite-difference time-domain algorithm for the analysis of electromagnetic propagation, University of Delaware, Newark, M.E.E. Thesis, 2002.

30

Nanoparticles for Drug Delivery

30.1 Introduction...**30**-1
30.2 Synthesis of Solid Nanoparticles......................................**30**-2
30.3 Processing Parameters **30**-5
      Surfactant/Stabilizer • Type of Polymer • Polymer Choice
      • Polymer Molecular Weight • Collection Method
30.4 Characterization..**30**-11
      Size and Encapsulation Efficiency • Zeta Potential • Surface
      Modification
30.5 Nanoparticulate Delivery Systems.....................................**30**-14
      Liposomes • Polymeric Micelles • Worm-like Micelles
      • Polymersomes
30.6 Targeted Drug Delivery Using Nanoparticles...............**30**-16
      Oral Delivery • Brain Delivery • Arterial Delivery
      • Tumor Therapy • Lymphatic System and Vaccines
      • Pulmonary Delivery
30.7 Drug Release **30**-19
      Mechanisms • Release Characteristics
Acknowledgments **30**-21
References...**30**-21

Meredith L. Hans
Anthony M. Lowman
Drexel University

30.1 Introduction

Polymer nanoparticles are particles of less than 1 μm diameter that are prepared from natural or synthetic polymers. Nanoparticles have become an important area of research in the field of drug delivery because they have the ability to deliver a wide range of drugs to different areas of the body for sustained periods of time. The small size of nanoparticles is integral for systemic circulation. Natural polymers (i.e., proteins or polysaccharides) have not been widely used for this purpose since they vary in purity, and often require cross-linking that could denature the embedded drug. Consequently, synthetic polymers have received significantly more attention in this area. The most widely used polymers for nanoparticles have been poly-ε-caprolactone (PCL), poly(lactic acid) (PLA), poly(glycolic acid) (PGA), and their co-polymers, poly(lactide-co-glycolide) (PLGA) [1–3]. In addition, block co-polymers of PLA and poly(ethylene glycol) (PEG) and poly(amino acids) have been used to make nanoparticles and micelle-like structures [4,5]. These polymers are known for both their biocompatibility and resorbability through natural pathways. Additionally, the degradation rate and accordingly the drug release rate can be manipulated by varying the ratio of PLA or PCL, increased hydrophobicity, to PGA, and increased hydrophilicity.

During the 1980s and 1990s, several drug delivery systems were developed to improve the efficiency of drugs and minimize toxic side effects [6]. The early nano- and microparticles were mainly formulated from poly(alkylcyanoacrylate) [6]. Initial promise of microparticles was dampened by the fact that there was a size limit for the particles to cross the intestinal lumen into the lymphatic system following oral delivery. Likewise, the therapeutic effect of drug-loaded nanoparticles was relatively poor due to rapid clearance of the particles by phagocytosis postintravenous administration. In recent years, headway has been made in solving this problem by the addition of surface modifications to nanoparticles. Nanoparticles, such as liposomes, micelles, worm-like micelles, polymersomes, and vesicles have also been proposed recently in the literature as promising drug delivery vehicles because of their small size and hydrophilic outer shell.

In recent years, significant research has been done on nanoparticles as oral drug delivery vehicles. For this application, the major interest is in lymphatic uptake of the nanoparticles by the Peyer's patches in the gut-associated lymphoid tissue (GALT). There have been many reports on the optimum size for Peyer's patch uptake ranging from <1 to <5µm [7,8]. It has been shown that microparticles remain in the Peyer's patches while nanoparticles are disseminated systemically [9].

Nanoparticles have a further advantage over larger microparticles because they are better suited for intravenous (IV) delivery. The smallest capillaries in the body are 5 to 6µm in diameter. The size of particles being distributed into the bloodstream must be significantly smaller than 5 µm, and should not form aggregates, to ensure that the particles do not form an embolism.

Clearly, a wide variety of drugs can be delivered using nanoparticulate carriers via a number of routes. Nanoparticles can be used to deliver hydrophilic drugs, hydrophobic drugs, proteins, vaccines, biological macromolecules, etc. [10–13]. They can be formulated for targeted delivery to the lymphatic system, brain, arterial walls, lungs, liver, spleen, or made for long-term systemic circulation. Therefore, numerous protocols exist for synthesizing nanoparticles based on the type of drug used and the desired delivery route. Once a protocol is chosen, the parameters must be tailored to create the best possible characteristics of the nanoparticles. Four of the most important characteristics of nanoparticles are their size, encapsulation efficiency, zeta potential (surface charge), and release characteristics. In this chapter, we intend to summarize many of the types of nanoparticles available for drug delivery, the techniques used for preparing polymeric nanoparticles, including the types of polymers and stabilizers used, and how these techniques affect the structure and properties of the nanoparticles. Additionally, we will discuss advances in surface modifications, targeted drug delivery applications, and release mechanisms and characteristics.

30.2   Synthesis of Solid Nanoparticles

As stated previously, there are several different methods for preparing nanoparticles. Additionally, numerous methods exist for incorporating drugs into the particles. For example, drugs can be entrapped in the polymer matrix, encapsulated in a nanoparticle core, surrounded by a shell-like polymer membrane, chemically conjugated to the polymer, or bound to the particle's surface by adsorption. Many of the previously mentioned nanoscale carriers such as micelles and polymersomes are synthesized via self-assembly mechanisms. The following section will deal primarily with the production of solid nanoparticles. A summary of these methods including the types of polymer, solvent, stabilizer, and drugs used is given in Tables 30.1 and 30.2.

The most common method used for the preparation of solid, polymeric nanoparticles is the emulsification–solvent evaporation technique. This technique has been successful for encapsulating hydrophobic drugs, but has had poor results in incorporating bioactive agents of a hydrophilic nature. Briefly, solvent evaporation is carried out by dissolving the polymer and the compound in an organic solvent. Frequently, dichloromethane is used for PLGA copolymers. The emulsion is prepared by adding water and a surfactant to the polymer solution. In many cases, nano-sized polymer droplets are induced by sonication or homogenization. The organic solvent is then evaporated and the nanoparticles are usually collected by centrifugation and lyophilization [7,14–17].

A modification on this procedure has led to the protocol favored for encapsulating hydrophilic compounds and proteins, the double or multiple emulsion technique. First, a hydrophilic drug and a stabilizer are dissolved in water. The primary emulsion is prepared by dispersing the aqueous phase into an organic

TABLE 30.1 Comparison of Methods for Nanoparticle Preparation

Method	Polymer	Solvent	Stabilizer	Size	References
Solvent diffusion	PLGA	Acetone	Pluronic F-127	200 nm	[88]
	PLGA	Acetone/DCM	PVA	200–300 nm	[58]
	PLA-PEG	MC	PVA/PVP	~130 nm	[114]
	PHDCA	THF	—	150 nm	[98]
	PLGA	Acetone	Sodium cholate	161 nm	[96]
	PLGA	Propylene carbonate	PVA or DMAB	~100 nm	[53]
Solvent displacement	PLA	Acetone/MC	Pluronic F68	123 ± 23 nm	[106]
	SB-PVA-g-PLGA	Acetone/Ethyl acetate	Poloxamer 188	~110 nm	[87]
Nanoprecipitation	PLGA/PLA/PCL	Acetone	Pluronic F68	110–208 nm	[59]
	PLGA	Acetonitrile	—	157.1 ± 1.9 nm	[57]
Solvent evaporation	PLA-PEG-PLA	DCM	—	193–335 nm	[129]
	PLGA	DCM	PVA	800 nm	[47]
	PLGA	DCM	TPGS	>300 nm	[67]
Multiple emulsion	PLGA	Ethyl acetate	—	>200 nm	[97]
	PEG-PLGA	DCM	PVA	~300 nm	[93]
	PLGA	Ethyl acetate/MC	PVA	335–743 nm	[92]
	PLGA-mPEG	DCM	—	133.5 ± 3.7–163.3 ± 3.6	[130]
	PLGA	DCM	PVA	213.8 ± 10.9 nm	[51]
	PLGA	DCM/Acetone	PVA	100 nm	[46]
	PLGA	Ethyl acetate	PVA	192 ± 12 nm	[49]
	PLGA	Ethyl acetate	PVA	300–350 nm	[55]
	PLGA	DCM	PVA	380 ± 40–1720 ± 110 nm	[50]
Salting out	PLA	Acetone	PVA	300–700 nm	[60]
Ionic gelation	Chitosan	TPP	—	278 ± 03 nm	[106]
Interfacial deposition	PLGA	Acetone	—	135 nm	[103]
Phase inversion nanoencapsulation	PLGA	MC	—	>5 μm	[52]
Polymerization	CS-PAA	—	—	206 ± 22 nm	[64]
	PECA	—	Pluronic F68	320 ± 12 nm	[66,85]
	PE-2-CA	—	—	380 ± 120 nm	[84]
Modified microemulsion	PolyoxStyl 20-stearyl ether	—	Emulsifying wax	~67 nm	[106]

DCM, dicloromethane; MC, methylene chloride; PVP, polyvinylpyrrolidone; PHDCA, poly(hexadecylcyanoacrylate); THF, tetrahydrofuran; SB-PVA-g-PLGA, sulfobutylated PVA-graft-PLGA; PCL, poly(epsilon-caprolactone); TPP, sodium tripolyphosphate; PAA, poly(acrylic acid); PECA, polyethylcyanoacrylate; PE-2-CA, polyethyl-2-cyanoacrylate.

solvent containing a dissolved polymer. This is then re-emulsified in an outer aqueous phase also containing a stabilizer [8,9,15,18–20]. From here, the procedure for obtaining the nanoparticles is similar to the single emulsion technique for solvent removal. The main problem with trying to encapsulate a hydrophilic molecule like a protein or a peptide drug is the rapid diffusion of the molecule into the outer aqueous phase during the emulsification. This can result in poor encapsulation efficiency, i.e., drug loading. Therefore, it is critical to have an immediate formation of a polymer membrane during the first water-in-oil emulsion. Song et al. [15] were able to accomplish this by dissolving a high concentration of high-molecular-weight PLGA in the oil phase consisting of 80/20 wt% dichloromethane/acetone solution. Additionally, the viscosity of the inner aqueous phase was increased by increasing the concentration of the stabilizer, bovine serum albumin (BSA). The primary emulsion was then emulsified with Pluronic F68 resulting in drug-loaded particles of approximately 100nm [15].

Another method that has been used to encapsulate insulin for oral delivery is phase inversion nanoencapsulation (PIN) [15,21]. In one example, Zn–insulin is dissolved in Tris-HCl and a portion of that is recrystallized by the addition of 10% $ZnSO_4$. The precipitate is added to a polymer solution of PLGA in

TABLE 30.2 Comparison of Nanoparticle Size for Different Drug-Loaded Particles

Polymer	Drug	Size	References
PLGA	Doxorubicin	200 nm	[88]
PLGA/PLA/PCL	Isradipine	110–208 nm	[59]
PLGA	U-86983	144 ± 37–88 ± 41 nm	[46]
PLGA	Rose bengal	150 nm	[103]
PLGA	Triptorelin	335–743 nm	[92]
PLGA	Procaine hydrochloride	164 ± 1.1–209.5 ± 2.7 nm	[57]
PLGA-mPEG	Cisplatin	133.5 ± 3.7–163.3 ± 3.6 nm	[130]
PLGA	Insulin	>1 μm	[52]
PLGA	Hemagglutinin	~250 nm	[6]
PLGA	Haloperidol	800 nm	[47]
PLGA	Estrogen	~100 nm	[53]
PEO-PLGA	Paclitaxel	150 ± 25 nm	[45]
PLA	Tetnus toxoid	>200 nm	[97]
PLA	Savoxepine	~300–700 nm	[60]
PLA	PDGFRβ tyrphostin inhibitor	123 ± 23 nm	[106]
PLA-PEG-PLA	Progesterone	193–335 nm	[129]
PECA	Amocicillin	320 ± 12 nm	[85]
Poly(butyl cyanoacrylate)	Dalargin	250 nm	[83]
Chitosan	Cyclosporin A	283 ± 24–281 ± 05 nm	[119]
PLGA	Paclitaxel	>300 nm	[67]
PLGA	Paclitaxel	<200 nm	[94]
PLA	N^6-cyclopentyladenosine	210 ± 50–390 ± 90 nm	[91]

methylene chloride. This mixture is emulsified and dispersed in 1L of petroleum ether, which results in the spontaneous formation of nanoparticles [21].

All of the techniques mentioned previously use toxic, chlorinated solvents that could degrade certain drugs and proteins if they come into their contact during the process. Consequently, an effort has been made to develop other techniques in order to increase drug stability during the synthesis. One such technique is the emulsification–diffusion method. This method uses a partially water-soluble solvent like acetone or propylene carbonate. The polymer and bioactive compound are dissolved in the solvent and emulsified in the aqueous phase containing the stabilizer. The stabilizer prevents the aggregation of emulsion droplets by adsorbing the surface of the droplets. Water is added to the emulsion to allow the diffusion of the solvent into the water. The solution is stirred leading to the nanoprecipitation of the particles. They can then be collected by centrifugation or the solvent can be removed by dialysis [22,23].

One problem with this technique is that water-soluble drugs tend to leak out of the polymer phase during the solvent diffusion step. To improve this process for water-soluble drugs, Takeuchi et al. [23] changed the dispersing medium from an aqueous solution to a medium chain triglyceride and added a surfactant, Span® 80, to the polymer phase. The nanoparticles were collected from the oily suspension by centrifugation. A double emulsification solvent diffusion technique has also been demonstrated to increase encapsulation efficiency of water-soluble drugs and maintenance of protein activity [24]. Protein activity during the fabrication process is a delicate balance between energy input (mechanical stirring, homogenization, and sonication) and particle size. A synergistic effect between mechanical stirring and ultrasound was shown to produce nanoparticles of 300 nm in diameter, while maintaining 85% of the starting activity of the cystatin protein [10]. Table 30.3 shows the changes in particle size with varied stirring rates. The addition of protein protectants, such as BSA or sugars, was also found to be essential in maintaining the biologically active, three-dimensional structure of cystatin [10]. These protectants may serve to shield proteins from interfaces during nanoparticle formation and lyophilization.

Several parameters can also be changed to benefit the encapsulation of hydrophilic molecules. Govender et al. [2] found that increasing the aqueous phase pH to 9.3 and incorporating pH-responsive

TABLE 30.3 The Effect of Combination of Stirring and Bath Sonication on Size of Particles Made from Polymer RG® 503H (Mean ± S.D., $n = 3$)

Organic Solvent	10,000 rpm[a]	Particle Size (nm) 7,500 rpm[a]	5,000 rpm[a]
Ethyl acetate	254 ± 16	254 ± 30	331 ± 25
Dichloromethane/acetone	235 ± 19	318 ± 14	314 ± 28

[a]Stirring rate.

Source: Reprinted from Cegnar, M. et al., *Eur. J. Pharm. Sci.*, 22, 357–364, 2004.

excipients, such as poly(methyl methacrylate-co-methacrylic acid) (PMMA-MAA) and lauric and caprylic acid, increased hydrophilic drug encapsulation without affecting the particle size, morphology, or yield. Murakami et al. [25] effectively modified the solvent diffusion technique by using two water-miscible solvents, one with more affinity for PLGA and the other with more affinity for the stabilizer, PVA, such as acetone and ethanol.

Nanoparticles can also be synthesized by the nanoprecipitation method. Briefly, the polymer and drug are dissolved in acetone and added to an aqueous solution containing a surfactant/stabilizer. The acetone is evaporated under reduced pressure and the nanoparticles remain in the suspension resulting in particles from 110 to 208 nm [26]. The salting-out process is another method, which does not require the use of chlorinated solvents. Using this technique, a water-in-oil emulsion is formed containing polymer, acetone, magnesium acetate tetrahydrate, stabilizer, and the active compound. Subsequently, water is added until the volume is sufficient to allow for diffusion of acetone into water, which results in the formation of nanoparticles. This suspension is purified by cross-flow filtration and lyophilization [27]. However, one disadvantage to this procedure is that it uses salts that may be incompatible with many bioactive compounds.

In most published techniques, nanoparticles are synthesized from biocompatible polymers. However, it is possible to make biodegradable nanoparticles from monomers or macromonomers by polycondensation reactions [28,29]. These processes also result in sizes ranging from 200 to 300 nm. Nanoparticles can also be made from hydrophilic polysaccharides like chitosan (CS). CS nanoparticles can be formed by the spontaneous ionic gelatin process [18,30]. CS-poly(acrylic acid) nanoparticles have also been made by polymerization of acrylic acid and the "dropping method" [31]. The resulting nanoparticles have small sizes and positive surface potentials. This technique is promising as the particles can be prepared under mild conditions without using harmful organic solvents.

30.3 Processing Parameters

The method of producing polymeric nanoparticles has several independent variables. Consequently, total drug loading, nanoparticle stability, and release characteristics may vary with slight changes in processing parameters. First, one must consider the selection of the components used in the nanoparticle production, including the polymer, molecular weight of the polymer, the surfactant, the drug, and the solvent [32]. For example, different surfactants may produce particles of different sizes [33]. An increase in polymer molecular weight will cause an impact on the release rate from the particles causing slower pore formation within the particles and therefore slower release [9,34]. Other processing variables include the time of emulsification, the amount of energy input, and the volume of the sample being emulsified. As energy input into an emulsion increases, the resulting particle size decreases [35]. In addition, there are four separate concentrations that can be altered: the polymer, drug, surfactant, and solvent. Often, a low concentration of surfactants will result in a high degree of polydispersity and aggregation [36]. Finally, the recovery of the particles can be changed depending on the method of lyophilization or centrifugation.

30.3.1 Surfactant/Stabilizer

One key parameter is the type of surfactant/stabilizer to be used. A wide range of synthetic and natural molecules with varying properties has been proposed to prepare nanoparticles. Feng and Huang [17] have investigated the use of phospholipids as natural emulsifiers. In their study, dipalmitoyl-phos-phatidylcholine (DPPC) improved the flow and phagocytal properties due to a denser packing of DPPC molecules on the surface of the nanoparticles leading to a smoother surface than particles made with the synthetic polymer, poly(vinyl alcohol) (PVA). DPPC also improved the encapsulation efficiency compared to PVA using the emulsification–solvent evaporation method. In a different study conducted by Kwon et al. [22], PLGA nanoparticles prepared using didodecyl dimethyl ammonium bromide (DMAB) were smaller than particles prepared with PVA. Lemoine and Preat [8] found that the presence of PVA in the inner aqueous phase produced smaller particles than Span® 40 [7]. When Pluronic is used as a stabilizer, the grade used can have a distinct effect on the size of the nanoparticles. For example, particles prepared with Pluronic F68 were smaller than particles prepared with Pluronic F108 [37].

A promising stabilizer for nanoparticles is the amphiphile D-α-tocopheryl polyethylene glycol 1000 succinate vitamin E (TPGS). TPGS has high emulsification efficiency, can increase incorporation efficiency when used as a matrix component, and can be used as a cellular adhesion enhancer. TPGS can be used at a concentration as low as 0.015% (w/v); in fact a lower concentration decreases particle size and polydispersity [38]. Figure 30.1 shows scanning electron micrographs (SEM) of PLGA particles made with PVA and TPGS. Moreover, the addition of TPGS dramatically reduced the release rate of paclitaxel from PLGA nanoparticles compared to those made with PVA (Figure 30.2). Using TPGS as an emulsifier, uptake by Caco-2 cells was greater than that of PVA-coated particles [33]. In addition, only TPGS-coated particles were found to be taken up in the nucleus [33].

The amount of stabilizer used will also have an effect on the properties of the nanoparticles. Most importantly, if the concentration of the stabilizer is too low, aggregation of the polymer droplets will occur and little if any nanoparticles will be recovered. Alternatively, if too much of the stabilizer is used, the drug incorporation could be reduced due to interaction between the drug and the stabilizer. However, when the stabilizer concentration is between the "limits," adjusting the concentration can be a means of controlling nanoparticle size. For example, in using the solvent evaporation technique, increasing the PVA concentration will decrease the particle size [8,15]. However, when using the emulsification–diffusion method, Kwon et al. [22] found that a PVA concentration from 2 to 4% was ideal for creating smaller nanoparticles, ~100 nm in diameter.

30.3.2 Type of Polymer

Biodegradable polymers retain their properties for a limited period of time *in vivo* and then gradually degrade into materials that can become soluble or are metabolized and excreted from the body. In order

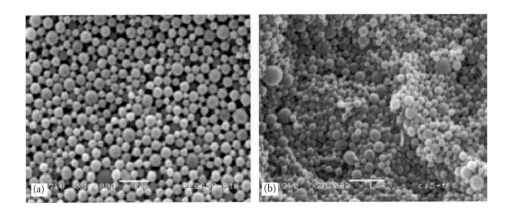

FIGURE 30.1 SEM images of coumarin 6-loaded PLGA particles coated with PVA (a) and vitamin E TPGS (b) (scale bar = 1 μm). (Reprinted from Yin Win, K. and Feng, S. S., *Biomaterials*, 26, 2713–2722, 2005. With permission.)

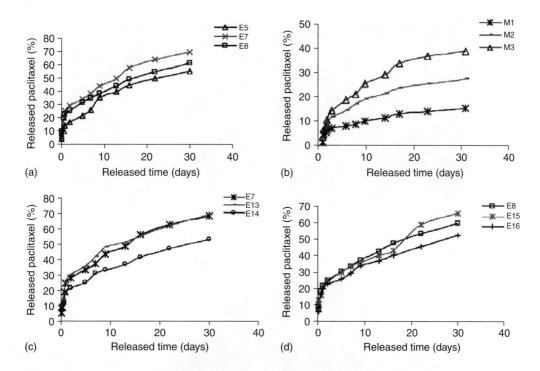

FIGURE 30.2 *In vitro* release curves of paclitaxel-loaded nanoparticles prepared under various experiment param-eters (a) E5: PLA, E7: PLGA (75:25), E8: PLGA (50:50); (b) ratio for PLGA-TPGS: M1 (2:1), M2 (1:1), M3 (1:2); (c) PLGA (75:25) concentration — E7: 0.125, E13: 0.188, E14: 0.25; (d) PLGA (50:50) concentration — E8: 0.125, E15: 0.188, E16: 0.25. (Reprinted from Mu, L. and Feng, S. S., *J. Control. Release*, 86, 33–48, 2003. With permission.)

to be used for *in vivo* applications, the polymers used for such systems must have favorable properties for biocompatibility, processability, sterilization capability, and shelf life. In the past, polystyrene or gold nanoparticles were used to investigate particle distribution and uptake. However, biodegradable polymer particles have several properties, such as hydrophobicity, surface charge, particle size distribution, density, or protein adsorption, which are different from polystyrene and gold that might have an impact on results of these studies [39]. Figure 30.3 and Figure 30.4 show examples of physical internalization of PLGA nanoparticles by vascular smooth muscle cells (VSMCs).

30.3.2.1 Poly(lactide-co-glycolide)

Many biodegradable systems rely on the random co-polymers of PLGA. These classes of polymers are highly biocompatible and have good mechanical properties for drug delivery applications [40]. In addi-tion, PLA and PLGA have been approved by the U.S. FDA for numerous clinical applications, such as sutures, bone plates, abdominal mesh, and extended-release pharmaceuticals. PLGA degrades chemically by hydrolytic cleavage of the ester bonds in the polymer backbone. Its degradation products, lactic acid and glycolic acid, are water-soluble, nontoxic products of normal metabolism that are either excreted or further metabolized to carbon dioxide and water in the Krebs cycle [41,42]. The composition of the PLGA random co-polymer, that is the relative amount of lactic acid and glycolic acid monomeric units, determines the degradation rate [34,43]. Since PGA is more hydrophilic than PLA a higher proportion of PGA incorporated into the co-polymer will increase the degradation rate by allowing more biological fluids to penetrate and swell the polymer matrix.

30.3.2.2 Poly(lactic acid)

Poly(lactic acid) occurs naturally as the pure enantiomeric poly(L-lactic acid) (LPLA) with a semicrys-talline structure. However, most types of PLA used for biological applications exist in the racemic

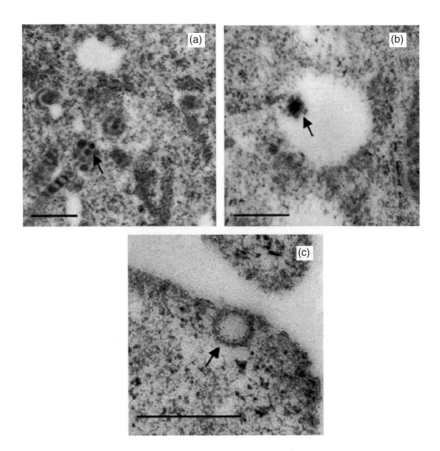

FIGURE 30.3 TEM images showing (a) nanoparticles (*indicated by arrow*) present in cytoplasm (×16,900), (b) a nanoparticle (*indicated by arrow*) interacting with vesicular membrane (×21,000), and (c) control vascular smooth muscle cells (VSMCs) (untreated cells) with nanoparticle-like vesicles of approximately 100 nm (*indicated by arrow*) (×38,000). Scale bars, 500 nm. (Reprinted from Panyam, J. et al., *Int. J. Pharm.*, 262, 1–11, 2003. With permission.)

D,L form (DLPLA) and are amorphous polymers. PLGA is also an amorphous polymer; both DLPLA and PLGA have glass transition temperatures above body temperature. The biomedical uses of PLA have been reported since the 1960s [44]. Numerous systems already utilize PLGA and PLA, including several micro- and nanoparticle systems as well as devices to control thyrotropin-releasing hormone in controlling metabolism [42], L-dopa to treat Parkinson's disease [45], and naltrexone in treating narcotic addiction [46] to successfully achieve long-term delivery. Several intraocular systems, including Vitrasert® (Bausch and Lomb), offer biocompatible delivery systems with controlled release drug therapy for periods ranging from several days up to 1 year [47].

30.3.2.3   Poly-ε-caprolactone

Poly-ε-caprolactone is another biodegradable and nontoxic polyester. PCL is polymerized similarly to PLA and PLGA, by ring-opening polymerization [48,49]. PCL is a semicrystalline polymer owing to its regular structure. The melting temperature of PCL is above body temperature, but its T_g is −60°C; so in the body, the semicrystalline structure of PCL results in high toughness, because the amorphous domains are in the rubbery state [50]. Hydrolysis of PCL yields 6-hydroxycaproic acid, which enters the citric acid cycle and is metabolized. Degradation of PCL occurs at a slower rate than PLA. PCL has also been used in blends and co-polymers with other biodegradable polymers [51]. Combinations of polymers allow

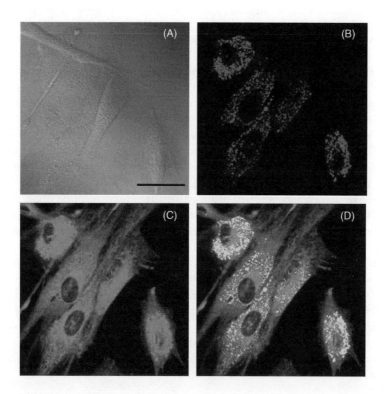

FIGURE 30.4 Confocal microscopic images demonstrating the intracellular distribution of 6-coumarin-loaded nanoparticles in VSMCs. (A) Differential interference contrast image showing the outline of the cells. (B) Cells stained with LysoTracker® Red and visualized using a RITC filter. (C) Uptake of green fluorescent 6-coumarin-loaded nanoparticles in VSMCs visualized using a FITC filter. (D) Overlay of (B, C) showing the co-localization of nanoparticles with endolysosomes. Scale bar, 25μm. (Reprinted from Panyam, J. et al., *Int. J. Pharm.*, 262, 1–11, 2003. With permission.)

the user to tailor mechanical properties and degradation kinetics, among other characteristics, to suit the needs of a specific application.

The amorphous regions of a semicrystalline polymer degrade prior to the crystalline domains [52]. This can lead to a change in the release profile. Thus, polymers that have a higher percent crystallinity are more impervious to water and therefore degrade at a slower rate. The drug entrapped in the amorphous region is released first and at a faster rate than the drug entrapped by the crystalline domains. The percent crystallinity in a polymer depends on the type of polymer used in the application, the polymer's composition, and the processing conditions for the polymer system. PCL typically has the highest percent crystallinity and the slowest degradation rate. This is evident by comparing PCL and LPLA versus PGA and PEG.

30.3.3 Polymer Choice

The polymer chosen to formulate the nanoparticles will strongly affect the structure, properties, and applications of the particles. As stated previously, PLGA has been the most common polymer used to make biodegradable nanoparticles, however, these are clearly not the optimal carrier for all drug delivery applications. For each application and drug, one must evaluate the properties of the system (drug and particle) and determine whether or not it is the optimal formulation for a given drug delivery application. For example, poly(butyl cyanoacrylate) nanoparticles have been successful in delivering drugs to the brain [53]. Other cyanoacrylate-based nanoparticles, such as polyalkylcyanoacrylate (PACA) and polyethylcyanoacrylate (PECA), have also been prepared. They are considered to be promising drug delivery systems due to their muco-adhesive properties and ability to entrap a variety of biologically active compounds. These polymers are biodegradable, biocompatible, as well as compatible with a wide range of compatible drugs [37,54]. Furthermore, these polymers have a faster degradation rate than PLGA, which in some cases may be more desirable. PECA nanoparticles have been prepared by emulsion polymerization in the presence and absence of different molecular weights PEGs, using Pluronic F68 as the stabilizer [55].

pH-sensitive nanoparticles made from a poly(methylacrylic acid and methacyrlate) co-polymer can increase the oral bioavailability of drugs like cyclosporin A by releasing their load at a specific pH within the gastrointestinal tract. The pH sensitivity allows this to happen as close as possible to the drug's absorption window through the Peyer's patches [56].

Other groups have successfully prepared nanoparticles from functionalized PLGA polymers. In one study, Jung et al. [57] synthesized nanoparticles made of a branched, biodegradable polymer, poly (2-sulfobutyl-vinyl alcohol)-g-PLGA. The purpose of using sulfobutyl groups attached to the hydrophilic backbone was to provide a higher affinity to proteins by electrostatic interactions that would favor adsorptive protein loading. Adjustments can be made to the characteristic nanoparticles by differing degrees of substitution of sulfobutyl groups. In another case, a carboxylic end group of PLGA was conjugated to a hydroxyl group of doxorubicin and formulated into nanoparticles [58]. This modification produced a sustained release of the drug that was approximately six times longer than the unconjugated drug [59]. The presence of the carboxylic end group on PLGA may also help in preserving cystatin activity in PLGA particles [10]. However, the carboxylic acid end group may also increase the overall release rate of drug from particles, so in some cases, a methyl-capped PLGA carboxylic acid end group can be utilized.

30.3.4 Polymer Molecular Weight

Polymer molecular weight, being an important determinant of mechanical strength, is also a key factor in determining the degradation rate of biodegradable polymers. Low-molecular-weight polymers degrade faster than high-molecular-weight polymers thereby losing their structural integrity more quickly. As chain scission occurs over time, the small polymer chains that result become more soluble in the aqueous environment of the body. This introduces "holes" into the polymer matrix. Consequently, lower-molecular-weight polymers release drug molecules more quickly [9,15]. This can be used to further engineer a system to control the release rate. A combination of molecular weights might be used to tailor a system to meet the demands of specific release profiles.

30.3.5 Collection Method

Another factor that can affect the properties of nanoparticles is the final freeze-drying process. It has been reported that additives such as saccharides are necessary for cryoprotection of nanoparticles in the freeze-drying process [60]. These saccharides may act as a spacing matrix to prevent particle aggregation. Because of the possibility of aggregation, the freeze-drying procedure can affect the "effective" nanoparticle size, and consequently, their release behavior and accordingly the drug pharmacokinetics [61].

Nanoparticles can also be collected by dialysis, ultracentrifugation, and gel filtration. While gel filtration has shown decreased encapsulation efficiency and total drug incorporation, it is thought that this collection method may remove drug adsorbed onto the particle surface, which may cause a dangerous release of drug immediately upon immersion in body fluid [62]. Figure 30.5 shows the difference in particle size and morphology in particles collected by gel filtration and ultracentrifugation. This burst effect will be discussed later in more detail.

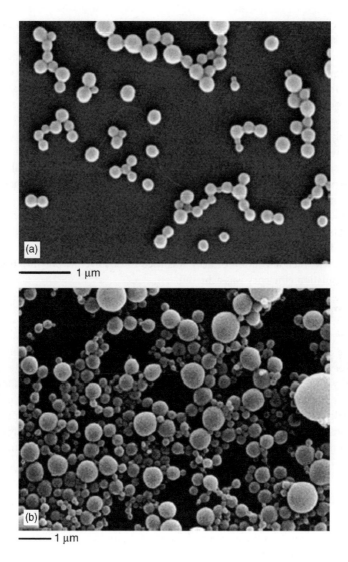

FIGURE 30.5 SEM of Oct-CPA-loaded nanospheres prepared by the nanoprecipitation method: (a) II-Oct-CPA sample recovered by gel filtration and (b) IV-Oct-CPA sample recovered by ultracentrifugation. (Reprinted from Dalpiaz, A. et al., *Biomaterials*, 26, 1299–1306, 2005. With permission.)

30.4 Characterization

30.4.1 Size and Encapsulation Efficiency

When considering a particular polymeric nanoparticle for a given drug delivery application, particle size and encapsulation efficiency are two of the most important characteristics. It is necessary to determine first what the goal of the nanoparticle delivery system is before determining the size desired. For example, if the goal is rapid dissolution in the body or arterial uptake, then the size of the nanoparticles should be ~100 nm or less. If prolonged dissolution is required, or targeting the mononuclear phagocytic system (MPS), larger particles around 800 nm, or particles engineered to have a stealth quality would be preferable. A comparison of various drugs encapsulated and the resulting sizes of the particles are summarized in Table 30.2. From examination of these data, it appears that the encapsulation efficiency increases with the diameter of the nanoparticles. In one study, the encapsulation efficiency was maximized in the double-emulsion solvent evaporation technique when the pH of the internal and the external aqueous phases were brought to the isoelectric point of the peptide being encapsulated, methylene chloride was used as a solvent, and the PLGA was rich in free carboxylic end groups [63]. It has also been found that by adding a freeze-thaw step during the primary emulsion of a double-emulsion process, there is an increase in overall protein incorporation by inducing the polymer phase to precipitate around the primary emulsion [35].

The molecular weight of the polymer has opposite effects on nanoparticle size and encapsulation efficiency. Smaller size nanoparticles, ~100 nm, can be prepared with lower-molecular-weight polymer, however, at the expense of reduced drug encapsulation efficiency. On the other hand, an increase in polymer concentration increases encapsulation efficiency and the size of the nanoparticles [9,15,22].

The synthesis method can also have a profound effect on the encapsulation efficiency. In loading paclitaxel into PLGA nanoparticles using the nanoprecipitation method, when the drug and polymer were mixed first and then solubilized in the organic solvent prior to fabrication, encapsulation efficiency was 15% [3]. However, when a solution of drug was used to dissolve the polymer prior to fabrication, nearly 100% encapsulation efficiency was achieved.

30.4.2   Zeta Potential

Another characteristic of polymeric nanoparticles that is of interest is zeta potential. The zeta potential is a measure of the charge of the particle, the relation being that the larger the absolute value of the zeta potential, the larger the amount of charge of the surface. In a sense, the zeta potential represents an index for particle stability. In the case of charged particles, as the zeta potential increases, the repulsive inter-actions will be larger leading to the formation of more stable particles with a more uniform size distri-bution. A physically stable nanosuspension solely stabilized by electrostatic repulsion will have a minimum zeta potential of ±30mV [64]. This stability is important in preventing aggregation. When a surface modification like PEG is added, the negative zeta potential is lowered, increasing the nanoparticle's stability [18].

30.4.3   Surface Modification

Before deciding which of the techniques to be used for synthesizing nanoparticles, one must consider the nature of the drug as well as the means and duration desired for the delivery. That will determine not only how the particles are synthesized, but also what the nature of the particles should be. In particular, the body recognizes hydrophobic particles as foreign and thus they are rapidly taken up by the MPS. However, if sustained systemic circulation is required then the surface of the hydrophobic nanoparticles must be modified in order to prevent phagocytosis [65].

Following intravenous administration, hydrophobic nanoparticles are rapidly cleared from the systemic circulation by the MPS, ending in the liver or the spleen [27]. If the goal is to treat a condition in the liver, then the proper choice for the application would be a hydrophobic nanoparticle. While it would appear that the hydrophobic nature of most biodegradable particles would limit the applicability of these carriers in many drug delivery applications, one may overcome concerns of clearance by the MPS through surface modification techniques. The goal of these modification techniques is to produce a particle that is not recognized by the MPS due to the hydrophilic nature of the surface [65,66].

Several types of surface-modified nanoparticles that have been described in recent literature are summarized in Table 30.4. The most common moiety used for surface modification is PEG [65,66]. PEG is a hydrophilic, non-ionic polymer that has been shown to exhibit excellent biocompatibility. PEG molecules can be added to the particles via a number of different routes including covalent bonding, mixing in during nanoparticle preparation, or surface adsorption [4,20,65–68]. The presence of a PEG brush on the surface of nanoparticles can serve other functions besides increasing residence time in the systemic circulation. For one, PEG tethers on the particle surface can reduce protein and enzyme adsorption

TABLE 30.4 Nanoparticle Sizes after Surface Modifications

Polymer	Surface Modification	Size (nm)	References
PLGA	Poloxamine 908	~160	[103]
PLGA	Poloxamer 407	~160	[103]
PLGA	Chitosan	500 ± 29	[49]
PLGA-mPEG	mPEG	133.5 ± 3.7 – 163.3 ± 3.6	[130]
PLGA-mPEG	mPEG	113.5 ± 14.3	[100]
PLGA-PEG	PEG	198.1 ± 11.1	[51]
PLA	PEG	164 – 270	[96]
PLA	PEG 6000	295	[60]
PLA-PEG	PEG	>200	[97]
PLA-PEG	PEG	~130	[114]
PHDCA	PEG	~150	[98]
PECA	PEG	220 ± 10 – 280 ± 8	[85]
PBCA	Polysorbate 80	250	[83]
PEG-PLGA	PEG	~300	[93]
Polyoxyl 20-stearyl ether	Thiamine	67	[105]
PEG-PACA	Transferrin	101.4 ± 7.2	[108]
PLA	Polysorbate 80	~160	[104]
PEI-b-PLGA	PEG	~100	[24]
PLA	Neutravidin™	~270	[111]
Glycoprotein-liposomes	Lectin	100	[19]

on the surface, which for PLGA-based particles will retard degradation [68]. The degree of protein adsorption can be minimized by altering the density and molecular weight of PEG on the surface [65]. The stability of PLA particles has been shown to increase in simulated gastric fluid (SGF) with the addition of PEG on the particle surface. After 4 h in SGF, 9% of the PLA nanoparticles converted into lactate versus 3% conversion for PEG–PLA particles [68]. PEG is also believed to facilitate transport through the Peyer's patches of the GALT [18].

As stated previously, the primary reason for interest in preparing PEG-functionalized particles is to improve the long-term systemic circulation of the nanoparticles [65,66]. The PEG-functionalized particles are not seen as a foreign body and in combination with the nanoscale size of the particles, are not taken up by the body, allowing them to circulate longer providing for a sustained systemic drug release. Because of their behavior, these PEG-functionalized nanoparticles are often called "stealth nanoparticles" [66]. Furthermore, it has been determined that PEG MW is important with respect to MPS uptake. For example, Leroux et al. [27] showed that an increase in PEG molecular weight in PLGA nanoparticles was associated with less interaction with the MPS and longer systemic circulation. Also, PEG-containing PLGA nanoparticles synthesized by Li et al. [20] were able to extend the half-life of BSA in a rat model from 13.6 min to 4.5 h [69]. Another study compared the dosages of PLGA nanoparticles versus PEG–PLGA nanoparticles. The PLGA nanoparticle pharmacokinetics seemed to depend on MPS saturation. However, the pharmacokinetics of PEG–PLGA dosages did not exhibit the same dependence on dosage/MPS saturation due to their stealth nature [70]. PEG may also benefit nanoparticle interaction with blood constituents. PLGA nanoparticles were shown to cause damage to red blood cells; PEGylated nanoparticles caused less damage [71]. It should be noted that the red blood cell damage was also concentration dependent.

Another way to prevent nanoparticles from becoming sequestered and eliminated by the spleen is by noncovalent adhesion of particles on to erythrocyte membranes [72]. This adhesion is controlled by van der Waals, electrostatic, hydrophobic, and hydrogen-bonding forces. The adhesion did not change the morphology of the red blood cells, and the particles were retained on their membranes for 24 h and beyond. This technology improved the circulation time of nanoparticles 10-fold.

Poloxamer and poloxamines have also been shown to reduce capture by macrophages and increase the time for systemic circulation. Similarly, PLGA particles coated with poloxamer 407 and poloxamine

908 extended the half-life of rose bengal, a hydrophilic model drug, with ~30% left in the bloodstream after 1 h postnanoparticle administration, as opposed to 8% present after 5 min post-free drug administration [73].

Another polymer used for surface modification is CS. The addition of CS to the surface of PLGA nanoparticles resulted in increased penetration of macromolecules in mucosal surfaces [30]. CS-coated PLGA particles were able to increase the positive zeta potential of the particles and increase the efficiency of tetanus toxoid protein encapsulation. Radiolabeled tetanus toxoid was used to show enhanced transport across nasal and intestinal epithelium using CS-coated particles versus uncoated particles, with a higher percentage of ^{125}I present in the lymph nodes for CS-coated particles [18].

30.5   Nanoparticulate Delivery Systems

30.5.1   Liposomes

Lipid bilayers occur throughout science and nature — cells use them to regulate chemical species within and outside of cells. Liposomes serve as excellent mimics of naturally occurring cell membranes [74]. Investigations with liposomes mimicking natural cell functions, especially those involving chemical transport, has led to their use as vehicles for drug delivery [75–78]. A primary advantage of liposomes is their high level of biocompatibility. Liposomes now constitute a mainstream technology for drug delivery; clinical approval has been given to liposomal formulations of anticancer drugs, doxorubicin (Doxil®/Caelyx® and Myocet®) and daunorubicin (Daunosome®) [79]. Another advantage of liposomes is their ability to transport a wide diversity of drugs that can be hydrophilic, lipophilic, or amphiphilic. This relates to the amphiphilic nature of phospholipid molecules themselves, which self-assemble in water to form bilayers that enclose an aqueous interior. Hydrophilic drugs can therefore be entrapped within the aqueous core, whereas hydrophobic drugs partition into the hydrocarbon-rich region of the bilayer. Loading techniques, such as the ammonium sulfate method [80] and the pH gradient method [81], can be used to place amphiphilic drugs (e.g., doxorubicin and vincristine, respectively) at the inner-phospholipid-monolayer/water interface [79].

A major disadvantage of liposomes as drug delivery vehicles is their rapid clearance from blood via the MPS, or reticuloendothelial system. This limitation was overcome by the advancement of "PEGylated Stealth®" liposomes, which avoid protein adsorption, and hence subsequent recognition and uptake, via a surface coating of PEG [80–82]. Incorporation of PEG into ordinary liposomes increases their circulation half-life from minutes to hours [79,83]. In addition to conveying this stealth quality, PEG also increases the susceptibility of liposomes to ultrasound-induced leakage [84]. This quality might prove useful in the development of a targeted, localized drug delivery system using external ultrasound as a remote mechanical stimulus to trigger drug release. Liposomes can also be decorated with glycoproteins and sugar chains to target specific cells [85].

30.5.2   Polymeric Micelles

Polymeric micelles can be assembled from block co-polymers composed of hydrophilic and hydrophobic segments. The hydrophobic segment creates the inner core of the micelle, while the hydrophilic segment creates the outer shell in an aqueous media [86]. Polymeric micelles can be used as a drug delivery device by either physically entrapping the drug in the core (i.e., hydrophobic drugs can be trapped inside the micelle by hydrophobic interactions), or by chemically conjugating the drug to the hydrophobic block prior to micelle formation [87,88]. Drugs either physically or chemically trapped in the hydrophobic core are protected from chemical degradation, causing unwanted side effects [87,88]. Micelles have several benefits as drug delivery vehicles. Their hydrophilic outer shell and small size (<100 nm) renders these particles nearly invisible to the reticuloendothelial system, allowing for long-term circulation in the bloodstream. Furthermore, micelle stability is relatively high due to the fact that, unlike other aggregates for drug delivery such as vesicles, they are a thermodynamically equilibrium aggregate. Micelles spontaneously form when the concentration of the amphiphile composing them (lipids, surfactants, or block copolymers) is higher than a critical concentration called the critical micelle concentration (CMC). The CMC is significantly lower for higher molecular weight polymers as opposed to surfactants or lipids. Micelle stability is further enhanced by crystallization or rigidity in the polymer core, which prevents rapid dissolution of block co-polymer micelles *in vivo*.

Triblock copolymers such as poly(ethylene oxide)-block-poly(propylene oxide)-block-poly(ethylene oxide) (PEO-b-PPO-b-PEO) or Pluronics® have frequently been studied to create block co-polymer micelles. Pluronics® conjugated to brain-specific antibodies or insulin have been used to solubilize haloperidol and other small compounds, such as FITC [89].

Polyethylenimine (PEI)-block-PLGA has been used to form micelles for cellular uptake [90]. Figure 30.6 shows size, polydispersity, and morphology of these micelles using transmission electron microscopy. Using confocal laser scanning microscopy (CLSM), PEI–PLGA micelles were shown to be absorbed onto human keratinocyte cell surfaces and translocated into the cytoplasm. These particles were not cytotoxic to the cells for up to 2 days at a concentration of 50 µg/mL. By comparison there was limited cellular internalization of PLGA particles [90].

Poly(ethylene oxide)-Poly(amino acid) (PEO-PAA) block co-polymers have received a great deal of attention, because the biodegradable amino acid core has free functional groups for chemical modification [88]. To achieve this, PEO-PBLA (Poly(β-benzyl L-aspartate)) is synthesized from β-benzyl N-carboxy L-aspartate anhydride (BLA-NCA) and α-methyl-ω-aminopoly(oxyethylene). PEO-P(Asp) is then prepared by debenzylation under alkaline conditions of the PEO-PBLA. Following this step, ADR is conjugated via amide bond formation [91]. Micelles prepared from drug conjugates have the added advantage that extended drug release will still be possible in the event of dissociation of the micelle as the drug must still be cleaved from the polymer chain. Stability of these PEO-P(Asp(ADR)) micelles has been shown

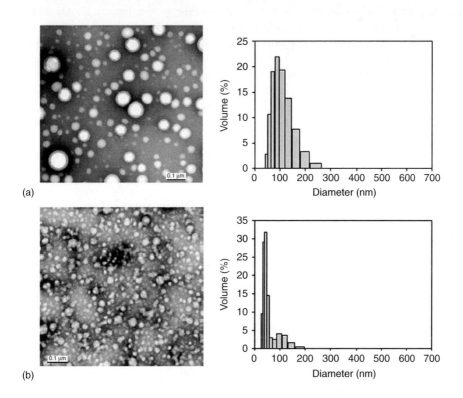

(a)

(b)

FIGURE 30.6 TEM pictures and size distributions of PEI–PLGA aggregates formed in (a) pure water, (b) 30 mM HCl (aqueous solution) (scale bar = 100 nm). (Reprinted from Nam, Y.S. et al., *Biomaterials*, 24, 2053–2059, 2003. With permission.)

to be dependent on a longer chain of PEO in comparison to the chain length of P(Asp) and ADR content. PEO-PBLA could also be synthesized with free hydroxyl groups on the outer core of the micelle for further conjugation to a targeting moiety (i.e., antibody, glucose) [87].

30.5.3 Worm-like Micelles

Cylindrical worm micelles [92] are a promising new class of supermolecular drug/dye carriers [93] to explore for a number of reasons. First, even if microns long, they can "worm" through small pores, including gels, and circulate for perhaps weeks. Second, targeted worms can cooperatively zip up — binding with high affinity — to surfaces or cells that bear suitable receptors. And third, once bound, internalization by the cell leads to delivery of a relatively large amount of drug all at once [94].

It is of central importance that several system characteristics be elucidated, including polymer molecular weight versus worm diameter, worm stability, and flexibility [93]. The first issue has already been addressed with vesicles [95]. Nanoscale worm micelles can be very stable; they appear similar to filamentous phages that have been used with great success *in vivo* for phage display of targeting ligands (including tumors) [96]. Unlike phages that carry nucleic acids, worm micelles can carry lipophilic drugs.

30.5.4 Polymersomes

Recently, interest has been focused on polymeric vesicles, composed of hydrophobic–hydrophilic diblock co-polymers, as drug delivery vehicles [95,97–100]. The advantages of these polymersomes, as compared to liposomes, include enhanced mechanical stability and greater flexibility to tailor bi-layer characteristics, such as thickness and chemical composition [98,101–104]. Moreover, it has been speculated that protein interactions with the polymeric bilayers will greatly differ from their interactions with lipid ones, thereby affecting drug delivery characteristics, such as circulation time *in vivo*. Indeed, Photos et al. [105] have shown that the circulation time of polymersomes *in vivo* increases with the bilayer thickness. Pata and Dan [106] have found that the characteristics of the polymeric bi-layer, when compared to the lipid bi-layer, are quite different and can be used for tailoring the carrier properties. Recently, Meng et al. [107] have synthesized biodegradable polymersomes from block co-polymers of PEG-PLA.

Finally, polymeric vesicles have been synthesized from multiblock co-polymers of Pluronic F27 with a PLA block on either end [108]. The vesicle structure is characterized by a hydrophilic core with hydrophobic layers, and can form several conformations such as bilayer or onion-like vesicles. The PLA-F27-PLA co-polymers exhibit a decrease in T_g and T_m values, which indicate an increased permeability and chain mobility in aqueous solutions. This may explain the high burst release from these vesicles, but does not eliminate this class of particles as a drug delivery system [108].

30.6 Targeted Drug Delivery Using Nanoparticles

30.6.1 Oral Delivery

Oral delivery of nanoparticles has focused on uptake via the Peyer's patches in the GALT. Peyer's patches are characterized by M cells that overlie the lymphoid tissue and are specialized for endocytosis and transport into intraepithelial spaces and adjacent lymphoid tissue. There have been several differing opinions as to the ease of nanoparticle transport through the M cells, and the method by which this occurs [109,110]. One theory is that nanoparticles bind the apical membrane of the M cells, followed by a rapid internalization and a "shuttling" to the lymphocytes [110,111]. The size and surface charge of the nanoparticles are crucial for their uptake. However, there have only been two published phase 1 clinical trials examining the oral uptake of PLGA nanoparticles encapsulating *Escherichia coli* antigens, with no clear benefit determined [109]. There is some promise in identifying M cell receptors and targeting them on the surface of nanoparticles. The carbohydrate epitope, sialytated Lewis antigen A (SLAA), has been identified on human M cells [109]. This application of targeted nanoparticles in oral delivery holds tremendous promise for the development of oral vaccines and in cancer therapy.

30.6.2 Brain Delivery

Another exciting application of surface-modified particles is targeted drug delivery to cells or organs. Kreuter et al. [53] were able to deliver several drugs successfully through the blood–brain barrier using polysorbate 80-coated poly(butylcyanoacrylate) nanoparticles. It is thought that after administration of the polysorbate 80-coated particles, apolipoprotein E (ApoE) adsorbs onto the surface coating. The ApoE protein mimics low-density lipoprotein (LDL) causing the particles to be transported via the LDL receptors into the brain. The effects of polysorbate-80 on endocytosis by the blood–brain barrier were confirmed by Sun et al. [112] with PLA nanoparticles. Nanoparticles made from emulsifying wax and polyoxyl 20-stearyl ether linked to a thiamine surface ligand, fabricated using microemulsion precursors and had an average final diameter of 67 nm [113]. These particles are able to associate with the blood–brain barrier thiamine transporters and thereby increase the unidirectional transfer coefficient for the particles into the brain.

30.6.3 Arterial Delivery

There are other specific areas, where nanoparticle administration may have an advantage over microparticulate-based drug delivery systems. One area that has been of recent interest is in prevention of restenosis [15,114]. Restenosis is a major postoperative concern following arterial surgery. In order to inhibit VSMC proliferation, drugs must be delivered at a high concentration over a long period of time. Nanoparticles offer an advantage, because the medication would not have to be delivered systemically as they are small enough for cellular internalization and connective tissue permeation. Several types of drugs including antiproliferative agents have been used to test this method of delivery. PLA nanoparticles were loaded with platelet-derived growth factor receptor β tyrphostin inhibitor and delivered intra-lumenally to an injured rat carotid artery [115]. The drug had the desired effect of preventing restenosis, but of significance was the absence of drug in other areas of the arteries and systemic circulation. Song et al. [15] found that specific additives after nanoparticle formation, such as heparin, DMAB, or fibrinogen, could enhance arterial retention of the particles. Suh et al. [14] created PEO–PLGA nanoparticles, which had an initial burst release of 40% of the antiproliferative drug in the first 3 days. However, a total of 85% of the drug was released after 4 weeks. This shows that nanoparticles have a great potential for long-term arterial drug delivery.

30.6.4 Tumor Therapy

One of the most common targeting applications for nanoparticles is of chemotheraputic agents against cancerous cells and tumor sites. Figure 30.7 shows the cellular internalization of PLGA nanoparticles into MCF-10A neoT cells. Oral delivery of chemotherapy is of high interest, because chemotheraputic agents are eliminated by the first-pass effect with cytochrome p450 [33]. Another reason that it has attracted attention is due to the phenomena known as the enhanced permeation and retention (EPR) effect. The vasculature around tumor sites is inherently leaky due to the rapid vascularization necessary to serve fast-growing tumors [116]. Much attention has also been given to lymphatic targeting using nanoparticles. In addition, poor lymphatic drainage at the site prevents elimination of the particles from the tumor tissue. Another benefit of tumor targeting is the fact that many cancer cells over-express specific antigens. Coating or binding the ligands on nanoparticulate surfaces can exploit this property. One example is covalently attaching sugar chains to nanoparticles. Yamazaki et al. [85] used sugar chain-remodeled glycoprotein–liposome conjugates for binding to E-selectin. These particles showed uptake by solid tumor tissue. Cancer cells over-express transferrin, which is normally used for iron uptake. Paclitaxel-loaded PACA nanoparticles with a PEG linker chain conjugated to transferrin were shown to have a decreased burst release over nontargeted particles and were able to increase the lifespan in tumor-bearing mice with decreased weight loss compared to mice given conventional paclitaxel [117]. The decreased burst may be due to the removal of surface-associated drug during the activation process and the decreased weight loss may point to a reduction in paclitaxel-associated side effects.

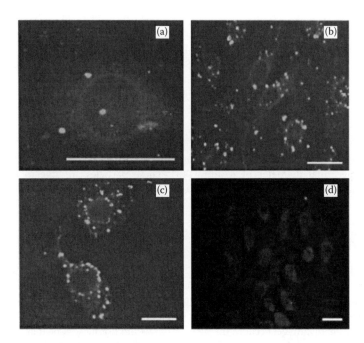

FIGURE 30.7 Internalization of NPs loaded with Alexa Fluor 488-labeled cystatin into MCF-10A neoT cells. (a) 10 min incubation, concentration = 100 μg NP/mL; (b) 30 min incubation, concentration = 100 μg NP/mL; (c) 45 min incubation, concentration = 100 μg NP/mL; (d) free Alexa Fluor 488-labeled cystatin, 45 min incubation, concentration = 0.1 μ*M*. Scale bar, 20 μm. All the images are representative of at least two independent experiments. (Reprinted from Cegnar, M. et al., *Exp. Cell Res.*, 301, 223–231, 2004. With permission.)

Yokoyama et al. [91,118–121] have conducted numerous studies on the chemical conjugation as well as physical entrapment of adriamycin (ADR, also known as doxorubicin), an anticancer drug, in PEO–Poly(aspartic acid) micelles (PEO–P(Asp)). It was found that polymer micelles with conjugated ADR alone or physically entrapped ADR alone did not have anti-tumor activity. However, high amounts of both chemically conjugated and physically entrapped ADR were necessary for anti-tumor activity [121]. Other anticancer drugs have been used in block co-polymer micelles. Cisplatin has been bound to P(Asp) using the same block co-polymer PEO-P(Asp) [122]. Methotrexate (Mtx) esters of PEO-block-poly (2-hydroxyethyl-L-aspartimide) (PEO-PHEA) were found to form stable micelles with a sustained release profile, dependent on the amount of Mtx substitution [123].

One method for targeting that is attracting attention is by exploiting the high noncovalent binding of biotin and avidin. Each avidin molecule binds four biotin molecules. Using the multifunctionality of this technology, drugs and homing molecules for cancer cells can be combined [124]. By using biotinylated PEG–PCL and associating that with avidin-bound lectin, the amount of nanoparticles associated with Caco-2 cells is increased dramatically [1]. NeutrAvidin™ has been covalently attached to the surface of PLA nanoparticles via sulfhydryl groups [125]. These thiol groups are bound via a carbodiimide reaction with cystamine. Using this technology, biotinylated antibodies or ligands can be attached to the particle surface. Biotinylated worm-like micelles have also been found to be stable in an aqueous solution for at least a month, and have the potential for the delivery of large quantities of hydrophobic drugs or dyes [94].

30.6.5 Lymphatic System and Vaccines

The lymphatic absorption of a drug via the GALT has an advantage over a portal blood route since it avoids any liver presystemic metabolism, known as the first-pass effect. This could be beneficial for anticancer treatment, mucosal immunity, as well as having the potential for staining the lymph nodes prior to surgery [126]. Nanoparticles can also be used to carry antisense oligonucleotides and plasmid DNA that can be used to treat some forms of cancer and viral infections, as well as a new vaccination approach. Antisense oligionucleotides normally have poor stability and cannot easily penetrate cells, but are easily encapsulated in nanoparticles [13,127]. High doses of antibiotics and antiparasitics are given to treat gastrointestinal bacteria and parasites because only 10 to 15% of the drug administered is absorbed [64]. The increased muco-adhesivity of nanoparticles could be effective in treating these pathogens with lower doses of drugs.

Nanoparticles have also had some success as a new delivery vehicle for vaccines. CS nanoparticles have been successful as a nasal vaccine in some animal studies producing significant IgG serum responses and superior IgA secretory responses when used in influenza, pertussis, and diphtheria vaccines [128]. Another option for vaccine delivery is the delivery of PLGA nanoparticles to dendritic cells [12]. Dendritic cells initiate an antigen-specific immune response. The co-delivery of antigens and immunomodulators in the same particles helps to overcome peripheral tolerance against self-antigens. This technology can potentially be used for cancer vaccines or hepatitis antigens.

30.6.6 Pulmonary Delivery

The lungs, owing to their large surface area, good mucosal permeation, well-developed vascular system, thin alveolar walls, and low activity of drug-metabolizing enzymes are another target area for nanoparticles. Direct administration to the lungs is beneficial because of the decreased systemic side effects, the availability for increased dose levels at the site, and the lack of first-pass metabolism. The tracheo-bronchial region is protected by a mucosal barrier, which can be cleared by nanoparticle technology. One obstacle for the inhalation of nanoparticles is their inappropriate mass median aerodynamic diameter. They do not have sufficient size for sedimentation or impaction deposition mechanisms and would be exhaled postinhalation [129]. Carrier particles have been proposed, which after deposition in the carrier matrix dissolve, while the nanoparticles are released.

When CS-coated nanoparticles were administered via the lungs, there were detectable blood levels of the drug 24 h after administration, as opposed to 8 h for the noncoated particles [23]. PACA nanoparticles have been found to be cytotoxic to airway epithelial cells, but gelatin and human serum albumin nanoparticles of ~200 nm in diameter were taken up by bronchial epithelial cells with no evidence of cytotoxicity or inflammation [130].

Nanoparticles have been used to target other mucosal surfaces. Long-term extra-ocular (cornea and conjunctiva) drug delivery with nanoparticles provides an improvement in conventional drug delivery in this region. It was possible to deliver drug-loaded CS nanoparticles to the extra-ocular structures over the course of 48 h at higher levels than with free drug, without exposing the inner ocular structures (iris and aqueous humour) to the drug [131].

Nanoparticles may be used for skin delivery to prolong the residence time of sunscreen agents in the stratum corneum or to deliver vitamin A to the upper layers of the skin [132]. CLSM demonstrates nanoparticles accumulating in the follicular openings, with higher percentages of smaller diameter nanoparticles accumulating.

30.7 Drug Release

30.7.1 Mechanisms

Biodegradable polymers release drug in one of two ways: erosion and diffusion. Release from biodegradable polymers *in vivo* is governed by a combination of both mechanisms, which depends on the relative rates of erosion and diffusion. Erosion is defined as the physical dissolution of a polymer as a result of its degradation [133]. Most biodegradable polymers used for drug delivery are degraded by hydrolysis. Hydrolysis is a reaction between water molecules and bonds in the polymer backbone, typically ester bonds, which repeatedly cuts the polymer chain until it is returned to monomers. Other biodegradable polymers are enzymatically degradable, which is also a type of chain scission. As water molecules break chemical bonds along the polymer chain, the physical integrity of the polymer degrades and allows drug to be released.

There are two possible mechanisms of erosion. When water is confined to the surface of the matrix, as in the case of a hydrophobic polymer, chain scission will occur only on the surface and the drug will be released as the surface of the polymer matrix erodes. If the water penetrates the polymer matrix faster than it hydrolyzes the bonds on the surface, then erosion will occur throughout the entire material — this is also called bulk erosion. In many cases, the erosion of a polymer matrix *in vivo* is some combination of these mechanisms. Degradation by surface erosion alone may be preferred, because the degradation rate can be controlled through the surface area of the matrix [134].

In the case of diffusion-controlled release, the drug's concentration gradient in the polymer matrix is the driving force for the molecules to diffuse into the surrounding medium. The diffusion of a drug molecule through the polymer matrix is dependent upon the solubility of the drug in the polymer matrix and the surrounding medium, the diffusion coefficient of the drug molecule, the molecular weight of the drug, its concentration throughout the polymer matrix, and the distance necessary for diffusion. Drugs can be either distributed evenly throughout the matrix or encapsulated as a reservoir [135]. The release rate for the reservoir system also factors in the membrane thickness and area. Practically, reservoir systems often have a lag period after placement *in vivo*, as opposed to the burst release present for most other systems. However, these systems need to be carefully engineered to prevent premature membrane rupture that might release a toxic amount of drug into the body.

When a drug is dissolved in the matrix and the mechanism for delivery is diffusion, then the driving force for release is the concentration gradient and release predictions can be made based on Fick's laws of diffusion [134]. Cumulative release from diffusion-controlled matrix devices is inversely proportional to the square root of time [136]. This presents an engineering challenge because surface area becomes smaller due to degradation, with a resulting decrement in the release rate.

Frequently, diffusion-controlled release is important in the early stages of drug release. For many of the polymeric delivery systems, there is some concentration of drug molecules entrapped near the surface of the matrix and adsorbed onto the surface of the matrix. Upon immersion into a medium, the release of these drug molecules is controlled by the rate of diffusion of the drug into the surrounding environment. This can cause a problem referred to as the "burst effect," which can potentially release a toxic amount of drug in some geometries (frequently 50% or more of the incorporated drug) into the body within the first 24 h [137]. This burst release is part of what is frequently referred to as a biphasic release profile [3]. During the first phase, the burst release, the structural integrity of the nanoparticles is maintained. The second phase, or the linear release, is characterized by pore formation, particle deformation, and fusion [138]. One method for eliminating the burst release is by tailoring the collection of the nanoparticles to remove the drug attached to the surface. By collecting PLA nanoparticles through gel filtration, the antiischemic drug N^6-cyclopentyladenosine near the surface was removed, which decreased the overall encapsulation efficiency, and eliminated the burst release from the particles [62].

The specific chemical and biological characteristics of the drug and the polymer are crucial in designing a polymeric delivery system. For example, drugs with greater hydrophilicity can increase the overall release rate by promoting polymer swelling and degradation which in turn increases drug diffusion, and certain drug molecules may potentially react with the polymer matrix [139]. The drug's molecular weight, solubility in biological fluids as well as its miscibility in the polymer matrix will influence the drug's diffusivity from the system and the concentration profile of the drug throughout the matrix. Since polymeric delivery systems are rarely homogenous throughout the entire matrix, the drug's diffusivity, and therefore release rate, can change with the local polymer composition and structure. One example of this phenomenon showing the effect of drug polymer interaction is the release of the prodrug, 3-methoxyxanthone and its active form xanthone, from PLGA nanocapsules [140]. Incorporation into nanocapsules was significantly higher than that of solid nanoparticles. The release profile of these drugs suggest a physical interaction of the drug with the polymer when given the nanocapsule structure [140].

30.7.2 Release Characteristics

The release characteristics of polymeric nanoparticles are among the most important features of the drug/polymer formulations because of the possible applications in sustained drug delivery. There are several factors that affect the release rate of the entrapped drug. Larger particles have a smaller initial burst release and longer sustained release than smaller particles. In addition, the greater the drug loading, the greater the burst and faster the release rate. For example, PLA nanoparticles containing 16.7% savoxepine released 90% of their drug load in 24 h, as opposed to particles containing 7.1% savoxepine, which released their content over 3 weeks [27]. The initial burst release is thought to be caused by poorly entrapped drug, or drug adsorbed onto the outside of the particles. When using polymers, which interact with a drug, like PLGA with a free COOH group and proteins, the burst release is lower and in some cases absent, and drug release is prolonged [9,63].

The addition of other polymers to PLA-based polymers can also be used to control drug release. For example, PEG has been polymerized into a PLA homopolymer creating a PLA-PEG-PLA co-polymer [141]. The amount of the drug (in this case progesterone) released increased with the PEG content and the molecular weight of the co-polymers. The drug release continued to increase as the total molecular weight of the co-polymers decreased. The initial burst was decreased in the absence of lower molecular weight polymers. The content of PEG in the co-polymer affected the size of the particles as well as the degradation of the polymers. Similar effects were seen with PLGA-mPEG nanoparticles loaded with cisplatin [142]. Consequently, it would be possible to alter the release rate of the drug by changing the amount of PEG in the co-polymer as well as the molecular weights of the polymers.

Acknowledgments

We thank the Nanotechnology Institute of Southeastern Pennsylvania and IGERT Grant # DGE-0221664 for support.

References

1. R. Gref, P. Couvreur, G. Barratt, and E. Mysiakine, Surface-engineered nanoparticles for multiple ligand coupling, *Biomaterials*, 24, 4529–4537, 2003.
2. T. Govender, S. Stolnik, M. C. Garnett, L. Illum, and S. S. Davis, PLGA nanoparticles prepared by nanoprecipitation: drug loading and release studies of a water soluble drug, *J. Control. Release*, 57, 171–185, 1999.
3. C. Fonseca, S. Simoes, and R. Gaspar, Paclitaxel-loaded PLGA nanoparticles: preparation, physio-chemical characterization and *in vitro* anti-tumoral activity, *J. Control. Release*, 83, 273–286, 2002.
4. Y. Dong and S. Feng, Methoxy poly(ethylene glycol)-poly(lactide) (MPEG-PLA) nanoparticles for controlled drug delivery of anticancer drugs, *Biomaterials*, 25, 2843–2849, 2004.
5. X. Zhang, Y. Li, X. Chen, X. Wang, X. Xu, Q. Liang, J. Hu, and X. Jing, Synthesis and characterization of the paclitaxel/MPEG-PLA block copolymer conjugate, *Biomaterials*, 26, 2121–2128, 2005.
6. M. Aprahamian, C. Michel, W. Humbert, J.-P. Devissaguet, and C. Damge, Transmucosal passage of polyalkylcyanoacrylate nanocapsules as a new drug carrier in the small intestine, *Biol. Cell*, 61, 69–76, 1987.
7. A.-M. Torche, H. Jouan, P. L. Corre, E. Albina, R. Primault, A. Jestin, and R. L. Verge, *Ex vivo* and *in situ* PLGA microspheres uptake by pig ileal Peyer's patch segment, *Int. J. Pharm.*, 201, 15–27, 2000.
8. D. Lemoine and V. Preat, Polymeric nanoparticles as delivery system for influenza virus glycopro-teins, *J. Control. Release*, 54, 15–27, 1998.
9. M. D. Blanco and M. J. Alonso, Development and characterization of protein-loaded poly(lactide-co-glycolide) nanospheres, *Eur. J. Pharm. Biopharm.*, 43, 287–294, 1997.
10. M. Ccgnar, J. Kos, and J. Kristl, Cystatin incorporated in poly(lactide-co-glycolide) nanoparticles: development and fundamental studies on preservation of its activity, *Eur. J. Pharm. Sci.*, 22, 357–364, 2004.

11. J. Barichello, M. Morishita, K. Takayama, and T. Nagai, Encapsulation of hydrophilic and lipophilic drugs in PLGA nanoparticles by the nanoprecipitation method, *Drug Dev. Indus. Pharm.*, 25, 471–476, 1999.

12. P. Elamanchili, M. Diwan, M. Cao, and J. Samuel, Characterization of poly(D,L-lactic-co-glycolic acid) based nanoparticulate system for enhanced delivery of antigens to dendritic cells, *Vaccine*, 22, 2406–2412, 2004.

13. C. Perez, A. Sanchez, D. Putnam, D. Ting, R. Langer, and M. J. Alonso, Poly(lactic acid)-poly(ethylene glycol) nanoparticles as new carriers for the delivery of plasmid DNA, *J. Control. Release*, 75, 211–224, 2001.

14. H. Suh, B. Jeong, R. Rathi, and S. W. Kim, Regulation of smooth muscle cell proliferation using paclitaxel-loaded poly(ethylene oxide)-poly(lactide/glycolide) nanospheres, *J. Biomed. Mater. Res.*, 42, 331–338, 1998.

15. C. X. Song, V. Lanhasetwar, H. Murphy, X. Qu, W. R. Humphrey, R. J. Shebuski, and R. J. Levy, Formulation and characterization of biodegradable nanoparticles for intravascular local drug delivery, *J. Control. Release*, 43, 197–212, 1997.

16. Y.-H. Cheng, L. Illum, and S. S. Davis, A poly(D,L-lactide-co-glycolide) microsphere depot system for delivery of haloperidol, *J. Control. Release*, 55, 203–212, 1998.

17. S.-S. Feng and G. Huang, Effects of emulsifiers on the controlled release of paclitaxel (Taxol) from nanospheres of biodegradable polymers, *J. Control. Release*, 71, 53–69, 2001.

18. A. Vila, A. Sanchez, M. Tobio, P. Calvo, and M. J. Alonso, Design of biodegradable particles for protien delivery, *J. Control. Release*, 78, 15–24, 2002.

19. H. Rafati, A. G. A. Coombes, J. Adler, J. Holland, and S. S. Davis, Protein-loaded poly(DL-lactide-co-glycolide) microparticles for oral administration: formulation, structural and release characteristics, *J. Control. Release*, 43, 89–102, 1997.

20. Y.-P. Li, Y.-Y. Pei, X.-Y. Zhang, Z.-H. Gu, Z. H. Zhou, W.-F. Yuan, J.-J. Zhao, J.-H. Zhu, and X.-J. Gao, PEGylated PLGA nanoparticles as protein carriers: synthesis, preparation and biodistribution in rats, *J. Control. Release*, 71, 203–211, 2001.

21. G. P. Carino, J. S. Jacob, and E. Mathiowitz, Nanosphere based oral insulin delivery, *J. Control. Release*, 65, 261–269, 2000.

22. H.-Y. Kwon, J.-Y. Lee, S.-W. Choi, Y. Jang, and J.-H. Kim, Preparation of PLGA nanoparticles containing estrogen by emulsification–diffusion method, *Colloids Surfaces*, 182, 123–130, 2001.

23. H. Takeuchi, H. Yamamoto, and Y. Kawashima, Mucoadhesive nanoparticulate systems for peptide drug delivery, *Adv. Drug Del.*, 47, 39–54, 2001.

24. M. Cegnar, A. Premzl, V. Zavasnik-Bergant, J. Kristl, and J. Kos, Poly(lactide-co-glycolide) nanoparticles as a carrier system for delivering cysteine protease inhibitor cystatin into tumor cells, *Exp. Cell Res.*, 301, 223–231, 2004.

25. H. Murakami, M. Kobayahi, H. Takeuchi, and Y. Kawashima, Preparation of poly(DL-lactide-co-glycolide) nanoparticles by modified spontaneous emulsification solvent diffusion method, *Int. J. Pharm.*, 187, 143–152, 1999.

26. M. L.-L. Verger, L. Fluckiger, Y. Kim, M. Hoffman, and P. Maincent, Preparation and characterization of nanoparticles containing an antihypertensive agent, *Eur. J. Pharm. Biopharm.*, 46, 137–143, 1998.

27. J.-C. Leroux, E. Allemann, F. D. Jaeghere, E. Doelker, and R. Gurny, Biodegradable nanoparticles — from sustained release formulations to improved site specific drug delivery, *J. Control. Release*, 39, 339–350, 1996.

28. S. Sakuma, M. Hayashi, and M. Akashi, Design of nanoparticles composed of graft copolymers for oral peptide delivery, *Adv. Drug Del.*, 47, 21–37, 2001.

29. N. Behan, C. Birkinshaw, and N. Clarke, Poly *n*-butyl cyanoacrylate nanoparticles: a mechanistic study of polymerisation and particle formation, *Biomaterials*, 22, 1335–1344, 2001.

30. K. A. Janes, P. Calvo, and M. J. Alonso, Polysaccharide colloidal particles as delivery systems for macromolecules, *Adv. Drug Del.*, 47, 83–97, 2001.

31. Y. Hu, X. Jiang, Y. Ding, H. Ge, Y. Yuan, and C. Yang, Synthesis and characterization of chitosan–poly(acrylic acid) nanoparticles, *Biomaterials*, 23, 3193–3201, 2002.

32. H. Rafati, E. C. Lavelle, A. G. Coombes, S. Stolnik, J. Holland, and S. S. Davis, The immune response to a model antigen associated with PLG microparticles prepared using different surfactants, *Vaccine*, 15, 1888–1897, 1997.

33. K. Yin Win and S. S. Feng, Effects of particle size and surface coating on cellular uptake of polymeric nanoparticles for oral delivery of anticancer drugs, *Biomaterials*, 26, 2713–2722, 2005.

34. M. J. Dorta, O. Manguia, and M. Llabres, Effects of polymerization variables on PLGA properties: molecular weight, composition, and chain structure, *Int. J. Pharm.*, 100, 9–14, 1993.

35. T. D. Dziubla, A. Karim, and V. R. Muzykantov, Polymer nanocarriers protecting active enzyme cargo against proteolysis, *J. Control. Release*, 102, 427–439, 2005.

36. S. K. Sahoo, J. Panyam, S. Prabha, and V. Labhasetwar, Residual polyvinyl alcohol associated with poly (D,L-lactide-co-glycolide) nanoparticles affects their physical properties and cellular uptake, *J. Control. Release*, 82, 105–114, 2002.

37. G. Fontana, G. Pitarresi, V. Tomarchio, B. Carlisi, and P. L. S. Biagio, Preparation, characterization and *in vitro* antimicrobial activity of ampicillin-loaded polyethylcyanoacrylate nanoparticles, *Biomaterials*, 19, 1009–1017, 1998.

38. L. Mu and S. S. Feng, A novel controlled release formulation for the anticancer drug paclitaxel (Taxol): PLGA nanoparticles containing vitamin E TPGS, *J. Control. Release*, 86, 33–48, 2003.

39. J. Panyam, S. K. Sahoo, S. Prabha, T. Bargar, and V. Labhasetwar, Fluorescence and electron microscopy probes for cellular and tissue uptake of poly(D,L-lactide-co-glycolide) nanoparticles, *Int. J. Pharm.*, 262, 1–11, 2003.

40. J. P. Kitchell and D. L. Wise, Poly(lactic/glycolic acid) biodegradable drug-polymer matrix systems, *Methods Enzymol.*, 112, 436–448, 1985.

41. H. Curtis, *Biology*, Worth Publishers Inc., New York, 1983.

42. H. Okada and H. Toguchi, Biodegradable microspheres in drug delivery, *Crit. Rev. Ther. Drug Carrier Syst.*, 12, 1–99, 1995.

43. K. Sung, R. Han, O. Hu, and L. Hsu, Controlled release of nalbuphine prodrugs from biodegradable polymeric matrices: influence of prodrug hydrophilicy and polymer composition, *Int. J. Pharm.*, 172, 17–25, 1998.

44. R. K. Kulkarni, K. C. Pani, C. Neuman, and F. Leonard, Polylactic acid for surgical implants, *Arch. Surg.* 93, 839–843, 1966.

45. B. A. Sabel, P. Dominiak, W. Hauser, M. J. During, and A. Freese, Levodopa delivery from controlled-release polymer matrix: delivery of more than 600 days *in vitro* and 225 days of elevated plasma levels after subcutaneous implantation in rats, *J. Pharmacol. Exp. Ther.*, 255, 914–922, 1990.

46. A. C. Sharon and D. L. Wise, Development of drug delivery systems for use in treatment of narcotic addiction, *NIDA Res. Monogr.* 28, 194–213, 1981.

47. S. J. Siegel, K. I. Winey, R. E. Gur, R. H. Lenox, W. B. Bilker, D. Ikeda, N. Gandhi, and W.-X. Zhang, Surgically implantable long-term antipsychotic delivery systems for the treatment of schizophrenia, *Neuropsychopharmacology*, 26, 817–823, 2002.

48. H. Qian, J. Bei, and S. Wang, Synthesis, characterization and degradation of ABA block copolymer of -lactide and -caprolactone, *Polym. Degrad. Stab.*, 68, 423–429, 2000.

49. P. Dubois, M. Krishnan, and R. Narayan, Aliphatic polyester-grafted starch-like polysaccharides by ring-opening polymerization, *Polymer*, 40, 3091–3100, 1999.

50. C. G. Pitt, Poly-ε-caprolactone and its copolymers, in *Biodegradable Polymers as Drug Delivery Systems*, Vol. 2, M. Chasin and R. Langer, Eds., Marcel Dekker, New York, 1990, pp. 71, chapter 3.

51. Y. Lemmouchi, E. Schacht, and C. Lootens, In vitro release of trypanocidal drugs from biodegradable implants based on poly(e-caprolactone) and poly(D,L-lactide), *J. Control. Release*, 55, 79–85, 1998.

52. M. Miyajima, A. Koshika, J. Okada, M. Ikeda, and K. Nishimura, Effect of polymer crystallinity on papaverine release from poly(lactic acid) matrix, *J. Control. Release*, 49, 207–215, 1997.

53. J. Kreuter, Nanoparticulate systems for brain delivery of drugs, *Adv. Drug Del.*, 47, 65–81, 2001.
54. J. L. Arias, V. Gallardo, S. A. Gomez-Lopera, R. C. Plaza, and A. V. Delgado, Synthesis and characterization of poly(ethyl-2-cyanoacrylate) nanoparticles, *J. Control. Release*, 77, 309–321, 2001.
55. G. Fontana, M. Licciardi, S. Mansueto, D. Schilaci, and G. Giammona, Amoxicillin-loaded poly-ethylcyanoacrylate nanoparticles: influenece of PEG coating on the particle size, drug release rate and phagocytic uptake, *Biomaterials*, 22, 2857–2865, 2001.
56. J. Dai, T. Nagai, X. Wang, T. Zhang, M. Meng, and Q. Zhang, pH-sensitive nanoparticles for improving the oral bioavailability of cyclosporine A, *Int. J. Pharm.*, 280, 229–240, 2004.
57. T. Jung, A. Breitenbach, and T. Kissel, Sulfobutylated poly(vinyl alcohol)-graft-poly(lactide-co-glycolide)s facilitate the preparation of small negatively charged biodegradable nanospheres, *J. Control. Release*, 67, 157–169, 2000.
58. H. S. Yoo, J. E. Oh, K. H. Lee, and T. G. Park, Biodegradable nanoparticles containing doxorubicin-PLGA conjugate for sustained release, *Pharm. Res.*, 16, 1114–1118, 1999.
59. H. S. Yoo, K. H. Lee, J. E. Oh, and T. G. Park, *In vitro* and *in vivo* anti-tumor activities of nanoparticles based on doxorubicin-PLGA conjugates, *J. Control. Release*, 68, 419–431, 2000.
60. M. Chacon, J. Molpeceres, L. Berges, M. Guzman, and M. R. Aberturas, Stability and freeze-drying of cyclosporine loaded poly (D,L lactide-glycolide) carriers, *Eur. J. Pharm. Sci.*, 8, 99–107, 1999.
61. A. Saez, M. Guzman, J. Molpeceres, and M. R. Aberturas, Freeze-drying of polycarolactone and poly(D,L-lactic-glycolic) nanoparticles induces minor particle size changes affecting the oral pharmacokinetics of loaded drugs, *Eur. J. Pharm. Biopharm.*, 50, 379–387, 2000.
62. A. Dalpiaz, E. Leo, F. Vitali, B. Pavan, A. Scatturin, F. Bortolotti, S. Manfredini, E. Durini, F. Forni, B. Brina, and M. A. Vandelli, Development and characterization of biodegradable nanospheres as delivery systems of anti-ischemic adenosine derivatives, *Biomaterials*, 26, 1299–1306, 2005.
63. S. Nicoli, P. Santi, P. Couvreur, G. Couarraze, P. Colombo, and E. Fattal, Design of triptorelin loaded nanospheres for transdermal iontophoretic administration, *Int. J. Pharm.*, 214, 31–35, 2001.
64. R. H. Muller, C. Jacobs, and O. Kayser, Nanosuspensions as particulate drug formulations in therapy. Rationale for development and what we can expect for the future, *Adv. Drug Del.*, 47, 3–19, 2001.
65. R. Gref, M. Luck, P. Quellec, M. Marchand, E. Dellacherie, S. Harnisch, T. Blunk, and R. H. Muller, Stealth corona-core nanoparticles surface modified by polyethylene glycol (PEG): influences of the corona (PEG chain length and surface density) and of the core composition on phagocytic uptake and plasma protein adsorption, *Colloids Surfaces B: Biointerfaces*, 18, 301–313, 2000.
66. M. T. Peracchia, E. Fattal, D. Desmaele, M. Besnard, J. P. Noel, J. M. Gomis, M. Appel, J. d'Angelo, and P. Couvreur, Stealth PEGylated polycyanoacrylate nanoparticles for intravenous administration and splenic targeting, *J. Control. Release*, 60, 121–128, 1999.
67. P. Calvo, B. Gouritin, I. Brigger, C. Lasmezas, J. Deslys, A. Williams, J. P. Andreux, D. Dormont, and P. Couvreur, PEGylated polycyanoacrylate nanoparticles as vector for drug delivery in prion diseases, *J. Neurosci. Methods*, 11, 151–155, 2001.
68. M. Tobio, A. Sanchez, A. Vila, I. Soriano, C. Evora, J. L. Vila-Jato, and M. J. Alonso, The role of PEG on the stability in digestive fluids and *in vivo* fate of PEG-PLA nanoparticles following oral administration, *Colloids Surfaces B: Biointerfaces*, 18, 315–323, 2000.
69. J. E. Blackwell, N. M. Daglia, B. Dickerson, E. L. Berg, and D. J. Goetz, Ligand coated nanosphere adhesion to E- and P-selectin under static and flow conditions, *Ann. Biomed. Eng.*, 29, 523–533, 2001.
70. Z. Panagi, A. Beletsi, G. Evangelatos, E. Livaniou, D. S. Ithakissios, and K. Avgoustakis, Effect of dose on the biodistribution and pharmacokinetics of PLGA and PLGA-mPEG nanoparticles, *Int. J. Pharm.*, 221, 143–152, 2001.
71. D. Kim, H. El-Shall, D. Dennis, and T. Morey, Interaction of PLGA nanoparticles with human blood constituents, *Colloids Surfaces B Biointerfaces*, 40, 83–91, 2005.

72. E. Chambers and S. Mitragotri, Prolonged circulation of large polymeric nanoparticles by non-covalent adsorption on erythrocytes, *J. Control. Release*, 100, 111–119, 2004.

73. H. M. Redhead, S. S. Davis, and L. Illum, Drug delivery in poly(lactide-co-glycolide) nanoparticles surface modified with poloxamer 407 and poloxamine 908: *in vitro* characterization and *in vivo* evaluation, *J. Control. Release*, 70, 353–363, 2001.

74. E. Kisak, B. Coldren, and J. Zasadinski, Nanocompartments enclosing vesicles, colloids, and macromolecules via interdigitated lipid bilayers, *Langmuir*, 18, 284–288, 2002.

75. T. Allen, Liposomal drug delivery, *Curr. Opin. Colloid Interfaces Sci.*, 1, 645–651, 1996.

76. D. Lasic, *Liposomes: from Physics to Applications*, Elsevier, Amsterdam, 1993.

77. D. Lasic and D. Papahadjopoulos, Liposomes and biopolymers in drug and gene delivery, *Curr. Opin. Solid State Mater. Sci.*, 1, 392–400, 1996.

78. D. Papahadjopoulos, T. Allen, A. Gabizon, E. Mayhew, K. Matthay, S. Huang, K. Lee, M. Woodle, D. Lasic, and C. Redemann, Sterically stabilized liposomes: improvements in pharmacokinetics and antitumor therapeutic efficacy, *Proc. Natl. Acad. Sci.*, 88, 11460–11464, 1991.

79. P. Sapra and T. Allen, Ligand-targeted liposomal anticancer drugs, *Prog. Lipid Res.*, 42, 439–462, 2003.

80. T. Allen and A. Chonn, Large unilamellar liposomes with low uptake into the reticuloendothelial system, *FEBS Lett.*, 223, 42–46, 1987.

81. A. Gabizon and D. Papahadjopoulos, Liposome formulations with prolonged circulation time in blood and enhanced uptake by tumors, *Proc. Natl. Acad. Sci. U.S.A.*, 85, 6949, 1988.

82. Papahadjopoulos and A. Gabizon, Liposomes designed to avoid the reticuloendothelial system, *Prog. Clin. Biol. Res.*, 343, 85–93, 1990.

83. T. Allen and C. Hansen, Pharmacokinetics of stealth versus conventional liposomes: effect of dose, *Biochim. Biophys. Acta*, 1068, 133–141, 1991.

84. H. Lin and J. Thomas, PEG-lipids and oligo(ethylene glycol) surfactants enhance the ultrasonic permeabilizability of liposomes, *Langmuir*, 19, 1098–1105, 2003.

85. N. Yamazaki, S. Kojima, and H. Yokoyama, Biomedical nanotechnology for active drug delivery systems by applying sugar-chain molecular functions, *Curr. Appl. Phys.*, 5, 112–117, 2005.

86. G. S. Kwon and K. Kataoka, Block copolymer micelles as long-circulating drug vehicles, *Adv. Drug Del. Rev.*, 16, 295–309, 1995.

87. S. Cammas-Marion, T. Okano, and K. Kataoka, Functional and site-specific macromolecular micelles as high potential drug carriers, *Colloids Surfaces B: Biointerfaces*, 16, 207–215, 1999.

88. A. Lavasanifar, J. Sammuel, and G. S. Kwon, Poly(ethylene oxide)-block-poly(L-amino acid) micelles for drug delivery, *Adv. Drug Del.*, 54, 169–190, 2002.

89. A. V. Kabanov, E. V. Batrakova, N. S. Melik-Nubarov, N. A. Fedoseev, T. Y. Dorodnich, V. Y. Alakhov, V. P. Chekhonin, I. R. Nazarova, and V. A. Kabanov, A new class of drug carriers: micelles of poly(oxyethylene)-poly(oxypropylene) block copolymers as microcontainers for drug targeting from blood in brain, *J. Control. Release*, 22, 141–158, 1992.

90. Y. S. Nam, H. S. Kang, J. Y. Park, T. G. Park, S. H. Han, and I. S. Chang, New micelle-like polymer aggregates made from PEI-PLGA diblock copolymers: micellar characteristics and cellular uptake, *Biomaterials*, 24, 2053–2059, 2003.

91. M. Yokoyama, G. S. Kwon, T. Okano, Y. Sakurai, T. Seto, and K. Kataoka, Preparation of micelle-forming polymer-drug conjugates, *Bioconjugate Chem.*, 3, 295–301, 1992.

92. Y. Won, H. Davis, and F. Bates, Giant wormlike rubber micelles, *Science*, 283, 960–963, 1999.

93. P. Dalhaimer, F. Bates, H. Aranda-Espinoza, and D. Discher, Synthetic cell elements from block copolymers — hydrodynamic aspects, *Comptes Rendus d'Academie Sci. — Serie IV Phys.*, 4, 251–258, 2003.

94. P. Dalhaimer, A. J. Engler, R. Parthasarathy, and D. E. Discher, Targeted worm micelles, *Biomacromolecules*, 5, 1714–1719, 2004.

95. H. Bermudez, A. Brannan, D. Hammer, F. Bates, and D. Discher, Molecular weight dependence of polymersome membrane structure, elasticity, and stability, *Macromolecules*, 35, 8203–8208, 2002.

96. W. Arap, R. Pasqualini, and E. Ruoslahti, Chemotherapy targeted to tumor vasculature, *Curr. Opin. Oncol.*, 10, 560–565, 1998.

97. Z. G. Gao, K. H. Oh, and C. K. Kim, Preparation and characterization of sustained-release microspheres of chlorpromazine, *J. Microencapsul.*, 15, 75–83, 1998.

98. B. Discher, Y.-Y. Won, D. Ege, J. Lee, D. Discher, and D. Hammer, Polymersomes: tough vesicles made from diblock copolymers, *Science*, 284, 1143–1146, 1999.

99. C. Dufes, A. Schatzlein, L. Tetley, A. Gray, D. Watson, J. Olivier, W. Couet, and I. Uchegbu, Niosomes and polymeric chitosan based vesicles bearing transferrin and glucose ligands for drug targeting, *Pharm. Res.*, 17, 1250–1258, 2000.

100. B. Discher, D. Hammer, F. Bates, and D. Discher, Polymer vesicles in various media, *Curr. Opin. Colloids Interfaces Sci.* 5, 125–131, 2000.

101. J. C.-M. Lee, H. Bermudez, B. M. Discher, M. A. Sheehan, Y.-Y. Won, F. S. Bates, and D. E. Discher, Preparation, stability, and *in vitro* performance of vesicles made with diblock coploymers, *Biotechnol. Bioeng.*, 73, 135–145, 2001.

102. D. Discher and A. Eisenberg, Polymer vesicles, *Science*, 297, 967–973, 2002.

103. R. Dimova, U. Seifert, B. Pouligny, S. Forster, and H. Dobereiner, Hyperviscous diblock copolymer vesicles., *Eur. Phys. J.*, 7, 241–250, 2002.

104. D. Discher, P. Photos, F. Ahmed, R. Parthasarathy, and F. Bates. Polymersomes: a new platform for drug targeting, in *Biomedical Aspects of Drug Targeting*, V. Torchilinand and V. Muzykantov, Eds., Springer, Philadelphia, 2002.

105. P. Photos, B. Discher, L. Bacakova, F. Bates, and D. Discher, Polymersomes in vivo: increased stealth with both PEG density and molecular weight, *J. Control. Release*, 90, 323–334, 2003.

106. V. Pata and N. Dan, The effect of chain length on protein solubilization in polymer-based vesicles (polymersomes), *Biophys. J.*, 85, 2111–2118, 2003.

107. F. Meng, C. Hiemstra, G. Engbers, and J. Feijen, Biodegradable polymersomes, *Macromolecules*, 36, 3004–3006, 2003.

108. X. Xiong, K. Tam, and L. Gan, Release kinetics of hydrophobic and hydrophilic model drugs from pluronic F127/poly(lactic acid) nanoparticles, *J. Control. Release*, 103, 73–82, 2005.

109. D. J. Brayden and A. W. Baird, Apical membrane receptors on intestinal M cells: potential targets for vaccine delivery, *Adv. Drug Deliv. Rev.*, 56, 721–726, 2004.

110. A. T. Florence, The oral absorption of micro- and nanoparticulates: neither exceptional nor unusual, *Pharm. Res.*, 14, 259–266, 1997.

111. A. T. Florence and N. Hussain, Transcytosis of nanoparticle and dendrimer delivery systems: evolving vistas, *Adv. Drug Deliv. Rev.*, 50, S69–S89, 2001.

112. W. Sun, C. Xie, H. Wang, and Y. Hu, Specific role of polysorbate 80 coating on the targeting of nanoparticles to the brain, *Biomaterials*, 25, 3065–3071, 2004.

113. P. R. Lockman, M. O. Oyewumi, J. M. Koziara, K. E. Roder, R. J. Mumper, and D. D. Allen, Brain uptake of thiamine-coated nanoparticles, *J. Control. Release*, 93, 271–282, 2003.

114. C. Song, V. Labhasetwar, X. Cui, T. Underwood, and R. J. Levy, Arterial uptake of biodegradable nanoparticles for intravascular local drug delivery: results with an acute dog model, *J. Control. Release*, 54, 201–211, 1998.

115. I. Fishbein, M. Chorny, L. Rabinovich, S. Banai, I. Gati, and G. Golomb, Nanoparticulate delivery system of a tyrphostin for the treatment of restenosis, *J. Control. Release*, 65, 221–229, 2000.

116. L. Brannon-Peppas and J. O. Blanchette, Nanoparticle and targeted systems for cancer therapy, *Adv. Drug Deliv. Rev.*, 56, 1649–1659, 2004.

117. Z. Xu, W. Gu, J. Huang, H. Sui, Z. Zhou, Y. Yang, Z. Yan, and Y. Li, *In vitro* and *in vivo* evaluation of actively targetable nanoparticles for paclitaxel delivery, *Int. J. Pharm.*, 288, 361–368, 2005.

118. M. Yokoyama, A. Satoh, Y. Sakurai, T. Okano, Y. Matsumura, T. Kakizoe, and K. Kataoka, Incorporation of water-insoluble anticancer drug into polymeric micelles and control of their particle size, *J. Control. Release*, 55, 219–229, 1998.

119. M. Yokoyama, T. Okano, Y. Sakurai, and K. Kataoka, Improved synthesis of adriamycin-conjugated poly (ethylene oxide)-poly(aspartic acid) block copolymer and formation of unimodal micellar structure with controlled amount of physically entrapped adriamycin, *J. Control. Release*, 32, 269–277, 1994.

120. M. Yokoyama, G. S. Kwon, T. Okano, Y. Sakurai, M. Naito, and K. Kataoka, Influencing factors on *in vitro* micelle stability of adriamycin-block copolymer conjugates, *J. Control. Release*, 28, 59–65, 1994.

121. M. Yokoyama, S. Fukushima, R. Uehara, K. Okamoto, K. Kataoka, S. Sakurai, and T. Okano, Characterization of physical entrapment and chemical conjugation of adriamycin in polymeric micelles and their design for *in vivo* delivery to a solid tumor, *J. Control. Release*, 50, 79–92, 1998.

122. M. Yokoyama, T. Okano, Y. Sakurai, S. Suwa, and K. Kataoka, Introduction of cisplatin into polymeric micelles, *J. Control. Release*, 39, 351–356, 1996.

123. Y. Li and G. S. Kwon, Methotrexate esters of poly (ethylene oxide)-block-poly(2-hydroxyethyl-L-aspartamide). Part I: Effects of the level of methotrexate conjugation on the stabiliy of micelles and on drug release, *Pharm. Res.*, 17, 607–611, 2000.

124. T. Ouchi, E. Yamabe, K. Hara, M. Hirai, and Y. Ohya, Design of attachment type of drug delivery system by complex formation of avidin with biotinyl drug model and biotinyl saccharide, *J. Control. Release*, 94, 281–291, 2004.

125. L. Nobs, F. Buchegger, R. Gurny, and E. Allemann, Poly(lactic acid) nanoparticles labeled with biologically active neutravidin for active targeting, *Eur. J. Pharm. Biopharm.*, 58, 483–490, 2004.

126. Y. Nishioka and H. Yoshino, Lymphatic targeting with nanoparticulate system, *Adv. Drug Deliv.*, 47, 55–64, 2001.

127. G. Lambert, E. Fattal, and P. Couvreur, Nanoparticulate systems for the delivery of antisense oligonucleotides, *Adv. Drug Deliv.*, 47, 99–112, 2001.

128. L. Illum, I. Jabbal-Gill, M. Hinchcliffe, A. N. Fisher, and S. S. Davis, Chitosan as a novel nasal delivery system for vaccines, *Adv. Drug Deliv. Rev.*, 51, 81–96, 2001.

129. J. O. Sham, Y. Zhang, W. H. Finlay, W. H. Roa, and R. Lobenberg, Formulation and characterization of spray-dried powders containing nanoparticles for aerosol delivery to the lung, *Int. J. Pharm.* 269, 457–467, 2004.

130. M. Brzoska, K. Langer, C. Coester, S. Loitsch, T. O. Wagner, and C. Mallinckrodt, Incorporation of biodegradable nanoparticles into human airway epithelium cells — in vitro study of the suitability as a vehicle for drug or gene delivery in pulmonary diseases, *Biochem. Biophys. Res. Commun.*, 318, 562–570, 2004.

131. A. DeCampos, A. Sanchez, and M. Alonso, Chitosan nanoparticles: a new vehicle for the improvement of the delivery of drugs to the ocular surface. Application to cyclosporin A, *Int. J. Pharm.*, 224, 159–168, 2001.

132. R. Alvarez-Roman, A. Naik, Y. N. Kalia, R. H. Guy, and H. Fessi, Skin penetration and distribution of polymeric nanoparticles, *J. Control. Release*, 99, 53–62, 2004.

133. U. Edlund and A. C. Albertsson, Degradable polymer microspheres for controlled drug delivery, *Degrad. Aliph. Polyest.: Adv. Polym. Sci.*, 157, 67–112, 2002.

134. M. Maeda, S. Moriuchi, A. Sano, and T. Yoshimine, New drug delivery system for water-soluble drugs using silicone and its usefulness for local treatment: application of GCV-silicone to GCV/HSV-tk gene therapy for brain tumor, *J. Control. Release*, 84, 15–25, 2002.

135. J. J. Kim and K. Park, Modulated insulin delivery from glucose-sensitive hydrogel dosage forms, *J. Control. Release*, 77, 39–47, 2001.

136. D. S. Hsieh, W. D. Rhine, and R. Langer, Zero-order controlled-release polymer matrices for micro- and macromolecules, *J. Pharm. Sci.*, 72, 17–22, 1983.

137. X. Huang and C. S. Brazel, On the importance and mechanisms of burst release in matrix-controlled drug delivery systems, *J. Control. Release*, 73, 121–136, 2001.

138. J. Panyam, M. M. Dali, S. K. Sahoo, W. Ma, S. S. Chakravarthi, G. L. Amidon, R. J. Levy, and V. Labhasetwar, Polymer degradation and in vitro release of a model protein from poly(D,L-lactide-co-glycolide) nano- and microparticles, *J. Control. Release*, 92, 173–187, 2003.

139. F. I. Liu, J. H. Kuo, K. C. Sung, and O. Y. Hu, Biodegradable polymeric microspheres for nalbuphine prodrug controlled delivery: *in vitro* characterization and *in vivo* pharmacokinetic studies, *Int. J. Pharm.*, 257, 23–31, 2003.

140. M. Texeira, M. Alonso, M. Pinto, and C. Barbosa, Development and characterization of PLGA nanospheres and nanocapsules containing xanthone and 3-methoxyxanthone, *Eur. J. Pharm. Biopharm.*, 59, 491–500, 2005.

141. J. Matsumoto, Y. Nakada, K. Sakurai, T. Nakamura, and Y. Takahashi, Preparation of nanoparticles consisted of poly (L-lactide)-poly(ethylene glycol)-poly(L-lactide) and their evolution *in vitro*, *Int. J. Pharm.*, 185, 93–101, 1999.

142. K. Avgoustakis, A. Beletsi, Z. Panagi, P. Klepetsanis, A. G. Karydas, and D. S. Ithakissios, PLGA-mPEG nanoparticles of cisplatin: *in vitro* nanoparticle degradation, *in vivo* drug release and *in vivo* drug residence in blood properties, *J. Control. Release*, 79, 123–135, 2002.

Index

3-D assembly
  proposed schematic process flow, **17**-30
  using DNA and nanoparticles, **17**-16
2-D distributed feedback laser generator, **25**-17
3-D force microscopes (3DFMs), **15**-2. *See also* Magnetic
      manipulation
  chromatin manipulation with, **15**-18
  cilia manipulation with, **15**-17
  cilia tracking with first-generation system, **15**-16
  force directionality data for second-generation system,
      15-14
  force *vs.* distance for second-generation system,
      15-13
  generated forces from first-generation system, **15**-12
  optical system diagram for second-generation system,
      15-15
  spherical plots for directionality of forces, **15**-14
  tracking system response to motion, **15**-11
  user interface, **15**-10
2-D molecular assembly, steps for nanotubes, **20**-7
2-D parabolic potential, schematic representation,
      13-11

A

Ab initio CNT simulation methods, **23**-28–**23**-32
Abacus, **7**-4, **7**-5
Academy of Sciences of Ukraine, xii
Accelerated design tools
  analysis of negative (left-handed) materials,
      29-11–**29**-15
  Celerity Hardware Accelerator, **29**-2
  coupled-resonator optical waveguides, **29**-5–**29**-11
  Mie theory comparison, **28**-3–**28**-5
  for nanophotonics devices and applications, **29**-1–**29**-2
  subwavelength optical systems, **29**-15–**29**-20
Accumulated polarization, as function of current in
      degenerate/nondegenerate semiconductors,
      9-29
Action potentials
  in human brain, **7**-21
  information transmission by, **7**-22
  in neurons, **7**-25
Activating carriers, **7**-24
Active matrix addressing, in memory devices, **4**-12
Active nanostructures products, **3**-17–**3**-18
Adamantane, in supramolecular design, **18**-27
Adaptive defect-tolerant molecular presenting-and-
      memory platforms, **7**-52–**7**-55

Aedomers, **18**-20–**18**-21, **19**-23–**19**-25
  chemical structure examples, **19**-24
  directed stacking in water, **18**-22
  structural examples, **19**-25
AIA materials, **2**-3
Air-bridge microcavity, **25**-17–**25**-18
Air Force Office of Scientific Research, ix
Alignment, in nanodevice manufacturing, **20**-11
Alkane chain, current-voltage characteristics of, **9**-18
Alkyl-free space-filling model, **19**-4
Alligator clips, **5**-17, **5**-21
  oligo derivatives with, **5**-18
  synthesis from formamide precursor, **5**-21
Alpha-peptides, **19**-27
Alpha relaxation, **16**-11
Amidocryptands, **19**-17
Amino acids
  alpha-helix and beta-sheet from, **18**-46
  structural features, **18**-45
Amylopectin topology, **24**-6
Analytic potential functions, quantum basis of, **28**-6–**28**-11
Analytical Engine, **7**-6
Analytical techniques, **11**-2
AND/OR logic gates, **5**-6, **7**-37, **7**-75, **7**-76
  from QCA cells, **5**-5
  in QCA structures, **6**-4
  simulating with SPICE/HSPICE software, **5**-23
Angiogenesis, inhibition by dendrimers, **24**-27
Angular equation, **7**-73
Angular wave function, **7**-73
Annulene-based muscles, redox-controlled, **11**-19
Antibody attachment, **17**-10
Antisense oligonucleotides, **30**-19
Antiviral agents, dendrimers as, **24**-26–**24**-27
Applications, training of workers for emerging, **3**-20
Applied information theory, biomolecular processing
      applications, **7**-25–**7**-30
Architecture changes, **2**-5–**2**-6
Army Research Office, ix
Arterial drug delivery, using nanoparticles, **30**-17
Artificial atoms
  many-body Hamiltonian of, **13**-10
  modeling of, **13**-8–**13**-10
  quantum modeling of, **13**-16–**13**-19
Artificial micro- and nanostructures
  biological and chemical-mediated self-assembly,
      17-1–**17**-2
  miniaturization and development of, **17**-2–**17**-3
Artificial muscles, **11**-16–**11**-17

Asymmetrical lens simulation, **29**-18–**29**-19

Atom relays, **6**-10

Atomic force microscopy, **11**-3, **15**-2

Atomic layer deposition, **4**-11

Atomic rearrangement, **1**-8, **3**-11

Atoms, and quantum mechanics, **1**-9

Au-(1)-Au device, I(V) characteristics, **5**-18

Au-(2)-Au device, I(V) characteristics of, **5**-19

Aviram-Ratner Donor-insulator-Acceptor construct, **9**-2

Axles, in molecular rotors, **11**-26

Axo-dendritic organelles, **7**-22

   input-output representation, **7**-25

Azobenzene valves, light-regulated, **11**-22–**11**-24

B

Babbage, Charles, **7**-6

Bandwidth, and memory system performance, **7**-46

Barrier properties

   barrier enhancement model, **21**-54

   biodegradability and bio-related properties, **21**-59–**21**-60

   electrical and thermal properties, **21**-58–**21**-59

   of nanoadditives, **21**-53–**21**-54

   optical properties, **21**-54–**21**-56

   surface modification, **21**-56–**21**-58

Base pairing, hydrogen bond examples, **18**-11

Beam models, for carbon nanotubes, **23**-7–**23**-9

Beam splitters, **25**-15–**25**-16

   amplitude propagation plots, **29**-20

   initial and optimized intensities in focal plane, **29**-20

   lens configuration, **29**-18

   optimized 1-to-3, **29**-19

   simulation results, **29**-19

Bengal Rose formation, in dendrimer center, **18**-60

Benzene-dithiolate molecule

   effects of tilting on I-C curve of, **9**-5

   measured conductance through, **12**-9

   schematic representation, **9**-5

Benzoic acid dimers, **18**-15

Beta-amino acid residues, structural examples, **19**-26

Beta-peptides, **19**-26–**19**-27

   chemical structure examples, **19**-28

   main helices of, **19**-27

Bimolecular processing

   and applied information theory and information estimates, **7**-25–**7**-30

   and fluidic molecular platforms, **7**-30–**7**-32

   and neuromorphological reconfigurable molecular processing platforms, **7**-32–**7**-33

   and neuroscience, **7**-20–**7**-25

Binary logics, bioprocessing features as, **7**-30

Binding angles, in ligand and metal complexes, **18**-35

Bio-inspired active device assembly, **17**-30–**17**-31

Bio-inspired self-assembly, **17**-3–**17**-5

   biological entity-mediated, **17**-12–**17**-21

   with chemically treated surfaces, **17**-22–**17**-25

   electrically mediated, **17**-25–**17**-27

   fluidics mediated, **17**-27–**17**-30

motivation for, **17**-5

state of the art, **17**-12

Biocomponent fibers

   partially soluble, **21**-10–**21**-13

   segmented pie fiber, **21**-7

   splittable, **21**-6–**21**-10

   tipped trilobal and segmented ribbon fibers, **21**-7

Biodegradability

   of nanoadditives, **21**-59–**21**-60

   photographic evidence, **21**-61

Biolinkers

   DNA as, **17**-6–**17**-9

   future developments in, **17**-31

   protein complexes as, **17**-9–**17**-10

   in self-assembly, **17**-6

   strategies for assembly with, **17**-10–**17**-12

Biological detection/protection, **3**-22

Biological entity-mediated self-assembly, **17**-12

   DNA-directed micro- and nanowires, **17**-16–**17**-18

   DNA-inspired self-assembly of electronic devices, **17**-18–**17**-19

   DNA-mediated nanostructure assembly, **17**-12–**17**-16

   future directions, **17**-29–**17**-30

   nanostructures mediated by DNA, **17**-12

   protein complex-mediated assembly of micro- and nanostructures, **17**-19–**17**-22

Biological information processing, real-time, **7**-7–**7**-8

Biological nanovalves, **11**-26

Biological structure control strategy, **24**-4

Biology

   convergence of engineering with, **17**-3

   convergence of nanotechnology with, **3**-9

   nanoscale examples in, **1**-5

Biomedical sciences, magnetic manipulation for, **15**-1–**15**-2

Biomimetic structures approach, **18**-43–**18**-44

   building new designs from old subunits, **18**-47–**18**-48

   building old designs from new subunits, **18**-48–**18**-53

   DNA, **18**-44

   limiting cases of MBB flexibility in, **18**-51

   primary and secondary structures, **18**-50–**18**-51

   proteins, **18**-44–**18**-47

   synthetic bis-ortho-carborane MBB for, **18**-52

Biomolecular cell trapping, **26**-4

Biomolecular electromechanical switches, **7**-23

Biomolecular processing and communication, **7**-7

   applied information theory and information estimates, **7**-25–**7**-30

Biomolecules, as information and routing carriers, **7**-24

BionomiX Inc., xi

Biotechnology, convergence with nanotechnology, **3**-4

Bipolar devices, three-terminal, **4**-3

Bipolarons, **12**-16

bis-amino acids, **19**-29–**19**-31

   building block approach, **19**-30

   rod-like and curved structures formed by, **19**-31

Bistable nanotubes, device potential, **5**-12

Bistable [2]rotaxanes, **11**-8–**11**-10, **11**-20

   folded conformation of, **11**-11, **11**-12

   interactions with metal electrodes, **11**-13

LB studies of, **11**-10–**11**-12

molecular structure and thermodynamic properties, **11**-16

redox-controlled switching behavior of, **11**-14

Bistable saturation

in QCA arrays, **6**-3

in single-domain magnets, **6**-7

BLOCK approach

examples of molecular structural synthesis, **19**-29

to large structures, **19**-27–**19**-29

Block copolymers, nanoscale patterning with, **20**-9–**20**-10

Boltzmann formula, **7**-11

as approximation to transport equation, **12**-5

Bond breakage, due to electric current, **12**-21

Bonding

covalent, **18**-3–**18**-5

mechanisms in carbon nanotubes, **23**-2–**23**-3

Books. *See also Encyclopedia Britannica*

fitting on credit card, **2**-1

Boraadarnantane, **18**-8

Bose-Einstein statistics, **7**-13

Bottle brush molecule, **28**-15, **28**-16

Bottom-up fabrication, **7**-3, **7**-9, **7**-18, **7**-19, **11**-4

in computer circuitry fabrication, **5**-2

convergence with top-down method, **10**-2–**10**-3

limited progress in, **7**-4

natural models of, **11**-4–**11**-5

in nature, **5**-3

for organic nanostructures, **24**-4–**24**-6

Bragg reflectors, **25**-16

Brain delivery, using nanoparticles, **30**-17

Brain modeling, based on neuron-to-neuron interactions, **3**-17

Branch cell construction, **24**-11

Branch cell structural parameters, **24**-10

Branched junction, **17**-13

Breit-Wigner formula, **9**-7

Brenner, Donald W., xii

Broadening

effect on I-V characteristics, **14**-12

MATLAB one-level model, **14**-21

MATLAB unrestricted one-level model, **14**-22–**14**-23

schematic diagram, **14**-5

Brownian dynamics

of fluidic molecules and ions, **7**-32

and limitations on minimum particle trapping size, **16**-2–-**16**-23

role at molecular level, **11**-3

Buckling behavior, of SWCNTs, **23**-10

Buckyball, **27**-5

Building block approaches. *See also* Molecular building blocks

A- and B-type building block examples, **19**-30

to linear macromolecules, **19**-1–**19**-3, **19**-17–**19**-31

to nonlinear macromolecules, **19**-1–**19**-17

Building block reactivity patterns, multiscale, **24**-35

Bush, Vannevar, **7**-6

Bush administration

and nanotechnology opportunities, **3**-20

support for NNI, **3**-9, **3**-10

C

Cache memory

associative *vs.* content-addressable, **7**-46–7-46

integration of data memory and tag memory via, **7**-47

CAD-supported SLSI design, **7**-19

hardware-software co-design and, **7**-55

limitations of current, **7**-38

proof-of-concept tools in, **7**-40

California Institute of Technology, xi

Clinton's announcement of NNI at, **3**-7

Calixarenes, **19**-14–**19**-16

synthesis strategies, **19**-14

Cancer detection, **3**-13

Cancer treatment, lymphatic system nanoparticle drug delivery for, **30**-18

Capacitor charging time, **4**-6–**4**-7

Capillary electrophoresis

erythrocyte determination by, **26**-13

parallel intracellular study by, **26**-14

Car-Parrinello molecular dynamics, **11**-3

Carbon, covalent, atomic, and van der Waals radii, **7**-12

Carbon black, **27**-4

crystallographic arrangements, **21**-49

as nanoadditive, **21**-48

typical furnace process, **21**-48

Carbon fibers

2D cross-section, **21**-49

as nanoadditives, **21**-48–**21**-49

Carbon nanofibers

growth process, **22**-18

as nanoadditives, **21**-50

replicate stress-strain behavior, **21**-51

Carbon nanostructures/nanocomposites, **27**-1–**27**-2

diamond-like structures in, **27**-19–**27**-21

extended carbon family, **27**-3–**27**-5

fullerene-based, **27**-9

modern history of, **27**-2–**27**-3

nanoscale diagram, **27**-4

nanotube-based, **27**-9–**27**-19

remaining challenges, **27**-21

structure, notation, and synthesis, **27**-5–**27**-9

Carbon nanotube experimental techniques, **23**-20

instruments for mechanical study, **23**-21–**23**-22

methods and tools for mechanical measurement, **23**-22–**23**-28

Carbon nanotube mechanics, **23**-1–**23**-2

experimental techniques, **23**-20–**23**-29

mechanical applications, **23**-39–**23**-48

mechanical properties of CNTs, **23**-2–**23**-20

simulation methods, **23**-29–**23**-38

Carbon nanotube tips, **22**-21–**22**-22

Carbon nanotubes, **5**-14, **10**-4, **21**-27, **22**-1–**22**-2

5-7-7-5 dislocations, **23**-15

application development, **22**-20–**22**-26

armchair, zig-zag, and chiral morphologies, **21**-28, **23**-4

axial compression of, **22**-6

beam and shell elastic models, **23**-7–**23**-9

bonding mechanisms of, **23**-2–**23**-3

computational modeling and simulation, **22**-3–**22**-14

cross-linking to polyethylene matrix, **27**-17

defect-mediate two-terminal junction in, **22**-12
deposition by AC electrophoresis, **23**-46
double-walled, **22**-7
elastic buckling and local deformation of, **23**-9–**23**-13
elastic properties based on crystal elasticity, **23**-17–**23**-20
elastic properties of, **23**-4–**23**-7
electronic structure, metallic armchair nanotube and
    semiconductor zig-zag nanotube, **13**-23
electrostatics of, **13**-27–**13**-28
field emissions applications, **22**-25
force balance for, **13**-25
forces in shell-sliding experiments, **23**-17
formation of PE layers around, **22**-9
four-level dendritic neural tree, **22**-14
future outlook, **13**-30
graphene sheet lattice, **22**-2
from graphene sheet to single-walled nanotubes,
    23-3–**23**-4
growth of, **22**-14–**22**-19
hexagonal bonding structure for graphite layer, **23**-3
honeycomb lattice structure, **13**-21
interlayer connections upon collapse, **23**-12–**23**-13
kink formation in, **27**-17
lowest conduction sub-band and highest valence
    sub-band, **13**-23
material development, **22**-19–**22**-20
mechanical applications, **23**-39–**23**-48
methyl radical chemical chemiabsorption into,
    27-17
microelectronics applications, **22**-22–**22**-24
microscopy applications, **22**-20–**22**-22
modeling as elastic materials, **22**-4–**22**-13
modeling of, **13**-20–**13**-24
modern history of, **27**-2
molecular structure of, **23**-2–**23**-4
multi-walled structures, **23**-4
multiscale analysis, **23**-39
as nanoadditives, **21**-49–**21**-50
from nanoscale to macroscale, **21**-29–**21**-30
nanotube mechanics, **13**-25–**13**-27
nanotube nanotweezers device, **13**-24
as new building block for molecular-scale circuits,
    6-9
ON and OFF states of, **13**-24
physical cross-bar junctions in, **22**-12
polymer/nanotube composites, **21**-30–**21**-31
polypropylene/nanotube composites, **21**-31–**21**-34
production methods, **21**-27
properties, **21**-28–**21**-29
radial distribution function, **22**-9
schematic diagram, **27**-8
scroll-like structures, **23**-4
self-assembly of, **10**-8
self-consistent charge density of armchair nanotube,
    13-27
sensor applications, **22**-24–**22**-25
simulated-device geometry, **13**-27
specific composite conductivity as function of weight
    fraction, **21**-59
Stone-Wallace bond rotation defect, **22**-7
strength characteristics, **23**-13–**23**-17

structure, **21**-27
    notation, and synthesis, **27**-5–**27**-9
structure and properties, **22**-2–**22**-3
system geometry with quantum connection to van der
    Waals forces, **13**-26
TBA electronic structure of graphite monolayer valence
    bonds, **13**-22
thermal conductivity as function of temperature for,
    22-11
three-terminal junctions in, **22**-13
Carbon protrusions, SEM image, **27**-6
Carbonaecous materials
    carbon black, **21**-48
    carbon nanofibers, **21**-48–**21**-50
    carbon nanotubes, **21**-49–**21**-50
    graphite, **21**-48
    mechanical reinforcement, **21**-50–**21**-52
    as nanoadditives, **21**-48
    properties and areas of application, **21**-50
Carboranes, **18**-25
Carrier hopping, in monolayers of saturated molecules,
    9-20
Carrier transport, in Tour wires, **9**-3
Casted films, **21**-60
Catalytic chemistry, **12**-21
    dendrimers in, **24**-29
Catenanes, **5**-12, **5**-13, **5**-23, **18**-19–**18**-20
    acetylcholine-mediated formation, **19**-7
    building block approaches to, **19**-3–**19**-9
    cyclic bis[2]catenane formation, **19**-7
    effects of metal, temperature, and concentration on
        formation of, **18**-20
    ion-templated formation, **19**-6
    metal-templated formation, **19**-5
    tristable, **19**-8
CBPQT recognition system, **11**-7–**11**-8, **11**-11
    complexation by nanopores, **11**-24
    demonstration of mechanical movements, **11**-15
    mechanical movement between DNP and TTF
        recognition sites, **11**-13
    rate of electron transfer to, **11**-14
Celerity Hardware Accelerator, **29**-2
Cell biology nanofluidics, **26**-1–**26**-2
    biomolecular cell trapping, **26**-4
    cell sorting in microdevices, **26**-5–**26**-7
    cellular nucleic acid isolation, **26**-8–**26**-10
    devices for monitoring single-cell physiology,
        26-7–**26**-20
    electrokinetic cell trapping, **26**-4–**26**-5
    electroosmotic and dielectrophoretic cell sorting,
        26-5–**26**-6
    electroporation schemes, **26**-8
    manipulation and mobilization of cells in microdevices,
        26-2–**26**-5
    mechanical cell trapping, **26**-2–**26**-4
    microchannel geometry and cell culture, **26**-17–**26**-20
    nanofluidic electrophoresis, **26**-12
    nucleic acid amplification/detection, **26**-10–**26**-12
    optical cell sorting, **26**-6–**26**-27
    pressure-based cell sorting, **26**-6
    proteomics applications, **26**-12–**26**-15

quantifying intracellular protein and small molecule detection, 26-15
single-cell electroporation, **26**-7–**26**-8
and surface chemistry in microdevices, **26**-15–**26**-17
Cell culture
  microchannel geometry and, **26**-17–**26**-20
  in microdevices, 26-15
  surface chemistry and, **26**-15–**26**-17
Cell manipulation and immobilization, 26-2
  biomolecular cell trapping, 26-4
  electrokinetic cell trapping, **26**-4–**26**-5
  mechanical cell trapping, **26**-2–**26**-4
Cell sorting
  electroosmotic and dielectrophoretic, **26**-5–**26**-6
  in microdevices, 26-5
  optical methods, **26**-6–**26**-7
  pressure-based, 26-6
Cell-to-cell interactions, in quantum dots, **6**-2, **6**-3, **6**-5
Cellular automata, 10-5
Cellular neural networks (CNNs), **2**-4, **2**-5
Cellular Nonlinear Networks, 10-5
Cellular nucleic acid isolation, **26**-8–**26**-10
Center of Functional Engineered Architectonics, 10-9
Centers of excellence, NNI support for, 3-22
Central processing unit (CPU), 7-43
Channel capacitance, 7-28, 7-30
  for analog channels, 7-29
  for Poisson processes, 7-29
Channel drop filters, 25-15
  frequency spectrum of time varying electric field, 25-16
  in photonic crystals, 25-15
Characterization tools, 20-11
Charge mobility
  and circuit speed, 4-6
  and signal delay time, 4-10
Charge movement
  in diffuse double layer, **16**-9–**16**-11
  in Stern layer, **16**-8–**16**-9
Charge transport, 4-12
Charging effects, **14**-8–**14**-11
Charging time, 4-7
  reducing with dual-gate designs, 4-11
Chelating agents, dendrimers as, **24**-27–**24**-28
Chemical detection/protection, 3-22
Chemical process engineering, dendrimers in, **24**-29–**24**-30
Chemical synthesis, nanoscale, 1-9
Chemically treated surfaces, self-assembly on, **17**-22–**17**-25
Chirality effects, in molecular conductance junctions, 12-22
Chromatin manipulation, with **3**-D force microscopes (3DFMs), **15**-17–**15**-19
Chronic illness
  detection by subcellular interventions, 3-3
  reducing through nanotechnology, 3-17
Cilia manipulation, with **3**-D force microscopes (3DFMs), **15**-15–**15**-17
Circuit design, challenges in molecular engineering, 4-12
Circuit speed
  in biological systems, 7-8
  and transistor drive current, 4-6
Clausius-Mossotti factor, **16**-3, **16**-5, **16**-6
  with crossover frequencies, 16-18

Clinton administration, announcement of NNI at Caltech, 3-7
CMOL architecture, **6**-11, **10**-16
CMOL circuits, 10-15
CMOL FPGA, **10**-17, **10**-18
CMOL memory, 10-17
  defect tolerance of, **10**-16
CMOS circuits
  advances beyond, **10**-5–**10**-6
  continuing viability of, 7-10
  current landscape, **10**-3–**10**-5
  decision diagram derivation for, 7-37
  future device roadmap, 10-5
  signal delay determinants in, 4-10
  technical challenges of, 10-1
CMOS device parameters and scaling, 4-6
  alternate device structures, 4-11
  constant field scaling, **4**-8–**4**-9
  interconnects and parasitics, 4-10
  mobility and subthreshold slope, **4**-6–**4**-8
  power dissipation, **4**-8–**4**-9
  reliability, **4**-10–**4**-11
CMOS device roadmap, **10**-4–**10**-5
CMOS devices
  alternate device structures for, 4-11
  inverter circuit approach, 4-5
  lithographic approaches to nanoscale, 4-13
  manufacturing practices, 4-1
  parameters and scaling, **4**-6–**4**-11
  reduced power consumption in, 4-4
  three-terminal, 4-3
  typical gain values, 4-5
CMOS inverters, 4-5
  from QCA cells, 5-5
CNT-based FETs, **10**-5, **22**-23
CNT films, selective defect removal from, **20**-13–**20**-14
CNT-reinforced nanocomposites, **27**-9–**27**-19
CNT simulation methods, 23-29
  ab initio and tight binding methods, **23**-29–**23**-32
  classical molecular dynamics, **23**-33–**23**-36
  continuum and multiscale models, **23**-36–**23**-38
  density functional theory (DFT), **23**-31–**23**-32
  Hartree-Fock approximation, **23**-30–**23**-31
  interlayer potentials, **23**-34–**23**-36
  tight-binding method, 23-32
Code converter, in hierarchical finite-state machines, 7-52
Cognitive technologies, **3**-4, **7**-20
  in biological systems, **7**-20–**7**-21
  convergence with nanotechnology, 3-9
Coherence, loss of, and inelasticity, 12-12
Coherence length, 12-5
Coherent transport, 12-4
  generalized Landauer formula for, 12-4
  and Landauer transmission model of conductance, **12**-5–**12**-10
  and length scales in mesoscopic systems, **12**-4–**12**-5
Combinatorial logic optimization, 7-51, 7-52
Combined top-down and bottom-up nanomanufacturing, 20-5
  directed self-assembly approach, **20**-6–**20**-7
  directed self-assembly using nanotemplates, **20**-7–**20**-9

nanoscale patterning, **20**-5–**20**-6
nanoscale patterning using block copolymers,
    20-9–**20**-10
Commercialization timelines, **3**-14
Compilers, **7**-57
Complementary metal oxide semiconductor (CMOS)
    devices, **4**-1. *See also* CMOS devices
Complex molecular devices, characterization, **5**-23–**5**-24
Complex paralleling, **7**-47, **7**-48
Complex permittivity, **16**-12
Complex spheroids, multishell model of, **16**-11–**16**-13
Composite nano-structured materials, **3**-16, **27**-1–**27**-2. *See
    also* Carbon nanostructures/nanocomposites
composite nanofibers via electrospinning, **21**-18–**21**-20
Computation
    based on coupling-induced ordering phenomena, **6**-7
    harnessing nanodevices for, **6**-1
    physical limits on, **10**-10
    with QCA arrays, **6**-5
Computational modeling
    of carbon nanotubes, **22**-3–**22**-5
    molecular electronics with nanotube junctions,
        22-11–**22**-13
    of molecular movement, **11**-2, **11**-12
    of molecular rotors, **11**-34
    and nanomechanics of nanotubes and composites,
        22-5–**22**-10
    nanotube sensors and actuators, **22**-14
    vibrational characteristics and thermal transport in
        nanotubes, **22**-10–**22**-11
Computer architecture, **7**-44
Computers, miniaturizing, **1**-5
Condensed phase movement, **11**-13–**11**-16
Conditional probability distribution, **7**-28
Conductance, **12**-1–**12**-3. *See also* Molecular conductance
    junctions
    defined, **12**-5
    and electron transfer rate constant, **12**-14
    as function of voltage for thiol-gold, **12**-8
    Landauer transmission model of, **12**-5–**12**-10
    and limitations on minimum particle trapping size,
        16-20–**16**-23
    quantum of, **12**-5
    schematic for molecular wires, **14**-2
    tilting mechanism in extrinsic molecular switching,
        9-4–**9**-5
Conductance gap, factors determining, **14**-9–**14**-10
Conductance ratio
    improving in SPINFETs, **8**-7
    SPINFET limitations, **8**-7
Conductance switching, in simulated nanomolecules, **5**-23
Conductivity states
    in memory devices, **5**-17
    of nanopore devices, **5**-16
Confined fluids, nanometer-scale properties simulation,
    28-17–**28**-19
Constant field scaling, **4**-8–**4**-9
Constant voltage scaling, **4**-8
Consultative Boards for Advancing Nanotechnology
    (CBAN), **3**-15
    objectives, **3**-16

Contact time, **12**-15
Contacts
    engineering challenges, **4**-14
    improving resistance, **4**-12
    molecular conformational effects at, **4**-14
Continuity equation, and spatial distribution of electron
    density, **9**-27
Control unit, in hierarchical finite-state machines, **7**-49
Controllable wettability, **11**-5
    surfaces with, **11**-35
Controlled molecular motion, issues in, **11**-3–**11**-4
Controlled organic nanostructures, importance in biology,
    24-3
Controlled-release drug molecules, **11**-21
Convection, and limitations on minimum particle trapping
    size, **16**-20–**16**-23
Converging technologies, **3**-3, **3**-4
    role of nanoscale in, **3**-3–**3**-5
    transforming tools, **3**-5
Coordinate vectors, **18**-37
Coordination complexes, **18**-5–**18**-7
    examples in metal complexes, **18**-5
Coordination geometries, for metal ions, **18**-36
Copper complexation structures, **18**-39
Core-shell patterns, **24**-12
    influence on molecular reactivity of dendrimers, **24**-34
    symmetry properties of, **24**-32
Cost issues, in circuit density development, **5**-2
Coulomb repulsion, **4**-14, **14**-8
    broadening and, **14**-12
    between CBPQT and TTF dication, **11**-9
    and charging effects, **14**-8–**14**-11
    current-voltage characteristics diagram, **14**-13
    by electrode material defects, **9**-16
    in QCA technology, **5**-5, **6**-3
    for QCA *vs.* metallic dots, **6**-7
    suppression through broadening, **14**-13
    unrestricted model, **14**-11–**14**-12
Coulomb staircase, for quantum dots, **13**-13
Coumarin-loaded particles
    PLGAs, **30**-6
    in VSMCs, **30**-9
Coumarin valves, light-regulated, **11**-22–**11**-24
Coupled-resonator optical waveguides, **29**-5, **29**-9–**29**-11
    delay time and amplitude as function of number of
        rings, **29**-11
    five-ring simulation, **29**-7–**29**-8
    numerical simulations, **29**-5
    propagation through five-ring CROW structure,
        29-9
    propagation through two-ring CROW structure,
        29-7
    pulse propagation through five-ring CROW, **29**-9
    pulse propagation through ten-ring CROW, **29**-10
    pulse propagation through twenty-ring CROW, **29**-10
    pulse propagation through two-ring CROW, **29**-8
    ten- and twenty-ring simulations, **29**-9
    transmission spectra at throughput and output ports,
        29-7, **29**-8
    two-ring simulation, **29**-6–**29**-7
Coupling-induced ordering phenomena, **6**-7

Covalent bonds, **18**-3–**18**-5
  challenges to SWNT technologies, **5**-14
  molecular simulations of, **28**-4–**28**-5
  and molecular Tinkertoy approach, **18**-22–**18**-31
  rotation in, **18**-4
Cow pea mosaic virus (CPMV) capsid organization, **10**-8
Cross-talk
  isolating molecular interconnects to avoid, **4**-14
  in nanomagnets, **9**-22
  with passive addressing, **4**-12
  zero, **25**-16–**25**-17
Crossbar arrays, **5**-8–**5**-15, **10**-5
  in Teramac computer, **6**-10
Crossed NW arrays. *See also* Nanowires (NW)
  layer-by-layer assembly and transport measurements, **5**-13
Crossover frequencies, **16**-8, **16**-19
Cryptands, as building blocks, **19**-17, **19**-18
Crystal elasticity, of carbon nanotubes, **23**-17–**23**-20
Crystal engineering, **18**-15–**18**-16
Crystal growth techniques, **2**-4
Cubic scaffolding, **18**-29–**18**-31
Current, as function of time for molecular junctions, **12**-21
Current density, **7**-68
Current flow
  as balancing act, **14**-6–**14**-8
  modification by molecular binding, **12**-10
Current/voltage characteristics, for molecular wire with six gold atoms, **12**-9
Current-voltage (I-V) characteristics
  for broadened one-level model, **14**-12
  broadening effects on, **14**-12
  and coulomb blockade, **14**-8–**14**-12
  with Coulomb blockade, **14**-13
  of molecules, **14**-1–**14**-3
  for one-level model, **14**-8
  and potential drop for **6**-atom QPC, **14**-19
  and potential drop for asymmetrically connected QPC, **14**-19
  for restricted and unrestricted one-level models, **14**-11
  schematic for molecular wires, **14**-2
  with suppression of Coulomb blockade by broadening, **14**-13
  for two-level model, **14**-10
  for two-terminal devices, **7**-59
Current *vs.* voltage curve, **4**-3
Cyclic molecules, AND/NAND gates in, **7**-76
Cylindrical systems, electron transport solution for, **7**-73
Cystatin-loaded nanoparticles, **30**-18

D

3D devices, Poisson equation for, **7**-74
3D directly interconnected molecular electronics, **7**-43
3D image processing, in biological systems, **7**-7–**7**-8
3D integrated circuits, **7**-37
3D molecular integrated circuits, **7**-5, **7**-18, **7**-20
  in biological systems, **7**-20
  design of, **7**-34
  design results, **7**-40
  proof-of-concept, **7**-39

3D molecular signal/data processing
  by fluidic processing devices, **7**-31
  and memory platforms, **7**-43–**7**-49
3D nanosystems, **3**-18. *See also* Third dimension
  heat dissipation problems, **4**-16
1D potential energy profile, **7**-76
3D problems, in molecular device modeling and analysis, **7**-71–**7**-75
3D synthetic fluidic device/module, **7**-31
Dangling bond, **9**-18, **9**-20
Data acquisition, for experimental magnetic manipulation systems, **15**-9–**15**-11
Data processing platforms, molecular, **7**-4–**7**-8
Data structures, for 3D ^{M}ICs, **7**-34
Datapath
  control by micro instruction set, **7**-50
  In hierarchical finite-state machines, **7**-49
  ICs in, **7**-43
  Implementing through pipelining, **7**-48
Dative bonds, **18**-7–**18**-9
  delivery pathway for, **18**-8
De Broglie conjecture, **7**-14, **7**-61
  and Schrödinger equation, **7**-64
Decision diagrams, **7**-34
  dimensions of, **7**-38
  and logic design of ^{M}ICs, **7**-37–**7**-41
  mapping into logical networks, **7**-43
Defect control
  defect removal due to micro-/nanoscale contamination, **20**-11–**20**-14
  in nanomanufacturing, **20**-11
Defect databases, storing street blockages in, **5**-8
Defect generation, **4**-10
  limitations to current technology, **5**-2
  as obstacles to CNN technology, **2**-5
  in silicon-based inorganic materials, **4**-14
Defect precursor distribution
  increasing effect at nanoscale, **13**-4
  reliability from, **13**-4–**13**-5
Defect removal, **20**-11–**20**-12
  selective removal from CNT films, **20**-13–**20**-14
  selective removal in assembly applications, **20**-12–**20**-13
Defect tolerant designs, **4**-14
  adaptive defect-tolerant molecular presenting-and-memory platforms, **7**-52–**7**-55
  CMOL memory, **10**-16, **10**-17
  with CMOL technology, **10**-15
  coexistence with high performance and acceptable power consumption, **10**-17
  with crossbar arrays, **5**-8
  engineering challenges, **5**-14
  at Hewlett-Packard, **5**-4
  hybrid CMOS/nanodevice circuits, **10**-14
  in Teramac computer, **5**-8, **5**-9, **6**-10
  vs. current microcircuit technology, **5**-2–**5**-3
Defects
  role in extrinsic molecular switching, **9**-9–**9**-12
  role in molecular transport, **9**-16–**9**-20
Dendrimer-carbohydrate conjugates, for polyvalent recognition, **24**-25
Dendrimer core-shell architecture, **24**-12

Dendrimers, **18**-53–**18**-57, **24**-1–**24**-2, **24**-10–**24**-11
 amplification and functionalization of surface groups,
   24-20
 as angiogenesis inhibitors, **24**-27
 as antiviral agents, **24**-26–**24**-27
 Bengal Rose formation, **18**-60
 as catalysts and redox active nanoparticles, **24**-29
 as chelating agents for cations, **24**-27–**24**-28
 in chemical and environmental process engineering,
   24-29–**24**-30
 comparison with biological building blocks spectrum,
   24-5
 core-shell patterns influencing reactivity, **24**-34–**24**-35
 de Gennes dense packing, **24**-15–**24**-19
 dendritic state, **24**-6–**24**-19
 divergent and convergent synthesis methods,
   24-11–**24**-13
 dynamics in solution, **24**-24
 features of interest to nanoscientists, **24**-13–**24**-15
 framework, **18**-54
 as functional nanomaterials, **24**-27–**24**-30
 as genetic material transfer agents, **24**-24–**24**-25
 guest-host interactions in, **18**-57–**18**-59
 historical technology periods and, **24**-2–**24**-3
 from hydrogen bonding, **18**-59
 and importance of controlled organic nanostructures in
   biology, **24**-3–**24**-4
 as ligands for anions, **24**-28
 as magnetic resonance imaging contrast agents,
   24-26
 as megamers, **24**-32–**24**-34
 microenvironments of, **18**-59–**18**-60
 molecular simulation, **28**-25–**28**-26
 as nanopharmaceuticals and nanomedical devices,
   24-24–**24**-27
 nanoscale container/scaffolding properties, **24**-19–**24**-20
 nanoscale dimensions and shape-mimicking proteins in,
   24-20
 nanoscale monodispersity in, **24**-19
 PAMAM, **24**-16
 polycells from inclusion of heterogeneous monomers,
   18-56
 and potential bottom-up synthesis structures for organic
   nanostructures, **24**-4–**24**-6
 as reactive molecules for synthesis of complex
   nanoarchitectures, **24**-30–**24**-34
 saturated-shell architecture approach, **24**-32–**24**-33
 shape, conformation, and solvation of, **24**-21–**24**-24
 shape changes in, **24**-15
 size and structure in solution, **24**-21
 structural and physical properties in solution,
   24-20–**24**-24
 as targeted drug-delivery agents, **24**-25
 as unimolecular micelle mimics, **24**-28–**24**-29
 unique properties of, **24**-19–**24**-20
 unsaturated-shell-architecture approach, **24**-34
 and wet-dry world of nanotechnology, **24**-4
Dendritic architecture, **24**-6
 comparing dendritic molecular chemistry with organic
   and polymer chemistries, **24**-6–**24**-8
 comparison with polymer properties, **24**-9

dendritic polymers, **24**-8–**24**-10
dendrons/dendrimers, **24**-10–**24**-19
Dendritic branching mathematics, **24**-14
Dendron construction, **24**-11
Density functional theory (DFT), **5**-17, **11**-2
 calculated electrostatic potential from, **12**-20
 calculation for I(V) molecule characteristics, **5**-23, **9**-8
 calculations at Si-SO$_2$ interface, **13**-3
 in CNT simulation, **23**-31–**23**-32
Density matrix, time evolution of, **12**-15
Density problems, **2**-2
 CMOS device advantages for power consumption, **4**-4
 financial roadblocks to solving, **5**-2
 for multiterminal structures, **4**-16
 with thin-film electronics, **4**-16
Department of Agriculture (USDA), **3**-8
Department of Defense (DOD), **3**-1, **3**-7, **3**-8
Department of Energy (DOE), ix, **3**-1, **3**-7
Department of Justice (DOJ), **3**-8
Department of Labor, anticipatory training of workers for
   nanotechnology applications, **3**-20
Desalinization, with nanotechnology, **3**-18
Design tools
 Verilog, **7**-57
 VHDL, **7**-57
Dethreading process, **11**-8
Detonation nanodiamonds, **27**-8
Device-device interactions, **6**-2
Device parameter optimization, difficulties with aggressive
   scaling of gate length, **10**-3
Device performance, analyzing for 1D case, **7**-64
Device switching speed, for molecular computing and
   processing platforms (^MPPs), **7**-16
Diagnostics applications, **3**-22
Diamond nanorods, **27**-2, **27**-4
 schematic illustration, **27**-8
Diamondoid scaffolding, **18**-26–**18**-27, **27**-3
Diazonium chemistry, **5**-24
Diazonium containing molecular electronics candidate,
   synthesis, **5**-21
Dielectric deflector, in PBG networks, **25**-27
Dielectrophoresis, **16**-2–**16**-3, **26**-4
 biomolecular applications, **16**-26
 frequency-conductivity diagram, **16**-7
 nanoscale applications, **16**-23–**16**-26
 schematic diagram with polarizable particle, **16**-2
 single nanoparticle trapping with, **16**-24–**16**-25
 in solid particles, **16**-3–**16**-7
 trapping of palladium sphere by positive, **16**-28
 traveling wave, **16**-13–**16**-16
Dielectrophoretic cell sorting, **26**-5–**26**-6
Diffractive optical elements (DOEs), **29**-15
 DOE lens with subwavelength features, **29**-16
Diffuse double layer, charge movement in, **16**-9–**16**-11
Diffusion-based drug release, **30**-19–**30**-20
Digital calculating machine, **7**-4, **7**-5
Digital revolution, convergence with nanotechnology, **3**-9
Dinitro-containing derivatives, synthesis, **5**-20
Diode-transistor logic (DTL), **7**-35
Diodes, **12**-4
 gating and control of junctions by, **12**-10–**12**-11

Dip-pen nanolithography, **11**-3, **20**-5
Directed self-assembly, **9**-3. *See also* Self-assembly
   of conductive polymers using nanoscale templates,
     20-10–**20**-11
   of nanoelements, **20**-6–**20**-7
   using nanotemplates, **20**-7–**20**-9
Discrete event time simulation, in VHDL design, **7**-58
Discretization spacing, **7**-70
Dispersibility, in carbon nanotubes, **27**-10
Dispersion-based nonchannel waveguiding, **25**-30
Dispersion-based routing structure, **25**-30–**25**-31
Dispersion-based tunable beam splitter, **25**-32
   steady-state result of Hz field, **25**-32
Dispersion-based variable beamsplitter, **25**-31–**25**-33
Distribution statistics, for molecular computing and
     processing platforms (^MPPs), **7**-12–**7**-13
DNA
   bases of, **17**-7
   as biomimetic structure, **18**-44
   connectivity and hydrogen bonding in, **18**-45
   denaturing and hybridization of, **17**-8
   structural components of nucleotide bases, **18**-44
DNA/Au colloidal particles, **17**-14
DNA biolinkers, **17**-6–**17**-9
   attachment to gold surfaces, **17**-8–**17**-9
DNA building blocks, rigid and unpaired regions, **18**-47
DNA-directed micro- and nanowires, **17**-16–**17**-18
   process flow for silver nanowires, **17**-17
DNA hybridization, **17**-15
DNA junctions, from rigid and sticky engineering, **18**-48
DNA/organic hybrids, **18**-49
Doping densities
   controlling, **4**-12
   in silicon chips, **2**-3
Double-barrier single-wall heterojunction transistors, **7**-76
Double layer effects, **16**-7–**16**-8
   alpha relaxation, **16**-11
   charge movement in diffuse double layer, **16**-9–**16**-11
   charge movement in Stern layer, **16**-8–**16**-9
Drain current, **7**-69
Drain source voltage, **7**-69
DRAM challenges, **10**-3
Dresselhaus spin-orbit interactions, **8**-3–**8**-4
Drift-diffusion models
   application to real ion channels, **13**-35
   of ion channels, **13**-34–**13**-35
Droplets, guided motion across surfaces, **11**-39
Drug delivery
   arterial delivery, **30**-17
   brain delivery, **30**-17
   coumarin-loaded PLGA particles, **30**-6
   drug release characteristics, **30**-21
   drug release mechanisms, **30**-19–**30**-20
   internalization of cystatin-loaded nanoparticles, **30**-18
   lymphatic system delivery, **30**-18–**30**-19
   nanoparticle characterization, **30**-11–**30**-14
   nanoparticle size comparisons, **30**-4
   nanoparticles for, **30**-1–**30**-2
   nanoparticulate delivery systems, **30**-14–**30**-16
   oral delivery, **30**-16
   processing parameters for nanoparticles, **30**-5–**30**-11

   pulmonary delivery, **30**-19
   synthesis of solid nanoparticles for, **30**-2–**30**-5
   targeting with dendrimers, **24**-25
   TEM images of nanoparticle-cellular interactions, **30**-8
   tumor therapy, **30**-17–**30**-18
   vaccines, **30**-18–**30**-19
Drug release
   characteristics of, **30**-21
   mechanisms of, **30**-19–**30**-20
Dual-gate designs, **4**-11
Dynamic RAM (DRAM), **4**-11–**4**-12
   molecular ICs implementing, **7**-34
Dynamical analysis, in molecular conductance junctions,
     12-22

E

E-beam lithography, **5**-2
Edge-driven computation, in QCA, **6**-6
Education and outreach, early NNI achievements, **3**-13
Effective cell size, in MOSFETs, **7**-8
Effective channel length, for FETs, **7**-8
Effective latency, **7**-46
Effective memory speed, as key performance parameter in
     design, **7**-45
Effective nuclear charge, **7**-15
Effective tunneling barrier, **12**-12
Einstein relation, for nondegenerate semiconductors, **9**-32
Elastic buckling, in carbon nanotubes, **23**-9–**23**-13
Elastic constants, **23**-4
Elastic properties, of carbon nanotubes, **23**-4–**23**-7
Electric circuits, nanoscale, **1**-8
Electric current, **7**-8
Electrical barrier properties, **21**-58–**21**-59
Electrical breakdown methods, MWNT and SWNT
     production using, **5**-14
Electrical engineering circuit simulation programs, **5**-22–**5**-23
Electrically addressable molecules, **9**-13–**9**-14
Electrically mediated self-assembly, **17**-25–**17**-27
Electrodeposition, **7**-2
Electrodes, as continuum structures, **12**-2
Electrografting, **11**-3
Electrokinetics
   analysis of key spectrum events, **16**-18–**16**-20
   cell trapping by, **26**-4–**26**-5
   dielectrophoresis, **16**-2–**16**-3
   dielectrophoretic behavior of solid particles, **16**-3–**16**-7
   double layer effects, **16**-7–**16**-13
   limitations on minimum particle trapping size,
     16-20–**16**-23
   multishell model of complex spheroids, **16**-11–**16**-13
   phase-related effects, **16**-13–**16**-16
   theoretical aspects, **16**-2
Electron charge
   in organic and semiconductor nanodevices, **9**-1–**9**-2
   as primary state variable, **10**-5
   and spin-torque domain wall switching in nanomagnets,
     9-22–**9**-23
Electron-electron scattering, **12**-11, **12**-12
Electron energy calculations, **7**-15
Electron/hole transport, in mesoscopic systems, **12**-5

Electron hopping, **12**-11, **12**-14–**12**-17
    at long distances, **12**-17
    thermally activated, **12**-16
Electron microscope, need for improvements to, **1**-4
Electron-photon scattering, **12**-11, **12**-12
Electron projection lithography, **10**-3
Electron transfer
    as fundamental process of molecular junction transport,
        12-4
    rate constant for, **12**-16
Electron transport
    generic problems related to, **9**-4
    magnetic effects on, **12**-22
    measuring in individual molecules, **12**-23
    in molecular devices, **9**-2–**9**-4
Electron transport ^{ME}devices, **7**-58
    3D examination, **7**-71–**7**-75
    in time-and spatial-varying metastable potentials, **7**-75
Electron traversal time, **7**-64
Electron tunneling, **7**-69, **9**-2, **12**-14
    dimensionless transmission probability of, **7**-59
    at short distances, **12**-17
    through finite potential barriers, **7**-60, **7**-70
    in ultra thin oxides, **13**-5–**13**-6
Electron velocity, **7**-16
    time derivative of, **7**-62
Electronic devices, DNA-inspired self-assembly of,
        17-18–**17**-19
Electronic mean free path length, **12**-5
Electronic Numerical Integrator and Computer, **7**-5, **7**-6
Electronic transport, **11**-12–**11**-13
Electronics, nanoscale modeling, **13**-1–**13**-2
Electroosmotic cell sorting, **26**-5–**26**-6
Electroporation schemes, **26**-8
Electrorotation, **16**-13–**16**-16
    schematic diagram, **16**-16
Electrospinning
    of nanofibers, **21**-2–**21**-3
    of polymeric nanofibers, **21**-17–**21**-18
Electrospun nanofibers
    composite nanofibers via electrospinning, **21**-18–**21**-20
    magnetic characterization of, **21**-22–**21**-24
    with magnetic domains for smart tagging, **21**-17
    magnetic nanoparticles synthesis and functionalization,
        21-20–**21**-22
    PEO nanofibers, **21**-18
    technological implications, **21**-25
Electrostatic forces
    biomolecular applications, **16**-26
    nanoparticle manipulation by, **16**-1–**16**-2
    nanoscale dielectrophoresis applications, **16**-23–**16**-26
    and particle separation, **16**-26–**16**-28
    theoretical aspects of AC electrokinetics, **16**-2–**16**-23
Electrostatic potentials, **12**-10
    applied and self-consistent field potentials, **14**-18
    bond breaking due to, **12**-21
    calculation from density functional theory, **12**-20
    for carbon nanotubes, **13**-27–**13**-28
    in molecular conductance junctions, **12**-18–**12**-20
    schematic diagram for two geometries, **14**-3
Electrostatically addressable templates, **20**-10

Elliptical particles, modeling and diagram, **16**-13
Embedded Atom Method (EAM) potentials for metals,
        28-4
Embedded memories, **10**-16
Emerging memory devices, **10**-12
Emerging research devices, **10**-10. *See also* Nanodevices
    examples of, **10**-11–**10**-13
Emitter-coupled logic (ECL), **7**-35
Encapsulation efficiency, in nanoparticles for drug delivery,
        30-11–**30**-12
Encyclopedia Britannica, fitting on head of pin, **1**-3, **2**-1
Energy barriers, and logical states, **7**-35
Energy conversion, **3**-22
    as objective of nanotechnology development, **3**-18
Energy level broadening, **14**-12
    schematic diagram, **14**-5
Energy levels, for molecular computing and processing
        platforms (^{M}PPs), **7**-13–**7**-15
Energy relaxation time, **10**-11
Engineering, convergence with biology, **17**-3
Engineering challenges
    for defect-tolerant approaches, **5**-14
    feasibility of molecular electronics, **7**-7
    interconnects, contacts, and interfaces, **4**-14
    material patterning and tolerances, **4**-13
    for molecular circuits, **4**-12–**4**-13
    in molecular electronics, **4**-1–**4**-2
    for QCA method, **5**-7–**5**-8
    self-assembly of large component numbers, **6**-9
    signal delay/decay, **5**-14
Enhancement/depletion mode circuits, **4**-4
Entropy
    applications for molecular computing and processing
        platforms (^{M}PPs), **7**-11–**7**-12
    differential in neurons, **7**-27
    in directed self-assembly, **9**-3
    as function of window size and time binwidth, **7**-29
    Renyi measure, **7**-28
Entropy difference, dependence on number of possible
        microscopic states, **7**-11–**7**-12
Environmental, health, and safety (EHS), nanotechnology
        implications, **3**-6
Environmental improvement, nanoscale processes for, **3**-22
Environmental process engineering, dendrimers in,
        24-29–**24**-30
Equiaxed nanomaterials, **21**-44
    metals, **21**-44–**21**-45
    oxides, **21**-45–**21**-46
    platelet nanomaterials, **21**-46–**21**-47
Equilibrium energy diagram
    for strongly coupled molecule, **14**-4
    for weakly coupled molecule, **14**-4
Equipment costs, future CMOS challenges, **10**-1
Erosion-based drug release, **30**-19–**30**-20
Euler approximation, **7**-70
European Union, estimated nanotechnology expenditures,
        3-12
Evaporation, miniaturization by, **1**-6
Evolution matching
    between ideal and fabricated molecular platform, **7**-53
    schematic diagram, **7**-54

Exchange engineering, **13**-16–**13**-19

Executing information carriers, **7**-24

Executive Office of the President (EOP), **3**-7

Expectation values, **7**-65, **7**-66

Experimental physics, **1**-1

techniques for molecular junction transport, **12**-3–**12**-4

Explosives detection/protection, **3**-22

Extended carbon family, **27**-3–**27**-5

Extraction tools, applications of computer modeling to, **28**-13

Extreme ultraviolet lithography (EUV), **5**-2, **10**-3, **10**-6

microexposure tool (MET), **10**-6–**10**-7

Extrinsic molecular switching, **9**-9, **9**-36

due to molecule tilting, **9**-4–**9**-5

in molecular ferroelectric PVDF, **9**-13

role of defects and molecular reconfigurations in, **9**-9–**9**-12

F

F-values, in electron microscopes, **1**-4

Fab 22, construction costs, **5**-2

Fabrication error analysis, with photonic crystals, **25**-29

Fan-out, in neurons, **7**-25

Federal agencies. *See* Government agencies

Femtoscopic particles, **7**-3

Fermi-Dirac distribution function, **7**-13, **7**-14, **7**-77

application to electron tunneling, **7**-59

Fermi energy, **14**-3–**14**-5

equilibrium energy diagrams, **14**-4

Fermi wavelength, **12**-5

Ferroelectric random access memories (FERAM), **9**-2

Ferromagnetic contacts, in SPINFETs, **8**-1

Ferromagnetic semiconductors (FMSs)

avoiding Schottky barrier problems with, **9**-2

electron density with spin at FM-S junction, **9**-25

energy diagram, **9**-24

spin injection in junctions, **9**-23

with transparent Schottky barrier, **9**-24

Feynman, Richard P., **1**-1

as father of nanotechnology, **2**-1

Feynman challenge, **1**-1–**1**-7, **2**-1–**2**-2

Feynman ratchet, **11**-3, **11**-4

Fiber-based composites, **27**-1

Field-effect transistor (FET) devices, **10**-5

based on nanowires, **5**-24

carbon nanotube-based, **6**-10

with molecular channel regions, **4**-15

schematic diagram, vertical, **4**-11

transit time for, **7**-68

Field emission scanning electron microscopy (FESEM), **17**-28

spin-coated thin film, **20**-9

Finite potential barriers, electron tunneling through, **7**-70

First-pass effect, avoiding through nanoparticle drug delivery via GALT, **30**-18

Flash memory, **4**-11–**4**-12

Flexing cantilevers, with palindromic [3]rotaxanes, **11**-20–**11**-21

Floating body DRAM, **10**-12

Fluid geometries, **12**-20

Fluidic molecular electronics, **7**-5, **7**-7, **7**-10, **7**-16–**7**-18

emulation of brain neurons by, **7**-31, **7**-32

networking and interconnects, **7**-26

and neurological information processing and memory postulate, **7**-20–**7**-25

Fluidic molecular platforms, **7**-30–**7**-32

Fluidic self-assembly, **17**-27–**17**-30

Fluidics-mediated self-assembly, **17**-27–**17**-29

Force generation, ubiquity in biological systems, **15**-1

Fortune 500 companies, nanotechnology activities, **3**-14

Forward and reverse bias voltages, estimating for molecular quantum dot rectifiers, **9**-6

Frequency multipliers, spintronic devices as, **9**-35

Fringing fields, **4**-12

Fullerene-based nanocomposites, **27**-9

Fullerene nanotubes

kink formation in, **28**-25

molecular simulation, **28**-24–**28**-25

Fullerene nanowires

spacing of, **20**-6

as zero-dimensional structures, **27**-3

Functional achiral foldamers, heparin-binding, **19**-22–**19**-23

G

GaA materials, **2**-3

Gain, **4**-15

importance in silicon-based devices, **4**-5–**4**-6

molecular devices capable of producing, **4**-15

in multiterminal structures, **4**-15–**4**-16

for SETs *vs.* MOS transistors, **6**-8

in two-terminal devices, **4**-14–**4**-15

Gate dielectric breakdown, **4**-10

reduction challenges, **4**-12

Gate leakage, **4**-9, **10**-3

challenges in, **4**-8

Gate length

in MOSFETs, **7**-8

scaling challenges for CMOS technology, **10**-3

Gate source voltage, **7**-69

Gated nanowire devices, **4**-15

Gatekeepers

molecular, **11**-24–**11**-25

supramolecular, **11**-24

Gates. *See* Logic gates

Gating mechanisms, **12**-3

in biological ion channels, **2**-6

and control of molecular junctions, **12**-10–**12**-11

Gears, applications of computer modeling to, **28**-13

Generalized Landauer formula, for coherent transport, **12**-4–**12**-10

Genetic material transfer, dendrimers as agents of, **24**-24–**24**-25

Geometry

in molecular wire junctions, **12**-20–**12**-21

UCLA-Hewlett Packard collaboration, **12**-10

Giant magnetoresistance (MR), **9**-20

Glass transition temperature, of CNT-PE composites, **22**-8

Global government investments, **3**-13

Global interconnects, and signal delay issues, **4**-10

Goddard, William A. III, xi
Gold
  biocompatibility of, **11**-22
  DNA attachment to, **17**-8–**17**-9
Gold impurities
  current-voltage characteristics of alkane chain with,
    9-18
  equilibrium position of, **9**-17
  local density of states and transmission for, **9**-17
Government agencies
  nanotechnology R&D at, **3**-23–**3**-24
  orchestration of international, **3**-5
Government expenditures, international comparisons,
  3-12
Graphite, **27**-3
  comparison of EOS for, **23**-35
  crystallographic arrangements, **21**-49
  as nanoadditive, **21**-48
Gravity, at micron scale, **10**-8
Green's function, **7**-58, **12**-6, **12**-7, **12**-9, **12**-10
  NEGF formalism and resistance, **14**-12–**14**-15
  schematic diagram, **14**-17
Grey-goo scenarios, risk of, **3**-6
Ground state
  computing in QCA with, **6**-5–**6**-6
  natural system tendency to assume, **6**-5
  and quantum dot arrangements, **6**-3
Ground state coconformation (GSCC), **11**-8
Group velocity, **7**-61, **7**-62
Guest-host interactions
  from dendrimer engineering, **18**-58
  in dendrimers, **18**-57–**18**-59
Guest molecules, **19**-20
  encapsulation of, **18**-42
Gut-associated lymphoid tissue (GALT), **30**-2
  nanoparticle absorption via, **30**-18–**30**-19

H

H-vacancy, **9**-19, **9**-20
Hairpin DNA compounds, rate behavior as function of
  distance for, **12**-19
Half metals
  injection efficiency of, **10**-12
  limits of spin injection efficiency, **8**-8
Hamiltonian, **12**-7
  many-body, **13**-10
  and multiterminal quantum-effect devices, **7**-75–**7**-75
  in nonballistic SPINFETs, **8**-4–**8**-5
  single-particle, **13**-10–**13**-12
  two-terminal device with, **7**-58
Hard semiconductor junctions, **12**-3
Hardware acceleration platforms, **29**-2
Hardware and software co-design, **7**-55–**7**-58
  hierarchical finite-state machines in, **7**-49–**7**-52
Hardware description languages (HDLs), **7**-57
Hartree-Fock approximation, **23**-30–**23**-31
  for quantum dots, **13**-12–**13**-13
Healthcare applications, **3**-22
Heat deflection temperature, **21**-53
Heat generation problems, **2**-2, **2**-3, **4**-8, **4**-12

and organic materials' temperature sensitivity, **4**-9
  reduction through QCA approach, **5**-4
  in two-terminal devices, **4**-15
Heisenberg uncertainty principle, **7**-35
  in molecular electronics modeling and analysis,
    7-60–**7**-61
Heparin, antithrombin III-binding pentasaccharide of,
  19-22
Hepatocyte self-assembly, **10**-8
Heterogeneous simulations, vii
Heterojunction bipolar transistors, **7**-75, **7**-77
Heterostructures, **9**-32–**9**-34
  additional degree of freedom, **9**-34
  as multifunctional devices, **9**-34
Hewlett-Packard, **7**-3
  Teramac computer, **5**-4, **6**-10
Hierarchic graphs (HGs), **7**-49
  control algorithm represented by, **7**-50
  multiple-level sequential, **7**-51
  transformation to state transition table, **7**-51
Hierarchical fine-state machines, use in hardware and
  software design, **7**-49–**7**-52
Hierarchical self-assembling, **3**-11
High circuit density, **10**-14
High-fidelity modeling, vii
High-frequency spin-valve effect, **9**-33–**9**-35
High-K/metal-gate gate stacks, **10**-4
High-level programming languages, **7**-57
High-performance high-yield ^{M}ICs, **7**-53
High-performance software, MATLAB, **7**-70
High-Q microcavities, **25**-13–**25**-14
High-quality filters, **25**-15
High-resolution mass spectrometry, **11**-2
High-resolution transmission electron microscopy
  (HRTEM), of MWCNT, **23**-5
High school competition, Feynman's challenge, **1**-9
Hole transfer efficiency, **12**-19
Homeland Security, **3**-8
HOMO-LUMO gap, **9**-16, **11**-16, **14**-10
  with gold impurities, **9**-17
  in molecular quantum dot switching, **9**-14
  for strongly coupled molecules, **14**-4
Homo-oligomers, Ndi and Dan, **19**-24
Homologous recombination reaction, **17**-18
Honeycomb lattices
  hexagonal moleculars from benzene junctions,
    18-29
  and molecular Tinkertoy approach, **18**-28–**18**-29
Hopping transport, **12**-14–**12**-17
  in polymeric structures, **12**-16
Horizontal multidisciplinary R&D, **3**-5
Hot-electron interface trap generation, for submicron
  nMOSFETs, **13**-4
HSPICE software, **5**-22
Human brain
  action potentials in, **7**-21
  neuronal communication in, **7**-26
  number of neurons, **7**-17
Hund's rules, for quantum dots, **13**-12–**13**-13
Hybrid circuit approach, **10**-14
Hybrid microscopic/macroscopic systems, **12**-5

Hybrid photonic crystal structures, **25**-21–**25**-23
  band diagrams, **25**-22
  transmission spectra, **25**-23
Hybrid silicon/molecular electronics, **4**-16–**4**-17
Hydrogen atoms, energy level calculations, **7**-14
Hydrogen bonding
  aromatic/carboxylic acid assemblies, **18**-15
  benzamide dimers, **18**-16
  crystal engineering, **18**-15–**18**-16
  dendrimer formation from, **18**-59
  in isophthalic acid derivatives, **18**-17
  in melamine and cyanuric acid, **18**-17
  as molecular building block, **18**-10–**18**-12
  orientational specificity in, **18**-17
  in rotaxane templation, **19**-11
  in self-assembly, **17**-5
  in supramolecular design, **18**-13–**18**-18
  in supramolecular structures, **18**-16–**18**-18
Hydrophilicity, manipulating, **11**-35
Hydrophobicity
  manipulating, **11**-35
  and self-assembly, **17**-6
Hypercell design, **7**-41–**7**-43
Hypercells, **7**-19, **7**-20, **7**-34
  design and aggregation in functional molecular ICs, **7**-37
  design of, **7**-41–**7**-43
  logic functions performed by, **7**-34
  LWDDs embedded in, **7**-38
  molecular NAND gates in, **7**-36
  multiplexer-based, **7**-42
  as reduced decision tree, **7**-37

I

Iafrate, Gerald J., xiii
Icosahedral viruses, self-assembly of, **10**-9
Image charges, in molecular conductance junctions, **12**-18–**12**-20
Imprint technology, **10**-3
Incoherence, in molecular junction transport, **12**-14–**12**-17
Industrial prototyping and commercialization, **3**-3
Inelasticity, **12**-3
  onset of, **12**-11–**12**-13
Information, microrepresentation of, **2**-2
Information carriers, **7**-24, **7**-26, **7**-32
  biomolecules as, **7**-24
  in fluidic devices, **7**-30
  in human brain, **7**-21, **7**-22
  molecules and ions as, **7**-31
  in neurons, **7**-43–**7**-44
Information estimates, biomolecular processing applications, **7**-25–**7**-30
Information processing
  in biomolecules, **7**-24
  and neurobiomimetics, **7**-20–**7**-25
Information Technology Research (ITR), **3**-4
Informationally irreversible gates, **10**-10
Informationally reversible gates, **10**-10
Infrastructure, **3**-22
  for nanotechnology user capabilities, **3**-15
Input voltage, ratio of output voltage change to, **3**-5

Instrumentation, nanoscale, **3**-22
Insulator resistance change memory, **10**-12
Integrated circuits (ICs), **7**-2
  design problems, **4**-12
  power dissipation issues, **4**-2
  trends in miniaturization, **17**-2
Integrated-injection logic (ILL), **7**-35
Intel Systems
  EUV microexposure tool (MET), **10**-6, **10**-7
  Fab 22 facility costs, **5**-2
Interagency Working Group on Nanoscale Science, Engineering and Technology (IWGN), **3**-9
Interatomic forces
  molecular simulations of, **28**-3
  in two-atom system, **23**-34
Interconnect line widths, and metal substitutions, **5**-1
Interconnects, **4**-12
  in CMOS devices, **4**-10
  as engineering challenge for molecular circuits, **4**-14, **10**-1
  isolating to avoid cross-talk and signal decay, **4**-14
  loading of transmission lines and power dissipation, **4**-9
  in nanotransistors, **2**-3
  scaling problems, **4**-10
Interface trap generation, failure function for, **13**-5
Interfaces, issues in molecular circuits, **4**-14
Interfacial deposition method, **30**-3
Interlayer potentials simulations, **23**-34–**23**-36
Interlocked dimer
  contraction of, **11**-18
  templated synthesis of, **11**-17
International Business Machines, **7**-6
International collaboration, **3**-5
International Dialog on Responsible Nanotechnology R&D, **3**-19–**3**-20
International Risk Governance Council, **3**-20
International Technology Roadmap for Semiconductors, **4**-1, **4**-8
Intracellular protein detection, **26**-15
Intramolecular electron transfer, **12**-13
Intrinsic molecular switching, **9**-13
Intrinsic polarization, in molecular ferroelectric PVDF, **9**-13
Inverters
  CNT-based, **22**-23
  from QCA cells, **5**-5
Investment
  five NNI modes of, **3**-22
  national government level, **3**-11
  by NNI in 1999, **3**-9
  by U.S. federal departments and agencies, **3**-8
Ion channels
  application of drift-diffusion model to real, **13**-35
  comparison of measured and computed current-voltage curves for, **13**-36
  as example of natural nanotechnology, **2**-6
  gramicidin channel geometric representation, **13**-37
  hierarchical approach to modeling, **13**-33–**13**-34
  mesh representation of *ompF* trimer, **13**-35
  modeling ion-water interactions in, **13**-36–**13**-37

Monte Carlo simulations, **13**-36
   resolving single-ion dynamics, **13**-36
   role in biological systems, **7**-22
Ion-water interactions, **13**-36–**13**-37
Ionic bonds
   molecular simulations, **28**-5–**28**-6
   in self-assembly, **17**-6
Ionic channel simulation, **13**-32–**13**-33
   drift-diffusion models, **13**-34–**13**-35
   molecular structure of porin channel from *E. coli*, **13**-33
Ionic gelation method, **30**-3
Ionic strength, manipulation of nanoporous materials by, **11**-25
Irregular dispersive phenomena
   dispersion-based nonchannel waveguiding, **25**-30
   dispersion-based routing structure, **25**-30–**25**-31
   dispersion-based variable beamsplitter, **25**-31–**25**-33
   in photonic crystals, **25**-29
   self-collimation, **25**-30
   spot-size converters, **25**-30
   super-prism phenomenon, **25**-29–**25**-30
Irreversible gates, **10**-10
Irreversible thin film regulators, **11**-22
Isophthalic acid derivatives, **18**-17
ITRS 2004 Emerging Technology Sequence, **10**-6

J

Japan, estimated nanotechnology expenditures, **3**-12
Journal articles, increase in nanotechnology subjects, **3**-10

K

Kalil, Tom, **3**-7
Keldysh formalism, **12**-7
Kiev Polytechnic Institute, xii
Kinesins, **11**-4, **11**-5
Kinetic equation, **14**-6

L

Lambda rule, **7**-8
Landauer Buttiker time, **12**-15
Landauer coherent conductance, **12**-7, **12**-8, **12**-11
Landauer expression, **7**-59
   generalized formula for coherent transport, **12**-4
   obtaining current for molQD rectifiers from, **9**-7
Landauer transmission model of conductance, **12**-5
   extensions to molecular wires, **12**-5–**12**-10
Lane, Neal, **3**-7
Langevin equation, **11**-28
Langmuir-Blodgett technique, **9**-6, **9**-16, **11**-3
   formation of hot spots using, **9**-9
   for self-assembly of molecules on water, **9**-3
   studies of molecular movement, **11**-10–**11**-12
Laplace equation, **7**-75
Large-scale integration, favoring of silicon technology by, **2**-4
Large systems, tyranny of, **2**-2
Latching switches, **10**-15

Latency
   decreases with molecular electronics, **7**-10
   and memory system performance, **7**-46
Layered composites, **27**-1
Layered silicates, schematic of, **21**-46
LB monolayers, switching of, **11**-14
Left-handed materials analysis, **29**-11–**29**-12
   demonstrated negative refractive index, **29**-15
   LHM prism model, **29**-14
   LHM structure, **29**-12
   modeling and simulation, **29**-12–**29**-14
   numerical characterization setup, **29**-13
   numerical demonstration of negative refractive index, **29**-14–**29**-15
   transmission spectra, **29**-14
Legendre function, **7**-73, **7**-74
Length scales, in mesoscopic systems, **12**-4–**12**-5
Leucippus of Miletus, **7**-2
Ligands, dendrimers as, **24**-28
Light-driven single-molecule force measurements, **11**-20
Light-emitting diodes (LEDs), **2**-4
Light regulation
   of azobenzene and coumarin valves, **11**-22–**11**-24
   of unidirectional rotation, **11**-30-**11**-33
Light-responsive surfaces, wettability of, **11**-36–**11**-37
Linear-connecting pi-systems, **18**-25
Linear macromolecules
   aedomers, **19**-23–**19**-25
   beta-peptides, **19**-26–**19**-27
   bis-amino acids, **19**-29–**19**-31
   BLOCK approach to large structures, **19**-27–**19**-29
   building-block approaches to, **19**-17–**19**-18
   functional heparin-binding achiral foldamers, **19**-22–**19**-23
   m-Phenylene ethynylene foldamers, **19**-19–**19**-22
Linear word-level decision diagrams (LWDDs), **7**-38
   as departure from existing logic design tools, **7**-39
Liposomes, as nanoparticulate delivery systems, **30**-14
Liquid-vapor interface, **11**-35
Lithographic process, **4**-12, **11**-4
   creating CNNs with, **2**-5
   nanoimprint lithography, **20**-3–**20**-4
   for nanoscale CMOS devices, **4**-13
   progress in, **10**-7
   technological limitations of current, **5**-2
Living cell array, **26**-11
Local conductance, **9**-11
   experimental setup for mapping, **9**-10
   on gold surfaces, **11**-34
Local deformation, of carbon nanotubes, **23**-9–**23**-13
Local interconnects, signal transfer rates for, **4**-19
Localization length, **12**-5
Logic gates, **7**-20
   constructing with crossed NW, **5**-24
   device and circuits perspective, **7**-34–**7**-36
   limitations of current CAD-supported design, **7**-38
   and molecular electronics, **7**-34–**7**-36
   schematic, **7**-33
Logic routable molecular universal logic gates, **7**-32
Logic truth table, for QCA structures, **6**-4
Logical states, and energy barriers, **7**-35

Long interconnects, need to minimize, **4**-10
Long-term R&D vision, **3**-5
Lotus leaf surface, SEM image, **21**-57
Low-loss optical waveguides, **25**-12–**25**-13
Low power operation, **7**-10
    CMOS device advantages, **4**-16
    with high capacitance, **4**-8
    in human brain, **7**-17
    of nanodevices, **6**-1
Lowest unoccupied molecular orbit (LUMO), **5**-17
Lubrication, nanoscale problems, **1**-6–**1**-8
Lymphatic system drug delivery, using nanoparticles,
    30-18–**30**-19
Lyshevski, Sergey Edward, xii

M

m-Phenylene ethynylene foldamers, **19**-19–**19**-22
    structure of, **19**-19
Macroscopic valves, **11**-21
Magic ring catenation, **19**-4
Magnetic coupling phenomena, **6**-7
Magnetic forces, **15**-3
Magnetic implementations, **6**-7
Magnetic manipulation
    of biological specimen, **15**-15
    biomedical science applications, **15**-1–**15**-2
    chromatin manipulation, **15**-17–**15**-19
    of cilia, **15**-15–**15**-16
    computer control and data acquisition, **15**-9–**15**-11
    description of magnetic system, **15**-3–**15**-7
    experimental results, **15**-16–**15**-17
    implementation of proposed system, **15**-9
    magnetic circuit topology, **15**-4
    magnetic force calibration, **15**-11–**15**-12
    magnetic force directionality, **15**-12–**15**-15
    magnetic forces in proposed system, **15**-3
    pole materials and methods, **15**-7
    position detection and, **15**-8–**15**-9
    system performance, **15**-11–**15**-19
    tracking resolution characterization, **15**-11
Magnetic nanoparticles
    electrospinning of, **21**-20–**21**-22
    TEM image, **21**-19
Magnetic resonance imaging, dendrimers as contrast
    agents in, **24**-26
Magnetic systems
    electric circuit analog, **15**-8
    engineering drawing for second-generation system, **15**-6
    experimental, **15**-4–**15**-7
    first-generation system engineering drawing, **15**-5
    pole materials and methods, **15**-7
    three-pole design, **15**-8
Magnetic tunnel junctions (MTJs), **9**-21
Majority gates
    operation demonstration, **5**-8
    from QCA cells, **5**-5, **5**-6
    in QCA structures, **6**-4
Manufacture, **7**-19. *See also* Nanomanufacturing
    decrease in IC feature sizes, **7**-2
    development of new chip techniques, **5**-1

inherent nanodevice difficulties, **6**-2
    at nanoscale level, **3**-17, **3**-22
    nine grand challenges of, **20**-2
Manufacturing costs, as final frontier, **10**-2
Manufacturing time, reducing with molecular technology,
    5-3
Mark I project, **7**-6
Mask-less lithography, **10**-3
Material patterning, **4**-12, **4**-13
Material structures, size range in nanotechnology, **3**-2
Material tolerances, **4**-13
    statistical analysis and control of, **4**-13
Materials of construction, and density achievements in
    computing, **5**-1
Materials Research Source LLC, xi
MATLAB, **7**-70
    broadened one-level model, **14**-21
    discrete one-level model, **14**-20
    discrete two-level model, **14**-20–**14**-21
    models of resistance, **14**-20
    unrestricted broadened one-level model, **14**-22–**14**-23
    unrestricted discrete one-level model, **14**-22
Matter
    gap in fundamental knowledge of, **3**-2
    systematic control at nanoscale level, **3**-11
Maximal polarization, condition for, **9**-31
Maximum absorbance, **7**-16
Maxwell-Boltzmann statistics, **7**-14
    for distinguishable particles, **7**-13
Mechanical applications
    of carbon nanotubes, **23**-39
    nanoelectromechanical systems (NEMS), **23**-44–**23**-46
    nanoropes, **23**-39–**23**-42
    nanotube filling, **23**-42–**23**-44
    nanotube-reinforced polymer, **23**-46–**23**-48
Mechanical cell trapping, **26**-2–**26**-4
Mechanical measurement
    based on nanomanipulation, **23**-26–**23**-28
    of carbon nanotubes, **23**-22
    challenges and directions, **23**-28–**23**-29
    mechanical resonance method, **23**-22–**23**-23
    scanning-force microscopy method, **23**-24–**23**-26
Mechanical properties
    approach to probing, **23**-25
    of polypropylene/nanotube combinations, **21**-33–**21**-34
Mechanical resonance method, **23**-22–**23**-23
Megamers, **24**-31
    approaches to synthesis, **24**-32–**24**-34
Melacine stator unit, **11**-27
Meltblowing process
    in nanofiber manufacture, **21**-3–**21**-5
    schematic representation, **21**-4
Memory density, of molecular ICs, **7**-34
Memory devices, **4**-11
    conductivity states, **5**-17
    3D, **7**-43–**7**-49
    DRAM, **4**-11–**4**-12
    flash memory, **4**-11–**4**-12
    neuroscience analogies, **7**-20–**7**-25
    passive and active matrix addressing for, **4**-12
    SRAM, **4**-11–**4**-12

Memory hierarchies, and decreased latency, 7-46
Memory-processor interface, 7-46
MEMs comb drive, 20-12
Merged-transistor logic (MTL), 7-35
Mesoscopic systems, length scales in, 12-4–12-5
Metal contamination, and chip yield problems, 5-3
Metal ions
    coordination geometries for, 18-36
    and defects in organic thin films, 9-16
    recovery from solution by dendrimer enhanced
        filtration, 24-30
Metal-ligand bonding, 18-6
    coordination cavities and channels, 18-33
    ladders, rods, racks formed from, 18-41
    structural motifs, 18-32
Metal-oxide-semiconductor field effect transistor
        (MOSFET) structures, 4-3
    effective cell areas and gate length, 7-8
    vs. single-electron circuits, 6-8
Metallic split-ring resonator, in LHM model, 29-12
Metals
    as equiaxed nanomaterials, 21-44–12-45
    molecular simulations, 28-4
Metastable state coconformation (MSCC), 11-8
Metastable states, 7-75
    potential in QCA arrays, 6-6
Micelle mimics, dendrimers as, 24-28–24-29
Microchannel geometry, and cell culture, 26-17–26-20
Microchips
    lithographic production techniques, 2-3
    storage capacities of current, 2-1
    superiorities to natural systems, 2-7
Microcraft, 3-22
Microelectronic technology, 5-1–5-2
    accomplishments of, 7-2
    2D, 7-18
    future trends, 7-10
    limitations of current, 5-2–5-3
    retrospect, 7-8–7-11
    in signal processing and computing platforms, 7-7
Microelectronics Advanced Research Corporation
        (MARCO), 10-1
Microenvironments, dendrimers and, 18-59–18-60
Microfabricated quadrupole DEP trap, 26-5
Microfluidics, 11-40
    microfluidic array device channel, 26-3
Microrobotics, 3-22
Microscopic behavior, 7-3
Microscopic reversibility, 11-4
    law of, 11-3
Microscopic states
    calculating transitions between, 7-15
    number of, 7-11–7-12
Microtubule-associated proteins (MAPs), 7-22
    quantum information processing in, 7-23
Mie theory comparison, 28-3–28-5
    hardware *vs.* analytic solvers, 29-4
Miniaturization, 6-1, 17-2–17-3
    of computers, 1-5
    by evaporation, 1-6
Minimal feature size, 7-8

Minimum allowable separation, 7-8
Mixed-signal neuromorphic networks, 10-16
Mobility, in CMOS devices, 4-6–4-8
Modeling and analysis
    of carbon nanotube nanoelectromechanical systems,
        13-20–13-24
    3D problem in quantum mechanics, 7-71–7-75
    Heisenberg uncertainty principle in, 7-60–7-61
    to interpret and predict molecular wire junction
        experiments, 12-2
    of left-handed materials, 29-12–29-14
    modeling concepts, 7-58–7-59
    of molecular electronic devices, 7-58
    multi-terminal quantum-effect devices, 7-75–7-78
    particle velocity, 7-61–7-64
    Schrödinger equation in, 7-64–7-71
Modified microemulsion method, 30-3
Modulation doping, 2-4
Molecular AND gates, 7-37
Molecular architectonics, 7-11, 7-38
    and hybridization states of carbon, 24-7
Molecular assembling, 3-11
Molecular beam epitaxy, 7-2
Molecular building blocks, 18-1–18-3
    approaches to, 18-13
    biomimetic structures approach, 18-43–18-53
    bonding and connectivity, 18-3–18-12
    coordination complexes, 18-5–18-7
    covalent architectures and molecular Tinkertoy
        approach, 18-22–18-31
    dative bonds, 18-7–18-9
    dendrimers, 18-53–18-60
    hydrogen bonds, 18-10–18-12
    pi-interactions, 18-9–18-10
    supramolecular chemistry approach, 18-13–18-21
    transition metals and coordination complexes,
        18-32–18-43
Molecular circuits, 6-9–6-11
    hybrid silicon/molecular electronics, 4-16–4-17
    interconnects, contacts, and interfaces, 4-14
    material patterning and tolerances, 4-13
    opportunities and challenges, 4-12–4-13
    power dissipation and gain issues, 4-14–4-16
    reliability issues, 4-13
    and thin-film electronics, 4-16
Molecular compasses, 11-28–11-30
Molecular compensator, 7-53, 7-54
Molecular complexity, as function of covalent synthesis
        strategies and dimensions, 24-5
Molecular computers. *See also* Molecular electronic
        computing architectures
    nanocell approach, 5-15–5-19
Molecular computing and processing platforms (MPPs),
        7-1–7-4
    adaptive defect-tolerant molecular presenting-and-
        memory platforms, 7-52–7-55
    bimolecular processing, 7-20–7-33
    3D molecular signal/data processing and memory
        platforms, 7-43–7-49
    data and signal processing platforms, 7-4–7-9
    design of 3D molecular integrated circuits, 7-34–7-43

device switching speed, **7**-16

energy levels, **7**-13–**7**-15

entropy and its applications, **7**-11–**7**-12

fluidic molecular electronics, **7**-20–**7**-33

hardware and software co-design for, **7**-55–**7**-58

hierarchical finite-state machines, **7**-49–**7**-52

micro- and nanoelectronics retrospect and prospects, **7**-8–**7**-11

modeling and analysis, **7**-58–**7**-78

performance estimates, **7**-11–**7**-18

photon absorption and transition energetics, **7**-16

processing performance estimates, **7**-16–**7**-18

synthesis taxonomy in molecular integrated circuit design, **7**-18–**7**-20

Molecular conductance junctions

advanced theoretical challenges in, **12**-17–**12**-22

chirality effects in, **12**-22

coherent transport theory, **12**-4

dynamical analysis challenges, **12**-22

electrostatic potentials and image charges in, **12**-18–**12**-20

experimental techniques for molecular junction transport, **12**-3–**12**-4

gating and control of, **12**-10–**12**-11

generalized Landauer formula for, **12**-4

geometry in, **12**-20–**12**-21

and nonadiabatic electron transfer, **12**-13–**12**-14

onset of incoherence and hopping transport, **12**-14–**12**-17

and onset of inelasticity, **12**-11–**12**-12

photo-assisted transitions in, **12**-22

reactions in, **12**-21

resistance in, **14**-1–**14**-3

theory and modeling progress report, **12**-1–**12**-3

Molecular containers, from building blocks, **19**-13–**19**-17

Molecular dynamics simulations, **11**-3, **23**-33

bonding potentials, **23**-33–**23**-34

Molecular electronic computing architectures, **5**-1, **5**-4

complex molecular device characterization, **5**-23–**5**-24

crossbar arrays, **5**-8–**5**-15

functional blocks to nanocell approach, **5**-19–**5**-23

nanocell approach, **5**-15–**5**-19

and present microelectronic technology, **5**-1–**5**-2

quantum cellular automata (QCA), **5**-4–**5**-8

switch characterization, **5**-23–**5**-24

Molecular electronic devices, **7**-58

Molecular electronics, vii, **5**-3

CMOS device parameters and scaling, **4**-6–**4**-11

computer architectures based on, **5**-4–**5**-23

departure from Moore's conjectures, **7**-9

electron transport in, **9**-2–**9**-4

engineering challenges in, **4**-1–**4**-2

envisioned advancements, **7**-10

extrinsic molecular switching in, **9**-4–**9**-5

gates and, **7**-24–**7**-36

goal of, **4**-2

memory devices, **4**-11–**4**-12

modeling and analysis of devices, **7**-58–**7**-78

modeling with nanotube junctions, **22**-11–**22**-13

molecular quantum dot rectifiers, **9**-5–**9**-8

molecular switches in, **9**-8–**9**-15

opportunities and challenges for molecular circuits, **4**-12–**4**-17

role of defects in molecular transport, **9**-16–**9**-20

technological feasibility and soundness, **7**-10

vs. silicon-based electrical devices and logic circuits, **4**-2–**4**-6

Molecular EXOR gates, **7**-39

Molecular ferroelectric PVDF, intrinsic polarization and extrinsic conductance switching in, **9**-13

Molecular gatekeepers, **11**-24–**11**-25

Molecular gear simulation, **28**-15

Molecular gyroscopes, **11**-28–**11**-30

molecular structure of proposed, **11**-29

Molecular integrated circuits (^MICs)

application-specific, **7**-53

decision diagrams and logic design of, **7**-37–**7**-41

design of 3D, **7**-34–**7**-43

designing high-performance/yield through defect tolerance, **7**-53

logic design, **7**-37

synthesis taxonomy in design of, **7**-18–**7**-20

Molecular junction transport

experimental techniques for, **12**-3–**12**-4

vs. nonadiabatic electron transfer, **12**-13

Molecular library model, **18**-34

metal complex and ligand classification, **18**-35

tetrahedra from, **18**-42

Molecular lithography, **17**-18

Molecular machines, **11**-1–**11**-2

biological nanovalves, **11**-26

bistable [2]rotaxanes, **11**-8–**11**-10

bottom-up assembly of, **11**-4–**11**-5

CBPQT recognition system case study, **11**-7–**11**-8

comparison of molecular environments, **11**-6–**11**-7

computational modeling of, **11**-12

and controlled molecular motion quandary, **11**-3–**11**-4

electronic transport in, **11**-12–**11**-13

enabling technologies, **11**-2–**11**-3

flexing cantilevers with palindromic [3]rotaxanes, **11**-20–**11**-21

irreversible thin film regulators, **11**-22

LB studies, **11**-10–**11**-12

light-driven unidirectional rotation in, **11**-30–**11**-33

light-regulated azobenzene and coumarin valves, **11**-22–**11**-24

light-responsive surfaces, **11**-36–**11**-37

models from nature, **11**-4–**11**-5

molecular compasses and gyroscopes, **11**-28–**11**-30

molecular gatekeepers, **11**-24–**11**-25

molecular muscle systems, **11**-16–**11**-21

molecular nanovalves, **11**-21–**11**-26

molecular rotors, **11**-26–**11**-35

nanofluidics devices, **11**-39–**11**-40

observing condensed phase movement in, **11**-13–**11**-16

photo-controlled polymeric muscles, **11**-19–**11**-20

polymeric valves, **11**-25

potential of artificial muscles, **11**-16–**11**-17

redox-controlled annulene-based muscles, **11**-19

supramolecular gatekeepers, **11**-24

surface-rotor interactions in, **11**-33–**11**-35

surfaces with controllable wettability, **11**-35–**11**-40

taxonomy of, **11**-5–**11**-6

transferring molecular movement to solid state in, **11**-6–**11**-16

transition metal controlled actuators, **11**-17–**11**-18

typical valve systems, **11**-21–**11**-22

Molecular memory, **10**-12

Molecular modeling

  of analytic potential functions, **28**-6–**28**-11

  applications of, **28**-13–**28**-27

  considerations in molecular dynamics simulations, **28**-12–**28**-13

  contributions to nanometer-scale science and technology, **28**-1–**28**-2

  of dendrimers, **28**-25–**28**-26

  first-principles approaches, **28**-12–**28**-13

  of fullerene nanotubes, **28**-24–**28**-25

  of interatomic potential energy functions and forces, **28**-3

  molecular simulations, **28**-2–**28**-11

  of nanometer-scale indentation, **28**-19–**28**-23

  of nanometer-scale properties of confined fluids, **28**-17–**28**-19

  of nanostructured materials, **28**-26–**28**-27

  of new materials and structures, **28**-23–**28**-27

  pump, gear, motor, valve, and extraction tool applications, **28**-13–**28**-17

  of workhorse potential energy functions, **28**-4–**28**-6

Molecular motors, schematic diagram, **19**-9

Molecular movement

  in bistable [2]rotaxanes, **11**-8–**11**-10

  and CBPQT recognition system case study, **11**-7–**11**-8

  comparison of molecular environments, **11**-6–**11**-7

  computational modeling of, **11**-12

  condensed phase movement, **11**-13–**11**-16

  and electronic transport, **11**-12–**11**-13

  LB studies of, **11**-10–**11**-12

  transferring to solid state, **11**-6

Molecular muscle systems, **11**-6, **11**-16

  flexing cantilevers with palindromic [3]rotaxanes, **11**-20–**11**-21

  photo-controlled polymeric muscles, **11**-19–**11**-20

  potential of artificial muscles, **11**-16–**11**-17

  redox-controlled annulene-based muscles, **11**-19

  transition metal controlled actuators, **11**-17–**11**-18

Molecular NAND gates, **7**-36, **7**-37, **7**-39, **7**-40

Molecular nanosystems products, **3**-18

Molecular nanovalves, **11**-21

  biological nanovalves, **11**-26

  comparison with typical valve systems, **11**-21–**11**-22

  irreversible thin film regulators, **11**-22

  light-regulated azobenzene and coumarin valves, **11**-22–**11**-24

  molecular gatekeepers, **11**-24–**11**-25

  polymeric valves, **11**-25

  supramolecular gatekeepers, **11**-24

Molecular NOR gates, **7**-36, **7**-39

Molecular optoelectronics, **12**-1

Molecular orbitals (MOs), **9**-3

Molecular processing platforms (^MPPs), **7**-1–**7**-4, **7**-44. *See also* Molecular computing and processing platforms (^MPPs)

  and progress in device physics, **7**-45

  synthesis taxonomy in design of, **7**-18–**7**-20

Molecular pump and gear, **28**-14

Molecular quantum dot rectifiers, **9**-5–**9**-8, **9**-14–**9**-15, **9**-37

  degeneration requirements for switching, **9**-15

  schematic diagram, **9**-14

  transmission coefficient *vs.* energy for, **9**-7

Molecular random access memory (MRAM), measured logic diagram, **5**-20

Molecular reconfigurations, role in extrinsic molecular switching, **9**-9–**9**-12

Molecular rotors, **11**-5, **11**-6, **11**-26, **18**-6

  altitudinal rotors on gold surface, **11**-34

  light-driven unidirectional rotation, **11**-30–**11**-33

  molecular compasses and gyroscopes, **11**-28–**11**-30

  molecular structure of, **11**-33

  and rotation as fundamental molecular motion, **11**-26–**11**-27

  rotation by design in, **11**-27–**11**-28

  second-generation, **11**-31, **11**-32

  surface-rotor interactions, **11**-33–**11**-35

Molecular-scale electronics, **5**-3

Molecular shape change, **24**-16

Molecular shuttle switches, **4**-14, **11**-5, **11**-6

Molecular simulations, **28**-2–**28**-3

  covalent bonding, **28**-4–**28**-5

  interatomic potential energy functions and forces, **28**-3

  ionic bonding, **28**-5–**28**-6

  of metals, **28**-4

Molecular switches, **9**-8–**9**-9, **11**-5, **11**-6

  atom relay model, **6**-10

  biomolecular, **7**-23

  catenanes and rotaxanes as, **5**-12

  characterization, **5**-23–**5**-24

  at cross junctions of SWNT arrays, **5**-21

  electrically addressable molecules as, **9**-13–**9**-14

  extrinsic switching in organic molecular films, **9**-9–**9**-12

  intrinsic polarization and extrinsic conductance switching in molecular ferroelectric PVDF, **9**-13

  measuring electrical characteristics of, **5**-4

  quantum dot switchings, **9**-14–**9**-15

  role of defects and molecular reconfigurations, **9**-9–**9**-12

Molecular Tinkertoy approach, **18**-22–**18**-24

  chemical precedent for, **18**-26

  cubic scaffolding and, **18**-29–**18**-31

  diamondoid scaffolding in, **18**-26–**18**-27

  engineering kit for, **18**-23

  honeycomb lattices and, **18**-29–**18**-30

  size control in, **18**-24–**18**-26

  stability features, **18**-24

Molecular transistor, conceptual model, **14**-2

Molecular transport, role of defects in, **9**-16–**9**-20

Molecular-type field effect transistors, **12**-11

Molecular wire junctions

  construction of, **12**-2

  dependence on electrical transduction, **12**-1

  rate behavior, **12**-17

resistance of, **14**-1–**14**-3
schematic diagram, **12**-2
with six gold atoms, **12**-9, **14**-16
Molecular wires, extension of Landauer transmission
model of conductance to, **12**-5–**12**-10
Molecular zippers, **18**-20
from amide oligomers, **18**-21
Molecule-electrode contact, role in organic and
semiconductor nanodevices, **9**-4–**9**-5
Molecule placement, difficulties in QCA method, **5**-7
Molecule/solid interfaces, **4**-14
Molecule tailoring, **3**-11
Molecule tilting, extrinsic molecular switching due to,
9-4–**9**-5
Molecules
as fundamental computing units, **5**-3
using as devices, **3**-18
Moletronics, and spintronics, **9**-1–**9**-2
Momentum
average per electron, **7**-68
expectation values, **7**-65
root-mean-square, **7**-67
solving for, **7**-71
of wave packet, **7**-63
Monte Carlo simulations, for ion channels, **13**-36
Moore's conjectures, **7**-9
Moore's law, **5**-1, **10**-2, **10**-3, **10**-18
and scaling challenges, **4**-1
Mortal programming, in molecular electronics, **5**-22–**5**-23
MOS devices
bipolar, **4**-3
operating characteristics, **4**-7
transistor cross-section, **4**-4
MOSFETs, **4**-3. *See also* Metal-oxide-semiconductor field
effect transistor (MOSFET) structures
comparison with SPINFETs, **8**-8, **8**-10–**8**-11
fundamental limits on speed and power, **8**-8–**8**-9
limits of switching delay in, **8**-10
poly-Si gate depletion, **10**-4
slowing of channel length scaling, **10**-4
Motors, applications of computer modeling to, **28**-13
mRNA isolation/first strand synthesis device, **26**-9
MSTJs, Langmuir layer of bistable [2]rotaxanes, **11**-16
Multilevel memory device, **10**-13
Zhou's, **10**-12
Multiple data streams, **7**-45
Multiple emulsion method, **30**-3
Multiple instruction stream, **7**-45
Multiple-valued logics, **7**-30, **7**-31
design of, **7**-38
I-V characteristics, **7**-78
Multiprocessor architecture, **7**-49
Multiscale models
of carbon nanotubes, **23**-39
for CNT simulation, **23**-36–**23**-38
Multiscale self-assembling, **3**-18
Multiscale theory
and modeling of carbon nanotube
nanoelectromechanical systems, **13**-20–**13**-24
operation of NEM switches, **13**-24–**13**-25
Multishell model of complex spheroids, **16**-11–**16**-13

Multiterminal structures
challenges for gain production, **4**-6
power dissipation and gain issues, **4**-15–**4**-16
with input, control, output terminals, **7**-58
multi-terminal quantum-effect molecular electronic
devices, **7**-75–**7**-78
Multiwalled nanotubes (MWNT), **5**-14
as-grown bamboo section, **23**-28
bending and buckling behaviors, **23**-24
clamping in place for tensile loading experiment, **23**-27
deformability in, **23**-12
electric field-driven resonance in, **23**-23
grown by thermal CVD, **22**-17
growth on alumina template, **22**-18
HRTEM images, **23**-5, **23**-10
molecular structure, **23**-4
schematic diagram, **21**-27
SEM images of electric field-induced resonance,
23-9
sideways buckling and strain accumulation, **22**-6
stress-strain curves, **23**-6
TEM image, **22**-3
tensile loading of, **23**-14
Muscle contraction, models of, **11**-4

N

NAND gates, **7**-75, **7**-76, **10**-10
Nano-bio-info-cogno (NBIC) tools, **3**-3, **3**-5
transforming effect on society, **3**-3
Nano-floating gate memory, **10**-12
Nano-motors, **3**-11
Nanoadditives
barrier properties, **21**-53–**21**-60
carbonaceous materials, **21**-48–**21**-52
classification of, **21**-44
equiaxed nanomaterials, **21**-44–**21**-47
in textiles, **21**-42–**21**-44
thermal stability and FR, **21**-52–**21**-53
Nanoarchitectonics, **10**-1–**10**-2
nanoarchitectures, **10**-13–**10**-18
nanodevices/emerging research devices,
10-10–**10**-13
top-down *vs.* bottom-up manufacture, **10**-2–**10**-3
Nanocars, **19**-3
rolling on gold surface, **19**-4
STM manipulation of, **19**-4
Nanocells, **4**-15
functional blocks to, **5**-19–**5**-23
in molecular computing, **5**-15–**5**-19
schematic diagram of proposed, **5**-22
synthesis, **5**-15
Nanocircuits, **6**-1
Nanodevices, **10**-10
advantages of, **6**-1
examples of emerging, **10**-11–**10**-13
power dissipation in, **10**-10–**10**-11
Nanoelectromechanical systems (NEMS), **2**-7,
23-44
fabrication methods, **23**-44–**23**-46
Nanoelectronic circuit architectures, **6**-1–**6**-2

molecular circuits, **6**-9–**6**-11
quantum-dot cellular automata (QCA), **6**-2–**6**-8 (*See also* Quantum cellular automata (QCA))
single-electron circuits, **6**-8–**6**-9
Nanoelectronic self-assembly, future, **10**-7–**10**-9
Nanoelectronics, **3**-22
advances in, **10**-1–**10**-2
and current CMOS landscape, **10**-3–**10**-5
nanostructure fabrication, **10**-6–**10**-9
retrospect and prospects, **7**-8–**7**-11
top-down *vs.* bottom-up fabrication, **10**-2–**10**-3
Nanofibers, **21**-2
biocomponent islands-in-the-sea fiber, **21**-10
cone calorimeter data, **21**-53
effects of number of islands and polymer composition on fiber diameter, **21**-10
electrospinning of, **21**-2–**21**-3, **21**-17–**21**-25
meltblowing process, **21**-3–**21**-5
partially soluble biocomponent fibers, **21**-10–**21**-13
segmented pie fiber cross-sections, **21**-9
splittable biocomponent fibers, **21**-6–**21**-10
stress-strain behavior, **21**-52
TEM micrograph, **21**-47
TGA analysis, **21**-52
Nanofluidic electroporesis, **26**-12
Nanofluidics
for cell biology, **26**-1–**26**-2
and wettability, **11**-39–**11**-40
Nanographs, Mihail C. Roco, **3**-25
Nanoimprint lithography, for nanoscale devices, **20**-3–**20**-4
Nanoinformatics, **3**-19
Nanomachines, **1**-6
Nanomagnetics, **3**-22, **6**-7
spin-torque domain wall switching in, **9**-22–**9**-23
Nanomanipulation testing stage, **23**-42
Nanomanufacturing, **20**-1–**20**-2
bottom-up approach to, **20**-4–**20**-5
challenges in, **20**-2–**20**-3
combined top-down and bottom-up approaches, **20**-5–**20**-11
in nanoscale trenches, **20**-8
redistribution of homopolymers, **20**-9
registration and alignment issues, **20**-11
reliability and defect control, **20**-11–**20**-14
top-down approach to, **20**-3–**20**-4
Nanomaterials manufacturing
future of, **21**-60–**21**-61
safety issues, **3**-15
Nanomedical devices, dendrimers as, **24**-24–**24**-27
Nanometer-scale forces, **10**-8
Nanometer-scale indentation simulation, **28**-19–**28**-23
Nanoparticle characterization, **30**-11
size and encapsulation efficiency, **30**-11–**30**-12
surface modification, **30**-12–**30**-14
zeta potential, **30**-12
Nanoparticle preparation methods, **30**-3
nanoprecipitation, **30**-11
Nanoparticle processing parameters, **30**-5–**30**-6
collection method, **30**-10–**30**-11
poly-e-caprolactone, **30**-8–**30**-9
poly(lactic acid), **30**-7–**30**-8

poly(lactide-co-glycolide), **30**-7
polymer choice, **30**-9–**30**-10
polymer molecular weight, **30**-10
polymer type, **30**-6–**30**-9
Nanoparticle size
after surface modifications, **30**-13
comparisons for drug delivery, **30**-4
effects of stirring and bath sonication on, **30**-5
and encapsulation efficiency, **30**-11–**30**-12
Nanoparticles
assembly with/without cleaning control, **20**-13
bright colors of, **21**-57
silver TEM image, **21**-45
trapping single, **16**-24–**16**-26
uses in textile applications, **21**-44
Nanoparticulate delivery systems, **30**-14
liposomes, **30**-14
polymeric micelles, **30**-14–**30**-16
polymersomes, **30**-16
worm-like micelles, **30**-16
Nanopharmaceuticals, dendrimers as, **24**-24–**24**-27
Nanophotonics, **3**-22
accelerated design tools for, **29**-1–**29**-2
Nanopore devices, **5**-15, **5**-16
Nanoprecipitation, **30**-3, **30**-11
Nanoproducts, marketing of existing, **3**-14
Nanoropes, **23**-39–**23**-42
before and after untangling, **23**-45
multiscale image of materials reinforced by, **23**-48
twisting mechanisms, **23**-41
Nanoscale
ability to measure and transform at, **3**-2
knowledge development and education originating from, **3**-17
negligible gravity and inertia at, **10**-8
systematic control and manufacture at, **3**-17
technical challenges, **3**-16
Nanoscale Center for Learning and Teaching, **3**-13
Nanoscale designed catalysts, **3**-16
Nanoscale dimensions, as function of atoms, monomers, branch cells, dendrimers, megamers, **24**-31
Nanoscale information, **1**-3–**1**-4
Nanoscale information Science Education network, **3**-13
Nanoscale instrumentation, **3**-22
Nanoscale manipulation, **1**-2
Nanoscale metrology, **3**-22
Nanoscale modeling, **13**-1–**13**-2
of carbon nanotube nanoelectromechanical systems, **13**-20–**13**-30
of quantum dots and artificial atoms, **13**-8–**13**-19
of Si-SiO$_2$ interface, **13**-2–**13**-8
simulation of ionic channels with, **13**-32–**13**-37
Nanoscale monodispersity, in dendrimers, **24**-19
Nanoscale patterning, **20**-5–**20**-6
using block copolymers, **20**-9–**20**-10
Nanoscale scaffolding, in dendrimers, **24**-19–**24**-20
Nanoscale Science, Engineering and Technology (NSET) subcommittee, **3**-9, **10**-1
discussion of unexpected consequences, **3**-15
Nanoscale science/engineering, NSF centers for, **3**-24
Nanoscale steric phenomena, **24**-15–**24**-19

Nanoscale templates, directed self-assembly with, **20**-7–**20**-10
Nanoscale trenches, **20**-8
Nanoscaled electronics, vii
Nanosilver-coated polyurethane, optical micrograph, **21**-58
Nanostructure fabrication, **10**-6
   control beyond dendrimers, **24**-30–**24**-32
   future nanoelectric self-assembly, **10**-7–**10**-9
   progress in lithography processes, **10**-6–**10**-7
   self-assembly on DNA scaffolds, **17**-12
Nanostructured materials
   by design, **3**-22
   molecular simulation, **28**-26–**28**-27
Nanosurgery, visionary concepts, **1**-7
Nanotechnology, vii
   awareness of potential unexpected consequences, **3**-15
   challenges per NNI, **3**-22
   defined, **3**-2–**3**-5
   development of foundational knowledge in, **3**-11
   estimated international government expenditures, **3**-12
   historical forerunners, vii
   as key component of converging technologies, **3**-3–**3**-5
   and long-term R&D vision, **3**-5
   overcoming of scepticism about, **3**-6
   potential economic impacts, **3**-3
   search for relevance of, **3**-6
   society implications, **3**-6
   vertical industrial development, **3**-5
   wet and dry world of, **24**-4
   worldwide market estimates, **3**-9
Nanotechnology education and outreach, **3**-13
Nanotechnology products, four generations of, **3**-17–**3**-18
Nanotemplates, **20**-7–**20**-8
   for guided self-assembly of polymer melts, **20**-8–**20**-9
Nanotube application development, **22**-20
   CNT-based nanoelectronics, **22**-22–**22**-24
   CNTs in microscopy, **22**-20
   field emissions, **22**-25
   miscellaneous applications, **22**-27
   nanotube-polymer composites, **22**-25–**22**-26
   sensor applications, **22**-24–**22**-25
Nanotube filling, **23**-42–**23**-44
Nanotube growth, **22**-14–**22**-15
   arc process and laser ablation, **22**-15
   catalyst preparation, **22**-16–**22**-19
   chemical vapor deposition, **22**-15–**22**-16
Nanotube junctions, in molecular electronics, **22**-11–**22**-13
Nanotube material development, **22**-19
   characterization step, **22**-19–**22**-20
   functionalization step, **22**-20
   purification step, **22**-19
Nanotube-polymer composites, application development, **22**-25–**22**-26
Nanotube-reinforced polymer, **23**-46–**23**-48
Nanotube sensors and actuators, modeling of, **22**-14
Nanotube strength, **23**-13
   due to bond breaking or plastic yielding, **23**-13–**23**-16
   due to interlayer sliding, **23**-16–**23**-17

Nanotubes, **5**-14. *See also* Carbon nanotubes; Single-walled carbon nanotubes (SWNT)
   assembly diagram, **20**-4
   bridging of crack in polystyrene film, **23**-47
   device potential, **5**-12
   modeling of C_{60} inside, **23**-43
   morphologies, **21**-28
   random orientation, **23**-47
   steps in 2-D molecular assembly, **20**-7
Nanovalves, biological, **11**-26
Nanowire crossbar, **10**-15
Nanowire spacing, **20**-6
Nanowires (NW), **5**-9
   assembly and transport measurements, **5**-13
   Au-nucleated, **5**-11
   biotemplated synthesis of, **10**-9
   constructing logic gates using, **5**-24
   guidance problems, **5**-12
   imprinted self-assembled, **6**-11
   three-terminal transport measurements, **5**-10
Naphthalene rectifiers, **9**-6
   I-V curves for, **9**-9
NASA-Ames Research Center, ix
NASA-Langley Research Center, ix
National Aeronautical and Space Administration (NASA), **3**-1, **3**-7, **3**-8
   establishment of academic nanotechnology centers, **3**-15
National Institute for Occupational Safety and Health (NIOSH), **3**-8
National Institute of Standards and Technology (NIST), **3**-1, **3**-7
National Institutes of Health (NIH), **3**-1, **3**-7, **3**-8
National Nanotechnology Initiative (NNI), **3**-1–**3**-2, **10**-1
   contributions of key federal agencies to, **3**-8
   five-year review, **3**-10–**3**-16
   goals in second strategic plan, **3**-23
   historical origins, **3**-6–**3**-10
   as inclusive process, **3**-6
   industry collaborative boards, **3**-16
   key factors n establishing, **3**-5–**3**-6
   membership, **3**-23–**3**-24
   modes of support in first strategic plan, **3**-22
   new governance approach, **3**-19–**3**-20
   new science and engineering approach, **3**-19–**3**-20
   recognition of contributions, **3**-24–**3**-25
   specific per capita expenditure to 2005, **3**-13
   technical challenges in current phase, **3**-16–**3**-19
National Nanotechnology Research, **3**-4
National Science Foundation, ix, **3**-1, **3**-7, **3**-8, **10**-1
   centers for nanoscale science and engineering, **3**-24
   nanotechnology definitions by, **3**-2
Natural resources, reducing consumption through molecular technologies, **5**-3
Naturally occurring structures, size and scale of, **17**-2
Nature
   bottom-up fabrication in, **10**-2
   information processing in, **7**-7–**7**-8
   models of bottom-up assembly, **11**-4–**11**-5
   multiple instruction streams in, **7**-45
   nanoscales in, **2**-6–**2**-7

self-assembly in, **10**-9
teaching unity of, **3**-17
Naval Warfare Centers, xii
Near-field optical coupling, **6**-8
Negative bias, **14**-6
Negative differential resistance (NDR) devices, **4**-2–**4**-3
with gold impurities, **9**-19
with load resistor, **4**-3
nitro analine derivatives, **5**-17
Negative refractive index, numerical demonstration of,
    29-14–**29**-15
Neilson's tortuous path model, **21**-54
NEM switches, **13**-24–**13**-25
analytical consideration of pull-in, **13**-28–**13**-29
pull-in gap as function of initial device gap, **13**-29
pull-in voltage as function of gap, **13**-29
Network on Nanotechnology in Society, **3**-13
NeuroBioMimetics, **7**-20, **7**-21
Neuromorphic circuits, **10**-15
Neuromorphic Networks, **10**-5
Neuromorphological reconfigurable molecular processing
    platforms, **7**-32–**7**-33
Neurons
communication in, **7**-26
as devices *vs.* systems, **7**-21
emulation by fluidic molecular processing devices, **7**-31,
    7-32
excitatory and inhibitory inputs in, **7**-27
fan-out, **7**-25
input-output representation, **7**-25
signal/data processing between, **7**-43
as switching devices, **7**-25, **7**-27
as systems (processing modules), **7**-30
Neuroscience, information processing and memory
    postulates, **7**-20–**7**-25
Neurotransmitter molecules, information processing in, **7**-22
New applications, promise of, **3**-3
Nitro aniline oligo derivatives, **5**-15
with alligator clips, **5**-18
conductivity of, **5**-16
NDR behavior of, **5**-17
Nitrogen lone-pair donors, **18**-33
NMOS devices, **4**-4
in CMOS circuits, **4**-9
inverter circuit approach, **4**-5
Nonadiabatic electron transfer, molecular junction
    conductance and, **12**-13–**12**-14
Nonballistic SPINFETs, **8**-4–**8**-8
Nonequilibrium Green's function (NEGF) formalism,
    14-12–**14**-15
Nonfunctionalized electrode, **17**-29
Nonlinear macromolecules
building block approaches to, **19**-1–**19**-3
catenanes, **19**-3–**19**-9
molecular containers, **19**-13–**19**-17
nanocars, **19**-3
rotaxanes, **19**-9–**19**-13
Nonoligomers, as molecular building blocks, **19**-2
Nonvolatile memory, **4**-12, **10**-3
SWNT-based devices, **5**-11–**5**-12
Normalized angular wave function, **7**-73

North Carolina State University, xiii
Notre Dame NanoDevices group, **6**-2
Novel devices, **7**-9, **7**-18
discovery of, **7**-3
NSF awards, in nanoscale science, **3**-10
Nuclear magnetic resonance (NMR), **11**-2
Nucleic acid amplification/detection, **26**-10–**26**-12
Numerical aperture (NA) microscopy, **15**-1
Numerov three-point-different expression, **7**-70, **7**-71

O

Object-oriented programming, **7**-57
Off-resonance tunneling electrons, **12**-12
Office of Management and Budget (OMB), **3**-7
Office of Naval Research, ix
Oligomers, as molecular building blocks, **19**-2
Omnipotent programming, in molecular electronics,
    5-22–**5**-23
Omniscient programming, in molecular electronics,
    5-22–**5**-23
On-off ratios, for carbon nanotube SETs, **6**-9
Open biocomponent spunbond process, **21**-8
Operating voltage, in molecular devices, **4**-15
Optical applications, inefficiency of silicon for, **2**-4
Optical barrier properties, of nanoadditives, **21**-54–**21**-56
Optical beam splitters, **25**-23
with hetero-structure lattice, **25**-24, **25**-25
output *vs.* radius of air holes, **25**-33
Optical cell sorting, **26**-6–**26**-7
Optical limiters, **25**-15
Optical networks
cross-sectional view, **25**-26
FDTD simulations, **25**-27, **25**-28
schematic diagram of on-chip, **25**-26
Optical pattern generation, **2**-4, **10**-7
Optical spectrometers, **25**-18
creation in PBG or rectangular lattice, **25**-20
multiple-channel, **25**-19–**25**-21
single-channel, **25**-18–**25**-19
Optical switches, using PBG, **25**-25
Oral drug delivery, using nanoparticles, **30**-16
Orbital magnetic dipole moment, **7**-15
Orchestrated objective reduction model, **7**-24
Organic chemistry, comparison with dendritic molecular
    chemistry, **24**-6–**24**-8
Organic materials
with long spin-relaxation time, **9**-37
sensitivity to temperature and heat generation, **4**-9
spin injection into, **9**-36
Organic nanodevices, electron charge and spin transport
    in, **9**-1–**9**-2
Organic nanostructures, bottom-up synthesis strategies
    for, **24**-4–**24**-6
Organic thin films, characterization of defects in, **9**-16
Organization, in silicon *vs.* organic molecules, **4**-2
Orientational specificity, in hydrogen-bonded structures,
    18-17
Oscillatory dependence, **9**-35
Output impedance, **4**-2
Output node capacitance, **4**-8

Output states, 7-24
Output voltage, ratio to input voltage, **4**-5
Oxides, as equiaxed nanomaterials, **21**-45–**21**-46

P

p/n junction diode, **4**-3
Packing density, of nanodevices, **6**-1
Pair potential, in two-atom system, **23**-34
Pairwise cell-cell interactions, PDMS device for analysis of, **26**-18
Palindromic [3]rotaxanes, **11**-20–**11**-21
    structure and graphical representation, **11**-21
Palladium sphere, trapping by positive dielectrophoresis, **16**-28
PAMAM dendrimers, **24**-16, **24**-33
    comparison of molecular weights an shell filling as function of generation, **24**-17
    inner cavities and channels, **24**-23
    periodic properties, **24**-20
    radii of gyration, **24**-22
    water domains in, **24**-23
Pantographs, **1**-7
Parallel memories
    higher bandwidth with, **7**-46
    integrating multiple MIC memory bands as, **7**-47
Parallel molecular NOR arrays, **7**-35
Parallelism, **2**-5
Parasitics, **4**-12
    and alternate designs for device-device coupling, **6**-2
    in CMOS devices, **4**-10
    concerns in hybrid silicon/molecular systems, **4**-17
    in ultra-shallow junction formation, **10**-3
Partially soluble biocomponent fibers, **21**-10–**21**-13
Particle removal forces, **20**-13
Particle separation, **16**-26–**16**-28
Particle velocity, **7**-16
    modeling and analysis, **7**-61–**7**-64
Pascal, Blaise, **7**-4
Passive matrix addressing, in memory devices, **4**-12
Passive nanostructure products, **3**-17–**3**-18
Patents
    decrease in 2005, **3**-15
    increased in number of nanotechnology-related, **3**-10
    molecular computer, **7**-7
    U.S. share of, **3**-13, **3**-14
Pattern generation, **2**-3
    challenges in molecular circuits, **4**-12, **10**-3
    convergence between top-down and bottom-up methods, **10**-2
    optical, **2**-4
PCR devices, nanofluidics applications, **26**-10
PDMS bacterial chemostat, **26**-19
PE oligomer 59, **19**-21
PEI-PLGA aggregates, TEM pictures and size distributions of, **30**-15
Pentium 4 chip, **7**-5
    density achievements, **5**-1
PEO nanofibers, **21**-18
    out-of-phase component of AC susceptibility, **21**-24
    room temperature equilibrium magnetization of, **21**-22

TEM electron diffraction pattern in, **21**-21
TEM image, **21**-20
Peptide bonds, **17**-9
Performance estimates, **7**-11
    device switching speed, **7**-16
    distribution statistics, **7**-12–**7**-13
    energy levels, **7**-13–**7**-15
    entropy and its applications, **7**-11–**7**-12
    photon absorption and transition energetics, **7**-16
    processing speed, **7**-16–**7**-18
Periodic Table of Elements, **7**-2
Peripheral substitutions, **18**-30
Perpetual motion, unidirectional, **11**-3, **11**-4
Perturbation methods, calculating responses in molecular junctions with, **12**-23
Pharmaceutical synthesis, gains from nanotechnology, **3**-17
Phase-change memories (PCM), **9**-2–**9**-3
Phase change memory, **10**-12
Phase inversion nanoencapsulation, **30**-3
Phase-related effects, electrorotation and traveling wave dielectrophoresis, **16**-13–**16**-16
Phase velocity, **7**-61, **7**-62
Photo-assisted transitions, in molecular conductance junctions, **12**-22
Photo-controlled polymeric muscles, **11**-19–**11**-20
Photochromic molecules, **9**-3
Photon absorption and transition energetics, performance estimates, **7**-16
Photonic bandgap applications, **25**-11–**25**-12
    2-D distributed feedback laser generator, **25**-17
    air-bridge microcavity, **25**-17–**25**-18
    beam splitters, **25**-15–**25**-16
    Bragg reflectors, **25**-16
    channel drop filters in photonic crystals, **25**-15
    control of spontaneous emission, **25**-18
    enhancing patch antenna performance, **25**-18
    fabrication error analysis, **25**-29
    high-Q microcavities, **25**-13–**25**-14
    high-quality filters, **25**-15
    hybrid photonic crystal structures, **25**-21–**25**-23
    low-loss optical waveguides, **25**-12–**25**-13
    optical limiters, **25**-15
    optical spectrometers, **25**-18–**25**-21
    optical switch using PBG, **25**-25
    photonic crystal fibers, **25**-17
    photonic crystal optical networks, **25**-25–**25**-28
    second harmonic generation, **25**-17
    surface-emitting laser diode, **25**-18
    tunable photonic crystals, **25**-23–**25**-25
    waveguide bends, **25**-13
    zero cross-talk, **25**-16–**25**-17
Photonic bandgap (PBG), **25**-2
    microcavity built in, **25**-14
    sharp wave-guide built on, **25**-14
Photonic bandgaps, structure analysis, **25**-5–**25**-10
Photonic crystal fibers, **25**-17
Photonic crystals, **25**-1–**25**-3
    analogy with semiconductor crystals, **25**-4–**25**-5
    applications using irregular dispersive phenomena in, **25**-29–**25**-33

bandgap applications, **25**-11–**25**-29
comparative dispersion diagrams, **25**-8, **25**-9
doping of, **25**-10–**25**-11
electromagnetic localization in, **25**-10
enhancing patch antenna performance with, **25**-18
future applications, **25**-33
mechanism of action, **25**-3–**25**-4
microcavities in, **25**-11
optical networks using, **25**-25–**25**-29
PhC slab and dispersion diagram, **25**-3
and photonic bandgap structure analysis, **25**-5–**25**-10
with point defect, **25**-11
steady-state field solution through PBG waveguide,
    25-13
tunable, **25**-23–**25**-25
unit cell for band structure calculations, **25**-8
waveguide created in, **25**-12
Photoresist pattern, in carbon nanotubes, **22**-21
Pi-interactions, **18**-9–**18**-10, **18**-18–**18**-19
aedemers, **18**-20–**18**-21
catenanes, **18**-19–**18**-20
molecular zippers, **18**-20
pi-cation interactions, **18**-9
staggered pi-stacking, **18**-9
in supramolecular chemistry, **18**-18–**18**-21
Piezoresponse force microscopy, **3**-25
Pipelining technique, **7**-48
Platelet nanomaterials, **21**-46–**21**-47
PMOS devices, **4**-4
in CMOS circuits, **4**-9
Point-contact transistors, difficulties in scaling, **2**-2
Point defects, in photonic crystals, **25**-11
Poisson equation, **7**-75, **7**-77
in 3D topology devices, **7**-74
Polarons, **12**-16
Poly-e-caprolactone, **30**-8–**30**-9
Poly(lactic acid), **30**-7–**30**-8
Poly(lactide-co-glycolide), **30**-7
Polymer chemistry
comparison with dendritic molecular chemistry,
    24-6–**24**-8
polymer architectures and state transitions, **24**-6
Polymer/clay composites
morphologies, **21**-47
permeability coefficients, **21**-55
UV-VIS transmittance spectra, **21**-56
Polymer melts, guided self-assembly of, **20**-8–**20**-9
Polymer/nanotube composites, **21**-30
matrix material of, **21**-31
SWNTs *vs.* MWNTs, **21**-30–**21**-31
Polymer technology, convergence with nanotechnology,
    24-9
Polymeric micelles, as nanoparticulate delivery systems,
    30-14–**30**-16
Polymeric muscles, photo-controlled, **11**-19–**11**-20
Polymeric structures, hopping mechanism in, **12**-16
Polymeric valves, **11**-25
Polymerization method, **30**-3
Polymers, surface activation of, **21**-39–**21**-41
Polymersomes, as nanoparticulate delivery systems, **30**-16
Polynomial electrode array, **16**-4

electric fields and forces generated by, **16**-23
fluorescence photo of beads collecting in, **16**-5
separation of herpes virus by, **16**-25
Polypropylene/nanotube composites, **21**-31
crystallization of, **21**-31–**21**-32
mechanical properties, **21**-33–**21**-34
morphology, **21**-32–**21**-33
polypropylene, **21**-31
Polyurethane, optical micrograph, **21**-58
Porphyrin squares, **18**-30
metal-ligand coordination stacking design from,
    18-31
Position-momentum-uncertainty relation, **7**-60
deriving for electrons, **7**-61
Positive bias, **14**-6
Potential barriers, electron tunneling through, **7**-70
Potential energy profile, 1D, **7**-76
Power consumption, **10**-1
challenge of controlling, **4**-9
of defect-tolerant devices, **10**-17
low in QCA devices, **4**-15
reduced in CMOS devices, **4**-4
Power dissipation, **4**-2, **4**-8–**4**-9, **4**-12, **4**-15
advantages of silicon device engineering, **4**-9
and alternatives to silicon technology, **9**-36
avoiding with active addressing, **4**-12
as bottleneck of silicon technology, **9**-1
challenges for CMOS technology, **10**-3
comparison between SPINFETs and MOSFETs, **8**-10
as function of generation node, **10**-4
increase with speed, **4**-9
and interconnect loading, **4**-9
by molecular ICs, **7**-34
in multiterminal structures, **4**-15–**4**-16
in nanodevices, **10**-10–**10**-11
reducing by eliminating current flow, **8**-11
and reliability, **4**-11
in Tour wires, **9**-3
in two-terminal devices, **4**-15
Power loss, **4**-9
Presidential Council of Advisors in Science and Technology
    (PCAST), **3**-7
Pressure-based cell sorting, **26**-6
Pretzelane formation, **19**-6
Probability current density, **7**-67, **7**-68
Probability density, **7**-68
Probe microscopies, **11**-3
Procedural programming, **7**-57
Processing and memory platforms, **7**-56
Processing speed
microchip advantages, **2**-7
nanodevice advantages, **6**-1
performance estimates, **7**-16–**7**-18
Processing steps, reducing number through molecular
    technology, **5**-3
Professional societies, **3**-15
Programmable diodes, **10**-15
Programmable gate arrays (PGAs)
in conventional ICs, **7**-52
in reconfigurable MPPs, **7**-53
Programmable interconnect blocks (PIBs), **7**-53

Programmable logic blocks (PLBs), **7**-53
Programming logic, **7**-57
Protein complexes
 attachment to surfaces, **17**-9–**17**-10
 as biolinkers, **17**-9
 as biomimetic structures, **18**-44–**18**-47
 comparisons with dendrimers as function of generation,
  24-21
 formation of, **17**-9
 mediated self-assembly of nano- and microstructures by,
  17-19–**17**-22
Proteoglycan topology, **24**-6
Proteomics applications, of nanofluidics, **26**-12–**26**-15
Prototyping, **7**-20
Pseudorotaxanes, ring-closing metathesis strategy for
  clipping, **19**-6
Pseudoscientific claims, **3**-6
Pseudotaxanes, **11**-7, **11**-8, **11**-10
 as gateposts, **11**-24
Pulmonary drug delivery, using nanoparticles, **30**-19
Pumps, applications of computer modeling to, **28**-13
Push-pull dynamics, **3**-19
 in nanotechnology development, **3**-4–**3**-5
pyrolyzed photoresist film (PPF), **5**-24

Q

QCA arrays, computing schematic, **6**-8
QCA fan-out, **5**-6, **6**-3, **6**-4
QCA logic, **6**-3–**6**-5
QCA shift register, **6**-6
Quadrupolar electrode array, **16**-3, **16**-4, **16**-23
 electric fields and forces generated by, **16**-23
 fluorescence photograph of beads collecting in, **16**-5
 separation of herpes simplex virus by, **16**-25
Quantum cellular automata (QCA), **4**-15, **5**-4–**5**-8, **6**-2,
  10-5, **10**-10
 binary wire, **6**-4
 computing with, **6**-5–**6**-6
 edge-driven computation in, **6**-6
 engineering challenges, **5**-7
 four-dot QCA cell, **5**-5
 ground state computing, **6**-5–**6**-6
 implementations, **6**-6–**6**-8
 majority cell logic table, **5**-7
 majority gates from, **5**-5, **5**-7, 506
 proposed quantum-dot cell, **6**-2–**6**-3
 QCA fan-out structures, **5**-6, **6**-4
 QCA logic, **6**-3–**6**-5
 schematic cell diagram, **6**-2
 shift register, **6**-6
Quantum computing, **5**-4, **7**-55, **10**-10, **10**-11
Quantum density functional theory (DFT) techniques,
  5-17
Quantum dots, **2**-4
 addition energy of single vertical, **13**-14
 as artificial atoms, **5**-4
 Calculated singlet triplet energy separation, **13**-15
 Coulomb staircase for, **13**-13
 discovering switchable, **9**-15

double dot spin for occupation numbers in two coupling
  regimes, **13**-18
exchange interaction as function of magnetic field in
  coupled dot system, **13**-19
full-scale simulation of, **13**-13–**13**-16
Hartree-Fock approximation and Hund's rules, **13**-12
modeling of, **13**-8–**13**-10
molecular quantum dot switching, **9**-14–**9**-15
Notre Dame cell proposal, **6**-2–**6**-3
planar coupled device schematic, **13**-17
with protein attachment scheme, **17**-20
schematic representation, **13**-9
in SET parametron, **6**-9
single-particle Hamiltonian and shell structures,
  13-10–**13**-12
Quantum entanglement, **2**-4–**2**-5
Quantum mechanics, **1**-9
 breakthroughs in, **10**-2
 3D problem in molecular electronic devices, **7**-71–**7**-75
Quantum physics, for solid molecular devices, **7**-9
Quantum point contact (QPC), **14**-15–**14**-19
Quantum states, energy differences between, **7**-14
Quantum theory, **7**-2
Quantum transport, **4**-14
Quantum tunneling, **7**-35
Quantum-well resonant tunneling diodes, **7**-75
Quantum wells, **2**-4
Qubits, obstacles to implementation, **2**-6
Quinone molecular electronics candidate, synthesis of,
  5-20

R

Radial equation, **7**-73
Radiological detection/protection, **3**-22
Raman spectra, of SWCNT, **22**-20
Rashba interaction, in SPINFETs, **8**-3
Rate behavior, as function of distance, **12**-19
Reactions, in molecular conductance junctions, **12**-21
Read/write speed, of molecular ICs, **7**-34
Real-time processing, in biological systems, **7**-7–**7**-8
Reconfigurable architecture, **7**-32–**7**-33
 with defect-tolerant components, **6**-10–**6**-11
 programmable logic units in, **7**-53
 routing and networking, **7**-33
Recursive cell, HG stack memory with, **7**-51
Redox active nanoparticles, dendrimers as, **24**-29
Redox-controlled annulene-based muscles, **11**-9, **11**-19
Redox-controlled mechanical switching, **11**-13
Redundant concept, applying to defect-tolerant devices,
  7-53
Register-level subsystems, **7**-49, **7**-52
Registration, in nanodevice manufacturing, **20**-11
Reinforcement scales, variation in, **21**-43
Relaxed nanotube bundles, **23**-40
Reliability issues, **4**-12
 hot-electron interface trap generation for submicron
  nMOSFETs, **13**-4
 increasing effect of defect precursor distribution at
  nanoscale, **13**-4
 in molecular circuits, **4**-13–**4**-14

in nanomanufacturing, **20**-11
and power dissipation, **4**-11
reliability and characterization tools, **20**-11
reliability from defect precursor distribution, **13**-4–**13**-5
Renyi entropy measure, **7**-28
Research and development (R&D)
    estimated international government expenditures,
        3-12
    horizontal multidisciplinary stage, **3**-5
    importance to nanotechnology development, **3**-15
    and innovation results, **3**-13
    national government investments in, **3**-11
    NNI expenditure per capita, **3**-11
    NNI programs for, **3**-1
Resistance
    broadening effects on, **14**-12
    and charging effects, **14**-8–**14**-11
    and Coulomb blockade, **14**-8–**14**-12
    and current flow as balancing act, **14**-6–**14**-8
    and Fermi energy, **14**-3–**14**-5
    MATLAB code models, **14**-20–**14**-23
    of molecular conductors, **14**-1–**14**-3
    and nonequilibrium Green's function (NEGF)
        formalism, **14**-12–**14**-15
    qualitative discussion, **14**-3–**14**-8
    quantum point contact (QPC) example, **14**-15–**14**-19
Resistor-transistor logic (RTL), **7**-35
Resonant threshold, current above, **9**-7
Resonant tunneling, **2**-4, **7**-75
    in spin-selective barriers, **8**-8
Resonant tunneling diode (RTD), **4**-2
Responsible development, **3**-19
Reversible gates, **10**-10
Reversible switching cycle, I-V characteristics of, **9**-10
Risk governance, **3**-19, **3**-20
Roco, Mihail C., **3**-24–**3**-25
Roll-up vector, **23**-3
Room temperature
    exceeding thermal energy at, **2**-4
    magnetic implementations at, **6**-7
    as obstacle to CNNs, **2**-5
    QCA implementations at, **6**-7
Root-mean-square momentum, **7**-67
Rotation
    in covalent bonds, **18**-4
    as fundamental molecular motion, **11**-26–**11**-27
Rotational barriers, **11**-34
Rotational freedom, **18**-51
Rotators, **11**-26
Rotaxanes, **5**-12, **5**-14, **11**-5, **11**-7, **11**-8
    as basis of reversible nanovalves, **11**-25
    bistable [2]rotaxanes, **11**-8–**11**-10
    building block approaches, **19**-9–**19**-13
    conductance spectrum, **11**-13
    hydrogen-bond templating of, **19**-11
    mechanically interlocking, **11**-6
    molecular structure of switchable, **11**-39
    near-IR squaraine dye in, **19**-13
    palindromic [3]rotaxanes, **11**-20–**11**-21
    reversible construction through imine bond formation,
        19-10

rotaxane 35 schematic, **19**-11
threading-followed-by-shrinking mechanism, **19**-12
Rotors
    difficulty of distinguishing from stators at molecular
        level, **11**-26
    with dipoles in molecular compasses, **11**-30
    factors influencing performance, **11**-28
    three main components of, **11**-26
Routing carriers, **7**-24, **7**-32
    biomolecules as, **7**-25
    in human brain, **7**-22
    molecules and ions as, **7**-31

S

Safety, of nanomaterial manufacturing, **3**-15
Salting out method, **30**-3
Sandia National Laboratory, discovery instrumentation
    platform at, **3**-17
Sarcomeres, **11**-17
Saturated hydrocarbons, **18**-25
Saturated-shell architecture, dendrimers and, **24**-32–**24**-33
Scalability, **10**-1
    and CMOS device parameters, **4**-6–**4**-11
    compromises with retention time, cell isolation, reading
        sensitivity, **10**-3
    economics as determining factor, **2**-3
    Moore's Law challenges, **4**-1
Scanning electron microscopy
    coumarin-loaded PLGA particles, **30**-6
    cylindrical display, **17**-24
    lotus leaf surface, **21**-57
    measurement of nanowires with, **5**-12
    three-terminal device image, **5**-10
Scanning-force microscopy, in CNT measurement,
    23-24–**23**-26
Scanning probe microscopy (SPM), **3**-25, **5**-23, **12**-3
Scanning tunneling microscopy (STM), **2**-1, **5**-23, **9**-4, **11**-3
    blinking of molecules in, **9**-12
    of blinking styrene molecules, **9**-11
    inelastic electron scattering effects in, **9**-12
    in SWNT samples, **5**-12
Scattering behavior, off dielectric sphere, **29**-3
Schottky barrier, **9**-24, **9**-33
    avoiding with ferromagnetic semiconductors (FMSs),
        9-3
Schottky-gated resonant tunneling, **7**-75
Schrödinger equation, **7**-2, **7**-3, **7**-16, **7**-77
    for electron tunneling problem, **7**-69
    in modeling and analysis, **7**-64–**7**-71
    modeling and analysis based on, **7**-58
    obtaining wave function via, **7**-59
    partial differential, **7**-72
    reduction to three independent 1D equations, **7**-72
    solution with quantum-mechanical model Hamiltonian,
        6-3
    solving through discretization, **7**-70, **7**-71
Second-generation molecular rotors, **11**-31, **11**-32
Second harmonic generation, **25**-17
Second law of thermodynamics, and Feynman ratchet,
    11-4

Self-assembled monolayer (SAM), **9**-4, **11**-3, **17**-10
  process flow for hydrophilic/hydrophobic surfaces using, **17**-23
Self-assembly, **3**-16, **3**-18, **4**-13, **6**-1, **7**-3, **10**-2
  of (16-mercapto)hexadecanoic acid, **11**-38
  at atomic, micro, and nanoscales, **10**-8
  bio-inspired, **17**-3–**17**-5, **17**-12
  biolinkers in, **17**-6–**17**-12
  biological- and chemical-mediated, **17**-1–**17**-2
  categorization of, **17**-4–**17**-5
  of charged device on oppositely charged regions, **17**-25
  with chemically treated surfaces, **17**-22–**17**-25
  and component numbers in molecular circuits, **6**-9
  defined, **17**-3–**17**-4
  development of, **12**-3
  of DNA-conjugated gold nanoparticles, **17**-14
  of DNA-inspired electronic devices, **17**-18–**17**-19
  effects of hydrophobic and hydrophilic regions on, **11**-35
  electrically mediated, **17**-25–**17**-27
  fluidics mediated, **17**-27–**17**-29
  forces and interactions of, **17**-5–**17**-6
  four possible strategies of, **17**-11
  of future nanoelectronic devices, **10**-7–**10**-9
  hydrogen bonding in, **17**-5
  hydrophobic interactions in, **17**-6
  ionic bonds in, **17**-6
  Langmuir-Blodgett technique, **9**-3
  of molecular-scale switches, **5**-4
  of nanostructures on DNA scaffolds, **17**-12
  of nanowires, **5**-14
  process flow for, **17**-27
  strategies for assembly with biolinkers, **17**-10–**17**-12
  of unidirectional rotor on gold, **11**-32
  van der Waals interactions in, **17**-5
Self-collimation, **25**-30
Self-consistent field (SCF) problem, **14**-8–**14**-9
Self-energy matrix, **12**-6, **14**-17
Self-organization, **3**-16
  in three-dimensional nature, **2**-6
Semiconductor Industry Association (SIA)
  forecast for worldwide microchip sales, **10**-1
  International Technology Roadmap for Semiconductors, **4**-1, **4**-8
Semiconductor laser diodes, **2**-4
Semiconductor nanodevices, electron charge and spin transport in, **9**-1–**9**-2
Semiconductor Research Corporation (SRC), **10**-1
Semiconductors
  spin injection/extraction into/from, **9**-23–**9**-24
  using for spin transport, **9**-2
Sensor applications, for CNTs, **23**-45
Separation of variables, **7**-73
Series molecular NAND arrays, **7**-35
SET parametron, **6**-8, **6**-9
Sharvin limit, **12**-6
Shell models, for carbon nanotubes, **23**-7–**23**-9
Si-SO$_2$ interface
  ball-and-stick model of heterojunction, **13**-6
  density functional calculations, **13**-3
  hot-electron interface trap generation for submicron nMOSFETs, **13**-4

nanostructure studies of, **13**-2
  reliability considerations, **13**-4–**13**-5
  Si-H bonds at, **13**-2–**13**-3, **13**-3
  tunneling in ultra thin oxides at, **13**-5–**13**-7
Sidewall functionalization methods, with SWNT, **5**-11
Signal assignment statement, in VHDL, **7**-58
Signal delay/decay
  determining factors in CMOS circuits, **4**-10
  isolating molecular interconnects to avoid, **4**-14
  problems with defect-tolerant NW approach, **5**-14
  in QCA units, **4**-15
Signal processing platforms, molecular, **7**-4–**7**-8
Signal propagation, in QCA structures, **6**-4
Silicon
  bottlenecks of standard technology, **9**-1
  covalent, atomic, and van der Waals radii, **7**-12
  extending current technology toward low-power, high-speed applications, **9**-37
  inappropriateness for optical applications, **2**-4
  innovations in, **10**-4
  STM images of blinking styrene molecules on, **9**-11
  technical limitations of, **9**-36
  transistor scaling, **3**-16
  unlikelihood of displacement by molecular technology, **9**-4
Silicon-based electrical devices, **4**-2
  basic three-terminal logic circuits, **4**-4–**4**-5
  importance of gain in, **4**-5–**4**-6
  negative differential resistance devices, **4**-2–**4**-3
  three-terminal bipolar, MOS, and CMOS devices, **4**-3
  two-terminal diode devices, **4**-2–**4**-3
Silicon doping, practical limitations of, **2**-2–**2**-3
Silicon nanowires, **5**-9
  assembly using fluidics, **17**-31
Silicon-on-insulator (SOI) technology, **2**-3
Simple paralleling, **7**-47, **7**-48
Single-bit full adder, QCA-based, **6**-4
Single-cell electroporation, **26**-6–**26**-7
Single-cell monitoring, **26**-7
  cellular nucleic acid isolation, **26**-8–**26**-10
  intracellular protein and small molecule detection, **26**-15
  nanofluidic electroporesis, **26**-12
  nucleic acid amplification/detection, **26**-10–**26**-12
  proteomics applications, **26**-12–**26**-15
  single-cell electroporation, **26**-7–**26**-8
Single channel optical spectrometers, **25**-18–**25**-19
  FDTD simulation results, **25**-20
  spectral results, **25**-21
Single data stream, **7**-45
Single-domain magnets, **6**-7
Single-electron circuits, **6**-8–**6**-9
Single-electron memory, **10**-12
Single electron transistor (SET), **4**-15
  carbon nanotube-based, **6**-9
Single-electron transistors, **6**-8
Single instruction stream, **7**-45
Single-molecule analytical tools, **11**-2
Single-walled carbon nanotubes (SWNT), **5**-4, **5**-14, **10**-8, **10**-9, **21**-30–**21**-31. *See also* Carbon nanotubes
  AFM image of bundle, **23**-25
  assembly in nanotrench-based templates, **20**-8

bending to create quantum dots, **5**-5
buckling behavior, **23**-10
change in cross-section as function of twist angle, **23**-41
chemical self-assembly diagram, **17**-24
fluorination advantages and problems, **5**-11
functional switches at array cross-junctions, **5**-21
growth by thermal CVD, 18, **22**-21
indentation simulation, **28**-23
lack of solubility, **5**-9
lateral force as function of AFM tip position, **23**-26
as nanowires in molecular computers, **5**-9–**5**-10
Raman spectra, **22**-20
scanning tunneling microscopy of, **5**-12
schematic diagram, **21**-27
selective breakdown, **5**-15
sideways buckling and strain accumulation, **22**-6
stress-strain curves, **23**-6
TEM image, **22**-3
TEM image of tensile-loaded bundle, **23**-14
tensile loading experiment setup, **23**-27
twisting mechanisms, **23**-40
vibration in, **23**-22
Six-terminal devices, **7**-78
Size-dependent novel properties/phenomena, **3**-2–**3**-3
Skeletal muscle contraction, by sarcomeres, **11**-17
SLSI software, for proof-of-concept 3D molecular ICs,
    7-40, **7**-41
Small-molecule delivery, experimental setup, **26**-17
Small molecule detection, **26**-15
Smart surfaces
    for controlling wettability, **11**-35
    smart tagging of textile products, **21**-17–**21**-25
Societal implications, **3**-15, **3**-19, **3**-22
Soft lithography, **20**-2–**20**-3
Soft molecular junctions, **12**-3, **12**-4
Solid molecular electronics, **7**-5, **7**-16, **7**-18
    energy band structures in, **7**-13
Solid particles, dielectrophoretic behavior of, **16**-3–**16**-7
Solid-state operation, challenge for polymeric muscles,
    11-19
Solid-state rectifiers, **7**-6
Solitons, **12**-16
Solution phase, *vs.* attachment to surfaces, **11**-7
Solvent diffusion, **30**-3
Solvent displacement, **30**-3
Solvent evaporation method, **30**-3
Spatial derivatives, **7**-66
Specific composite conductivity, as function of CNT weight
    fraction, **21**-59
Spectroscopic techniques, **11**-2
Speed, bottlenecks in silicon technology, **9**-1
Speed advantages, in MOSFETs *vs.* SPINFETs, **8**-10
Spherical systems, **7**-72–**7**-73, **27**-1
    electron transport solution, **7**-72
SPICE software, **5**-22
Spin, as state variable in nanodevices, **10**-11
Spin accumulation, **9**-24–**9**-32
    proportional to current density through drain magnet,
    9-22
Spin channel current, drift-diffusion approximation of,
    9-26

Spin extraction, **9**-24–**9**-31
    conditions for efficient, **9**-31–**9**-33
    from semiconductors, **9**-23–**9**-24
Spin field effect transistors (SPINFETs), **8**-1–**8**-3
    differences from MOSFETs, **8**-1
    Dresselhaus spin-orbit interaction-based, **8**-3–**8**-4
    eigenenergies and eigenspinors of Hamiltonian,
        8-5
    eigenspin components, **8**-6
    energy dispersion relations of spin-split subbands,
        8-6
    half metals and limits of spin injection efficiency, **8**-8
    limitations in comparison with MOSFETs, **8**-10–**8**-11
    magnetic field strength and carrier velocity, **8**-2
    maximum ratio of on-to-off conductance for, **8**-7
    motivations for research, **8**-8–**8**-10
    non-idealities, **8**-4
    nonballistic type, **8**-4–**8**-8
    precession angle of, **8**-3, **8**-4
    precession angle with Dresselhaus-type, **8**-3
    Rashba interaction type, **8**-3
    rate of spin precession in space, **8**-2
    spin frequency of injected carriers, **8**-2
    spin injection efficiency at source contact, **8**-7
    spin precession time bottleneck, **8**-10
    spin relaxation time bottleneck, **8**-10
    structure of, **8**-2
    total effective magnetic field for, **8**-4
Spin injection
    boundary condition, **9**-27
    characteristic current density, **9**-27
    conditions for efficient, **9**-31–**9**-33
    control of in heterostructures, **9**-32, **9**-33
    high-frequency spin-valve effect, **9**-33–**9**-35
    into organic materials, **9**-36
    processes and devices, **9**-20–**9**-21
    at reverse bias, **9**-26
    into semiconductors, **9**-23–**9**-24
    spin accumulation and extraction, **9**-24–**9**-31
    and spin-torque domain wall switching in nanomagnets,
        9-22–**9**-23
    tunnel magnetoresistance in, **9**-21–**9**-22
Spin injection efficiency, **9**-21
    conditions for, **9**-30
    at FM-S interface, **9**-28
    at forward *vs.* reverse bias voltage, **9**-31
    limits at room temperature, **8**-7–**8**-8
    at room temperature, **9**-29
Spin orbit interaction, and gate voltage in spin transistors,
    8-10
Spin polarization, **9**-21
    of current, **9**-30
Spin precession time, as bottleneck in SPINFETs, **8**-10
Spin relaxation time, **9**-36
    as bottleneck in SPINFETs, **8**-10
Spin-relaxation time, **9**-31
Spin-torque domain wall switching, in nanomagnets,
    9-22–**9**-23
Spin transistors, **10**-11
Spin transport, in organic and semiconductor nanodevices,
    9-1–**9**-2

SPINFETS. *See* Spin field effect transistors (SPINFETs)
Spintronics, 2-4–2-5
  devices, **9**-35–**9**-36
  and moletronics, **9**-1–**9**-2
  and quantum entanglement, **2**-5
  use of spin degrees of freedom in, **9**-37
Spiropyans, photoinduced isomerization of, **11**-37
Splittable biocomponent fibers, **21**-6–**21**-10
Spontaneous emission, control through photonic bandgap
     technology, **25**-18
Spot-size converters, **25**-30
Square formation, in solution, **18**-40
Square-law detectors
  spintronic devices as, **9**-35–**9**-36
  and variation of emitter current, **9**-36
Stack memory, **7**-51, **7**-52
Staggered pi-stacking, **18**-9
Standards, development of, **3**-22
Standby power
  in CMOS devices, **4**-15
  nanocell problems, **4**-15
  in power loss analysis, **4**-9
Startup companies, U.S. share of, **3**-13
State encoding, **7**-51
State variables, **10**-5
Static RAM (SRAM), **4**-11–**4**-12
  molecular ICs implementing, **7**-34
Stationary scattering theory, **12**-6
Statistical analysis, **7**-12–**7**-13
Stators, in molecular rotors, **11**-26
Step and flash imprint lithography, **20**-3
Stern layer, charge movement in, **16**-8–**16**-9
Stimuli, biosystem detection of, **7**-20
Stone-Wallace bond rotation defect, **22**-7
Storage, new forms of, **2**-3–**2**-5
Strain energy, in carbon nanotubes, **23**-4–**23**-7
Strained-Si channels, **10**-4
Stray charges, **6**-8
Street blockages, in Teramac computer, **5**-8
Stress-strain curves, for SWCNT bundles, **23**-6
Styrene molecules
  blinking behavior of, **9**-11
  variation of current on silicon with time, **9**-12
Subthreshold slope, in CMOS devices, **4**-5–**4**-8
Subwavelength optical systems, **29**-15–**29**-16
  asymmetrical lens simulation and iterative design,
     29-18–**29**-19
  focusing light, **29**-17
  intensity *vs.* distance data, **29**-18
  numerical validation in hardware, **29**-16–**29**-18
  subwavelength lens geometry, **29**-17
Super-large-scale integration (SLSI), **7**-4
  CAD-supported design of, **7**-19
  decision diagrams/trees in design of, **7**-39
Super-prism phenomenon, **25**-29–**25**-30
Superconductivity, **1**-8
Superscalar processing, **7**-49
Supramolecular chemistry, **18**-13–**18**-14
  as enabling technology, **11**-2
  hydrogen bonding in, **18**-14–**18**-18
  pi-interactions in, **18**-18–**18**-21

Supramolecular gatekeepers, **11**-24
Supramolecular structures, hydrogen bonding in,
     18-16–**18**-18
Surface activation
  polymers or large molecules, **21**-39–**21**-41
  small molecules, **21**-38–**21**-39
  of textile nanofibers, **21**-38
Surface area/volume ratios, for reinforcement filler
     geometries, **21**-43
Surface chemistry, and cell culture, **26**-15–**26**-17
Surface-emitting laser diodes, using photonic bandgap
     crystal cavity, **25**-18
Surface-liquid interface, **11**-35
Surface modification
  as barrier property, **21**-56–**21**-58
  nanoparticle size after, **30**-13
  in nanoparticles for drug delivery, **30**-12–**30**-14
Surface properties
  effects on electronic properties, **11**-12
  electrostatic control of, **11**-37–**11**-39
Surface-rotor interactions, **11**-33–**11**-35
Surface tension gradient, in nanofluidics, **11**-30
Surface-vapor interface, **11**-35
Switch capacitance, and signal delay time, **4**-10
Switching delay, in MOSFETs *vs.* SPINFETs, **8**-10
Switching devices
  neurons as, **7**-25, **7**-27
  new forms of, **2**-3–**2**-5
Switching energy, **7**-15
Switching function, for hypercells, **7**-42
Switching speed, **4**-6, **7**-16
  in biological systems, **7**-8
  fundamental limitations in MOSFETs, **8**-8–**8**-9
  for molecular computing and processing platforms
     (^{M}PPs), **7**-16
  and power consumption in CMOS devices, **4**-4
Symmetry interaction model, **18**-34–**18**-38
  helicate formation from, **18**-42
  structural examples, **18**-37
Synaptic cleft, in biosystems, **7**-21
Synaptic transmission, **7**-25
Synthesis taxonomy, in design of ^{M}ICs and processing
     platforms, **7**-18–**7**-20
Synthetic chemistry, **11**-1
Synthetic fluidic molecular processing module, **7**-31
System states, time evolution of, **7**-65
Systems of nanosystems products, **3**-18

T

Targeted delivery, in pharmaceutics, **11**-22
Technological innovation
  improving through nanotechnology, **3**-19
  technology periods and paradigm shifts in, **24**-2–**24**-3
Temperature
  and coherence, **2**-6
  manipulation of nanoporous materials by, **11**-25
Temperature dependence
  of conductance in simple organic films, **9**-20
  of electron transport through organic molecular films,
     9-16

of nanopore junction current at forward and backward
     biases, **12**-18
in thiophene devices, **9**-20
Template-directed slipping, **11**-8
Tensile strength
     comparison of common engineering materials, **21**-29
     of double-walled CNT, **22**-7
Teramac computer, **5**-4, **5**-8, **6**-10, **6**-11
Terminal velocity, **16**-21
Tersoff test, **28**-4
Tetrahedral model, of diamondoid scaffolding, **18**-26
Tetraureas, as building blocks for multicatenances,
     19-14
Textile nanotechnologies, **21**-1–**21**-2
     carbon nanotubes, **21**-27–**21**-34
     electrospun nanofibers with magnetic domains,
          21-17–**21**-25
     nanoadditives in textiles, **21**-42–**21**-61
     nanofibers, **21**-2–**21**-13
     open biocomponent spunbond process, **21**-8
     spunbonded fiber diameter as function of number of
          segments, **21**-9
     surface activation, **21**-38–**21**-41
Theoretical challenges, in molecular conductance
     junctions, **12**-17–**12**-22
Thermal coefficient
     of CNT-PE composites, **22**-8
     dependence on MMT content in nanofibers, **21**-53
Thermal conductivity
     in carbon nanotubes, **22**-11
     comparisons among nanofibers, **21**-59
Thermal energy
     barrier properties, **21**-58–**21**-59
     exceeding, **2**-4
Thermal stability, of nanoadditives, **21**-52–**21**-53
Thermal transport, in nanotubes, **22**-10–**22**-11
Thin-film electronics, **4**-16
     in vertical structures, **4**-11
Thiol-gold molecules, schematic, **17**-8
Thiol group, as alligator clips, **5**-17
Thiols, on gold, **11**-3
Thiophene molecules, switching behavior in, **9**-19–**9**-20
Third dimension, **2**-7. *See also* 3D nanosystems
     nature's use of, **2**-6
     and transistor packing density, **2**-3
     using in nanotechnology, **2**-2
     visualization and numerical simulation of, **3**-17
Thomas Arithmometer, **7**-5
Threading process, **11**-8
Three-dimensional structures, transition metal based,
     18-40–**18**-43
Three-terminal devices, **4**-3, **10**-14
     from cyclic molecule with carbon interconnecting
          framework, **7**-76
     DRAM, **4**-12
     logic circuits, **4**-4–**4**-5
     SEM image, **5**-10
     transport measurements, nanowire devices, **5**-10
Thymine, change in surface hydrophobicity/hydrophilicity,
     11-37
Tight-binding simulation method, **23**-32

Tilting angle, **9**-36–**9**-37
     effects on I-V curves of benzene-dithiolate molecules, **9**-5
Time derivatives, **7**-66
Tiny hands analogy, **1**-8
TMV-templated nanowires, **10**-9
Toffoli gate, **10**-10
Tolerance control, **4**-12
Top-down fabrication method, **20**-3
     convergence with bottom-up method, **10**-2–**10**-3
     limitations in computer circuitry, **5**-2
     nanoimprint lithography for nanoscale devices,
          20-3–**20**-4
     in silicon industry, **5**-3
Top-down synthesis taxonomy, **7**-19
Total energy
     expectation values, **7**-65
     using Schrödinger equation, **7**-66
Total polarization, **9**-32
Tour, extrinsic telegraph switching in, **9**-37
Tour wires, **9**-4
     carrier transport in, **9**-3
Tradeoffs, between molecular stability and molecular
     switching, **9**-13
Transforming development, **3**-19
Transistor drive current, and circuit speed, **4**-6
Transistor-transistor logic (TTL), **7**-35
Transistors
     number per logic chip, **5**-1, **5**-2
     scalability history, **2**-2–**2**-3
Transition metal controlled molecular actuators,
     11-17–**11**-18
Transition metals, **18**-32–**18**-34
     molecular library model and, **18**-34
     symmetry interaction model, **18**-34–**18**-38
     three-dimensional structures, **18**-40–**18**-43
     two-dimensional structures, **18**-38–**18**-40
Transmission coefficient, **7**-74, **12**-6
Transmission probability, **9**-8
Traveling wave dielectrophoresis, **16**-13–**16**-16
     schematic diagram, **16**-17
Triangle formation, in solution, **18**-40
Tridymite-based oxides, integrated transmission *vs.*
     incident electron energy, **13**-7
Triodes, gating and control of junctions by, **12**-10–**12**-11
Tryptycene rotor unit, **11**-27
     in molecular compasses, **11**-30
Tumor therapy, using nanoparticles, **30**-17–**30**-18
Tunable photonic crystals, **25**-23–**25**-25
     dispersion-based tunable beam splitter, **25**-32
Tunnel magnetoresistance (TMR), **9**-21
     in FM-insulator-FM structures, **9**-20
     nonvolatile magnetic memory based on, **9**-22
     in spin injection, **9**-21–**9**-22
Two-dimensional structures, transition metal based,
     18-38–**18**-40
Two-impurity channel conductance, **9**-19
Two-terminal diode devices, **4**-2–**4**-3, **10**-13
     with Hamiltonian, **7**-58
     molecules as, **7**-7
     power dissipation and gain issues, **4**-14–**4**-15
Tyranny of large systems, **2**-2–**2**-3

U

UCLA-Hewlett Packard collaboration, on switching with geometric change, **12**-10
Ultra-shallow junction formation, challenges to, **10**-3
Ultra thin oxides, tunneling in, **13**-5–**13**-6
Ultrafast spin-injection devices, **9**-35
Unidirectional rotation, **11**-28, **11**-30
   on gold surfaces, **11**-32
   light-driven, **11**-30–**11**-33
   molecular structure of first-generation rotor, **11**-31
   second-generation rotor, **11**-31
Unique capabilities, **7**-19
   in molecular electronics, **7**-3
United States
   estimated nanotechnology expenditures, **3**-12
   share of global nanotechnology investments, **3**-13–**3**-14
Unsaturated-shell architecture approach, dendrimers in, **24**-34
U.S. Air Force Research Laboratories, xii
U.S. government investment, in nanotechnology, **3**-6

V

Vaccine delivery, using nanoparticles, **30**-18–**30**-19
Valve systems
   applications of computer modeling to, **28**-13
   in nature, **11**-21–**11**-22
Van der Waals attractions, **1**-8, **7**-12
   geometry of nanotube systems with, **13**-26
   in self-assembly, **17**-5
Velocity modulation transistor (VMT), similarities to SPINFETs, **8**-9
Verilog, **7**-57
Vertical cavity surface emitting laser diodes (VCSELs), **2**-4
Vertical field-effect transistors, **4**-11
Very High Speed Integrated Circuit Hardware Description Language (VHDL), **7**-57
Vibrational characteristics, of carbon nanotubes, **22**-10–**22**-11
Vibronic coupling, **12**-16
Virtual control unit design, **7**-49–**7**-52
Vitamin E-loaded nanoparticles, **30**-6
VLSI design, logic in, **7**-35
Voltage inverter trace, for CMOS inverter, **4**-5
Voltage requirements, for flash memory, **4**-12
Von Neumann architecture, **10**-5

W

Water, as example of molecular assembly, **10**-9
Water filtration/desalinization, with nanotechnology, **3**-18
Water immersion lithography, **10**-7
Wave function
   in 3D problem, **7**-77
   for free particle, **7**-66
   obtaining with Schrödinger equation, **7**-59
   and time evolution of system states, **7**-65
Wave packet
   energy and momentum, **7**-63
   magnitude, **7**-61
Waveguide bends, **25**-13
Wavelength, as core element of lithography, **10**-6
Wettability, **11**-5
   and electrostatic control of surface properties, **11**-37–**11**-39
   energetic contributors to, **11**-35
   light-responsive surface and, **11**-36–**11**-37
   and nanofluidics, **11**-39–**11**-40
   surfaces with controllable, **11**-35
   and switching surface design, **11**-37
White House Office of Science and Technology Policy (OSTP), **3**-6
Wire splitting, in QCA structures, **6**-4
Workers, anticipatory training for nanotechnology applications, **3**-20
Worldwide market, estimates, **3**-9
Worm-like micelles, as nanoparticulate delivery systems, **30**-16
Writing small, **1**-2–**1**-3

X

X-ray lithography, **5**-2

Y

Young's modulus
   CNT-PE composite, **22**-8
   CNT properties, **23**-4–**23**-7

Z

Zero cross-talk, **25**-16–**25**-17
Zeta potential, in nanoparticles for drug delivery, **30**-12
Zhou multilevel memory device, **10**-12, **10**-13